Vascular Plants
of New Mexico

Vascular Plants of New Mexico

Kenneth D. Heil & Steve L. O'Kane, Jr.

 MISSOURI BOTANICAL GARDEN PRESS

ISBN 978-1-935641-30-8
Library of Congress Control Number 2023949248
Monographs in Systematic Botany from the Missouri Botanical Garden, Volume 140
ISSN 0161-1542

Managing Editor: Allison M. Brock
Editor: Lisa J. Pepper
Press Coordinator: Amanda Koehler
Copyeditor: Laurel Anderton
Cover and interior design: Paula Catalano
Indexer: Franz Stadler
Typesetting and layout: Integrated Publishing Solutions

This monograph was printed on 18 February 2025.

Front cover painting and frontispiece: *Rio Grande - River of Dreams* © Rod Hubble.
Back cover: photos by Steve L. O'Kane, Jr. (Ken at Pup Canyon, NM) and Russ Kleinman (Steve at Kneeling Nun Overlook, NM).

CONTENTS

Color plates of rare, endangered, and endemic plants of New Mexico follow page 596
Oil paintings by Rod Hubble follow color plates

PREFACE

Throughout New Mexico new species of plants and new plant distributions continue to be discovered, and molecular genetic information continues to modify our understanding of the phylogeny and classification of vascular plants. Further, climate change obviously and continually alters plant migration and distributions. *Vascular Plants of New Mexico* is meant to be a field-portable identification guide to our current knowledge of all vascular plant species, subspecies, and varieties in the state, about 4200, as documented by herbarium specimens. It is a systematic, multidecade effort involving over 60 contributing botanists and artists such as Rod Hubble, a renowned New Mexico painter. Major herbaria contributing to this effort were those of San Juan College (SJNM), the University of New Mexico (UNM), and New Mexico State University (NMSU). Consultations with numerous other herbaria were made when additional information was needed. Collections and distributions documented by SEINet, especially the Arizona–New Mexico Portal (https://swbiodiversity.org/seinet/) and the Biota of North America Program (BONAP), were especially helpful. BONAP data and maps were generously supplied by John Kartesz and greatly aided our fieldwork. Photographs of the rarest species, mostly by us, are included to illustrate the fascinating diversity of the New Mexican flora.

The diverse flora of New Mexico found among its 33 counties is distributed among five major river systems (Gila, San Juan, Pecos, Rio Grande, and Canadian), six major mountain ranges (Sangre de Cristo, San Juan, Peloncillo, Gila, Sacramento, and Guadalupe), two national parks (Carlsbad Caverns and White Sands), a world heritage site (Chaco Culture National Historical Park), nine national monuments (Aztec Ruins, Capulin Volcano, Gila Cliff Dwellings, Bandelier, El Malpais, El Morro, Fort Union, Petroglyph, and Salinas Pueblo Missions), 295 Spanish land grants of varying size, and 27 wilderness areas.

Although our intention is to include all known taxa of vascular plants in New Mexico, no flora is ever complete, and continued exploration will yield new plants and new distributions. We'd have it no other way.

Kenneth D. Heil and Steve L. O'Kane, Jr., 2022

ACKNOWLEDGMENTS

Many people including colleagues, students, coworkers, friends, and family aided in the production of *Vascular Plants of New Mexico*. The list of those who made this project possible is vast; however, our limited and aging memories make it impossible to thank everyone who contributed.

Thanks to Michael Ottinger and Toni Pendergrass of San Juan College. Laurie Gruel aided us immensely with managing grant money. We are especially thankful to Mike Howard and Zoe Davidson of the New Mexico Bureau of Land Management, Gregory Silsby and David Anderson of White Sands Missile Range, and the University of Northern Iowa for summer salaries. Also, thanks to Brian Essex of the Melrose Bombing Range.

Funding was generously provided by the Bureau of Land Management, Native Plant Society of New Mexico, Ecosphere Environmental Services (Mike Fitzgerald), Nelson Consulting (Steve Nelson), and San Juan College. Publication of this book would not have been possible without the funding provided by the Environmental Department of White Sands Missile Range (WSMR–U.S. Army) and the botanical contributions of employees Gregory Silsby and David Anderson. As a tribute to their contributions, a brief history of WSMR and a checklist of the flora of WSMR are presented in the appendices.

Collecting permits for botanical explorations were provided by David Anderson and Gregory Silsby (WSMR); Fort Bliss Environmental Department; Daniela Roth and Jeff Cole (Navajo Nation); Frank Simms (Jicarilla Apache Nation); Mike Howard, Lilis Urban, and Zoe Davidson (Bureau of Land Management); and the U.S. Forest Service (permits, Albuquerque Office). Private landholders graciously allowed us to collect plants and/or provided valuable information, including Glenn A. Briscoe (Briscoe Ranch), Danielle J. Smith (Sabinoso area), Lillian Lester (Sabinoso area), Mary and Mauricio Gomez, Salso Gomez[†], and Richard Gooding. Arnold Clifford, of the Navajo Nation, was our guide and liaison to the wonders of the northwestern part of the state.

Thank you to those individuals and institutions who helped with collection and plant information: U-Trail Pecos Wilderness Horseback Packers, Rio Costilla Cooperative Livestock Association, White Sands Missile Range, and Sabinoso Wilderness Area. We also thank those who have allowed us access to herbarium collections and those who provided assistance, identification, herbarium work, and/or fieldwork beyond what was necessary for their particular taxonomic treatments: Adam Martinez, Steven Perkins, Dave Schleser, Don Hyder, John Anderson, Les Lundquist, Ron Hartman[†], Arnold Clifford, Chick Keller, Gene Jercinovic, Donna Hobbs, Guy Nesom, Kelly Allred, Mark Porter, Wayne Mietty, Ilyse Gold, C. Barre Hellquist, Ihsan Al-Shehbaz, David Anderson, Mike Heil, Mark Heil, Nate Heil, Adriano Tsinigine, Jeremy McClain, Gregory Silsby, Ross McCauley, Ron Coleman[†], Turner Collins, Gregory Penn, Glenn Rink, David Morin, Fred and Pam Romero, and Ben Legler. A special tip o' the hat to Jennifer Ackerfield, who graciously provided her expertise in "filling in the cracks" by writing several orphan family treatments that we had neither the time nor expertise to complete.

Plant distributions would have been impossible without the help of John Kartesz of the Biota of North America Program (BONAP) and Edward Gilbert of SEINct. A very special thanks to them.

Our deepest thanks go to Rod Hubble for his beautiful paintings that capture the heart of our botanical landscapes. Artistic contributions like these are not common in modern scientific publications. In addition to efforts by O'Kane and Heil to photograph all the rarest species in the state, Patrick Alexander, especially, contributed several important photographs of rare plants found in southern New Mexico.

As in our Flora of the Four Corners project, students and herbarium assistants tackled the daily herbarium work of gluing specimens, entering data, producing labels, and filing with good humor and positive attitudes. We thank Starla White, Cynthia Holmes, Lonnie Herring, Rosanna Flores, Sande Burr, Brean Reed, Aleisha Shevokas, Michelle Szumlinski, Gregory Penn, Les Lundquist, Laverne Clark, and Nick Moore. A special thanks to the staff of the Missouri Botanical Garden Press, especially Allison Brock, managing editor, who gave us positive encouragement and feedback. We are indebted to Dieter Wilken and Peter Lesica, whose reviews (at that time anonymous) significantly improved this publication. Remaining errors and shortcomings we claim as our own. No doubt we have omitted the names of people we should have included, and we thank them as well.

A very special dedication and thanks to our dear late friend Dave Jamieson, who was indefatigable in the field, a source of botanical inspiration, and, most importantly, an endless source of gut-busting humor. He is sorely missed.

Lastly and most importantly we are grateful for the forbearance and encouragement of our spouses, Marilyn Heil and Arlene Prather-O'Kane, for putting up with and encouraging the *Vascular Plants of New Mexico* project and for providing counsel so graciously and for so long.

CONTRIBUTORS

Jennifer Ackerfield Denver Botanical Garden Herbarium

Robert P. Adams Baylor University

Jason A. Alexander University and Jepson Herbaria, University of California, Berkeley

Kelly W. Allred New Mexico State University

David Anderson White Sands Missile Range

John L. Anderson Bureau of Land Management

Tina I. Ayers Deaver Herbarium, Northern Arizona University

Gary I. Baird Brigham Young University–Idaho

Susan C. Barber Oklahoma City University

Fred R. Barrie Missouri Botanical Garden

Paul E. Berry University of Michigan Herbarium

David E. Boufford Harvard University Herbaria

Josh M. Brokaw Abilene Christian University

John Burris San Juan College

Anita F. Cholewa University of Minnesota, Bell Museum

Ronald A. Coleman[†] University of Arizona

L. Turner Collins Evangel University

Mihai Costea Wilfrid Laurier University

Robert D. Dorn Mountain West Environmental Services

J. Mark Egger University of Washington

Barbara Ertter University and Jepson Herbaria, University of California, Berkeley

David J. Ferguson Rio Grande Botanic Garden

Mark Fishbein Oklahoma State University

Craig C. Freeman R. L. McGregor Herbarium, University of Kansas

Sherel Goodrich U.S. Forest Service

Margaret M. Hanes Eastern Michigan University

Kristin Huisinga Harned[†] Gila Hot Springs

Ronald L. Hartman[†] Rocky Mountain Herbarium, University of Wyoming

Kenneth D. Heil San Juan College

Mark Heil Geospatial scientist

C. Barre Hellquist Massachusetts College of Liberal Arts

Peter C. Hoch Missouri Botanical Garden

Rod Hubble www.rodhubble.com (paintings)

Larry Hufford Washington State University

Don Hyder San Juan College

Nicholas J. Jensen Conservation analyst

Diana D. Jolles Plymouth State University

John Kartesz Biota of North America Program (BONAP)

Nancy R. Khan Smithsonian Institution

Ben S. Legler University of Washington Herbarium, Burke Museum

Walter A. Lewis† Missouri Botanical Garden

Max H. Licher Consultant

Timothy K. Lowrey University of New Mexico Herbarium, University of New Mexico

Robert L. Mathiasen Northern Arizona University

Ross A. McCauley Fort Lewis College

Angela McDonnell St. Cloud State University

James McGrath Consultant

Lynn M. Moore Consultant

Sandra M. Namoff Rancho Santa Ana Botanic Garden

Mare Nazaire Rancho Santa Ana Botanic Garden

Guy L. Nesom Academy of Natural Sciences, Drexel University

William R. Norris Western New Mexico University

Steve L. O'Kane, Jr. Grant Herbarium, University of Northern Iowa

Steven R. Perkins Chevron Corporation, Environmental Department

J. Mark Porter Rancho Santa Ana Botanic Garden

Richard K. Rabeler University of Michigan

R. M. Rhode University of California, Davis

Glenn R. Rink Northern Arizona University

John J. Schenk Georgia Southern University

Randall Wheeler Scott Deaver Herbarium, Northern Arizona University

Robert C. Sivinski University of New Mexico Herbarium

John R. Spence National Park Service, Glen Canyon National Recreation Area

Thomas R. Stoughton Plymouth State University

Debra K. Trock California Academy of Sciences

Warren L. Wagner Smithsonian Institution

Elizabeth F. Wells George Washington University

Alan T. Whittemore U.S. National Arboretum

Michael D. Windham Duke University

INTRODUCTION

Vascular Plants of New Mexico is a field-portable guide to identifying the vascular plants spontaneously growing in the state. As such, it contains taxonomic keys, short plant descriptions, plant community types, elevation ranges, and general plant distributions. All known and documented species, subspecies, and varieties of vascular plants are included. We designed this book to be useful to plant enthusiasts, amateur botanists, professional biologists, consultants, academics, ecologists, and land management personnel.

As with our *Flora of the Four Corners Region*, expert volunteer authors were solicited and were asked to complete brief treatments following a readable and consistent format. Beside descriptions, also included in the book are dichotomous keys, paintings by Rod Hubble, and photographs of over 200 rare plants found throughout New Mexico. Synonyms provided for taxa are only those that would be commonly encountered. A more complete synonymy can be found in Allred (2020). Measurements are metric, except elevations are given in feet because we have found that most users in the United States are more familiar with the vegetation and climate at elevations given in those units.

Due to having numerous contributors, there is a diversity of taxonomic philosophies and opinions. Editors Kenneth D. Heil and Steve L. O'Kane, Jr. were responsible for raising funds, editing manuscripts, writing numerous treatments, taking photographs of habitat types and rare plant species, applying for permits, collecting plants, and performing many and varied day-to-day duties. The editors attempted to maintain accuracy and consistency throughout the book. Individual authors, however, had the final say about taxonomic treatments, recognition of taxa, and length of descriptions. Neither complete uniformity among treatments nor strictly parallel descriptions is to be expected. However, authors were instructed to provide characters that would best confirm identification in the fewest words possible.

Constraints to the project were ever present and included limited funding for fieldwork and herbarium workers, extended drought throughout the state, the sheer size and remoteness of the study area, and the inaccessibility of private land. However, between 2007 and 2022, over 10,000 New Mexico collections were made by the editors alone, focusing on areas that were the least collected and explored. We have attempted to treat every species, subspecies, and variety of plant known to occur in New Mexico and to accurately describe their morphology, general distribution, habitat, and New Mexico elevations. Identification keys have been constructed to be as simple and user-friendly as possible.

Four large watersheds occur in New Mexico: (1) the Rio Grande Basin: Rio Grande and Pecos River; (2) the Arkansas-White-Red Basin: Canadian River; (3) the Upper Colorado Basin: San Juan River; and (4) the Lower Colorado Basin: Gila River. The entire area encompasses 121,697 square miles. The highest point is 13,161 feet at Wheeler Peak in Taos County, in the Sangre de Cristo Mountains within the Carson National Forest. The lowest point is Red Bluff Reservoir on the Pecos River in Eddy County, at 2842 feet. Because of the large elevation gradient, vegetation throughout New Mexico varies from alpine tundra to coniferous forest to montane grassland, mountain shrubland, sagebrush, shortgrass prairie, and Chihuahuan desert scrub.

Older floras are available for New Mexico. These are far out of date in their nomenclature and the taxa included and are extremely difficult to obtain: *Flora of New Mexico* (Wooton & Standley, 1915); *A Flora of Arizona and New Mexico* (Tidestrom & Kittell, 1941); and *A Flora of New Mexico* (Martin &

Hutchins, 1980–1981). The recent *Flora Neomexicana III: An Illustrated Identification Manual* (Allred et al., 2020) is an excellent reference tool and identification guide, but it doesn't lend itself to heavy field use, especially as it consists of two large paperback volumes. Those volumes in conjunction with our *Vascular Flora of New Mexico* will provide users with two independent ways to identify the plants of New Mexico.

NOTEWORTHY HISTORICAL COLLECTORS

Listed below are historical collectors and authors whose work was especially noteworthy within New Mexico. Many of these botanists collected throughout the Southwest and periodically visited New Mexico (Dickerman, 1985; Ewan, 1950). Many of their names are reflected in the names of our plants—for example, *Quercus gambelii*—or as the authors of these names.

Baker, Charles F. 1872–1927. Entomologist and botanist. Collected at Aztec, San Juan County.

Barneby, Rupert C. 1911–2000. Specialist in Fabaceae, notably *Astragalus* and *Oxytropis*; collected extensively throughout the American West.

Bigelow, John Milton. 1804–1878. Botanist and U.S. Army surgeon with the Mexican Boundary Survey. In 1853 he made numerous collections, mostly east of Albuquerque.

Castetter, Edward Franklin. 1896–1978. Ethnobotanist, professor, and vice president of the University of New Mexico; collected throughout New Mexico.

Emory, William. 1811–1887. Engineer, major, U.S. Army Corps of Topographical Engineers. Identified with the Mexican Boundary Survey, where he collected and preserved plants.

Fendler, Augustus. 1813–1883. German American botanical collector, native of eastern Prussia. Collected from Bent's Fort to Santa Fe in 1846.

Gambel, William. 1821–1849. Friend of renowned botanist Thomas Nuttall; he was encouraged to explore the West. He spent time in the southern Rocky Mountains and collected in the upper San Juan River drainage on an expedition from Albuquerque to Los Angeles. Gambel spent at least two weeks in Santa Fe.

Greene, Edward Lee. 1843–1915. First professor of botany at the University of California and botanical explorer in New Mexico and Colorado.

Kittell, Sister M(ary) Teresita. 1892–1990. Coauthor, with I. F. Tidestrom, of *A Flora of Arizona and New Mexico* (1941).

Newberry, John Strong. 1822–1892. Paleontologist, geologist, botanist; educated in medicine (M.D.); collected in Largo Canyon.

Parry, Charles C. 1823–1890. British American explorer and botanist. He was with the Mexican Boundary Survey from 1849 to 1852, collected with E. Hall and J. P. Harbour in 1861, and served as botanist to the Kansas Pacific Railway Survey along the present route across northern New Mexico.

Ripley, Harry D. D. 1908–1973. Collecting companion to Rupert C. Barneby.

Smith, Charles Piper. 1877–1955. High school teacher in San Jose, California, and later professor of botany at the Agricultural College of Utah; collected in Rio Arriba and San Juan Counties.

Standley, Paul Carpenter. 1884–1963. Coauthor, with E. O. Wooton, of *Flora of New Mexico* (1915).

Thurber, George. 1821–1878. Botanist, naturalist, author, and editor; quartermaster and commissary on the Mexican Boundary Survey, 1850–1853; made numerous collections around Silver City.

Tidestrom, Ivar Frederick. 1864–1956. Studied under E. L. Greene; coauthor, with Sister Teresita Kittell, of *A Flora of Arizona and New Mexico* (1941).

Wislizenus, Frederick A. 1810–1889. Naturalist, explorer, physician from Germany; collected along the Rio Grande from near Santa Fe to Mesa del Norte, Mexico.

Wooton, Elmer Ottis. 1865–1945. Taught chemistry and botany at New Mexico College of Agricultural and Mechanic Arts; coauthor, with P. C. Standley, of *Flora of New Mexico* (1915).

Wright, Charles. 1811–1885. Botanical explorer. Collected mostly in Ciudad Juárez, Chihuahua, Mexico, and El Paso, Texas. May have made some collections in New Mexico near the Texas state line.

GEOLOGY
John Burris

New Mexico is a land of stunning variation in its landscapes, ranging from alpine forests to arid badlands. This diversity is in large part a result of its long and complex geologic history. To understand New Mexico's geology is, in fact, a lesson in the history of the state, but on a scale lasting billions of years. With this depth of time, having and sharing a complete understanding of the entirety of the state's geologic history is impossible, though some volumes attempt to do so (e.g., Mack & Giles, 2004). The following is a brief overview of the geologic history of New Mexico, and how these past events have shaped the state as it is today.

The included geologic time scale (Table 1), based on the International Chronostratigraphic Chart (Cohen et al., 2021), provides context for the names of the time periods discussed here. The oldest rocks in New Mexico are from the Proterozoic eon, approximately 1800 million years old (Myr) (Karlstrom et al., 2004). While the earth has a long history preceding the age of these rocks, older rocks either did not form during that time or have been buried, metamorphosed, or eroded in New Mexico. These ancient crystalline basement rocks range from 1800 to approximately 1000 Myr and are generally found in mountain uplifts, including the Sandia, San Andres, Zuni, and Sangre de Cristo Mountains. Rocks of this age are generally intrusive igneous and metamorphic and record the tectonic assembly of the North American continent from smaller landmasses and later assembly of the pre-Pangaean supercontinent Rodinia.

Toward the end of the Proterozoic eon, Rodinia began to break apart, and some stability came to the New Mexico area. No sedimentary rocks were preserved during this time, however, leaving a massive

Table 1. Geologic times exposed in the rocks of New Mexico.

Eon	Era	Period	Millions of Years BP	Events and Environments	Orogenies
Phanerozoic	Cenozoic	Quaternary	2.58–0.00	Valles Caldera	
Phanerozoic	Cenozoic	Neogene	23–2.58	Rio Grande rifting, volcanism along the Jemez Lineament	
Phanerozoic	Cenozoic	Paleogene	66–23	K-Pg mass extinction, explosive volcanism	Laramide Orogeny
Phanerozoic	Mesozoic	Cretaceous	145–66	Western Interior Seaway, Pangaea breakup	Laramide Orogeny
Phanerozoic	Mesozoic	Jurassic	201–145	Sundance Sea and terrestrial basins	
Phanerozoic	Mesozoic	Triassic	251–201	Arid conditions continue, terrestrial basins	
Phanerozoic	Paleozoic	Permian	299–251	Arid conditions, Pangaea assembly, P-Tr mass extinction	Ancestral Rockies
Phanerozoic	Paleozoic	Pennsylvanian	323–299	Great Unconformity between Proterozoic and Cambrian, New Mexico periodically covered by ocean	Ancestral Rockies
Phanerozoic	Paleozoic	Mississippian	359–323	Great Unconformity between Proterozoic and Cambrian, New Mexico periodically covered by ocean	Transcontinental Arch
Phanerozoic	Paleozoic	Devonian	419–359	Great Unconformity between Proterozoic and Cambrian, New Mexico periodically covered by ocean	Transcontinental Arch
Phanerozoic	Paleozoic	Silurian	444–419	Great Unconformity between Proterozoic and Cambrian, New Mexico periodically covered by ocean	Transcontinental Arch
Phanerozoic	Paleozoic	Ordovician	485–444	Great Unconformity between Proterozoic and Cambrian, New Mexico periodically covered by ocean	Transcontinental Arch
Phanerozoic	Paleozoic	Cambrian	541–485	Great Unconformity between Proterozoic and Cambrian, New Mexico periodically covered by ocean	Transcontinental Arch
Precambrian	Proterozoic		2500–541	Crystalline basement records assembly and breakup of Rodinia	
Precambrian	Archean		4000–2500	Not known in New Mexico	
Precambrian	Hadean		4600–4000	Not known in New Mexico	

unconformity, and in some cases, there are hundreds of millions of years of missing rocks because of the unroofing of the basement rock. This "Great Unconformity" is capped by Cambrian-Ordovician sediments, recording a worldwide rise in sea level and bringing marine conditions to southern New Mexico (Mack, 2004). The entire Paleozoic era was marked by fluctuating sea levels as New Mexico remained warm and near the equator, and various sandstones, shales, and limestones were deposited. The sea would rise and cover parts of the state, followed by sea level falling and erosion removing some of the previous deposits. Invertebrate and vertebrate fossils are well represented in New Mexico's Paleozoic strata, including superb Permian trace fossils (Lucas & Heckert, 1995). Complicating the history was the influence of an upland area centered on New Mexico called the Transcontinental Arch during the Silurian through Mississippian, and the formation of the Ancestral Rockies uplifts and basins from the late Mississippian to early Permian (Kues & Giles, 2004). While massive reef deposits are found in southern New Mexico, the assembly of the supercontinent Pangaea was nearing completion, leading to increasing aridity and the presence of evaporite deposits.

As New Mexico entered the Mesozoic era, Pangaea completed its assembly and stayed mostly intact through the Triassic and into the Jurassic. New Mexico continued to be close to the equator and was generally above sea level. At times, vast deserts covered North America, creating massive wind-blown dune deposits. Rivers controlled much of the deposition through the Triassic and Jurassic as well (Lucas, 2004). Marine conditions returned to northern New Mexico during the Jurassic, creating the Sundance Sea. An arm of this sea became completely isolated, and as evaporation concentrated the dissolved minerals in the water, beds of gypsum were deposited. As Pangaea was breaking up in the Cretaceous, the planet warmed, and the sea rose to cover a broad basin that formed east of the growing Rocky Mountains. A broad inland sea called the Western Interior Seaway flooded North America from the Gulf of Mexico to the Arctic Ocean. Its western shoreline moved back and forth through New Mexico with both minor and major fluctuations in sea level. At times, New Mexico was a muddy to sandy seafloor, but as the sea shifted, delta, river, and poorly drained swamp environments returned (Nummedal, 2004). Both the marine and terrestrial fossil records are excellent for the Mesozoic, including Triassic dinosaurs near Ghost Ranch and Abiquiú, and marine invertebrates and dinosaurs from the San Juan Basin.

The regression of the Western Interior Seaway to the east occurred in part because of the beginning of the Laramide Orogeny, an unusual mountain-building event that created many of the mountains seen in New Mexico today. The Laramide began in the Cretaceous, though the bulk of this event crossed into the Paleogene. The Laramide Orogeny was different from other mountain-building events in a few noteworthy ways. For one, the mountains formed were far from the margin of the continent. For another, the faults that lifted the mountains were nearly vertical, instead of the typical compression folding and thrust faulting seen along convergent boundaries. Adjacent to these highlands were deep sedimentary basins that collected the sediments weathered and transported from the mountains. At the time of the Laramide Orogeny, the Farallon Plate was being overridden by North America in a process called subduction. The subducting plate usually descends into the mantle at approximately a 45° angle. The Farallon Plate, however, descended at a much shallower angle, perhaps as little as 5°–10°. As a result, the Farallon Plate scraped beneath North America much like a spatula beneath a pancake, pushing mountain building far inland from the continental margin. The results of the Laramide Orogeny are dominant features in New Mexico today, including the Hidalgo Uplift, Potrillo Uplift, Defiance Uplift, Nacimiento Uplift, Sangre de Cristo Mountains, Raton Basin, and San Juan Basin to name just a few, though many Laramide features have been diminished by 40 million years of erosion. As North America continued consuming the Farallon Plate, only small remnants of the once massive plate remained adjacent to Central America and the Pacific Northwest, and the Laramide Orogeny ceased (Cather, 2004).

New Mexico records a major biological event—the Cretaceous-Paleogene mass extinction that ended the non-avian dinosaurs. As a result of this mass extinction, mammals were poised to fill empty ecological niches and explode in diversity. The Paleogene sedimentary rocks of the San Juan Basin superbly preserve this mammal fossil record in muds and sands in vast river deposits (Williamson, 1996).

New Mexico had two major pulses of volcanic activity in the Cenozoic era. Explosive volcanism tied to Laramide tectonics was occurring in the southwestern part of the state, in the Mogollon-Datil and Boot Heel volcanic fields, at the end of the Paleogene into the Neogene (Chapin et al., 2004). These volcanic fields record the changing tectonism as the Laramide came to an end and the Rio Grande Rift began to form, creating one of New Mexico's most striking features. The Rio Grande Rift is a zone of extension in the center of the state, wider and older in the south and narrower and younger in the north. It formed as the Colorado Plateau pulled away from the Great Plains, thinning the crust. The rift is bordered by uplifts like the Sandia Mountains on both the east and west sides. The Rio Grande today is constrained by these resulting topographic features.

The second pulse of volcanism began in the Pliocene, centered on northern and central New Mexico. This volcanism primarily followed a zone of crustal weakness known as the Jemez Lineament and resulted in the Zuni-Bandera volcanic field (and El Malpais National Monument), Mount Taylor, the Taos Plateau volcanic field, Albuquerque volcanoes, and the Raton-Clayton volcanic field. The Valles Caldera is found at the intersection of the Jemez Lineament and the Rio Grande Rift, formed by the collapse of the magma chamber after an immense and violent eruption around 1.2–0.6 Myr (Goff & Gardner, 2004).

Four physiographic provinces converge in New Mexico today (Fig. 1). To the east are the Great Plains, part of North America's ancient and stable continental interior. The Colorado Plateau is found in the northwest of the state, a large, uplifted region centered on the Four Corners area, left relatively untouched by previous mountain-building events. It is marked by deep canyons formed as rivers directed their erosive powers downward in a gravitational attempt to reach sea level. In the north-central part of the state are the Sangre de Cristo Mountains, the southern extension of the Rocky Mountains. Finally, there is the Rio Grande Rift in the southwestern and central part of New Mexico. The rift is an extension of the Basin and Range Province, part of the continent that has been stretched into narrow mountains and broad, sediment-filled basins. The presence of all four of these provinces has added to the diversity of the ecosystems and environments in New Mexico.

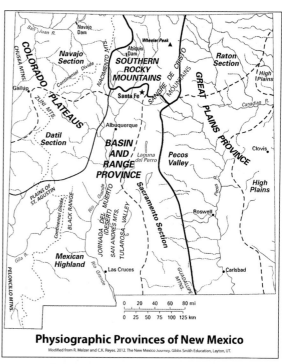

Physiographic Provinces of New Mexico
Modified from R. Melzer and C.K. Reyes. 2012. The New Mexico Journey. Gibbs Smith Education, Layton, UT.

Figure 1. Physiographic provinces.

VEGETATION

Steve L. O'Kane, Jr.

When pondering a landscape, one is presented with a puzzle. Why does this landscape look different from that one? Why is this landscape dominated by trees and that one by grasses? Why does this canyon wall have scattered shrubs while that one, just across the road or stream, is blanketed with Douglas-fir trees? People have contemplated these puzzles for as long as we have interacted with the land, either for aesthetic or scientific reasons or for reasons of necessity, such as intuiting where water, plant foods, or game might be. In New Mexico and other western states, striking geologic features may first catch the eye, but it is the vegetation—the local assemblage and arrangement of plants—that most informs what we mean by landscape. What is it, then, that determines not only the combination of plant species that inhabit an area, but how they are aggregated and arranged?

Factors Affecting Vegetation

If one had to reduce all of the possible variables that determine where a particular species or community of species resides on the landscape, it has long been known that temperature and the availability and seasonal timing of moisture are the dominant factors, and that these two factors are often inextricably connected. For example, the upper elevational limit of species is typically determined by temperature, while the lower elevational limit is determined by precipitation (Daubenmire, 1942, 1943). Soil nutrients and the vegetational history of an area are also important determinants of current vegetation.

Across the state, temperature and precipitation are most affected by latitude, weather patterns, elevation, and local and regional topography. In general, more southerly and lower areas are warmer, and more northerly and higher areas are cooler. Southern areas receive more direct sunlight because they are closer to the equator. We have all experienced the air cooling as we ascend a mountain This phenomenon is mainly a product of air expanding and cooling with increasing elevation. Conversely, air compresses and warms with decreasing elevation.

Elevation - The rate of temperature change with elevation is called the adiabatic lapse rate. Dry air changes at the dry adiabatic lapse rate, about 5.4°F per 1000 feet of elevation (3°C/1000 ft.). Moist air changes, on average, at about half that rate, or 2.7°F per 1000 feet of elevation (1.5°C/1000 ft.), the moist adiabatic lapse rate. This difference in the rate of change for dry versus cool air explains why the phenomenon of rain shadows can be so profound. Rain shadows occur when air moves across a mountain or range of mountains. As the air ascends it cools; at some point the cooling will cause precipitation to form on the windward side of the mountain. The leeward side will be drier because of this loss of moisture. However, because of the difference in adiabatic lapse rates, the leeward side of the mountain will also be warmer.

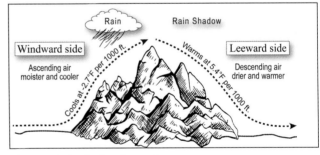

For example (see Fig. 2), say a moist body of air at 80°F at 4000 feet moves from west to east across a mountain range that has an average height of 9000 feet and then descends the other side. At the top of the mountain the air temperature, having decreased at the *moist* lapse rate, will be 80°F – (2.7°F × 5) = 66.5°F. On the leeward side of the mountain the temperature, increasing at the *dry* lapse rate, will be 66.5°F + (5.4°F × 5) = 93.5°F. The air on the leeward side is *both* drier and warmer. Windward sides of mountains are both cooler and moister.

Figure 2. Rain shadow formation.

Slope Position - Locally, the position on a slope has a strong effect on temperature. The air is not directly heated by the sun. Rather, solar radiation heats the ground, which then heats the air by conduction, convection, and the reradiation of sunlight by longwave radiation. Because warm air rises, during the daytime air tends to move upslope, leaving valley bottoms warmer than the tops of local hills or mountains (discounting the effect of the more regional adiabatic lapse rate). In the evening, airflow shifts, again locally, from higher to lower elevations because cool air is denser and settles in valleys. That is why it seems as if every time you sit around a campfire the smoke seems to be constantly changing direction. We tend to sit around campfires for breakfast and supper, just when shifts in wind direction are occurring as a result of daily heating or cooling. Interestingly, midslope positions, on average, have the warmest overall temperatures. Firefighters know that nighttime trouble spots are usually at midslope. Midslopes, then, often have different vegetation than local upper or lower slopes.

Local slope position also affects the amount of moisture available from precipitation, particularly from the heavy summer storms that are common in the state. At the tops of hills, ridges, and mountains, precipitation that does not soak into the ground has nowhere to go but downhill. Consequently, other things being equal, higher local topographic positions, even very small hilltops or mounds, are drier than midslopes or bottom slopes. Bottom slopes have the most moisture and are sometimes even adjacent to running or standing water.

Aspect - Aspect (also called exposure) refers to the direction a slope faces: north, northeast, south, etc. In the Northern Hemisphere south-facing slopes (SE through SW) receive significantly more solar radiation than do north-facing slopes (NE through NW). East-facing (ENE, E, ESE) and west-facing (WSW, W, WNW) slopes receive intermediate amounts of solar input. South-facing slopes, then, are warmer and consequently drier because of increased evaporation. West-facing slopes tend to be drier than east-facing slopes even though they receive the same amount of solar radiation. This is because west-facing slopes receive their highest level of solar radiation during the warmer parts of the day, while east-facing slopes receive their strongest illumination during the cooler parts of the day. So, for example, WSW slopes are warmer and drier than ESE slopes.

As a result of the effects of aspect, it has long been recognized that the zonation of major vegetation types varies noticeably from north-facing to south-facing slopes. Bands of vegetation are situated at lower elevations on north-facing slopes than they are on south-facing slopes. The following illustration (Fig. 3) shows the effects of latitude, altitude, and aspect on vegetation.

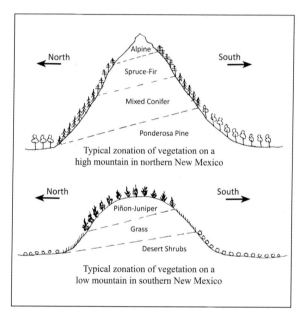

Figure 3. Effects of latitude, altitude, and aspect on New Mexico vegetation zonation. Modified with permission from Dick-Peddie (1993).

Soil Texture - The texture of a local soil, which is the proportion of sand (large particles), silt (medium-sized particles), and clay (very small particles) present, is primarily a result of geology and the time the soil has had to form, with geology being the larger determinant in younger soils. Coarse-textured soils with a lot of sand, gravel, or pebbles is said to be a light soil. Light soils allow for good aeration but rapid loss of water and nutrients into the ground. Conversely, heavy soils, those high in clay, allow little air penetration, have little ability for water to flow through the soil, and tenaciously bind water and nutrients such that they are not available to plants. Clay soils, even if water is present, are essentially deserts, with desert-adapted plants often the only inhabitants. Soils that are intermediate in texture, the loams, have balanced mixtures of sand, silt, and clay and have good aeration and optimal water and mineral retention.

Plants Themselves - Plants on a site affect their own environment. As time goes by, plants change the soil by adding organic matter and by aiding in the weathering of the parent geologic material. These two factors alone mean that soils tend to become moister over time. Further, as taller plants replace shorter plants, a new factor is added: shade. Shade means that the understory is cooler and therefore moister. Trees and tall shrubs also block drying winds. As a result of these changes, plants tend to change their environment in such a way that a different set of species is then able to inhabit a site. This process of one set of plants giving way to a different set of plants is called *succession*. A typical example is seen when a spruce-fir (*Picea-Abies*) forest is burned, blown down by high winds, wiped out by a landslide, or killed by bark beetles (Schimpf et al., 1980) (Fig. 4A). Aspens (*Populus tremuloides*), whose seedlings and saplings cannot tolerate shade, eventually invade the now open and sunny site. Once these are mature, they are replaced in turn by a spruce-fir forest whose seedlings can tolerate shade. Spruce-fir forests can persist for long periods because they can reproduce under their own shade. Aspen forests typically have an understory rich in species; spruce-fir forests, because of their dense shade and deep layers of accumulated organic material from dropped needles, are low in species richness. In this example, then, there is a cycle not only of aspen and spruce-fir, but also of species abundance. Another example of succession following disturbance is one that ultimately leads to piñon-juniper woodland (Fig. 4B).

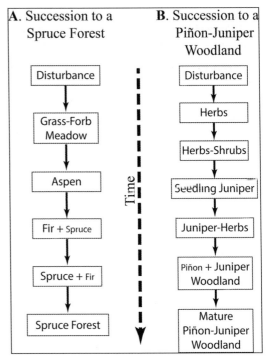

Figure 4. Examples of vegetation succession in New Mexico. Modified with permission from Dick-Peddie (1993).

Some plants, rather than making their local environments more hospitable to other species, exclude other species by depositing "anti-other," or allelopathic, chemicals on and in the soil. Examples include the invasive alien weeds Russian knapweed (*Rhaponticum repens*) and spotted knapweed (*Centaurea stoebe*) (Alford et al., 2009), and the abundant and native Utah juniper (*Juniperus osteosperma*) and big sagebrush (*Artemisia tridentata*) (Salls & Bannister, 2014).

Regional Climate – The following is taken from the New Mexico Climate Center website at New Mexico State University (https://weather.nmsu.edu/climate/about/, retrieved 14 July 2020) with slight modification.

Average annual precipitation ranges from less than 10 inches over much of the southern desert and the Rio Grande and San Juan Valleys to more than 20 inches at higher elevations in the State. [Most precipitation falls] during the warmest 6 months of the year, May through October. Much of the winter precipitation falls as snow in the mountain areas, but it may occur as either rain or snow in the valleys. Average annual snowfall ranges from about 3 inches at the Southern Desert and Southeastern Plains stations to well over 100 inches at Northern Mountain stations. It may exceed 300 inches in the highest mountains of the north.

Summer rains fall almost entirely during brief, but frequently intense thunderstorms. The general southeasterly circulation from the Gulf of Mexico brings moisture for these storms into the State, and strong surface heating combined with orographic lifting as the air moves over higher terrain causes air currents and condensation [see **Elevation** above]. July and August are the rainiest months in most of the State, with from 30 to 40 percent of the year's total moisture falling at that time. The San Juan Valley area is least affected by this summer circulation, receiving about 25 percent of its annual rainfall during July and August.

Winter precipitation is caused mainly by frontal activity associated with the general movement of Pacific Ocean storms across the country from west to east. As these storms move inland, much of the moisture is precipitated over the coastal and inland mountain ranges of California, Nevada, Arizona, and Utah [the rain shadow effect]. Much of the remaining moisture falls on the western slope of the

Continental Divide and over northern and high central mountain ranges. Winter is the driest season in New Mexico except for the portion west of the Continental Divide. This dryness is most noticeable in the Central Valley and on eastern slopes of the mountains.

Nutrients and Toxic Compounds – The availability of nutrients and the presence of inhibitory heavy metals or other chemicals is the result of an area's geology, with the age of the soil having only a slight effect in New Mexico. Some soils, such as those derived from some kinds of limestone, are low in available nutrients, making it difficult for plants to survive. Other soils, like those derived from lava, can form rich soils high in nutrients. Some soils are high in heavy metals, such as magnesium and nickel, that interfere with the growth of all but a few specially adapted species. Classic cases of heavy metal toxicity come from serpentine and dolomitic soils, both of which are high in magnesium. Gypsum-derived soils, composed mainly of calcium sulfate dihydrate, support relatively few species but include a number of rare "gyp" taxa endemic to New Mexico, such as *Anulocaulis leiosolenus* var. *howardii*, *Linum allredii*, *Physaria newberryi* var. *yesicola*, and *Nerisyrenia hypercorax*. The unique chemical constitution of unusual or rare geologic formations underlies the habitat of most of the state's rare plant species.

Vegetation History – That which came before in large part determines what is present now. Prior to the late Cretaceous, New Mexico was covered by a continental sea. Once that sea retreated, vegetation began to occupy the land surface, with plant colonists coming from three main sources: (1) the Arcto-Tertiary Geoflora of the northerly parts of the world; (2) the Neotropical-Tertiary Geoflora of more southern latitudes; and (3) the Madro-Tertiary Geoflora of drier regions of Mexico and adjacent areas of what is now the United States. This latter flora was in large part derived from an admixture of the Arcto-Tertiary and Neotropical-Tertiary Geofloras.

The much more recent glacial-interglacial events of the Pleistocene and Holocene have also had a major effect on vegetation. For example, vegetation that is now found at high elevations was found at lower elevations during warmer periods, and vice versa. These climatic pulses can lead to species becoming isolated in patches of suitable habitat where they can accumulate differences that are sometimes recognized as separate species, subspecies, or varieties.

Vegetation Classification

Many ways have been devised to classify vegetation into recognizable, presumably natural, units (summarized in Küchler, 1980; see also Bailey, 1983; Omernik, 1987; USNVC, 2021). The methodology of Omernik (1987, 1995) and his collaborators (Omernik et al., 2000) is widely accepted. His ecoregion concept is "based on the premise that ecological regions can be identified through the analysis of the patterns and the composition of biotic and abiotic phenomena that affect or reflect differences in ecosystem quality and integrity" (Omernik, 2003). Our map is modified from that of Griffith et al. (2006; public domain). This large-scale view of the intersection of major vegetation types and regional physiography helps put our plants into ecological context. See the New Mexico Ecoregion III map (Fig. 5).

A finer-scale classification of New Mexico's vegetation is that of Dick-Peddie (1993). His careful assessment of the state's vegetation is based on what he calls "major types," which are then divided into series. Unlike the ecosystem approach, Dick-Peddie's approach is focused almost solely on the dominant species and their physiognomy (herb, shrub, tree). Major vegetation types are delineated on the New Mexico vegetation map (**inside back cover**, modified from Dick-Peddie [1993] and used with permission). Brief comments about each of the major vegetation types follow. It should be remembered that each of these types can include several to many more finely described subordinate types. Some types, such as riparian vegetation, cannot be shown on a map of this scale but are ecologically important and very diverse.

Alpine Tundra – Alpine tundra is very limited in New Mexico and is confined to elevations above 11,500 feet. The growing season is brief, and strong winds, extreme cold, and high solar insolation are present year-round. Tundra is typically found adjacent to spruce, fir, or spruce-fir forests, and although

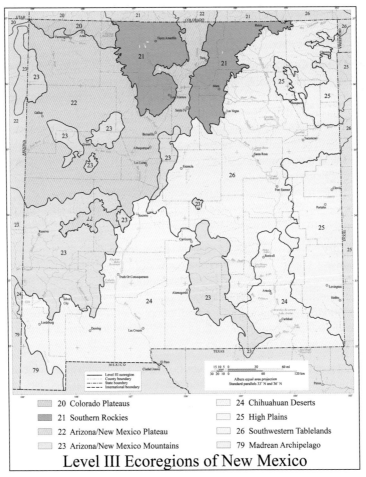

Figure 5. Ecoregion map.

the transition from forest to tundra is often abrupt, it can consist of a narrow ecotone of stunted and twisted trees in a picturesque zone called krummholz. Especially on steep or south-facing slopes, Subalpine-Montane Grasslands can grade into Alpine Tundra, for example on Sierra Blanca. The short vegetation is often dominated by a turf of *Carex* spp. and *Kobresia myosuroides*. Cushion plants like alpine forget-me-not (*Eritrichium nanum*), alpine clovers (*Trifolium* spp.), moss-pink (*Silene acaulis*), and alpine sandwort (*Cherleria obtusiloba*) are found where the soil is rocky or gravelly (Fig. 6).

Subalpine Coniferous Forest – These forests occur from about 9500 ft. up to the Alpine Tundra. Like the Alpine Tundra, this vegetation type experiences a short growing season because of cold winter temperatures. Unlike in tundra, heavy snow accumulations are common as the trees block the wind near ground level. These forests are often called "spruce-fir forests" because they are dominated by Engelmann spruce (*Picea engelmannii*) and corkbark fir (*Abies arizonica*), either alone or in combination. Subdivisions, called series, of this vegetation type are named for the particular combination of spruce and fir as well as of bristlecone pine (*Pinus aristata*), limber pine (*P. flexilis*), and, at lower elevations, Douglas-fir (*Pseudotsuga menziesii*). Subalpine Coniferous Forests often have dark understories because their dense tree canopies block sunlight and skylight. This lack of light and a heavy accumulation of spent needles preclude the forest floor from being diverse. Species of low-growing *Vaccinium* are common. In more open forests, species richness is much greater, with several shrubs and many species of grasses, sedges, and forbs. Small meadows are often encountered, as are small riparian areas. In these places, diversity can be quite high. Near the Colorado border, subalpine fir (*Abies bifolia*) is to be expected (Fig. 7).

Figure 6. Alpine Tundra. Culebra Range near the Colorado border. (O'Kane)

Figure 7. Subalpine Coniferous Forest. East of Chama. (O'Kane)

Montane Coniferous Forest - In montane forests, found below subalpine forests at about 8000–10,000 ft., Douglas-fir (*Pseudotsuga menziesii* var. *glauca*) becomes a dominant tree, typically in combination with various mixtures of white fir (*Abies concolor*), blue spruce (*Picea pungens*), aspen (*Populus tremuloides*), southwestern white pine (*Pinus strobiformis*), and limber pine (*P. aristata*). At lower elevations Gambel oak (*Quercus gambelii*) is an important component, as is ponderosa pine (*Pinus ponderosa*), especially in areas that were burned in the recent past. As such, they are often referred to as Mixed Conifer Forests. These midelevation forests are highly productive because of their relatively long growing season and the input of both winter and summer precipitation, which allow the formation of more fertile soils. The species diversity of Montane Coniferous Forests far exceeds that of Subalpine Coniferous Forests (Fig. 8).

Dick-Peddie (1993) broadly divides Montane Coniferous Forests into Upper Montane Coniferous Forests generally above 8000 ft. and Lower Montane Coniferous Forests generally below 8500 ft., down to where forests gradually give way to the scattered trees of woodlands. Lower Montane Coniferous Forests are typically dominated by ponderosa pine or Chihuahua pine (*Pinus leiophylla*), often in combination with piñon pine and various species of oak, such as Gambel oak (*Quercus gambelii*), silverleaf oak (*Q. hypoleucoides*), and gray oak (*Q. grisea*). Especially in the lower montane, grasses such as blue grama (*Bouteloua gracilis*), Arizona fescue (*Festuca arizonica*), mountain muhly (*Muhlenbergia montana*), mutton grass (*Poa fendleriana*), and pine dropseed (*Muhlenbergia tricholepis*) can be dominant in the understory.

Subalpine-Montane Grassland - Grasslands occurring within a matrix of Subalpine or Montane Coniferous Forests are often referred to as meadows or parks, or sometimes by the Spanish name *valles*. These typically occur from about 8900 to 11,500 ft., where they may grade into Alpine Tundra. Most meadows are found in valley bottoms on smooth terrain with little slope. Exceptions are those on steeper, usually south-facing slopes. In valley bottoms, research shows that trees are excluded by cold air drainage coming from above and by finer soil textures. This phenomenon has been called a reversed

Figure 8. Montane Coniferous Forest. Note Pinus ponderosa *in the lower part of view. (O'Kane)*

tree line because trees are locally excluded by cold temperatures at both higher and lower temperatures (Coop & Givnish, 2007). On steeper slopes trees may be excluded because of drier soil, especially when south facing. Commonly encountered grasses are Thurber fescue (*Festuca thurberi*), Arizona fescue (*F. arizonica*), Parry oatgrass (*Danthonia parryi*), mountain muhly (*Muhlenbergia montana*), pine drop-seed (*M. tricholepis*), tufted hairgrass (*Deschampsia caespitosa*), bluegrass species such as *Poa interior*, *P. fendleriana*, and *P. longiligula*, and sedge species such as *Carex pellita* and *C. aquatilis*. Grass species are often spread along elevational gradients. For example, in the Sierra Blanca region sheep fescue (*Festuca ovina*) occurred at the highest elevation, with Thurber fescue next lowest, and Arizona fescue the lowest (Moir, 1967). Mountain meadows can be quite diverse and home to spectacular displays of wildflowers. Overgrazing, however, can radically alter these communities. In some places dandelion (*Taraxacum officinale*) and Kentucky bluegrass (*Poa pratensis*) can cause dense infestations (Fig. 9).

Coniferous and Mixed Woodland - In woodlands, trees are relatively widely spaced, and their canopies do not touch. Individual trees are also shorter than are typically found in forests. Woodlands are usually found lower in elevation than forests and are drier and usually warmer. In New Mexico, the trees of our woodlands are piñon pines (*Pinus* spp.) or junipers (*Juniperus* spp.). Most often only a single species of pine is present, typically *Pinus edulis*, along with one or more species of juniper. One-seed juniper (*Juniperus monosperma*) is by far the most common and widespread species, but Utah juniper (*J. osteosperma*) and Rocky Mountain juniper (*J. scopulorum*) can dominate in northwestern New Mexico. Alligator juniper (*J. deppeana*) can dominate in the south-central and southwest, and redberry juniper (*J. pinchotii*) or Mexican juniper (*J. ashei*) in the southeast. Areas with a significant admixture of non-coniferous trees are considered Mixed Woodlands. Typical broadleaf trees include Gambel oak, Texas madrone (*Arbutus xalapensis*), pointleaf manzanita (*Arctostaphylos pungens*), Wright's siltassel (*Garrya wrightii*), Arizona white oak (*Quercus arizonica*), gray oak (*Q. grisea*), and Emory oak (*Q. emoryi*) (Fig. 10).

Juniper Savanna – Juniper Savanna forms an ecotone between the Coniferous and Mixed Woodlands and the Grasslands. Savanna vegetation consists mainly of grasses with widely scattered trees,

Figure 9. Subalpine-Montane Grassland. Carson National Forest, Lagunitas area. (O'Kane)

Figure 10. Coniferous and Mixed Woodland. Here is a mixture of Pinus edulis *and* Juniperus monosperma. *Elsewhere, the species of pine and juniper are different. (O'Kane)*

typically fewer than 130 per acre (320 per hectare). The trees of nearly all of the Juniper Savanna in New Mexico are one-seed juniper (*Juniperus monosperma*). In southern New Mexico, redberry juniper (*J. pinchotii*) can be the dominant species. Oak is rarely the tree of our savannas, but Arizona white oak (*J. arizonica*) forms small areas of savanna in the Animas Mountains. Large shrubs such as pointleaf manzanita (*Arctostaphylos pungens*), Bigelow sagebrush (*Artemisia bigelovii*), shadscale (*Atriplex confertifolia*), and wavyleaf oak (*Quercus undulata*) can sometimes form a conspicuous part of the vegetation (Fig. 11).

   Plains-Mesa Grassland - At lower, and therefore drier, elevations, Juniper Savanna gives way to the grasslands of the Great Plains and similar grasslands of mesa tops. These grasslands are the dominant vegetation of much of the eastern half of the state as well as large areas in the west-central portion. In yet lower or drier areas Plains-Mesa Grassland grades into either Desert Grassland or Chihuahuan Desert Scrub. The extensive grasslands of North America can be broadly divided into the more mesic (30-40 inches of precipitation) Tallgrass Prairies of the eastern Great Plains, the central (20-30 inches of precipitation) Mixed Grass Prairies (also called Midgrass Prairies), and the drier (< 15-25 inches of precipitation) Shortgrass Prairies (sometimes referred to as Steppes) of the western Great Plains. Our Plains-Mesa Grasslands are of the latter kind (Fig. 12).

   Both Plains Grasslands and Mesa Grasslands, when undisturbed and in their climax condition, are made up almost entirely of grasses, with blue grama (*Bouteloua gracilis*) dominant or a major component of the grass community. In most of the Plains Grasslands, buffalograss (*B. dactyloides*) is codominant. Other common grasses include Indian ricegrass (*Eriocoma hymenoides*), New Mexico feathergrass (*Hesperostipa neomexicana*), needle-and-thread grass (*Hesperostipa comata*), sideoats grama (*B. curtipendula*), and three-awn grasses (*Aristida fendleriana* and *A. longiseta*). In swales of southern New Mexico, tobosa (*Hilaria mutica*) forms local grasslands.

Figure 11. Juniper Savanna. Manzano Mountains. (Heil)

Figure 12. Plains-Mesa Grassland. Plains of St. Agustin at the Very Large Array. (Heil)

Desert Grassland – These dry grasslands typically occur between the Plains-Mesa Grasslands at higher, somewhat more mesic elevations and the Chihuahuan Desert Scrub or Great Basin Desert Scrub at lower, drier elevations. Many areas of Desert Grassland were probably once Plains Grassland before being disturbed, mainly by grazing. Common grasses include black grama (*Bouteloua eriopoda*), blue grama (*B. gracilis*), alkali sacaton (*Sporobolus airoides*), Indian ricegrass (*Eriocoma hymenoides*), and tobosa (*Hilaria mutica*). The diversity of nongrass species in these grasslands is considerably higher than in more mesic grasslands, and forbs often make up a significant portion of the plant community. Shrubs like creosote bush (*Larrea tridentata*), broom snakeweed (*Gutierrezia sarothrae*), species of *Ephedra*, *Agave*, mariola (*Parthenium incanum*), and mesquite (*Prosopis*) can cause different areas of the Desert Grassland to be ecologically and visually quite different (Fig. 13).

Chihuahuan Desert Scrub – This shrub-dominated vegetation type extends into New Mexico from its primary extent in Mexico. Creosote bush (*Larrea tridentata*) and tarbush (*Flourensia cernua*) are the major shrubs in this area that otherwise closely resembles Desert Grassland. In New Mexico most stands of Chihuahuan Desert Scrub have creosote bush as the sole dominant. Areas that have not been disturbed in some way for 100 or more years have a smaller component of tarbush. In some smaller areas whitethorn (*Vachellia constricta* or *V. vernicosa*) may be codominant with creosote bush (Fig. 14).

Plains-Mesa Sand Scrub – Dry areas of deep sand in the old floodplains of the Pecos River, Rio Grande, the lower reaches of the Rio Chama, and the western edges of the Llano Estacado caprock host shrub communities adapted to dry, unstable conditions. Sand sagebrush (*Artemisia filifolia*) is typically the dominant shrub. Other common shrubs include four-wing saltbush (*Atriplex canescens*), Torrey jointfir (*Ephedra torreyana*), and skunkbush sumac (*Rhus trilobata*). In the east-central and southeastern part of the state, sand scrublands sometimes have shinnery oak (*Quercus havardii*) as a dominant. Grasses make up a major portion of plant communities. Common sand-adapted grasses include sand bluestem (*Andropogon gerardi* subsp. *hallii*), lovegrasses (*Eragrostis intermedia* and *E. secundiflora*), purple three-awn (*Aristida purpurea*), sand paspalum (*Paspalum stramineum*), bristlegrasses (*Setaria*

Figure 13. Desert Grassland. Near the border with Trans-Pecos Texas. (O'Kane)

Figure 14. Chihuahuan Desert Scrub. White Sands and the San Andres Mountains in the background. (O'Kane)

leucopila and *S. texana*), sandhill muhly (*Muhlenbergia pungens*), and dropseeds (*Sporobolis cryptandrus* and *S. airoides*) (Fig. 15).

Great Basin Desert Scrub – The Great Basin Desert Scrub, often referred to as simply the Great Basin Desert, is found in the northwestern quarter of the state. This desert, mainly because of its elevation, is considered a cold desert. Unlike in the other desert scrublands of the state, much of the precipitation, 50%–60%, comes as winter snow rather than summer thunderstorms. Big sagebrush (*Artemisia tridentata*) is the dominant shrub nearly everywhere the vegetation type is found. Rabbitbrushes (*Chrysothamnus* spp. and *Ericameria* spp.) are often codominant or nearly so. Locally, Mormon tea (*Ephedra* spp.), Bigelow's sagebrush (*Artemisia bigelovii*), four-wing saltbush (*Atriplex canescens*), shadscale (*A. confertifolia*), or, where especially saline, greasewood (*Sarcobatus vermiculatus*), may be dominant or codominant. The diversity of nonwoody plants is high because of varying elevations, moisture availability, slopes, and geologic substrates (Fig. 16).

Montane Scrub – These scrublands occur in areas that would normally host more mesic communities and are typically found embedded in them. Usually the surrounding vegetation matrix is Montane Coniferous Forest. The name Mixed Mountain Shrub, often applied to this vegetation type, is apt because wherever it is found it consists of some mixture of several shrub species. The presence of this community type is determined by local conditions that increase dryness such as southwest-facing slopes, shallow and rocky or gravelly soil, steep slopes, and escarpment rims. The list of shrubs that are found in myriad combinations is extensive. Just a few of the most common are mountain mahogany (*Cercocarpus montanus*, the most abundant and widespread), antelope bitterbrush (*Purshia tridentata*), snowberry (*Symphoricarpos* spp.), rabbitbrush (*Chrysothamnus* and *Ericameria* spp.), oceanspray (*Holodiscus dumosus*), skunkbush (*Rhus* spp.), California buckthorn (*Frangula californica*), oaks (*Quercus* spp.), chokecherry (*Prunus virginiana*), gooseberry and currant (*Ribes* spp.), serviceberry (*Amelanchier utahensis*), buckbrush (*Ceanothus* spp.), and Stansbury cliffrose (*Purshia stansburyana*) (Fig. 17).

Figure 15. Plains-Mesa Sand Scrub. Between the Rio Grande and the Manzano Mountains. (O'Kane)

Figure 16. Great Basin Desert Scrub. Almost a monoculture of Artemisia tridentata *because of over-grazing. (O'Kane)*

Figure 17. Montane Scrub. This example is dominated by Quercus gambelii *and* Amelanchier utahensis. *Various other mixtures of shrubs are possible, mainly on a latitudinal gradient. (O'Kane)*

Closed Basin Communities – These communities are found in areas that either do not drain by surface flow to any river system or have topography so flat that surface water spreads out rather than forming gullies and arroyos. Precipitation mostly evaporates in place, causing soils to become saline, often strongly so, and often gypsiferous. Because of their waterlogged status at least part of the time, these areas are frequently classified as riparian areas. Closed Basin Communities can be classified as Closed Basin Scrub, Playas, or Alkali Sinks depending on vegetational composition, drainage regime, or the concentration of various salts, mostly sodium chloride (Fig. 18).

Closed Basin Scrub communities are dominated by shrubs, with four-wing saltbush (*Atriplex canescens*) being the typical dominant. Pale wolfberry (*Lycium pallidum*) is sometimes codominant, as is shadscale (*Atriplex confertifolia*) or greasewood (*Sarcobatus vermiculatus*). Forbs may be abundant, with most of these from the salt-loving plant family Amaranthaceae. When gypsum is abundant, the grasses gyp grama (*Bouteloua breviseta*) and gyp dropseed (*Sporobolus nealleyi*) can be common.

Playas are areas of especially flat topography in which water stands for long enough periods to disallow the establishment of shrubs, at least in their centers. These areas generally occur in the lower portions of arid basins and usually have soils with a clay or caliche hardpan. They have internal drainage, periodically flood, and accumulate sediment (Neal, 1975). Vegetation is scant but may be more abundant in areas that might be called salt marshes.

Alkali Sinks are areas that, because of their closed hydrology, topographic position, and surrounding geology, have extremely high concentrations of alkali salts and a basic pH. Water is lost from the system entirely by evaporation because percolation is precluded by the presence of a hardpan soil horizon of accumulated clay or caliche. Shrubs able to tolerate these conditions include iodinebush (*Allenrolfea occidentalis*), Trans-Pecos false clapdaisy (*Pseudoclappia arenaria*), and species of *Suaeda* (*S. calceoliformis* and *S. nigra*). Saltgrass (*Distichlis spicata*) is common and often abundant.

Figure 18. Closed Basin Communities. West of Lordsburg, here with the rare Atriplex griffithsii. *This is but one of several possible communities of this type. (O'Kane)*

USING A DICHOTOMOUS KEY

Dichotomous keys present a set of choices for the plant you want to identify. For each pair of choices, choose the couplet that best fits the plant you have in hand. Continue working through the key, moving through numbered pairs of questions until you arrive at the plant you are attempting to identify. For example, say you want to identify a species of *Corispermum* using the following key. You measure the fruits and find that they are 4.5 mm long and have clearly visible wings that are 0.5 mm wide and that they are arranged in a dense inflorescence. You should arrive at *C. welshii* as your identification. Now go on to read the description of that species, checking to be sure you are confident in your identification.

1.   Fruits 1.8–3.2 mm, wingless or with a scarcely visible wing…**C. villosum**
1'.   Fruits 2.5–4.6 mm, with a visible wing 0.2–0.6 mm wide…(2)
2.   Wings of fruits 0.4–0.6 mm wide; inflorescence usually compact and dense…**C. welshii**
2'.   Wings of fruits 0.2–0.4 mm wide; inflorescence usually lax and interrupted…**C. americanum**

NOMENCLATURE

Each plant is given a two-element name, or binomial. The first element of a scientific name is the genus (plural genera). The second element is the specific epithet. The genus and specific epithet together form the species name. The binomial is followed by the abbreviation of the name of the person or persons who first applied that name to the plant. For example, in *Convolvulus arvensis* L., the "L." stands for Linnaeus. The genus name is always capitalized, and the specific epithet is lowercase. Below the species level, two widely used taxonomic categories are subspecies and variety. Each has its own authority, as in *Opuntia polyacantha* Haw. var. *hystricina* (Engelm. & J. M. Bigelow) B. D. Parfitt. George Engelmann and J. M. Bigelow first described the variety *hystricina* as a full species. Bruce Parfitt subsequently reduced it to a variety of *Opuntia polyacantha*.

RARITY, ENDEMISM, AND ENDANGERMENT

Whether a species, subspecies, or variety is rare, endemic, or endangered is indicated by one or more of the following codes:

R – Rare according to the New Mexico Rare Plant Technical Council (NMRPTC)
E – Endemic to New Mexico
FT – Federally Threatened
FE – Federally Endangered

Taxa considered rare (**R**) are those listed as such by the NMRPTC, which is made up of professional botanists and land managers familiar with the plants of New Mexico and their current and potential threats. Rare taxa, based on NMRPTC criteria, are narrowly endemic to a specific geographic (e.g., mountain range, drainage basin) or geologic feature (geologic substrate), or a small area of a broader phytogeographic region, such as the southern Rocky Mountains or the northern Chihuahuan Desert. Taxa can be locally abundant within their narrow range, more widespread but numerically rare and never locally common (e.g., *Peniocereus greggii*), or numerically abundant only in a few small, widely scattered habitats (e.g., *Puccinellia parishii*, *Helianthus paradoxus*) (New Mexico Rare Plant Technical Council, New Mexico Rare Plants Home Page, https://nmrareplants.unm.edu/about, accessed 27 January 2022).

Biologists consider rarity itself to be a form of endangerment (Harnik, 2012). Rare taxa are vulnerable to loss of genetic diversity, devastating local diseases, accidental or purposeful habitat destruction, loss of pollinators or dispersal agents, human collection (see Stuart et al., 2006) or intentional destruction, long-term climate change, short-term local climate anomalies like droughts and floods, invasive species, and other phenomena that would not threaten more common and widespread taxa.

Endemic taxa (**E**) are those that occur only in New Mexico, a list of which can be found in Allred et al. (2020: 47–48). A few taxa that are very nearly endemic are also listed as **E** (or **R**) with the appropriate caveat.

Taxa protected by the Endangered Species Act, as administered by the U.S. Fish and Wildlife Service, are listed as either Federally Threatened (**FT**) or, more critically, as Federally Endangered (**FE**). These taxa are protected by force of federal law.

CONVENTIONS USED IN SPECIES DESCRIPTIONS

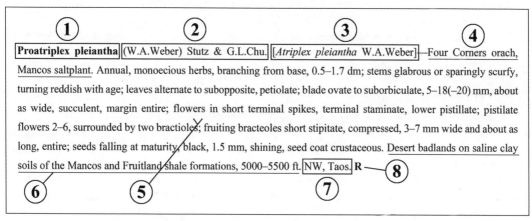

Figure 19. An example of a species entry. Explanations for numbered parts are given below.

1. **Currently accepted scientific name.**
2. **Author or authors of the current name**. In this example, W. A. Weber first described the spe-

cies, and Stutz and Chu transferred it from the genus *Atriplex* to the genus *Proatriplex*. Abbreviations for author names are those given by the International Plant Names Index (https://www.ipni.org).

3. **Synonym or synonyms of the scientific name**. The synonyms provided in this book are only those that users are likely to encounter using internet sources or recent floras published in about the last 75 years. To reduce the size of this book, the provided synonymy is far from complete. Users wishing a more exhaustive list of synonyms should consult Allred's (2020) praiseworthy Annotated Checklist.

4. **Common, or vernacular, name**. Most "common" names have been created by various government agencies to provide a name that they think people will more readily remember or be able to pronounce. The common names chosen by us and our contributing authors are typically those found on websites such as the U.S.D.A. PLANTS Database (https://plants.sc.egov.usda.gov/home).

5. **Botanical description**. We and our contributing authors have attempted to provide descriptions that are minimal and focused on verifying identifications made using the keys. Authors' descriptions, not unexpectedly, vary in length and detail. A **Glossary** with most of the descriptive words used is provided at the back of the book. Descriptions typically start with the gross attributes of a species, such as annual-perennial, herb-shrub-tree, etc.; they then move on to stems and leaves, and then to inflorescences/infructescences. Next flowers are described from the outside toward the center: sepals, petals, stamens, and pistils. Finally, fruits and sometimes seeds are described. Unobservable details, like chromosome number, are not given. All measurements are in metric units. Unless otherwise specified, a measurement refers to length or height. Where two measurements are given, the first is length and the second is width, as in 12 × 5 mm or 5 cm × 6 mm.

6. **Brief habitat description**. Descriptions typically indicate associated vegetation, but especially when species grow in nearly barren areas, geology and soil type are mentioned. Elevation range is given in feet, rather than meters, because users in the United States are more likely to be familiar with the ecological conditions at elevations in those units. Feet are easily converted to meters by multiplying by 0.3048 or by dividing by 3.2808.

7. **Distribution in New Mexico**. We have divided the state into nine broad regions (**inside front cover**). For example, **NW** refers to the northwestern division of the state, **SC** is south-central, **NCS** refers to the central north–south portion of the state, etc. In addition to these abbreviations, a county or counties are also or only mentioned, especially when species are found in fewer than three counties. If a plant is found more or less throughout the state, it is listed as **widespread**. Counties disjunct from the main regions are listed; for example, the distribution of *Carex nebrascensis* is given as NC, NW, WC, Sierra.

8. **Rare, endemic, and/or endangered**. These categories are described above. **A taxon with one of these codes (R, E, FT, FE) is illustrated by a photograph**.

In larger treatments with a main author, if select genera were authored by or coauthored with another, those names are listed under the genus only.

As a space-saving measure, if a family has only one species in New Mexico, only the species is described, but with a somewhat expanded description. Similarly, if only one genus in a family is present, both the genus and the included species are described, but not the family. If a family has two or more genera present, the family is fully described. Occasionally, taxa that do not occur in the flora area are included in the keys and/or given a brief treatment. Their names are in roman instead of boldface, and those taxa that are expected to occur in New Mexico are listed as "[to be expected]" in the keys.

KEY TO THE
VASCULAR PLANT FAMILIES
OF NEW MEXICO

With the kind permission of the late William A. Weber, who passed away in 2020, most of this key has been adapted from his excellent family key appearing in *Colorado Flora: Western Slope* (Weber & Wittmann, 2001). Bill's personality, wealth of stories, and that twinkle in his piercing, pale blues eyes will be sorely missed and never forgotten. The ferns and lycophytes key was written by Michael D. Windham, and the woody "dicot" key was adapted by Heil from Carter (2012).

KEY TO MAIN GROUPS
1.   Plants reproducing by spores, not producing seeds, flowers, or cones; the main plant body the haploid (*n*) gametophyte…**FERNS AND LYCOPHYTES**
1′. Plants reproducing by seeds; plants producing flowers or cones; the main plant body the sporophyte (2*n*) (SEED PLANTS)…(2)
2.   Flowers never present; seeds on the open face of a scale or bract, usually in a cone (fleshy and berrylike in *Juniperus*); leaves needlelike or scalelike; plants evergreen; trees and shrubs…**GYMNOSPERMS**
2′. Flowers present; seeds contained in a fruit; leaves seldom needlelike or scalelike; plants rarely evergreen; herbs, shrubs, and trees…**ANGIOSPERMS**

FERNS AND LYCOPHYTES
(Families start on page 45)
1.   Stems with 3–30 equally spaced ridges and grooves their entire length, with sharply defined, hollow internodes; leaves fused into cylindrical sheaths around nodes, with a fringe of unfused teeth distally (these often deciduous)…**Equisetaceae**
1′. Stems not ridged and grooved their entire length, with poorly defined, solid internodes; leaves not fused into cylindrical sheaths around nodes…(2)
2.   Plants aquatic to amphibious, occurring in or around standing water…(3)
2′. Plants terrestrial, often on rocks (occasionally growing along edges of fast-moving streams)…(5)
3.   Plants with linear, grasslike leaf blades, rooted in shallow water around margins of natural ponds and lakes at elevations > 8700 ft.; sporangia embedded in widened leaf bases near attachment to corm… **Isoëtaceae**
3′. Plants with oblong to orbicular leaf blades, free-floating or rooted in soil periodically inundated by water at elevations < 8700 ft.; sporangia contained within specialized structures (sporocarps) clearly distinct from leaves…(4)
4.   Plants rooted in soil periodically inundated by water; petioles usually much > 1 cm; blades 1-pinnate into 4 fan-shaped, pseudopeltate pinnae…**Marsileaceae**
4′. Plants free-floating or stranded on shore, any rootlike structures not embedded in soil; leaves sessile or the petioles < 1 cm, blades entire or bilobed…**Salviniaceae**
5.   Leaves < 1 cm, arranged in pseudowhorls on aboveground stems; sporangia attached to adaxial base of specialized leaves (sporophylls) clustered in terminal strobili…(6)
5′. Leaves > 1 cm, spirally arranged on underground stems; sporangia attached to abaxial leaf surfaces or clustered on green axes (modified leaf segments) resembling stems…(7)
6.   Leaf blades (the part not fused to the stem) 4–8 mm, the central vein either not apparent or marked by an obscure longitudinal ridge abaxially; megasporangia absent, all sporangia containing thousands of spores 30–50 µ diam.…**Lycopodiaceae**
6′. Leaf blades (the part not fused to the stem) < 4 mm, the central vein marked by a shallow groove abax-

ially; megasporangia present in some strobili, containing 4 (rarely 2 or 1) spores 250–500 μ diam.…
Selaginellaceae

7. Sporangia > 0.3 mm, each containing hundreds of spores; fertile leaves with very distinct fertile and sterile portions; leaves arising singly (occasionally in pairs) from short, erect stems…**Ophioglossaceae**

7'. Sporangia < 0.3 mm, each containing no more than 64 spores; fertile leaves lacking distinct fertile and sterile portions; leaves clustered or, if arising singly, the stems long-creeping and horizontal…(8)

8. Blade margins usually recurved, often forming false indusia partially or fully protecting young sporangia; sporangia distributed along leaf margins or following veins, usually not clustered into discrete sori; spores trilete…(9)

8'. Blade margins not recurved, young sporangia protected by membranous or filamentous indusium arising from soral receptacle or fully exposed; sporangia clustered into discrete (sometimes confluent at maturity) sori ± centered between segment margins and midrib; spores monolete…(10)

9. Blades > 20 cm wide, sparsely to moderately hairy abaxially; leaves arising singly from long-creeping stems; stems hairy to glabrescent, lacking scales; petioles with > 3 vascular bundles at base…
Dennstaedtiaceae

9'. Blades < 20 cm wide or, if wider, then glabrous with leaves clustered on short, suberect stems; stems with many (rarely sparse) scales; petioles with 1 vascular bundle at base (2 in *Astrolepis*)…**Pteridaceae**

10. Leaves regularly pinnatifid, the segments all connected by a strip of leaf tissue along rachises; sori lacking indusia…**Polypodiaceae**

10'. Leaves 1–3-pinnate (rarely alternately forked or simple); sori usually with indusia but these often inconspicuous at maturity and occasionally deciduous or absent…(11)

11. Leaves 1-pinnate (pinnae often toothed but not pinnatifid) or alternately forked (very rarely simple), blades ≤ 5 cm wide; sori and indusia oblong to linear; stem scales black throughout or strongly clathrate with thick, black radial walls and clear luminae…**Aspleniaceae**

11'. Leaves 1-pinnate-pinnatifid to 3-pinnate or, if simply pinnate, then blades > 5 cm wide and sori round; stem scales brown throughout or with darker central stripe, not entirely black or strongly clathrate…(12)

12. Stem and petiole base scales (at least some) > 5 mm; indusia (where persistent) round, cordate, or elongate, arising within sori or extending along vein on 1 side…(13)

12'. Stem and petiole base scales < 5 mm; indusia (where persistent) ovate-lanceolate, scalelike or filamentous, narrowly attached at proximal edge of sori or completely surrounding them…(14)

13. Rachises and lower pinna midribs with many lance-attenuate scales abaxially; blades 1–2-pinnate; petioles with > 3 vascular bundles at base (seen in cross-section)…**Dryopteridaceae**

13'. Rachises and lower pinna midribs with few or no lance-attenuate scales; blades usually 2-pinnate-pinnatifid; petioles with 2 vascular bundles at base…**Athyriaceae**

14. Veins usually obscure, terminating before reaching segment margins; indusia consisting of 4–16 filamentous or scalelike lobes emerging around entire circumference of sori, always present but often inconspicuous; stem scales (at least some) bicolored, with dark central stripe and brown margins…
Woodsiaceae

14'. Veins conspicuous, intersecting segment margins; indusia consisting of single hoodlike, ovate-lanceolate membrane emerging at proximal edge of sori, this often deciduous at maturity (absent in *Gymnocarpium*); stem scales all concolored, tan to brown (weakly clathrate in some *Cystopteris*)…**Cystopteridaceae**

GYMNOSPERMS
(Families start on page 71)

1. Shrubs with jointed green stems, the leaves represented by small black triangular scales in whorls at the joint; male cones yellow with stamenlike sporangia; female (ovulate) cones green, of a few loose thin scales…**Ephedraceae**

1'. Trees or shrubs with either needlelike or scalelike overlapping leaves, producing berrylike gray cones or cones with woody scales…(2)

2. Fruit a berrylike cone, the scales fused together and detectable only by their protruding tips; shrubs or small trees with decussately arranged scalelike leaves or flat sharp needles…**Cupressaceae**

2'. Fruit a woody cone with spirally arranged scales; small or large trees with needle leaves…**Pinaceae**

ANGIOSPERMS
(Families start on page 79)

1. Plants parasitic or saprophytic (carnivorous and aquatic in Lentibulariaceae), often brightly colored but not deep green (Orchidaceae or Viscaceae epiphytes may be yellowish green)…**KEY 1: PARASITES AND SAPROPHYTES**

1'. Plants autotrophic, not parasitic, producing chlorophyll and therefore with green leaves or stems…(2)

2. Mature leaves reduced to spines arising from areoles; stems succulent…**Cactaceae**

2'. Plants not as above, without areoles; stems various…(3)

3. Plants growing from water, submersed or emersed, with or without floating leaves, some floating and unrooted in the substrate; plants sometimes stranded on drying edges of retreating bodies of water… **KEY 2: AQUATIC PLANTS**

3'. Plants terrestrial or in moist or saturated soil, neither submerged nor with floating leaves…(4)

4. Vines (herbaceous) or lianas (woody), climbing or twining among other plants, often possessing suckers or tendrils…**KEY 3: VINES AND LIANAS**

4'. Plants not vines; herbaceous or woody…(5)

5. Flower parts principally 3-merous; leaves mostly parallel-veined; seeds with 1 cotyledon; herbaceous except some Agavaceae and Ruscaceae…**KEY 4: MONOCOTS**

5'. Flower parts 4- or 5-merous (rarely 2- or 3-merous); leaves mostly net-veined; seeds usually with 2 cotyledons; herbaceous or woody…(6)

6. Trees or shrubs…**KEY 5: WOODY "DICOTS"**

6'. Herbaceous, sometimes woody at the base…**KEY 6: HERBACEOUS "DICOTS"**

KEY 1: PARASITES AND SAPROPHYTES

1. Internal parasites on roots and stems, the vegetative body penetrating the host plant and only the flowers showing…**Apodanthaceae**

1'. Plants not as above…(2)

2. Plants epiphytic herbs, with short stems; ball-shaped clumps on oaks, not growing in the soil; bootheel region…**Bromeliaceae**

2'. Plants saprophytic or parasitic herbs; not ball-shaped clumps, aquatic or terrestrial; widespread…(3)

3. Plants insectivorous, aquatic; leaves finely dissected, bearing small bladderlike traps…**Lentibulariaceae**

3'. Plants parasitic or saprophytic; often lacking chlorophyll…(4)

4. Plants attached to the bark of trees, or by suckers to the aboveground stems of herbs…(5)

4'. Plants without obvious attachments to the aboveground parts of their host…(6)

5. Plants not threadlike, variously colored, including shades of green, attached to the branches of trees… **Viscaceae** (mistletoe)

5'. Plants threadlike and orange or yellow, attached by suckers to aerial parts of herbs…**Convolvulaceae** (*Cuscuta*)

6. Flowers actinomorphic…(7)

6'. Flowers zygomorphic…(8)

7. Stems green, glabrous, not thick and stout; petals absent; sepals petaloid, white to pink; stamens 5… **Comandraceae**

7'. Stems reddish to purplish brown, or yellowish to white, glabrous or glandular, often thick and stout; sepals and petals present; stamens 6–12…**Ericaceae** (*Monotropa, Pterospora*)

8. Flowers without a labellum; stamens 4, epipetalous but free of one another; petals united below; ovary superior…**Orobanchaceae**

8'. Flowers with the lowest petal enlarged, markedly different from the others (forming a labellum); anthers united with the gynoecium to form a column; petals distinct; ovary inferior…**Orchidaceae**

KEY 2: AQUATIC PLANTS

1. Plants free-floating, not rooted in the soil, small and the leaves or plants usually not > 1 cm or leaves and plants larger…(2)

1'. Plants not free-floating, rooted in the soil, the leaves generally larger…(3)

2. Plants disklike or thalluslike, without true stems and leaves; flowers small and inconspicuous, white, extremely rarely seen…**Araceae**

2'. Plants not disklike or thalluslike, with true stems and oblong or ovate to round leaves; flowers showy, yellow or purple…**Pontederiaceae**

3. Leaves trifoliately compound with sheathing bases; flowers white and densely hairy within, in a bracteate raceme…**Menyanthaceae**

3'. Leaves simple to dissected or pinnately compound; flowers unlike the above…(4)

4. Leaves lobed or finely dissected…(5)

4'. Leaves simple and undivided, the margins entire or toothed…(8)

5. Leaves bearing small bladderlike traps; flowers showy, yellow, spurred, on racemes projecting above the water level…**Lentibulariaceae**

5'. Leaves not bearing bladders; flowers not spurred…(6)

6. Leaves alternate, palmately divided; flowers with white or yellow petals…**Ranunculaceae**

6'. Leaves whorled; flowers greenish, inconspicuous…(7)

7. Leaves dichotomously dissected, the margins finely serrulate; perianth absent; ovary unicarpellate… **Ceratophyllaceae**

7'. Leaves pinnately dissected, the margins entire; perianth of 4 sepals; ovary 4-carpellate…**Haloragaceae**

8. Stems short or lacking, the leaves attached to the bottom, elongate-linear, the tips floating on the sur-

face; flowers clustered in dense balls, lower clusters carpellate and the upper staminate…**Typhaceae** (*Sparganium*)

8'.   Plants with definite stems and otherwise not as above…(9)

9.   Plants mostly 1 m+; shoots and flowering stems stiffly erect; cattails…**Typhaceae**

9'.   Plants < 1 m; shoots and flowering stems often lax or limp…(10)

10.   Leaves simple, entire or slightly toothed…(11)

10'.   Leaves distinctly lobed or compound…(24)

11.   Leaves arranged in whorls, linear or oblong…(12)

11'.   Leaves not whorled, variously shaped…(13)

12.   Leaves lax, translucent (only 2 cell layers thick); flowers, when present, sessile (carpellate) or long-pedicelled (staminate)…**Hydrocharitaceae**

12'.   Leaves rather rigid, unless submerged, opaque (> 2 cell layers thick); flowers sessile in leaf axils…**Plantaginaceae** (*Hippuris*)

13.   Leaves peltate; < 6 cm across; petiole attached near the middle of the blade…**Araliaceae** (*Hydrocotyle*)

13'.   Leaves not peltate; > 6 cm across; petiole not as above…(14)

14.   Leaves nearly orbicular, cordate, thick, and leathery, floating; flowers large, yellow, solitary…**Nymphaeaceae**

14'.   Leaves narrower, not cordate, floating or not; flowers not as above…(15)

15.   Leaves linear or filiform…(16)

15'.   Leaves with distinctly broadened blades…(20)

16.   Flowers in terminal or axillary spikes…**Potamogetonaceae**

16'.   Flowers in the leaf axils, sessile or on slender, often coiled peduncles…(17)

17.   Fruit minute, blackish, on an elongate, often coiled peduncle; leaves filiform, > 3 cm…**Ruppiaceae**

17'.   Flowers and fruits sessile in the leaf axils; leaves < 3 cm…(18)

18.   Fruit rounded or emarginate, oblong or wider, not beaked…**Plantaginaceae** (*Callitriche*)

18'.   Fruit narrowly cylindrical, tapered to a beak…(19)

19.   Fruit flattened, slightly curved with a stout beak; leaves filiform…**Potamogetonaceae**

19'.   Fruit terete, straight, the beak whitish, not rigid; leaves linear, flat, the margins finely toothed…**Hydrocharitaceae** (*Najas*)

20.   Leaves alternate…(21)

20'.   Leaves opposite…(22)

21.   Floating leaves pinnately veined; flowers with showy pink perianth parts…**Polygonaceae** (*Persicaria*)

21'.   Floating leaves with parallel veins, or floating leaves absent; flowers greenish, not showy…**Potamogetonaceae**

22.   Petals absent or 4–5, flowers actinomorphic; annuals or perennials…**Onagraceae** (*Ludwigia*)

22'.   Petals absent, minute, or showy in Alismataceae; perennials…(23)

23.   Calyx and corolla absent; ovary 4-locular; stipules lacking; floating and submerged leaves often different…**Plantaginaceae** (*Callitriche*)

23'.   Calyx and corolla often present; ovary 3- or 5-locular; stipules present; leaves not dimorphic…**Elatinaceae**

24.   Fruit a silicle or silique; stamens 6, leaves pinnately divided with ovate segments, or terete and linear and in a densely packed basal rosette…**Brassicaceae** (*Nasturtium*)

24'.   Fruit various, not a silicle or silique; stamens 4 or 6–9; leaves various, not pinnately divided…(25)

25.   Leaves hastate, sagittate, linear, linear-lanceolate, narrowly elliptic, or ovate; fruit an achene; carpels distinct and numerous; stamens 6–9; flowers 3-merous, not tightly grouped at the base of the plant…**Alismataceae**

25'.   Leaves elliptic, mostly 2 cm; fruit a capsule; carpels united; stamens 4; flowers 5-merous, small and solitary, on short peduncles tightly grouped at the base of the plant…**Scrophulariaceae** (*Limosella*)

KEY 3: VINES AND LIANAS

1.   Plants parasitic, twining on the aerial stems of various herbs, yellowish and lacking chlorophyll, the stems thin and threadlike; leaves minute and reduced to scales…**Convolvulaceae** (*Cuscuta*)

1'.   Plants green, not parasitic; leaves usually present and conspicuous…(2)

2.   Leaves opposite…(3)

2'.   Leaves alternate…(6)

3.   Leaves entire, sagittate to hastate; flowers greenish to brown, in umbels…**Apocynaceae** (*Sarcostemma*, *Matelea*, *Metastelma*)

3'.   Leaves palmately lobed or pinnately compound to dissected; flowers not as above…(4)

4.   Leaves palmately 3–5-lobed, with toothed margins; flowers imperfect, the pistillate flowers subtended by papery bracts and aggregated into a conelike structure…**Cannabaceae** (*Humulus*)

4'.   Leaves pinnately compound or dissected, the leaflets toothed or entire; flowers perfect or imperfect, not aggregated into a conelike structure…(5)

5.   Flowers zygomorphic; sepals orange-red and thick; petals orange to orange-red, united, tubular; sta-
      mens 5, these markedly unequal; fruit a loculicidal capsule…**Bignoniaceae** (*Campsis*)
5'.   Flowers actinomorphic; sepals purple, bluish, or white; petals absent; stamens numerous; fruit an aggre-
      gate of cypselas with long feathery styles…**Ranunculaceae** (*Clematis*)
6.   Leaves palmately, pinnately, or ternately compound…(7)
6'.   Leaves simple, sometimes lobed (leaves deeply 3-lobed in *Solanum*, not truly compound)…(10)
7.   Leaves palmately compound…(8)
7'.   Leaves pinnately or ternately compound…(9)
8.   Tendrils with adhesive cups; fruit a berry…**Vitaceae**
8'.   Tendrils without adhesive cups; fruit an ovoid pepo…**Cucurbitaceae**
9.   Flowers zygomorphic with a banner, wing, and keel, variously colored but not greenish; leaves stipulate,
      pinnately compound or ternately compound with the terminal leaflet modified into a tendril, the leaflets
      not shiny; stamens usually 10, usually diadelphous; fruit a legume or a loment…**Fabaceae**
9'.   Flowers actinomorphic, greenish; leaves exstipulate, ternately compound with shiny leaflets; stamens 5,
      alternate with the petals, equal in size; fruit a white or cream-colored drupe…**Anacardiaceae** (*Toxico-
      dendron*—poison ivy!)
10.   Tendrils present…(11)
10'. Tendrils absent…(12)
11.   Vines woody; ovary superior; fruit a berry…**Vitaceae**
11'. Vines herbaceous; ovary inferior; fruit a pepo…**Cucurbitaceae**
12.   Stems with a sheathing membranous stipule (ocrea) at the petiole base; perianth of tepals, all parts
      similar, petaloid, greenish or whitish…**Polygonaceae** (*Fallopia*, *Polygonum*)
12'. Stipules absent; perianth in 2 distinct series (sepals and petals), the petals purple, white, pink, or red…
      (13)
13.   Flowers purple; fruit a berry; plants climbing but not twining…**Solanaceae** (*Solanum dulcamara*)
13'. Flowers white, pink, or red to reddish orange; fruit a capsule; plants twining and climbing on other
      plants…**Convolvulaceae**

KEY 4: MONOCOTS
(Aquatics, parasites, and saprophytes are found in the appropriate keys above.)
1.   Plants epiphytic on oaks in the bootheel region of the state…**Bromeliaceae** (*Tillandsia*)
1'.   Plants not epiphytic…(2)
2.   Plants woody, at least at the base, low shrubs to small trees, flowering peduncles ± woody, sometimes
      massive…(3)
2'.   Plants herbaceous…(6)
3.   Leaves a little soft when pinched between thumb and forefinger, somewhat fleshy and succulent, thick
      in cross-section, usually > 6 mm thick, the leaf base expanded in width; ovary inferior…**Agavaceae**
      (*Agave*)
3'.   Leaves firm or rigid and fibrous, not succulent, thin in cross-section, usually < 6 mm thick, the leaf base
      expanded only slightly; ovary superior…(4)
4.   Leaf margins with prominent curved or hooked prickles or teeth, leaves rigid, the leaf base forming a
      shallow "spoon"; the stem short, mostly subterranean, the leaves appearing as a ball at ground level;
      plants dioecious; flowers unisexual; fruit a 1-celled winged capsule; seeds 3-angled; staminate flowers in
      dense flexible spikes…**Ruscaceae** (*Dasylirion*)
4'.   Leaf margins lacking prominent curved or hooked prickles, sometimes with fine teeth or fibrous; leaves
      not rigid, the leaf base flat; plants with perfect flowers…(5)
5.   Leaves rigid, daggerlike, and spine-tipped; flowers > 2 cm, in looser racemes or panicles, complete and
      bisexual; sepals and petals > 10 mm, each with several veins; fruit with many flat seeds…**Agavaceae**
      (*Yucca*)
5'.   Leaves firm but not rigid, flexible, grasslike, and not spine-tipped; flowers < 1 cm, numerous in dense ra-
      cemes or panicles, incomplete and unisexual, but also occasionally bisexual; sepals and petals < 5 mm,
      each with a single vein; fruit with 3 or fewer globose seeds in each carpel…**Ruscaceae** (*Nolina*)
6.   Plants tall, fernlike; leaves minute, triangular, papery, subtending clusters of filiform green cladodes;
      flowers small, yellowish; fruit a red berry…**Asparagaceae** (*Asparagus*)
6'.   Plants not as above…(7)
7.   Flowers minute, enclosed in chaffy bracts; typical perianth lacking; flowers arranged in spikes or spike-
      lets (grasses and sedges)…(8)
7'.   Flowers larger, not enclosed in chaffy bracts; perianth of 3 or 6 parts (rarely absent), these may be papery
      or chaffy…(9)
8.   Leaves 2-ranked (in 2 rows on the stem); sheaths usually open, or at least the margins not fused at the
      summit (with a few exceptions); stem cylindrical or flattened and almost always hollow; anthers attached
      to filaments at their middle…**Poaceae**

8'. Leaves 3-ranked, sometimes absent; sheaths usually closed, the margins fused; stems typically triangular in cross-section and solid (a few cylindrical and hollow); anthers attached at one end…**Cyperaceae**

9. Flowers minute, in headlike clusters, cymose, or in narrow spikelike racemes, scalelike; fruit a capsule or a schizocarp…(10)

9'. Flowers not minute, inflorescence various; tepals or petals various; fruit various…(11)

10. Flowers in headlike clusters or cymose; fruit a loculicidal capsule…**Juncaceae**

10'. Flowers in a spikelike raceme; fruit a schizocarp with 3 or 6 mericarps…**Juncaginaceae**

11. Ovary wholly inferior, the floral parts attached to the top of the ovary; flowers zygomorphic or actinomorphic…(12)

11'. Ovary superior or only partly inferior; flowers actinomorphic…(15)

12. Flowers zygomorphic; stamens united with gynoecium to form a column; lower petal forming a distinctive lip (labellum)…**Orchidaceae**

12'. Flowers actinomorphic; not as above…(13)

13. Foliage fleshy-thickened, glabrous or nearly so…**Amaryllidaceae** (*Habranthus*)

13'. Foliage not as above…(14)

14. Leaves equitant; tepals deep blue to purple or yellow, rarely white, in similar or dissimilar whorls, not hairy along their undersides; stamens 3; plants from rhizomes or fibrous roots…**Iridaceae**

14'. Leaves not equitant; tepals yellow, all similar in appearance, the outer ones usually hairy along their undersides; stamens 6; plants from small corms…**Hypoxidaceae** (*Hypoxis*)

15. Flowers in an umbel subtended by papery bracts…(16)

15'. Flowers variously arranged but not in an umbel subtended by papery bracts…(17)

16. Plants with onion odor (lacking in *Nothoscordum*); inflorescence bracts 2 (sometimes fused), wholly enclosing immature inflorescence, but generally not formed among pedicels; stems and leaves arising from bulb…**Alliaceae**

16'. Plants without onion odor; inflorescence bracts (2)3+, not wholly enclosing immature inflorescence, and some formed among pedicels; stems and leaves arising from corm…**Themidaceae**

17. Inflorescence a large, densely flowered panicle 20–80 cm; plants to 2 m; fruit a capsule…**Melanthiaceae** (*Veratrum*)

17'. Plants not as above…(18)

18. Tepals differentiated into two 3-merous series, one petaloid and colorful and the other sepaloid and green…(19)

18'. Tepals all ± similar in color and texture…(20)

19. Flowers subtended by a decidedly leafy bract (spathe); stem arising from thickened fleshy roots… **Commelinaceae**

19'. Flowers subtended by a not especially leafy bract; stem arising from a bulb…**Liliaceae** (*Calochortus*)

20. Ovary subterranean or arising from near ground level; plants acaulescent or subacaulescent… **Agavaceae**

20'. Ovary evidently well above the ground and/or plants evidently caulescent…(21)

21. Flowers mostly solitary or 2–3(4) together, borne at the stem tops or in the leaf axils…(22)

21'. Flowers usually several to numerous in umbels, well-developed racemes, or panicles…(28)

22. Leaves basal or absent, none borne on the flowering stems…(23)

22'. Leaves borne on the flowering stems…(24)

23. Leaves 1–2, elliptic, mottled with purple; flowers bright yellow, nodding…**Liliaceae** (*Erythronium*)

23'. Leaves few to several, linear, not mottled; flowers white, erect…**Themidaceae** (*Milla*)

24. Flowers borne in leaf axils…(25)

24'. Flowers borne at stem tips…(26)

25. Perianth segments united to near the tips…**Ruscaceae** (*Polygonatum*)

25'. Perianth segments separate to near the base…**Liliaceae** (*Streptopus*)

26. Leaves whorled at the upper nodes, alternate below, lanceolate; flowers orange-red…**Liliaceae** (*Lilium*)

26'. Leaves alternate throughout, lanceolate to ovate; flowers whitish or yellowish…(27)

27. Stems simple, unbranched; tepals neither swollen nor slightly inflated above the base; flowers erect to spreading…**Ruscaceae** (*Maianthemum*)

27'. Stems branched below; tepals weakly gibbous above the base; flowers nodding…**Liliaceae** (*Prosartes*)

28. Flowers borne on well-developed and evident pedicels…(29)

28'. Flowers sessile or nearly so and borne in dense clusters, pedicels and branches absent or scarcely evident…(30)

29. Perianth 5–12 mm, white to pale pink…**Asphodelaceae** (*Asphodelus*)

29'. Perianth 4.5–6 cm, yellow basally with darker tawny orange zones and stripes, veins reticulate [reported as escaped along roadsides in San Miguel County]…Hemerocallidaceae (*Hemerocallis*)

30. Flowers blue-purple; perianth segments united into an urn-shaped tube…**Hyacinthaceae** (*Muscari*)

30'. Flowers white to greenish to pale yellow; perianth segments separate…**Melanthiaceae**

KEY 5: WOODY "DICOTS"

(Subkeys include most genera in addition to families.)

1. Leaves scalelike, < 5 mm, overlapping and appressed to the stem…**Tamaricaceae**
1'. Leaves larger and otherwise not as above…(2)
2. Leaves and twigs covered by silvery or brownish peltate scales…**Elaeagnaceae**
2'. Leaves and twigs not covered by peltate scales…(3)
3. Flowers in heads and surrounded by an involucre of bracts (phyllaries); pappus often present surrounding the petals of disk and/or ray florets; stamens united by the anthers; ovary inferior…**Asteraceae**
3'. Flowers not in heads and surrounded by an involucre; pappus not present; stamens not united by the anthers; ovary superior or inferior…(4)
4. Leaves 3–9-cleft from the base, the segments narrowly linear and spinulose; flowers cream to yellowish and often tinged purplish, the corolla salverform and united into a tube below, 12–25 mm; fruit a loculicidal capsule with 3 valves…**Polemoniaceae** (*Linanthus pungens, Ipomopsis congesta*)
4'. Leaves and flowers not as above…(5)
5. Leaf arrangement opposite or whorled…(6)
5'. Leaf arrangement alternate…(7)
6. Leaves simple, may be divided or lobed (sometimes deeply so), but not separated into leaflets…**Key 5A**
6'. Leaves compound, separated into leaflets…**Key 5B**
7. Leaves usually absent; stems greenish, branching and ending in thorns…**Koeberliniaceae**
7'. Leaves and stems not as above…(8)
8. Leaves simple, may be divided or lobed (sometimes deeply so), but not separated into leaflets…**Key 5C**
8'. Leaves compound, separated into leaflets…**Key 5D**

KEY 5A: LEAVES OPPOSITE OR WHORLED; SIMPLE

1. Leaves palmately lobed…**Sapindaceae** (*Acer*)
1'. Leaves not palmately lobed…(2)
2. Leaves > 15 × 12 cm; large tree…**Bignoniaceae** (*Catalpa*)
2'. Leaves < 15 × 12 cm; trees or shrubs…(3)
3. Leaves whorled, 3–6 leaves per node; flowers red, white, or yellow…**Rubiaceae** (*Bouvardia*)
3'. Leaves opposite, sometimes fascicled; flowers variously colored…(4)
4. Leaves averaging < 1 cm wide…(5)
4'. Leaves averaging > 1 cm wide…(27)
5. Leaves averaging < 4 cm…(6)
5'. Leaves averaging > 4 cm…(11)
6. Leaves in fascicles or clusters of 2+ leaves at each node…(7)
6'. Leaves 2 to a node, 1 on each side of the stem, not in fascicles or clusters…(11)
7. Flowers in heads; ovary inferior; fruit a cypsela…**Asteraceae** (*Carphochaete, Thymophylla*)
7'. Flowers solitary or in simple cymes, not in heads; ovary inferior or superior; fruit variable…(8)
8. Sepals 4; petals 4; stamens 4 or 8…(9)
8'. Sepals 5; petals 5; stamens 5 or 10…(10)
9. Widely branched shrub, 1–2 m; stamens 8; flower solitary with 4 white, clawed petals…**Hydrangeaceae** (*Fendlera*)
9'. Small shrub, 20–55 cm; stamens 4; flowers consisting of a small cyme with 4 white petals…**Rubiaceae** (*Hedyotis*)
10. Shrub to 1+ m; leaves linear-oblanceolate to elliptic, 4–12 mm; flowers borne in small compound cymes; limestone, piñon-juniper and sagebrush communities…**Hydrangeaceae** (*Fendlerella*)
10'. Shrub to 40 cm; leaves small, leathery, and heathlike; flowers solitary or in small clusters of 3–5; gypsum soils and margins of salt lakes…**Frankeniaceae**
11. Leaves filiform, linear or awl-shaped, usually averaging < 0.5 cm wide; sessile or with a minute petiole…(12)
11'. Leaves broader, not linear or awl-shaped, usually averaging 0.5–1 cm wide…(16)
12. Flowers imperfect (dioecious) or occasionally monoecious, not brightly colored, lacking petals, subtended by distinctive bracts; plants reaching 1–2 m, often with some rigid branches becoming spinescent…**Amaranthaceae** (*Atriplex*)
12'. Flowers perfect, often showy, white, pink, red, yellow, lavender, blue, or purple, with or without bracts; plants reaching 60 cm, with or without spinescent branches…(13)
13. Fruit of 2 or 3 winged samaras; flowers yellow to yellow orange–yellow, in axillary clusters of two…**Malpighiaceae**
13'. Fruit a capsule or nutlet; flowers never yellow; inflorescence a raceme or a panicle…(14)
14. Fruit a follicle; flowers solitary, axillary or terminal, actinomorphic; sepals 4; petals 4, white; stamens 8…**Crossosomataceae**
14'. Fruit a capsule or nutlet; flowers in axillary spikes or loose to open panicles, seldom solitary, actinomorphic; sepals and petals 4 or 5, color various…(15)

15. Stems square in cross-section; ovary divided into 4 lobes; fruit a nutlet...**Lamiaceae** (*Poliomintha, Salazaria*)
15′. Stems round in cross-section; ovary 2-locular, not lobed; fruit a capsule...**Plantaginaceae** (*Penstemon*)
16. Stems square in cross-section; plants aromatic; fruit a nutlet; stamens 2 or 4...(17)
16′. Stems round or terete in cross-section; fruit a drupe; stamen number variable...(18)
17. Corolla purple to bluish white, zygomorphic, < 1.5 cm; stamens 2...**Lamiaceae** (*Salvia*)
17′. Corolla white, slightly zygomorphic; stamens 4...**Verbenaceae** (*Aloysia*)
18. Fruit a drupe; flowers bluish to green or white or tubular and white to pink...(19)
18′. Fruit an achene or capsule; flowers red to orange, lavender to blue, yellow, or white...(21)
19. Corolla tubular, pink to reddish white; flowers solitary, in pairs or in several-flowered racemes; fruit white...**Caprifoliaceae**
19′. Corolla not tubular, may be absent or if present, white; flowers in terminal panicles or lateral clusters; fruit dark red to purple or black...(20)
20. Flowers inconspicuous, appearing before the leaves in axillary clusters; corolla absent; fruit purple to black; leaves oblanceolate to elliptic; widespread, moist soils along rivers and streams...**Oleaceae** (*Forestiera*)
20′. Flowers small, appearing with the leaves in terminal panicles; sepals and petals 5; fruit black; leaves elliptic to obovate or oblong; Hidalgo Canyon, steep banks of an intermittent stream...**Rhamnaceae** (*Sageretia*)
21. Flowers in dense, sessile, or slightly stalked clusters in the axils of the upper leaves or terminal cymes; flowers < 5 mm, crowded; corolla cream-colored to white or light blue; leaf margins crenate to dentate, entire or with a few teeth...(22)
21′. Flowers borne at the apex of the stems or axillary, solitary or in few-flowered panicles, cymes, or corymbs; flowers > 10 mm; corolla showy, red-orange, lavender, blue, yellow, or white; leaf margins entire...(23)
22. Flowers 4-merous, in dense, tomentose, ball-like clusters in leaf axils, greenish yellow to yellow; leaves tomentose, sessile, narrowly oblong to linear-cuneate, margins crenate to dentate-crenate; fruit a pod-like capsule...**Scrophulariaceae**
22′. Flowers 5-merous, in terminal or axillary clusters, mostly white or bluish or pinkish; leaves thick and leathery, elliptic to obovate, margins entire or slightly toothed; fruit a 3-lobed capsule...**Rhamnaceae** (*Ceanothus*)
23. Flowers bright yellow; leaves oblanceolate or lanceolate, alternate above and opposite below...**Oleaceae** (*Menodora*)
23′. Flowers white, red-orange, blue, or lavender; leaves ovate to elliptic, oblanceolate, or lanceolate, opposite...(24)
24. Flowers in heads surrounded by an involucre of bracts; fruit a cypsela...**Asteraceae**
24′. Flowers not as above; fruit not a cypsela...(25)
25. Petals showy, red, purple-red, orange-red, lavender, bright blue, or white; forming a tube; stamens 2–4...**Acanthaceae** (*Anisacanthus, Ruella*)
25′. Petals none or white to pink; stamens 13–90 or 8...(26)
26. Petals none; stamens 8; fruit bright red, a drupe...**Elaeagnaceae** (*Shepherdia*)
26′. Petals white or pink, forming a hypanthium; stamens 13–90; fruit a capsule...**Hydrangeaceae**
27. Fruit a capsule...(28)
27′. Fruit a drupe, berry, or nutlet...(29)
28. Leaves lanceolate; flowers showy, bright red to orange-red...**Acanthaceae** (*Anisacanthus*)
28′. Leaves oval, ovate, or elliptic; flowers not showy, small, greenish yellow...**Simmondsiaceae**
29. Fruit a nutlet; stamens 2 or 4 per flower; stems square in cross-section...(30)
29′. Fruit a drupe or berry; stamens variable in number, mostly 4; stems not square in cross-section (young tips of branches may be square in *Garrya*)...(31)
30. Ovary 4-lobed, 4-loculed, producing 4 nutlets; corolla blue, zygomorphic; stamens 2...**Lamiaceae** (*Salvia*)
30′. Ovary with 2 locules, producing 2 nutlets; corolla white, only slightly zygomorphic; stamens 4...**Verbenaceae** (*Aloysia*)
31. Flowers reduced in size, green, and obscure, corolla absent or much reduced; fruit a drupe, purple to black; leaves leathery...**Garryaceae**
31′. Flowers larger, pink to red, yellow, or white, corolla well developed; fruit a berrylike drupe, color variable; leaves not leathery...(32)
32. Flowers zygomorphic (often weakly so); corolla tubular, salverform, or funnelform, often paired in the leaf axils; fruit a white, black, red, or orange berry...**Caprifoliaceae** (*Lonicera, Symphoricarpos*)
32′. Flowers actinomorphic; corolla various in shape; not paired in the leaf axils; fruit a follicle, berry, drupe, or capsule...(33)
33. Fruit a winged samara; flowers appearing before the leaves, petals absent; leaves ovate to oval, often with rounded tips...**Oleaceae** (*Fraxinus*)

33'. Fruit a drupe or berry; flowers white, pink, yellow, lavender, or small and greenish; leaves various… (34)

34. Young branches and leaves densely pubescent with golden-brown or silvery stellate or peltate hairs; fruit covered with silvery-gray hairs; sepals 4, united into a hypanthium; petals absent; stems dark gray to black…**Elaeagnaceae**

34'. Young branches and leaves without golden-brown or silvery stellate or peltate hairs; flowers numerous in cymes; fruit not covered with silvery-gray hairs, red; petals white; older stems reddish…**Cornaceae**

KEY 5B: LEAVES OPPOSITE OR WHORLED; COMPOUND

1. Leaflets 2, connate basally, evergreen, leathery, resinous; leaves appearing simple and 2-lobed…**Zygophyllaceae** (*Larrea*)

1'. Leaflets 3+, plants not as above…(2)

2. Leaflets mostly 3; fruit a double samaroid schizocarp; pith round in cross-section; bundle scars 3…**Sapindaceae** (*Acer*)

2'. Leaflets 5–7; fruit a single samara, capsule, or berry; otherwise, not as above…(3)

3. Fruit a berry; pith of older stems > 1/2 the diameter of the stem…**Viburnaceae** (*Sambucus*)

3'. Fruit a capsule or single samara; pith of older stems < 1/2 the diameter of the stem…(4)

4. Leaflets bright green on both sides; flowers bright yellow, 4–6 cm, corolla tubular; fruit a capsule with comose seeds…**Bignoniaceae** (*Tecoma*)

4'. Leaflets dull green, at least on 1 side; flowers with petals lacking or light yellow, with 4 narrow corolla lobes; fruit a single samara with a single seed…**Oleaceae** (*Fraxinus*)

KEY 5C: LEAVES ALTERNATE AND SIMPLE

1. Leaves lobed or notched at the apex…(2)

1'. Leaves not lobed, or only slightly notched; in some species leaves absent or scalelike…(11)

2. Plants with spines or thorns on stems…(3)

2'. Plants without spines or thorns on stems…(4)

3. Spines 1.5+ cm; leaves villous, 0.5–2 cm, finely divided into 5–7-parted lobes; stems white-tomentose; inflorescence a head of 6–20 small yellow-green flowers…**Asteraceae** (*Artemisia*)

3'. Spines < 1.5 cm; leaves glabrous to glandular-pubescent, 1–10 cm, lobed or notched; stems whitish to dark brown, not tomentose; inflorescence a raceme or flowers solitary, white to cream, dull pink, or red or orange…**Grossulariaceae** (*Ribes*)

4. Leaves pinnately lobed with 1 main vein (midrib) from the base or 2 or 3 small lobes at leaf apex…(5)

4'. Leaves palmately lobed and/or with > 1 main vein (midrib) from near the base…(7)

5. Inflorescence a head of many flowers surrounded by an involucre of bracts; fruit a cypsela; sepals absent or forming a pappus of hair at the top of an inferior ovary…**Asteraceae** (*Artemisia*, *Viguiera*, *Senecio*, *Isocoma*, *Parthenium*, *Brickellia*)

5'. Inflorescence a catkin or flowers borne solitarily; fruit a nut, an acorn, or an achene; sepals and petals reduced and inconspicuous or fused only at the base or from a hypanthium and conspicuous…(6)

6. Fruit a nut or an acorn; leaves 3–15 cm; sepals and petals reduced and inconspicuous; stamens 4 to several…**Fagaceae**

6'. Fruit an achene with an elongate and exserted style; leaves 0.5–2 cm; sepals and petals conspicuous; stamens numerous, usually 20–44…**Rosaceae**

7. Fruit a several-seeded berry…**Grossulariaceae**

7'. Fruit an aggregate, multiple achene or capsule…(8)

8. Shrubs, often with shredding bark; flowers and fruit borne solitarily, 1 or 2 on a leafy branch, or in corymbs; flowers perfect and complete…**Rosaceae** (*Physocarpus*, *Rubus*)

8'. Mostly trees; flowers borne in catkins, a drooping spike, or in a ball or head forming achenes; floral parts absent or reduced in size, plants often monoecious or dioecious…(9)

9. Inflorescence a pendulous, globose head, green in flower, becoming brown in fruit and producing many achenes; plants monoecious; bark on older branches and trunk scaling off to expose white or greenish inner bark…**Platanaceae**

9'. Inflorescence a pendulous elongate catkin, green in flower and producing either a number of yellow-brown capsules with long silky hairs, or a fleshy multiple fruit of numerous red, black, or white drupes; bark smooth or forming fissures or ridges…(10)

10. Inflorescence a pendulous elongate catkin producing tomentose capsules containing many seeds with long silky hairs; leaves deltoid-ovate, 4–8 cm wide, 3–5-lobed, toothed, upper surface dark green to blue-green, lower surface densely white-tomentose…**Salicaceae** (*Populus*)

10'. Inflorescence a pendulous catkin producing fleshy multiple fruits with red, black, or white drupes; leaves 1–6-lobed, obliquely based, ovate or deeply notched, with 3 main veins from the base…**Moraceae**

11. Plants with spines or thorns on stems…(12)

11'. Plants without spines or thorns on stems…(36)

12. Shrubs, a heavily branched, stiff, smooth, green mass with thick thorns; leaves absent or scalelike; fruit a black berry…**Koeberliniaceae**

12′. Plants with spines and thorns on stems and leaves or forming slender branches, not appearing to make up the entire plant; some true leaves, not scalelike; fruit various…(13)

13. Leaf margins entire…(14)

13′. Leaf margins toothed, serrate, dentate, or crenate…(29)

14. Leaves averaging < 4 mm wide…(15)

14′. Leaves averaging > 4 mm wide…(21)

15. Leaf blades glabrous or sparsely pubescent, 1–4 cm, succulent…**Sarcobataceae**

15′. Leaf blades not as above…(16)

16. Fruit a fleshy 1-seeded drupe, borne from a single flower; flowers inconspicuous in a small axillary cyme or solitary…**Rhamnaceae** (*Condalia*)

16′. Fruit a 1-seeded legume, follicle, achene, or utricle; flowers borne in heads, racemes, panicles, or spikes…(17)

17. Fruit a legume with several seeds, or a legumelike round pod with 1 seed and armed with sharp spines; inflorescence a raceme or solitary; flowers purple to purple-pink to red…(18)

17′. Fruit an achene, utricle, or follicle; inflorescence a head, panicle, spike, or solitary; flowers white or yellowish…(19)

18. Shrubs to 1 m, glabrous; fruit a loment, not spiny…**Fabaceae** (*Alhagi*)

18′. Low shrubs to 70 cm, herbage gray-pubescent, sometimes glandular; fruit a 1-seeded pod, thick-walled and spiny…**Krameriaceae**

19. Inflorescence a head of 5–9 flowers, 4–6 phyllaries; fruit a cypsela, 6–8 mm; flowers yellow; leaves linear to oblanceolate, often fascicled…**Asteraceae** (*Tetradymia*)

19′. Inflorescence a cyme, spike, or flowers borne solitarily; fruit a utricle or follicle…(20)

20. Young twigs tan to brown or gray, rigid, becoming spinescent; leaves linear, fleshy, 15–40 mm; flowers imperfect and incomplete, perianth lacking, mostly dioecious; fruit a reddish winged utricle…**Amaranthaceae** (*Atriplex*)

20′. Young twigs green, slender, stiff, angled, glabrous or glaucous, terminally spinescent; leaves small, 5–12 × 1–2.5 mm; flowers perfect and complete, solitary, sepals 5, petals 5; fruit a follicle…**Crossosomataceae** (*Glossopetalon*)

21. Fruit a large, dense, green aggregate, 7–13 cm diam., made up of many 1-seeded drupelets, containing sticky white juice…**Moraceae** (*Maclura*)

21′. Fruit a much smaller berry, drupe, capsule, utricle, or winged seed, all < 2 cm diam.; not containing sticky white juice…(22)

22. Fruit a berry, borne singly, in pairs or in small fascicles…(23)

22′. Fruit a drupe, capsule, utricle, or winged seed; inflorescence a cyme, raceme, panicle, or spike, solitary or in axillary clusters…(24)

23. Leaves oblong to obovate, mostly fascicled on short branches, leathery, glabrate, dark green above and white or tan, woolly-villous beneath, 2–6 × 1.4–2.5 cm; fruit a lustrous black berry; extreme SW New Mexico…**Sapotaceae** (*Sideroxylon*)

23′. Leaves linear to elliptic or oblanceolate, light or glaucous green, 1–6 × 0.3–1.2 cm; fruit a globose to ovoid red berry; widespread…**Solanaceae** (*Lycium*)

24. Fruit a capsule; flowers and fruit borne in terminal racemes or narrow panicles, mostly at ends of branches…(25)

24′. Fruit a drupe, utricle, or samaralike seed; flowers seldom terminal…(26)

25. Stems numerous, grooved, long and canelike, 1.5–5 m, radiating whiplike from a common base; flowers appearing first followed by leaves; flowers bright crimson; fruit a capsule producing many winged seeds…**Fouquieriaceae**

25′. Stems not as above; flowers and leaves often present at the same time; flowers small, white; fruit a 3-lobed, small dry capsule…**Rhamnaceae** (*Ceanothus*)

26. Fruit a utricle; fruiting bracts in pairs and orbicular or obovate, formed on each side of the mature seed and enclosing it; plants dioecious or monoecious; inflorescence a spike, panicle, or axillary cluster…(27)

26′. Fruit a 1-seeded fleshy drupe, not surrounded by paired bracts; flowers perfect or sometimes monoecious; inflorescence a cyme or flowers solitary…(28)

27. Shrubs to 1 m; leaves orbicular-ovate, 10–20 × 10–20 mm, densely scurfy…**Amaranthaceae** (*Atriplex*)

27′. Shrubs to 1.5 m; leaves oblanceolate to spatulate, 2–3 cm × 2–12 mm, ca. 4 times longer than wide, scurfy-pubescent when young, becoming glabrate…**Amaranthaceae** (*Grayia*)

28. Spines commonly paired at the nodes, stipular, averaging < 2 mm; leaves with 3 main veins, leaf base oblique…**Cannabaceae** (*Celtis*)

28′. Spines commonly borne singly, not paired, 1 spine borne at each node or formed at tips of branches, averaging > 2 cm; leaves with 1 main vein, leaf base attenuate, cuneate, or obtuse, not oblique…**Rhamnaceae** (*Ziziphus*)

29. Spines 2 or 3 at the nodes, commonly stipulate, averaging < 2 cm; fruit a drupe or berry…(30)

29'. Spines borne singly at the nodes or on exserted modified stems averaging > 2 cm; fruit a drupe or pome…(32)

30. Spines 3 at the nodes and stipular; leaves fascicled, 1 main vein from the base, some leaves spine-tipped; fruit a several-seeded berry…**Berberidaceae**

30'. Spines paired at the nodes and stipules, with some stems exserted and forming spines; leaves alternate, 3 main veins from the base; fruit a several-seeded berry…(31)

31. Leaf bases oblique, some leaves entire and some 3-toothed toward the apex; fruit a 1-seeded, yellow or orange, fleshy drupe…**Cannabaceae** (*Celtis*)

31'. Leaf bases obtuse, cuneate to slightly oblique, most leaves serrate and glandular-tipped; fruit a 1- or 2-seeded, orange to brown, fleshy drupe…**Rhamnaceae** (*Ziziphus*)

32. Leaves averaging < 2.5 × 1 cm; fruit a drupe with 1 or 3 stones…(33)

32'. Leaves averaging > 2.5 × 1 cm; fruit a drupe with 1 stone or a pome…(34)

33. Fruit with 1 stone, blue to black; flowers in small axillary cymes of 2–7 flowers; twigs grayish green with stout leaf-bearing spines…**Rhamnaceae** (*Ziziphus*)

33'. Fruit with 3 stones, dark blue to black and juicy; flowers in terminal panicles or spikes…**Rhamnaceae** (*Sageretia*)

34. Fruit a 1-seeded plumlike drupe; carpel 1; lenticels horizontal and prominent…**Rosaceae** (*Prunus*)

34'. Fruit an applelike pome; carpels several, commonly 5; lenticels oval, sometimes prominent or obscure…(35)

35. Fruit averaging > 2 cm diam.; leaves permanently pubescent to tomentose on lower surface, ovate-oblong to broadly elliptic, 6–9 × 3–6 cm, coarsely serrate; flowers showy, rose-colored…**Rosaceae** (*Malus*)

35'. Fruit averaging < 2 cm diam.; leaves various; flowers white to pink…**Rosaceae** (*Crataegus*)

36. Inflorescence a head of several to many flowers; heads may contain all disk florets, all ray florets, or a combination of disk and ray florets…**Asteraceae**

36'. Inflorescence various, not a head…(37)

37. Fruit an acorn…**Fagaceae**

37'. Fruit variable, not an acorn…(38)

38. Buds enclosed by a single hoodlike scale; individual seeds covered with long hairs; flowers unisexual, borne in catkins, usually erect or slightly pendulous; fruit a capsule…**Salicaceae**

38'. Buds enclosed with several overlapping scales or bud scales lacking; seeds not covered with long hairs; flowers of various inflorescence types, if catkins, usually pendulous rather than erect…(39)

39. Plants and stems leafless or with small scalelike green leaves, never with obvious broad leaves…(40)

39'. Plants and stems with obvious leaves…(42)

40. Cedarlike shrubs or small trees, 7–8 m, quickly spreading over large areas; red bark when young…**Tamaricaceae**

40'. Shrubs not cedarlike, to 2 m, usually growing in isolated clumps, not large areas…(41)

41. Stems distinctly jointed and swollen, not waxy, woody at base, 0.3–1.5 m, succulent, glabrous, green; leaves scalelike; fruit a utricle…**Amaranthaceae** (*Allenrolfea*)

41'. Stems not distinctly jointed, swollen, or waxy, to 80 cm, mostly greenish or gray; leaves usually absent, if present linear and quickly deciduous; fruit a small capsule…**Euphorbiaceae** (*Euphorbia*)

42. Fruit a utricle; leaves scurfy, covered with gray to white-mealy scales on both sides or densely stellate-pubescent…**Amaranthaceae** (*Krascheninnikovia*, *Atriplex*)

42'. Fruit various, not a utricle; leaves various, not mealy…(43)

43. Leaves thick to leathery, evergreen, glabrous; flowers cup-shaped or urn-shaped, waxy, < 1 cm, white to pink or greenish white; inflorescence a raceme or panicle; sepals and petals 4 or 5; stamens 8 or 10; fruit a berry, drupe, or berrylike capsule…**Ericaceae**

43'. Leaves not as above; flowers never cup-shaped or urn-shaped…(44)

44. Perianth of 6 petaloid tepals, in 2 whorls of 3, white, yellow, or pink; plants only weakly woody…**Polygonaceae**

44'. Perianth not of 6 petaloid tepals, flower color various; plants clearly shrubs or trees…(45)

45. Plants covered with stellate or forked hairs or small, scurfy, peltate scales; flowers unisexual; plants monoecious or dioecious; fruit a 2- or 3-celled capsule…**Euphorbiaceae**

45'. Plants not covered with stellate or forked hairs or scurfy, peltate scales; flowers bisexual or unisexual; inflorescence variable; fruit variable…(46)

46. Leaf bases oblique or distinctly asymmetric or unequal on most leaves; fruit a drupe, samara, or multiple; inflorescences pendulous catkins, spikes, cymes, small clusters or solitary…(47)

46'. Leaf bases evenly cordate, rounded-attenuate, cuneate, or symmetric, not oblique or unequal; fruits variable; inflorescences variable…(48)

47. Leaves with 1 main midvein from the base, leaf margins double-dentate; fruit a 1-seeded samara with the wing surrounding the seed…**Ulmaceae**

47'. Leaves with 3–5 veins from the base; fruit a drupe or multiple…(49)

48. Fruit a succulent, cylindrical multiple, consisting of many small fleshy nutlets, 1.5–2.5 cm, bright red and becoming black to purple or white when mature, sweet, juicy, edible; leaves mostly deeply lobed, margins serrate to crenate…**Moraceae** (*Morus*)

48'. Fruit a globose fleshy drupe, never a fleshy multiple or catkin; leaves never lobed…**Cannabaceae** (*Celtis*)

49. Margins of leaves entire, not toothed, lobed, or divided, rarely slightly undulate or with a few teeth… (50)

49'. Margins of leaves toothed, serrate, dentate, or crenate, not distinctly cleft or lobed…(59)

50. Leaves averaging mostly > 1 cm wide…(51)

50'. Leaves averaging < 1 cm wide…(52)

51. Fruit a legume; leaves broadly cordate to reniform, bases cordate, palmately veined; petioles > 3 cm… **Fabaceae** (*Cercis*)

51'. Fruit a drupe or capsule; leaves lanceolate or elliptic to obovate, bases rounded to cuneate to attenuate, not cordate, pinnately veined; petioles < 2.5 cm…**Rhamnaceae** (*Rhamnus, Ceanothus*)

52. Leaves 7–17 cm × 5–10 mm, linear to linear-lanceolate; large shrubs or trees to 10 m…**Bignoniaceae** (*Chilopsis*)

52'. Leaves 0.3–5 cm × 1–9 mm, leaf shape various; shrubs or subshrubs, seldom reaching 3 m…(53)

53. Plants true shrubs, erect, 50 cm to 3 m…(54)

53'. Plants perennial subshrubs, prostrate to erect, 10–50 cm…(57)

54. Leaves 0.5–1.9 × 1–2 mm, linear to linear-spatulate, gland-dotted on abaxial surface; fruit a legume… **Fabaceae** (*Psorothamnus*)

54'. Leaves averaging 4 mm wide, not gland-dotted on abaxial surface; fruit a capsule or pome…(55)

55. Leaves 1.5–4 cm × 4–9 mm; in fascicles at the ends of branches, sessile or nearly so, narrowly oblanceolate; fruit a pome…**Rosaceae** (*Peraphyllum*)

55'. Leaves 0.3–1.8 cm, scattered along the stem or a few in small cluster, not in fascicles, oval to elliptic to obovate; fruit a capsule…(56)

56. Flowers borne toward the ends of the branches in small clusters, 2–4 mm, white to cream-colored; leaves small, scabrous to sparsely hispid, leathery…**Celastraceae** (*Mortonia, Parnassia*)

56'. Flowers axillary along the stem, usually solitary, occasionally in pairs, pale violet to purple or pink; leaves silvery-canescent, greenish, ovate-spatulate to obovate, not leathery…**Scrophulariaceae** (*Leucophyllum*)

57. Leaves linear, 1.5 mm or less wide, glandular-hispid to hirsute and revolute; small shrubs to 40 cm; lobes densely plumose-hairy…**Boraginaceae**

57'. Leaves various, not linear, > 2 mm wide, glabrous to pubescent to tomentose; plants various, calyx lobes not densely plumose-hairy…(58)

58. Plants erect, small, clumped perennials to 35 cm; woody below and herbaceous above; flowers < 3 cm; leaves lanceolate to obovate or oblong to ovate, 0.9–3.2 cm × 5–9 mm; inflorescence a cyme…**Oleaceae** (*Menodora*)

58'. Plants low, densely cespitose subshrubs, forming large flat mats, often over rock faces; leaves appearing fascicled, mostly spatulate to lanceolate or narrowly oblanceolate, 5–15 × 2–5 mm; inflorescence a raceme…**Rosaceae** (*Petrophytum*)

59. Plants low shrubs or suffrutescent perennials from a woody base, < 80 cm or prostrate; fruit a 5-celled, warty, globose capsule…**Malvaceae** (*Ayenia*)

59'. Plants tall shrubs or trees, 1+ m, woody throughout, never prostrate; fruit variable…(60)

60. Inflorescence a catkin; plants monoecious or dioecious; flowers imperfect and incomplete, not showy; fruit a nutlet or capsule…(61)

60'. Inflorescence various; flowers perfect; fruit various, not a catkin…(62)

61. Plants monoecious trees or large shrubs; buds not distinctly resinous and sticky to the touch; staminate catkins elongate and pendulous, pistillate catkins pinecone-like; sepals or petals present, not both; leaves often double-serrate; fruit a nutlet…**Betulaceae**

61'. Plants dioecious trees or large shrubs; buds often resinous and sticky to the touch; catkins both staminate and pistillate, elongate and pendulous, never pinecone-like; sepals and petals none; leaves never double-serrate; fruit a capsule…**Salicaceae**

62. Fruit a drupe, 1-seeded or berrylike with 2–4 seeds; flowers white to pink or green…(63)

62'. Fruit a capsule, achene, follicle, or pome; flower color variable…(64)

63. Flowers small, inconspicuous, greenish, not showy, sepals and petals in 4s or 5s or absent, stamens 4 or 5; fruit a berrylike drupe with 2–4 seeds…**Rhamnaceae** (*Rhamnus*)

63'. Flowers showy, white or pink, perfect, sepals and petals in 5s, stamens 10+; fruit a 1-seeded drupe… **Rosaceae** (*Prunus*)

64. Fruit a pome or achene; widespread…**Rosaceae**

64'. Fruit a capsule or woody follicle; S counties…(65)

65. Leaves narrowly lanceolate, 3.8–8 × 0.6–1.4 cm, margins serrate, glabrous to puberulent on the adaxial surface and puberulent to tomentose on the abaxial surface; petals white…**Rosaceae** (*Vauquelinia*)

65'. Leaves ovate to ovate-cordate, 1.5–4 × 0.7–1.5 cm, margins serrate to crenate, densely pubescent with stellate or simple hairs on the adaxial and abaxial surfaces; petals white, yellow, or pink…**Malvaceae** (*Abutilon*)

KEY 5D: LEAVES ALTERNATE AND COMPOUND

1. Leaves palmately compound or trifoliate…(2)
1'. Leaves pinnately compound…(11)
2. Plants with spines or thorns on leaves or stems…(3)
2'. Plants without spines or thorns on leaves or stems…(6)
3. Leaflets spine-tipped, leathery, and sessile, to 6 cm, 3–7 lobes on a trifoliate, hollylike leaf… **Berberidaceae**
3'. Leaflets not spine-tipped, spines and thorns confined to the stems of the plant…(4)
4. Leaflets consistently trifoliate, the 3 leaflets entire, commonly triangular with a truncate base, terminal petiolule several times longer than the 2 lateral petiolules; petals scarlet; fruit a legume…**Fabaceae** (*Erythrina*)
4'. Leaflets predominantly trifoliate with some leaves occasionally palmately or pinnately compound, the leaflets serrate or biserrate, ovate to lanceolate or slightly cordate, not triangular, the terminal petiolule longer but not several times longer than the lateral petiolules; petals variable in color, not scarlet; fruit a drupe, cluster of drupelets, or an aggregate…(5)
5. Flowers borne singly at the tips of the branches; petals 5, pink to red, sepals 5, numerous stamens; fruit a large globose hip, red at maturity…**Rosaceae** (*Rosa*)
5'. Flowers borne in corymbs or racemes; petals 5, white, sepals 5, numerous stamens; fruit an aggregate, red to black…**Rosaceae** (*Rubus*)
6. Plants becoming woody vines with tendrils; leaflets commonly 5–7…**Vitaceae** (*Parthenocissus*)
6'. Plants shrubs that occasionally become small trees to 5 m, or woody vines that do not produce tendrils; leaflets mostly 3–5, seldom 6 or 7…(7)
7. Plants producing both small compound and smaller simple leaves, leaves and leaflets linear to filiform, <1.5 cm × 1 mm; flowers blue…**Fabaceae** (*Psorothamnus*)
7'. Plants producing only compound leaves, leaflets variable, not filiform, mostly 1+ cm; flowers white, cream, light green, or yellow…(8)
8. Leaflets 3–7 per leaf, obovate to lanceolate, silky-white on the abaxial surface, somewhat revolute, entire; flowers 5-merous, with bright yellow petals; fruit an achene…**Rosaceae** (*Dasiphora*)
8'. Leaflets 3 per leaf, shape and pubescence variable; flower color variable; fruit variable…(9)
9. Fruit a dry, straw-colored, orbicular samara…**Rutaceae** (*Ptelea*)
9'. Fruit a dry, 1-seeded drupe or a legume…(10)
10. Fruit a legume; inflorescence a dense spike of yellow flowers…**Fabaceae** (*Dalea*)
10'. Fruit a dry, 1-seeded drupe, white, cream-colored, red, or orange, inflorescence a panicle…**Anacardiaceae** (*Toxicodendron*)
11. Leaves odd-pinnately compound…(12)
11'. Leaves even-pinnately compound…(26)
12. Plants with spines or thorns on leaves or stems…(13)
12'. Plants without spines or thorns on leaves or stems…(16)
13. Leaflets spine-tipped, leathery, on a hollylike leaf; stems without spines or thorns…**Berberidaceae** (*Berberis*)
13'. Leaflets not spine-tipped; spines and thorns confined to the stems of the plant…(14)
14. Leaflets 9–23 per leaf, margins entire; fruit a legume; flowers zygomorphic, borne in many-flowered racemes, petals yellow, white, rose, or bluish purple…**Fabaceae** (*Robinia*)
14'. Leaflets 3–11 per leaf, margins serrate or dentate, not entire; fruit a hip or drupelet; flowers actinomorphic, borne singly, few to a cluster or simple raceme; petals pink, rose, or white…(15)
15. Fruit a cluster of drupelets forming an aggregate fruit or raspberry; leaves glaucous or strongly whitened beneath, leaflets commonly 3–5; petals white…**Rosaceae** (*Rubus*)
15'. Fruit a hip or a berrylike structure formed from a hypanthium enclosing several seeds; leaves green to pale green, not strongly glaucous, leaflets 3–11; petals pink, rose, or rarely white…**Rosaceae** (*Rosa*)
16. Fruit a legume, 1–10 cm, 1- to several-seeded…**Fabaceae**
16'. Fruit various, not a legume…(17)
17. Leaflets averaging <1 cm wide…(18)
17'. Leaflets averaging >1 cm wide…(20)
18. Flowers several in solitary heads, ray florets yellow; leaflets filiform, averaging > 3 cm…**Asteraceae** (*Sidneya*)
18'. Flowers not in solitary heads; flower color variable; leaflet shape variable, not filiform…(19)

19. Leaf rachis not winged; fruit an achene; leaflets 3–7, silky-white on the abaxial surface, appearing trifoliate to palmate; twigs never spinescent; flowers yellow…**Rosaceae** (*Dasiphora*)

19.' Leaf rachis winged; fruit a red to orange drupe; leaflets 5–9, oblong-elliptic to obovate, some branches appearing spinescent; flowers greenish white…**Anacardiaceae** (*Rhus*)

20. Fruit a drupe or nut…(21)

20.' Fruit a pome, capsule, berry, or samara…(23)

21. Fruit a nut; pith chambered; inflorescence a catkin…**Juglandaceae** (*Juglans*)

21.' Fruit a drupe; pith continuous; inflorescence a panicle or thyrse…(22)

22. Leaves bipinnate or twice compound, to 60 cm, with up to 45 leaflets; flowers perfect…**Meliaceae** (*Melia*)

22.' Leaves 1-pinnate, to 40 cm, with up to 25 leaflets; flowers polygamous…**Anacardiaceae** (*Rhus*)

23. Fruit a red berrylike pome; flowers white; inflorescence a flat-topped compound cyme; hypanthium urn-shaped or turbinate…**Rosaceae** (*Sorbus*)

23.' Fruit a samara, berry, or capsule; flowers yellow, yellow-green, or bright pink; inflorescence an extended panicle…(24)

24. Fruit a narrow samara; leaves 25–60 cm…**Simaroubaceae** (*Ailanthus*)

24.' Fruit a capsule, berry, or drupe; leaves 10–30 cm…(25)

25. Fruit a globose berry; leaflets entire, both odd- and even-pinnate; flowers white…**Sapindaceae** (*Sapindus*)

25.' Fruit a papery or leathery capsule; leaflets toothed, odd-pinnate; flowers bright yellow or rose-pink…**Sapindaceae** (*Ungnadia*)

26. Spines or thorns absent on stems and leaves…**Fabaceae** (*Senna, Caesalpinia, Calliandra, Acaciella, Vachellia*)

26.' Spines or thorns present on stems or leaves…(27)

27. Leaflets 2.5–4 cm; plant a large tree reaching 15–20 m, the trunk often thorny, thorns often reaching 5+ cm; fruit 15–40 × 1.4–3.5 cm…**Fabaceae** (*Gleditsia*)

27.' Leaflets averaging < 2 cm; plant a shrub or small tree, reaching 5–8 m; thorns rarely reaching 4 cm; fruit < 18 × 1.5 cm…(28)

28. Stamens numerous, > 10 per flower and free; shrubs…(29)

28.' Stamens 10 per flower or fewer; shrubs or small trees…(30)

29. Spines curved or hooked; flowers cream-colored or white; fruit 4–14 × 1–2 cm…**Fabaceae** (*Senegalia*)

29.' Spines straight or absent; flowers yellow; fruit 4–10 × 0.2–0.5 cm…**Fabaceae** (*Vachellia*)

30. Flowers in spikes or racemes, 3.5–8 cm, yellow to yellow-green or pale green; stems with straight, needle-like, solitary or paired stipular spines…**Fabaceae** (*Prosopis*)

30.' Flowers in globose heads or short spikes, 3–5 cm, white or pink to pale red; stems with short, straight to recurved, scattered, flattened prickles…**Fabaceae** (*Mimosa*)

KEY 6: HERBACEOUS "DICOTS"

1. Cauline leaves in whorls…(2)

1.' Cauline leaves alternate or opposite, not in whorls…(4)

2. Leaves in a single whorl of broad leaf blades at the top of the stem…**Cornaceae** (*Cornus*)

2.' Leaves in several whorls along the stem…(3)

3. Stems square; plants with hairs, often with recurved hooks…**Rubiaceae**

3.' Stems not square; plants glabrous, tall, stout…**Gentianaceae** (*Frasera*)

4. Flowers several to many, sessile in heads that often resemble a single "flower," each flower cluster surrounded or subtended by an involucre…(5)

4.' Flowers not as above…(6)

5. Involucre papery or umbrellalike, consisting of a single individual cup; flowers obviously separate and not tightly confined by the involucre…**Nyctaginaceae**

5.' Involucre not papery or umbrellalike, the flowers in a dense head…**Asteraceae**

6. Perianth none or of a single whorl (tepals or sepals), these all much alike in color and texture…**Key 6A**

6.' Perianth present, of 2 whorls, the outer whorl (sepals) and inner whorl (petals) usually conspicuously different in texture, color, or both…(7)

7. Petals separate…**Key 6B**

7.' Petals united (at least at the base)…(8)

8. Corolla radially symmetric…**Key 6C**

8.' Corolla bilaterally symmetric…**Key 6D**

KEY 6A: PERIANTH NONE OR OF A SINGLE WHORL

1. Plants dioecious…(2)

1.' Plants not dioecious…(4)

2. Leaves simple, not lobed or compound…**Amaranthaceae**

2'.   Leaves lobed or compound…(3)

3.   Staminate flowers in racemes, carpellate in clusters; fruit nutlike; leaves digitately compound with narrow serrate leaflets…**Cannabaceae** (*Cannabis*)

3'.   All flowers in open panicles; cypselas ribbed, in clusters; leaves ternately compound with small palmately lobed leaflets…**Ranunculaceae** (*Thalictrum*)

4.   Perianth completely absent…(5)

4'.   Perianth present, consisting of a single whorl…(6)

5.   Leaves deeply lobed, divided, or compound…**Ranunculaceae**

5'.   Leaves entire…**Saururaceae**

6.   Ovary inferior…(7)

6'.   Ovary superior…(18)

7.   Plants with tendrils…**Cucurbitaceae**

7'.   Plants without tendrils…(8)

8.   Ovary with 2 locules, 1 ovule in each; fruit 2-seeded…(9)

8'.   Ovary with 1 locule, this with 1–2 ovules (or ovary with 1–3 locules but only 1 locule containing an ovule); fruit 1-seeded…(10)

9.   Petals united at the base; leaves opposite or whorled; flowers often in cymes or paniculate, headlike, clustered or solitary, never in umbels…**Rubiaceae**

9'.   Petals separate; leaves alternate or basal; flowers in umbels…**Apiaceae**

10.   Leaves alternate, glaucous; flowers greenish, white; fruit a drupe…**Comandraceae** (*Comandra*)

10'. Leaves opposite; flowers white or pink; fruit an achene…(11)

11.   Leaves pinnately lobed or divided; fruits with a pappus of plumose bristles…**Caprifoliaceae** (*Valeriana*)

11'. Leaves not as above; fruits not as above…(12)

12.   Leaves simple, entire; flowers pinkish or flesh-colored; fruits hard and bony or with papery wings…**Nyctaginaceae**

12'. Plants not as above…(13)

13.   Leaves markedly fleshy and succulent; plants usually prostrate; fruit a circumscissile capsule…**Aizoaceae**

13'. Leaves not succulent; plants erect to prostrate; fruit not a circumscissile capsule…(14)

14.   Plants semiaquatics of wet meadows, ponds, and marshes; flowers solitary…**Onagraceae** (*Ludwigia*)

14'. Plants of dry, terrestrial habitats…(15)

15.   Leaves simple, not separated into leaflets; flowers shaped like a Dutch pipe…**Aristolochiaceae**

15'. Leaves compound, separated into leaflets; flowers not shaped like a Dutch pipe…(16)

16.   Flowers in spikes; stipules large and conspicuous…**Rosaceae** (*Alchemilla, Sanguisorba*)

16'. Flowers in umbels or panicles; stipules absent or inconspicuous…(17)

17.   Styles 2; flowers in umbels or umbel-like heads…**Apiaceae**

17'. Styles 1; flowers in panicles…**Araliaceae**

18.   Pistils 2 to many in each flower (if 1, the fruit a fleshy, several-seeded berry); stamens usually numerous…(19)

18'. Pistil 1 in each flower; stamens 1 to many (but usually not > 10 in most families)…(21)

19.   Tepals greenish, shorter than the stamens; cypselas ribbed; plants with compound leaves, the leaflets ovate, lobed…**Ranunculaceae** (*Thalictrum*)

19'. Plants not as above…(20)

20.   Carpels enclosed in a 4-angled calyx cup; leaves with stipules; flowers in a dense head…**Rosaceae**

20'. Carpels not enclosed in the calyx; leaves lacking stipules; flowers not in a dense head, often showy…**Ranunculaceae**

21.   Plants with milky juice…**Euphorbiaceae**

21'. Plants with or without milky juice…(22)

22.   Flowers unisexual, ovary on a stalk (in this group individual flowers are reduced to single stamens or a single pistil; the individual, reduced flowers are surrounded by a cuplike involucre that resembles a perianth)…**Euphorbiaceae**

22'. Flowers perfect…(23)

23.   Stamens > twice as many as perianth segments…(24)

23'. Stamens twice as many as perianth segments or fewer…(28)

24.   Perianth small, inconspicuous…(25)

24'. Perianth well developed, colored…(26)

25.   Leaves compound, alternate; flowers in terminal racemes; fruit a red or white berry…**Ranunculaceae** (*Actaea*)

25'. Leaves simple, opposite; flowers in axillary clusters; fruit a capsule…**Aizoaceae** (*Sesuvium*)

26.   Leaves entire, simple…**Talinaceae**

26'. Leaves toothed, lobed, dissected, or compound…(27)

27.   Perianth segments 4 or 8; leaves dissected or compound…**Papaveraceae**

27'. Perianth segments 5; leaves twice pinnately compound…**Fabaceae** (*Mimosa*)
28.   Leaves opposite or whorled…(29)
28'. Leaves alternate or crowded at the base of the stem, usually toothed; flowers in terminal spikes or racemes…(46)
29.   Ovary with 2+ locules…(30)
29'. Ovary with 1 locule…(36)
30.   Flowers unisexual; plants with milky juice…**Euphorbiaceae**
30'. Flowers bisexual; plants lacking milky juice…(31)
31.   Leaves often fleshy; plants not mealy…(32)
31'. Leaves various (fleshy in some *Amaranthus* and *Talinum*); plants mealy (some *Amaranthus*) or not…(34)
32.   Exocarp and endocarp not separating…**Montiaceae**
32'. Exocarp and endocarp separating…(33)
33.   Fruit dehiscence basal; seeds usually black and glossy with a strophiola; embryo curved…**Talinaceae**
33'. Fruit dehiscence apical; seeds usually pale, without a strophiola; embryo rather straight… **Anacampserotaceae**
34.   Flowers in a terminal, pedunculate cluster…**Talinaceae** (*Talinum*)
34'. Flowers axillary and sessile…(35)
35.   Leaves opposite; fruit circumscissile…**Aizoaceae**
35'. Leaves whorled; fruit splitting lengthwise…**Molluginaceae**
36.   Ovary with several ovules; fruit a several-seeded capsule…(37)
36'. Ovary with 1 ovule; fruit a 1-seeded achene or utricle…(39)
37.   Perianth of united tepals, pink; leaves oblong, glaucous…**Primulaceae** (*Glaux*)
37'. Perianth of separate tepals; leaves otherwise…(38)
38.   Perianth segments 5…**Caryophyllaceae**
38'. Perianth segments 4…**Lythraceae** (*Ammannia*)
39.   Leaves with stipules either papery or sheathing the stem…(40)
39'. Stipules none or if present, herbaceous and not sheathing the stem…(41)
40.   Leaves opposite; stipules not united around the stem…**Caryophyllaceae**
40'. Leaves alternate; stipules united around the stem in a sheath just above the node…**Polygonaceae**
41.   Style and stigma 1…(42)
41'. Styles and stigmas 2+…(44)
42.   Perianth segments petal-like, colored, fused into a tube; leaves entire or slightly lobed…**Nyctaginaceae**
42'. Perianth segments sepal-like, green, small and inconspicuous; leaves toothed or entire…(43)
43.   Leaves entire; herbage thickly covered with woolly or silky branched hairs…**Amaranthaceae** (*Tidestromia*)
43'. Leaves entire or toothed; herbage glabrous or sparsely pilose, often with stinging hairs…**Urticaceae**
44.   Leaves reduced to small fleshy scales; stems succulent, jointed; inflorescence a fleshy spike…**Amaranthaceae** (*Salicornia*)
44'. Leaves, stems, and inflorescences other than above…(45)
45.   Perianth and subtending bracts dry and scarious; herbage without a mealy or powdery surface; habitat various, generally not saline, often weedy…**Amaranthaceae**
45'. Perianth and subtending bracts herbaceous or membranous; herbage often fleshy, often with a mealy or powdery surface; often in saline soils…**Amaranthaceae** ("chenopods")
46.   Leaves with stipules, generally well developed and obvious…(47)
46'. Leaves without stipules, or stipules reduced to scalelike bracts…(52)
47.   Style and stigma 1…(48)
47'. Styles and stigmas 2+…(49)
48.   Ovule 1 per ovary; fruit an achene…**Rosaceae**
48'. Ovules several to many; fruit a capsule…**Violaceae**
49.   Leaves palmately compound or lobed…**Polygonaceae**
49'. Leaves simple, not lobed…(50)
50.   Stipules fused into a sheath around the stem…**Polygonaceae**
50'. Stipules distinct, not fused into a sheath…(51)
51.   Locules with 1 ovule…**Euphorbiaceae**
51'. Locules with 2 ovules…**Phyllanthaceae**
52.   Perianth segments 6 (petal-like tepals)…**Polygonaceae**
52'. Perianth segments 5 or fewer, generally not petal-like…(53)
53.   Ovule 1 per ovary, inflorescence various…(54)
53'. Ovules 2 to many per ovary; inflorescence a raceme…(55)
54.   Perianth and subtending bracts dry and scarious; herbage without a mealy or powdery surface; habitat various, generally not saline, often weedy…**Amaranthaceae**

54'. Perianth and subtending bracts herbaceous or membranous; herbage often fleshy, often with a mealy or powdery surface; often in saline soils…**Amaranthaceae** ("chenopods")

55. Perianth segments 4…(56)

55'. Perianth segments 5…(57)

56. Fruit a fleshy berry; stamens 4…**Petiveriaceae**

56'. Fruit a dry silique or silicle; stamens 6…**Brassicaceae**

57. Flowers actinomorphic; stamens free; fruit a berry…**Phytolaccaceae** (*Phytolacca*)

57'. Flowers zygomorphic; stamens fused; fruit a capsule…**Polygalaceae**

KEY 6B: PETALS PRESENT, SEPARATE

1. Stamens alternating with branched staminodes having terminal, yellow, antherlike glands…**Celastraceae** (*Parnassia*)

1'. Stamens lacking alternating branched staminodes…(2)

2. Ovary inferior, or at least the lower 1/2 fused to the hypanthium or calyx tube…(3)

2'. Ovary superior (if hypanthium present, the ovary may seem to be inferior, but it is not embedded in the hypanthium tissues, as in a rose hip)…(10)

3. Prostrate succulent herbs with rounded, oblanceolate, thick leaves; stamens variable in number, petals yellow…**Portulacaceae** (*Portulaca*)

3'. Not prostrate, succulent herbs…(4)

4. Styles 2+, separate…(5)

4'. Style 1, may be lobed or cleft…(7)

5. Stipules present; fruit a capsule; sepals 5; petals 5; flowers borne 2+ together or not showy… **Saxifragaceae**

5'. Stipules none; fruit various; flowers various…(6)

6. Locules 4–6; fruit a several-seeded berry; leaves basal, ternately compound…**Araliaceae**

6'. Locules 2; fruit dry, separating into two 1-seeded mericarps…**Apiaceae**

7. Flowers in involucrate heads, the involucre simulating the calyx; floral tube constricted above the ovary and enclosing the 1-seeded fruit (thus appearing inferior)…**Nyctaginaceae**

7'. Flowers not in involucrate heads; both calyx and corolla present; fruit 2- to several-seeded…(8)

8. Foliage sandpapery, with minutely barbed hairs…**Loasaceae**

8'. Plants lacking barbed hairs…(9)

9. Stamens 2, 4, or 8; petals 2 or 4; locules usually 4 (2 in *Circaea*)…**Onagraceae**

9'. Stamens 4–5; petals 4–5; locules usually 2…**Cornaceae**

10. Sepals 2…(11)

10'. Sepals > 2…(12)

11. Sepals deciduous; plants not succulent…**Papaveraceae**

11'. Sepals persistent; plants somewhat succulent…**Portulacaceae**

12. Stamens united in a column around the styles…**Malvaceae**

12'. Stamens not united in a column around the styles…(13)

13. Flowers zygomorphic…(14)

13'. Flowers actinomorphic…(21)

14. Plants annual; leaves lanceolate, entire; flowers small, inconspicuous; inflorescence a slender terminal spike…**Resedaceae**

14'. Plants perennial; leaves various; flowers large, conspicuous; inflorescence various…(15)

15. Leaves palmately lobed or parted; flowers orange, large, conspicuous and borne in a terminal raceme… **Cochlospermaceae**

15'. Plants not as above…(16)

16. Petals 3, usually red, sometimes green; flowers tiny…**Cistaceae**

16'. Plants not as above…(17)

17. Leaves pinnately or palmately compound…(18)

17'. Leaves simple, entire to deeply lobed, not truly compound…(19)

18. Ovary with 1 placenta; petals 5 (a banner, 2 wings, and a keel that consists of 2 partly united petals enclosing the stamens and style); flowers usually shaped like those of a sweet pea…**Fabaceae**

18'. Ovary with 2 placentae on opposite sides of the ovary; petals 4; stamens exserted…**Brassicaceae**

19. Stamens many; carpels > 1; fruit a follicle (dry and dehiscing along 1 suture)…**Ranunculaceae**

19'. Stamens 10 or fewer; ovary of a single or several united carpels; fruit a capsule (dry and dehiscing along > 1 suture)…(20)

20. Flowers not spurred, but with a large upper petal (banner)…**Polygalaceae**

20'. Flowers spurred…**Violaceae**

21. Stamens of the same number as the petals and opposite them…(22)

21'. Stamens fewer or more numerous than the petals, or if the same number, then alternate with them…(24)

22. Sepals, petals, and stamens each 6, 3 of the sepals petal-like; leaf margins spiny…**Berberidaceae**

22'.   Sepals, petals, and stamens 2–5 (sepals rarely 6); spineless…(23)
23.   Style and stigma 1; sepals usually 5…**Primulaceae**
23'.   Styles and stigmas 2+; sepals usually 2…**Portulacaceae**
24.   Ovary 1 (a single unit), with 1 locule…(25)
24'.   Ovaries > 1 (several units), or if 1, then with 2+ locules…(38)
25.   Stamens 13+…(26)
25'.   Stamens 12 or fewer…(32)
26.   Ovary simple (of a single carpel having 1 placenta, 1 style, 1 stigma; many such ovaries may be present in a single flower)…**Ranunculaceae**
26'.   Ovary compound (2+ placentae, styles, or stigmas)…(27)
27.   Placenta free-central or basal…**Portulacaceae**
27'.   Placenta parietal…(28)
28.   Ovary with 2 parietal placentae; plants usually viscid and ill-smelling…**Cleomaceae**
28'.   Ovary with 3+ placentae; plants not viscid or ill-smelling…(29)
29.   Leaves alternate, toothed or lobed; flowers white or cream; juice milky…**Papaveraceae**
29'.   Leaves opposite or alternate; flowers yellow or white; juice not milky…(30)
30.   Leaves simple, entire, opposite or whorled, not fleshy, with minute translucent dots…**Hypericaceae**
30'   Leaves simple or compound, alternate, fleshy, without minute translucent dots…(31)
31.   Leaves opposite; mostly with 2 fleshy sessile leaflets; Doña Ana County…**Zygophyllaceae** (*Zygophyllum*)
31'.   Leaves alternate; blades divided into irregular linear lobes; widespread in S…**Nitrariaceae** (*Peganum*)
32.   Pistil simple (with 1 placenta, style, and stigma)…(33)
32'.   Pistil compound (with > 1 placenta, style, or stigma)…(34)
33.   Stamens and petals attached to the rim of the calyx tube (hypanthium)…**Rosaceae**
33'.   Stamens and petals not attached to the calyx tube…**Fabaceae**
34.   Ovules attached to base of ovary or to a free-central placenta…(35)
34'.   Ovules attached to 2+ parietal placentae…(37)
35.   Calyx of united sepals; petals with claws…**Caryophyllaceae**
35'.   Calyx of separate sepals; petals not stalked…(36)
36.   Sepals 2 or numerous; stamens commonly opposite the petals, sometimes fewer than the petals, sometimes numerous; leaves commonly basal and succulent…**Portulacaceae**
36'.   Sepals usually 5; stamens not opposite the petals, usually 5 or 10 (rarely 3); leaves usually opposite, not especially succulent…**Caryophyllaceae**
37.   Ovary with 2 parietal placentae; sepals and petals 4 each…**Brassicaceae**
37'.   Ovary with 3–5 parietal placentae; sepals and petals 5…**Hypericaceae**
38.   Plants with milky juice or stinging hairs; ovary stipitate, exserted from the cyathium…**Euphorbiaceae**
38'.   Plants without milky juice or stinging hairs…(39)
39.   Perianth parts in 2s (rarely) or 4s…**Brassicaceae**
39'.   Perianth parts in 5s or numerous…(40)
40.   Leaves trifoliate, acrid-tasting…**Oxalidaceae**
40'.   Leaves not trifoliate or acrid-tasting…(41)
41.   Leaves linear or oblong, succulent…**Crassulaceae**
41'.   Leaves not as above…(42)
42.   Stamens numerous…(43)
42'.   Stamens not > 10…(45)
43.   Leaves elliptic with translucent dots and often minute, black, marginal dots on leaves and petals; stamens tending to be in 5 groups; fruit a capsule…**Hypericaceae**
43'.   Leaves and stamens not as above; fruit an achene…(44)
44.   Stipules lacking; hypanthium not developed…**Ranunculaceae**
44'.   Stipules present; hypanthium well developed…**Rosaceae**
45.   Petals waxy; anthers opening by terminal pores; leaves often leathery…**Ericaceae**
45'.   Petals not waxy; anthers opening by slits…(46)
46.   Fruit 5-carpellate, separating at maturity into five 1-seeded segments (mericarps)…(47)
46'.   Fruit not 5-carpellate, or if so, not separating into mericarps…(48)
47.   Flowers pink or white; mericarps not spiny…**Geraniaceae**
47'.   Flowers yellow; mericarps stoutly spiny (*Tribulus*) or not…**Zygophyllaceae**
48.   Petals yellow, copper, or blue, falling within a few hours; capsule 10-locular (1 ovule per locule)…**Linaceae**
48'.   Petals white, yellow, or pink-purple, not falling soon after flowering; capsule not as above…**Saxifragaceae**

KEY 6C: PETALS UNITED; FLOWERS RADIALLY SYMMETRIC

1.   Plants with milky juice…(2)
1'.   Plants without milky juice…(4)

2. Corolla rotate, with a central structure (gynostegium) consisting of fused stigmas and stamens; corona present, consisting of erect tubular hoods, these usually enclosing fine, projecting toothlike structures (horns)...**Apocynaceae** (*Asclepias*)
2'. Corolla bell-shaped, without a gynostegium; corona absent...(3)
3. Leaves opposite...**Apocynaeae**
3'. Leaves alternate...**Campanulaceae**
4. Ovary superior...(5)
4'. Ovary inferior or 1/2 inferior...(28)
5. Stamens more numerous than corolla lobes, 6 to many...(6)
5'. Stamens as many as corolla lobes or fewer...(14)
6. Leaves compound...(7)
6'. Leaves simple...(8)
7. Leaflets 3; petals 5; stamens 10...**Oxalidaceae**
7'. Leaflets many; petals 4; stamens 6...**Papaveraceae**
8. Flower parts in 3s; fruit a triangular achene...**Polygonaceae**
8'. Flower parts in 5s; fruit a capsule or berry...(9)
9. Leaves succulent...**Crassulaceae**
9'. Leaves not succulent...(10)
10. Leaves opposite...(11)
10'. Leaves alternate...(12)
11. Leaf pairs usually unequal in size; style 1...**Nyctaginaceae**
11'. Leaf pairs usually equal in size; styles 3–5...**Caryophyllaceae**
12. Stamens numerous, > 10, united into a tube around the style...**Malvaceae**
12'. Stamens 6–10, not united into a tube around the style, separate and distinct...(13)
13. Stamens 10 (5 fertile alternating with 5 sterile staminodes); petals with long, coiled threadlike stalks and united at the broadened tips over the stamens...**Malvaceae**
13'. Stamens 6–10, separate and distinct; anthers opening by pores at the basal end; petals waxy...**Ericaceae** (*Pyrola*)
14. Fertile stamens of the same number as corolla lobes, usually 5...(15)
14'. Fertile stamens (having well-developed anthers) 2 or 4, usually fewer in number than the obvious corolla lobes, 1+ sterile stamens (staminodes) sometimes present...(23)
15. Stamens opposite the petals, sometimes with additional sterile stamens (staminodes) alternate with the petals...(16)
15'. Stamens alternate with the petals; sterile stamens none...(20)
16. Plants annual or perennial; usually succulent, glabrous herbs, often with swollen nodes...**Montiaceae** (*Montia*, *Claytonia*, *Lewisia*, *Phemeranthus*)
16'. Plants and/or leaves otherwise...(17)
17. Styles and stigmas 4–5; fruit an achene or a utricle; ovule 1...**Plumbaginaceae**
17'. Style and stigma 1; fruit a capsule; ovules several to numerous...(18)
18. Leaves borne on the stem...**Primulaceae** (*Lysimachia*)
18'. Leaves all basal...(19)
19. Inflorescence a raceme or panicle...**Primulaceae** (*Samolus*)
19'. Inflorescence an umbel...**Primulaceae** (*Primula*, *Androsace*)
20. Flowers yellow, in dense spikes or racemes > 20 cm; filaments hairy...**Scrophulariaceae** (*Verbascum*)
20'. Flowers variously colored, never in spikes or elongate racemes...(21)
21. Leaves cauline, all opposite or whorled, simple, entire...**Gentianaceae**
21'. Leaves not as above...(22)
22. Ovary not 4-lobed; fruit a capsule or berry, usually several-seeded...(23)
22'. Ovary 4-lobed, developing into 4 (or by abortion fewer) 1-seeded nutlets...(24)
23. Leaves alternate; stem not square in cross-section; not aromatic...(27)
23'. Leaves opposite; stem square in cross-section; mostly aromatic...**Lamiaceae**
24. Ovary 4-lobed or -grooved and 4-chambered; fruit of 4 nutlets or mericarps or 1–3 of the 4 nutlets abortive; style 1; flowers usually in 1-sided cymes...(25)
24'. Ovary entire, 1–3-chambered; fruit a capsule or berry; styles and flowers various, sometimes coiled...(27)
25. Ovary 4-lobed, subdivided into 4 uniovulate lobes separated by deep folds and dehiscing as individual nutlets (may be fewer by abortion and sometimes the nutlets not all alike); style gynobasic and passing through the 4 lobes of the ovary...**Boraginaceae**
25'. Ovary entire or 4-grooved in flower; fruit schizocarpic, separating into 2–4 individual mericarps (nutlets), may be fewer by abortion; style apical from the ovary...(26)
26. Style undivided, with a conic or discoid stigmatic head...**Heliotropiaceae**
26'. Style with 2 branches, each with a capitate stigma...**Ehretiaceae**

27. Ovary with 1 locule; leaves basal; flowers solitary, scapose; stoloniferous plants rooted in mud...**Plantaginaceae** (*Limosella*)

27'. Ovary with 2+ locules; plants not as above...(28)

28. Stigma 3-lobed or style 3-branched; ovary with 3 locules...**Polemoniaceae**

28'. Stigma entire or 2-lobed, or style 2-cleft; ovary usually with 2 locules...(29)

29. Style 2–5-cleft apically...(30)

29'. Style entire, not cleft apically...(32)

30. Style 3–5-cleft; small trees, shrubs, or perennial herbs; fruit a drupe...**Viburnaceae**

30'. Style 2-cleft or 2-branched below the stigmas; perennial herbs; fruit a capsule...(31)

31. Leaves basal and cauline, rarely all cauline; simple or variously divided to bipinnate; flowers usually in coiled 1-sided cymes; style 1 with stigmatic branches...**Hydrophyllaceae**

31'. Leaves all cauline, entire; flowers solitary and axillary, or in terminal nonscorpioid cymes; stylodia 2, distinct to the base...**Namaceae**

32. Style 1, the stigma entire or 2-lobed; fruit a capsule or berry...**Solanaceae**

32'. Plants not as above...(33)

33. Stamens numerous, > 10 per flower...**Loasaceae**

33'. Stamens 10 or fewer per flower...(34)

34. Stamens borne opposite the midveins of the petals...**Primulaceae**

34'. Plants not as above...(35)

35. Stems prickly...**Caprifoliaceae** (*Dipsacus*)

35'. Stems not prickly...(36)

36. Leaves alternate or basal...**Campanulaceae**

36'. Leaves opposite or whorled...(37)

37. Stems creeping, slightly woody; leaves opposite, crenate; flowers 2, pink, pendent from an erect stalk...**Caprifoliaceae** (*Linnaea*)

37'. Stems erect or sprawling, herbaceous; leaves opposite or whorled, entire; flowers white, minute, in cymes...**Rubiaceae**

KEY 6D: PETALS UNITED; FLOWERS BILATERALLY SYMMETRIC

1. Ovary inferior...**Caprifoliaceae** (*Valeriana*)

1'. Ovary superior...(2)

2. Stems 4-angled; leaves opposite...(3)

2'. Stems not 4-angled; leaves opposite, alternate, or basal...(5)

3. Corolla not 2-lipped, with a narrow tube and flaring lobes...**Verbenaceae**

3'. Corolla 2-lipped, usually strongly so, the tube not especially narrow...(4)

4. Corolla crimson, brownish, green, or yellow; style gynoecium not lobed or divided; foliage never with a minty odor...**Scrophulariaceae**

4'. Corolla color various; style arising from the base of the lobed gynoecium; foliage usually with a minty or otherwise aromatic odor...**Lamiaceae**

5. Calyx tubular, ribbed or pleated; fruit a capsule, loculicidal...**Phrymaceae**

5'. Calyx various, not ribbed or pleated; fruit a septicidal capsule or nutlet...(6)

6. Corolla papery; leaves basal; inflorescence a spike; fruit a circumscissile capsule...**Plantaginaceae** (*Plantago*)

6'. Corolla not papery; plants otherwise not as above...(7)

7. Calyx split to the base along the lower side; corolla 3–5 cm and nearly as wide; fruit woody, with long claws...**Martyniaceae**

7'. Calyx not split to the base on the lower side; corolla various; fruit not as above...(8)

8. Corolla not spurred or saclike; fruit loculicidal...**Orobanchaceae**

8'. Corolla base spurred or saclike; fruit dehiscent by terminal slits or pores...(9)

9. Plants annual...(10)

9'. Plants perennial...(14)

10. Flowers with a fifth sterile stamen or fertile stamens 4 or 2 (then with 2 staminodes); corolla gibbous, saclike, or bilabiate...(11)

10'. Flowers not with sterile stamens; corolla not gibbous or saclike...(12)

11. Flowers with a fifth sterile stamen; corolla gibbous or saclike...**Plantaginaceae** (*Collinsia*)

11'. Flowers with fertile stamens 4 or 2 (then with 2 staminodes); corolla bilabiate...**Linderniaceae**

12. Calyx strongly 5-angled or 5-pleated...**Phrymaceae** (*Erythranthe*)

12'. Calyx not 5-angled or 5-pleated...(13)

13. Leaves pinnatifid, lobes usually toothed; corolla with yellow tube and violet limb...**Plantaginaceae** (*Schistophragma*)

13'. Leaves usually entire; corolla mostly purple or pinkish purple...**Orobanchaceae** (*Agalinis*)

14. Sterile stamens absent; leaves palmately or pinnately veined; calyx strongly 5-angled or 5-pleated… **Phrymaceae** (*Erythranthe*)

14'. Sterile stamen present; leaves rarely palmately veined; calyx not 5-angled or 5-pleated…(15)

15. Corolla urn-shaped, 6–20 mm; sterile stamens flattened, scalelike; stems 4-angled…**Scrophulariaceae**

15'. Corolla campanulate to tubular, 10–40 mm; sterile stamen slender, mostly elongate-filiform; stems usually round…**Plantaginaceae** (*Penstemon*)

FERNS AND LYCOPHYTES

Michael D. Windham

(Pteridophyte Phylogeny Group, 2016; Smith, 1993)

ISOËTACEAE – QUILLWORT FAMILY

ISOËTES L. – Quillwort

Isoëtes bolanderi Engelm.—Bolander's quillwort. Plants rooted aquatics (rarely amphibious or up-rooted and stranded on shore by wave action); stems barely subterranean, erect, globose, ± glabrous corms, bilobed but otherwise not ridged or grooved, internodes poorly defined, not hollow; stem scales absent; leaves monomorphic, all potentially fertile, spirally arranged, closely spaced, to 20 cm; petioles lacking; blades linear, entire, < 0.5 cm wide, glabrous, abruptly tapering to fine tip, margins rounded (flattened and winged proximally but not recurved); veins apparent or obscure, 1 per leaf, terminating at apex; sori absent, each fertile leaf with single sporangium embedded in center of widened base; indusia absent; sporangia > 0.3 mm; spores dimorphic, microspores thousands per sporangium, 20–30 μ, mono-lete, megaspores usually 50-100 per sporangium, 300–500 μ, trilete, rugose or tuberculate. In shallow water (rarely emergent) around margins of high-elevation ponds and lakes, 8900-10,700 ft. NC, NW.

LYCOPODIACEAE – CLUBMOSS FAMILY

Plants terrestrial; stems superficial, horizontal, spreading over ground surface and rooting at nodes, gla-brous, ± ridged by adnate leaf bases but these short and discontinuous, internodes poorly defined, not hollow; stem scales absent; leaves dimorphic (obscurely so in *Huperzia*), the fertile leaves usually shorter, broader, and paler than the sterile leaves, spirally pseudowhorled, closely spaced, to 1 cm; petioles lack-ing; blades linear, entire to denticulate, < 0.2 cm wide, glabrous, the margins flat to rounded; veins ob-scure or marked by a longitudinal ridge abaxially, 1 per leaf, terminating before reaching apex; sori absent, each fertile leaf (sporophyll) with single sporangium attached adaxially at center of widened base adja-cent to stem, sporophylls aggregated to form terminal strobili or (in *Huperzia*) interrupted fertile zones on upright stems; indusia absent; sporangia > 0.3 mm; spores monomorphic, hundreds per sporangium, 30–50 μm, trilete (Field et al., 2016).

Two genera and three species are reported for New Mexico, but only *Lycopodium annotinum* is confirmed.

1. Fertile and sterile stems very similar, both rooting along short-creeping base and becoming almost erect distally; leaves subtending sporangia (sporophylls) very similar to sterile leaves, aggregated into nonterminal zones along otherwise undifferentiated stems [to be expected]…Huperzia lucidula (Michx.) Trevis.
1'. Fertile stems arising as upright (nonrooting) branches from long-creeping, horizontal stems; leaves subtending sporangia (sporophylls) strongly differentiated from sterile leaves, aggregated into terminal stro-bili…**Lycopodium**

LYCOPODIUM L. – Clubmoss

1. Strobili sessile on undifferentiated sterile leafy branches; leaves lacking hairlike tips…**L. annotinum**
1'. Strobili terminating on differentiated stalks (peduncles) with highly reduced leaves; stem leaves with hair-

like tips (these deciduous on older leaves but always present near apices of sterile shoots) [to be expected]…L. clavatum L.

Lycopodium annotinum L. [*Spinulum annotinum* (L.) A. Haines]—Bristly clubmoss. Stems of 2 types: horizontal, rooting stems giving rise to upright, nonrooting branches periodically producing sporangia; leaves of upright branches lacking hairlike tips, dimorphic, the sterile leaves with linear-lanceolate blades mostly > 4 mm, fertile leaves (sporophylls) ovate, < 4 mm; sporophylls aggregated into terminal strobili sessile on sterile leafy branches; strobili solitary. In moist duff on forest floor, 8600–10,700 ft. NC. Although some authors assign this species to the segregate genus *Spinulum* A. Haines, DNA analyses by Field et al. (2016) indicate that it forms a monophyletic group with *Lycopodium* s.s.

SELAGINELLACEAE – SPIKEMOSS FAMILY

SELAGINELLA P. Beauv. - Spikemoss

Plants usually growing on or around rocks; stems superficial (rarely with subterranean rhizomes), horizontal (fertile stems erect in some species), often spreading parallel to ground surface and rooting at nodes, glabrous (rarely sparsely hispid), often ± ridged by adnate leaf bases but these short and discontinuous, internodes poorly defined, not hollow; stem scales absent; leaves dimorphic depending on fertility (fertile leaves shorter and broader than sterile) and often orientation (dorsal vs. ventral, median vs. lateral), spirally pseudowhorled, closely spaced, to 0.4 cm; petioles lacking; blades linear to ovate, denticulate to ciliate, < 0.1 cm wide, glabrous or less often minutely hispid, the margins flat, with hispid teeth, cilia, and/or terminal setae; veins marked by shallow longitudinal groove abaxially, 1 per leaf, terminating before reaching apex; sori absent, each fertile leaf (sporophyll) with single sporangium attached adaxially at center of widened base adjacent to stem, sporophylls aggregated to form terminal strobili; indusia absent; sporangia > 0.3 mm; spores dimorphic, microspores hundreds per sporangium, 30–60 µ, trilete, megaspores 4 (rarely 2 or 1) per sporangium, 250–500 µ, trilete.

Eleven species are reported for New Mexico, with nine confirmed.

1. Stems strongly flattened dorsiventrally, forming rosettes of frondlike structures radially arranged on localized root cluster; leaves clearly dimorphic, green leaves much more prevalent on dorsal surfaces and lateral and median leaf rows well differentiated…(2)
1'. Stems not strongly flattened, forming compact to loose mats proliferating laterally by creeping surficial or subterranean branches with diffuse roots; leaves monomorphic, or the dorsal and ventral leaves somewhat different, but median and lateral rows not clearly distinct…(3)
2. Leaf apices with hairlike, often twisted tip > 0.5 mm…**S. pilifera**
2'. Leaf apices acuminate, the scarious, flattened tip < 0.5 mm…S. lepidophylla (Hook. & Grev.) Spring
3. Fertile aboveground stems mostly erect or ascending, rhizophores produced only near bases (proximal 1/2), the distal unrooted portions usually > 3 cm…(4)
3'. Fertile aboveground stems creeping along substrate, rhizophores produced most of their length, the distal unrooted portions < 3 cm…(6)
4. Leaf bases partially fused to main stems, forming low decurrent ridges; blade margins short-bristly to denticulate, the projections < 0.1 mm; N New Mexico…**S. weatherbiana**
4'. Leaf bases abruptly attached to main stems, not forming low ridges; blade margins ciliate proximally or throughout, at least some cilia > 0.1 mm; S New Mexico…(5)
5. Leaf apices with bristle tip > 0.5 mm; leaves > 2.8 mm; sporangia and spores mostly well-formed… **S. rupincola**
5'. Leaf apices with bristle tip < 0.5 mm; leaves < 2.8 mm; sporangia and/or spores abortive… **S. ×neomexicana**
6. Leaves ± dimorphic, the ventral leaves longer, broader, yellowish to tan, and curved upward; stems (distal portion) curling dorsally when dormant to expose largely nonphotosynthetic ventral surface…(7)
6'. Leaves monomorphic, or if slightly dimorphic, then green leaves ± equally prevalent on dorsal and ventral surfaces; stems (distal portion) usually not curling dorsally when dormant…(9)
7. Leaves on ventral side of main stems lanceolate, ± papery; leaf apices with flattened, scarious bristle tip usually < 0.2 mm…S. arizonica Maxon
7'. Leaves on ventral side of main stems linear-lanceolate, ± fleshy; leaf apices with rounded to slightly flattened, stiff bristle tip > 0.2 mm (bristle often deciduous with age in *S. wrightii*)…(8)

8. Leaves on ventral side of main stems < 2.8 mm, partially fused to stems, forming low decurrent ridges; leaves clearly dimorphic, green leaves much more prevalent on dorsal surfaces; cilia on blade margins mostly < 0.15 mm…**S. peruviana**
8'. Leaves (the largest) on ventral side of main stems > 2.8 mm, usually abruptly attached, not forming low ridges; leaves weakly dimorphic, green leaves only slightly more prevalent on dorsal surfaces; cilia on blade margins mostly > 0.15 mm…**S. wrightii**
9. Leaves on ventral side of main stems < 2 mm, abruptly attached, not forming low ridges on stems… **S. mutica**
9'. Leaves on ventral side of main stems > 2 mm, the bases partially fused to stems, forming low decurrent ridges…(10)
10. Plants forming compact, closely branched decumbent mats; strobili often closely spaced; leaves appressed and closely spaced such that stem surfaces are rarely visible; leaves of main stems weakly dimorphic, the ventral slightly larger and curved upward…**S. densa**
10'. Plants forming diffuse, open-branched prostrate mats; strobili scattered; leaves ascending to loosely appressed, spaced such that stem surfaces are often visible; leaves of main stems monomorphic… **S. underwoodii**

Selaginella densa Rydb.—Rocky Mountain spikemoss. Plants forming compact, closely branched mats; stems not strongly flattened dorsiventrally, spreading laterally by decumbent branches, all stems creeping along substrate surface, rhizophores produced nearly throughout, the distal unrooted portion of stems < 3 cm, usually not curling dorsally when dormant; leaves of main stems weakly dimorphic, green leaves slightly more prevalent on dorsal surfaces, ventral leaves slightly larger and curved upward, linear-lanceolate, the largest mostly > 2.8 mm, ± fleshy, appressed and closely spaced such that stem surfaces are rarely visible; ventral leaf bases partially fused to main stems, forming low decurrent ridges; blade margins usually short-ciliate proximally, the largest cilia < 0.15 mm; leaf apices with a rounded, stiff, mostly persistent bristle tip > 0.5 mm; strobili often closely spaced; sporangia and spores mostly well formed. New Mexico plants represent **var. scopulorum** (Maxon) R. M. Tryon or intermediates between this and var. *densa* [*S. scopulorum* Maxon]. Usually on igneous rocks, 6200–13,000 ft. NE, NC, C, WC, SC, SW.

Selaginella mutica D. C. Eaton ex Underw.—Blunt-leaf spikemoss. Plants forming diffuse, open-branched mats; stems not flattened dorsiventrally, spreading laterally by prostrate branches, all stems creeping along substrate surface, rhizophores produced nearly throughout, the distal unrooted portion of stems < 3 cm, not curling dorsally when dormant; leaves of main stems monomorphic, lanceolate, 1–2 mm, ± fleshy, tightly appressed but spaced such that stem surfaces are occasionally visible; ventral leaf bases abruptly attached to main stems, not forming low ridges; blade margins ciliate to short-bristly, the largest cilia < 0.15 mm; leaf apices blunt, minutely mucronate or with a rounded, ± flexible, mostly persistent bristle tip < 0.5 mm; strobili scattered; sporangia and spores mostly well formed. Two infraspecific taxa have been recognized in New Mexico: the more southern **var. limitanea** Weath. with short-bristly leaf margins and a longer apical bristle, and the widespread **var. mutica** with long-ciliate leaf margins and a highly reduced apical bristle. These appear to intergrade and more work is needed to determine whether they represent distinct taxa. On a variety of substrates including both limestone (common) and granite (rare), 4400–9500 ft. All but NW and likely there as well.

Selaginella ×neomexicana Maxon—New Mexico spikemoss. Plants forming ascending clumps or loose decumbent mats; stems not strongly flattened, spreading laterally by creeping subterranean and surficial branches, rhizophores confined to lower 1/2 of upright stems or nearly throughout on rhizomatous stems, fertile aboveground stems usually erect or ascending with distal unrooted portion > 3 cm, not curling when dormant; leaves of aboveground stems monomorphic, linear-lanceolate, the largest < 2.8 mm, ± fleshy, tightly appressed and closely spaced such that stem surfaces are rarely visible; leaf bases abruptly attached to main stems, not forming low ridges; blade margins ciliate, the largest cilia < 0.15 mm; leaf apices with a rounded, stiff, mostly persistent bristle tip < 0.5 mm; strobili scattered; sporangia and/or spores abortive. Usually on igneous rocks, 6000–6300 ft. SC, SW. This sterile hybrid has arisen repeatedly where the ranges of *S. mutica* and *S. rupincola* overlap.

Selaginella peruviana (Milde) Hieron. [*S. sheldonii* Maxon]—Peruvian spikemoss. Plants forming compact to diffuse mats; stems not strongly flattened dorsiventrally, spreading laterally by decumbent branches, all stems creeping along substrate surface, rhizophores produced nearly throughout, distal unrooted portion of stems < 3 cm, curling dorsally when dormant to expose largely nonphotosynthetic ventral surface; leaves of main stems clearly dimorphic, green leaves much more prevalent on dorsal surfaces, the ventral leaves longer, broader, yellowish to tan, and curved upward, linear-lanceolate, mostly < 2.8 mm, ± fleshy, appressed and closely spaced such that stem surfaces are rarely visible; ventral leaf bases partially fused to main stems, forming low decurrent ridges; blade margins ciliate to short-bristly, the largest cilia < 0.15 mm; leaf apices with a rounded, ± flexible, mostly persistent bristle tip 0.3–0.8 mm; strobili scattered or closely spaced; sporangia and spores mostly well formed. On a variety of rock types, 3400–7100 ft. NE, EC, C, SE, SC.

Selaginella pilifera A. Braun—Hairy resurrection spikemoss. Plants forming rosettes; stems strongly flattened dorsiventrally, forming clusters of frondlike structures radially arranged around central root cluster, not spreading laterally, stems decumbent when hydrated, strongly incurved when dry, rhizophores restricted to rosette base, distal unrooted portion of stems > 3 cm, curling dorsally when dormant to expose nonphotosynthetic ventral surface; leaves strongly dimorphic, green leaves much more prevalent on dorsal surfaces and lateral and medial leaf rows strongly differentiated, lateral leaves elliptic to elliptic-ovate, 2–3.5 mm, ± fleshy, tightly appressed and closely spaced such that stem surfaces are very rarely visible; lateral leaf bases cordate, abruptly attached to main stems, not forming low ridges; lateral blade margins short-ciliate proximally, denticulate distally, the largest cilia < 0.1 mm; lateral leaf apices with a proximally flattened, often twisted, mostly persistent hairlike tip > 0.5 mm; strobili scattered; sporangia and spores mostly well formed. Usually on limestone, 4400–5200 ft. SE.

Selaginella rupincola Underw.—Rock-loving spikemoss. Plants forming upright clumps; stems not flattened dorsiventrally, spreading laterally by creeping subterranean (rarely surficial) branches, fertile aboveground stems erect or ascending, rhizophores confined to lower 1/2 of upright stems, distal unrooted portion of stems > 3 cm, not regularly curling when dormant; leaves of aboveground stems monomorphic, linear-lanceolate, the largest > 2.8 mm, ± fleshy, tightly appressed and closely spaced such that stem surfaces are rarely visible; leaf bases abruptly attached to main stems, not forming low ridges; blade margins ciliate proximally or throughout, the largest cilia > 0.15 mm; leaf apices with a rounded, stiff, mostly persistent bristle tip > 0.5 mm; strobili scattered; sporangia and spores mostly well formed. Usually on igneous rocks, 4500–7400 ft. SC, SW.

Selaginella underwoodii Hieron.—Underwood's spikemoss. Plants usually forming diffuse, open-branched mats; stems not strongly flattened dorsiventrally, spreading laterally by prostrate branches, all stems creeping along substrate surface, rhizophores produced nearly throughout, distal unrooted portion of stems < 3 cm, not regularly curling dorsally when dormant; leaves of main stems monomorphic, linear-lanceolate, 2–3.5 mm, ± fleshy, ascending to loosely appressed, spaced such that stem surfaces are often visible; ventral leaf bases partially fused to main stems, forming low decurrent ridges; blade margins denticulate or short-bristly, the largest projections < 0.1 mm; leaf apices with a rounded, ± flexible, mostly persistent bristle tip 0.3–0.8 mm; strobili scattered; sporangia and spores mostly well formed. On a variety of rock types, 4900–10,400 ft. NE, NC, NW, C, WC, SC, SW.

Selaginella weatherbiana R. M. Tryon—Weatherby's spikemoss. Plants forming upright clumps; stems not strongly flattened, spreading laterally by creeping subterranean (rarely surficial) branches, fertile aboveground stems erect or ascending, rhizophores confined to lower 1/2 of upright stems, distal unrooted portion of stems > 3 cm, usually not curling when dormant; leaves of aboveground stems monomorphic, linear-lanceolate, the largest < 2.8 mm, ± fleshy, appressed and spaced such that stem surfaces are occasionally visible; leaf bases partially fused to main stems, forming low decurrent ridges; blade margins short-bristly to denticulate, the largest projections < 0.1 mm; leaf apices with a rounded, ± flexible, mostly persistent bristle tip, the largest > 0.5 mm; strobili scattered; sporangia and spores mostly well formed. Usually on igneous rocks, 7700–11,700 ft. NC.

Selaginella wrightii Hieron.—Wright's spikemoss. Plants forming mostly compact mats; stems not strongly flattened dorsiventrally, spreading laterally by decumbent branches, all stems creeping along substrate surface, rhizophores produced nearly throughout, distal unrooted portion of stems < 3 cm, curling dorsally when dormant to expose ventral surface; leaves weakly dimorphic, green leaves slightly more prevalent on dorsal surfaces, the ventral leaves somewhat longer, broader, yellowish to tan, and curved upward, linear-lanceolate, the largest > 2.8 mm, ± fleshy, appressed and closely spaced such that stem surfaces are rarely visible; ventral leaf bases abruptly attached to main stems, not forming low ridges; blade margins ciliate proximally, the largest cilia > 0.15 mm; leaf apices with a slightly flattened, stiff, often deciduous bristle tip < 0.5 mm; strobili usually closely spaced; sporangia and spores mostly well formed. Usually on limestone, 3200–4600 ft. SE.

ASPLENIACEAE – SPLEENWORT FERN FAMILY

ASPLENIUM L. – Spleenwort ferns

Plants growing on or around rocks; stems subterranean, ± erect, compact to short-creeping, scaly rhizomes, not continuously ridged and grooved, internodes poorly defined, not hollow; stem scales ≤ 5 mm, black throughout or strongly clathrate with thick, black radial walls and clear luminae; leaves monomorphic or weakly dimorphic with the fertile longer and erect, spirally arranged, clustered, to 25 cm; petioles mostly > 1 cm, with 2 vascular bundles at base and few scales like those on stems; blades linear to lanceolate, 1-pinnate or alternately forked (rarely simple), < 5 cm wide, with rare scales or hairs on rachises and midribs or glabrous, margins flat; veins mostly obscure, pinnate, terminating before reaching margins; sori borne abaxially on veins, ± centered between segment midrib and margins, discrete, oblong to linear; indusia oblong to linear, attached along vein and overlapping sorus with free edge facing midrib, persistent but often inconspicuous at maturity; sporangia < 0.3 mm; spores monomorphic, (32)64 per sporangium, 25–50 μ, monolete.

1. Blades simple, > 2 cm wide…**A. scolopendrium**
1'. Blades 1-pinnate or alternately forked, or if simple, then much < 2 cm wide…(2)
2. Rachises green; petioles longer than blades; blades alternately forked, with ≤ 5 linear pinnae…**A. septentrionale**
2'. Rachises reddish brown to blackish; petioles shorter than blades; blades clearly 1-pinnate, with > 20 elliptic to triangular pinnae…(3)
3. Pinnae (most) in subopposite pairs ("alas de ángeles"); rachises blackish to dark purplish brown…(4)
3'. Pinnae (most) alternately arranged, or if subopposite, then rachises reddish brown…(5)
4. Pinna margins entire to shallowly crenate; blade apices tapering to lobed terminal pinna, rachis not extending beyond pinnae to form rooting tip…**A. resiliens**
4'. Pinna margins prominently dentate-serrate; blade apices with rachis extending beyond pinnae to form whiplike rooting tip [to be expected]…A. palmeri Maxon
5. Blades < 1.2 cm wide; rachis pinnae lacking acroscopic auricles or these ± inconspicuous and usually not overlapping rachises…**A. trichomanes**
5'. Blades (largest) > 1.2 cm wide; pinnae with well-developed acroscopic auricles often overlapping rachises…**A. platyneuron**

Asplenium platyneuron (L.) Britton, Sterns & Poggenb.—Ebony spleenwort fern. Leaves ± dimorphic, erect fertile leaves usually arising from cluster of shorter, spreading sterile leaves; petioles shorter than blades; blades linear to narrowly oblanceolate, 1-pinnate, > 1.2 cm wide, apices tapering to lobed terminal pinna, not rooting at tips; rachises reddish brown to purplish; pinnae 52–60, alternate (at least proximally), elliptic to broadly triangular, articulate, base abruptly tapered, often oblique with well-developed, usually serrulate acroscopic auricle (occasionally basiscopic as well), often overlapping rachis; margins serrate-dentate (rarely crenate); largest fertile pinnae 8–25+ mm, with 10–15+ sori. On sandstone and igneous rocks, 6000–7000 ft. NE.

Asplenium resiliens Kunze—Black-stemmed spleenwort fern. Leaves monomorphic; petioles shorter than blades; blades ± linear, 1-pinnate, > 1.2 cm wide, apices tapering to lobed terminal pinna, not rooting at tips; rachises blackish; pinnae 36–62, subopposite, elliptic to broadly triangular, articulate, base abruptly tapered, usually oblique with well-developed, entire acroscopic auricle (occasionally basiscopic as well), occasionally overlapping rachis; margins entire to shallowly crenate; largest fertile pinnae 6–12 mm, with 8–14 sori. Usually on calcareous rocks, 4500–7100 ft. C, SE, SC, SW.

Asplenium scolopendrium L.—Hart's-tongue spleenwort fern. Leaves monomorphic; petioles shorter than blades; blades linear-oblong, simple, 2–5 cm wide, apices acuminate to rounded, not rooting at tips; rachises brown proximally, straw-colored distally; pinnae absent, the blade margins entire; blades with 1–4 sori. On basalt in a sheltered lava cave; ca. 7600 ft. New Mexico plants presumably represent the East Asian/American tetraploid subsp. *japonicum* (Kom.) Rasbach, Reichst. & Viane, distinguished from its diploid European counterpart (subsp. *scolopendrium*) primarily by its larger (> 31 µ) spores. **R**

Asplenium septentrionale (L.) Hoffm.—Forked spleenwort fern. Leaves monomorphic; petioles longer than blades; blades lanceolate or linear, alternately forked or rarely simple, 0.2–3 cm wide, apices attenuate to terminal pinna, not rooting at tips; rachises green; pinnae 1–5, alternate, linear, not articulate, base gradually and symmetrically tapered, not overlapping rachis, margins entire except for 1–3 ± appressed, linear lobes distally; largest fertile pinnae 8–26 mm, with 1–4 sori. New Mexico plants represent the widespread tetraploid **subsp. septentrionale**, distinguished from its diploid Eurasian counterpart (subsp. *caucasicum* Fraser-Jenk. & Lovis) primarily by its larger (> 37 µ) spores. On sandstone and igneous rocks, 5600–10,200 ft. NE, NC, WC, SW.

Asplenium trichomanes L.—Maidenhair spleenwort fern. Leaves monomorphic; petioles shorter than blades; blades linear, 1-pinnate, < 1.2 cm wide, apices tapering to lobed terminal pinna, not rooting at tips; rachises reddish brown; pinnae 28–44, alternate to subopposite, elliptic, broadly triangular or fan-shaped, articulate, base abruptly tapered, occasionally oblique with ± inconspicuous acroscopic auricle, usually not overlapping rachis; margins crenate to shallowly dentate; largest fertile pinnae 3–6 mm, with 4–8 sori. Our plants represent the diploid **subsp. trichomanes**, distinguished from its autotetraploid derivative (subsp. *quadrivalens* D. E. Mey.) primarily by its smaller (< 33 µ) spores and preference for noncalcareous substrates. On sandstone and igneous rocks, 5500–9500 ft. NE, NC, C, WC, SC, SW.

ATHYRIACEAE – LADY FERN FAMILY

ATHYRIUM Roth – Lady ferns

Athyrium filix-femina (L.) Roth ex Mertens—Lady fern. Plants terrestrial; stems subterranean, ± erect, short-creeping, scaly rhizomes, not continuously ridged and grooved, internodes poorly defined, not hollow; stem scales (at least some) > 5 mm, concolored, brown; leaves monomorphic, spirally arranged, clustered, to 75 cm; petioles > 1 cm, with 2 vascular bundles at base and many scales like those on stems; blades narrowly elliptic to oblanceolate, 1-2-pinnate-pinnatifid, mostly 10–20 cm wide, with scattered scales and glands on rachises and midribs, margins flat; veins apparent, pinnate, terminating at translucent border just short of segment margin; sori borne abaxially on veins, ± centered between segment midrib and margins, discrete, ± elongate, usually hooked toward distal end and often J- or U-shaped; indusia same shape as sori, attached along vein and overlapping sori usually with free edge facing midrib, often persistent but inconspicuous at maturity; sporangia < 0.3 mm; spores monomorphic, 64 per sporangium, 30–50 µ, monolete. New Mexico plants represent **subsp. californicum** (Butters) Hultén, distinguished by its combination of erose-dentate to sparsely fimbriate (rather than long-ciliate) indusia and brown (rather than yellowish) spores. Near streams or springs in montane forests, 6000–10,400 ft. NE, NC, C, WC, SC, SW.

CYSTOPTERIDACEAE – FRAGILE FERN FAMILY

Plants usually growing on or around rocks, occasionally terrestrial; stems subterranean, horizontal, short- to long-creeping, scaly rhizomes, not continuously ridged and grooved, internodes poorly defined, not hollow; stem scales < 5 mm, concolored or weakly clathrate with thick, radial walls and clear luminae; leaves monomorphic or weakly dimorphic with the fertile longer and narrower, spirally arranged, clustered (widely spaced in *Gymnocarpium*), to 40 cm; petioles > 1 cm, with 2 vascular bundles at base and rare scales like those on stems; blades 1–3-pinnate-pinnatifid, usually < 10 cm wide, with scattered hairs (rarely hairlike scales) on rachises, otherwise glabrous or sparsely glandular, margins flat; veins conspicuous, pinnate, intersecting margins; sori borne abaxially on veins, ± centered between segment midrib and margins, mostly discrete, round; indusia consisting of lanceolate membrane narrowly attached at proximal edge of and initially covering sori, mostly deciduous at maturity; sporangia < 0.3 mm; spores monomorphic, 64 per sporangium, 30–50 μ, monolete.

1. Blades oblong-lanceolate to narrowly triangular, petioles shorter than blades; leaves clustered near growing tip of creeping (but not cordlike) stem; indusia consisting of lanceolate membrane narrowly attached at proximal edge of sori, this often deciduous at maturity…**Cystopteris**
1'. Blades broadly triangular, petioles usually longer than blades; leaves arising singly from long-creeping, cordlike stems; indusia absent…**Gymnocarpium**

CYSTOPTERIS Bernh. – Fragile ferns

Stems creeping but not cordlike; leaves clustered near growing tip; blades elliptic-lanceolate to narrowly triangular, petioles shorter than blades; indusia consisting of lanceolate membrane narrowly attached at proximal edge of sori, this often deciduous at maturity

1. Blades elliptic to lanceolate, usually widest near middle, nearly glabrous; petioles mostly lustrous brown at maturity (at least proximally); veins usually intersecting margins at tips of prominent lobes or teeth (occasionally in shallow notches at apices of such teeth); widespread in shaded montane habitats… **C. fragilis**
1'. Blades triangular to broadly lanceolate, widest at or near base, with sparse to many minute, glandular hairs; petioles straw-colored to tan at maturity, brown only near point of attachment; veins (most) intersecting margins in shallow notches between teeth; rare on limestone or sandstone outcrops near Arizona and Texas borders…(2)
2. Blades of fertile leaves narrowly triangular, often widest at base with long-attenuate apices, commonly with many minute, glandular hairs concentrated at pinna/rachis junctions; lowermost pinnae usually attached to rachises by narrow acroscopic wing of green tissue; rachises often producing bulblets abaxially on distal portion, the bulblets plump, ± glabrous or with sparse, pale scales…**C. bulbifera**
2'. Blades of fertile leaves triangular to broadly lanceolate, widest at or near base with short-attenuate to acute apices, glandular hairs present but usually sparse; lowermost pinnae not attached to rachises by narrow wing of green tissue; rachises occasionally producing bulblets abaxially on distal portion, the bulblets usually small and misshapen, with clathrate scales exhibiting thick dark brown radial walls and clear luminae…**C. tennesseensis**

Cystopteris bulbifera (L.) Bernh.—Bulblet fragile fern. Stems short-creeping, the scales lanceolate, tan to brown or somewhat clathrate; leaves often slightly dimorphic, the early-season sterile leaves shorter and less attenuate; petioles often pinkish when young, mostly straw-colored to tan at maturity; blades of fertile leaves narrowly triangular, 2-pinnate to 2-pinnate-pinnatifid, often widest at base with long-attenuate apices; lowermost pinnae usually attached to rachises by narrow acroscopic wing of green tissue, often with many minute glandular hairs concentrated at pinna/rachis junctions; rachises often with bulblets abaxially on distal portion, the bulblets plump and sparsely scaly to ± glabrous; veins intersecting margins in notches between teeth; indusia with minute glandular hairs. Rare on limestone outcrops, 6000–7500 ft. WC, SE.

Cystopteris fragilis (L.) Bernh. [*C. reevesiana* Lellinger]—Fragile fern. Stems creeping, scales ovate to lanceolate, uniformly tan or brown; leaves monomorphic; petioles often dark brown (at least proximally) at maturity; blades elliptic to lanceolate, 2–3-pinnate, usually widest near middle with short-attenuate to acute apices; lowermost pinnae not attached to rachises by narrow wing of green tissue, nearly glabrous; rachises not producing bulblets; veins intersecting margins at tips of prominent lobes or teeth (occasionally in shallow notches at apices of such teeth); indusia lacking gland-tipped hairs. The vast majority of collections from New Mexico represent **subsp. tenuifolia** Clute [*C. reevesiana* Lellinger], a diploid taxon with more dissected blades (often 3-pinnate) and long-creeping stems. A few specimens with less divided blades and short-creeping stems from rocky, high-elevation habitats in the N counties belong to the tetraploid **subsp. fragilis**. Widespread in shaded montane habitats, 6000–12,600 ft. NE, NC, NW, C, WC, SC, SW.

Cystopteris tennesseensis Shaver—Tennessee fragile fern. Stems creeping, scales lanceolate, ± clathrate with thick dark brown radial walls and clear luminae; leaves monomorphic; petioles green to straw-colored throughout or darker at base; blade triangular to broadly lanceolate, 2-pinnate-pinnatifid, widest at or near base with short-attenuate to acute apices; lowermost pinnae not attached to rachises by narrow wing of green tissue, with sparse glandular hairs, rachises occasionally producing inconspicuous bulblets abaxially on distal portion, the bulblets usually small and misshapen, with clathrate scales; veins intersecting margins in both teeth and notches; indusia with sparse glandular hairs. Our plants represent **subsp. utahensis** (Windham & Haufler) Windham, which arose through hybridization between *C. bulbifera* and *C. fragilis* subsp. *tenuifolia*. Rare on limestone or sandstone outcrops near the Arizona and Texas borders, 6800–7600 ft. WC, SE.

GYMNOCARPIUM Newman - Oak fern

Gymnocarpium dryopteris (L.) Newman—Common oak fern. Stems long-creeping and cordlike; leaves arising singly, often well separated from stem growing tip; petioles usually longer than blades; blades broadly triangular; indusia absent. In soil among igneous rocks, 7800–9600 ft. NC.

DENNSTAEDTIACEAE - BRACKEN FERN FAMILY

PTERIDIUM Gled. ex Scop. - Bracken ferns

Pteridium aquilinum (L.) Kuhn—Northern bracken fern. Plants terrestrial; stems subterranean, horizontal, long-creeping, sparsely hairy rhizomes, not continuously ridged and grooved, internodes poorly defined, not hollow; stem scales absent; leaves monomorphic, spirally arranged, widely spaced, mostly 40–100 cm; petioles > 10 cm, with > 3 vascular bundles at base, sparsely hairy to glabrescent, lacking scales; blades broadly triangular, usually 3-pinnate, 20–80 cm wide, the surfaces hairy to glabrescent, margins ± recurved; veins mostly obscure, pinnate, terminating before reaching margins; sori borne abaxially on veins closely paralleling segment margins, confluent, forming ± continuous linear band; indusia linear, attached along submarginal vein and overlapping sorus with free edge facing midrib, persistent but usually inconspicuous, covered by modified, recurved leaf margins (false indusia); sporangia < 0.3 mm; spores monomorphic, 64 per sporangium, 25–40 µ, trilete. Forming extensive clones in and around montane meadows, 6500–11,200 ft. NE, NC, NW, C, WC, SC, SW. Although assigned to **var.** (now subsp.) **pubescens** Underw. in earlier references, some collections from New Mexico appear to represent **subsp. latiusculum** (Desv.) Hultén. This is not surprising given that all specimens from the Madrean sky islands just across the border in northern Mexico were assigned to var. *latiusculum* by Mickel and Smith (2004). The primary distinguishing features involve the distribution of hairs on abaxial blade surfaces. In subsp. *latiusculum*, the inner indusium is glabrous and hairs are confined to costae and major veins; in subsp. *pubescens* the inner indusium is ciliate and hairs are distributed over the entire abaxial blade surface. The two subspecies are known to intergrade where their ranges overlap.

DRYOPTERIDACEAE - WOOD FERN FAMILY

Plants growing on or around rocks; stems subterranean, ± erect, short-creeping, scaly rhizomes, not continuously ridged and grooved, internodes poorly defined, not hollow; stem scales (at least some) > 5 mm, concolored, brown; leaves monomorphic or weakly dimorphic, spirally arranged, clustered, to 70 cm; petioles > 1 cm, with > 3 vascular bundles at base and many scales like those on stems; blades elliptic to ovate-lanceolate, 1–2-pinnate, 2–30 cm wide, with many lance-attenuate and hairlike scales abaxially on rachises and proximal pinna midribs, margins flat; veins mostly obscure, pinnate, forming expanded, submarginal hydathodes or extending to translucent border; sori borne abaxially on veins, ± centered between segment midrib and margins or closer to margins, mostly discrete, round; indusia round or cordate, emerging within sori, often conspicuous and persistent at maturity; sporangia < 0.3 mm; spores monomorphic, 64 per sporangium, 30–50 μ, monolete.

1. Blades 1-pinnate; pinnae 8–20 per fertile leaf…**Phanerophlebia**
1'. Blades 1-pinnate-pinnatifid to 2-pinnate; pinnae 28–90 per fertile leaf…(2)
2. Blades ovate-lanceolate, 10–30 cm wide…**Dryopteris**
2'. Blades linear-elliptic to narrowly lanceolate, 3–6 cm wide…**Polystichum**

DRYOPTERIS Adans. – Wood ferns

Dryopteris filix-mas (L.) Schott—Male fern. Blades 1-pinnate-pinnatifid to 2-pinnate, 10–30 cm wide; pinnae 28–52 per fertile leaf, the marginal teeth lacking prominent apical projections; sori in single row ± centered between segment margin and midrib; indusia cordate, attached at proximal edge of sori but usually enveloped by maturing sporangia. Usually on igneous rocks or soils derived from them, 6000–9400 ft. NE, NC, C, WC, SC, SW.

PHANEROPHLEBIA C. Presl. – Mexican holly ferns

Phanerophlebia auriculata Underw.—Auriculate Mexican holly fern. Blades 1-pinnate, 8–18 cm wide; pinnae 8–20 per fertile leaf, the marginal teeth with prominent apical projections ("spines") to 1 mm; sori in 2–4 rows between segment (pinna) margins and midrib; indusia round-peltate, attached at center of sori. Usually in crevices of igneous rocks, 4800–7300 ft. SC, SW.

POLYSTICHUM Roth – Holly ferns

Polystichum scopulinum (D. C. Eaton) Maxon—Mountain holly fern. Blades 1-pinnate-pinnatifid proximally, 3–6 cm wide; pinnae 46–90 per fertile leaf, the marginal teeth with or without apical projections ("spines") to 0.7 mm; sori in single row ± centered between segment margin and midrib; indusia round, peltate, attached at center of sori. Crevices of basalt lava flow, ca. 7000 ft. WC.

EQUISETACEAE – HORSETAIL FERN FAMILY

EQUISETUM L. – Horsetail ferns

Plants rooted near water or terrestrial; stems dimorphic, with subterranean, ± horizontal, long-creeping, hairy to glabrescent rhizomes giving rise to ± erect, nonrooting, glabrous aerial stems showing fertile/sterile dimorphism in some species, aerial stems branched or unbranched, with 3–30 equally spaced ridges and grooves their entire length, internodes sharply defined, hollow in main stems; stem scales absent; leaves monomorphic, whorled, closely spaced, to 2 cm (including nodal sheath); petioles lacking; blades fused into nodal sheaths proximally, with ring of teeth distally (these often deciduous), < 0.3 cm wide, glabrous, margins flat to slightly incurved; veins mostly obscure, dichotomously 1-branched, terminating before reaching apex; sori absent, sporangia borne on abaxial surface of peltate stem outgrowths (sporangiophores), these aggregated into terminal strobili; indusia absent; sporangia > 0.3 mm; spores monomorphic, hundreds per sporangium, 30–80 μ, alete.

1. Aerial stems dimorphic, the pinkish brown, unbranched fertile stems quickly wilting, the green, regularly branched sterile stems persisting through 1 growing season; stem branches prominently 3–4-ridged…
   E. arvense
1'. Aerial stems monomorphic, all green, unbranched (with some irregular branching in injury forms), and persisting 1+ growing seasons; stem branches (when present) with > 4 ridges…(2)
2. Leaf sheaths mostly grayish, regularly framed by dark bands near base and apex, rectangular in side view (not flaring distally); aerial stems persisting > 1 year, the ridges tuberculate with ± conic silica deposits…
   E. hyemale
2'. Leaf sheaths (at least those on upper part of stem) greenish, not regularly framed by 2 dark bands, ± flared

distally in side view; aerial stems completely or distally deciduous at end of growing season, the ridges rugulate with ± elongate ("cross-bar") silica deposits…(3)

3. Proximal leaf sheaths generally not banded; strobili of main stems usually blunt at apex; sporangia dehiscing at maturity to release greenish spores with filamentous elaters…**E. laevigatum**

3'. Proximal leaf sheaths often irregularly banded; strobili of main stems with sharp point at apex; sporangia not dehiscing at maturity, containing malformed, whitish spores lacking elaters…**E. ×ferrissii**

Equisetum arvense L.—Common horsetail fern. Aerial stems dimorphic, the pinkish-brown, un-branched fertile stems deciduous within a week, the green, regularly branched sterile stems persisting through 1 growing season, the ridges of main stems smooth proximally, sawtooth-tuberulate distally; leaf sheaths greenish to straw-colored at base, the ridges dark brown to black distally; stem branches prominently 3–4-ridged; strobili blunt at apices; sporangia dehiscing at maturity to release greenish spores with filamentous elaters. Forming dense clones adjacent to springs and flowing montane streams, 5000–11,200 ft. Widespread except E, SW.

Equisetum ×ferrissii Clute—Hybrid scouring-rush fern. Aerial stems monomorphic, green, un-branched (with some irregular branching in injury forms), distal portion deciduous at end of growing season, proximal portion often persisting > 1 year, the ridges rugulate with ± elongate ("cross-bar") silica deposits; leaf sheaths usually greenish on distal part of stem, irregularly banded proximally, slightly flared distally in side view; stem branches (when present) with > 4 ridges; strobili with sharp point at apices; sporangia not dehiscing at maturity, containing malformed, whitish spores lacking elaters. Usually forming dense clones adjacent to springs and flowing streams, 3700–7300 ft. NC, NW, EC, C, WC, SC. This very successful hybrid between *E. hyemale* and *E. laevigatum* spreads vegetatively by water transport of uprooted rhizome segments and often occurs in the absence of either parent.

Equisetum hyemale L.—Common scouring-rush fern. Aerial stems monomorphic, green, un-branched (with some irregular branching in injury forms), persisting > 1 year, the ridges tuberculate with ± conic silica deposits; leaf sheaths mostly grayish, regularly bordered by dark bands at base and apex, rectangular in side view (not flaring distally); stem branches (when present) with > 4 ridges; strobili with sharp point at apices; sporangia dehiscing at maturity to release greenish spores with filamentous elaters. Our plants represent **subsp. affine** (Engelm.) Calder & Roy L. Taylor. Forming dense clones adjacent to springs and flowing montane streams, 5000–8700 ft. NE, NC, NW, C, WC, SC, SW.

Equisetum laevigatum A. Braun—Smooth scouring-rush fern. Aerial stems monomorphic, green, unbranched (with some irregular branching in injury forms), usually deciduous at end of growing season, proximal portion occasionally persisting > 1 year, the ridges rugulate with ± elongate ("cross-bar") silica deposits; leaf sheaths greenish (rarely irregularly banded proximally), noticeably flared distally in side view; stem branches (when present) with > 4 ridges; strobili usually blunt at apices (at least on main stems); sporangia dehiscing at maturity to release greenish spores with filamentous elaters. Usually forming clones adjacent to flowing or standing water but often extending into surrounding drier soils, 3700–9600 ft. Widespread except SE.

MARSILEACEAE – PEPPERWORT FERN FAMILY

MARSILEA L. – Water clover ferns

Two species reported for New Mexico but only *M. vestita* confirmed.

1. Distal tooth of sporocarps > 0.3 mm, acute; sporocarps > 3 mm wide…**M. vestita**

1'. Distal tooth of sporocarps < 0.3 mm, blunt or absent; sporocarps < 3 mm wide…M. mollis B. L. Rob. & Fernald

Marsilea vestita Hook. & Grev.—Hairy water clover fern. Plants rooted aquatics, often amphibious in dry season; stems superficial or barely subterranean, horizontal, creeping, sparsely hairy rhizomes, not continuously ridged and grooved, internodes poorly defined, not hollow; stem scales absent; leaves often dimorphic, those of emergent plants smaller and more pubescent, spirally arranged, closely to widely spaced, to 25 cm; petioles > 1 cm, with 1 vascular bundle at base, hairy to glabrescent, lacking scales; blades orbicular, 1-pinnate into 4 fan-shaped, pseudopeltate pinnae, < 5 cm wide, hairy to glabrescent, the margins flat; veins apparent (in aquatic plants) to obscure (in amphibious plants), dichotomously branched, mostly obscure, extending to translucent pinna border; sori enclosed within hardened protective structures (sporocarps) attached by short stalks to basal portion of petioles, sporocarps elliptic to globose, > 3 mm wide, with acute distal tooth > 0.3 mm, containing both microsporangia and megasporangia; indusia membranous, enclosing sorus and partially fused to indusia of adjacent sori; sporangia mostly < 0.5 mm; spores dimorphic, microspores 32(16 or 32)64, 30–80 μ, trilete, megaspores 1 per sporangium, 400–525 μ, alete. In periodically flooded soil in and around standing water, 4000–8700 ft. NE, NC, EC, C, WC, SE, SC, SW.

OPHIOGLOSSACEAE – ADDER'S TONGUE FAMILY

Ben S. Legler

Perennial herbs, glabrous or sparsely pubescent; stems erect, subterranean, producing 1(2) leaves each year; leaves normally divided into a sterile leaflike segment (trophophore) and a fertile spore-bearing axis (sporophore), these joined below to form a common stalk; trophophore simple or pinnately to ternately lobed or compound, rarely absent; sporophore unbranched or branched, occasionally absent; sporangia globose, 2-valved, embedded in sporophore axis or exposed in grapelike clusters; spores numerous; gametophytes subterranean. Measurements, descriptions, and key characters given here apply specifically to New Mexico plants.

1. Trophophore simple, entire, veins anastomosing, not reaching trophophore margin; sporophore unbranched, with sporangia forming 2 rows sunken into opposing sides of sporophore axis…**Ophioglossum**
1′. Trophophore usually lobed or compound (rarely entire), veins free, extending to trophophore margin; sporophore usually branched, with sporangia exposed and mostly in irregular clusters…(2)
2. Trophophore 1–2(3) times pinnate or ternate (rarely simple or absent), narrowly oblong to lanceolate or deltate, rarely > 5 cm wide; plants glabrous; sporophore nearly always present…**Botrychium**
2′. Trophophore 3–4 times ternately or pinnately compound, deltate, (2–)4–20 cm wide; plants sparsely hairy when young, becoming glabrate in age; sporophore often absent in small plants…(3)
3. Trophophore and sporophore diverging well aboveground on an elongate common stalk; trophophore sessile, thin, deciduous, with acute lobes…**Botrypus**
3′. Trophophore and sporophore diverging near or below ground level on a short common stalk; trophophore long-stalked, thick, often persisting through winter, with rounded to broadly acute lobes…**Sceptridium**

BOTRYCHIUM Sw. – Moonwort

Plants glabrous; common stalk short or elongate; trophophore sessile or stalked, pinnately to ternately compound, rarely simple or absent; pinnae linear, forked, fan-shaped, or pinnately lobed, margins entire, toothed, or lobed, veins extending to margin; sporophore usually branched; sporangia exposed in clusters. *Botrychium* is a difficult group of cryptic taxa containing both diploids and morphologically intermediate allopolyploids (see Dauphin et al., 2018). Not all specimens can be reliably identified based on morphology. Carefully flattened and pressed trophophores are important for identification when collecting specimens; belowground parts are not needed. Plant color in life should be noted, as color is usually lost in dried specimens. *Botrychium* species often co-occur; where one species is found others are likely to be present. Most of our taxa prefer mesic, well-drained, subalpine meadows and meadowlike areas, including old disturbances, with a low, lawnlike cover of grasses, *Carex*, and perennial, herbaceous species of Rosaceae (e.g., *Fragaria*, *Potentilla*) and Asteraceae (e.g., *Achillea*, *Antennaria*, *Taraxacum*), often with regenerating young conifers (usually *Picea* and *Abies*). All are obligately mycorrhizal.

1. Trophophore and sporophore joined at or below ground level on a very short common stalk; middle and upper pinnae fan-shaped to wedge-shaped; basal pinnae similar to the upper or, on large plants, greatly elongate and again pinnately lobed; trophophore apex broadly rounded…(2)

1'. Trophophore and sporophore joined well above ground level on an elongate, exposed common stalk; pinnae and trophophore apex various, sometimes as above…(3)

2. Upper pinnae broadly confluent (or trophophore unlobed in very small plants); basal pinnae often much elongate and spoon-shaped or again lobed, if short and fan-shaped then usually with a smoothly rounded junction between side and outer margins; plants of moist streamside meadows and seeps…**B. simplex**

2'. Upper pinnae nearly distinct, not broadly confluent; basal pinnae short and fan-shaped with an angular junction between side and outer margins, or occasionally elongate and again lobed; plants of mesic to gravelly meadows or old disturbance areas…**B. minganense**

3. Trophophore pinnate-pinnatifid or 2-pinnate, the middle and basal pinnae ovate, lanceolate, or oblong, broadest near or below middle, narrowed to tip, pinnately lobed with pinnate venation (except in small plants); trophophore oblong-deltate to triangular, apex obtuse to acute; common stalk with or without red tinge…(4)

3'. Trophophore 1-pinnate, the middle (and usually also basal) pinnae fan-shaped, wedge-shaped, or linear, broadest at outer margin or parallel-sided, entire or palmately lobed with palmate or parallel venation; trophophore lanceolate to oblong, apex generally rounded; common stalk never red-tinged…(7)

4. Trophophore broadly triangular or pentagonal in outline, as broad as long; pinnae lobes ± acute, not overlapping; sporophore usually ternately branched, stalk 1/2 or less length of trophophore at spore release   **B. lanceolatum**

4'. Trophophore ovate to narrowly triangular in outline, clearly longer than broad; pinnae lobes rounded to acute, sporophore stalk < or > 1/2 length of trophophore at spore release…(5)

5. Sporophore stalk 1/2 or less length of trophophore at spore release, the sporophore usually pinnately branched; pinnae narrow, well spaced, with well-spaced lobes; basal pinnae often mitten-shaped with an enlarged lower basal lobe; trophophore sessile; plants yellow-green in life, usually shiny, often also glaucous…**B. echo**

5'. Sporophore stalk (1/3–)1/2–2 times as long as trophophore at spore release, the sporophore pinnately or ternately branched; pinnae ovate to broadly lanceolate, often approximate to overlapping, with approximate to overlapping lobes; basal pinnae usually not mitten-shaped; trophophore sessile or stalked; plants rarely yellow-green…(6)

6. Plants gray-green to dull green and glaucous in life; pinnae usually asymmetrically lobed (lobes enlarged on lower side), the basal pinnae often much enlarged; trophophore usually short-stalked; common stalk with a distinct reddish stripe below trophophore; sporophore often ternately branched…
B. hesperium

6'. Plants lustrous grass-green, not glaucous in life; pinnae symmetrically lobed, evenly reduced in size upward; trophophore usually sessile; common stalk green or with a light maroon tinge or stripe; sporophore usually pinnately branched…**B. pinnatum**

7. Sporophore stalk at spore release < 1/3 length of trophophore; unlobed pinnae (or lobes of larger pinnae) linear to narrowly wedge-shaped, spanning an angle of < 45°(–60°); trophophore strictly sessile…
B. campestre

7'. Sporophore stalk at spore release 1/2–2 times as long as trophophore; pinnae wedge-shaped, fan-shaped, or mushroom-shaped, at least the lower pinnae usually spanning an angle of > 60°; trophophore sessile or stalked…(8)

8. Plants glossy green to glossy yellow-green in life; pinnae in 2–4(5) pairs, spreading at nearly a right angle to rachis, mostly broadly fan-shaped, outer margin smoothly rounded, often with a broadly rounded junction to side margins; basal pinnae usually asymmetrically enlarged on lower side; sporophore ca. equal in length to trophophore, branches often widely spreading to descending…**B. tunux**

8'. Plants grass-green to pale green in life, sometimes glaucous or yellowish but not or scarcely glossy; pinnae in (2)3–7 pairs, the middle and upper pinnae generally ascending, outer margin smooth to cleft or lobed, often with an angular junction to side margins; basal pinnae rarely asymmetrically enlarged on lower side; sporophore stalk often longer than trophophore, branches usually ascending…(9)

9. Pinnae broadly fan-shaped, often overlapping, the lower forming a symmetric half circle spreading at ca. right angle to rachis, the middle and upper transitioning to a quarter circle with straight margins, the upper margin erect (parallel to rachis) and lower margin descending to spreading, outer margins usually smooth; trophophore usually sessile…**B. neolunaria**

9'. Pinnae narrowly to broadly fan-shaped, overlapping or well spaced, the lower sometimes forming a half circle but then often either symmetrically trilobed or asymmetrically enlarged on upper side; upper pinnae not shaped as above; trophophore usually stalked…(10)

10. Plants deep green to grass-green (occasionally yellow-green) in life; pinnae narrowly to broadly fan-shaped, entire to symmetrically 3- or 5-lobed, the middle lobe usually largest; sporophore and trophophore stalks ± straight, forming a narrow V shape…**B. minganense**

10'. Plants pale green to whitish green in life; pinnae narrowly fan-shaped to irregularly mushroom-shaped, entire to asymmetrically 2–8-lobed, the upper lobe usually largest; sporophore and trophophore stalks often sigmoidal…**B. furculatum**

Botrychium campestre W. H. Wagner & Farrar—Prairie moonwort, slender moonwort. Plants 2–15 cm, whitish green to light green, glaucous or not; common stalk elongate aboveground, not red-tinged; trophophore sessile, oblong to oblong-lanceolate, 1-pinnate; pinnae ascending, linear to narrowly wedge-shaped, unlobed or with narrow lobes; sporophore stalk < 1/3 length of trophophore at spore release. Many high-elevation populations in the S Rocky Mountains encompass the genetic and morphological diversity of both varieties, while those of lower elevations usually display more restricted genotypes; not all can be confidently assigned to variety.

1. Pinnae narrowly wedge-shaped, unlobed or with shallow, nonspreading lobes; basal pinnae generally no larger than middle pinnae; plants of lower-elevation grasslands, prairies, or open woodlands…**var. campestre**
1′. Pinnae linear to narrowly wedge-shaped, unlobed or often with deep, spreading lobes; basal pinnae often larger than middle pinnae; plants usually of subalpine areas…**var. lineare**

var. campestre—Prairie moonwort. Trophophore oblong, basal pinnae generally no larger than middle pinnae; pinnae narrowly wedge-shaped, unlobed or with shallow, nonspreading lobes. Well-drained grasslands, prairies, and open woodlands, 8850 ft. In New Mexico reported only from the Chuska Mountains in McKinley County; these plants have not been confirmed genetically and their varietal status remains uncertain.

var. lineare (W. H. Wagner) Farrar [*B. lineare* W. H. Wagner]—Slender moonwort. Trophophore narrowly oblong-lanceolate, basal pinnae often larger than middle pinnae; pinnae linear to narrowly wedge-shaped, often deeply divided into several spreading, linear to narrowly wedge-shaped lobes. Lightly vegetated subalpine scree slopes and grassy subalpine meadows, 11,000–11,800 ft. In New Mexico known from only 1 site near Wheeler Peak in Taos County; to be expected elsewhere.

Botrychium echo W. H. Wagner—Echo moonwort. Plants 3–10(–15) cm, lustrous to dull yellow-green, often glaucous; common stalk elongate aboveground, often faintly red-tinged; trophophore sessile, ovate to narrowly triangular, (1)2 times pinnate; pinnae ascending, oblong to elliptic or nearly linear, asymmetrically lobed, lobes acute or obtuse, basal pinnae often mitten-shaped with an enlarged lower basal lobe; sporophore stalk < 1/2 as long as trophophore at spore release. Mesic, gravelly subalpine meadows, gravelly road shoulders, old clearcuts, ski slopes, 9800–11,800 ft. NC. Tetraploid derived from *B. lanceolatum* s.l. × *B. campestre* s.l.

Botrychium furculatum S. J. Popovich & Farrar—Forked moonwort. Plants 3–15 cm, pale green to whitish green, often glaucous; common stalk elongate aboveground, not red-tinged; trophophore stalked, narrowly oblong, 1-pinnate; pinnae spreading-ascending, narrowly wedge-shaped to fan-shaped or irregularly mushroom-shaped, entire or asymmetrically lobed with upper lobe often the largest, side margins often concave; sporophore stalk (1/2–)1–2 times as long as trophophore at spore release. Mesic, usually gravelly or grassy subalpine meadows, gravelly road shoulders, old clearcuts, 9500–11,800 ft. Colfax, Rio Arriba, Taos. Tetraploid derived from *B. pallidum* × *B.* sp.

Botrychium hesperium (Maxon & R. T. Clausen) W. H. Wagner & Lellinger—Western moonwort. Plants 5–20 cm, dull gray-green, usually glaucous; common stalk elongate aboveground, with a reddish stripe below trophophore; trophophore short-stalked, ovate to narrowly triangular, (1)2 times pinnate; pinnae spreading-ascending, often overlapping, ovate to broadly lanceolate, asymmetrically lobed, lobes usually obtuse, basal pair of pinnae often much enlarged relative to the next pair; sporophore stalk (1/3–)1/2–2 times as long as trophophore at spore release. Mesic subalpine meadows, ski slopes, gravelly road shoulders, 9800–11,800 ft. NC. Tetraploid derived from *B. lanceolatum* s.l. × *B.* sp.

Botrychium lanceolatum (Gmel.) Ångstr.—Triangle moonwort. Plants 5–20 cm, glossy green or dull yellow-green, not glaucous; common stalk elongate aboveground, green or with a red stripe below trophophore; trophophore sessile, broadly deltate or triangular, (1)2 times pinnate, the basal pinnae usually as large as rest of trophophore; pinnae spreading-ascending, narrowly to broadly lanceolate, lobes acute or obtuse; sporophore stalk shorter than trophophore at spore release. Mesic subalpine meadows and forest openings, ski slopes, gravelly road shoulders, old meadowlike disturbance areas, 9800–12,000 ft.

NC. Diploid. Two genotypes are present in New Mexico. The green genotype has a green common stalk, a lustrous green trophophore, and basal pinnae straight and symmetrically lobed. The red genotype has a red stripe on the common stalk, a dull yellow-green trophophore, and basal pinnae curved upward with much-enlarged lower basal lobes.

Botrychium minganense Vict.—Mingan moonwort. Plants 2–15 cm, grass-green or occasionally pale green to yellowish green, glaucous or not; common stalk usually elongate aboveground, not red-tinged; trophophore usually stalked, lanceolate to oblong, 1-pinnate or sometimes the basal or lower pinnae elongate and pinnately lobed (then lobes fan-shaped); pinnae spreading-ascending, broadly to narrowly fan-shaped, outer margins rounded, entire to symmetrically 3- or 5-lobed; sporophore stalk (1/2–)1–2 times as long as trophophore at spore release. Mesic subalpine meadows and forest openings, avalanche chutes, ski slopes, road shoulders, old clearcuts, 10,000–12,500 ft. NC, WC, SC.

Tetraploid derived from *B. neolunaria* × *B.* sp. The most common species in New Mexico, extremely variable and easily confused with other species.

Botrychium neolunaria Stensvold & Farrar [*B. lunaria* L., misapplied]—Common moonwort. Plants 3–20 cm, grass-green, not glaucous; common stalk elongate aboveground, not red-tinged; trophophore usually sessile, oblong, 1-pinnate; pinnae spreading, usually overlapping, broadly fan-shaped, the lowest pinnae forming a spreading half circle, the upper pinnae forming an ascending quarter circle, outer margins broadly rounded; sporophore stalk 1–2 times as long as trophophore at spore release. Mesic subalpine meadows, scree slopes, road shoulders, 9900–11,800 ft. NC, WC, SC. Diploid.

Botrychium pinnatum H. St. John—Northwestern moonwort. Plants 3–15 cm, lustrous green to grass-green, not glaucous; common stalk elongate aboveground, green or uniformly red-tinged; trophophore sessile or rarely short-stalked, oblong-deltate, (1)2 times pinnate; pinnae spreading to ascending, lanceolate to broadly ovate, symmetrically lobed, lobes obtuse to broadly acute; sporophore stalk 0.5–1.5 times as long as trophophore at spore release. Mesic subalpine meadows, ski slopes, gravelly road shoulders, old clearcuts, 10,400–11,100 ft. Colfax, Santa Fe, Taos. Tetraploid derived from *B. lanceolatum* s.l. × *B. neolunaria*.

Botrychium simplex E. Hitchc.—Least moonwort. Plants 2–8 cm, grass-green to gray-green, usually glaucous; common stalk very short, the trophophore and sporophore diverging very near or below ground level except sometimes in very small plants; trophophore stalked, oblong, and 1-pinnate, or ternate with the basal pinnae greatly elongate and spoon-shaped to again lobed, or occasionally unlobed in very small plants; pinnae spreading-ascending, rounded to fan-shaped, the upper pinnae usually broadly confluent, lobe margins usually broadly rounded; sporophore stalk 1–2(3) times as long as trophophore at spore release. New Mexico material belongs to **var. simplex**. Moist subalpine streamside meadows, margins of seeps, 9200–10,500 ft. Colfax, Rio Arriba, Taos. Diploid.

Botrychium tunux Stensvold & Farrar—Moosewort. Plants 2–6(–10) cm, green to yellow-green, glossy, not glaucous; common stalk elongate aboveground, not red-tinged; trophophore sessile or short-stalked, oblong, 1-pinnate; pinnae spreading at nearly a right angle to rachis, broadly fan-shaped (except in very small plants), the lowest pinnae usually asymmetrically enlarged on lower side, outer margins broadly rounded with a rounded junction to side margins, usually entire; sporophore stalk 1/2–1 times as long as trophophore at spore release, branches often widely spreading to descending. Gravelly, sparsely vegetated, subalpine scree slopes, 11,800 ft. Diploid. In New Mexico known from only 1 site near Wheeler Peak in Taos County; to be expected elsewhere in high mountains.

BOTRYPUS Michx. – Rattlesnake fern

Botrypus virginianus (L.) Michx. [*Botrychium virginianum* (L.) Sw.]—Virginia rattlesnake fern. Plants sparsely pubescent when young, becoming glabrate, to 40 cm; common stalk elongate, the trophophore and sporophore diverging well aboveground; trophophore sessile, deltate, blade to 30 cm long and wide, 3–4 times ternately compound, thin and herbaceous, deciduous; pinnae to 12 pairs, ultimate lobes lanceolate to ovate, tips acute, margins usually toothed, veins extending to margin; sporophore 1–2 times as long as trophophore, pinnately branched; sporangia exposed in clusters. Seep in a montane, forested canyon, 7150 ft. In New Mexico known from 1 site, in Los Alamos County.

OPHIOGLOSSUM L. – Adder's tongue

Ophioglossum engelmannii Prantl—Limestone adder's tongue. Plants glabrous, to 25 cm; common stalk short; trophophore sessile or nearly so, erect to spreading, unlobed, margins entire, veins anastamosing, not extending to trophophore margin, blade ca. 10 × 4.5 cm, ovate to ovate-lanceolate, often folded, narrowed abruptly to base, apiculate at tip, veins forming secondary areoles within primary areoles; sporophore 1.5–2.5 times as long as trophophore, sporangia forming 2 rows sunken into opposing sides of sporophore axis. Moist areas over limestone. In New Mexico known only from Doña Ana County. White Sands Missile Range, 4000–4500 ft.

SCEPTRIDIUM Lyon – Grapefern

Sceptridium multifidum (S. G. Gmel.) M. Nishida [*Botrychium multifidum* (S. G. Gmel.) Trevis]— Leather grapefern. Plants pubescent (sparsely so with age), 5–20 cm; trophophore and sporophore diverging near or below ground level; trophophore stalk 1–5 cm, blade 2–10 cm long and wide, thick-textured, pinnae lobes obliquely ovate, often overlapping, tips obtuse to rounded, margins entire or shallowly crenate, often curled, pinnae to 10 pairs, ultimate lobes lanceolate to ovate or fan-shaped, veins extending to margin; sporophore pinnately branched, (1–)1.5–2.5 times as long as trophophore, sporangia exposed in clusters. Moist to wet meadows, lakeshores, drained ponds, 9800–10,000 ft. In New Mexico known only from the Sangre de Cristo Mountains in Taos County.

POLYPODIACEAE – POLYPODY FERN FAMILY

1. Blades with ovate scales concentrated near sori and scattered over abaxial surfaces; stem scales strongly clathrate with thick, black radial walls and clear luminae [to be expected]…Pleopeltis riograndensis (T. Wendt) E. G. Andrews & Windham
1'. Blades nearly glabrous or with sporadic lanceolate scales along rachises on abaxial surfaces; stem scales not strongly clathrate…**Polypodium**

POLYPODIUM L. – Polypody ferns

Plants growing on or around rocks; stems subterranean, horizontal, creeping, scaly rhizomes, not continuously ridged and grooved, internodes poorly defined, not hollow; stem scales < 5 mm, concolored (tan to brown) or weakly bicolored with darker central stripe and brown margins; leaves monomorphic, spirally arranged, closely to widely spaced, to 25 cm; petioles > 1 cm, with 3 vascular bundles at base and few scales like those on stems; blades narrowly oblong to triangular, regularly pinnatifid with segments all connected by strip of leaf tissue along rachises, < 5 cm wide, glabrous except for rare scales or hairs on rachises and midribs, margins flat; veins mostly obscure, pinnate, terminating before reaching margins; sori borne abaxially on veins ± centered between segment midrib and margins or closer to margins, usually discrete, round to oval; indusia absent, sori fully exposed; sporangia < 0.3 mm; spores monomorphic, 64 per sporangium, 50–70 µ, monolete.

1. Stem scales linear-lanceolate, dark brown, usually coarsely toothed and contorted distally; rachises (proximal part on abaxial surface) with sporadic lance-ovate scales > 6 cells wide; immature sori circular, submarginal; sporangiasters (sterile sporangia with dark bulbous heads) present…**P. saximontanum**
1'. Stem scales narrowly ovate, tan to brown, entire to denticulate, usually not contorted distally; rachises (proximal part on abaxial surface) with sporadic linear-lanceolate scales < 6 cells wide; immature sori oval, situated midway between pinna midrib and margins; sporangiasters absent…**P. hesperium**

Polypodium hesperium Maxon—Western polypody fern. Stem scales narrowly ovate, 25–40 cells wide just above point of attachment, ± symmetrically tapered, entire to denticulate, usually not contorted distally, tan or brown throughout, rarely with a few indurated cells medially; blades elliptic-lanceolate to triangular, mostly widest near base; rachises (proximal part on abaxial surface) with sporadic linear-lanceolate scales < 6 cells wide; sori oval when immature, situated midway between pinna midrib and margins; sporangiasters absent. On a variety of acidic and mildly basic rocks, 5400–8700 ft. NC, C, SC, SW.

Polypodium saximontanum Windham—Rocky Mountain polypody fern. Stem scales linear-lanceolate, 20–30 cells wide just above point of attachment, asymmetrically tapered, often coarsely toothed and contorted distally, mostly dark brown with indurated cells commonly forming a faint medial stripe; blades linear-elliptic, usually widest near middle; rachises (proximal part on abaxial surface) with sporadic lance-ovate scales > 6 cells wide; sori round when immature, submarginal; sporangiasters present. On granitic or gneissic rocks, ca. 8200 ft. NC.

PTERIDACEAE – BRAKE FERN FAMILY

Plants growing on or around rocks; stems subterranean, horizontal, compact to long-creeping, scaly rhizomes, not continuously ridged and grooved, internodes poorly defined, not hollow; stem scales < 5 mm, concolored or bicolored with dark central stripe and brown margins; leaves monomorphic or fertile and sterile leaves dimorphic, spirally arranged, clustered (rarely widely spaced), usually < 50 cm; petioles > 1 cm, with 1 vascular bundle (rarely 2) at base, often hairy and with few to many scales like those on stems; blades linear to pentagonal, 1–6-pinnate, 1–25 cm wide, the surfaces variously hairy, scaly, farinose (with whitish or yellowish powder) or glabrous, margins recurved (rarely flat); veins mostly obscure, pinnate, terminating before reaching margins (conspicuous, dichotomously branched, and intersecting margins in *Adiantum*); sori borne abaxially along veins, diffuse, often confluent with age and forming interrupted, submarginal bands; indusia absent, the young sporangia often covered by modified or unmodified, recurved leaf margins (false indusia); sporangia < 0.3 mm; spores monomorphic, 64 or 32 (rarely 16) per sporangium, 30–70 µ, trilete.

1. Veins of ultimate segments conspicuous, dichotomously branched, intersecting margins; sporangia borne directly on reflexed marginal lobes scattered along acroscopic edges of ultimate segments…
   Adiantum
1'. Veins of ultimate segments inconspicuous, or if apparent, then pinnately branched and terminating before reaching margins; sporangia borne on abaxial leaf surfaces around periphery of ultimate segments…(2)
2. Main axes of blades green or straw-colored abaxially, glabrous; leaves dimorphic, the ultimate segments of sterile leaves regularly toothed or lobed, each tooth with a prominent hydathode adaxially…
   Cryptogramma
2'. Main axes of blades brown to blackish abaxially, or if lighter, then glandular, hairy, or scaly; leaves mostly monomorphic, the ultimate segments of any sterile leaves not regularly toothed or lobed…(3)
3. Blades pinnate-pinnatifid or simply pinnate throughout; abaxial surfaces densely covered with scales…
   Astrolepis
3'. Blades 2–6-pinnate proximally, or if less divided, then abaxial surfaces hairy or farinose, not densely scaly…(4)
4. Blades with glands producing whitish or yellowish powder (farina) intermixed with sporangia or covering entire abaxial surface…(5)
4'. Blades without glands or powdery farina on abaxial surface…(8)
5. Ultimate segments (most) narrowed at base, the well-defined stalk dark brown or blackish…(6)
5'. Ultimate segments broadly attached, or if narrowed at base, then the stalk poorly defined and dull brown…(7)
6. Stem scales (at least some) bicolored with a dark, sclerotic central stripe and thin, brown margins; farina confined to narrow submarginal bands associated with sporangia…**Pellaea** (in part)
6'. Stem scales concolored; farina entirely covering abaxial surfaces of ultimate segments…**Argyrochosma** (in part)
7. Sporangia following veins for most of their length [to be expected]…Pentagramma triangularis (Kaulf.) Yatsk., Windham & E. Wollenw. subsp. maxonii (Weath.) Yatsk., Windham & E. Wollenw.
7'. Sporangia confined to modified vein tips near blade margins…**Notholaena**
8. Blades conspicuously hairy or scaly abaxially…(9)

8'.   Blades sparsely pubescent to glabrous abaxially…(10)
9.   Blades deeply pinnate-pinnatifid and pentagonal…**Bommeria**
9'.   Blades 2–4-pinnate, or if pinnate-pinnatifid, then linear or lanceolate…**Myriopteris** (in part)
10.   Stem scales bicolored with a dark, central stripe, or if concolored, then largest unlobed ultimate segments > 7 mm…**Pellaea** (in part)
10'. Stem scales concolored to weakly bicolored; largest unlobed ultimate segments < 7 mm…(11)
11.   Ultimate segments broadly attached, or if narrowed at base, then the stalk entirely green or with greenish wings…**Myriopteris** (in part)
11'. Ultimate segments (most) abruptly narrowed at base, the stalk dark brown throughout…**Argyrochosma** (in part)

ADIANTUM L. – Maidenhair ferns

Adiantum capillus-veneris L.—Southern maidenhair. Stems short-creeping, the scales concolored, brown; leaves slightly dimorphic, clustered; blades lanceolate to ovate, 3–5-pinnate, abaxially and adaxially glabrous; main axes of blades dark brown to blackish abaxially, glabrous; ultimate segments > 7 mm, narrowed to well-defined dark brown or blackish stalks; fertile segment margins with strongly differentiated, reflexed lobes along acroscopic edges forming discrete false indusia; veins conspicuous, dichotomously branched, intersecting segment margins; sporangia borne along veins on abaxial surfaces of prominent false indusia. Moist, usually calcareous habitats, 3500–6200 ft. EC, SE, SC, SW.

ARGYROCHOSMA (J. Sm.) Windham – False cloak ferns

Stems compact, the scales concolored, tan to dark brown; leaves monomorphic, clustered; blades ovate-triangular to lanceolate, 3–6-pinnate, abaxially with abundant glands producing whitish farina (glabrous in *A. microphylla*), adaxially sparsely glandular to glabrous; main axes of blades brown to blackish abaxially, glabrous or with farinose glands; ultimate segments < 7 mm, narrowed to well-defined brown or blackish stalks; fertile segment margins flat or more often recurved and forming poorly differentiated false indusia extending entire length of segment; veins mostly inconspicuous, pinnately branched, terminating before reaching segment margins; sporangia borne along distal 1/2 of secondary veins (nearly their full length in *A. incana*) on abaxial surfaces of ultimate segments, often submarginal and partially protected by false indusia (*Notholaena* p.p.; *Pellaea* p.p.).

1.   Abaxial surfaces of blades glabrous, lacking glands or whitish farina…**A. microphylla**
1'. Abaxial surfaces of blades with minute, glandular hairs producing whitish farina…(2)
2.   Main axes of blades distinctly flexuous, sharply bent at regular intervals…**A. fendleri**
2'. Main axes of blades straight or nearly so…(3)
3.   Ultimate segments not articulate, dark color of stalks extending into segment bases abaxially; margins of ultimate segments usually recurved, often partially covering the mostly submarginal sporangia… **A. limitanea**
3'. Ultimate segments articulate, dark color of stalks stopping abruptly at segment bases; margins of ultimate segments flat, the sporangia fully exposed and extending along veins for most of their length… **A. incana**

Argyrochosma fendleri (Kunze) Windham—Fendler's false cloak fern. Stem scales brown; leaves to 25 cm; petioles dark brown; blades triangular, 4–6-pinnate proximally, with whitish farina abaxially, glabrous or sparsely glandular adaxially; main axes of blades rounded adaxially, distinctly flexuous; ultimate segments not articulate, dark color of stalks continuing into segment bases abaxially; segment margins recurved to flat; sporangia submarginal, borne on distal 1/4 of secondary veins. On igneous rocks, 5400–8000 ft. NC, C, SC, SW.

Argyrochosma incana (C. Presl) Windham—Hairy or gray false cloak fern. Stem scales brown; leaves to 20 cm; petioles black; blades ovate, 3–4-pinnate proximally, with dense, whitish farina abaxially, glabrous adaxially; main axes of blades rounded to slightly flattened adaxially, straight or nearly so; ultimate segments articulate, dark color of stalks stopping abruptly at segment bases abaxially; segment margins flat; sporangia following secondary veins for most of length. On volcanic rocks, 4400–4500 ft. SW.

Argyrochosma limitanea (Maxon) Windham—Southwestern false cloak fern. Stem scales brown; leaves to 30 cm; petioles reddish brown to black; blades lanceolate to triangular, 3–5-pinnate proximally, with dense white farina abaxially, glabrous or sparsely glandular adaxially; main axes of blades rounded to slightly flattened adaxially, straight to slightly flexuous; ultimate segments not articulate, dark color of stalks continuing into segment bases abaxially; segment margins recurved; sporangia submarginal, borne on distal 1/2 of secondary veins. Two infraspecific taxa have been recognized: **subsp. limitanea** with lowermost pinnae at least 1/2 as long as the blades, and **subsp. mexicana** (Maxon) Windham with lowermost pinnae 1/4–1/3 as long as the blades. Their ranges overlap broadly in New Mexico and more work is needed to determine whether they represent distinct taxa. Usually on calcareous or volcanic rocks, 4200–7500 ft. SE, SC, SW.

Argyrochosma microphylla (Mett. ex Kuhn) Windham [*Notholaena parvifolia* Tryon]—Small-leaf false cloak fern. Stem scales brown; leaves to 25 cm; petioles brown; blade triangular to ovate, 3–4-pinnate proximally, glabrous abaxially and adaxially; main axes of blades flattened or shallowly grooved adaxially, straight to somewhat flexuous; ultimate segments articulate, dark color of stalks stopping abruptly at segment bases; segment margins recurved; sporangia submarginal, borne on distal 1/3 of secondary veins. On calcareous rocks, 3400–6200 ft. SE, SC.

ASTROLEPIS D. M. Benham & Windham – Star-scaled cloak ferns

Stems compact to short-creeping, the scales mostly concolored, tan to brown; leaves monomorphic, clustered; blades mostly linear, pinnate-pinnatifid to simply pinnate, abaxially covered by ciliate or erose scales, adaxially with scattered stellate or ciliate scales, but often glabrescent when mature; main axes of blades (rachises) brown to tan abaxially, with scales throughout; ultimate segments (pinnae) > 7 mm except in *A. cochisensis*, narrowed to poorly defined, dull brown stalks; fertile segment margins flat, not forming differentiated false indusia; veins inconspicuous, pinnately branched, terminating before reaching segment margins; sporangia borne along veins near margins on abaxial surfaces of pinnae (often clustered near notches between pinna lobes), not protected by false indusia (*Cheilanthes* p.p.; *Notholaena* p.p.).

1. Abaxial scales ovate with rounded to broadly acute apices; most pinnae < 7 mm…**A. cochisensis**
1'. Abaxial scales lanceolate with ± attenuate apices; most pinnae > 7 mm…(2)
2. Adaxial scales mostly persistent, ciliate with prominent, usually plane central axes; pinnae asymmetrically 3–8-lobed, the lobes usually more prominent on convexly curved acroscopic edge…**A. integerrima**
2'. Adaxial scales mostly deciduous at maturity, highly dissected or ciliate with narrow, often twisted central axes; pinnae mostly symmetrically 7–12-lobed…(3)
3. Adaxial pinna surfaces glabrescent, most scales deciduous with age; central axes of adaxial scales 1–2 cells wide; pinnae usually deeply lobed…**A. sinuata**
3'. Adaxial pinna surfaces sparsely scaly, at least some scales persistent; central axes of adaxial scales 2–4 cells wide; pinnae usually shallowly lobed…**A. windhamii**

Astrolepis cochisensis (Goodd.) D. M. Benham & Windham—Cochise star-scaled cloak fern. Leaf blades shallowly 1-pinnate-pinnatifid, the largest pinnae < 7 mm, symmetrically or asymmetrically 3–5-lobed, the lobes broadly rounded, separated by very shallow sinuses; abaxial scales ovate with ± rounded apices, usually < 1 mm, erose-dentate to coarsely ciliate; adaxial scales sparse, mostly deciduous at maturity, orbicular to ovate-elliptic, coarsely ciliate with prominent, usually plane central axes > 5 cells wide. Two subspecies have been recognized in New Mexico; the 64-spored sexual diploid **subsp. chihuahuensis** D. M. Benham is confined to the SE part of the state; the 32-spored apomictic triploid **subsp. cochisensis** is more widespread. On limestone and other calcareous substrates, 3100–7000 ft. C, SE, SC, SW.

Astrolepis integerrima (Hook.) D. M. Benham & Windham—Upright star-scaled cloak fern. Leaf blades shallowly 1-pinnate-pinnatifid, the largest pinnae > 7 mm, asymmetrically 3–8-lobed, the lobes usually more prominent on convexly curved acroscopic edge, broadly rounded, separated by mostly shallow sinuses; abaxial scales lanceolate with ± attenuate apices, usually > 1 mm, coarsely ciliate; adaxial scales abundant, mostly persistent, lanceolate, ciliate with prominent, usually plane central axes often > 5 cells wide. Usually on limestone or other calcareous substrates, 3400–6900 ft. C, EC, SE, SC, SW.

Astrolepis sinuata (Lag. ex Sw.) D. M. Benham & Windham—Wavy-margined star-scaled cloak fern. Leaf blades 1-pinnate-pinnatifid, the largest pinnae > 7 mm, symmetrically 7–12-lobed, the lobes narrowly rounded to acute, separated by deep sinuses; abaxial scales lanceolate with ± attenuate apices, usually > 1 mm, denticulate to short-ciliate with delicate marginal projections; adaxial scales sparse, deciduous, highly dissected with very narrow, often twisted central axes 1–2 cells wide. Two subspecies have been recognized in New Mexico: the 64-spored sexual diploid **subsp. mexicana** D. M. Benham is currently known only from Eddy County; the 32-spored apomictic triploid **subsp. sinuata** is more widespread. On a variety of substrates but rarely on limestone, 4200–6800 ft. C, WC, SE, SC, SW.

Astrolepis windhamii D. M. Benham—Windham's star-scaled cloak fern. Leaf blades 1-pinnate-pinnatifid, the largest pinnae > 7 mm, usually symmetrically 7–10-lobed, the lobes rounded, separated by shallow sinuses; abaxial scales lanceolate with ± attenuate apices, usually > 1 mm, ciliate; adaxial scales sparse, some persistent, highly dissected with narrow, often twisted central axes 2–4 cells wide. On calcareous or volcanic rocks, 3400–7500 ft. SE, SC, SW.

BOMMERIA E. Fourn. – Bommeria fern

Bommeria hispida Underw.—Copper fern. Stems long-creeping, the scales concolored, brown; leaves monomorphic, scattered; blades pentagonal, deeply pinnatifid, abaxially with coiled and needle-like hairs plus scattered scales, adaxially with appressed, needlelike hairs; main axes of blades brown abaxially, with hairs and scales throughout; ultimate segments often > 7 mm, sessile and broadly attached to axes; fertile segment margins flat to slightly recurved, not forming differentiated false indusia; veins inconspicuous, pinnately branched, terminating before reaching segment margins; sporangia borne along distal 2/3 of secondary veins on abaxial surfaces of ultimate segments, not protected by false indusia. On acidic to neutral rocks, 5000–7600 ft. WC, SC, SW.

CRYPTOGRAMMA R. Br. – Parsley fern

Cryptogramma acrostichoides R. Br.—American parsley fern. Stems short-creeping, the scales bicolored with dark central stripe and lighter brown margins; leaves clustered, dimorphic, the shorter sterile leaves with ultimate segments symmetrically toothed or lobed, each tooth with prominent hydathode adaxially; blades triangular to ovate-lanceolate, 2–3-pinnate proximally, abaxially with yellowish, farina-producing glands among the sporangia, adaxially glabrous or sparsely puberulent; main axes of blades green to straw-colored abaxially, glabrous; ultimate segments mostly < 7 mm, usually narrowed to poorly defined stalks, these greenish throughout; fertile segment margins recurved when young to form false indusia extending entire length of segments; veins usually inconspicuous, pinnately branched, terminating before reaching segment margins; sporangia borne along distal 1/2 of secondary veins on abaxial surfaces of ultimate segments, protected by false indusia when young. Acidic to neutral rock outcrops, 8200–12,500 ft. NC.

MYRIOPTERIS Fée – Lip ferns

Stems compact to long-creeping, the scales bicolored with dark central stripe and lighter brown margins or concolored in some species; leaves monomorphic to weakly dimorphic, clustered to scattered; blades lanceolate, linear-oblong, or ovate-triangular, pinnate-pinnatifid to 4-pinnate proximally, abaxially hairy and/or scaly, rarely glabrous, adaxially hairy to glabrous; main axes of blades brown, blackish or rarely greenish abaxially, hairy, scaly, or rarely almost glabrous; ultimate segments < 7 mm (except in *M. aurea*), often narrowed to stalks, these greenish or dark-colored; fertile segment margins recurved (except in *M. aurea*) to form somewhat differentiated false indusia usually extending entire length of segment; veins usually inconspicuous, pinnately branched, terminating before reaching segment margins; sporangia borne near submarginal vein tips on abaxial surfaces of ultimate segments, often partially or fully covered by false indusia (*Cheilanthes* p.p.; see Grusz & Windham, 2013).

1. Rachises shallowly grooved adaxially, glabrous except for widely scattered hairlike scales...**M. wrightii**
1′. Rachises rounded adaxially, hairy and/or scaly...(2)

2.   Ultimate segments scabrous adaxially, covered with glassy, sharp-tipped, broad-based hairs…**M. scabra**
2′.   Ultimate segments smooth adaxially, hairs often present but not glassy, sharp-tipped, or broad-based…(3)
3.   Rachises lacking scales, the indument entirely of abundant to sparse hairs…(4)
3′.   Rachises with narrow to broad scales (inconspicuous in *M. tomentosa*), hairs usually present as well…(6)
4.   Ultimate segments inconspicuously hairy to glabrescent abaxially…**M. alabamensis**
4′.   Ultimate segments conspicuously hairy abaxially…(5)
5.   Blades 1-pinnate-pinnatifid…**M. aurea**
5′.   Blades 3-pinnate proximally…**M. gracilis**
6.   Stems compact, the leaves clustered; stem scales ± linear, at least some strongly bicolored with well-defined dark, sclerotic central stripe and thin brown margins…(7)
6′.   Stems creeping, the leaves ± scattered; stem scales lanceolate to ovate, concolored, brown to blackish, or if bicolored, the dark, sclerotic central portion and pale margins not well defined…(9)
7.   Blade scales inconspicuous, linear, the largest ≤ 0.4 mm wide…**M. tomentosa**
7′.   Blade scales usually conspicuous, lanceolate to ovate, the largest > 0.4 mm wide…(8)
8.   Ultimate segments with scattered coarse hairs adaxially; blade scales concealing much of abaxial surface, ovate-lanceolate, the whitish body usually strongly contrasting with the darker region of attachment…**M. windhamii**
8′.   Ultimate segments finely tomentose to glabrescent adaxially, coarse hairs absent; blade scales concealing only the costae, lanceolate, the brownish body not strongly contrasting with the darker region of attachment…**M. rufa**
9.   Blade scales conspicuously ciliate, the many fine cilia usually extending entire length of scales and apparent adaxially…(10)
9′.   Blade scales lacking cilia or cilia relatively inconspicuous, ± confined to proximal 1/2 of scales and not apparent adaxially…(11)
10.   Cilia of blade scales very conspicuous adaxially, entangled, individual scales difficult to remove; stem scales mostly concolored, light brown, thin…**M. lindheimeri**
10′.   Cilia of blade scales apparent adaxially but usually not entangled, individual scales ± easily removed; stem scales often bicolored with poorly defined, dark sclerotic central portion and narrow pale margins…**M. yavapensis**
11.   Blade scales lacking cilia, the margins entire to erose-dentate throughout…**M. fendleri**
11′.   Blade scales with relatively coarse cilia ± confined to proximal 1/2 of scales…**M. wootonii**

Myriopteris alabamensis (Buckley) Grusz & Windham [*Cheilanthes alabamensis* (Buckley) Kunze]—Alabama lip fern. Stems short-creeping to compact; stem scales mostly concolored, brown, thin; leaves clustered; blades 2-pinnate to 2-pinnate-pinnatifid proximally, lacking abaxial scales; rachises rounded, adaxially conspicuously hairy, abaxially sparsely hairy to glabrescent; ultimate segments oblong, > 1 mm, with sparse, inconspicuous hairs adaxially and abaxially. Usually on limestone, 3900–6200 ft. SE, SC, SW.

Myriopteris aurea (Poir.) Grusz & Windham [*Cheilanthes bonariensis* (Willd.) Proctor]—Golden lip fern. Stems short-creeping to compact; stem scales mostly bicolored, with dark sclerotic central stripe and thin brown margins; leaves clustered; blades 1-pinnate-pinnatifid proximally, lacking abaxial scales; rachises rounded, hairy; ultimate segments elongate-triangular, > 1 mm, adaxially sparsely hairy, abaxially densely tomentose. On a variety of substrates though rarely observed on limestone, 4200–6200 ft. WC, SC, SW.

Myriopteris fendleri (Hook.) E. Fourn. [*Cheilanthes fendleri* Hook.]—Fendler's lip fern. Stems creeping; stem scales usually concolored, brown, thin; leaves scattered; blades 3–4-pinnate proximally, abaxially with conspicuous, lance-ovate scales, the largest 0.4–1.2 mm wide, often concealing ultimate segments, entire to denticulate, not ciliate; rachises rounded adaxially, lacking hairs, with scattered scales; ultimate segments round to oblong, > 1 mm, adaxially glabrous, abaxially glabrous or rarely a few small scales near base. On a variety of acidic and mildly basic substrates, 4600–8600 ft. NE, NC, EC, C, WC, SC, SW.

Myriopteris gracilis Fée [*Cheilanthes feei* T. Moore]—Slender lip fern. Stems compact to short-creeping; stem scales a mix of concolored (brown) and bicolored (with dark sclerotic central stripe); leaves clustered; blades 3-pinnate proximally, lacking abaxial scales; rachises rounded adaxially, hairy, lacking scales; ultimate segments round to oblong, the largest > 1 mm, adaxially sparsely hairy to glabrescent, abaxially hairy. On limestone or other calcareous substrates, 3500–10,000 ft. Widespread.

Myriopteris lindheimeri (Hook.) J. Sm. [*Cheilanthes lindheimeri* Hook.]—Lindheimer's lip fern. Stems long-creeping; stem scales mostly concolored, light brown, thin; leaves scattered; blades 4-pinnate proximally, abaxially with conspicuous, lance-ovate scales, the largest 0.4–1 mm wide, mostly concealing ultimate segments, long-ciliate throughout, cilia very conspicuous adaxially, fine, curly, strongly entangled; rachises rounded adaxially, with hairs and scattered linear-lanceolate scales; ultimate segments round to slightly oblong, mostly < 1 mm, adaxially and abaxially with sporadic branched hairs. On a variety of acidic to mildly basic substrates, 4100–7000 ft. WC, SC, SW.

Myriopteris rufa Fée [*Cheilanthes eatonii* Baker]—Reddish lip fern. Stems compact; stem scales mostly bicolored with dark, sclerotic central stripe and thin brown margins; leaves clustered; blades 3–4-pinnate proximally, abaxially with conspicuous, lanceolate scales, the largest 0.4–0.7 mm wide, not concealing ultimate segments, erose-dentate, not ciliate; rachises rounded adaxially, with scattered hairs and narrowly lanceolate scales; ultimate segments oval to round, the largest > 1 mm, adaxially with fine, unbranched hairs or glabrescent, abaxially densely tomentose. On a variety of substrates including limestone and granite, 3900–9000 ft. Widespread.

Myriopteris scabra (C. Chr.) Grusz & Windham [*Cheilanthes horridula* Maxon]—Prickly lip fern. Stems short-creeping; stem scales mostly concolored, brown, thin; leaves clustered; rachises rounded adaxially, with scattered hairs and linear-lanceolate scales; blades 1-pinnate-pinnatifid to 2-pinnate at base, abaxially with mostly inconspicuous, lanceolate scales, the largest 0.4–0.6 mm wide, not concealing ultimate segments, erose, not ciliate; ultimate segments oblong to elongate-triangular, > 1 mm, adaxially scabrous with glassy, sharp-tipped, broad-based hairs, these present but less apparent abaxially. On limestone at 3700 ft. SE.

Myriopteris tomentosa Fée [*Cheilanthes tomentosa* Link]—Woolly lip fern. Stems compact; stem scales mostly bicolored with dark sclerotic central stripe and thin brown margins; leaves clustered; blades 3–4-pinnate proximally, abaxially with inconspicuous, linear scales, the largest 0.1–0.4 mm wide, not concealing ultimate segments, entire, not ciliate; rachises rounded abaxially, with hairs and scattered linear scales; ultimate segments oblong, the largest > 1 mm, adaxially with fine, unbranched hairs, abaxially densely tomentose. On a variety of substrates including limestone and granite, 5500–6700 ft. SC, SW.

Myriopteris windhamii Grusz [*Cheilanthes villosa* Davenp. ex Maxon]—Villous lip fern. Stems compact; stem scales mostly bicolored with dark sclerotic central stripe and thin brown margins; leaves clustered; blades 3–4-pinnate proximally, abaxially with conspicuous, ovate to lanceolate scales, the largest 0.4–1.5 mm wide, often concealing ultimate segments, erose-dentate, not ciliate; rachises rounded adaxially, lacking hairs, with scattered linear to lanceolate scales; ultimate segments round to oval, the largest > 1 mm, adaxially with scattered, coarse, twisted hairs, abaxially with rare hairs and scales. Usually on limestone, 4700–7100 ft. C, SE, SC, SW.

Myriopteris wootonii (Maxon) Grusz & Windham [*Cheilanthes wootonii* Maxon]—Wooton's lip fern. Stems long-creeping; stem scales concolored, brown, thin or rarely weakly bicolored with poorly defined, dark sclerotic central stripe; leaves scattered; blades 3–4-pinnate proximally, abaxially with conspicuous, lanceolate scales, the largest 0.4–0.8 mm wide, often concealing ultimate segments, with coarse cilia mostly confined to proximal 1/2, the cilia usually not apparent adaxially and not entangled; rachises rounded adaxially, with sparse hairs and scattered linear-lanceolate scales; ultimate segments round to oblong, the largest 1–3 mm, adaxially glabrous, abaxially glabrous or with a few small scales near base. Usually on igneous substrates, 4400–7500 ft. NE, NC, EC, C, WC, SC, SW.

Myriopteris wrightii (Hook.) Grusz & Windham [*Cheilanthes wrightii* Hook.]—Wright's lip fern. Stems creeping; stem scales mostly concolored, brown, thin; leaves clustered to somewhat scattered; blades 2-pinnate-pinnatifid proximally, lacking abaxial scales; rachises shallowly grooved adaxially, glabrous except for occasional hairlike scales; ultimate segments oblong to linear, not beadlike, the largest 3–7 mm, adaxially and abaxially glabrous. Rocky slopes and ledges usually on igneous substrates, 4500–7100 ft. SC, SW.

Myriopteris yavapensis (T. Reeves ex Windham) Grusz & Windham [*Cheilanthes yavapensis* T. Reeves ex Windham]—Yavapai lip fern. Stems creeping; stem scales mostly bicolored with dark sclerotic central stripe and thin brown margins; leaves scattered; blades 4-pinnate proximally, abaxial scales conspicuous, lanceolate, the largest 0.4–1 mm wide, often concealing ultimate segments, with cilia distributed entire length of scale, these not strongly entangled but usually apparent adaxially; rachises rounded adaxially, with sparse hairs and scattered linear-lanceolate scales; ultimate segments round to oblong, the largest usually 1–2 mm, adaxially with rare branched hairs, abaxially glabrous or with a few small scales near base. Usually on igneous substrates, 4500–6800 ft. SC, SW.

NOTHOLAENA R. Br. – Cloak ferns

Stems compact to short-creeping, the scales concolored or bicolored with dark central stripe and lighter brown margins; leaves monomorphic, clustered; blades linear-lanceolate to pentagonal, deeply pinnatifid to 2-pinnate-pinnatifid proximally, abaxially covered with glands producing whitish or yellowish farina, adaxially sparsely glandular or glabrous; main axes of blades brown or blackish abaxially, with hairs, scales, farinose glands, or rarely glabrous; ultimate segments mostly 2–15 mm, broadly attached to axes or rarely narrowed to a poorly defined, dull brown stalk, fertile segment margins often slightly recurved to form poorly defined false indusia extending entire length of segment; veins usually inconspicuous, pinnately branched, terminating before reaching segment margins; sporangia borne on submarginal vein tips on abaxial surfaces of ultimate segments, rarely protected by false indusia.

1. Blades pentagonal, 1–2 times longer than wide; abaxial surfaces with dense farina-producing glands, lacking scales...**N. standleyi**
1'. Blades narrowly lanceolate, 3–6 times longer than wide; abaxial surfaces with scattered scales in addition to glands and farina [to be expected in SW]...N. grayi Davenp.

Notholaena standleyi Maxon—Standley's cloak fern. Stem scales (most) bicolored with dark central stripe and broad, brown margins; blades pentagonal, 1–2 times longer than wide, deeply 2–3-pinnatifid (distal pinnae connected to basal pinnae by narrow wing of green tissue), abaxially with dense glands producing yellowish farina, adaxially nearly glabrous; basal pinnae larger than adjacent pair, the basiscopic pinnules greatly enlarged. On various substrates including granite and limestone, 4600–7000 ft. NE, NC, C, WC, SC, SW.

PELLAEA Link – Cliff-brake ferns

Stems compact to long-creeping, the scales concolored or bicolored with dark, central stripe and lighter brown margins; leaves monomorphic to weakly dimorphic, clustered to scattered; blades linear-lanceolate to ovate-triangular, usually 2-pinnate proximally, abaxially glabrous, hairy, or with glands among the sporangia producing submarginal band of whitish farina, adaxially usually glabrous; main axes of blades brownish or blackish, with scattered hairs or more often glabrous; ultimate segments often > 7 mm, narrowed to a well-defined stalk, this usually brown or blackish; fertile segment margins recurved to form weakly differentiated false indusia extending entire length of segment; veins usually inconspicuous, pinnately branched, terminating before reaching segment margins; sporangia borne along distal 1/2 of secondary veins on abaxial surfaces of ultimate segments, usually partially or fully protected by false indusia.

1. Stems long-creeping, the leaves scattered; rachises straw-colored, tan, or grayish...**P. intermedia**
1'. Stems compact, the leaves clustered; rachises dark brown to blackish...(2)
2. Stem scales concolored, uniformly reddish brown or tan; rachises rounded adaxially...(3)
2'. Stem scales (most) bicolored, with dark central stripe and lighter margins; rachises ± grooved adaxially...(4)
3. Rachises glabrous or with very few hairlike scales...**P. glabella**
3'. Rachises with abundant hairs adaxially...**P. atropurpurea**
4. Rachillae shorter than or equal to ultimate segments; sporangia intermixed with sparse farina-producing glands...**P. wrightiana**
4'. Rachillae longer than ultimate segments; sporangia usually intermixed with abundant farina-producing glands...(5)

5. Pinnae (largest on fertile leaves) usually with > 10 ultimate segments; rachillae (longest) > 2 times longer than ultimate segments; spores well formed…**P. truncata**

5'. Pinnae (largest on fertile leaves) usually with < 10 ultimate segments; rachillae (longest) < 2 times longer than ultimate segments; spores malformed…**P. ×wagneri**

Pellaea atropurpurea (L.) Link—Purple cliff-brake fern. Stems compact; stem scales concolored, reddish brown or tan; leaves clustered; blades 2–18 cm wide; rachises reddish purple, rounded adaxially, densely pubescent adaxially with short, curly, appressed hairs; pinnae with 3–15 ultimate segments; rachillae 10–100 mm, often longer than ultimate segments, the latter 10–75 mm, abaxially sparsely villous near midrib; margins weakly recurved to flat on mature fertile segments, usually partially covering the sporangia, farina-producing glands absent; spores mostly well formed. On limestone or other basic to neutral substrates, 3900–8000 ft. NE, C, WC, SE, SC, SW.

Pellaea glabella Mett. ex Kuhn—Smooth cliff-brake fern. Stems compact; scales concolored, reddish brown; leaves clustered; blades 1–8 cm wide; rachises brown, rounded adaxially, glabrous or with very few hairlike scales; pinnae with 3–7 ultimate segments; rachillae 1–50 mm, often shorter than fertile ultimate segments, the latter 5–20 mm, glabrous; margins recurved on mature fertile segments, covering the sporangia, farina-producing glands absent; spores mostly well formed. New Mexico plants are **subsp. simplex** (Butters) Á. Löve & D. Löve. On sandstone and limestone, 7400–7600 ft. WC.

Pellaea intermedia Mett. ex Kuhn—Intermediate cliff-brake fern. Stems creeping; scales mostly bicolored with black central stripe and brown margins; leaves scattered; blades 4–16 cm wide; rachises straw-colored, tan, or grayish, rounded or slightly flattened adaxially, ± pubescent; pinnae usually with 7–21 ultimate segments; rachillae 20–100 mm, longer than fertile ultimate segments, the latter 5–15 mm, abaxially puberulent or rarely glabrous; margins recurved on mature fertile segments, covering the sporangia, farina-producing glands absent; spores mostly well formed. Usually on limestone, 5000–7300 ft. C, SE, SC, SW.

Pellaea truncata Goodd. [*P. longimucronata* Hook.]—Spiny cliff-brake fern. Stems compact; scales bicolored with dark central stripe and brown margins; leaves clustered; blades 4–18 cm wide; rachises brown, shallowly grooved adaxially, nearly glabrous; pinnae with 9–25 ultimate segments; rachillae 12–70 mm, > 2 times longer than fertile ultimate segments, the latter 4–12 mm, glabrous; margins recurved on mature fertile segments, covering the sporangia, the latter intermixed with abundant farina-producing glands; spores mostly well formed. On various substrates but rarely on limestone, 4400–7000 ft. NC, C, WC, SE, SC, SW. A strongly outcrossing sexual diploid commonly hybridizing with *P. wrightiana* to produce intermediate plants with malformed spores.

Pellaea ×wagneri Windham—Wagner's hybrid cliff-brake fern. Stems compact; scales bicolored with dark central stripe and brown margins; leaves clustered; blades 2.5–10 cm wide; rachises brown, shallowly grooved adaxially, nearly glabrous; pinnae with 7–11 ultimate segments; rachillae 7–37 mm, 1–2 times longer than fertile ultimate segments, the latter 6–18 mm, glabrous; margins recurved on mature fertile segments, covering the sporangia, the latter intermixed with abundant farina-producing glands; spores mostly malformed. On various substrates but rarely on limestone, 5200–7000 ft. C, WC, SC, SW. This triploid hybrid appears sporadically in habitats where its parents co-occur; spores are mostly nonfunctional in these otherwise vigorous plants.

Pellaea wrightiana Hook.—Wright's cliff-brake fern. Stems compact; scales bicolored with dark central stripe and brown margins; leaves clustered; blades 1.5–5 cm wide; rachises brown, shallowly grooved adaxially, nearly glabrous; pinnae usually with 3–9 ultimate segments; rachillae 2–20 mm, shorter than or equal to ultimate segments, the latter 5–20 mm, glabrous; margins recurved on mature fertile segments, covering the sporangia, the latter intermixed with sparse farina-producing glands; spores mostly well formed. On a variety of acidic to mildly basic substrates, 5200–7500 ft. NE, C, WC, SC, SW. A sexual tetraploid commonly hybridizing with *P. truncata* to produce intermediate plants with malformed spores.

SALVINIACEAE – MOSQUITO FERN FAMILY

Plants free-floating aquatics or stranded on shore, any rootlike structures not embedded in soil; stems surficial, horizontal, short-creeping, glabrous to hairy, not continuously ridged and grooved, internodes poorly defined, not hollow; stem scales absent or hairlike; leaves dimorphic with dorsal leaves or lobes photosynthetic and those on ventral surfaces modified as floats or rootlike structures, spirally arranged, closely spaced, to 5 cm; petioles absent (*Azolla*) or to 1 cm (*Salvinia*), with 1 vascular bundle and scattered hairs when present, scales lacking; blades oblong to orbicular, entire, bilobed, or some highly dissected and rootlike, < 2 cm wide, with prominent columnar projections (complex, multicellular "hairs") on 1 surface (*Salvinia*) or easily overlooked domelike or papillate hairs (*Azolla*), margins flat; veins obscure to apparent, 1 per leaf (*Azolla*) or pinnate-reticulate (*Salvinia*), terminating at translucent border just short of margin; sori enclosed within membranous protective structures (sporocarps) attached by short stalks to basal portion of petioles, the sporocarps globose to elliptic, < 2 mm wide, containing either microsporangia or megasporangia; indusia modified to form sporocarp wall; sporangia < 0.5 mm, the microsporangia containing 64 or 32 spores, the megasporangia containing a single functional spore; spores dimorphic, microspores 15–35 μ, trilete, megaspores 225–375 μ, trilete.

1. Blades > 5 mm, with prominent columnar projections (complex, multicellular "hairs") on 1 surface; true roots absent but submerged leaves dissected to produce highly branched, rootlike structures...**Salvinia**
1′. Blades < 2 mm, lacking prominent columnar projections, the largest hairs domelike and 1–3-celled; true roots present, unbranched...**Azolla**

AZOLLA Lam. – Mosquito ferns

True roots present, unbranched; stems glabrous; blades < 2 mm, lacking prominent columnar projections, the largest hairs domelike or papillate and 1–3-celled; megaspore surfaces pitted or granulate or with raised angular bumps. Several species reported for New Mexico but taxonomy very confused (see Evrard & Van Hove, 2004).

1. Leaves with at least some 2–3-celled hairs on surfaces of upper leaf lobes; plants usually < 1.7 cm; megaspores pitted or granulate but plants very rarely fertile...**A. cristata**
1′. Leaves with strictly 1-celled hairs on surfaces of upper leaf lobes; plants (the largest) usually > 1.7 cm; megaspores with raised angular bumps and plants commonly fertile...**A. filiculoides**

Azolla cristata Kaulf [*A. caroliniana* auct. non Willd.; *A. mexicana* Schltdl. & Cham. ex Kunze; *A. microphylla* auct. non Kaulf.]—Small-leaf mosquito fern. Plants usually < 1.7 cm; leaves with at least some 2–3-celled hairs on surfaces of upper leaf lobes; megaspores pitted or granulate but the plants very rarely fertile. Forming floating colonies on standing or slow-moving water, 3700–5000 ft. C, SC, SW.

Azolla filiculoides Lam.—Pacific mosquito fern. Plants (the largest) usually > 1.7 cm; leaves with strictly 1-celled hairs on surfaces of upper leaf lobes; megaspores with raised angular bumps and the plants commonly fertile. Forming floating colonies on standing or slow-moving water, ca. 5350 ft. SW.

SALVINIA Ség. – Floating ferns

Salvinia minima Baker—Water spangles. True roots absent but submerged leaves dissected to produce highly branched, rootlike structures; stems with abundant flattened hairs; blades > 5 mm, with prominent columnar projections (complex, multicellular "hairs") on 1 surface; megaspore surfaces ridged and perforate. Floating on anthropogenic pond, ca. 4000 ft. SC.

WOODSIACEAE – CLIFF FERN FAMILY

WOODSIA R. Br. – Cliff ferns

Plants growing on or around rocks; stems subterranean, horizontal, compact to short-creeping scaly rhizomes, not continuously ridged and grooved, internodes poorly defined, not hollow; stem scales < 5 mm, concolored or bicolored with dark central stripe and brown margins; leaves monomorphic, spirally arranged, clustered, to 30 cm; petioles > 1 cm, with 2 vascular bundles at base, ± glandular-hairy and with few scales like those on stems; blades narrowly elliptic to oblanceolate, 1–2-pinnate-pinnatifid, usually < 5 cm wide, with scattered scales and sparse to abundant hairs on rachises, surface indument primarily of short-stalked glands (plus flattened multicellular hairs in *W. scopulina*) but indusial segments easily mistaken for hairs or scales, margins flat; veins usually obscure, pinnate, terminating before reaching margins; sori borne abaxially on veins, ± centered between segment midrib and margins, mostly discrete, round; indusia consisting of filamentous or scalelike lobes attached around circumference of sori, always present but often inconspicuous at maturity; sporangia < 0.3 mm; spores monomorphic, 64 per sporangium, 30–50 μ, monolete. All New Mexico species hybridize when they co-occur, blurring the distinctions among taxa. Most hybrids can be identified as such based on their malformed spores, but some populations thought to be *W. cathcartiana* × *W. plummerae* appear to be fertile and possibly worthy of species recognition.

1. Blades with flattened, multicellular hairs concentrated along rachises and midribs on both surfaces; petioles usually dark reddish brown, relatively brittle and easily shattered [reported but no specimen available]…W. scopulina D. C. Eaton
1′. Blades lacking flattened, multicellular hairs along rachises and midribs; petioles light brown to straw-colored, or if dark reddish brown, then somewhat pliable and resistant to shattering…(2)
2. Indusia composed of multicellular, scalelike segments proximally, these often branched or dissected into hairlike tips…(3)
2′. Indusia composed of mostly uniseriate, hairlike segments…(4)
3. Proximal portion of mature petioles reddish brown to purplish; blades with abundant, bulbous-tipped, glandular hairs, often ± sticky; vein tips not enlarged, barely visible adaxially; segment margins not thickened or lustrous adaxially…**W. plummerae**
3′. Petioles light brown or straw-colored proximally (sometimes darker at very base); blades sparsely to moderately glandular with short, ± columnar glands, never sticky; vein tips usually enlarged to form whitish hydathodes visible adaxially; segment margins usually thickened, lustrous adaxially…**W. cochisensis**
4. Segment margins (viewed abaxially) smooth to ± ragged but with few translucent projections; petioles reddish brown or dark purple proximally; indusial filaments usually inconspicuous, concealed by or slightly surpassing mature sporangia…**W. cathcartiana**
4′. Segment margins (viewed abaxially) with many translucent projections or filaments on teeth; petioles usually light brown or straw-colored proximally (sometimes darker at very base); indusial filaments often conspicuous, usually surpassing mature sporangia…(5)
5. Translucent projections on segment margins mostly 1–2-celled, occasionally filamentous; largest pinnae divided into 3–7 pairs of closely spaced segments, pinna apices usually abruptly tapered to rounded…**W. neomexicana**
5′. Translucent projections on segment margins mostly multicellular, often prolonged to form twisted filaments; largest pinnae with 7–14 pairs of discrete segments, pinna apices narrowly acute to attenuate…**W. phillipsii**

Woodsia cathcartiana B. L. Rob.—Cathcart's cliff fern. Petioles reddish brown to dark purple proximally, somewhat pliable and resistant to shattering; blades lacking flattened, multicellular hairs along rachises and midribs, sparsely to moderately glandular with clear, ± columnar glands, never sticky, margins thin, not lustrous, with occasional glands, minutely denticulate and often appearing somewhat ragged, rarely with 1–2-celled translucent projections; vein tips slightly (if at all) enlarged, barely visible adaxially; indusia composed primarily of uniseriate, hairlike segments, concealed by or slightly surpassing mature sporangia. Limestone cliffs and ledges, 6800–11,000 ft. NE, NC, NW, C, WC.

Woodsia cochisensis Windham—Cochise cliff fern. Petioles light brown or straw-colored, occasionally darker at very base, relatively brittle and easily shattered; blades lacking flattened, multicellular hairs along rachises and midribs, sparsely to moderately glandular with short, ± columnar glands, never sticky; pinnae abruptly tapered to rounded or broadly acute apex, occasionally attenuate, the largest with 4–9 pairs of ultimate segments, abaxial and adaxial surfaces glandular, lacking flattened, multicellular hairs; segment margins lustrous and somewhat thickened adaxially, sparsely glandular with occasional 1–2-celled translucent projections; vein tips enlarged to form whitish hydathodes visible adaxially; indusia of multiseriate, scalelike segments, usually ± lacerate and hairlike distally, often surpassing mature sporangia. Often near creeks and springs; cliffs and ledges, ca. 6000 ft. SC, SW.

Woodsia neomexicana Windham—New Mexico cliff fern. Petioles light brown or straw-colored, occasionally darker at very base, relatively brittle and easily shattered; blades lacking flattened, multi-cellular hairs along rachises and midribs, glabrescent to sparsely glandular with clear, ± columnar glands, never sticky; pinnae abruptly tapered to a rounded or broadly acute apex, the largest with 3–7 pairs of closely spaced ultimate segments, margins thin, not lustrous, with rare glands and abundant 1–2-celled translucent projections on teeth; vein tips occasionally enlarged to form whitish hydathodes visible adaxially; indusia composed almost exclusively of uniseriate, hairlike segments, usually surpassing mature sporangia. Mostly in cracks of igneous rock; cliffs and ledges, 5800–8800 ft. NE, NC, NW, C, WC, C, SC, SW.

Woodsia phillipsii Windham—Phillips' cliff fern. Petioles light brown or straw-colored, occasionally darker at very base, relatively brittle and easily shattered; blades lacking flattened, multicellular hairs along rachises and midribs, sparsely to moderately glandular with clear, ± columnar glands, never sticky; pinnae often attenuate to a narrowly acute apex, the largest with 7–14 pairs of widely spaced ultimate segments, margins slightly thickened, often lustrous adaxially, with occasional glands, appearing ciliate because multicellular translucent projections on teeth often prolonged to form twisted filaments; vein tips usually enlarged to form whitish hydathodes visible adaxially; indusia composed primarily of uniseriate, hairlike segments, often greatly surpassing mature sporangia. Cliffs, 5900–7600 ft. WC, SC, SW.

Woodsia plummerae Lemmon—Plummer's cliff fern. Petioles reddish brown to dark purple, ± pliable and resistant to shattering; blades lacking flattened, multicellular hairs along rachises and midribs, copiously glandular with clear, bulbous-tipped hairs, often ± sticky; pinnae abruptly tapered to rounded or broadly acute apex, occasionally attenuate, the largest with 5–11 pairs of closely spaced ultimate segments, margins thin, not lustrous, glandular, rarely with 1–2-celled translucent projections on teeth; vein tips rarely enlarged or visible adaxially; indusia of multiseriate, scalelike segments, these often ± lacerate distally, occasionally surpassing mature sporangia. Often on granite; cliffs and ledges, 5400–8600 ft. NE, NC, C, WC, SC, SW.

GYMNOSPERMS

KEY TO FAMILIES

1. Shrubs with jointed green stems, the leaves represented by small black triangular scales in whorls at the joint; male cones yellow with stamenlike sporangia; female (ovulate) cones green, of a few loose thin scales…**Ephedraceae**
1'. Trees or shrubs with either needlelike or scalelike overlapping leaves, producing berrylike gray cones or cones with woody scales…(2)
2. Cone berrylike, the scales fused together and detectable only by their protruding tips; shrubs or small trees with decussately arranged scalelike leaves or flat sharp needles…**Cupressaceae**
2'. Cone woody with spirally arranged scales; small or large trees with needle leaves…**Pinaceae**

CUPRESSACEAE – CYPRESS FAMILY

Robert P. Adams

Trees or shrubs; adult leaves scalelike, juvenile leaves decurrent, bark fibrous and furrowed or exfoliating in plates; seed cones persistent in *Hesperocyparis* and deciduous at maturity in *Juniperus*, terminal (except axillary in *J. communis*), scales fused in *Juniperus*, abutting in *Hesperocyparis* but appearing spherical to ovoid, scales woody to fleshy (resembling a fruit in some *Juniperus*), brown to reddish brown to blue when ripe, opening at maturity in *Hesperocyparis* and the entire seed cone fused in *Juniperus*; seeds 1–20 per scale or 1 to numerous (~100) per cone (Adams, 1993).

1. Seed cones opening at maturity, seeds shed, persistent on trees, cone scales woody, monoecious…**Hesperocyparis**
1'. Seed cones berrylike, remaining closed at maturity, seeds retained, cones shed after maturity, cone scales generally fleshy or fibrous, monoecious or dioecious…**Juniperus**

HESPEROCYPARIS Bartel & R. A. Price – Cypress

Hesperocyparis arizonica (Greene) Bartel [*Cupressus arizonica* Greene]—Arizona cypress. Trees monoecious, to 30 m, usually single-stemmed; bark exfoliating in strips or small flakes, but not smooth; leaves opposite in 4 ranks, scale (adult) leaves with serrate margins, abaxial gland generally visible, often ruptured, whip (juvenile) leaves serrate; seed cones gray to brown at maturity, persisting on tree, globose or oblong, 2–3 cm, scales persistent; seeds 8–15(–20) per scale, 4–6(–8) mm. Dry to mesic slopes and along streams, 5800–7360 ft. Doña Ana, Grant, Luna.

JUNIPERUS L. – Juniper, cedar

Trees, shrubs, or prostrate; leaves opposite in 4 ranks or in whorls of 3; adult leaves scalelike to subulate; seed cones globose to ovoid, 2–20 mm, remaining closed, usually glaucous, fleshy or fibrous to obscurely woody; seeds 1–6 per cone.

1. Leaves all acicular (subulate, jointed at the base) and spreading; seed cones sessile, axillary; low shrubs with ascending branchlet tips (or occasionally spreading shrubs)…**J. communis**
1'. Leaves decurrent (not jointed at the base), both whiplike and scalelike; seed cones terminal; trees to upright shrubs…(2)
2. Whip- and scale-leaf margins entire (×20) or with irregular teeth (×40)…**J. scopulorum**

2'. Whip- and scale-leaf margins denticulate (×20)…(3)
3. Seed cones with (3)4–5(6) seeds, fibrous to obscurely woody; trunk bark exfoliating in square or quadrangular plates…**J. deppeana**
3'. Seed cones with 1–2(3) seeds, fleshy to fibrous (when mature and fresh); trunk bark exfoliating in strips…
(4)
4. Mature seed cones orange, reddish orange, red, bronze, or reddish brown, appearing pink or rose-colored
if covered with bloom…(5)
4'. Mature seed cone dark blue or dark bluish black to bluish brown, with a light to heavy coat of bloom (waxy
glaucous) appearing light blue…(7)
5. Mature seed cones orange to red, with light bloom appearing pink or rose-colored; whip-leaf ventral side
white-glaucous, glands on whip-leaves visible, raised, elongate and divided (often 3 glands); often single-stemmed shrub-trees with stocky, clumpy foliage…(6)
5'. Mature seed cones copper to reddish brown, with no bloom; whip-leaf ventral side not white-glaucous,
glands on whip-leaves visible, raised, oval, not divided; shrubs with elongate terminal whips…**J. pinchotii**
6. Shorter whip-leaf glands 1/2 or less as long as the associated sheath…**J. arizonica**
6'. Longer whip-leaf glands > 1/2 as long as the associated sheath…**J. coahuilensis**
7. Plants dioecious, glands on scale-leaves visible (conspicuous) and ruptured; seed cones 1-seeded, with
a soft, juicy pericarp, 6–8 mm diam., reddish blue to brownish blue, glaucous but not appearing white…
J. monosperma
7'. Plants monoecious, glands on scale-leaves not conspicuous (embedded in leaves, not visible), not ruptured;
seed cones 1- or 2-seeded, dry, often woody, 8–9(–13) mm diam., bluish brown, very glaucous, appearing
white…**J. osteosperma**

Juniperus arizonica (R. P. Adams) R. P. Adams [*J. coahuilensis* (Martínez) Gaussen ex. R. P. Adams
var. *arizonica* R. P. Adams]—Arizona juniper. Shrub to small tree, dioecious, 3–8 m, often with a single
stem; bark brown, thin, exfoliating in long ragged strips; leaves decurrent (whip) and scale, whip- and
scale-leaf margins denticulate (×20), white-glaucous on adaxial leaf surface, at least 1/4+ of whip-leaf
glands with a white-crystalline exudate; seed cones rose to pinkish but yellow-orange, orange, or dark
red beneath the white-blue glaucous coating, soft and juicy, 6–7 mm, 1(2)-seeded, seeds 4–5 mm.
Bouteloua grasslands and adjacent rocky slopes, 4680–8675 ft. WC, SC, SW.

Juniperus coahuilensis (Martínez) Gaussen ex R. P. Adams [*J. erythrocarpa* Cory var. *coahuilensis*
Martínez]—Rose-fruited juniper. Shrub to small tree, dioecious, 3–8 m, often with a single stem; bark
brown, thin, exfoliating in long ragged strips; leaves both whip and scale, whip- and scale-leaf margins
denticulate (×20), white-glaucous on adaxial leaf surface, at least 1/4+ of whip-leaf glands with a white-
crystalline exudate; seed cones rose to pinkish but yellow-orange, orange, or dark red beneath the
white-blue glaucous coating, soft and juicy, 6–7 mm, 1(2)-seeded, seeds 4–5 mm. *Bouteloua* grasslands
and adjacent rocky slopes, 4400–8675 ft. SW, SC.

Juniperus communis L.—Common juniper. Shrubs monoecious, prostrate or low with ascending
branchlet tips; leaves upturned, to 15 × 1.6 mm, glaucous stomatal band about as wide as each green
marginal band, apex acute and mucronate to acuminate; seed cones 6–9 mm, shorter than leaves. Our
material belongs to **var. depressa** Pursh. Rocky soil, slopes, summits, 7000–11,500 ft. NC, NW, C, WC,
Lincoln.

Juniperus deppeana Steud. [*J. deppeana* var. *pachyphlaea* (Torr.) Martínez]—Alligator juniper.
Trees dioecious, to 10–15(–30) m; bark brown, exfoliating in rectangular plates; leaves green with abaxial
gland ovate to elliptic, exudate absent or present, margins denticulate (×20), whip-leaves 3–6 mm, not
glaucous adaxially, scale-leaves 1–2 mm, not overlapping, keeled, apex acute to mucronate; seed cones
8–15 mm, reddish tan to dark reddish brown, glaucous, fibrous to obscurely woody, with (3)4–5(6) seeds;
seeds 6–9 mm. Our material belongs to **var. deppeana**. Rocky soils, slopes, mountains, 6000–8000 ft.
C, WC, SE, SC, SW.

Juniperus monosperma (Engelm.) Sarg.—One-seed juniper. Shrubs or small trees, dioecious, to
7(–12) m; bark gray to brown, exfoliating in thin strips; leaves with abaxial glands elongate, with an evident white-crystalline exudate, margins denticulate (×20), whip-leaves 4–6 mm, scale-leaves 1–3 mm,
not overlapping, or if so, by < 1/4 their length, keeled, apex acute to acuminate, spreading; seed cones

6-8 mm, reddish blue to brownish blue, glaucous, fleshy and resinous, with 1(-3) seeds; seeds 4-5 mm. Dry, rocky soils and slopes, 4000-8000 ft. Widespread.

Juniperus osteosperma (Torr.) Little [*J. utahensis* (Engelm.) Lemmon]—Utah juniper. Trees or shrubs, monoecious, to 6(-12) m; bark exfoliating in thin gray-brown strips; leaves with abaxial glands inconspicuous and embedded, exudate absent, margins denticulate (×20), whip-leaves 3-5 mm, scale-leaves 1-2 mm, most not overlapping, keeled, apex rounded, acute or occasionally obtuse, appressed; seed cones with straight peduncles, (6-)8-9(-12) mm, bluish brown, often almost tan beneath glaucous coating, fibrous, with 1(2) seeds; seeds 4-5 mm. Dry, rocky soil and slopes, 4800-7500 ft. NW, WC.

Juniperus pinchotii Sudw. [*J. erythrocarpa* Cory]—Copper berry juniper. Shrub to small shrubby tree, dioecious; 1-6 m; bark thin, ashy-gray, exfoliating in long strips; leaves both decurrent (whip) and scalelike, whip- and scale-leaf margins denticulate (×20), yellow-green, with a white-crystalline (mostly camphor) exudate; seed cones copper to copper-red, not glaucous, 6-8(-10) mm, soft and juicy, 1(2)-seeded; seeds 4-5 mm. Gravelly soils on rolling hills and ravines, limestone, gypsum, 3000-5000 ft. SE, SC, Chaves, Quay.

Juniperus scopulorum Sarg.—Rocky Mountain juniper. Trees dioecious, to 20 m, single-stemmed (rarely multistemmed); bark brown, exfoliating in thin strips, that of small branchlets (5-10 mm diam.) smooth; leaves with exudate absent, margins entire (×20 and ×40), whip-leaves 3-6 mm, scale-leaves 1-3 mm, mostly not overlapping, keeled to rounded, apex obtuse to acute, appressed or spreading; seed cones 6-9 mm, appearing light blue when heavily glaucous, but dark blue-black beneath glaucous coating when mature, resinous to fibrous, with (1)2(3) seeds; seeds 4-5 mm. Rocky soils on slopes and eroded hillsides, 5700-9000 ft. Widespread except EC.

EPHEDRACEAE - EPHEDRA FAMILY

Kenneth D. Heil

Dioecious shrubs; branches green to olive-green, opposite or whorled, striate; leaves scalelike, opposite or whorled, ± connate; male cones compound, borne at the nodes or terminal, with 2-8 microsporophylls, these free or with stalks united, with a calyxlike involucre surrounding the stalks; female cones solitary or whorled, sessile or peduncled, subtended by firm or scarious bracts; seeds 1-3, hard, somewhat angled to almost terete (Stevenson, 1993 and refs. therein).

EPHEDRA L. - Ephedra, Mormon tea

Stems simulate those of an *Equisetum*, especially in being green and striate, with a solid black pith; leaves scalelike, either opposite or in whorls of 3; cones laterally produced.

1. Leaves and bracts mostly 3 per node…(2)
1'. Leaves and bracts mostly 2 per node…(3)
2. Cones always sessile; terminal buds acute at apex…**E. torreyana**
2'. Cones usually with short, scaly peduncles; terminal buds spinelike…**E. trifurca**
3. Twigs viscid…**E. cutleri**
3'. Twigs not viscid…(4)
4. Leaf bases shredding brown, turning gray with age…**E. aspera**
4'. Leaf bases persistent, forming a black, thickened collar…(5)
5. Twigs with smooth ridges; seed cones sessile or on short, scaly peduncles; inner bracts membranous with a yellow center and base…**E. viridis**
5'. Twigs with slightly scabrous ridges; seed cones usually on long, smooth peduncles; inner bracts often fleshy and orange…**E. coryi**

Ephedra aspera Engelm. ex S. Watson [*E. nevadensis* S. Watson var. *aspera* (Engelm. ex S. Watson) L. D. Benson]—Boundary ephedra. Erect shrubs, 0.5-1.5 m, bark gray; branches opposite or whorled, pale to dark green, yellow with age, terminal buds conic; leaves mostly opposite, 1–3, connate to 1/2–7/8 their length; male cones 2 at the nodes, 4–7 mm, mostly sessile, bracts opposite, 6–10 pairs, yellow to red-brown, membranous; female cones mostly 2 at the nodes, 6–10 mm, sessile or on short, scaly peduncles, bracts opposite, 5–7 pairs, membranous, red-brown. Chihuahuan desert scrub, piñon-juniper woodland communities, 3440–6250 ft. EC, C, SE, SC, SW.

Ephedra coryi E. L. Reed—Cory's ephedra. Rhizomatous shrubs forming clumps, 0.25–1.5 × 3–5 m; bark red, branches alternate or twigs whorled, bright green, becoming yellow-green with age, terminal buds conic; leaves opposite, 2–5 mm, connate to 1/2–3/4 their length; male cones 2 to several at the nodes, 4–6 mm, on very short, scaly peduncles, bracts opposite, 5–9 pairs, light yellow, membranous; female cones 2 to several at the nodes, 7–15 mm, on smooth peduncles, bracts opposite, 3–4 pairs, inner pairs becoming fleshy, orange. Sandy sites, 3700–4050 ft. SE.

Ephedra cutleri Peebles [*E. viridis* Coville var. *viscida* (H. C. Cutler) L. D. Benson]—Navajo ephedra. Rhizomatous shrubs forming clumps, 0.25–1.5 m diam., bark reddish brown; branches alternate or whorled, twigs bright green, becoming yellow-green with age, viscid, terminal buds conic; leaves opposite, 2–5 mm, connate 1/4–1/2 their length; male cones 2 to several at the nodes, 4–6 mm, on very short, scaly peduncles, bracts opposite, 5–9 pairs, light yellow, membranous; female cones 2 to several at the nodes, 7–15 mm, on peduncles 5–25 mm, bracts opposite, 3 or 4 pairs, membranous. Mostly in sandy areas with blackbrush, mixed desert shrub, mixed grass, rabbitbrush, and piñon-juniper communities, 5860–6435 ft. NW, Cibola.

Ephedra torreyana S. Watson—Torrey's ephedra. Erect shrubs, 2–10 m; branches blue-green to olive-green, appearing smooth, rigid, terete, to 3.5 mm thick, solitary or whorled at the nodes; leaves ternate or whorled, 2–5 mm, connate for nearly 2/3 their length, at maturity somewhat persistent; male cones solitary to several in a whorl, ovate, sessile, 6–8 mm; female cones solitary to several at the nodes, ovoid, 9–13 mm, sessile, bracts in 3s in whorls of 5 or 6; seeds solitary or 2, pale brown to yellow-green, 7–10 mm. Dry, sandy or rocky hillsides in blackbrush, salt desert scrub, mountain brush, and piñon-juniper communities, 3500–7700 ft. Widespread except NE.

Ephedra trifurca Torr. ex S. Watson—Mexican tea. Erect shrubs, 0.5–5 m, bark gray; branches alternate or whorled, twigs pale green, gray with age, terminal buds spinelike; leaves in whorls of 3.5–15 mm, connate to 1/2–3/4 their length; male cones 1 to several at the nodes, 6–10 mm, bracts in whorls of 3, reddish brown; female cones 1 to several at the nodes, 10–15 mm, on short, scaly peduncles, bracts in 6–9 whorls of 3. Chihuahuan desert scrub, sandy areas, 3500–6000 ft. SC, SW, Socorro.

Ephedra viridis Coville—Mormon tea. Shrubs 1–15 dm, spreading to erect; branches rigid to flexible, bright green to yellow-green or 3/4 fastigiate and broomlike; leaves opposite, 1.5–4 mm; male cones 2+, sessile, 5–7 mm, bracts opposite, 2–4 mm, membranous, pale yellow; female cones 6–10 mm, sessile, with 4–8 pairs of ovate bracts 4–7 mm; seeds paired, brown, 5–8 mm. Blackbrush, salt desert scrub, sagebrush, mountain brush, piñon-juniper, and rabbitbrush communities, 5000–9730 ft. NW, C, SC.

PINACEAE – PINE FAMILY

Kenneth D. Heil

In our area, evergreen; usually excurrent (with undivided main trunk); resinous, aromatic; monoecious; leaves needlelike or narrowly linear, arranged spirally on the branches or clumped (fascicled) on short lateral shoots; pollen cones (male) small, soft, with spirally arranged pollen-bearing scales (microsporophylls), drying after shedding pollen (anthesis) and soon deciduous; seed cones (female) mostly large, becoming woody or papery, maturing and usually releasing seeds in 1 or 2 growing seasons, scales overlapping, spirally arranged; seeds 2 per scale, borne on upper (adaxial) side, each with attached membranous wing (lacking in *Pinus edulis*) (Thieret, 1993).

1. Leaves (needles) in bundles of 2+…**Pinus**
1'. Leaves solitary, not in bundles…(2)
2. Seed cones with conspicuous 3-pronged bracts extending beyond the end of each scale…**Pseudotsuga**
2'. Seed cones without 3-pronged bracts between the cone scales…(3)
3. Leaves attached directly to twig, forming scars flush with twig surface or nearly so; seed cones growing upright on upper branches, disintegrating and releasing seeds in autumn…**Abies**
3'. Leaves attached to tips of peglike projections (sterigmata) from twig that remain after leaves drop; seed cones ± pendent from branches, falling intact after seeds are shed at the end of summer…**Picea**

ABIES Mill. – Fir

Tree with crown, usually spirelike, leaf scars prominent, flush with twig surface or slightly raised or depressed all around; bark thin, smooth, with resin blisters when young; leaves (needles) borne singly, arranged spirally; seed cones maturing by late summer of the first year, growing upright on upper branches, disintegrating and falling scale by scale when mature.

1. Leaves on lower branches (1.5–)2–3 cm, dark bluish green; seed cones dark brown to purplish; common tree of subalpine forests; bark often whitish, spongy to the touch…**A. arizonica**
1'. Leaves on lower branches (2.5–)3–6 cm, pale grayish green; seed cones grayish green; common tree in montane forests; bark gray and smooth, not spongy to the touch…**A. concolor**

Abies arizonica Merriam [*A. bifolia* A. Murray var. *arizonica* (Merriam) O'Kane & K. D. Heil]—Rocky Mountain subalpine fir. Tree to ca. 30 m; trunk 0.7 m wide; bark gray and smooth on young trees, thickening (to 18 cm); leaves 1–2.5 cm, light green to bluish green; seed cones dark purple-blue to grayish purple, 5–10 × 3–3.5 cm. This species, along with *Picea engelmannii*, dominates moist subalpine forests up to timberline in our area, 8000–11,800 ft. *Abies lasiocarpa* is a species of the Pacific Northwest and was long thought to be a species of the high mountains in New Mexico. NC, NW, C, WC.

Abies concolor (Gordon & Glend.) Lindl. ex Hildebr.—White fir. Tree to ca. 40 m, trunk to 1 m; bark gray and smooth on young trees, becoming thick with age; leaves 1.5–6 cm, pale grayish green; seed cones olive-green, 7–12 × 3–5 cm. Dry, sunny slopes, 6200–8500 ft. NC, NW, C, WC, SC. *Abies concolor* is probably the most drought-tolerant North American *Abies*.

PICEA A. Dietr. – Spruce

Tree, crown broadly conic to spirelike; bark gray to reddish brown; leaves relatively rigid, borne sessile on a persistent peglike base, apex usually sharp-pointed; seed cones borne mostly on upper branches, maturing in 1 growing season, pendent.

1. Young twigs finely pubescent; leaf tips usually acute, but not sharp; seed cones mostly 3–7 cm…**P. engelmannii**
1'. Young twigs essentially glabrous; leaf tips usually attenuate into rigid, sharp spines; seed cones mostly 6–11 cm…**P. pungens**

Picea engelmannii Engelm.—Engelmann spruce. Tree with stems to 30 × 1 m, crown narrowly conic, young twigs finely pubescent, with peglike leaf bases; leaves 1.6–3 cm, blue-green, 4-angled in cross-section, flexible; apex acute but not usually very sharp; seed cones 4–7(–8) cm. Moist upper montane and subalpine forests, primarily on N-facing slopes, often with *Abies arizonica* in dense stands, 8000–12,000 ft. NC, NW, C, WC, SC.

Picea pungens Engelm.—Colorado blue spruce. Tree with stems to 25 × 0.6 m; crown broadly conic, young twigs glabrous, with peglike leaf bases; leaves 1.6–3 cm, rigid, blue-green to whitish, the latter giving the tree a silvery appearance, 4-angled in cross-section, apex spine-tipped; seed cones (5–)6–11(–12) cm. Usually upper montane to lower subalpine, often along streams, especially in the lower portion of its altitudinal range, 7000–11,000 ft. NC, NW, C, WC, SC.

PINUS L. – Pine

Trees, some shrubby at timberline, crown often open when mature; bark thin, scaly, and continuously shed, or thick, woody, deeply fissured, and persistent; leaves needlelike, in fascicles (bundles) of (1)2–5, reaching 2–18 cm; seed cones becoming woody and brown.

1. Leaves (needles) mostly 1–3 per cluster…(2)
1'. Leaves (needles) mostly (4)5 per cluster…(6)
2. Needles mostly 1–2 per cluster, 2–4 cm…**P. edulis**
2'. Needles mostly 3 per cluster, 3–40 cm…(3)
3. Leaf sheaths early-deciduous…(4)
3'. Leaf sheaths persistent…(5)
4. Needles mostly 6–12 cm; plants monoecious…**P. leiophylla**
4'. Needles mostly 3–6 cm; plants nearly dioecious…**P. cembroides**
5. Needles mostly 10–22 cm…**P. scopulorum**
5'. Needles mostly 25–40 cm…**P. engelmannii**
6. Needles mostly 10–22 cm…**P. arizonica**
6'. Needles mostly 3–8 cm…(7)
7. Needles conspicuously resinous, mostly strongly curved; cone scales with prickles or bristles…**P. aristata**
7'. Needles not resinous, straight or nearly so; cone scales lacking prickles or bristles…(8)
8. Cones 6–15 cm; terminal exposed portion of cone scales not recurved; mostly above 8850 ft.…**P. flexilis**
8'. Cones 15–25 cm; terminal exposed portion of cone scales recurved; mostly below 8850 ft.…**P. strobiformis**

Pinus aristata Engelm.—Rocky Mountain bristlecone pine, Colorado bristlecone pine. Trees to 15(–20) m, shrublike and twisted toward timberline; bark whitish and smooth on young trees, dark and fissured on old; leaves 2–4 cm, mostly 5 per bundle; seed cones dark purplish when young, becoming brown when mature, 6–11 cm, each scale tipped with a slender incurved prickle to 7 mm. Mostly upper subalpine to lower fringes of alpine zone, typically on relatively dry, S-facing slopes, 8500–12,500 ft. NC.

Pinus arizonica Engelm. [*P. ponderosa* P. Lawson & C. Lawson var. *arizonica* (Engelm.) Shaw]—Arizona pine. Trees to 30 m; bark dark brown when young, rough, scaly, and becoming deeply furrowed, cinnamon-brown with age; leaves (3–4)5 per fascicle, 7–17 cm; pollen cones 6–9 cm. Mesas, slopes, canyons, canyon rims, 7250–8050 ft. Hidalgo.

Pinus cembroides Zucc.—Mexican piñon. Shrubs or trees to 15 m, strongly tapering, much branched, crown rounded; bark red-brown to dark brown, furrowed; leaves (2)3(4) per fascicle, 2–6 cm; pollen cones to 10 mm. Piñon-juniper woodlands, mesas, foothills, 4400–8200 ft. Hidalgo.

Pinus edulis Engelm.—Piñon. Trees to 15 m, usually shorter, with dense conic crown when young, becoming open and rounded or irregular with age, trunk often dividing into 2+ major branches near base and becoming twisted with age; bark grayish to reddish brown, irregularly and shallowly furrowed; leaves 1–2(3) per fascicle, 2–4(–5) cm; seed cones (2–)3–5 cm when open. Mesas, canyons, mountain slopes, typically forming open woodlands with *Juniperus* spp. and smaller shrubs, 4200–9000 ft.

1. Leaves mostly 2 per fascicle…**var. edulis**
1'. Leaves mostly 1 per fascicle…**var. fallax**

var. edulis—Piñon pine. Tree or shrub to 20 m; needles 2(3); mostly 4500–7500 ft. Widespread.

var. fallax Little—Single needle piñon. Tree or shrub to 15 m; needles 1(2); < 4500 ft. SW.

Pinus engelmannii Carrière—Apache pine. Trees to 35 m, crown straight, irregularly rounded; bark dark brown, deeply furrowed; leaves 3(–5) per fascicle, (20–)25–45 cm; seed cones 11–14 cm. Dry mountain ranges and plateaus, 6500–7000 ft. Grant, Hidalgo.

Pinus flexilis E. James—Limber pine. Trees in our area are narrowly pyramidal when young, crown broadening and opening with age, 10–15(–20) m; bark smooth and whitish on younger branches, becoming dark gray and divided into scaly rectangular plates on older trunk(s); leaves 5 per fascicle, (3–)3.5–

6(–9) cm; seed cones 6–15 cm, mostly yellowish brown. Mostly subalpine, usually in exposed, windy, rocky, thinly vegetated sites, 7500–10,500 ft. NC.

Pinus leiophylla Schiede ex Schltdl. & Cham.—Chihuahua pine. Trees to 25 m, crown conic, becoming rounded; bark brown to red-brown, narrowly furrowed; leaves (2)3(4) per fascicle, 6–15 cm; seed cones 3.5–5(–9) cm. Our material belongs to **var. chihuahuana** (Englem.) Shaw. Hillsides; pine, oak, juniper woodlands, 4600–6600 ft. Grant, Hidalgo.

Pinus scopulorum (Engelm.) Lemmon [*P. ponderosa* P. Lawson & C. Lawson var. *scopulorum* Engelm.]—Ponderosa pine. Trees to ca. 40 m; mature bark orange to deep reddish brown, with irregular fissures; leaves mostly 3 per fascicle; seed cones (5–)10(–12) cm. Sheltered canyons at its lowest limits to exposed ridges at its highest localities; usually the dominant tree at the lowest forested elevations, 5600–10,000 ft. Widespread except EC.

Pinus strobiformis Engelm. [*P. reflexa* (Engelm.) Engelm.]—Southwestern white pine. Trees to 30 m, conic when young, becoming more rounded and irregular with age; bark gray, becoming more deeply furrowed with age; leaves 5 per fascicle; seed cones 12–25 cm. Montane, valley bottoms or relatively moist slopes with some soil development, 7800–10,500 ft. NC, C, WC, SC, SW.

PSEUDOTSUGA Carrière – Douglas-fir

Pseudotsuga menziesii (Mirb.) Franco—Douglas-fir. Tree, crown broadening and flattening with age, to 40 m in our area, crown narrow to broadly conic; bark gray, smooth with resin blisters when young, becoming dark, deeply fissured in age; leaves borne singly, alternate, short-stalked, 15–30 mm, flattened, usually bluish green; seed cones 4–7 cm, bracts spreading, exserted and clearly visible between scales, 3-pronged; seeds winged. Our material belongs to **var. glauca** (Mayr) Franco. Mostly in the mountains and occasionally in protected areas at lower elevations, 6500–10,000 ft. NE, NC, NW, C, WC, SE, SC, SW.

ANGIOSPERMS

ACANTHACEAE – ACANTHUS FAMILY

Kenneth D. Heil

Herbs or small shrubs (ours); leaves opposite, simple, entire; flowers bisexual, bilateral (to nearly regular), often subtended by large, often colorful bracts and bractlets; corolla sympetalous, 5-lobed and ± 2-lipped; calyx connate, with 4 or 5 segments; stamens (2)4 and didynamous, epipetalous, anthers often asymmetric; pistil 2-carpellate, ovary superior, 2-locular, placentation axile; fruit a capsule, usually dehiscing explosively (Yatskievych & Fischer, 1984).

1. Stamens 2…(2)
1'. Stamens 4…(7)
2. Inflorescence a dense spike with overlapping bracts…(3)
2'. Inflorescence not a dense spike with overlapping bracts; flowers solitary in axils, in small axillary cymes, or in racemes…(4)
3. Spikes borne on scaly peduncles, bracts appressed scales; flowers pale blue or lavender…**Elytraria**
3'. Spikes not borne on scaly peduncles, bracts leafy, curved away from axis; flowers yellowish, upper lobe with a violet patch…**Tetramerium**
4. Stems hexagonal; corolla deeply 2-lipped; flowers covered by green, heart-shaped bracts…**Dicliptera**
4'. Stems terete or 4-sided; corolla 2-lipped or not; flowers not covered by green, heart-shaped bracts…(5)
5. Shrubs usually > 1 m; flowers reddish to orange, 3–5 cm…**Anisacanthus**
5'. Perennial herbs < 1 m; flowers purple, lavender, or yellowish, < 3 cm…(6)
6. Flowers blue or white with maroon veins; stamens exserted…**Carlowrightia**
6'. Flowers lavender to purple; stamens included…**Justicia**
7. Plants low, < 10 cm, forming mats of small rosettes without apparent leafy stems; flowers pink with white streaks…**Stenandrium**
7'. Plants usually > 10 cm, with obvious leafy stems; flowers pale lavender to purple…(8)
8. Plants ± procumbent, spreading from a central rootstock, flowers sessile, 1.8–2.5 cm, in axillary clusters…**Dischoriste**
8'. Plants not procumbent, shrubs to 4 dm; flowers solitary, pedicelled, 3–3.5 cm, in upper nodes…**Ruellia**

ANISACANTHUS Nees – Desert honeysuckle

Anisacanthus thurberi (Torr.) A. Gray—Thurber's desert honeysuckle. Branched shrubs with whitish exfoliating bark to ca. 1 m; leaves opposite, lanceolate, puberulent to glabrate, 2–5 cm; corolla purplish red to orange, axillary, tube slender, limb bilabiate, upper lip entire or 2-cleft, lower lip 3-lobed; calyx 4–5 mm, lobes longer than the tube; stamens 2, included in corolla; fruit a capsule, ca. 2 cm, stipelike base longer than body. Rocky washes and canyons, Chihuahuan desert scrub to mixed oak woodlands, 3000–6250 ft. WC, SW, sw of NW.

CARLOWRIGHTIA A. Gray – Wrightwort

Herbs or shrubs, stems slender, often diffuse; leaves opposite, linear to ovate or orbicular; flowers in axillary clusters, corolla blue, purple, or white; stamens 2; fruit an ovoid, flattened capsule; seeds flat, few (Thomas, 1983).

1. Leaves sessile, linear; corolla blue; widespread…**C. linearifolia**
1'. Leaves petiolate, ovate to orbicular; corolla white with maroon veins; Eddy County…**C. texana**

Carlowrightia linearifolia (Torr.) A. Gray—Heath wrightwort. Shrubs to 30 cm; leaves linear to linear-filiform, 10–20 mm; calyx 5-parted; corolla blue, ca. 5 mm, twice as long as tube. Rocky soils, often in canyon bottoms, Chihuahuan desert scrub and grasslands, 3800–5900 ft. W, C, SC.

Carlowrightia texana Henrickson & T. F. Daniel—Texas wrightwort. Small shrubs to 3.5 cm; leaves broadly ovate, mostly 6–16 mm; calyx 3–6 mm, lobes 2–5 mm; corolla white with maroon veins, yellow eye on upper lip, 5.5–7 mm; capsules 7.5–12.5 mm. Limestone flats and hills, Chihuahuan desert scrub and mesquite woodlands, 3000–3500 ft. Known from 1 collection in Eddy County.

DICLIPTERA Juss. – Foldwing

Dicliptera resupinata (Vahl) Juss.—Arizona foldwing. Herbaceous perennial; stems branched, leafy, 30–70 cm, divaricately branched; leaves opposite, lanceolate to ovate, 2–8 cm, slender-petioled; flowers axillary or few flowered, subtended by a pair of bractlets, these cordate to slightly cuneate at the base; corolla rose-purple, 15–20 mm, the lips about as long as the tube; capsules 5 mm, stipitate. Piñon pine, juniper, and oak woodland, 4500–6000 ft. One old specimen, labeled "Guadalupe Pass." Either Hidalgo County or adjacent Arizona.

DYSCHORISTE Nees – Snakeherb

Dyschoriste schiedeana (Nees) Kuntze—Schied's snakeherb. Herbaceous, caulescent; stems 10–30 cm; leaves numerous, 1–4 cm, oblong to obovate or spatulate; flowers axillary, sessile, 1 to few; calyx lobes subulate, 8–15 mm; corolla purple, puberulent to pubescent, 15–20 mm wide, lobes as long as throat; capsules 10–12 mm.

1. Stems procumbent, sprawling from a central rootstock; calyx lobe margins with hairs < 0.3 mm…**var. decumbens**
1'. Stems erect-ascending, sometimes with lateral decumbent stems; calyx lobe margins with hairs 0.2–1.2 mm…**var. cinerascens**

var. **cinerascens** Henrickson & Hilsenb.—Schied's snakeherb. Stems mostly erect, 10–30 cm; leaves 2–4 cm; calyx lobes 8–10 mm; corolla blue or purple. Rocky slopes in grassland communities, 3000–3500 ft. SW, SE.

var. **decumbens** (A. Gray) Henrickson—Spreading snakeherb. Stems mostly procumbent, sprawling, to 30 cm; leaves 1–3 cm; calyx lobes 12–15 mm; corolla purple. Flats, hillsides, canyon bottoms, grasslands and oak, piñon pine woodlands, 3000–6000 ft. SW, SE.

ELYTRARIA Michx. – Scaly-stem

Elytraria imbricata (Vahl) Pers. [*Justicia imbricata* Vahl]—Purple scaly-stem. Plants acaulescent herbs with numerous green bracts; basal leaves oblong to elliptic, 5–12 cm, entire; spikes 1–3 cm; flowers borne in dense peduncled spikes, both spikes and peduncles bearing imbricate bracts; floral bracts 6–8 mm; corolla 5–6 mm, blue; capsule ca. 6 mm, constricted at the base. Steep, rocky sites; oak, juniper, *Rhus* communities, 4500–5575 ft. Hidalgo, ?Sandoval.

JUSTICIA L. – Water-willow

Herbs or shrubs; leaves opposite, petiolate; flowers solitary or in spikes or panicles; bracts imbricate; flowers white, pink, or purple, tube narrow, limb 2-lipped, upper lip 2-lobed, lower lip 3-lobed; stamens 2, often slightly exserted; capsules clavate.

1. Corolla tube slender and cylindrical…**J. pilosella**
1'. Corolla tube short, not cylindrical…**J. wrightii**

Justicia pilosella (Nees) Hilsenb. [*Siphonoglossa pilosella* (Nees) Torr.]—Gregg's tube-tongue. Plants suffrutescent, low, to 3 dm; stems hirsute; leaves subsessile, to 4 cm; flowers solitary in axils of uppermost leaves; bractlets subtending calyx; calyx to 2.5 mm; corolla lavender, to 28 mm, tube cylindrical; capsules clavate, 8–9 mm. Rocky soils, Chihuahuan desert scrub, 3675–4800 ft. Doña Ana, Eddy.

Justicia wrightii A. Gray [*Ecbolium wrightii* (A. Gray) Kuntze]—Wright's justicia. Suffrutescent perennial, 8–15 cm; leaves sessile, linear to oblanceolate, 10–15 mm, margins flat; flowers solitary, sessile, in leaf axils; branchlets 2.5–11 mm; calyces 4-lobed, 2.5–4.5 mm; corollas 8.5–12 mm, purple; capsules 7–8 mm. Limestone benches, Chihuahuan desert scrub, 3500–4500 ft. Eddy. **R**

RUELLIA L. - Wild petunia

Ruellia parryi A. Gray—Parry's wild petunia. Shrub to 4 dm, usually profusely branched; stems branched above, aging to gray; leaves with petioles, obovate to lanceolate, 1–2 cm, mainly entire; flowers solitary, in axils of leaves; calyx with 5 lobes, 7–10 mm; corolla pale lavender, 3–3.5 cm, the tube ca. 15 mm; capsule 3 mm. Limestone hills and ledges, canyon bottoms, arroyos, montane scrub and piñon-juniper woodland communities, 4000–5500 ft. SC, SE.

STENANDRIUM Nees - Shaggy tuft

Stenandrium barbatum Torr. & A. Gray—Early shaggy tuft. Low perennial herbs, hirsute with shaggy whitish hairs; roots stout; stems short, scapiform; leaves crowded, entire, hirsute; flowers bracteate, terminal spikes; calyx segments nearly equal; corolla pink with white streaks, the tube slender, limb 5-lobed; stamens 4, included in corolla. Limestone slopes and ridges, conglomerate gravel, Chihuahuan desert scrub and black grama and mesquite, juniper communities, 3600–6000 ft. S.

TETRAMERIUM C. F. Gaertn. - Tetramerium

Tetramerium nervosum Nees [*T. hispidum* Nees]—Hairy fournwort. Plants suffruticose, decumbent to ascending, to 1 m; stems several, hispid; leaves pubescent to glabrate, petioled, lanceolate to lance-ovate, to 5 cm; flowers in spikes; bracts lanceolate to lance-ovate; corollas white to cream or pale yellow, upper lobe with a violet patch; stamens 2, included; capsules 4–5 mm. Disturbed sites in canyons, washes, and hillsides. Known from only 1 collection in Socorro County, 4000–5000 ft. Weedy in Mexico.

AGAVACEAE - AGAVE FAMILY

Steve L. O'Kane, Jr.

Perennial or monocarpic, usually woody to some degree at least at the very base, some forming trees (herbaceous in *Leucocrinum*); stems often thickened; leaves persistent, usually fibrous, often spiny or filiferous and separating into elongate fibers, in whorls or rosettes or alternate; flowers bisexual, actinomorphic, hypogynous or epigynous; tepals 6, usually entirely petaloid, distinct to connate; stamens 6, distinct, free or adnate to a floral tube; pistil with 3 locules (or 6-locular by false septa); fruit a capsule or spongy or fleshy berry (baccate) (Utech, 2002).

1. Flowers rose-red to salmon-colored, and present most of the growing season; leaves with slender threads on margins; known from cultivation in New Mexico, but may escape…**Hesperaloe**
1′. Flowers yellow or ± white, sometimes tinged red, present only during the flowering period; leaves with or without threads on margins (some of these then cultivated)…(2)
2. Leaves ± herbaceous, narrowly to very narrowly linear, < 9 mm wide, flexuous…(3)
2′. Leaves leathery, hard, or succulent, wider than narrowly linear; typically > 9 mm wide (if < 9 mm wide, leaves then with fibrous margins), firm to stiff…(4)
3. Tepals yellow; flowers above ground level, ovary aboveground…**Echeandia**
3′. Tepals white; flowers at ground level, ovary subterranean…**Leucocrinum**

4. Flowers yellow to yellow-green; ovary inferior and leaves with spiny margins (except in *Agave schottii* with entire or fibrous leaves)…**Agave**

4'. Flowers white or ochroleucous, sometimes tinged red; ovary superior and leaves lacking spiny margins (entire, fibrous, or ± dentate or denticulate)…**Yucca**

AGAVE L. – Agave

Perennial, some monocarpic, usually nearly acaulescent (or caulescent), scapose; leaves in robust, succulent rosettes, evergreen, linear-lanceolate to ovate, firm to rigid, with marginal spines (fibrous-margined in *A. schottii*) and a sharp-pointed apical spine; inflorescences much exceeding foliage, atop a semiwoody stalk, spicate, racemose, or paniculate, open to dense; flowers funnelform to tubular, showy, mostly yellow, infrequently whitish or reddish, tepals connate basally into tube atop a typically constricted neck; ovary inferior, greenish at anthesis (Reveal & Hodgson, 2002).

1. Inflorescences spicate, subspicate, or rarely narrowly racemose-paniculate; flowers on peduncles or lateral branches < 5 cm (subgenus *Littaea*)…(2)

1' Inflorescences paniculate; flowers on peduncles > 10 cm (subgenus *Agave*)…(3)

2. Leaf margins entire or filiferous; perianth lobes no more than twice the length of perianth tube…
A. schottii

2'. Leaf margins conspicuously armed; perianth lobes more than twice the length of perianth tube…
A. lechuguilla

3. Leaves of rosettes usually open; leaf blades linear-lanceolate to lanceolate or oblanceolate, 35–92 cm; inflorescences open, usually with 8–26(–32) lateral branches…**A. palmeri**

3'. Leaves of rosettes somewhat open to dense; leaf blades linear to lanceolate to broadly lanceolate, 18–65 cm; inflorescences dense, usually with (10–)20–40 lateral branches (sometimes fewer in *A. parryi*)…(4)

4. Perianth tube 6–18 mm, as long as or shorter than limb lobes (13–27 mm); filaments inserted above middle of perianth tube to just below rim; apical leaf spine 1.5–4 cm; flowering spring to summer; SW New Mexico…**A. parryi**

4'. Perianth tube 4–7 mm, much shorter than limb lobes (14–18 mm); filaments inserted just below rim of perianth tube; leaf apical spine 2.5–5 cm; flowering summer to early fall; SC and SE New Mexico…
A. gracilipes

Agave gracilipes Trel.—Slimfoot century plant. Rosettes usually solitary, 3–4 × 7–8 dm, dense; leaves ascending, 18–30 × 4.5–7 cm, glaucous and yellow- or gray-green, lanceolate to broadly lanceolate, concavo-convex apically, marginal teeth single, well defined, 2–8 mm, apical spine reddish brown to gray, acicular, 2.5–5 cm; scape (1.8–)4–5 m; inflorescences narrowly to somewhat broadly paniculate, dense; perianth red in bud, yellow to yellow-green at anthesis, tube broadly campanulate, 4–7 × 9–15 mm, lobes spreading to ascending, slightly unequal, 14–18 mm. Gravelly to rocky, often calcareous places in grasslands, desert scrub, and piñon-juniper woodlands, 4600–5750 ft. SE, SC.

Agave lechuguilla Torr.—Lechuguilla. Plants frequently suckering, rosettes 3–4 × 5–6 dm; leaves ascending to erect, 25–50 × 2–5 cm, light green to yellowish green, sometimes checkmarked, linear-lanceolate, concavo-convex apically, margins easily detached, teeth 2–6 mm, apical spine grayish, conic to subulate, 1.5–4.5 cm; scape 2–3.5 m; inflorescences spicate, densely flowered on distal 1/2; perianth yellow, frequently tinged with red or purple, tube campanulate, 1.5–4 × 6–12 mm, lobes ascending, subequal, 11–20 mm. Gravelly to rocky calcareous places in desert scrub, lower elevations to ca. 4400 ft. SE, SC.

Agave palmeri Engelm.—Palmer's century plant. Rosettes usually solitary, 4–13 × 7–13 dm, open; leaves ascending to spreading, 35–92 × 3.5–19 cm, pale to glaucous-green or green, sometimes tinged reddish, linear-lanceolate to lanceolate, concavo-convex apically, marginal teeth 3–6 mm, apical spine reddish brown to brown, acicular, 3–6 cm; scape 2–7 m; inflorescences broadly paniculate, open; perianth cream to pale yellow or light green, tube urceolate, 10–18 × 10–16 mm, lobes erect, strongly unequal, (6–)9–18 mm, apex often flushed with maroon. Sandy to gravelly places on limestone in oak woodlands and grassy plains, lower elevations to 5950 ft. WC, Luna.

Agave parryi Engelm.—Parry's century plant. Plants freely suckering, rosettes 3.5–7.5 × 4–8.5 dm, somewhat open to dense; leaves ascending to erect, 7–65 × 4–20 cm, glaucous-gray to light green, linear-lanceolate to lanceolate or broadly ovate, rigid, adaxially nearly plane to concave toward apex, abaxially convex; marginal teeth 3–8 mm (rarely absent), apical spine dark brown to gray, subulate or acicular, 1.5–4 cm; scape 2–6 m; inflorescences broadly paniculate, somewhat open to dense, perianth pink to red or red to orange in bud, yellow to yellowish green at anthesis, tube campanulate, 6–18 × 11–21 mm, limb lobes erect to ascending, unequal, 13–27 mm. Two varieties in our area.

1. Rosettes flat-topped; scape 2–4.5 m; perianth tube 12–18 mm, lobes 3–4 mm wide; capsules 2.5–3.5 cm... **var. neomexicana**
1'. Rosettes globose; scape 4–6 m; perianth tube 6–12 mm, lobes 4–7 mm wide; capsules 3.5–5 cm...**var. parryi**

var. neomexicana (Wooton & Standl.) McKechnie [*A. neomexicana* Wooton & Standl., *A. parryi* subsp. *neomexicana* (Wooton & Standl.) B. Ullrich]—Gravelly to rocky places in grasslands and desert scrub, 4500–6000 ft. WC, SC.

var. parryi [*A. parryi* subsp. *parryi*]—Gravelly to rocky places in grasslands, desert scrub, chaparral, piñon-juniper, oak woodlands, 4500–8000 ft. WC, SE, SC, SW.

Agave schottii Engelm.—Schott's century plant. Plants freely suckering, rosettes 3–6 × 6–12 dm; leaves mostly erect, widest near base, 20–40(–50) × 0.7–2.5 cm, yellowish green or deep green, sometimes with conspicuous white bud-prints on both surfaces, linear, firm, adaxially plane or somewhat concave toward apex, abaxially convex to deeply convex toward base, margins filiferous or not, unarmed; apex acuminate to long-acuminate, spine brown or grayish, acicular, 0.8–1.9 cm, scape 1.6–4 m; inflorescences spicate or subspicate to narrowly racemose-paniculate; perianth yellow, tube funnelform, 8–14 × 5–8(–13) mm, lobes erect to incurved, subequal, (7–)11–16 mm. Our plants are **var. schottii**. Gravelly to rocky places, mostly in desert scrub, grasslands, juniper and oak woodlands, 4500–4900 ft. Hidalgo.

ECHEANDIA Ortega – Echeandia, crag-lily

Echeandia flavescens (Schult. & Schult. f.) Cruden [*Anthericum flavescens* Schult. & Schult. f., *A. torreyi* Baker]—Torrey's crag-lily. Herbs, scapose, from corms with enlarged storage roots; basal leaves 3–15, very narrowly linear to narrowly linear, 8–40 cm × 0.5–9 mm, surrounded by fibrous leaf bases from previous year; cauline leaves 0–3, long-acuminate base, 1–9(–13) cm; inflorescence 0–1(–3)-branched, 21–60(–84) cm; tepals yellow, strongly reflexed to spreading. Open montane forests, mainly *Pinus* but sometimes mixed with *Pseudotsuga* and *Abies*, and grassy areas at forest margins, 5400–10,200(–11,800) ft. C, WC, SE, SC, SW, McKinley.

HESPERALOE Engelm. – False yucca

Hesperaloe parviflora (Torr.) J. M. Coult.—Red-flower false yucca. Cespitose, acaulescent, semi-succulent, short- to long-rhizomatous; leaves in basal rosettes, linear, thick and striate-ridged abaxially, 5–12.5 dm × 1–2.5 cm, margins with slender threads; inflorescences loose, 3–8-branched, panicles 1–2.5 m; perianth narrowly tubular, rose-red to salmon, 2.5–3.5 cm. Native of Texas, widely cultivated and expected to escape.

LEUCOCRINUM Nutt. ex A. Gray – Star-lily

Leucocrinum montanum Nutt. ex A. Gray—Star-lily. Herbs, acaulescent, from short, deeply buried, fleshy roots; leaves few, tufted, each tuft surrounded basally by membranous sheaths, linear, 10–20 cm × 2–8 mm, distal-most occasionally fibrous; inflorescences umbel-like, at ground level; flowers showy, fragrant; tepals connate below middle, white, narrowly oblong, equal; perianth tube long, slender, 4–8(–10) cm; limb lobes spreading. Shortgrass prairie, montane meadows, open montane forests, 5250–8000 ft. NE, NC.

YUCCA L. - Yucca, Spanish bayonet

Perennial, acaulescent or caulescent, sometimes a tree; leaves in robust rosettes, linear-lanceolate, usually rigid and leathery, occasionally fleshy, margins entire or denticulate, often with elongate fibers, apex sharp-pointed; inflorescences paniculate or racemose, peduncle sometimes scapelike; perianth campanulate or globose; tepals fleshy, distinct or connate at base, whitish to cream or tinged slightly with green or purple; ovary superior, usually green (Hess & Robbins, 2002; Sivinski, 2008).

1. Fruits indehiscent, pendent, fleshy and succulent at maturity (section *Yucca*)...(2)
1'. Fruits dehiscent, ± ascending to erect, dry at maturity (section *Chaenocarpa*)...(5)
2. Tepals connate basally for 1+ mm; pistil 2.8–8 cm...(3)
2'. Tepals distinct, or connate basally for < 1 mm; pistil 1.5–4 cm...(4)
3. Mature plants > 2.5 m; tepals 3.9–10.8 cm...**Y. faxoniana**
3'. Mature plants < 2.5 m; tepals 4.5–13 cm...**Y. baccata**
4. Leaf blade thin, flexible, margins entire, occasionally with slender fibers...**Y. madrensis**
4'. Leaf blade thick, rigid, margins entire, denticulate, or with slender or coarse fibers...**Y. treculeana**
5. Inflorescences of population predominantly paniculate, sometimes upper 1/3 of inflorescence racemose and lower 2/3 branched; plants acaulescent or caulescent with erect stems to 5 m...(6)
5'. Inflorescences of population predominantly racemose, sometimes with a few branches in lowest nodes of racemes; most plants in population acaulescent, some may have short stems usually < 0.5 m...(7)
6. Population caulescent with stems 1–5 m, often treelike; peduncles long, the lowest panicle branches at least 3 dm above leaf tips; S New Mexico, mostly W of the Pecos River...**Y. elata**
6'. Population mostly acaulescent (stems rarely to 1.5 m); peduncles generally short, the lowest panicle branches within leaf rosette or just above leaf tips; plains of the EC and SE New Mexico...**Y. campestris**
7. Leaf upper surfaces convex, the blade narrowly lanceolate, usually 1–2 cm wide...(8)
7'. Leaf upper surfaces flat, the blade linear or linear-lanceolate, usually < 1 cm wide...(9)
8. Peduncles long in most of the population, holding the lowest flowers of the racemes ≥ 10 cm above the leaf tips; styles pale green or ochroleucous; rocky ridges and hillsides in high plains of NE New Mexico...**Y. neomexicana**
8'. Peduncles short in most of the population, holding the lowest flowers of the racemes within the leaves or near the leaf tips; styles green; rare in NW New Mexico in mountains and on sandstone slickrock...**Y. harrimaniae**
9. Peduncles long in most of the population, holding the lowest flowers of the raceme ≥ 10 cm above the leaf tips of the rosette; racemes (excluding peduncle) long on most plants, often > 1.5 times longer than leaves; capsules usually deeply constricted near the middle, plains of McKinley and San Juan Counties...**Y. angustissima**
9'. Peduncles short in most of the population, holding the lowest flowers of the racemes within the leaves or near the leaf tips; racemes long or short; capsules constricted or not...(10)
10. Racemes usually loosely flowered; styles short, somewhat swollen, dark or medium green; high plains of NE New Mexico...**Y. glauca**
10'. Racemes densely flowered; styles on most plants ± elongate, terete or oblong-cylindrical, usually ochroleucous, rarely pale green...**Y. baileyi**

Yucca angustissima Engelm. ex Trel.—Narrow-leaf yucca. Plants solitary or colonial, ± acaulescent, rarely with stems to 40 cm, rosettes to 3 m diam.; leaves spreading, linear, lanceolate, concavo-convex or plano-keeled, widest near middle, 20–80(–150) × 0.4–2 cm, rigid or flexible, not glaucous, margins becoming filiferous, white aging darker, apex long-acuminate, spine acicular, 3–7 mm; inflorescences racemose, some paniculate proximally, arising well beyond rosettes, (4–)8–20 dm; perianth campanulate to globose, tepals distinct, white to cream or greenish white, often tinged pink or brown, broad to narrowly elliptic to lanceolate-elliptic or orbiculate, 3–6.5 × 1.3–2.5 cm; style white to pale green, 3–13 mm; stigmas lobed. Our plants are **var. angustissima**. Sandy places, sandstone outcrops, piñon-juniper, ponderosa pine, pine-oak, 5250–8750. McKinley, San Juan.

Yucca baccata Torr.—Banana yucca. Plants often forming open colonies of rosettes, < 2.5 m; stems, if present, decumbent, to 2 m; leaves bluish green, concavo-convex, 30–100 × 2–6 cm, rigid, scabrous or glaucous, margins brown; inflorescences paniculate, dense, completely within to mostly extending beyond rosettes, ovoid, 6–8.2 dm, peduncle scapelike, to 0.8 m; perianth campanulate, tepals connate basally to form shallow floral cup 7–12 mm, usually cream-colored, occasionally tinged with purple, 4.5–

13 cm; filaments connate proximally into collarlike structure; style 5-7 mm; fruit baccate, indehiscent, fleshy, succulent, elongate. Two varieties in our area.

1. Plants acaulescent or caulescent; stems when present 1-6, aerial or subterranean, < 0.3 m; leaf margins coarse, curling; peduncle 0.6-0.8 m…**var. baccata**
1'. Plants caulescent; stems 1-24, aerial, often branched, some reaching 2 m; leaf margins filiferous; peduncle ≤ 0.3 m…**var. brevifolia**

var. baccata—Habitats various, rocky slopes, oak, grasslands, oak–ponderosa pine, piñon-juniper woodlands, 4900-9050 ft. Widespread except E.

var. brevifolia L. D. Benson & Darrow [*Y. arizonica* McKelvey]—Sonoran Desert, desert grasslands, oak woodlands, 4400-5200 ft. Hidalgo.

Yucca baileyi Wooton & Standl.—Bailey's bouncing yucca, Navajo yucca. Plants solitary or forming colonies 1.3-2 m diam., acaulescent, to 2 dm; rosettes usually small, symmetric; leaves yellowish green, plano-convex or plano-keeled, occasionally falcate, widest near middle, 25-50 cm × 6-9 mm, rigid, margins recurved, filiferous, whitish, apex spinose, spine acicular, to 3.2 mm; inflorescences racemose, arising within or just beyond rosettes, 2.5-4.5(-8.5) dm; perianth campanulate, tepals distinct, ovate to obovate or elliptic, 5-6.5 × 1.5-3.2 cm; style white, 7 mm.

1. Racemes (excluding scape) of the population usually short and < 1.5 times the length of the leaves; capsules not constricted or only slightly so; usually in the mountains…(2)
1'. Racemes longer, most plants of the population with racemes ~1.5 times and sometimes up to 2.5 times the length of the leaves; capsules often deeply constricted near the middle; usually on plains and foothills…**var. intermedia**
2. Plants forming dense, compact colonies of rosettes…**var. navajoa**
2'. Plants in loose colonies of scattered rosettes…**var. baileyi**

var. baileyi [*Y. baileyi* var. *navajoa* (J. M. Webber) J. M. Webber; *Y. navajoa* J. M. Webber]—Mountains, adjacent woodlands and grasslands, 5400-8200 ft. NC, NW.

var. intermedia (McKelvey) Reveal [*Y. intermedia* McKelvey var. *ramosa* McKelvey]—Piñon-juniper woodlands to adjacent grasslands, 4700-7800 ft. NC, C. **E**

var. navajoa (J. M. Webber) J. M. Webber—Four Corners region, especially in the Zuni and Chuska Mountains. Considered synonymous with var. *baileyi* by some authors. Mountains, piñon-juniper, piñon-oak communities, adjacent grassy areas, 6400-8800 ft. NW.

Yucca campestris McKelvey—Plains yucca. Plants forming small or large, open colonies, acaulescent or occasionally caulescent and arborescent; rosettes usually small, stems 0.6-1 m; leaves linear, plano-convex or plano-keeled, widest near middle, 40-65 × 0.3-0.7(-1.5) cm, rigid, margins filiferous, white, apex spinose, spine acicular, 7 mm; inflorescences paniculate, arising within or occasionally beyond rosettes, narrowly ellipsoid, 6-10 dm; perianth globose, tepals connate, dull green, sometimes tinged pink, 4.1-6.5 × 1.5-2.5 cm; style bright green. Deep sands, grassy, often with sand sage, 3150-4700 ft. NE, EC, SE.

Yucca elata (Engelm.) Engelm.—Soaptree yucca. Plants solitary or forming small colonies, caulescent or rarely acaulescent, arborescent, mostly few-branched, 1.2-4.5 m; rosettes usually large, symmetric or asymmetric; stems erect, thick, 1-2.5 m; leaves pale green, linear, widest near middle, 25-95 × 0.2-1.3 cm, flexible, margins curled, filiferous, whitish, apex tapering to short spine; inflorescences mostly paniculate, sometimes distally racemose, arising beyond rosettes, mostly narrowly ovoid to ovoid, 7-15 × 2.5-6.5 dm; perianth campanulate or globose, tepals distinct, creamy white, often tinged green or pink, narrow to broadly elliptic or ovate, 4.5-5.7 × 1.3-3.2 cm; style white or pale green, 6-11 mm. Our plants are **var. elata**. Grasslands, desert scrub, creosote, mesquite, juniper, sagebrush, 3300-6800 ft. C, WC, SE, SC, SW.

Yucca faxoniana Sarg.—Eve's-needle, Faxon yucca, Spanish dagger. Plants solitary, erect, arborescent, 2.5-6.9 m, including inflorescence; stems simple or with 2-4 branches, to 5.1 m, average diam.

32 cm; leaves yellowish green, 43–115 × 3.1–8.4 cm, rigid, margins conspicuous, curling, filiferous, brown; inflorescences paniculate, often with proximal branches arising beyond rosettes, broadly ovoid, 5.5–25.5 dm; perianth campanulate, tepals connate basally into floral cup 1–32 mm, white to greenish white, ovate, 3.9–10.8 cm; style 4.5 mm; fruit baccate, indehiscent, elongate, fleshy, succulent. Rocky slopes, flat plains, 3250–4600 ft. SE, SC.

Yucca glauca Nutt. [*Y. angustifolia* Pursh]—Soapweed yucca. Plants forming small to moderate colonies, acaulescent or caulescent and arborescent, occasionally branched; rosettes 1–15 per colony, usually small; stems erect, to 0.4 m; leaves linear to linear-lanceolate, concave to concavo-convex, widest near middle, 40–60 × 0.8–1.2 cm, rigid, margins entire, filiferous, white, apex blunt to acicular; inflorescences racemose, occasionally paniculate proximally, arising within or just beyond rosettes, 5–10 dm; tepals distinct, greenish white to white, elliptic, 5–5.3 × 2.6–3.5 cm, apex acute; style dark green, 10 mm. Prairies, open oak-brush, and waste areas in sandy or limestone soils, 4250–8400(–9450) ft. NE, NC.

Yucca harrimaniae Trel.—Spanish bayonet. Plants in dense to open colonies, acaulescent or short-caulescent; rosettes usually small; stems < 0.3 m; leaves pale green, linear- or spatulate-lanceolate, concavo-convex, widest near middle, 30–50 × 1.8–4.3 cm, rigid, margins filiferous, white or brown, apex pungent; inflorescences racemose, rarely paniculate proximally, arising within or just beyond rosettes, 3.5–7 dm; perianth broadly campanulate; tepals distinct, pale yellow or greenish yellow, usually tinged purple, broadly lanceolate, 4–5.3 × 1.6–3.4 cm; style pale to bright green, 9–13 mm. Our plants are **var. harrimaniae**. Piñon-juniper, clay slopes, sandstone outcrops, pine-oak communities, 5250–8150 ft. San Juan.

Yucca madrensis Gentry—Mountain yucca. Plants solitary, caulescent, < 3 m; stems simple, unbranched, < 2 m; leaves erect to reflexing with age, bluish glaucous or green to yellow-green, thin, flat to conduplicate, flexible, margins entire, serrulate, or occasionally filiferous with slender fibers; inflorescences erect, paniculate, somewhat open, arising 1/4–1/2 within rosettes, elongate-ovoid, to 8 dm; perianth ovoid, tepals distinct, or barely connate basally, white, ovate-lanceolate to lanceolate, ca. 3 × 1.4 cm; fruit baccate, fleshy, succulent. Plants previously identified as *Y. schottii* Engelm. will key here (see Hess & Robbins [2002] for a discussion). Madrean pine-oak forest, rocky slopes, 3900–5400 ft. Hidalgo.

Yucca neomexicana Wooton & Standl.—New Mexico yucca. Plants single or in open colonies, acaulescent or rarely caulescent; rosettes usually small, mostly asymmetric; stems, when present, < 1 m; leaf blades spreading, including distal leaves, spatulate-lanceolate, concavo-convex, thin, widest near middle, 15–46 × 0.7–2 cm, flexible, rather glaucous, margins filiferous; inflorescences racemose, rarely paniculate proximally, arising within or more often 0–20 cm beyond rosettes, 4–7 dm; perianth campanulate, tepals distinct, white to somewhat greenish white, usually tinged pink or purple, broadly lanceolate, 3–4.7 × 1.5–3 cm; style pale green or rarely white, 9–13 mm. Exposed rocky areas, ponderosa pine, ponderosa pine–piñon, piñon-juniper woodlands, sagebrush-shadscale, adjacent grasslands, 4650–8050 ft. NE, NC.

Yucca treculeana Carrière [*Y. torreyi* Shafer]—Don Quixote's lace. Plants often forming colonies, arborescent, to 7 m; rosettes with an overall ragged appearance; stems 1–8, 14–15 cm diam.; leaves erect, yellowish to bluish green, usually U- or V-shaped in cross-section, thick, 36–128 × 1.6–7 cm, rigid, scabrous, margins filiferous with straight, coarse fibers; inflorescences erect, paniculate, arising mostly within rosettes, variable in shape, usually ovoid, 18 dm; perianth globose, tepals distinct, cream-colored, occasionally tinged with purple, ovate, 2.7–8.1 × 1–3.4 cm, apex rounded or acute; style 2–8 mm; fruit baccate, indehiscent, fleshy, succulent. Grassy or rocky slopes or mesas, brushland, chaparral, 3050–5300 ft. SE, SC, Hidalgo.

AIZOACEAE – CARPETWEED FAMILY

Kenneth D. Heil

Herbs (ours), shrubs, or subshrubs, annual or perennial, often succulent, glabrous, hairy, or scaly; roots mostly fibrous; stems underground, prostrate and mat-forming to erect; leaves cauline or basal, alternate or opposite, sessile or petiolate; inflorescences axillary or terminal, flowers solitary or in cymes; flowers bisexual (ours), inconspicuous to showy; calyx colored (ours); stamens 1–700; pistil 1; ovary superior or 1/2 inferior; fruit a capsule with several to numerous seeds (Vivrette et al., 2003).

1. Styles 2–5; leaves of each pair equal; stipules absent; plants perennial…**Sesuvium**
1'. Styles 1–2; leaves of each pair unequal; stipules present; plants annual…**Trianthema**

SESUVIUM L. – Sea purslane

Sesuvium verrucosum Raf.—Western sea purslane. Plants perennial, herbaceous, succulent, papillate with crystalline globules, abundant; stems prostrate, forming mats to 2 m diam., branched from base; leaves opposite, blades linear to widely spatulate, to 4 cm; flowers solitary in leaf axils; calyx lobes rose or orange adaxially, 2–20 mm; ovary 1/2 inferior; capsules ovoid-globose, 4–5 mm, dehiscence circumscissile. Margins of mostly saline, alkaline, and gypseous habitats, 3150–5000 ft. C, EC, SC, SE.

TRIANTHEMA L. – Horse purslane

Trianthema portulacastrum L.—Desert horse purslane. Annual herbs (ours); roots fibrous; stems prostrate; leaves subopposite, stipules attached to margin of petiole, blades terete, flat, linear to orbiculate; flowers solitary or in cymes; calyx lobes 5, adaxial surface colored; stamens 5–10(–20); ovary superior, 1–2-loculed; fruit a capsule, dehiscence circumscissile near base; seeds 1–12. Moist or dry alkaline flats, playas, banks of rivers and creeks, gardens, irrigated soils, 4325–5940 ft. C, EC, SW, SC.

ALISMATACEAE – WATER-PLANTAIN, ARROWHEAD FAMILY

C. Barre Hellquist

Annual or perennial; rhizomatous, stoloniferous, or cormose; herbs; roots septate or nonseptate; leaves basal, sessile or petiolate, submersed, floating; basal lobes present or absent, veins parallel from base of blade to apex; inflorescences scapose racemes, panicles, occasionally umbels, erect, floating, or decumbent; flowers bisexual or unisexual, if unisexual plants dioecious; hypogynous; sepals 3, petals 3; stamens (0)6–9(–30); pistils 0–6 and up to 1500 in some; ovules 1 or 2; fruits cypselas or follicles; mature seeds with endosperm absent (Haynes & Hellquist, 2000a).

1. Pistils and cypselas in a single, flat-topped ring, flowers all bisexual, stamens 6…**Alisma**
1'. Pistils and cypselas in a dense globose head; flowers bisexual or unisexual, stamens > 6…(2)
2. Roots nonseptate; all flowers bisexual; cypselas thickened, ridged, not winged; leaf blades never sagittate…**Echinodorus**
2'. Roots septate; upper flowers unisexual, usually staminate, lower flowers pistillate, occasionally bisexual (petals with basal spot) or sometimes dioecious; cypselas flattened, winged; leaf blades often sagittate…**Sagittaria**

ALISMA L. – Water-plantain

Perennial herbs, submersed or emersed, some floating-leaved; often rhizomatous, stolons, corms, and tubers absent. Roots nonseptate; leaves sessile or petiolate, petioles triangular, blades linear to ovate, bases attenuate to rounded, apex obtuse to acute; inflorescence a panicle of 2–10 whorls, erect, emersed; flowers bisexual, pedicels ascending, bracts subtending pedicels; receptacle flattened; sepals erect; petals pink or white, stamens 6–9; pistils 15–20 in a single flat-topped ringed receptacle; ovule 1; fruit a laterally compressed achene.

1. Leaves submersed and ribbonlike, or if emersed, with lanceolate to narrowly elliptic blades; cypselas with 2 dorsal grooves and a central ridge…**A. gramineum**
1'. Leaves emersed, with ovate to elliptic blades, occasionally submersed to floating; cypselas with a single dorsal groove…(2)
2. Flowers 3–3.5 mm wide, sepals at anthesis 1.5–2.5 mm; petals 1–3 mm; fruiting heads 2–4 mm diam.; cypselas 1.5–2 mm…**A. subcordatum**
2'. Flowers 7–13 mm wide, sepals at anthesis 3–4(–6) mm; petals 3.5–6 mm; fruiting heads 4–7 mm diam.; cypselas 2.2–3 mm…**A. triviale**

Alisma gramineum Lej.—Grass-leaved water-plantain. Perennial to 50 cm; leaves submersed, floating, or emersed; submersed leaves sessile and linear, 0.2–2(–3) mm wide, blade present or absent; emersed leaves petiolate, rarely sessile, blade linear-lanceolate, lanceolate, or narrowly elliptic, 0.4–1.5 cm wide; flowers purplish white, 2–4 mm; fruiting head 3–6 mm diam. Ponds and lakes, stranded or shallow water, 6800–7575 ft. Colfax, McKinley, Rio Arriba.

Alisma subcordatum Raf.—Southern water-plantain. Perennial herb to 60 cm; leaves emersed, petiolate, blades ovate to elliptic; sepals 1.5–2.5 mm; petals white, 1–3 mm; fruiting head 2–4 mm diam.; achene obliquely ovoid with 1 dorsal groove, 1.5–2.2 mm; beak erect. Shallow ponds, stream margins, marshes, ditches, 4425–9375 ft. Cibola, Rio Arriba, San Juan.

Alisma triviale Pursh—Northern water-plantain. Perennial herbs to 1 m; leaves emersed, petiolate; blade linear-lanceolate to broadly elliptic or oval; sepals 3, 3–4(–6) mm; petals 3, white, 3.5–6 mm: fruiting head 4–7 mm diam.; achene ovoid with 1 dorsal groove, 2.1–3 mm, beak erect or nearly erect. Shallow ponds, stream margins, marshes, ditches, 4575–8290 ft. NC, NW, Catron, Socorro.

ECHINODORUS Rich. ex Engelm. – Burhead

Plants annual or perennial, emersed, floating-leaved, rarely submersed; rhizomes present or absent, roots nonseptate; leaves sessile or petiolate, petioles triangular, blades with translucent dots or lines present or absent, margins entire or undulating, apex obtuse to acute; inflorescences racemes or panicles of 1–18 whorls, erect or decumbent; flowers bisexual, pedicels ascending to recurved, petals white, stamens 9–25, pistils 15–250; fruits plump, lateral wings absent, glands often present.

Echinodorus berteroi (Spreng.) Fassett—Upright burhead. Annual or perennial; rhizomatous; leaves emersed or rarely submersed; petiole terete to triangular, 2–26 cm; blade with transparent distinct lines, elliptic, lanceolate, or ovate, 2.6–15.5 × 0.5–20 cm, base truncate, rarely cordate or tapering; inflorescence a raceme, rarely a panicle of 1–9 whorls; bracts distinct, lanceolate, spreading or ascending, 0.6–2.8 cm; flowers with sepals spreading to recurved, 9–13-veined; pistils 45–200; fruits plump, 3–5-ribbed, glands 1 or 2, beak terminal. Ditches, streams, shallow water, 4400 ft. Roosevelt.

SAGITTARIA L. – Arrowhead

Perennial herbs, rarely annual, submersed, floating-leaved, or emersed; rhizomes usually present, often with tubers, or stolons; roots septate; leaves sessile or petiolate; blades sagittate, hastate, cuneate, lanceolate, elliptic, or obovate; inflorescences racemes, panicles, rarely umbels, of 1–7 whorls; flowers unisexual (rarely bisexual); staminate flowers pedicellate, distal to pistillate flowers; pistillate flowers mostly pedicellate, occasionally sessile; bracts subtending pedicels; petals white, rarely with pink spot or tinge, stamens 7–30; filaments linear to dilated, glabrous or pubescent; pistils to 1500; fruit compressed, wings often present (Bogin, 1955).

1. Fruiting pedicels recurved, rarely spreading; pistillate sepals mostly erect, closely enclosing flower and fruiting head…(2)
1'. Fruiting pedicels spreading, ascending, or absent; pistillate sepals spreading or recurved, not enclosing flower or fruit…(3)
2. Leaves emersed and submersed, submersed leaves linear, sessile, emersed leaves petiolate, blade hastate to sagittate; inflorescence of 1–15 whorls; fruiting heads 1.2–2.1 cm diam.…**S. calycina**
2'. Leaves submersed, blade phyllodial; inflorescence of 2–7 whorls; fruiting heads 0.4–0.6 cm diam.… **S. demersa**

3. Emersed leaves sagittate…(4)
3′. Emersed leaves linear or linear-oblanceolate, not sagittate…**S. graminea**
4. Cypselas with beak horizontal, 1-2 mm; bracts 3-8…**S. latifolia**
4′. Cypselas with beak ascending or erect, 0.1-0.6 mm; bracts 0.7-4 cm…(5)
5. Emersed plants with recurved petioles, blades linear to sagittate; submersed leaves phyllodial, floating leaves cordate to sagittate…**S. cuneata**
5′. Emersed plants with ascending to erect petioles, blades sagittate with basal lobes longer than rest of blade; submersed and floating leaves absent…**S. longiloba**

Sagittaria calycina Engelm. [*S. montevidensis* Cham. & Schltdl.]—Hooded arrowhead. Annual or perennial herbs to 100 cm; rhizomes present; submersed leaves sessile, emersed leaves petiolate, blade hastate to sagittate; inflorescence of 1-15 whorls, floating or emersed; bracts distinct; flowers with a ring of sterile stamens. New Mexico material belongs to **subsp. calycina**. Shore of lakes and ponds, 3800-5875 ft. Catron, Doña Ana, Socorro.

Sagittaria cuneata E. Sheld.—Northern arrowhead. Perennial herbs to 40 cm, to 80 cm submersed, stolons present; leaves emersed or submersed, rosettes often with floating leaves, or submersed, flaccid, flattened, linear; petiole triangular, blade cordate, sagittate, or linear; inflorescence a raceme, rarely a panicle of 2-10 whorls, emersed; peduncles 10-50 cm; bracts connate, 7-40 mm; fruiting pedicels ascending; flowers to 2.5 cm diam.; sepals recurved; stamens 10-24; fruiting heads 0.8-1.5 cm diam.; cypselas obovoid, 1.8-2.6 × 1.3-2.5 mm, beaked. Lakes, ponds, rivers, 4400-9800 ft. NE, NC, NW, C, WC.

Sagittaria demersa J. G. Sm.—Chihuahuan arrowhead. Annual herbs to 60 cm; stolons and corm present; leaves submersed, phyllodial, 12-53 cm; inflorescence a raceme of 2-7 whorls, floating or emersed; peduncles 13.5-28 cm; bracts ovate to lanceolate, 1.5-2 mm; fruiting pedicels 1.5-6.5 cm; flowers 1.5-5 cm diam.; sepals mostly erect, often enclosing flowering and fruiting heads; fruiting heads 0.4-0.6 cm diam.; cypselas oblanceoloid to obovoid, 1.5-1 mm, beak lateral, erect, ca. 1.1 mm. Streams and lakes, 8075-8225 ft. Colfax, San Miguel.

Sagittaria graminea Michx.—Grassy arrowhead. Perennial herbs to 100 cm; rhizomes coarse; leaves submersed, phyllodial, to 1 cm wide, flattened on upper surface, 6.4-3.5 × 0.5-4 cm, or emersed, blade linear to linear-oblanceolate, 2.5-17.4 × 0.2-4 cm; inflorescence a raceme of 1-12 whorls; peduncles 6.5-29.7 cm; bracts 20-50 mm; pistillate pedicels spreading, 0.5-3 cm; flowers to 2.3 cm diam.; sepals recurved to spreading, not enclosing flower, filaments pubescent; fruiting heads 0.6-1.5 cm diam.; cypselas oblanceolate, lacking keel, 1.5-2.8 × 1.1-1.5 mm, beak erect, 0.2 mm. Lakes and ponds, 8000-8100 ft. Colfax.

Sagittaria latifolia Willd.—Broadleaf arrowhead. Perennial herbs to 150 cm; rhizomes absent, stolons present, bearing corms; leaves emersed, petiole triangular, erect, blade sagittate, hastate, or lanceolate; inflorescence a raceme, rarely a panicle of 3-90 whorls, emersed; peduncles 10-59 cm; bracts connate for at least 1/4 of total length, 3-8 mm; fruiting pedicels spreading; flowers to 4 cm diam.; sepals recurved to spreading, stamens 21-40; fruiting head 1-1.7 cm diam.; cypselas oblanceolate, 2.5-3.5 × to 2 mm, beak lateral, horizontal, 1-2 mm. Pond and lake shores, streams, marshes, 3925-8400 ft. NC, NW, Roosevelt, Sierra, Socorro.

Sagittaria longiloba Engelm. ex J. G. Sm.—Longbarb arrowhead. Perennial to 100 cm, producing stolons and corms; leaves emersed, erect; blade sagittate, 11.5-26.5 × 0.8-15 cm, basal lobes longer than rest of blade; inflorescence a raceme, rarely a panicle, 5-17 whorls; peduncles 25-96 cm; bracts 6.5-15 mm, delicate; fruit pedicels spreading, 1.5-4.4 cm; flowers to 3 cm diam.; sepals recurved to spreading; fruiting heads 0.9-1.5 cm diam.; cypselas oblanceoloid, 1.2-2.5 × 0.8-1.6 mm, beak lateral, erect, 0.1-0.6 mm. Ditches and margins of streams, lakes, ponds, 3500-5375 ft. Catron, Doña Ana, Socorro.

ALLIACEAE – ONION FAMILY

Steve L. O'Kane, Jr.

Biennial or perennial herbs, usually with a distinctive onionlike odor, acaulescent from a bulb with a papery or fibrous covering, rhizomes sometimes present; leaves ± basal, acicular, linear, or linear-lanceolate, flat, channeled, or terete; inflorescence a terminal umbel with membranous and spathelike bracts; flowers bisexual, actinomorphic; perianth petaloid, of 2 whorls of 3 tepals, campanulate, urceolate, tubular, to nearly rotate; tepals distinct to connate, entire (sometimes denticulate or obscurely toothed); stamens 6, epitepalous; ovary superior, 3-locular, placentation axile; fruit a loculicidal capsule (Simpson, 2019).

1. Plants with an onion odor; tepals white, pink, lilac, or rose-purple; pedicels within the umbels all of equal length, umbels therefore symmetric…**Allium**
1'. Plants lacking an onion odor; tepals white to yellowish white, midvein typically green proximally, at least the outer tepals abaxially with the midvein pale to deep red or purplish red; pedicels within the umbels of different lengths, making umbels asymmetric…**Nothoscordium**

ALLIUM L. – Onion, garlic, wild leek

Bulbs with onion odor and taste, solitary or clustered, bulb coat papery-membranous or fibrous, rhizomes present or not; leaves ± basal, 1–12, linear, terete, channeled, or flat; scape usually persistent, terete or flattened; umbels symmetric, sometimes replaced totally or partially by bulbils, bracts conspicuous, ± fused, usually 3+-veined; tepals white, pink, rose-purple, or red, rarely greenish yellow; ovary crested with processes or not (McNeal & Jacobson, 2002; Sivinski, 1998).

1. Outer bulb coat persisting as a conspicuous covering of coarse, anastomosing fibers; rhizomes lacking…(2)
1'. Outer bulb coat without fibers or with parallel fibers, typically papery, never fibrous-reticulate; with or without rhizomes…(6)
2. Bracts of involucre 2–5-nerved (occasionally coalescing into what appears to be a single wide nerve in *A. macropetalum*)…(3)
2'. Bracts of involucre mostly 1-nerved…(4)
3. Ovary conspicuously crested with 3 pairs of short, flat projections; leaves usually 2 per scape; desert and plains, W and C New Mexico…**A. macropetalum**
3'. Ovary lacking a crest; leaves usually 3 per scape; desert and plains of E New Mexico…**A. perdulce**
4. Perianth spreading-rotate; epidermal cells of inner bulb coats (under outer reticulum) intricately contorted (use lens); portions of outer bulb coat fused into irregular, solid pieces except along the ragged top and bottom edges of bulb; common on hills and plains of SE New Mexico…**A. drummondii**
4'. Perianth urceolate; epidermal cells of innermost bulb coats rectangular and vertically elongate (use lens); entire outer bulb coats a reticulate fabric of coarse fibers with open interstices…(5)
5. Leaves usually 2 per scape; spring-flowering; rare in NW and NE New Mexico…**A. textile**
5'. Leaves usually 3+ per scape; summer-flowering; widespread…**A. geyeri**
6. Bulbs attached to stout, dark, irislike rhizomes; leaves flat, strap-shaped, (4–)5–8 mm wide; mountains in W and SC New Mexico…**A. gooddingii**
6'. Bulbs with or without rhizomes, if rhizomes present, then slender and pale; leaves linear-channeled or broadly U-shaped in cross-section, usually < 5 mm wide (occasionally flat and > 5 mm wide in *A. cernuum*)…(7)
7. Umbel nodding from a bend in the scape below the involucral bracts; tepals obtuse; stamens exserted from perianth; in all New Mexico mountain ranges and on NE plains…**A. cernuum**
7'. Umbel erect; tepals acute or acuminate; stamens shorter than perianth…(8)
8. Inner whorl of tepals long-acuminate with recurved tips, margins minutely serrulate-dentate; outer tepals similar, but conspicuously broader and usually entire; outer bulb coat cells relatively square with thick, wafflelike walls; rare in W New Mexico…**A. acuminatum**
8'. Inner and outer tepals entire and not conspicuously wider or narrower; other characters never combined as above…(9)
9. Ovary and capsule conspicuously crested…(10)
9'. Ovary and capsule not crested…(11)
10. Scape usually < 10(–12) cm; perianth (8–)10–14 mm; rhizomes lacking; outer bulb coat dark brown, cells elongating vertically; desert species of SW New Mexico…**A. bigelovii**
10'. Scape 10–30(–40) cm; perianth 6–10 mm; slender rhizomes present at the base of the bulb; outer bulb coat grayish, cells elongating horizontally; rare in mountains of W New Mexico…**A. bisceptrum**

11. Bulb subspherical, often proliferating from the base by slender, scaled rhizomes; perianth campanulate-spreading; tepals white (often drying pinkish), with a dark red-purple midrib on outer surface; anthers red-purple (drying brown); igneous ridges and canyons in SW New Mexico...**A. rhizomatum**

11'. Bulb ovoid, rhizomes absent; perianth spreading-rotate; tepals white to pale pink (drying pink), outer midrib absent or vague; anthers yellow; calcareous ridges and canyons in S New Mexico...**A. kunthii**

Allium acuminatum Hook.—Tapertip onion. Bulb coat membranous, ± yellow-brown, prominently cellular-reticulate; leaf blades subterete or ± channeled, 7–30 cm × 1–3 mm; umbel loose, hemispherical; spathe bracts 2, 3–7-veined, lanceolate to ovate; flowers campanulate, 8–15 mm; tepals erect, bright pink to rose-purple (or white), lanceolate to lance-ovate, becoming rigid and keeled in fruit, margins finely denticulate, apex acuminate, outer longer and wider than inner, spreading to recurved at tip, inner with strongly recurved tips; anthers yellow; pollen yellow; ovary crested; processes 3, central, 2-lobed, rounded, minute, margins entire. Dry slopes and plains, piñon-juniper, ponderosa pine, 5700–7250 ft.

Allium bigelovii S. Watson—Bigelow's onion. Bulb coat membranous, dark brown, prominently reticulate; leaves subterete to channeled, 16–21 cm × 2–4 mm; umbel loose to ± compact, hemispherical; spathe bracts 2, 2–11-veined, lance-ovate to ovate; flowers campanulate, 8–14 mm; tepals erect, white to pale pink, outer with red midribs, inner with red tips, lanceolate, becoming papery and ± rigid in fruit, apex acute; stamens included; anthers purple; pollen yellow; ovary crested; processes 6, prominent, flat, triangular, margins entire to coarsely toothed. Open, rocky, gravelly foot-slopes of desert mountain ranges, 4350–6350 ft.

Allium bisceptrum S. Watson [*A. palmeri* S. Watson]—Twincrest onion, aspen onion. Bulbs producing a cluster of basal bulbils or filiform rhizomes terminated by bulbils; bulb coat membranous, light brown to gray, obscurely cellular-reticulate; leaf blades flat, broadly channeled, 8–30 cm × 1–13 mm; umbel loose, 15–40-flowered, globose; spathe bracts persistent, 2, 3–4-veined, ovate to lanceolate; flowers stellate, 7–10 mm; tepals spreading, lilac to white, lanceolate, ± equal, becoming papery in fruit, apex acuminate; anthers purple; pollen yellow; ovary conspicuously crested; processes 6, central, distinct, flattened, triangular, margins papillose-denticulate. Meadows, aspen groves, occasionally ponderosa pine, less commonly open slopes in mountains, 7200–10,150 ft.

Allium cernuum Roth [*A. neomexicanum* Rydb.]—Nodding onion. Bulb coat membranous, enclosing 1+ bulbs, grayish or brownish, minutely striate; leaf blades flat, channeled to broadly V-shaped in cross-section, 10–25 cm × 1–6 mm, margins entire or denticulate; umbel nodding from the bent scape, loose, hemispherical; spathe bracts 2, 3-veined, lanceolate; flowers campanulate, 4–6 mm; tepals ± erect, pink or white, elliptic-ovate, withering in fruit, apex ± obtuse, at least outer tepals strongly incurved, midribs not thickened; stamens exserted; anthers yellow; pollen yellow; ovary conspicuously crested; processes 6, flattened, ± triangular, margins entire or toothed. Various habitats of mountains, meadows, oak woodland, mixed conifer, ponderosa pine, piñon-juniper, grassy openings, 5000–7400 ft.

Allium drummondii Regel—Regel's onion. Bulb coat fibrous, light reddish brown, reticulate, portions fused, with no openings between fibers; leaf blades flat, channeled, 10–30 cm × 1–5 mm, margins entire; umbel compact to ± loose, hemispherical-globose, rarely replaced by bulbils; spathe bracts persistent, 2–3, 1-veined, ovate; flowers campanulate to ± stellate, 6–9 mm; tepals spreading, white, pink, or red, rarely greenish yellow, ovate to lanceolate, becoming papery and rigid in fruit, apex obtuse or acute, midribs somewhat thickened; anthers yellow; pollen light yellow; ovary crestless. Plains, hills, prairies, particularly in limestone soils, 3000–8600(–10,000) ft.

Allium geyeri S. Watson—Geyer's onion. Bulb coat fibrous, gray or brown, reticulate, cells coarse-meshed; leaf blades flat, channeled, (6–)12–30 cm × 1–5 mm, margins entire or denticulate; umbel compact, hemispherical to globose or 0–5-flowered and flowers mainly replaced by bulbils; spathe bracts 2–3, mostly 1-veined, ovate to lanceolate; flowers urceolate-campanulate, 4–10 mm; tepals erect or spreading, pink to white, ovate to lanceolate, permanently investing the fruit or withering if fruit not produced, apex obtuse to acuminate; anthers yellow; pollen yellow; ovary when present inconspicuously crested; processes 6, central, low, distinct or connate in pairs across septa, ± erect, rounded, margins entire, developed or obsolete in fruit. Two varieties in our area.

1. Most flowers replaced by bulbils, the remaining flowers sterile; scattered in several W and N New Mexico mountain ranges…**var. tenerum**
1'. Flowers not replaced by bulbils, flowers all fertile; widespread in most New Mexico mountain ranges… **var. geyeri**

var. geyeri—Geyer's onion. Seeds produced. In mountains, moist, open slopes, meadows, tundra, forest openings, stream banks, 4800–12,500 ft.

var. tenerum M. E. Jones—Bulbil onion. Seeds rarely produced and then only in non-bulbil-producing flowers. Moist meadows and along streams, 8200–12,450 ft.

Allium gooddingii Ownbey—Gooding's onion. Bulb coat membranous, brownish, minutely striate; plants with a thick irislike rhizome; leaf blades flat, 8–25 cm × 4–8 mm; umbel loose, conic; spathe bracts 2, 3–5-veined, narrowly lanceolate; flowers campanulate, 8–10 mm; tepals erect, pink, elliptic, withering in fruit, apex obtuse; anthers white or purple; pollen white; ovary crestless. Steep, rocky slopes, canyon bottoms, coniferous forests, 7850–12,450 ft. **R & E**

Allium kunthli G. Don   Kunth's onion. Bulb coat membranous, grayish or brownish, with or without obscure, delicate cellular markings, sometimes striate; leaf blades flat, channeled, 10–21 cm × 1–3 mm, margins and veins sometimes denticulate; umbel loose, conic; spathe bracts 2, 3–5-veined, lanceolate; flowers rotate-spreading (to campanulate), 4–8 mm; tepals ± spreading, white or pale pink (particularly on midribs), lanceolate, becoming papery and withering in fruit, apex acute to acuminate; anthers yellow or purple; pollen yellow; ovary crestless. Dry, rocky, calcareous substrates, hills and mountains, typically in piñon-juniper woodlands, sometimes with ponderosa pine–juniper or grassy areas, 7050–8800 ft.

Allium macropetalum Rydb.—Large-flower onion. Bulb coat fibrous, brown, reticulate, cells usually coarse-meshed; leaf blades channeled, semiterete, 8–20 cm × 1–3 mm; umbel compact to loose, hemispherical to globose; spathe bracts 2–3, 3–5-veined, ovate to lanceolate; flowers campanulate, 8–12 mm; tepals spreading, pink with deeper pink or reddish midveins, lanceolate, becoming papery in fruit, not investing capsule, apex obtuse to acuminate, midrib scarcely thickened; anthers yellow or purple; pollen yellow; ovary usually conspicuously crested; processes 6, central, usually connate in pairs across septa, ± erect, flattened, triangular, to 2 mm, margins entire, mostly well developed in fruit. Desert plains and hills, sandy washes, juniper, piñon-juniper, shadscale, oak shrub, desert scrub, 3450–8350 ft.

Allium perdulce S. V. Fraser—Plains onion. Bulb coat fibrous, dark brown, reticulate, cells coarse-meshed; leaf blades flat, channeled, 8–30 cm × 1–2(–3) mm, margins entire; umbel loose, hemispherical-globose; spathe bracts 2–3, 3–7-veined, ovate; flowers urceolate, 7–10 mm; tepals erect, white or pale pink with deep pink midribs to deep rose, lanceolate, ± equal, becoming callous-keeled and permanently investing capsule, margins entire, apex obtuse or acute; anthers yellow or purple; pollen yellow; ovary crestless. Our plants are **var. perdulce**. Usually sandy soils, desert and grassland plains, 3550–5700 (–6200) ft.

Allium rhizomatum Wooton & Standl.—Gland onion. Bulbs replaced yearly by new bulbs at tip of rhizome; bulb coats enclosing parent bulbs membranous, grayish, lacking cellular reticulation; leaf blades flat, 20–35 cm × 2–3 mm; umbel loose, globose to hemispherical; spathe bracts 2, 3-veined, ovate to lance-ovate; flowers stellate, 6–9 mm; tepals erect, pink with purplish or pinkish midveins, oblong to lanceolate, slightly carinate basally, becoming papery in fruit, margins entire, apex acute to acuminate; anthers yellow or pink; pollen yellow or white; ovary crestless, 3-grooved with thickened ridge on either side of groove. Dry, usually grassy areas, canyon bottoms, rocky areas, piñon-juniper, open woodlands, 5150–8600 ft.

Allium textile A. Nelson & J. F. Macbr. [*A. geyeri* S. Watson var. *textile* (A. Nelson & J. F. Macbr.) B. Boivin]—White wild onion. Bulb coat fibrous, gray or brown, reticulate, fine-meshed; leaf blades ± straight, channeled, semiterete, 10–40 cm × 1–3(–5) mm, margins entire or denticulate; umbel compact to ± loose, hemispherical; spathe bracts 3, usually 1-veined, ovate; flowers urceolate to campanulate, 5–7 mm; tepals erect, white or rarely pink, with red or reddish-brown midribs; outer whorl broader and permanently covering the capsule, apex obtuse to acuminate; inner whorl narrower, apex distinctly spreading; anthers yellow; pollen yellow; ovary ± conspicuously crested; processes 6, central, distinct or connate in pairs across septa, ± erect, rounded, developed or not in fruit. Dry plains and hills, sagebrush, piñon-juniper, 5000–8500(–9400) ft.

NOTHOSCORDUM Kunth – Crow poison, false garlic

Nothoscordum bivalve (L.) Britton [*Allium bivalve* (L.) Kuntze]—Crow poison. Bulbs lacking onion odor or taste; outer bulb coats membranous, brown; leaves basal, 1–4, blades filiform to linear, to 30 cm × 1–4(–5) mm; umbels usually asymmetric; spathe bracts 2, membranous; flowers on pedicels ascending to ± erect, withering-persistent; tepals connate in lower 1/3, subequal, whitish to cream, at least outer ones with red or purplish-red midvein, elliptic, 8–15 × 3–4.5 mm, apex acute or acuminate; anthers yellow; ovary crestless. Open woodlands, grasslands, barrens, 4750–5650 ft. (Jacobsen & McNeal, 2002).

AMARANTHACEAE – AMARANTH FAMILY

Ross A. McCauley

Annual or perennial, hermaphroditic, dioecious, monoecious, or polygamous herbs or shrubs (ours); leaves simple, spiral or opposite, exstipulate, succulent or reduced in some taxa; inflorescence of solitary flowers or a spike, panicle, cyme, or thyrse with subtending bracts and bracteoles generally bristlelike; flowers small, bisexual or unisexual, usually actinomorphic; perianth uniseriate, consisting of (0–2)3–5(6–8) distinct or basally connate tepals; stamens (1–2)3–5(6–8), generally the same number as the tepals, distinct or basally connate into a tube, staminodia and pseudostaminodia common in some taxa; anthers di- or monothecal; gynoecium with a superior ovary, 1–3(–5) carpels, 1 locule; fruit a nutlet, berry, dehiscent capsule, or utricle (Powell & Worthington, 2018; Robertson & Clemants, 2003; Welch et al., 2003; Welsh & Atwood, 2013).

1. Stems jointed, fleshy; leaves reduced to scales; flowers in dense cylindrical spikes, obscure among bracts or in stems; plants of saline environments…(2)
1'. Stems not jointed, fleshy or not; leaves usually well developed; plants of various environments…(3)
2. Shrubs 30–150 cm, ± glaucous, generally with a blackish hue; branches and leaves alternate… **Allenrolfea**
2'. Herbs or rhizomatous shrubs 15–30 cm; leaves opposite, scalelike and connate into a sheath surrounding the stem, with scarcely projecting distinct tips, stems succulent, plants of saline sites…**Salicornia**
3. Stem leaves alternate or rarely opposite proximally…(4)
3'. Stem leaves opposite…(24)
4. Herbs, flowers imperfect, subtended by 3 bracts (1 bract and 2 bracteoles) (sometimes modified)… **Amaranthus**
4'. Herbs, subshrubs, or shrubs; plants monoecious or dioecious; flowers perfect or imperfect, flowers (pistillate if flowers imperfect) subtended by 1–5 bracts…(5)
5. Leaves or bracts of inflorescence bristle- or spine-tipped…(6)
5'. Leaves or bracts of inflorescence not bristle- or spine-tipped (7)
6. Fruiting perianth abaxially winged; leaves linear to sublate, herbaceous or fleshy, spine-tipped or not; bracts of inflorescence ovate-lanceolate, spine-tipped…**Salsola**
6'. Fruiting perianth apically winged; leaves terete, fleshy-succulent, bristle-tipped; bracts of inflorescence similar to leaves…**Halogeton**
7. Leaves cylindrical to linear, generally fleshy or semisucculent…(8)
7'. Leaves with flattened blades and/or not fleshy or succulent…(10)
8. Shrubs, armed with thorny branchlets; staminate flowers in spikes, pistillate flowers solitary and axillary [this genus is now treated in Sarcobataceae]…**Sarcobatus**
8'. Shrubs or herbs, not armed; flowers perfect or perfect and pistillate…(9)
9. Herbage villous-tomentose; plants low subshrubs (to 50 cm)…**Neokochia**
9'. Herbage glabrous, glaucous, or puberulent; plants annual or perennial, or if subshrubs, then tall (to 150 cm)…**Suaeda**
10. Plants densely white-hairy with at least some dendritic hairs, these becoming golden-brown in age; shrubs…**Krascheninnikovia**

10'.   Plants variously hairy or glabrous, but not as above; shrubs or herbs of various distributions…(11)

11.   Flowers imperfect, the pistillate enclosed in 2 accrescent or connate bracteoles…(12)

11'.   Flowers perfect or some also pistillate, all with sepals and not enclosed by paired bracteoles…(16)

12.   Leaf blades orbicular or suborbicular, flabellate to broadly obtuse, with conspicuous reddish veins; fruit flask-shaped in outline…**Suckleya**

12'.   Leaves mainly ovate to lanceolate, acute to obtuse basally…(13)

13.   Leaves exhibiting Kranz anatomy (leaf veins are dark green, indicating high concentrations of chloroplasts in the bundle-sheath cells surrounding the veins; best viewed with ×10 hand lens)…**Atriplex** (in part)

13'.   Leaves not exhibiting Kranz anatomy (leaf veins not dark green due to even dispersal of chloroplasts across the mesophyll)…(14)

14.   Erect shrubs; paired fruiting bracteoles present and connate, samaralike…**Grayia**

14'.   Sprawling, prostrate, or erect herbaceous annuals; fruiting bracteoles present, connate or distinct…(15)

15.   Bracteoles enclosing 2–6 pistillate flowers, slightly connate in proximal 1/2; pistillate flowers with perianths…**Proatriplex**

15'.   Bracteoles enclosing a single pistillate flower, connate; pistillate flowers without perianths…**Atriplex** (in part)

16.   Perianth horizontally winged in fruit…(17)

16'.   Perianth not horizontally winged in fruit…(18)

17.   Leaf blade margins sinuate-dentate; plants villous or tomentulose, becoming glabrous with maturity… **Cycloloma**

17'.   Leaf blade margins entire; plants ± pubescent but not villous or tomentulose…**Bassia** (*B. scoparia*)

18.   Perianth segments developing spiniform, hooked, or conic appendages…**Bassia**

18'.   Perianth segments rounded or keeled abaxially, lacking spines or appendages…(19)

19.   Calyx lobes 1–3, the fruit largely exposed; stamens 1–3; plants of sandy environments…**Corispermum**

19'.   Calyx lobes 5, largely concealing to exposing the fruit, stamens usually 5…(20)

20.   Plants aromatic, leaves and perianth with stalked glandular hairs and/or subsessile glands…**Dysphania**

20'.   Plants nonaromatic (but sometimes fetid), vesicular-hairy (farinose) or glabrous…(21)

21.   Stems unbranched or sparingly branched; basal leaves often forming a rosette; perianth often changing to succulent or hardened in fruit, sometimes reduced to 1 lobe; stigmas 2–4; seeds vertical…**Blitum**

21'.   Stems usually branched; basal leaves not in a rosette; perianth unchanged in fruit, not reduced; stigmas 2(3), seeds vertical and/or horizontal…(22)

22.   Flowers often dimorphic, in lateral flowers perianth segments 3(–5), seeds mostly vertical or sometimes horizontal; stamens 1–3…**Oxybasis**

22'.   Flowers not dimorphic, perianth segments 5, seeds exclusively horizontal; stamens almost always 5…(23)

23.   Young stems and leaves densely covered with vesicular globose trichomes becoming cup-shaped when dry and mostly persistent at maturity; perianth segments without prominent midvein visible inside; seeds smooth or striate and somewhat rugulose, sometimes pitted…**Chenopodium**

23'.   Young stems and leaves with vesicular trichomes becoming totally collapsed when dry, mostly caducous or rarely present at maturity; perianth segments with prominent midvein visible inside; seeds distinctly pitted to sometimes rugulose or almost smooth…**Chenopodiastrum**

24.   Herbs, subshrubs, or shrubs; plants monoecious or dioecious; flowers imperfect, pistillate flowers subtended by 2 bracts…**Atriplex** (in part)

24'.   Herbs (sometimes becoming woody at base); plants monoecious; flowers perfect or imperfect, subtended by 3 bracts (1 bract and 2 bracteoles)…(25)

25.   Inflorescences of axillary and terminal panicles (to 10+ cm wide); flowers mostly imperfect…**Iresine**

25'.   Inflorescences of terminal panicles with spicate branches, or axillary single flowers, flower clusters, spikes, or terminal spikes or headlike clusters; flowers perfect…(26)

26.   Erect or ascending herbs; inflorescences mostly terminal with spicate branches on elongate peduncles, spikes appearing cottony collectively because of woolly perianths; tepals united in fruit into an indurated tube…**Froelichia**

26'.   Prostrate to decumbent herbs, or stems ascending or erect, or erect with a woody base; inflorescences in few-flowered axillary clusters, spikes densely aggregated at nodes, or subglobose to short-cylindrical heads; tepals separate, united only at the base, or campanulate…(27)

27.   Plants densely stellate-pubescent; leaves gray-green; flowers in small axillary clusters, inconspicuous, lacking pinkish or white bracts…**Tidestromia**

27'.   Plants pubescent or glabrous but without stellate pubescence; leaves green and glossy to gray-green or yellowish green; flowers in axillary clusters or in heads or short-cylindroid spikes…(28)

28.   Flowers in relatively large (1–2.8 cm diam.), globose or short-cylindrical heads; scarious bracts and tepals pinkish or whitish; plants erect to procumbent…**Gomphrena**

28'.   Flowers in relatively small (usually < 1 cm diam.) axillary clusters; bracts and tepals whitish or stramineous, or subtending hairs whitish; plants prostrate to slightly ascending…(29)

29. Leaf blades spatulate to orbicular, dark green to yellowish green and glossy above, glabrous or sparsely pubescent (hairs, if present, mostly underneath, at lower margins, and in petiole area); flower clusters mostly to entirely glabrous outside to occasionally villous; bracts and tepals with short-spinose tips… **Alternanthera**

29'. Leaf blades lanceolate or oval to rounded-obovate, pubescent, densely so underneath; flower clusters pubescent outside; bracts and tepals lacking short-spinose tips…(30)

30. Leaves of basal rosette persistent at time of flowering; filament tube free from perianth; tepals distinct, 3-nerved, green along nerves…**Gossypianthus**

30'. Leaves of basal rosette early-deciduous and not persisting at time of flowering; filament tube adnate to perianth, tepals connate proximally, with a single nerve, nerve not extending to tip of free lobe, not green along nerve…**Guilleminea**

ALLENROLFEA Kuntze – Pickleweed

Allenrolfea occidentalis (S. Watson) Kuntze [*Halostachys occidentalis* S. Watson]—Iodine bush, chamiso verde. Shrubs 3–15 dm, ± glaucous, generally with a blackish hue; stems erect or decumbent, much branched, woody proximally, fleshy distally; branches articulated into small joints (2–)3–5(–10) × 1–4.5 mm; leaves alternate, deciduous, reduced to scales, blade 2–4 × 2–3 mm; inflorescences terminal spikes 6–25 × 2.5–4 mm; flowers spirally arranged in axils of deciduous, peltate, fleshy bracts; stamens 1–2; stigmas 2(3), usually distinct; fruit a utricle enclosed by perianth. Alkaline flats, riparian areas, salt desert scrub communities, 3400–6900 ft. NC, C, EC, SC, SE.

ALTERNANTHERA Forssk. – Chaff-flower

Herbs (ours), annual or perennial; stems prostrate, decumbent, ascending, or erect, indumentum of simple trichomes; leaves opposite, blades lanceolate to ovate, ovate-rhombic, or obovate-rhombic, margins entire; inflorescences axillary or terminal, sessile or pedunculate, several-flowered cylindrical spikes or globose heads, without immediately subtending leaves; bracts and bracteoles scarious; flowers bisexual; tepals 5, distinct; stamens 3–5; filaments connate basally into tube or short cup; pseudostaminodes 5, alternating with stamens; ovule 1; style 1, ca. 0.2 mm; stigma capitate or rarely 2-lobed; utricles compressed, ovoid or obovoid, indehiscent.

1. Tepals 3–5 mm, densely villous, generally soft to the touch; leaf blades longer than broad; pseudostaminode margins usually entire…**A. caracasana**

1'. Tepals 5–7 mm, sparsely villous, generally very stiff to the touch; leaf blades usually as broad as long; pseudostaminode margins dentate…**A. pungens**

Alternanthera caracasana Kunth—Mat chaff-flower. Perennial herbs; stems prostrate to procumbent, villous to glabrate; leaves rhombic-ovate to obovate, 0.5–2.5 × 0.3–1.5 cm, apex rounded, apiculate, sparsely villous; inflorescences axillary, sessile; heads white to stramineous, ovoid, 0.5–0.8 × 0.4–0.6 cm; bracts shorter than tepals, apex long-attenuate, aristate; tepals whitish to stramineous, lanceolate, 3–5 mm, apex acuminate, spinose-tipped, densely villous; stamens 5; pseudostaminode margins usually entire, rarely dentate. Waste ground, open disturbed sites, lawns, 3100–5680 ft. SE, SC, SW. Introduced.

Alternanthera pungens Kunth [*A. repens* (L.) Link]—Khakiweed, khaki joyweed. Perennial herbs; stems prostrate, villous; leaves oval to obovate, 1.3–3 × 1.1–1.7 cm, apex rounded, pilose, glabrate; inflorescences axillary, sessile; heads stramineous, globose to ovoid, 0.6–1 × 0.6–0.7 cm; bracts equaling tepals, apex attenuate; tepals stramineous, lanceolate, 5–7 mm, apex acuminate, spinose-tipped, sparsely villous; stamens 5; pseudostaminode margins dentate. Waste ground, yucca grasslands, 3670–5800 ft. Doña Ana, Eddy, Grant.

AMARANTHUS L. – Amaranth, pigweed

Annual or perennial herbs, monoecious or dioecious; leaves alternate; inflorescences terminal and/or axillary, dichasia subtended by persistent bracts; bracteoles absent or 1–2; flowers unisexual; pistillate flowers with 1 pistil; ovule 1; style 0.1–1 mm or absent; stigmas 2–3(–5); staminate flowers with 3–5 stamens, filaments distinct, anthers 4-locular, pseudostaminodes absent; fruit a utricle loosely enclosed by inner tepals; seed 1, subglobose or lenticular, usually smooth, shiny.

1. Plants dioecious; inflorescences terminal spikes, thyrses, or panicles…(2)
1'. Plants monoecious; inflorescences terminal spikes and panicles or axillary glomerules or clusters…(9)
2. Plants with pistillate flowers…(3)
2'. Plants with staminate flowers (pistillate flowers are usually required for positive identification)…(6)
3. Floral bracts deltate or rhombic-deltate, leaflike, margins crenate or denticulate, completely enfolding flower; leaf blades linear or narrowly linear-lanceolate, margins crispate or erose, or irregularly undulate …**A. acanthochiton** (in part)
3'. Bracts ovate to narrowly lanceolate, not leaflike, margins entire, not enfolding flower; leaf blades variable in shape, margins entire to slightly undulate…(4)
4. Pistillate flowers with tepals absent or 1–2, usually < 2 mm; leaf blades variable, narrowly to broadly ovate, obovate, elliptic, usually > 1 cm wide…**A. tuberculatus** (in part)
4'. Pistillate flowers usually with 5 tepals, at least outer tepals > 2 mm; utricle dehiscence usually circumscissile…(5)
5. Floral bracts 4–6 mm, longer than tepals, outer tepals acuminate or acute-acuminate at apex…**A. palmeri** (in part)
5'. Floral bracts 1.5–3(–4) mm, shorter than or equaling tepals, outer tepals rounded at apex…**A. arenicola** (in part)
6. Leaf blades linear or linear-lanceolate, margins crispate or erose…**A. acanthochiton** (in part)
6'. Leaf blades variable in shape, margins entire or sometimes slightly undulate…(7)
7. Outer tepals with apex acute or obtuse, apiculate, dark midribs not excurrent…**A. arenicola** (in part)
7'. Outer tepals with apex acuminate, midribs excurrent as rigid spines…(8)
8. Bracts 2 mm, shorter than outer tepals, apex acuminate to short-subulate…**A. tuberculatus** (in part)
8'. Bracts 4 mm, equaling or exceeding outer tepals, apex usually long-subulate…**A. palmeri** (in part)
9. Inflorescences axillary clusters or glomerules, distal nodes sometimes condensed into leafy spikes…(10)
9'. Inflorescences terminal spikes and/or panicles, leafless or almost leafless at least in the distal part, axillary spikes or clusters usually also present…(16)
10. Pistillate flowers usually with only 1 well-developed tepal, sometimes with 1–3 distinctly unequal tepals; plants prostrate…**A. californicus**
10'. Pistillate flowers with 3–5 equal or subequal tepals, at least 2 tepals well developed; plants various…(11)
11. Tepals of pistillate flowers fan-shaped, margins fimbriate or denticulate; utricles dehiscent…**A. fimbriatus** (in part)
11'. Tepals of pistillate flowers spatulate or narrowly ovate to oblanceolate, lanceolate, or linear, margins entire to minutely erose; utricles indehiscent or dehiscent…(12)
12. Pistillate flowers usually with 3 tepals; fruits usually regularly dehiscent…**A. albus**
12'. Pistillate flowers usually with (4)5 tepals; fruits usually indehiscent or tardily dehiscent (regularly dehiscent in *A. blitoides* and *A. torreyi*)…(13)
13. Inflorescence axes thickened, becoming indurate at maturity; plants low, spreading, procumbent to decumbent…**A. crassipes**
13'. Inflorescence axes not thickened, not indurate at maturity; plants various…(14)
14. Utricles indehiscent or tardily dehiscent…**A. polygonoides**
14'. Utricles with dehiscence regularly circumscissile…(15)
15. Tepals narrowly ovate to broadly linear; leaf blades usually obovate to elliptic-spatulate…**A. blitoides**
15'. Tepals spatulate; leaf blades lanceolate, oblanceolate, or ovate-lanceolate…**A. torreyi** (in part)
16. Tepals of pistillate flowers fan-shaped to spatulate, base contracted into claw; terminal spikes unbranched or nearly so, usually interrupted, narrow and slender; leaf blades linear to ovate-lanceolate…(17)
16'. Tepals of pistillate flowers spatulate-obovate, oblanceolate, ovate-elliptic, or elliptic to lanceolate-linear, base never contracted into claw; terminal inflorescences variable, usually branched and ± dense; leaf blades usually rhombic-ovate to elliptic…(19)
17. Utricles indehiscent…**A. obcordatus**
17'. Utricles with dehiscence regularly circumscissile…(18)
18. Tepals of pistillate flowers fan-shaped, margins fimbriate or denticulate…**A. fimbriatus** (in part)
18'. Tepals of pistillate flowers spatulate, margins entire, rarely minutely erose…**A. torreyi** (in part)
19. Utricles indehiscent; tepals of pistillate flowers usually 2–3; inflorescence bracts shorter than tepals… **A. viridis**
19'. Utricles dehiscent; tepals of pistillate flowers usually 5 (or 3–5 on the same plant in *A. powellii*); inflorescence bracts exceeding tepals (sometimes equal to or shorter than tepals in *A. hypochondriacus*)…(20)
20. Fully developed inflorescences large and robust, usually brightly colored, red, purple, deep beet-red, occasionally white or yellowish, rarely green in some forms; bracts equal to or exceeding style

branches at maturity; seeds white, ivory, reddish, brown, or black; plants cultivated and rarely escaped ...**A. hypochondriacus**

20'. Inflorescences moderately large, usually green, occasionally silvery-green, sometimes with reddish tint; bracts in most species exceeding style branches and tepals, almost equal to tepals in some rare forms of *A. retroflexus*; seeds brown to black; plants wild, often weedy...(21)

21. Plants densely viscid-pubescent; inflorescences usually unbranched...**A. viscidulus**
21'. Plants not viscid (occasionally slightly viscid in some forms of *A. retroflexus*); inflorescences branched... (22)

22. Tepals of pistillate flowers obtuse, rounded, or emarginate at apex...(23)
22'. Tepals of pistillate flowers acute or acuminate to aristate at apex...(24)

23. Plants glabrous or nearly so; tepals of pistillate flowers 1.5-2 mm...**A. wrightii**
23'. Plants densely to moderately pubescent; tepals of pistillate flowers (2-)2.5-3.5(-4) mm...**A. retroflexus**

24. Bracts 2-4 mm; inflorescences variable, usually soft and lax, with spreading branches...**A. hybridus**
24'. Bracts 4-7 mm; inflorescences usually stiff, with erect branches...**A. powellii**

Amaranthus acanthochiton J. D. Sauer [*Acanthochiton wrightii* Torr.]—Greenstripe amaranth. Plants annual, glabrous or glabrescent; stems erect, much branched, 0.1-0.8 m; leaf blades narrowly linear-lanceolate to linear, 2-8 × 0.2-1.2(-1.7) cm, margins erose, crispate, or irregularly undulate; inflorescences of erect terminal spikes; bracts completely enfolding flower; seeds dark reddish brown to brown, 1-1.3 mm. Sandy areas, sand dunes, riverbanks, acacia scrubland, disturbed habitats, 3800-6750 ft. NW, C, SW, SC.

Amaranthus albus L.—White or tumbleweed amaranth, quelite. Plants annual, glabrous or glabrescent or viscid-pubescent, stems usually erect, rarely prostrate, much branched, bushy, 0.1-1 m; leaf blades obovate to narrowly spatulate, mostly 0.5 × 0.5-1.5 cm (early proximal leaves to 8 cm), margins entire, plane (or ± distinctly undulate); inflorescences axillary glomerules; seeds dark reddish brown to black, lenticular, 0.6-1 mm. Introduced in disturbed sites in vacant areas, riparian communities, roadsides, woodlands, agricultural fields, 3800-8700 ft. Widespread.

Amaranthus arenicola I. M. Johnst.—Sand amaranth. Plants annual, ± glabrous; stems erect, usually branched, 0.4-1.5(-2) m; leaf blades mostly narrowly ovate, obovate, elliptic, or lanceolate, 1.5-8 × 0.5-3 cm, thin and soft, margins entire, plane or irregularly undulate; inflorescences mostly terminal, spikes to panicles, erect to nodding, rarely with proximal axillary clusters; pistillate tepals with terminal mucro; seeds dark reddish brown, (0.9-)1-1.2 mm. Sandy habitats, sand hills, riverbanks, sand prairies, disturbed areas, agricultural fields, 2970-8760 ft. C, SW, SC, SE, Harding, Sandoval, Union.

Amaranthus blitoides S. Watson—Prostrate or mat amaranth. Plants annual, glabrous; stems prostrate or ascending, much branched from base, (0.1-)0.2-0.6(-1) m; leaf blades obovate, elliptic, or spatulate, 1-2(-4) × 0.5-1(-1.5) cm, margins usually entire, plane, rarely slightly undulate; inflorescences axillary glomerules; seeds black, lenticular to broadly plumply lenticular, 1.3-1.6 mm. Disturbed sites in woodlands, desert scrub, arroyo bottoms, fields, waste places, sandy flats, 3320-8900 ft. Widespread. Introduced.

Amaranthus californicus (Moq.) S. Watson—California amaranth. Plants annual, glabrous; stems prostrate, whitish or tinged with red, much branched from base, 0.1-0.5 m; leaf blades pale green, veins prominent, obovate, spatulate, or oblanceolate to linear, 0.3-2(-3) × 0.2-1.5 cm, margins entire, plane or slightly undulate; inflorescences axillary clusters; seeds very dark reddish brown, lenticular, (0.6-)0.7-1 mm. Seasonally moist flats, wash bottoms, 5400-6700 ft. S, Otero, Sandoval.

Amaranthus crassipes Schltdl.—Warnock's amaranth, clubfoot amaranth. Plants annual, glabrous; stems prostrate or weakly ascending, branched from base, 0.1-0.6 m; leaf blades broadly elliptic, obovate, orbiculate, or oblanceolate, (0.5-)1-3(-4.5) × 0.3-2(-2.5) cm, margins entire, plane to undulate; inflorescences axillary clusters; axes much thickened, appearing inflated, becoming indurate at maturity; bracts of pistillate flowers keeled; seeds dark brownish or reddish black to black, compressed-ovoid to broadly lenticular, 1-1.4 mm. Our plants belong to **var. warnockii** (I. M. Johnst.) Henrickson. Disturbed lowlands and hillsides, 3500-7700 ft. SW, SC, SE.

Amaranthus fimbriatus (Torr.) Benth. ex S. Watson—Fringed amaranth. Plants annual, glabrous; stems erect or with lateral branches ascending, branched from base, 0.3–0.7(–1) m; leaf blades linear to narrowly lanceolate, (1–)2–6(–10) × 0.1–0.5(–1) cm, margins entire, plane; inflorescences mostly axillary clusters distally condensed in lax, unbranched, almost leafless, slender, terminal spikes; seeds black or dark reddish brown, lenticular to broadly lenticular, 0.8–1 mm. Sandy, gravelly slopes, dunes, mesquite community, 3900–4400 ft. Doña Ana, Socorro.

Amaranthus hybridus L.—Smooth amaranth. Plants annual, glabrous or glabrescent; stems erect, green or sometimes reddish purple, 0.3–2(–2.5) m; leaf blades ovate, rhombic-ovate, or lanceolate, (2–)4–15 × (1–)2–6 cm, margins entire; inflorescences terminal and axillary, erect or reflexed, occasionally nodding, green or olive-green, occasionally with silvery or reddish-purple tint, leafless at least distally; seeds black to dark reddish brown, lenticular to lenticular-globose, 1–1.3 mm. Introduced in waste places, agricultural and fallow fields, disturbed woodlands, roadsides, riparian areas, 3800–10,600 ft. Widespread. Introduced.

Amaranthus hypochondriacus L.—Prince's feather. Plants annual, glabrous or moderately pubescent in distal parts; stems usually erect, green or reddish purple, branched, 0.4–2(–2.5) m; leaf blades rhombic-ovate to broadly lanceolate, 4–12 × 2–7 cm, larger in robust plants, margins entire; inflorescences predominantly terminal, often with few spikes at distal axils, stiff, erect, dark red, purple, or deep beet-red, less commonly yellowish or greenish, leafless at least in distal part; bracts subspinescent; seeds white, ivory, pinkish white, or black to dark reddish brown, subglobose to lenticular, 1–1.4 mm. Escape from cultivation. Doña Ana, Sandoval. Introduced.

Amaranthus obcordatus (A. Gray) Standl.—Trans-Pecos amaranth. Plants annual, glabrous; stems erect or ascending, branched, 0.1–0.5 m; leaf blades oblong-lanceolate to lanceolate-linear, 1–3 × 0.2–1 cm, margins entire, plane or sometimes undulate; inflorescences mostly axillary, but at apex flowers also condensed in terminal spikes or spicate panicles, usually leafy proximally or nearly leafless distally; seeds dark reddish brown to nearly black, lenticular or broadly lenticular, 0.6–0.8 mm. Semideserts, naturally disturbed habitats. Known from only 1 collection in New Mexico, from 1851. Sierra.

Amaranthus palmeri S. Watson—Palmer's amaranth, quelite. Plants annual, glabrous; stems erect, branched, (0.3–)0.5–1.5(–3) m; leaf blades obovate or rhombic-obovate to elliptic proximally, sometimes lanceolate distally, 1.5–7 × 1–3.5 cm, margins entire, plane; inflorescences terminal, linear spikes to panicles, usually drooping, occasionally erect, with few axillary clusters, uninterrupted or interrupted in proximal part of plant; bracts of pistillate flowers with long-excurrent midrib; seeds dark reddish brown to brown, 1–1.2 mm. Disturbed open sites in woodlands, grasslands, roadsides, floodplains, agricultural fields, 3570–8850 ft. Widespread.

Amaranthus polygonoides Zoll.—Tropical or smartweed amaranth. Plants annual, glabrescent proximally, pubescent distally, becoming glabrous at maturity; stems erect-ascending to prostrate, branched mostly at base, 0.1–0.5 m; leaf blades ovate or obovate-rhombic to narrowly ovate, sometimes lanceolate, 1.5–3(–4) × 0.5–1.5(–2) cm, margins entire to undulate-erose; inflorescences axillary, congested clusters; seeds dark reddish brown to black, lenticular, 0.8–1 mm diam. Hillsides, disturbed habitats, ca. 4395 ft. Known from only 1 collection in New Mexico. Luna.

Amaranthus powellii S. Watson—Powell's amaranth. Plants annual, glabrous or moderately pubescent toward inflorescences, becoming glabrescent at maturity; stems usually erect, stiff, green or sometimes reddish purple, branched to nearly simple, 0.3–1.5(–2) m; leaf blades rhombic-ovate to broadly lanceolate, 4–8 × 2–3 cm, margins entire; inflorescences mostly terminal, usually with spikes at distal axils, erect and rigid, green to silvery-green, occasionally tinged red, leafless at least distally; seeds black, subglobose to lenticular, 1–1.4 mm. Disturbed habitats, agricultural fields, roadsides, woodlands, waste areas, riparian zones, 3800–10,000 ft. Widespread.

Amaranthus retroflexus L.—Red-root pigweed. Plants annual, densely to moderately pubescent; stems erect, reddish near base, branched in distal part, 0.2–1.5(–2) m; leaf blades ovate to rhombic-ovate,

2–15 × 1–7 cm, margins entire, plane or slightly undulate; inflorescences terminal and axillary, erect or reflexed at tip, green or silvery-green, often with reddish or yellowish tint, branched, leafless at least distally, usually short and thick; bract apex acuminate with excurrent midrib; seeds black to dark reddish brown, lenticular to subglobose-lenticular, 1–1.3 mm. Introduced in riparian zones, desert grasslands, shortgrass prairies, openings in woodlands, agricultural fields, roadsides, waste areas, 3700–9400 ft. Widespread. Introduced.

Amaranthus torreyi (A. Gray) Benth. ex S. Watson—Torrey's amaranth. Plants annual, glabrescent to sparsely pubescent; stems erect or ascending proximally, much branched especially near base, 0.1–0.7 m; leaf blades oblanceolate or lanceolate to ovate-lanceolate, (1.2–)1.5–5(–7) × 0.3–2 cm, margins entire, plane or slightly undulate; inflorescences axillary clusters, toward apex aggregated in spikes; seeds black, subglobose to broadly lenticular, 1 mm. Sandy, rocky, and gravelly areas, slopes, arroyos, canyon bottoms, washes, 3900–7800 ft. NW, C, WC, SW, SC, Roosevelt.

Amaranthus tuberculatus (Moq.) J. D. Sauer [*Acnida tuberculata* Moq.]—Rough-fruit amaranth, tall water-hemp. Plants annual, glabrous; stems erect to sometimes ascending or rarely prostrate, branched, rarely simple, (0.5–)1–2(–3) m; leaf blades ovate or obovate proximally, oblong or elliptic to narrowly lanceolate distally, 1.5–15 × 0.5–3 cm, margins entire, plane; inflorescences terminal, linear spikes to panicles, occasionally interrupted-moniliform, remote, globose glomerules; seeds dark reddish brown to dark brown, 0.7–1 mm. Typically in disturbed moist areas, 3500–4000 ft. NW, C, WC, SW, SC, Roosevelt. Introduced.

Amaranthus viridis L.—Slender amaranth. Plants annual to short-lived perennial, glabrous; stems erect, simple or with lateral branches, 0.2–1 m; leaf blades rhombic-ovate or ovate, 1–7 × 0.5–5 cm, margins entire, plane; inflorescences slender spikes aggregated into elongate terminal panicles, also from distal axils, green, leafless at least distally; seeds black or dark brown, subglobose to thick-lenticular, 1 mm. Introduced in fields, lawns, gardens, waste areas, other disturbed habitats, 4200–7700 ft. WC, SW, Doña Ana, Otero, San Miguel.

Amaranthus viscidulus Greene—Sticky amaranth. Plants annual, densely viscid-pubescent (especially distal parts), becoming glabrescent proximally; stems erect or ascending, often whitish or tinged with red, branched to nearly simple, 0.2–1 m; leaf blades rhombic-ovate, ovate, obovate, or elliptic, 1–4.5 × 0.5–2.5 cm, usually somewhat fleshy, margins entire; inflorescences terminal, unbranched, stout spikes and axillary clusters, erect, usually greenish or reddish, leafless at least distally; bract apex spinescent; seeds black, lenticular to subglobose-lenticular, 1–1.2 mm. Open dry slopes, 6500–7000 ft. Lincoln, Rio Arriba, San Miguel.

Amaranthus wrightii S. Watson—Wright's amaranth. Plants annual, glabrous or nearly so; stems erect or ascending, often whitish or tinged with red, simple to sparingly branched distally, or occasionally basally, 0.2–1 m; leaf blades rhombic-ovate to elliptic-lanceolate, 1.5–6 × 0.5–3 cm, margins entire, plane to slightly undulate; inflorescences terminal and axillary in distal part of plant, erect, usually reddish green, branched, leafless at least distally, short and thick; bract apex spinescent; seeds dark reddish brown to nearly black, lenticular to subglobose-lenticular, 1 mm. Disturbed habitats in woodlands, stream banks, canyons, semideserts, 3800–9000 ft. NW, C, WC, SW, SC.

ATRIPLEX L. – Saltbush

Herbs or shrubs, annual or perennial, monoecious or dioecious; leaves persistent or tardily deciduous, alternate to opposite; blades entire, serrate, or lobed, with venation either of Kranz type or normal dicotyledonous type; inflorescences axillary or terminal; flowers borne in axillary clusters or glomerules, or in terminal spikes or spicate panicles; staminate flowers with 3–5-parted calyx, ebracteate; stamens 3–5; pistillate flowers lacking perianth, or in few species with (1–)3–5-lobed perianth, commonly enclosed within pair of bracteoles; fruiting bracteoles enlarged in fruit, fruit tightly enclosed in fruiting bracteoles; seeds flattened.

1.   Plants perennial shrubs…(2)
1'.   Plants annual or perennial herbs…(9)
2.   Leaves sinuate to sinuate-dentate to hastate, or strongly undulate-crisped and appearing lobed; herbage silvery white…**A. acanthocarpa**
2'.   Leaves entire or merely hastately lobed, or if sinuate-dentate or denticulate, herbage green or gray…(3)
3.   Bracteoles conspicuously longitudinally 4-winged, or with tubercles aligned in 4 parallel rows…(4)
3'.   Bracteoles lacking lateral wings (tubercles sometimes aligned in 4 rows)…(5)
4.   Leaves mainly 0.3–0.8 cm wide; bract tip lacking lateral teeth; shrubs mainly 8–20 dm; widespread in New Mexico…**A. canescens**
4'.   Leaves often > 8 mm wide; bract tip with or without lateral teeth; shrubs mainly 2–8 dm; NW New Mexico…**A. garrettii**
5.   Plants definitely spiny, branches terminating in thorns; bracteoles foliose, entire, united only at base, surfaces lacking appendages, shrub to 8 dm…**A. confertifolia**
5'.   Plants not definitely spiny, or if somewhat so bracteoles at least 1/3 united, surfaces appendaged or not; shrubs to 20 dm…(6)
6.   Leaf blades typically some subhastate; shrubs to 10 dm; branchlets conspicuously and sharply angled; bootheel region, SW New Mexico…**A. griffithsii**
6'.   Leaf blades attenuate to rounded basally, seldom some subhastate; shrubs mainly < 20 dm; not restricted to SW New Mexico…(7)
7.   Leaves 2–4+ mm wide; bracteoles with appendages on basal 1/3; staminate flowers in spikes; plants prostrate…**A. corrugata**
7'.   Leaves often > 4 mm wide; bracteoles with appendages various; staminate flowers mainly in panicles; plants not or seldom prostrate; distribution various or other…(8)
8.   Leaves oblong-ovate to orbicular, > 1 cm wide, proximalmost alternate; stems stiffly erect; staminate glomerules very numerous…**A. obovata**
8'.   Leaves linear to oblong, mainly < 1 cm wide, or if wider, proximal opposite; stems prostrate to ascending, or less commonly erect; staminate glomerules numerous…**A. cuneata**
9.   Leaves usually green on both surfaces, glabrous or sparingly powdery or finely scurfy…(10)
9'.   Leaves white to gray, densely and finely scurfy, especially adaxially…(14)
10.   Pistillate flowers of 2 kinds: some with calyx 3–5-lobed and seed horizontal, others lacking perianth, enclosed in a pair of bracteoles, seed vertical; fruiting bracteoles samaralike, strongly compressed, oval to orbicular or ovate…**A. hortensis**
10'.   Pistillate flowers all alike or, if dimorphic, both kinds lacking perianth, enclosed within bracteoles, and seed vertical; fruiting bracteoles variously compressed (orbicular in *A. micrantha*)…(11)
11.   Bracteoles ± thickened with spongy tissue, especially toward base…(12)
11'.   Bracteoles not thickened…(13)
12.   Lower leaves ovate-lanceolate or linear, or triangular or triangular-hastate, typically thickened and ± scurfy, even in age; seeds ellipsoid, wider than long…**A. dioica** (in part)
12'.   Lower leaves (sometimes all or most) triangular and thin textured; seeds usually distinctly dimorphic, mostly small and glossy black, but also some larger, and dull brown…**A. prostrata**
13.   Bracteoles ovate to widely triangular, surfaces often 2-tuberculate, margin toothed; leaves usually thickened, ± scurfy, even at maturity…**A. dioica** (in part)
13'.   Bracteoles orbiculate-ovate, surfaces smooth, margin entire; leaves usually thin, green on both sides, not or scarcely scurfy…**A. micrantha**
14.   Plants perennial, monoecious…(15)
14'.   Plants annual, dioecious…(16)
15.   Seeds dimorphic: black, 1.5–1.7 mm, or brown, 2 mm; bracteoles fleshy; plants low-growing, many-stemmed; leaf margin irregularly dentate…**A. semibaccata**
15'.   Seeds monomorphic, either black or brown; bracts not fleshy; plants not low-growing; leaf margin entire to dentate…**A. argentea** (in part)
16.   Leaf blades all or most of them dentate or sinuate-dentate, leaves alternate…**A. rosea**
16'.   Leaf blades not all dentate, some, or all, of them entire, leaves commonly opposite or subopposite at proximal nodes, alternate or opposite…(17)
17.   Bracteoles on stipes 2–6 mm, body 5–6 mm thick, globose, with hornlike appendages on both faces…**A. saccaria** (in part)
17'.   Bracteoles variously sessile or, if stipitate, not otherwise as above…(18)
18.   Bracteoles samaralike, orbicular, often > 10 mm, margin 2–4 times as wide as body; staminate flowers in deciduous terminal panicles…**A. graciliflora**
18'.   Bracteoles commonly < 10 mm, margin usually little if at all wider than body; staminate flowers in panicles or some or all in axillary glomerules…(19)
19.   Fruiting bracteoles orbicular, finely and regularly radiately dentate to base, strongly compressed…**A. elegans**

19′. Fruiting bracteoles not orbicular, or if so, never radiately dentate to base and not at once strongly compressed…(20)
20. Bracteoles ovate-oblong, broadest at or near base, fused to near summit, tridentate apically, occasionally with marginal or less commonly with facial appendages; leaves small, sessile, blade typically entire, ovate to linear…**A. powellii** (in part)
20′. Bracteoles broadest near or above middle, usually rounded or truncate at apex, or if not so, leaves dentate; leaves usually large and often some petiolate…(21)
21. Leaf blades, most of them, lance-ovate to deltoid, broadest at or near base, margins entire…**A. argentea** (in part)
21′. Leaves linear to lanceolate or oblong, margins sinuate-dentate to less commonly entire…**A. wrightii**

Atriplex acanthocarpa (Torr.) S. Watson—Tubercled saltbush, burscale. Shrubs or subshrubs, unarmed, dioecious, evergreen, 2–10 × 4–10+ dm; leaves opposite, becoming alternate distally, blades oblong to oblong-lanceolate, ovate, obovate, or spatulate, 12–40(–50) × 5–25 mm, margins entire or sinuate-dentate to undulate-crisped; staminate flowers in interrupted or crowded glomerules in paniculate spikes; pistillate flowers few to solitary, in axillary clusters or in crowded or interrupted, spicate racemes or racemose panicles; fruiting bracteoles with flattened to hornlike tubercles to 8 mm; seeds brown, 1.5–2 mm. Our plants belong to **var. acanthocarpa**. Alkaline soils in desert scrub, playas, mesquite woodlands, 3800–6600 ft. SW, Catron.

Atriplex argentea Nutt.—Silver-scale saltbush. Perennial herbs, simple to branched, 0.5–6 dm; leaves alternate to opposite proximally, blades lance-ovate, lanceolate, deltoid, or cordate, 5–75 × 4–50(–75) mm, margins entire to closely repand-dentate; flowers in axillary glomerules and terminal, interrupted spikes; fruiting bracteole faces smooth, tuberculate, or crested, processes sometimes again toothed, teeth then aligned with axis of process; seeds brown, 1.5–2 mm. Five varieties, three of which occur in the flora. NW, NC, C, WC, SC, Chaves, Quay.

1. Leaf blades elliptic to oval, attenuate to a cuneate base; NW New Mexico…**var. rydbergii**
1′. Leaf blades triangular-ovate to oval, base broadly obtuse to acute or less commonly cuneate…(2)
2. Distal leaves short-petiolate, proximal leaves alternate; plants mostly < 4 dm; widespread in New Mexico…**var. argentea**
2′. Distal leaves sessile, proximal leaves opposite; plants mostly > 4 dm; C to S New Mexico…**var. mohavensis**

var. argentea—Stems erect, 3–4 dm; leaves proximally alternate; fruiting bracteoles smooth or sparsely to densely tuberculate or cristate. Alkaline sandy to clay soils in woodlands, salt desert scrublands, riparian zones, and roadsides, 4400–7000 ft.

var. mohavensis (M. E. Jones) S. L. Welsh—Stems erect or decumbent, much branched, 3–12(–20) dm, forming clumps 3–10(–30) dm broad; fruiting bracteoles with few irregular, green projections or crests, or unappendaged. Alkaline flats, sandy soils of floodplains, 3800–4900 ft.

var. rydbergii (Standl.) S. L. Welsh—Stems ascending, erect, (0.2–)1–5 dm; fruiting bracteoles shallowly or deeply and coarsely dentate, sides smooth or with 1 to few thickened processes, and these sometimes again appendaged. Saline substrates derived mainly from Mancos Shale and Morrison Formations, alluvium, salt desert scrub, floodplain communities, ca. 6100 ft.

Atriplex canescens (Pursh) Nutt.—Four-wing saltbush. Shrubs, dioecious or rarely monoecious, 8–20 dm, as wide or wider; leaves persistent, alternate, blades linear to oblanceolate, oblong, or obovate, mainly 10–40 × 3–8 mm; margins entire; staminate flowers yellow to brown in clusters 2–3 mm wide, borne in panicles 3–15 cm; pistillate flowers borne in panicles 5–40 cm; fruiting bracteoles with 4 prominent, dentate to entire wings extending length of bract, united throughout, mainly 8–12 mm wide and about as long; seeds 1.5–2.5 mm. Our plants belong to **var. canescens.** Sandy or gravelly sites, commonly nonsaline, sometimes in saline soils, wide range of community types, 3000–7700 ft. Widespread.

Atriplex confertifolia (Torr. & Frém.) S. Watson—Shadscale. Shrubs, spinescent, dioecious, 3–8 dm; leaves persistent, alternate, blades orbiculate to ovate, elliptic, or oval, 9–25(–45) × 4–20(–25) mm, margins entire; flowers in clusters 2–4 mm wide or in spikes to 1 cm, axillary, in branched panicles; fruiting bracteole faces smooth, lacking appendages; seeds 1.5–2 mm. Gravelly to fine-textured alkaline soils, salt desert scrub, sagebrush, piñon-juniper, grassland, 5100–7450 ft. NW, NC, C, WC, SW, Union.

Atriplex corrugata S. Watson—Mat saltbush. Shrubs dioecious or rarely monoecious, low-spreading, 0.3–1.5 × 3–15 dm; leaves persistent, opposite proximally, alternate distally, blades linear to linear-oblanceolate, or oblong, 3–18 × 2–6 mm, margins entire, staminate flowers borne in spikes 1–8 cm; pistillate flowers in leafy bracteate spikes 5–15 cm; fruiting bracteole margins entire or undulate, densely tuberculate to smooth; seeds brown, 1.5 mm. Alkaline soils, usually fine-textured, derived from various shale formations, vegetative badlands, salt desert scrub, 4600–6200 ft. NW.

Atriplex cuneata A. Nelson [*A. gardneri* var. *cuneata* (A. Nelson) S. L. Welsh]—Castle Valley saltbush. Subshrubs, dioecious or sparingly monoecious, 1–4.5 × 4–20 dm; stems decumbent; leaves opposite to subalternate proximally, alternate distally, blades grayish green, obovate to spatulate, orbiculate or oblong, 10–50 × (5–)10–25 mm, margins entire; staminate flowers in glomerules 3–4 mm thick, borne in terminal panicles 2–11 cm; pistillate flowers in leafy unbranched spikes or spicate panicles 5–23 cm; fruiting bracteole face densely tuberculate to smooth and free of tubercles; seeds 2–2.5 mm. Our plants belong to **var. cuneata** (A. Nelson) S. L. Welsh. Saline, fine-textured substrates on Mancos Shale and other formations of similar texture and salinity, salt desert scrub, 4900–6800 ft. NW.

Atriplex dioica Raf.—Thickleaf orach. Perennial herb, monoecious, erect and branching, 3–15 dm; stems green or striped; leaves opposite or subopposite at least proximally, blades thickened, strongly 3-veined from near base, lanceolate to linear-lanceolate or often triangular-ovate or lance-ovate, 30–125 × 25–60(–80) mm, basal lobes spreading to mainly antrorse, margins entire to sparingly dentate; flowers terminal on lateral branches, in spiciform spikes, 2–9 cm; fruiting bracteoles green, blackening in age, faces smooth or with 2 tubercles; seeds ellipsoid, wider than long, dimorphic: brown, 1.5–3 mm, or black, 1–2 mm. Alkaline soils in riparian communities and swales, 5280–7600 ft. NW, C, Socorro.

Atriplex elegans (Moq.) D. Dietr.—White-scale saltbush. Herbs, annual to perennial, monoecious; stems procumbent to erect, stramineous or whitish, simple to much branched at base, 0.5–4.5 dm; leaf blades elliptic to spatulate, oblanceolate, oblong, or obovate, 5–30(–35) × 2–8(–12) mm, margins entire or irregularly dentate, densely scurfy abaxially, green and glabrate adaxially; fruiting bracteole margins dentate to incised, face smooth; seeds brown, 1–1.5 mm. Our plants belong to **var. elegans.** Fine-textured saline soils, disturbed sites, roadsides, 3000–5400 ft. SW, SC, Lea.

Atriplex garrettii Rydb.—Garrett's saltbush. Shrubs or subshrubs, dioecious (rarely monoecious), 2–8(–10) × 5–10+ dm; leaves opposite or subopposite proximally, or some alternate distally, blades yellow-green, ovate to obovate, elliptic, or orbiculate, 8–55 × 6–32 mm, margins entire or repand-dentate, sparingly scurfy; staminate flowers in clusters 2–4 mm wide, on panicles 2–10(–15) cm; pistillate flowers in spikes or spicate panicles 4–30 cm; fruiting bracteole surface smooth, reticulate, or with flattened processes, wings 3–4 mm wide; seeds brown, 2 mm. Fine-textured saline soils, talus slopes, 4800–5600 ft. San Juan.

Atriplex graciliflora M. E. Jones—Slender-flower saltbush. Annual herbs, monoecious, branching from base, 1–3 dm; stems often red-purple, sparingly farinose when young; leaves mainly alternate, blades cordate-ovate to deltoid, (5–)8–20(–25) mm; staminate flowers in loose, deciduous, terminal panicles overtopping foliage, rachis and branches filiform, glomerules beadlike in alternate position along rachis; fruiting bracteoles winged, wings undulate or entire, surfaces smooth; seeds white, 3 mm. Edge of water bodies in badland communities on saline, often salt-encrusted and semibarren substrates derived from Mancos Shale, Entrada, and other fine-textured formations, ca. 7000 ft. Known from 1 collection in New Mexico. Sandoval.

Atriplex griffithsii Standl. [*A. torreyi* (S. Watson) S. Watson) var. *griffithsii* (Standl.) G. D. Br.]—Griffith's saltbush. Shrubs, dioecious, forming broad clumps, 3–10 dm; branchlets slender; leaves persistent, alternate, blades grayish or greenish, ovate to deltate, rhombic, oval, or lanceolate, typically subhastate, (5–)12–32 × 4–16(–20) mm, margins entire (rarely toothed); staminate flowers in clusters 1 mm wide, borne in panicles 10–30 cm; pistillate flowers borne in panicles 5–20 cm; fruiting bracteole margins crenate, apex rounded; seeds brown, 1.4–1.9 mm. Saline playa margins, 4150–4320 ft. Hidalgo. **R**

Atriplex hortensis L.—Garden orach. Annual herbs, glabrous, green to yellowish or reddish, 5–15 (–25) dm; stems erect, mostly branched; leaves opposite or alternate, blades green, ovate or ovate-lanceolate to cordate-hastate at base, 15–180 × 8–135 mm, margins entire or more rarely irregularly toothed or lobed, inflorescences of spikes disposed in leafless panicles; fruiting bracteoles samaralike, orbicular to oval or ovate, faces smooth; seeds black, 1–2 mm, or olivaceous brown, 3–4.5 mm. Cultivated and occasionally escaped. San Juan, Taos. Introduced.

Atriplex micrantha Ledeb. [*A. heterosperma* Bunge]—Russian atriplex, two-seed orach. Annual herbs, monoecious, erect, branching from base, 5–15 dm; leaves alternate except proximal-most, blades green, triangular to lance-triangular, 30–120 × 12–90 mm, hastate or subcuneate, margins subentire or irregularly dentate; flowers in terminal or axillary pyramidal panicles 6–25+ cm; fruiting bracteole margins entire, surfaces smooth; seeds dimorphic, yellowish brown or black, 1.5–3 mm. Riparian habitats, 5300–6800 ft. Mora, Sandoval, San Juan, Socorro. Introduced.

Atriplex obovata Moq. [*A. greggii* S. Watson]—Ovate-leaf or New Mexico saltbush. Subshrubs, dioecious, clump-forming, 2–8 dm; stems stiffly erect; leaves tardily deciduous, alternate or proximal-most subopposite, blades gray-green, oblong-ovate to elliptic or orbiculate, 8–30(–35) × 6–20 mm, margins entire or rarely dentate; staminate flowers in clusters 2–3 mm wide, in panicles 6–30 cm; pistillate flowers in small, numerous glomerules in axils of elongate, terminal, leafy-bracteate spikes; fruiting bracteole margins sharply toothed, apical tooth subtended by 2–6 equal or smaller teeth, faces smooth or rarely tuberculate; seeds brown, 2.4–2.8 mm. Fine-textured, generally alkaline substrates, salt desert scrub, desert grassland, piñon-juniper communities, floodplains, 3800–6900 ft. NW, C, NC, SW.

Atriplex powellii S. Watson—Powell's orach. Annual herbs, dioecious or monoecious; stems slender to stout, 1–5(–7) dm, branching throughout; herbage pubescent with scurfy and arachnoid hairs; leaves alternate, blades 3-veined, ovate to rhombic or orbiculate to elliptic, 0.4–5 × 0.2–3 cm, margins entire, fruiting bracteole surfaces with thickened processes or smooth; seeds greenish, yellowish, or brown, 0.9–2 mm. NW, NC.

1. Leaves mostly conspicuously 3-veined; plants mostly dioecious; fruiting bracteoles easily discerned at bract bases...**var. powellii**
1'. Leaves inconspicuously, if at all, 3-veined; plants evidently monoecious; fruiting bracteoles apparent only by removing the subtending bract...**var. minuticarpa**

var. minuticarpa (Stutz & G. L. Chu) S. L. Welsh—Smallbract orach. Fine-textured saline and seleniferous silts and clays, ca. 6700 ft.

var. powellii—Fine-textured saline and seleniferous silts and clays, riparian zones, 5400–6400 ft.

Atriplex prostrata Boucher ex DC.—Thinleaf orach. Annual herbs, monoecious, erect, decumbent or procumbent, branching, 1–10 dm; stems green or striped; leaves opposite or subopposite, blades triangular-hastate, lobes spreading, 20–100 mm, margins entire, serrate, dentate, or irregularly toothed; flowers in spiciform naked spikes 2–9 cm, sometimes forming terminal panicles; glomerules tight, contiguous or irregularly spaced; fruiting bracteoles green, becoming brown to black at maturity, faces smooth or with 2 tubercles; seeds dimorphic: brown, 1–2.5 mm, or black, 1–1.5 mm. Introduced in riparian communities, gallery bosque, particularly on saline soils, 4000–5700 ft. NW, WC, SE, San Miguel.

Atriplex rosea L.—Tumbling orach. Annual herbs, erect, 1–10(–20) dm; stems simple or divaricately branching throughout; herbage whitish scurfy to glabrate; leaves alternate, blades prominently 3-veined, ovate to lanceolate, 12–80 × 6–50 mm, margins irregularly sinuate-dentate and often subhastately lobed or rarely some entire; flowers in axillary glomerules or interrupted terminal spikes; fruiting bracteoles conspicuously dentate, sharply tuberculate to almost smooth on faces; seeds dimorphic: brown, 2–2.5 mm, or black, 1–2 mm. Riparian communities, shale badlands, 4000–7800 ft. NW, WC, SC. Introduced.

Atriplex saccaria S. Watson—Stalked orach, medusa-head orach. Annual herbs, erect, forming rounded clumps, 0.5–4(–5) dm; stems usually branched from base, herbage scurfy; leaves alternate to subopposite, blades cordate-ovate or subreniform to ovate to deltoid-ovate or oval, 6–40 × 4–30 mm, margins entire or subhastately lobed or sometimes undulate-dentate, staminate glomerules in distal axils or in short, naked, terminal (early-deciduous) panicles; pistillate flowers in fascicles of 1–3 in proximal axils; fruiting bracteoles irregularly dentate with flat, cristate, or hornlike appendages or smooth; seeds brownish to whitish, 1.5–2.3 mm. Two co-occurring varieties are found in the flora: var. **saccaria**, with leaves ovate to cordate and fruiting bracteoles with stipes to 10 mm but seldom exceeding 6 mm; and **var. cornuta** (M. E. Jones) S. L. Welsh, with leaves rhombic to oval or triangular and fruiting bracteoles with flattened processes, borne on stipes mainly 2–15 mm. Salt desert scrub communities, washes, piñon-juniper woodlands, roadsides, 3700–7250 ft. NW, WC.

Atriplex semibaccata R. Br.—Australian, creeping, or berry saltbush. Perennial, monoecious, herbs or subshrubs, 0.5–8 dm, spreading to 15+ dm wide, stems decumbent-prostrate, white-scurfy when young; leaves alternate, blades 1-veined, spatulate or obovate to oblong or elliptic, 5–30(–40) × 2–9(–12) mm, margins remotely dentate to subentire; staminate flowers in small, terminal, leafy-bracteate glomerules 1.5 mm wide; pistillate flowers solitary or in few-flowered clusters; fruiting bracteoles red-fleshy at maturity, margins toothed; seeds dimorphic: black, 1.5–1.7 mm, or brown, 2 mm. Saline waste places, roads, floodplains, disturbed communities, 3800–5500 ft. SW, SC, Santa Fe. Introduced.

Atriplex wrightii S. Watson—Wright's saltbush. Annual herbs, stems erect and ascending, sparsely branched or simple, 1.5–10(–15) dm, scurfy when young; leaf blades white abaxially, green adaxially, linear to lanceolate, elliptic, or oblong, 15–75 × (1–)3–25 mm, thin, margins coarsely sinuate-dentate or entire; staminate flowers in glomerules, forming slender, usually dense, naked, terminal, narrowly paniculate spikes; pistillate flowers in few-flowered axillary clusters; fruiting bracteole faces 3-veined, usually unappendaged, rarely obscurely tuberculate; seeds pale brown, 1 mm. Disturbed sandy soils, roadsides, old fields, vacant lots, 3700–4750 ft. SW.

BASSIA All. – Smother-weed, summer cypress

Annual herbs, densely pubescent; stems erect to prostrate, branched or simple; leaves alternate, sessile (or sometimes narrowed to pseudopetiole); blades linear, lanceolate, or lanceolate-elliptic, flat or semiterete (semicylindrical in transverse section, ± fleshy), bases cuneate, margins entire, apex obtuse to acute; inflorescences terminal spikes, flowers (1)2–3 in axils; flowers bisexual, sessile, ebracteolate; perianth segments 5, with spiniform, hooked, or conic appendages; stamens 5; styles and stigmas 2(3); fruiting bracts absent; utricles ovate-compressed; seeds lenticular, brownish, smooth (Kadereit & Freitag, 2011).

1. Perianth segments at maturity developing a spine (each of the 5 perianth members with an apically hooked or coiled spine); utricle with a narrow transverse wing...**B. hyssopifolia**
1'. Perianth segments at maturity horizontally winged (each of the 5 perianth members with a fanlike segment); utricle not winged...**B. scoparia**

Bassia hyssopifolia (Pall.) Kuntze [*Salsola hyssopifolia* Pall.]—Aromatic smother-weed. Plants 5–100 cm; stems divaricately branched or simple; inflorescences with ± straight axes; perianth segments with thin, hooked spine adaxially at maturity. Alkaline playas, saltgrass and riparian communities, degraded farmland, 3400–8200 ft. Scattered. Introduced.

Bassia scoparia (L.) A. J. Scott [*B. sieversiana* (Pall.) W. A. Weber; *Kochia scoparia* (L.) Schrad.]— Mexican fireweed, summer cypress. Plants to 1 m; stems erect, bushy-branched, often hairy above; leaves stalkless, 2–5 × 3–7 mm, linear to narrowly lance-shaped with a rounded or slightly pointed tip, ± hairy; inflorescences 5–10 cm, hairy; utricles enclosed in the persistent, incurved sepals, with a short, horizontal wing on the outer side of each sepal. Roadsides, waste ground, disturbed areas, 3200–8200 ft. Widespread. Introduced.

BLITUM L. – Poverty-weed

Annual or perennial, nonaromatic herbs, glabrous or sometimes with stipitate vesicular hairs and sticky when young; stems erect to prostrate, several from the base; leaves alternate, petiolate, basal ones often long-petiolate; blades thin or thickish and ± succulent, triangular to triangular-hastate or triangular-lanceolate, or spatulate; margins entire to dentate; inflorescences of spicately arranged glomerules, ebracteate or axillary; flowers bisexual or pistillate; perianth segments 0–5, connate only at base or close to the middle, not keeled; stamens 1–5; stigmas 2–4; fruit a utricle, seeds vertical, broadly ovate to orbicular, margins slightly acute to rounded or truncate, dark brown to black (Fuentes-Bazán et al., 2012).

1. Perianth lobe 1; fruits largely exposed; stamens 0, 1, or 2…**B. nuttallianum**
1'. Perianth lobes 3–4, largely enfolding and concealing to exposing fruits; stamens 1–3…(2)
2. Glomerules not subtended by leaflike bracts in distal 1/2 of spike; flowers maturing uniformly from apex of plant to base…**B. capitatum**
2'. Glomerules subtended by leaflike bracts throughout the inflorescence; flowers maturing from base of plant to apex…**B. virgatum**

Blitum capitatum L. [*Chenopodium capitatum* (L.) Ambrosi]—Strawberry-spinach. Stems erect to ascending or decumbent, 1.5–10 dm; leaf blades lanceolate, ovate, triangular, or triangular-hastate, 2.5–10 × 1–9 cm; spikes 5–20 cm; glomerules globose, 3–10 mm diam.; perianth segments 3, connate only at base, often becoming fleshy and red in fruit. Widespread except E plains.

1. Flower clusters often > 6(–12) mm wide, the calyx becoming red and fleshy at maturity…**subsp. capitatum**
1'. Flower clusters commonly < 6(–8) mm wide, the calyx not fleshy, though sometimes reddish at maturity… **subsp. hastatum**

subsp. capitatum—Clearings in mixed conifer communities, alpine meadows, streamsides, 7200–11,200 ft.

subsp. hastatum (Rydb.) Mosyakin [*Chenopodium capitatum* var. *parvicapitatum* S. L. Welsh]—Clearings in mixed conifer and aspen communities, montane grasslands, 7180–10,500 ft.

Blitum nuttallianum Schult. [*Monolepis nuttalliana* (Schult.) Greene]—Nuttall's poverty-weed. Stems prostrate to ascending, 0.5–2(–5) dm; principal leaf blades hastately lobed near base, narrowly triangular, lanceolate, narrowly elliptic, or linear, 1–3(–4) cm × 2–15(–25) mm at lobes; perianth segment 1, utricles 1.1–1.5 mm. Open sites in blackbrush, shadscale, sagebrush, montane forest communities, 3980–10,800 ft. NW, NC, C, WC, SW.

Blitum virgatum L. [*Chenopodium foliosum* (Moench) Asch.]—Leafy goosefoot. Stems erect to ascending, branched, 1.4–6(–8) dm; leaf blades narrowly triangular or oblong-triangular to almost deltate, 1.7–7.5(–9) × 0.8–3.5 cm, glomerules globose, 3–8 mm diam.; bracts leaflike throughout inflorescence; flowers maturing from base to apex; perianth segments 3(4), becoming red, enlarged, and fleshy in fruit. Open areas in mixed conifer and aspen communities, subalpine, 7400–11,200 ft. NW, WC, Lincoln, Sierra. Introduced.

CHENOPODIASTRUM S. Fuentes, Uotila & Borsch – Mock goosefoot

Annual, nonaromatic herbs, young stems and leaves glabrescent, with vesicular trichomes; stems erect, branched; leaves alternate, petiolate; blades thickish, triangular, ovate, or rhombic-ovate to lanceolate; margins irregularly dentate to lobed, or pinnatifid; inflorescences axillary and terminal, leafy to leafless, with flowers in small dense glomerules; flowers bisexual or pistillate; perianth segments 5, basally connate, with strong midrib enclosing or spreading in fruit; stamens 5; stigmas 2; fruit a utricle; seeds lenticular, round in outline, margins acute to fairly obtuse, black (Fuentes-Bazán et al., 2012).

1. Leaf blades usually 1.5–5.5 × 0.8–3 cm, margins irregularly dentate, teeth usually shallow; perianth segments distinct nearly to the base; fruits 1–1.5 mm across…**C. murale**
1'. Leaf blades usually 3.5–15 × 2–9 cm, margins with 2–5 prominent teeth per side; perianth segments connate basally; fruits 1.3–2 mm across…**C. simplex**

Chenopodiastrum murale (L.) S. Fuentes, Uotila & Borsch [*Chenopodium murale* L.]—Nettle-leaf mock goosefoot. Stems erect, branched, 1–6(–10) dm; proximal branches decumbent; leaf blades triangular, ovate, or rhombic-ovate, 0.8–4(–8) × 0.4–3(–5) cm, glomerules in terminal and lateral panicles, 6–7 × 4–5 cm; glomerules subglobose, 2–4 mm diam., or some flowers not in glomerules. Roadsides and open areas, 6100–8800 ft. WC, SW, Santa Fe. Introduced.

Chenopodiastrum simplex (Torr.) S. Fuentes, Uotila & Borsch [*Chenopodium simplex* (Torr.) Raf.]— Maple-leaf goosefoot, giant-seed mock goosefoot. Stems erect, branched, 3–15 dm; leaf blades ovate to triangular, 3.5–15 × 2–9 cm, margins sinuate-dentate with 1–5 coarse, acute teeth per side; glomerules in terminal and lateral spikes and panicles, 6–15 cm; glomerules irregularly globose, flowers in different stages of development, 0.5–2 mm diam. Riparian woodlands and riverbanks, 5100–9800 ft. NC, NE, SC, Eddy, Torrance.

CHENOPODIUM L. – Goosefoot, lamb's-quarters, pigweed

Annual or perennial, nonaromatic (but sometimes fetid) herbs (ours); young stems and leaves often densely farinose, monoecious or (rarely) dioecious; stems erect to prostrate, branched, branches alternate or the lowermost ones sometimes subopposite; leaves alternate, petiolate; sometimes fleshy, linear to trullate, rhombic, or triangular-hastate; margins entire to dentate or lobed; inflorescences terminal and lateral, ebracteate or with bractlike leaves, with flowers in compact or loose glomerules; flowers in monoecious plants dimorphic, bisexual or pistillate; segments (4)5, connate near the base or close to the middle; stamens 5; stigmas 2; fruit a utricle, seeds depressed-globular to lenticular, margins rounded to subacute, black, almost smooth to finely striate, rugulose or variously pitted (Fuentes-Bazán et al., 2012; Benet-Pierce & Simpson, 2014, 2017).

1. Primary leaves linear, linear-lanceolate, or occasionally narrowly oblong-ovate, 2–3 times longer than broad or longer; usually without teeth or lobes or occasionally with a pair of basal lobes…(2)
1'. Leaves ovate, rhombic, triangular, or lanceolate, to 2 times longer than broad; usually with basal lobes and often with additional teeth on margin…(9)
2. Leaves with 1 vein, blades linear, usually somewhat fleshy, margins entire…(3)
2'. Leaves with 3 veins from base, blades linear, lanceolate, or occasionally narrowly oblong-ovate or triangular-rhombic, usually without teeth or lobes or occasionally with a pair of basal lobes…(5)
3. Sepals connate to or above widest part of seed; seeds 1.3–1.5 mm diam., margins acute…**C. cycloides**
3'. Sepals connate only at very base; seeds 0.8–1.5 mm diam., margins rounded…(4)
4. Seeds 1.4–1.6 mm diam.; well-grown plants > 4 dm…**C. pallescens**
4'. Seeds 0.9–11.1 mm diam.; plants chiefly < 4 dm…**C. leptophyllum**
5. Leaves 2–3 times longer than broad or longer, oblong or oblong-lanceolate…(6)
5'. Leaves 2–3 times longer than broad, usually narrowly oblong-ovate, broadly oblong, or deltoid-rhombic …**C. atrovirens**
6. Fruit with pericarp adherent, minutely granular-roughened…**C. hians**
6'. Fruit with pericarp separable…(7)
7. Plants spreading or erect; leaf blades usually unlobed…(8)
7'. Plants strictly erect; primary leaves usually with basal lobes…**C. pratericola**
8. Plants usually spreading; seeds flat on both sides, seed coat finely tessellate and very shiny…**C. nitens**
8'. Plants usually erect and branching early; seeds conic on 1 side, seed coat not tessellate or very shiny… **C. desiccatum**
9. Seeds honeycomb-pitted…(10)
9'. Seeds smooth or areolate…(14)
10. Lower leaves entire or with only 1–2 teeth or lobes near base…(11)
10'. Lower leaves serrate and usually lobed…**C. berlandieri**
11. Seeds subglobose, entirely covered by perianth segments at maturity; pericarp slightly to prominently whitened; leaves broadest very near base…**C. watsonii**
11'. Seeds ± flattened, partially exposed at maturity; pericarp black; leaves usually prominently lobed above the base [*C. neomexicanum* complex]…(12)
12. Fruit ~1 mm diam., margin thick in side view (~0.15–0.2 mm)…**C. arizonicum**
12'. Fruit 1.2–1.6 mm diam., margin thin in side view (~0.1 mm)…(13)
13. Plants unbranched or sparsely branched from base, branches strongly ascending; leaf blades campanulate; fruit margin narrow in top view, pericarp broadly pitted, grayish…**C. neomexicanum**

13'. Plants heavily branched from above base, branches spreading, ascending; leaf blades deltoid to rhombic-ovate; fruit margin wide in top view; pericarp papillate, mottled brown…**C. lenticulare**
14. Leaves triangular…(15)
14'. Leaves ovate to broadly ovate, rhombic, or lanceolate, variously lobed or toothed…(16)
15. Plants simple below, branching above; seeds 1–1.3 mm diam.…**C. fremontii**
15'. Plants profusely branching from base; seeds 0.9–1.1(–1.2) mm diam.…**C. incanum**
16. Leaf blades without teeth except for often present basal lobes or teeth…**C. atrovirens**
16'. Leaf blades with lateral teeth and often basal lobes…**C. album**

Chenopodium album L.—Lamb's-quarters, pigweed. Erect to sprawling annual, 1–30 dm; sparsely to densely farinose; leaf blades ovate-lanceolate to rhombic-lanceolate or broadly oblong, 1–5.5(–12) × 0.5–3.8(–8) cm, margins sinuous-dentate to shallowly serrate or entire; glomerules subglobose, 3–4 mm diam., utricles depressed-ovoid; pericarp nonadherent, occasionally adherent, smooth to papillate; seeds 0.9–1.6 mm, seed coat smooth, indistinctly granulate and/or radially grooved, or with faint reticulate-rugose ridges. Disturbed soils in open habitats, 3800–9800 ft. Widespread. Introduced.

Chenopodium arizonicum Standl.—Erect annual, 3–6 dm, ill-scented; profusely branched above, reddish; leaf blades triangular to rhombic-ovate, 0.7–1.6 cm, central lobe large, margins entire; utricles ca. 1 mm diam.; pericarp adherent, with medium-sized papillae, seed coat reticulate. Generally moist soils, piñon-juniper and coniferous woodlands, desert scrub, desert grasslands, alluvium, roadsides, 4300–7900 ft. Sierra.

Chenopodium atrovirens Rydb.—Piñon goosefoot, mountain goosefoot. Erect annual, usually much branched, 0.7–6.5 dm; farinose; leaf blades ovate, broadly oblong, oval, or occasionally triangular, 3-veined, 1–3 × 0.4–2.2 cm, thick, margins entire or occasionally with basal lobe; glomerules in terminal and axillary paniculate spikes, 2–8 × 1–1.5 cm; utricles ovoid; pericarp adherent or nonadherent, smooth; seeds 0.9–1.3 mm, rugulate. Open dry sandy areas, meadows in oak and mixed conifer forests, other disturbed sites, 4900–10,400 ft. Widespread except E plains.

Chenopodium berlandieri Moq.—Pit-seed goosefoot. Erect annual, much branched to simple, 1–10.5 dm, farinose; leaves nonaromatic; blades lanceolate, rhombic, ovate, or triangular, 1.2–12(–15) × 0.5–7.5(–9) cm, margins serrate, irregularly dentate, or entire, generally with 2 basal lobes; glomerules in compound spikes, 5–17 cm; glomerules irregularly rounded, 4–7 mm; utricles depressed-ovoid; pericarp adherent or nonadherent; seeds round, 1–1.3 mm, honeycomb-pitted. Two weakly differentiated varieties are occasionally recognized for the state: **var. sinuatum** (Murr) Wahl, with yellow-tinted style bases; and **var. zschackei** (Murr) Murr ex Graebn., with style bases lacking a yellow tint. Disturbed woodlands, shrublands, grasslands, riparian zones, roadsides, 3800–9400 ft. Widespread except SE.

Chenopodium cycloides A. Nelson—Sandhill goosefoot. Erect annual, branched, 3–8 dm, glabrous to sparsely farinose; leaves nonaromatic; blades linear, 1-veined, 1–3 × 0.1–0.2 cm, somewhat fleshy, margins entire; glomerules in terminal and axillary panicles of interrupted spikes, 10–20 × 4–6 cm; utricles ovoid; pericarp adherent, red, minutely tuberculate; seeds round, 1.3–1.5 mm, rugulate. Open sandy areas especially around blowouts on sand dunes and in grassland and shinnery oak communities, 3600–4280 ft. EC, SE, Doña Ana, Rio Arriba.

Chenopodium desiccatum A. Nelson—Arid-land goosefoot. Spreading annual, ca. 1 dm, white-mealy throughout, leaf blades oblong to linear, 1.5–2.5 × 0.4–0.6, margins entire; glomerules in dense panicles; utricles ovoid; pericarp nonadherent; seeds ovoid, warty. Much reported for New Mexico. Recent work by Benet-Pierce and Simpson (2014) has reinterpreted this taxon in the light of consistent seed characters and has placed its range N of New Mexico extending S to C Colorado. They do suggest it may be possible for *C. desiccatum* to occur within the state. Widespread.

Chenopodium fremontii S. Watson—Fremont's goosefoot. Erect to spreading annual, 1–8 dm, farinose; leaf blades usually broadly triangular to ovate or elliptic, 0.7–6 cm, thick, margins entire to dentate or with a pair of basal teeth or lobes; glomerules in terminal and axillary interrupted spikes, 16–22 × 4–5 cm; glomerules 2–5 mm diam.; utricles ovoid; pericarp nonadherent, warty-smooth; seeds round,

1–1.3 mm; seed coat ± smooth. Two overlapping varieties can be recognized for the state: **var. fremontii** exhibits broadly triangular-hastate leaves with entire margins, while **var. pringlei** (Standl.) Aellen exhibits broadly triangular leaf blades with margins coarsely and irregularly sinuate-dentate or shallowly repand-dentate. A variety of habitats in desert, cliffs, talus, and moist shaded areas under aspen, juniper, or piñon, often in riparian habitats, 4400–10,500 ft. Widespread except EC, SE.

Chenopodium hians Standl.—Hians goosefoot. Erect annuals, 2–3(–4) dm, farinose; leaf blades elliptic-oblong or narrowly lanceolate, 3-veined, 1–2.5 × 0.3–0.6(–0.8) cm, margins entire; glomerules in lateral spikes or panicles, 5–7 cm; utricles ovoid; pericarp adherent, with minute brown and white papillae; seeds round, 1–1.4 mm, rugulate. Open, dry or moist areas including prairies, pastures, sand hills, roadsides, lakeshores, 5400–8700 ft. NC, NE, WC, SW.

Chenopodium incanum (S. Watson) A. Heller—White-mealy goosefoot. Erect annual, branched from base, 0.6–7.5 dm, farinose; leaf blades triangular to ovate, 1–1.5(–2.3) × 0.5–1.6 cm, thin or thick, margins usually with 2 basal teeth or lobes; glomerules in terminal and lateral panicles, 3.5–12 × 1–7 cm; utricles ovoid; pericarp nonadherent, smooth; seeds round, 0.9–1.25 mm, wrinkled. Two weakly differentiated varieties occur across the state with overlapping ranges. **var. incanum** consists of plants that are densely branched from the base, with leaf blades exhibiting acute basal teeth and apices; and **var. elatum** D. J. Crawford consists of plants with erect or spreading stems with leaf blades exhibiting rounded basal teeth and apices. Open sandy soils, hillsides, savannas, prairies, roadsides, woodlands, 3000–9500 ft. Widespread.

Chenopodium lenticulare Aellen—Erect annual, branched from base to 7 dm; plants with faint or no odor; leaf blades triangular or ovate, often with obtuse or rounded basal lobes, or often the blades broadly rhombic, rhombic-ovate, or ovate to ovate-lanceolate, 0.8–3 cm, margins entire; utricles 1.5 mm diam., equatorial margin or rim prominent (at least 0.2 mm wide); pericarp semiadherent, mottled with narrow, elongate papillae, seeds smooth. Generally moist soils, piñon-juniper and coniferous woodlands, desert scrub, desert grasslands, alluvium, roadsides, 5900–8200 ft. Colfax, Eddy, Sandoval, San Miguel.

Chenopodium leptophyllum (Moq.) Nutt. ex S. Watson—Slimleaf goosefoot. Erect annual, usually branching from base, 1–4 dm, farinose; leaf blades linear, 1-veined, 0.7–2.6(–3) × 0.1–0.3 cm, ± fleshy, margins entire; glomerules in terminal and axillary panicles; utricles ovoid; pericarp adherent, smooth; seeds ovoid, 0.9–1.1 mm, finely rugulate. Open, often disturbed sandy areas and fields, woodlands, roadsides, 3700–9100 ft. Widespread.

Chenopodium neomexicanum Standl.—New Mexico goosefoot. Erect annual, 2.5–7 dm, sparsely farinose; plants ill-scented; leaf blades campanulate or broadly rhombic-ovate, 0.7–2.9 × 0.4–2.1 cm, subhastate with low, rounded or acutish lobes, margins entire above lobes; glomerules in paniculate spikes, 21–24 × 4–9 cm; utricles ovoid; pericarp adherent, broadly pitted; seeds lenticular or round, 1–1.3 mm, reticulate. Generally moist soils, piñon-juniper and coniferous woodlands, desert scrub, desert grasslands, alluvium, roadsides, 4300–8600 ft. Widespread except N and E plains.

Chenopodium nitens Benet-Pierce & M. G. Simpson [*C. desiccatum* A. Nelson, in part]—Shiny-seed goosefoot. Prostrate to occasionally erect annual, forming mats to ca. 3 dm; leaf blades narrowly elliptic-oblong to lanceolate, very rarely basally lobed, 6–14(–18) × 2–4 mm; glomerules in axillary and terminal spikes, 1–7 cm; utricles ovoid; pericarp nonadherent; seeds ovoid. Open, undisturbed soil in shortgrass prairies and sandy stabilized dunes, 4200–9800 ft. Eddy.

Chenopodium pallescens Standl.—Slim-leaf goosefoot. Erect annual, much branched from base, 3–6.5 dm, glabrate or sparsely and finely farinose; leaf blades linear, 1-veined, 1–3.5 × 0.1–0.3 cm, somewhat fleshy, margins entire; glomerules in terminal and axillary panicles or cymes of spikes, 4–9 × 5–9 cm; utricles ovoid; pericarp adherent, finely tuberculate; seeds round, 1.4–1.6 mm, rugulate. Open, stony or sandy grasslands, woodlands, 3800–7000 ft. SE, San Miguel.

Chenopodium pratericola Rydb. [*C. desiccatum* A. Nelson var. *leptophylloides* (Murr) Wahl]—Narrowleaf goosefoot. Erect annual, 2–8 dm, farinose; leaf blades linear to narrowly lanceolate or oblong-elliptic, (1-)3-veined, 1.5-4.2(-6) × 0.4-1(-1.4) cm, thick and ± fleshy, margins entire or with pair of lobes near base; glomerules in terminal and axillary panicles, 1-13 × 0.15-0.5 cm; utricles ovoid; pericarp non-adherent, smooth; seeds round, 0.9-1.3 mm, rugulate. Open sandy soils, piñon woodlands, sagebrush, often in saline or alkaline habitat, 3640-9700 ft. Widespread.

Chenopodium watsonii A. Nelson—Watson's goosefoot. Erect annuals, branched from base, 1-4.5 dm, farinose; leaves aromatic, blades broadly rounded-triangular to rounded-rhombic or ovate, 1-2.6 × 0.5-2.9 cm, margins entire or with 1-2 teeth on each side at base; glomerules in paniculate spikes, 14-24 × 2-3 cm; utricles ovoid; pericarp adherent, honeycombed; seeds subglobose, 0.9-1.3 mm, coarsely honeycombed. Open sites in woodlands, shrublands, and savannas, roadsides, 4000-8100 ft. Widespread except EC, SE.

CORISPERMUM L. - Tickseed, bugseed

Annual herbs with dendroid, sometimes almost stellate hairs, occasionally glabrous, or becoming glabrous at maturity, stems erect or ascending, rarely prostrate; leaves alternate, sessile; blades lanceolate, linear-lanceolate, linear, or filiform, flat or convolute at maturity, margins entire, apex acute; inflorescences terminal spikes; flowers solitary in axils of leaflike bracts; flowers bisexual; perianth segments 0-1(-3), scalelike; stamens 1-3(-5); ovary superior; stigmas and styles 2; cypselas largely exposed, vertical, sessile, wing (if present) with entire, undulate, or minutely erose-denticulate margins; pericarp strongly accrescent to seed or small portions not accrescent, forming small whitish bladders ("warts"); seeds vertical.

1. Fruits 1.8-3.2 mm, wingless or with a scarcely visible wing, style base protruding beyond wing… **C. villosum**
1′. Fruits 2.5-4.6 mm, with a visible wing 0.2-0.6 mm wide…(2)
2. Wings of fruits 0.4-0.6 mm wide; inflorescence usually compact and dense…**C. welshii**
2′. Wings of fruits 0.2-0.4 mm wide; inflorescence usually lax and interrupted…**C. americanum**

Corispermum americanum (Nutt.) Nutt.—American tickseed. Plants branched near base, 10-35 (-50) cm, sparsely pubescent; leaf blades linear or narrowly linear (occasionally linear-lanceolate or almost filiform), usually plane or occasionally folded, 1.5-3.5(-4) × 0.1-0.3 cm; inflorescences usually lax, interrupted, rarely ± condensed distally; fruits yellowish brown, greenish brown, light brown, or brown, often with reddish-brown spots and whitish warts; (2.3-)2.5-4.5 × 2-3.5 mm, wings translucent, thin, margins entire or rarely indistinctly erose. NW, NC, WC, SE, Doña Ana, Quay.

1. Fruits 2.5-3.5 mm, wings usually 0.2-0.3 mm wide, occasionally almost wingless…**var. americanum**
1′. Fruits 3.5-4.5 mm, wings usually 0.3-0.4 mm wide…**var. rydbergii**

var. americanum—Sandy soils in piñon-juniper communities, dune communities, disturbed soils, 3050-6200 ft.

var. rydbergii Mosyakin—Sandy soils in ponderosa pine, piñon-juniper communities, 5500-7800 ft.

Corispermum villosum Rydb.—Hairy bugseed. Plants usually branched from base, (5-)10-30(-35) cm, densely or sparsely pubescent; leaf blades linear-oblanceolate, linear, or rarely narrowly linear, usually plane, (1-)1.5-3.5 × (0.1-)0.2-0.3 cm; inflorescences rather compact, dense, condensed in distal 1/2, occasionally interrupted in proximal 1/2, fruits yellowish brown, light brown, or dark brown, usually with reddish-brown spots and occasionally whitish warts, 1.8-3(-3.2) × 1.5-2 mm, dull; wings absent or to 0.1(-0.15) mm wide, margins entire. Piñon-juniper shrublands, ca. 6900 ft. NW, Doña Ana.

Corispermum welshii Mosyakin—Welsh's bugseed. Plants branched near base, 10-35 cm, densely or sparsely pubescent; leaf blades linear-lanceolate or linear, usually plane, 1-6 × 0.2-0.5 cm; inflorescences usually compact and dense, rarely ± lax, and condensed only at apex, fruits yellowish brown, light brown, or brown, usually with reddish-brown spots and whitish warts, (3.3-)3.7-4.6 × (2.7-)3-3.6 mm, wings translucent, thin, (0.3-)0.4-0.6 mm wide, margins entire or irregularly minutely erose-denticulate. Sandy sites, dry valleys, 4200-7200 ft. NW, WC, SC, Quay.

CYCLOLOMA Moq. – Winged-pigweed

Cycloloma atriplicifolium (Spreng.) J. M. Coult.—Winged-pigweed, tumble-ringwing. Annual herbs, villous or tomentulose, becoming glabrous at maturity, stems erect, much branched, 5–80 cm; leaves alternate, petiolate or almost sessile, blades oblong-ovate, oblong, or lanceolate, 2–7(–8) × 0.5–2 cm, margins sinuate-dentate, inflorescences diffusely branched, paniculate, interrupted, linear spikes; flowers solitary or few in axils of short bracts; flowers bisexual or pistillate; perianth segments 5, connate to above middle; stamens 5; stigmas 3; utricles with membranous, circular wing; seeds ovate-lenticular or globose-lenticular; seed coat black, smooth or indistinctly sculptured. Sand dunes, shinnery oak communities, sandy riparian zones, 3050–8300 ft. NW, NE, EC, C, SC, SE.

DYSPHANIA R. Br. – Glandular goosefoot

Annual (ours) or short-lived perennial, aromatic herbs, with glandular hairs and subsessile glands; stems erect-ascending, decumbent, or prostrate, branched; leaves alternate, petiolate, blades thin, lanceolate, oblanceolate, ovate, or elliptic, often pinnately lobed; margins entire, dentate, or serrate; inflorescences terminal and axillary, ebracteate, of loose, compound ebracteate cymes, or glomerules arranged spicately and often subtended by reduced leaflike bracts; flowers bisexual or rarely unisexual; perianth segments 1–5, free near to the base and later loosely covering the fruit, or fused to form a sac surrounding the fruit; fruit a utricle.

1. Inflorescence of small sessile glomerules arranged in spikes; leaves entire to sinuate-toothed, > 4 cm in length; sepals not glandular; stigmas 3…**D. ambrosioides**
1'. Inflorescence open with cymose branching; leaves pinnatifid, < 4 cm long; sepals glandular-hairy; stigmas 2…(2)
2. Cymes with the central flower developed, the lateral flowers abortive with spinelike pedicels; sepals with golden sessile glands and a single hornlike appendage…**D. graveolens**
2'. Cymes with all flowers developed; sepals with stalked glandular hairs and lacking a hornlike appendage… **D. botrys**

Dysphania ambrosioides (L.) Mosyakin & Clemants [*Chenopodium ambrosioides* L.]—Mexican-tea. Stems erect to ascending, much-branched, 3–10(–15) dm; leaf blades ovate to oblong-lanceolate or lanceolate, 2–8(–12) × 0.5–4(–5.5) cm, copiously gland-dotted (rarely glabrous); inflorescences lateral spikes, 3–7 cm; glomerules globose, 1.5–2.3 mm diam.; utricles ovoid; pericarp nonadherent, rugose to smooth, seeds reddish brown, ovoid, 0.6–1 × 0.4–0.5 mm; seed coat rugose to smooth. River bottoms, dry lake beds, waste areas, 5400–8800 ft.

Dysphania botrys (L.) Mosyakin & Clemants [*Chenopodium botrys* L.]—Jerusalem-oak. Stems erect to ascending, branched at base to ± simple, 1–6(–10) dm, leaf blades ovate to elliptic, 1.3–4 × 0.6–2.7 cm, margins lyrate-sinuate, pinnatifid, or occasionally entire, glandular-pubescent abaxially; inflorescences axillary cymes, often arranged in terminal thyrses, 12–24 cm; utricles subglobose; pericarp adherent, papillose, becoming rugose, usually white-blotchy; seeds globose to subglobose, (0.5–)0.6–0.8 × 0.5–0.7 mm, seed coat rugose. Sandy or gravelly soils, dry rocky slopes, piñon-juniper woodlands, roadsides, 6200–8850 ft.

Dysphania graveolens (Willd.) Mosyakin & Clemants [*Chenopodium graveolens* Willd.]—Incised or fetid goosefoot. Stems erect, 2.3–5.2 dm, leaf blades lanceolate to ovate or elliptic, 1.7–4.5 × 0.7–2.6 cm, margins pinnatifid or entire; inflorescences terminal compound cymes, 8.5–22 cm; utricles subglobose; pericarp adherent, papillose, becoming rugose, usually white-blotchy; seeds subglobose, 0.6–0.9 × 0.5–0.7 mm, seed coat rugose. Piñon-juniper woodlands, mixed conifer communities, roadsides, 5200–10,200 ft.

FROELICHIA Moench – Snake-cotton, cottonweed

Herbs (ours), annual or perennial; stems erect or procumbent, simple to much branched, usually richly pubescent; leaves opposite, most abundant on proximal 1/2 of plant; blades linear, lanceolate, oblance-

olate, oblong, or orbiculate, fulvous abaxially, margins entire; inflorescences terminal, erect, pedunculate, spiciform, mostly compound; flowers arranged in 3- or 5-ranked spirals; bracteoles enclosing and
falling with the flowers; flowers bisexual; tepals 5, connate at least to middle, becoming indurate in fruit
and developing lateral wings or crests and in some species, facial tubercles or spines; stamens 5; filaments connate into a tube, pseudostaminodes 5, alternating with stamens; ovule 1; style 1, short or elongate, stigma capitate; fruit a utricle.

1. Plants perennial with a woody taproot; pubescence of tepals of mature flowers bright white, appearing as
    small cotton balls; flowers 3.5–5.5 mm…**F. arizonica**
1'. Plants annual or short-lived perennial, taproots semiwoody; pubescence of tepals of mature flowers gray-
    ish white, not appearing as small cotton balls; flowers 2.4–3.8 mm…**F. gracilis**

   Froelichia arizonica Thornber ex Standl.—Arizona snake-cotton. Plants perennial; taproot woody;
stems ascending or decumbent, sericeous-tomentose with white hairs, 3–10 dm; leaves usually crowded
at base, blades lanceolate, proximal leaves 3–12 × 0.5–2.5 cm, scaberulous or canescent adaxially, sericeous-tomentose with bright whitish hairs abaxially; spikes stout, flowers in a 3-ranked spiral; flowers
(3.5–)4–5.5 mm; tepals with dense, bright white pubescence; utricles winged laterally, wing margins
irregularly dentate, 1 or both faces with 1+ basal tubercles or spines. Open rocky or gravelly hillsides,
3450–5950 ft. WC, SW. *Froelichia floridana* (Nutt.) Moq. may just enter New Mexico at the eastern
edge. While often reported for the state, specimens identified to this taxon represent *F. gracilis*. This
taxon differs from *F. arizonica* in being a robust annual with brownish hairs and flowers arranged in a
5-ranked spiral. It is restricted to sandy soils of plains.

   Froelichia gracilis (Hook.) Moq.—Slender snake-cotton. Plants annual or short-lived perennial;
taproots semiwoody; stems several (rarely 1), erect or ascending, sometimes procumbent, usually much
branched from base, slender, villous-tomentose with grayish white hairs, 1–5(–10) dm; leaves predomi-
nant on proximal 1/3 of plant, blades linear to lanceolate or lance-elliptic, largest leaves 1.6–9(–13.5) ×
0.2–0.9(–1.2) cm, canescent or sericeous adaxially, sericeous-tomentose with white or gray hairs abaxially; spikes sparsely branched, flowers in a 3-ranked spiral; flowers 2.4–3.8 mm; tepals with grayish pubescence; utricles with irregularly and deeply cut ("spiny") lateral wings, both faces with distinct spines
or tubercles. Open areas on rocky hillsides, roadsides, stream beds, desert scrub communities, woodlands, shortgrass prairies, 4000–8000 ft. NE, EC, C, SW, SC, SE, Cibola.

GOMPHRENA L. – Globe-amaranth

Herbs, annual or perennial; stems ascending, decumbent, prostrate, or erect; leaves opposite, blades
ovate to obovate, not fleshy, margins entire, long-pilose abaxially and sometimes adaxially; inflorescences terminal and/or axillary, sessile, subglobose heads, often subtended by involucres of sessile
leaves, bracts and bracteoles thin; flowers bisexual; tepals 5, connate proximally; stamens 5; filaments
connate basally into tube; pseudostaminodes absent; ovule 1; style 1, 1.5–4 mm; stigmas 2(3), filiform;
utricles included in tepals, stramineous, ovoid or oblong, somewhat compressed, membranous, usually
indehiscent.

1. Plants cespitose, perennial, from a woody, branched caudex…**G. caespitosa**
1'. Plants not cespitose, stems simple or branched, annual or perennial…(2)
2. Plants annual, roots fibrous, bractlets with laciniate crests, often tinged red or pink…**G. nitida**
2'. Plants perennial, roots woody, fibrous, or fusiform, bractlets with or without laciniate crests…(3)
3. Bractlets not cristate along keel, roots fibrous to woody, plants 4–60 cm tall…**G. sonorae**
3'. Bractlets cristate along keel, roots fleshy, fusiform, plants 20–70 cm tall…G. haageana

   Gomphrena caespitosa Torr. [*G. viridis* Wooton & Standl.]—Tufted globe-amaranth. Plants perennial, cespitose, 0.3–1.5 dm; roots woody; stems ascending, sparsely villous to glabrate; leaves mostly
basal; blades green to gray-green, obovate to elliptic-oblong, 3.5–7.5 × 2–3 cm, apex rounded or obtuse,
glabrous to densely grayish appressed-pilose; inflorescence heads white, subglobose to cylindrical, 35–
70 × 8–20 mm; bractlets not crested; tepals white to yellow-pilose; utricles ovoid, 2 mm, apex acute.
Open woodlands, especially oak-juniper woodlands, desert grasslands, hillsides, 4200–7000 ft. WC, SW.

Gomphrena haageana Klotzsch—Rio Grande globe-amaranth. Plants perennial; roots fusiform, fleshy; stems erect, pilose, 2-7 dm; leaves green, oblanceolate to oblong-linear, 3-10 × 0.3-1 cm, apex acute to acuminate, mucronate, pilose; inflorescence heads stramineous, globose to short-cylindrical, 20-28 mm diam.; bractlets crested along keel; tepals densely lanate; utricles ovoid, 2.2 mm, apex acute. Rocky banks and hillsides. No documented specimens for New Mexico but to be expected in the far S.

Gomphrena nitida Rothr.—Pearly globe-amaranth. Plants annual; roots fibrous; stems usually erect, pilose or strigose, 2-7 dm; leaves green, obovate or oblong, 1.5-6 × 0.4-2.5 cm, apex obtuse or acute, pilose; inflorescence heads yellowish white to reddish, subglobose, 12-16 mm diam.; bractlets with laciniate crests; tepals lanate; utricles ovoid, 1.5 mm, apex truncate. Open woodlands, grasslands, hillsides, 4900-7500 ft. WC, SW.

Gomphrena sonorae Torr.—Sonoran globe-amaranth. Plants perennial or rarely annual; roots woody or fibrous; stems erect or ascending, sparsely pilose, glabrate, 1.5-6 dm; leaves green, elliptic, oblong, oblanceolate, or linear, 1-3 × 0.5-1.2 cm, sparsely pilose; inflorescence heads white, globose, 0.8 1.3 mm diam., subtended by 2 leaves; bractlets white, not crested; tepals densely lanose proximally; utricles ovoid, 2.2 mm, apex acute. Sandy slopes, open woodlands, scrub, dry streambeds, 5100-5800 ft. Grant, Hidalgo.

GOSSYPIANTHUS Hook. – Cottonflower

Gossypianthus lanuginosus (Poir.) Moq.—Cottonflower. Perennial herbs; stems prostrate to decumbent, sparsely pilose to strongly woolly-villous with whitish, straight to wavy-crinkled hairs; leaves opposite, basal and cauline, blades linear to spatulate, 1.6-9.7 × 0.2-1(-1.7) cm; glabrous or sparsely to densely pilose to strigose; inflorescences axillary, sessile, few-flowered glomerules, 1-10 × 5-8 mm, white; bracts and bracteoles membranous; flowers bisexual; tepals 5, distinct; stamens 5, filaments connate basally into tube, pseudostaminodes absent; ovule 1; style 1, ca. 0.2 mm; stigma 2-lobed, capitate; utricles broadly ovoid, indehiscent, included within tepals, 1.4-2 mm. Our plants belong to **var. lanuginosus**. Rocky knolls, shortgrass prairies, 4000-9000 ft. Lincoln, Roosevelt, San Miguel.

GRAYIA Hook. & Arn. – Hopsage

Erect shrubs or subshrubs, 1.5-15 dm, dioecious or monoecious. Stems branched, younger branches ribbed/striate, older bark gray-brown; leaves solitary or fasciculate, alternate, sessile; blades elliptic, ovate, obovate, spatulate, or linear-oblanceolate, 6-80 × 1.5-42 mm, succulent or coriaceous, margins entire, surfaces green to grayish; inflorescences with staminate flowers glomerulate (2-5 flowers per glomerule), in interrupted axillary or terminal spikes or panicles, sometimes mixed with pistillate flowers, pistillate flowers in terminal and axillary interrupted panicles; perianth lobes 4-5, united in proximal 1/2; stamens 4-5, pistillate flowers borne singly within 2 opposite bracteoles; perianth lacking; stigmas 2; fruiting bracteoles folded along midribs and completely connate except for apex, 4-14 × 3-15 mm; utricles 2-4 × 1.5-2 mm, orbicular to obovoid, laterally compressed-lenticular; pericarps membranous, free or slightly adherent to seeds, brown to yellowish brown, dull; seed coats thin (Zacharias & Baldwin, 2010).

1. Branches divaricate, often thorny; older stems with longitudinal reddish striations; pubescence of branched hairs…**G. spinosa**
1′. Branches erect, not thorny; stems without longitudinal striations; pubescence of scurfy or unbranched hairs…(2)
2. Leaves mainly < 6 mm broad…**G. brandegeei**
2′. Leaves mainly > 6 mm broad (to 25+ mm)…**G. plummeri**

Grayia brandegeei A. Gray—Spiny hopsage. Plants monoecious; stems 1-5 dm; leaves of main stem 13-80 × 1.5-6 mm; blades green; fruiting bracteoles 2(-4)-winged, samaralike, 3.4-9 mm diam. Fine-textured, often saline and seleniferous substrates, juniper woodlands, 5900-6000 ft. McKinley, San Juan.

Grayia plummeri (Stutz & S. C. Sand.) E. H. Zacharias—Plummer's siltbush. Plants monoecious; stems 1-5 dm; leaves of main stem usually 13-80 × 6-25 mm; blades green; fruiting bracteoles 2(-4)-winged, samaralike, 3.4-9 mm diam. Alkaline soils, shale slopes, 6000-6100 ft. San Juan.

Grayia spinosa (Hook.) Moq.—Spiny hopsage. Plants dioecious (rarely monoecious); stems 3-10 (-15) dm, becoming reddish brown with whitish ribs exfoliating in strips; leaves of main stems 1-2.5(-4.2) cm × 1.5-6(-10) mm; blades green; fruiting bracts wholly connate, sessile, orbicular to broadly elliptic, 7.5-14 × 6-12 mm; wings thickened near margin, yellowish green, whitish, or pink to red-tinged, smooth, glabrous. Valleys, foothills, dry, alkaline or scarcely alkaline soils, sagebrush communities, ca. 5900 ft. San Juan.

GUILLEMINEA Kunth – Matweed

Guilleminea densa (Humb. & Bonpl. ex Schult.) Moq.—Small matweed. Perennial herbs; stems much branched, procumbent, villous to densely woolly-villous; leaves basal and cauline, opposite, petiolate, blades oblong-elliptic to lanceolate, (8-)20-40 × (1-)6-9(-14) mm, glabrous adaxially, villous abaxially; inflorescences axillary, sessile, few-flowered glomerules, 2-3 × 3-5 mm, white; flowers bisexual; tepals 5, connate proximally; stamens 5, filaments connate basally into tube, pseudostaminodes absent; ovule 1; style 1, ca. 0.2 mm; stigma 2-lobed, capitate; utricles broadly ovoid, included within tepals, 1.1-1.4 mm. Our plants belong to **var. aggregata** Uline & W. L. Bray. Dry waste ground, creek beds, roadsides, gravelly soil, riparian scrub, piñon-juniper communities, 3700-7700 ft. EC, C, WC, SC, SW.

HALOGETON C. A. Mey. ex Ledeb. – Saltlover

Halogeton glomeratus (M. Bieb.) C. A. Mey.—Saltlover. Annual herbs, polygamous, herbage glaucous, glabrous except for axillary tufts of long white hairs; stems erect or spreading, much branched, 1-4 dm, somewhat fleshy; leaves alternate, sessile, ± succulent; blades terete, linear, 4-14(-17) mm, base expanded, clasping, apex obtuse with caducous, flexible bristle; inflorescences axillary glomerules; bracts leaflike; flowers bisexual and unisexual; perianth persistent, deeply 5-parted, not imbricate, apically winged, often indurate; stamens 3-5; stigmas 2; fruit a utricle, orbiculate; pericarp somewhat adherent to seed; seeds orbiculate, seed coat brown or brownish black. Alkaline or sandy soils in salt desert scrub communities, badlands, piñon-juniper forest edges, 4070-7120 ft. NW, NC, Grant, Quay, Sierra. An introduced noxious and toxic weed.

IRESINE P. Browne – Bloodleaf

Iresine heterophylla Standl.—Standley's bloodleaf. Perennial herbs; stems erect to ascending or sprawling, sparsely pubescent, 5-10 dm; leaves opposite, blades broadly rhombic-ovate, 3-6 × 2-4 cm, pubescent abaxially; inflorescence a panicle, 1.5-15 cm; bracts and bracteoles 1/3-1/2 as long as tepals; flowers unisexual; tepals deeply 5-parted, whitish to stramineous, ovate, 1-1.3 mm, apex acute, densely lanate; pseudostaminodes absent or short; ovule 1; style 1, ca. 0.2 mm; stigmas branched, slender or capitate; utricles included in tepals, greenish, ovoid, 0.8 mm, apex acute. Rocky soil in sheltered ravines and canyons, 4100-7000 ft. Doña Ana, Grant, Socorro.

KRASCHENINNIKOVIA Gueldenst. – Winter fat

Krascheninnikovia lanata (Pursh) A. Meeuse & A. Smit [*Ceratoides lanata* (Pursh) J. T. Howell; *Eurotia lanata* (Pursh) Moq.]—Winter fat. Subshrubs, monoecious or dioecious, herbage densely white to brownish tomentose, hairs stellate, plants 1.5-5 dm; stems erect, basal branches woody, flowering branches herbaceous; leaves alternate, petiolate, those of main stem with blades linear to narrowly lanceolate, 1-3(-4) × 1.5-3.5(-5) cm, not fleshy, margins revolute; inflorescences axillary clusters or small spikes; flowers unisexual; staminate flowers with bractlets absent, perianth 4-parted, stamens 4, deciduous after anthesis; pistillate flowers enclosed in 2 partially connate, slightly keeled, densely hirsute bractlets with free tips hornlike, perianth absent, stigmas 2; fruiting bracts ovate, 4-7.5 mm, margins connate proximally, fruit a utricle, 0.5–3.5 mm, densely pubescent; seeds ovate, seed coat brown, covered with white hairs. Foothills, flats, grasslands, shrublands, open piñon-juniper woodlands, usually in relatively low-alkaline soils, 3500–9500 ft. Widespread.

NEOKOCHIA (Ulbr.) G. L. Chu & S. C. Sand. – Green molly

Neokochia americana (S. Watson) G. L. Chu & S. C. Sand. [*Bassia americana* (S. Watson) A. J. Scott; *Kochia americana* S. Watson]—Green molly. Subshrubs with woody bases, monoecious, plants whitish gray or grayish green, 5–35(–50) cm, tomentose-sericeous or almost glabrous, stems erect or ascending, simple or branched only at base; branches ± erect; leaves sessile, blades linear, terete or semiterete, 4–25 × 0.5–2 mm, fleshy, sericeous or almost glabrous; inflorescences spicate; 1–4-flowered in axils of bracts; flowers perfect or pistillate, fruiting perianth 5-winged, the wings 1.5–2.5 mm, scarious, tomentose (rarely almost glabrate). Desert scrub communities, generally on alkaline soils, 3800–5700 ft. Bernalillo, Quay, Roosevelt, San Juan (Kadereit & Freitag, 2011).

OXYBASIS Kar. & Kir. – Goosefoot

Annual, monoecious, nonaromatic herbs, ± glabrous, sometimes leaves densely farinose below; stems erect to ascending or prostrate, branched; leaves alternate, petiolate; blades somewhat fleshy, triangular to narrowly triangular, hastate or rhombic or lanceolate; margins entire to dentate; inflorescence axillary and terminal, usually largely leafy or bracteate, sometimes ebracteate, flowers in compact glomerules; flowers usually dimorphic; terminal flowers bisexual, perianth segments 3–5, stamens 1(–5); stigmas 2(–3); lateral flowers usually pistillate, perianth segments 3(4), connate; stamen 0–1; stigmas 2; fruit a utricle with membranous pericarp, free or loosely attached to seed; seeds oval to orbicular, seed coat brownish to black, almost smooth to finely reticulate or minutely pitted (Fuentes-Bazán et al., 2012).

1. Leaf blades lanceolate or oblong, glaucous abaxially…**O. glauca**
1'. Leaf blades triangular or rhombic, green abaxially…**O. rubra**

Oxybasis glauca (L.) S. Fuentes, Uotila & Borsch [*Chenopodium glaucum* L.]—Oak-leaved goosefoot. Stems erect to prostrate, branched from base, 0.5–2.5(–4) dm, farinose; leaf blades lanceolate to oblong or ovate, 0.5–4 × 0.3–1.5 cm, margins with acute teeth densely farinose adaxially, glaucous abaxially; inflorescences glomerules in terminal or lateral spikes, 5–10 cm; bracts leaflike in inflorescence, elliptic, 0.2–1 × 0.1–0.5 cm, or absent at least in terminal 1/2 of inflorescence; utricles ovoid; pericarp nonadherent, smooth; seeds 0.6–1.1 mm. Our plants belong to **subsp. salina** (Standl.) Mosyakin. Arroyo bottoms, shorelines, riparian communities, 3300–9090 ft. NW, NC, EC, C, WC, SC, SW. Introduced.

Oxybasis rubra (L.) S. Fuentes, Uotila & Borsch [*Chenopodium rubrum* L.]—Red Eurasian goosefoot. Stems erect to ascending or prostrate, much branched, 0.1–6(–8) dm, glabrous; leaf blades triangular to rhombic, 1–9 × 1–6 cm, margins dentate or entire; inflorescences lateral glomerules sessile on lateral branched spikes; bracts linear, 0.4–2 cm; utricles ovoid; pericarp nonadherent, reticulate-punctate; seeds 0.6–1(–1.2) mm. Two varieties, both occurring sympatrically in New Mexico, are sometimes recognized: **var. humilis** (Hook.) Mosyakin, with stems decumbent or spreading and leaf margins entire or shallowly dentate; and **var. rubra**, with stems erect or ascending and leaf margins deeply dentate. Moist open areas, riparian communities, dry lake beds, open woodlands, occasionally in weedy sites, 4100–9800 ft. NW, NC, EC, C, WC, SW, Doña Ana, Eddy. Introduced.

PROATRIPLEX (W. A. Weber) Stutz & G. L. Chu – Proatriplex

Proatriplex pleiantha (W. A. Weber) Stutz & G. L. Chu [*Atriplex pleiantha* W. A. Weber]—Four Corners orach, Mancos saltplant. Annual, monoecious herbs, branching from base, 0.5–1.7 dm; stems glabrous or sparingly scurfy, turning reddish with age; leaves alternate to subopposite, petiolate; blades ovate to suborbiculate, 5–18(–20) mm, about as wide, succulent, margins entire; flowers in short terminal spikes, terminal staminate, lower pistillate; pistillate flowers 2–6, surrounded by 2 bracteoles; fruiting bracteoles short-stipitate, compressed, 3–7 mm wide and about as long, entire; seeds falling at maturity, black, 1.5 mm, shining, seed coat crustaceous. Desert badlands on saline clay soils of the Mancos Shale and Fruitland Formations, 5000–5500 ft. NW, Taos. **R**

SALICORNIA L. – Glasswort

Herbs or shrubs, monoecious, fleshy, glabrous; stems prostrate to erect, simple to many-branched, apparently jointed and fleshy when young; leaves opposite, connate basally, sessile, decurrent portions forming fleshy segments enclosing stem, fleshy; blades reduced to fleshy scales, margins entire, narrow, scarious; inflorescences spikes, terminal or lateral, apparently jointed, each joint (fertile segment) consisting of 2 axillary, opposite, usually 3(–5)-flowered cymes embedded in fleshy tissue of distal internode; flowers bisexual or unisexual, ± radially symmetric; perianth segments persistent in fruit, usually 3–4, connate except for extreme tips, fleshy; stamens (0)1–2; styles 2–3; fruits utriclelike; seeds ellipsoid, seed coat yellowish brown, thin, membranous.

1. Plants annual herbs; all stems terminated by an inflorescence…**S. rubra**
1'. Plants perennial shrubs; many stems entirely vegetative…**S. utahensis**

   Salicornia rubra A. Nelson—Red glasswort, marshfire pickleweed. Stems usually erect, green with red or purple at base and apex of segments and around flowers, often becoming completely red in fruit, simple or with primary and secondary branches, (1–)5–25 cm, spikes weakly torulose, 0.5–3(–5) cm, with 4–10(–19) fertile segments; fertile segments (second to fourth in main spikes) 2.1–4.4 × 1.8–3.2 mm. Saline and alkaline marshes and flats, 4100–6000 ft. NC, EC, Otero.

   Salicornia utahensis Tidestr. [*Sarcocornia utahensis* (Tidestr.) A. J. Scott]—Utah swampfire. Woody stems procumbent to erect, rhizomatous (rhizomes often long-creeping), 10–30 cm; young branches with fleshy segments 5–20 × 2–3 mm; larger terminal spikes with 3–20 fertile segments, 10–40 mm; larger fertile segments 2.5–4 × 2.5–4 mm; seeds 1.3–1.5 mm, smooth except for straight conic papillae on edge. Saline and alkaline marshes, flats, riparian zones, 3400–5000 ft. C, SC, SE.

SALSOLA L. – Tumbleweed

Herbs, monoecious, annual (ours), glabrous or ± pubescent; stems erect, ascending, or prostrate; leaves alternate, sessile, blades lanceolate, linear, or filiform to subulate, semiterete, margins entire basally, apex obtuse, soft and subspinescent or narrowed to spine or soft bristle; inflorescences spicate, flowers solitary in axils of bracts or reduced distal leaves; bracts ovate-lanceolate, spine-tipped; flowers bisexual, with 2 bracteoles; perianth segments persistent, 5, covering utricle at maturity, often developing transverse, dorsal, membranous or ± coriaceous wing; stamens 5; styles and stigmas 2(3); fruits utricles; pericarp adherent; seeds orbicular, seed coat black or brown.

1. Bracts appressed; inflorescences relatively long, slender, spikelike; flowers usually gall-like below…
    S. collina
1'. Bracts spreading; inflorescences of axillary flowers on the branches; flowers usually not gall-like below…
    (2)
2. Branchlets often papillate; leaves yellow-green; sepal wings in fruit 2.5–4.5 mm…**S. paulsenii**
2'. Branchlets often hispidulous (short, stiff hairs); leaves greenish; sepal wings in fruit 0.5–2.5 mm…
    S. tragus

   Salsola collina Pall.—Slender Russian thistle. Herbs, 10–100 cm, sparsely to densely papillose or hispid (rarely subglabrous); leaf blades filiform to narrowly linear, 1–2 mm wide, apex with soft bristle; inflorescences not interrupted, dense, 1-flowered (rarely 2–3-flowered), often also in axils of proximal leaves and branches, lower ones tightly enclosed in bracts and bracteoles, forming gall-like caducous balls at maturity; bracts alternate, strongly imbricate and appressed at maturity. Waste places, roadsides, cultivated fields, disturbed natural and seminatural plant communities, often in sandy soils, 5400–8000 ft. NW, NC, C, WC, SC. Introduced.

   Salsola paulsenii Litv.—Barbwire Russian thistle. Herbs, 10–80(–100) cm, glabrous or sparsely papillose to hispid, profusely branched from or near base; leaf blades filiform to narrowly linear, usually

< 1 mm wide, fleshy or not, not swollen at base, apex subspinose or spinescent; inflorescences distinctly interrupted at maturity; perianth segments prominently winged; fruiting perianth 7–12 mm. Sandy soils, disturbed natural and seminatural plant communities, semideserts, deserts, 4000–6700 ft. McKinley, Sandoval. Introduced.

Salsola tragus L.—Tumbleweed. Herbs, (5–)10–100 cm, sparsely papillose to hispid or glabrous; profusely branched from or near base; leaf blades filiform or narrowly linear, not fleshy, apex subspinescent (spine < 1.5 mm); inflorescences interrupted at maturity (at least proximally); perianth segments with prominent, membranous wing at maturity; fruiting perianth ca. 4–10 mm. Disturbed areas, roadsides, cultivated fields, riparian sands, desert scrublands, shortgrass prairies, 3000–9400 ft. Widespread. An introduced noxious weed.

SUAEDA Forssk. ex J. F. Gmel. – Seablite, seepweed

Herbs, annual or perennial, or subshrubs or shrubs, monoecious, glabrous or pubescent; stems prostrate to erect, leaves usually alternate, sometimes opposite, sessile to short-petiolate, fleshy, blades linear to linear-lanceolate, terete or flat, margins entire; inflorescences clustered in the axils and crowded into dense, terminal and lateral spikes; flowers usually bisexual, clusters subtended by a bract; perianth segments 5, often hooded, keeled, winged, or horned; stamens (1–2)5; ovary 1-celled, stigmas 2–3(–5); fruits utricles.

1. Herbs, annual, glabrous; perianth segments of different sizes (1–3 larger), horned, somewhat keeled or wing-margined; fresh leaf cross-sections uniformly green…**S. calceoliformis**
1'. Subshrubs or shrubs (or sometimes appearing annual), usually densely short-villous, less often moderately short-pubescent or puberulent distally to nearly glabrous in some populations; perianth segments all the same size, rounded on the back; fresh leaf cross-sections with dark green peripheral ring…**S. nigra**

Suaeda calceoliformis (Hook.) Moq.—Paiuteweed, horned seablite. Annual herbs, prostrate to erect, green to dark red, 0.5–8(–10) dm, glaucous; stems usually striped; leaves tightly ascending, sometimes ± spreading; blades linear-lanceolate, adaxial surface flat, (5–)10–40 × 0.2–15 mm; glomerules usually crowded in 1–6 cm compound spikes, 3–5(–7)-flowered; bracts subtending branches leaflike, often slightly broader than leaves; perianth segments transversely winged proximally. Saline or alkaline wetland soils, riparian communities, edge of sagebrush, playas, 3400–8500 ft. NW, NC, NE, EC, C, WC, SC, SE.

Suaeda nigra (Raf.) J. F. Macbr. [*S. moquinii* (Torr.) Greene]—Bush seepweed. Shrubs, subshrubs, or facultative annuals, ± erect, 2–15 dm; stems brown to gray-brown, herbaceous stems green to dark red; leaves ascending to widely spreading, blades linear to narrowly lanceolate, subcylindrical to flattened, (5–)10–30 × 1–2 mm; glomerules confined to distal stems and branches, 1–12-flowered; bracts usually shorter than leaves, 3–15 mm. Alkaline soils, roadsides, lakeshores, riparian communities, salt desert scrub communities, 3250–8100 ft. Widespread except NC, EC.

SUCKLEYA A. Gray – Suckleya

Suckleya suckleyana (Torr.) Rydb.—Poison suckleya. Herbs, annual, monoecious; stems prostrate or ascending, usually purplish red, 5–30 cm, diffusely branched; leaves alternate, petiolate; blades rhombic-ovate to suborbicular, 1–3 × 0.5–2 cm, margins repand-dentate, sparsely covered with inflated unicellular trichomes (scurfy when dry); inflorescences of staminate and pistillate flowers in mixed clusters in axils of nearly all leaves; staminate flower perianth segments usually 4; stamens usually 4; rudimentary ovary present; pistillate flower perianth segments becoming marginally connate, 4-lobed; stigmas 2; bracteoles enlarged in fruit, winged, triangular-ovate or rhombic-ovate, 5–6 mm; wings 6, longitudinal, narrow, margins crenate-denticulate; utricles compressed. Agricultural lands, riparian areas, salt desert scrub, piñon-juniper woodlands, 4260–9000 ft. NE, NC, NW, SC, Hidalgo.

TIDESTROMIA Standl. – Honeysweet, espanta vaquero

Herbs or subshrubs, annual or perennial, glabrous to densely candelabriform-pubescent or with random barbed projections; stems prostrate to ascending; leaves opposite to alternate proximally, sessile or petiolate, blades lanceolate to orbicular; inflorescences axillary, subtended by 2 subopposite, involucral leaves becoming indurate and connate in age; flowers bisexual; tepals 5, distinct, keeled; stamens 5; filaments connate at base into low cups; pseudostaminodes absent or triangular short lobes; staminodes present or absent; ovule 1; styles absent or short; stigmas bilobed; utricules subglobose, indehiscent.

1. Plants subshrubby perennials; stem bases usually with buds…**T. suffruticosa**
1′. Plants herbaceous annuals; stem bases without buds…**T. lanuginosa**

Tidestromia lanuginosa (Nutt.) Standl.—Woolly tidestromia. Annual herbs; stems yellowish green to rarely reddish, to 50 cm, glabrous to lanuginose; trichomes completely candelabriform; buds absent on stem bases; leaves widely obovate to lanceolate, 0.8–3.2 × 0.9–3 cm; inflorescences 1–3-flowered; flowers 1.5–3 mm; tepals yellowish, glabrous or lanuginose; pseudostaminodes absent or short-triangular lobes; utricles 1.3–1.6 × 1–1.3 mm. Widespread except NC, NE.

1. Terminal cells of trichomes with irregular or spreading projections; pollen with microspines on tectum… **subsp. eliassoniana**
1′. Terminal cells of trichomes with only irregular projections; pollen lacking microspines on tectum…**subsp. lanuginosa**

subsp. eliassoniana Sánch. Pino & Flores Olv.—Rocky slopes, ca. 4400 ft.

subsp. lanuginosa—Rocky or sandy slopes, roadsides, dry washes, desert scrublands, shortgrass prairies, juniper-oak grasslands, 3140–6800 ft.

Tidestromia suffruticosa (Torr.) Standl.—Shrubby tidestromia. Subshrubs; stems gray-green to reddish, to 60 cm, glabrate to densely lanuginose, trichomes completely or partially candelabriform, buds commonly present, rarely absent on stem bases; leaves lanceolate to widely ovate or reniform, 0.6–4.5 × 0.4–2.7 cm; inflorescences 1–3(4)-flowered; flowers 1.7–3 mm; tepals yellowish or yellowish brown, densely lanuginose to glabrous; pseudostaminodes absent or triangular lobes; utricles 1–1.8 × 0.8–1.6 mm. Our plants belong to **var. suffruticosa**. Rocky slopes and arroyos, often on limestone, 1130–5600 ft. SW, SC, Chaves.

AMARYLLIDACEAE – AMARYLLIS FAMILY

Steve L. O'Kane, Jr.

In our area a family of one genus and one species. In APG (2016), this is a much larger family that includes Alliaceae s.s. Here we follow Simpson (2019).

HABRANTHUS Herb. – Copper-lily

Habranthus longifolius (Hemsl.) Flagg, G. Lom. Sm. & Meerow [*Zephyranthes longifolia* Hemsl.]— Longleaf copper-lily. Herbs, perennial, scapose, from bulbs; leaves sessile, blades linear, dull green, to 1(–2) mm wide; inflorescence 1-flowered, with a spathe; flowers actinomorphic; perianth yellow, funnelform, 2.1–2.8 cm, connate basally into green tube ca. 1/4 perianth length; stamens 6, of 2 unequal sets; filaments filiform, 0.7–1.1 cm; anthers in 2 nonoverlapping sets, 3–6 mm; style longer than perianth tube; stigma trifid, usually among anthers. Sandy, gravelly, calcareous, alkaline soils, piñon-juniper, desert scrub, desert grassland, bajadas, 3250–7300(–?8350) ft. NE, EC, WC, SE, SC, SW (Flagg et al., 2010).

ANACAMPSEROTACEAE – ANACAMPSEROS FAMILY

David J. Ferguson

TALINOPSIS A. Gray – Arroyo fameflower

Talinopsis frutescens A. Gray [*Grahamia frutescens* (A. Gray) G. D. Rowley]—Pink arroyo fame-flower. Suffrutescent to woody subshrub to 1.5 m; stems erect, glabrous; leaves succulent, terete, in whorls of (2-)4-6(-8) per node, sometimes in axillary fascicles on older stems; inflorescences terminal and lateral, cymose, with ovoid to terete mostly paired foliaceous bracts; flowers opening in afternoon, sessile, 10-19 mm diam.; sepals 2; petals usually 5, (ours) pink, magenta, or white; stamens mostly 20-25; capsule ovoid-conic, 9-18 mm, dehiscent; seeds brown, 2 mm. Broken terrain on limestone substrates, Chihuahuan desert scrub, 4250-6000 ft. (Nyffeler & Eggli, 2010; Packer, 2003).

ANACARDIACEAE – SUMAC FAMILY

Kenneth D. Heil

Woody shrubs or vines; leaves alternate, pinnately or ternately compound; inflorescence a terminal or axillary thyrse, or panicle, eventually cymose; flowers small, actinomorphic, perfect or imperfect and then with reduced parts of the opposite sex; fruit a drupe, often with a waxy or oily mesocarp (Acker-field, 2015; Anderson, 2006).

1. Leaflets 3, mostly 5-11 cm; berries white to cream…**Toxicodendron**
1'. Leaflets 3-9, mostly < 5 cm; berries red to orange…**Rhus**

RHUS L. – Sumac

Shrubs; sap usually acrid or resinous; leaves simple or odd-pinnately compound; flowers mostly 5-mer-ous, inconspicuous, greenish, yellowish, or whitish; ovary surrounded by a flattened, lobed disk; stamens inserted; fruit a small, dry, yellow or red drupe.

1. Leaves simple, rarely trifoliate; Peloncillo and Big Hatchet Mountains…**R. ovata**
1'. Leaves with mostly 1-5+ leaflets; more widespread…(2)
2. Leaves with 1-3 leaflets…**R. trilobata**
2'. Leaves with 5+ leaflets…(3)
3. Rachis of leaves winged…(4)
3'. Rachis of leaves not winged…(5)
4. Flowers appearing before the leaves; stems densely branched; leaflets < 2 cm…**R. microphylla**
4'. Flowers appearing after the leaves; stems not highly numerous, not tangled; leaflets > 3 cm…
   R. lanceolata
5. Leaflets 3-5…**R. virens**
5'. Leaflets 11-25…**R. glabra**

Rhus glabra L.—Smooth sumac. Mostly shrubs, 1-3 m, bark dark gray; leaflets 11-30, narrowly lanceolate to oblong, 2-8 cm, serrate, sessile, dark green above, paler beneath; inflorescence a dense thyrse, 12-25 cm; flower petals to 3 mm, white; fruit subglobse, 4-5 mm, dark red, glandular-pubescent. Rich soil in oak and ponderosa pine woodlands, often in canyons and riparian zones, 5350-8700 ft. NC, C, WC, SC, SW, San Miguel.

Rhus lanceolata (A. Gray) Britton [*R. copallinum* L. var. *lanceolata* A. Gray]—Prairie sumac. Shrubs to 3 m, branches grayish puberulent; leaves deciduous, leaflets 11-21, oblong-lanceolate to lanceolate, 25-75 mm, acuminate, upper surfaces dark green, lower surfaces paler, the rachis winged; flowers in terminal panicles, petals yellowish green to white, 3 mm; stamens exserted; fruit subglobose, 4-5 mm, dark red, glandular-hairy. Limestone, riparian woodlands with oak, juniper, walnut, 4000-6000 ft. SE, SC.

Rhus microphylla Engelm.—Little leaf sumac. Densely branched shrubs to 2 m, old bark dark gray, stiff and spinescent, puberulent to glabrate; leaves 12–20 mm, deciduous, odd-pinnately compound, 5–9-foliate, with a winged rachis; leaflets sessile, elliptic, 6–9 mm; inflorescences small, dense spikes, terminal and axillary, 8–12 mm; flower petals appearing early, before or with the first leaves, to 3 mm, cream; fruit ovoid, 5–7 mm diam., dark red to orange, glandular-hairy. Gravelly mesas and rocky hillsides, often on limestone, Chihuahuan desert scrub, semidesert grassland, oak, dry washes, riparian woodlands, 3446–6500 ft. NE, EC, SE, SC, SW. Bernalillo.

Rhus ovata S. Watson—Sugar bush. Evergreen shrubs or small trees to 5 m; old bark shaggy; leaves simple, entire, mostly ovate, 4–8.5 × 3–5 cm, leathery, bright green, glabrous, petioles 10–20 mm; inflorescence of dense panicles, 2.5–3.5 cm; flowers to 5 mm; sepals magenta; petals cream to pinkish; fruit lenticular-orbicular, 5–7 mm diam., dark reddish, glandular-pubescent, viscid. Rocky hillsides with interior chaparral; Peloncillo and Big Hatchet Mountains, ca. 5900 ft. Hidalgo.

Rhus trilobata Nutt. [*R. aromatica* Aiton]—Threeleaf sumac, lemonade sumac. Shrubs with spreading branches, sometimes forming thickets, to 3 m; bark gray, lenticular; leaves trifoliate or palmately lobed to simple and unlobed, leaflets sessile, ovate to rhombic, crenate to deeply lobed, glabrous to puberulent, terminal leaflet 15–35 mm, foliage thin, deciduous and dark red in the fall; inflorescences short dense panicles of compound spikes, arising from lateral branches, and in most varieties appearing early before the leaves in the spring, 10–15 mm; flowers to 3 mm, petals obovate, pale yellow, glabrous; fruit lenticular-orbicular, 6–8 mm diam., dull orange to dark reddish, villous and/or short-glandular-pubescent, viscid. Widespread.

1. Leaves unifoliate, or if trifoliate, then the lateral leaflets much smaller than the terminal; NW New Mexico…**var. simplicifolia**
1'. Leaves always trifoliate, the lateral leaflets not conspicuously smaller than the terminal…(2)
2. Young twigs densely velvety-pilose, the hairs usually yellowish, leaves tomentose beneath…**var. pilosissima**
2'. Young twigs glabrescent to puberulent, the hairs not yellowish, leaves not tomentose beneath…(3)
3. Terminal leaflet much longer than wide…**var. trilobata**
3'. Terminal leaflet slightly longer than wide…(4)
4. Flowers appearing after the leaves mature; pedicels 5–8 mm…**var. racemulosa**
4'. Flowers appearing before or with the leaves; pedicels 2–4 mm…(5)
5. Terminal leaflet deeply lobed; leaflets somewhat pubescent…**var. quinata**
5'. Terminal leaflet not deeply lobed; leaflets mostly glabrate…**var. anisophylla**

var. anisophylla (Greene) Jeps.—Terminal leaflet slightly longer than wide, leaflets glabrous above, minutely pubescent or glabrate beneath. Rocky slopes, canyon bottoms, riparian areas, 5300–7400 ft.

var. pilosissima Engl.—Terminal leaflet slightly longer than wide; young branches densely soft-pilose, tomentose beneath. Canyons and hillsides, shortgrass prairie, juniper savanna, piñon-juniper woodlands, 3900–8000 ft.

var. quinata Jeps.—Terminal leaflet slightly longer than wide, deeply lobed, leaflets persistently pubescent. Canyons, hillsides, mesas, chaparral and piñon-juniper woodlands, 5500–7300 ft.

var. racemulosa (Greene) F. A. Barkley—Terminal leaflet slightly longer than wide; flowering after maturation of the leaves. Canyons, hillsides, piñon-juniper woodlands, pine and Douglas-fir communities, 5185–9500 ft.

var. simplicifolia (Greene) F. A. Barkley—Leaves unifoliate or if trifoliate, then the lateral leaflets much smaller than the terminal; pubescent. Salt desert scrub, slickrock and juniper communities, 5000–6500 ft.

var. trilobata—Terminal leaflet much longer than wide, glabrous above, minutely pubescent or glabrate beneath. Geographically and ecologically wide-ranging. Rocky ledges and slopes to canyon and river bottoms, deserts, grasslands, piñon-juniper woodlands, ponderosa pine forests, riparian zones, 5000–7500 ft.

Rhus virens Lindh. ex A. Gray [*R. virens* var. *choriophylla* (Wooton & Standl.) L. D. Benson]—Evergreen sumac. Sparsely branched shrubs or small trees to 3 m, bark gray, lenticular; leaves evergreen, petiolate, odd-pinnately compound, 5-9-foliate, leaflets lanceolate or elliptic to ovate, 25-50 mm, leathery, dull green above, paler and puberulent to glabrate beneath; inflorescence an open panicle, to 8 cm, terminal and axillary; flower petals to 5 mm, cream; fruit lenticular-orbicular, to 6 mm diam., orange, glandular-pubescent. Rocky hillsides, steep slopes, canyons, upper edge of Chihuahuan desert to semidesert grassland, 4000-7500 ft. SE, SC, SW.

TOXICODENDRON Mill. – Poison ivy, poison oak

Toxicodendron rydbergii (Small ex Rydb.) Greene—Allergenic, rhizomatous shrubs; stems pubescent, red-brown, resinous, weakly branched; leaves 3-foliate (rarely 4- or 5-foliate), leaflets shiny green, glabrous or sometimes minutely strigose, ovate, 2-12 cm, lateral leaflets smaller than terminal leaflet; inflorescence an open raceme or panicle; flowers dioecious, 5-merous, small, petals 2-3 mm, yellowish green, 3-lobed; fruit a drupe, fleshy, globose, shiny white with yellow or green tints, 4-6 mm. Moist, shaded canyons, stream banks, around springs, 3500-8500 ft. Widespread except EC, SE, Doña Ana, Luna.

APIACEAE (UMBELLIFERAE) – PARSLEY FAMILY

Sherel Goodrich and Kenneth D. Heil

Annual, biennial, or perennial acaulescent or caulescent herbs from taproots, fibrous or tuberous roots, or rhizomes; leaves simple to decompound, petioles typically sheathing at base or long-petiolate or the upper leaves reduced to dilated sheaths; inflorescence of compound umbels, the primary umbels with or without a subtending involucre of bracts, the secondary umbels (umbellets) with or without a subtending involucel of bractlets; flowers mostly regular; sepals 5 or absent; petals 5, small, white, yellow, or purple; stamens 5, small, alternate with the petals; pistil 1, ovary inferior, bicarpellate, 2-loculed, with 1 ovule per locule, 2 styles with or without a conic base (stylopodium); fruit a schizocarp of 2 mericarps (Ackerfield, 2015; Allred & Ivey, 2012; Hartman et al., 2013; Martin & Hutchins, 1981c; New Mexico Rare Plant Technical Council, 1999).

1.   At least lower leaves simple…(2)
1'.   Leaves compound…(5)
2.   Schizocarps in burlike heads…**Eryngium**
2'.   Schizocarps not in burlike heads…(3)
3.   Leaves peltate…**Hydrocotyle**
3'.   Leaves not peltate…(4)
4.   Leaves entire, linear to narrowly oblong…**Bupleurum**
4'.   Leaves palmately lobed…**Bowlesia**
5.   Plants annual…(6)
5'.   Plants perennial…(14)
6.   Lower leaflets palmately lobed…**Coriandrum**
6'.   Leaflets not palmately lobed…(7)
7.   Leaves 1-pinnate, lower leaves with broad leaflets, upper leaves with linear lobes…**Eurytaenia**
7'.   Leaves not as above…(8)
8.   Fruit glabrous…**Cyclospermum**
8'.   Fruit scaberulous to hairy or with prickles…(9)
9.   Involucre well developed, the bracts 1+ times pinnate…(10)
9'.   Involucre lacking or of a single linear bract; fruit various…(11)
10.   Involucral bracts similar to the leaves; fruit with uncinate prickles on the ribs…**Yabea**
10'. Involucre pinnate-dissected but much different from the leaves; fruit with barbed prickles…**Daucus**
11.   Stems retrorsely and leaves antrorsely rough-scabrous; primary leaflets with ± lanceolate lobes or teeth; fruit with glochidiate and minutely barbellate prickles…**Torilis**
11'. Stems glabrous or scabrous on ridges distally; leaves glabrous or scabrous on margins and midribs; leaflets divided into linear, oblong, or spatulate, narrow segments; fruit various…(12)

12. Fruit without secondary ribs; stems sparsely retrorsely hispid below…**Chaerophyllum**

12'. Fruit with or without secondary ribs; stems glabrous…(13)

13. Leaflet and bractlet margins and midribs prominently scabrous; fruit ovoid-oblong to urceolate-ovoid or broadly ellipsoid, ribs sparsely to densely scaberulous with single-celled papillalike projections… **Ammoselinum**

13'. Herbage glabrous; fruit broadly ovoid to ellipsoid or elliptic-ovoid, ribs and intervals variously hairy or at least tuberculate…**Spermolepis**

14. Axils of upper leaves bearing bulblets; umbels rarely bearing fruit; plants of wet places…**Cicuta**

14'. Axils of leaves without bulblets; umbels bearing fruit; plants not restricted to wet places…(15)

15. Leaves palmately cleft into 5–9 toothed segments; flowers yellow; fruits covered with stout, hooked prickles…**Sanicula**

15'. Leaves various; fruits not covered by hooked prickles, or if so, then flowers white (*Daucus*)…(16)

16. Plants caulescent; pseudoscape absent; peduncles few to several, mostly shorter than the leafy stem; styles rarely > 1 mm; stylopodium present; petals white…(17)

16'. Plants acaulescent, the leaves sometimes whorled atop a pseudoscape, or if subcaulescent, the usually solitary peduncle longer than the short leafy stem, and lateral umbels (if any) typically borne on the lower 1/3 of the plant; styles often > 1 mm; stylopodium absent or present; petals yellow, white, or purple…(18)

17. Leaves pinnate or ternate; leaflets mostly sessile…**KEY 1**

17'. Leaves various, often pinnately decompound; leaflets usually petiolulate, at least the primary ones (sessile in *Carum*)…**KEY 2**

18. Leaves ternate or biternate with 3–9 leaflets or rarely a few simple, usually only 2–3 per plant; petals white…**Orogenia**

18'. Leaves and leaflets not as above, or if so, plants mostly taller and/or petals yellow…**KEY 3**

KEY 1: PLANTS CAULESCENT; PEDUNCLES AND UMBELS MOSTLY SHORTER THAN THE STEM; STYLOPODIUM PRESENT (EXCEPT *CYMOPTERUS*); LEAVES PINNATE OR TERNATE; LEAFLETS SESSILE

1. Leaflets entire, linear or linear-elliptic, mostly 2–5 cm; petals yellow when fresh; fruit 6–8 mm… **Cymopterus**

1'. Leaflets toothed and/or lobed, not linear; petals yellow or white; fruit various…(2)

2. Leaves ternate, the upper ones sometimes simple, the leaflets 8–36 cm, about as wide; plants 9 dm–2+ m…(3)

2'. Leaves pinnate, the leaflets < 8 cm and much narrower; plants shorter or not villous-woolly…(4)

3. Involucre and involucels well developed, sometimes spreading or deflexed, the bractlets (2–)4–12; fruit 1.5–3 mm, the ribs not winged; plants of very wet places; often growing in water from fibrous roots…(4)

3'. Involucre lacking or infrequently of 1 or 2 bracts; involucels often lacking; fruit often 3 mm or else the ribs winged; plants of various habitats, from a taproot or tuberous root…(5)

4. Stems often sprawling, sometimes stoloniferous; leaves with (3–)5–15 opposite pairs of leaflets, these 0.3–4(–6.5) cm; margins, especially of the upper leaves, usually irregularly toothed or cleft; rays 4–16; ribs of the fruit obscure…**Berula**

4'. Stems erect, not stoloniferous; leaves with 4–6 opposite pairs of leaflets, these 2–8(–15) cm, margins mostly evenly serrate; rays 11–24; ribs of the fruit prominently corky…**Sium**

5. Umbels often > 7 per stem; fruit strongly flattened dorsally, 5–8 × 3–6 mm, the lateral ribs slightly winged, the dorsal ribs filiform; petals greenish yellow or reddish; plants adventive or cultivated…**Pastinaca**

5'. Umbels < 7 per stem; fruit not strongly flattened dorsally, or if so, 3–5 mm; petals white or greenish; plants native…(6)

6. Fruit linear to narrowly clavate, > 10 mm; leaves 1–3-ternate or ternate-pinnate; peduncles mostly not subtended by dilated, bladeless sheaths, or these greatly reduced…**Osmorhiza**

6'. Fruit elliptic to broadly oblong, 3–6 mm; leaves mostly 1-pinnate; peduncles often with subtending dilated sheaths…(7)

7. Fruit strongly flattened, the dorsal ribs filiform, the lateral ribs conspicuously winged; plants with tuberous roots…**Oxypolis**

7'. Fruit cross-section rounded in outline, the dorsal and lateral ribs prominently thickened wings; plants from taproots…**Angelica**

KEY 2: PLANTS CAULESCENT; PEDUNCLES AND UMBELS MOSTLY SHORTER THAN THE STEMS; STYLOPODIUM PRESENT (EXCEPT *LOMATIUM*); LEAVES > 1-COMPOUND; PRIMARY LEAFLETS NOT SESSILE

1. Ultimate leaf segments, in part, > 2 cm, toothed or lobed, but not entire or pinnatifid…(2)

1'. Ultimate leaf segments < 2 cm, or if longer, entire or pinnatifid…(5)

2. Involucels of mostly 6 bractlets, 1–4 mm; umbels 6–20+ per stem, the rays 15–26, 1.5–4 cm; fruit 1.5–4 mm, the ribs corky...**Cicuta**

2'. Involucels mostly absent, umbels often < 6 per stem and/or the rays either fewer or longer than above or both; fruit various...(3)

3. Fruit < 2 mm...**Apium**

3'. Fruit > 2 mm...(4)

4. Fruit linear to clavate, (10–)12–25 mm, bristly-hispid (except in *Osmorhiza occidentalis*), the dorsal ribs not prominent; leaflets often hirtellous; dilated sheaths seldom subtending the peduncles...**Osmorhiza**

4'. Fruit oblong to elliptic, 4–5 mm, not bristly-hispid, the dorsal ribs with small wings; leaflets glabrous; peduncles often subtended by dilated bladeless or nearly bladeless sheaths...**Angelica**

5. Fruits and ovaries with bristly hairs; involucre often of pinnatifid or compound bracts...**Daucus**

5'. Fruits and ovaries without bristly hairs; involucre mostly of entire bracts...(6)

6. Involucel and involucre absent or much reduced...(7)

6'. Involucel and involucre present...(10)

7. Petals yellow; plants introduced; ultimate segments of leaves filiform, 1–40 × ± 0.5 mm wide... **Foeniculum**

7'. Petals white or yellow (in *Lomatium*) and plants native; ultimate segments various, often > 0.5 mm wide. (8)

8. Petals yellow, or if white then plants pubescent; fruit 6–20 mm, dorsally flattened, the dorsal ribs filiform, the lateral ribs winged; stylopodium absent...**Lomatium**

8'. Petals white; plants glabrous; fruit 3–8 mm, rounded, the dorsal and lateral ribs narrowly winged; stylopodium low-conic...(9)

9. Primary leaflets sessile...**Carum**

9'. Primary leaflets on petiolules...**Ligusticum**

10. Annuals, 9–12 dm tall; leaves 10 × 11 cm, triangular-ovate; fruit 3–4 mm...**Daucosma**

10'. Biennials or perennials, to 3 m tall; leaf length various; fruit (2)4–12 mm...(11)

11. Plants up 1–3 m tall; leaf blades 2–5 × 2–5 dm; fruit 8–12 mm...**Heracleum**

11'. Plants without the above characters...(12)

12. Stems usually purple-spotted, usually much branched, mostly with 10–30+ umbels; plants 5–30 dm; involucre of 2–6 bracts, 2–6(–15) mm; naturalized, weedy in disturbed mesic sites...**Conium**

12'. Stems not purple-spotted, few-branched, with (1–)3–7(–12) umbels; plants to 10 dm; involucre absent or seldom as above; native, often montane...(13)

13. Ultimate leaf segments < 10, at least some well > 3 cm...**Perideridia**

13'. Ultimate leaf segments > 10, < 3 cm...(14)

14. Involucels with 2, rarely 3 bractlets...**Ligusticum**

14'. Involucels usually with > 3 bractlets...(15)

15. Fruit glabrous, stems glabrous...**Conioselinum**

15'. Fruit granular-roughened, stems hirtellous just below the umbel...**Harbouria**

KEY 3: PLANTS TYPICALLY ACAULESCENT; STYLES OFTEN > 1 MM; STYLOPODIUM ABSENT (EXCEPT *PODISTERA*)

1. Fruit strongly flattened dorsally, dorsal ribs filiform, not winged, the lateral ribs winged, body 8–18(–20) mm, or if shorter, usually pubescent; involucre absent...**Lomatium**

1'. Fruit not strongly flattened, or if so, the dorsal ribs winged, at least in part, the body usually < 8 mm, the wings sometimes to 12(–15) × 2–2.5 mm, especially in plants with conspicuous involucres; plants glabrous to hirtellous...(2)

2. Stylopodium conic; leaves 1-pinnately compound with palmatifid leaflets; bractlets of involucels, in part, with 2 or 3+ apical teeth...**Podistera**

2'. Stylopodium absent; leaves either more than 1-compound or leaflets not palmatifid; bractlets entire... (3)

3. Bractlets of involucel large, showy, scarious, white to purple, ± connate below, tending to form a cup around the umbellet...**Vesper**

3'. Bractlets of involucel (and bracts of involucre, if present) relatively small and inconspicuous, or larger but herbaceous or coriaceous...(4)

4. At least some dorsal ribs of fruit ± winged, the body flattened dorsally...**Cymopterus**

4'. Dorsal ribs of fruit prominent but not winged, the body slightly flattened laterally...(5)

5. Leaves 1-pinnate with sessile, linear leaflets...**Neoparrya**

5'. Leaves not as above...**Aletes**

ALETES J. M. Coult. & Rose – Indian parsley

Acaulescent or caulescent perennial herbs, with taproot surmounted by a caudex; involucel of narrow bractlets; flowers yellow; stylopodium lacking; fruit compressed laterally; ribs obscure or corky-winged.

1. Leaflets sessile or nearly so, ultimate segments not filiform…**A. acaulis**
1'. Primary leaflets with petiolules, ultimate leaf segments filiform or nearly so…**A. filifolius**

Aletes acaulis (Torr.) J. M. Coult. & Rose [*Deweya acaulis* Torr.]—Stemless Indian parsley. Plants 5–35 cm, scabrous; leaflets 4–15 mm, lanceolate to orbicular or fan-shaped, pinnately divided or some essentially palmately divided, dentate; involucel bractlets 2–3 mm, lanceolate or linear; rays 8–15, 19–20 mm, subequal, spreading to reflexed; fruit 4–7 mm, with corky wings. Rocky areas, mixed conifer and aspen forests, 5950–10,680 ft. C, WC, SE, SW.

Aletes filifolius Mathias, Constance & W. L. Theob.—Trans-Pecos Indian parsley. Plants to 40 cm, caulescent; leaves ternate-pinnate with linear to filiform ultimate segments, blades 2.5–20 cm; bracts none; rays 4–21, 6–20 mm; bractlets linear to lanceolate, 2–5 mm, free to slightly connate; fruit oblong to ovoid-oblong, 2.4–8 mm, with corky wings. Rocky places, bluffs, mountain summits, stream and canyon bottoms, mostly on limestone, piñon-juniper and mixed conifer communities, 4500–9500 ft. C, SE, SC, SW.

AMMOSELINUM Torr. & A. Gray – Sandparsley

Ammoselinum popei Torr. & A. Gray—Plains sandparsley. Annual, odorless or faintly aromatic, taprooted herbs; plants 8–35(–60) cm; leaves alternate, the basal 3-ternately compound, cauline 2–3-ternately compound; leaf blades oblong in outline, 1–4 cm, ultimate segments 2–10 mm; leaflet and bractlet margins and midribs prominently scabrous; umbels compound; rays (0–)6–25 mm; pedicels (0–)2–5 mm; involucel bractlets linear and entire or less commonly 2–3 times divided; sepals obsolete or greatly reduced; fruit 3–5 mm, dorsal ribs rounded, densely and coarsely papillate-scaberulous with translucent, multicelled, convex, apically rounded, papillalike hairs, lateral ribs corky-thickened. Calcareous soils in SE plains and Chihuahuan desert scrub, 3110–4675 ft. EC, C, SE, SC.

ANGELICA L. – Angelica

Plants perennial, caulescent, single-stemmed herbs from a stout taproot; leaves pinnately to ternately 1–3 times compound, with broad leaflets; lower blades on elongate petioles, the middle ones often arising directly from a dilated sheath, the upper ones often much reduced or absent and the leaves reduced to a dilated sheath; inflorescence an open umbel; involucre and involucel absent or of narrow, scarious or foliaceous bracts or bractlets; calyx teeth minute or obsolete; petals white, seldom pink or yellow; fruit elliptic-oblong to orbicular, strongly compressed dorsally, the lateral and dorsal ribs with small but obvious wings, or the ribs all corky-thickened and scarcely winged.

1. Fruit 7–8 mm; umbels globose; plants mostly > 1 m…**A. ampla**
1'. Fruit 4–5 mm; umbels ± flat-topped; plants mostly < 1 m…(2)
2. Ovaries and immature fruit glabrous; bractlets of involucels linear-lanceolate to lanceolate, usually > 1 mm wide; flowers usually purplish brown…**A. grayi**
2'. Ovaries and mature fruit scabrous to hispidulous; bractlets of involucels absent, filiform, or narrowly linear, not > 1 mm wide; flowers white to pink…**A. pinnata**

Angelica ampla A. Nelson—Giant angelica. Plants usually > 1 m; leaves ternate then bipinnate, leaflets 3–20 mm, ovate, serrate; involucel bractlets few, filiform; rays 30–40; umbels globular; petals white; fruit 7–8 mm, oblong-oval, the ribs narrowly winged. Montane; moist places, especially along streams, 6000–9000 ft. Colfax, Otero.

Angelica grayi J. M. Coult. & Rose—Gray's angelica. Plants 20–60 cm, stout, mostly > 1 cm thick at the base; leaves pinnate to bipinnate or ternate-pinnate, the middle division larger; leaflets 1–5 cm, ovate to lanceolate, serrate or sometimes lobed; involucre wanting or of foliaceous bracts; rays many;

involucel of bractlets 5-18 mm, usually > 1 mm wide, linear-lanceolate to lanceolate; pedicels 2-6 mm; flowers purplish brown; fruit 4-5 mm, glabrous, oval, dorsal ribs narrowly winged, lateral ribs broader-winged. Alpine scree slopes and mountain meadows, 9800-13,000 ft. NE, NC, NW.

Angelica pinnata S. Watson [*A. leporina* S. Watson]—Small-leaved angelica. Plants 4.5-10(-15) dm, glabrous or nearly so, except scabrous to hirtellous in the inflorescence, without persistent leaf bases, from a taproot and sometimes branched crown; leaves pinnate or partly bipinnate; rays 7-14, 2-8.5 cm, scabrous to hirtellous; involucels absent or very rarely of 1+ green to scarious, linear or nearly linear bractlets, 3-13 mm; pedicels 3-7 mm; petals white; fruit 4-5 mm, the lateral wings ca. 1 mm wide, the dorsal wings ca. 0.5 mm wide. Oak, maple, aspen, Douglas-fir, spruce-fir, willow, wet meadow communities, very often along streams or around seeps, 7100-12,000 ft. NC, NW.

APIUM L. - Celery

Apium graveolens L.—Wild celery. Annual to perennial caulescent glabrous herbs from fibrous roots; plants mostly 5-15 dm, freely branched; leaves pinnately 1-compound to ternately compound, the ultimate segments filiform to suborbicular; basal leaves long-petiolate, becoming smaller upward, pinnately compound with 1-4 pairs of primary pinnae, these deeply cleft into 3 coarsely few-toothed segments; umbels compound, short-pedunculate or some lateral ones sessile; involucre lacking or of a few foliose bracts; involucel lacking; pedicels ca. 5 mm; calyx teeth minute or obsolete; styles short, spreading, the stylopodium short-conic; carpophore entire to deeply bifid; fruit flattened laterally, ca. 1.5 mm, about as wide, the ribs strongly raised. Moist places in S foothills and mountains, 3640-6600 ft. SE, SC, SW, Guadalupe.

BERULA W. D. J. Koch - Water parsnip

Berula erecta (Huds.) Coville—Cutleaf water parsnip. Perennial, acaulescent, glabrous herbs, often stoloniferous; plants 5-10 dm, from numerous fibrous roots; leaves pinnately compound with (3-)5-15 opposite pairs of lateral leaflets, or submerged leaves (if present) often with filiform-dissected blades; leaf blades 2-31 cm; leaflets 0.3-4(-6.5) cm, sessile, toothed to incised or occasionally a few entire; inflorescence an open umbel; bracts of involucre 1-6, 2-15(-25) mm, linear or elliptic, entire, toothed, or rarely pinnatifid; rays 4-16, 0.5-2.5(-4) mm; flowers white; fruit mostly 2 mm, the ribs obscure. In mud and water of streams, seeps, springs, marshes, swamps, margins of ponds and lakes, wet hanging gardens, 3800-8380 ft. Widespread. Introduced.

BOWLESIA Ruiz & Pav. - Bowlesia

Bowlesia incana Ruiz & Pav. [*B. septentrionalis* J. M. Coult. & Rose]—Hoary bowlesia. Slender branching annuals with stellate pubescence; plants delicate with weak trailing stems, 1-5 dm, dichotomously branched; petioles slender, 2-8 cm; leaves opposite; blades reniform to cordate, 0.5-3 cm broad, 5-9-lobed; umbels with 1-6 flowers, yellowish green; fruit 1-1.5 mm, ovate to round, inflated, stellate-hairy, prickly or nearly glabrous. Shade of rocks, trees, shrubs; canyons in SW New Mexico, 4500-5100 ft. Grant, Luna.

BUPLEURUM L. - Hare's ear

Bupleurum americanum J. M. Coult. & Rose [*B. triradiatum* Adams ex Hoffm. subsp. *arcticum* (Regel) Hultén]—American thorow-wax. Caulescent perennials from a branching caudex and taproot, the caudex sheathed with dark brown leaf bases, the dead leaves often coiled upon drying; stems few to several, mostly simple, 1.5-5(-7) dm; leaves simple, basal and cauline, 2-25 cm, narrowly oblong to linear, with 3-5 prominent, parallel veins; rays 1-8(-14), 0.5-5 cm; involucre of 1 to several unequal, lanceolate to ovate bracts; involucel of 5-8 ovate to lanceolate bractlets; petals yellow, greenish, or purple; ovaries glabrous; fruit glabrous, 3-4 mm, the ribs raised but wingless. Piñon-juniper woodlands in Lincoln County, elevation unknown.

CARUM L. - Carum

Carum carvi L.—Caraway. Perennial, caulescent, glabrous herbs from taproots; plants 3–6(–10) dm; leaves 2–3 times pinnate, blades 5–16 cm, oblong in outline, ultimate segments 2–8(–15) × 0.5–2 mm, linear and entire or obovate and toothed; involucre lacking or inconspicuous; rays 1.5–8 cm; involucels lacking or of minute scarious teeth; pedicels (5–)8–20 mm; petals white; filaments white, anthers pale green or whitish; fruit 3–4 mm, the ribs filiform. An escape from gardens in N New Mexico, 7120–9400 ft. Colfax, Taos. Introduced.

CHAEROPHYLLUM L. - Chervil

Chaerophyllum tainturieri Hook.—Chervil. Weedy annuals; stems erect, solitary, usually branched near base, 1.5–9 dm, densely retrorsely hispid below to glabrate, sparsely hispid above; leaves ternate-pinnately decompound, lobes linear to ovate, glabrous to ± hispid; umbels simple or with 2 or 3 rays; involucel of several conspicuous ovate, rounded to acute, ciliate-margined bractlets; flowers white, 3–10 per umbellet; calyx teeth obsolete; fruit 3–10 on clavate pedicels, ribs wider than spaces between ribs, glabrous. A single collection in the Guadalupe Mountains; cobbles in a dry channel, 4880 ft. Eddy.

CICUTA L. - Water Hemlock

Cicuta maculata L. [*C. douglasii* (DC.) J. M. Coult. & Rose, misapplied]—Spotted water hemlock. Perennial, caulescent, glabrous, violently poisonous herbs; plants 6–21+ dm, stems 5–15+ mm diam., hollow, with clusters of fibrous roots surmounted by a thickened crown; leaves pinnate or ternate-pinnate, leaflets 2–11 × 3–25 mm, lanceolate to linear, serrate; involucre absent or of 1 or few linear bracts to 1 cm; rays 15–26, 1.5–4 cm; bractlets of involucels ca. 6, 1–4 mm, linear or narrowly deltoid; petals white; stamens white; fruit 2–4 mm, oval to globose, the ribs prominent, ± corky, green. This plant is extremely poisonous. You need mature fruits to distinguish between varieties.

1. Style usually < 1 mm; fruit usually about as wide as long, constricted at commissure; principal stem leaflets > 5 times longer than wide…**var. angustifolia**
1'. Style usually > 1 mm; fruit longer than wide, constricted or not constricted at commissure; principal stem leaflets < 5 times longer than wide…**var. maculata**

var. angustifolia Hook.—Spotted water hemlock. Widespread in New Mexico. Marshes, stream banks, ditches, 5300–10,300 ft. NC, NW, SC.

var. maculata—Spotted water hemlock. Marshes, edges of ponds, stream banks, ditches, wet prairie depressions. Lincoln, San Miguel, Santa Fe.

CONIOSELINUM Hoffm. - Hemlock-parsley

Conioselinum scopulorum (A. Gray) J. M. Coult. & Rose [*Ligusticum scopulorum* A. Gray]—Rocky Mountain hemlock-parsley. Perennial, ± caulescent herbs; plants 3–10 dm, glabrous except in inflorescence, from a fusiform taproot with simple or very sparingly branched crown; leaves pinnate or ternate-pinnate, leaf blades 3.5–19 cm, ovate in outline, ultimate segments 2–15 × 1–5 mm; involucre absent or of 1 or few linear bracts to 1 cm; rays 9–15, 1.5–5 cm; involucels of 3–6 linear or linear-filiform bractlets, 2–8 mm; petals white; stamens white; fruit 4–6 mm, lateral ribs narrowly corky-winged, dorsal ones not winged. Streamsides and meadows in the mountains, 6700–12,000 ft. NE, NC, NW, C, WC, SC.

CONIUM L. - Poison hemlock

Conium maculatum L.—Poison hemlock. Biennial, caulescent, glabrous herbs from stout taproots with purple-spotted, freely branching hollow stems; plants 5–30 dm; leaves pinnate or ternate-pinnately decompound, the lower ones usually 2+-pinnate and then pinnatifid, the upper 1-pinnate, pinnatifid, sessile; ultimate leaflets pinnatifid, the lobes entire or toothed, the widest confluent portions 2–5(–10) mm wide; involucral bracts 2–6, 2–6(–15) mm, entire and ovate or deltoid; rays 9–16, 1–4 cm; bractlets of involucels 4–6, 1–3 mm, shaped like the involucral bracts; pedicels 2–6 mm; petals white; stamens white; fruit 2–2.5 mm, the ribs prominently ridged, narrower than the intervals. Roadside weed of ditches, 4400–8860 ft. NE, NW, C, WC, SC.

CORIANDRUM L – Coriander

Coriandrum sativum L.—Coriander. Plants annual, caulescent, 2–7 dm, glabrous; lower leaves ternately or pinnately divided into ovate or obovate segments 1–2 cm, variously toothed or incised; cauline leaves decompound into linear segments; rays 2–8, 1–2.5 cm; inflorescence of loose compound umbels; petals white to rose; fruit 3–5 mm, subglobose, hard, with slender ribs. Disturbed areas, 3610–7145 ft. Scattered. Introduced.

CYCLOSPERMUM Lag. – Marsh parsley

Cyclospermum leptophyllum (Pers.) Sprague ex Britton & P. Wilson—Marsh parsley. Annual herbs; plants glabrous, 15–70 cm; leaves 3–4 times pinnately dissected, the lower petiolate, the upper sessile, the ultimate segments filiform, 1.5–7 x 0.5–1 mm; petals white; fruit ribbed, carpophore shortly 2-cleft, 1.2–3 mm, glabrous. Weed of lawns, parks, ballfields, 3935–4445 ft. C, SC, SW, Cibola, Lea.

CYMOPTERUS Raf. – Spring parsley

Perennial, acaulescent or subcaulescent, glabrous or scabrous herbs from slender to greatly enlarged and tuberlike taproots to branching woody caudices; leaves all basal or basal and 1 to few cauline mostly on lower 1/2 of stems, ternate to pinnate or ternate-pinnately compound, rarely simple and ternately cleft; inflorescence an open to congested umbel, solitary to several; involucel of separate or united bractlets; rays few to several; involucral bracts spreading to reflexed or absent; calyx teeth obsolete to conspicuous; petals white, yellow, or purple; stylopodium absent; carpophore absent; fruit ovoid to oblong, terete to somewhat flattened dorsally, the lateral and usually 1+ of the dorsal ribs with corky-thickened to papery wings, usually prominent.

1. Bractlets of the involucel large, showy, scarious, white to purple, ± connate below, tending to form a cup around the umbellet; bracts of the involucre similar to the bractlets of the involucel, or smaller and less conspicuous [see below]…**Vesper multinervatus**
1'. Bractlets of the involucel (and bracts of the involucre, if present) relatively small and inconspicuous, or larger but herbaceous or coriaceous, the involucel generally asymmetric…(2)
2. Plants low, mat-forming perennial herbs with a taproot and often rhizomatous; caudex branching and basal leaves only; herbage glabrous to hirtellous-scabrous, not glandular; alpine and subalpine…(3)
2'. Plants of various habit, sometimes densely tufted but not mat-forming…(4)
3. Bractlets obovate, toothed at the apex, usually purplish…**C. bakeri**
3'. Bractlets linear or narrowly elliptic, entire, acute to acuminate…**C. alpinus**
4. Stems evidently scabrous-hirtellous or puberulent just below the umbel, sometimes also just below the nodes, otherwise glabrous or nearly so…(5)
4'. Stems or scapes glabrous or sometimes short-hairy or scabrous, but not with the pubescence pattern described above…(8)
5. Ultimate leaf segments 5–70 mm, 0.8–1(–2) mm wide, entire; plants acaulescent…**C. spellenbergii**
5'. Ultimate leaf segments various, but commonly not as above; plants acaulescent or caulescent…(6)
6. Stems ± leafy; young fruit puberulent…**C. davidsonii**
6'. Stems with (0)1 to few leaves; young fruit glabrous…(7)
7. Rays usually 1–2 cm, ascending; fruit 3–5 mm; widespread in the mountains…**C. lemmonii**
7'. Rays usually 1.8–5.5 cm, spreading; fruit 6–9 mm; C to SE New Mexico…**C. longiradiatus**
8. At least some leaves merely pinnatifid to sub-bipinnatifid…(9)
8'. Leaves mostly 2–3+ times pinnately or ternately dissected…(10)
9. Rays mostly 5–8, to 6 mm…**C. sessiliflorus**
9'. Rays mostly 8–15, sometimes > 6 mm…**Aletes**
10. Plants from a branched, ± woody caudex, mostly clothed at the base with marcescent leaf bases, often of rocky places; sepals 0.5–2 mm…(11)
10'. Plants from fibrous taproots with simple or sparingly branched crowns, without or with few marcescent leaf bases, not specifically of rocky places; sepals 0.1–0.4 mm…(15)

11. Plants strongly caulescent [see above]…**Aletes filifolius**
11'. Plants acaulescent or nearly so…(12)
12. Plants weakly or not aromatic…(13)
12'. Plants strongly aromatic…(14)
13. Ultimate leaf segments 5-70 mm, entire…**C. spellenbergii**
13'. Ultimate leaf segments 1-12 mm…**C. sessiliflorus**
14. Lowest pair of primary leaflets (1/4-)1/3-3/4+ the length of the leaf blade, mostly 3-9 cm, several times longer than the upper pairs, on petiolules 1-4 cm; plants mostly at 4600-6850 ft.…**C. terebinthinus**
14'. Lowest pair of primary leaflets 1/4 or less the length of the leaf blade, to 2.7 cm, often not more than twice as long as some of the upper pairs, sessile or on petiolules to 1 cm; plants mostly at 9500+ ft.…
     C. longilobus
15. Bracts of involucre conspicuous, scarious [to be expected in extreme E NM]…C. macrorhizus (Mathias, Constance & W. L. Theob.) B. L. Turner
15'. Bracts of involucre inconspicuous or absent, herbaceous, never scarious…(16)
16. Involucels rarely wholly green, not foliose, often scarious-margined and/or bractlets linear or narrowly elliptic and not > 1.5 mm wide; plants not viscid and not with adhering sand grains…**C. purpureus**
16'. Involucels green and foliose, seldom scarious-margined, bractlets 1.5-4 mm wide; plants obscurely viscid and with adhering grains of sand…**C. glomeratus**

Cymopterus alpinus A. Gray [*Oreoxis alpina* (A. Gray) J. M. Coult. & Rose]—Alpine oreoxis. Plants acaulescent, mat-forming to tufted, with taproot and branched caudex clothed with persistent leaf bases; leaf blades lanceolate to narrowly ovate in outline, 1-4 × 0.3-2 cm, mostly pinnate-pinnatifid to bipinnate-pinnatifid below; involucre absent; rays 4-11, 0.5-4 mm; involucel of 5-9 bractlets, linear to ovate, entire, 0.5-6 mm; petals yellow; fruit 4-6 mm, broadly elliptic, wings 3-5, 0.5-1 mm wide, straight, smooth, corky. Forb-grass, spruce, alpine communities, raw escarpments, barren ridge communities, 8700-12,600 ft. NC, Cibola, Sandoval.

Cymopterus bakeri (J. M. Coult. & Rose) M. E. Jones [*Oreoxis bakeri* J. M. Coult. & Rose]—Baker's oreoxis. Plants 1-12 cm, acaulescent, cespitose with taproot and branched caudex clothed with persistent leaf bases; leaf blades lanceolate, 1-5 × 0.5-3 cm, mostly pinnate-pinnatifid to bipinnate, leaflets sessile, ultimate leaf segments 2-8 × 0.5-1.2 mm; involucre absent; rays 5-14, 2-8 mm; bractlets of involucel 4-7, 3-5(6)-lobed, 3.5-5 mm; petals yellow; fruit 3.5-5 mm, broadly elliptic, wings usually 5, 0.2-1 mm wide, straight, smooth, corky. Alpine communities, raw escarpments, barren ridge communities, 9700-13,160 ft. NC.

Cymopterus davidsonii (J. M. Coult. & Rose) R. L. Hartm. [*Pteryxia davidsonii* (J. M. Coult. & Rose) Mathias & Constance]—Davidson's wavewing. Similar to *C. lemmonii* but with leafier stems and fruits averaging slightly smaller and puberulent. Rocky places in piñon-juniper woodland and lower montane coniferous forest, 6500-8000 ft. Catron, Grant. **R**

Cymopterus glomeratus (Nutt.) DC. [*C. acaulis* Raf.; *C. fendleri* A. Gray]—Plains spring parsley. Plants 7-18(-27) cm; herbage often ± viscid and dotted with sand grains; leaves basal, or more often whorled, 2-3 times pinnate, blades to 7 cm, the confluent portions 1-7(-12) mm wide; involucres absent; rays 6-9, 1-13 mm; bractlets of involucel 3-8(-11) × ca. 1.5-4 mm; petals yellow when fresh, sooner or later fading to white or cream when dried; fruit 5-10 mm, wings to 2 mm wide, straight or wavy, slightly corky. Desert shrub, sagebrush, piñon-juniper communities, often on sandy soil, 3800-7500 ft. Widespread except EC, SW.

Cymopterus lemmonii (J. M. Coult. & Rose) Dorn [*Pseudocymopterus montanus* (A. Gray) J. M. Coult. & Rose]—Lemmon's spring parsley. Plants acaulescent to caulescent, (2-)8-50 cm from taproot and simple or branched caudex clothed with persistent leaf bases; leaves basal and sometimes 1-3 cauline ones, blades narrowly lanceolate to broadly ovate or triangular in outline, 2-15 × 1.5-12 cm, pinnate-pinnatifid to tripinnate, in part, ultimate leaf segments mostly 2-40 × 0.5-4 mm, linear to lanceolate; involucre absent or bracts 1 or 2; rays 8-18; involucel of 5-10 bractlets, linear to lanceolate; petals bright yellow or rarely reddish; fruit 3-6 mm. Grass-forb, aspen, Douglas-fir, spruce-fir communities, windswept ridges and raw escarpments, 7500-12,500 ft. Widespread except EC, SE.

Cymopterus longilobus (Rydb.) W. A. Weber [*C. hendersonii* (J. M. Coult. & Rose) Cronquist]— Wavewing. Plants 7–35 cm, acaulescent, tufted, glabrous, strongly aromatic, with taproot and branched caudex clothed with persistent leaf and peduncle bases; leaf blades oblong to lanceolate or ovate in outline, 1–10 × 0.8–2.5 cm, mostly bipinnate or tripinnate in part; involucre absent or rarely of 1–3 linear to filiform bracts; rays 6–16, 0.5–2.4 cm; involucel of 2–6 bractlets, 2–10 mm; pedicels 1–3 mm, to 6 mm in fruit; petals yellow, often fading to whitish; fruit 4–8 mm, broadly elliptic, lateral wings 0.3–1.3 mm wide. Gravelly slopes, rock outcrops, alpine meadows, 6100–13,030 ft. Rio Arriba, Taos.

Cymopterus longiradiatus (Mathias, Constance & W. L. Theob.) B. L. Turner [*Pseudocymopterus longiradiatus* Mathias, Constance & W. L. Theob.]—Trans-Pecos false mountainparsley. Stems to 90 cm, 1–3 stem leaves; leaves ovate to oblong-ovate, ultimate divisions of leaflets linear-oblong to ovate, acute or acuminate; peduncles to 50 cm; branchlets of involucel linear-lanceolate, to 11 mm; fruit 6–9 mm, the commissure with 2 oil tubes. Limestone, moist, shaded areas in canyons, piñon-juniper and oak communities, 4700–7600 ft. WC, SE, SC.

Cymopterus purpureus S. Watson—Purple spring parsley. Plants acaulescent, tufted, simple or branched caudex clothed with persistent leaf bases, leaves leathery to somewhat fleshy, blades broadly ovate to triangular in outline, 1.5–13 × 1.5–10 cm, pinnate-pinnatifid to bipinnate-pinnatifid or ternate; ultimate segments 2–25 × mostly 1–15 mm; involucre absent; rays 5–22, 5–15 mm, to 95 mm in fruit; involucel of (1–)4–8 bractlets, linear to lanceolate or triangular, 2–4 mm; petals yellow, sometimes becoming purplish; fruit 4–8 mm. Sandy to heavy clay soils, desert shrub, sagebrush, piñon-juniper, mountain brush, ponderosa pine communities, 5660–7250 ft. NW.

Cymopterus sessiliflorus (W. L. Theob. & C. C. Tseng) R. L. Hartm. [*Aletes macdougalii* J. M. Coult. & Rose subsp. *breviradiatus* W. L. Theob. & C. C. Tseng]—Sessile-flower Indian parsley. Plants acaulescent, 6–20 cm, tufted to cespitose, branched caudex clothed with persistent leaf bases; leaves glabrous, blades lanceolate to broadly ovate, 1.5–8 × 8–3.5 cm, pinnate to pinnate-pinnatifid or nearly bipinnatifid; involucre absent; rays 4–9, 2–15 mm; involucel of 3–8 bractlets, 2–5 mm; petals yellow; fruit 5–6 mm, broadly elliptic, wings 3–5. Rocky slopes, ledges, and crevices of badlands and canyonlands, sandy ground in piñon-juniper communities, 5000–7200 ft. NW, C, WC, SC.

Cymopterus spellenbergii R. L. Hartm. & J. E. Larson—Taos spring parsley. Plants acaulescent, weakly or not aromatic, from taproot and simple or branched caudex clothed with persistent leaf bases; leaves 1- or 2+-pinnate, with filiform or linear ultimate segments 5–70 mm; peduncles sparsely to moderately puberulent at the apex with peglike hairs (×20); rays 6–24, 3–8 mm; involucel of 5–7 entire bractlets, 1–3 mm; pedicels 1–1.5 mm; flowers yellow; fruit 4.8–5.2 mm, the wings < 1 mm wide; carpophore spilt to the base. Basaltic boulders on the Taos Plateau; open piñon-juniper and Douglas-fir communities, 6200–8000 ft. Rio Arriba, Taos. **R & E**

Cymopterus terebinthinus (Hook.) Torr. & A. Gray [*C. petraeus* M. E. Jones]—Rock-parsley. Plants acaulescent, strongly aromatic; leaf blades lanceolate to ovate in outline, 1.5–14 × 2–10 cm, mostly pinnate to pinnate-pinnatifid, ultimate segments 0.2–7 × 0.2–1.5 mm; involucre absent; rays 7–13, involucel of (0)1–5 bractlets, linear to linear-subulate, entire, 2–5 mm; petals bright yellow, often fading to whitish; fruit 5–8 mm, ovoid-oblong, wings 2–5. Our material belongs to **var. petraeus** (Jones) Goodrich. Desert shrub and piñon-juniper communities, often in talus, colluvium, crevices of rock outcrops, sandy to clayey soil, 5500–6000 ft. McKinley, San Juan.

DAUCOSMA Engelm. & A. Gray – Meadow parasol

Daucosma laciniata Engelm. & A. Gray—Meadow parasol. Plants erect, annual, 9–12 dm; leaves triangular-ovate, ternate-pinnately dissected to 10 × 11 cm, ultimate divisions lanceolate; inflorescence of compound umbels; involucre of pinnately parted bracts; involucel of bractlets like the bracts; rays 14–17, 2–5 cm; flowers white, calyx teeth subulate, the stylopodium conic; carpophore divided to the base; fruit ovoid-oblong, 3–4 mm, compressed laterally, glabrous. Hidalgo.

DAUCUS L. – Carrot

Annual and biennial, caulescent herbs from taproots; leaves pinnately dissected; inflorescence of open umbels; involucre of pinnatifid bracts; involucel of toothed or entire bracts or absent; calyx teeth evident to obsolete; petals white or those of the central flower of the umbel or umbellet often purple, or rarely all the flowers pink or yellow; stylopodium conic; carpophore entire or bifid at the apex; fruit oblong to ovoid, slightly compressed and evidently ribbed dorsally, with 2 ribs on the commissure, with stout, spreading bristles, in part somewhat uncinate on ribs.

1. Plants biennial; involucral bracts widely spreading…**D. carota**
1'. Plants annual; involucral bracts appressed to the umbel in fruit…**D. pusillus**

Daucus carota L.—Wild carrot, Queen Anne's lace. Stems 6–10 dm, from a taproot; leaves in rosettes and cauline, mostly 1–2 times pinnate and then pinnatifid, basal and lower blades 5–15+ cm, ultimate segments 1–10 × 0.5–2 mm, elliptic, narrowly deltoid, or linear; involucre of pinnatifid bracts, 1–5 cm, the segments linear and narrow; rays 15–60+, (0.5–)1–6 cm; involucels similar to the involucre but smaller; petals white to yellowish; fruit 3–4 mm, bristly-hirsute in rows, the hairs or bristles 2 mm. Roadsides, irrigation ditches, other moist places, 5900–9150 ft. Scattered. Introduced.

Daucus pusillus Michx.—American wild carrot. Plants annual, variously hispid; leaves 3–4 times pinnate with linear ultimate segments; involucre leaflike, equaling or exceeding the rays; involucel bracts linear, about equaling the pedicels; petals white; fruit oblong, 3–5 mm, the primary ribs bristly, the secondary ribs with a single row of prominent barbed bristles. Rocky sites in SW New Mexico, 3900–6000 ft. SC, SW.

ERYNGIUM L. – Eryngo

Perennial, usually glabrous herbs; leaves entire or toothed to deeply cleft and often spinose-toothed; inflorescence of dense, bracteolate heads in cymes or racemes; involucre of 1+ series of entire or variously toothed or cleft bracts subtending the head; flowers sessile, white to blue or purple; calyx lobes well developed, firm, sometimes spinescent; stylopodium wanting; carpophore wanting; fruit globose to obovoid, variously covered with scales or tubercles, the ribs obsolete.

1. Leaves linear, mostly entire, parallel-veined…**E. sparganophyllum**
1'. Leaves lanceolate to oblanceolate, toothed or serrate; reticulate-veined…(2)
2. Lower leaves pinnatifid or bipinnatifid; bracts yellowish above; plants from a taproot…**E. heterophyllum**
2'. Lower leaves spinose-serrate; bracts silvery-white above; plants from fascicled roots…**E. lemmonii**

Eryngium heterophyllum Engelm.—Wright's eryngo. Stems 30–40 cm; leaves pectinately toothed to pinnatifid or bipinnatifid, reticulate-veined; heads ovoid; floral bracts linear to lanceolate, entire or with 1 or 2 pairs of spines near the middle, usually yellowish above; flowers pale blue; fruit ovoid, 2–3 mm. Seeps, springs, canyon bottoms, 4200–6500 ft. SW, Hidalgo.

Eryngium lemmonii J. M. Coult. & Rose—Chiricahua Mountain eryngo. Stems to 40 cm, from a fascicle of roots; leaves lanceolate to oblanceolate, reticulate-veined, lower cauline leaves spinose-serrate; inflorescence successively 3-branched; heads ovoid; bracts broadly lanceolate to oblanceolate, spinose-serrate, with 2 or 3 pairs of teeth, silvery-white above; fruit ovoid, laterally flattened. Animas Mountains; springs, wet ground, 7000–8500 ft. Hidalgo.

Eryngium sparganophyllum Hemsl.—Arizona eryngo. Plants to 1.5 m, from fibrous roots; leaves basal and on lower 1/3 of stem, to 1 m, simple, linear, entire or rarely with 1 or 2 spinose teeth; involucre of ovate or lanceolate bracts; heads burlike, 12–25 mm, with bractlets subtending the flowers; petals cream or bluish purple; fruit ovoid, 3–4 mm. Marshy ground, 4520 ft. Hidalgo (at Playas). **R**

EURYTAENIA Torr. & A. Gray – Spreadwing

Eurytaenia texana Torr. & A. Gray [*E. hinckleyi* Mathias & Constance]—Texas spreadwing. Slender, erect, caulescent, branching, scaberulous, herbaceous annuals from slender taproots; basal leaves petiolate, lobed or pinnatifid with obtuse, crenate-serrate lobes; cauline leaves pinnately or ternate-pinnately dissected with narrow, often elongate, entire or serrate divisions; inflorescence of compound umbels; involucre of 1-2-ternate, reflexed bracts shorter than the rays, the bractlets like the bracts, or entire; rays few, spreading-ascending; pedicels spreading-ascending, scaberulous; flowers white; stylopodium depressed; carpophore divided to the base; fruit 4-6 × 4-6 mm, oblong-oval to orbicular, strongly flattened dorsally, scaberulous to glabrous, the lateral wings thin, lighter colored than the body. Sandy soil, plains, 3300-3800 ft. SE.

FOENICULUM Mill. - Fennel

Foeniculum vulgare Mill.—Sweet fennel. Short-lived caulescent perennial herb, 0.5-2 m, with solitary taproot, with strong odor of anise; stems glabrous, glaucous, branched above, from a taproot; leaves to 3 times ternate-pinnately compound with 6-9 opposite pairs of lateral primary leaflets, larger blades 30-40 cm, ovate in outline, finely and completely dissected, the elongate, filiform ultimate segments 4-40 × <1 mm, the lowest pair of primary leaflets on petiolules often > 2 cm; umbels several; rays 10-40, 2-8 cm; sepals obsolete, petals yellow; fruit 3.5-4 mm. Roadsides and waste places, 4785-6860 ft. Doña Ana, Socorro, Torrance. Introduced.

HARBOURIA J. M. Coult. & Rose - Harbouria

Harbouria trachypleura (A. Gray) J. M. Coult. & Rose—Harbouria. Perennial, caulescent herb from a taproot; plants 8-50 cm, caulescent but sometimes with a single cauline leaf, hirtellous just below the umbel and often in the umbel, otherwise glabrous; leaves mostly basal, ovate-oblong in outline, ultimate segments 2-30 mm; inflorescence of compound umbels; rays 8-30; pedicels 1-4 mm; petals yellow; fruit 3-6 mm, granular-roughened. Rocky areas in piñon-juniper, ponderosa pine, Douglas-fir communities, 6000-8000 ft. NC, NW, C, WC, Doña Ana, Lincoln, Roosevelt.

HERACLEUM L. - Cowparsnip

Heracleum lanatum Michx. [*H. sphondylium* L. subsp. *lanatum* (Michx.) Á. Löve & D. Löve; *H. maximum* W. Bartram]—Common cowparsnip. Perennial herbs, 8-25 dm, glabrate or thinly to densely villous; leaves ternate or upper ones simple, leaf blades to 40 cm, ovate to orbicular, leaflets 8-36 cm, ovate to orbicular, usually with 3 major lobes that are again lobed and coarsely toothed; involucre absent or of few, entire bracts to 2 cm; rays 12-25, 3.5-12 cm; involucels of 3-5 linear, subulate, or caudate bractlets to 15 mm; petals white, (2-)4-8.5 mm; fruit 8-12 mm, strongly flattened, the lateral ribs with wings. Aspen, tall forb, fir, oak-maple, willow, streamside, wet meadow communities, 6880-12,120 ft. NC.

HYDROCOTYLE L. - Water pennywort

Hydrocotyle verticillata Thunb.—Water pennywort. Perennial herbs; plants glabrous, with slender creeping stems; leaves peltate, suborbicular, 0.5-6 cm wide, shallowly lobed and often crenate; petioles slender, 3-20+ cm; peduncles slender, axillary; flowers verticillate in few to several well-separated whorls; petals greenish or yellowish white to purplish; fruits subsessile, subtruncated at the base, 1.5-2 mm. Marshes, slow-moving streams, and edges of ponds, 4000-4750 ft. Doña Ana, Eddy, Socorro.

LIGUSTICUM L. - Licorice-root

Perennial, caulescent or acaulescent, taprooted herb; leaves ternately or ternate-pinnately compound or dissected; inflorescence of open umbels; involucre and involucel absent or of a few narrow bracts or bractlets; calyx teeth evident or obscure; petals white; stylopodium low-conic; carpophore divided to the base; fruit oblong to ovate or suborbicular, subterete or slightly compressed laterally, the ribs evident, often winged.

1. Ultimate leaf segments (at least some) elliptic or broader, some usually > 3 mm wide, sometimes toothed or lobed…**L. porteri**
1'. Ultimate leaf segments ± linear or elliptic, mostly 0.7–3 mm wide…**L. filicinum**

Ligusticum filicinum S. Watson—Fernleaf ligusticum. Similar to *L. porteri* but different as listed in the key and often with fewer umbels, most of these alternate, less often opposite, rarely whorled. Colfax, San Miguel.

Ligusticum porteri J. M. Coult. & Rose. [*L. brevilobum* Rydb.]—Southern ligusticum. Robust plants with 1 to several stems from base, mostly 5–12 dm; blades orbicular in outline, ternate-pinnately dissected with broader ultimate segments, these (1.5–)3–8 mm wide; inflorescence a terminal umbel often subtended by a whorl of 3–8 lateral umbels; petals white; fruit oblong, 5–8 mm. Sagebrush, oak, aspen, Douglas-fir, spruce-fir, occasionally open forb-grass communities, 7200–12,000 ft. Widespread except EC, SE, SW.

LOMATIUM Raf. – Desert parsley, biscuit-root

Plants perennial, acaulescent or caulescent, glabrous or pubescent, from a slender taproot with sometimes 1+ tuberlike segments, or from a thickened, woody, branching caudex, sometimes clothed at the base with marcescent leaf bases; stems simple or rarely branched; leaves pinnate or pinnately to ternate-pinnately compound, inflorescence of open umbels; involucre absent or inconspicuous; rays few to many; involucel mostly separate or partly united bractlets, rarely absent; petals small, yellow, white, greenish yellow, or purplish; calyx teeth obsolete or small, or conspicuous in some species; fruit oblong to orbicular or obovate, flattened dorsally.

1. Ultimate leaf segments mostly < 30, some usually > 1 cm; leaves not lacelike…**L. triternatum**
1'. Ultimate leaf segments numerous, mostly < 1 cm; leaves lacelike…(2)
2. Larger mature leaves with blades (10–)15–30 cm, ternate-pinnately compound, the larger ultimate segments 2–3 mm wide; plants 3–13 dm; peduncles fistulose, (3–)4–6(–10) mm thick at the base… **L. dissectum**
2'. Larger mature leaves with blades 2–11 cm, or if longer, either not at all ternate or with ultimate segments not > 1 mm wide; plants rarely > 50 cm; peduncles fistulose or not, often < 4 mm thick…(3)
3. Plants glabrous or at most scabrous; petals yellow when fresh, fading whitish when dried…**L. grayi**
3'. Plants pubescent; petals white or yellow…(4)
4. Petals yellow…**L. foeniculaceum**
4'. Petals white…(5)
5. Ovaries and fruits pubescent; lowest pair of leaflets sessile or nearly so…**L. nevadense**
5'. Ovaries and fruits glabrous; lowest pair of primary leaflets petiolulate…**L. orientale**

Lomatium dissectum (Nutt.) Mathias & Constance [*L. multifidum* (Nutt.) R. P. McNeil & Darrach]— Giant lomatium. Plants 30–130 cm, mostly short-caulescent, bushy, puberulent, rarely glabrous; leaf blades ovate to orbicular, 10–30 × 6–30 cm, cauline leaves reduced, ternate- to biternate- or pinnate-pinnatifid, ultimate segments numerous, 1–12 × 0.5–3 mm; involucre lacking or occasionally of 1–3 bracts; rays 9–27; involucel of 3–7 bractlets, mostly 3–6 mm; petals yellow, yellow-green, or purplish; fruit 9–15(–20) mm, lateral wings 1–2 mm wide, corky. Our material belongs to **var. eatonii** (J. M. Coult. & Rose) Cronquist. Mountain brush, piñon-juniper, ponderosa pine communities, 6500–7500 ft. Rio Arriba.

Lomatium foeniculaceum (Nutt.) J. M. Coult. & Rose—Desert parsley. Plants mostly 6–30 cm, from somewhat fleshy root sometimes capped with a few-branched caudex, densely hirtellous-puberulent throughout; leaves ternate-pinnate, dissected into numerous ultimate segments 1–3 mm; rays of umbel elongating unequally, to 6 cm; fruiting pedicels 3–11 mm; petals yellow; fruit 5–11 mm, wings ca. 1/2 as wide as the body. New Mexico plants belong to **var. macdougalii** (J. M. Coult. & Rose) Cronquist. Rocky hillsides, foothills to ponderosa pine communities, 4600–5500 ft. SW, Grant, Hidalgo, Rio Arriba.

Lomatium grayi (J. M. Coult. & Rose) J. M. Coult. & Rose—Milfoil lomatium. Plants 8–40(–80) cm, acaulescent or subcaulescent, tufted, strongly aromatic, glabrous; leaf blades ovate to lanceolate in outline, 7–16 × 3–5 cm, ternate-bipinnately to tripinnately dissected, ultimate segments 1–3(–6) × 0.2–0.3 mm; involucre lacking; rays 10–26; involucels of 3–6 bractlets, 3–5 mm; petals yellow, fading to whitish with age; fruit 8–12 mm, lateral wings 1–2 mm wide. Our material belongs to **var. grayi**. Sagebrush, piñon-juniper, mountain brush, ponderosa pine, Douglas-fir communities, 6000–8000 ft. Rio Arriba, San Juan.

Lomatium nevadense (S. Watson) J. M. Coult. & Rose—Nevada lomatium. Plants 10–36 cm, acaulescent or subcaulescent, ± pubescent; leaf blades lanceolate to oblong, 7–24 cm, pinnate-bipinnatifid or partly tripinnatifid, ultimate segments mostly 50–150, 1–15 × 1–2 mm, acute; involucre absent; rays 8–13, 1–5 cm, to 6 cm in fruit; involucel of 3–8 bractlets, 3–10 mm; petals white; fruit 6–20 × 5–10 mm, glabrous, lateral wings 1–3 mm wide.

1. Ovary and usually fruit pubescent…**var. nevadense**
1'. Ovary and fruit glabrous…**var. parishii**

var. nevadense—Nevada lomatium. Ovary pubescent. Foothills and piñon-juniper communities, 5200–6100 ft. Grant.

var. parishii (J. M. Coult. & Rose) Jeps.—Parish's biscuit-root. Ovary glabrous. Desert scrub and piñon-juniper communities, 3925–6200 ft. NW, WC, SW.

Lomatium orientale J. M. Coult. & Rose—Eastern desert parsley. Plants 10–40 cm, puberulent; leaves tripinnate, 3–8 cm, the ultimate divisions linear, 2–5 mm; rays 1.2–5.5 cm, subequal; involucel bractlets linear-lanceolate, 2–4 mm; petals white; anthers red; fruit 6–9 mm, lateral wings 0.5–1 mm wide. Rocky areas, alluvial fans in SW, 4515–8500 ft. WC, SW, Torrance, Union.

Lomatium triternatum (Pursh) J. M. Coult. & Rose—Nineleaf biscuit-root. Plants 20–70 cm, acaulescent or subcaulescent, glabrous to hirtellous; leaf blades ovate to broadly obovate, 4–20 × 2–10 cm, ternate- to biternate-pinnately compound, ultimate leaflets or segments linear to lanceolate; rays 4–20; involucel of 6–10 bractlets, 1.5–8 mm, linear to lanceolate, entire; petals bright yellow, fading to white with age; fruit broadly elliptic, 8–15 × 4–11 mm, glabrous, lateral wings 2–2.5(–4) mm. New Mexico material is **subsp. platycarpum** (Torr.) Cronquist. Sagebrush-grass, piñon-juniper woodland, mountain brush, ponderosa pine, dry meadow communities, 7040–9050 ft. NW.

NEOPARRYA Mathias – Neoparrya

Neoparrya lithophila Mathias [*Aletes lithophila* (Mathias) W. A. Weber]—Bill's neoparrya. Plants perennial, acaulescent, glabrous, to ca. 15 cm; leaves pinnately compound with linear leaflets; leaf blades 8–10 cm, pinnate with linear leaflets 5–20 mm; rays 5–15 mm; involucel bractlets narrowly lanceolate, ca. 3 mm; flowers yellow; fruit 3–5 mm, compressed laterally, ribs prominent but not winged. Volcanic rock, piñon pine, sagebrush, 7830–7900 ft. N Taos.

OROGENIA S. Watson – Indian potato

Orogenia linearifolia S. Watson—Great Basin Indian potato. Perennial plants from tuber, low, glabrous, acaulescent or very short-caulescent; plants 4–10+ cm, from a globose, edible root, leaves mostly 2 or 3, 1- or 2-ternate or rarely simple, the leaflets narrow, elongate, the blades only slightly elevated above ground level, the ultimate segments mostly 1–4.5 × 0.5–4 mm; inflorescence a loose umbel, compact at anthesis, 1–2 cm, becoming more open in fruit, pedicels < 2 mm; petals white; fruit 3–4 mm, oblong-elliptic, with evident dorsal ribs. Mountain brush and ponderosa pine communities. A single collection from near Lumberton, 7120 ft. Rio Arriba.

OSMORHIZA Raf. – Sweetroot

Perennial, caulescent, usually pubescent herbs from taproots with simple or branched crowns; leaves ternately or pinnately 1–3 times compound; inflorescence an open or loose umbel; involucre absent or of

1 to few narrow foliaceous bracts; involucel absent or of several foliaceous reflexed bractlets; petals and stamens white, greenish white, yellow, pink, or purple; fruit linear or clavate, somewhat compressed laterally, bristly-hispid to glabrous, the ribs narrow.

1. Stylar beak 2–3.5 mm…**O. longistylis**
1'. Stylar beak 2 mm or less…(2)
2. Mature fruit including tails mostly 16–25 mm, the apex concavely pointed into a beak 1–2 mm; fruiting pedicels mostly ascending-spreading…**O. chilensis**
2'. Mature fruit including tails mostly 13–18 mm, the apex convex and obtuse; fruiting pedicels horizontally divergently spreading to sometimes ascending…**O. depauperata**

Osmorhiza chilensis Hook. & Arn. [*O. nuda* Torr.; *O. berteroi* DC.]—Chile sweet cicely. Stems often solitary, 18–75 cm; leaves basal and 2–3 cauline, biternate, usually with 9 distinct leaflets; leaf blades 5–15 cm, the lateral primary leaflets nearly as long as the central one, blades of leaflets 1–4(–5.5) cm, elliptic to ovate; involucre absent; rays 3–7, 2.5–9(–13) cm; involucels absent; fruit including tails 16–25 mm, linear-clavate, bristly-hispid, the beak concavely pointed, 1–2 mm. Mountain brush, aspen, Douglas-fir, white fir, narrowleaf cottonwood, riparian communities, 7300–9200 ft. NC, NW.

Osmorhiza depauperata Phil.—Blunt-fruit sweet cicely. Stems mostly solitary, 14–63(–77) cm, often with a slight ring of hairs at the nodes; leaves basal and 1–3 cauline, biternate, usually with 9 distinct leaflets, or the upper cauline ones 1-ternate, leaf blades (2–)4–11 cm; umbels 3–6; involucre absent, or rarely of a solitary bract to 12 mm; rays 3–5, 1.5–8.5 cm; involucels absent or infrequently of 1 or 2 separate ciliolate bractlets to 3 mm; petals greenish white; fruit including tails (11–)13–18 mm, linear-clavate, bristly-hispid, the beak convex-obtuse. Mountain brush, aspen, ponderosa pine, Douglas-fir, spruce-fir, riparian communities, 6900–12,500 ft. NC, NW, C, WC, SC.

Osmorhiza longistylis (Torr.) DC.—Longstyle sweetroot. Similar to *O. chilensis*, with the main difference stylar beak 2–3.5 mm. Cool canyons and ravines in the foothills, 7050–8100 ft. Colfax, Union.

OXYPOLIS Raf. – Cowbane

Oxypolis fendleri (A. Gray) A. Heller—Fendler's cowbane. Perennial, caulescent herbs from fascicled tuberous roots; stems 6–8 dm; leaves pinnate with 2–5 pairs of opposite lateral leaflets, the upper ones sometimes reduced to bladeless or nearly bladeless sheaths, upper blades sessile on a dilated sheath; blades 7–17 cm, oblong in outline, the leaflets sessile, 2–5 cm, ovate to orbicular, those of the upper leaves lanceolate to linear and sometimes entire; umbels usually 4+ per stem; involucre absent; inflorescence an open umbel; rays 5–14, 1–5(–7) cm, ascending; involucels absent; petals and stamens white; fruit 3–5 mm, oblong to oval, strongly flattened dorsally, with dorsal ribs filiform and lateral ribs broadly winged. Stream banks, 6950–12,960 ft. NC, NW, C, WC, SC.

PASTINACA L. – Parsnip

Pastinaca sativa L.—Garden parsnip. Plants 8–15 dm, caulescent, aromatic; leaves pinnate or partly bipinnate in some lower leaflets; leaf blades 12–35+ cm, oblong in outline, leaflets sessile and sometimes confluent or the lower ones sometimes on petiolules, blades 2.5–12 cm, lanceolate to ovate; umbels 6–15+; involucre absent or of 1 to few linear bracts to 2(–4) cm; inflorescence an open compound umbel; rays 9–25, 0.8–8.5 cm; involucels absent or infrequently of 1 to few linear to 2(–4) bracts; petals greenish yellow or reddish; fruit 5–8 × 3–6 mm, elliptic to obovate, strongly flattened dorsally, the dorsal ribs filiform, the lateral ones narrowly winged. Weedy, of ditch banks, roadsides, fence lines, gardens, fields, margins of ponds and lakes, 6850–9600 ft. NC, NW, C, Cibola, Otero. Introduced.

PERIDERIDIA Rchb. – Yampa

Plants perennial, glabrous, from 1 or a cluster of tuberous-thickened, edible roots; leaves cauline and basal, pinnately or ternate-pinnately compound or dissected with mostly narrow ultimate segments; inflorescence of compound umbels; involucre of few to many small, narrow, ± scarious bracts; involucel mostly of scarious or scarious-margined or colored bractlets, or obsolete; flowers white or pink; fruit glabrous.

1.   Basal leaves 1–2-ternate or 1–2-pinnate with 1–3 pairs of primary leaflets…**P. parishii**
1'. Basal leaves ± pinnate with 3–5 pairs of primary leaflets, lower leaflets sometimes lobed or ternately dissected…**P. gairdneri**

Perideridia gairdneri (Hook. & Arn.) Mathias—Yampa. Plants 30–140 cm; stems erect, branched; basal leaves 1-pinnate, basal blades 20–35 cm, oblong to ovate, leaflets 2–12 cm; cauline leaves 1–2-pinnate or 1–2–3-ternate; inflorescence of compound umbels; rays 7–16, 1.5–7 cm; flowers small, many, petals white; fruit with 2 dry, 1-seeded halves, glabrous, not winged. Moist soil of meadows, streamsides, grasslands. One record from the Mogollon Mountains, 8500 ft. Catron.

Perideridia parishii (J. M. Coult. & Rose) A. Nelson & J. F. Macbr.—Parish's yampa. Plants 20–90 cm; stems solitary; basal leaves 1-pinnate or 1-ternate, the ultimate segments lance-linear, 3–10 cm × 3–7 mm; cauline leaves smaller; umbels 1 to several, mostly 2–4 cm wide at anthesis, flat or concave with central rays shorter than the outer; involucre lacking or of 1–2 setaceous bracts, bractlets of involucel lance-linear, 1–2 mm; petals white; fruit oblong or elliptic to seldom orbicular, 3–4.5 mm, ribs threadlike. Our plants are **subsp. parishii**. Moist meadows and woods, 7000–8000 ft. Catron, Grant, Socorro.

PODISTERA S. Watson – Podistera

Podistera eastwoodiae (J. M. Coult. & Rose) Mathias & Constance—Eastwood's podistera. Plants 7–30 cm, acaulescent, tufted; leaves pinnate with deeply lobed leaflets; leaf blades narrowly to broadly oblong in outline, 2–9 × 0.8–3 cm, pinnate with 4–6 opposite (sometimes appearing whorled) pairs of lateral leaflets, leaflets sessile, 5–20 × mostly 5–25 mm; involucre absent; inflorescence a solitary, open to congested umbel; rays 5–8, 2–8 mm; involucel of 3–6 bractlets, linear to broadly spatulate, usually 3(5–7)-lobed apically; petals greenish yellow, often purplish with age; fruit 3–4 mm, broadly ovoid, ribs prominent, not winged. Alpine meadows and spruce-fir forest communities, 8400–13,000 ft. NC.

SANICULA L. – Black snakeroot

Sanicula marilandica L.—Maryland sanicle. Stems 4–12 dm, single, often branched above; leaves alternate, the lowermost cauline leaves well developed, blades 6–16 cm wide, palmately 5–7-parted or palmately compound, cauline leaves usually several, gradually reduced upward and becoming sessile; ultimate umbels ca. 1 cm wide or less at anthesis, subtended by a few minute narrow bractlets, mostly 15–25-flowered; calyx teeth well developed, mostly connate; fruit ovoid, 4–6 × 3–5 mm wide, covered with numerous uncinate prickles. Lightly wooded slopes and flats from foothills to moderate elevations in the mountains, 7000–8675 ft. Colfax, Rio Arriba, San Miguel.

SIUM L. – Water parsnip

Sium suave Walter—Water parsnip. Perennial, caulescent herbs; stems 5–10 dm; leaves pinnate or occasionally partly bipinnate, with 4–6 opposite pairs of sessile lateral leaflets, lower blades 14–32 cm, the upper ones reduced, leaflets 2–8(–15) × (1–)3–8(–20) mm, linear to lanceolate, sharply and uniformly serrate to pinnatifid with linear segments; umbels 3–11+ per stem; involucre of 1–6 separate, often reflexed bracts 2–9 mm; inflorescence an open umbel; rays 11–24, 1.5–3 cm; involucels of (2–)5–12 separate bractlets 2–5 mm; petals and stamens white; fruit 2–3 mm, the ribs prominent and corky but hardly winged. Mudflats, marshlands, wet meadows, along streams and shorelines, ponds and lakes, 5200–9650 ft. Rio Arriba, Sandoval.

SPERMOLEPIS Raf. – Scaleseed

Slender, erect or spreading, caulescent, branching, glabrous annuals from slender taproots; leaves ternately or ternate-pinnately decompound, the ultimate divisions linear to filiform; inflorescence of compound umbels; involucre lacking, the involucel of a few linear bractlets; rays few, erect to divaricate;

petals white; stylopodium low-conic; carpophore 2-cleft; fruit ovoid, flattened laterally, smooth, tuber-culate, or echinate, the ribs filiform, rounded.

1. Fruit densely echinate-bristly with sharp-pointed, apically hooked hairs…(2)
1′. Fruit tuberculate with multicellular trichome bases…(3)
2. All umbels distinctly pedunculate…**S. echinata**
2′. Distal umbels sessile, proximal umbels sometimes pedunculate…**S. lateriflora**
3. Tubercles irregularly scattered, some with short, erect hairs; peduncles 0.9–3.5 cm…**S. organensis**
3′. Tubercles densely arranged, without hairs; peduncles 2–7 cm…**S. inermis**

Spermolepis echinata (Nutt. ex DC.) A. Heller—Prickly scaleseed. Plants 5–40 cm; leaf blades divided into 3 segments, each segment again divided into 3 segments, blade 7–25 mm wide, ovate in outline; inflorescence in axillary and terminal compound umbels, rays 5–14, 1–15 mm; flowers white; fruit a capsule, splitting into 2 single-seeded mericarps, ovate, 1.5–2 mm wide, with short bristles and hooked tips. Rocky slopes, Chihuahuan desert scrub, roadsides, 4430–5000 ft. Doña Ana.

Spermolepis inermis (Nutt. ex DC.) Mathias & Constance—Red River scaleseed. Plants 8–80 cm; leaf blades oblong-ovate, 3–5 cm, 3-pinnately compound, ultimate segments filiform, 3–30 × 0.1–1 mm; petioles 4–15 mm; umbels terminal and axillary, all pedunculate; involucel bractlets 1–4, linear to linear-lanceolate, 2–5 mm, margins scabrous-toothed; fruiting rays 5–11, unequal, 1–13 mm; umbellets 2–7-flowered; fruiting pedicels 0–6 mm; fruit 1.2–2 mm, tuberculate without hairs. Clayey, sandy, and limestone soils, prairies, oak-juniper woodlands, ditch banks, roadsides, 3500–4500 ft. SE, Chaves, Eddy.

Spermolepis lateriflora G. L. Nesom—Sessile-umbel scaleseed. Plants 5–35 cm; leaf blades broadly ovate, mostly 1–5 cm, finely ternately dissected, ultimate segments linear to oblong, 4–12 mm; peduncles absent or 20–70 cm; umbels usually axillary only, usually sessile at all nodes, always sessile at distal nodes; involucel bractlets 2–4, linear, 1–2 mm; fruiting rays 4–5 per node, 10–14 mm; umbellets 3–8-flowered; fruiting pedicels 1–6 mm; fruit 1.2–2.2 mm, densely echinate-bristly with apically hooked hairs. Sandy to rocky soils, desert grassland, desert scrub, mesquite, oak savanna, oak juniper woodlands, 3850–6000 ft. SW.

Spermolepis organensis G. L. Nesom—Organ Mountain scaleseed. Similar to *S. inermis* in its strictly pedunculate, terminal and axillary umbels; different in its shorter fruiting peduncles and its corky fruit surface with vaguely formed tubercles, some tubercles producing short, straight hairs, some with-out hairs. Known from the type collection of 1995 and rediscovered 2.5 miles N of the type location in 2016 by Patrick Alexander and Gregory Penn. Foothills of the Organ Mountains, 4700–5500 ft. Doña Ana. **R & E**

TORILIS Adans. – Hedge-parsley

Torilis arvensis (Huds.) Link—Field hedge-parsley. Plants 3–10 dm; stems retrorsely strigose; leaves antrorsely strigose or partly glabrous; leaves ovate or lance-ovate in outline, 2–3 times pinnate, or the upper only 1-pinnate; involucre lacking or of a single slender bract; rays 0.5–2.5 cm; involucel of narrow bracts; pedicels 1–4 mm; petals white; fruit 3–5 mm, with widely spreading upcurved prickles ca. 1 mm. Pinos Altos Mountains, mixed conifer, 6000–7000 ft. Eddy, Grant. Introduced.

VESPER R. L. Hartm. & G. L. Nesom – Spring parsley

Perennial, acaulescent or subacaulescent herbs with development of a pseudoscape, from thick taproots; leaves pinnately, bipinnately, or ternate-bipinnately compound, somewhat fleshy, thus often minutely wrinkled on drying, glabrous or margins rarely scaberulous, not viscid, usually pallid-glaucous, the ulti-mate segments often confluent and overlapping; inflorescence compact; involucral bracts basally con-nate, prominently nerved, all white or purplish scarious or with broad white-scarious margins; mericarps dorsally compressed with 4–5 thin, broad, dorsal and lateral wings, with 3–9 oil tubes per interval.

1. Fruiting peduncles shorter than or equaling the leaves; mericarp wings conspicuously enlarged at the base…**V. montanus**

1'. Fruiting peduncles equaling or longer than the leaves; mericarp wings not conspicuously enlarged at the base…(2)
2. Involucel bractlets with lacerate-fringed distal margins…**V. macrorhizus**
2'. Involucel bractlets with entire or irregularly toothed or lobed margins…(3)
3. Involucre mostly a low hyaline sheath; involucel bractlets commonly purplish to rosy, 5–8-nerved; pedicels 0–1 mm…**V. multinervatus**
3'. Involucre of 1–8 oblong to obovate, often variously lobed bracts; involucel bracts greenish white to white, 1–3(–5)-nerved; pedicels 1–12 mm…(4)
4. Umbels in fruit tightly globose, rays 1–4(–8) mm, pedicels 1–4 mm; carpophores absent; fruit orbicular, 10–12 mm…**V. purpurascens**
4'. Umbels in fruit relatively open, ± flat-topped, rays 10–50 mm, pedicels 5–12 mm; carpophores well developed; fruit oblong, 8 mm…(5)
5. Involucel bractlets connate for 1/3–2/3+ of length, the free portion usually abruptly enlarged distally, broadly ovate to orbicular, with mostly 1 vein, occasionally with 1–2 pairs of shorter lateral veins, parallel to divergent or branched…**V. bulbosus**
5'. Involucel bractlets connate to 1/3 of length, the free portion gradually expanding distally, obovate to spatulate, with mostly 3 veins arising from the base, parallel below, gradually flaring distally, equal or nearly so…**V. constancei**

Vesper bulbosus (A. Nelson) R. L. Hartm. & G. L. Nesom [*Cymopterus bulbosus* A. Nelson]—Bulbous spring parsley. Plants 8–27 cm; leaf blades lanceolate to broadly ovate in outline, 2–8 × 1.5–5 cm, pinnate-pinnatifid to bipinnate-pinnatifid below, ultimate leaf segments 0.3–5 × 0.1–2.5 mm, oblong to elliptic; involucre lacking or reduced to a ring or cup, or the bracts to 13 mm; rays 3–8, 5–30 mm; involucel bracts 3–10 mm, with a green or purple midrib and 1 or 2 lateral nerves ca. 1/2 as long as the midrib; petals and stamens white or purplish in age; fruit 6–17 mm, wings 1.7–4 mm wide. Desert shrub and juniper communities, 4900–8500 ft. NE, NC, NW, C.

Vesper constancei (R. L. Hartm.) R. L. Hartm. & G. L. Nesom [*Cymopterus constancei* R. L. Hartm.]—Constance's spring parsley. Plants 3–18 cm; leaf blades lanceolate to broadly ovate in outline, 2.5–8 cm, ultimate leaf segments 0.2–2.5 × 0.5–1 mm, oblong to elliptic; inflorescence of 1–8+ umbels, in fruit loose; involucre of 1–8 bracts 4–10 mm; rays 3–6, 3–5 mm; involucel of 4–6 bractlets, obovate to spatulate, 4.5–7 mm, fused in lower 20%–30%, white, scarious with usually 3 dark green to purple veins arising from base, parallel below; pedicels 1–3 mm, to 12 mm in fruit; petals white or cream to purple; fruit 7–14 mm, wings 3–4 mm wide. Grasslands, sagebrush, piñon-juniper, ponderosa pine communities, 5000–8700 ft. NW, C.

Vesper macrorhizus (Buckley) R. L. Hartm. & G. L. Nesom [*Cymopterus macrorhizus* Buckley]—Bigroot spring parsley. Plants 5–25 cm; leaf blades 1.5–6.5 cm; leaflets entire to pinnately lobed, the ultimate segments 1–3 mm; involucre lacking, or of 1+ linear bracts; rays 1–6, spreading, 5–30 mm; involucel of conspicuous, subcuneate bractlets, lacerate-fringed at the apex, white with a dark central nerve; pedicels 0–2 mm; petals pinkish; fruit ovoid to ovoid-oblong, 4–9 mm, 3–8 mm wide including wings. Limestone ridges and hillsides, gypsum exposures, sandy prairies, mesquite-grassland communities, 4050–6235 ft. Quay, Roosevelt, San Miguel.

Vesper montanus (Nutt. in Torr. & A. Gray) R. L. Hartm. & G. L. Nesom [*Cymopterus montanus* Nutt. ex Torr. & A. Gray]—Mountain spring parsley. Plants 5–30 cm; leaf blades ovate-oblong, 1.5–8 cm, pinnate, bipinnate, or occasionally ternate-bipinnate, ultimate segments mucronate, 1–2 mm; involucre lacking or of a low inconspicuous sheath, or of conspicuous linear-oblong bracts; rays 3–6, 0.5–20 mm; involucel of conspicuous, ovate-oblong, mostly acute bractlets, white with a conspicuous green central nerve; rays 3–6, 0.5–20 mm; petals white or purple; fruit 5–12 mm, the wings about twice as wide as the body. Sandy soils, E plains and mountain hillsides, 4000–8600 ft. Widespread except NW, NC, SC, SW.

Vesper multinervatus (J. M. Coult. & Rose) R. L. Hartm. & G. L. Nesom [*Cymopterus multinervatus* (J. M. Coult. & Rose) Tidestr.]—Purplenerve spring parsley. Plants 7–15 cm; leaf blades 1–7 cm, with 3–5 pairs of lateral leaflets, ultimate segments to 4(–7) mm; primary umbel with several rays to ca. 2 cm; involucre virtually obsolete or of a few irregularly developed scarious, whitish or somewhat anthrocyanic, basally connate bracts; rays (3–)5–11, 3–10 mm; involucel of several rounded bractlets 5–10 mm, with several nerves; flowers purplish; fruit 10-16 mm, broadly multiwinged, the wings commonly 4-7 mm wide. Rocky soils, desert scrub, desert grassland, piñon-juniper communities, 4100-7250 ft. SW, Cibola, Grant, Hidalgo.

Vesper purpurascens (A. Gray) R. L. Hartm. & G. L. Nesom [*Cymopterus purpurascens* (A. Gray) M. E. Jones]—Wide-wing spring parsley. Plants acaulescent, tufted; stem pseudoscapes 1 or 2; leaves somewhat fleshy, blades lanceolate to broadly ovate in outline, 1.2-7 × 1.5-5 cm, pinnate-pinnatifid to bipinnate-pinnatifid below, with 3-6 opposite pairs of lateral leaflets; inflorescence of 1-8 obscurely distinct umbels; involucre usually of 8-10 bracts 8-15 mm, bracts fused into a lobed to variously parted cup; rays 0-8, 1-8 mm; involucels of 4-6 bractlets; petals white or purplish; fruit 7-15 mm, broadly elliptic to suborbicular, tan to purplish, wings 5, 2-4 mm. Desert shrub, sagebrush, piñon-juniper, ponderosa pine communities, on eolian sand to heavy clay, 5020-8075 ft. NW, Cibola.

YABEA Koso-Pol. – Yabea

Yabea microcarpa (Hook. & Arn.) Koso-Pol.—California hedge-parsley. Annual, caulescent herbs; plants 8-40 cm, pubescent with spreading-hispid hairs, from slender taproot; leaves 2-3(4) times pinnate or ternate-pinnate, the blades 1-5 cm, oblong or ovate in outline, the ultimate segments 1-8 × 0.5-2 mm; involucre resembling the upper leaves or a little smaller; umbels compound; rays 1.5-10 cm; involucels pinnatifid or entire; pedicels 5-15 mm; petals and stamens white; fruit 3-7 mm, oblong or ovoid, somewhat compressed laterally, with spreading uncinate prickles along alternating ribs and bristly-hairy on the other ribs. Desert foothills, 4285-6000 ft. SW, Grant, Hidalgo, Luna.

APOCYNACEAE – MILKWEED OR DOGBANE FAMILY

Mark Fishbein

Perennial herbs, subshrubs, or vines, usually with milky latex; perennating organs tubers, caudices, or rhizomes; stems prostrate to erect, twining or not twining; leaves simple, opposite, subopposite, or alternate, petiolate or sessile, margins entire, venation pinnate, pinnipalmate, or a single vein, stipules absent (but colleters usually present in a stipular position); inflorescences cymes (often umbelliform or racemiform); flowers radially symmetric; perianth biseriate, 5-merous; coronas usually present (absent in *Amsonia, Haplophyton, Mandevilla, Vinca*); stamens 5, epipetalous, filaments free or fused, anthers free, connivent or fused to styles forming a gynostegium, pollen shed singly, in tetrads or in pollinia; pistil 2-carpellate, usually connate through styles with ovaries free, placentation parietal; fruit 1 or 2 follicles; seeds few to numerous, with or without a coma (Krings et al., 2023, and refs. within).

1. Stems twining…(2)
1'. Stems not twining…(4)
2. Leaves linear, 1-2.5 mm wide; corollas purely cream, campanulate, lobes conspicuously pubescent adaxially with downward-pointed hairs; follicles 0.3-0.5 cm broad…**Metastelma**
2'. Leaves linear to ovate, 1-45 mm wide; corollas green, brown, purple, or cream, rotate or campanulate, if cream, then tinged or lined with red, purple, pink, or green and rotate, lobes inconspicuously pubescent with straight hairs or glabrous adaxially; follicles 0.6-3.5 cm broad…(3)
3. Stems and leaves with sparse, non-glandular pubescence; corona white, of 2 parts: a pentagonal ring in corolla throat plus 5 inflated knobs on backs of stamens…**Funastrum**
3'. Stems and leaves with dense glandular pubescence mixed with longer, straight, non-glandular trichomes; corona green, of a single series of fused, lobed segments arising from junction of stamens and corolla… **Matelea** (in part)
4. Leaves alternate…(5)
4'. Leaves opposite or subopposite…(7)
5. Corolla rotate with reflexed or ascending lobes; corona present; seeds comose at one end…**Asclepias** (in part)
5'. Corolla salverform; corona absent; seeds bare or comose at both ends…(6)
6. Corolla bright yellow, some leaves opposite, seeds comose at both ends…**Haplophyton** (in part)
6'. Corolla blue or purplish to white, all leaves alternate, seeds bare…**Amsonia**

7.   Corolla salverform or funnelform with a narrow tube…(8)
7'. Corolla rotate, urceolate, cylindrical, or campanulate, lacking a narrow tube…(10)
8.   Corolla yellow, tube 6-8 mm; seeds with coma at both ends…**Haplophyton** (in part)
8'. Corolla white, pink, blue, or violet, tube (including throat) 12-40 mm; seeds with coma at one end or bare…(9)
9.   Corolla white to pink, tube (including throat) 30-40 mm; follicles 5.5-12 cm; seeds comose; stems erect…**Mandevilla**
9'. Corolla blue to pale violet (cultivars may be white or red-violet), tube (including throat) 12-15 mm; follicles 2-4 cm long, seeds lacking a coma; stems trailing…**Vinca**
10. Gynostegium absent, filaments separate, anthers connivent; corona of minute scales on corolla below sinuses; pollen in tetrads; seeds fusiform…**Apocynum**
10'. Gynostegium present, stamens fused their entire length; corona of 5 segments, free or united, attached to staminal column or point of insertion of filaments on corolla; pollen in pollinia; seeds flat, ovate…(11)
11. Stems erect to decumbent; corona segments attached to staminal column above insertion of filaments; leaf bases truncate to cuneate, not deeply cordate…**Asclepias** (in part)
11'. Stems prostrate; corona segments attached at insertion point of filaments; leaf bases cordate to truncate…**Matelea** (in part)

AMSONIA Walter – Bluestar

Plants erect herbs; latex white; leaves alternate, stipular colleters present, laminar colleters absent; inflorescences terminal and subterminal cymes; corollas salverform, estivation contorted; gynostegium absent; corona absent; pollen in monads; follicles usually paired, erect; seeds cylindrical, coma absent.

1.   Stems tomentose; branches exceeding inflorescence; calyx tube tomentose; corolla lobes white to blue; follicles constricted between seeds; seeds 14-19 mm…**A. arenaria**
1'. Stems glabrous to pubescent; branches at most slightly exceeding inflorescence; calyx tube glabrous to pubescent; corolla lobes purple to magenta, maroon, or lavender; follicles smooth; seeds 5-11 mm…(2)
2.   Corolla tube 28-41 mm; leaf blades linear, 0.1-0.3 cm wide; seeds 5-8 mm; stems branched throughout…**A. longiflora**
2'. Corolla tube 7.5-23 mm; leaf blades elliptic, oblong, lanceolate, or linear, 0.2-1.8 cm wide; seeds 6-11 mm; stems branched only (or nearly so) above the middle…(3)
3.   Corolla tube 6-12 mm, lobes 3-7 mm; stems glabrous to sparsely pubescent; Grant, Hidalgo, Luna, and San Juan counties…(4)
3'. Corolla tube 12-23 mm, lobes 6-11 mm, stems pubescent; Eddy and Socorro counties…(5)
4.   Calyx glabrous; corolla lobes 5-8.5 mm; main stem leaf blades ovate to lanceolate; San Juan County…**A. jonesii**
4'. Calyx pubescent; corolla lobes 3-5 mm; main stem leaf blades oblong, elliptic, lanceolate, or linear; Grant, Hidalgo, and Luna counties…**A. palmeri**
5.   Corolla tube 12-17 mm, lobes 6-9 mm; stems 10-20 cm; Eddy County…**A. tharpii**
5'. Corolla tube 18-23 mm, lobes 8-11 mm; stems 15-50 cm; Socorro County…**A. fugatei**

Amsonia arenaria Standl.—Woolly bluestar, sand bluestar. Stems erect, 15-70 cm, branched throughout; leaf blades on main stem linear, 3-6 × 0.1-0.4 cm; inflorescences exceeded by vegetative branches, flowers 5-20; calyx tomentose at base, glabrous above; corollas purple to magenta, maroon, or lavender, tubes 9-11 mm, lobes 6-8 mm; follicles 3-10.5 × 0.4-0.7 cm, constricted between seeds; seeds 14-19 × 3.5-5 mm. Dunes, sandhills, bajadas, sandy and gravelly soils, desert scrub, 3900-5100 ft. SC.

Amsonia fugatei S. P. McLaughlin—Fugate's bluestar. Stems erect, 18-50 cm, branched above middle; leaf blades on main stem narrowly elliptic to lanceolate, 3-5.5 × 0.5-0.7 cm; inflorescences slightly exceeded by vegetative branches, flowers 5-15; calyx lobes ciliate; corolla tubes blue to purplish, 18-23 mm, lobes white to cream, 8-11 mm; follicles 3.5-6 × 0.2-0.4 cm, smooth; seeds 8-10 × 2-3 mm. Dunes, bluffs, slopes, limestone, gypsum, conglomerate, sandy, silty, and rocky soils, caliche, desert scrub, 3700-5300 ft. Socorro. **R**

Amsonia jonesii Woodson—Jones' bluestar. Stems ± erect, 20-50 cm; leaf blades ovate-lanceolate, 3-6.5 × 12-27 mm, glabrous, rather thick and leathery; inflorescences a condensed corymb; calyx 1-3 mm, lobes lanceolate to narrowly lanceolate; corolla tubes whitish to powder-blue or dark bluish, 4.5-8 mm; follicles 6-9 mm, narrowly cylindrical. Sandy, gravelly, or clay soils in sagebrush and piñon-juniper woodlands, 5300-5750 ft. San Juan.

Amsonia longiflora Torr.—Tubular bluestar. Stems erect, 15-60 cm, branched; leaf blades on main stem linear, 3-6.5 × 0.1-0.3 cm; inflorescences exceeding vegetative branches, flowers 3-15; calyx glabrous or pubescent; corolla tubes blue to lavender or purplish, 28-41 mm, lobes white to cream or pale pinkish or bluish, 9-13.5 mm; follicles 6-11 × 0.3-0.4 cm, smooth; seeds 5-8 × 1.5-3 mm. Canyons, arroyos, slopes, hills, bajadas, valleys, limestone, gypsum, sandstone, rocky, clay, silty, and sandy soils, alluvium, desert scrub, 3400-5100 ft. SE, SC, SW. In addition to the typical variety, **var. salpignantha** (Woodson) S. P. McLaughlin, which differs by pubescent (vs. glabrous) stems, leaves, and calyces, is found in New Mexico. The varieties do not appear to be geographically or ecologically segregated.

Amsonia palmeri A. Gray—Palmer's bluestar. Stems erect, 20-65 cm, branched mostly above middle; leaf blades on main stem oblong or elliptic to narrowly lanceolate or linear, 4.5-7.5 × 0.2-1.8 cm; inflorescences exceeding or only slightly exceeded by vegetative branches, flowers 5-20; calyx lobes pubescent, ciliate, or glabrous; corolla tubes white or bluish, 7.5-12 mm, lobes white to blue or yellowish, 3-5 mm; follicles 2-10 × 0.2-0.4 cm, smooth; seeds 6-10 × 1-2.5 mm. Slopes, canyons, arroyos, granite, basalt, sandy, clay, and gravelly soils, desert grassland, desert scrub, riparian woodland, 3800-4500 ft. Grant, Hidalgo, Socorro.

Amsonia tharpii Woodson—Tharp's bluestar. Stems erect, 10-20 cm, branched mostly above middle; leaf blades on main stem oblong or elliptic to lanceolate, 2.5-5 × 0.4-1 cm; inflorescences slightly exceeded by vegetative branches, flowers 5-20; calyx glabrous; corolla tubes purplish, 12-17 mm, lobes white to pale blue or greenish, 6-9 mm; follicles 2-7 × 0.2-0.5 cm, smooth; seeds 7-11 × 2-3 mm. Hills, slopes, arroyos, flats, limestone, gypsum, dolomite, sandy soils, desert scrub, 3000-3800 ft. Known in New Mexico only from Eddy County. **R**

APOCYNUM L. – Dogbane

Plants rhizomatous herbs; latex white; leaves opposite, stipular colleters present, laminar colleters absent; inflorescences terminal and subterminal cymes; corollas urceolate, cylindrical, or campanulate, lobes erect to reflexed, estivation contorted; gynostegium absent; corona of 5 minute scales below corolla sinuses; pollen in tetrads; follicles usually paired; seeds cylindrical, comose at 1 end. Plants intermediate between the 2 species present in the state are assigned to **Apocynum ×floribundum** Greene.

1. Leaves drooping; inflorescences borne above vegetative branches; corolla campanulate, 4.5-9 mm, white or pink (often with pink stripes), lobes reflexed…**A. androsaemifolium**
1'. Leaves spreading to ascending; inflorescences overtopped by some vegetative branches; corolla urceolate or cylindrical, 2.5-5 mm, white, cream, yellowish, or greenish, lobes erect to ascending…**A. cannabinum**

Apocynum androsaemifolium L. [*A. androsaemifolium* var. *pumilum* A. Gray; *A. pumilum* (A. Gray) Greene]—Spreading dogbane. Stems erect to spreading, lacking a central, dominant stem, 20-70 cm; leaves drooping; blades ovate to elliptic, 2.5-10 × 1-6 cm; inflorescences exceeding vegetative branches, flowers 5-30; corollas campanulate, white or pink (often pink-striped), 4.5-9 mm, lobes reflexed; follicles erect to pendulous, 6-15 cm, smooth; seeds 1.5-2.5 mm, coma 1-2 cm. Slopes, canyons, streamsides, rhyolite, granite, basalt, limestone, sandstone, talus, silty, clay, sandy, and rocky soils, juniper and piñon-juniper woodlands, pine, mixed conifer, spruce-fir, and riparian forests, 7000-9500 ft. NE, NC, NW, C, WC, SC.

Apocynum cannabinum L. [*A. cannabinum* var. *pubescens* (R. Br.) A. DC.; *A. hyperificifolium* Aiton; *A. sibiricum* Jacq.]—Indian hemp, common dogbane, hemp dogbane. Stems erect, central stem dominant, 30-150 cm; leaves spreading to ascending; blades ovate to oblong or elliptic, 3.5-14 × 0.5-7 cm; inflorescences exceeded by vegetative branches, flowers 10-70; corollas urceolate to cylindrical, white, cream, yellowish, or greenish, 2.5-5 mm, lobes erect to ascending; follicles pendulous, 6-20 cm, smooth; seeds 3.5-6 mm, coma 1.5-3 cm. Arroyos, canyons, slopes, cliffs, stream banks, marshes, seeps, pond and lake shores, dunes, limestone, rhyolite, volcanic ash, silty, rocky, and sandy soils, shortgrass prairie, oak and juniper grasslands, piñon-juniper and oak woodlands, pine-oak, pine, mixed conifer, and riparian forests, 3900-9000 ft. Widespread except SE.

ASCLEPIAS L. – Milkweed

Plants herbs, subshrubs, or shrubs; latex white (clear in *A. tuberosa*); leaves opposite or alternate, stipular colleters present (rarely absent), laminar colleters present or absent; inflorescences extra-axillary or terminal, cymose, umbelliform; corollas rotate with reflexed to spreading lobes (campanulate in *A. asperula*), estivation valvate; gynostegium present; corona of 5 cavitate, laminar, or clavate segments, usually with an adaxial subulate or falcate appendage arising from the cavity, if present; pollinia present; follicles usually solitary; seeds flat, comose at 1 end.

1.   Leaves alternate (sometimes a few subopposite)…(2)
1ʹ. Leaves opposite, subopposite, or whorled…(14)
2.   Leaves filiform to narrowly linear, 0.1–0.3 cm wide…(3)
2ʹ. Leaves linear, elliptic, oblong, lanceolate, ovate, oblanceolate, or obovate, 0.3–0.9 cm wide…(8)
3.   Stems 30–160 cm…(4)
3ʹ. Stems 5–30 cm…(6)
4.   Stems woody; leaves needlelike, 1.5–4 cm; corona segments with rod-shaped, exserted appendages… **A. linaria**
4ʹ. Stems herbaceous; leaves not needlelike, 5–19 cm; corona segment appendages absent or at most a low crest…(5)
5.   Gynostegial column (measured from corolla to base of anthers) 0.5–1.5 mm; fused anthers barrel-shaped, anther wings evenly crescent-shaped and separate throughout, apical anther appendages narrowly pandurate, conduplicate, not obscuring corpuscula…**A. engelmanniana**
5ʹ. Gynostegial column 0–0.5 mm; fused anthers obconic, anther wings connivent, except at the wider base, apical anther appendages deltoid, obscuring corpuscula…**A. rusbyi**
6.   Corolla lobes 4.5–6 mm; corona segments 3.5–4.5 mm; follicles 1.5–2 cm wide, rugose…**A. involucrata** (in part)
6ʹ. Corolla lobes 3–5 mm; corona segments 1–2.5 mm; follicles 0.5–1.5 cm wide, smooth…(7)
7.   Corolla red-violet; corona segment appendages ligulate, barely exserted; fruiting pedicels upcurved… **A. uncialis** (in part)
7ʹ. Corolla pink or cream; corona segment appendages acicular, well exserted; fruiting pedicels straight… **A. pumila**
8.   Corolla campanulate, lobes ascending and exceeding corona segments, 7–10 mm; corona segments clavate-tubular, deflexed at base, ascending to incurved at apex, margins connivent…**A. asperula**
8ʹ. Corolla reflexed, lobes sometimes spreading, not overtopping corona segments, 3–8 mm; corona segments cupulate, conduplicate, or tubular, margins separate…(9)
9.   Stems 15–100 cm; corolla lobes 7–8 mm…(10)
9ʹ. Stems 5–15 cm; corolla lobes 3–6 mm…(11)
10. Stems, leaves, peduncles, and follicles hirsute; corolla and corona red, orange, or yellow; anthers yellow to yellowish green; seeds 8–9 mm; widespread…**A. tuberosa**
10ʹ. Stems, leaves, peduncles, and follicles with curved trichomes, pilosulous, tomentulose, or glabrate; corolla pink, red, or green; corona cream, segments with reddish or purplish dorsal stripes; anthers brown; seeds 6–7 mm; rare, known only historically from Colfax County…**A. hallii** (in part)
11. Corolla green, sometimes tinged red or pink; corona segment appendages well exserted, sharply inflexed toward or over style apex; follicles rugose…(12)
11ʹ. Corolla red-violet; corona segment appendages included or barely exserted from segment; follicles smooth…(13)
12. Leaves 0.2–0.8 cm wide; anthers 1.5–2 mm; corona segments cream, usually with a pink or red dorsal stripe, conduplicate, 3.5–4.5 mm; seeds 7–8 × 5–6 mm…**A. involucrata** (in part)
12ʹ. Leaves 0.5–2 cm wide; anthers 1–1.5 mm; corona segments yellow to yellowish cream, tubular, 2–3 mm; seeds 8–12 × 6–8 mm…**A. macrosperma** (in part)
13. Corona segment appendages falcate, included; Colorado Plateau, San Juan County…**A. sanjuanensis** (in part)
13ʹ. Corona segment appendages ligulate, barely exserted; NE to SW New Mexico, not in San Juan County… **A. uncialis** (in part)
14. Leaves whorled…**A. subverticillata**
14ʹ. Leaves opposite or subopposite…(15)
15. Plants subshrubs; corona segment apex long-caudate…**A. macrotis**
15ʹ. Plants herbaceous; corona segment apex truncate to acute…(16)
16. Stems and leaves glabrous, glaucous…**A. elata**
16ʹ. Stems or leaves usually with trichomes, or if glabrous, never glaucous…(17)
17. Stems prostrate or decumbent…(18)

17'. Stems spreading to erect…(23)

18. Leaves 1.5–5 cm; corollas red-violet; corona segments 1–2 mm; follicles 3–5 × 0.5–1.5 cm…(19)

18'. Leaves 1.5–15 cm; corollas green, sometimes tinged red, pink, purple, or brown; corona segments 2–12 mm; follicles 4.5–10 × 1.2–3 cm…(20)

19. Corona segment appendages falcate, included; Colorado Plateau, San Juan County…**A. sanjuanensis** (in part)

19'. Corona segment appendages ligulate, barely exserted; NE to SW New Mexico, not in San Juan County… **A. uncialis** (in part)

20. Leaves 1.2–7.5 cm wide, apex obtuse to acute; umbels extra-axillary; corolla lobes 7–13 mm; anthers 1.7–2.5 mm; corona segments green, cream, or green and cream, sinuous-tubular, 5–11 mm…(21)

20'. Leaves 0.2–2 cm wide, apex acute to attenuate; umbels terminal; corolla lobes 4.5–6.5 mm; anthers 1–2 mm; corona segments cream with pink or red dorsal stripe or yellow, conduplicate or tubular, but not sinuous, 2–4.5 mm…(22)

21. Leaves ovate or lanceolate to oblong or elliptic, base cuneate to obtuse, apex obtuse to rounded, sometimes conduplicate; pedicels 10–20 mm; gynostegial columns 1–1.5 mm; anthers 2–2.5 mm; corona segments relatively slender, green, sometimes tinged bronze, apex white, apex deeply emarginate; widespread in New Mexico…**A. oenotheroides** (in part)

21'. Leaves ovate to lanceolate, base cuneate or obtuse to truncate or subcordate, apex obtuse to acute, rarely conduplicate; pedicels 17–30 mm; gynostegial columns 0.3–0.5 mm; anthers 1.7–2 mm; corona segments relatively stout, cream or green with cream apex, apex truncate; Catron, Grant, Hidalgo Counties…**A. nyctaginifolia** (in part)

22. Leaves 0.2–0.8 cm wide; anthers 1.5–2 mm; corona segments cream, usually with a pink or red dorsal stripe, conduplicate, 3.5–4.5 mm; seeds 7–8 × 5–6 mm…**A. involucrata** (in part)

22'. Leaves 0.5–2 cm wide; anthers 1–1.5 mm; corona segments yellow to yellowish cream, tubular, 2–3 mm; seeds 8–12 × 6–8 mm…**A. macrosperma** (in part)

23. Corolla lobes 3–5 mm…(24)

23'. Corolla lobes 5–15 mm…(26)

24. Stems, leaves, peduncles, pedicels, and calyces puberulent with curved trichomes to glabrate; umbels pendent…**A. quinquedentata** (in part)

24'. Stems, leaves, peduncles, pedicels, and calyces tomentose, at least initially…(25)

25. Stems 20–40 cm; leaves linear-lanceolate, 5–15 × 0.3–1.3 cm; peduncles 0.2–1.5 cm; abaxial corolla lobes minutely pilosulous; anthers 2–2.5 mm; corona segments tubular, appendage ligulate; follicles ribbed… **A. brachystephana** (in part)

25'. Stems 6–15 cm; leaves orbicular to ovate or oblate, 1.5–3.2 × 1.7–4 cm; peduncles 2.7–6 cm; abaxial corolla lobes glabrous; anthers 1–1.5 mm; corona segments conduplicate, appendage falcate; follicles smooth…**A. nummularia**

26. Stems, leaves, peduncles, or pedicels persistently tomentose or tomentulose…(27)

26'. Stems, leaves, peduncles, and pedicels variously indumented, sometimes ephemerally tomentose or tomentulose…(32)

27. Corona segments 7–15 mm…(28)

27'. Corona segments 2–7 mm…(29)

28. Petioles 2–6 mm; leaves strongly bicolored (adaxially dark green, abaxially white or gray, tomentose); corolla lobes deep maroon to green adaxially, glabrous; anthers 2–2.5 mm; coronas deep maroon to yellowish green, segments 7–9 mm; appendage absent or merely a low crest; follicles fusiform, 1.2–1.4 cm wide…**A. hypoleuca**

28'. Petioles 4–12 mm; leaves uniformly green on both surfaces; corolla lobes dark pink (rarely pale) adaxially, hirtellous at base; anthers 2.5–3 mm; coronas pale pink to nearly cream, segments 9–15 mm; appendage subulate, sharply inflexed over style apex; follicles lance-ovoid, 2–3 cm wide…**A. speciosa**

29. Corolla lobes minutely pilosulous abaxially; flowers 4–15 per umbel; corona segments 1.5–2 mm; follicles ribbed…**A. brachystephana** (in part)

29'. Corolla lobes glabrous abaxially, flowers 9–60 per umbel; corona segments 3.5–6.5 mm; follicles smooth…(30)

30. Corolla pink or red or pinkish green or reddish green; anthers brown; corona segments cream with red, pink, or purple dorsal stripe, apex acute, appendage glabrous; rare, known only historically from Colfax County…**A. hallii** (in part)

30'. Corolla green to yellowish green, sometimes tinged red or purple; anthers green; corona segments cream, greenish cream, or yellowish cream, not striped red, pink, or purple, apex truncate to rounded, appendage papillose…(31)

31. Petioles 0–4 mm, leaf bases cordate; anthers 3–3.5 mm; corona segment apices truncate, not emarginate, oblique, appendage apex not upturned; seeds 7–8 × 5–6 mm…**A. latifolia** (in part)

31'. Petioles 7–17 mm, leaf bases rounded, truncate, or subcordate; anthers 2–2.5 mm; corona segment apices truncate, emarginate, not oblique, appendage apex upturned; seeds 9–12 × 6–8 mm…**A. arenaria**

32. Stems, leaves, peduncles, and pedicels hirsute, hirsutulous, or hirtellous…(33)
32′. Stems, leaves, peduncles, and pedicels variously indumented, but trichomes not both straight and spreading…(34)
33. Leaves ovate or lanceolate to oblong or elliptic, base cuneate to obtuse, apex obtuse to rounded, sometimes conduplicate; pedicels 10–20 mm; gynostegial columns 1–1.5 mm; anthers 2–2.5 mm; corona segments relatively slender, green, sometimes tinged bronze, apex white, apex deeply emarginate; widespread in New Mexico…**A. oenotheroides** (in part)
33′. Leaves ovate to lanceolate, base cuneate or obtuse to truncate or subcordate, apex obtuse to acute, rarely conduplicate; pedicels 17–30 mm; gynostegial columns 0.3–0.5 mm; anthers 1.7–2 mm; corona segments relatively stout, cream or green with cream apex, apex truncate; Catron, Grant, Hidalgo Counties…**A. nyctaginifolia** (in part)
34. Corona appendage crestlike and barely exserted or absent…(35)
34′. Corona appendage falcate or acicular, well exserted…(36)
35. Leaves linear to broadly oval or nearly orbicular, 0.8–6 cm wide; flowers 22–60 per umbel; anthers 3–4 mm; corona segments laminar, appressed to gynostegium, appendage absent or obscure; follicles 1.5–2 cm wide…**A. viridiflora**
35′. Leaves linear, 0.2–0.6 cm wide; flowers 4–10 per umbel; anthers 1.5–2 mm; corona segments conduplicate, separate from gynostegium, appendage crestlike, barely exserted; follicles 0.5–1 cm wide… **A. quinquedentata** (in part)
36. Corolla green; corona segment apex truncate, appendage papillose…**A. latifolia** (in part)
36′. Corolla pink, red, white, pinkish green, or reddish green; corona segment apex acute to obtuse, appendage glabrous…(37)
37. Interpetiolar ridges present; pedicels 10–15 mm; corolla lobes 5–6 mm; corona segments 2–2.5 mm; corona segment appendages arching above style apex; fruiting pedicels straight…**A. incarnata**
37′. Interpetiolar ridges absent; pedicels 16–28 mm; corolla lobes 7–8 mm; corona segments 5.5–6.5 mm; corona segment appendages sharply inflexed over style apex; fruiting pedicels upcurved; rare, known only historically from Colfax County…**A. hallii** (in part)

Asclepias arenaria Torr.—Sand milkweed. Plants decumbent to erect, rhizomatous herbs, rarely branched, 20–100 cm; latex white; leaves opposite, stipular colleters present, blades oblong or obovate to ovate or oval, 4.2–11.5 × 2.5–7.5 cm, laminar colleters present; inflorescences extra-axillary (sometimes appearing terminal), flowers 14–51; corollas green to yellowish green, reflexed, lobes spreading, 7–8 mm, glabrous; corona segments cream to greenish or yellowish, conduplicate, 3.5–4 mm, apex truncate to rounded, appendage falcate, sharply incurved over style apex, papillose; follicles erect on upcurved pedicels, lance-ovoid, 5.5–10 × 2–2.8 cm, smooth; seeds 9–12 × 6–8 mm, coma 2–3 cm. Sand hills, dunes, sandy soil, prairies, pastures, grasslands, oak scrub, riparian areas, 3300–4600 ft. NE, EC, SE, Otero.

Asclepias asperula (Decne.) Woodson—Antelope horns, spider milkweed, spider antelope horns. Plants decumbent to erect herbs, unbranched above base, 15–60 cm; latex white; leaves alternate (rarely a few subopposite), stipular colleters present, blades linear-lanceolate to linear, 7.7–17 × 0.4–3.6 cm, laminar colleters absent; inflorescences terminal, flowers 10–60; corollas pale green (sometimes tinged red abaxially), campanulate, lobes ascending, 7–10 mm, puberulent abaxially at apex; corona segments reddish purple with white upper margin, clavate-tubular, 4.5–7 mm, apex rounded, appendage an internal, low crest, papillose; follicles erect on upcurved pedicels, lance-ovoid, 6–11.5 × 1–2.5 cm, ribbed (sometimes also muricate at apex); seeds 5–8 × 4–6 mm, coma 2.5–4 cm. Our plants are **subsp. asperula**. Slopes, hillsides, mesas, ridgetops, bajadas, canyons, arroyos, streamsides, lakesides, rhyolite, basalt, granite, limestone, shale, sandstone, alluvium, talus, rocky, gravelly, clay, sandy, and silty soils, grasslands, piñon-juniper, oak, juniper, pine, pine-oak, mixed conifer, and mesquite woodlands, chaparral, riparian areas, desert scrub, 4000–8500 ft. Widespread.

Asclepias brachystephana Engelm. ex Torr.—Short-crowned milkweed. Plants multistemmed, erect herbs, unbranched above base, 20–40 cm; latex white; leaves opposite to subopposite, stipular colleters present, blades linear-lanceolate, 5–15 × 0.3–1.3 cm, laminar colleters absent; inflorescences extra-axillary, flowers 4–15; corollas red-violet (rarely green with red tinge), lobes reflexed, 4–6 mm, pilosulous; corona segments red-violet to pink at base, white at apex, tubular, 1.5–2 mm, apex truncate, oblique, appendage ligulate, sharply inflexed toward gynostegium, glabrous; follicles erect on upcurved pedicels, lance-ovoid, 5–7 × 1.2–1.8 cm, ribbed; seeds 6–7 × 4–6 mm, coma 2–2.5 cm. Plains, bajadas, pastures, arroyos, stream banks, riparian areas, limestone, igneous substrates, alluvium, gravelly, clay, silty, and sandy soils, desert grassland, desert scrub, oak-juniper, juniper, 3200-6200 ft. EC, C, WC, SE, SC, SW.

Asclepias elata Benth. [*A. glaucescens* Kunth subsp. *elata* (Benth.) E. Fourn.]—Nodding milkweed. Plants erect, glaucous herbs, unbranched, 30-80 cm; latex white; leaves opposite, stipular colleters present, blades oval or elliptic to oblong or lanceolate, 5.5-14 × 1.5-6 cm, laminar colleters absent; inflorescences terminal and extra-axillary, flowers 7-20; corollas green, reflexed, lobes spreading, 11-14 mm, glabrous; corona segments white, yellow to tan dorsally, conduplicate, 4-6 mm, apex truncate, appendage an included crest, glabrous; follicles erect on upcurved pedicels, fusiform to lance-ovoid, 8-13 × 1-2 cm, smooth; seeds 6-7 × 3-6 mm, coma 4-4.5 cm. Canyons, arroyos, stream banks, slopes, igneous substrates, limestone, rocky, sandy, and clay soils, piñon-juniper and oak woodlands, pine-oak and riparian forests, meadows, 4600-5500 ft. Known in New Mexico from only a few collections from Hidalgo and Sierra Counties.

Asclepias engelmanniana Woodson—Engelmann's milkweed. Plants erect herbs, occasionally branched, 40-160 cm; latex white; leaves alternate, drooping, stipular colleters present, blades linear, conduplicate, 5-19 × 0.15-0.3 cm, laminar colleters absent; inflorescences extra-axillary, flowers 14-23; corollas pale green to cream, yellow, tan, or russet, reflexed, lobes spreading, 4-5 mm, glabrous; corona segments cream to tan or yellow, chute-shaped, 2-3 mm, apex retuse to nearly truncate, appendage absent; follicles erect on upcurved pedicels, lance-ovoid, 6-10 × 1.2-2 cm, smooth; seeds 8-9 × 5-6 mm, coma 2-2.5 cm. Slopes, plains, valleys, arroyos, canyons, streamsides, dunes, sandstone, limestone, gypsum, sandy, gravelly, clay, calcareous, and rocky soils, prairies, shrubby and juniper grasslands, 3600-7700 ft. Widespread except NW, SE, SC.

Asclepias hallii A. Gray—Hall's milkweed. Plants erect to ascending, rhizomatous herbs, unbranched below inflorescence, 30-70 cm; latex white; leaves alternate to subopposite, stipular colleters present, blades narrowly lanceolate to ovate, 5-16 × 1.5-9 cm, laminar colleters present; inflorescences extra-axillary, flowers 9-29; corollas pale pink to red or greenish pink, reflexed, lobes spreading, 7-8 mm, minutely papillose in throat; corona segments cream with red or pink to purple dorsal stripe, conduplicate, 5.5-6.5 mm, apex acute, appendage falcate, sharply inflexed over style apex, glabrous; follicles erect on upcurved pedicels, lance-ovoid, 8-12 × 0.7-1.5 cm, smooth; seeds 6-7 × 4 mm, coma 2.5-3.5 cm. Known in New Mexico from a single, historical specimen, 8500 ft.; possibly extirpated. Colfax.

Asclepias hypoleuca (A. Gray) Woodson—Mahogany milkweed, talayote. Plants erect herbs, unbranched, 25-100 cm; latex white; leaves opposite, stipular colleters present, blades ovate or lanceolate to oblong, elliptic, or oval, 5.5-11.5 × 1-5 cm, laminar colleters present; inflorescences terminal and extra-axillary, flowers 12-35; corollas deep maroon to green, reflexed, lobes usually spreading, 8-10 mm, pilosulous abaxially; corona segments deep maroon to green, conduplicate, 7-9 mm, apex truncate and long-tapering distally, appendage absent or a low, included crest; follicles erect on upcurved pedicels, fusiform, 9-11.5 × 1.2-1.4 cm, smooth; seeds 6-7 × 4-5 mm, coma 3-3.5 cm. Slopes, rocky soils, pine and oak forests, 7500-9000 ft. Known in New Mexico only from Catron and Grant Counties.

Asclepias incarnata L.—Swamp milkweed. Plants erect herbs, unbranched below inflorescence, 30-150 cm; latex white; leaves opposite, stipular colleters and interpetiolar ridge present, blades lanceolate to linear-lanceolate, 5-15 × 0.5-4 cm, laminar colleters present; inflorescences extra-axillary, branched, flowers 10-31; corollas bright pink to white, reflexed, lobes spreading, 5-6 mm, minutely papillose in throat; corona segments pink to white, paler than corolla, tubular, 2-2.5 mm, apex obtuse, appendage acicular, arching over style apex; follicles erect on straight pedicels, fusiform, 6-9 × 0.8-1.2 cm, smooth; seeds 8-9 × 5-6 mm, coma 1.5-2 mm. Our plants are **subsp. incarnata**. Canyons, hillsides, calcareous soils, marshes, seeps, wet depressions, wet meadows, riparian woods, 3100-7100 ft. EC, C, SW, SC, Colfax.

Asclepias involucrata Engelm. ex Torr.—Dwarf milkweed. Plants decumbent herbs, unbranched above base, 5-18 cm; latex white; leaves opposite to alternate, stipular colleters present, blades linear to narrowly lanceolate, 1.5–12 × 0.2–0.8 cm, laminar colleters absent; inflorescences terminal, flowers 6–35; corollas green, sometimes tinged pink or red, lobes reflexed, 4.5–6 mm, glabrous; corona segments cream, usually with pink or red dorsal stripe, conduplicate, 3.5–4.5 mm, apex truncate with spreading tip, appendage falcate, sharply inflexed toward style apex; follicles erect on upcurved pedicels, ovoid, 4.5–5.5 × 1.5–2 cm, rugose; seeds 7–8 × 5–6 mm, coma 1.5–2 cm. Hills, slopes, ridges, mesas, arroyos, flats, dunes, limestone, sandstone, rhyolite, rocky, gravelly, sandy, gypsum, and clay soils, short-grass prairies, shrubby and desert grasslands, oak, juniper, and piñon-juniper woodlands, 4200–7600 ft. NC, NW, EC, C, WC, SW.

Asclepias latifolia (Torr.) Raf.—Corn-kernel milkweed, broad-leaf milkweed. Plants erect, rhizomatous herbs, unbranched, 25–100 cm; latex white; leaves opposite, stipular colleters present, blades oval or oblong to ovate or orbicular, 5.5–14 × 3–14 cm, laminar colleters present; inflorescences extra-axillary, flowers 20–59; corollas green, lobes reflexed, 7–9 mm, papillose in throat; corona segments cream to yellow, conduplicate, 3–5.5 mm, apex truncate, oblique, appendage falcate, sharply inflexed over style apex, papillose; follicles erect on upcurved pedicels, ovoid, 6.5–9.5 × 2–3 cm, smooth; seeds 7–8 × 5–6 mm, coma 3–4 cm. Plains, hills, streamsides, dunes, arroyos, limestone, shale, sandstone, silty, clay, sandy, gypsum, rocky, and gravelly soils, prairies, shrubby grasslands, desert scrub, piñon-juniper, juniper, and riparian woodlands, 3700–7100 ft. NE, NC, NW, EC, C, SE, SC, SW.

Asclepias linaria Cav.—Pine-needle milkweed, hierba del cuervo. Plants shrubs with rounded crowns, branched, 30–70 cm; latex white; leaves alternate, stipular colleters present, blades linear, 1.5–4 × 0.1–0.15 cm, laminar colleters absent; inflorescences extra-axillary, flowers 9–30; corollas green to cream, often tinged red or purple, reflexed, lobes spreading, minutely hirsutulous in throat; corona segments cream, sometimes with green or purple dorsal stripe, cupulate, 2.5–3 mm, apex obtuse to rounded, appendage rod-shaped, slightly exserted, glabrous; follicles erect on upcurved pedicels, ovoid, 3.5–5 × 0.6–1 cm, smooth; seeds 5–6 × 3–4 mm, coma 1.5–2 cm. Canyon slopes, 4300–4800 ft. Known in New Mexico only from the Peloncillo Mountains, Hidalgo County.

Asclepias macrosperma Eastw.—Large-seed milkweed, dwarf milkweed, Eastwood's milkweed. Plants decumbent, unbranched above base, 6–15 cm; latex white; leaves opposite or alternate, stipular colleters present, blades linear-lanceolate to lance-ovate, 2.5–7 × 0.5–2 cm, laminar colleters absent; inflorescences terminal, flowers 12–40; corollas green, tinged red, lobes reflexed, 4.5–5 mm, glabrous; corona segments yellow to yellowish cream, tubular, 2–3 mm, apex truncate, appendage falcate, sharply inflexed toward or over style apex, glabrous; follicles erect on upcurved pedicels, ovoid, 5–6.5 × 1.2–2 cm, rugose; seeds 8–12 × 6–8 mm, coma 1.5–2 cm. Arroyos, canyons, slopes, sandstone, gypsum, sandy, silty, and clay soils, juniper and piñon-juniper woodlands, shrubby grasslands, 4900–6300 ft. San Juan.

Asclepias macrotis Torr.—Long-hood milkweed. Plants cespitose, rhizomatous subshrubs, branched at base, 10–35 cm; latex white; leaves opposite, stipular colleters present, blades filiform, 2.5–7 × 0.05–0.15 cm, laminar colleters absent; inflorescences extra-axillary, flowers 2–7; corollas greenish cream, tinged red, reflexed, lobes spreading, 4–5 mm; corona segments cream to greenish cream, reddish brown at base or as a dorsal stripe, conduplicate, 4–5 mm, apex truncate, long-caudate, appendage falcate, inflexed toward style apex, hirtellous; follicles erect on upcurved pedicels, fusiform, 4.5–6 × 0.5–0.7 cm, smooth; seeds 6–8 × 2.5–4 mm, coma 2–2.5 cm. Mesas, slopes, cliffs, canyons, arroyo margins, limestone, dolomite, sandstone, shale, rhyolite, gypsum, talus, rocky, clay, and sandy soils, alluvium, oak, juniper, and piñon-juniper woodlands, chaparral, shrubby grasslands, prairies, 4500–7500 ft. Scattered.

Asclepias nummularia Torr.—Tufted milkweed. Plants erect herbs, unbranched, 6–15 cm; latex white; leaves opposite, stipular colleters present, blades orbicular to obovate or oblate, 1.5–3.2 × 1.7–4 cm (larger on nonreproductive shoots), laminar colleters present or absent; inflorescences mostly terminal, flowers 5–28; corollas pinkish violet to tan, lobes reflexed, 4–5 mm, minutely papillose in throat; corona segments white, red-violet at base, conduplicate, 2.5–3 mm, apex truncate, appendage falcate, sharply inflexed over style apex, glabrous; follicles erect on upcurved pedicels, lance-ovoid, 4–7.5 × 1.2–1.5 cm, smooth; seeds 6–7 × 4–5 mm, coma 1.5–2.5 cm. Slopes, rocky, volcanic soils, grasslands, piñon-juniper and oak-juniper woodlands, 5600-6000 ft. Known in New Mexico only from Grant, Hidalgo, and Sierra Counties.

Asclepias nyctaginifolia A. Gray—Mohave milkweed. Plants decumbent to erect herbs, unbranched above base, 15-40 cm; latex white; leaves opposite, stipular colleters present, blades ovate to lanceolate, 4.5-15 × 1.5-7.5 cm, laminar colleters present or absent; inflorescences extra-axillary, flowers 5-28; corollas green, sometimes tinged red or purple, lobes reflexed, 9-13 mm, sometimes hirsutulous near apex abaxially; coronas cream or green with cream apex, tubular, slightly sinuous, 8-11 mm, apex truncate, slightly flared, appendage ligulate, sharply incurved, greatly exceeded by segment margin, minutely papillose; follicles erect on upcurved pedicels, lance-ovoid, 6.5-10 × 1.5-3 cm, smooth; seeds 6-8 × 4.5-6.5 mm, coma 2-4 cm. Slopes, streamsides, limestone, dolomite, rhyolite, sandy soils, pine, pine-oak, and riparian forests, mesquite grassland, 3900-6700 ft. Known in New Mexico only from Catron, Grant, Hidalgo, and Sierra Counties.

Asclepias oenotheroides Schltdl. & Cham.—Zizotes milkweed, hierba de zizotes, longhood milkweed, sidecluster milkweed. Plants decumbent to erect herbs, unbranched above base, 10-50 cm; latex white; leaves opposite, stipular colleters present, blades ovate or lanceolate to oblong or elliptic, 4-11 × 1.2-6.5 cm, laminar colleters absent; inflorescences extra-axillary, flowers 8-32; corollas green, sometimes tinged red or brown, lobes reflexed, 9-12 mm, sometimes hirsutulous near apex abaxially; corona segments green or bronze, white at apex, tubular, slightly sinuous, 7-10 mm, apex deeply emarginate, flared, appendage ligulate, sharply incurved, exceeded by segment margin, sometimes minutely papillose; follicles erect on upcurved pedicels, lance-ovoid, 4.5-9.5 × 1.2-2.5 cm, smooth; seeds 6-8 × 5-6 mm, coma 2-3 cm. Hills, slopes, ridges, mesas, bajadas, dunes, arroyos, valleys, limestone, sandstone, shale, granite, volcanic ash, caliche, alluvium, sandy, clay, gravelly, and rocky soils, desert scrub, mesquite grasslands, prairies, piñon-juniper and juniper woodlands, 3300-6300 ft. NC, EC, C, SE, SC, SW.

Asclepias pumila (A. Gray) Vail—Plains milkweed, low milkweed, dwarf milkweed. Plants erect, rhizomatous herbs, sometimes branched, 10-30 cm; latex white; leaves alternate, stipular colleters present, blades linear, 2.5-5.5 × 0.05-0.1 cm, laminar colleters absent; inflorescences extra-axillary, flowers 3-13. corollas pink to cream, reflexed, lobes spreading, 3.5-4.5 mm, minutely papillose in throat; corona segments cream, sometimes tinged or striped pink, tubular, 2-2.5 mm, apex obtuse, appendage acicular, arching toward style apex, glabrous; follicles erect on straight pedicels, narrowly fusiform, 2.5-9.5 × 0.5-1 cm, smooth; seeds 5-6 × 3-5 mm, coma 2-3 cm. Plains, bajadas, basalt, sandy, rocky, gravelly, and clay soils, shortgrass prairies, oak scrub, desert and shrubby grasslands, 4100-7400 ft. Scattered throughout.

Asclepias quinquedentata A. Gray—Slim-pod milkweed. Plants erect to ascending herbs, unbranched above base, 10-60 cm; latex white; leaves opposite, stipular colleters present, blades linear, 6-14 × 0.2-0.6 cm, laminar colleters present or absent; inflorescences extra-axillary, flowers 4-10; corollas green, sometimes tinged red, lobes reflexed, 4-6 mm, pilose abaxially; corona segments pink to red-violet at base, white at apex, conduplicate, 3-4 mm, apex truncate, appendage a barely exserted crest, glabrous; follicles erect on upcurved pedicels, fusiform, 8.5-16 × 0.5-1 cm, smooth; seeds 4-5 × 3-4 mm, coma 2-2.5 cm. Hillsides, canyons, rhyolite, basalt, rocky soil, pine-oak forest, piñon-juniper woodland, 6500-8000 ft. Catron, Grant, Hidalgo, Socorro.

Asclepias rusbyi (Vail) Woodson [*A. engelmanniana* Woodson var. *rusbyi* (Vail) Kearney]—Rusby's milkweed. Plants erect herbs, sometimes branched, 50-100 cm; latex white; leaves alternate, drooping, stipular colleters present, blades linear, conduplicate, 9-15 × 0.2-0.3 cm, laminar colleters absent; inflorescences extra-axillary, sometimes branched, flowers 7-28; corollas green or brown, reflexed, lobes spreading, 4-6 mm, glabrous; corona segments yellowish green to yellow, chute-shaped, 1.5-2.5 mm, apex truncate, appendage absent; follicles erect on upcurved pedicels, lance-ovoid, 9-12.5 × 1.5-2.5 cm, smooth; seeds 6-8 × 4-6 mm, coma 2-2.5 cm. Canyons, slopes, lakesides, pine-oak forest, riparian woods, 6000-7400 ft. NW, Grant, Lincoln, Quay.

Asclepias sanjuanensis K. D. Heil, J. M. Porter & S. L. Welsh—San Juan milkweed. Plants decumbent, rhizomatous herbs, unbranched above base, 4-13 cm; latex white; leaves opposite or alternate, stipular colleters present or absent, blades lanceolate to ovate, 2–5 × 0.5–2 cm, laminar colleters absent; inflorescences terminal and extra-axillary, flowers 3–8; corollas red-violet, reflexed, lobes spreading, 4–5 mm, glabrous; corona segments red-violet at base, white to orange at apex, cupulate, 1.5–2 mm, apex truncate, appendage falcate, barely exserted, glabrous; follicles erect on upcurved pedicels, ovoid, 3–5 × 0.7–1 cm, smooth; seeds 8–9 × 6–7 mm, coma 1–1.5 cm. Slopes, dunes, hills, ridges, sandstone, sandy and clay soils, alluvium, juniper and piñon-juniper woodlands, shrubby grassland, 5300–6800 ft. McKinley, San Juan. **R & E**

Asclepias speciosa Torr.—Showy milkweed. Plants erect, rhizomatous herbs, rarely branched, 30–125 cm; latex white; leaves opposite, stipular colleters present, blades lanceolate or ovate to oblong, 6–20 × 2–14 cm, laminar colleters present; inflorescences extra-axillary, flowers 3–34; corollas dark pink (rarely pale), reflexed, lobes spreading, 9–12 mm, densely pilose abaxially; corona segments pale pink to nearly cream, scoop-shaped, 9–15 mm, apex truncate and long-attenuate, appendage subulate, sharply inflexed over style apex, glabrous; follicles erect on upcurved pedicels, lance-ovoid, 9–12 × 2–3 cm, muricate; seeds 7–9 × 4–5 mm, coma 2.5–3 cm. Slopes, flats, valleys, canyons, streamsides, marshes, lake edges, ditches, seeps, limestone, basalt, clay, sandy, silty, gravelly, rocky, and saline soils, pine, spruce, and mixed conifer forests, riparian woods, grasslands, prairies, meadows, old fields, 4500–9400 ft. Scattered.

Asclepias subverticillata (A. Gray) Vail—Western whorled milkweed, horsetail milkweed, poison milkweed. Plants erect, rhizomatous herbs, sometimes branched, 20–90 cm; latex white; leaves whorled (may be opposite on branches), interpetiolar ridge and stipular colleters present, blades linear, 3–13 × 0.1–0.4 cm, laminar colleters absent; inflorescences extra-axillary, flowers 9–25; corollas pale green to cream, sometimes tinged pink or tan, reflexed, lobes spreading, 3.5–4.5 mm, minutely papillose in throat; corona segments cream, sometimes tinged green or pink, cupulate, 1.5–2 mm, apex obtuse, appendage acicular, arching over style apex, glabrous; follicles erect on straight pedicels, narrowly fusiform, 6–8.5 × 0.5–0.9 cm, smooth; seeds 5–8 × 3.5–5 mm, coma 2–2.5 cm. Mesas, valleys, slopes, flats, depressions, marshes, canyons, streamsides, arroyos, pond and lake margins, seeps, playas, limestone, sandstone, granite, clay, sandy, silty, gravelly, and rocky soils, alluvium, shortgrass prairie, desert scrub, juniper and desert grasslands, pine savannas, chaparral, oak, juniper, piñon-juniper, and riparian woodlands, pine, mixed conifer, and spruce-fir forests, 3000–8200 ft. Widespread.

Asclepias tuberosa L.—Butterflyweed, butterfly milkweed, pleurisy root. Plants erect to ascending herbs, unbranched below inflorescence, 15–90 cm; latex clear; leaves alternate, stipular colleters present, blades narrowly elliptic or lanceolate to oblong or linear, 2–12 × 0.5–3 cm, laminar colleters present or absent; inflorescences corymbs of extra-axillary umbels on branches, flowers 5–27; corollas reddish orange to yellow, reflexed, lobes spreading, 6–8 mm, minutely papillose in throat; corona segments reddish orange to yellow, conduplicate, 5.5–7 mm, apex acute, appendage subulate, arching above style apex, glabrous; follicles erect on upcurved pedicels, fusiform, 7–14 × 1.2–2 cm, smooth; seeds 8–9 × 4–5 mm, coma 3–5 cm. Our plants are **subsp. interior** Woodson. Slopes, streamsides, arroyos, canyons, limestone, granite, sandy, silty, clay, gravelly, and rocky soils, alluvium, meadows, oak and piñon-juniper woodlands, pine, pine-oak, mixed conifer, spruce-fir, and riparian forests, 4600–8500 ft. NC, C, WC, SE, SC, SW.

Asclepias uncialis Greene—Wheel milkweed, dwarf milkweed. Plants decumbent, rhizomatous herbs, unbranched above base, 4–10 cm; latex white; leaves opposite or alternate, stipular colleters present, blades linear to lanceolate, 1.7–5 × 0.2–1 cm, laminar colleters absent; inflorescences terminal and extra-axillary, flowers 3–7; corollas red-violet, reflexed, lobes spreading, 3–5 mm, glabrous; corona segments red-violet at base, white to orange at apex, cupulate, 1–2 mm, apex truncate, appendage ligulate, barely exserted, glabrous; follicles erect on upcurved pedicels, ovoid, 3–5 × 0.8–1.5 cm, smooth; seeds 7–8 × 5–6 mm, coma 1.5–2 cm. Rolling hills, bajadas, arroyo banks, sandstone, rhyolite, sandy, clay, and calcareous soils, shortgrass prairie, juniper and shrubby grasslands, piñon-juniper woodland, 4200–7100 ft. NE, C, Grant.

Asclepias viridiflora Raf.—Green milkweed, green comet milkweed. Erect to ascending herbs, rarely branched, 20–125 cm; latex white; leaves opposite to subopposite, stipular colleters present, blades linear to nearly orbicular, 2–13 × 0.8–6 cm, laminar colleters absent; inflorescences extra-axillary, flowers 22–60; corollas yellowish green, sometimes tinged red, lobes reflexed, 5–7 mm, abaxially inconspicuously pilosulous at apex; corona segments green to cream, sometimes tinged red, laminar, 3–4 mm, apex obtuse, appendage absent or obscure; follicles erect on upcurved pedicels, fusiform to lance-ovoid, 6–10 × 1.5–2 cm, smooth; seeds 7–8 × 4–5 mm, coma 2.5–3 cm. Plains, hills, mesas, slopes, dunes, valleys, playas, shale, sandstone, limestone, basalt, sandy, silty, clay, and calcareous soils, alluvium, shortgrass prairie, shrubby, oak, and desert grasslands, piñon-juniper and pine-oak woodlands, pine forest, 4100–7500 ft. NE, NC, NW, EC, Doña Ana, Grant, Torrance.

FUNASTRUM E. Fourn. – Twinevine, milkweed vine

Plants vines, woody and corky at base; latex white; leaves opposite, stipular colleters present, laminar colleters present or absent; inflorescences extra-axillary, cymose, umbelliform; corollas rotate-campanulate, estivation contorted; gynostegium present; corona in 2 parts, a raised, fleshy ring at insertion of staminal column and a whorl of 5 inflated segments on staminal column; pollinia present; follicles usually solitary; seeds flat, comose at 1 end.

1. Corolla glabrous adaxially; inflated corona segments oblong with medial constriction; leaf margins usually crisped; follicles 10–13 cm…**F. crispum**
1'. Corolla inconspicuously hispidulous adaxially; inflated corona segments ovoid without a constriction; leaf margins plane; follicles 6–9 cm…(2)
2. Leaf blades linear to linear-lanceolate, 0.1–1.5 cm wide, bases truncate to hastate or sagittate, rarely cordate…**F. heterophyllum**
2'. Leaf blades lanceolate to ovate, 0.7–4.5 cm wide, bases cordate, rarely sagittate or hastate… **F. cynanchoides**

Funastrum crispum (Benth.) Schltr. [*Sarcostemma crispum* Benth.]—Wavy twinevine. Leaf blades lanceolate to nearly linear, 3–10 × 0.2–3 cm, bases cordate, hastate, or truncate; laminar colleters present or absent; flowers 4–13; corollas green, red, or brown, lobes 4–9 mm, glabrous adaxially; corona ring green to pink, inflated corona segments white to pink, green to magenta at base, oblong with medial constriction, 1.5–2.5 mm; follicles lance-ovoid, 10–13 × 0.6–1.2 cm, smooth; seeds 8–9 × 3–4 mm, coma 3–3.5 cm. Hills, mesas, canyons, limestone, shale, sandstone, granite, basalt, sandy, gravelly, clay, and rocky soils, juniper and desert grasslands, chaparral, desert scrub, riparian woods, 4500–6900 ft. NE, SE, WC, SC, SW, Sandoval.

Funastrum cynanchoides (Decne.) Schltr. [*Sarcostemma cynanchoides* Decne.]—Fringed twinevine. Leaf blades ovate to lanceolate, 1.2–8 × 0.7–4.5 cm, bases cordate, hastate, or sagittate; laminar colleters present; flowers 6–31; corollas cream, often tinged pink at lobe tips or throughout, lobes 3–6 mm, hispidulous adaxially, sometimes just at lobe tips; corona ring green to pink, inflated corona segments white, ovoid, 1.5–2.5 mm; follicles ovoid to lance-ovoid, 6–9 × 0.8–1.6 cm, smooth; seeds 5–6 × 2–3 mm, coma 2.5–3 cm. Streamsides, arroyos, sandy, clay, and gravelly soils, riparian woodland, desert grassland, 3800–6300 ft. SE, SC, SW, Guadalupe.

Funastrum heterophyllum (Engelm. ex. Torr.) Standl. [*Sarcostemma cynanchoides* Decne. subsp. *hartwegii* R. W. Holm]—Fringed twinevine. Leaf blades linear to linear-lanceolate, 1.5–8 × 0.1–1.5 cm, bases hastate, truncate, or sagittate (sometimes cordate); laminar colleters present or absent; flowers 5–13; corollas pink to purple or cream with pink or purple blotches, lobes 3.5–5 mm, hispidulous adaxially, sometimes just at lobe tips; corona ring green to pink, inflated corona segments white at apex, green at base, ovoid, 1.5–2 mm; follicles ovoid to lance-ovoid, 6–9 × 0.6–1 cm, smooth; seeds 5–6 × 2–3 mm, coma 2.5–3 cm. Canyons, arroyos, mesas, bajadas, limestone, sandy, gravelly, and rocky soils, alluvium, desert scrub, oak and riparian woodlands, 4100–5000 ft. Scattered.

HAPLOPHYTON A. DC. – Cockroach plant, hierba de la cucaracha

Haplophyton cimicidum A. DC. [*H. crooksii* (L. D. Benson) L. D. Benson]—Cockroach plant, hierba de la cucaracha. Plants sparsely to densely branched erect herbs; latex clear; leaves opposite below or throughout, sometimes alternate above, stipular colleters present, blades narrowly lanceolate, 1.1–3.5 × 0.3–1.2 cm, laminar colleters absent; inflorescences terminal, cymose, flowers 1–3; corollas bright yellow, salverform, tubes 6–8 mm, glabrous, estivation contorted; gynostegium absent; corona absent; pollen in monads; follicles usually paired, very narrowly cylindrical, 6–10 × 0.2–0.3 cm; seeds fusiform, 6–8 mm, comose at both ends; coma 0.7–1 cm. Our plants are **var. crooksii** L. D. Benson. Arroyos, canyon slopes, igneous, limestone, and sandstone substrates, pine-oak-juniper woodland, desert grassland, 4200–5200 ft. Grant, Hidalgo, Luna.

MANDEVILLA Lindl. – Rocktrumpet, mandevilla

Mandevilla brachysiphon (Torr.) Pichon [*Macrosiphonia brachysiphon* (Torr.) A. Gray]—Huachuca Mountain rocktrumpet. Plants rhizomatous subshrubs, unbranched above base, 20–40 cm; latex white; leaves opposite (rarely whorled), stipular colleters present, blades narrowly ovate to oblong or nearly orbicular, 1.4–3 cm, laminar colleters absent; inflorescences terminal, flowers 1–2; corollas white, tinged red or pink, salverform, tube and throat 30–40 mm, lobes 15–25 mm, glabrous, estivation contorted; gynostegium absent; corona absent; pollen in monads; follicles usually paired, cylindrical, 5.5–12 × 0.4–0.5 cm, smooth; seeds linear, 5–7 mm, comose at 1 end, coma 1–1.2 mm. Hills, canyons, slopes, limestone, igneous, and metamorphic substrates, rocky soils, piñon-juniper woodland, desert grassland, desert scrub, 4700–5900 ft. Hidalgo, Luna.

MATELEA Aubl. – Matelea, milkvine

Angela McDonnell and Mark Fishbein

Plants vines or decumbent to prostrate herbs; latex white; leaves opposite, stipular colleters present or absent, laminar colleters present; inflorescences extra-axillary, cymose, umbelliform; corollas rotate, campanulate, or tubular, estivation contorted; gynostegium present; corona of 5 fleshy, fused segments; pollinia present; follicles usually solitary; seeds flat, comose at 1 end (Fishbein & McDonnell, 2023).

1. Plants vines; leaves inconspicuously puberulent; corollas tubular-campanulate, green, lobes spreading to reflexed; follicles fusiform, smooth, mottled green and gray…**M. producta**
1'. Plants prostrate to decumbent herbs; leaves densely hirsute; corollas rotate to rotate-campanulate, deep maroon, deep purple, brown, or cream with green or pink venation, lobes spreading to ascending; follicles ellipsoid to ovoid or lance-ovoid, muricate, green…(2)
2. Corolla deep maroon or purple to brown, densely pilose to hirsute adaxially; corona segments deep maroon, purple, or brown, glabrous; follicles 4.5–8.5 cm; seeds 8–11 × 7–10 mm; Lea County…**M. biflora**
2'. Corolla white with green to pinkish veins adaxially, greenish to pinkish abaxially, inconspicuously hirsute adaxially only at base of lobes; corona segments white, translucent, hirsutulous; follicles 5–6 cm; seeds 7–9 × 5–7 mm; Hidalgo County…**M. chihuahuensis**

Matelea biflora (Raf.) Woodson—Two-flowered milkvine, star milkvine. Plants prostrate to decumbent herbs; latex white; stems not twining; leaves opposite, stipular colleters present, blades ovate to deltoid, 0.8–5 × 0.6–3.5 cm, base truncate to cordate, laminar colleters present or absent; inflorescences extra-axillary, flowers 1–2; corollas deep maroon to brown, rotate-campanulate, lobes spreading, 3–6 mm, adaxially pilose to hirsute; corona of 5 fused segments with medial lobe incumbent on anthers, deep maroon or purple to brown, glabrous; follicles ellipsoid to ovoid, 4.5–8.5 × 1.5–3.5 cm, muricate; seeds 8–11 × 7–10 mm, coma 2.5–4 cm. Limestone, clay, and rocky soils, shortgrass prairie, 4000–4100 ft. Lea.

Matelea chihuahuensis (A. Gray) Woodson—Chihuahuan hairy milkvine. Plants prostrate to decumbent herbs; latex white; stems not twining; leaves opposite, stipular colleters present, blades deltoid, 0.7–3.5 × 0.7–2.5 cm, base truncate to cordate, laminar colleters present; inflorescences extra-axillary, flowers 2–5; corollas white with green to pinkish veins adaxially, greenish to pinkish abaxially, rotate-campanulate, lobes spreading to ascending, 2.4–5 mm, inconspicuously hirsute adaxially only at base of lobes; corona of 5 fused segments, apically connivent and incumbent on anthers, translucent white, hirsutulous; follicles ovoid to lance-ovoid, 5–6 × 1.5–2.5 cm, muricate; seeds 7–9 × 5–7 mm, coma 1.5–3 cm. Valleys, bajadas, slopes, arroyo banks, sandy and clay soils, desert grassland, 4100–5400 ft. Hidalgo.

Matelea producta (Torr.) Woodson—Texas milkvine. Plants vines, weakly twining at first; latex white; stems herbaceous; leaves opposite, stipular colleters present or absent, blades ovate, 2–8 × 1–4.5 cm, base deeply cordate, laminar colleters present; inflorescences extra-axillary, umbelliform to racemiform, flowers 1–6; corollas green to yellowish, tubular-campanulate, 7–12 mm, lobes spreading to recurved, abaxially pubescent, adaxially glabrous; corona a shallowly lobed ring of 5 fused segments, green, glabrous; follicles fusiform, 7–9 × 1.5–2 cm, smooth; seeds 7–8 × 4–5 mm, coma 2–3 cm. Canyons, slopes, bajadas, valleys, arroyos, granite, rocky, sandy, and clay soils, desert scrub, mesquite and desert grasslands, oak, oak-juniper, and piñon-juniper woodlands, 3800–6600 ft. EC, C, SW, Chaves.

METASTELMA R. Br. – Swallow-wort, milkweed vine

Metastelma mexicanum (Brandegee) Fishbein & R. A. Levin [*Cynanchum wigginsii* Shinners]—Wiggins' swallow-wort. Plants intricately twining vines, herbaceous; latex white; stems woody and corky at base; leaves opposite, stipular colleters present, blades linear, 1.5–4 × 0.1–0.3 cm, base cuneate, laminar colleters present; inflorescences extra-axillary, umbelliform, flowers 2–6; corollas cream, campanulate, 2–3 mm, lobes with recurved tips, adaxially pubescent, estivation valvate; gynostegium present; gynostegial corona of 5 free, laminar, linear segments slightly shorter than gynostegium; pollinia present; follicles solitary, fusiform, 4–6 × 0.3–0.5 cm, smooth; seeds flat, 4–6 × 0.3–0.5 mm, comose at 1 end; coma 1–2 cm. Hills and canyons, rocky slopes, igneous substrates, piñon-juniper woodlands, 4500–5000 ft. Hidalgo.

VINCA L. – Periwinkle, vinca

Vinca major L.—Greater periwinkle. Plants trailing, rhizomatous herbs or subshrubs; latex clear or nearly so; leaves opposite, stipular colleters present, blades elliptic to broadly ovate, 2–9 × 2–6 cm, laminar colleters present near top of petiole; inflorescences axillary, flower 1; corolla blue or light purple (may be reddish or white in cultivars), funnelform (nearly salverform), tube and throat 12–15 mm, limb 30–50 mm across, glabrous, estivation contorted; gynostegium absent; corona absent; pollen in monads; follicles usually paired, narrowly cylindrical, falcate, 2–4 cm, smooth; seeds ellipsoid, 7–8 mm, coma absent. Roadsides, abandoned homesites, valleys, 4900–7300 ft. Scattered, cultivated but not yet known to be naturalized.

APODANTHACEAE – STEM-SUCKER FAMILY

Kenneth D. Heil

PILOSTYLES Guill. – Stem-sucker

Pilostyles thurberi A. Gray [*P. covillei* Rose]—Thurber's stem-sucker. Herbaceous, dioecious stem parasites, achlorophyllous; roots not produced; bracts 4–7, imbricate, circular to ovate, 1–1.5 mm; flowers brown to maroon, 1.5–2 mm; sepals similar to bracts; stylar column with apex expanded to 1 mm diam., papillose along margins; anthers 15–21, in ring of ca. 3 rows, 1-locular, connate to distal portion of central column, not produced in pistillate flowers, column expanded at apex, sometimes into knob or disk; in pistillate flowers ovary partially inferior, indistinctly 4- or 5-carpellate, 1-locular; placentation parietal, ovules mostly 120–200. Parasitic on species of *Dalea* and *Psorothamnus*. Desert scrub, 4015–6200 ft. Scattered (Yatskievych, 2015).

ARACEAE – ARUM FAMILY

C. Barre Hellquist

Perennial, few annual, terrestrial, wetland, emergent, or floating, sometimes reduced to small floating green leaves, usually with milky or watery latex in terrestrial species; rhizomes, corms, and stolons present in nonfloating species, or with short roots from floating fronds, stems absent; leaves or fronds distinct, mostly alternate or clustered, some not differentiated into petiole; inflorescence a spadix subtended by a spathe or solitary in *Lemma*; flowers monoecious or unisexual, perianth absent, stamens 1–12; fruit a berry or follicle; seeds 1–40.

1. Plants with roots; fronds (leaves) ovate, obovate, flat, gibbous, 1–21-veined…(2)
1'. Plants lacking roots; fronds (leaves) boat-shaped, lacking veins…**Wolffia**
2. Roots 1 per leaf; fronds (leaves) with 1(–7) veins…**Lemna**
2'. Roots 2–22 per leaf, leaves with 7–21 veins…**Spirodela**

LEMNA L. – Duckweed

Perennial, aquatic; roots 1 per leaf; fronds (leaves) floating or submersed, 1–20, often in clusters, lance-ovate, flat, or gibbous, 1–15 mm, upper surfaces often with papillae along the veins, veins 1–7; flowers 1(2) per frond, surrounded by a small inflated scale; stamens 2; seeds 1–4, longitudinally ribbed (Landolt, 1986, 2000; Landolt & Kandeler, 1987).

1. Plants submersed, fronds long-petiolate, 6–15 mm, with lateral fronds usually remaining attached to the parent plant…**L. trisulca**
1'. Plants floating, fronds elliptic to linear-oblong, 1–6 mm, forming single plants or in clusters…(2)
2. Leaves 1-veined…(3)
2'. Leaves 3–5(–7)-veined…(4)
3. Vein mostly prominent, longer than extension of air spaces, at least 3/4 length between node and apex; fronds 1–5 mm…**L. valdiviana**
3'. Vein sometimes indistinct, very rarely longer than extension of air spaces, not longer than 2/3 distance between node and apex; fronds 1–4 mm…**L. minuta**
4. Root sheath winged at base; root usually with acute tip; fronds with 1+ distinct papillae near apex…(5)
4'. Root sheath not winged at base; root usually rounded at tip; fronds with or without distinct papillae near apex…(6)
5. Fronds 1–1.7 times as long as broad; root sheath wing 2–3 times as long as wide…**L. perpusilla**
5'. Fronds 1–3 times as long as broad; root sheath wing 1–2.5 times as long as wide…**L. aequinoctialis**
6. Plants forming small brown turions; fronds flat with line of papillae along upper surface…**L. turionifera** (in part)
6'. Plants usually not forming turions; fronds usually gibbous with distinct papillae near apex…(7)
7. Largest air spaces longer than 0.3 mm, if frond red on lower surface, color originating from margins and red spots on lower surface beginning near apex…**L. gibba**
7'. Largest air space 0.3 mm or less, if frond red, originating from point of root attachment…(8)
8. Fronds not reddish on the lower surface; distance between lateral veins near or proximal to the middle, upper surface lacking a distinct row of papillae…**L. minor**
8'. Fronds often reddish on lower surface; distance between lateral veins near or proximal to center, upper surface with distinct papillae…(9)
9. Fronds flat, with papillae on midline of upper surface; plants producing olive to brown rootless turions… **L. turionifera**
9'. Fronds usually gibbous with papillae above node and near apex on upper surface, not between node and apex; plants not producing turions…**L. obscura**

Lemna aequinoctialis Welw.—Lesser duckweed. Roots to 3 mm, tip acute; sheath winged at base, wing 1–2.5 times as long as broad; fronds floating, 1–22 in clusters, ovate-lanceolate, flat, 1–6 mm, 1–3 times as long as wide, veins 3, distance between lateral veins near or proximal to middle; 1 distinct papilla near apex on upper surface and 1 small above node, lower surface lacking reddish purple pigment; turions not formed. Quiet waters, 7600–7700 ft. NW, WC.

Lemna gibba L.—Inflated duckweed. Roots to 15 cm, tip mostly rounded, sheathed, not winged; fronds floating, 2–5+, orbicular-obovate, often gibbous (bulging) with distinct papillae near apex, 1–8 mm, 1–1.5 times as long as wide, veins mostly 4–5 originating from node, red spots on lower surface beginning near apex. Quiet waters, 4000–7500 ft. WC, SW, SC.

Lemna minor L.—Common duckweed. Roots to 15 cm, tip mostly rounded, wingless fronds floating, 1 or 2–5+ in clusters, ovate, slightly gibbous, flat, 1–8 mm, veins 3(–5), distance between lateral veins near or proximal to the middle, upper surface green, papillae not always distinct. Quiet waters. 3600–9600 ft. NC, NW, WC, SC, SW.

Lemna minuta Kunth—Least duckweed. Roots to 15 cm, tip rounded or pointed; sheaths wingless; fronds floating, 1 or 2 often clustered, obovate, flat, 1–4 mm, longer than broad, vein not extending beyond aerenchyma tissue, vein 1. Quiet waters, 4000–10,480 ft. NW, C, WC.

Lemna obscura (Austin) Daubs—Little duckweed. Roots to 15 cm, tip usually rounded; sheath wingless; fronds floating, 1 or 2–5+ in clusters, obovate, nearly symmetric at base, flat or gibbous with papillae above node or near apex on upper surface, not between node and apex, 1–3.5 mm, veins 3, greatest distance between lateral veins near middle; papillae near apex, lower surface often slightly red-colored, upper surface often red-spotted, veins 3. Quiet waters, 6600–8660 ft. NC, WC, SW.

Lemna perpusilla Torr.—Minute duckweed. Roots to 3.5 cm, tip usually sharp-pointed; sheath laterally winged at base, 2–3 times as long as wide; fronds floating, laterally 1 or 2 to few in clusters, ovate-obovate, flat, 1 mm, 1–1.7 times as long as wide, 1 distinct papilla near apex of upper surface and a median line of small papillae, lacking reddish-purple pigmentation, veins 3. Quiet waters, 5570–7300 ft. Catron, Rio Arriba, Socorro.

Lemna trisulca L.—Star duckweed. Roots lacking or to 2.5 cm, tips pointed, sheaths wingless often deciduous, fronds vegetative submersed, floating, usually attached, 3–25 mm, often connected by 2–20 mm stalk, apical margin dentate to erose, veins (1–)3, papillae absent. Still, often alkaline waters, 4575–8450 ft. Catron, Rio Arriba, Socorro.

Lemna turionifera Landolt—Turion duckweed. Roots shorter than 15 cm, tip usually rounded, sheath wingless; fronds floating, 1 or 2 often clustered, obovate, slightly gibbous to flat with line of papillae on midline of upper surface, 1–4 mm; veins 3, olive to brown rootless turions sometimes formed. Quiet waters, 5500–9875 ft. WC, Sandoval.

Lemna valdiviana Phil.—Pale duckweed. Roots to 1.5 cm, tip rounded to pointed, sheath not winged; fronds floating, sometimes submerged, 1 or 2 or in groups, ovate-lanceolate, flat or slightly convex, pale green or green to translucent, thin 1–5 mm, 1.3–3 times as long as wide, vein 1 mostly prominent extending beyond aerenchyma tissue. Quiet waters, 4360–7890 ft. NC, WC, NC.

SPIRODELA Schleid. – Great duckweed

Spirodela polyrhiza (L.) Schleid.—Giant duckweed, common duck-meat. Perennial floating aquatic; roots 7–21, to 3 cm from lower surface of frond; fronds often cohering in groups of 1–20, round-obovate, 3–10 mm diam., 1–1.5 times as long as wide, upper surface with red spot in center, lower surface reddish purple, (5–)7–21 veins, flat, rarely gibbous; flowers with 1 or 2 ovules; fruit 1–1.5 mm, laterally winged; seeds with 12–20 ribs. Quiet waters, 4500–7850 ft. Hidalgo, Socorro.

WOLFFIA Horkel ex Schleid. – Water-meal

Wolffia brasiliensis Wedd. [*W. punctata* Griseb.]—Brazilian water-meal. Perennial, floating and submersed; roots absent; fronds boat-shaped, 0.5–2.6 mm, rounded at apex, papilla prominent in center of upper surface, veins lacking, upper surface bright green, with 50–100 stomates; turions light green, globular, smaller than fronds. Quiet water, 8800–8900 ft. San Juan.

ARALIACEAE – GINSENG FAMILY

Kenneth D. Heil and Steve L. O'Kane, Jr.

ARALIA L. – Spikenard

Aralia racemosa L.—American spikenard. Perennial shrubs, caulescent; leaves alternate, pinnately or ternately compound, with 3 primary divisions, with serrate margins; leaflets toothed; inflorescence a panicle of numerous umbels; flowers small, actinomorphic, in umbels, greenish white; sepals 5; petals 5, usually distinct; stamens usually 5, alternate with petals; pistil 1; ovary inferior, with 2–5 locules with 1 ovule per locule; styles connate or distinct, usually swollen at the base to form a stylopodium; drupe purplish black, 4–6 mm diam. Riparian, marshland, canyon bottoms, mixed conifer and subalpine communities, 5585–10,500 ft. NE, NC, NW, C, WC, SC.

ARISTOLOCHIACEAE – DUTCHMAN'S-PIPE FAMILY

Kenneth D. Heil

ARISTOLOCHIA L. – Dutchman's-pipe

Perennial herbs; stems procumbent; leaves alternate, 2-ranked, true stipules absent, blades membranous to leathery; calyx (ours) mostly brown-purple, bilaterally symmetric, tubular, usually bent or curved, 1- or 3-lobed, base with utricle (basal, inflated portion of calyx surrounding or containing gynostemium); corolla absent; stamens 5–6, adnate to styles and stigmas, forming gynostemium; ovary inferior; capsule dry, dehiscent; seeds flattened or rounded, sometimes winged (Barringer & Whittemore, 1997).

1. Leaf venation palmate; leaf blades as wide as long or wider…**A. wrightii**
1'. Leaf venation pinnate or palmate-pinnate; leaf blades longer than wide…**A. watsonii**

Aristolochia watsonii Wooton & Standl. [*A. porphyrophylla* Pfeifer]—Indianroot. Herbs, procumbent, to 0.5 m; young stems smooth, glabrous; leaf blades lanceolate, 8 × 5 cm, apex acuminate to acute, surfaces abaxially tomentose to tomentulose; calyx brown-purple, straight to curved; utricle angled upward, ovoid to narrowly ellipsoid, 0.6–0.7 × 0.5 cm; capsules ovoid to obovoid, 1.5 × 1–2 cm; seeds flat, triangular, 0.3–0.4 × 0.4 mm. Sandy and rocky sites, Chihuahuan desert scrub, piñon-juniper, oak, 4140–5655 ft. SC, SW.

Aristolochia wrightii Seem. Bot. [*A. brevipes* Benth. var. *wrightii* (Seem.) Duch.]—Wright's Dutchman's-pipe. Herbs, procumbent, to 0.4 m; young stems slightly striate, velutinous with yellowish trichomes; leaf blades ovate to hastate, (0–)3-lobed, 1–4 × 2–4 cm, apex obtuse to acute, surfaces velutinous with yellowish trichomes; calyx brown-purple, straight; utricle angled upward, ovoid, to 0.7 × 0.4 cm; capsules ± globose, 1.2 × 1–3 cm; seeds flat, triangular, 0.4 × 0.5 cm. Igneous and limestone substrates, 4400–5500 ft. Luna.

ASPARAGACEAE – ASPARAGUS FAMILY

Steve L. O'Kane, Jr.

In APG (2016), this is a much larger family that includes Agavaceae, Ruscaceae, and Themidaceae. Here we follow Simpson (2019) (Straley & Utech, 2002).

ASPARAGUS L. – Asparagus

Asparagus officinalis L.—Asparagus. Herbs, perennial, erect, 1–2.5 m; annual stems from fibrous rhizomes, densely branched distally, finely dissected, ascending to perpendicular; cladophylls in clusters at nodes, filiform, straight or curved, 1–3 cm; leaves scalelike, 3–4 mm, blades lanceolate; inflorescences in axillary racemes, these 1–3-flowered; pedicels with a conspicuous joint; some flowers unisexual; perianth campanulate, greenish white; tepals 6, 3–8 × 1–2 mm, connate for 1–2 mm; ovaries superior, 3-locular; style 3-branched distally; berries red, 6–10 mm. Escaped from cultivation, now found in riparian areas, wet or moist areas near old habitations, and other moist sites, especially somewhat disturbed ones, 3850–7900 ft. NC, NW, C, SC, Guadalupe, Roosevelt. Introduced.

ASPHODELACEAE – ASPHODEL FAMILY

Steve L. O'Kane, Jr.

ASPHODELUS L. – Onionweed

Asphodelus fistulosus L.—Onionweed. Herbs, annual or short-lived perennial, scapose, root crowns thickened with many fibrous roots; leaves numerous, basal; blades 5–35 cm × 2–4 mm, cylindrical, sheathing, margins entire; scapes hollow; inflorescences racemose or paniculate, bracteate; bracts persistent, narrowly lanceolate, scarious; flowers diurnal; tepals 6, erect to spreading, distinct or barely connate basally, equal, white to pale pink, each with single prominent vein, vein dark pink or brown; stamens 6, distinct, shorter than tepals; filaments expanded at base; anthers dorsifixed; ovary 3-locular; septal nectaries present; style 1; stigma weakly 3-lobed; fruits capsular, globose, hard, dehiscence loculicidal. Waste places in Doña Ana and Luna Counties; can be an aggressive escape. Introduced.

ASTERACEAE – SUNFLOWER FAMILY

Family description by Kenneth D. Heil and Steve L. O'Kane, Jr.

Timothy Lowrey kindly provided the genus key (see Lowrey, 2020), which was, with permission, reformatted and slightly modified.

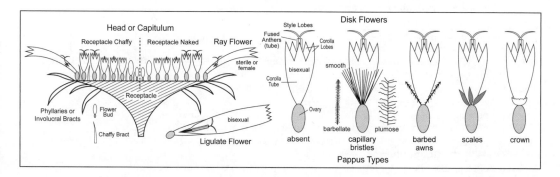

Annual, biennial, or perennial; herbs or shrubs; leaves alternate, opposite, or whorled, simple, pinnatifid, palmatifid, or compound; inflorescence of involucrate multiflowered heads, these solitary or several in corymbose, racemose, paniculate, or cymose clusters; flowers within a head few to numerous, inserted on a common receptacle, surrounded by green bracts forming a cup-shaped, cylindrical, or urn-shaped involucre enclosing the flowers in bud; heads of three possible types: (1) entirely of tubular, actinomorphic disk florets, (2) entirely of flattened, zygomorphic ligulate florets, or (3) with tubular, actinomorphic corollas forming a central disk, and an outer radiating series of strap-shaped ray corollas; receptacles flat, convex, conic, or cylindrical, naked or bearing chaffy bracts, scales, or hairs; florets with calyx lacking or usually present as a pappus of capillary bristles, scales, or awns and crowning the summit of the ovary; stamens alternate with the corolla lobes, filaments free (rarely connate), anthers united and forming a tube (rarely separate) surrounding the style; ovary inferior, of 2 carpels, 1 locule with a single ovule; styles 1, 2-cleft, exserted through the anther tube; fruit a cypsela (Welsh, 2016).

KEY TO ASTERACEAE

1. Involucres conspicuously armed with hooked prickles, stout spines, or prominent wings…**KEY A**
1'. Involucres unarmed and not as above, lacking prickles, spines, or wings…(2)
2. Leaves and/or phyllaries obviously dotted with translucent oil glands…**KEY B**
2'. Leaves and phyllaries not as above, occasionally glandular-pitted, but these tiny and not translucent…(3)
3. Corollas all raylike or bilabiate, tubular disk florets absent…**KEY C**
3'. Corollas not all raylike, tubular disk florets present…(4)
4. Corollas all tubular; ray florets absent, or the rays vestigial, minute, and scarcely evident…(5)
4'. Corollas not all tubular, ray florets present and evident…(6)
5. Pappus of capillary bristles, wholly or in part, sometimes plumose…**KEY D**
5'. Pappus of scales, awns, very short chaffy bristles, or absent, not capillary and not plumose…**KEY E**
6. Pappus of capillary bristles, at least in part…**KEY F**
6'. Pappus of awns, scales, or absent…(7)
7. Pappus of awns or scales…**KEY G**
7'. Pappus absent…**KEY H**

KEY A: INVOLUCRES WITH PRICKLES, SPINES, FRINGED APPENDAGES, OR WINGS

1. Involucres covered with numerous hooked prickles…(2)
1'. Involucres lacking hooked prickles…(3)
2. Bur (involucre) completely enclosing the flowers, no flowers protruding or visible at the apex; lower leaves not large, not cordate at base…**Xanthium**
2'. Bur (involucre) vaselike, the flowers exposed at the apex; lower leaves large, resembling rhubarb, cordate at base…**Arctium**
3. Plants shrubs with filiform leaves; fruiting involucres with conspicuous hyaline wings…**Ambrosia**
3'. Plants herbaceous, the leaves not filiform; fruiting involucres lacking wings, the phyllaries with fringed appendages, modified into spines or with prominent spine tips…(4)
4. Phyllaries fringed with slender, spinelike teeth…(5)
4'. Phyllaries modified into sharp spines or with prominent spine tips…(6)
5. Heads radiate; peripheral florets elongate and expanded…**Plectocephalus**
5'. Heads discoid…**Centaurea**
6. Heads unisexual and of 2 kinds, the staminate unarmed and borne in terminal racemes, the pistillate spinose and borne below in the leaf axils…**Ambrosia**
6'. Heads bisexual, all essentially the same on a single plant…(7)
7. Leaves lacking spiny margins…**Centaurea**
7'. Leaves with spiny margins…(8)
8. Flowers yellow to red…(9)
8'. Flowers white, purple, or pink…(10)
9. Pappus absent or of narrow overlapping scales; receptacle chaffy; cultivated and escaping…**Carthamus**
9'. Pappus of plumose bristles; receptacle bristly; weedy…**Cirsium**
10. Leaves variegated, with white veins or mottlings…**Silybum**
10'. Leaves lacking white veins or mottlings…(11)
11. Pappus plumose, the bristles feathery; receptacle densely bristly…**Cirsium**
11'. Pappus not plumose, the bristles simple; receptacle bristly or not…(12)
12. Receptacles with dense bristles or narrow chaffy scales between disk florets, fleshy or honeycombed; heads erect…**Onopordum**
12'. Receptacles densely bristly, not fleshy or honeycombed; heads nodding…**Carduus**

KEY B: LEAVES AND/OR PHYLLARIES OBVIOUSLY DOTTED WITH TRANSLUCENT OIL GLANDS

1. Leaves simple, bristly-ciliate at the base; style branches of disk florets very short, much < 1 mm…**Pectis**
1'. Leaves pinnately parted, or if simple, not bristly-ciliate at the base; style branches of disk florets ca. 1 mm…(2)
2. Phyllaries separate to the base or nearly so…(3)
2'. Phyllaries united at least 1/3 their length…(6)
3. Involucres subtended to an additional series of tiny bracts (calyculate); pappus with at least some small scales in addition to bristles…(4)
3'. Involucres lacking an additional basal series of tiny bracts (not calyculate); pappus lacking any scales, entirely of separate bristles…**Dyssodia**

4. Ray florets present…**Chrysactinia**
4'. Ray florets absent…**Porophyllum**
5. Involucres lacking an additional basal series of tiny bracts; pappus of 2 awns and 2 scales…**Tagetes**
5'. Involucres subtended by an additional series of tiny bracts; pappus of several awns and scales…(6)
6. Plants perennial…**Thymophylla**
6'. Plants annual…(7)
7. Plants mostly 5–30 cm; involucres 4–6 × 27 mm; receptacle glabrous or nearly so; phyllaries strongly united 2/3+ of their length…**Thymophylla**
7'. Plants mostly 30–70 cm; involucres 7–18 × 5–12 mm; receptacle with fine bristles; phyllaries weakly united ca. 1/2 their length…**Adenophyllum**

KEY C: COROLLAS ALL RAYLIKE OR BILABIATE

1. Corollas all bilabiate, the outer lobe 3-toothed, the inner lobe 2-toothed; juice watery, not milky…(2)
1'. Corollas ligulate, not bilabiate; juice usually milky…(5)
2. Plants shrubs, woody at least in the lower 1/2; corollas yellow…**Trixis**
2'. Plants herbaceous; corollas whitish or purplish…(3)
3. Flowering stems evidently leafy; leaves spiny-toothed or spinulose-dentate; caudices brown-woolly… **Acourtia**
3'. Flowering stems scapose, lacking leaves; leaves entire, not spiny-toothed…(4)
4. Heads nodding in bud and fruit, erect in flower; outer florets creamy-white, rarely purple-tinged… **Chaptalia**
4'. Heads erect in bud, flowering, and fruiting; outer florets pinkish to purplish, inner florets white… **Leibnitzia**
5. Pappus of plumose bristles, at least in part…(6)
5'. Pappus of simple bristles, awns, scales, or lacking…(10)
6. Florets white, pink, or lavender…(7)
6'. Florets yellow or purple…(8)
7. Basal leaves not withered at flowering, cauline leaves well developed; florets 15–30, white, sometimes with rose or purple veins abaxially; annual…**Rafinesquia**
7'. Basal leaves withered at flowering (except in forms of *S. thurberi* flowering in late spring/early summer), cauline leaves mostly reduced to subulate scales; florets 4–14, pink, lavender, or white; annual or perennial…**Stephanomeria**
8. Leaves all basal; phyllaries 20–30 in 3–4 series; plants 10–60 cm tall…**Hypochaeris**
8'. Leaves basal and cauline; phyllaries 18–30 in 3–5 series or 5–16 in 1 series; plants to 150 cm tall…(9)
9. Basal leaves with margins entire, not lobed; phyllaries 5–16, in 1 series…**Tragopogon**
9'. Basal leaves usually pinnately lobed or toothed; phyllaries 18–30, in 3–5 series…**Scorzonera**
10. Flowering stems scapose, lacking leaves or bracts, terminated by a single head…(11)
10'. Flowering stems with leaves or bracts and/or bearing 2+ heads…(18)
11. Cypselae obviously beaked at summit, short in *Uropappus*…(12)
11'. Cypselae not beaked at summit…(17)
12. Pappus of bristle-tipped scales on all florets or of scales on outer florets and bristles on inner florets… (13)
12'. Pappus of slender capillary bristles…(14)
13. Leaves oblanceolate or oblong; pappus of scales on outer florets and plumose bristles on inner florets… **Leontodon**
13'. Leaves linear to narrowly lanceolate, grasslike; pappus of 5–6 bifid scales tipped with bristles 2–6 mm… **Uropappus**
14. Corollas whitish to purplish…**Chaptalia**
14'. Corollas yellow to orange…(15)
15. Phyllaries in 3–4 graduated series; cypselae 10-ribbed, not at all spinulose…**Agoseris**
15'. Phyllaries in 2 unequal series, the lower very short and usually reflexed, the upper longer and erect; cypselae 4- or 5-ribbed, spinulose-roughened on the upper 1/2…(16)
16. Calyculi 8 in number, 3–8 mm; pappus bristles 10–12 mm…**Pyrrhopappus**
16'. Calyculi 12–18 in number, 6–12 mm; pappus bristles 5–6(–8) mm…**Taraxacum**
17. Leaves oblanceolate or oblong…**Leontodon**
17'. Leaves linear-lanceolate…**Nothocalais**
18. Pappus absent…**Lapsana**
18'. Pappus present, of bristles, scales, or both…(19)
19. Peduncles inflated distally; phyllaries enfolding outer cypselae…**Hedypnois**
19'. Peduncles not inflated; phyllaries not enfolding outer cypselae…(20)
20. Pappus of numerous unawned scales, lacking bristles entirely; florets blue, rarely white…**Cichorium**
20'. Pappus of bristles, at least in part; florets other than blue (bluish in some *Lactuca*)…(21)

21.   Receptacle chaffy or bristly…(22)
21'.   Receptacle naked…(24)
22.   Florets yellow…**Malacothrix**
22'.   Florets reddish, pinkish, or whitish…(23)
23.   Annuals; upper stems and heads with tacklike, stalked glands; receptacle bristly…**Calycoseris**
23'.   Perennials; upper stems and heads lacking glands; receptacle chaffy…**Pinaropappus**
24.   Cypselae obviously beaked at the summit…(25)
24'.   Cypselae not beaked, occasionally narrowed at the apex…(26)
25.   Cypselae flattened in cross-section; beak lacking a ring of reflexed hairs at the summit, just beneath the pappus…**Lactuca**
25'.   Cypselae terete or scarcely flattened; beak with a ring of reflexed hairs at the summit, just beneath the pappus…**Pyrrhopappus**
26.   Cypselae flattened in cross-section…**Sonchus**
26'.   Cypselae not flattened…(27)
27.   Florets white, pinkish, or purplish when fresh…(28)
27'.   Florets yellow or orange when fresh (sometimes white in *Malacothrix*)…(30)
28.   Rays white to cream-colored; cypselae ca. 10-ribbed…**Hieracium**
28'.   Rays pinkish or purplish; cypselae ca. 5-ribbed…(29)
29.   Plants annual; involucres 4–5 mm…**Prenanthella**
29'.   Plants perennial; involucres 10–25+ mm…**Lygodesmia**
30.   Pappus composed of an outer series of small scales and an inner series of bristles…**Krigia**
30'.   Pappus composed entirely of bristles, lacking scales…(31)
31.   Pappus bristles ± united at the base and falling as a unit, leaving 1–8 persistent stiffer bristles on the achenes…**Malacothrix**
31'.   Pappus bristles not united, all persistent on the achenes…(32)
32.   Plants fibrous-rooted; pappus mostly tan to brown; phyllaries not thickened; heads often nodding…**Hieracium**
32'.   Plants taprooted; pappus whitish; phyllaries somewhat thickened at the base or on the midrib; heads seldom nodding…**Crepis**

KEY D: RAY FLORETS ABSENT; PAPPUS OF CAPILLARY BRISTLES, WHOLLY OR IN PART, SOMETIMES PLUMOSE

1.   Plants obvious shrubs or subshrubs…(2)
1'.   Plants herbaceous or woody only at the base…(17)
2.   Heads unisexual, the plants dioecious and the sexes borne on different plants…**Baccharis**
2'.   Heads bisexual, the plants perfect with both sexes in the same head…(3)
3.   Florets yellow…(4)
3'.   Florets bluish to purplish, or white to cream…(12)
4.   Phyllaries 4–6, in a single series; spinescent or unarmed…**Tetradymia**
4'.   Phyllaries more numerous, in 2+ series; unarmed…(5)
5.   Phyllaries tending to be aligned in vertical ranks, the midrib of one ± overlapping the midrib of the next…(6)
5'.   Phyllaries not aligned vertically…(7)
6.   Disk florets 4–15, cypselae cylindrical…**Lorandersonia**
6'.   Disk florets 2–7, cypselae oblong or top-shaped…**Chrysothamnus**
7.   Pappus of plumose bristles…**Bebbia**
7'.   Pappus of smooth or merely barbellate bristles…(8)
8.   Florets 3; stems with silvery hairs and glandular blisters…**Lepidospartum**
8'.   Florets > 3; stems glabrous or variously hairy, glandular blisters absent…(9)
9.   Stems densely tomentose, without glandular hairs…**Ericameria**
9'.   Stems not tomentose, but with glandular dots or stalked glandular hairs…(10)
10.   Leaves entire, never toothed; stems resinous…**Ericameria**
10'.   Leaves usually toothed, sometimes entire; stems never resinous…(11)
11.   Plants tufted, often mound-forming; heads mostly 1 per branch tip, not in corymbiform clusters…**Xanthisma**
11'.   Plants not tufted, stems elongate; heads in terminal corymbiform clusters…**Isocoma**
12.   Flowers blue or purple…**Pluchea**
12'.   Flowers white or cream…(13)
13.   Phyllaries 4–6; florets 5…**Stevia**
13'.   Phyllaries 8–45; florets 10–25…(14)
14.   Cypselae 8–10-ribbed…(15)
14'.   Cypselae 4–5-ribbed…(16)

15. Leaves linear; leaf margins always entire…**Asanthus**
15'. Leaves not linear, leaves deltoid, lanceolate, or ovate; leaf margins toothed or lobed…**Brickellia**
16. Phyllaries equal in length…**Ageratina**
16'. Phyllaries unequal in length, the outer shorter…**Brickelliastrum**
17. Receptacles paleate (some or all florets subtended by a palea, a bract on the receptacle)…(18)
17'. Receptacles lacking paleae…(19)
18. Pappus bristles 1–10, hidden in head…**Stylocline**
18'. Pappus bristles 13–28+, visible in heads…**Logfia**
19. Leaves opposite or whorled with 3+ leaves per node…(20)
19'. Leaves alternate…(30)
20. Corollas yellow to orange…**Arnica**
20'. Corollas white or pink to purplish…(21)
21. Phyllaries 5–6 in 1–2 series…**Stevia**
21'. Phyllaries 7–45 in 2–8 series…(22)
22. Cypselae 8–10-ribbed…**Brickellia**
22'. Cypselae 4–5-ribbed…(23)
23. Plants annual; pappus bristles plumose…**Carminatia**
23'. Plants perennial; pappus bristles not plumose…(24)
24. Phyllaries ± equal in length…(25)
24'. Phyllaries unequal in length, the outer shorter…(27)
25. Receptacles conic…**Conoclinium**
25'. Receptacles flat or convex…(26)
26. Phyllaries 7–16 in 1–2 series; florets 3–13…**Koanophyllon**
26'. Phyllaries 30 in 2–3 series; florets 10–60…**Ageratina**
27. Leaves whorled…**Eutrochium**
27'. Leaves opposite…(28)
28. Florets 10–25…**Fleischmannia**
28'. Florets 25–50…(29)
29. Corollas white to yellowish white; phyllaries 2–4-nerved…**Brickelliastrum**
29'. Corollas blue, lavender, or pinkish (rarely white); phyllaries 3-nerved…**Koanophyllon**
30. Phyllaries in 1–2 series, equal in length and often subtended by smaller calyxlike bracts (calyculi)…(31)
30'. Phyllaries in 3–10 series, unequal in length, calyculi lacking…(32)
31. Corollas white or purplish, sometimes yellowish; leaves 3–4 times pinnately compound…**Psacalium**
31'. Corollas yellow; leaves at most 1–2-pinnatifid and not compound…**Packera/Senecio**
32. Phyllaries striate with prominent nerves, generally 5–6 in number, sometimes more…**Brickellia**
32'. Phyllaries not obviously striate…(33)
33. Corollas white, blue, pink, or purple…(34)
33'. Corollas cream, yellow, or orange…(39)
34. Pappus of plumose bristles…(35)
34'. Pappus not plumose…(36)
35. Leaves gland-dotted; heads in spikes or racemes…**Liatris**
35'. Leaves not gland-dotted; heads in panicles or corymbs…**Rhaponticum**
36. Phyllaries wholly scarious or with the margins obviously scarious…**Gnaphalium**
36'. Phyllaries not scarious or scarious-margined…(37)
37. Leaves and stem densely arachnoid-tomentose; plants 2–25 cm…**Gamochaeta**
37'. Leaves and stems puberulent or glandular-pubescent, not arachnoid-tomentose; plants 30–200+ cm… (38)
38. Heads discoid, all florets similar and bisexual; plants perennial…**Vernonia**
38'. Heads disciform, florets of 2 kinds: the outer florets filiform and pistillate, the inner florets expanded and staminate; plants annual and perennial…**Pluchea**
39. Phyllaries wholly scarious…(40)
39'. Phyllaries not wholly scarious…(43)
40. Plants not dioecious; all heads with a similar number of florets…(41)
40'. Plants dioecious; male and female heads with different number of florets…(42)
41. Florets yellowish or reddish…**Pseudognaphalium**
41'. Florets purplish…**Gamochaeta**
42. Basal leaves present at flowering; plants 4–25 cm…**Antennaria**
42'. Basal leaves withered at flowering; plants 20–80 cm…**Anaphalis**
43. Plants annual or biennial…(44)
43'. Plants perennial…(45)
44. Leaves gland-dotted or with stalked glandular hairs…**Laennecia**
44'. Leaves not gland-dotted or with glandular hairs, variously hairy otherwise…**Conyza**

45. Leaves mostly basal, stem scapose…**Psathyrotopsis**
45'. Leaves basal and cauline, stem not scapose…(46)
46. Phyllary midnerves translucent and swollen…**Solidago**
46'. Phyllary midnerves not translucent or swollen…(47)
47. Heads 1–3, not in flat-topped clusters; plants 2–20 cm…**Erigeron**
47'. Heads > 5, in flat-topped clusters; plants 20–120 cm…**Isocoma**

KEY E: RAY FLORETS ABSENT; PAPPUS OF SCALES, AWNS, VERY SHORT CHAFFY BRISTLES, OR ABSENT, NEITHER CAPILLARY NOR PLUMOSE

1. Receptacles paleate (some or all inner florets subtended by a palea, a receptacular bract)…(2)
1'. Receptacles without paleae…(20)
2. Pappus absent…(3)
2'. Pappus present, of awns or scales…(13)
3. Leaves opposite throughout or at least on lower stem…(4)
3'. Leaves alternate…(9)
4. Stems, leaves, and phyllaries villous with stipitate-glandular black or yellow hairs; annual; malodorous… **Madia**
4'. Stems, leaves, and phyllaries glabrous or variously hairy but without stipitate-glandular hairs; annual or perennial; not malodorous…(5)
5. Florets without showy corollas; plants wind-pollinated…(6)
5'. Florets with showy corollas, plants not wind-pollinated…(8)
6. Cypselae strongly flattened, with corky wings…**Dicoria**
6'. Cypselae not strongly flattened, corky wings absent…(7)
7. Heads in racemes or spikes…**Iva**
7'. Heads in panicles…**Cyclachaena**
8. Phyllaries strongly united; leaves simple, pinnately lobed…**Thelesperma**
8'. Phyllaries not or only weakly united; leaves compound with 3–5 leaflets…**Bidens**
9. Plants annual…(10)
9'. Plants perennial, biennial, or annual…(12)
10. Leaves 1–3-pinnately lobed…**Hedosyne**
10'. Leaves entire…(11)
11. Stems, leaves, and heads villous with stipitate-glandular black or yellow hairs; malodorous…**Madia**
11'. Stems, leaves, and heads lanuginose, whitish or grayish, not glandular; not malodorous…(12)
12. Leaves subulate to lanceolate; outer female florets enclosed by saccate paleae…**Stylocline**
12'. Leaves oblanceolate to obovate; outer female florets not enclosed by paleae…**Diaperia**
13. Leaves deeply pinnately lobed, lobes linear or filiform; subshrubs or shrubs…**Oxytenia**
13'. Leaves entire, or if lobed, lobes not linear; shrubs, perennial, biennial, or annual…(14)
14. Involucres with distinct calyculi (a separate outer set of bracts subtending the main phyllaries, resembling a calyx)…(15)
14'. Involucres without calyculi…(16)
15. Phyllaries fused 1/5–7/8 of their length; pappus of scales or smooth awns…**Thelesperma**
15'. Phyllaries free or fused only to 1/10 of their length; pappus of barbellate or ciliate awns…**Bidens**
16. Phyllaries falling together with an outer cypsela and adjacent 2 disk florets…**Parthenium**
16'. Phyllaries persistent, not falling with the achenes…(17)
17. Pappus plumose, of bristlelike scales…**Bebbia**
17'. Pappus not plumose, the scales not bristlelike…(18)
18. Plants woody shrubs…**Flourensia**
18'. Plants herbaceous…(19)
19. Corolla brownish red, brownish purple, or red; pappus scales aristate…**Gaillardia**
19'. Corolla white, pinkish, cream, or pale yellow; pappus scales not aristate…**Chaenactis**
20. Pappus absent or nearly so…(21)
20'. Pappus present…(39)
21. Leaves mostly or all opposite…(22)
21'. Leaves alternate…(29)
22. Corollas yellow…(23)
22'. Corollas white, blue, lavender, pink, or purple…(25)
23. Florets 1–5 per head; heads in tightly packed clusters…**Flaveria**
23'. Florets 20–100 per head; heads borne singly or in open clusters, not in headlike or tightly packed arrays…(24)
24. Leaf blades usually 3-lobed or sometimes up to 5-lobed, not triangular-hastate, apices not long-tailed; phyllaries 8–16 in 2–3 series, not fused; S New Mexico…**Perityle**

24′. Leaf blades triangular-hastate, entire, dentate, or shallowly lobed, apices long-tailed; phyllaries 15–21, fused together in 1 series…**Pericome**

25. Fruits flattened, margins with corky wings; NW New Mexico…**Dicoria**

25′. Fruits mostly prismatic or columnar, margins without corky wings…(26)

26. Florets 20–125 per head…**Ageratum**

26′. Florets 5–15 per head…(27)

27. Involucres cylindric; heads in flat-topped corymbs…**Stevia**

27′. Involucres not cylindrical, but campanulate or hemispherical, heads in spikes, racemes, or panicles…(28)

28. Heads in spikes or racemes…**Iva**

28′. Heads in panicles…**Cyclachaena**

29. Corolla mostly white, sometimes blue, lavender, pink, or purple…(30)

29′. Corolla yellow…(33)

30. Plants annual…**Chaenactis**

30′. Plants perennial…(31)

31. Phyllaries in 6+ series, with fringed appendages…**Centaurea**

31′. Phyllaries in 1–3 series, without fringed appendages; cypselae various…(32)

32. Involucres cylindric; heads in flat-topped corymbs…**Stevia**

32′. Involucres campanulate or hemispherical; heads in elongate panicles…**Leuciva**

33. Stems winged by decurrent leaf bases; phyllary margins herbaceous…**Helenium**

33′. Stems not winged; phyllary margins scarious…(34)

34. Plants annual or biennial…(35)

34′. Plants perennial…(37)

35. Plants 30–80 cm; usually aromatic (sage)…**Artemisia**

35′. Plants 20–30 cm…(36)

36. Foliage aromatic, with pineapple odor when bruised; florets all with corollas…**Matricaria**

36′. Foliage not aromatic; peripheral florets lacking corollas…**Cotula**

37. Plants herbaceous…**Tanacetum**

37′. Plants shrubs or subshrubs…(38)

38. Heads in panicles, racemes, or spikes; usually aromatic (sage)…**Artemisia**

38′. Heads borne singly or in flat-topped corymbs…**Pentzia**

39. Leaves all basal, ovate to suborbicular…**Chamaechaenactis**

39′. Leaves various, not all basal…(40)

40. Leaves mostly opposite or whorled, the upper cauline leaves may be alternate…(41)

40′. Leaves alternate throughout…(49)

41. Corollas yellow…(42)

41′. Corollas white to cream or blue, lavender, pink, or purple…(45)

42. Corollas 5-lobed; fruits not flattened, strongly angled…(43)

42′. Corollas 4-lobed; cypselae strongly flattened or weakly 3–4-angled…(44)

43. Phyllaries hairy (hirsutulous) and gland-dotted; disk florets 15–30…**Picradeniopsis**

43′. Phyllaries gland-dotted, otherwise glabrous; disk florets 2–8…**Schkuhria**

44. Leaf blades usually 3-lobed or sometimes up to 5-lobed, not triangular-hastate, apices not long-tailed; phyllaries 8–16 in 2–3 series, not fused…**Perityle**

44′. Leaf blades triangular-hastate, entire, dentate, or shallowly lobed, apices long-tailed; phyllaries 15–21 fused together in 1 series…**Pericome**

45. Corollas 4-lobed…**Perityle**

45′. Corollas 5-lobed…(46)

46. Phyllaries 5, in 1 series…**Stevia**

46′. Phyllaries 8–45, in 2–8+ series…(47)

47. Cypselae 4-angled, not ribbed, densely hairy…**Palafoxia**

47′. Cypselae 4–10-ribbed, not densely hairy…(48)

48. Cypselae 8–10-ribbed…**Carphochaete**

48′. Cypselae 4–5-ribbed…**Ageratum**

49. Corollas mostly white or blue, lavender, pink, or purple…(50)

49′. Corollas yellow to orange…(55)

50. Phyllaries toothed or fringed…**Centaurea**

50′. Phyllaries not toothed or fringed…(51)

51. Phyllaries 35–70 in 3–8 series…**Vernonia**

51′. Phyllaries 5–21 in 1–2 series…(52)

52. Phyllary margins membranous or scarious…(53)

52′. Phyllary margins herbaceous throughout…(54)

53. Pappus scales aristate…**Hymenothrix**

53'. Pappus scales rounded, not aristate...**Hymenopappus**
54. Phyllaries 5 in 1 series; florets 5...**Stevia**
54'. Phyllaries 5-21 in 1-2 series; florets 8-70...**Chaenactis**
55. Primary leaves forming recurved spines...**Tetradymia**
55'. Primary leaves not spiny...(56)
56. Phyllary margins scarious or membranous...(57)
56'. Phyllary margins not scarious...(59)
57. Foliage not aromatic when crushed; pappus of orbicular scales or absent...**Hymenopappus**
57'. Foliage aromatic when crushed; pappus coroniform or absent...(58)
58. Plants annual; stems 4-40 cm...**Matricaria**
58'. Plants perennial; stems 40-150 cm...**Tanacetum**
59. Phyllary apices usually looped, hooked, or curved at anthesis; involucres notably resinous...**Grindelia**
59'. Phyllary apices erect at anthesis; involucres not resinous...(60)
60. Stems winged by decurrent leaf bases...**Helenium**
60'. Stems not winged...(61)
61. Pappus of outer florets of scales, pappus of inner florets with longer bristles...**Erigeron**
61'. Pappus wholly of scales...(62)
62. Corollas white, cream, or pinkish; receptacles without stout bristles (setae)...**Chaenactis**
62'. Corollas brown-purple or red-brown; receptacles with stout bristles...**Gaillardia**

KEY F: RAY FLORETS PRESENT; PAPPUS OF CAPILLARY BRISTLES, AT LEAST IN PART

1. Ray florets white, pink, or purple...(2)
1'. Ray florets yellow, orange, or red...(27)
2. Shrubs and subshrubs...(3)
2'. Annuals, biennials, or herbaceous perennials...(6)
3. Plants ± thorny, thorns green; leaves reduced; branches often wandlike; weedy...**Chloracantha**
3'. Plants not thorny; leaves not reduced, branches not wandlike; not particularly weedy...(4)
4. Leaves cordate and clasping the stem, margin spinulose-serrate...**Herrickia**
4'. Leaves not cordate or clasping the stem, margin not spinulose-serrate, or if serrate, then teeth bristle-tipped...(5)
5. Cypselae dimorphic (ray cypselae 3-sided, disk cypselae compressed), each with 6-18 ribs...**Xanthisma**
5'. Cypselae all similar, not dimorphic, with 2-3 ribs...**Ionactis**
6. Plants annual or biennial...(7)
6'. Plants perennial...(16)
7. Heads solitary, sessile or pedunculate...(8)
7'. Heads in clusters, either in panicles or corymbs...(12)
8. Cypselae turbinate or cylindrical, not compressed, sometimes slightly flattened...(9)
8'. Cypselae oblanceolate or oblong, compressed or clearly flattened...(11)
9. Leaves deeply 1-2-pinnatifid, lobes bristle-tipped...**Machaeranthera**
9'. Leaves entire or toothed, or if pinnatifid, then lobes not bristle-tipped...(10)
10. Ray florets with prominent pappus; leaves entire or toothed...**Dieteria**
10'. Ray pappus absent or present; if pappus present then leaves pinnatifid or bipinnatifid throughout...**Leucosyris**
11. Phyllaries usually equal in height; phyllary nerves golden-resinous...**Erigeron**
11'. Phyllaries strongly unequal in height; phyllary nerves not golden-resinous...**Townsendia**
12. Ray florets with reduced lamina 0.5-1 mm, or lamina nearly absent...(13)
12'. Ray florets with lamina > 1 mm...(14)
13. Leaf faces and cypselae stipitate-glandular or gland-dotted; phyllaries lacking orange to brown mid-nerves...**Laennecia**
13'. Leaf faces and cypselae not glandular; phyllaries with orange to brown midnerves...**Conyza**
14. Pappus of ray florets absent...**Psilactis**
14'. Pappus of ray florets present, composed of bristles similar to those of disk florets...(15)
15. Stems and leaves usually hairy and sometimes glandular, but glabrous in *Dieteria canescens* var. *glabra*; plants of grasslands, woodlands, or dry streambeds...**Dieteria**
15'. Stems and leaves usually glabrous; plants usually of marshy habitats, moist soils, wet swales, and stream banks...**Symphyotrichum**
16. Stems thorny (thorns green), or if not thorny, then wandlike with reduced leaves; weedy...**Chloracantha**
16'. Stems not thorny or wandlike; not weedy...(17)
17. Cypsela margins ribbed; cypsela faces 1-2-nerved or nerves absent...(17)
17'. Cypsela margins not ribbed; cypsela faces 3-12-nerved...(20)
18. Phyllaries keeled...**Ionactis**
18'. Phyllaries not keeled...(19)

19. Phyllaries unequal in length; pappus of 12–35 narrow scales, sometimes bristlelike…**Townsendia**
19'. Phyllaries equal in length; pappus of outer shorter bristles or scales plus 5–40 inner longer bristles, bristles sometimes absent…**Erigeron**
20. Pappus of relatively coarse bristles, bases flattened; cypselae dimorphic, ray cypselae 3-sided, disk cypselae flattened…(21)
20'. Pappus of fine bristles, not basally flattened; ray and disk cypselae similar…(22)
21. Subshrubs; pappus bristles coarsely barbed…**Xanthisma**
21'. Herbaceous perennials; pappus bristles finely barbed…**Leucosyris**
22. Phyllaries equal or subequal in length; leaf blades linear or narrowly lanceolate…**Almutaster**
22'. Phyllaries unequal in length; leaf blades lanceolate or broader…(23)
23. Plants taprooted…**Dieteria**
23'. Plant rhizomatous…(24)
24. Leaf bases clasping the stems…(25)
24'. Leaf bases not clasping the stems…(26)
25. Pappus of yellowish to cinnamon or tawny stiff bristles…**Herrickia**
25'. Pappus of white or brownish soft bristles…**Symphyotrichum**
26. Heads borne singly and terminally on branches; cauline leaves densely overlapping, coriaceous…**Chaetopappa**
26'. Heads in corymbs, panicles, or racemes (except in *Symphyotrichum foliaceum*, occurring in alpine or subalpine meadows); cauline leaves not densely overlapping or coriaceous…**Symphyotrichum**
27. Leaves opposite or subopposite, or if alternate, then leaves mostly basal…(28)
27'. Leaves alternate…(30)
28. Leaves succulent, filiform to linear…**Haploësthes**
28'. Leaves not succulent, triangular, oblanceolate, elliptic, or cordate-ovate…(27)
29. Plants perennial; widespread throughout New Mexico…**Arnica**
29'. Plants annual; SW New Mexico…**Bartlettia**
30. Phyllaries in 1–2 series, equal in length, often subtended by smaller calyculi…(31)
30'. Phyllaries in 3+ series, unequal in length, calyculi absent…(34)
31. Annuals, herbaceous perennials or low subshrubs, woody only at the base…(32)
31'. Shrubs, woody well above the base…(35)
32. Leaves, at least the larger, (7–)8–17 cm wide and suborbicular to ovate; SW New Mexico…**Roldana**
32'. Leaves < 8 cm wide, not suborbicular or ovate…**Packera/Senecio**
33. Leaves linear and evenly distributed on stem…**Senecio**
33'. Leaves lance-elliptic, lanceolate, or lance-linear, clustered at ends of stems…**Barkleyanthus**
34. Shrubs or subshrubs…(35)
34'. Annuals, biennials, or herbaceous perennials…(39)
35. Phyllaries in obvious vertical ranks…(36)
35'. Phyllaries in spirals, not in vertical ranks…(37)
36. Leaves with 3–5 raised parallel veins; leaf blades gland-dotted…**Petradoria**
36'. Leaves without raised parallel veins, 1-nerved; leaf blades not gland-dotted…**Lorandersonia**
37. Basal leaves pinnatifid, lobes bristle-tipped; pappus bristles flattened at base…**Xanthisma**
37'. Basal leaves not pinnatifid, entire or shallowly toothed; pappus bristles not flattened at base…(38)
38. Plants rhizomatous; stems glaucous, woody only at base; moist to wet soils in streambeds, lakeshores, or marshes…**Euthamia**
38'. Plants not rhizomatous; stems not glaucous, obviously woody; dry habitats…**Ericameria**
39. Receptacles with scales…**Xanthisma**
39'. Receptacles naked, without scales…(40)
40. Pappus of ray and disk florets of small outer scales and larger inner bristles; ray floret pappus sometimes absent…**Heterotheca**
40'. Pappus of ray and disk florets entirely of bristles; ray floret pappus always present…(41)
41. Lamina of ray florets, when present, 0.5–1 mm, otherwise disciform…**Laennecia**
41'. Lamina of ray florets 2+ mm, never disciform…(42)
42. Plants annual; cypselae dimorphic, ray cypselae 3-angled, disk cypselae compressed…**Rayjacksonia**
42'. Plants perennial; cypselae all similar…(43)
43. Pappus brownish…**Pyrrocoma**
43'. Pappus white…(44)
44. Heads generally 1 per stem, occasionally 2–6…(45)
44'. Heads numerous on stems…(46)
45. Peduncles 10–130 mm; phyllaries unequal in length, the outer not foliaceous…**Stenotus**
45'. Peduncles 3–8 mm; phyllaries equal or subequal in length, the outer foliaceous…**Tonestus**
46. Cauline leaves clasping or subclasping; stems and leaves obviously stipitate-glandular; cypselae 12–16-nerved, nerves whitish and raised…**Oreochrysum**

46'.   Cauline leaves not clasping; stems and leaves generally not stipitate-glandular but leaves sometimes stipitate-glandular; cypselae 5–8-nerved, nerves not whitish or raised…**Solidago**

KEY G: RAY FLORETS PRESENT; PAPPUS OF AWNS OR SCALES

1.   Receptacles paleate…(2)
1'.   Receptacles without paleae…(39)
2.   Phyllaries, stems, and leaves with black glandular hairs; phyllaries in 1 series…**Layia**
2'.   Phyllaries, stems, and leaves lacking black glandular hairs; phyllaries in 2–7 series…(3)
3.   Phyllaries and paleae nearly transparent and striate with brown stripes…(4)
3'.   Phyllaries and paleae not transparent and striate…(5)
4.   Ray florets 1–3, pale yellow to orange; pappus absent or consisting of 2–3 retrorsely barbed awns… **Heterosperma**
4'.   Ray florets 3–6, yellow; pappus of 2 awns not retrorsely barbed…**Dicranocarpus**
5.   Calyculi present, 1–8+ bractlets subtending phyllaries…(6)
5'.   Calyculi absent…(10)
6.   Phyllaries united > 1/5 their length…**Thelesperma**
6'.   Phyllaries free or united < 1/10 their length…(7)
7.   Cypselae compressed…(8)
7'.   Cypselae not compressed, 4-angled or terete…(9)
8.   Pappus of barbellate awns; cypselae not winged except for *Bidens polylepis*…**Bidens** (in part)
8'.   Pappus of scales or bristly cusps; cypselae winged…**Coreopsis**
9.   Cypselae with 1 groove on each face; ray florets pink, purple, rose-pink, violet, or white…**Cosmos**
9'.   Cypselae without grooves, or if grooves present, 2 on each face; ray florets yellow or white…**Bidens** (in part)
10.   Phyllaries usually falling with ray cypselae and adjacent fruit, not persistent in fruit…(11)
10'. Phyllaries persistent in fruit…(12)
11.   Leaves alternate…**Parthenium**
11'. Leaves opposite…**Galinsoga**
12.   Receptacles obviously columnar or cone-shaped…(13)
12'. Receptacles not columnar or strongly cone-shaped…(16)
13.   Ray florets purple or pink…**Echinacea**
13'. Ray florets yellow, maroon, or whitish…(14)
14.   Ray floret lamina persistent and becoming papery in fruiting heads…**Sanvitalia**
14'. Ray floret lamina not persistent and papery in fruiting heads…(15)
15.   Phyllaries equal or subequal in length; cypselae 4-angled, not compressed…**Rudbeckia**
15'. Phyllaries unequal in length, outer much longer than inner; cypselae strongly compressed…**Ratibida**
16.   Ray florets persistent in fruit, becoming papery…(17)
16'. Ray florets not persistent in fruit and not papery…(19)
17.   Leaf margins serrate to coarsely toothed…**Heliopsis**
17'. Leaf margins entire…(18)
18.   Subshrubs, obviously woody at base; ray floret lamina 7–18 mm…**Zinnia**
18'. Annuals or herbaceous perennials; ray floret lamina 1.5–2.5 mm…**Sanvitalia**
19.   Inner phyllaries broadly obovate or orbicular…**Berlandiera**
19'. Inner phyllaries not broadly obovate or orbicular…(20)
20.   Ray florets 2–4; leaves linear to filiform…**Pseudoclappia**
20'. Ray florets 5+; leaves not linear to filiform, broader…(21)
21.   Ray florets sterile, not producing fruits…(22)
21'. Ray florets fertile…(26)
22.   Cypselae flattened, with thin margins…**Simsia** (in part)
22'. Cypselae convex or 3–4-angled…(23)
23.   Shrubs; leaves often lobed…**Sidneya**
23'. Annuals or herbaceous perennials; leaves not lobed…(24)
24.   Pappus falling readily, not persistent in fruit…**Helianthus**
24'. Pappus persistent in fruit…(25)
25.   Petioles < 1 cm; phyllary apices gradually narrowed…**Aldama**
25'. Petioles 1–2 cm; phyllary apices abruptly narrowed…**Viguiera**
26.   Ray florets white…(27)
26'. Ray florets yellow, orange, or brown…(28)
27.   Ray florets 8; leaves pinnately lobed or compound, alternate…**Hymenopappus**
27'. Ray florets 20–40; leaves entire or serrate, opposite…**Eclipta**
28.   Disk florets sterile (female-sterile), only ray florets producing fruits…(29)
28'. Disk florets bisexual and fertile…(30)

29. Ray florets 8–9…**Engelmannia**
29'. Ray florets 12–38…**Silphium**
30. Leaves alternate, basal and/or cauline…(31)
30'. Leaves mostly opposite, all cauline…(35)
31. Cypselae 3–4-angled…**Wyethia**
31'. Cypselae compressed or flattened, not 3–4-angled…(32)
32. Cypselae winged…(33)
32'. Cypselae not winged…(34)
33. Pappus of 2+ subulate awns or scales only; cypsela margins glabrous…**Verbesina**
33'. Pappus of 2+ subulate scales plus up to 4 shorter scales; cypsela margins usually ciliate…**Helianthella** (in part)
34. Plants glabrous; outer phyllaries longer than inner phyllaries (*Flourensia pringlei*)…**Flourensia**
34'. Plants variously pubescent; outer phyllaries shorter than inner phyllaries…**Encelia**
35. Cypselae winged, wings membranous or corky…(36)
35'. Cypselae not winged…(37)
36. Leaves gland-dotted; pappus in 2 series, of 2–3 scales or awns plus 2–8 shorter scales and awns…**Jefea**
36'. Leaves not gland-dotted; pappus in 1 series of 2–3 awns or scales…**Verbesina**
37. Phyllaries 5 in 1 series; involucres 3–8 mm diam.…**Calyptocarpus**
37'. Phyllaries 12–35 in 2–5 series; involucres 10–50 mm diam.…(38)
38. Leaf margins entire…**Helianthella** (in part)
38'. Leaf margins coarsely serrate…**Lasianthaea**
39. Leaves all opposite, or opposite below and alternate above…(40)
39'. Leaves alternate…(49)
40. Cypselae compressed, with ciliate margins…**Perityle**
40'. Cypselae not compressed, 4–5-angled or 10–15-ribbed, margins not ciliate…(41)
41. Cypselae 10–15-ribbed, not 4–5-angled…(42)
41'. Cypselae 4–5-angled, not ribbed…(44)
42. Heads borne singly…**Pseudoclappia** (in part)
42'. Heads in compact flat-topped clusters…(43)
43. Ray florets 3–5, pappus 10, of 5 scales and 5 bristles…**Sartwellia**
43'. Ray florets (0)1, pappus of 2–4 scales…**Flaveria**
44. Ray florets 1–8…(45)
44'. Ray florets 6–13…(47)
45. Plants perennial…**Picradeniopsis** (in part)
45'. Plants annual or biennial…(46)
46. Ray florets (0)1–2; ray corollas yellow or white…**Schkuhria**
46'. Ray florets 3–8; ray corollas pinkish to purplish…**Palafoxia**
47. Ray florets white with red veins…**Eriophyllum**
47'. Ray florets yellow…(48)
48. Plants annual…**Lasthenia**
48'. Plants perennial or if annual, ray flowers absent…**Picradeniopsis** (in part)
49. Pappus mixed, of scales and bristles…(50)
49'. Pappus wholly of scales…(56)
50. Ray florets white, sometimes blue, purple, lilac, maroon, or pink…(51)
50'. Ray florets mostly yellow to orange…(54)
51. Phyllary margins prominently white-scarious-margined…**Chaetopappa**
51'. Phyllary margins not white-scarious although they may be scarious or not…(52)
52. Phyllaries equal in height, generally not imbricate…**Erigeron**
52'. Phyllaries unequal in height, imbricate…(53)
53. Pappus of scales subtending an inner set of longer bristles, bristles terete, not flattened…**Ionactis**
53'. Pappus of lanceolate, subulate, or setiform scales (flattened bristles)…**Townsendia**
54. Pappus of bristles subtending subulate scales…**Grindelia**
54'. Pappus of scales subtending bristles…(55)
55. Cypselae 2-ribbed, with thickened margins; heads borne singly or in 2s or 3s…**Erigeron**
55'. Cypselae 4–12-ribbed; heads in compound clusters, rarely borne singly…**Heterotheca**
56. Ray florets white, sometimes blue, purple, or violet…(57)
56'. Ray florets yellow to orange…(60)
57. Ray florets 5–8…**Hymenopappus**
57'. Ray florets 10–67…(58)
58. Cypselae strongly compressed to flattened…**Townsendia**
58'. Cypselae not strongly flattened, often terete…(59)
59. Leaves gland-dotted, glabrous, or minutely hairy…**Gutierrezia**

59'. Leaves not gland-dotted, leaves obviously hairy, strigose, or hirsute…**Aphanostephus**
60. Phyllaries united 1/2–3/4 their length; ray florets with dark basal blotch or spot on upper surface… **Gazania**
60'. Phyllaries not united, or united < 1/2 their length; rays without dark basal blotch or spot…(61)
61. Ray corollas becoming reflexed, dry and persisting past flowering…(62)
61'. Ray corollas withering and falling after flowering…(65)
62. Heads in flat-topped or spherical clusters…**Psilostrophe** (in part)
62'. Heads borne singly at tips of stems…(63)
63. Shrubs or subshrubs…**Psilostrophe** (in part)
63'. Herbaceous annuals or perennials…(64)
64. Leaves and stems woolly, not gland-dotted…**Baileya**
64'. Leaves glabrous or hairy, gland-dotted, not woolly…**Tetraneuris**
65. Disk florets female-sterile, not producing fruits…(66)
65'. Disk florets bisexual, producing fruits…(67)
66. Annuals; ray florets 5–15…**Amphiachyris**
66'. Perennials; ray florets 1–5…**Hymenoxys** (in part)
67. Disk florets brown-purple or red-brown or tipped with brown-purple or red-brown…(68)
67'. Disk florets mostly yellow or cream…(69)
68. Stems winged by decurrent leaf bases…**Helenium**
68'. Stems not winged…**Gaillardia**
69. Phyllaries mostly unequal in length, imbricate…**Gutierrezia**
69'. Phyllaries mostly equal to subequal in length, imbricate…(70)
70. Cypselae not strongly 4-angled, length 3+ times diam.…**Hymenoxys** (in part)
70'. Cypselae strongly 4-angled, length usually 1–2 times diam.…**Platyschkuhria**

KEY H: RAY FLORETS PRESENT; PAPPUS ABSENT

1. Receptacles paleate, receptacular bracts or paleae present…(2)
1'. Receptacles not paleate, without receptacular bracts…(29)
2. Phyllaries with scarious margins…(3)
2'. Phyllaries without scarious margins, herbaceous or margins narrowly membranous…(4)
3. Disk florets 5–8; heads in compact, flat-topped clusters…**Achillea**
3'. Disk florets 10–15, sterile; heads borne singly or in loose clusters…**Anthemis**
4. Heads with calyculi, 1–8+ bractlets subtending phyllaries…(5)
4'. Heads without calyculi…(9)
5. Phyllaries united > 1/5 their total length…**Thelesperma**
5'. Phyllaries free or united < 1/10 their total length…(6)
6. Cypselae 3–4-angled or linear-fusiform…(7)
6'. Cypselae compressed…(8)
7. Cypselae with 1 groove on each face; ray corollas pink, purple, rose-pink, violet, or white…**Cosmos**
7'. Cypselae without grooves, or if grooves present, 2 on each face; ray corollas yellow or white, never pink, purple, or violet…**Bidens**
8. Inner cypselae beaked; ray floret lamina 1–2 mm…**Heterosperma**
8'. Inner cypselae not beaked; ray floret lamina 4–30+ mm…**Coreopsis**
9. Ray floret corollas white or pale yellow fading to white…(10)
9'. Ray floret corollas yellow or orange…(14)
10. Plants annual, occasionally perennial in *Eclipta*…(11)
10'. Plants perennial…(12)
11. Ray florets 5–8…**Galinsoga**
11'. Ray florets 20–40…**Eclipta**
12. Leaves basal, alternate…**Hymenopappus**
12'. Leaves cauline, opposite…(13)
13. Ray corollas persistent in fruit, becoming papery; phyllaries persistent in fruit…**Zinnia**
13'. Ray corollas not persistent in fruit, not becoming papery; phyllaries shed together with ray achenes… **Melampodium**
14. Inner phyllaries broadly ovate or orbicular…**Berlandiera**
14'. Inner phyllaries not broadly ovate or orbicular, narrower…(15)
15. Ray corollas persistent in fruit, becoming papery…(16)
15'. Ray corollas not persistent in fruit, not papery…(17)
16. Leaves petiolate, margins serrate or toothed…**Heliopsis**
16'. Leaves sessile, margins entire…**Zinnia**
17. Phyllaries enfolding ray florets shed together with ray cypselae…**Melampodium**
17'. Phyllaries not enfolding ray florets, persistent in fruit…(18)

18.   Receptacles columnar or cone-shaped, 8–20 mm high…(19)
18'. Receptacles flat to convex, 0–5 mm high…(20)
19.   Phyllaries equal or subequal in length; cypselae 4-angled, not compressed…**Rudbeckia**
19'. Phyllaries unequal in length, outer much longer than inner; cypselae strongly compressed…**Ratibida**
20.   Ray florets sterile, not producing fruit…(21)
20'. Ray florets fertile, producing fruit…(26)
21.   Cypselae flattened, thin-margined…**Simsia**
21'. Cypselae biconvex or 3–4-angled, not strongly flattened…(22)
22.   Plants annual…**Heliomeris**
22'. Plants perennial…(23)
23.   Plants shrubs…(24)
23'. Plants perennial…(25)
24.   Leaves petiolate, petioles 2–7 mm…**Encelia**
24'. Leaves sessile or subsessile, petioles to 1 mm if present…**Viguiera**
25.   Leaves sessile…**Heliomeris**
25'. Leaves petiolate…**Zaluzania**
26.   Disk florets female-sterile, not producing fruit…**Silphium**
26'. Disk florets bisexual, producing fruit…(27)
27.   Cypselae 3–4-angled…**Wyethia**
27'. Cypselae compressed to strongly flattened…(28)
28.   Cypselae winged…**Verbesina**
28'. Cypselae not winged…**Helianthella**
29.   Shrubs with thorny stems; disk cypselae winged…**Osteospermum**
29'. Annual or perennial herbs, if woody, only at base; thorns absent; disk cypselae not winged…(30)
30.   Phyllaries with prominent scarious margins…(31)
30'. Phyllaries herbaceous, without prominent scarious margins…(34)
31.   Phyllaries equal or subequal; perennials or biennials…**Hymenopappus**
31'. Phyllaries in 2–5 series; annuals…(32)
32.   Perennials with rhizomes; cypselae 10-ribbed…**Leucanthemum**
32'. Annuals, never rhizomatous; cypselae 3–5-ribbed…(33)
33.   Leaves 2–3 times pinnately lobed, lobes filiform…**Tripleurospermum**
33'. Leaves entire or with a pinnatifid margin, lobes not filiform…**Aphanostephus**
34.   Leaves opposite; ray florets 1…**Flaveria**
34'. Leaves alternate; ray florets 2+…(35)
35.   Shrubs or subshrubs…**Gymnosperma**
35'. Annuals, biennials, or herbaceous perennials…(36)
36.   Ray florets 3–5…**Hymenoxys**
36'. Ray florets 10–55…(37)
37.   Leaves densely white-woolly, not gland-dotted…**Baileya**
37'. Leaves not white-woolly, green, usually gland-dotted…**Hymenothrix**

ACHILLEA L. – Yarrow

Jennifer Ackerfield

Achillea millefolium L.—Common yarrow. Perennial herbs; plants 0.6–6.5 dm, densely tomentose to glabrate; leaves alternate, lanceolate to oblong, 3.5–35 × 0.5–3.5 cm, finely 1–2 times pinnately dissected; heads radiate, 10–100 in a corymbiform array; involucral bracts in 3 series, dry with scarious or hyaline margins and a green midrib; receptacle chaffy; ray florets pistillate and fertile, rays white or light pink, 1.5–3 mm. Common in gravelly soil of roadsides, prairies, open slopes, mountain meadows; a highly variable species occurring from the high plains to the alpine, 3800–13,000 ft. Widespread except EC, SE (Mulligan & Bassett, 1959; Tryl, 1975).

ACOURTIA D. Don – Desertpeony

Kenneth D. Heil

Perennials, 2.5–50(–150) cm; stems glabrate or resinous-punctate; leaves basal, cauline, or both, oblong, lanceolate, ovate, or rhombic, spinulose-toothed, sessile or clasping at the base; heads solitary or corymbose or paniculate; phyllaries in 1–7 series; corollas pink to lavender or white, zygomorphic, bilabiate, lower lip 3-toothed, upper lip 2-lobed; cypselae ± fusiform or terete to cylindrical, 4–10 mm, not beaked, usually ± ribbed; pappus tan or white, ± barbellate to nearly smooth (Simpson, 2006).

1. Plants 2.5–30 cm; leaf blades rhombic-orbiculate to suborbiculate, hollylike…**A. nana**
1'. Plants 30–150 cm; leaf blades ovate to ovate-elliptic or oblong-lanceolate to elliptic-oblong, not holly-like…(2)
2. Leaf blades oval to ovate-elliptic; phyllaries acuminate; florets 3–6…**A. thurberi**
2'. Leaf blades oblong-lanceolate to elliptic-oblong; phyllaries obtuse to acute; florets 8–12…**A. wrightii**

Acourtia nana (A. Gray) Reveal & R. M. King [*Perezia nana* A. Gray]—Desert holly. Plants 2.5–30 cm; leaves sessile, blades rhombic-orbiculate to suborbiculate, 10–50 mm, margins coarsely and irregularly prickly-dentate; heads borne singly; involucres campanulate, 14–17 mm; florets 15–24; corollas lavender-pink or white, 10–17 mm; cypselae subcylindrical, 3–7.5 mm, densely stipitate-glandular; pappus white or tawny. Sandy, silty, or caliche soils, desert scrub communities, 3150–8500 ft. EC, C, SE, SC, SW, Cibola, Quay.

Acourtia thurberi (A. Gray) Reveal & R. M. King [*Perezia thurberi* A. Gray]—Thurber's desertpeony. Plants 40–150 cm; leaves sessile, blades ovate to ovate-elliptic, 1.5 cm (cauline) to 18 cm (basal), margins acerose-denticulate, faces densely glandular-puberulent; heads in subcongested corymbiform arrays; involucres obconic to campanulate, 7–9 mm; florets 3–6; corollas lavender-pink (purple), 7–12 mm; cypselae subcylindrical to subfusiform, 3–7 mm, glandular; pappus bright white, 8–9 mm. Gravel and caliche soils, desert scrub communities, 4200–6175 ft. Grant, Hidalgo.

Acourtia wrightii (A. Gray) Reveal & R. M. King [*Perezia wrightii* A. Gray]—Brownfoot. Plants 30–120 cm; leaves sessile, blades oblong-lanceolate to elliptic-oblong, 2.5–13 cm, margins dentate to denticulate, faces minutely stipitate-glandular and/or hirtellous; heads in corymbiform arrays; involucres turbinate, 5–8 mm; florets 8–12; corollas pink or purple, 9–20 mm; cypselae linear-fusiform, 2–6 mm, glandular-puberulent; pappus bright white, 9–12 mm. Gravel, sandy, or caliche soils, desert scrub communities, 6000–7300 ft. SE, SC, SW.

ADENOPHYLLUM Pers. – Dogweed

Don Hyder

Adenophyllum wrightii A. Gray [*Dyssodia neomexicana* (A. Gray) B. L. Rob.]—Wright's dogweed. Annuals, to 70 cm; leaves linear or pinnately lobed, 25–35 mm, lobes 5–9, linear (each with oil gland at base); phyllaries 8–13, narrowly to broadly oblanceolate, remaining connate in fruit; ray florets usually 8, corollas yellow-orange; disk florets 30–50, corollas pale yellow, often tipped with crimson, 4–5 mm; cypselae 3.5–4.5 mm; pappus of 10 outer, erose scales 0.7–1 mm, plus 10 inner, unequally 3-aristate scales 5–6.5 mm. Swales and drainages in piñon-juniper woodland communities, 7000–7200 ft. Catron, Grant, Sierra (Nesom, 2006a).

AGERATINA Spach – Snakeroot

Kenneth D. Heil

Herbs or shrubs; leaves mostly opposite; heads discoid; involucral bracts imbricate or subequal, green to chartaceous or weakly coriaceous; receptacle naked, flat or conic; disk florets blue, pink, purple, or white; ray florets absent; pappus of capillary bristles; cypselae prismatic, usually 5-angled (Ackerfield, 2015; Nesom, 2006b).

1. Shrubs…**A. wrightii**
1'. Perennials or subshrubs…(2)
2. Leaves mostly sessile; heads in open, loose arrays…**A. lemmonii**
2'. Leaves petiolate; heads in compact clusters…(3)
3. Leaves yellow-green or grayish yellow-green, blades triangular to lance-ovate, or ovate; phyllaries mostly granular-puberulent…**A. herbacea**
3'. Leaves mostly green, blades lanceolate to lance-ovate; phyllaries glabrous or glabrescent…**A. rothrockii**

Ageratina herbacea (A. Gray) R. M. King & H. Rob. [*Eupatorium herbaceum* (A. Gray) Greene]—Fragrant snakeroot. Plants (2-)3-6(-8) dm, from a woody caudex; leaves opposite, deltoid to widely lanceolate or ovate, 2-5(-7) cm, dentate, abaxially hispidulous to glabrate; involucral bracts 4-5 mm; disk florets white; cypselae 2-3 mm. Rocky hillsides, forest openings, meadows, 4250-10,000 ft. Widespread except EC, SE.

Ageratina lemmonii (B. L. Rob.) R. M. King & H. Rob. [*Eupatorium lemmonii* B. L. Rob.]—Lemmon's snakeroot. Plants 20-40(-70) cm; stems commonly purple, erect, puberulous; leaves opposite, petioles usually 0, blades ovate-lanceolate, 2-4.5 × 0.5-2.5 cm, margins shallowly serrate, abaxial faces glabrous or glabrate, gland-dotted; heads in loose open arrays; involucres 4-5.5 mm; corollas white; cypselae sparsely and finely hispidulous. Rocky sites, open ponderosa pine-oak woodlands; 1 record from the Gila National Forest, 8845 ft. Catron.

Ageratina rothrockii (A. Gray) R. M. King & H. Rob. [*Eupatorium rothrockii* A. Gray]—Rothrock's snakeroot. Plants (20-)40-70(-150) cm; leaves opposite, blades lanceolate to lance-ovate, mostly 3-6 × (1.5-)2-3 cm, margins serrate to crenate, sparsely puberulent abaxially, mostly along nerves; heads clustered; involucres 5-7 mm; corollas white; cypselae sparsely and finely strigose-hirsute. Rocky slopes and ledges, oak-juniper-pine woodlands, mixed conifer, spruce-fir forests, 5700-10,200 ft. C, WC, SC, Grant.

Ageratina wrightii (A. Gray) R. M. King & H. Rob. [*Eupatorium wrightii* A. Gray]—Wright's snakeroot. Shrubs, 50-150 cm; leaves opposite proximally, alternate on distal 1/3-1/2 of stem; blades ovate to deltate-ovate, mostly 1-2 × 0.5-1.5(-2) cm, margins entire or shallowly crenate, abaxial faces gland-dotted; heads clustered; involucres 3.5-4.5 mm; corollas pinkish white to pink; cypselae sparsely hispidulous. Limestone slopes and ledges, desert mountains from desert scrub to piñon-juniper woodland communities, 4200-7200 ft. SE, SC, SW, Lincoln.

AGERATUM L. – Whiteweed

Kenneth D. Heil

Ageratum corymbosum Zuccagni [*A. corymbosum* var. *jaliscense* B. L. Rob.; *A. salicifolium* Hemsl.]—Flat-top whiteweed. Perennials or subshrubs, 30-100 cm; stems often decumbent, puberulent to minutely strigose-hispid; leaves (ours) cauline, opposite; leaf blades ovate to rhombic-lanceolate, 3-8 × 1-3.5 cm, margins toothed, abaxial faces densely gland-dotted; heads discoid, in dense to open, cymiform to corymbiform arrays; involucres 5-6 mm; florets 20-125, corollas usually blue to lavender, sometimes white; cypselae 4-5-ribbed, glabrous; pappus usually crowns of connate scales. Ledges and cliffs, along streams, in desert grasslands, oak-agave, oak, oak-juniper, and pine-oak woodlands, 4400-6800 ft. Hidalgo (Nesom, 2006c).

AGOSERIS Raf. – Fleecetop

Gary I. Baird

Annual or perennial herbs, sap milky; stems scapose, glabrous or distally hairy; leaves all basal, blades oblanceolate to lance-linear, simple, margins entire to pinnately lobed, faces glabrous and glaucous or pubescent; heads ligulate, solitary; involucres cylindrical to conic; phyllaries 10-30+ in 3-5 series; florets 10-100+, bisexual, matutinal, corollas yellow (outer abaxially red-purple, this still evident when dried) or orange (often drying purplish), rarely pink; cypselae uniform or dissimilar (peripheral differing from central), fusiform to narrowly conic or obconic, 10-ribbed, flaxen to brownish or purplish, glabrous or scabridulous to hirtellous, apically beaked; beak 1-25 mm; pappus of numerous bristles, white, persistent (Baird, 2006).

Hybrids between the various taxa can be expected wherever they occur together.

1. Plants annual; ligules typically 2-3 mm; anthers to 1 mm...**A. heterophylla**
1'. Plants perennial; ligules typically 5-20 mm; anthers 2-6 mm...(2)

2. Corollas orange (rarely pinkish but in either case usually drying purple)…**A. aurantiaca** (in part)
2'. Corollas yellow (drying yellow or pale except red-purple stripe on outer florets remaining evident)…(3)
3. Mature cypsela beak 1–4 mm (usually < 0.5 times length of cypsela body)…(4)
3'. Mature cypsela beak 5–10 mm (usually 1–2.5 times length of cypsela body)…(5)
4. Leaves entire, glabrous and (usually) glaucous, but if toothed or lobed (lobe pairs 2–3), then mostly pubescent (uncommonly glabrous); peduncles glabrous or sparsely to densely villous distally; phyllaries glabrous or faces villous to glandular-villous…**A. glauca**
4'. Leaves toothed or lobed (lobe pairs 2–5+), glabrous and (usually) glaucous (rarely sparsely pubescent); peduncles glabrous or slightly pubescent distally; phyllaries ciliate, sometimes also pubescent on midvein, eglandular (uncommonly glabrous)…**A. ×agrestis** (in part)
5. Ligules 4–12 mm; leaves entire or with 2–4 pairs of spreading to antrorse lobes; mostly of higher, mesic habitats (montane forests and meadows to alpine tundra)…**A. aurantiaca** (in part)
5'. Ligules 10–20 mm; leaves toothed or lobed with 2–7 pairs of spreading to retrorse lobes (uncommonly entire); mostly of lower and/or drier habitats (sagebrush, grassland, open forests, alpine meadows)…(6)
6. Leaf blades mostly < 15 cm; peduncles mostly 10–25 cm; base of head usually villous (rarely glabrate); cypsela beak mostly 5–10 mm, slender…**A. parviflora**
6'. Leaf blades mostly > 15 cm; peduncles mostly 35–55 cm; base of head glabrous or scantily pubescent; cypsela beak mostly 1–7 mm, stout or slender…**A. ×agrestis** (in part)

Agoseris ×agrestis Osterh. (pro sp.) [*A. glauca* (Pursh) Raf. var. *agrestis* (Osterh.) Q. Jones ex Cronquist]—Foothill fleecetop. Perennials; scapes 35–55 cm, glabrous or scant-pubescent distally; leaf bases pale (or slightly purplish), blades 15–25 cm × 5–15 mm (excluding lobes), margins toothed or lobed (lobes 2–5+ pairs, mostly spreading), lobules mostly lacking, faces glabrous and glaucous (rarely scant-pubescent); phyllaries green or medially red-purple, faces glabrous or pubescent on midvein, margins ciliate (rarely glabrous), inner phyllaries elongating (or not) in fruit; florets 15–100, peripheral exceeding involucre; corolla yellow, ligules 10–18 mm; cypselae 6–9 mm, ribs uniform (sometimes obscure), apex gradually tapering to beak; beak stout to slender, 1–7 mm (about equal to length of body or less); pappus 10–15 mm. Meadows and open woodlands, 6560–8200 ft.

Agoseris aurantiaca (Hook.) Greene—Mountain fleecetop. Perennials; scapes 10–40 cm, villous to woolly, sometimes glabrous or nearly so; leaf bases red-purple (rarely pale), blades 10–30 cm × 5–15 mm (excluding lobes), margins entire to lobed (lobes 2–4 pairs, spreading to antrorse), lobules often present, faces glabrous and glaucous to sparsely villous; phyllaries green to red-purple, often black-purple-spotted or -blotched, faces glabrous to villous, margins ciliate (rarely glabrous), inner phyllaries elongating in fruit; florets 15–50+, peripheral ± equal to involucre; corolla orange or yellow (rarely pink), ligules 5–10 mm; cypselae 6–9 mm, ribs often thickened distally, apex abruptly to gradually tapering to beak; beak slender, 5–8 mm (about equal to length of body); pappus 9–15 mm. Montane to alpine meadows and forests.

1. Phyllaries mostly lanceolate, margins ciliate proximally (mainly), faces predominantly villous; corollas orange (pink); cypsela apex (usually) abruptly narrowing to beak, ribs often thickened distally…**var. aurantiaca**
1'. Phyllaries mostly ovate or obovate, margins ciliate distally (mainly), faces predominantly glabrous; corollas orange or yellow; cypsela apex (usually) gradually tapering to beak, ribs mostly uniform…**var. purpurea**

var. aurantiaca—Scapes usually shorter than leaves at flowering, villous to tomentose when young, somewhat glabrate with age; outer phyllaries mostly lanceolate, margins ciliate (especially proximally), faces sparsely to densely pubescent (rarely glabrous); florets mostly orange (occasionally pinkish, rarely yellow); ligules mostly < 9 mm, ± equal to phyllaries; cypselae ribs often thickened distally, minutely pubescent (rarely glabrous), apex often abruptly narrowed to beak. Alpine and subalpine communities, > 10,500 ft. in the Sangre de Cristo and San Juan Mountains. NC, NW, WC. New Mexico is the S limit of this taxon and all specimens seen from the state appear to be ± intermediate with var. *purpurea*.

var. purpurea (A. Gray) Cronquist—Colorado fleecetop. Scapes usually equal to or longer than leaves at flowering, tomentose to woolly when young, somewhat glabrate with age; outer phyllaries mostly ovate to obovate, margins ciliate to pubescent (especially distally), faces glabrous to sparsely pubescent; florets orange or yellow (rarely pinkish); ligules often > 9 mm, subequal to exceeding phyllaries (especially if yellow); cypselae ribs usually uniform, glabrous (mostly) to minutely scabrous, apex often gradually tapered to beak. Montane and alpine meadows and forests, 6900–11,800 ft. in mountains throughout New Mexico. NC, WC, SC. This taxon forms hybrids with both *A. glauca* and *A. parviflora*.

Agoseris glauca (Pursh) Raf.—Prairie fleecetop. Perennials; scapes 5–70 cm, glabrous to densely pubescent or glandular-pubescent; leaf bases pale (or slightly purplish), blades 5–30 cm × 3–50 mm (excluding lobes), margins entire, toothed, or lobed (lobes 2–3 pairs, spreading to antrorse), lobules lacking, faces glabrous and glaucous to densely villous; phyllaries green or red-purple, often with a darker midstripe and/or with blackish speckles or spots, tip often blackish, faces glabrous to villous, glandular or not, margins glabrous or ciliate (often glandular-ciliate), inner not (or little) elongating in fruit; florets 15–150+, peripheral much exceeding involucre; corolla yellow, ligules 10–20 mm; cypselae 5–9 mm, ribs uniform, apex gradually tapering to beak; beak stout (or lacking), 1–4 mm (mostly < 1/2 length of body); pappus 8–18 mm. Wet to mesic prairies and sloughs to montane forests and alpine meadows.

1. Plants mostly glabrous and glaucous, or slightly pubescent distally; leaves mostly entire; involucres glabrous (rarely slightly pubescent at base), nonglandular; phyllaries mostly lanceolate, margins ± straight; mostly of wet (alkaline) meadows, riparian habitats, or other wet areas…**var. glauca**
1'. Plants mostly pubescent, especially at base of head; leaves mostly toothed or lobed; involucres pubescent or glandular-pubescent; phyllaries mostly ovate to obovate or oblong, margins ± wavy or undulate; mostly of montane or alpine meadows and forests…**var. dasycephala**

var. dasycephala (Torr. & A. Gray) Jeps.—Arctic fleecetop. Scapes 5–40 cm, glabrate but remaining densely pubescent at base of head; leaf blades oblanceolate, 5–35 cm × 5–35 mm, margins entire or with 2–3 pairs of teeth or lobes, lobules lacking, apices rounded to acuminate, faces mostly pubescent, occasionally glabrous; outer phyllaries ovate to obovate, medially red-purple to blackish (rarely all green), villous, often glandular-pubescent, margins often wavy or undulate. Montane to alpine meadows and forests, 9515–11,800 ft. NC (currently unverified from New Mexico but occurs in mountains of S Colorado and is expected here).

var. glauca—Scapes 15–70 cm, glabrous; leaf blades linear-elliptic, 5–30 cm × 3–10 mm, margins entire, apices acuminate to attenuate, faces glabrous and glaucous; outer phyllaries lanceolate, green or medially red-purple, often with a darker midstripe, glabrous, margins straight. Wet (often alkaline) meadows, 8200–11,800 ft. NC, NW, WC.

Agoseris heterophylla (Nutt.) Greene—Arizona fleecetop. Annuals; scapes 2–20 cm, villous to tomentose (sometimes glabrate proximally); leaf bases pale, blades 2–12 cm × 3–12 mm (excluding lobes), margins toothed to lobed (lobes 2–3 pairs, spreading to antrorse), lobules lacking, abaxial face glabrous, adaxial face pubescent (rarely all glabrous); phyllaries green or medially red-purple, faces villous-lanate, usually some hairs glandular, margins ciliate, inner elongating in fruit; florets 5–50, peripheral ± equal to involucre; corollas yellow, ligules 2–3 mm; cypselae 3–4 mm, ribs uniform, apex gradually tapering to beak; beak slender, 5–6 mm (1.5–2.5 times length of body); pappus 4–9 mm. New Mexico material belongs to **var. quentinii** G. I. Baird. Desert grasslands, 4920–5575 ft. WC, SW.

Agoseris parviflora (Nutt.) D. Dietr. [*A. glauca* (Pursh) Raf. var. *laciniata* (D. C. Eaton) Kuntze]—Steppe fleecetop. Perennials; scapes 10–25 cm, glabrate to villous, mostly remaining villous distally; leaf bases pale (or slightly purplish), blades 4–15 cm × 1–15 mm (excluding lobes), margins lobed or toothed, rarely entire (lobes 3–8 pairs, retrorse to spreading), lobules often present, faces glabrous, sometimes glaucous, to (uncommonly) densely villous (sometimes with a tuft of hair at base of lobes); phyllaries red-purple, usually with a darker midstripe, rarely spotted or all green; faces glabrous to loosely pubescent, eglandular; margins ciliate to lanate, inner elongating in fruit; florets 30–50, peripheral exceeding involucre; corollas yellow, ligules 10–20 mm; cypselae 5–9 mm, ribs mostly uniform, apex often abruptly tapered; beak slender, mostly 5–10 mm, rarely shorter (generally 1–2 times length of body); pappus 10–20 mm. Sagebrush, piñon-juniper, and ponderosa pine to high-elevation meadows, 6230–11,150 ft. NC, NW. Cibola.

ALDAMA La Llave – Aldama

Kenneth D. Heil

Aldama cordifolia (A. Gray) E. E. Schill. & Panero [*Viguiera cordifolia* A. Gray]—Heartleaf golden-eye. Erect perennial; stems 50–100 cm, shrubby, woody, straight and held stiffly erect; leaves opposite, blades ovate to deltate-ovate or lance-ovate (distal), 2–10 × 1.3–7 cm, margins serrate or serrulate, faces strigose; inflorescences open corymbiform cymes; involucre campanulate to subhemispherical; heads borne singly or 2–9 in ± corymbiform arrays; involucres 11–15 × 7–15 mm; ray florets 6–8, yellow to or-ange; disk florets yellow, 40+; corollas 5.5–6.6 mm; cypselae biconvex to slightly trigonous, black, 5–6.5 mm, ± strigose; pappus of 2 lacerate, aristate scales. Limestone soils, pine forests, 4250–8750 ft. Catron, Grant, Hidalgo (Schilling & Panero, 2011).

ALMUTASTER Á. Löve & D. Löve – Alkali marsh aster

Kenneth D. Hcil

Almutaster pauciflorus (Nutt.) Á. Löve & D. Löve [*Aster pauciflorus* Nutt.]—Few-flowered aster. Perennial herbs; plants 3–12 dm, with lengthy rhizomes; leaves alternate, simple, linear, 6–10 cm, gla-brous; heads radiate; involucral bracts in 3–4 series, subequal, densely short-glandular-stipitate, 4–7 mm; receptacle naked; disk florets yellow, 4–6 mm; ray florets white to purple; pappus of a single series of barbellate bristles; cypselae hairy, 7–10-nerved. Meadows or moist places, especially in alkaline soil, 3400–7600 ft. NC, NW, C, WC, Chaves.

AMBROSIA L. – Ragweed

Kenneth D. Heil

Annuals, perennials, or shrubs, 10–400+ cm; stems branched; leaves usually cauline, sessile or petiolate; blades usually pinnately, sometimes palmately lobed, margins entire or toothed, faces hairy or glabrate, gland-dotted or stipitate-glandular; heads discoid, usually in racemiform to spiciform arrays; pistillate heads: phyllaries 12–30(–80+), usually with free tips forming tubercles, spines, or wings, florets 1(–5+), corollas 0; staminate heads: phyllaries 5–16+ in ± 1 series; receptacles ± flat or convex; florets 5–60+, corollas whitish or purplish, lobes 5; cypselae black, ± ovoid or fusiform (Strother, 2006a).

1. Shrubs…**A. monogyra**
1′. Annuals or perennials…(2)
2. Annuals…(3)
2′. Perennials…(5)
3. Leaves mostly opposite, usually with some blades palmately 3–5-lobed…**A. trifida**
3′. Leaves opposite and alternate…(4)
4. Involucres of staminate heads (2–)3–5(–7) mm diam., usually each with 1–5+ black nerves; burs ± fusiform to obpyramidal, 3–5 mm, spines 8–18+, 2–4(–5) mm…**A. acanthicarpa**
4′. Involucres of staminate heads 2–3+ mm diam., usually without black nerves; globose to pyriform, 2–3 mm, spines or tubercles 3–5+, 0.1–0.5+ mm…**A. artemisiifolia**
5. Leaves whitish-tomentose or pubescent…(6)
5′. Leaves greenish above and below, sparsely hairy…(7)
6. Leaves finely dissected, blades dark green above, whitish-tomentose below…**A. tomentosa**
6′. Leaves broadly lobed, blades and stems uniformly silvery-gray pubescent or irregularly glabrate in age…
    A. grayi
7. Bur spines (1–)8–13+, subulate, tips usually uncinate…**A. confertiflora**
7′. Bur spines or tubercles 0 or 1–6, stoutly conic to ± acerose, tips straight…**A. psilostachya**

Ambrosia acanthicarpa Hook. [*Franseria acanthicarpa* (Hook.) Coville]—Flatspine bur ragweed. Annuals, 10–80+ cm; stems erect; leaves opposite (proximal) and alternate; blades ± deltate, 15–40 (–85+) × 12–35(–80+) mm, 1-2-pinnately lobed, margins entire or toothed, hispid, and strigillose to seri-ceous, gland-dotted; pistillate heads: florets 1; staminate heads: peduncles 0.5–2 mm; involucres shal-lowly cup-shaped, usually with 1–5+ black nerves; burs fusiform to obpyramidal, 3–5 mm, spines 8–18+, straight or uncinate. Sandy soils, disturbed sites, 3800–7170 ft. Widespread.

Ambrosia artemisiifolia L. [*A. elatior* L.]—Annual ragweed. Annuals 10-60(-150+) cm; stems erect; leaves opposite (proximal) and alternate; blades deltate to lanceolate or elliptic, 25-55(-90+) × 20-30(-50+) mm, 1-2-pinnately lobed, margins entire or toothed, faces pilosulous to strigillose, gland-dotted; pistillate heads: florets 1; staminate heads: peduncles 0.5-1.5 mm; involucres shallowly cup-shaped; burs ± globose to pyriform, 2-3 mm, spines or tubercles 3-5+, 0.1-0.5+ mm, tips straight. Disturbed sites, 5900-7330 ft. Scattered.

Ambrosia confertiflora DC. [*Franseria confertiflora* (DC.) Rydb.]—Weakleaf bur ragweed. Perennials, 20-80(-150+) cm; leaves mostly alternate, blades lanceolate to ovate, 40-85(-150) × 20-35(-55) mm, laciniately lobed, margins entire, faces strigillose to sericeous, gland-dotted; pistillate heads: florets 1(2); staminate heads: peduncles 0.5-2 mm; involucres cup-shaped, 1.5-3+ mm diam., burs pyramidal to pyriform, 1-2 mm, spines (1-)5-12+, tips uncinate. Disturbed sites, 4640-6325 ft. NE, NC, NW, C, WC, SC, SW, Eddy.

Ambrosia grayi (A. Nelson) Shinners—Gray's ragweed. Perennial, 1-4 dm; leaves mostly alternate, elliptic to ovate, 4-6(-10) cm, 1-2-pinnately lobed, coarse-hairy to sericeous; pistillate heads with 2 florets, spines in several series on involucre; staminate heads with tomentose, 5-9-lobed involucral bracts. E plains, fields, roadsides, 4300-7300 ft. Colfax, Curry, Union.

Ambrosia monogyra (Torr. & A. Gray) Strother & B. G. Baldwin [*Hymenoclea monogyra* Torr. & A. Gray]—Singlewhorl burrobush. Shrubs, 30-150(-400) cm; leaves mostly alternate, blades mostly filiform, 5-30(-65+) × 0.5-1.5 mm, abaxial faces glabrous or glabrate, adaxial faces densely scabrellous; pistillate heads: floret 1; staminate heads: peduncles 0-0.5 mm; involucres ± cup-shaped, 2-4 mm diam.; burs fusiform to pyriform, 4-5 mm, wings 7-12+, mostly around middle, 2-3 × 1-2 mm. Desert washes and ravines, 3450-6850 ft. WC, SE, SC, SW, Bernalillo.

Ambrosia psilostachya DC.—Cuman ragweed. Perennials, 10-60(-100+) cm; leaves proximally opposite, distally alternate, blades deltate to lanceolate, 20-60(-140) × 8-35(-50+) mm, pinnately toothed to 1-pinnately lobed, margins entire or toothed, abaxial and adaxial faces hirsutulous to strigose, gland-dotted; pistillate heads: floret 1; staminate heads: peduncles 0.5-2 mm; involucres obliquely cup-shaped, 2-4(-5) mm diam.; burs ± obpyramidal to globose, 2-3 mm, spines or tubercles 0 or 1-6, stoutly conic to acerose, (0.1-)0.5-1 mm, tips straight. Disturbed sites, alkaline, clay soils, 3800-8700 ft. Widespread.

Ambrosia tomentosa Nutt.—Skeleton-leaf bur ragweed. Perennial, 1-4 dm; leaves mostly alternate, elliptic, 5-10 cm, 1-3-pinnately lobed, silvery-gray; pistillate heads with 2 florets, spines in several series on involucre; staminate heads with short-hairy involucral bracts. Roadsides, streams, weedy, disturbed sites, plains to mixed conifer communities, 5525-9840 ft. NE, NC, SC.

Ambrosia trifida L.—Giant ragweed. Annuals, 30-150(-400+) cm; leaves mostly opposite, blades rounded-deltate to ovate or elliptic, 40-150(-250+) × 30-70(-200+) mm, usually some blades palmately 3(-5)-lobed, margins usually toothed, rarely entire, abaxial and adaxial faces ± scabrellous and gland-dotted; pistillate heads: floret 1; staminate heads: peduncles 1-3+ mm; involucres ± saucer-shaped, 2-4 mm diam., often with 1-3 black nerves; burs ± pyramidal, 3-5(-7+) mm, spines 4-5, 0.5-1 mm, tips straight. Disturbed sites, 3650-8055 ft. NE, NC, NW, Grant, Lea, Luna.

AMPHIACHYRIS (DC.) Nutt. – Broomweed

Kenneth D. Heil

Amphiachyris dracunculoides (DC.) Nutt. [*Brachyris dracunculoides* DC.; *Gutierrezia dracunculoides* (DC.) S. F. Blake]—Prairie broomweed. Annuals, plants 30-100(-200) cm; stems usually branched distally, 0.3-1(-2) mm diam.; leaves linear to lanceolate, margins entire, faces gland-dotted; leaf blades narrowly to broadly lanceolate, 5-60 × 0.5-6 mm; heads radiate, in crowded corymbiform arrays; receptacles deeply pitted, glabrous; involucres narrowly campanulate to turbinate, 2-4 mm diam.; ray florets 7-12, corollas yellow, sometimes drying orange-tinged; disk florets 10-21, corollas yellow; cypselae 1.2-

2.2 mm, 7-9-ribbed, short-setulose. Calcareous clay or sandy soils, disturbed habitats, 3200-4670 ft. SE, Quay, Socorro (Nesom, 2006d).

ANAPHALIS DC. – Pearly everlasting

Jennifer Ackerfield

Anaphalis margaritacea (L.) Benth. & Hook.—Pearly everlasting. Caulescent, dioecious or polygamo-dioecious, perennial herbs; plants from slender rhizomes, 2-8(-12) dm; leaves alternate, simple, entire, linear to widely lanceolate, 3-10(-15) cm, tomentose or glabrate below, green and glabrate above; heads discoid; involucral bracts imbricate in several series, outer series ovate, white, and scarious, inner series less scarious; involucre 5-7 mm; receptacle naked; disk florets numerous; ray florets absent; pappus of capillary bristles; cypselae 0.5-1 mm, glabrous. Common in mountain meadows and forest openings from the montane to subalpine, 7000-12,500 ft. NC, NW, C, WC, SC, Grant (Ackerfield, 2015).

ANTENNARIA Gaertn. – Pussy-toes, everlasting

Jennifer Ackerfield and Kenneth D. Heil

Dioecious, white-woolly, perennial herbs; leaves simple, mostly entire; heads unisexual, discoid; involucral bracts imbricate in several series, scarious and dry at least at the tip, white or sometimes colored; receptacle naked; disk florets numerous; ray florets absent; pappus of capillary bristles; cypselae round or compressed (Ackerfield, 2015; Allred et al., 2020; Bayer, 2006).

1. Heads solitary and terminal; plants low, mostly < 10 cm…(2)
1'. Heads few to many in capitate or corymbose clusters; plants usually well < 10 cm…(4)
2. Plants lacking stolons; involucral bracts blackish green or brownish and often with a colorless margin and apex…**A. dimorpha**
2'. Plants with stolons (*careful - these short in A. rosulata!*); involucral bracts (at least inner ones) white or occasionally pink…(3)
3. Leaves 5-10 mm; stems usually < 1 cm; heads scarcely elevated above basal leaves…**A. rosulata**
3'. Leaves 10-35 mm; stems usually > 1 cm; heads usually elevated well above basal leaves…**A. parvifolia**
4. Leaves glabrous and green above, or distinctly less hairy than the lower side…(5)
4'. Leaves white-tomentose on both sides, with ± uniform hairs…(6)
5. Involucral bracts widest above middle, tips abruptly acute; plants generally smaller, usually to 15 cm…**A. marginata**
5'. Involucral bracts narrow and pointed; plants generally more robust and taller, 10-40 cm…**A. neglecta**
6. Plants small and low, to 3 cm, heads scarcely elevated above basal leaves…**A. rosulata**
6'. Plants > 3 cm, heads well above basal leaves…(7)
7. Tips of scarious involucral bracts white or pink (sometimes with a dark spot at base); plants generally from foothills to montane, only seldom alpine…(8)
7'. Tips of scarious involucral bracts (at least outer and middle ones) brown (sometimes streaked with pink) to blackish, rarely tips all white or yellowish; plants generally subalpine and alpine…(11)
8. Involucral bracts with a conspicuous dark spot at base of scarious portion; basal leaves narrowly oblanceolate…**A. corymbosa**
8'. Involucral bracts lacking a conspicuous dark spot at base of scarious portion; basal leaves usually broader, cuneate-oblanceolate or spatulate…(9)
9. Involucre 7-11 mm (heads larger); plants 2-15 cm…**A. parvifolia**
9'. Involucre usually < 8 mm (heads smaller); plants 10-45 cm…(10)
10. Involucral bracts rose or pinkish at least near apex…**A. rosea**
10'. Involucral bracts whitish or yellowish green at apex…**A. microphylla**
11. Outer involucral bracts blackish or blackish green throughout, usually sharp-pointed at apex…**A. media**
11'. Outer involucral bracts brown, or rarely white or yellowish, usually blunt at apex…**A. umbrinella**

Antennaria corymbosa E. E. Nelson—Flat-top pussy-toes. Plants 15-30 cm, with stolons; leaves 3-3.5 cm, narrowly linear, densely white-tomentose; heads few in a small corymb; involucre 4-5 mm, bracts with a brown spot at base with white tips. Common along stream banks, moist meadows, open woods, 8770-12,000 ft. NC, NW.

Antennaria dimorpha (Nutt.) Torr. & A. Gray—Low pussy-toes. Plants 1-3 cm, lacking stolons; leaves linear to oblanceolate, 1-3 cm, white-tomentose; heads solitary; involucre 5-7 mm, bracts brown or blackish green distally. Dry, rocky, open places at lower elevations, often in sagebrush communities, 6000–7650 ft. Rio Arriba, San Juan.

Antennaria marginata Greene [*A. fendleri* Greene]—White-margin pussy-toes. Plants 2–15(–25) cm, with stolons; leaves oblanceolate to spatulate, 1–3 cm, green and glabrous above and white-tomentose below; heads numerous in a compact cluster; involucre 6–9 mm, bracts brownish at base and white above. Found in dry places, often with ponderosa and piñon pine, 5500–10,000 ft. Widespread except E.

Antennaria media Greene—Rocky Mountain pussy-toes. Plants 5–15 cm, with stolons; leaves 0.6–2.5 cm, oblanceolate to spatulate, tomentose; heads numerous in a subcapitate cyme; involucre 4–7 mm, bracts blackish green with white-scarious tips. Common on rocky alpine slopes, 8100–13,000 ft. NC.

Antennaria microphylla Rydb.—Little-leaf pussy-toes. Plants 10–40 cm, with stolons; leaves spatulate, 0.5–1.5 mm, tomentose; heads 3–10, in a corymbose array; involucre 4–8 mm, bracts brown or brownish green at base and white above. Common on open slopes, forest openings, stream banks, sagebrush meadows, 5900–13,000 ft. NC.

Antennaria neglecta Greene [including *A. howellii* Greene subsp. *neodioica* (Greene) R. J. Bayer]—Field pussy-toes. Plants 10–40 cm, with stolons; leaves oblanceolate to spatulate, 1.5–6.5 cm, becoming glabrous above and white-tomentose below; heads several in a dense cyme; involucre 5–9 mm, bracts white above. Open meadows and ponderosa pine forests, 7200–9200 ft. Colfax.

Antennaria parvifolia Nutt.—Small-leaf pussy-toes. Plants 2–15 cm, with stolons; leaves oblanceolate to spatulate, 1–3.5 cm, white-tomentose; heads 3–8 or rarely solitary, in a corymbiform array; involucre 7–11 mm, bracts white or occasionally pinkish. Common in open meadows, rocky slopes, 4600–12,000 ft. Widespread except EC.

Antennaria rosea Greene—Rosy pussy-toes. Plants 10–45 cm, with stolons; leaves oblanceolate to spatulate, 15–20 cm, tomentose; heads 3–20 in corymbiform arrays; involucre 4–7(–10) mm, bracts pink or reddish below and white above. Common in open meadows, rocky slopes, forests, 5000–12,800 ft. NC, NW, C.

Antennaria rosulata Rydb.—Kaibab pussy-toes. Plants 1–3 cm, with short stolons; leaves spatulate, 0.5–1 cm, white-tomentose; heads solitary or 2–3; involucre 5–9 mm, outer bracts green and inner white-scarious. Infrequent on open rocky slopes and in dry meadows, often associated with sagebrush communities, 6500–11,000 ft. NE, NC, NW, Cibola.

Antennaria umbrinella Rydb.—Umber pussy-toes. Plants 7–20 cm, with stolons; leaves spatulate to narrowly oblanceolate, 1–2 cm, white-tomentose; heads 3–8, in a dense cyme; involucre 5–6 mm, outer bracts brown or greenish and inner with white tips. Common on rocky slopes, dry meadows, open forests, 7000–12,400 ft. NC, C, SC.

ANTHEMIS L. – Chamomile, dog fennel

Jennifer Ackerfield

Anthemis cotula L.—Dog fennel. Annuals, often ill-scented, glabrous or somewhat hairy above; leaves 1–2-pinnately lobed into linear or filiform ultimate segments, 25–55 × 15–30 mm; heads radiate (ours); involucral bracts numerous in 2 to several series, imbricate, scarious-margined; involucre 5–9 mm diam., villous; receptacle chaffy throughout or only near center; disk florets 2–3 mm, yellow, sparsely gland-dotted; ray florets 5–15 mm, white, sterile; pappus absent; cypselae 1.3–2 mm, ribs usually tuberculate. Locally common in disturbed places such as roadsides and fields, 8200–8600 ft. Colfax, Sandoval. Introduced (Ackerfield, 2015).

APHANOSTEPHUS DC. - Lazydaisy

Kenneth D. Heil

Annuals or perennials, 5–50 cm; stems erect to decumbent or ascending; leaves basal and cauline, alternate, broadly oblanceolate to linear-lanceolate, margins entire or pinnatifid, faces loosely strigose to hirsute; heads radiate, borne singly; involucres broadly to depressed-hemispherical, ray florets white with purplish midstripe; disk florets yellow with orange resin ducts; cypselae usually 4-angled, faces usually sparsely strigose; pappus 0 or with scales (Nesom, 2006e).

1. Plants perennial…**A. riddellii**
1'. Plants annual or biennial…(2)
2. Cypselae hairs coiled…**A. skirrhobasis**
2'. Cypselae hairs not coiled, straight…**A. ramosissimus**

Aphanostephus ramosissimus DC.—Plains lazydaisy. Annuals; stem hairs spreading to deflexed, 0.2–0.6 mm; phyllary apices acute; ray florets 20–41; disk floret corolla bases not conspicuously swollen and indurate; cypselae hairs apically straight; pappus minutely ciliate crowns 0.1–0.2 mm or absent or nearly so.

1. Involucres 4.7–7.5 mm; cypselae mostly 1.4–1.6 mm; pappus 0 or with minutely ciliate rings, often nearly smooth at apices…**var. humilis**
1'. Involucres 3.2–4.8 mm; cypselae mostly 1.2–1.4 mm; pappus 0.2–0.6 mm, usually minutely ciliate-tipped… **var. ramosissimus**

var. humilis (Benth.) B. L. Turner & Birdsong [*A. arizonicus* A. Gray]—Plains dozedaisy. Plants 6–15 cm; involucres 4.7–7.5 mm; ray florets 26–35; cypselae (1.1–)1.4–1.6 mm; pappus 0 or minutely ciliate rings 0.1–0.2 mm, sometimes nearly smooth. Limestone, sand and gravel bars, creosote and mesquite savannas, disturbed sites, 3500–7025 ft. NC, EC, C, SE, SC, SW,

var. ramosissimus—Plains dozedaisy. Plants 20–45 cm; involucres 3.2–4.8 mm; ray florets 20–41; cypselae (1–)1.2–1.4 mm; pappus coroniform, indurate, cupuliform, 0.2–0.6 mm, usually minutely ciliate. Loam soils, caliche, limestone, oak-juniper openings, mesquite pastures, disturbed sites, 3450–5350 ft. EC, C, SE, SC.

Aphanostephus riddellii Torr. & A. Gray—Riddell's lazydaisy. Perennials, 10–40(–50) cm, caudices usually woody; stem hairs spreading to deflexed or antrorse, 0.2–0.5 mm; phyllary apices long-acuminate; ray florets 30–75; disk floret corolla bases not conspicuously swollen; cypselae hairs apically glochidiate; pappus coroniform, minutely ciliate or lacerate, 0.1–0.3 mm. Calcareous soils, often with oaks, 3500–6200 ft. SE, SW, Doña Ana, Lincoln.

Aphanostephus skirrhobasis (DC.) Trel. ex Coville & Branner [*Keerlia skirrhobasis* DC.]—Arkansas lazydaisy. Annuals, 5–45 cm; stem hairs spreading to deflexed; phyllary apices acute to short-acuminate; ray florets 18–45; disk floret corolla bases (mature) bulbous-swollen; cypselae hairs coiled; pappus usually coroniform, uneven, rarely of acute to awn-tipped scales. New Mexico material belongs to **var. skirrhobasis**. Gravelly, sandy, or silty soils, oak woodlands, mesquite savannas, disturbed sites, 3050–4400 ft. EC, SE, Lincoln.

ARCTIUM L. - Burdock

Jennifer Ackerfield

Arctium minus (Hill) Bernh.—Burdock. Biennial herbs; plants 5–30 dm; basal leaves long-petiolate, petioles hollow, blades 3–6 dm, thinly gray-tomentose below and green above, the margins coarsely dentate; heads in racemose clusters, sessile or on short peduncles to 4 cm, 1.5–4 cm wide; involucral bracts minutely hairy, often glandular, imbricate in several series, each tip with an inward-pointing hooked bristle; receptacle densely bristly; disk florets pink or purplish, rarely white; ray florets absent; pappus 1–3.5 mm; cypselae oblong, slightly compressed, with many nerves. Common in waste places, roadsides, fields, 4500–7500 ft. NE, NC, NW, C, SC. Introduced (Keil, 2006a).

ARNICA L. – Arnica

Jennifer Ackerfield

Perennial herbs; leaves mostly opposite, simple; heads radiate or rarely discoid; involucral bracts herbaceous, ± in 2 series, subequal; receptacle fimbrillate or hirsute; disk florets yellow to orange; ray florets yellow to orange; pappus of numerous white to tawny, barbellate to subplumose, capillary bristles; cypselae cylindrical to fusiform, 5-10-nerved (Ackerfield, 2015).

1. Heads lacking ray florets, or rarely with some very short ray florets…**A. parryi**
1'. Heads with conspicuous ray florets (1-3 cm)…(2)
2. Stem leaves of mostly 5-12 pairs, gradually reduced upward; heads generally 5-20 (rarely 3) per stem…
   A. chamissonis
2'. Stem leaves of mostly 2-4 (rarely 5) pairs (not counting basal rosette if present), often strongly reduced upward; heads < 5 per stem…(3)
3. Leaf blades broad, mostly 1-2.5 times as long as wide, cordate, elliptic, or ovate…**A. cordifolia**
3'. Leaf blades narrow, mostly 3-10 times as long as wide, lanceolate, oblanceolate, or narrowly elliptic…(4)
4. Pappus subplumose, tawny…**A. mollis**
4'. Pappus merely barbellate, white or nearly so…(5)
5. Leaves mostly basal, with tufts of long brown woolly hairs in the axils of basal leaves (very soft to the touch, even if not clearly visible); plants of lower elevations, from foothills to montane…**A. fulgens**
5'. Leaves cauline, 2-4 pairs per stem, without long brown woolly hairs in the axils of basal leaves; montane to alpine…**A. latifolia**

Arnica chamissonis Less.—Chamisso arnica. Plants 2-10 dm, hairy or subtomentose; leaves all cauline, with 5-10 pairs per stem, lanceolate to oblanceolate, with entire to denticulate margins, 5-30 × 1-4 cm, upper sessile and lowermost petiolate; heads 5-15; involucre 8-13 mm, bracts with a subapical tuft of long white hairs internally; rays 1.5-2 cm; pappus barbellate, tawny. Common in mountain meadows and moist places, 8000-9950 ft. NC, NW, Cibola.

Arnica cordifolia Hook.—Heartleaf arnica. Plants 1-6 dm, glandular-hairy to loosely white-hairy; leaves cauline and basal, with 2-4 pairs per stem, cordate, with dentate margins, 4-12 × 3-9 cm, long-petiolate; heads 1-3(-7); involucre 13-20 mm, with spreading white hairs; rays 1.5-3 cm; pappus barbellate, white or stramineous. Common in moist, shaded forests and along streams from foothills to subalpine, 6375-11,500 ft. NC, NW, Sierra, Socorro.

Arnica fulgens Pursh—Foothill arnica. Plants 2-6 dm, glandular-stipitate and often also hairy; leaves cauline and basal, with 2-4 pairs per stem, oblanceolate to narrowly elliptic, entire or nearly so, 3-12 × 1-4 cm, sessile above; heads 1-3; involucre 10-15 mm, glandular and hairy; rays 1.5-2.5 cm; pappus barbellate, white or stramineous. Common in open meadows, rocky slopes, 8300-9350 ft. McKinley, San Juan.

Arnica latifolia Bong.—Broadleaf arnica. Plants 1-6 dm, glandular; leaves cauline, 2-4 pairs per stem, elliptic, usually toothed, 2-14 × 1.5-8 cm, sessile or nearly so; heads 1-3; involucre 10-18 mm, glandular; rays 1-2.5 cm; pappus barbellate, white. Locally common in moist forests, meadows, open places in the mountains, 8800-11,600 ft. Rio Arriba, Taos.

Arnica mollis Hook.—Hairy arnica. Plants 15-70 cm; leaves (2)3(4) pairs, mostly cauline, blades broadly elliptic or narrowly to broadly lanceolate, 4-20 × 1-4 cm, margins entire or irregularly denticulate, faces sparsely to moderately hairy; heads 1 or 3-7; involucres hemispherical to campanulate; rays 10-22; pappus tawny, bristles plumose. Moist meadows surrounded by spruce-fir forest; 1 record in Sangre de Cristo Mountains, 11,000 ft. Taos.

Arnica parryi A. Gray—Parry's arnica. Plants 2-6 dm, glandular above; leaves cauline and basal, 2-4 pairs per stem, lanceolate, entire to toothed, 5-20 × 1.5-6 cm, lowermost petiolate; heads 3-9; involucre 10-14 mm, minutely hairy and glandular; rays absent; pappus barbellate, tawny. Common in open forests and meadows, 8600-10,250 ft. McKinley, Rio Arriba.

ARTEMISIA L. – Sagebrush, wormwood

Kenneth D. Heil

Annuals, biennials, perennials, or shrubs, 3–350 cm, mostly aromatic; stems 1–10, usually erect and branched, glabrous or hairy; leaves basal or basal and cauline, alternate; inflorescence paniculate, racemose, or spicate; heads usually discoid; involucres campanulate, globose, ovoid, or turbinate; phyllaries 2–20, in 4–7 series; receptacles flat, convex, or conic, glabrous or hairy; ray florets absent (subradiate in *A. bigelovii*); disk florets 2–20(–32) per head, bisexual and fertile, or functionally staminate, corollas usually pale yellow, rarely red; cypselae fusiform, glabrous or hairy, often gland-dotted; pappus absent in ours (Ackerfield, 2015; Porter & Clifford, 2013; Shultz, 2006).

1.   Plants shrubs or subshrubs, woody well above base…(2)
1'. Plants herbaceous…(10)
2.   Heads with both ray (subradiate) and disk florets, ray florets 2-lipped; branchlets of inflorescence spreading to reflexed; plants of rimrock areas and other rocky sites…**A. bigelovii**
2'   Heads with disk florets only; branchlets of inflorescence variously disposed; distribution various…(3)
3.   Leaves 1–3-pinnately or ternately dissected, segments linear…(4)
3'. Leaves entire or toothed, or if lobed, lobes oblong or broader, or if linear, tall shrubs of sandy areas at low elevations…(7)
4.   Plants silvery-canescent; receptacle hairy; commonly on windswept ridges, but not always so restricted… **A. frigida**
4'. Plants green to gray-green; receptacle glabrous, or if hairy, plants of low elevations…(5)
5.   Branches spreading, spinescent; flowering in spring…**A. spinescens**
5'. Branches erect or ascending, not spinescent; flowering in late summer and autumn…(6)
6.   Plants commonly 0.5–2 dm; leaves 0.3–1 cm; on clay soils with igneous, calcareous, or dolomitic gravels… **A. pygmaea**
6'. Plants usually 5–15 dm; leaves mainly 0.6–8 cm…**A. filifolia** (in part)
7.   Leaves linear-filiform, < 1 mm wide, entire or 3-pinnately divided; tall plants of sandy low-elevation sites…**A. filifolia** (in part)
7'. Leaves broader, entire or segments > 1 mm wide; habitat and elevation various…(8)
8.   Persistent leaves entire or with 1 or 2 teeth (ephemeral ones often tridentate); heads borne in slender panicles; plants of middle to high elevations…**A. cana**
8'. Persistent and ephemeral leaves toothed or lobed at the apex; heads borne in slender-spicate to broad panicles; plants of low to moderate elevations…(9)
9.   Plants usually < 3 dm; leaves usually < 1 cm, with small glandular swellings showing through tomentum as scattered green dots; foliage dull yellowish to lead-gray or rarely silvery…**A. nova**
9'. Plants mainly > 3 dm; leaves usually > 1 cm, nonglandular; foliage silvery-canescent…**A. tridentata**
10. Involucral bracts with distinct dark brown or black margins; heads few, arranged in a spike or narrow raceme; subalpine and alpine…(11)
10'. Involucral bracts without distinct dark margins; heads and inflorescences various; plants variously distributed…(12)
11. Basal leaves 1-pinnately parted or cleft at the apex; stem leaves entire; heads relatively few, usually 1–5, these larger (6–11 mm broad); disk florets 30–130 per head; corollas glabrous…**A. pattersonii**
11'. Basal leaves bipinnately compound or ternate-pinnatifid; stem leaves pinnatifid to entire; heads 5–12, these smaller (3–6 mm broad); disk florets 10–30 per head; corollas usually hairy…**A. scopulorum**
12. Receptacles with long woolly hairs between the florets…**A. frigida**
12'. Receptacles naked, without long woolly hairs between the florets…(13)
13. Leaves glabrous, not at all tomentose or occasionally sparingly and loosely villous…(14)
13'. Leaves tomentose (at least below), conspicuously villous, or resinous-glandular…(16)
14. Leaves entire to rarely a few cleft; disk florets functionally staminate and sterile, only marginal ones pistillate and fertile…**A. dracunculus**
14'. Leaves variously pinnatifid; disk florets all potentially fertile, marginal ones pistillate and inner ones perfect…(15)
15. Plants small, 0.8–3 dm; perennials from woody caudices or rootstock with well-developed basal leaves; heads larger, 4–7 mm broad…**A. parryi**
15'. Plants tall, 3–30 dm; biennials from well-developed taproots, without well-developed basal leaves; heads 1–2 mm broad…**A. biennis**
16. Stem leaves and a basal cluster of leaves present (these sometimes withered or lacking), pinnatifid (at least at base); stems often reddish-tinged…(17)
16'. Leaves all cauline with no basal cluster, entire to variously pinnatifid; stems green to gray or rarely reddish…(18)

17. Heads mostly 4-5 mm broad with 45-90 disk florets per head; leaves generally bicolored, green and glabrous above, gray-tomentose below…**A. franserioides** (in part)
17'. Heads mostly 1-3 mm broad with 5-30 disk florets per head; leaves not bicolored, green and villous to pilose on both sides…**A. campestris**
18. Leaves 2-3-pinnatifid…**A franserioides** (in part)
18'. Leaves entire, toothed or 1-pinnatifid, or incised…(19)
19. Leaves mostly under 1.5 cm in length, dissected to midrib into slender 0.5-1-mm-wide segments… **A. carruthii**
19'. Leaves mostly over 2 cm in length (if shorter, then inflorescence broad and 3 cm+ wide, not narrow), entire or lobed to dissected in wider (1.5-4 mm at the narrowest) segments…**A. ludoviciana**

Artemisia biennis Willd.—Biennial wormwood. Annuals or biennials, (10-)30-80(-150) cm; stem 1, glabrous; leaves cauline, sessile, blades broadly lanceolate to ovate, 4-10(-13) × 1.5-4 cm, 1-2-pinnately lobed, coarsely toothed, glabrous; heads erect, in leafy paniculate to spicate capitulescences; involucres 2-4 × 2-4 mm; receptacles naked; phyllaries green, glabrous; disk florets 15-40, corollas pale yellow. Disturbed sites, lake beds, pond margins, mudflats, 5800-8900 ft. NC, NW, WC. Introduced.

Artemisia bigelovii A. Gray [*A. petrophila* Wooton & Standl.]—Bigelow sagebrush. Shrubs, 20-40 (-60) cm, mildly aromatic; stems silvery, canescent; leaves persistent, light gray-green, blades narrowly cuneate, 0.5-3 × 0.2-0.5 cm, entire or 3(-5)-lobed; heads nodding, capitulescences, 6-25 × 1-4 cm; involucres globose, 2-3 × 1.5-2.5 mm; phyllaries canescent or tomentose; receptacles naked; disk florets 3-4, usually with 2 perfect florets and 1 marginal pistillate floret. Rimrock, rocky and sandy soils, desert scrub, piñon-juniper woodlands, 3500-8300 ft. Widespread except S.

Artemisia campestris L.—Western sagewort, sand wormwood. Biennials or perennials, 30-100 cm, faintly aromatic; stems 3-5, turning reddish brown, tomentose or glabrous; leaves often with persistent basal rosettes, basal blades 4-12 cm, cauline 2-4 × 0.5-1.5 cm, 2-3-pinnately lobed, surfaces green and glabrous or gray-green and sparsely hairy; heads paniculate, in mostly leafless panicles, 10-22 × 1-3(-7) cm; involucres turbinate, 2-3 × 2-3 mm; phyllaries glabrous or villous-tomentose; receptacles naked; disk florets 5-30.

1. Plants biennial with mostly 1 stem; basal leaves withered at time of flowering…**var. caudata**
1'. Plants perennial with several stems; basal leaves present and not withered at time of flowering…**var. pacifica**

var. caudata (Michx.) E. J. Palmer & Steyerm. [*A. caudata* Michx.; *Oligosporus caudatus* (Michx.) Poljakov]—Field sagewort. Rocky and sandy soils, open meadows and roadsides, (3600-)5900-9800 ft. NC, EC, C, WC.

var. pacifica (Nutt.) M. Peck [*A. pacifica* Nutt.; *Oligosporus pacificus* (Nutt.) Poljakov]—Field sagewort. Sandy soils, grasslands and open meadows, 4500-10,500 ft. NE, NC, NW, C, WC.

Artemisia cana Pursh—Silver sagebrush. Shrubs, 50-90(-150) cm, aromatic; stems (ours) white and sparsely tomentose or brown and glabrous; leaves deciduous, (ours) bright green or dark gray-green, blades (ours) linear to narrowly lanceolate, (1.5-)2-3 × 0.2-0.4 cm, ours irregularly lobed, sparsely hairy or glabrescent, viscid; heads in sparsely paniculate capitulescences; involucres narrowly to broadly campanulate, 3-4 × 2-3(-4) mm; receptacles naked; disk florets 8-16(-20). New Mexico material belongs to **subsp. viscidula** (Osterh.) Beetle. Rocky and sandy sites, arroyo benches, moist meadows, piñon-juniper woodlands, ponderosa pine forests, 5900-8400 ft. Rio Arriba, Sandoval.

Artemisia carruthii Alph. Wood ex J. H. Carruth [*A. bakeri* Greene; *A. coloradensis* Osterh.]— Carruth wormwood. Plants 15-40(-70) cm, faintly aromatic, rhizomatous; stems brown to gray-green, bases somewhat woody, sparsely to densely tomentose; leaves cauline, bicolored, blades narrowly elliptic, 0.1-2.5(-3) × 0.5-1 cm, deeply pinnatifid with 3-5 lobes, densely tomentose; heads usually nodding, in leafy paniculate capitulescences, 10-30 × 3-9 cm; involucres campanulate, 2-3 × 1.5-3 mm; phyllaries gray-tomentose; receptacles naked; disk florets 6-30. Sandy soil, sagebrush, ponderosa pine, mixed conifer forests, 5000-10,600 ft. Widespread except EC, SE.

Artemisia dracunculus L. [*A. aromatica* A. Nelson]—Wild tarragon. Perennials, often subshrubs, 50–120(–150) cm, strongly tarragon-scented or not aromatic, rhizomatous; stems green to brown or reddish brown; cauline leaf blades bright green, linear, lanceolate, or oblong, 1–7 × 0.1–0.5(–0.9) cm, mostly entire; heads in terminal or lateral leafy, paniculate capitulescences, 15–45 × 6–30 cm; involucres globose, 2–3 × 2–3.5(–6) cm; phyllaries light brown; receptacles naked; disk florets 10–30. Desert scrub, piñon-juniper woodlands, ponderosa pine and mixed conifer forests, 3500–10,100 ft. Widespread.

Artemisia filifolia Torr. [*A. plattensis* Nutt.; *Oligosporus filifolius* (Torr.) Poljakov]—Sand sage. Shrubs, 60–180 cm, faintly aromatic; stems green or gray-green, glabrous or sparsely hairy; leaves gray-green, blades linear to obovate, 1.5–5(–6) × 0.1–2.5 cm, entire to 3-lobed, lobes filiform, < 1 mm wide; heads in paniculate capitulescences, 8–15(–17) × 2–5 cm; involucres globose, 1.5–2 × 1.5–2 mm; phyllaries densely hairy; receptacles naked; disk florets 1–6. Sandy soils, dunes, prairies, desert scrub, piñon-juniper woodlands, ponderosa pine forests, 3500–7500(–9500) ft. Widespread.

Artemisia franserioides Greene—Bursage. Biennials or perennials, 30–100 cm, faintly aromatic, rhizomatous; stems reddish brown, leafy, glabrous or glabrate; leaves in basal rosettes and cauline, bicolored, abaxially white and adaxially green, blades ovate, 3–7(–20) × 2–4(–6) cm, 2–3-pinnately lobed; heads nodding, in paniculate to racemose capitulescences, 10–35 × 2–4 cm; involucres broadly ovate, 3–5 × 4–5(–6) mm; phyllaries sparsely hairy; receptacles naked; disk florets 45–90. Talus slopes and cliffs, mixed conifer and spruce-fir forests, 5500–11,000 ft. NC, C, WC, SC, SW.

Artemisia frigida Willd.—Fringed sage. Perennials, 10–40 cm, forming silvery mats or mounds, strongly aromatic; stems gray-green, glabrescent; leaves silvery-gray, blades ovate, 0.5–1.5(–2.5) cm, 1–2-ternately lobed, surfaces densely whitish pubescent; heads in leafy paniculate capitulescences, 0.5–2(–4) × 4–15(–20) cm; involucres globose, 3–5 × (2–)5–6 mm; phyllaries gray-green, densely tomentose; receptacles long-hairy; disk florets 10–50. Rocky sites, desert scrub, piñon-juniper woodlands, ponderosa pine and mixed conifer forests, 4000–9950 ft. Widespread except EC, SE, SW.

Artemisia ludoviciana Nutt.—Silver wormwood, white sage. Perennials, 20–80 cm, rhizomatous, aromatic; stems gray-green, hairy; leaves cauline, gray-green, green, white, or bicolored, blades linear to broadly elliptic, 1.5–11 × 0.5–4 cm, lobed or deeply pinnatifid, hairy; heads erect or nodding, in congested to open paniculate capitulescences; involucres campanulate or turbinate, (1–)2–5 × 2–5(–8) mm; phyllaries gray-green; receptacles naked; disk florets 8–13. Numerous varieties have been described. We recognize 5 intergrading taxa.

1. Leaves deeply lobed, nearly to the midrib…**var. incompta**
1'. Leaves entire or shallowly lobed…(2)
2. Heads in paniculiform arrays, (4–)8–30 cm diam.; leaves mostly 1.5–2 cm…**var. albula**
2'. Heads in paniculiform or racemiform arrays, 1–6 cm diam.; leaves 1.5–11 cm…(3)
3. Leaf margins plane…**var. ludoviciana**
3'. Leaf margins revolute…(4)
4. Leaves bicolored, margins mostly entire, abaxial faces glabrous…**var. redolens**
4'. Leaves gray-green, margins usually lobed, abaxial faces hairy…**var. mexicana**

var. albula (Wooton) Shinners [*A. albula* Wooton]—White wormwood. Leave whitish green, blades entire, margins revolute. Sandy soils, desert scrub, piñon-juniper woodlands, ponderosa pine forests, 4100–9100 ft. Widespread except EC.

var. incompta (Nutt.) Cronquist [*A. incompta* Nutt.]—Mountain wormwood. Leaves bicolored, white or gray-green abaxially and green adaxially, irregularly lobed. Montane meadows, mixed conifer and spruce-fir forests, (5200–)6500–8200 ft. NC, NW, Otero.

var. ludoviciana—Louisiana wormwood. Leaves gray, margins plane. Disturbed sites, meadows, rocky slopes, sagebrush, piñon-juniper woodlands, ponderosa pine and mixed conifer forests, (4900–)5900–10,200 ft. Widespread except SE.

var. mexicana (Willd. ex Spreng.) Fernald [*A. revoluta* Rydb.]—Mexican wormwood. Leaves bicolored, white or gray-green abaxially and green adaxially, margins revolute. Rocky soils, mountain slopes, piñon-juniper woodlands, ponderosa pine forests, 4700–10,300 ft. Widespread except E.

var. redolens (A. Gray) Shinners [*A. redolens* A. Gray]—White sagebrush. Leaves bicolored, white and green, lobed. Rocky soils, ponderosa pine, mixed conifer, and spruce-fir forests, 6200–10,000 ft. NW, C, WC, SC.

Artemisia nova A. Nelson [*A. arbuscula* Nutt. subsp. *nova* (A. Nelson) G. H. Ward]—Black sage. Shrubs, 10–30(–50) cm, aromatic; stems dark brown with age; leaves bright green to dark green or gray-green, 0.5–2 × 0.2–1 cm, apex 3-lobed, surfaces sparsely hairy, gland-dotted; heads arranged in paniculate capitulescences, 4–10 × 0.5–3 cm; involucres turbinate, 2–3 × 2 mm; phyllaries straw-colored or light green; receptacles naked; disk florets (2)3–5(6). Shallow soils and mountain slopes, shadscale, piñon-juniper woodlands, mountain brush communities, 5550–9025 ft. NC, NW.

Artemisia parryi A. Gray [*A. laciniata* Willd. subsp. *parryi* (A. Gray) W. A. Weber]—Parry's sagebrush. Perennials; stems 10–40 cm, mildly aromatic; stems reddish brown, glabrous; leaves mostly basal, greenish, 2–3-pinnate, cauline 1.5–0.8 × 0.2 mm; heads spreading to nodding, spiciform arrays; involucres globose, 3–4 × 4–5 mm; phyllaries greenish or yellowish, glabrous or sparsely hairy; receptacles naked; disk florets 40–50. Rocky soils, subalpine and alpine meadows, 9350–11,000 ft. Colfax, Sandoval, Taos.

Artemisia pattersonii A. Gray [*A. monocephala* (A. Gray) A. Heller]—Patterson sagewort. Perennials, 8–20 cm, mildly aromatic; stems gray-brown, glabrate or finely pubescent; leaves gray-green, blades broadly spatulate, 2–4 × 0.5 cm, pinnately lobed, faces silky-hairy; heads borne singly or 2–5, spreading to nodding, in paniculiform or racemiform arrays; involucres broadly hemispherical, 5–8 × 5–8(–10) mm; phyllaries gray, villous; receptacles long-hairy; disk florets 32–100. Alpine meadows, 11,400–12,700 ft. Santa Fe, Taos.

Artemisia pygmaea A. Gray [*Seriphidium pygmaeum* (A. Gray) W. A. Weber]—Pygmy sage. Shrubs, 5–10 cm, slightly aromatic; stems stiffly erect, pale to light brown, sparsely tomentose; leaves bright green, blades oblong to ovate, 0.3–0.5 × 0.2–0.3 cm, pinnately lobed, glabrous or sparsely tomentose, resinous; heads sessile, in paniculate or racemose capitulescences; involucres narrowly turbinate, 2–3 × 3–4 mm; phyllaries green; receptacles naked; disk florets 3–5. Fine-textured soils of gypsum or shale, sagebrush, piñon-juniper woodlands, 6375–7250 ft. McKinley, Rio Arriba, Taos.

Artemisia scopulorum A. Gray—Alpine sagebrush. Perennials, 10–25 cm, mildly aromatic; stems gray-green, glabrate; leaves gray-green, blades of basal leaves oblanceolate, 2–7 × 0.1 cm, 2-pinnately lobed, silky-canescent; heads 5–22, in spicate capitulescences; involucres globose or subglobose, 3.7–4.4 × 4–7 mm; phyllaries green; receptacles long-hairy; disk florets 15–30. Rocky sites, meadows in subalpine and alpine communities, 11,350–13,160 ft. NC.

Artemisia spinescens D. C. Eaton [*Picrothamnus desertorum* Nutt.]—Budsage. Aromatic subshrubs and shrubs, 5–30(–50) cm; stems erect, some lateral branches persistent, forming thorns, villous to cobwebby; leaves mostly cauline, 15–20 × 1–5(–20), blades orbiculate, simple or 1–2-pedately lobed, surfaces villous and gland-dotted; heads 2–12, mostly in leafy, racemose or spicate capitulescences; involucres obconic, 2–3(–5) mm diam.; phyllaries 5–8, in 2 series; receptacles naked; disk florets 5–20. Sand, clay, and saline soils, desert scrub, greasewood, shadscale, black sagebrush communities, 5100–6950 ft. NW.

Artemisia tridentata Nutt.—Big sage. Aromatic shrubs, 40–200(–300) cm, herbage canescent; stems glabrate, bark gray; leaves gray-green, blades cuneate, (0.4–)0.5–3.5 × 0.1–0.7 cm, 3-lobed, surfaces densely hairy; heads usually erect, arranged in paniculate capitulescences; involucres lanceolate (1–)1.5–4 × 1–3 mm; phyllaries densely tomentose; receptacles naked; disk florets 3–9(–11).

1. Shrubs, 100–200(–300) cm; heads in relatively broad, paniculate capitulescences; involucres 1.5–2.5 × 1–2 mm; usually in sandy soils of valley bottoms, lower montane slopes, and along drainages...**subsp. tridentata**

1'. Shrubs, 30–150 cm; heads in relatively narrow, paniculate capitulescences; mountains, cold desert basins, high plateaus, foothills…(2)
2. Shrubs, 60–80(–150) cm; crowns flat-topped; capitulescences 10–15 cm; involucres 2–3 × 1.5–3 mm; mountains…**subsp. vaseyana**
2'. Shrubs, 30–50(–150) cm; crowns rounded; capitulescences 2–6(–8) cm; involucres (1–)1.5–2 × 1.5–2 mm; cold desert basins, high plateaus, sometimes foothills…**subsp. wyomingensis**

subsp. tridentata—Big sage, Great Basin sagebrush. Plants 100–200(–300) cm; stems with vegetative branches subequal to flowering branches; leaves cuneate or lanceolate, 0.5–1.2(–2.5) × 0.2–0.3 (–0.6) cm, 3-lobed; involucres 1.5–2.5 × 1–2 mm; florets 4–6. Deep, well-drained (often sandy) soils in valley bottoms, lower montane slopes, and along drainages, 5375–7650 ft. NC, NW, Chaves, Grant, Lincoln.

subsp. vaseyana (Rydb.) Beetle [*A. tridentata* var. *pauciflora* Winward & Goodrich]—Mountain sagebrush. Plants 60–80(–150) cm, strongly aromatic, crowns flat-topped; leaves broadly cuneate, 1.2–3.5 × 0.3–0.7 cm, regularly 3-lobed to irregularly toothed; involucres 2–3 × 1.5–3 mm, florets 3–9. Rocky soils, montane meadows, sometimes forested areas, 7050–8700 ft. NC, NW.

subsp. wyomingensis Beetle & A. M. Young—Wyoming sagebrush. Plants 30–50(–150) cm, rounded in outline; leaves narrowly to broadly cuneate, (0.4–)0.7–1.1(–2) × (0.1–)0.2–0.3 cm, 3-lobed; involucres (1–)1.5–2 × 1.5–2 mm; florets 4–8. Rocky or fine-grained soils, cold desert basins to high plateaus, foothills, 5200–8500 ft. NC, NW.

ASANTHUS R. M. King & H. Rob. – Brickellbush

Kenneth D. Heil

Asanthus squamulosus (A. Gray) R. M. King & H. Rob. [*Brickellia squamulosa* A. Gray]—Mule Mountain false brickellbush. Perennials or subshrubs, 30–60(–100) cm; leaves mostly opposite, mostly linear to filiform, blades 3–7 cm × 2–6 mm, faces glabrous or puberulent, usually gland-dotted; heads discoid; florets 8–14, corollas whitish; cypselae 2–3 mm, scabrellous on ribs; pappus of 20–100 barbellate bristles. Pine-oak woodlands, 4200–6500 ft. SW (Nesom, 2006f).

BACCHARIS L. – Groundsel-tree

Kenneth D. Heil

Perennials, subshrubs, or shrubs, 10–400 cm, dioecious (rarely monoecious), usually glabrous, often resinous; bases woody; stems 1–20+; margins entire or coarsely serrate, faces usually glabrous; heads discoid; receptacles flat; functionally staminate florets 10–50, corollas white to pale yellow; pistillate florets 20–150, corollas whitish; cypselae light brown, obovoid to cylindrical, ± compressed; pappus persistent or falling, of 25–50 whitish to tawny, rarely brownish, minutely barbellate bristles (Ackerfield, 2015; Allred & Ivey, 2012; Sundberg & Bogler, 2006).

1. Stems hispidulous, at least distally among heads…**B. brachyphylla**
1'. Stems glabrous or glabrate…(2)
2. Perennials or subshrubs; stems 10–80 cm, much branched from base…(3)
2'. Shrubs; stems mostly 60+ cm…(5)
3. Pistillate involucres 7–9 mm; phyllaries keeled; leaves 10–40 × 1–4 mm; margins finely undulate…
   B. texana
3'. Pistillate involucres 4–5 or 9–14 mm; phyllaries not keeled; leaves 5–30 × 1–2(–7) mm; margins entire, finely serrate or irregularly dentate, not undulate…(4)
4. Pistillate involucres 4–4.5 mm; pappus ca. 4 mm, tawny; leaf margins often irregularly toothed…
   B. havardii
4'. Pistillate involucres 9–14 mm; pappus 15–20 mm, brownish; leaf margins entire or finely serrate…
   B. wrightii
5. Heads in compact racemiform arrays; leaves mostly 5–15 mm, clustered in fascicles…**B. pteronioides**
5'. Heads in ± terminal corymbiform or paniculiform arrays; leaves not as above…(6)
6. Plants broomlike, densely stemmed, branches ± parallel; leaves sparse or absent at flowering; heads borne singly at tips of branches or in paniculiform arrays…**B. sarothroides**

6'.   Plants bushy or sparingly branched, not broomlike; leaves present at flowering; heads in corymbiform or paniculiform arrays…(7)
7.   Leaves obovate or oblanceolate…(8)
7'.   Leaves lanceolate, elliptic, or oblong to narrowly oblanceolate…(9)
8.   Leaves not thickened, margins coarsely serrate or 2-serrate; apical leaves reduced to bracts; pappus 3-4.5 mm…**B. bigelovii**
8'.   Leaves thickened, margins entire or coarsely dentate; apical leaves not reduced to bracts; pappus 6-9 mm…**B. pilularis**
9.   Leaves oblong to narrowly oblanceolate, margins entire or coarsely and irregularly serrate; pappus 8-12 mm…**B. salicina**
9'.   Leaves lanceolate-elliptic or linear-oblanceolate to narrowly oblong, margins finely serrate; pappus 4-6 mm…(10)
10.   Leaves 30-150 mm, margins entire or finely serrate from bases to apices, teeth blunt-tipped… **B. salicifolia**
10'. Leaves 20-40(-80) mm, margins evenly serrate, teeth spinulose…**B. thesioides**

Baccharis bigelovii A. Gray—Bigelow's false willow. Shrubs, 3-10 dm, branched from bases; leaves obovate to oblanceolate, 20-35 × 3-15 mm, margins irregularly incised to coarsely serrate, gland-dotted; heads 20-50 in corymbiform arrays; involucres campanulate, staminate 4-5 mm, pistillate 4-5 mm; staminate florets 15-20, corollas 3.5-4 mm; pistillate florets 25-30, corollas 2-2.6 mm; pappus 3-4.5 mm. Rocky sites in coniferous forests, 5400-7400 ft. SW, Eddy, Torrance.

Baccharis brachyphylla A. Gray—Shortleaf false willow. Shrubs or subshrubs, 2-10 dm, densely branched from bases, stems hispidulous distally near heads; leaves linear to linear-lanceolate, 5-17 × 1-2 mm, margins entire, glabrous; heads in paniculiform arrays; involucres funnelform or campanulate, staminate 3-5.2 mm, pistillate 4-6 mm; staminate florets mostly 12-18, corollas 3.3-4.2 mm; pistillate florets 8-18, corollas 2-2.8 mm; pappus 4.5-7 mm. Canyons, washes, desert scrub, 4400-6300 ft. SW.

Baccharis havardii A. Gray—Havard's false willow. Subshrubs, 15-70 cm, much branched from base; leaves narrowly spatulate to linear, 20-40 × 2-3 mm, margins entire or toothed, gland-dotted; heads in broad paniculiform arrays; involucres cylindro-campanulate, staminate 3 mm, pistillate 4-4.5 mm; staminate florets 12-15, corollas 3-4 mm; pistillate florets 15-20, corollas 3 mm; pappus 4 mm. Calcareous gravels, desert scrub, 5700-7250 ft. Eddy, Torrance.

Baccharis pilularis DC.—Dwarf chaparral false willow. Shrubs, 1.5-45 dm, prostrate and mat-forming to erect, much branched; leaves oblanceolate to obovate, 5-40 × 2-15 mm, entire or coarsely dentate, gland-dotted; heads 100-200+, in leafy paniculiform arrays; involucres hemispherical to campanulate, staminate 3.2-5 mm, pistillate 3-6 mm; staminate florets 20-34, corollas 3-4 mm; pistillate florets 19-43, corollas 2.5-3.5 mm; pappus 6-9 mm. New Mexico material belongs to **subsp. consanguinea** (DC.) C. B. Wolf. Oak woodlands and coniferous forest openings, 3900-4600 ft. Grant.

Baccharis pteronioides DC. [*B. ramulosa* (DC.) A. Gray]—Yerba de pasmo. Shrubs, 3-10 dm, diffusely and evenly branched; leaves linear to lanceolate or spatulate, 5-25 × 1-6 mm, margins sharply serrate, gland-dotted; heads 10-20+ in spreading racemiform arrays; involucres campanulate, staminate 4-5 mm, pistillate 5-6(-7) mm; staminate florets 15-20, corollas 4-5 mm; pistillate florets 15-20, corollas 4-5 mm; pappus 8-10 mm. Oak woodlands and grassland communities, 3800-8900 ft. S, Harding, Quay, Sandoval, San Miguel.

Baccharis salicifolia (Ruiz & Pav.) Pers. [*B. glutinosa* Pers.; *B. viminea* DC.]—Mule's seepwillow. Shrubs, 3-40 dm, stems spreading to ascending; leaves lanceolate-elliptic, willowlike, 30-150 × 3-20 mm, margins usually finely serrate, gland-dotted; heads in terminal, compound corymbiform arrays; involucres hemispherical, staminate 3-6 mm, pistillate 3-6 mm; staminate florets mostly 17-48, corollas 4-6 mm; pistillate florets 50-150, corollas 2-3.5 mm; pappus 3-6 mm. Stream banks, washes, floodplains, riparian woodlands, 3500-6900 ft. WC, SC, SW, Guadalupe, Harding, Quay, San Juan.

Baccharis salicina Torr. & A. Gray—Great Plains false willow. Shrubs, 10-30 dm; leaves oblong to oblanceolate, 2.5-7 cm, margins usually irregularly toothed above, finely gland-dotted; heads 100-200+, in a paniculate array; involucres 4-9 mm, bracts gland-dotted, resinous, otherwise glabrous; staminate florets 20-25, corollas 3-5 mm; pistillate florets 25-30, corollas 3-4 mm; pappus 8-12 mm, whitish. Sandy or saline soil along streams and in montane meadows and grasslands, 3300-9125 ft. Widespread.

Baccharis sarothroides A. Gray—Groundsel desertbloom. Shrubs, 10-40 dm, broomlike; leaves linear-lanceolate, 5-15 × 1-2 mm, margins entire, often revolute, gland-dotted; heads borne singly in dense paniculiform arrays; involucres cylindrical to hemispherical, staminate 4-5.2 mm, pistillate 3-8 mm; staminate florets 18-35, corollas 4.2-5 mm; pistillate florets 19-31, corollas 2.5-3.5 mm; pappus 7-12 mm. Washes, roadsides, mesquite flats, 4000-6200 ft. SC, SW.

Baccharis texana (Torr. & A. Gray) A. Gray—Texas prairie false willow. Perennials or subshrubs, 2.5-6 dm, erect or procumbent; leaves linear to narrowly lanceolate, 10-40 × 1-4 mm, margins minutely undulate, gland-dotted; heads on short peduncles, in loose corymbiform arrays; involucres campanulate, staminate 4-7 mm, pistillate 7-9 mm; staminate florets 15-20, corollas 4-5 mm; pistillate florets 20-30, corollas 3.5-4 mm; pappus 11-14 mm. Dry prairies, mesas, brushy flats; 1 record reported at Laguna Colorada, 7500 ft. Needs verification.

Baccharis thesioides Kunth [*B. alamosana* S. F. Blake]—Arizona baccharis. Shrubs, 10-20 dm, openly branched from base; leaves linear-oblanceolate to narrowly oblong, 20-40(-80) × 4-8 mm, margins evenly serrate, teeth spinulose, gland-dotted; involucres campanulate, staminate 3-6 mm, pistillate 3-6 mm; staminate florets 20-30, corollas 3 mm; pistillate florets 30, corollas 2.2-3 mm; pappus 4-6 mm. Oak-pine forests, 5000-8000 ft. WC, SW.

Baccharis wrightii A. Gray—Wright's false willow. Shrubs or subshrubs, 1-8 dm; leaves oblanceolate to narrowly oblong, 0.5-1(-2.5) cm, with entire or finely toothed margins, not resinous; heads solitary and terminal on each branch; involucres 5-14 mm, glabrous, eglandular; staminate florets 20-30, corollas 4.5-6 mm; pistillate florets 20-30, corollas 3-5 mm; pappus 15-20 mm, tawny or reddish. Sandy, dry soil in plains, rocky places of canyons and mesas, 3475-7375 ft. Widespread.

BAILEYA Harv. & A. Gray ex Torr. – Desert marigold

Kenneth D. Heil

Baileya multiradiata Harv. & A. Gray [*B. pleniradiata* Harv. & A. Gray var. *multiradiata* (Harv. & A. Gray) Kearney; *B. thurberi* Rydb.]—Desert marigold. Biannual or perennial (ours), plants mostly 20-100 cm; stems erect, branched from base, ± throughout, floccose-woolly; leaves alternate, blades lance-linear to broadly ovate, usually pedately to pinnately lobed, margins entire, faces floccose-woolly; heads radiate, borne singly; involucres hemispherical, mostly 5-10 × 10-25 mm; ray florets mostly 34-55, corollas yellow; disk florets 100+, corollas yellow; phyllaries mostly 21-34, floccose-tomentose; cypselae 4 mm, obpyramidal, weakly ribbed or striate, glandular-pubescent; pappus usually 0. Widespread on stony slopes, mesas, sandy plains, 3040-7000 ft. Widespread except N, EC (Turner, 2006).

BARKLEYANTHUS H. Rob. & Brettell – Willow ragwort

Kenneth D. Heil

Barkleyanthus salicifolius (Kunth) H. Rob. & Brettell [*Cineraria salicifolia* Kunth]—Willow ragwort. Shrubs, 100-200(-400+) cm; stems abundantly branched, arching distally; leaves tapering to weakly defined petioles, cauline, alternate, lance-elliptic or lanceolate to lance-linear, margins obscurely dentate to subentire or entire; blades 3-10(-15) × 0.5-1.5 cm; heads radiate, in cymiform or paniculiform arrays, (3-)8-16 in each array, arrays usually clustered; phyllaries 4-6(-8) mm, tips green, obtuse; ray corollas yellow, 8-12+ mm; disk corollas yellow, (5-)7-8+ mm; cypselae ± prismatic to obpyramidal, 5-nerved, (1.5-)2-3 mm; pappus 5-6+ mm. Open rocky and disturbed sites, 4500-4805 ft. Hidalgo (Barkley, 2006a).

BARTLETTIA A. Gray – Bartlettia

Kenneth D. Heil and Steve L. O'Kane, Jr.

Bartlettia scaposa A. Gray—Bartlettia daisy. Annuals, mostly 5–25 cm; leaves mostly basal, sub-opposite or alternate, leaf blades 8–25 × 9–23 mm, petiolate, faces sparsely hirsute or glabrate; heads radiate, borne singly; involucres turbinate to hemispherical, mostly 7–9 mm diam.; receptacles knobby, epaleate; phyllaries 8–22 in 2–3 series; ray florets 5–13, pistillate, fertile, corollas golden yellow, tubes 3–3.5 mm, laminae 7–8 × 3–4 mm; disk florets 30–90, 4–7.5 mm, bisexual, fertile, corollas yellow, sometimes suffused with purple; cypselae 2.5–4 mm, obcompressed (each shed with subtending, linear, membranous scale). Gravelly or adobe soils, grasslands, playas, desert scrub, 4260–4410 ft. Hidalgo, Luna (Strother, 2006b). **R**

BEBBIA Greene – Sweetbush

Kenneth D. Heil

Bebbia juncea (Benth.) Greene [*Carphephorus junceus* Benth.]—Sweetbush. Subshrubs or shrubs, 50–140(–300) cm; stems erect or spreading, glabrous or rough-hairy; leaves cauline, opposite or alternate, sessile (or petiolate), blades linear to narrowly elliptic, 5–65 × 1–6(–12) mm, faces rough-hairy to glabrate; heads usually discoid, borne singly or in corymbiform arrays; involucres ± campanulate or broader, 5–15 mm diam.; phyllaries persistent, 15–30(–40+) in 3–5 series, ovate or elliptic to lanceolate or lance-linear, strongly unequal; receptacles convex, paleate, paleae persistent, lanceolate to linear-lanceolate or linear-elliptic; ray florets 0; disk florets usually 20–50, corollas yellow to orange, 4.5–9 mm, lobes 5; cypselae ± compressed, ± 3-angled; pappus of 15–25+ plumose bristles. Our plants belong to **var. aspera** Greene. Rocky and sandy soils, dry hillsides, slopes, desert washes, canyons, 4500–6000 ft. Doña Ana, Luna (Whalen, 2006).

BERLANDIERA DC. – Greeneyes

Kenneth D. Heil

Perennials or subshrubs, 8–120 cm; stems decumbent to erect, usually branched; leaves basal and/or cauline, alternate; heads radiate, borne singly or in paniculiform to corymbiform arrays; involucres 12–30 mm diam.; phyllaries usually 14–22 in 2–3+ series; receptacles ± turbinate; ray florets (2–)8(–13), corollas pale yellow to orange-yellow, abaxially green or red to maroon; disk florets 80–200+, corollas yellow or red to maroon; cypselae obcompressed or obflattened, mostly obovate; pappus 0 (Pinkava, 2006).

1. Leaf blades lyrate-pinnatifid, faces velvety; peduncles hairy with many reddish, wartlike hairs surpassing appressed white hairs…**B. lyrata**
1'. Leaf blades elongate-deltate, lanceolate, or ovate; peduncles without wartlike hairs, hairs hirsute, grayish to reddish, erect or spreading…(2)
2. Leaves mostly basal; disk florets yellow…**B. macvaughii**
2'. Leaves cauline; disk florets red to maroon…**B. texana**

Berlandiera lyrata Benth. [*B. incisa* Torr. & A. Gray]—Chocolate flower. Plants 10–60(–120) cm; stems erect to decumbent, usually branched; leaf blades oblanceolate or obovate to spatulate, often lyrate, sometimes ± pinnatifid, faces ± velvety; heads in corymbiform arrays; peduncles hairy with some hairs reddish, bulbous-based, wartlike; involucres 13–17 mm diam.; ray corollas deep yellow to orange-yellow; disk corollas red to maroon (rarely yellow); cypselae 4.5–6 × 2.7–3.7 mm. Dry, sandy loams, rocky, limestone soils, roadsides, grasslands with mesquite, oak, and juniper, 3250–8000 ft. Widespread except NW.

Berlandiera macvaughii B. L. Turner—Plants 15–35 cm; leaves mostly basal, cauline leaves mostly absent, blades ovate to linear-oblanceolate, sparsely pubescent above, more so below, margins crenulate; heads 1–2 per peduncle; ray florets 8, corollas yellow; pappus absent; cypselae glabrous dorsally, pubescent ventrally, 5–6 mm. Limestone, Guadalupe Mountains; 1 record, 7200 ft. Eddy. **R**

Berlandiera texana DC. [*B. longifolia* Nutt.]—Texas greeneyes. Plants to 120 cm; stems (erect, usually suffrutescent) much branched distally; leaf blades elongate-deltate to lanceolate, margins dentate, serrate, or 2-serrate, faces hirsute to ± scabrous; heads in paniculiform or corymbiform arrays; peduncles densely hirsute; involucres 18–27 mm diam.; ray corollas deep yellow to orange-yellow; disk corollas red to maroon; cypselae ovate, 4.5–6 × 3–4.8 mm. Dry, rocky, calcareous and sandy soils, 3600–4050 ft. Chaves, Quay, Roosevelt.

BIDENS L. – Beggar-ticks

Kenneth D. Heil and Jennifer Ackerfield

Herbs; leaves opposite; heads discoid or radiate; involucral bracts in 2 distinct series, the outer herbaceous and the inner usually striate; receptacles chaffy; disk florets yellow or orange; ray florets (if present) neutral or pistillate, yellow to white or pink; pappus of 2–4 awns or teeth, these usually retrorsely barbed; cypselae flattened or sometimes 4-angled, hairy (Ackerfield, 2015; Allred & Ivey, 2012; Sherff, 1937; Strother & Weedon, 2006). Key adapted from Strother and Weedon (2006) and Allred and Ivey (2012).

1. Leaves simple, entire or merely toothed…(2)
1′. Leaves pinnately dissected or pinnately or trifoliately compound…(5)
2. Leaves petiolate, petioles 3–25 mm and wing-margined…**B. tripartita** (in part)
2′. Leaves sessile…(3)
3. Bractlets subtending heads 7–35+ mm…**B. tripartita** (in part)
3′. Bractlets subtending heads mostly 3–12 mm, sometimes longer…(4)
4. Ray florets 2–15 mm; chaff of receptacle yellowish at tips, margins of cypselae noticeably thickened or winged…**B. cernua**
4′. Ray florets 15–25 mm; chaff of receptacle reddish at tips, margins of cypselae not noticeably thickened or winged…**B. laevis**
5. Cypselae 4-angled, linear-fusiform, thickened near the middle…(6)
5′. Cypselae flattened, oblanceolate, thickest toward the tips…(13)
6. Heads narrow, 1–3(–4) mm wide, disk florets 5–10…(7)
6′. Heads broader, (3–)4–8 mm wide; disk florets 10–50…(10)
7. Leaf lobes 5–15 mm wide…**B. bigelovii** (in part)
7′. Leaf lobes 0.5–5 mm wide…(8)
8. Bractlets subtending heads spatulate to linear, sometimes lobed and foliaceous, 3–20 mm, usually surpassing phyllaries…**B. lemmonii**
8′. Bractlets subtending heads linear, 1–4 mm, not surpassing phyllaries…(9)
9. Ultimate leaf lobes 2–3+ mm wide; cypselae hairy, at least toward apex…**B. leptocephala**
9′. Ultimate leaf lobes 0.5–2 mm wide; cypselae usually glabrous…**B. heterosperma**
10. Ultimate leaf lobes linear, 5–15 × 1–3 mm…**B. tenuisecta**
10′. Ultimate leaf lobes larger, 10–30+ × 5–25+ mm…(11)
11. Leaves 1-pinnately lobed, lobes or leaflets mostly 10–40 mm wide…**B. pilosa**
11′. Leaves mostly 2–3-pinnately lobed, ultimate lobes mostly 5–20 mm wide…(12)
12. Outer cypselae 6–7 mm, inner 10–14 mm; pappus of 2–3 awns…**B. bigelovii** (in part)
12′. Outer cypselae 7–15 mm, inner 12–18 mm; pappus of 3–4 awns…**B. bipinnata**
13. Ray florets 8, rays 10–25 mm…**B. polylepis**
13′. Ray florets absent or 1–5, rays (when present) 2–4 mm…**B. frondosa**

Bidens bigelovii A. Gray—Bigelow's beggar-ticks. Annuals, 2–8 dm, glabrous or nearly so; leaves 2.5–9 cm, 2–3-pinnatifid; involucre 2.5–7 mm, outer bracts linear and inner lanceolate; rays 0 or 1–3(–7) mm, whitish; pappus of 2 retrorsely barbed awns 2–4 mm; cypselae narrowly linear in middle, outer ones linear-cuneate. Wet soil along streams and pond edges, drier soil of canyons and hillsides, 4000–8200 ft. S, San Miguel, Union.

Bidens bipinnata L.—Spanish needles. Annuals, (1.5–)3–10(–15) dm; leaves 20–50 mm, rounded-deltate to ovate or lanceolate; involucres ± campanulate, 5–7 × 3–4(–5) mm; rays 0 or 3–5+, yellowish or whitish; pappus of 2–4 erect to divergent, retrorsely barbed awns 2–4 mm; outer cypselae weakly obcompressed, inner ± 4-angled, linear to linear-fusiform. Disturbed wettish sites, fields, forests, 4300–8500 ft. C, WC, SC, Grant, Union. Introduced.

Bidens cernua L.—Nodding beggar-ticks. Annuals, 2-10(-40) dm, glabrous or stems scabrous-hispid; leaves lanceolate to lance-ovate, sessile, margins usually toothed, 4-20 cm; involucres of lance-linear to ovate bracts 2-10 mm; rays 2-15 mm, yellow or sometimes absent; pappus of (2-)4 retrorsely barbed awns 2-4 mm; cypselae 4-angled, narrowly cuneate. Common along streams, ditches, or in other low, wet places or disturbed areas, 4000-8800 ft. Widespread except SE, SC. Introduced.

Bidens frondosa L.—Devil's beggar-ticks. Annuals, 2-12(-18) dm, glabrous or nearly so; leaves pinnately compound into 3-5 lanceolate or lance-ovate segments, petiolate, 3-20 cm; involucres of linear-spatulate outer bracts and ovate to lanceolate inner bracts; rays 0 or 2-3.5 mm, yellow; pappus of 2 antrorsely or retrorsely barbed awns 2-5 mm; cypselae narrowly cuneate, flat, with 1 nerve on each face. Common in disturbed areas, along ditches, especially in wet soil, 3850-7500 ft. NC, NW, C, WC, Doña Ana, Grant, Harding. Introduced.

Bidens heterosperma A. Gray—Rocky Mountain beggar-ticks. Annuals, (5-)10-30(-60+) cm; leaves with petioles 3-20 mm, blades usually rounded-deltate, 1-2-pinnatisect; involucres ± cylindrical; rays 0 or 1-3+, yellowish; pappus 0 or of 2-3 spreading to divergent, retrorsely barbed awns 1-2 mm; outer cypselae ± obcompressed and linear, inner similar, ± equally 4-angled, linear-fusiform. Seeps on rocky slopes, 5500-8550 ft. SC, SW, Torrance.

Bidens laevis (L.) Britton, Sterns & Poggenb.—Smooth beggar-ticks. Annuals, (1-)2-6(-12+) dm; leaves sessile, blades obovate or elliptic to lanceolate or linear; involucres turbinate to hemispherical; rays mostly 7-8, rarely 0, orange-yellow; pappus of 2-4 ± erect, retrorsely barbed awns 3-5 mm; cypselae obcompressed, flattened or unequally 3-4-angled, outer 6-8 mm, inner 8-10 mm. Meadows, marshes, margins of ponds and streams, 3900-5500 ft. Doña Ana, Grant, Hidalgo, Lincoln, San Juan, Socorro.

Bidens lemmonii A. Gray—Lemmon's beggar-ticks. Annuals, (1-)1.5-2.5(-3+) dm; leaves with petioles 10-20 mm, blades oblanceolate to linear or rounded-deltate; involucres campanulate to cylindrical; rays 0 or 1-3, whitish; pappus of 2-3 erect, retrorsely barbed awns 1-3 mm; outer cypselae ± equally 4-angled, linear-fusiform, inner similar. Wettish spots on rocky slopes, 4000-7300 ft. Lincoln.

Bidens leptocephala Sherff—Few-flower beggar-ticks. Annuals, 1-2.5(-5) dm; leaves with petioles 5-40 mm, blades rounded-deltate; rays 0 or 2-3, yellowish or whitish; involucres ± cylindrical; pappus of 2-3 erect, retrorsely barbed awns 1-3 mm; outer cypselae flat or unequally 4-angled, linear; inner similar, linear-fusiform. Along streams, 4075-6900 ft. SC, SW, Lincoln.

Bidens pilosa L.—Hairy beggar-ticks. Annuals, (1-)3-6(-18-25) dm; leaves with petioles mostly 10-30 mm, blades either ovate to lanceolate or 1-pinnately lobed, ovate to lanceolate; involucres turbinate to campanulate; rays 0 or 3-8+, whitish to pinkish or yellowish; pappus of 2 ± erect, antrorsely barbed or smooth awns 1-2.4 mm; cypselae flattened, linear to narrowly cuneate. Ponds, swamps, other wet sites, 5100-7000 ft. WC, Bernalillo, San Miguel. Introduced.

Bidens polylepis S. F. Blake—Bearded beggar-ticks. Annuals or biennials, 3-10 dm; leaves with petioles 5-30 mm, blades deltate to ovate; involucres turbinate to hemispherical; rays 8, golden yellow; pappus 0, or 2 divergent, retrorsely barbed awns or ± deltate scales; cypselae flattened, obovate or oblanceolate to cuneate. Marshes, bogs, floodplains, disturbed sites; 1 record, 3955 ft. Eddy.

Bidens tenuisecta A. Gray—Slimlobe beggar-ticks. Annuals, 1-4 dm, glabrous or nearly so; leaves pinnately 2-3 times dissected into small, linear segments 2-5 cm, petiolate; involucres of lance-linear bracts, hispid-hirsute; rays 4-6 mm, yellow; pappus of 2 retrorsely barbed awns 1-3 mm; cypselae weakly 4-angled, coarsely ribbed. Wet soil along streams and ditches, roadsides, disturbed places, 5900-10,000 ft. Scattered except SE.

Bidens tripartita L. [*B. comosa* (A. Gray) Wiegand]—Straw-stem beggar-ticks. Annuals, 2-15 dm, glabrous or nearly so; leaves elliptic-lanceolate, 3-15 cm, sessile or petiolate; involucres of ovate bracts; rays 0 or 4-8 mm, yellow; pappus absent or of 2-3 retrorsely barbed awns 2-3(-6) mm; cypselae very flat, usually smooth. Along streams, ditches, moist disturbed areas, 4900-9885 ft. Scattered except SC, SE.

BRICKELLIA Elliott – Brickellbush

Randall W. Scott

Shrubs, subshrubs, or perennials (annuals); leaves cauline, opposite or alternate, simple; heads discoid; involucres cylindrical to obconic or campanulate; phyllaries persistent in mostly 4–8 unequal series; receptacles naked; corollas usually white to cream, sometimes greenish, purplish, or yellowish; cypselae narrowly prismatic, 10-ribbed; pappus of 10–80 smooth or barbellulate to barbellate, sometimes plumose or subplumose bristles in a single series (Scott, 2006a).

1.   Pappus bristles plumose or subplumose, rarely barbellate…(2)
1'. Pappus bristles smooth or barbellulate to barbellate, sometimes subplumose…(3)
2.   Phyllaries puberulent, often densely gland-dotted as well…**B. eupatorioides**
2'. Phyllaries glabrous or puberulent (not gland-dotted)…**B. brachyphylla**
3.   Florets (40–)45–90…(4)
3'. Florets 3–30(–50)…(5)
4.   Petioles 10–70 mm; peduncles 4–30 mm; involucres 7–12 mm; phyllaries 30–40; corollas 6.5–7.5 mm…
     B. grandiflora
4'. Petioles 3–20 mm; peduncles 20–50 mm; involucres 12–14 mm; phyllaries 40–60; corollas 9–10 mm…
     B. simplex
5.   Florets 3–7…B. longifolia
5'. Florets 8–30(–50)…(6)
6.   Petioles 4–70 mm…(7)
6'. Petioles 0–5 mm…(14)
7.   Leaf apices acuminate or long-acuminate to attenuate…(8)
7'. Leaf apices acute to obtuse or rounded…(10)
8.   Outer phyllary apices obtuse to acute (cypselae 2–2.5 mm)…**B. rusbyi**
8'. Outer phyllary apices acute, acuminate, or mucronate…(9)
9.   Shrubs; leaf margins subentire or ± toothed (usually with 1–3 sets of sharp teeth near base); phyllaries 17–22…**B. coulteri**
9'. Perennials; leaf margins crenate, dentate, or serrate; phyllaries 28–40…**B. grandiflora**
10. Perennials; stems and peduncles stipitate-glandular…(11)
10'. Shrubs; stems and peduncles pubescent or glandular-pubescent…(13)
11. Leaves opposite, leaf bases cordate to cordate-clasping…**B. amplexicaulis** (in part)
11'. Leaves alternate, leaf bases cordate to cuneate…(12)
12. Leaf bases truncate to cordate; petioles 20–35 mm; outer phyllaries ovate to lanceolate, shorter than inner, glandular-hirtellous…**B. floribunda**
12'. Leaf bases rounded to cuneate; petioles 5–10 mm; outer phyllaries ovate to lance-ovate or lanceolate, equaling or surpassing inner, glabrous or sparsely glandular-hirtellous…**B. chenopodina**
13. Leaf bases cuneate, margins laciniate-dentate; peduncles 5–40 mm…**B. baccharidea**
13'. Leaf bases cordate to truncate, margins crenate to serrate; peduncles 1–5 mm…**B. californica**
14. Leaves mostly opposite…(15)
14'. Leaves mostly alternate…(19)
15. Stems and leaves puberulent to pubescent, rarely gland-dotted (not glandular-pubescent)…(16)
15'. Stems and leaves puberulent to pubescent and gland-dotted (glandular-pubescent) or stipitate-glandular…(17)
16. Perennials, 12–30 cm; leaves deltate-ovate to lanceolate, 7–30 × 4–28 mm…**B. parvula**
16'. Perennials, mostly 30–90 cm; leaves oblong, obovate, or ovate, (15–)30–80 × 10–35 mm…
     B. betonicifolia
17. Leaf blades oblong to linear, 2–10 mm wide, margins entire or dentate; peduncles (20–)30–150 mm…
     B. venosa
17'. Leaf blades elliptic, lanceolate, oblong, obovate, ovate, or subdeltate, mostly 10–75 mm wide, margins crenate to serrate; peduncles 0–15(–25) mm…(18)
18. Leaf bases rounded to cordate-clasping; peduncles densely stipitate-glandular…**B. amplexicaulis** (in part)
18'. Leaf bases ± cuneate; peduncles densely tomentose, often gland-dotted…**B. lemmonii** (in part)
19. Florets 25–50; leaf blades elliptic, oblong, or lance-linear, margins entire…**B. oblongifolia**
19'. Florets 8–28(–34); leaf blades ovate to suborbiculate; margins entire or dentate…(20)
20. Leaves mostly cordate or ovate to suborbiculate (if oblong, peduncles viscid or glandular-villous)…(21)
20'. Leaves elliptic, lanceolate, lance-ovate, obovate, ovate, or subdeltate…**B. lemmonii** (in part)
21. Peduncles (bracteate) 2–12 mm, viscid to glandular-villous…**B. microphylla**
21'. Peduncles 0.5–2.5 mm, glandular-viscid…**B. laciniata**

Brickellia amplexicaulis B. L. Rob.—Earleaf brickellbush. Perennial; leaves opposite, blades lance-olate to oblong, 40–160 × 10–75 mm, faces densely to sparsely pubescent and stipitate-glandular; heads erect, in paniculiform arrays; involucres cylindrical to campanulate, 9–13 mm; phyllaries 18–24 in 4–5 unequal series, outer bracts ovate; corollas pale yellow or cream, often purple or red-tinged, 6–8 mm; cypselae 2.8–4 mm; pappus of 35–60 white, barbellulate bristles. A single record on a rocky, rhyolitic slope in an oak community, 5550 ft. Hidalgo.

Brickellia baccharidea A. Gray—Resinleaf brickellbush. Shrubs; leaves alternate, blades rhomboid-ovate, 10–40 × 10–20 mm, faces glabrate or glandular; heads mostly erect, in dense paniculiform arrays; involucres narrowly campanulate, 7–11 mm; phyllaries 35–40 in 5–7 unequal series, outer bracts ovate, inner lanceolate; corollas cream-colored, 6–7 mm; cypselae 2–3 mm; pappus of 18–24 white, barbellu-late bristles. Sandy, granitic soils, limestone slopes, granitic cliffs in desert scrub communities, 4250–6600 ft. Doña Ana, Hidalgo, Luna.

Brickellia betonicifolia A. Gray—Betonyleaf brickellbush. Perennials; leaves opposite, blades ovate, oblong, or obovate, 15–80 × 10–35 mm, faces sparsely to densely pubescent and glandular; heads erect, usually in loose paniculiform arrays, involucres cylindrical to narrowly campanulate, 8–12 mm; phyllaries 17–20 in 3–5 unequal series; outer bracts lance-ovate, sparsely pubescent, inner narrowly lanceolate, glabrous; corollas purplish, 7–9 mm; cypselae 3–4 mm; pappus of 40–52 white, barbellu-late bristles. Open, limestone hillsides and creek drainages, piñon-juniper, ponderosa pine-oak to mixed conifer communities, 5275–8500 ft. SW, Catron, Lincoln.

Brickellia brachyphylla (A. Gray) A. Gray—Plumed or resinleaf brickellbush. Perennials; leaves mostly alternate, occasionally subopposite; blades lanceolate or lance-ovate, 10–60 × 4–20 mm, faces sparsely to densely pubescent, glandular; heads erect, in racemiform to paniculiform arrays, rarely soli-tary; involucres cylindrical to campanulate, 8–11 mm; phyllaries 15–20 in 4–5 unequal series, 5–9 mm; outer bracts lance-ovate, puberulent, inner bracts narrowly lanceolate, glabrous; corollas pale yellow-green, often purple-tinged, 4.5–6 mm; cypselae 2.5–5.3 mm; pappus of 27–32 white, plumose, rarely barbellate, bristles. Limestone cliffs, rocky ridges, canyon walls, hillsides, riparian areas, piñon-juniper, oak, ponderosa pine–oak, mixed conifer communities, 3930–9600 ft. Widespread.

Brickellia californica (Torr. & A. Gray) A. Gray—California brickellbush. Shrubs; leaves alternate, blades ovate to deltate, faces puberulent to glabrate, often glandular; heads mostly erect, borne in leafy paniculiform arrays; involucres cylindrical to obconic, 7–12 mm; phyllaries 21–35 in 5–6 unequal series; outer bracts ovate to lance-ovate, glabrous to sparsely pubescent-glandular, inner bracts lanceolate, glabrous; corollas pale yellow-green, 5.5–8 mm; cypselae 2.5–3.5 mm; pappus of 24–30 white, barbel-late bristles. Dry, rocky hillsides, arroyos, canyons, desert scrub, grasslands, piñon-juniper, oak, ponder-osa pine–oak, mixed conifer communities, 3600–10,000 ft. Widespread.

Brickellia chenopodina (Greene) B. L. Rob.—Chenopod brickellbush. Perennial; leaves alternate, blades rhomboid-ovate to lanceolate, 20–50 × 10–40 mm, faces glabrous or sparsely glandular; heads erect, borne singly, terminating elongate lateral branches; involucres cylindrical to campanulate, 8–9 mm; phyllaries 22–26 in 5–6 subequal or unequal series; outer bracts lanceolate to lance-ovate, inner lanceolate to linear-lanceolate; corollas pale yellow-green or greenish white, 5–6 mm; cypselae 2.5–3 mm; pappus of 30–35 white, smooth to barbellulate bristles. Rare; known only from Gila River bottom near Cliff, Grant County, 4500 ft. Listed as Extinct in the Wild in IUCN Red Book 1997 (Gillett & Walter, 1998). *Brickellia chenopodina* is very similar to *B. floribunda* and may be little more than a shade form of the latter, with which it is sympatric.

Brickellia coulteri A. Gray—Coulter's brickellbush. Shrub; leaves opposite, blades broadly ovate to deltoid, 20–60 × 10–30 mm, faces minutely pubescent; heads erect, in loose, paniculiform arrays; invo-lucres cylindrical to campanulate, 8–12 mm; phyllaries 17–22 in 4–6 unequal series, outer lance-ovate to narrowly lanceolate, inner narrowly lanceolate to linear, corollas pale yellow-green, often purple-tinged, 7–8.2 mm; cypselae 3–5 mm; pappus of 28–40 barbellulate to smooth bristles.

1. Midstem leaves mostly 20-40(-50) mm; peduncles 5-25 mm, mostly lacking glandular hairs...**var. brachiata**
1'. Midstem leaves mostly 40-60 mm; peduncles mostly 15-40 mm, mostly glandular-pubescent...**var. coulteri**

var. brachiata (A. Gray) B. L. Turner—Leaf blades 10-40(-50) × 10-20 mm, midstem blades mostly 20-40(-50) mm; peduncles 5-25 mm, pubescent, mostly lacking glandular hairs. Among rocks and cracks in cliffs; desert scrub, 3200-5000 ft. SC, SW.

var. coulteri—Leaf blades 20-60 × 10-30 mm, midstem blades mostly 40-60 mm; peduncles mostly 15-40 mm, mostly glandular-pubescent. Desert scrub, sandy soils, among rocks and cracks in cliff faces, 4200-4600 ft. Luna.

Brickellia eupatorioides (L.) Shinners [*Kuhnia eupatorioides* L.]—False boneset. Perennials; leaves mostly opposite, blades lanceolate, lance-linear, lance-ovate, linear, oblong, or lance-rhombic, 25-100 × 0.5-40 mm, faces glandular-pubescent; heads erect, in paniculiform or corymbiform arrays; involucres cylindrical to narrowly campanulate, 7-15 mm; phyllaries 22-26 in 4-6 unequal series, outer bracts ovate to lance-ovate, inner lanceolate; corollas pale yellow, yellow-green, pinkish lavender, or maroon, 4.5-6 mm; cypselae 2.7-5.5 mm; pappus of 20-28 white or tawny, usually plumose or subplumose bristles.

1. Leaves linear or linear-lanceolate, 0.5-3 mm wide, usually 1-nerved from base; petioles 0-1 mm; heads solitary or in loose paniculiform clusters of 2-3 heads...**var. chlorolepis**
1'. Leaves linear-lanceolate, lanceolate, lance-ovate, or lance-rhombic, 1-40 mm wide, 1-3-nerved from base; petioles 0-10 mm; heads mostly in dense corymbiform clusters of 3-8 heads...**var. corymbulosa**

var. chlorolepis (Wooton & Standl.) B. L. Turner [*B. chlorolepis* (Wooton & Standl.) Shinners; *Kuhnia chlorolepis* Wooton & Standl.]—Leaves opposite or alternate, blades linear to linear-lanceolate, single-nerved from base, 5-90 × 1-3.5 mm; heads in paniculiform arrays; involucres 7-14 mm; cypselae 4-6 mm. Open and wooded areas on a wide range of soil types; desert scrub, grasslands, piñon-juniper, oak, ponderosa pine-oak, mixed conifer communities, 3500-9000 ft. Widespread except NW.

var. corymbulosa (Torr. & A. Gray) Shinners [*Kuhnia eupatorioides* L. var. *corymbulosa* Torr. & A. Gray]—Leaves opposite, blades linear-lanceolate to lanceolate, 1- or 3-nerved from base, 5-90 × 2-15 mm; heads in paniculiform or corymbiform arrays; involucres 8-15 mm; cypselae 2.7-5 mm. Open areas, wide range of soil types; desert scrub, piñon-juniper, 5000-7500 ft. Colfax, Grant. A widespread taxon that intergrades with var. *chlorolepis* in E New Mexico.

Brickellia floribunda A. Gray—Chihuahuan brickellbush. Perennials, bases woody; leaves alternate, blades deltoid-ovate or rhomboid-ovate, 20-140 × 10-100 mm, faces glandular-pubescent; heads nodding, in paniculiform arrays; involucres cylindrical to campanulate, 7.5-8.5 mm; phyllaries 22-26 in 5-6 unequal series, outer bracts ovate to lanceolate, inner linear-lanceolate; corollas pale yellow-green or greenish white, 5-6.5 mm; cypselae 2-3 mm; pappus of 30-35 white, barbellulate bristles. Near streams, dry arroyos, and in canyon bottoms on limestone rocks; piñon-juniper-oak and ponderosa pine-oak communities, 4200-7500 ft. NW, SC, SW, Sandoval, Torrance.

Brickellia grandiflora (Hook.) Nutt.—Tasselflower brickellbush. Perennials; leaves opposite or alternate; blades deltoid-ovate, lance-ovate, or subcordate, 15-120 × 20-70 mm, faces glandular, puberulent; heads nodding, borne in loose corymbiform or paniculiform arrays; involucres cylindrical or obconic, 7-12 mm; phyllaries 30-40 in 5-7 unequal series, outer bracts lance-ovate to lanceolate, inner lanceolate to lance-linear; corollas pale yellow-green, 6.5-7.5 mm; cypselae 4-5 mm; pappus of 20-30 white, barbellate bristles. Rocky hillsides, forests, dry slopes, canyons, riparian areas, piñon-juniper-oak, ponderosa pine-oak, mixed conifer, spruce-bristlecone communities, 4770-12,200 ft. Widespread except EC.

Brickellia laciniata A. Gray—Splitleaf brickellbush. Shrubs; leaves alternate, blades oblong to broadly ovate, 6-15 × 5-9 mm, faces sparsely pubescent and glandular-punctate; heads mostly erect, in dense, leafy, racemose or paniculiform arrays; involucres cylindrical, 6-8 mm; phyllaries 15-25 in 4-6 unequal series, outer bracts ovate, inner bracts lanceolate; corollas 4.8-5.7 mm, pale yellow-green, often purple-tinged; cypselae 2.5–3 mm; pappus of 24–28 tawny, barbellate bristles. Dry slopes and stream-beds (arroyos), riparian areas, desert scrub communities, 3400–6000 ft. SE, SC, Grant, Quay, Sierra.

Brickellia lemmonii A. Gray [*B. viejensis* Flyr]—Lemmon's brickellbush. Perennials; leaves alternate, occasionally opposite, blades elliptic to lanceolate, 20–70 × 7–20 mm, faces pubescent-glandular with raised reticulate venation; heads erect, in paniculiform arrays; involucres narrowly cylindrical, 10–13 mm; phyllaries 18–20 in 3–4(5) uneven series, outer bracts ovate, inner bracts narrowly lanceolate to linear-oblong; corollas pale yellow-green, 6–8.2 mm; cypselae 2.5–4 mm; pappus of 32–43 white, barbellate to subplumose bristles.

1. Leaf blades lanceolate to ovate, length mostly 2–5 times width; inner phyllaries mostly acute...**var. lemmonii**
1′. Leaf blades broadly ovate to subdeltoid, length mostly 1–2 times width, inner phyllaries mostly obtuse to nearly rounded [to be expected]...var. conduplicata

var. conduplicata (B. L. Rob.) B. L. Turner [*B. conduplicata* B. L. Rob.]—Leaf blades broadly ovate to subdeltate, 20–60 × 7–30 mm, length 1–2 times width; peduncles 2–22 mm; phyllaries (inner) with apices mostly obtuse to nearly rounded. Igneous soils, montane and canyon slopes, 4500–7000 ft. No records seen, to be expected.

var. lemmonii—Leaf blades elliptic to lanceolate, 20–70 × 7–20 mm, length mostly 2–5 times width; peduncles 2–15 mm; phyllaries (inner) with apices mostly acute. Igneous soils, montane and canyon slopes and bottoms, desert scrub and juniper-oak communities, 4500–6800 ft. SW, Socorro.

Brickellia longifolia S. Watson—Longleaf brickellbush. Shrubs, 20–200(–250) cm; leaves alternate, blades 3-nerved from bases, lance-elliptic, lance-ovate, lance-linear, or linear (folded or falcate), 10–130 × 2–9 mm, margins entire or nearly so, faces gland-dotted, often puberulent (shiny); heads in (leafy) paniculiform arrays; phyllaries 10–24 in 6–8 series, pale green to stramineous, 3–5-striate, unequal, (glutinous) margins scarious; outer bracts ovate (apices acute), inner bracts lanceolate (apices obtuse); corollas cream, 3.5–4.5 mm; cypselae 1.8–2.5 mm, scabrous; pappus of 30–40 barbellulate bristles. Washes, stream margins, seeps, hanging gardens. To be expected along the New Mexico–Arizona state lines.

Brickellia microphylla (Nutt.) A. Gray—Littleleaf brickellbush. Shrubs; leaves alternate; blades ovate to suborbicular, 3–20 × 1–15 mm, faces glandular-villous or hispidulous; heads erect, in loose paniculiform arrays, often clustered at ends of branches; involucres cylindrical to narrowly campanulate, 8–12 mm; phyllaries 30–48 in 5–8 unequal series, outer bracts obovate to suborbicular, inner bracts linear-oblong; corollas pale yellow, often purple-tinged, 5.5–7 mm; pappus of 18–24 white, minutely barbellate bristles. Our material belongs to **var. scabra** A. Gray. Dry, rocky places, canyon walls, sand dunes, washes, desert scrub and piñon-juniper communities, 4500–7200 ft. NW, Grant, Luna, Sierra.

Brickellia oblongifolia Nutt.—Narrowleaf brickellbush. Perennials or subshrubs; leaves alternate, sometimes subopposite, blades lance-linear to oblong or elliptic, 9–40 × 1–15 mm, faces pubescent to villous, often stipitate-glandular; heads erect, solitary or in corymbiform arrays; involucres cylindrical to campanulate, 10–20 mm; phyllaries 25–35 in 4–6 uneven series, outer bracts ovate to lanceolate, inner bracts linear-lanceolate to linear, corollas pale yellow-green or cream, often purple-tinged, 5–10 mm; cypselae 3–7 mm; pappus of 18–25 white, barbellate to subplumose bristles. New Mexico material belongs to **var. linifolia** (D. C. Eaton) B. L. Rob. Deserts, grasslands, dry rocky hillsides, piñon-juniper and ponderosa pine–oak communities, 4500–7550 ft. NW.

Brickellia parvula A. Gray—Small brickellbush. Perennials; leaves opposite, blades deltoid-ovate to lanceolate, 7–30 × 4–28 mm, faces puberulent, glandular; heads erect, borne singly or in corymbose-paniculiform arrays; involucres cylindrical to campanulate, 8–12 mm; phyllaries 12–15 in 3–5 uneven series, outer bracts ovate to lance-ovate, inner bracts narrowly lanceolate; corollas pale-green or yellow, 5.5–6.5 mm; cypselae 4.5–5 mm; pappus of 24–30 white, barbellate, sometimes subplumose bristles. Granitic cliffs or boulders, canyon drainages, desert scrub, 6000–6500 ft. Doña Ana, Luna.

Brickellia rusbyi A. Gray—Rusby's brickellbush. Perennials; leaves alternate, blades deltoid or rhombic-ovate, 5–10 × 2–5 cm, upper faces glabrous to sparsely pubescent, often punctate-glandular, lower faces sparsely villous, glandular; heads nodding, in paniculiform arrays; involucres cylindrical to campanulate, 7–10 mm; phyllaries 22–26 in 4–5 uneven series, outer bracts ovate to lanceolate, often purple-tinged, inner bracts linear to oblong; corollas yellowish, 5.3–6.5 mm; cypselae 2–2.5 mm; pappus of 26–32 white, barbellate bristles. Rocky slopes, limestone hillsides, shaded areas, 5650–8710 ft. NC, C, SC, SW, San Miguel.

Brickellia simplex A. Gray—Perennials; leaves opposite or alternate, blades deltoid to lanceolate, 10–60 × 7–50 mm, faces sparsely to densely pubescent, punctate-glandular; heads borne singly or in open paniculiform arrays; involucres cylindrical to broadly campanulate, 12–14 mm; phyllaries 40–60 in 5–6 uneven series, outermost bracts narrowly lanceolate, inner bracts lanceolate; corollas pale yellow, often purple-tinged, 9–10 mm; cypselae 4–5 mm; pappus of 32–36 white, barbellate bristles. Oak woodlands, canyons, dry mountain slopes, 5200–7000 ft. Eddy, Hidalgo.

Brickellia venosa (Wooton & Standl.) B. L. Rob.—Perennials; leaves opposite, blades linear to oblong, 20–60 × 2–10 mm, faces pubescent, glandular; heads erect, in open paniculiform arrays, borne singly, often terminal on lateral branches; involucres cylindrical to campanulate, 9–10 mm; phyllaries 25–35 in 6–9 uneven series, outer bracts broadly ovate to orbicular, inner bracts lanceolate to linear-lanceolate; corollas pale yellow, often purple-tinged, 5.5–7 mm; cypselae 3–3.5 mm; pappus of 30–40 white, barbellate bristles. Dry hills, canyon walls, mesas, limestone or granitic outcrops, desert scrub, juniper-oak, ponderosa pine-oak, 4300–6500 ft. WC, SC, SW.

BRICKELLIASTRUM R. M. King & H. Rob. – Brickellbush

Randall W. Scott

Brickelliastrum fendleri (A. Gray) R. M. King & H. Rob. [*Brickellia fendleri* A. Gray; *Steviopsis fendleri* (A. Gray) B. L. Turner]—Fendler's brickellbush. Perennials; leaf blades 15–80 × 10–60 mm; heads erect to nodding; involucres 5–7 mm, outer bracts acute to acuminate, inner acute; florets 25–35; corollas 3.5–6 mm; cypselae 2.3–3.5 mm; pappus bristles 4–5 mm. Rock outcrops, cliff faces, canyon bottoms, roadsides, piñon-juniper, ponderosa pine, mixed conifer, aspen-spruce, montane meadows, 5000–11,000 ft. Widespread except NW, EC, SE (Scott, 2006b).

CALYCOSERIS A. Gray – Tack-stem

Kenneth D. Heil and Steve L. O'Kane, Jr.

Calycoseris wrightii A. Gray—White tack-stem. Annuals, 5–30 cm; stems 1–3, stems, distal branches, and phyllaries gland-dotted, glands straw-colored and resemble tacks; leaves basal and cauline, alternate, sessile, basal blades pinnately lobed, distal reduced to linear; heads borne singly or in open, cymiform arrays; involucres campanulate, 10–12+ mm diam.; receptacles flat, smooth, bristly; ligules white, often with fine reddish veins abaxially, 2–3 cm; phyllaries 12–20 in 1 series; florets ca. 25; corollas white or yellow, showy; cypselae dark tan to brown, shallowly grooved between ribs, faces weakly cross-rugulose. Gravels derived from limestone, volcanics, granites, and caliche soils, often in creosote or mesquite communities, 3900–6100 ft. C, WC, SC, SW (Gottlieb, 2006a).

CALYPTOCARPUS Less. – Straggler daisy

Kenneth D. Heil

Calyptocarpus vialis Less.—Straggler daisy. Perennials, to 30 cm; stems prostrate to decumbent, branched throughout; leaves cauline, opposite, petiolate, blades deltate to ovate or lanceolate; heads radiate, borne singly (in axils); involucres obconic, 3–8 mm diam.; phyllaries persistent, 5 in 1(2) series; receptacles convex; ray florets 3–8, pistillate, corollas pale yellow; disk florets 10–20, corollas yellow; cypelae compressed or flattened, cuneate, pappus persistent. Weedy, in lawns and dry disturbed areas; known only from Doña Ana County, 3701–4341 ft. (Allred & Ivey, 2012; Strother, 2006c).

CARDUUS L. – Musk thistle

Jennifer Ackerfield

Annual or biennial spiny herbs; leaves alternate, bases strongly decurrent; heads discoid; involucral bracts imbricate in several series, tips spreading to reflexed, midribs excurrent as spines; receptacles bristly; disk florets purplish or reddish or rarely white; ray florets absent; pappus of barbellate bristles; cypselae with longitudinal lines, glabrous, basifixed (Ackerfield, 2015).

1. Heads 1–2.5 cm diam.; outer involucral bracts narrowly lanceolate, 1–2 mm wide at the base, without a constriction in the middle…**C. acanthoides**
1'. Heads 3–7 cm diam.; outer involucral bracts ovate-lanceolate to lanceolate, 2–10 mm wide, with a shallow constriction in the middle…**C. nutans**

Carduus acanthoides L.—Plumeless thistle. Biennials, 3–15 dm, with strongly spiny-winged stems; leaves 10–30 cm, deeply 1–2-pinnately lobed or spiny-toothed, sparsely hairy; involucre 1–2.5 cm diam.; disk florets 13–20 mm, purple; pappus 11–13 mm; cypselae 2.5–3 mm. Disturbed places, open fields, roadsides, 6560–7000 ft. Otero. Introduced.

Carduus nutans L. [*C. nutans* subsp. *macrolepis* (Peterm.) Kazmi]—Musk thistle, nodding thistle. Annuals or biennials, 4–20+ dm, with spiny-winged stems; leaves 10–40 cm, 1–2-pinnately lobed; involucre 3–7 cm diam.; disk florets 15–28 mm, purple; pappus 13–25 mm; cypselae 4–5 mm. Common in disturbed places, open fields and meadows, roadsides, 4005–9835 ft. NC, NW, EC, C, SC. Introduced.

CARMINATIA Moc. ex DC. – Plumeweed

Kenneth D. Heil

Carminatia tenuiflora DC. [*Brickellia tenuiflora* (DC.) D. J. Keil & Pinkava]—Plumeweed. Annuals, 10–100+ cm; stems erect, usually puberulent and/or villous to pilose, often in lines; leaves mostly cauline, usually opposite, blades triangular to broadly ovate, 2–8 × 1.5–6 cm; heads discoid, in spiciform or narrow, ± paniculiform arrays, 1–10+ per principal node; involucres cylindrical, 9–12 mm; phyllaries persistent, (12–)17–22+ in 3–4+ series, 3-nerved, glabrous; corollas 5–6 mm, greenish white to cream-colored; cypselae gray to nearly black, 3.5–5 mm; usually 5-ribbed, minutely puberulent; pappus white, 5–8 mm, of 8–13 plumose bristles or setiform scales in 1 series. Oak–ponderosa pine–juniper woodlands, grasslands, riparian areas, 5200–7800 ft. WC, SC, SW (Keil, 2006b).

CARPHOCHAETE A. Gray – Bristlehead

Kenneth D. Heil

Carphochaete bigelovii A. Gray—Bigelow's bristlehead. Subshrubs or shrubs, (8–)20–45(–120+) [very rarely to 300] cm; stems usually stiff, erect; leaves cauline, opposite, sessile, often fascicled on older growth; blades 5–35 mm, narrowly elliptic or oblanceolate to linear, faces glabrate or puberulent, gland-dotted (in pits); heads discoid, usually borne singly or in pairs, terminal or in leaf axils; involucres 12–20 mm; phyllaries persistent, 8–12+ in 3–4+ series, gland-dotted, sometimes puberulent; receptacles flat to slightly convex; corollas 15–20 mm, throats usually purplish, lobes creamy-white; cypselae yellow-green to golden-brown, 11–14 mm; pappus scales mostly brownish to purplish with colorless, hyaline margins. Sandy soils, rock outcrops in grasslands, chaparral, pine-oak woodlands, 4800–7200 ft. WC, SC, SW (Keil, 2006c).

CARTHAMUS L. – Safflower

Kenneth D. Heil

Carthamus tinctorius L.—Safflower. Plants 3–10 dm; stems striate; leaves sessile, clasping, elliptic to ovate or oblong, spine-tipped; involucres 2–4 cm diam.; disk corollas 2–3 cm, yellow to red; cypselae white, 6–9 mm. Often seen along roadsides, 4000–5000 ft. Bernalillo. Introduced cultivar (Keil, 2006d).

CENTAUREA L. – Knapweed, star-thistle

Kenneth D. Heil

Annuals, biennials, or perennials, 20–300 cm, glabrous or tomentose; stems erect, ascending; leaves basal and cauline, petiolate or sessile, proximal blade margins often ± deeply lobed, faces glabrous or ± tomentose, sometimes also villous, strigose, or puberulent, often glandular-punctate; heads discoid, disciform, or radiate; involucres cylindrical or ovoid to hemispherical; phyllaries many in 6 to many series; receptacles flat, epaleate, bristly; corollas white to blue, pink, purple, or yellow, bilateral or radial; cypselae ± barrel-shaped, ± compressed, smooth or ribbed, glabrous or with fine, 1-celled hairs; pappus 0 or ± persistent, of 1–3 series of smooth or minutely barbed, stiff bristles or narrow scales. All species introduced (Keil & Ochsmann, 2006).

1. Corollas yellow…(2)
1′. Corollas white to pink, blue, or purple…(3)
2. Central spines of principal phyllaries 5–10 mm…**C. melitensis**
2′. Central spines of principal phyllaries 10–25 mm…**C. solstitialis**
3. Principal phyllaries tipped with spines…(4)
3′. Principal phyllaries not spine-tipped…(5)
4. Central spines of principal phyllaries 10–25 mm…**C. calcitrapa**
4′. Central spines of principal phyllaries 1–5 mm…**C. diffusa**
5. Annual…**C. cyanus**
5′. Perennial…**C. stoebe**

Centaurea calcitrapa L.—Purple star-thistle. Annuals, biennials, or short-lived perennials, 20–100 cm; stems 1 to several, often forming rounded mounds, puberulent to loosely tomentose; leaves puberulent to loosely gray-tomentose, becoming ± glabrous, minutely resin-gland-dotted, blades 10–20 cm, 1–3 times pinnately dissected, rosette with central cluster of spines; heads disciform, borne singly or in leafy cymiform arrays; involucres ovoid, 15–20 × 6–8 mm; principal phyllary bodies greenish or stramineous, spiny-fringed at base, each tipped by a stout spreading spine 10–25 mm; florets 25–40, corollas purple, all ± equal, 15–24 mm; cypselae 2.5–3.4 mm, glabrous. Pastures, fields, roadsides, 4250–7800. Chaves, Otero. Introduced.

Centaurea cyanus L. [*Leucacantha cyanus* (L.) Nieuwl. & Lunell]—Bachelor's-button. Annuals, 20–100 cm; leaves ± loosely gray-tomentose; basal leaf blades linear-lanceolate, 3–10 cm, margins entire or with remote linear lobes, apices acute; cauline linear; heads radiant, in open, rounded or ± flat-topped cymiform arrays; corollas blue (white to purple), those of sterile florets raylike, enlarged, 20–25 mm, those of fertile florets 10–15 mm; cypselae stramineous or pale blue, 4–5 mm, finely hairy; pappus of many unequal stiff bristles, 2–4 mm. Disturbed sites, forests, grasslands, roadsides, ca. 3900 ft. Doña Ana. Introduced.

Centaurea diffusa Lam. [*Acosta diffusa* (Lam.) Soják]—Diffuse knapweed. Annuals or perennials, 20–80 cm, stems 1 to several, much branched throughout, puberulent and ± gray-tomentose; leaves hispidulous and ± short-tomentose, blades 10–20 cm, margins bipinnately dissected into narrow lobes; midcauline sessile, bipinnately dissected; heads disciform, in open paniculiform arrays; involucres narrowly ovoid or cylindrical, 10–13 × 3–5 mm; principal phyllary bodies pale green, each phyllary tipped by spine 1–3 mm; florets 25–35; corollas cream-white (rarely pink or pale purple), 12–13 mm; cypselae dark brown, ca. 2–3 mm; pappus 0 or < 0.5 mm, only rudimentary. Disturbed sites, grasslands, coniferous forests, 5350–6330 ft. NC, NW. Introduced.

Centaurea melitensis L.—Maltese star-thistle. Annuals, 10–100 cm, herbage loosely gray-tomentose and villous with jointed multicellular hairs, sometimes minutely scabrous, minutely resin-gland-dotted; stems 1 to few, few- to many-branched distally; leaf blades linear to oblong or oblanceolate, 1–15 cm, margins entire to dentate or pinnately lobed; heads disciform; involucres ovoid, 10–15 mm, loosely cobwebby-tomentose or becoming glabrous; principal phyllary bodies ± stramineous, each tipped by slender spine 5–10 mm; florets many, corollas yellow; cypselae ca. 2.5 mm, finely hairy; pappus of many white, unequal, stiff bristles. Roadsides, fields, woodlands, 3115–4600 ft. SE, SC, SW. Introduced.

Centaurea solstitialis L.—Yellow star-thistle. Annuals, 10–100 cm; stems simple or often branched from base, forming rounded bushy plants, gray-tomentose; leaves gray-tomentose and scabrous to short-bristly, blades 1–15 cm, margins pinnately lobed or dissected; cauline long-decurrent; heads disciform, borne singly or in open leafy arrays; involucres ovoid, 13–17 mm, loosely cobwebby-tomentose or becoming glabrous; principal phyllary bodies pale green, radiating cluster of spines, stout central spine 10–25 mm; florets many, corollas yellow, all ± equal, 13–20 mm; cypselae 2–3 mm, glabrous; pappus of many white, unequal bristles 2–4 mm, fine. Disturbed sites, roadsides, fields, pastures, 4500–6800 ft. Grant, San Miguel. Introduced.

Centaurea stoebe L. [*Acosta maculosa* (Lam.) Holub; *Centaurea biebersteinii* DC.; *C. maculosa* Lam.]—Spotted knapweed. Perennials, 30–150 cm; gray-tomentose; upper leaves sessile, gray-tomentose and gland-dotted; blades 10–15 cm, margins divided; heads discoid; phyllaries ovate with several prominent parallel veins, lower portion green to pink-tinged, appendages dark brown or black, scarious and fringed with slender teeth; florets 30–40, corollas pink to purple, rarely white; cypselae brown, finely pubescent; pappus of many stiff white bristles. New Mexico material belongs to **subsp. micranthos** (S. G. Gmel. ex Gugler) Hayek. Disturbed sites, roadsides, 4540–8700 ft. NC, NW, WC, Otero. Introduced.

CHAENACTIS DC. – Dustymaiden, pincushion

Kenneth D. Heil

Annuals or perennials; stems erect, 4–60 cm; leaves basal and cauline, alternate, usually petiolate; heads discoid; phyllaries in 1–2 series; receptacles naked (ours); ray florets 0; disk florets white or pale cream, tubes shorter than cylindrical or funnelform throats, lobes 5; cypselae clavate, ± cylindrical or compressed; pappus of scales (Morefield, 2006a).

1. Annuals; farinose, not arachnoid; largest leaf blades (2)3–4 times pinnately lobed; Hidalgo County... **C. carphoclinia**
1'. Annuals or perennials; arachnoid to thinly lanuginose; largest leaf blades 1–2 times pinnately lobed; widespread...(2)
2. Biennials or perennials; leaf blades gland-dotted beneath indumenta...**C. douglasii**
2'. Annuals; leaf blades not gland-dotted...**C. stevioides**

Chaenactis carphoclinia A. Gray [*C. carphoclinia* var. *attenuata* (A. Gray) M. E. Jones]—Pebble or straw-bed pincushion. Annuals; proximal indument predominantly ± farinose, not arachnoid; plants (5–)10–30(–40) cm; leaves basal and cauline, longest 1–6(–7) cm, deltate to ± ovate, (2)3–4-pinnately lobed, not gland-dotted; involucres mostly 5–10 mm; longest phyllaries 7–10 mm; heads discoid, mostly 3–20+ per stem; corollas white to pinkish, 4–6 mm; cypselae ± terete or compressed; pappus 0. New Mexico material belongs to **var. carphoclinia**. Pyramid Mountains, bajadas; around a spring at 4330 ft. Hidalgo.

Chaenactis douglasii (Hook.) Hook. & Arn.—Douglas's dustymaiden. Biennials or perennials; stems erect to spreading, (1–)5–50(–60) cm; indument mostly arachnoid-sericeous to thinly lanuginose; leaves basal or basal and ± cauline, (1–)2–12(–15) cm, largest blades elliptic or slightly lanceolate to ovate, primary lobes (4)5–9(–12) pairs; involucres obconic to hemispherical, the bracts 9–17 mm; heads 1–25+ per stem, disk florets 5–8 mm; longest phyllaries 9–17 mm; corollas 5–8 mm; cypselae 5–8 mm; longest pappus scales 3–6 mm. Our plants belong to **var. douglasii**. Sagebrush communities and piñon-juniper woodlands, 5000–7800 ft. NC, NW.

Chaenactis stevioides Hook. & Arn.—Esteve's dustymaiden. Annuals; 5–30(–45) cm; proximal indumenta grayish, ± arachnoid-sericeous; stems 1–12; leaves basal, usually withering, ± cauline, 1–10 cm, mostly 1–2- pinnately lobed; heads mostly 3–20+ per stem; involucres obconic to hemispherical; longest phyllaries 5.5–10 mm; corollas white to pinkish or cream, 4.5–6.5 mm; cypselae (3–)4–6.5 mm; pappus of (1–)4(5) scales, usually in 1 series. Sandy or clay soils, desert scrub, piñon-juniper woodlands, 3900–6400 ft. SC, SW, San Juan.

CHAETOPAPPA DC. – Leastdaisy

Jennifer Ackerfield and Kenneth D. Heil

Herbs; leaves alternate, simple, entire; heads radiate; involucral bracts imbricate in 2–6 series, green, rather dry, the margins scarious; receptacles naked; disk florets yellow; ray florets white to pinkish or drying purple; pappus of a hyaline crown or scales, capillary bristles, a few awnlike bristles, a combination of alternating scales and bristles, or reduced to an inconspicuous crown and appearing absent; cypselae terete to strongly compressed, 2–10-ribbed (Ackerfield, 2015; Nesom, 2010a).

1. Stems 1–2(–5) cm; pappus in 2 series, inner of 4–6 bristles plus outer of minute scales…**C. hersheyi**
1'. Stems 6–12 cm; pappus in 1 series, usually of 20–30 bristles…**C. ericoides**

Chaetopappa ericoides (Torr.) G. L. Nesom [*Leucelene ericoides* (Torr.) Greene]—Rose heath. Perennials, 6–12 cm, from rhizomes; leaves narrowly oblanceolate to lanceolate, 4–12 mm, usually glandular-stipitate, coriaceous; involucres 6–7.5 mm; pappus of 1 series of barbellate bristles; cypselae 2.5–3.5 mm. Shale, sandstone, igneous, gypsum substrates, creosote, juniper, piñon-juniper, ponderosa pine communities, 3000–8700 ft. Widespread.

Chaetopappa hersheyi S. F. Blake—Guadalupe leastdaisy. Perennials, 1–2(–5) cm, taprooted; leaves linear to oblanceolate, 3–8 mm, faces glabrous; involucres 4.5–5 mm; pappus in 2 series, outer of minute scales, inner of ca. 5 bristles; ray cypselae 2–3-nerved, disk cypselae 5-nerved. Limestone cliffs in the Guadalupe Mountains, piñon-juniper and mixed conifer communities, ca. 5000 ft. **R**

CHAMAECHAENACTIS Rydb. – Chamaechaenactis

Kenneth D. Heil

Chamaechaenactis scaposa (Eastw.) Rydb. [*Chaenactis scaposa* Eastw.]—Fullstem. Densely cespitose, ± pulvinate perennials; caudices thickly branched; leaves basal, 3–40 mm, length mostly 1–2 times width; largest blades 4–15(–18) × 3–13(–15) mm; longest phyllaries 9–17 mm; disk florets white or cream, 5–9 mm; ray florets absent; cypselae 5–8 mm, densely hairy; longest pappus scales 4–7 mm. Mostly in barren silty to clay soils, rarely in sandstone or limestone, mostly in piñon-juniper woodlands, 5600–7000 ft. McKinley, San Juan (Morefield, 2006b).

CHAPTALIA Vent. – Silverpuff

Kenneth D. Heil

Chaptalia texana Greene [*C. nutans* (L.) Pol. var. *texana* (Greene) Burkart]—Silverpuff. Perennials, 3–40(–100) cm; leaves petiolate, blades obovate to ovate or elliptic or sublyrate, 3–21 cm, abaxial faces usually covered with dense wool, margins lobed to denticulate; heads nodding in bud and fruit, erect in flowering; phyllaries in 2–5+ series; receptacles flat to convex; florets in 1–2 series; outer pistillate, corollas evenly cream-colored, turning crimson, inner florets bisexual; cypselae 11.5–13 mm, fusiform, often slightly flattened, ribs mostly 4–12, faces glabrous or hairy. Rocky limestone soils, oak woodlands, 7500–8000 ft. Otero, Sierra (Nesom, 2006g).

CHLORACANTHA G. L. Nesom, Y. B. Suh, D. R. Morgan, S. D. Sundb. & B. B. Simpson – Mexican devilweed

Kenneth D. Heil

Chloracantha spinosa (Benth.) G. L. Nesom [*Aster spinosus* Benth.]—Mexican devilweed, spiny aster. Perennials or subshrubs, 50–150 cm, usually thorny proximally, some populations producing thornless, wandlike branches; leaves oblanceolate, mostly 10–40 mm; heads borne singly in loose corymbo-paniculiform arrays; involucres 4–6 mm; ray florets 20–33 in 1–2 series, corollas white; disk florets mostly 20–70, corollas yellow with orange resin ducts; cypselae fusiform-cylindrical, slightly compressed; pappus of 30–60 barbellate bristles. Riverbanks, saline flats, ditches, irrigation channels, 3000–6000 ft. EC, C, SE, SC, SW. Introduced (Nesom, 2006h).

CHRYSACTINIA A. GRAY – Damianita

Kenneth D. Heil

Chrysactinia mexicana A. Gray—Damianita daisy. Shrubs or subshrubs, evergreen, plants twiggy; stems erect; leaves cauline, mostly alternate, blades 5–10(–23) × 1–2 mm, simple, linear to clavate, peduncles 15–75 mm; heads radiate, borne singly; involucres turbinate to hemispherical; phyllaries persistent; receptacles convex to hemispherical; ray florets pistillate, fertile, corollas bright yellow, laminae 6–12 × 2–4 mm; disk florets 4.5–7 mm, bisexual, fertile, corollas yellow to orange; cypselae 3–4 mm, cylindrical to fusiform; pappus 3–5.5 mm, with persistent bristles. Limestone soils in S desert scrub habitats, 3904–8802 ft. C, SE, SC, SW, Roosevelt (Allred & Ivey, 2012; Strother, 2006d).

CHRYSOTHAMNUS Nutt. – Rabbitbrush

Gary I. Baird and Kenneth D. Heil

Shrubs or subshrubs, 10–120 cm, often scopiform; stems erect to spreading or decumbent, bark white to beige, glabrous or puberulent, sometimes glandular or resinous; leaves cauline, alternate, sessile, blades linear-filiform to lanceolate or oblanceolate, flat to keeled, straight to twisted, margins entire, faces glabrous or puberulent, sometimes glandular or resinous; heads discoid in corymbiform arrays; involucres cylindrical to obconic; phyllaries 12–25 in 3–6 imbricate series, spiraled to strongly aligned in vertical ranks, faces glabrous or hairy, often glandular or resinous; ray florets 0; disk florets 4–14 (mostly ~5), bisexual, fertile, corollas yellow; cypselae cylindrical to obconic, often prismatic, brownish, faces glabrous or hairy, seldom glandular; pappus of 15–50+ slender bristles in 1 series, beige to buff (Urbatsch et al., 2006a).

1. Cypselae glabrous or obscurely glandular…(2)
1′. Cypselae moderately to densely hairy, eglandular…(3)
2. Involucres 6–8 mm; phyllaries rounded to acute, weakly keeled or aligned…**C. vaseyi**
2′. Involucres 9–15 mm; phyllaries acuminate to cuspidate, strongly keeled and aligned…**C. depressus**
3. Leaves 0.5–2 mm wide; phyllaries acuminate to cuspidate, weakly aligned; cypselae pilose…**C. greenei**
3′. Leaves 0.5–10 mm wide; phyllaries obtuse to acute or rounded, ± aligned; cypselae puberulent…
   C. viscidiflorus

Chrysothamnus depressus Nutt.—Longflower rabbitbrush. Shrubs, 7–27 cm; stems densely puberulent; leaves flat to keeled, not twisted but often arching, blades narrowly oblanceolate, 7–30 × 1.5–7 mm, faces glabrous or puberulent, sometimes stipitate-glandular; heads few in compact clusters; involucres fusiform-obconic, 9–15 × 3–5 mm; phyllaries 20–25 in 4–6 series, strongly aligned, faces puberulent; disk florets 5–6, corollas 7–11 mm; cypselae tan, 5–6.5 mm, glabrous or sparsely glandular; pappus 5.5–7.5 mm. Dry, clay to sandy or rocky soils of plains, mesas, and ridges in sagebrush, piñon-juniper, and ponderosa pine communities, 5800–9000 ft. NW, C, WC.

Chrysothamnus greenei (A. Gray) Greene—Threadleaf rabbitbrush. Shrubs, 14–38 cm; stems glabrous, resinous; leaves sulcate, often twisted, sometimes arching; blades linear-filiform, 10–40 × 0.5–2 mm, faces glabrous, resinous; heads few to many in loose to compact, corymbiform arrays; involucres cylindrical, 5–8 × 1.5–2.5 mm; phyllaries 15–20 in 3–4 series, weakly or not aligned, faces glabrous, resinous; disk florets 4–5, corollas 4–5.5 mm; cypselae fulvous-tan, 3–4 mm, hairy; pappus 4–5 mm. Dry, sandy to rocky, sometimes saline soils of washes, flats, and slopes in grassland, sagebrush, piñon-juniper, and ponderosa pine communities, 5600–8300 ft. NC, NW, C, WC.

Chrysothamnus vaseyi (A. Gray) Greene—Vasey's rabbitbrush. Shrubs, 8–26 cm; stems glabrous or puberulent, resinous; leaves flat, straight or slightly twisted, blades linear-oblanceolate, 10–40 × 1–2.5 mm, faces glabrous or puberulent, often gland-dotted; heads few to many in crowded corymbiform arrays; involucres cylindrical to obconic, 5–8 × 2–4 mm; phyllaries 12–18 in 3–4 series, weakly or not aligned, faces glabrous or gland-dotted; disk florets 5–7, corollas 4.5–6.5 mm; cypselae fulvous-tan, 4–5 mm, glabrous; pappus 3.5–5 mm. Mesic to dry, clay to rocky, often disturbed soils of drainages, flats, and slopes in sagebrush, piñon-juniper, ponderosa pine, and montane meadows, 7300–9700 ft. NC, Socorro.

Chrysothamnus viscidiflorus (Hook.) Nutt.—Twistleaf rabbitbrush. Shrubs, 20–120 cm; stems glabrous or puberulent, often resinous; leaves flat or sulcate, weakly to strongly twisted, blades linear to lanceolate, 10–75 × 0.5–10 mm, faces glabrous or abaxial faces puberulent, often resinous; heads in congested corymbiform arrays; involucres cylindrical to obconic, 5–7 × 1.5–2.5 mm; phyllaries 12–24 in 3–4 series, weakly or not aligned, faces glabrous or puberulent, often resinous; disk florets 4–14 (mostly ~5), corollas 5–6.5 mm; cypselae pale to fulvous-tan, 2.5–4 mm; pappus 3.5–6 mm.

1.   Herbage ± puberulent; leaves only weakly twisted, if at all…**var. lanceolatus**
1'. Herbage ± glabrous; leaves simply twisted to corkscrewed…**var. viscidiflorus**

var. lanceolatus (Nutt.) Greene [*C. lanceolatus* Nutt.]—Shrubs mostly < 50 cm; stems mostly densely puberulent; leaf blades 15–45 × 2–6 mm, obscurely twisted or with a single twist, faces mostly puberulent, rarely glabrous; involucres obconic, 5–6.5 mm; phyllary faces mostly puberulent; disk florets usually 5, corollas 5.5–6 mm. Sagebrush and piñon-juniper communities, 7185–9500 ft. NC, NW.

var. viscidiflorus [*C. viscidiflorus* var. *stenophyllus* (A. Gray) H. M. Hall]—Shrubs mostly > 50 cm; stems mostly glabrous and viscid-resinous; leaf blades 20–75 × 1–10 mm, often strongly twisted, sometimes weakly so, faces mostly glabrous; involucres cylindrical to obconic, 5.5–7 mm; phyllary faces mostly glabrous; disk florets 4–14 (mostly ~5), corollas 5–6.5 mm. Piñon-juniper, ponderosa pine, mixed conifer communities, 4600–9500 ft. NC, NW, C, WC, SC.

CICHORIUM L. – Chicory

Gary I. Baird

Cichorium intybus L.—Common chicory. Perennial herbs, taprooted, sap milky; stems usually 1, branched; stem 1, 30–100 cm tall, hairy (proximally) to glabrous (distally); leaf blades oblanceolate to lance-linear, basal 10–25 × 1–7 cm, margins runcinate to dentate or entire, abaxial faces setose-hispid on midvein, adaxial face glabrous, cauline reduced upward; heads sessile in glomerules of 1–3; involucres cylindric, the phyllaries 10–15+ in 2 series, phyllary margins scabrous-hispidulous, faces often with a few glandular setae; ligulate florets ca 10–20, outer exceeding involucre, ligule 10–15 × 4–5 mm; cypselae 2–3 mm; pappus scales to 0.5 mm. Weedy and invasive, roadsides, ditchbanks, and other ruderal habitats, 4900–7100 ft. (Strother, 2006).

CIRSIUM Mill. – Thistle

Jennifer Ackerfield

Spiny herbs; leaves alternate, spiny-margined; heads discoid; involucral bracts imbricate in several series or subequal, the midribs usually strong and excurrent as spines; receptacles densely bristly; disk florets perfect and fertile (except in *C. arvense*, which is dioecious), purplish, pink, reddish, yellowish, or white; ray florets absent; pappus of plumose bristles; cypselae glabrous, nerveless (Ackerfield et al., 2020; Barlow-Irick, 2002; Keil, 2006e; Sivinski, 2016).

1.   Plants acaulescent with no true stem; heads clustered at ground level…**C. tioganum**
1'. Plants with a conspicuous leafy stem; heads elevated on a stem…(2)
2.   Flowers yellow to yellowish green…(3)
2'. Flowers pink, purple, red, or white…(4)
3.   Leaves shallowly lobed to nearly entire; involucral bracts not obscured by dense covering of woolly hairs; involucral bract margins often fringed or erose…**C. parryi**
3'. Leaves deeply lobed; involucral bracts obscured by dense covering of woolly hairs; involucral bract margins not fringed or erose…**C. funkiae**
4.   Upper leaf surface densely covered with short, rigid bristles…**C. vulgare**
4'. Upper leaf surface glabrous or glabrate to tomentose, but never with short, rigid bristles…(5)
5.   Corolla lobes at least twice as long as corolla throat…(6)
5'. Corolla lobes < twice as long as corolla throat…(9)
6.   Flowers bright red or reddish pink…**C. arizonicum**
6'. Flowers purple…(7)
7.   Leaves conspicuously decurrent on the stem; stems and abaxial leaf midveins with septate hairs; involu-

cral bract spines plentiful, usually present on every phyllary row, to 20 mm; all involucral bracts similar in appearance…**C. chellyense**

7'. Leaves not conspicuously decurrent on the stem; stems and abaxial leaf midveins usually lacking septate hairs; involucral bract spines usually present only on the bottom few rows of phyllaries, to 6 mm; upper involucral bracts usually purplish or rose-colored and lower bracts yellowish green to green…(8)

8. Style tips 1.2–3 mm; corolla lobes 13–18 mm…**C. calcareum**

8'. Style tips 2–4.5 mm; corolla lobes 8–15 mm…**C. pulchellum**

9. Involucral bracts densely arachnoid- or woolly-pubescent (hairs obscuring bracts); heads numerous in a dense terminal cluster; alpine…**C. culebraense**

9'. Involucral bracts glabrous, or if cobwebby-pubescent then bracts evidently visible, not obscured by hairs; heads not numerous in a dense terminal cluster; variously distributed…(10)

10. Heads nodding; involucral bracts purple, at least the lower ones reflexed; leaves green and glabrous on both sides; stems dark maroon…**C. vinaceum**

10'. Heads erect; involucral bracts green, yellowish, or purple-tinged, all ascending to erect; leaves often white-tomentose below; stems usually green…(11)

11. Middle stem leaves long-decurrent on stem for > 1 cm…(12)

11'. Middle stem leaves sessile and clasping, or shortly decurrent on stem for < 1 cm…(14)

12. Leaves glabrous or nearly so, thick and leathery; heads 1–2 cm wide; wetlands…**C. wrightii**

12'. Leaves white-tomentose below, green and glabrous above, not thick or leathery; heads 1.5–4.5 cm wide; dry places…(13)

13. Heads 2.5–4.5 cm wide; involucral bract spines 4–12 mm, stout; plants usually forming clumps of several stems…**C. ochrocentrum**

13'. Heads 1.5–2.5 cm wide; involucral bract spines 1–5 mm, slender; plants not forming clumps of several stems…**C. texanum**

14. Heads in dense terminal clusters or racemiform arrays, closely subtended by leafy bracts overtopping the heads (these usually in a whorl below heads); stems usually fleshy and thick; flowers white… **C. coloradense**

14'. Heads in loose clusters or racemiform arrays but not closely subtended by leafy bracts overtopping the heads; stems not fleshy and thick; flowers white, pink, or purple…(15)

15. Middle and inner involucral bracts with erose or fringed scarious margins [to be expected]… C. centaureae

15'. Middle and inner involucral bracts with entire margins…(16)

16. Plants dioecious (sometimes imperfectly so); pistillate flowers shorter than pappus in fruit; from horizontal creeping roots…**C. arvense**

16'. Plants monoecious; flowers longer than pappus; from taproots…(17)

17. Involucral bracts mostly dark purple…**C. grahamii**

17'. Involucral bracts mostly green or grayish green…(18)

18. Involucral bracts densely tomentose on the outer surface or cobwebby; outer involucral bracts strongly reflexed, with reflexed tip including a significant portion of green bract…**C. neomexicanum**

18'. Involucral bracts glabrous or thinly tomentose, mostly on the margins, sometimes weakly cobwebby; outer involucral bracts not as above…(19)

19. Leaves deeply pinnatifid nearly to the midrib; ultimate segments narrowly lanceolate, dark green, and nearly glabrous above, whitish-tomentose below; pappus 15–20 mm…**C. wheeleri**

19'. Leaves usually not as deeply pinnatifid, often gray-tomentose above and below (especially younger leaves); pappus 20–38 mm…(20)

20. Heads smaller, the involucre 18–30 mm; involucral bracts mostly ovate, in 3–4(5) rows…**C. tracyi**

20'. Heads larger, the involucre 25–50 mm; involucral bracts lanceolate to ovate, in 5–7 rows…**C. undulatum**

Cirsium arizonicum (A. Gray) Petr. [*Cnicus arizonicus* A. Gray]—Arizona thistle. Plants 3–15 dm; leaves unlobed and spinulose to shallowly lobed or pinnately divided nearly to the midrib, thinly tomentose to glabrous and green above, usually thinly tomentose below, 3–45 cm, sessile or bases decurrent to 1 cm; involucres 2–3.2 × 1.5–3 cm, dorsal ridge absent, glabrous to loosely cobwebby-pubescent, spines erect or lower reflexed, margins entire to minutely ciliolate; disk florets red or reddish pink, 24–34 mm; pappus 17–28 mm; cypselae 3.5–6 mm.

1. Leaves tomentose abaxially…**var. arizonicum**

1'. Leaves glabrous or rarely with septate trichomes abaxially on the midveins…**var. rothrockii**

var. arizonicum [*C. nidulum* (M. E. Jones) Petr.; *C. arizonicum* var. *nidulum* (M. E. Jones) S. L. Welsh]—Arizona thistle. Piñon-juniper, ponderosa pine forests, mixed conifer, spruce-fir communities, 3900–11,800 ft.

var. rothrockii (A. Gray) D. J. Keil [*Cnicus rothrockii* A. Gray]—Rothrock's thistle. Rocky slopes, embankments, pine-oak-juniper, 7000–8000 ft.

Cirsium arvense (L.) Scop. [*Breea arvensis* (L.) Less.]—Canada thistle. Plants dioecious (sometimes imperfectly so), 3–20 dm; leaves 5–18 cm, shallowly to deeply pinnately lobed or occasionally unlobed, white-tomentose below, green and glabrate above; involucres 10–20 mm, glabrous to arachnoid-hairy, bracts imbricate, spines ca. 1 mm; disk florets pink-purple or rarely white, 12–24 mm; pappus 13–32 mm; cypselae 2–4 mm. Disturbed places, roadsides, ditches, fields, meadows, 4000–10,500 ft. NC, NW, WC, Eddy, Otero. Introduced.

Cirsium calcareum (M. E. Jones) Wooton & Standl.—Cainville thistle. Plants 5–15 dm; leaves to 45 cm, glabrate above and tomentose below, deeply pinnatifid, decurrent to 1 cm; involucres 2.2–3.3 × 1.5–3 cm, lacking a glutinous dorsal ridge, spines 1.5–6 mm; disk florets purple, 24–32 mm, lobes 13–18 mm; style tips 1.2–3 mm; pappus 19–28 mm; cypselae 3.5–6 mm. Canyon bottoms, rocky outcrops. NC, NW, C.

Cirsium centaureae (Rydb.) K. Schum. [*C. latcrifolium* (Osterh.) Petr.]—Centaury thistle. Plants 2–10 dm; leaves pinnatifid into broad segments, thinly tomentose below, 10–30 cm, decurrent to 2 cm; involucres 15–25 mm, dorsal ridge absent, glabrous to densely woolly-tomentose, spines not reflexed, bracts imbricate, margins erose or fringed; disk florets usually white, 16–25 mm; pappus 10–20 mm; cypselae 4.5–7 mm. Mountain meadows, forest clearings. No records seen but to be expected.

Cirsium chellyense R. J. Moore & Frankton—Queen thistle. Plants 5–15 dm; leaves to 40 cm, glabrate above and tomentose below, deeply pinnatifid, decurrent 1.5–2 cm, usually with multicellular hairs on the underside along the midvein; involucres 2.3–2.8 × 1.5–2.5 cm, lacking a glutinous dorsal ridge, spines 6–12 mm; disk florets pink or purple, 24–32 mm, lobes 12–16 mm; style tips 2.2–3.5 mm; pappus 20–27 mm; cypselae 3.5–6 mm. Canyon bottoms, rocky outcroppings, ponderosa pine forests, 6500–9500 ft. Chuska Mountains. McKinley, San Juan.

Cirsium coloradense Cockerell ex Daniels [*C. scariosum* Nutt. var. *coloradense* (Rydb.) D. J. Keil]—Colorado thistle. Plants acaulescent or 3–10 dm; leaves pinnatifid, with a broad central midstripe, 10–40 cm, tomentose below, glabrate and green above, sessile; involucres 20–35 mm, dorsal ridge absent, bracts imbricate, often erose on the margins, spines 2–5 mm, spreading to ascending; disk florets white to lavender or pinkish purple, 20–35 mm; pappus 15–30 mm; cypselae 4–6 mm. Forest clearings, along streams and roadsides, meadows, 7000–10,000 ft. NC.

Cirsium culebraense Ackerf.—Culebra Mountains thistle. Plants 5–10 dm; leaves pinnatifid, 10–35 cm, the margins usually strongly undulate, glabrous to finely arachnoid-tomentose below, sessile; involucres 20–25 mm, densely tomentose, heads densely compacted in an erect terminal cluster; disk florets pink to pale purple, 14–25 mm; pappus 8–20 mm; cypselae 5.5–7 mm. Alpine meadows, scree slopes, 10,900–12,500 ft. Taos.

Cirsium funkiae Ackerf.—Funky thistle. Plants 2–8 dm; leaves pinnatifid, 3–35 cm, tomentose below and glabrate above, sometimes decurrent to 4 cm; involucres 30–35 mm, densely tomentose, heads densely compacted into a nodding terminal cluster, spines 10–18 mm; disk florets yellow, 18–25 mm; pappus 10–15 mm; cypselae 4–6 mm. Alpine meadows, scree slopes, 11,400–12,500 ft. Mora, Santa Fe.

Cirsium grahamii A. Gray—Graham's thistle. Plants 5–10 dm; leaves entire to coarsely dentate or deeply pinnatifid, 20–30 cm, white-tomentose below and glabrate above, sessile or shortly decurrent to 1 cm; involucres 2–3 × 2–4 cm, purplish, bracts with a well-developed glutinous dorsal ridge, spines 1.5–2.5 mm, erect to ascending; disk florets 22–30 mm, deep purple; pappus 22–30 mm; cypselae 4–5.5 mm. Rare in canyon bottoms and moist meadows, 5500–8700 ft. Catron.

Cirsium neomexicanum A. Gray—New Mexico thistle. Plants 5–20 dm; leaves pinnatifid with a broad central midstripe, 10–35 cm, thinly to densely tomentose on both sides, decurrent 1–3 cm; involucres 20–35 mm, tomentose, the outer reflexed from the middle, weakly spine-tipped; disk florets white or pale purple, 22–35 mm; pappus 20–25 mm; cypselae 5–6 mm. Dry hillsides or along roadsides, often in clay or shale soil, 3000–9500 ft. Widespread except NE, EC, SE.

Cirsium ochrocentrum A. Gray—Yellowspine thistle. Plants 2–10 dm; leaves pinnatifid with a broad central midstripe, densely white-tomentose below, green above, 8–22 cm, decurrent; involucres 25–45 mm, with a prominent dorsal ridge, spines 4–12 mm, the outer reflexed; disk florets 30–45 mm, purple, pink, red, or rarely white; pappus 17–30 mm; cypselae 7–8 mm. Grasslands and dry, open slopes, 3200–10,300 ft. Widespread.

1. Stems with numerous leaves and crowded nodes; flowers white to pale pink or lavender; widespread… **var. ochrocentrum**
1'. Stems with well-separated nodes; flowers reddish purple to rose-pink or red, rarely white; WC, SW, Rio Arriba…**var. martinii**

var. martini (Barlow-Irick) D. J. Keil—Martin's thistle. Roadsides, desert grasslands, and piñon-juniper woodlands. Catron, Grant, Hidalgo.

var. ochrocentrum—Yellowspine thistle. Roadsides, grasslands, sagebrush flats; widespread.

Cirsium parryi (A. Gray) Petr. [*C. gilense* Wooton & Standl.; *C. pallidum* (Wooton & Standl.) Wooton & Standl.; *C. inornatum* (Wooton & Standl.) Wooton & Standl.]—Parry's thistle. Plants 3–20 dm; leaves pinnatifid with a broad central midstripe, glabrate above and arachnoid below, 10–30 cm, not decurrent; involucres 20–30 mm, densely arachnoid-pubescent, lacking a glutinous dorsal ridge, the margins often erose, spines 4–7 mm; disk florets greenish yellow, 11–17 mm; pappus 9–15 mm; cypselae 4–6 mm. Forests and meadows, 6500–12,500 ft. NC, NW, C, WC, SC.

Cirsium pulchellum Wooton & Standl. [*C. arizonicum* (A. Gray) Petr. var. *bipinnatum* (Eastw.) D. J. Keil]—Cainville thistle. Plants 3–15 dm; leaves to 50(–60) cm, glabrate above and tomentose below, deeply pinnatifid, decurrent 1.5–2 cm; involucres 1.8–3 × 1.5–2.7 cm, lacking a glutinous dorsal ridge, spines 1–6 mm; disk florets reddish purple to purple, 22–34 mm, lobes 8–15 mm; style tips 2–4.5 mm; pappus 18–25 mm; cypselae 3.5–6 mm. Canyon bottoms, 5000–8500 ft.

Cirsium texanum Buckley—Texas thistle. Plants 2–20 dm; leaves shallowly to deeply pinnatifid, 7–30 cm, white-tomentose below, glabrate above, decurrent 1–3 cm; involucres 1.5–2 × 1.5–2 cm, with a well-developed glutinous dorsal ridge, spines 1–5 mm, spreading to ascending; disk florets white to pink-purple, 20–25 mm; pappus 15–16 mm; cypselae 3–5 mm. Roadside ditches, riparian valley bottoms, 3000–4000 ft. Black River and lower Pecos River basins. Chaves, Eddy.

Cirsium tioganum (A. Gray) Petr. [*C. americanum* (A. Gray) Daniels; *C. acaulescens* K. Schum.]—Dinnerplate thistle. Plants acaulescent; leaves pinnatifid with a broad central midstripe, 10–40 cm, tomentose below, glabrate and green above, sessile; involucres 15–30 mm, dorsal ridge absent, bracts imbricate, spines 2–12 mm, spreading to ascending; disk florets white to pink, rarely lavender, 22–30 mm; pappus 20–25 mm; cypselae 4.5–6 mm. Forest clearings, wet meadows; known from a single collection, elevation unknown. Rio Arriba.

Cirsium tracyi (Rydb.) Petr. [*C. undulatum* (Nutt.) Spreng. var. *tracyi* (Rydb.) S. L. Welsh]—Tracy's thistle. Plants 3–15 dm; leaves pinnately lobed, sometimes shallowly so, 10–40 cm, white-tomentose below, glabrate above, sessile or very shortly decurrent (< 1 cm); involucres 18–30 mm, with a well-developed glutinous dorsal ridge, bracts imbricate, spines 3–7 mm, reflexed to spreading; disk florets pink-purple or sometimes white, 22–30 mm; pappus 18–20 mm; cypselae 5–8 mm. Hillsides, canyons, dry soil, disturbed areas, 6000–9700 ft. NW, Cibola.

Cirsium undulatum (Nutt.) Spreng.—Wavyleaf thistle. Plants 3–15 dm; leaves coarsely dentate to pinnatifid, 10–45 cm, white-tomentose below and glabrate above, sessile or shortly decurrent; involucres 25–50 mm, with a well-developed glutinous dorsal ridge, bracts imbricate, spines 3–5 mm, outer reflexed to spreading; disk florets 25–36 mm, purple, pink, or rarely white; pappus 22–40 mm; cypselae 4–7 mm. Sandy or gravelly soil, grasslands, open hillsides, 3500–10,300 ft. Widespread.

Cirsium vinaceum (Wooton & Standl.) Wooton & Standl.—Sacramento Mountains thistle. Plants 10–20 dm; leaves 1–2 times pinnately lobed or divided, 10–50 cm, glossy green and glabrous, sessile to auriculate-clasping; involucres 2–3 × 2–3 cm, dorsal ridge poorly developed (inconspicuous), spines 1–3 mm, lower bracts strongly reflexed; disk florets rose-purple, 20–26 mm; pappus 18–20 mm; cypselae ca. 5 mm. Springs, seeps, wet meadows, 7500–9500 ft. Sacramento Mountains. Otero. **FE & E**

Cirsium vulgare (Savi) Ten.—Bull thistle. Plants 5–15 dm; leaves pinnatifid, white-tomentose below and with short, rigid bristles above, 12–30(–50) cm, decurrent, leaf bases extending nearly from 1 node to the next; involucres 25–40 mm, dorsal ridge absent, spines 2–5 mm, spreading; disk florets dark purple, 25–35 mm; pappus 20–30 mm; cypselae 3–5 mm. Disturbed places, forest openings, roadsides, fields, 4500–9500 ft. NC, NW, EC, C, WC, SE, SC. Introduced.

Cirsium wheeleri (A. Gray) Petr. [*C. perennans* (Greene) Wooton & Standl.]—Wheeler's thistle. Plants 3–10 dm; leaves shallowly to deeply pinnatifid, 10–25 cm, white-tomentose below and dark green and glabrate above, sessile; involucres 17–30 mm, dorsal ridge absent, spines 3–7 mm, spreading to ascending; disk florets purple, pink, or rarely white, 21–30 mm; pappus 15–25 mm; cypselae 6.5–7 mm. Piñon-juniper woodlands, canyon bottoms, rocky soil, 5500–10,300 ft. NW, C, WC, SC, SW,

Cirsium wrightii A. Gray—Wright's thistle. Plants 10–30 dm; leaves unlobed or shallowly to deeply pinnatifid, 10–60 cm, glabrous or nearly so, sessile; involucres 1–2 × 1–2 cm, dorsal ridge present, spines ca. 1 mm, spreading to ascending; disk florets pink-purple or white, 19–21 mm; pappus 15–16 mm; cypselae ca. 4.5 mm. Spring-fed wet meadows, salt marshes, 3200–7900 ft. EC, C, SE, SC. **FT**

CONOCLINIUM DC. – Thoroughwort

Kenneth D. Heil

Conoclinium dissectum A. Gray [*C. greggii* (A. Gray) Small; *Eupatorium greggii* A. Gray]—Palm-leaf mistflower. Perennials, 50–200 cm; stems erect (often from knotty crowns, sometimes basally lignescent); leaves cauline, opposite, leaf blades ovate-deltate to ovate, 1.5–4 cm, bases attenuate, margins dissected or lobed, apices pointed or rounded; heads discoid, in tight, corymbiform arrays; phyllaries 3.5–5 mm, persistent, in 2–3 series; involucres hemispherical, 3–6 mm diam.; receptacles conic; florets 35–70+, corollas blue, lavender, or purple, 2.5–3.5 mm; cypselae 1.8–2.5 mm, hispidulous; pappus bristle tips not dilated. Sandy or rocky soils, depressions and ditches; mesquite and creosote communities, 4300–5650 ft. SC, SW (Patterson & Nesom, 2006).

COREOPSIS L. – Tickseed

Don Hyder

Annual or perennial herbs; leaves opposite; heads radiate; involucral bracts in 2 series and dimorphic, fused at the base, the outer bracts shorter, narrower, and more herbaceous than the striate inner bracts; receptacle chaffy; disk florets 4- or 5-lobed or toothed; ray florets usually yellow; pappus absent or of 2 short awns or teeth or a minute crown; cypselae flattened (Allred & Ivey, 2012; Strother, 2006e).

1. Leaves mostly entire or with a few lobes…**C. lanceolata**
1´. Leaves pinnatifid…(2)
2. Ray florets pistillate, with styles…**C. californica**
2´. Ray florets neuter, lacking styles…(3)
3. Plants perennial; ray and disk florets usually yellow throughout, the face of the head thus appearing yellow, or at least the rays yellow…**C. grandiflora**
3´. Plants annual; rays usually with a reddish spot or band toward the base; disk florets purplish or reddish brown; the face of the head thus appearing yellow toward the perimeter and reddish toward the center…(4)
4. Disk corollas with 4 lobes…**C. tinctoria**
4´. Disk corollas with 5 lobes [record needs verification]…C. basalis

Coreopsis basalis (A. Dietr.) S. F. Blake—Goldenmane tickseed. Annuals, 10–50+ cm; leaves basal and cauline, lance-ovate, 7–9+ mm; ray florets yellow, 15–20+ mm, laminae usually with proximal, red-brown to purple spots or bands; disk corollas 3-4 mm, apices red-brown to purple; cypselae 1.2-1.8 mm, wingless. Our plants would belong to **var. wrightii** (A. Gray) S. F. Blake. Sandy soils in open, often disturbed places; occurrence needs verification.

Coreopsis californica (Nutt.) H. Sharsm.—California tickseed. Annuals, 5-20(-30+) cm; leaf blades simple or 1(2) times pinnately lobed; peduncles 5-15(-30+) cm; phyllaries 5-8, obovate to oblanceolate, 4-6(-7+) mm; ray florets yellow; disk florets yellow; cypselae ± oblong, marked adaxially with red dots or dashes, wings corky-thickened. New Mexico material belongs to **var. newberryi** (A. Gray) E. B. Sm. Openings in desert scrub; 1 record near Virden, 4200 ft. Hidalgo.

Coreopsis grandiflora Hogg ex Sweet—Big-flower coreopsis. Perennials, 4-6+ dm; leaves usually pinnately lobed into linear or lance-linear segments, 1.5-5(-9) cm; ray florets 1.2-2.5+ cm, yellow; disk florets 3-5 mm, 5-lobed, the tips yellow; cypselae 2-4 mm, winged. Introduced, seen only in cultivation but could escape onto nearby roadsides, 7497-8035 ft. C, SC, Eddy.

Coreopsis lanceolata L.—Lanceleaf tickseed. Perennials, 1-3(-6) dm; leaves simple and entire or with 1-2 lateral lobes, 5-12 cm; ray florets 15-35 mm, yellow; disk florets 6-7.5 mm, 5-lobed; cypselae 2.5-4 mm, winged. Introduced, occasionally escaping from cultivation, 4452-8281 ft. NE, C, SC, Eddy.

Coreopsis tinctoria Nutt.—Plains coreopsis. Annuals, (1-)3-8(-15) dm; leaves 1-2-pinnate, rarely simple, the ultimate segments linear, 1-5 cm; ray florets yellow with a red spot at the base, 1.2-2 cm; disk florets 2-4 mm, 4-lobed; cypselae 1.5-4 mm, wings absent or not. Common along roadsides, occasionally escaping from cultivation at higher elevations, 5052-8793 ft. EC, C, Cibola, Grant, McKinley.

COSMOS Cav. – Cosmos

Kenneth D. Heil

Annuals (perennials or subshrubs), 30-250 cm; stem usually 1, erect or ascending, branched distally or ± throughout; leaves mostly cauline, opposite, blades usually 1-3 times pinnately lobed; heads radiate, borne singly or in corymbiform arrays; involucres hemispherical or subhemispherical (cylindrical), 3-15 mm diam.; phyllaries persistent, (5-)8 in ± 2 series; receptacles flat; ray florets (0 or 5) or 8, corollas white to pink or purple, or yellow to red-orange; disk florets 10-20(-80+), bisexual, corollas yellow (orange) (at least distally); cypselae dark brown or black, relatively slender, quadrangular-cylindrical or -fusiform; pappus persistent (falling), of 2-4(-8) retrorsely (antrorsely) barbed awns, sometimes 0 (Kiger, 2006).

1. Ray laminae 15-50 mm...**C. bipinnatus**
1'. Ray laminae 5-9 mm...**C. parviflorus**

Cosmos bipinnatus Cav.—Garden cosmos. Plants 30-200 cm, glabrous or sparsely puberulent; leaf blades 6-11 cm, ultimate lobes to 1.5 mm wide, margins entire; peduncles 10-20 cm; involucres 7-15 mm diam.; phyllaries erect, 7-13 mm; ray corollas white, pink, or purplish, 15-50 mm; disk corollas 5-7 mm; cypselae 7-16 mm, glabrous, papillose; pappus 0, or of 2-3 ascending to erect awns, 1-3 mm. Disturbed sites, roadsides, 3800-7000 ft. Grant, Hidalgo, Otero.

Cosmos parviflorus (Jacq.) Pers. [*Coreopsis parviflora* Jacq.]—Southwestern cosmos. Plants 30-90 cm, glabrous or sparsely pubescent; leaf blades 2.5-6.5 cm, ultimate lobes to 1 mm wide, margins usually spinulose-ciliate; peduncles 10-30 cm; involucres (3-)5-10(-15) mm diam.; phyllaries erect, 5-8 mm; ray corollas white to rose-pink or violet, 5-9 mm; disk corollas 4-5 mm; cypselae 9-16 mm, setulose; pappus of 2-4 erect awns, 2-3 mm. Open or forested slopes and canyons, sometimes in disturbed and cultivated areas, 4500-9320 ft. All zones except NE, NW, EC, SE.

COTULA L. – Waterbuttons

Kenneth D. Heil and Steve L. O'Kane, Jr.

Cotula australis (Sieber ex Spreng.) Hook. [*Anacyclus australis* Sieber ex Spreng.]—Australian waterbuttons. Annuals, 2-25 cm; stem usually 1, erect or prostrate to decumbent or ascending, sometimes rooting at nodes; leaves mostly cauline, mostly 2–3 cm, obovate to spatulate, margins entire or irregularly toothed; involucres broadly hemispherical to saucer-shaped, 3–6 mm diam.; heads disciform or radiate, borne singly; receptacles flat to convex, epaleate; phyllaries 13–22+ in 2–3 series; ray florets 0; disk corollas ochroleucous to pale yellow, 0.5–0.8 mm; outer cypselae stalked, 1–1.2 mm, ± winged; inner cypselae sessile, 0.8–1 mm, not winged. Two records, disturbed areas, golf courses, 4000–7215 ft. Otero, Socorro. Introduced (Watson, 2006).

CREPIS L. – Hawksbeard

Gary I. Baird

Annual or perennial herbs, taprooted, often with a stout, branched caudex, sap milky; stems erect, leafy or scapiform, glabrous or hairy, often hispid or setose, sometimes stipitate-glandular; leaves basal and cauline, petiolate or sessile, basal blades lanceolate, ovate-elliptic, oblanceolate, obovate, or spatulate, cauline reduced, lance-linear, margins entire to dentate, or laciniate to runcinate-pinnatifid; heads liguliflorous, 2–60 in loose, cymosely corymbiform-paniculiform arrays (rarely solitary); involucres cylindrical to campanulate; calyculi of 5–12 subulate to lance-deltate bractlets, unequal; phyllaries 5–18 in 1–2 series, lanceolate, ± equal, bases often keeled, faces glabrous, tomentose, or setose, sometimes stipltate-glandular; florets 5–70, bisexual, fertile, ligules yellow; cypselae monomorphic, yellowish or reddish brown to nearly black, cylindrical to fusiform, often curved, distally tapered to obscurely beaked, ribs 10–18, faces glabrous or hispidulous; pappus of 80–150 bristles, white to flaxen, barbellulate, persistent (Bogler, 2006a).

As a whole, *Crepis* is taxonomically challenging. Our native species generally consist of sexually reproducing diploid populations whose boundaries are compromised by asexually reproducing (apomictic) polyploid hybrids. The line separating some of the apomicts sometimes appears both artificial and arbitrary, as they can combine various characteristics of the different parental taxa, making the identification of individual specimens difficult.

1. Plants annual; cypselae red-brown to nearly black...**C. tectorum**
1'. Plants perennial; cypselae yellow-brown to dark brown...(2)
2. Stems and leaves glabrous and glaucous (rarely hispidulous); basal leaves entire to dentate or ± pinnately lobed; phyllaries mostly 12–16 (sometimes as few as 10)...**C. runcinata**
2'. Stems and leaves tomentose, sometimes setose and/or stipitate-glandular; basal leaves laciniate or runcinate-pinnatifid (rarely pinnately lobed or dentate); phyllaries mostly 5–10 (sometimes as many as 13)...(3)
3. Basal leaves mostly runcinate-pinnatifid, lobes dentate or lobulate; principal phyllaries villous-tomentose, often stipitate-glandular or glandular-setose; calycular bractlets to 1/2 length of phyllaries... **C. occidentalis**
3'. Basal leaves mostly pinnately lobed, lobes entire or weakly dentate; principal phyllaries glabrous (or tomentulose distally); calycular bractlets mostly < 1/4 length of phyllaries...**C. acuminata**

Crepis acuminata Nutt.—Longleaf hawksbeard. Plants perennial, bases usually pale (not yellow), caudex ± shallow, not clad with persistent leaf bases; stems 20–65 cm; leaves basal and cauline, basal leaves lance-elliptic, 8–40 × 1–6+ cm, faces ± tomentulose, margins laciniately lobed or pinnatifid, apices long-acuminate, cauline leaves similar but reduced, distal-most filiform and bractlike; heads 30–70+ per stem; involucres cylindrical-campanulate, 8–16 mm; calyculi of 5–7 deltate bractlets, 1–2 mm; phyllaries 5–8, lanceolate, 8–12 mm, faces mostly glabrous, sometimes sparsely tomentulose distally; florets 5–10 (–15), ligules 10–18 mm; cypselae 6–9 mm, pale yellowish brown to flaxen; pappus 6–9 mm. Piñon-juniper, mountain brush communities, 7875–9525 ft. NW, Cibola.

Reports of limestone hawksbeard (*C. intermedia* A. Gray) from New Mexico (Rio Arriba) have not been substantiated. This species is a polyploid, apomictic hybrid that exhibits features intermediate between *C. acuminata* and *C. occidentalis*. In general appearance, it is similar to *C. acuminata* but can be (often arbitrarily) separated from that species in the following ways: leaf lobes usually denticulate or lobate (vs. entire); cauline leaves usually pinnately lobed (vs. usually entire); phyllaries ± villous-tomentulose (vs. pubescent only at the tip) and with (at least a few) blackish, nonglandular setae distally on midrib (vs. setae lacking); and fruits yellow to golden (vs. pale yellowish brown to flaxen).

Crepis occidentalis Nutt.—Western hawksbeard. Plants perennial, bases often yellow, caudex often deep-seated and clad with long-persistent leaf bases; stems 10–40 cm, branched throughout, tomentose and usually stipitate-glandular; leaves basal and cauline, basal leaves elliptic, 8–20 × 2–5 cm, faces tomentose, mostly stipitate-glandular, margins runcinate-pinnatifid, apices acute-acuminate, cauline leaves similar but reduced, distal-most sometimes bractlike; heads 4–20 per stem; involucres cylindrical, 11–19 mm; calyculi of 6–8 lance-linear bractlets, 3–6 mm; phyllaries 7–10, lanceolate, 12–15 mm, faces tomentose, usually glandular-setose (setae black or greenish); florets 10–40, ligules 18–22 mm; cypselae 6–10 mm, golden-brown to dark brown; pappus 10–12 mm. Four varieties (or subspecies) are traditionally recognized, but their separation is difficult and often arbitrary. All of our material appears assignable (±) to **var. occidentalis**. Sagebrush, grassland, piñon-juniper, ponderosa pine–oak communities, 6200–8200 ft. NE, NC, NW.

Crepis runcinata (E. James) Torr. & A. Gray—Fiddleleaf hawksbeard. Plants perennial, bases pale or green, caudex shallow, not clad with persistent leaf bases; stems 15–65 cm, scapiform, mostly branching distally, glabrous or glabrate, rarely pubescent distally; leaves mostly basal (rosulate), basal leaves linear-oblanceolate to elliptic-obovate, or spatulate, 3–30 × 0.5–4 cm, faces glabrous, often glaucous, margins entire or weakly dentate to shallowly laciniate-runcinate-lobed, apices acute to rounded, cauline leaves filiform-linear, bractlike; heads 2–16 per stem (rarely solitary); involucres obconic-campanulate, 8–20 mm; calyculi of 5–12 lance-deltate bractlets, 1–3 mm; phyllaries 10–16, linear-lanceolate, 8–10 mm, green to sordid-green, faces glabrous or tomentulose, sometimes stipitate-glandular (rarely setose); florets 20–50, ligules 7–18 mm; cypselae flaxen to golden-brown or dark brown, 4–8 mm; pappus 4–9 mm. Saline seeps, springs, ciénegas, other wet meadow and riparian habitats, 4250–10,825 ft.

A number of poorly defined varieties (or subspecies) have been proposed. The following key can be used to separate the 3 variants occurring in our area.

1. Phyllaries stipitate-glandular, often strongly sordid-green throughout (especially when dry)...**var. runcinata**
1'. Phyllaries eglandular, mostly green but shading to sordid-green distally...(2)
2. Leaves obovate to spatulate, 5–10+ × 1–3+ cm...**var. glauca**
2'. Leaves narrowly oblanceolate, < 5 cm, < 1 cm wide...**var. barberi**

var. barberi (Greenm.) B. L. Turner—Barber's hawksbeard. Leaf blades narrowly oblanceolate, 2–4 cm × 3–10 mm, faces glaucous; heads 2–4 per stem; calycular bractlets much shorter than phyllaries; phyllaries mostly green, glabrous (apex sometimes pubescent), eglandular; pappus 7–8 mm. 4600–8875 ft. NC, C, Guadalupe.

var. glauca (Nutt.) B. Boivin—Smooth hawksbeard. Leaf blades oblanceolate to spatulate, 5–15 cm × 10–30 mm, faces glaucous; heads 2–10 per stem; calycular bractlets to ca. 1/2 length of phyllaries; phyllaries mostly green, glabrous (apex sometimes pubescent), eglandular; pappus 5–6 mm. 4250–7225 ft. NE, NC, NW, WC.

var. runcinata—Leaf blades oblanceolate, 5–25 cm × 5–25 mm, faces sometimes glaucous; heads 2–16 per stem; calycular bractlets to ca. 1/2 length of phyllaries; phyllaries usually black-green, stipitate-glandular; pappus 4–6 mm. 4900–10,825 ft. NC, C, WC.

Crepis tectorum L.—Narrowleaf hawksbeard. Plants annual, bases pale or green (not yellow), caudex lacking; stems 10–100 cm, tomentulose and/or hispid; leaves basal and cauline, basal leaves lanceolate to oblanceolate, 5–15 × 1–4 cm, faces glabrous or abaxial faces tomentose, margins entire or dentate to pinnately or irregularly lobed, apices acute to acuminate; heads 5–20+ per stem; involucres cylindro-campanulate, 6–9 × 7–8 mm; calyculi of ~12 subulate bractlets 2–5 mm; phyllaries 12–15, lanceolate, 5–9 mm, faces tomentose to hispidulous; florets 30–70, ligules 10–13 mm; cypselae reddish or purplish brown, 3–4 mm; pappus 4–5 mm. Disturbed areas, 7200–10,170 ft. NC. Introduced.

CYCLACHAENA Fresen. – Carelessweed

Kenneth D. Heil

Cyclachaena xanthiifolia (Nutt.) Fresen. [*Iva xanthiifolia* Nutt.]—Carelessweed. Annuals, 30–120 (–200) cm; stems erect; leaves cauline, mostly opposite, deltate, ovate, or rhombic, blades 6–12(–20+) × 5–12(–18+) cm; margins usually toothed, faces densely to sparsely scabrellous to strigillose, usually gland-dotted; heads ± disciform, in paniculiform arrays; involucres 2–3 × 3–5 mm; phyllaries 10–12+ in 2+ series; pistillate florets 5, corollas whitish; functionally staminate florets 5–10(–20+), corollas whitish; cypselae 2–2.5(–3) mm; pappus 0. Disturbed sites; abandoned fields, floodplains, stream banks, 4000–8700(–10,150) ft. NE, NC, NW, C, Grant (Strother, 2006f).

DIAPERIA Nutt. – Rabbit-tobacco, dwarf cudweed

Kenneth D. Heil

Annuals, 3–25 cm; stems 1, erect, or 2–10, ascending to ± prostrate; leaves basal and cauline, alternate, blades oblanceolate to obovate; heads borne singly or in glomerules; involucres inconspicuous; phyllaries (2–)4–6, ± equal (similar to paleae); receptacles pulvinate to conic (height 0.2–2.4 times diam.), glabrous; pistillate palea bodies with 5+ nerves; staminate or bisexual paleae readily falling (coherent with pistillate); pistillate florets 13–35+; functionally staminate or bisexual florets 2–5; corolla lobes mostly 4; cypselae light to dark brown, faces glabrous, minutely papillate, dull or ± shiny; pappus absent (Morefield, 2006c).

1. Heads in subdichasiform arrays, ± campanulate to spherical, 2–3.3 mm, height ± equal to diam....**D. verna**
1′. Heads in strictly dichasiform or pseudo-polytomous arrays, ellipsoid to ± cylindrical, 3.5–4.5 mm, heights 2–3 times diam....**D. prolifera**

Diaperia prolifera (Nutt. ex DC.) Nutt. [*Evax prolifera* Nutt. ex DC.]—Big-head rabbit-tobacco. Plants grayish green to silvery, 3–15 cm, sericeous to lanuginose; stems mostly 2–10; largest leaves 7–15 × 2–4 mm; heads in strictly dichasiform or pseudo-polytomous arrays, height 2–3 times diam.; receptacles 0.4–0.6 mm or ± 0.9–1.1 mm, height 0.5–0.7 or 2–2.4 times diam.; pistillate paleae imbricate, longest 2.5–4 mm; staminate paleae ± 3, apices erect to somewhat spreading, ± plane; cypselae ± angular, obcompressed, mostly 0.9–1.2 mm.

1. Plants grayish to greenish, loosely lanuginose; heads 4–40+ in glomerules; receptacle height mostly 0.5–0.7 times diam....**var. prolifera**
1′. Plants silvery-white, tightly sericeous; heads borne singly or 2–3 in largest glomerules; receptacle height mostly 2–2.4 times diam....**var. barnebyi**

var. barnebyi Morefield—Barneby rabbit-tobacco. Plants silvery-white, 3–9 cm, tightly sericeous; stems mostly 1–5; branches equal to unequal, distal strictly ascending to erect; largest leaves 7–11 × 2–3 mm; heads borne singly, or 2–3 in largest glomerules; receptacles narrowly conic, 0.9–1.1 mm, height mostly 2–2.4 times diam. Open, dry, shallow, rocky or gravelly soils, usually over limestone or gypsum, sometimes with extra moisture (dry drainages, disturbed places), 4450–4725 ft. Eddy, Lincoln.

var. prolifera—Big-head rabbit-tobacco. Plants mostly grayish green, 3–15 cm, loosely lanuginose; stems mostly 2–10; branches ± equal, distal mostly spreading to ascending; largest leaves 9–15 × 2–4 mm; heads 4–40+ in largest glomerules; receptacles broadly conic, 0.4–0.6 mm, height mostly 0.5–0.7 times diam. Dry, open, often disturbed silty to clay soils, barren to grassy, brushy, or wooded slopes, plains, prairies, usually over carbonate (limestone, chalk). Union.

Diaperia verna (Raf.) Morefield [*Evax verna* Raf.]—Spring or many-stem rabbit-tobacco. Plants greenish to grayish, 2–15(–25) cm, ± lanuginose; stems mostly 2–10; largest leaves 7–13 × 2–4 mm; heads mostly distal, in subdichasiform arrays, 2–3.3 mm, height ± equal to diam.; receptacles 0.3–0.6 mm, height ± 0.2–0.5 times diam.; pistillate paleae scarcely imbricate, longest 1.9–2.7 mm; staminate paleae mostly 3–5, apices somewhat spreading, ± plane; cypselae rounded, ± terete, mostly 0.7–0.9 mm. New Mexico material belongs to **var. verna**. Dry grassy areas, Chihuahuan desert scrub, open, barren areas to plains, often in disturbed urban areas, 4200–4850 ft. EC, C, SE, SC, SW.

DICORIA Torr. & A. Gray – Twinbugs

Kenneth D. Heil

Dicoria canescens A. Gray [*D. brandegeei* A. Gray]—Desert twinbugs. (Ours) annuals or short-lived perennials, 2–8 dm; stems erect, branched; leaves proximally opposite, otherwise alternate, blades lance-linear to lanceolate or deltate, 10–30(–120) × 3–20(–30+) mm, margins entire or toothed, faces sericeous to strigillose, usually gland-dotted; inflorescence a large paniculate cluster; heads disciform with 1–4 marginal pistillate flowers and 5–20 central staminate flowers, 2.5–3 mm; cypselae 3–8 mm; pappus 0. Sandy, alkaline soils, desert washes and dunes, 5300–6700 ft. McKinley, Sandoval, San Juan (Strother, 2006g).

DICRANOCARPUS A. Gray – Dicranocarpus

Kenneth D. Heil

Dicranocarpus parviflorus A. Gray [*Heterosperma dicranocarpum* A. Gray]—Pitchfork. Annuals, 10–70 cm; stems erect, branched dichotomously; leaves opposite, blades 1(2) times pedately or pinnately lobed, 2–8 cm, margins entire; heads radiate, borne singly or (2–4) in cymiform arrays; involucres cylindrical to obconic, 1.5–3 mm diam.; receptacles slightly convex; ray florets 3–6, 2–3 mm, corollas yellow; disk florets 3–4+, 1–3 mm, corollas yellow; cypselae (3–)5–10 mm; pappus persistent, of 2 spreading to recurved, papillate awns, 2–5 mm. Alkaline or gypseous soils, Chihuahuan desert scrub, 3300–6000 ft. (Allison, 2006). EC, C, WC, SE, SC, SW.

DIETERIA Nutt. – Tansy-aster

Kenneth D. Heil and Jennifer Ackerfield

Annuals, biennials, or perennials; leaves alternate, simple; heads radiate or discoid; involucral bracts imbricate in 3–12 series, the base hard and tannish and the apex herbaceous, often bristle-tipped; receptacles naked; ray florets white, pink, purple, or blue, or absent; disk florets yellow; cypselae flattened, smooth or with 8–12 ribs; pappus of barbellate bristles in 1–3 series, persistent (Morgan, 2006a).

1. Phyllaries and peduncles variously hairy, sometimes stipitate-glandular, rarely both prominently stipitate-glandular…(2)
1'. Phyllaries and peduncles both prominently stipitate-glandular…(3)
2. Phyllaries usually hairy throughout, on both indurate bases and herbaceous apices; apices acute to long-acuminate, 1–6 mm; midstem leaf blades 6–20 mm wide…**D. asteroides** (in part)
2'. Phyllaries hairy only on herbaceous apices; apices acute to acuminate, 1–3 mm; midstem leaf blades 1.5–6(–8) mm wide…**D. canescens** (in part)
3. Leaves stiffly stipitate-glandular…**D. asteroides** (in part)
3'. Leaves glabrous or hairy, but not stiffly stipitate-glandular…(4)
4. Midstem leaf blades lanceolate to oblanceolate and 5–15 mm wide, or phyllary apices long-acuminate (2–6 mm), or both…**D. bigelovii**
4'. Midstem leaf blades linear-lanceolate to linear or linear-oblanceolate, 1.5–5 mm wide; phyllary apices acute to acuminate (1–3 mm)…**D. canescens** (in part)

Dieteria asteroides Torr. [*Machaeranthera asteroides* (Torr.) Greene]—Fall tansy-aster. Biennials or short-lived perennials; leaves lanceolate to oblanceolate, midstem blades 20–100 × (2–)6–20 mm, margins entire to irregularly serrate, puberulent or canescent, often sparsely short-stipitate-glandular; involucres broadly turbinate to hemispherical; ray florets white to purple, 10–20 mm.

1. Stems glabrous or stipitate-glandular; leaves stiffly stipitate-glandular…**var. glandulosa**
1'. Stems hairy to canescent, sometimes sparsely glandular; leaves puberulent or canescent, sometimes sparsely stipitate-glandular…**var. asteroides**

var. asteroides [*Aster canescens* Pursh var. *tephrodes* A. Gray]—Fall tansy-aster. Stems hairy to canescent, sometimes sparsely glandular; leaves: midstem blades 6–20 mm wide, margins usually serrate or serrulate, puberulent or canescent. Grasslands, creosote bush scrublands, 4250–9500 ft. NC, NW, C, WC, SC, SW.

var. glandulosa (B. L. Turner) D. R. Morgan & R. L. Hartm. [*Machaeranthera asteroides* (Torr.) Greene var. *glandulosa* B. L. Turner]—Fall tansy-aster. Stems glabrous or densely stipitate-glandular; leaves: mid blades 6–20 mm wide, margins usually serrate, faces sometimes puberulent, stiffly long-stipitate-glandular. Grasslands, creosote bush scrublands, pine-oak woodlands; 1 record at Emory Pass, 8230 ft. Sierra.

Dieteria bigelovii (A. Gray) D. R. Morgan & R. L. Hartm. [*Machaeranthera pattersonii* (A. Gray) Greene]—Bigelow's tansy-aster. Biennials or perennials, 1–10 dm; leaves lanceolate to oblanceolate, margins denticulate to nearly entire, 4–20 cm, glabrous to sparsely hairy and often glandular-stipitate; involucres 8–15 mm, glandular, mostly with reflexed tips; ray florets 10–25 mm, lavender-blue to purple. New Mexico material belongs to **var. bigelovii**. Common along roadsides, on open slopes, and in meadows and forest clearings, 5400–12,000 ft. Widespread, EC, SE.

Dieteria canescens (Pursh) Nutt. [*Aster canescens* Pursh; *Machaeranthera canescens* (Pursh) A. Gray]—Hoary tansy-aster. Biennials or perennials, 1–5 dm; leaves linear-oblanceolate to spatulate, toothed to nearly entire, finely canescent-puberulent, sometimes also glandular-stipitate to nearly glabrous, to 10 cm; involucres 5.5–12 mm, canescent-puberulent and/or glandular, the bracts appressed to reflexed; ray florets blue-purple to pink, 5–15 mm.

1. Phyllaries spreading to reflexed, rarely appressed…**var. aristata** (in part)
1'. Phyllaries usually appressed, sometimes spreading, rarely reflexed…(2)
2. Phyllaries densely appressed-hairy, usually eglandular, sometimes sparsely stipitate-glandular, then obscured by appressed hairs…**var. ambigua**
2'. Phyllaries glabrous or stipitate-glandular, rarely appressed-hairy or canescent…(3)
3. Stems stipitate-glandular, sometimes also canescent, rarely glabrous…**var. aristata** (in part)
3'. Stems glabrous, puberulent, or canescent, sometimes sparsely stipitate-glandular…**var. glabra**

var. ambigua (B. L. Turner) D. R. Morgan & R. L. Hartm. [*Machaeranthera canescens* (Pursh) A. Gray var. *ambigua* B. L. Turner]—Hoary tansy-aster. Annuals, biennials, or short-lived perennials; stems 1+, ascending to erect, canescent, sometimes sparsely stipitate-glandular; mid leaf blades lanceolate to linear; phyllaries in 5–10 series, usually appressed, sometimes spreading, apices densely appressed-hairy, usually eglandular, sometimes sparsely stipitate-glandular, then obscured by appressed hairs; cypselae glabrous or glabrate. Grasslands and pine forests, 4900–8350 ft. NC, NW, C, WC.

var. aristata (Eastw.) D. R. Morgan & R. L. Hartm. [*Aster canescens* Pursh var. *aristatus* Eastw.; *Machaeranthera canescens* (Pursh) A. Gray var. *aristata* (Eastw.) B. L. Turner; *M. rigida* Greene]—Hoary tansy-aster. Annuals or biennials; stems 1+, erect, stipitate-glandular, sometimes also canescent, rarely glabrous; midstem and distal leaf blades linear-lanceolate to linear; phyllaries in 3–6 series, appressed, spreading, or reflexed, apices sparsely to moderately stipitate-glandular, rarely canescent; cypselae sparsely to moderately appressed-hairy. Grasslands, sagebrush scrublands, piñon-juniper woodlands, pine-oak woodlands, 3800–9550 ft. NC, NW, C.

var. glabra (A. Gray) D. R. Morgan & R. L. Hartm. [*Machaeranthera canescens* (Pursh) A. Gray var. *glabra* A. Gray]—Hoary tansy-aster. Annuals or biennials, rarely short-lived perennials; stems usually 1 (sometimes more), erect, glabrous, puberulent, or canescent, sometimes sparsely stipitate-glandular; mid leaf blades lanceolate to linear; phyllaries in 4–8 series, usually appressed, sometimes spreading, rarely reflexed; apices usually glabrous or moderately stipitate-glandular, rarely canescent; cypselae sparsely appressed-hairy. Grasslands, creosote bush scrublands, piñon-juniper woodlands, pine forests, 3500–9100 ft. Widespread except SW.

DYSSODIA Cav. – Dyssodia

Jennifer Ackerfield

Dyssodia papposa (Vent.) Hitchc. [*Tagetes papposa* Vent.]—Fetid marigold. Annuals or perennials, plants 10–30(70+) cm, malodorous; leaves (1)2–3-pinnate, 1.5–5 cm, glabrous or minutely hairy, dotted with oil glands, the uppermost often alternate; heads radiate; involucres 6–10 mm, bracts in 2 series but equal, the bracts with 1-7 oil glands; receptacles naked; ray florets 1.5-2.5 mm, yellow to orange; disk florets ca. 3 mm, yellow to orange; cypselae striate-angled, 3-3.5 mm; pappus of blunt or awned scales. Common in disturbed places, along roadsides, and in fields and gardens, 4100-10,200 ft. Once you smell this plant, you will understand how it got its name "fetid marigold"! Widespread (Ackerfield, 2015).

ECHINACEA Moench – Purple coneflower

Kenneth D. Heil and Steve L. O'Kane, Jr.

Echinacea angustifolia DC.—Narrow-leaved purple coneflower. Perennial herbs; plants to 70 cm; leaves alternate, elliptic to lanceolate, 7-30 × 0.5-4 cm, margins entire; heads radiate; involucral bracts in 3 or 4 series, imbricate, narrow, herbaceous and intergrading into the chaffy bracts; phyllaries lanceolate to ovate; ray corollas pink to purplish, laminae reflexed; disk corollas 5-7 mm, lobes usually purple; cypselae 4-angled, 4-5 mm. Dry prairies, barrens, rocky to sandy-clay soils, 4385-7000 ft. Quay, Torrance, Union (Ackerfield, 2015).

ECLIPTA L. – False daisy

Don Hyder

Eclipta prostrata (L.) L.—False daisy. Annual or perennial herbs, 10-50(-70+) cm; stems erect or decumbent, branched from bases and/or distally (sometimes rooting at proximal nodes); leaves cauline, petiolate or sessile, opposite, blades 2-10 cm × 4-30+ mm; heads radiate, in loose, corymbiform arrays or borne singly; ray florets 20-40 (in 2-3+ series), corollas white or whitish, ray laminae ca. 2 mm; disk corollas ca. 1.5 mm; cypselae ca. 2.5 mm. Wet places, often ruderal, 3799-10679 ft. SC, SW, Chaves, Socorro (Allred & Ivey, 2012; Strother, 2006h).

ENCELIA Adans. – Brittlebush

Kenneth D. Heil

Perennials, subshrubs, or shrubs, (10-)30-150 cm; stems erect, usually branched from base, often throughout; leaves usually cauline, sometimes basal (ours), alternate, blades oblanceolate to linear or ovate to deltate, margins usually entire, rarely toothed, faces hirtellous to scabrellous or sparsely canescent; heads radiate or discoid, borne singly; involucres ± hemispherical or broader, 9-22 mm diam.; phyllaries persistent, 18-30(-50+) in 2-3+ series; ray florets 8-40, corollas yellow; disk floret corollas yellow; cypselae strongly compressed, obovate to cuneate, 5-8 mm; pappus usually 0, or bristlelike awns (Clark, 2006).

1. Perennial herbs; leaves mostly basal [to be expected]...E. scaposa
1'. Shrubs; leaves cauline...**E. virginensis**

Encelia scaposa (A. Gray) A. Gray [*E. scaposa* var. *stenophylla* Shinners]—Onehead brittlebush. Perennials, 10-30(-60) cm; leaves mostly basal, blades greenish to cinereous, mostly narrowly oblanceolate to linear, 30-100 mm, mostly 1-8 mm wide, faces ± hirtellous to scabrellous; heads borne singly; involucres 12-22 mm; ray florets 20-40; disk corollas yellow, ca. 5 mm; cypselae cuneate to obovate, ca. 5 mm, faces ± villous; pappus readily falling, of 2 bristlelike awns. Rocky desert slopes, 4250-5575 ft. No vouchers seen; to be expected.

Encelia virginensis A. Nelson [*E. frutescens* (A. Gray) A. Gray var. *virginensis* (A. Nelson) S. F. Blake]—Virginia River brittlebush. Shrubs, 50-150 cm; leaves cauline, blades gray-green, narrowly ovate to deltate, 12-25 mm, faces sparsely canescent and strigose; heads borne singly; involucres 9-13 mm; ray florets 11-21; disk corollas yellow, 5-6 mm; cypselae 5-8 mm; pappus usually 0, rarely of 1-2 bristlelike awns. Desert flats, rocky slopes, roadsides, 4150-4750 ft. Grant, Hidalgo, Luna.

ENGELMANNIA A. Gray ex Nutt. – Engelmann's daisy

Kenneth D. Heil

Engelmannia peristenia (Raf.) Goodman & C. A. Lawson [*E. pinnatifida* Nutt.]—Engelmann's daisy, chocolate flower. Perennials, 20–50(–100) cm; stems erect, strigose, hispid, or hirsute; leaves basal and cauline, alternate, oblong to lanceolate, usually 1(2) times pinnately lobed, basal blades (5–)10–30 cm, margins lobed or toothed; involucres 6–10 mm; disk florets perfect, sterile, partly enclosed by the receptacle chaff; ray florets 8–10, 10–16 mm, yellow; pappus an irregular crown of ciliate scales, 1–2 mm; cypselae 3–4 mm. Desert scrub, grasslands, piñon-juniper woodlands, 3480–8000 ft. (Keil, 2006f).

ERICAMERIA Nutt. – Rabbitbrush, goldenbush

Gary I. Baird and Kenneth D. Heil

Shrubs 10–500 cm, often scopiform; stems erect to spreading, glabrous with tan bark, or densely to-mentose to felty, concealing the stem surface, often gland-dotted or resinous; leaves cauline, alternate, petiolate or sessile, blades filiform to spatulate, flat to sulcate, mostly straight, margins entire, flat or wavy, faces glabrous or tomentose, sometimes stipitate-glandular or punctate-glandular and resinous-viscid; heads radiate or discoid, solitary or in loose to dense clusters, forming racemiform to cymiform or paniculiform arrays; involucres subcylindrical to campanulate; phyllaries 10–60 in 3–7 series, subequal to imbricate, spiraled to strongly aligned, faces glabrous or tomentose, sometimes resinous; ray florets 0 or 3–6, pistillate, corollas yellow; disk florets 5–20, bisexual, corollas yellow; cypselae cylindrical to turbinate, often prismatic, tan to fulvous, glabrous or hairy; pappus of 20–60 subequal, slender bristles, beige to tan (Urbatsch et al., 2006b).

Reports of *E. linearifolia* from the study area have not been verified and would be far outside its normal range. It shares similar features with *E. laricifolia* and the two may be mistaken for each other, but *E. linearifolia* blooms in late spring to early summer, while *E. laricifolia* blooms from late summer into fall.

1. Stems and leaves glabrous, usually gland-dotted and often viscid…(2)
1'. Stems and leaves tomentose, this sometimes tightly matted and appearing smooth, glandular or not…(3)
2. Leaves spatulate, 4–16 mm wide; heads discoid…**E. cuneata**
2'. Leaves filiform to narrowly oblanceolate, 1–2 mm wide; heads radiate…**E. laricifolia**
3. Phyllaries subequal, not or weakly aligned, outer leaflike, often exceeding inner; disk florets 5–20…
   E. parryi
3'. Phyllaries unequal, often strongly aligned, outer bractlike, shorter than inner; disk florets 4–6…
   E. nauseosa

Ericameria cuneata (A. Gray) McClatchie [*Haplopappus cuneatus* A. Gray]—Cliff goldenbush. Plants 10–100 cm; stems glabrous, resinous; leaves short-petiolate, blades spatulate, 10–25 × 4–16 mm, faces glabrous, glandular-viscid; heads discoid in compact to loose cymiform arrays, or solitary, involu-cres cylindrical-turbinate, 8–12 × 5–7 mm; phyllaries 20–60, lanceolate to linear-oblong, unequal, spi-rally arranged; ray florets 0; disk florets 7–15, corollas 5–6 mm; cypselae 2.5–3 mm, villous; pappus 6.5–8 mm. Our material belongs to **var. spathulata** (A. Gray) H. M. Hall. Rocky hillsides and canyon walls, 4450–6000 ft. Hidalgo.

Ericameria laricifolia (A. Gray) Shinners [*Haplopappus laricifolia* A. Gray]—Turpentine bush. Plants 30–100 cm; stems glabrous, gland-dotted, resinous; leaves sessile, blades linear, subterete, 10–20 × 1–2 mm, faces glabrous, glandular-viscid; heads radiate, in cymiform arrays; involucres turbinate, 3–5 × 3–5 mm; phyllaries 12–20, linear to lanceolate, unequal, spirally arranged; ray florets 3–6, laminae 4–5 × 1–2 mm; disk florets 6–16, corollas 5–6 mm; cypselae 3.5–4 mm, villous; pappus 3.5–5 mm. Mesas, canyons, rock walls, Chihuahuan desert scrub, oak, juniper, and piñon pine woodlands, 4490–8000 ft. WC, SC, SW.

Ericameria nauseosa (Pursh) G. L. Nesom & G. I. Baird [*Chrysothamnus nauseosus* (Pursh) Britton]— Feltstem rabbitbrush, chamisa. Plants 10–250 cm; stems whitish, greenish, or yellowish green, surface concealed by a tightly matted, feltlike tomentum; leaves sessile, blades linear-filiform to linearly oblan-ceolate, 10–90 × 0.5–3 mm, faces densely gray-tomentose to subglabrous and greenish, obscurely glandular; heads discoid, in conic to flat-topped, cymiform arrays; involucres obconic to subcylindrical, 6–20 × 2–4 mm; phyllaries 10–30, 1.5–14 × 0.7–1.5 mm, unequal, aligned in strong to weak vertical ranks; ray florets 0; disk florets (4)5(6), corollas 6–14 mm; cypselae 3–8 mm, glabrous or hairy; pappus 4–13 mm.

1. Involucres 15–20 mm; corollas 12–14 mm...**var. arenaria**
1'. Involucres 6–12 mm; corollas 6–11 mm...(2)
2. Cypselae glabrous...(3)
2'. Cypselae hairy...(5)
3. Stems yellow-green, tomentum relatively compact; leaves 30–50 mm...**var. nitida** (in part)
3'. Stems gray-white, tomentum relatively loose; leaves 10–35 mm...(4)
4. Plants 2–12 dm; leaves 0.5–1 mm wide; pappus 7.5–11 mm...**var. bigelovii**
4'. Plants 2–5 dm; leaves 1–1.5 mm wide; pappus 5–7.5 mm...**var. texensis**
5. Pappus 9–10 mm; corollas 9.5–11 mm...**var. nitida** (in part)
5'. Pappus 4–7.5 mm; corollas 6–9.5 mm...(6)
6. Stems grayish to whitish, tomentum relatively loose; outer phyllaries tomentose...(7)
6'. Stems yellow-green to green or gray-green, tomentum relatively compact; outer phyllaries glabrous or puberulent...(8)
7. Plants 4–12 dm; inner phyllaries mostly linear-lanceolate (widest at or below the middle), acute, tomentose...**var. hololeuca**
7'. Plants 13–20 dm; inner phyllaries mostly lance-ovate to oblanceolate (widest at or above the middle), obtuse, glabrous...**var. latisquamea**
8. Leaves to 1 mm wide, 1-veined; corolla lobes 1.5–2.5 mm; style appendages equaling or shorter than stigmatic portion...**var. oreophila**
8'. Leaves 1–3 mm wide, mostly 3-veined; corolla lobes 0.5–1.5 mm; style appendages longer than stigmatic portion...**var. graveolens**

var. arenaria (L. C. Anderson) G. L. Nesom & G. I. Baird [*Chrysothamnus nauseosus* (Pall. ex Pursh) Britton var. *arenarius* (L. C. Anderson) S. L. Welsh]—Navajo rabbitbrush. Plants 70–120 cm; stems grayish green, moderately leafy, ± compactly tomentose; leaves grayish green, blades linear, 40–75 × 0.8–1.1 mm, faces tomentose; involucres 16–19 mm; phyllaries (19–)22–27; corollas 12–14 mm, tubes glabrous, lobes 1.2–2 mm, glabrous; style appendages longer than stigmatic portions; cypselae hairy; pappus 9.5–12.5 mm. Sand hills and sandstone cliffs, 4920–6560 ft. Sandoval, San Juan.

var. bigelovii (A. Gray) G. L. Nesom & G. I. Baird [*Chrysothamnus nauseosus* (Pall. ex Pursh) Britton var. *bigelovii* (A. Gray) H. M. Hall]—Bigelow's rabbitbrush. Plants 15–120 cm; stems whitish, nearly leafless at flowering, loosely tomentose; leaves mostly grayish white, blades filiform, 15–30 × 0.5–1 mm, faces tomentulose; involucres 10.5–12.5 mm; phyllaries 14–18(–25); corollas 9–11 mm, tubes glabrous or puberulent, lobes 0.8–1.5 mm, glabrous; style appendages longer than stigmatic portions; cypselae glabrous; pappus 7–11 mm. Desert grasslands, piñon-juniper, ponderosa pine, mixed conifer communities, 4900–8150 ft. NC, NW, C, WC.

var. graveolens (Nutt.) Reveal & Schuyler [*Chrysothamnus nauseosus* (Pall. ex Pursh) Britton var. *graveolens* (Nutt.) H. M. Hall]—Plains rabbitbrush. Plants 50–160 cm; stems yellowish green to greenish white, leafy, compactly tomentose; leaves greenish, blades broadly linear, 30–90 × 1–3 mm, faces weakly tomentulose; involucres 6–8 mm; phyllaries 14–18, abaxial faces usually glabrous, outer sometimes sparsely hairy; corollas 7–9+ mm, tubes puberulent or glabrous, lobes 0.5–1.5 mm, glabrous; style appendages longer than stigmatic portions; cypselae densely hairy; pappus 5–6.5 mm. Plains, desert grasslands, piñon-juniper, ponderosa pine, mixed conifer communities, 3700–11,000 ft. NE, NC, NW, EC, C, WC, SE.

var. hololeuca (A. Gray) G. L. Nesom & G. I. Baird [*Chrysothamnus nauseosus* (Pall. ex Pursh) Britton var. *hololeucus* (A. Gray) H. M. Hall]—White rabbitbrush. Plants 40–120 cm (pleasantly fragrant); stems whitish, leafy, densely to loosely tomentose; leaves whitish, blades filiform to linear, 10–35 × 1–2 mm, faces densely tomentose; involucres 7–9 mm; phyllaries 14–16, abaxial faces tomentose; corollas 7–9.5 mm, tubes puberulent to cobwebby, lobes 0.5–1 mm, glabrous; style appendages shorter than stigmatic portions; cypselae densely hairy; pappus 5–7 mm. Gravelly or sandy slopes, piñon-juniper and ponderosa pine communities, 5720–8660 ft. NC, NW, C, WC.

var. latisquamea (A. Gray) G. L. Nesom & G. I. Baird [*Chrysothamnus nauseosus* (Pall. ex Pursh) Britton var. *latisquameus* (A. Gray) H. M. Hall]—Madrean rabbitbrush. Plants 130–200 cm; stems white, leafy, loosely tomentose; leaves whitish, blades filiform, 15–45 × 1–2 mm, faces loosely tomentose; involucres 7.5–10.5 mm; phyllaries 15–17, obtuse, outer abaxial faces densely tomentulose, inner glabrous; corollas 7–9.5 mm, tubes glabrous or puberulent, lobes 0.6–1 mm, glabrous; style appendages equaling or shorter than stigmatic portions; cypselae densely hairy; pappus 5–7.5 mm. Plains, desert grasslands, piñon-juniper, ponderosa pine, mixed conifer communities, 5300–10,680 ft. NC, C, WC, SC, SW.

var. nitida (L. C. Anderson) G. L. Nesom & G. I. Baird [*Chrysothamnus nauseosus* (Pall. ex Pursh) Britton var. *nitidus* (L. C. Anderson) S. L. Welsh]—Arroyo rabbitbrush. Plants 60–150 cm; stems yellowish green, leafy at flowering, densely tomentose; leaves yellowish green, blades linear, 30–50 × 1–1.5 mm, faces glabrate; involucres 10–12.5 mm; phyllaries 13–19, outer abaxial faces scurfy-tomentulose, inner glabrous; corollas 9.5–11 mm, tubes mostly glabrous, lobes 0.7–1 mm, glabrous or villous; style appendages longer than stigmatic portions; cypselae usually glabrous, sometimes hairy; pappus 9–10 mm. Sandy gravels of dry streambeds, *Atriplex* and piñon-juniper communities, 4900–6600 ft. Cibola, Rio Arriba, San Juan, Taos.

var. oreophila (A. Nelson) G. L. Nesom & G. I. Baird [*Chrysothamnus nauseosus* (Pall. ex Pursh) Britton var. *oreophilus* (A. Nelson) H. M. Hall; *C. nauseosus* var. *consimilis* (Greene) H. M. Hall]—Great Basin rabbitbrush. Plants 70–250 cm; stems greenish to grayish, usually leafy, compactly tomentose; leaves yellowish green, blades filiform to linear, 20–50 × 0.8–1 mm, faces tomentose to glabrate; involucres 6.5–10 mm; phyllaries 11–20, apices erect, acute, glabrous; corollas 6–9 mm, tubes glabrous, lobes 1.5–2.5 mm, glabrous; style appendages shorter than or equaling stigmatic portions; cypselae hairy; pappus 4–7.5 mm. Alkaline valleys and plains, piñon-juniper, ponderosa pine, mixed conifer communities, 5500–9500 ft. NE, NC, NW, WC.

var. texensis (L. C. Anderson) G. L. Nesom & G. I. Baird [*Chrysothamnus nauseosus* (Pall. ex Pursh) Britton subsp. *texensis* L. C. Anderson]—Guadalupe rabbitbrush. Plants 20–40(–50) cm; stems whitish, leafy, loosely tomentose; leaves grayish green, blades linear, 10–35 × 1(–1.5) mm, faces loosely tomentulose; involucres 8–11.5 mm; phyllaries 16–23 (outer margins ciliate), outer abaxial faces tomentose, glabrescent; corollas 8.5–11 mm, tubes glabrous or sparsely puberulent, lobes 0.6–1.3 mm, glabrous; style appendages shorter than stigmatic portions; cypselae glabrous; pappus 5–7.45 mm. Guadalupe Mountains; limestone cliffs and among boulders, less common on gravel alluvium of streambeds; Chihuahuan desert scrub, piñon-juniper woodlands, 4900–7000 ft. Eddy, Otero. **R**

Ericameria parryi (A. Gray) G. L. Nesom & G. I. Baird [*Linosyris parryi* A. Gray]—Parry's rabbitbrush. Plants 10–100 cm; stems ascending to erect, greenish when young becoming tan and gray; leaves erect to ascending or spreading; blades linear to spatulate 10–80 × 0.5–8(–14) mm, faces glabrous or gray, greenish, or yellowish tomentose; heads usually in congested, racemiform or cymiform clusters; involucres subcylindric, 9–18 × 4–8 mm; phyllaries 10–20 in 3–6 series, tan, ovate to lanceolate or elliptic, 5–11+ × 0.7–2 mm; ray florets 0; disk florets 5–20; corollas 8–12.5 mm; cypselae tan, narrowly ellipsoid to subturbinate; pappus off-white to brown, 3.3–7.5 mm.

1. Distalmost leaves overtopping arrays…(2)
1'. Distalmost leaves usually shorter than or equaling arrays…(3)
2. Leaf faces minutely glabrous or puberulent, often minutely stipitate glandular; florets 8–20…**var. parryi**
2'. Leaf faces tomentulose to tomentose, sometimes gland-dotted; florets 5–7(8)…**var. howardii**
3. Leaf faces tomentulose; corolla lobes 0.7–1 mm…**var. affinis**
3'. Leaf faces glabrous or sparsely hairy, sometimes viscid; corolla lobes 1–2 mm…**var. attenuata**

var. affinis (A. Nelson) G. L. Nesom & G. I. Baird [*Chrysothamnus affinis* A. Nelson; *C. parryi* (A. Gray) Greene subsp. *affinis* (A. Nelson) L. C. Anderson; *C. parryi* var. *affinis* (A. Nelson) Cronquist]—Parry's rabbitbrush. Plants 15–60 cm; leaves crowded, greenish yellow; blades 1-nerved, linear, 25–40 × 1–3 mm, faces tomentulose, eglandular; distalmost shorter than arrays; heads 5–10+ in racemiform arrays; involucres 10.5–13 mm; phyllaries 12–15, outer herbaceous or herbaceous-tipped, otherwise chartaceous;

florets 5–8; corollas pale yellow, 8–9.1 mm, tubes glabrous or puberulent, throats abruptly dilated, lobes 0.7–1 mm. Mountain slopes, 6400–7425 ft. Rio Arriba, Sandoval, Taos.

var. attenuata (M. E. Jones) G. L. Nesom & G. I. Baird [*Bigelowia howardii* (Parry ex A. Gray) A. Gray var. *attenuata* M. E. Jones; *Chrysothamnus parryi* (A. Gray) Greene subsp. *attenuatus* (M. E. Jones) H. M. Hall & Clements]—Narrow-bract rabbitbrush. Plants 20–60 cm; leaves moderately crowded, green; blades 1-nerved, linear, 20–40 × ca. 1 mm, faces glabrous or sparsely hairy, eglandular, somewhat viscid; distalmost shorter than arrays; heads 5–10 in compact, racemiform arrays; involucres 11–12.5 mm; phyllaries 13–22, apices erect, attenuate; florets 5–7; corollas clear yellow, 10–11 mm, tubes glabrous, throats gradually dilated, lobes 1.5–2 mm. Mountain slopes, dry, rocky soils, sagebrush, piñon-juniper, ponderosa pine, and aspen communities, 6200–9200 ft. NW.

var. howardii (Parry ex A. Gray) G. L. Nesom & G. I. Baird [*Linosyris howardii* Parry ex A. Gray; *Chrysothamnus parryi* (A. Gray) Greene subsp. *howardii* (Parry ex A. Gray) H. M. Hall & Clements]—Howard's rabbitbrush. Plants 30–60 cm; leaves crowded, gray; blades 1-nerved, linear, 20–40 × ca. 1 mm, faces tomentose, eglandular; distalmost overtopping arrays; heads 5–10+ in compact, racemiform arrays, or terminal glomerules; involucres 9.5–12 mm; phyllaries 12–20, mostly chartaceous, apices spreading, acuminate; florets 5–7; corollas pale yellow, 8–10.7 mm, tubes hairy, throats gradually dilated, lobes 1.2–1.7 mm. Sagebrush, piñon-juniper, ponderosa pine, 6890–9000 ft. NW, NC.

var. parryi—Plants 30–100 cm; leaves usually crowded, green; blades 3-nerved (only midnerves prominent), linear, 30–80 × 2–3 mm, faces glabrous or puberulent, often minutely stipitate-glandular, ± resinous; distalmost overtopping arrays; heads usually 5–12+ in racemiform arrays; involucres 9–12 mm; phyllaries 10–15, mostly chartaceous, outermost sometimes herbaceous-tipped, apices erect, attenuate; florets 8–20; corollas yellow, 7.9–10 mm, tubes sparsely hairy, throats gradually dilated, lobes 1.4–2.1 mm. Sagebrush, piñon-juniper, ponderosa pine, mixed conifer, 5600–10395 ft. NW, NC.

ERIGERON L. – Fleabane

Guy L. Nesom

Herbs, annual, biennial, or perennial, glabrous or hairy, commonly glandular; taprooted, fibrous-rooted, or rhizomatous and fibrous-rooted, sometimes stoloniferous; leaves basal and/or cauline, alternate; heads usually radiate, sometimes discoid or disciform, single or in a loose corymb or panicle; phyllaries in 2–5 subequal series; ray florets in 1(2+) series, sometimes 0, corollas white to bluish or purplish to pink, ligules sometimes absent; cypselae oblong to oblong-obovoid, compressed to flattened with 2(–4) ribs or subterete with 5–14 ribs, glabrous or strigose or sericeous, eglandular; pappus of persistent or deciduous barbellate bristles, usually with outer setae or scales, sometimes only on ray or only on disk cypselae, sometimes absent (Nesom, 2006i).

Records of *E. consimilis* Cronquist and *E. engelmannii* A. Nelson from New Mexico are based on misidentifications of *E. pulcherrimus* A. Heller and *E. abajoensis* Cronquist. Records of *E. caespitosus* Nutt. are based mostly on collections of *E. abajoensis*. *Erigeron strigosus* Muhl. ex Willd. may occur along roadsides or in disturbed sites, but no vouchers from New Mexico have been seen for this account.

1. Ray florets either absent or present but with ligules absent or extremely short…(2)
1'. Ray florets present, with prominent ligules…(7)
2. Plants perennial; ray florets absent…**E. aphanactis**
2'. Plants annual; ray florets present but ligules absent or extremely short…(3)
3. Phyllaries loose, ca. equal in length, 4–6 mm; heads on bracteate or ebracteate peduncles…(4)
3'. Phyllaries rigid, graduated in length, 1.5–3.5 mm; heads on leafy or prominently bracteate peduncles…(5)
4. Heads usually in loose, racemiform arrays; pistillate florets all with a filiform lamina; pappus bristles not accrescent…**E. lonchophyllus**
4'. Heads usually in corymbiform arrays; pistillate florets in 2 series, outer with nearly filiform ligules, inner essentially without a ligule; pappus bristles accrescent…**E. acris**
5. Plants branched throughout and spreading, without a main axis, 5–25 cm…**E. divaricatus**
5'. Plants erect, branched mostly distally, with a main axis, (3–)10–200 cm…(6)

6. Stems hispidulous; receptacles 3–5 mm diam. in fruit; involucres hispidulous to loosely strigose; pistillate florets mostly 60–150…**E. bonariensis**

6'. Stems glabrous to very sparsely strigose; receptacles 1–1.5(–3) mm diam. in fruit; involucres glabrous; pistillate florets 20–30(–45)…**E. canadensis**

7. Annual…(8)

7'. Perennial…(13)

8. Plants producing long, sparsely leafy, aboveground runners or stolons…(9)

8'. Plants without aboveground runners or stolons…(10)

9. Stems sparsely strigose with closely appressed hairs; stolons often rooting at nodes…**E. flagellaris** (in part)

9'. Stems densely hirsute with distinctly deflexed hairs; stolons not rooting at nodes…**E. tracyi** (in part)

10. Ray and disk cypselae without pappus bristles…**E. versicolor**

10'. Ray and disk cypselae with pappus bristles…(11)

11. Fibrous-rooted; cauline leaves clasping…**E. philadelphicus**

11'. Taprooted; cauline leaves not clasping…(12)

12. Stems hirsutulous with upcurved hairs, eglandular; outer pappus a cartilaginous crown; some ray florets positioned among the inner phyllaries…**E. bellidiastrum**

12'. Stems hirsute with patent hairs, minutely glandular; outer pappus of minute scales and setae; all ray florets inside the phyllaries…**E. divergens**

13. Taprooted…(14)

13'. Rhizomatous, without a taproot…(35)

14. Plants producing long, sparsely leafy, aboveground runners or stolons…(15)

14'. Plants without aboveground runners or stolons…(16)

15. Stems sparsely strigose with closely appressed hairs; stolons often rooting at nodes…**E. flagellaris** (in part)

15'. Stems densely hirsute with distinctly deflexed hairs; stolons not rooting at nodes…**E. tracyi** (in part)

16. Stems 10–30 cm, brittle, subintricately branched, stiffly hispidulous, eglandular…**E. bigelovii**

16'. Stems not as above…(17)

17. Early-season plants; stems basally reddish; basal leaves large, toothed or lobed; single, large heads on ebracteate peduncles (later-season plants have branching stems with more numerous, much smaller, diffusely arranged heads)…**E. modestus**

17'. Not early-season plants; stems, leaves, and heads not as above…(18)

18. Caudex multicipital, sometimes irregularly branched; basal and lower cauline leaves long-petiolate, usually with toothed or lobed blades; upper portions of plant like *E. divergens*…**E. incomptus**

18'. Caudex and leaves not as above…(19)

19. Petioles prominently ciliate with spreading hairs…(20)

19'. Petioles not ciliate with spreading hairs…(21)

20. Disk corollas hirsute-strigose with sharp-pointed hairs; outer pappus of scales 0.2–0.5 mm, inner of (7–)10–14(15) bristles…**E. concinnus**

20'. Disk corollas glabrous or slightly puberulent with glandular-viscid, blunt-tipped hairs; outer pappus of setae or subulate scales, 0.1–0.3 mm, inner of 12–27 bristles…**E. pumilus**

21. Leaves deeply lobed or pinnatisect…(22)

21'. Leaves entire…(25)

22. Stems (8–)20–90 cm; heads (1–)5–25(–50); ray ligules white, reflexing; pappus bristles (8–)10–12, basally deciduous…(23)

22'. Stems 4–11 cm; heads 1; ray ligules white to blue, pink, or purplish, reflexing or not; pappus bristles 12–30, persistent…(24)

23. Stems, leaves, and phyllaries eglandular or glands minute and noncapitate, otherwise usually strigose… **E. neomexicanus**

23'. Stems, leaves, and phyllaries densely glandular (glands relatively large, capitate), otherwise glabrous or stems sparsely hirsute-villous…**E. oreophilus**

24. Leaves ternately lobed or dissected; ray ligules white to pink or blue, not reflexing…**E. compositus**

24'. Leaves pinnatisect; ray ligules blue to purplish, reflexing…**E. pinnatisectus**

25. Stems densely and prominently glandular…**E. vetensis**

25'. Stems eglandular…(26)

26. Stems glabrous or sparsely strigose; leaves glabrous to glabrate; 9850–11,500 ft.…**E. subglaber**

26'. Stems distinctly strigose; leaves strigose or hirsute; 4100–9300 ft. (–11,400 ft. in *E. abajoensis*)…(27)

27. Phyllaries predominantly glandular, sometimes also sparsely hairy…(28)

27'. Phyllaries predominantly strigose or hirsute to villous-hirsute, sometimes also obscurely glandular…(30)

28. Leaves linear; ray ligules coiling at maturity; cypsela faces glabrous…**E. sivinskii**

28'. Leaves oblanceolate to spatulate; ray ligules not coiling or reflexing at maturity; cypsela faces strigose…(29)

29. Stems 3.5–8 cm; basal leaves 8–23, 1.5–3 mm wide, cauline linear; phyllaries narrowly elliptic-lanceolate, herbaceous without a broad medial line, purplish at least distally, glandular and hirsute…**E. acomanus**
29′. Stems 8–15 cm; basal leaves 20–32 mm, 5–7 mm wide, cauline oblanceolate; phyllaries narrowly oblong-lanceolate with a broad, orange medial line, minutely glandular, otherwise glabrate…**E. spellenbergii**
30. Cypselae glabrous, (8–)10–14-nerved; heads 1…**E. canus**
30′. Cypselae strigose, (2–)4(5)-, 3–4-, or 6–8-nerved; heads 1 or few…(31)
31. Stems and leaves loosely strigose to hirsute; phyllaries hirsute; rays reflexing…**E. abajoensis**
31′. Stems and leaves strigose; phyllaries strigose or villous-hirsute; rays not reflexing…(32)
32. Stems green or often purple at the base, leaves green, basal 5–11(–19) cm; rays not coiling or reflexing; cypselae 2-nerved…**E. eatonii**
32′. Stems and leaves gray to silvery, closely strigose, basal 1–7 cm; rays coiling; cypselae (2–)4(5)-, 3–4-, or 6–8-nerved…(33)
33. Basal leaves usually deciduous by flowering; heads (1–)3–10; ray ligules 1–1.5 mm wide; cypselae 3–4-nerved…**E. sparsifolius**
33′. Basal leaves persistent; heads 1; ray ligules 1.5–2.8 mm wide…(34)
34. Ray ligules 1.5–2.2 mm wide; phyllaries strigose; cypselae 6–8-nerved…**E. argentatus**
34′. Ray ligules 1.6–2.8 mm wide; phyllaries villous-hirsute; cypselae (2–)4(5)-nerved…**E. pulcherrimus**
35. Hairs of involucres with prominently colored crosswalls…(36)
35′. Hairs of involucres with colorless crosswalls…(38)
36. Crosswalls reddish to purple…**E. elatior**
36′. Crosswalls black…(37)
37. Stems 10–70 cm; leaves basal and cauline, broadly oblanceolate to elliptic or oblong-lanceolate, 4–12 (–15) cm; 8900–12,100 ft.…**E. coulteri**
37′. Stems 3–12(–21) cm; leaves basal, spatulate to oblanceolate, (1–)2–5(–15) cm; 11,600–13,000 ft.…
   E. melanocephalus
38. Plants mainly fibrous-rooted, without distinct rhizomes or a caudex…**E. glabellus**
38′. Plants distinctly rhizomatous, with or without a caudex…(39)
39. Stems arising directly from rhizomes, without a caudex…(40)
39′. Stems arising from a thickened, lignescent caudex…(42)
40. Stems 25–45 cm; cauline leaves essentially all linear, little reduced from the proximal; cypselae 3.5–4.5 mm, 5–6-nerved, nearly glabrous…**E. rhizomatus**
40′. Stems 0.5–3.5 cm or 4–12(–15) cm; cauline leaves strongly reduced from basal or bracteate; ray ligules reflexing; cypselae 1.5–2.2 mm, 2(3)-nerved, sparsely strigose…(41)
41. Stems 4–12(–15) cm; basal leaves 1.5–7 cm; phyllaries in 2–3 series…**E. leiomerus**
41′. Stems 0.5–3.5 cm; basal leaves 0.5–1.2 cm; phyllaries in 3–4(5) series…**E. scopulinus**
42. Leaves all cauline (basal absent or deciduous)…(43)
42′. Leaves basal and cauline…(45)
43. Stems, leaves, and involucres densely stipitate-glandular with long-stiped hairs…**E. vreelandii**
43′. Stems, leaves, and involucres eglandular or minutely glandular…(44)
44. Leaves and involucres glabrous to sparsely strigose-hirsute, often minutely glandular distally…
   E. speciosus
44′. Leaves and involucres hirsute to strigose-hirsute, eglandular…**E. subtrinervis**
45. Rhizomes slender, producing scale-leaved runners or stoloniform rhizomes…(46)
45′. Rhizomes relatively thick, sometimes woody, fibrous-rooted…(47)
46. Stems usually glabrous, minutely glandular to nearly eglandular; basal leaves persistent; phyllaries usually glabrous, minutely glandular…**E. eximius**
46′. Stems hirsute to hirtellous with deflexed hairs, eglandular or sparsely glandular; basal leaves usually withering by flowering; phyllaries pilose-hirsute, sometimes minutely glandular…**E. rybius**
47. Stems strigillose with loosely appressed, slightly crinkled hairs (most densely so immediately below heads); phyllaries usually glabrous, densely and evenly stipitate-glandular, without other vestiture; ray ligules 1.5–3 mm wide; cypselae 2.5–2.8 mm, (4)5(–7)-nerved…**E. glacialis**
47′. Stems mostly hirsute to strigose with straight hairs; phyllaries glandular, hairy, or mixed; ray ligules 0.5–1.5 mm wide; cypselae 1.1–2.1 mm, 2-nerved…(48)
48. Heads 1; phyllaries woolly-villous; 10,200–12,900 ft.…**E. grandiflorus**
48′. Heads 1–6; phyllaries pilose-hirsute, strigose-hirsute, hirsute-villous, or glabrate; 7000–11,600 ft.…(49)
49. Rhizomes with short, slender branches; habit cespitose; petiole and proximal blade margins spreading-ciliate…**E. hessii**
49′. Rhizomes unbranched; habit not cespitose; petiole margins eciliate…(50)
50. Stems erect from the base; ray florets 25–80; ligules 1–1.5 mm wide…**E. arizonicus**
50′. Stems decumbent-ascending to ascending from the base; ray florets 75–150; ligules 0.5 mm wide…
   E. formosissimus

Erigeron abajoensis Cronquist [*E. awapensis* S. L. Welsh]—Abajo fleabane. Taprooted perennials with relatively thick, short caudex branches retaining old, fibrous leaf bases; stems 5–25(–30) cm, ascending to decumbent, loosely strigose to hirsute or hirtellous, eglandular; leaves basal and cauline, basal narrowly oblanceolate to oblong-oblanceolate, narrowly spatulate-oblanceolate or nearly linear, entire, (1–)3-nerved, 1.5–9(–12) cm, hirtellous, eglandular, cauline linear to linear-oblong or narrowly oblanceolate, gradually reduced distally; heads 1–4 on branches from distal 1/2 of stem; phyllaries hirsute-villous to hirtellous, minutely glandular; ray ligules blue to pink or white, 5–6 mm, reflexing. Rocky or gravelly slopes, talus, cliffs, crevices, sandstone outcrops, Gambel oak, sagebrush, piñon-juniper, ponderosa pine, subalpine meadows, (6800–)8700–11,400 ft. San Juan (Nesom, 2020a).

Erigeron acomanus Spellenb. & P. J. Knight—Acoma fleabane. Taprooted perennials with lignescent rhizomes and caudex branches, forming mats; stems 3.5–8 cm, strigillose with stiff, whitish hairs, eglandular; leaves mostly basal in rosettes, narrowly oblanceolate to obovate or spatulate, entire, 8–23 mm, blades 1.5–3 mm wide, strigillose, cauline reduced, linear; heads 1; involucres 6–8 mm wide (pressed), phyllaries loosely strigose, minutely glandular; ray florets 10–12, ligules white, not coiling or reflexing. Shaded, sandy slopes at the bases of cliffs of Entrada Sandstone, piñon-juniper, 6550–7550 ft. Cibola, McKinley. **R & E**

Erigeron acris L.—Bitter fleabane. Annuals, biennials, or short-lived perennials, usually fibrous-rooted, sometimes apparently taprooted, slender caudices simple or branched; stems 20–60(–80) cm; leaves basal and cauline, oblanceolate to spatulate, entire, rarely serrate-dentate, 2–14 cm, cauline gradually reduced; heads 1 or 3–35, usually corymbiform, on curved-ascending peduncles; ray (pistillate) florets in 2 series; outer 150–250, ligules white to pink or purplish, filiform, 3–4.5 mm, erect, inner florets fewer than outer and eligulate; pappus bristles accrescent. New Mexico material belongs to **var. kamtschaticus** (DC.) Herder. Roadsides, ridges, volcanic cliffs, talus slopes, depressions among boulders, 9600–10,900 ft. Colfax, Sandoval. New Mexico collections previously identified as *E. nivalis* Nutt. are *E. acris*.

Erigeron aphanactis (A. Gray) Greene [*E. concinnus* (Hook. & Arn.) Torr. & A. Gray var. *aphanactis* A. Gray]—Rayless shaggy fleabane. Taprooted perennials with branching caudices; stems 2–20(–30) cm, canescent-hirsute, densely stipitate-glandular; leaves basal and cauline, petioles prominently ciliate with thick-based, spreading hairs, basal linear-oblanceolate to spatulate, 20–80 cm, canescent-hirsute, densely stipitate-glandular, cauline gradually or abruptly reduced; heads 1–4, disciform; ray florets 30–45, corollas without ligules or ligules shorter than involucres; disk corolla throats white-indurate, inflated, conspicuously puberulent. Our material belongs to **var. aphanactis**. Clay, sandy, or rocky soil, talus, washes, riparian, desert shrub, sagebrush, piñon-juniper, ponderosa pine forests, 5250–7700 ft. San Juan.

Erigeron argentatus A. Gray—Silver fleabane. Taprooted perennials from branched caudices retaining old leaf bases; stems mostly 15–40 cm, densely gray-green to silvery-strigose with white, closely appressed hairs, eglandular; leaves basal and cauline, basal spatulate to oblanceolate or narrowly oblanceolate, entire, silvery-strigose, eglandular, cauline linear to linear-oblanceolate; heads 1; phyllaries silvery-strigose, minutely glandular; ray florets 20–50(–75), ligules blue to lavender to pink or white, 9–15 mm, coiling; cypselae terete to slightly compressed, 2.8–3.4 mm, 6–8-nerved. Ridges and slopes in dry, sandy or gravelly soil, desert shrub, sagebrush, piñon-juniper woodlands, 5700–6200 ft. San Juan.

Erigeron arizonicus A. Gray—Arizona fleabane. Perennials from fibrous-rooted rhizomes, caudices or rhizomes relatively thick, simple or branched, sometimes resembling a taproot; stems 8–40(–60) cm, strigose to hirsute, eglandular; leaves basal and cauline, basal spatulate, obovate to ovate, or elliptic, often 3-nerved, entire to shallowly serrate, loosely strigose, rarely minutely glandular, cauline slightly smaller, relatively even-sized, bases subclasping, becoming nonclasping distally; heads 1–4(–6) on distal 1/3–1/2 of stem, peduncles 1.5–9 cm; phyllaries pilose-hirsute, minutely glandular; ray ligules white, coiling at tips. Openings, woods edges, roadsides, open slopes, ponderosa pine forests, mixed conifer communities, 7500–9000 ft. Catron, Grant.

Erigeron bellidiastrum Nutt.—Sand fleabane. Taprooted annuals; stems 3.5–30(–50) cm, loosely strigose-hirsutulous with upcurved hairs, usually eglandular, sometimes minutely glandular (var. *arenarius*); leaves cauline or mostly so (basal deciduous), linear to oblanceolate or spatulate, entire, lobed, or pinnately dissected, 1–6(–8) cm; phyllaries hispidulous, minutely glandular; rays relatively few, ligules white, often with abaxial lilac midstripe, 4–7.5 mm, not coiling or reflexing, some ray florets consistently produced between the phyllaries, the mature cypselae of these held in place as the phyllaries reflex at maturity; pappus 1-seriate, the outer series absent or a cartilaginous crown. Dunes, loose sand, creek sides, washes, floodplains, desert shrub, sagebrush, juniper, piñon-juniper, 4400–5750 ft.

1. Basal and proximal cauline leaf margins deeply dentate to pinnately lobed…**var. arenarius**
1'. Basal and proximal cauline leaf margins entire or rarely with pair of shallow teeth…(2)
2. Proximal stems mostly 1–2(–2.5) mm wide; basal and proximal cauline leaf blades linear to linear-oblanceolate, 10–15(–30) × 1–2.5(–3) mm…**var. bellidiastrum**
2'. Proximal stems mostly (2–)2.5–5 mm wide; basal and proximal cauline leaf blades oblanceolate, 20–40 (–60) × 3–5(–15) mm…**var. robustus**

var. arenarius (Greene) G. L. Nesom [*E. arenarius* Greene]—Sandwort daisy fleabane. Stems moderately to intricately branched; proximal mostly 1–2(–2.5) mm wide; basal and proximal cauline leaf blades oblanceolate, 20–40 × 3–8 mm, margins deeply dentate to pinnately lobed. Open habitats in deep, loose sand, sometimes with oak, 2850–4265 ft. SE, SC, SW.

var. bellidiastrum—Western daisy fleabane. Stems intricately branched; proximal mostly 1–2(–2.5) mm wide; basal and proximal cauline leaf blades linear to linear-oblanceolate, 10–15(–30) × 1–2.5(–3) mm, margins entire or rarely with a pair of shallow teeth. Open habitats in deep, loose sand, desert shrub, sagebrush, juniper, piñon-juniper woodlands, 2900–6235 ft. Widespread.

var. robustus Cronquist—Western daisy fleabane. Stems moderately branched, usually simple proximally; proximal mostly (2–)2.5–5 mm wide; basal and proximal cauline leaf blades oblanceolate, 20–40(–60) × 3–5(–15) mm, margins entire or rarely with a pair of shallow teeth. Open habitats in deep, loose sand, 4900–9000 ft. Doña Ana, Grant.

Erigeron bigelovii A. Gray—Bigelow's fleabane. Taprooted perennials, often short-lived with woody caudices; stems 10–30 cm, brittle, subintricately branched, stiffly hispidulous, eglandular; leaves mostly cauline, basal withering, linear-oblong, rarely oblanceolate to obovate, 4–10(–20) mm, entire, hispid, minutely glandular, reduced distally; heads 1(–3) per main branch, 20–30 in later season; phyllaries glabrous or outer sparsely hispid, minutely glandular; ray ligules white to pink or purplish, without abaxial midstripe, not coiling or reflexing. Cliff bases, limestone faces and outcrops, sandstone slopes, 4000–5700 ft. EC, SC, SW.

Erigeron bonariensis L. [*Conyza bonariensis* (L.) Cronquist]—Sidewalk conyza. Taprooted annuals; stems 10–100(–150+) cm, branched mostly distally; proximal leaves oblanceolate, 3–8(–12) cm, distally becoming narrowly oblanceolate to linear; heads mostly paniculiform to racemiform, rarely corymbiform; fruiting receptacles (2–)2.5–4(–5) mm wide; phyllaries hispidulous; receptacles 3–5 mm diam. in fruit; pistillate florets 60–150; corollas ± equaling or surpassing styles, laminae 0 or to 0.3 mm; disk florets 8–12; cypselae 1–1.5 mm. Lawns, disturbed places, 3800–4200 ft. Doña Ana, McKinley, San Juan.

Erigeron canadensis L. [*Conyza canadensis* (L.) Cronquist]—Old field conyza. Taprooted annuals; stems (3–)50–200(–350) cm, branched mostly distally; leaves oblanceolate to linear, 2–5(–10) cm, glabrate; heads paniculiform or corymbiform; fruiting receptacles 1–1.5 mm wide; phyllaries usually glabrous, sometimes sparsely strigose; receptacles 1–1.5(–3) mm diam. in fruit; pistillate florets 20–30 (–45+), ligules 0.3–1 mm, equaling or slightly longer than styles; disk florets 8–30+; cypselae 1–1.5 mm. Roadsides, disturbed sites, old fields, recent burns, riparian areas, irrigation ditches, 3600–7700(–9200) ft. Widespread.

Erigeron canus A. Gray—Hoary fleabane. Taprooted perennials, often with thick caudex branches retaining old fibrous leaf bases, stems and leaves evenly and closely white strigose-canescent, eglandular; stems 5–35 cm; leaves mostly basal, linear-oblong to linear-oblanceolate, 2–10 cm; heads 1(–4) on leafy-bracteate peduncles; phyllaries densely hirsute to strigose-hirsute, minutely glandular; ray ligules white to light blue, reflexing; cypselae nearly terete, 2.8–3.5 mm, (8–)10–14-nerved, glabrous. Sandstone mesas and flats, gravelly loam, basalt breaks, prairie, ponderosa pine, piñon-juniper, 5400–8200(–9300) ft. NE, NC, NW, WC, SW.

Erigeron compositus Pursh—Dwarf mountain fleabane. Taprooted perennials, caudices simple or branches usually relatively thick and short, rarely slender and rhizomelike, covered with persistent leaf bases; stems 5–15(–25) cm, sparsely hispid-pilose, minutely glandular; leaves mostly basal, (1)2–3(4) times ternately lobed or dissected, 0.5–5(–7) cm, cauline bractlike, mostly entire; heads 1; ray ligules white to pink or blue, usually 6–12 mm, often reduced to tubes (heads disciform), not coiling or reflexing. Known only from N Taos County in the Latir Peak Wilderness Area; alpine tundra, 11,500–12,500 ft.

Erigeron concinnus (Hook. & Arn.) Torr. & A. Gray [*E. pumilus* Nutt. var. *concinnoides* Cronquist]— Navajo fleabane. Taprooted perennials, caudices simple or branched, branches sometimes rhizomelike; stems mostly 10–30 cm, hispid-pilose, minutely glandular; leaves mostly basal or basal and cauline, narrowly oblanceolate to linear-oblong, entire, 10–50(–80) cm, petioles ciliate with thick-based, spreading hairs, cauline reduced distally; heads (1)2–3(–5); phyllaries hirsute to hirsute-villous, minutely glandular, midvein region orange or yellowish; ray florets 50–100(–125), ligules white, 6–15 mm, reflexing; disk corolla throats indurate and inflated, hirsute-strigose with biseriate, sharply pointed hairs; pappus of (7–) 10–15 bristles and an outer series of scales 0.2–0.5 mm. Sagebrush-juniper, piñon-juniper, 3600–6300 ft. *Erigeron concinnus* and *E. pumilus* in New Mexico approach each other closely in geographic range without intergradation. McKinley, San Juan.

Erigeron coulteri Porter—Coulter's fleabane. Perennials from fibrous-rooted rhizomes, sometimes with branched caudices and scale-leaved stolons; stems 10–70 cm, sparsely hispid-villous (hair crosswalls black) to glabrate, eglandular; leaves basal and cauline, broadly oblanceolate to elliptic or oblong-lanceolate, 4–12(–15) cm, entire or with 1–5 pairs of shallow teeth, bases usually clasping, cauline becoming elliptic-ovate to lanceolate, gradually reduced distally; heads 1(–4); phyllaries hirsute-villous, hair crosswalls black, minutely glandular; ray ligules white, 9–25 mm, coiling. Meadows and openings in spruce-fir, 8900–12,100 ft. NC, Socorro.

Erigeron divaricatus Michx. [*Conyza ramosissima* Cronquist]—Plains conyza. Taprooted annuals; stems 5–25+ cm, branched throughout; leaves narrowly spatulate to linear, 5–15 mm; heads solitary and scattered or loosely corymbiform; fruiting receptacles 0.5 0.75 mm wide; phyllaries glabrous or glabrate; receptacles 0.7–1 mm diam. in fruit; pistillate florets 20–30, ligules 0.3–0.8 mm, equaling or slightly longer than styles; disk florets 3–8; cypselae 1–1.5 mm. Lawns, disturbed sites, 3600–7000 ft. Doña Ana, Eddy, Grant.

Erigeron divergens Torr. & A. Gray—Spreading fleabane. Taprooted annuals; stems (7–)12–40(–70) cm, usually single from the base but sometimes with congested multicipital caudex, freely branching above the base, puberulous-hirsutulous (hairs not evenly deflexed), minutely glandular; leaves basal (usually deciduous) and cauline, obovate-spatulate, entire or with 2–3 pairs of teeth or lobes; heads 5–100+, paniculate or loosely corymboid; phyllaries hirsute, minutely glandular; ray florets 75–150, ligules white, without abaxial midstripe, not coiling or reflexing. Juniper-grassland, riparian, piñon-juniper, ponderosa pine-oak, mixed conifer forests, meadows, openings, roadsides and other disturbed sites, canyon bottoms, (4600–)5100–9400 ft. An early-spring form (*E. wootonii* Rydb.) has a slightly fleshy taproot, dense, persistent basal leaves, and multiple stems ascending from the base. Roadsides and gravelly sites, 4100–4900 ft. Widespread.

Erigeron eatonii A. Gray—Eaton's fleabane. Taprooted perennials; stems decumbent-ascending from the base, often proximally purple, simple or few-branched above the middle, loosely short-strigose, eglandular; basal leaves persistent, linear-oblanceolate, 3-veined, 5–11(–19) cm, base long-attenuate and petiolelike, cauline mostly linear; heads 1–3 on long peduncles; ray ligules white, not reflexing or coiling. In New Mexico known only from the Beautiful Mountain area. Open areas in piñon-juniper woodlands, ponderosa pine-oak, aspen, and Douglas-fir forests, 6400–9000 ft. Rio Arriba, San Juan.

Erigeron elatior (A. Gray) Greene—Fuzzy-head fleabane. Perennials from woody, relatively thick, fibrous-rooted rhizomes; stems 20–60 cm, villous, minutely glandular; leaves mostly cauline (basal not persistent), 1-nerved, obovate, becoming ovate to ovate-lanceolate distally, entire, (2–)3–9 cm, nearly equal-sized and evenly distributed distally, bases clasping; heads 1(–6); phyllaries linear, commonly with spreading-reflexing apices, minutely glandular and densely villous-lanate, hair crosswalls reddish to purple; rays 75–150, ligules pink to rose-purple, 10–20 mm, coiling. Talus, rocky sites, meadows, lakeshores, mixed conifer communities, 8500–12,300 ft. Mora, Rio Arriba, Taos.

Erigeron eximius Greene—Spruce-fir fleabane. Perennials from fibrous-rooted rhizomes, caudices or primary rhizomes slender, simple or branched, usually producing slender, scale-leaved rhizomes or stolons bearing terminal leaf tufts; stems 15–60 cm, glabrous or sparsely hirsute-pilose near the heads, densely minutely glandular to nearly eglandular; leaves basal (persistent) and cauline, basal spatulate to elliptic-spatulate (to oblanceolate-obovate), entire or serrulate to mucronulate, 3–15 cm, cauline usually clasping to subclasping, even-sized and widely spaced, glabrous, distal often glandular; heads 1–5(–15), phyllaries minutely glandular and without other vestiture; ray ligules white to bluish or lavender, 12–20 mm, coiling; cypselae 1.8–2.5 mm, 2(–4)-nerved. Subalpine meadows, openings in fir, spruce-fir, and aspen forests, (7500–)8800–11,500 ft. Widespread except E, S.

Erigeron flagellaris A. Gray—Trailing fleabane. Biennials or short-lived perennials, fibrous-rooted or sometimes taprooted, caudices lignescent, rarely branched; stems strigose with consistently antrorsely appressed hairs, first erect, 3–15 cm, then producing herbaceous, leafy, prostrate runners, usually with rooting plantlets at tips, the populations often becoming clonal mats; leaves basal and cauline; basal broadly oblanceolate to elliptic, entire or dentate, 20–55 mm, cauline abruptly reduced distally; heads 1(–3, on proximal branches); phyllaries strigose to loosely hirsute, minutely glandular; ray ligules white, often with an abaxial midstripe, not coiling or reflexing. Grasslands, piñon-juniper woodlands, ponderosa pine–oak, mixed conifer, spruce-fir forests, (5800–)7000–10,500 ft. Widespread.

Erigeron formosissimus Greene [*E. formosissimus* var. *viscidus* (Rydb.) Cronquist]—Beautiful fleabane. Perennials from fibrous-rooted, variably thickened rhizomes; stems 10–40(–55) cm, ascending, densely hirsute to hirsutulous or glabrous, minutely glandular to stipitate-glandular; leaves basal and cauline, oblanceolate to oblanceolate-spatulate, entire, 2–10(–15) cm, cauline with clasping bases, becoming ovate to lanceolate, gradually reduced; heads 1–6; phyllaries glabrous or hirsute-villous, densely minutely glandular to stipitate-glandular; ray florets 75–150, ligules usually blue to purple, 8–15 mm, sometimes coiling. Stream banks, meadows, roadsides, piñon-juniper woodlands, ponderosa pine–oak, mixed conifer, spruce-fir forests, 7000–11,600 ft. Widespread except EC, SE, SW.

Erigeron glabellus Nutt.—Streamside fleabane. Perennials, fibrous-rooted, without rhizomes, caudex often multicipital; stems erect from the base, branching at midstem and above, hirsute-strigose to softly hirsutulous, eglandular; basal leaves persistent, mostly oblanceolate, cauline usually strongly reduced distally, not clasping; heads loosely corymboid on long, bracteate peduncles; rays numerous, ligules filiform (like *E. divergens*), white to light blue, not coiling. Streamsides and wet meadows, 7000–9000 ft. NC, C, Cibola. Collections identified as *E. glabellus* are commonly some other species, especially *E. formosissimus* or *E. speciosus*.

Erigeron glacialis (Nutt.) A. Nelson [*E. peregrinus* Greene var. *callianthemus* (Greene) Cronquist]—Subalpine fleabane. Perennials from fibrous-rooted rhizomes, caudices usually simple, thick; stems 5–55 (–70) cm, densely strigillose with loosely appressed, slightly crinkled hairs (most densely so immediately proximal to heads), eglandular; leaves basal and cauline, linear-oblanceolate to broadly lanceolate or spatulate, entire, mostly 3–16(–20) cm, glabrous or glabrate, eglandular, cauline with subclasping bases, gradually reduced distally; heads 1(–8); phyllaries usually glabrous, rarely sparsely villous, densely and evenly stipitate-glandular; ray ligules blue to purple or pink, less commonly white to pale blue, coiling; cypselae 2.5–2.8 mm, (4)5(–7)-nerved. New Mexico material belongs to **var. glacialis**. Subalpine meadows, bogs, openings and edges of spruce-fir forests, 11,000–12,100 ft. NC.

Erigeron grandiflorus Hook. [*E. simplex* Greene]—Rocky Mountain alpine fleabane. Perennials from fibrous-rooted rhizomes, caudices or rhizomes crownlike or branches relatively short and thick; stems 2-25 cm, erect to decumbent-ascending, sparsely to moderately pilose to villous-hirsute, stipitate-glandular; leaves basal and cauline, oblanceolate to obovate or spatulate, 1-6(-9) cm, entire, cauline abruptly or gradually reduced; heads 1; phyllaries moderately to densely woolly-villous, hairs flattened and often with reddish crosswalls, minutely glandular; ray ligules blue to pink or purplish, rarely white, coiling. Spruce openings, krummholz, alpine tundra, 10,200-12,900 ft. NC, NW.

Erigeron hessii G. L. Nesom—Hess's fleabane. Perennials from fibrous-rooted rhizomes, often cespitose, caudices or rhizomes thickened, branched, lignescent; stems 5-16 cm, loosely strigose to hirsute, eglandular; leaves mostly basal, oblanceolate to oblanceolate-spatulate or elliptic-ovate, 2-5(-7) cm, usually entire, cauline subclasping, reduced in size; heads 1(-3); phyllaries coarsely strigose-hirsute along midribs, sparsely to densely minutely glandular; ray ligules white to light lavender, 8-13 mm, slightly coiling at maturity. Known only from the Whitewater Baldy area. Andesitic dikes in rhyolite, bedrock cracks; open areas, upper montane to subalpine conifer forests, 9500-10,200 ft. Catron. **R & E**

Erigeron incomptus A. Gray—Hybrid fleabane. Perennials from a woody taproot, caudex woody, often irregularly branched; stems 15-40(-70) cm, freely branching above, puberulous-hirsutulous, hairs not stiffly deflexed, minutely glandular; leaves basal and cauline, basal narrowly oblanceolate, often deeply lobed to pinnatifid but sometimes subentire with long narrow petioles, crowded along basal portion of the stem, petioles persistent; heads 5-100+, paniculate or loosely corymboid; phyllaries hirsute, minutely glandular; ray ligules white, without abaxial midstripe, not coiling or reflexing, ligules barely longer than the involucre in scattered forms. Gravelly soil; creosote bush-grassland, piñon-juniper woodland, 4700-7800 ft. Scattered throughout. Similar to *E. divergens* and often identified as such but with a woody base and distinctly perennial. Intermediates are formed with *E. tracyi*.

Erigeron leiomerus A. Gray—Rockslide fleabane. Perennials from a diffuse system of relatively long, slender, rhizomelike branches; stems 4-12(-15) cm, decumbent to ascending or erect, often purplish, glabrous or sparsely strigillose, eglandular; leaves basal and cauline, basal oblanceolate to obovate or spatulate, entire, 1.5-7 cm, cauline reduced distally, not subclasping, glabrous or sparsely strigose, eglandular; heads 1; phyllaries often narrowly elliptic, purplish, glabrous or sparsely strigose, minutely glandular; ray ligules white to blue or purple, 6-11 mm, reflexing. Crevices, fell-fields, talus, rocky meadows, 8100-11,500 ft. NC, Socorro.

Erigeron lonchophyllus Hook.—Lance-leaf fleabane. Biennials or short-lived perennials, sometimes appearing annual, fibrous-rooted, caudices simple; stems 2-45(-60) cm; leaves basal and cauline, basal oblanceolate to spatulate, entire, 1.3-8(-15) cm, cauline mostly linear, sometimes longer than basal, usually erect; heads 1 or 3-12 in a loose raceme on erect peduncles; ray (pistillate) florets 70-130 in 1 series, ligules filiform, 2-3 mm, erect, white to light pink; pappus bristles not accrescent. Wet and marshy meadows and hillsides, lakeshores, streamsides, 7500-9900(-10,800) ft. NC, NW.

Erigeron melanocephalus (A. Nelson) A. Nelson—Black-head fleabane. Perennials from fibrous-rooted rhizomes, caudices decumbent, often branched, rhizomelike, sometimes taprootlike; stems 3-12 (-21) cm, villous (hairs with black crosswalls), minutely glandular near heads; leaves mostly basal, spatulate to oblanceolate, entire, (1)2-5(-15) cm, cauline linear, bractlike; heads 1; phyllaries villous-sericeous (hairs flattened, crosswalls black, imparting a distinct black color to involucres), glandular; ray ligules white to purple, tardily coiling. Fell-fields, scree, rocky soil, receding snowbanks, clearings in spruce-fir, krummholz and alpine tundra communities, 11,600-13,000 ft. NC.

Erigeron modestus A. Gray—Modest fleabane. Perennials from a woody or lignescent taproot, early-season plants, fibrous-rooted and appearing annual or biennial, caudices usually woody, sometimes branched; stems 8-40 cm, branches at first from midstem or proximal, later more distal, proximally reddish in early season, sometimes woody or lignescent, loosely strigose to villous, often glandular; leaves basal (withering after early season) and cauline, oblanceolate to spatulate, entire or with 1-2(3) pairs of teeth, 2-5(-10) cm, gradually reduced distally; heads 1(-3, -50 in later season); phyllaries sparsely to moderately hirsute, minutely glandular; ray florets 24-65(-170), ligules white, often with abaxial mid-stripe, not coiling or reflexing. Limestone slopes, canyon bottoms, grasslands, juniper-scrub, moist to wet disturbed sites, 4100-7000 ft. SE, SC, SW. *Erigeron modestus* is apparently a polyploid complex incorporating genes from *E. flagellaris*, *E. tracyi*, elements of *E. divergens*, and the Mexican *E. pubescens* Kunth; arbitrary identifications may be necessary. Early-season forms (March-April) often have stems red at the bases; large, toothed or lobed basal leaves 30-60(-100) × 5-10(-18) mm; single, large heads (involucres 9-12 mm diam.) on ebracteate peduncles 5-15 cm; and ray florets 50-170. Later-season plants have branching stems with more numerous, much smaller, diffusely arranged heads.

Erigeron neomexicanus A. Gray—New Mexico fleabane. Taprooted perennials with woody caudices; stems 20-70 cm, moderately to densely strigose, eglandular or glands minute; leaves basal (sometimes withering) and cauline, usually deeply pinnatifid, sometimes dentate to entire, strigose, eglandular, cauline gradually reduced distally; heads (1-)5-15(-30), loosely corymbiform; phyllaries strigose to hirsute mostly along the midregion, usually minutely glandular, rarely eglandular; ray florets 70-150, ligules white, reflexing; pappus of (8-)10-12 fragile, basally deciduous bristles. Piñon-juniper woodlands, ponderosa pine-oak forests, mixed conifer forests, grassy slopes, (4900-)6000-8500(-10,400) ft. C, WC.

Erigeron oreophilus Greenm.—Chaparral fleabane. Taprooted perennials with woody caudices; stems (8-)25-90 cm, sparsely hirsute-villous (hairs 0.6-2 mm) or glabrous, densely stipitate-glandular (at least on distal 1/2, glands relatively large, capitate); leaves basal (sometimes withering) and cauline, usually deeply pinnatifid, usually glabrous, stipitate-glandular, cauline gradually reduced distally; heads (1-)5-25(-50), loosely corymbiform; phyllaries glabrous or sparsely hirsute, densely stipitate-glandular; ray florets 75-130, ligules white, reflexing; pappus of (8-)10-12 fragile, basally deciduous bristles. Piñon-juniper woodlands, ponderosa pine-oak forests, grassy slopes, (4900-)6000-8500(-10,400) ft. WC, SW, Lincoln.

Erigeron philadelphicus L.—Philadelphia fleabane. Annuals, biennials, or short-lived perennials, fibrous-rooted, caudices simple; stems 4-80 cm, loosely strigose to sparsely hirsute distally, minutely glandular; leaves basal (persistent or withering) and cauline, basal oblanceolate to obovate, (1.5-)3-11 (-15) cm, shallowly crenate to coarsely serrate or pinnately lobed, cauline oblong-oblanceolate to lanceolate, gradually reduced distally, bases clasping to auriculate-clasping; heads (1-)3-35, usually in corymbiform arrays; ray florets 150-250(-400), ligules usually white, sometimes pinkish, 5-10 mm, sometimes tardily coiling. New Mexico plants are **var. philadelphicus**. Canal and river margins, ditch banks, seeps, roadsides, riparian areas, piñon-juniper woodlands, 4800-8200 ft. Scattered.

Erigeron pinnatisectus (A. Gray) A. Nelson—Feather-leaf fleabane. Taprooted perennials, caudex branches usually relatively thick, woody; stems 4-11 cm, sparsely to moderately hispidulous and minutely glandular, sometimes merely glandular; leaves mostly basal and proximal, pinnatifid with linear lobes, often folding along the middle, 20-40(-50) mm, glabrate to sparsely hirsute, usually minutely glandular, cauline abruptly reduced distally or nearly absent, entire; heads 1; ray ligules light blue to purplish, reflexing. Scree, talus, gravelly slopes, among rocks, rocky ridges, tundra, 11,400-12,700 ft. NC.

Erigeron pulcherrimus A. Heller [*E. bistiensis* G. L. Nesom & Hevron]—Basin fleabane. Taprooted perennials with relatively short and thick caudex branches; stems (5-)7-30(-35) cm, gray-green, strigose with stiff white hairs, eglandular; leaves basal and cauline, basal linear to narrowly oblanceolate, 1-7 cm, entire, cauline little reduced for 1/2-3/4 of the stems, strigose with stiff white hairs, eglandular; heads 1; phyllaries hirsute to villous-hirsute, minutely glandular; ray ligules broad, white to pink, usually with a prominent pink or purplish midstripe, tardily coiling; cypselae 2.5-3 mm, (2-)4(5)-nerved, faces and margins densely strigose-sericeous. Clay, gypsum-clay, shale, sand and gravelly sand; Moenkopi, Chinle, and Fruitland Formations, often on eroded slopes; sandy draws and stream bottoms; desert scrub, piñon-juniper, 5800-8800 ft. NC, NW.

Erigeron pumilus Nutt.—Shaggy fleabane. Taprooted perennials, caudices with relatively short, thick branches; stems 5-30(-50) cm, hirsute to hispid-hirsute with slightly deflexed hairs, glandular; leaves basal and cauline, basal oblanceolate to narrowly oblanceolate, entire, 2-8 cm, petioles prominently ciliate with spreading, thick-based hairs, cauline becoming linear-lanceolate, little reduced distally; heads 1-5(-50); phyllaries hirsute to hispid-hirsute, minutely glandular, midvein region orange to yellowish; ray ligules white, reflexing; disk corolla throats indurate and inflated, indurate portion glabrous, shiny; pappus of 15-27 bristles and a short series of setae or scales. Our material belongs to **var. pumilus**. Sagebrush, piñon-juniper woodlands, ponderosa pine–oak forests, 5200-7500 ft. NC, NW, WC.

Erigeron rhizomatus Cronquist—Zuni fleabane. Perennials from fibrous-rooted, slender, scale-leaved, creeping rhizomes or rhizomelike caudex branches without well-defined central axes, plants growing in clumps to ca. 3 dm diam.; stems 25-45 cm, sparsely branching from near the base, sparsely strigose to strigose-hirsutulous, sometimes minutely glandular; leaves all cauline, narrowly oblong to oblong-oblanceolate, linear and relatively even-sized distally, glabrous except for marginal cilia; heads 1 or rarely 2-3 from proximal branches; involucres 13-16 mm wide; phyllaries sparsely strigose, sometimes sparsely minute-glandular; rays 25-45, ligules white or tinged with blue-violet, 6-7 mm, not coiling or reflexing; cypselae 3.5-4.5 mm, 5-6-nerved, nearly glabrous. Barren detrital clay hillsides or benches, shale-derived soils of the Chinle or Baca Formations, piñon-juniper woodlands, ponderosa pine and Douglas-fir forests, 6550-8200 ft. Catron, McKinley, San Juan. **FT**

Erigeron rybius G. L. Nesom—Sacramento Mountain fleabane. Perennials from fibrous-rooted rhizomes, primary rhizomes slender, with systems of lignescent, branched rhizomes and slender, herbaceous, scale-leaved stolons bearing terminal leaf tufts; stems 15-35 cm, hirsute to hirtellous with retrorsely spreading hairs, eglandular or sometimes sparsely minutely glandular; leaves basal and cauline, elliptic-ovate to spatulate-obovate, entire to mucronulate or shallowly serrate, 1.8-13.5 cm, cauline clasping to subclasping, becoming lanceolate and entire, nearly even-sized distally; heads 1-6 from distal branches; phyllaries pilose-hirsute, usually eglandular, sometimes minutely glandular; rays 47-99, ligules 11-20 mm, white, drying white or lilac-tinged, coiling. Meadows, grassy forest openings, disturbed areas, ponderosa pine–oak, mixed conifer, and spruce-fir forests, 5900-9200(-10,100) ft. Lincoln, Otero. **R & E**

Erigeron scopulinus G. L. Nesom & V. D. Roth—Cliff-face fleabane. Perennials from a system of relatively slender, basal offsets and slender, woody rhizomes 1-15 cm, mat-forming; stems 0.5-3.5 cm, decumbent to ascending, sparsely strigose, eglandular; leaves basal, blades spatulate, entire, often folding, 5-12 mm, glabrous to sparsely strigose, eglandular; heads 1 on an erect, scapiform peduncle; involucres 4-4.5 × 4-7 mm; phyllaries glabrous or sparsely strigose, sometimes minutely glandular; rays 10-20; ligules white, 5.5-9 mm, reflexing. Crevices and soil pockets of porphyritic and rhyolitic cliffs, 6550-9500 ft. Catron, Sierra, Socorro. **R**

Erigeron sivinskii G. L. Nesom—Sivinski's fleabane. Perennials from a thick taproot and short caudex branches with persistent old leaf bases; stems 5-8 cm, unbranched, short-strigose with white, stiff, closely appressed trichomes, eglandular; leaves basal and cauline, linear, 1-3.5 cm, cauline strictly ascending and continuing relatively unreduced in size at least halfway up the stems; heads 1; phyllaries granular-glandular, the outer also sparsely pilose with a few crisped-spreading hairs arising centrally; ray ligules white, drying pink to purple, sometimes as a midstripe, coiling with maturity; cypselae 2(3)-nerved, faces glabrous, margins sparsely ciliate. Red clay slopes of Summerville Formation and eroded shale slopes of Chinle Formation, desert scrub, piñon-juniper woodlands, 5600-7700 ft. McKinley. **R**

Erigeron sparsifolius Eastw. [*E. utahensis* A. Gray var. *sparsifolius* (Eastw.) Cronquist]—Bracted Utah fleabane. Taprooted perennials from branched caudices; stems 10-55 cm, sparsely strigose, eglandular; leaves basal (usually withering by flowering and cauline, basal and proximal-most oblanceolate-spatulate, entire, 2-5 cm, densely and closely strigose, eglandular, cauline abruptly reduced to erect linear bracts, relatively even-sized; heads (1-)3-10 from distal branches; phyllaries sparsely strigose to glabrous, minutely glandular; ray ligules white to blue, 4-8 mm, coiling; disk corollas 2-3.5 mm, viscid-puberulent with blunt, multicellular hairs; cypselae 1.5-2.5 mm, 3-4-orange-nerved. Rocky or sandy soil, soil pockets and crevices in sandstone, hanging gardens, canyon bottoms, river terraces, 4100-5600 ft. San Juan.

Erigeron speciosus (Lindl.) DC. [*E. speciosus* var. *macranthus* (Nutt.) Cronquist]—Showy fleabane. Perennials from short, thick, fibrous-rooted rhizomes, caudices relatively thick; stems 30–80(–100) cm, branching distally, glabrous or sparsely hirsute-pilose, often minutely glandular distally, often reddish; leaves basal (usually withering) and cauline, cauline ovate to ovate-lanceolate, oblong-lanceolate, or lanceolate, entire, base usually clasping to subclasping, nearly even-sized or sometimes largest at midstem, continuing to immediately below heads; heads (2–)4–20 in corymbiform arrays; phyllaries linear, usually glabrous, sometimes sparsely hirsute-pilose, minutely glandular; ray florets 75–150, ligules blue to lavender, rarely whitish, 8–16 mm, 1 mm wide, slightly coiling. Canyon bottoms, meadows, riparian communities, piñon–ponderosa pine–oak forests, mixed conifer, (5900–)7100–9900 ft. Widespread except E, S.

Erigeron spellenbergii G. L. Nesom—Spellenberg's fleabane. Taprooted perennials with lignescent rhizomes and caudex branches, colonial; stems 8–15 cm, strigillose with stiff, whitish hairs, eglandular; leaves basal and cauline, basal in rosettes, oblanceolate to narrowly obovate or spatulate, entire, 1-nerved, 20–32 × 5–7 mm wide, even-sized, base attenuate to a narrow petiole ca. 1/2 the leaf length, strigillose, cauline oblanceolate, slightly reduced from the basal, even-sized and evenly placed along the stems; heads 1; phyllaries narrowly oblong-lanceolate with a broad, orange medial line, glabrate except for minute glandularity; ray floret ligules white, not coiling or reflexing. Steep N slopes in accumulating sand at base of Zuni Sandstone cliffs, 6500–7000 ft. Known only from Blue Water Canyon in Cibola County. **R** [not pictured in plates]

Erigeron subglaber Cronquist—Pecos fleabane. Taprooted perennials with relatively short, thick caudex branches; stems 3–7 cm, glabrous or sparsely strigose, eglandular; leaves mostly basal, broadly oblanceolate to subspatulate, mostly 2–5(–6) cm, glabrous or glabrate, eglandular, entire, cauline gradually or abruptly reduced; heads 1; phyllaries glabrous, eglandular; ray ligules purplish to lavender, 6–9 mm, not coiling or reflexing. Rocky open meadows and slopes, openings in spruce-fir forests, 9850–11,500 ft. Mora, San Miguel. **R & E**

Erigeron subtrinervis Rydb. ex Porter & Britton [*E. speciosus* (Lindl.) DC. var. *mollis* (A. Gray) S. L. Welsh]—Three-nerve fleabane. Perennials, rhizomatous to subrhizomatous, fibrous-rooted, caudices usually branched, woody, thick; stems 15–90 cm, branching distally, moderately to densely hirsute, eglandular; leaves basal (usually withering) and cauline, lanceolate to oblong, oblong-ovate, or broadly ovate, evenly hirsute to strigose-hirsute, base usually clasping to subclasping, nearly even-sized or sometimes largest at midstem, continuing to immediately below heads; heads 1–6(–21) in corymbiform arrays; phyllaries linear, moderately to densely hirsute, minutely glandular; ray florets 100–150, ligules blue to lavender, 7–18 × 1 mm, coiling. Ponderosa pine–oak forests, mixed conifer, 7000–9500(–10,400) ft. NC, C, WC, SC. *Erigeron subtrinervis* is variable in vestiture.

Erigeron tracyi Greene [*E. colomexicanus* A. Nelson.; *E. divergens* Torr. & A. Gray var. *cinereus* A. Gray]—Running fleabane. Biennials to short-lived perennials, taprooted or fibrous-rooted, caudices simple or branched, woody; stems first erect, then producing long, leafy, prostrate runners sometimes with rooting plantlets at tips, often branched proximally, densely and evenly hirsutulous with deflexed hairs, sparsely minutely glandular; leaves mostly basal in early season, then withering, oblanceolate to spatulate, 1–3(–6) cm, long-petiolate, entire to dentate or lobed, hirsute, cauline linear to linear-oblanceolate; heads 1(–3 from midstem or proximal branches) on long, ebracteate peduncles; ray ligules white, often purplish abaxially, sometimes with an abaxial midstripe, not coiling or reflexing. Rocky slopes, canyon bottoms, lava flows, grassland, piñon-juniper woodlands, ponderosa pine–oak, mixed conifer, and spruce-fir forests, 4200–8700 ft. Widespread. Similar in habit to *E. flagellaris*, but the stem pubescence is different, the stolons much less commonly produce rooting plantlets at the tips, and the plants tend to produce woody or lignescent caudices.

Erigeron versicolor (Greenm.) G. L. Nesom [*E. geiseri* Shinners; *E. gilensis* Wooton & Standl.; *E. mimegletes* Shinners]—Bald-fruit fleabane. Annuals or rarely short-lived perennials, taprooted or rarely fibrous-rooted; stems 12–50(–80) cm; leaves basal (usually not persistent) and cauline, narrowly lanceolate to oblanceolate, entire or shallowly crenate to serrate, (0.5–)2–5 cm, cauline gradually reduced;

heads (1–)5 to ca. 100; ray ligules white, not reflexing or coiling; disk corollas 1.2–2 mm; cypselae 0.8–1.3 mm; pappus only of blunt scales or a fimbriate crown to 0.15 mm, without bristles. Wet meadows, mixed conifer, 8100–8900 ft. Catron, Rio Arriba, Taos.

Erigeron vetensis Rydb.—Early blue-top fleabane. Taprooted perennials, caudices with short, thick branches, retaining old leaf bases; stems 5–25 cm, glabrous to sparsely hirsute or villous, densely glandular; leaves basal and cauline, linear to narrowly oblanceolate, entire, 2–9(–15) cm, glabrous or sparsely hispidulous, minutely glandular, petioles ciliate with thick-based spreading hairs, cauline gradually reduced; heads 1; phyllaries hispid to hispid-villous, densely glandular; ray ligules bluish to purplish, sometimes white, reflexing. Subalpine meadows, barren summits, ridges, crevices and ledges, gravelly slopes, rocky meadows, 7300–10,600 ft. NE, NC, NW.

Erigeron vreelandii Rydb. [*E. platyphyllus* Greene; *E. rudis* Wooton & Standl.]—Vreeland's fleabane. Perennials, rhizomatous to subrhizomatous, fibrous-rooted, caudices usually relatively short and few-branched, thick, woody; stems 30–80 cm, sometimes sparsely pilose, densely stipitate-glandular; leaves basal (usually withering) and cauline, densely stipitate-glandular, cauline narrowly lanceolate to oblong-, ovate-, or elliptic-lanceolate, entire, base usually clasping to subclasping, nearly even-sized or sometimes largest at midstem, continuing to immediately below heads; heads 1–15(–22) in corymbiform arrays; phyllaries linear, densely stipitate-glandular; ray florets 75–150, ligules blue to lavender, 9–17 × 0.8–1 mm, coiling. Ponderosa pine–oak, mixed conifer, 7400–9500 ft. NC, NW, C, WC, SC, SW. Intermediates are apparently formed with *E. speciosus*.

ERIOPHYLLUM Lag. – Woolly sunflower

Kenneth D. Heil

Eriophyllum lanosum (A. Gray) A. Gray—White easter-bonnets, woolly daisy. Annuals, 3–15 cm; stems decumbent to ascending; leaf blades oblanceolate to linear, 5–20 mm, rarely lobed, ultimate margins, usually entire, faces sparsely woolly; heads borne singly; involucres campanulate to obconic, 3–5 mm diam.; phyllaries 8–10, distinct; ray florets 8–10, laminae white with red veins, 3–5 mm; disk florets 10–20, corollas 2–3 mm; cypselae 2.5–4.5 mm; pappus of 5 subulate scales. Sandy or gravelly openings, desert scrublands, 4000–5000 ft. Hidalgo.

EUTHAMIA (Nutt.) Cass. – Goldentop

Jennifer Ackerfield

Perennial herbs or subshrubs; leaves alternate, entire, glandular-punctate; heads radiate; involucral bracts imbricate in several series, chartaceous, yellowish- or greenish-tipped, ± glutinous; receptacles ± fimbrillate; ray florets usually more numerous than disk florets, yellow; cypselae with several nerves, nearly terete; pappus of capillary bristles (Ackerfield, 2015; Taylor & Taylor, 1983).

1. Inflorescence elongate or rounded and interrupted, with lateral corymbiform clusters arising from the axils of well-developed leafy bracts; plants often > 1 m…**E. occidentalis**
1'. Inflorescence broad and flat-topped, without lateral corymbiform clusters; plants usually < 1 m… **E. graminifolia**

Euthamia graminifolia (L.) Nutt. [*Solidago graminifolia* (L.) Salisb.]—Flat-top goldenrod. Perennials, 3–15 dm; leaves 3–5-nerved, linear to lanceolate, 3.5–13 cm, with scabrous margins, little or obscurely gland-dotted; heads in a flat-topped array; involucres 3–5.5 mm; disk florets 2.5–3.5 mm. Usually along streams and ditches in sandy soil, although sometimes in drier sites; reported from the Burro Mountains, Gila Bird Area, ca. 4375 ft. Grant.

Euthamia occidentalis Nutt. [*Solidago occidentalis* (Nutt.) Torr. & A. Gray]—Western goldentop. Perennials or subshrubs, 4–20 dm; leaves 3–5-nerved, linear, 8–10 cm, with scabrous margins, gland-dotted and often sparsely hairy; heads in narrow, elongate arrays; involucres 3.5–5 mm, the bracts yellowish brown, lanceolate to linear; disk florets 3–4.2 mm. In moist soil and along rivers and ditches, 3850–9500 ft. NC, NW, EC, C, SC, SW.

EUTROCHIUM Raf. – Joe pye weed

Kenneth D. Heil

Eutrochium maculatum (L.) E. E. Lamont [*Eupatorium maculatum* L.]—Spotted joe pye weed. Perennials, 60–150 cm; stems erect, unbranched, usually purple-spotted, densely puberulent throughout, glandular-puberulent distally; leaves mostly cauline, mostly whorled, in (3s)4s–5s(6s); blades lance-elliptic to lanceolate or lance-ovate, usually 6–17 × 1.5–5(–7) cm, margins sharply serrate or doubly serrate, abaxial faces gland-dotted and densely pubescent, adaxial faces sparingly hairy or glabrous; heads in flat-topped, corymbiform arrays; involucres cylindrical, often purplish, 6.5–9 × 3.5–7 mm; phyllaries 10–22 in 5–6 series, often gland-dotted, glabrous or densely pubescent; florets (8)9–20(–22), corollas purplish, 4.5–7.5 mm, throats funnelform; cypselae 3–5 mm, dark brown to black or yellowish brown, 5-ribbed, usually gland-dotted. New Mexico material belongs to **var. bruneri** (A. Gray) E. E. Lamont [*Eupatorium bruneri* A. Gray]. Stream and canal banks, wet meadows, bogs, seeps, calcareous soils, 6000–7900 ft. C, SC, Colfax, Grant (Lamont, 2006).

FLAVERIA Juss. – Yellowtops

Kenneth D. Heil and Steve L. O'Kane, Jr.

Herbs; leaves opposite, sessile, simple; heads small, numerous, crowded into small glomerules aggregated into a terminal cluster, radiate (ours); involucral bracts subequal; receptacles naked, small; disk florets yellow; ray floret usually 1 per head, yellowish, inconspicuous; pappus absent or of a few scales; cypselae small, black, 8–10-ribbed (Ackerfield, 2015; Yarborough & Powell, 2006a).

1. Leaves connate-perfoliate, margins entire; pappus scales present…**F. chlorifolia**
1'. Leaves not connate-perfoliate, margins weakly serrate to spinose-serrate; pappus scales absent…(2)
2. Heads in tight, axillary glomerules…**F. trinervia**
2'. Heads in corymbiform arrays, seldom in tight, axillary glomerules…**F. campestris**

Flaveria campestris J. R. Johnst.—Alkali yellowtops. Annuals, 18–90 cm; stems usually erect; leaves linear-lanceolate to oblanceolate, 30–90 × 6–22 mm, margins strongly to weakly serrate or spinulose-serrate; phyllaries 3, elliptic, obovate, or oblong-obovate; ray florets 0 or 1, laminae yellow, 1.5–2.5 mm; disk florets 5–8, corolla tubes 0.8–1.3 mm; cypselae 2.8–3.6 mm. Saline soils, lake, pond, and stream margins, floodplains, disturbed pastures, 3550–5860 ft. Scattered throughout.

Flaveria chlorifolia A. Gray—Clasping yellowtops. Perennials; stems erect, to 200 cm, glaucous; leaves sessile, blades oblong-ovate to lanceolate or hastate, (20–)30–100 × 10–40(–50) mm, margins entire; heads 25–150+ in paniculiform arrays; involucres 5–7 mm; ray florets 0; disk florets 9–14; cypselae narrowly oblanceolate to linear, 2.5–3 mm. Pecos River drainage; marshes, springs, rivers, irrigation canals, roadside ditches, 3350–6400 ft. Chaves, Eddy, Guadalupe.

Flaveria trinervia (Spreng.) C. Mohr—Clustered yellowtops. Annuals, to 200 cm; stems erect; leaves lanceolate or oblanceolate to elliptic or subovate, 30–150 × 7–40 mm, margins serrate, serrate-dentate, or spinulose-serrate; phyllaries usually 2, oblong; ray florets 0–1, laminae pale yellow or whitish, 0.5–1 mm; disk florets 0–2, corolla tubes 0.5–1.4 mm; cypselae 2–2.6 mm. Near water, saline and gypseous areas, 3435–5900 ft. C, SE, SC.

FLEISCHMANNIA Sch. Bip. – Slender-thoroughwort

Kenneth D. Heil

Fleischmannia sonorae (A. Gray) R. M. King & H. Rob. [*Eupatorium sonorae* A. Gray]—Sonoran slender-thoroughwort. Plants 30–80 cm (ours); stems usually puberulent, hairs curled erect to ascending-erect, not sprawling-scandent; cauline leaves opposite, petioles 0.5–3 cm, blades deltate-ovate, 1.5–5 × 0.6–2 cm, bases broadly cuneate to subtruncate, margins irregularly crenate-serrate, apices acute to acuminate; heads discoid, in loose, corymbiform arrays; involucres obconic to hemispherical, (4–)5–6 mm; phyllaries persistent, 20–30 in 2–4 series, 2–3-nerved; receptacles flat or slightly convex; corollas usually pale purple, rarely white; cypselae prismatic, 5(-8)-ribbed, 1-1.2(-1.5) mm, glabrous or glabrate. Rich soils, along streams, rocky slopes; 1 old record. Hidalgo (most likely) (Nesom, 2006j).

FLOURENSIA DC. – American tarwort

Don Hyder

Subshrubs or shrubs; stems erect; leaves cauline, alternate; heads discoid or radiate; involucres campanulate to hemispherical; receptacles flat to conic-ovoid; corollas yellow; disk florets yellow; cypselae compressed or flattened (Allred & Ivey, 2012; Strother, 2006i).

1. Shrubs; leaves 10-30 × 4-18 mm; ray florets absent…**F. cernua**
1'. Subshrubs; leaves 30-100 × 10-40 mm; ray florets present…**F. pringlei**

Flourensia cernua DC.—American tarwort. Shrubs; stems to 2 m; branches puberulent; leaves 10-30 × 4-18 mm, blades elliptic to ovate, petioles 1-2 mm; heads 10-12 mm; pappus a pair of slender unequal awns. Usually in limestone or alkaline or clay soils and gravelly sites; desert scrub, 3500-6800 ft. C, WC, SE, SC, SW, Quay. Plants have a tarry odor.

Flourensia pringlei (A. Gray) S. F. Blake.—Pringle's tarwort. Subshrubs; stems to 40 cm; leaves somewhat resinous, cuneate at the base, blades elliptic to lance-oblong, petioles short; heads 2.5-4 cm wide; ray florets 13-21, yellow. Rocky and disturbed slopes, 5125-5185 ft. Hidalgo.

GAILLARDIA Foug. – Blanketflower

Jennifer Ackerfield and Kenneth D. Heil

Herbs; leaves alternate or basal; heads radiate; involucral bracts in 2 or 3 series, with chartaceous bases and herbaceous tips; receptacles alveolate with long stiff setae, or the setae absent; disk florets woolly-villous; ray florets yellow to partly or wholly red or purple, 3-toothed; cypselae partly or wholly covered with long hairs; pappus of scales, usually with an awn (Biddulph, 1944; Stoutamire, 1977; Strother, 2006j; Turner & Whalen, 1975).

1. Plants annual; ray corollas usually bicolored, brown-purple to red proximally, tipped with yellow or orange…**G. pulchella**
1'. Plants perennial; ray corollas various…(2)
2. Leaves mostly entire, linear…**G. multiceps**
2'. Leaves mostly pinnatifid…(3)
3. Receptacular setae mostly 1-3 mm; cypselae hairy, hairs inserted at bases and on faces…**G. pinnatifida**
3'. Receptacular setae mostly 3-6 mm; cypselae hairy, hairs inserted at bases and not on faces…**G. aristata**

Gaillardia aristata Pursh—Blanketflower. Perennials, 3-6 dm; leaves linear-oblong to lance-ovate, entire to toothed or somewhat pinnatifid, 5-20 cm, hirsute and resinous-glandular; ray florets 1-3.5 cm, yellow or with purple at the base; disk florets 4.5-5.5 mm, brownish purple or rarely yellow; pappus scales 5-6 mm, the awn twice as long as the body. Very common in dry to moderately moist, open places such as in meadows and along roadsides, often in disturbed habitats, 5800-12,200 ft. Scattered throughout.

Gaillardia multiceps Greene—Onion blanketflower. Perennials, 30-45+ cm; leaves narrowly spatulate or linear, 2-6 cm × 3-5(-8+) mm, margins entire, faces sparsely and minutely hispidulous or glabrate; ray florets 13-20 mm, corollas yellow; disk florets 4-4.5 mm, yellow, distally purplish, pappus of 10 lanceolate, aristate scales 6-9 mm. Gypseous soils, including dunes, 3100-6200 ft. EC, C, SE.

Gaillardia pinnatifida Torr.—Red dome blanketflower. Perennials, 1-5 dm; leaves deeply pinnatifid, to 12 cm, villous and resinous-glandular; ray florets 1-2 cm, yellow; disk florets reddish to brownish purple; pappus scales 3-7 mm, the awn no more than 1/2 as long as the body. Common in dry, rocky, sandy, or clay soil of open places, 3150-8600 ft. Widespread.

Gaillardia pulchella Foug.—Indian blanketflower. Annuals or short-lived perennials, 1-6 dm; leaves lanceolate or oblong, entire to toothed or somewhat pinnatifid, 2-8 cm, glandular-villous; ray florets 1-2 cm, usually reddish purple with a yellow tip; disk florets brownish purple; pappus scales 5–6 mm, the awn about equal in length to the body. Scattered on dry, open prairie, 3300–9700 ft. Widespread.

GALINSOGA Ruiz & Pav. – Quickweed, gallant-soldier

Kenneth D. Heil and Steve L. O'Kane, Jr.

Galinsoga parviflora Cav.—Quickweed. Annual herbs; plants 4–35 cm; leaves opposite, lanceolate to ovate, blades 7–80 × 3–20 mm, margins entire or serrulate, sometimes serrate; congested arrays of radiate heads; axillary peduncles shorter than subtending bractlets; involucral bracts few in 1–3 poorly defined series, several-nerved; phyllaries persistent; receptacles chaffy; disk florets 15–50, yellow to greenish yellow; ray florets usually dull white to pink, laminae 0.5–1.8 × 0.7–1.5 mm. Our material belongs to **var. semicalva** A. Gray [*Galinsoga semicalva* (A. Gray) H. St. John & D. White]. Rocky slopes, 5365–9460 ft. C, WC, SC, SW, San Miguel (Ackerfield, 2015).

GAMOCHAETA Wedd. – Everlasting

Kenneth D. Heil

Gamochaeta stagnalis (I. M. Johnst.) Anderb. [*Gnaphalium stagnale* I. M. Johnst.]—Desert cud-weed. Annuals, 2.5–20(–35) cm; stems erect to decumbent-ascending, loosely arachnose-tomentose; leaves mostly cauline, alternate, sessile, 1–2.5(–3) cm × 2–6 mm, blades mostly oblanceolate to oblong-oblanceolate, margins entire, faces concolored or weakly bicolored, both loosely tomentose or adaxial glabrescent and greener; heads in capitate clusters or interrupted, usually in spiciform arrays; involucres campanulate, 2.5–3 mm, bases sparsely arachnose; bisexual florets (2)3(4); all corollas purplish distally; corollas all yellow or purplish-tipped; cypselae tan, oblong, 0.3–0.5 mm. Sandy, often moist soils, riparian areas, desert grasslands, juniper-grasslands, creosote bush-mesquite-cholla, oak woodlands; 1 record from the Peloncillo Mountains, 4500 ft. Hidalgo (Nesom, 2006k).

GAZANIA Gaertn. – Treasure flower

Kenneth D. Heil

Gazania linearis (Thunberg) Druce [*Gorteria linearis* Thunberg; *Gazania longiscapa* DC.]—Treasure flower. Perennials; plants 10–35 cm; stems sometimes proximally woody; leaves mostly basal, blades either linear to lanceolate and not lobed 10–20(–38) cm × 6–10 mm, or oblanceolate to oblong and pin-nately lobed 10–20 cm × 25–50 mm, or both, margins usually entire, sometimes prickly, abaxial faces white-villous, adaxial faces glabrate to arachnose; heads 3.5–8 cm diam.; phyllaries: outer lanceolate, margins prickly-ciliate, the inner with margins undulate, ciliate with submarginal dark stripe; ray florets 13–18, corolla laminae yellow or orange, usually each with a dark abaxial stripe and adaxial basal blotch or spot; cypselae 1–2 mm; pappus of 7–8 scales. One record in Las Cruces; found in a vacant lot at 4000 ft.

GNAPHALIUM L. – Cudweed

Jennifer Ackerfield

Annual herbs, generally white-woolly; leaves alternate, entire, sessile; heads in dense glomerules at the ends of branches and in leaf axils, disciform; involucral bracts slightly imbricate, usually white at the tip; receptacles naked; ray florets absent; disk florets yellowish or whitish; cypselae small, nerveless, round or flat; pappus of capillary bristles (Ackerfield, 2015).

1. Leaves lanceolate to oblanceolate or oblong, generally wider, 2–10 mm wide; leaves subtending the heads not much longer than the inflorescence…**G. palustre**
1'. Leaves linear or linear-oblanceolate, narrow, 0.5–3(–5) mm wide; leaves subtending the heads much lon-ger than the inflorescence…(2)
2. Heads in axillary spiciform glomerules; leaves linear to linear-oblanceolate…**G. exilifolium**
2'. Heads in capitate glomerules terminating the branches or sometimes with axillary glomerules; leaves oblanceolate…**G. uliginosum**

Gnaphalium exilifolium A. Nelson—Slender cudweed. Annuals, 3–25 cm; leaves linear to linear-oblanceolate, 0.4–5 cm, loosely tomentose; involucres 2.5–3.5 mm, the bracts brownish with woolly bases, the inner with white, acute tips. Streams and pond margins, 5600–10,900 ft. NC, NW, WC, SW, Otero.

Gnaphalium palustre Nutt.—Diffuse cudweed. Annuals, 1–30 cm; leaves oblanceolate to lanceolate or oblong, 1–3.5 cm, woolly-tomentose; involucres 2.5–4 mm, the bracts brownish with woolly bases, the inner with white, blunt tips. Sandy soil of moist places along streams and pond margins, sometimes in alkaline soil, 3800–10,400 ft. Grant, Rio Arriba, Sandoval.

Gnaphalium uliginosum L.—Marsh cudweed. Annuals, 3–25 cm; leaves oblanceolate, 1–5 cm, loosely tomentose; involucres 2–4 mm, the bracts brownish with woolly bases, the inner with white, acute tips. Streams and pond margins, swampy places, sometimes disturbed areas, 7000–9150 ft. Catron, Colfax, Sandoval, Sierra. Introduced.

GRINDELIA Willd. · Gumweed

Jennifer Ackerfield and Kenneth D. Heil

Herbs; leaves alternate, sessile and often clasping, the margins entire to sharply toothed; heads radiate or discoid; involucral bracts imbricate, sticky and gummy, firm, with herbaceous, reflexed tip; receptacles naked; ray florets yellow to orange; disk florets yellow; cypselae smooth, compressed to quadrangular; pappus of 2 to several firm awns (Ackerfield, 2015; Strother & Wetter, 2005).

1. Pappus persistent or tardily falling, 25–40 barbellate bristles subtending 8–15+ barbellate, setiform awns or subulate scales; E New Mexico…**G. ciliata**
1'. Pappus readily falling, 1–8+ smooth or barbellulate to barbellate bristles, setiform awns, or subulate scales; widespread…(2)
2. Stems variously hairy and/or glandular…(3)
2'. Stems glabrous…(5)
3. Leaf margins crenate, teeth blunt, resin-tipped; Lincoln and Eddy Counties…**G. havardii**
3'. Leaf margins serrate to dentate, the teeth sharp, apiculate to setose; various locations…(4)
4. Pappus of 2–4 ± straight, usually barbellulate bristles; S New Mexico…**G. scabra**
4'. Pappus of 2–4(–6) usually contorted to curled, sometimes nearly straight, setiform awns or scales; widespread…**G. hirsutula** (in part)
5. Ray florets present…(6)
5'. Ray florets absent, heads with only disk florets…(8)
6. Middle and upper involucral bracts erect and appressed…**G. arizonica**
6'. Middle and upper involucral bracts reflexed…(7)
7. Stem leaves closely and evenly crenulate-serrate or closely serrulate to only remotely closely serrulate, the teeth mostly 1–2(–2.5) mm apart, if entire then the involucre surface conspicuously and abundantly resinous…**G. squarrosa**
7'. Stem leaves sharply and coarsely and more remotely dentate to serrate, or the upper and middle sometimes entire or serrate near the apex only, the teeth mostly 2.5–5 mm apart, or if entire then the involucre surface only moderately resinous…**G. hirsutula** (in part)
8. Middle and outer involucral bracts with a loosely or moderately reflexed tip, not as tightly curled as the next; heads depressed-hemispherical or broadly bowl-shaped, usually much broader than high; cypselae 2–3 mm, smooth to striate on the angles or deeply furrowed or ribbed, the ridges sometimes wrinkled… **G. nuda**
8'. Middle and outer involucral bracts with a strongly revolute, tightly curled tip; heads deeply bowl-shaped, usually as high as or higher than broad; cypselae 3–5.5 mm, slightly nerved or smooth…(9)
9. Stigmas linear-lanceolate; leaves conspicuously resinous-punctate…**G. fastigiata**
9'. Stigmas oblong-lanceolate or oblong; leaves less conspicuously resinous-punctate…**G. inornata**

Grindelia arizonica A. Gray [including *G. laciniata* Rydb.]—Arizona gumweed. Plants 4–8 dm; leaves oblong to oblanceolate or ovate-lanceolate, with entire to serrate or denticulate margins, 2–7 cm; involucres 7–10 mm, the bracts erect and appressed; ray florets 12–20; cypselae 2.3–3.5 mm; pappus of 2–4 straight or weakly contorted, smooth setiform awns or subulate scales.

1. Heads hemispherical; phyllary apices subulate…**var. neomexicana**
1'. Heads campanulate; phyllary apices lance-acuminate to deltate…**var. arizonica**

var. arizonica—Arizona gumweed. Piñon-juniper woodlands, ponderosa pine communities, 6600-8300 ft. SW.

var. neomexicana (Wooton & Standl.) G. L. Nesom [*G. neomexicana* Wooton & Standl.]—New Mexico gumweed. Piñon-juniper woodlands, ponderosa pine, mixed conifer communities, 5485-9000 ft. WC, SW. **R & E**

Grindelia ciliata (Nutt.) Spreng. [*G. papposa* G. L. Nesom & Y. B. Suh]—Spanish gold. Stems erect, glabrous; leaves oblong to obovate; margins dentate, teeth apiculate to setose, 38 cm; involucres 10-15 mm, the bracts recurved to straight; ray florets 25-45; cypselae 2-4 mm; pappus persistent or tardily falling, 25-40 barbellate bristles. Disturbed sites, prairies, railroads, roadsides, 3870-4090 ft. EC, SE, Quay.

Grindelia fastigiata Greene—Pointed gumweed. Plants 3-15 dm; leaves oblanceolate to oblong-spatulate, 1.5-15 cm, the margins entire to dentate or closely serrate; involucres 8-11 mm, the upper tips closely and strongly revolute; ray florets absent; cypselae 3.5-5 mm; pappus of 2-3(-6) contorted or curled, sometimes straight, mostly smooth scales or setiform awns. Dry, open places, often in sandy soil; 1 record, 6825 ft. Rio Arriba.

Grindelia havardii Steyerm.—Havard's gumweed. Plants 3-15 dm; leaves ovate to oblong, mostly 1.5-5.5 cm; the margins crenate, blunt, resin-tipped; involucres 8-13 mm, recurved to straight; ray florets 18-25; cypselae 2-3.5 mm; pappus straight, usually smooth setiform awns, 4-7 mm, 3560-4400 ft. Eddy, Lincoln.

Grindelia hirsutula Hook. & Arn. [including *G. acutifolia* Steyerm.; *G. decumbens* Greene; *G. revoluta* Steyerm.]—Hirsute gumweed. Plants 1.5-8 dm; leaves oblong to oblong-lanceolate or ovate-lanceolate, 2-8.5 cm, the margins entire to dentate; involucres 8-15 mm, the upper bracts strongly reflexed or revolute; ray florets 12-37; cypselae 3-5 mm; pappus of 2-3(-6) contorted or curled, sometimes straight, mostly smooth scales or setiform awns. Dry, open hillsides and canyon slopes, 5200-8800 ft. Scattered.

Grindelia acutifolia and *G. revoluta* are included within *G. hirsutula* by Strother and Wetter (2005). We have also included *G. decumbens* within the broadly defined *G. hirsutula* because all of the characters for *G. decumbens* and *G. revoluta* overlap. The main delimiting character supporting maintaining these 3 as separate species seems to be geographic locality. The rayless *G. inornata* and *G. fastigiata* are also included within the *G. hirsutula* complex by Strother and Wetter. We have chosen to use a varietal ranking to categorize the morphological variation and geographic differences between the following forms of *G. hirsutula* and choose to keep the rayless forms as separate species.

1. Lower leaves with small glandular hairs on the lower sides along the main vein, the margins mostly conspicuously scabridulous…**var. acutifolia**
1'. Lower leaves without glandular hairs, the margins mostly scarcely scabridulous…**var. decumbens**

var. acutifolia (Steyerm.) Ackerf. [*G. acutifolia* Steyerm.]—Raton gumweed. Uncommon in dry, open places; found near Trinidad in the Raton Mesa region, 6700-8100 ft. NE, NC.

var. decumbens (Greene) Ackerf. [*G. decumbens* Greene]—Reclined gumweed. Dry, open hillsides and canyon slopes, 7350-9150 ft. NW.

Grindelia inornata Greene—Colorado gumweed. Plants 2.5-8 dm; leaves obovate-oblong, 2-5.5 cm, the margins dentate; involucres 8-12 mm, the upper bracts strongly reflexed; ray florets absent; cypselae 3-5 mm; pappus of 2-3(-6) contorted or curled, sometimes straight, mostly smooth scales or setiform awns. Locally common on dry hillsides and slopes, 4430-5250 ft. Colfax, Mora, San Miguel, Union.

Grindelia nuda Wood—Curlycup gumweed. Plants 1.5-6 dm; leaves oval to broadly oblong, closely and evenly crenate, 1.5-4.5 cm; involucres 8-15 mm, with loosely or moderately reflexed tips; ray florets absent; cypselae 2-3 mm; pappus of 2-3(-8) straight or contorted to curled, smooth or barbellulate to barbellate scales or awns. Dry, open places. Widespread.

Strother and Wetter (2005) group *G. nuda* and *G. aphanactis* within a more broadly defined *G. squarrosa*. However, we choose to keep the discoid plants as separate varieties based on the presence of ray florets as well as several additional morphological characteristics.

1. Stem leaves oval to ovate or broadly oblong, 1.5-3 times longer than wide, closely and evenly crenulate-serrate or closely serrulate to only remotely closely serrulate, the teeth mostly 1 mm apart, rarely wider; cypselae smooth to striate on the angles; pappus awns usually entire or remotely serrulate; stems often greenish...**var. nuda**
1'. Stem leaves oblanceolate to oblong, mostly 5-10 times longer than wide, entire or finely to coarsely dentate or the lower sometimes pinnatifid, the teeth mostly 2+ mm apart; cypselae deeply furrowed or ribbed, the ridges sometimes wrinkled; pappus awns closely serrulate to setulose-serrulate; stems usually reddish...**var. aphanactis**

var. aphanactis (Rydb.) G. L. Nesom—Moist to dry fields, meadows, roadsides, sandy places along streams, 3170-8900 ft.

var. nuda [*G. pinnatifida* Wooton & Standl.]—Uncommon in dry, open places on plains, 4500-8200 ft.

Grindelia scabra Greene—Rough gumweed. Plants 1-7 dm; leaves ovoid-oblong or spatulate, 2-8.5 cm, the margins serrate to dentate, teeth apiculate to setose; involucres 7-11 mm, slightly recurved to nearly straight; ray florets 17-30; cypselae 2.5-3 mm; pappus of 2-4 straight, usually barbellulate to barbellate, sometimes smooth bristles. Dry, rocky slops and mesas, 5200-9500 ft. SC, SW, Bernalillo, Valencia.

Grindelia squarrosa (Pursh) Dunal—Curlycup gumweed. Plants 1-10 dm; leaves oblong to oblanceolate, 3-7 cm, the margins closely and evenly crenulate-serrate, entire, or remotely serrulate; involucres 6-11 mm, the upper strongly reflexed; ray florets 22-36; cypselae 2-3.5 mm. Very common in dry, open places, 3500-8500 ft.

Strother and Wetter (2005) combine some of the rayless *Grindelia* species (*G. nuda* and *G. aphanactis*) with *G. squarrosa* and eliminate the following varieties. However, even though the following varieties intergrade, there are significant morphological differences between them. We choose to keep a narrower circumscription of the group as completed by Steyermark (1937). Widespread.

1. Leaves relatively narrow, the middle and upper ones mostly 5-8 times as long as wide, linear-oblong to oblanceolate; 6000-7320 ft....**var. serrulata** (Rydb.) Steyerm.
1'. Leaves relatively broad, the middle and upper ones mostly 2-4 times as long as wide, ovate or oblong; 4000-9500 ft....**var. squarrosa**

GUTIERREZIA Lag. – Snakeweed

Kenneth D. Heil

Annuals, perennials, or subshrubs, 10-150(-200) cm; stems erect to ascending, usually branched; leaves alternate, sessile or petiolate, blades linear to lanceolate or spatulate, margins entire, faces glabrous or minutely hairy, gland-dotted, resinous; heads radiate; involucres cylindrical to campanulate; phyllaries 4-40 in 2-4 series; receptacles flat to conic, pitted; ray florets 1-30, corollas yellow or white; disk florets 1-150, corollas yellow or white; cypselae light tan to purplish black, clavate or cylindrical, not compressed; pappus persistent or readily falling, coroniform or of 5-10 whitish, often erose-margined scales in 1-2 series (Nesom, 2006l).

1. Annuals...(2)
1'. Perennials...(4)
2. Cauline leaves mostly 3-5-nerved, proximal usually persistent at flowering; cypsela faces pebbly or warty with raised oil cavities, glabrous...**G. wrightii**
2'. Cauline leaves mostly 1-nerved (sometimes 3-nerved in *G. texana*), proximal usually absent at flowering; cypsela faces moderately to densely hairy, never with oil cavities...(3)
3. Stems smooth, glabrous; cypsela hairs with acute apices; pappus of scales or essentially absent... **G. texana**

3'. Stems papillate-scabrous, often sparsely so; cypsela hairs with clavate apices; pappus of scales…
    G. sphaerocephala
4. Involucres cylindrical, 1–1.5 mm diam.; ray florets 1(2)…**G. microcephala**
4'. Involucres campanulate to cylindrical, 1.5–2(–3) mm diam.; ray florets (2)3–8…**G. sarothrae**

Gutierrezia microcephala (DC.) A. Gray—Threadleaf snakeweed. Subshrubs, 2–14 dm; leaves linear to lanceolate, glabrous or somewhat hispid; heads sessile or subsessile, 2–6 in compact, flat-topped arrays; involucres 1–1.5 mm diam.; ray florets 1 or rarely 2; disk florets 1 or rarely 2; cypselae 1–2.5 mm. Gravelly, gypsum, and limestone substrates, grasslands, oak and oak-pine woodlands, 3100–8700 ft. Widespread except NE, NC.

Gutierrezia sarothrae (Pursh) Britton & Rusby—Broom snakeweed. Subshrubs, 1–6(–10) dm; leaves linear to lanceolate, glabrous to hispidulous; heads sessile or subsessile in dense, flat-topped arrays; involucres 1.5–2(–3) mm diam.; ray florets (2)3–5(–8), 3–5.5 mm; disk florets (2)3–5(–9); cypselae 0.8–2.2 mm. Abundant in overgrazed sites, rocky, open slopes, grasslands, montane meadows, 3200–10,000 ft. *Gutierrezia longifolia* Greene is usually placed in synonymy with *G. sarothrae*. It is recognized by longer, wider, 3-nerved leaves and is found at middle to higher elevations, 6200–9575 ft. Widespread.

Gutierrezia sphaerocephala A. Gray—Roundleaf snakeweed. Annuals, 2–5 dm; leaves linear, proximal usually absent at flowering, 1-nerved, 0.5–1.2 mm wide; heads in loose arrays; ray florets 10–15, 3–5.5 mm; disk florets 18–37; cypselae 0.9–1.4 mm. Alkaline flats with *Atriplex* and gypsum or limestone substrates, grasslands, lake edges, oak-pine-juniper woodlands, 3250–7050 ft. EC, C, SE, SC, SW.

Gutierrezia texana (DC.) Torr. & A. Gray [*Xanthocephalum texanum* (DC.) Shinners]—Texas snakeweed. Annuals, 1–10 dm; leaves linear, 1- or 3-nerved, proximal usually absent at flowering; heads in loose arrays; involucres 2–4.5 mm diam.; ray florets 5–23, 3–6 mm; disk florets 7–48; cypselae 1.3–1.8 mm. New Mexico material belongs to **var. glutinosa** (S. Schauer) M. A. Lane. Grasslands and pine-oak-juniper woodlands, 3600–5600 ft. EC, SE, Doña Ana, Grant, Socorro.

Gutierrezia wrightii A. Gray [*Xanthocephalum wrightii* (A. Gray) A. Gray]—Wright's snakeweed. Annuals, 3–20 dm; leaves mostly narrowly elliptic-lanceolate, 40–70 × 2–4 mm, proximal usually persistent at flowering, mostly 3- or 5-nerved; heads usually in open arrays; involucres campanulate, 4.5–7.5 mm diam.; ray florets 8–19, 5–12 mm; disk florets 30–60; cypselae 0.7–2 mm, faces pebbly or warty with raised blisterlike oil cavities. Meadows and clearings in pine, pine-oak, and pine-fir woodlands, 5500–8800 ft. EC, C, WC, SC, SW.

GYMNOSPERMA Less. – Jackass-clover

Kenneth D. Heil

Gymnosperma glutinosum (Spreng.) Less. [*Selloa glutinosa* Spreng.]—Gumweed. Shrubs, 5–15 (–20) dm, glabrous, heavily resinous; leaf blades linear to narrowly lanceolate, oblanceolate, or elliptic, mostly 25–50 mm, margins entire, faces gland-dotted (in pits); heads radiate, sessile to subsessile, in compact glomerules, in terminal, corymbiform arrays; involucres 3–3.8 mm; receptacles flat, pitted; ray florets 4–9, 2–3 mm, corollas yellow; disk florets 4–6, corollas orange-yellow; cypselae columnar or fusiform, terete or slightly compressed, 1–1.4 mm; pappus essentially 0. Sandy or rocky slopes, crevices, and ledges, creosote bush, piñon-juniper, pine-oak-maple, 4400–8600 ft. SE, SC, SW (Nesom, 2006m).

HAPLOËSTHES A. Gray – False broomweed

Don Hyder

Haploësthes greggii A. Gray—False broomweed. Perennials or subshrubs, stems erect to sprawling, branched from bases or throughout, to 60 cm; leaves cauline, opposite, linear to filiform, 2–4 cm, blades 2–8 cm × 0.5–4 mm; heads radiate; receptacles convex; ray florets 3–6, 2–4 mm, corollas yellow;

disk florets yellow; cypselae 15-25 mm; pappus 2.5-3.5 mm. Our plants belong to **var. texana** (J. M. Coult.) I. M. Johnst. Limestone and gypsum outcrops and rubble, 4600-5575 ft. EC, C, SC (Allred & Ivey, 2012; Strother, 2006k).

HEDOSYNE Strother - Marsh-elder

Don Hyder

Hedosyne ambrosiifolia (A. Gray) Strother [*Iva ambrosiifolia* (A. Gray) A. Gray]—Ragged marsh-elder. Annuals; stems erect, strictly branched; leaves cauline, blades 3-5(-9) × 4-5(-8) cm, lobes 1-3 mm wide; petioles 5-12(-45) mm; involucres 2-3+ mm; outer phyllaries 5, sparsely strigose or glabrous; heads usually in loose, paniculiform arrays; receptacles hemispherical; corollas whitish, funnelform, 1.5-2 mm; cypselae 1.4-1.7 mm. Disturbed sites (roadsides, washes, etc.), in sandy, gypseous, or calcareous soils, 3700-7300 ft. C, SE, SC, SW (Allred & Ivey, 2012; Strother, 2006l).

HEDYPNOIS Mill. - Cretanweed

Don Hyder

Hedypnois cretica (L.) Dum. Cours.—Cretanweed. Annuals, taprooted; stems usually 1; leaves basal and cauline, margins entire or dentate, blades 5-150(-250) × 25(-35+) mm; peduncles 2-5(-15+) cm; heads borne singly or in loose, corymbiform arrays; abaxial faces of phyllaries glabrous or scabrous to hispid; pistillate florets 5-10, corollas 0; functionally staminate florets 5-10+, corollas whitish, funnelform, lobes 5, soon reflexed; cypselae 5-7.5. Known only from a lawn in Las Cruces, 3900 ft. Doña Ana. Introduced (Allred & Ivey, 2012; Strother, 2006m).

HELENIUM L. - Sneezeweed

Jennifer Ackerfield and Kenneth D. Heil

Annual or perennial herbs; leaves simple, alternate; heads usually radiate (ours); involucral bracts subequal in 2-4 series, ± herbaceous, the outer reflexed; receptacles naked or with a few scattered bristles; ray florets 3-lobed at the apex, or absent; disk florets perfect and fertile; cypselae obpyramidal, 4-5-angled; pappus of 5-10 scarious or hyaline, awn-tipped thin scales (Ackerfield, 2015; Bierner, 2006a).

1. Stems not winged; annual…**H. amarum**
1'. Stems ± winged by decurrent leaf bases; annual or perennial…(2)
2. Disk florets yellow; perennial with fibrous roots or rhizomes…**H. autumnale**
2'. Disk florets reddish brown; annual from a taproot…**H. microcephalum**

Helenium amarum (Raf.) H. Rock—Sneezeweed. Annuals, 1-8 dm; basal leaf blades linear to ovate, entire or pinnately toothed or lobed, distal leaf blades linear, entire; involucres 5-9 mm; ray florets 4.5-14 mm; disk florets 75-250+, corollas yellow proximally, yellow to yellow-brown distally; cypselae 0.7-1.3 mm, moderately to densely hairy; pappus of 6-8 entire, aristate scales 1-1.8 mm. New Mexico material belongs to **var. badium** (A. Gray ex S. Watson) Waterf. Shortgrass prairie, 4100-4500 ft. Curry, Roosevelt.

Helenium autumnale L.—Mountain sneezeweed. Perennials, 5-13 dm; leaves lanceolate to obovate, entire to dentate or weakly lobed, hairy; involucres 8-20 mm, the bracts hairy; ray florets 10-23 mm; disk florets yellow, 2.4-4 mm; cypselae 1-2 mm, hairy; pappus of 4-7 aristate scales 0.5-1.5 mm. Around ponds, lakes, streams, fields, seepage areas, 3800-8150 ft. NC, Chaves, Guadalupe, Otero.

Helenium microcephalum DC.—Smallhead sneezeweed. Annuals, 2-12 dm; leaves narrowly elliptic to oblong-elliptic, serrate to deeply toothed or lobed, glabrous to hairy; involucres 4-8 mm, the bracts hairy; ray florets 2.6-9 mm; disk florets mostly reddish brown, 1.2-2.5 mm; cypselae 0.7-1.5 mm, hairy; pappus of 6 scales 0.3-0.7 mm. Our material belongs to **var. microcephalum**. Slickrock, sandy or clay soils, around potholes, ponds, lakes, and streams, 3200-4900 ft. SE, SC.

HELIANTHELLA Torr. & A. Gray – Dwarf sunflower

Kenneth D. Heil

Perennials, 20–150 cm; stems erect, leaves basal and cauline, opposite and alternate, simple, entire; heads radiate; involucral bracts subequal or imbricate; receptacles chaffy; bracts persistent and clasping the achenes; disk florets yellow, purple, or brownish purple; ray florets yellow; pappus of persistent slender awns or scales, or absent; cypselae ± compressed, margins usually ± winged; pappus 0 or of persistent scales (Ackerfield, 2015; Weber, 2006).

1. Heads 5–15+; ray laminae 8–14+ mm; disk corollas purple to brown…**H. microcephala**
1'. Heads 1(–3+); ray laminae (11–)15–30(–45+) mm; disk corollas yellow…(2)
2. Heads erect…**H. uniflora**
2'. Heads ± nodding…(3)
3. Plants 20–50 cm; leaf blades broadest distal to middle, not leathery, faces rough-pubescent; involucres 15–20 mm diam.; ray florets 8–14, corollas pale yellow…**H. parryi**
3'. Plants 50–150 cm; leaf blades broadest near middle, leathery, faces sparsely hirsute or glabrous; involucres 40–50 diam.; ray florets (13–)21, corollas bright yellow…**H. quinquenervis**

Helianthella microcephala (A. Gray) A. Gray [*Encelia microcephala* A. Gray]—Purpledisk little sunflower. Plants 20–60(–80) cm; cauline leaves largest proximal to midstem; blades linear-spatulate to oblong-spatulate, 6–25 cm, faces hispidulous to scabrellous or glabrate; heads (3–)5–15+ in corymbiform arrays; disk florets purple to brown, ray florets 6–12 × 2–4 mm. Semidesert sites, often with piñon-juniper and scattered ponderosa pine, 6100–7075(–8400) ft. NW, Catron.

Helianthella parryi A. Gray—Parry's little sunflower. Plants 20–50 cm; cauline leaves largest proximal to midstem; blades oblanceolate or spatulate, faces rough-pubescent; heads usually borne singly, ± nodding; disk florets yellow; ray florets 8–14, corollas pale yellow. Mixed conifer forests to timberline, (5175–)7000–12,335 ft. NC, NW, C, WC, Eddy, Otero.

Helianthella quinquenervis (Hook.) A. Gray [*Helianthus quinquenervis* Hook.]—Fivenerve little sunflower. Plants (30–)50–150 cm; cauline leaves largest proximal to midstem; blades 3- or 5-nerved, elliptic or ovate-lanceolate to lanceolate, 10–50 cm, faces sparsely hirsute or glabrous; heads usually borne singly, ± nodding; disk florets yellow; ray florets 25–40 mm, corollas bright yellow. Mountain meadows and aspen forests, 7600–12,050 ft. NC, C, WC, SC.

Helianthella uniflora (Nutt.) Torr. & A. Gray [*Helianthus uniflorus* Nutt.]—Common little sunflower. Plants 40–120 cm; cauline leaves largest at midstem; blades lanceolate to elliptic, rarely ovate, 12–25 cm, faces puberulent, hirsute, or scabrous; heads usually borne singly, sometimes 2–3+, erect; disk florets yellow; ray florets 15–45 mm. Grasslands, meadows, sagebrush, mixed conifer and spruce-fir forests; 1 record, 8700 ft. Rio Arriba, San Juan.

HELIANTHUS L. – Sunflower

Kenneth D. Heil

Annuals or perennials, taprooted or rhizomatous, some tuberous; stems mostly 1; leaves mostly cauline, petiolate or sessile, opposite or alternate; heads radiate, rarely discoid, borne singly or few to many; involucres campanulate to hemispherical; phyllaries mostly 10–40+ in 2–4 series; disk florets mostly 30–150+, corollas yellow or reddish purple; ray florets 5–30, corollas yellow; cypselae ± 4-angled, glabrous, glabrate, or pubescent; pappus of 2 primary ± subulate scales (Ackerfield, 2015; Schilling, 2006a).

1. Plants annual from a taproot; leaves mostly alternate (or proximal opposite); paleae prominently 3-toothed; disk corollas reddish purple, at least distally, rarely yellowish…(2)
1'. Plants perennial from rhizomes or widely creeping roots; leaves opposite or alternate; paleae weakly 3-toothed; disk corollas yellow or reddish purple…(5)
2. Plants 100–300 cm; leaf blades (largest) 10–40 cm wide, abaxial faces gland-dotted; phyllaries ovate to lance-ovate, apices narrowed abruptly…**H. annuus**

2'.   Plants mostly 25–200 cm; leaf blades (largest) < 12 cm wide, abaxial faces sometimes gland-dotted; phyllaries usually lanceolate to lance-ovate, apices narrowed gradually…(3)
3.   Palea apices glabrous, not bearded…**H. paradoxus**
3'.   Palea apices (at least central palea) bearded with tufts of whitish hairs…(4)
4.   Leaf blades usually lanceolate to deltate-lanceolate, bases truncate to cuneate; phyllary apices relatively short-attenuate…**H. petiolaris**
4'.   Leaf blades usually lanceolate to deltate-ovate or ovate, bases truncate to cordate; phyllary apices long-attenuate…**H. neglectus**
5.   Leaves (at flowering) mostly basal, cauline leaves abruptly smaller…**H. pauciflorus** (in part)
5'.   Leaves (at flowering) mostly cauline, not abruptly smaller distally…(6)
6.   Disk corolla lobes reddish…(7)
6'.   Disk corolla lobes yellow…(9)
7.   Leaves petiolate to subsessile, margins subentire or serrate; ray laminae 15–37 mm…**H. pauciflorus** (in part)
7'.   Leaves sessile or nearly so, margins often irregularly toothed or lobed; ray laminae 8–11 mm…(8)
8.   Plants 50–120 cm; stems glabrate to strigose or hispid; leaves green or grayish, not bluish green…**H. laciniatus**
8'.   Plants 40–70 cm; stems glabrous or glabrate (glaucous); leaves often bluish green, glabrous or glabrate…**H. ciliaris**
9.   Stems glabrous or glabrate, at least proximal to arrays of heads…(10)
9'.   Stems hairy ± throughout…(11)
10.   Plants 20–30 cm; stems ascending to erect; ray laminae 7–9 mm; disk corollas 3–3.5 mm…**H. arizonensis**
10'. Plants 100–400 cm; stems erect; ray laminae 20–25 mm; disk corollas 5–7 mm…**H. nuttallii**
11.   Heads in a spiciform or racemiform inflorescence; upper and sometimes middle leaves folded lengthwise…**H. maximiliani**
11'. Heads few or solitary; leaves not folded…**H. tuberosus**

Helianthus annuus L.—Common sunflower. Annuals, 10–30 dm; leaves ovate to ovate-oblong, hispid, 10–40 × 5–35 cm, margins serrate; involucres 15–40(–200) mm diam., bracts 13–25 mm, margins usually ciliate, hispid; disk florets 5–8 mm, reddish purple or rarely yellow; ray florets 25–50 mm; pappus of scales, 2–3.5 mm. Disturbed sites, especially fields and roadsides, 3275–9320 ft. Widespread.

Helianthus arizonensis R. C. Jacks.—Arizona sunflower. Perennials, 20–30 cm; leaves lanceolate, glabrous, 2–7 × 0.5–1.3 cm, margins entire or undulate; involucres 9–18 mm diam.; phyllaries 16–19; disk florets 3–3.5 mm, lobes yellow, anthers reddish brown; ray florets 7–9 mm; pappus of 2 deltate, erose scales. W of Quemado in open pine communities, 4000–7000 ft. Catron. **R**

Helianthus ciliaris DC.—Texas blueweed. Perennials, 40–70 cm, rhizomatous or with creeping roots; leaves sessile, blades often bluish green, linear to lanceolate, 3–7.5 × 0.5–2.2 cm, margins entire or serrate, usually ciliate, faces glabrous or glabrate to hispid; involucres 12–25 mm diam.; phyllaries 16–19; disk florets 4–6 mm, reddish; ray florets 8–9 mm; pappus of scales, 1.2–1.5 mm. Dry to moist sites, usually disturbed areas, 3500–8600 ft. Widespread except N. Introduced.

Helianthus laciniatus A. Gray [*H. crenatus* R. C. Jacks.]—Alkali sunflower. Perennials, 50–120 (–200) cm; leaves sessile, green or grayish, lanceolate, 5–9 × 0.5–3.5 cm, margins entire or irregularly toothed to lobed, faces strigose, gland-dotted; involucres 10–24 mm diam.; phyllaries 16–21; disk florets 40+, corollas 4.8–5.8 mm, lobes reddish; ray florets 8–11 mm; pappus of scales, 1.4–2.5 mm. Alkaline soils, 3950–5400 ft. SC, SW, Torrance.

Helianthus maximiliani Schrad. [*H. dalyi* Britton]—Maximilian sunflower. Perennials, 50–300 cm, rhizomatous; leaves mostly alternate, blades light green to gray green, lanceolate, 10–30 × 2–5.5 cm, margins entire or serrulate, abaxial faces scabrous to hispid, gland-dotted; involucres 13–28 mm diam.; phyllaries 30–40; disk florets 5–7, yellow; ray florets 15–40 mm; pappus of 2 aristate scales, 3–4.1 mm. Moist prairie of the E plains, gardens where it escapes, 4000–6700 ft. NE, NW, C.

Helianthus neglectus Heiser—Neglected sunflower. Annuals, 80–200 cm; leaves petiolate, 7–14 × 7.5–12.3 cm, margins subentire to serrulate, abaxial faces strigose, not gland-dotted; involucres 23–28 × 10–14 mm; phyllaries 25–35; disk florets 150+, corollas 6–6.5 mm, lobes reddish; ray florets 21–31; pappus of 2 aristate scales, 2.8–3.2 mm. Sandy soil, shinnery oak communities and dunes, 3078–4195 ft. Eddy, Lea, Roosevelt.

Helianthus nuttallii Torr. & A. Gray—Nuttall's sunflower. Perennials, 100–400 cm, rhizomatous; leaves opposite to alternate, light to dark green, lanceolate to ± ovate, 4–20 × 0.8–4 cm, margins entire or serrate, abaxial faces hispid to villous-tomentose, gland-dotted; involucres 10–20 mm diam.; phyllaries 30–38; disk florets 5–7, yellow; ray florets 15–40 mm; pappus of scales, 3–4 mm. Moist meadows, stream and pond margins, 5600–7800(–9570) ft. NC, WC, SE.

Helianthus paradoxus Heiser—Pecos or paradox sunflower. Annuals, 130–200 cm; leaves mostly cauline, opposite or alternate, blades lanceolate to lance-ovate, 7–17.5 × 1.7–8.5 cm, with 3 prominent veins, margins entire or toothed, abaxial faces ± scabrous, not gland-dotted; involucres 15–20 mm diam.; phyllaries 15–25; disk florets 50+, corollas 5–5.5 mm, lobes reddish; ray florets 12–20; pappus of 2 lanceolate scales, 2.5–2.9 mm. Saturated saline soils, desert wetlands, 3300–6600 ft. Chaves, Cibola, Guadalupe, Socorro, Valencia. **FT**

Helianthus pauciflorus Nutt. [*H. subrhomboideus* Rydb.; *H. rigidus* Desf. subsp. *subrhomboideus* (Rydb.) Heiser]—Stiff sunflower. Perennials, 50–120 cm; leaves mostly basal, opposite, blades rhombic-ovate to lance-linear, 5–12 cm, apices acute or obtuse, margins serrate to subentire, abaxial faces sparsely hispid, gland-dotted; involucres 15–23 mm diam.; phyllaries 25–35; disk florets 6.5–7 mm, red to reddish purple; ray florets 20–35 mm; pappus of scales, 4–5 mm. Our material belongs to **subsp. subrhomboideus** (Rydb.) O. Spring & E. E. Schill. Grasslands, montane meadows, open ponderosa pine forests, 6000–9320 ft. NE, NC, C, Eddy, Lincoln.

Helianthus petiolaris Nutt.—Prairie sunflower. Annuals, 40–200 cm; leaves mostly cauline, mostly alternate, blades often bluish green, lanceolate to deltate-ovate or ovate, 4–15 × 1–8 cm, margins entire or ± serrate, abaxial faces hispid to villous; involucres 10–24 mm diam.; phyllaries 14–25; disk florets reddish purple, 4.5–6 mm; ray florets 15–20 mm; pappus of scales, 1.5–3 mm.

1. Leaves densely canescent on abaxial and adaxial surfaces…**var. canescens**
1'. Leaves glabrous, scabrous, or short-hispid, not canescent…(2)
2. Stems usually hispidulous to strigillose; peduncles usually bractless; phyllaries 3–5 mm wide…**var. petiolaris**
2'. Stems usually ± hispid; peduncles usually each with a leafy bract subtending heads; phyllaries 2–3.5 mm wide…**var. fallax**

var. canescens A. Gray [*H. niveus* (Benth.) Brandegee subsp. *canescens* (A. Gray) Heiser]—Prairie sunflower. Stems densely canescent; leaves with abaxial and adaxial faces canescent, densely gland-dotted; peduncles usually ebracteate; phyllaries 1–2 mm wide; disk corolla throats gradually narrowed distal to slight, not densely hairy basal bulges. Desert scrub, prairies, piñon-juniper woodlands, ponderosa pine forests, 4900–5725(–8500) ft. C, SE, SW.

var. fallax (Heiser) B. L. Turner—Prairie sunflower. Stems usually ± hispid or rarely hirsute to glabrate; leaves with abaxial faces sparsely, if at all gland-dotted; peduncles usually with leafy bracts subtending heads; phyllaries 2–3.5 mm wide; disk corollas with throats gradually narrowed distal to slight, not densely hairy basal bulges. Sandy soils, dry and disturbed sites, 4250–8600 ft. Widespread.

var. petiolaris—Prairie sunflower. Stems usually hispidulous to strigillose, rarely ± hirsute to glabrate; leaves with abaxial faces sparsely, if at all gland-dotted; peduncles usually bractless; phyllaries 3–5 mm wide; disk corollas with throats abruptly narrowed distal to densely hairy basal bulges. Sandy soils, dry and disturbed sites, 3050–6750 ft. Widespread.

Helianthus tuberosus L.—Jerusalem artichoke. Perennials, 50–200+ cm, rhizomatous, producing tubers; leaves mostly cauline, opposite or alternate proximally, blades lanceolate to ovate, 10–23 × 7–15 cm, margins entire or serrate, abaxial faces hirsute to scabrous, gland-dotted; involucres 10–25 × 8–12 mm; phyllaries 22–35; disk florets 6–7 mm, yellow; ray florets 25–40 mm; pappus of scales, 2–3 mm. Disturbed sites; 1 record for New Mexico, elevation unknown. Rio Arriba. Introduced.

HELIOMERIS Nutt. – Goldeneye

Kenneth D. Heil

Annuals or perennials, (2–)10–90(–120+) cm; stems erect, branched; leaves mostly cauline, opposite or alternate, sessile or subsessile; heads radiate, borne singly or in cymiform to paniculiform arrays; involucres 6–14 mm diam.; disk florets 25–50+, 3–4 mm; ray florets 5–14, 7–20 mm, gland-dotted below; cypselae black or gray, weakly 4-angled, glabrous; pappus 0 (Schilling, 2006b). *Heliomeris* has often been submerged within *Viguiera*.

1. Annuals; leaf margins conspicuously ciliate at least 3/4+ their length (hairs often 0.5+ mm)…**H. hispida**
1'. Annuals or perennials; leaf margins ciliate to 1/4 their length (hairs to 0.5 mm)…(2)
2. Perennials with woody caudices; proximal and midstem leaves relatively broad, length 2–8 times width…**H. multiflora** (in part)
2'. Annuals (taprooted) or perennials (woody caudices); proximal and midstem leaves relatively narrow, length 6–30+ times width…(3)
3. Annuals (taprooted); proximal and midstem leaves (40–)80–160 × 4–8(–12) mm…**H. longifolia** (in part)
3'. Annuals (taprooted); proximal and midstem leaves 10–70(–85) × 1.5–5 mm, or perennials (woody caudices); proximal and midstem leaves 10–90 × 2–20 mm…(4)
4. Annuals (taprooted); proximal and midstem leaves 10–70(–85) × 1.5–5 mm; heads relatively small; involucres 6–9 mm diam.…**H. longifolia** (in part)
4'. Perennials (woody caudices); proximal and midstem leaves 10–90 × 2–20 mm; heads relatively large; involucres 6–14 mm diam.…**H. multiflora** (in part)

Heliomeris hispida (A. Gray) Cockerell [*H. multiflora* Nutt. var. *hispida* A. Gray; *H. hispida* var. *ciliata* (B. L. Rob. & Greenm.) Cockerell]—Hairy goldeneye. Annuals, 10–90 cm (taprooted); stems hispid; leaves mostly alternate, blades lance-linear, 20–100 × 1–3 mm, margins conspicuously ciliate at least 3/4 their length, hairs often 0.5+ mm, faces hispid, abaxial sometimes gland-dotted; involucres 9–14 mm diam.; phyllaries 5.5–10 mm; ray florets 9–15, 6–13 mm; disk florets 50+, corollas 2–3 mm; cypselae black, 1.5–2.5 mm. Saline marshes and meadows, 5000–6000 ft. Grant, Hidalgo.

Heliomeris longifolia (B. L. Rob. & Greenm.) Cockerell [*Viguiera longifolia* (B. L. Rob. & Greenm.) S. F. Blake]—Longleaf false goldeneye. Annuals, 7–15 cm (taprooted); stems strigose; leaves mostly opposite proximally, sometimes alternate distally, blades lance-linear to linear, 10–160 × 1.5–8(–12) mm, margins ciliate to 1/4 their length, hairs mostly < 0.5 mm; faces strigose to strigillose, abaxial often gland-dotted; involucres 6–14 mm diam.; phyllaries 3–6 mm; ray florets 12–14, laminae ± 5–17 mm; disk florets 50+, corollas 2.5–4 mm; cypselae black or mottled, 1.8–2 mm.

1. Stems branching mostly in distal 1/2, bases usually 7+ mm diam.; leaves (40–)80–160 × 4–8(–12) mm; involucres 9–14 mm diam.…**var. longifolia**
1'. Stems branching ± throughout, bases usually < 5 mm diam.; leaves 10–70(–85) × 1.5–5 mm; involucres 6–9 mm diam.…**var. annua**

var. annua (M. E. Jones) W. F. Yates—Annual longleaf false goldeneye. Stems branching ± throughout, bases usually < 5 mm diam.; leaves 10–70(–85) × 1.5–5 mm; involucres 6–9 mm diam. Desert scrub, piñon-juniper woodlands, ponderosa pine forests, 4310–7500 ft. SC, SW.

var. longifolia—Longleaf false goldeneye. Stems branching mostly in distal 1/2, bases usually 7+ mm diam.; leaves (40–)80–160 × 4–8(–12) mm; involucres 9–14 mm diam. Desert scrub, piñon-juniper woodlands, ponderosa pine forests, mixed conifer forests, 4200–8800 ft. NC, C, WC, SC.

Heliomeris multiflora Nutt. [*Viguiera multiflora* (Nutt.) S. F. Blake]—Showy goldeneye. Perennials, 20–120+ cm (woody caudices); stems strigose or puberulent to glabrate; leaves opposite or alternate, blades elliptic, lance-linear, lance-ovate, linear, or ovate, 10–90 × 2–28 mm, ciliate to 1/4 their length, hairs mostly < 0.5 mm, faces strigose, abaxial often gland-dotted; involucres 6–14 mm diam.; phyllaries 4–8 mm; ray florets 5–14, laminae 7–20 mm; disk florets 50+, corollas 3–4 mm; cypselae black or gray-striate, 1.2–3 mm.

1. Leaves mostly alternate, blades elliptic to ovate, apices obtuse, often mucronate…**var. brevifolia**
1'. Leaves mostly opposite, blades lance-ovate, lance-linear, or linear, apices acute…(2)

2. Leaves lance-ovate to lance-linear, 5–20 mm wide, margins flat...**var. multiflora**
2'. Leaves lance-linear to linear, 2–5 mm wide, margins usually strongly revolute...**var. nevadensis**

var. brevifolia (Greene ex Wooton & Standl.) W. F. Yates [*H. brevifolia* (Greene ex Wooton & Standl.) Cockerell]—Showy goldeneye. Leaves mostly alternate; blades elliptic to ovate, 8–28 mm wide, margins usually flat or obscurely revolute, apices obtuse (often mucronate). Canyon bottoms, piñon-juniper woodlands, ponderosa pine and mixed conifer forests, 6500–11,000 ft. Socorro.

var. multiflora—False goldeneye. Leaves mostly opposite; blades lance-ovate to lance-linear, 5–20 mm wide, margins flat, apices acute. Moist habitats in piñon-juniper woodlands, ponderosa pine and mixed conifer forests, (4800–)6500–11,000 ft. Widespread except E.

var. nevadensis (A. Nelson) W. F. Yates [*H. nevadensis* (A. Nelson) Cockerell]—Showy goldeneye. Leaves mostly opposite; blades lance-linear to linear, 2–5 mm wide, margins usually strongly revolute, apices acute. Dry rocky sites, piñon-juniper woodlands, ponderosa pine and mixed conifer forests (4700–)6550–10,000(–11,000) ft. NC, NW, C, WC, SC, SW.

HELIOPSIS Pers. – Oxeye, sunflower everlasting

Kenneth D. Heil

(Ours) perennials, 30–150 cm; stems erect or trailing; leaves cauline, opposite, blades ± 3-nerved from base, deltate or ovate to lanceolate, margins serrate or coarsely toothed, faces glabrous or hairy; heads radiate, borne singly; involucres turbinate to hemispherical, 8–14 mm diam.; phyllaries persistent, 12–20 in 2–3 series; receptacles convex to conic; ray florets [0–]5–20, pistillate, fertile; corollas yellow to orange; disk florets 30–150+, bisexual, corollas yellow or brown to purple; cypselae brown to black-brown, subterete or obscurely 3(ray)- or 4(disk)-angled (not winged); pappus 0, or persistent (Smith, 2006a).

1. Leaf blades (1.5–)3–6 × 0.8–3 cm, sparsely pubescent to glabrescent; cypselae rugulose to subtuberculate...**H. parvifolia**
1'. Leaf blades 6–12(–15) × 2.5–6(–12) cm, glabrous, pubescent, or scabrous; cypselae smooth...
    H. helianthoides

Heliopsis helianthoides (L.) Sweet [*Buphthalmum helianthoides* L.]—False sunflower, oxeye, smooth oxeye. Perennials, (40–)80–150 cm; stems stramineous to reddish brown, glabrous or hairy; leaf blades deltate to narrowly ovate-lanceolate, 6–12(–15) × 2.5–6(–12) cm, faces glabrous, sparsely pubescent, moderately to densely scabrellous, or scabrous; heads 1–15+; involucres 12–25 mm diam.; phyllaries glabrescent to densely pubescent on margins; ray florets 10–18, corollas golden yellow; disk florets 10–75+, corollas yellowish to brownish yellow; cypselae 4–5 mm, glabrous or pubescent on angles, smooth; pappus 0 or of 2–4 minute, toothlike scales. New Mexico material belongs to **var. scabra** (Dunal) Fernald [*H. scabra* Dunal; *H. helianthoides* var. *occidentalis* (T. R. Fisher) Steyerm.]. Open woods, prairies, old pastures, edges of fields, meadows, road banks, ditches, 5250–9640 ft. NE, NC, C, SC.

Heliopsis parvifolia A. Gray—Mountain oxeye. Perennials, 30–50(–80) cm; stems stramineous to reddish, glabrous or sparingly hairy; leaf blades deltate-lanceolate, (1.5–)3–6 × 0.8–3 cm, margins irregularly dentate to subentire, faces sparsely pubescent to glabrescent; heads 1–10+; involucres 12–20 mm diam.; phyllaries densely pubescent on margins and apices; ray florets 9–11, corollas yellow; disk florets 8–50+, corollas greenish yellow to yellow-brown; cypselae 4.5–5 mm, glabrous, rugulose to subtuberculate; pappus 0. Open, rocky mountain slopes, canyons, 5800–7250 ft. Hidalgo, San Miguel.

HERRICKIA Wooton & Standl. – Herrickia

Jennifer Ackerfield

Perennial herbs or subshrubs; leaves alternate; heads radiate; involucral bracts subequal to imbricate in 3–6 series, conspicuously keeled, the tips herbaceous and green and the bases chartaceous; receptacles naked; ray florets white to purple; disk florets yellow but sometimes drying purplish; cypselae compressed, striate; pappus simple, strongly barbellate bristles (Ackerfield, 2015).

1.   Leaves with entire margins; involucral bracts appressed…**H. glauca**
1'. Leaves with sharply spinulose-serrate margins; involucral bracts spreading…**H. horrida**

Herrickia glauca (Nutt.) Brouillet [*Aster glaucodes* S. F. Blake; *Eucephalus glaucus* Nutt.; *Eurybia glauca* (Nutt.) G. L. Nesom]—Gray aster. Perennials, 2–7 dm; leaves oblong to lanceolate, 4–12 cm, glabrous, margins entire; involucres 6–9 mm diam.; ray florets 8–18 mm, purple; disk florets 6.5–7.5 mm, yellow; pappus of bristles, 6–7 mm. Our material is **var. glauca**. Along roadsides and in open, rocky places in the mountains, often in association with piñon-juniper communities, 5300–10,000 ft. NW.

Herrickia horrida Wooton & Standl. [*Aster horridus* (Wooton & Standl.) S. F. Blake; *Eurybia horrida* (Wooton & Standl.) G. L. Nesom]—Horrid herrickia. Perennials or subshrubs, 3–6 dm; leaves oblong to nearly orbiculate, 1–5 cm, thick and rigid, the margins sharply spinose-serrate, glandular-stipitate and scabrous; involucres 8–12 mm diam.; ray florets 15–25 mm, purple; disk florets 8–9 mm, yellow; pappus of bristles, 7–8 mm. Uncommon on rocky hillsides and the sides of canyons, often with oak, 4600–9850 ft. Colfax, Harding, San Miguel.

HETEROSPERMA Cav. – Heterosperma

Jennifer Ackerfield

Heterosperma pinnatum Cav.—Wingpetal. Annual herbs, 1–4(–7) dm; leaves opposite, leaves usually 1–4 cm, the ultimate divisions 0.5–1(–3) mm wide; heads radiate; involucral bracts in 2 series, the outer 3–5 bracts linear and herbaceous, the inner series oval, striate, membranous; receptacles chaffy; ray florets 1–3(–8), ca. 1–2 mm, yellow; disk florets ca. 2.5 mm, yellow; pappus of 2 or 3 deciduous awns or absent; cypselae of 2 different kinds, the outer with corky wings and the inner tapering to barbellate beaks, 5–18 mm, glabrous. Rocky hills and dry grasslands, widespread in scattered localities, 4325–8400 ft. Widespread except NW, EC (Ackerfield, 2015; Allred & Ivey, 2012).

HETEROTHECA Cass. – Goldenaster

Guy L. Nesom

Herbaceous annuals and perennials, taprooted, sometimes with elongate, rhizomelike caudex branches; stems and leaves mostly sericeous to strigose, hirsute, or hispid, often sessile- or stipitate-glandular; leaves basal and cauline, basal often not persistent, margins entire; heads radiate, solitary or in a corymboid arrangement; rays yellow, pistillate, fertile; disk florets hermaphroditic, fertile; pappus of a long series of barbellate bristles and a short series of thick bristles or scales; cypselae sparsely strigose (ray cypselae glabrous and epappose in *H. subaxillaris* and *H. psammophila*) (Harms, 1963, 1974; Nesom, 2019, 2020b; Semple 1996, 2006, 2008).

1.   Annual or biennial; ray cypselae without pappus…(2)
1'. Perennial; ray cypselae with pappus…(3)
2.   Cauline leaves mostly ovate to ovate-deltate, about as long as wide, the proximal usually petiolate and nonclasping; innermost phyllaries 5–6(–7) mm…**H. subaxillaris**
2'. Cauline leaves oblong to oblong-obovate, oblong-ovate, or triangular-ovate, commonly 3–4 times longer than wide, usually epetiolate and clasping to subclasping nearly to the base of the stem; innermost phyllaries 6–8(–9) mm…**H. psammophila**
3.   Heads not immediately subtended by prominent herbaceous bracts…(4)
3'. Heads immediately subtended by prominent herbaceous (capitular) bracts…(15)
4.   Heads solitary, long-peduncled…(5)
4'. Heads in terminal clusters, short-peduncled…(7)
5.   Stems densely strigose with closely appressed hairs, eglandular…**H. pedunculata** (in part)
5'. Stems sparsely strigose to hirsute, prominently sessile-glandular…(6)
6.   Leaves oblong to oblong-lanceolate to oblanceolate, distal lanceolate to triangular-lanceolate, mostly 2–6 mm wide…**H. polothrix** (in part)
6'. Leaves broadly elliptic-oblanceolate to obovate or obovate-spatulate, 8–14 mm wide…**H. viscida**
7.   Heads few, loosely arranged; stems and leaves greenish, sparse vestiture not obscuring the surfaces; leaf surfaces sessile-glandular at least adaxially…**H. polothrix** (in part)

7'.   Heads in a corymboid cluster; stems and leaves grayish to silvery from denser vesture; leaf surfaces eglandular or glandular beneath nonglandular hairs…(8)

8.   Stems hirsute, usually glandular…(9)

8'.   Stems strigose, eglandular…(11)

9.   Heads loosely arranged from branches originating midstem to distally; midstem axillary buds prominent…**H. loboensis**

9'.   Heads clustered from branches originating distally; midstem axillary buds usually not produced…(10)

10.   Leaf surfaces hirsute; phyllaries loosely strigose to hirsute; ray corollas 9–11 × 1–1.3 mm…**H. hirsutissima**

10'.   Leaf surfaces strigose; phyllaries closely and finely strigose; ray corollas 10–14 × 1.5–3 mm… **H. sierrablancensis**

11.   Plants usually producing elongate, rhizomelike caudex branches…(12)

11'.   Plants without rhizomes or elongate caudex branches…(13)

12.   Leaf margins ciliate; stems densely leafy, with axillary buds and axillary tufts of leaves; leaf surfaces sessile-glandular beneath strigose vesture…**H. canescens** (in part)

12'.   Leaf margins eciliate; stems sparsely leafy, without axillary buds and axillary tufts of leaves; leaf surfaces eglandular…**H. pedunculata** (in part)

13.   Heads on short, ebracteate peduncles…**H. zionensis**

13'.   Heads on peduncles evenly leafy or bracteate to immediately beneath the involucre…(14)

14.   Leaves rigid, margins narrowly revolute; axillary tufts of small leaves absent; vesture sparse to moderately dense, not obscuring the green epidermis, trichome bases strongly pustulate-enlarged; leaf surfaces and phyllary apices eglandular; plants cespitose, not rhizomatous…**H. angustifolia**

14'.   Leaves not rigid, margins flat; axillary tufts of small leaves abundant and prominent; vesture dense and silvery-sericeous, obscuring the epidermis, trichome bases usually wider than the extended portion above but not strongly pustulate; leaf surfaces and phyllary apices sessile-glandular; plants colonial, rhizomatous…**H. canescens** (in part)

15.   Middle and upper leaves narrow, > 5 times longer than broad…**H. scabrifolia**

15'.   Middle and upper leaves < 5 times longer than broad…(16)

16.   Leaves 1–5 mm wide, midrib thickened and raised on abaxial surface, apex often with a thick-indurate, sharp-pointed, recurving terminal mucro; involucres 6–10 mm wide…**H. arizonica**

16'.   Leaves 5–35 mm wide, midrib not thickened or raised, apex rounded to obtuse or acute, without a thick, sharp mucro; involucres (5– in *E. fulcrata*) 9–15 mm wide…(17)

17.   Nonglandular vesture absent or greatly reduced except along leaf and bract margins… **H. cryptocephala**

17'.   Nonglandular vesture prominent on stems and leaf and bract surfaces…(18)

18.   Phyllaries hirsute-villous with long, fine hairs; alpine…**H. pumila**

18'.   Phyllaries strigillose to strigose or strigose-hirsute; stems erect to ascending; habitats below alpine… (19)

19.   Heads in panicles from branches from midstem and above; involucres 5–8 mm wide; capitular bracts no longer than the involucre; rays 6–8 mm…**H. paniculata**

19'.   Heads few from distal branches; involucres mostly 10–14 mm wide; capitular bracts usually longer than the involucre; rays 10–12 mm…(20)

20.   Stems and leaves strigillose with extremely thin, slightly flexuous, closely appressed hairs, without enlarged basal cells; plants usually with basal offsets and short rhizomes…**H. nitidula**

20'.   Stems and leaves hirsute or hirsutulous to hirsute-villous, pilose-strigose, or loosely strigose, hairs usually with enlarged basal cells; plants without basal offsets and rhizomes…**H. fulcrata**

Heterotheca angustifolia Shinners—Prairie candles. Cespitose, from a thick, woody taproot, without rhizomatous caudex branches; stems stiffly erect; stems and leaves loosely strigose with stiff, thick-based hairs not obscuring the green surfaces, eglandular; leaves rigid, oblanceolate to narrowly oblanceolate, ascending, margins narrowly revolute; axillary tufts of small leaves absent or not dense and prominent; heads mostly short-pedunculate in a closely clustered capitulescence; involucres 5–8 mm wide. Grassland, brushland, piñon-juniper, shinnery oak, counties immediately bordering Texas, 4000–5200 ft. NE, EC, SE.

Heterotheca arizonica (Semple) G. L. Nesom [*H. fulcrata* (Greene) Shinners var. *arizonica* Semple]—Arizona goldenaster. Stems and leaves densely and prominently sessile-glandular to short-stipitate-glandular, essentially green because of the relatively sparse, hirsutulous vesture of nonglandular hairs; cauline leaves elliptic to elliptic-oblanceolate or elliptic-lanceolate, 5–15 mm, usually relatively even-sized, spreading or ascending, often with only the midvein visible and distinctly thickened and raised on abaxial surface, with a thick-indurate, sharp-pointed, often recurving terminal mucro; heads usually immediately subtended by small, linear to oblanceolate capitular bracts; phyllaries glabrous to sparsely minutely glandular to sparsely strigose. Streambeds, roadsides, base of canyon walls, 4800–6500 ft. Hidalgo, Luna.

Heterotheca canescens (DC.) Shinners—Hoary goldenaster. Colonial, commonly with long, rhizomelike, node-rooting caudex branches; stems densely leafy, with axillary buds and axillary tufts of leaves; stems, leaves, and phyllaries silvery-sericeous, densely strigose, sessile-glandular beneath the strigose layer; leaves oblanceolate, margins ciliate; heads corymboid on short peduncles, rarely with small capitular bracts. Ponderosa pine–Gambel oak, pine woodland edges, piñon-juniper, meadows, prairies, usually in sandy soil, 4800–7500 ft. NE, EC.

Heterotheca cryptocephala (Wooton & Standl.) G. L. Nesom—Hidden head goldenaster. Stems, leaves, bracts, and phyllaries sessile-glandular, distinctly green in aspect because of reduced nonglandular vestiture; leaves oblong, sessile, clasping to subclasping; capitular bracts large and foliaceous, ciliate-margined, 3–4(5) per head; involucres 9–12 mm wide; ray florets 10–18. Ponderosa pine–oak, ponderosa pine-fir-spruce, Gambel oak-ash, cliffs and talus, rocky road embankments; Sacramento Mountains and N San Andres Mountains, 6200–9000 ft. SC. **R**

Heterotheca fulcrata (Greene) Shinners [*H. fulcrata* var. *senilis* (Wooton & Standl.) Semple]—Ciliate-bracted goldenaster. Stems lax; stems and leaves hirsute-villous to hirsute or hirsutulous, often also sessile-glandular; leaves oblong to oblong-lanceolate or -oblanceolate, basally truncate to subauriculate, often subclasping, midcauline 15–55 mm; heads solitary or on leafy-bracteate peduncles (3–)4–8 cm, capitular bracts usually densely long-ciliate at least along the proximal margins; involucres (5–)12–16 mm wide; phyllaries sparsely strigose to sparsely or densely hirsute, eglandular to (less commonly) glandular; ray florets 16–24. Plants in Bernalillo, San Miguel, Socorro, Taos, and Torrance Counties have relatively small heads, eciliate or short-ciliate leaf and bract margins, and "wiry"-thin stems from the base. Piñon-juniper, ponderosa pine, ponderosa pine–Douglas-fir, mixed conifer, open rocky meadows, steep slopes, cliffs and cliff bases, talus, rocky road embankments, 5200–10,500 ft. Bernalillo, Eddy, Hidalgo, Luna, Otero, San Miguel, Socorro, Taos, and Torrance.

Heterotheca hirsutissima (Greene) G. L. Nesom [*H. horrida* (Rydb.) V. L. Harms; *H. villosa* (Pursh) Shinners var. *nana* (A. Gray) Semple]—Bristly goldenaster. Stems hirsute, vestiture (1)2-storied, eglandular, overstory hairs deflexed-spreading, 1–3(–4.5) mm; leaves mostly oblanceolate-oblong to lanceolate-oblong, subpetiolate, not clasping or subclasping, sessile-glandular on both surfaces; heads in a loosely corymboid capitulescence, without capitular bracts; involucres 8–18 mm wide (pressed); phyllaries loosely strigose, eglandular, inner 5–9 mm; ray florets 14–18, corollas 9–11 × 1–1.3 mm. Short plants with small heads and exceptionally small leaves occur in central counties (Lincoln, Torrance, and Valencia); exceptionally tall plants (25–50 cm to 100+ cm) occur in NC counties. Pine-fir, ponderosa pine–oak, juniper, juniper-*Opuntia*, piñon pine, rabbitbrush, grassland, roadsides, fencerows, rock outcrops, rocky hillsides, gravel, sand, and clay soil, 4600–9200 ft. NE, NC, NW, EC, C, WC.

Heterotheca loboensis G. L. Nesom—Lobo goldenaster. Stems erect from the base, 20–60 cm, unbranched until above the middle, sparsely hirsute to strigose-hirsute with thick-based hairs, glandular, with prominent axillary fascicles of small leaves; cauline leaves even-sized and evenly distributed, strigose-hirsute to hirsute-strigose, glandular adaxially; heads usually numerous in a corymboid arrangement, on bracteate peduncles, without capitular bracts; phyllaries in 4–5(6) series strongly graduated in length. W side of Guadalupe Mountains; rocky hills, mesas, canyon bottoms, crevices, roadsides, ca. 5800 ft. Eddy.

Heterotheca nitidula (Wooton & Standl.) G. L. Nesom—Strigillose goldenaster. Stems, leaves, bracts, and phyllaries minutely glandular and sparsely to moderately strigillose with extremely thin, slightly flexuous, closely appressed hairs without enlarged basal cells (more reduced vestiture in the Zuni Mountains area of Cibola and McKinley Counties); leaves mostly narrowly oblanceolate, 3–5.5 cm, basally attenuate; heads solitary or usually in clusters of 2–6, with prominent, lanceolate capitular bracts;

ray florets 10–12 mm. Ponderosa pine, also ponderosa pine–mixed conifer, piñon-juniper, and montane grassland, 7400–10,400 ft. NW, WC.

Heterotheca paniculata G. L. Nesom—Panicled goldenaster. Stems numerous from the base, often thin and wiry, hirsute to loosely hirsute-strigose, 20–50 cm, ascending-erect to decumbent or declining from the base and often sprawling downward, often producing many lateral branches from nodes midstem upward; leaves mostly 1–2.5 cm, minutely glandular and otherwise sparsely hirsute, margins coarsely short-ciliate; heads in a loose panicle (from lateral branches), short-pedunculate, each subtended by 1–5 capitular bracts usually no longer than the involucre; involucres 5–8 mm wide (pressed), inner phyllaries 5–7 mm; ray florets 7–18. Rocky slopes, talus, cliff crevices and ledges, ravines and canyon bottoms, gravelly soil, 5200–10,500 ft. Doña Ana, San Miguel, Socorro, Torrance.

Heterotheca pedunculata (Greene) G. L. Nesom [*H. villosa* (Pursh) Shinners var. *pedunculata* (Greene) V. L. Harms ex Semple]—Elegant goldenaster. Rhizomatous, with ascending stems; stems and leaves strigose with short, thin-based, closely appressed hairs, eglandular; leaves usually oblanceolate, narrowed to a subpetiolar base; heads on long peduncles, without capitular bracts. Grassland, sagebrush, piñon-juniper, ponderosa pine, pine-fir, aspen-spruce, 7000–10,100 ft. NC, NW.

Heterotheca polothrix G. L. Nesom [*H. villosa* (Pursh) Shinners var. *scabra* (Eastw.) Semple]—Wand-flowered goldenaster. Stems erect, usually little branched until distal 1/3, sessile-glandular and loosely strigose to hirsute, sometimes essentially without nonglandular hairs; leaves cauline, relatively widely spaced, grading gradually into peduncular bracts, adaxial and abaxial surfaces often distinctly different in vestiture, the adaxial more densely glandular and less densely hirsute-strigose, moderately to densely sessile-glandular, nonglandular hairs often less dense adaxially, often restricted to midvein area (middle 1/3) of adaxial surfaces, or sometimes nearly absent on both surfaces; heads usually solitary on long, sparsely reduced-bracteate peduncles or few in a loosely corymboid arrangement; involucres 10–13 mm wide; phyllaries sparsely to moderately strigose, eglandular or minutely sessile-glandular. Sandy hills, alluvium, 5500–7700 ft. McKinley, San Juan.

Heterotheca psammophila B. Wagenkn.—Arizona camphorweed. Annual, thick-stemmed, to 1.5(–2) m; stems and leaves sparsely to densely hirsute to hispid-pilose, herbage drying green, without a dark cast; cauline leaves oblong to oblong-obovate, oblong-ovate, or triangular-ovate, commonly 3–4 times longer than wide, usually epetiolate and clasping to subclasping nearly to the base of the stem, vestiture reduced adaxially; involucres 6–8(–9) mm; phyllaries densely glandular, without nonglandular hairs, often purple-tipped; ray cypselae epappose; disk cypselae with strongly developed outer pappus. Disturbed sandy, gravelly soil, dry washes, river and creek alluvium, roadsides, urban sites, 4200–5800 ft. Hidalgo.

Heterotheca pumila (Greene) Semple—Alpine goldenaster. Stems erect or ascending-erect to decumbent-ascending, hirsute; leaves oblong to oblanceolate, lower leaves subpetiolate, often becoming sessile distally, mostly even-sized to the heads; heads solitary to clustered on short (5–20 mm, rarely to 35 mm) peduncles; capitular bracts longer than the involucre; involucres 10–15 mm wide; phyllaries prominently hirsute-villous with long, fine hairs, glandular or eglandular, often purplish along the margins. Talus, scree, and rocky slopes, alpine; in New Mexico known from only a single collection at 12,500 ft. in the Culebra Range in Taos County, the S extremity of its range.

Heterotheca scabrifolia (A. Nelson) G. L. Nesom—Rough-leaf goldenaster. Taprooted; stems numerous from the base; stems, leaves, and bracts hispid-hirsute, surfaces not obscured, trichomes with distinctly pustulate bases, phyllaries densely sessile-glandular and without other hairs, or sessile-glandular and sparsely strigose; leaf and bract margins eciliate or short-ciliate along proximal 1/5-1/4; leaves linear to narrowly oblanceolate, rigid, 0.5–2(–4) mm wide; heads 5–8 mm wide (pressed), in cymiform to subcymiform clusters on leafy-bracteate peduncles (0.5–)1–2.5 cm; capitular bracts with revolute, ciliate margins; ray florets 14–20. Two collections have been cited from E counties of New Mexico, as *H. stenophylla* (A. Gray) Shinners. Roosevelt, Union.

Heterotheca sierrablancensis (Semple) G. L. Nesom [*H. villosa* (Pursh) Shinners var. *sierrablancensis* Semple]—Sierra Blanca goldenaster. Stems mostly 20–35(–46) cm, erect from the base, hirsute; cauline leaves sessile (epetiolate), basally subclasping, mostly oblanceolate-oblong to lanceolate-oblong, densely strigose and with densely glandular surfaces; heads without capitular bracts; phyllaries closely and finely strigose; ray corollas 9–14 mm, with broad ligules. Mostly in the White Mountains; rocky slopes, crevices, outcrops, roadcuts, gravelly loam, mixed conifer-oak, spruce-fir, 7200–11,500 ft. Lincoln, Otero. **R**

Heterotheca subaxillaris (Lam.) Britton & Rusby [*H. latifolia* Buckley; *H. subaxillaris* (Lam.) Britton & Rusby subsp. *latifolia* (Buckley) Semple]—Camphorweed. Annual or biennial; upper cauline leaves usually at least slightly subclasping, becoming reduced in size but mostly remaining ovate; involucres 5–8 mm; phyllaries densely glandular, usually without nonglandular hairs; ray cypselae epappose; disk cypselae with strongly developed outer pappus. Roadsides, disturbed areas, pastures, woodland clearings, grassland, riparian areas, 3300–5000(–6900) ft. Widespread.

Heterotheca viscida (A. Gray) V. L. Harms—Sticky goldenaster. Stems, leaves, and phyllaries prominently sessile-glandular and often viscid, coarsely hirsute (sparsely so to nearly glabrate except for marginal cilia and hairs along veins), green at least partly from the low density of nonglandular hairs; stems mostly decumbent; leaves mostly (1–)2–5 cm, elliptic-oblanceolate or oblong-oblanceolate to broadly oblanceolate, proximal with subpetiolate bases but distally becoming sessile and auriculate to subauriculate, clasping to subclasping; heads mostly solitary (a second head occasionally produced on a very short branch), capitular bracts absent. Rocky slopes, cliff faces and ledges, 5800–8900 ft. Doña Ana, Luna, Socorro.

Heterotheca zionensis Semple—Zion goldenaster. Taprooted; stems in a cluster from the caudex, erect, to 50 cm; stems, leaves, and phyllaries closely strigose-sericeous, usually silvery but sometimes darker, surfaces eglandular or less commonly sessile-glandular, stem vesture usually 1-storied but sometimes with a longer, spreading overstory; leaves ascending to spreading to deflexed, axillary fascicles of small leaves often evident; heads in a corymboid cluster, without capitular bracts. Roadsides, arroyos and canyon bottoms, clay and sandy soil, piñon-juniper, ponderosa pine–Gambel oak, grassland, 4000–8900 ft. NC, NW, SE, SC, SW. Occasional plants (sometimes perhaps grazed or otherwise damaged) are sometimes reduced in height and leaf size but are still distinct (vs. *H. canescens* and *H. pedunculata*, which are similar in vesture).

HIERACIUM L. - Hawkweed

Gary I. Baird

Perennial herbs, taprooted with a branched, fibrous-rooted caudex, sap milky; stems 1+ per caudex branch, leafy or scapiform, leafy stolons present or lacking, glabrous or hairy with a mix of diverse hairs; leaves basal and/or cauline, petioles distinct to obscure, blades lance-linear to spatulate, reduced distally, margins entire to dentate, faces glabrous to hairy; heads ligulate, arrays racemiform to paniculiform; involucres ± campanulate; calyculi of 3–13 bractlets, unequal, erect; phyllaries 10–20 in ~2 series, lance-linear, subequal, faces glabrous or diversely hairy, margins not (or slightly) scarious; florets 12–60+, bisexual or apomictic, corollas white, yellow, or pink, 5–15 mm; cypselae uniform, 2–7 mm, columnar or tapered, ± 10-ribbed, red-brown to black, glabrous, beakless; pappus of few to numerous bristles in 1–2 series, white, sordid, or flaxen, persistent.

The genus *Hieracium* has a reputation for being taxonomically challenging. Species boundaries are not always well defined, and some specimens may exhibit phenotypic inconsistencies, making identification difficult. Vesture has been commonly but cautiously utilized to separate species. Three main trichome types occur, and the surfaces of stems, leaves, peduncles, and phyllaries may bear one or more (or none) of them in diverse patterns. The following terms are used here to distinguish them: (1) "pilose" refers to hairs, 1–15 mm, ± smooth and straight (if curly then the term "lanate" is used), and white, yellow, or brown; (2) "floccose" refers to very short hairs, < 0.5 mm, dendritically branched, and white or sordid;

(3) "glandular" refers to gland-tipped hairs, ~0.5-1 mm, which are often black or dark at the base but yellow or pale distally (Strother, 2006n).

1. Stems scapose or subscapose, leaves mostly basal, cauline leaves few and rapidly reduced in size, bract-like…(2)
1'. Stems ± leafy, leaves cauline or basal and cauline (rarely nearly all basal), gradually reduced in size upward and becoming bractlike distally…(4)
2. Cypselae 5-9 mm, distal portion tapered; involucres 10-15 mm; basal leaves often purplish abaxially… **H. fendleri**
2'. Cypselae 1-4 mm, columnar, not tapering; involucres 5-10 mm; basal leaves ± green on both surfaces…(3)
3. Cypselae 2-4 mm; plants lacking leafy stolons (or these rare)…**H. triste**
3'. Cypselae 1-2 mm; plants with leafy stolons usually present…**H. ×floribundum**
4. Ligules pinkish white (often drying rosy-purple); basal leaves usually smaller and pilose, cauline leaves larger and glabrous-glaucous…**H. carneum**
4'. Ligules yellow to ochroleucous or whitish (drying yellowish or whitish); basal leaves ± equal to or larger than cauline leaves, both usually ± the same in vestiture (usually more densely hairy proximally)…(5)
5. Cypselae columnar, 2-4 mm (rarely more); involucres 6-10 mm…(6)
5'. Cypselae tapered distally, 4-6 mm (rarely less); involucres 9-11 mm (rarely less)…(7)
6. Basal portion of stem sparsely to densely pilose; pappus flaxen (rarely white); phyllaries mostly 14-18; heads relatively many (5-60) in thyrsoid-paniculiform arrays…**H. abscissum**
6'. Basal portion of stem densely lanate; pappus white; phyllaries mostly 18-24; heads relatively few (5-20) in cymose arrays…H. pringlei A. Gray
7. Involucres glandular, lacking pilose hairs…**H. crepidispermum**
7'. Involucres with both pilose and glandular hairs…**H. brevipilum**

Hieracium abscissum Less. [*H. rusbyi* Greene]—Rusby's hawkweed. Stems leafy, 40-75 cm, proximally sparsely to densely pilose, not or obscurely glandular, becoming floccose and glandular distally; stolons absent; leaves basal and/or cauline, ± clasping, blades oblanceolate to lanceolate, 35-150 × 7-28 mm, faces pilose, glandular; heads 5-60 in paniculiform arrays, peduncles floccose and/or glandular; involucres 6-10 mm, floccose, glandular (pilose hairs lacking); florets 20-24, corollas ~9 mm, yellow; cypselae 2-3 mm, columnar; pappus 4-5 mm, flaxen (rarely white). Openings in coniferous woodlands; 2 records, ca. 8500 ft. Catron.

Hieracium brevipilum Greene [*H. fendleri* Sch. Bip. var. *mogollense* A. Gray]—Mogollon hawkweed. Stems 25-65 cm, proximally pilose and glandular, becoming more sparse distally; stolons absent; leaves basal and cauline, ± clasping, blades oblanceolate to lanceolate, 35-120 × 10-18 mm, faces pilose (glabrous); heads 6-10 in loose, paniculiform arrays, peduncles floccose, glandular, ± pilose; involucres 10-11 mm, pilose, glandular, sometimes floccose; florets 15-25, corollas ~8 mm, yellow; cypselae ~5 mm, ± columnar but tapered distally; pappus 5-6 mm, white. Meadows in montane coniferous forests, 7700-10,300 ft. Catron. **R**

Hieracium carneum Greene—Huachuca hawkweed. Stems 30-60 cm, proximally pilose (rarely glabrous), becoming quickly glabrous and glaucous distally; stolons absent; leaves basal (withered by anthesis) and cauline, blades oblanceolate to lance-linear, 40-170 × 3-18 mm, basal leaves pilose, cauline leaves glabrous, glaucous; heads 5-70 in corymbiform to paniculiform arrays, peduncles ± short-pilose and glandular; involucres 7-10 mm, glabrous or short-pilose (long hairs lacking), obscurely floccose proximally, sometimes glandular; florets ~20, corollas ~8 mm, pink to rosy-white (drying roseate); cypselae 3-4.5 mm, columnar or slightly tapered; pappus 4-6 mm, white. Openings in pine-oak, ponderosa pine, and Douglas-fir woodlands, 5600-10,200 ft. SE, SC, SW, Lincoln, San Miguel.

Hieracium crepidispermum Fr. [*H. lemmonii* A. Gray]—Mexican hawkweed. Stems 20-100 cm, proximally densely pilose, becoming moderately pilose and floccose distally, stolons absent; leaves mostly cauline, cordate-clasping, blades oblanceolate to linear-elliptic, 30-180 × 15-40 mm, faces pilose; heads 8-35 in compact, paniculiform arrays, peduncles ± pilose, floccose, glandular; involucres 7-12 mm, proximally floccose, glandular, distally ± glandular (rarely, if ever, pilose); florets 25-40, corollas 8-9 mm, whitish to pale yellow; cypselae 4-6 mm, columnar proximally but tapered distally; pappus 5-6 mm, white or sometimes flaxen. Meadows in pine-oak and montane coniferous forest, 9000 ft. WC, Lincoln.

Hieracium fendleri Sch. Bip.—Fendler's hawkweed. Stems 10-75 cm, base with tufts of long-pilose hairs, proximally pilose, becoming sparsely pilose distally; stolons mostly absent; leaves basal (cauline 0-2 and bractlike), blades elliptic to oblanceolate, 20-100 × 12-45 mm, faces ± pilose, often purple abaxially; heads 2-10 (rarely more) in paniculiform arrays, peduncles pilose and glandular, sometimes also floccose; involucres 10-15 mm, ± floccose, sometimes also sparsely glandular and/or pilose; florets 15-30, corollas 10-15 mm, yellow; cypselae ± columnar, 5-7 mm; pappus 5-9 mm, flaxen (rarely white). Open, moist areas, often in sandy soil, in piñon-juniper-oak, ponderosa pine, and montane coniferous forest, 5700-11,300 ft. Widespread except E.

Hieracium ×floribundum Wimm. & Grab. [*Pilosella ×floribunda* (Wimm. & Grab.) Fr.]—Yellow devil hawkweed. Stems 20-80 cm, pilose, ± glaucous; leafy stolons usually present; leaves mostly basal (cauline 0-2, bractlike), blades elliptic to oblanceolate, 30-140 × 6-20 mm, abaxial faces pilose, adaxial faces ± glabrous and glaucous; heads 3-50, glomerate in congested to open, ± paniculiform arrays, peduncles ± pilose, floccose, glandular; involucres 5-9 mm, glandular, ± pilose, floccose; florets 30-60, corollas yellow; cypselae columnar, 1-2 mm; pappus white. Moist meadows in montane coniferous forest; 1 record, ca. 9170 ft. Sandoval.

Hieracium triste Willd. ex Spreng. [*H. gracile* Hook.]—Slender hawkweed. Stems 10-30 cm, proximally glabrous or inconspicuously floccose, becoming floccose distally; leaves mostly basal (cauline 0-2, bractlike), blades obovate to spatulate, 15-40 × 5-15 mm, faces glabrous, sometimes glandular; heads (1-)2-8 in racemiform to corymbiform arrays, peduncles floccose, sparsely pilose, and/or glandular; involucres 6-10 mm, pilose, floccose, glandular; florets 20-60, corollas 5-6 mm, yellow; cypselae columnar, 2-3 mm; pappus 4-5 mm, white or sordid. Moist alpine or subalpine meadows, 9600-12,500 ft. NC.

HYMENOPAPPUS L'Hér. – Hymenopappus

Kenneth D. Heil

Biennials or perennials, to 120 cm; stems erect, 1 stem per crown in biennials, usually 3+ in perennials; leaves mostly basal, blades usually 1-2+ times pinnately lobed, lobes usually ± filiform, ultimate margins usually entire, often tomentose, usually gland-dotted; heads radiate or discoid, in ± corymbiform arrays; involucres obconic to hemispherical, 4-15+ mm diam.; phyllaries 5-13+ in 2-3+ series; receptacles flat or convex; ray florets 0 or 8, corollas white to ochroleucous; disk florets 12-70+, corollas usually yellow to ochroleucous or whitish to purplish; cypselae obpyramidal, 4-, sometimes 5-angled; pappus 0 or of 12-22 orbiculate to spatulate scales (Allred & Ivey, 2012; Strother, 2006o).

1. Corollas white…(2)
1'. Corollas usually yellow…(4)
2. Biennials; heads 20-40…**H. biennis**
2'. Perennials; heads 3-8…(3)
3. Basal leaves 12-25 cm; peduncles 6-15 cm…**H. newberryi**
3'. Basal leaves 8-14 cm; peduncles 3-8 cm…**H. radiatus**
4. Disk corollas whitish…(5)
4'. Disk corollas yellowish…(6)
5. Perennials, usually 3+ stems from crown…**H. filifolius** (in part)
5'. Biennials, usually 1 stem from crown…**H. tenuifolius**
6. Perennials, usually 3+ stems from crown…(7)
6'. Biennials, usually 1 stem from crown…**H. flavescens**
7. Leaf blades simple or 1-pinnate; cypselae glabrous or sparsely hirtellous…**H. mexicanus**
7'. Leaf blades 2-pinnate; cypselae hirtellous to villous…**H. filifolius** (in part)

Hymenopappus biennis B. L. Turner—Biennial woollywhite. Biennials, 60-100 cm; leaves basal, 2-pinnate, 6-16 cm; heads 20-40 per stem; ray florets 9, corollas white; disk florets 32-50+, corollas yellow; cypselae glabrous, 4 mm. Limestone soils, piñon-juniper, ponderosa pine, and mixed conifer communities, 5100-10,300 ft. C, WC, SC.

Hymenopappus filifolius Hook.—Fineleaf hymenopappus. Perennials, 0.5-10 dm; leaves bipinnately dissected, the ultimate segments linear to filiform, tomentose to glabrate, 3-20 cm; heads 1-30

(-60) per stem; involucres 3-14 mm, white or yellowish-membranous; ray florets absent; disk florets 2-7 mm, densely glandular to glabrate; cypselae densely hairy, 3-7 mm.

1. Terminal lobes of basal leaves 2-6 mm; phyllaries 3-7 mm; corollas 2-3 mm…**var. pauciflorus**
1'. Terminal lobes of basal leaves (2-)3-50 mm; phyllaries 5-14 mm; corollas 2.5-7 mm…(2)
2. Corollas 4-7 mm; anthers 3-4 mm; cypselae 5-7 mm…**var. lugens**
2'. Corollas 2.5-4.5 mm; anthers 2-3 mm; cypselae 4-6 mm…(3)
3. Cauline leaves 2-8; heads 5-60; corolla throats 1.3-1.8 mm…**var. polycephalus**
3'. Cauline leaves (0-)2-4; heads 1-6; corolla throats 1.5-2.5 mm…**var. cinereus**

var. cinereus (Rydb.) I. M. Johnst. [*H. cinereus* Rydb.]—Stems 15-40 cm; basal leaf axils ± densely tomentose; heads 1-6; florets 25-40, corollas yellowish or whitish; cypselae 4-6 mm. Piñon-juniper to ponderosa pine communities, 3900-8750 ft. SE, SW.

var. lugens (Greene) Jeps. [*H. lugens* Greene]—Stems 20-60 cm; basal leaf axils ± densely tomentose; heads 3-8; florets 20-70, corollas yellowish or whitish; cypselae 5-6 mm. Juniper-oak, oak, piñon-juniper, ponderosa pine communities, 4800-8600 ft. NC, NW, C, WC, SW.

var. pauciflorus (I. M. Johnst.) B. L. Turner [*H. pauciflorus* I. M. Johnst.]—Stems 20-35 cm; basal leaf axils ± densely tomentose; heads 2-15(-30); florets 10-30, corollas yellowish; cypselae 3-4.5 mm. Sandy soils, piñon-juniper to ponderosa pine communities, 4000-7200 ft. NC, NW, WC, SC.

var. polycephalus (Osterh.) B. L. Turner [*H. polycephalus* Osterh.]—Stems (20-)30-60 cm; basal leaf axils ± densely tomentose; heads 5-60; florets 20-50, corollas yellowish; cypselae 4-5 mm. Sandy and clay soils, prairies, 4700-5100 ft. NE.

Hymenopappus flavescens A. Gray—College-flower. Biennials, 3-9 dm; basal leaves 2-pinnate, 6-15 cm; heads 15-100; ray florets 0; disk florets 20-40, corollas yellow; cypselae 3-4.5 mm, ± villous.

1. Lobes of basal leaves 1-6 mm wide; abaxial faces glabrous or less hairy than adaxial…**var. flavescens**
1'. Lobes of basal leaves 1-2 mm wide; abaxial and adaxial faces ± equally hairy…**var. canotomentosus**

var. canotomentosus A. Gray—Lobes of basal leaves 1-2 mm wide; abaxial and adaxial faces ± equally hairy; pappus 1-1.5 mm. Sandy, gravelly, and limestone soils, prairie, piñon-juniper, and ponderosa pine communities, 3500-7500 ft. NE, NC, C, WC, SE, SC, SW.

var. flavescens—Lobes of basal leaves (1-)2-6 mm wide; abaxial faces glabrous or less hairy than adaxial; pappus 0.5-1(-1.2) mm. Sandy soils, prairie communities, 4000-6200 ft. NE, EC, C.

Hymenopappus mexicanus A. Gray—Mexican woollywhite. Perennials, 2-9 dm; basal leaves simple or 1-pinnate, 6-20 cm, cauline leaves 0 or 3-6; heads 1-8(-20+) per stem; ray florets 0; disk florets 20-40, corollas yellow, 3-4.5 mm; cypselae 4-6 mm, glabrous or sparsely hairy. Montane meadows in ponderosa pine and mixed conifer communities, 5700-10,000 ft. WC, McKinley.

Hymenopappus newberryi (A. Gray ex Porter & J. M. Coult.) I. M. Johnst. [*Leucampyx newberryi* A. Gray ex Porter & J. M. Coult.]—Newberry's hymenopappus. Perennials, 20-60 cm; basal leaves 2-pinnate, 12-25 cm, cauline leaves 1-3(-5); heads 3-8 per stem; ray florets 8, corollas white or pinkish, 14-20 mm; disk florets 60-150+, corollas yellowish; cypselae 3.5-4 mm, glabrous. Clay and igneous soils, montane meadows in ponderosa pine, mixed conifer, and spruce-fir communities, 6700-11,500 ft. NC, WC, SE.

Hymenopappus radiatus Rose—Ray hymenopappus. Perennials, 3-4.5 dm; basal leaves 2-pinnate, 8-14 cm, cauline leaves 2-3; heads 6-8 per stem; ray florets 8, corollas white, 14-16 mm; disk florets 30-50, corollas yellowish, 3-4 mm; cypselae 4-4.5 mm, ± hairy. Sandy soils, pine woodlands, 6300-9600 ft. NC, Catron, Grant, Otero.

Hymenopappus tenuifolius Pursh—Chalk Hill hymenopappus. Biennials, 4-15 dm; basal leaves 2-pinnate, 8-15 cm, cauline leaves 8-30; heads 20-200 per stem; ray florets 0; disk florets 25-50, corollas whitish, 2.5-3 mm; cypselae 3.5-4.5 mm. Sandy, gravelly, silty, and limestone soils, grasslands, juniper, and piñon-juniper communities, 3100-8200 ft. NE, NC, C, WC, SE, Hidalgo.

HYMENOTHRIX A. Gray – Thimblehead

Kenneth D. Heil

Annuals, biennials, or perennials, 10–70(–150) cm; stems erect, branched; leaves basal and cauline; mostly alternate, blades 2 times ternately (or pinnately) lobed, ultimate margins entire, faces usually ± scabrellous, sometimes gland-dotted; heads radiate or discoid, in corymbiform to paniculiform arrays; involucres obconic to hemispherical, 4–8+ mm diam.; phyllaries persistent, 8–16 in 2–3 series; receptacles flat or convex, knobby or smooth, epaleate; ray florets 0, or 3–8+, corollas yellowish; disk florets 10–30+, corollas yellowish or whitish or pinkish to purplish; cypselae obpyramidal, 4–5-angled; pappus persistent, of 12–18 narrowly lanceolate to subulate scales in 1 series, some or all ± aristate (Allred et al., 2020; Strother, 2006p).

1. Ray florets 0; disk florets white or pinkish to purplish; anthers pinkish to purplish; phyllaries oblong to ovate or obovate...**H. wrightii**
1'. Ray florets present; disk florets yellow or cream-colored; anthers yellow...(2)
2. Pappus mostly absent, rarely with lance-linear scales...**H. dissecta**
2'. Pappus of ovate, oblanceolate, or lanceolate scales...(3)
3. Leaf lobes oblong or ovate to oblanceolate, 2–8 mm wide; corolla ligules 5–6 mm...**H. pedata**
3'. Leaf lobes filiform to linear, sometimes oblong, 0.5–2.5 mm wide; corolla ligules 2–4 mm...(4)
4. Ray florets 3–8; pappus scales all aristate...**H. wislizeni**
4'. Ray florets 8–13; pappus scales of only inner florets aristate...**H. biternata**

Hymenothrix biternata (A. Gray) B. G. Baldwin [*Amauriopsis biternata* (A. Gray) B. L. Turner]—Slim-lobe sunray daisy. Annuals, 8–40 cm; leaf blades simple or 1–2 times ternately lobed, lobes filiform to linear, 0.5–2.5 mm; corolla yellow; cypselae of ovate scales 0.5–1 mm. Gravelly soils, desert scrub, roadsides, 4000–5850 ft. SW, Chaves.

Hymenothrix dissecta (A. Gray) B. G. Baldwin [*Amauria dissecta* A. Gray]—Thimblehead. Annuals or biennials; leaf blades 1–2 times ternately lobed, 2–6(–12) mm wide, lobe margins entire or toothed, gland-dotted; corolla yellow; cypselae 2.5–4 mm. Piñon-juniper-ponderosa pine, mixed conifer, and spruce-fir forests, (5600–)6460–11,000 ft. Widespread except E.

Hymenothrix pedata (A. Gray) B. G. Baldwin [*Bahia pedata* A. Gray]—Bluntscale bahia. Annuals or biennials, 15–70(–120+) cm; stems mostly erect; leaves all or mostly alternate, blades ovate or obovate to lanceolate or oblanceolate, 2–25 × 1–6(–8) mm, faces sparsely scabrellous, usually gland-dotted; involucres 4–6+ × 8–12+ mm; ray florets 10–15; corolla laminae 5–6(–10+) mm; disk florets 40–80+; corollas 3–3.5 mm; cypselae 2.5–3.5+ mm, faces ± hirtellous; pappus of ± spatulate to oblanceolate, apically ± muticous scales 1–1.5 mm. Openings in grasslands and piñon-juniper woodlands, 3300–8000 ft. EC, C, SE, SC, San Miguel.

Hymenothrix wislizeni A. Gray—Trans-Pecos thimblehead. Annuals, 30–70 cm; ray florets 3–8; disk florets 15–30, corollas creamy to bright yellow, mostly 5–7 mm; anthers yellowish; cypselae 3–5 mm; pappus scales 1.5–6 mm. Open slopes and washes, sandy or gravelly soils, 3500–6100 ft. SC, SW.

Hymenothrix wrightii A. Gray—Wright's thimblehead. Annuals or perennials, 3–6 dm; ray florets 0; disk florets 15–30; corollas white or pinkish to purplish, 5–6 mm; anthers pinkish to purplish; cypselae 4–5 mm; pappus scales 4–6 mm. Rocky places, often with piñon and juniper, 4100–8000 ft. SC, SW.

HYMENOXYS Cass. – Rubberweed, bitterweed

Kenneth D. Heil and Jennifer Ackerfield

Annual, biennial, or perennial herbs; leaves alternate; heads radiate (ours); involucral bracts subequal in 2–4 series or unequal in 2 series, herbaceous; receptacles naked; ray florets yellow, yellow-orange, or orange; disk florets yellow; cypselae mostly 5-angled; pappus of awned or aristate scales (Bierner, 2006b).

1. Annuals from slender taproots...**H. odorata**
1'. Biennials or perennials...(2)

2. Disk florets 6-15, functionally staminate; receptacles flat…**H. ambigens**
2'. Disk florets 25-400+, bisexual; receptacles not flat…(3)
3. Phyllaries 22-50 in 2-3 series, subequal…(4)
3'. Phyllaries (11-)16-30(-40) in 2 series, unequal…(5)
4. Plants 30-100 cm; leaf blades simple, phyllaries 36-50 in 2 series; ray corollas usually yellow-orange to orange…**H. hoopesii**
4'. Plants 8-24 cm; leaf blades simple and/or lobed; phyllaries 22-40+ in 2-3 series; ray corollas yellow…
   H. brandegeei
5. Outer phyllaries basally connate to 1/5 their length; inner phyllaries narrowly lanceolate to oblanceolate, 8.5-12.6 mm, aristate…**H. bigelovii**
5'. Outer phyllaries basally connate 1/4-2/3 their length; inner phyllaries obovate to oblanceolate, 2.8-8 mm, apices usually mucronate, sometimes acuminate…(6)
6. Stems 1-20(-30+) cm; plants usually with highly branched, woody caudices; basal leaf bases densely long-villous-woolly…**H. richardsonii**
6'. Stems 1-10(-20) cm; plants often with sparingly or moderately branched, woody caudices; basal leaf bases sparsely, if at all, long-villous-woolly…(7)
7. Involucres 4-8 mm diam.…(8)
7'. Involucres (8-)10-18 mm diam.…(9)
8. Perennials; midblades simple or lobed…**H. rusbyi**
8'. Biennials or perennials; midblades lobed…**H. brachyactis**
9. Leaf blades lobed; outer phyllaries basally connate 1/2-2/3 their length…**H. vaseyi**
9'. Leaf blades simple or lobed; outer phyllaries basally connate 1/4-1/2 their length…**H. helenioides**

Hymenoxys ambigens (S. F. Blake) Bierner [*Plummera ambigens* S. F. Blake]—Pinaleño Mountains rubberweed. Perennials, 30-40 cm; leaves simple or lobed (lobes 3-19+), glabrous or sparsely hairy, gland-dotted; heads 35-100+ per plant; involucres 5-7.5 mm; ray florets 3.4-4.2 mm, yellow; disk florets 3-4 mm; cypselae 1.7-2.3 mm, hairy. New Mexico material belongs to **var. neomexicana** W. L. Wagner. Open woodlands in the Animas and Peloncillo Mountains, 5250-6250 ft. Hidalgo. **R & E**

Hymenoxys bigelovii (A. Gray) K. F. Parker [*Actinella bigelovii* A. Gray]—Bigelow's rubberweed. Perennials, 2-7 dm; leaves usually simple, rarely lobed, glabrous or ± hairy, eglandular or sparsely gland-dotted, basal leaf bases ± long-villous-woolly; heads usually borne singly; involucres 13 mm; ray florets 13-26 mm, corollas yellow; disk florets 5.7-7.4 mm; cypselae 4.2-4.7 mm. Roadsides, edges of juniper-pine and pine forests, 6000-7750 ft. WC, SW, McKinley.

Hymenoxys brachyactis Wooton & Standl.—East View rubberweed, tall bitterweed. Biennials or perennials, 3-6 dm; leaves simple or lobed, glabrous, gland-dotted; heads (7-)40-150(-250) per plant; ray florets 7-8.5 mm, yellow; disk florets 3.1 mm; cypselae 2.1-2.5 mm. Roadsides, open areas, edges of pine forests, 5850-8200 ft. Lincoln, Socorro, Torrance, Valencia. **R & E**

Hymenoxys brandegeei (Porter ex A. Gray) K. F. Parker [*Tetraneuris brandegeei* (Porter ex A. Gray) K. F. Parker]—Brandegee's rubberweed. Perennials, to 2 dm; leaves simple or usually with 2-3 linear lobes, glabrate to villous; heads solitary, the peduncles densely tomentose under the heads; involucres 7-13 mm, tomentose; ray florets 10-20 mm, yellow; disk florets 4-5.5 mm; cypselae 2.5-3 mm. Alpine in N mountains, 11,000-13,260 ft. NC, Lincoln, McKinley, Otero.

Hymenoxys helenioides (Rydb.) Cockerell [*Picradenia helenioides* Rydb.]—Intermountain rubberweed. Perennials, 1-5(-10) dm; leaves simple or 3-lobed, glabrous to sparsely hairy, 7-20 cm; heads 5-50+; involucres 5-7 mm, sparsely to moderately hairy and gland-dotted; ray florets 17-30 mm, yellow to yellow-orange; disk florets 3.5-5.5 mm; cypselae 2.5-3.5 mm. Uncommon in open meadows and along roadsides; 1 collection, 8070 ft. San Juan. Some botanists feel *H. helenioides* is a hybrid between *H. richardsonii* and *H. hoopesii*.

Hymenoxys hoopesii (A. Gray) Bierner [*Dugaldia hoopesii* (A. Gray) Rydb.; *Helenium hoopesii* A. Gray]—Orange sneezeweed, owl's claws. Perennials, 2-10 dm; leaves simple, oblanceolate, loosely villous-tomentose to glabrate, 10-30 cm; heads solitary to 12; involucres 6-10 mm; ray florets 15-30 mm, orange or orange-yellow; disk florets 4-5.5 mm; cypselae 3.5-4.5 mm. Common in moist places along streams, in meadows, and on open slopes, 6000-11,800 ft. Widespread except NE, EC, SE, SW.

Hymenoxys odorata DC. [*Picradenia odorata* (DC.) Britton]—Bitter rubberweed. Annuals, 1–8 dm; leaves deeply divided into 3–5 linear lobes, hairy to minutely hairy, mostly 2–5 cm; heads 15–300+; involucres 3–5 mm, gland-dotted; ray florets 5–10 mm, yellow; disk florets 2.5–4 mm; cypselae 1.5–2.5 mm. Dry, open places, dry lake beds, 3500–8100 ft. Widespread except NC, SE.

Hymenoxys richardsonii (Hook.) Cockerell—Colorado rubberweed. Perennials, 0.7–3.5 dm; leaves cleft into 3–5 linear or linear-filiform segments, sparsely hairy to glabrate, usually punctate; heads 5–300+; involucres 4–8 mm, often glandular-punctate; ray florets 7–15 mm, yellow; disk florets 3–5 mm; cypselae 2–3 mm. Common in dry, open or sparsely wooded places, often in rocky or sandy soil, 5400–10,500 ft.

1. Heads relatively few and larger, usually 1–4 per stem and involucres 5–8 mm; rays 8–14; plants 0.7–2.4 dm...**var. richardsonii**.
1'. Heads relatively more numerous and smaller, usually > 4 per stem and involucres 4–5 mm; plants 1.9–3.4 dm...**var. floribunda**

var. floribunda (A. Gray) K. F. Parker [*Actinella richardsonii* (Hook.) Nutt. var. *floribunda* A. Gray]—Stems glabrous or ± hairy; leaves glabrous or sparsely hairy; involucres 7–8 × 7–9 mm; ray florets 7–11 mm; disk florets 3–4 mm; cypselae 2 mm. Roadsides, open edges of forests, 4050–11,000 ft. Widespread except E.

var. richardsonii—Stems ± hairy; leaves ± hairy; involucres 8–11 × 9–14 mm; ray florets 11–17 mm; disk florets 3–5 mm; cypselae 2–3 mm. Roadsides, open areas, edges of forests, 5500–10,700 ft. NC, C.

Hymenoxys rusbyi (A. Gray) Cockerell [*Actinella rusbyi* A. Gray]—Rusby's rubberweed or bitterweed. Perennials, 3–15 dm; leaves simple or lobed, glabrous, gland-dotted; heads 50–250+ per plant; ray florets 4.5–8 mm, yellow; disk florets 2.7–3.2 mm; cypselae 2–3 mm. Roadsides, open areas, edges of forests, 5600–8800 ft. Catron, Grant.

Hymenoxys vaseyi (A. Gray) Cockerell [*Actinella vaseyi* A. Gray]—Vasey's rubberweed or bitterweed. Perennials, 2–6 dm; leaves usually lobed, glabrous or sparsely hairy, gland-dotted; heads 10–50+ per plant; ray florets 9–15 mm, yellow; disk florets 3.5 mm; cypselae 2–2.7 mm. Open areas, edges of forests, 6900–8200 ft. Doña Ana, Grant, Sierra, Socorro. **R & E**

HYPOCHAERIS L. – Cat's ear

Kenneth D. Heil

Hypochaeris radicata L.—Hairy cat's ear. Perennials, 10–60 cm; stems 1–15, erect, usually branched, glabrous or coarsely hirsute proximally; leaves all basal, blades oblanceolate or lyrate to slightly runcinate, 50–350 × 5–30 mm, margins coarsely dentate to pinnatifid, faces ± hirsute; heads usually 2–7 in loose arrays, sometimes borne singly; involucres cylindrical or campanulate, 10–25 × 10–20 mm; phyllaries 20–30, margins scarious, green to darkened; florets 10–15 mm, surpassing phyllaries at flowering; corollas bright yellow or grayish green; cypselae all beaked, beaks 3–5 mm; bodies golden brown; pappus of whitish bristles in 2 series. One collection along a fence in Animas, 4400 ft. Hidalgo. Introduced (Bogler, 2006b).

IONACTIS Greene – Ankle-aster

Kenneth D. Heil

Ionactis elegans (Soreng & Spellenb.) G. L. Nesom [*Chaetopappa elegans* Soreng & Spellenb.]—Sierra Blanca cliff daisy. Perennials, plants 3–5(–9) cm, with relatively short caudex branches, cespitose; stems erect, proximally herbaceous or slightly woody, eglandular; cauline leaves alternate, sessile, densely clustered proximally, internodes not visible, reduced in size distally; middle and distal leaves 5–17 mm; blades spatulate (proximally) or linear to narrowly oblong or elliptic-lanceolate, margins green, faces glabrous or sparsely puberulent, eglandular; heads radiate, 1(–3); involucres broadly turbinate to campanulate, 4.5–6.5 mm; phyllaries 20–60 in 2–6 series; receptacles flat, pitted, epaleate; ray florets nearly white or pale pink to lavender; disk floret corollas yellow, 3.8–5 mm; cypselae 2–2.5 mm, faces sessile- to stipitate-glandular. Granite outcrops, mixed conifer forests, 7600–9500 ft. Lincoln (Nesom, 2006n). **R & E**

ISOCOMA Nutt. – Jimmyweed, goldenweed

Kenneth D. Heil

Perennials or subshrubs, (4–)20–120(–150) cm, bases often woody; stems usually gland-dotted, often resinous; leaves linear to oblanceolate or obovate, margins entire or toothed to pinnatifid, faces glabrous or pubescent, usually gland-dotted (in pits); heads discoid; involucres obconic to turbinate or campanulate; phyllaries 15–30 in (3)4–6 series; receptacles flat, pitted; ray florets absent; disk florets 8–34, corollas yellow with dark orange-resinous veins; cypselae brownish, obpyramidal, terete or subterete, 5–11-ribbed, sometimes thick and resinous, faces sericeous; pappus of 40–50 barbellate bristles. Previously included in *Haplopappus* (Nesom, 2006o).

1. Leaves pinnatifid…(2)
1'. Leaves entire or shallowly toothed, not pinnatifid…(3)
2. Herbage minutely hispidulous or sparsely puberulous…**I. tenuisecta**
2'. Herbage glabrous…**I. azteca**
3. Involucres 3.2–5.5 mm; florets (8–)11–17(–21)…**I. pluriflora**
3'. Involucres 5.5–7.5 mm; florets 19–25…**I. rusbyi**

Isocoma azteca G. L. Nesom—Apache jimmyweed. Herbage glabrous, sometimes sparsely stipitate-glandular, not resinous; leaf blades narrowly oblong to narrowly oblanceolate, 20–50 mm, margins shallowly to deeply pinnatifid; involucres 7–8 × 5–7.5 mm; florets 18–25, corollas 5–6 mm. Sandy, clay, gypseous, and saline soils, desert scrub, piñon-juniper woodlands, commonly with *Atriplex*, 5100–7250 ft. NW.

Isocoma pluriflora (Torr. & A. Gray) Greene [*Linosyris pluriflora* Torr. & A. Gray; *Haplopappus pluriflorus* (Torr. & A. Gray) H. M. Hall]—Southern jimmyweed. Herbage usually glabrous or sparsely hispidulous; leaves sometimes stipitate-glandular, never resinous; leaf blades oblanceolate to narrowly oblong-oblanceolate or nearly linear, mostly 10–40(–50) mm, margins usually entire, sometimes shallowly toothed; involucres 3.2–5.5 × 2.5–4 mm; florets (8–)11–17(–21), corollas 5–6 mm. Igneous, calcareous, gypsum, sandy, and clay soils, grasslands, desert scrub, juniper-oak, piñon-juniper, ponderosa pine communities, 3100–7800 ft. Widespread except NE, NC, NW.

Isocoma rusbyi Greene [*Haplopappus rusbyi* (Greene) Cronquist]—Rusby's jimmyweed. Herbage glabrous, not resinous; leaf blades narrowly elliptic-oblong to elliptic-obovate, 20–50 mm, margins entire; involucres (5.5–)6–9.5 × 5–7.5 mm; florets 19–25, corollas 5–6.5 mm. Rocky, sandy to clay soils, usually in saline habitats, desert scrub and juniper communities, 5100–6900 ft. Rio Arriba, San Juan.

Isocoma tenuisecta Greene [*Haplopappus tenuisectus* (Greene) S. F. Blake ex L. D. Benson]—Shrine jimmyweed, burroweed. Herbage minutely hispidulous to hirtellous or sparsely puberulous, not resinous; leaf blades oblong-oblanceolate, 20–35 mm, margins pinnatifid; involucres 4–6.5 × 2–2.8 mm; florets 8–12(–15), corollas 4.5–6 mm. Sandy and gravelly flats, grasslands, juniper-oak, often with creosote bush, 3800–7250 ft. NW, WC, SW.

IVA L. – Marsh elder

Kenneth D. Heil

Annuals and perennials, (10–)50–100(–150+); stems usually erect, sometimes decumbent; cauline leaves mostly opposite, margins entire or toothed, faces glabrous or scabrellous, often gland-dotted; heads discoid, racemiform, or spiciform arrays; involucres 3–4(–5) mm; phyllaries 3–5; ray florets absent; pistillate flowers 1–5; staminate flowers 5–22; often ± obcompressed, usually gland-dotted; pappus 0 (Strother, 2006q).

1. Annuals; leaf blades deltate or ovate to elliptic or lanceolate; heads in ± spiciform arrays; peduncles mostly 0.5-1 mm...**I. annua**
1'. Perennials; leaf blades narrowly ovate or obovate to elliptic or spatulate; heads in ± racemiform arrays; peduncles 1-2 mm...**I. axillaris**

Iva annua L.—Povertyweed. Annuals, (10-)50-100(-150+) cm; stems erect; leaf blades deltate or ovate to elliptic or lanceolate, 30-100(-150+) × 8-45(-80) mm, margins ± toothed, faces ± scabrellous, gland-dotted; heads in ± spiciform arrays; involucres hemispherical, 3-4(-5) mm; outer phyllaries 3-5; pistillate flowers 3-5, corollas 0.5-1 mm; functionally staminate flowers 8-12+, corollas 2-2.5 mm; cypselae 2-3 mm. Disturbed sites, moist soils; Bosque del Apache National Wildlife Refuge, ca. 4500 ft. Socorro.

Iva axillaris Pursh—Povertyweed. Perennials, 10-40(-60) cm (rhizomatous); stems erect; leaf blades narrowly ovate or obovate to elliptic or spatulate, 15-25(-45) × 3-8(-15) mm, margins rarely toothed, faces usually ± strigose to scabrellous, gland-dotted; heads in ± racemiform arrays; involucres 2-3 mm; phyllaries 3-5; pistillate flowers 3-5(-8), corollas 0.5-1.5 mm; functionally staminate flowers 4-8(-20+), corollas 2-2.5 mm, cypselae 2.5-3 mm. Seasonally wet, saline habitats, 4025-8200 ft. NE, NC, NW, EC, C, Chaves, Luna.

JEFEA Strother – Jefea

Don Hyder

Jefea brevifolia (A. Gray) Strother—Boss daisy. Subshrubs or shrubs; stems erect; leaves cauline, blades 8-25+ × 4-20 mm, sometimes obscurely 3(-5)-lobed; heads radiate, borne singly; involucres campanulate to hemispherical; ray florets 5-13(-20), pistillate, fertile, corollas yellow to orange; disk florets 30-60(-100+), bisexual, fertile, corollas yellow to orange, tubes shorter than or about equaling funnelform throats, lobes 5, yellow to orange; cypselae 3-angled (peripheral) or strongly compressed, pappus fragile or persistent. Limestone, desert scrub, 5000-6000 ft. Doña Ana (Allred & Ivey, 2012; Strother, 2006r).

KOANOPHYLLON Arruda – Thoroughwort

Kenneth D. Heil

Perennials or subshrubs, 5-15 dm (ours); stems erect, branched; leaves mostly opposite, sometimes subopposite or alternate on distal 1/3 of stems, blades mostly triangular-lanceolate or lanceolate, margins entire or dentate, faces glabrous or sparsely strigose, sometimes sparsely gland-dotted; heads discoid; involucres 2-3 mm diam.; flowers 3-13(-20), corollas whitish to pinkish; cypselae prismatic, 5-ribbed, sparsely hairy (Nesom, 2006p).

1. Leaves usually opposite, sometimes subopposite to alternate on distal 1/3 of stems, blade apices acuminate, adaxial faces glabrous; involucres 4.5-5.5 mm; outermost phyllaries narrowly lanceolate, glabrous, margins hyaline on proximal 2/3, eciliate [to be expected]...K. solidaginifolium
1'. Leaves opposite, blade apices acute, adaxial faces sparsely strigose to hispidulous; involucres 3-5.4 mm; outermost phyllaries ovate-elliptic or obovate to narrowly oblong-lanceolate, puberulent, margins herbaceous, usually weakly ciliate...**K. palmeri**

Koanophyllon palmeri (A. Gray) R. M. King & H. Rob. [*Eupatorium palmeri* A. Gray]—Palmer's umbrella thoroughwort. Perennials or subshrubs, 5-15 dm; leaves opposite, blades lanceolate to triangular-lanceolate, (2-)3-6 × 0.8-2 cm, margins shallowly serrate to subentire, apices acute, abaxial faces eglandular or minutely gland-dotted, adaxial sparsely strigose to hispidulous; involucres (3-)3.5-4 mm; corollas white, (2-)2.4-2.8 mm; cypselae 1.6-1.8 mm, sparsely hispidulous. Shaded rocks along streams, crevices, oak woodlands, 4000-6800 ft. Hidalgo.

Koanophyllon solidaginifolium (A. Gray) R. M. King & H. Rob. [*Eupatorium solidaginifolium* A. Gray]— Shrubby umbrella thoroughwort. Perennials or subshrubs (shrubs), 6-10 dm; leaves mostly opposite, sometimes subopposite to alternate on distal 1/3 of stems, blades lanceolate, mostly 3-8(-11) × 1-3 cm, margins entire or shallowly crenate, apices acuminate, abaxial faces eglandular, adaxial glabrous; involucres 4.5–5.5 mm; corollas usually white, sometimes purple-tinged to yellowish, 2.5–3 mm; cypselae 1.8–2.8 mm, sparsely hispidulous. Canyon walls, ledges, slopes, talus, limestone hills and ridges, on and among rocks, in woods along streams. No records seen. To be expected in Hidalgo County.

KRIGIA Schreb. – Dwarf dandelion

Kenneth D. Heil

Krigia biflora (Walter) S. F. Blake [*Hyoseris biflora* Walter]—Orange dwarf dandelion. Perennials, 10–70 cm; stems 1–5+, erect; leaves mostly basal (rosettes), some cauline (proximal), blades oblanceolate to obovate or spatulate, 5–25 cm; heads borne singly, (2)3–20+; involucres 7–11 mm; phyllaries 8–18, in 1–2 series, reflexed in fruit, lanceolate; receptacles flat or low-convex, pitted; florets 25–60, corollas orange or yellow-orange, 15–25 mm; cypselae reddish brown, columnar, 2–2.5 mm, 12-ribbed; pappus of ca. 10 outer scales 0.3–0.5 mm, plus 20–40 barbellulate inner bristles 4.5–5.5 mm. Shaded areas, often near streams; meadows, ponderosa pine, mixed conifer, and aspen communities, 7320–9600 ft. NC, Catron, Otero (Chambers & O'Kennon, 2006).

LACTUCA L. – Lettuce

Gary I. Baird

Annual, biennial, or perennial herbs, sap milky; stems erect or decumbent; leaves basal and/or cauline, alternate, simple, sessile or petiolate, bases often auriculate, blades linear-lanceolate to orbiculate, faces glabrous-glaucous or pubescent to prickly (on midrib), entire to pinnately lobed, ultimate margins entire to dentate or prickly; heads liguliflorous, arrays racemiform to paniculiform or corymbiform; involucres cylindrical to obclavate; phyllaries imbricate, herbaceous; florets bisexual, ligules blue, whitish, or yellow (then often red-purple or bluish abaxially); cypselae ellipsoid, flattened, brownish, grayish, or blackish, often mottled, glabrous or scaberulous, ribbed, beakless or beaked, beaks stout to filiform, green, white, or pale; pappus of bristles, barbellulate, white to brownish, persistent (Strother, 2006s).

1. Plants rhizomatous perennials...**L. tatarica**
1'. Plants taprooted annuals or biennials...(2)
2. Corollas blue or whitish (rarely yellowish)...(3)
2'. Corollas yellow (sometimes red-purple or bluish abaxially and drying bluish)...(4)
3. Cypsela beaks stout, to 0.5 mm, faces 4–6-nerved; pappus brownish or sordid...**L. biennis**
3'. Cypsela beaks slender, 2–4 mm, faces 1(–3)-nerved; pappus white...**L. graminifolia**
4. Cypsela faces 5–9-nerved (rarely fewer)...(5)
4'. Cypsela faces 1(–3)-nerved...(7)
5. Cypsela beaks 5–6 mm (rarely shorter); heads in racemiform to spiciform arrays...**L. saligna**
5'. Cypsela beaks 2.5–5 mm; heads in paniculiform to corymbiform arrays...(6)
6. Blades (of undivided cauline leaves) oblong to obovate; leaves often twisted at base so blade is held in a vertical plane (in full sun), margins and midribs prickly (rarely lacking)...**L. serriola**
6'. Blades (of undivided cauline leaves) ovate to orbiculate; leaves held in a horizontal plane (not twisted at base), margins and midribs glabrous (rarely with a few prickles)...**L. sativa**
7. Cypsela bodies 2.5–3.5 mm; involucres mostly 10–12 mm; florets mostly 15–20...**L. canadensis**
7'. Cypsela bodies 4.5–5 mm; involucres mostly 12–15 mm; florets mostly 20–50...**L. ludoviciana**

Lactuca biennis (Moench) Fernald—Tall blue lettuce. Biennials (annuals), 0.5–3 m; undivided leaf blades ovate to lanceolate, midribs glabrous or sparsely pilose, ultimate margins entire or denticulate; heads in paniculiform arrays; involucres 7–12 mm; florets mostly 20–30; corollas pale blue or whitish (rarely yellowish); cypsela bodies 4–5 mm, brown, faces (4)5–6-nerved, beaks stout, to 0.5 mm; pappus brownish, 4–6 mm. Along drainages, riparian areas, creek banks, 6000–8700 ft. NC, NW.

Lactuca canadensis L.—Canada lettuce. Biennials, 0.5–2 m; undivided leaf blades mostly oblong to obovate or spatulate (rarely lance-linear), midribs glabrous or sparsely pilose, ultimate margins entire or denticulate; heads in paniculiform to corymbiform arrays; involucres 10–12 mm; florets mostly 15–20; corollas yellowish (rarely bluish); cypsela bodies 2.5–3.5 mm, brown, faces 1(–3)-nerved, beaks slender,

1-3 mm; pappus white, 5-6 mm. Riparian sites, canyon bottoms, margins of creeks and rivers, wet meadows, roadsides, 5500-10,500 ft. NC, NW, C, WC, SE.

Lactuca graminifolia Michx.—Arizona lettuce. Biennials, 0.25-1.5 m; undivided leaf blades spatulate to lance-linear, midribs sometimes setose but not prickly, ultimate margins entire or denticulate; heads in paniculiform arrays; involucres 12-20 mm; florets mostly 15-20; corollas bluish; cypselae bodies 5-6 mm, brown, faces 1(-3)-nerved, beaks slender, 2-4 mm; pappus white, 5-9 mm. New Mexico material belongs to **var. arizonica** McVaugh. Sandy ridges, pine forests, canyon bottoms, 4300-9400 ft. NC, C, WC, SE, SW.

Lactuca ludoviciana (Nutt.) Riddell—Prairie lettuce. Biennials, 0.2-1.5 m; undivided leaf blades oblanceolate to obovate or spatulate, midribs pilose-setose, ultimate margins denticulate (pilose-ciliate); heads in paniculiform arrays; involucres 12-15 mm; florets mostly 20-50; corollas yellow (rarely bluish); cypsela bodies 4.5-5 mm, brown to black, faces 1(-3)-nerved, beaks slender, 2.5-4.5 mm; pappus white, 5-10 mm. Uncommon; openings in pine forests, stream and river banks, prairies, 5000-7550 ft. Bernalillo, Grant, San Juan.

Lactuca saligna L.—Willowleaf lettuce. Annuals, 0.2-1 m; undivided leaf blades lance-linear, midribs usually prickly-setose, ultimate margins entire or denticulate; heads in racemiform to spiciform arrays; involucres 6-13 mm; florets mostly 6-20; corollas yellow; cypsela bodies 2.5-3.5 mm, pale brown, faces 5-7-nerved, beaks slender, (2-)5-6 mm; pappus white, 5-6 mm. Disturbed sites, 3100-4950 ft. Socorro. Introduced.

Lactuca sativa L.—Garden lettuce. Annuals (biennials), 0.2-1 m; undivided leaf blades ovate to orbiculate, midribs glabrous (rarely prickly-setose), ultimate margins entire or denticulate; heads in paniculiform to corymbiform arrays; involucres 8-13 mm; florets mostly 7-30; corollas yellow (sometimes streaked with violet); cypsela bodies 3-4 mm, gray-white or tan (black), faces 5-9-nerved, beaks slender, 3-5 mm; pappus white, 3.5-4 mm. Disturbed sites and abandoned plantings, 4500-7600 ft. Bernalillo, Colfax, McKinley. Introduced.

Lactuca serriola L.—Prickly lettuce. Annuals, 0.2-1 m; undivided leaf blades oblong to obovate, midribs prickly (rarely glabrous), ultimate margins denticulate-prickly; heads in paniculiform arrays; involucres 9-12 mm; florets mostly 12-20; corollas yellow; cypsela bodies 2.5-3.5 mm, gray-white to tan, faces (3-)5-9-nerved; beaks slender, 2.5-4 mm; pappus white, 3-5 mm. Disturbed sites and roadsides, 3100-10,100 ft. Widespread. Introduced.

Lactuca tatarica (L.) C. A. Mey. [*L. oblongifolia* Nutt.]—Blue lettuce. Perennials, rhizomatous, 0.2-1 m, undivided leaf blades oblong-elliptic to ovate, sometimes lance-linear, midribs glabrous, ultimate margins entire to dentate; heads in paniculiform to corymbiform arrays (sometimes solitary); involucres 12-20 mm; florets mostly 10-50; corollas blue; cypsela bodies 4-5 mm, red-brown to blackish, faces 4-6-ribbed, beakless or with a stout beak to 1 mm; pappus white, 7-12 mm. Our material belongs to **subsp. pulchella** (Pursh) Stebbins. Calcareous sites, clearings in pine forests or shrublands, meadows, roadsides, stream banks, other wet areas, 3900-10,700 ft. NC, NW, C, WC, SE.

LAENNECIA Cass. – Horseweed

Jennifer Ackerfield and Kenneth D. Heil

Annual herbs; leaves alternate, sessile and clasping, glandular; heads disciform or inconspicuously radiate; involucral bracts not much imbricate, scarcely herbaceous; receptacles flat, naked; pistillate flowers numerous, corollas slender, rayless or with a short, inconspicuous, white or pinkish ray; disk florets few, yellow; cypselae 2-nerved; pappus of capillary bristles (Nesom, 1990).

1. Leaf blades mostly 1-2 times pinnately lobed, pappus 2-3 mm…**L. sophiifolia**
1'. Leaf blades coarsely toothed or entire; pappus 3.5-4 mm…(2)
2. Leaf blades entire or distally toothed, loosely arachnoid-tomentose; pistillate flowers with a short ligule (ca. 0.5-1 mm); heads usually in an elongate, narrow, thyrsoid inflorescence…**L. schiedeana**

2'. Leaf blades coarsely toothed from near the base or entire, leaves not arachnoid-tomentose; pistillate flowers without a ligule; heads in an elongate, wider, thyrsoid-paniculiform inflorescence…**L. coulteri**

Laennecia coulteri (A. Gray) G. L. Nesom [*Conyza coulteri* A. Gray]—Coulter's horseweed. Plants 1-15 dm; leaves spatulate to oblong, 2-5 cm, villous, glandular-viscid, the margins lobed to coarsely toothed or entire; involucres 2.5-3.5 mm; pistillate flowers 60-100, lacking a ligule; disk florets 5-20; cypselae 0.5-1 mm. Along ditches, dry stream beds, pine forests, 3200-9300 ft. Widespread except N, EC.

Laennecia schiedeana (Less.) G. L. Nesom [*Conyza schiedeana* (Less.) Cronquist]—Pineland marestail. Plants 1-5(-10) dm; leaves spatulate to oblong or linear, 2-5 cm, tomentose, glandular, the margins usually entire or sometimes lobed distally; involucres 4-6 mm; pistillate flowers 60-100+, the ligules 0.5-1 mm; disk florets 5-10; cypselae 1-1.5 mm. Uncommon in open pine forests, 3600-10,000 ft. Widespread except NE, NW, EC, SE, SW.

Laennecia sophiifolia (Kunth) G. L. Nesom [*Conyza sophiifolia* Kunth]—Leafy horseweed. Plants 1-5(-15) dm; leaves ovate to spatulate, pinnately lobed, hirtellous to villous, at least on the margins; involucres 2.5-3.5 mm; pistillate flowers 20-40+; disk florets 2-6; cypselae 0.6-0.8 mm, glabrous. Disturbed sites, among igneous rocks, 5450-9550 ft. WC, SC, SW.

LAPSANA L. – Nipplewort

Jennifer Ackerfield

Lapsana communis L.—Common nipplewort. Lactiferous annuals, 1.5-10(-15) dm, hirsute to nearly glabrous; leaves alternate, entire to variously dentate or pinnatifid, petiolate, ovate, or rounded-obtuse, lowermost lyrate, 1-15(-30) cm; heads small, ligulate; involucres 5-8 mm, the bracts in 2 series, 3-9 mm; receptacles naked; ray florets 8-15, 1-10 mm, yellow; disk florets absent; cypselae 3-5 mm, narrow, glabrous, with many nerves. Weed in disturbed forests; 1 record at Ruidoso, 6440 ft. Lincoln. Introduced (Ackerfield, 2015).

LASIANTHAEA DC. – Lasianthaea

Don Hyder

Lasianthaea podocephala (A. Gray) K. M. Becker—San Pedro daisy. Shrubs, rootstocks bearing tubers; stems erect; leaves cauline, opposite, blades 2-8+ cm; heads radiate, borne in corymbiform arrays, involucres hemispherical, phyllaries persistent, herbaceous to papery, receptacles convex, ray florets (4-)11-14(-30), pistillate; corollas yellow to orange (purplish); disk florets yellow to orange (purplish); cypselae 3-angled, 2.5-4.5 mm, compressed or flattened, all with sharp edges, not truly winged; pappus persistent, 1-4 mm. Open or shaded sites, pine-oak woodlands, 6000-8500 ft. Hidalgo (Allred & Ivey, 2012; Strother, 2006t).

LASTHENIA Cass. – Goldfields

Kenneth D. Heil and Steve L. O'Kane, Jr.

Lasthenia gracilis (DC.) Greene [*Baeria gracilis* (DC.) A. Gray]—Common goldfields. Annuals, to 40 cm; stems erect or decumbent, ± hairy; leaves mostly cauline, opposite, usually linear, 8-70 × 1-3(-6) mm, margins entire or with 3-5+ teeth, faces hairy (ours); heads radiate; involucres campanulate or hemispherical, 5-10 mm; receptacles hemispherical to narrowly conic or subulate; phyllaries 4-13, in 1 series, ovate-lanceolate; ray florets 6-13, corollas yellow to orangish, sometimes white; anther appendages deltate; cypselae ± linear, to 3 mm, black to gray. Desert grasslands, ca. 4150 ft. Grant, Hidalgo (Chan & Ornduff, 2006).

LAYIA Hook. & Arn. ex DC. - Tidytips

Kenneth D. Heil

Layia glandulosa (Hook.) Hook. & Arn. [*Blepharipappus glandulosus* Hook.]—Whitedaisy tidytips. Annuals, plants 3-60 cm, glandular; stems often uniformly dark purple; leaves mostly alternate, sessile, 6-100 mm, obovate to linear, margins toothed to lobed, faces glabrous or hirsute to strigose; involucres ± campanulate to hemispherical, 4-11 mm; phyllaries 3-14; heads usually radiate, borne singly or in ± corymbiform arrays; receptacles flat to convex; ray florets 3-14, often white, sometimes yellow, 3-22 mm; disk florets 17-100+, 3.5-6.5 mm; ray cypselae obcompressed, glabrous; disk pappus of 10-15 scales, 2-5 mm, each plumose and often adaxially woolly. Openings in scrub, woodlands, forests, grasslands, and meadows, 4000-6250 ft. Grant, Hidalgo (Baldwin et al., 2006).

LEIBNITZIA Cass. - Seemann's sunbonnet

Kenneth D. Heil

Leibnitzia lyrata (Sch. Bip.) G. L. Nesom [*Gerbera lyrata* Sch. Bip.; *Leibnitzia seemannii* (Sch. Bip.) G. L. Nesom]—Seemann's sunbonnet. Perennials, 5-60+ cm; leaves basal, 3-19 cm, elliptic to obovate, oblanceolate, or lyrate, margins usually sinuately lobed to dentate, abaxial faces thinly gray-tomentose, adaxial faces glabrous or glabrescent; involucres 9-20+ mm; heads vernal-chasmogamous, autumnal-cleistogamous; receptacles flat to convex; florets with outer corollas usually pinkish, inner corollas usually whitish; cypselae ± fusiform, tan to purplish, 6-10 mm; pappus of 50-80+ stramineous, barbellulate bristles. Open areas in pine and pine-oak woodland communities, 7000-9650 ft. WC, SC, Colfax, San Miguel (Nesom, 2006q).

LEONTODON L. - Hairy hawkbit

Kenneth D. Heil

Leontodon saxatilis Lam.—Hairy hawkbit. (Ours) perennials, 10-40 cm; stems 1-15+, ascending, simple, glabrous or coarsely hispid; leaf blades oblanceolate to oblong, 2-15 × 0.5-2.5 cm, margins entire, dentate, or deeply lobed, faces hispid or hirsute; heads borne singly; involucres campanulate, 6-13 × 4-9 mm; phyllaries 16-20; florets 20-30; corollas bright yellow to grayish yellow (outer faces), 8-15 mm; cypselae fusiform, 4-5.5 mm, outer curved, thick, not beaked, often enclosed by phyllaries, inner beaked, beaks 1 mm. Our plants belong to **var. saxatilis**. Ca. 5200 ft. San Miguel, Union (Bogler, 2006c).

LEPIDOSPARTUM (A. Gray) A. Gray - Scalebroom

Steven R. Perkins

Lepidospartum burgessii B. L. Turner—Burgess' broomsage. Shrubs 10-150 cm, broomlike; stems erect, many-branched, with silvery hairs with glandular blisters; leaves cauline, filiform to acerose, alternate, 3-20 mm; heads discoid, terminal on stems; involucres turbinate to cylindrical, 7-8 mm; phyllaries 8-13, persistent, unequal (outer shorter), ovate to elliptic, pannose-tomentose; ray florets 0; disk florets 3, yellow; cypselae ca. 4 mm, 5-nerved, densely pilose; pappus 5-8 mm, persistent, with ca. 150 barbellulate bristles. Stabilized gypsum dunes, 3600-4000 ft. Otero (Allred & Ivey, 2012; New Mexico Rare Plant Technical Council, 1999; Strother, 2006u). **R**

LEUCANTHEMUM Mill. - Oxeye daisy

Jennifer Ackerfield

Leucanthemum vulgare Lam. [*Chrysanthemum leucanthemum* L.]—Oxeye daisy. Perennial herbs; plants 1-10 dm; leaves alternate, simple, obovate to spatulate or oblanceolate, 4-15 cm, the upper leaves auriculate-clasping, margins usually pinnately lobed or toothed; involucres 10-20 mm diam., bracts imbricate in 2-4 series, the margins and tips scarious; receptacles convex, naked; ray florets white, 12-30 mm; disk florets yellow; cypselae 1.5-2.5 mm, 10-ribbed; pappus lacking. Widespread, roadsides, disturbed areas, old mining camps, mountain meadows, 4400-10,450 ft. NC, NW, C, WC, SE. Introduced (Ackerfield, 2015).

LEUCIVA Rydb. – Marsh elder

Steven R. Perkins

Leuciva dealbata (A. Gray) Rydb.—Woolly marsh elder. Perennial herbs, 20-60+ cm; stems erect, branching from bases or distally; leaves cauline, mostly alternate, petioles 1-5+ mm, blades lanceolate to oblanceolate, usually gland-dotted; heads discoid or disciform in paniculiform arrays; involucres ± hemispherical, 2-3+ mm; phyllaries persistent, 5+ in 1(2) series, outer 5 sparsely strigose or glabrous abaxially; pistillate florets 0 or 3-5, corollas whitish and tubular if present; functionally staminate florets 5-12+, corollas whitish, funnelform, lobes 5; receptacles convex, epaleate; cypselae 1.5-2 mm, pyriform, densely gland-dotted; pappus 0. Roadsides, drainages, calcareous plains, primarily (but not exclusively) in the S half of New Mexico, 3900-6500 ft. EC, C, WC, SE, SC, SW (Allred & Ivey, 2012; Strother, 2006v).

LEUCOSYRIS Greene – Desert tansy-aster

Kenneth D. Heil

Annuals or perennials, 10-60 cm; stems glabrous or gland-dotted; leaves entire or pinnatifid; heads radiate or discoid; involucres turbinate to depressed-hemispherical, 3-12 × 4-16 mm; receptacles naked or with scales; ray florets blue, purple, violet, or white; disk florets yellow; pappus persistent, of 20-40 white or whitish (tawny in *L. blepharophylla*), barbellulate, apically attenuate bristles in 2-3 series (Hartman & Bogler, 2006; Pruski & Hartman, 2012). Previously included in *Arida* and *Machaeranthera*.

1. Leaves (at least some) pinnatifid to 2-pinnatifid…**L. parviflora**
1′. Leaves entire or toothed…(2)
2. Perennials; leaf margins entire with 8-20 cilia per side; involucres broadly turbinate…**L. blepharophylla**
2′. Annuals; leaf margins entire, eciliate or with 1-8 cilia per side; involucres hemispherical…**L. riparia**

Leucosyris blepharophylla (A. Gray) Pruski & R. L. Hartm. [*Aster blepharophyllus* A. Gray; *Machaeranthera gypsitherma* G. L. Nesom, Vorobik & R. L. Hartm.; *Arida blepharophylla* (A. Gray) D. R. Morgan & R. L. Hartm.]—La Playa Springs tansy-aster. Perennials, 4-35 cm; leaves linear-oblanceolate, 10-40 × 1.5-3 mm, succulent, margins entire, with 8-20 cilia per side; ray florets blue; disk florets yellow or sometimes purple-tinged; pappus tawny, setose. Alkaline springs and seeps in Chihuahuan desert scrub, 3900-4600 ft. Hidalgo. **R**

Leucosyris parviflora (A. Gray) Pruski & R. L. Hartm. [*Machaeranthera parviflora* A. Gray; *Arida parviflora* (A. Gray) D. R. Morgan & R. L. Hartm.]—Smallflower tansy-aster. Annuals or short-lived perennials, 1-5 dm; leaves lanceolate to oblong, 1-3 cm, glabrous or sparsely stipitate-glandular; ray florets 6-8 mm, blue to purple; disk florets 4-5 mm; cypselae 1.5-2 mm. Alkaline soil, often in association with greasewood, 3600-7300 ft. SE, SC, SW, Socorro, Torrance.

Leucosyris riparia (Kunth) Pruski & R. L. Hartm. [*Aster riparius* Kunth; *Machaeranthera riparia* (Kunth) A. G. Jones; *Arida riparia* (Kunth) D. R. Morgan & R. L. Hartm.]—Chiricahua Mountain tansy-aster, alkali aster. Annuals, 25-60 cm; leaves oblanceolate, 20-30 × 3-5 mm, eciliate or with 1-8 cilia per side, glabrous; ray florets bluish purple to lavender; disk florets yellow; cypselae 2.5-3 mm. Low valleys, saline soils on mudflats, edges of playas, 4090-4400 ft. Hidalgo.

LIATRIS Gaertn. ex Schreb. – Blazing star, gay-feather

Jennifer Ackerfield and Kenneth D. Heil

Perennial herbs; leaves alternate, simple, entire, generally conspicuously punctate; heads discoid; involucral bracts imbricate, herbaceous, the margins scarious; receptacles naked; disk florets purple, pink, or rarely white; pappus of plumose or barbellate bristles, in 1+ series; cypselae ca. 10-ribbed, somewhat cylindrical but pointed at the base (Nesom, 2006r).

1. Involucral bract tips rounded, scarious, and usually broadly lacerate or fringed; heads broadly campanulate to hemispherical; stems reddish, glabrous below with white appressed hairs above...**L. ligulistylis**
1'. Involucral bract tips acute to acuminate, ciliate-margined or narrowly scarious-margined; heads cylindrical to weakly campanulate; stems green to yellowish brown or the upper sometimes reddish, glabrous or nearly so...(2)
2. Pappus barbellate; corolla ca. 6 mm, glabrous inside; leaves lax, the lower 5-veined...**L. lancifolia**
2'. Pappus plumose; corolla 9-12 mm, pilose or pubescent inside; leaves rigid, often coarsely ciliate-margined, 1-3-veined (very rarely a couple of leaves 5-veined)...**L. punctata**

Liatris lancifolia (Greene) Kittell—Lanceleaf blazing star. Plants (2-)4-8 dm; leaves broadly linear, the lower 5-veined, glabrous, 10-30 cm; involucres 7-11 mm, the bracts erect, acute; disk florets 12-15, ca. 6 mm, purple, glabrous; pappus barbellate, 5 mm. Prairies, moist banks of spring-fed springs, 3800-6800 ft. Chaves, Doña Ana, Lincoln, Otero.

Liatris ligulistylis (A. Nelson) K. Schum.—Rocky Mountain blazing star. Plants 1-6 dm; leaves broadly linear, glabrous to hispid on the midvein to densely hairy on both sides, 8-25 cm; involucres 13-20 mm, the bracts scarious and sometimes broadly lacerate, purple; disk florets 30-70, 9-11 mm, purple, glabrous; pappus barbellate, 8-10 mm. Wet meadows and along streams, 7800-9515 ft. Colfax, San Miguel, Union.

Liatris punctata Hook.—Dotted blazing star. Plants 1.5-8 dm; leaves linear, 1-3-veined, glabrous with ciliate margins, 8-15 cm; involucres 15-20 mm, the bracts ciliate or membranous-margined, rounded; disk florets 4-8, 9-12 mm, purple or rarely white, pilose inside; pappus plumose, 9-11 mm. Open grasslands and meadows, especially common on E plains.

1. Heads in dense spiciform arrays, closely spaced; stems usually obscured by heads...**var. punctata**
1'. Heads in loose spiciform arrays, widely spaced; stems evident...**var. mexicana**

var. mexicana Gaiser—Mexican blazing star. Leaves 50-120 × 2-7 mm; heads in loose spiciform arrays; involucres 10-15 mm; florets 4-6. Canyon bottoms, grassy areas, mesquite, 4300-7200 ft. Eddy.

var. punctata—Dotted blazing star. Leaves 100-140 × 1-5 mm; heads in dense spiciform arrays; involucres 10-14 mm; florets 4-8. Prairies and sagebrush communities, 4300-8400 ft. NE, NC, EC, C, WC, SE, SC.

LOGFIA Cass. - Cottonrose

Kenneth D. Heil

Annuals, 1-50(-70) cm; stems 1, erect, or 2-10+, ascending to prostrate; leaves cauline, alternate, blades subulate to obovate; heads usually in glomerules of 2-10(-14) in racemiform to paniculiform or dichasiform arrays; involucres 0 or inconspicuous; phyllaries 0, vestigial, 1-4 (unequal), or 4-6 (equal); receptacles fungiform to obovoid, glabrous; pistillate florets 14-45+; functionally staminate florets 0; bisexual florets 2-10; corolla lobes 4-5, ± equal; cypselae brown, dimorphic, faces glabrous; outer pistillate pappus 0, inner pistillate and bisexual pappus of (11-)13-28+ bristles (Morefield, 2006d).

1. Stems prostrate; outer phyllaries papery; corollas yellow to brownish, 5-lobed; inner cypselae smooth...**L. depressa**
1'. Stems mostly erect; outer phyllaries cartilaginous; corollas reddish to purplish, 4-lobed; inner cypselae papillate...**L. filaginoides**

Logfia depressa (A. Gray) Holub [*Filago depressa* A. Gray]—Dwarf cottonrose. Plants 1-5(-10) cm; stems typically ± prostrate; leaves elliptic to obovate, largest 6-8(-10) × 1-2 mm; heads in glomerules of 2-5 in ± dichasiform arrays; receptacles obovoid, 0.9-1.2 mm; pistillate paleae (except innermost) 7-13 in 2(3) series, chartaceous; innermost paleae ± spreading in 1 series; corollas 1.3-2 mm, lobes mostly 5, yellowish to brownish; cypselae compressed, 0.7-0.9 mm, mostly smooth. Gravelly soils, desert flats and alluvial slopes; recently found near Lordsburg and Silver City, 4480-5800 ft. Grant, Hidalgo.

Logfia filaginoides (Hook. & Arn.) Morefield [*Gnaphalium filaginoides* Hook. & Arn.]—California cottonrose, fluffweed. Plants 1-30(-55) cm; stems 1(-7), typically ± erect; leaves mostly oblanceolate, largest 10–15(–20) × 2–3(–4) mm; heads mostly in glomerules of 2–4 in racemiform or paniculiform arrays, 3.5–4.5 × 2.5–3 mm; phyllaries 0, vestigial, or 1–4, unequal; outer pistillate florets 7–13, epappose, inner 14–35, pappose; bisexual florets 4–7; corollas 1.9–2.8 mm, bright reddish to purplish; outer cypselae nearly straight, mostly 0.9–1 mm; inner mostly papillate; pappus of 17–23+ bristles. Disturbed areas in Chihuahuan desert scrub communities, 3910–4700 ft. Doña Ana, Luna.

LORANDERSONIA Urbatsch, R. P. Roberts & Neubig – Candlebrush

Gary I. Baird and Kenneth D. Heil

Shrubs or subshrubs, 5–350 cm, mostly scopiform; stems erect to spreading, green to white or gray-tan, glabrous to puberulent or scabridulous, glandular or not; leaves basal and/or cauline, alternate, sessile or petiolate, blades linear-lanceolate to spatulate, flat to concave, margins entire, faces glabrous to puberulent or scabridulous, eglandular or glandular-resinous; heads discoid or radiate in cymiform to corymbiform arrays; involucres cylindrical to obconic; phyllaries 13–30 in 3–6 series, chartaceous with subapical green spot or with a caudate herbaceous apex, spiraled to strongly aligned; ray florets 0 or 1–20, pistillate, fertile, ligules yellow; disk florets 4–15, bisexual, fertile, corollas yellow; cypselae cylindrical to obconic, often prismatic, ribbed, brownish, faces sparsely to densely hairy (or glabrous); pappus of 20–80+ slender bristles in 1 series, beige to buff (Urbatsch & Neubig, 2006).

1. Heads radiate; suffrutescent perennial herbs, 5–20 cm...**L. microcephala**
1'. Heads discoid (rarely with 1 to few raylike florets); shrubs, 25–350 cm...(2)
2. Involucres 10–15 mm; disk corollas 9–14 mm; cypselae 4.5–7 mm...(3)
2'. Involucres 4–7 mm; disk corollas 4–5.5 mm; cypselae 2–3.5 mm...(4)
3. Leaf margins ciliolate...**L. baileyi**
3'. Leaf margins eciliolate...**L. pulchella**
4. Plants soboliferous; cypselae ribs 10–12, faces densely hairy; pappus bristles 4.5–7 mm...**L. linifolia**
4'. Plants not soboliferous; cypselae ribs 4–6, faces sparsely hairy; pappus bristles 3–4 mm...**L. spathulata**

Lorandersonia baileyi (Wooton & Standl.) Urbatsch, R. P. Roberts & Neubig [*Chrysothamnus baileyi* Wooton & Standl.]—Bailey's candlebrush. Shrubs, 25–70 cm, not soboliferous; stems single, glabrous or puberulent; leaves cauline, blades linear-oblong, 4–35 × 0.5–3 mm, margins ciliolate, faces glabrous; phyllaries 20–25 in 4–5 series, vertically aligned; ray florets 0; disk florets 5, corollas 9–14 mm; cypselae 5–7 mm, 5–8-ribbed, faces sparsely or rarely densely hairy; pappus 9–12. Sandy to rocky soils in grassland, open oak, mesquite, sagebrush, and piñon-juniper communities, 3500–8200 ft. Widespread except NC, SW.

Lorandersonia linifolia (Greene) Urbatsch, R. P. Roberts & Neubig [*Chrysothamnus linifolius* Greene]—Alkali candlebrush. Shrubs, 70–350 cm, soboliferous; stems single, glabrous or scabridulous; leaves cauline, blades broadly linear to lanceolate or oblong, 20–75 × 3–8 mm, margins scabridulous, faces glabrous, gland-dotted; phyllaries 15–18 in 3–4 series, vertically aligned; ray florets 0; disk florets 4–6, corollas 4–5.5 mm; cypselae 2.5–3.5 mm, 10–12-ribbed, faces densely hairy; pappus 4.5–7 mm. Moist, alkaline soils in riparian habitats, 4270–7425 ft. NW, Lincoln, Socorro.

Lorandersonia microcephala (Cronquist) Urbatsch, R. P. Roberts & Neubig [*Tonestus microcephalus* (Cronquist) G. L. Nesom & D. R. Morgan]—Crevice candlebrush. Perennials, suffrutescent (caudices woody), 5–20 cm, not soboliferous; stems 3–20+, sparsely stipitate-glandular, resinous; leaves basal and cauline, blades elliptic to narrowly oblanceolate, 20–40 × 2–3.5 mm, margins and faces glabrous or stipitate-glandular; phyllaries 24–30 in 3–4 series, not aligned; ray florets 5–8; disk florets 10–15, corollas 4.3–4.7 mm; cypselae 1.5–2.3 mm, 5–7-ribbed, faces hairy; pappus 3–5 mm. Crevices and clefts in granite outcrops in ponderosa pine woodlands, 8000–8500 ft. Rio Arriba, Taos. **R & E**

Lorandersonia pulchella (A. Gray) Urbatsch, R. P. Roberts & Neubig [*Chrysothamnus pulchellus* (A. Gray) Greene]—Wright's candlebrush. Shrubs, 80–120 cm, not soboliferous; stems single, glabrous; leaves cauline, blades linear or linear-oblong, 10–40 × 1–2.5 mm, margins and faces glabrous; phyllaries 24–30 in 4–5 series, vertically aligned; ray florets 0; disk florets 5, corollas 9–12 mm; cypselae 4.5–6.3 mm, 5–8-ribbed, faces glabrous or sparsely hairy; pappus 9–11 mm. Sandy to rocky soils in grassland, open sagebrush, piñon-juniper, and mountain brush communities, 3360–7565 ft. Widespread except NE, NC, SW.

Lorandersonia spathulata (L. C. Anderson) Urbatsch, R. P. Roberts & Neubig [*Chrysothamnus spathulatus* L. C. Anderson]—Mescalero candlebrush. Shrubs, 70–150 cm, not soboliferous; stems single, scabridulous; leaves cauline, blades oblanceolate to spatulate, 25–50 × 1–3 mm, margins and faces scabridulous; phyllaries 13–16 in 3–4 series, weakly aligned; ray florets 0 (rarely 1 to few); disk florets 4–5, corollas 4–5 mm; cypselae 2–3 mm, 4–6-ribbed, faces sparsely hairy; pappus 3–4 mm. Piñon-juniper and oak woodlands, 6000–8050 ft. C, WC, SE.

LYGODESMIA D. Don – Skeletonweed, rushpink

Gary I. Baird

Perennial herbs, taprooted and rhizomatous, 5–70 cm, sap milky, white to yellowish; stems mostly erect, simple to much branched; leaves basal and cauline, sessile, blades linear-subulate to lance-linear, reduced upward to scales in some, faces glabrous, margins entire; heads liguliflorous, 1–50, solitary at the ends of branches, forming open to bunched arrays, peduncles obscure to conspicuous, not inflated distally, bracteate; involucres cylindrical; calyculi of 8–10 scalelike bractlets, ovate to subulate, unequal; phyllaries 5–12 in ± 1 series, linear, equal, gray-green, faces glabrous or puberulent, margins scarious, apex keeled or flat; florets 5–12, ligules pink-purple, rarely white; cypselae cylindrical-fusiform, terete to angled, beakless, tan, faces glabrous, smooth or roughened and grooved; pappus of 60–80 smooth bristles in 1–2+ series, sordid-white, persistent (may be connate basally and falling as a unit) (Bogler, 2006d).

1. Involucres 10–16 mm; ligules 3–4 mm wide; cypselae 6–10 mm; pappus 6–9 mm…**L. juncea**
1′. Involucres 18–25 mm; ligules 5–10 mm wide; cypselae 10–17 mm; pappus 10–15 mm…(2)
2. Leaves primarily basal, forming rosettes, cauline leaves reduced to scales…**L. texana**
2′. Leaves primarily cauline, well developed or scalelike, basal rosettes lacking…(3)
3. Phyllaries 8–12, apically keeled; florets 6–12; cypselae smooth, not grooved; plants mostly > 15 cm (sometimes less)…**L. grandiflora**
3′. Phyllaries 5–6, apically flat; florets 5–7; cypselae rugose, grooved; plants mostly < 15 cm (rarely more)…
    L. arizonica

Lygodesmia arizonica Tomb [*L. grandiflora* (Nutt.) Torr. & A. Gray var. *arizonica* (Tomb) S. L. Welsh]— Navajo skeletonweed. Plants 5–15 cm (rarely more); stems sparsely branched from base; leaves not forming basal rosettes, cauline present at flowering, blades of proximal leaves 40–100 × 2–6 mm, margins entire; distal leaves smaller but not reduced to scales; heads mostly solitary, involucres 18–25 mm; main phyllaries 5–6, flat apically; florets usually 5, ligules 18–25 × 6–10 mm; cypselae 10–13 mm, abruptly narrowed apically, faces roughened and grooved; pappus 10–13 mm. Sandy soils, desert grasslands, desert shrub, 4600–6000 ft. McKinley, San Juan.

Lygodesmia grandiflora (Nutt.) Torr. & A. Gray—Canyonlands skeletonweed. Plants 5–40 cm; stems simple or sparingly branched from base; leaves not forming basal rosettes, cauline present at flowering, blades of proximal leaves 50–150 × 1–3 mm, margins entire (rarely lobed), distal leaves reduced to scales; heads mostly in open arrays; involucres 18–21 mm; main phyllaries 8–12, keeled apically; florets 6–12, ligules 16–19 × ca. 6 mm; cypselae 10–13 mm, tapered, faces smooth, striate; pappus 10–13 mm. Sandy or gravelly soils, piñon, juniper, and sagebrush communities, 5000–6500 ft. NW, Chaves, DeBaca.

Lygodesmia juncea (Pursh) D. Don ex Hook.—Prairie skeletonweed. Plants 10–35(-70) cm; stems much branched from base and distally; leaves not forming basal rosettes, cauline absent at flowering, blades of proximal leaves 5–30(-60) × 1–2(-4) mm, margins entire, distal leaves reduced to scales; heads mostly in bunched arrays; involucres 10–16 mm; main phyllaries 5–7, flat apically; florets usually 5, ligules 10–12 × 3–4 mm; cypselae 6–10 mm, faces obscurely striate (rarely sets seed); pappus 6–9 mm. Silty to sandy soils, disturbed sites, roadsides, desert grasslands, shortgrass prairies, montane grasslands, 4000–7800 ft. NE, NC, EC, C, Eddy, Roosevelt.

Lygodesmia texana (Torr. & A. Gray) Greene ex Small—Texas skeletonweed. Plants 25–65 cm; stems branched from base and distally; leaves forming basal rosettes, cauline sometimes withering before flowering, blades of proximal leaves 100–200 × 1–8 mm, margins usually lobed, distal leaves reduced to scales; heads mostly solitary; involucres 18–25 mm; main phyllaries 8–10, keeled apically; florets 8–12; ligules 20–24 × 6–8 mm; cypselae 11–17 mm, tapered, faces smooth, striate; pappus 10–15 mm. Sandy, calcareous, alkaline soils, oak-juniper woodlands, mesquite shrublands, open grasslands, 3450–5500 ft. EC, SE, Union.

MACHAERANTHERA Nees – Tansy-aster

Jennifer Ackerfield and Kenneth D. Heil

Herbs; leaves alternate, simple, deeply pinnatifid or lobed, glandular-stipitate; heads radiate; involucral bracts imbricate, spreading or reflexed, often bristle-tipped; receptacles naked or with a few scales; ray florets white, pink, purple, or blue; disk florets yellow; cypselae usually obovate, smooth or with ribs; pappus of filiform bristles (Ackerfield, 2015; Morgan & Hartman, 2006).

1. Involucres hemispherical; phyllary apices spreading to reflexed; disk corolla lobes 0.3–0.7 mm, glabrous or glabrate…**M. tanacetifolia**
1'. Involucres broadly turbinate; phyllary apices appressed; disk corolla lobes 0.7–1 mm, hairy…**M. tagetina**

Machaeranthera tagetina Greene [*Aster tagetinus* (Greene) S. F. Blake]—Mesa tansy-aster. Annuals, 5–50 cm; leaves 5–40 mm; involucral bracts broadly turbinate, 5–10 mm; ray florets blue, violet, or purple, 2–4 mm; pappus 2–8 mm. Grasslands, creosote desert scrub, pine-oak woodlands, roadsides, 4400–7650 ft. Grant, Hidalgo, Otero, Socorro.

Machaeranthera tanacetifolia (Kunth) Nees [*Aster tanacetifolius* Kunth]—Tansyleaf tansy-aster. Taprooted annuals or biennials, 5–100 cm; leaves 3–12 cm, 1-2-pinnate; involucral bracts in 3–6 series, 4–11 mm, tips spreading to reflexed (appressed); ray florets blue or purple; cypselae 2–3.5(–4) mm; pappus 2–8 mm. Common in sandy or rocky soil on plains and dry, open places in valleys, 3000–9500 ft. Widespread.

MADIA Molina – Tarweed

Jennifer Ackerfield

Madia glomerata Hook.—Mountain tarweed. Herbs, ± glandular and strong-scented; plants 5–120 cm; leaves simple, linear to narrowly linear, 2–10 cm; heads radiate or rarely discoid, usually in tightly grouped glomerules; involucral bracts equal, pilose and glandular-hairy; receptacles with a single series of chaff between the ray and disk florets; ray florets 1–3 mm, yellow or purplish or absent; disk florets 3–4.5 mm, hairy, yellow; cypselae black, usually radially compressed; pappus usually absent. Moist montane meadows surrounded by ponderosa pine–Gambel oak and mixed conifer communities, 8000–10,250 ft. NW, Cibola, Rio Arriba (Ackerfield, 2015).

MALACOTHRIX DC. – Desert dandelion

Kenneth D. Heil

(Ours) annuals, 3–60 cm; stems 1–15, mostly erect; leaves usually basal and cauline, sessile, margins entire or dentate, faces usually glabrous; heads usually in corymbiform to paniculiform arrays; outer phyllaries intergrading with inner or 3–30+ in 1–2 series; involucres broadly to narrowly campanulate, sometimes hemispherical; florets 15–270, corollas yellow or white; cypselae of capillary bristles; pappus 0 (Davis, 2006).

1. Involucres hemispherical; phyllaries orbiculate (outer) to oblong or linear, hyaline margins 1–2.5 mm wide…**M. coulteri**
1'. Involucres usually ± campanulate to hemispherical in *M. sonchoides*; bractlets of calyculi and/or phyllaries lanceolate to linear, hyaline margins 0.05–0.3(–1) mm wide…(2)

2. Corollas 4–10 mm; outer ligules exserted 1–4 mm…(3)
2'. Corollas (7–)10–23+ mm; outer ligules exserted 5–15+ mm…(4)
3. Corollas white or pale yellow; cypselae 1.7–2 mm (bases slightly expanded, distal 0.3 mm smooth)… **M. sonorae**
3'. Corollas usually yellow, sometimes white; cypselae 1.7–2.3 mm (bases not expanded, distal 0.1–0.2 mm smooth)…**M. stebbinsii**
4. Proximal cauline leaves pinnately lobed (lobes 3–8 pairs, ± oblong to triangular, ± equal, apices obtuse or acute), ± fleshy, ultimate margins usually dentate…**M. sonchoides**
4'. Proximal cauline leaves pinnately lobed (lobes 2–6+ pairs, filiform or triangular to oblong, subequal to unequal, apices acute), not fleshy, ultimate margins dentate or entire…(5)
5. Proximal cauline leaves usually pinnately lobed, lobes filiform; receptacles bristly…**M. glabrata**
5'. Proximal cauline leaves sometimes pinnately lobed, lobes relatively broad, triangular to deltate; receptacles not bristly…**M. fendleri**

Malacothrix coulteri Harv. & A. Gray [*Malacolepis coulteri* (Harv. & A. Gray) A. Heller]—Snake's head desert dandelion. Annuals, 10–60 cm; stems 1–6, ascending or erect; proximal cauline leaves linear to obovate, sometimes pinnately lobed, ultimate margins entire or dentate, faces glabrous; involucres hemispherical, 10–22+ × 6–22+ mm; phyllaries (25–)40–60+ in 4–6+ series, hyaline margins 1–2.5 mm wide; receptacles densely bristly; florets usually pale yellow, sometimes white, 8–12 mm; cypselae 1.6–3.2 mm. Sandy soil, desert scrub, 4100–4760 ft. Grant, Hidalgo.

Malacothrix fendleri A. Gray—Fendler's desert dandelion. Annuals, 3–15(–25+) cm; stems 3–8+, ± decumbent or spreading-ascending; proximal cauline leaves elliptic to oblong-oblanceolate, sometimes pinnately lobed (lobes 2–4+ pairs, oblong to triangular, unequal, apices acute), not fleshy, ultimate margins usually dentate, faces glabrous; involucres ± campanulate, 7–10 × 5–6+ mm; phyllaries 13–25+ in 2–3 series, hyaline margins 0.05–0.3 mm wide; receptacles not bristly; florets yellow, usually with red or purplish abaxial stripes, 6–14 mm; cypselae 1.8–2.4 mm. Desert scrub, grasslands, piñon-juniper woodlands, 3850–7700 ft. NW, C, WC, SC, SW.

Malacothrix glabrata (A. Gray ex D. C. Eaton) A. Gray [*M. californica* DC. var. *glabrata* A. Gray ex D. C. Eaton]—Smooth desert dandelion. Annuals, (5–)10–40+ cm; stems (1–)3–5+, ascending to erect; proximal cauline leaves usually pinnately lobed (lobes 3–6+ pairs, usually filiform), ultimate margins entire, faces glabrous or ± hairy; involucres campanulate to hemispherical, 9–17 × 4–7 mm; phyllaries 20–25+ in 2–3 series, hyaline margins 0.05–0.3 mm wide; receptacles bristly; florets usually pale yellow, sometimes white, 15–23+ mm; cypselae 2–3.3 mm. Desert scrub, 4600–5650 ft. Grant, Hidalgo, Luna.

Malacothrix sonchoides (Nutt.) Torr. & A. Gray [*M. runcinata* A. Nelson]—Sow-thistle desert dandelion. Annuals, (5–)10–25(–50) cm; stems 1–5, ascending to erect; proximal cauline leaves narrowly oblong to elliptic, pinnately lobed (lobes 3–8+ pairs, oblong to triangular), ultimate margins dentate to denticulate, faces glabrous; involucres ± campanulate to hemispherical, 7–13 × 4–6(–12+) mm; receptacles bristly; florets lemon-yellow, 10–14(–16) mm; cypselae 1.8–3 mm. Fine sand, dunes, arroyos, grasslands, 4600–7100 ft. NW, C, Grant, Hidalgo.

Malacothrix sonorae W. S. Davis & P. H. Raven—Sonoran desert dandelion. Annuals, 10–35 cm; stems 1(–9), erect; proximal cauline leaves narrowly oblanceolate to obovate, usually pinnately lobed (lobes oblong to triangular), not fleshy, ultimate margins ± dentate, faces glabrous; involucres ± campanulate, 6–9 × 4–6.6 mm; phyllaries 12–15+ in 2(3) series, hyaline margins 0.05–0.2 mm wide; receptacles not bristly; florets white or pale yellow, 6–10+ mm; cypselae 1.7–2 mm. Desert scrub and pine-oak communities, 4750–5400 ft. Grant, Luna.

Malacothrix stebbinsii W. S. Davis & P. H. Raven [*M. clevelandii* A. Gray var. *stebbinsii* (W. S. Davis & P. H. Raven) Cronquist]—Stebbins' desert dandelion. Annuals, 5–60 cm; stems 1–5+, erect; proximal cauline leaves obovate, pinnately lobed (lobes 3–5+ pairs, triangular to oblong or linear), not fleshy, ultimate margins entire or dentate, faces glabrous; involucres ± campanulate, 7–10 × 3–6 mm; phyllaries 16–20+ in 2–3 series, margins 0.05–0.1 mm wide, faces glabrous; receptacles not bristly; florets yellow, sometimes white, 6–7 mm; cypselae 1.7–2.3 mm. Gravelly soils, desert scrub and sagebrush communities, 4400–6500 ft. WC, SC.

MATRICARIA L. – Chamomile

Jennifer Ackerfield

Matricaria discoidea DC.—Pineapple weed. Annual herbs; plants (1-)4-40(-50) cm, sweet-scented, branching from base; leaves alternate, pinnatifid, to 6.5(-8.5) cm; heads discoid, usually solitary; involucres 2.5-3.8 mm, bracts subequal in (2)3-4(5) series, usually with hyaline-scarious margins; receptacles naked; ray florets absent; disk florets 1-2 mm, green-yellow; cypselae laterally compressed to subcylindrical, light brown, nerves white, with narrow brown glands extending to or at least near the bottom of the achene. Roadsides, disturbed places, 5480-9750 ft. NC, NW. Introduced (Ackerfield, 2015).

MELAMPODIUM L. – Blackfoot

Steven R. Perkins

Annuals, perennials, or subshrubs, 5-50(-150+) cm; stems erect or prostrate; leaves cauline, opposite, faces usually hairy, usually gland-dotted; heads radiate, borne singly; involucres mostly hemispherical; phyllaries persistent (outer) or falling with cypselae; 8-20+ in 2 series; receptacles flat or convex to conic; ray florets (3-)5-13+, pistillate, fertile; disk florets (3-)45-70(-100+), functionally staminate, lobes 5; pappus 0 (Allred & Ivey, 2012; Strother, 2006w).

1. Plants perennial; ray florets cream to white; disk florets 25-50...**M. leucanthum**
1'. Plants annual; ray florets yellow; disk florets usually 10 or fewer...(2)
2. Peduncles 4-30+ mm; ray florets 7-12; disk florets 8-10+...**M. longicorne**
2'. Peduncles 0-3(-11+) mm; ray florets 5-8; disk florets 5-8...**M. strigosum**

Melampodium leucanthum Torr. & A. Gray—Plains blackfoot. Perennials or subshrubs, 12-40 (-60) cm; leaves lanceolate, linear-oblong, or linear, pinnately lobed with 1-6 lobes; peduncles 3-7 cm; outer phyllaries 5; ray florets 8-13, corollas usually cream to white, sometimes purplish; disk florets 25-50; fruits 1.5-2.6 mm. Grasslands, piñon-juniper communities, desert scrub, (1400-)3200-7700(-9200) ft. Widespread except NW.

Melampodium longicorne A. Gray—Arizona blackfoot. Annuals, 10-60+ cm; leaves lance-elliptic or oblanceolate to lance-linear; peduncles 4-30+ mm; outer phyllaries 5; ray florets 7-12, corollas yellow; disk florets 8-10+; fruits 3-3.5+ mm. Mountain foothills and oak woodlands, 5000-6300 ft. Grant, Hidalgo.

Melampodium strigosum Stuessy—Shaggy blackfoot. Annuals, 5-35+ cm; leaves oblanceolate to oblong-linear; peduncles 0-3(-11+); outer phyllaries 5; ray florets 5-8, corollas yellow; disk florets 5-8+; fruits 2.2-3 mm. Woodlands and pine forest openings, 5600-6800 ft. Grant, Hidalgo, Otero.

NOTHOCALAIS (A. Gray) Greene – Whitehead

Gary I. Baird

Nothocalais cuspidata (Pursh) Greene [*Microseris cuspidata* (Pursh) Sch. Bip.]—Prairie whitehead. Perennial herbs, sap milky; stems lacking; leaves all basal, blades 7-30 cm × 3-20 mm, lance-linear, simple, pale, margins undulate, ciliate; heads liguliflorous, solitary, erect; peduncles scapose, ebracteate or with a subapical bract; involucres ovoid to campanulate; receptacles epaleate; phyllaries 10-30+ in 3-5 series; ray florets exceeding involucres, fertile, corolla yellow (outermost usually abaxially red-purple, this still evident when dried); disk florets lacking; cypselae 7-10 mm, monomorphic, cylindrical-fusiform, brown, glabrous or scaberulous distally, 10-ribbed, not beaked; pappus of 40-80 intergrading setiform scales and barbellulate bristles, subequal, white, persistent (Chambers, 2006a). Dry prairies and open piñon-juniper woodlands, 6550-8200 ft. Potential habitat in Colfax, Union.

ONOPORDUM L. – Scotch-thistle

Kenneth D. Heil

Onopordum acanthium L.—Scotch cotton thistle. Biennials, spiny herbs; plants to 30 dm; leaves 1–5 dm, lobed, decurrent-winged on the stems, dentate to shallowly lobed, the ultimate segments triangular, with spines 5–20 mm, tomentose-woolly; heads discoid; involucres tomentose, with spines 2–5 mm, bracts imbricate, spine-tipped; disk florets 20–25 mm, purple or pinkish white; ray florets absent; pappus 6–9 mm, of barbellate bristles; cypselae 4–5 mm. Disturbed places, roadsides, pipelines, pastures, 4035–7800 ft. NC, NW, EC, SE. Introduced (Keil, 2006g).

OREOCHRYSUM Rydb. – Goldenrod

Kenneth D. Heil

Oreochrysum parryi (A. Gray) Rydb. [*Haplopappus parryi* A. Gray]—Parry's goldenrod. Perennials, 15–60(–100) cm; rhizomatous; stems erect, usually simple, often purple proximally; leaves basal and cauline, alternate, lower leaves mostly 4–15 cm, oblanceolate to spatulate, 1-nerved, clasping or sub-clasping; heads radiate, distinctly flat-topped, tightly corymbiform; involucres 10–11 × 6–8 mm; phyllaries 15–24 in 3–4 series; ray florets 12–20, 6–10 mm, corollas yellow; disk florets 25–37, 7–9 mm; cypselae fusiform; pappus of capillary bristles. Moist to dry meadows, wooded slopes, often in partial shade, (5600–)6600–12,450 ft. Widespread except N (Nesom, 2006s).

OSTEOSPERMUM L. – Daisybush

Kenneth D. Heil

Osteospermum spinescens Thunb.—Daisybush. Shrubs, 10–30+ cm, stems becoming thorns; leaf blades cuneate to clavate, 3–12(–20+) × 1–4+ mm, margins entire or toothed, faces arachnose, glabrescent, often stipitate-glandular distally; phyllaries 5–8(–12+), 4–8 mm; ray florets 5–8(–13+), corollas yellow to orange; disk florets 12–20+, corollas yellow to orange; cypselae 8–10 mm, usually 3-winged. Known from a single plant that has persisted near Silver City, ca. 5900 ft. Grant (Strother, 2006x).

OXYTENIA Nutt. – Copperweed

Gary I. Baird

Oxytenia acerosa Nutt.—Copperweed. Perennials, becoming woody proximally; leaves cauline, simple, alternate, blades pinnatifid, 20–150 × 1–2 mm; heads ca. 2 mm; inflorescence in paniculiform clusters with 1–5 heads per node, nodding, sessile; involucres 2–3 mm; phyllaries ca. 10–15 in ± in overlapping series; pistillate flowers 5, corollas lacking, ovaries densely white-villous; staminate flowers 10–25, corollas whitish or pale yellow, gland-dotted; cypselae blackish, obovoid, ± obcompressed and angled; pappus lacking. Riparian communities, often in alkaline soils, 4400–4900 ft. Sandoval, San Juan (Baird, 2013).

PACKERA Á. Löve & D. Löve – Ragwort, groundsel

Debra K. Trock

Herbs, annuals, biennials, or perennials; leaves alternate, mostly basal; inflorescences of compact or open corymbs, cymes, or subumbelliform, rarely single, terminal; heads radiate or discoid; calyculi 0 or 1–5+ bractlets; involucres campanulate to cylindrical, phyllaries persistent; ray florets 0 or 5–13, pale yellow to orange-red; disk florets 20–80+, corollas pale yellow to orange-red; stamens 5; ovary inferior; cypselae tan to dark brown, cylindrical, usually 5–10-ribbed, glabrous or pubescent; pappus of white, barbellulate bristles (New Mexico Rare Plant Technical Council, 2020; Trock, 2006).

1. Plants usually glabrous or glabrate (at most arachnose to tomentose at bases of stems, in axils of leaves, and/or at bases of heads); basal leaves usually entire or dentate, seldom pinnately lobed, pinnatifid, or pinnatisect…(2)

1. Plants usually tomentose, sometimes glabrous, glabrate, or glabrescent and/or basal leaves pinnately lobed, pinnatifid, or pinnatisect…(12)
2. Plants 1-3 dm; heads usually 1-3; subalpine to alpine…**P. dimorphophylla**
2′. Plants 2-10 dm; heads (2-)6-20(-100+); never alpine…(3)
3. Bases of basal leaves (and proximal cauline) abruptly contracted or truncate to strongly cordate or sub-cordate…(4)
3′. Bases of basal leaves (and cauline) gradually tapering (sometimes obtuse)…(8)
4. Basal leaf blades broadly lanceolate to subhastate…**P. pseudaurea**
4′. Basal leaf blades cordate, obovate, orbiculate, ovate, or reniform…(5)
5. Basal leaf blades cordate or reniform…**P. cardamine**
5′. Basal leaf blades orbiculate, orbiculate-ovate, or obovate…(6)
6. Stems glabrous or sparsely hairy (at bases); peduncles ebracteate or inconspicuously bracteate, glabrous; calyculi 0 or inconspicuous…**P. crocata** (in part)
6′. Stems tomentose (at bases and in leaf axils); peduncles conspicuously bracteate, glabrous or sparsely tomentose; calyculi conspicuous…(7)
7. Plants stoloniferous; cauline leaves gradually reduced, clasping, margins pinnatisect or pinnately lobed…**P. obovata** (in part)
7′. Plants fibrous-rooted; cauline leaves gradually to abruptly reduced, not clasping, margins entire or sub-entire…**P. streptanthifolia** (in part)
8. Basal (and proximal cauline) leaf blades obovate, orbiculate, ovate, or suborbiculate, length usually 1-2 times width…(9)
8′. Basal (and proximal cauline) leaf blades elliptic, elliptic-ovate, lanceolate, lyrate, oblanceolate, orbiculate, ovate, or spatulate, length usually 2-3 times width…(10)
9. Stems and leaf axils usually glabrous, sometimes tomentose; heads 8-15+; peduncles conspicuously bracteate, glabrous or proximally tomentose; calyculi conspicuous…**P. obovata** (in part)
9′. Stems and leaf axils usually floccose-tomentose, sometimes glabrate; heads 3-9+; peduncles inconspicuously bracteate, loosely tomentose or glabrescent; calyculi 0…**P. hartiana**
10. Plants taprooted (caudices woody)…**P. tridenticulata**
10′. Plants fibrous-rooted, rhizomatous, or taprooted (caudices not woody)…(11)
11. Basal leaves relatively thick and turgid, blades oblanceolate, orbiculate, ovate, or spatulate; cauline leaves gradually or abruptly reduced, entire or subentire; calyculi conspicuous; disk florets 35-60+…**P. streptanthifolia** (in part)
11′. Basal leaves relatively thin, not turgid, blades lanceolate to narrowly elliptic; cauline leaves gradually reduced, margins dissected, incised, lacerate, or lobed; calyculi 0 or inconspicuous; disk florets 50-80+…**P. crocata** (in part)
12. Basal and cauline leaves usually pinnately lobed, pinnatifid, or pinnatisect (herbage usually glabrous or glabrate, sometimes persistently tomentose…(13)
12′. Basal leaves usually entire or ± dentate or pinnately lobed (herbage usually tomentose or glabrescent)…(19)
13. Stems and leaves usually persistently tomentose, sometimes glabrescent…(14)
13′. Stems and leaves usually glabrous, sometimes tomentose (mostly bases and leaf axils)…(16)
14. Basal (and proximal cauline) leaf blades broadly elliptic, obovate, orbiculate, ovate, or spatulate…**P. plattensis** (in part)
14′. Basal (and proximal cauline) leaf blades narrowly elliptic, lanceolate, lyrate, oblanceolate, or sublyrate…(15)
15. Stems and leaves tomentose; basal leaf margins evenly pinnatifid or pinnatisect; ray florets 6-8, corolla laminae 5-7 mm; disk florets 30-40+; cypselae glabrous…**P. fendleri** (in part)
15′. Stems densely tomentose at bases, otherwise sparsely tomentose; abaxial faces of basal leaves sometimes sparsely hairy, margins pinnately lobed to irregularly pinnatisect; ray florets 8-10, corolla laminae 9-10 mm; disk florets 60-70+; cypselae hirtellous on ribs…**P. plattensis** (in part)
16. Plants taprooted…(17)
16′. Plants rhizomatous or fibrous-rooted…**P. sanguisorboides**
17. Basal leaf margins pinnately lobed, lateral lobes 6+ pairs or lobing irregular, terminal lobes smaller than laterals…(18)
17′. Basal leaf margins pinnately lobed, lateral lobes 1-6 pairs, terminal lobes larger than laterals…**P. quercetorum**
18. Basal leaf blades narrowly lanceolate; peduncles inconspicuously bracteate or ebracteate; calyculi inconspicuous; pappus 3.5-4.5 mm [to be expected]…P. millelobata
18′. Basal leaf blades obovate, oblanceolate, or spatulate; peduncles bracteate; calyculi conspicuous; pappus 5-6 mm…**P. multilobata**
19. Plants 3-15 cm; heads 1-6(-15+)…(20)
19′. Plants 10-50+ cm; heads (1-)6-30+…(21)

20. Plants 3-5 cm; shortgrass prairie; NE New Mexico...**P. spellenbergii**
20'. Plants 4-15 cm; piñon-juniper woodlands and mixed conifer forests; NW New Mexico...**P. cliffordii**
21. Plants ± rhizomatous, rhizomes branched; stems canescent, floccose-tomentose, lanate, or lanate-tomentose...(22)
21'. Plants fibrous-rooted (caudices slender) or rhizomatous (rhizomes creeping); stems glabrous, glabrate, arachno-tomentose, sparsely lanate, or sparsely tomentose (then sometimes glabrescent)...(23)
22. Cauline leaves gradually reduced (blades similar to basal); cypselae 2.5-3.5 mm; plains and sagebrush communities...**P. cana** (in part)
22'. Cauline leaves abruptly reduced (bractlike); cypselae 1.5-2 mm; krummholz to alpine communities...**P. werneriifolia**
23. Plants taprooted (caudices often branched)...**P. neomexicana** (in part)
23'. Plants fibrous-rooted or rhizomatous...(24)
24. Basal leaf blades usually broadly oblanceolate to spatulate, sometimes lanceolate to elliptic (then cypselae 1-1.5 mm and hirtellous on ribs)...**P. streptanthifolia** (in part)
24'. Basal leaf blades elliptic, lanceolate, oblanceolate, or obovate (not spatulate)...(25)
25. Basal leaves relatively leathery, or thick and turgid; blades narrowly elliptic to lanceolate or oblanceolate, margins entire, subentire, or wavy (26)
25'. Basal leaves not leathery, not thick and turgid; blades elliptic, oblanceolate, orbiculate, ovate, or sublyrate, margins entire, crenate, dentate, serrate, pinnately lobed, or pinnatisect...(27)
26. Cauline leaves relatively leathery, gradually reduced (proximal and middle nearly as large as basal); heads 10-30+ in open or congested, cymiform arrays; peduncles 0 or ebracteate, densely tomentose (usually relatively short)...**P. cynthioides**
26'. Cauline leaves relatively thick and turgid, not leathery, abruptly or gradually reduced; heads 4-15+ in open, corymbiform, cymiform, or subumbelliform arrays; peduncles bracteate, sparsely tomentose or glabrate...**P. streptanthifolia** (in part)
27. Basal leaf blades orbiculate to broadly ovate...(28)
27'. Basal leaf blades narrowly elliptic, narrowly ovate, or oblanceolate...(29)
28. Basal leaf margins entire, weakly crenate, dentate, or lobed; phyllaries glabrous; cypselae glabrous...**P. streptanthifolia** (in part)
28'. Basal leaf margins dentate to pinnatisect; phyllaries densely tomentose proximally; cypselae usually hirtellous, sometimes glabrous...**P. plattensis** (in part)
29. Basal leaf margins crenate or serrate-dentate to pinnately lobed, pinnatifid, or pinnatisect...(30)
29'. Basal leaf margins entire, subentire, or dentate (toward apices)...(31)
30. Plants rhizomatous and/or fibrous-rooted, sometimes stoloniferous (mostly E populations); basal leaf margins dentate or pinnately lobed to pinnatisect; disk florets 60-70+; cypselae usually hirtellous, sometimes glabrous; pappus 6.5-7.5 mm...**P. plattensis** (in part)
30'. Plants rhizomatous (rhizomes branched, plants not fibrous-rooted or stoloniferous); basal leaf margins wavy or evenly pinnatifid to pinnatisect; disk florets 30-40+; cypselae glabrous; pappus 4-5 mm...**P. fendleri** (in part)
31. Stems densely lanate-tomentose or canescent; calyculi inconspicuous; cypselae glabrous...**P. cana** (in part)
31'. Stems lanate- or arachno-tomentose or glabrescent; calyculi conspicuous; cypselae usually hirtellous on ribs, sometimes glabrous...**P. neomexicana** (in part)

Packera cana (Hook.) W. A. Weber & Á. Löve [*Senecio canus* Hook.]—Woolly groundsel. Herbage densely lanate-pubescent or canescent; stems mostly 1-3 + dm, single and unbranched; basal leaves ovate or elliptic to lanceolate, margins entire, undulate or weakly dentate; cauline leaves reduced, sessile or weakly clasping, elliptic to lanceolate, margins entire to dentate; inflorescence a corymb of 8-15+ heads; heads radiate; phyllaries 13-21, green; ray florets 8, 10, or 13, laminae 8-10 mm; disk florets 35-50+; cypselae 2.5-3.5+ mm, glabrous. Open high plains, sagebrush associations, rocky dry slopes, crevices in granitic and limestone outcrops, 5875-9000 ft. Colfax, Grant, Harding, Sandoval.

Packera cardamine (Greene) W. A. Weber & Á. Löve [*Senecio cardamine* Greene]—Bittercress ragwort. Herbage glabrous; stems 2-6+ dm, single and unbranched; basal leaves orbiculate, ovate, obovate, or subreniform, margins crenate, dentate, or wavy; cauline leaves abruptly reduced, sessile and clasping (sometimes auriculate), broadly lanceolate to hastate, margins irregularly dentate, incised, or crenate; inflorescence a cyme of 3-8 heads; heads radiate; phyllaries 13, light green; ray florets usually 8 (rarely 0), laminae 8-11 mm; disk florets 30-45+; cypselae 1.5-2 mm, glabrous. Canyons, meadows, spruce forests, 8000-10,000 ft. In New Mexico, known only from the Mogollon Mountains. Catron. **R**

Packera cliffordii (N. D. Atwood & S. L. Welsh) O'Kane—Clifford's groundsel. Tufted perennial, 4-15 cm, forming clumps to 2 dm across, covered with white, felty or woolly tomentulose trichomes; basal leaves linear or linear-lanceolate, 2-4 cm × 1-3 mm, flat or with margins rolled inward; stem leaves reduced to small scales; flower stems taller than basal leaves and terminated by 1 or 3 heads; flower heads ca. 10 mm; involucral bracts in 1 or 2 series, green or becoming purplish at maturity; disk florets yellow; ray florets absent; cypselae sparsely short-hairy along the longitudinal angles. Limy mudstones and sandy soils in piñon-juniper woodlands to mixed conifer communities, 7380-7700 ft. McKinley. **R**

Packera crocata (Rydb.) W. A. Weber & Á. Löve [*Senecio crocatus* Rydb.; *S. pyrrhochrous* Greene]— Saffron ragwort. Herbage glabrous; stems 2-6+ dm, single and unbranched or rarely 2-3 clustered; basal leaves narrowly lanceolate or ovate to oblong-ovate, margins subentire to crenate-dentate; cauline leaves reduced, sessile and weakly clasping, lanceolate to oblong, margins sublyrate or lobed; inflorescence a corymb of 7-15+ heads; heads radiate; phyllaries 13 or 21, light green with reddish tips; ray florets 8 or 13, laminae 6-8+ mm, deep yellow to orange-red; disk florets 60-80+; cypselae 1-1.5 mm, glabrous. Wet meadows, along trails, on rocky outcrops at midelevations to subalpine, 8500-10,000 ft. Rio Arriba, Sandoval, Taos.

Packera cynthioides (Greene) W. A. Weber & Á. Löve [*Senecio cynthioides* Greene; *S. wrightii* Greenm.]—White Mountain ragwort. Herbage densely lanate-tomentose or canescent; stems 2-4+ dm, single and unbranched or rarely 2-3 clustered; basal leaves thick and leathery, narrowly lanceolate or oblanceolate, margins entire, subentire, dentate, or wavy; cauline leaves gradually reduced, sessile, lanceolate to oblanceolate, margins entire or wavy; inflorescence a congested cyme of 10-30+ heads; heads radiate; phyllaries 8 or 13, green, often with reddish tips; ray florets 5-8, laminae 8-10+ mm; disk florets 35-45+; cypselae 1-1.5 mm, glabrous. Loose, rocky soils on steep slopes in subalpine and pine-juniper forests, 6000-10,400 ft. C, WC, Eddy, Otero. This species has more well-developed cauline leaves and blooms later in the season than other groundsels in New Mexico.

Packera dimorphophylla (Greene) W. A. Weber & Á. Löve [*Senecio dimorphophyllus* Greene]— Splitleaf groundsel. Herbage glabrous; stems 1-3+ dm, single and unbranched or rarely 2-3 clustered; basal leaves thick and fleshy, ovate to oblong-lanceolate, margins entire, wavy, or crenate; cauline leaves gradually reduced, some as large as basal leaves, sessile, auriculate, distinctly clasping, subentire to irregularly dissected; inflorescence a congested corymb or subumbellate with 1-6+ heads; heads radiate; phyllaries (8-)13-21, green, often with reddish tips; ray florets 8 or 13, laminae 5-8+ mm, deep yellow to orange; disk florets 46-60+; cypselae 0.75-1.5 mm. New Mexico material belongs to **var. dimorphophylla**. Wet or drying open alpine and subalpine meadows or steep wet slopes, 7200-12,100 ft. Rio Arriba, Taos.

Packera fendleri (A. Gray) W. A. Weber & Á. Löve [*Senecio fendleri* A. Gray]—Fendler's ragwort. Herbage floccose-tomentose or glabrescent; stems 1-4+ dm, single and unbranched or in dense clumps; basal leaves lanceolate to oblanceolate; margins evenly pinnatifid to pinnatisect or wavy, lower surface sometimes glabrescent; cauline leaves gradually reduced, sessile, lanceolate to oblanceolate, margins pinnatisect to wavy; inflorescence an open or compact corymb with 6-25+ heads; heads radiate; phyllaries 13, green; ray florets 6-8+; laminae 5-7 mm; disk florets 30-40+; cypselae 2.5-3 mm, glabrous. Steep slopes in loose, dry, rocky or gravelly soils along streams, open forests, disturbed sites, 5175-11,800 ft. Widespread except EC, SE, SW.

Packera hartiana (A. Heller) W. A. Weber & Á. Löve [*Senecio hartianus* A. Heller; *S. quaerens* Greene]—Hart's ragwort. Herbage loosely tomentose, sometimes glabrate; stems 4-7 dm, single and unbranched or 2-3 clustered; basal leaves ovate to obovate or suborbiculate, margins serrate-dentate to weakly crenate; cauline leaves gradually reduced, lower ones petiolate, sublyrate, margins weakly pinnatisect, upper ones sessile and pinnatisect to subentire; inflorescence an open corymb with 3-9+ heads; heads radiate; phyllaries (13-)21, green; ray florets 8 or 13, laminae 5-8 mm; disk florets 50-65+; cypselae 1-2 mm, glabrous. Meadows or open areas, woodlands, along streams, 5000-10,135 ft. NC, NW, C, WC, SE, SC.

Packera millelobata (Rydb.) W. A. Weber & Á. Löve [*Senecio millelobatus* Rydb.]—Uinta ragwort. Herbage glabrous or sparsely tomentose in leaf axils; stems 3-5+ dm, single and unbranched or occasionally branched or 2-5+ loosely clustered; basal leaves narrowly lanceolate, margins pinnatifid with lateral lobes in 6+ pairs, terminal lobes smaller than lateral ones; cauline leaves gradually reduced, sessile and pinnatifid; inflorescence an open corymb with 6-20+ heads; heads radiate; phyllaries (13-)21, green with yellow tips; ray florets (8-)13, laminae 8-10+ mm; disk florets 40-55+; cypselae 1-1.5 mm, hirtellous on the ribs. Streambeds, openings in wooded areas on limestone-derived soils or on igneous-derived soils that are damp during the growing season. The only known collections of this species in New Mexico are the type specimens, which were collected from "Hills on the Limpia."

Packera multilobata (Torr. & A. Gray) W. A. Weber & Á. Löve [*Senecio multilobatus* Torr. & A. Gray]— Lobeleaf groundsel. Herbage glabrous or glabrescent, rarely sparsely tomentose throughout; stems 2-5+ dm, single and unbranched or 2-5 clustered; basal and proximal cauline leaves obovate, oblanceolate, spatulate, or lyrate to sublyrate; margins ± pinnately lobed, lateral lobes 3-6 pairs and smaller than terminal lobes; ultimate margins incised or dentate; cauline leaves gradually reduced, sessile; inflorescence an open corymb or subumbellate with 10-30+ heads; heads radiate; phyllaries 13-21, green (tips often yellow); ray florets 8-13, laminae 7-10 mm; disk florets 40-50+; cypselae 2-3 mm, glabrous or hirtellous along the ribs. Dry, rocky or sandy soils in sagebrush, woodlands, subalpine areas, 5000-10,500 ft. NW, SE, SC, Cibola, San Miguel.

Packera neomexicana (A. Gray) W. A. Weber & Á. Löve [*Senecio neomexicanus* A. Gray]—New Mexico groundsel. Herbage lanate to arachnoid-tomentose or glabrescent; stems 2-5+ dm, 1 or 2-5+ clustered, branched or unbranched; basal and lower cauline leaves petiolate, blades ovate, lanceolate, or narrowly lanceolate, margins subentire or denticulate to serrulate; cauline leaves abruptly or gradually reduced, upper leaves sessile, lanceolate, margins entire; inflorescence open or compact, corymbiform or subumbellate, with 3-20+ heads (subtended by smaller arrays arising from leaf axils); heads radiate; phyllaries 13 or 21, green or yellowish; ray florets (5-)8 or 13, laminae 4-10 mm; disk florets 40-60+; cypselae 1.5-2.5 mm, usually hirtellous along the ribs.

1. Plants loosely tomentose or glabrate; basal leaf blades narrowly lanceolate, margins subentire or irregularly dentate…**var. mutabilis**
1'. Plants usually densely lanate or woolly tomentose, seldom glabrate; basal leaf blades ovate or broadly lanceolate, margins dentate to deeply dentate or dissected…(2)
2. Plants taprooted; cauline leaves conspicuous…**var. neomexicana**
2'. Plants rhizomatous or fibrous rooted; cauline leaves inconspicuous…**var. toumeyi**

var. mutabilis (Greene) W. A. Weber & Á. Löve [*Senecio neomexicanus* (A. Gray) var. *mutabilis* (Greene) T. M. Barkley; *S. mutabilis* Greene; *S. neomexicanus* (A. Gray) var. *metcalfei* (Greene ex Wooton & Standl.) T. M. Barkley]—New Mexico groundsel. Herbage loosely tomentose or glabrate; stems 1-3; basal leaves narrow, bases tapering, margins subentire or irregularly dentate; cauline leaves conspicuous and abruptly reduced. Rocky, well-drained soils, meadows, coniferous woodlands, 5125-10,500 ft. NE, NC, NW, C, WC, SC.

var. neomexicana—New Mexico groundsel. Herbage usually densely lanate or woolly-tomentose, often glabrescent on lower leaf faces; stems 1; basal leaves ovate to obovate, margins deeply dentate or dissected; cauline leaves conspicuous. Rocky soils, oak-conifer or chaparral associations, 5000-9500 ft. NC, NW, C, WC, SC, SW.

var. toumeyi (Greene) Trock & T. M. Barkley [*Senecio neomexicanus* A. Gray var. *toumeyi* (Greene) T. M. Barkley, *S. toumeyi* Greene]—Toumey's groundsel. Herbage densely lanate or woolly-tomentose; stems 1, plants scapiform; basal leaves ovate to obovate, margins irregularly dentate, lower surface glabrate; cauline leaves inconspicuous and bractlike. Well-drained, rocky soils in coniferous woodlands, 7600-8500 ft. Known only from the Animas and Black Mountains. WC, SW.

Packera obovata (Muhl. ex Willd.) W. A. Weber & Á. Löve [*Senecio obovatus* Muhl. ex Willd.]— Roundleaf ragwort. Herbage glabrous, sometimes tomentose at the base and in leaf axils; stems 2-5+ dm, 1 or several loosely clustered; basal and lower cauline leaves petiolate, orbiculate, ovate, or obovate, margins crenate, dentate, or serrate; cauline leaves gradually reduced, sessile, clasping, margins pinnatisect or sublyrate; inflorescences open or congested corymbs of 6-15+ heads; heads radiate; phyllaries 13 or 21, green (rarely with reddish tips); ray florets 8-13(-21), laminae 7-10 mm; disk florets 40-50+; cypselae 1-1.5 mm, glabrous or rarely hirsute along the ribs. Meadows in deciduous woodlands, wet ditches, stream banks, rocky hillsides, 6800-7215 ft. Eddy.

Packera plattensis (Nutt.) W. A. Weber & Á. Löve [*Senecio plattensis* Nutt.]—Prairie groundsel. Herbage sparsely tomentose or glabrescent, floccose-tomentose at the base of stem and in leaf axils; stems 2-6+ dm, 1 or 2-3 clustered, unbranched; basal leaves narrowly elliptic to elliptic-ovate or oblanceolate to suborbiculate or sublyrate, margins subentire to crenate, serrate-dentate, or pinnately lobed; lower cauline leaves gradually reduced, petiolate, margins sublyrate or pinnatisect; upper cauline leaves sessile, margins subentire or irregularly dissected; inflorescence open or congested corymbs of 6-20+ heads; heads radiate; phyllaries 13 or 21, green (tips occasionally reddish); ray florets 8-10, laminae 9-10 mm; disk florets 60-70+; cypselae 1.5-2.5 mm, usually hirtellous. Prairies, meadows, open wooded areas, along highways and railroads, usually on limestone, 6450-8650 ft. Catron, Otero.

Packera pseudaurea (Rydb.) W. A. Weber & Á. Löve [*Senecio pseudaureus* Rydb. var. *flavulus* (Rydb.) Greenm.; *S. flavulus* Greene]—False-gold groundsel. Herbage glabrous or sparsely tomentose proximally and in leaf axils; stems 2-4 dm, slender and unbranched; basal leaves broadly ovate, margins bluntly dentate or denticulate, rarely sharply dentate; cauline leaves gradually reduced, sessile and sometimes clasping; inflorescence a congested cyme of 5-12 heads; heads radiate (rarely eradiate); phyllaries (13-)21(-30+), light green; ray florets 0, 8, or 13, laminae 6-10+ mm; disk florets 70-80+; cypselae 1-1.5 mm, glabrous. Our material belongs to **var. flavula** (Greene) Trock & T. M. Barkley. Damp soils, stream banks, woodlands, meadows, 5850-11,000 ft. NE, NC, NW, SC.

Packera quercetorum (Greene) C. Jeffrey [*Senecio quercetorum* Greene; *S. macropus* Greenm.]— Oak Creek ragwort. Herbage glabrous or tomentose at the base and in leaf axils; stems 6-10+ dm, 1 or 2-4 loosely clustered, deeply purple-tinged near the base, lightly tinged distally, unbranched; basal (and proximal cauline) leaves obovate or lyrate, pinnately lobed, lateral lobes in 2-6+ pairs, terminal lobes larger than laterals, midribs narrowly winged, ultimate margins sharply dentate, crenate-dentate, or irregularly incised; upper cauline leaves sessile or short-petiolate and shallowly lobed; inflorescence an open cyme of 15-40+ heads; heads radiate; phyllaries (13-)21, green with yellow tips; ray florets (8-)13, laminae 6-10+ mm; disk florets 60-70+; cypselae 1.5-2 mm, glabrous or slightly scabrellous. Rocky soils, open areas, scrub oak and piñon forests, chaparral, 5175-6875 ft. Catron, Grant.

Packera sanguisorboides (Rydb.) W. A. Weber & Á. Löve [*Senecio sanguisorboides* Rydb.]—Burnet ragwort. Herbage glabrous, tomentose in leaf axils; stems 1 or 2-3 clustered, unbranched; basal (and proximal cauline) leaves broadly oblanceolate, pinnately lobed, lateral lobes 2-3+ pairs, terminal lobes larger than laterals, lobes ovate to reniform, midribs not winged, ultimate margins crenate to dentate; cauline leaves sessile or short-petiolate, lyrate to sublyrate, midribs winged, terminal lobes weakly distinct, margins shallowly dentate; inflorescence a subumbel or compound cyme of 2-4+ clusters with 2-5+ heads each; heads radiate; phyllaries 13, bright green with lighter green or yellowish tips; ray florets 8, laminae 6-12 mm; disk florets 35-50+; cypselae 1.5-2 mm, glabrous. Damp, open meadows, spruce-aspen forests, 7800-12,500 ft. NC, C, Lincoln, Otero. **E**

Packera spellenbergii (T. M. Barkley) C. Jeffrey [*Senecio spellenbergii* T. M. Barkley]—Carrizo Creek ragwort. Herbage densely tomentose, becoming glabrate distally; stems 3-6+ cm, 1 (rarely 2), unbranched; basal leaves leathery, sessile, linear, margins entire, revolute; cauline leaves abruptly reduced, linear, bractlike; inflorescence of 1 (rarely 2) heads; heads radiate or eradiate; phyllaries 13, purple to deep reddish, 6-9+ mm; ray florets 0, 5, or 8, laminae 4-7 mm; disk florets 30-40+; cypselae 3-3.5 mm, hirtellous. Chalky, sandy limestone, shortgrass prairie and juniper savanna communities, 5400-5800 ft. Harding, Union. **R & E**

Packera streptanthifolia (Greene) W. A. Weber & Á. Löve [*Senecio streptanthifolius* Greene]— Rocky Mountain groundsel. Herbage glabrous or sometimes sparsely tomentose in leaf axils; stems 1-5+

dm, 1 or 2-5 clustered, unbranched; basal leaves thick and turgid, spatulate to oblanceolate, or ovate to orbiculate, margins entire, crenate, dentate, or weakly lobed; cauline leaves gradually or abruptly reduced, short-petiolate or sessile, margins entire; inflorescence a loose corymb of 2-20+ heads; heads radiate; phyllaries (8-)13 or 21, green (tips sometimes reddish); ray florets 8 or 13, laminae 5-10 mm; disk florets 35-60+; cypselae 1.5-2.5 mm, glabrous. Forests, open meadows, valleys, dry to damp and loamy soils, 6000-12,450 ft. NE, NC, WC.

Packera tridenticulata (Rydb.) W. A. Weber & Á. Löve [*Senecio tridenticulatus* Rydb.]—Threetooth ragwort. Herbage glabrous, rarely sparsely tomentose; stems 1-3+ cm, 1 or many clustered, unbranched; basal leaves relatively thick, lanceolate to narrowly oblanceolate, margins usually entire or rarely pinnatisect; cauline leaves gradually reduced, short-petiolate or sessile; inflorescence a corymb of 4-15+ heads; heads radiate; phyllaries 13 or 21, green; ray florets 8-10(-13), laminae 5-8+ mm; disk florets 45-60+; cypselae 1.5-2.5 mm, glabrous or sparsely hirtellous along the ribs. Open, dry areas, roadsides, gravelly or sandy slopes, shortgrass prairies, sagebrush scrub, 4600-10,200 ft. NE, NC.

Packera werneriifolia (A. Gray) W. A. Weber & Á. Löve [*Senecio werneriifolius* A. Gray; *S. saxosus* Klatt]—Hoary groundsel. Herbage tomentose, canescent, or glabrate; stems 7-15+ cm, 1 or 3-5 clustered, unbranched; basal leaves sessile or with very short petioles, blades narrowly lanceolate to elliptic, margins entire or dentate toward the apex, occasionally revolute; or in some populations leaf blades distinctly petiolate with blades ovate to orbicular; cauline leaves reduced to mere bracts and plants with a scapose aspect; inflorescence of 1-5 or rarely 8 heads; heads radiate; phyllaries 13 or 21, green or with reddish tips; ray florets 8 or 13, laminae 5-10 mm, or occasionally absent; disk florets 30-50+; cypselae 1.5-2 mm, glabrous. Rocky talus slopes, sandy soils in forest openings near or above timberline, 8000-11,850 ft. NC, WC, SC.

PALAFOXIA Lag. - Palafox

Steven R. Perkins

Annuals or perennials, 10-80(-150+) cm; stems erect, branched; leaves cauline, opposite or alternate; leaf blades with 1 or 3 nerves, broadly lanceolate to linear, margins entire; heads radiate, discoid, disciform, or radiate; involucres narrowly cylindrical or turbinate to hemispherical, 3-15+ mm diam.; phyllaries persistent or falling; ray florets 0 or 3-13, corollas white or pinkish to purplish; disk florets 12-40(-90), corollas pinkish to purplish or whitish, lobes 5; receptacles flat, pitted or knobby, epaleate; cypselae 4-angled, densely to sparsely hairy; pappus usually persistent, 4-10 scales in 1-2 series (Allred & Ivey, 2012; Strother, 2006y).

1. Ray florets absent; disk corollas 7-10 mm...**P. rosea**
1'. Ray florets present; disk corollas 10-14 mm...**P. sphacelata**

Palafoxia rosea (Bush) Cory [*Othake roseum* Bush; *Palafoxia rosea* var. *macrolepis* (Rydb.) B. L. Turner & M. I. Morris]—Rosy palafox. Annuals, 10-50 cm; stems scabrous to glabrate; leaf blades linear-lanceolate; involucres ± turbinate; ray florets 0; disk florets 5-30, corollas actinomorphic; cypselae 5-8 mm. Sandy soils and grasslands in E and S portions of New Mexico, 3300-5900 ft. NE, EC, SE, Hidalgo, Mora.

Palafoxia sphacelata (Nutt. ex Torr.) Cory [*Stevia sphacelata* Nutt. ex Torr.]—Othake. Annuals, 10-90 cm; stems usually proximally hispid, distally stipitate-glandular; leaf blades broadly to narrowly lanceolate; involucres broadly to narrowly turbinate; ray florets 3-5, corollas 15-25 mm; disk florets 15-35, corollas ± actinomorphic; cypselae 6-9 mm. Sandy soils in desert scrub, grassland, roadsides, 3300-7500 ft. Widespread except WC, SW.

PARTHENIUM L. - Feverfew

Steven R. Perkins

Annuals, biennials, perennials, subshrubs, or shrubs, 1-120(-400+) cm; stems ± erect, generally branched; leaves cauline or (sometimes) rosettes, alternate, faces usually hairy; heads radiate or (sometimes) ±

disciform; involucres ± hemispherical; phyllaries falling, 10(-16) in 2 series; receptacles flat to conic, scarious or membranous; pistillate florets 5(-8), fertile, corollas ochroleucous; disk florets 12-60+, functionally staminate, corollas ochroleucous, lobes 5; pappus 0 (Allred & Ivey, 2012; Strother, 2016z).

1. Plants short (1-2 cm), cespitose or forming mats; heads borne singly…**P. alpinum**
1'. Plants taller, not cespitose or forming mats; heads in clusters…(2)
2. Plants shrubs or woody…**P. incanum**
2'. Plants herbaceous…**P. confertum**

Parthenium alpinum (Nutt.) Torr. & A. Gray [*Bolophyta alpina* Nutt.; *P. tetraneuris* Barneby; *P. alpinum* var. *tetraneuris* (Barneby) Rollins]—Alpine feverfew. Perennials, 1-2 cm, cespitose or forming mats; leaves oblanceolate to spatulate, margins entire, faces strigillose-sericeous and obscurely gland-dotted; heads ± disciform, borne singly; outer phyllaries 5-8, ± linear, inner 5-8, ± orbiculate; pistillate florets 5-8, corolla laminae 0; disk florets 18-28+; cypselae oblanceoloid, 4 mm. Open, calcareous slopes and ridges, 5500-7700 ft. Harding, McKinley.

Parthenium confertum A. Gray [*P. confertum* var. *divaricatum* Rollins; *P. confertum* var. *microcephalum* Rollins]—Gray's feverfew. Biennials, 10-30(-60+) cm; leaves ovate or rounded-deltate to elliptic, faces strigillose with (usually) erect hairs 1-2 mm, gland-dotted; heads disciform or radiate in open ± paniculiform arrays; peduncles 2-8(-12+) mm; outer phyllaries 5, lance-ovate to elliptic, inner 5, ± orbiculate; pistillate florets 5; disk florets 20-30+; cypselae ± obovoid, 2-3 mm. New Mexico material belongs to **var. lyratum** (A. Gray) Rollins. Piñon-juniper woodlands, grasslands, desert scrub, canyon bottoms, 3000-6900(-8000) ft. C, SE, SC, SW, San Miguel.

Parthenium incanum Kunth—Mariola. Shrubs 30-100+ cm; leaves oval-elliptic to obovate, faces tomentose, gland-dotted; heads radiate, forming compound corymbiform arrays; peduncles 1-3+ mm; outer phyllaries 5, oblong, inner 5, orbiculate; pistillate florets 5, corolla laminae ovate; disk florets 8-20(-30+); cypselae obovoid, 1.5-2 mm. Often on limestone, rocky and sandy soils, piñon-juniper woodlands, desert scrub, 3200-7600 ft. NE, EC, C, SE, SC, SW.

PECTIS L. – Chinchweed

Lynn M. Moore

Annual or perennial herbs, often lemon- or spicy-scented; leaves simple, margins usually setose-ciliate near bases, faces glabrous or hairy, often dotted with oil glands, opposite; heads radiate, phyllaries in 1 series, usually distinct, bracts often with oil glands on margins, tips, and/or faces; ray florets yellow, often reddened abaxially; disk florets yellow, regular to slightly bilabiate; pappus of awns, bristles, or scales, or reduced to an irregular crown, that of ray florets differing from that of disk florets; receptacles naked; cypselae cylindrical, many-ribbed, hairy (Allred & Ivey, 2012; Keil, 2006h).

1. Ray florets 3-5…(2)
1'. Ray florets 8-15…(4)
2. Phyllaries distinct, spreading and falling individually, each with a ray achene…**P. filipes**
2'. Phyllaries coherent at least at bases, falling together enclosing all cypselae of a head…(3)
3. Ray florets 3(4)…**P. cylindrica**
3'. Ray florets 5…**P. prostrata**
4. Plants perennial; heads solitary, peduncles 3-16 cm…**P. longipes**
4'. Plants annual; heads in congested cymes, peduncles 1-4 cm…(5)
5. Pappus of disk florets 16-24 subplumose bristles; cypselae strigillose to short-pilose, hair tips curled and bulbous…**P. papposa**
5'. Pappus of disk florets a reduced crown and/or 0-7 awns or bristles; cypselae strigillose, hair tips straight and forked…**P. angustifolia**

Pectis angustifolia Torr.—Limoncillo, lemonscent, crownseed pectis. Plants scented annuals, 1-20 cm, stems hairy or glabrous; leaves 10-45 mm, margins with 2-5 pairs of setae; heads congested in cymes, peduncles 1-20 mm; phyllaries falling separately, 1 or 2 swollen oil glands at tip plus 2-5 pairs of smaller oil glands along margins; ray florets 8, 3-5(-7) mm, pappus reduced crown and/or 0-7 awns or bristles, 1-2 mm; cypselae 2.5-4 mm, strigillose, hair tips straight, forked.

1. Bases of upper leaves noticeably expanded, crowded toward tips of branches, often partly concealing peduncles; pappus usually a reduced crown (ours)...**var. angustifolia**
1'. Bases of upper leaves seldom expanded; pappus mostly 1-7 awns or bristles...**var. tenella**

var. angustifolia—Narrowleaf pectis. Plants lemon-scented, distal bases of leaves conspicuously flared, crowded toward tips of branches, often concealing peduncles; phyllaries linear, 2.5-5.5 mm, width ± uniform, pappus a reduced crown (ours). Piñon-juniper, shortgrass prairie, canyon bottoms, desert scrub, roadsides, loose sandy soils, 3325-8200 ft. Widespread.

var. tenella (DC.) D. J. Keil—Lemonscent. Plants strongly spicy-scented, distal bases of leaves not expanded, peduncles scarcely concealed by bases; phyllaries linear, 3-5 mm, widest near middle, pappus of 0-7 awns or bristles 1-3 mm and/or reduced crown. Deserts, shortgrass prairies, dry woodlands, roadsides, volcanics, sandy soils, 3000-3700 ft. Grant.

Pectis cylindrica (Fernald) Rydb.—Three-rayed or Sonoran chinchweed. Plants annuals, often mat-forming, 1-20 cm, stems hairy or glabrate; leaves 10-30 mm, margins with 2-5 pairs of setae; heads solitary or in congested cymes, peduncles 1-5 mm; phyllaries falling together, densely dotted with scattered, circular oil glands; ray florets 3(4), 3-4 mm, scarcely surpassing phyllaries, pappus of 2 (rays) or 5 (disks) lanceolate scales 1.5-3.5 mm; cypselae 4-5.5 mm, hairy, glandular hairy toward tip. Desert shrub, piñon-juniper woodlands, grasslands, roadsides, volcanics, 4330-6300 ft. SC, SW.

Pectis filipes Harv. & A. Gray—Threadstalk chinchweed. Plants annuals, strong-scented, 5-40 cm, stems hairy or glabrous; leaves 10-60 mm, margins with 1-4 pairs of setae; heads solitary or in cymes, peduncles 20-65 mm; phyllaries falling separately, 0-2 oil glands at tip plus 0-3 pairs of smaller oil glands along margins; ray florets 5, 4-9 mm, pappus of 0-3 awns, 3-4 mm, usually with a shorter crown; cypselae 2.5-4 mm, strigillose to short-pilose. Our plants are **var. subnuda** Fernald. Desert shrub, piñon-juniper woodlands, montane coniferous forests, volcanics, gravelly soils, 3200-6500 ft. WC, SC, SW.

Pectis longipes A. Gray—Longstalk chinchweed. Plants rhizomatous perennials, lemon- or spicy-scented, 8-25 cm, stems glabrous; leaves 10-15 mm, margins with 1-4 pairs of setae; heads solitary, peduncles (30-)50-160 mm; phyllaries falling separately, 1-3 swollen oil glands at tip plus 1-3 pairs of narrow oil glands along margins; ray florets (8-)13(-15), 8-12 mm, pappus of rays 1-2 awns, 3-3.5 mm, pappus of disks 2-30 unequal bristles, 3-5 mm; cypselae 2.5-4.5 mm, strigillose, hair tips acute or blunt. Grasslands, juniper-oak, ponderosa pine–juniper woodlands, roadsides, gravelly sandy soils, 5200-7500 ft. WC, SW.

Pectis papposa Harv. & A. Gray—Common chinchweed. Plants annuals, strong-scented, 1-30 cm, stems hairy or glabrous; leaves 10-60 mm, margins with 1-3 pairs of setae; heads congested in cymes, peduncles 3-40 mm; phyllaries falling separately, 1-5 oil glands at tip plus 2-5 pairs of oil glands along margins; ray florets (7)8(-10), 3-8 mm, pappus of rays a reduced crown, rarely 1+ awns or bristles, 1-4 mm, pappus of disks 16-24 subplumose bristles, 1.5-4 mm, rarely a crown; cypselae 2-5.5 mm, strigillose to short-pilose, hair tips curled, bulbous.

1. Phyllaries 3-5 mm; disk florets 6-15...**var. papposa**
1'. Phyllaries 5-8 mm; disk florets 12-24...**var. grandis**

var. grandis D. J. Keil—Common chinchweed. Peduncles 10-40 mm; phyllaries 5-8 mm; ray florets 5-8 mm; disk florets 12-24(-34); disk pappus 2.5-4 mm; cypselae 3-5.5 mm. Desert shrub, piñon-juniper woodlands, roadsides, coarse, loose, sandy soils, 3850-4940 ft. SC.

var. papposa—Common chinchweed. Peduncles 3-10(-25) mm; phyllaries 3-5 mm; ray florets 3-6 mm; disk florets 6-15(-18); disk pappus 1-2.5 mm; cypselae 2-4.5 mm. Deserts, grasslands, open areas, roadsides, 4000-5000 ft. C, WC, SE, SC, SW, San Juan.

Pectis prostrata Cav.—Spreading chinchweed. Plants annuals, often mat-forming, 1-30 cm, stems hairy; leaves 10-40 mm, margins with 4-12 pairs of setae; heads solitary or in congested cymes, peduncles 1-2 mm; phyllaries falling together, often dotted with oil glands along margins and sometimes along midribs; ray florets 5, 1–3 mm, pappus of 2 (rays) or 5 (disks) lanceolate scales 1.5–2.5 mm; cypselae 2.5–4.5 mm, strigillose. Deserts, grasslands, piñon-juniper and ponderosa pine woodlands, arroyos, canyon bottoms, bajadas, gravelly loose soils, 3975–6500 ft. WC, SC, SW.

PENTZIA Thunb. – African sheepbush

Kenneth D. Heil

Pentzia incana (Thunb.) Kuntze [*Chrysanthemum incanum* Thunb.]—African sheepbush. Shrubs, mostly 15–35 cm; stem branches ascending, finely appressed-tomentulose; leaves often fascicled, blades ± obovate to cuneate, 3–9 mm, 1(2) times divided, ultimate lobes linear to narrowly spatulate, revolute, inconspicuously appressed-puberulent to tomentulose or glabrate, minutely gland-dotted; heads discoid, single or in corymbiform arrays; involucres 3–4 × 4–5 mm; outer phyllaries with relatively narrow scarious margins; ray florets 0; disk florets 40–100, corollas ± goblet-shaped, yellow, 1.8–2+ mm; cypselae gray-brown, 1.2–1.8 mm. San Andres Mountains, Ropes Spring, 5675 ft. Doña Ana. Introduced (Keil, 2006i).

PERICOME A. Gray – Mountain tail-leaf

Kenneth D. Heil

Pericome caudata A. Gray—Mountain tail-leaf. Perennials, (20–)50–150+ cm; stems densely puberulent to tomentulose, sometimes gland-dotted; leaves mostly cauline, mostly opposite, mostly deltate-hastate, ovate, or cordate, margins entire, cleft, or toothed, usually gland-dotted; heads 3–30+, in corymbiform arrays; involucres 4–10 mm diam.; phyllaries 16–24 in 1 series; ray florets absent; disk florets 30–70, corollas yellow; cypselae flattened, 3–5 mm; pappus a crown of hyaline scales, sometimes with 1–2 bristles. Open, rocky hillsides, ponderosa pine and mixed conifer forests, (4250–)5575–10,700 ft. Widespread except EC, SE (Yarborough & Powell, 2006b).

PERITYLE Benth. – Rock daisy

Kenneth D. Heil

(Ours) perennial herbs or subshrubs; stems erect to pendent, 6–40 cm, glabrous, hairy, often gland-dotted; leaves mostly cauline, proximally opposite, distally alternate; blades usually 3-lobed, margins entire, toothed, or lobed; heads radiate or discoid, borne singly in corymbiform arrays; involucres glabrous or hairy; ray florets 0 or (1–)3–18; corollas cream, yellow, or white; disk florets 5–150, yellow; stamens 4; cypselae flatted to subcylindrical; pappus of (1–2)6–35 bristles (Yarborough & Powell, 2006c).

1. Pappus mostly of 25–35 bristles; leaves serrate-dentate, subentire, or shallowly lobed…(2)
1'. Pappus of 1–6 bristles; leaves 3–5-lobed or pinnately divided…(3)
2. Disk florets 20–150; heads borne singly (nodding or erect)…**P. cernua**
2'. Disk florets 5–6(–8); heads in corymbiform arrays…**P. quinqueflora**
3. Pappus of 2(3) bristles; subshrubs; widespread in SW New Mexico…**P. coronopifolia**
3'. Pappus of 2–3(–6) bristles; perennials or subshrubs; San Andres, Big Hatchet, or Peloncillo Mountains…(4)
4. Disk florets 20–40; ray florets 0; heads 1(2–3); phyllaries 10–16; Big Hatchet and Peloncillo Mountains…**P. lemmonii**
4'. Disk florets 40–50; ray florets 0 or 4–8; phyllaries 13–16(–22); San Andres Mountains…**P. staurophylla**

Perityle cernua (Greene) Shinners [*Laphamia cernua* Greene]—Organ Mountain rock daisy. Plants 3–12(–20) cm; leaf blades ovate-reniform to ovate-cordate, 10–40 × 8–40 mm, margins unevenly serrate-dentate; heads borne singly, nodding or erect, 10–12 × 12–14 mm; involucres broadly campanulate; phyllaries 18–28, 6–9 × 1.2–2 mm; disk florets 20–150, corollas yellow; cypselae 2.1–3 mm; pappus of 25–35 bristles. Cliffs of fractured rhyolite, 5000–8800 ft. Doña Ana. **R & E**

Perityle coronopifolia A. Gray [*Laphamia coronopifolia* (A. Gray) Hemsl.]—Crow-foot rock daisy. Subshrubs, 6–36 cm; leaf blades pedately 3-lobed, lobes spatulate or linear, or 2–3-pinnatifid, lobes linear-filiform, 4–30 × 4–20 mm, margins entire; heads 2–5, in corymbiform arrays, 5–6.5 × 5–6 mm; involucres campanulate; phyllaries 2–16, 3.5–5 × 0.5–1.5 mm; ray florets 8–12, corollas white, 3–7 × 2–3 mm; disk florets 30–40, corollas yellow, often purple-tinged; cypselae 1.8–2.5 mm; pappus of 2(3+) barbellulate bristles. Cliff faces, 4600–8800 ft. WC, SC, SW.

Perityle lemmonii (A. Gray) J. F. Macbr. [*Laphamia lemmonii* A. Gray]—Lemmon's rock daisy. Perennials or subshrubs, 6–17(–23) cm; leaf blades ovate to ovate-deltate or 3–5-lobed or pinnately divided, 6–18 × 7–20 mm; heads borne singly or (2–3) in corymbiform arrays, 7–10 × 5–9 mm; involucres campanulate; phyllaries 10–16, 4–6.4(–8) × 1–1.7 mm; ray florets 0; disk florets 20–40, corollas yellow, often tinged with purple; cypselae 2.5–3.2(–3.6) mm; pappus of 1(2) delicate bristles. Crevices of granitic boulders and cliffs, 5000–6000 ft. Hidalgo.

Perityle quinqueflora (Steyerm.) Shinners [*Laphamia quinqueflora* Steyerm.]—Five-flower rock daisy. Plants 7–30 cm; leaf blades reniform to cordate, 8–20(–33) × 8–20(–33) mm; heads in corymbiform arrays, 7–8(–9) × 2–3 mm; involucres cylindrical to narrowly campanulate; phyllaries 5–6(–9), (1–) 1.3–1.7 mm; disk florets 5–6(–8), corollas yellow; cypselae 1.9–2(–2.9) mm; pappus of 25–30 bristles. Crevices of limestone (rarely igneous) rock, canyons, bluffs, caprock, 5000–6000 ft. Eddy. **R**

Perityle staurophylla (Barneby) Shinners [*Laphamia staurophylla* Barneby]—New Mexico rock daisy. Perennials or subshrubs, 15–40 cm; leaf blades usually deeply divided, 5–30 × 7–25 mm, lobes 3, relatively broad to linear, usually secondarily lobed, cleft, or parted, sometimes almost entire; heads usually in corymbiform arrays, sometimes borne singly, 6–7.5 × 4.5–6 mm; involucres campanulate; phyllaries 13–16(–22), 3.4–4 × 0.7–1.1(–1.8) mm; ray florets 0, or 4–8, corollas yellow, 3.6–4.8 × 1.2–2.2 mm; disk florets 40–50, corollas yellow; cypselae 1.8–2.3 mm; pappus of 2–3(–6) bristles.

1. Leaf lobes usually linear to filiform; ray florets 0; N end of San Andres Mountains…**var. homoflora**
1'. Leaf lobes relatively broad to filiform; ray florets 4–8; not at N end of San Andres Mountains…**var. staurophylla**

var. homoflora Todsen—San Andres rock daisy. Leaf lobes usually linear to filiform; ray florets 0; San Andres Mountains, N of Rhodes Canyon. Crevices in limestone, 6400–7000 ft. Sierra, Socorro. **R & E**

var. staurophylla—New Mexico rock daisy. Leaf lobes relatively broad to filiform; ray florets 4–8; SC mountains. Crevices in limestone, 4900–7000 ft. Doña Ana, Sierra, Socorro. **R & E**

PETRADORIA Greene - Rock goldenrod

Kenneth D. Heil

Petradoria pumila (Nutt.) Greene—Rock goldenrod. Perennials, stems 1–20+; leaves basal and cauline, stiffly erect to ascending, alternate, blades linear to lanceolate or oblanceolate, 20–120 × 1–12 mm, faces glabrous or scabrous, gland-dotted, margins entire; heads radiate in densely crowded, corymbiform arrays; involucres 1–7 × 0.5–1+ mm; phyllaries 10–21, in 3–6 series; ray florets (1)2–3, 2.5–4.5 mm, corollas yellow; disk florets 4.5–6.2 mm, corollas yellow; cypselae 4–5 mm; pappus of capillary bristles (Urbatsch et al., 2006c).

1. Leaves usually 1(–3)-nerved, 1–2 mm wide; involucres 1.3–2 mm wide; ray florets usually 1, laminae 0.7–1.5 mm wide; disk florets 2–3…**var. graminea**
1'. Leaves usually 3–5-nerved, 2–12 mm wide; involucres 1.9–3 mm wide; ray florets usually (1)2–3, laminae 1–2.4 mm wide; disk florets 2–4…**var. pumila**

var. graminea (Wooton & Standl.) S. L. Welsh [*P. graminea* Wooton & Standl.]—Grass-leaved rock goldenrod. Stems 10–25 cm; leaf blades usually 1(–3)-nerved, linear to narrowly lanceolate, 30–90 × 1–2 mm; heads 10–25; involucres 5–8 × 1.3–2 mm; phyllaries 11–15; ray florets usually 1, corollas 5–7.5 mm; disk florets 2–3, corollas 4–5.8 mm. Rocky habitats, piñon-juniper woodlands, ponderosa pine forests, 6500–7875 ft. NC, NW.

var. pumila—Rock goldenrod. Stems 0.8–3 cm; leaf blades usually 3–5-nerved, broadly linear to oblanceolate, rarely lanceolate, 30–120 × 2–12 mm; heads 25+; involucres 6–9.5 × 1.9–3 mm; phyllaries

10-21; ray florets (1)2-3, corollas 4.5-9 mm; disk florets 2-4, corollas 4.5-6.2 mm. Dry, rocky places, 5200-8725 ft. NC, NW, Cibola.

PICRADENIOPSIS Rydb. - Bahia

Jennifer Ackerfield

Annual and perennial herbs; leaves opposite, palmately to pinnately dissected with the lobes often further divided; heads radiate; involucral bracts subequal in 2 ± distinct series, the outer keeled; receptacles naked; ray florets yellow; disk florets yellow; cypselae narrowly obpyramidal; pappus a crown of ca. 8 scales (Ackerfield, 2015; Ellison, 1964; Stuessy et al., 1973).

1. Annuals; ray florets absent...**P. multiflora**
1'. Perennials; ray florets present...(2)
2. Ray florets 9-15; disk florets 60-120; distal leaves opposite or alternate...**P. absinthifolia**
2'. Ray florets 3-8; disk florets 25-60; distal leaves opposite...(3)
3. Cypselae conspicuously gland-dotted; pappus scales ± ovate, with a short midrib becoming obscured upward and rarely exserted as a bristle...**P. oppositifolia**
3'. Cypselae hairy, not gland-dotted; pappus scales lanceolate, with a well-developed midrib that often extends into a prominent bristle at the tip...**P. woodhousei**

Picradeniopsis absinthifolia (Benth.) B. G. Baldwin [*Bahia dealbata* A. Gray]—Hairyseed bahia. Perennials, 10-40 cm; stems spreading to erect; leaves all or mostly opposite, lobes lanceolate to oblong, 10-25(-55+) × 2-5(-20+) mm, faces ± densely scabrello-canescent, usually gland-dotted; involucres 5-7+ × 9-14+ mm; ray florets 8-13+; corolla laminae 6-15+ mm; disk florets 60-80(-120+); corollas 3-4 mm; cypselae 3-4.5+ mm, faces hirtellous to ± strigose; pappus of ± spatulate to obovate, apically ± muticous scales 1-1.5 mm. Calcareous soils, mesquite and desert scrub, 3300-8800 ft. C, SE, SC, SW.

Picradeniopsis multiflora (Hook. & Arn.) B. G. Baldwin [*Bahia neomexicana* (A. Gray) A. Gray; *Schkuhria multiflora* Hook. & Arn.]—Many-flower false threadleaf. Plants mostly 3-12(-25+) cm; stems ± decumbent-ascending to erect; leaves mostly 1-3 cm; blades linear or lobed, faces puberulent and gland-dotted; involucres ± obconic, 5-6 mm; phyllaries 7-10+, green to purple, weakly ± hirsutulous and gland-dotted; ray florets 0; disk florets 15-30+, corollas yellowish, 1-2 mm; cypselae 3 mm; pappus of 8 white to tawny or purplish scales, 1-2 mm. Roadsides, sandy slopes, washes, piñon-juniper and ponderosa pine communities, 6500-8500 ft. Widespread except NE, EC, SE.

Picradeniopsis oppositifolia (Nutt.) Rydb. [*Bahia oppositifolia* (Nutt.) DC.]—Oppositeleaf bahia. Plants 0.3-2 dm; leaves linear to lanceolate, 1-3 cm, hairy and gland-dotted; involucres 5-7 mm; ray florets 3-5 mm; disk florets 3.5-5 mm; cypselae 3-5 mm, gland-dotted; pappus ovate with a short midrib becoming obscured upward and rarely exserted as a bristle, 0.5-1.5 mm. Common in sandy or gravelly soil of open places on the plains and outer foothills, 4200-9000 ft. NE, NC, EC, C, McKinley.

Picradeniopsis woodhousei (A. Gray) Rydb. [*Bahia woodhousei* (A. Gray) A. Gray]—Woodhouse's bahia. Plants 0.3-2 dm; leaves linear to lanceolate, 0.8-2.5 cm, hairy and gland-dotted; involucres 5-7 mm; ray florets 2-5 mm; disk florets 3-3.5 mm; cypselae 3-4 mm, usually hairy and not gland-dotted; pappus lanceolate or linear, with a well-developed midrib that is often extended into a prominent bristle. Sandy soils, disturbed sites, roadsides, open areas in piñon-juniper to ponderosa pine communities, montane meadows, 3850-9700 ft. NE, NC, NW, EC, WC, Otero.

PINAROPAPPUS Less. - Rocklettuce

Kenneth D. Heil

Perennials, 3-40 cm; stems glabrous; basal leaf blades linear to lanceolate, margins entire, toothed, or pinnately lobed (faces glabrous), cauline leaves foliaceous or reduced to minute bracts; heads borne singly; involucres cylindrical to campanulate, 3-20 mm diam.; phyllaries 18-22 in 3-5 series, margins scarious; florets mostly 20-40, corollas pink, purple, lavender, or nearly white; cypselae golden or yellowish brown, ribs 5-6 (Bogler, 2006e).

1. Plants 3–7 cm, forming dense mats; involucres cylindrical, 8–10 × 3–5 mm; phyllaries purplish, margins white…**P. parvus**
1′. Plants 10–30 cm, forming individual rosettes or clumps; involucres 10–15 × 12–20 mm; phyllaries pale green, margins pink…**P. roseus**

Pinaropappus parvus S. F. Blake—Small rocklettuce. Perennials, 3–7 cm, forming dense clumps and mats; stems 3–10+; leaf blades linear-oblanceolate, 2–5 cm × 1–3 mm; phyllaries purplish, margins white, 6–8 mm, apices purple to dull brown; florets 20–30, corollas pink, 6–8 mm; cypselae 4–5 mm. Rocky ledges, limestone cliffs, 5200–7500 ft. SE, SC.

Pinaropappus roseus (Less.) Less. [*Achyrophorus roseus* Less.]—White dandelion. Perennials, 10–40 cm, forming individual rosettes or clumps; stems 1–20+; leaf blades narrowly oblanceolate, 4–12 cm × 2–15 mm; phyllaries pale green, margins pink, 2–14 mm, apices dark brown; florets 20–40, corollas pale pink abaxially, white to yellow adaxially; cypselae 5–6 mm. Limestone areas, roadsides, cliffs, open grassy flats, 3150–5860 ft. Eddy, Otero.

PLATYSCHKUHRIA (A. Gray) Rydb. – Basin daisy

Jennifer Ackerfield

Platyschkuhria integrifolia (A. Gray) Rydb. [*Bahia nudicaulis* A. Gray]—Basin daisy. Perennial herbs; plants to 5 dm; leaves alternate or sometimes opposite above; lanceolate or sometimes ovate, 2–10 cm × 5–35 mm, margins entire, gland-dotted and sparsely to densely scabrous; heads radiate; involucres 9–12 × 12–30 mm, bracts equal or subequal in 2 series, ± herbaceous; receptacles naked; disk florets 3–7 mm, yellow; ray florets 6–15 mm, yellow; cypselae (1–)3–5(–8) mm, narrowly obpyramidal; pappus 0.5–4 mm, of 8–16 scales with lacerate margins and midribs sometimes excurrent as a short awn. New Mexico material is **var. oblongifolia** (A. Gray) W. L. Ellison. Shale slopes, clay soil, sandy mesas, 5000–7250 ft. NW (Ackerfield, 2015).

PLECTOCEPHALUS D. Don – Basketflower

Kenneth D. Heil

Annuals, 30–200 cm, not spiny; stems erect, branched; leaves basal and cauline, blade margins entire or dentate, faces puberulent, minutely glandular-punctate; heads radiate, borne singly or in open cymiform arrays; involucres ovoid to hemispherical or campanulate, 30–60 mm diam.; phyllaries many in 8–10+ series; receptacles flat, epaleate, bristly; corollas pink to purple; cypselae obovoid or ± barrel-shaped, ± compressed, weakly ribbed, glabrous or puberulent with 2-celled hairs; pappus readily falling, of 1–3 series of stiff, minutely barbed bristles. Previously in the genus *Centaurea* (Keil, 2006j).

1. Midphyllaries with 4–8 pairs of lobes, distally light to dark-stramineous…**P. americanus**
1′. Midphyllaries with 9–15 pairs of lobes, distally medium brown to dark brown…**P. rothrockii**

Plectocephalus americanus (Nutt.) D. Don [*Centaurea americana* Nutt.]—American basketflower. Plants 50–200 cm; stems usually 1, erect, glabrous, minutely scabrous and glandular; leaves scabrous; basal leaf blades oblanceolate to narrowly obovate, 10–20 cm, margins entire or sparingly denticulate; cauline leaves sessile, blades ovate to lanceolate, mostly 5–10 cm, entire or serrulate; phyllary apices with appendages erect to spreading, whitish to stramineous, fringed with 9–15 slender spinelike teeth; corollas of neutral florets pink-purple (rarely white), 35–50 mm; cypselae grayish brown to black, 4–5 mm, glabrous or with white hairs near bases; pappus bristles unequal, stiff, 6–14 mm. Prairies, fields, open woods, grasslands, roadsides, other disturbed sites, 3025–8100 ft. NE, EC, C, WC, SE, SC.

Plectocephalus rothrockii (Greenm.) D. J. N. Hind [*Centaurea rothrockii* Greenm.]—Rothrock's basketflower. Plants 30–150 cm; stems usually 1, erect, loosely tomentose, sometimes glabrate; leaves loosely tomentose, soon glabrate; basal leaf blades oblanceolate, margins entire or denticulate; cauline leaves sessile, often clasping, blades ovate to lanceolate or oblong, margins entire or serrulate to denticulate; phyllary appendages erect to spreading, ± brown, fringed with 15–25 slender, ciliate, spinelike teeth 2-4 mm; corollas of neutral florets pink to purple (rarely white), 30-70 mm; cypselae dark brown, ca. 4 mm, glabrous; pappus bristles unequal, stiff, 5-6 mm. Damp soil near streams, roadsides, open pine-oak woodlands and forests, 5415-8600 ft. WC.

PLUCHEA Cass. – Camphor-weed

Kenneth D. Heil

Annuals, perennials, or shrubs, (20-)150-300(-500) cm; leaves cauline, alternate, blades mostly elliptic, lanceolate, oblanceolate, obovate, or ovate, margins entire or dentate, faces glabrate to sparsely to densely silvery-hairy; heads disciform; involucres 4-6 mm diam.; phyllaries in 3-6+ series; peripheral (pistillate) flowers pinkish, lavender, purplish, or rosy; inner (functionally staminate) flowers 2-40+, corollas pinkish, lavender, or purplish (ours); cypselae oblong-cylindrical, ribs 4-8; pappus persistent barbellate bristles in 1 series (Nesom, 2006t).

1. Shrubs (ours); leaves and stems sericeous, not glandular…**P. sericea**
1′. Annuals or perennials; leaves and stems glandular…**P. odorata**

Pluchea odorata (L.) Cass. [*Conyza odorata* L.]—Sweetscent, shrubby camphor-weed. Annuals or perennials, 2-20 dm; stems glandular; leaves petiolate or sessile; blades lance-ovate to ovate, mostly 4-15 cm, faces glabrate to densely pubescent, margins shallowly serrate; involucres 5-6 mm; phyllaries cream; corollas pink to rosy or purple. New Mexico material belongs to **var. odorata**. Saline habitats, freshwater springs, moist drainages, 3320-4600 ft. EC, C, SE, SC.

Pluchea sericea (Nutt.) Coville [*Polypappus sericeus* Nutt.]—Arrowweed. Shrubs, not aromatic, 15-30(-50) dm; stems sericeous; leaves sessile, blades lanceolate or narrowly lanceolate, mostly 1-5 cm, faces silvery-sericeous, not glandular, margins entire; involucres 4-6 mm; phyllaries pink to purplish; corollas pink to purplish. Floodplains, stream banks, dry lake beds, dunes, 3000-5400 ft. C, SC.

POROPHYLLUM Guett. – Poreleaf

Steven R. Perkins

Annuals, biennials, perennials, subshrubs, or shrubs, 10-120(-200+) cm; strongly pungent-scented; stems erect, usually branched; leaves cauline, margins crenate or entire, faces usually glabrous, with scattered oil glands; heads discoid, borne singly or in ± corymbiform arrays; involucres cylindrical to campanulate; phyllaries 5-10 in ± 2 series; receptacles convex to conic; ray florets 0; disk florets (5-)10-80(-100+), corollas yellow, whitish, greenish, or purplish, lobes 5; pappus persistent, 25-50(-100) bristles in 1-2+ series (Allred & Ivey, 2012; Keil, 2012; Strother, 2006aa).

1. Plants herbaceous annuals; leaves elliptic to ovate or obovate…**P. ruderale**
1′. Plants perennial shrubs or subshrubs; leaves linear to filiform…(2)
2. Corollas whitish or purplish; phyllaries 5…**P. gracile**
2′. Corollas yellowish; phyllaries 7-10…**P. scoparium**

Porophyllum gracile Benth.—Slender poreleaf. Subshrubs or shrubs, 20-70 cm; internodes mostly 10-30 mm; leaves linear to filiform; phyllaries 5, oblong to linear, 9-15 mm; florets (5-)12-30, corollas whitish to purplish; cypselae 6-10 mm; pappus bristles 7-9 mm. Rocky slopes and flats, desert scrub, 4400-5600 ft. SC, SW, Lincoln.

Porophyllum ruderale (Jacq.) Cass.—Yerba porosa. Annual, usually 10-50 cm; stems glabrous, branches ascending; leaves 1-5 cm, narrowly elliptic to ovate or obovate; heads 1 to few, peduncle expanded below head; phyllaries 5, linear, dotted or streaked with glands; florets 30-50+, corollas ± purple or ± white; pappus 6-7 mm. New Mexico material belongs to **var. macrocephalum** (DC.) Cronquist [*Kleinia ruderalis* Jacq.]. Moist canyons, 4600-6000 ft. Hidalgo.

Porophyllum scoparium A. Gray—Trans-Pecos poreleaf. Subshrubs or shrubs, 20-60+ cm, internodes usually 10-30 mm; leaves filiform, 10-40 × 1-2 mm; phyllaries 7-10, oblanceolate to linear; florets usually 40–80+, corollas yellow; pappus bristles 6–7.5 mm. Desert scrub in rocky or gravelly sites, foothills, arroyos, limestone substrates, 4300–6000 ft. C, SE, Hidalgo.

PRENANTHELLA Rydb. – Brightwhite

Gary I. Baird

Prenanthella exigua (A. Gray) Rydb. [*Lygodesmia exigua* (A. Gray) A. Gray]—Brightwhite. Annual herbs, sap milky; stems 1–5+, much branched, 7–20 cm, stramineous and stiff proximally, flexuous distally; leaves basal and cauline, alternate, simple, sessile to petiolate, blades 1–3 cm, runcinate-oblanceolate, margins dentate to lobed, reduced distally to minute bracts; heads liguliflorous, numerous in dense paniculiform arrays, erect; involucres 4–5 mm, cylindrical-ovoid; phyllaries lance-linear, equal, green to purple (at least medially), faces glabrous to glandular-puberulent, margins scarious; florets 3–4, white or pinkish; cypselae 3–5 mm; pappus 2–3 mm. Sandy, rocky, or clay soils; sagebrush, shadscale, and piñon-juniper communities, 5300–5700 ft. Hidalgo, San Juan (Chambers, 2006b).

PSACALIUM Cass. – Indianbush

Kenneth D. Heil and Steve L. O'Kane, Jr.

Psacalium decompositum (A. Gray) H. Rob. & Brettell [*Odontotrichum decompositum* (A. Gray) Rydb.]—Desert Indianbush. Perennials, caudices thick, tough; stems densely hairy proximally, sparsely hairy to subglabrous distally; leaves mostly basal, ovate or elliptic, 20–40 cm, deeply pinnatisect; heads discoid, in corymbiform to paniculiform arrays; involucres cylindrical to weakly turbinate, 1.5–10+ mm diam.; receptacles flat, epaleate; phyllaries mostly 5–7 mm; corollas 7–9 mm, mostly white or ochroleucous; cypselae 4–5 mm, ± ellipsoid, ± compressed, 10–18-ribbed. Shady sites, open woodlands, 5635–8000 ft. Grant, Hidalgo (Barkley, 2006b).

PSATHYROTOPSIS Rydb. – Turtleback

Steven R. Perkins

Psathyrotopsis scaposa (A. Gray) H. Rob. [*Psathyrotes scaposa* A. Gray; *Pseudobartlettia scaposa* (A. Gray) Rydb.]—Naked turtleback. Perennial forbs 8–40 cm; stems 1–3+, erect; leaves mostly basal, alternate, petiolate, faces ± tomentose and stipitate-glandular; heads discoid, borne singly or in corymbiform arrays; involucres campanulate to hemispherical, 4–6(–10) mm diam.; phyllaries persistent, 8–21 in 2–3 series; ray florets 0; disk florets 16–50, corollas ochroleucous to bright yellow, lobes 5; receptacles flat, epaleate; cypselae narrowly obconic; pappus 70–90 unequal bristles in 2–3 series, 1–3 mm, persistent. Creosote bush communities in Doña Ana County, 3800–4000 ft. (Allred & Ivey, 2012; Strother, 2006bb).

PSEUDOCLAPPIA Rydb. – False clapdaisy

Steven R. Perkins

Pseudoclappia arenaria Rydb.—Trans-Pecos false clapdaisy. Subshrubs or shrubs, 5–40+ cm; stems mostly erect, branched from base; leaves cauline, mostly alternate, sessile, blades 10–35 mm, linear to filiform (terete), margins entire, faces glabrous; heads radiate, borne singly; involucres obconic, 4–6 mm diam.; phyllaries persistent, 12–16 in 2+ series; ray florets 2–4+, corollas yellowish, laminae 5–11 mm; disk florets 20–40+, corollas 6–7 mm, yellowish; receptacles convex, usually epaleate; cypselae ± columnar, ca. 3 mm; pappus ca. 50 bristles, 4–6 mm, 3–4 series. Often in gypseous soils, grasslands, alkaline seeps, lake edges, 3300–6600 ft. C, SE, SC, Sandoval (Allred & Ivey, 2012; Strother, 2006cc).

PSEUDOGNAPHALIUM Kirp. – Rabbit-tobacco

Kenneth D. Heil

Annuals, biennials, or perennials, (ours) 15–90(–100); stems 1+, ± woolly-tomentose; leaves basal and cauline, alternate, usually sessile, blades mostly narrowly lanceolate to oblanceolate, margins entire, faces bicolored or concolored, tomentose to velutinous; heads disciform, usually in glomerules in cor-ymbiform or paniculiform arrays; involucres mostly campanulate; phyllaries in (2-)3-7(-10) series; pistil-late flowers (15-)25-200, corollas yellowish; bisexual flowers 5-30, corollas yellowish; pappus of 10-12 barbellate bristles (Nesom, 2006u).

1. Leaf faces strongly to weakly bicolored (abaxial gray to white, tomentose, adaxial green, not tomentose, sometimes glandular)…(2)
1.' Leaf faces concolored or weakly bicolored (both usually gray to gray-green or greenish, tomentose, adaxial sometimes glandular beneath tomentum)…(4)
2. Leaves crowded, blades linear to linear-lanceolate, margins strongly revolute…**P. leucocephalum**
2.' Leaves not crowded, blades lanceolate to oblanceolate or spatulate, margins flat or slightly revolute…(3)
3. Involucres 4.5-5.5 mm; phyllaries in 4-5 series; bisexual flowers 7-12…**P. macounii**
3.' Involucres 3.5-4 mm; phyllaries in 2-3 series; bisexual flowers (1)2-6…**P. pringlei**
4. Leaf bases clasping to subclasping…(5)
4.' Leaf bases not clasping or subclasping…(6)
5. Involucres 3-4 mm; bisexual flowers 5-10, corollas red-tipped; cypselae with papilliform hairs…**P. luteoalbum**
5.' Involucres 4-6 mm; bisexual flowers mostly 18-28, corollas evenly yellowish, not red-tipped; cypselae gla-brous…**P. stramineum**
6. Leaf bases not decurrent…**P. canescens**
6.' Leaf bases decurrent…**P. jaliscense**

Pseudognaphalium canescens (DC.) Anderb. [*Gnaphalium canescens* DC.; *G. wrightii* A. Gray]—Wright's rabbit-tobacco. Annuals or perennials, 20-70(-100+) cm; stems tomentose; leaf blades narrowly to broadly oblanceolate, mostly 2-4(-5) cm × 2-8(-15) mm, bases not clasping, not decurrent, margins flat, tomentose; heads in loose corymbiform arrays; involucres 4-5 mm; phyllaries in 3-4 series; pistil-late flowers (16-)24-44; bisexual flowers (1)2-5(6). Rocky sites, grasslands, oak-pine-juniper woodlands, ponderosa pine forests, 3800-7500(-8800) ft. NE, C, WC, SE, SC, SW.

Pseudognaphalium jaliscense (Greenm.) Anderb. [*Gnaphalium jaliscense* Greenm.]—Jalisco rabbit-tobacco. Annuals or biennials, 30-70 cm; stems densely and persistently loosely woolly-tomentose-sericeous, not glandular; leaf blades narrowly lanceolate to nearly linear, 3-10 cm × 3-6 mm, bases not clasping, decurrent, margins flat or slightly revolute, faces concolored, tomentose-sericeous, sessile-glandular beneath tomentum; heads in corymbiform arrays; involucres 5-6 mm; phyllaries in 5-6(7) series; pistillate flowers (80-)115(-180); bisexual flowers (6-)8-12(-30). Disturbed sites, openings in oak-pine-juniper and ponderosa pine forests, 6000-7400 ft. NE, Grant, Hidalgo, Lincoln.

Pseudognaphalium leucocephalum (A. Gray) Anderb. [*Gnaphalium leucocephalum* A. Gray]—White rabbit-tobacco. Biennials or short-lived perennials, 30-60 cm; stems densely and persistently white-tomentose, usually with stipitate-glandular hairs; leaf blades linear-lanceolate, 3-7 cm × 1-5(-6) mm, bases subclasping, not decurrent, margins strongly revolute, faces bicolored, abaxial densely white-tomentose, adaxial green, stipitate-glandular; heads in corymbiform arrays; involucres broadly campanulate, 5-6 mm; phyllaries in 5-7 series; pistillate flowers 66-85; bisexual flowers mostly 6-14. Sandy and gravelly sites, stream bottoms, arroyos, riparian communities, oak-pine woodlands, (4260-) 6000-8500 ft. Catron, Doña Ana, Hidalgo.

Pseudognaphalium luteoalbum (L.) Hilliard & B. L. Burtt [*Gnaphalium luteoalbum* L.]—Jersey cudweed. Annuals, 15-40 cm; stems white-tomentose, not glandular; leaf blades narrowly obovate to subspatulate, 1-3(-6) cm × 2-8 mm, bases subclasping, usually decurrent, margins weakly revolute, faces mostly concolored to weakly bicolored, abaxial gray-tomentose, adaxial usually gray-tomentose, sometimes glabrescent, neither glandular; heads in terminal glomerules; involucres 1-3(-6) cm × 2-8 mm; phyllaries in 3-4 series; pistillate flowers 135-160; bisexual flowers 5-10. Disturbed sites, fields, pastures, arroyos, stream banks, gardens, 3850-6580 ft. NE, C, SC, SW.

Pseudognaphalium macounii (Greene) Kartesz [*Gnaphalium macounii* Greene]—Macoun's cud-weed. Annuals or biennials, often sweetly fragrant, 40-90 cm; stems stipitate-glandular, usually white-tomentose; leaf blades lanceolate to oblanceolate, 3-10 cm × 3-13 mm, bases not clasping, decurrent,

5-10 mm, margins flat to slightly revolute, faces weakly bicolored, abaxial tomentose, adaxial stipitate-glandular, otherwise glabrescent or glabrous; heads in corymbiform arrays; involucres 4.5-5.5 mm; phyllaries in 4-5 series; pistillate flowers 47-101(-156); bisexual flowers 5-12(-21). Dry, open sites, piñon-juniper woodlands, ponderosa pine and mixed conifer forests, (5250-)6800-10,400 ft. Widespread except E, SE, SW.

Pseudognaphalium pringlei (A. Gray) Anderb. [*Gnaphalium pringlei* A. Gray]—Pringle's rabbit-tobacco. Annuals or perennials, 30-80 cm; stems lightly white-tomentose and/or glabrescent; minutely glandular beneath indument; leaf blades oblanceolate-spatulate to obovate, 5-10 cm × 10-20 mm, margins slightly revolute, faces bicolored, abaxial thinly white-tomentose, adaxial minutely glandular; heads in loose, corymbiform arrays; involucres 3.5-4 mm; phyllaries in 2-3 series; pistillate flowers 15-40(-64); bisexual flowers (1)2-6. Rocky sites, outcrops, slopes, crevices and thin soil on cliffs, oak and oak-pine woodlands, 6075-7550 ft. WC, SW.

Pseudognaphalium stramineum (Kunth) Anderb. [*Gnaphalium stramineum* Kunth; *G. chilense* Spreng.]—Cotton-batting-plant. Annuals or biennials, 30-60(-80) cm; stems loosely tomentose, not glandular; leaf blades oblong to narrowly oblanceolate or subspatulate, 2-8(-9.5) cm × 2-5(-10) mm, bases subclasping, usually not decurrent, margins flat or slightly revolute, faces concolored, not glandular; heads in terminal glomerules; involucres 4-6 mm; phyllaries in 4-5 series; pistillate florets 160-200; bisexual florets (8-)18-28. Disturbed sites, fields, streamsides, washes, dunes, roadsides, 3800-9500 ft. Widespread.

PSILACTIS A. Gray – Tansy-aster

Kenneth D. Heil

Annuals or biennials, 15-150 cm; stems erect or weakly ascending, simple, glabrate to densely hairy and glandular; leaves basal and cauline, basal blades 1-nerved, obovate to linear-oblanceolate, proximal cauline blades lanceolate, elliptic, or obovate to linear-oblanceolate, margins entire, coarsely toothed or pinnately lobed, faces appressed-hairy, distal blades ovate to lanceolate to linear, smaller, margins entire, faces stipitate-glandular and/or appressed-hairy; heads radiate, borne in loosely corymbiform arrays; involucres 2-9 × 4-18 mm; phyllaries 30-45 in 2-3 series, herbaceous, margins scarious; ray florets 10-70, pistillate, fertile, corollas white to blue or purple; disk florets 15-150, bisexual, corollas yellow, lobes often purple (Morgan, 2006b).

1. Involucres 2-4 mm; ray laminae 1-4 mm...**P. brevilingulata**
1'. Involucres 4-8 mm; ray laminae 4-15 mm...**P. asteroides**

Psilactis asteroides A. Gray [*Machaeranthera boltoniae* (Greene) B. L. Turner & D. B. Horne]—New Mexico tansy-aster. Annuals, 25-150 cm; stems and branches often sparsely appressed-hairy, stipitate-glandular distally, occasionally proximally; distal leaf blades lanceolate to narrowly elliptic or linear, smallest 2.5-5 × 0.7-1.5 mm, ray florets 20-40, laminae 5-9 × 0.5-1 mm; disk florets 40-75, corollas 2-3.5 mm; ray cypselae 1-1.8 mm, sparsely appressed-hairy; disk cypselae 1.2-2.2 mm, sparsely appressed-hairy. Wet or occasionally flooded habitats, stream banks, lakeshores, ditches, fields, disturbed areas, 3900-8180 ft. NW, NC, SC, SW.

Psilactis brevilingulata Sch. Bip. ex Hemsl. [*Machaeranthera brevilingulata* (Sch. Bip. ex Hemsl.) B. L. Turner & D. B. Horne]—Trans-Pecos tansy-aster. Annuals, 15-75 cm; stems and branches stipitate-glandular; distal leaf blades sessile, lanceolate to linear-lanceolate, smallest 2-10 × 0.5-1.5 mm; ray florets 15-40, laminae 1-4 × 0.2-0.5 mm; disk florets 15-35, corollas 1.5-2.5 mm; ray cypselae 1-1.5 mm, appressed-hairy; disk cypselae 1.3-2 mm, appressed-hairy; 1 collection from the Hachita Valley, Chihuahuan desert scrub, 4350 ft. Hidalgo.

PSILOSTROPHE DC. – Paperflower

Lynn M. Moore

Biennials, perennials, subshrubs, or shrubs; leaves simple, basal and/or cauline, alternate; heads radiate in corymbs, clusters, or solitary, involucres cylindrical to campanulate or obconic, 2–7 mm diam., phyllaries persistent, 5–12 in 1–2 series; receptacles naked; ray florets 1–8, yellow to orange, marcescent, spreading or reflexed in fruit; disk florets 5–25+, yellow to orange; cypselae striate-ribbed, usually glabrous, sometimes gland-dotted; pappus of 4–8 entire scales, 2–3 mm (Allred et al., 2020; Strother, 2006dd).

1. Plants shrubs or subshrubs; stems white; heads solitary…**P. cooperi**
1'. Plants herbaceous biennials or perennials; stems gray to gray-green; heads in corymbs…(2)
2. Stems with mostly appressed hairs, greenish; rays reflexed in fruit…**P. sparsiflora**
2'. Stems with cobwebby-villous hairs, gray to gray-green; rays spreading horizontal in fruit…(3)
3. Peduncles of flowering heads mostly 5–40 mm; rays 5–14 mm…**P. tagetina**
3'. Peduncles of flowering heads mostly 1–5 mm; rays 3–6 mm…**P. villosa**

Psilostrophe cooperi (A. Gray) Greene—White-stem paperflower. Plants shrubs or subshrubs, (15–)25–30(–50+) cm, stems pannose, white; peduncles 35–60(–80+) mm; heads solitary; ray florets 3–6, (8–)12–20 mm, spreading to reflexed in fruit, disk florets (6–)10–17(–25); cypselae usually glabrous, sometimes gland-dotted. Oak-juniper woodlands, roadsides, open rocky areas, 4000–6400 ft. NE, WC, SW, Chaves.

Psilostrophe sparsiflora (A. Gray) A. Nelson—Green-stem paperflower. Plants biennial or perennial herbs, 10–40+ cm, stems ± strigillose, greenish; peduncles 8–12(–25+) mm; heads in corymbs; ray florets (1)2–3(4), (6–)8–10+ mm, reflexed in fruit, disk florets 7–9; cypselae usually glabrous, sometimes gland-dotted. Piñon-juniper, ponderosa pine woodlands, desert grasslands, arroyos, canyon bottoms, sandy loose soils, 4000–7800 ft. NW, C, WC, SC, SW, Roosevelt.

Psilostrophe tagetina (Nutt.) Greene—Woolly paperflower. Plants biennial or perennial herbs, 10–30(–60+) cm, stems cobwebby-villous, gray to gray-green; peduncles (3–)12–20(–40) mm; heads in corymbs; ray florets 3–4(–6), (3–)7–14 mm, spreading in fruit, disk florets 6–9(–12); cypselae usually glabrous, sometimes hairy and/or gland-dotted. Piñon-juniper, ponderosa pine woodlands, shortgrass prairie, roadsides, sandy, gypseous, or limestone soils, 3100–7400 ft. Widespread.

Psilostrophe villosa Rydb. ex Britton—Woolly paperflower. Plants biennial or perennial herbs, 10–45(–60+) cm, stems cobwebby-villous, gray to gray-green; peduncles (0.5–)1–3(–5+) mm; heads in corymbs; ray florets (2)3(–5), 3–4(–6) mm, spreading in fruit, disk florets 5–8(–12); cypselae usually glabrous, sometimes gland-dotted. Grasslands, alkaline, gypseous soils, open disturbed areas, roadsides, 3890–6440 ft. EC, C, SE, SC.

PYRRHOPAPPUS DC. – Desert-chicory

Steven R. Perkins

Annuals or perennials, 5–100+ cm, taprooted or rhizomatous; stems 1(2–5+), erect; leaves basal or basal and cauline, basal leaves ± petiolate, distal leaves usually sessile; heads single or in loose, corymbiform arrays; peduncles not inflated distally, sometimes bracteate; involucres cylindrical, 4–5(–8+) mm diam.; phyllaries 8–21+ in ± 2 series; florets (20–)30–150+, corollas yellow to whitish; cypselae reddish brown to stramineous, grooves (or broad ribs) 5; pappus persistent (Allred & Ivey, 2012; Strother, 2006ee).

1. Plants perennials; roots with tuberiform swellings 1–15 cm below soil surface; stems usually scapiform; most or all leaves basal; anthers 4.5–5 mm [likely present but no specimens seen]…P. grandiflorus
1'. Plants annuals or perennials; roots lacking tuberiform swellings; stems rarely scapiform; cauline leaves usually 3–9+; anthers 2.5–4 mm…(2)
2. Plants annuals; stems usually sparsely to densely pilosulous proximally, sometimes glabrous; distal cauline leaves usually (3–)5–7(–9+) times pinnately lobed…**P. pauciflorus**
2'. Plants annuals or perennials; stems usually glabrous proximally, sometimes pilosulous; blades of distal cauline leaves usually entire or with 1–2 lobes near bases; sometimes 3–5(–7+) times pinnately lobed…(3)

3. Plants perennials; involucres 12-15(-20+) mm; florets (20-)30-60; cypselae stramineous; pappus 6-7 mm...**P. rothrockii**
3'. Plants annuals; involucres 17-24+ mm; florets 50-150+; cypselae reddish brown; pappus 7-10 mm...**P. carolinianus**

Pyrrhopappus carolinianus (Walter) DC.—Carolina desert-chicory. Annuals, (5-)20-50(-100+) cm; stems usually branching from base and/or distally, usually glabrous proximally; cauline leaves (1-)3-9+, proximal mostly lanceolate, margins usually dentate, distal margins entire or with 1-2 lobes near base; heads (1-)3-5+ in loose corymbiform arrays; involucres ± cylindrical to campanulate, 17-24+ mm; phyllaries 16-21+; florets 50-150+; cypselae bodies reddish brown, 4-6 mm, beaks 8-10 mm; pappus 7-10+ mm. Sandy soils, disturbed sites, 4900-5900 ft. The last documented collection in New Mexico was in 1935. Bernalillo, Doña Ana.

Pyrrhopappus grandiflorus (Nutt.) Nutt. [*Barkhausia grandiflora* Nutt.]—Tuberous desert-chicory. Occasionally reported in New Mexico, but no valid specimens have been documented.

Pyrrhopappus pauciflorus (D. Don) DC. [*Chondrilla pauciflora* D. Don; *Pyrrhopappus geiseri* Shinners]—Smallflower desert-chicory. Annuals, 5-40(-80+) cm; stems branching from base and/or distally, usually pilosulous proximally; cauline leaves 1-3(-5+), proximal mostly oblanceolate to lanceolate, margins usually pinnately lobed; heads (1-)3-7+ in loose corymbiform arrays; involucres ± campanulate to cylindrical, 16-22 mm; phyllaries 13-21; florets 50-60; cypselae bodies reddish brown, 4-5 mm, beaks 7-9 mm; pappus 7-10 mm. Often associated with disturbed sites, prairies, and clay soils, 2900-7400 ft. Scattered.

Pyrrhopappus rothrockii A. Gray—Smallflower desert-chicory. Perennials, 15-40 cm; stems branching from base and/or distally; cauline leaves (1-)3-9+, proximal mostly spatulate or oblanceolate to linear; heads (1-)3-5+ in loose corymbiform arrays; involucres ± cylindrical, 12-15(-20+) mm; phyllaries 13-16; florets (20-)30-60+; cypselae bodies stramineous, 3-4 mm, beaks 6-7 mm; pappus 6-7 mm. Associated with meadows, stream banks, and floodplains, 3800-7000 ft. NC, WC, SC. Some taxonomists include *P. rothrockii* within *P. pauciflorus*.

PYRROCOMA Hook. – Goldenweed

Jennifer Ackerfield

Pyrrocoma crocea (A. Gray) Greene [*Haplopappus croceus* A. Gray]—Curlyhead goldenweed. Perennial herbs; plants 1-8 dm; leaves chiefly basal, simple, oblanceolate to narrowly elliptic, 8-45 cm, entire, lacking cilia, glabrous or rarely sparsely hairy; heads radiate, usually solitary; involucres 10-20 mm, the bracts large and herbaceous, nearly equal or subequal in 2 to several series; receptacles naked; disk florets 7-13 mm, yellow; ray florets 9-35 mm, yellow; pappus of capillary bristles, generally rigid, unequal, tawny, 6-12 mm; cypselae compressed, 5-8 mm, 3-4-angled. New Mexico material belongs to **var. crocea**. Ponderosa pine, mixed conifer, and subalpine forests, (6200-)7400-11,350 ft. NC, Catron (Ackerfield, 2015).

RAFINESQUIA Nutt. – Rafinesque's chicory

Kenneth D. Heil

Rafinesquia neomexicana A. Gray—Desert chicory. Annuals, 15-150 cm; plants 15-60 cm; stems 1-3, erect, glabrous; leaves basal and cauline, blades oblong to oblanceolate, pinnately lobed (lobes broad or narrow); cauline sessile, sometimes auriculate-clasping; heads borne singly or in open, paniculiform arrays; involucres cylindroconic, (4-)6-15+ mm diam.; calyculi of 8-14, spreading to reflexed, unequal bractlets; phyllaries 7-20 in 1 series; receptacles flat, smooth, glabrous; florets 15-30, corollas white, sometimes with rose or purplish veins; cypselae tan to mottled grayish brown, fusiform, 9-18(-20) mm, pappus ± persistent, of 5-21 white or sordid, ± plumose barbs in 1 series. Open sites, sandy and gravelly soils in *Larrea*, *Opuntia*, *Nolina*, *Quercus*, and *Juniperus* communities, 3875-6200 ft. SC, SW (Gottlieb, 2006b).

RATIBIDA Raf. – Prairie coneflower, Mexican-hat

Jennifer Ackerfield and Kenneth D. Heil

Perennials, 15–120 cm; stems 1–12+, erect; leaves basal and cauline, alternate, blades lanceolate to ovate or oblanceolate, pinnately lobed to 1–2-pinnatifid, margins entire or serrate, faces usually strigose-hirsute, usually gland-dotted; heads radiate, in corymbiform arrays; involucres 8–16 mm diam.; phyllaries 5–15 in 2 series; receptacles subspherical to columnar; ray florets 3–15+, corollas yellow, maroon, or bicolored; disk florets 50–400+, corollas yellowish green and purplish; cypselae strongly compressed; pappus of connate scales (Ackerfield, 2015; Urbatsch & Cox, 2006a).

1. Receptacles long-columnar, to 4.5 cm; rays 7–35 mm; pappus of 1 or 2 teeth or rarely absent; heads solitary or several at the ends of long peduncles…**R. columnifera**
1'. Receptacles globular, 0.8–1.5 cm; rays 4–8 mm; pappus a thick crown; heads on short peduncles and often closely clustered…**R. tagetes**

Ratibida columnifera (Nutt.) Wooton & Standl.—Prairie coneflower. Plants 3–10 dm; leaves 2–15 cm, 1–2-pinnatifid, the ultimate segments linear to narrowly ovate, hirsute and gland-dotted; disk florets greenish yellow and often purplish distally, 1–2.5 mm; ray florets yellow, purplish yellow, or reddish maroon, sometimes bicolored, 7–35 mm; pappus of 1–2 teeth or absent. Common on the plains, in fields, along roadsides, and in open places in the foothills, ponderosa pine and mixed conifer forests, 3055–10,300 ft. Widespread.

Ratibida tagetes (E. James) Barnhart—Short-ray prairie coneflower. Plants to 6 dm; leaves 0.5–9 cm, 1–2-pinnatifid, the ultimate segments linear to narrowly obovate, hirsute and gland-dotted; disk florets greenish yellow, sometimes purplish distally, 1.2–2.5 mm; ray florets yellow or purplish, 3–8 mm; pappus a thick crown. Sandy soil, most common in grasslands of E plains; open areas in piñon-juniper woodlands and ponderosa pine forests, 2900–9200 ft. Widespread.

RAYJACKSONIA R. L. Hartm. & M. A. Lane – Camphor daisy

Kenneth D. Heil

Rayjacksonia annua (Rydb.) R. L. Hartm. & M. A. Lane [*Haplopappus annuus* (Rydb.) Cory; *Machaeranthera annua* (Rydb.) Shinners]—Viscid camphor daisy. Annuals, 20–100 cm, herbaceous or suffrutescent; leaf blades oblanceolate to oblong or oblong-lanceolate, (1–)4–15(–25) mm wide; heads on short, sometimes bracteate peduncles, not surpassed by distal leaves; involucres (4.5–)6–9 × 9–15(–20) mm; phyllaries in 3–4 series, strongly unequal; ray florets (13–)20–36, corollas 6–11(–14) mm, corollas yellow; disk floret corolla tubes ± equaling limbs, corollas yellow; pappus of 30–40 unequal barbellate, attenuate bristles. Lea, Quay (Nesom, 2006v).

RHAPONTICUM Ludw. – Russian knapweed

Kenneth D. Heil

Rhaponticum repens (L.) Hidalgo [*Centaurea repens* L.; *C. picris* Pall. ex Willd.; *Acroptilon repens* (L.) DC.]—Russian knapweed. Creeping roots usually dark brown or black; stems ± cobwebby-tomentose; basal and proximal cauline leaves often deciduous by flowering, blades oblong, 4–15 cm; middle and distal leaves linear to linear-lanceolate or oblong, 1–7 cm; involucres 9–17 mm, loosely cobwebby; apices of inner phyllaries acute or acuminate, densely short-pilose; corollas 11–14 mm, blue, pink, or white; cypselae ivory to grayish or brown, 2–4 mm; pappus bristles white, 6–11 mm. Disturbed sites. roadsides, cultivated ground, ditch banks, 3600–8500 ft. NC, NW, EC, C, WC, SE, SC. Introduced (Allred et al., 2020).

ROLDANA La Llave – Roldana

Kenneth D. Heil

Roldana hartwegii (Benth.) H. Rob. & Brettell [*Senecio hartwegii* Benth.]—Roldana. Perennials or subshrubs, mostly (20–)60–100(–300) cm; stems single, often marked with purplish spots or striae;

leaves mostly 3-7 cm; leaves mostly cauline, alternate, blades suborbiculate to ovate or oblong; phyllaries 4-6 mm; heads radiate; phyllaries mostly 4-6 mm; ray florets (0 or 3)5-8, 3-4 × 1.5-2 mm, corollas yellow; disk florets 6-8 mm, corollas yellow or yellow-orange; cypselae (1-)2.5 mm; pappus 4-6 mm, 40-80 white barbellulate bristles. Likely in pine-oak forests in the bootheel region. Reported from New Mexico (Funston, 2006), but no records seen.

RUDBECKIA L. – Coneflower, black-eyed Susan

Kenneth D. Heil

Annuals, biennials, or perennials, mostly 50-300 cm; stems 1-15+, erect, branched distally, glabrous or hairy, sometimes glaucous; leaves basal and cauline; alternate, petiolate or sessile, blades elliptic, lanceolate, linear, oblanceolate, ovate, or spatulate, often pinnate, ultimate margins entire to toothed, faces glabrous or hairy; heads radiate or discoid; involucres hemispherical to rotate, 15-30+ mm diam.; phyllaries persistent, 5-20; receptacles subspherical to ovoid, or conic to columnar; ray florets 0 or 5-25+, usually yellow to yellow-orange or bicolored or maroon; disk florets 50-800+, bisexual, fertile; corollas yellow, yellowish green, or brown-purple; cypselae black, ± obpyramidal and 4-angled (Allred & Ivey, 2012; Urbatsch & Cox, 2006b).

1. Pappus a short crown; leaves deeply 3-lobed or -parted or pinnatifid…**R. laciniata**
1'. Pappus absent; leaves not pinnatifid or lobed…(2)
2. Cauline leaves sessile, bases clasping…**R. amplexicaulis**
2'. Cauline leaves pubescent with spreading hairs; stem leaves sessile to petiolate, bases not clasping…
   R. hirta

Rudbeckia amplexicaulis Vahl—Clasping coneflower. Leaf blades 3-15 × 0.5-4 cm; involucres 1-4 cm diam.; phyllaries spreading to reflexed, green, linear to lanceolate, herbaceous; ray laminae spreading, eventually reflexed, elliptic to obovate, 12-30 × 7-15 mm, abaxially hirsute; disk corollas 2.8-3.5 mm; cypselae with each face 4-5-striate and minutely cross-rugose, glabrous; pappus 0 (cypselae each with ring of tan tissue at apex, ca. 0.1 mm). One old record from the Mesilla Valley, 3900 ft. Doña Ana.

Rudbeckia hirta L. [*R. serotina* Nutt.]—Black-eyed Susan. Annuals or perennials, to 100 cm; stems hispid to hirsute, branched mostly beyond midheight, leafy throughout; basal leaf blades lanceolate to oblanceolate, 1-2.5(-5) cm wide (length 3-5 times width); cauline blades spatulate, oblanceolate, or broadly linear; margins entire or serrulate; faces scabrous to hirsute; heads borne singly or 2-5 in loose, corymbiform arrays; phyllaries to 3 cm; ray florets 8-16, corollas yellow to yellow-orange, sometimes maroon; disk florets 250-500+; corollas yellowish green, distally brown-purple; cypselae 1.5-2.7 mm; pappus 0. New Mexico material belongs to **var. pulcherrima** Farw. Canyon bottoms, wet areas in meadows, along streams, 6000-10,800 ft. NC, WC, Lincoln.

Rudbeckia laciniata L.—Cutleaf coneflower. Perennials, 50-300 cm; leaf blades broadly ovate to lanceolate, mostly 1-2-pinnatifid or pinnately compound, margins entire or dentate, faces glabrous or hairy; heads 2-25 in loose, corymbiform arrays; phyllaries to 2 cm; ray florets 8-12, laminae 15-50 × 4-14 mm; disk florets 150-300+, corollas yellow to yellowish green; cypselae 3-4.5 mm; pappus coroniform or of 4 scales, to 1.5 mm.

1. Receptacles ovoid; disks (17-)20-30 mm…**var. ampla**
1'. Receptacles globose or hemispherical; disks 10-20 mm…**var. laciniata**

var. ampla (A. Nelson) Cronquist—Rocky Mountain cutleaf coneflower. Proximal and midcauline blades 5-9-lobed; paleae 4.5-6.5 mm; cypselae 4-5.5 mm; pappus mostly > 0.7 mm. Wet sites, along streams and meadows, 6300-10,000 ft. NE, NC, NW, C, WC, SC.

var. laciniata L.—Cutleaf coneflower. Proximal cauline leaves pinnatifid, midcauline leaves 5-9-lobed; paleae 4.4-6.1 mm; cypselae 4.2-6 mm; pappus 0.1-0.7 mm. Wet areas, meadows in mixed conifer and spruce-fir forests, 7000-8200 ft. NC, Catron.

SANVITALIA Lam. - Creeping zinnia

Gary I. Baird

Sanvitalia abertii A. Gray—Abert's creeping zinnia. Annuals, 3–30 cm, taprooted, herbage pubescent; stems 1, erect, branched from base or throughout; leaves cauline, simple, blades lanceolate to linear, usually 3-nerved, margins mostly entire; heads radiate, broadly campanulate; involucres mostly 5–10 mm diam.; phyllaries ca. 10 in 2 series, subequal receptacles ± conic; ray florets mostly 5–20, pistillate, corollas whitish or yellowish; disk florets 15–60, bisexual, corollas yellow to orange; cypselae dimorphic, glabrous or pubescent. Piñon-juniper and desert grassland communities, 4200–8150 ft. Widespread except E (Strother, 2006ff).

SARTWELLIA A. Gray - Glowwort

Steven R. Perkins

Sartwellia flaveriae A. Gray—Threadleaf glowwort. Perennial forbs, 10–30 cm, ± succulent; stems ascending to erect; leaves cauline, opposite, sessile; leaf blades linear to filiform, 20–50 × 1–2.5 mm, margins entire; heads radiate, in corymbiform arrays; involucres campanulate, 2–3 mm diam.; phyllaries persistent, 5 in 1 series, yellowish (usually) or greenish; ray florets 3–5, corollas yellow; disk florets 5–15, corollas ca. 2 mm, yellowish, lobes 5; receptacles flat, epaleate; cypselae ± cylindrical, 1.5–2 mm. Usually associated with gypseous soils and outcrops, 3300–6700(–8100) ft. C, SE, SC (Allred & Ivey, 2012; Strother, 2006gg).

SCHKUHRIA Roth - False threadleaf

Kenneth D. Heil

Schkuhria pinnata (Lam.) Kuntze ex Thell. [*S. wislizeni* A. Gray]—Pinnate false threadleaf. Annuals; plants (10–)25–40(–70+) cm; stems usually strictly erect; leaves mostly 10–25(–40) mm; leaves mostly cauline, mostly opposite, blades linear or lobed, lobes linear to filiform, faces scaberulous and gland-dotted; involucres obconic or obpyramidal, 4–6+ mm; phyllaries 4–6, green to purple, weakly carinate, gland-dotted, otherwise usually glabrous; ray florets usually 1(2), sometimes 0; corollas yellow to white; disk florets 2–6(–8+); corollas yellow (sometimes with purple), 1.5–2 mm; cypselae 3–4 mm; pappus of 8 white to tawny or purplish scales 1–2.5 mm. Roadsides, pastures, wooded slopes, 6400–7040 ft. NC, NW, EC, C, WC, SC, SW. Plants that are hirsutulous and gland-dotted, that lack ray florets, and that have 15–30 disk florets are *Picradeniopsis multiflora* (Hook. & Arn.) B. G. Baldwin (Strother, 2006hh).

SCORZONERA L. - Vipergrass

Gary I. Baird

Scorzonera laciniata L.—Cutleaf vipergrass. Annual, biennial, or perennial herbs, sap milky; stems proximally white-nerved and often purplish; basal leaf bases broad, clasping, white-nerved, blades 6–26 cm, faces sparsely pubescent or glabrate, midrib prominent, white, lobes 1–3+ pairs, subopposite, linear, 2–13 mm, cauline leaf blades mostly entire; involucres 7–30 mm; phyllaries lance-ovate, imbricate, mostly green; ray florets 25–100+, ligules 4–7 mm; cypselae 9–12 mm, 5-angled, 10-ribbed, proximal 1/3 hollow and enlarged; pappus 8–17 mm. Ruderal weed of disturbed sites and waste places, often in alkaline soils, 3125–8750(–10,900) ft. Widespread. Introduced (Strother, 2006ii and refs. therein).

SENECIO L. - Groundsel, ragwort, butterweed

Debra K. Trock

Herbs (ours), 5–200+ cm; herbage glabrous or hairy; leaves basal and/or cauline, petiolate or sessile, blades linear (filiform) to ovate or triangular, margins entire to serrate or toothed, sometimes with callous denticles; heads usually erect, sometimes nodding, usually radiate, occasionally discoid, arranged in corymbiform, paniculiform, or racemiform arrays; involucres campanulate or cylindrical; ray florets 5, 8,

13, or 21 (sometimes 0), pistillate and fertile, corollas yellow; disk florets with 5-lobed yellow corollas; cypselae cylindrical, usually 5-ribbed or -angled, glabrous or hairy; pappus usually persistent, of white barbellulate or smooth bristles (Barkley, 2006c).

1. Annuals…**S. vulgaris**
1'. Shrubs, subshrubs, perennials, or biennials…(2)
2. Shrubs to subshrubs, leaves evenly distributed along stems…(3)
2'. Biennials or perennials; leaves mostly basal and midcauline, abruptly to only gradually reduced distally… (7)
3. Herbage lanate-tomentose or unevenly glabrescent…(4)
3'. Herbage glabrous or at most sparsely tomentose in leaf axils and among heads…(5)
4. Leaves evenly distributed along stems, narrowly linear-filiform, sometimes pinnatifid with segments linear or filiform; heads radiate, ray florets with laminae 10–18 mm; cypselae hairy…**S. flaccidus** (in part)
4'. Leaves mostly on upper ends of stems, narrowly linear-filiform, thick, turgid, and recurved; heads radiate or rays lacking entirely, ray florets with laminae 5–10 mm; cypselae glabrous…**S. warnockii**
5. Leaves evenly distributed along stems, lower either smaller or withering early; involucres narrowly campanulate to cylindrical, calyculi < 1/3 the length of phyllaries; rays 5 or 8, laminae 8–12 mm…(6)
5'. Leaves ± evenly distributed along stems; involucres campanulate; calyculi of 3–5+ bractlets (or lacking), usually > 1/2 the length of phyllaries; rays 8 or 13, laminae 10–18 mm…**S. flaccidus** (in part)
6. Lower leaves withering early; involucres narrowly campanulate to broadly cylindrical; calyculi of 3–8 bractlets, < 1/3 the length of phyllaries; ray florets 8…**S. riddellii**
6'. Lower leaves smaller than distal leaves; involucres cylindrical; calyculi 0 or 1–3 minute (< 2 mm) bractlets; ray florets 5…**S. spartioides**
7. Heads usually nodding, at least in bud…(8)
7'. Heads usually erect or nearly so…(13)
8. Heads radiate; herbage glabrous or glabrescent or at most sparsely tomentose in leaf axils…(9)
8'. Heads eradiate; herbage glabrous to heavily tomentose…(11)
9. Plants 10–60 cm; leaves lanceolate to oblanceolate…**S. amplectens**
9'. Plants 5–20 cm; leaves oblanceolate to ovate…(10)
10. Herbage glabrous, usually purple-tinged; leaves mostly basal…reported as S. soldanella, but probably **S. amplectens var. holmii**
10'. Herbage tomentose, persistent on lower surfaces, glabrescent elsewhere; leaves evenly distributed along stem…**S. taraxacoides**
11. Herbage glabrous; leaves narrowly lanceolate to oblanceolate…**S. pudicus**
11'. Herbage tomentose; leaves lanceolate, oblanceolate, triangular, or ovate…(12)
12. Herbage floccose-tomentose to glabrous; leaves ovate to lanceolate; heads 3–20, phyllaries 13 or 21… **S. bigelovii**
12'. Herbage sparsely to heavily tomentose; leaves broadly lanceolate to triangular; heads 6–12; phyllaries 8…**S. sacramentanus**
13. Leaves ± evenly distributed along stem, basal leaves withering early (or upper leaves only slightly reduced)…(14)
13'. Leaves mostly basal, petiolate; middle and distal cauline leaves reduced, sessile or clasping…(16)
14. Herbage glabrous or at most lightly tomentose when young…(15)
14'. Herbage pubescent with sticky glandular hairs [to be expected]…S. parryi
15. Lower leaves withering early, blades narrowly lanceolate to ovate; calyculi of 3–5+ bractlets up to 3/4 the length of phyllaries; ray florets with laminae 5–10 mm…**S. eremophilus**
15'. Lower leaves well developed, only slightly reduced distally, blades narrowly triangular; calyculi of 2–6 minute (< 2 mm) bractlets; ray florets with laminae 9–15 mm…**S. triangularis**
16. Herbage canescent, densely tomentose, or glabrescent…(17)
16'. Herbage glaucous, glabrous, or glabrate at flowering…(18)
17. Leaves abruptly reduced distally, margins subentire to denticulate; inflorescence of 1 or rarely 2–3 heads; phyllaries 21, tips greenish with white bristles; ray florets 13, laminae 8–10 mm; cypselae hairy…**S. actinella**
17'. Leaves progressively reduced distally, margins dentate with dark callous denticles; inflorescence a corymbiform array of 20–60+ heads; phyllaries 5 or 8, tips black; ray florets 5, laminae 5–8 mm; cypselae glabrous…**S. atratus**
18. Leaves progressively reduced distally; upper leaves sessile and bractlike; involucres with calyculi < 3 mm; phyllaries 13 or 21, tips green, brownish (not black), or at most minutely black…(19)
18'. Leaves gradually reduced distally, midcauline leaves often larger than basal leaves; upper leaves clasping and small; involucres with calyculi 1/3 the length of phyllaries; phyllaries 13 or 21 (rarely 8), tips black with short hairs…**S. crassulus**

19. Leaf blades elliptic, lanceolate, or oblanceolate, margins entire or dentate; growing on open prairies or plains at 7000–9000 ft....**S. integerrimus**
19'. Leaf blades ovate to obovate; margins subentire to wavy with callous denticles; growing in rocky or disturbed areas at 6560–11,500 ft....**S. wootonii**

Senecio actinella Greene [*S. mogollonicus* Greene; *S. actinella* var. *mogollonicus* (Greene) Greenm.]— Flagstaff ragwort. Perennials; herbage densely tomentose to glabrescent; stems 10–35+ cm; basal leaves petiolate, blades narrowly ovate, bases tapered, margins subentire to denticulate; cauline leaves bractlike; heads 1 (rarely 2 or 3); heads radiate; calyculi of 1–8+ bractlets, 1/3–3/4 the length of phyllaries; ray florets 13; cypselae hairy. Rocky woodlands, especially in pine-dominated forests, 6800–9000 ft. WC, SC, San Miguel.

Senecio amplectens A. Gray [*Ligularia amplectens* (A. Gray) W. A. Weber]—Showy ragwort. Perennials; herbage glabrous or sparsely hairy in leaf axils and among heads, often purple-tinged; stems mostly 10–60 cm; leaves petiolate, often clasping, progressively reduced distally, blades lanceolate to oblanceolate, bases tapered, margins dentate to denticulate, distal leaves bractlike; inflorescences of 1–10 nodding heads; heads radiate; ray florets 13, rarely fewer; cypselae glabrous.

1. Plants sparsely hairy; stems 30–60 cm; phyllaries with black or dark purple tips; growing at or below timberline...**var. amplectens**
1'. Plants glabrous to glabrate;, stems 1–30 cm; phyllaries with green or light purple tips; growing at or above timberline...**var. holmii**

var. amplectens—Showy alpine ragwort. Herbage sparsely hairy; stems 30–60+ cm; leaves mostly proximal (basal withering before flowering); phyllaries with black or purplish tips, often with black hairs on abaxial faces. Damp or drying rocky sites in conifer associations and near timberline, 9600–12,500 ft. NC.

var. holmii (Greene) H. D. Harr.—Holm's ragwort. Herbage glabrous or glabrate at flowering; stems 5–30 cm; leaves mostly basal, midcauline and distal leaves reduced; phyllaries green or light purplish, glabrous. Loose, rocky soil at or above timberline, 9700–13,000 ft. NC.

Senecio atratus Greene [*S. milleflorus* Greene; *S. atratus* var. *milleflorus* (Greene) Greenm.]—Tall blacktip ragwort. Perennials; herbage densely tomentose to canescent, sometimes unevenly glabrescent; stems 20–80+ cm; leaves petiolate, progressively reduced distally; basal leaves petiolate, blades oblong-ovate to oblanceolate, bases tapering, margins dentate with dark, callous denticles; inflorescences in arrays of 20–60+ heads; heads radiate; calyculi of 2–5 bractlets, < 1/3 length of phyllaries; ray florets usually 5; cypselae glabrous. Dry or drying, rocky or sandy sites in coniferous forests, disturbed sites, 9200–13,125 ft. NC.

Senecio bigelovii A. Gray [*Ligularia bigelovii* (A. Gray) W. A. Weber]—Perennials; herbage floccose-tomentose, glabrescent, or sometimes glabrous; stems 20–120 cm; leaves progressively reduced distally; lower leaves petiolate, blades ovate to lanceolate, bases contracted or tapered, margins subentire or serrate to dentate; middle and upper cauline leaves sessile, becoming bractlike, often clasping; inflorescences in arrays of 3–20 nodding heads; heads eradiate; calyculi of 4–10 bractlets, 1/3–1/2 the length of phyllaries.

1. Plants glabrous or sparsely hairy; middle and upper cauline leaves clasping; phyllaries 13, 6–8 mm...**var. bigelovii**
1'. Plants thickly tomentose or unevenly glabrescent; middle and upper cauline leaves sessile but not clasping, phyllaries 21, 8–12 mm...**var. hallii**

var. bigelovii [*S. rusbyi* Greene]—Nodding ragwort. Herbage sparsely tomentose or glabrous; lower leaf blades with bases contracted or obtuse; middle and upper leaves clasping; heads 6–20. Meadows, rocky, open sites in conifer-dominated forests, 5900–9850 ft. NC, NW, C, WC, SC.

var. hallii A. Gray [*S. accedens* Greene]—Hall's ragwort. Herbage woolly-tomentose, especially among the heads and on the lower surfaces of leaves, becoming unevenly glabrescent; lower leaf blades tapered at the base; middle and upper cauline leaves sessile and not clasping; heads 7–12. Grassy, damp or drying sites in open coniferous forests, 5575–11,500 ft. NC.

Senecio crassulus A. Gray [*S. lapathifolius* Franch.; *S. semiamplexicaulis* Rydb.]—Thickleaf ragwort. Perennials; herbage glabrous; stems 15–65 cm; leaves gradually reduced distally, middle leaves occasionally larger than lower ones; lower and middle leaves petiolate, blades subelliptic to broadly lanceolate, bases tapered, margins subentire to sharply dentate with callous teeth, upper leaves often clasping and smaller; inflorescences in arrays of 3–12 heads; heads radiate, calyculi of 1–5 bractlets to 1/3 the length of phyllaries; ray florets 8 or 13, laminae 5–12 mm; cypselae glabrous. Meadows, open places in forests, moist to drying hillsides, 7200–12,150 ft.

Senecio eremophilus Richardson—Perennials; herbage glabrous to glabrate; stems 20–120+ cm, single or loosely clustered, unbranched; leaves evenly distributed, lower ones withering before flowering, sessile or short-petiolate, blades narrowly lanceolate, lanceolate, or ovate, bases tapered, margins pinnate, lacerate, or rarely dentate; arrays of 10–60+ heads (involucres campanulate to cylindrical); heads radiate; calyculi of 3–5+ bractlets, often to 3/4 the length of phyllaries; ray florets 8, laminae 5–10 mm; cypselae glabrous, rarely hairy.

1. Heads 10–35; involucres campanulate, 7–10 mm wide; phyllaries 6–8 mm, with green (minutely black) tips; ray florets with laminae 5–10 mm…**var. kingii**
1'. Heads 20–60; involucres cylindrical, 4–5 mm wide; phyllaries 3–5 mm, with black tips; ray florets with laminae 5–6 mm…**var. macdougalii**

var. kingii Greenm. [*S. kingii* Hook.; *S. ambrosioides* Mart. ex Baker]—King's ragwort. Heads 10–35; involucres campanulate, 7–10 mm wide; phyllaries 6–8 mm, tips green or minutely if at all black; ray floret laminae 5–10 mm. Rocky, open sites in grassy areas, 7200–11,300 ft. NC, NW, C, WC, SC.

var. macdougalii (A. Heller) Cronquist [*S. macdougalii* A. Heller]—MacDougal's ragwort. Heads 20–60; involucres cylindrical, 4–5 mm wide; phyllaries 3–5 mm, tips black; ray floret laminae 5–6 mm. Coniferous forest associations in damp and drying soils, 7875–11,800 ft. NC, NW, C, WC, SC.

Senecio flaccidus Less.—Subshrubs or shrubs, 30–120 cm; herbage glabrous or lanate-tomentose; stems usually multiple, slightly branched; leaves ± evenly distributed, sessile or short-petiolate; blades narrowly linear or filiform; ultimate margins entire or slightly toothed; heads 2–20 in corymbiform clusters sometimes gathered into larger showy arrays; heads radiate; calyculi of 3–5+ bractlets (sometimes 0), usually > 1/2 the length of phyllaries; ray florets 8 or 13, laminae 10–18 mm; cypselae hairy.

1. Plants tomentose; calyculi with bractlets minute or 0; involucres cylindrical or weakly campanulate; phyllaries 5–8 mm…**var. flaccidus**
1'. Plants glabrous to glabrate; calyculi with bractlets to 1/2 the length of phyllaries; involucres campanulate; phyllaries 7–10 mm…**var. monoensis**

var. flaccidus [*S. longilobus* Benth.; *S. douglasii* DC. var. *longilobus* (Benth.) L. D. Benson]—Threadleaf ragwort. Herbage tomentose with whitish hairs; calyculi with bractlets minute or lacking; involucres cylindrical to weakly campanulate; phyllaries 5–8 mm. Plains, streambeds, dry, open, sandy or rocky sites, 3275–10,500 ft. Widespread.

var. monoensis (Greene) B. L. Turner & T. M. Barkley [*S. monoensis* Greene]—Smooth threadleaf ragwort. Herbage glabrate or glabrous; calyculi with bractlets well developed to 1/2 the length of phyllaries; involucres campanulate; phyllaries 7–10 mm. Desert basins and washes, open rocky or sandy sites, 3400–6200 ft. Scattered.

Senecio integerrimus Nutt.—Lamb's tongue ragwort. Perennials; stems 15–70 cm; leaves progressively reduced distally, lower and middle leaves petiolate, blades elliptic, lanceolate, or oblanceolate, bases tapered or truncate, margins entire or dentate; upper leaves sessile and bractlike; inflorescences of 4–20 heads; heads radiate; calyculi of 1–5+ bractlets, usually < 2 mm; ray florets 5–8; laminae 6–8 mm; cypselae usually glabrous, rarely hirtellous on the ribs. New Mexico material belongs to **var. integerrimus**. Open prairies and plains, 7000–9100 ft. NC, Catron, Otero.

Senecio parryi A. Gray—Mountain ragwort. Perennials; herbage pubescent with sticky-glandular hairs; stems 30–90 cm; leaves evenly distributed along stem, lower ones often withering before flowering, short-petiolate, blades ranging from ovate to obovate, spatulate, or lanceolate, margins dentate; middle and upper leaves narrower and clasping; inflorescences in arrays of 12–30 heads; heads radiate; calyculi of 5–15+ prominent bractlets, often to nearly as long as phyllaries; ray florets 13, laminae 8–12 mm; cypselae hairy. Desert mountains in disturbed and rocky sites. To be expected at 4300–8600 ft.

Senecio pudicus Greene [*S. cernuus* A. Gray; *Ligularia pudica* (Greene) W. A. Weber]—Bashful ragwort. Perennials or biennials; herbage glabrous; stems 50–80 cm; leaves progressively reduced distally, lower leaves petiolate, blades narrowly lanceolate to oblanceolate, bases tapered, margins entire or dentate; midcauline and distal leaves sessile and bractlike; inflorescences in arrays of 2–35 nodding heads; heads eradiate; calyculi of 2–6 bractlets; cypselae glabrous. Coniferous and aspen woodlands; 2 records, ca. 7675–8085 ft. Colfax.

Senecio riddellii Torr. & A. Gray [*S. spartioides* Torr. & A. Gray var. *riddellii* (Torr. & A. Gray) Greenm.; *S. filifolius* Nutt. var. *fremontii* Torr & A. Gray]—Riddell's ragwort. Subshrubs, 30–100 cm; herbage glabrous; stems multiple, branching upward; leaves evenly distributed (lower often withering before flowering), sessile or short-petiolate, blades simple and linear-filiform or pinnately divided into linear-filiform lobes, margins entire; inflorescences in arrays of 5–20 heads; heads radiate; calyculi of 3–8+ bractlets, < 1/3 the length of phyllaries; ray florets 8, laminae 8–10 mm; cypselae hirtellous. Drying, open floodplains in sandy, rocky soils, 3500–8700 ft. Widespread.

Senecio sacramentanus Wooton & Standl.—Sacramento ragwort. Perennials; herbage sparsely to heavily tomentose; stems 20–50 cm, single or clustered, unbranched; leaves progressively reduced distally, lower leaves petiolate, blades broadly lanceolate to triangular, margins dentate or serrate; midcauline and distal leaves sessile and bractlike; inflorescences of 6–12 nodding heads; heads eradiate; calyculi of 2–5+ bractlets, < 1/2 the length of phyllaries; cypselae glabrous. Montane meadows, 5900–10,600 ft. Lincoln, Otero. **R & E**

Senecio soldanella A. Gray [*Ligularia soldanella* (A. Gray) W. A. Weber]—Colorado ragwort. Perennials; herbage glabrous; stems 5–20 cm, single, often purple-tinged, unbranched; leaves mostly basal, with winged petioles, blades broadly ovate to ovate-orbiculate, margins weakly dentate or denticulate to subentire, cauline leaves bractlike; inflorescence of a single nodding head (rarely 2); heads radiate; calyculi of 3–5+ bractlets, 1/2–2/3 the length of phyllaries; ray florets 13, laminae 10–14 mm; cypselae glabrous. Talus slopes, 11,700–12,625 ft. Specimens keying here are most likely *S. amplectens* var. *holmii* (q.v.).

Senecio spartioides Torr. & A. Gray [*S. multicapitatus* Greenm. ex Rydb.]—Broom-like ragwort. Subshrub; stems 20–120 cm, multiple-clustered, branching and arching upward; leaves evenly distributed, with lower ones often smaller, sessile or short-petiolate, blades narrow-linear or filiform, margins entire; inflorescences in compound arrays of 10–60 heads; heads radiate; calyculi 0 or 1–3 minute bractlets; ray florets 5, rarely 13, laminae 8–12 mm; cypselae usually mostly hirtellous. Stream banks, hillsides, open, dry, disturbed sites, 3510–3500 ft. Widespread.

Senecio taraxacoides (A. Gray) Greene [*S. amplectens* A. Gray var. *taraxacoides* A. Gray; *Ligularia taraxacoides* (A. Gray) W. A. Weber]—Dandelion ragwort. Perennials; herbage tomentose, persistent on lower leaf surfaces, glabrescent elsewhere, sometimes with a purplish tinge; stems 5–14 cm; leaves evenly distributed along stem, petiolate, blades oblanceolate to ovate, bases tapered, margins dentate or incised to subpinnatifid; upper leaves bractlike; inflorescence of 1(2–5) heads; heads radiate; calyculi of 2–5 bractlets usually > 1/2 the length of phyllaries; ray florets usually 13, rarely fewer or 0, laminae 8–12 mm; cypselae glabrous. Alpine peaks and tundra, 8700–13,170 ft. NC.

Senecio triangularis Hook. [*S. triangularis* var. *angustifolius* G. N. Jones]—Arrowleaf ragwort. Perennials; herbage glabrous or lightly tomentose when young; stems 30–180 cm; leaves evenly distributed along stem, upper ones smaller and petiolate, becoming sessile, blades narrowly triangular, margins dentate or rarely subentire; inflorescences in arrays of 15–60 heads; heads radiate; calyculi of 2–6 minute bractlets; ray florets 8, laminae 9–15 mm; cypselae glabrous. Coniferous forests and rocky stream banks in open and damp places, 7930–12,000 ft. NC, SW, SC.

Senecio vulgaris L.—Common groundsel. Annuals, 10–60 cm; herbage glabrous or sparsely and unevenly tomentose when young; stems usually 1; leaves evenly distributed, blades ovate to oblanceolate, 2–10 × 0.5–2(–4) cm, margins lobulate to dentate; heads 8–20 in loose corymbiform arrays; phyllaries ± 21, 4–6 mm, tips usually green, sometimes black; ray florets 0; cypselae usually sparsely hairy or nearly glabrous. Disturbed sites, 3900–9075 ft. NC, NW, C, SC.

Senecio warnockii Shinners—Warnock's ragwort. Subshrubs; herbage unevenly lanate-tomentose, becoming glabrescent; stems 20–40 cm; leaves mostly on upper ends of stems, thick and turgid, often recurved, sessile or short-petiolate with blades filiform to narrowly linear, margins entire; inflorescences in arrays of 3–10 heads; heads radiate or eradiate; calyculi 0 or of 3–5 bractlets, < 1/2 the length of phyllaries; ray florets 8 or rarely 0; laminae 5–10 mm; cypselae glabrous. Gypsiferous soils, 3350–3650 ft. Eddy, Otero. **R**

Senecio wootonii Greene [*S. anacletus* Greene]—Wooton's ragwort. Perennials; herbage glaucous or glabrous; stems 15–50 cm; leaves progressively reduced distally, lower leaves petiolate, blades ovate to obovate or lanceolate, bases tapered, margins subentire to wavy with callous denticles; midcauline and upper leaves sessile and bractlike; inflorescences in arrays of 3–24 heads; heads radiate; calyculi of 1–3 bractlets, < 3 mm; ray florets 8–10; laminae 4–10 mm; cypselae glabrous. Rocky, disturbed areas in damp or drying sites, 6500–11,500 ft. C, WC, SC, SW.

SIDNEYA E. E. Schill. & Panero – Goldeneye

Kenneth D. Heil

Sidneya tenuifolia (A. Gray) E. E. Schill. & Panero [*Heliomeris tenuifolia* A. Gray; *Viguiera tenuifolia* Gardner; *V. stenoloba* S. F. Blake]—Skeletonleaf goldeneye. Shrubs, 50–150 cm; leaves opposite or alternate, sessile or subsessile, blades ovate, shallowly to deeply lobed or linear, 1.5–10.8 × 1–9 cm, margins entire, abaxial faces strigillose and gland-dotted, adaxial densely strigose to glabrate; heads mostly borne singly; involucres 5–12 × 7–9 mm; ray florets 13–18, laminae 7–12, yellow; disk florets 100+, corollas 3.3–4.2 mm, yellow; cypselae 2–3 mm; pappus 0. Chihuahuan desert scrub, 3000–7800 ft. SE, SC, Colfax, Quay (Schilling, 2006c; Schilling & Panero, 2011).

SILPHIUM L. – Rosinweed

Kenneth D. Heil

Perennials, 20–250+ cm; stems usually erect, usually branched, often vernicose with resinous exudates; leaves basal and cauline, whorled, opposite, or alternate, blades deltate, elliptic, linear, ovate, or rhombic, margins entire or toothed; heads radiate, in paniculiform or racemiform arrays; involucres campanulate to hemispherical, 10–30 mm diam.; phyllaries persistent, 11–45 in 2–4 series; receptacles flat to slightly convex, ray florets 8–35+ in 1–4 series, corollas yellow or white; disk florets 20–200+, corollas yellow or white; cypselae black to brown, obflattened; pappus 0, or persistent, of 2 awns (Clevinger, 2006).

1. Leaf blades lanceolate, linear-ovate, or rhombic, 4–60 × 1–30 cm; bases attenuate to truncate…**S. laciniatum**
1'. Leaf blades lanceolate to ovate, 2–23 × 0.1–11 cm; bases round to caudate…**S. integrifolium**

Silphium integrifolium Michx.—Wholeleaf rosinweed. Plants caulescent, 40–200 cm; stems terete to slightly square; basal leaves caducous, cauline leaves opposite, sessile, blades lanceolate to ovate, 2–23 × 0.1–11 cm, faces glabrous; phyllaries 17–37 in 2–3 series, outer appressed, abaxial faces glabrous; ray florets 20–36, corollas yellow; disk florets 130–225, corollas yellow; cypselae 9–14 × 6–10 mm; pappus 1–4 mm. New Mexico material belongs to **var. laeve** Torr. & A. Gray [*Silphium speciosum* Nutt.]. Roadsides, 6850–6900 ft. Mora, San Miguel.

Silphium laciniatum L.—Compass plant. Plants scapiform, (40-)100-300 cm; stems terete, hirsute, hispid, or scabrous; basal leaves persistent, petiolate or sessile, cauline leaves petiolate or sessile; blades lanceolate, linear, ovate, or rhombic, 4-60 × 1-30 cm, faces hirsute, hispid, or scabrous; phyllaries 25-45 in 2-3 series, outer reflexed or appressed, abaxial faces hispid to scabrous, ± stipitate-glandular; ray florets 27-38, corollas yellow; disk florets 100-275, corollas yellow; cypselae 10-18 × 6-12 mm; pappus 1-3 mm. Roadsides, 6425-7200 ft. Bernalillo, McKinley, San Miguel, Santa Fe.

SILYBUM Adans. – Milk thistle

Kenneth D. Heil and Steve L. O'Kane, Jr.

Silybum marianum (L.) Gaertn. [*Carduus marianus* L.]—Blessed milk thistle, chardon. Annuals or biennials, 15-300 cm; stems erect, usually simple, glabrous or slightly tomentose; leaves basal and cauline, wing-petioled, margins dentate, teeth and lobes spine-tipped, blades 15-60+ cm, margins coarsely lobed, cauline leaves clasping, bases spiny; heads discoid, borne singly; involucres ovoid to spherical, 15-60 mm diam.; receptacles flat, epaleate, covered with whitish bristles; phyllary appendages spreading, ovate, 1-4 cm including long-tapered spine tips; corollas 26-35 mm, pink to purple; cypselae ovoid, slightly compressed, not ribbed, brown- and black-spotted, 6-8 mm. Roadsides, pastures, waste areas, 4400-6200 ft. Doña Ana, Grant. Introduced (Keil, 2006k).

SIMSIA Pers. – Bush sunflower

Kenneth D. Heil

Annuals, perennials, or subshrubs (shrubs), 20-400 cm; stems erect or ascending; leaves cauline, opposite (proximal) or alternate (whorled), blades mostly deltate to ovate (linear), sometimes 3(-5)-lobed (pinnatifid), ultimate margins entire or toothed, faces often gland-dotted or ± stipitate-glandular to glandular-puberulent; heads radiate (discoid), borne singly or in 2s or 3s, or in tight to loose, corymbiform (paniculiform) arrays; involucres 5-16(-22) mm diam.; phyllaries persistent, in 2-4 series; ray florets (0-)5-21(-45), corollas orange-yellow (lemon-yellow, pink, purple, or white); disk florets (12-)13-154 (-172), bisexual, corollas concolorous with rays (usually turning purple apically); cypselae flattened, thin-margined; pappus 0, or fragile or readily falling (Spooner, 2006).

1. Perennials or subshrubs; ray florets 8-21; disk florets (26-)90-154; anthers usually yellow, rarely black… **S. calva**
1'. Annuals; ray florets 5-10; disk florets 13-27; anthers yellow proximally, usually purple to bronze distally… **S. lagasceiformis**

Simsia calva (A. Gray & Engelm.) A. Gray [*Barrattia calva* A. Gray & Engelm.]—Awnless bush sunflower. Perennials or subshrubs, 30-150 cm; leaf blades ovate, 2-8 × 1.5-6 cm, sometimes 3-lobed; heads usually borne singly, sometimes in 2s or 3s; involucres 10-12 × 7-16 mm; phyllaries 21-43, subequal to unequal; ray florets 8-21; corollas light orange-yellow; disk florets (26-)90-154; anthers usually yellow, rarely black; cypselae 3.5-5.7 mm; pappus 0 or to 4 mm. One collection from the Big Hatchet Mountains; clay soil in a small canyon surrounded by shrub-grassland, 5870 ft. Hidalgo.

Simsia lagasceiformis DC. [*Simsia exaristata* A. Gray]—Annual bush sunflower. Annuals, 20-400 cm; leaf blades ovate to deltate, 2-21 × 1-16 cm, rarely 3-lobed; heads in tight to loose corymbiform arrays; involucres 8-12 × 5-10 mm; phyllaries 13-19, unequal; ray florets 5-10, corollas orange-yellow; disk florets 13-27, anthers yellow proximally, usually purple to bronze distally; cypselae 4.2-6 mm; pappus usually 2.5-4.6 mm, rarely 0. Chihuahuan desert scrub; 1 collection from Crow Flats, 3800 ft. Otero.

SOLIDAGO L. – Goldenrod

Kenneth D. Heil

Perennial herbs; leaves alternate, simple; heads radiate; involucral bracts imbricate in several series or not much imbricate, ± chartaceous at the base, with green tips; receptacles naked; ray florets yellow;

disk florets yellow; cypselae subterete or angled, short-pubescent to glabrous; pappus of capillary bristles (Ackerfield, 2015; Allred et al., 2020; Keller, 1999; Semple & Cook, 2006).

(Key adapted from Allred et al. [2020])

1. Inflorescence flat-topped; phyllaries multiveined; basal leaf petiole bases persisting on rootstock... **S. rigida**
1'. Inflorescence narrow-paniculiform to secund-conic, or if ± flat-topped then phyllaries single-veined and basal leaf petiole bases not persisting...(2)
2. Lower stem leaves deciduous and not the largest; basal rosette leaves not present; inflorescence secund-conic; lower and midstem leaves and sometimes upper leaves triple-nerved...(3)
2'. Lower stem leaves often persisting and the largest; basal rosette leaves often present; inflorescence narrow to broadly paniculiform; leaves not as above...(6)
3. Middle to lower stems densely short-strigose-villous...(4)
3'. Lower to middle-upper stems glabrous or lower stems glabrous to very sparsely strigose and becoming more so distally...(5)
4. Leaves lanceolate; upper stem leaves triple-nerved; inflorescences from about as tall as wide to 2 times as tall as wide...**S. altissima**
4'. Leaves linear-lanceolate; upper stem leaves not triple-nerved; inflorescences usually much taller than wide...**S. altiplanities**
5. Inflorescence parts not glandular, or very rarely sparsely so; wetter locations on prairies and at bases of mountains in N counties...**S. gigantea**
5'. Inflorescence parts sparsely to moderately stipitate-glandular; lower elevations to high mountain meadows...**S. lepida**
6. Outer phyllaries 1/2+ as long as inner; inflorescence rounded if small, to paniculiform if lower branches developed; upper leaves broadest at the base, margins with long-ciliate hairs; middle to higher elevations...**S. multiradiata**
6'. Outer phyllaries 1/4–1/3 as long as inner; inflorescence narrow to broadly paniculiform; upper leaf margins not long-ciliate; elevations various...(7)
7. Phyllaries and sometimes stems glutinous-resinous...**S. glutinosa** (in part)
7'. Phyllaries not glutinous-resinous, but sometimes stipitate-glandular...(8)
8. Inflorescence narrow to broadly paniculiform, not secund-conic...(9)
8'. Inflorescence secund-conic, either elongate or compressed and pseudocorymbiform...(14)
9. Large lower stem leaves usually present; stems glabrous below to strigose in inflorescences...(10)
9'. Large lower stem leaves usually absent; stems usually short-strigose-villous...(11)
10. Leaves pale green, somewhat glaucous; cypselae glabrous...**S. pallida**
10'. Leaves bright green, not pale, not glaucous; cypselae finely strigose...**S. glutinosa** (in part)
11. Basal rosette and low stem leaves narrowly oblanceolate, petioles 1–4 cm; middle and upper stem leaves lanceolate-elliptic to linear-lanceolate-elliptic; phyllaries resinous, glabrate; cypselae sparsely strigose; Guadalupe Mountains...**S. correllii**
11'. Basal rosettes nearly always absent, lower stem leaves winged-petiolate; middle and upper stem leaves narrowly to broadly elliptic; phyllaries glandular, resinous, or strigose; cypselae glabrous to moderately densely strigose; not limited to the Guadalupe Mountains...(12)
12. Cypselae glabrous, rarely glabrate to very sparsely strigose, rarely strigose; arrays of heads often narrow-elongate; cauline leaves entire; Union County...**S. petiolaris**
12'. Cypselae very sparsely to moderately densely strigose (rarely glabrate); locations various...(13)
13. Cypselae moderately densely strigose; arrays rounded-corymbiform to paniculiform on older shoots; phyllaries from densely strigose to densely glandular, with mixed hairs and glands common; widespread...**S. wrightii**
13'. Cypselae very sparsely strigose; arrays congested, narrow-paniculiform; phyllaries lanceolate, sparsely glandular and moderately strigose distally; cauline leaves grayish green, moderately short-strigose; stems densely short villous-canescent; Union County...**S. capulinensis**
14. Stems glabrous; rhizomatous...**S. missouriensis**
14'. Stems sparsely to densely strigillose; short-branched caudices to rhizomatous...(15)
15. Heads secund, in rounded, pseudocorymbiform, compressed-paniculiform arrays; leaves soft-pubescent; N Rio Arriba County...**S. nana**
15'. Heads in secund arrays (sometimes weakly so), not compressed-pseudocorymbiform; locations various...(16)
16. Plants with short-branched caudices; heads in softly canescent arrays, secund to apically recurved; prairies and foothills; NE New Mexico...**S. nemoralis**
16'. Plants with short to long-creeping rhizomes; heads in thyrsiform to secund-pyramidal, paniculiform arrays...(17)
17. Heads in paniculiform arrays, usually compact, branches broadly thyrsiform to somewhat secund-

pyramidal; proximal branches reflexed-recurved distally; basal leaves withering by flowering; prairies…
S. mollis
17′.   Heads in cone-shaped arrays with branches narrowly secund, or open, lax, pyramidal; basal leaves often present at flowering; lower to midmountain elevations…**S. velutina**

Solidago altiplanities C. E. S. Taylor & R. John Taylor—Goldenrod. Plants 30–100 cm; leaves elliptic to linear-lanceolate, 40–90 × 4–5 mm, 3-nerved, faces finely sparsely to moderately strigose; ray florets 4–5, ligules 1.5–2.5 mm; disk florets 6–8, corollas 3–3.5 mm; pappus 3 mm. Gypsum, lava flows, shale soils, grasslands. Elevation unknown. Colfax, Union.

Solidago altissima L.—Canada goldenrod. Plants 5–20 dm; leaves oblanceolate to lanceolate, middle 4.5–10 cm, strongly 3-nerved, entire to serrate, scabrous above and moderately strigillose below; involucres 2–3 mm, bracts imbricate; ray florets (laminae) 0.7–1.5(–2) × 0.1–0.4(–0.5) mm; disk florets 2.3–3.6 mm; cypselae 0.5–1.5 mm; pappus 2.5–3.5 mm. New Mexico material belongs to **var. gilvocanescens** (Rydb.) Semple. Dry to moist soils, disturbed sites, grasslands, near streams and ponds, roadsides, 3500–9800 ft. NE, SC, Sandoval.

Solidago capulinensis Cockerell & D. M. Andrews—Capulin goldenrod. Plants (4–)8–12 dm; leaves elliptic-lanceolate to elliptic-oblanceolate, 4–7 cm, surface puberulent and minutely stipitate-glandular; involucres 3–5 mm; ray florets 11–15, 5–7 × 1–1.2 mm; disk florets 5–6 mm; cypselae 2.5–3 mm, glabrous; pappus of apically acute barbellate bristles. Among boulders of basalt and sandstone, hillsides with juniper, savannas, grasslands, 5000–7500 ft. Union. **R**

Solidago correllii Semple [*S. wrightii* A. Gray var. *guadalupensis* G. L. Nesom]—Plants 3–7 dm, ascending to erect, upper stems sparsely to densely hirsute; leaves alternate, lanceolate, 3–6 cm × 4–7 mm, narrowly lanceolate to linear-lanceolate; inflorescences subcorymboid; heads 9–12 mm; rays ca. 4 × 2 mm; phyllaries glabrous to minutely sessile-glandular; cypselae glabrous, rarely strigose. Limestone, desert scrub, oak-maple, ponderosa pine, 4300–7100 ft. SC, SE. **R**

Solidago gigantea Aiton [*S. serotina* Aiton]—Giant goldenrod. Plants (5–)10–15+ dm; leaves narrowly elliptic to lanceolate, 6–17 cm, glabrous to somewhat hairy on the nerves, 3-nerved, sharply serrate; involucres 2.5–4 mm, bracts in 3–4 series; ray florets (8–)10–18, 1–3 mm; disk florets 2.5–4 mm; cypselae 1.3–1.5 mm; pappus 2–2.5 mm. Common in moist places, especially on E plains. Colfax, San Miguel, Union.

Solidago glutinosa Nutt. [*S. spathulata* DC; *S. simplex* Kunth]—Mt. Albert goldenrod. Plants 0.5–6 dm; leaves oblanceolate to spatulate, to 15 cm, glabrous, toothed to crenate (mostly in the upper 1/2 to upper 1/3); involucres 4–7 mm; disk florets 4–5 mm; ray florets (6–)8–12, mostly 3–4 mm; pappus 2–5 mm; cypselae 2–3.2 mm. Common in mountain meadows, forests, and the alpine, 6000–13,000 ft. NC, NW, C, WC, SC.

Solidago lepida DC. [*S. canadensis* L. var. *lepida* (DC.) Cronquist]—Western goldenrod. Plants 2.5–15 dm; leaves narrowly to broadly oblanceolate, 10–15 cm × 15–23 mm, glabrate to short-villous, margins subentire to coarsely serrate; involucres campanulate, mostly 2.5–4 mm; ray florets 7–22, 1–2 mm; disk florets mostly 5–9, 2–5 mm; cypselae 0.6–1.2 mm, strigillose.

1.   Heads in thyrsiform arrays; proximal branches ascending…**var. lepida**
1′.   Heads in pyramidal paniculiform arrays; proximal branches arching to recurved…**var. salebrosa**

var. lepida DC.—Goldenrod. Fields, mesas, piñon-juniper woodlands, ponderosa pine forests, streams, rivers, 5000–7000 ft. Catron, Grant.

var. salebrosa (Piper) Semple [*S. serotina* Aiton var. *salebrosa* Piper; *S. canadensis* L. subsp. *salebrosa* (Piper) D. D. Keck]—Canada goldenrod. Meadows, streams, rivers, 3000–10,100 ft. NC, C, WC, SC.

Solidago missouriensis Nutt.—Missouri goldenrod. Plants 1.5–9 dm; leaves linear-lanceolate to narrowly elliptic, 4–25 cm, glabrous, entire to serrate, 3-nerved (often weakly so); involucres 3–5 mm; ray florets 5–11, 2–3 mm; disk florets 2–4 mm; cypselae 1–2 mm; pappus 2.5–3 mm. Common in open meadows, forests, plains, 3700–10,000 ft. NW, C, WC.

Solidago mollis Bartl.—Velvety goldenrod. Plants 1–5(–7) dm; leaves elliptic to ovate, 3–8 cm, 3-nerved, hirsute, toothed on the upper 1/3; involucres 3.5–6 mm; ray florets 6–10, 1–2 mm; disk florets 2.5–4 mm; cypselae 1.5–2 mm; pappus 2–3 mm. Common in sandy or rocky soil of plains and outer foothills, 3400–7000 ft. Mora, Quay, San Miguel, Union.

Solidago multiradiata Aiton—Rocky Mountain goldenrod. Plants 0.5–5 dm; leaves oblanceolate to spatulate, 2–10 cm, petiole margins strongly ciliate, entire to remotely toothed, glabrous; involucres 4–7 mm; ray florets 12–18, 3–4 mm; disk florets 3–5 mm; cypselae 1.5–4 mm; pappus 3–4 mm. Common in rocky soil of mountain meadows, forests, and the alpine, 6500–13,000 ft. NC, WC, Lincoln. Can hybridize with *S. glutinosa* at high elevations.

Solidago nana Nutt.—Baby goldenrod. Plants 1–5 dm; leaves oblanceolate to spatulate, 2–10 cm, weakly or scarcely 3-nerved, short-hairy, entire to remotely toothed, involucres 4–6 mm; ray florets (5)6–10, ca. 3 mm; disk florets 8–20; cypselae 2–3 mm; pappus 3.5–4 mm. Common in meadows and on open or slightly forested slopes, 5000–9800 ft. NC, C.

Solidago nemoralis Aiton [*S. nemoralis* Aiton var. *longipetiolata* (Mack. & Bush) E. J. Palmer & Steyerm.]—Gray goldenrod. Plants 1–10(–13) dm; leaves oblanceolate to narrowly ovate, 5–20 cm, 1-nerved or obscurely 3-nerved, densely short-hairy, margins subentire to crenate-dentate; involucres 4–6 mm; ray florets 5–11, 2.8–5.5 mm; disk florets 2.5–4.5 mm; cypselae 0.5–2 mm; pappus 2–4 mm. New Mexico material belongs to **var. decemflora** (DC.) Fernald. Locally common on open slopes, meadows, 5000–7500 ft. NE, C, SC.

Solidago pallida (Porter) Rydb. [*S. speciosa* Nutt.]—Showy goldenrod. Plants 2–15(–20) dm; leaves ovate to lanceolate or oblong, 5–15(–30) cm, usually entire, 1-nerved, glabrous; involucres 3–6 mm; disk florets 2.5–4 mm; ray florets 4–8(–11), 3–4 mm; pappus 3–4.5 mm; cypselae 1.5–2.5 mm. Uncommon or locally common in open places in the foothills in sandy or rocky soil, 5400–9000 ft. Bernalillo, Sandoval, San Miguel.

Solidago petiolaris Aiton [*S. angusta* Torr. & A. Gray; *S. squarrulosa* (Torr. & A. Gray) Alph. Wood]—Downy ragged goldenrod. Plants 40–150 cm; leaves basal, absent at flowering; blades mostly lanceolate-elliptic or ovate, sometimes linear-lanceolate, 30–150 × 5–30 mm, margins entire or few-toothed; abaxial faces sometimes resinous and shiny, glabrous or strigillose, adaxial glabrous or scabrous; heads in mostly paniculiform arrays; involucres campanulate, 4.5–7.5 mm; ray florets (5–)7–9; laminae 3–7 × 1–2 mm; disk florets (8–)10–16, corollas 4–5 mm; cypselae 3–4 mm, glabrous; pappus ca. 4 mm. Sandy soils, prairie communities, 6000–7000 ft. Union.

Solidago rigida L. [*S. canescens* (Rydb.) Friesner]—Stiff goldenrod. Plants 2–16 dm; leaves elliptic-oblong to lanceolate, 5–15(–25) cm, 1-nerved, densely hairy, entire or obscurely toothed; involucres 5–9 mm, bracts conspicuously striate; ray florets 6–13, 1.4–5.5 mm; disk florets 4–6 mm; cypselae 0.8–2 mm; pappus 3–4 mm. Our material belongs to **var. humilis** Porter [*Oligoneuron rigidum* (L.) Small var. *humilis* (Porter) G. L. Nesom]. Common along roadsides, open meadows, 3500–8800 ft. NC, SC.

Solidago velutina DC. [*S. sparsiflora* A. Gray]—Three-nerve goldenrod. Plants 1.5–8(–15) dm; leaves elliptic to spatulate or oblanceolate, 5–12 cm, usually entire, short-hairy, 3-nerved; involucres 3–4.5 mm; disk florets 3.5–6 mm; ray florets 6–10, 3–6 mm; pappus 2.5–4.7 mm; cypselae 0.7–3 mm. New Mexico material belongs to **subsp. sparsiflora** (A. Gray) Semple. Common on open slopes, mountain meadows, forest clearings, 5500–10,500 ft. Widespread except E.

Solidago wrightii A. Gray—Wright's goldenrod. Plants 2–6 dm; leaves elliptic to oblong, to 8 cm, short-hairy, mostly 1-nerved, margins entire; involucres 5–6 mm; disk florets 3–4 mm; ray florets 6–10, 3–5 mm; pappus 3–4 mm; cypselae 1.5–2.5 mm.

Two weak varieties exist:

1. Foliage and stems scabrous-pubescent, with stipitate glands…**var. adenophora**
1'. Foliage and stems lacking stipitate glands…**var. wrightii**

var. adenophylla S. F. Blake—NC, WC, SC.

var. wrightii—Widespread except NW.

SONCHUS L. – Sow-thistle

Gary I. Baird

Herbs, 3–350+ cm, taprooted, rhizomatous, or stoloniferous; stems erect, usually glabrous, sometimes stipitate-glandular; leaves alternate, margins usually 1(2) times pinnately lobed, ultimate margins usually dentate (teeth often ± prickly), sometimes entire; heads ligulate; involucres campanulate to urceolate, 5–15+ mm diam.; phyllaries 27–50 in 3–5+ series, unequal; receptacles flat to convex, glabrous; corollas yellow to orange; cypselae stramineous or reddish to dark brown, beaks 0, ribs usually 2–4(5+) on each face; pappus white, smooth or barbellulate bristles (Hyatt, 2006).

1. Perennials with extensive roots systems; heads 3–5 cm wide…**S. arvensis**
1'. Annuals from taproots; heads 1.5–2.5 cm wide…(2)
2. Leaf base auricles usually recurved or curled, rounded; cypselae usually wrinkled…**S. oleraceus**
2'. Leaf base auricles usually straight, sometimes curved; cypselae not wrinkled…**S. asper**

Sonchus arvensis L.—Field sow-thistle. Perennials, 0–150(–200) cm, usually rhizomatous or stoloniferous; blades of midcauline leaves oblong to lanceolate, (3–)6–40 × 2–15 cm, bases auriculate, margins usually pinnately lobed, dentate, or entire; peduncles sessile- or stipitate-glandular; involucres 10–17+ mm; phyllaries sessile- or stipitate-glandular; cypselae dark brown, oblanceoloid to ellipsoid, 2.5–3.5 mm, ribs 4–5(+) on each face, pappus 8–14 mm. New Mexico material belongs to **subsp. uliginosus** (M. Bieb.) Nyman [*S. uliginosus* M. Bieb.]. Irrigation ditches and wet meadows, 3270–7565 ft. NC, NW, C, Eddy. Introduced.

Sonchus asper (L.) Hill [*S. oleraceus* L. var. *asper* L.]—Spiny-leaf sow-thistle. Annuals or biennials, 10–120(–200+) cm; blades of midcauline leaves spatulate or oblong to obovate or lanceolate, 6–30 × 1–15 cm, bases auriculate, margins often pinnately lobed, usually prickly-dentate; peduncles usually stipitate-glandular, sometimes glabrous; involucres 9–13+ mm; phyllaries usually stipitate-glandular; cypselae stramineous to reddish brown, mostly ellipsoid, strongly compressed, ± winged, 2–3 mm, ribs 3(–5) on each face; pappus 6–9 mm. Disturbed sites along roadsides and streams, 3900–9000 ft. Widespread. Introduced.

Sonchus oleraceus L.—Common sow-thistle. Annuals or biennials, 10–140(–200) cm; blades of midcauline leaves spatulate or oblong to obovate or lanceolate, 6–35 × 1–15 cm, bases auriculate, margins usually pinnately (often runcinately) lobed, entire, or dentate; peduncles usually glabrous, sometimes stipitate-glandular; involucres 9–13+ mm; phyllaries usually glabrous, sometimes tomentose and/or stipitate-glandular; cypselae dark brown, mostly oblanceoloid, 2.5–3.5+ mm, ribs 2–4 on each face, faces transversely rugulose or tuberculate across and between ribs; pappus 5–8 mm. Disturbed sites such as roadsides, gardens, and along streams, 3900–8610 ft. NC, NW, C, SC, SW, Roosevelt. Introduced.

STENOTUS Nutt. – Mock goldenweed

Jennifer Ackerfield

Stenotus armerioides Nutt.—Thrift mock goldenweed. Perennial herbs, 1–3 dm; leaves chiefly basal, crowded at branch tips, alternate, simple, entire, narrowly oblanceolate to spatulate, 1.7–9 cm, margins sometimes ciliate, gland-dotted; heads radiate, 1–2(–4); involucral bracts subequal in 2–3 series, with a greenish center or tip and scarious margins; receptacles naked; ray florets 5.5–19 mm, yellow; disk florets 4–9 mm, yellow; cypselae 2–6 mm; pappus 3.5–7.5 mm, compressed, densely hairy, of white capillary bristles. New Mexico material belongs to **var. armerioides**. Sandy or rocky soil, grasslands, badlands, sagebrush, piñon-juniper woodlands, 5500–9100 ft. NW (Ackerfield, 2015; Morse, 2006).

STEPHANOMERIA Nutt. – Wirelettuce, skeletonweed

Gary I. Baird

Annuals, taprooted, or perennials, herbage (mostly) glabrous or pubescent, sometimes glandular, sap milky; leaves basal and (mostly) cauline, simple, alternate, blades oblanceolate to linear, entire to dentate or pinnatifid; heads liguliflorous, borne singly or clustered; involucres cylindrical-conic, calyculi of 3–7 bractlets, unequal; phyllaries 5–8 in 1–2 series, equal; receptacles flat, epaleate; ray florets matutinal, corollas ligulate, pink, lavender, or white; disk florets lacking; cypselae ± monomorphic, light to dark brown or stramineous, ribs 5, broad and forming corners, separated by a narrow groove on each face (ours), smooth to tuberculate, beakless; pappus of bristles, persistent, white to tawny, plumose (at least partly so) (Gottlieb, 2006c).

1. Plants annual, taprooted…**S. exigua**
1'. Plants perennial, rhizomatous (deeply) and/or with a woody caudex…(2)
2. Florets 8–16…**S. thurberi**
2'. Florets 4–6…(3)
3. Pappus bristles tan or tawny (white only at base), plumose on distal 1/2–3/4; peduncles glabrous…**S. pauciflora**
3'. Pappus bristles white, plumose throughout; peduncles glandular-pubescent…**S. tenuifolia**

Stephanomeria exigua Nutt.—Small wirelettuce. Annuals; stems mostly 1, 8–64 cm, usually diffusely branched distally; leaves basal and lower cauline, blades linear to narrowly oblanceolate, 2–9 cm × 1–14 mm, margins dentate to laciniately pinnatifid or subentire; heads terminal or axillary; involucres 7–9 mm; ray florets 6, exceeding involucres; corollas pink-lavender; cypselae 3–4 mm, light to dark tan or stramineous; pappus 5–6 mm, white to stramineous. Our material is all assignable to **var. exigua**. Sagebrush, desert scrub, piñon-juniper communities, 3935–8100 ft. NC, NW, C, WC, SC, SW.

Stephanomeria pauciflora (Torr.) A. Nelson [*Prenanthes pauciflora* Torr.; *Lygodesmia pauciflora* (Torr.) Shinners]—Prairie wirelettuce. Perennials; stems 1–5+, 10–50 cm, diffusely branched distally; leaves basal and lower cauline, blades linear, 2–6 cm × 2–8 mm, margins entire to dentate or irregularly pinnately lobed; heads terminal and axillary; involucres 6–10 mm; ray florets 5–9, exceeding involucres; corollas pink or lavender; cypselae 3–5 mm, stramineous; pappus 5–7 mm, white and not plumose, or stramineous and plumose throughout. Juniper and desert shrub communities, in sandy to clay soils, often in wash bottoms and alkaline habitats, 3340–8660 ft. Widespread.

Stephanomeria tenuifolia (Raf.) H. M. Hall. [*Prenanthes tenuifolia* Torr.]—Slender wirelettuce. Perennials; stems numerous, slender, persistent, 30–75 cm, erect, branched proximally or throughout; leaves basal and lower cauline, blades linear to filiform, 2–10 cm × 1 mm or less (excluding lobes), margins entire or dentate to lobed; heads terminal and axillary; involucres 7–11 mm; phyllaries 4–5 in 1–2 series; ray florets 4–5, exceeding involucres; corollas lavender or white; cypselae 3–4 mm, stramineous; pappus 5–6 mm, white, not abruptly widened proximally, plumose. Riparian areas and cliff habitats or canyon bottoms in piñon-juniper, ponderosa pine, and Douglas-fir communities, 3250–7500 ft. EC, NW, C, WC, SE, SC, SW.

Stephanomeria thurberi A. Gray—Thurber's wirelettuce. Perennials, 20–50 cm; stems single, branched on distal 1/3–1/2; basal leaf blades oblanceolate to spatulate, runcinate, 4–7 cm, margins pinnately lobed, cauline reduced, scalelike to linear and threadlike; heads borne singly on branch tips; involucres 9–11(–12) mm; phyllaries 6–8, glabrous; florets 10–16(–20); cypselae tan, 5–6 mm; pappus of 30–40 white bristles (persistent), wholly plumose. Sandy sites in juniper-mesquite grasslands and yellow pine forests, sometimes growing as a weed along roadsides, 4265–10,190 ft. WC, SC, SW, Lincoln.

STEVIA Cav. – Candyleaf

Kenneth D. Heil

Annuals, perennials, or shrubs, 10–100 cm; leaves cauline; all or mostly opposite or mostly alternate; petiolate or sessile; blades lanceolate, lance-linear, or ovate to ovate-deltate, margins entire or serrate to serrulate, faces hirtellous, puberulent, glabrescent, or glabrous (sometimes shiny), sometimes gland-dotted; heads discoid, in loose to dense, corymbiform arrays; involucres ± cylindrical, (1-)2–3 mm diam.; phyllaries persistent, 5(6) in ± 1 series; corollas purple to pink or white, throats narrowly funnelform; cypselae columnar to prismatic or fusiform, 5-ribbed, gland-dotted and/or scabrellous; pappus of 5 distinct scales (Nesom, 2006w).

1. Annuals...**S. micrantha**
1'. Perennials or shrubs...(2)
2. Shrubs...**S. salicifolia**
2'. Perennials...(3)
3. Leaves mostly opposite; petioles 0-3(-5) mm...**S. plummerae**
3'. Leaves mostly alternate; petioles 0...**S. serrata**

Stevia micrantha Lag.—Annual candyleaf. Annuals, 1–3.5(-4) dm; leaves mostly opposite, blades ovate to ovate-deltate, 3–3.5(-5) cm, margins serrate; involucres 6–8 mm; corollas white. Grassy openings, shady sites, oak, oak-pine, piñon-juniper, and ponderosa pine woodlands, 5000–7800 ft. Grant, Hidalgo, Sierra.

Stevia plummerae A. Gray [*S. plummerae* var. *alba* A. Gray]—Plummer's candyleaf. Perennials, 30–80 cm; leaves mostly opposite, petioles 0-3(-5) mm, blades lanceolate, lance-ovate, or oblanceolate, (2-)3–10 cm, margins coarsely serrate; involucres 5.5–8.5 mm; corollas whitish, pale rose, pink, or red. Canyon walls, oak, ponderosa pine to Douglas-fir woodlands. Uncommon, 7200–8300 ft. Catron, Grant, Sierra.

Stevia salicifolia Cav.—Willow-leaf candyleaf. Shrubs, 10–50(-80) cm; leaves opposite, blades lance-linear to lanceolate or narrowly elliptic, 3–9 cm, margins entire or serrate to serrulate; corollas white. Rocky sites, crevices, among boulders, oak and oak-pine woodlands; 1 record near Cloudcroft, 8267 ft. Otero.

Stevia serrata Cav.—Saw-tooth candyleaf. Perennials, 40–100 cm; leaves mostly alternate, petioles 0; blades narrowly lanceolate to lance-linear, 1.5–4 cm, margins serrulate; corollas white or pink. Disturbed sites, roadsides, oak-grasslands, oak-pine grasslands, conifer-oak, ponderosa pine–Douglas-fir, mixed conifer, 6000–10,000 ft. WC, SC, SW.

STYLOCLINE Nutt. – Desert woollyhead

Kenneth D. Heil

Annuals, 2–10(-20) cm; leaves cauline, mostly alternate, blades oblanceolate to lanceolate; heads in paniculiform to cymiform arrays; involucres 0 or inconspicuous; phyllaries 0 or vestigial; receptacles cylindrical, 2–3 mm; pistillate paleae ca. 3.4–4.5 mm; staminate paleae readily falling, mostly 2–4; pistillate flowers 12–25+; functionally staminate florets 3–6; bisexual florets 0; cypselae compressed, faces glabrous; pappus usually of 1–10(-13) bristles (Morefield, 2006e).

1. Receptacles clavate, height 2.8–3.5 times diam.; proximal leaves blunt...**S. sonorensis**
1'. Receptacles ± cylindrical, height 4–8 times diam.; proximal leaves acute...**S. micropoides**

Stylocline micropoides A. Gray—Desert woollyhead. Plants 2–14(-20) cm; longest leaves 8–20 mm, mostly subulate to lanceolate; longest pistillate paleae 3.4–4.5 mm, winged distally; functionally staminate flowers 3–6; ovaries vestigial, 0–0.3 mm; corollas 1.2–1.9 mm; cypselae 1–1.4 mm, compressed; staminate pappus of 2–5(-10) smooth to barbellulate bristles 1.1–2 mm. Sandy or gravelly soils, desert scrub, 4100–5200 ft. SC, SW.

Stylocline sonorensis Wiggins—Sonoran or mesquite neststraw. Plants 2–10(–15) cm; leaves blunt (proximal) or acute (median and distal), mucronate, longest 6–13 mm; largest capitular leaves ± elliptic to narrowly ovate, 3–10 × 2–3 mm; phyllaries 0; receptacles clavate, 1.2–2.2 mm, height 2.8–3.5 times diam.; longest pistillate paleae 1.9–3.1 mm, winged distally; functionally staminate florets 2–5; ovaries partially developed, 0.3–0.6 mm; corollas 0.9–1.4 mm; cypselae 0.6–0.8 mm, slightly compressed; staminate pappus of (1–)3–8 barbellate bristles 0.9–1.3 mm. Grassy hillsides, sandy drainages with mesquite, 4400–4800 ft. Grant, Hidalgo.

SYMPHYOTRICHUM Ness – Aster

Jennifer Ackerfield and Kenneth D. Heil

Herbs; leaves alternate, simple; heads usually radiate; involucral bracts in 2+ series, imbricate, green at the tip and chartaceous below; receptacles naked; ray florets white, blue, purple, or pink, sometimes the ligule very short and inconspicuous and the heads appearing discoid; disk florets yellow, purple, pink, or occasionally white; cypselae usually several-nerved; pappus of capillary bristles or double with an inner series of capillary bristles and an outer series of short hairs (Ackerfield, 2015; Allred et al., 2020; Brouillet et al., 2006). Previously included in *Aster*.

This is a very complex genus with many species exhibiting weak reproductive barriers. Some species are allopolyploids formed from hybridization between two other species and can backcross to one or more parents, resulting in plants with traits from one or more species. Other species hybridize where their ranges overlap, again resulting in a hybrid with traits of one or more species. In such cases, determinations can be quite difficult, and specimen annotation labels often read "with influence of other species."

1. Annuals from taproots; rays inconspicuous (< 2 mm) and barely surpassing disk florets or essentially absent…(2)
1'. Perennials with woody caudices or creeping rhizomes; rays conspicuous and well developed, far surpassing disk florets…(4)
2. Ray florets 16–30(–54), in 1–3 series, laminae 0.2–1.3 mm wide; phyllaries unequal…**S. subulatum**
2'. Ray florets (14–)75–110+, in 2–5 series, laminae 0 or 0.1–0.2 mm wide…(3)
3. Rays 1.5–2 mm, pinkish; involucral bracts oblong to narrowly oblanceolate…**S. frondosum**
3'. Rays essentially absent; involucral bracts acute to acuminate…**S. ciliatum**
4. Stems and/or involucral bracts with glandular hairs…(5)
4'. Stems and involucral bracts without glandular hairs, instead glabrous to variously hairy…(9)
5. Leaves linear or narrowly oblong, > 7 times as long as wide, usually < 4 mm wide…(6)
5'. Leaves lanceolate, oblong, or elliptic, < 5 times as long as wide and principal leaves > 4 mm wide…(8)
6. Leaf margins with coarse bristly-ciliate hairs, these spaced approximately 1 mm apart…**S. fendleri**
6'. Leaf margins without ciliate hairs, or if present then these not long and coarse and very closely spaced…(7)
7. Involucral bracts hairy, not glandular [likely present]…S. ×amethystinum
7'. Involucral bracts glandular…**S. campestre**
8. Leaf bases not auriculate but sometimes slightly clasping; involucral bracts oblanceolate, obtuse, or rarely acute, the tips green, not at all purple or purple-margined; stems not reddish…**S. oblongifolium**
8'. At least some leaf bases auriculate or cordate-clasping; involucral bracts acute or acuminate, usually with purple margins or tips; upper stems often reddish or purplish…**S. novae-angliae**
9. Involucral bracts and most leaves with a white spinulose tip; ray florets usually white or sometimes violet or pink…(10)
9'. Involucral bracts and leaves without a white spinulose tip; ray florets various…(13)
10. Stems and leaves glabrous, although often ciliate-margined; basal leaf cluster usually present at flowering time; stem leaves usually a mix of long leaves and short leaves…**S. porteri**
10'. Stems and leaves hairy; basal leaf cluster not present at flowering time or rarely so; stem leaves all generally the same length…(11)
11. Ray florets violet, blue, pink, or very seldom white; involucral bracts with a small spinulose tip, the green portion ovate, nearly orbicular; heads usually larger (involucres 6–15 mm wide)…**S. ascendens** (in part)
11'. Ray florets white; involucral bracts with a conspicuous spinulose tip, the green portion lanceolate; heads usually smaller (involucres 4.5–9 mm wide)…(12)
12. Heads secund (situated on 1 side of the stem; this is easiest to see on unpressed specimens) and relatively small; involucres 2.5–5 (average 4) mm; disk florets 5–25 per head…**S. ericoides**

12'. Heads usually not secund, and larger; involucres 4.5–8 (average 5.5) mm; disk florets 13–36 per head…
      S. falcatum
13.   Outer involucral bracts equaling or surpassing inner involucral bracts in length, mostly foliaceous…(14)
13'. Outer involucral bracts not equaling or surpassing inner involucral bracts in length, foliaceous or not…(16)
14.   Inflorescence a long panicle with numerous leaves and many heads; ray florets pink or white (sometimes
      drying purple)…**S. eatonii** (in part)
14'. Inflorescence with few heads, or if many then inflorescence a cymose panicle with reduced leaves; ray
      florets pink, purple, blue, or violet…(15)
15.   Middle stem leaves narrowly lanceolate, mostly < 1 cm wide; involucral bracts narrow, never leafy,
      sometimes purple-tipped but rarely purple-margined; plants below 10,500 ft. elevation; stems erect…
      S. spathulatum (in part)
15'. Middle stem leaves lanceolate to ovate, mostly > 1 cm wide; involucral bracts wider and leafy (if bracts
      narrow and linear and leaves < 1 cm wide, then bracts usually purple-tipped and -margined, and plants
      alpine, typically above 10,000 ft., with decumbent to erect stems)…**S. foliaceum**
16.   Outer involucral bracts obtuse (the green portion ovate to nearly orbicular)…**S. ascendens** (in part)
16'. Outer involucral bracts acute…(17)
17.   Stem leaves not or little clasping…(18)
17'. Stem leaves clasping…(23)
18.   Stem leaves ovate to broadly elliptic…(19)
18'. Stem leaves linear to narrowly elliptic…(21)
19.   Plants tufted, with short rhizomes; inflorescences racemose to narrowly paniculate, the branches as-
      cending; ray corollas pink…**S. eatonii** (in part)
19'. Plants with long rhizomes; inflorescence paniculate to corymblike, the branches usually open; ray corol-
      las usually violet to blue…(20)
20.   Leaves thin, the margins flat; bracts of peduncles 5–12+; outer phyllaries linear-lanceolate…**S. lanceo-
      latum** (in part)
20'. Leaves firm, the margins often revolute; bracts of peduncles 1–3; outer phyllaries oblong-lanceolate…
      S. praealtum
21.   Plants tufted, with short rhizomes; inflorescences racemose to narrowly paniculate; ray corollas pink…
      S. eatonii (in part)
21'. Plants colonial, with long rhizomes; inflorescences paniculate to corymblike; ray corollas usually violet to
      blue…(22)
22.   Basal leaves persistent at flowering time; cypselae not compressed, 2.5–3.5 mm…**S. spathulatum** (in
      part)
22'. Basal leaves withering by flowering time; cypselae ± compressed, 1.5–2 mm…**S. lanceolatum** (in part)
23.   Bracts on peduncles large and foliaceous; peduncles ± hispid-pilose…**S. lanceolatum** (in part)
23'. Bracts on peduncles narrow to subulate; peduncles glabrous or nearly so…**S. laeve**

Symphyotrichum ×amethystinum (Nutt.) G. L. Nesom [*Aster ×amethystinus* Nutt.]—A hybrid
between *S. novae-angliae* and *S. ericoides* occurring where their ranges overlap, 5000–6000 ft. No
records seen.

Symphyotrichum ascendens (Lindl.) G. L. Nesom [*Aster ascendens* Lindl.; *Virgulaster ascendens*
(Lindl.) Semple]—Western aster. Perennials, 1–12 dm; leaves linear or narrowly lanceolate, 2–10 cm, entire, glabrous to hairy; involucres (4–)5–7 mm, tips of bracts rounded or sometimes minutely mucronate;
ray florets 25–65(–80), 6–10 mm, pink or violet to blue, occasionally white. Common in a wide variety of
habitats including along roadsides, in meadows, along streams, and in sagebrush to mixed conifer,
5500–11,650 ft. NC, NW, SC. *Symphyotrichum ascendens* is an allopolyploid derived from the hybridization of *S. spathulatum* and *S. falcatum* (Allen, 1985). It is a variable species, probably from the result
of backcrosses to the original parents or even to another, closely related species. It tends to resemble
S. spathulatum morphologically more than *S. falcatum*.

Symphyotrichum campestre (Nutt.) G. L. Nesom [*Aster campestris* Nutt.; *Virgulus campestris*
(Nutt.) Reveal & Keener]—Western meadow aster. Perennials, 1–4 dm; leaves linear-oblanceolate, 1–8
cm, glabrous to sparsely scabrous to glandular-stipitate, mucronate or white-spinulose; involucres 5.5–8
mm, sparsely to densely glandular-stipitate; ray florets 15–30, 5–15 mm, purple. Infrequent in dry meadows, dry lake beds, lake margins, open places in the mountains, 5800–9500 ft. NC, NW, Cibola.

Symphyotrichum ciliatum (Ledeb.) G. L. Nesom [*Aster brachyactis* S. F. Blake; *Brachyactis ciliata*
(Ledeb.) Ledeb. subsp. *angusta* (Lindl.) A. G. Jones]—Rayless alkali aster. Annuals, 0.7–7 dm; leaves linear-

oblanceolate to spatulate, 3–8(–15) cm, glabrous, margins ciliate to scabrous; involucres 5–7(–11) mm, bracts acute, glabrous; ray florets absent. Usually along the borders of lakes or streams in moist, saline soil, 4600–9600 ft. NC, NW, Cibola, Otero.

Symphyotrichum eatonii (A. Gray) G. L. Nesom [*Aster bracteolatus* Nutt.; *A. eatonii* (A. Gray) Howell]—Eaton's aster. Perennials, (2.7–)6–15 dm; leaves linear to narrowly elliptic or oblanceolate, 0.8–15 cm, glabrous to minutely hairy, margins ciliate; involucres 4.5–8(–10) mm, bracts mucronate; ray florets 20–40, 5–12 mm, pink or white. Infrequent along streams, moist meadows, open slopes in the mountains, 6000–7500 ft. NC.

Symphyotrichum ericoides (L.) G. L. Nesom [*Aster ericoides* L.; *Virgulus ericoides* (L.) Reveal & Keener]—White aster. Perennials, 3–10 dm; leaves linear to lanceolate, 5–7 cm, hirsute, white-spinulose at the tips; involucres 2.5–5 mm, bracts white-spinulose-tipped; ray florets 10–18, 3–8 mm, white.

1. Plants colonial, rhizomatous; involucres cylindro-campanulate when fresh…**var. ericoides**
1'. Plants cespitose, with cormoid caudices, not strongly rhizomatous; involucres broadly campanulate when fresh…**var. pansum**

var. ericoides—White aster. Plants densely to diffusely colonial; strongly rhizomatous, not cormoid; stems 1, decumbent to erect; involucres cylindro-campanulate. Desert scrub, piñon-juniper to ponderosa pine and montane grassland communities, 3800–8200 ft. NC, EC, WC, SC.

var. pansum (S. F. Blake) G. L. Nesom [*Aster multiflorus* Aiton var. *pansus* S. F. Blake; *A. ericoides* L. var. *pansus* (S. F. Blake) B. Boivin]—Many-flowered aster. Plants cespitose, with cormoid caudices, not strongly rhizomatous; stems 1–10+, decumbent to ascending or erect to arching; involucres broadly campanulate when fresh. Sacaton-saltgrass community. San Juan.

Stem hair type has been used as a criterion for separating *S. ericoides* (hairs appressed or ascending) from *S. falcatum* (hairs spreading). However, Jones (1978) notes that in populations of *S. ericoides*, colonies of plants with both pubescence types occurred side by side and that the plants were otherwise indistinguishable from each other. In addition, stem pubescence can vary between both types on a single specimen. Although *S. ericoides* and *S. falcatum* can cross and produce hybrid plants, these are uncommon, indicating a strong reproductive barrier between the two species.

Symphyotrichum falcatum (Lindl.) G. L. Nesom [*Aster falcatus* Lindl.; *Virgulus falcatus* (Lindl.) Reveal & Keener]—White prairie aster. Perennials, 8 dm; leaves linear to narrowly lanceolate, 2–8 cm, white-spinulose-tipped; involucres 4.5–8 mm, bracts white-spinulose-tipped; ray florets 17–35, 6–8 mm, white. Common in fields, along roadsides, open plains and meadows, 3500–10,000 ft.

1. Plants cespitose, 1–5(–10)-stemmed, usually sparsely appressed-strigose, with cormoid caudices; new shoots developing near bases of old stems; plants sparsely hairy…**var. falcatum**
1'. Plants colonial, usually 1-stemmed, sometimes clumped; new shoots developing at ends of elongate rhizomes, rhizomes entangled; plants usually densely hairy…**var. commutatum**

var. commutatum (Torr. & A. Gray) G. L. Nesom [*Aster commutatus* (Torr. & A. Gray) A. Gray]—White prairie aster. Plants colonial, usually densely hairy, long-rhizomatous; phyllaries ± unequal, apices strongly squarrose. Prairies, roadsides, stream banks, 5300–8300 ft. Widespread except EC, SE.

var. falcatum—White prairie aster. Plants cespitose, usually sparsely appressed-strigose, with cormoid caudices; phyllaries subequal, apices not strongly squarrose. Stream banks, alkali lakes and flats, prairies, mesic montane habitats, 3900–8200 ft. NE, NC, C, WC.

Symphyotrichum fendleri (A. Gray) G. L. Nesom [*Aster fendleri* A. Gray]—Fendler's aster. Perennials, 6–30 cm; basal leaves linear-oblanceolate, 20–40 × 5–30 mm, cauline linear to linear-lanceolate, 10–40 × 5–20 mm, margins entire; heads in racemiform to paniculiform arrays; involucres campanulate, 4–7 mm; ray florets 10–20, corollas light to dark lavender to purple; disk florets (7–)10–30, corollas yellow, becoming reddish purple. Soils eroded from limestone and sandstone outcrops; prairies, pastures, roadsides, 5660–7300 ft. Colfax, Mora, Rio Arriba.

Symphyotrichum foliaceum (Lindl. ex DC.) G. L. Nesom [*Aster foliaceus* Lindl. ex DC.]—Alpine leafybract aster. Perennials, 0.5–10 dm; leaves oblanceolate, lanceolate, or ovate, 1.8–16 cm; involucres 6–12 mm, bracts usually large and leafy, acute to rounded at the tips; ray florets 15–60, 9–20 mm, pink, purple, or blue.

There are 3 varieties of *S. foliaceum* in New Mexico. These can intergrade, but for the most part they are distinct:

1. Plants shorter, < 2 dm; involucral bracts with purple tips and margins; middle stem leaves mostly < 1 cm wide…**var. apricum**
1'. Plants taller, > 2 dm; involucral bracts typically without purple tips and margins; middle stem leaves mostly > 1 cm wide…(2)
2. Involucral bracts oblong to ovate with an obtuse to acutish tip, the outer foliaceous ones broadly lanceolate to ovate with a rounded apex…**var. canbyi**
2'. Involucral bracts linear with an acute to acuminate tip, the outer foliaceous ones linear with a very acute apex…**var. parryi**

var. apricum (A. Gray) G. L. Nesom—Alpine leafy aster. Alpine and subalpine slopes and meadows and along streams, 9300–10,250 ft. NC.

var. canbyi (A. Gray) G. L. Nesom—Canby's aster. Typically drier habitats such as woods and open slopes, usually at lower elevations than var. *apricum*, 6200–10,700 ft. NC, NW, C, WC.

var. parryi (D. C. Eaton) G. L. Nesom—Parry's aster. Moist habitats such as along streams and in wet meadows, usually at lower elevations than var. *apricum*, 6100–10,500 ft. NC.

Symphyotrichum frondosum (Nutt.) G. L. Nesom [*Aster frondosus* (Nutt.) Torr. & A. Gray; *Brachyactis frondosa* (Nutt.) A. Gray]—Short-rayed alkali aster. Annuals, 0.5–14 dm; leaves oblanceolate to linear, 1–11 cm, glabrous; involucres 5–9 mm, bracts glabrous; ray florets 90–110, 1.5–2 mm, pink to pinkish white. Usually along the borders of lakes or ponds in moist, saline soil, 5300–9000 ft. NW, Cibola, Mora.

Symphyotrichum laeve (L.) Á. Löve & D. Löve [*Aster laevis* L. var. *geyeri* A. Gray]—Smooth blue aster. Perennials, 4–12 dm; leaves linear-oblanceolate to lanceolate or elliptic, 0.8–14 cm, glabrous; involucres 5–8 mm, bracts glabrous; ray florets 15–30, 6–9 mm, purple or blue. Our material belongs to **var. geyeri** (A. Gray) G. L. Nesom. Common in open meadows, along streams, and in forest openings, 5000–10,170 ft. NC, NW, C, SC, Cibola, Eddy.

Symphyotrichum lanceolatum (Willd.) G. L. Nesom [*Aster hesperius* A. Gray]—Western lined aster. Perennials, 3–15(–20) dm; leaves linear, lance-ovate, oblanceolate, or obovate, 4–15 cm, glabrous; involucres 3–8 mm, outer bracts sometimes foliaceous; ray florets 16–50, white to pinkish or pale purple. New Mexico material belongs to **subsp. hesperium** (A. Gray) G. L. Nesom. Along streams and ditches and in moist meadows, 4100–10,350 ft. NE, NC, NW, C, WC, SC.

Symphyotrichum foliaceum specimens with influence of other *Symphyotrichum* species may key out here because the outer involucral bracts are not equal to or longer than the inner involucral bracts in these hybrids. However, this species can be separated from *S. lanceolatum* by the presence of fewer, larger, and showier heads (involucre width 10–20 mm as opposed to 7–10(–12) mm in *S. lanceolatum*).

Symphyotrichum novae-angliae (L.) G. L. Nesom [*Aster novae-angliae* L.; *Virgulus novae-angliae* (L.) Reveal & Keener]—New England aster. Perennials, 3–12 dm; leaves spatulate to oblanceolate, oblong, or lanceolate, 2–10 cm, hairy and glandular-stipitate; involucres 7–10(–15) mm, bracts glandular-stipitate; ray florets 40–75(–100), 9–15 mm, dark pink to purple, rarely white. Roadsides and open meadows; can also escape from cultivation and is sometimes intentionally introduced, 5000–7800 ft. NC, Grant, Otero.

Symphyotrichum oblongifolium (Nutt.) G. L. Nesom [*Aster oblongifolius* Nutt.; *Virgulus oblongifolius* (Nutt.) Reveal & Keener]—Aromatic aster. Perennials, 1–10 dm; leaves linear-lanceolate to oblong or oblanceolate, 2–10 cm, glandular-stipitate; involucres 5–9 mm, bracts hairy and glandular-stipitate;

ray florets 25–35, 9–15 mm, purple or rose-purple. Dry, rocky, open sites and mesas, 5250–8700 ft. NE, Grant.

Symphyotrichum porteri (A. Gray) G. L. Nesom [*Aster porteri* A. Gray]—Smooth white aster. Perennials, 2–4 dm; leaves linear to linear-oblanceolate, 5–10 cm, white-spinulose-tipped; involucres 4–5 mm, bracts white-spinulose-tipped; ray florets 4–8 mm, white. Open fields, meadows, roadsides, ponderosa pine forests, 5250–7000 ft. Harding, Mora, San Miguel.

Symphyotrichum praealtum (Poir.) G. L. Nesom [*Aster praealtus* Poir.]—Willowleaf aster. Perennials, (10–)50-150(-200) cm, with fleshy long rhizomes; basal leaf blades spatulate, 40–70 × 10–25 mm; cauline blades elliptic or lanceolate to oblanceolate or linear-lanceolate, 40–100(-150) × 3–18 mm; involucres campanulate, (4–)5-7(-8) mm; ray florets (6–)20–35, corollas pale blue-violet to lavender or rose-purple, rarely white; disk florets 20–30(-35+), corollas cream or light yellow turning pinkish purple. Wet loamy soils, prairies, meadows, stream banks, ditches, 4500–7900 ft. NC, WC, Otero.

Symphyotrichum spathulatum (Lindl.) G. L. Nesom [*Aster occidentalis* (Nutt.) Torr. & A. Gray; *A. spathulatus* Lindl. var. *spathulatus*]—Western mountain aster. Perennials, 1–8.5 dm; leaves oblanceolate to lanceolate or elliptic, 1–15 cm, glabrous; involucres 5–8 mm; ray florets 15–50, 6–15 mm, purple to blue. Infrequent along streams and in moist meadows in the mountains, 6770–10,600 ft. NW. *Symphyotrichum spathulatum* freely hybridizes with *S. foliaceum*, especially var. *parryi* and sometimes var. *apricum* where their habitats overlap.

Symphyotrichum subulatum (Michx.) G. L. Nesom [*Aster subulatus* Michx.]—Annual saltmarsh aster. Annuals, (1–)3-15 dm; leaves lanceolate or ovate to oblanceolate, 1–9 cm, glabrous; involucres 5–7 mm; ray florets 17–45, 2–7 mm, white to pink or lavender to blue.

1. Ray florets lavender to blue, 3.5–7 mm…**var. ligulatum**
1′. Ray florets white to pink, 1.9–3 mm…**var. parviflorum**

var. ligulatum (Shinners) S. D. Sund. [*Aster subulatus* Michx. var. *ligulatus* Shinners; *Symphyotrichum divaricatum* (Nutt.) G. L. Nesom]—Saltmarsh aster. Plants (1–)6-20 dm; involucres (5.2–)6-6.9 mm; ray florets lavender to blue, (1.5–)4.5-7 × 0.9–1.3 mm. Marshy habitats, often weedy, roadsides, lawns, disturbed sites, 3900–6550 ft. Widespread except NE.

var. parviflorum (Nees) S. D. Sundb. [*Tripolium subulatum* (Michx.) Nees var. *parviflorum* Nees; *Symphyotrichum expansum* (Poepp. ex Spreng.) G. L. Nesom]—Southwestern annual saltmarsh aster. Plants (1–)7-15 dm; involucres 5–6.2 mm; ray florets white, sometimes pink, 1.9–3 × 0.2–0.5 mm. Marshy habitats, roadsides, disturbed sites, 3300–6900 ft. NC, EC, WC, SE, SC.

TAGETES L. – Marigold

Kenneth D. Heil

Tagetes micrantha Cav.—Licorice marigold. Annuals, mostly 3–15(-35+) cm (ours); leaves cauline; mostly opposite, petiolate or sessile, blades 10–25(-35+) mm overall, lobes or leaflets 0 or 3–5, mostly linear to filiform overall, usually 1–3-pinnate, oil glands scattered and/or submarginal; involucres 9–12+ mm diam.; phyllaries persistent, 3–21+ in 1–2 series; receptacles convex to conic; ray florets 0–1+; ochroleucous to yellowish, ± 1.5–2.5 mm; disk florets 5–6+, corollas 2.5–3.5 mm. Piñon-juniper woodlands, ponderosa pine forests, NC, WC, SC, SW (Strother, 2006jj).

TANACETUM L. – Tansy

Jennifer Ackerfield

Tanacetum vulgare L.—Tansy. Perennial herbs, 4–15 dm; leaves alternate, oblong to elliptic or oval, 4–20 cm, pinnatifid, the margins toothed, gland-dotted, glabrous to sparsely hairy; heads in a corymbiform inflorescence, discoid or disciform; involucral bracts imbricate in 2–3 series, the inner at least with scarious margins and tips; receptacles naked; ray florets usually lacking or < 1 cm; ray florets absent; disk florets 2–3 mm, yellow, corollas 5-lobed; cypselae 3–10-ribbed; pappus a short crown. Commonly culti-vated in gardens and occasionally escaping along roadsides, ditches, and meadows, 6700–8400 ft. Rio Arriba, Taos. Introduced (Ackerfield, 2015).

TARAXACUM F. H. Wigg. – Dandelion

Gary I. Baird

Perennial herbs, taprooted, 5–30 cm, sap milky; stems lacking; leaves all basal, simple, obscurely petio-late, blades oblanceolate to obovate, dentate to runcinately pinnatifid (rarely entire), ultimate margins entire to dentate, faces glabrous, glaucous, or pubescent; heads liguliflorous, solitary; involucres cylindro-campanulate to urceolate; calycular bractlets 6–18+, erect to squarrose; phyllaries 8–24+ in 1–2 series, equal, erect, herbaceous, faces glabrous, sometimes glaucous; florets bisexual, ligules yellow (peripheral often with a purplish or grayish stripe abaxially, this still evident when dried); cypselae monomorphic, flaxen, brownish, reddish, olivaceous, or grayish, mostly obovoid, coarsely muricate (especially distally), mostly 5–15-ribbed, apex slenderly beaked; pappus of 50–100+ bristles in 1 series, barbellulate, white to cream, persistent (Brouillet, 2006).

1. Phyllaries and calycular bractlets ± corniculate…(2)
1'. Phyllaries and calycular bractlets not corniculate…(3)
2. Calycular bractlets erect to spreading, corniculate; cypselae flaxen to brownish or olivaceous; native plants of alpine habitats (mostly > 10,000 ft.)…**T. ceratophorum**
2'. Calycular bractlets soon recurved or reflexed, corniculate or flat; cypselae dull red to reddish brown; intro-duced plants of often weedy areas (mostly < 8000 ft.)…**T. erythrospermum** (in part)
3. Plants < 5 cm; leaves mostly 10 or fewer, dentate, the teeth spreading and with entire margins; calycular bractlets 10 or fewer, ovate to deltate, erect to spreading (ultimately reflexed); native plants of alpine habitats (mostly > 11,500 ft.)…**T. scopulorum**
3'. Plants > 5 cm; leaves mostly > 10, subentire to runcinately pinnatifid; calycular bractlets > 10, lanceolate, soon recurved or reflexed; introduced, ruderal weeds of mostly disturbed areas (mostly < 10,000 ft.)…(4)
4. Cypsela bodies dull red to reddish brown…**T. erythrospermum** (in part)
4'. Cypsela bodies flaxen, olivaceous, tan, or grayish…**T. officinale**

Taraxacum ceratophorum (Ledeb.) DC.—Horned dandelion. Leaf blades 2–9 cm × 5–25 mm (in-cluding teeth), margins subentire to runcinately dentate; scapes 1.5–12.5 cm; involucres 8–18 mm; calycu-lar bractlets 10–16, 4–8 mm, lance-ovate, erect to spreading; phyllaries 12–14, 8–14 mm, lanceolate, cor-niculate (rarely not); florets ~50, peripheral exceeding phyllaries, ligules 8–11 × 1–2 mm; cypsela bodies 4–5 mm, flaxen to brownish or olivaceous, principal ribs 5, smaller ribs 10–15, beaks 5–14 mm; pappus 5–7 mm. Alpine meadows and slopes, 7875–12,750 ft. NC, Otero, Sierra.

Taraxacum erythrospermum Andrz. [*T. laevigatum* DC.]—Redseed dandelion. Leaf blades 5–11 cm × 5–35 mm (including lobes), margins runcinately pinnatifid; scapes 3–24 cm; involucres 10–15 mm; calycular bractlets 10–16, ovate to lanceolate, recurved to reflexed; phyllaries 12–18, 10–15 mm, lanceo-late, corniculate or not; florets ~75–150, peripheral ± exceeding phyllaries, ligules 5–9 × 1–2 mm; cypsela bodies 2.5–4 mm, dull red to red-brown, ribs ~15, beaks 5–7 mm; pappus 5–8 mm. Riparian areas, mead-ows, sagebrush, piñon-juniper-oak, ponderosa pine, and mixed conifer–aspen communities, often in dis-turbed sites, 5700–10,230 ft. NC, NW, C, WC, Otero.

Taraxacum officinale F. H. Wigg.—Common dandelion. Leaf blades 5–38 cm × 10–60 mm (includ-ing lobes), margins runcinately dentate, lobed, or pinnatifid; scapes 6–50 cm; involucres 10–20 mm; calycular bractlets 12–24, ovate to lanceolate, recurved to reflexed; phyllaries 12–22, 10–20 mm, lanceo-late, not corniculate; florets ~50–150, peripheral exceeding phyllaries, ligules 7–12 × 1–2 mm; cypsela bodies 2.5–4 mm, flaxen to brownish or olivaceous, larger ribs 5, smaller ribs 10, beak 5–9 mm; pappus 5–8 mm. Riparian areas, meadows, sagebrush, piñon-juniper-oak, ponderosa pine, mixed conifer–aspen, and alpine communities, often in disturbed or cultivated areas, 3100–12,150 ft. Widespread. Introduced.

Taraxacum scopulorum (A. Gray) Rydb. [*T. lyratum* (Ledeb.) DC.]—Alpine dandelion. Leaf blades oblanceolate, 2–5 cm × 5–6 mm (including lobes), margins entire to dentate-lobed; scapes 1–4 cm; invo-lucres 6–8 mm; calycular bractlets ~8, ovate, erect; phyllaries 8–12, 6–8 mm, lanceolate, not corniculate;

florets 25–50, peripheral just exceeding phyllaries, ligules 5–6 × ~1 mm; cypsela bodies ~3 mm, reddish brown to flaxen, larger ribs ~5, smaller ribs ~10, beak 3–4 mm; pappus 4–5 mm. Uncommon in alpine meadows, 11,800–12,150 ft. Mora, Taos.

TETRADYMIA DC. – Horsebrush

Lynn M. Moore

Shrubs, mostly 30–200 cm; leaves simple, usually with fascicles of secondary leaves in axils, alternate; heads discoid, involucres turbinate to cylindrical or hemispherical, phyllaries persistent, 4–6 in 1–2 series, distinct, equal or subequal; receptacles naked; ray florets 0, disk florets 4–9, cream to bright yellow, tubes longer than throats, pappus 0 or of 70–150 bristles or 20–30 scales; cypselae prismatic to obconic or fusiform, obscurely 5-ribbed (Allred & Ivey, 2012; Strother, 2006kk).

1. Plants spinose, primary leaves forming spines…**T. spinosa**
1'. Plants lacking spines…(2)
2. Leaf blades linear-filiform…**T. filifolia**
2'. Leaf blades spatulate to lanceolate…**T. canescens**

Tetradymia canescens DC.—Spineless horsebrush. Shrubs, 10–80 cm; stems unarmed, pannose but for floccose or glabrescent streaks; leaf blades 5–40 mm; primary leaves lanceolate to spatulate, tomentose to sericeous; heads 3–8; involucres 6–12 mm; disk florets 4, cream to bright yellow; pappus 100–150 bristles, 6–11 mm; cypselae 3–5 mm, glabrous to hirsute. Piñon-juniper, oak, ponderosa pine woodlands, sagebrush, grasslands, 5000–9200 ft. NC, NW, C, WC, Chaves, Luna.

Tetradymia filifolia Greene—Threadleaf horsebrush. Shrubs, 10–50 cm; stems unarmed, tomentose to glabrescent; leaf blades 5–20 mm; primary leaves linear-subulate, floccose, sometimes glabrate; heads 3–7; involucres 7–10 mm; disk florets 4, cream to bright yellow; pappus 100–130 bristles, 6–8 mm; cypselae 3–5 mm, hirsute. Piñon-juniper woodlands, grassland, limestone, sandy gypseous soils, 5000–7000 ft. NC, C, WC, SC. **E**

Tetradymia spinosa Hook. & Arn.—Shortspine horsebrush. Shrubs, 10–100 cm; stems spiny, evenly pannose; primary leaves forming recurved spines, secondary leaf blades linear-filiform to spatulate, 3–25 mm, glabrous or glabrescent; heads 1–2 in axils of spines; involucres 8–12 mm; disk florets 5, pale to bright yellow; pappus ± 25 subulate scales, 6–9 mm; cypselae 6–8 mm, copiously pilose, hairs 9–12 mm. Piñon-juniper woodlands, grasslands, sandy soils, 5000–6600 ft. NW.

TETRANEURIS Greene – Four-nerve daisy

Lynn M. Moore

Plants annuals or perennials, 2–50+ cm; leaves basal and/or cauline, simple, alternate; heads radiate or discoid, involucres hemispherical to campanulate, phyllaries 11–60+ in 3 series; receptacles naked; ray florets 0 or 7–27, yellow, usually 3-lobed, disk florets 20–250+, yellow proximally, yellow or purplish distally; cypselae ± obpyramidal, moderately to densely hairy; pappus persistent, of 4–8 usually aristate scales (Ackerfield, 2015; Allred et al., 2020; Bierner & Turner, 2006).

1. Plants annual…**T. linearifolia**
1'. Plants perennial, with caudices…(2)
2. Caudex branches not notably thickened distally; basal leaves not tightly clustered, internodes often evident…**T. scaposa**
2'. Caudex branches notably thickened distally; basal leaves tightly clustered, internodes usually not evident…(3)
3. Leaves all basal…(4)
3'. Leaves both basal and cauline…(5)
4. Midribs of leaves distinct; outer phyllaries 4–8, margins conspicuously scarious, 0.5–1.2 mm wide… **T. torreyana**
4'. Midribs of leaves indistinct; outer phyllaries 6–12, margins not or slightly scarious, 0–0.4 mm wide… **T. acaulis**

5. Leaf blades ± densely canescent, hairs tightly appressed and silvery looking…**T. argentea**
5'. Leaf blades glabrous to moderately sericeous…**T. ivesiana**

Tetraneuris acaulis (Pursh) Greene [*Actinella eradiata* (A. Nelson) A. Nelson; *Hymenoxys acaulis* (Pursh) K. F. Parker]—Stemless four-nerve daisy. Plants perennial, with branched caudices, branches notably thickened distally, 2-30+ cm; leaves all basal, new leaves tightly clustered, blades spatulate to linear-oblanceolate, glabrous or ± hairy, sometimes gland-dotted; heads 1-35(-60); involucres 7-12 mm, outer phyllaries 6-12, 3.9-9(-11.5) mm, margins slightly scarious, abaxial faces ± hairy; ray florets 8-15 (-21), corollas 9-19 mm; pappus of 5-8 aristate scales; cypselae 2-4 mm.

1. Leaves densely gland-dotted, not strigose-canescent or sericeous, often glabrous…**var. arizonica**
1'. Leaves sparsely to densely gland-dotted, usually densely hairy and strigose-canescent or sericeous…(2)
2. Flowering stems rarely > 7 cm; peduncles densely lanate below the heads…**var. caespitosa**
2'. Flowering stems often > 7 cm; peduncles not densely lanate below the heads…**var. acaulis**

var. acaulis—Stemless four-nerve daisy. Plants (3-)10-20(-30+) cm; peduncles (2-)8-20(-30) cm, leaves usually densely, sometimes sparsely, strigose-canescent, sparsely to densely gland-dotted; outer phyllaries 6-10, 4-7 mm; pappus 2.2-2.9 mm; cypselae 2.5-3 mm. Shortgrass prairie, piñon-juniper, ponderosa pine woodlands, sagebrush, oak, dry open areas, roadsides, loose sandy soils, 4450-10,400 ft. NE, NC, C, SE, SC,

var. arizonica (Greene) K. F. Parker—Arizona four-nerve daisy. Plants (2-)6-15(-30+) cm; peduncles (1-)5-15(-30) cm, leaves glabrous or moderately to densely hairy (not strigose-canescent or sericeous), densely gland-dotted; outer phyllaries 7-10, 3.9-6.5 mm; pappus 2-3.5 mm; cypselae 3-3.7 mm. Shortgrass prairie, piñon-juniper woodland, dry open areas, roadsides, mesa tops, 4500-6700 ft. NE, NC, NW, Torrance.

var. caespitosa A. Nelson—Cespitose four-nerve daisy. Plants 2-8(-12+) cm; peduncles 0.5-8(-12) cm, leaves sparsely to densely hairy, often lanate, sometimes sericeous, sparsely to sometimes densely gland-dotted; outer phyllaries 6-12, 6.8-9(-11.5) mm; pappus 2.5-3.5 mm; cypselae 2-3(-4) mm. Alpine tundra, windswept ridges at high elevations, piñon-juniper woodland, 7380-12,600 ft. NC, NW, C, Cibola.

Tetraneuris argentea (A. Gray) Greene—Perkysue. Plants perennial, with branched caudices, branches notably thickened distally, 6-25(-42+) cm; leaves basal and cauline, basal leaves tightly clustered, blades spatulate to oblanceolate, usually densely silvery strigose-canescent, ± gland-dotted; heads 1-10(-30) per plant; outer phyllaries 8-11, 4-6.5 mm, usually scarious, abaxial faces densely hairy; ray florets 8-14, corollas 11.8-17 mm; pappus of 5-6 aristate scales, 2.1-3.5 mm; cypselae 2.4-3.8 mm. Piñon-juniper, ponderosa pine woodlands, riparian areas, dry washes, roadsides, loose limestone or gypseous soils, 5300-9200 ft. NC, NW, C, WC, SC.

Tetraneuris ivesiana Greene—Ives' four-nerve daisy. Plants perennial, with branched caudices, branches notably thickened distally, 10-26+ cm; leaves basal and cauline, basal leaves tightly clustered, blades linear-oblanceolate, glabrous or sparsely to moderately lanate to sericeous, usually densely gland-dotted; heads (1-)5-30(-40) per plant; outer phyllaries 7-12, 5-6.4 mm, usually scarious, abaxial faces ± hairy; ray florets 7-10, corollas 10-20 mm; pappus of 5-7 aristate scales, 2.7-4.5 mm; cypselae 3-4.1 mm. Piñon-juniper, ponderosa pine, mixed conifer woodlands, open areas, loose rocky substrates, gypseous soils, 5300-9100 ft. NC, NW, C, Cibola.

Tetraneuris linearifolia Greene—Fineleaf four-nerve daisy. Plants taprooted annuals, 16-50+ cm; leaves basal and cauline, blades oblanceolate to linear, with 1-2 teeth or lobes, ± hairy, ± gland-dotted; heads 8-50(-80) per plant; outer phyllaries 8-21, 2.4-5.5 mm, scarious or not, abaxial faces ± hairy; ray florets 9-25, corollas 8.2-16.8 mm; pappus of 4-8 often aristate scales, 1-2.5 mm; cypselae 1.5-2.6 mm. Our plants are **var. linearifolia**. Piñon-juniper woodlands, arid rocky arroyos, loose gypseous soils, 3280-7200 ft. SE.

Tetraneuris scaposa (DC.) Greene—Stemmy four-nerve daisy. Plants perennial, with branched caudices, branches not thickened distally, 14-40+ cm; leaves all basal, new leaves not tightly clustered, blades spatulate to linear, sparsely to densely hairy, ± gland-dotted; heads 1–50 per plant; outer phyllaries 8–16, 3.8–6.6 mm, sometimes to often slightly scarious, abaxial faces ± hairy; ray florets 12–26, corollas 7.4–22 mm; pappus of 5–7 aristate scales, 1.6–2.3 mm; cypselae 2–3 mm. Our plants are **var. scaposa**. Shortgrass prairie, piñon-juniper woodland, dry open areas, roadsides, slopes, 3300–8800 ft. Widespread except N, WC.

Tetraneuris torreyana (Nutt.) Greene [*Actinella torreyana* Nutt.; *Hymenoxys depressa* (Torr. & A. Gray) S. L. Welsh & Reveal; *H. torreyana* (Nutt.) K. F. Parker]—Torrey's four-nerve daisy. Perennials, 2–15+ cm; caudices branched, branches notably thickened distally; stems 1–40, erect; leaves all basal, blades spatulate or oblanceolate to linear-oblanceolate, entire, glabrous or usually moderately to densely hairy, densely gland-dotted; heads 1–10(–40) per plant; outer phyllaries 4–8, 6–8.5 mm; ray florets 7–14; corollas 11–17 mm; disk florets 25–150+, corollas yellow, 4–4.7 cm; cypselae 3–4 mm. Open areas in piñon-juniper woodlands and ponderosa pine forests, 7100–8040 ft. NW, Harding, San Miguel.

THELESPERMA Less. – Greenthread

Kenneth D. Heil

Annuals, perennials, or subshrubs, 10–80+; stems usually 1, erect; leaves mostly basal, basal and cauline, or mostly cauline, mostly opposite, blades usually 1(–3) times pinnately lobed, faces usually glabrous; heads radiate or discoid, borne singly or in loose, corymbiform arrays; involucres hemispherical to urceolate; phyllaries 5–8 in ± 2 series; receptacles flat to convex; ray florets 0 or ca. 8, corollas yellow or red-brown; disk florets 20–100+, corollas yellow or red-brown; cypselae dark red-brown or stramineous, ± obcompressed; pappus 0 or persistent, of 2 retrorsely ciliate, subulate scales or awns (Strother, 2006ll).

1. Throats of disk corollas equal to or longer than lobes; pappus usually 0, rarely of 2 awns at 0.1–0.3(–0.5) mm…(2)
1'. Throats of disk corollas shorter than lobes; pappus usually of 2 awns or scales at (0.5–)1–3 mm, rarely 0… (4)
2. Cauline leaves ± scattered over proximal 3/4+ of plant height, internodes 45–95 mm…**T. simplicifolium**
2'. Cauline leaves ± crowded over proximal 1/4–1/2 of plant height, internodes mostly 5–25+ mm or 1–5(–35) mm…(3)
3. Leaf lobes mostly linear to filiform, 5–25(–45+) × 0.5(–1) mm; cypselae 2–3 mm…**T. longipes**
3'. Leaf lobes mostly oblanceolate to linear, (5–)10–35(–45+) × (1–)2–3(–5) mm; cypselae 5–7 mm…**T. subnudum**
4. Annuals…**T. filifolium**
4'. Perennials…(5)
5. Plants 10–30(–50) cm; ray florets usually 8, rarely 9; disk corollas red-brown…**T. ambiguum**
5'. Plants (20–)30–80+; ray florets 0; disk corollas yellow…**T. megapotamicum**

Thelesperma ambiguum A. Gray [*T. megapotamicum* (Spreng.) Kuntze var. *ambiguum* (A. Gray) Shinners]—Colorado greenthread. Perennial herbs or subshrubs, 10–30(–50) cm; cauline leaves crowded to ± scattered over proximal 1/3–1/2 of plant height, internodes mostly 5–50(–80+) mm, (5–)15–75 × 0.5–1(–2) mm; ray florets usually 8, rarely 0, laminae yellow, 4–8(–12+) mm; disk corollas red-brown, throats shorter than lobes; cypselae 4–5+ mm; pappus usually 1.5–2+ mm, rarely 0. Disturbed sites on sands or clays, meadows, pastures, 3350–7170 ft. SW, Quay, Socorro.

Thelesperma filifolium (Hook.) A. Gray [*T. filifolium* var. *intermedium* (Rydb.) Shinners]—Stiff greenthread. Annuals; stems 10–40(–70+) cm; cauline leaves crowded to ± scattered over proximal 1/2–3/4 of plant height, internodes mostly 10–35(–50+) mm, lobes mostly linear to filiform, sometimes oblanceolate, 5–30(–55+) × 0.5–1(–3+) mm; ray florets 8, laminae yellow to golden yellow, sometimes proximally suffused with red-brown, 12–20+ mm; disk corollas red-brown or yellow with red-brown nerves; cypselae 3.5–4+ mm; pappus 0.5–1(–2+) mm. Disturbed sites on clays or sandy soils, desert scrub, piñon-juniper woodlands, ponderosa pine forests, 3350–7950 ft. NW, EC, C, SE, SC, Hidalgo.

Thelesperma longipes A. Gray—Longstalk greenthread. Perennial herbs or subshrubs; stems 20–40+ cm; cauline leaves mostly crowded over proximal 1/4–1/2 of plant height, internodes mostly 5–25+ mm; lobes mostly linear to filiform, 5–25(–45+) × 0.5(–1) mm; ray florets 0; disk corollas yellow, sometimes with red-brown nerves, throats equal to or longer than lobes; cypselae 2–3 mm; pappus usually 0, rarely 0.1–0.3+ mm. Often on limestone ridges, openings in desert scrub, piñon-juniper woodlands, ponderosa pine forests, 3000–8100 ft. EC, C, WC, SE, SC, SW.

Thelesperma megapotamicum (Spreng.) Kuntze [*T. gracile* (Torr.) A. Gray]—Hopi tea, Navajo tea, greenthread. Perennial herbs or subshrubs, (20–)30–80+ cm; cauline leaves ± scattered over proximal 1/2–3/4 of plant height, internodes 40–100 mm, lobes mostly linear to filiform, sometimes oblanceolate, 20–40(–50+) × 0.5–1(–2.5) mm; ray florets 0, disk corollas yellow, often with red-brown nerves, throats shorter than lobes; cypselae 5–8 mm; pappus 1–2(–3) mm. Sandy or clay soil, disturbed sites in desert scrub, oak-piñon-juniper woodlands, ponderosa pine forests, 3050–10,200 ft. Widespread.

Thelesperma simplicifolium (A. Gray) A. Gray [*T. curvicarpum* Melchert]—Slender greenthread. Perennial herbs or subshrubs, (20–)30–70+ cm; cauline leaves ± scattered over proximal 3/4+ of plant height, internodes mostly 45–95 mm, lobes mostly linear to filiform, sometimes oblanceolate, (5–)15–45(–60) × 0.5–1(–2) mm; ray florets usually 8, rarely 0, laminae yellow, 9–15(–20+) mm; disk corollas yellow with red-brown nerves, throats equal to or longer than lobes; cypselae 3–4 mm; pappus usually 0, rarely 0.1–0.3+ mm. Often on limestone, openings in desert scrub and oak-juniper woodlands, 4460–7175 ft. Chaves, San Miguel, Socorro.

Thelesperma subnudum A. Gray [*T. caespitosum* Dorn; *T. subnudum* var. *alpinum* S. L. Welsh]—Navajo tea. Perennial herbs, stems 10–30(–40+) cm; cauline leaves mostly crowded over proximal 1/4–1/2 of plant height, internodes mostly 1–5(–35) mm, lobes mostly oblanceolate to linear, (5–)10–35 (–45+) × (1–)2–3(–5) mm; ray florets 0 or 8, laminae yellow, (6–)12–20+ mm; disk corollas yellow, sometimes with red-brown nerves, throats equal to or longer than lobes; cypselae 5–7 mm; pappus usually 0, rarely 0.1–0.5 mm. Often on talus, openings in piñon-juniper woodlands or ponderosa pine forests, 5000–9700 ft. NE, NC, NW, Lincoln.

THYMOPHYLLA Lag. – Dogweed

Lynn M. Moore

Plants annuals, perennials, subshrubs, or shrubs; leaves cauline, opposite or alternate, tips spinulose; heads radiate or rarely discoid, involucres campanulate to obconic, 2–7 mm diam., bractlets subtending heads 0 or 1–8, deltate to subulate, bearing oil glands, phyllaries persistent, 8–13(–22) in ± 2 series, sometimes strongly connate, usually bearing oil glands; receptacles naked; ray florets 5, 8, 13, or 21, yellow to orange, rarely white; disk florets 16–100+, yellow to orange; pappus coroniform or of 10(–20) scales, each scale erose, or 1–5-awned, or a fascicle of 5–9 basally connate bristles; cypselae obpyramidal, obconic, or cylindro-clavate, glabrous or sparsely strigillose (Allred & Ivey, 2012; Strother, 2006mm).

1. Leaves entire, not lobed…**T. acerosa**
1'. Leaves mostly pinnatifid or lobed…(2)
2. Plants annual; leaves mostly alternate…(3)
2'. Plants perennial; leaves mostly opposite…(4)
3. Bractlets subtending heads 3–8; disk florets 50–100+…**T. tenuiloba**
3'. Bractlets subtending heads 0–2; disk florets 25–45…**T. aurea**
4. Plants green, usually puberulent to canescent, sometimes glabrous; bractlets 0 or 1–5…**T. pentachaeta**
4'. Plants ashy-white, tomentose; bractlets 1–3…**T. setifolia**

Thymophylla acerosa (DC.) Strother—Pricklyleaf dogweed. Plants subshrubs or shrubs, ca. 25 cm, green, usually hairy, sometimes glabrous; leaves opposite, blades linear to acerose, entire; heads campanulate to cylindrical, 5–7 mm, peduncles 0–10 mm; phyllaries ca. 13, abaxial faces hairy or glabrous, bractlets 5, lance-linear; ray florets 7–8, lemon-yellow; disk florets 18–25+, pale yellow; pappus scales ca. 20, each a fascicle of 3–5 bristles, 3–4 mm; cypselae 3–3.5 mm. Desert scrub, dry rocky slopes and outcrops, disturbed areas, roadsides, 3250–7600 ft. Widespread except N, WC.

Thymophylla aurea (A. Gray) Greene ex Britton & A. Br.—Many-awn pricklyleaf. Plants annuals, ca. 20(–30) cm, green, glabrous or hairy; leaves alternate, blades linear, lobes 5–13; heads obconic to campanulate, 5-6 mm, peduncles 10-70 mm; phyllaries 12-15, abaxial faces hairy or glabrous, bractlets 0 or 1-2, subulate; ray florets 8-12, bright yellow; disk florets 30-45, yellow; pappus scales erose or awned; cypselae 3 mm.

    1.   Peduncles 10-30 mm; pappus scales 8-10, erose, 0.3-0.6 mm…**var. aurea**
    1'.   Peduncles 20-70 mm; pappus scales 18-20, aristate, 2-3 mm…**var. polychaeta**

    var. aurea—Many-awn pricklyleaf. Peduncles 10-30 mm; pappus scales 8-10, erose, 0.3-0.6 mm. Desert scrub, swales in grasslands, silty clay flats, 4000-4400 ft. SC, SW.

    var. polychaeta (A. Gray) Strother—Many-awn pricklyleaf. Peduncles 20-70 mm; pappus 18-20 aristate scales, 2-3 mm. Desert scrub, grasslands, silty clay flats, roadsides, 3950-6000 ft. Lea, Luna.

    Thymophylla pentachaeta (DC.) Small—Five-needle pricklyleaf. Plants perennials or subshrubs, ca. 15(-25) cm, ± grayish to green, usually hairy, sometimes glabrous; leaves mostly opposite, blades mostly pinnately lobed, lobes 3-11; heads obconic to campanulate or hemispherical, 4-6 mm, peduncles 20-100 mm; phyllaries 12-21, abaxial faces puberulent or glabrous, bractlets 0 or 1-5, deltate; ray florets (8-)12-21, yellow to orange-yellow, disk florets 16-80, yellow, sometimes irregular; pappus scales 10, erose and/or awned, 1-3 mm; cypselae 2-3 mm.

    1.   Heads cylindrical, 2-3.5 mm wide; disk florets 16-40…**var. hartwegii**
    1'.   Heads campanulate to hemispherical, mostly 4-5 mm wide; disk florets 50-80…**var. belenidium**

    var. belenidium (DC.) Strother—Five-needle pricklyleaf. Leaf lobes 3-7, subequal, heads obconic to campanulate, phyllaries ca. 13, margins of outer distinct almost to base, abaxial faces glabrous, margins ciliate, bractlets 3-5; disk florets 50-70; pappus usually of 10, 3-awned scales, sometimes of 5 blunt scales alternating with 5, 3-awned scales. Desert scrub and grasslands, dry slopes and hills, calcareous soils, 3200-6200 ft. C, SE, SC, SW.

    var. hartwegii (A. Gray) Strother—Hartweg's pricklyleaf. Leaf lobes 3-7, unequal, heads cylindrical, phyllaries ca. 13, margins of outer distinct 1/3 or less their length, abaxial faces sparsely hairy, often glabrescent, bractlets 3-5; disk florets 16-40, pappus of 5 erose scales alternating with 5, 1-3-awned scales, 3500-5400 ft. SE, SC, SW.

    Thymophylla setifolia Lag.—Texas pricklyleaf. Plants perennials, ca. 15 cm, ashy-white, tomentose; leaves opposite, lobes 3-7, from near base; heads obconic to campanulate, 3.5-4 mm, peduncles 30-40 mm; phyllaries 9-15, margins of outer distinct 1/3 or less their length, abaxial faces glabrous, bractlets 1-3, subulate; ray florets 7-10, bright yellow; disk florets 20-40, dull yellow; pappus scales connate, 0.3-0.5 mm; cypselae 1.5-2.2 mm. Our plants are **var. greggii** (A. Gray) Strother. Desert pavement, rocky arroyos, canyon bottoms, calcareous soils, 4300-6800 ft. SE, SC, Grant, Socorro.

    Thymophylla tenuiloba (DC.) Small—Bristle-leaf pricklyleaf. Plants annuals, ca. 30 cm, green, glabrous or sparsely hairy; leaves mostly alternate, lobes 7-15; heads obconic, 5-7 mm, peduncles 30-80 mm; phyllaries 12-22, margins of outer distinct 1/5 or less their length, abaxial faces glabrous or sparsely hairy, bractlets 3-8, deltate to subulate; ray florets 10-21, yellow-orange; disk florets 50-100+, yellow; pappus scales 10-12, each 3-5-awned; cypselae 2-3.5 mm. Our plants are **var. tenuiloba.** Basin flats, roadsides, 3900 ft. Luna. Introduced.

TONESTUS A. Nelson – Goldenweed

Jennifer Ackerfield

    Tonestus pygmaeus (Torr. & A. Gray) A. Nelson [*Haplopappus pygmaeus* (Torr. & A. Gray) A. Gray]—Pygmy goldenweed. Perennials; plants 1-9 cm; leaves alternate and basal, linear to spatulate or oblong, 1-5(-9.5) cm, eglandular or sparsely glandular-stipitate, entire; heads radiate; involucres 8-20 mm, the outer bracts ovate to elliptic, with rounded or shortly mucronate tip; ray florets 6.5-8.5 mm, yellow; disk florets 4.5-7.5 mm, yellow; cypselae 2-5 mm, narrowly oblong to subcylindrical, somewhat compressed, usually ribbed, villous. Locally common in rocky soil of the alpine, 9000–13,100 ft. NC (Ackerfield, 2015; Morse, 2010).

TOWNSENDIA Hook. – Townsendia

Timothy K. Lowrey

Annuals, biennials, or perennials, mostly taprooted, sometimes rhizomatous; stems decumbent to erect, glabrous or variously hairy; leaves basal and/or cauline, alternate, petiolate but petioles grading into blades; leaf blades 1-nerved, spatulate, oblanceolate, lanceolate, or linear, margins entire, sometimes dentate, faces glabrous or hairy; heads radiate, borne singly on scapes, leafy stems, or sessile in rosette leaves; phyllaries 16–80 in 3–7 series, unequal, margins somewhat scarious, faces glabrous or hairy; ray florets 4–67, pistillate, fertile, corollas white, blue, pinkish, or lavender; disk florets 16–100, bisexual, fertile, corollas yellow, 5-lobed; pappus of 12–35 lanceolate, subulate, or setiform scales (flattened bristles); cypselae compressed, obovate to oblanceolate, 2–3-nerved, faces glabrous or hairy, some species with glochidiate hairs, variable (Strother, 2006nn).

1. Plants 1–3 cm, pulvinate; heads sessile, nestled among rosette of basal leaves…(2)
1'. Plants 3–35 cm; heads at tips of stems…(3)
2. Leaves mostly 1–2 mm wide; phyllaries lanceolate, mostly 7–9 mm, 2–5 times longer than wide; rays 5–10 mm…**T. leptotes**
2'. Leaves mostly 2–6 mm wide; phyllaries linear, 10–17 mm, 6+ times longer than wide; rays 12–18 mm… **T. exscapa**
3. Pappus of all florets < 1 mm; plants rhizomatous or stoloniferous…**T. formosa**
3'. Pappus of disk florets 1–7 mm; plants taprooted…(4)
4. Phyllaries mostly 16–30, in 3–4 series…(5)
4'. Phyllaries 30–60, in 4–8 series…(8)
5. Plants annual; pappus of disk florets 1–2.5 mm…**T. annua**
5'. Plants biennial to perennial; pappus of disk florets mostly 2.5–8 mm…(6)
6. Stems only moderately hairy and the surface seldom hidden; disk pappus shorter than disk corolla… **T. fendleri**
6'. Stems densely hairy so the surface is hidden by the hairs; disk pappus equaling or longer than disk corolla…(7)
7. Rays 5–12 mm; pappus of disk florets 4–6 mm…**T. incana**
7'. Rays 3–6 mm; pappus of disk florets 2.5–3 mm…**T. gypsophila**
8. Pappus of disk florets 8–12, most scales 0.5–1 mm, only 1–2 bristles 1–4 mm; phyllaries 10–14 mm… **T. eximia**
8'. Pappus of disk florets 15–30, 4–6 mm; phyllaries 8–10 mm…**T. grandiflora**

Townsendia annua Beaman—Annual Townsend daisy. Annuals; stems erect to decumbent, 2–16 cm, strigose; leaves mostly cauline in adult plants; leaf blades spatulate, 8–25 × 1–5 mm, faces strigillose; heads borne at tips of stems; phyllaries 26–30 in 3 series, lance-ovate to oblanceolate, faces glabrate to strigillose; ray florets 12–30, corollas white or pinkish, ligules 5–8 mm; disk florets 60–120; pappus dimorphic; cypselae 2–3 mm, faces hairy, hairs glochidiate. Disturbed ground, sandy soils, piñon-juniper woodlands, arroyo bottoms, grassland, desert scrub, 3600–7650 ft. NC, NW, C, WC, SC, SW, Lea.

Townsendia eximia A. Gray—Tall Townsend daisy. Biennials or perennials; stems 6–15(–30) cm, erect, strigose; leaves basal and cauline; leaf blades spatulate to oblanceolate, 15–60(–120) × 2–10 mm, margins ciliate, faces mostly glabrous; heads borne at leafy stem tips; phyllaries 50–80 in 5–6 series, lanceolate, apices acute, hardened, glabrous; ray florets 15–55, corollas blue, ligules 12–20 mm; disk florets 100–150; pappus dimorphic; cypselae 3–4 mm, faces hairy, hairs glochidiate or with forked tips. Piñon-juniper woodlands and mixed conifer woodlands, C and NC mountains, 6000–11,800 ft. NC, C, Socorro.

Townsendia exscapa (Richardson) Porter [*Aster exscapus* Richardson; *Townsendia sericea* Hook.]— Easter daisy. Perennials, pulvinate; stems 1–3 cm, hirsute to strigose; leaves in basal rosettes; leaf blades spatulate, oblanceolate, or linear, 12–75 × 2–6 mm, faces strigillose; heads sessile, nestled in rosette leaves; phyllaries 30–60 in 4–7 series, lanceolate to linear, 10–17 mm, 6+ times longer than wide, herbaceous, faces sparsely hairy or glabrate; ray florets 11–40, corollas white, sometime tinged pink, ligules 12–18 mm; disk florets 100–150; cypselae 4–6 mm, faces hairy, hairs glochidiate, ray cypselae orange when mature and fresh. Gravelly hills, piñon-juniper and ponderosa forests, 4100–9915 ft. Widespread. One of the first species to flower in the spring.

Townsendia fendleri A. Gray—Fendler's Townsend daisy. Perennials; stems 3–20 cm, decumbent to erect, pilose to strigose; heads borne at leafy stem tips; phyllaries 22–40 in 4–5 series, lance-ovate to lanceolate, faces strigose; ray florets 10–25, corollas white or pinkish, ligules 5–10 mm; disk florets 40–80; pappus dimorphic; cypselae 2–3 mm, faces hairy, hairs glochidiate. Sandy soils and gypsum substrates, desert scrub, piñon-juniper woodlands, 4300–7800 ft. NC, NW, C, WC.

Townsendia formosa Greene—Smooth Townsend daisy. Perennials, rhizomatous and sometime stoloniferous; stems 10–35(–75) cm, strigillose; leaves basal and cauline, basal leaves forming a rosette; leaf blades spatulate to oblanceolate, 15–75 × 2–20 mm, reduced upward, becoming subulate distally, stems appearing almost scapose; heads borne at stem tips; phyllaries 30–45 in 4–6 series, lance-ovate to lanceolate, apices acute to apiculate, bodies hardened, glabrous; ray florets 20–34, corollas blue to white, sometimes pinkish, ligules 10–25 mm; disk florets 8–150; pappus a 2-toothed corona; cypselae 3–5 mm, faces glabrous or stipitate-glandular. Montane slopes and meadows, mixed conifer woodlands, 6800–9000 ft. WC, SC, San Miguel.

Townsendia grandiflora Nutt.—Largeflower Townsend daisy. Biennials or perennials; stems 3–30 cm, strigose; leaves basal and cauline; leaf blades spatulate to oblanceolate, sometimes nearly linear, 20–40(–90) × 1–10 mm, margins ciliate, faces mostly glabrous, sometimes sparsely hairy; heads borne at stem tips; phyllaries 30–40 in 4–5 series, lance-ovate to lanceolate, apices acute, hardened, and apiculate, faces mostly glabrous; ray florets 20–40, corollas blue, sometimes nearly white, ligules 7–20 mm; disk florets 80–120; pappus dimorphic; cypselae 3.5–4 mm, faces hairy, hairs forked or glochidiate. Shale and limestone substrates, piñon-juniper woodlands, ponderosa pine forests, grasslands, 5650–8800 ft. NC, Harding, Socorro.

Townsendia gypsophila Lowrey & P. J. Knight—Gypsum-loving Townsend daisy. Perennials, some plants with branching caudices; stems 3–12(–25) cm, pilose to strigose; leaves basal and cauline; leaf blades spatulate to oblanceolate, some nearly linear, 12–50 × 2–9 mm, faces strigose; heads borne at stem tips; phyllaries 20–24 in 3–4 series, lance-ovate to lanceolate, faces strigose; ray florets 8–30, corollas white to pinkish, ligules 5–14 mm; disk florets 20–100; cypselae 3–4 mm, faces hairy, hairs glochidiate. Gypsum soils and outcrops, salt desert scrub with scattered juniper, 5500–6000 ft. Sandoval. **R**

Townsendia incana Nutt.—Hoary Townsend daisy. Perennials; stems decumbent to erect, 2–12 cm, densely pilose-hirsute, epidermis hidden by dense hairs; leaves basal and cauline; leaf blades spatulate to oblanceolate, 3–40 × 1–4 mm, faces strigose; heads borne at stem tips; phyllaries 16–28 in 3–4 series, lance-ovate to lanceolate, faces strigose; ray florets 8–34, corollas white to pinkish, ligules 5–12 mm; disk florets 60–80; cypselae 3.5–5 mm, faces hairy, hairs glochidiate. Soils derived from shale and sandstone, piñon-juniper woodlands, rock benches, 4600–8350 ft. NC, WC, Guadalupe.

Townsendia leptotes (A. Gray) Osterh.—Common Townsend daisy. Perennials, pulvinate; stems 1–3 cm, strigose; leaves basal and cauline, appearing to be in rosettes because of very short internodes; leaf blades spatulate to obovate, 8–40 × 1–2 mm, faces strigose to glabrate; heads usually sessile; phyllaries 40–50 in 4–5 series, lanceolate, herbaceous, mostly 7–9 mm, 2–5 times longer than wide; faces strigose or glabrous; ray florets 13–34, corollas white to pinkish, ligules 5–10 mm; disk florets 40–80; cypselae 3 mm, glabrous or sparsely hairy proximally, hairs glochidiate. Piñon-juniper-oak and ponderosa woodlands, canyon bottoms, 6000–8225 ft. NW, Cibola.

TRAGOPOGON L. – Milkgrass

Gary I. Baird

Biennial herbs, 15–150 cm, sap milky; stems erect; leaves basal (first year) and cauline (second year), alternate, simple, sessile, base often dilated-clasping, blades linear to ovate-lanceolate, faces glabrous or tomentose to loosely lanate, margins entire; heads liguliflorous, solitary (terminal), peduncles sometimes inflated and fistulose; involucres cylindrical to obclavate; phyllaries 5–15 in 1–2 series, subequal, herbaceous, faces often keeled, elongating in fruit; florets 10–150+, bisexual, ligules yellow and/or purple; cypselae monomorphic or somewhat dimorphic, fusiform to cylindrical, brownish to stramineous or whitish, muricate, tomentulose immediately below pappus, otherwise glabrous, 10-ribbed, beaked; pappus of 12–20 bristlelike, plumose scales in 1 series, 20–30 mm, whitish to flaxen, persistent (Soltis, 2006).

1. Corollas purple…**T. porrifolius**
1'. Corollas yellow…(2)
2. Peduncles inflated and fistulose distally; corollas usually pale yellow, the outer ones mostly much shorter than the involucres; leaf apices erect…**T. dubius**
2'. Peduncles not inflated or fistulose; corollas usually bright yellow, the outer ones mostly surpassing the involucres; (at least upper) leaf apices ± recurved or coiled…**T. pratensis**

Tragopogon dubius Scop.—Desert milkgrass. Leaf blades lance-linear, 10–40 cm × 2–4 mm (cauline progressively shorter but basally wider upward), erect to ascending, apices erect or straight; peduncles distally inflated and fistulose; involucres 20–60 mm; florets much shorter than phyllaries; corollas pale yellow, ligules 5–20 mm; cypselae 20–30 mm, gradually tapered to beak, beak and body about equal in length. Disturbed sites, typically in drier sites than *T. pratensis*, 3135–11,370 ft. Widespread. Introduced.

Tragopogon porrifolius L.—Salsify. Leaf blades lance-linear to lance-ovate, 15–30 cm × 5–15 mm (cauline progressively shorter but not much wider at base), erect to ascending, apices erect or straight; peduncles distally inflated and fistulose; involucres 35–70 mm; florets shorter than to equaling phyllaries; corollas purple, ligules 10–25 mm; cypselae 25–40 mm, rapidly tapered to beak, beak ± longer than body. Occasionally cultivated, shady and moist sites, (4325–)6000–9575 ft. NC, C, SC, Grant. Introduced.

Tragopogon pratensis L.—Meadow milkgrass. Leaf blades lanceolate to ovate-lanceolate, 10–30 cm × 5–20 mm (cauline progressively shorter but not much wider at base), erect to ascending, apices arched to recurved or coiled (at least distal leaves); peduncles not inflated or fistulose; involucres 10–35 mm; florets equaling or surpassing phyllaries; corollas bright yellow, ligules 5–20 mm; cypselae 15–25 mm, gradually tapered to beak, beak shorter than body. Disturbed sites, 5300–10,660 ft. Widespread except SE, SW. Introduced.

TRIPLEUROSPERMUM Sch. Bip. – Mayweed

Jennifer Ackerfield

Tripleurospermum inodorum (L.) Sch. Bip. [*Matricaria maritima* L.; *M. perforata* Mérat; *Tripleurospermum perforatum* (Mérat) M. Laínz]—Wild chamomile. Usually annuals, (5–)30–60(–80) cm; leaves alternate, 2–8 cm, the ultimate lobes filiform; heads radiate; involucral bracts in 2–5 series, bracts oblong, the margins colorless to light brown; receptacles naked; ray florets 10–25, (4–)10–13(–20) mm, white; disk florets 1–2.5 mm, yellow or greenish; cypselae light brown, laterally compressed to subcylindrical. Uncommon in disturbed places, 6800–8700 ft. Colfax, Rio Arriba, Taos. Introduced (Ackerfield, 2015).

TRIXIS P. Browne – Threefold

Kenneth D. Heil

Trixis californica Kellogg—American threefold. Plants 20–200 cm; leaves usually ascending, almost parallel with stems, blades linear-lanceolate to lanceolate, 2–11 cm, margins entire or denticulate (often revolute), abaxial and adaxial faces glandular; heads usually in corymbiform or paniculiform arrays, rarely borne singly; phyllaries usually 8, 8-14 mm; florets 11-25, corollas yellow (aging white), zygomorphic; cypselae 6-10.5 mm, papillalike double hairs producing mucilage when wetted; pappus 7.5-12 mm. New Mexico material belongs to **var. californica**. Dry, rocky slopes, washes, desert flats, thorn scrub, 3500-7800 ft. WC, SC, SW, Chaves (Keil, 2006l).

UROPAPPUS Nutt. – False silverpuffs

Gary I. Baird

Uropappus lindleyi (DC.) Nutt. [*Microseris lindleyi* (DC.) A. Gray; *Uropappus linearifolius* Nutt.]—Lindley's silverpuffs. Annual herbs, sap milky; stems obscure, branched near base; leaves all basal (or appearing so), simple, leaf blades 5-20 cm, bases clasping, margins often villous-ciliate, faces glabrous to villous-puberulent, lobes 0-5+ pairs, subopposite, linear-lanceolate, 3-14 mm; heads liguliflorous, solitary, erect; involucres 15-25 mm, fusiform-ovoid; phyllaries 5-25 in 3-4 series, lanceolate, imbricate, green or medially rosy-purple, faces glabrous, margins entire, scarious, inner elongating with fruit; ray florets 5-150, ligules 4-10 mm, pale yellow; disk florets lacking; cypselae 7-17 mm; pappus scales 5-15 mm, bristles 4-6 mm. Sandy to rocky soils, often of roadsides, in desert, grassland, piñon-juniper, and riparian communities, 3800-6725 ft. WC, SC, SW, Rio Arriba (Chambers, 1964; Strother, 2006oo).

VERBESINA L. – Crownbeard

Kenneth D. Heil

Annuals or perennials, 7-15+ cm or mostly 30-150+ cm; stems usually erect; leaves basal and/or cauline, opposite or alternate; blades sometimes pinnately or palmately lobed, margins subentire or toothed, faces glabrous or hairy; heads radiate or discoid; phyllaries persistent; receptacles flat to convex or ± conic; ray florets 0 or (1-)5-30, either pistillate and fertile or styliferous and sterile, or neuter, corollas yellow to orange; disk florets 8-150(-300+), corollas usually concolorous; cypselae ± flattened; pappus usually of 2 subulate scales or awns (Strother, 2006pp).

1. Leaves all or mostly alternate...**V. encelioides**
1'. Leaves all or mostly opposite...(2)
2. Plants mostly 7-15+ cm; abaxial faces of leaves mostly strigose-sericeous; cypselae ± strigillose...**V. nana**
2'. Plants mostly 30-100(-150+) cm; faces of leaves, at least abaxial, scabrellous to hirtellous or hirsutulous, not strigose-sericeous; cypselae glabrous...(3)
3. Leaf blades ± lance-linear, length 10-15 times width...**V. longifolia**
3'. Leaf blades lance-elliptic, ovate-deltate, or rhombic; length 1.5-2.5 times width...(4)
4. Phyllaries 12-16 in ± 2 series, lance-linear, lance-ovate, or linear, 4-7 mm...**V. oreophila**
4'. Phyllaries 18-30+ in 3-4 series, oblong to elliptic...**V. rothrockii**

Verbesina encelioides (Cav.) Benth. & Hook. f. ex A. Gray [*V. encelioides* var. *exauriculata* B. L. Rob. & Greenm.]—Golden crownbeard. Plants 10-50(-120+) cm, annuals; leaves all or mostly alternate, blades deltate-ovate or rhombic to lanceolate, 3-8(-12+) × 2-4(-6+) cm, margins coarsely toothed to subentire, faces scabrellous to sericeous; heads usually borne singly, sometimes 2-3+ in loose, cymiform or corymbiform arrays; involucres 10-20+ mm diam.; phyllaries 12-18+ in 1-2 series, lance-ovate or lance-linear to linear; ray florets (8-)12-15+; disk florets 80-150+, corollas yellow; cypselae 3.5-5+ mm, faces strigillose. Swales and disturbed sites, 3000-8900(-10,000) ft. Widespread. Introduced.

Verbesina longifolia (A. Gray) A. Gray—Longleaf crownbeard. Plants 60-150+ cm; leaves all or mostly opposite, blades ± lance-linear, length 10-15 times width, 9-12(-20+) × 0.6-1.2(-1.5+) cm, margins obscurely toothed to subentire; heads borne singly or 3-5+ in loose, corymbiform arrays; involucres 11-15+ mm; phyllaries 18-24+ in 2-3 series, linear to lance-linear; ray florets 12-13; disk florets 60-100+, corollas yellow; cypselae 7 mm, faces glabrous. Rocky slopes, grassy sites, open areas in oak-pine communities, 5500-6275 ft. Hidalgo.

Verbesina nana (A. Gray) B. L. Rob. & Greenm.—Dwarf crownbeard. Plants mostly 7-15+; leaves all or mostly opposite; blades ± deltate to rhombic, 3-8+ × 2-5+ cm, margins toothed, faces strigose; heads borne singly or 2-3 together; involucres 15-25+ mm diam.; phyllaries 28-35+ in 2-3 series, linear to lance-ovate; ray florets 12–16+; disk florets 80–150+, corollas yellow to orange; cypselae 6.5–8 mm, faces ± strigillose. Silty flats, desert scrub communities, 3500–4025 ft. Chaves, Eddy.

Verbesina oreophila Wooton & Standl.—Mountain crownbeard. Plants 50–100+ cm; leaves all or mostly opposite (distal sometimes alternate), blades ovate-deltate to lance-elliptic, 3–8(–11) × 1–4(–6) cm, margins coarsely toothed to subentire, faces ± hirsutulous to scabrellous; heads borne singly or 2–6 in loose, corymbiform arrays; involucres ± hemispherical, 8–10 mm diam.; phyllaries 12–16 in 2 series, ± erect, linear to lance-ovate or lance-linear, 4–7 mm; ray florets 13; laminae 9–12+ mm; disk florets 60–100+, corollas yellow; cypselae dark brown, obovate, 4–5 mm, faces glabrous; pappus 0–0.5 mm. Limestone outcrops and soils, 6900–8300 ft. Eddy, Lincoln, Otero.

Verbesina rothrockii B. L. Rob. & Greenm.—Rothrock's crownbeard. Plants 30–60+ cm; leaves all or mostly opposite (distal sometimes alternate); blades ovate-deltate or rhombic to lance-elliptic, 3–5 × 1–3 cm, bases subtruncate to cuneate, margins coarsely toothed to subentire, faces ± hirtellous to scabrellous; heads borne singly; involucres ± hemispherical, 10–15+ mm diam.; phyllaries 18–30+ in 3–4 series, ± erect, oblong to elliptic, 6–10+ mm; ray florets 8–13; laminae 15–25+ mm; disk florets 60–100+, corollas yellow; cypselae dark brown, ± elliptic, 10 mm, faces glabrous; pappus 0–0.5 mm. Rocky slopes and limestone substrates, 4300–8500 ft. Grant, Hidalgo, Luna.

VERNONIA Schreb. – Ironweed

Steven R. Perkins

Perennials, 20–200(–300) cm; leaves mostly cauline (rarely basal), margins toothed (rarely entire), bases usually ± cuneate, apices acute to attenuate, abaxial faces usually resin-gland-dotted or pitted; heads discoid, ± pedunculate; involucres ± campanulate to obconic or hemispherical, 3–8(–11+) mm diam.; phyllaries 18–70+ in 4–7+ series, margins entire; florets 9–30(–65+), corollas purplish or pink (rarely white); cypselae 8–10-ribbed, columnar or arcuate; pappus persistent (Allred et al., 2020; Strother, 2006qq).

1. Leaves 8–12 mm wide; blades narrowly lanceolate to lance-linear; phyllaries 35–40+…**V. marginata**
1'. Leaves 18–50 mm wide; blades elliptic to lance-ovate or lanceolate; phyllaries 50–70+…**V. missurica**

Vernonia marginata (Torr.) Raf. [*V. altissima* Nutt. var. *marginata* Torr.]—Plains ironweed. Perennials, 30–50(–80+) cm; stems puberulent to glabrescent; leaves mostly cauline, blades narrowly lanceolate to lance-linear, 5–15+ cm × (2.5–)8–12+ mm, abaxially glabrate, adaxially puberulent to glabrescent; heads in corymbiform arrays; involucres narrowly campanulate, (7–)9–11 mm; phyllaries 35–40+; florets 10–25+; cypselae 4–5 mm; pappus stramineous to purplish. E New Mexico in river bottoms, stream banks, ditches, grassy swales, plains, 3600–4900 ft. NE, EC, Chaves.

Vernonia missurica Raf.—Missouri ironweed. Perennials, 60–120(–200+) cm; stems puberulent; leaves mostly cauline, blades elliptic to lance-ovate or lanceolate, 6–16(–20+) cm × 18–50 mm, abaxially usually puberulent to tomentose or pannose, adaxially scabrellous, glabrescent; heads in corymbiform-scorpioid arrays; involucres broadly campanulate to urceolate, (6–)7–10+ mm; phyllaries 50–70+ in 6–7 series; florets 30–55+; pappus stramineous to whitish. Associated with loamy to sandy soils; roadsides, drainages, grassy swales, prairies, 3000–6400 ft. Lea, Otero, Torrance.

VIGUIERA Kunth – Goldeneye

Kenneth D. Heil and Steve L. O'Kane, Jr.

Viguiera dentata (Cav.) Spreng. [*Helianthus dentatus* Cav.]—Sunflower goldeneye, toothleaf goldeneye. Perennials, 100–200 cm; leaves opposite or alternate, petioles 10–55 mm, blades ovate or rhombic-ovate to lance-ovate or lanceolate, 3.5–12.5 × 1–8 cm, margins serrate or serrulate; heads mostly 3–9+ in ± corymbiform arrays; involucres 11–18 × 7–10 mm; phyllary apices abruptly narrowed; ray florets 10–14; disk florets 50+, corollas 3–4 mm, staminal filaments hairy; cypselae 3.5–3.8 mm, ± strigose. Dry slopes and canyons, fields, roadside ditches, 3560–8300 ft. C, WC, SE, SC, SW (Allred & Ivey, 2012; Schilling, 2006c).

WYETHIA Nutt. – Mule's ears

Jennifer Ackerfield

Perennial herbs; leaves alternate, simple; heads radiate (ours); involucral bracts subequal in several series, herbaceous or coriaceous, the outer often long and foliaceous; receptacles chaffy; ray florets yellow (ours); disk florets yellow; pappus a crown of unequal laciniate scales, these often prolonged into awns; cypselae of disk florets radially compressed-quadrangular (Ackerfield, 2015).

1. Leaves narrowly oblong to lanceolate or nearly linear, narrower (0.5–2 cm wide), the lowermost reduced; older stems whitish…**W. scabra**
1'. Leaves elliptic to lance-ovate, wider (3–16 cm wide), the basal ones the largest; older stems not whitish…
   W. arizonica

Wyethia arizonica A. Gray—Arizona mule's ears. Plants 2–3(–10) dm; leaves lanceolate to elliptic, 12–30 cm, scabrous to hirsute, sometimes glabrate, the cauline leaves usually short-petiolate; involucres 18–30 mm diam.; ray florets (25–)35–50 mm; cypselae 9–10 mm. Dry hills and slopes, often with sagebrush, 5400–9000 ft. Rio Arriba, San Juan.

Wyethia scabra Hook. [*Scabrethia scabra* (Hook.) W. A. Weber]—Whitestem sunflower. Plants 2–6 dm; leaves narrowly oblong to lanceolate or nearly linear, hispid to scabrous or sometimes glabrescent; involucres (15–)20–35 mm diam.; ray florets 15–50 mm; cypselae 7–9 mm. Uncommon on dry, open slopes in sandy soil, 5000–6500 ft. NW, Catron.

XANTHISMA DC. – Sleepy daisy

Kenneth D. Heil

Annuals, biennials, perennials, or subshrubs, 3–100 cm, taprooted, caudices woody; stems erect, spreading, or sprawling, glabrous or hispid to hispidulous, villous, or stipitate-glandular; leaves basal and cauline, short-petiolate or sessile, blades 1-nerved, lanceolate to oblanceolate or spatulate, margins entire, serrate, dentate, pinnatifid, or 2-pinnatifid, faces usually glabrous, hispid, hispidulous, or villous, sometimes also stipitate-glandular; heads radiate or discoid, borne singly or in corymbiform arrays; involucres turbinate, campanulate, or hemispherical, 4–10 × 6–25 mm; phyllaries 26–80+ in 2–8 series; receptacles flat to convex, pitted; ray florets 0 (in *X. grindelioides*) or 12–60+, corollas white, pink, red-purple, purple, or yellow; disk florets 15–200+, bisexual, corollas yellow; cypselae ellipsoid to obovoid, oblong, or obscurely cordate, ± 3-sided (Allred, 2020; Hartman, 2006).

1. Ray florets absent…**X. grindelioides**
1'. Ray florets present…(2)
2. Rays white, pink, or purple…(3)
2'. Rays yellow…(4)
3. Leaves usually finely or obscurely serrate or serrulate, usually with 12–25 teeth per side; peduncles hispid or hispidulous…**X. blephariphyllum**
3'. Leaves serrate, often coarsely, with 5–14 teeth per side; peduncles stipitate-glandular…**X. gypsophilum**
4. Inner phyllaries with proximal portion stalklike, abruptly enlarged into ovate to orbiculate or elliptic blade, 2–5 mm wide, apices acuminate to obtuse or broadly rounded, not bristle-tipped…**X. texanum**
4'. Phyllaries not markedly expanded distally, linear to broadly oblong or lanceolate, 1–2 mm wide, apices narrowly obtuse to long-attenuate, usually bristle-tipped…(5)
5. Leaf teeth terminating in a stiff callus, not bristle-tipped…**X. viscidum**
5'. Leaf teeth terminating in a bristle 1.5–3 mm…(6)
6. Annuals, taprooted…**X. gracile**
6'. Perennials, stems with a woody base…**X. spinulosum**

Xanthisma blephariphyllum (A. Gray) D. R. Morgan & R. L. Hartm. [*Haplopappus blephariphyllus* A. Gray; *Machaeranthera blephariphylla* (A. Gray) Shinners]—Tansy aster. Subshrubs, 15–40 cm; stems 5–15+; cauline leaves oblong to oblanceolate, 15–45 × 4–15 mm, margins evenly, finely or obscurely serrate or serrulate, teeth 12–25 per side, faces densely hairy, occasionally stipitate-glandular; involucres depressed-hemispherical, 4–8 × 8–15 mm; phyllaries in 5–6 series, broadly linear or narrowly lanceolate; ray florets 20–45, corollas white, laminae 9–13.5 × 2–3.1 mm; disk florets 60–120+, corollas 4.6–6.3 mm;

cypselae oblong to obovoid, 2–2.8 mm, faces moderately whitish to tawny-hairy. Calcareous soils, piñon-juniper woodlands, 4600–7600 ft. C, SE, SC, Hidalgo.

Xanthisma gracile (Nutt.) D. R. Morgan & R. L. Hartm. [*Dieteria gracilis* Nutt; *Haplopappus gracilis* (Nutt.) A. Gray; *Machaeranthera gracilis* (Nutt.) Shinners]—Slender goldenweed. Annuals, 5–45 cm; stems 1–15+, leaf blades 20–60 × 10–25 mm, pinnatifid to 2-pinnatifid, obovate to oblanceolate (proximal), oblong to linear (distal), 4–10(–20) × 1–3 mm, margins evenly serrate to serrulate, teeth 3–6 per side; involucres hemispherical, 0.6–0.8 × 0.8–1.3 cm; phyllaries in 4–5 series; ray florets 12–26, corollas yellow; disk florets 50–100+, corollas 4–5 mm; cypselae narrowly obovoid to oblanceoloid, 1.5–2.8 mm. Rocky to sandy washes, slopes, roadsides, disturbed sites, 4000–8600(–9800) ft. Widespread except NE.

Xanthisma grindelioides (Nutt.) D. R. Morgan & R. L. Hartm. [*Haplopappus nuttallii* Torr. & A. Gray; *Machaeranthera grindelioides* (Nutt.) Shinners]—Goldenweed, gumweed aster. Subshrubs, 3–35 cm, often mound-forming; stems 10–30+, simple or branched; leaf blades oblong, obovate, or lanceolate to spatulate or narrowly ovate to obovate, 7–60 × 2–13 mm, margins usually evenly coarsely serrate to serrulate, teeth 4–16 per side, faces sparsely to densely hairy, often minutely stipitate-glandular; heads 1–10+; involucres hemispherical to campanulate, 0.5–1 × 0.6–1.5 cm; phyllaries in 3–5 series; ray florets 0; disk florets 15–50+, corollas 5–8.5 mm; cypselae narrowly obovoid, 2–3.5 mm. Gypsum, sandy, and clay badlands, scattered piñon-juniper communities, 4500–7925 ft. NC, NW, Catron, Harding, Lincoln.

Xanthisma gypsophilum (B. L. Turner) D. R. Morgan & R. L. Hartm. [*Machaeranthera gypsophila* B. L. Turner]—Gypsum sleepy daisy. Subshrubs, 15–50 cm; stems 5–10+, hispid or minutely stipitate-glandular; leaf blades narrowly to broadly oblong to obovate, 5–60 × 2–10 mm, margins unevenly, often coarsely serrate, teeth 5–14 per side, involucres depressed-hemispherical, 0.4–0.7 × 0.9–1.5 cm; phyllaries in 4–6 series; ray florets 12–28, corollas usually white, often pink- or purple-tinged abaxially, rarely purple; disk florets 50–120+, corollas 4–6.1 mm; cypselae oblong to obovoid, 1.5–2.6 mm. Gypseous, calcareous, or sandy plains in Doña Ana County, 6560–7775 ft.

Xanthisma spinulosum (Pursh) D. R. Morgan & R. L. Hartm. [*Haplopappus spinulosus* (Pursh) DC.; *Machaeranthera pinnatifida* (Hook.) Shinners]—Spiny goldenweed. Perennials or subshrubs, 10–100 cm; stems 1–30+; leaf blades 1.5–3 × 0.8–1.3 cm, pinnatifid to 2+-pinnatifid, blades oblong to lanceolate, 0.2–8 × 0.1–3 cm, margins deeply lobed to coarsely dentate or ± entire, teeth 4–18+ per side; involucres hemispherical to cupulate, 6–10 × 8–25 mm; phyllaries in 5–6 series; ray florets 14–60, corollas yellow; disk florets 30–150+, corollas 4–5 mm; cypselae narrowly obovoid, 1.8–2.5 mm.

1. Midinternodes 5–20 mm; heads mostly borne singly on peduncles 2–15 cm; involucres 15–25 cm wide… **var. paradoxum**
1′. Midinternodes 2–5 mm; heads on relatively terminal and later peduncles 0.5–2 cm; involucres 8–15 mm wide…(2)
2. Stems stiffly erect (var. *chihuahuanum* also), branched in distal 1/3; leaves ascending, margins pinnatifid, lobes oblong-lanceolate to triangular, faces mostly glabrous, sometimes lightly tomentose, eglandular; heads subsessile, usually clustered…**var. glaberrimum**
2′. Stems spreading to sprawling or stiffly erect (var. *chihuahuanum*), usually much branched in proximal 1/2–2/3; leaves mostly spreading to loosely ascending, margins pinnatifid or 2-pinnatifid, lobes linear to lanceolate, faces hairy, rarely glabrous, usually stipitate-glandular; heads pedicellate in diffuse arrays…(3)
3. Stems usually spreading to sprawling, rarely stiffly erect, 10–40 cm, rarely woody at bases; leaves not much reduced distally, margins pinnatifid to 2-pinnatifid; involucres 8–12 mm wide…**var. spinulosum**
3′. Stems stiffly erect, 30–50 cm, often woody at bases; proximal leaves with margins pinnate to pinnatifid, distal often much reduced, coarsely dentate to ± entire; involucres 12–15 mm wide…**var. chihuahuanum**

var. chihuahuanum (B. L. Turner & R. L. Hartm.) D. R. Morgan & R. L. Hartm. [*Haplopappus spinulosus* (Pursh) DC. var. *chihuahuanus* (B. L. Turner & R. L. Hartm.) Gandhi]—Perennials, 30–50 cm, often woody at base; stems stiffly erect, usually much branched in proximal 1/2–2/3; midinternodes 2–5 mm; leaves numerous, concentrated in proximal ± 1/2, spreading to loosely ascending, margins of proximal pinnatifid to 2-pinnatifid, of distal coarsely dentate to ± entire, lobes linear to lanceolate, 3–5 mm, faces usually hairy, rarely glabrous, often densely stipitate-glandular; heads 2–10+ in diffuse arrays; peduncles relatively terminal and lateral, 0.5–2 cm; involucres 12–15 mm wide. Clay to rocky washes, Chihuahuan desert scrub, 3110–7900 ft. EC, C, SE, SC, SW.

var. glaberrimum (Rydb.) D. R. Morgan & R. L. Hartm. [*Haplopappus spinulosus* (Pursh) DC. subsp. *glaberrimus* (Rydb.) H. M. Hall; *Machaeranthera pinnatifida* (Hook.) Shinners var. *glaberrima* (Rydb.) B. L. Turner & R. L. Hartm.]—Perennials, 20–50 cm, often woody at base; stems stiffly erect, branched in distal 1/3; midinternodes 2–5 mm; leaves numerous, strictly ascending, not reduced distally, margins pinnatifid, lobes oblong-lanceolate to triangular, 2–7 mm, faces usually glabrous, sometimes lightly tomentose, eglandular; heads 2–10+, subsessile, usually clustered; peduncles relatively terminal and lateral, 0.5–2 cm; involucres 8–15 mm wide. Gypsum and sandy to gravelly soils, prairies, hillsides, roadsides, 3275–7280 ft. NE, NC, EC, C, SE.

var. paradoxum (B. L. Turner & R. L. Hartm.) D. R. Morgan & R. L. Hartm. [*Macheranthera pinnatifida* (Hook.) Shinners var. *paradoxa* B. L. Turner & R. L. Hartm.; *Haplopappus spinulosus* (Pursh) DC. var. *paradoxus* (B. L. Turner & R. L. Hartm.) Cronquist]—Perennials, (15–)30–60 cm, often woody proximally or at crowns; stems erect to spreading, branching in proximal 1/3–2/3; midinternodes 5–20+ mm; leaves relatively sparse, crowded proximally, mostly spreading to loosely ascending, often reduced distally, margins pinnatifid to 2-pinnatifid, lobes linear, 2–5 mm, often toothed, faces glabrous or puberulent, stipitate-glandular; heads mostly borne singly; peduncles short (sometimes elongate), 2–15 cm, leafy; involucres 15–25+ mm wide. Rocky to sandy slopes, washes, disturbed areas, ca. 5400 ft. NW.

var. spinulosum [*Haplopappus texensis* R. C. Jacks.]—Perennials, 10–40 cm, rarely woody at base; stems spreading to sprawling, rarely stiffly erect, usually much branched in proximal 1/2–2/3; midinternodes 2–5 mm; leaves numerous, equally spaced, mostly spreading to loosely ascending, not much reduced distally, margins pinnatifid or 2-pinnatifid, lobes linear to lanceolate, mostly 2–5 mm, faces usually hairy, rarely glabrous, usually stipitate-glandular; heads 2–10+ in diffuse arrays; peduncles relatively terminal and lateral, 0.5–2 cm; involucres 8–12 mm wide. Mostly rocky to sandy soil in a variety of habitats, 3000–10,200 ft. Widespread.

Xanthisma texanum DC.—Texas sleepy daisy. Annuals, rarely biennials; stems 1–3, mostly glabrous; basal leaves (if persisting) 50–80 × 15–25 mm, pinnatifid, rarely 2-pinnatifid, cauline leaves spatulate, distal 2/3 narrowly to broadly lanceolate, 5–35 × 8–12 mm, margins of proximal pinnatifid to coarsely serrate, of distal minutely, evenly serrulate or entire, teeth or cilia 50–90+ per side, faces glabrous; involucres hemispherical, 5–10 × 11–20 mm; phyllaries in 3–4 series, abruptly expanded distal to stalks, bodies mostly ovate (3.5–8 × 2.5–5 mm distal to widest point), apices acute to acuminate, flared distally, ovate portion 3.5–8 × 2.5–5 mm; ray florets 12–34, corollas yellow; disk florets 50–200+, corollas 4.5–5 mm; cypselae 1.6–1.8 mm. New Mexico plants belong to **subsp. drummondii** (Torr. & A. Gray) Semple. Sandy soils; roadsides and other disturbed sites, 4015–4500 ft. EC, SE, Sandoval.

Xanthisma viscidum (Wooton & Standl.) D. R. Morgan & R. L. Hartm. [*Haplopappus havardii* Waterf.; *H. viscidus* (Wooton & Standl.) S. F. Blake]—Viscid sleepy daisy. Annuals, 25–60 cm; stems stipitate-glandular; leaves obovate to oblanceolate or oblong, 5–30 × 2–15 mm, margins coarsely serrate to dentate, teeth 4–7 per side, entire, faces densely stipitate-glandular; involucres hemispherical, 8–10 × 13–18 mm; phyllaries in 5–7 series, ray florets 14–18, corollas yellow; disk florets 20–30+, corollas 5.5–6 mm; cypselae oblong to narrowly ellipsoid, 2.3–2.7 mm. Sandy soils; roadsides and other disturbed sites, 3000–4800 ft. SE, Socorro.

XANTHIUM L. – Cocklebur

Jennifer Ackerfield

Annual herbs; leaves alternate; heads unisexual, discoid; involucral bracts few, distinct, and subherbaceous, or absent in staminate heads, fused into a 2-chambered prickly bur in pistillate heads; receptacles chaffy; ray florets absent; disk florets staminate or pistillate, pistillate flowers lacking a corolla; pappus absent (Ackerfield, 2015).

1. Leaves densely silvery-sericeous below, lanceolate and tapering to the base; nodes with a 3-forked axillary spine…**X. spinosum**
1'. Leaves not silvery below, ovate to suborbicular and cordate to truncate at the base; nodes unarmed…
X. strumarium

Xanthium spinosum L. [*Acanthoxanthium spinosum* (L.) Fourr.]—Spiny cocklebur. Plants 1–6 (–15) dm; stems with a 3-forked axillary spine at each node; leaves lanceolate, 4–12 cm, densely silvery-sericeous below. Uncommon in disturbed places, fields, and along roadsides, 3900–7100 ft. NE, EC, C, WC, SE, SC, SW. Introduced.

Xanthium strumarium L.—Common cocklebur. Plants 1–10(–20) dm; stems unarmed; leaves deltate to ovate or suborbiculate or palmately lobed, 4–15(–20) cm, not silvery-sericeous below. Common in disturbed places, fields, and along roadsides, 3275–7940 ft. Widespread. Introduced.

ZALUZANIA Pers. – Yellow streamers

Steven R. Perkins

Zaluzania grayana B. L. Rob. & Greenm.—Yellow streamers. Perennial forbs or subshrubs, 30–80 (–250+) cm; stems erect, branched; leaves cauline, mostly opposite, petiolate; leaf blades deltate to cordate (ovate to lanceolate), 2–8 × 1–6 mm, margins coarsely toothed, faces sparsely strigillose and gland-dotted; heads radiate (discoid), in loose corymbiform (paniculiform) arrays; involucres hemispherical, 4–8 mm diam.; phyllaries persistent, 10–25 in 2–3+ series; ray florets (0 or 4–)8–10, corollas yellow, laminae 15–20 mm; disk florets 30–100+, corollas yellow, lobes 5; cypselae ± compressed, 2–3 mm; pappus 0. Limestone slopes, 5900–6500 ft. Hidalgo (Allred et al., 2020; Strother, 2006rr).

ZINNIA L. – Zinnia

Kenneth D. Heil

Annuals or subshrubs, 8–22 cm; stems greenish to gray; leaf blades 1- or 3-nerved, linear, 8–30 × 1–3 mm, glabrescent, scabrous, or strigose; involucres campanulate or narrowly campanulate to cylindrical, 3–8 × 5–8 mm; phyllaries oblong, glabrous or appressed-hairy, scarious; ray florets 3–6, corollas bright yellow or white, 4–18 mm; disk florets 8–24, corollas red, green, or yellow; cypselae 2.4–5 mm, 3-angled (Smith, 2006b).

1. Flowers bright yellow to orange; leaves 3-veined…**Z. grandiflora**
1'. Flowers white; leaves 1-veined…**Z. acerosa**

Zinnia acerosa (DC.) A. Gray [*Diplothrix acerosa* DC.]—Desert or shrubby zinnia. Subshrubs, to 16 cm; stems greenish to gray; leaf blades 1-nerved, linear to acerose, 8–20 × 1–2 mm, scabrous to glabrescent; involucres campanulate, 3–5 × 5–7 mm; ray florets 4–7, corollas mostly white, 7–10 mm; disk florets 8–13, corollas yellow or tinged with purple, 3–6 mm; cypselae 2.4–4 mm. Rocky, open slopes, calcareous soils, 3560–6100 ft. EC, SE, SC, SW.

Zinnia grandiflora Nutt.—Rocky Mountain zinnia. Subshrubs, 8–22 cm; stems greenish; leaf blades 1- or 3-nerved, linear, 10–30 × 2–3 mm, strigose to scabrous; involucres narrowly campanulate to cylindrical, 5–8 × 5–8 mm; ray florets 3–6, corollas bright yellow, mostly 10–18 mm; disk florets 18–24, corollas red or green, to 10 mm; cypselae 4–5 mm. Calcareous soils, mesas, shortgrass prairies, piñon-juniper woodlands, 3150–8000 ft. Widespread.

BERBERIDACEAE – BARBERRY FAMILY

Kenneth D. Heil

BERBERIS L. – Barberry

Erect to sprawling evergreen or deciduous shrubs, 1–3 m; stems dimorphic, composed of elongate primary shoots and short secondary shoots, bark glabrous, gray to red or brown, spines simple or 1–5-branched; inflorescence a lax, open raceme or umbel; flowers 3-merous, subtended by bracts; sepals 3 or 6, yellowish; petals 6, yellow, stamens 6; fruit a berry. Species with spiny stems have traditionally been included in *Berberis* s.s. while those lacking spiny stems have been included in *Mahonia*. Most of the latter, except *B. repens*, have recently been transferred to *Alloberberis* (Yu & Chung, 2017). *Berberis*, however, as recognized here is monophyletic (Whetstone, 1997).

1. Stems spiny; leaves simple; plants deciduous…(2)
1′. Stems not spiny; leaves compound; plants evergreen…(3)
2. Older branches gray; racemes 10–20-flowered…**B. vulgaris**
2′. Older branches purple; racemes 3–15-flowered…**B. fendleri**
3. Plants low-growing, 0.2–2 dm; racemes dense, 25–50-flowered; berries blue…**B. repens**
3′. Plants tall, > 2 dm; racemes loose, 3–7-flowered; berries yellow, red, purple, or brownish…(4)
4. All leaves trifoliate; terminal leaflet sessile…**B. trifoliolata**
4′. Leaves 5–11-foliate; terminal leaflet stalked…(5)
5. Racemes dense, 25–70-flowered; bracteoles rounded or obtuse; berries blue…**B. wilcoxii**
5′. Racemes loose, 1–11-flowered; bracteoles acuminate; berries yellow, purple, or red to brown…(6)
6. Terminal leaflet 1–2.5 times as long as wide; berries dry, inflated, 12–18 mm, yellow or red to brown…
    B. fremontii
6′. Terminal leaflet 2–5 times as long as wide; berries juicy, solid, 5–8 mm, purple or red…**B. haematocarpa**

Berberis fendleri A. Gray—Fendler barberry. Erect to sprawling shrubs, 1–2 m; older bark brown to purple, spines simple or 1–2-branched; leaves deciduous, simple, lanceolate, or oblanceolate to narrowly elliptic from a long-attenuate base, 1–6 × 0.5–1.7 cm; inflorescence a lax raceme of 5–12 flowers; flower filaments lacking, 2 teeth near apex; fruit red, ovoid, 5–8 mm. Mesic to riparian woodlands and around springs, high-elevation piñon-juniper woodlands to coniferous forest communities, 6000–10,170 ft. NC, NW, C, WC.

Berberis fremontii Torr. [*Mahonia fremontii* (Torr.) Fedde]—Frémont barberry. Erect to sprawling shrubs to 3 m, not stoloniferous; bark on second-year stems gray-brown to brown-purple; leaves evergreen, odd-pinnate, 2–10 cm, leaflets (3–)5–9(–11), dull gray-green, 10–30 × 5–15 mm, margins serrate with 3–6 pairs of sharp teeth; inflorescence racemose or umbellate, open, 3–10-flowered, 2 smaller red bractlets present below flower; flower petals 3–4 mm; fruit dry, inflated, yellowish, brown-red to purple, ellipsoid, 2–4 mm. Widespread on rocky slopes and washes in desert scrub, riparian zones, and piñon-juniper woodlands, 5500–7600 ft. NC, C, WC, SE.

Berberis haematocarpa Wooton [*Mahonia haematocarpa* (Wooton) Fedde]—Algerita. Shrubs, evergreen, 1–4 m; bark of second-year stems grayish purple, spines absent; leaves 3–9-foliate, leaflet blades thick and rigid, surfaces abaxially dull, papillose, adaxially dull, glaucous, terminal leaflet blade 1.5–3.8 × 0.5–1.1 cm, 2–5 times as long as wide, margins undulate or crispate, toothed or lobed, with 2–4 teeth; inflorescence racemose, 3–7-flowered, bracteoles membranous, apex acuminate; flower anther filaments without distal pair of recurved lateral teeth; fruit purplish red, spherical or short-ellipsoid, 5–8 mm, juicy, solid. Slopes and flats in Chihuahuan desert shrubland, desert grassland, and dry oak woodland, 4000–10,200 ft. NE, C, WC, SE, SC, SW.

Berberis repens Lindl. [*Mahonia repens* (Lindl.) G. Don]—Oregon grape. Low, sprawling or creeping shrubs to 0.3 m, stoloniferous; stems monomorphic, mostly lacking short secondary shoots, bark of second-year stems gray-purple; leaves odd-pinnate, 10–30 cm, dull dark green above, paler green below, leaflets (3–)5–7, ovate, 18–70 mm, margins serrate with 6–10 pairs of sharp teeth; inflorescence racemose, dense, 10–30-flowered, bractlets absent; flower petals 5–7 mm, cleft at apex and bilobed; fruit dark blue to blue or purple-black, ovoid, 6–10 mm. Shaded sites under trees on mountain slopes, in ponderosa pine, mixed conifer, and subalpine forest communities, 6500–10,500 ft. Widespread except E, SW.

Berberis trifoliolata Moric. [*Mahonia trifoliolata* (Moric.) Fedde]—Shrubs, evergreen, 1–3.5 m; stems with gray or grayish-purple bark, spines absent; leaves 3-foliate, blades thick and rigid, terminal leaflet sessile, blade 2.3–5.8 × 0.9–2 cm, 1.6–3.1 times as long as wide; toothed or lobed, with 1–3 teeth or lobes 3–7 mm tipped with spines; inflorescence racemose, 1–8-flowered; bracteoles membranous, apex acuminate; flowers without distal pair of recurved lateral teeth; fruit red, spherical, 6–11 mm, juicy, solid. Slopes and flats in grassland, Chihuahuan desert scrub, and piñon-juniper woodlands, 3380–5000 ft. C, SE, SC, SW.

Berberis vulgaris L.—Common barberry. Erect shrubs, 1–3 m; older bark gray, spines simple or 3-branched; leaves deciduous, obovate to oblanceolate or sometimes elliptic, from a short- to long-attenuate base, 2–7 × 0.6–3 cm, margins strongly bristle-tipped; inflorescence a somewhat open raceme of 10–20 flowers; flower filaments lacking 2 teeth near apex; fruit purple to red, ellipsoid, 9–11 mm. Often cultivated as an ornamental, sometimes escapes locally but never becomes common or invasive in the Southwest, 5200–7300 ft. San Juan, Taos. Introduced.

Berberis wilcoxii Kearney—Wilcox's barberry. Shrubs, evergreen, 0.3–2 m; stems ± monomorphic, bark of older stems purple or brown, glabrous, spines absent; leaves 5–9-foliate, leaflet blades thick and rigid, surfaces glossy, green, terminal leaflet stalked, blade 2.6–6.6 × 1.7–4.4 cm, 1–2.5 times as long as wide, margins plane to crispate, toothed, each with 3–5 teeth; inflorescences racemose, dense, 30–50-flowered, bracteoles membranous, apex rounded or obtuse; flower anther filaments with distal pair of recurved lateral teeth; fruit blue, glaucous, oblong-ovoid, 6–11 mm, juicy, solid. Rocky slopes and canyons in the bootheel area, 4920–6560 ft. Hidalgo.

BETULACEAE – BIRCH FAMILY

Kenneth D. Heil

Trees and shrubs, deciduous; leaves alternate, simple; stipules deciduous, leaf blades sometimes lobed, margins toothed or serrate to nearly entire; inflorescences unisexual; staminate catkins pendulous, elongate, cylindrical, conspicuously bracteate, consisting of crowded, reduced, 1–3-flowered clusters; pistillate inflorescences either of erect to pendulous bracteate catkins, or of compact 2–3-flowered clusters subtended by leafy involucres; pistillate flowers small, highly reduced, pistil 1, 2(3)-carpellate, ovary inferior; fruits nuts, nutlets, or 2-winged samaras, 1-seeded, often subtended or enclosed by foliaceous hull developed from 2–3 bracts (Heil, 2013a).

1. Infructescences mostly > 4 cm, not conelike, not woody, bracts forming inflated bladders and completely enclosing fruits; fruits small nutlets; rare in New Mexico…**Ostrya**
1'. Infructescences 1–4 cm, conelike, woody; bracts not forming inflated bladders; fruits tiny samaras; widespread…(2)
2. Pistillate catkins conelike, with persistent woody bracts; pith of twigs triangular in cross-section…**Alnus**
2'. Pistillate catkins with firm, not woody, deciduous scales; pith of twigs round in cross-section…**Betula**

ALNUS Mill. – Alder

Trees or shrubs; leaves alternate, simple, serrate to lobed; staminate catkins 1–3, bracts subtending 3–6 flowers; flowers with 2–4 stamens; pistillate catkins conelike, woody bracts subtending 2 flowers; fruit a nutlet.

1. Leaves ovate to oblong-ovate, rounded or cuneate to subcordate at the base; stamens 4; widespread…
   A. incana
1'. Leaves elliptic or oblong-ovate, acute or short-cuneate at the base; stamens mostly (1)2(3); mountains in WC, SW…**A. oblongifolia**

Alnus incana (L.) Moench [*A. tenuifolia* Nutt.]—Thinleaf alder. Large shrubs or small trees to 10 m, bark thin, red, twigs pubescent and glandular; leaves with petioles 5–30 mm, blades 2–11 × 1–10 cm, ovate or oblong-ovate, obtuse, acute, or short-acuminate at apex, rounded or broadly cuneate to subcordate at base, doubly serrate, usually hairy along the veins beneath. New Mexico material belongs to **subsp. tenuifolia** (Nutt.) Breitung. Riparian areas, stream banks, seeps, springs, 6500–12,120 ft. NE, NC, NW, C, WC, SW.

Alnus oblongifolia Torr.—New Mexico alder. Large shrubs or trees to 30 m; bark of older trees grayish brown; leaf blades elliptic or oblong-ovate to lanceolate, mostly 3–9 × 1.5–7 cm, bases acute to short-cuneate, margins doubly serrate; staminate flowers with (1)2(3) stamens. Mountain canyons along streams, 4300–8600 ft. NW, C, WC, SW.

BETULA L. – Birch

Trees and shrubs, outer bark smooth, often separable in sheets; leaves alternate, simple, serrate to crenate or doubly serrate; staminate catkins 1–4 per bud, pendulous or spreading in flower; staminate flowers in clusters of 3, stamens 2; pistillate catkins usually solitary, erect in flower, pistillate flowers 2–3 per bract; fruit a samara.

1. Low to moderately sized shrub, 0.5–1.2 m; leaf blades 0.5–2.5 cm, oval to orbicular, usually 10 or fewer teeth per side; rare, Sandoval County...**B. glandulosa**
1'. Tall shrubs or trees, mostly 4–8 m; leaf blades 1.5–8 cm, ovate to deltoid, usually 10–40 teeth per side; widespread in N New Mexico...**B. occidentalis**

Betula glandulosa Michx.—Glandular birch, swamp birch. Low shrubs, 0.5–1.2 m, twigs and branches glabrous, densely resinous-glandular; leaves with petioles 2–10 mm, blades 1–2.5 cm, suborbicular to obovate, apex rounded, base rounded or broadly cuneate, margins crenate-serrate, 10 or fewer teeth per side, not hairy in lower vein axils; pistillate catkins 7–20 × 3–8 mm; bracts commonly glabrous dorsally, ciliate; fruit a samara with wings narrower than 1/2 the body width. Beaver ponds, fens, iron bogs in subalpine communities, 8460–9200 ft. Sandoval.

Betula occidentalis Hook.—Water birch. Shrubs or small trees, mostly 3–6 m, with several trunks, to 2.5+ dm thick; bark not exfoliating, reddish or yellowish brown, shining, with pale horizontal lenticels, twigs pubescent to glabrous with reddish crystalline resin glands; leaves with petioles 5–15 mm, blades 1–5 × 0.7–4 cm, ovate, acute or abruptly acuminate apically, obtuse to rounded, sharply and often doubly serrate, mostly 15–25 teeth per side; pistillate catkins 15–40 mm thick, the bracts puberulent and ciliate; fruit a samara with wings subequal to width of the nutlet. Riparian communities, along streams, seeps, and springs, 5400–8540 ft. NE, NC, NW.

OSTRYA Scop. – Hop-hornbeam

Ostrya knowltonii Coville—Western hop-hornbeam. Small trees, mostly 2–6 m, trunks 3–18 cm thick, branchlets spreading-hairy, ± stipitate-glandular, becoming glabrous; leaves with blades 0.8–8 × 0.8–5 cm, ovate to lance-ovate or elliptic, doubly serrate, acute apically, rounded to obtuse basally; staminate catkins 1.5–3 cm; pistillate catkins 0.7–1 cm, in fruit to 4.5 cm, about as wide; fruit with aments 2–5 cm, the individual sacs 10–25 mm, greenish white to brownish. Shaded defiles and hanging garden communities, 4000–7200 ft. SE, SW.

BIGNONIACEAE – BIGNONIA FAMILY

Kenneth D. Heil

Trees, shrubs, or lianas; leaves compound or simple, exstipulate; flowers perfect, zygomorphic; calyx 5-lobed, the lobes unequal but not bilabiate; corolla 5-lobed, bilabiate or actinomorphic, imbricate; stamens 5, unequal, distinct; ovary superior, 2-carpellate; fruit a loculicidal or septicidal capsule (Ackerfield, 2015; Allred et al., 2020).

1. Leaves compound, toothed…(2)
1'. Leaves simple, entire…(3)
2. Woody vines; flowers orange or reddish…**Campsis**
2'. Shrubs; flowers yellow…**Tecoma**
3. Leaves ovate, strongly petioled…**Catalpa**
3'. Leaves linear, nearly sessile…**Chilopsis**

CAMPSIS Lour. – Trumpet creeper

Campsis radicans (L.) Bureau—Trumpet vine. Lianas to 10 m; leaves opposite, to 30 cm; leaflets 5–13, widely lanceolate, serrate; flowers orange to reddish; calyx orange-red, to 2.5 cm, tubular; corolla zygomorphic, tubular-funnelform; capsules 1–2 dm. Escaped ornamental, 3450–6075 ft. Scattered. Introduced (Ackerfield, 2015; Allred et al., 2020).

CATALPA Scop. – Catalpa tree

Catalpa speciosa Warder—Northern catalpa. Trees to 30 m; leaves opposite, ovate, 15–30+ cm, glabrous adaxially, densely hairy abaxially; flowers white with purple spots, pink, or yellowish; calyx of 2 distinct lobes, 1 cm; corolla with 2 smaller upper lobes and 3 larger lower lobes, (1–)5(–7) cm; capsules 25–45(–60) cm, brown; seeds 3–4 cm. Cultivated as an ornamental, escapes along roadsides and vacant lots, 3980–6800 ft. Scattered. Introduced (Ackerfield, 2015; Allred et al., 2020).

CHILOPSIS D. Don – Desert willow

Chilopsis linearis (Cav.) Sweet—Desert willow. Small trees and large shrubs to 4 m; stems often several, glabrous to puberulent; leaves mostly alternate, simple, entire, linear or linear lanceolate, 10–15 cm; flowers showy, in short terminal racemes; calyx inflated, upper lip 3-toothed, lower lip 2-toothed; corolla funnelform, 5-lobed, 25–35 mm, lavender to nearly white, often with purplish lines; capsules 6–30 cm; seeds numerous, the wings dissected into many hairs. Often cultivated as an ornamental. Integration exists between the 2 varieties (Allred et al., 2020).

1. Leaves strongly arcuate…**var. arcuata**
1'. Leaves straight or slightly curved…**var. linearis**

var. arcuata Fosberg—Desert willow. Deserts, near arroyos and foothills, 3110–6000 ft. C, SC, SW.

var. linearis—Desert willow. Deserts, near arroyos and foothills, 3600–6550 ft. C, SE, SW.

TECOMA Juss. – Trumpet bush

Tecoma stans (L.) Juss. ex Kunth—Yellow trumpet. Shrubs to 2+ m; leaves opposite, 6–12 cm, 5–13-foliate; leaflets incised-serrate, linear to lanceolate, petiolate, adaxial surface glabrous, abaxial glabrous to pubescent; flowers in racemes or panicles; calyx tubular, glabrous, lobes triangular; corolla funnelform-campanulate, showy, bright yellow, slightly bilabiate; capsules 6–12 cm, linear. New Mexico material belongs to **var. angustata** Rehder. Well-drained soils; hillsides and stream beds, 4075–5800 ft. SC, SW (Allred et al., 2020).

BORAGINACEAE – FORGET-ME-NOT FAMILY

Robert C. Sivinski

Herbs or rarely shrubs, often with stiff hairs; leaves usually alternate, simple, entire; stipules absent; inflorescence of cymes (branches often helicoid) or thyrses; flowers actinomorphic, perfect; sepals 5;

petals 5, connate into salverform, funnelform, or tubular corollas, faucal appendages (fornices) usually at the throat of salverform corollas; stamens 5; pistil superior, of 2 united carpels, 4-lobed, style basal from a gynobase below and through the ovary, entire or rarely branched near the apex; stigma capitate or bilobed; fruit of 4 homomorphic or heteromorphic nutlets (or fewer by abortion), smooth or commonly roughened with bumps, wrinkles, or prickles or rarely winged. Boraginaceae are rendered here in the narrow sense, so as not to include the families Hydrophyllaceae, Heliotropiaceae, Ehretiaceae, and Namaceae (Allred et al., 2020; Boraginales Working Group, 2016).

1.   Nutlets armed with hooked or barbed prickles or bristles…(2)
1'. Nutlets unarmed (toothed or lacerate in *Eritrichium*, but not hooked or barbed)…(5)
2.   Bristles of nutlets merely hooked at the tips, not glochidiate with several barbs; nutlets widely spreading when mature…**Pectocarya**
2'. Bristles of nutlets glochidiate at the tips with several barbs; nutlets spreading or erect when mature…(3)
3.   Nutlets covered over the entire surface with numerous short barbs; nutlets spreading when mature… **Cynoglossum**
3'. Nutlets barbed only on the angles or dorsal side, the entire surface not covered as above…(4)
4.   Pedicels erect in fruit; plants annual…**Lappula**
4'. Pedicels reflexed in fruit; plants biennial or perennial…**Hackelia**
5.   Ovary entire or shallowly lobed, the style terminal on the ovary…(6)
5'. Ovary deeply 4-lobed, the style basal…(7)
6.   Style distinctly cleft; stigmas 2, not subtended by a ring or disk (*Tiquilia*)…go to **EHRETIACEAE**
6'. Style not divided, simple; stigma subtended by a ring or disk…go to **HELIOTROPIACEAE**
7.   Flowers large, 2.5–8 cm, hairy…**Lithospermum** (in part)
7'. Flowers smaller, < 2.5 cm, glabrous or hairy…(8)
8.   Corolla blue, rarely white or pinkish…(9)
8'. Corolla white, greenish white, cream-colored, yellow, or orange…(12)
9.   Plants pulvinate-cespitose, flowering stems to 10 cm; foliage conspicuously villous to strigose… **Eritrichium**
9'. Plants not at all pulvinate-cespitose, the flowering stems nearly always > 10 cm; foliage glabrous to obscurely or lightly pubescent…(10)
10. Corolla rotate-salverform, the lobes spreading at nearly right angles and about the same length as the short tube…**Myosotis**
10'. Corolla tubular-funnelform, the lobes erect to ascending and usually shorter than the tube…(11)
11. Nutlet attachment scar surrounded by a thick ring or collar…**Symphytum**
11'. Nutlet attachment scar not surrounded by a thick ring or collar…**Mertensia**
12. Gynobase low, not at all pyramidal…(13)
12'. Gynobase raised, ± pyramidal…(15)
13. Corolla yellow, orange-yellow, or greenish yellow…**Lithospermum** (in part)
13'. Corolla greenish white or creamy white…(14)
14. Leaves with 5–7 raised veins beneath; corolla tubular, hairy, lobes erect or folded tightly inward over the throat; style strongly exserted…**Lithospermum** (in part)
14'. Leaves without obvious veins except midrib; corolla broadly rotate-salverform (saucer-shaped), glabrous; style included in the tube…**Antiphytum**
15. Corolla orange or bright yellow, the throat open and not crested (lacking fornices); plants annual… **Amsinckia**
15'. Corolla pale yellow to white, the throat usually crested (with fornices); plants perennial, or if annual then not yellow…(16)
16. Nutlets with a keel on the ventral surface, the attachment scar not elongate but raised and wartlike… **Plagiobothrys**
16'. Nutlets with a groove or open triangular scar running most of the length of the ventral surface, this often expanded at the base (upper part of scar sometimes with overlapping edges)…(17)
17. Plants biennial or perennial; corolla limb 4–14 mm wide…**Oreocarya**
17'. Plants annual; corolla limb 1–5 mm wide…(18)
18. Stigma terminating in a short style; stems not wiry; roots not purple-dye-stained…**Cryptantha/ Johnstonella**
18'. Stigma sessile on an elongate gynobase; stems wiry; roots charged with a purple dye…**Eremocarya**

AMSINCKIA Lehm. – Fiddleneck

Annual herbs; stems erect, generally bristly; leaves pustulate-hirsute, basal and cauline, alternate, sessile or lower shortly petioled, linear-lanceolate or oblong; inflorescence of spikelike helicoid cymes elongating at maturity; calyx lobes 5 or 3–4 when fused; corolla orange or yellow, throat generally open, fornices absent, limb generally with 5 red interior spots; nutlets 4, erect, triangular, generally with rounded or sharp tubercles, attachment scar elliptic.

1. Corolla tube 20-nerved below the attachment of the stamens; calyx lobes unequal in width, commonly 3 or 4, 1 or 2 of them 2-lobed; nutlets rounded-tuberculate…**A. tessellata**
1'. Corolla tube 10-nerved below the attachment of the stamens; calyx lobes 5, distinct, ± equal; nutlets sharply tuberculate…**A. intermedia**

Amsinckia intermedia Fisch. & C. A. Mey.—Common fiddleneck. Stems 3–10 dm, spreading-hispid; leaves oblong to lance-linear; cymes naked or a few bracts on lowest flowers; calyx lobes 5, ± equal, not fused above the base; corolla yellow-orange with 5 red spots, 7–11 mm; tube 10-veined near base; nutlets usually 4 maturing per calyx, 2–3.5 mm, sharp-tubercled, often with a dorsal keel. Rocky slopes and bajadas in desert scrub, 4525–5350 ft. Infrequently collected in the N Peloncillo Mountains. Hidalgo.

Amsinckia tessellata A. Gray—Bristly fiddleneck. Stems 1–6 dm, spreading-hispid; leaves lance-linear or more often lance-oblong or lance-ovate, to 10 cm; calyx lobes unequal, commonly only (2–)4 by lateral fusion, the broader one(s) often apically bidentate; corolla yellow or orange, 7–14 mm, tube 20-nerved near the base; nutlets 2.5–3.5 mm, surface cobblestonelike or round-tubercled, ridged or not. Desert scrub, 4000–5100 ft. Doña Ana, Grant, Hidalgo.

ANTIPHYTUM DC. ex Meisn. – Saucerflower

Antiphytum floribundum (Torr.) A. Gray [*Amblynotopsis floribunda* (Torr.) J. F. Macbr.; *Eritrichium floribundum* Torr.]—Texas saucerflower. Biennial or short-lived perennial with a dense basal leaf rosette; stems stiffly erect, 3–10 dm, with several ascending branches especially above the middle strigose and spreading-hirsute; basal leaves linear-oblanceolate, 3–10 cm; proximal cauline leaves sometimes opposite, upper leaves alternate, conduplicate, linear, acute, usually < 3 cm; inflorescence of bracteate cymules; calyx segments lanceolate, unequal, ca. 5 mm in fruit; corolla white or creamy-white, not or barely surpassing the calyx lobes; fornices absent. Igneous rocky slopes, oak woodlands; known from only a single collection in the Animas Mountains. Hidalgo (Sivinski et al., 1994).

CRYPTANTHA Lehm. ex G. Don – Cryptantha

Annual herbs, strigose, hirsute, or hispid-setose; leaves narrowly oblanceolate, spatulate, or rarely linear; inflorescence of helicoid, naked or bracteate cymules (branched false racemes or false spikes) that usually elongate with maturity; calyx cleft to the base or nearly so; corolla white, salverform with a spreading or nearly closed limb, faucal appendages (fornices) at the throat small, often yellow; homostylous; stigma capitate; ovules usually 4 (*C. recurvata* is exceptional with only 2 ovules); nutlets erect, 4 or 1–3 by abortion, triangular-ovate, lanceolate, or winged, smooth or variously roughened, heteromorphic or all similar, affixed to an elongate gynobase, the ventral scar closed, narrowly open, or forming a triangular areola. Similar segregate genera *Eremocarya* and *Johnstonella* are included in the following *Cryptantha* key to species (Hasenstab-Lehman & Simpson, 2012).

1. Nutlet margins decidedly winged…**C. pterocarya**
1'. Nutlet margins rounded or sharply angled, never winged…(2)
2. Taproot charged with red-purple dye; gynobase elongate, surpassing the nutlets and terminated by a sessile stigma, without a differentiated style; fruiting calyx persistent (*E. micrantha*)…go to **EREMOCARYA**
2'. Taproot without red dye (sometimes slightly dye-stained in *C. recurvata*); gynobase shorter than the nutlets and topped by a definite style that may or may not surpass the nutlets; fruiting calyx deciduous…(3)
3. Usually a solitary nutlet matured in each calyx…(4)
3'. Nutlets normally 4 per calyx (often fewer by abortion)…(5)
4. Calyx lobes and nutlet decidedly recurved or deflexed; nutlet muricate…**C. recurvata**
4'. Calyx and nutlet not curved or bent; nutlet smooth…**C. gracilis**
5. Nutlets in each calyx all smooth-surfaced…**C. fendleri**
5'. Nutlets all rough…(6)
6. Nutlets decidedly heteromorphic, one larger and/or differently ornamented than the others…(7)

6'. Nutlets all alike in size and surface ornamentation…(9)
7.   Odd nutlet < 1.5 mm; nutlet margins angled or rounded; style surpassing odd nutlet; midrib of fruiting calyx lobes moderately thickened but not noticeably expanded and hard (*J. angustifolia*)…go to **JOHN-STONELLA** (in part)
7'. Odd nutlet 2–3 mm; nutlet margins rounded; style subequal to odd nutlet; midrib of fruiting calyx lobes conspicuously thickened and bony…(8)
8.   Cymules bracteate (most flowers subtended by small, leafy bracts)…**C. minima**
8'. Cymules naked (may have 1 or 2 small bracts near the base)…**C. crassisepala**
9.   Cymules bractless or nearly so…(10)
9'. Cymules bracteate throughout…(13)
10. Fruiting calyx < 3 mm (*J. pusilla*)…go to **JOHNSTONELLA** (in part)
10'. Fruiting calyx > 5 mm…(11)
11. Stems spreading-hirsute, the branches erect or ascending…**C. barbigera**
11'. Stems strigose, erect or often flexuous and laxly branched…(12)
12. Stems slender, flexuous-sprawling; nutlets narrowly lanceolate and long-acuminate…**C. nevadensis**
12'. Stems rigid, stiffly erect; nutlets lance-ovate and narrowly acute…**C. juniperensis**
13. Plants low (5–15 cm), stems dichotomously branching from the base outward; flowering in spring… **C. mexicana**
13'. Plants usually taller (15–30 cm), stems initially straight and erect, forming a short central axis and producing dichotomously branching laterals; flowering in late summer…**C. albida**

Cryptantha albida (Kunth) I. M. Johnst.—New Mexico cryptantha. Stems 1 to few from the base, usually forming a central axis, 15–40 cm, antrorsely strigose and sparingly hispid; leaves spatulate to linear-spatulate, usually folded, abundant along the stems; cymules bracteate, paired; fruiting calyx to 3 mm; corolla limb < 2.5 mm wide; style surpassing mature nutlets; nutlets usually 4, all alike, triangular-ovate, ca. 1 mm, tuberculate, scar triangular, excavated. Calcareous soils, desert scrub, arid grassland, 4200–5800 ft. Chaves, Eddy, Otero.

Cryptantha barbigera (A. Gray) Greene—Bearded cryptantha. Stems erect, 1–4 dm, bristly-hirsute with some shorter appressed hairs; leaves oblong to lance-linear; cymules paired (rarely solitary or ternate), naked; fruiting calyx 5–8 mm, segments lance-linear with white-villous margins; corolla inconspicuous, 1–2 mm wide; style nearly equal to mature nutlet tips; nutlets 1–4 maturing per calyx, lanceolate, all alike, 1.5–2.5 mm, densely verrucose-muricate, scar closed above, gradually dilated toward the basal triangular areola. Rocky slopes, Chihuahuan desert scrub, 4300–5800 ft. SC, SW.

Cryptantha crassisepala (Torr. & A. Gray) Greene—Thicksepal cryptantha. Stems branched from the base, 5–15 cm, spreading or erect without a strong central axis; leaves linear to narrowly oblanceolate; cymules naked, solitary (rarely paired); fruiting calyx 4–6.5 mm; sepal midribs thick and bony at maturity; corolla limb 1–6 mm wide; style surpassed by odd nutlet; nutlets lance-ovate, usually 4, heteromorphic, one evidently larger (2–2.5 mm) and more firmly attached to the gynobase, minutely granulate or muricate; the 1–3 consimilar nutlets readily deciduous, 1.2–1.8 mm and granulate-tuberculate, scar open and commonly excavated. We have 2 weakly differentiated varieties:

1.   Corolla limb 3.5–6 mm diam.…**var. crassisepala**
1'. Corolla limb < 3.5 mm diam.…**var. elachantha**

var. crassisepala—Arid grasslands, 3000–3600 ft. NE, EC, C, SE, SC, Luna.

var. elachantha I. M. Johnst.—Desert scrub to piñon-juniper woodland, 3000–6600 ft. Widespread.

Cryptantha fendleri (A. Gray) Greene—Fendler's cryptantha. Stems solitary or occasionally branching from the base, 1–4 dm, erect; leaves mostly cauline, linear-acute; inflorescence broad, cymules naked; fruiting calyx 4–6 mm, corolla minute (1 mm wide); style subequal to mature nutlets; nutlets all alike, usually 4 maturing, narrowly lanceolate, smooth and shinning, scar closed except at the small basal areola. Deep sandy soils in piñon-juniper woodlands and ponderosa pine forests, 5000–7800 ft. NC, NW, C, WC, Sierra.

Cryptantha gracilis Osterh.—Slender cryptantha. Stems branching when well developed, 1–3 dm; leaves basal and scattered along the stem, linear or narrowly spatulate; cymules naked, compact and not much elongating at maturity; fruiting calyx 2-3 mm; corolla minute (ca. 1 mm wide); mature nutlet surpassing the style; nutlets lanceolate, 1.4-2 mm, smooth and shining, usually 1 maturing per calyx (rarely 2-3), scar closed except for small basal areola. Desert scrub, piñon-juniper woodlands, 5850-6050 ft. NW.

Cryptantha juniperensis M. G. Simpson & R. B. Kelley [*Cryptantha nevadensis* A. Nelson & P. B. Kenn. var. *rigida* I. M. Johnst.]—Woodland cryptantha. Stems erect, 1-4 dm, strigose; leaves linear-lanceolate; cymules paired (rarely solitary), naked; fruiting calyx 4.5-7.5 mm; corolla limb 2-5 mm wide; style nearly equal to mature nutlet tips; nutlets usually 4 maturing (or fewer by abortion), lance-ovate, all alike, 1.8-2.2 mm, tuberculate, scar closed above, gradually dilated toward the basal triangular areola. Barely entering the state; foothills, dry arroyos in the Peloncillo Mountains, 4220 ft. Hidalgo.

Cryptantha mexicana (Brandegee) I. M. Johnst.—Mexican cryptantha. Stems few to several, branching from the base, lax or ascending, 5-20 cm; cymules paired, completely bracteate; fruiting calyx 3-4 mm; corolla minute, 1-2 mm wide; style barely surpassing the nutlets; nutlets all alike, usually 4, triangular-ovate, ca. 1 mm, tuberculate, scar triangular and occupying much of the ventral surface, excavated. Calcareous soils, Chihuahuan desert scrub, 3550-5600 ft. SE, SC, SW.

Cryptantha minima Rydb.—Little cryptantha. Stems several, branching from the base, 5-15 cm; leaves basal and cauline, narrowly oblanceolate; cymules continuously or interruptedly bracteate and terminating the branches; fruiting calyx 4-7 mm; sepal midribs becoming thick and bony at maturity; corolla inconspicuous, 1-2 mm wide; style surpassed by odd nutlet; nutlets usually 4, lance-ovate, heteromorphic, odd nutlet 2-3 mm, finely granulate, consimilar nutlets 1.2-1.5 mm, tuberculate, scar open especially at the base. Arid grasslands and juniper savanna, 4150-8260 ft. NE, NC, EC, C, SE, SC.

Cryptantha nevadensis A. Nelson & P. B. Kenn.—Nevada cryptantha. Stems 1-4 dm, slender and lax, branches often sinuous or flexuous and sprawling, strigose; leaves linear or nearly so; cymules naked, usually paired (often ternate or single); fruiting calyx 6-10 mm, segments with recurved tips; corolla minute, 1-2 mm wide; style about equal to or barely surpassing nutlet tips; nutlets 1-4 matured per calyx, all alike, narrowly lanceolate, 2-2.4 mm, sharply tuberculate-muricate, scar closed above, gradually dilated toward the basal triangular areola. Chihuahuan desert scrub, 4000-4800 ft. Grant, Hidalgo.

Cryptantha pterocarya (Torr.) Greene—Wing-nut cryptantha. Stems freely branching when well developed, erect, 1-4 dm; leaves linear or nearly so; cymules naked, paired (rarely solitary or ternate); fruiting calyx 4-5 mm, sepals lance-ovate; corolla minute (0.5-2 mm wide); style subequal to nutlet tips; nutlets usually 4, all alike or heteromorphic, 2.2-3.2 mm, all (or only 3) conspicuously wing-margined, body surface verrucose-muricate, scar narrowly open above and dilated below. We have 2 morphologically distinct but geographically sympatric varieties:

1. Nutlets decidedly heteromorphic, 3 winged and 1 wingless...**var. pterocarya**
1'. Nutlets all alike, all 4 winged...**var. cycloptera**

var. cycloptera (Greene) J. F. Macbr.—Desert scrub, piñon-juniper woodlands, 4000-5850 ft. NW, C, SC, SW.

var. pterocarya—Desert scrub, piñon-juniper woodlands, 5400-6050 ft. NW, SC, SW.

Cryptantha recurvata Coville—Curvenut cryptantha. Stems freely branching when well developed, lax and spreading or ascending, 1-4 dm; leaves scattered, linear to lance-oblong; cymules naked, usually paired; fruiting calyx 2.5-3.5 mm, asymmetric, bent and recurved; corolla minute, ca. 1 mm wide; style surpassed by nutlet tip; nutlets only 1 matured per calyx, somewhat recurved-bent in alignment with the calyx, finely granulate-muricate, scar closed or narrowly open. Desert scrub, piñon-juniper woodlands, 5380-5700 ft. San Juan.

CYNOGLOSSUM L. - Hound's tongue

Cynoglossum officinale L.—Hound's tongue. Biennial with a thick taproot; stem single, leafy, soft-villous, 3-12 dm, with several ascending branches; leaves alternate, entire, mostly basal, those petiolate, oblanceolate or narrowly elliptic, 10–30 cm; cauline leaves sessile, lanceolate, gradually reduced up-ward; sepals broad, blunt; inflorescence of numerous single or paired, ebracteate, helicoid cymules (false racemes) in upper leaf axils and terminating the branches; corolla dull red, aging violet-purple, fornices prominent and broadly rounded; nutlets 1–4, ovate, 5–7 mm, obovate, dorsally flattened or convex, at-tachment scar extending from middle to point of the ventral surface. Noxious weed in disturbed mon-tane sites and riparian areas, 6060–10,100 ft. NE, NC, NW, Lincoln. Introduced.

EREMOCARYA Greene – Eremocarya

Eremocarya is often included in the genus *Cryptantha* s.l. (Hasenstab-Lehman & Simpson, 2012).

Eremocarya micrantha (Torr.) Greene [*Cryptantha micrantha* (Torr.) I. M. Johnst.]—Redroot ere-mocarya. Annual herb from a taproot charged with purple-red dye; stems strigose, slender, 3–10 cm; leaves sparse, < 1 cm, mostly cauline, linear, strigose to ciliate-hirsute near the base; inflorescence of many small, biseriate, bracteate cymules; fruiting calyx tardily deciduous, segments lanceolate, hirsute; corolla white, 1 mm or less wide, salverform; fornices inconspicuous or absent; nutlets evidently sur-passed by an elongate gynobase capped by a sessile stigma; nutlets narrowly lanceolate, 1 mm, smooth or finely tuberculate, usually 4, all alike or somewhat heteromorphic with an odd nutlet that is slightly larger and more firmly attached to the gynobase, scar narrowly open or closed and slightly dilated at the base. Chihuahuan desert scrub, with sagebrush and saltbush in the Four Corners region, 4000–6750 ft. NC, SC, SW, Sandoval, San Juan.

ERITRICHIUM Schrad. ex Gaudin – Forget-me-not

Eritrichium nanum (L.) Schrad. ex Gaudin [*Eritrichium aretioides* (Cham.) DC. var. *elongatum* Rydb.]—Alpine forget-me-not. Cushionlike, long-lived perennial, densely hairy; acaulescent or stems short and slender, to 1 dm; leaves mostly basal, silvery-villous, entire, narrowly ovate to oblong, 5–10 mm, imbricate on flowering stems; inflorescence of compact cymose clusters, somewhat elongate into naked or bracteate false racemes; flowers with blue, rarely white, corolla limbs 4–8 mm wide; corolla tube equal to the calyx, fornices in the throat well developed; nutlets 1–4 maturing, dorsal margins entire or with a few short teeth on the upper margin angles. Our plants are the North American **var. elongatum** (Rydb.) Cronquist. Rocky ledges and slopes, alpine tundra at or above timberline, 11,400–13,160 ft. NC.

HACKELIA Opiz – Stickseed

Biennial taprooted herbs; stems erect or ascending from a basal rosette of leaves, strigose-hirsute; stem leaves alternate, short-petioled or sessile; inflorescence an indeterminate, cymosely branched false raceme; pedicels decurved in fruit; calyx deeply cleft, sepals nearly distinct, all alike; corolla white or blue, rotate-salverform or funnelform, yellow or white faucal appendages (fornices) nearly closing the throat; stigma capitate; nutlets usually 4 maturing (sometimes fewer by abortion), mostly homomor-phic, lanceolate or ovate, attached to a pyramidal gynobase by a medial areola, margins armed with stout, usually flattened glochidiate prickles, dorsal surface verrucose-hispidulous and sometimes with small intramarginal glochidiate prickles.

1. Corolla limb white to cream-colored; inflorescence bracteate…**H. ursina**
1′. Corolla limb blue, rarely pale violet or pinkish; inflorescence bracteate or not…(2)
2. Leaves hispid-hirsute, pustulate bases of coarse hairs evident…(3)
2′. Leaves soft-hirsute or strigose, pustulate bases of hairs inconspicuous or absent…(4)
3. Corolla limb inconspicuous, 1–2.5 mm across; nutlet margin prickles < 1.5 mm…**H. besseyi**
3′. Corolla limb conspicuous, 4–8 mm across; nutlet margin prickles > 1.5 mm…**H. hirsuta**
4. Intramarginal prickles 1–4 on dorsal nutlet surface between the larger marginal prickles…**H. pinetorum** (in part)
4′. Intramarginal prickles absent…(5)
5. Inflorescence mostly elongate and narrow, racemose branches usually 3–6 cm in fruit; midstem leaf blades narrowly oblanceolate or lance-linear, narrowly acute or acuminate…**H. floribunda**
5′. Inflorescence open and spreading, racemose branches 4–10 cm in fruit; midstem leaf blades elliptic or lanceolate, acute or obtuse…**H. pinetorum** (in part)

Hackelia besseyi (Rydb.) J. L. Gentry [*Hackelia grisea* (Wooton & Standl.) I. M. Johnst.]—Bessey's stickseed. Stems 1 to several, 1–6 dm; leaves oblanceolate-oblong farther up the stem, hispid-hirsute or hirsute-strigose, usually obtuse; flowers in cymose false racemes, 5–12 cm in fruit, irregularly bracteate for most of their length; pedicels 4–8 mm in fruit; corolla blue, 1–2.5 mm wide; nutlets 2–2.5 mm, marginal prickles usually < 1.5 mm, dorsal surface without intramarginal prickles. Slopes in piñon-juniper woodlands, ponderosa pine and mixed conifer forests, 6560–10,500 ft. NC, NW, C, SC.

Hackelia floribunda (Lehm.) I. M. Johnst.—Manyflower stickseed. Stems 1 to few, 5–12 dm; stem leaves narrowly oblanceolate or lance-linear, narrowly acute to acuminate, soft-strigose or velutinous; flowers in cymose false racemes, 5–6 cm in fruit, bracts absent or sparsely bracteate; pedicels short, 3–6 mm in fruit; corolla blue, 3–6 mm wide; nutlets 2.5–3.5 mm, marginal prickles usually 2–3 mm, dorsal surface without intramarginal prickles or very rarely with 1 small prickle. Forest openings and valley bottoms in high mountain ranges, 5950–10,180 ft. NC, NW, C, WC, SC.

Hackelia hirsuta (Wooton & Standl.) I. M. Johnst.—New Mexico stickseed. Stems 1 to several, 1–6 dm; leaves oblanceolate or oblong farther up the stem, hispid-hirsute and hirsute-strigose, usually obtuse; flowers in cymose false racemes, 5–12 cm in fruit, regularly bracteate for entire length; pedicels 6–8 mm in fruit, corolla blue, rarely pale pink, 4–8 mm wide; nutlets 2.5–3.5 mm, marginal prickles usually 2–3.5 mm, dorsal surface without intramarginal prickles. Endemic to N mountains in pine-oak and mixed conifer forests, 7500–10,800 ft. NE, NC. **R & E**

Hackelia pinetorum (Greene ex A. Gray) I. M. Johnst.—Livermore stickseed. Stems 1 to few, 2–8 dm; stem leaf blades lanceolate or elliptic, acute or obtuse, hirsute-strigose or hispidulous; flowers in cymose false racemes, 4–10 cm in fruit, bracts absent or a few at the base; pedicels 4–10 mm in fruit; corolla blue, pale violet, rarely white, 5–7 mm wide; nutlets 2–3 mm, marginal prickles usually 2–3 mm, dorsal surface with or rarely without intramarginal prickles.

1. Small intramarginal prickles 1, 2, or 3(4) on dorsal nutlet surface…**var. pinetorum**
1'. Small prickles on dorsal nutlet surface absent…**var. jonesii**

var. jonesii J. L. Gentry—S New Mexico. Sierra Madrean; variation occurs in the Organ and White Mountains, 6500–7500 ft. C, SC.

var. pinetorum—Pine-oak and mixed conifer forests, 6850–10,340 ft. C, WC.

Hackelia ursina (Greene ex A. Gray) I. M. Johnst.—Chihuahuan stickseed. Stems 1 to several, 1–6 dm; leaves oblanceolate or oblong farther up the stem, hispid-hirsute and hirsute-strigose, usually obtuse; flowers in cymose false racemes, 5–12 cm in fruit, regularly bracteate for entire length; pedicels 6–8 mm in fruit, corolla white or cream-colored, 5–10 mm wide; nutlets 1–2 mm, marginal prickles usually 1.5–2 mm, dorsal surface with or without intramarginal prickles. We have 2 varieties:

1. Small intramarginal prickles on dorsal nutlet surface absent…**var. ursina**
1'. Small intramarginal prickles present on at least some nutlets in the inflorescence…**var. pustulata**

var. pustulata (J. F. Macbr.) J. L. Gentry—Pine-oak woodland in bootheel region, 5250–7550 ft. Hidalgo.

var. ursina—Pine-oak woodlands and mixed conifer forest, 6060–8850 ft. WC.

JOHNSTONELLA Brand - Pick-me-not

Annual herbs, strigose, hirsute, or hispid-setose; leaves linear-oblanceolate; inflorescence of helicoid cymules (false racemes or false spikes), naked or bracteate at base; calyx cleft to the base or nearly so; corolla white, salverform with a spreading or nearly closed limb and small fornices at the throat; homostylous; stigma capitate; nutlets 4, or 1–3 by abortion, triangular-ovate or lanceolate, tuberculate, heteromorphic or all similar, affixed to an elongate gynobase. Segregation of the genus *Johnstonella* from *Cryptantha* is based (for the most part) on molecular evidence and lacks consistent morphological distinction (Hasenstab-Lehman & Simpson, 2012).

1. Nutlets lanceolate, decidedly heteromorphic with 3 small consimilar nutlets and 1 larger odd nutlet…
   J. angustifolia
1'. Nutlets triangular-ovate, all alike…**J. pusilla**

Johnstonella angustifolia (Torr.) Hasenstab & M. G. Simpson [*Cryptantha angustifolia* (Torr.) Greene]—Narrowleaf pick-me-not. Stems diffusely branched from the base, 5–20 cm; leaves scattered, linear; cymules usually paired, naked or sparsely bracteate at the base; fruiting calyx hispid, 2.5–4 mm, segments narrowly lanceolate, tardily deciduous; corolla limb < 2.5 mm wide; style usually surpassing all the nutlets; nutlets lanceolate, usually 4, heteromorphic, the nutlet in the abaxial position more firmly attached and slightly larger (> 1.5 mm) than the other 3 (ca. 1 mm), all finely tuberculate, odd nutlet margins angled or rounded, scar subulate, closed or narrowly open. Bajadas, desert scrub, 3500–4650 ft. SC, SW.

Johnstonella pusilla (Torr. & A. Gray) Hasenstab & M. G. Simpson [*Cryptantha pusilla* (Torr. & A. Gray) Greene]—Little pick-me-not. Stems few to several from the base, slender, 3–15 cm; leaves mostly basal, scattered above, linear-spatulate; cymules solitary or paired with few (if any) minute bracts; fruiting calyx 2–2.5 mm, broadly ovate, segments lance-ovate; corolla minute (< 1 mm wide); style usually surpassing nutlets; nutlets all alike, usually 4 maturing, ca. 1 mm, triangular-ovate, bent, tuberculate, margins sharply angled, scar subulate and dilated at base into a triangular areola. Rocky, arid slopes and hills, desert scrub, 4260–5250 ft. SC, SW.

LAPPULA Moench – Sheepbur

Annual taprooted herbs; strigose-hirsute, puberulent; stems erect or ascending; leaves alternate, narrowly oblanceolate or linear-elliptic; inflorescence an indeterminate, bracteate, cymosely branched false raceme; pedicels ascending in flower and fruit; calyx deeply cleft, sepals nearly distinct, all alike; corolla inconspicuous, white or blue, funnelform, yellow or white faucal appendages (fornices) nearly closing the throat; nutlets usually 4 maturing (sometimes fewer by abortion), homomorphic or heteromorphic, attached to an erect gynobase along the length of a median ventral keel, margins armed with stout glochidiate prickles (Allred, 1999).

1. Nutlets with 2+ rows of slender marginal prickles that are not confluent at their bases; corolla 3–4 mm wide…**L. squarrosa**
1'. Nutlets with a single row of marginal prickles that are distinct or confluent at their bases; corolla 1–2 mm wide…**L. occidentalis**

Lappula occidentalis (S. Watson) Greene—Sheepbur. Stems 1 to several, 1–6 dm; leaves oblanceolate where crowded at the base, becoming linear or narrowly oblong farther up the stem; cymes greatly elongate in fruit; pedicels 1–3 mm, corolla 1–2 mm wide; nutlets homomorphic or heteromorphic, lanceolate, margins with a single row of glochidiate prickles, 2–2.5 mm, dorsal surface tuberculate or papillate. Introduced.

1. Marginal prickles distinct or scarcely confluent at their bases, not forming a wing or cuplike structure…
   var. occidentalis
1'. Marginal prickles confluent at their bases to form a conspicuous wing or cuplike, sometimes swollen, structure on 2, 3, or all 4 of the nutlets…**var. cupulata**

var. cupulata (A. Gray) L. C. Higgins [*Lappula texana* (Scheele) Britton]—Desert scrub to piñon-juniper woodlands, 3800–8200 ft. Widespread.

var. occidentalis [*Lappula redowskii* (Hornem.) Greene]. Desert scrub, piñon-juniper woodlands, ponderosa pine forests, 4260–9500 ft. Widespread.

Lappula squarrosa (Retz.) Dumort. [*Lappula echinata* Gilib.]—Bristly sheepbur. Stems 1 to few, 1–6 dm; leaves oblanceolate near the base, becoming narrowly oblong or lanceolate farther up the stem; cymes elongate in fruit; pedicels 1–3 mm, ascending in fruit; corolla blue, 3–4 mm wide; nutlets homomorphic, lanceolate, margins with 2 or 3 rows of glochidiate prickles, 2–2.5 mm, dorsal surface papillate.

Disturbed openings in ponderosa pine and mixed conifer forests, 5740–8200 ft. NC, Catron, Otero, San Miguel. Introduced.

LITHOSPERMUM L. – Gromwell

Perennial herbs from a branching caudex or taproot; root crowns and stem bases sometimes purple-dye-stained; stems usually erect, simple below, hirsute or strigose, often branched above; leaves alternate, strigose; inflorescence of racemose, often helicoid branches; flowers bracteate, homostylous or heterostylous, usually chasmogamous, (but *L. incisum* and *L. parksii* can have cleistogamous flowers later in the season); calyx deeply lobed, segments usually narrow; corolla yellow or pale green, salverform, funnelform, or tubular, throat with or without faucal appendages (fornices); nutlets 1–4, erect, basally attached, ovoid or angular, lustrous, bone-white, gray, or brownish, smooth, pitted, or roughened (Cohen & Davis, 2009).

1. Corolla yellow, rarely greenish; upper stem leaves linear, narrowly lanceolate, or oblong, only midvein prominent…(2)
1'. Corolla pale green or greenish yellow; upper stem leaves broadly lanceolate, midvein and lateral veins prominent…(6)
2. Corolla lobes erose or fimbriate on early chasmogamous flowers; later (lower) cleistogamous flowers smaller, entire…**L. incisum**
2'. Corolla lobes entire; cleistogamous flowers present or absent…(3)
3. Nutlets distinctly roughened; faucal appendages (fornices) present at corolla throat; cleistogamous flowers sometimes present…**L. parksii**
3'. Nutlets smooth or slightly pitted; faucal appendages at corolla throat absent; cleistogamous flowers absent…(4)
4. Stems 1 or few arising from a crowded rosette of basal leaves that are larger than the middle and upper stem leaves…**L. cobrense**
4'. Stems 1 to several arising from buds on a stout root crown or caudex; lowest stem leaves usually poorly developed and smaller than the middle stem leaves…(5)
5. Corolla definitely yellow, tube 9–15 mm; flowers heterostylic, heteromorphic…**L. multiflorum**
5'. Corolla pale yellow, often tinged with green, tube 4–7 mm; flowers homostylic; anthers in the corolla throat surpassing the style…**L. ruderale**
6. Corolla tube > 3 cm…**L. macromeria**
6'. Corolla tube < 3 cm…(7)
7. Corolla lobes spreading-reflexed…**L. viride**
7'. Corolla lobes erect, like closed valves when fresh…**L. onosmodium**

Lithospermum cobrense Greene—Smooththroat gromwell. Stems single or few from a leafy rosette, erect, simple or sparsely branched, 1–5 dm; rosette leaves usually withering before anthesis, oblanceolate, 5–10 cm × 5–15 mm; stem leaves numerous, crowded, narrowly oblong to linear, revolute, obtuse, 1–3 × 2–5 mm; inflorescence a helicoid raceme terminating the stems; corolla funnelform, pale yellow, fornices absent; homostylous; nutlets 2.5–3 mm, plump, smooth. Openings in pine-oak forest to mixed conifer forest, 5080–10,000 ft. C, WC, SC, SW, Sandoval.

Lithospermum incisum Lehm.—Fringed gromwell. Stems few to several from a stout, usually purple-dye-stained taproot, erect or spreading, branched, 0.5–3 dm; stem leaves narrowly lanceolate or linear, acute to acuminate; flowers in small cymes in the leaf axils; chasmogamous corolla yellow, salverform, tube 15–30 mm, lobes erose to fimbriate, fornices present but not prominent; style heteromorphic, 5–30 mm; cleistogamous corolla smaller with entire lobes or failing to open; nutlets 3–3.5 mm with a ventral keel, smooth and sparsely pitted. Desert scrub, piñon-juniper woodlands, 3450–8200 ft. Widespread.

Lithospermum macromeria J. I. Cohen [*Macromeria viridiflora* DC.]—Giant trumpets. Stems single or few from a woody caudex, erect, to 1.2 m; leaves cauline, sessile, obovate-elliptic below, lanceolate above, veins sunken; flowers in terminal helicoid cymes; corolla narrowly funnelform, canescent-hairy, pale grayish green to yellowish green, tube 4–6 cm, lobes erect to spreading, fornices absent; nutlets smooth, 3–4 mm, ovoid, apex acute. Ponderosa pine forests, oak woodlands, 5840–9340 ft. C, SC, SW, Mora, San Miguel.

Lithospermum multiflorum Torr. ex A. Gray—Manyflowered gromwell. Stems single or few from a woody caudex charged with purple dye, erect, to 60 cm; leaves cauline, sessile, narrowly lanceolate, oblong, or linear, often revolute; flowers in terminal and axillary scorpioid cymes; corolla tubular-funnelform, yellow, tube 9-15 mm, fornices absent; style heteromorphic; nutlets smooth or sparsely pitted, 2.5-3.5 mm, ovoid, white or brownish. Oak woodlands, ponderosa pine and mixed conifer forests, 5900-10,000 ft. Widespread except EC, SE, SW.

Lithospermum onosmodium J. I. Cohen [*Onosmodium bejariense* DC.; *O. molle* Michx.; *O. occidentale* Mack.]—Marleseed. Stems single or few from a woody caudex, erect, to 1 m; leaves cauline, sessile, obovate-oblanceolate below, lanceolate above, the veins sunken; flowers in terminal racemes; corolla narrowly tubular, canescent-hairy, greenish white or yellowish green, tube 8-15 mm, lobes erect, like closed valves when fresh, fornices absent; style long-exserted; nutlets smooth, 3.5-4 mm, ovoid, white. Moist draws in plains and foothills, 4920-8100 ft. Colfax, Mora, Union.

Lithospermum parksii I. M. Johnst.—Parks' gromwell. Stems single or few from a stout taproot, erect or spreading, 0.5-3 dm; stem leaves narrowly lanceolate or linear, acute or obtuse; flowers in the leaf axils; chasmogamous corolla yellow, salverform, tube 8-15 mm, lobes entire, fornices present; cleistogamous corolla (when present) much smaller and failing to open; nutlets 3 mm, white or gray, verrucose or rugulose. Limestone, slopes and canyon bottoms, 4300-4950 ft. Eddy, Otero.

Lithospermum ruderale Douglas ex Lehm.—Western gromwell. Stems few to several from a stout taproot, 2-6 dm; leaves cauline, narrowly lanceolate or linear; flowers borne in leafy-bracteate clusters within the densely leafy upper stem; corolla pale yellow, often tinged with green, salverform, tube 4-7, lobes entire, fornices absent; homostylous; nutlets 3.5-5 mm, plump with a ventral keel, gray. Piñon-juniper woodland, mountain brush. Colfax.

Lithospermum viride Greene—Green gromwell. Stems single or few from a woody caudex, 3-7 dm, erect; leaves cauline, sessile or on 1-2 mm petioles, elliptic to ovate-lanceolate, lateral veins obvious, often sunken; flowers in terminal racemes; corolla greenish yellow, tubular, tube 14-25 mm, externally villous, lobes reflexed, fornices absent; homostylous; nutlets 3.5-4 mm; smooth or sparsely pitted, white, gray, or tan. Piñon-juniper-oak woodlands, 5580-8200 ft. SE, SC, SW, Lincoln.

MERTENSIA Roth – Bluebell, lungwort

Mare Nazaire

Herbs, perennial, glabrous, strigose, hispid, or hirsute, hairs often with pustulate bases; plants erect, ascending, or decumbent; leaves basal and cauline, ovate, elliptic, linear, or lanceolate, thin or subcoriaceous and somewhat fleshy, lateral veins conspicuous or not, margins entire; inflorescences ebracteate cymes, compound, paniclelike, few- to many-flowered; flowers pale blue, blue, or dark blue, occasionally white, throat/limb dilated, funnelform, or campanulate; fornices alternating with stamens; stamens included; ovary 2-celled, each cell deeply 2-lobed; style gynobasic; stigma entire, capitate; nutlets (3)4, attached laterally to gynobase, off-white, pale brown, or dark brown, usually rugulose and papillate, ovate to tetrahedral, not compressed (Nazaire & Hufford, 2014).

1. Plants relatively tall and robust, 40-150 cm; leaves thin, lateral venation conspicuous…(2)
1'. Plants relatively short, seldom exceeding 40 cm; leaves usually thick, lateral venation inconspicuous…(3)
2. Stems, leaves, and pedicels usually glaucous; calyx lobes oblong to ovate, 1.1-2 mm; apex of calyx lobes obtuse to subacute…**M. ciliata**
2'. Stems, leaves, and pedicels not glaucous; calyx lobes lanceolate, linear-lanceolate, or linear-oblong, 2-3.4 mm; apex of calyx lobes acuminate…**M. franciscana**
3. Corollas with nearly salverform, subrotate, or sometimes broadly funnelform throat/limb; filaments stipelike; anthers not surpassing fornices…(4)
3'. Corollas with dilated to funnelform throat/limb; filaments ligular, rarely stipelike; anthers surpassing fornices…(5)
4. Calyces cleft 1/2-3/4 to base, strigose to sericeous; plants montane to subalpine…**M. brevistyla**
4'. Calyces cleft 3/4-4/5 to base, glabrous, occasionally strigose at base; plants subalpine to alpine…**M. tweedyi**

5. Calyx cleft 3/4+ to base; plants usually of montane to alpine habitats at middle to high elevations...
M. ovata
5'. Calyx cleft up to 3/4 to base; plants usually of prairies, foothills, and woodlands at low to middle elevations...(6)
6. Calyx cleft 1/3-1/2 to base...**M. fendleri**
6'. Calyx cleft 1/2-3/4 to base...(7)
7. Taproot fusiform; leaf blades oblong-ovate to ovate-elliptic, apex rounded to obtuse...**M. fusiformis**
7'. Taproot slender to stout, not fusiform; leaf blades linear-lanceolate, lanceolate, or oblanceolate, apex acute to subacute...**M. lanceolata**

Mertensia brevistyla S. Watson—Short-styled bluebell. Plants 12-30 cm; stems erect to ascending; leaves oblanceolate to obelliptic, thin, green, lateral venation inconspicuous, abaxially glabrous, adaxially strigose; pedicels strigose; calyx 2-4.2 mm, cleft 1/2-3/4 to base, strigose to sericeous, lobes deltoid-lanceolate, 1.4-2.5(-3.2) mm, apex acute or subacute; corolla blue, throat/limb broadly funnelform to nearly salverform, 7-10.5 mm, glabrous within; filaments inserted below fornices, stipelike, 0.2-0.6 mm; stigma not surpassing anthers or fornices. Montane to subalpine meadows, open slopes, rocky outcrops, aspen, oak, sagebrush communities, 4920-11,480 ft. Rio Arriba.

Mertensia ciliata G. Don—Tall fringed, mountain, or streamside bluebell. Plants 40-100(-150) cm; stems erect; leaves ovate, lanceolate, or elliptic, thin, green, lateral venation conspicuous, often glaucous, abaxially glabrous or strigose, adaxially pustulate, otherwise glabrous; pedicels usually sparsely pustulate; calyx 1.6-3(-4) mm, cleft 2/3-3/4 to base, glabrous, occasionally sparsely strigose at base, calyx lobes oblong to ovate, 1.3-2 mm, apex subacute to obtuse; corolla blue, occasionally white, throat/limb dilated to funnelform, 9.6-17 mm, glabrous or sparsely pubescent within; filaments inserted below fornices, ligular, 1.3-2.7 mm. Banks of streams and rivers, subalpine meadows, aspen woods, coniferous forests, seeps and marshy areas, 3900-12,470 ft. NC.

Mertensia fendleri A. Gray [*M. lanceolata* (Pursh) DC. var. *fendleri* (A. Gray) A. Gray]—Fendler's bluebell. Plants 15-40(-50) cm; stems erect; leaves ovate or lanceolate to linear-lanceolate or oblanceolate, somewhat thick, green, lateral venation inconspicuous; calyx 3.5-6 mm, cleft 1/3-1/2 to base, strigose, calyx lobes deltoid to deltoid-lanceolate, 1.1-2.6 mm, apex acute; corolla pale blue, throat/limb dilated, 7.5-11.5 mm, densely pubescent within; filaments inserted below fornices, ligular, 1.9-3 mm.

1. Basal and cauline leaves abaxially glabrous; pedicels strigose...**var. fendleri**
1'. Basal and cauline leaves abaxially densely strigose to hairs spreading; pedicels densely strigose...**var. pubens**

var. fendleri—Leaves abaxially glabrous; pedicels strigose. Stream banks, canyons, moist meadows, woodlands, 8200-9840 ft. NC, NW, C, Cibola.

var. pubens J. F. Macbr. [*M. lanceolata* (Pursh) DC. var. *pubens* (J. F. Macbr.) L. O. Williams]—Leaves abaxially densely strigose to hairs spreading; pedicels densely strigose. Stream banks, canyons, moist meadows, woodlands, 8200-9840 ft. NC.

Mertensia franciscana A. Heller [*M. pratensis* A. Heller]—Franciscan bluebell. Plants 40-100(-150) cm; stems erect to ascending; leaves broadly elliptic to ovate or lanceolate, thin, green, lateral venation conspicuous, abaxially glabrous to sparsely strigose, adaxially strigose; pedicels densely strigose to strigillose; calyx (2.2-)2.8-4.4 mm, cleft 2/3-4/5 to base, sparsely strigose to glabrous, calyx lobes linear-lanceolate to linear-oblong, (1.3-)2.2-3.4 mm, apex acuminate; corolla blue to pale blue, throat/limb dilated to funnelform, (8.7-)9.6-15 mm, pubescent within, occasionally glabrous; filaments inserted below fornices, ligular, 1.4-2.4 mm. Subalpine and montane meadows, aspen woods, coniferous forests, banks of streams and rivers, canyon bottoms, 6890-12,470 ft. Widespread except E, SW.

Mertensia fusiformis Greene—Spindle-rooted bluebell. Plants 12-30 cm, taproot fusiform; stems erect to ascending; leaves oblong-ovate to ovate-elliptic, somewhat thick, green to gray-green, lateral venation inconspicuous, abaxially glabrous, adaxially strigose; pedicels densely strigose; calyx 2.5-4.3 mm, cleft 2/3-3/4 to base, strigose, densely so at base, calyx lobes deltoid-lanceolate, 1.7-3.4 mm, apex acute; corolla blue, throat/limb dilated to funnelform, (5–)7–14 mm, pubescent within; filaments inserted below fornices, ligular, 1.3–2.4(–3.8) mm. Montane meadows, open slopes, rocky outcrops, aspen, oak, sagebrush communities, 5570–10,500 ft. NC.

Mertensia lanceolata (Pursh) DC.—Narrow-leaf or prairie bluebell. Plants 15–40 cm, taproot slender to stout, not fusiform; stems erect to ascending; leaves linear-lanceolate, lanceolate, or oblanceolate, somewhat thick, green, lateral venation inconspicuous, abaxially glabrous, adaxially strigose; pedicels sparsely strigose to pustulate, occasionally glabrous; calyx 2.9–6 mm, cleft 2/3–3/4 to base, short-strigose to pustulate, otherwise glabrous, calyx lobes lanceolate to linear-lanceolate, 1.9–4.9 mm, apex acute to acuminate; corolla blue to pale blue, throat/limb dilated to funnelform, 8.6–15.4 mm, densely pubescent within; filaments inserted below fornices, ligular, 1.7–3.1 mm. Grasslands, prairies, meadows, sagebrush slopes, stream banks, 2600–8530 ft. NE, NC, NW, C, WC.

Mertensia ovata Rydb.—Plants 8–30 cm; stems decumbent to ascending; leaves broadly ovate to elliptic or ovate, somewhat thick, coriaceous, glaucous, bluish green to green, lateral venation inconspicuous, abaxially glabrous, adaxially glabrous or short-strigose; pedicels glabrous, occasionally strigillose; calyx 3.6–6 mm, cleft 3/4–9/10 to base, glabrous, calyx lobes linear-lanceolate, 2.8–5.3 mm, apex acute to acuminate; corolla dark blue, occasionally with whitish tube, throat/limb dilated to funnelform, 10–13.3 mm, pubescent within; filaments inserted at level of fornices, ligular, 1.7–2.5 mm. Our plants are **var. caelestina** (A. Nelson & Cockerell) Nazaire & L. Hufford. Rocky outcrops in alpine and subalpine, slopes above timberline, 11,480–12,800 ft. NE, NC.

Mertensia tweedyi Rydb. [*M. alpina* G. Don misapplied]—Alpine bluebell. Plants 8–25 cm; stems decumbent to ascending; leaves ovate, lanceolate, or oblanceolate, somewhat thick, green, lateral venation inconspicuous, abaxially glabrous, adaxially strigose; pedicels strigose; calyx 2.2–3.4 mm, cleft 3/4–4/5 to base, glabrous, occasionally strigose at base, calyx lobes lanceolate to oblong, 1.8–2.8 mm, apex acute to subacute; corolla blue, throat/limb funnelform to nearly salverform, (5.5–)6–10.1 mm, glabrous within; filaments inserted below fornices, stipelike, 0.2–0.5 mm; stigma not surpassing anthers or fornices. Alpine fell-fields, subalpine meadows, steep rocky slopes above timberline, 8850–13,167 ft. Colfax, Santa Fe, Taos.

MYOSOTIS L. – Forget-me-not

Myosotis scorpioides L.—Common forget-me-not. Perennial with fibrous roots; stems leafy, 2–6 dm, often stoloniferous; leaves alternate, oblanceolate to elliptic, 2–8 cm; inflorescence of single or paired helicoid racemes, elongate in fruit; calyx tube strigose; corolla salverform, blue, fornices prominent, limb 5–10 mm wide; nutlets 4, blackish, smooth and shiny, attachment scar small, basilateral with a raised margin. Occasional in shallow water and wet soil of mountain streams, 5900–7880 ft. Lincoln, San Miguel. Introduced (Hutchins, 1974).

OREOCARYA Greene – Cat's-eye

Perennial or biennial herbs, strigose (rarely glabrate) and often hirsute or hispid-setose, larger hairs often with pustulate bases; leaves alternate; flowers borne in terminal or lateral modified cymes (cymules) that are scorpioid or aggregated into a thyrse or terminal subcapitate cluster; calyx cleft to the base or nearly so; corolla white or yellow, salverform with a spreading or rotate limb and prominent fornices at the throat (these usually yellow); anthers included in the corolla tube; homostylous or heterostylous; stigma capitate; nutlets 4, or 1–3 by abortion, ovate to lanceolate, smooth to variously roughened, homomorphic, affixed to elongate gynobase, ventral scar closed or narrowly open (Hasenstab-Lehman & Simpson, 2012).

1. Dorsal surface of mature nutlets smooth and shiny…(2)
1′. Dorsal nutlet surface roughened with tubercles, murications, or wrinkles…(4)
2. Entire corolla yellow; nutlets straight-lanceolate, usually maturing 1 per calyx (rarely 2)…**O. flava**
2′. Corolla limb white (tube and fornices often yellowish); nutlets ovate-lanceolate, decidedly curved inward toward the style, usually maturing 4 per calyx (sometimes fewer by abortion)…(3)

3. Corolla tube lacking basal scales; plants biennials or short-lived perennials…**O. palmeri**
3'. Interior base of corolla tube ringed with small (< 1 mm) antrorse scales; plants evidently perennial…**O. suffruticosa**
4. Corolla tube elongate, usually exceeding the calyx by at least 2 mm…(5)
4'. Corolla tube about equal to the calyx…(9)
5. Nutlets muricate (murications sometimes setulose-tipped), usually maturing 1 per calyx (sometimes 2)…**O. fulvocanescens**
5'. Nutlets rugose or tuberculate; usually maturing 4 per calyx (sometimes fewer by abortion)…(6)
6. Nutlets lance-ovate, straight, scar narrowly open for nearly entire length, NW New Mexico…(7)
6'. Nutlets ovate, decidedly curved toward the style, scar closed for entire length, S and EC New Mexico…(8)
7. Inflorescence subcapitate, < 5 cm; corolla tube 10–12 mm; plants usually < 15 cm…**O. paradoxa**
7'. Inflorescence elongate, 5–30 cm; corolla tube 7–10 mm; plants 10–40 cm…**O. flavoculata**
8. Flowers heterostylous; corolla limb 10–14 mm diam., fornices bright yellow…**O. paysonii**
8'. Flowers homostylous; corolla limb 6–10 mm diam., fornices white or pale yellow…**O. oblata** (in part)
9. Nutlet margins conspicuously papery-winged; plants coarse, 4–10 dm…**O. setosissima**
9'. Nutlet margins not papery-winged; plants smaller, < 5 dm…(10)
10. Corolla tube 6–10 mm; nutlets decidedly bent toward the style…(11)
10'. Corolla tube 6 mm or less; nutlets erect…(12)
11. Midcauline leaves narrowly oblanceolate, > 2 mm wide; mature inflorescence usually elongate-interrupted, rarely subcapitate; nutlets tuberculate-rugose, 2.4–3 mm…**O oblata**
11'. Midcauline leaves linear or linear-oblanceolate, < 2 mm wide; mature inflorescence terminal and subcapitate, occasionally with a reduced cymule at 1 or 2 additional nodes; nutlets finely tuberculate 2–2.4 mm…**O. worthingtonii**
12. Mature inflorescence densely flowered and broad (> 1 dm broad)…**O. thyrsiflora**
12'. Inflorescence fewer-flowered and narrower (< 1 dm broad)…**O. bakeri**

Oreocarya bakeri Greene [*Cryptantha bakeri* (Greene) Payson]—Baker's cat's-eye. Biennial or short-lived, taprooted perennial; stems 1 to few from the base, 1–3 dm; leaves oblanceolate or spatulate, spreading-hirsute and sericeous-strigose; inflorescence elongate-cylindrical in flower; flowers homostylic; fruiting calyx strigose-hirsute, 6–9 mm; corolla tube equal to the calyx at anthesis; corolla limb white, 7–10 mm wide; nutlets ovate-lanceolate, usually all 4 maturing, rugose-tuberculate on both surfaces; scar closed, with conspicuously raised margins. Sandstones or sandy clay soils, sagebrush, piñon-juniper woodlands, ponderosa pine forests, 4500–8700 ft. Rio Arriba, San Juan.

Oreocarya flava A. Nelson [*Cryptantha flava* (A. Nelson) Payson]—Yellow cat's-eye. Cespitose perennial; stems few to several from a branching woody caudex, 1–4 dm; leaves mostly basal; linear-oblanceolate and silvery-strigose; inflorescence elongate and thyrsoid-cylindrical, conspicuously yellow-setose; flowers heterostylic; calyx 9–12 mm in fruit, lobes linear; corolla yellow throughout, tube surpassing the calyx at anthesis, limb 7–11 mm wide; nutlets lance-ovate, usually only 1 maturing, sometimes 2, 3.4–4.2 mm, smooth and glossy. Sandy soil, desert scrub, sagebrush, piñon-juniper woodlands, 5250–7550 ft. NW, NC, C, Sandoval, Socorro.

Oreocarya flavoculata A. Nelson [*Cryptantha flavoculata* (A. Nelson) Payson]—Roughseed cat's-eye. Cespitose perennial; stems few to several from a branching caudex, 0.5–2.5 dm; leaves oblanceolate to spatulate, obtuse, rarely acute, coarsely strigose to sericeous-puberulent; inflorescence narrowly cylindrical, rarely somewhat open and spreading; flowers heterostylic; calyx segments lanceolate to ovate, 8–10 mm in fruit; corolla tube surpassing the calyx at anthesis; corolla limb white, rotate, 8–12 mm wide; nutlets lanceolate to lance-ovate, usually 4 maturing, 2.5–3.8 mm, tuberculate-muricate, scar open, constricted near the middle and surrounded by elevated margins. Sagebrush, piñon-juniper woodlands. San Juan.

Oreocarya fulvocanescens (S. Watson) Greene—Tawny cat's-eye. Cespitose perennial; stems few to several from a branching woody caudex, 0.5–3 dm; leaves mostly basal, narrowly oblanceolate, acute or obtuse, uniformly silvery-strigose, cauline leaves usually narrower and spreading-hispid; inflorescence narrowly cylindrical to somewhat open at maturity, rarely subcapitate, conspicuously tawny-setose or silvery-strigose; flowers heterostylic; fruiting calyx 6–13 mm, lobes linear, hispid to strigose; corolla tube surpassing the calyx at anthesis; corolla limb white, 6–9 mm wide, reflexed after anthesis; nutlets lance-ovate, 3.1–4.4 mm, usually only 1 or sometimes 2 maturing per calyx, both surfaces muricate, often with sharp, setose tips terminating some or all of the murications; scar closed or only slightly open.

1. Calyx densely hispid-strigose; interior calyx lobe faces strigulose, the green surface partly visible…**var. fulvocanescens**
1'. Calyx densely strigose and sparsely hispid; interior calyx lobe faces obscured by dense, silvery pubescence…**var. nitida**

var. fulvocanescens [*Cryptantha fulvocanescens* (S. Watson) Payson; *C. fulvocanescens* (S. Watson) Payson var. *echinoides* (M. E. Jones) L. C. Higgins]—Shale, clayey sand, or gypseous soil, desert scrub, sagebrush, piñon-juniper woodlands, 4000–8040 ft. NC, NW, C, WC, SW, Otero.

var. nitida (Greene) R. B. Kelley [*Cryptantha fulvocanescens* (S. Watson) Payson var. *nitida* (Greene) Sivinski; *Oreocarya nitida* Greene]—Sandstone outcrops in mixed desert scrub and piñon-juniper; uncommon in San Juan County, 5580–6080 ft. The inflorescence of this variety is more silvery-strigose and less hispid than that of var. *fulvocanescens*. On average, it also has shorter pedicels, longer calyces, and more flowers per cymule and is less likely to have setose tips on the nutlet murications.

Oreocarya oblata (M. E. Jones) J. F. Macbr. [*Cryptantha oblata* (M. E. Jones) Payson]—Rough cat's-eye. Cespitose perennial; stems few to several from the base, 1–3 dm, simple; leaves oblanceolate, rarely linear-oblanceolate, strigose and coarsely appressed-setose; inflorescence short-cylindrical to subcapitate; flowers homostylic; fruiting calyx 8–10 mm, segments lance-linear, densely setose; corolla tube exceeding or (rarely) subequal to the calyx tips, lacking interior basal scales; corolla limb white, 6–10 mm wide; nutlets 2.5–3 mm, usually 4 maturing, dorsal surface narrowly ovate, bowed outward from the base and inward to the tip; dorsal surface rugose-tuberculate, scar closed. Gravelly limestone or caliche soil, desert scrub, 4400–6200 ft. C, SE, SC, SW.

Oreocarya palmeri (A. Gray) Greene [*Cryptantha coryi* I. M. Johnst.; *C. palmeri* (A. Gray) Payson]—Palmer's cat's-eye. Biennial or short-lived perennial; stems 1 to few from the base, 0.5–3 dm; leaves oblanceolate to linear-oblanceolate, strigose and subtomentose; inflorescence cylindrical to broadly ovate; flowers homostylic; fruiting calyx 8–10 mm, segments lanceolate; corolla tube equal to calyx, lacking interior basal scales; corolla limb white, 6–9 mm wide; nutlets 2.5–2.8 mm, usually 4 maturing, narrowly ovate, bowed outward from the base and inward to the tip, smooth and glossy, margins sharply angled, scar closed. Limestone hills,; desert scrub, 3400–4350 ft. Chaves, Eddy.

Oreocarya paradoxa A. Nelson [*Cryptantha paradoxa* (A. Nelson) Payson]—Handsome cat's-eye. Cespitose perennial; stems several from a branching caudex, 0.5–1.5 dm; leaves mostly basal, with fine-strigose pubescence and appressed-pustulate bristles on the lower surface, cauline leaves more loosely villous; inflorescence subcapitate, 1–4 cm; flowers heterostylic; fruiting calyx 6–8 mm; corolla tube surpassing the calyx tips; corolla limb white, rotate, 9–12 mm wide; nutlets lance-ovate, usually 4 maturing, 2–2.8 mm, rugose-tuberculate; scar narrowly open, constricted below the middle, margins elevated. Alkaline soil, saltbush scrub, 4820–5220 ft. San Juan.

Oreocarya paysonii J. F. Macbr. [*Cryptantha paysonii* (J. F. Macbr.) I. M. Johnst.]—Payson's cat's-eye. Cespitose perennial; stems few to several from branching woody caudex, 1–3 dm; leaves oblanceolate, strigose and appressed-setose; inflorescence short-cylindrical or subcapitate; flowers heterostylic; fruiting calyx 8–10 mm, segments linear-lanceolate, densely setose; corolla tube surpassing the calyx; corolla limb white, 10–14 mm wide, tube lacking basal scales; nutlets usually 4 maturing, dorsal surface narrowly ovate, 2.5–3 mm, bowed outward from base and inward to tip, finely rugose-tuberculate on both surfaces, scar closed. Limestone or caliche, desert scrub, juniper savanna, 4660–6080 ft. C, SC, DeBaca, Eddy.

Oreocarya setosissima (A. Gray) Greene [*Cryptantha setosissima* (A. Gray) Payson]—Bristly cryptantha. Biennial; stem simple, usually solitary from a stout taproot, 4–10 dm, finely puberulent and coarsely spreading-setose with long (2–4 mm) bristles; leaves basal and cauline, gradually reduced above, oblanceolate, villous-tomentose; inflorescence densely hispid, narrowly cylindrical and interrupted below when in flower; cymules elongating in fruit; flowers homostylic; fruiting calyx 6–13 mm; corolla tube equal to calyx; corolla limb 7–10 mm wide; nutlets usually 4 maturing, 4.5–6 mm, margins decidedly papery-winged, dorsal body surface finely muricate. Sporadic and rare in pine-oak forests, 7860–9020 ft. NW, NC.

Oreocarya suffruticosa (Torr.) Greene—Bownut cat's-eye. Perennial; stems 1 to several from a branching, often woody caudex, 1–4 dm; leaves oblanceolate to lance-linear, obtuse to acute, strigose to villous-puberulent or glabrate; inflorescence a cymule often elongating at maturity; calyx segments ovate-lanceolate, 5–7 mm in fruit; flowers homostylic; corolla limb white, 4–8 wide, tube equal to calyx and ringed with short (< 1 mm) antrorse scales at the interior base; nutlets 1–4 maturing, 1.8–2.5 mm, smooth and shiny, deeper than wide, dorsal surface narrowly ovate, bowed outward from the base and inward to the tip, scar closed with a small basal pocket. A polymorphic species with several named and unnamed, morphologically and geographically inconsistent variations.

1. Upper leaf surface pubescent; plants not usually on gypsum…**var. suffruticosa**
1'. Upper leaf surface glabrous; plants entirely restricted to gypsum habitats…**var. pustulosa**

var. pustulosa (Rydb.) R. B. Kelley [*Cryptantha cinerea* (Greene) Cronquist var. *pustulosa* (Rydb.) L. C. Higgins; *Oreocarya pustulosa* Rydb.]—Gypsum strata of the Todilto and Yeso Formations, 4600–6600 ft. NC, C. This glabrous or glabrate variety occupies only gypsum substrates in New Mexico but is in deep sand habitats in adjacent Arizona, Colorado, and Utah.

var. suffruticosa [*Cryptantha cinerea* (Greene) Cronquist; *C. cinerea* (Greene) Cronquist var. *jamesii* Cronquist; *C. jamesii* Payson]—Desert scrub to piñon-juniper woodlands, 4900–8600 ft. Widespread.

Oreocarya thyrsiflora Greene [*Cryptantha thyrsiflora* (Greene) Payson]—Calcareous cat's-eye. Biennial or short-lived perennial; stems stout, 1 to few from basal rosette, 2–4 dm; leaves oblanceolate, hispid and pustulate; inflorescence broad at maturity, 1–2.5 dm wide, dense and diffuse with numerous elongating cymules, hispid, foliar bracts evident; fruiting calyx 7–9 mm; corolla limb white, 4–8 mm wide, tube equal to calyx, interior basal scales present; nutlets 1–4 maturing, lanceolate, 2.5–3.5 mm, dorsal surface rugose-tuberculate, margins angled, scar narrowly open. Limestone and caliche soil, short-grass prairie, 5180–8040 ft. NE, NC.

Oreocarya worthingtonii Sivinski—Worthington's cat's-eye. Cespitose perennial from a branching caudex; stems slender, 1–2 mm wide, 0.5–2 dm; basal leaves linear-oblanceolate, 2–7 cm, 2–3(4) mm wide, strigulose on upper surface, pustulate-strigose on lower surface; cauline leaves linear to linear-oblanceolate, 1–2.5 cm, strigose, spreading hirsute on the margins; inflorescence a short, subcapitate thyrse of 2–3 cymules; bracts similar to cauline leaves; calyx accrescent, segments lanceolate, 5–7 mm at anthesis 7–9 in fruit; corolla salverform, tube usually exceeding the sepal tips by 1–2 mm, 6–9 mm, yellow, fornices yellow; corolla limb white, rotate, 5–8 mm diam.; anthers 1.5–2 mm; styles 4–5 mm; fruit globose, usually maturing 4 nutlets; nutlets ovoid, margins acute, 2–2.4 × 1.9–2.3 mm, dorsally convex, bowed outward from the base then inward at the tip, dorsal surface finely tuberculate, ventral surface rugulose-papillose, scar closed with a small basal pocketlike areole. Grows on pockets of fine sand, containing small amounts gypsum in Big Dog Canyon, Upper Dog Canyon, and the Brokeoff Mountains at 4100–6235 ft. Eddy, Otero (Sivinski, 2023).

PECTOCARYA DC. ex Meisn. – Combseed

Annual taprooted herbs; strigose; stems slender; leaves linear or narrowly elliptic, lowermost sometimes opposite, all others alternate; inflorescence an indeterminate, stemlike, irregularly bracteate false raceme that may or may not be forked; calyx deeply cleft, sepals nearly distinct, all alike or dissimilar in fruit; corolla minute, white, salverform or funnelform, faucal appendages (fornices) nearly closing the throat, often yellow; nutlets radially spreading, homomorphic or heteromorphic, uncinate-bristly at least toward the apex, margins often raised or winged, usually 4 maturing (sometimes fewer by abortion).

1. Nutlets obovate or rounded, both the body and the very thin, conspicuous wing with slender uncinate bristles...**P. setosa**
1'. Nutlets oblong to linear, with bristles only on the margins...(2)
2. Nutlet margins mostly entire or undulate, bristly only near the distal end...**P. heterocarpa**
2'. Nutlet margins lacerate or toothed most of their length, as well as at the end...(3)
3. Nutlets conspicuously recurved, the margins narrow, with nearly distinct teeth...**P. recurvata**
3'. Nutlets nearly straight, the margins broad and conspicuous, with confluent teeth...**P. platycarpa**

Pectocarya heterocarpa (I. M. Johnst.) I. M. Johnst.—Chuckwalla combseed. Stems prostrate to spreading, 3–15 cm; leaves linear, 1–3 cm; flowers on recurving pedicels in fruit, calyx bilaterally symmetric in fruit; corolla inconspicuous, 1–1.5 mm wide; nutlets usually 4, sometimes fewer, heteromorphic, often in dissimilar pairs, linear-oblong, 1.8–3 mm, margins usually entire, uncinate bristles mostly at the apex, apex slightly incurved. Desert scrub, 4400–4600 ft. Hidalgo, Luna.

Pectocarya platycarpa (Munz & I. M. Johnst.) Munz & I. M. Johnst.—Broad combseed. Stems widely ascending, 3–15 cm; leaves linear, 1–3 cm; flowers on ascending or slightly curved pedicels in fruit, calyx bilaterally symmetric In fruit; corolla inconspicuous, 1–1.5 mm wide; nutlets not all alike, 1 or 2 smaller and less strongly wing-margined than the others, all oblong, 2.5–3.5 mm, margins winged with several uncinate bristles confluent at their swollen triangular bases. Desert scrub, 4300–5200 ft. Doña Ana, Grant, Hidalgo.

Pectocarya recurvata I. M. Johnst.—Bent combseed. Stems widely spreading or ascending, 3–15 cm; leaves narrowly elliptic or linear, 1–3 cm; flowers on recurving pedicels in fruit, calyx bilaterally symmetric in fruit; corolla inconspicuous, 1–1.5 mm wide; nutlets usually 4, rarely fewer, conspicuously recurved, slightly heteromorphic, 1 or 2 smaller than the others, linear-oblong, 2–3 mm, margins lined with several uncinate bristles. Desert scrub, 4300–5025 ft. Grant, Hidalgo, Luna.

Pectocarya setosa A. Gray—Bristly combseed. Stems erect to ascending, 3–20 cm; leaves linear or narrowly oblanceolate, 1–3 cm; flowers on ascending or slightly curved pedicels in fruit, calyx radially symmetric in fruit; sepals linear, strigose and armed with a few stiff bristles; corolla inconspicuous, 1–2 mm wide; nutlets usually 4, sometimes fewer, heteromorphic, 2 of them scarious-wing-margined and 2 wingless, orbicular, 2–3 mm across, slender uncinate bristles scattered over the dorsal surface. Arid brushy slopes; an 1884 collection near Acoma Pueblo at 6300 ft. is the only one from New Mexico. Cibola (Veno, 1979).

PLAGIOBOTHRYS Fisch. & C. A. Mey. – Popcorn-flower

Annual herbs, strigose-hirsute or hispid; leaves alternate above, often opposite on the lower stem; inflorescence of helicoid, naked or bracteate modified cymes appearing as racemes or spikes, usually elongating at maturity; corolla white, salverform with yellow faucal appendages (fornices) at the throat; anthers included in the corolla tube; homostylous; stigma capitate; nutlets homomorphic, 4, or fewer by abortion, ventral keel usually evident, attachment scar elevated, carunclelike, basilateral or near the middle.

1. Fruiting calyx circumscissile; leaf veins, roots, and basal part of stem charged with a red-purple dye; stems 10–50 cm, ascending to erect...**P. arizonicus**
1'. Fruiting calyx not circumscissile; leaf veins, roots, and basal part of stem not dye-stained; stems 2–15 cm, prostrate to ascending...**P. scouleri**

Plagiobothrys arizonicus (A. Gray) Greene ex A. Gray [*Eritrichium canescens* (Benth.) A. Gray var. *arizonicum* A. Gray]—Arizona popcorn-flower. Stems several, spreading to ascending, 10–50 cm; basal part of stem, roots, and leaf veins charged with red-purple dye; leaves all alternate, oblanceolate to nearly linear, spreading-hirsute; inflorescence terminating the branches, irregularly leafy-bracteate below; flowers short-pedicellate, calyx 2.5–4 mm in fruit, circumscissile; corolla limb 1.5–3 mm wide; nutlets lance-ovate, 1.5–2.5 mm, finely tuberculate between dorsal ridges, attachment scar near the middle. Desert scrub, 4300–6000 ft. WC, SC, SW.

Plagiobothrys scouleri (Hook. & Arn.) I. M. Johnst. [*Plagiobothrys scopulorum* (Greene) I. M. Johnst.]—Meadow popcorn-flower. Stems several, usually prostrate or somewhat ascending, 2–15 cm; leave essentially all cauline, lower 1–4 pairs opposite, alternate above, linear to narrowly oblanceolate, strigose, 1–7 cm; inflorescence terminating the branches, irregularly bracteate with leaflike bracts; flowers short-pedicellate, calyx 2–4 mm in fruit; corolla inconspicuous, limb 1–2 mm wide; nutlets lance-ovate, 1.5–2 mm, variously roughened with rugae or tuberculations, sometimes with short setose projections, the attachment scar small, basilateral. Our plants belong to **var. hispidulus** (Greene) Dorn [*P. scouleri* (Hook. & Arn.) I. M. Johnst. var. *penicillatus* (Greene) Cronquist]. Drying mud of low, seasonally wet areas in N mountains, 8060–10,180 ft. NE, NC, NW, Sandoval.

SYMPHYTUM L. – Comfrey

Symphytum officinale L.—Common comfrey. Perennial herb from a thick, carrotlike root; stems ascending to erect, 5–10 dm; leaves mostly cauline, alternate, entire, lanceolate to ovate, 5–30 cm, with decurrent blades on the upper stem, short spreading-hairy; inflorescence cymose with paired helicoid, ebracteate false racemes; calyx deeply lobed, bristly, to 9 mm in fruit, lobes lanceolate; corolla 15–19 mm, pink or bluish, rarely white, funnelform or urceolate with throat expanded above tube, lobes erect; stamens included; style included or slightly exserted; nutlets 1–4, blackish, 4–5 mm, ovoid, tip somewhat incurved, ventrally keeled, attachment scar basal with a ringlike, minute-toothed rim. Garden escape in disturbed montane sites and riparian areas; 1 record, 7220 ft. Santa Fe. Introduced.

BRASSICACEAE (CRUCIFERAE) – MUSTARD FAMILY

Steve L. O'Kane, Jr.

Herbs, subshrubs, and small shrubs; glabrous, glaucous, glandular, or variously pubescent with simple, 2- to many-branched, stellate, or peltate trichomes; leaves alternate and/or in a basal rosette; flowers hypogenous, actinomorphic to somewhat bilateral, perfect; sepals and petals 4, distinct; stamens 6 (rarely 2 or 4), the outer 2 shorter than the inner 4, rarely of equal length or in 3 pairs of different lengths; fruit a bivalved capsule, elongate (> 3 times longer than wide; a silique) or not (< 3 times longer than wide; a silicle), in cross-section terete and uncompressed, quadrangular, or compressed parallel (latiseptate) or perpendicular (angustiseptate) to the septum, the septum rimmed by a persistent replum; seeds with cotyledons accumbent, incumbent, or conduplicate (see glossary). Pedicel measurements refer to fruiting pedicels. Citations for individual genera are mainly not given, as treatments by Al-Shehbaz (2010), Holmgren (2005), O'Kane (2013), and Rollins (1993) were heavily relied on and contain an extensive list of citations.

KEY BASED ON FLOWERS

1. Plants entirely glabrous or with simple or glandular trichomes only (scan the entire plant at ×10)…(2)
1'. Plants with at least some trichomes branched, malpighian, forked, peltate, or stellate…(3)
2. Cauline leaves absent, or when present, petioles or leaf bases mostly auriculate, clasping, or amplexicaul…**FLOWER KEY 1**
2'. Cauline leaves present, petioles or leaf bases not auriculate, clasping, or amplexicaul…**FLOWER KEY 2**
3. Ovaries and young fruits linear (fruits generally siliques)…**FLOWER KEY 3**
3'. Ovaries and young fruits broader than linear (fruits generally silicles)…**FLOWER KEY 4**

KEY BASED ON FRUITS

1. Fruit a silicle (< 3 times longer than wide)…(2)
1'. Fruit a silique (> 3 times longer than wide)…(3)
2. Plants entirely glabrous or with simple or glandular trichomes only (scan the entire plant at ×10)…**FRUIT KEY 1**
2'. Plants with at least some trichomes branched, malpighian, forked, peltate, or stellate…**FRUIT KEY 2**
3. Plants entirely glabrous or with simple or glandular trichomes only (scan the entire plant at ×10)…**FRUIT KEY 3**
3'. Plants with at least some trichomes branched, malpighian, forked, peltate, or stellate…**FRUIT KEY 4**

FLOWER KEY 1

Plants entirely glabrous or with simple or glandular trichomes only. Cauline leaves absent, or when present, petioles or leaf bases often auriculate, clasping, or amplexicaul.

1.   Plants scapose, cauline leaves entirely absent…(2)
1'.   Plants not scapose; cauline leaves present…(3)
2.   Petals yellow…**Diplotaxis**
2'.   Petals white, lavender, or purple…**Cardamine** (in part)
3.   Cauline leaves petiolate…(4)
3'.   Cauline leaves sessile…(5)
4.   Plants terrestrial, not rooting from proximal nodes…**Cardamine** (in part)
4'.   Plants aquatic, rooting from proximal nodes…**Nasturtium**
5.   Stamens 2…**Lepidium**
5'.   Stamens 6, rarely 4…(6)
6.   Fruits on stalks (gynophores); stamens all exserted from the flowers…(7)
6'.   Fruits sessile; all or some stamens included in the flowers…(9)
7.   Stigmas evidently 2-lobed; stamens tetradynamous…**Thelypodiopsis** (in part)
7'.   Stigmas entire; stamens ± equal in length…(8)
8.   Sepals spreading to reflexed, petals white or yellow, 12–25 mm…**Stanleya**
8'.   Sepals erect to ascending; petals white or purple, 6–12 mm (rarely to 18 mm)…**Thelypodium** (in part)
9.   Petal margins crisped and/or channeled; stamens typically in 3 pairs of unequal length…(10)
9'.   Petal margins not or rarely crisped, never channeled; stamens tetradynamous…(11)
10.   Fruit terete; seeds not winged…**Caulanthus** (in part)
10'. Fruit latiseptate; seeds often winged…**Streptanthus** (in part)
11.   Petals yellow…(12)
11'. Petals white, lavender, pink, purple, or violet…(19)
12.   Ovules 1 or 2 per ovary…(13)
12'. Ovules > 2 per ovary…(14)
13.   Basal and lower leaf blade margins 2- or 3-pinnatifid or -pinnatisect…**Lepidium** (in part)
13'. Basal and lower leaf blade margins entire, repand, or dentate…**Isatis**
14.   Calyx urceolate; petal margins crisped…see key at either **Mostacillastrum** or **Thelypodiopsis** (in part)
14'. Calyx not urceolate; petal margins not crisped…(15)
15.   Fruiting pedicels reflexed; petal claws not attenuate…**Caulanthus** (in part)
15'. Fruiting pedicels divaricate to erect; petal claws attenuate or absent…(16)
16.   Fruit with a sterile beak; petals usually 8–30 mm, claw evident…**Brassica**
16'. Fruit lacking a sterile beak; petals usually < 9 mm, claw absent…(17)
17.   Cauline leaf base cordate-amplexicaul…**Conringia**
17'. Cauline leaf base auriculate…(18)
18.   Ovary and young fruit linear; petals generally 4–9 mm; stems angular distally…**Barbarea**
18'. Ovary and young fruit globose, ovoid, or oblong; petals generally < 4 mm; stems rarely angular distally…
      Rorippa
19.   Ovary and young fruit ovate, orbicular, or cordate…(20)
19'. Ovary and young fruit linear…(21)
20.   Ovules 4(–6) per ovary; plants usually with trichomes…**Lepidium** (in part)
20'. Ovules (4–)6–10 per ovary; plants glabrous…**Noccaea**
21.   Petals not differentiated into blade and claw, gradually narrowing to base…**Boechera**
21'. Petals differentiated into blade and claw…(22)
22.   Fruit reflexed; petal claw not attenuate…**Caulanthus** (in part)
22'. Fruit erect to divaricate; petal claw attenuate…(23)
23.   Petals magenta or purple, with a deeper purple center…**Streptanthus** (in part)
23'. Petals white or lavender, or if purple, without a deeper purple center…(24)
24.   Fruit with an evident sterile beak; petals generally > 15 mm…**Brassica**
24'. Fruit lacking a sterile beak; petals generally < 15 mm…(25)
25.   Stigmas 2-lobed…**Thelypodiopsis** (in part)
25'. Stigmas entire…**Thelypodium** (in part)

FLOWER KEY 2

Plants entirely glabrous or with simple or glandular trichomes only. Cauline leaves present, petioles or leaf bases not auriculate, clasping, or amplexicaul.

1.   Racemes bracteate throughout…(2)
1'.   Racemes ebracteate or only the proximal-most flowers bracteate…(4)
2.   Petals pink or purple; cauline leaves 3–5-foliate…**Cardamine** (in part)
2'.   Petals yellow; cauline leaves simple…(3)

3. Basal leaf blade margins 1- or 2(3)-pinnatisect; petals 4–20 mm…**Selenia**
3'. Basal leaf blade margins dentate, sinuate- to runcinate-pinnatifid; petals 1.5–2 mm…**Sisymbrium** (in part)
4. Petals absent…(5)
4'. Petals present…(6)
5. Sepals 3.7–1.2 mm; ovules 6–22 per ovary…**Cardamine** (in part)
5'. Sepals 1.2–3 mm; ovules > 70 per ovary…**Rorippa** (in part)
6. Flowers zygomorphic…(7)
6'. Flowers actinomorphic…(8)
7. Ovaries and young fruits linear; ovules > 10 per ovary; stamens in 3 unequal-length pairs…**Streptan-thus** (in part)
7'. Ovaries and young fruits ovate; ovules 2 per ovary; stamens tetradynamous…**Iberis**
8. Stamens 2 or 4…(9)
8'. Stamens 6…(12)
9. Stamens 2…**Lepidium** (in part)
9'. Stamens 4…(10)
10. Ovaries linear…**Cardamine** (in part)
10'. Ovaries wider, not linear…(11)
11. Plants very delicate; ovules > 10 per ovary; generally moist habitats…**Hornungia**
11'. Plants not especially delicate; ovules 2 or 4 per ovary; generally dry habitats…**Lepidium** (in part)
12. Ovaries wider than linear (or rarely very narrowly lanceolate)…(13)
12'. Ovaries linear, rarely very narrowly lanceolate…(17)
13. Ovules > 24 per ovary…**Rorippa** (in part)
13'. Ovules < 14 per ovary…(14)
14. Ovaries and young fruits with 2 dissimilar segments…**Rapistrum**
14'. Ovaries and young fruits unsegmented…(15)
15. Ovules > 6 per ovary; basal leaves dentate, cauline leaves deeply pinnatifid…**Armoracia**
15'. Ovules 2(–4) per ovary; leaves not as above…(16)
16. Fruit angustiseptate; ovules 2(–4) per ovary…**Lepidium** (in part)
16'. Fruit latiseptate; ovules 1 per ovary…**Thysanocarpus**
17. Cauline leaves compound…(18)
17'. Cauline leaves simple (if divided, the blade not interrupted between lobes)…(19)
18. Plants terrestrial, not rooting from proximal stem nodes…**Cardamine** (in part)
18'. Plants aquatic, rooting from proximal stem nodes…**Nasturtium**
19. Plants with glandular papillae…**Chorispora**
19'. Plants without glandular papillae…(20)
20. Stamens ± equal in length, well exserted…(21)
20'. Stamens tetradynamous or in 3 pairs of unequal length, exserted or not…(23)
21. Petals yellow; claws and/or filaments usually pubescent, rarely both glabrous…**Stanleya**
21'. Petals not yellow; claws and filaments glabrous…(22)
22. Fruit subsessile on a gynophore as broad as the fruit…**Caulanthus** (in part)
22'. Fruit stipitate on a gynophore narrower than the fruit…**Thelypodium**
23. Stamens in 3 pairs of unequal length…(24)
23'. Stamens tetradynamous…(26)
24. Fruit terete in cross-section; seeds not winged…**Caulanthus** (in part)
24'. Fruit slightly to evidently latiseptate; seeds winged all around or only distally…(25)
25. Sepals 5–12 mm; cotyledons accumbent…**Streptanthus** (in part)
25'. Sepals 2–4(–5) mm; cotyledons incumbent…**Streptanthella**
26. Petal blades with purple or brown veins darker than the rest of the blade…(27)
26'. Petal blades with veins of the same color intensity as the rest of the blade…(30)
27. Perennial, glaucous or glabrous; leaf blade margins entire or dentate…(28)
27'. Annual, not glaucous, often pubescent; leaf blade margins pinnately lobed…(29)
28. Stigma evidently 2-lobed; ovules > 76 per ovary…**Caulanthus** (in part)
28'. Stigma obscurely 2-lobed; ovules < 62 per ovary…**Hesperidanthus** (in part)
29. Fruit indehiscent, corky, valvular segments rudimentary and seedless…**Raphanus**
29'. Fruit dehiscent, not corky, valvular segments well developed and seeded…**Eruca**
30. Cauline leaf blade margins entire…(31)
30'. Cauline leaf blade margins pinnately divided, sinuate, or dentate…(35)
31. Plants biennial…**Boechera** (in part)
31'. Plants perennial from a caudex, rhizomes, or small tubers…(32)
32. Plants from rhizomes or small tubers…**Cardamine** (in part)
32'. Plants from a caudex, rhizomes or tubers absent…(33)

33. Petals yellow, lavender, or purple…**Hesperidanthus** (in part)
33'. Petals white…(34)
34. Ovules < 18 per ovary; the mature style appearing as an extension of the broad replum…**Cardamine** (in part)
34'. Ovules > 28 per ovary; the mature style not appearing as an extension of the broad replum…**Boechera** (in part)
35. Petals yellow…(36)
35'. Petals white, pink, lavender, or purple…(40)
36. Ovaries and fruits 2-segmented, the terminal segment forming a sterile beak…(37)
36'. Ovaries and fruits unsegmented…(38)
37. Fruit valves 1-veined; sepals usually erect or ascending, rarely spreading…**Brassica**
37'. Fruit valves 3–7-veined; sepals spreading to reflexed…**Sinapis**
38. Stigma entire…**Rorippa** (in part)
38'. Stigma 2-lobed…(39)
39. Seeds biseriate; fruit valves 1-veined; cotyledons conduplicate…**Diplotaxis** (in part)
39'. Seeds uniseriate; fruit valves 3-veined; cotyledons incumbent…**Sisymbrium** (in part)
40. Ovary and fruit 2-segmented, with a sterile beak…**Diplotaxis** (in part)
40'. Ovary and fruit unsegmented…(41)
41. Petals usually purple, or if not, then margins crisped and/or channeled…**Caulanthus** (in part)
41'. Petals white or lavender, margins neither crisped nor channeled…(42)
42. Leaf blade margins pinnatisect or pinnatifid throughout…**Cardamine** (in part)
42'. Leaf blade margins entire or dentate…(43)
43. Fruit terete in cross-section; seeds not winged; cotyledons incumbent…**Caulanthus** (in part)
43'. Fruit latiseptate; seeds winged; cotyledons accumbent…**Streptanthus** (in part)

FLOWER KEY 3

Plants with at least some trichomes branched, malpighian, forked, peltate, or stellate. Ovaries and young fruits linear (fruit generally siliques).

1. Plants scapose, lacking cauline leaves entirely…**Draba** (in part)
1'. Plants not scapose, at least 1 cauline leaf present…(2)
2. Cauline leaves sessile, blade bases auriculate, sagittate, or amplexicaul…(3)
2'. Cauline leaves petiolate, or (rarely) if sessile, blade bases not auriculate, sagittate, or amplexicaul…(6)
3. Trichomes stalked, stellate, rarely with a few simple ones…**Arabis**
3'. Trichomes not as above or a mixture of more than one kind…(4)
4. Annual; stigma 2-lobed; petals with purple veins darker than the blades…**Caulanthus**
4'. Biennial or perennial; stigmas ± entire; petals with veins not purple, their color intensity the same as the blade…(5)
5. Petals yellow or cream-white, rarely pink; young fruits appressed to rachises…**Turritis**
5'. Petals white, pink, lavender, or purple; young fruits not or rarely appressed to rachises…**Boechera** (in part)
6. Trichomes sessile, medifixed, malpighian, or stellate…**Erysimum**
6'. Trichomes stalked or simple, forked, or dendritic…(7)
7. Stigmas 2-lobed…(8)
7'. Stigmas entire…(11)
8. Sepals spreading to reflexed…**Nerisyrenia**
8'. Sepals erect, sometimes connivent…(9)
9. Plants with multiseriate glands; stigmas 2-horned; petal margins crisped…**Matthiola** (in part)
9'. Plants lacking multiseriate glands; stigmas not horned; petal margins smooth…(10)
10. Plants tomentose, trichomes dendritic; sepals 10–15 mm…**Matthiola** (in part)
10'. Plants not tomentose, trichomes simple and/or forked; sepals 5–8 mm…**Hesperis**
11. Cauline leaf blades 1–3-pinnatisect, pectinate, or rarely pinnatifid…(12)
11'. Cauline leaf blades entire, dentate, sinuate, or lyrate…(13)
12. Petals yellow; trichomes dendritic…**Descurainia**
12'. Petals white; trichomes simple or forked…**Arabidopsis** (in part)
13. Flowers cup-shaped; petals white with purple tips…**Pennellia**
13'. Flowers not cup-shaped; petals various colors, including white, but without purple tips…(14)
14. Proximal-most flowers bracteate…**Draba** (in part)
14'. Proximal-most flowers ebracteate…(15)
15. Leaf blade margins, at least some, lobed, sinuate, or lyrate…(16)
15'. Leaf blade margins entire or dentate…(17)
16. Trichomes forked; fruit terete or latiseptate, glabrous…**Arabidopsis** (in part)
16'. Trichomes subdendritic; fruit terete, pubescent…**Halimolobos**

17.   Petals yellow…**Draba** (in part)
17′.   Petals white, lavender, pink, or purple…(18)
18.   Ovary and young fruit pubescent…**Boechera** (in part)
18′.   Ovary and young fruit glabrous…(19)
19.   Petals (4-)5-18 mm; seeds winged…**Boechera** (in part)
19′.   Petals 2-4(-5) mm; seeds not winged…(20)
20.   Fruit latiseptate; seeds in 2 rows in each locule; cotyledons accumbent…**Draba** (in part)
20′.   Fruit terete, rarely slightly latiseptate; seeds in 1 row in each locule; cotyledons incumbent…**Arabidopsis** (in part)

FLOWER KEY 4
Plants with at least some trichomes branched, malpighian, forked, peltate, or stellate. Ovaries and young fruits broader than linear (fruit generally silicles).
1.   Plants scapose, cauline leaves entirely absent…**Draba** (in part)
1′.   Plants not scapose, at least 1 cauline leaf present…(2)
2.   Cauline leaves sessile, blade bases auriculate, sagittate, or amplexicaul…(3)
2′.   Cauline leaves usually petiolate, or if sessile then blade bases neither auriculate nor amplexicaul…(4)
3.   Petals white; at least some branched trichomes sessile, stellate…**Capsella**
3′.   Petals usually yellow, rarely cream-white or white with yellow claws; branched trichomes stalked, not stellate…**Camelina**
4.   Petal apex deeply bilobed…**Berteroa**
4′.   Petal apex rounded, retuse, or emarginate…(5)
5.   Cauline leaf blade trichomes all malpighian…(6)
5′.   Cauline leaf blade trichomes not all malpighian…(7)
6.   Petals yellow; ovules 14-18 per ovary…**Draba** (in part)
6′.   Petals white or purplish violet; ovules 2 per ovary…**Lobularia**
7.   Some filaments winged, dentate, or appendaged…**Alyssum** (in part)
7′.   No filaments winged, dentate, or appendaged…(8)
8.   Ovules 1 or 2 per ovary…(9)
8′.   Ovules 4-100 per ovary…(10)
9.   Ovaries and young fruits angustiseptate, didymous with suborbicular lobes; petals 4-10 mm… **Dimorphocarpa**
9′.   Ovaries and young fruits latiseptate, orbicular to obovate; petals 1.5-3(-4) mm…**Alyssum** (in part)
10.   Leaf blades 1-3-pinnatisect, -pinnatifid, or palmately 3-7-lobed…(11)
10′.   Leaf blades usually entire or dentate, rarely sinuate…(12)
11.   Petals yellow; most or all trichomes dendritic…**Descurainia**
11′.   Petals white; trichomes simple or forked…**Hornungia** (in part)
12.   Trichomes all stellate or lepidote…**Physaria**
12′.   Trichomes usually not stellate or lepidote, at least never exclusively so…(13)
13.   Ovaries and young fruits angustiseptate…(14)
13′.   Ovaries and young fruits latiseptate…(15)
14.   Petals 1.5-2.5 mm; styles 0.3-0.8…**Halimolobos**
14′.   Petals 0.6-1.2 mm; styles ± obsolete…**Hornungia** (in part)
15.   Plants perennial…**Draba** (in part)
15′.   Plants annual…(16)
16.   Petals usually yellow, 1.4-8.5 mm (if white, late-season flowers often cleistogamous and apetalous)… **Draba** (in part)
16′.   Petals usually white, 0.6-5 mm…(17)
17.   Fruit glabrous; flowers 0.6-1.2 mm…**Hornungia**
17′.   Fruit pubescent; flowers 2-5 mm…**Tomostima**

FRUIT KEY 1
Fruit a silicle (< 3 times longer than wide). Plants entirely glabrous or with simple or glandular trichomes only (look closely and scan the entire plant).
1.   Fruit 2-segmented…**Rapistrum**
1′.   Fruit unsegmented…(2)
2.   Fruit angustiseptate…(3)
2′.   Fruit terete or latiseptate…(10)
3.   Cauline leaves usually petiolate, or if sessile, blade bases not auriculate, sagittate, or amplexicaul…(4)
3′.   Cauline leaves sessile, blade bases auriculate, sagittate, or amplexicaul…(7)
4.   Seeds 8-24 per fruit; plants of moist to wet habitats…(5)
4′.   Seeds 2(-4) per fruit; plants of drier habitats or an escaped ornamental…(6)

5.   Perennial, 5–20 dm; roots fleshy…**Armoracia**
5'. Annual, 0.2–3 dm; roots not fleshy…**Hornungia**
6.   Racemes not or slightly elongate in fruit; seeds winged; petals of 2 lengths; escaped ornamental…**Iberis**
6'. Racemes elongate in fruit; seeds wingless; petals of equal length; not an escaped ornamental…**Lepid-ium** (in part)
7.   Fruit indehiscent, samaroid, 1-seeded…**Isatis**
7'. Fruit dehiscent, not samaroid, 2–16-seeded…(8)
8.   Seeds usually 2 per fruit…**Lepidium** (in part)
8'. Seeds 4–16 per fruit…(9)
9.   Seeds blackish brown, concentrically striate or alveolate; plants with foul odor…**Thlaspi**
9'. Seeds yellow-brown or brown, usually smooth, rarely minutely reticulate; plants without a foul odor…**Noccaea**
10. Racemes bracteate throughout…**Selenia**
10'. Racemes ebracteate…(11)
11. Fruit dehiscent; seeds 8–90 per fruit…**Rorippa**
11'. Fruit indehiscent; seeds 1 or 2(–4) per fruit…(12)
12. Fruit strongly latiseptate, winged with radiating rays…**Thysanocarpus**
12'. Fruit terete, not winged…**Lepidium** (in part)

FRUIT KEY 2
Fruit a silicle (< 3 times longer than wide). Plants with at least some trichomes branched, malpighian, forked, peltate, or stellate.
1.   Plants scapose; cauline leaves entirely absent…**Draba** (in part)
1'. Plants not scapose; at least 1 cauline leaf present…(2)
2.   Cauline leaves sessile, blade bases auriculate, sagittate, or amplexicaul…(3)
2'. Cauline leaves usually petiolate, or if sessile then blade bases not auriculate, sagittate, or amplexicaul…(4)
3.   Fruit obdeltoid to obdeltoid-cordiform, strongly angustiseptate…**Capsella**
3'. Fruit pyriform, obovoid, or depressed-globose, ± terete…**Camelina**
4.   Fruit angustiseptate…(5)
4'. Fruit latiseptate, rarely terete or 4-angled…(8)
5.   Seeds 2 per fruit; fruit didymous, breaking into 1-seeded segments, never inflated…**Dimorphocarpa**
5'. Seeds 4–100 per fruit; fruit not didymous, or if so (some *Physaria*), inflated and not breaking into 1-seeded segments…(6)
6.   Annual or biennial; styles 0–1 mm; cotyledons incumbent…(7)
6'. Perennial (rarely annual); styles 1.8–7 mm; cotyledons accumbent…**Physaria** (in part)
7.   Fruit valves glabrous; fruit 10–24-seeded…**Hornungia**
7'. Fruit valves pubescent; fruit 60–100-seeded…**Halimolobos**
8.   Cauline leaf blades 1- or 2-pinnatifid or -pinnatisect…**Descurainia**
8'. Cauline leaf blades entire, crenate, dentate, or sinuate…(9)
9.   Seeds 1 or 2 per fruit…(10)
9'. Seeds 4+ per fruit…(11)
10. Trichomes stellate when branched…**Alyssum**
10'. Trichomes malpighian…**Lobularia**
11. Seeds narrowly wing-margined; lateral pair of filaments with a basal toothlike appendage; petals deeply bifid apically…**Berteroa**
11'. Seeds not winged (if narrowly wing-margined then trichomes all stellate); filaments not appendaged; petals ± entire apically…(12)
12. Trichomes ± exclusively stellate to stellate-lepidote; septa with an apical midvein…**Physaria** (in part)
12'. Trichomes mixed, not exclusively stellate to stellate-lepidote; septa lacking an apical midvein…**Draba** (in part)

FRUIT KEY 3
Fruit a silique (> 3 times longer than wide). Plants entirely glabrous or with simple or glandular trichomes only.
1.   Plants scapose, cauline leaves entirely absent…(2)
1'. Plants not scapose, at least 1 cauline leaf present…(3)
2.   Petals yellow; plants annuals or short-lived perennials; cotyledons conduplicate; ovules, and typically seeds, ≥ 20 per fruit…**Diplotaxis** (in part)
2'. Petals white, pink, purple, or lilac; plants perennial with rhizomes or small tubers; cotyledons accumbent; ovules, and typically seeds, ≤ 20 per fruit…**Cardamine**
3.   Fruit indehiscent or breaking into 1-seeded segments…(4)
3'. Fruit dehiscent and not breaking into 1-seeded segments…(6)
4.   Plants with glandular papillae or tubercles…**Chorispora**

4'.   Plants eglandular…(5)
5.   Fruit segmented, usually with > 1 seed, not samaroid…**Raphanus**
5'.   Fruit not segmented, 1-seeded, samaroid…**Isatis**
6.   Fruit usually segmented; cotyledons conduplicate…(7)
6'.   Fruit unsegmented; cotyledons incumbent or accumbent…(11)
7.   Cauline leaf blade bases auriculate to amplexicaul…**Brassica** (in part)
7'.   Cauline leaf blade bases neither auriculate nor amplexicaul…(8)
8.   Seeds biseriate…(9)
8'.   Seeds uniseriate…(10)
9.   Terminal fruit segment beaklike, tubular, or clavate; petal veins the same color as the blade…**Diplotaxis** (in part)
9'.   Terminal fruit segment flattened and tapering from the body of the fruit to the apex; petal veins darker than the rest of the blade…**Eruca**
10.   Fruit valves each prominently 3–5(–7)-veined…**Sinapis**
10'. Fruit valves each obscurely veined or only the midvein prominent…**Brassica** (in part)
11.   Fruit latiseptate…(12)
11'. Fruit terete or 4-angled…(20)
12.   Seeds winged at least distally…(13)
12'. Seeds not winged…(17)
13.   Cotyledons incumbent…**Streptanthella**
13'. Cotyledons accumbent…(14)
14.   Stigmas usually prominently 2-lobed, or if entire, stamens in 3 pairs of unequal length and/or calyces urceolate…**Streptanthus** (in part)
14'. Stigmas entire; stamens tetradynamous; calyces never urceolate…(15)
15.   Cauline leaf bases not auriculate or sagittate…**Boechera** (in part)
15'. Cauline leaf bases auriculate or sagittate…(16)
16.   Fruit 6–11.7 cm; seeds < 28 per fruit…**Boechera** (in part)
16'. Fruit 1.5–4.7 cm; seeds > 50 per fruit…**Dryopetalon**
17.   Plants tomentose; gynophores 10–20 mm…**Stanleya** (in part)
17'. Plants glabrous or sparsely pubescent; gynophores 0–11 mm…(18)
18.   Leaf blade margins dentate or pinnately lobed…(19)
18'. Leaf blade margins entire in distal-most leaves…**Streptanthus** (in part)
19.   Cauline leaves sessile, auriculate or amplexicaul; petals yellow…**Barbarea** (in part)
19'. Cauline leaves petiolate, neither auriculate nor amplexicaul; petals white, lavender, or purple…**Thelypodium** (in part)
20.   Cauline leaves pinnately compound; stems rooting from proximal nodes…**Nasturtium**
20'. Cauline leaves simple, sometimes pinnately lobed; stems not rooting from proximal nodes (except some *Rorippa*)…(21)
21.   Cauline leaf blade bases auriculate, sagittate, or amplexicaul…(22)
21'. Cauline leaf blade bases not auriculate, sagittate, or amplexicaul…(28)
22.   Stigmas prominently 2-lobed…**Thelypodiopsis** (in part)
22'. Stigmas usually entire, rarely obscurely 2-lobed…(23)
23.   Gynophores < 25 mm; petals never purple…**Stanleya** (in part)
23'. Gynophores 0–5 mm, or if longer then petals purple…(24)
24.   Cauline leaf blade margins dentate, serrate, or pinnately lobed…(25)
24'. Cauline leaf blade margins entire or repand…(26)
25.   Seeds biseriate, surfaces colliculate or foveolate…**Rorippa** (in part)
25'. Seeds uniseriate, surfaces reticulate or papillate…**Barbarea** (in part)
26.   Fruit 4-angled; seeds papillose, mucilaginous when wetted…**Conringia**
26'. Fruit terete; seeds reticulate, not mucilaginous when wetted…(27)
27.   Gynophores 0.6–8.5 mm; anthers often coiled after dehiscence…**Thelypodium** (in part)
27'. Gynophores < 0.4 mm; anthers not coiled after dehiscence…**Mostacillastrum** (in part)
28.   Stigmas 2-lobed…**Sisymbrium**
28'. Stigmas entire…(29)
29.   Gynophores (0.5–)1.2–28 mm; anthers coiled after dehiscence…(30)
29'. Gynophores 0–1 mm; anthers not coiled after dehiscence…(31)
30.   Petals yellow, claws and/or filaments pubescent or papillate; sepals spreading to reflexed…**Stanleya** (in part)
30'. Petals white, lavender, or purple, claws and filaments glabrous; sepals erect or ascending…**Thelypodium** (in part)
31.   Perennial from a caudex…**Hesperidanthus**
31'. Annual or rarely biennial, not from a developed caudex…(32)

32.   Petals pinnatifid; filaments and petal claws papillate basally…**Dryopetalon**
32'. Petals entire; filaments and petal claws not papillate basally…(33)
33.   Seeds biseriate; cotyledons accumbent…**Rorippa** (in part)
33'. Seeds uniseriate; cotyledons incumbent…**Caulanthus**

FRUIT KEY 4

Fruit a silique (> 3 times longer than wide). Plants with at least some trichomes branched, malpighian, forked, peltate, or stellate.

1.   Plants scapose, cauline leaves entirely absent…**Draba** (in part)
1'.   Plants not scapose, at least 1 cauline leaf present…(2)
2.   Cauline leaves sessile, blade bases auriculate, sagittate, or amplexicaul…(3)
2'.   Cauline leaves petiolate, or if sessile then bases not auriculate, sagittate, or amplexicaul…(5)
3.   Fruit latiseptate…(4)
3'.   Fruit subterete-quadrangular…**Turritis**
4.   Fruit straight and erect; branched trichomes stellate, 3–5-rayed…**Arabis**
4'.   Fruit straight, curved, or arcuate and spreading, reflexed, or pendent; branched trichomes forked with 2–14 rays, or dendritic, not stellate…**Boechera** (in part)
5.   Trichomes sessile, malpighian and sometimes mixed with sessile 3–5-rayed ones, simple trichomes always absent…**Erysimum**
5'.   Trichomes stalked, cruciform, dendritic, stellate, submalpighian, or forked, simple trichomes sometimes present…(6)
6.   Stigmas conic, 2-lobed, lobes connivent and often decurrent…(7)
6'.   Stigmas capitate, entire…(10)
7.   Seeds winged; stigmas 2-horned; cotyledons accumbent…**Matthiola**
7'.   Seeds not winged; stigmas without horns; cotyledons incumbent…(8)
8.   Fruit torulose, glabrous; trichomes simple and forked; herbage often with glandular trichomes…**Hesperis**
8'.   Fruit not torulose, pubescent; some trichomes dendritic; herbage not glandular…(9)
9.   Annual; fruit 4-angled; sepals erect; style obsolete…**Strigosella**
9'.   Perennial herbs or subshrubs; fruit angustiseptate or terete; sepals spreading; styles 0.9–4.3 mm…**Nerisyrenia**
10.   Cauline leaf blades pinnatifid to pinnate…**Descurainia**
10'. Cauline leaf blades entire or dentate, rarely sinuate…(11)
11.   Seeds winged…(12)
11'. Seeds not winged…(13)
12.   Fruiting racemes secund; styles obsolete; flowers cup-shaped…**Pennellia** (in part)
12'. Fruiting racemes usually not secund; styles evident; flowers not cup-shaped…**Boechera** (in part)
13.   Seeds biseriate…(14)
13'. Seeds uniseriate…(16)
14.   Fruiting pedicels pendent…(15)
14'. Fruiting pedicels erect, ascending or spreading…**Draba** (in part)
15.   Seeds 8–90 per fruit; cotyledons accumbent…**Boechera** (in part)
15'. Seeds 150–250 per fruit; cotyledons incumbent…**Pennellia** (in part)
16.   Fruit latiseptate; cotyledons accumbent…**Boechera** (in part)
16'. Fruit terete or 4-angled; cotyledons incumbent…(17)
17.   Fruit pubescent with a mixture branched trichomes, some fine and sessile, some stalked and coarser…**Halimolobos**
17'. Fruit glabrous or slightly pubescent with trichomes all of ± the same kind, variously branched…(18)
18.   Plants with dendritic trichomes; fruit pubescent…**Pennellia** (in part)
18'. Plants glabrous or with simple and/or forked trichomes; fruit glabrous…**Arabidopsis**

ALYSSUM L. – Alyssum, madwort

Annual herbs, pubescent with appressed-stellate trichomes, these sometimes mixed with simple ones; leaves mostly cauline, entire, spatulate to narrowly oblanceolate, narrowing to a sessile base; petals pale yellow, aging to cream or white; silicles elliptic to orbicular, inflated in the center with a flattened margin, strongly latiseptate with 2 ovules.

1.   Silicles glabrous…**A. desertorum**
1'.   Silicles pubescent with stellate hairs…(2)
2.   Sepals persistent in fruit; silicles 3–4 mm wide…**A. alyssoides**
2'.   Sepals deciduous after anthesis; silicles 4–5.2 mm wide…**A. simplex**

Alyssum alyssoides (L.) L.—Pale alyssum, pale madwort. Leaves to 3(-4) cm; pedicels 2-3 mm; sepals persistent; petals 2.5-4 mm; filaments slender, neither winged nor appendaged; silicles ovate to orbicular, 3-4 mm wide, covered with minute stellate trichomes, emarginate at the apex; seeds ovoid-obovoid, plump, 1.2-2 mm, narrowly winged. Disturbed sites, roadsides, wash bottoms, 6250-9100 ft. N. Introduced.

Alyssum desertorum Stapf—Desert alyssum, madwort. Leaves to 2.5(-3.5) cm; pedicels 1.5-4 mm; sepals deciduous; petals 1.8-3 mm; median pairs of filaments narrowly winged at base, lateral filaments with a broadly winged appendage apically notched into 2 teeth; silicles orbicular, 3-4 mm wide, glabrous, rarely with a few scattered stellate trichomes, emarginate at the apex; seeds broadly ovate, 1.2-1.5 mm, narrowly winged. Disturbed areas, especially on bare ground, 6200-10,200 ft. N. Introduced.

Alyssum simplex Rudolphi—Wild alyssum, small alyssum, field alyssum. Leaves 4-25 mm; pedicels 2.5-5 mm; sepals deciduous; petals 2.7-3.6 mm; median pairs of filaments broadly winged and apically 1- or 2-toothed, lateral filaments with a broadly winged appendage apically; silicles broadly obovate to suborbicular, 4-5.2 mm wide, covered with coarse, stellate trichomes, emarginate at the apex; seeds broadly ovate, 1.8-2.1 mm wide, narrowly winged. Disturbed sites, roadsides, wash bottoms, meadows, sagebrush, rabbitbrush, piñon-juniper communities, 4900-10,000 ft. N, WC, C. Introduced.

ARABIDOPSIS Heynh. – Rockcress

Arabidopsis thaliana (L.) Heynh.—Thale cress. Annuals; glabrous or pubescent, trichomes simple, sometimes mixed with stalked, forked ones; basal leaves shortly petiolate, obovate, spatulate, ovate, or elliptic, 0.8-3.5(-4.5) cm, margins entire, repand, or dentate, apex obtuse, adaxial surface with predominantly simple and stalked, 1-forked trichomes, cauline leaves subsessile, blades lanceolate, linear, oblong, or elliptic, 0.5-2.2 cm, margins usually entire, rarely toothed; petals white, spatulate, 2-4 mm; siliques linear, terete, 0.8-1.7 cm, valves each with distinct midvein. Disturbed, usually sandy areas within its range, often a greenhouse weed, not known in the wild in New Mexico. Introduced.

ARABIS L. – Rockcress

Arabis pycnocarpa M. Hopkins [*A. hirsuta* DC. var. *pycnocarpa* (M. Hopkins) Rollins]—Hopkins' rockcress. Biennial or short-lived perennials; stems from basal rosette, 2-5.5 dm; rosulate leaves oblanceolate or obovate, the largest 11-17 mm wide, entire to dentate, apex obtuse, sparsely hirsute with simple and stalked, forked hairs; cauline leaves 5-23 mm, entire or dentate, the upper lanceolate to ± linear, auricles absent or to 0.5 mm, sparsely ciliate, sporadically hirsute to nearly glabrous; pedicels erect to erect-ascending, 4-13 mm, glabrous; petals white, 3-5.5 mm; siliques erect to erect-ascending, often appressed to rachis, glabrous, latiseptate. Our plants belong to **subsp. pycnocarpa**. Riparian habitats and open forest floor in mountainous areas, 6800-10,500 ft. Widespread, except southernmost tier of counties and SE, EC.

ARMORACIA G. Gaertn., B. Mey. & Scherb. – Armoracia

Armoracia rusticana G. Gaertn., B. Mey. & Scherb.—Horseradish. Roots fleshy or woody; stems 5-15 dm; basal leaf petioles to 60 cm and broadly expanded basally, broadly oblong, oblong-lanceolate, or ovate, to 60 cm, margins coarsely crenate, rarely pinnatifid, proximal cauline leaves short-petiolate, lobed, oblong to linear-oblong, smaller than basal, margins pinnatifid or pinnatisect, distal cauline leaves sessile or short-petiolate, linear to linear-lanceolate, bases cuneate or attenuate, margins serrate or crenate, rarely entire; pedicels ascending, 8-20 mm; petals white, obovate or oblanceolate, 5-8 mm, claw to 1.5 mm; siliques 4-6 mm, rarely produced. Disturbed, moist places, escaped from cultivation. Reported from the state with no locality. Introduced.

BARBAREA W. T. Aiton – Yellowrocket

Biennial or perennial herbs from a taproot, glabrous or sparsely pubescent with simple trichomes; stems erect, branched, usually stiff, angled; basal leaves often withering, reduced upward, auriculate-clasping, pinnatifid or pinnately compound; cauline leaves auriculate at base, entire to pinnatifid; flowers rather showy; sepals ascending, lateral pair often gibbous-based; petals yellow or yellowish, obovate to oblong; siliques linear, terete, quadrangular, or latiseptate, often torulose or weakly torulose, erect to divaricate.

1. Venation of upper leaves pinnate; upper leaves pinnatifid at least to some degree; siliques relatively thick, 1.1–1.5 mm wide…**B. orthoceras**
1'. Venation of upper leaves palmate-pinnate; upper leaves coarsely dentate; siliques narrow, 0.5–1.1 mm thick…**B. vulgaris**

Barbarea orthoceras Ledeb.—American yellowrocket. Glabrous or sparsely hirsute; stems freely branched, 2–6(–7.5) dm; basal leaves often withering, long-petioled, 3–12 cm, blades lyrate-pinnatifid to pinnate, terminal lobe ovate, margins nearly entire to crenate or dentate, lower petiole bases sometimes ciliate with simple hairs; petals bright to pale yellow, 3–6 mm, oblanceolate or spatulate; siliques 1.5–4 cm × 1.1–1.5 mm, slightly angustiseptate but 4-angled, ascending to divaricately ascending. Damp or moist places, riverbanks, aspen groves, lake margins, cottonwood groves, valley bottoms, vacant lots, meadows, (3600–)5700–9500 ft. Widely scattered. Introduced.

Barbarea vulgaris W. T. Aiton—Garden yellowrocket. Glabrous or petiole bases sometimes ciliate with simple hairs; stems usually branched above, 1.5–10 dm; leaves petiolate, 5–13 cm, blades oblanceolate in outline, lyrate-pinnatifid, (entire or) crenate-dentate, the terminal lobe larger and broadly ovate to suborbicular; petals bright yellow, 4.5–7 mm, oblanceolate or spatulate; siliques 1.5–3 × 0.5–1.1 mm, ± terete, divaricately ascending, prominently nerved. Damp or moist places, especially when disturbed, 5500–9400 ft. NC. Introduced.

BERTEROA DC. – Berteroa

Berteroa incana (L.) DC. [*Alyssum incanum* L.]—Hoary false madwort. Annual or perennial, grayish green with fine, dense, appressed-stellate trichomes; stems to 10 dm; basal leaves usually forming a rosette, entire to pinnatifid, slenderly petiolate, 3–5 cm; cauline leaves appressed-ascending, entire, rarely dentate, obtuse apically, 2–5.5 cm, short-petiolate below, sessile and reduced above, basal and lower cauline leaves withering by anthesis; inflorescence dense, narrow; petals bilobed, white, deeply bifid, 4–6 mm; silicles 4–8 mm, pubescent, somewhat inflated, valves with 1 faint vein near the base, erect. Disturbed areas of streamsides and roadsides, 7600–8350 ft. NC. Introduced.

BOECHERA Á. Löve & D. Löve – Rockcress

Michael D. Windham

Perennial (biennial) herbs with branched trichomes occasionally mixed with simple ones; stems 1–9 per caudex branch, arising terminally or laterally, erect to ascending, occasionally branched; basal leaves rosulate or forming sterile upright clusters, usually petiolate, dentate to entire; cauline leaves ± sessile, usually auriculate, entire to dentate; sepals variously pubescent, occasionally glabrous; petals white, pale lavender, or purple, narrowly spatulate to obovate, not clawed; siliques linear, latiseptate, straight or curved, pendent to loosely ascending or erect, glabrous or pubescent. As circumscribed here, the flora includes four sexual diploid species and a complex array of apomictic hybrids with intermediate morphologies. All taxa previously in *Arabis*.

1. Plants glabrous except for appressed, longitudinally aligned, sessile, 2-rayed trichomes on basal leaves and lower stems (occasionally with few spreading, simple trichomes on petioles and lower stems); fruiting pedicels erect…**B. stricta**
1'. Plants hirsute or pubescent (at least proximally) with simple and/or stalked, branched trichomes; fruiting pedicels pendent to ascending, not erect…(2)
2. Plants pubescent throughout, the ovaries, fruits, and rachises with abundant, ± evenly distributed trichomes…**B. pulchra** s.l.

2'. Plants often lacking trichomes distally, the ovaries, fruits, and rachises glabrous or with sparse, unevenly distributed trichomes…(3)

3. Sepals glabrous or with mostly simple and/or 2-rayed trichomes…(4)

3'. Sepals pubescent with at least some (often abundant) 3–7-rayed trichomes…(5)

4. Fruiting pedicels strongly ascending, the fruit bases usually < 7 mm from rachises…**B. austromontana**

4'. Fruiting pedicels slightly ascending to horizontal, the fruit bases usually > 7 mm from rachises…**B. perennans** s.l.

5. Pedicels with minute, dendritic trichomes distally, these always including 5-, 6-, and 7-rayed forms comprising > 50% of pubescence…(6)

5'. Pedicels glabrous, or if pubescent, then usually lacking minute, dendritic trichomes dominated by 5–7-rayed forms…(7)

6. Young ovaries glabrous; caudices simple or sparingly branched, usually nonwoody, with leaf clusters at ground level; pollen mostly ellipsoid and symmetrically tricolpate…**B. crandallii** s.l. (in part)

6'. Young ovaries (ca. 90%) with sparse trichomes distally, a few of these retained on ca. 50% of fruits; caudices mostly branched, often woody, with leaf clusters ± elevated above ground level; pollen ovoid-spheroid and asymmetrically multicolpate or malformed…**B. quadrangulensis**

7. Mature fruits ascending, the apices of > 50% extending above point of pedicel attachment to rachis…(8)

7'. Mature fruits horizontal, descending, or pendent, the apices of > 50% positioned below pedicel attachment to rachis…(9)

8. Main stem axes usually with < 7 cauline leaves; sepals always with 5- and 6-rayed trichomes, together comprising > 50% of pubescence; pollen mostly ellipsoid and symmetrically tricolpate…**B. crandallii** s.l. (in part)

8'. Main stem axes usually with > 7 cauline leaves; sepals without 5- and 6-rayed trichomes or these comprising < 50% of pubescence; pollen mostly ovoid-spheroid and asymmetrically multicolpate or malformed…**B. gracilenta** s.l. (in part)

9. Flowering/fruiting stems usually 1 per caudex branch and arising from center of basal leaf rosette, sterile upright leaf clusters rare…(10)

9'. Flowering/fruiting stems usually > 1 per caudex branch and/or arising laterally below sterile upright leaf clusters…(13)

10. Lower stem trichomes simple and 2-rayed; basal leaves with marginal cilia extending to middle of petiole…**B. selbyi**

10'. Lower stem trichomes 3–7-rayed; basal leaves without marginal cilia or these confined to petiole bases…(11)

11. Mature fruits mostly < 1.5 mm wide; largest trichomes of lower stems < 0.5 mm [to be expected]…B. lignifera (A. Nelson) W. A. Weber (in part)

11'. Mature fruits mostly > 1.5 mm wide; largest trichomes of lower stems usually > 0.5 mm…(12)

12. Mature fruits closely pendent, with bases of lowermost fruits positioned 1–6 mm from rachises; sepals 3.3–5.1 mm at full anthesis; fruiting pedicels mostly (> 90%) pubescent with 20–140 trichomes attached to visible surfaces…**B. consanguinea**

12'. Mature fruits widely pendent, with bases of lowermost fruits positioned 6.5–15 mm from rachises; sepals 2.5–3.5 mm at full anthesis; lowermost fruiting pedicels glabrous, or if pubescent (< 25%), then with < 20 trichomes attached to visible surfaces…**B. pseudoconsanguinea**

13. Basal leaves entire; lower stem trichomes mostly appressed, < 0.5 mm…(14)

13'. Basal leaves mostly dentate (but often obscurely so); lower stem trichomes spreading, the largest often > 0.5 mm…(15)

14. Fruiting pedicels usually ascending proximally; lower stem trichomes loosely appressed, the largest often spreading and less branched, with 2–4 rays; pollen mostly ellipsoid and symmetrically tricolpate; seed wing nearly continuous, > 0.1 mm wide…**B. crandallii** s.l. (in part)

14'. Fruiting pedicels usually horizontal or descending proximally; lower stem trichomes appressed, the largest usually the most branched, with 5–7 rays; pollen mostly ovoid-spheroid and asymmetrically multicolpate or malformed; seed wing ± discontinuous, < 0.1 mm wide…B. lignifera (A. Nelson) W. A. Weber (in part)

15. Plants of S 1/2 of New Mexico; caudices mostly branched, often woody, with leaf clusters ± elevated above ground level; pollen mostly ellipsoid and symmetrically tricolpate…**B. perennans** s.s.

15'. Plants of N 1/2 of New Mexico; caudices simple or sparingly branched, usually nonwoody, with leaf clusters at ground level; pollen mostly ovoid-spheroid and asymmetrically multicolpate or malformed… **B. gracilenta** s.l. (in part)

Boechera austromontana Windham & Allphin [*B. divaricarpa* (A. Nelson) Á. Löve & D. Löve misapplied]—Southern mountain rockcress. Stems 1–4 per caudex branch, arising from center of basal rosette or below sterile upright leaf clusters, glabrous or with rare 2-rayed and simple trichomes; basal leaves entire (dentate), margins and surfaces with subsessile, 2–3-rayed trichomes; cauline leaves 7–28;

pedicels ascending, straight or slightly upcurved; sepals glabrous (with rare 2-rayed and simple trichomes); ovaries glabrous; fruits ascending; seeds in 2 rows or irregular. Apomictic diploid and triploid hybrids containing genomes from *B. perennans* s.l. and *B. stricta*. Dry montane meadows and forest margins, 7700–9000 ft. NW, NC.

Boechera consanguinea (Greene) Windham & Al-Shehbaz—Kindred rockcress. Stems 1 per caudex branch, usually arising from center of basal rosette, with 3–6-rayed trichomes proximally; basal leaves dentate (entire), margins and surfaces with 5–7-rayed trichomes; cauline leaves 15–36; pedicels slightly descending to horizontal, curved downward; sepals with 3–6-rayed trichomes; ovaries usually glabrous; fruits closely pendent; seeds in 2 rows or irregular. Apomictic triploid hybrids containing genomes from *B. exilis* (A. Nelson) Dorn, *B. perennans* s.l., and *B. retrofracta* (Graham) Á. Löve & D. Löve. Rocky places in piñon-juniper woodland, mountain brush, and ponderosa pine forest, 6000–7600 ft. NW.

Boechera crandallii (B. L. Rob.) W. A. Weber—Crandall's rockcress. Stems usually 2–5 per caudex branch, arising laterally below sterile, upright leaf clusters, with 2–6-rayed trichomes proximally; basal leaves entire to dentate, margins and surfaces with mostly 4–7-rayed trichomes (simple and 2–3-rayed in subsp. *villosa*); cauline leaves 3–16; pedicels ascending to divaricate, straight or gently downcurved; sepals with mostly 4–7-rayed trichomes; ovaries glabrous; fruits ascending to widely pendent; seeds in 1 row; New Mexico plants mostly sexual diploids. Rocky or sandy habitats in piñon-juniper woodland, 6200–7500 ft. Three subspecies in New Mexico:

1. Basal leaves with mostly simple, 2- and 3-rayed trichomes…**subsp. villosa**
1'. Basal leaves with mostly 4–7-rayed trichomes…(2)
2. Mature fruits ascending to horizontal; basal leaves oblanceolate to obovate, the largest usually > 6 mm wide, often shallowly dentate, with slightly overlapping, 4–7-rayed trichomes…**subsp. thompsonii**
2'. Mature fruits widely pendent; basal leaves narrowly oblanceolate to linear, the largest usually < 6 mm wide, entire, with dense 5–7-rayed trichomes…**subsp. kelseyana**

subsp. kelseyana (Windham & Allphin) Windham [*B. kelseyana* Windham & Allphin]—Ann's rockcress. Sexual diploids. Sandy soil under piñon or juniper, 6200–7500 ft. NW.

subsp. thompsonii (S. L. Welsh) Windham [*B. pallidifolia* (Rollins) Windham & Al-Shehbaz misapplied]—Thompson's rockcress. New Mexico plants mostly sexual diploids. Sandy places, usually under piñon or juniper, 6300–7500 ft. NW, NC.

subsp. villosa (Windham & Al-Shehbaz) Windham [*B. villosa* Windham & Al-Shehbaz]—Villous rockcress. Sexual diploid known only from the type collection. Basalt outcrop in piñon-juniper woodland, ca. 6900 ft. NC.

Boechera gracilenta (Greene) Windham & Al-Shehbaz—Graceful rockcress. Stems 1–8 per caudex branch, usually arising laterally below sterile, upright leaf clusters, with 2–5-rayed trichomes proximally; basal leaves dentate (entire), margins and surfaces with 2–9-rayed trichomes; cauline leaves 5–20; pedicels slightly ascending to horizontal, curved downward; sepals with 2–7-rayed trichomes; ovaries glabrous; fruits divaricate, descending to horizontal (pendent); seeds in 2 rows or irregular (1 row). Apomictic diploid and triploid hybrids containing genomes from *B. crandallii* s.l. and *B. perennans* s.l. Rocky habitats in desert scrub, piñon-juniper woodland, and mountain brush, 4650–7800 ft. NW, NC, WC, C.

Boechera perennans (S. Watson) W. A. Weber—Perennial rockcress. Stems 1–9 per caudex branch, arising laterally below sterile, upright leaf clusters or from center of basal rosette, with simple and/or 2–3-rayed trichomes proximally; basal leaves dentate (entire), surfaces with simple and/or 2–5-rayed trichomes; cauline leaves 6–38; pedicels horizontal to slightly ascending or descending, curved downward; sepals with simple and/or 2–3-rayed trichomes or glabrous; ovaries glabrous; fruits divaricate, descending to pendent (horizontal); seeds in 2 rows or irregular (1 row in subsp. *perennans*). Geologically diverse rocky habitats, 4400–9200 ft. Ten subspecies (8 confirmed, 2 others likely):

1. Sepals (100%) and lower portion of most flowering stems (> 70%) glabrous…(2)
1'. Sepals (> 90%) and lower portion of most flowering stems (> 95%) hirsute with sparse to abundant trichomes…(5)

2. Basal leaf margins with simple cilia extending from petiole base to leaf apex (rarely absent from distal portion of blade); basal leaf surfaces often (ca. 50%) glabrous, or if hirsute then without 3-rayed trichomes or these comprising < 10% of hairs…**subsp. carrizozoensis**

2′. Basal leaf margins with simple cilia confined to petiole, rarely extending to proximal portion of blade; basal leaf surfaces mostly (> 90%) hirsute with 3-rayed trichomes comprising 10%–90% of hairs…(3)

3. Basal leaf surfaces with 2-rayed trichomes comprising < 20% of hairs…**subsp. zephyra**

3′. Basal leaf surfaces with 2-rayed trichomes comprising 20%–80% of hairs…(4)

4. Petals aging pale lavender; well-formed pollen ellipsoid, or if spheroid, averaging < 21 microns wide; flowering stems terminal, arising from center of basal rosette [to be expected]…subsp. texana (Windham & Al-Shehbaz) Windham

4′. Petals aging lavender-purple; well-formed pollen mostly spheroid, averaging > 21 microns wide; flowering stems often (ca. 50%) lateral, these arising at base of sterile, upright leaf cluster…**subsp. porphyrea**

5. Sepals with 3-rayed trichomes comprising 10%–80% of hairs; seeds forming a single row at widest point of fruit; mature caudices often woody, usually branched (to 13 cm), often elevating leaf rosettes above ground level…**subsp. perennans**

5′. Sepals without 3-rayed trichomes or these comprising < 10% of hairs; seeds forming 2 parallel rows (occasionally irregular) at widest point of fruit; mature caudices not woody, rarely branched (to 3 cm) or elevating leaf rosettes above ground level…(6)

6. Basal leaf margins with simple cilia confined to petiole or (in subsp. *centrifendleri*) occasionally extending to proximal portion of blade; basal leaves mostly (> 90%) dentate, the largest teeth > 0.7 mm; mature seeds (100%) with wing > 0.1 mm wide…(7)

6′. Basal leaf margins with simple cilia extending from petiole base to apex (rarely absent from distal portion of blade); basal leaves entire to shallowly dentate, the teeth < 0.7 mm; mature seeds usually (> 90%) without wing or the wing < 0.1 mm wide…(8)

7. Basal leaf margins with simple cilia confined to base of petiole; basal leaf surfaces with 3-rayed trichomes usually comprising > 30% of hairs; mature fruits (largest) usually > 2 mm wide…**subsp. gracilipes**

7′. Basal leaf margins with simple cilia extending from petiole to proximal portion of blade; basal leaf surfaces without 3-rayed trichomes or these comprising < 30% of hairs; mature fruits (largest) usually < 2 mm wide…**subsp. centrifendleri**

8. Flowering stems usually (> 90%) lateral, arising at base of sterile, upright leaf cluster; petals > 0.9 mm wide at full anthesis, most (> 90%) aging pale lavender…**subsp. fendleri**

8′. Flowering stems mostly (> 80%) terminal, arising from center of basal rosette; petals < 0.9 mm wide at full anthesis, most (> 90%) remaining whitish with age…(9)

9. Mature fruits pendent; well-formed pollen ellipsoid, averaging < 18 microns wide [to be expected]…subsp. spatifolia (Rydb.) Windham

9′. Mature fruits divaricate-descending; well-formed pollen spheroid, averaging > 18 microns wide…**subsp. sanluisensis**

subsp. carrizozoensis (P. J. Alexander) Windham [*B. carrizozoensis* P. J. Alexander]—Carrizozo rockcress. Apomictic diploid hybrids containing genomes from subsp. *texana* (Windham & Al-Shehbaz) Windham and an unknown sexual diploid. Rocky places in desert scrub and piñon-juniper-oak woodland, 4700–6900 ft. WC, C, SC. **E** [not pictured in plates]

subsp. centrifendleri (P. J. Alexander) Windham [*B. centrifendleri* P. J. Alexander]—Plateau rockcress. Apomictic diploid hybrids containing genomes from subsp. *fendleri* and *gracilipes*. Rocky places in ponderosa pine forest and piñon-juniper-oak woodland, 6600–8850 ft. NW, NC, WC, C, SW.

subsp. fendleri (S. Watson) Windham [*B. fendleri* (S. Watson) W. A. Weber]—Fendler's rockcress. Sexual and apomictic diploids. Rocky places in ponderosa pine forest and piñon-juniper-oak woodland, 6400–9200 ft. NW, NC, WC, C, SW.

subsp. gracilipes (Greene) Windham [*B. gracilipes* (Greene) Dorn]—Flagstaff rockcress. Sexual diploids. Rocky places in ponderosa pine forest and piñon-juniper-oak woodland, 6500 ft. SW.

subsp. perennans—Perennial rockcress. Sexual and apomictic diploids. Rocky places in desert scrub and pine-oak-juniper woodland, 4600–6000 ft. WC, SW, SC.

subsp. porphyrea (Wooton & Standl.) Windham [*B. porphyrea* (Wooton & Standl.) Windham, Al-Shehbaz & P. J. Alexander]—Porphyrea rockcress. Apomictic triploid hybrids containing genomes from subsp. *gracilipes*, *perennans*, and *texana* (Windham & Al-Shehbaz) Windham. Rocky places in desert scrub and piñon-juniper-oak woodland, 4600–6900 ft. NC, C, SC, SE.

subsp. sanluisensis (P. J. Alexander) Windham [*B. sanluisensis* P. J. Alexander]—San Luis rockcress. Apomictic diploid hybrids containing genomes from subsp. *fendleri* and *spatifolia* (Rydb.) Windham. Rocky places in open ponderosa pine and mixed conifer forests, 6750–9800 ft. NW, NC. **R**

subsp. zephyra (P. J. Alexander) Windham [*B. zephyra* P. J. Alexander]—Wind Mountain rockcress. Apomictic diploid hybrids containing genomes from subsp. *perennans* and *texana* (Windham & Al-Shehbaz) Windham. Rocky places in desert scrub and piñon-juniper-oak woodland, 4450–6400 ft. SW, SC, SE. **R**

Boechera pseudoconsanguinea Windham & Allphin [*B. consanguinea* (Greene) Windham & Al-Shehbaz misapplied]—San Juan rockcress. Stems 1(2) per caudex branch, usually arising from center of basal rosette, with 2–6-rayed trichomes proximally; basal leaves mostly entire, surfaces with 3–8-rayed trichomes; cauline leaves 9–38; pedicels horizontal to slightly descending, gently downcurved; sepals with 4–6-rayed trichomes; ovaries glabrous; fruits widely pendent; seeds in 2 rows or irregular. Apomictic triploid hybrid containing genomes from *B. crandallii* s.l., *B. exilis*, and *B. perennans* s.l. Rocky habitats in piñon-juniper woodland and mountain brush, 5800–7500 ft. NW.

Boechera pulchra (M. E. Jones ex S. Watson) W. A. Weber [*B. formosa* (Greene) Windham & Al-Shehbaz]—Beautiful rockcress. Stems 1 per caudex branch, usually arising from center of basal rosette, with 2–7-rayed trichomes proximally; basal leaves entire, surfaces with mostly 4–7-rayed trichomes; cauline leaves 10–17; pedicels horizontal to slightly descending, downcurved; sepals with mostly 4–7-rayed trichomes; ovaries densely pubescent; fruits pendent; seeds in 2 rows. Our plants belong to the white-flowered Colorado Plateau endemic **subsp. pallens** (M. E. Jones) W. A. Weber. Rocky places in desert scrub and piñon-juniper woodland, 4650–6500 ft. NW.

Boechera quadrangulensis Windham & Allphin [*B. duchesnensis* (Rollins) Windham & Al-Shehbaz misapplied]—Four Corners rockcress. Stems usually 1 per caudex branch, arising from center of basal rosette or laterally below sterile, upright leaf clusters, with 2–6-rayed trichomes proximally; basal leaves entire (slightly dentate), surfaces with mostly 4–7-rayed trichomes; cauline leaves 6–15; pedicels slightly ascending to horizontal, straight or gently downcurved; sepals with mostly 4–7-rayed trichomes; ovaries (ca. 90%) with sparse trichomes distally; fruits horizontal to slightly descending; seeds in 2 rows or irregular. Apomictic triploid hybrid containing genomes from *B. crandallii* s.l., *B. perennans*, and *B. pulchra* s.l. Rocky places in desert scrub, piñon-juniper woodland, and mountain brush, 4650–7500 ft. NW.

Boechera selbyi (Rydb.) W. A. Weber—Selby's rockcress. Stems 1(2) per caudex branch, usually arising from center of basal rosette, with simple and 2-rayed trichomes proximally; basal leaves slightly dentate (entire), surfaces with 2–4-rayed trichomes; cauline leaves 22–50; pedicels ± horizontal, curved downward; sepals with 2–4-rayed trichomes; ovaries glabrous; fruits pendent; seeds in 2 rows or irregular. Apomictic diploid hybrids containing genomes from *B. perennans* s.l., and *B. retrofracta* (Graham) Á. Löve & D. Löve. Rocky places in open pine-oak or mixed conifer forest, 7000–8500 ft. NC.

Boechera stricta (Graham) Al-Shehbaz [*Arabis drummondii* A. Gray]—Erect rockcress. Stems mostly 1 per caudex branch, usually arising from center of basal rosette, glabrous or with sessile, 2-rayed (simple) trichomes; basal leaves entire (slightly dentate), margins and often abaxial surfaces with sessile, longitudinally aligned, 2-rayed trichomes; cauline leaves 5–23; pedicels erect, usually straight; sepals glabrous; ovaries glabrous; fruits erect and often appressed to rachises; seeds in 2 rows. New Mexico plants are sexual diploids. Widespread in W North America, this species reaches the S limit of its range in N New Mexico. Montane meadows and open aspen-conifer forests, 7000–12,700 ft. NW, NC, WC.

BRASSICA L. – Mustard

Annual (mostly), biennial, or perennial herbs; glaucous, glabrous, or pubescent with simple hairs; stem usually single and freely branched above; basal leaves usually rosulate, petiolate, pinnate or lyrate-pinnatifid; cauline leaves short-petiolate, sessile, or auriculate to amplexicaul; petals yellow, broadly to narrowly obovate; siliques oblong to linear, terete to latiseptate or rarely quadrangular, erect to divaricate, usually divided into 2 dissimilar portions, a usually seedless beak and a basal seed-bearing segment. Several species are important crops and weeds.

1. Cauline leaves sessile, blade bases auriculate and/or amplexicaul…(2)
1.' Cauline leaves petiolate or sessile, blade bases tapered, not auriculate or amplexicaul…(4)
2. Biennials or perennials; petals >16 mm; beak of fruits 3–11 mm…**B. oleracea**
2.' Annuals or biennials; petals 6–16 mm; beak of fruits 7–22 mm…(3)
3. Flowers usually not overtopping buds, rarely at same level, when open; petals pale yellow, 10–16 mm; beak of fruits 5–16 mm…**B. napus**
3.' Flowers overtopping or equaling buds when open; petals deep yellow, 6–12 mm; beak of fruits 8–22 mm… **B. rapa**
4. Fruits and pedicels erect, ± appressed to rachises; fruits 10–27 mm, not torulose; fruiting pedicels 2–6 mm…**B. nigra**
4.' Fruits and pedicels spreading to ascending, not appressed to rachises; fruits often 2+ cm, torulose; fruiting pedicels 7–20 mm…(5)
5. Basal leaves persistent, blades with 4–10 lobes each side, surfaces hirsute; petals 4–7 × 1.5–2.5 mm… **B. tournefortii**
5.' Basal leaves deciduous, blades with 1–3(4) lobes each side, surfaces glabrous or nearly so; petals 9–13 × 3–7.5 mm…**B. juncea**

Brassica juncea (L.) Czern.—Chinese, brown, Indian, or leaf mustard, mustard greens. Glaucous and glabrous; basal leaves early-deciduous, pinnatifid to pinnately lobed, 1–3 lobes per side; cauline leaves short-petiolate, oblong or lanceolate, reduced in size distally, base tapered or cuneate, margins dentate to lobed; pedicels slender, spreading to divaricately ascending, 10–15 mm; petals pale yellow, ovate to obovate, 9–13 mm, apex rounded or emarginate; siliques spreading to nearly erect, torulose, subcylindrical or somewhat flattened, 3–5 cm × 2–5 mm. Escape from cultivation, 6000–8500 ft. Cibola, Grant, San Miguel, Santa Fe.

Brassica napus L.—Canola, oilseed rape, rape, rapeseed, rutabaga. Glaucous and glabrous, glabrescent, or pubescent with coarse trichomes; basal leaves (rosulate when biennial) with blades lyrate-pinnatifid, ± pinnately lobed, 0–6 lobes each side, these smaller than terminal; cauline leaves sessile; base auriculate or amplexicaul, margins entire; petals golden or creamy to pale yellow, broadly obovate, 10–16 mm, apex rounded; siliques spreading to ascending, smooth or slightly torulose, terete, 5–10 cm × 3.5–5 mm. Escape from cultivation, reported at 7800–8000 ft. Sierra.

Brassica nigra (L.) W. D. J. Koch—Black mustard. Sparsely to densely hirsute-hispid, proximally rarely subglabrate; basal leaves lyrate-pinnatifid to sinuate-lobed, lobes 1–3 each side, these smaller than terminal ovate or obtuse lobe; cauline leaves sessile or subsessile, ovate-elliptic to lanceolate, similar to basal, reduced distally and less divided, base tapered, not auriculate or amplexicaul, margins entire to sinuate-serrate; pedicels erect, 3–5 mm; petals yellow, ovate, 7–11 mm, apex rounded; siliques erect-ascending, ± appressed to rachis, smooth, ± 4-angled, 1–2.5(–2.7) cm × 2–3 mm. Escape from cultivation. San Juan, Socorro.

Brassica oleracea L.—Cabbage, kale, collard, cauliflower, broccoli, kohlrabi, Brussels sprouts. Glaucous and glabrous; basal leaves pinnatifid or margins dentate; distal cauline leaves sessile, oblong to lanceolate, base auriculate and amplexicaul, margins entire; pedicels spreading to ascending, 14–25 mm; petals yellow, white, or lemon-yellow, ovate or elliptic, 18–25 mm, apex rounded; siliques spreading to ascending, smooth, ± 4-angled or subterete, 5–8 cm × 3–4 mm. Escape from cultivation. Socorro.

Brassica rapa L.—Field mustard, rape mustard, turnip. Green to slightly glaucous, glabrous or sparsely hairy; basal leaves ± lyrate-pinnatifid to pinnate to pinnatisect, lobes 2–4(–6) each side, terminal lobe oblong-obovate, obtuse, large, blade surfaces usually setose; cauline leaves sessile, base auriculate to amplexicaul, margins subentire; petals deep yellow to yellow, obovate, 6–11 mm; siliques ascending to somewhat spreading, torulose, terete, 3–8 cm × 2–4 mm. Escape from cultivation, reported at 4250–7500 ft. Catron, Doña Ana, Hidalgo, Mora. Native to Europe but now established in much of North and Central America. The source of canola (rapeseed) oil, bok choy, and turnips.

Brassica tournefortii Gouan—Sahara mustard. Densely hirsute proximally, glabrescent distally; rosettes persistent, basal leaves lyrate to pinnatisect, 4–10 lobes each side; cauline leaves sessile, reduced in size distally, distal-most bractlike, base tapered, not auriculate or amplexicaul; fruiting pedicels widely spreading, 8–15 mm; petals pale yellow, fading, or sometimes white, oblanceolate, 4–7 mm; siliques on a gynophore to 1 mm, widely spreading to ascending, torulose, cylindrical, 3–7 cm × 2–4 mm. Disturbed sites, 3900–6000 ft. Bernalillo, Doña Ana, Sierra. Introduced.

CAMELINA Crantz – False flax

Annual or biennial, usually pubescent with dendritic to stellate (rarely simple) hairs; stem single and simple or branched above; leaves both basal and cauline, simple, entire to dentate, cauline leaves sessile, sagittate or amplexicaul, lanceolate to linear; sepals erect to ascending, the inner slightly gibbous-based; petals yellow to white, obovate or spatulate; silicles obovoid, generally a little inflated but ± latiseptate and reticulate, ascending to divaricate or erect and appressed to the rachis, valves firm, 1-nerved.

    1. Silicles 7–10+ mm; plants ± glabrous (rarely sparsely pubescent)…**C. sativa**
    1'. Silicles < 7 mm; plants densely pubescent, especially below…(2)
    2. Petals < 5 mm, yellow; basal leaves withering by flowering time…**C. microcarpa**
    2'. Petals 6–9 mm, white outside, pale yellow inside; basal leaves present at flowering time…**C. rumelica**

Camelina microcarpa Andrz. ex DC.—Littlepod false flax. Annual, hirsute, at least below, with simple (rarely forked) trichomes with a lower level of smaller stellate trichomes; leaves primarily cauline, the basal ones withering by anthesis, short-petiolate to sessile, sagittate-auriculate above, 2–8 cm, lanceolate to oblanceolate, entire or nearly so; pedicels spreading-ascending; petals pale yellow, fading white, 3–6 mm; silicles obovoid, obtuse at apex, 5–7 mm, inflated but firm, glabrous. Weedy in desert shrub, piñon-juniper, slickrock, roadsides, irrigation ditches, mountain shrub, sagebrush, pond margins, 5400–9000 ft. N. Introduced.

Camelina rumelica Velen.—Graceful false flax. Annual, hirsute with ± dense, simple, long white trichomes, nearly glabrous above, especially in the inflorescence; basal leaves forming a loose rosette, lanceolate to oblong, entire to irregularly dentate, cauline leaves oblong, irregularly toothed, with acute auricles; fruiting pedicels horizontally spreading; petals pale yellow-white to white, oblong-spatulate with a very short claw, 6–9 mm; silicles obovate, 5–7.5 mm. Disturbed areas of piñon-juniper communities, 7100–7950 ft. Colfax, San Juan. Introduced.

Camelina sativa (L.) Crantz—Annual (biennial), ± glabrous, if trichomes present these simple and branched; basal leaves oblong-lanceolate, sinuately and irregularly toothed, cauline leaves linear-oblong, base sagittate or strongly auriculate, margins entire or remotely denticulate; fruiting pedicels ascending; petals yellow, spatulate with a narrow claw; silicles obovate, 7–10 mm, typically truncate apically. Disturbed areas, especially roadsides, 6150–8400 ft. McKinley. Introduced.

CAPSELLA Medik., nom. cons. – Shepherd's purse

Capsella bursa-pastoris (L.) Medik.—Shepherd's purse. Annual or winter annual, pubescent with a mix of simple and stellate trichomes, ± glabrous above, 1–5 dm; basal leaves rosette-forming, oblanceolate, subentire to toothed to lyrate-pinnatifid, 3–6 cm on a winged petiole; cauline leaves becoming lanceolate-oblanceolate, remotely serrate-dentate, sessile and auriculate-clasping above; petals white, oblanceolate to spatulate, 2–3 mm; silicles triangular-obcordate with a cuneate base, strongly angustiseptate, 4.5–7 mm. Disturbed, usually mesic areas of many habitats, 3600–10,500 ft. Widespread. Introduced.

CARDAMINE L. – Bittercress

Annual, biennial, or perennial herbs, often rhizomatous, rhizomes sometimes tuberous; glabrous or pubescent with simple hairs; stems erect to ascending, simple or branched; basal leaves pinnately compound, trifoliate, pinnatifid, or entire; cauline leaves petiolate or sessile, sometimes auriculate, simple or compound; sepals erect or ascending, petals white to purplish, obovate to spatulate; siliques linear to subulate, somewhat latiseptate to subterete, with coiling, elastically dehiscent valves.

1. Leaves all simple (a few may be slightly pinnatifid); petals showy, 7–12 mm…**C. cordifolia**
1′. Leaves all pinnate (some may be pinnatifid); petals < 7 mm…(2)
2. Basal leaves rosulate, persistent to anthesis…**C. hirsuta**
2′. Basal leaves not rosulate, often withered by anthesis…**C. pensylvanica**

Cardamine cordifolia A. Gray—Heartleaf bittercress, large-mountain bittercress. Perennial with slender rhizomes, ± glabrous, sometimes densely pubescent near the base with simple, short, spreading trichomes; plants to 6(–8) dm; leaves cauline, lower petioles to 8 cm, decreasing in size upward, blades reniform to ovate-cordate to deltate-cordate, simple, somewhat fleshy, the lowermost sometimes with a pair of small ovoid lobes; petals white, with a broad obovate blade emarginate or truncate-rounded at the tip, 7–12 mm; siliques linear, straight, very slightly latiseptate, 2–3.7 mm. Wet meadows, open or forested riparian areas, 6600–12,000 ft. Widespread.

Cardamine hirsuta L.—Hairy bittercress. Annual, rhizomes absent, ± sparsely hirsute basally, often glabrous distally; plants to 4.5 dm; basal leaves rosulate, ca. 8–15-foliate, 3.5–17 cm, lateral leaflets oblong, ovate, obovate, or orbicular, smaller than terminal, margins entire, repand, crenate, or 3-lobed; terminal leaflet 0.4–2 cm, entire, repand, dentate, or 3- or 5-lobed, cauline leaves 1–4(–6), compound, 1.2–7 cm; petals (sometimes absent) white, spatulate, 2.5–5 mm; siliques linear, torulose, 1.5–2.8 cm. Reported at 3950 ft. Doña Ana. *Cardamine flexuosa* With. will key here and is known from flower beds in Doña Ana and Eddy Counties. It differs from *C. hirsuta* in lacking conspicuous basal rosettes of leaves, having raceme rachises that zigzag, and having ascending to divaricate siliques rather than erect to ascending.

Cardamine pensylvanica Muhl. ex Willd.—Pennsylvania bittercress. Biennial or short-lived perennial from a fibrous root system, glabrous, sometimes hirsute near the base; leaves pinnate (-pinnatifid) with 5–11(–19) leaflets, lateral leaflets broadly elliptic to obovate, sessile, cuneate at base; terminal leaflets more broadly obovate and slightly larger, margins entire to dentate; petals white, oblanceolate to narrowly spatulate, not clawed, 2–4 mm; siliques slender, 5–10 mm, terete. Wet areas and irrigation canals, 5300–5400 ft. San Juan.

CAULANTHUS S. Watson – Wild-cabbage

Caulanthus lasiophyllus (Hook. & Arn.) Payson [*Streptanthus lasiophyllus* (Hook. & Arn.) Hoover]—California mustard. Annual, sparsely to densely hispid or hirsute, rarely subglabrate; erect, 2–14 dm; basal leaves soon withered, cauline leaves petiolate, lanceolate to oblong or oblanceolate, pinnatifid, 2–12 cm × 5–50 mm, margins of lateral lobes dentate or entire; pedicels strongly reflexed or spreading; petals white, narrowly oblanceolate, 2.5–6.5 mm, not channeled or crisped, claw undifferentiated from blade; siliques ascending or descending, sometimes subtorulose, terete, 2–5 cm × 0.7–1.2 mm, glabrous or sparsely pubescent. Elsewhere known from dry habitats; our report is from 4200 ft. Hidalgo.

CHORISPORA R. Br. ex DC., nom. cons. – Chorispora

Chorispora tenella (Pall.) DC.—Blue mustard, crossflower. Annual from a taproot, with stipitate-glandular trichomes and often a few simple nonglandular trichomes; branched throughout, to 5 dm; leaves elliptic-oblong to (ob)lanceolate, deeply sinuate-dentate, the long petioles reduced apically and leaves becoming sessile, main leaf blades 3–8 cm; lower flowers subtended by leafy bracts; petals blue-purple or magenta with darker veins (rarely white), often with a yellowish throat, long-clawed, oblanceolate, 9–12 mm; siliques woody, with a stout, tapering beak, torulose, terete, 3–4.5 mm, curved upward to less commonly ± straight, indehiscent and breaking into 1-seeded segments. Disturbed sites of canyon bottoms, roadsides, sagebrush, piñon-juniper, cottonwood, ponderosa pine communities, 4000–9100 ft. Widespread except SE. Introduced (Eurasian).

CONRINGIA Heist. ex Fabr. – Hare's ear

Conringia orientalis (L.) Dumort—Annual or winter annual, glabrous and glaucous; 3–8 dm including the overtopping fruits; basal leaves obovate to oblanceolate and tapering to a winged petiole, (sub) entire, 5–9 cm, cauline leaves oblong-lanceolate, sessile, cordate-clasping, entire; inflorescence corymbiform, few-flowered; petals cream-white to pale yellow, long-clawed, narrowly oblanceolate, 9–12 mm; siliques ascending to nearly erect, linear, quadrangular but sometimes nearly terete, somewhat torulose, tapering to a slender tip with a thick style, 8–13 mm. Disturbed sites of roadsides, canyon bottoms, piñon-juniper, grass-dominated communities, 6200–8500 ft. N, Torrance. Introduced (Eurasian).

DESCURAINIA Webb & Berthel., nom. cons. – Tansymustard

Annual or biennial herbs with dendritic trichomes, sometimes mixed with glandular or simple ones; stem single, erect, usually branched; leaves petiolate, basal leaves rosulate, withering early and often not seen, cauline 2- or 3-pinnatisect to pinnate, bipinnate, or tripinnate, often finely divided, smaller distally; petals yellow to more rarely cream-white, spatulate or obovate to oblong, weakly clawed; fruit a silique or silicle, the septum sometimes perforated or lacking. Fully mature fruits needed for accurate identification. Expect intermediate morphologies, especially between *D. incisa* and *D. pinnata* and their subspecies (Goodson & Al-Shehbaz, 2010).

1. Fruits sparsely to densely pubescent at least when young…(2)
1'. Fruits glabrous…(3)
2. Seeds biseriate; ovules 48–64 per ovary; fruits 1–1.3 mm wide; fruiting pedicels 13–31 mm…**D. adenophora**
2'. Seeds uniseriate; ovules 16–40 per ovary; fruits 0.7–1 mm wide; fruiting pedicels 6–15 mm…**D. obtusa**
3. Fruits evidently fusiform, obovate, clavate, or broadly ellipsoid, rarely broadly linear (wider distally)…(4)
3'. Fruits linear (sometimes oblong in *D. brevisiliqua*), sometimes shortly tapering at both ends…(5)
4. Fruits usually clavate, rarely broadly linear (wider distally); seeds biseriate; ovules 16–40 per ovary; valves each with a distinct midvein…**D. pinnata**
4'. Fruits fusiform, obovate, or broadly ellipsoid, appearing plump; seeds uniseriate; ovules 4–12 per ovary; valves each with an obscure midvein…**D. californica**
5. Fruits usually ± strictly appressed to rachises, at least apically; septa with a distinct midvein; fruiting pedicels erect to erect-ascending…**D. incana**
5'. Fruits not appressed to rachises; septa with a distinct midvein or not; fruiting pedicels horizontal, divaricate, or ascending, occasionally some nearly erect…(6)
6. Septa with a distinct midvein (sometimes faint, check several fruits); ovules (4–)8–10 per ovary (check several fruits); plants short-lived perennials [to be expected at high elevations in the San Juan Mountains]…D. kenheilii
6'. Septa lacking a distinct midvein (except *D. sophia* with a midvein); ovules (10–)14–48 per ovary; plants annual or biennial…(7)
7. Leaf blades 2- or 3-pinnate; fruit septa appearing 2- or 3-veined…**D. sophia**
7'. Leaf blades usually 1-pinnate; fruit septa not veined…(8)
8. Plants eglandular; fruits < 8(–10) mm; petals < 1.2 mm; seeds < 0.8 mm…**D. brevisiliqua**
8'. Plants glandular or eglandular; fruits > 8 mm; petals > 1.7 mm; seeds > 0.9 mm…(9)
9. Plants canescent or not; distal segments of cauline leaf blades oblong to lanceolate, or linear, margins dentate, denticulate, or entire; fruits straight or strongly curved inward…**D. incisa**
9'. Plants not canescent; distal segments of cauline leaf blades linear or oblong, margins entire; fruits straight or slightly curved inward…**D. longepedicellata**

Descurainia adenophora (Wooton & Standl.) O. E. Schulz [*D. obtusa* (Greene) O. E. Schulz subsp. *adenophora* (Wooton & Standl.) Detling]—Blunt tansymustard. Biennials; glandular (at least distally), finely pubescent, often canescent, trichomes dendritic, sometimes mixed with simple ones; stems branched distally, 4.5–13 dm; basal leaves pinnate, 2–10 cm, with 2–5 pairs of lobes; cauline leaves densely pubescent; pedicels divaricate, straight, 13–31 mm; petals oblanceolate, 1.8–2.6 mm; siliques divaricate to erect, linear, slightly torulose, 8–16(–20) × 1–1.3 mm, abruptly acute at both ends. Piñon-juniper, ponderosa pine, mixed woodlands, montane conifers, 5500–11,000 ft. WC, Grant.

Descurainia brevisiliqua (Detling) Al-Shehbaz & Goodson—Biennials; eglandular, finely pubescent, often canescent, trichomes dendritic, rarely also mixed with simple ones; stems branched distally,

6–11 dm; basal leaves pinnate, 1–5 cm, with 2–5 pairs of lobes, linear, margins entire or serrate to incised (apex obtuse); cauline leaves densely pubescent; pedicels divaricate-ascending, straight, 4–9 mm; petals oblanceolate, 0.7–1 mm; siliques divaricate to erect, linear to oblong, not torulose, 3–9 × 1–1.2 mm. Generally sandy areas, grassy areas, piñon-juniper, ponderosa pine, 6450–8200 ft. WC, C, NC.

Descurainia californica (A. Gray) O. E. Schulz—Sierra tansymustard. Annuals or biennials; eglandular, usually pubescent, sometimes glabrous distally, trichomes dendritic; stems branched distally, 1.3–10.5+ dm; basal leaves pinnate, 1.5–6 cm, with 2–4(5) pairs or lateral lobes; cauline leaves sparsely pubescent; pedicels divaricate to ascending or suberect, often straight, 3–11 mm; petals oblanceolate, 1.1–1.8 mm; silicles divaricate to erect, fusiform, not torulose, 2–6 × 0.8–1.3 mm, long-acute at both ends. Disturbed areas, piñon-juniper, dry hillsides, decomposed granite slopes, sagebrush, moist roadsides, open woods, fir-spruce or aspen communities, gravel and talus slopes, 3450–10,200 ft. Widespread except the southernmost and easternmost counties.

Descurainia incana (Bernh. ex Fisch. & C. A. Mey.) Dorn [*D. richardsonii* O. E. Schulz]—Mountain tansymustard. Biennials; rarely glandular, finely pubescent, sometimes canescent, trichomes dendritic; stems often many-branched distally, 1.5–12 dm; basal leaves pinnatifid, 1.5–11 cm; cauline leaves with (2)3–5 pairs of lateral lobes; pedicels erect to erect-ascending, straight, 2–11 mm; petals oblanceolate, 1.2–2 mm; siliques erect, often strictly appressed to rachis, linear, slightly torulose, 4–12 × 0.7–1.3 mm, acute at both ends. Mainly alpine and subalpine areas, gravel and sand bars, scree, grassy slopes, spruce-fir, pine, aspen, sagebrush communities, 6800–10,700 ft. Widespread except southeasternmost and easternmost counties.

Descurainia incisa (Engelm. ex A. Gray) Britton—Mountain tansymustard. Annuals; glandular or eglandular, densely to sparsely pubescent, glabrous or pubescent distally, sometimes canescent, trichomes dendritic; stems branched distally or sometimes throughout, 1.3–10.7 dm; basal leaves pinnate, with 2–9(–10) pairs of lateral lobes, cauline leaves pubescent or glabrous; pedicels ascending to divaricate or horizontal, straight, 4–30 mm; petals narrowly oblanceolate, 1.7–2.8 mm; siliques erect to ascending, linear, slightly torulose, 8–20 × 0.9–1.3 mm, straight or slightly to strongly curved inward. A difficult and variable species that cannot always be reliably distinguished from *D. pinnata* and its subspecies. Moist areas, meadows, streamsides.

1. Fruiting pedicels ascending to divaricate, 3–10(–12) mm; lateral lobes of basal and proximal cauline blades (3–)5–9 pairs, margins usually coarsely dentate to incised, rarely crenate or pinnatifid; lobes of distal cauline blades oblong to lanceolate, margins dentate to denticulate; fruits straight or curved inward... **subsp. incisa**
1'. Fruiting pedicels horizontal to divaricate, (10–)13–30 mm; lateral lobes of basal and proximal cauline blades 2 or 3(4) pairs, margins usually entire; lobes of distal cauline blades linear, margins entire; fruits curved inward...**subsp. paysonii**

subsp. incisa [*D. richardsonii* O. E. Schulz subsp. *incisa* (Engelm. ex A. Gray) Detling]—Mountain tansymustard. Herbage subglabrous to moderately pubescent; inflorescence rachis and branches lacking stipitate-glandular hairs; pedicels equaling or shorter than the mature siliques. Alpine, stream banks, disturbed sites, roadsides, meadows, sagebrush and juniper communities, open woods, rocky cliffs, sandy areas, talus slopes, 5200–11,000 ft. Widespread except easternmost counties.

subsp. paysonii (Detling) Rollins—Payson's tansymustard. Herbage canescent; inflorescence rachis and branches lacking stipitate-glandular hairs; pedicels longer than the mature siliques. Piñon-juniper, sagebrush, desert shrub, and bitterbrush communities, 6050–7250 ft. McKinley, Rio Arriba, Socorro.

Descurainia kenheilii Al-Shehbaz—Heil's tansymustard. Biennial or short-lived perennials; eglandular, glabrescent to sparsely pubescent throughout with dendritic trichomes; stems erect, branched distally, mostly 1.4–11 dm, but when dwarfed, as in the holotype, unbranched and as little as 0.1 dm; basal leaves pinnate, oblanceolate to obovate in outline, lateral leaflets 2–5 per side, (1–)1.2–6.5 cm; cauline leaves smaller distally, sparsely pubescent to glabrescent; pedicels ascending to erect, not appressed to the stem, straight, 1–7 mm; petals narrowly oblanceolate, 1–1.5 mm; siliques ascending to erect, glabrous, all or most linear, sometimes some curved, 5–12 × 1–1.3 mm, smooth or somewhat torulose, tapered at both ends, the ends rounded-acute to acute, valves each with a distinct midvein; septum with midvein, this often faint. Subalpine moist and wet meadows, alpine tundra, talus slopes, rock outcrops, grassy streamsides, in both undisturbed and disturbed areas, 9260–9310 ft. (higher and lower in adjacent Colorado). Rio Arriba (O'Kane & Heil, 2022).

Descurainia longepedicellata (E. Fourn.) O. E. Schulz [*D. pinnata* (Walter) Britton subsp. *brachycarpa* (Richardson) Detling; *D. pinnata* (Walter) Britton var. *filipes* (A. Gray) M. Peck; *D. incisa* (Engelm. ex A. Gray) Britton var. *filipes* (A. Gray) N. H. Holmgren]—Western tansymustard. Annuals; rarely glandular, moderately to sparsely pubescent, often glabrous distally, not canescent, trichomes dendritic; stems often branched distally, 2–7 dm; basal leaves pinnate, 1.5–7 cm; cauline leaves usually glabrous, rarely pubescent; pedicels horizontal to divaricate, straight, 9–18 mm; petals narrowly oblanceolate, 1.7–2.6 mm; siliques erect, linear, not torulose, 9–17 × 0.8–1.1 mm, straight or slightly curved inward. Sandy plains and banks, dry washes, open hillsides, sagebrush and juniper or pine communities, grasslands, 3800–9300(–10,300) ft. NW, Curry, Lincoln.

Descurainia obtusa (Greene) O. E. Schulz [*Sophia obtusa* Greene]—Blunt tansymustard. Biennials; glandular or eglandular, finely pubescent, often canescent, trichomes dendritic, sometimes mixed with simple ones; stems 4–12(–15) dm; basal leaves pinnate, 1–6 cm, with 2–5 pairs of lateral lobes, oblanceolate to linear or narrowly lanceolate, entire or serrate, rarely incised; cauline leaves densely pubescent; pedicels ascending to divaricate, straight, 6–15 mm; siliques divaricate to suberect, linear, slightly torulose, 10–20(–23) × 0.7–1 mm, acute at both ends. Gravelly ground, sandy areas, disturbed sites, open forests, plateaus, abandoned mine areas, dry streams and washes, 5000–9700 ft. NW, NC, WC, C.

Descurainia pinnata (Walter) Britton—Western tansymustard. Annuals; glandular or eglandular, sparsely to densely pubescent, sometimes glabrous distally, canescent or not, trichomes dendritic; stems branched basally and/or distally, 0.8–8 dm; basal leaves with 4–9 lateral lobes; cauline leaves densely pubescent; pedicels usually ascending to divaricate or horizontal, rarely descending, straight or slightly recurved, 4–23 mm; petals yellow or pale yellow-white, narrowly oblanceolate, 1–3 mm; siliques erect to ascending, usually clavate and wider distally, rarely broadly linear, not torulose, 4–15 × 1.2–2.2 mm. A difficult and variable species that cannot always be reliably distinguished from *D. incana* and its subspecies. Elevations are for all subspecies due to misidentifications and specimens often not identified to subspecies: 3300–9050 ft. Three subspecies in our flora. Widespread.

1. Rachises glabrous, eglandular…**subsp. glabra**
1'. Rachises sparsely to densely pubescent, glandular or eglandular…(2)
2. Plants canescent, often eglandular; stems branching basally or distal to base; sepals purple or rose…**subsp. ochroleuca**
2'. Plants not canescent, often glandular; stems unbranched basally; sepals yellow…**subsp. brachycarpa**

subsp. brachycarpa (Richardson) Detling [*D. intermedia* (Rydb.) Daniels]—Plants usually glandular, rarely eglandular, usually not canescent; stems unbranched basally, branched distally; raceme rachises sparsely to densely pubescent, often glandular, fruiting pedicels divaricate to ascending, forming 20°–60°(–80)° angle, 8–20 mm; sepals yellow, 1.5–2.6 mm; petals 1.7–3 × 0.6–1 mm. Sagebrush, piñon-juniper, disturbed sites, limestone ledges, foothills, canyon margins, gravel washes, dry slopes, cliffs, streamsides, prairies. NC.

subsp. glabra (Wooton & Standl.) Detling—Plants eglandular, not canescent; stems unbranched basally, branched distally; raceme rachises glabrous, eglandular, fruiting pedicels divaricate to horizontal or descending, forming 70°–90°(–100)° angle, 4–13 mm; sepals rose (at least apically), 0.8–1.5 mm; petals 1–1.8 × 0.3–0.7 mm. Sandy areas, scrub and bush communities, juniper, oak, and pine woodlands, sagebrush, limestone outcrops. S half of state.

subsp. ochroleuca (Wooton) Detling—Plants usually eglandular, rarely glandular, distinctly canescent; stems branched basally or just distal to base, branched distally; raceme rachises often densely pubescent, eglandular; fruiting pedicels divaricate to ascending, forming 30°-60° angle, 5-12 mm; sepals purple or rose, 1-2 mm; petals 1.5-2 × 0.3-0.5 mm. Gravelly and stony hills, desert grasslands, roadsides. Widespread.

Descurainia sophia (L.) Webb ex Prantl—Northern tansymustard. Annuals; eglandular, sparsely to densely pubescent, sometimes glabrous distally, trichomes dendritic; stems unbranched or branched distally, 1-9 dm; basal leaves 2- or 3-pinnate; cauline leaves often glabrous; pedicels divaricate to ascending, straight, 6-18 mm; petals narrowly oblanceolate, 2-3 mm; siliques divaricate-ascending to erect, narrowly linear, torulose, 12-30 × 0.5-1 mm, straight or curved upward. Waste places and disturbed sites, hillsides, mountain slopes, canyon bottoms, stream banks, deserts, sagebrush, piñon-juniper, ponderosa pine, 3750-9800 ft. Widespread. Introduced.

DIMORPHOCARPA Rollins – Spectaclepod

Annual, biennial, or perennial herbs, pubescent with dendritic trichomes; stem erect or decumbent, generally single, branched or not; leaves basal and cauline, petiolate (or sessile above), entire, dentate, or lobed; sepals spreading to reflexed; petals white to lavender, obovate; silicles strongly didymous and angustiseptate, strongly flattened, spectacle-shaped, divaricate, glabrous or pubescent, valves winged and dehiscing with the enclosed seed; seed 1 per locule.

1. Fruit valves 8-10 mm; petals (7-)8-12 mm, claws not expanded basally; distal cauline leaves sessile, blades usually ovate to narrowly oblong, rarely lanceolate, bases obtuse to truncate…**D. candicans**
1'. Fruit valves 4-5.5 mm; petals 4-7(-8) mm, claws expanded basally; distal cauline leaves shortly petiolate or subsessile, blades linear to lanceolate, bases cuneate…**D. wislizeni**

Dimorphocarpa candicans (Raf.) Rollins [*D. palmeri* (Payson) Rollins; *Dithyrea wislizeni* Engelm. var. *palmeri* Payson]—Palmer's spectaclepod. Annual or biennial; basal leaves lanceolate to oblong or ovate, 4-9 cm, base cuneate to obtuse, margins dentate; cauline leaves usually ovate to narrowly oblong, rarely lanceolate, base obtuse to truncate, margins entire, sometimes repand; pedicels divaricate; petals 7-12 mm, claw 2-3 mm, not expanded basally; silicles with each valve suborbicular or orbicular, 7-10 × 6-10 mm, base rounded. Sandy hills and plains, prairies, sand dunes, mesquite, ca. 4000 ft. Few definitive records, as this may not be separable from *D. wislizeni*.

Dimorphocarpa wislizeni (Engelm.) Rollins [*Dithyrea wislizeni* Engelm.]—Spectaclepod, tourist-plant. Annual; basal leaves lanceolate to linear-lanceolate, 2-79 cm, base cuneate to attenuate, margins pinnately lobed to coarsely dentate; cauline leaves linear to narrowly lanceolate, base cuneate, margins usually entire, rarely dentate, or repand; pedicels divaricate to slightly reflexed; petals 4-7.5 mm, claw 1-1.5 mm, expanded basally; silicles with valves ovoid-oblong, rarely suborbicular, 4-6 × 4-7.5 mm, base slightly rounded. Sandy areas of roadsides, knolls, hills and dunes, streambeds and dry washes, desert flats, hanging gardens, sandy pockets in slickrock, piñon-juniper communities, disturbed areas, 3100-8150 ft. Widespread.

DIPLOTAXIS DC. – Wallrocket

Annual or perennial; scapose or not; strongly scented, glabrous, glabrescent, or pubescent; stems erect or ascending, branched; basal leaf margins sinuate to deeply pinnatifid; pedicels ascending, divaricate, or reflexed, stout to slender; petals yellow, obovate (apex rounded or truncate); siliques dehiscent, sessile or stipitate, with a terminal beak.

1. Perennial, with adventitious buds on roots; stems frequently foliose, glabrescent or sparsely pubescent basally; gynophores 0.5-3 mm…**D. tenuifolia**
1'. Annual or short-lived perennial, without buds on roots; stems frequently scapose, moderately pubescent; gynophores absent or to 0.5 mm…**D. muralis**

Diplotaxis muralis DC.—Annual or stinking wallrocket. Annual or short-lived perennial, frequently scapose or subscapose, taprooted; stems ascending to suberect, moderately pubescent, spreading basally, retrorse distally; basal leaves with 2-4(-6) lobes each side; petals 5-9 × 3-5 mm; gynophore obsolete or to 0.5 mm; siliques erect-spreading, 1.5–4 cm × 1.5–2.5 mm. Waste ground, disturbed sites, road-sides, railroads, around buildings, grazed grasslands, 6550–7050 ft. Grant, Lincoln. Introduced.

Diplotaxis tenuifolia DC.—Perennial wallrocket. Perennials, usually suffrutescent, roots with shoots from adventitious buds, glaucescent; stems erect, 2–9 dm, glabrescent or sparsely pubescent basally; basal leaves with 2–5 lobes each side; pedicels 8–35 mm; petals yellow, 7–12 × 5–8 mm; gyno-phore 0.5–3 mm; siliques usually erect, rarely ascending, 2–5 cm × 1.5–2.5 mm. Waste places, disturbed areas, wharf and railroad ballast, sandy beaches, muddy shores, wet woods, mountain slopes, 4000–7600(–8800) ft. Doña Ana, Lincoln, Otero, Sierra. Introduced.

DRABA L. – Draba

Plants perennial, less frequently annual or biennial; caudex simple or branched, ± absent in some; tri-chomes stalked or sessile, simple, forked, cruciform, stellate, malpighian, or dendritic, often > 1 kind pres-ent, stems unbranched or branched (usually distally); leaves basal and cauline, basal usually rosulate, usu-ally petiolate, rarely sessile, blade margins usually entire or toothed, rarely pinnately lobed; lateral pair of sepals not saccate or subsaccate basally; petals yellow, white, pink, or purple; all flowers petalous and chasmogamous; fruits silicles or siliques, latiseptate (ours) or terete; ovules 4–70 per ovary; style obso-lete to evident (Al-Shehbaz, 2012a, 2013; Al-Shehbaz et al., 2010).

Species of *Tomostima* (formerly in *Draba*) might key here and can be distinguished as follows:

A. Plants annual; flowers of 2 types: early-season flowers petalous and chasmogamous and late-season flowers of the lateral branches apetalous and cleistogamous; petals white; plants occurring below 7700 ft. elevation…go to **TOMOSTIMA**

A'. Plants perennial, occasionally biennial or annual in some species; flowers all of 1 type: petalous and chas-mogamous; petals yellow or white; plants usually occurring above 7700 ft. elevation, or if lower, then petals yellow…**DRABA** [key follows]

1. Plants annual or biennial; lacking an evident caudex (root crown)…(2)
1'. Plants perennial; caudex usually present and evident…(7)
2. Styles well developed, 0.8–3 mm…(3)
2'. Styles ± obsolete to minute, < 0.2 mm…(4)
3. Cauline leaves > 8; rachises pubescent; fruiting pedicels mainly 4–10 mm…**D. helleriana** (in part)
3'. Cauline leaves 1–3; rachises usually glabrous, rarely sparsely pubescent; fruiting pedicels mainly 9–18 mm…**D. mogollonica** (in part)
4. Rachises pubescent   **D. rectifructa**
4'. Rachises glabrous…(5)
5. Cauline leaves 4–15; fruiting pedicels 1.5–7 times longer than the fruit [reports from New Mexico are doubtful and are most likely misidentifications]…D. nemorosa L.
5'. Cauline leaves 1–3(–5); fruiting pedicels subequal to or shorter than the fruit…(6)
6. Lower surface of basal leaves glabrous or sparsely pubescent with simple and 2-rayed trichomes; stems usually glabrous, rarely sparsely pubescent basally with simple trichomes…**D. crassifolia** (in part)
6'. Lower surface of basal leaves pubescent with 2–4-rayed trichomes; stems basally pubescent with 2–4-rayed (rarely simple) trichomes…**D. albertina** (in part)
7. Flowering stems leafless…(8)
7'. Flowering stems with (1)2+ leaves…(9)
8. Stems branched distally, pubescent basally; cauline leaf blades pubescent…**D. albertina** (in part)
8'. Stems unbranched throughout (rarely branched distally), glabrous throughout (rarely pubescent ba-sally); cauline leaf blades (when present) glabrous…**D. crassifolia** (in part)
9. Lower leaf surfaces glabrous or with simple trichomes…(10)
9'. Lower leaf surfaces pubescent with only branched trichomes…(16)
10. Styles 1–3 mm; fruits usually twisted, rarely plane; petals 4–7.5 mm…(11)
10'. Styles < 1 mm; fruits plane, rarely slightly twisted; petals 1.5–3 mm (to 4.5 mm in D. grayana)…(13)
11. Fruits twisted up to 3 turns, rarely plane; stems basally with trichomes 0.4–2.2 mm; ovules 20–34 per ovary…**D. streptocarpa**
11'. Fruits twisted up to 1 full turn, rarely plane; stems basally with trichomes 0.1–0.7 mm; ovules 12–24 per ovary…(12)
12. Basal leaves not differentiated into a blade and petiole, 0.5–0.8 cm, entire leaf persistent, imbricate, margins with setiform trichomes…**D. heilii**

12'. Basal leaves differentiated into a blade and petiole, 1.2–8.5 cm, only petiole persistent, leaves not imbricate, margins lacking setiform trichomes…(13)

13. Fruits lanceolate to ovate-lanceolate, (7–)8–14 × 3–5 mm; rachises pubescent, the trichomes crisped… **D. crassa**

13'. Fruits linear-elliptic to elliptic, 5–10(–13) × 1.5–2.5 mm; rachises usually glabrous, rarely pubescent, the trichomes non-crisped… **D. standleyi**

14. Plants perennial, densely cespitose; fruiting pedicels 1.5–6 mm [reports from New Mexico are doubtful and are most likely misidentifications]…D. grayana (Rydb.) C. L. Hitchcock

14'. Plants annual, biennial, or perennial, not cespitose; fruiting pedicels 5–14+ mm…(15)

15. Lower leaf surface pubescent with 2–4-rayed trichomes; stems pubescent, trichomes simple and 2-rayed… **D. albertina** (in part)

15'. Lower leaf surface glabrous or with some simple and 2-rayed trichomes; stems glabrous, rarely pubescent basally with simple trichomes… **D. crassifolia** (in part)

16. Fruit valves glabrous…(17)

16'. Fruit valves pubescent, at least on the margins…(19)

17. Stem and leaf trichomes sessile, 2 rays parallel to the long axis of the stem or midvein, some malpighian… **D. spectabilis** (in part)

17'. Stem and leaf trichomes stalked, rays not parallel to the long axis of the stem or midvein, none malpighian…(18)

18. Styles evident, 1–3 mm; petals 4–6 mm… **D. abajoensis**

18'. Styles nearly obsolete, 0.01–0.12 mm; petals 2–3.2 mm… **D. albertina** (in part)

19. Lower surface of leaf with at least some 7–15-rayed trichomes…(20)

19'. Lower surface of leaf with 2–4(–6)-rayed trichomes…(23)

20. Petals yellow…(21)

20'. Petals white…(22)

21. Cauline leaves 5–20+; racemes basally bracteate, 18–52+-flowered; fruits (6–)9–14(–17) mm, often subappressed to the rachis; ovules 28–38+ per ovary… **D. aurea** (in part)

21'. Cauline leaves 1–4; racemes ebracteate throughout, 4–10(–18)-flowered; fruits 3–10 mm, ascending; ovules 10–18 per ovary… **D. streptobrachia**

22. Styles evident, 0.8–2.3 mm; fruits twisted; plants forming tangled mats… **D. smithii**

22'. Styles nearly obsolete, 0.1–0.6 mm; fruits plane or only slightly twisted; plants not forming tangled mats… **D. cana**

23. Fruit trichomes mainly 2-rayed and/or simple…(24)

23'. Fruit trichomes all or nearly all 4-rayed…(27)

24. Petals white… **D. henrici**

24'. Petals yellow…(25)

25. Stem and leaf trichomes sessile, 2 rays parallel to the long axis of the stem or midvein, some trichomes malpighian… **D. spectabilis** (in part)

25'. Stem and leaf trichomes stalked, rays not parallel to the long axis of the stem or midvein, not malpighian…(26)

26. Fruit valves puberulent, at least along margins, with simple trichomes; lower leaf surfaces with short-stalked cruciform trichomes; margins of cauline leaves entire or denticulate… **D. petrophila**

26'. Fruit valves puberulent with 2(–4)-rayed trichomes; lower leaf surfaces with stalked, cruciform, and fewer 3–5-rayed trichomes; margins of cauline leaves usually dentate, rarely subentire… **D. helleriana** (in part)

27. Cauline leaves 1–3… **D. mogollonica** (in part)

27'. Cauline leaves 5–30+…(28)

28. Styles mainly 0.5–1.2 mm; petals 3.5–5 mm; ovules ≥ 28 per ovary; fruits often subappressed to rachis… **D. aurea** (in part)

28'. Styles mainly 1.5–3.5 mm; petals 5–7 mm; ovules 14–28 per ovary; fruits not subappressed to rachis… **D. helleriana** (in part)

Draba abajoensis Windham & Al-Shehbaz [*D. spectabilis* Greene var. *glabrescens* O. E. Schulz]— Abajo Mountains draba. Perennial; caudex simple or few-branched; stems unbranched, pubescent proximally; basal leaves rosulate, oblanceolate to spatulate, 1–4 cm, pubescent with (2–)4-rayed trichomes, sometimes adaxially with simple trichomes; cauline leaves 4–10, broadly ovate to lanceolate or oblong, pubescent; racemes ebracteate, glabrous or pubescent; petals yellow, 4–6 mm; fruits elliptic to elliptic-lanceolate, 5–10 × 2–3 mm, plane, glabrous; style 1–3 mm; ovules 10–18 per ovary. Piñon and ponderosa pine with Gambel oak, 6400–8760 ft. Chuska Mountains. San Juan. **R**

Draba albertina Greene [*D. crassifolia* Graham var. *albertina* (Greene) O. E. Schulz]—Slender draba. Annual, biennial or short-lived perennial; stems simple or branched above, pubescent proximally; basal leaves rosulate, obovate to oblanceolate or linear-lanceolate, mainly 10–30 mm, ciliate with simple trichomes; pubescent beneath with 2–4-rayed trichomes; upper surfaces with simple trichomes sometimes mixed with 2-rayed ones, rarely glabrous; cauline leaves 1–3(–5), rarely absent, lanceolate to elliptic or ovate; racemes ebracteate, glabrous; petals yellow, 2–3 mm; fruits lanceolate to narrowly elliptic or linear, 5–13 × 1–2.1 mm, plane, glabrous or rarely puberulent; style 0.01–0.12 mm; ovules 20–40 per ovary. Moist meadows and grassy areas near conifers, 9000–10,700 ft. NW, NC.

Draba aurea Vahl ex Hornem.—Golden draba. Perennial; caudex simple or branched; stems simple or branched above, pubescent throughout; basal leaves rosulate, oblanceolate to obovate, mainly 10–40 mm, both surfaces pubescent with stalked, mainly 4–7-rayed trichomes, petiole ciliate; cauline leaves 5–25, oblong to lanceolate or ovate, pubescent; racemes bracteate along lowermost flowers, very rarely ebracteate, pubescent; petals yellow, 3.5–5 mm; fruits lanceolate to linear-lanceolate, 8–15 × 2–3.5 mm, flattened, slightly twisted or plane, pubescent with a mixture of simple and short-stalked, 2–4-rayed trichomes; style 0.5–1.5 mm; ovules 28–40 per ovary. Meadows, rockslides, crevices and ledges, coniferous forests and woodlands, below snowbanks, alpine slopes, turf, tundra, 6560–12,960 ft. W, NC, S.

Draba cana Rydb. [*D. breweri* S. Watson var. *cana* (Rydb.) Rollins]—Hoary draba. Perennial; caudex simple or short-branched; stems simple or branched above, pubescent; basal leaves rosulate, linear to oblanceolate or oblong, 6–20+ mm, both surfaces pubescent with short-stalked, 4–12-rayed trichomes, base ciliate; cauline leaves 3–15, lanceolate to ovate or oblong, pubescent; racemes basally bracteate; petals white, oblanceolate to spatulate, 2.3–4.5 mm; fruits narrowly lanceolate to linear or very rarely ovate-oblong, 5–10 × 1.5–2.5 mm, flattened, slightly twisted or plane, pubescent with short-stalked, 3–7-rayed trichomes; style 0.1–0.6 mm; ovules 28–48 per ovary. Alpine tundra, open areas in subalpine forests, 10,600–12,700 ft. NC.

Draba crassa Rydb.—Thickleaf draba. Perennial; cespitose; caudices well developed, simple to many-branched; stems simple, glabrous proximally, pubescent distally; basal leaves rosulate, oblanceolate, 2–7 mm, glabrous but petiole and margin ciliate; cauline leaves 2–4(–6), ovate to oblong, ciliate with simple and stalked, 2-rayed trichomes; racemes ebracteate, pubescent; petals yellow, spatulate to sub-obovate, 3.5–6 mm; fruits lanceolate to ovate-lanceolate, 8–14 × 3–5 mm, flattened, slightly twisted, glabrous; style (0.4–)0.7–1.5 mm; ovules 16–20 per ovary. Alpine areas; in New Mexico known from W Colfax County.

Draba crassifolia Graham [*D. parryi* Rydb.]—Snowbed draba. Biennial to short-lived perennial; rarely with a few-branched caudex; stems often scapose, simple or rarely branched above, glabrous or rarely pubescent proximally; basal leaves rosulate, oblanceolate to obovate, mainly 5–25 mm, glabrous or ciliate to sparsely pubescent with simple and 2-rayed trichomes; cauline leaves absent or rarely 1; racemes ebracteate, glabrous; petals yellow or cream, often fading white, oblanceolate, 1.5–3 mm; fruits narrowly elliptic to lanceolate or rarely linear-lanceolate, (3–)5–10 × 1.5–2.5 mm, plane, glabrous; style 0.02–0.1 mm; ovules mainly 16–25 per ovary. Alpine tundra, talus slopes, bare snowmelt areas, 9000–12,960 ft. NC, WC.

Draba heilii Al-Shehbaz—Heil's draba. Perennial; densely cespitose from a few-branched caudex; stems unbranched or branched distally, glabrous throughout, rarely pilose; basal leaves rosulate, linear-lanceolate, 0.5–0.8 cm, glabrous or rarely pilose with simple and stalked, 2-rayed trichomes; cauline leaves 6–8, linear-oblong, margins ciliate; racemes ebracteate, usually glabrous, rarely pubescent; petals yellow, oblong-oblanceolate, 4–6; fruits narrowly lanceolate, twisted 1/2 turn or plane, flattened, 7–10 × 1.3–1.6 mm, glabrous; style 0.7–1.3 mm; ovules 16–20 per ovary. Alpine tundra and tree line, 11,800–12,400 ft. Mora, Rio Arriba. **R & E**

Draba helleriana Greene [*D. neomexicana* Greene]—Heller's draba. Biennial or short-lived perennial; caudices simple to many-branched; stems usually branched throughout, hirsute proximally, glabrous or pubescent above; basal leaves rosulate, oblanceolate to obovate, 9–41+ mm, pubescent with stalked, (3)4(5)-rayed trichomes; cauline leaves mainly 10–35, ovate to lanceolate or oblong, pubescent with mostly 4-rayed trichomes mixed with fewer or more simple or 2-rayed ones; racemes ebracteate or rarely lowermost 1–3 flowers bracteate, pubescent; petals yellow, oblanceolate, 5–7 mm; fruits lanceolate to ovate or oblong-lanceolate, 5–15 × 2–3.5 mm, flattened, slightly to strongly twisted or plane, puberulent; style 1–3.5 mm; ovules 14–28 per ovary. Tundra, rocky alpine and subalpine meadows, rocky areas in pine-oak and spruce-fir woodlands, aspen groves, 6200-12,890 ft. W, NCS.

Draba henrici Al-Shehbaz—Henk's draba. Perennial; caudex few- to several-branched; stems simple throughout or branched distally, pubescent near base with a mixture of simple trichomes and smaller 4-6-rayed trichomes; basal leaves rosulate, narrowly oblanceolate, 5–12 mm, both surfaces moderately pubescent with stalked, 4(-7)-rayed trichomes; cauline leaves 8-20, overlapping, lanceolate to oblong or narrowly ovate, both surfaces with 4(-7)-rayed trichomes, usually mixed above with fewer simple trichomes; racemes ebracteate, pubescent; petals white, oblanceolate, 3–3.5 mm; fruit (immature) narrowly oblong, 6–8 × ca. 1.5 mm, slightly twisted, puberulent with simple and 2(-4)-rayed trichomes; style 0.3-0.8 mm, ovule number not known. Gravelly alpine tundra. Taos. **R & E**

Draba mogollonica Greene—Mogollon Mountains draba. Annual, biennial, or rarely short-lived perennial; stems usually branched, pubescent proximally, glabrous distally; basal leaves rosulate, spatulate to obovate or oblanceolate, (1.5-)3–10 cm, pubescent with cruciform, and fewer 2- or 3-rayed, trichomes; cauline leaves 1-3, oblong to ovate, pubescent; racemes ebracteate, glabrous, rarely basally with a few trichomes; petals yellow, spatulate, 5–8.5 mm; fruits linear-elliptic to elliptic-lanceolate, slightly twisted, flattened, 6–19 × 2.5-3.5 mm, glabrous or pubescent with simple and 2(-4)-rayed trichomes; style (0.8-)1–2.2 mm; ovules 24-36 per ovary. Areas of ponderosa pine or mixed conifers in rocky or gravelly spots, 4770-8000 ft. WC, SW. **R & E**

Draba petrophila Greene [*D. helleriana* Greene var. *blumeri* C. L. Hitchc.; *D. helleriana* Greene var. *petrophila* (Greene) O. E. Schulz]—Santa Rita Mountains draba. Perennial; cespitose; caudex simple or branched; stems usually unbranched, hirsute throughout; basal leaves rosulate, oblanceolate, 1-6 cm, pubescent, lower surface with cruciform trichomes, adaxially often similar, sometimes with fewer, simple and 2-rayed trichomes; cauline leaves 3-10, ovate to lanceolate or oblong, pubescent; racemes ebracteate, pubescent; petals yellow, oblanceolate, 3.5–6 mm; fruits lanceolate to elliptic, often strongly twisted, flattened, 5-11 × 2-3 mm, puberulent at least along margin, with simple trichomes; style 0.8-1.8(-2.5) mm, ovules 14-24 per ovary. Open, often rocky areas in ponderosa pine, mixed conifers, spruce and spruce-fir forests, WC, SW, SC (reports from NC doubtful).

Draba rectifructa C. L. Hitchc.—Mountain draba. Annual; stems simple or few-branched above, pubescent; basal leaves rosulate, oblanceolate to obovate, 1–2(–3) mm, lower surface pubescent with a mixture of simple and 2–4-rayed trichomes, upper surfaces with simple and 2-rayed trichomes; cauline leaves 3–10(–17), lanceolate to ovate or oblong, pubescent; racemes ebracteate, mainly 20–50+-flowered, pubescent; petals yellow, sometimes fading to white, oblanceolate, 1.5–3 × 0.5–1 mm; fruits lanceolate, 6–10 × 1.3–2.3 mm, flattened, not twisted, pubescent; style 0.01–0.1 mm; ovules 32–60 per ovary. Meadows and other grassy areas of ponderosa pine or mixed conifer forest openings, 6500–9620 ft. NC, C (doubtful in SC, Sacramento Mountains).

Draba smithii Gilg & O. E. Schulz—Smith's draba. Perennial; matted with many slender, prostrate caudex branches; stems usually simple, pubescent throughout; basal leaves rosulate, obovate to narrowly oblanceolate, 5–15+ mm, pubescent with subdendritic, 5–12-rayed trichomes; cauline leaves (2–)3–8, oblong to lanceolate, pubescent; racemes ebracteate, pubescent; petals white, spatulate, 4–6 mm; fruits ovate-lanceolate, 5–9 × 2–3 mm, flattened, twisted, pubescent; style 0.7–2.3 mm; ovules 16–20 per ovary. Rock crevices, ledges, moist areas. Our one report is from Taos County at 6900 ft. in ponderosa pine-Gambel oak. **R & E**

Draba spectabilis Greene—Showy draba. Perennial; caudices simple or few-branched; stems simple, pubescent proximally, usually glabrous or sparsely pubescent distally; basal leaves rosulate, oblanceolate to obovate, mainly 15–50+ mm, both surfaces pubescent with sessile and malpighian or cruciform trichomes or with minutely stalked 2- or 4-rayed ones; cauline leaves (3–)4–14, broadly ovate to lanceolate or oblong, pubescent; racemes ebracteate, glabrous or pubescent; petals yellow, oblanceolate, 4–6.5 mm; fruits lanceolate to oblong or elliptic, 6–13 × 2–3.5 mm, flattened, not twisted, glabrous or puberulent; style (0.5–)1–2.7 mm; ovules 12–24 per ovary. Mixed conifers, spruce-fir, tundra, 7200–9760 ft. NC.

Draba standleyi J. F. Macbr. & Payson—Standley's draba. Perennial; densely pulvinate from a branched caudex; stems unbranched, usually glabrous throughout or sparsely pubescent proximally, rarely sparsely pubescent distally; basal leaves rosulate, narrowly oblanceolate to linear-lanceolate, mainly 1.5–7 cm, surfaces glabrous or pubescent with simple, rarely 2-rayed, trichomes; cauline leaves 1–8, lanceolate to narrowly oblong, pubescent; racemes ebracteate, usually glabrous, rarely pubescent; petals yellow, oblanceolate, 4–6 mm; fruits linear-elliptic to elliptic, twisted or plane, flattened, 5–10(–13) × 1.5–2.5 mm, usually glabrous, rarely puberulent; style 0.7–1.4(–1.8) mm; ovules 12–24 per ovary. Igneous rock outcrops, stabilized talus slopes, 5700–10,000 ft. SW, SC. **R & E**

Draba streptobrachia R. A. Price—Alpine tundra draba. Perennial; caudex branched; stems unbranched, pubescent throughout; basal leaves rosulate, oblanceolate to linear-oblanceolate, 0.4–3(–4) cm, pubescent with 3–8-rayed trichomes; cauline leaves (1)2–4(5), oblong to ovate or linear, pubescent; racemes ebracteate, pubescent; petals yellow, spatulate, 3–5 mm; fruits ovate to elliptic or lanceolate, slightly twisted or plane, flattened, (3–)5–10 × 2–4 mm; valves often pubescent, occasionally glabrous, trichomes simple and minutely stalked, 2–4-rayed, 0.03–0.25 mm; style 0.3–0.8(–1.2) mm; ovules 10–16(–18) per ovary. Alpine tundra, scree, and slopes, crevices in rock ledges, 12,270–12,750 ft. Taos.

Draba streptocarpa A. Gray—Pretty draba. Perennial; caudex simple or branched; stems usually unbranched, hirsute proximally, sparsely pubescent or glabrous distally or throughout; basal leaves rosulate, oblong-oblanceolate to linear-oblanceolate, 0.5–3.8 cm, strigose to hirsute beneath with long-stalked, 2(–4)-rayed trichomes, usually with simple ones, strigose above with simple trichomes; cauline leaves 2–15, oblong to lanceolate, pubescent; racemes ebracteate or lowermost 1–3 flowers bracteate, sparsely pubescent; petals yellow, oblanceolate to spatulate, (4–)5–7.5 mm; fruits ovate to linear-lanceolate, usually strongly twisted (to 3 turns), rarely plane, flattened, 5–16 × 2–3 mm, puberulent along margin with simple trichomes; style 0.8–2.5 mm; ovules 20–34 per ovary. Various habitats, streamsides, meadows, alpine tundra, rock crevices, oak, mixed conifer, spruce-fir, 6500–12,650 ft. NC.

DRYOPETALON A. Gray – Rock-mustard

Dryopetalon runcinatum A. Gray—Rock-mustard. Annual or biennial; usually hirsute proximally; stems erect, unbranched or branched basally, branched distally; basal leaves spatulate to obovate in outline, lyrate, pinnatifid, or bipinnately lobed, (2–)4–20(–25) cm × 30–80 mm; cauline leaf base not auriculate, similar to basal; fruiting pedicels divaricate-ascending; petals white to purplish, spatulate, 6–9(–11) × 2.5–3.5(–4.5) mm, claw 2.5–4(–5) mm, margins pinnatifid, 5–7(–11)-lobed, papillate basally; silicles linear, straight or arcuate, 2–6 cm × 0.5–1.2 mm; ovules 60–110 per ovary; style 0.1–0.7(–1) mm. Ledges, shade of boulders and cliffs, foothills, canyons, scrub woodlands, streambeds, rocky basalt, crevices, 4000–7400 ft. SW. Notable for its pinnatifid to runcinate leaf and petal margins.

ERUCA Mill. – Garden-rocket

Eruca vesicaria (L.) Cav.—Garden-rocket, arugula. Stems branched basally, 1–10 dm, glabrous, hirsute, or hispid; basal leaves often withered by fruiting, widely oblanceolate or pinnatisect, 3–17 cm, lobes 3–9 on each side, lobe margins entire or dentate; cauline leaves lobed or not, similar to basal; pedicels subappressed to rachis, 2–10 mm; petals broadly obovate, 12–26 mm; siliques 1.1–4 cm × 2.5–5 mm; terminal segment 5-veined, as long as or slightly shorter than valves. Our plants are **subsp. sativa** (Mill.) Thell. Roadsides, disturbed areas, waste places, cultivated fields, dry ditches, rocky outcrops, gravelly slopes, sandy plains, open rangelands, lawns, 3700–5200(–7000) ft. Doña Ana, Grant, Hidalgo. Introduced.

ERYSIMUM L. – Wallflower

Annual, biennial, or perennial, herbs or rarely subshrubs or shrubs, pubescent with 2- to few-branched malpighian or stellate trichomes; stems ribbed, single to multiple from base, branched or not; lower leaves petiolate, entire to dentate, rarely pinnatifid; upper short-petiolate or sessile; petals yellow to orange, less frequently white, magenta, or purple, orbicular or obovate to oblong, long-clawed; siliques linear, rarely oblong, smooth or torulose, terete or 4-angled, rarely a little compressed, valves sometimes ± keeled.

1. Petals > 12.5 mm, showy; style mostly > 1.5 mm; plants biennial or perennial…(2)
1'. Petals < 12 mm, not especially showy; style mostly < 1.5 mm; plants annual, biennial, or perennial…(3)
2. Valves densely pubescent between midvein and replum with 2-rayed trichomes; fruits widely spreading or divaricate, usually straight, rarely curved upward…**E. asperum**
2'. Valves pubescent with 2–6-rayed trichomes; fruits divaricate or ascending to erect, straight or curved upward…**E. capitatum**
3. Siliques widely spreading and strongly torulose at maturity; pedicels thick, ± the same diameter as the siliques…**E. repandum**
3'. Siliques ascending to erect, not noticeably torulose at maturity; pedicels slender, narrower than the siliques…(4)
4. Plants annual; petals < 5(–6) mm; siliques ± quadrangular at maturity, the "facets" flat; flower buds not especially notched at apex; seeds twisted…**E. cheiranthoides**
4'. Plants biennial or short-lived perennial; petals (4–)5–7 mm; siliques subterete (but may be 4-ribbed at maturity, and then the "facets" rounded); flower buds noticeably notched at apex; seeds not twisted… **E. inconspicuum**

Erysimum asperum (Nutt.) DC.—Western wallflower. Biennial; trichomes of leaves 2- or 3-rayed; stems to 8 dm; basal leaves often withered by fruiting, oblanceolate, 2–10 cm, base attenuate, margins dentate, apex acute; cauline leaves entire or denticulate; pedicels horizontal to divaricate; petals yellow, obovate to suborbicular, 13–22 × 4–9 mm, claw 8–15 mm; siliques widely spreading or divaricate, narrowly linear, usually straight, rarely curved upward, not torulose, 4–13 cm × 1.2–2.7 mm, 4-angled, strongly longitudinally 4-striped, densely pubescent outside, trichomes 2-rayed between midvein and replum, glabrous inside. Prairies, sand dunes, roadsides, bluffs, sandhills along stream banks, knolls, open plains, 6500–9500(–10,800) ft. NC, NE, EC.

Erysimum capitatum (Douglas ex Hook.) Greene—Western wallflower, sand dune wallflower. Biennial or short-lived perennial (sometimes long-lived at high elevations) from a thick, leaf-covered caudex, variously pubescent with 2-branched trichomes with short stalks and/or with 3-(7)-branched trichomes mixed in; stems to 8(–10) dm; basal leaves linear-lanceolate to spatulate, entire to dentate, gradually narrowing to the base; pedicels divaricately ascending to ascending; petals 13–28 mm, the claw exceeding the blade, showy, yellow, orange-yellow, cream-yellow, or lavender (high elevations); siliques quadrangular to slightly flattened in cross-section, 3–15 × 1–3 mm.

1. Upper leaf surfaces with mostly 3-rayed trichomes; some plants to 7+ dm; seeds > 1.2 mm wide, often distally winged…**var. capitatum**
1'. Upper leaf surfaces with mostly 2-rayed trichomes; some plants to 5.5 dm; seeds < 1.2 mm wide, essentially wingless…**var. purshii**

var. capitatum [*E. capitatum* var. *argillosum* (Greene) R. J. Davis; *E. wheeleri* Rothr.]—Sepals green; petals yellow or yellow-orange, sometimes fading to orange; seeds typically > 1.2 mm wide. Hillsides, meadows, valley bottoms, alpine areas, deserts, woodlands, sandy mesas, 3900–12,800 ft. Widespread.

var. purshii (Durand) Rollins—Sepals green but sometimes fading purple at upper elevations; petals typically yellow or yellow-orange, occasionally copper or burnt-orange especially when fading, or deep blue-purple at higher elevations; seeds typically < 1.2 mm wide. Meadows, dry slopes, hillsides, 4300–12,600 ft. Individuals generally higher than 10,000 ft. that are long-lived from a stout caudex and that have comparatively massive petals might be better recognized as **var. nivale** (Greene) N. H. Holmgren, if not at the species level.

Erysimum cheiranthoides L.—Wormseed wallflower. Annual (or biennial), pubescent with mostly 2-branched trichomes, stems to 10 dm; leaves linear to narrowly lanceolate to elliptic-oblong, 3–11 × to 2.2 cm, entire to denticulate; pedicels divaricately ascending to erect, 1.5–3 mm; petals pale yellow, 3.5–5 mm; siliques 1.5–3 × 1–1.3 mm, terete-quadrangular, pubescent with 3- or 4-branched trichomes. Piñon-juniper communities, clay hills, wash bottoms with tamarisk, < 7000–8700 ft. Rio Arriba. Introduced.

Erysimum inconspicuum (S. Watson) MacMill. [*E. asperum* (Nutt.) DC. var. *inconspicuum* S. Watson]—Shy wallflower, lesser wallflower. Biennial or short-lived perennial, pubescent with mostly 2-branched trichomes except leaf surfaces sometimes with 3-branched trichomes; stems from an often swollen caudex, to 6 dm; leaves linear to linear-oblanceolate, entire to sparsely dentate, 3–8 cm × 2–6 mm; pedicels divaricately ascending to ascending; petals 5–12 mm, pale yellow; siliques 3–6 × 1–1.7 mm, subterete, pubescent. Our plants are **var. inconspicuum**. Roadsides and ponderosa pine, piñon-juniper, Gambel oak, and mountain shrub communities, 5300–9000(–11,800) ft. NC.

Erysimum repandum L.—Spreading wallflower, treacle wallflower. Annual, pubescent with 2- or 3-branched trichomes; stems to 4.5 dm; leaves narrowly oblanceolate to linear, sinuate-dentate to almost entire, 1–8 cm; pedicels horizontally spreading to slightly spreading; petals 6–8 mm, pale yellow; siliques 4–8 cm × 1.1–1.5 mm, on pedicels ± the same diameter as the fruit, torulose. Disturbed areas, tamarisk, saltbush, desert shrub, roadsides, saline areas, (3450–)4800–8000 ft. NE, Chaves, Grant, Lincoln, Torrance. Introduced.

HALIMOLOBOS Tausch – Fissurewort

Halimolobos diffusa (A. Gray) O. E. Schulz—Spreading fissurewort. Perennial; stems erect to ascending, often paniculately branched distally, to 12 dm, trichomes ± sessile; basal leaves absent on older plants, cauline leaves oblanceolate or lanceolate to oblong or elliptic, 2–7 cm × 5–20+ mm, base cuneate, margins sinuately lobed or dentate, surfaces with minutely stalked to subsessile trichomes; pedicels usually divaricate, rarely slightly descending, 2–7.5 mm; petals spatulate, 1.8–2.5 mm, claw distinctly differentiated from blade; siliques straight, subtorulose, linear, terete, 0.6–1.7 cm × 0.5–0.8 mm. Shaded talus, ravines, granite outcrops, rock crevices, bluffs, steep canyons, limestone slopes, oak-juniper communities, igneous slopes, 4200–8000 ft. SE, WC, Eddy, Lincoln, San Miguel.

HESPERIDANTHUS (B. L. Rob.) Rydb. – Plains-mustard

Hesperidanthus linearifolius (A. Gray) Rydb. [*Schoenocrambe linearifolia* (A. Gray) Rollins; *Streptanthus linearifolius* A. Gray; *Thelypodium linearifolium* (A. Gray) S. Watson; *Sisymbrium linearifolium* (A. Gray) Payson; *Thelypodiopsis linearifolia* (A. Gray) Al-Shehbaz]—Slimleaf plains-mustard. Perennial, glabrous and glaucous; stems 1 or a few, erect, branched upward, to 15 dm; first-year rosette leaves early-deciduous; cauline oblanceolate and becoming linear or linear-lanceolate upward, thick, short-petiolate to cuneate, entire to weakly dentate; pedicels divaricately ascending to spreading; petals showy, lavender to purplish with darker veins, 9–20 mm, claw linear, blade obovate to suborbicular and spreading at right angle to the claw, apically finely crenulate; siliques straight, 3.5–9 mm, terete. Ponderosa pine variously with juniper, piñon pine, Gambel oak, and cliffrose, 3800–10,950 ft. Widespread.

HESPERIS L. – Dame's rocket

Hesperis matronalis L.—Dame's rocket, mother-of-the-evening. Biennial or perennial, hirsute with simple and forked trichomes, sometimes with shorter forked ones mixed in; stems 1 (to several), erect, simple or sparingly branched; leaves ± all lanceolate to ovate-lanceolate, ± serrate to dentate, cuneate basally, 7–12 cm, larger below, smaller above, petiolate, the lower on long petioles; flowers unusually fragrant, especially at night; petals lilac to purple (rarely white), blade broadly ovate, 17–23 mm including the long claw; siliques terete, torulose, glabrous, longitudinally ribbed, 4–12 cm, narrow. Horticultural escape, especially near old mining towns, 7750–8100 ft. Colfax, Sandoval, San Miguel. Introduced.

HORNUNGIA Rchb. - Hornungia

Hornungia procumbens (L.) Hayek—Ovalpurse. Delicate annual, glabrous or sparsely pubescent with simple or branched trichomes; stems freely branched, to 15 cm; leaves 1–3 cm, obovate to lyrate-pinnatifid below, reduced in size upward and becoming obovate to narrowly lanceolate, cuneate at base; pedicels divaricate; petals white, spatulate, about equaling the sepals; siliques elliptic to slightly obovate, 2.3–4.3 × 1.4–2.4 mm, valves reticulate-veined, standing erect on pedicels. Seeps, moist areas in and below hanging gardens, shaded areas on benches above wash bottoms, < 5600 ft. San Juan.

IBERIS L. - Candytuft

Iberis umbellata L.—Globe candytuft. Annual; glabrous, stems erect, branched distally, 1–6 dm; basal leaves absent, cauline leaves petiolate or distally sessile, linear-oblanceolate, 2–6 cm, margins entire or subentire; pedicels divaricate to ascending, 4–11 mm; petals white or pink to purple, abaxial pair 10–16 × 5–7 mm, adaxial pair 6–10 × 2–5 mm; silicles ovate, 7–10.5 × 5–8 mm, apically notched, the valves extending into subacuminate wing; style 2–4.5 mm, included or exserted beyond apical notch. Garden escape, stream banks, abandoned gardens and lawns, urban areas, docks. San Miguel. Introduced.

ISATIS L. - Woad

Isatis tinctoria L.—Dyer's woad. Biennial, glaucous, usually glabrous, sometimes pubescent proximally; stems to 15 dm; basal leaves oblong or oblanceolate, to 20 cm, margins entire, repand, or dentate, apex obtuse; cauline leaves oblong or lanceolate, rarely linear-oblong, base sagittate or auriculate; pedicels 5–10 mm; petals 2.5–4 × 0.9–1.5 mm, base attenuate; siliques black or dark brown, often broader distal to middle, 0.9–2.7 cm × 3–8 mm, sometimes slightly constricted, apex subacute or rounded. Roadside noxious weed; a single record, 6300 ft. Sandoval. Introduced.

LEPIDIUM L. - Peppergrass, pepperwort, peppercress

Herbs or subshrubs, annual, biennial, or perennial; trichomes absent or simple; basal leaves rosulate or not, simple, entire or variously toothed or divided and 1–3-pinnatisect; cauline leaves petiolate or sessile, bases various, margins entire, dentate, or dissected; petals white, yellow, pink, or purple, sometimes rudimentary or absent, entire or emarginate; stamens 2 or 4 and equal, or 6 and tetradynamous; silicles strongly flattened parallel to the septum or inflated and terete, sessile, unsegmented, glabrous or pubescent, thin or strongly thickened and ornamented; style absent or evident, included or exserted from apical notch of fruit.

1. Cauline leaves (at least some) sessile, blade bases auriculate, sagittate, or amplexicaul…(2)
1'. Cauline leaves usually petiolate, or if sessile, bases not auriculate, sagittate, or amplexicaul…(7)
2. Plants glabrous or sparsely pubescent proximally; uppermost cauline leaf blade bases cordate-amplexicaul; basal leaf blades with margins 2- or 3-pinnatifid or pinnatisect; petals yellow…**L. perfoliatum**
2'. Plants usually puberulent or hirsute proximally, rarely glabrate; uppermost cauline leaf blade bases auriculate or sagittate; basal leaf blades with margins usually entire, subentire, dentate, denticulate, lyrate, or sinuate, rarely 1- or 2-pinnatifid; petals (when present) white…(3)
3. Stamens 2; petals absent or rudimentary; basal leaf blades with margins 1- or 2-pinnatifid; styles usually obsolete (rarely to 0.1 mm)…**L. oblongum** (in part)
3'. Stamens 6; petals present; basal leaf blades with margins usually entire, subentire, dentate, denticulate, lyrate, or sinuate, rarely pinnatifid; styles evident, ≥ 0.2 mm…(4)
4. Plants not rhizomatous; fruits dehiscent, apically broadly winged, notch present; racemes much elongated in fruit…**L. campestre**
4'. Plants rhizomatous; fruits indehiscent, apically not winged, notch absent; racemes not much elongated (corymbose panicles) in fruit…(5)
5. Fruits flattened, cordate to subreniform, valves reticulate-veined…**L. draba**
5'. Fruits inflated, globose, subglobose, obovoid, or obcompressed-globose, valves not reticulate-veined…(6)
6. Fruits usually globose, rarely subglobose, 2.5–4.4(–5) mm wide, valves puberulent; styles 0.5–1.5 mm…**L. appelianum**

6'. Fruits obovoid to subglobose, or obcompressed-globose, (3.5-)4–7 mm wide, valves glabrous; styles (0.8-)1.2–2 mm…**L. chalepense**

7. Valves rugose or rugose-verrucose…**L. didymum**

7'. Valves usually smooth…(8)

8. Subshrubs or perennials, with a woody caudex, sometimes with persistent petiolar remains…(9)

8'. Annuals or biennials, lacking a woody caudex, without persistent petiolar remains…(12)

9. Basal and proximal-most cauline leaf blade margins usually entire, crenate, or dentate, rarely palmately 3–5-lobed at apex…(10)

9'. Basal and proximal-most cauline leaf blade margins pinnately lobed…(11)

10. Plants rhizomatous; fruits (1.6-)1.8–2.4(–2.7) mm, apically not winged, notch usually absent; styles 0.05–0.15 mm…**L. latifolium**

10'. Plants not rhizomatous; fruits (2.5-)3–4(–4.4) mm, apically winged, notch present; styles 0.2–1 mm… **L. crenatum**

11. Fruits usually ovate to suborbicular, rarely oblong; basal leaf blades with margins 1- or 2-pinnatifid to -pinnatisect; cauline leaf blade margins often pinnately lobed…**L. montanum** (in part)

11'. Fruits broadly ovate; basal leaf blades with margins pinnately lobed; cauline leaf blade margins usually entire, rarely dentate…(12)

12. Perennials or subshrubs; stems usually 1–5.5 dm; middle cauline leaf blades 0.7–2.5 mm wide…**L. alyssoides**

12'. Perennials; stems usually 4.5–17 dm; middle cauline leaf blades 3–10 mm wide…**L. eastwoodiae** (in part)

13. Stamens 2…(14)

13'. Stamens 4 or 6…(21)

14. Fruits 1.7–2.1 × 1.2–1.6 mm; plants puberulent with clavate trichomes…**L. sordidum**

14'. Fruits 1.8–7 × 1.5–5.5 mm; plants puberulent, hirsute, or hispid with cylindrical trichomes…(15)

15. Fruiting pedicels often strongly flattened, 0.2–0.7 mm wide; fruit valves hirsute to hispid, sometimes only on margins…**L. lasiocarpum**

15'. Fruiting pedicels terete or only slightly flattened, 0.1–0.3(–4) mm wide; fruit valves glabrous or puberulent at least on margins…(16)

16. Fruits elliptic…(17)

16'. Fruits obovate, suborbicular, or orbicular…(18)

17. Basal leaf blades pinnatifid; racemes slightly elongated in fruit, rachises with curved trichomes; fruiting pedicels usually puberulent adaxially, rarely throughout…**L. ramosissimum**

17'. Basal leaf blades (1)2- or 3-pinnatisect; racemes considerably elongated in fruit, rachises with straight trichomes; fruiting pedicels puberulent throughout…**L. ruderale**

18. Plants hirsute; basal leaf blade margins pinnatifid…(19)

18'. Plants puberulent or glabrous; basal leaf blade margins dentate, serrate, lyrate, or pinnatifid…(20)

19. Stems often simple from base; rachises pubescent, trichomes curved, with fewer longer, straight ones… **L. austrinum**

19'. Stems often several from base; rachises hirsute, trichomes mostly straight…**L. oblongum** (in part)

20. Fruits obovate to obovate-suborbicular, widest beyond middle; rachises with straight, slender to sub-clavate trichomes; petals absent or rudimentary, 0.3–0.9 mm…**L. densiflorum**

20'. Fruits orbicular, widest at middle; rachises usually with curved, cylindrical trichomes, rarely glabrous; petals usually present, rarely rudimentary, 1–2.5 mm…**L. virginicum**

21. Annuals; rachises pilose, trichomes straight…**L. thurberi**

21'. Annuals or biennials; rachises puberulent, trichomes straight or curved…(22)

22. Petals suborbicular, 1.5–2.5 mm wide; cauline leaf blades lanceolate or oblanceolate to linear; stems 4–17 dm, often simple from base…**L. eastwoodiae** (in part)

22'. Petals spatulate to oblanceolate, 1.3–1.8 mm wide; cauline leaf blades often pinnatifid to pinnatisect, sometimes linear; stems 0.4–5(–7) dm, simple or few to several from base…**L. montanum** (in part)

Lepidium alyssoides A. Gray [*L. alyssoides* var. *angustifolium* (C. L. Hitchc.) Rollins]—Mesa pepperwort. Subshrubs or herbs, perennial, woody base often elevated, glabrous or minutely puberulent; basal leaves often not rosulate, blades 1.5–10 cm, pinnately lobed; cauline leaves sessile, linear, 1–9 cm, not auriculate; sepals ovate to oblong, 1–2 mm; petals white, suborbicular, 2–3 mm; stamens 6; silicles ovate, 2–4 × 1.8–3 mm, apically winged, apical notch 0.1–0.3 mm deep, glabrous; style 0.2–0.6 mm, exserted beyond apical notch; seeds brown, ovate, 1.5–1.8 mm. Piñon-juniper, sagebrush, grasslands, sandstone outcrops, gypsum flats, sand dunes, dry flats and river bottoms, roadsides, 3050–9500 ft. Widespread.

Lepidium appelianum Al-Shehbaz [*Cardaria pubescens* (C. A. Mey.) Jarm.]—Hairy whitetop, globe-podded hoarycress. Herbs, perennial, rhizomatous, densely hirsute; basal leaves basal, not rosulate, often withered by anthesis, blades obovate to oblanceolate, 2-7 cm; cauline leaves sessile, blades oblong or lanceolate, 1-7 cm, base sagittate; sepals oblong, 1.4-2 mm; petals white, broadly obovate, 2.2-4 mm; stamens 6; silicles globose to subglobose, 2-5 mm diam., inflated, apically wingless, apical notch absent, densely puberulent; style 0.5-1.5 mm. Roadsides, sagebrush communities, alkaline meadows, waste ground, ditches, streamsides, fields, pastures, 4900-8000 ft. Bernalillo, Lincoln, Rio Arriba, San Juan. Introduced.

Lepidium austrinum Small—Southern pepperwort. Herbs, annual or biennial; basal leaves densely hirsute, rosulate, later withered, blades pinnatifid, 2-8.3 cm; cauline leaves shortly petiolate, blades oblanceolate to nearly linear, 1-6 cm, base attenuate to cuneate, not auriculate; sepals oblong, 0.8-1 mm; petals (sometimes absent) white, oblanceolate, 0.4-1.6; stamens 2; silicles elliptic-obovate to obovate-orbicular, 2.4-3.2 × 1.8-2.5 mm, apically winged, apical notch 0.2-0.5 mm deep, sparsely puberulent, trichomes sometimes restricted to margin; style 0.05-0.1 mm, included in apical notch. Disturbed ground, railroad tracks and embankments, fields, knolls, stream banks, waste areas, open banks, roadsides, sandy terraces. The presence of this species in the state needs verification.

Lepidium campestre (L.) W. T. Aiton [*Thlaspi campestre* L.; *Neolepia campestris* (L.) W. A. Weber]—Fieldcress, field pepperwort. Herbs, annual, densely hirsute; basal leaves rosulate, blades oblanceolate or oblong, 1.5-7 cm; cauline leaves oblong, lanceolate, or narrowly deltoid-lanceolate, 1-6 cm, base sagittate or auriculate; sepals oblong, 1-1.8 mm; petals white, spatulate, 1.5-2.5 mm; stamens 6; silicles broadly oblong to ovate, 4.5-6.5 × 3.5-5 mm, curved adaxially, apically broadly winged, apical notch (0.2-)0.4-0.6 mm deep, papillate except for wing; style 0.2-0.6 mm, slightly exserted to included in apical notch. Roadsides, grasslands, pine woodlands, rocky slopes, forests, disturbed areas, 5400-7800 ft. Rio Arriba, San Juan, Taos. Introduced.

Lepidium chalepense L. [*Cardaria draba* (L.) Desv. subsp. *chalapensis* (L.) O. E. Schulz]—Whitetop. Herbs, perennial, rhizomatous, densely hirsute to subglabrous; basal leaves not rosulate, withered early, blades obovate, spatulate, or ovate, 2-12 × 1-3.7 cm; cauline leaves sessile, blades obovate to oblong or lanceolate to oblanceolate, 2-11 × 1.2-4 cm, pubescent or glabrous, base sagittate-amplexicaul or auriculate; sepals oblong to ovate, 1.7-3 mm; petals white, obovate, 3-5 mm; stamens 6; silicles obovoid to subglobose or obcompressed-globose, 3.5-6 × 4-6.5 mm, apex wingless, apical notch absent, glabrous; style 1-2 mm. Mountain slopes, roadsides, fields, agricultural lands, riverbanks, pastures, waste areas, 4750-8200 ft. Colfax, Rio Arriba, Sandoval, San Juan, Taos, Valencia. Introduced.

Lepidium crenatum (Greene) Rydb.—Alkaline pepperwort. Subshrubs, perennial with woody portion elevated, puberulent; basal leaves rosulate on sterile shoots, blades oblanceolate to spatulate, 3-8 cm; cauline leaves short-petiolate to sessile, blades oblong to oblanceolate, 1-3.5 cm, base cuneate; sepals oblong to ovate, 1.3-1.8 mm; petals white, suborbicular to broadly obovate, 2-3 mm; stamens 6; silicles broadly ovate, 3-4 × 2-2.8 mm, apically winged, apical notch 0.1-0.2 mm deep; style 0.2-0.6 mm, exserted beyond apical notch. Piñon-juniper and brush communities, clay bluffs of sandstone mesas, arroyo banks, 3500-6700 ft. San Juan.

Lepidium densiflorum Schrad.—Prairie pepperwort, common peppergrass. Herbs, annual or biennial, puberulent or glabrous; basal leaves rosulate, withered early, blades oblanceolate, spatulate, or oblong, 2-10 cm, cauline leaves shortly petiolate, blades narrowly oblanceolate or linear, 1-7 cm, base attenuate to cuneate; sepals oblong, 0.5-1 mm wide, deciduous; petals absent or rudimentary and white, filiform, 0.3-0.9 mm; stamens 2; silicles obovate to obovate-suborbicular, 2-3.5 × 1.5-3 mm, widest above middle, apically winged, apical notch 0.2-0.4 mm deep, glabrous or sometimes sparsely puberulent at least on margin; style 0.1-0.2 mm, included in apical notch. Disturbed sites, grasslands, chaparral, meadows, sagebrush flats, floodplains, gravelly hillsides, roadsides, 3650-9850 ft. No varieties recognized here. Widespread. Introduced.

Lepidium didymum L. [*Coronopus didymus* (L.) Sm.]—Lesser swinecress. Annual, fetid, glabrous or pilose; basal leaves soon withering, not rosulate, blades 1- or 2-pinnatisect, 1-7 cm, cauline leaves short-petiolate to subsessile, blades similar to basal, smaller and less divided distally, 1.5-4 cm, base not auriculate; sepals ovate, 0.5–0.8 mm; petals white, elliptic to linear, 0.4–0.5; stamens 2; silicles schizocarpic, didymous, 1.3–1.7 × 2–2.5 mm, apically not winged, apical notch 0.2–0.4 mm deep, rugose, strongly veined, glabrous; style absent or obsolete, included in apical notch. Disturbed areas; reported from Doña Ana County. Introduced.

Lepidium draba L. [*Cardaria draba* (L.) Desv.]—Hoarycress, whitetop. Herbs, perennial, rhizomatous, hirsute or subglabrous; basal leaves not rosulate, withered early, blades obovate, spatulate, or ovate, 2–12 cm, cauline leaves sessile, blades ovate, elliptic, oblong, lanceolate, oblanceolate, or obovate, 2–12 cm, pubescent or glabrous, base sagittate-amplexicaul or auriculate; sepals oblong to ovate, 1.5–2 mm; petals white, obovate, 3–4.5 mm; stamens 6; silicles cordate to subreniform, 2–4 × 3.7–5.5 mm, apex obtuse to subacute, wingless, apical notch absent, glabrous, reticulate-veined; style 1–2 mm. Mountain slopes, roadsides, fields, agricultural lands, riversides, disturbed ground, pastures, waste areas, 3850–7900 ft. Widespread except SE. Introduced.

Lepidium eastwoodiae Wooton [*L. alyssoides* A. Gray var. *eastwoodiae* (Wooton) Rollins; *L. moabense* S. L. Welsh]—Eastwood's pepperwort. Herbs, annual, biennial, or perennial with woody base, glabrous or pubescent; basal leaves not rosulate, soon deciduous, blades 3–8 cm, pinnatifid; cauline leaves short-petiolate or sessile, blades narrowly lanceolate or oblanceolate to linear, 3–7 cm, base attenuate to cuneate; sepals suborbicular to oblong, 0.8–1.5 mm; petals white, suborbicular, 2.2–3.5 mm; stamens 6; silicles broadly ovate, 2–3.5 × 1.8–2.6 mm, apically winged, apical notch 0.1–0.2 mm deep, not veined; style 0.3–0.6 mm, exserted beyond apical notch. Piñon-juniper, sagebrush, mixed desert shrub communities, 4800–9000 ft. NCS.

Lepidium lasiocarpum Nutt. ex Torr. & A. Gray—Hairy-pod pepperwort. Herbs, annual, hirsute or hispid; basal leaves not rosulate, later withered, blades spatulate to oblanceolate, 1.5–6 cm, lyrate-pinnatifid, pinnatisect, or 2-pinnatifid, very rarely dentate; cauline leaves subsessile or petiolate, blades lanceolate to oblanceolate, 1–4 cm, base cuneate; sepals oblong, 1–1.3 mm; petals absent or present and white, oblanceolate to linear, 0.5–1.5 mm; stamens 2; silicles dehiscent, ovate to ovate-orbicular, 2.8–4.5 × 2.4–3.6 mm, base broadly cuneate to rounded, apically winged, apical notch 0.3–0.6 mm deep, hirsute to hispid on surface or along margin, not veined; style obsolete or to 0.1 mm, included in apical notch. Piñon-juniper, sagebrush, shrub communities, open deserts, dry washes and flats, waste places, stream beds, roadsides, sandy areas, rockslides, stony slopes, 3200–6600 ft. Widespread with no geographical pattern to the subspecies.

1. Fruit valves hirsute or fringed on margin, trichomes not pustular-based; fruiting pedicels 0.2–0.4(–0.6) mm wide, usually < 3 times as wide as thick; nectar glands toothlike, to 0.2 mm…**subsp. lasiocarpum**
1'. Fruit valves hispid, trichomes pustular-based; fruiting pedicels (0.4–)0.5–0.7 mm wide, often > 3 times as wide as thick; nectar glands subulate, not toothlike, 0.3–0.5 mm…**subsp. wrightii**

subsp. lasiocarpum—Piñon-juniper woodlands, sagebrush and other shrub communities, open deserts, dry washes and flats, waste places, streambeds, roadsides, sandy areas, rockslides, stony slopes.

subsp. wrightii (A. Gray) Thell.—Roadsides, rocky draws, dry washes, loose sand, stony or gravelly areas, clay flats.

Lepidium latifolium L. [*Cardaria latifolia* (L.) Spach]—Broadleaf pepperwort, tall whitetop. Herbs, perennial, with thick rhizomes, glabrous or pubescent; basal leaves not rosulate, blades leathery, elliptic-ovate to oblong, 3–20 cm; cauline leaves sessile or short-petiolate, blades oblong to elliptic-ovate or lanceolate, 2–11 cm, base cuneate; sepals suborbicular to ovate, 1–1.4 mm; petals white, obovate, 1.8–2.5 mm; stamens 6; silicles oblong-elliptic to broadly ovate or suborbicular, 1.8–2.5 × 1.3–1.8 mm, apically wingless, apical notch absent or rarely to 0.1 mm, glabrous or sparsely pilose, not veined; style 0.05–0.15 mm, if present exserted beyond apical notch. Disturbed places, 3750–8000 ft. Most of the state except SE. Introduced.

Lepidium montanum Nutt. [*L. albiflorum* A. Nelson & P. B. Kenn.; *L. alyssoides* A. Gray var. *jonesii* (Rydb.) Thell.; *L. alyssoides* A. Gray var. *stenocarpum* Thell.; *L. brachybotryum* Rydb.; *L. corymbosum*

Hook. & Arn.; *L. crandallii* Rydb.; *L. integrifolium* Nutt. var. *heterophyllum* S. Watson; *L. jonesii* Rydb.]—Mountain pepperwort. Herbs or subshrubs, woody at the base, annual, biennial, or perennial, cespitose or not, glabrous or pubescent; basal leaves rosulate or not, blades 1- or 2-pinnatifid to -pinnatisect, rarely undivided, 1.5–6 cm; cauline leaves shortly petiolate, blades similar to basal leaves or undivided and linear, base cuneate to attenuate; sepals oblong to broadly ovate, 1.2–2 mm; petals white, spatulate to oblanceolate, 2.2–4 mm; stamens 6; silicles ovate to suborbicular or rarely oblong, 2–5 × 1.8–4 mm, apically winged, apical notch 0.1–0.3 mm deep, glabrous or rarely puberulent, not veined; style 0.2–0.9 mm, exserted beyond or rarely subequaling apical notch. Piñon-juniper, sagebrush, shrub communities, rocky hillsides and crevices, bajadas, washes, gypsum areas, sandstone cliffs, limestone gravel, playas, knolls, clay hills, sandy areas, alkaline flats and lowlands, roadsides, 2950–9200 ft. Widespread except most of NE. The synonymy of this species and its purported number of subspecific taxa are enormous (Al-Shehbaz, 2010), and only a few of the most commonly encountered are given above. No subspecific taxa are recognized here.

Lepidium oblongum Small—Veiny pepperweed. Annual; basal leaves not rosulate, blades 1- or 2-pinnatifid, 0.7–3.5 cm; cauline leaves usually sessile, rarely shortly petiolate, blades obovate to oblanceolate, 0.8–2 cm, base cuneate, auriculate or not, margins dentate to laciniate or pinnatifid; sepals ovate to broadly oblong, 0.7–1 mm; petals (absent or rudimentary), white, linear-oblanceolate, 0.1–0.7 mm; stamens 2; siliques orbicular to broadly obovate or elliptic, 2.2–3.5 × 2–3 mm, apically winged, apical notch 0.2–0.3 mm deep, glabrous or sparsely puberulent along margin; style to 0.1 mm, included in apical notch. Prairies, pastures, floodplains, waste ground, llanos, disturbed areas, roadsides, flats, calcareous sand, alluvial terraces, 3550–5900 ft. SE, SW.

Lepidium perfoliatum L.—Clasping pepperwort. Herbs, annual or occasionally biennial, glaucous, glabrous or sparsely pubescent below; basal leaves rosulate, 2- or 3-pinnatifid or -pinnatisect, 3–8(–15) cm, lobes linear to oblong, entire; cauline leaves sessile, ovate to cordate or suborbicular, 1–4 cm, base deeply cordate-amplexicaul, margins entire; sepals oblong, 0.8–1.3 mm; petals pale yellow, narrowly spatulate, 1–2 mm; stamens 6; silicles orbicular to rhombic or broadly obovate, 3–5 × 3–4.1 mm, apically winged, apical notch 0.1–0.3 mm deep, glabrous, not veined; style 0.1–0.4 mm, subequaling or slightly exserted beyond apical notch. Waste places, piñon-juniper, sagebrush flats, open deserts, grasslands, alkaline flats, fields, 5000–8000 ft. NW, Taos. Introduced.

Lepidium ramosissimum A. Nelson—Branched pepperwort. Herbs, biennial, puberulent; basal leaves not rosulate, soon withering, blades oblanceolate, pinnatifid, 2–5 cm; cauline leaves shortly petiolate to sessile, blades oblanceolate, 1.2–5.5 cm, base attenuate to cuneate, uppermost leaves linear, entire; sepals oblong, 0.6–1 mm; petals absent or rudimentary and white, linear, 0.2–1 mm; stamens 2; silicles elliptic, 2.2–3.2 × 1.7–2.1 mm, apically winged, apical notch 0.1–0.3 mm deep, glabrous or puberulent at least along margin, not veined; style obsolete or rarely to 0.1 mm, included in apical notch. Sagebrush, pine woodlands, waste ground, roadsides, alkaline flats, abandoned fields, 3600–10,000 ft. Scattered but mainly in N.

Lepidium ruderale L.—Annual or biennial, fetid, puberulent; basal leaves rosulate, blades (1)2- or 3-pinnatisect (lobes oblong), 2–7 cm, usually entire, rarely dentate; cauline leaves sessile, blades linear, 1–2 cm, base cuneate, entire; sepals oblong, 0.5–1 mm; petals absent or rudimentary, white, linear, 0.2–0.5; stamens 2; silicles elliptic, 1.5–3 × 1.5–2.3 mm, apically winged, apical notch 0.1–0.2 mm deep, glabrous; style obsolete or to 0.1 mm, included in apical notch. Fields, pastures, waste places, roadsides, gardens, 4300–4600 ft. Grant, Guadalupe, Hidalgo.

Lepidium sordidum A. Gray—Sordid pepperweed. Annual, puberulent; basal leaves soon withered, not rosulate, blades 1- or 2-pinnatifid, 3–5.6 cm; cauline leaves petiolate, blades pinnatifid, 0.7–2 cm, base attenuate to cuneate; sepals oblong, 0.5–0.7 mm; petals absent or rudimentary, white, linear, 0.2–0.4; stamens 2; silicles ovate-elliptic, 1.7–2.1 × 1.2–1.6 mm, apically winged, apical notch 0.1–0.2 mm deep; valves thin, smooth, not or weakly veined, glabrous; style 0.1–0.15 mm, included in or equaling apical notch. Alluvial fans, sandy flats, rocky hillsides, grassy valleys, canyons, 4000–4300 ft. SE.

Lepidium thurberi Wooton—Thurber's pepperweed. Annual, pubescent, simple trichomes to 1 mm along with shorter, clavate ones; basal leaves often withered at anthesis, rosulate, blades pinnatifid, 2-9 cm; cauline leaves shortly petiolate, 1.5-6 cm, base not auriculate; sepals suborbicular to broadly ovate, 1-1.6 mm; petals white, broadly obovate to suborbicular, 3-4 mm; stamens 6; silicles broadly ovate to orbicular, 2-2.9 × 2-2.8 mm, apically winged, apical notch 0.1-0.2 mm deep, not veined, glabrous; style 0.3-0.8 mm, exserted beyond apical notch. Salt flats, mesquite and creosote bush communities, playas, stream banks, sandy deserts, washes, clay bottoms, bluffs, gravelly granitic sand, grasslands, alluvial fans, roadsides, silty terraces, washes, gravelly flats, 3800-6300(-8400) ft. WC, C, SW, SC.

Lepidium virginicum L.—Virginia pepperweed. Herbs, annual, puberulent; basal leaves not rosulate, withered by anthesis, blades obovate or spatulate to oblanceolate, 2-13 cm, pinnatifid to lyrate or dentate; cauline leaves shortly petiolate, oblanceolate or linear, 1-6 cm, base attenuate to subcuneate; sepals oblong to ovate, 0.7-1 mm; petals white, spatulate to oblanceolate, 1-2.5 mm, rarely rudimentary; stamens 2; silicles ± orbicular, 2.5-3.5 mm diam., apically winged, apical notch 0.2-0.5 mm, glabrous, not veined; style 0.1-0.2 mm, included in apical notch, 4200-9500 ft.

1. Fruiting pedicels terete, 0.15-0.2 mm wide; silicles glabrous…**subsp. virginicum**
1'. Fruiting pedicels flattened at least below apex, (0.2-)0.3-0.4 mm wide; silicles glabrous or puberulent… **subsp. menziesii**

subsp. menziesii (DC.) Thell. [*L. virginicum* var. *medium* (Greene) C. L. Hitchc.; *L. virginicum* var. *pubescens* (Greene) Thell.; *L. medium* Greene]—Menzies' pepperweed. Fruiting pedicels terete, 0.15-0.2 mm wide, puberulent adaxially or glabrous; silicles glabrous or puberulent. Roadsides, bottomlands, gravelly and sandy shores, waste ground, riverbanks, grassy meadows, dry flats and creek beds, abandoned fields, woods, cliffs, plains, pastures, sagebrush and other desert shrub communities, dry mountain slopes. NW, NC, SW, SC.

subsp. virginicum—Fruiting pedicels terete, 0.15-0.2 mm wide, puberulent adaxially or glabrous; silicles glabrous. Fields, roadsides, waste places, disturbed sites, fields, grassy areas. W, NCS.

LOBULARIA Desv. - Lobularia

Lobularia maritima (L.) Desv.—Sweet alyssum. Plants suffruticose (when subshrubs); stems 0.5-3 dm; leaf blades linear or lanceolate-oblanceolate, 1.5-4 cm, base attenuate, apex acute; sepals often tinged purplish, 1.4-2 mm; petals broadly obovate, 2-3 × 1.5-2.5 mm, abruptly contracted into claw; silicles broadly ovate, obovate, or suborbicular, 2-4 × 1.5-2.5 mm, thin, sparsely pubescent. Cultivated ornamental escaped to roadsides, waste places, vacant lots, cultivated fields, walls; a coastal species, ephemeral. Doña Ana, Valencia. Introduced.

MATTHIOLA L. - Stock

Matthiola longipetala (Vent.) DC. [*M. bicornis* (Sibth. & Sm.) DC.]—Night scented stock. Annual, sparsely to moderately pubescent, glandular papillae present or not; cauline leaves petiolate, distally sessile, blades linear-lanceolate to lanceolate or oblanceolate, 3-10 cm, pinnatisect to sinuate or dentate, rarely entire or subentire, lateral leaflets sessile or petiolulate, rachis not winged, blade smaller than terminal; petals purple, pink, yellow, or brown, rarely white, oblong to linear-lanceolate, 16-25 mm, margin crisped; siliques ascending to divaricate or, rarely, descending, straight, terete, 3-9 cm × 1-2 mm, pubescent and often glandular; stigma with 2 straight or upcurved, sometimes reflexed horns. Garden escape, roadsides, disturbed areas, waste ground, fields, (3900-)4900-6050 ft. Doña Ana, Grant. Introduced.

MOSTACILLASTRUM O. E. Schulz - False tumblemustard

Herbs, annual (ours); trichomes absent (ours), herbage often glaucous; basal leaves petiolate, rosulate or not, simple, entire or dentate to pinnately lobed; cauline leaves petiolate or sessile, sometimes auriculate to sagittate, entire or dentate, rarely pinnately lobed to pinnatisect; racemes ebracteate, corymbose; petals white to purplish, apex obtuse; ovules 30-80(-140) per ovary; siliques linear, terete, unsegmented;

valves with a prominent midvein and often distinct marginal veins, smooth or rarely torulose; stigma capitate, entire or slightly 2-lobed, narrower in diameter than style; seeds uniseriate, wingless; cotyledons incumbent. Our species recently segregated from *Thelypodiopsis* (Al-Shehbaz, 2012a).

1. Sepals and petals yellow; siliques on stipes 2–8 mm…**Thelypodiopsis aurea**
1′. Sepals and petals white to pale lavender; siliques on stipes < 0.4 mm…(2)
2. Proximal-most cauline leaf blades with pinnatifid to sinuate-dentate margins, distal ones with subamplexicaul or auriculate bases…**M. purpusii**
2′. Proximal-most cauline leaf blades usually with entire or repand margins, rarely denticulate, distal ones with auriculate bases…**M. subauriculatum**

Mostacillastrum purpusii (Brandegee) Al-Shehbaz [*Thelypodiopsis purpusii* (Brandegee) Rollins]— Purpus' tumblemustard. Annual from a taproot, glabrous and leaves glaucous; leaves somewhat fleshy, basal leaves few and early-withering, these tapering to a winged petiole, pinnatifid to shallowly sinuate-dentate; cauline leaves sessile, auriculate, ovate to elliptic-lanceolate, entire; sepals green or purplish. 3.5–4.5 mm; petals oblanceolate, 4–4.5 mm, white; siliques ascending, terete, narrowly linear, sessile to subsessile on a short stipe < 0.3 mm, body 4–5.5 cm. Duff beneath trees and shrubs in piñon-juniper and mountain shrub communities, 4200–8500 ft. Much of the state except NE and the E tier of counties.

Mostacillastrum subauriculatum Al-Shehbaz [*Thelypodiopsis vaseyi* (S. Watson ex B. L. Rob.) Rollins]—Las Vegas tumblemustard. Annual; glabrous and often glaucous; cauline leaves sessile and auriculate, proximally oblong to lanceolate or oblanceolate, distally linear to narrowly oblong or lanceolate, usually entire, rarely denticulate; sepals whitish or purplish, 1.8–2.5 mm; petals obovate to spatulate, 2.5–4.5, white; silicles erect to ascending, straight or curved, strongly torulose, 1.5–2.5 cm × 1–1.2 mm. Open wooded slopes, mixed conifer forests, canyons, 6500–10,650 ft. NCS. Nearly **R**

NASTURTIUM W. T. Aiton, nom. cons. – Watercress

Perennial, rhizomatous, aquatic or semiaquatic herbs, glabrous or sparsely pubescent with simple trichomes; stems rooting at lower nodes, typically prostrate or decumbent, less frequently erect; leaves all cauline, petiolate (to sessile), the petioles often with small auricles; blades usually pinnate with 1–6(–12) pairs of lateral leaflets, base obtuse, cuneate, or subcordate, margins entire or repand, apex obtuse; pedicels divaricate or descending, straight or occasionally recurved; petals white, obovate or narrowly spatulate, not clawed; siliques linear, terete, smooth or somewhat torulose, obscurely veined.

1. Siliques (1.8–)2–3 mm wide, latiseptate; seeds biseriate, coarsely reticulate, with 25–50(–60) areolae on each side…**N. officinale**
1′. Siliques 1–1.5(–1.8) mm wide, ± terete; seeds uniseriate, moderately reticulate, with (75–)100–150(–175) areolae on each side…**N. microphyllum**

Nasturtium microphyllum Boenn. ex Rchb.—Onerow yellowcress. Leaf blades (3–)5–9(–11)-foliate, 2–10(–15) cm; terminal leaflet (or simple blade) suborbicular or oblong, 1–4 cm; sepals 2.5–4 × 1–1.5 mm; petals white or pink, spatulate or obovate, 4.5–6 × 1.5–3 mm, apex rounded; siliques 1.4–2.2(–2.7) cm × 1–1.5(–1.8) mm; style 0.5–1.5(–2) mm; seeds uniseriate, moderately reticulate with (75–)100–150(–175) areolae on each side. Lake margins, streams, ponds, springs, river shores, seeps, swales, wet meadows. Reported. Introduced.

Nasturtium officinale R. Br. [*Rorippa nasturtium-aquaticum* (L.) Hayek; *Sisymbrium nasturtium-aquaticum* L.]—Watercress. Leaf blades 3–9(–13)-foliate, 2–20 cm; terminal leaflet (or simple blade) suborbicular to ovate, or oblong to lanceolate, (0.4–)1–4(–5) cm; sepals 2–3.5 × 0.9–1.6 mm; petals white or pink, spatulate or obovate, 2.8–4.5(–6) × 1.5–2.5 mm, apex rounded; siliques (0.6–)1–1.8(–2.5) cm × (1.8–)2–2.5(–3) mm; style 0.5–1(–1.5) mm; seeds biseriate, coarsely reticulate, with 25–50(–60) areolae on each side. Flowing streams, ditches, lake margins, swamps, marshes, seeps, 4000–9500 ft. Widespread. Introduced.

NERISYRENIA Greene – Fanmustard

Perennial subshrubs with a woody caudex; usually pubescent, sometimes glabrous or glabrate, trichomes long-stalked to subsessile, dendritic; leaves cauline, simple, petiolate or sessile, often fleshy, margins entire, dentate, or repand; pedicels divaricate to ascending, or rarely recurved, slender; petals white to cream-white, often fading lavender, rarely lavender, obovate to spatulate or broadly elliptic; siliques or silicles sessile, linear to oblong or obovoid, straight or curved, angustiseptate or terete.

1. Leaves linear, all < 5 mm wide…**N. linearifolia**
1'. Leaves oblanceolate, spatulate, or obovate, the larger ones > 5 mm wide…(2)
2. Fruits crispate, < 15 mm; infructescences < 7 cm; petals < 5 mm wide, remaining white…**N. hypercorax**
2'. Fruits not crispate, > 15 mm; infructescences > 7 cm; petals > 5 mm wide, fading lavender…**N. camporum**

Nerisyrenia camporum (A. Gray) Greene—Bicolor fanmustard. Plants moderately to densely pubescent or glabrate; leaves attenuate at base, blades obovate to spatulate or oblanceolate, rarely elliptic, 1 4.5 cm × (4-)7-30 mm, not fleshy, margins dentate, repand, or entire; racemes to 3.5 dm; pedicels (7-)10-20 mm; sepals 5-9 × 1-2 mm; petals obovate, (8-)10-15 × 5-9 mm, claw often flattened, margin dentate, to 2 mm; siliques angustiseptate, 15-35 × 1.5-3(-4) mm. Clay flats, gypseous clay, gravelly knolls, hillsides, sandy washes, 2900-6700 ft. S half of state.

Nerisyrenia hypercorax P. J. Alexander & M. J. Moore—Crow Flats fanmustard. Moderately pubescent; leaves attenuate at base, blades oblanceolate to spatulate, 1.8-4.5 cm × (4-)6-14 mm, fleshy, margins entire to weakly sinuate, or rarely obscurely sinuate-dentate; racemes to 4.5(-6.5) dm; pedicels 3-9(-13) mm; sepals 3-6 × 1-1.5 mm; petals obovate to spatulate, 7-9 × 3.5-4.5 mm; siliques angustiseptate, 5-12(-16) × 2.2-3 mm. Sparsely vegetated exposures of gypseous clay of the Yeso Formation, 4250-3600 ft. Chavez, Otero. **R & E**

Nerisyrenia linearifolia (S. Watson) Greene—White Sands fanmustard. Moderately to densely pubescent; leaves attenuate at base, blades linear to linear-oblanceolate, 1.5-7 cm × 1-4.5 mm, fleshy, margins entire, rarely dentate; racemes to 3.5 dm; pedicels 6-14 mm; sepals 4.8-7.5 × 1-2 mm; petals obovate to spatulate, 8-13 × 5-8.5 mm; siliques terete to slightly angustiseptate, 0.9-3 cm × 1-2.2 mm. Gypsum soils in knolls, bluffs, open flats, 3000-6900 ft. S half of state except westernmost tier of counties.

NOCCAEA Moench – Pennycress

Noccaea fendleri (A. Gray) Holub [*Thlaspi montanum* L. misapplied]—Wild candytuft, Fendler's pennycress. Perennial, glabrous, often glaucous; leaves simple, basal rosulate, petiolate, 0.4-3 cm, entire, denticulate, or dentate; cauline leaves sessile, ovate or suboblong, 0.4-2.8 cm, auriculate to sub-amplexicaul at base, entire or dentate; petals white to pinkish purple, spatulate, 3.5-13 mm; silicles divaricately ascending to horizontal-spreading, obovate to obcordate, cuneate at base, winged (rarely wingless), 5-10 × 2.5-5 mm. Mountain brush, Gambel oak, ponderosa pine, spruce-fir, aspen, subalpine meadows, alpine tundra, 5000-12,800 ft. Widespread. Two weak subspecies in our area.

1. Petals pinkish purple or occasionally white, (6-)6.5-13 mm; styles (1.8-)2.5-4.2 mm; fruits 7-12(-16) mm; racemes often compact…**subsp. fendleri**
1'. Petals white or occasionally pinkish purple, 3.4-7(-8.5) mm; styles 0.4-2.2(-3) mm; fruits 2.5-8(-12) mm; racemes often lax…**subsp. glauca**

subsp. fendleri—Fendler's pennycress. Petals pinkish purple or occasionally white, (6-)6.5-13 mm; styles (1.8-)2.5-4.2 mm; fruits 7-12(-16) mm; racemes often compact. Mountain brush, ponderosa pine, mixed conifer, aspen, subalpine meadows, alpine tundra, 7000-13,000 ft. Widespread in the mountains.

subsp. glauca (A. Nelson) Al-Shehbaz & M. Koch—Pennycress. Petals white or occasionally pinkish purple, 3.4-7(-8.5) mm; styles 0.4-2.2(-3) mm; fruits 2.5-8(-12) mm; racemes often lax. Talus slopes and montane meadows, 7000-12,000 ft. Widespread in the mountains.

PENNELLIA Nieuwl. – Mock thelypody

Biennial or perennial herbs, pubescent with simple, forked, or 4–5-branched dendritic hairs, often basally glabrescent; stems single or few, erect, simple or branched; basal leaves rosulate, oblanceolate, entire or sinuate-dentate to shallowly lobed, obtuse at apex; cauline leaves petiolate to sessile, entire or dentate to shallowly lobed; sepals oblong and often purplish; petals purple to white, subequaling or rarely exceeding the sepals; siliques linear, terete or latiseptate, erect to pendent.

1. Siliques pendent on downward-arching pedicels; flowers zygomorphic, at least slightly so…**P. longifolia**
1′. Siliques erect or ascending on erect or divaricately ascending pedicels; flowers actinomorphic…**P. micrantha**

Pennellia longifolia (Benth.) Rollins—Long-leaf mock thelypody. Pubescent with coarse, simple or forked, spreading trichomes; basal leaves entire to sinuate-dentate, apically obtuse; cauline leaves linear, few, entire or lower shallowly dentate, densely pubescent; inflorescence strongly elongate in fruit, secund; petals purplish, often white at the base, veins darker, narrowly oblanceolate, 4–6 mm; siliques pendent, straight, linear, slightly latiseptate, 5–20 cm. Ponderosa pine–Gambel oak communities, 5100–10,700 ft. Widespread except easternmost counties.

Pennellia micrantha (A. Gray) Nieuwl.—Mountain mock thelypody. Glabrous above and pubescent below with forked or dendritic trichomes; basal leaves sinuate-dentate to shallowly lobed, rarely entire; cauline leaves ± like the basal ones except fewer, becoming much reduced and glabrous above; inflorescence narrow and elongate; petals white or more rarely purplish, spatulate to narrowly ligulate, gradually narrowing to base, 2–8 mm; siliques erect or nearly so, slender, terete, 2–5.8 cm. Slopes and talus in spruce-fir-aspen zone, 5050–10,000 ft. Widespread except easternmost counties.

PHYSARIA (Nutt. ex Torr. & A. Gray) A. Gray – Bladderpod

Annual, biennial, or perennial herbs, densely pubescent with stellate, webbed-stellate, or lepidote trichomes; basal leaves usually rosulate, entire to pinnatifid, petiolate or the blade tapering into the petiole; cauline leaves entire to dentate, petiolate, the blade tapering into the petiole, or ± sessile; petals yellow, in some cream-yellow, white, white with purple veins, or purplish; silicles of two forms: (1) didymous, latiseptate, with inflated, papery or leathery valves, or (2) not didymous, terete to latiseptate or angustiseptate, with valves not independently inflated, ± chartaceous or firm. Includes most species previously in *Lesquerella* (Al-Shehbaz & O'Kane, 2002; O'Kane, 2010).

1. Fruits strongly to moderately inflated, "double" (didymous); valves retaining seeds after dehiscence, basal sinus usually present; replum narrower than fruit (traditionally the genus *Physaria*)…(2)
1′. Fruits not or only slightly inflated, "single" (not didymous); valves not retaining seeds after dehiscence, basal sinus absent (or nearly so); replum usually as wide as or wider than fruit (traditionally the genus *Lesquerella*)…(4)
2. Fruit valves with flat or somewhat concave sides…**P. newberryi**
2′. Fruit valves with rounded (convex) or rounded-irregular sides…(3)
3. Fruit valves smoothly rounded and strongly inflated; basal leaf margins entire, terminally rounded or rounded-apiculate…**P. acutifolia**
3′. Fruit valves rounded-irregular and moderately inflated; basal leaf margins usually dentate or pinnatifid, rarely subentire, terminally acute or obtuse, not rounded…**P. floribunda**
4. Fruits glabrous externally…(5)
4′. Fruits pubescent externally…(12)
5. Plants forming tight, hard mounds, mats, or tufts…**P. navajoensis**
5′. Plants loose or dense, but not as above…(6)
6. Trichome rays fused ca. 1/2 their length (observe at ×10 or more); veins of petals obviously orange…**P. fendleri**
6′. Trichomes rays not at all fused or fused only at the very base; veins of petals sometimes slightly darker, but not orange…(7)
7. Petals white, sometimes purple-veined…**P. purpurea**
7′. Petals yellow or yellow-orange…(8)
8. Fruiting pedicels recurved; fruits ± pendent…**P. aurea** (in part)
8′. Fruiting pedicels sigmoid, straight, or upcurved; fruits not pendent…(9)

9. Basal leaves with clearly differentiated petioles and blades…(10)
9'. Basal leaves with the blade tapering into the petiole, petiole and blade not differentiated…(11)
10. Racemes subumbellate to densely corymbiform; cauline leaf blades elliptic to obovate…**P. ovalifolia**
10'. Racemes elongate; cauline leaf blades obovate to rhombic…**P. pruinosa**
11. Plants annual; cauline leaves strongly overlapping…**P. gordonii**
11'. Plants perennial; cauline leaves not or loosely overlapping…**P. pinetorum**
12. Basal and cauline leaves essentially alike except sometimes in size, basal leaves mostly narrow, with blades < 5 mm wide and tapering to the petiole…(13)
12'. Basal and cauline leaves dissimilar in shape, basal leaf blades usually > 5 mm wide and ± differentiated from the petiole…(16)
13. Fruiting pedicels recurved in an arch…**P. ludoviciana**
13'. Fruiting pedicels straight, sigmoid, or curved-ascending…(14)
14. Basal leaf blades involute (rarely some nearly flat) and linear to linear-oblanceolate or narrowly spatulate…**P. intermedia**
14'. Basal leaf blades flat and linear, linear-oblanceolate, spatulate to nearly rhombic, oblanceolate, or elliptic…(15)
15. Cauline leaf blades secund; basal leaf blades linear; plants compact…**P. calcicola**
15'. Cauline leaf blades not secund; basal leaf blades narrowly oblanceolate to broadly elliptic; plants not compact…**P. rectipes**
16. Fruits strongly compressed parallel to the replum (latiseptate)…**P. gooddingii**
16'. Fruits not or only slightly compressed, and then perpendicular to the replum (angustiseptate)…(17)
17. Fruiting pedicels recurved; fruits pendent…**P. aurea** (in part)
17'. Fruiting pedicels sigmoid, straight, or ascending; fruits horizontal to erect…(18)
18. Fruits elliptic to lanceolate in outline, apex ± acute, often slightly compressed…**P. montana**
18'. Fruits subglobose to ovoid, apex rounded, truncate, or obtuse, not or slightly compressed…(19)
19. Caudices thickened; fruit apex somewhat compressed…**P. valida**
19'. Caudices not thickened; fruit apex not compressed or the fruit slightly compressed throughout…**P. lata**

Physaria acutifolia Rydb.—Rydberg's twinpod, sharpleaf twinpod (a misnomer). Perennial; caudex branched; silvery-pubescent; stems 0.4–2 dm; basal leaves with a slender, often narrowly winged petiole, blades obovate to orbicular or rhombic-orbicular, 2–9 cm, abruptly narrowed to petiole, entire, rarely with few scattered teeth; cauline leaves spatulate to oblanceolate, 1–3 cm, entire; pedicels divaricate, slightly sigmoid or nearly straight, 6–12 mm; petals spatulate, 6–11 mm; silicles erect, didymous, suborbicular, inflated, 5–15 × 6–20 mm, papery, basal and apical sinuses similar, pubescent, trichomes appressed. Our plants are **var. acutifolia**. Piñon-juniper, ponderosa pine, sagebrush, rabbitbrush, blackbrush, and saltbush communities, especially in clayey substrates, 5450–7300 ft. Lincoln, Rio Arriba, San Juan, Taos.

Physaria aurea (Wooton) O'Kane & Al-Shehbaz [*Lesquerella aurea* Wooton]—Golden bladderpod. Biennial or short-lived perennial; caudex branched; densely pubescent; stems to ca. 2.5 cm, shallowly dentate or sometimes lyrate-pinnatifid; cauline leaves obovate to rhombic or oblanceolate, 2–4(–6) cm, entire or shallowly and remotely dentate; pedicels strongly recurved, to 20 mm; petals obovate to spatulate, 4.5–7.5 mm; silicles ± pendent, ovoid and angustiseptate or globose, 4–6(–8) mm. Open sites and bare areas in rocky limestone soil in mountains, road banks, open woods, 6500–9150 ft. Lincoln, Otero. **R**

Physaria calcicola (Rollins) O'Kane & Al-Shehbaz [*Lesquerella calcicola* Rollins]—Rocky Mountain bladderpod. Perennial; caudex branched; densely silvery-pubescent; stems 1–3 dm; basal leaves linear, 2–7(–10) cm, entire, repand, or shallowly dentate; cauline leaves spatulate to linear, (1–)2–3(–4.5) cm, entire, sometimes involute; pedicels spreading, sharply sigmoid, 8–15 mm; petals spatulate, 7–9(–11) mm; silicles ovate to oblong, not compressed at distal margins or apex, 5–9 mm, sparsely pubescent. Shale bluffs, limestone hillsides, gypsiferous knolls and ravines, various calcareous substrates, 5600–7300 ft. NE, Taos.

Physaria fendleri (A. Gray) O'Kane & Al-Shehbaz [*Lesquerella fendleri* (A. Gray) S. Watson]—Fendler's bladderpod. Perennial; caudex branched, sometimes woody at base; densely silvery-pubescent, trichome rays webbed, fused ca. 1/2 their length; stems 0.4–2.5(–4) dm; basal leaves linear to somewhat elliptic, 1–4(–8) cm, entire or coarsely dentate; cauline leaves linear to narrowly oblanceolate, rarely elliptic to rhombic, 0.5–2.5 cm, entire or remotely dentate, sometimes involute; pedicels divaricatespreading to erect, usually straight or slightly curved, occasionally sigmoid, 8–20+ mm; petals yellow and usually orange or orange-yellow at junction of blade and claw, sometimes also with orange guidelines, obdeltate to obovate, 8–12 mm; silicles globose, broadly ellipsoid, or ovoid, not or slightly inflated, 5–8 mm, glabrous. Piñon-juniper and desert shrub communities, slickrock, 3050–7850 ft. Throughout.

Physaria floribunda Rydb.—Point-tip twinpod. Perennial; caudex branched; silvery-pubescent throughout; stems 1–2 dm; basal leaves broadly oblanceolate, 3–8 cm, dentate or pinnatifid, rarely subentire; cauline leaves spatulate to linear-oblanceolate, 1–3 cm, entire, rarely toothed; pedicels recurved, 6–15 mm; petals spatulate, 9–11 mm; silicles pendent on arching pedicels, less frequently widely divergent, irregular in shape, base obtuse or slightly cordate, apex deeply and broadly notched, not strongly inflated, 8–11 × 8–12 mm, papery. Steep hillsides, decomposed granite, rocky banks, shale hills, rocky ravines, sagebrush and piñon-juniper areas, 7350–8900 ft. NC.

Physaria gooddingii (Rollins & E. A. Shaw) O'Kane & Al-Shehbaz [*Lesquerella gooddingii* Rollins & E. A. Shaw]—Goodding's bladderpod. Annual or biennial; without caudex; densely pubescent, trichomes sometimes with U-shaped notch on 1 side; basal leaves obovate or elliptic, to ca. 3 cm, sinuate or shallowly dentate; cauline leaves obovate to broadly elliptic, 1–3 cm, sinuate or shallowly toothed; pedicels recurved or sigmoid, somewhat expanded apically; petals cuneate, 6.5–8 mm, slightly expanded at base; silicles oblong or broadly elliptic, strongly latiseptate, 0.5–0.8 cm, pubescent, trichomes spreading, sparsely pubescent inside. Mountainous areas, open areas in piñon-juniper and ponderosa pine forests, 5750–7700 ft. Catron, Grant, Sierra. **R & E**

Physaria gordonii (A. Gray) O'Kane & Al-Shehbaz [*Lesquerella gordonii* (A. Gray) S. Watson; *L. gordonii* (A. Gray) S. Watson var. *densifolia* Rollins; *P. gordonii* subsp. *densifolia* (Rollins) O'Kane & Al-Shehbaz]—Gordon's bladderpod. Annual, biennial, or short-lived perennial, with a fine taproot; usually densely pubescent; basal leaves obovate to broadly oblong, 1.5–5(–8) cm, lyrate-pinnatifid, dentate, or entire; cauline leaves linear to oblanceolate, often falcate, 1–4(–7) cm, entire, repand, or shallowly dentate; pedicels divaricate-ascending, sigmoid or sometimes nearly straight, 5–15(–25) mm; petals yellow to orange, claw sometimes whitish, 5–8(–10) mm; silicles shortly stipitate, subglobose, not or slightly compressed, 0.3–0.8 cm, firm, glabrous. Sandy or light soils, rocky plains, caprock ledges, gravelly brushland, sandy desert washes, stream bottoms, pastures, roadsides, abandoned fields, 2900–6900(–7800) ft. S half of state, Mora, Valencia. Reports of *P. tenella* from New Mexico belong here,

Physaria intermedia (S. Watson) O'Kane & Al-Shehbaz [*Lesquerella intermedia* (S. Watson) A. Heller]—Mid bladderpod, Watson's bladderpod. Perennial, from a usually branched underground caudex clothed in old leaf bases; silvery or gray from dense crust of trichomes; basal leaves clustered on stem base, entire, linear to linear-oblanceolate, usually involute, sometimes somewhat flattened, 2–5 cm; cauline leaves linear-oblanceolate to linear, usually involute, 1–4 cm; petals oblanceolate, yellow, 6.5–10.5 mm; silicles subglobose to slightly ovoid, rarely a little compressed, sparsely pubescent, 3.5–5.5 mm. Piñon-juniper, sometimes with scattered Douglas-fir, 5100–7900 ft. N, Catron.

Physaria lata (Wooton & Standl.) O'Kane & Al-Shehbaz [*Lesquerella lata* Wooton & Standl.]—Lincoln County bladderpod. Perennial; densely pubescent; basal leaves elliptic to obovate, 3–4 cm; cauline leaves elliptic to obovate, 1–2 cm, entire; pedicels sigmoid, 5–8 mm; petals narrowly spatulate, 7–8 mm; silicles substipitate, globose, ellipsoid, or obovoid, not or slightly compressed, 0.3–0.4 cm, sparsely pubescent, sometimes few trichomes inside. Limestone soils and rocky places, piñon-juniper-oak woodland, montane coniferous forest, 6000–8200 ft. Lincoln. Possibly better considered a variant form of *P. pinetorum*. **R & E**

Physaria ludoviciana (Nutt.) O'Kane & Al-Shehbaz [*Lesquerella ludoviciana* (Nutt.) S. Watson]—Silver bladderpod, foothill bladderpod. Perennial; densely pubescent; basal leaves tufted, basal and cauline linear, entire, rarely shallowly dentate, some simple hairs present on leaf bases, no differentiation between blade and petiole, flat or often involute, to 8(–12) cm; pedicels recurved; petals yellow, erect, linear or narrowly oblong to lanceolate, 6.5–8.2 mm; silicles subglobose or shortly obovoid, terete to slightly latiseptate, densely pubescent, pendulous. Piñon-juniper communities in deep sand and wash bottoms, 6200–7400 ft. NW.

Physaria montana (A. Gray) Greene [*Lesquerella montana* (A. Gray) S. Watson]—Mountain bladderpod. Perennial, caudex branched, sometimes somewhat woody; pubescent; basal leaves tufted or rosette-forming, suborbicular or obovate to elliptic, entire to sinuate or shallowly dentate; cauline leaves often secund, linear to obovate or rhombic, entire or shallowly dentate; petals yellow to orange-yellow, sometimes fading purple, narrowly spatulate, 7.5–12 mm; silicles ellipsoid or ovoid, slightly angustiseptate, densely pubescent, 6–12 mm. Piñon-juniper communities, sometimes with Gambel oak, 5600–8600 ft. N, Catron.

Physaria navajoensis (O'Kane) O'Kane & Al-Shehbaz [*Lesquerella navajoensis* O'Kane]—Navajo bladderpod. Perennial, forming small hemispherical cushions; silvery-gray from a dense crust of trichomes; leaves essentially all cauline, linear-oblanceolate, tapering to an indistinct petiole, entire, 3–8(–13) mm; petals yellow, spatulate, deep yellow, faintly orange at junction of blade and claw, 5.2–6.5 mm; silicles becoming reddish or copper-colored at maturity, glabrous, ovate, often a little compressed at margins apically, acute, 3–4.9 mm. Piñon-juniper communities on limestone. A very rare, mat- and mound-forming species limited to windswept outcrops of Todilto limestone, 7000–8050 ft. McKinley. **R**

Physaria newberryi A. Gray—Newberry's twinpod. Perennial, cespitose from a simple or branched caudex; densely pubescent with silvery trichomes; basal leaves in a dense tuft or rosette, obovate to oblanceolate, entire or few-toothed, to 8(–12) cm; cauline leaves entire, oblanceolate, smaller than the basal, entire, lanceolate to narrowly lanceolate; petals yellow, narrowly oblanceolate, 9.5–12 mm; silicles strongly didymous, inflated but with angular valves with concave sides, the apical sinus deep and V-shaped, the base nor or barely notched.

1. Styles < 4 mm, shorter than fruit sinuses…**subsp. newberryi**
1′. Styles 5–9 mm, longer than fruit sinuses…**subsp. yesicola**

subsp. newberryi—Plants usually not mound-forming; caudex simple or few-branched (some populations on shifting substrates with elongate caudex branches); styles < 4 mm, shorter than fruit sinus. Piñon-juniper, Gambel oak, ponderosa pine communities, 5400–7800 ft. W half of state.

subsp. yesicola (Sivinski) O'Kane [*P. newberryi* var. *yesicola* Sivinski]—Plants mound-forming; caudex diffusely branched; styles 5–9 mm, longer than fruit sinus. Soils of the Yeso Formation in the Sierra Lucero Range and Mesa Lucero, sandy and silty soil, associated with piñon pine and one-seed juniper, 5700–6950 ft. Cibola, Valencia. **R & E**

Physaria ovalifolia (Rydb.) O'Kane & Al-Shehbaz [*Lesquerella ovalifolia* Rydb.]—Roundleaf bladderpod. Perennial; caudex branched, thickened by persistent leaf bases; densely pubescent; basal leaves suborbicular to elliptic, ovate, or deltate, 0.5–2(–6.5) cm, entire or shallowly dentate; cauline leaves narrowly elliptic or obovate, 0.5–2.5(–4) cm, entire; pedicels spreading at right angles, sometimes nearly erect, ± straight, 5–15(–20) mm, stout; petals yellow or white, suborbicular to obovate or obdeltate, 6.5–15 mm, apex sometimes emarginate; silicles sessile or shortly stipitate, < 1 mm, subglobose to broadly ellipsoid, inflated or slightly compressed, terete or subterete, 0.4–0.9 cm. Bare limestone flats, rocky knolls and slopes, limestone and gypsiferous outcrops, rock crevices, exposed caprock, 3800–7150 ft.

1. Caudices usually simple; petals white, (9–)11–15 mm, often 2 times as long as sepals; racemes usually elongate…**subsp. alba**
1′. Caudices branched and well developed; petals yellow, rarely whitish, 6.5–12(–14) mm, usually 1.5 times or less as long as sepals; racemes not elongate, subumbellate…**subsp. ovalifolia**

subsp. alba (Goodman) O'Kane & Al-Shehbaz—Stems usually few from base, erect or outer decumbent, 1.5–2.5 dm; basal leaves often broadly elliptic, base narrowing gradually to petiole, margins sinuate to dentate, or deltate, < 1 cm, and base abruptly narrowed to petiole. Limy and gravelly knolls, grassland hills, limestone hillsides and breaks, gypsum, shale, rocky calcareous soils, stony areas, prairie pastures, limestone roadcuts. Rarely encountered in NE.

subsp. ovalifolia—Caudex branched and well developed; stems several from base, stiffly erect, 1–2 dm; basal suborbicular or ovate to elliptic, base narrowing abruptly to petiole, margins entire. Bare limestone flats, rocky knolls and slopes, limestone chip, gypseous outcrops, rock crevices, exposed caprock. E half of state, Socorro.

Physaria pinetorum (Wooton & Standl.) O'Kane & Al-Shehbaz [*Lesquerella pinetorum* Wooton & Standl.]—White Mountain bladderpod. Perennial, caudex simple or branched; cespitose to densely so and matted; densely pubescent; basal leaves rhombic to elliptic and irregularly angular, sometimes spatulate to oblanceolate, 1.5–7.5(–10) cm, entire, undulate, or lyrate, or with 2 to few weak teeth; cauline leaves not or very loosely overlapping, spatulate to oblanceolate, 1–4 cm, entire; pedicels sigmoid or curved-ascending, 4.5–15(–20) mm; petals narrowly spatulate to broadly cuneate, 6–13 mm; silicles substipitate, globose or obovoid to ellipsoid, sometimes slightly obcompressed, 3.4–9 cm, firm, glabrous. Two subspecies.

1. Infructescences subumbellate, barely or not at all exceeding the basal leaves, or if exceeding them and somewhat elongate, the stems short, prostrate or nearly so, and the fruits mainly subumbellate at the ends of the stems; plants matlike, typically dense but looser where protected; ca. 10,700 ft. elevation, limestone-derived soils and rock crevices, Sandia Crest, Bernalillo County…**subsp. iveyana**
1'. Infructescences racemose throughout, elongate and evidently much exceeding the basal leaves; plants cespitose but never forming hard, dense mats; more widely distributed, 4750–10,650 ft. elevation, coarse, mainly granitic, but some limy soils…**subsp. pinetorum**

subsp. iveyana (O'Kane, K. N. Sm. & K. A. Arp) O'Kane—Ivey's bladderpod. Wind-swept, high-elevation, barren gray Madera Formation limestone escarpment on the W-facing summit of Sandia Peak, 10,700 ft. Bernalillo. **R & E**

subsp. pinetorum—White Mountain bladderpod. Steep slopes with scrub oak, piñon-juniper, open woodlands, sandy ridges, exposed rock crevices, rocky hillsides, coarse granitic or limy soil, 4750–10,650 ft. W, NCS.

Physaria pruinosa (Greene) O'Kane & Al-Shehbaz [*Lesquerella pruinosa* Greene]—Frosty bladderpod, Pagosa bladderpod. Perennial, from a simple or branched caudex; densely pubescent; basal leaves suborbicular or obovate to rhombic, entire to sinuate or shallowly dentate, abruptly narrowed to the petiole, 4–8 cm; cauline leaves obovate to rhombic, entire to shallowly toothed; petals yellow, spatulate, ca. 9 mm; silicles subglobose or ellipsoid, inflated and thin-walled, glabrous, often becoming coppery at maturity, terete, 6–9 mm. Gray clay, Mancos Shale hills with ponderosa pine and Gambel oak or sagebrush and grasses surrounded by ponderosa pine; 1 record barely in New Mexico, 6850 ft. Rio Arriba. **R**

Physaria purpurea (A. Gray) O'Kane & Al-Shehbaz [*Lesquerella purpurea* (A. Gray) S. Watson]—Rose bladderpod. Perennial, from a usually woody, simple caudex; densely pubescent; basal leaves elliptic or obovate to oblong, 4–15 cm, entire, dentate, or lyrate-pinnatifid; cauline leaves broadly elliptic to obovate or rhombic, 0.5–3(–5) cm; pedicels spreading-recurved or loosely sigmoid, 5–25 mm; petals white, often purple-veined, fading purplish, suborbicular to obovate, obdeltate, or cuneate, 4.5–10(–12) mm, apex ± emarginate; silicles subglobose to broadly ellipsoid, not or slightly inflated, 0.4–0.8 cm, firm, glabrous. Rocky draws and canyons, moist and shady areas of ridges, rock crevices on limestone ledges, shady lava cliffs, sand and gravel of ephemerally dry stream beds, rocky slopes, talus, shade of vegetation, 3450–5950 ft. S except Lea and Roosevelt.

Physaria rectipes (Wooton & Standl.) O'Kane & Al-Shehbaz [*Lesquerella rectipes* Wooton & Standl.]—Straight bladderpod. Perennial; from a simple or branched caudex; pubescent and gray-green; basal leaves ± rosulate, entire to weakly toothed, oblanceolate on the outside and often becoming narrower toward the middle and often involute when young; cauline leaves oblanceolate, obtuse, ± entire; petals yellow, erect to spreading, oblanceolate, 6.5–10 mm; silicles ovoid to ellipsoid to subglobose, pubescent, terete or a little latiseptate at the apex, 4–7 mm. Wash bottoms, piñon-juniper, ponderosa pine, Gambel oak, Douglas-fir communities, (3800–)5200–8750 ft. NW, NC, WC, C.

Physaria valida (Greene) O'Kane & Al-Shehbaz [*Lesquerella valida* Greene]—Strong bladderpod. Perennial, from a thickened branched caudex; densely pubescent; basal leaves elliptic to lanceolate or obovate, 3–8 cm, entire; cauline leaves elliptic or obovate, to 2 cm, entire; pedicels divaricate-ascending to horizontal, straight to loosely curved, to 15 mm; petals bright yellow, ligulate or broadly obovate, 7.5–8.5 mm; silicles suborbicular to broadly ovate or ellipsoid, slightly compressed, 0.6–0.8 cm, pubescent. Limestone soils, steep slopes, roadcuts, open woods, 5300–8000 ft. Eddy, Lincoln, Otero.

RAPHANUS L. – Radish

Annuals or biennials, from slender or fleshy roots (size, shape, and color variable in cultivated forms); glabrous or pubescent; stems erect; leaves basal and cauline, basal not rosulate, margins lyrately lobed or pinnatifid to pinnatisect, cauline dentate or lobed; pedicels divaricate, ascending, spreading, or reflexed; petals white, creamy-white, yellow, pink, or lilac (usually with darker veins), suborbicular; siliques sessile, glabrous, with 2 segments, cylindrical, fusiform, oblong, or ellipsoid, smooth to strongly moniliform.

1. Petals pale or creamy-white; fruits strongly constricted between seeds and usually breaking into units, strongly ribbed, beak narrowly conic…**R. raphanistrum**
1'. Petals usually purple or pink, sometimes white; fruits not or rarely slightly constricted between seeds and usually not breaking into units, not ribbed, beak narrowly to broadly conic to linear…**R. sativus**

Raphanus raphanistrum L.—Wild radish, jointed charlock. Annual, roots not fleshy; sparsely to densely pubescent; basal leaf petioles 1–6 cm, blades oblong, obovate, or oblanceolate in outline, lyrate or pinnatifid, sometimes undivided, 3–15(–22) cm × 10–50 mm, dentate, lobes 1–4 each side, oblong or ovate, to 4 cm × to 20 mm; petals yellow or creamy-white, veins dark brown or purple, 15–25; siliques cylindrical or narrowly lanceolate, strongly constricted between seeds, strongly ribbed, valvular segment 1–1.5 mm; terminal segment 2–11(–14) cm, beak narrowly conic. Disturbed waste places, cultivated fields, roadsides, orchards, hill slopes. Santa Fe, Socorro. Introduced.

Raphanus sativus L.—Cultivated radish. Annual or biennial, roots often fleshy in cultivated forms; often sparsely scabrous or hispid, sometimes glabrous; basal leaf petioles 1–30 cm, blades oblong, obovate, oblanceolate, or spatulate in outline, lyrate or pinnatisect, sometimes undivided, to 60 cm × to 200 mm, dentate, lobes 1–12 each side, oblong or ovate, to 10 cm × 50 mm; petals usually purple or pink, sometimes white, veins often darker, 15–25 mm; siliques usually fusiform or lanceolate, sometimes ovoid or cylindrical, smooth or rarely slightly constricted between seeds, not ribbed, valvular segment 1–3.5 mm; terminal segment 3–15(–25) cm, beak narrowly to broadly conic to linear. Roadsides, disturbed areas, waste places, cultivated fields, gardens, orchards. Otero, Roosevelt, Valencia. Introduced.

RAPISTRUM Crantz – Wild-turnip, turnipweed

Rapistrum rugosum (L.) All.—Bastard-cabbage. Hispid proximally, glabrous distally; basal leaf blades with 1–5 lobes each side, 2–25 cm, irregularly dentate, lateral lobes oblong or ovate, terminal lobe suborbicular or ovate, larger than lateral; pedicels erect and appressed to rachis, 1.5–5 mm; petals pale yellow, 6–11 × 2.5–4 mm; fruits with valvular segment ellipsoid, 0.7–3 × 0.5–1.5 mm, terminal segment globose or ovoid, 1.5–3.5 × 1–2.8 mm, usually rugose or ribbed, rarely smooth. Roadsides, disturbed sites, waste places, fields, grassy banks, ballast, 3000–4200 ft. Eddy, Roosevelt. Introduced.

RORIPPA Scop. – Yellowcress

Annual, biennial, or perennial herbs; glabrous or pubescent with simple trichomes; basal leaves, if present, usually rosulate, simple or 1–3-pinnatisect; cauline leaves petiolate or auriculate to amplexicaul, progressively reduced in size upward, entire to pinnatisect; pedicels often with 2 minute basal glands; petals yellow, white, or purplish, rarely absent, ovate to oblanceolate, rarely clawed; siliques or silicles 2(rarely 3–6)-loculed and -valved (very unusual in the family), globose, obovoid, or oblong to linear, terete or somewhat latiseptate, smooth or torulose, the false septum occasionally perforated.

1. Fruits silicles, < 3 times longer than wide…(2)
1'. Fruits siliques, > 3 times longer than wide…(7)
2. Fruit valves papillate, strigose, or pilose…**R. tenerrima** (in part)
2'. Fruit valves glabrous…(3)
3. Fruits globose or subglobose, 1.2–3.2 mm…(4)
3'. Fruits lanceolate, oblong, ovoid, ellipsoid, or pyriform, 2–10 mm…(5)
4. Perennial with thickened rhizomes; fruiting pedicels 4–15 mm; petals 3–5 mm; styles 1–2 mm…**R. austriaca**
4'. Annual or biennial without thickened rhizomes; fruiting pedicels 1.5–3.7(–4.3) mm; petals 0.6–1.2 mm; styles 0.1–0.7(–7) mm…**R. sphaerocarpa**
5. Perennial with a well-developed caudex; cauline leaf blade bases not auriculate, not amplexicaul…**R. alpina** (in part)
5'. Annual or, rarely, short-lived perennial without a caudex; cauline leaf blade bases sometimes auriculate or amplexicaul…(6)
6. Stems ascending, decumbent, or prostrate, few to several from base; sepals 0.8–1.8 mm; petals 0.5–1.8 mm…**R. curvipes** (in part)
6'. Stems erect, often simple from base; sepals 1.5–2.6 mm; petals (1.5–)1.8–3 mm…**R. palustris**
7. Fruits oblong to lanceolate, (1.5–)9–10 mm…(8)
7'. Fruits linear or oblong-linear, (8–)10–40 mm…(11)
8. Cauline leaf blade bases not auriculate…(9)
8'. Cauline (distal) leaf blade bases auriculate…(10)
9. Perennial (with caudex); fruit valves glabrous; petals (1.3–)1.5–2 mm…**R. alpina** (in part)
9'. Annual; fruit valves papillate; petals 0.5–0.8 mm…**R. tenerrima** (in part)
10. Perennial, glabrous or pubescent, trichomes hemispherical, vesicular; petals 2.7–6 mm; styles (0.8–)1–3.5 mm…**R. sinuata** (in part)
10'. Annual or, rarely, perennial, glabrous or hirsute, trichomes cylindrical; petals 0.5–1.8 mm; styles 0.3–1 mm…**R. curvipes** (in part)
11. Plants pubescent (trichomes vesicular, hemispherical or clavate)…(12)
11'. Plants usually glabrous, rarely pubescent (trichomes not vesicular)…(13)
12. Perennial with rhizomes; fruiting pedicels sigmoid or recurved; petals 2.7–6 mm; ovules < 90 per ovary…**R. sinuata** (in part)
12'. Annual or biennial without rhizomes; fruiting pedicels straight or curved-ascending; petals 1–2 mm; ovules > 100 per ovary…**R. teres**
13. Perennial; distal cauline leaf blades pinnatisect; petals 1.5–2.5 mm wide; seeds rarely produced…**R. sylvestris**
13'. Annual or biennial; cauline leaf blades not pinnatisect; petals 0.7–1.2 mm wide; seeds usually produced…**R. microtitis**

Rorippa alpina (S. Watson) Rydb. [*R. curvipes* Greene var. *alpina* (S. Watson) Stuckey]—Alpine yellowcress. Perennial, caudex simple or usually branched; herbage glabrous; stems slender, weak, several to many, prostrate to decumbent, freely branching; leaves basal and cauline, somewhat reduced in size upward, lanceolate to obovate, crenate to pinnatifid to nearly entire, blade tapering to the petiole; inflorescence narrow, elongating in fruit, terminal; petals yellow, fading to pale yellow, 2–2.5 mm, spatulate to ovate; siliques (or silicles) oblong, tapering toward apex, 3–5(–8) mm, straight or slightly curved upward, terete, usually slightly torulose, glabrous; style 0.5–1 mm. Lake, pond, and stream margins, alpine meadows, talus slopes, moist open forests, 6550–12,200 ft. NW, NC.

Rorippa austriaca (Crantz) Besser—Austrian yellowcress. Perennial; rhizomes thickened, short; usually glabrous, rarely pubescent proximally; stems erect, much branched distally; basal leaves not rosulate, margins pinnatifid; cauline leaves auriculate to amplexicaul, entire or serrate; inflorescence elongate; petals yellow, obovate, 3–5 mm; silicles (rarely produced) straight, globose or subglobose, 2.5–3.2 mm; style 1–1.5(–2) mm. Mudflats, floodplains, fields, roadsides, lakeshores, marshes, ditches, stream banks, wet grasslands, waste ground; 1 record, 5330 ft. San Juan. Introduced.

Rorippa curvipes Greene—Bluntleaf yellowcress. Annual or short-lived perennial from a taproot; glabrous or sparsely pubescent; stems prostrate to erect, branched; basal leaves withering early, entire to pinnately divided, margins entire or slightly and irregularly toothed; cauline leaves auriculate or not, oblong to oblanceolate, pinnatifid to near the midrib, the terminal lobe similar to the others, entire to crenate; inflorescence axillary and terminal, elongating in fruit; petals spatulate to narrowly oblanceolate, yellow, 0.5–1.5(–2.8) mm, shorter than the sepals; siliques (or silicles) pyriform to short-cylindrical, usually slightly curved upward, 1.4–8 mm, glabrous. Disturbed sites of lakeshores and floodplains, 4050–9350(–10,600 doubtful) ft. W, NCS.

Rorippa microtitis (B. L. Rob.) Rollins—Chihuahuan yellowcress. Annual (or biennial); usually glabrous throughout, rarely sparsely pubescent proximally; stems erect, branched distally; basal leaves often not rosulate, pinnatifid; cauline leaves petiolate, pinnatifid to pinnatisect or pectinate, lobes 4–10 on each side, base auriculate, entire, dentate, or distally pinnately lobed; inflorescence subumbellate and elongate; petals yellow, spatulate, 2.5–4 mm; siliques usually curved inward, rarely straight, linear, 8–20 mm; style 0.5–1.2 mm. Wet pastures, moist fields, ponds, ditches, meadows, 7800–8600 ft. Catron.

Rorippa palustris (L.) Besser [*R. islandica* (Oeder ex Murray) Borbás misapplied]—Bog yellowcress. Annual, biennial, or short-lived perennial; glabrous or hirsute below and becoming sparingly hirsute to glabrous above; stems stout to more rarely decumbent at the base or even prostrate, branched; leaves basal and cauline, short-petiolate or auriculate, oblong to oblanceolate, deeply pinnatifid, the lobes irregularly dentate to crenate-sinuate, less frequently entire, the terminal lobe larger, ovate, rounded to obtuse apically; inflorescences axillary and terminal, elongating in fruit; petals oblanceolate or spatulate, yellow, 0.8–2.7 mm; silicles or siliques (2.2–)3–14 mm, obtuse or rounded on each end, the shortest sometimes 3- or 4-valved (very unusual for members of the family), globose to elongate-oblong, glabrous, 4400–10,900 ft. Mainly throughout except SE.

1. Stems and abaxial leaf blade surfaces usually glabrous, rarely sparsely pubescent proximally…**var. palustris**
1'. Stems and abaxial leaf blade surfaces often densely hirsute…**var. hispida**

var. hispida (Butters & Abbe) Stuckey—Hispid yellowcress. Margins of ponds and streams, lakeshores, gravelly beaches, roadside ditches, mudflats, wet meadows, stream banks, springy ledges.

var. palustris [*R. palustris* var. *fernaldiana* (Butters & Abbe) Stuckey]—Bog yellowcress. Marshlands, pastures, prairies, meadows, swales, flats, sandbars, wet ground, stream banks, moist depressions, ditches, estuaries, waste ground, roadsides, sloughs, shores of lakes and ponds, bogs, thickets, grasslands.

Rorippa sinuata (Nutt.) Hitchc. [*Nasturtium sinuatum* Nutt.]—Spreading yellowcress. Perennial from a taproot, often with creeping roots and adventitious shoots; sparsely to densely pubescent with vesicular trichomes; stems prostrate to decumbent, rarely erect and then rhizomatous; leaves all cauline, somewhat fleshy, short-petiolate below and slightly auriculate above, deeply pinnatifid, the lobes rounded, entire to coarsely sinuate or crenate to toothed, upper surfaces glabrous; inflorescences axillary and terminal; petals oblanceolate, yellow, fading to pale yellow, 2.5–4.5(–6) mm; siliques or silicles short- to long-cylindrical or oblong to lanceolate, terete, glabrous to densely covered with vesicular trichomes. Lake, pond, and stream margins, irrigation ditches, wash bottoms, wet meadows, 4000–9100 ft. Nearly throughout.

Rorippa sphaerocarpa (A. Gray) Britton [*R. obtusa* (Nutt.) Britton var. *sphaerocarpa* (A. Gray) Cory]—Roundfruit yellowcress. Biennial from a taproot; glabrous or hirsute below (rarely hirsute throughout); stems decumbent to erect, usually branching throughout; leaves basal and cauline, short-petiolate, basal ones withering by flowering time, sessile to slightly auriculate, oblong to oblanceolate, pinnately divided to the midrib, the lobes irregularly serrate, terminal lobe larger than the others; inflorescence axillary and terminal, elongating in fruit; petals often shorter than the sepals, 0.6–1.2 mm, yellow; silicles globose to ovoid, < 1.5 times as long as wide, 1–3 mm, terete. Ephemeral creek banks, lake margins, marshy areas, hanging gardens, ephemeral ponds, 4000–10,850(–11,400) ft. NW, NC, WC, C.

Rorippa sylvestris (L.) Besser—Creeping yellowcress. Perennial; glabrous or sparsely pubescent; stems prostrate, decumbent, ascending, or suberect, branched mainly basally; basal leaves not rosulate, similar to cauline; cauline leaves petiolate or distally subsessile, deeply pinnatisect, lobes 3–6 on each side, dentate, serrate, subentire, or distally pinnatisect; inflorescence elongate; petals yellow, spatulate or obovate, 2.2–6 mm; siliques straight, usually linear, rarely oblong-linear, 10–20(–25), glabrous. Ditches, damp areas, shores of ponds and lakes, sandy beaches, waste ground, ditches, wet roadsides, meadows, washes, fields, gardens, 4250–8000 ft. Cibola, Curry, Guadalupe, San Miguel. Introduced.

Rorippa tenerrima Greene—Southern marsh yellowcress. Annual from a taproot; glabrous; stems prostrate to decumbent, noticeably light-colored; leaves basal and cauline, short-petiolate, deeply and very regularly pinnatifid, narrowly oblong to narrowly elliptic, lobes entire or sinuate-margined, the terminal lobe larger; inflorescence axillary and terminal, elongating in fruit; petals oblanceolate or spatulate, 0.5–0.8 mm, yellow; siliques (or silicles) sparsely to densely papillate-pubescent, cylindrical to narrowly lanceolate, (2–)3–7(–9) mm. Wash bottoms, moist disturbed areas, sagebrush, ephemeral ponds, floodplains, hanging gardens, 5850–8200 ft. NW.

Rorippa teres (Michx.) Stuckey—Southern marsh yellowcress. Annual (or biennial); puberulent, at least proximally, or glabrous, trichomes, at least in part, clavate to hemispherical, vesicular; stems usually prostrate or decumbent, rarely erect, branched distally; basal leaves rosulate, pinnatifid; cauline leaves short-petiolate to auriculate, oblong, oblanceolate to obovate, or lyrate-pinnatisect, lateral lobes oblong to ovate, pinnatifid to pinnatisect, rarely 2-pinnatifid; petals yellow, spatulate, 1–2 mm; siliques straight or curved, linear to oblong-linear, 8–14(–21), glabrous or pubescent. Wet areas, muddy ground, edges of canals and ditches, sandy fields, margins of ponds, streamsides, peat, 5500–8800(–12,200 doubtful) ft. NW, WC, C.

SELENIA Nutt. – Selenia

Selenia dissecta Torr. & A. Gray—Texas selenia. Winter annuals, often nearly acaulescent; stems often inflated into a thick crown, usually ascending, rarely decumbent; basal leaves rosulate, blade margins usually 2-, rarely 3-pinnatisect, lobes 5–10(–15) on each side, smaller than terminal; cauline leaves similar to basal, smaller distally; pedicels usually from basal leaf axil; petals broadly spatulate to obovate, 12–20 × 5–9 mm, apex rounded; siliques oblong to elliptic, latiseptate, 1.4–4 cm × 8–17 mm, slightly fleshy when green, thick, papery, base and apex acute, reticulate-veined. Grassy banks, pastures, salt draws, gypseous llanos, roadsides, sandy alluvium, limestone or sandy areas, creosote bush, 3800–7000 ft. NCS except far N.

SINAPIS L. – Charlock

Annual (perennial) herbs (or subshrubs), glabrous or pubescent with simple trichomes; stems erect, usually branched; leaves usually all cauline, petiolate, lyrate-pinnatifid or pinnatisect, sometimes dentate, progressively reduced in size upward, ± sessile and entire (except lobing); petals yellow (rarely white), obovate; siliques oblong to linear, often divided into a body and a beak, terete or strongly latiseptate, sometimes ± corky, often torulose.

1. Fruiting pedicels divaricate; fruits lanceolate, hispid, with trichomes of 2 types (subsetiform mixed with shorter, slender ones); terminal segment of fruit flattened, equal to or longer than valves, seedless…**S. alba**
1'. Fruiting pedicels ascending to suberect; fruits linear, glabrous or pubescent with only 1 type of trichome; terminal segment or fruit terete, much shorter than valves, seedless or 1-seeded…**S. arvensis**

Sinapis alba L.—White mustard. Basal leaf blades oblong, ovate, or lanceolate, lobes 1–3 each side, dentate or repand, rarely pinnatifid; cauline leaves (distal) shortly petiolate, blades ovate or oblong-ovate, margins coarsely dentate, rarely subentire; petals pale yellow, 7–14 mm; siliques lanceolate, 1.5–4.2(–5) cm, body terete or slightly flattened, 0.5–2 cm, 2–5-seeded per locule, beak flattened, 1–2.5 cm, equal to or longer than valves, seedless, hispid. Escape from cultivation, roadsides, waste places, disturbed areas, grain fields, cultivated areas, gardens, orchards. McKinley. Introduced.

Sinapis arvensis L.—Charlock, wild mustard, corn-mustard. Leaves all cauline, petiolate below, sessile above, lyrate-pinnatifid with coarsely toothed lobes, the terminal lobe broadly ovate; petals yellow, 7–11 mm; siliques 2–3.5 cm, differentiated into a body and a beak, the body 1.5–2.5 cm, linear, subterete, torulose, the valves distinctly 5–7-nerved, the beak 0.7–1.6 cm, narrowly conic, slightly flattened distally. Moist disturbed areas of alkali seeps, pond margins, ponderosa pine, meadows, 5800–7900 ft. Bernalillo, Lea, McKinley, Otero, Sandoval. Introduced.

SISYMBRIUM L. – Hedgemustard

Annual or perennial herbs, glabrous or pubescent with simple, rarely branched, trichomes; stems usually erect and branched; leaves cauline, petiolate, runcinate, lyrate-pinnatisect, or pinnatifid, the upper progressively reduced upward and becoming short-petiolate to sessile; petals yellow, rarely white or lavender, clawed; siliques linear or lanceolate, terete or rarely ± latiseptate, smooth or torulose, straight to slightly arched.

1. Fruits subulate-linear, 0.7–1.4(–1.8) cm; fruiting pedicels appressed to rachises…**S. officinale**
1'. Fruits narrowly linear, 2–13 cm; fruiting pedicels not appressed to rachises…(2)
2. Upper leaves entire; perennials; glabrous and usually glaucous…**S. linifolium**
2'. Upper leaves all or mostly entire; annuals; glabrous or pubescent…(3)
3. Upper and lower leaves distinctly different, the lower with oblong and dentate lobes, the upper with narrow, linear or filiform lobes; fruiting pedicels thick and stout, almost as thick as the siliques…**S. altissimum**
3'. Upper and lower leaves essentially the same, differing only in size; fruiting pedicels slender, narrower than the siliques…(4)
4. Fruiting pedicels nearly as wide as fruit…**S. orientale**
4'. Fruiting pedicels evidently narrower than fruit…(5)
5. Flowers showy, sepals 3–4 mm, petals 5.2–6.5 mm, bright yellow…**S. loeselii**
5'. Flowers not especially showy, sepals 1.7–2.5 mm, petals 2–4 mm, pale yellow…**S. irio**

Sisymbrium altissimum L.—Tall tumblemustard. Annual from a taproot; hispid below, usually glabrous above; leaves broadly lanceolate to oblong or oblanceolate, to 20 cm, lower runcinate to pinnatifid, upper pinnately divided into narrowly linear segments; sepals 3–5.5 mm, the outer 2 with a short horn at the apex; petals oblanceolate, 5.5–8.5 mm, pale yellow, fading to white; siliques linear, about the same diameter as the pedicel, terete, 5.5–8.5 cm, valves 3-nerved. Disturbed areas of desert grassland, washes, roadsides, townsites, piñon-juniper, desert shrub, and sagebrush communities, 3800–9200 ft. Nearly throughout. Introduced.

Sisymbrium irio L.—London rocket. Annual from a taproot; glabrous or sparsely hispid above; leaves pinnately lobed with a larger hastate terminal lobe, becoming reduced and sometimes simple above; sepals 1.7–2.5 mm; petals oblanceolate, 2–4 mm, pale yellow; siliques straight or slightly incurved, terete, glabrous to sparsely pubescent, slender, 3–5 cm, valves 3-nerved. Disturbed areas of piñon-juniper communities, wash bottoms, roadsides, and campgrounds, 3100–8900 ft. Nearly throughout except NE. Introduced.

Sisymbrium linifolium (Nutt.) Nutt. ex Torr. & A. Gray [*Schoenocrambe linifolia* (Nutt.) Greene]—Flaxleaf plainsmustard. Perennial, rhizomatous; glabrous and glaucous, rarely sparsely pilose below; leaves all cauline, lower ones to 7 cm, somewhat fleshy, linear to linear-oblanceolate, entire to pinnatifid, becoming entire, narrower, and linear above as well as nearly terete; sepals 4–6.5 mm; petals narrowly oblanceolate or spatulate, claw lacking, 7–12 mm, yellow; siliques linear, terete, 4–7 cm, wider than the pedicel. Wash bottoms and hanging gardens, 6050–8150 ft. NW, C.

Sisymbrium loeselii L.—Small tumbleweed mustard. Annual from a taproot; sparsely to densely hispid below, usually glabrous above; leaves to 15 cm below, smaller above, broadly deltoid-lanceolate to lanceolate, runcinate-pinnatifid, typically with a large, acuminate, irregularly serrate to dentate terminal lobe; sepals 3–4 mm; petals obovate, 5.2–6.5 mm, bright yellow; siliques straight or slightly incurved, terete, valves prominently 3-nerved, 2–3.4(–4) cm, evidently wider than the pedicel. Piñon-juniper communities, disturbed areas, 3900–8150 ft. N, Hidalgo. Introduced.

Sisymbrium officinale (L.) Scop.—Hedgemustard. Annual; glabrous or pubescent; basal leaves usually rosulate, broadly oblanceolate or oblong-obovate 2–10(–15) cm, lyrate-pinnatifid, pinnatisect, or runcinate, lobes (2)3 or 4(5) on each side, oblong or lanceolate, smaller than terminal lobe, entire, dentate, or lobed, terminal lobe suborbicular or deltate; cauline leaves similar to basal; sepals 2–2.5 mm;

petals spatulate, 2.5–4 mm, yellow; siliques erect, subulate-linear, straight, slightly torulose or smooth, stout, 0.7–1.8 cm, glabrous or pubescent. Roadsides, fields, pastures, waste ground, deserts. Curry, Lea, Lincoln, McKinley, Rio Arriba, Roosevelt. Introduced.

Sisymbrium orientale L.—Indian hedgemustard. Annual; glabrous or pubescent; basal leaves rosulate, broadly oblanceolate to oblong-oblanceolate, 3–8(–10) cm, runcinate-pinnatipartite, lobes 2–5 on each side, oblong or lanceolate, much smaller than terminal lobe, subentire or dentate, terminal lobe lanceolate, deltate, or often hastate; cauline leaves similar to basal; sepals 3.5–5.5 mm; petals spatulate, 6–10 mm; siliques narrowly linear, straight, smooth, stout, 5–10(–13) cm, glabrous or pubescent. Waste ground, roadsides, disturbed sites, 4050–5950 ft. Colfax, Lea, Quay. Introduced.

STANLEYA Nutt. – Prince's plume

Biennial or perennial herbs, often woody at the base; glabrous or pubescent with simple (rarely some branched in upper portions of the plant) trichomes, often glaucous; plants often with a basal rosette of petiolate, entire, toothed, or pinnatifid leaves; cauline leaves entire, toothed, or pinnatifid, petiolate or occasionally auriculate or amplexicaul; petals bright yellow or white to very pale yellow; stamens not tetradynamous, all ± the same length and coiling as they dehisce from the top downward; siliques glabrous, on a long stipe of 1–3 cm at maturity, linear, subterete or latiseptate, ± torulose, on widely spreading pedicels.

1. Cauline leaves sessile, bases auriculate to sagittate…**S. viridiflora**
1'. Cauline leaves petiolate, bases not auriculate to sagittate…(2)
2. Petal blade 2.5–5 mm wide, white to pale yellow, obovate to orbicular, hairy only at the inner summit of the claw…**S. albescens**
2'. Petal blade 1–2.5 mm wide, bright yellow, oblanceolate to obovate, the claw hairy nearly its full length on the inner side…**S. pinnata**

Stanleya albescens M. E. Jones—White prince's plume. Biennial from a taproot, glabrous and glaucous; leaves ± fleshy, petiolate, essentially all cauline, lyrate-pinnatifid, runcinate, or the uppermost rarely entire, to 20 cm; sepals 9–13 mm; petals hairy on inner side near the junction of blade and claw, 2.5–5 mm broad, blade obovate to orbicular, white to very pale yellow; siliques erect to spreading, (sub)terete, gently curved upward, 3–6 cm, on a long stipe 15–26 mm. Desert scrub in saline soils, 6100–6500 ft. Reported for McKinley but no specimens seen.

Stanleya pinnata (Pursh) Britton—Prince's plume. Perennial subshrub, glabrous or sparsely pubescent, ± glaucous; lower leaves entire to deeply pinnately lobed, broadly lanceolate, to 15 cm, upper leaves linear-lanceolate to ovate, entire or pinnatifid, smaller than lower leaves; sepals 8–16 mm; petals hairy nearly the entire length of the inner side, 1–2.5 mm broad, blade oblanceolate to obovate, yellow (rarely white in some); siliques glabrous, widely spreading, (sub)terete, arcuate to nearly straight, 4–8(–12) cm, on a long stipe 12–21 mm, 4550–8250 ft.

1. Leaves all entire (sometimes some lower ones dentate or shallowly pinnatifid); infrequent…**var. integrifolia**
1'. Leaves (at least the lower ones) pinnately divided; common…**var. pinnata**

var. integrifolia (E. James) Rollins [*S. integrifolia* E. James; *S. pinnatifida* Nutt. var. *integrifolia* (E. James) B. L. Rob.]—Golden prince's plume. Piñon-juniper, piñon-shrub, desert scrub, clay hills, shaly slopes. San Juan.

var. pinnata—Desert prince's plume. Selenium- and often uranium-bearing soils of desert scrub, piñon-juniper, sagebrush, badlands, and (rarely) hanging gardens. Northernmost and westernmost counties.

Stanleya viridiflora Nutt.—Green prince's plume. Perennial from a short-branched caudex covered with old leaf bases surmounting a thick, woody taproot, glabrous, glaucous; leaves ± fleshy, petiolate, basal and cauline, the basal leaves in a well-developed rosette, lanceolate to ovate, narrowly cuneate at the base, entire or occasionally coarsely toothed or lyrate-pinnatifid, cauline leaves crowded; sepals

12-18 mm; petals linear-oblong, erose at apex, 13-20 mm, lemon-yellow to pale yellow to nearly white; siliques on stout, widely spreading pedicels, glabrous, arcuate, (sub)terete, 3-7 cm, on a long stipe, 10-20(-25) mm. Clayey soils of mat saltbush and desert scrub communities, 5200-5600 ft. San Juan.

STREPTANTHELLA Rydb. – Fiddle-mustard

Streptanthella longirostris (S. Watson) Rydb.—Longbeak fiddle-mustard, longbeak twistflower. Annual herbs; glabrous and usually glaucous; leaves all cauline (basal ones, if present, very early-deciduous), narrowly (ob)lanceolate, obtuse, entire, dentate, or rarely pinnatifid, becoming linear and entire above; petals white or yellowish, sometimes purple-tinged, narrowly spatulate to narrowly oblanceolate, crispate, 3.5-6.5 mm; siliques linear, latiseptate, narrow, 3-5 cm, pendent or pendent-appressed, on strongly reflexed pedicels. Piñon-juniper, juniper-shrub, slickrock, desert scrub, alcoves, hanging gardens, 4600-7050 ft. NW, Socorro.

STREPTANTHUS Nutt.   Jewelflower

Annual (or perennial) herbs; often glaucous, glabrous to pubescent; basal leaves petiolate; cauline sessile and auriculate or amplexicaul, entire or dentate (or pinnatifid); petals lavender to purple or magenta, obovate or elliptic to linear or lanceolate, with crispate or involute-rolled blades; siliques sessile or stipitate, linear, latiseptate or subterete, smooth or torulose.

1. Perennials, with woody caudex; calyces ± campanulate; adaxial pair of filaments distinct; anthers 2.5-5 mm, adaxial pair fertile…**S. cordatus**
1'. Annuals or biennials, without woody caudex; calyces often urceolate, sometimes campanulate; adaxial pair of filaments often connate, sometimes distinct; anthers 1-6 mm, adaxial pair often sterile, sometimes fertile…(2)
2. Petals creamy-white with purple veins or purple with white margins, 14-18 mm, margins crisped, 1-1.5 mm wide; fruit valves each with obscure midvein…**S. carinatus**
2'. Petals lavender or purplish lavender, 16-27 mm, margins not crisped, 6-9 mm wide; fruit valves each with prominent midvein…**S. platycarpus**

Streptanthus carinatus C. Wright ex A. Gray—Lyreleaf jewelflower. Annuals or biennials; glaucous and glabrous; leaf margins runcinate-pinnatifid, dentate, or entire, cauline blades ovate to lanceolate, 2.5-14 cm, smaller distally, base auriculate to amplexicaul; fruiting pedicels straight or curved upward, divaricate-ascending, 7-22(-35) mm; calyx urceolate or campanulate, sepals purple, or ochroleucous to yellowish, 8-11 mm, keeled; petals white with purplish veins or purple with white margins, recurved, 14-18 mm, margins crisped, claw about as wide as blade; silicles ascending, smooth, straight, strongly flattened, 3-8 cm × 4.5-6 mm. S half of state except easternmost.

1. Basal and proximal cauline leaf blade margins runcinate-pinnatifid; sepals purple; petals purple with white margins…**subsp. carinatus**
1'. Basal and proximal cauline leaf blade margins entire, dentate, or runcinate-pinnatifid; sepals ochroleucous to yellowish; petals white with purplish veins…**subsp. arizonicus**

subsp. arizonicus (S. Watson) Kruckeb., Rodman & Worth. [*Disaccanthus luteus* Greene]—Well-drained sandy or rocky soils, gravelly bajada slopes, canyons, scrub or open woodland, gravelly roadsides, desert scrub, sandy grasslands, 4000-7000 ft.

subsp. carinatus [*Streptanthus validus* (Greene) Cory]—Cliff bases, talus, limestone, gravelly slopes, brushy areas, canyons, isolated in washes, 2950-7000 ft.

Streptanthus cordatus Nutt.—Heartleaf twistflower. Perennial, from a simple or closely branched caudex, sometimes rhizomatous; glaucous and glabrous except the ciliate margins of some lower leaves; leaf margins dentate, cauline blades broadly oblong to nearly orbicular, dentate to entire, to 10 cm, base auriculate and sagittate; pedicels straight, divaricate-ascending, 3-11(-14) mm; calyx flask-shaped, sepals greenish brown to purple, the outer pair strongly gibbous-based, scarious-margined, 7.5-12 mm; petals involute along margins, dark purple to brownish, often with white margins, 9.5-16 mm, blade wider than claw; siliques linear to curved, strongly latiseptate, 5-10 cm × to 7 mm. Our plants are **var. cordatus**.

Piñon-juniper, piñon–Douglas-fir, and Douglas-fir–ponderosa pine communities, often in wash bottoms, 5500–8450 ft. Lincoln, Rio Arriba, Sandoval, San Juan, Taos.

Streptanthus platycarpus A. Gray [*S. sparsiflorus* Rollins]—Broadpod jewelflower. Annual; glaucous, usually glabrous throughout; basal leaf margins sinuate to lyrate or pinnatifid, oblanceolate; cauline blades ovate, 1.5–15 cm, base auriculate to amplexicaul, entire or sinuate-dentate; pedicels divaricate to ascending, straight, 7–35 mm; calyx campanulate; sepals purple to deep maroon, 8–12 mm; petals lavender or purplish lavender, 16–27 mm, blade 6–9 mm wide, margins not crisped, claw crisped; siliques erect to divaricate-ascending, smooth, straight, strongly flattened, 4–9.5 cm × 4.5–6 mm. Cliff ledges, often on limestone, rocky sites, roadsides, rocky creek beds, 4700 ft. Eddy. If *S. sparsiflorus* is recognized separately from *S. platycarpus*, it too is considered a rare plant. **R**

STRIGOSELLA Boiss. – Strigosella

Strigosella africana (L.) Botsch. [*Malcolmia africana* (L.) W. T. Aiton]—African mustard, African adder's-mouth. Annual; densely pubescent with small, coarse, freely branched trichomes; leaves oblanceolate, petiolate, remotely and coarsely dentate, 3–7 cm, smaller and sessile distally; inflorescence few-flowered, elongate, lower flowers in leaf axils; petals pink to rose-violet, clawed with an obovate blade, 6–10 mm; siliques terete to subquadrangular, straight, slightly torulose, divaricately ascending to divaricate, 4–6.5 cm on stout pedicels about the same thickness as the siliques. Desert scrub, saltbush, piñon-juniper communities, roadsides, disturbed areas, 4900–5750 ft. San Juan. Introduced intrusive.

THELYPODIOPSIS Rydb. – Tumblemustard

Only one species in our area. Other species have recently been reassigned to *Mostacillastrum* (Al-Shehbaz, 2012a).

1. Sepals and petals yellow; siliques on stipes 2–8 mm…**T. aurea**
1′. Sepals and petals white to pale lavender; siliques on stipes < 0.4 mm…**go to MOSTACILLASTRUM**

Thelypodiopsis aurea (Eastw.) Rydb.—Durango tumblemustard. Biennial or short-lived perennial from a taproot; glabrous or rarely sparsely pubescent at the base; leaves somewhat fleshy, basal leaves oblanceolate, tapering to a winged petiole, irregularly dentate; cauline leaves sessile, auriculate, oblong to lanceolate, acute-tipped; sepals yellow, 5–7(–9) mm; petals narrowly oblong with a slight constriction at the junction of the blade and claw, 7–13 mm, yellow; siliques erect to slightly divaricate, slender, nearly straight to more commonly incurved, somewhat torulose, stipe 2–8 mm, body 4.5–9 mm. Ledges, alkali flats, clay flats and hills, desert shrub, piñon-juniper, badlands, saltbush communities, 4600–7450 ft. NW.

THELYPODIUM Endl. – Thelypody

Annual, biennial, or rarely short-lived perennial, herbs; glaucous and glabrous, rarely pubescent; basal leaves ± rosulate, usually withering early, petiolate, relatively broad, entire, toothed or pinnately lobed; cauline leaves petiolate to sessile, often auriculate to amplexicaul, entire, toothed or pinnately lobed; petals white to purple, spatulate or linear to obovate; siliques on indistinct to long stipes, linear, terete or ± latiseptate, usually torulose.

1. Cauline leaves sessile…**T. integrifolium**
1′. Cauline leaves petiolate…(2)
2. Stems hirsute or glabrous basally; fruits terete, submoniliform to strongly torulose (replum constricted between seeds)…**T. laxiflorum**
2′. Stems glabrous basally; fruits flattened, torulose (replum not constricted between seeds)…(3)
3. Annuals; styles conic or subconic in fruit; petals spatulate; distal-most cauline leaf blade margins pinnately lobed (pectinate)…**T. texanum**
3′. Biennials; styles clavate to subclavate in fruit; petals oblong to linear; distal-most cauline leaf blade margins usually entire or dentate, rarely lobed…**T. wrightii**

Thelypodium integrifolium (Nutt.) Endl.—Entire-leaved thelypody. Biennial from a taproot; glabrous, glaucous; fleshy; leaves ± fleshy, basal oblong, oblanceolate, or spatulate, entire, less frequently denticulate or repand, to 7 cm; cauline leaves oblanceolate to lanceolate below and becoming linear-lanceolate to linear above, sessile, entire; inflorescence dense and little elongating; petals lavender to purplish, rarely white, 4.5-10 mm; siliques straight to strongly incurved, terete to a little flattened, torulose to submoniliform, divaricately ascending to horizontal (rarely reflexed), 1-3.5(-6) cm on a stipe 0.5-4 mm. Our plants are **subsp. gracilipes** (B. L. Rob.) Al-Shehbaz. Moist areas of canyon bottoms, floodplains, cottonwood groves, alcoves, hanging gardens, piñon-juniper communities, willow and shrub communities, 4800-7400 ft. NW.

Thelypodium laxiflorum Al-Shehbaz—Droopflower thelypody. Biennial from a taproot; glabrous or sparsely to densely hirsute below; basal leaves oblanceolate, often pinnately to lyrately lobed, laciniate, or sinuate; cauline leaves linear to linear-lanceolate or lanceolate, sometime oblanceolate or oblong especially below, sinuate to dentate; inflorescence elongating and lax in fruit; petals white to rarely lavender, 5.5-9 mm; siliques slender, straight or somewhat variously curved, terete, divaricate to somewhat reflexed, submoniliform to strongly torulose, 2.5-6 cm, on a stipe 0.3-3 mm. Wet areas in canyon bottoms and Gambel oak communities; a single specimen from San Juan County, ca. 5750 ft.

Thelypodium texanum (Cory) Rollins [*Sibara grisea* Rollins]—Texas thelypody. Annual; slightly glaucous, glabrous; basal leaves oblanceolate or spatulate, pinnately lobed, lateral lobes often oblong, sometimes ovate; cauline leaves pectinate with linear lobes, similar to basal, much smaller; inflorescence somewhat lax, considerably elongate in fruit; petals white, spatulate to oblanceolate, 3.5-7 mm; siliques torulose, straight or slightly recurved, flattened, 2-7 cm, on a stipe 0.5-1(-2) mm. Barren hillsides, creek beds, stream banks, 4700-7500 ft. Eddy, Otero.

Thelypodium wrightii A. Gray—Wright's thelypody. Biennial from a taproot; glabrous and glaucous; basal leaves oblanceolate, pinnatifid to coarsely toothed; cauline leaves reduced in size upward, lanceolate to narrowly lanceolate, tapering to the petiole, entire (usually) to dentate to sinuate; inflorescence dense and becoming strongly elongate; petals white to rarely lavender, blade (ob)lanceolate, 4.5-7 mm; siliques latiseptate to subterete, straight to somewhat curved, torulose, horizontal to reflexed, 5-7(-9) cm. Talus and rock outcrops, 4000-9500 ft. Throughout except E counties.

THLASPI L. – Pennycress

Thlaspi arvense L.—Field pennycress. Annual or winter annual from a taproot; glabrous; leaves basal and cauline, the basal ones early deciduous and not usually seen, decreasing in size upward, oblanceolate, irregularly dentate to sinuate, obtuse to rounded apically, petiolate below, becoming sessile and sagittate-clasping above, to 4.5(-6) cm; petals spatulate, 2.5-3.7 mm, white; silicles strongly angustiseptate, oval to oblong-obcordate, strongly wing-margined and deeply emarginate, 7-20 mm. Disturbed areas of roadsides, canal banks, stock ponds, sagebrush, piñon-juniper, spruce, 5000-9250(-10,000) ft. Mainly throughout except below 5000 ft. Introduced.

THYSANOCARPUS Hook. – Fringepod, lacepod

Thysanocarpus curvipes Hook.—Fringepod. Annual; often hirsute, sometimes glabrous; stems 1-6(-8) dm; basal leaves oblanceolate to obovate, 1-6(-13) cm, subentire to sinuate-dentate; cauline leaves lanceolate, widest at base, auriculate-clasping, auricles extending around stem on at least some leaves; pedicels smoothly recurved; silicles flat or plano-convex, obovate to nearly orbicular, 3-6(-9) mm wide, pubescent or glabrous, trichomes clavate and 0.2-0.4 mm, or pointed and ± 0.2 mm, strongly winged, the wing entire, perforate, or incised. Rocky slopes, washes, oak woodlands, streamsides, meadows, sometimes serpentine soils, 4000-6200 ft. Grant, Hidalgo, Luna.

TOMOSTIMA Raf. – Tomostima

Plants annual, without caudices; trichomes simple and 2–7-rayed; stems erect to ascending, simple or branched basally; usually leafless, leaves proximal when present (except leafy to the inflorescence in *T. platycarpa*); leaves basal and not or only loosely forming rosettes, entire to toothed; lateral pair of sepals not saccate or subsaccate basally; petals white; later in the season chasmogamous flowers are replaced by apetalous cleistogamous flowers; fruits silicles or siliques, latiseptate; ovules (12–)24–88; styles obsolete or nearly so, < 0.1(–0.4) mm; otherwise much like, and recently segregated from *Draba*. *Tomostima platycarpa* (Torr. & Gray) Al-Shehbaz, M. Koch & Jordon-Thaden, found in adjacent Arizona and Texas, is likely present in the S part of the state and is included in the key.

1. Stems leafy to the inflorescence [to be expected]…T. platycarpa (Torr. & Gray) Al-Shehbaz, M. Koch & Jordon-Thaden
1'. Stems scapose (rarely with 1 or a few leaves well below the inflorescence)…(2)
2. Rachises and fruiting pedicels densely pubescent…**T. cuneifolia**
2'. Rachises and fruiting pedicels glabrous (rarely sparsely pubescent)…**T. reptans**

Tomostima cuneifolia (Nutt. ex Torr. & A. Gray) Al-Shehbaz, M. Koch & Jordon-Thaden [*Draba cuneifolia* Nutt. ex Torr. & A. Gray]—Wedge-leaf stonecress. Annual; taprooted; stems simple or branched at base, erect to ascending, hirsute proximally, pubescent throughout with stalked, 2–4(–5)-rayed trichomes; lower leaves rosulate, spatulate or broadly obovate, 8–35 mm, often ciliate proximally with simple trichomes; both surfaces uniformly pubescent with stalked, 2–4(–5)-rayed trichomes; upper surfaces usually with a few simple trichomes; cauline leaves absent or 1–6 along lower 1/3 of stem, often dentate on distal 1/2, pubescent like basal leaves; racemes ebracteate, 10–50-flowered; petals white (lacking in late-season flowers); fruits oblong to linear, 6–16 × 1.7–3 mm, not twisted, puberulent with simple (rarely some 2-rayed) trichomes or glabrous; style 0.01–0.1 mm; ovules 24–66(–72) per fruit. Rocks and ledges, sparsely vegetated clay to gravelly soil, canyon bottoms, piñon-juniper, piñon-sage, 3400–8100 ft. Throughout except NE.

Tomostima reptans (Lam.) Al-Shehbaz, M. Koch & Jordon-Thaden [*Draba reptans* (Lam.) Fernald]—Carolina stonecress. Annual; taprooted; stems often branched at base, erect to ascending, sparsely to densely pubescent proximally with 2–3-rayed trichomes 0.1–0.6 mm, these sometimes mixed with simple or spurred trichomes to 0.9 mm, usually glabrous distally; lower leaves rosulate, elliptic to suborbicular, 5–23 mm, often ciliate proximally with simple trichomes 0.4–1 mm; lower surfaces uniformly pubescent with stalked, 2–4-rayed trichomes; upper surfaces with larger simple and smaller 2-rayed trichomes; cauline leaves absent or rarely 1–3 just above the base, entire, pubescent like basal leaves; racemes ebracteate, 3–16-flowered, usually subumbellate; petals white (lacking in late-season flowers); fruits linear to linear-oblong, 5–20 × 1.2–2.3 mm, not twisted, glabrous or with antrorse, simple (rarely spurred or 2-rayed) trichomes; style 0.01–0.1 mm; ovules 32–88 per fruit. Canyon bottoms, sandy to gravelly soil, typically in piñon-juniper or piñon-juniper-oak communities, 3500–7750 ft. N, C.

TURRITIS L. – Tower mustard

Turritis glabra L. [*Arabis glabra* (L.) Bernh.]—Tower mustard, tower cress. Biennial, rarely short-lived perennial; sparsely to densely pilose basally, glabrous distally; basal leaves rosulate, oblanceolate or oblong, the largest 10–30 mm wide, crenate to repand (rarely entire), hirsute with 2–4-rayed (and a few simple) hairs; cauline leaves overlapping and often concealing stem proximally, entire or dentate, the upper lanceolate with prominent auricles, glabrous; petals pale yellow to cream-white, sometimes aging lilac, 3–5.5 mm; silicles erect, appressed to rachis, glabrous, subterete-quadrangular. Meadows, forest margins, and riverbanks, often in disturbed habitats, 7000–10,950 ft. N, Lincoln.

BROMELIACEAE – BROMELIAD OR PINEAPPLE FAMILY

Kenneth D. Heil

TILLANDSIA L. – Ball moss

Tillandsia recurvata (L.) L.—Ball moss. Herbs, perennial, usually epiphytic or attached to various objects; plants in dense spherical clusters, short; leaves 4–10, 2-ranked, recurving, gray, 6–12 × 0.2–0.3 cm, densely pruinose-scaly, blades subulate, terete distally, succulent, margins involute to nearly tubular; inflorescence a scape, conspicuous, erect, 2-5 cm, ± 1 mm diam.; flowers usually 2, conspicuous; sepals free, lanceolate, not keeled, 6-8 mm, surfaces glabrous; corolla tubular, petals violet, elliptic, 0.7-1 cm, fruit to 3 cm. One record from rocky slopes at bottom of Indian Creek Canyon, 5800 ft. Hidalgo (Luther & Brown, 2000).

CACTACEAE – CACTUS FAMILY

Kenneth D. Heil

Perennial, stem succulents, some becoming woody; stems globose, cylindroid, columnar, or flattened, leafless (at least ultimately), with areoles (nodes) where spines and sometimes glochids are produced; spines variable, may be alike or diverse, long or short, straight, curved, or both; flowers near or below apex of stem, epigynous, with hypanthium forming a floral tube, perianth of numerous tepals grading from sepaloids to petaloids, stamens numerous, carpels 3-24, fused, style 1, stigma lobes free, as many as the carpels, ovary inferior; fruit fleshy or dry at maturity, a many-seeded berry. Descriptions closely follow those in Parfitt and Gibson (2003), although the taxonomy sometimes departs from that treatment (see also Boissevain & Davidson, 1940; Weniger, 1969).

1. Areoles bearing minute sharp-pointed barbed bristles (glochids) as well as spines; seeds enveloped in bony aril; stems made up of a series of cylindroid, club-shaped, or flattened joints…(2)
1′. Areoles not bearing glochids; seeds without bony covering; stems generally not composed of a series of joints…(4)
2. Joints of stem approximately circular in cross-section, cylindrical to club-shaped, often prominently tubercled (if not, thin-cylindrical); main spines with a thin, paperlike, eventually deciduous sheath (may be restricted to tip of spine and may be absent due to age)…(3)
2′. Joints of stem usually flattened (may be circular in cross-section in *O. fragilis*, but then thick-ovoid to globose in shape), without prominent tubercles on mature stems; spines with no sheath (pricklypears)… **Opuntia**
3. Joints of stem cylindroid; spines entirely enveloped with relatively persistent, translucent, papery sheath (chollas)… **Cylindropuntia**
3′. Joints of stem obovate or club-shaped; spines with ephemeral apical sheath… **Grusonia**
4. Stems < 2 cm diam.; roots turnip-shaped or tuberlike… **Peniocereus**
4′. Stems > 2 cm diam.; roots diffuse…(5)
5. Flowers and fruits always appearing below the current stem apex on stems 1+ years old (hedgehog)… **Echinocereus**
5′. Flowers and fruits produced on new growth of the current season…(6)
6. Spines, at least some, hooked at the ends like fishhooks…(7)
6′. Spines straight to curving or arching, not fishhooked at the ends (sometimes hooked in *Coryphantha robustispina*)…(11)
7. Stems 20+ cm wide, large and barrel-like; central spine cross-ridged… **Ferocactus** (in part)
7′. Stems < 15 cm wide, not barrel-like; central spines not cross-ridged…(8)
8. Tubercles nipplelike projections, distinct, not forming longitudinal ridges…(9)
8′. Tubercles united for at least 1/2 their length and forming confluent longitudinal ridges…(15)
9. Hooked spines 6-10 cm… **Ferocactus** (in part)
9′. Hooked spines 2-3 cm…(10)
10. Some radial spines hooked… **Glandulicactus**
10′. Radial spines straight… **Sclerocactus** (in part)
11. Central spines absent (*S. mesae-verdae*)… **Sclerocactus** (in part)
11′. Central spines 1-5…(12)
12. Central spines curved, strongly cross-ribbed; fruits woolly and bearing spine-tipped sepaloids…(13)
12′. Central spines straight, not cross-ribbed; fruits glabrous…(14)
13. Central spines 30-70, 3-9 mm wide at the base; longer radial spines usually > 3 cm; ribs usually > 13; stem epidermis pubescent… **Homalocephala**
13′. Central spines 18-43, 1-2.5 mm wide at the base; longer radial spines usually < 3 cm; ribs usually < 13; stem epidermis glabrous… **Echinocactus**
14. Flowers 3.5-6 cm; radial spines 8-17; Eddy County… **Thelocactus**
14′. Flowers 2-3 cm; radial spines 13-25; widespread… **Echinomastus**
15. Flowers borne remote from the stem apex, forming a ring of flowers around the stem; glands absent from mature areoles of stems; fruits 5-30(-40) mm… **Mammillaria**

15'. Flowers borne at stem apex, in axillary end of a narrow, linear, areolar groove extending along adaxial surface of the tubercle from spine-cluster to tubercle axil; glands present or absent in mature areoles of stem; fruits 3–50 mm…(16)

16. Fruit dehiscent along 1 vertical suture, dry at maturity, short-cylindrical; seeds papillate…**Pediocactus**

16'. Fruit indehiscent, generally juicy at maturity, spherical, clavate, ellipsoid, or ovoid to obovoid; seeds reticulate or pitted…(17)

17. Flowers minute, nearly hidden in a depression at the tip of the stem; stems spherical, obscured by a dense covering of spines; fruits bright red…**Epithelantha**

17'. Flowers easily visible; stems easily seen or partially covered by dense spines; fruits green or red…(18)

18. Seeds smooth or weakly raised-reticulate, brown when dry, red or yellow when fresh; outer tepals entire, denticulate, or minutely fringed…**Coryphantha**

18'. Seeds pitted, black or brown when dry, black, brown, red, or yellow when fresh; outer tepals fringed… **Pelecyphora**

CORYPHANTHA (Engelm.) Lem. – Pincushion cactus

Erect, spherical, and unbranched, or if branched, then ultimately forming low clumps or small mats; stems single to numerous, hemispherical or globose to ovoid or cylindroid, tuberculate, 5–20 × 4–8.5 cm; spines 7–12 per areole, color various, needlelike, differentiated into radial and central spines; radial spines straight or curved; central spines straight, curved, or hooked, terete, 2–17 mm; flowers borne at or near stem apex on new growth of current year and/or last-produced areoles of preceding year, tepals 3–5.5 × 3.5–7 cm, variously colored, stigma lobes 6–13; fruits indehiscent, green, 1.4–2.5 × 1.2–1.8 mm, usually juicy; seeds usually reddish brown, 1.2–3.5 mm diam.

1. In mature plants, areolar groove extending 1/2–3/4 of distance from spine cluster to axil of tubercle; stem cortex mucilaginous; inner tepals rose-pink to magenta; stems often forming large mounds…**C. macromeris**

1'. In mature plants, areolar groove connecting spine clusters and axils of tubercles; stem cortex not mucilaginous; inner tepals yellowish; stems mostly singular…**C. robustispina**

Coryphantha macromeris (Engelm.) Lem.—Nipple beehive cactus. Plants forming low mats or hemispherical mounds to 100 cm diam.; stems hemispherical to short-cylindrical, 5–20 × (1.5–)4–8 cm; tubercles unusually large; spines mostly 7–12 per areole, radial spines mostly 9–15, central spines (1–)3–8; flowers apical or nearly so, 30–50 × 40–70 mm; outer tepals heavily fringed, bright rose-pink or magenta; fruits dark green; seeds reddish brown. Chihuahuan desert scrub, usually sandy alluvium, gypsum, or clay, rarely crevices or steep slopes, 4265–6070 ft. SE, SC, Chaves, Luna.

Coryphantha robustispina (Schott ex Engelm.) Britton & Rose [*C. muehlenpfordtii* (Poselger) Britton & Rose]—Pineapple cactus. Stems mostly single, spherical or ovoid, 5–10 × 5.5–8.5 cm; spines 7–20+, variously colored, radial spines 6–16(–20), 11–35 mm, central spines 1–4 per areole, porrect or slightly ascending; flowers nearly apical, mostly 45–64 × 50–73 mm; tepals minutely fringed, dark golden yellow; fruits green; seeds reddish brown. New Mexico material belongs to **subsp. scheeri** (Lem.) N. P. Taylor. Oak-juniper savannas to *Larrea* and *Atriplex* associations, grassy hills and valley floors, deep, sandy or silty soils, 3000–5200 ft. SE, SC, SW. **R**

CYLINDROPUNTIA (Engelm.) F. M. Knuth – Cholla

David J. Ferguson and Kenneth D. Heil

Shrubs to occasionally small trees, spreading-procumbent to erect, usually much branched, with primary branches indeterminate; stems glabrous, approximately cylindrical, weakly to more often strongly tubercled, mostly approximately straight, lateral branches randomly alternate to whorled, often segmented, terminal segments sometimes easily dislodged, 0.5–40(–50) × 0.3–5.5 cm; spines with eventually deciduous enveloping translucent papery sheath; major spines flattened to terete; flowers perfect (ours) or dioecious, with inner tepals white, green, yellow, orange, red, or magenta, sometimes brownish; fruit at maturity dry and tan to brown, or (ours) fleshy and green or yellow to red or purplish, strongly tubercled to smooth; seeds pale yellowish to tan, roughly lenticular, with encircling rim narrow and inconspicuous. Rare hybrids between species are intermediate in character to their parents and will not key out properly (Baker et al., 2009; Barnett & Barnett, 2016; Ferguson, 1999).

1. Stems mostly < 1.3 cm diam., not or weakly tubercled, side branches mostly whorled; spines 0–4 per areole; fruit not to weakly tubercled when mature, often reddish or red when ripe…(2)
1'. Stems mostly > 1.3 cm diam., prominently tubercled, side branches mostly alternate; spines numerous; fruit prominently tubercled, mostly ripening greenish to yellow, only rarely tinged reddish…(3)
2. Stems mostly < 1 cm thick; spines 0–1 per areole; flowers yellow green to yellow; fruit mostly < 1.5 cm, bright red at maturity…**C. leptocaulis**
2'. Stems mostly 1+ cm thick; spines 2–4 per areole; flowers variable in color, usually pink; fruit mostly > 1.5 cm, usually greenish, orange, or pinkish when mature…**C. kleiniae**
3. Mature plants usually > 1 m; flowers bright rose-pink to magenta (very rarely other colors)…**C. imbricata**
3'. Mature plants usually < 1 m; flowers not bright rose-pink to magenta…(4)
4. Roots tuberous; joints loosely attached; spine sheaths loose and baggy, yellow (becoming gold-brown in age), dominating color of plant from a distance; flowers green to coppery-green, sometimes flushed reddish; fruit mostly sterile, green at maturity; seeds rare; occasional in grasslands, E of the Rio Grande Valley, rare in SW…**C. davisii**
4'. Roots not tuberous; joints firmly attached; fruit usually fertile with abundant seeds, usually yellow at maturity; W of the Rio Grande Valley…(5)
5. Spine sheaths mostly loose and baggy, of varied colors but typically white to yellow and dominating color of plant; flowers greenish yellow; NW quarter of state…**C. whipplei**
5'. Spine sheaths not noticeably baggy, mostly pinkish, not dominating appearance of plant; flowers usually light orange, often with green highlights, rarely varying to red; Santa Fe to Cochiti and Española (rare)…**C. viridiflora**

Cylindropuntia ×anasaziensis D. J. Barnett & Donnie Barnett—Anasazi cholla. Hybrid between *C. imbricata* and *C. whipplei* and keys out here. Usually found near archaeological sites where one parent was prehistorically cultivated within the range of the other. These plants exhibit traits of their parents in spination and coloration, including flower color, and may appear similar to *C. viridiflora*. However, *C. viridiflora* behaves as a stable, sexually reproducing species, whereas *C. ×anasaziensis* is apparently sterile. Found in a few widely scattered localities. NW, C, WC.

Cylindropuntia davisii (Engelm. & J. M. Bigelow) F. M. Knuth [*Opuntia davisii* Engelm. & J. M. Bigelow]—Davis cholla. Shrubs, densely branched, 0.2–0.6 m; stem segments easily detached, whorled, obscuring trunks, light green or brownish/purplish in winter, 4–6 × 0.8–1.2 cm; spines 7–13(–21) per areole, spreading, obscuring stems, tan- to red-brown to nearly black, with sheaths yellow (usually ours) to tan or brownish, the longest 1.5–5 cm; tepals yellow-green or green to greenish bronze, sometimes red-tinged; fruits sometimes in chains of 2, yellow, top-shaped, often bearing few short rigid spines on basal 1/2; seeds if present few at most, tan to yellowish, 3–4 mm diam. Sandy grasslands, rarely into bordering habitats, mostly E of the Rio Grande Valley, 2900–6000 ft. SC, C, SW, Rio Arriba.

Cylindropuntia imbricata (Haw.) F. M. Knuth [*Opuntia arborescens* Engelm.]—Tree cholla, cane cholla. Vertical shrubs to small trees, with usually several short trunks, widely branching, 1–2.5(–4) m; stem segments firmly attached, whorled or subwhorled, green to gray-green, often purplish to purple in dormancy, cylindrical to weakly clavate, (5–)10–40 × (1.5–)2–3.5(–5) cm; spines (0–)6–18(–30) per areole, usually pale brownish to pinkish or red-brown, with sheaths white, yellowish, tan, pinkish, red-brown, or brown; tepals bright rose-pink to magenta (rarely other colors), to 35 mm; fruits yellow at maturity, sometimes tinged slightly reddish, fleshy, tuberculate, spineless or with few thin bristlelike spines to 1.5 cm; seeds pale tan to yellow, 3–5 mm diam. Desert shrub, grasslands, piñon-juniper woodlands. Two subspecies in New Mexico intergrade where they meet (Black Range and Cookes Range).

1. Tubercles of stems usually 2–5 cm, widely spaced (typically 4 longitudinal rows visible in side view); spines usually not obscuring stems; fruit with tubercles mostly > 2 times as long as wide and usually prominently raised and riblike; roughly E of Continental Divide…**subsp. imbricata**
1'. Tubercles of stems usually 0.5–1.5 cm, crowded (typically 5–6 longitudinal rows visible in side view); spines commonly interwoven so as to largely obscure stems; fruits with tubercles mostly < 2 times as long as wide, mostly more domelike and less prominently raised; W of Black Range and Cookes Range…**subsp. spinosior**

subsp. imbricata—Cane cholla. Desert grasslands, desert scrub, piñon-juniper woodlands, mesquite grasslands, 3600-8300 ft. Widespread except SW.

subsp. spinosior M. A. Baker, Cloud-H. & Majure [*Opuntia spinosior* (Engelm.) Toumey; *C. spinosior* (Engelm.) F. M. Knuth]—Spiny cholla. Desert grasslands and desert scrub, 4200-7200 ft. SW.

Cylindropuntia kleiniae (DC.) F. M. Knuth [*Opuntia kleiniae* DC.]—Pencil cholla, Klein cholla. Shrubs to small trees, often scraggly, often thicket-forming, openly branched, 0.5-2.5 m; stem segments usually alternate, green or tinged purplish in winter, 4-20 × 0.6-1.2 cm; spines (0-)2-4 per areole, usually in most areoles, straight to slightly curved, deflexed to erect, not obscuring stem, tan to red-brown, often yellowish apically, sheath white to yellowish or brass-orange, often of deeper color distally, acicular, the longest 1-3(-4) cm; tepals usually pink, often greenish basally, varying to yellow, tan, brownish, or purplish, 15-25 mm; fruits mostly greenish or yellowish to orange, sometimes nearly red, fleshy; seeds tan to yellowish, 3-4 mm. Limestone rocky slopes, desert grasslands, desert scrub, often creosote bush-mesquite flats and washes, 3000-6000 ft. SE, SC.

Cylindropuntia leptocaulis (DC.) F. M. Knuth [*Opuntia leptocaulis* DC.]—Desert Christmas cactus. Shrubs or small trees, occasionally thicket-forming, sparingly to densely branched, 0.5-2.5 m; stem segments usually alternate, green, gray-green, often blotched or tinged purplish, 2-8 × 0.3-30 cm; spines 0-1 (very rarely 2-3) per areole, varying from spineless to present in most areoles, deflexed to erect, flexible, straight to slightly curved, tan or grayish to red-brown, often yellowish apically; sheath mostly white to yellowish, brassy, or brownish, often of deeper color distally, the longest (0.4-)1.4-5 cm; tepals pale yellow-green to yellow, sometimes tipped red or tinged pink, 5-10 mm; fruit bright red, sometimes partly green (rarely orange or yellow instead), fleshy; seeds tan to pale yellow, 3-4.5 mm diam. Deserts, grasslands, shrublands, oak-juniper woodlands, flats, bajadas, and slopes, sandy, loamy to gravelly substrates, 2850-6000 ft. EC, C, WC, SE, SC, SW.

Cylindropuntia viridiflora (Britton & Rose) F. M. Knuth [*Opuntia viridiflora* Britton & Rose]—Santa Fe cholla. Shrubs, densely branched, 0.6-1(-2) m; stem segments moderately firmly attached, dull green to brownish/purplish in winter, 0.5-1.5 × 1.3-2(-2.5) cm; spines 2-8(-10) spines per areole, not much interlacing with spines of adjacent areoles, tan or pinkish to brown, sheaths mostly pale brown; tepals light orange, rarely to dull red, often tinged green basally, 18-30 mm; fruits tubercled, yellow, fleshy, ± 2 cm; seeds tan to pale yellowish, ± 3 mm. Sandy-gravelly to rocky alluvium slopes, open piñon-juniper woodlands, 5600-7200 ft. Known within area bounded roughly by Chimayo, Cochiti, Española, and Santa Fe. **R & E**

Cylindropuntia whipplei (Engelm. & J. M. Bigelow) F. M. Knuth [*Opuntia whipplei* Engelm. & J. M. Bigelow]—Whipple cholla, rat-tail cholla. Low to upright shrub, sparingly to densely branched, sometimes almost matlike, 0.1-0.6(-2) m; stem segments firmly attached, green, 3-9(-15) × 0.5-1.5(-2.2) cm; spines (1-)3-8(-10) per areole, in all but basal-most areoles, interlacing with spines of adjacent areoles, whitish or pale yellow, pale red-brown; tepals yellow to green-yellow, 15-25(-30) mm; fruits yellow to greenish yellow, fleshy; seeds pale yellow, 3-3.5 × 2.5-3.5 mm. Desert scrub, plains grasslands, juniper woodlands, oak, piñon, and pine forests, sagebrush, 4800-7500. NW.

ECHINOCACTUS Link & Otto – Barrel cactus

Echinocactus horizonthalonius Lem.—Blue barrel cactus, Turk's cap. Plants normally unbranched; stems pale gray-green to bright gray-blue, flat-topped or hemispherical, 4-25(-45) × 8-15(-20) cm; ribs (7)8(9), vertical to helically curving around stem; spines (5-)8(-10) per areole, pink, gray, tan, or brown, not hiding stem surface; radial spines 5(-8) per areole, similar to central spines, central spines 1(-3) per areole, 18-43 × 1-2.5(-3) mm; flowers 5-7 × 5-6.5(-9.5) cm, inner tepals bright rose-pink or magenta; fruits indehiscent or weakly dehiscent, pink or red, usually quickly drying to tan shell, 10-30 mm; seeds black or gray, 2-3 mm. Arid rocky slopes, primarily limestone, 4000-7565 ft. EC, C, SE, SC.

ECHINOCEREUS Engelm. – Strawberry cactus, hedgehog cactus, pitayita

Plants usually erect or ascending, branched or unbranched, sometimes forming dense mounds to 500 branches; stems unsegmented, green, spherical to long-cylindrical, 3–40(–70) × (2.5-)4–15 cm; spines (0-)4–36 per areole, white, yellow, reddish, brown, or black, (1-)4.5–100 × 0.1–2.5 mm, radial spines (1-)4–38(–45) per areole, straight or curved, 2–40(–50) mm; central spines 0–17 per areole, straight, curved, or twisted, never hooked; flowers broadly to narrowly funnelform, 2.5–120 × (10-)15–150 mm, tepals pink, red, magenta, orange, yellow, brownish, or greenish (rarely white); fruits green, purplish brown, pink, or red, juicy, drying quickly; seeds black or dark reddish brown, 0.8–2 mm (Taylor, 1985).

1. Tepals both crimson/scarlet/carmine and > 15 mm; tips stiff, strong enough for hummingbirds to perch; flowers remaining fully open at night and in cold temperatures...(2)
1'. Tepals yellow to brownish or pink to purple, 10–69(–77) mm and 20–35 mm or less in some populations of *E. viridiflorus*); flowers partly or completely closing at night and in cold temperatures...(4)
2. Largest spines usually terete (rarely somewhat flat); inner tepals usually crimson or scarlet, or orange-red...**E. coccineus**
2'. Largest spines angular or terete in cross-section; inner tepals usually carmine or crimson...(3)
3. Spines 3–11 per areole, angled and/or papillate (use lens, ×15); ribs 5–8; central spines 0–1(–4) per areole; widespread...**E. triglochidiatus**
3'. Spines (8-)9–18 per areole, angled to terete; ribs 8–13; central spines 1–8 per areole; SW New Mexico...**E. arizonicus**
4. Tepals usually 15 mm or less, greenish, brownish, or rarely carmine...**E. viridiflorus**
4'. Tepals 20+ mm, showy, pink to purple or yellow...(5)
5. Central spines shorter than the longest radial spines, or absent, 2–17(–30) mm; stems usually spherical to short-cylindrical; plants usually 0–10-branched...(6)
5'. Central spines mostly longer than the longest radial spines, usually 20–80 mm; stems elongate, usually short- to long-cylindrical; plants usually branched with age (except in *E. fendleri*), branches (4-)12–500...(8)
6. Spines of flower tube thick, more spinelike than bristlelike; young flower buds appearing relatively naked or merely spiny, never completely hidden by vesture; flowers succulent and durable, reopening for 2 to several days...**E. dasyacanthus**
6'. Spines of flower tube unusually slender, relatively bristlelike; young flower buds hidden by their own long wool and bristles; flowers ephemeral, often wilting after only a few hours...(7)
7. Central spines always absent; inner tepals white-banded proximally; plants from far W of the Rio Grande, mostly W of the Continental Divide...**E. rigidissimus**
7'. Central spines 0–7 per areole, usually 1–6 mm; inner tepals silvery-pink to magenta, usually white, crimson, green, or multicolored proximally, plants from E of the Rio Grande...**E. reichenbachii**
8. Plants unbranched or sometimes forming loose clumps of < 20 stems; radial spines 4–10; central spines mostly 1...**E. fendleri**
8'. Plants branched, forming clumps or compact mounds of 20–100(–500) branches; radial spines 7–14; central spines mostly 2–4...**E. stramineus**

Echinocereus arizonicus Rose ex Orcutt [*E. coccineus* Engelm. var. *arizonicus* (Rose ex Orcutt) D. J. Ferguson]—Arizona hedgehog cactus. Plants few- to many-branched; stems usually erect, cylindrical, 10–40 × 5–10 cm; spines (8-)9–18 per areole, straight or contorted; radial spines 7–14 per areole, yellowish to brownish, becoming gray, 5–25 mm; central spines 1–4 or 3–8 per areole, brownish yellow to reddish black, becoming gray, 15–50 mm; flowers 5.5–7 × 3.5–5 cm; tepals bright orange-red to dark red, tips thick and rigid; fruits green, 20–30 mm. Chihuahuan desert scrub, interior chaparral, desert grasslands, steep walls of canyons, limestone hills, among granite boulders, 4480–5640 ft. Grant, Hidalgo.

Echinocereus coccineus Engelm. [*E. triglochidiatus* Engelm. var. *melanacanthus* (Engelm.) L. D. Benson]—Claret-cup cactus. Plants commonly 20–100(–500)-branched, loosely aggregated into clumps or tightly packed into rounded mounds, to 100 cm diam.; stems erect, cylindrical or spherical, 5–40 × 4–15 cm; spines (1-)5–16(–22) per areole, mostly straight, ashy-white to gray, brown, yellowish, reddish, or black; radial spines (1-)4–13(–18) per areole, (3-)5–40(–49) mm; central spines 0–6 per areole, spreading to projecting outward, mostly (5-)10–80 mm; flowers (2.5-)3.8–8(–9) × (1.5-)3–7 cm; tepals crimson or scarlet, less often orange-red; fruits greenish or yellowish to pinkish, bright red. Chihuahuan desert scrub, piñon-juniper and oak woodlands, grasslands, montane forest, bajadas, rocky slopes, cliffs, 3980–7800 ft. W, NCS.

Echinocereus dasyacanthus Engelm. [*E. pectinatus* (Scheidw.) Engelm. var. *dasyacanthus* (Engelm.) W. Earle ex N. P. Taylor]—Texas rainbow cactus. Plants unbranched or few-branched; stems erect, ovoid, becoming cylindrical with age, 11–23(–40) × (4.5–)5.5–7(–10) cm; spines 19–28 per areole, straight, appressed or spreading in all directions, mostly pink to pale yellow, white, or tan, radial spines mostly 17–25 per areole, 5–15(–25) mm; central spines mostly 8–12 per areole, terete, largest spines 4.5–9.5(–14) mm; flowers 7–8.5(–10) × 7–12 cm, tepals yellow, salmon, or rose-pink; fruits dark dull purplish to maroon, 2–3.5 cm. Chihuahuan desert scrub, valleys to rocky canyon sides, limestone, 3545–7260 ft. NE, EC, SE, SC, Cibola, Hidalgo.

Echinocereus fendleri (Engelm.) Sencke ex J. N. Haage—Fendler's hedgehog cactus. Plants unbranched or sometimes in small clumps; stems erect or slightly decumbent, ovoid to cylindrical with age, 7.5–17(–30) × (3.3–)3.8–7.5(–10) cm; spines (2–)4–12(–16) per areole, straight or curved, radial spines mostly 4–10 per areole, white or with contrasting black or brown spines in same areoles, central spines (0)1(–3) per areole, (10–)25–42(–62) mm, like others in color or darker; flowers 5–11 × 5–11 cm, tepals magenta to nearly white; fruits red to purple.

1. Central spine straight and porrect; stems 1–10...**var. rectispinus**
1'. Central spine curving upward; stems 1–4...(2)
2. Central spine mostly 1, to 3.8 cm...**var. fendleri**
2'. Central spine mostly 0, if present, to 2.9 cm...**var. kuenzleri**

var. fendleri—Fendler's hedgehog cactus. Stems to 15 × 6 cm; spines 8–10, central spine mostly 1, to 3.8 cm, curved upward. Shortgrass prairie and piñon-juniper woodland communities, 4860–7430 ft. NC, NW, EC, C, WC, SC.

var. kuenzleri (Castetter, P. Pierce & K. H. Schwer.) L. D. Benson [*E. kuenzleri* Castetter, P. Pierce & K. H. Schwer.]—Kuenzler's hedgehog cactus. Stems to 25 × 10 cm; spines 5–10, central spine mostly 0. Gravelly to rocky slopes and benches of limestone or limy sandstone, Great Plains grassland, oak woodland, piñon-juniper woodland, 5200–6600 ft. Chaves, Eddy, Lincoln, Otero. This taxon is often included with var. *fendleri*. The complex needs further study. **FE & E**

var. rectispinus (Peebles) L. D. Benson [*E. rectispinus* Peebles]—Pink-flower hedgehog cactus. Stems to 25 × 10 cm; spines 8–10, central spine 1, straight and porrect. Sandy and gravelly slopes and benches, Chihuahuan scrub, oak woodland, piñon-juniper woodland, 4800–6000 ft. C, SC, SW. This taxon is often included with var. *fendleri*.

Echinocereus reichenbachii (Terscheck ex Walp.) J. N. Haage—Lace hedgehog cactus. Plants unbranched or sometimes 12-branched; stems erect, cylindrical or short-cylindrical, 7.5–30(–40) × (2.5–)4–10 cm; spines 15–36 per areole, white to tan, dull pink, dark brown, or purplish black, radial spines 12–36 per areole, pectinately arranged or nearly so, 2–8(–25) mm, central spines 0–7 per areole, 1–6(–15) mm; flowers mostly 4.5–8 × 5–10 cm, tepals silvery-pink to magenta; fruits green or olive-green, 15–28 mm. Chihuahuan desert scrub, grasslands, oak-juniper woodlands, 3000–4830 ft. NE, EC, SE.

Echinocereus rigidissimus (Engelm.) F. Haage—Rainbow hedgehog cactus. Plants unbranched (very rarely few-branched); stems erect, short-cylindrical, mostly 6–18 × 9–11 cm; spines 15–23 per areole, all radial, straight or slightly curved toward stem, pectinately arranged, bright pink or pink-and-white in alternating bands of color around stem, 6–10 mm; flowers 6–8 × 6–9 cm; tepals bright rose-pink or magenta, conspicuously white-banded proximally; fruits greenish or dark purplish brownish. Gravelly hills, steep canyon sides, semidesert grasslands, oak woodlands, interior chaparral, igneous substrates, 4950–5575 ft. SW.

Echinocereus ×roetteri (Engelm.) Rümpler [*E. lloydii* Britton & Rose]—Lloyd's hedgehog cactus is basically an intermediate in vegetative and floral characters to *E. coccineus* and *E. dasyacanthus* or in backcross plants. Grant, Hidalgo.

Echinocereus stramineus (Engelm.) F. Seitz [*E. enneacanthus* Engelm. var. *stramineus* (Engelm.) L. D. Benson]—Strawberry hedgehog cactus. Plants branched, forming clumps or compact mounds of

20-100(-500) branches, clumps 15-60(-100) cm; stems erect, long-ovoid, < 30 × 4.5-11 cm; spines 9-14 (-16) per areole, straight, straw-colored, darkest spines tan or brown, radial spines 7-14 per areole, 15-40 mm, central spines 2-4(5) per areole; flowers 8-12 × 10-12.5(-15) cm, tepals rose-pink to magenta; fruits bright pinkish brown. Chihuahuan desert scrub, rocky slopes, rarely flats, igneous and sedimentary substrates, 3295-4930 ft. Doña Ana, Eddy, Otero.

Echinocereus triglochidiatus Engelm.—Claret-cup cactus. Plants unbranched or 1-76-branched, forming large mounds of branches to 300; stems usually erect or nearly so, cylindrical (spherical), (2-) 5-70 × (3-)5-13 cm; spines (0-)3-11 per areole, straight to curved or contorted, white to yellow, gray, or black, radial spines (0)1-10 per areole, (0-)15-90 mm, central spines 0-1(-4) per areole, angular, (0-)50-120 mm; flowers (4-)5-10 × 3-7 cm; tepals bright orange-red to dark red, mostly 25-40 × 10-15 mm; fruits green to yellow-green or pink (rarely red). Plants of *E. triglochidiatus* in NW portion of New Mexico have been called *E. triglochidiatus* var. *mojavensis* (Engelmann & J. M. Bigelow) L. D. Benson. That taxon includes straight-spined plants. Plants with the fewest and largest spines, called *E. triglochidiatus* var. *triglochidiatus*, occupy the C and E portion of New Mexico. The largest spines, whether central or radial, of var. *triglochidiatus* are sharply angular in cross-section and 1-2 mm thick.

1. Stems with 5-7 ribs; areoles widely spaced so that green of stem not obscured; spines stout and thick, often > 1.5 mm thick, prominently angular in cross-section; radial spines usually 8 or fewer…**var. triglochidiatus**
1'. Stems with 7+ (most often 9) ribs; areoles more closely spaced and spines often obscuring green of stem; spines more slender, usually < 1 mm thick, less angular in cross-section, and usually nearly terete; radial spines usually > 8…**var. mojavensis**

var. mojavensis (Engelm. & J. M. Bigelow) L. D. Benson [*Cereus mojavensis* Engelm. & J. M. Bigelow var. *zuniensis* Engelm.]—Mojave claret-cup cactus. Stems 2.5-7.5 cm diam., clusters to 30 × 60 cm; central spines 0-2(-4) per areole, 1-5 cm, usually angular in cross-section, often curved and twisted, radial spines 5-12 per areole, 5-30 mm. Ephedra, sagebrush, and piñon-juniper communities, 5000-8000 ft. McKinley, San Juan.

var. triglochidiatus [*E. triglochidiatus* var. *gonacanthus* (Engelm. & J. M. Bigelow) Boissev.]—Thick-spine claret-cup cactus. Stems 5-12 cm diam., clusters to 90 cm across and 60 cm tall; central spines 0-1(2) per areole, 1-5 cm, angular in cross-section, often curved and twisted, radial spines (0-)5-12 per areole, 5-30 mm, angular in cross-section, often curved and twisted. Sagebrush, piñon-juniper, and mountain brush communities, 5000-7500 ft. This variety blends with var. *mojavensis* to the NW, and plants of the Four Corners region may be intermediate in character. Widespread except E.

Echinocereus viridiflorus Engelm.—Small-flowered hedgehog cactus. Plants unbranched or few-branched; stems erect, spherical to short-cylindrical, mostly 8-30 × 2.8-9 cm; spines mostly 18-30 per areole, stiff (flexible) and straight, red, white, and yellow, radial spines 12-38(-45) per areole, (2-)4-18 mm, central spines 0-17 per areole, spreading, 35-40 mm; flowers 2-3.5 × (1-)1.5-3 cm, tepals shades of yellow or brown, frequently tending toward yellowish green or brick-red; fruits green, 6-17 mm.

1. Stems ovoid to elongate-ovoid, mostly 2.5-5(-12.5) cm; flowers green to yellowish green, lemon-scented…**var. viridiflorus**
1'. Stems cylindrical, to 20 cm; flowers green, reddish brown, or red to reddish purple, not lemon-scented…(2)
2. Central spines 3-12, lower central spine usually white or white on the surface, (1.5-)2.5-4.3 cm, directed downward but usually curving upward…**var. chloranthus**
2'. Central spines (0)1-3, lower central, when present, colored like the other central spines, 0.7-1.4 cm, directed downward or porrect, straight or slightly curved…**var. cylindricus**

var. chloranthus (Engelm.) Backeb. [*E. chloranthus* (Engelm.) Haage]—Western green-flowered hedgehog cactus. Stems cylindrical, to 20 × 8 cm; spines 17-29, central spines 3-12, red, brown, or red and white, radial spines 15-23, to 12 mm; flowers dark green to yellowish green or greenish brown, 2-3.5 cm. Chihuahuan desert scrub, oak woodland, and piñon-juniper woodland communities, 3900-6000 ft. SE, SW.

var. cylindricus (Engelm.) Rümpler [*E. chloranthus* (Engelm.) Haage var. *cylindricus* (Engelm.) N. P. Taylor]—Small-flowered hedgehog cactus. Stems cylindrical, to 25 × 5-7.5 cm; spines 14-33, central spines 0–2(3), gray or yellowish at their bases with reddish distal halves, or the main central spine may be whitish to yellow, radial spines 14–23(–33), ca. 10 mm; flowers amber or sulfur-yellow to greenish brown, reddish brown, or carmine, 2–3.3 cm. Grassland, piñon-juniper, and montane communities, 4625–8600 ft. C, SE.

var. viridiflorus—Stems ovoid to elongate-ovoid, 2.5–5(–12.5) cm; spines 9–25, central spines 0–1, red, brownish, white, or pale gray, radial spines 8–24, ca. 10 mm; flowers green to yellowish green, 2–2.5 cm. Grassland, sagebrush, and piñon-juniper woodland communities, 4600–7500 ft. NE, NC, C, EC, SC.

ECHINOMASTUS Britton & Rose – Fish hook cactus

Echinomastus intertextus (Engelm.) Britton & Rose—Woven-spine pineapple cactus. Plants erect, usually unbranched, not deep-seated in substrate; stems green, spherical or ovoid to short-cylindrical, 5–17(–20) × 3–10 cm, glabrous; spines 20–29 per areole (ours), dull to pale gray, pinkish, or reddish brown, radial spines 13–25 per areole, straight or curved, shortest 5–20 mm, central spines (3)4 per areole straight, 0.5–14(–20) mm, others 10–18 mm; flowers near stem apex, 2.5–3 × 2.5–3 cm; tepals silvery-white to pale lavender-pink, rarely white with pale pink midstripes; fruits circumscissile near base, green, brown, or dull pink; seeds black or nearly so, spherical or hemispherical, 1.8–2.5 mm (Anderson, 1986).

1. Radial spines and adaxial central spines closely appressed; abaxial, porrect central spine 0.5–4(–5) mm… **var. intertextus**
1'. Radial and adaxial central spines diffusely spreading; abaxial, porrect central spine 4–15(–20) mm… **var. dasyacanthus**

var. dasyacanthus (Engelm.) Backeb. [*Neolloydia intertexta* (Engelm.) L. D. Benson var. *dasyacantha* (Engelm.) L. D. Benson]—Stems spherical, obovoid, or ovoid to cylindrical, appearing bristly; spines slightly appressed to spreading, except for the porrect adaxial central spine, which is 4–20 mm. Desert grasslands and plains grasslands, upper edge of Chihuahuan desert scrub, grassy hills, bajadas, sometimes with oak and juniper, igneous substrates (rarely limestone), 3610–6870 ft. Doña Ana.

var. intertextus—Woven-spine pineapple cactus. Stems spherical when young, ovoid to ovoid-cylindrical with age, appearing smooth; spines closely appressed, except for the porrect abaxial central spine, which is 0.5–4(–5) mm. Desert grasslands and plains grasslands, grassy hills, bajadas, sometimes with juniper or oak, igneous substrates (rarely limestone), 4000–5800 ft. C, SC, SW, Cibola.

EPITHELANTHA F. A. C. Weber ex Britton & Rose – Button cactus

Epithelantha micromeris (Engelm.) F. A. C. Weber ex Britton & Rose—Button cactus. Plants miniature, erect, usually unbranched, not deep-seated in substrate; stems mostly spherical, often flat-topped, 1–4(–6) × 2–4(–6) cm, surface completely obscured by spines; spines 20–40 per areole, in 1–3 series, white to ashy-gray, straight, terete, slender, not distinguishable as radial and central spines; flowers borne at adaxial margins of spine clusters deep within woolly stem apex, inconspicuous, 0.6–0.9 × 0.3–0.5 cm, tepals pink to white or pale yellow, (1–)2–6(–9) × 1–2.3(–3) mm; fruits indehiscent, bright red, weakly succulent, soon drying and papery; seeds blackish, 1.2–1.4 × 1 mm. Rocky substrates of limestone and igneous origin, hills and ridges, desert grasslands, Chihuahuan desert scrub communities, 3900–6400 ft. C, SE, SC, SW.

FEROCACTUS Britton & Rose – Barrel cactus

Plants short-cylindroid, globose, or ovoid; stems unbranched or few-branched from the base, mostly 15–100 × 10–45(–60) cm; central spines mostly 4, reddish to salmon, light gray or yellowish to brown, cross-ribbed, straight, curved, or hooked, lower one the largest and hooked, mostly 3–11 × 0.5–3.5 mm, flattened, radial spines mostly 8–15 per areole, brown, reddish, stramineous, or grayish, needlelike or slender, straight or curving, 1.5–7 cm; flowers on new growth, borne in a ring near the apex; 4.5–7.5 cm diam., tepals yellow, orange-yellow, or red; fruit fleshy, green to reddish green, globular or short-cylindroid, 3–5 cm; seeds black, 1.4–1.6 × 1–2.5 mm (Taylor, 1979).

1. Mature stems mostly 15-200 cm (1-3 m); central spines strongly cross-ribbed, mostly 1.5-3.5 mm wide at base; fruit yellow at maturity, the rind thick, pulp dry…**F. wislizeni**
1'. Mature stems mostly 10-30 cm; central spines not cross-ribbed, mostly 1-2 mm wide at base; fruit green or reddish tinged, the rind thin, pulp juicy…**F. hamatacanthus**

Ferocactus hamatacanthus (Muehlenpf.) Britton & Rose—Giant fish hook cactus. Stems erect, spherical to short-cylindrical, 10-63 × 7.5-30 cm; spines 12-16(-28) per areole, pinkish brown and/or straw-colored, often imparting appearance of dried grass clump, central spines 4(-8) per areole, curved or strongly hooked, principal central spine (40-)60-165 × 1-5 mm; flowers yellow, 5.5-8(-10) × 6.5-9.5 cm; fruits indehiscent, green or maroon, juicy; seeds 1.4-1.6 mm. Chihuahuan desert and semidesert grasslands, 4000-5500 ft. Otero. This taxon is known tentatively from the Cornudas Mountains area. No vouchers have been seen.

Ferocactus wislizeni (Engelm.) Britton & Rose—Candy barrel cactus. Stems usually leaning southward, depressed-spherical to ovoid-cylindrical, 19-100(-300) × (20-)36-65(-100) cm; spines 16-25(-29) per areole, central spines and larger radial spines dull pink, gray, or tan; smallest spines per areole white, central spines (1)2-4, often with several subulate subcentral spines, rigid; principal central spine strongly hooked, 36-120 mm from curve of hook to base of spine; flowers similar in color inside and out, 4-8.5 × 4-6.5 cm, tepals orange, red, or yellow; fruits ± readily dehiscent through basal pore, bright yellow, fleshy; seeds mostly 2-2.5 mm. Desert scrub, grasslands, oak woodlands, flats, bajadas, usually relatively deep soils of limestone and igneous origin, 4000-5500 ft. SC, SW.

GLANDULICACTUS Backeb. – Eagle-claw cactus

Glandulicactus uncinatus (Galeotti ex Pfeiff.) Backeb. [*Ancistrocactus uncinatus* (Galeotti) L. D. Benson var. *wrightii* (Engelm.) L. D. Benson]—Eagle-claw cactus. Plants erect, usually unbranched, not deep-seated; stems bluish green or grayish green, spherical to cylindrical, with ribs, mostly 7-15 × 5-7.5 cm; spines straw-colored to pale gray or some pink to reddish, radial spines (5-)8-10 per areole, abaxial 3 hooked, mostly 20-35 mm, central hooked spine 50-90(-130) × 0.5-1(-1.7) mm; flowers at stem apex, orange-red or reddish to purplish or brownish purple; fruit indehiscent, brilliant red (green), very succulent, conspicuously scaly; seeds black, 1.3-1.5 × 0. 8-1 mm. Our material belongs to **var**. **wrightii** (Engelm.) Backeb. Chihuahuan desert scrub, semidesert grasslands, limestone outcrops, sometimes igneous substrates or alluvium, 4500-6560 ft. SE, SC.

GRUSONIA F. Rchb. ex Britton & Rose – Club cholla

David J. Ferguson and Kenneth D. Heil

Mostly trailing shrubs forming mats or clumps; stems segmented, segments determinate, firmly attached to easily dislodged, subequal in length, cylindrical to roughly globular, often obovoid or ovoid to clavate, prominently tubercled (ours) or ribbed, glabrous; spines with sheath only at apices, early-deciduous; major spines mostly angular-flattened to ribbonlike, occasionally terete, mostly swollen at base; flowers with ovary and floral tube bearing prominent tufts of white to brown wool and often spines; fruit obconic to ellipsoid, obovoid, or globose, tuberculate, fleshy, usually yellow at maturity but in most species soon drying, often with numerous prominent glochids and often bearing spines; seeds tan to pale yellowish, approximately lenticular, with marginal rim inconspicuous (Fenstermacher, 2016; Pinkava, 1999). The genus *Corynopuntia* F. M. Knuth is included here.

1. Swollen tuberous roots sometimes present; stem segments easily dislodged; spines relatively slender, terete or flattened; near S state line E from Rio Grande Valley…(2)
1'. Roots diffuse-fibrous; stem segments very firmly attached; main spines stout, angularly flattened; from elsewhere…(3)
2. Often with greatly swollen tuberous central roots; spines terete or nearly so…**G. grahamii**
2'. Usually without greatly swollen tuberous roots; main spines distinctly flattened; rare in New Mexico… **G. schottii**
3. Stem segments 2.5-5(-7.5) cm; tubercles prominent, narrow, 4-6 times longer than wide, mostly ob-

scured by interlacing spines; Rio Grande and Pecos drainages, N from about Truth or Consequences and Roswell; rare in Four Corners…**G. clavata**

3'. Stem segments 7–19 cm; tubercles very prominent, broad, 1–3.5 times longer than wide, little obscured by interlacing spines; Gila drainage, SW…**G. emoryi**

Grusonia clavata (Engelm.) H. Rob. [*Corynopuntia clavata* (Engelm.) F. M. Knuth; *Opuntia clavata* Engelm.]—Club cholla. Shrubs, mat-forming, 5–15 cm; stem segments short-clavate, often curved, 2.5–5(-7.5) × 2–4 cm; areoles circular, wool white to gray; spines spreading, 7–15, white or tinged tan, yellowish, or pinkish, main spines angularly flattened, daggerlike, 12–40 mm; glochids yellow, ± 4 mm; tepals bright yellow, to 3 cm; fruits yellow, obconic to ellipsoid or narrowly barrel-shaped, fleshy, prominently covered by dense clusters of long glochids; seeds tan to yellowish, smooth, 3–5 mm wide. Desert and grassland communities, 3600–6500 ft. NE, NC, NW, EC, C, WC, SC. A New Mexico endemic within the Rio Grande and Pecos drainages; rare in the Four Corners. **E**

Grusonia emoryi (Engelm.) Pinkava [*Opuntia emoryi* Engelm.; incl. *G. stanlyi* (Engelm. ex B. D. Jacks.) H. Rob.]—Devil club cholla. Shrubs, clump- or mat-forming, 7–30 cm; stem segments elongate-clavate, usually curved, 7–19 × 3–6 cm; areoles with white to gray wool; spines 10–30, spreading, main spines angularly flattened, yellow or tan to red-brown or nearly black, often tipped yellow, longest 3–7 cm; New Mexico plants have several stiff terete spines per areole, to 10 mm; glochids yellow to brown; tepals yellow, 3–5 cm, style cream; fruits yellow, obconic to ellipsoid, fleshy, pale glochids prominent; seeds pale tan to yellowish, 3.5–6 mm, smooth. Chihuahuan and Sonoran Deserts, sandy or gravelly desert flats, washes, hills, 3700–4300 ft. SW.

Grusonia grahamii (Engelm.) H. Rob. [*Opuntia grahamii* Engelm.]—Graham club cholla. Compact, clumping, low shrubs, 2–12 cm; stem segments usually easily disarticulating, mostly ovoid to ellipsoid or sometimes obovoid, mostly uncurved, 1.5–7 × 1.3–3.5 cm; areoles with white to yellowish or brownish wool; spines (6–)8–15, terete to subulate, white, tan, red-brown to near black, often 2-toned or paler near tip, longest 2–6 cm; glochids white to yellowish, to 6 cm × 1 mm; tepals bright yellow to pink, 20–25 mm; fruits greenish to yellow, narrowly obconic to ellipsoid, fleshy, with prominent glochids and few thin spines; seeds tan to pale yellowish, 3–5 mm, smooth. Chihuahuan desert scrub, sandy, gravelly, or rocky bajadas, flats, and hills, 3600–5250 ft. SC, SE. Mostly sandy areas between El Paso and Las Cruces, E to W base of Guadalupe Mountains.

Grusonia schottii (Engelm.) H. Rob. [*Opuntia schottii* Engelm.]—Schott club cholla. Clump- or mat-forming shrubs, 2–12 cm; stem segments usually easily disarticulating, mostly obovoid to clavate, often curved, 2–7 × 1.3–3.5 cm; areoles with white to yellowish or brownish wool; spines 8–15, main spines flattened, white, yellowish, tan, red-brown, to 6 cm × 2(-2.5) mm; glochids white to yellowish, tan, or reddish, to ± 6 mm; tepals bright yellow, 20–25 mm; fruits greenish to yellow, narrowly obconic to ellipsoid, fleshy, with prominent glochids and few thin spines; seeds tan to pale yellowish, 3–5 mm, smooth. Chihuahuan Desert, mostly sandy, silty, or rocky flats, and low hills, 3600–4000 ft. SC, SE. Mostly in vicinity of Dell City and Cornudas Mountains. Intergrades freely with *G. grahamii* in New Mexico, and our populations are variable and somewhat atypical.

HOMALOCEPHALA Britton & Rose – Horse-crippler

Homalocephala texensis (Hopffer) Britton & Rose [*Echinocactus texensis* Hopffer]—Horse-crippler, devil's-head. Plants unbranched (very rarely branched); stems pale gray-green, aboveground portion flat-topped, deep-seated, flush with soil surface, 10–30 × 10–30 cm; spines (6–)7–8 per areole, mostly decurved or 1 porrect and straight, pale tan, pink, or reddish to gray, radial spines (5)6–7 per areole; central spine 1 per areole; (20–)40–60(–80) × 1.5–4(–8) mm; flowers 5–6 × 5–6 cm; inner tepals bright rose-pink to pale silvery-pink; fruits indehiscent, fleshy; seeds black; 2.5–3 mm. Chihuahuan desert scrub, grasslands, openings in oak woodlands, thorn scrub, deep soils, saline flats, limestone hills, 3340–4425 ft. EC, SE.

MAMMILLARIA Haw. – Pincushion cactus

Plants branched or unbranched, deep-seated in substrate or not, roots diffuse or taproots; stems unsegmented, green to gray-green, sometimes purplish under stress, spherical to cylindrical or turbinate, often flat-topped, 0-16(-30) × 2-12(-30) cm, firm or flaccid; tubercles distinct, not confluent into ribs, cortex and pith containing latex (absent in *M. grahamii*); spines (6)7-60(-90) per areole, of every color that cactus spines can be, mostly 2-25 × 0.01-0.6 mm, radial spines (5-)8-35(-80) per areole, mostly straight to curved, central spines 0-4(-7); flowers in a ring distant from stem apex, tepals yellow, white, rose-pink, magenta, or maroon, 4.5-19 × 1.5-8 mm; fruits indehiscent, usually pink, bright red, or greenish, usually juicy; seeds black, brown, or reddish, 0.8-1.5 × 0.7-1.4 mm.

1. Central spines hooked…(2)
1′. Central spines, when present, straight or curved throughout their length…(4)
2. Outer tepals entire or short-fringed; seeds small, 0.8-1.2 mm; cortex and pith not mucilaginous, cut surfaces almost dry to the touch; stems relatively firm…**M. grahamii** (in part)
2′. Outer tepals long-fringed; seeds large, 1.3-2.4 mm; cortex and pith mucilaginous (slimy); stems soft and flabby…(3)
3. Inner tepals rose or magenta…**M. wrightii**
3′. Inner tepals white, pale greenish, or pale rose-pink…**M. viridiflora**
4. Central spines 0 or indefinitely numerous (depending on interpretation), not differentiated from radial spines; spines (26-)40-60(-90) per areole; stems (1.4-)2-4(-7) cm diam.…**M. lasiacantha**
4′. Central spines (0)1-12 per areole, at least 1 strongly differentiated from radial spines; if central spines absent, then spines 10 or fewer per areole and mature stems mostly > 7.5 cm diam.…(5)
5. Stems spherical to cylindrical, (2.3-)3.5-6.8 cm diam., latex absent; inner tepals bright rose-pink or rose-purple; spines (19-)26-33(-38) per areole; seeds black…**M. grahamii** (in part)
5′. Stems low, flat-topped, usually 7.5-25 cm diam., latex present, sticky, white; inner tepals greenish yellow, white, cream, or pale pink, with midstripes of pink, lavender, or brown; spines (6-)7-22(-26) per areole; seeds orange or yellow when fresh, becoming reddish brown…(6)
6. Fruits brilliant scarlet, carmine, or crimson; radial spines often > 12 per areole…**M. heyderi**
6′. Fruits relatively pale purplish pink, whitish, or pale green; radial spines (5)6-12 per areole…**M. meiacantha**

Mammillaria grahamii Engelm. [*M. microcarpa* Engelm.]—Graham's fish hook cactus. Plants branched or unbranched, roots diffuse; stems spherical to cylindrical, usually 5-16 × 3.5-6.8 cm, firm, latex absent; spines (19-)26-33(-38) per areole, radial spines 17-35 per areole, whitish or pale tan, 6-12 × 0.1-0.15 mm, central spines (2)3-4 per areole, 1-3(4) spines hooked, reddish to purplish brown to almost black, subcentral spines 1-3 per areole; flowers 2 × 1.8-3.5(-4.5) cm, bright rose-pink or rose-purple; fruits green, turning bright red, scarlet, or carmine; seeds black, 0.8-1 × 0.7-0.9 mm. Chihuahuan desert scrub, grasslands, oak woodlands, alluvial slopes, hills, canyons, silty, sandy, gravelly, or rocky soils of igneous or calcareous origin, 3500-5780 ft. SE, SC, SW.

Mammillaria heyderi Muehlenpf.—Little nipple cactus. Plants unbranched, protruding relatively little above soil; roots obconic taproots; stems top-shaped, flat-topped, protruding aboveground, 0-2 × (4-)7.5-15 cm, latex abundant, sticky, white; spines 8-22 per areole, usually brownish, radial spines mostly 10-22 per areole, white to white-and-brown or brown, 6-15 mm, central spines (0)1(-4), 2-8 × 0.15-0.45 mm; flowers 1.9-3.8 × 1.5-3 cm; white, greenish, or cream to pale pink; fruits brilliant red, scarlet, carmine, or crimson; seeds reddish brown, sometimes yellowish, 1-1.2 mm.

1. Central spines 0.15-0.35 mm diam.; radial spines (7-)13-17(-26) per areole, abaxial radial spines 6-11(-16) mm…**var. heyderi**
1′. Central spines 0.35-0.45 mm diam.; radial spines 10-14 per areole, abaxial radial spines 9-15 mm…**var. bullingtoniana**

var. bullingtoniana Castetter, P. Pierce & K. H. Schwer.—Cream cactus. Stems 5-11 × 5-12(-15) cm, tubercles 11-25 × 4-8 mm; spines usually 10-14 per areole, radial spines usually 10-14 per areole, central spines, when present, 0.35-0.45 mm diam.; flowers 2.2-3.3 × 1.7-3 cm. Chihuahuan desert scrub, desert grasslands, lower edge of oak zone, 3935-6890 ft. SC, SW.

var. heyderi [*M. gummifera* Engelm. var. *applanata* (Engelm.) L. D. Benson]—Heyder's nipple cactus. Stems 4-9 × 4-10(-15) cm, tubercles 9-20 × 3-7 mm; spines (8-)14-18(-27) per areole, radial spines

(7-)13-17(-26) per areole, central spines stiff, 0.15-0.35 mm diam.; flowers 1.8-2.3 × 1.8-1.9 cm. Chihuahuan desert scrub and thorn scrub, usually limestone and alluvial substrates, 3295-5200 ft. WC, SE, SC.

Mammillaria lasiacantha Engelm.—Golf ball cactus. Plants unbranched, usually deep-seated in substrate and inconspicuous; stems depressed-spherical to short-cylindrical, (1-)2-3.5 × (1.4-)2-4(-7) cm, firm, latex clear or slightly milky; spines mostly 40-60 per areole, in several series but all equally thin, mostly appressed, white or very pale pink, bristlelike, 0.6-5(-6) × 0.05-0.1 mm, central spines 0; flowers 0.9-1.5(-2) × 0.8-1.3 cm, tepals white or cream, usually with sharply defined midstripes; fruits scarlet, cylindrical or clavate; seeds black. Chihuahuan desert scrub with *Agave lechuguilla*, rocky hills, gravelly slopes, usually on limestone, 3295-6300 ft. Eddy, Lincoln, Otero.

Mammillaria meiacantha Engelm. [*M. heyderi* Muehlenpf. var. *meiacantha* (Engelm.) L. D. Benson]—Nipple cactus. Plants unbranched; stems flat-topped, 10 × 8-10(-30) cm, firm; tubercles 8-17 × 4-11 mm; latex abundant, sticky, white; spines (6)7-8(-10) per areole, needlelike, 6.5-13.5 mm, white, reddish brown, gray, or yellowish, central spines (0)1 per areole, inconspicuous against radial spines, mostly 5-12; flowers 2.5-3.5 × 1.9-3.5 cm, white to pale pink, often with pink or lavender midstripes; fruits purplish pink; seeds reddish brown, 1.1-1.2 mm. Great Plains grasslands, Chihuahuan desert scrub, pine-oak woodlands, 4860-7000 ft. NE, NC, C, SE, SC.

Mammillaria viridiflora (Britton & Rose) Boed.—Green flower nipple cactus. Plants usually unbranched; stems flat-topped or spherical to short-cylindrical, ± flaccid, latex absent; spines 19-31 per areole, usually white or brown-and-white, radial spines 13-27 per areole, central spines 1-2 per areole, porrect or strongly projecting, all hooked, 7-20 mm; flowers 2-3.5 × 1.8-3 cm, tepals usually white, cream, pale tan, greenish white, or pale rose-pink; fruits green or purple; seeds chocolate-brown, 1.3 × 0.9 mm. Semidesert grasslands, interior chaparral, piñon-juniper and oak woodlands, crevices, boulders, canyon sides, gravelly igneous substrates, 5800-6400 ft. Grant.

Mammillaria wrightii Engelm.—Wright's nipple cactus. Plants usually unbranched; stems flat-topped or spherical, 4-8 × 4-8 cm, ± flaccid; latex absent; spines 9-31(-34) per areole, white, usually tipped brown, radial spines 8-30 per areole, lateral spines bristlelike, 7-11 mm, central spines 1-4(-7) per areole, 1 or all hooked, (5-)12-14(-21) mm; flowers 2.5-3.5 × 2.2-4.5 cm, tepals rose-pink or magenta (white), margins often paler; fruit green or dull purple, juicy; seeds black, 1.3-1.5 mm.

1.   Radial spines usually 8-15 per areole; fruits 12.5-26 mm diam....**var. wrightii**
1'. Radial spines usually 16-30 per areole; fruits 6-15 mm diam....**var. wilcoxii**

var. wilcoxii (Toumey ex K. Schum.) W. T. Marshall [*M. wilcoxii* Toumey ex K. Schum.]—Wilcox's fish hook cactus. Stem tubercles 6-21 mm; radial spines usually 16-30 per areole; flowers 2.2-5.1 cm diam.; fruits 6-15 mm diam. Semidesert grasslands, Madrean pine-oak woodlands, steep, rocky slopes, canyons, valleys, usually on alluvial or igneous substrates, 4600-6500 ft. SC, SW.

var. wrightii—Wright's fishhook cactus. Stem tubercles 8-24 mm; radial spines usually 8-15 per areole; flowers 2.5-7.5 cm diam.; fruits 12.5-26 mm diam. Semidesert grasslands, plains grasslands, piñon-juniper woodlands, gentle slopes, mesas, valleys, usually on alluvial or igneous substrates, 5270-7500 ft. C, WC, McKinley.

OPUNTIA Mill. - Pricklypear

David J. Ferguson

Low shrubs to small trees, clumping or trailing to erect, usually multibranched; stems nonwoody to woody, initially composed of series of cladodes (pads), green, sometimes bluish or reddish to purple, usually flattened, 1-60(-120) × 1.2-40 cm, nearly smooth to tuberculate, glabrous or pubescent, areoles arranged diagonally across face of cladodes or "pads," bearing wool, glochids, and usually spines; spines 0-15+ per areole, may be barbed, varied in color from white, gray, or yellow to deep red-brown or black, often fading to white, gray, or black with age; tepals and filaments matching in color or not, white, yellow, or orange to pink, red, or magenta; stigma usually white or yellowish to green; fruits if dry, tan to brown (fading to gray), if fleshy, green, yellow, orange, or red to deep purple, variously cylindroid or barrel-shaped, spineless or spiny; usually with glochids; seeds roughly lenticular to thick-discoid, 2-10 mm diam. (Crook & Mottram, 1995-2005; Ferguson, 1987, 1988, 1999; Griffiths, 1908-1911, 1914-1916; Pinkava, 2003).

In keying *Opuntia*, it is best to look at a combination of traits. Cladodes too young or old and edge areoles may not properly show the characteristics. The keys leading to *O. phaeacantha* and *O. engelmannii* must be used with some allowance for overlapping traits. Every trait may be affected by the environment and should be used tentatively and in combination with other traits. The fact that these plants may grow together and with similar *Opuntia* species but do not intergrade shows their distinctness from one another. They are often similar to one another, and a reliable key is difficult to construct.

1.   Low plants with creases on older cladodes; wrinkled and often becoming prostrate in winter; fruit without juicy pulp, dry and brittle at maturity, often with numerous spines…(2)
1'.   Plants of various habit; fruit fleshy, with juicy pulp at maturity, with at most a few inconspicuous spines… (6)
2.   Cladodes usually in tight clumps from common base, grayish to bluish, mostly broad and flat above, tapering down to approximately cylindrical at base; spineless with areoles mostly sunken in dimplelike depressions, glochids often short and inconspicuous; flowers pink to magenta…**O. basilaris**
2'.   Cladodes usually stacked or chaining along ground, glabrous; flowers various, but often not pink to magenta; stigma rich green…(3)
3.   Very small, cladodes mostly < 5 cm and proportionately thick, may be little flattened and obovoid to nearly globose, easily detached from plant; mostly in sandy areas…(4)
3'.   Plants with cladodes larger, mostly at least 5 cm, broad and distinctly flattened, not easily detached; plants of various habitats…(5)
4.   Plants forming colonies interconnected by rhizomes, bearing areoles and glochids, ± 6 mm diam.; usually with majority of cladodes at least somewhat flattened; spines mostly deflexed, not strongly barbed; rare; SC and SW…**O. arenaria**
4'.   Plants mostly lacking rhizomes; most plants with majority of cladodes rounded and little to not flattened; spines mostly spreading outward, barbed and easily/firmly attaching to flesh or cloth; NW…**O. fragilis**
5.   Plants mostly compact, little-spreading, procumbent to upright clumps of few-segmented stems from a central base; cladodes with areoles typically < 2 mm on young mature cladodes, and numerous; spines in all areoles numerous, white to yellow, slender, flexible, often elongating greatly with age; spines on fruit similarly colored, thin and flexible; flowers green-yellow or yellow (may fade to orange before closing), without red bases; filaments often red; steep rocky slopes and gypsiferous sites…**O. trichophora**
5'.   Plants mostly spreading mats of chaining cladodes, rooting as they progress; cladodes with areoles often large and few in number; spines variable in number, length, and thickness, normally stiff, sometimes with some slender and flexible along lower portion of older cladodes; flowers and filaments variable in color from yellow to magenta, often with red tepal bases; various habitats…**O. polyacantha**
6.   Sprawling plants with long, narrow cladodes mostly 0.5-1(-2) m × 10-20 cm; slender yellow spines (0)1-2(-4)…**O. lindheimeri** (in part)
6'.   Plants various, pads variously shaped but never long and narrow as above; spines various…(7)
7.   Plants spineless, low, broad, many-stemmed, shrublike, mostly 30-60 cm; cladodes green to purplish, minutely pubescent, oval to circular; glochids in prominent tight clumps, increasing in length and number on older trunks; flowers pale yellow (very rarely orange or red) with deep green stigmas…**O. microdasys**
7'.   Growth habit various, not as above; cladodes spreading or clumping low to ground, glabrous, usually bearing at least some spines; flower color various…(8)
8.   Plants low, nonwoody, creeping or sometimes tightly clumping, < 30 cm, becoming prostrate and mostly strongly wrinkled and soft in winter; often developing permanent creases across older cladodes…(9)
8'.   Plants low to upright, woody-stemmed, often > 30 cm, rarely becoming noticeably wrinkled during winter or drought; without permanent creases across midcladodes…(14)
9.   Main spines slender, 1-2; upper areoles occasionally with 1(2) short, fine, deflexed spines; flowers usually with pale stigmas and often nearly white…(10)
9'.   Main spines thicker than above, 3+ spines in most areoles; stigmas usually green…(11)
10.   Plants spreading mats of chaining, mostly obovoid cladodes, rooting where they touch the soil; often with fusiform or nodulelike tubers along adventive roots; flowers usually pale yellow, often red in center, opening wide, mostly 6+ cm diam.; sandy soils in grasslands, piñon-juniper woodlands, ponderosa pine forests…**O. macrorhiza**
10'.   Plants not spreading mats of obovoid chains; cladodes in clumps from swollen tuberous taproot, not rooting when contacting ground; often narrowed near base (stipitate); flowers of varied colors, mostly not opening widely and remaining roughly "tulip-shaped," < 5 cm diam.; not in sandy sites…**O. pottsii**

11. Cladodes mostly rhombic or broadly elliptic, becoming lavender in winter; main spines 2–4, stout, angular in cross-section, 4–5 cm, white to yellowish-tinged; lower bristlelike spines often missing or sometimes 1 or 2 in few areoles, white; flowers yellow (rarely to orange), without red center; fruit orange to red; scattered colonies on sandy soils; W from Rio Grande Valley, N from Belen and Fence Lake…**O. zuniensis**

11′. Cladodes rounded to obovate; spines 5+ in most areoles; with both stout longer main spines and thinner, shorter, spines conspicuous in most areoles; flowers yellow, orange, pink, red, or magenta; fruit purplish red and/or brownish-tinged…(12)

12. Older cladodes with creases; main spines terete, erect with 2–5 mixed-size flattened lower deflexed spines; flowers yellow without red center; petaloids curved upward distally (flowers roughly open "tulip-shaped") and rolled down along margins, giving ruffled or crisped appearance; fruit dull brownish-tinted and sweet, with a few spines; seeds irregular with wide margins, 6+ mm diam.…**O. cymochila**

12′. Older cladodes without creases; main spines often > 1 mm thick, often angular in cross-section, not as above; flowers yellow, with orange or red center; petaloids usually widely spreading and not noticeably rolled under at margins; fruit dark purplish red, orange, or pink, not brownish-tinted when ripe, bland; seeds usually smaller than above, with narrow rims and more regular in shape…(13)

13. Cladodes < 10 cm, rounded and nearly as broad as long; with > 5 areoles in diagonal row across midcladode; main spines 4+…**O. tortispina**

13′. Cladodes > 10 cm, obovate and longer than broad; with < 6 areoles in diagonal row across midcladode; main spines < 4…**O. phaeacantha**

14. Plants treelike in habit, typically 2(–2.5) m, developing 1(–5) near-erect trunks; flowers yellow with stigma usually pale…**O. chlorotica**

14′. Plants not treelike in habit, either low and spreading or bushy, rarely developing vertical trunks, and then usually only as part of a larger multistemmed bush; fruit juicy and usually entirely eaten by animals upon ripening; flowers various…(15)

15. Cladodes and fruits mostly 7+ per diagonal row across midcladode; areoles on sides of most recent mature cladodes rather small, with glochids arranged in an inner clump and outer ring, usually of different lengths; spines mostly few and relatively slender…(16)

15′. Cladodes and fruits usually 7 or fewer per diagonal row across midcladode; areoles and spines various…(18)

16. Plants < 60 cm wide and tall, branches mostly erect or ascending; cladodes < 17 cm, narrowed toward base and obdeltoid to obovate, or circular; dull grayish or bluish, turning purple to light magenta (often intensely so) in dormancy or under stress; spines 0–4, mostly only in upper areoles, with main spines usually slender, to 12 cm, black-brown to black, white at tip, or occasionally all white; flowers yellow with deep red center; stigma white to pale greenish; fruit typically of a dull brown-tinted pink to light purple, narrowed at the top with acute flangelike rim…**O. macrocentra**

16′. Plants well > 60 cm wide, branches often lying on ground; cladodes > 17 cm, broadly ovate to circular, typically dark to rich green or slightly bluish when young, more yellowish in age; spines 1–3 (occasionally more, or none) per areole, < 3 cm, relatively pale in color and not white or black; flowers yellow, may fade to orange, rarely with distinct orange or red centers, (if so, not sharply defined); fruit usually deep purplish when ripe, usually barrel-shaped or obovate to globose and with top curved inward to rim…(17)

17. Low-spreading bushes, 30–60(–90) cm tall and much wider; cladodes obovate to ovate or circular, wavy; becoming pale purplish in winter (sometimes only blotches); areoles on sides of mature cladodes < 3 mm; spines creamy or pale golden to gold-brownish or dark red-brown, darkening toward base, stoutest terete; flowers typically pale yellow, often slightly brassy in center, with pale stigma; fruit 4–5 cm; seeds ca. 4–5 mm diam.; seedlings not hairy…**O. gilvescens**

17′. Large, robust shrubs, 1–2 m tall and somewhat to much wider; cladodes broadly ovate or circular, or sometimes broadly oval, not narrower in basal 1/2 than apical 1/2; not or only faintly becoming purplish in winter; areoles on sides of most mature cladodes 3–4 mm; spines pale whitish, yellowish, or brownish, little darkened toward base, 1 usually stouter, somewhat flattened, and curving down; flowers bright yellow with large dark green stigma; fruit > 5 cm; seeds 3–4 mm; seedlings hairy…**O. orbiculata**

18. Areoles of mature cladodes < 3 mm, not elevated, with glochids evenly and compactly arranged in 2 distinct series, an inner clump and an outer ring of different lengths; flowers yellow, with orange to red centers, "veins" of petaloids not noticeably darker than surrounding color; stigma pale; fruit without tubercles, light-colored seedlings not hairy…(19)

18′. Areoles of mature cladodes > 3 mm, often with stem slightly elevated around areole, with glochids of mixed sizes scattered through areole; flowers yellow, fading orange, with fine, faintly orange veining, flushed reddish toward center, but rarely with defined orange or red centers; stigma robust, rich green; fruit large, with slight tubercles, deep purple when ripe; seedlings often hairy…(21)

19. Small, spreading, ground-hugging plants, not > 2–3 cladodes high, not > 30 cm (taller in protected locations); cladodes not > 17 cm; flowers ± 5–7 cm diam., not opening widely, yellow, with orange to red center; stigma white to pale green; fruit red to deep purplish red and paler, 4–5 cm; seeds 4–5 mm diam.…**O. camanchica**

19′. Large, wide, bushy plants, often several cladodes high; cladodes > 17 cm to considerably larger; flowers > 7 cm diam., opening widely, often with orange to red center; stigma usually stout, green; fruit > 5–6 cm, dark red to magenta when mature; seeds 3–4 mm diam….(20)

20. Joints often glaucous, broadly obovate to nearly circular; spines > 3 per areole, with longest porrect and to 5 cm; main spine whitish with orange-brown base, sometimes entirely white or considerably darker, rarely yellowish; flowers rich or pale yellow, orange or red in center, not fading orange before closing; fruit barrel-shaped to roughly egg-shaped, usually sweet; lower mountains, hills, canyons…**O. dulcis**

20′. Joints not glaucous, narrowly to broadly obovate; spines usually 3 or fewer per areole, with longest somewhat deflexed and < 3 cm; main spine pale to distinctly yellowish or pale brassy, with a translucent "smooth" appearance, reddish or brownish toward base; flowers rich yellow or orangy-yellow and fading to orange before closing; fruit distinctly pyriform, bitter/sour; shallow breaks in shortgrass prairies; S from Roswell and Tatum, E of the Pecos Valley…**O. pyrocarpa**

21. Plants compact bushes < 1 m tall and wide; main stems ascending to erect; younger cladodes appearing noticeably grayish/bluish even when old, broadly obovate to circular, < 20 cm; spine arrangement dimorphic, either with 3–5 main spines, often angular in cross-section, radiating outward and downward and with dark bases and subtended by few small white spines, or with only 1–3 shorter white spines deflexed downward from somewhat smaller areoles…**O. confusa**

21′. Plants diffuse, spreading, or considerably larger; coloring and dimensions various, not in above combination; spination not noticeably dimorphic…(22)

22. Plants spreading and < 1 m; cladodes rounded, averaging < 25 cm, though often with some pads larger; fruit broad, broadly obovate to near globose, ca. 1–1.3 times as long as wide, < 5 cm…(23)

22′. Plants > 1 m; cladodes various, averaging > 25 cm; fruit 5+ cm…(24)

23. Areoles pale; spines white to yellow, without dark base, appearing translucent, main spines terete to subulate 1–3(–5), to ± 3 cm, with 1 sometimes porrect, but mostly deflexed or spreading downward, some curving downward; canyons and mesas, prairies, desert scrub, piñon-juniper woodlands…**O. cyclodes**

23′. Areoles dark; spines white or cream to pale yellowish with brown to nearly black base, appearing opaque, main spines variously terete, flattened or angular in cross-section, longest > 4 cm, spreading outward and downward, or some curving slightly upward; desert scrub, piñon-juniper woodlands…**O. arizonica**

24. Cladodes obovate to circular or broadly ovate, not > 1.5 times as long as wide; areoles light in color; spines not increasing in number with age, oldest stems bare of spines; fruit rounded, not much longer than wide to ca. 1.5 times as long as wide…(25)

24′. Cladodes more elongate, > 1.5 times as long as wide, oval or obovate, tending toward rhombic; areoles dark (often blackish) brown; spines tending to increase in number on older stems; fruit more elongate, about twice as long as wide…(27)

25. Cladodes circular or broadly obovate, often wavy or curved (potato chip–like), somewhat bluish or grayish; main spines stout, similar in most areoles, angular in cross-section, < 3 cm, 3–5 arranged in a bird's-foot pattern, opaque chalky-white (rarely yellowish), base sometimes darkened, and if so chocolate-brown, not yellowish, with 1–2 much smaller spines randomly placed, and 1 of those darker in color; glochids yellow to yellow-brown; bajadas and rocky slopes, S of Belen and W of the Pecos River…**O. discata**

25′. Cladodes longer than wide, obovate, broadly rounded apically, not wavy, less bluish or grayish; spination highly variable, 0–6 main spines well > 3 cm, 1–2 erect and terete with 1–3 flattened to angular, spreading down and laterally, 0–3 small white lower spines, spines unevenly distributed on cladode, missing from many areoles; areoles yellow to red-brown; distribution various…(26)

26. Spines longest and more numerous in distal 1/2 of cladode, spines opaque, white, cream, tan, or somewhat yellowish, bases often entirely dark red-brown, often with "annular" patterning; 0–3 smaller white lower spines; fruit sweet; S of Albuquerque and throughout S New Mexico…**O. engelmannii**

26′. Spines longest and most numerous toward lower side margin and toward distal end of vertical cladodes; spines translucent yellow, rarely almost white, sometimes reddish or brownish near base; fruit sour; near Pecos River and near E base of Guadalupe Mountains…**O. lindheimeri** (in part)

27. Cladodes dark yellowish green, with new growth distinctly dark and brownish at first; spines 1–5 per areole, relatively slender, greatly varied in length from plant to plant, 1–10 cm, pale yellowish apically, grading through white and yellow to a nearly black base; flowers deep yellow-tinged orange, fading to orange before closing; fruit sour; locally common at lower levels on W side of Sandia, Manzano, and the Los Pinos Mountains…**O. wootonii**

27′. Cladodes distinctly bluish or grayish green, especially younger ones, new growth green to blue-green; spines several per areole and stout, angular in cross-section; annular-patterned, white to cream-colored with deep brown bases; flowers light yellow, not fading to orange; fruit sweet; Sandia Mountains and San Ysidro S into Trans-Pecos, Texas…**O. valida**

Opuntia arenaria Engelm. [*O. polyacantha* Haw. var. *arenaria* (Engelm.) B. D. Parfitt]—Sand pricklypear. Small clumping plants mostly < 10 cm; cladodes easily detached, small, 1–5(–7) cm, proportionately thick, circular to flattened in cross-section, variable in shape, from elongate to circular or globose, main spines 1–3, mostly limited to distal 1/2 of cladode, reflexed to porrect, usually terete, mostly whitish to brown or yellowish, longest exceptionally to 4 cm, smaller spines 3–7 usually in all areoles, deflexed, white, mostly < 1 cm; glochids prominent, most often yellow; flowers yellow, often with red in center, fading to peachy-pink or orange before closing; stigma green. Stable eolian sand areas, mostly in dune fields, 3650–4500 ft. Doña Ana, Luna. **R**

Opuntia arizonica Griffiths—Arizona pricklypear. Much-branched shrub mostly < 1 m tall and wider than tall; main branches usually lying along ground; cladodes broadly obovate to nearly circular, ca. 20 cm (exceptionally to 27 cm); spines mostly in all areoles, 3–6(–8), with 3 or 4 largest to 5 cm; flower petaloids yellow, faint reddish veining often visible; stigma stout, green; fruit broadly obovate to nearly globose, to 4–5 cm, often slightly "lumpy," deep purplish red inside and out (sometimes partly greenish inside), very juicy and sweet; seeds ± 4 mm; seedlings not hairy. Desert scrub, piñon-juniper woodlands, 4100–7000 ft. NC, EC, C, WC, SE, SC, SW. Similar to *O. engelmannii* and *O. discata*, but smaller, lower, more spreading, and spinier on average.

Opuntia basilaris Engelm. & J. M. Bigelow—Beavertail cactus. Tightly clumping plants with cladodes arising from base of older cladodes and only occasionally stacking to 2 cladodes high, oval, obovate, obdeltoid, spatulate, dull gray- or blue-green, sometimes purplish in winter, spineless, minutely pubescent; flowers early (usually first species of *Opuntia* to flower), pink to magenta (very rarely white), stigma white to faintly yellowish; fruit quickly dry at maturity, tan to grayish; seeds large, ± 6–7 mm, creamwhite. Our plants are **var. basilaris**. Occasional adventive in Chihuahuan desert scrub, 3700–4100 ft. Doña Ana.

Opuntia camanchica Engelm. & J. M. Bigelow. [*O. phaeacantha* Engelm. var. *brunnea* Engelm.; *O. chihuahuensis* Rose]—Comanche pricklypear. Low-spreading to creeping plants typically not > 2 cladodes (± 30 cm) high, spreading to 2+ m, obovate to circular, 12–16(–20) cm; spines 1–6(–8), main upper spines erect to spreading or deflexed, angular in cross-section, to 5–10 cm, white or yellow to red-brown or black, usually dark basally, small lower spines mostly deflexed and white; glochids mostly in 2 even concentric compact groupings; flowers yellow, often with well-differentiated orange to red center; stigma white to pale green; fruit barrel-shaped to obovate, dull red to deep purplish red, greenish, pinkish, or magenta inside, bland; seeds 4–5 mm; seedlings not hairy. Bajadas and flats in desert scrub, prairies, piñon-juniper and oak woodlands, ponderosa pine forests, 2900–7500 ft. Widespread.

Opuntia chlorotica Engelm. & J. M. Bigelow [*O. palmeri* Engelm. ex J. M. Coult.]—Plants erect, treelike, to 2(–2.5) m with 1(–4) well-defined vertical trunks to 20(–30) cm diam.; cladodes circular to obovate or oval, 15–20 cm; spines 0–3(–6) in only distal or all areoles of younger cladodes, increasing to 30+ per areole on old stems and trunks; flower petaloids yellow, sometimes with faint reddish veining or longitudinal streaks, without saturated red bases; fruit broad, with numerous prominent areoles, bland; seeds ± 3(–4) mm; seedlings not hairy. Mostly on steep rocky slopes among open scrub or woodland, occasionally scattered in more gentle terrain of grasslands or desert scrub communities, 4000–7200 ft.

1. Stems green or somewhat yellowish; spines mostly < 4 cm, stiff; spines and glochids yellow; fruit globose, bright red, to 5 cm…**var. chlorotica**
1'. Stems mostly grayish to bluish or purplish, often becoming strongly purple in drought or winter; spines 3–7 cm, flexible; spines and glochids yellow to red-brown; fruit barrel-shaped, pink to purplish, to 4 cm…**var. santa-rita**

var. chlorotica—Pancake pricklypear, clock-face pricklypear. WC, SC, SW.

var. santa-rita Griffiths & Hare [*O. gosseliniana* F. A. C. Weber var. *santa-rita* (Griffiths & Hare) L. D. Benson; *O. violacea* Engelm. var. *santa-rita* (Griffiths & Hare) L. D. Benson]—Santa Rita pricklypear. Hidalgo.

Opuntia confusa Griffiths—Tucson pricklypear. Compact bushes 30–60 cm (–1 m), semiupright with branches ascending to erect, to 3–4 cladodes high; cladodes obovate to circular, gray-green or bluish, 15–20 cm; areoles (5+ mm) bearing several stout, spreading, long, angular spines; spines dimorphic, to ± 5 cm, pale cream, basally deep chocolate-brown to black-brown, other cladodes with few deflexed, mostly white, slender spines to ± 3 cm from smaller areoles; glochids orange-brown to red-brown; flowers yellow, often with ill-defined orange to red center becoming pink or orange before closing; stigma green; fruit barrel-shaped to obovate, ± 5 cm, deep purplish red inside and out, sweet; seeds ± 4 mm; seedlings not hairy. Rocky bajadas and hillsides, Chihuahuan desert scrub, 4000-5000 ft. Doña Ana, Otero.

Opuntia cyclodes (Engelm. & J. M. Bigelow) Rose [*O. engelmannii* Salm-Dyck ex Engelm. var. *cyclodes* Engelm. & J. M. Bigelow]—Rio Gallinas pricklypear. Much-branched shrub mostly < 1 m tall and wider than tall; main branches usually lying along ground; cladodes broadly obovate to nearly circular, 20-27 cm, spines usually in upper 1/2 to all areoles; spines 1-3(-5), to 3 cm, yellowish to yellow, small lower spine(s) white; flower petaloids yellow, with very faint reddish veining; stigma stout, green; fruit broadly obovate to nearly globose, 4-5 cm, often slightly "lumpy," deep purplish red, sweet; seeds ± 4 mm; seedlings not hairy. Mesas and canyons, prairies, oak scrub, piñon-juniper woodlands, 3700-7000 ft. NE, EC, SE, SC, SW. Similar to *O. lindheimeri* but smaller, lower, and more spreading.

Opuntia cymochila Engelm. & J. M. Bigelow [*O. mesacantha* Raf. var. *oplocarpa* Engelm. ex J. M. Coult.]—Plains pricklypear. Low-creeping plants typically not > 2 cladodes high, rarely to 25 cm, spreading to 1+ m; cladodes broadly obovate to circular or even wider than long, 6-10(-15) cm, green to dark green, but varying to purplish in winter; spines in most areoles, 1(-3), terete, ± erect, often only in upper 1/2 or less, mostly white or yellowish or brown, with darker base, to 5(-8) cm; smaller spines white, 1-5, deflexed/spreading downward; flowers yellow, occasionally brassy in center but rarely orange to red, petaloids usually curved upward distally, flower roughly open "tulip-shaped" and rolled under along margins, giving ruffled or crisped appearance; stigma green; fruit barrel-shaped or obovoid to fusiform, often slightly tubercled, with few bristlelike spines to 12 mm, usually narrowed abruptly at top to acute, sometimes flangelike rim; seeds irregular in shape with wide margins, usually (4-)5-7 mm; seedlings not hairy. Grasslands and piñon-juniper woodlands. Widespread.

Opuntia discata Griffiths [*O. engelmannii* Salm-Dyck ex Engelm. var. *discata* (Griffiths) C. Z. Nelson]—Bird's foot pricklypear. Bushy plants 0.9-1.5 m, stems ascending to erect; cladodes bluish gray-green, becoming greener with age, broadly ovate to circular or broadly obovate, 20-25(-32) cm; spines (1-)3-7 per areole, white to cream or faintly pinkish, rarely slightly yellowish, often with brownish bases, angularly flattened, divergent to recurved in "bird's-foot" pattern, 2-2.5(-4) cm; flowers light yellow, sometimes with faint fine reddish veining, sometimes fading to orange or red before closing; fruits mostly broadly obovate to nearly globose, dark red to purple throughout, 5-7 cm, sweet; seeds 3-4 mm; seedlings hairy. Desert scrub, grasslands, oak woodlands, open piñon-juniper woodlands, 4000-6000 ft. EC, C, WC, SE, SC.

Opuntia dulcis Engelm. [*O. phaeacantha* Engelm. var. *major* Engelm.; *O. eocarpa* Griffiths; *O. expansa* Griffiths]—Sweet pricklypear. Compact bushes 30-60 cm (-1 m), semiupright, mostly 3-4 cladodes high, cladodes obovate to approximately circular, dull gray to somewhat grayish, mostly 17-23 cm; spines 1-6, to ± 3(-6) cm with up to 3 main upper spines, main upper spine usually erect to porrect and terete in cross-section, others angular in cross-section and deflexed to spreading; spines white or yellow to red-brown or black, most often pale white to brownish with brownish base; glochids in 2 even, concentric, compact groupings; flowers yellow, with well-differentiated orange to red center, often fading to orange before closing; stigma pale to rich green; fruit barrel-shaped to obovate, 1.3-6 cm, deep purplish red inside and out, sweet and very juicy; seeds 3.5-5 mm; seedlings not hairy. Prairies, desert scrub, piñon-juniper woodlands, 3000-7000 ft. NE, NC, C, WC, SE, SC, SW.

Opuntia engelmannii Salm-Dyck ex Engelm. [*O. gregoriana* Griffiths; *O. procumbens* Engelm. & J. M. Bigelow; *O. recurvospina* Griffiths]—Engelmann's pricklypear. Large, bushy plants mostly 1-1.5(-2) m, main stems/trunks procumbent to erect; cladodes oval to broadly obovate or sometimes circular, 20-30 cm; areoles mostly < 8 in a diagonal row across midcladode; spines in all or only upper areoles, usually longest in upper areoles, (0)1-6(-8) per areole, white to cream, often with darker brownish bases, varying to yellowish and shades of reddish brown to sometimes dark red-brown, upper main spines porrect and longest, remaining spines spreading outward or downward, flattened; flowers light yellow, with fine reddish veining, not often fading to orange before closing; fruits obovate to nearly globose, occasionally broadly pyriform, dark red to purple throughout, 6–9 cm; seeds 2.5–3.5 mm; seedlings hairy. Desert scrub, piñon-juniper woodlands, open ponderosa pine forests, 3400–7000 ft. EC, C, WC, SE, SC, SW.

Opuntia fragilis (Nutt.) Haw. [*Cactus fragilis* Nutt.; *Opuntia brachyarthra* Engelm. & I. M. Bigelow]—Brittle pricklypear. Shrubs low, mat-forming, 2–10 cm tall; stem segments subspheric to cylindric to flattened, dark green, glabrous, the terminal ones easily detached, elliptic-obovate, (1.5–)2–5.5 × (1–)1.5–3 cm, tubercles low, becoming pronounced when dried; spines gray with brown tips, terete, straight, spreading, 3–8 per areole, the longest 3.5 cm; flowers with inner tepals yellow, sometimes basally red, 2–2.6 cm; fruit tan, dry, 1–3 × 0.8–1.5; seeds 5–6 mm diam. Barren areas in grasslands, piñon-juniper woodlands, ponderosa pine forests, 5800–8030 ft. McKinley, San Juan.

Opuntia gilvescens Griffiths—Bajada pricklypear. Diffuse bushes, 30–60 cm, 2–3 cladodes high; cladodes broadly ovate to circular or obovate, often somewhat wavy, dark dull green, often reddish to purplish in winter, 15–20(–25) cm; spines 0–3(–5), to ± 2.5(–4) cm, terete or flattened in cross-section, with 1–2 main upper spines porrect at first and later often becoming deflexed, dirty-white or pale yellowish to brassy or orangy-brown and somewhat darker toward base, lower spines mostly deflexed or curving down, white or pale brownish; flowers light yellow, often with somewhat brassy center, rarely somewhat orange in center; stigma pale to rich green; fruit mostly broadly rounded, barrel-shaped to sometimes globose or pyriform, 4–6(–7) cm, deep red to purplish red, sweet; seeds 4–5 mm; seedlings not hairy. Mostly on bajada terrain near bases of mountains; grasslands, desert scrub, piñon-juniper woodlands, 3000–7000 ft. Widespread.

Opuntia lindheimeri Engelm.[*O. engelmannii* Salm-Dyck ex Engelm. var. *lindheimeri* (Engelm.) B. D. Parfitt & Pinkava; *O. ferruginispina* Griffiths; *O. linguiformis* Griffiths; *O. texana* Griffiths; etc.]—Texas pricklypear. Bushy plants 1–1.5(–2) m tall and usually wider; main stems/trunks procumbent to rarely erect, usually becoming spineless; cladodes green to dark green, often yellowish, oval to broadly obovate, 20–30 cm; glochids yellow to red-brown; spines in all or only some areoles, occasional plants nearly spineless, with 1 short spine in 0 to few areoles, 0–3(–6) per areole, usually translucent-yellow, sometimes pale yellowish or varying to rarely red-brown, upper main spine usually porrect and longest, remaining spines spreading outward or downward; flowers light yellow, with fine reddish veining, fading to orange or red before closing; fruits mostly broadly pyriform to broadly obovate, smooth, dark red to purple throughout, 6–9 cm, sour; seeds 2.5–3.5 mm; seedlings hairy. New Mexico material belongs to **var. subarmata** (Griffiths) Elizondo & J. A. Wehbe. Rare in New Mexico. Desert scrub, piñon-juniper, mostly near the Pecos River, canyons along E base of Guadalupe Mountains, 3600–5000 ft. Eddy, Otero. Plants of the nominate var. *lindheimeri* are commonly grown in gardens, usually with red flowers or as the mutant 'Cow's Tongue' cultivar 'Linguiformis.' These are often discarded into the wild or may establish from seed.

Opuntia macrocentra Engelm. [*O. violacea* Engelm. ex B. D. Jacks.; *O. violacea* Engelm. ex B. D. Jacks. var. *macrocentra* (Engelm.) L. D. Benson]—Purple pricklypear. Semierect to ascending to 60 cm; cladodes dull gray-green to blue-green or purple, most plants purple in drought or winter, broadly obovate to obdeltoid or spatulate, 10–20 cm; glochids mostly red-brown; spines 0–4, mostly only in upper areoles, with main spines usually slender, to 12 cm, black-brown to black, often white at tip, or occasionally all white; flowers bright yellow with well-defined contrasting red center; stigma white to pale greenish; fruit dull brown-tinted pink to light purple, narrowed at the top with acute, often almost flangelike rim; seeds ± 4–5 mm; seedlings occasionally hairy. Our material belongs to **var. macrocentra**. Chihuahuan desert scrub, but habitats vary widely, 3295–5500 ft. NE, NC, EC, C, WC, S.

Opuntia violacea Engelm. ex B. D. Jacks. var. *castetteri* L. D. Benson (Hueco Mountain purple pricklypear) also keys here. It differs in being smaller and more compact, with cladodes often wider than long and curved, usually not tapered down toward base; also, the areoles are more numerous, and spines are few, long, slender, and white. Hueco and Cornudas Mountains. Otero.

Opuntia macrorhiza Engelm. [*O. rafinesquei* Engelm.]—Prairie pricklypear. Low-creeping plants not > 2 cladodes high, rarely to 20 cm, spreading to 1+ m; roots mostly adventitious from lower side of branches, often prominently swollen and tuberlike; cladodes wrinkling, often creasing, becoming prostrate in winter, broadly oval to obovate, 7–12(–15) cm, green to dark; spines in only upper areoles, main spines 0–2(3) with 1 ± erect and the others deflexed, mostly off-white tinted yellowish or orange-brown at base, 2–5 cm, with 0–3 smaller, white, deflexed spines; flowers pale yellow, often with well-defined orange to red centers; fruit narrowly obovoid to pyriform, without flangelike rim, greenish, pink, deep red, or deep purplish red, bland to sour; seeds 3–4 mm; seedlings not hairy. Sandy soil, grasslands, open piñon-juniper woodlands, ponderosa pine forests, 3200–7000 ft. Widespread.

Opuntia microdasys (Lehm.) Pfeiff. [*O. macrocalyx* Griffiths]—Bunny ears cactus. Distinctive spineless species with small pubescent pads bearing prominent clumps of glochids; glochids in tight clumps, white or yellow to dark red-brown; flowers mostly pale yellow with deep green stigma (very rarely orange or red); fruit bright red, rounded, with glochids matching those of stems. Occasional adventive in Chihuahuan Desert in S New Mexico; from Mexico, native primarily to Chihuahuan Desert region, ± 4000 ft. Doña Ana, Eddy.

Opuntia orbiculata Salm-Dyck ex Pfeiff. [*O. crinifera* Salm-Dyck ex Pfeiff.; *O. dillei* Griffiths; *O. microcarpa* Engelm. ex B. D. Jacks.]—Round pricklypear. Compact bushy plants 1–2 m, branches procumbent to erect; pads ovate to circular or occasionally broadly obovate; spines 1–3(–5), to ± 2.5(–4) cm, mostly flattened in cross-section, often curving downward, with 1–2 main upper spines porrect at first, later often becoming deflexed, mostly dirty-white to yellowish, orange-brown, or red-brown, somewhat darker toward base, lower spines mostly spreading or curving down or deflexed and paler; glochids yellow to red-brown; flowers bright yellow, often with faint orange centers, often fading to orange or red before closing; fruit mostly broadly rounded, barrel-shaped to broadly obovate, 5–8 cm, deep red to purplish red, sweet; seeds 2–4 mm; seedlings hairy. NC, EC, C, WC, SE, SC, SW. This pricklypear is often confused with *O. engelmannii* and *O. lindheimeri*.

Opuntia phaeacantha Engelm. [*O. phaeacantha* var. *piercei* Fosberg; *O. tenuispina* Engelm.]—New Mexico pricklypear. Spreading to creeping plants, typically not > 2 cladodes (± 30 cm) high, spreading to 2+ m; cladodes obovate to rarely circular, 12–16(–20) cm, green to glaucous-green, but varying to distinctly bluish and often purplish in winter; spines 1(–9) cm with up to 4 main upper spines, erect to porrect and terete in cross-section, white or yellow to red-brown or black, small lower spines deflexed and white; glochids yellow to red-brown; flowers yellow, with orange to red center; stigma white to rich green; fruit mostly barrel-shaped to obovate, most often red to deep purplish red, flavor bland; seeds 4–5 mm, seedlings not hairy. Grasslands, desert scrub, oak woodlands, piñon-juniper woodlands, sagebrush, ponderosa pine, mixed conifer forests, 2900–8000 ft. Widespread.

Opuntia polyacantha Haw. [*O. ferox* Haw.; *O. media* Haw.; *O. missouriensis* DC.]—Creeping plants mostly 1–2 cladodes high, not > 20 cm; cladodes green to gray-green, with numerous areoles; spination highly variable from nearly spineless to extremely spiny, erect to porrect and smaller, mostly deflexed secondary spines, mostly all rigid but sometimes with lower spines thin, elongating and flexible, white or yellow to deep red-brown or nearly black, often paler apically, secondary spines white or at least paler; petaloids yellow, orange, red, pink, magenta, or 2-tone combinations; stigma green; fruit dry at maturity, tan to grayish, usually spiny; seeds large, irregular, with sharp rims, (3–)5–8 mm. Varieties of *O. polyacantha* intergrade where they meet, and many plants are not typical of any one variety.

1. Plants dwarf, cladodes usually < 5 cm; spines variable, but mostly short; mostly of montane woodlands and sagebrush communities; > 7000 ft....**var. schweriniana**
1'. Plants larger; cladodes mostly (5–)7–12(–17) cm; spines variable, mostly longer than above; various plant communities; generally < 7000 ft....(2)
2. Spines few, mostly short and limited to upper part of cladodes; fruit sometimes spineless; sagebrush and open piñon-juniper woodlands...**var. juniperina**
2'. Numerous spines on entire cladode; fruit spiny; various plant communities...(3)
3. Areoles typically < 1 cm apart; central spine typically 1 per areole and < 3 cm; fruit nearly globose, not red-

dish; Great Plains and Rocky Mountain grassland basins and mesas mostly E of the Sangre de Cristo Mountains…**var. polyacantha**
3'. Areoles typically > 1 cm apart; central spines often 2+ and mostly well > 3(–12) cm; fruit elongate, often turning reddish before drying; desert scrub and grassland areas W of the Great Plains…**var. hystricina**

var. hystricina (Engelm. & J. M. Bigelow) B. D. Parfitt—Porcupine pricklypear. Grasslands, piñon-juniper woodlands, 4500–7500 ft. NC, NW, C, WC, SC.

var. juniperina (Britton & Rose) L. D. Benson [*O. sphaerocarpa* Engelm. & J. M. Bigelow]—Juniper pricklypear. Sagebrush communities, piñon-juniper woodlands, ponderosa pine forest openings, 4900–9500 ft. NE, NC, NW, C, WC.

var. polyacantha [*O. polyacantha* var. *rufispina* (Engelm. & J. M. Bigelow) L. D. Benson; *O. polyacantha* Haw. var. *albispina* (Engelm. & J. M. Bigelow) J. M. Coult.]—Plains pricklypear, starvation pricklypear. Prairies, sagebrush communities, piñon-juniper woodlands, 4300–8000 ft. Widespread.

var. schweriniana (K. Schum.) Backeb.—Dwarf mountain pricklypear, dwarf cactus. Montane grasslands, sagebrush, piñon-juniper woodlands, ponderosa pine forests, mixed conifer forests, 6900–10,000 ft. NC, NW, C.

Opuntia pottsii Salm-Dyck [*O. macrorhiza* Engelm. var. *pottsii* (Salm-Dyck) L. D. Benson]—Small plants forming clumps of cladodes from a central swollen, tuberous rootstock, < 3 dm wide, exceptionally spreading to 2 dm; cladodes mostly dark to glaucous gray- or blue-green, 5–8(–11) cm; spines 0–4 per areole in upper areoles, slender, to 1 mm diam., 1–2 porrect upper spines per areole, 3–5(–8) cm, white or yellowish to reddish or sometimes blackish, often darker near base, smaller spines white to paler and mostly reflexed to deflexed; flowers 3.5–5 cm diam., pale or bright yellow, orange, red, pink, or purplish red, or combinations, often with darker orange to red center; stigma mostly white to pale greenish; fruit greenish or pinkish to deep purplish red, bland; seeds 3–4(–5) mm. Desert grasslands, Chihuahuan desert scrub, oak woodlands, 3360–6800 ft. SW, SC, SE

Opuntia pyrocarpa Griffiths—Pear-fruit pricklypear. Diffuse bushes 30–60(–100) cm, branches mostly spreading, with secondary branches and ends ascending to erect, to roughly 3–4 cladodes high, often thicket-forming; cladodes narrowly to broadly obovate, rounded or obtuse apically, 18–25(–30) cm; areoles with 1–2 main upper spines, porrect at first and later becoming deflexed, pale yellowish to pale brassy or orangy-brown or dirty-white to gray-brown, mostly slightly darker toward base, 1–2 lower main spines spreading or curving down, somewhat lighter in color; 0–3 secondary small white spines; flowers dark yellow, often orange toward center, fading to orange before closing; stigma pale to rich green; fruit obovate to more often pyriform, to 7 cm, deep purplish red, sour; seeds 3.5–5 mm; seedlings not hairy. Shallow breaks in shortgrass prairie communities, 3000–7000 ft. EC, SE.

Opuntia tortispina Engelm. & J. M. Bigelow—Desert grassland pricklypear. Creeping plants typically not > 2 cladodes high, to 25 cm, spreading to 1.5+ m; cladodes broadly obovate to circular, 8–12(–16) cm; areoles with up to 6 main upper spines erect to spreading, deflexed, or porrect, angular in cross-section, spreading, white or yellow to red-brown or black; flowers yellow, with orange to center; stigma green; fruit barrel-shaped to obovate, to 4 cm, red to deep purplish red, bland; seeds 4–5 mm; seedlings not hairy. Grasslands, sagebrush, piñon-juniper woodlands, oak woodlands, ponderosa pine forests. Widespread. Easily confused with *O. phaeacantha* and *O. cymochila*.

Opuntia trichophora (Engelm. & J. M. Bigelow) Britton & Rose [*O. polyacantha* Haw. var. *trichophora* (Engelm. & J. M. Bigelow) J. M. Coult.]—Grizzlybear cactus, woolly cactus. Clumping, often semi-erect plants to 25 × 30 cm or wider; cladodes green to gray-green, with numerous small areoles all spine-bearing; spines numerous, white to yellowish, sometimes reddish or brownish in the south, thin and flexible on cladodes and fruits, often curving variously, almost hairlike, longest to 5(–12) cm; flowers yellow, occasionally fading orange; filaments orange or red; fruit dry at maturity, tan to grayish; seeds large, ± 6–7 mm. Steep rocky slopes, sandy soil, gypsiferous soil, desert scrub, piñon-juniper woodlands,

3800–8000 ft. Widespread. Similar to spinier variants of *O. polyacantha* but differs in ploidy and is often sympatric without intermediate plants.

Opuntia valida Griffiths—Valida pricklypear, San Antonio pricklypear. Broad shrub with main branches horizontal to mostly ascending to ca. 1 × 2 m; cladodes ovate, elliptic, obovate, or often rhomboid, 20–30(–40) cm; spines (1)2–7(10), with 2–5 main spines stout, angularly flattened, divergent, to 5(–7) cm × 1.5 mm thick, chalky-white, redbrown to nearly black at base, with 2–5 smaller, pale, recurved spines below; flowers light yellow with fine reddish veining, fading to orange before closing; stigma green; fruit reddish purple, narrowly oblong to obovate, slightly "lumpy" especially near rim, deep purple, very juicy, sweet; seeds 4–5 mm diam.; seedlings hairy. Prairies, desert scrub, oak woodlands, piñon-juniper woodlands, 3500–6500 ft. NC, EC, C, SE, SC.

Opuntia wootonii Griffiths—Wooton pricklypear. Shrub, main branches horizontal to ascending, 1 × 2 m; cladodes ovate, elliptic, obovate, or often rhomboid, 20–30(–40) cm; areoles with (1–)4–6 spines, 2–4 divergent erect main spines to 5–7(–11) cm, red-brown, yellow, whitish, or combinations of these, with 1–3 smaller pale recurved spines below; flowers deep yellow with fine reddish veining, sometimes tinged reddish toward center, fading to orange before closing; stigma green; fruit narrowly oblong to obovate or pyriform, slightly "lumpy" especially near rim, purplish red, very juicy, sour/tart; seeds ± 4 mm diam.; seedlings sparsely hairy or not. Grasslands, piñon-juniper woodlands, oak woodlands, 4000–7600 ft. NC, EC, C, WC, SE, SC.

Opuntia zuniensis Griffiths—Zuni pricklypear. Low-spreading to creeping plants typically not > 2 cladodes (± 20 cm) high, spreading to 2+ m; cladodes obovate or broadly elliptic to rhomboid, 10–16 (–20) cm, often glaucous to bluish green, but varying to distinctly reddish lavender in winter; main spines 2–4(–6), usually all similar, angular in cross-section, cream-white to yellowish-tinged, lower white bristlelike spines 0–2 in few areoles; flowers yellow (rarely to orange), without red center; stigma green; fruit ellipsoid to obovoid, ± 1.5 times as long as wide, yellowish to pink or red, bland; seeds ± 4 mm; seedlings not hairy. Sandy areas, grasslands and openings in piñon-juniper woodlands, 4700–7500 ft. NC, NW, C, WC.

PEDIOCACTUS Britton & Rose – Plains cactus

Plants unbranched or few-branched, deep-seated in substrate or rising 1–15 cm above substrate, stems unsegmented, green or gray-green, globular to short-cylindrical or depressed-ovoid to ovoid or globose, 0.7–15 × 1–15 cm; spines 18–46, reddish tan, pink, or white, needle-shaped, 5–21 × 0.3 mm, radial spines 18–35 per areole, 1–13 mm; central spines 0–11, needlelike; flowers borne at adaxial margins of areoles at stem apex, funnelform, 1–3.5 × 1–2.5 cm; tepals yellow, peach, pink, magenta, cream, or white, 8–25 × 3–8 mm; fruits dehiscent along 1 vertical suture, green, often turning reddish brown, becoming dry at maturity; seeds brown, black, or gray (Heil et al., 1981).

1. Stems 3–30 cm; central spines 4–12 per areole…**P. simpsonii**
1'. Stems 0.7–6.8 cm; central spines 0(–2) per areole…**P. knowltonii**

Pediocactus knowltonii L. D. Benson—Knowlton's cactus. Plants branched or unbranched; stems globular to short-cylindrical, 0.7–5.5 × 1–3 cm; spines smooth, relatively hard, all radial, mostly 18–26 per areole, spreading, recurved, or somewhat pectinate, reddish tan, pink, or white, 1–1.5 mm; flowers 1–3.5 × 1–2.5 cm, tepals pink; fruits green, drying reddish tan, turbinate; seeds black, 1.5 × 1–1.2 mm. Gravel pavements in piñon-juniper woodlands with mixed sagebrush, 6560 ft. San Juan. A New Mexico endemic. **FE & E**

Pediocactus simpsonii (Engelm.) Britton & Rose [*Echinocactus simpsonii* Engelm. var. *minor* Engelm.]—Mountain cactus. Plants only occasionally branched; stems depressed-ovoid to ovoid or globose, 2.5–15(–25) × 2.5–15 cm; spines smooth, hard, radial spines 15–35 per areole, widely spreading, white, slender, 3–13 mm, central spines 4–11 per areole, reddish brown (rarely black), with basal 1/2 cream or yellow, 5–21 × 0.3 mm; flowers 1.2–3 × 1–2.5 cm, tepals white, pink, magenta, yellow, or yellow-green;

fruits green tinged with red, drying reddish brown; seeds gray to black, 2–3 × 1.5–2 mm. Piñon-juniper woodlands, sagebrush, montane, prairie grasslands, coniferous forests, 6100–10,200 ft. NC, NW. *Pedio-cactus simpsonii* is an exceedingly variable species.

PELECYPHORA C. Ehrenb. – Escobaria

Plants low-growing, solitary or clustered; stems depressed-globose to cylindrical, tubercles often becoming corky and deciduous with age; spines usually short, fine, straight, densely covering the plant; flowers arising on the upper edge of the areolar groove, often not fully opening; fruits globose or oblong, mostly naked, usually red but sometimes green or pink; seeds broadly oval to nearly circular, 1–1.7 mm diam., brown or blackish brown.

1. Floral remnant on fruit deciduous; spines 6–20 per areole; stems deep-seated in substrate, often nearly subterranean except in growing season…**P. missouriensis**
1′. Floral remnant on fruit persistent; spines (10–)15–76(–95) per areole; stems usually not deep-seated, > 1/2 aboveground (sometimes deep-seated and flat-topped in winter in *P. vivipara*)…(2)
2. Central spines erect, appressed and therefore inconspicuous against radial spines…(3)
2′. Central spines radiating in every direction, porrect and conspicuous…(4)
3. Spines (21–)31–44(–55) per areole; stigma lobes dark green to bright yellow; seeds black; fruit bright red…**P. duncanii**
3′. Spines (12–)19–23(–25) per areole; stigma lobes white (to pink or purple); seeds brown; fruit green to dull reddish…**P. vivipara** (in part)
4. Ripe fruit green or maroon to dull reddish pink (sometimes blood-red or crimson in *E. sneedii*); stigma lobes pink, purple, or white; widespread throughout New Mexico…(5)
4′. Ripe fruit red; stigma lobes white; S New Mexico…**P. tuberculosa**
5. Branches 0–30; inner tepals pale rose-pink to intense pink; fruits green, dull pink, or brownish red; widespread…**P. vivipara** (in part)
5′. Branches 0–250; inner tepals generally white, cream, pale tan, greenish white, or pale rose-pink; fruits either red or green; SE New Mexico, W Texas…(6)
6. Central spines with a yellowish cast or darkly pigmented…(7)
6′. Central spines white…(8)
7. Central spines with a yellowish cast; individual stems to 3 cm diam., solitary or 10–20+ stems; igneous substrates; Organ Mountains…**P. organensis**
7′. Central spines darkly pigmented; individual stems mostly > 3 cm diam., often forming small clumps of 2–12 stems; limestone substrates; Sacramento Mountains…**P. villardii**
8. Stems densely clustered and forming large clumps; flowers white with pink or magenta midveins or brownish pink; Franklin and Guadalupe Mountains…(9)
8′. Stems solitary or loosely clustered and not forming large clumps; flowers yellow, cream, or pink; Florida, Peloncillo, Little Hatchet, Big Hatchet, San Andres and Guadalupe Mountains…(10)
9. Spines appressed against the plant, sometimes with 1 to a few short porrect centrals; Guadalupe Mountains…**P. leei**
9′. Spines spreading to erect; Bishop Cap area, Franklin Mountains and limestones of the southern Organ Mountains…**P. sneedii**
10. Stems ovoid, mostly solitary; flowers pale yellow, cream, or pink; Guadalupe Mountains…**P. guadalupensis**
10′. Stems cylindroid, solitary to densely clustered; flowers pinkish; Florida, Peloncillo, Little Hatchet, Big Hatchet, and San Andres Mountains…(11)
11. Central spines mostly 9–11 mm; stems mostly solitary or a few loosely clustered; Florida, Peloncillo, Little Hatchet and Big Hatchet Mountains…**P. orcuttii**
11′. Central spines to 25 mm; stems 20+, loosely clustered; San Andres Mountains…**P. sandbergii**

Pelecyphora duncanii (Hester) D. Aquino & Dan. Sánchez [*Coryphantha duncanii* (Hester) L. D. Benson]—Duncan's pincushion cactus. Plants unbranched (rarely to 8 mature branches), white-bristly spines obscuring stem; stems deep-seated, inconspicuous, obovoid, ovoid, or spherical, 2.5–6 × 1–3.4 cm; spines mostly 31–44 per areole, all snowy-white, with dark tips on medium and large spines, radial spines 18–41 per areole, 4–9 × 0.03–0.2 mm, outer central spines mostly 3–9 per areole; flowers nearly apical, 15–30, 13–19 mm; white or cream to light pink, stigma lobes green to bright yellow; fruits bright red, not very succulent; seeds black, ±1.2 mm. Limestone slopes, mostly in crevices of massive limestone outcrops, 5100 ft. Sierra. **R**

Pelecyphora guadalupensis (S. Brack & K. D. Heil) J. M. Porter & K. D. Heil, **comb. nov.** [*Escobaria guadalupensis* S. Brack & K. D. Heil (Brack & Heil, 1986)]—Guadalupe pincushion cactus. Stems mostly solitary or sometimes up to 5 in a clump, globose to obovoid, 4–10 × 3–5 cm; spines ca. 50–60 per areole, spreading, typically white, often darker at the tip, fading to gray, mostly 7–14 mm; flowers to 2 cm wide (usually smaller), tepals pale yellowish to pinkish or nearly white, usually with midribs darker, stigmas white to pink; fruits green to somewhat reddish; seeds ca. 1 mm, brown. Cracks in limestone and rocky soils of broken mountainous terrain in open oak and piñon-juniper woodland, lower montane coniferous forest, and Chihuahuan desert scrub, 5200–7200 ft. Endemic to the Guadalupe Mountains. Chaves, Eddy, Otero. **R**

Pelecyphora leei (Rose ex Boed.) J. M. Porter & K. D. Heil, **comb. nov.** [*Escobaria leei* Rose ex Boed. (Boedeker, 1933); *Coryphantha sneedii* (Britton & Rose) A. Berger var. *leei* (Rose ex Boed.) L. Benson; *Escobaria sneedii* Britton & Rose var. *leei* (Rose ex Boed.) D. R. Hunt]—Lee's pincushion cactus. Plants forming small dense clusters of up to 250+, dominated by immature stems, cylindric to clavate stems, 3.5–10 × 1.3–3 cm; spines mostly 62–95 per areole, 1–2.5 mm, appressed to the tubercle or spreading in southern populations, white to cream-colored or gray in age; flowers 1–2 cm, campanulate to funnel-form, tepals 5–12 mm, rose-purple to tan-pink midribs darkest; stigmas white; fruits ellipsoid, green to dull brownish-red; seeds 1–1.5 mm, reddish-brown. Cracks in limestone in broken terrain and steep slopes of Chihuahuan desert scrub, 4000–5000 ft. Endemic to the Guadalupe Mountains, New Mexico. Eddy. **FT & E**

Pelecyphora missouriensis (Sweet) D. Aquino & Dan. Sánchez—Missouri foxtail cactus. Plants (ours) unbranched; spine-bearing areoles with short white wool; stems deep-seated in substrate, becoming flat-topped and nearly subterranean in winter, spherical to obconic, 2–8 × 1.8–10 cm; spines 6–21 per areole, bright white, pale gray, or pale tan, radial spines 6–20, 4–16 mm, central spines 0–2 per areole, 9–20 mm; flowers 18–50 × 15–50 mm; tepals fringed pale greenish yellow to yellow-green or pinkish, stigma lobes 3–7, green or yellowish; fruits orange-red to scarlet; seeds black, 1.4–2.2 mm. Plains, stony shortgrass prairies, woodlands of ponderosa pine, piñon, juniper, or *Quercus gambelii*, often restricted to sedimentary rocks, 5500–6500 ft. Bernalillo, Sandoval.

Pelecyphora orcuttii (Rose ex Boed.) J. M. Porter & K. D. Heil, **comb. nov.** [*Escobaria orcuttii* Rose ex Boed. (Boedeker, 1933); *Coryphantha orcuttii* (Rose ex Boed.) Zimmerman]—Orcutt pincushion cactus. Stems solitary or commonly branched to form small dense clusters, the individual stems to 15 × 6 cm; spines whitish, usually with purplish or brownish tips, often brittle, 45–60 per areole, spreading, 8–11 mm; flowers to 2 cm wide, tepals pale yellowish to pinkish or nearly white, usually with midribs darker, stigmas white to pink; fruits green to somewhat reddish; seeds ca. 1 mm, brown. In cracks of limestone or in rocky soils of broken mountainous terrain in Chihuahuan desert scrub, desert grassland, and oak woodland, 5200–6000 ft. Endemic to the Florida, Little Hatchet, and Big Hatchet Mountains. Hidalgo, Luna. **R**

Pelecyphora organensis (Zimmerman) J. M. Porter & K. D. Heil, **comb. nov.** [*Coryphantha organensis* Zimmerman (Zimmerman, 1972); *Escobaria organensis* (Zimmerman) Castetter, P. Pierce & K. H. Schwer.]—Organ Mountain pincushion cactus. Stems solitary or commonly branched to form small dense clusters, the individual stems mostly 15 × 2–4 cm; spines with a yellowish cast, often reddish brown at the tips, 20–60 per areole, slender and bristlelike, 1–2.5 cm; flowers to 2 cm wide, pale yellowish to pinkish or nearly white; fruits green to somewhat reddish; seeds ca. 1 mm, brown. Andesite, quartz-monzonite, and to a lesser extent rhyolite and limestone in broken mountainous terrain; Chihuahuan desert scrub, open oak and piñon-juniper woodlands, 4400–8530 ft. Endemic to the Organ and northern Franklin Mountains. Doña Ana. **R & E**

Pelecyphora sandbergii (Castetter, P. Pierce & K. H. Schwer.) J. M. Porter & K. D. Heil, **comb. nov**. [*Escobaria sandbergii* Castetter, P. Pierce, K. H. Schwer. (Castetter et al., 1975a)]—Sandberg pincushion cactus. Stems solitary or commonly branched to form small dense clusters of 20+ stems, the individual stems mostly 15 × 2–4 cm; spines 20–60 per areole, white, often darker at tip, 1–2.5 cm; flowers to 2 cm wide; tepals pale yellowish to pinkish or nearly white, stigmas white to pink; fruits green to somewhat reddish; seeds ca. 1 mm, brown. Rocky, igneous and limestone soils in Chihuahuan desert scrub and open oak and piñon-juniper woodland in mountainous terrain, 4200-7400 ft. Endemic to the San Andres Mountains. Doña Ana, Sierra. **R & E**

Pelecyphora sneedii (Britton & Rose) D. Aquino & Dan. Sánchez [*Coryphantha sneedii* (Britton & Rose) A. Berger; *Escobaria sneedii* Britton & Rose]—Sneed's pincushion cactus. Plants growing in clumps, or commonly branched to form small dense clusters, usually not dominated by immature stems; stems mostly 10 × 1-2 cm; spines mostly 31-68 per areole, not or only slightly appressed, slender, mostly 1-5 mm, 15 ± central, stouter spines to ca. 10 mm, the spines typically white but varying to yellowish, pinkish, or pale brownish, often darker at tip; flowers to 1.5 cm wide (usually smaller); tepals pale yellowish to pinkish or nearly white, stigmas white to pink; fruits green to somewhat reddish; seeds ca. 0.8 mm, brown. Primarily cracks in limestone in broken terrain and steep slopes of Chihuahuan desert scrub, 3900-6000 ft. Bishop Cap area and Franklin Mountains. Doña Ana. **FE**

Pelecyphora tuberculosa (Engelm.) D. Aquino & Dan. Sánchez—Cob cactus. Plants usually branched and small-stemmed (to 50 branches), sometimes unbranched, corncoblike on or below ground portion, only on oldest plants; stems ovoid to cylindrical, 4-16 × mostly 3-6 cm; spines 21-41 per areole, ashy-white, gray, or pale tan, radial spines 15-41 per areole, gray, 6-13.5 mm, central spines usually 5, outer central spines mostly 3 per areole, longest spines 10-15 mm; flowers mostly apical, 20-30 × 20-45 mm, tepals white, pale rose-pink, or pale lavender-pink, stigma lobes 4-6, white; fruits green, maroon, or bright red, not very succulent; seeds reddish brown. Stony grasslands, oak-juniper savannas, *Larrea* scrub, often with *Agave lechuguilla*, limestone mountainsides or igneous rocks, 3600-6500 ft. SE, SC, SW.

Pelecyphora villardii (Castetter, P. Pierce & K. H. Schwer.) J. M. Porter & K. D. Heil, **comb. nov.** [*Escobaria villardii* Castetter, P. Pierce & K. H. Schwer. (Castetter, 1975b)]—Villard pincushion cactus. Stems solitary, commonly branched to form small dense clusters, mostly 15 × 2-4 cm; spines ca. 20-50 per areole, radial spines 20-36, 5-12 mm, white, central spines 8-10, the longest 12-20 mm, ashy-white with dark tips; flowers to 2.5 cm wide, pale yellowish to pinkish or nearly white; fruit green to somewhat reddish; seeds ca. 1 mm. Loamy soils of desert grassland with Chihuahuan desert scrub on broad limestone benches in mountainous terrain, 4500-6500 ft. Franklin and Sacramento Mountains. Doña Ana, Otero. **R & E**

Pelecyphora vivipara (Nutt.) D. Aquino & Dan. Sánchez—Pincushion cactus. Plants usually unbranched or with age in some populations to 30 branches; stems spherical, ovoid, obovoid, or cylindrical with age, 2.5-20 × 3-11 cm; spines 11-55 per areole, radial spines 10-40 per areole, appressed, pectinately arranged, white, tan, pale pinkish, or reddish brown, central spines 3-14 per areole, often strongly projecting in a "bird's-foot" arrangement or radiating like spokes, longest spines 9-25 × 0.2-0.7 mm; flowers subapical, 20-57 × 25-67 mm, pale rose-pink to reddish pink or magenta, stigma lobes 5-13, white to magenta; fruits green, slowly turning dull brownish red, juicy; seeds bright reddish brown. Four weak varieties are found in New Mexico.

1. Central spines mostly 4(-11), the lower one turned downward…(2)
1'. Central spines 5-7, the lower one not turned downward…(3)
2. Spines 13-29 per areole, not obscuring the tubercles; central spines lustrous, reddish brown or orange… **var. vivipara**
2'. Spines 23-55 per areole, obscuring the tubercles; central spines drab, white or brown…**var. neomexicana**
3. Central spines 1.2-1.5 cm; radial spines white to brown; Hidalgo, Catron, Grant Counties…**var. bisbeeana**
3'. Central spines 1.5-2 cm; radial spines white; widespread…**var. arizonica**

var. arizonica (Engelm.) J. M. Porter & K. D. Heil, **comb. nov.** [*Mammillaria arizonica* Engelm. (Engelmann, 1876); *Coryphantha vivipara* (Nutt.) Britton & Rose var. *arizonica* (Engelm.) W. T. Marshall]—Arizona beehive cactus. Stems ovoid, 5-10 × 5-6.2 cm; spines 25-37 per areole, central spines 5-7, 1.5-1.9 cm, red, white basally, color mingling with the mass of radials, radial spines 20-30, 1.2-1.5 cm; flowers deep pink, mostly 2.5-3 × mostly 4.5 cm; fruits green, to 2 cm. Piñon-juniper woodlands and montane forest communities, 4700-7200 ft. NW, WC, SW, Otero.

var. bisbeeana (Orcutt) J. M. Porter & K. D. Heil, **comb. nov.** [*Coryphantha bisbeeana* Orcutt (Orcutt, 1926)]—Bisbee beehive cactus. Stems ovoid, 5–7.5 × 6.5–7 cm; spines 25–36 per areole, central spines 5–6, 13–16 mm, brown or gray with pink or brown tips, seeming to mingle with the radial spines, radial spines 20–30, 10–12 mm; flowers pink, mostly 2.5–5 cm; fruits green, mostly 2 cm. Desert grassland communities, 4500–5500 ft. Catron, Grant, Hidalgo.

var. neomexicana (Engelm.) J. M. Porter & K. D. Heil, **comb. nov.** [*Mammillaria vivipara* (Nutt.) Haw. subvar. *neo-mexicana* Engelm. (Engelmann, 1856); *Coryphantha radiosa* (Engelm.) Rydb.]—New Mexico beehive cactus. Stems globose, ovoid to cylindroid, mostly 6–20 × 3.5–7.5 cm; spines 25–54 per areole, central spines 5–11, often in a "bird's-foot" arrangement, 1 descending, porrect, the longest 1.4–2 cm, tan, reddish brown, or dark reddish brown, radial spines mostly 25–34, 0.8–1.4(–2) cm; flowers magenta, mostly 2.5–5 × 3 cm; fruits green. Grasslands, desert scrub to mountain woodlands, 3250–9000 ft. Widespread.

var. vivipara—Eastern beehive cactus. Stems globose to ovoid, unbranched to profusely branched, 2.5–20 × 3–10.5 cm; spines mostly 13–29 per areole, central spines 4, often in a "bird's-foot" arrangement, 1 descending, porrect, and 3 ascending, longest 2–2.3 cm, reddish brown, radial spines 10–26, 0.7–2 cm, appressed; flowers magenta, rose-pink, or pale rose, mostly 2.5–5 × 3 cm; fruits green. Grasslands, mesquite and piñon-juniper woodland communities, 4000–6500 ft. NE, EC, SE.

PENIOCEREUS (A. Berger) Britton & Rose – Night-blooming cereus

Peniocereus greggii (Engelm.) Britton & Rose [*Cereus greggii* Engelm.]—Desert night-blooming cereus. Shrubs, erect to sprawling, usually inconspicuous, sparingly branched; roots turnip-shaped, usually 15–30 × 5–12 cm (much larger ones known); stems gray to gray-green, columnar, distally terete or 3–5-angled, ribs 4–6, simple or with 2–5 branches, 40–120(–300) cm, often narrowed toward base; spines (9–)11–15(–17) per areole, usually in 3 vertical rows, radial and central spines similar, yellowish white and black; flowers nocturnal, usually fragrant, salverform, with long tube flaring abruptly near apex, 15–17 × 5–6 cm; tepals white or lightly tinged cream or pink; fruits fleshy, bright red, darkening in age; seeds 3–4 × 2–2.5 mm. Our material belongs to **var. greggii.** Chihuahuan desert scrub, sandy or gravelly loams, along washes, on creosote bush flats or gentle slopes, 3935–5250 ft. Doña Ana, Grant, Hidalgo, Luna. **R**

SCLEROCACTUS Britton & Rose – Eagle-claw cactus

Plants erect, usually unbranched, sometimes deep-seated in substrate in winter but never flat-topped; roots diffuse; stems pale to dark green or bluish green, ovoid, spherical, depressed-spherical, depressed-hemispherical, cylindrical, or elongate-cylindrical, 1–40(–45) × 1.8–15(–20) cm, tubercles usually coalescent into ribs; spines 2–17(–29), gray, white, yellow, straw-colored, red, reddish brown, brown, pink to purplish pink, or black, radial spines 2–11(–18) per areole, central spines (0–)1–6(–11) per areole, usually of 2–3 distinct types, 1+ hooked (rarely none hooked), acicular or subulate or both (ribbonlike and papery in *S. papyracanthus*), longest spines 7–15 mm; flowers 1–6.7 × 1–6(–7) cm, borne at or near stem apex, tepals white, cream, yellow, or pink to purplish; fruits dehiscent, turning tan, pink, or red; seeds brown or black, 1.5–3 × 1.9–4.5 mm (Heil & Porter, 1994).

1. Central spines (0)1(–4) per areole, ribbonlike, flexible, twisting or curled, papery, with obscure adaxial midrib, lacking hook, if central spine absent, then radial spines strongly flattened…**S. papyracanthus**
1'. Central spines (0)1–6(–11) per areole, terete to angled, or if all flattened and papery, then some hooked, if central spine absent, then radial spines not flattened…(2)
2. Abaxial central spine nearly always absent, if present, then erect and/or hooked, 7–15 mm, brown; flowers yellow or cream (rarely pink)…**S. mesae-verdae**
2'. Abaxial central spines nearly always present, usually hooked (sometimes straight), 10–100 mm, variable in color; flowers variable in color…(3)
3. Flowers 2.2–3.5 × 1.5–3.5 cm; ovary minutely papillate, appearing smooth; adaxial central spine 1.5–2 mm wide, or if < 1.5 mm wide, then radial spines 4–6 per areole; radial spines 4–12 per areole…**S. cloverae**
3'. Flowers 3–5.7 × 2–5.5 cm; ovary papillate, appearing granular; adaxial central spine 0.5–1.5 mm wide; radial spines 5–17 per areole…**S. parviflorus**

Sclerocactus cloverae K. D. Heil & J. M. Porter [*S. cloverae* subsp. *brackii* K. D. Heil & J. M. Porter]—Clover eagle-claw cactus. Stems unbranched (occasionally branched near base), green, ovoid to elongate-cylindrical, 2.9–25(–35) × 2.8–12.5(–20) cm; ribs usually (11–)13(–15), well developed; spines obscuring stems; radial spines 4–6 per areole, 19 × 1.3–2 mm; central spines 6–9 per areole, usually 8, straw-colored to brown, usually hooked, (15–)30–46 × 1.5 mm, lateral central spines 5–8 per areole; flowers 2.5–3.5(–4) × 1.6–3.1(–3.6) cm, tepals purple, sometimes suffused with brown; fruits dehiscent, green to tan, sometimes suffused with pink; seeds brown or black, 1.2–2.5 × 1.9–3.5 mm. Sandy or clay soils, desert grasslands, desert scrub, piñon-juniper to ponderosa pine communities, 5000–7200 ft. Rio Arriba, Sandoval, San Juan.

Sclerocactus mesae-verdae (Boissev. & C. Davidson) L. D. Benson [*Coloradoa mesae-verdae* Boissev. & C. Davidson]—Mesa Verde cactus. Stems usually unbranched (occasionally 2–3-branched), pale green, depressed-spherical to ovoid, 1–11(–18) × 1–8(–10) cm; spines not obscuring stem, radial spines 7–14 per areole, spreading, straw-colored, 6–13 mm, central spines 0(–4) per areole, very rarely 1 central spine per areole, hooked, 7–15 mm, brown; flowers 1–3.5 × 1–3 cm, tepals yellow to cream (rarely pink); fruits indehiscent to irregularly dehiscent, green, becoming tan at maturity; seeds black, 2.5–3 × 3–4 mm.

1. Stem length 1–2.5 cm; stem diam. 1–4.5 cm; fruit 5–7 mm; Menefee Formation...**subsp. depressus**
1'. Stem length 3.2–11 cm or rarely greater; stem diam. 3.8–8 cm; fruit 8–10 mm; Mancos Shale and Fruitland Shale...**subsp. mesae-verdae**

subsp. depressus O'Kane, K. D. Heil & A. Clifford—Mesa Verde cactus. Stems mostly solitary, depressed-globose, 10–25 × 10–45 mm; central spines 0(–2), reddish tan, spreading to erect, 7–10 mm; radial spines 7–12, straw-colored, spreading, 3–7 mm. Desert pavement and clay hillsides, 5730–6050 ft. NW. **[?FT] & E**

subsp. mesae-verdae—Mesa Verde cactus. Stems usually unbranched (occasionally 2–3-branched), pale green, depressed-spherical to ovoid, 3.2–11(–18) × 3.8–8(–10) cm; spines not obscuring stem, radial spines 7–14 per areole, spreading, straw-colored, 6–13 mm, central spines 0(1) per areole, very rarely 1 central spine per areole, hooked, 7–15 mm, brown. Salt desert scrub, 4900–5900 ft. NW. **FT**

Sclerocactus papyracanthus (Engelm.) N. P. Taylor [*Toumeya papyracantha* (Engelm.) Britton & Rose]—Grama-grass cactus. Stems unbranched, cylindrical or obconic-cylindrical, 2–7.5(–8) × 1.2–2.5 cm, tubercles prominent; spines dense, obscuring stems, radial spines 5–10 per areole, white, straight, flat, (2–)3–5 × 0.3–0.6 mm, central spines 1(–4) per areole, whitish to tan or gray, flat, flexible, papery; flowers 2–2.5 × 1–2.5 cm, tepals white with brown midstripes; fruits indehiscent or irregularly dehiscent, green, dry at maturity; seeds black, 2.5–3 × 2–2.5 mm. Desert grasslands, piñon-juniper woodlands, Chihuahuan desert scrub, 4920–7215 ft. NC, C, SC, Cibola, Grant. Nearly **E**

Sclerocactus parviflorus Clover & Jotter [*Echinocactus parviflorus* (Clover & Jotter) L. D. Benson]—Devil's-claw cactus. Stems unbranched or branched near base, depressed-spherical, spherical, cylindrical, or elongate-cylindrical, 4.5–45 × 3.5–14.5 cm; spines frequently obscuring stems, radial spines 8–17 per areole, usually white, 6–36 mm; central spines mostly 4–6 per areole, 1–5 hooked, white, straw-colored, pink, purple, or black, 15–72 × 0.6–1 mm; flowers (2–)3–5.7(–7) × 2.5–5.5(–8) cm, tepals rose to purple, pink, or yellow (rarely white); fruits irregularly dehiscent, green turning reddish pink; seeds dark brown to black, 2.5–3.5 × 1.5–3 mm. Our material belongs to **subsp. intermedius** (Peebles) K. D. Heil & J. M. Porter. Sandy, gravelly, or clay hills, mesas, and washes, desert grasslands, saltbush, sagebrush, rabbitbrush, piñon-juniper woodlands, 4865–5575 ft. NW, WC.

THELOCACTUS (K. Schum.) Britton & Rose – Thelocactus

Thelocactus bicolor (Galeotti ex Pfeiff.) Britton & Rose [*Echinocactus bicolor* Galeotti ex Pfeiff.]—Glory of Texas. Plants few-branched or unbranched; stems ovoid to cylindrical, 7.5–18(–38) × 5–10(–13) cm; ribs 8–13, tubercles confluent into rounded ribs; spines red and white; radial spines 9–18 per areole, longest spines 12–30 mm; adaxial spines 16–75(–100) × 1.5 mm; central spines (1–)4, longest spines terete to angular, (13-)20-45(-60) × 0.4-1.5 mm, adaxial 3 like radials but larger. New Mexico material belongs to **var. bicolor**. Limestone, Chihuahuan desert scrub, 4200-4800 ft. Eddy.

CAMPANULACEAE – BELLFLOWER FAMILY

Jennifer Ackerfield

Herbs or shrubs; leaves usually alternate or sometimes opposite or whorled, petiolate, usually simple or rarely pinnately compound, exstipulate; inflorescence a cyme, panicle, raceme, or spike, or flowers solitary, terminal or axillary; flowers perfect, actinomorphic and often campanulate, or zygomorphic and often bilabiate; sepals 5, distinct; petals 5, usually connate; stamens 5, alternating with the petals, the anthers or filaments sometimes fused into a tube; pistil 1; stigmas 2, 3, or 5; ovary inferior or superior, with axile or parietal placentation; fruit usually a capsule (Ackerfield, 2015; McVaugh, 1936, 1945).

1. Corolla zygomorphic…(2)
1'. Corolla actinomorphic…(3)
2. Corolla 1.3-3 mm, white with reddish tips on the upper lobes…**Nemacladus**
2'. Corolla 15-35 mm, red, blue, or purple…**Lobelia**
3. Flowers pedicellate, the lowermost flowers not cleistogamous; perennials…**Campanula**
3'. Flowers sessile in leaf axils, the lowermost flowers cleistogamous; annuals…**Triodanis**

CAMPANULA L. – Bellflower, harebell

Perennial herbs (ours); leaves alternate, petiolate, simple; flowers actinomorphic, campanulate, funnelform, or rotate; corolla blue to purple; stamens distinct; ovary inferior, 3-5-carpellate; fruit a poricidal capsule.

1. Flowers numerous in long, 1-sided racemes; leaves mostly with serrate margins, only the very upper leaves entire; plants 4-15 dm…**C. rapunculoides**
1'. Flowers solitary or few in loose racemes or panicles; leaves entire or a few remotely serrate or obscurely toothed, or just the basal leaves dentate; plants 0.4-5(-7.5) dm…(2)
2. Corolla lobes 1/4-1/3 the length of the corolla, not extending to the midpoint of the corolla; capsule nodding, opening by pores near the base; flowers usually in a loose raceme or panicle, occasionally solitary… **C. rotundifolia**
2'. Corolla lobes 1/2 the length of the corolla; capsule erect, opening by pores near the apex; flowers solitary…(3)
3. Corolla 17-22 mm; calyx mostly 8-12 mm…**C. parryi**
3'. Corolla 5-8.5 mm; calyx mostly 2.5-5.5 mm…**C. uniflora**

    Campanula parryi A. Gray—Rocky Mountain bellflower. Plants 0.7-2.5 dm; leaves lanceolate to narrowly elliptic, 1-3.5 × 0.1-1.2 cm, the basal petiolate and upper sessile, glabrous with ciliate margins; inflorescence solitary; calyx (5.5-)8-15(-19) mm; corolla 17-23 mm, blue to blue-purple, the lobes ca. 1/2 the corolla length. Our plants are **var. parryi**. Grassy meadows and along streams, 6880-11,500 ft. NC, NW, WC.

    Campanula rapunculoides L.—Rover bellflower. Plants 4-15 dm, with deep-seated rhizomes; leaves lanceolate to ovate, 3-15 × 1-5 cm, the lower petiolate and upper sessile, glabrous above and usually hispid below, the margins irregularly double-serrate; inflorescence a 1-sided raceme; calyx 5-8 mm; corolla 15-30 mm, blue or blue-purple, the lobes ca. 1/3-1/2 the corolla length. Cultivated garden plant sometimes escaping and found along roadsides. Colfax. Introduced.

    Campanula rotundifolia L. [*C. gieseckeana* Vest ex Roem. & Schultes; *C. groenlandica* Berl.]—Bluebell of Scotland; harebell. Plants (0.6-)1-5(-7.5) dm; leaves oblanceolate to orbicular below and linear to lanceolate above, 0.7-3 × 0.4-1.2 cm, the lower petiolate and upper sessile, glabrous, the margins of basal leaves dentate and of cauline leaves entire; inflorescence solitary or a loose raceme; calyx 4-8 mm; corolla 10-20 mm, blue or blue-purple, the lobes ca. 1/4-1/3 the corolla length. Dry, rocky soil of slopes and meadows, 5900-12,500 ft.

Campanula uniflora L.—Arctic bellflower. Plants 0.4–1 dm; leaves elliptic to oblanceolate, 1–3.5 × 0.2–0.8 cm, the basal petiolate and upper sessile, glabrous, the margins shallowly lobed or obscurely toothed; inflorescence solitary; calyx 2.5–6 mm; corolla 5–9(–12) mm, blue to blue-purple, the lobes ca. 1/2 the corolla length. Alpine meadows and scree slopes, 9500–12,800 ft. Taos.

LOBELIA L. – Lobelia

Herbs or shrubs; leaves alternate, simple; inflorescence a terminal, spikelike raceme; flowers zygomorphic, bilabiate; corolla blue, red, yellow, or white; stamens connate by the anthers and filaments, distinct from the petals; ovary inferior, 2-carpellate; fruit a capsule splitting by 2 valves near the tip.

1. Flowers deep red…**L. cardinalis**
1'. Flowers blue, blue-purple, or purple…(2)
2. Perennials; leaves not clasping the stem, the margins remotely toothed, sometimes appearing nearly entire…**L. anatina**
2'. Annuals; leaves often clasping the stem, the margins regularly and conspicuously toothed…**L. fenestralis**

Lobelia anatina E. Wimm.—Apache lobelia. Perennials to 8 dm; leaves 4–8 cm, linear to lanceolate or oblanceolate, the margins remotely toothed; corolla blue with a white center, 1.5–2.5 cm; capsules 6–9 mm. Moist meadows, along streams, and on floodplains, 6800–8900 ft. Catron, Grant, Sierra.

Lobelia cardinalis L.—Cardinal flower. Perennials to 15 dm; leaves 2.5–20 cm, lanceolate to ovate, the margins with gland-tipped teeth; corolla red, 2–3.5 cm; capsules 5–9 mm. Moist places, along creeks, and on floodplains, 3600–8530 ft. NC, C, WC, SE, SC, SW.

Lobelia fenestralis Cav.—Fringeleaf lobelia. Annuals to 7 dm; leaves 2–7 cm, lanceolate to oblanceolate, the margins regularly toothed; corolla blue, blue-purple, or purple, with a white center, 1.5–2.5 cm; capsules 3–8 mm. Moist meadows and swales, 5080–5750 ft. Hidalgo.

NEMACLADUS Nutt. – Threadplant

Nemacladus orientalis (McVaugh) Morin [*N. glanduliferus* Jeps. var. *orientalis* McVaugh]—Glandular threadplant. Annuals 5–25 cm, stems branched from the base; leaves alternate, oblanceolate to narrowly elliptic, 3–16 mm, hairy, the margins toothed or pinnately lobed; calyx 0.8–2.3 mm; corolla zygomorphic, white or the upper lobes reddish-tipped, 1.3–3 mm; capsules 2–4 mm. Rocky soil of desert shrub communities, 4150–5085 ft. Hidalgo, Luna.

TRIODANIS Raf. – Venus' looking-glass

Triodanis perfoliata (L.) Nieuwl.—Clasping Venus' looking-glass. Annuals, 1–10 dm; leaves alternate, the upper sessile and clasping, cordate to broadly ovate, 0.5–3 × 0.8–2.5 cm, the margins crenate to serrate; corolla actinomorphic, purple or sometimes white, the tube 1–2.5 mm and lobes 4.5–7 mm; capsules 3.5–6(–8) mm, with pores at the middle or below, these 0.5–1.5 mm wide. Sandy or rocky soil of open places, often in disturbed areas, 3900–7550 ft.

1. Capsules with pores at the apex; floral bracts longer than wide…**var. biflora**
1'. Capsules with pores at the middle or below; floral bracts about as wide as long, sometimes wider than long…**var. perfoliata**

var. biflora (Ruiz & Pav.) T. R. Bradley—Grant, Hidalgo.

var. perfoliata—EC, C, WC, SE, SC, SW.

CANNABACEAE – HEMP FAMILY

Kenneth D. Heil

Herbs, annual or perennial, taprooted or rhizomatous, erect or twining, aromatic, pubescent with small glands and hairs, hairs with or without cystoliths; stems usually branched, usually ridged or furrowed; leaves decussate proximally, often alternate distally, simple to palmately lobed or compound, petiolate; stipules persistent, triangular; blade margins serrate; flowers unisexual, staminate and pistillate usually on different plants; staminate flowers 20–200+, pedicellate; sepals 5, stamens 5; pistillate flowers 10–50, perianth a thin undivided layer adhering to ovary, obscure; pistil 1, usually 2-carpellate; ovary superior, fruit fleshy; seed an achene (Whittemore, 2012).

1. Trees or shrubs; leaves simple or ± pinnate…**Celtis**
1'. Herbs or vines; leaves simple to compound, palmate…(2)
2. Herbs; leaves palmately compound…**Cannabis**
2'. Trailing vines; leaves simple with palmate lobes…**Humulus**

CANNABIS L. – Hemp, marijuana

Cannabis sativa L.—Hemp, marijuana, pot, grass, maryjane, cannabis. Stems 0.2–6 m; leaf blades mostly 3–9, linear to linear-lanceolate, 3–15 × 0.2–1.7 cm, margins coarsely serrate; surfaces abaxially whitish green with scattered yellowish brown, resinous dots, strigose, adaxially darker green with large, stiff, bulbous-based conic hairs; flowers unisexual, stamens caducous after anthesis, somewhat shorter than sepals, filaments 0.5–1 mm; cypselae white or greenish, mottled with purple, 2–5 mm. Disturbed sites such as roadsides and vacant lots, occasionally in fallow fields, 4900–8000 ft. NW, Bernalillo, Otero, Sierra. Introduced.

CELTIS L. – Hackberry

Trees or rarely shrubs, to 30 m; crowns spreading; bark usually gray, smooth or often fissured and conspicuously warty; branches slender, glabrous or pubescent, with or without thorns; leaves with stipules falling early; blades deltate to ovate to oblong-lanceolate, base oblique or cuneate to rounded, margins entire or serrate-dentate; venation 3(-5)-pinnate; flowers usually unisexual, staminate and pistillate on same plants, along with a few bisexual flowers, stamens 4–5; fruit a drupe; seed within a stone.

1. Branches with thorns; leaf blade usually < 2 cm wide…**C. pallida**
1'. Branches without thorns; leaf blade usually > 2 cm wide…(2)
2. Leaf blade typically elliptic-lanceolate to ovate-lanceolate, apex sharply acute to acuminate, margins mostly entire…**C. laevigata**
2'. Leaf blade typically broadly to narrowly ovate to oblong-lanceolate, apex blunt or obtuse to abruptly long-acuminate, acute, or short-acuminate, margins variable…(3)
3. Leaf blade typically 4.5 cm or less, margins entire or somewhat serrate above the middle…**C. reticulata**
3'. Leaf blade mostly 5+ cm, margins coarsely serrate for at least part of length…**C. occidentalis**

Celtis laevigata Willd. [*C. laevigata* var. *anomala* Sarg.]—Sugarberry, palo blanco. Trees, to 30 m; trunks to 1 m diam., crowns broad, spreading; bark light gray, smooth or covered with corky warts; branches without thorns; leaf blades typically elliptic-lanceolate to ovate-lanceolate, (4-)6–8(-15) × (2-)3–4 cm, thin and membranous to leathery, margins entire or rarely with a few long teeth; fruit a drupe, orange to brown or red; stones 4.5–7 × 5–6 mm. Rich bottomlands along streams, floodplains, 4188–4265 ft. Chaves, Sierra.

Celtis occidentalis L. [*C. pumila* Pursh]—Hackberry. Trees or shrubs, crowns rounded; bark gray, deeply furrowed, warty with age; branches without thorns; leaf blades lance-ovate to broadly ovate or deltate, 5–12 × 3–6(-9) cm, leathery, margins conspicuously serrate to well below middle; fruit dark orange to purple or blue-black, stones cream-colored, 7–9 × 5–8 mm, reticulate. Moist soil along streams, floodplains, 5100–5500 ft. NE.

Celtis pallida Torr. [*C. spinosa* Spreng. var. *pallida* (Torr.) M. C. Johnst.]—Desert hackberry. Shrubs, to 3 m; crowns rounded; bark gray, smooth; branches spreading, flexuous, whitish gray, with thorns, puberulent; thorns single or in pairs, 3-25 mm; leaf blades ovate to ovate-oblong, to 2-3 × 1.5-2 cm, thick, margins entire or crenate-dentate; flowers mostly staminate on proximal branches, terminal flower bisexual; fruit orange, yellow, or red. Deserts, canyons, mesas, washes, foothills, thickets, brushland, grassland, near gravelly or well-drained sandy soil, 4250-6000 ft. Hidalgo, Luna.

Celtis reticulata Torr. [*C. brevipes* S. Watson]—Netleaf hackberry, palo blanco. Trees or shrubs, (1-)7(-16) m; trunks rarely 6 dm diam.; crowns ± rounded, bark gray with corky ridges; branches without thorns, upright, villous when young; leaf petioles 3-8 mm; blades ovate, 2-4.5(-7) × 1.5-3.5 cm, thick, rigid, margins entire or somewhat serrate above middle, surfaces pubescent, abaxially yellow-green, adaxially gray-green; fruit reddish or reddish black. Dry hills, often on limestone, ravine banks, rocky outcrops, occasionally in sandy soils, 3460-7500 ft. Widespread.

HUMULUS L. – Hops

Humulus lupulus L.—Hops. Herbs, perennial, rhizomatous, 1-6(-7) m; stems branched, pubescent at nodes; leaves simple, with blades ± cordate, palmately 5-7-lobed, 5-15 cm, surfaces abaxially resin-dotted and/or gland-dotted, with soft-pubescent veins, margins dentate-serrate; flowers solitary or paired, subtended by bracts and bracteoles; staminate and pistillate flowers usually on different plants; cypselae yellowish, ovoid, compressed, glandular. Our material belongs to **var. neomexicanus** A. Nelson & Cockerell. Slopes, riverbanks, woods, 5000-9800 ft. Widespread except EC, SE, SW.

CAPRIFOLIACEAE – HONEYSUCKLE FAMILY

Fred R. Barrie

[Dipsacaceae; Valerianaceae]. Small trees, shrubs, subshrubs, or woody vines; leaves opposite, simple; flowers zygomorphic or nearly actinomorphic, perfect or imperfect, borne in pairs or in dense involucrate heads; calyx usually small and 5-lobed; corolla usually 5-lobed, often bilabiate; stamens 1-5; ovary inferior, 2-5-carpellate; fruit an achene, capsule, berry, or dry. (Ackerfield, 2015; Allred et al., 2020; Barrie, 2013; Cronquist, 1984; Judd et al., 2016).

1. Stems and leaves prickly; heads elongate, ovoid to cylindrical…**Dipsacus**
1'. Stems and leaves not prickly; heads not elongate or ovoid to cylindrical…(2)
2. Plants herbaceous; stamens 1-3; ill-scented…**Valeriana**
2'. Plants shrubs or subshrubs; stamens 4-5; not ill-scented…(3)
3. Subshrubs, creeping with prostrate stems; stamens 4; flowers in terminal pairs on long peduncles…
    Linnaea
3'. Shrubs, erect; stamens 5; flowers in axillary pairs or terminal clusters, on shorter peduncles…(4)
4. Flowers slightly to strongly bilabiate, yellow, yellowish red, pink, or white; axillary pairs of flowers subtended by a pair of bracts; berries black, orange, or red with many seeds…**Lonicera**
4'. Flowers not bilabiate, flowers pink or white; flowers not subtended by a pair of bracts; berries white…
    Symphoricarpos

DIPSACUS L. – Teasel

Dipsacus fullonum L. [*D. sylvestris* Huds.]—Common teasel. Biennial herbs with prickly stems, 5-20(-30) dm; leaves sessile, lanceolate to oblanceolate, to 4 dm × 4-10 cm, often with prickly margins and nerves, entire or crenate-serrate; heads 3-10 × 3-5 cm; involucral bracts linear, spine-tipped, subtending dense heads, upturned and as long as or longer than the heads; receptacle chaffy; calyx 4-lobed; corolla 4-lobed, white to purple; cypselae 3-8 mm. Moist locations, ditch banks, river margins, roadsides, 5600-9700 ft. NC, Lincoln, Otero. Native of Europe, now a widespread weed.

LINNAEA L. - Twinflower

Linnaea borealis L.—Twinflower. Evergreen subshrubs with stems trailing and rooting at the nodes; leaves opposite, exstipulate, elliptic to obovate to nearly orbicular, 0.5-2 cm × 3-15 mm, entire or few-toothed, glabrous to hairy; peduncles glandular-stipitate, often hairy; flowers actinomorphic or nearly so; calyx 5-lobed; corolla 5-lobed, 6-12 mm, pink or pinkish, borne in pairs; stamens 4; ovary 3-carpellate; fruit a dry drupe, 1.5-3 mm. New Mexico material belongs to **var. longiflora** Torr. Moist forests, mixed conifer and spruce-fir communities, 7500-11,435 ft. NC, NW.

LONICERA L. - Honeysuckle

Shrubs or woody vines; leaves opposite, the margins mostly entire; flowers actinomorphic or nearly so to zygomorphic and bilabiate, borne in axillary pairs or in terminal clusters; calyx 5-lobed; corolla 5-lobed; stamens 5; ovary 2-3-carpellate; fruit a berry.

1. Uppermost pair of leaves connate-perfoliate; flowers in whorled clusters at the ends of the stems, absent from the axils along the stem…(2)
1'. Uppermost pair of leaves distinct, not connate-perfoliate; flowers in pairs in the axils along the stem and sometimes crowded at the stem tips…(3)
2. Flowers strongly bilabiate, white to cream; blades glabrous or hairy below…**L. albiflora**
2'. Flowers mostly regular, orange, pink, red, or purplish; blades glabrous to very sparsely hairy below…**L. arizonica**
3. Stems twining, trailing; corolla 3-5 cm, strongly bilabiate; fruit black…**L. japonica**
3'. Stems not twining or trailing, mostly upright or bushy-branched; corolla mostly < 3 cm, not strongly bilabiate; fruit red, orange, or black…(4)
4. Flowers yellow or yellowish red, hairy externally, each pair subtended by large (> 5 mm wide), green and foliaceous, glandular bracts; fruit red, becoming black at maturity…**L. involucrata**
4'. Flowers pink, white, ochroleucous, or light yellow, glabrous, each pair subtended by narrow (to 1 mm wide), eglandular bracts; fruit red or orange…(5)
5. Flowers pale yellow to ochroleucous; leaves rounded to obtuse at the tips…**L. utahensis**
5'. Flowers rose-pink or white; leaves acute or acuminate at the tips…(6)
6. Plants glabrous; flowers rose-pink…**L. tatarica**
6'. Plants thinly hairy beneath; flowers deep to light pink…**L. ×bella**

Lonicera albiflora Torr. & A. Gray—White honeysuckle. Shrubs to 2.5 m; stems sometimes twining; leaves ovate or obovate to nearly orbicular, ca. 5 cm; corolla white, 10-20 m, slightly gibbous; berries orange or red. S mountains, piñon-juniper and ponderosa pine communities, 4000-7300 ft. C, WC, SC, SW, Eddy.

Lonicera arizonica Rehder—Arizona honeysuckle. Viny shrubs; leaves ovate, 2-7 cm, glabrous adaxially, pale and usually pubescent abaxially, margins ciliate; corollas nearly actinomorphic, bright red, 34-45 mm; berries red. S and W mountains, piñon-juniper, ponderosa pine–oak, and mixed conifer communities, 5000-9600 ft. C, WC, SW, Eddy, McKinley, Otero.

Lonicera ×bella Zabel—Showy fly honeysuckle. Shrubs to 3 m; stems hollow, the new growth usually sparsely pubescent; leaves ovate-oblong, blades 3-6 × 1.5-3.5 cm, sparsely pubescent on the underside; bracts linear, 5-15 mm; sepals ca. 0.5 mm; corollas pink or white, becoming yellow; berries red or yellow, ca. 1 cm diam. A hybrid between *L. tatarica* L. and *L. morrowii* A. Gray. Floodplains and irrigation canals, 5380-7220 ft. San Juan, Santa Fe.

Lonicera involucrata (Richardson) Banks ex Spreng.—Black twinberry. Shrubs 0.5-2(3) m; leaves pubescent, blades ovate to elliptic or obovate, 5-12 × 2-6 cm; bracts (8-)10-15(-20) mm, leafy, forming a showy involucre surrounding the flowers and fruits, red or purple; calyx reduced; corolla yellow or with a touch of red, hairy, 1-1.5 cm; berries red, becoming glossy-black, ca. 1 cm diam. Streamsides and other moist sites, aspen, Douglas-fir, and spruce-fir communities, 8000-12,000 ft. NC, NW, C, WC, Otero.

Lonicera japonica Thunb. ex Murray—Japanese honeysuckle. Trailing or climbing vines; stems, petioles, and inflorescences pubescent; leaf blades ovate to elliptic, 3-8 × 1.5-4 cm; bracts similar in form to the leaves, to 1 cm; sepals lanceolate, ca. 1 mm; corolla white or yellow, bilabiate, 2–5 cm; berries black, 1–1.5 cm diam. Widely planted as an ornamental and locally naturalized, 3900–5900 ft. NW, Doña Ana. Introduced.

Lonicera tatarica L.—Tatarian honeysuckle. Shrubs to 3 m; leaves elliptic to oblong, hirsute below; bracts minute; corolla rose-pink, 11–20 mm; berries orange or reddish-orange. 4025–6900 ft. Few collections, scattered.

Lonicera utahensis S. Watson—Utah honeysuckle. Shrubs 0.5–2 m, glabrous or with a few hairs; leaves oblong or ovate, blades 1.5–5 × 1–3 cm; bracts linear, 2–3 mm; sepals reduced and nearly obsolete; corolla white or yellow, 1–2 cm; fruit red or salmon-pink, ca. 1 cm diam. Mixed conifer and spruce-fir forests, 8800–10,000 ft. Catron, Grant, Sierra.

SYMPHORICARPOS Duhamel – Snowberry

Shrubs; leaves simple; inflorescence terminal or axillary, racemose, with tightly clustered flowers or the flowers solitary or paired in the leaf axils; flowers actinomorphic, perfect, 4- or 5-merous; corollas pink or white, tubular, bell-shaped, or funnelform; stamens included to weakly exserted; ovary 4-locular; fruit a 2-seeded berry (Ackerfield, 2015; Heil et al., 2013; Jones, 1940).

1. Corolla densely hairy at the level of insertion of the filaments (near the top of the corolla tube), campanulate, with lobes nearly as long as or longer than the tube…(2)
1'. Corolla glabrous densely hairy within below the level of insertion of the filaments (near the base of the corolla tube), campanulate to salverform, with lobes shorter than the tube…(3)
2. Stamens and style exserted from the corolla; flowers several, 6–15 in dense, spicate clusters in the leaf axils; corolla lobes evidently spreading; leaves mostly 2.5–8 cm…**S. occidentalis**
2'. Stamens and style included in the corolla; flowers generally fewer, 2–3(–5) in the leaf axils; corolla lobes scarcely spreading; leaves 1–3(–5) cm…**S. albus**
3. Anthers sessile or nearly so; style 4–7 mm, stiffly hairy above the middle; corolla salverform, the lobes abruptly spreading; leaves 0.6–1.5(–2) cm, oblanceolate or narrowly elliptic although sometimes broadly elliptic and to 9 mm wide…**S. longiflorus**
3'. Anthers on short but evident filaments; style 2–4 mm, glabrous; corolla elongate-campanulate or sub-salverform, the lobes slightly spreading; leaves 0.7–3.5(–5) cm, elliptic to ovate, 3–25 mm wide…**S. rotundifolius**

Symphoricarpos albus (L.) S. F. Blake [*S. pauciflorus* (J. W. Robbins) Britton]—White snowberry. Shrubs 1–2 m; stems pubescent; leaves ovate to elliptic, 1–5 × 0.8–2.5 cm, short-pilose along the veins; corollas 5–8 mm, the lobes shallow and scarcely spreading, densely hairy within; berries white, 7–9 mm. Ponderosa pine–Gambel oak and mixed conifer communities, 7380–9400 ft. NE, C, Catron, Lincoln, Sierra.

Symphoricarpos longiflorus A. Gray—Desert snowberry. Shrubs 0.5–2 m; stems branched; leaves elliptic or ovate, 5–20 × 2–8 cm, uniformly pubescent; inflorescence terminal with 5–6 flowers or 1–2 flowers in the leaf axils; bracts ovate, 2 mm; calyx lobes triangular, connate, 1–2 mm; corollas tubular, light pink; fruit white, 6–10 mm. Piñon-juniper woodland and ponderosa pine communities, 5000–6675 ft. SC, SW, Eddy.

Symphoricarpos occidentalis Hook. [*S. fragrans* A. Nelson & P. B. Kenn.]—Wolfberry. Shrubs to 1 m, forming thickets, the twigs puberulent; leaves ovate, 2.5–11 × 2–6 cm, entire or with a few irregular teeth; flowers 6–15 in dense, spicate clusters in the leaf axils; corolla 5–8 mm, the lobes as long as or longer than the tube, spreading, densely hairy within; berries white, 6–9 mm. Near streams, lakes, meadows, and on mountain slopes, 6300–8200 ft. Colfax.

Symphoricarpos rotundifolius A. Gray [*S. oreophilus* A. Gray; *S. utahensis* Rydb.]—Mountain snowberry. Shrubs 0.5–2 m; stems red or tan; leaves elliptic to ovate or broadly ovate; inflorescence terminal, few-flowered or 1–2 flowers in the leaf axils; bracts ovate; corollas funnel-shaped, pink or white, 7–14 mm; fruit white, 6–15 mm. Gambel oak woodlands, ponderosa pine, Douglas-fir, and spruce-fir communities, 4750–11,000 ft. NC, NW, C, WC, SC, SW.

VALERIANA L. – Valerian

Annual, biennial, or perennial herbs, rhizomatous or taprooted; stems erect; flowers with a calyx of 6–30 plumose setae, 2–8 mm; enrolled at anthesis, unfurling in mature fruit to form a persistent, pappuslike structure; corolla rotate to infundibular or tubular, gibbous near the base, the corolla of female flowers commonly 1/3–1/2 the size of that of perfect flowers; anthers 3, exserted or included in perfect flowers, vestigial in female flowers; ovary with 2 sterile locules reduced or vestigial; fruit with pappose calyx persistent (Cronquist, 1984).

1.　Plants annual or possibly biennial, the roots irregularly globose…**V. sorbifolia**
1'.　Plants perennial…(2)
2.　Plants with a taproot and branching caudex…(3)
2'.　Plant rhizomatous, roots fibrous…(4)
3.　Leaves 10–50 cm, narrowly elliptic to oblanceolate, the margins typically ciliate…**V. edulis**
3'.　Leaves 4–15 cm, obovate, glabrous…**V. texana**
4.　Leaf bases truncate; corollas 10–20 mm,,,**V. arizonica**
4'.　Leaf bases rounded to cuneate; corollas 2–9 mm…(5)
5.　Basal leaves 3–14 cm; corollas of perfect flowers infundibular, 5–9 mm…**V. acutiloba**
5'.　Basal leaves 10–25 cm; corollas of perfect flowers rotate, 2–5 mm…**V. occidentalis**

Valeriana acutiloba Rydb.—Sharpleaf valerian. Plants rhizomatous, the stems, flowers, and/or fruits usually lightly pubescent; stems 20–80 cm; leaves mostly basal, with 1–3 cauline pairs, the basal leaves 3–14 cm, simple or less commonly pinnatifid with 1–3 pairs of reduced lateral lobes, elliptic to narrowly elliptic or oblanceolate; inflorescence capitate in flower or with 1–2 pairs of reduced lateral branches; perfect flowers infundibular, calyx 12–16-fid; corolla white or pink, 5–9 mm; stamens exserted; female flowers 3–5 mm; fruit lance-ovate to narrowly oblong, 3–5 mm. Pine forests and wet meadows, 6500–12,000 ft. NC, NW, C, WC, SC.

Valeriana arizonica A. Gray—Arizona valerian. Plants rhizomatous, glabrous or pubescent at leaf and inflorescence nodes; stems 10–30 cm; leaves basal and cauline; basal leaves ovate or deltate, 0.8–6.5 × 1.5–3 cm; cauline leaves in 2 or 3 pairs, reduced, simple or pinnatifid; inflorescence capitate, expanding in fruit; flowers perfect; calyx 10–12-fid; corolla 10–20 mm, infundibular, white or pink; stamens exserted; fruits 2–5 mm, ovate-lanceolate. Ponderosa pine and mixed conifer forests, 6000–9000 ft. NC, NW, C, WC, SC.

Valeriana edulis Nutt.—Tobacco root. Plants taprooted; stems 1–4 from a branching caudex, glabrous, 0.5–1.5 m; leaves basal or with 1–3 cauline pairs, 10–50 × 0.5–5 cm, narrowly elliptic to oblanceolate or obovate-spatulate, margins entire or with 2–10 linear to oblanceolate lobes, ciliate, rarely glabrous; surfaces glabrous or with scattered simple hairs restricted to the veins; inflorescence paniculiform, 7–60 cm; flowers male (perfect in appearance) or female; calyx 8–14-fid; corolla white to greenish white or ivory, rotate, corollas 1–5 mm; stamens weakly to strongly exserted; fruit ovate to elliptic, 2–4.5 mm. Common in meadows and forest openings, 6000–12,850 ft. Widespread except EC, SE, SW.

Valeriana occidentalis A. Heller—Western valerian. Plants rhizomatous, glabrous; stems 40–100 cm; basal leaves 10–25 cm, simple or pinnatifid with 1–4 pairs of lateral leaflets, the blade or terminal lobe elliptic, 4–8 cm, the lateral lobes 1/4–1/2 the size of the terminal lobe and similar in shape; cauline leaves in 1–4 pairs, the lowermost 1–2 pairs often well developed, 5–15 cm, pinnatifid, the uppermost pairs usually reduced; inflorescence capitate to somewhat conic; flowers perfect or female; calyx 10–14-fid; corollas white, rotate, 2–5 mm; stamens exserted; fruits lanceolate, 3–5 × 1–2 mm. Mixed conifer forests to wet alpine tundra, 7800–11,000 ft. NC, Catron, Grant.

Valeriana sorbifolia Kunth—Pineland valerian. Annual or possibly biennial, 1–12 cm; stem pubescent basally or at the nodes only; leaves cauline, odd-pinnate, 3–20 cm, blades 2–18 × 1–8 cm, elliptic to obovate, glabrous or pubescent along the rachis only; lateral leaflets in 1–6 opposite or subequal pairs, 2–45 × 1–30 mm, elliptic or ovate; margins serrate; inflorescence 12–50 × 3–20 cm wide in late flower and fruit; calyx 6-fid; corollas white or pink; bisexual corollas 1.4–2.5 × 0.4–0.9 mm, tubular; female corollas 0.9–1 × 0.5–0.7 mm, infundibular; stamens included; fruit 1–2.2 × 0.5–1.4 mm, ovate or elliptic, uniformly pubescent or adaxially only. Pine forests in the Animas Mountains, 6700 ft. Hidalgo.

Valeriana texana Steyerm.—Texas valerian. Plants perennial, taprooted; stems 1–4 from a branching caudex, 7–30 cm; leaves simple, basal with 1 or 2 cauline pairs, blades 1.8–11.5 × 0.6–3.8 cm, obovate, margins entire, surfaces glabrous; inflorescence 3–6 × 2–5 cm in flower, 9–16 × 5–12 in fruit; calyx 6–8-fid; corollas white, rotate, gibbous; bisexual corollas 3–3.8 × 1.3–1.5 mm, the lobes 1 mm; female corollas 1.1–1.9 × 0.7–1.2 mm; stamens and style exserted; fruit 2.1–2.6 × 0.6–1.1 mm, narrowly elliptic to elliptic, glabrous. Narrow fissures in exposed limestone and granite rock faces and canyon walls in the Guadalupe Mountains, 6230–8200 ft. Eddy, Lincoln, Otero. **R**

CARYOPHYLLACEAE – PINK FAMILY

Richard K. Rabeler and Ronald L. Hartman†

Annual to perennial herbs; plants erect to prostrate; leaves simple, opposite, or rarely appearing whorled, petioles present or absent, stipules present or absent; flowers actinomorphic, usually bisexual; sepals 4 or 5, distinct or connate below; petals 4 or 5 or absent, distinct, often clawed, limb apex entire, notched, or 2(–4)-fid; stamens (0)1–10; pistil 1, ovary superior, locules 1 or 3–5, usually with free-central placentae, styles (0)1–5, stigmas often linear along adaxial surface of styles (or branches); fruit a capsule with carpel valves splitting apically to throughout, or more rarely an indehiscent utricle (Hartman & Rabeler, 2013; Madhani et al., 2018; Rabeler & Hartman, 2005).

We follow Dillenberger and Kadereit (2014) in their treatment of *Minuartia* s.s. as being a European genus; our species formerly included in *Minuartia* are now placed in *Cherleria* (*C. biflora* and *C. obtusiloba*) and *Sabulina* (*S. macrantha*, *S. rubella*, and *S. stricta*).

Two additional species reported from New Mexico are not treated here. For *Agrostemma githago* L., the voucher (*Ellis 382* [cited in Wooton & Standley, 1915]) has not been located (K. Allred, pers. comm.); and for *Arenaria benthamii* Fenzl ex Torr. & A. Gray, the voucher (*Thurber* in 1881, GH), although labeled "New Mexico," was collected near El Paso, Texas (W. Kittridge, pers. comm.).

1. Leaves with stipules, ovate to triangular, filiform to subulate, or bristlelike, mostly scarious, connate…(2)
1′. Leaves without stipules…(5)
2. Fruit a utricle; seed 1; petals absent; staminodes filiform…**Paronychia**
2′. Fruit a capsule; seeds 3–150+; petals present, sometimes rudimentary or absent; staminodes absent…(3)
3. Petal blade apex divided into 2 or 4 lobes; styles (2)3, connate proximally…**Drymaria**
3′. Petal blade apex entire or rudimentary or petals absent; styles 3, distinct…(4)
4. Flowers in terminal cymes, sometimes solitary; stipules lanceolate to ovate or broadly triangular; sepals lacking lateral spurs…**Spergularia**
4′. Flowers axillary, 1(2); stipules bristlelike; sepals often with 2 lateral spurs…**Loeflingia**
5. Sepals connate 1/2+ of length, into a cup or (usually) a prominent tube; petals white to pink, red, or purple; perianth hypogynous…(6)
5′. Sepals distinct or essentially so; petals usually white; perianth hypogynous or sometimes perigynous…(10)
6. Styles 3(4) or 5 (absent in staminate flowers); fruit valves 3–5 or splitting into 6–10 teeth; flowers perfect (except *S. latifolia*)…**Silene**
6′. Styles 2(3); fruit valves usually 4; plants with perfect flowers…(7)
7. Flower or flowers subtended or enclosed by 2–6 involucral bracts…**Dianthus**
7′. Flower or flowers not subtended by involucral bracts…(8)
8. Calyx 2–4 mm, cup-shaped, commissures between adjacent sepals veinless, scarious…**Gypsophila** (in part)
8′. Calyx 7.5–25 mm, oblong-cylindrical to ovoid, commissures between adjacent sepals absent…(9)
9. Plants perennial; calyx tube terete; coronal appendages 2…**Saponaria**
9′. Plants annual; calyx tube 5-angled or -keeled; coronal appendages absent…**Gypsophila** (in part)
10. Petals absent or rudimentary…(11)
10′. Petals present…(12)

11. Capsule cylindrical, often curved, opening by 8 or 10 teeth (*C. glomeratum*, *C. nutans*)…**Cerastium** (in part)
11'. Capsule ovoid to globose, symmetric, opening by 6(8 or 10) valves or teeth…**Stellaria** (in part)
12. Petal apices 2-lobed, often divided nearly to base, or 4-lobed…(13)
12'. Petal apices entire, emarginate, or notched…(16)
13. Plants prostrate or sprawling, glabrous, glaucous, and succulent; leaves appearing whorled; petals 4-lobed; inflorescences umbelliform…**Drymaria**
13'. Plants, if prostrate, not glabrous, glaucous, and succulent; leaves opposite; petals 2-lobed; inflorescences rarely umbelliform…(14)
14. Petals bilobed for 1/10–1/5 their length; fruit spherical, opening by 6 2–3-times recoiled valves; rhizomes usually with tuberous thickenings…**Torreyostellaria**
14'. Petals bilobed or bifid nearly to base; fruit ovoid or globose to cylindrical, opening by teeth or valves, valves not recoiled; rhizomes, when present, without tuberous thickenings…(15)
15. Capsule cylindrical, often curved, opening by 10 teeth…**Cerastium** (in part)
15'. Capsule ovoid to globose, opening by 6(8 or 10) valves or teeth…**Stellaria** (in part)
16. Capsule valves and styles equal in number…(17)
16'. Capsule valves or teeth 2 times number of styles…(19)
17. Sepals (4)5; styles (4)5; capsule valves (4)5…**Sagina**
17'. Sepals 5; styles 3; capsule valves or teeth 3…(18)
18. Sepals hooded, apex obtuse, margins incurved; seeds obscurely sculptured…**Cherleria**
18'. Sepals not hooded, apex acute to acuminate, sometimes obtuse, margins not incurved; seeds tuberculate or papillate…**Sabulina**
19. Styles 5; capsule cylindrical, opening by 10 teeth…**Cerastium** (in part)
19'. Styles 3; capsule ovoid to urceolate or globose, opening by 6 valves or teeth…(20)
20. Leaf blades filiform to subulate, congested at or near base of flowering stem, apex narrowly blunt to usually apiculate or spinose…**Eremogone**
20'. Leaf blades ovate to lanceolate (sometimes narrowly so), not congested at or near base of flowering stem, apex acute to obtuse…(21)
21. Seeds 1–2.2 mm, appendage present, elliptic, white, spongy; plants perennial…**Moehringia**
21'. Seeds 0.4–0.8 mm, appendage absent; plants annual or perennial…**Arenaria**

ARENARIA L. – Sandwort

Plants annual or perennial, with taproot or slender rhizomes, prostrate to ascending or erect, sometimes densely matted; stems simple or branched, terete to ellipsoid; leaves opposite, not congested at or near base of flowering stem, without stipules, blades linear-lanceolate to broadly ovate or rarely orbiculate; inflorescence an open cyme or flowers solitary, flower(s) not subtended by involucral bracts; flower(s) hypogynous; sepals 5, distinct or nearly so, apex acute to acuminate, not hooded, awn absent; petals 5, white, not clawed, auricles absent, coronal appendages absent, apex entire; stamens 10, arising from ovary base, staminodes 0; ovary 1-locular; styles 3; capsule ovoid to cylindrical-ovoid, opening by 6 ascending to recurved teeth; seeds 0.4–0.8 mm, black, shiny or dull, smooth or tuberculate, marginal wing absent, appendage absent.

1. Plants perennial; leaves linear-lanceolate to narrowly elliptic or oblanceolate, 1-veined; seeds suborbicular, black, shiny, smooth…**A. lanuginosa**
1'. Plants annual; leaves elliptic to broadly ovate or rarely orbiculate, 3–5-veined; seeds reniform, ashy-brown or ashy-black, not shiny, tuberculate…**A. serpyllifolia**

Arenaria lanuginosa (Michx.) Rohrb. [*A. saxosa* A. Gray; *A. confusa* Rydb.; *A. lanuginosa* subsp. *saxosa* (A. Gray) Maguire; *A. saxosa* A. Gray var. *cinerascens* B. L. Rob.; *Spergulastrum lanuginosum* Michx. subsp. *saxosum* (A. Gray) W. A. Weber]—Spreading sandwort. Plants perennial, erect or ascending to procumbent; stems 5–60 cm, internodes retrorsely puberulent throughout or in lines; leaves linear-lanceolate to narrowly elliptic or oblanceolate, 3–35 mm, 1-veined, margins scarious and shiny, ciliate, apex obtuse or acute to apiculate; cymes often much branched, 1–80+-flowered; sepals often appearing prominently keeled proximally, to 6 mm in fruit, apex acute to acuminate, not pustulate; petals narrowly spatulate to obovate, 1.5–6 mm; capsule ± loosely to tightly enclosed by calyx, ovoid, 3–6 mm, 0.8–1.5 times sepal length; seeds black, suborbiculate, slightly compressed, 0.7–0.8 mm, shiny, smooth. Our plants are **var. saxosa** (A. Gray) Zarucchi, R. L. Hartm. & Rabeler. Understory of mixed coniferous forests, sandstone outcrops, stream banks, meadows, gravelly soils, 4000–11,800 ft. Widespread except E. This taxon is extremely polymorphic and deserves further study. Patrick Alexander has informally recognized three phases that intergrade; see Allred and Ivey (2012) for a key to these phases and Keller et al. (2017) for comments on forms found in the Jemez Mountains.

Arenaria serpyllifolia L.—Thymeleaf sandwort. Plants annual, erect to ascending; stems 3-40+ cm, internodes uniformly puberulent; leaves elliptic to broadly ovate or rarely orbiculate, 2-7 mm, 3-5-veined, margins herbaceous and dull, ciliate, apex acute to acuminate; cymes 3-50+-flowered; sepals not keeled, to 4 mm in fruit, apex narrowly acute to acuminate, ± minutely pustulate; petals oblong, 0.6-2.7 mm; capsule loosely enclosed by the calyx, ovoid to cylindrical-ovoid, 3-3.5 mm, 0.8-1.2 times sepal length; seeds ashy-black, reniform, plump, 0.5-0.6 mm, not shiny, with low-elongate, prominent tubercles. Our plants are **var. serpyllifolia**. Lawns, disturbed sites, 4350 ft. Luna. Introduced.

CERASTIUM L. – Mouse-ear chickweed

Plants annual or perennial, with a taproot or slender rhizomes, ascending to erect, sometimes prostrate; stems simple or branched, terete; leaves opposite, not congested at or near base of flowering stem, without stipules, blades linear to oblanceolate, elliptic, or spatulate; inflorescence an open cyme or flowers borne in axils of herbaceous bracts along the stem, flowers not subtended by involucral bracts; flowers hypogynous or weakly perigynous; sepals 5, distinct or nearly so, apex acute or acuminate to obtuse, not hooded, awn absent; petals 5 or rarely 0, white, clawed, auricles absent, coronal appendages absent, apex bilobed 1/5-1/2 of length; stamens (5 or) 10, arising from ovary base, staminodes absent; ovary 1-locular; styles 5; capsule cylindrical, straight or usually curved, opening by 10 erect or spreading, convolute or revolute teeth; seeds 0.4-1.2 mm, pale to dark or reddish brown, tuberculate, marginal wing absent, appendage absent.

1. Capsules straight, tips of teeth rolled outward; leaves broadly spatulate, petiolate…**C. texanum**
1'. Capsules usually curved, tips of teeth erect or spreading, may roll inward; leaves otherwise…(2)
2. Plants annual, all stems producing flowers…(3)
2'. Plants perennial, often with nonflowering stems…(8)
3. Inflorescence bracts, at least uppermost, with scarious margins…**C. fontanum** (in part)
3'. Inflorescence bracts all completely herbaceous…(4)
4. Sepals with long hairs, some exceeding sepal tip…**C. glomeratum**
4'. Sepals with shorter hairs, none exceeding sepal tip…(5)
5. Flowers borne in axils of herbaceous bracts along the stem…**C. axillare**
5'. Flowers in terminal, dichotomous cymes or clusters…(6)
6. Pedicels equaling or shorter than capsules, often deflexed at base of pedicel…**C. brachypodum**
6'. Pedicels longer than capsules, sharply deflexed at summit of pedicel…(7)
7. Leaves and stems with long, woolly hairs, especially below; inflorescence of mature plants not bushy, not equaling 1/2 plant height; sepals ovate-lanceolate, apex broadly acute to obtuse, inner with broad scarious margins (ca. as wide as herbaceous center); midstem leaves lanceolate to narrowly elliptic, (3-)6-16 mm wide; capsules (9-)10-13 mm…**C. nutans**
7'. Leaves and stems with short, erect or spreading hairs; inflorescence of mature plants bushy, at least 1/2 plant height; sepals narrowly lanceolate, apex sharply acute to acuminate, inner with narrow scarious margins (narrower than herbaceous center); midstem leaves linear-lanceolate, 1.5-6 mm wide; capsules 5-10 (-11) mm…**C. fastigiatum**
8. Petals about equaling the sepals; stems erect to spreading…**C. fontanum** (in part)
8'. Petals usually 1.5-2 times the sepal length; stems matted with old leaves persisting…(9)
9. Bracts herbaceous or only the uppermost slightly scarious-margined; leaves narrowly oblong to oblong-lanceolate or oblanceolate, usually obtuse, those of the flowering stems without axillary clusters of secondary leaves; plants strictly alpine…**C. beeringianum**
9'. Bracts scarious-margined or only the lowermost completely herbaceous; leaves linear to narrowly lanceolate, usually acute, those of the flowering stems with axillary clusters of secondary leaves; plants generally in disturbed areas…**C. arvense**

Cerastium arvense L.—Field or prairie mouse-ear. Plants perennial, clumped and taprooted, or mat-forming and long-creeping, ascending to erect; stems 5-30 cm, pilose-subglabrous proximally, glandular-pubescent distally, nonflowering shoots present; leaves of flowering stems lanceolate or oblanceolate to linear, 2-25 × 1-5 mm, apex acute, rarely obtuse, subglabrous to hairy, axillary clusters of leaves present below; flowers in terminal cymes, inflorescence bract margins narrowly scarious (or only the lowest completely herbaceous), glandular-hairy, pedicels curved at summit, longer than capsules; sepals narrowly lanceolate to lance-elliptic, 3.5–6(–7) mm, with narrowly scarious margins, apex acute, softly hairy, hairs not exceeding sepal tips; petals obovate, 7.5–12.5 mm, ca. 2 times sepal length; stamens 10; capsule cylindrical, curved, < 1.5 times sepal length, teeth erect, convolute; seeds brown, 0.6–1.1 mm. Our plants are **subsp. strictum** Gaudin. Roadsides, meadows, riparian areas, mixed conifer woodlands, alpine meadows and slopes, disturbed areas, 7000–13,000 ft. NC, NW, C, WC, SC, SW.

Cerastium axillare Correll—Trans-Pecos mouse-ear chickweed. Plants annual, taprooted, usually erect; stems 6–40 cm, simple or branched below, glandular-pilose, hairs in midstem region equaling or longer than stem diam., nonflowering shoots absent; leaves oblanceolate to spatulate (lower) or linear-lanceolate, lanceolate, or narrowly elliptic (upper), 7–25 × 1–6 mm, apex acute, rarely obtuse, glandular-pilose, axillary clusters of leaves absent; flowers borne, often singly, in axils of herbaceous bracts along the stem, inflorescence bracts herbaceous, glandular-hairy, pedicels sharply curved at summit, shorter than to 2 times as long as capsules; sepals lanceolate, 3 5 mm, with broad (inner) or very narrow (outer) scarious margins, apex acute, glandular-hispid, hairs not extending beyond sepal tips; petals oblanceolate, 2–3 mm, shorter than sepals; stamens 5; capsule narrowly cylindrical, curved, ca. 2 times sepal length, teeth erect, convolute; seeds light brown, 0.4–0.7 mm. Rocky canyons, grassy sinkholes, mountain slopes, 4400–8500(9840) ft. SE, SC, SW, Socorro.

Cerastium beeringianum Cham. & Schltdl.—Bering mouse-ear chickweed. Plants perennial, taprooted but often matted, somewhat erect; stems 10–25 cm, glabrous to sparsely pubescent, especially above, hairs widely spreading to slightly deflexed, often glandular; nonflowering, prostrate shoots present; leaves narrowly oblong to oblong-lanceolate or oblanceolate, 5–20 × 2–5 mm, apex usually obtuse, hairy, axillary clusters of leaves absent; flowers in terminal cymes, inflorescence bracts herbaceous (or only uppermost sometimes with a slightly scarious margin), hairs glandular and not glandular, pedicels angled at base in fruit, 1–5 times as long as capsules; sepals lanceolate to lance-elliptic, 3–7 mm, with broad scarious margins, apex acute, densely glandular-hairy, hairs not extending beyond sepal tips; petals broadly oblanceolate, 6–12 mm, usually equaling (rarely 2 times) sepal length; stamens 10; capsules cylindrical, curved, ca. 2 times sepal length, teeth erect, convolute; seeds pale to dark brown, 0.7–1.1 mm. Subalpine riparian areas, alpine tundra meadows and slopes, 11,200–12,620 ft. NC, C, SC, Cibola.

Cerastium brachypodum (Engelm. ex A. Gray) B. L. Rob. [*Cerastium nutans* Raf. var. *brachypodum* Engelm. ex A. Gray]—Short-stalked mouse-ear chickweed. Plants annual, taprooted, erect; stems 5–20 cm, glandular-pubescent, nodes without long, woolly hairs, nonflowering shoots absent; leaves lanceolate to narrowly elliptic or oblanceolate (midstem) or ovate to spatulate (lower leaves), 5–30 × 2–8 mm, apex ± acute (or obtuse in lowermost leaves), usually sparsely, softly hairy, axillary clusters of leaves absent; flowers in terminal cymes, inflorescence bracts herbaceous, glandular-puberulent, pedicels deflexed at base in fruit, as long as capsules; sepals broadly lanceolate, 3–4.5 mm, with broad (inner) or narrow (outer) scarious margins (or outer may be entirely herbaceous), apex broadly acute, glabrate or hairs sparse, short, glandular, not extending beyond sepal tips; petals ovate-elliptic, 3–4 mm, equaling or shorter than sepals; stamens 10; capsules cylindrical, curved, 2–2.25 times sepal length, teeth erect, convolute; seeds golden-brown, 0.4–0.7 mm. Grasslands, wet meadows, open mixed conifer woods, 5000–10,400 ft. NC, C, WC, SC.

Cerastium fastigiatum Greene—Fastigiate mouse-ear chickweed. Plants annual, taprooted, erect; stems branched at base, the branches ascending (creating a bushy appearance), 10–50 cm, with stiff, stipitate-glandular, spreading or slightly reflexed hairs shorter than diam. of stem, soft woolly hairs absent, nonflowering shoots absent; leaves linear-lanceolate (midstem) or narrowly oblanceolate to spatulate (lower leaves), 20–70 × 1.5–6 mm, apex acute to acuminate, hairs short, stiff, spreading, glandular, axillary clusters of leaves absent; flowers in very lax cymes (at least 1/2 plant height), inflorescence bracts herbaceous, glandular-hairy, pedicels bent near summit, longer than the capsules; sepals narrowly lanceolate, 4–5 mm, with narrow (narrower than the herbaceous center of inner sepals) scarious margins, apex sharply acute to acuminate, hairs sparse or not so, usually short, glandular-hispid, not extending beyond sepal tips; petals oblanceolate, 4–5 mm, equaling the sepals; stamens 10; capsules cylindrical, curved, 5–10(–11) mm, ca. 2 times sepal length, teeth erect to slightly spreading, convolute; seeds golden-brown, 0.5–0.8 mm. Sandy canyons, grassy or rocky openings in mixed conifer woodlands, rhyolitic out-crops, and wet meadows, 4500–10,660 ft. C, WC, SE, SC, San Miguel. In following Morton (2005a) in segregating *C. fastigiatum* from *C. nutans*, we found that *C. fastigiatum* is likely to be more widespread than previously assumed.

Cerastium fontanum Baumg. [*C. vulgare* Hartm.; *C. fontanum* subsp. *triviale* (Link) Jalas]—Common mouse-ear chickweed. Plants perennial, tufted to mat-forming, often rhizomatous, erect; stems 10–45 cm, softly pubescent with eglandular, straight hairs, nonflowering, decumbent shoots sometimes present, leaves elliptic to ovate-oblong or oblanceolate to spatulate (leaves on sterile shoots), 10–25 × 3–8 mm, apex subacute (obtuse on sterile shoots), hairs dense, nonglandular, flowers in terminal cymes, inflorescence bract margins herbaceous, uppermost often with narrow scarious margins, hairs not glan-dular, pedicels slightly curved near summit, about equaling the capsules; sepals ovate-lanceolate, 9–13 mm, with narrow scarious margins, apex acute, hairs nonglandular, rarely glandular, not extending be-yond sepal tips; petals oblanceolate, 5–7 mm, 1–1.5 times sepal length; stamens 10, occasionally 5; cap-sule narrowly cylindrical, ca. 2 times sepal length, teeth erect, convolute; seeds reddish brown, 0.4–1.2 mm. Our plants are **subsp. vulgare** (Hartm.) Greuter & Burdet. Riparian areas, stream banks, meadows in mixed forests, canyon bottoms, roadsides, disturbed areas, 5100–10,880 ft. NC, C, WC, Lincoln, Sierra. Introduced.

Cerastium glomeratum Thuill.—Sticky mouse-ear chickweed. Plants annual, taprooted, erect or ascending; stems branched, 5–45 cm, hairy, glandular at least above, rarely entirely eglandular, non-flowering shoots absent; leaves broadly ovate or elliptic-ovate (midstem) or oblanceolate or obovate, sometimes spatulate (lower leaves), 5–20 × 2–8 mm, apex apiculate, hairs long, white, and spreading, axillary clusters of leaves absent; flowers in dense, cymose clusters or somewhat open cymes, inflores-cence bracts herbaceous, hairs long, mainly not glandular, pedicels erect, often slightly arched near sum-mit, shorter than capsules; sepals lanceolate, 4–5 mm, with narrow scarious margins, apex acute, hairs glandular, long white hairs extending beyond sepal tips; petals oblanceolate (rarely absent), 3–5 mm, shorter than sepals; stamens 10; capsule narrowly cylindrical, ca. 2 times sepal length, teeth erect, con-volute; seeds pale brown, 0.5–0.6 mm. Wet coniferous forests, stream banks, rhyolite slopes, 5500–8000 ft. Catron, Doña Ana.

Cerastium nutans Raf.—Nodding mouse-ear chickweed. Plants annual, taprooted, erect; stems simple or branched (but not bushy), 10–50 cm, softly hairy, often with a few long, woolly hairs at lower-most nodes, glandular above, nonflowering shoots rarely present; leaves lanceolate to narrowly lanceo-late-triangular (midstem) or spatulate (lower leaves), 25–32 × (3–)6–16 mm, apex mostly acute, softly hairy and glandular or woolly-hairy, especially below, sterile axillary clusters of leaves absent; flowers in terminal cymes (not equaling 1/2 plant height), inflorescence bracts herbaceous, glandular-hairy, pedicels sharply bent near summit, 1.5–5 times as long as capsules; sepals ovate-lanceolate, 4–6 mm, with broad (inner as wide as the herbaceous center) or narrow (outer) scarious margins (or outer may be entirely herbaceous), apex acute to obtuse, hairs sparse, short, glandular, not extending beyond sepal tips; petals oblanceolate, 3–6 mm, 1–1.5 times sepal length; stamens 10; capsule cylindrical, curved, (9–)10–13 mm, 2–3 times as long as sepals, teeth erect, convolute; seeds golden-brown, 0.5–0.8 mm. Yellow pine and mixed conifer woods, gravelly soils, 4600–10,400 ft. WC, SW. This species, along with *C. fastigiatum* and *C. brachypodum*, deserves further study in our area. While most plants of *C. nutans* are likely **var. ob-tectum** Kearney & Peebles, plants of **var. nutans** (leaves not marcescent; long woolly hairs only at lower stem nodes) may occur in New Mexico. See Morton (2005a) for further information about these taxa.

Cerastium texanum Britton [*C. sordidum* B. L. Rob.; *Stellaria montana* Rose]—Chihuahuan mouse-ear chickweed. Plants annual, taprooted, erect; stems sparingly branched below, 15–35 cm, sparsely glandular-pilose, nonflowering shoots absent; leaves linear-lanceolate to narrowly oblanceolate (mid-stem) or broadly spatulate-petiolate (lower leaves), 7–55 × 3–16 mm, apex acute or obtuse, sometimes short-acuminate, softly pilose, small axillary tufts of leaves absent; flowers in open and loose cymes;

inflorescence bracts herbaceous, pilose, pedicels sharply deflexed at base in fruit, 1–2 times as long as capsules; sepals lanceolate to ovate, 3–6 mm, with narrow, scarious margins, apex acute, hairs short, glandular, not extending beyond sepal tips; petals oblanceolate, 5–8 mm, 1.5–2 times sepal length; stamens 5; capsule cylindrical, straight, 1.5–2 times as long as sepals, becoming outwardly coiled; seeds red-brown, 0.4–0.7 mm. Canyon bottoms, oak woodlands, on igneous substrates, 4800–5500 ft. SC, SW, Catron, Lincoln.

CHERLERIA L. – Sandwort

Plants perennial with woody taproot, cespitose or mat-forming, flowering stems erect; stems simple or branched, terete; leaves opposite, not congested at or near base of flowering stem, without stipules, blades needlelike to subulate; inflorescence an open cyme or flowers solitary, flower(s) not subtended by involucral bracts; flower(s) perigynous; sepals 5, distinct, apex obtuse, hooded via incurved margins, awn absent; petals 5, white or pink, not clawed, auricles absent, coronal appendages absent, apex entire; stamens 10, arising from hypanthium, staminodes absent; ovary 1-locular; styles 3; capsule narrowly ellipsoid, opening by 3 incurved to recurved valves; seeds 0.6–0.8 mm, red-tan or brown, smooth or obscurely sculptured, marginal wing absent, appendage absent.

1. Leaves straight, ± flat in cross-section; sepals (or just tips) glabrous to sparsely hairy, distinctly less hairy than the pedicels, recurved in fruit; petals 1.2–1.5 times sepal length; moist areas…**C. biflora**
1'. Leaves ± outwardly curved, strongly keeled, triangular in cross-section; sepals and pedicels equally stipitate-glandular, erect in fruit; petals 2+ times sepal length; drier areas…**C. obtusiloba**

Cherleria biflora (L.) A. J. Moore & Dillenb. [*Arenaria biflora* L.; *Minuartia biflora* (L.) Schinz & Thell.]—Mountain sandwort. Flowering stems erect, 2–10 cm, glabrous or rarely retrorsely hairy; leaves needlelike to subulate, 5–10 mm, straight, ± flat in cross-section; margins glabrous, apex rounded to acute; sepals oblong to narrowly lanceolate, 3.5–4.5 mm, sepals (or just tips) glabrous to sparsely hairy, distinctly less hairy than the pedicels, recurved in fruit; petals 1.2–1.5 times sepal length; capsule 5.5 mm, longer than sepals; seeds brown, 0.7–0.8 mm, smooth to obscurely sculptured. Subalpine lakes, spruce-fir forests, alpine meadows, 11,500–12,700 ft. NC.

Cherleria obtusiloba (Rydb.) A. J. Moore & Dillenb. [*Arenaria obtusiloba* (Rydb.) Fernald; *Minuartia obtusiloba* (Rydb.) House]—Alpine stitchwort. Flowering stems erect, 1–12 cm, stipitate-glandular; leaves needlelike to subulate, 1.5–8 mm, ± outwardly curved, strongly keeled, triangular in cross-section, margins glabrous, apex often apiculate; sepals narrowly ovate to oblong, 2.9–6.5 mm, sepals and pedicels equally stipitate-glandular, erect in fruit; petals 2+ times sepal length; capsule 3.5–6 mm, equaling sepals; seeds reddish tan, 0.6–0.7 mm, obscurely sculptured. Alpine tundra and dry, rocky subalpine meadows and talus slopes, (8800–)9800–13,025 ft. NC, Cibola, Lincoln, Otero.

DIANTHUS L. – Pink, carnation

Dianthus armeria L.—Deptford pink. Plants erect; stems 4–88 cm, often pubescent; leaves linear to narrowly oblanceolate, 2–10.5 cm, green, margins basally ciliate, apex rarely obtuse (below) or acute; inflorescence bracts linear, equaling or longer than calyx, herbaceous, apex acute; calyx 20–25-veined, 11–20 mm, hairy, lobes acuminate; petals bearded, 3–8(–10) mm; capsule 10–16 mm, slightly shorter than calyx; seeds shield-shaped. Our plants are **subsp. armeria**. Grasslands in ponderosa pine forests, roadsides, granite outcrops, 6000–9875 ft. NC, Catron, Grant. Introduced.

DRYMARIA Willd. ex Schult. – Drymary

Plants annual with taproot, prostrate or ascending to erect; stems simple to often branched distally, terete; leaves opposite or appearing whorled, not congested at or near base of flowering stem, stipules filiform to subulate, herbaceous to indurate, or absent, blades suborbiculate to spatulate or linear to oblong; inflorescence a terminal cyme or umbel-like cluster, flower(s) not subtended by involucral bracts; flower(s) hypogynous; sepals 5, distinct, apex acute to acuminate, obtuse, or blunt to rounded, hooded, awn absent; petals 5, white, clawed, auricles absent, coronal appendages absent, apex divided into 2 or 4 lobes; stamens 5, arising from base of ovary, staminodes 0; ovary 1-locular; styles (2)3; capsule ellipsoid to globose, opening by (2)3 spreading to recurved valves; seeds 0.5–0.9(–1.3) mm, tan, reddish brown, dark brown to purplish, or olive-green to black, tuberculate, marginal wing absent, appendage absent.

Specimens from New Mexico labeled *Drymaria laxiflora* Benth., a species native to Mexico and southern Texas and included in Allred and Ivey (2012), are *D. glandulosa*.

1. Blades of cauline leaves ovate or reniform to suborbicular, 4–25 mm wide, base truncate to cordate or obtuse to rounded…(2)
1'. Blades of cauline leaves linear to oblong or lanceolate, 0.2–3 mm wide, base briefly obtuse to attenuate… (3)
2. Leaves opposite, base truncate to cordate, stipules present; stems erect or ascending; flowers in terminal cymes; petals 2-fid, lobe apex not notched; seeds < 1 mm…**D. glandulosa**
2'. Leaves appearing whorled, base obtuse to rounded, stipules absent; stems prostrate or sprawling; flowers in axillary umbel-like clusters; petals 4-fid, lobe apex distinctly notched; seeds 1–1.5 mm…**D. pachyphylla**
3. Leaves appearing whorled, at least in part; flowers in a terminal cyme becoming elongate and racemose; petals 4-lobed; seeds 0.8–0.9 mm…**D. molluginea**
3'. Leaves opposite; flowers in a congested to open terminal cyme; petals 2-lobed; seeds 0.5–0.7 mm…(4)
4. Herbaceous portion of sepals ± oblong, apex blunt or rounded, veins ± parallel, apically confluent… **D. depressa**
4'. Herbaceous portion of sepals lanceolate, apex acute to acuminate, veins with lateral pair distinctly arching outward…**D. leptophylla**

Drymaria depressa Greene [*D. effusa* A. Gray var. *depressa* (Greene) J. A. Duke]—Pinewoods drymary. Plants annual, ascending to erect; stems generally branching at base, 0.5–5 cm, glabrous or minutely puberulent, not glaucous; leaves opposite, stipules (often deciduous) entire, subulate, 0.5–1.2 mm, leaves orbiculate to spatulate (basal leaves) or oblong (cauline leaves), 0.3–1 cm × 0.2–3 mm, base gradually tapered, apex rounded to acute; cymes terminal, congested to open, 3–25+-flowered; sepals subequal, lanceolate, oblong, or ovate (herbaceous portion ± oblong), 1.8–2.3 mm, with 3 prominent, ± parallel, apically confluent veins, apex blunt to rounded, hood at ± right angle to apex, formed in part by scarious margins, glabrous; petals 2-lobed for ca. 1/2 their total length, 1.5–2.8 mm, 0.75–1 times sepal length; seeds light reddish brown to tan, with snail-shell or teardrop shape, 0.5–0.6 mm, tubercles minute, rounded. Open glades, rocky or gravelly slopes in coniferous woodlands, 8200–10,700 ft. Catron, Grant, Rio Arriba.

Drymaria glandulosa Bartl. [*D. fendleri* S. Watson]—Fendler's drymary. Plants annual or perennial, erect or ascending; stems simple or sparingly branched, 5–35 cm, pubescent to stipitate-glandular (especially above), not glaucous; leaves opposite, stipules persistent, divided into 2 filiform segments, 0.5–1 mm, leaves ovate to reniform, 0.5–1.5(–2) cm × 4–17(–25) mm, base truncate to cordate, apex acute to cuspidate; cymes terminal, usually congested, 3–15-flowered; sepals subequal or outer sepals shorter than inner, lanceolate (herbaceous portion similar), 3–4.8 mm, with 3 distinct, usually prominent veins arcing outward at midsection and ± confluent apically, apex acute to setaceous-acuminate (herbaceous portion similar), not hooded, glabrous or stipitate-glandular; petals 2-lobed for 1/2+ their length, 1.2–3.2 mm, equaling or shorter than sepals; seeds tan to reddish brown, with snail-shell shape, 0.5–0.7 mm, tubercles prominent, rounded. Our plants are **var. glandulosa**. Rocky and gravelly areas, piñon-juniper and pine-oak woodlands, 4000–8000 ft. NC, NW, C, WC, SC, SW.

Drymaria leptophylla (Cham. & Schltdl.) Fenzl ex Rohrb. [*D. tenella* A. Gray]—Canyon drymary. Plants annual, erect; stems sparingly branched, 8–25 cm, glabrous, not glaucous; leaves mostly opposite, stipules (often deciduous) divided into 2 filiform segments, 0.3–1 mm, leaves linear to narrowly oblong, 0.5–2.5 cm × 0.2–1.2 mm, base gradually tapered, apex rounded to apiculate; cymes terminal, open, (5–)15–75+-flowered; sepals unequal, with outer 2 often shorter (± 0.5 mm) than inner, lanceolate to ovate (herbaceous portion lanceolate), 1.8–3.5 mm, with 3 prominent veins arcing outward at midsection and confluent apically, apex acute to acuminate (herbaceous portion similar), hood oblique, present at least on outer 2 sepals, glabrous or with a few sessile glands; petals 2-lobed for 1/2 or less their length, 1.3–2.6 mm, 0.5–1 times sepal length; seeds tan, with snail-shell shape, 0.6–0.7 mm, tubercles minute, rounded. Our plants are **var. leptophylla**. Piñon-juniper and oak-pine woodlands, disturbed roadsides, gravelly and rocky soils, 5400–8400(–10,000) ft. NC, NW, C, WC, SW, Eddy.

Drymaria molluginea (Ser.) Didr. [*D. sperguloides* A. Gray]—Slimleaf drymary. Plants annual, erect; stems simple or dichotomously branched, 3–25(–30) cm, glabrous or sparsely glandular, not glaucous; leaves mostly appearing whorled, stipules ± persistent, simple, filiform to subulate, 0.5–2 mm, leaves linear, 1–3(–3.5) cm × 0.5–1.5(–2) mm, base gradually tapered, apex rounded or sometimes apiculate; cymes terminal, open and becoming elongate and racemose, 3–30+-flowered; sepals subequal, broadly ovate to ± orbiculate (herbaceous portion elliptic to oblong or lanceolate), 2–3.5 mm, with midvein prominent, lateral pair often evident, then arcing outward at midsection and confluent apically, apex obtuse (herbaceous portion similar), hood oblique or at right angles to apex, formed in part by scarious margins, glabrous; petals 4-lobed for 1/2 or less their length, 1.7–2.5 mm, subequal to sepals; seeds dark brown to purplish, horseshoe-shaped, 0.8–0.9 mm, tubercles minute, rounded. Gravelly soils, montane grasslands, oak and piñon-juniper woodlands, 5000–8835 ft. NC, NW, C, WC, SW. *Drymaria arenarioides* Willd. ex Schult., a federally listed noxious weed, is not yet known in New Mexico but is native to areas just south of the Mexican border. Plants would resemble *D. molluginea* but with stipitate-glandular (vs. glabrous) stems and leaves.

Drymaria pachyphylla Wooton & Standl.—Thickleaf drymary, inkweed. Plants annual, nearly prostrate; stems nearly prostrate, radiating pseudoverticillately from base, 10–20 cm, succulent, glabrous, glaucous; leaves appearing whorled, stipules absent, leaves ovate to suborbiculate, (0.2–)0.5–1.3 cm × 4–10 mm, base obtuse to rounded, apex ± obtuse; umbel-like clusters axillary, congested, 3–12-flowered; sepals subequal, oblong to broadly elliptic (herbaceous portion similar), 2–3.5 mm, with 3 or 5 obscure veins usually not confluent apically, apex obtuse (herbaceous portion generally acute), not hooded, glabrous; petals 4-lobed for 1/2 or less their length, 2.5–3 mm, 0.6–1 times sepal length; seeds olive-green to black, teardrop-shaped, 1.1–1.3 mm, tubercles marginal, minute, elongate. Clay soils, adobe flats, 3750–4740 ft. SC, Luna.

EREMOGONE Fenzl – Sandwort

Plants perennial, usually with branched, woody base, erect; stems simple or branched, terete; leaves opposite, congested at or near base of flowering stems, without stipules, blades needlelike or filiform to subulate; inflorescence an open or congested cyme, flower(s) not subtended by involucral bracts; flower(s) weakly perigynous; sepals 5, distinct or nearly so, apex acute or acuminate, hooded or not, awn absent; petals 5, white or sometimes yellowish white or brownish to reddish pink, clawed or not, auricles absent, coronal appendages absent, apex entire or emarginate; stamens 10, arising from hypanthium, staminodes 0; ovary 1-locular; styles 3; capsule ovoid, opening by 6 ascending to recurved teeth; seeds 1.2–2 mm, brown or black, papillate or tuberculate, marginal wing absent, appendage absent.

Specimens we have seen labeled *Eremogone aculeata* (S. Watson) Ikonn., included in Allred and Ivey (2012), are either *E. eastwoodiae* or *E. fendleri*.

1. Flowers in dense, congested, capitate cymes…**E. hookeri**
1'. Flowers in open cymes…(2)
2. Sepals glabrous throughout or essentially so…**E. eastwoodiae**
2'. Sepals moderately to densely stipitate-glandular…**E. fendleri**

Eremogone eastwoodiae (Rydb.) Ikonn. [*Arenaria eastwoodiae* Rydb.]—Eastwood's sandwort. Plants perennial, erect, densely tufted, green; stems 10–25 cm, glabrous or stipitate-glandular; basal leaves spreading to recurved, needlelike, 1–3 cm, flexible to rigid, cauline leaves 2–4 pairs, reduced upward, margins glabrous to puberulent, not glaucous, apex spinose; cymes open, 3–17-flowered; sepals lanceolate to ovate-lanceolate, (3.5–)4–6.5 mm, apex narrowly acute to acuminate, glabrous; petals yellowish white or sometimes brownish to reddish pink, broadly oblong-elliptic to oblanceolate, 0.9–1.1 times sepal length; capsules 4–6 mm; seeds brown, 1.2–1.7 mm, papillate, subechinate. Piñon-juniper woodlands, desert scrub, sandy soils, 5000–7500 ft.

1. Stems and pedicels stipitate-glandular…**var. adenophora**
1′. Stems and pedicels glabrous…**var. eastwoodiae**

var. adenophora (Kearney & Peebles) R. L. Hartm. & Rabeler—Stems and pedicels stipitate-glandular. Piñon-juniper woodlands, desert scrub, sandy soils, 5500–7435 ft. NC, NW.

var. eastwoodiae—Stems and pedicels glabrous. Piñon-juniper woodlands, desert scrub, sandstone soils and cliffs, 5000–7500 ft. NC, NW.

Eremogone fendleri (A. Gray) Ikonn. [*Arenaria fendleri* A. Gray]—Fendler's sandwort. Plants perennial, erect, mostly tufted, bluish green; stems 10–30(–40) cm, densely stipitate-glandular; basal leaves ascending or recurved, filiform, 1–10 cm, flexible, cauline leaves (4)5+ pairs, reduced upward or not, margins glabrous to puberulent, not glaucous, apex apiculate to spinose; cymes open, 3–35-flowered; sepals linear-lanceolate, 4–7.5 mm, apex acuminate, moderately to densely stipitate-glandular; petals white, oblong-elliptic to spatulate, 0.9–1.3 times sepal length; capsules 5–7 mm; seeds black, 1.5–1.9 mm, tuberculate. Grasslands, sagebrush slopes, oak and mixed conifer woodlands, rock outcrops, stony slopes, meadows, roadsides, alpine tundra, 5990–13,025 ft. Widespread except EC, SE, Hidalgo, Luna.

Eremogone hookeri (Nutt.) W. A. Weber [*Arenaria hookeri* Nutt.]—Hooker's sandwort. Plants perennial, erect, densely or loosely matted, green; stems 1–15(–20) cm, scabrid-puberulent; basal leaves persistent, straight to recurved, subulate to needlelike, 0.3–4 cm, cauline leaves 1–4 pairs, often larger than basal leaves, margins minutely puberulent, often glaucous, apex spinose; cymes dense, congested, capitate, 3–30+-flowered; sepals linear-lanceolate to lanceolate, 5–8(–9) mm, apex narrowly acute or acuminate, glabrous or pubescent; petals white, oblanceolate, about equaling sepals; capsules to 4 mm; seeds black, 1.8–2 mm, tuberculate. Plains, exposed slopes and ridges, granitic outcrops, 6000–6200 ft.

1. Basal leaves 0.3–1.5 cm, straight or recurved, rigid; sepals 5–8(–9) mm…**var. hookeri**
1′. Basal leaves 2–4 cm, straight, rigid or flexible; sepals (7–)8–10 mm…**var. pinetorum**

var. hookeri—Basal leaves 0.3–1.5 cm, straight or recurved, rigid; sepals 5–8(–9) mm. Plains, exposed slopes and ridges, 6000 ft. Guadalupe, Union.

var. pinetorum (A. Nelson) Dorn—Basal leaves 2–4 cm, straight, rigid or flexible; sepals (7–)8–10 mm. Granitic outcrops, 6200 ft. NE, EC.

GYPSOPHILA L. – Baby's-breath

Plants annual or perennial, with woody base or taproot, erect; stems simple or usually branched, terete; leaves opposite, not congested at or near base of flowering stem, without stipules, blades oblong to narrowly lanceolate or ovate-lanceolate; inflorescence a dense to open cyme or thyrse, flower(s) not subtended by involucral bracts; flower(s) hypogynous; sepals 5, connate proximally into cup or tube, sometimes 5-angled or -keeled, apex obtuse, acute, or acuminate, not hooded, awn absent; petals 5, white or light pink to purplish, clawed or barely so, auricles absent, coronal appendages absent; apex entire or shallowly bilobed; stamens 10, arising with petals from low nectariferous disk or adnate to petals; staminodes absent; ovary 1-locular or 2-locular proximally; styles 2; capsule globose or oblong, opening by 4 slightly recurving valves or spreading teeth; seeds 1–2.5 mm, brown to black, tuberculate or papillose, marginal wing absent, appendages absent.

1. Plants perennial; calyx 2–4 mm, cup-shaped, commissures between adjacent sepals veinless, scarious… **G. scorzonerifolia**
1′. Plants annual; calyx 7.5–17 mm, tubular, 5-angled or -keeled, commissures between adjacent sepals absent…**G. vaccaria**

Gypsophila scorzonerifolia Ser.—Glandular baby's-breath. Plants perennial, erect; stems 50–200 cm, glabrous below, glandular-puberulent above; leaves oblong-lanceolate to narrowly ovate, 2–15 cm, margins glabrous or minutely roughened, apex obtuse to acute; calyx cup-shaped, 2.5–4 mm, commissures between adjacent sepals veinless, lobe apex obtuse; petals white with pink tinge to light purplish pink (drying darker), barely clawed, 4–6 mm; capsule globose, 3–4 mm, about equaling calyx; seeds reniform, 1–1.3 mm, coarsely tuberculate. Roadsides and disturbed areas, 5300–5725 ft. Bernalillo, Grant. Introduced.

Gypsophila vaccaria (L.) Sm. [*Vaccaria hispanica* (Mill.) Rauschert]—Cow herb, cow-cockle. Plants annual, erect; stems 20–100 cm, glabrous, glaucous; leaves oblong or lanceolate to ovate-lanceolate, 2–10 cm, margins glabrous, apex acute; calyx tubular, 5-angled or -keeled, 7.5–17 mm, commissures between adjacent sepals absent, lobe apex acute or acuminate; petals pink to purplish, claw 8–14 mm, limb 3–8 mm; capsule oblong to subglobose, 8–9(–10) mm, shorter than calyx; seeds subglobose, 2–2.5 mm, papillose. Cultivated fields and waste ground, 7200–7500 ft. NC. Introduced. The most recent specimen seen was collected in 1932.

LOEFLINGIA L. – Pygmy-leaf

Loeflingia squarrosa Nutt.—Plants annual, prostrate to erect; stems 1–12 cm, covered with stalked glands, somewhat fleshy; leaves with filamentous to spinose stipules, 0.4–1.5 mm, leaves awl-like to oblong, 0.4–5.5 mm, margins bristly, apex blunt to spine-tipped; sepals erect to squarrose, resembling leaves, usually with 2 filamentous to stiff lateral spurs, 1.8–6.5 mm, apex acute to spinose; capsule 1.5–3.7 mm, 0.5–0.8 times sepal length. Mesquite and oak shrublands, 3180–4100 ft. Eddy, Roosevelt.

MOEHRINGIA L. – Sandwort

Plants perennial, with slender rhizomes, ascending or decumbent to erect; stems simple or branched, terete or angled; leaves opposite, not congested at or near base of flowering stem, without stipules, blades oblanceolate to elliptic; inflorescence an open cyme or flowers solitary, flower(s) not subtended by involucral bracts; flower(s) weakly perigynous; sepals 5, distinct, apex obtuse to acute or acuminate, not hooded, awn absent; petals 5, white, not clawed, auricles absent, coronal appendages absent, apex entire; stamens 10, arising from hypanthium, staminodes absent; ovary 1-locular; styles 3; capsule ovoid to subglobose, opening by 6 revolute teeth; 1–2.2 mm, reddish brown to blackish, shiny, smooth to minutely tuberculate, marginal wing absent, appendage elliptic, white, spongy.

1. Stem pubescence retrorse; sepals (herbaceous portion) oblong to elliptic, apex mostly rounded or obtuse, 1.7–2.8(–3) mm; petals ca. 2 times sepal length…**M. lateriflora**
1'. Stem pubescence peglike, spreading; sepals (herbaceous portion) lanceolate, apex acute to acuminate, (2.8–)3–6 mm; petals 0.75–1.5 times sepal length…**M. macrophylla**

Moehringia lateriflora (L.) Fenzl [*Arenaria lateriflora* L.]—Grove or blunt-leaf sandwort. Plants ascending or decumbent; stems 5–30 cm, terete, uniformly retrorsely pubescent; leaves broadly elliptic to oblong-elliptic or oblanceolate, 6–35 mm, margins granular to minutely serrulate-ciliate, apex obtuse or rounded; sepals not keeled, ovate or obovate, herbaceous portion oblong to elliptic, 1.7–3 mm, apex mostly obtuse or rounded; petals 1.7–2.8(–3) mm, ca. 2 times sepal length; capsule subglobose, 3–5 mm, 1.5–2 times sepal length; seeds reniform, 1 mm, smooth. Wet areas in mixed conifer forests, 8200–9865 ft. NC, San Miguel.

Moehringia macrophylla (Hook.) Fenzl [*Arenaria macrophylla* Hook.]—Bigleaf sandwort. Plants ascending to erect; stems 2–18 cm, angled or grooved, hairs minute, spreading, peglike; leaves lanceolate to elliptic, 15–70 mm, margins smooth to minutely granular, often ciliate below, apex acute; sepals at least somewhat keeled, ovate, herbaceous portion lanceolate, 2.8–6 mm, apex acute to acuminate; petals (2.8–)3–6 mm, 0.75–1.5 times sepal length; capsule ovoid, 5 mm, about as long as sepals; seeds oval, 1.5–2.2 mm, tuberculate. Rocky summits, canyons, riparian areas in spruce-fir and birch-maple forests, 7100–10,700 ft. NC, C, Cibola, Socorro.

PARONYCHIA Mill. – Nailwort, whitlow-wort

Plants perennial, with branched, woody base, prostrate, ascending, or erect; stems simple or branched, terete or angled; leaves opposite, not congested at or near base of flowering stem, stipules ovate or lanceolate to subulate, scarious, blades linear, linear-subulate, or narrowly elliptic to narrowly elliptic-oblanceolate; inflorescence an open or congested cyme or flowers solitary, flower(s) not subtended by involucral bracts; flower(s) perigynous; sepals 5, connate basally, apex obtuse to rounded, hooded, awned subapically on abaxial surface, awn filiform to stout (often thickened-conic basally or weakly spinulose apically); petals absent; stamens 5, arising from hypanthium, staminodes 5; ovary 1-locular; style 1; utricle ovoid; seeds brown, smooth, marginal wing absent, appendage absent.

1. Leaves linear or linear-subulate…(2)
1'. Leaves narrowly elliptic to narrowly elliptic-oblanceolate or lanceolate…(4)
2. Cymes terminal, 3–6-flowered, congested or flowers solitary; flowers 3.6–5 mm; awns erect or somewhat spreading, white, 1–1.5(–2) mm…**P. sessiliflora**
2'. Cymes terminal; 3–70-flowered, open or congested; flowers 1.8–3.5 mm; awns divergent, yellowish, 0.4–0.9 mm…(3)
3. Stems 8–15 cm, prostrate and mat-forming; stipules 2–8 mm, apex acuminate; cymes congested, with 3–7 flowers; sepals 1.7–2 mm…**P. depressa**
3'. Stems 10–35 cm, erect to ascending; stipules 5–15 mm, apex long-acuminate; cymes open, with 20–70 flowers; sepals 1.3–1.8 mm…**P. jamesii**
4. Leaves narrowly elliptic to narrowly elliptic-oblanceolate, fleshy, apex obtuse to subacute; stipules ovate, apex subobtuse, entire; awns erect, white, 0.3–0.6(–1) mm…**P. pulvinata**
4'. Leaves lanceolate, leathery, apex sharply mucronate; stipules lanceolate, 5–8 mm, apex acuminate, often deeply cleft, awns moderately divergent, bright white, 1.4–2.1 mm…**P. wilkinsonii**

Paronychia depressa Nutt. ex Torr. & A. Gray—Spreading nailwort. Plants perennial, often matted, caudex branched, woody; stems 8–15 cm, prostrate to sprawling, much branched; stipules lanceolate, 2–8 mm, apex acuminate, entire; leaves linear, 8–15 mm, leathery, apex shortly cuspidate; cymes terminal, 3–7-flowered, congested in clusters 7–25 mm wide; flowers ± ovate, 2.3–3.5 mm; sepals green to purple-brown, oblong to lanceolate-oblong, 1.7–2 mm, margins whitish to translucent, 0.05–0.1 mm wide, scarious, awns divergent, yellowish, conic in basal 1/2–2/3, 0.7–0.9 mm; utricle smooth, glabrous. Calcareous grasslands, 5700 ft. Harding, Union.

Paronychia jamesii Torr. & A. Gray—James' nailwort. Plants perennial, caudex branched; stems 10–35 cm, erect to ascending, much branched; stipules lanceolate, 5–15 mm, apex long-acuminate, entire; leaves linear, 7–25 mm, leathery, apex obtuse to subacute or submucronate; cymes terminal, 20–70-flowered, open, clusters 10–20 mm wide; flowers short-campanulate, 1.8–2.8 mm; sepals green to red-brown, oblong, 1.3–1.8 mm, margins whitish to translucent, 0.05–0.1 mm wide, scarious, awns widely divergent, yellowish, conic in basal 1/2–2/3, 0.4–0.8 mm; utricle smooth, glabrous. Limestone soils, prairie and dune areas, rocky slopes, pine-juniper, desert scrub, 2990–8060 ft. Widespread except NW, Catron, Taos.

Paronychia pulvinata A. Gray—Rocky Mountain nailwort. Plants perennial, cushion-forming, caudex much branched; stems 5–10 cm, prostrate, much branched; stipules ovate, 3–6 mm, apex subobtuse, entire; leaves narrowly elliptic to narrowly elliptic-oblanceolate, 2–5 mm, fleshy, obscurely nerved, apex obtuse to subacute; cymes none, flowers solitary at end of shoots, almost concealed; flowers elliptic-oblong, 2.5–2.8 mm; sepals whitish to green, narrowly oblong to ovate-oblong, 1.5–1.7 mm, margins white, 0.2–0.3 mm wide, papery, awns erect, white, weakly spinulose apically, 0.3–0.6(–1) mm; utricle smooth, glabrous. Alpine tundra, subalpine meadows, rocky areas, 10,200–12,930 ft. NC.

Paronychia sessiliflora Nutt.—Creeping nailwort. Plants perennial, densely cespitose and mat-forming, caudex branched; stems 5–25 cm, erect to ascending, branched proximally; stipules lanceolate to subulate, 2–3 mm, apex long-acuminate, often deeply cleft; leaves linear-subulate, 4–7.5 mm, leathery, prominently nerved, apex acute to shortly cuspidate-mucronate; cymes terminal, 3–6-flowered, congested or flowers solitary; flowers ± ovate, 3.6–5 mm; sepals green to red-brown, lanceolate to oblong, 1.5–2 mm, margins whitish to translucent, 0.1–0.2 mm wide, scarious, awns erect or somewhat spreading, white, narrowly conic below, 1–1.5(–2) mm; utricle densely pubescent in distal 1/2. Limestone soils, dry prairies, piñon-juniper woodlands, 4535–7300 ft. NE, NC, NW, C, EC, Sierra.

Paronychia wilkinsonii S. Watson—Wilkinson's nailwort. Plants perennial, cushion-forming, caudex much branched, stems 4–10 cm, erect to ascending; stipules lanceolate, 5–8 mm, apex acuminate, often deeply cleft; leaves lanceolate, 5–9 mm, leathery, apex sharply mucronate; cymes terminal or subterminal, 3–7(–10)-flowered, densely congested in clusters 5–15 mm wide; flowers short-campanulate, 3.3–4.5 mm; sepals red-brown, lanceolate, 1.7–2 mm, margins white, 0.2–0.3 mm wide, papery to scarious, awns moderately divergent, bright white, conic below, 1.4–2.1 mm; utricle papillate distally. Limestone cobble, 4900 ft. Otero. Rare in New Mexico. **R**

SABULINA Rchb. – Sandwort

Plants perennial with taproot, often mat-forming or ascending to erect; stems simple or branched, terete; leaves opposite, not congested at or near base of flowering stem, without stipules, blades linear-filiform or needlelike to subulate or lanceolate-oblong; inflorescence an open cyme or flowers solitary, flower(s) not subtended by involucral bracts; flower(s) perigynous; sepals 5, distinct, apex acute to acuminate or sometimes obtuse, not hooded, awn absent; petals 5, white, not clawed, auricles absent, coronal appendages absent, apex entire; stamens 10, arising from hypanthium, staminodes absent; ovary 1-locular; styles 3; capsule ovoid or ellipsoid, opening by 3 incurved to recurved valves, seeds 0.4 1 mm, brown or black, tuberculate or papillate, marginal wing absent, appendage absent.

1. Pedicels and sepals stipitate-glandular; seeds reddish brown, 0.4–0.5 mm; capsule 4.5–5 mm, longer than sepals…**S. rubella**
1'. Pedicels and sepals glabrous; seeds black, 0.7–1 mm; capsule 3–4 mm, usually shorter than sepals…(2)
2. Leaf apex rounded; cymes 2–5-flowered or flowers solitary…**S. macrantha**
2'. Leaf apex blunt to sharply pointed; cymes 5–30+-flowered…**S. stricta**

Sabulina macrantha (Rydb.) Dillenb. & Kadereit [*Minuartia macrantha* (Rydb.) House; *Arenaria macrantha* (Rydb.) A. Nelson]—Large-flower sandwort. Plants erect to procumbent, cespitose or mat-forming; stems 2–15 cm, glabrous, leaves subulate to linear, 5–10 mm, margins glabrous, apex rounded; cymes 2–5-flowered or flowers solitary, pedicels glabrous; sepals ovate to lanceolate, 3.5–5 mm, to 5.5 mm in fruit, apex green to purple, sharply acute to acuminate, glabrous; petals oblong to obovate, 0.7–1.8 times sepal length; capsule 3–3.8 mm, shorter than sepals; seeds black, 0.7–1 mm, tuberculate; tubercles low, rounded. Limestone rocks, openings in spruce-fir forests, 10,000–10,490 ft. Bernalillo, Cibola.

Sabulina rubella (Wahlenb.) Dillenb. & Kadereit [*Arenaria rubella* (Wahlenb.) Sm.; *Minuartia rubella* (Wahlenb.) Hiern]—Reddish sandwort. Plants ascending to erect, cespitose or mat-forming; stems 2–8(–18) cm, usually moderately to densely stipitate-glandular, leaves subulate, 1.5–10 mm, margins glabrous, apex acute to apiculate; cymes 3–7+-flowered or rarely flowers solitary, pedicels stipitate-glandular; sepals ovate to lanceolate, 2.5–3.2 mm, not enlarging in fruit, apex green to purple, acute to acuminate, stipitate-glandular; petals elliptic, 0.8–1.3 times sepal length; capsule 4.5–5 mm, longer than sepals; seeds reddish brown, 0.4–0.5 mm, tuberculate; tubercles low, elongate, usually rounded. Alpine tundra, talus slopes, snowmelt springs, meadows, rock fields, 9866–12,584 ft. NC, Bernalillo.

Sabulina stricta (Sw.) Rchb. [*Minuartia michauxii* (Fenzl) Farw.; *Arenaria stricta* Michx.]—Michaux's stitchwort. Plants erect to ascending, cespitose; stems 8–40 cm, glabrous, leaves linear-filiform to linear-lanceolate, 8–30 mm, margins glabrous, apex blunt to sharply pointed; cymes 5–30+-flowered, pedicels glabrous; sepals ovate to lanceolate, 3–6 mm, not enlarging in fruit, apex green, acute to mostly acuminate, glabrous; petals oblong-obovate, 1.3–2 times sepal length; capsule 3–4 mm, usually shorter than sepals; seeds black, suborbiculate, compressed, 0.8–0.9 mm, tuberculate; tubercles elongate. Dry, calcareous soils in grasslands, 5500–6400 ft. (*Heil 29580*, collected at 12,400 ft. in tundra from Mora County, appears to be this taxon).

SAGINA L. – Pearlwort

Sagina saginoides (L.) H. Karst.—Arctic pearlwort. Plants ascending to sometimes procumbent, tufted to cespitose; stems 1–5 cm, glabrous; leaves linear, 10–20 mm, margins glabrous, apex usually apiculate; pedicels frequently recurved after flowering, erect in fruit; flowers mostly 5-merous, sepals elliptic, 2–2.5 mm, glabrous, margins hyaline, white to purple, apex obtuse to rounded, remaining appressed to capsule; petals elliptic, 1.5–2 mm; capsule 2.5–3.5 mm, 1.5–2 times sepal length; seeds obliquely triangular with distinct adaxial groove, 0.3–0.4 mm. Meadows, riparian areas in mixed conifer forests, 8200–11,435 ft. NC, NW.

SAPONARIA L. – Soapwort, bouncing bet

Saponaria officinalis L.—Bouncing bet. Plants perennial, erect, colonial; stems 30–90 cm, glabrous; leaves ovate to elliptic, strongly 3(–5)-veined, 3–15 cm, margins minutely roughened, apex acute or rounded; calyx green or reddish, often cleft, 15–25 mm, lobe apex acute or acuminate; petal limb 8–15 mm, often drying to dull purple; capsule 15–20 mm. Roadsides in ponderosa pine forests, disturbed areas, escaping from cultivation, 5400–8000 ft. N, C, SW, Catron, Curry, Roosevelt. Introduced.

SILENE L. – Campion, catchfly

Plants annual or perennial, with taproot or branched, woody caudex, ascending to erect. Stems simple or branched, terete to somewhat angled; leaves opposite, not congested at or near base of flowering stem, without stipules, blades linear to spatulate, lanceolate or broadly elliptic; inflorescence a congested cyme, or frequently flowers few or solitary, flower(s) not subtended by involucral bracts; flower(s) hypogynous; sepals 5, connate proximally into tube, terete, apex triangular to linear-acuminate, not hooded, awn absent; petals 5, white, greenish white, yellowish, pink, scarlet, or dark red, clawed, auricles 2, coronal appendages 2, apex entire or 2–4(–8)-lobed; stamens 10, adnate with petals to carpophore, staminodes absent; ovary 1- or 3–5-locular; styles usually 3 or 5, rarely 4; capsule ovoid to cylindrical, opening by 3–5 valves, usually splitting into 6, 8, or 10 teeth; seeds 0.5–2.5 mm, reddish to black, tuberculate or papillate, marginal wing rarely present, appendage absent.

Silene csereii Baumg. has been reported near Pecos but has not yet been collected (K. Allred, pers. comm.). It would key near *S. vulgaris*, differing in having purple (vs. white) filaments and a calyx that is barely (vs. conspicuously) inflated and constricted above (vs. not) in fruit.

1.   Plants annual, taproots generally slender…(2)
1'. Plants perennial, arising from a rhizome or a branched, generally woody caudex (rarely a taproot)…(5)
2.   Calyx tubes clearly 20–30-veined; flowers pistillate; styles (4)5…**S. latifolia** (in part)
2'. Calyx tubes clearly (8–)10-veined; flowers perfect and styles 3, or staminate…(3)
3.   Calyx tubes 5–6 mm, lobes 1–2 mm; calyx glabrous…**S. antirrhina**
3'. Calyx tubes 15–20 mm, lobes 3–13 mm; calyx hairy…(4)
4.   Calyx tube veins not clearly netted above middle of tube, lobes 3–6 mm, obtuse to acuminate; flowers staminate…**S. latifolia** (in part)
4'. Calyx tube veins netted above middle of tube, lobes 6–13 mm, acuminate; flowers perfect…**S. noctiflora**
5.   Petals scarlet, lobes 2, obconic, or 4–6, linear to oblong; calyx (in flower) 19–32 mm, cylindrical…(6)
5'. Petals white to greenish white or pink, entire or lobes 2–4(–8); calyx (in flower) 5–20 mm, campanulate to cylindrical…(7)
6.   Plants 10–20 cm; leaves 1–5 mm wide; flowers usually solitary; petal lobes 2, obconic…**S. plankii**
6'. Plants 30–60 cm; leaves 15–30 mm wide; flowers 3–5 in a cyme; petal lobes 4–6, linear to oblong…**S. laciniata**
7.   Calyx entirely green or reddish, pale commissures between obscure veins absent…(8)
7'. Calyx with pale commissures between usually prominent veins…(10)
8.   Calyx puberulent to stipitate-glandular…**S. menziesii**
8'. Calyx glabrous…(9)
9.   Plants densely matted and forming a cushion; stems 3–6(–10) cm; flowers solitary, petals bright pink or rarely white; plants of alpine areas…**S. acaulis**
9'. Plants not densely matted; stems 20–40 cm; flowers in open cymes, petals white; plants mostly of lower elevations…**S. vulgaris**
10. Flowers unisexual; styles (4)5 in pistillate flowers…**S. latifolia**
10'. Flowers perfect; styles 3(–5)…(11)
11. Plants 2–20 cm; cauline leaves 1–3 pairs; seeds winged, wing 0.25–1 times width of body of seed; plants of alpine areas…**S. hitchguirei**
11'. Plants mostly 20–80 cm; cauline leaves 1–12 pairs; seeds not winged; plants mostly of lower elevations…(12)
12. Styles (4)5; capsule teeth (8–)10; calyx lobes 1–2 mm…**S. drummondii**

12'. Styles 3(4); capsule teeth 6(-8); calyx lobes 3-7 mm…(13)
13. Petals white to pale yellowish, limb 2-lobed, each lobe cleft into 3-4 smaller lobes, 5-8 mm; calyx lobes narrowly lanceolate, 5-7 mm…**S. wrightii**
13'. Petals white to greenish white or pink, limb 2- or 4-lobed, each often with a small lateral tooth, 2.5-8 mm; calyx lobes lanceolate, 2-5 mm…(14)
14. Stems simple or seldom branching; cymes narrow, flowers 2-8+ per flowering node; pedicels shorter than or equaling the calyx tube; calyx teeth not recurved…**S. scouleri**
14'. Stems often branched; cymes open, pedicels longer than the calyx tube; calyx teeth often recurved… **S. thurberi**

Silene acaulis (L.) Jacq. [*S. acaulis* subsp. *subacaulescens* (F. N. Williams) Hultén]—Moss campion. Plants perennial, densely matted and forming cushions, flowering stems erect; stems 3-6(-10) cm, internodes glabrous; leaves linear, 0.4-1.5 cm × 0.5-1.2 mm, 3-6 pairs per stem, apex acute; flowers solitary, pedicels shorter than or equaling the calyx tube; flowers perfect or unisexual; calyx tubular to campanulate, 10-veined, not netted above, reddish, pale commissures between veins absent, 6-9 × 2-3 mm (in fruit to 13 mm and 5 mm, respectively), glabrous, lobes lanceolate to ovate, 1-1.5 mm; petals bright pink (rarely white), 1.2-1.5 times calyx length, limb entire (rounded) to 2-lobed, lobes rounded, 2.5-3.5 mm, appendages poorly developed; stamens exserted 0-1 mm in staminate flowers; styles 3, exserted 2-3 mm; capsule opening by 6 teeth; seeds light brown, 0.8-1 mm, margins rugose, not winged. Subalpine and alpine tundra meadows, spruce forests, talus slopes, 9800-13,000 ft. NC.

Silene antirrhina L.—Sleepy catchfly. Plants annual, erect, taproot slender; stems 15-70 cm, upper internode(s) with a glutinous band accumulating debris; leaves narrowly oblanceolate to linear, 1-7 cm × 2-12 mm, 7(-11) pairs per stem, apex acute to obtuse; cymes open, 1-25+-flowered, pedicels mostly longer than calyx tube; flowers perfect; calyx ovoid to ellipsoid, 10-veined, not netted above, pale commissures between veins present, 5-6 × 2-3 mm (in fruit to 8 mm and 5 mm, respectively), glabrous, lobes (teeth) narrowly triangular, 1-2 mm; petals mostly dark red throughout or in part, 1.1-1.2 times calyx length, limb 2-lobed, lobes rounded, 2.5 mm, appendages 2, 0.1-0.4 mm; stamens included; styles 3, included; capsule opening by 6 teeth; seeds gray-brown, 0.5-0.8 mm, margins finely papillate, not winged. Desert shrub and grasslands, piñon-juniper woodland communities, limestone or igneous rocky slopes, riparian areas, disturbed sites, 4100-8700 ft. C, SE, SC, SW, widely scattered in N.

Silene drummondii Hook.—Drummond's catchfly, forked catchfly. Plants perennial, erect, caudex 1- to few-branched; stems 20-70 cm, internodes uniformly stipitate-glandular, without glutinous bands; leaves narrowly lanceolate to oblanceolate or linear, 3-10 cm × 4-12 mm, 2 5 pairs per stem, apex acute to rounded; cymes narrow, 1-10+-flowered, pedicels shorter or longer than calyx tube; flowers perfect; calyx cylindrical or broadly tubular to ellipsoid, 10-veined, not netted above, pale commissures between veins present, 12-18 × 4-8 mm (in capsule to 14-18 mm and 5-8 mm, respectively), glandular-pubescent, lobes triangular, 1-1.5 mm; petals white, often dark red at least in part, 1-1.5 times calyx length, limb 2-lobed, lobes rounded, 1-3 mm, appendages 2, 0.1-0.4 mm; stamens included; styles (4) mostly 5, included; capsule opening by (4) mostly 5 teeth; seeds dark brown, 0.7-1.2 mm, margins minutely papillate, not winged. Ponderosa pine, mixed aspen-conifer, spruce-fir forests, dry meadows, riparian areas, rocky slopes, roadsides, 7000-11,800 ft.

While most of our plants can be recognized as one of two varieties, intermediates are known in the southern Rocky Mountains as noted by Morton (2005c).

1. Petals not exceeding calyx, white; fruiting calyces mostly cylindrical or broadly tubular, 2.25-3 times as long as broad; seeds 0.7-0.8 mm; inflorescence mostly 3-10-flowered…**var**. **drummondii**
1'. Petals exserted 4-6 mm, dark red to purple apically; fruiting calyces narrowly ovoid to ellipsoid (pressed), 2 times as long as broad; seeds 1-1.2 mm; inflorescence mostly 1-4-flowered…**var**. **striata**

var. **drummondii**—Inflorescence mostly 3-10-flowered; calyx in fruit mostly cylindrical or broadly tubular, 2.25-3 times as long as broad; petals not exceeding calyx, white. Ponderosa pine, mixed aspen-conifer forests, spruce-fir forests, dry meadows, riparian areas, rocky slopes, roadsides, 7000-11,800 ft. NC, NW, WC.

var. striata (Rydb.) Bocquet [*S. drummondii* subsp. *striata* (Rydb.) J. K. Morton]—Inflorescence typically 1–4(–12)-flowered; calyx in fruit narrowly ovoid to ellipsoid, 2 times as long as broad; petals exserted 4–6 mm, dark red to purple apically. Mixed conifer forests and spruce-fir forests, meadows, riparian areas, rocky areas, 8300–10,000 ft. NC, NW, C.

Silene hitchguirei Bocquet [*Lychnis apetala* L. var. *montana* (S. Watson) C. L. Hitchc.; *Silene uralensis* (Rupr.) Bocquet subsp. *montana* (S. Watson) McNeill]—Mountain campion. Plants perennial, erect, caudex 1- to few-branched, with tuft of basal leaves; stems 2–10 cm, internodes uniformly stipitate-glandular; leaves narrowly oblanceolate and long-spatulate, 5–8 cm × 5–10 mm, 1–3 pairs per stem, apex acute; flowers solitary or cymes 2–4-flowered, pedicels equaling or longer than calyx tube; flowers perfect; calyx elliptic, 10-veined, netted above, pale commissures between veins present, 6–12 × 5–8 mm, enlarged little in fruit, densely pubescent, hairs purple-septate, lobes triangular, 1.5–4 mm; petals white or pink, 1–1.5 times calyx length, limb emarginate, 3 mm, appendages 2, 0.2–0.3 mm; stamens equaling calyx; styles 5, equaling calyx; capsule opening by 5 or 10 teeth; seeds brown, 0.7–1.2 mm, margins not papillate, wing < 1/4 width of body of seed. Spruce forests, 11,500–12,850 ft. Taos.

Silene laciniata Cav.—Gregg's Mexican pink or campion. Plants perennial, erect, caudex 1- to few-branched; stems 20–60 cm, internodes uniformly stipitate-glandular, without glutinous bands; leaves lanceolate or elliptic to oblanceolate, 3–6 cm × 8–30 mm, 4–7 pairs per stem, apex acute; cymes open, 3–5-flowered, pedicels shorter than to (mostly) longer than calyx tube; flowers perfect; calyx cylindrical, 10-nerved, netted above, pale commissures between veins present, 19–23 × 5–6 mm (in fruit to 25 mm and 8 mm, respectively), stipitate-glandular, hairs with clear cross-walls, lobes triangular, 3–3.5 mm; petals scarlet, 1.4–1.6 times calyx length, limb 4–6-lobed, lobes linear to oblong, 6–15 mm, appendages 2, 1–2 mm; stamens exserted 8–10 mm; styles 3, exserted 9–12 mm; capsule opening by 6 teeth; seeds reddish brown, 1.2–2.5 mm, margins long-papillate, not winged. Our material belongs to **var. greggii** (A. Gray) S. Watson. Grassy slopes in piñon-oak woodlands, ponderosa pine and spruce forests, cliffs, rocky soils, 4000–11,000 ft. NW, C, WC, SE, SC, SW, San Miguel.

Silene latifolia Poir. [*Lychnis alba* Mill.; *Silene pratensis* (Rafn) Gren. & Godr.]—White campion or cockle. Plants annual or briefly perennial, erect or decumbent at base, taproot woody; stems 30–80+ cm, internodes uniformly pubescent, often stipitate-glandular; leaves lanceolate to elliptic, 2–12 cm × 6–25 mm, 3–8 pairs per stem, apex acute; cymes open, 3–30-flowered, pedicels shorter than calyx tube; flowers unisexual; calyx in staminate flowers tubular, 10-veined, not netted above, pale commissures between veins present, 15–20 × 5–8 mm, in "fruit" to 22 mm and 9 mm, respectively, in pistillate flowers elliptic, 20-veined, not netted above, pale commissures between veins present, 15–25 × 5–9 mm (in fruit to 23 mm and 15 mm, respectively), lobes narrowly to broadly triangular, 3–6 mm; petals white, ca. 2 times calyx length, limb mostly deeply 2-lobed, lobes rounded, 2.5–5 mm, appendages 2, 1–1.5 mm; stamens exserted 5–7 mm; styles (4) mostly 5, exserted 7–10 mm; capsule opening by (4) mostly 5 or 10 teeth; seeds dark gray-brown, 1.2–1.5 mm, margins coarsely tuberculate, not winged. Roadsides, disturbed sites, riparian areas, mixed aspen-conifer woodlands, 5100–9840 ft. NC. Introduced.

Silene menziesii Hook. [*S. menziesii* var. *viscosa* (Greene) C. L. Hitchc. & Maguire]—Menzies' catchfly. Plants perennial, shoots decumbent to matted at base, taproot slender, flowering stems decumbent to erect; stems mostly 5–30 cm, internodes uniformly stipitate-glandular, without glutinous bands; leaves 2–4(–6) cm × 3–20 mm, broadly elliptic to oblanceolate, 5–20 pairs per stem, apex acute to acuminate; flowers solitary or in leafy cymes, 3–7+-flowered, pedicels longer than calyx tube; flowers functionally perfect or unisexual; calyx campanulate, obscurely 10-nerved, not netted above, pale commissures between veins absent, 5–7 × 3–6 mm, enlarged little in fruit, puberulent to stipitate-glandular, lobes lanceolate, 1.5–3 mm; petals white, 1–1.7 times calyx length, limb 2-lobed, lobes oblong, 1.5–3 mm; appendages 2, 0.1–0.3 mm; stamens equaling calyx in staminate flowers; styles mostly 3(4), exserted 1–3 mm; capsule opening by 3 or 6(–8) teeth; seeds brown, 0.6–1 mm, margins obscurely reticulate, not winged. Riparian areas, slopes, ponderosa pine-oak forests, 6400–10,150 ft. NW, NC.

Silene noctiflora L.—Night-flowering catchfly. Plants annual, erect, taproot slender; stems 30–70+ cm, internodes uniformly stipitate-glandular, without glutinous bands; leaves 1–12 cm × 3–35 mm, broadly elliptic to lanceolate, 5–9 pairs per stem, apex acute to short-acuminate; cymes 3–9-flowered, pedicels shorter than to sometimes longer than calyx tube; flowers perfect; calyx ovate-elliptic, 10-veined, netted above middle of tube, pale commissures between veins present, 17–20 × 5–8 mm (in fruit to 22 mm and 12 mm, respectively), densely pilose, often stipitate-glandular, lobes linear-lanceolate to subulate, 6–13 mm; petals white or pinkish, 1.2–1.3 times calyx length, limb deeply 2-lobed, lobes usually oblong, 4–6 mm, appendages 2, 0.5–1.5 mm; stamens exserted 1–3 mm; styles 3, exserted 3–4 mm; capsule opening by 6 teeth; seeds grayish brown, 1–1.6 mm, margins tuberculate, not winged. Riparian areas, roadsides in pine-oak forest, 7200–7700 ft. Mora, Sandoval, Socorro. Introduced.

Silene plankii C. L. Hitchc. & Maguire—Rio Grande fire pink. Plants perennial, ascending; caudex many-branched, woody; stems 10–20 cm, internodes finely retrorse gray-puberulent; leaves linear to narrowly lanceolate or oblanceolate, 1–4 cm × 1–5 mm, 5–9 pairs per stem, apex sharply acuminate; flowers usually solitary; pedicels shorter than calyx tube, flowers perfect; calyx tubular, 10-veined, not netted above, green, pale commissures between veins absent, 20–30 × 3–6 mm, glandular-puberulent, lobes lanceolate, 2–4 mm; petals scarlet, about equaling calyx, limb shallowly 2-lobed, lobes rounded, 7–10 mm, appendages 2, 1–1.5 mm; stamens exserted, ± equaling corolla lobes; styles 3, exserted, ± equaling corolla lobes; capsule opening by 6 teeth; seeds brown, 1.5 mm, margins papillate, not winged. Crevices in rhyolite or granite cliffs, 5800–8910 ft. NC, C, SC. **R & E**

Silene scouleri Hook.—Hall's catchfly, Pringle's catchfly. Plants perennial, erect, caudex 1- to several-branched; stems mostly 25–70 cm, internodes uniformly stipitate-glandular, without glutinous bands; leaves ovate-lanceolate to lanceolate, oblanceolate, or rarely linear, 6–25 cm × 4–30 mm, 1–12 pairs per stem, apex acute; cymes narrow, 2–8+ flowers per node, flowering nodes 4–6+; pedicels shorter than to equaling calyx tube; flowers perfect; calyx campanulate, 10-veined, netted above, pale commissures between veins present, 8–13 × 3–7 mm (in fruit to 18 mm and 8 mm, respectively), densely stipitate-glandular, lobes lanceolate, 2–5 mm, apex not recurved; petals white, greenish white, or pink, 1.2–1.3 times calyx length, limb deeply 2–4 lobed, often each with a lateral tooth, lobes broadly oblong, 2.5–8 mm, appendages 2, 1–3 mm; stamens equaling calyx; styles mostly 3(4), exserted 4–6 mm; capsule opening by mostly 6(–8) teeth; seeds grayish brown, 0.9–1.5 mm, margins papillate, not winged. Meadows and riparian areas in piñon-juniper, pine-oak, mixed conifer, and spruce-fir forests, rocky slopes, roadsides, 6000–12,500 ft. Widespread except E, SW.

Silene scouleri in our area likely requires further study. While Morton (2005c) and Allred and Ivey (2012) both noted the presence of two subspecies (**subsp. hallii** (S. Watson) C. L. Hitchc. & Maguire and **subsp. pringlei** (S. Watson) C. L. Hitchc. & Maguire), we found segregating the subspecies more difficult than expected and here refrain from trying to distinguish them.

Silene thurberi S. Watson [*S. plicata* S. Watson]—Thurber's catchfly. Plants perennial, erect, caudex branched, woody; stems 30–80 cm, internodes scabrid-puberulent and glandular-viscid; leaves oblanceolate (basal and lower cauline) to lanceolate (upper cauline), 5–18 cm × 5–30 mm, 6–8+ pairs per stem, apex acute; cymes open, 1–10+-flowered; pedicels longer than calyx tube; flowers perfect; calyx tubular, 10-veined, not netted, pale commissures between veins present, 8–12 × 2–4 mm, becoming campanulate and 5–7 mm wide in fruit, viscid glandular-hairy, especially on veins, lobes erect, narrowly lanceolate, 3–4 mm, apex recurved; petals greenish white, about equaling calyx, limb 2-lobed, usually each with a lateral tooth, ca. 3 mm, appendages 0.3–1 mm; stamens exserted, 1 mm; styles 3, equaling stamens; capsule opening by 6 teeth; seeds blackish, ca. 1 mm, margins coarsely papillate, not winged. Rocky places in juniper-oak woodlands, 5300–6000 ft. Hidalgo. Rare in New Mexico. **R**

Silene vulgaris (Moench) Garcke [*S. cucubalus* Wibel]—Bladder campion. Plants perennial, erect, caudex 1- to few-branched; stems 20–45 cm, upper internodes glabrous; leaves 2–8 cm × 5–30 mm, broadly oblong to oblanceolate or lanceolate, 7–12 pairs per stem, apex acute to acuminate; cymes open, 5–50-flowered, pedicels shorter than or equaling calyx tube; flowers perfect; calyx campanulate, 20–

veined, netted (especially in fruit), green or reddish, pale commissures between veins absent, 10-16 × 7-11 mm (in fruit to 20 mm and 15 mm, respectively), glabrous, lobes (teeth) broadly triangular, 2-3 mm; petals white, 1.3-1.5 times calyx length, limb mostly 2-lobed, lobes broadly rounded, 1.5-3.5 mm, appendages absent or essentially so; stamens exserted 1-3 mm; styles 3, exserted 5-12 mm; capsule opening by 6 teeth; seeds black-brown, 1-1.3 mm, margins finely tuberculate, not winged. Roadsides, ~7000 ft. Cibola, Rio Arriba, San Miguel. Introduced.

Silene wrightii A. Gray—Wright's catchfly. Plants perennial, spreading to ascending, caudex branched, woody; stems 10-30 cm, internodes densely hairy, glandular; leaves elliptic-lanceolate to oblanceolate (upper cauline), 1.5-6 cm × 3-14 mm, 5-9 pairs per stem, apex sharply acuminate; flowers axillary or in few-flowered cymes; pedicels shorter than to equaling calyx tube; flowers perfect; calyx tubular to narrowly obconic, 10-veined, not netted, pale commissures between veins present, 16-20 × 4-5 mm, becoming clavate and to 7 mm in fruit, coarsely glandular-hairy and viscid, lobes narrowly lanceolate, 5-7 mm; petals white to pale yellowish, ca. 2 times calyx length, limb 2-lobed, each lobe cleft into 3-4 smaller lobes, lanceolate to oblong, 5-8 mm, appendages 2, very short; stamens exserted, shorter than petal limbs; styles 3, exserted, shorter than petal limbs; capsule opening by 3 teeth that later split into 6; seeds brown, ca. 1.5 mm, margins papillate, not winged. Andesite and rhyolite cliffs, 6200-8600 ft. WC. **R**

SPERGULARIA (Pers.) J. Presl & C. Presl – Sand-spurry

Plants annual or short-lived perennial with taproot, erect to sprawling; stems simple to freely branched above or throughout, terete; leaves opposite, not congested at or near base of flowering stem, stipules lanceolate or broadly triangular, scarious, blades filiform to linear; inflorescence a simple or compound cyme or flowers solitary, flower(s) not subtended by involucral bracts; flower(s) weakly perigynous; sepals 5, briefly connate, apex acute to rounded or obtuse, not hooded, awn absent; petals 5, white to pink or rosy, not clawed, auricles absent, coronal appendages absent, apex entire; stamens 1-10, arising from rim of hypanthial disk, staminodes absent; ovary 1-locular; styles 3; capsule ovoid, opening by 3 spreading valves with recurved tips; seeds 0.4-0.8 mm, brown or reddish brown, smooth to papillate, submarginally grooved, marginal wing present or absent, appendage absent.

Spergularia media (L.) C. Presl has been reported from New Mexico; see Allred and Ivey (2012). Although it is known from Utah and Colorado, we did not locate a voucher of this species from New Mexico. It is a coarser plant with larger flowers, and most seeds are winged.

1. Stipules conspicuous, usually shiny white, lanceolate-acuminate; leaves scarcely fleshy; stamens 6-10; nonsaline areas in mountains > 8000 ft....**S. rubra**
1'. Stipules inconspicuous, usually dull white to tan, broadly triangular to reniform, not acuminate or scarcely so; leaves fleshy; stamens 2-5; saline areas in basins and drainages < 5500 ft....**S. marina**

Spergularia marina (L.) Griseb. [*S. salina* J. Presl & C. Presl]—Salt-marsh sand-spurry, lesser seaspurry. Plants annual, erect to ascending or prostrate; stems 8-30 cm, with stalked glands at least above; leaves with inconspicuous stipules, dull white to tan, broadly triangular, 1.2-3.5 mm, apex acute to short-acuminate, leaves linear, (0.8-)1.5-4 cm, fleshy, axillary leaf clusters usually absent, apex blunt to apiculate; sepals ovate to elliptic, 2.5-4.5 mm, to 4.8 mm in fruit, apex rounded to acute; petals white or pink to rosy, ovate to elliptic-oblong, 0.75-1 times sepal length; stamens 2-3, rarely 5; capsule 2.8-6.4 mm, 1-1.5 times sepal length; seeds light brown to reddish brown, broadly ovate, mostly plump, 0.5-0.7(-0.8) mm, dull, ± smooth, often with gland-tipped papillae; wing usually absent or incomplete. Riparian areas in canyons, weedy areas, 5000-7300 ft. Bernalillo, Cibola, Los Alamos, Sandoval. Introduced.

Spergularia rubra (L.) J. Presl & C. Presl—Red sandspurry. Plants annual or short-lived perennial, erect to ascending or prostrate; stems 4-25 cm, with stalked glands at least above; leaves with conspicuous stipules, shiny white, lanceolate, 3.5-5 mm, apex long-acuminate, leaves filiform to linear, 0.4-1.5 cm, scarcely fleshy, axillary clusters of 2-4+ per node present, apex apiculate to spine-tipped; sepals lanceolate, 2-3.2 mm, to 4 mm in fruit, apex obtuse to acute; petals pink, obovate to ovate, 0.9-1 times sepal length; stamens 6–10; capsule 3.5–5 mm, 1–1.2 times sepal length; seeds red-brown to dark brown, broadly ovate or ± truncate, plump, 0.4–0.6 mm, sculpturing of parallel, wavy lines, margins with peglike papillae; wing absent. Our plants belong to **var. rubra.** Roadsides and trailheads, 8200–9400 ft. Rio Arriba. Introduced.

STELLARIA L. – Chickweed, starwort

Plants annual or perennial, with taproot or slender rhizomes, prostrate to ascending or erect; stems simple or branched, usually 4-angled; leaves opposite, not congested at or near base of flowering stem, without stipules, blades linear or lanceolate to broadly ovate or elliptic; inflorescence an open or umbellate cyme or flowers solitary, flower(s) not subtended by involucral bracts; flower(s) hypogynous or weakly perigynous; sepals (4)5, distinct, apex acute, acuminate, or obtuse, not hooded, awn absent; petals (0–)5, white, not clawed, auricles absent, coronal appendages absent, apex 2-fid for 2/3–4/5 of length; stamens (1–)5–10, arising from nectariferous disk, staminodes 0; ovary 1-locular; styles usually 3(–5); capsule ovoid to globose, opening by 6(8 or 10) ascending to recurved valves; seeds 0.5–1.3 mm, pale or reddish brown to dark brown, papillate, tuberculate, or rugose, marginal wing absent, appendage absent.

Stellaria crassifolia Ehrh. was reported from New Mexico by Allred and Ivey (2012); specimens identified as such that we have seen are instead *S. longipes*.

1. Plants annual, from a taproot…(1)
1'. Plants perennial, from slender or elongate rhizomes…(4)
2. Inflorescence bracts reduced, mostly scarious; cauline leaves linear-lanceolate, rarely ovate…**S. nitens**
2'. Inflorescence bracts resembling leaves; cauline leaves ovate to elliptic or deltate…(3)
3. Leaf bases cordate to truncate; seeds covered with prominent stalked glands…**S. cuspidata**
3'. Leaf bases rounded or tapered; seeds with rounded or flat-topped tubercles…**S. media**
4. Inflorescence bracts scarious or scarious-margined, reduced; flowers in a terminal cyme…(5)
4'. Inflorescence bracts resembling leaves; flowers in a leafy terminal cyme or solitary in upper leaf axils…(7)
5. Cyme umbel-like, with 3–5+ pedicels appearing to radiate from 1 point; pedicels deflexed below, often curved apically in fruit; petals absent…**S. irrigua**
5'. Cyme not umbel-like; pedicels spreading or erect to ascending, straight to arcuate in fruit; petals present…(6)
6. Leaf margins minutely tuberculate-scaberulous as viewed under a strong (×20) lens, dull; cymes open, axillary or terminating axillary shoots with spreading pedicels; petals 2-fid to near base…**S. longifolia**
6'. Leaf margins smooth, lustrous, often coriaceous; cymes narrow, terminal with erect to ascending pedicels; petals deeply notched…**S. longipes** (in part)
7. Petals equaling or longer than sepals…(8)
7'. Petals < 0.8 times the length of sepals or absent…(9)
8. Plants not matted; leaves linear to linear-lanceolate, (20–)27–35(–42) mm, widest at or near middle, often arching, pairs mostly not pointed upward; flowers solitary; axillary sepals 4.5–6 mm in flower…**S. porsildii**
8'. Plants often matted; leaves lanceolate to ovate-triangular, 8–26(–40) mm, widest near base, pairs usually pointed upward; flowers solitary or often in narrow cymes; sepals 3–4 mm in flower…**S. longipes** (in part)
9. Mature plants purplish throughout; leaves fleshy, margin transparent to white or purple; flowers solitary, axillary, often concealed by upper leaves…**S. sanjuanensis**
9'. Mature plants green; leaves not fleshy, margins green; flowers 3 to many in a leafy cyme…**S. calycantha**

Stellaria calycantha (Ledeb.) Bong.—Northern starwort. Plants perennial, erect, sometimes trailing, forming clumps, green, rhizomes slender; stems branched, 15–35 cm, glabrous or pilose; leaves ovate to elliptic, 5–25 mm, widest over lower 1/2, not fleshy, petiole absent or essentially so, margins glabrous or rarely ciliate, herbaceous to thinly scarious, base tapered, apex acute; cymes terminal, flowers 3–11+, bracts resembling leaves; pedicels ascending in fruit; sepals 5, ovate, 1.2–1.5 mm in flower to 2.5 mm in fruit, veins obscure, glabrous, margins scarious, broad, apex broadly acute; petals 1–5 or absent, 2-lobed to near base, 0.4–0.6 times sepal length; stamens 5; styles 3–5; capsule dark purple, ± globose, 3–5 mm, valves 6, 8, or 10; seeds brown, ovate, 0.5–0.9 mm, smooth or shallowly tuberculate. Meadows, stream banks, spruce-fir communities, scree slopes in alpine communities, 9680–10,340 ft. NC, NW, C, Cibola.

Stellaria cuspidata Willd. ex D. F. K. Schltdl.—Mexican chickweed. Plants annual, decumbent, green, taprooted; stems much branched, 15–70 cm, softly glandular-hairy; leaves ovate to deltate, 10–45 mm, widest near middle, not fleshy, petiole present (basal) or sessile (upper), margins entire, glabrous, herbaceous, rarely ciliate, base cordate or truncate, apex acuminate; cymes terminal, flowers (3–)5 to ca. 35, bracts resembling leaves, pedicels ascending to spreading, sometimes deflexed in fruit; sepals 5, lanceolate to ovate-lanceolate, 4–5 mm, to 8 mm in fruit, veins obscure, hairy on midrib, ± ciliate on margins, margins scarious, narrow, apex acute or acuminate; petals 4–5, 2-lobed to near base, 2–8 mm, shorter than to 2 times sepal length; stamens 3–8; styles 3; capsule green, transparent, ovoid, 4–6 mm, ± equaling sepals, valves 6, recurved; seeds reddish brown, round, 1–1.2 mm, surface covered with prominent, stalked glands. Shaded ravines, under boulders, 5100–8000 ft. Doña Ana.

1. Petals 5–8 mm, 1.5–2 times sepal length; sepals narrowly lanceolate, apex acuminate, 7–8 mm in fruit; higher elevations…**subsp. cuspidata**
1'. Petals 2–4 mm, equaling or shorter than sepals; sepals ovate-lanceolate, 4–5 mm in fruit; lower elevations…**subsp. prostrata**

subsp. cuspidata—Petals 5–8 mm, 1.5–2 times sepal length; sepals narrowly lanceolate, to 7–8 mm in fruit. Under boulders, 8000 ft.

subsp. prostrata (Baldwin) J. K. Morton [*S. prostrata* Baldwin]—Petals 2–4 mm, equaling or shorter than sepals; sepals ovate-lanceolate, to 4–5 mm in fruit. Shaded ravines, 5100 ft.

Stellaria irrigua Bunge [*S. umbellata* Turcz.]—Umbellate starwort. Plants perennial, erect, clumped or matted, green, rhizomes slender; stems branched at base, 5–30 cm, glabrous; leaves elliptic to lanceolate, 30–90 mm, widest near middle, somewhat fleshy, petiole ± absent, margins glabrous, herbaceous, base round to tapering, apex acute; cymes terminal or subterminal, umbelloid (3–5+ pedicels appearing to radiate from 1 point), (2–)5- to many-flowered, bracts mostly scarious, reduced, pedicels deflexed below, often curved apically in fruit; sepals 5, lanceolate, 1.5–2 mm in flower, to 3 mm in fruit, veins 3, glabrous, margins scarious, narrow, apex obtuse; petals absent; stamens 5; styles 3; capsule tan, conic, 3–4.5 mm, valves 6; seeds brownish, round, 0.5–0.7 mm, shallowly rugose. Spruce-fir forests, stream banks, moist meadows, alpine tundra and talus slopes, 8800–12,700 ft. Colfax, Taos. See Sharples and Tripp (2019) for an explanation of the application of this name to what has most recently been known as *S. umbellata*.

Stellaria longifolia Muhl. ex Willd.—Long-leaved starwort. Plants perennial, straggling to erect, forming loose clumps, green, rhizomes elongate; stems diffusely branched, 10–35 cm, glabrous except angles minutely roughened; leaves linear to very narrowly elliptic, 8–40 mm, widest near middle, not fleshy, petiole absent, margins ciliate near base, minutely tuberculate-scaberulous (×20), dull, herbaceous, base tapering, apex acuminate to acute; cymes open, axillary or terminating axillary shoots, flowers 2 to many, bracts mostly scarious, reduced, pedicels straight to arcuate in fruit; sepals 5, ovate-elliptic, 2.5–3 mm in flower, to 4 mm in fruit, veins 3, obscure, glabrous, margins scarious, broad, apex acute; petals 5, 2-lobed to near base, 2–3.5 mm, 0.9–1.3 times sepal length; stamens 5–10; styles 3; capsule blackish purple or tan, ovoid-conic, 3–6, valves 6; seeds brown, broadly reniform, 0.7–0.8 mm, slightly rugose. Wet montane meadows, riparian areas in mixed conifer woodlands, 7050–10,160 ft. NC, NW, WC.

Stellaria longipes Goldie—Goldie's starwort. Plants perennial, ascending to erect, forming small to large clumps or (often) mats, or diffuse, green, rhizomes slender; stems often branched, 3–32 cm, glabrous or softly hairy; leaves linear-lanceolate to ovate-triangular, 8–26(–40) mm, widest near base, pairs usually pointed upward, not fleshy, petiole absent, margins glabrous or ciliate, smooth, lustrous, often coriaceous, base blunt, apex acute to acuminate; cyme narrow, terminal, flowers 1–20, bracts resembling leaves or scarious, reduced, pedicels ascending to erect in fruit; sepals 5, lanceolate to ovate-lanceolate, 3.5–4 mm in flower, to 5 mm in fruit, veins 3, midrib prominent, glabrous or hairy, margins scarious, narrow, sometimes ciliate, apex acute; petals 5, deeply notched, 3–8 mm, 1.1–1.6 times sepal length; stamens 5–10; styles 3; capsule purplish black, ovoid to lanceoloid, 4–6 mm, valves 6; seeds brown, reniform to globose, 0.6–0.9 mm, shallowly tuberculate to smooth. Our plants are **subsp. longipes**.

Riparian areas, stream edges, aspen, mixed conifer and spruce-fir forests, alpine and subalpine meadows and talus slopes, 6240–12,700 ft. NC, NW, C, WC, SC.

Stellaria media (L.) Vill.—Common chickweed. Plants annual, creeping to ascending, green, tap-rooted; stems diffusely branched, 5–40 cm, villous in a line between nodes; leaves ovate to broadly ellip-tic, 5–40 mm, widest near middle, not fleshy, petiole present (especially below) or not (upper), margins often ciliate, herbaceous, base rounded or tapered, apex acute or shortly acuminate; cyme terminal, flowers 5 to many, bracts resembling leaves, pedicels ascending, deflexed at base in fruit; sepals 5, ovate-lanceolate, 4–5 mm in flower, to 6 mm in fruit, veins obscure, usually glandular-hairy, margins scarious, narrow, apex obtuse; petals 5 or absent, 2-lobed to near base, 0.5–0.9 times sepal length; sta-mens 3–5(–8); styles 3; capsule green to tan, ovoid-oblong, 3–5 mm, valves 6; seeds reddish brown, broadly reniform to round, 0.9–1.3 mm, with rounded or flat-topped tubercles. Lawns, moist disturbed areas, 3900–8850 ft. Apetalous plants, such as *Jercinovic 1202* (NMC, SNM, UNM), can be difficult to segregate from *Stellaria pallida* (Dumort.) Crép. (lesser chickweed). This species resembles a smaller *S. media*, an often yellowish-green plant with sepals 2–3 mm and seeds that are 0.7–0.8 mm and yellow-brown. Scattered. Introduced.

Stellaria nitens Nutt.—Shining starwort. Plants annual, erect, green, taprooted; stems sparingly branched below inflorescence, 3–25 cm, glabrous or sparsely hairy; leaves oblanceolate to obovate (below) or linear-lanceolate (above), 5–15 mm, widest near middle, not fleshy, petiole absent or rarely present (below); margins often ciliate, herbaceous, base round, apex acuminate; cyme open, terminal, flowers 3–21+, bracts reduced, herbaceous below, scarious above, pedicels ascending to erect, straight in fruit; sepals 5, very narrowly lanceolate, 2.8–4.2 mm, veins 3, prominent and ridged, glabrous, margins scarious, broad, apex acuminate; petals 5 or absent, 2-lobed, 1–3 mm, shorter than sepals; stamens 3–5; styles 3; capsule green or tan, narrowly ovoid, 2–3 mm, valves 3, splitting into 6; seeds brown, round, 0.5–0.7 mm, minutely tuberculate. Rocky, grassy slopes, 4400 ft. Luna.

Stellaria porsildii C. C. Chinnappa—Porsild's starwort. Plants perennial, erect, rarely straggling, rarely clumped, never compact and mat-forming, green, rhizomes slender; stems diffusely branched, mainly at base, 9–20 cm, glabrous; leaves linear to linear-lanceolate, (20–)27–35(–42) mm, widest at or near middle, often arching, pairs mostly not pointed upward, not fleshy, petiole absent, margins entire, sparsely ciliate at base, base cuneate, apex acute to somewhat acuminate; flowers solitary in upper leaf axils, bracts absent, pedicels erect; sepals 5, ovate-lanceolate to lanceolate, 4.5–6 mm, veins 3, midrib prominent, glabrous, margins membranous, narrow, apex acute; petals 5, deeply 2-lobed, (4–)5–6 mm, equaling or slightly longer than sepals; stamens 10; styles 3; capsule black, oblong, 6–8 mm, valves 6; seeds dark brown, broadly ovate, 0.8–1 mm, shallowly tuberculate. See Morton (2005b) for a discussion of difficulties in distinguishing *S. porsildii* from *S. longipes*. Shaded slopes in open, mixed aspen-conifer woodlands, 8000–8920 ft. Grant. **R**

Stellaria sanjuanensis M. T. Sharples & E. A. Tripp [*Alsine polygonoides* Greene ex Rydb.]—Altai chickweed, Altai starwort. Plants perennial, ascending to spreading, forming mats or low cushions, ma-ture plants purplish throughout, rhizomes elongate; stems somewhat branched, 2–10 cm, glabrous; leaves elliptic or lanceolate to oblanceolate, 1–10 mm, widest near middle, fleshy, petiole absent; mar-gins glabrous, transparent, white to purple, base tapered to truncate, apex acute; flowers solitary, often concealed by upper leaves, bracts absent, pedicels curved to deflexed in fruit; sepals 5, lanceolate, 2.5–3 mm in flower, to 4 mm in fruit, veins 3, glabrous, margins membranous, apex acute; petals 5, 2-lobed to near base, 0.4–0.5 times sepal length; stamens 5; styles 3; capsule green to tan, ovoid-obtuse, 2.5–3 mm, valves 6; seeds pale brown, reniform, 1–1.2 mm, sides smooth to shallowly rugose, margins thickened with shallow longitudinal ridges. Steep talus and scree slopes, 11,500–12,600 ft. Colfax, Taos.

TORREYOSTELLARIA Gang Yao, B. Xue & Z. Q. Song – Sticky starwort

Torreyostellaria jamesiana (Torr.) Gang Yao, B. Xue & Z. Q. Song [*Pseudostellaria jamesiana* (Torr.) W. A. Weber & R. L. Hartm.; *Stellaria jamesiana* Torr.]—Tuber starwort. Plants perennial, ascending to erect; stems 12–60 cm, 4-angled, glabrous proximally or stipitate-glandular throughout or at least in inflorescence; leaves linear to linear-lanceolate or broadly lanceolate, 2–10(–15) cm, margins smooth or serrulate, glabrous or stipitate-glandular, apex rarely obtuse (below) or acute; pedicels recurved to reflexed in fruit, uniformly stipitate-glandular; sepals lanceolate to narrowly ovate, 3–6 mm, stipitate-glandular, apex acute to acuminate; petals 7–9.5 mm, apex with V-shaped notch 1–2 mm deep, lobes broadly rounded; capsule 4.5–5 mm; seeds 1–3, reddish brown, broadly elliptic, ± plump, 2–3.4 mm; tubercles conic to elongate, rounded. Meadows, canyons, riparian areas, roadsides in oak or mixed conifer forests, 6600–11,800 ft. NC, NW, C, WC, SC. We follow Xue et al. (2023) who placed *Pseudostellaria jamesiana* in a monotypic genus, showing it was most closely related to, but morphologically distinct from, *Hesperostellaria americana* (Porter ex B. L. Rob.) Gang Yao, B. Xue & Z. Q. Song (*Stellaria americana* Porter ex B. L. Rob.), a species of the northern Rockies.

CELASTRACEAE – BITTERSWEET FAMILY

John R. Spence

Herbs, shrubs, or vines, annual to perennial, deciduous to evergreen; leaves simple, stipules present or absent, alternate or opposite to fascicled, margins serrate, dentate, spiny, or entire; flowers bisexual or unisexual, actinomorphic or weakly zygomorphic; perianth biseriate, (3)4–5(–7)-merous; stamens 3–5 (–10), staminodes present or absent; pistil 1, 1–5-carpellate, ovary superior to partially inferior, locules 1–5, placentation axile or parietal, style 0–3, stigmas 2–5; fruit a loculicidal capsule or nutlike drupe, seeds 1–2 per locule (Ma et al., 2016). *Canotia holacantha* Torr. ex A. Gray should be looked for near the Arizona–New Mexico border, as it occurs within a few miles to the west at midelevations on limestone.

1. Herbaceous perennials; staminodes present, opposite petals...**Parnassia**
1'. Shrubs or trees; staminodes lacking...(2)
2. Leaves alternate; fruit nutlike...**Mortonia**
2'. Leaves opposite; fruit a capsule...**Paxistima**

MORTONIA A. Gray – Saddlebush

Mortonia scabrella A. Gray—Scabrid mortonia, Chihuahuan mortonia. Erect shrubs; branches rounded, rough-scabrous, stems to 2.5 m; leaves evergreen, alternate, crowded, petiole absent or very short, margins entire, revolute, elliptic, or ovate to suborbiculate, (4–)5–6(–7) × (3–)4–5(–6) mm, length 1.1–1.3 times width, stiff, surface finely scabrous; inflorescences terminal, cymose; flowers actinomorphic, perigynous; sepals 5, round-deltate, to 1 mm; petals 5, obovate to oblong, 1–1.3 mm, white, nectary intrastaminal, stamens 5, adnate to nectary; staminodes 0; pistil 5-carpellate; ovary superior, 5-locular, placentation axile, ovules 2 per locule; style 1, stigmas 5; fruit a hard indehiscent nutlike drupe, beaked, 3–5 mm; seeds 1 per fruit. Rocky slopes, hillsides, ledges, rock outcrops, 4055–6930 ft. Doña Ana, Hidalgo, Otero.

PARNASSIA L. – Grass of Parnassus, bog star

John R. Spence and Kenneth D. Heil

Herbs, perennial; leaves mostly basal, alternate, simple, petiole present in basal leaves, usually absent in cauline leaf, blade margins entire or fimbriate; inflorescences terminal, flowers solitary, actinomorphic; sepals 5; petals 5, white, the margins entire or fimbriate, stamens 5; staminodes 5, opposite petals, usually deeply divided; pistil (3)4(5)-carpellate; ovary superior to 1/2 inferior; fruit a capsule (Ball, 2016). Previously included in Parnassiaceae or Saxifragaceae; recent molecular work shows the genus to be basal to the remainder of the Celastraceae.

1. Petal margins fimbriate proximally; leaf blades cordate at the base, cauline leaves predominantly on middle or distal portions of stem...**P. fimbriata**
1'. Petal margins entire; leaf blades cuneate to rounded at the base, cauline leaves predominantly on proximal to middle portions of stem...**P. palustris**

Parnassia fimbriata K. D. Koenig—Fringed grass of Parnassus. Stems 10-35 cm; leaves mostly basal in rosettes, blades of larger leaves orbicular-reniform to reniform, 10-50 × 15-60 mm, base cordate, cauline leaves mostly in upper 1/2 of stem; flower sepals 4-6 mm, petals 8-14 × 4-8 mm, length 2 times sepals, fimbriate proximally; staminodes irregularly divided into 5-10+ oblong, obtuse lobes; capsules 8-12 mm. Stream banks, wet meadows, fens, seeps in forest glades, alpine ravines, 6900-12,200 ft. NC, San Miguel.

Parnassia palustris L. [*P. montanensis* Fernald & Rydb.; *P. palustris* var. *montanensis* (Fernald & Rydb.) C. L. Hitchc. & Ownbey]—Mountain grass of Parnassus. Stems 8-35(-50) cm; basal leaves in rosettes, blades ovate to suborbiculate, 6-40 × 4-30 mm, base rounded to cordate, cauline leaves mostly in lower 1/2 of stem; flower sepals 4-11 mm; petals (7-)8-17(-20) × 5-12 mm, margins entire; staminodes divided distally into (7-)9-27 gland-tipped filaments; capsules (6-)8-10 mm. Wet meadows, fens, ditches, stream margins, 6000-10,850 ft. NC, NW, Grant, Otero.

PAXISTIMA Raf. – Mountain-lover

Paxistima myrsinites (Pursh) Raf.—Mountain-lover, myrtle boxwood. Low-growing to prostrate shrubs; branches 4-angled; leaves short, evergreen, opposite, petiolate, blades ovate to elliptic, 7-35 × 3-12 mm, margins serrate, apex rounded to acute; inflorescences axillary, cymose, or flowers single; flowers actinomorphic, perigynous; sepals 4, obtuse-deltate, to 1 mm; petals 4, ovate, to 1.2 mm, red-brown to yellow, nectary intrastaminal, stamens 4, adnate to nectary; staminodes 0; pistil 2-carpellate; ovary superior, 2-locular, placentation axile, ovules 2 per locule; style 1, stigmas 2; fruit a capsule, not beaked; seeds brown to black, 1 per fruit with fleshy aril. Shaded and forested sites in mountains, ravines, 6100-12,305 ft. Widespread except E, SW.

CERATOPHYLLACEAE – HORNWORT FAMILY

C. Barre Hellquist

CERATOPHYLLUM L. – Hornwort, coontail

Ceratophyllum demersum L.—Hornwort, coontail. Perennial, rootless, submersed, floating, olive-green, lacking rhizomes; fragments generate into new plants; stems elongate, branching, brittle, cord-like; leaves whorled, finely dissected into capillary to linear and flattened serrate divisions, as many as 12 in a verticil; leaf divisions toothed, ca. 15 mm; flowers small, unisexual, lacking a perianth, solitary, sessile in leaf axil; staminate flowers in pairs on opposite sides of the stem, short-stalked, with a calyxlike involucre of 8-12 short, stout, oblong segments, each 2-3-toothed at apex, stamens 12-20, filaments short, the large anthers terminating in 2-3 sharp points; pistillate flowers usually solitary with a unilocular ovary and long style; fruit 4-6 mm, with a terminal persistent style 4-6 mm, with 2 recurved basal, lateral spines; fruit a 1-seeded, ovoid-oblong achene (Les, 1997). Still water of lakes, ponds, and streams, 3050-9400 ft. Scattered.

CISTACEAE – ROCKROSE FAMILY

Don Hyder

LECHEA L. – Pinweed

Herbs and shrubs; leaves alternate, subopposite, opposite, or whorled, margins revolute; flowers 3-5; sepals 5, the outer 2 smaller than the inner 3; petals 3, maroon or green; stamens (3-)5-15(-25), filaments distinct; ovary superior, unilocular, 2-, 3-, 5-, or 6-12-carpellate; fruit a capsule (Allred et al., 2020; Hodgdon, 1938).

1. Flowering stems spreading-villous…**L. mucronata**
1'. Flowering stems with appressed hairs…**L. mensalis**

Lechea mensalis Hodgdon—Narrowleaf pinweed. Herbs, perennial; stems 15-25 cm, sericeous; leaves of flowering stems alternate, blades linear, 8-15 × to 1.5 mm, with appressed hairs; flower sepals ca. 2 mm, outer sepals equaling or longer than inner; capsules 1.5-2 × 1.2-1.3 mm. Sandstone; Guadalupe Mountains, ca. 6800 ft. Eddy. **R**

Lechea mucronata Raf. [*L. minor* L.; *L. villosa* Elliott]—Hairy pinweed. Herbs, biennial or perennial; basal stems produced; flowering stems erect, (15-)30-90 cm, densely spreading-villous; leaves of flowering stems opposite or whorled; blades elliptic to ovate, 10-30 × 3-4 mm, apex acute to obtuse, mucronate, abaxial surface villous, adaxial glabrous; flower sepals 1.4-2 mm, outer sepals shorter than inner; capsules subglobose, 1.4-1.7 × 1.3-1.6 mm. Shin-oak communities, 4100-4300 ft. Roosevelt.

CLEOMACEAE – CLEOME FAMILY

Steve L. O'Kane, Jr.

Annual or perennial, ill-smelling herbs; plants erect; leaves palmately 3-, 5-, or 7-compound; flowers actinomorphic or zygomorphic; perianth biseriate, 4-merous, parts distinct; stamens 6 (to 8-27 in *Polanisia*); pistil 1 and 2-carpellate, ovary superior, with a replum (but no false septum as in Brassicaceae), stipitate or subsessile, locule 1, with 2 parietal placentae, style 1, stigma 2-lobed; fruit a 2-valved septicidal capsule or a didymous schizocarp forming a pair of 2(-4)-seeded mericarps (Tucker & Vanderpool, 2010). Previously included in Capparaceae.

1. Stamens 8-20(-27); plants sticky, glandular-puberulent; petals white to cream; ovary subsessile… **Polanisia**
1'. Stamens 6; plants not sticky, glabrous, or if puberulent not glandular; petals colored, not or very rarely white or cream; fruit evidently stipitate…**Cleomella**

CLEOMELLA DC. – Stinkweed, cleomella

Plants annual or perennial, generally ascending to erect, usually branched from the base, glabrous or spreading-hispidulous, often red-tinged; stems 6-30 cm; leaves palmately compound, leaflets 3 or 5(7); flowers actinomorphic or zygomorphic, petals yellow to yellow-orange, or pink to rose or purple (rarely white in a few), upper 2 often recurved, stamens 6, anthers coiled when dry, ovary stipitate; fruit obdeltoid, rhomboid, deltoid, or ovoid to linear or oblong; seeds 2-12. In order to recognize monophyletic genera in the family, *Cleomella* was recently expanded to include *Peritoma* and *Wislizenia* (Roalson et al., 2015).

1. Fruit didymous, a schizocarp forming 2 indehiscent mericarps (traditionally *Wislizenia*)…**C. refracta**
1'. Fruit not didymous, a dehiscent 2-valved capsule…(2)
2. Fruits clearly longer than broad, many-seeded, somewhat constricted between the seeds (traditionally *Peritoma*)…(3)
2'. Fruits short and broad, as wide as long or wider, (1)2-16-seeded, not constricted between the seeds (traditionally *Cleomella*)…(5)
3. Petals yellow; leaflets 3 or 5 (or 7)…**C. lutea**
3'. Petals pink, rose, or purple (rarely white); leaflets 3…(4)
4. Leaflets 6-15 mm wide; petals 7-12 mm; capsules 15-40 mm…**C. serrulata**
4'. Leaflets ca. 1 mm wide; petals 4-5 mm; capsules 15-25 mm…**C. multicaulis**
5. Leaf surface spreading-hispidulous; styles 2-3(-5) mm; plants usually spreading or matlike, sometimes erect…**C. obtusifolia**
5'. Leaf surface glabrous; styles 0.7-2 mm; plants ± erect…(6)
6. Gynophore 2.5-6(-8) mm in fruit; bracts ± rudimentary…**C. palmeriana**
6'. Gynophore 6-17 mm in fruit; bracts unifoliate…**C. longipes**

Cleomella longipes Torr.—Chiricahua Mountain stinkweed. Plants erect-ascending, glabrous; stems sparsely branched (central stem dominant), 30-80 cm; leaflets oblanceolate to oblong-lanceolate, 1.5-3 ×

0.4–1 cm; bracts unifoliate, 5–12 mm; pedicels 5–8 mm; petals yellow, oblong, 6–9 mm; stamens yellow, 8–12 mm, somewhat exceeding the petals; gynophore ascending, 6–17 mm in fruit; fruit rhomboid, 4–8 × 6–10 mm. Saline and/or alkaline flats, 2850–5525 ft. Hidalgo, Sierra.

Cleomella lutea (Hook.) Roalson & J. C. Hall [*Cleome lutea* Hook.; *Peritoma lutea* (Hook.) Raf.]—Yellow beeplant. Herbage glabrous or slightly hairy; stems mainly 3–15 dm, branched or simple; leaflets (3)5(7), oblong to lanceolate or elliptic, 0.8–5 cm × 2–15 mm; flowers with pedicels 3–10 mm; sepals and petals yellow; stamens with yellow filaments; fruit 1.5–3(–4) cm, spreading, surface striate, on stipes 5–17 mm. Clayey or sandy flats, desert scrub, piñon-juniper to ponderosa pine communities, 4300–7100 ft. McKinley, San Juan.

Cleomella multicaulis (DC.) J. C. Hall & Roalson [*Cleome multicaulis* DC.; *Cleome sonorae* A. Gray; *Peritoma multicaulis* (DC.) Iltis]—Slender spiderflower. Herbage glabrous; stems 2–6 dm, unbranched or sparsely branched; leaflets 3, linear to very narrowly elliptic, 1–2 × 0.1 cm; flowers with pedicels 7–15 mm; sepals yellow; petals white, pink, or rose; stamens with yellow filaments; fruit 1.5–2.5 cm, spreading and pendulous, surface striate, on stipes 10–20 mm. Saline and/or alkaline soils of sinks, meadows, and old lake beds, ca. 4500 ft. In New Mexico known from only two 1851 collections in the bootheel region. **R**

Cleomella obtusifolia Torr. & Frém.—Mojave stinkweed. Plants usually spreading or matlike (rarely erect), moderately to densely hairy (sometimes spreading-hispidulous); stems 10–40(–120) cm; leaflets elliptic to obovate, 0.5–1.5 cm; bracts unifoliate; pedicels 3–12 mm; petals yellow or orange, oblong, 3.5–6 mm; stamens yellow, 8–14 mm; anthers 1.5–2.2 mm; gynophore reflexed, 4–5 mm in fruit; ovary ± rhomboid, 1–1.5 mm; fruit rhomboid, 3.5–4 × 7–10 mm. Dry, open, alkaline, gravelly or sandy flats; 1 report from Luna County, ca. 4250 ft.

Cleomella palmeriana M. E. Jones—Palmer cleomella. Plants erect-ascending, glabrous; stems branching from the base, 6–30 cm; leaflets 9–20 × 2–9 mm, elliptic to oblong or lance-oblong; upper bracts reduced to setae; pedicels 5–6 mm; petals 3–4 mm and tipped with red in bud; stamens 7–9 mm, well exserted; fruit obtuse apically, base triangular-acute, 2–5 × 5–9 mm on an often recurved gynophore 2.5–7 mm. Our plants are **var. palmeriana**. Dry, open, alkaline, gravelly, sandy, or clayey flats and desert shrub communities, 3900–6100 ft.

Cleomella refracta (Engelm.) J. C. Hall & Roalson [*Wislizenia refracta* Engelm.]—Jackass-clover. Plants annual or perennial, glabrous-puberulent; stems branched from the base, 0.5–24 dm; leaves palmately compound with usually 3 leaflets, petiole 3–25 mm; inflorescence a dense, terminal raceme to 20 cm in fruit; flowers actinomorphic; pedicels 5–10 mm; petals elliptic, yellow, tapered to the base; stamens 8–14 mm; ovary of 2 nearly separate lobes, each usually with 1 ovule; fruit of 2 nutlets with deciduous valves on a reflexed gynophore 2–12 mm; seeds usually 1 per nutlet. Our plants belong to **var. refracta**. Alkaline soils in valleys and washes, 3000–4800 ft. and possibly higher. WC, C, SC, SW, San Miguel.

Cleomella serrulata (Pursh) Roalson & J. C. Hall [*Cleome serrulata* Pursh; *Peritoma serrulata* (Pursh) DC.]—Rocky Mountain beeplant. Herbage glabrous or slightly hairy; stems mainly 3–20+ dm, simple or branched above; leaflets 3, lanceolate, elliptic, or oblanceolate, 1.5–6(–7) cm × 6–15 mm, entire; flowers with pedicels 12–23 mm; sepals purple to green; petals purple (rarely white); stamens with purple filaments; fruit 2.5–7.5 cm, descending to pendulous, surface smooth, on stipes 1–15(–25) mm. Prairies, piñon-juniper woodlands, desert scrub, disturbed sites, often along roads and stream banks, 2950–9200 ft. Widespread, except SE, EC, and southernmost counties.

POLANISIA Raf. – Clammyweed

Plants viscid-puberulent with stalked glands and strongly rank-smelling; leaves palmately compound, leaflets 3; inflorescence a dense terminal raceme; flowers zygomorphic, petals white, stamens 8–20, long-exserted, purple, and unequal; fruit capsular, elongate, erect, subsessile, glandular, somewhat flattened, valves persistent and dehiscing apically, seeds many.

1. Leaflets linear or narrowly elliptic, 0.1–0.2 cm wide; styles persistent in fruit, 2.5–4.5 mm…**P. jamesii**
1'. Leaflets obovate to oblanceolate or broadly elliptic, 0.5–2(–3) cm wide; styles deciduous in fruit, 5–40 mm…(2)
2. Plants annual (rarely short-lived perennial); stamens 10–20; styles 5–17 mm; seeds roughened or tuberculate-rugose, 2–2.3 mm…**P. dodecandra**
2'. Plants perennial; stamens 20–27; styles 20–40 mm; seeds smooth, 0.7–2 mm…**P. uniglandulosa**

Polanisia dodecandra (L.) DC.—Clammyweed. Plants annual; stems 1.5–8 dm; leaflets oblanceolate to obovate or broadly elliptic, 1.9–4.5(–6.5) × 0.5–2(–3) cm; pedicels 10–21 mm; sepals purplish; petals white to cream, emarginate; stamens 10–20; styles 5–17 mm, deciduous in fruit; gynophore 0–2 mm in fruit; stigma purple. Our plants belong to **var. trachysperma** (Torr. & A. Gray) Iltis. Gravelly, sandy, or shaly-clayey soil, sunny places along streams, open woodlands, grasslands, roadsides, 3475–6575 ft. Widespread.

Polanisia jamesii (Torr. & A. Gray) Iltis—James' clammyweed. Plants annual; stems 15–40(–60) cm; leaflets linear to narrowly elliptic, 1.5–2.8 × 0.1–0.2 cm; pedicels 5–15 mm; sepals pale yellow; petals white, emarginate to lacerate; stamens 6–9; gynophore 2–4 mm in fruit; style persistent in fruit, 2.5–4.5 mm; stigma red. Sandy soil, often in blowouts in prairies and stabilized dunes, 3800–4900 ft. E.

Polanisia uniglandulosa (Cav.) DC. [*Cleome uniglandulosa* Cav.; *P. dodecandra* (L.) DC. subsp. *uniglandulosa* (Cav.) Iltis]—Mexican clammyweed. Plants perennial, stems 40–80 cm; leaflets broadly elliptic to oblanceolate, 2–4 × 1–2 cm; pedicels 10–25(–40) mm; sepals purple; petals white, emarginate to lacerate; stamens 20–27; gynophore 0–2 mm in fruit; style deciduous in fruit, 20–40 mm; stigma purple. Piñon, juniper, and oak woodlands, arroyos, riverbeds, roadsides, pastures, 4500–8000 ft. WC, C, SW, SC.

COCHLOSPERMACEAE – YELLOWSHOW FAMILY

Kenneth D. Heil

AMOREUXIA DC. – Yellowshow

Amoreuxia palmatifida DC.—Mexican yellowshow. Perennial herbs; stems 20–60 cm; leaves alternate, 3–10 cm, sparsely lanulose-puberulent, glandular, 7(9)-lobed, 2–7.5 × 2.5–10(–14) cm, margins serrate (crenate); flowers actinomorphic, perfect, 4–6.5 cm diam.; sepals 5, 15–20 × 3–5 mm; petals 5, yellow to orange, distal petals each with 2 conspicuous red marks at base; stamens numerous; pistil single, superior, of 3–5 united carpels; capsules broadly ovoid, 2–4 cm, sparsely pubescent. Rocky sites in deserts, grasslands, oak forests, 4500–5240 ft. Grant, Hidalgo (Johnson-Fulton & Watson, 2017).

COMANDRACEAE – BASTARD TOADFLAX FAMILY

John R. Spence

COMANDRA Nutt. – Bastard toadflax

Comandra umbellata (L.) Nutt.—Bastard toadflax. Herbaceous perennial root hemiparasites, evergreen to deciduous, arising from rhizomes; stems to 40 cm, usually dying or not dying back in winter; rhizome cortex blue when fresh, blackish when dry; leaves alternate, simple, late-deciduous, lanceolate to ovate, 5–50 × 2–15 mm, somewhat fleshy, glaucous-green, gray-green often fading yellow-green; inflorescence of 3–6 thyrses; flowers bisexual, sessile to short-pedicellate, subtended by 1–3 bracts; sepals 0; petals 4–7; perianth funnelform, with (3–)5(–7) lobes, white, pink, or purple; pistil 1, ovary inferior, 1-carpellate, 1-locular, placentation free-central; style 1, filiform, stigma 1, capitate; fruit an ovoid to globose drupe, somewhat fleshy, 5–9 mm, purple to red-brown (Nickrent, 2016a).

1. Leaf blade lateral veins obscure on abaxial surface; proximal part of aerial stems not overwintering; herbs, 5–33 cm…**subsp. pallida**
1'. Leaf blade lateral veins apparent on abaxial surface; proximal part of aerial stems overwintering; herbs to subshrubs, 15–40 cm…subsp. californica

subsp. californica (Eastw. ex Rydb.) Piehl [*Comandra californica* Eastw. ex Rydb.]—Subshrubs, 15–40 cm; rhizome cortex blue, drying blackish; aerial stems often much branched; proximal portions overwintering; leaf blades light green to bluish or grayish green, ovate, lanceolate, or linear, 1.7–5.3 cm. Foothills, open coniferous forests, oak woodlands, chaparral margins. No records seen but to be looked for in SW.

subsp. pallida (A. DC.) Piehl [*Comandra pallida* A. DC.]—Herbs, 5–33 cm; rhizome cortex blue, drying blackish; aerial stems much branched; proximal portions not overwintering; leaf blades blue-green to grayish green, glaucous, linear to lanceolate, elliptic, or ovate, 0.9–4.3 cm. Sandy and rocky slopes, sagebrush communities, coniferous forests, 3180–9850 ft. Widespread.

COMMELINACEAE – SPIDERWORT FAMILY

Kenneth D. Heil

Perennial (ours) or annual herbs, usually somewhat succulent; leaves basal or cauline, with sheathing bases, alternate, simple, margins entire, venation parallel; inflorescence terminal or axillary, cymose, thyrsiform, or umbel-like; flowers actinomorphic or zygomorphic, perfect or both perfect and staminate on the same plant, sepals 3, petals 3, stamens 6, all fertile or some staminodial or absent, ovary superior, 2–3-locular, stigma 1, simple; fruit a capsule (Faden, 2000).

1. Flowers zygomorphic; stamens of 2 types, filaments naked; inflorescence subtended by 1 spathelike bract…**Commelina**
1'. Flowers actinomorphic; stamens all alike, filaments bearded below; inflorescence subtended by several leaflike bracts…**Tradescantia**

COMMELINA L. – Dayflower, widow's-tears

Leaves 2-ranked or spirally arranged, not glaucous; blades sessile or petiolate; inflorescence terminal, leaf-opposed; cymes 1 or 2, enclosed in spathes; flowers both perfect and staminate, zygomorphic, sepals distinct or proximal 2 connate, petals distinct, all the same color or the proximal petal a different color and smaller than the distal 2; stamens 5 or 6, 3 fertile, 2 or 3 sterile (staminodial), filaments glabrous; fruit 2- or 3-valved, 2- or 3-locular.

1. Spathe with a long, narrow tip, ca. 1.5 times the length of the body of the spathe, the margins distinct to the base; petals all alike, blue…**C. dianthifolia**
1'. Spathe without a long, narrow tip, the margins connate basally; petals unlike, the upper 2 large and blue, the lower smaller and white…**C. erecta**

Commelina dianthifolia Delile—Birdsbill dayflower. Perennial plants with tuberous-thickened roots; stems simple or branched, erect or decumbent; leaves linear-lanceolate, acuminate, 4–15 × 0.4–1 cm; inflorescence a distal cyme, usually 1-flowered, exserted, spathe 3–6 cm, not connate at base, very long-acuminate or caudate, tip as long as or longer than the body, glabrous or short-pubescent; flower petals blue, 10–12 mm, proximal petal smaller. Rocky soils in shaded or open ground, ponderosa pine and oak habitat, 5500–10,250 ft. Widespread except EC.

Commelina erecta L. [*C. angustifolia* Michx.; *C. crispa* Wooton]—Whitemouth dayflower. Herbs, perennial, roots fleshy, stout, tufted; stems cespitose, usually erect to ascending; leaves linear to lanceolate, leaf sheath auriculate at apex, blades sessile or petiolate, linear to lanceolate, 5–15 × 0.3–4 cm, acuminate; inflorescence a distal cyme, spathes solitary or clustered, 1–2.5(–4) × 0.7–1.5(–2.5) cm, margins connate, glabrous except along connate edge, apex acute to acuminate; flowers bisexual and staminate, 1.5–4 cm wide; proximal petal minute, white, distal petals blue (rarely lavender or white). Rocky areas, hillsides, oak woodlands, pine woods, sand dunes, shale barrens, roadsides, 3180–8020 ft. NC, WC. Widespread.

TRADESCANTIA L. – Spiderwort

Leaves arranged spirally or 2-ranked, sessile, linear to oblong, produced in a perfoliate sheath; inflorescence terminal or terminal with axillary pairs of cymes, cymes sessile, umbel-like, subtended by a spathe-like bract, either similar to leaves or different; flowers perfect, actinomorphic, sepals distinct, petals distinct, white to pink, blue, or violet, stamens 6, all fertile, filaments bearded or glabrous, ovary with 3 locules; fruit a dry loculicidal capsule, 3-valved, 3-locular; seeds usually 2 per locule.

1.   Sepals glabrous…(2)
1'. Sepals covered with eglandular hairs…(3)
2.   Petals rose to magenta or purple; leaf blades 4–10 cm × 2–5 mm…**T. wrightii**
2'. Petals blue; leaf blades 5–6.2 cm × 4–9 mm…**T. occidentalis** (in part)
3.   Roots thick but not tuberlike, rootstock none; stems stout, often branched; sheaths glabrous; corolla usually > 2 cm diam.…**T. occidentalis** (in part)
3'. Roots partly tuberous-thickened, fascicled at base of stem or borne on a creeping rootstock; stems slender; usually unbranched; sheaths pubescent or puberulent, especially on the margins; corolla not > 2 cm diam. (all flower parts smaller than above)…**T. pinetorum**

Tradescantia occidentalis (Britton) Smyth—Western spiderwort. Stems typically 30–60 cm, erect, branching, glabrous; leaves linear-lanceolate, long-acuminate, glabrous, 5–30 × 0.2–2 cm; inflorescence an umbel-like cyme, terminal or axillary, pedicels erect to spreading or reflexed, glandular-puberulent; flowers broadly campanulate, sepals elliptic, acute to acuminate, 4–10 mm, greenish or suffused with rose or purple, glandular-puberulent, petals broadly ovate, 7–16 mm, magenta, purple, rose-purple, or bluish, stamens erect, filaments densely pilose.

1.   Sepals glabrous…**var. scopulorum**
1'. Sepals pubescent…**var. occidentalis**

var. occidentalis—Stems 5–90 cm; leaves 5–50 × 0.2–3 cm; flower pedicels 0.8–3 cm, glandular-puberulent, rarely nearly glabrous; sepals 4–11 mm, glandular-puberulent, rarely nearly glabrous; petals bright blue to rose or magenta, 1.2–1.6 cm. Sandy and rocky soils, prairies, piñon-juniper, ponderosa pine, Douglas-fir communities, 3150–9400 ft. NE, NC, NW, EC, C.

var. scopulorum (Rose) E. S. Anderson & Woodson [*T. scopulorum* Rose]—Prairie spiderwort. Stems 14–35 cm; leaves 5–45 × 0.4–0.9 cm; flower pedicels 1–2 cm, glabrous; sepals 5–8 mm, glabrous; petals bright blue, 0.7–1 cm. Moist canyons and stream banks, 4300–8400 ft. SE, SC, SW.

Tradescantia pinetorum Greene—Pinewoods spiderwort. Stems slender, sparsely branched, 8–39 cm, usually scabridulous; leaves linear-lanceolate, glabrous, glaucous, 1–10 × 0.15–0.8 cm; inflorescence terminal, solitary or axillary, pedicels glandular-puberulent; flower sepals 4–6 mm, frequently suffused with red or purple, glandular-puberulent, petals 9–12 mm, magenta, purple, rose-purple, or bluish, filaments pilose. Moist canyons and stream banks, usually in pine woods, 5500–8600 ft. NW, WC, SC, SW.

Tradescantia wrightii Rose & Bush—Wright's spiderwort. Stems unbranched, 5–18 cm; leaves linear-lanceolate, 4–10 × 0.2–0.5 cm, firmly membranous to subsucculent, glaucous or glaucescent, glabrous; inflorescences terminal, solitary, bracts foliaceous; flower pedicels 1.2–1.7 cm, with few to many minute glandular hairs (or glabrous), sepals glaucous or glaucescent, 0.5–0.6 cm, glabrous or with a few minute glandular hairs at base, petals distinct, rose to magenta or purple, filaments bearded. Limestone, desert scrub, moist canyons, stream banks, 4000–7500 ft. SE, SC, Sierra, Socorro.

CONVOLVULACEAE – MORNING GLORY FAMILY

Sandra M. Namoff and Nicholas J. Jensen

Plants annual or perennial; vines, herbs, or subshrubs, often with milky sap, glabrous or hairy; leaves alternate, simple, entire to pinnately or palmately compound, usually petiolate, exstipulate; inflorescences solitary, cymose, rarely paniculate; bracts often opposite, 0–2, foliaceous or scalelike; flowers radial, bisexual, usually showy but sometimes inconspicuous; sepals 5, usually free or connate only at base, equal or unequally imbricate, sometimes persistent; petals 5, usually united; corolla funnelform, rotate, or campanulate, sometimes lobed; stamens 5, distinct, usually epipetalous; gynoecium 2 united carpels; ovules usually 4, styles 1 or 2, stigmas variable; fruits generally dehiscent capsules; seeds papillate or pubescent.

1. Plants orange or yellow; not photosynthetic; parasitic; leaves lacking…**Cuscuta**
1'. Plants green; photosynthetic; leaves present…(2)
2. Styles 2, free or united only at base…(3)
2'. Style 1…(5)
3. Leaves 2-lobed, reniform, orbicular, cordate, auriculate, truncate, conspicuously pedicellate; flowers mostly green, 5 mm or less; plants stoloniferous and mat-forming…**Dichondra**
3'. Leaves entire, unlobed, shape varying from linear to ovate, bases acute to rounded, usually sessile or subsessile; flowers varying in color but not greenish, 7+ mm; plants rhizomatous, more erect…(4)
4. Stigmas 2, capitate; flowers rotate…**Cressa**
4' Stigmas 4, linear to club-shaped; flowers widely funnelform to rotate…**Evolvulus**
5. Stigma capitate (headlike), unlobed or 2-3-lobed, flowers purple, blue, red, orange, rarely white…**Ipomoea**
5'. Stigma bifid (forked), lobes 2, cylindrical to linear-filamentous, flowers usually white…(6)
6. Stigma lobes cylindrical, tips obtuse; floral bracts large and surrounding calyx; pedicel short or absent; calyx 7–25 mm…**Calystegia**
6'. Stigma lobes linear to narrowly spoon-shaped, tips acute; floral bracts scalelike; pedicel long; calyx 3–10 mm…**Convolvulus**

CALYSTEGIA R. Br. – False bindweed

Plants perennial, herbs or vines, perennating from rhizomes or caudex; stems twining to climbing; glabrous or hairy; leaves alternate, sagittate to hastate, petiolate, usually > 1 cm; inflorescences solitary or cymose; flowers conspicuous; sepals free, ovate to oblong, ± equal, usually glabrous; corolla funnelform, white or pink; stigma lobes 2, terete, tips blunt; fruits capsular, 4-valved, surrounded by enlarged sepals and sometimes bracts; seeds 1–4, glabrous, smooth to verrucose.

1. Floral bracts strongly overlapping at least 1/2 their length, inflated at the base; leaf sinuses broad and almost square-sided; flowers sometimes in pairs in the axils (subsp. *fraterniflora*)…**C. silvatica**
1'. Floral bracts not or only slightly overlapping, flat and mostly keeled, not or only slightly inflated at the base; leaf sinuses acute to rounded; flowers always single in the axils…(2)
2. Plants finely pubescent; basal lobes of leaves without teeth; bract bases flat or slightly auriculate, bract margins overlapping to enclose the sepals…**C. macounii**
2'. Plants glabrous or glabrate; basal lobes of leaves with 2 teeth per lobe; bract bases keeled, bract margins not overlapping (hard to see on specimens)…**C. sepium**

Calystegia macounii (Greene) Brummitt [*Convolvulus macounii* Greene]—Macoun's bindweed. Plants perennial vines; pubescence fine; stems trailing to weakly twining; leaf blades deltate to ovate, bases lobed, sagittate, hastate, auriculate, 20–60 mm; inflorescences solitary; bracts oval to ovate-oblong; pedicel shorter than bracts or absent; sepals equal, elliptic to ovate; corolla white, funnelform; stamens included; fruits capsular, 4-valved, surrounded by enlarged sepals. Riparian areas, moist places, ditches, disturbed places, 6400–7000 ft. Cibola, San Miguel.

Calystegia sepium (L.) R. Br. [*Convolvulus sepium* L.]—Hedge bindweed. Plants perennial vines, glabrous or glabrate; stems trailing to climbing; leaf blades broadly triangular, bases lobed, hastate with 2 teeth per lobe, 36–50(–150) mm; bracts oval or oblong; pedicel shorter than bracts or absent; sepals equal, lanceolate; corolla white, funnelform; stamens included; fruits capsular. The taxon in New Mexico is **subsp. angulata** Brummitt. Disturbed riparian sites, edges of marshes, roadsides, stream banks, tidal swamps, 3200–7200 ft. NC, NW, Doña Ana, Eddy, Lincoln.

Calystegia silvatica (Kit.) Griseb. [*Convolvulus silvaticus* Kit.]—Shortstalk false bindweed. Plants strongly climbing; herbage glabrous to pubescent; leaves 5–8 cm, broadly lanceolate to ovate or triangular, basal lobes rounded to strongly angled; flowers solitary or in pairs in the axils; floral bracts 13–30 mm, strongly overlapping and enfolding the calyx; sepals 14–16 mm; corolla white, rarely pinkish, 4–8 cm. New Mexico material belongs to **subsp. fraterniflora** (Mack. & Bush) Brummitt. Moist roadsides, meadows, along creeks and streams, 3850–7060 ft. Colfax, Doña Ana, Grant.

CONVOLVULUS L. – Bindweed

Plants perennial herbs or vines; stems decumbent to twining, glabrous to pubescent; leaves simple, entire to dissected, 1–10 cm; inflorescence solitary to cymose, 2–5 flowers per peduncle; sepals usually unequal, 3–12 mm; corolla usually white or pink, sometimes tinged or striped with blue or pink, funnelform to ± rotate, 12–30 mm; limb 5-angled to 5-lobed; fruits capsular, dehiscence valvate; seeds 1–4, trigonous or rounded.

1. Leaf blade length ± equal to width; calyx < 5 mm; rhizomatous, forming large patches; wet, ruderal, disturbed areas…**C. arvensis**
1'. Leaf blade length > width; calyx > 5 mm; caudex sometimes divided, but not forming large patches; intact habitat…**C. equitans**

Convolvulus arvensis L.—Field bindweed. Stems trailing, rarely twining to 1 m, forming large mats; leaf blades deltate-ovate, ovate-lanceolate; bases sagittate, hastate, auriculate; inflorescences solitary or cymose, with 2–3 flowers; bracts scalelike; sepals unequal, 3.5–5 mm; corolla white to pink, funnelform, 12–30 mm; fruits capsular, 5–7 mm diam.; seeds 1–4. Common noxious weed of disturbed sites, 3110–9325 ft. Widespread. Introduced.

Convolvulus equitans Benth.—Texas bindweed. Stems trailing, rarely twining, not forming large mats; leaf blades elliptic to triangular-lanceolate or linear; bases lobed, sagittate to hastate; inflorescences solitary or cymose, with 2–3 flowers; bracts scalelike; sepals elliptic to ovate, 6–12 mm; corolla pink or white, funnelform, 25–30 mm; fruits capsular, 7–8 mm diam. Grasslands, hills, plains, 3000–7100 ft. Widespread.

CRESSA L. – Alkaliweed

Cressa truxillensis Kunth—Alkaliweed. Plants perennial herbs; hairs short, stout to long-silky; stems erect to trailing to 0.5 m; leaves sessile; margins entire, unlobed; blades elliptic to lance-ovate; bases acute to obtuse; inflorescences solitary, flowers concentrated in upper leaf axils; bracts scalelike or absent; pedicels absent or to 2 mm; flowers small; sepals elliptic to obovate; 3–4.5 × 2–3 mm; corolla white, salverform, 5–6.5 mm; lobes narrowly acute, villous abaxially; stamens exserted; styles exserted; fruits unilocular, 4-valvate, apically pubescent; seeds often 1 by abortion, glabrous. Moist saline and/or alkaline sites, 3500–4625 ft. SE, SC, Bernalillo, Socorro.

CUSCUTA L. – Dodder

Mihai Costea

Holoparasitic plants, annual or sometimes perennating inside the host stems; stems filiform, yellow or orange, trailing or twining and attached to the host by numerous small haustoria, glabrous; leaves reduced, scalelike; inflorescences cymose; flowers (3)4- or 5-merous, actinomorphic, fleshy, glabrous or papillate; calyx gamosepalous; corolla gamopetalous, white, interior with a corona of fimbriate appendages called infrastaminal scales; ovary superior, styles 2, stigmas capitate or linear (ours); fruits capsules, circumscissile, dehiscent or indehiscent; seeds 1–4 per capsule, usually with seed coat cells pitted when dry and papillate when hydrated (Costea et al., 2015; Yuncker, 1932).

Although a few species can be easily recognized in the field using a good magnifier (×30 or higher), the identification of most dodders requires the dissection of flowers and examination under a stereomicroscope. *Cuscuta epithymum*, *C. glomerata*, *C. gronovii*, *C. salina*, and *C. suaveolens* are mentioned from New Mexico by various sources, but no herbarium specimens were found and they are not included in the flora. Also, a number of other species—*C. azteca*, *C. cuspidata*, *C. cephalanthi*, *C. coryli*, *C. leptantha*, *C. tuberculata*, and *C. umbrosa*—are known from only one or very few old herbarium collections, and their occurrence and conservation status in New Mexico should be reassessed.

1. Styles equal; stigmas linear (subgenus *Cuscuta*)…**C. approximata**
1′. Styles unequal; stigmas capitate (subgenus *Grammica*)…(2)
2. Bracts 2–11 at base of pedicels and/or flowers…(3)
2′. Bracts 0–1 at base of pedicels and/or flowers…(4)
3. Pedicels 2–5 mm; inflorescences loose, paniculiform…**C. cuspidata**
3′. Pedicels 0–1 mm; inflorescences dense, glomerulate or short-spiciform…**C. squamata**
4. Capsules circumscissile near the base (the dehiscence line is also readily detectable at the base of ovaries)…(5)
4′. Capsules indehiscent…(11)
5. Corolla tube cylindrical; infrastaminal scales 1/4–1/2 the length of corolla tube…(6)
5′. Corolla tube campanulate (in *C. chinensis* and *C. azteca* becoming globose-depressed or urn-shaped in fruit); infrastaminal scales equaling or exceeding corolla tube…(8)
6. Flowers 5-merous; calyx lobes carinate; corolla lobes erect…**C. tuberculata**
6′. Flowers (3)4-merous; calyx lobes not carinate; corolla lobes spreading to reflexed…(7)
7. Calyx equaling corolla tube; infrastaminal scales 1/4–1/3 length of corolla tube…**C. liliputana**
7′. Calyx 1/3–1/2 length of corolla tube; infrastaminal scales ca. 1/2 length of corolla tube…**C. leptantha**
8. Flowers sessile or pedicels to 2 mm; inflorescences glomerulate or densely paniculiform…(9)
8′. Flowers pedicellate (pedicels 2–10 mm); inflorescences umbelliform…(10)
9. Calyx lobes rounded or obtuse, overlapping, carinate; corolla lobes rounded…**C. chinensis**
9′. Calyx lobes acute, not overlapping, not carinate but with a few multicellular projections along midveins; corolla lobes acute…**C. azteca**
10. Flowers 4–5.5(–6) mm; calyx lobes acuminate…**C. legitima**
10′. Flowers 2–3 mm; calyx lobes obtuse to acute…**C. umbellata**
11. Corolla lobes rounded or obtuse…(12)
11′. Corolla lobes acute to acuminate…(13)
12. Flowers mostly 3–4-merous; capsules depressed-globose, not risen around the interstylar aperture, capped by withered corolla…**C. cephalanthi**
12′. Flowers mostly 5-merous; capsules globose or ovoid, risen around the interstylar aperture, surrounded by withered corolla…**C. umbrosa**
13. Multicellular protuberances present on the calyx (do not mistake for the unicellular papillae)…(14)
13′. Multicellular protuberances absent…(15)
14. Each calyx lobe with distal conic projection, 0.5–0.75 mm; infrastaminal scales reduced, with a few distal teeth; styles 0.2–0.4 mm…**C. warneri**
14′. Each calyx lobe with 1–2(–5) hornlike or domelike, multicellular protuberances 0.1–0.3 mm; infrastaminal scales well developed, with numerous fimbriae…**C. draconella**
15. Flowers (3)4-merous; infrastaminal scales with 1–3 fimbriae on each side of filament attachment or with denticulate wings; withered corolla capping the capsule…**C. coryli**
15′. Flowers mostly 5-merous; infrastaminal scales well developed with numerous fimbriae; withered corolla surrounding or at the base of capsule…(16)
16. Perianth fleshy; capsules globose to ovoid, risen around the interstylar aperture…**C. indecora**
16′. Perianth membranous; capsules globose to globose-depressed to depressed, not risen around the interstylar aperture…(17)
17. Calyx lobes overlapping at base; papillae absent; withered corolla enveloping 1/3 or less of capsule base…**C. campestris**
17′. Calyx lobes not or only slightly overlapping at base; papillae present on the perianth, ovaries/capsules, and sometimes on the pedicels; withered corolla enveloping 1/2+ of the capsule…**C. glabrior**

Cuscuta approximata Bab.—Alfalfa dodder. Inflorescences spherical-glomerulate; pedicels absent to 2 mm; flowers 5-merous, 3–4.2 mm, papillae absent or rarely present on corolla lobes; calyx nearly as long as corolla tube, lobes basally overlapping, broadly triangular to obovate-rhombic, distally with a fleshy appendage; apex acute to acuminate; corolla 2.5–4 mm, the tube 2–2.2 mm, campanulate but becoming globose or urceolate in fruit, lobes spreading, shorter than to equaling the tube, ovate-orbicular, apex obtuse, straight; stamens exserted; infrastaminal scales shorter than to equaling corolla tube, oblong to ovate, shallowly fimbriate in the distal 1/2; styles plus stigma 0.5–1.5 mm, as long as or longer than the ovary; stigmas linear; capsules circumscissile, depressed-globose, 1.5–2.3 × 1.6–2 mm, not risen around the small interstylar aperture, translucent, capped by the withered corolla; seeds 3–4 per capsule, 0.8–1.1 × 0.6–0.7 mm. Known from only 1 specimen cited by Yuncker (1932) from San Juan County; may not have persisted in the flora. Most common hosts in North America are herbaceous genera from the Fabaceae, Asteraceae, and Lamiaceae. San Juan.

Cuscuta azteca Costea & Stefanović [*C. potosina* W. Schaffn. ex Engelm. var. *globifera* Yunck.]— Globe dodder. Inflorescences glomerulate; pedicels 0.4–1.3 mm; flowers 5-merous, 2–2.6 mm, papillae absent; calyx equaling corolla tube, lobes not overlapping, ovate, not carinate but with a few multicellular protuberances along midveins, apex acute; corolla 1.5–2.1 mm, the tube 0.8–1.2 mm, campanulate but becoming globose in fruit, lobes erect to spreading, shorter than to equaling the tube, ovate-triangular, apex acute, straight to slightly incurved; stamens exserted; infrastaminal scales equaling corolla tube or slightly longer, oblong, uniformly fimbriate in the distal 1/2; styles 0.4–0.7 mm, shorter than to as long as the ovary; stigmas capitate; capsules circumscissile, depressed-globose, 1.5–2.5 × 1–1.8 mm, not risen around the small interstylar aperture, translucent, loosely surrounded or capped by the withered corolla; seeds 3–4 per capsule, 0.8–1 × 0.7–0.9 mm. Known only from Caballero Canyon and Mogollon Mountains, ca. 6500 ft.; hosts are herbaceous Fabaceae (especially species of *Dalea*) and other herbaceous plants. Grant, Otero.

Cuscuta campestris Yunck. [*C. pentagona* Engelm. var. *calycina* Engelm.]—Field dodder. Inflorescences dense, corymbiform or glomerulate; pedicels 0.3–2.5(–3.5) mm; flowers 5-merous, 1.9–3.6 mm, papillae absent; calyx ca. as long as corolla tube, lobes overlapping at base, ovate-triangular, not carinate, apex obtuse to rounded; corolla 2–3.5 mm, the tube campanulate, lobes spreading, triangular-lanceolate, ca. as long as the tube, apex acute to acuminate, inflexed; stamens exserted; infrastaminal scales equaling or exceeding corolla tube, oblong-ovate to spatulate, densely fimbriate, styles 0.5–1.6 mm, shorter than to ca. as long as the ovary; stigmas capitate; capsules indehiscent, globose to globose-depressed, 1.3–2.8 × 1.9–3.8 mm; not risen around the large interstylar aperture, not translucent, persistent corolla enveloping 1/3 or less of the capsule base; seeds 2–4 per capsule, 1.12–1.54 × 0.9–1.1 mm. Margins of roads, disturbed sites, 2800–7220 ft.; potential weed of forage legumes, tomatoes, and other horticultural crops; hosts are herbaceous plants from numerous families and genera. NW, C, WC, SC, SW.

Cuscuta cephalanthi Engelm.—Buttonbush dodder. Inflorescences paniculiform; pedicels 0.4–2 mm; flowers 3–4-merous (rarely some 5-merous), 2–3 mm, papillae absent; calyx shallowly cupulate, ca. 1/2 the corolla tube, lobes slightly overlapping basally, oblong-ovate, not carinate, apex obtuse; corolla 1.8–2.8 mm, the tube cylindrical-campanulate, lobes spreading, 1/3–1/2 the length of the tube, ovate, apex obtuse, straight; stamens enclosed to slightly exserted; infrastaminal scales somewhat shorter than to equaling corolla tube, oblong, fimbriate distally; styles (0.6–)1–2 mm, ca. as long as or longer than ovary; stigmas capitate; capsules indehiscent, depressed-globose to globose, 2.5–3.2(–4) × 2–4 mm, not risen around the small interstylar aperture, not translucent, capped by the withered corolla; seeds 1–2 per capsule, 1.49–2 × 1.3–1.45 mm. Known from only 1 collection made in 1899 in Sacramento Mountains. Hosts outside New Mexico include both woody and herbaceous plants (e.g., *Achillea, Boehmeria, Cephalanthus, Decodon, Hypericum, Justicia, Lythrum, Physostegia, Polygonum, Pycnanthemum, Saururus, Symphyotrichum, Solidago, Teucrium, Vernonia,* and *Vicia*). Doña Ana, Otero.

Cuscuta chinensis Lam. [*C. applanata* Engelm.]—Gila River dodder. Inflorescences glomerulate to densely paniculiform; pedicels 0.4–2 mm; flowers 5-merous, 2.5–3.5 mm, papillae absent; calyx ca. as long as the corolla tube, lobes basally overlapping, broadly triangular-ovate, carinate and with multicellular protuberances along midveins, apex rounded or obtuse; corolla 2–2.4 mm, the tube campanulate, becoming globose or urn-shaped, lobes spreading, ca. as long as the tube, ovate-lanceolate, apex rounded, ± incurved (but not inflexed); stamens exserted; infrastaminal scales 1.2–1.8 mm, equaling or longer than corolla tube, obovate, fimbriate in the distal 1/2; styles 0.9–1.7 mm, equaling or longer than the ovary; stigmas capitate; capsules circumscissile, depressed-globose, 1.8–2.5 × 0.8–1.6 mm, not risen around the small interstylar aperture, translucent, surrounded by the withered corolla; seeds (1–)3–4 per capsule, 0.85–1.2 × 0.8–1.1 mm. New Mexico material belongs to **var. applanata** (Engelm.) Costea & Stefanović. Clay or sandy flats, margins of roads, 3200–5500 ft.; hosts are herbaceous, including species of *Amaranthus, Ambrosia, Anisacanthus, Bahia, Baileya, Boerhavia, Chamaecrista, Chamaesaracha, Coldenia, Croton, Dalea, Flaveria, Ipomoea, Parthenium, Sanvitalia, Solanum, Tragia, Viguiera*. Doña Ana, Sandoval, Sierra.

Cuscuta coryli Engelm.—Hazel dodder. Inflorescences paniculiform; pedicels 0.5–3 mm; flowers (3)4-merous, 1.7–2.6(–3) mm, papillae usually absent but perianth cells fleshy, domelike; calyx cupulate, equaling or somewhat longer than corolla tube, lobes not or only slightly overlapping at base, triangular-ovate, ± carinate, apex acute; corolla 1.5–2.5 mm, the tube campanulate to suburceolate, lobes spreading to erect, 1/3 to equaling corolla tube, triangular-ovate, apex acute, inflexed; stamens enclosed; infrastaminal scales equaling or exceeding corolla tube, oblong, bifid with 1–3 fimbriae on each side of filament attachment or with denticulate wings; styles 0.7–1.8 mm, ca. as long as or sometimes longer than the ovary; stigmas capitate; capsules indehiscent, globose to globose-depressed, 1.8–2.5 × 3.5–5 mm, thickened around the large interstylar aperture, not translucent, capped by the withered corolla; seeds 2–4 per capsule, 1.32–1.65 × 1.25–1.4 mm. Known from only 1 specimen collected from Bluff Creek in 1847. Riverbanks, mesophyllous places; hosts are herbaceous and woody plants, including *Corylus, Helianthus, Monarda, Rubus, Solidago, Symphyotrichum*. Doña Ana.

Cuscuta cuspidata Engelm.—Cusp dodder. Inflorescences loosely paniculiform; bracts 2–4; pedicels 2–5 mm; flowers 5-merous, 3.5–4.2 mm; papillae absent; calyx ca. 1/2 corolla tube, lobes basally overlapping, ovate, not carinate, apex acute to cuspidate; corolla 3.3–4 mm, the tube cylindrical-campanulate, lobes reflexed 1/3–1/2 the corolla length, ovate-oblong to ovate-triangular, apex cuspidate-acute to obtuse, straight; stamens exserted; infrastaminal scales 3/4–4/5 of the corolla tube, oblong, ± uniformly densely fimbriate; styles 2–2.6 mm, longer than the ovary; stigmas capitate; capsules indehiscent, globose to globose-depressed, 2.5–3 × 2.8–3.2 mm, with a thickened and risen ridge or collar around the small interstylar aperture, translucent, capped by the withered corolla; seeds 2–4 per capsule, 1.1–1.4 × 1–1.1 mm. Sandy soils, prairies, ca. 5000 ft.; hosts are Asteraceae: *Ambrosia, Amphiachyris, Baccharis, Croptilon, Eclipta, Helianthus, Heterotheca, Iva, Liatris*. Chaves, Curry, Doña Ana, Union.

Cuscuta draconella Costea & Stefanović—Dragon dodder. Inflorescences glomerulate; pedicels 0.1–0.8 mm; flowers 5-merous, 2.5–3.6(–4) mm, papillae present or absent on pedicels, calyx, and corolla; calyx equaling corolla tube, lobes overlapping basally, triangular-ovate or broadly ovate to subround, with 1–2(–5) hornlike or domelike, multicellular protuberances 0.1–0.3 mm distributed along midveins, apex subacute, rounded, or obtuse; corolla 2–3.5 mm, tube campanulate, lobes spreading to reflexed, triangular-ovate, equaling corolla tube, apex ± cucullate, acute, or truncate, cuspidate, inflexed; stamens exserted; infrastaminal scales spatulate to obovate, 3/4 corolla tube length, with numerous fimbriae; styles 0.5–1.2 mm, shorter than to equaling the ovary; stigmas capitate; capsules indehiscent, obovoid, apically thickened and raised into a collar (no mature capsules or seeds seen). New Mexico specimens belong to a form with a papillate perianth and acute calyx lobes. Rocky arroyos, 5250–7000 ft.; hosts are *Atriplex, Gutierrezia*, and *Thelesperma*. Probably rare but conservation status not yet assessed. Catron, Doña Ana, Grant, Socorro. **R**

Cuscuta glabrior (Engelm.) Yunck.—Bushclover dodder. Inflorescences glomerulate or densely corymbiform; pedicels 0.8–4(–5) mm; flowers 5-merous, 2.5–3.8 mm; papillae present or absent on pedicels, perianth, and ovary; calyx cupulate, ± equaling corolla tube, lobes not overlapping, ovate-triangular, midvein not carinate, apex obtuse to subacute; corolla white, 1.4–3.4 mm, tube campanulate, later globose and saccate between lines of stamen attachments, lobes spreading to reflexed, triangular, equaling corolla tube, apex acute to acuminate, inflexed; stamens exserted; infrastaminal scales ovate to spatulate, equaling or longer than corolla tube, uniformly densely fimbriate; styles 0.9–1.6 mm, equaling or longer than ovary; stigmas capitate; capsules indehiscent, globose to depressed-globose, 1.5–2.8 × 2.1–3.5 mm, not raised around the midsized to large interstylar aperture, not translucent, 1/2–2/3 enveloped by withered corolla; seeds 2–4, 0.9–1.1 × 0.8–1 mm. Known from only 3 specimens; 1 near the border with

Texas (Lea County) and 2 collected by C. Wright in 1851 with no locality indicated, ca. 3000 ft. Hosts outside New Mexico are numerous herbaceous plants. Chaves, Curry, Doña Ana, Lea.

Cuscuta indecora Choisy—Large-seed dodder. Inflorescences paniculiform or corymbiform; pedicels 0.5–6 mm; flowers 5-merous, 3–4.5 mm, papillae usually present on pedicels, perianth, and ovary, or if absent, perianth cells domelike, fleshy; calyx 1/2–3/4 the length of corolla tube, lobes overlapping basally, triangular-ovate, not carinate, apex acute; corolla 2.5–4(–5) mm, tube campanulate but becoming subglobose or urceolate in fruit, lobes spreading, triangular-ovate, 1/3 to equaling tube, apex acute, inflexed; stamens barely exserted; infrastaminal scales obovate to spatulate, equaling corolla tube, uniformly fimbriate; styles 1–2.5 mm, equaling the ovary; stigmas capitate; capsules indehiscent, globose to ovoid, 2–3.5 × 1.9–4(–5) mm, raised around the medium-sized interstylar aperture, translucent, surrounded or capped by withered corolla; seeds 2–4, 1.4–1.8 × 1.2–1.6 mm. New Mexico plants are **var. indecora**. Flats, outcrops, bluffs, slopes on sandy or gypsum substrates, 3500–5000 ft.; hosts are numerous herbaceous plants; potentially a weed of forage legume crops. WC, SE, SC, Quay, San Juan, Union.

Cuscuta legitima Costea & Stefanović [*C. umbellata* Kunth var. *reflexa* (J. M. Coult.) Yunck.]— Large-flowered flat-globe dodder. Inflorescences umbelliform; pedicels 2–10 mm; flowers 5-merous, 4–5.5(–6) mm, papillae absent; calyx campanulate, longer than corolla tube, lobes not basally overlapping, ovate-lanceolate, not carinate, apex acuminate; corolla 3.8–5.2(–5.6) mm, tube campanulate, lobes reflexed, longer than the tube, linear-lanceolate, apex acuminate, straight; stamens exserted; infrastaminal scales equaling or slightly longer than corolla tube, spatulate to obovate, rounded, uniformly fimbriate; styles 0.9–2.5 mm, longer than the ovary; stigmas capitate; capsules circumscissile, globose-depressed, 2–3 × 1–2 mm, slightly risen around the inconspicuous interstylar aperture, translucent, surrounded or capped by the withered corolla; seeds 2–4 per capsule, 0.9–1.2 × 0.8–0.9 mm. Flats, arroyos in sandy substrates, 3000–4500 ft.; hosts are herbaceous and include *Allionia*, *Amaranthus*, *Chamaesaracha*, *Evolvulus*, *Kallstroemia*, *Salsola*, *Solanum*, *Tidestromia*, *Trianthema*, *Tribulus*. Doña Ana.

Cuscuta leptantha Engelm.—Slender dodder. Inflorescences umbelliform; pedicels 0.75–7 mm; flowers 4-merous, 3.5–4.5(–5) mm; papillae present on pedicels and perianth; calyx campanulate, 1/3–1/2 the corolla tube, lobes not basally overlapping, triangular-ovate, not carinate, apex acute; corolla 3–4 mm, tube 1.5–2.5 mm, cylindrical, lobes spreading or reflexed, as long as the tube, lanceolate, apex acute, ± cucullate; stamens short-exserted; infrastaminal scales ca. 1/2 the corolla tube, oblong, short-fimbriate; styles 1.2–2.1 mm, longer than the ovary; stigmas capitate; capsules circumscissile, globose, 1.5–2 × 1.6–1.9 mm, slightly risen around the inconspicuous interstylar aperture, translucent, capped by the withered corolla; seeds 2–4 per capsule, 0.75–0.9 × 0.7–0.8 mm. Known from only 1 specimen collected at El Morro National Monument that could not be verified (*Rink 5370*, ASC), ca. 7300 ft.; host is *Euphorbia*. Cibola.

Cuscuta liliputana Costea & Stefanović—Lilliputian dodder. Inflorescences umbelliform; pedicels 1–5 mm; flowers (3)4-merous, 2.8–4 mm, papillae present on pedicels, calyx, and corolla; calyx equaling the corolla tube, lobes ovate-triangular, sometimes with multicellular protuberances on midveins, not overlapping, apex acute to acuminate; corolla 3–3.6 mm, tube 1.5–2 mm, cylindrical; lobes spreading or reflexed, lanceolate, apex acute; stamens exserted; infrastaminal scales 1/4–1/3 of the corolla tube, oblong to slightly obovate, distally fimbriate; styles 0.8–2.5 mm, longer than the ovary; stigmas capitate; capsules circumscissile, 1.5–2.2 × 0.75–1.5 mm, globose to globose-depressed, slightly risen around the small interstylar aperture, translucent, capped by the withered corolla; seeds 2–4 per capsule, 0.8–1.15 × 0.7–0.85 mm. Shortgrass prairies, roadsides, 4265–5500 ft.; host is *Euphorbia*. Probably rare but conservation status not yet assessed. Sierra.

Cuscuta squamata Engelm.—Scale-flower dodder. Inflorescences glomerulate or short-spiciform; bracts 4–5(–10) at base of clusters and flowers; pedicels 0–1 mm; flowers 5-merous, 5–6 mm, papillae absent; calyx 1/2–2/3 length of corolla tube, lobes broadly overlapping basally, ovate, not carinate, apex acute to cuspidate; corolla 4–5.5 mm, tube 2.4–3.5 mm, cylindrical, lobes spreading to reflexed, ovate-lanceolate to oblong-ovate, 1/3–4/5 length of tube, apex acute, sometimes cuspidate, straight; stamens barely exserted; infrastaminal scales oblong, equaling corolla tube, densely fimbriate; styles 2.5–3.3 mm, longer than ovary; stigmas capitate; capsules indehiscent, subglobose or ovoid to subconic, 3.4–4.5 × 2.2–3 mm, ± raised around small interstylar aperture, not translucent, capped by withered corolla; seeds 2–4, 1.5–1.7 × 1.1–1.3 mm. Desert scrub, roadsides, 3600–5000 ft. Hosts are herbaceous Asteraceae. Chaves, Doña Ana, Socorro.

Cuscuta tuberculata Brandegee—Tubercle dodder. Inflorescences umbelliform; pedicels 2–3(–5) mm; flowers 5-merous, 2.5–4 mm, papillae present especially at base of corolla tube; calyx 1/3–1/2 length of corolla tube, lobes not overlapping basally, triangular to lanceolate, carinate and/or with multicellular protuberances on midveins, apex acute to acuminate; corolla 2–3.5 mm, the tube cylindrical, 1.5–2.2 mm, lobes erect, triangular-lanceolate, equaling the tube, apex acute, straight; stamens barely exserted; infrastaminal scales ovate to oblong, 1/3–1/2 length of corolla tube, short-fimbriate; styles 1.5–3 mm, longer than the ovary; stigmas capitate; capsules circumscissile, globose, 1.3–2.2 × 1–2.3 mm, slightly raised around small interstylar aperture, translucent, capped by withered corolla; seeds 2–4, 0.6–0.9 × 0.3–0.5 mm. Hosts are *Boerhavia*, more rarely *Amaranthus*, and genera from the Nyctaginaceae and Euphorbiaceae. Grant, Otero.

Cuscuta umbellata Kunth [*C. fasciculata* Yunck.]—Flatglobe dodder. Inflorescences umbelliform; pedicels 2–10 mm; flowers 5-merous, 2–3 mm; papillae sometimes present on corolla lobes; calyx equaling corolla tube, lobes not overlapping basally, triangular-ovate, not carinate, apex obtuse to acute; corolla 2–2.5 mm, tube campanulate, 0.6–1.2 mm, lobes reflexed, oblong to lanceolate, longer than the tube, apex obtuse to acute, straight; stamens exserted; infrastaminal scales subspatulate to obovate, equaling or slightly longer than corolla tube, densely fimbriate; styles 0.8–1.7 mm, longer than ovary; stigmas capitate; capsules circumscissile, depressed-globose, 1–2.5 × 0.7–1.2 mm, slightly raised around medium-sized interstylar aperture, translucent, surrounded or capped by withered corolla; seeds 3–4 per capsule, 0.8–1.2 × 0.6–0.8 mm. Sandy, gravelly, or alkaline soils, savanna, meadow, desert scrub, piñon-juniper communities, 3000–7200 ft.; hosts are *Allionia*, *Alternanthera*, *Amaranthus*, *Atriplex*, *Boerhavia*, *Gilia*, *Iresine*, *Kallstroemia*, *Phyloxera*, *Portulaca*, *Salsola*, *Selinocarpus*, *Sesuvium*, *Suaeda*, *Tidestromia*, *Trianthema*, *Tribulus*. NC, C, NW, EC, SC, Union.

Cuscuta umbrosa Beyr. ex Hook. [*C. curta* (Engelm.) Rydb.; *C. megalocarpa* Rydb.]—Big-fruit dodder. Inflorescences paniculiform (becoming globular in fruit); pedicels 0.9–7 mm; flowers 5-merous (rarely some 4-merous), 2–3.5(–4.4) mm, papillae absent; calyx ca. 1/2 the length of corolla tube, lobes overlapping basally, ovate, not carinate, apex rounded or obtuse; corolla 2–4 mm, tube campanulate, 1.7–2.3(–2.7) mm, lobes spreading to reflexed, ovate to broadly triangular-ovate, 1/4–1/3 length of tube, apex rounded to obtuse; stamens exserted; infrastaminal scales broadly oblong, (1/3–)1/2 length of corolla tube, fimbriation mostly in distal 1/2; styles 0.3–0.9 mm, 1/4 length of ovary; stigmas capitate; capsules indehiscent, ovoid to globose, 3.5–6.5(–7) × 3–5(–6) mm, raised around the relatively small interstylar aperture, not translucent, withered corolla surrounding capsule; seeds 3–4, 1.8–2.5 × 1.5–1.6 mm. Known from a few old herbarium specimens collected in the White Mountains, ca. 7400 ft. Hosts are both woody and herbaceous plants, including *Ampelopsis*, *Clematis*, *Convolvulus*, *Humulus*, *Rubus*, *Salix*, *Solidago*, *Symphoricarpos*. Otero.

Cuscuta warneri Yunck.—Warner's dodder. Inflorescences glomerulate or corymbiform; pedicels 0.5–1 mm; flowers 5-merous, 2.1–4 mm; papillae present on pedicels, corolla, and ovary/capsule; calyx campanulate-cupulate, ca. 1/2 the corolla length, lobes not overlapping, triangular-ovate, carinate, each apically enlarged to form a large, divergent, hornlike fleshy projection, 0.5–0.75 mm; corolla 1.8–3.5 mm, the tube 1.5–2.5 mm, campanulate-urceolate, lobes erect, connivent, 1/4–1/3 the tube length, triangular-ovate, apex acute, inflexed; stamens included; infrastaminal scales ca. 4/5 the corolla tube, oblong, truncate, irregularly dentate distally; styles 0.2–0.4 mm, much shorter than the ovary; stigmas capitate; capsules indehiscent, globose, raised in a collar around the styles, 1.8–2.5 × 1.9–3 mm, translucent, surrounded or capped by the withered corolla; seeds 2–4 per capsule, 1.32–1.65 × 1.25–1.4 mm. Mesophyllous areas, 4700–4800 ft. Hosts are *Phyla*. Roosevelt, Sierra. **R**

DICHONDRA J. R. Forst. & G. Forst. – Pony's foot

Plants perennial herbs, glabrate to pubescent; stems stoloniferous, mat-forming; leaf blades cordate, reniform, or orbiculate, conspicuously petiolate, 3–51 mm wide; inflorescences solitary, rarely paired; bracts reduced or absent; flowers inconspicuous; calyx deeply 5-lobed; corollas rotate or broadly campanulate, with a short tube, lobes 5-parted, white or greenish or greenish yellow; ovary emarginate to deeply 2-lobed; styles 2, free or united only at base; stigma capitate; fruits capsular or utricular, entire or 2-lobed; seeds 1–2(–4).

1. Corollas ca. 2 mm; stems sparsely appressed-pubescent; fruit indehiscent; disturbed sites, lawns, gardens…**D. micrantha**
1′. Corollas 3–5 mm; stems pilose to tomentose; fruit dehiscent; foothills, canyon bottoms, plains…(2)
2. Plants with appressed, silvery pubescence; peduncles usually < 6 mm, recurved near the attachment to the calyx…**D. argentea**
2′. Plants with silky, green pubescence; peduncles usually > 6 mm, recurved near the attachment to the calyx…**D. brachypoda**

Dichondra argentea Humb. & Bonpl. ex Willd.—Silver pony's foot. Plants with hairs silvery, long, silky; stems trailing, to 0.5 m; leaves petiolate, petiole 1–5 cm; margins shallowly to deeply 2-lobed; blades circular; bases reniform to orbicular; pubescence same density on both surfaces; inflorescences solitary; bracts scalelike or absent; flowers inconspicuous; sepals connate at base, 2–2.6 mm; corolla cream to greenish, rotate to broadly campanulate, 3.4–4 mm; fruit a capsule. Desert scrub, oak woodlands, 4590–5800 ft. SC, SW.

Dichondra brachypoda Wooton & Standl.—New Mexico pony's foot. Plants with hairs green, long, silky; stems trailing, to 0.5 m; leaves petiolate, petiole 1.5–15 cm; blades circular; bases reniform, truncate, orbicular; shallowly to deeply 2-lobed; indumentum more dense abaxially; inflorescences solitary; bracts scalelike or absent; flowers inconspicuous; sepals connate at base, 2.5–4 mm, corolla greenish cream, rotate to broadly campanulate; 3.5–5 mm; fruit a capsule. Oak woodlands, lower ponderosa pine zones, 3175–7200 ft. SE, SC, SW.

Dichondra micrantha Urb.—Pony's foot. Plants with herbage ± pubescent with appressed hairs; leaf blades reniform to suborbicular, 8–16 × 9–21 mm; bases broadly cordate; calyx broadly campanulate, 1.5–2 mm; corollas campanulate to funnelform, subequal to the calyx; anthers purplish. Disturbed and weedy sites, lawns, sidewalks, gardens. Chaves, Doña Ana. Introduced.

EVOLVULUS L. – Dwarf morning glory

Plants perennial herbs or small suffrutescent shrubs; stems not twining but sometimes creeping; leaves entire, unlobed, linear to ovate, 2–35 mm, usually small, sessile or subsessile, surfaces glabrate, glabrous, or hairy; inflorescences solitary or 2–3-flowered cymes; bracts opposite, usually scalelike; sepals equal, lance-linear to ovate, 2–6 mm; corolla usually blue, lavender, purple, or white, rarely violet, widely funnelform to rotate, 3–15+ mm; fruit a capsule; seeds 1–4, small, smooth or minutely verrucose.

1. Peduncles plus pedicels filiform, longer or shorter than subtending leaves…(2)
1′. Peduncles plus pedicels stout or absent, always shorter than subtending leaves…(3)
2. Corollas < 10 mm wide; sepals < 3 mm; trichomes on stems appressed…**E. alsinoides**
2′. Corollas > 10 mm wide; sepals > 3 mm; trichomes on stems long, spreading, rarely appressed…**E. arizonicus**
3. Distal leaf arrangement distichous…**E. sericeus**
3′. Distal leaf arrangement pentastichous, not distichous…(4)
4. Pubescence same density on upper and lower surfaces; outermost sepals lanceolate; corolla not turning yellow with age; plants in deep sand and other habitats…**E. nuttallianus**
4′. Pubescence denser on adaxial (upper) side; outermost sepals ovate; corolla yellowing with age; plants only in deep sand environments…**E. arenarius**

Evolvulus alsinoides (L.) L. [*E. linifolius* L.]—Ojo de víbora. Plants subshrubs, 6–50 cm, hairs sparse or densely matted; leaves ± sessile, blades elliptic, lanceolate, oblong, or ovate; inflorescences solitary or cymose with 2 flowers; sepals glabrous to pilose, 2–2.5 mm; corolla pale blue, sometimes white, rotate to broadly funnelform. New Mexico material belongs to **var. angustifolius** Torr. Disturbed rocky sites, desert scrub and piñon-juniper communities, 4500–5850 ft. SE, SC, SW.

Evolvulus arenarius R. T. Harms—High plains dwarf morning glory. Plants perennial herbs, sub-shrubs, 8–40 cm, hairs sparse, silver-gray; leaves ± sessile; blades linear to narrowly elliptic; sepals un-equal, 3.1–4.1 mm; corolla lavender, aging to yellow, rotate to broadly funnelform, 9–14 mm. Deep sand, grasslands, desert scrub, 3000–3300 ft. Chaves.

Evolvulus arizonicus A. Gray [*E. laetus* A. Gray]—Blue eyes, false flax. Plants perennial, suffrutes-cent subshrubs, 10–30 cm, hairs sparsely or densely matted; leaves ± sessile, blades lanceolate, lance-linear, or linear; inflorescences solitary or 2–3-flowered; sepals lanceolate to lance-linear, pilose to to-mentose, 3–3.5 mm; corolla blue, rotate to broadly funnelform, (10–)12–22 mm. Disturbed rocky sites, 4500–5575 ft. Doña Ana, Hidalgo, Luna.

Evolvulus nuttallianus Schult.—Shaggy morning glory. Plants perennial subshrubs, 1–1.5 dm, hairs densely matted; leaves ± sessile, blades usually elliptic, sometimes linear-oblong or narrowly lanceolate to oblanceolate; inflorescences solitary, flowers sessile or on peduncles; sepals unequal, 4–5 mm, corolla blue or lavender, rotate to broadly funnelform. Oak woodlands, ponderosa pine zones, rocky prairies, juniper-piñon woodlands, chaparral, 3400–7200 ft. Widespread except WC.

Evolvulus sericeus Sw.—Silvery morning glory. Plants perennial herbs, subshrubs, 10–30 cm, hairs long, straight, appressed; leaves ± sessile, blades elliptic, lanceolate, oblong, or ovate; inflorescences solitary; sepals unequal, 3–5 mm; corolla pale blue, lavender, or white, aging to yellow, rotate to broadly funnelform, 7–12 mm. Chaparral, pine woodlands, desert grasslands, 3450–7100 ft. C, SE, SC, SW, San Miguel.

IPOMOEA L. – Morning glory

Plants annual or perennial herbs, vines, or shrubs; stems decumbent, erect, trailing, twining, or climbing, glabrous or hairy; leaves simple or compound, usually cordate, lanceolate, linear, ovate, reniform, sagit-tate, or palmately lobed; inflorescences usually solitary or in 2–3-flowered cymes, rarely in panicles; bracts foliaceous or scalelike; corollas purple, red, pink, orange, rarely white, shape usually funnelform, some-times campanulate or salverform, 2–8 cm; styles 1; stigmas entire or 2–3-lobed, capitate or globose; fruits capsular, dehiscence irregular or valvate; seeds 1–4(–6).

1.   Leaves entire (not lobed or palmately dissected) and linear; stems erect...**I. leptophylla**
1′. Leaves usually lobed or palmately dissected, or if entire, not linear; stems usually trailing, often twining...(2)
2.   Leaves palmately compound or dissected, margins not entire...(3)
2′. Leaves shallowly to deeply lobed, or if dissected, only on distal portion, margins sometimes entire...(7)
3.   Corollas > 65 mm, white; flowers nocturnal...**I. tenuiloba**
3′. Corollas < 65 mm, not white; flowers diurnal...(4)
4.   Corollas 10–12 mm...**I. costellata**
4′. Corollas > 12 mm...(5)
5.   Plants annual; sepals glabrous or hirsute but without muricate or tuberculate texture...**I. ternifolia**
5′. Plants perennial; sepals glabrous, with muricate or tuberculate texture...(6)
6.   Peduncles 5–7 mm, leaves 5–15 mm; sepals prominently muricate-tuberculate, especially along mar-gins...**I. plummerae** (in part)
6′. Peduncles 15–30 mm, leaves 10–30 mm; sepals muricate but only along margins...**I. capillacea**
7.   Leaves wedge-shaped, divided into linear segments distally...**I. plummerae** (in part)
7′. Leaves of various shapes (not wedge-shaped), often lobed but without distal linear segments...(8)
8.   Corollas red to orange...**I. cristulata**
8′. Corollas various colors including blue, lavender, purple, or white (rarely red in *I. purpurea*)...(9)
9.   Sepals hairy...(10)
9′. Sepals glabrous, some with muricate surface texture or occasionally ciliate along margins...(15)
10. Sepal tips abruptly acute...**I. purpurea**
10′. Sepal tips long-acuminate, or if acute, not abruptly so...(11)

11. Leaves and stems with long, silky white hairs…**I. pubescens**
11′. Leaves and stems without long, silky white hairs…(12)
12. Corollas > 45 mm; plants perennial…(13)
12′. Corollas < 45 mm; plants annual…(14)
13. Sepals 11–14 mm; flowers nocturnal…**I. gilana**
13′. Sepals 18–30 mm; flowers diurnal…**I. lindheimeri**
14. Leaves and stems hairy; leaves shallowly lobed, without glandular dots on abaxial surface…**I. hederacea**
14′. Leaves and stems glabrous; leaves deeply lobed, with glandular dots on abaxial surface…**I. barbatisepala**
15. Sepals smooth, without surface texture, < 14 mm…**I. cordatotriloba**
15′. Sepals with rugose or muricate surface texture, > 14 mm…(16)
16. Corollas > 26 mm; sepal midveins rugose, otherwise surface texture glabrous…**I. cardiophylla**
16′. Corollas < 26 mm; sepal surface texture muricate…**I. dumetorum**

Ipomoea barbatisepala A. Gray—Canyon morning glory. Plants annual; stems twining, glabrous; leaves orbiculate to ovate in outline, 30–80 × 15–85 mm, palmately 5–7-lobed, lobes lanceolate or rhombic, bases cordate, glabrous, often gland-dotted on abaxial surface; sepals lance-linear, 10–12 × 1–2 mm, hispid-pilose except at base, tips elongate and narrowly linear; corollas blue to purple or white, funnelform, 16–20(–25) × 18–20 mm. Desert scrub, canyon bottoms, rocky habitats, 4100–5900 ft. SE, SC, SW.

Ipomoea capillacea (Kunth) G. Don [*I. muricata* (L.) Jacq.; *I. patens* (A. Gray) House]—Purple morning glory. Plants perennial from tubers; stems ascending to erect, sometimes trailing, glabrous; leaves sessile, palmately compound, 5–7-lobed, appearing fascicled, lobes (3–)5–15(–25) × 0.2–1 mm; sepals unequal, elliptic, oblong, or ovate, 5–6 × 2–3 mm, muricate to tuberculate at least on midvein; corollas funnelform, lavender to red-purple, tubes lighter in color, 30–40 × 20–25 mm. Coniferous forests, oak woodlands, grassland, washes, (4450–)5700–8800 ft. WC, Lincoln.

Ipomoea cardiophylla A. Gray—Dutchman's-pipe morning glory. Plants annual from fibrous roots; stems twining, glabrous; leaves simple, cordate, 20–60 × 14–38 mm, bases cordate, glabrous; sepals triangular, 6 × 3–4 mm, tips acute, midveins rugose, glabrous; corollas funnelform, blue (drying pink or purple), 26–27 × 30–35 mm. Desert scrub, piñon-juniper woodland, coniferous forest, oak woodland, canyon bottoms, 4400–6500(–8500) ft. SC, SW.

Ipomoea cordatotriloba Dennst.—Torrey's tievine. Plants perennial; stems twining; leaves cordate-ovate, lance-ovate, or ovate, 10–90 × 10–90 mm, sometimes 3–5(–7)-lobed, lobes rounded, glabrous; sepals lanceolate to ovate, 8–14 mm, margins ciliate or not, surfaces glabrous; corollas funnelform, pale rose-violet, tubes darker in color, violet, 20–30 mm. New Mexico material belongs to **var. torreyana** (A. Gray) D. F. Austin. Collected once in Doña Ana County. A weed twining in shrubs, 4100 ft.

Ipomoea costellata Torr.—Crested morning glory. Plants annual from taproot; stems erect at first, then trailing, twining near tips, glabrous; leaves sessile or on petioles 10–30 mm, deeply palmately dissected, lobes 5–9, segments lance-linear, linear, oblanceolate, or spatulate, 7–28 × 0.5–3(–8) mm, glabrous to sparsely hispidulous; sepals oblong to lanceolate, unequal, outer 3–5 × 1–2 mm, inner 4–6 × 2–3 mm, tips acute, inner rugose along veins, glabrous; corollas funnelform, pale lavender to pink, 10–12 mm. Desert scrub, wash bottoms, grasslands, piñon-juniper and oak woodlands, coniferous forests, 3800–7350 ft. WC, SE, SC, SW, Cibola, Sandoval, San Miguel.

Ipomoea cristulata Hallier f. [*I. coccinea* auct. non L. var. *coccinea*; *I. coccinea* auct. non L. var. *hederifolia* (L.) A. Gray]—Trans-Pecos morning glory. Annual; stems twining, glabrous, sometimes pilose at nodes; leaves ovate and entire or partially to fully 3–7-lobed, 15–100 × 10–70 mm, glabrous or rarely pilose on abaxial surfaces; sepals oblong, subequal, outer 3–3.5 × 2–2.5 mm, inner 4–5.5 × 3–3.5 mm, subterminal appendages 3–5 mm, glabrous; corollas red to red-orange, salverform, 18–26 × 10–15 mm. Disturbed areas, roadsides, desert scrub, canyon bottoms, oak woodlands, coniferous forests, 4000–8350 ft. NW, C, WC, SE, SC, SW.

Ipomoea dumetorum Willd.—Railway creeper. Plants annual, twining to trailing; leaves deltate, ovate, or ovate-elongate, 24–80 × 8–87 mm, margins sometimes 3-toothed, glabrous; sepals elongate-ovate to ovate, 3.5–8 mm, tips acute to obtuse, muricate; corollas dark lavender to pink, rarely white, funnelform, 15–18 mm. Known from only 2 collections, 1 each in the White and Organ Mountains, open dry to wet habitats, washes, 7050–7400 ft. Doña Ana, Lincoln.

Ipomoea gilana K. Keith & J. A. McDonald—Black Range morning glory. Plants perennial; stems twining; leaves entire to slightly or deeply 5–7-lobed, 30–80 × 30–70 mm, lobes 10–60 × 5–20 mm, trichomes present along veins, otherwise glabrous; sepals ovate-elongate, 11–14(–16) × 3–5(–7) mm; margins hyaline, tips acute or acuminate, outer sepals sparsely pilose, inner sepals glabrous; corollas pale blue with irregular pink blushes, narrowly funnelform, 60–70 × 60–75 mm. Endemic to the Black Range. Open woodlands of oak and piñon-juniper, 6580–6700 ft. Sierra. **R & E**

Ipomoea hederacea Jacq.—Ivy-leaf morning glory. Plants annual; stems twining, densely to sparsely pubescent; leaves ovate to orbicular, entire or 3–5-lobed, 50–120 × 50–120 mm, pubescent; petioles to 12 mm, hairy with retrorse trichomes; sepals lanceolate, proximally ovate, abruptly narrowed to ± curved or spreading, distal portion notably longer than dilated base, 12–24 × 4–5 mm, densely long-hirsute at least on proximal 1/3; corollas funnelform, light blue, tube white or pale yellow inside, 2–37 (–45) × 17–35 mm. Disturbed areas, oak, piñon-juniper woodlands, 4400–7160 ft. C, SC, SW.

Ipomoea leptophylla Torr.—Bush morning glory. Plants perennial from a large taproot; stems erect, sometimes trailing; leaves simple and sessile, lance-linear to linear, 10–80(–150) × 2–8(–10) mm, base cuneate, glabrous; sepals elliptic, orbiculate, 5–10 mm, tips obtuse; corollas funnelform, lavender or pink to red, throats darker, 50–90 mm. Desert scrub, desert grasslands, piñon-juniper and oak woodlands, canyon bottoms, roadsides, 3600–7125 ft. N, EC, SE, Doña Ana, Hidalgo.

Ipomoea lindheimeri A. Gray—Lindheimer's morning glory. Plants perennial; stems twining; leaves broadly ovate to reniform, 50–60 × 50–80 mm, usually with 3–5(–7) lanceolate lobes, hirsute or sericeous; sepals lanceolate to lance-ovate, 18–30 × 3–5(–9) mm, dilated at base, hirsute; corollas funnelform, blue to violet, 55–90 mm, limb 60–70 mm diam. Carbonate ridges and slopes, grasslands, piñon-juniper and oak woodlands, 3500–7115 ft. SE, SC, SW.

Ipomoea plummerae A. Gray—Huachuca Mountain morning glory. Plants perennial from tuber-like roots; stems trailing, sometimes ascending; leaves orbicular to broadly elliptic, 1–3 cm, often divided into 5–7 filiform or narrowly linear segments 1–4 mm wide; sepals unequal, oblong to ovate, outer 5–8 × 2–3 mm, inner 7–9 × 3–4 mm, muricate; corollas funnelform, purple, 25–31 mm, limb 18–22 mm diam.

1. Blades compound, divided into filiform segments…**var. plummerae**
1'. Blades simple, lobed, the apices laciniate-lobed…**var. cuneifolia**

var. cuneifolia (A. Gray) J. F. Macbr. [*I. egregia* House]—Cuneate-leaved Huachuca Mountain morning glory. Known from a single collection near Bear Mountain, NW of Silver City. Coniferous forests, ca. 6500 ft. Grant.

var. plummerae—Huachuca Mountain morning glory. Piñon-juniper and oak woodlands, coniferous forests, 7015–8800 ft. C, WC, SC.

Ipomoea pubescens Lam.—Silky morning glory. Perennial from a large, oblong root; stems twining, hirsute with appressed trichomes; leaves cordate to ovate, deeply 3–5-lobed, 20–80 × 20–90 mm, hirsute or coarsely sericeous, trichomes appressed; sepals unequal, lance-ovate to ovate, broadly ovate at base and narrowing gradually to an acuminate apex, bases truncate, 9–21 × 2–11 mm, hispid to sericeous; corollas funnelform, blue to violet with white throat, 55–80 × 60–70 mm. Montane scrub, piñon-juniper woodlands, grasslands, 5400–7300 ft. SC, SW.

Ipomoea purpurea (L.) Roth [*I. desertorum* House; *I. mexicana* A. Gray]—Common morning glory. Annuals; stems vining, loosely pubescent to tomentose with short-appressed, retrorse trichomes mixed with long-spreading trichomes; leaves cordate to ovate, entire to 3–5-lobed, 10–110(–180) × 10–120(–160) mm, hairy with retrorse trichomes; sepals elliptic, broad at base, tapering abruptly to acute or acuminate apex, 8–15 × 2.4–4.5 mm, bases with spreading trichomes; corollas funnelform, blue, red, or purple, white within tube, (25–)40–60 × 24–48(–70) mm. Moist, disturbed habitats, 3800–7600 ft. Widespread except NE. Introduced.

Ipomoea tenuiloba Torr.—Spiderleaf. Plants perennial; stems trailing, twining near tips, or prostrate, glabrous; leaves orbiculate, palmately 5–9-lobed, lobes linear, 10–70 × 0.5–2 mm, glabrous; sepals unequal, outer oblong-lanceolate, inner obovate, to 14 mm, margins scarious, muricate along midrib to nearly smooth; corollas salverform, generally white, occasionally pink or purple, 65–100 × 3–36 mm. New Mexico material belongs to **var. tenuiloba**. Canyon bottoms, oak and piñon-juniper woodlands, 5850–8900 ft. C, SE, SW, Otero.

Ipomoea ternifolia Cav. [*I. leptotoma* Torr.]—Tripleleaf morning glory. Plants annual with fibrous roots; stems trailing, twining near tips, glabrous; leaves palmately divided with 5–7 filiform or linear segments, 15–30 × 15–30 mm, glabrous to scarcely setose; sepals unequal, 9–14 mm, hirsute along veins or glabrous; corollas funnelform, purple, 23–45 × 23–46 mm. Our plants belong to **var. leptotoma** (Torr.) J. A. McDonald. Grasslands, desert scrub, 4900–5420 ft. Hidalgo.

CORNACEAE – DOGWOOD FAMILY

Kenneth D. Heil

CORNUS L. – Dogwood

Shrubs or herbaceous perennial plants; leaves opposite or whorled, simple, entire, deciduous, 3–9 cm, lanceolate, elliptic, or ovate to obovate, acute to short-acuminate; inflorescence a terminal cyme with many flowers or capitate; flowers perfect, regular, small, sepals 4–5, 0.4–0.5 mm; petals 4–5, distinct, 1.5–3 mm, yellow to purple or white; stamens the same number as the petals and alternate to them, pistil 1, ovary inferior, mostly 2-loculed, 1 ovule per locule, style 1; fruit a 1–2-seeded drupe, 7–9 mm diam. (Heil & Porter, 2013a).

1. Plants herbaceous, not > 25 cm; leaves 4–6 in a single whorl; inflorescence capitate, subtended by 4 petaloid bracts; fruit red…**C. canadensis**
1'. Woody shrubs > 25 cm; leaves many and opposite; inflorescence a flat-topped cyme, petaloid bracts not present; fruit white…**C. sericea**

Cornus canadensis L.—Bunchberry. Herbaceous plants from woody rootstocks; stems 5–20 cm; leaves 4–6, whorled in 1 series near apex, often a pair of smaller or scalelike leaves below, 3–6 cm, ovate to obovate, acute to short-acuminate, cuneate at base, sparsely strigose above, glabrous and lighter in color below; inflorescence a capitate cyme with many flowers subtended by 4 large white or yellowish bracts, these ca. 10–15 mm, nearly as wide; fruit red. Spruce-fir and aspen communities, 7200–11,500 ft. Sandoval, Taos.

Cornus sericea L.—Red-osier dogwood. Woody shrubs to 4 m, often as broad; branches red to purplish or yellowish, subglabrous to strigulose, older stems grayish green, mostly glabrous; leaves mostly 5–9 cm, lanceolate, elliptic to ovate, acute to acuminate, cuneate, glabrous or nearly so above, strigillose below with spreading hairs along the veins; inflorescence a flat-topped cyme with many flowers; flowers with white petals; fruit white. Common along streams and in moist ravines, 5000–9800 ft. Widespread except E.

CRASSULACEAE – STONECROP FAMILY

Kenneth D. Heil and Steve L. O'Kane, Jr.

Herbs (ours), often succulent; leaves alternate, opposite, or whorled, simple, exstipulate, flowers perfect, actinomorphic; sepals 4 or 5, distinct; petals 4 or 5, distinct, white, yellow, pink, reddish, or purple;

stamens 4, 5, 8, or 10, epipetalous or not; pistils 4 or 5, with a nectariferous appendage near the base; ovary superior; fruit a follicle (Ackerfield, 2015; Moran, 2009).

1. Annuals; stems usually prostrate; flowers minute and inconspicuous, solitary in leaf axils…**Crassula**
1'. Perennials; stems erect; flowers larger and conspicuous, aggregated in cymes…(2)
2. Leaves persistent in dense, basal rosettes; flowers pale yellow, irregularly dotted and red-banded in distal 1/2…**Graptopetalum**
2'. Leaves not in dense, basal rosettes; flowers various but not irregularly dotted…(3)
3. Leaves often clustered in persistent rosettes or not; not dying back in winter; rootstock slender, lacking scalelike leaves…**Sedum**
3'. Leaves cauline, not forming rosettes; dying back in winter; rootstock stout, with scalelike leaves… **Rhodiola**

CRASSULA L. – Pygmyweed

Crassula aquatica (L.) Schönland [*Tillaea aquatica* L.]—Water pygmyweed. Plants annual, aquatic, sometimes stranded; leaves opposite, oblanceolate to linear, 2–6 mm; inflorescences lax; flowers 1 per node; flowers 4-merous; sepals 0.5–1.5 mm; petals 1–2 mm, whitish; stamens 4; pistils 4, distinct. Shallow water and moist soil around vernal pools; 1 record, 7415 ft. Cibola.

GRAPTOPETALUM Rose – Leatherpetal

Graptopetalum rusbyi (Greene) Rose [*Echeveria rusbyi* (Greene) A. Nelson & J. F. Macbr.]—San Francisco River leatherpetal. Perennial herbs; stems succulent, ascending, 3–10 mm thick; leaves 2–6 (–10) cm diam., rosettes densely cespitose, blades green or reddish, rhombic-obovate to oblanceolate, 1.5–5 × 0.8–1.8 cm; inflorescences mostly flat cymes; flowers (5)6–7(8)-merous; corolla 14–21 diam., pale yellow, irregularly dotted and red-banded in distal 1/2; stamens as many or 2 times as many as sepals; pistils erect, connate basally; fruits mostly erect. Rock crevices, especially on N slopes and shaded cliffs. No records seen; known from Grant County.

RHODIOLA L. – Stonecrop

Perennials; leaves alternate, entire to toothed, fleshy; flowers perfect or rarely imperfect; in terminal or axillary cymes; sepals 4 or usually 5; petals 4 or usually 5; stamens twice the number of sepals; pistils 4 or usually 5.

1. Petals deep red to purple, 2.4–4 mm; sepals 1.6–3 mm… **R. integrifolia**
1'. Petals pink to rose or white, 7–12 mm; sepals 4–7 mm… **R. rhodantha**

Rhodiola integrifolia Raf. [*Sedum integrifolium* (Raf.) A. Nelson]—King's crown. Stems 0.3–4.5 dm; leaves flat, oblanceolate to ovate or elliptic, green or glaucous, scalelike below, 4–40 × 2–15 mm, entire to toothed; sepals 1.6–3 mm, dark purple; petals 2.4–4 mm, dark purple to red or yellow and red at the apex.

1. Petals mostly dark red or purple throughout…**subsp. integrifolia**
1'. Petals yellow with red tips…**subsp. neomexicana**

subsp. integrifolia [*Sedum integrifolium* (Raf.) A. Nelson subsp. *procerum* R. T. Clausen]—Ledge stonecrop. Stems 3–15(–50) cm; leaf blades mostly green, sometimes glaucous, ovate to elliptic or oblanceolate, 0.5–3(–5) × 0.2–1.5(–2) cm; petals mostly dark red throughout; seeds 1.4–2 mm. Cliffs and rocky slopes, alpine meadows, alpine tundra, 7965–13,160 ft. NC, C, WC.

subsp. neomexicana (Britton) H. Ohba [*R. neomexicana* Britton; *Sedum integrifolium* (Raf.) A. Nelson subsp. *neomexicanum* (Britton) R. T. Clausen]—New Mexico stonecrop. Stems 15–25 cm; leaf blades green, not glaucous, linear-oblanceolate, 2–3.5 × 0.3–0.7 cm; petals yellow, red at apex and on keel; seeds 1.1–1.9 mm. Porphyritic rocks, 8600–11,950 ft. Lincoln, Otero. **R & E**

Rhodiola rhodantha (A. Gray) H. Jacobsen [*Sedum rhodanthum* A. Gray]—Rose crown. Stems 10–60 × 0.2–0.6 cm; leaf blades green, not glaucous, linear-oblong to linear-oblanceolate, 1–4.5 × 0.2–

0.7 cm; sepals linear-lanceolate, 3–9 mm; petals nearly erect with tips outcurved, greenish or white to rose. Wet meadows, stream banks, rock fields, rock crevices, 9050–12,595 ft. NC, Lincoln.

SEDUM L. – Stonecrop

Perennials; leaves opposite or alternate, entire to toothed, fleshy; flowers perfect or rarely imperfect, in terminal or axillary cymes; sepals 4 or usually 5; petals 4 or usually 5; stamens 8 or 10; pistils 4 or usually 5.

  1. Petals yellow…(2)
  1'. Petals white…(3)
  2. Leaves opposite; inflorescences 2–7-flowered…**S. debile**
  2'. Leaves alternate; inflorescences 5–25-flowered…**S. lanceolatum**
  3. Leaves terete or nearly so…**S. stelliforme**
  3'. Leaves flattened, or rounded only on the back…(4)
  4. Flowers with unique musky odor; leaves immediately deciduous…**S. wrightii**
  4'. Flowers without musky odor; leaves more firmly attached…**S. cockerellii**

Sedum cockerellii Britton—Cockerell's stonecrop. Stems rarely branched, erect; leaf blades green or yellow-green, obovate or oblong, spatulate, 9.5–15 × 1.5–3.5 mm; petals white, streaked with pink. Ponderosa pine and mixed conifer forests, shallow soils, usually in shade, 5645–10,200 ft. Widespread except E, San Juan.

Sedum debile S. Watson [*Gormania debilis* (S. Watson) Britton]—Orpine stonecrop. Stems decumbent, branched; leaves opposite, sessile, blades pale green, speckled with pink, usually elliptic, oblanceolate, or obovate, subterete, 4.2–7.2 × 2.8–4.3 mm; petals yellow, 6–9 mm. Open, rocky places in the mountains. Colfax.

Sedum lanceolatum Torr. [*S. stenopetalum* Pursh]—Spearleaf stonecrop. Plants mostly 0.4–2.5 dm; leaves alternate, linear to lanceolate, subterete, 3–20 mm, not densely imbricate; petals yellow, 5–8 mm. Rocky soils and dry slopes, 8000–13,160 ft. NC, NW.

Sedum stelliforme S. Watson—Huachuca Mountain stonecrop. Stems erect, sometimes bearing rosettes; leaves alternate, linear-oblanceolate, terete to subterete, 4–9(–15) × 1–2 mm; flowering branches and sepals with glistening patches; petals white, tinged with purple, 4–7 mm. Moist cliffs in ponderosa pine and mixed conifer woodlands, moist areas in grasslands, 5475–10,700 ft. WC, SW.

Sedum wrightii A. Gray—Wright's stonecrop. Plants mostly 2–20 cm; leaves alternate, easily detached from stem, elliptic to oblanceolate, terete, 5–12 mm; petals white, 5–9 mm.

  1. Leaf blades of flowering shoots 10+ mm; sepals 4–9 mm; petals 5–9 mm…**var. wrightii**
  1'. Leaf blades of flowering shoots 6–9 mm; sepals ca. 4 mm; petals 5–6 mm…**var. priscum**

var. priscum (R. T. Clausen) H. Ohba—Wright's stonecrop. Flowering shoots: leaf blades 6–9 mm; inflorescences 2–14-flowered or flowers solitary; pedicels 1.1–3.7 mm; flower sepals ca. 4 mm; petals 5–6 mm; seeds ca. 0.5 mm. Rocky slopes and cliffs, piñon pine to Douglas-fir woodlands, 7875–8200 ft. Catron.

var. wrightii—Wright's stonecrop. Flowering shoots: leaf blades 10+ mm; inflorescences 10–30-flowered; pedicels to 1 mm; flower sepals 4–9 mm; petals 5–9 mm; seeds 0.6 mm. Cliffs and other rocky sites, 5000–8300 ft. C, WC, SC.

CROSSOSOMATACEAE – ROCKFLOWER FAMILY

John R. Spence

Shrubs, intricately branched, often spinescent, bark often exfoliating in thin longitudinal plates; leaves deciduous or drought-deciduous, simple, alternate or opposite or sometimes fasciculate on short branches, stipules absent or very small, petioles absent or short, margins smooth to apically lobed; inflorescence of 1 to few flowers on short terminal shoots; flowers bisexual, actinomorphic, perigynous, hypanthium cup-shaped; perianth biseriate, sepals and petals distinct, (3)4–5(6)-merous; stamens 4–50; pistils 1–9, free, superior, style short or indistinct, stigma discoid to linear, ovules 1–20+; fruit a follicle, seeds 1–22 per follicle, arillate, arils white to tan, discoid to fimbrillate.

1. Leaves alternate, entire; sepals and petals mostly 5; stigmas discoid…**Glossopetalon**
1'. Leaves opposite, entire to often 3-lobed; sepals and petals 4; stigmas linear…**Apacheria**

APACHERIA C. T. Mason – Apacheria

Apacheria chiricahuensis C. T. Mason—Chiricahua apachebush. Low-growing, intricately branched shrubs, to 50 cm, branches tan to weakly orange-tinged, not strongly spinescent; leaves opposite, entire to 3-lobed, oblanceolate to spatulate, 3–8 × 1–2.5 mm, apiculate, glabrous; flowers single, sessile to short-pedunculate, sepals 4, 3–4 mm; petals 4, white, 4–5 mm; stamens 8, carpels (1–)4, ± distinct, sessile; follicles (1–)4, obovoid to turbinate, longitudinally striate; seeds 1–2 per follicle, brown to yellow. Volcanic (rhyolite) and gypsum rock outcrops and ledges, 5250–7450 ft. Hidalgo, Sierra, Socorro. Nearly **E**

GLOSSOPETALON A. Gray – Greasebush

Glossopetalon spinescens A. Gray—Spiny greasebush. Tall, often upright shrubs, to 3 m; secondary branchlets ascending-divaricate, green when young, becoming brown to orange-brown with age, older branches gray, tips strongly spinescent; leaves small, 3–17 × 1–6 mm, oblanceolate to elliptic, entire, tip smooth or mucronate; inflorescences of 1(–3) flowers, sepals 4–5(–6), ovate, 1–3 mm, sometimes unequal, petals 4–5(6), white, 4–9 mm, elliptic to oblanceolate; stamens 5–10, in 1 or 2 unequal series, longer series opposite petals; follicles 1–2, ovoid to ellipsoid, 3–5 mm, ± striate; seeds 1–2 per follicle, brown, yellow, or cream.

1. Stipules absent…**var. spinescens**
1' Stipules present, thick, reddish purple where attached to stem…**var. planitierum**

var. planitierum (Ensign) Yatsk.—Shrubs, low-growing to somewhat erect, to 1.2 m; leaves slowly deciduous, often some remaining at flowering; stamens mostly 8 in 2 unequal series. Dry prairies, ledges, cliff tops, drainages, on limestone, gypsum, or sandstone; prairie communities, scattered juniper, 4312–4830 ft. Curry, Quay, Santa Fe.

var. spinescens—Shrubs, to 3 m; stipules absent, leaves early-deciduous, generally before flowering; stamens 6–10, in 1 equal or 2 unequal series. Rocky washes, hillslopes, ledges, and cliffs, mostly on limestone, 4000–7000 ft. C, SE, SC, SW, Cibola, McKinley.

CUCURBITACEAE – CUCUMBER FAMILY

Kenneth D. Heil and Guy L. Nesom

Herbs or rarely shrubs, monoecious or dioecious; stems prostrate, procumbent, sprawling, trailing, or climbing; tendrils usually present, unbranched or branched; leaves alternate, petiolate, simple or compound, often palmately lobed; inflorescences paniculate, racemose, umbellate to subumbellate, fasciculate, corymbose, or solitary flowers; flowers imperfect, usually actinomorphic; sepals 5, distinct; petals 5, distinct or connate, usually yellow, orange, or white, sometimes green; stamens (2)3 or 5, with 4 mostly connate in pairs; anthers connate or distinct; fruit a berry, capsule, or pepo; seeds mostly compressed, sometimes winged (Nesom, 2015).

1. Inflorescences 1–16…(2)
1'. Inflorescences 1(–3)…(3)
2. Plants annual; pistillate flowers sessile to subsessile in umbelliform clusters at peduncle apex; seeds 1…
    Sicyos
2'. Plants perennial; pistillate flowers in axillary fascicles or in axillary racemoid to corymboid panicles; seeds many…**Apodanthera**

3. Fruits echinate, muricate to muriculate, subaculeate, or tuberculate…(4)
3'. Fruits usually smooth, ribbed, warty, or furrowed, rarely tuberculate…(9)
4. Staminate flowers solitary…(5)
4'. Staminate flowers (1-)2-200 in racemes, racemoid panicles, panicles, or fascicles…(6)
5. Seeds red-arillate; filaments inserted near hypanthium rim; fruit surface minutely tuberculate, muriculate in longitudinal rows…**Momordica**
5'. Seeds not arillate; filaments inserted near hypanthium base; fruit surface netted, warty, or scaly…**Cucumis** (in part)
6. Leaves 3-7-foliate; fruits fleshy-capsular…**Cyclanthera**
6'. Leaves 3-7-lobed; fruits dry, thin-walled…(7)
7. Plants perennial; roots tuberous; stamens 3(4)…**Marah** (in part)
7'. Plants annual; taprooted or roots slender-fibrous; stamens 3-5…(8)
8. Petals (5)6, narrowly triangular to linear-lanceolate or oblong-lanceolate; spinules glabrous or slightly scabrous; leaf surfaces glabrous or scabrous; stamens 3…**Echinocystis**
8'. Petals 5, triangular-ovate or oblong to deltate; spinules with stipitate-glandular hairs or pubescent-eglandular; leaf surfaces hispid to hispidulous; stamens (4)5…**Echinopepon**
9. Leaf lobes each pinnately lobed to shallowly sinuate-lobulate…**Citrullus**
9'. Leaf lobes not pinnately lobed to shallowly sinuate-lobulate (margins lobulate in *Ibervillea*)…(10)
10. Petals usually white to cream, yellow to pale green, rarely greenish…(11)
10'. Petals greenish yellow, cream, yellow to orange-yellow, or orange…(12)
11. Individual flowers and fruits loosely enclosed within pair of basally cordate bracts…**Sicyosperma**
11'. Individual flowers and fruits not enclosed within pair of bracts…**Marah** (in part)
12. Seeds arillate…**Ibervillea**
12'. Seeds not arillate…(13)
13. Tendrils usually unbranched, rarely unbranched or 2-branched, sometimes absent; petals 2-25 mm…**Cucumis** (in part)
13'. Tendrils 2-7-branched or absent; petals 25-90 mm…**Cucurbita**

APODANTHERA Arn. – Melon loco

Apodanthera undulata A. Gray—Melon loco. Vines fetid; branches to 3 m, harshly strigose; tendrils usually branched from 1-5 cm beyond base; leaf blades 8-15 cm wide, surfaces moderately to densely white-sericeous abaxially, glabrous adaxially; staminate flowers in long-pedunculate racemes; pistillate flowers in loose fascicles; hypanthium 10-18 mm; sepals erect, not recurving; pepos 6-10 cm. New Mexico material belongs to **var. undulata**. Stream terraces, dunes, grasslands, Chihuahuan desert scrub, mesquite, oak-juniper woodlands, 4000-6000 ft. SE, SC, SW, Chaves.

CITRULLUS Schrad. – Watermelon

Citrullus lanatus (Thunb.) Matsum. & Nakai [*C. lanatus* subsp. *vulgaris* (Schrad.) Fursa]—Watermelon. Vines, annual; stems climbing or trailing, 50-200 cm, villous; roots fibrous; tendrils 2-3-branched; leaf blades ovate to lance-ovate or ovate-triangular, mostly 8-20 cm, 3-5-lobed, lobes pinnately shallowly sinuate-lobulate, margins denticulate, surfaces hirsute abaxially, hispid on veins and veinlets, glabrous or scabrous adaxially with translucent dots; hypanthium broadly campanulate; sepals 3-5 mm; petals obovate-oblong to widely oblanceolate, 7-16 mm; pepos green, mottled with paler green and yellowish to whitish stripes, globose to oblong-ellipsoid, 12-35+ cm diam.; rind tough, not durable, mesocarp red to orange, yellow, or greenish, juicy, sweet. Our material belongs to **subsp. lanatus**. Gardens, fields, vacant lots, roadsides, gravel bars, stream banks, riparian thickets, 3445-5300 ft. Scattered. Introduced.

CUCUMIS L. – Cantaloupe, honeydew, muskmelon

Cucumis melo L.—Cantaloupe, honeydew, muskmelon. Plants monoecious or andromonoecious; leaf blades broadly ovate, unlobed or palmately 3-5-lobed, 2-14(-26) × 2-15(-26) cm, margins entire or weakly serrate; staminate flowers 2-7(-18) in fascicles or panicles; pistillate flower calyx lobes 1.5-3(-8) mm, petals 3.5-9(-20) mm, corolla tube 1-1.6(-3) mm; pepos mostly ellipsoid to subglobose, surface netted, warty, scaly, ridged, or smooth, flesh orange to yellowish, green, or whitish. Ditches along the Rio Grande, 4000-5500 ft. Doña Ana, San Juan. Introduced from Africa.

CUCURBITA L. – Squash, gourd

Coarse, monoecious perennials; stems prostrate, tendrilled from enlarged roots, pubescence rough and appressed; flowers relatively few and large, solitary, axillary; hypanthium of staminate flowers campanulate or tubular; hypanthium of pistillate flowers subglobose; stamens 3, filaments distinct; anthers coherent and contorted; staminodia of pistillate flowers 3; ovary 1-celled, 3–5-carpellate; fruit globose, smooth (Rhodes et al., 1968).

1.   Leaf blades unlobed or shallowly 2-lobed, longer than broad…**C. foetidissima**
1'. Leaf blades palmately, deeply 5-lobed, about as wide as long…(2)
2.   Lobes of leaf blades lanceolate to linear, narrow and reaching nearly to the midrib…**C. digitata**
2'. Lobes of leaf blades broad, ovoid-elliptic, often hardly reaching to the midrib…**C. pepo**

Cucurbita digitata A. Gray—Finger-leaf or coyote gourd. Perennial; stems usually sprawling, harshly pubescent; tendrils 2–5-branched; leaves about as wide as long, the blades to ca. 25 cm, palmately 5-lobed, sinuses nearly or completely to petiole, margins coarsely toothed or remotely sinuate-dentate to serrate, surfaces hispid to hispidulous; hypanthium cylindrical to narrowly campanulate; sepals linear-subulate, 3–5 mm; corolla bright yellow, narrowly campanulate, 4–7 cm; pepos dark green with 10 whitish stripes, mostly globose to depressed-globose, smooth. Grasslands, mesquite, juniper-scrub, disturbed sites, 3895–5650 ft. SW, Eddy.

Cucurbita foetidissima Kunth [*Pepo foetidissima* (Kunth) Britton]—Buffalo or foetid gourd. Perennial; stems prostrate, puberulent to scabrous with pustulate-based hairs; tendrils 3–7-branched 3–6 cm above base; leaf blades mostly triangular-ovate, (10–)12–30(–40) × (6–)8–20(–30) cm, longer than broad, margins coarsely and widely mucronulate to denticulate, surfaces with short pubescence; flower sepals 10–25 mm; corolla golden yellow, 6–10 cm; pepos depressed-globose to globose or oblong-globose, 5–10 cm, smooth. Rocky to clay soils, prairies, mesquite scrub, piñon-juniper woodlands, oak-pine forests, disturbed sites, 3220–7800 ft. Widespread.

Cucurbita pepo L.—Field or jack-o'-lantern pumpkin. Annual; stems creeping or climbing, hispid with persistent, strongly pustulate-based hairs; tendrils 2–7-branched 1–5 cm above base; leaf blades broadly ovate-cordate to triangular-cordate or reniform, shallowly to deeply palmately (3–)5-7-lobed, 20–30 × 20–35 cm, usually broader than long or equally so, margins denticulate to serrate-denticulate, surfaces hirsute, villous-strigose, or hispidulous-scabrous; flower sepals 8–25 mm; corolla yellow to golden yellow or orange, 4–10 cm; pepos globose or depressed-globose to ovoid, 5–10(–25) cm, usually smooth. Disturbed sites, 4000–7010 ft. San Juan.

CYCLANTHERA Schrad. – Bur cucumber

Cyclanthera naudiniana Cogn.—Cut-leaf cyclanthera. Stems glabrous except for minutely villosulous nodes; tendrils mostly 2-branched; leaves 3-foliate, lateral pair of leaflets deeply divided, terminal leaflet 3–4.5 cm, blade broadly lanceolate, leaflet margins coarsely serrate to shallowly or deeply lobed; flowers solitary, in fascicles of 2–4, or along short axes; staminate corollas 3.5–4.9 mm diam.; capsules narrowly ovoid, slightly oblique, short-beaked, 15–25 mm, spinules 3–4 mm. Streamsides and piñon-juniper woodlands, 5800–7400 ft. Mora, San Miguel, Union.

ECHINOCYSTIS Torr. & A. Gray – Wild cucumber

Echinocystis lobata (Michx.) Torr. & A. Gray [*Sicyos lobatus* Michx.]—Concombre grimpant. Annual, monoecious, climbing or trailing; stems glabrate; tendrils 3-branched; leaf blades depressed-orbiculate to suborbiculate or ovate, 2–8(–12) cm, palmately (3–)5(–7)-lobed, margins entire or serrulate-cuspidate, surfaces glabrous or slightly scabrous, hair bases pustulate; staminate racemes 8–14 cm; pistillate peduncles 2–5 cm; flower sepals (5)6, filiform, almost pricklelike; petals (5)6, white to greenish white, 3–6 mm, corolla rotate; pepos 3–5 cm, spinules 4–6 mm, glabrous or slightly scabrous. Marshes, irrigation ditches, fencerows, 5600–7000 ft. NC, NW, San Miguel.

ECHINOPEPON Naudin – Wild balsam apple

Monoecious annuals, stems climbing; mostly hairy; tendrils (2)3-branched; leaf blades reniform to orbiculate, deeply to shallowly palmately 3-5-lobed or dissected, surfaces mostly hispid, eglandular; staminate flowers 50-100 in branched or unbranched axillary racemes; pistillate flowers solitary, from same or different axils as staminate; flower sepals 5(6); petals 5(6); staminate flowers: stamens (4)5, filaments inserted at base of corolla; pistillate flowers: ovary 2-locular, ovoid, ovules 2-5 per locule; style 1, inserted at base of hypanthium; fruit ovoid, beaked, hispid or echinate, opening by an apical lid; seeds 4-10, ovoid to quadranguloid.

1. Leaf blades 4-5(-7) cm wide; corollas 8-12 mm diam., petal apices emarginate; capsule surfaces and prickles eglandular-pubescent, prickles mostly 3-5 mm...**E. coulteri**
1'. Leaf blades 5-8(-15) cm wide; corollas 6-8 mm diam., petal apices acute; capsule surfaces and prickles hirsute, hairs stipitate-glandular, prickles 10-20 mm...**E. wrightii**

Echinopepon coulteri (A. Gray) Rose [*Elaterium coulteri* A. Gray]—Coulter's wild balsam apple. Stems 0.5-4 m, sparsely short-villous, hairs eglandular; leaves mostly orbiculate, deeply to shallowly lobed, 4-5(-7) cm wide, margins entire or slightly sinuate-denticulate, surfaces sparsely hispid; staminate flowers in simple racemes 5-14(-20) cm; flower sepals linear to subulate; petals oblong to deltate, 4-6 mm; capsules 2-3 × 1.2 cm, surface and prickles eglandular-pubescent, beak 5-8 mm, prickles mostly 3-5 mm. Rocky hillsides, washes, montane riparian sites, 6460-6600 ft. WC, SC, SW.

Echinopepon wrightii (A. Gray) S. Watson [*Elaterium wrightii* A. Gray]—Wright's wild balsam apple. Stems 0.5-4 m, hirsute, hairs stipitate-glandular; leaves reniform to orbiculate, angular or undulate to shallowly lobed, 5-8(-15) cm wide, margins entire or denticulate, surfaces hispid to hispidulous; staminate flowers in simple or compound racemes 6-12 cm; flower sepals narrowly triangular; petals triangular-ovate, 3-4 mm, corolla rotate, 6-8 mm diam.; capsules 2.5-3 × 1.4-1.8 cm, surface and prickles hirsute, hairs stipitate-glandular, beak 10 mm, prickles 10-20 mm. Canyon bottoms, desert scrub, desert grasslands, woodlands, riparian sites, 5280-8000 ft. SW.

IBERVILLEA Greene – Globeberry

Plants perennial, dioecious, trailing and climbing; stems glabrous; roots tuberous; tendrils simple; leaf blades ovate to suborbiculate, broadly ovate, or reniform, usually palmately 5-lobed or pedately 3-lobed, margins lobulate or coarsely toothed, surfaces eglandular; staminate flowers solitary or 4-12(-24) in axillary racemes; pistillate flowers solitary; flower sepals 5; petals 5, distinct, greenish yellow to yellow, 3-7 mm; staminate flowers: stamens 3, filaments inserted near hypanthium rim; pistillate flowers: ovary 3(-5)-locular; fruits berrylike, dark orange to red, globose to ellipsoid, smooth.

1. Leaf lobes 10-25 mm wide; petals 5-7 mm; fruits 2-3.5 cm...**I. lindheimeri**
1'. Leaf lobes 2-5 mm wide; petals 3-4 mm; fruits 1-1.5 cm...**I. tenuisecta**

Ibervillea lindheimeri (A. Gray) Greene [*I. tenella* (Naudin) Small]—Lindheimer's globeberry. Stems to 3 m; leaf blades usually deeply palmately 5-lobed or pedately 3-lobed, 6-12 cm wide, often slightly succulent, surfaces glabrous or sparsely scabridulous; staminate flowers usually 4-12(-24) in racemes or terminal corymboid clusters 1-3 cm, rarely solitary; petals 5-7 mm; fruits dark orange to red, globose to ellipsoid, 2-3.5 cm. Piñon-juniper woodlands, ponderosa pine and riparian communities, 6110-8000 ft. Eddy.

Ibervillea tenuisecta (A. Gray) Small [*I. lindheimeri* (A. Gray) Greene var. *tenuisecta* (A. Gray) M. C. Johnst.]—Deer-apples. Stems to 1 m; leaf blades deeply pedately 3-lobed, 2-6 cm wide, slightly succulent, surfaces glabrous; staminate flowers 4-7(-20) in racemes; petals 3-4 mm; fruits bright red, globose, 1-1.5 cm. Rocky hills, arroyos, flats, Chihuahuan desert scrub, mesquite-oak woodlands, 3285-5835 ft. EC, SE, SC, SW, Guadalupe.

MARAH Kellogg – Manroot

Marah gilensis (Greene) Greene—Gila manroot. Plants perennial, monoecious, trailing or climbing; stems usually glabrous; roots tuberlike; tendrils unbranched or 2–3-branched; leaf blades suborbiculate, shallowly to deeply palmately 5(–7)-lobed, 4–10 cm wide, margins entire or remotely and coarsely dentate-lobulate, surfaces eglandular; staminate flowers in axillary or racemoid panicles; pistillate flowers solitary, in same axils as staminate flowers; sepals 5, petals 5, usually white or greenish; stamens 3(4); filaments inserted near hypanthium base; pistillate flowers: ovary (2–)4(–8)-locular; fruits capsular, yellowish green to orange-yellow. Desert flats and washes, stream beds, canyon bottoms and slopes, desert scrub, chaparral, oak, riparian areas, 4000–5000 ft. Grant.

MOMORDICA L. – Balsam-apple

Momordica balsamina L.—Balsam-apple. Plants annual, monoecious, climbing or trailing; stems pubescent to glabrescent; leaf blades broadly ovate or reniform to orbiculate, palmately 3–5-lobed, 1–9(–12) cm, margins sinuate-dentate, surfaces glabrous or sparsely hairy; staminate and pistillate flowers solitary; flower sepals 5; petals 5, distinct, yellow to bright yellow, 8–15 mm; fruits orange-red, broadly ovoid, 2.5–4(–7) cm. Disturbed areas, roadsides, fencerows; a single record from Doña Ana County, 3905 ft. Introduced.

SICYOS L. – Bur cucumber

Plants annual, monoecious, climbing or trailing; stems glabrous or hairy, often viscid-pubescent when young; tendrils 2–5-branched; leaf blades ovate or orbiculate to suborbiculate or reniform, deeply to shallowly palmately 3–5-angular-lobed, margins usually serrate to denticulate, surfaces eglandular; staminate flowers 3–22(–34) in axillary racemes or panicles; pistillate flowers 4–16, sessile to subsessile in umbelliform clusters; sepals 5; petals 5, white to greenish white, yellowish green, or yellow; pepos dark green to gray, fusiform to ovoid; seeds 1.

1. Pepos not echinate…**S. glaber**
1′. Pepos echinate…(2)
2. Staminate inflorescences 10–16-flowered; stigmas 3-lobed; mature stems glabrate to sparsely minutely stipitate-glandular; leaf blades deeply lobed, sinuses (1/3–)1/2–2/3 to base…**S. microphyllus**
2′. Staminate inflorescences 3–10 flowered; stigmas 2-lobed; mature stems glabrescent but remaining villous and stipitate-glandular; leaf blades shallowly lobed to angulate, sinuses 1/5–1/4 to base…**S. laciniatus**

Sicyos glaber Wooton—Smooth bur cucumber. Stems sparsely villous, becoming sparsely hirsutulous to glabrous; leaf blades pentagonal-angulate to reniform-angulate or shallowly 5-angulate, sinuses 1/4–1/3 to base, (3–)4–7 × 5–9 cm; staminate inflorescences 6–22-flowered; pistillate 4–10-flowered; staminate corollas greenish yellow to yellow, 2–4 mm; pistillate corollas 2–3 mm including tube; pepos ovoid, 4–6 mm, not echinate, glabrous or sparsely villosulous. Canyon bottoms, juniper–mountain mahogany woodlands, 4495–8045 ft. Doña Ana, Hidalgo. **R**

Sicyos laciniatus L. [*S. ampelophyllus* Wooton & Standl.]—Cut-leaf bur cucumber. Stems densely villous and stipitate-glandular, glabrescent; leaf blades broadly ovate-angulate to reniform-angulate or shallowly 5-lobed, sinuses 1/5–1/4 to base, 3–5(–8) × 4–11(–13) cm; staminate inflorescences 3–10-flowered; pistillate 4–12-flowered; staminate corollas white, 2 mm; pistillate corollas 1.5–2 mm including tube; pepos ovoid, 5–6 mm, echinate, spinules retrorsely barbellate and glabrous. Rocky slopes, cliff faces, streamsides, floodplains, cottonwood-willow, juniper, oak, pine-oak, 4380–7600 ft. C, WC, SC, SW.

Sicyos microphyllus Kunth [*S. deppei* G. Don]—Little-leaf bur cucumber. Young stems sparsely puberulent, stipitate-glandular, hairs viscid, glabrescent, becoming glabrate; leaf blades orbiculate-pentagonal to deeply 5-lobed, sinuses (1/3–)1/2–2/3 to base, 4–10 × (4–)6–12 cm; staminate inflorescences 10–16-flowered; pistillate 4–10-flowered; staminate corollas greenish white, 2–3 mm; pistillate corollas 1–2 mm (essentially without a tube); pepos ovoid, 5–8 mm, echinate and sparsely, finely villous, spinules retrorsely barbellate. Rocky slopes and ridges, cliff bases, streamsides, ponderosa pine woodlands, 5600–7700 ft. C, WC, SC.

SICYOSPERMA A. Gray – Climbing arrowheads

Sicyosperma gracile A. Gray—Climbing arrowheads. Plants annual, monoecious, climbing; stems glabrous or sparsely puberulent; tendrils 2-branched; leaf blades broadly to shallowly triangular, usually shallowly palmately 3(–5)-lobed to hastate, 2–7 cm wide, margins entire or denticulate, surfaces eglandular; staminate flowers 20–40 in axillary, compound racemes; flower sepals 5; petals 5, white, corollas salverform; pepos whitish, ellipsoid-ovoid, 4–5 mm, smooth, glabrous, indehiscent; seeds 1, ovoid, compressed. Streamsides, riparian sites. Reported for New Mexico but no specimens seen. Potential habitat in the bootheel region.

CYPERACEAE – SEDGE FAMILY

Max Licher and Glenn R. Rink

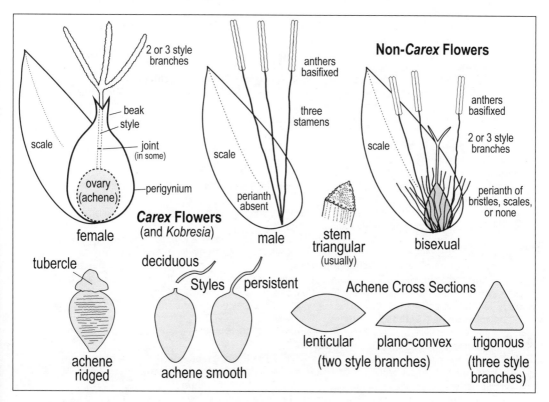

Grasslike herbs, cespitose or rhizomatous; culms usually triangular, occasionally round, or rarely compressed; leaves simple, basal and/or cauline, usually 3-ranked; blades linear, or leaves reduced to bladeless sheaths; sheaths fused; inflorescence composed of 1 to many spikelets arranged in spikes, panicles, umbels, or racemes, in varying orders of complexity, sometimes compressed into a compact head, usually subtended by a leaflike or scalelike bract; spikelets composed of 1 to many florets, usually spirally arranged, occasionally 2-ranked along the axis of the spikelet (reduced to a single floret in *Carex* and some *Kobresia*); flowers bisexual (unisexual in *Carex* and *Kobresia*), each usually subtended by a single scale, with perianth lacking or reduced to bristles and/or scales; ovary superior (in 2 genera, the pistillate flower/achene is partially [*Kobresia*] or completely [*Carex*] enclosed in a saclike bract called a perigynium); stamens usually 3; anthers basifixed; style 1, with either 2 or 3 stigmas; fruit an achene, either biconvex or trigonous, corresponding to the number of stigmas (Ball et al., 2002 and refs. therein; Bauters et al., 2014; Dai et al., 2010; Goetghebeur & Van den Borre, 1999; Goodrich, 2013; Harms, 1972; Hartman, 1973; Kral, 1971; Larridon et al., 2014; Larson & Catling, 1996; Martin & Hutchins, 1980a; Rothrock, 2009a–d; Sivinski et al., 1995 [1996]; Smith, 2001; Smith et al., 2002; Tucker, 1994; Whitmore & Schuyler, 2002; Wilson et al., 2014).

1. Achene completely or partially enclosed in a saclike bract (perigynium); perianth absent; flowers all unisexual, with staminate and pistillate flowers in separate spikes or at opposite ends of the same spike, or rarely randomly mixed in a spike…(2)
1'. Achene not enclosed in a saclike bract (perigynium); perianth absent or present, when present consisting of bristles or scalelike structures; flowers usually bisexual, spikes and/or spikelets usually with flowers similarly arranged…(3)
2. Perigynium open, with margin unsealed on 1 side, merely wrapping around the achene…**Kobresia**
2'. Perigynium closed, with sealed margins, completely enclosing the achene except for an apical opening for the style…**Carex**
3. Fruiting spikelets or heads resembling a dense tuft of cotton due to the numerous elongate, hairlike perianth bristles that obscure the flowers and scales; montane wetland plants of high elevations… **Eriophorum**
3'. Fruiting spikes or heads not cottony as above; plants of various habitats and elevations…(4)
4. Spikelets with floral scales arranged in 2 opposite ranks on either side of the rachis, compressed to terete in cross-section…(5)
4'. Spikelets with floral scales arranged spirally around the rachis, terete in cross-section…(6)
5. Proximal scales of the spikelet fertile or the first one empty; perianth absent; spikelets compressed, subterete, or quadrangular; leaf sheaths not blackish; widespread…**Cyperus** (in part – *Cyperus* s.s.)
5'. Proximal 2+ scales of the spikelet sterile (empty); perianth of bristles usually present; spikelets compressed; leaf sheaths blackish; known in New Mexico from 1 location…**Schoenus**
6. Inflorescence consisting of a single terminal spikelet, without an involucral bract…**Eleocharis**
6'. Inflorescence consisting of multiple spikelets, or if a single spikelet, appearing lateral with a bractlike extension of the culm surpassing the spikelets, or with a noticeable involucral bract subtending the spikelets…(7)
7. Perianth present, of bristles and/or spatulate scales (do not confuse these with the remnants of filaments after anthers have fallen)…(8)
7'. Perianth absent…(11)
8. Perianth dimorphic, of 3 stipitate-based, spatulate scales, alternating with 3 much shorter bristles… **Fuirena**
8'. Perianth monomorphic, of bristles only…(9)
9. Inflorescence bract apparently single, appearing as a continuation of the culm so that the inflorescence appears lateral (smaller bracts occasionally present, but scalelike and not green); leaves either all basal or confined to the bottom 1/3 of the culm; leaves with obvious blades, or significantly reduced to little more than basal sheaths…**Schoenoplectus**
9'. Inflorescence bracts 2+, leafy and spreading and not resembling the culm; at least some leaf blades occurring above the middle of the culm; all leaves with obvious blades…(10)
10. Spikelets large (4–10 mm wide), commonly 3–40; cypselae 2.3+ mm; culms with cormlike bases; leaf ligules lacking…**Bolboschoenus**
10'. Spikelets relatively small (< 4 mm wide), commonly > 40; cypselae 1.8 mm or less; culms without cormlike bases; leaf ligules present…**Scirpus**
11. Large perennial plants usually 1.5–2 m; leaves basal and cauline (above the lower 1/4 of the culm), blades scabrid with fine sawtooth margins; inflorescences terminal and often lateral (from upper leaf axils), conspicuously branched and rebranched; spikelets 100–1000…**Cladium**
11'. Small annuals to larger perennial plants (to 1 m in *Fimbristylis*); leaves all basal (from the lower 1/4 of the culm), blades without sawtooth margins, at most moderately scabrid; inflorescences terminal only, simple to branched; spikelets 50 or fewer…(12)
12. Style base not enlarged in fruit; inner transparent scale behind the thicker primary floral scale present or lacking…**Cyperus** (in part – *C. michelianus*)
12'. Style base enlarged in fruit; inner transparent scale always lacking…(13)
13. Style base persistent as a tubercle in fruit; stigmas 3; plants of dry habitats…**Bulbostylis**
13'. Style base deciduous in fruit; stigmas 2 in our species; plants of wet habitats…**Fimbristylis**

BOLBOSCHOENUS (Asch.) Palla – Tuberous bulrush

Bolboschoenus maritimus (L.) Palla—Alkali bulrush. Plants perennial, colonial from stout rhizomes with tubers; culms 50–150 cm, (1–)3–8 mm wide, with cormose bases; leaves basal and cauline, ligules absent, blades 2–12 mm wide; inflorescences of (1)2–40 sessile spikelets in a terminal cluster, or up to 1/2 the spikelets on 1–4 rays; inflorescence bracts 1–4, surpassing the inflorescences, leaflike; spikelets ovoid to lanceoloid, 7–40 × (4–)7–10 mm; scales 5–8 mm, apex acute, bifid with slender awn; stigmas 2(3); ovary superior; cypselae usually biconvex, 2.3–4.1 mm; stamens 3. Our plants belong to **subsp. paludosus** (A. Nelson) T. Koyama [*Scirpus paludosus* A. Nelson]. Watercourses and marshes, ponds and lakeshores, often brackish or alkaline, 3000–8700 ft. Widespread.

BULBOSTYLIS Kunth – Hairsedge

Plants annual or less frequently perennial; cespitose with fibrous roots only; culms wiry, terete to angulate, ribbed; leaves basal, blades linear to filiform, involute, with scabrous-ciliate margins; sheaths glabrous and usually fimbriate-ciliate apically, ligules absent; inflorescence terminal, of 1 to many spikelets arranged in simple to compound anthelae (cymose corymbs); inflorescence bracts (1)2–8, leaflike to scalelike; spikelets ovoid to lanceoloid, with 6–50 florets spirally arranged; flowers bisexual, subtended by a scale; perianth absent; stamens 1–3; ovary superior, style with 3 stigmas (in our species), style base persistent as a small articulated tubercle on the top of the achene; achene trigonous or 3-lobed, 3-ribbed.

1. Spikelets 1 per culm; culms to 10(–15) cm; leaves 1/2 to slightly exceeding culms; basal spikelets usually present…(2)
1′. Spikelets usually > 1 per culm (often 1 in depauperate individuals); culms 10–30 cm; leaves 1/4–1/3 culm length; basal spikelets only occasionally present…**B. capillaris**
2. Anthers 2; culms to 15 cm; leaves 1/2 to exceeding culms; involucral bracts 1 or 0; cypselae of the basal spikelets larger (1.5 mm) than those on culms (1 mm); basal spikelets dissimilar to those on culms…**B. funckii**
2′. Anthers 3; culms to 7 cm; leaves exceeding culms; involucral bracts 2; cypselae of the basal spikelets the same size as those on culms (1 mm); basal spikelets similar to those on culms…**B. schaffneri**

Bulbostylis capillaris (L.) Kunth ex C. B. Clarke—Densetuft hairsedge. Plants annual; culms 30 cm; leaf blades 1/4–1/3 the length of the culm, ca. 0.5 mm wide, involute; inflorescence terminal (occasionally basal and terminal), of 1–3(–7) spikelets in a subcapitate cluster to an open cyme; bracts 1–2, filiform, proximal bracts shorter to longer than the inflorescence; spikelets 3–5 mm, ovoid to lanceoloid; scales 1.5–2 mm, red-brown with a green or tan midrib, apex acute; stigmas 3, stamens (1)2; cypselae trigonous-obovoid, 1 mm. Sandy or gravelly soils, rocky slopes, roadsides, semiriparian desert scrub, shallow depressions in piñon-juniper woodland, 5000–7600 ft. WC, SC, SW.

Bulbostylis funckii (Steud.) C. B. Clarke—Funck's hairsedge. Plants annual; culms 5–10(–15) cm; leaf blades 1/2 to slightly exceeding the culm, 0.5 mm wide, flat or involute; inflorescence terminal and basal; terminal inflorescence mostly sterile with a single spikelet, bract when present appearing as an extension of the culm; basal inflorescence of individual fertile florets in the sheath axils at the culm base; terminal spikelet 5–7 mm, lanceoloid, apex acuminate; scales 2–2.5 mm, reddish brown with a green or tan midrib excurrent as a mucro; stigmas 3, stamens 2; cypselae broadly trigonous-obovoid, 1 mm in terminal spikelets, 1.5 mm in basal spikelets. Sandy clearings, road banks, fields, disturbed areas, 7500 ft. Catron.

Bulbostylis schaffneri (Boeckeler) C. B. Clarke—Schaffner's hairsedge. Plants annual, diminutive; culms 5–7 cm; leaf blades exceeding the culms, 0.5 mm wide, filiform, flat to involute; inflorescence a single spikelet, either terminal and/or sub-basal on very short scapes; bracts 2, proximal leaflike and exceeding the spikelet; basal spikelets similar to the terminal ones; spikelets 5 mm, ovoid; scales 2–2.5 mm, pale brown with a greenish or yellowish midrib excurrent as a mucro; stigmas 3, stamens 3; cypselae trigonous-obovoid, 1 mm. Sandy or gravelly clearings, in pine or oak woodlands, 5100–8400 ft. Catron, Hidalgo, Socorro.

CAREX L. – Sedge

Plants grasslike, perennial, mostly monoecious, but some species dioecious, densely cespitose to colonial from long-creeping rhizomes; culms mostly trigonous, occasionally subterete; leaves 3-ranked, blades simple, linear, ligule present and attached to blade, sheaths usually closed, the fronts often hyaline and ripping or shredding in age; inflorescences simple or compound, of 1+ sessile or pedunculate spikes; spikes entirely staminate, entirely pistillate, androgynous (containing staminate and pistillate flowers, with the staminate flowers above the pistillate ones), gynecandrous (containing staminate and pistillate flowers, with the pistillate flowers above the staminate ones), or more rarely with staminate and pistillate flowers intermixed in no obvious pattern; bracts leaflike or reduced to a scale; flowers with perianth absent, staminate flowers consisting of 3 stamens, each subtended by a scale ("staminate scale"), pistillate flowers consisting of a pistil with 2 or 3 stigmas, each surrounded by a perigynium (a saclike bract open only at the top, with the style and/or stigmas exposed), the characters of the mature perigynia essential for species differentiation, each perigynium subtended by a scale ("pistillate scale"); fruit an achene, 1 per flower, enclosed in the perigynium, biconvex (lenticular) or plano-convex in flowers with 2 stigmas, trigonous in flowers with 3 stigmas.

1. Spikes solitary per culm...**KEY A**
1′. Spikes multiple per culm...(2)
2. Perigynia hairy (at least some hairs present on the upper 1/2)...**KEY B**
2′. Perigynia glabrous...(3)
3. Spikes generally of 2 types, the terminal spike(s) staminate (rarely androgynous) or gynecandrous; the lower spikes predominantly pistillate or androgynous, often pedunculate...(4)
3′. Spikes similar to each other in shape and/or gender arrangement, sessile, subsessile, or short-pedunculate ...(6)
4. Stigmas 3, cypselae trigonous...(5)
4′. Stigmas 2, cypselae biconvex...**KEY E**
5. Terminal spike staminate (rarely androgynous)...**KEY C**
5′. Terminal spike gynecandrous...**KEY D**
6. Terminal spike androgynous or staminate or pistillate; lateral spikes androgynous or staminate or pistillate; stigmas 2 (3 in *C. muriculata*)...**KEY F**
6′. Terminal spike gynecandrous (sometimes appearing wholly pistillate after anthers have fallen); lateral spikes gynecandrous or pistillate; stigmas 2 or 3...(7)
7. Stigmas 3; spikes sessile to pedunculate...**KEY D**
7′. Stigmas 2; spikes sessile...(8)
8. Perigynia not winged...**KEY G**
8′. Perigynia winged...**KEY H**

KEY A: SINGLE SPIKE PER CULM

1. Plants cespitose, growing in dense to loose clumps...(2)
1′. Plants rhizomatous, with single stems to small clusters of stems arising from obvious rhizomes...(6)
2. Perigynia apices rounded to slightly retuse, without beaks; plants soft, lax, of wet/moist habitats within high-elevation (8500–10,500 ft.) coniferous forests, loosely cespitose from short rhizomes...**C. leptalea**
2′. Perigynia apices contracted, beaked; plants more stiffly upright, habitat and elevation various, densely tufted, rhizomes lacking or inconspicuous...(3)
3. Perigynia with stipitate base, glabrous, ellipsoid to narrowly ovoid or lanceoloid; spike with staminate portion shorter than pistillate portion; pistillate scales ovate to lanceolate, dark brown with narrow hyaline margins; plants of dry to moist alpine habitats...**C. micropoda**
3′. Perigynia lacking stipitate base, glabrous to finely pubescent, ovoid to obovoid; spike with staminate portion longer than pistillate portion (rarely equal); pistillate scales mostly broadly ovate, or broadly obovate to suborbiculate, light tan to greenish, yellowish, or reddish brown, with broad hyaline margins; plants of dry habitats in piñon-juniper woodland, mountain meadows, grasslands, or alpine...(4)
4. Leaf blades folded or channeled, wider blades 0.8–1.5(–2.8) mm wide near base, culms sometimes scabrous below the inflorescence; plants of mountain meadows, dry slopes, and grasslands, not alpine habitats...**C. oreocharis**
4′. Leaf blades involute-cylindrical, filiform, wider blades 0.2–0.8 mm wide near base, culms mostly smooth below the inflorescence; plants of piñon-juniper woodland or alpine habitats...(5)
5. Perigynia glabrous or sparsely short-hirsute/ciliate only on the upper portion near the base of the beak; plants of alpine rock fields and meadows above timberline, 10,500–12,800 ft....**C. elynoides**
5′. Perigynia usually short-pubescent all over, at least on the distal portion; plants of desert scrub and open areas within piñon-juniper woodland, 5900–8200 ft....**C. filifolia**
6. Spikes with pistillate and staminate sections separated by a section of bare rachis, pistillate portion of 1–2(3) separated perigynia; perigynia (4.4–)4.9–6.4(–8.4) mm, obovoid, rounded at top with a minute beak or beakless; culms to 50 cm; plants of dry woodlands and forests, open slopes and meadows... **C. geyeri**

6'. Spikes with pistillate and staminate sections contiguous, pistillate portion usually with > 3 adjacent perigynia; perigynia to 4.7 mm (appearing longer in *C. microglochin* due to exserted rachilla), widest usually at the middle or below, beaked; culms to 25 cm; plants of wet or dry habitats...(7)

7. Plants of wet habitats; leaf blades 0.2–1 mm wide...(8)

7'. Plants of drier habitats; leaf blades 0.5–3(–3.8) mm wide...(9)

8. Perigynia lance-subulate, 3.4–4.7 mm, > 4 times as long as wide, strongly deflexed at maturity; rachilla present, exserted from orifice to 2.2 mm (appearing like an extension of the beak); stigmas 3...**C. microglochin**

8'. Perigynia ovoid, oblong, or plumply elliptic, 2.6–3.4 mm, < 3 times as long as wide, spreading to slightly deflexed at maturity; rachilla absent; stigmas 2...**C. gynocrates**

9. Perigynia uniformly deep red-brown to dark brown, shiny; sheath fronts with red-purple dots; leaf blades 0.5–1.6(–2) mm wide, tips generally straight or curved, not curling; rhizomes often reddish purple; plants of montane grasslands, open forests, ponderosa pine savannas, and rocky ridges up to 10,500 ft.... **C. obtusata**

9'. Perigynia cream or pale green with darker tip, dull; sheath fronts without red-purple dots; leaf blades 1.4–3(–3.8) mm wide, tips often curling; rhizomes brown; planls of alpine summits, dry meadows, and rocky ridges at or above 11,500 ft....**C. rupestris**

KEY B: MULTIPLE SPIKES PER CULM, PERIGYNIA PUBESCENT OR MURICATE-WARTY

1. Inflorescence a dense head of indistinguishable similar androgynous spikes; plants single-stemmed to occurring in small clumps connected by long rhizomes; dry habitats...**C. duriuscula** (in part)

1'. Inflorescence elongate with spikes of different types, terminal spike staminate and lateral spikes mostly pistillate, or if all spikes similar and androgynous, then separated along the rachis and easily distinguishable; plants cespitose or colonial-rhizomatous; varying habitats (note that some species have both cauline and basal inflorescences, the latter having a simpler, compact structure)...(2)

2. Pistillate spikes 1–7 cm, each with 40+ perigynia; plants colonial from long rhizomes; wetland habitats, usually pond and lake margins, along streams, wet meadows, or seasonally flooded wetlands...(3)

2'. Pistillate (androgynous in *C. muriculata*) spikes 1.2 cm or less (to 2.2 cm in *C. muriculata*), each with < 20 perigynia; plants loosely to densely cespitose; dry to mesic woodland, forest, or prairie habitats...(4)

3. Perigynia occasionally with a few bristles along the nerves, otherwise glabrous; base of leaf blades and summit of leaf sheaths usually pubescent; perigynia 6–8.2(–9) mm, beak bidentate with spreading teeth 1.6–2.2(–3.6) mm...**C. atherodes** (in part)

3'. Perigynia densely pubescent over entire surface; base of leaf blades and summit of leaf sheaths glabrous; perigynia 2.5–3.8(5.2) mm, beak bidentate with straight teeth 0.4–0.6 mm...**C. pellita**

4. Surface of perigynia muricate-warty; lateral spikes 2–5, androgynous or sometimes appearing pistillate due to cryptic male flowers...**C. muriculata** (in part)

4'. Surface of perigynia ± pubescent (proximally glabrous in *C. planostachys*) but not warty; lateral spikes 1–3(4), pistillate...(5)

5. Plants with cauline inflorescences only, usually with both staminate and pistillate spikes...(6)

5'. Plants with both basal and cauline inflorescences, occasionally basal only, the cauline inflorescences with staminate and pistillate spikes, the basal spikes on short peduncles, usually all pistillate...(7)

6. Culms to 50 cm, longer to much longer than the leaves; perigynia prominently 10–25-nerved at least to midbody; proximal inflorescence bract as long as or longer than the inflorescence, less frequently shorter; staminate scales pale green throughout, sometimes with a darker green midstripe; SW mountains only...**C. leucodonta**

6'. Culms to 35 cm, shorter to longer than the leaves; perigynia nerveless except for 2 strong marginal nerves; proximal inflorescence bract usually shorter than the inflorescence, rarely a bit longer; staminate scales with brown body and pale or green midstripe; found primarily in the N half of the state (hybrids with basal inflorescences are occasional)...**C. heliophila**

7. Perigynia with many prominent nerves, hispidulous distally, glabrous proximally, or nearly glabrous throughout; beak sometimes abruptly bent; S mountains only...**C. planostachys** (in part)

7'. Perigynia nerveless or with few fine nerves between the 2 prominent marginal nerves, uniformly pubescent; beak slightly bent or straight; of various distributions...(8)

8. Old leaf bases persisting as coarse fibers, tan to brown, usually without any red or orange; cauline pistillate spike bract usually shorter than the inflorescence; pistillate scales shorter than to as long as the perigynia; staminate spikes often thick with many strongly overlapping scales...**C. geophila**

8'. Old leaf bases only slightly fibrous, with some red or orange; cauline pistillate spike bract shorter to longer than the inflorescence; pistillate scales usually shorter than the perigynia; staminate spikes usually slender, with fewer, less overlapping scales, with more of the scale length exposed...(9)

9. Perigynia 2.3–3.4(–4.2) mm; beaks 0.5–1(–1.5) mm; plants loosely cespitose, often lax; alpine rock fields and tundra at or above timberline...**C. brevipes**

9′. Perigynia 3–4.9 mm; beaks 0.7–2 mm; plants densely cespitose, strict or lax; piñon-juniper woodland to ponderosa pine or mixed conifer forests at or below timberline (rarely above in *C. rossii*)…(10)

10. Cauline pistillate spikes 3–6(–15)-flowered; proximal inflorescence bract wide, flat, longer to much longer than the inflorescence; culms 7–25(–40) cm; staminate spikes (6–)13–16.5 mm, the scales pale to slightly tinged with brown or red (not deep purple)…**C. rossii**

10′. Cauline pistillate spikes 1–4-flowered; proximal inflorescence bract narrow, folded or involute, shorter to slightly longer than the inflorescence; culms 3–15 cm; staminate spikes 4.8–13 mm, the scales usually darker, with purple-black tinge…(11)

11. Cauline inflorescence with 2–4 pistillate spikes, each with (1)2–3 perigynia, the perigynia thus appearing in clusters along the axis of the inflorescence; staminate spike 7.5–13 mm, on a peduncle of variable length, often with a pistillate flower close to its base; leaves flat, lax…**C. aff. rossii**

11′. Cauline inflorescence with 1–2(3) pistillate spikes, each with 1(2) perigynium, the perigynia thus often appearing separated along the axis of the inflorescence; staminate spike 4.8–8.3(–11.5) mm, usually on a long peduncle equal to or much longer than spike, rarely with a pistillate flower at its base; leaves often folded or involute, stiffer than the above…**C. pityophila**

KEY C: MULTIPLE SPIKES PER CULM, STIGMAS 3, PERIGYNIA GLABROUS, TERMINAL SPIKE STAMINATE (RARELY ANDROGYNOUS), MOSTLY WETLAND PLANTS

(Note: Only C. capillaris, C. geyeri, C. hystericina, and C. utriculata are known from more than a few locations in the state)

1. Mature pistillate spikes uniformly dark; perigynia and scales at least partially suffused with dark purple or blackish coloration; plants rare in New Mexico…(2)

1′. Mature pistillate spikes green to brown or straw-colored but not uniformly dark; either perigynia and scales both pale **OR** dark scales contrasting with pale-colored perigynia **OR** pale scales contrasting with darker perigynia; plants of varying frequency and distribution…(3)

2. Stigmas 2 (some flowers occasionally with 3); perigynia 2 times as long as wide, often shiny, ovate to ovate-lanceolate or broadly elliptic in profile, inflated (often flattened in pressing), the beaks smooth; plants colonial-rhizomatous…**C. saxatilis** (in part)

2′. Stigmas 3; perigynia 2.5–3 times as long as wide, not shiny, narrowly elliptic to lanceolate, not inflated, the beaks ciliate-setose (at least at the base); plants cespitose to short-rhizomatous…**C. luzulina** (in part)

3. Plants with robust pistillate spikes; longer pistillate spikes 2–13 cm, each with > 50 perigynia, the perigynia densely packed on the spike; perigynia moderately to strongly inflated (though often flattened/collapsed in pressing or flat to slightly inflated in *C. ultra*), the cypselae not filling the perigynia; plants often robust; wetland habitats…(4)

3′. Plants with delicate pistillate spikes; longer pistillate spikes < 3 cm (rarely longer in *C. luzulina*, *C. microdonta*, and *C. sprengelii*), each with < 50 perigynia, the perigynia loosely packed on the spike (or if densely packed, then the spike < 2 cm); perigynia not or barely inflated, the cypselae more nearly filling the perigynia; plants robust or not; varying habitats…(10)

4. Perigynia beaks 0.2–0.6 mm, bodies with red-brown spotting or blotching, flattened to only slightly inflated; leaf blades usually glaucous, thick, coriaceous, with harsh scabrous margins; plants densely cespitose…**C. ultra**

4′. Perigynia beaks 0.7–4.2 mm, bodies without red-brown spotting or blotching (though with dark brownish coloration in some species), usually inflated; leaf blades pale to dark green but not glaucous, thin to thick but not coriaceous and without harsh scabrous margins; plants cespitose to rhizomatous…(5)

5. Pistillate scales with a long scabrous awn, the awn > 1 mm, often equaling or longer than the body, distinct from the body of the scale…(6)

5′. Pistillate scales without an awn, or with midrib extended as a short, smooth awn < 1 mm, shorter than the body…(8)

6. Base of leaf blades and summit of leaf sheaths usually pubescent; perigynia beak bidentate with teeth 1.6–2.2(–3.6) mm; plants rhizomatous, producing large colonies…**C. atherodes** (in part)

6′. Base of leaf blades and summit of leaf sheaths glabrous; perigynia beak bidentate with teeth (0.2–)0.3–1 mm; plants rhizomatous or cespitose…(7)

7. Perigynia bodies obconic, widest near the top, abruptly contracted to the beak; proximal pistillate spike ascending to spreading; sheath fronts papery-membranous, easily torn, not ladder-fibrillose; plants rhizomatous; extreme S New Mexico…**C. aureolensis**

7′. Perigynia bodies lanceoloid to ovoid to ellipsoid, widest at or near the middle, gradually tapering into the beak; proximal pistillate spike often pendent; sheath fronts not papery-membranous, becoming ladder-fibrillose; plants densely cespitose; widespread…**C. hystericina**

8. Perigynia bodies broadly obovoid to subspherical, 3–7-nerved, beaks 0.7–1.1 mm; widest leaf blades 8–20 mm wide…**C. amplifolia**

8'.   Perigynia bodies lanceoloid, ovoid, or ellipsoid, 6–12-nerved, beaks 0.8–2 mm; widest leaf blades 4.5–10 mm wide; widespread (*C. utriculata*) or rare (*C. vesicaria*)…(9)

9.   Ligule of lowest leaf blade no more than 1.5 times as wide as long, with rounded or emarginate apex; culms often thick, clothed in old leaf bases; sheaths thick and spongy-based, often with crosswalls between the veins, appearing like brickwork, the sheaths surrounding the thickest culms rarely reddish, the sheaths surrounding narrower culms occasionally reddish; widest leaf blades to 10 mm wide; plants colonial from long rhizomes; perigynia strongly spreading at maturity; common and widespread…**C. utriculata**

9'.   Ligule of lowest leaf blade at least twice as long as wide, with acute apex; culms often thin, not clothed in old leaf bases; sheaths thin, not spongy-based, with few or no crosswalls between the veins, not appearing like brickwork, the sheaths often reddish; widest leaf blades to 6(–8) mm wide; plants loosely cespitose from short rhizomes; perigynia ascending to spreading at maturity; rare…**C. vesicaria**

10.   Proximal pistillate spikes pendent on long, flexuous peduncles…(11)

10'. Proximal pistillate spikes erect or spreading on stiff peduncles, or sessile…(14)

11.   Perigynia bodies abruptly contracted to a long conspicuous beak, the beak 1.9–2.8 mm, subequal to or slightly longer than the body; sheath bases disintegrating into a persistent tuft of vertical "horsehair" fibers; plants to 90 cm; leaf blades to 40 cm…**C. sprengelii**

11'. Perigynia bodies gradually or abruptly tapered into a shorter beak, the beak 0.1–1.1 mm, < 1/2 the length of the body, or absent; sheath bases not disintegrating into a persistent tuft of vertical fibers; plants 5–40(–70) cm; leaf blades usually < 10 cm…(12)

12.   Perigynia surface smooth, beak 0.3–1.1 mm; inflorescence bract with sheath 4+ mm; roots without a covering of fine hairs; plants cespitose…**C. capillaris**

12'. Perigynia surface papillate, beak 0.1(–0.5) mm or absent; inflorescence bract with sheath 3 mm or less; roots with a covering of fine hairs; plants rhizomatous (but stems sometimes in small clumps)…(13)

13.   Pistillate scales (0.6–)0.9–1.9(–2) mm wide, much narrower than and thus not concealing the perigynia, conspicuously longer than the mature perigynia and giving the pistillate spike a "shaggy" appearance, deciduous; plants rhizomatous but sometimes growing in clumps; phyllopodic, culms usually with dead leaf remains at base…**C. magellanica**

13'. Pistillate scales (1.7–)1.8–2.5(–3.4) mm wide, as wide as the perigynia and concealing them, shorter to longer than the mature perigynia, not giving the pistillate spike a "shaggy" appearance, persistent; plants rhizomatous, colonial; aphyllopodic, culms without dead leaf remains at base…**C. limosa**

14.   Inflorescence with only 1–2 individual sessile perigynia; perigynia obovoid, (4.4–)4.9–6.4(–8.4) mm, beakless, with a narrow, spongy base; plants rhizomatous, of dry montane forest, woodland, and meadow habitats…**C. geyeri**

14'. Inflorescence with > 3 perigynia, on multiple spikes (sometimes a single basal spike in several species); perigynia of various shapes, without a narrow spongy base, 4.5 mm or less, with beak (except in *C. conoidea*); plant growth form and habitat various, but not both as above…(15)

15.   Perigynia 1.5–2.2 mm; inflorescence with the terminal staminate spike surpassed by 1+ pedunculate pistillate spikes; inflorescence bract reduced to a sheath only, without blade; leaf blades very fine, 0.2–0.5(–1) mm wide; known from a single location in the Sacramento Mountains…**C. eburnea**

15'. Perigynia 2.4+ mm (1.8–3 mm in *C. viridula*); inflorescence with terminal staminate spike(s) surpassing all lateral pistillate spikes; inflorescence bract with blade; leaf blades 1+ mm wide; distribution various…(16)

16.   Perigynia with impressed (sunken) nerves, beakless…**C. conoidea**

16'. Perigynia with raised nerves, beaked…(17)

17.   Proximal pistillate spike arising from the lower 1/3 of the culm; plants rhizomatous…**C. microdonta**

17'. Proximal pistillate spike arising in the upper 1/2 (usually upper 1/3) of the culm (separate basal inflorescences can be present in *C. planostachys*); plants cespitose…(18)

18.   Proximal inflorescence bract greatly exceeding the inflorescence, erect to widely divergent; pistillate spikes oblong to ovoid to globose, the perigynia very densely packed; the terminal group of spikes tightly clustered and sessile to subsessile to short-pedunculate…**C. viridula**

18'. Proximal inflorescence bract shorter than to slightly exceeding the inflorescence, usually erect to ascending; pistillate spikes cylindrical to cylindrical-oblanceoloid, the perigynia more loosely packed than in *C. viridula*; the terminal spikes not tightly clustered, or if clustered, then usually more loosely so and short-pedunculate…(19)

19.   Plants of dry habitats in S New Mexico, basal inflorescences often present; culms to 15 cm…**C. planostachys** (in part)

19'. Plants of wet to mesic habitats in N New Mexico; culms usually 15+ cm…(20)

20.   Perigynia beak evident, straight, 0.5–1.5 mm, ciliate-serrulate (at least at the base); culms with leaf blades crowded near the base, fewer above on the culm…**C. luzulina** (in part)

20'. Perigynia beak short, abruptly bent, 0.6 mm or less, glabrous; culms with leaf blades more evenly distributed along the culm…**C. blanda**

KEY D: MULTIPLE SPIKES PER CULM, STIGMAS (2)3, PERIGYNIA GLABROUS, TERMINAL SPIKE GYNECANDROUS

1. All spikes sessile in a compact head and closely overlapping one another (occasionally the basal spike slightly remote and subsessile)…(2)
1′. Lower spikes pedunculate, some shortly so; inflorescence usually moderately elongate, always with 1 to several lower spikes remote from the terminal cluster…(3)
2. Perigynia ovate to suborbicular, 1.3–2.5(–2.8) times longer than wide, contracted abruptly to a beak 0.2–0.5(–0.6) mm; spikes often not readily distinguishable…**C. nova**
2′. Perigynia narrowly elliptic to narrowly ovate, 2.8–3.9 times longer than wide, tapering to a beak (0.3–)0.5–0.8(–0.9) mm; spikes easily distinguishable…**C. nelsonii**
3. Pistillate scale apices typically with awns 0.5–2(–3.5) mm; lateral spikes sessile to short-pedunculate; plants rhizomatous; lower leaf sheath fronts light-colored, becoming ladder-fibrillose with age; rare…**C. buxbaumii**
3′. Pistillate scale apices acute to acuminate, without awns; lateral spikes short- to long-pedunculate; plants loosely to densely cespitose; lower leaf sheaths not becoming ladder-fibrillose with age; rare to common in New Mexico…(4)
4. Lowest lateral spike gynecandrous (though often with only a few staminate flowers at the base), on a peduncle longer than the upper spikes); spikes generally long and cylindrical or narrowly ellipsoid, often > 3 times as long as wide…**C. bella**
4′. Lowest lateral spike wholly pistillate (rarely gynecandrous in *C. chalciolepis*), short-pedunculate; spikes oblong, ovoid, or ellipsoid, generally < 2.5 times as long as wide…(5)
5. Perigynia 1.9–2.7 mm, usually bright green or tan, contrasting with the dark scales; pistillate scales shorter than the perigynia; plants of seeps, moist meadows, stream banks, or forests…**C. stevenii**
5′. Perigynia 2.8–4.8 mm, usually partially suffused with darker color, not contrasting strongly with the dark scales; pistillate scales usually as long as or longer than the perigynia; plants of rocky alpine or subalpine habitats…(6)
6. Pistillate scales lanceolate to ovate-lanceolate, exceeding the perigynia in the upper 1/2 of the spike, giving the inflorescence a "shaggy" appearance; achene filling approximately 1/2 of the perigynia; inflorescence often nodding in maturity…**C. chalciolepis**
6′. Pistillate scales broadly ovate, about equaling the perigynia in length, not giving the inflorescence a "shaggy" appearance; achene filling most of the perigynium; inflorescence usually remaining erect…**C. albonigra**

KEY E: MULTIPLE SPIKES PER CULM, STIGMAS 2(3), PERIGYNIA GLABROUS, TERMINAL SPIKE STAMINATE (ANDROGYNOUS) OR GYNECANDROUS

1. Plants relatively slight; culms thin and flexuous, generally < 40 cm; perigynia nearly beakless; lateral pistillate spikes to 2.8 cm; leaf blades < 3.5 mm wide; rhizomes 0.7–1(–1.8) mm thick; stigmas usually 2, but sometimes with a few flowers (up to 20%) with 3 stigmas; perigynia sometimes golden-yellow to orange…**C. aurea**
1′. Plants more robust or coarse; culms thick and upright, 7–120 cm; perigynia beaks 0.1–1.1 mm; lateral pistillate spikes 0.5–8 cm; leaf blades 1–10(–12) mm wide; rhizomes 1–6 mm thick; stigmas 2 (occasionally 3 in *C. saxatilis*); perigynia never golden-yellow to orange…(2)
2. Plants cespitose or culms densely clustered in clumps…(3)
2′. Plants with culms occurring singly or a few together connected by rhizomes…(5)
3. Basal sheaths red-brown, often shiny, ladder-fibrillose; uncommon in SW New Mexico; 4500–8500 ft.…**C. senta** (in part)
3′. Basal sheaths not red-brown, not shiny, not becoming ladder-fibrillose; N and W New Mexico (*C. kelloggii*) or nearly statewide except SW New Mexico (*C. emoryi*); 3400–12,500 ft.…(4)
4. Plants cespitose; terminal staminate spike usually 1; perigynia abruptly and narrowly stipitate; ligules never shorter than wide; free portion of ligules 0.5–2 mm; spikes dark; 7000–12,500 ft.…**C. kelloggii**
4′. Plants rhizomatous; terminal staminate spikes 2+; base of perigynia attenuate to broadly stipitate; ligules usually shorter than wide; free portion of ligules to 0.5 mm; spikes pale; 3400–8800 ft.…**C. emoryi** (in part)
5. Perigynia beaks (0.2–)0.3–1.1 mm; perigynia inflated, suborbicular in cross-section (often flattened in pressed specimens); pistillate scales dark brown to reddish black, often with a white-hyaline apex; mostly 2-styled with biconvex achenes, occasionally 3-styled with trigonous achenes; rare; over 10,000 ft.…**C. saxatilis** (in part)
5′. Perigynia beaks 0.1–0.6 mm; perigynia flattened (not inflated), not suborbicular in cross-section; pistillate scales light- to dark-colored, dark scales lacking a white-hyaline apex; 2-styled with biconvex achenes; uncommon to common; 3400–13,000 ft.…(6)
6. Pistillate scales usually with broad pale midportion with light brown marginal stripes, spikes light-colored;

ligule usually broadly U-shaped to horizontal, usually shorter than wide; sheaths pale, fronts usually red-spotted…**C. emoryi** (in part)

6'. Pistillate scales usually dark with narrow, light-colored midvein (sometimes broad in *C. aquatilis*); spikes dark-colored; ligule V- or U-shaped, as long as or longer than wide; sheaths often red-brown, fronts not red-spotted…(7)

7. Some lower pistillate scale midveins excurrent as a short awn; leaf blades 3–10(–12) mm wide; perigynia beaks often bidentate and ciliate at the apex…**C. nebrascensis**

7'. Pistillate scales lacking awns; leaf blades 1–6(–8) mm wide; perigynia beaks entire or minutely bidentate, not ciliate at the apex (sometimes ciliate in *C. scopulorum*)…(8)

8. Basal sheaths ladder-fibrillose; SW New Mexico; 4500–8500 ft.…**C. senta** (in part)

8'. Basal sheaths not ladder-fibrillose; N New Mexico; 7000–13,000 ft.…(9)

9. Inflorescence elongate, (4–)6–18(–21) cm; proximal bract longer to slightly shorter than the inflorescence; proximal spike 1–5(–7) cm…**C. aquatilis**

9'. Inflorescence congested, 2–8 cm; proximal bract shorter than the inflorescence; proximal spike 0.5–3.1 cm…**C. scopulorum**

KEY F: MULTIPLE SPIKES PER CULM, STIGMAS 2 (3 IN *C. MURICULATA*), PERIGYNIA GLABROUS, SPIKES SESSILE, TERMINAL SPIKE ANDROGYNOUS OR STAMINATE OR PISTILLATE

1. Delicate plants, culms arching, flexible; mostly < 5 perigynia per spike (6, or rarely up to 8 in *C. radiata*)…(2)

1'. Coarse plants, culms upright and stiff; mostly > 5 perigynia per spike (sometimes fewer in *C. jonesii* and *C. occidentalis*)…(3)

2. Rhizomes conspicuous; mature perigynia unequally biconvex to round in cross-section, plump, the faces conspicuously veined, the margins not serrulate, 1–4 per spike; spikes 2–4(–7)…**C. disperma**

2'. Rhizomes inconspicuous; mature perigynia plano-convex to biconvex in cross-section, not plump, the faces not or inconspicuously veined, the margins serrulate, 1–6(–8) per spike; spikes 3–5(–8)…**C. radiata**

3. Plants rhizomatous, with clear separation between small groups of 1 to several culms; distance between the culms along the rhizome mostly > 1 cm…(4)

3'. Plants mostly cespitose, forming clumps; distance between the culms along the rhizome mostly < 1 cm, rarely to 2 cm in *C. agrostoides*…(9)

4. Rhizomes flexible, 0.6–1.6 mm thick; plants of dry habitats…(5)

4'. Rhizomes stout, 1.2–3.6 mm thick; plants of wet to mesic habitats…(7)

5. Plants mostly dioecious; staminate inflorescence 7–20 mm wide, pistillate inflorescence 8–20 mm wide, appearing shaggy…**C. douglasii**

5'. Plants mostly monoecious; inflorescence 2–9 mm wide, not appearing shaggy…(6)

6. Culms to 30 cm; inflorescence compact, 0.7–2 cm; all spikes tightly overlapping and indistinguishable; perigynia 2.8–3.8 mm; beak 0.3–0.8; culms smooth just below the inflorescence…**C. duriuscula** (in part)

6'. Culms to 90 cm; inflorescence elongate, 1–3.5 cm; lowest spikes loosely overlapping and distinguishable; perigynia 3.6–6 mm; beak 1.2–2.5 mm; culms usually scabrous just below the inflorescence…**C. siccata**

7. Perigynia beaks 0.3–0.5 mm; perigynia 1.7–2.4 mm, dark reddish brown; culm bases typically pale; plants of wetter habitats frequently flooded throughout the growing season…**C. simulata**

7'. Perigynia beaks > 0.5 mm; perigynia 2–4.2 mm, brown to black at maturity; plants of springs or moist to wet habitats but usually not those that are flooded throughout the growing season…(8)

8. Culm bases typically dark brown; rhizomes typically not straight; inflorescences branched or spicate; plants usually of wetlands with saturated soils or moist areas that dry out seasonally, sometimes weedy…**C. praegracilis**

8'. Culm bases typically light brown; rhizomes typically straight; inflorescences always branched; plants typically of springs or creek banks, never weedy…**C. agrostoides** (in part)

9. Inflorescence 0.8–1.5(–2) cm, spikes and branches nearly indistinguishable…**C. jonesii**

9'. Inflorescence usually > 1.5 cm, spikes and branches more readily distinguishable…(10)

10. Beaks of perigynia light-colored and contrasting with the dark and shiny bodies; perigynia with 2 prominent raised ridges on either side of a dorsal central groove, otherwise without nerves, with a distinct hyaline flap at the top of the dorsal suture; known from a single high-elevation wetland in Rio Arriba County…**C. diandra**

10'. Beaks of perigynia the same color as the bodies; perigynia lacking 2 ridges, with or without nerves on at least 1 face, lacking a distinct hyaline flap at the top of the dorsal suture…(11)

11. Perigynia 3–5.5 mm, 2–3 mm wide; proximal leaf sheaths loose, longitudinally green-and-white-striped and with prominent green crossveins…**C. gravida**

11'. Perigynia 2.2–4.8 mm, 0.9–2 mm wide; proximal leaf sheaths tight, not or indistinctly striped, without prominent crossveins…(12)

12. Inflorescence branched, at least some spikes attached to the lower branches of the inflorescence; generally wet habitats…(13)

12'. Inflorescence unbranched, all spikes attached to the main axis of the inflorescence; generally dry habitats…(15)

13. Inflorescence bracts hairlike; perigynia 2.2–2.8 mm, sharp-margined basally…**C. vulpinoidea**

13'. Inflorescence bracts broader, not hairlike; perigynia 2.6–4.6 mm, round-margined basally…(14)

14. Culms 4–6 mm wide, with the angles winged; sheath fronts cross-rugulose; leaf blades 2–8(–11) mm wide; inflorescence robust, spiky with long, acute-tipped perigynia pointing in all directions; perigynia 3.6–4.6 mm…**C. stipata**

14'. Culms 1.5–3.5 mm wide, sharp-angled but not winged; sheath fronts not cross-rugulose; leaf blades 0.7–4(–4.4) mm wide; inflorescence softer, or if spiky then not so robust; perigynia 2.6–4.2 mm… **C. agrostoides** (in part)

15. Stigmas 3; cypselae trigonous; perigynia muricate-warty; limestone substrates; SE New Mexico… **C. muriculata** (in part)

15'. Stigmas 2; cypselae plano-convex to biconvex; perigynia smooth; various habitats, wide-ranging…(16)

16. Pistillate scales as long and wide as the perigynia, mostly concealing them; mature perigynia plano-convex but not plump, with marginal nerves usually on the margins of the body, rarely 1 nerve pushed over onto the ventral face, mostly unnerved on the faces; beak prominently (and doubly) serrulate on the margins, obviously bidentate; perigynia uniformly oriented in the spikes, giving the inflorescence a smooth appearance…**C. occidentalis**

16'. Pistillate scales usually shorter and narrower than the perigynia, the perigynia readily visible; mature perigynia plumply plano-convex, with at least 1 marginal nerve pushed over onto the ventral face, many-nerved dorsally and often ventrally; beak smooth or finely (and singly) serrulate on the margins, obliquely cleft or only slightly bidentate; some perigynia with random orientations in the spikes, giving the inflorescence a rough appearance…**C. vallicola**

KEY G: MULTIPLE SPIKES PER CULM, STIGMAS 2, PERIGYNIA GLABROUS AND UNWINGED, SPIKES SESSILE, TERMINAL SPIKE GYNECANDROUS

1. Inflorescence compact and capitate, spikes often indistinguishable…**C. illota**

1'. Inflorescence elongate, spikes easily distinguishable…(2)

2. Perigynia radiating in all directions (star-shaped) at maturity; perigynia widest near the base, with spongy, thickened tissue at the base apparent…(3)

2'. Perigynia spreading, ascending, and/or appressed, not radiating in all directions at maturity; perigynia widest well above the base, with spongy, thickened tissue at the base not apparent…(4)

3. Perigynia 2.1–3(–3.2) mm; beaks 0.3–1 mm…**C. interior**

3'. Perigynia (2.6–)2.8–3.5 mm; beaks 0.9–1.5 mm…**C. echinata**

4. Perigynia plano-convex, 2.5–3.8(–4) times longer than wide, margins with a definite edge, surfaces smooth; beaks 0.9–1.8(–2.5) mm…(5)

4'. Perigynia biconvex, (1.4–)1.6–2.7 times longer than wide, margins rounded, surfaces smooth to papillose; beaks to 1 mm…(6)

5. Ligule of distal leaf on culm about as long as wide, 0.7–2(–2.7) mm; cypselae 1.7–2.1 mm; inflorescence with 2–4(5) spikes…**C. deweyana**

5'. Ligule of distal leaf on culm much longer than wide, (2–)3.4–7.1 mm; cypselae 1.2–1.6(–2.1) mm; inflorescence with 4–6(–9) spikes…**C. bolanderi**

6. Spikes light-colored, the lower spikes well separated; pistillate scales hyaline throughout, sometimes tinged brown or brown in the middle with green midveins; perigynia dorsal sutures lacking dark coloration, or dark coloration present but not extending the entire length of the beak…(7)

6'. Spikes darker-colored, mostly strongly overlapping; pistillate scales reddish brown or light chestnut-brown with hyaline margins; perigynia dorsal sutures dark-colored for the entire length of the beak…(8)

7. Spikes (2–)3–8 mm, with 1–11 perigynia; perigynia with the dorsal suture well developed, often extending the length of the beak or more, sometimes with a white-hyaline flap…**C. brunnescens**

7'. Spikes 3–14 mm, with (2–)5–23(–30) perigynia; perigynia with the dorsal suture poorly developed, not extending the length of the beak, lacking a white-hyaline flap…**C. canescens**

8. Perigynia 1.5–2.3 mm, papillose; beaks 0.2–0.4(–0.5) mm; inflorescence with (3)4–5(6) spikes…**C. praeceptorum**

8'. Perigynia (2–)2.4–3.8 mm, smooth; beaks 0.5–1 mm; inflorescence with (1–)3(4) spikes…**C. lachenalii**

KEY H: MULTIPLE SPIKES PER CULM, SPIKES SIMILAR, STIGMAS 2, ALL SPIKES SESSILE AND GYNECANDROUS, PERIGYNIA WINGED
(section *Ovales*, excluding *C. illota*)

Sedges in section *Ovales* are among the most difficult to identify, and many species are highly variable. Hints: (A) When collecting, include at least 6 (preferably more) inflorescences to ensure good representation of the variation on a plant. (B) When determining perigynia characters, select mature perigynia from the middle 1/3 of the lower spikes and use measurements from the larger ones that are normal in shape. (C) Examine many perigynia from your specimen before deciding which way to go in the key regarding perigynia characters, but especially shape and beak characters. Many treatments discuss beaks as either terete and entire, or winged and serrulate to the tip. Tip characters have more variation than this implies, often on one spike. Some beaks that are terete are also serrulate to the tip (*C. ebenea*). (D) Discern whether the upper spikes are distinguishable by seeing if you can easily see the boundaries of the upper spikes without magnification or dissection, at arm's length. (E) Once you arrive at a tentative determination, make sure that your specimen matches the species description.

1.   Longest perigynia to 4 mm; wet areas…(2)
1'.   Longest perigynia > 4 mm; wet or dry areas…(5)
2.   Most inflorescences with 1–3 lower inflorescence bracts longer than the head (often broken off on dried specimens); broad (usually hyaline) margins of the basal portion of the lowest inflorescence bract > 1/2 the length of the lowest spike; perigynia beak tips terete and entire in the distal 0.3–0.8 mm… **C. athrostachya**
2'.   Few inflorescences with 1–3 lower inflorescence bracts longer than the head; if so, the broad (usually) hyaline margins of the basal portion of the lowest inflorescence bract < 1/2 the length of the lowest spike; perigynia winged and serrulate to the tips, or less commonly entire in distal portion (to 0.5 mm in *C. subfusca*)…(3)
3.   Upper spikes usually indistinguishable, usually tightly clustered; most perigynia apices acuminate to acute, more nearly terete, serrulate to the tip or sometimes as much as 0.5 mm of distal portion entire; widespread in N and SW… **C. subfusca** (in part)
3'.   Upper spikes usually distinguishable, clustered to moniliform; most perigynia apices acute and winged to the tip, rarely terete in the distal 0.2 mm, often serrulate to the tip; rare plants of N…(4)
4.   Lowest inflorescence internode 2–4.5 mm; upper spikes usually clustered; spikes (2–)5–9; presently known only from Sandoval County… **C. bebbii**
4'.   Lowest inflorescence internode (2–)4–9 mm; spikes moniliform; spikes 3–5(–8); presently known only from Rio Arriba County… **C. tenera**
5.   Longest perigynia to 5 mm (rarely to 5.2 mm in *C. phaeocephala*)…(6)
5'.   Longest perigynia > 5 mm…(17)
6.   Upper spikes distinguishable…(7)
6'.   Upper spikes congested and indistinguishable…(12)
7.   Perigynia < 2 times as long as wide, body suborbicular; spikes often clavate due to the large number of male flowers at the base, often rough-textured (due to prominent beaks on more widely spreading perigynia)… **C. brevior**
7'.   Perigynia > 2 times as long as wide, body various, but not orbicular; spikes not clavate (except in *C. tahoensis*), smooth-textured (perigynia more appressed than in the former)…(8)
8.   Plants found at or above tree line, of dry places; culms to (5–)15–45 cm; perigynia tips hyaline… **C. phaeocephala**
8'.   Plants found below tree line (rarely up to tree line in *C. tahoensis*), of wet to dry places; culms to 120 cm; perigynia tips hyaline or not…(9)
9.   Spikes 3–4(–6); cypselae nearly filling the middle 1/3 of the perigynia; culms loosely clustered, arising from short, linear rhizomes; dry places… **C. tahoensis** (in part)
9'.   Spikes (1–)3–10(–14); cypselae mostly filling the lower 1/2 of the perigynia; culms densely cespitose; wet places (*C. praticola* might be found in dry places)…(10)
10.   Most perigynia winged to the tip; pistillate scale apices acute to acuminate; moist areas < 8500 ft…. **C. scoparia** (in part)
10'.   Most perigynia terete at the tip, terete portion 0.4–1 mm; pistillate scale apices acute to obtuse; mostly wet areas > 8500 ft….(11)
11.   Perigynia 3.6–4.5 mm; spikes 5–9(–11) mm; inflorescence 0.6–2.2 cm… **C. pachystachya** (in part)
11'.   Perigynia (3.7–)4.5–6 mm; spikes 8.5–20 mm; inflorescence (1.7–)2.5–5 cm… **C. praticola**
12.   Culms to 25(–40) cm, often arching; distance from the beak tips to the tops of the cypselae (2.3–)2.6–3.8 mm; alpine tundra, > 11,800 ft…. **C. haydeniana**
12'.   Culms to 120 cm, rarely arching; distance from the beak tips to the tops of the cypselae (1.2–)1.5–3 mm; plants in forested areas or meadows < 11,800 ft., rarely higher…(13)
13.   Perigynia dorsal suture long-white-hyaline, tips hyaline; scales dark brown, sometimes with a metallic sheen, hyaline margins of middle scales 0.2–0.5 mm wide… **C. macloviana** (in part)
13'.   Perigynia dorsal suture not long-white-hyaline, tips brown; scales tan to brown, without a metallic sheen, hyaline margins of middle scales 0.1–0.3(–0.5) mm wide…(14)
14.   Perigynia plano-convex, narrowly wing-margined, 0.45–0.6 mm thick…(15)

14'. Perigynia flat except where distended over the achene to thinly plano-convex, broadly wing-margined, 0.3–0.5 mm thick…(16)

15. Perigynia 3.6–4.5 mm, 1.1–2 mm wide, slightly spreading when mature, veined or not, sometimes with coppery sheen, most perigynia beaks entire in distal 0.3–0.6 mm, sometimes a few serrulate to tips… **C. pachystachya** (in part)

15'. Perigynia 2.4–4(–4.3) mm, 0.9–1.2(–1.5) mm wide, ascending when mature, conspicuously veined, never with coppery sheen, most perigynia beaks serrulate to tips, sometimes up to 0.5 mm of distal portion entire…**C. subfusca** (in part)

16. Lowest inflorescence internode to 3 mm, inflorescence (0.7–)1.1–2.4, the base truncate to retuse; perigynia usually widest below the top of the achene…**C. microptera** (in part)

16'. Lowest inflorescence internode usually > 4 mm, inflorescence (1.5–)2–3.6 cm, the base tapered; perigynia usually widest at the top of the achene…**C. aff. microptera** (in part)

17. Upper spikes distinguishable; most perigynia winged to tip…(18)

17'. Upper spikes indistinguishable; perigynia not winged to tip (usually winged to tip in *C. wootonii*)…(23)

18. Perigynia not thick, flat except over the achene…(19)

18'. Perigynia thick, plano-convex, with cypselae more nearly filling the perigynia…(21)

19. Perigynia 3.3–5.4(–6.8) mm, (1–)1.2–2.2 mm wide; pistillate scale apices acuminate to acute; spikes (1–)3–9(–13), (5–)6–12(–16) mm; moist areas, 5000–8100 ft.… **C. scoparia** (in part)

19'. Perigynia (4.8–)5.2–7(–7.2) mm, (1.6–)1.8–3.5(–3.8) mm wide; pistillate scale apices acute; spikes 2–7(8), 8–21 mm; dry areas, 6700–11,000 ft.…(20)

20. Perigynia up to 2.3 times as long as wide, (2.2–)3–3.5(–3.8) mm wide; spikes 8–15 mm; proximal inflorescence internode (2–)4–5 mm…**C. egglestonii**

20'. Perigynia > 2.1 times as long as wide, (1.6–)1.8–2.9(–3) mm wide; spikes 9–21 mm; proximal inflorescence internode (2–)4–10 mm…**C. wootonii** (in part)

21. Perigynia usually nerveless ventrally, usually winged to the tips; spikes obovoid, with few staminate flowers at the base of each spike; hyaline margins on pistillate scales 0–0.5 mm wide…**C. wootonii** (in part)

21'. Perigynia usually nerved ventrally, winged to the tips or not; spikes narrowly fusiform, lanceoloid, to oblanceoloid, often with many staminate flowers at the base of each spike; hyaline margins on pistillate scales 0.2–0.8 mm wide…(22)

22. Perigynia 6–8.4 mm; distance from perigynia beak tip to top of the achene (1.8–)3.2–4.6 mm; pistillate scales shorter and narrower than the perigynia, or equal to the perigynia, pale green to light brown with a pale or green midvein…**C. petasata**

22'. Perigynia 4–4.8(–6.2) mm; distance from perigynia beak tip to top of the achene 1.7–2.2(–2.6) mm; pistillate scales usually covering the perigynia, bronze with a straw to tan midvein…**C. tahoensis** (in part)

23. Perigynia (1.6–)1.8–2.9(–3) mm wide; spikes 3–7(8), 9–21 mm…**C. wootonii** (in part)

23'. Perigynia 1.1–1.8 mm wide; spikes 5–12(13), 6–15 mm…(24)

24. Spikes 9–15 mm; proximal inflorescence internode usually 1–3 mm, rarely to 5 mm; perigynia (3.5–)5.3–7.1 mm; dry places, > 9000 ft.…**C. ebenea**

24'. Spikes 6–10 mm; proximal inflorescence internode usually at least (3–)4 mm; perigynia (3.5–)4–5.6 mm; wet places, 5700–9700 ft.…**C. aff. microptera** (in part)

Carex agrostoides Mack. [*C. chihuahuensis* Mack.]—Grassleaf sedge. Plants rhizomatous, often densely cespitose, rhizomes thick, straight, smooth, brown; culms 20–110 cm; leaf blades 0.7–4(–4.4) mm wide; inflorescence dense to elongate, of several upper spikes sessile on the rachis and on (1)2–9 lower multiple-spiked branches, spikes androgynous, pistillate scales similar in size to the perigynia, bronze with a green midvein and hyaline margins; perigynia ovate to lanceolate, plano-convex in cross-section, 2.6–4.2 × 1–1.8 mm, green or pale, turning tan or black at maturity, weakly to strongly nerved, margins rounded, base spongy and truncate to cordate, beak 0.6–0.8 mm; stigmas 2; cypselae 1.5–1.8 mm. Mountain springs and stream banks, and ciénegas in the plains, 4600–6200 ft. WC, SW.

Carex albonigra Mack.—Black-and-white sedge. Plants loosely cespitose from short rhizomes; culms 10–31 cm; leaf blades flat, 1.5–4.5(–5) mm wide, sheaths pale green to purplish red; inflorescence compact, of 2–3(4) similar-looking spikes, terminal spike gynecandrous, lateral spikes pistillate, short-pedunculate, erect; proximal bract shorter than or equal to the inflorescence; pistillate scales ovate, slightly shorter to longer than and as wide as the perigynia, brown to black, margins whitish-hyaline, tip obtuse; perigynia 2.8–3.3 × 1.1–1.9 mm, papillose, brown to black, sometimes light-colored proximally, ascending, nerveless, elliptic to obovate, beak 0.1–0.4 mm; stigmas 3; cypselae 1.6–1.7(–1.9) mm, nearly filling the body of the perigynia. Alpine, 11,400–12,600 ft. Colfax, Otero, Taos.

Carex amplifolia Boott—Bigleaf sedge. Plants forming large clumps from stout rhizomes; culms 50–90(–100) cm; leaf blades (5–)7–15(–20) mm wide, sheaths pale green to reddish, sometimes red-spotted; inflorescence elongate, of 1–2(3) terminal staminate spikes and 3–5(–7) lateral pistillate spikes, spikes subsessile to pedunculate; pistillate scales shorter and narrower than the perigynia, mucronate to awn-tipped; perigynia 2.4–3.1(–3.6) × 1.3–1.7 mm, subinflated, brownish green, with 3–7 nerves, obovoid to subspherical and contracted to a minutely bidentate beak, beak 0.7–1.1 mm, sometimes bent; stigmas 3; cypselae 0.9–1.9 mm. Wet areas in coniferous forests, 7800–8800 ft. Grant, Sierra.

Carex aquatilis Wahlenb.—Water sedge. Plants colonial from long rhizomes, rhizomes stout, scaly, 1.4–3.5 mm thick; culms 20–120 cm, arising singly or a few together; leaf blades 2.5–6 mm wide, sheaths brown or reddish, not ladder-fibrillose; inflorescence elongate, of 1–2(–4) terminal staminate spikes and 2–4(–7) lateral pistillate spikes; pistillate scales shorter to longer and often narrower than the perigynia, ovate to lanceolate, dark with pale midvein or none; perigynia elliptic or obovate, 2–3.6 × 1.3–2.3 mm, dull, papillose, unveined, beak 0.1–0.4 mm; stigmas 2; cypselae 1–1.8 mm. New Mexico material belongs to **var. aquatilis**. Stream banks, fens, wet meadows, 7000–12,100 ft. NE, NC, Catron.

Carex atherodes Spreng.—Wheat sedge. Plants colonial from long rhizomes, forming large stands; culms 30–110(–125) cm; leaf blades 3–9 mm wide, lower surface pubescent throughout or near the ligule, sheaths reddish purple to black, ladder-fibrillose; inflorescence elongate, of (1)2–8 terminal staminate spikes (rarely, 1–4 of these androgynous) and 1–4 remote, lateral pistillate pedunculate spikes, proximal bract leaflike, exceeding the inflorescence; pistillate scales shorter or longer than the perigynia, lanceolate to ovate, with a scabrous awn 1.2–6 mm; perigynia 6–8.2(–9) × (1–)1.3–1.8 mm, body lanceoloid, green to straw-colored, with many strong nerves, beak (2.1–)2.5–4.2 mm, bidentate with spreading teeth 1.6–2.2(–3.6) mm; stigmas 3; cypselae 2.8–3.1 mm. Floodplains and seasonally flooded wetlands, 5300–9300 ft. Rio Arriba, Sandoval.

Carex athrostachya Olney—Slenderbeak sedge. Plants cespitose; culms (5–)20–60(–80) cm; leaf blades 1.5–4(–5) mm wide; inflorescence a compact head of (4–)7–10(–15) sessile, similar, gynecandrous spikes, proximal bract much longer than the inflorescence; pistillate scales narrower and shorter than the perigynia, golden to reddish brown with pale center and hyaline margins; perigynia 2.8–4 × (0.8–)1–1.5(–1.8) mm, elliptic to ovate, tapering to a broad stipitate base, beak terete, entire in distal 0.3–0.8 mm, green to light brown, nerved dorsally and sometimes ventrally; stigmas 2; cypselae 1.2–1.6 mm. Wetlands, seasonally flooded ponds, depressions, streams, 6000–9900 ft. NC, NW, WC, Otero.

Carex aurea Nutt.—Golden sedge. Plants rhizomatous to loosely cespitose from long, slender rhizomes, rhizomes 0.7–1(–1.8) mm thick; culms 5–41 cm; leaf blades 1.3–3.5 mm wide, sheaths cream to brown; inflorescence of 1 terminal staminate (rarely gynecandrous) spike and 1–4(–7) lateral pistillate spikes, the proximal on long peduncles, sometimes also with pistillate spikes on long peduncles from basal or lower leaf axils; proximal bract leaflike, longer than the inflorescence, sheath (2–)3–15(–20) mm; pistillate scales about as wide and up to as long as the perigynia, brownish with pale midvein and hyaline margins, tip obtuse to cuspidate; perigynia globose to obovoid, (1.4–)2–2.8(–3.1) × 0.9–1.9 mm, pale green maturing to yellow-orange, veined or veinless, nearly beakless; stigmas 2(3); cypselae 1.3–1.8(–2) mm. Wet meadows, seepage slopes, springs, marginal to sluggish streams, 5500–11,600 ft. NC, NW, WC, SC, San Miguel.

Carex aureolensis Steud.—Goldenfruit sedge. Plants colonial from long rhizomes, forming large stands to loose clumps; culms (20–)30–65(–85) cm; leaf blades (2–)4–9 mm wide, scabrous, sheaths purplish red, septate-nodulose, sheath fronts papery-membranous; inflorescence elongate, composed of 1(2) terminal staminate spike(s) and 3–5 gynecandrous (rarely pistillate) lateral spikes, proximal bract much longer than the inflorescence; pistillate scales (0.4–)0.6–1.2 mm wide, shorter than the perigynia, hyaline with green midrib extending into a long, scabrous awn; perigynia (3.2–)4.3–5.4(–5.6) × 1.3–2(–2.5) mm, obconic, brown proximally, green to straw-colored distally, many-nerved, inflated, beak (1.2–)1.5–2.8 mm, bidentate with teeth 0.3–1 mm; stigmas 3; cypselae (1.2–)1.9–2.9 mm. Mountain springs and streams, 4300–5600 ft. Eddy, Luna.

Carex bebbii (L. H. Bailey) Olney ex Fernald—Bebb's sedge. Plants cespitose; culms (14-)20-67 (-90) cm; leaf blades 1-4 mm wide; inflorescence compact to elongate, of (2-)5-9 sessile, similar, gynecandrous spikes, proximal bract setaceous, shorter than the inflorescence; pistillate scales narrower and shorter than or equal to the perigynia, lanceolate to oblong-lanceolate, pale to reddish brown with pale or green midvein, sometimes with narrow hyaline margins; perigynia (1.9-)2.7-3.2 × 0.9-1.3(-1.6) mm, ovate or elliptic, light to dark reddish brown, nerveless or finely nerved dorsally, nerveless ventrally or with several nerves at the base only, beak winged and serrulate to the tip; stigmas 2; cypselae 1-1.4 mm. Wet meadows, stream banks, 7900-8800 ft. Sandoval.

Carex bella L. H Bailey—Southwestern showy sedge. Plants cespitose from short rhizomes; culms 15-64(-75) cm; leaf blades flat, (0.8-)1.2-4.1(-6) mm wide, sheaths green to dark reddish brown; inflorescence elongate, of (1)2-5 cylindrical gynecandrous spikes, lower spikes with progressively fewer staminate flowers and longer peduncles, mature lower spikes often drooping, proximal bract shorter than the inflorescence; pistillate scales ovate, shorter than the perigynia, dark reddish brown with a lighter midvein, tip acute, or extending as an awn to 0.6 mm; perigynia 2.4-3.2(-3.6) × 1.2-2(-2.4) mm, light green or beige, sometimes mottled, elliptic or obovate, ascending, nerveless or obscurely several-nerved, apex rounded, beak 0.1-0.5(-0.6) mm; stigmas 3; cypselae 1.8-2.6 mm, nearly filling the body of the perigynia. Montane coniferous forests to alpine, 7600-12,200 ft. NC, WC, SC, Lincoln.

Carex blanda Dewey—Eastern woodland sedge. Plants densely cespitose; culms 14-53 cm; leaf blades 1-10 mm wide; inflorescence elongate, of 1 terminal staminate spike and 3(4) lateral pistillate spikes, spikes sessile to pedunculate, proximal bract leaflike, longer than the inflorescence, sheathing > 4 mm; pistillate scales shorter than perigynia, hyaline with a green midvein extending as an obvious awn; perigynia 4-18 per spike, > 20-veined, obovoid to ellipsoid-obovoid, 2.5-3.8(-4.1) × 1.3-2.2 mm, beak 0.1-0.6 mm, abruptly bent; stigmas 3; cypselae obovoid, 2.1-3.2 mm. Wide range of habitats, often weedy; only 1 location in New Mexico, possible waif, 6400 ft. Rio Arriba.

Carex bolanderi Olney [*C. deweyana* Schwein. var. *bolanderi* (Olney) W. Boott]—Bolander's sedge. Plants cespitose; culms 20-75(-115) cm; leaf blades 0.6-3.5(-5.9) mm wide, ligule of distal leaf (2-)3.4-7.1 mm, much longer than wide; inflorescence elongate, of 4-6(-9) sessile, gynecandrous or pistillate spikes, proximal bract setaceous, shorter than the inflorescence; pistillate scales about as long as the body of the perigynia but beaks exposed, light-colored to hyaline with a greenish midvein, margins wide, hyaline, apex acuminate to short-awned; perigynia 2.8-3.8(-5.2) × 1.1-1.4 mm, 2.5-3.5(-4) times longer than wide, lanceolate, thin, translucent, beak 1.1-1.7(-2.5) mm, apex bidentate, teeth 0.1-0.6(-1) mm, stigmas 2; cypselae 1.2-1.6(-2.1) mm. Wetlands in coniferous forest, 6700-8800 ft. Catron, Sandoval, Sierra.

Carex brevior (Dewey) Mack.—Shortbeak sedge. Plants cespitose; culms 20-70 cm; leaf blades 1-2.8 mm wide; inflorescence compact to elongate, of 3-6(-10) sessile, similar, gynecandrous, clavate spikes, proximal bract scalelike or setaceous-prolonged, shorter than the inflorescence; pistillate scales narrower and shorter than the perigynia, pale yellow-brown with green midvein (that fades in time) and narrow hyaline margins; perigynia (2.6-)3.4-4.6 × (2-)2.2-3 mm, < 2 times as long as wide, body nearly orbicular, green to tan, nerveless ventrally and sometimes nerved dorsally; stigmas 2; cypselae 1.5-2.2 mm. Wet meadows, swales, springs, seeps, stream banks, lakeshores, 5400-8200 ft. NC, Doña Ana.

Carex brevipes W. Boott ex B. D. Jacks. [*C. deflexa* Hornem. var. *boothii* L. H. Bailey; *C. rossii* Boott var. *brevipes* (W. Boott ex B. D. Jacks.) Kük.]—Short sedge. Plants cespitose; culms 5-17(-31) cm; leaf blades 0.9-2.4(-3.2) mm wide, old sheaths weakly fibrous; inflorescence both basal and cauline, basal inflorescences of 1-2 pistillate spikes on a slender peduncle, cauline inflorescences of 1 terminal staminate spike and 1-4 lateral pistillate spikes, pistillate spikes each with 1-5 perigynia, the proximal bract leaflike, shorter to longer than the inflorescence; pistillate scales shorter than the perigynia; perigynia 2.3-3.4(-4.2) × 0.8-1.4 mm, with 2 marginal nerves, pubescent, obovoid to ellipsoid, abruptly contracted to the beak, beak often bent, with margins ciliate-serrulate; stigmas 3; cypselae 1.3-1.8 mm. Alpine rock fields and tundra at or above timberline, 11,800-12,400 ft. Mora, Taos.

Carex brunnescens (Pers.) Poir.—Brownish sedge. Plants cespitose; culms 10–60 cm; leaf blades 0.7–2.5 mm wide; inflorescence compact to elongate, of 3–10 sessile, easily distinguished gynecandrous spikes, proximal bract setaceous, shorter than the inflorescence; pistillate scales shorter than the perigynia, white-hyaline, often brown in the middle with a green midvein; perigynia 1.9–2.7 × (0.7–)0.9–1.2(–1.5) mm, 1.6–2.5 times longer than wide, smooth to papillose, biconvex, ovate to elliptic, nerved on both sides, marginal nerves prominent, smooth to papillose, green or brown, dark brown in age, beak 0.2–0.7 mm, serrulate, with a long dorsal suture usually the same color as or lighter than the perigynia, sometimes with a hyaline flap, usually extending the length of the beak; stigmas 2; cypselae 1.3–1.5 mm wide. Our material belongs to **subsp. brunnescens**. Boggy wetlands, 8400–8700 ft. Sandoval.

Carex buxbaumii Wahlenb.—Buxbaum's sedge. Plants rhizomatous, colonial; culms 25–45(–75) cm; leaf blades 1.5–3(–4) mm wide, sheaths reddish to dark brown, sheath fronts light-colored, ladder-fibrillose; inflorescence elongate, of 2–4(5) oblong-ovoid spikes, terminal spike gynecandrous, lateral spikes pistillate, sessile to short-pedunculate, proximal spike remote, proximal bract shorter to longer than the inflorescence; pistillate scales longer than the perigynia, lanceolate to ovate-lanceolate, brown to reddish black with a pale center, midvein raised, awned tip to 0.5–2(–3.5) mm; perigynia 2.5–3(–4) × 1.2–1.6(–2) mm, papillose, pale green, elliptic to obovate, smooth to strongly nerved on faces, rounded to tapered at top, beak to 0.2 mm; stigmas 3; cypselae 1.5–2.2 mm, nearly filling the body of the perigynia. Wet meadows and bogs, 10,300 ft. Rio Arriba.

Carex canescens L.—Silvery sedge. Plants cespitose; culms 15–40(–60) cm; leaf blades 1–3(–4) mm wide; inflorescence elongate, of 3–7(8) sessile, easily distinguished gynecandrous spikes, proximal bract setaceous above, shorter than the inflorescence; pistillate scales ovate, slightly shorter than and about the same width as the perigynia, hyaline, sometimes tinged brown or green, midvein green or light-colored; perigynia (1.6–)1.8–2.4(–2.7) × (0.7–)0.9–1.2(–1.3) mm, (1.4–)1.6–2.7 times longer than wide, papillose, elliptic-ovate, biconvex, usually pale green to yellowish green or brown, nerved on both faces, papillose, beak blunt, 0.1–0.5(–0.6) mm, dorsal suture inconspicuous; stigmas 2; cypselae 1–1.5 mm. New Mexico material belongs to **subsp. canescens**. Wetlands, 8000–12,000 ft. NC.

Carex capillaris L.—Hairlike sedge. Plants cespitose, culms 5–35(–60) cm; leaf blades 0.7–3 mm wide; inflorescence elongate, of 1 terminal staminate spike and 2–4 lateral pistillate spikes, staminate spike erect or flexuous, longer than or occasionally shorter than the upper lateral spikes, proximal pistillate spikes pendent on long, flexuous peduncles, proximal bract shorter to longer than the inflorescence, sheath 4+ mm; pistillate scales pale to medium brown with hyaline margins, shorter and narrower to wider than the perigynia; perigynia (2.3–)2.5–3(–3.5) × 0.6–1(–1.2) mm, ellipsoid to lanceoloid-ellipsoid, pale green to light or dark brown, sometimes shiny, 2-nerved, base stipitate, apex acute, beak 0.3–1.1 mm, smooth or serrulate; stigmas 3; cypselae (1.2–)1.3–1.7 mm. Wetlands, 8500–12,500 ft. NC, Otero.

Carex chalciolepis Holm [*C. heteroneura* W. Boott var. *chalciolepis* (Holm) F. J. Herm.; *C. atrata* L. var. *chalciolepis* (Holm) Kük.]—Holm's sedge. Plants cespitose from short rhizomes; culms 10–60(–75) cm, often drooping; leaf blades (0.8–)1.2–6 mm wide, sheaths tan to purplish red; inflorescence compact, often nodding, of (1)2–4(5) overlapping, similar-looking spikes, terminal spike gynecandrous, lateral spikes pistillate, lowest rarely gynecandrous, short-pedunculate, often remote; pistillate scales to 1+ mm longer than the perigynia, lanceolate to ovate-lanceolate, dark reddish brown to purplish black throughout or with a lighter midvein, often with hyaline margins, tip acute; perigynia 2.9–4.8 × 1.5–2.4 mm, papillose, flattened but distended by achene, dark red-brown to purplish black, or with marginal portions bright green or tan, broadly ovate, nerveless, beak 0.2–0.4(–0.5) mm; stigmas 3; cypselae (1.5–)1.8–2.1(–2.5) mm, filling proximal 1/2 or less of the perigynia. Subalpine and alpine meadows, rock fields, 10,300–13,000 ft. NC, Otero.

Carex conoidea Willd.—Openfield sedge. Plants cespitose from short rhizomes; culms (2–)8–35(–75) cm; leaf blades 1.8–3.2(–5.6) mm wide; inflorescence elongate, of 1 terminal staminate spike and 1–3 lateral pistillate spikes, sessile to pedunculate, rarely also with spikes on long peduncles from basal or lower leaf axils, lateral peduncles scaberulous, proximal bract longer than the inflorescence, sheath

4+ mm; pistillate scales silvery to reddish brown with a green midrib, shorter to slightly longer than the perigynia, entire, apex frequently awned, the awn (0.1–)0.7–2.2(–2.7) mm; perigynia 2.4–3.6(–4.3) × 1.2–1.8 mm, oblong-ovoid to oblong-obovoid, yellowish or brownish green, with 16–20(–25) impressed nerves, beak to 0.2 mm; stigmas 3; cypselae 1.8–2.6 mm. Wetlands, 8600–8800 ft. Mora, Sandoval.

Carex deweyana Schwein.—Dewey's sedge. Plants cespitose; culms (9–)14–52(–89) cm; ligule of distal leaf 0.7–2(–2.7) mm, about as long as wide; leaf blades 1.1–3.1(–4.2) mm wide; inflorescence elongate, of 2–4(5) sessile, easily distinguished, gynecandrous or pistillate spikes, proximal bract setaceous, shorter than the inflorescence; pistillate scales ovate to oblong-ovate, hyaline, midrib green, apex acuminate to short-awned; perigynia (3–)3.6–4.6 × 1.1–1.5(–1.7) mm, 2.5–3.8 times longer than wide, narrowly ovate to narrowly elliptic, pale, translucent, beak serrulate, 0.9–1.8 mm, apex bidentate with teeth 0.1–0.5(–0.7) mm; stigmas 2; cypselae 1.7–2.1 mm. Our plants are **var. deweyana**. Moist coniferous forest, 6500–9100 ft. NC, Catron.

Carex diandra Schrank –Lesser panicled sedge. Plants densely cespitose from short rhizomes; culms 20–100 cm; widest leaf blades 1–3.1 mm wide, sheath fronts red-spotted; inflorescence dense to elongate, of several upper spikes sessile on the rachis and on (0)1–7 lower multiple-spiked branches, spikes androgynous, proximal bract absent or short-cuspidate; pistillate scales similar in size to the perigynia, straw-colored to brownish with a pale midvein, the margins hyaline; perigynia deltoid-ovoid, (2–)2.3–2.8(–2.9) × 1–1.4 mm, green to dark chestnut-brown to black, with 2 raised ridges on either side of a dorsal central groove, shiny, with a hyaline flap at the top of the dorsal suture, beak 1–1.3 mm, often light-colored and contrasting with the dark body; stigmas 2; cypselae 1–1.4 mm. Wetlands, 8000–10,400 ft. Rio Arriba.

Carex disperma Dewey [*C. tenella* Schkuhr]—Softleaf sedge. Plants fine, rhizomatous, loosely cespitose; culms 6–60 cm, slender, arched or nodding; leaf blades 0.7–1.8(–2.5) mm wide, sheaths pale brown; inflorescence elongate, of 2–4(–7) sessile, globose, androgynous spikes, spikes with 1–4 perigynia, lowermost spikes well separated, uppermost more contiguous, proximal bract setaceous above, shorter than the inflorescence; pistillate scales shorter than to as long as the perigynia, hyaline with a prominent pale midvein; perigynia ellipsoid to obovoid, unequally biconvex in cross-section, 2.4–3.6 × 1–1.7 mm, pale green maturing to olive-green, becoming shiny dark brown, both faces many-nerved, broadly stipitate, beak 0.2–0.5 mm; stigmas 2; cypselae (1–)1.5–1.75(–2) mm. Wetlands, mossy and shady coniferous woods, 7400–11,600 ft. NC, Catron, Sierra.

Carex douglasii Boott—Douglas' sedge. Plants rhizomatous, forming dioecious colonies, rhizomes 0.8–1.4 thick, brown; culms 4–25 cm, arising singly or several from rhizomes, smooth below the inflorescence; leaf blades to 3 mm wide, sheaths brown; inflorescence compact, dense, unisexual, or with some opposite-gender flowers, the lower branch(es) usually with multiple sessile spikes, staminate inflorescence lanceoloid; pistillate inflorescence subglobose to fusiform, proximal bract shorter than the inflorescence; pistillate scales longer and wider than the perigynia; perigynia elliptic to ovate, plano-convex in cross-section, (3–)3.5–4.2(–4.8) × 1.2–2.1 mm, straw-colored to brown, nerveless, stipitate, beak 1.6–1.9 mm; stigmas 2; cypselae (1.4–)1.6–2 mm. Seasonally flooded depressions, grasslands to forests, 6600–10,000 ft. NC, C.

Carex duriuscula C. A. Mey. [*C. eleocharis* L. H. Bailey; *C. stenophylla* auct. non Wahlb.; *C. stenophylla* Wahlb. var. *eleocharis* (L. H. Bailey) Breitung]—Needleleaf sedge. Plants rhizomatous, colonial, the rhizomes 0.6–1.4 mm thick, brown, fibrillose; culms 6–30 cm, bluntly to sharply trigonous (though smooth) below the inflorescence; leaf blades 0.3–1.5(–2.3) mm wide, sheaths fibrillose; inflorescence compact, of 3–8 sessile androgynous spikes, or the plants unisexual in a clone, often infected with smut, then producing perigynia that are globose, smooth, and gray, spikes indistinguishable, proximal bract shorter than the inflorescence; pistillate scales as long as or longer than the perigynia, 2.4–4.1 mm, straw to dark reddish brown, margins hyaline, midvein pale, apex acuminate or with a protruding midvein; perigynia elliptic-ovate to nearly orbicular, 2.8–3.8 × 1.2–1.6 mm, plano-convex, greenish to dark reddish brown, black at maturity, shiny, often with stiff short hairs on the upper portion of the dorsal face, beak

0.3–0.8 mm; stigmas 2; cypselae 1.5–2.1 mm. Dry prairies, grasslands, openings in woodlands, 6300–11,500 ft. NC, NW, WC.

Carex ebenea Rydb.—Ebony sedge. Plants cespitose; culms 10–50 cm; leaf blades 2–3.5 mm wide; inflorescence a compact head of 5–12 similar, sessile, gynecandrous spikes, proximal bract shorter than the inflorescence; pistillate scales narrower and shorter than the perigynia, dark brown, sometimes with a green or gold midvein and narrow hyaline margins; perigynia (3.5–)5.3–7.1 × 1.1–1.5(–1.7) mm, lanceolate to ovate, green to coppery- or golden-brown, usually nerved on both sides, often serrulate to the tips; stigmas 2; cypselae 1.5–1.9 mm. Subalpine and alpine meadows, openings in coniferous forests, edges of wetlands and lakes, talus, 9000–13,300 ft. NC, Lincoln.

Carex eburnea Boott—Bristleleaf sedge. Plants cespitose; culms (12–)15–25(–30) cm; leaf blades wiry, involute, 0.2–0.5(–1) mm wide; inflorescence of 1 terminal staminate spike and 2–4 pistillate spikes, 1+ of the long-pedunculate pistillate spikes overtopping the short-peduncled central staminate spike, proximal bract bladeless, sheathing only; pistillate scales white-hyaline with green or brown midrib, shorter than or equal to the perigynia; perigynia light green becoming dark brown, 1.5–2.2 × 0.7–1.1 mm, beak 0.2–0.4(–0.5) mm; stigmas 3; cypselae 1.5 mm. Wet, spring-fed meadows in coniferous forests, 8800–8900 ft. Otero.

Carex echinata Murray—Star sedge. Plants cespitose; culms 9–35(–60) cm, wiry; leaf blades (0.4–)1.2–2(–2.5) mm wide; inflorescence elongate, of 2–6 sessile, easily distinguished gynecandrous spikes, terminal spike clavate with a prominent narrow staminate basal portion, lateral spikes sometimes pistillate, proximal internode 2–7(–8) mm, proximal bract scalelike or setaceous, shorter than the inflorescence; pistillate scales inconspicuous, ovate, mostly hyaline, brown in the middle and shorter than the perigynia; perigynia (2.6–)2.8–3.5 × 0.9–1.5 mm, (1.9–)2.1–2.9 times as long as wide, plano-convex, spreading in age, giving the spikes a starlike appearance, green to dark brown when mature, broadest slightly above the base, base plump with spongy tissue, beak serrulate, 0.9–1.5 mm, tip nearly entire to bidentate to 0.3 mm; stigmas 2; cypselae 1.3–1.7 mm. Our plants belong to **subsp. echinata**. Wetlands, 9000–10,800 ft. Rio Arriba, Sandoval, Santa Fe.

Carex egglestonii Mack.—Eggleston's sedge. Plants cespitose; culms 30–80 cm; leaf blades (1.5–)2.7–4(–6) mm wide; inflorescence compact, of 2–5(6) similar, distinguishable, sessile, gynecandrous spikes, proximal bract shorter than the inflorescence; pistillate scales narrower and shorter than the perigynia, ovate or ovate-lanceolate, gold to brown, sometimes with a green or gold midvein, the apex acute to acuminate; perigynia 5.2–5.9(–7.2) × (2.2–)3–3.5(–3.8) mm, 1.8–2.3 times as long as wide, broadly ovate, green to gold, not nerved or sometimes up to 6-nerved dorsally, beak winged to tip; stigmas 2; cypselae 1.5–2.4 mm. Subalpine and alpine meadows, steep grassy slopes, talus, 9700–9800 ft. Rio Arriba.

Carex elynoides Holm—Blackroot sedge. Plants densely cespitose, forming mats; culms 4–12(–20) cm, mostly smooth, bases crowded with old sheaths; leaf blades wiry, 0.2–0.6(–0.8) mm wide; inflorescence a single terminal androgynous spike, staminate portion mostly longer than or equal to the pistillate portion, bract absent; pistillate scales ovate or obovate, longer and wider than the perigynia, brown, with broad hyaline margins, midvein pale, tip obtuse; perigynia 2–3.5 × 1–2 mm, ovoid to obovoid, straw to yellowish green or cream-colored, uppermost part of perigynia brown to dark brown, nerveless, glabrous to sparsely hirsute and/or ciliate on the upper portion, beak hyaline, (0.2–)0.4–1 mm; stigmas 3; cypselae 1.8–2.6(–3) mm. Alpine, 11,100–13,200 ft. NC, SC, Lincoln.

Carex emoryi Dewey [*C. stricta* Lam. var. *emoryi* L. H. Bailey]—Emory's sedge. Plants in tufts of several culms connected by long rhizomes, rhizomes 1.5–3.5 mm thick; culms 12–120 cm; leaf blades 2–4(–6) mm wide, ligule usually U-shaped to horizontal, usually wider than long, free portion to 0.5 mm, sheaths straw to brown, not ladder-fibrillose; inflorescence elongate, of 2–3 terminal staminate spikes and 3–5 lateral pistillate spikes, short-peduncled, proximal bract leaflike, shorter than or equal to the inflorescence; pistillate scales shorter to longer and narrower than the perigynia, oblong, with light brown margins and a broad pale midportion, usually spring-flowering, with perigynia and pistillate scale deciduous by midsummer; perigynia ovate to elliptic, 1.7–3.2 × 1–2.1 mm, green to straw, papillose, veinless or up to 5-veined on each face; beak 0.1–0.4 mm, entire; stigmas 2; cypselae 1.5 mm. Banks of slow-moving rivers, streams, floodplain meadows, 3400–8800 ft. NE, NC, NW, C, DeBaca, Doña Ana.

Carex filifolia Nutt.—Threadleaf sedge. Plants densely cespitose, forming tufts; culms 6–30 cm, mostly smooth, bases crowded with old sheaths; leaf blades wiry, (0.2–)0.3–0.8 mm wide; sheaths straw to yellow-brown, sharply contrasting with the green leaves, ladder-fibrillose; inflorescence a single terminal androgynous spike, staminate portion mostly longer than the pistillate portion, bract absent; pistillate scales ovate to obovate, wider and mostly longer than the perigynia, reddish to yellowish brown, with broad hyaline margins, midvein pale, apex rounded or acute; perigynia 3.1–5 × 1.3–2.1(–2.4) mm, obovoid to obpyramidal, white to light brown at base, nerveless to 2(3)-nerved, pubescent on the upper 1/2, beak hyaline, (0.4–)0.7–1 mm; stigmas 3; cypselae 2.6–4.2 mm. Desert scrub to piñon juniper woodland, on sand and limestone, 5800–8300 ft. NC, NW, WC.

Carex geophila Mack.—Ground-loving sedge. Plants densely cespitose; culms 3–18(–22) cm; leaf blades 0.8–2.9 mm wide, sheaths tan to brown, persistent, forming a dense fibrous base; inflorescence both basal and cauline, basal inflorescences of 1–2 pistillate spikes on short, stout peduncles, cauline inflorescences of 1 reddish-brown terminal staminate spike and 1–3 lateral pistillate spikes, pistillate spikes each with 1–6 perigynia, the proximal bract leaflike, usually shorter than the inflorescence; pistillate scales shorter than to as long as the perigynia; perigynia 2.5–4.2 × 1.5–2 mm, with 2 thick marginal nerves and 0–15 lighter nerves, pubescent, obovoid, ellipsoid, or suborbicular, abruptly contracted to the beak, beak straight, rarely bent; stigmas 3; cypselae 1.6–2.8 mm. Mostly dry soil in ponderosa pine forests (occasionally piñon-juniper woodlands), (5450–)6000–8900 ft. NC, NW, C, WC, SW.

Carex geyeri Boott—Elk sedge. Plants rhizomatous, colonial or loosely clumping; culms 12–41(–50) cm; leaf blades flat, (1.2–)1.5–3.1(–3.5) mm wide; inflorescence a single terminal androgynous spike, the staminate portion longer than the pistillate portion and separated from it by a bare rachis, bract absent; pistillate scales narrowly oblong, ovate or lanceolate, longer than the perigynia, apex acuminate, often awned, green to brown, midvein pale, margin hyaline; perigynia (4.4–)4.9–6.4(–8.4) × (1.8–)2.2–2.8 mm, obovoid, greenish yellow to brown, glabrous, glossy, tapering to a spongy narrow base, apex rounded, beak minute or lacking; stigmas 3; cypselae 3.5–5(–6.2) mm. Dry forests and woodlands, open slopes, often in shade, 6300–10,200 ft. NC, NW.

Carex gravida L. H. Bailey—Heavy sedge. Plants cespitose; culms 20–120 cm, sheaths loose at the base, proximally often green-and-white-striped with crossveins on the dorsal surface, the ventral surface hyaline, occasionally cross-rugulose with fragile, often torn summit, widest leaf blades 3–6.5(–8) mm wide, ligules (1–)2–5(–7) mm; inflorescence compact to elongate, of (5–)8–19 androgynous spikes, the spikes usually sessile on the inflorescence axis (but inflorescence occasionally branched), proximal bract setaceous, shorter than the inflorescence; pistillate scales hyaline or brown with green center, body shorter or slightly longer than the perigynia, apex acuminate to short-awned; perigynia lanceolate to ovate, 3–5.5 × 2–3 mm, cream to brown with green margins, margins serrulate distally, beak 0.6–1.3(–1.6) mm, teeth 0.2–1 mm; stigmas 2; cypselae 1.8–2.5 mm. Rocky, sandy soil, 4900–6700 ft. Mora, Quay, Union.

Carex gynocrates Wormsk. ex Drejer [*C. dioica* L. subsp. *gynocrates* (Wormsk ex Drejer) Hultén; *C. dioica* L. var. *gynocrates* (Wormsk. ex Drejer) Ostenf.]—Northern bog sedge. Plants rhizomatous, rhizomes threadlike; culms 2–20 cm, arising singly or several together; leaf blades involute, 0.2–0.9 mm wide; inflorescence a single androgynous or unisexual spike, the staminate portion often shorter than or equal to the pistillate portion, bract absent; pistillate scales ovate, shorter and wider than the perigynia, light brown, margins hyaline, apex acute to mucronate, lowest scale sometimes awned; perigynia 2.6–3.4 × 1.1–1.7 mm, ovoid, oblong, or elliptic, biconvex, yellowish or green when young, dark brown and shiny when mature, nerved, beak 0.2–0.9 mm; stigmas 2; cypselae 1.2–1.7 mm. Bogs in coniferous forests, gravelly stream banks, 10,300–11,100 ft. Colfax, Taos.

Carex haydeniana Olney—Cloud sedge. Plants cespitose; culms 4–25(–40) cm; leaf blades 1–2.5 mm wide; inflorescence a compact head of 5–7(–9) erect to spreading, similar, sessile, gynecandrous spikes, proximal bract shorter than the inflorescence; pistillate scales narrower and shorter than the perigynia, reddish, coppery to dark brown, sometimes with a whitish or pale gold midvein; perigynia 4–5 × 1.5–2.2 mm, lance-ovate to broadly ovate, coppery to brown, often nerved on the dorsal side, sometimes nerved ventrally, flat except over the achene; stigmas 2; cypselae (1.2–)1.4–1.8 mm. Rocky or gravelly alpine slopes and clearings, 11,800–12,300 ft. Colfax, Taos.

Carex heliophila Mack. [*C. inops* L. H. Bailey subsp. *heliophila* (Mack.) Crins.]—Sun sedge. Plants in tufts on long rhizomes; culms 13–35 cm, scabrous distally; leaf blades (0.7–)1.1–2.8 mm wide, sheaths purplish red, fibrous; inflorescence cauline only, compact, of 1 reddish-brown, terminal, staminate or rarely androgynous spike on a 1–9(–14) mm peduncle, and 1–2(3) lateral pistillate spikes, pistillate spikes each with 1–10(–15) perigynia, proximal bract leaflike, shorter than the inflorescence; pistillate scales shorter to longer than the perigynia; perigynia 2.8–4(–4.9) × 1.2–2.2 mm, with 2 prominent marginal nerves, short-pubescent, obovoid to ovoid, abruptly contracted to the beak, beak straight to bent, the margin ciliate-serrulate or not, deeply bidentate; stigmas 3; cypselae 1.6–2.5 mm. Mixed conifer and ponderosa pine forests, juniper woodlands, montane grasslands, 6000–10,500 ft. NE, NC, NW, C, Sierra.

Carex hystericina Muhl. ex Willd.—Bottlebrush sedge. Plants densely clumping from short rhizomes; culms (15–)20–85(–100) cm; leaf blades (1.4–)2.5–8.5 mm wide, septate-nodulose, sheaths pale to reddish, ladder-fibrillose; inflorescence elongate, of 1 terminal staminate spike (rarely gynecandrous or mixed), and 2–4(–6) lateral pistillate spikes, these erect to spreading on flexuous, filiform peduncles, the proximal bract longer than the inflorescence; pistillate scales narrower and shorter than the perigynia, abruptly narrowed to scabrous awns 1.2–6 mm, often longer than the scale bodies; perigynia 4–7.3 × (0.8–)1.2–2.3 mm, nerved, ovoid to lanceoloid, inflated, tapering to a slender bidentate beak 1.5–2.7 (–3.5) mm, teeth 0.2–0.7(–0.9) mm; stigmas 3; cypselae 1.3–2.5 mm. Wetlands, 3900–7300 ft. NC, EC, C, WC, SE, SC.

Carex illota L. H. Bailey—Sheep sedge. Plants cespitose; culms (10–)16–32(–38) cm; leaf blades (1–) 1.4–2.5(–3) mm wide; inflorescence a compact head of 2–4(–6) gynecandrous spikes, spikes sessile, indistinct, prickly with protruding perigynia, proximal bract shorter than the inflorescence; pistillate scales dark brown, usually with pale or green midstripe, usually shorter than perigynia; perigynia ovate-lanceolate, plano-convex, 2.1–2.9(–3.2) × 0.8–1.3 mm, gold to dark brown, margin unwinged and smooth, beak poorly defined, brown, with prominent dark dorsal suture; stigmas 2; cypselae 1.2–1.5 mm. Wetlands, 10,400–12,000 ft. Mora, Rio Arriba, Taos.

Carex interior L. H. Bailey—Inland sedge. Plants densely cespitose; culms (10–)14–53 cm, wiry; leaf blades 0.4–2.9 mm wide; inflorescence compact to elongate, of 2–5 sessile, easily distinguished gynecandrous or pistillate spikes, terminal spike clavate with a prominent narrow staminate basal portion, proximal internode 2–9 mm, proximal bract scalelike or setaceous, shorter than the inflorescence; pistillate scales ovate, inconspicuous, mostly hyaline, brown in the middle, shorter than the perigynia; perigynia 2.1–3(–3.2) × 0.9–1.7, 1.5–2.5 times as long as wide, spreading in age, green to dark brown, serrulate above, beak 0.3–1 mm, nearly entire to bidentate to a depth of 0.3(–0.5) mm; stigmas 2; cypselae 1.2–1.6 mm. Wetlands, 6000–11,000 ft. NC, Grant, Otero.

Carex jonesii L. H. Bailey—Jones' sedge. Plants loosely cespitose; culms 20–40(–60) cm; leaf blades 1–3.5 mm wide; inflorescence a compact head of 4–10 indistinguishable androgynous spikes, mostly sessile on the axis, but sometimes sessile on obscure branches, proximal bract scalelike, shorter than the inflorescence; pistillate scales shorter than to as long as the perigynia, coppery to hyaline to dark brown, shiny; perigynia narrowly lanceolate to ovate-lanceolate, 2.5–2.9(–4.8) × 0.9–1.5 mm, green maturing to brown, 7–11-nerved dorsally, 5–9-nerved ventrally, widest near the slightly spongy, rounded base; stigmas 2; cypselae (1.1–)1.5(–2) mm. Subalpine wetlands, 9400–10,600 ft. Rio Arriba.

Carex kelloggii W. Boott [*C. lenticularis* Michx. var. *lipocarpa* (Holm) L. A. Standl.; *C. lenticularis* Michx. var. *pallida* (Boott) Dorn]—Kellogg's sedge. Plants densely cespitose; culms 15–80 cm; leaf blades 1–4 mm wide, sheaths straw to light brown, not ladder-fibrillose, free portion of the ligules 0.5–2 mm; inflorescence elongate, of 1(2) terminal staminate spike(s), and 3–4(–7) lateral pistillate spikes, lower spikes sessile or with peduncle to 11 cm, proximal bract leaflike, longer than the inflorescence; pistillate scales narrower and shorter than to equaling the perigynia, oblong-ovate, dark with a green midvein that does not reach the tip, tip obtuse to acute, awnless; perigynia elliptic to ovate, 1.8–3.2(–3.5) × 1–1.8 mm, dull white or green, nerveless or with 5–7 nerves on each face, abruptly stipitate, beak 0.1–0.5 mm, entire; stigmas 2; cypselae ca. 1 mm. Wetlands, 6900–12,500 ft. Catron, San Miguel, Santa Fe.

Carex lachenalii Schkuhr [*C. bipartita* All. var. *austromontana* F. J. Herm.]—Two-tipped sedge. Plants loosely cespitose, forming large turflike clumps; culms 9–30 cm; leaf blades 0.8–2 mm wide; inflorescence compact, of (1–)3(4) sessile gynecandrous spikes, proximal bract shorter than the inflorescence; spikes 4–8 mm, oblong, each containing 10–20 perigynia; pistillate scales oblong-ovate, obtuse, red-brown with yellowish-brown center, margins hyaline, subequal to perigynia; perigynia (2–)2.4–3.8 × 0.9–1.5 mm, 2.3–2.7 times longer than wide, obovate to ovate, biconvex, substipitate, smooth, pale yellow-green turning yellowish brown, dorsal suture extending the length of the beak and sometimes onto the perigynia body, darker than body, beak 0.5–1 mm, tip hyaline, either smooth or inconspicuously serrulate; stigmas 2; cypselae 1.2–1.5 mm. Alpine wetlands, 12,500–12,600 ft. Taos.

Carex leptalea Wahlenb.—Bristly stalked sedge. Plants cespitose, forming loose clumps; culms 9–60 cm, capillary, lax, usually longer than the leaves; leaf blades flat, 0.2–1.3 mm wide; inflorescence a single terminal androgynous spike, staminate portion usually shorter than the pistillate portion, often inconspicuous, bract absent; pistillate scales obovate to ovate to lanceolate, shorter than the perigynia, green to brown, midvein pale, apex acute to acuminate, sometimes awned, proximal scale awns sometimes prolonged as bristles, deciduous; perigynia 2.4–4.9(–5.4) × 0.8–1.3 mm, oval-elliptic with a pithy stipelike base, pale green to straw, nerved, apex rounded to retuse, beak lacking; stigmas 3; cypselae 1.3–1.9 mm. Bogs, meadows, streamsides in coniferous forests, 10,600–11,200 ft. Taos.

Carex leucodonta Holm—Huachuca Mountain sedge. Plants loosely cespitose; culms 15–50 cm; leaf blades 2–3.5(–4) mm wide, sheaths brown, persistent, fibrous; inflorescence cauline only, of 1 pale-colored terminal staminate or rarely androgynous spike and 2–3 lateral pistillate spikes, pistillate spikes each with 1–8(–10) perigynia, proximal pistillate bract leaflike, longer or shorter than the inflorescence; pistillate scales as long as or slightly longer than the perigynia; perigynia 3.4–4(–4.2) × (1.5–)1.7–2.2(–2.4) mm, 10–25-nerved, short-pubescent, obovoid to almost globose, abruptly contracted to a shallowly bidentate beak; stigmas 3; cypselae (1.6–)1.8–2.6 mm, globose. Open pine forests and oak-fir forests, 7500 ft. Grant.

Carex limosa L.—Mud sedge. Plants rhizomatous, colonial, roots covered with dense, pale yellow, felty pubescence; culms 15–40(–60) cm, arising singly or in small groups, sheaths often purplish red; leaf blades (0.8–)1–2.5(–3) mm wide, sheath fronts red-dotted; inflorescence elongate, of 1 terminal staminate spike and (0)1–2(3) lateral pistillate spikes, pistillate spikes pendent to ascending on flexuous peduncles, occasionally androgynous, proximal bract shorter to longer than the inflorescence, sheath to 3 mm; pistillate scales 3–4.8(–5.5) × (1.7–)1.8–2.5(–3.4) mm, ovate, shorter to longer than and as wide as or wider than the perigynia, yellowish brown to dark reddish brown, sometimes with a green midstripe, cuspidate to awn-tipped; perigynia 2.5–3.5(–4) × 1.2–2.3(–2.6) mm, ovoid, green to yellow-green, nerved, compressed-trigonous, papillose, beak 0.1(–0.5) mm, orifice entire or emarginate; stigmas 3; cypselae (1.5–)1.8–2.2(–2.7) mm. Bogs and fens, 10,200–11,700 ft. San Miguel, Taos.

Carex luzulina Olney—Woodrush sedge. Plants cespitose to short-rhizomatous; culms 15–60(–90) cm; leaf blades clustered near the base, 1.8–3.7(–9) mm wide, yellowish green; inflorescence elongate, of 1 terminal staminate spike and (2)3–5(6) lateral androgynous or pistillate spikes, lower spikes remote, sometimes pendent, proximal peduncle (5–)16–45 mm, proximal bract shorter than the inflorescence, sheath inflated; pistillate scales dark reddish brown to purple, margins hyaline, midvein scabrous, green or pale; perigynia (3–)3.5–5.5 × 0.9–1.6(–1.8) mm, 2.5–3 times as long as wide, narrowly elliptic to lanceolate, cream to green or purple, distal margins often ciliate-setose, beak 0.5–1.5 mm; stigmas 3; cypselae 1.3–2 mm. Our plants belong to **var. ablata** (L. H. Bailey) F. J. Herm. Wetlands, 9800 ft. Rio Arriba.

Carex macloviana d'Urv.—Thickhead sedge. Plants cespitose; culms 9–52(–60) cm; leaf blades

(1.7-)2-3.6 mm wide; inflorescence compact, of 3-5(-9) similar, sessile, gynecandrous spikes, proximal bract shorter than the inflorescence; pistillate scales narrower and shorter than the perigynia, gold to brown, sometimes with a paler midvein, hyaline margins 0.2-0.5 mm wide, lower scale tips usually obtuse; perigynia 3.5-4.5 × (1.1-)1.3-2 mm, ovate, straw, gold, red-brown, or coppery, sometimes with a metallic sheen, margins often brown, darker than perigynia body, nerved or not, usually with white-hyaline dorsal suture margins and white-hyaline tip; stigmas 2; cypselae (1.2-)1.4-1.9 mm. Wet areas and dry meadows in subalpine and alpine habitats, 8900-12,000 ft. Colfax, Rio Arriba, Taos.

Carex magellanica Lam. [*C. paupercula* Michx.]—Boreal bog sedge. Plants rhizomatous, roots covered with dense, pale yellow or whitish pubescence; culms arising singly, in small clusters, or sometimes forming larger, loose clumps, (10-)30-50(-70) cm; sheaths brown to reddish purple; leaf blades (1-)1.5-3(-4) mm wide; inflorescence elongate, of 1 terminal staminate spike and 1-3 lateral pistillate spikes, pistillate spikes pendent on flexuous peduncles, proximal inflorescence bract leaflike, shorter to longer than the inflorescence, with sheath to 3 mm; pistillate scales narrowly lanceolate to ovate-lanceolate, tapering to an acute tip, (2.8-)3.8-5.3(-7) × (0.6-)0.9-1.9(-2) mm, mostly narrower and conspicuously longer than the perigynia, reddish brown proximally to yellowish brown distally; perigynia 2-3.2(-3.8) × (1.1-)1.7-2.2(-2.5) mm, broadly ellipsoid to obovoid, compressed-trigonous, papillose, green turning deep brown to cream, nerved, coriaceous, apex rounded, beak 0.1(-0.2) mm or absent, orifice entire or emarginate; stigmas 3; cypselae 1.9-2 mm. New Mexico material belongs to **subsp. irrigua** (Wahlenb.) Hiitonen. Wetlands, 10,300-10,500 ft. Rio Arriba.

Carex microdonta Torr. & Hook.—Littletooth sedge. Plants colonial-rhizomatous; culms single or rarely 2-3 culms together, 7-48(-56) cm; leaf blades 2-6(-8) mm wide; sheaths brown; inflorescence elongate, of 1 erect terminal staminate spike and 2-4(5) erect lateral pistillate (rarely androgynous or staminate) spikes, the lowest spike pedunculate, arising in the lower 1/3 of the culm, sometimes almost basal, proximal bract shorter than the inflorescence; pistillate scales shorter to longer and narrower than the perigynia, green with hyaline margins, tips often awl-like; perigynia 2.6-4.1(-4.2) × 1-2 mm, narrowly ovoid to oblong-ovoid, green to light reddish brown, often rust-spotted, many-nerved, beak 0.3-0.9(-1) mm, nearly entire; stigmas 3; cypselae 1.7-2.7 mm. Limestone seeps in arid to mesic woodland sites, 6800-7100 ft. Eddy.

Carex microglochin Wahlenb.—Fewseeded bog sedge. Plants rhizomatous, colonial, culms (2-)4-25 cm, longer than the leaves; leaf blades involute, 0.3-1 mm wide; inflorescence a single terminal androgynous spike, the staminate portion shorter than to equal to the pistillate portion, bract absent; pistillate scales ovate-triangular, shorter and wider than the perigynia, deciduous, brown, apex obtuse to subacute; perigynia 3.4-4.7 × 0.6-1 mm, lance-subulate, beak ill-defined, pale green to straw, nerved, vestigial rachilla exserted 0.6-2.2 mm beyond the orifice; stigmas 3; cypselae 1.9-2.4 mm. Our material belongs to **subsp. microglochin**. Moist areas in alpine tundra, 10,800-11,700 ft. Colfax.

Carex micropoda C. A. Mey. [*C. crandalli* Gand; *C. pyrenaica* Wahlenb. subsp. *micropoda* (C. A. Mey.) Hultén]—Pyrenean sedge. Plants densely cespitose, forming tufts; culms 5-30 cm, shorter to longer than the leaves; leaf blades wiry, 0.25-1.5 mm wide, bases not fibrous, sheath fronts membranous; inflorescence a single terminal androgynous spike, staminate portion shorter than the pistillate portion, bract absent; pistillate scales ovate to lanceolate, mostly shorter than the perigynia, deciduous, brown, sometimes glossy, midvein pale, margins narrow, hyaline, apex obtuse to acute; perigynia 2.7-4.5 × 0.7-1.5 mm, elliptic to ovate or lanceolate with a narrow, cylindrical green stipe to ca. 0.4 mm, glabrous, light green to straw-colored, becoming dark brown at maturity, marginal nerves 2, beak hyaline, 0.1-0.5 mm; stigmas (2)3; cypselae 1.2-1.8 mm. Alpine, 11,500-12,600 ft. Taos.

Carex microptera Mack.—Smallwing sedge. Plants cespitose; culms 15-80(-110) cm; leaf blades 1-4 mm wide; inflorescence a compact head of 4-10(-14) similar, sessile, gynecandrous, mostly indistinguishable spikes, proximal bract usually shorter than the inflorescence; pistillate scales shorter and narrower than to as wide as the perigynia, brown, usually with a pale midvein, hyaline margin 0-0.2 mm wide, the tip acute; perigynia (2.5-)3.1-4.4(-4.8) × 1.1-2 mm, ovate, flat except at achene, usually widest below the top of the achene, green to tan or brown, dorsally nerved, ventrally nerved or not; cypselae 1.2–1.8 mm; stigmas 2. Moist to wet meadows and along streams, 5400–12,500 ft. Widespread except E.

Carex aff. microptera (This is currently an undescribed species with affinities to *C. microptera*.)—Apache sedge. Plants cespitose; culms 25–75 cm; leaf blades 1.4–4 mm wide; inflorescence compact, of 5–10(–13) similar, sessile, gynecandrous spikes, lower spikes somewhat remote and distinguishable, proximal bract shorter than the inflorescence; pistillate scales shorter and narrower than, to rarely as wide as the perigynia, deep brown with a pale or green midvein and hyaline margins; perigynia (3.5–)4–5.6 × 1.1–1.8 mm, ovate, flat except at achene, usually widest at the top of the achene, green to golden brown, usually many-nerved on both sides; stigmas 2; cypselae 1–1.4(–1.6) mm. Moist to wet meadows, along streams, 5700–9700 ft. Catron, Grant.

Carex muriculata F. J. Herm.—Schiede's sedge. Plants cespitose; culms 10–80 cm; leaf blades 1–2.7 mm wide, bases brown, fibrillose; inflorescence elongate, of 3–6 androgynous spikes, short-pedunculate to sessile, proximal bract longer to much longer than the inflorescence; pistillate scales oblong-lanceolate, usually exceeding the perigynia, tip acuminate, occasionally tapering to an awn, with green to brown center and whitish to pale brown translucent sides, strongly 3-veined; perigynia 3.4–4.9 × 1–1.4 mm, ellipsoid, green to dark brown, 8–16-veined, leathery, muricate-warty with tiny teeth, stipitate, trigonous in cross-section, beaks 0.6–1.3 mm, bidentate with teeth 0.2–0.6 mm; stigmas 3; cypselae 1.9–2.9 mm. Dry habitats on limestone, 4900–5900 ft. Eddy.

Carex nebrascensis Dewey—Nebraska sedge. Plants rhizomatous, often forming large colonies; culms 15–80 cm, arising singly or in small clusters; leaf blades 3–10(–12) mm wide, green to glaucous, sheaths dirty-white to dark brown, not ladder-fibrillose, the ligule U-shaped to truncate; inflorescence elongate, of 1–3 terminal staminate spikes and 2–4 lateral pistillate or androgynous spikes, lower spikes sessile or with peduncle to 2(–4) cm, proximal bract leaflike, shorter to longer than the inflorescence; pistillate scales equal to or longer than the perigynia, lanceolate, dark with a light-colored midvein, the midvein of at least the lower scales often excurrent and awned, the awn to 1(–3) mm; perigynia elliptic to obovate, (2.2–)2.5–4 × (1.3–)1.6–2.5 mm, dull, leathery, 3–7-nerved on each face, beak brown, 0.3–0.6 mm, bidentate with teeth to 0.2 mm, orifice often ciliate; stigmas 2; cypselae 0.3–2.5 mm. Wetlands, often in areas used by cattle, 5400–12,100 ft. NC, NW, WC, Sierra.

Carex nelsonii Mack.—Nelson's sedge. Plants densely cespitose; culms 9–19(–30) cm, longer than the leaves; leaf blades flat, (1.5–)2–3 mm wide, basal sheaths brown; inflorescence a compact head, of 3–4 similar-looking spikes, terminal spike gynecandrous, lateral spikes pistillate and sessile to short-pedunculate, proximal bract usually shorter than the inflorescence; pistillate scales ovate, shorter than or equal to and slightly narrower than the perigynia, dark brown to black to the margins, tip acute; perigynia 3.3–4 × 1–1.5 mm, 2.8–3.9 times longer than wide, papillose, narrowly elliptic to narrowly ovate, light-colored proximally, black distally, margins light-colored, flattened, nerved, beak (0.3–)0.5–0.8(–0.9) mm, dark, bidentate; stigmas 3; cypselae 1.3–1.7 mm, filling proximal 1/2 or less of the perigynia. Moist alpine meadows, springs, 12,200–12,400 ft. Taos.

Carex nova L. H. Bailey—Black sedge. Plants loosely cespitose; culms 14–48(–60) cm; leaf blades (0.9–)1.8–4 mm wide, sheaths light brown to dark reddish brown; inflorescence a compact head of 3–4 similar-looking spikes, terminal spike gynecandrous, lateral spikes pistillate, sessile to rarely short-pedunculate and remote, proximal bract shorter to longer than the inflorescence; pistillate scales ovate to lanceolate, mostly shorter and narrower to sometimes longer than the perigynia, dark brown to black, tip acute; perigynia 2.5–4 × 1.1–2(–3) mm, 1.3–2.5(–2.8) times longer than wide, papillose, pale yellow to dark brown or black distally, lighter-colored proximally, flattened, margins nerved, obovate to suborbicular, beak bidentate, 0.2–0.5(–0.6) mm; stigmas 3; cypselae 1.6–1.9 mm, filling proximal 1/2 or less of the perigynia. Moist subalpine and alpine meadows, springs, lake margins, 9400–12,500 ft. NC.

Carex obtusata Lilj.—Obtuse sedge. Plants rhizomatous, colonial, rhizomes slender, tough, dark brown to reddish purple; culms 4–25 cm, widely spaced, arising singly or several together; sheaths reddish purple, fronts red-dotted; leaf blades 0.5–1.6(–2) mm wide; inflorescence a single terminal androgynous spike, the staminate portion about as long as the pistillate portion, bract lacking; pistillate scales ovate to lanceolate, mostly shorter than and as wide as the perigynia, pale brown to reddish brown, midvein pale, margins hyaline, apex obtuse to acute to awned; perigynia 2.4–4.1 × 1–1.6(–1.8) mm, ellipsoid to obovoid, brown and shiny when mature, nerved, beak bidentate, 0.5–1.3 mm, tip hyaline; stigmas 3; cypselae 1.7–1.9 mm. Montane grasslands, open forests, rocky ridges, 6500–10,500 ft. NC, Cibola, Socorro.

Carex occidentalis L. H. Bailey—Western sedge. Plants densely cespitose or short-rhizomatous with culms crowded; culms 15–90 cm; leaf blades 1.4–2.8 mm wide; inflorescence elongate, of 4–12 sessile, androgynous, ascending spikes, unbranched, proximal bract setaceous or absent, shorter than the inflorescence; pistillate scales similar in size to the perigynia, brown with green center and hyaline margins; perigynia 1–8 per spike, elliptic to oblong-elliptic, plano-convex in cross-section, 3–4.8 × 1.4–1.9 mm, green maturing to straw, brown, or black, smooth, shiny, nerveless, sharp-margined, usually doubly serrulate above the middle, beak 0.6–1.3 mm; stigmas 2; cypselae (1.3–)1.9–2(–2.4) mm. Dry grasslands, woodlands, more mesic habitats at lower elevations, 5400–11,500 ft. Widespread except EC, SE, SW.

Carex oreocharis Holm—Grassyslope sedge. Plants densely cespitose, forming tufts; culms 6–30 (–50) cm, longer or sometimes shorter than the leaves, often scabrous below the inflorescence; bases crowded with brown sheaths that contrast sharply with the green leaves; leaf blades folded or channeled, (0.3–)0.8–1.5(–2.8) mm wide, often curled at the tips, inflorescence a single terminal androgynous spike, the staminate portion longer than the pistillate portion, bract absent; pistillate scales ovate or lanceolate to broadly orbicular, wider and mostly longer than the perigynia, yellowish green to brown, midvein pale, margins broad, hyaline, apex obtuse to awn-tipped; perigynia (2.7–)3–4.2 × 1.2–2.1 mm, broadly ovoid to obovoid, straw to pale yellowish green, short-pubescent, nerves 0–5(–8), margins ciliate near beak, beak hyaline, 0.2–1 mm; stigmas 3; cypselae 2.2–3 mm. Mountain meadows, dry slopes, 7500–10,900 ft. Colfax, Socorro, Taos.

Carex pachystachya Cham. ex Steud.—Chamisso sedge. Plants cespitose; culms 15–60(–120) cm; leaf blades 2.2–2.7 mm wide; inflorescence compact, of 4–10(–14) similar, sessile, gynecandrous spikes, proximal bract shorter than the inflorescence; pistillate scales shorter and narrower than the perigynia, ovate to ovate-lanceolate, brown and shiny, with a pale or green midvein, apex acute to obtuse; perigynia 3.6–4.5 × 1.1–2 mm, plano-convex, green or tan to brown, darker at margins, typically (5–)7–9(–11)-veined on both sides, sometimes nerveless ventrally, most perigynia entire in the distal 0.3–0.6 mm, a few sometimes serrulate to the tip, tapering to the terete beak; beak tip brown; stigmas 2; cypselae 1.2–1.4 mm. Montane canyons along streams, 9800 ft. Rio Arriba.

Carex pellita Muhl. ex Willd. [*C. lanuginosa* auct. non Michx., misapplied]—Woolly sedge. Plants colonial from long rhizomes; culms 15–100 cm; leaf blades 2–4.5(–7) mm wide, sheaths reddish purple, ladder-fibrillose; inflorescence elongate, of 1–3 terminal staminate spike(s), and 2–3(4) well-separated, ascending lateral pistillate spikes, lowest pistillate spike on a long, slender peduncle, proximal bract shorter to longer than the inflorescence; pistillate scales shorter or longer than the perigynia, apices acute to acuminate-awned; perigynia 2.5–3.8(–5.2) × 1.4–2.1(–2.8) mm, nerved on both sides, nerves hidden by uniform dense pubescence throughout, inflated, ovoid, rounded and spongy-thickened below, abruptly contracted to beak, beak (0.5–)0.9–1.2(–1.5) mm, deeply bidentate; stigmas 3, cypselae 1.5–2.1 mm, trigonous. A common obligate wetland sedge that occurs in a wide variety of habitats, 4700–10,900 ft. Widespread except EC, SE.

Carex petasata Dewey—Liddon sedge. Plants cespitose from short rhizomes; culms 28–85 cm, longer than the leaves; leaf blades 0.8–3 mm wide; inflorescence elongate, composed of (2)3–6(7) similar, sessile, gynecandrous spikes, proximal bract shorter than the inflorescence; pistillate scales shorter and narrower than, or equal to the perigynia, pale green to light brown with a pale or green midvein, with hyaline margins 0.2–0.8 wide; perigynia 6–8.4 × 1.7–2.4 mm, lanceolate to ovate, light olive-green to brown, usually many-nerved on both sides; stigmas 2; cypselae 2.2–3.3 mm. Dry to wet meadows, grasslands, open clearings in forests, 5100–11,900 ft. NC, WC, Union.

Carex phaeocephala Piper—Dunhead sedge. Plants cespitose; culms (5–)15–45 cm; leaf blades (0.4–)0.8–2.5 mm wide; inflorescence elongate, of (2)3–7 similar, sessile, gynecandrous spikes, proximal bract shorter than the inflorescence; pistillate scales equaling or slightly shorter than the perigynia with beaks exposed, often covering, or sometimes exposing the side of the perigynia, lanceolate to ovate, gold, red-brown, or dark brown with a pale midvein; perigynia (3.4–)3.6–4.4(–5.2) × (1–)1.1–1.5(–2.3) mm, ovate or elliptic-oblong, pale green to pale yellow to brown or red-brown, sometimes with green margins, veinless to few-nerved ventrally, several-nerved dorsally; stigmas 2; cypselae (1.3–)1.5–2 mm. Alpine or windswept high montane rocky areas, 10,100–12,700 ft. NC.

Carex pityophila Mack.—Loving sedge. Plants densely cespitose; culms 3–12 cm; leaf blades 0.8–1.5(–2.6) mm wide, folded or involute, wiry, sheaths reddish to purplish brown, slightly fibrous; inflorescences both basal and cauline, basal inflorescences a single spike of 1–2(3) perigynia on a filiform peduncle, cauline inflorescences of 1 terminal dark staminate spike on a peduncle as long as or longer than the spike, and 1–2(3) lateral pistillate spikes, each with a single perigynium or very rarely 2, separate along the rachis, proximal bract narrow to involute, subequal to slightly longer than the inflorescence; pistillate scales shorter than or equal to the perigynia; perigynia 3–4.6 × 1–1.7 mm, with 2 marginal nerves only, finely pubescent, ellipsoid to ovoid, abruptly contracted to beak; beak straight, smooth or weakly ciliate-serrulate, bidentate, stigmas 3; cypselae 1.9–2.1 mm, obtusely trigonous. Piñon-juniper woodland, ponderosa pine, mixed conifer forests, 7300–9900 ft. NC.

Carex planostachys Kunze [*C. lativena* S. D. Jones & G. D. Jones]—Cedar sedge. Plants densely cespitose; culms 4–15 cm; leaf blades 0.9–2 mm wide, sheaths tan to dark brown, scabrous, fibrous; inflorescences both basal and cauline, basal inflorescences of (1)2–3(4) pistillate spikes on a slender peduncle, cauline inflorescences of 1 pale-colored terminal staminate spike, and (0)1–2(3) lateral pistillate spikes, pistillate spikes each with 3–8 perigynia, proximal bract leaflike, shorter than or equal to the inflorescence; pistillate scales slightly shorter to slightly longer than the perigynia, apices acute, acuminate, or cuspidate; perigynia 3–4.9 × (0.7–)1–2.1 mm, with 14–18(–27) prominent nerves, glabrous proximally and hispidulous distally (or nearly glabrous throughout), obovoid, contracted to the beak, beak straight to strongly bent, bidentate; stigmas 3; cypselae (1.5–)1.9–2.5(–2.6) mm, trigonous, apices truncate to retuse or obtuse. Dry, rocky, oak-juniper woodland, 4200–6800 ft. Eddy, Hidalgo, Otero, Quay.

Carex praeceptorum Mack.—Teacher's sedge. Plants cespitose; culms 10–31 cm; leaf blades 1.5–2.5 mm wide; inflorescence compact to elongate, of (3)4–5(6) sessile gynecandrous spikes, proximal bract shorter than the inflorescence; pistillate scales light chestnut-brown with green or lighter center, margins hyaline, subequal to or as long as the perigynia; perigynia 1.5–2.3 × 1–1.2 mm, 1.9–2.1 times longer than wide, biconvex, widest near the middle, short-stipitate, papillose, faces nerved, nerves dark, raised, pale brown, often darker with age, beaks 0.2–0.4(–0.5) mm, smooth or serrulate, dorsal sutures about as long as the beaks, darker than perigynia body, sometimes extending onto the perigynia bodies; stigmas 2; cypselae 1.2–1.6 mm. Bogs and streamsides, 8400 ft. Rio Arriba.

Carex praegracilis W. Boott—Blackcreeper sedge. Plants rhizomatous with coarse, knotty, dark rhizomes, often in dense colonies, rhizomes 1.4–3.6(–4) mm thick; culms 5–65 cm, sharply triangular, scabrous below the heads, arising singly or a few together with internodes between culms to 2(–4) cm; leaf blades 1–3.5 mm wide, sheaths dark; inflorescence compact to elongate, of 4–12 upper spikes sessile on the main rachis and 0–5 multiple-spiked lower branches, proximal bract shorter than the inflorescence; spikes similar, androgynous or unisexual; pistillate scales similar in size to the perigynia, bronze with a green midvein, margins hyaline; perigynia elliptic to oblong, plano-convex, 2–3.8 × 1–1.8 mm, green or pale, maturing to brown or black, serrulate to smooth, nerveless ventrally, 0–7-nerved dorsally, base rounded to truncate, beak 0.7–1.3 mm; stigmas 2; cypselae 1.5–1.8 mm. Seasonal wetlands, often in agricultural areas, 3500–11,800 ft. Widespread except NE, EC, SE, SC.

Carex praticola Rydb.—Meadow sedge. Plants cespitose; culms (1–)30–70(–95) cm; leaf blades (1.5–)2–3 mm wide; inflorescence elongate, of 4–10 similar, sessile, gynecandrous spikes, proximal bract shorter than the inflorescence; pistillate scales equaling to longer and narrower than the perigynia, white, gold, coppery, or brown with a white or green to brown midvein, hyaline margins 0.1–0.3 mm wide, apex obtuse to acute; perigynia (3.7–)4.5–6 × 1.2–2 mm, length 2.8–3.1 times the width, often white, hyaline, green, gold, brown, or coppery over the achene, ovate to lanceolate, usually many-nerved on both sides, white-hyaline or not at tip; stigmas 2; cypselae 1.4–2.1(–2.7) mm. Moist to wet meadows, open dry woods, rocky areas, 9100–9300 ft. Rio Arriba.

Carex radiata (Wahlenb.) Small—Eastern star sedge. Plants densely cespitose, forming large clumps; culms 13–40(–80) cm; leaf blades 1.1–1.5(–1.9) mm wide; inflorescence elongate, of 3–5(–8) androgynous spikes, proximal internode 2–4+ times longer than the proximal spike, proximal bract threadlike, usually shorter than the inflorescence; pistillate scales ovate, shorter than the perigynia, hyaline with green or brown midvein; perigynia elliptic to ovate, plano-convex to biconvex, 2.2–2.8(–3.8) × 0.8–1.2 mm, green to pale, the faces weakly nerved at most, the margins serrulate distally, each green-nerved, base spongy, cuneate to stipitate, beak (0.1–)0.3–0.7 mm long; stigmas 2; cypselae (1.2–)1.3–1.6 × 0.9–1.2(–1.4) mm. Riparian, 7100–8500 ft. Sandoval, San Miguel.

Carex rossii Boott.—Ross' sedge. Plants cespitose; culms 7–25(–40) cm; leaf blades 0.8–3.2(–4) mm wide, lax, sheaths reddish brown, slightly fibrous; inflorescences both basal and cauline, basal inflorescences of 1–2 pistillate spikes on a slender, elongate peduncle, cauline inflorescences of 1 pale-colored terminal staminate spike, and 2–4 lateral pistillate spikes, pistillate spikes each with 3–6(–15) perigynia, proximal bract leaflike, usually much longer than the inflorescence; pistillate scales shorter than the perigynia; perigynia 3.2–4.9 × 1.1–1.7(–1.9) mm, obovoid to ellipsoid, with 2 marginal nerves, finely pubescent, abruptly contracted to the beak, beak straight (rarely slightly bent), with ciliate-serrulate margins, bidentate; stigmas 3; cypselae (1.9–)2–2.5 mm, obtusely trigonous. Ponderosa pine and mixed conifer forests, rarely above timberline, 6000–11,600(–12,300) ft. NC, C, WC. This elevation range includes the following new segregate species (which appears to be far more common and widespread than *C. rossii*). Once all collections have been reviewed, both geographic and elevation ranges for each species can be determined.

Carex aff. rossii D. B. Poindexter, ined. (This is currently an undescribed species with affinities to *C. rossii*.)—Plants cespitose; culms 7–15 cm; leaf blades 0.8–3.4(–3.7) mm wide, flat, lax, sheaths reddish to orangish brown, slightly fibrous; inflorescence both basal and cauline, basal inflorescences of 1–2 pistillate spikes on slender, elongate peduncles, cauline inflorescences of 1 dark-colored terminal staminate spike and 2–4 lateral pistillate spikes, pistillate spikes each with 1–4 perigynia, proximal bract narrow, sometimes involute, subequaling to slightly longer than the inflorescence; pistillate scales shorter than the perigynia; perigynia 3.2–4.5 × 1.2–1.8 mm, obovoid to ellipsoid, with 2 marginal nerves, finely pubescent, abruptly contracted to the beak, beak straight (rarely slightly bent), with ciliate-serrulate margins, bidentate; stigmas 3; cypselae (1.9–)2–2.5 mm, obtusely trigonous. Piñon-juniper woodland, ponderosa pine, mixed conifer forests. See *C. rossii* for elevation discussion. NC, C, WC.

Carex rupestris All.—Curly sedge. Plants rhizomatous, in loose tufts or forming large patches of sparse turf, rhizomes scaly, brown; culms 4–12(–15) cm, shorter to slightly longer than the leaves; leaf blades flat or channeled, 1.4–3(–3.8) mm wide, the tips often curling, lower sheath bases reddish purple to brownish; inflorescence a single terminal androgynous spike, staminate portion shorter than or as long as the pistillate portion, bract lacking; pistillate scales ovate to circular-ovate, longer and wider than the perigynia, brown, margins hyaline, midvein pale, lower scale apices acute to awn-tipped, upper scale apices obtuse; perigynia (2.5–)2.9–3.7(–4) × 0.9–1.6(–2) mm, ellipsoid to obovoid, cream to pale green, several-nerved, beak dark brown, 0.1–0.2 mm, entire, sometimes with a fringe of tiny, toothlike hairs; stigmas 3; cypselae 2.1–2.8 mm. Alpine, 11,800–13,000 ft. Colfax, Mora, Taos.

Carex saxatilis L.—Rock sedge. Plants rhizomatous, colonial, sometimes forming dense clumps; culms (8–)16–45(–80) cm, triangular and scabrous, angled above; leaf blades (1–)1.5–3(–5) mm wide, sheaths brown to reddish brown; inflorescence elongate, of 1–2(3) terminal staminate spikes and (1)2–3 lateral pistillate spikes, the lowest sometimes pendent, proximal bract shorter than or longer than the inflorescence; pistillate scales ovate to ovate-lanceolate, dark brown to reddish black, mostly shorter than to as long as the perigynia, apices often white-hyaline, acute or obtuse; perigynia 2.2–4.2(–5) × 1.1–2(–2.9) mm, 2 times as long as wide, elliptic to ovate, dark brown becoming reddish black distally, proximally dark yellowish, ventrally nerveless, dorsally 4–6-nerved, slightly inflated, loosely enclosing the achene; beak (0.2–)0.3–1.1 mm, smooth, bidentate with teeth to 0.2 mm; stigmas 2(3); cypselae (1.6–)1.8–2.6 mm. Wetlands, 10,300–11,600 ft. Rio Arriba, Taos.

Carex scoparia Schkuhr ex Willd.—Broom sedge. Plants cespitose, rarely short-rhizomatous; culms (17–)20–76(–100) cm; leaf blades 1.2–4(–4.2) mm wide; inflorescence compact to elongate, of (1–)3–9 (–13) similar, sessile, gynecandrous spikes, proximal bract shorter than the inflorescence; pistillate scales shorter and narrower than the perigynia, ovate with acuminate to acute apex, light brown, sometimes with a pale or pale green midvein; perigynia 3.3–5.4(–6.8) × (1–)1.2–2.2 mm, > twice as long as wide, ovate to narrowly ovate, cream to green to light brown, nerved or not; stigmas 2; cypselae (1–)1.2–1.7 mm. Our plants belong to **var. scoparia**. Wetlands, 5000–8100 ft. NC, C, WC, Doña Ana, Grant.

Carex scopulorum Holm—Mountain sedge. Plants rhizomatous, often colonial; culms 7–40(–65) cm; leaf blades 1.6–3.5(–6) mm wide, sheaths brown, not ladder-fibrillose; inflorescence dark, elongate, of 1–2 terminal staminate or androgynous spikes and 1–5 pistillate or androgynous lateral spikes, proximal bract shorter than the inflorescence; pistillate scales black, midvein pale, narrower and shorter or longer than the perigynia, awnless; perigynia elliptic or obovate, 2–3.1(–4) × 1.2–2.3 mm, pale brown, nerveless, beak 0.1–0.5(–0.6) mm, entire, sometimes ciliate at the apex; stigmas 2; cypselae 1.4 mm. New Mexico plants belong to **var. scopulorum**. Alpine and subalpine wetlands, 9600–13,000 ft. Taos.

Carex senta Boott—Swamp sedge. Plants rhizomatous to cespitose; culms 40–80(–100) cm; leaf blades 1–6(–8) mm wide, sheaths red-brown, often shiny, ladder-fibrillose, ligule U-shaped, longer than wide; inflorescence elongate, of 2–3 terminal staminate spikes and 2–4 lateral pistillate or androgynous spikes, each with a peduncle to 2(–4) cm, proximal bract shorter to longer than the inflorescence; pistillate scales narrower and shorter than to equaling the perigynia, oblong to lanceolate, dark with a greenish midvein, awnless; perigynia ovate to obovate, 3–3.5 × 1.5–2(–2.3) mm, 5–7-nerved on each face, beak brown, 0.2–0.3 mm, entire; stigmas 2; cypselae 1.5 mm. Streams, lakeshores, often forming tussocks on midstream rocks, 4500–8500 ft. Catron, Grant.

Carex siccata Dewey [*C. aenea* auct. non Fernald; *C. foenea* auct. non Willd.; *C. foenea* Willd. var. *tuberculata* F. J. Herm.]—Dryspike sedge. Plants rhizomatous, colonial; culms arising singly, 15–90 cm, scabrous-angled to rarely smooth above; leaf blades 1–3.5 mm wide, sheaths pale to brown; inflorescence elongate, unbranched, of 4–9 sessile spikes, upper spikes often indistinguishable, terminal spike androgynous or appearing pistillate, often subtended by 1+ staminate spikes, middle spikes staminate or androgynous, lower spikes staminate, pistillate, or androgynous, proximal bract setaceous, shorter than the inflorescence; pistillate scales shorter than to as long as the perigynia, brown with pale or green center, margins hyaline; perigynia elliptic to ovate, plano-convex, 3.6–6 × 1.2–2 mm, green to black, ~10-nerved dorsally, nerveless ventrally, sharp-margined, beak 1.2–2.5 mm; stigmas 2; cypselae 2 mm. Dry forests and grasslands, 6600–12,300 ft. NC, NW, C, WC, SE, SC.

Carex simulata Mack.—Analogue sedge. Plants rhizomatous, forming large, dense, unisexual stands; culms arising singly, 12–60 cm, sharply triangular and minutely scabrous above; leaf blades 1–3 mm wide, sheaths brown, fronts red-spotted, hyaline; inflorescence a narrow, compact, fusiform head, mostly unisexual, lowest branches often with 2 to several spikes, proximal bract usually shorter than the inflorescence; pistillate scales wider and longer than the perigynia, coppery to brownish or black, margins broad, hyaline; perigynia ovate to rhombic-orbicular, plano-convex to biconvex, 1.7–2.4 × 1.3–1.5 mm, chestnut, smooth, glossy, 2–6-nerved dorsally, nerveless ventrally, base rounded, spongy, beak 0.3–0.5 mm; stigmas 2; cypselae 1.2–1.5 mm. Wetlands with saturated soils, 6100–11,700 ft. NC, NW, C, Catron.

Carex sprengelii Dewey ex Spreng.—Sprengel's sedge. Plants loosely cespitose; culms 30–65(–90) cm, bases covered with dense, vertical, fibrous remains ("horse hair") of the previous year's leaves; leaf blades (1.8–)2–4 mm wide; inflorescence elongate, of 1 terminal staminate spike and (1–)3–5 lateral pistillate spikes, upper spike erect or drooping, lower spikes pendent on long, flexuous peduncles, proximal bract shorter than to equaling the inflorescence; pistillate scales ovate to oblong, apex acuminate, proximal scales sometimes awned, chestnut, hyaline-tinged, midrib green, mostly shorter than but sometimes as long as the perigynia; perigynia (4.5–)4.8–5.5(–6.5) × (1–)1.2–1.8(–2) mm, oblong, orbicular, tan to golden-green, with 2 marginal nerves, apex rounded, beak tubular, 1.9–2.8 mm, with bidentate teeth (0.3–)0.6–1 mm; stigmas 3; cypselae (2–)2.2–2.5 mm. Riparian areas in canyon bottoms, 7400–8800 ft. Colfax, Sandoval.

Carex stevenii (Holm) Kalela [*C. alpina* Lilj. var. *stevenii* Holm; *C. media* of NM authors; *C. media* R. Br. var. *stevenii* (Holm) Fernald; *C. norvegica* Retz. subsp. *stevenii* (Holm) D. F. Murray]—Steven's sedge. Plants cespitose; culms 10–60 cm, longer than the leaves; leaf blades 1.1–3.2(–4) mm wide, basal sheaths tan to purplish red, fronts hyaline; inflorescence compact, of 1 terminal gynecandrous spike, subtended by 1–3 pistillate spikes, lowermost spike sometimes remote, proximal bract shorter to longer than the inflorescence; pistillate scales shorter than the perigynia and as wide, broadly ovate, dark brown to black with hyaline margins, midvein absent or inconspicuous, the tip obtuse to acute; perigynia 1.9–2.7 × (0.8–) 1–1.4 mm, papillose, filled by the achene, pale green or pale yellow to brown, elliptic, margins nerved, serrulate distally, beak (0.1–)0.2–0.4 mm; stigmas 3; cypselae 1.2–2 mm. Meadows, springs, stream banks, partial shade in forests, (6200–)7700–12,300 ft. NC, Catron, Sierra.

Carex stipata Muhl. ex Willd.—Awl-fruit sedge. Plants cespitose; culms 20–75(–120) cm, sharply trigonous and broadly winged below the head; leaf blades 2–8(–11) mm wide, brown at the base, leaf sheaths ventrally cross-rugulose above; inflorescence dense to elongate, of 10–15 multiple-spiked branches, spikes androgynous, bristly, proximal bract setaceous, shorter than the inflorescence; pistillate scales shorter than to as long as the perigynia, hyaline to straw, midvein green; perigynia lance-triangular, plano-convex or biconvex, 3.6–4.6 × 1–1.6(–2) mm, base truncate or cordate, often bulbous, spongy, tapering to the tip, sometimes abruptly narrowing above the middle, forming a violin shape, green, sides nerved, beak 2–3.5 mm; stigmas 2; cypselae 1.3–2.3 mm. Our plants belong to **var. stipata**. Wetlands, 6000–9800 ft. NC, NW, WC, Lincoln.

Carex subfusca W. Boott—Brown sedge. Plants cespitose; culms 10–80(–105) cm; leaf blades 0.8–2 mm wide; inflorescence compact, of 4–12(–14) similar, sessile, gynecandrous spikes, the upper clustered and indistinguishable, the lower usually more remote, proximal bract shorter to occasionally longer than the inflorescence; pistillate scales shorter and narrower than to sometimes covering the perigynia, light green to brown, usually with a pale, straw, or green midvein and hyaline margins; perigynia 2.4–4(–4.3) × 0.9–1.2(–1.5) mm, plano-convex, narrowly to broadly ovate, green to tan or brown, dorsally nerved, ventrally nerved or not; stigmas 2; cypselae 1–1.6 mm. Moist to wet meadows, stream banks, forest margins, 5500–10,700 ft. NC, WC, SC, SW.

Carex tahoensis Smiley—Tahoe sedge. Plants cespitose; culms (15–)40–70 cm; leaf blades 1.5–2.2(–2.5) mm wide; inflorescence compact to more elongate, of 3–4(–6) similar, sessile, distinguishable, gynecandrous spikes, proximal bract shorter than the inflorescence; pistillate scales usually covering the perigynia, red-brown with a straw to tan midvein, apex generally acute to short-awned; perigynia 4–4.8(–6.2) × 1.4–2(–2.6) mm, lanceolate to ovate to oblanceolate, green to brown with age, many-nerved on both sides; stigmas 2; cypselae (1.7–)2.1–2.2(–2.5) mm. Grasslands, open rocky and sandy slopes, subalpine and alpine meadows, 9400–9900 ft. Rio Arriba, Taos.

Carex tenera Dewey—Quill sedge. Plants densely cespitose; culms 20–75(–90) cm; leaf blades 1.3–2.5 mm wide; inflorescence elongate, of 3–5(–8) similar, distinguishable, sessile, gynecandrous spikes, each separate along the rachis, proximal bract with a threadlike awn, much shorter than the inflorescence; pistillate scales shorter and narrower than the perigynia, ovate, brown-hyaline to pale brown, with a green to pale midvein; perigynia (2.1–)2.8–3.8(–4) × 1.4–2 mm, ovate to broadly ovate, green to straw-colored with a brown to red-brown tip, dorsal nerves (1–)4–9, ventral nerves 0–8; stigmas 2; cypselae 1.3–1.7 mm. Dry to wet open forests and meadows, seeps, 8000–8100 ft. Rio Arriba.

Carex ultra L. H. Bailey [*C. spissa* L. H. Bailey var. *ultra* (L. H. Bailey) Kük.]—Cochise sedge. Plants densely cespitose from stout rhizomes; culms 50–160 cm; leaf blades 6–15 mm wide, stiff, scabrous on margins and keel, sheaths brown or reddish, fronts densely red-spotted or blotched, ladder-fibrillose; inflorescence elongate, of 1–3 terminal staminate spikes and 3–6 lateral pistillate spikes, proximal bract leaflike, shorter to longer than the inflorescence; pistillate scales shorter than to as long as the perigynia, apex acute to acuminate, often ciliate-serrulate; perigynia 2–5 × 1.4–2.2 mm, lance-oblong or obovoid, pale green to reddish brown, with red-brown spots, nerved or not, compressed-trigonous, beak 0.2–0.6 mm; stigmas 3; cypselae 1.9–2.2 mm. Springs and stream banks, 4800–5200 ft. Hidalgo.

Carex utriculata Boott—Northwest Territory sedge. Plants rhizomatous, colonial, forming dense stands; culms 22–100 cm; leaf blades (1–)3–10 mm wide, sheaths spongy-thickened, brown, reddish in smaller plants, ligules of the lowest leaf blade shorter than wide, with rounded apex; inflorescence elongate, of (1)2–5(6) terminal staminate spikes and (1)2–3(4) ascending lateral pistillate spikes, proximal bract leaflike, usually longer than the inflorescence; pistillate scales lance-ovate to narrowly lanceolate, narrower and mostly shorter than the perigynia, apex acute to awn-tipped; perigynia 3–5(–5.8) × 1–2(–3) mm, ovoid to ellipsoid, densely packed, inflated, green to reddish brown, 6–10-veined; beak 0.8–1.8 mm, bidentate, teeth 0.2–0.5 mm; stigmas 3; cypselae 1.1–2 mm. Wet meadows, marshes, lake margins, stream banks, 5600–12,000 ft. NC, NW, WC, Otero.

Carex vallicola Dewey [*C. rusbyi* Mack.; *C. vallicola* var. *rusbyi* (Mack.) F. J. Herm.]—Valley sedge. Plants cespitose; culms 12–45 cm; leaf blades (0.5–)1–3 mm wide, sheaths green; inflorescence elongate, unbranched, of 5–8 sessile, androgynous, ascending to spreading spikes, proximal bract shorter than the inflorescence; pistillate scales shorter than the perigynia, hyaline with a green, 1–3-veined center; perigynia ovate, biconvex, (2.5–)3.1–4.3 × 1.6–2 mm, green, glossy, dorsally 7–17-nerved proximally, ventrally nerveless, marginal nerves pushed to the ventral face at maturity, marginal serrations none to scant, beak 0.5–1 mm; stigmas 2; cypselae 1.6–2.7 mm. Dry grasslands, woodlands, 6900–9400 ft. NC, C, WC.

Carex vesicaria L.—Blister sedge. Plants rhizomatous, forming clumps or dense stands; culms 30–95 cm; leaf blades 1.5–6(–8) mm wide, sheaths red to purplish brown, sometimes ladder-fibrillose, fronts often red-spotted, ligules of lowest leaf usually twice as long as wide, apex acute; inflorescence elongate, of 1–3(–5) terminal staminate spikes and (1)2–3 lateral pistillate spikes, proximal bract longer than the inflorescence; pistillate scales 2–6.5 mm, narrower and shorter than the perigynia, awnless or awned; perigynia 3.8–7.5 × (1.5–)2.2–3.4 mm, lanceoloid to ovoid-lanceoloid, inflated, 7–12-nerved, beak 1.1–2 mm, bidentate, teeth 0.3–1.2 mm; stigmas 3; cypselae 1.7–3 mm. Wet meadows, marshes, lake margins, stream banks, 7700–8400 ft. Cibola, Rio Arriba.

Carex viridula Michx. [*C. oederi* Retz. subsp. *viridula* (Michx.) Hultén]—Little green sedge. Plants cespitose; culms straight or arching, 10–35 cm; leaf blades 1.5–3(–4.5) mm wide; inflorescence compact, of 1 staminate (occasionally androgynous) spike, and 1–4(–7) lateral pistillate (rarely androgynous) spikes, proximal bract usually much longer than the inflorescence; pistillate scales much shorter than the perigynia, pale brown to golden-brown, midrib keeled, green, apex obtuse; perigynia 1.8–3 × 0.8–1.4 mm, yellow-green to green, inflated, nerved, beak 0.3–1.1 mm; stigmas 3; cypselae 1–1.6 mm. New Mexico plants belong to **subsp. viridula**. Wet meadows, alkaline seeps, springs, 7800–10,000 ft. Otero, Taos.

Carex vulpinoidea Michx.—Fox sedge. Plants cespitose; culms (20–)30–100 cm; leaf blades to 3.5 mm wide, leaf sheaths ventrally cross-rugulose near summit; inflorescence elongate, of 10–15 densely aggregated branches each with multiple androgynous spikes, inflorescence bracts multiple, hairlike, the proximal shorter than the inflorescence; spikes sessile, androgynous; pistillate scales shorter than to equaling the perigynia, hyaline to greenish brown; perigynia ovate, plano-convex, 2.2–2.8 × 1.3 mm, green to pale brown, dull, veinless or few-nerved dorsally, veinless ventrally, margins sharp, base spongy, truncate to cordate, beak 0.8–1.2 mm; stigmas 2; cypselae 1.2–1.4(–1.6) mm. Wetlands along streams and lakes, 5300–8200 ft. NC, WC, Catron, Grant.

Carex wootonii Mack.—Wooton's sedge. Plants cespitose, forming small clumps from short rhizomes; culms (15–)21–70(–75) cm; leaf blades 1–4 mm wide, inflorescence compact to elongate, of 3–7(8) similar, usually distinguishable, sessile, gynecandrous spikes, proximal bract shorter to occasionally longer than the inflorescence; pistillate scales shorter and narrower than the perigynia, reddish or golden-brown with a pale or green midvein; perigynia (4.8–)5.2–7(–7.2) × (1.6–)1.8–2.9(–3) mm, 2.1–3.1(–3.5) times longer than wide, narrowly ovate to ovate, forest-green when fresh, usually lacking nerves on either side; stigmas 2; cypselae 1.6–3 mm. Dry places in open meadows and slopes, clearings in forests, rocky areas, 6700–11,000 ft. C, WC, SC, Sandoval, San Miguel.

CLADIUM P. Browne – Saw-grass

Cladium californicum (S. Watson) O'Neill [*C. mariscus* (L.) Pohl var. *californicum* S. Watson]— California saw-grass. Plants perennial, robust, forming large clumps from stout (10 mm thick), short rhizomes; culms 1–2 m × 5–10 mm wide; leaf blades as long as the culms, 7–10 mm wide, scabrid with fine, sawtooth margins; inflorescence terminal and lateral, pedunculate, of many spikelets in a compound panicle with third- to fourth-order flexuous branches; inflorescence bracts 3–4+, leaflike; spikelets in clusters of (3)4–6, 3 mm, with 5–6 florets spirally arranged along the rachilla, ellipsoid to lanceoloid; scales 5–6 × 2.5–3 mm; styles with thickened base and 3 stigmas, stamens 2; cypselae ovoid, 1.5–2 mm. Alkaline marshes and springs, streamsides, 3200–5600 ft. SE, SC, Chaves, Guadalupe.

CYPERUS L. – Flatsedge

Plants grasslike, perennial or less frequently annual, cespitose to colonial from long rhizomes; culms trigonous to rounded in cross-section, solid, glabrous to less frequently scabrous; leaves 3-ranked, clustered at or near the base, blades well developed, linear, rarely reduced to sheaths only, ligules absent; inflorescence terminal, occasionally appearing pseudolateral when forced to the side by a strong upright bract, umbellate to capitate, of 1 to many spikes, the spikes in a sessile cluster at the top of the culm, and/or elevated on rays (most frequently unequal in length), second- and third-order umbellate branching present in some species; inflorescence bracts spirally arranged at the base of the first-order branching, leaflike; spikes composed of few to numerous (rarely 1) spikelets, arranged digitately, subdigitately, or cylindrically along an axis (the rachis); spikelets with few to many florets, usually distichously arranged along a rachilla, rarely spirally arranged, flattened (compressed) to quadrangular or terete in cross-section; flowers bisexual, subtended by scales, the scale bases often decurrent onto the rachilla, forming hyaline "wings"; perianth absent; stamens (1–2)3; style 1, stigmas 2 or 3; cypselae either lenticular or trigonous, corresponding with the number of stigmas, stipitate or not.

1. Stigmas 2; cypselae lenticular; spikelets highly compressed or terete in cross-section (if terete, the florets spirally arranged)…(2)
1′. Stigmas 3; cypselae trigonous; spikelets compressed, quadrangular, or terete in cross-section (not spirally arranged, except *C. michelianus*)…(7)
2. Spikelets with 1 fertile floret subtended by 1–3 scales; spikes 1–3(4), compact, ovoid to subspherical, sessile…(3)
2′. Spikelets with 1+ fertile floret, typically 3+; spikes 1 to many, compact to open, of varying shapes, sessile to pedunculate…(4)
3. Annual, diminutive; spikes 1–2(3), subequal, seemingly lateral; primary inflorescence bract ascending to vertical and often appearing to be an extension of the culm; floral scales 1–2 per spikelet, with outer scale opaque and inner scale membranous or sometimes absent; anthers 0.1–0.2 mm; cypselae terete… **C. subsquarrosus** (*Lipocarpha micrantha*)
3′. Perennial, cespitose; spikes 1–3(4), unequal, terminal, the primary spike vertical and larger than the others; inflorescence bracts spreading to reflexed; floral scales 2(3) per spikelet, similar in texture; anthers (0.4–)0.6–0.8(–1) mm; cypselae biconvex…**C. sesquiflorus** (*Kyllinga odorata*)
4. Florets spirally arranged on the spikelet rachilla (atypical for the genus); spikelets 100+, densely packed into a single ovoid to subglobose capitate head; stigmas 2 or 3…**C. michelianus** (in part)
4′. Florets distichously arranged on the spikelet rachilla (typical for the genus); spikelets 2–60, loosely distributed along the spike rachis, or digitately arranged in a radiating cluster; stigmas 2…(5)
5. Spikes loosely cylindrical, with an obvious rachis; floral scales widely spreading so that the spikelets have a sawtooth edge; plants annual…**C. flavicomus**
5′. Spikes subcapitate, lacking an obvious rachis; floral scales appressed so that the spikelets have a smooth edge; plants perennial or annual…(6)
6. Perennial with slender rhizomes, often producing dense clumps; spikes with (3–)5–25(–60) spikelets; floral scales 1.9–2.7 mm; anthers 0.6–0.8 mm…**C. niger**

6'.   Annual with fibrous roots; spikes with 3-5(-8) spikelets; floral scales 1.5-2.1 mm; anthers 0.4-0.5 mm...
      C. bipartitus

7.   Spikelets borne in digitate clusters or in umbellate heads, compressed in cross-section (except in *C. michelianus*)...(8)

7'.   Spikelets borne in linear spikes (sometimes the rachis so shortened that the spikes appear almost head-like, or with a few ascending spikes from a common terminus), compressed, quadrangular, or terete in cross-section...(13)

8.   Plants annual (occasionally biennial in *C. acuminatus*), lacking rhizomes or tuberous rootstocks...(9)

8'.   Plants perennial, with rhizomes or tuberous rootstocks...(11)

9.   Florets spirally arranged on the spikelet rachilla (atypical for the genus); spikes 100+, densely packed into a single ovoid to subglobose capitate head; stigmas 2 or 3...**C. michelianus** (in part)

9'.   Florets distichously arranged on the spikelet rachilla (typical for the genus); spikelets 2-75, digitate in a subspherical head, or subdigitate in a hemispherical cluster; stigmas 3...(10)

10.   Floral scales 2-keeled at the base, with 3-5 ribs visible on each side, ribbed almost to the margins; awned tip 0.5-1.0+ mm, strongly excurved...**C. squarrosus** (in part)

10'. Floral scales 1-keeled, with 0-2 ribs visible on each side, with a wide area smooth to the margins; cusp to 0.5 mm, slightly excurved...**C. acuminatus**

11.   Inflorescence a single dense spike (head), spikelets 20-60(-100), tightly radiating; inflorescence bracts horizontal to deflexed parallel to culm; floral scales often milky-white; culms 4-30 cm...**C. andinus**

11'. Inflorescence usually of > 1 spike; spikelets usually < 20 per spike, ascending in a loose cluster; inflorescence bracts strongly ascending; floral scales not milky-white; culms 15-50(-60) cm...(12)

12.   Floral scales (2.3-)2.8-3.2 mm, distal scales with apical cusp 0.3-1 mm; cypselae broadly ellipsoid; anthers 0.8-1.4 mm; upper culms usually scabrous on the angles; plants of sandy substrates...**C. schweinitzii** (in part)

12'. Floral scales (1.6-)1.8-2.4 mm, distal scales with apical cusp 0.1-0.3 mm; cypselae ovoid to obovoid; anthers 0.4-0.6 mm; upper culms usually smooth on the angles; plants primarily of rocky slopes...
      C. sphaerolepis (in part)

13.   Apex of floral scales with a definite cusp or awnlike tip > 0.2 mm...(14)

13'. Apex of floral scales obtuse to acute, lacking any extension or with a tiny cusp 0.1-0.2 mm...(19)

14.   Plants annual, small, slender, tufted, with fibrous roots; larger culms generally < 0.8 mm wide; floral scales with strongly excurved awns...**C. squarrosus** (in part)

14'. Plants perennial, robust, rhizomatous; larger culms generally > 1 mm wide; floral scales with a straight to slightly spreading cusp...(15)

15.   Floral scales deciduous (the empty spikelet rachilla generally remaining persistent on the spike); spikelets strongly compressed, > 2 times as wide as thick, mostly strongly ascending throughout the spike...
      (16)

15'. Floral scales persistent, with spikelets deciduous as a unit (including the rachilla); spikelets quadrangular to slightly compressed, < 2 times as wide as thick, mostly spreading at right angles to the rachis at mid-spike...(18)

16.   Most inflorescence bracts slightly ascending (at < 45 degrees above horizontal) to reflexed; spikes sessile (central one longest), forming an elongate head, rarely with 1 short ray...**C. fendlerianus**

16'. Most inflorescence bracts ascending at > 45 degrees above horizontal; most spikes usually on elongate rays, spikes of varying lengths...(17)

17.   Floral scales (2.3-)2.8-3.2 mm, distal scales with apical cusp 0.3-1 mm; cypselae broadly ellipsoid; anthers 0.8-1.4 mm; upper culms usually scabrous on the angles; plants of sandy substrates...**C. schweinitzii** (in part)

17'. Floral scales (1.6-)1.8-2.4 mm, distal scale with apical cusp 0.1-0.3 mm; cypselae ovoid to obovoid; anthers 0.4-0.6 mm; upper culms usually smooth on the angles; plants of rocky slopes...**C. sphaerolepis** (in part)

18.   Longest spikelets (2.2-)4-10(-18) mm, floral scales 1-5; distal scale with glabrous midrib, the mucro 0.1-0.3(-0.5) mm; anthers (0.3-)0.4-0.6 mm...**C. retroflexus**

18'. Longest spikelets 9-21 mm, floral scales 3-8(-13); distal scale with scabrid midrib, the mucro 0.6-1.9 mm; anthers 0.5-1.3 mm...**C. floribundus**

19.   Spikelets quadrangular to terete, sometimes slightly compressed in cross-section, < 1.5 times as wide as thick...(20)

19'. Spikelets compressed in cross-section, > 1.5 times as wide as thick...(24)

20.   Plants perennial, with short rhizomes; spikes < 1 cm wide, the rays usually with solitary spikes; spikelets with 1-5 floral scales...(21)

20'. Plants annual to short-lived perennial, with primarily fibrous roots (rarely producing rhizomes in *C. strigosus*); spikes significantly > 1 cm wide (except in depauperate specimens), the rays often with clusters of spikes at the apex; spikelets with 3-12(-30) floral scales...(22)

21.   Inflorescence a cluster of 3-6 sessile spikes, sometimes with 1-5 additional spikes on rays; floral scales

laterally pale green to whitish; cypselae (1.6–)1.8–2(–2.2) mm; bootheel of Hidalgo County…**C. pallidicolor**

21. Inflorescence of 1 sessile spike, with 4–11 additional spikes on rays; floral scales laterally tan to brown; cypselae 1.2–1.7 mm; plants known from 2 locations (Dona Ana and Rio Arriba Counties)…**C. retrorsus**

22. Floral scales < 3.2 mm; spikelets often appearing somewhat flexuous, with the floral scales deciduous from the rachilla or the rachilla disarticulating at each joint and falling with the scale; plants small to robust annuals without cormlike bases or rhizomes…(23)

22′. Floral scales 3.2–4.5(–6) mm; spikelets generally appearing straight and stiff, falling whole from the rachis of the spike; plants short-lived perennials with thickened cormlike bases (rarely producing rhizomes)… **C. strigosus** (in part)

23. Floral scales 1.3–1.5 mm, spreading at maturity, imbricate, deciduous from the persistent rachilla of the spikelet…**C. erythrorhizos** (in part)

23′. Floral scales 2–2.8(–3.2) mm, generally remaining appressed at maturity and not imbricate; spikelets disarticulating at each joint of the rachilla, each scale remaining attached to its rachis joint…**C. odoratus**

24. Floral scales 1–2 mm; spikelets densely packed on the spike rachis, the rachis either hidden completely or poorly visible between the spikelets; plants annual…(25)

24′. Floral scales (1.8–)2–4.5(–6) mm; spikelets loosely to moderately packed on the spike rachis, the rachis usually easily visible between the spikelets; plants annual or perennial…(26)

25. Floral scales 1–1.5 mm; cypselae (0.4–)0.7–1 mm; inflorescence bracts (3–)5–7(–11); plants native, from widely scattered locations in New Mexico…**C. erythrorhizos** (in part)

25′. Floral scales 1.5–2 mm; cypselae 1.3–1.4 mm; inflorescence bracts 3–4; plants introduced, known from the Rio Grande corridor in C New Mexico…**C. glomeratus**

26. Floral scales and spikelet rachilla persistent; plants perennial, colonial, with fine rhizomes bearing tubers in mature plants (often not collected)…(27)

26′. Floral scales or spikelets as a whole deciduous; plants annual or perennial, generally not colonial, the rhizomes if present not bearing tubers…(28)

27. Spikelets golden orange-brown, 10–20(–25) per spike…**C. esculentus**

27′. Spikelets purplish to reddish brown, (2)3–7(–12) per spike…**C. rotundus**

28. Floral scales 2.3–3.1 mm, often with some pink to reddish tinge; plants annual; culms 5–25 cm × 1–1.5 mm; widely scattered in SW mountains and river corridors…**C. parishii**

28′. Floral scales 3–4.5(–6) mm, tan to yellowish or reddish brown; plants perennial (sometimes flowering first year); culms 20–100 cm × 1–6 mm (shorter in depauperate *C. strigosus*); rare…(29)

29. Floral scales reddish brown at maturity, deciduous; plants robust perennials with long rhizomes; anthers 1.5–2 mm; EC New Mexico…**C. setigerus**

29′. Floral scales yellowish green to yellowish brown at maturity, persistent (the spikelet falling whole); plants short-lived perennials to annuals with hardened cormlike bases, rarely producing rhizomes; anthers 0.3–0.5 mm; known from only a few collections in the bootheel…**C. strigosus** (in part)

Cyperus acuminatus Torr. & Hook.—Tapertip flatsedge. Plants annual; culms (10–)20–30 cm, longer than the leaves; leaf blades 1–2(–4) mm wide; inflorescence of 1 terminal and 1–4 hemispherical to spherical heads (spikes) on rays, bracts 3–6, the longest strongly ascending to vertical, longer than the inflorescence; heads 7–12(–15) mm wide, with (15–)25–50(–75) digitate spikelets; spikelets compressed, of 8–20(–35) flowers; scales 1–2 mm, yellowish green to light reddish brown, apex acuminate, slightly excurved; stigmas 3; cypselae trigonous, ellipsoid, 0.8–1.1 mm, 0.3–0.4 mm wide. Wet shorelines and riverbanks, disturbed soils, 5100–6300 ft. Grant, Hidalgo.

Cyperus andinus Palla ex Kük.—Flatsedge. Plants perennial, cespitose from short, stout rhizomes; culms 4–30 cm, longer than the leaves; leaf blades 1–2.5(–3) mm wide; inflorescence of 1 terminal hemispherical to subspherical cluster of (5–)20–60(–100) sessile spikelets, bracts 2–5, horizontal to greatly reflexed and almost parallel to culm; spikelets compressed, of 8–12(–24) closely imbricate flowers; scales 1.7–3.2 mm, base greenish to tan, upper sides and apex whitish, apex acuminate to mucronate; stigmas 3; cypselae trigonous, obovoid, 0.8–0.9 × 0.5–0.8 mm. Clearings in woodlands and forests; known from 1 location in the Mogollon Mountains, 7600 ft. Sierra.

Cyperus bipartitus Torr. [*C. rivularis* Kunth]—Slender flatsedge. Plants annual; culms 3–25 cm, shorter to longer than the leaves; leaf blades 1–2 mm wide; inflorescence of 1 sessile and 1–4 additional hemispherical heads on rays, bracts 2–3, horizontal to ascending and longer than the inflorescence; heads 7–12(–15) mm wide, of 3–5(–8) digitate spikelets; spikelets compressed, of (6–)10–26(–32) closely imbricate flowers; scales 1.9–2.7 mm, light to dark brown, apex obtuse; stigmas 2; cypselae lenticular, obovoid, 1–1.3 × 0.6–0.8 mm. Emergent shorelines, stream banks, ditches, disturbed areas, 4000–6700 ft. NW, C, WC.

Cyperus erythrorhizos Muhl.—Redroot flatsedge. Plants annual, small to robust; culms 5–25(–100) cm, equal to slightly longer than the leaves; leaf blades 2–5(–11) mm wide; inflorescence of 1 terminal cluster of 1–3(–6) sessile spikes and 2–6(–12) more spikes or clusters of spikes on rays, bracts (3–)5–7 (–11), horizontal to slightly ascending; spikes cylindrical, 10–30(–55) × (6–)10–16(–23) mm; spikelets 20–40(–80), quadrangular to slightly compressed, of 6–16(–30) closely imbricate flowers; scales deciduous, 1.3–1.5 mm, medially green, laterally yellowish to reddish brown, apex obtuse with small mucro; stigmas 3; cypselae trigonous, ovoid, (0.4–)0.7–1 × 0.4–0.6 mm. Emergent shorelines, 3900–5600 ft. NW, C, SC, SW.

Cyperus esculentus L.—Yellow nutsedge. Plants perennial, rhizomatous, bearing edible tubers, stems single or in small clumps; culms (20–)30–40(–65) cm, longer than the leaves; leaf blades 2–6.5 mm wide; inflorescence of 5–10 spikes, 1 to several subsessile and the rest on rays, bracts (3)4–5(7), horizontal to ascending; spikes cylindrical, 18–30 × 18–35 mm; spikelets 10–20(–25), compressed, with 9–42 flowers; scales persistent, (1.8–)2–3.1 mm, medially brown to green, laterally yellow to orange-brown, apex obtuse to subacute; stigmas 3; cypselae trigonous, ellipsoid, (1.1–)1.3–1.6 × 0.3–0.6(–0.8) mm. Riparian areas, washes, ditches, croplands, roadside margins, 3600–8900 ft. Widespread.

1. Spikelets 2 mm wide or less; scales 1.8–2.7 mm…**var. leptostachyus** Boeckeler
1'. Spikelets 2–3 mm wide; scales 2.7–3.4 mm…**var. macrostachyus** Boeckeler

These varieties may be of questionable distinction, as the actual specimens seem to represent a continuous range, and the 2 characters are not consistently aligned within a given variety. Of those clearly identified, var. *leptostachyus* is far more common in New Mexico.

Cyperus fendlerianus Boeckeler—Fendler's flatsedge. Plants perennial, stems single or in small clumps from short, stout rhizomes; culms (7–)20–70(–85) cm, longer than the leaves; leaf blades (2–)3–5(–7) mm wide; inflorescence of (1–)3–4(–6) sessile spikes in a compact head, the longest erect, the others clustered at its base, bracts (2)3–6(–10), horizontal to reflexed; central spike cylindrical, (12–)18–30 × (6–)12–20 mm; spikelets (8–)15–30, slightly compressed, of 5–8(–10) flowers; floral scales decidu-ous, 2.4–2.8(–3.4) mm, ovate to orbiculate, medially brown to green, laterally yellow to green-brown, apex obtuse, mucronate; stigmas 3; cypselae trigonous, obovoid, 1.6–1.9 × 1.1–1.4 mm wide. Clearings in woodlands and forests, 4900–9500(–10,500) ft. Widespread except NE, EC.

Cyperus flavicomus Michx. [*C. albomarginatus* (Mart. & Schrad. ex Nees) Steud.]—Whitedge flatsedge. Plants annual; culms 30–75 cm, longer than the leaves; leaf blades 2–8 mm wide; inflorescence of 1 to several sessile spikes and (3–)5–11 spikes or clusters of spikes on rays, bracts 3–7, horizontal to ascending; spikes cylindrical, shorter than or equal to the width; spikelets 6–60, highly compressed, of 6–24 spreading flowers, the mature spikelet edge looking saw-toothed; scales (1.4–)1.7–2.3 mm, lat-erally ribbed, light to reddish brown with conspicuous clear margins, apex acute to obtuse; stigmas 2; cypselae lenticular, obovoid, 1.2–1.6 × 0.6–1.1 mm. Moist draws and grasslands in oak woodland; Animas and Peloncillo Mountains, 4000–5700 ft. Hidalgo.

Cyperus floribundus (Kük.) R. Carter & S. D. Jones. [*C. uniflorus* Torr. & Hook. var. *floribundus* Kük.]—Rio Grande flatsedge. Plants perennial, cespitose; culms 15–40 cm, slightly longer than the leaves; leaf blades 1–2 mm wide; inflorescence of 1 sessile spike and 4–14 single spikes on rays, bracts 3–6, ascending; spikes cylindrical, shorter than wide; spikelets (5–)20–35(–60), deciduous, quadrangular, of 3–8(–13) flowers; scales appressed, persistent, 2.6–4.8 mm, laterally whitish to reddish brown, medi-ally green, apex acute with mucro 0.3–0.5 mm, distal sterile scale involute with longer cusp; stigmas 3; cypselae trigonous, narrowly ellipsoid, 1.8–2.4 × 0.6–0.8 mm. Sandy, open areas; known from only 1 loca-tion, 4360 ft. Curry.

Cyperus glomeratus L.—Clustered flatsedge. Plants annual to biennial, single-stemmed to cespi-tose; culms 30–60(–90) cm, shorter to longer than the leaves; leaf blades 4–8 mm wide; inflorescence of a cluster of 1–5 sessile spikes and 3–8 spikes or clusters of spikes elevated on rays, bracts 3–4, horizontal to ascending; spikes cylindrical, dense; spikelets 50+, slightly compressed, of 8-16 slightly spreading flowers; scales 1.4-2 mm, reddish brown, apex obtuse; stigmas 3; cypselae trigonous, narrowly oblong, 1.3-1.4 × 0.5 mm. Riverbanks and ditches; Rio Grande floodplain near Albuquerque, 4700-5000 ft. Bernalillo. Introduced.

Cyperus michelianus (L.) Link—Michel's flatsedge. Plants annual; culms 2-25 cm, slightly shorter to longer than the leaves; leaf blades 1-2.5 mm wide; inflorescence a single dense terminal head of numerous sessile spikelets digitately aggregated on short, hidden rays, bracts 3-6, horizontal to slightly reflexed; spikelets of 10-20 flowers, spirally arranged on the rachis; scales 2 mm, green to yellowish white, apex with a slightly recurved mucro; stigmas 2(3); cypselae plano-convex or trigonous, narrowly oblong, ca. 1 mm. Wet areas along riverbanks, floodplains, pond margins; known from along the Gila River, 3900-4400 ft. Grant. Introduced.

Cyperus niger Ruiz & Pav.—Black flatsedge. Plants perennial, densely tufted from fine rhizomes; culms 5-40(-95) cm, longer than the leaves; leaf blades 1.5-3 mm wide; inflorescence of 1 terminal and rarely 1-2 hemispherical heads on rays, bracts 2-3, horizontal to reflexed; heads 7-20 mm wide, digitate; spikelets (3-)5-25(-60), highly compressed, of 4-18 flowers; scales 1.5-2.1 mm, chestnut-brown to blackish, apex obtuse; stigmas 2; cypselae lenticular, ellipsoid, 1.1-1.4 × 0.6-0.8 mm. Ciénegas, wet meadows, ditches, riverbanks, seeps, springs, 3900-7400 ft. NC, C, WC, SW.

Cyperus odoratus L.—Fragrant flatsedge. Plants annual to short-lived perennial, single-stemmed to cespitose; culms (5-)10-50(-130) cm, longer than the leaves; leaf blades 4-12 mm wide; inflorescence a cluster of 1-5(-12) sessile spikes and (0-)6-9(-12) spikes or clusters of spikes on rays, bracts 4-8(-10), horizontal to ascending; spikes cylindrical; spikelets 50+, terete to slightly flattened, of (4-)8-12(-30) appressed to slightly spreading flowers; scales 2-2.8(-3.2) mm, green medially, reddish brown to stramineous laterally, apex entire or emarginate with small mucro, each flower disarticulating with its rachis joint; stigmas 3; cypselae trigonous, narrowly ellipsoid to oblong, 1.2-1.5(-1.9) × 0.5-0.6(-0.75) mm. Riverbanks, lakeshores, emergent shorelines, disturbed wet areas, 2900-5800 ft. Scattered.

Cyperus pallidicolor (Kük.) G. C. Tucker [*C. subambiguus* Kük. var. *pallidicolor* Kük.]—Pale flatsedge. Plants perennial, stems cespitose from short rhizomes; culms (10-)30-50(-80) cm, longer than the leaves; leaf blades 2-5 mm wide; inflorescence of 3-6 sessile spikes in a compact head, occasionally 1-5 spikes elevated on rays, bracts (3)4-6(-8), horizontal to reflexed; spikes cylindrical; spikelets (10-)25-60(-100), terete, of 1-4 flowers, deciduous as a unit; scales appressed, 2.6-3.3 mm, broadly ellipsoid, medially green, laterally whitish to green, apex rounded; stigmas 3; cypselae trigonous, ellipsoid, (1.6-)1.8-2(-2.2) × 0.6-0.9 mm. Moist draws in oak-juniper woodland; found only in the Sky Island mountains, 5200-5300 ft. Hidalgo.

Cyperus parishii Britton ex Parish—Parish's flatsedge. Plants annual; culms 5-25 cm, longer than the leaves; leaf blades 2-4 mm wide; inflorescence of 1 terminal spike and 1-6 spikes elevated on rays, bracts 2-5, ascending; spikes cylindrical, 15-25 mm wide, shorter than wide; spikelets 5-30, compressed, of (4-)8-12 flowers; scales deciduous, 2.3-3.1 mm, medially green, laterally reddish purple to light reddish brown, apex acute to obtuse; stigmas 3; cypselae trigonous, brown to dark purplish brown, 1.1-1.3 × 0.6-0.9 mm. Stream banks, springs, desert washes, 3900-7600 ft. WC, SC, SW.

Cyperus retroflexus Buckley [*C. uniflorus* Thunb. var. *retroflexus* (Buckley) Kük.]—One-flower flatsedge. Plants perennial, stems tufted from short rhizomes; culms (5-)15-40(-80) cm, longer than the leaves; leaf blades (0.5-)1-3(-4) mm wide; inflorescence of 1 sessile spike and 3-6(-9) spikes elevated on rays, bracts 3-6, ascending, attenuate and often curling at maturity; spikes cylindrical; spikelets (5-)20-35(-65), quadrangular in cross-section, of 1-5 flowers, deciduous as a unit; scales appressed, 2.6-3.9 mm, narrowly ovate, medially green to tan, laterally whitish green with red spots or a deep red central area, apex rounded to acute and mucronate, terminal sterile scale involute and sometimes hooked; stigmas 3; cypselae trigonous, ellipsoid, (1.8-)2-2.4 × 0.6-0.9 mm. Damp to dry, sandy soils in desert scrub to piñon-juniper woodlands, 3400-5500 ft. NE, EC, SE, SC.

Cyperus retrorsus Chapm. [*C. cylindricus* (Elliott) Britton]—Pine-barren flatsedge. Plants perennial, cespitose from short rhizomes; culms 20–50(–85) cm, longer than the leaves; leaf blades 1.5–4(–5) mm wide; inflorescence of 1 sessile spike and 4–8(–11) spikes on rays, bracts 3–6(–10), ascending; spikes cylindrical, dense; spikelets 40–120, subterete, of 2–5 flowers, deciduous as a unit; scales appressed, 1.8–2.5 mm, oblong, medially green, laterally tan to brown, apex rounded to acute; stigmas 3; cypselae trigonous, oblong-ellipsoid, 1.2–1.7 × 0.5–0.6 mm. Damp to dry, sandy soils in open woods and thickets; Ghost Ranch and in Las Cruces, 3900–6400 ft. Doña Ana, Rio Arriba.

Cyperus rotundus L.—Purple nutsedge. Plants perennial, rhizomes bearing edible tubers, stems single or in small clumps; culms 10–35(–40) cm, longer than the leaves; leaf blades 2–6 mm wide; inflorescence of 4–10 spikes, 1 to several subsessile and the rest elevated on rays, bracts (2)3–5, horizontal to ascending; spikes cylindrical, 12–30 × 12–50 mm; spikelets (2)3–7(–12), compressed, of 6–36(–42) flowers; scales persistent, (1.8–)2.6–3.3 mm, ovate, medially green, laterally purple to reddish brown, apex obtuse; stigmas 3; cypselae trigonous, ellipsoid, 1.4–1.7(–1.9) × 0.8–1 mm. Croplands, lawns, ditches, roadside margins, 3600–4300 ft. Doña Ana, Eddy. Introduced.

Cyperus schweinitzii Torr.—Sand flatsedge. Plants perennial, stems single or in small clumps from short, stout rhizomes; culms (10–)20–50 cm, longer than the leaves, upper portion scabrous on the angles; leaf blades 2–6 mm wide; inflorescence of 1 loose cluster of sessile, ascending to upright spikelets, and 3–5 similar spikelet clusters on rays, bracts 3–7, ascending to erect; spikelets slightly compressed, of (1–)5–10(–14) flowers; scales deciduous, spreading, (2.3–)2.8–3.4 mm, broadly to oblong-ovate, laterally tan, medially green, apex rounded with prominent cusp; stigmas 3; cypselae trigonous, broadly ellipsoid to obovoid, 2–2.4 × 0.9–1.4 mm. Sandy soils, dunes, from piñon-juniper woodland to prairies, 4300–7500 ft. Widespread.

Cyperus sesquiflorus (Torr.) Mattf. & Kük. [*Kyllinga odorata* Vahl]—Fragrant spiked sedge. Plants perennial, cespitose; culms 10–25(–45) cm; leaf blades 2–3(–4) mm wide; inflorescence of 1–3(4) tightly clustered, sessile spikes, the central one upright, longer than the lateral ones, bracts (2)3–4, horizontal to slightly deflexed; central spike cylindrical, 6–12(–18) × 4–8 mm; spikelets (50–)75–100, compressed, 2–3 mm, of 2(3) scales, only the lower one bisexual and fertile; scales 2–2.5 mm, whitish, with 2–3 lateral ribs; stigmas 2; cypselae biconvex, broadly ovate, 2–2.5 × 0.7–0.9 mm. Damp grasslands; 1 collection at Ghost Ranch, 6400 ft. Rio Arriba.

Cyperus setigerus Torr. & Hook.—Plants perennial, rhizomatous; culms 50–100 cm; leaf blades 4–7 mm wide; inflorescence of 1–4 sessile spikes and 5–13 spikes or clusters of spikes on rays, second-order rays sometimes present, bracts 3–8, ascending; spikes cylindrical, 5–25 × 15–70 mm; spikelets 5–30, compressed, of 6–24 flowers; scales imbricate, appressed to somewhat spreading, deciduous, 3–4 mm, medially green, laterally reddish to yellowish brown, apex obtuse; stigmas 3; cypselae trigonous, oblong, 1.5 × 0.4–0.5 mm. Ditches, roadsides, croplands; 1 collection N of Santa Rosa, 5350 ft. Guadalupe.

Cyperus sphaerolepis Boeckeler [*C. rusbyi* Britton; *C. fendlerianus* Boeckeler var. *debilis* (Britton) Kük.]—Rusby's flatsedge. Plants perennial, stems single or in small clumps from short, stout rhizomes; culms (8–)15–40(–60) cm, longer than the leaves, upper culms smooth on the angles; leaf blades 1–3(–5) mm wide; inflorescence of 1–3 sessile ascending to upright spikelets, with (1)2–6(–11) similar spikelet clusters on rays, bracts 2–3(–7), ascending to erect; spikelets slightly compressed, of (1–)4–8(–14) flowers; scales deciduous, spreading, (1.4–)1.8–2.4 mm, broadly ovate to suborbiculate, medially green, laterally tan, apex obtuse, with short cusp; stigmas 3; cypselae trigonous, ovoid to obovoid, 1.4–1.6(–1.9) × 1–1.4 mm. Clearings in montane forests, pine-oak, piñon-juniper woodlands, rocky slopes, grasslands, 4800–8000 ft. C, WC, SE, SC, SW, Harding, San Miguel.

Cyperus squarrosus L. [*C. aristatus* Rottb.; *C. inflexus* Muhl.]—Awned flatsedge. Plants annual; culms 2–16 cm, longer than the leaves; leaf blades 0.5–2.5 mm wide; inflorescence of 1 sessile spike and infrequently (1–)3–6 spikes on rays, bracts (1)2–4, the longest ascending to vertical, longer than the inflorescence; spikes hemispherical, 9–20 mm wide, subdigitate; spikelets (2–)6–20(–40), compressed, of (4–)10–20(–34) spreading flowers; scales 1.2–1.8(–2.2) mm, deciduous, laterally green to tan to reddish brown, apex cuspidate and strongly excurved; stigmas 3; cypselae trigonous, obovoid, 0.7–0.8(–1.1) × 0.3–0.4(–0.5) mm. Receding shorelines, floodplains, road margins, soil pockets in bedrock, 3900–9100 ft. N, EC, C, WC, SE, SC, SW.

Cyperus strigosus L.—Straw-colored flatsedge. Plants perennial, tufted from short rhizomes; culms (1–)20–40(–90) cm, with cormlike bases; leaf blades 1–4(–8) mm wide; inflorescence of 1–4 sessile spikes and (1–)3–6(–8) spikes or clusters of spikes on rays, second-order rays sometimes present, bracts (3–)5–7(–10), ascending; spikes cylindrical, (6–)10–28(–50) × 10–40 mm; spikelets (5–)12–50, stiff and straight, slightly compressed, of 3–11 flowers, deciduous as a unit; scales appressed, 3.2–4.5(–6) mm, medially green, laterally yellowish green to yellowish brown, apex acute; stigmas 3; cypselae trigonous, narrowly oblong, (1.5–)1.8–2.4 × 0.5–0.6 mm. Pond shores, ciénegas, damp wash beds, 5000–5200 ft. Hidalgo.

Cyperus subsquarrosus (Muhl.) Bauters [*Lipocarpha micrantha* (Vahl) G. C. Tucker; *Scirpus micranthus* Vahl]—Small-flower halfchaff sedge. Plants annual, diminutive; culms 2–20 cm, longer than the leaves; leaves 0.5 mm wide or less, flat to involute; inflorescence pseudolateral, of 1–2(3) sessile spikes, bracts 1–2, the longest erect, appearing as an extension of the culm; spikes 1–3(–5) mm, ovoid; spikelets 50–150, spirally arranged, of 2 scales, only 1 scale fertile; main scale 0.9–1.6 mm, yellowish to reddish brown with a greenish midvein, oblanceolate to obovate, apex tapering to an acute point, spreading to slightly recurved; inner scale 0.1–0.2 mm or absent; stigmas 2; cypselae obovoid to terete, 0.4–0.6(–0.7) mm. Sandy soils, emergent shorelines, stream banks, pond margins; known from 1 location, 5000–5200 ft. Hidalgo.

ELEOCHARIS R. Br. – Spikerush

Plants perennial or less frequently annual, cespitose to colonial from long rhizomes, the rhizomes in some species terminating in tubers or bulbs, some species stoloniferous, fibrous-rooted in the annual species; culms terete to compressed in cross-section, spongy or hollow, glabrous; leaves 2 per culm, basal, blades absent, sheaths sometimes with a "tooth" or mucro projecting beyond the apex, ligules absent; inflorescence a single terminal spikelet, bract lacking; spikelet terete, ovoid to cylindrical, of 4–500+ florets spirally arranged along the rachilla; flowers bisexual, in the axil of a scale, the proximal 1–2(3) scales empty in some species, perianth of bristles, shorter to longer than the achene; stamens 1–3; pistil with 1 style and 2 or 3 stigmas, style base usually persistent as a tubercle on the top of the achene, often distinct in shape, color, and texture, sometimes similar and appearing as a beaklike continuation of the achene; achene biconvex, plano-convex, or trigonous to subterete.

1. Stigmas 2 (or a mix of 2 and 3, with up to 1/3 of flowers with 3 stigmas in several annual species); cypselae biconvex (or up to 1/3 compressed-trigonous)…(2)
1'. Stigmas all or mostly 3; cypselae trigonous…(9)
2. Plants perennial, colonial with obvious rhizomes in addition to fibrous roots; stigmas almost always 2…(3)
2'. Plants tufted annuals, with fibrous roots only (late-season plants may rarely develop rhizomes); up to 1/3 of flowers with 3 stigmas in some species…(5)
3. Proximal scale nearly orbicular, clasping 100% of the culm circumference; second proximal scale always with flower; culms 0.3–0.8(–1.4) mm wide, averaging 0.6 mm wide, terete; spikelet 3–18 mm; floral scales 15–50…**E. erythropoda**
3'. Proximal scale usually longer than wide, clasping 2/3–3/4 of the culm circumference; second proximal scale with or without flower; culms 0.5–5 mm wide, averaging 1.3 mm wide, terete or compressed; spikelet 5–40 mm; floral scales 30–100…(4)
4. Stem often strongly compressed; distal sheath summit subtruncate, sometimes with apical tooth present; spikelets 5–40 mm, the apex often sharply pointed; second proximal scale with or without flower …**E. macrostachya**
4'. Stem terete to slightly compressed; distal sheath summit oblique, often splitting, without apical tooth; spikelets 5–25 mm, the apex rounded to pointed; second proximal scale always without flower…**E. palustris**
5. Tubercles not strongly dorsoventrally compressed, cross-sectional shape similar to shape of achene, differentiated from the achene with a distinct change in color and texture or constriction; mature cypselae black; distal leaf sheath apex with an acute to acuminate tip, without apical tooth…(6)

5'. Tubercles strongly dorsoventrally compressed, cross-sectional shape proportionally much thinner than the cross-sectional shape of achene, distinct but ± confluent with the achene; mature cypselae stramineous to dark brown; distal leaf sheath apex oblique to acute, often toothed…(8)

6. Cypselae 0.3-0.5 mm, 0.3-0.4 mm wide; perianth bristles white or clear; tubercle usually 1/4 the width of the achene or less; culms 2-15 cm…**E. atropurpurea**

6'. Cypselae 0.5-1.1 mm, 0.3-0.7 mm wide; perianth bristles red-brown to whitish; tubercle usually > 1/4 as wide as the achene; culms 3-45 cm…(7)

7. Cypselae 0.5-0.65 mm; bristles 4-6, light brown to whitish, usually shorter than the achene; spikelets ellipsoid to ovoid, 2-4 mm, 10-26-flowered, the tip acute; floral scales brown with a conspicuous green or lighter midrib, 1 mm; culms 3-6(-9) cm, 0.2-0.3 mm thick…**E. microformis**

7'. Cypselae 0.7-1.1 mm; bristles (0-)6-8, brownish, usually exceeding the achene or absent; spikelets subglobose to ovoid, 3-7 mm, 28-50-flowered, the tip rounded; floral scales straw-colored with an inconspicuous lighter midrib, 1.5-2 mm; culms 5-25(-45) cm, 0.3-1 mm thick…**E. geniculata**

8. Larger spikelets lanceoloid to subcylindrical; tubercles not > 1/4 as high as achene; perianth bristles shorter than achene to equaling tubercle, or often absent…**E. engelmannii**

8'. Larger spikelets broadly ovoid to lanceoloid; tubercles 1/4-1/3 as high as the achene; perianth bristles equaling to exceeding the tubercle…**E. obtusa**

9. Floral scales cleft at the apex (at least on the lower scales); culms subterete to strongly compressed… **E. compressa**

9'. Floral scales entire; culms terete to slightly compressed…(10)

10. Cypselae with a network of strong vertical ridges interconnected by fine horizontal ridges; tubercle separated from the achene by a distinct constriction…(11)

10'. Cypselae without a regular pattern of ridges; tubercle either confluent with achene or with a distinct constriction…(12)

11. Plants annual, tufted, usually with many stems radiating from a central clump of fibrous roots, rarely producing rhizomes; culms 0.1-0.3 mm thick, often some curving upward, usually < 3 cm in our specimens; floral scales 1-1.5 mm, apex narrowly acute to acuminate; anthers 0.3-0.5 mm…**E. bella**

11'. Plants perennial, with fine rhizomes (hard to see), forming large, fine, irregular colonies; culms 0.2-0.7 mm thick, usually straight, 3-25 cm; floral scales 1.5-2.5(-3.5) mm, apex rounded to acute; anthers 0.6-1.5 mm…**E. acicularis**

12. Tubercle distinct from achene, with a definite constriction where it joins the rounded top of the achene; distal sheath apex usually with tooth on some culms; plants colonial from long rhizomes, 0.5-2 mm thick…(13)

12'. Tubercle confluent with achene, the achene tapering into an acute to acuminate tip; distal sheath apex without a tooth; large plants tufted from very short rhizomes, or small colonial plants from fine rhizomes, 0.1-1 mm thick…(14)

13. Spikelets narrowly lanceoloid to cylindrical, 3-20 mm; floral scales 15-40, 3-4 per mm of rachilla length… **E. parishii**

13'. Spikelets ovoid or ellipsoid to subcylindrical, 4-12 mm; floral scales 30-100, 6-10 per mm of rachilla length…**E. montevidensis**

14. Culms to > 1 m, tufted from short, stout, ascending or horizontal caudexlike rhizomes, often forming large, dense colonies; culms compressed, with some arching to decumbent and rooting at the tip; bulbs not present at the rhizome tips; floral scales 20-40…**E. rostellata**

14'. Culms < 35 cm, colonial from fine rhizomes; culms subterete to slightly compressed, erect, never rooting at the tips; bulbs or tubers often present at the rhizome tips; floral scales 3-25…(15)

15. Cypselae 1.5-2.7 mm; perianth bristles rudimentary to equaling the achene; floral scales 3-10, 2.5-6 mm; rhizomes 0.2-1 mm thick; culms 5-35 cm…**E. quinqueflora**

15'. Cypselae 0.7-1.1 mm; perianth bristles absent or rudimentary, < 1/2 achene length when present; floral scales 6-25, 1.7-2.5 mm; rhizomes 0.1-0.2 mm thick; culms 2-9 cm…**E. coloradoensis**

Eleocharis acicularis (L.) Roem. & Schult.—Needle spikerush. Plants perennial, colonial from fine rhizomes; culms 1-25(-60) cm, 0.2-0.5(-0.7) mm thick, terete to occasionally compressed; distal sheath tight to inflated, apex rounded; spikelets 2-8 × 1-2 mm, ovoid, lanceoloid, to subcylindrical; scales 4-25, 1.5-2.5 mm, apex blunt to acute; bristles usually absent, occasionally 2-4; anthers 0.7-1.5 mm; stigmas 3; cypselae trigonous to subterete, 0.7-1.1 × 0.3-0.6 mm, with longitudinal and fine horizontal ridges, tubercle pyramidal to depressed, with basal constriction. Wet shorelines and riverbanks, meadows, springs, disturbed soils, 6300-10,000 ft. NC, NW, WC, Harding, Hidalgo.

Eleocharis atropurpurea (Retz.) J. Presl & C. Presl—Purple spikerush. Plants annual, densely tufted; culms 2-12(-15) cm, 0.2-0.4 mm thick; distal sheath tight to the culm, apex acute; spikelets 2-6 × 1-2.5 mm, ovoid to ellipsoid; scales up to 100, 0.6-1.3 mm, apex rounded to acute; bristles (0-)4-6, typically 4, lacking or shorter than the achene; anthers 0.3–0.5 mm; stigmas 2; cypselae glossy black, biconvex, obovoid, 0.3–0.5 × 0.3–0.4 mm, tubercle whitish, flattened to subconic, with basal constriction. Floodplains, shorelines, riverbanks, stock ponds; known from 1 location in the White Sands area, 4000–4400 ft. Doña Ana.

Eleocharis bella (Piper) Svenson—Beautiful spikerush. Plants annual; culms 1–3(–7) cm, 0.2–0.3 mm thick, terete to 4-angled; distal sheath slightly inflated, apex oblique and acute; spikelets 1.5–2.5(–4) × 0.8–2 mm, ovoid; scales 4–6(–15), 1–1.5 mm, apex acute to rounded, sometimes slightly recurved; bristles absent; anthers 0.3–0.5 mm; stigmas 3; cypselae trigonous to subterete, 0.6–1.1 × 0.3–0.4 mm, with fine longitudinal and horizontal ridges, tubercle pyramidal to depressed, appressed to achene with or without much constriction. Drying soil on shorelines, riverbanks, and seasonally wet depressions (4200–) 7500–8000 ft. Catron, Sierra.

Eleocharis coloradoensis (Britton) Gilly [*E. parvula* (Roem. & Schult.) Link ex Bluff, Nees & Schauer var. *anachaeta* (Torr.) Svenson]—Colorado spikerush. Plants perennial, colonial from fine rhizomes often ending in 2.5–4 mm tubers with a terminal bud; culms 1.5–7(–9) cm, 0.2–0.5 mm thick, soft and spongy; distal sheath membranous, often disintegrating, apex rounded; spikelets 2–5(–6) × 1–2.5 mm, ovoid to ellipsoid; scales 5–10(–25), 1.7 mm, apex rounded to acute; bristles 0(–5), rudimentary to 1/2 the length of the achene; stigmas 3; anthers 0.6–1.2 mm; cypselae light brown, trigonous, obovoid, 0.75–1.1 × 0.5–0.7 mm, tubercle confluent with the achene. Floodplains, drying shorelines and riverbanks, ponds, ditches, 2900–6700 ft. NC, NW, SE, Chaves.

Eleocharis compressa Sull.—Flatstem spikerush. Plants perennial, colonial from rhizomes 2–3 mm thick; culms 8–45 cm, (0.2–)0.5–1 mm thick, 1–2 times as wide as thick; distal sheath membranous, apex obtuse to subtruncate, without tooth; spikelets 4–8 × 2–4 mm, ovoid; scales 20–60, spreading in fruit, 2–3(–4) mm, apex acute to acuminate, hyaline, bifid to shallowly cut in proximal scales, entire in upper scales; bristles 0–5, shorter than to as long as the achene; stigmas (2)3; anthers 0.7–2 mm; cypselae yellow to dark brown, trigonous, obovoid to obpyriform, 0.8–1.1 × 0.6–0.8 mm, tubercle with basal constriction, depressed or rudimentary. New Mexico material belongs to **var. acutisquamata** (Buckley) S. G. Sm. Seasonal seeps, depressions, meadows, woods; 1 location, 7600 ft. Colfax.

Eleocharis engelmannii Steud.—Engelmann's spikerush. Plants annual, densely tufted; culms 2–40 cm, 0.5–1.5(–2) mm thick, straight; distal sheath tight to culm, rarely flaring near apex, apex acute to rounded, sometimes with tooth to 0.3 mm; spikelets 5–10(–20) × 2–3(–4) mm, lanceoloid to subcylindrical to ovoid; scales 25–100(–200), 2–2.5 mm, apex rounded to acute; bristles (0–)5–8, vestigial to equaling the achene; anthers 0.3–0.7(–1) mm; stigmas 2–3; cypselae brown, biconvex or obovoid, 0.9–1.1(–1.5) × 0.7–1.1 mm, tubercle depressed, subdeltoid, confluent with and almost as wide as the achene, 2/5 or less as high as wide. Receding shorelines, riverbanks, seasonal wetlands, ponds, 7200–8400 ft. WC, Colfax, San Miguel.

Eleocharis erythropoda Steud.—Bald spikerush. Plants perennial, colonial from rhizomes; culms 8–55(–80) cm, 0.3–1.6 mm thick, terete; distal sheath thickened, tip obtuse to subacute, tooth rarely present; spikelets 3–18 × 2–3(–5) mm, ovoid or lanceoloid to subcylindrical; scales 15–50, 2–3.5 mm, apex rounded in proximate scales, acute in distal ones; bristles (0–)4; anthers 1–1.8 mm; stigmas 2; cypselae light yellow, biconvex, obovoid, 0.9–1.6 × 0.7–1.2 mm, tubercle pyramidal, higher to lower than wide, with basal constriction. Wet shorelines and riverbanks, meadows, springs, fens, ponds, (5000–)5600–8900 ft. NC, NW, EC, C, WC, SC, Luna.

Eleocharis geniculata (L.) Roem. & Schult.—Capitate spikerush. Plants annual, densely tufted; culms to 45 cm, 0.3–1 mm thick, often curving upward; distal sheath loose, apex acute to acuminate; spikelets (3–)4–7 × 1–4 mm, ovoid to subglobose, apex obtuse; scales 28–50, 1.5–2(–3) mm, apex rounded to acute; bristles (0–)6–8, lacking to exceeding the achene; stigmas 2; anthers 0.6–0.8 mm; cypselae glossy black, biconvex, obovoid, 0.7–0.9(–1.1) × 0.3–0.7 mm, tubercle whitish, depressed-umbonate, with basal constriction. Floodplains, shorelines, riverbanks, stock ponds, 4500–6500 ft. Doña Ana, Socorro, Valencia.

Eleocharis macrostachya Britton—Pale spikerush. Plants perennial, colonial from substantial rhizomes; culms 10–100 cm, 0.5–2.5(–3.5) mm thick, terete to compressed; distal sheath with apex obtuse to truncate, tooth present occasionally; spikelets 5–40 × 2–5 mm, ovoid to narrowly lanceoloid, tip usually acute; scales 30–80, 2.5–5.5 mm, apex rounded in proximate scales, acute in distal ones; bristles 4(5); stigmas 2; anthers 1.3–2.7 mm; cypselae yellow to dark yellowish brown, biconvex, obovoid, 1.1–1.9 × 0.8–1.5 mm, tubercle pyramidal, as high as to much higher than wide, with basal constriction. Wet shorelines and riverbanks, meadows, springs, ditches, pastures, ponds, 3100–11,600 ft. Widely scattered.

Eleocharis microformis Buckley—Buckley's spikerush. Plants annual, densely tufted; culms 3–6 (–9) cm, 0.2–0.3 mm thick, arching to straight; distal sheath tight, apex acute to acuminate; spikelets 2–3(–4) × 1–3 mm, ovoid to ellipsoid, the apex acute; scales 10–20(–26), 1 mm, apex obtuse; bristles (0–)4–6, retrorsely scabrous, shorter than the achene; stigmas 2(3); cypselae black, biconvex to thickly trigonous, obovoid, 0.5–0.65 × 0.4–0.5 mm, tubercle depressed-umbonate, distinct from achene with minor constriction. Wet pond margins and floodplains, springs, ephemeral swales, 4200–5600 ft. Doña Ana, Luna, Sierra (O'Kennon & Taylor, 2013).

Eleocharis montevidensis Kunth—Sand spikerush. Plants perennial, colonial from rhizomes; culms 25–50 cm, 0.5–1.2 mm thick, terete to elliptic; distal sheath tight to culm, apex truncate to rounded, apical tooth present on some to all of the culms; spikelets 4–12 × (1.5–)2–3 mm wide, ovoid to ellipsoid; scales 30–100, 1.5–2.5 mm, apex rounded to obtuse in proximate scales, sometimes acute in distal ones, sometimes horizontally wrinkled and recurved; bristles (0–)5–6(7); anthers 0.8–1.5 mm; stigmas 3; cypselae dark brown, compressed-trigonous, 0.7–1 × 0.6–0.8 mm, tubercle pyramidal, as high as wide, occasionally depressed, with basal constriction. Wet shorelines, riverbanks, meadows, springs, ponds, 3500–5800 ft. S.

Eleocharis obtusa (Willd.) Schult.—Blunt spikerush. Plants annual, densely tufted; culms 3–50(–90) cm, 0.2–2 mm thick, straight; distal sheath tight to slightly flaring at the apex, apex acute to obtuse, sometimes with tooth; spikelets (2–)5–13 × (2–)3–4 mm, broadly ovoid to less frequently lanceoloid or ellipsoid; scales 15–150, 1.5–2.5 mm, apex broadly rounded; bristles (5)6–7, exceeding the achene; anthers 0.3–0.6 mm; stigmas 2–3; cypselae brown, biconvex, obovoid, 0.9–1.2(–1.3) × 0.7–0.9 mm, tubercle deltoid, subconfluent with and almost as wide as the achene, 1/3–2/3 as high as wide. Receding shorelines and riverbanks, seasonal wetlands, ponds; known from 1 location, 6450 ft. San Juan, San Miguel.

Eleocharis palustris (L.) Roem. & Schult.—Common spikerush. Plants perennial, colonial from substantial rhizomes; culms 30–115 cm, 0.5–5 mm thick, terete to slightly compressed; distal sheath apex obtuse to truncate, tooth absent; spikelets 5–25 × 3–7 mm, ovoid to lanceoloid, apex obtuse to acute; scales 30–100, 3.5 mm, apex rounded in proximate scales, acute in distal ones; bristles (0–)4(5), tan to medium brown; anthers 1.5–2.2 mm; stigmas 2; cypselae tan to dark brown, biconvex, obovoid, 1.1–2 × 1–1.5 mm, tubercle pyramidal, as high as to much higher than wide, with basal constriction. Wet shorelines and riverbanks, meadows, springs, ditches, pastures, ponds, 3100–11,600 ft. Widespread.

Eleocharis parishii Britton—Parish's spikerush. Plants perennial, colonial from rhizomes; culms 10–50 cm, 0.2–0.7(–1) mm thick, terete to elliptic; distal sheath tight to culm, apex truncate to rounded, apical tooth present on some to all of the apices; spikelets 3–20 × 1.5–2.5 mm, narrowly lanceoloid to subcylindrical; scales 4–25, 1.5–2.5 mm, apex rounded to obtuse in proximate scales, acute in distal ones; bristles 3–7, usually present; anthers 1.1–2 mm; stigmas 3; cypselae yellowish to dark brown, compressed-trigonous, 0.8–1.4 × 0.5–0.7 mm, tubercle pyramidal, often higher than wide, with basal constriction. Wet shorelines and riverbanks, meadows, springs, ponds, 3900–8000 ft. NC, NW, C, WC, SC, SW.

Eleocharis quinqueflora (Hartmann) O. Schwartz. [*E. pauciflora* (Lightf.) Link]—Fewflower spikerush. Plants perennial, colonial from rhizomes 0.2–1 mm thick, often ending in buds, or buds found among culm bases; culms 5–35 cm, 0.2–0.5(–1.2) mm thick, subterete to compressed; distal sheath membranous, apex subtruncate to acute, tooth absent; spikelets 3–8 × 1.5–4 mm, ellipsoid to lanceoloid; scales 3–10, 2.5–6 mm, apex acute; bristles (0–)3–6, rudimentary to equaling the tubercle; anthers 2–2.5 mm; stigmas 3; cypselae tan to brown, equilateral to compressed-trigonous, obovoid, 1.6–2.3 × 0.7–1.3

mm, tubercle confluent, without basal constriction. Wet meadows, seeps, springs, fens in the Sangre de Cristo and Jemez Mountains, 8500–11,800 ft. SC, Cibola, Sandoval.

Eleocharis rostellata (Torr.) Torr.—Beaked spikerush. Plants perennial, tufted from short to ascending rhizomes, forming dense stands, rooting at the long, arching culm tips; culms 20–100 cm, 0.35–2 mm thick, strongly compressed, 1.5+ times as wide as thick; distal sheath tight to culm, apex rounded; spikelets 5–17 × 2.5–5 mm, ovoid; scales 20–40, 3.5–6 mm, apex rounded to subacute; bristles 5–6, subequal to the tubercle; anthers 2–2.4 mm; stigmas 3 (mix of 3 and 2); cypselae tan to brown, ovoid to obovoid, 1.5–2.5 × 1–1.2 mm, tubercle similar in color to the achene, without a basal constriction. Wet alkaline meadows, seeps, springs, fens, often dominant in its habitat, 3000–8800 ft. Widespread except SW.

ERIOPHORUM L. – Cottongrass, cottonsedge

Plants perennial; cespitose to colonial from long rhizomes; culms terete to trigonous; leaves basal and cauline, sometimes reduced to bladeless sheaths; blades flat to filiform, ligules present; inflorescence terminal, of a single upright spikelet to many spikelets arranged in a subcapitate to subumbellate panicle; inflorescence bracts 1 to several, leaflike to scalelike; spikelets with (10–)20–200 florets spirally arranged along the rachilla; flowers bisexual, subtended by a scale; perianth persistent, of (8–)10–25 hairlike, smooth, straight, elongate bristles, obscuring the scales in the spikelet; stamens 1–3; style with 3 stigmas, deciduous; fruit trigonous.

1. Spikelets solitary, erect…**E. scheuchzeri**
1'. Spikelets 2–10, spreading or nodding…**E. angustifolium**

Eriophorum angustifolium Honck. [*E. polystachion* L.]—Tall cottongrass. Plants perennial, colonial from rhizomes; culms to 1 m; leaves with trigonous tip, 1.5–6(–8) mm wide, distal blade longer than its sheath; inflorescence of (1)2–10 pendent spikelets in subumbels; inflorescence bract 1, leaflike, similar to distal leaf; spikelets 10–20 mm in flower, 20–50 mm in fruit, ovoid; scales 5–10 mm; perianth bristles 10+, white to pale yellow, 15–30 mm; stigmas 3; anthers 2–5 mm; cypselae oblanceoloid, 2–5 mm. Our plants belong to **subsp. angustifolium**. Marshes, bogs, fens, wet meadows, 10,400–11,900 ft. Rio Arriba, Santa Fe, Taos.

Eriophorum scheuchzeri Hoppe [*E. altaicum* Meinsh.]—White cottongrass. Plants perennial, colonial from rhizomes; culms 5–35(–70) cm; leaves with channeled to involute blades, 0.5–1.5 mm wide, distal blade reduced to a black-tipped sheath; inflorescence of 1 upright spikelet; inflorescence bract absent; spikelets 8–12(–40) mm in fruit, broadly obovoid to subglobose; scales 4–10 mm; perianth bristles 10+, bright white, 15–30 mm; stigmas 3; anthers 0.5–1.5 mm; cypselae narrowly oblong with a subulate beak, 0.4–2.5 mm. Tundra, marshes, peaty soils, riverbanks, pond shores, 12,500–12,600 ft. Taos. Rare.

FIMBRISTYLIS Vahl – Fimbry

Fimbristylis puberula (Michx.) Vahl—Hairy fimbry. Plants perennial, in small tufts from slender rhizomes; culms wandlike, to 100 cm, 1 mm thick; leaf blades 1–2 mm wide, involute, scabrid-ciliate; inflorescence of 2–20 spikelets in simple to compound anthelae; proximal inflorescence bract leaflike, shorter or longer than the inflorescence; spikelets 5–10 mm, ovoid, cylindrical, or ellipsoid; scales 2.5–3.5 mm, reddish brown with midrib excurrent as a mucro, apex obtuse; anthers 2–2.5 mm; style flat, fimbriate, stigmas 2; cypselae lenticular-obovoid, ca. 1 mm. New Mexico plants belong to **var. interior** (Britton) Kral. Moist sandy or silty soils in prairie swales or along stream banks; known from Blue Hole and Bitter Lake National Wildlife Refuge, 3400–4600 ft. Chaves, Guadalupe.

FUIRENA Rottb. – Umbrella sedge

Fuirena simplex Vahl—Western umbrella sedge. Plants annual or perennial; culms tufted or single, 10–100 cm, 0.5–1.5 mm thick; leaves cauline, blades 2–20 mm wide, margins pilose to hispid-ciliate;

inflorescence of 1–3(–5) clusters of spikelets; proximal inflorescence bract (when > 1 cluster of spikelets is present) leaflike, long-sheathing, shorter to longer than the inflorescence; spikelets 3–15(–20) mm, ovoid, lance-ovoid, or cylindrical; scales 1.5–2.5 mm, scabrid, midrib excurrent as a spreading to excurved awn, antrorsely scabrous, scale apex rounded; flowers bisexual, subtended by a scale; perianth of 2 types, the outer set of 3 sharp, short, retrorsely barbed bristles, the inner set of 3 stipitate, spatulate, longer scales, these apically barbed; stigmas 3, anthers 1–3, 0.4–1.4 mm; achene with stipe and beak, 1.2 mm. Moist soils at seeps and springs, along stream banks; known primarily from the Guadalupe and San Andreas Mountains, 4300–5900 ft. Chaves, Doña Ana, Eddy.

1. Anthers 0.9–1.4 mm; spikelets 1–5(–7), 6–20 mm; scales with 3–9 ribs, light-colored…**var. simplex**
1'. Anthers 0.4–0.7 mm; spikelets 1–12, 3–7 mm; scales with 3 principal ribs, occasionally with 1–2 less strong ribs, dark-colored…**var. aristulata**

var. aristulata (Torr.) Kral—Springs; known only from Doña Ana County, 5070 ft.

var. simplex Near springs, waterfalls, oak, mesquite, juniper; known only from Eddy County, 4340–5900 ft.

KOBRESIA Willd. – Bog sedge

Plants perennial, densely cespitose with short-ascending rhizomes usually not apparent; culms rounded-trigonous; leaves basal and cauline, basal sheaths persistent, blades involute to filiform, glabrous, ligules present; inflorescence terminal, spicate to tightly paniculate with 10–30 spikelets; proximal inflorescence bracts (when present) leaflike to scalelike; spikelets each with 1–4 sessile florets spirally arranged along the rachilla, the distal spikelets usually 1-flowered and staminate, the proximal spikelets 1-flowered and pistillate or 2–4-flowered with 1 pistillate flower proximally and 1–3 staminate flowers distally, sometimes with sterile scales, each spikelet enclosed within a scalelike bract (perigynium) that is open on 1 side; flowers unisexual, perianth absent; stamens 3; style with 3 stigmas, the style base persistent; cypselae trigonous, included within the perigynia.

1. Inflorescence a spike, unbranched; inflorescence bract 0; basal sheaths somewhat glossy, blades deciduous; plants of dry tundra habitats…**K. myosuroides**
1'. Inflorescence paniculate, with short branches; inflorescence bract present; basal sheaths dull, usually with withered remains of the blades attached; plants of mesic habitats…**K. simpliciuscula**

Kobresia myosuroides (Vill.) Fiori [*Carex myosuroides* Vill.; *K. bellardii* (All.) Degl. ex Loisel.]—Bellardi bog sedge. Plants perennial, densely cespitose; culms 5–20(–35) cm; leaves with filiform blades 0.2–0.5 mm wide, basal sheaths persisting, making a tuft below the current year's blades; inflorescence 10–30 × 2–3 mm, of many spikelets in a simple unbranched spike, proximal spikelets androgynous with 2 flowers (the lowest rarely 1-flowered and pistillate), the distal single-flowered and staminate, inflorescence bract absent; scales 2–3.5 mm, apex obtuse or cuspidate; stigmas 3; perigynia 2–3.5 mm, open on 1 side; cypselae trigonous, 2–2.8 mm. Alpine tundra and scree slopes; Sangre de Cristo Mountains, 10,200–12,800 ft. NC.

Kobresia simpliciuscula (Wahlenb.) Mack. [*Carex simpliciuscula* Wahlenb.]—Simple bog sedge. Plants perennial, densely cespitose; culms 5–35 cm; leaves with channeled blades 0.2–1.5(–2) mm wide, basal sheaths persisting with blades from the previous year; inflorescence (8–)10–35 × (2–)3–8 mm, of many spikelets in a panicle, proximal spikelets androgynous with 2 flowers or pistillate and 1-flowered, distal spikelets single-flowered, staminate; inflorescence bract present, shorter than the inflorescence; scales 2–3 mm, apex obtuse to subacute; stigmas 3; perigynia 2.5–3.2 mm, open on 1 side; cypselae trigonous, 2–3 mm. Wet montane to alpine meadows, bogs, 10,800–10,900 ft. Colfax.

SCHOENOPLECTUS (Rchb.) Palla – Naked-stemmed bulrush

Plants perennial (ours), colonial from long, tough rhizomes (ours); culms terete to trigonous, smooth; leaves all basal, rarely 1(2) cauline in the bottom 1/3 or less of the culm, blades rudimentary to well developed, flat to channeled, ligules present, membranous; inflorescences terminal, appearing lateral, of

1–100+ spikelets in capitate clusters to open panicles; inflorescence bracts 1–5, leaflike to stemlike, the proximal erect (ours), appearing as a continuation of the culm; spikelets terete, ovoid to lanceoloid or subcylindrical, with 8+ florets spirally arranged along the rachilla, scales deciduous, each subtending a flower or the proximal scale empty; flowers bisexual, perianth of 0–6(–8) retrorsely barbed bristles; stamens 3; style with 2–3 stigmas, deciduous, leaving an apiculus; achene biconvex to trigonous.

1. Inflorescence branches absent or scarcely developed, the spikelets borne in a tight cluster on the culm; culms trigonous…(2)
1'. Inflorescence branches well developed and evident, the spikelets borne on branches; culms cylindrical… (3)
2. Plants typically robust; spikelets light brown; scales often pale with brown dots, the apex notch 0.1–0.4 (–0.8) mm, with awn 0.2–0.6 mm; distal leaf blade much shorter than to equaling (rarely 1.5 times longer than) the sheath; proximal inflorescence bract 1–3(–6) cm, other bracts without blades; sides of the mid-culm deeply concave, rarely nearly flat…**S. americanus**
2'. Plants usually less robust; spikelets dark brown; scales often rich brown, the apex notch (0.3–)0.5–1 mm, with awn 0.5–1.5(–2.5) mm; distal leaf blade much longer than to nearly equaling the sheath; proximal inflorescence bract (1–)3–20 cm, other bracts with narrow blades sometimes exceeding the spikelets; sides of the midculm shallowly concave to flat or slightly convex…**S. pungens**
3. Spikelet scales uniformly orange-brown (sometimes with straw-colored streaks), the dorsal side of scales smooth to sparsely (rarely densely) scabrous; awns of the scales straight to bent, 0.2–0.8 mm; many spikelets solitary at the tips of the pedicels…**S. tabernaemontani**
3'. Spikelet scales wholly or partially pale gray to tan with prominent streaks, the dorsal side of the scales sparsely to often densely scabrous; awns of the scales usually contorted, 0.5–2 mm, most spikelets in sessile clusters of 2+…**S. acutus**

Schoenoplectus acutus (Muhl. ex Bigelow) Á. Löve & D. Löve [*Scirpus acutus* Muhl. ex Bigelow]— Hardstem bulrush. Plants perennial, colonial from stout rhizomes 5–15 mm thick; culms 1–4 m, 2–10 mm thick, cylindrical; leaf blades 1–2, distal blade shorter than the sheath; inflorescence appearing lateral, of 3–40(–190) spikelets in a compact panicle, most spikelets in clusters of 2–8; proximal inflorescence bract 1–9 cm; spikelets ovoid, 6–18(–24) mm; scales pale gray to light tan with prominent streaks, scabrid, 3–4 mm, apex notched 0.3–0.5 mm, awn 0.5–2 mm, often contorted (often broken off); perianth bristles (4–)6(–8); stigmas 2(3); cypselae dark gray-brown to almost white, obovoid, plano-convex to weakly trigonous, (1.5–)2–3 mm. Marshes, lakeshores, stream banks, often emergent in water to 1.5 m.

1. All styles with 2 stigmas; culms very firm, with air cavities in distal 1/4, mostly 0.5 mm thick…**var. acutus**
1'. Some styles with 3 stigmas; culms soft to firm, with larger air cavities in distal 1/4, mostly 1–2.5 mm thick… **var. occidentalis**

var. acutus—Hardstem bulrush. Calcareous to brackish marshes, fens, and slow streams, 3000–8000 ft. Widespread.

var. occidentalis (S. Watson) S. G. Sm.—Tule. Calcareous to brackish marshes, fens, and slow streams, 3850–9300 ft. Widespread.

Schoenoplectus americanus (Pers.) Volkart ex Schinz & R. Keller [*Scirpus americanus* Pers.; *Scirpus olneyi* A. Gray]—Olney's threesquare bulrush. Plants perennial, colonial from stout rhizomes 2–5 mm thick; culms 0.4–2.5 m, 3–10 mm thick, sharply trigonous-winged, sides deeply concave to flat; leaf blades 1–3, distal blade 0.2–1.5 times as long as the sheath; inflorescence appearing lateral, of 2–20 sessile spikelets in a capitate cluster; proximal inflorescence bract erect, 1–3(–6) cm; spikelets ovoid, 5–15 mm; scales orangish red to light tan with linear spots, smooth, margins ciliolate, 3–4 mm, apex notched 0.1–0.4 mm, awn straight, 0.2–0.6 mm; perianth bristles (2–)5–6(7); stigmas 2(3); cypselae brown, obovoid, plano-convex to compressed-trigonous, 1.8–2.8 mm. Marshes, ditches, stream banks, pond shores, 3400–7600 ft. Widespread except N.

Schoenoplectus pungens (Vahl) Palla [*Scirpus americanus* of NM authors, not Pers.; *Scirpus pungens* Vahl]—Common threesquare bulrush. Plants perennial, colonial from stout rhizomes 1–6 mm thick; culms 0.1–2 m, 1–6 mm thick, bluntly trigonous, sides convex to slightly concave; leaf blades 2–6, distal blade (1)2–5 times as long as the sheath; inflorescence appearing lateral, of 1–5(–10) sessile spikelets in a capitate cluster; proximal inflorescence bract erect, (1–)3–20 cm; spikelets ovoid, 5–23 mm; scales dark

orangish red to brown or tan with linear spots, smooth, margins ciliolate, 3–4 mm, apex notched (0.3–)
0.5–1 mm, awn irregularly bent, 0.5–1.5(–2.5) mm; perianth bristles 4–8; stigmas 3; cypselae brown,
obovoid, lenticular to compressed-trigonous, (2–)2.5–3.5 mm. Marshes, ditches, stream banks, lakeshores, often emergent in water to 0.7 m deep, 3000–8700 ft. Widespread.

Schoenoplectus tabernaemontani (C. C. Gmel.) Palla [*Scirpus tabernaemontani* C. C. Gmel.;
Scirpus validus Vahl]—Softstem bulrush. Plants perennial, colonial from stout rhizomes 3–10 mm thick;
culms 0.5–3 m, 2–10 mm thick, cylindrical; leaf blades 1–2, distal blade much shorter than the sheath;
inflorescence appearing lateral, of 15–200 spikelets in a panicle often with drooping branches, many
spikelets single, others in clusters of 2–4; proximal inflorescence bract erect, 1–8 cm; spikelets ovoid,
3–17 mm; scales dark to pale orange-brown with linear spots, smooth to lightly scabrid, 2–3.5 mm, apex
notched 0.2–0.3 mm, awn 0.2–0.8 mm, straight or bent (often broken off); perianth bristles 6; stigmas
2(3); cypselae dark gray-brown, obovoid, plano-convex, 1.5–2.8 mm. Marshes, lakeshores, stream banks,
often emergent in water to 1 m deep, 3600–8700 ft. Widespread except SE, SC.

SCHOENUS L. – Bog-rush

Schoenus nigricans L.—Black bog-rush. Plants perennial, cespitose, with caudexlike rhizomes;
culms 20–70 cm, 1 mm wide, wiry, terete; leaves basal only, with blackish sheaths, blades 0.8–2 mm wide,
involute; inflorescence of (1–)10–25 digitately clustered spikelets in an ovoid head; inflorescence bracts
1–2, spreading to erect, leaflike; spikelets compressed, distichous, green to dark brown or black, proximal
2–3 scales empty and the distal 3–8 containing a flower; perianth of 0–6 bristles, smooth or slightly
scabrous; stigmas 3; stamens 3, anthers 1–1.5 mm; cypselae whitish, glossy, ovoid to ellipsoid, 1–1.5 mm.
Alkaline marshes, springs, damp meadows; known from Karr Canyon, 6900 ft. Otero. Introduced. **R**

SCIRPUS L. – Bulrush

Plants perennial, cespitose or colonial from long rhizomes; culms trigonous to roundly trigonous, smooth,
glabrous; leaves basal and cauline, blades well developed, flat to V-shaped, prominently keeled, ligules
present; inflorescences terminal (sometimes lateral in the distal 1–3 leaf axils), of 50–500 spikelets in
subumbellate or corymbose panicles, inflorescence bracts usually 3, leaflike, ascending to spreading;
spikelets < 3.5(–5) mm diam., terete, ovoid to lanceoloid or subcylindrical, with 10–50 florets spirally
arranged along the rachilla; scales deciduous, each subtending a flower; flowers bisexual, perianth of
(0–)3–6 bristles; stamens 1–3; style with 2–3 stigmas, base not or scarcely enlarged, persistent; cypselae
biconvex, plano-convex, or trigonous. Many species formerly treated in *Scirpus* are now treated in *Bolboschoenus* and *Schoenoplectus*.

1. Terminal bract of the flowering stem single, resembling a prolongation of the culm so the inflorescence
   appears to be lateral rather than terminal (smaller bracts occasionally present but scalelike and not
   green)...go to **Schoenoplectus**
1′. Terminal bracts of the flowering stem 2+, leafy, spreading, not resembling a prolongation of the culm...(2)
2. Spikelets large, mostly 12–25 mm, few in number (commonly 3–40)...go to **Bolboschoenus**
2′. Spikelets small, mostly 3–6 mm, numerous (> 100)...(3)
3. Spikelets in open clusters, with all but the central spikelet long-pedicellate; perianth bristles smooth,
   strongly contorted, much longer than cypselae (sometimes not projecting beyond them because of their
   contortion)...**S. pendulus**
3′. Spikelets borne closely together in tight clusters with all spikelets sessile within each cluster; perianth
   bristles barbed, straight, or curved, shorter to longer than achenes...(4)
4. Stigmas predominantly 2; spikelets in many small clusters of 3–18; scales not (or only very shortly) awned;
   perianth bristles mostly 4 (sometimes up to 6), the teeth thick-walled, sharp-tipped, densely crowded
   over most of the bristle length...**S. microcarpus**
4′. Stigmas predominantly 3; spikelets in fewer, large clusters of 12–130, scales awned, awns often strongly
   spreading; perianth bristles 6, the teeth thin-walled, round-tipped, restricted mostly to distal 0.5 mm or
   less of the bristle length...**S. pallidus**

Scirpus microcarpus J. Presl & C. Presl—Panicled bulrush. Plants perennial, colonial from stout
rhizomes; culms 50–100(–150) cm; leaf blades 4–11, cauline, sheaths reddish, blades 5–15(–20) mm wide;

inflorescence of many dense clusters of 3-18 spikelets in an open, compound umbel; inflorescence bracts 2-5, leaflike, shorter to longer than the inflorescence; spikelets ovoid, 2-6(-8) × 3.5 mm; scales 1.1-3.4 mm, apex rounded to acute or apiculate to 0.2 mm; bristles (3)4(-6), stout, shorter to longer than the achene, retrorsely barbed; stigmas 2(3); cypselae almost white, ovate to obovate, biconvex to plano-convex, 0.7-1.6 mm. Marshes, moist meadows, stream banks, lakeshores, 5400-10,400 ft. NC, NW, C, WC.

Scirpus pallidus (Britton) Fernald [*S. atrovirens* Willd.]—Pale bulrush. Plants perennial, cespitose from short rhizomes; culms 60-110(-150) cm; leaf blades 5-10, cauline, sheaths green or whitish, blades 8-16 mm wide; inflorescence of a few dense clusters of 12-130 spikelets in a relatively compact, compound umbel; inflorescence bracts 2-5, leaflike, shorter to longer than the inflorescence; spikelets narrowly ovoid, 4-5 × 1.8-2.3 mm; scales 1.6-2.8 mm, apex with terete to flattened awn to 0.4-0.6(-1.2) mm; perianth bristles 6, stout, shorter than to equaling the achene, distally retrorsely toothed; stigmas (2)3; cypselae pale brown to almost white, elliptic to obovate in profile, plumply trigonous (plano-convex), 0.8-1.2 mm. Marshes, moist meadows, stream banks, 5200-8300 ft. NE, NC, NW, C, WC, SC.

Scirpus pendulus Muhl.—Rufous bulrush. Plants perennial, cespitose from short rhizomes; culms 50-150 cm; leaf blades basal and cauline, 5-7 cauline, sheaths whitish, blades 4-8(-12) mm wide; inflorescence of many individual spikelets in an open, compound cyme, rays often pendent; inflorescence bracts leaflike, shorter than the inflorescence; spikelets ovoid, lance-ovoid, or subcylindrical, 5-10(-12) × 2-3 mm; scales brown to red-brown with green midrib, 2 mm, apex mucronate; bristles 6, slender, longer than the achene, smooth; stigmas 3; cypselae pale to medium brown, elliptic, plumply trigonous, 1-1.2 mm. Marshes, moist meadows, stream banks, often on calcareous substrates, 5100-7500 ft. Harding, San Miguel.

EHRETIACEAE – EHRETIA FAMILY

Robert C. Sivinski

(Ours) leaves usually alternate, simple, entire; stipules absent; leaves solitary or clustered at the stem nodes, tomentose, strigose, or hispid, margins usually revolute; inflorescence of solitary axillary flowers, cymes, spikes, or corymbs with the ultimate inflorescence units cymose; flowers actinomorphic, perfect; sepals 5; petals 5, connate into salverform or funnelform corolla; stamens 5; pistil superior, of 2 united carpels with 2 ovules per locule, style apical, 2- or 4-branched; fruit a fleshy drupe, or nonfleshy schizocarp of 4 mericarps (or fewer by abortion) composed of nutlets. Genera in the Ehretiaceae have traditionally been included in the Boraginaceae s.l. (Boraginales Working Group, 2016; Richardson, 1977).

TIQUILIA Pers. – Crinklemat

1. Leaf blades broad, ovate to elliptic, 2-10 mm wide, villous-tomentose but not hispid-hirsute…(2)
1'. Leaf blades linear to narrowly obovate, 0.5-2 mm wide, strongly hispid-hirsute…(3)
2. Plants small erect shrubs; flowers aggregated into conspicuous, dense, plumose clusters…**T. greggii**
2'. Plants low and spreading; flowers solitary, not plumose…**T. canescens**
3. Upper blade surfaces usually green, glabrous to scabrous beneath the bristles, mostly linear and broadest at the middle or below; petioles glabrous or minutely scabrous, never villous, bristly on the margin; flowers 4-5 mm across; attachment on the nutlet open its entire length or at least below the middle… **T. hispidissima**
3'. Upper blade surfaces grayish, occasionally light green, usually puberulent-villous beneath the bristles, mostly oblanceolate and broadest above the middle; petioles usually villous, bristly on the margin; flowers 2-3 mm across; attachment scar on the nutlet entirely closed or open only above the middle…**T. gossypina**

Tiquilia canescens (DC.) A. T. Richardson [*Coldenia canescens* DC.]—Woody crinklemat. Suffrutescent herb from a woody caudex; stems prostrate, spreading, or rarely ascending; leaves shortly petiolate, ovate to elliptic, 7-15 × 4-8 mm, silvery-tomentose and strigose; flowers solitary; calyx persistent, lobes subulate or triangular; corolla pink or lavender, rarely white; nutlets 1-4, usually 4, 2.2-2.8 mm, brownish, minutely papillate, glabrous or puberulent apically. Calcareous substrates in desert scrub, 3600–5700 ft. C, EC, SE, SC, SW.

Tiquilia gossypina (Wooton & Standl.) A. T. Richardson [*Coldenia gossypina* (Wooton & Standl.) I. M. Johnst.]—Texas crinklemat. Mat-forming plant from a woody caudex; stems prostrate, spreading; leaves narrowly oblanceolate, to 8 × 1–2 mm, upper surface cinereous, occasionally green, usually puberulent-villous below the hispid bristles, in age detaching from a persistent petiole; flowers solitary; calyx persistent, lobes subulate or narrowly triangular; corolla purple to pink; nutlets 1–4, 1.1–1.3 mm, black with white pustules ventrally, attachment scar narrowly open only above the middle. Calcareous soils in desert scrub, 4200–4600 ft. Doña Ana.

Tiquilia greggii (Torr. & A. Gray) A. T. Richardson [*Coldenia greggii* (Torr. & A. Gray) A. Gray]— Plumed crinklemat. Low shrub, 2–7 dm; branches numerous and crowded; leaves shortly petiolate, ovate to elliptic, 4.5–8.5 × 2.1–4.2 mm, silvery-strigose; flowers clustered at the branch ends; calyx deciduous and enclosing the nutlet, segments lanceolate-caudate, the filiform ends plumose with thin stiff bristles; corollas pink to purple; nutlets 1 per calyx, 1.8–2.5 mm, dorsally shiny, purplish black, puberulent at the apex. Limestone in desert scrub, 3200–6000 ft. SE, SC, Lincoln.

Tiquilia hispidissima (Torr. & A. Gray) A. T. Richardson [*Coldenia hispidissima* (Torr. & A. Gray) A. Gray]—Hairy crinklemat. Mat-forming plant from a woody caudex; stems prostrate, spreading; leaves linear, to 8 × 1–1.8 mm, upper surface green, prickly with bristles, in age detaching from a persistent petiole; flowers solitary; calyx persistent, lobes subulate or narrowly triangular; corolla pink; nutlets 1–4, 1.1–1.5 mm, yellow to brown or sooty with white pustules ventrally, attachment scar open. Scattered locales on gypsum outcrops, 2850–5800 ft. C, EC, SE, SC.

ELAEAGNACEAE – OLEASTER FAMILY

Kenneth D. Heil

Shrubs or trees with lepidote or stellate trichomes; leaves alternate or opposite, simple, entire; flowers perfect or imperfect, regular, mostly 4-merous; perianth 4-lobed; stamens 4–8; pistil 1, ovary superior, 1-loculed; style 1; stigma 1; fruit a drupe or a berrylike, dry, indehiscent achene; seed mostly single, little or no endosperm (Heil, 2013b; Shultz & Varga, 2021).

1. Leaves alternate; stamens 4…**Elaeagnus**
1'. Leaves opposite; stamens 8…**Shepherdia**

ELAEAGNUS L. – Oleaster

Trees or shrubs, young twigs with dense lepidote-stellate trichomes; leaves alternate, small; flowers perfect or imperfect, actinomorphic, stamens 4; fruit drupelike, with persistent hypanthium base, dry and mealy.

1. Leaves mostly 3–8 times as long as wide; branchlets and leaves with silvery-peltate scales only; cultivated and escaped; widespread; mostly thorny…**E. angustifolia**
1'. Leaves mostly 1.5–3 times as long as wide; branchlets with both silvery- and brown-peltate scales; native, rare; no thorns…E. commutata

Elaeagnus angustifolia L.—Russian olive, oleaster. Rapid-growing, mostly thorny, small to large trees, to 12 m, trunks 1–5 dm thick; leaves lance-linear or narrowly elliptic, mostly 3–10 × 0.5–1.5 cm, silvery and densely stellate or lepidote, bicolored; flowers sickly fragrant, 8–12 mm, silvery and with yellow corolla lobes, slightly stellate-hairy; fruit ellipsoid, mostly 1 cm, at full maturity dull orange. Along most, if not all, New Mexico waterways, 3600–7420 ft. Widespread. Introduced.

Elaeagnus commutata Bernh. ex Rydb. [*E. argentea* Pursh, not Moench]—Silverberry. Unarmed shrub, mostly 2–5 m; young branches with brownish lepidote scales; leaves mostly 2–7 × 1–3 cm, elliptic to oblanceolate or lanceolate, acute, obtuse, or rounded apically, silvery on both sides but greener above than beneath; flowers 1–1.5 cm, corolla lobes yellowish; fruit ellipsoid, 9–14 mm, silvery, covered with scales. Stream banks with willow and Russian olive. To be expected in extreme NW New Mexico along the Animas River.

SHEPHERDIA Nutt. – Buffaloberry, soapberry

Shrubs, young twigs densely lepidote-stellate, brownish to silvery; leaves opposite; flowers perfect or imperfect, actinomorphic, sepals persistent in fruit, pistillate flowers with the hypanthium constricted at the summit; staminate flowers with 8 stamens alternating with the nectary glands; fruit berrylike.

1. Leaf blades mostly 3–7 × 1.5–4 cm, mostly acute at the base; fruit red, edible; plants often thorny...**S. argentea**
1'. Leaf blades mostly 2–5 × 0.5–2.5 cm, mostly rounded at the base; fruit yellow-red, unpalatable; plants unarmed...**S. canadensis**

Shepherdia argentea (Pursh) Nutt. [*Hippophae argentea* Pursh]—Silver buffaloberry. Often thorny, deciduous shrub or small tree, 2–5 m; branchlets covered with silvery-peltate scales; leaves short-petiolate, blades 0.5–6 × 3–14 mm, oblong or oblong-lanceolate to oblong-elliptic, mostly acute at the base, silvery with lepidote-stellate scales on both sides; flowers subsessile, 2.5–4 mm, corolla lobes yellowish; fruit 4–7 mm, red, edible. Stream banks and other moist sites, 5000–8140 ft. Rio Arriba, Sandoval, San Juan.

Shepherdia canadensis (L.) Nutt. [*Hippophae canadensis* L.]—Soapberry. Unarmed, highly branched, deciduous shrub, to 2 m; branchlets with brown-peltate scales; leaves short-petiolate, blades 0.5–1 × 1.5–4 cm, ovate to lanceolate, rounded apically and basally, green above, brownish lepidote beneath; flowers 2–3 mm, the lobes brownish; fruit ellipsoid, fleshy, bright red or yellow, bitter, 4–7 mm. Ponderosa pine, Douglas-fir, white fir, aspen, spruce-fir communities, 7620–10,400 ft. NC, NW.

ELATINACEAE – WATERWORT FAMILY

C. Barre Hellquist

Small annual or perennial herbs; leaves opposite, simple, with membranous stipules; flowers small, actinomorphic, axillary, persistent or withering, but persistent sepals and petals imbricate in the bud; stamens the same number as the petals, alternate with them or twice the number; ovary with 2–4 locules, placenta in the axis; fruit a capsule, valves alternate with ovary partitions; seeds oblong-cylindrical, straight or curved, usually with a reticulate surface pattern, the testa nearly filled by the cylindrical embryo (Razifard et al., 2016).

1. Plants glandular-hairy, 10+ cm, erect-ascending; flowers 5-merous; sepals cuspidate, scarious-margined; leaf margins serrulate...**Bergia**
1'. Plants glabrous, < 10 cm, creeping; flowers 2–4-merous; sepals obtuse, not scarious-margined; leaf margins entire...**Elatine**

BERGIA L. – Bergia

Bergia texana (Hook.) Seub.—Bergia. Plants prostrate to ascending, to 40 cm, base woody, ± glandular-puberulent throughout; flowers short-pedicelled, 1–3 in leaf axils; leaves elliptic-oblong to oblong-oblanceolate, tapering to base, serrulate to 3 cm; stipules lanceolate, deeply serrate; sepals 5, deltate, to 3.5 cm, acuminate, with thickened greenish midvein and scarious margins; petals 5, white, shorter than sepals; stamens 5 or 10; fruit ovoid, chambers 5, many-seeded; seeds oblong, somewhat curved, brown surface ± netlike. Drying mud, ponds, marshes, ditch banks; a single record from Doña Ana County, 4300 ft.

ELATINE L. – Waterwort

Annual, erect or prostrate, flaccid to succulent, to 10 cm, aquatic, amphibious, or terrestrial dwarf plants; leaves opposite, sessile or petiolate, with hyaline or toothed stipules; glabrous, blades linear-spatulate to oblong or orbicular-obovate, the margins obscurely and remotely crenate; inflorescence nodal, 1–2 flowers per node; flowers sessile or pedicelled, 2-, 3-, or 4-merous; sepals 2–3, obtuse, or 4-merous; stamens 8, unequal in size; petals 2–4; stamens as many as or twice as many as petals or reduced to 1; styles or capitate stigmas 2–4; fruit a membranous capsule, 2–4-celled, several- to many-seeded, 2–4-valved.

1. Capsule 4-celled; sepals and petals 4; stamens 8; seeds J- or U-shaped…**E. californica**
1'. Capsule 2- or 3-celled; sepals 2–3, petals 3; stamens 1–6; seeds straight or slightly curved…(2)
2. Stamens (1–)6, if 3, opposite the petals…**E. heterandra**
2'. Stamens 3, alternate to the petals…(3)
3. Seed pits 9–15 per row…**E. brachysperma**
3'. Seed pits 16–35 per row…(4)
4. Leaves rounded at apex; seed pits wider than long…**E. chilensis**
4'. Leaves blunt or notched at apex; seed pits ± as wide as long…**E. rubella**

Elatine brachysperma A. Gray—Short-seed waterwort. Small, low-spreading plants to 5 cm wide; leaves to 6 × 2 mm; flowers sessile, 3-merous; sepals 2 or with a third reduced; petals pinkish; capsule depressed, 3-celled; seeds oblong-ellipsoid with 9–15 pits in each triangular row. Muddy shores of vernal pools, ditches, and ponds, 5100–8315 ft. WC, SW.

Elatine californica A. Gray—California waterwort. Prostrate plants; leaves short-petioled to subsessile, to 6 × 2 mm; flowers on short pedicels, elongating in fruit to 1–2 times as long as fruit; sepals 4, oblong, uniting at base; petals 4, obovate; stamens 8; seeds J- or U-shaped, rounded at 1 end and truncate at the other. Muddy shores or shallow water; 1 record from S Chuska Mountains, 9025 ft. McKinley.

Elatine chilensis Gay—Chilean waterwort. Creeping aquatic or terrestrial plants to 10 cm, rooting at the nodes; leaves obovate to spatulate, rounded at apex, 3–4 × 1–3 mm, narrowed at base to a petiole with hyaline stipules; flowers solitary in leaf axils, sessile; sepals 2, sometimes with a third, much reduced, oblong; petals white to pink, orbicular; stamens 3, alternate to petals; seeds 20+ per cell, borne at the base of the placental axis, slightly curved, with 25–35 pits, wider than long. In shallow water or on muddy shores of lakes and ponds, 7100–9600 ft. Catron, San Juan, Taos. Introduced.

Elatine heterandra H. Mason—Mosquito waterwort. Decumbent plants 2–5 cm; leaves obovate to wide-oblong-elliptic; petiole 1/4 the length of the blade; flowers 1 per node; sepals 2; petals 3, ovate; stamens (1–)6 (if 3, opposite petals); fruit with 3 chambers; seeds wide-elliptic, curved, 12–16 pits per row, wider than long. Pond shores; 1 record from McKinley County, elevation unknown.

Elatine rubella Rydb.—Southwestern waterwort. Plants prostrate, sometimes erect, stems to 20 cm, rooting at the nodes; glabrous, often tinted red; leaves lance-oblong, 3–6(–12) mm, punctate above, margins entire, notched to blunt at apex; inflorescence with flowers 1–2 per node; sepals ca. 2 mm; petals 3; seeds narrowly oblong, straight or curved, 15–35 pits per row, as wide as long. Moist soil of vernal pools, pond margins, and mudflats, 8460–9000 ft. Catron, McKinley, Rio Arriba, Sandoval.

ERICACEAE – HEATH FAMILY

John L. Anderson and Jennifer Ackerfield

Shrubs, subshrubs, small trees, or perennial herbs, some without chlorophyll and mycoheterotrophic; leaves simple, evergreen or deciduous, alternate, entire or toothed, usually petiolate, lacking stipules, some scalelike; inflorescences usually bracteate terminal racemes or corymbs, or solitary and axillary; flowers usually perfect, actinomorphic; sepals 4 or 5, distinct or connate with short lobes; petals usually perfect, distinct or connate with short lobes; stamens as many or twice as many as the corolla lobes, anthers usually dehiscing by terminal pores, often with awnlike appendages called spurs; ovary usually superior (inferior in *Vaccinium*), pistil and style 1, mostly 4–5-locular with axile placentation; fruit usually a berry or drupe. Recent phylogenetic analysis supports the traditional classification of Ericaceae as including the Monotropoideae and Pyroloideae (Kron et al., 2002; Tucker, 2009).

1. Plants green with large leaves, autotrophic…(2)
1'. Plants not green, leaves absent or scalelike, mycoheterotrophic…(3)
2. Petals united; plants woody (suffrutescent in *Gaultheria*)…(4)
2'. Petals distinct; plants herbaceous (suffrutescent in *Chimaphila* and *Orthilia*)…(7)
3. Petals distinct; inflorescence not glandular, nodding…**Monotropa**
3'. Petals united; inflorescence glandular, erect…**Pterospora**
4. Leaves thin, deciduous; ovary inferior; fruit a fleshy berry…**Vaccinium**
4'. Leaves leathery, evergreen; ovary superior; fruit a rugulose berry, a capsule, or a mealy drupe…(5)
5. Plants suffrutescent to herbaceous, mat-forming; flowers solitary, axillary; fruit a capsule surrounded by the persistent, fleshy calyx forming a dry or mealy berrylike fruit…**Gaultheria**
5'. Plants shrubs or trees; flowers many in terminal inflorescences; fruit a berry or drupe…(6)
6. Trees of SW New Mexico; fruit a berry with rugulose surface…**Arbutus**
6'. Shrubs, widespread; fruit a berrylike drupe with smooth surface…**Arctostaphylos**
7. Flowers solitary, extremely fragrant…**Moneses**
7'. Flowers several on a stem…(8)
8. Flowers 4–10 in a terminal umbelliform corymb; stem leaves evident, oblanceolate or oblong-oblanceolate with sharply serrate margins; style straight and very short; stamens with dilated filaments that are ciliate on the margins, with maroon anthers…**Chimaphila**
8'. Flowers 4–25 in a terminal raceme; leaves basal or clustered at the base, elliptic to orbiculate; margins entire to serrate; style straight but > 2 mm or sigmoidally curved with an upturned end; stamens without dilated filaments…(9)
9. Flowers in a strongly 1-sided raceme; style straight, 3–5.5 mm; petals with 2 basal tubercles on the inner surface…**Orthilia**
9'. Flowers not in a 1-sided raceme; style sigmoidally curved with an upturned end or straight but 0.8–1.5 mm; petals without tubercles…**Pyrola**

ARBUTUS L. – Madrone

Trees to 15 m or sometimes shrubs, with checkered or exfoliating, smooth, reddish bark; leaves simple, alternate, evergreen, blades leathery, margins serrate or entire, petiolate; inflorescences terminal raceme or panicles; flowers perfect, actinomorphic, sepals 5-lobed, corolla white to pinkish, bell-shaped or urceolate; stamens 10; filaments dilated at the base and hairy; ovary superior; fruit a several-seeded, rugulose berry.

1. Leaves elliptic-lanceolate, tapered at the base; bark segments retained; fruit 8–10 mm diam.….**A. arizonica**
1'. Leaves elliptic to ovate, rounded at the base; bark segments exfoliating; fruit 6-8 mm diam.….**A. xalapensis**

Arbutus arizonica (A. Gray) Sarg.—Arizona madrone. Trees and sometimes shrubs; trunks with checkered gray or reddish bark; younger branches with smooth reddish bark; leaves lanceolate to elliptic, 5–12 × 1.5–3.5 cm, blades light green, glossy above, pale green below, glabrous, base tapered to acute; flowers pedicellate, sepals pale green, corollas to 6 mm; fruits 8–10 mm diam., reddish orange. Canyon bottoms and hillsides in oak-pine woodlands; Animas and Peloncillo Mountains, 4000–8200 ft. Hidalgo.

Arbutus xalapensis Kunth [*A. texana* Buckley]—Texas madrone. Trees to 8 m, with reddish, exfoliating bark; leaves elliptic to elliptic-ovate, 4–6 × 2–3 cm, blades pale to bright green, base tapered to slightly cordate; flowers pedicellate, sepals tan, corollas 5.1–5.5 mm; fruits 6–8 mm diam., dark red. Rocky hillsides, calcareous ledges; Guadalupe and Animas Mountains, 3600–7000 ft. SE, Hidalgo, Otero. Reported to hybridize with *A. arizonica* in the Animas Mountains (Sørensen, 2009).

ARCTOSTAPHYLOS Adans. – Manzanita, bearberry

Shrubs erect or prostrate and mat-forming; bark usually reddish brown, smooth, exfoliating; leaves simple, evergreen, leathery, alternate, usually entire; inflorescences terminal racemes or panicles; flowers perfect, actinomorphic, sepals 5, imbricate and distinct, corollas white to pink, urceolate; stamens 10, included; anthers spurred; filaments dilated, hairy; fruit a berrylike drupe with 2-10 nutlets.

1. Shrubs low, creeping and mat-forming, < 10 cm; leaves oblanceolate to obovate…**A. uva-ursi**
1'. Shrubs 1-5 m; leaves ovate to orbiculate or lanceolate-elliptic…(2)
2. Leaf blades 1.5-4 cm wide; petiole 7-15 mm [to be expected]…A. patula Greene
2'. Leaf blades 1-2 cm wide; petiole 4-8 mm…**A. pungens**

Arctostaphylos pungens Kunth—Point-leaf manzanita. Shrubs 1-2 m, erect with rigid, spreading stems with reddish-brown, smooth bark, branchlets densely pubescent; leaves bright green, shiny, petiole 4-9 mm, white-puberulent, blades 1.5-4 × 0.5-2 cm, base acute to rounded, tip acute and mucronate, margins entire (young leaves may be toothed); flowers pedicellate, pedicels 2-4 mm, sepals with reflexed lobes, corolla white to pinkish, urceolate, 5-8 mm, fruit a drupe, 5-11 mm wide, depressed-globose, orange to brownish red. Rocky hillsides, sometimes in dense stands, widespread, 5000-8200 ft. C, WC, SW, Sandoval, San Juan.

Arctostaphylos uva-ursi (L.) Spreng.—Bearberry, kinnikinnik. Low-growing prostrate shrub 0.1-0.2 m with branches trailing along the ground and rooting and forming mats; branchlets glabrous to puberulent; leaves dark green above, light green below, glabrous, blades 1-2.5 × 0.3-1 cm, base wedge-shaped, tip rounded and not mucronate, margins entire, petiole 2-5 mm, glandular; flowers pedicellate, pedicels 2-4 mm, glabrous, sepals with reflexed lobes, 1-2 mm, corollas white to pink, 4.5-8 mm, urceolate, ovary glabrous; fruit a drupe, depressed-globose, 6-12 mm, bright red, glabrous. Ground cover under coniferous forests, widespread in N and WC, 7000-10,000 ft. NC, NW, WC.

CHIMAPHILA Pursh – Pipsissewa

Chimaphila umbellata (L.) W. P. C. Barton—Plants to 3 dm; leaves whorled, lanceolate to oblanceolate, 3-7 cm, leathery, sharply toothed; flowers actinomorphic, in terminal corymbs or umbels; sepals persistent, distinct; petals pink or reddish pink, 5-7 mm; stamens 10, opposite and alternate with the petals; ovary 5-carpellate; style very short and inconspicuous; capsules 4-6 mm, wider than long. Mixed conifer and subalpine fir communities, 8500-11,000 ft. Santa Fe, Taos (Freeman, 2009a).

GAULTHERIA L. – Wintergreen

Gaultheria humifusa (Graham) Rydb. [*G. myrsinites* Hook.; *Vaccinium humifusum* Graham]— Alpine wintergreen, creeping wintergreen. Suffruticose shrub, rooting at the nodes and forming mats; leaves simple, evergreen, alternate, margins crenate or serrate, oval to elliptic, blades 0.8-2 × 5-15 mm, margins entire at base and serrulate near tip, base and tip rounded; flower sepals glabrous, corolla white to pink, 3-4 mm, campanulate, stamens 10, included; filaments glabrous; anthers dehiscing by terminal pores; ovary superior (ours); fruits berrylike, 5-7 mm, subglobose, red. Acidic, moist, mossy soil, meadows and stream banks, under subalpine forests, N, 10,000-12,500 ft. NC, WC, Sierra.

MONESES Salisb. ex Gray – Wood nymph

Moneses uniflora (L.) A. Gray—Wood nymph. Perennial herbs; stems solitary, 3-10(-12) cm, recurved above; leaves opposite or in clusters of 3, blades 1-2.5 cm × 6-20 mm, ovate-elliptic to obovate, crenate to finely serrate or subentire; inflorescence a solitary, nodding flower borne on a long peduncle; flower sepals mostly 15-25 mm, greenish white to yellowish; petals (4)5, fragrant, waxy-white, 7-12 mm; fruit depressed-globose, 5-8 mm diam. Moist, mossy areas, boggy stream banks, under coniferous forests, 8400-11,550 ft. NC, WC, Lincoln.

MONOTROPA L. – Indian-pipe

Monotropa hypopitys L. [*Hypopitys latisquama* Rydb.]—Indian-pipe. Plants perennial, lacking chlorophyll, white to reddish, arising from a cluster of matted roots; stems fleshy with reduced scalelike leaves, frequently clustered; inflorescence a raceme of nodding flowers, generally < 30 cm; flowers perfect, sepals 3–5, distinct, petals 4 or 5, 8–17 mm, distinct, yellow to orange, erect, often scalelike at the base; stamens 6–12, twice as many as the petals; anthers disklike, not awned; ovary superior; fruit an ovoid capsule, 6–10 × 4–8 mm. Understory in damp coniferous forests, 7000–10,000 ft. NC, NW, C, WC, SC, Eddy. This is an achlorophyllous, mycoheterotrophic species that obtains its nutrients through an ectomycorrhizal association with green plants (usually oaks or conifers).

ORTHILIA Raf. – Orthilia

Orthilia secunda (L.) House—Sidebells. Perennials, 0.5–2 dm; leaves basal, ovate to orbicular or elliptic, 1–3 cm, crenate-serrate; flowers in a strongly 1-sided terminal raceme; sepals persistent, distinct; petals 4–6 mm, white or greenish; stamens 10, alternate and opposite to the petals, ovary superior, 5-carpellate; style straight and well developed, exserted; stigma peltate and crateriform with 5 prominent spreading lobes. Common in moist coniferous forests, 8200–12,000 ft. NC, C, WC (Freeman, 2009b).

PTEROSPORA Nutt. – Pinedrops

Pterospora andromedea Nutt.—Pinedrops. Plants perennial, lacking chlorophyll, purplish brown to reddish purple; stems and leaves absent; inflorescence racemose, 30–150 cm; flowers perfect, actinomorphic, pendulous on recurved pedicels, calyx deeply 5-parted, corolla urceolate, petals united with short, spreading lobes, cream to yellowish; stamens 10, included, anthers awned, longitudinally dehiscent; fruit a 5-lobed capsule, 7–10 mm wide, opening from the base. Understory in litter of coniferous forests, 7000–9000 ft. Widespread except E, SW.

PYROLA L. – Wintergreen

Leaves basal, flowers actinomorphic or zygomorphic, in terminal racemes; sepals persistent, distinct; petals white, cream, greenish, or pink; stamens 10, alternate and opposite to the petals; ovary superior, 5-carpellate; stigma peltate or capitate (Freeman, 2009c).

1. Flowers actinomorphic; style short, (0.5–)0.8–1.5(–1.8) mm, straight and erect, without a collar beneath the stigma; anthers < 1.5 mm…**P. minor**
1'. Flowers zygomorphic; style elongate, 4–10 mm, sigmoidally curved with an upturned end and with an evident or inconspicuous collar just beneath the stigma; anthers 2–5 mm…(2)
2. Leaves dark green and prominently white- or gray-mottled along the midrib and main veins on the upper surface, the lower surface usually purplish red; flowers white to greenish white or sometimes pink-tinged… **P. picta**
2'. Leaves green, the veins not white-mottled, the lower surface sometimes purplish red, usually paler green; flowers white to pale green or pink to rosy…(3)
3. Petals pinkish to purplish; bracts of the scape 7–15 mm near the base of the scape…**P. asarifolia**
3'. Petals creamy-white to greenish white; bracts of the scape 2–9 mm near the base of the scape…(4)
4. Leaf blades mostly 1–3 cm and shorter than the petioles…**P. chlorantha**
4'. Leaf blades mostly 3–7 cm and longer than the petioles…**P. elliptica**

Pyrola asarifolia Michx.—Pink wintergreen. Plants 1.3–4 dm; leaves orbicular to ovate or broadly elliptic, mostly 3–7 × 3–6 cm, rounded at the tip, entire or serrulate; racemes mostly 10–25-flowered; petals pink to rose, 5–7 mm; anthers with terminal tubules 0.2–0.4 mm; stigma capitate, 0.7–1.5 mm wide; style exserted, 7–10 mm. Common along stream banks and in moist, shaded coniferous forests, 7950–10,900 ft. NC, NW.

Pyrola chlorantha Sw.—Green-flowered wintergreen. Plants 1–2.5 dm; leaves ovate to obovate, rounded, mostly 1–3 × 0.5–3 cm, entire to crenate-serrate; racemes mostly 2–10-flowered; petals white to cream or pale green, 5–7 mm; anthers with prominent terminal tubules 0.5–0.9 mm; stigma capitate, 1–1.5 mm wide; style exserted, (4–)5–7 mm. Common in moist, shaded coniferous forests, 7000–11,150 ft. NC, NW, WC.

Pyrola elliptica Nutt.—Shinleaf. Plants 1.1–2.7(–3) dm; leaves broadly elliptic to oblong or oblong-obovate, 1.2–8 × (0.8–)1.1–5.7 cm, crenate or obscurely denticulate; racemes 3–14(–21)-flowered; petals white to greenish white, 6–9 mm; anthers with terminal tubules 0.3–0.6 mm; stigma 0.7–1.2(–1.5) mm wide, lobes erect. Mixed conifer to subalpine forests, 7500–9100 ft. Catron, Rio Arriba, Sandoval.

Pyrola minor L.—Lesser wintergreen. Plants 0.8–2.5 dm; leaves oval to elliptic, rounded, 1–3 cm, entire to crenate; racemes mostly 5–13-flowered; petals pale pink or cream, 3.5–4.5 mm; anthers lacking terminal pores on the tubules; stigma peltate, with 5 short, spreading lobes; style included, straight, (0.5–)0.8–1.5(–1.8) mm. Common in moist, shaded coniferous forests, 8000–12,000 ft. NC, WC, McKinley.

Pyrola picta Sm.—White-veined wintergreen. Plants 1–2.5 dm; leaves oval to elliptic, dark green with white or gray mottling along the veins on the upper surface, 2–7 cm; racemes 2–7-flowered; petals greenish white to cream, 7–8 mm; anthers with terminal tubules; stigma capitate, 1–1.5 mm wide, with erect lobes; styles exserted, 4–9 mm. Moist, shaded coniferous forests, 7900–10,200 ft. NC, WC.

VACCINIUM L. – Blueberry, whortleberry

Shrubs openly branched; leaves deciduous, alternate, margins entire or serrulate, short-petiolate; inflorescence solitary in leaf axil of current year's growth; flowers perfect, actinomorphic; sepals 4–5 with small lobes or fused; corolla white to pinkish, shallowly 5-lobed, urceolate to campanulate; stamens 10, twice as many as the corolla lobes; anthers dehiscing terminally, with a pair of spurs; ovary inferior; fruit a berry with many seeds.

1. Branchlets of current year's growth terete or slightly angled, reddish green or yellowish…**V. cespitosum**
1'. Branchlets of current year's growth sharply angled, bright green…(2)
2. Berries blue; plants with flexuous branches; leaves 1–4 cm…**V. myrtillus**
2'. Berries red; plants with broomlike habit; leaves short (to 1.5 cm)…**V. scoparium**

Vaccinium cespitosum Michx.—Dwarf blueberry, bilberry, whortleberry. Low shrubs, 5–30 cm; stems densely branched, forming dense patches, branchlets terete or slightly angled, reddish green or yellowish green, not bright green; leaves thin, short-petiolate, glabrous, elliptic to obovate, blades 15–35 × 4.5–15 mm, base tapered to cuneate, tip obtuse to rounded; flowers pedicellate, pedicels 1.2–3.5 mm; sepals glabrous; corolla 4–6 × 3–5 mm, urceolate to globose; filaments glabrous; fruit a berry, subglobose, blue-glaucous, 6–8 mm wide. Mountain slopes and alpine tundra, edges of coniferous woods, meadows, streamsides, N, 8500–12,000 ft. Rio Arriba, Taos.

Vaccinium myrtillus L. [*V. oreophilum* Rydb.]—Blueberry, bilberry, whortleberry. Shrubs, 10–40 cm; stems openly branching and forming open colonies from woody rhizomes, branchlets sharply angled, bright green, flexuous; leaves thin, short-petiolate, ovate to elliptic, blades 1–4 cm × 7–16 mm, base obtuse, tip acute, glabrous; flowers pedicellate, 2–4 mm; sepals glabrous, corolla 4–5 × 5–7 mm, globose; filaments glabrous; fruit a berry, 5–9 mm wide, blue or blue-black. Mountain slopes and openings in mixed conifer forests, 8000–11,500 ft. NC, WC.

Vaccinium scoparium Leiberg ex Coville—Grouse whortleberry. Erect shrubs, 10–30 cm; stems narrowly branched with broomlike habit, forming colonies, branchlets sharply angled, green, glabrous, rigid; leaves thin, short-petiolate, ovate to elliptic, blades 0.5–1.5 × 3–7 mm, base obtuse, tip acute, glabrous; flowers pedicellate, 1.5–3.5 mm; sepals glabrous; corolla 2–4 × 3–4 mm, broadly urn-shaped, filaments glabrous; fruit a berry, 3–6 mm wide, bright red. Edges of subalpine spruce-fir forests and alpine meadows, N, 8500–12,000 ft. NC, Sierra.

EUPHORBIACEAE – SPURGE FAMILY

Kenneth D. Heil and Paul E. Berry

(Ours) shrubs or herbs, often with milky or watery latex; monoecious or dioecious; leaves alternate, opposite, or whorled, usually stipitate; flowers actinomorphic, imperfect, sometimes with a cyathium;

sepals 4–6 or absent; petals 4–6 or absent; stamens 1 to many; pistil 1; ovary superior, 3-carpellate; fruit a capsule (Levin & Gillespie, 2016).

1. Low, much-branched shrub to 1+ m…**Bernardia**
1′. Plants herbaceous, to 1 m or less…(2)
2. Leaves palmately 3–5-lobed (ours)…**Jatropha**
2′. Leaves entire or toothed, not lobed…(3)
3. Perianth absent or the calyx represented by a minute scale; flowers arranged in an often gland-margined cyathium…**Euphorbia**
3′. Perianth present, with 5 or 6 sepals, the flowers not arranged in cyathia…(4)
4. Herbage with stellate hairs that obscure the leaf surfaces…**Croton**
4′. Herbage glabrous to variously pubescent, not covered with stellate hairs…(5)
5. Plants with stinging hairs; styles undivided…**Tragia**
5′. Plants without stinging hairs; styles bifid or cleft…(6)
6. Plants with milky juice; glabrous…**Stillingia**
6′. Plants without milky juice; glabrous to pubescent…(7)
7. Petals present; styles 2-cleft…**Argythamnia**
7′. Petals absent; styles dissected into filiform segments…**Acalypha**

ACALYPHA L. – Three-seeded mercury

Plants monoecious; annual or perennial herbs; leaves alternate, entire or toothed; inflorescence an axillary or terminal spike; staminate flowers subtended by minute bractlets, sepals 4, stamens 8–16, united at the base; pistillate flowers subtended by a foliaceous bract, sepals 3–5, ovary 3-celled; fruit a 6-valved capsule.

1. Plants perennial; inflorescence terminal…**A. phleoides**
1′. Plants annual; staminate spikes axillary, pistillate spikes terminal…(2)
2. Upper leaves cordate at the base, 4–10 cm [to be expected]…A. ostryifolia
2′. Upper leaves acute to rounded at the base, 1–3 cm…**A. neomexicana**

Acalypha neomexicana Müll. Arg.—New Mexico copperleaf. Annual; stems erect, simple or branched, 2–5 cm, with sparse to dense, recurved hairs; leaves lanceolate to rhombic-ovate, mostly 1–3 cm; staminate spike axillary; pistillate spike terminal, flowers conspicuous, shallowly toothed; fruit mostly glabrous; seeds brown, with large deep pits in distinct rows. Moist areas in montane communities, often along streams, 3755–8000 ft. EC, C, WC, SE, SC, SW, McKinley.

Acalypha ostryifolia Riddell—Pineland three-seeded mercury. Annual; stems erect, branched, 35–70 cm, with sparse to dense, recurved, short, stiff hairs; leaves ovate, cordate or obtuse at the base, 30–80 × 15–55 mm, serrate, acuminate, petioles about equal to blade length; staminate spikes axillary, 5–35 mm; pistillate spikes terminal; flowers deeply cut into several linear lobes; capsules echinate with fleshy projections near the apex; seeds brown, tuberculate. Sandy soil in open piñon-juniper woodlands, 5000–6000 ft. No vouchers seen. However, potential habitat in the bootheel region.

Acalypha phleoides Cav.—Lindheimer's copperleaf. Perennial herb, mostly erect; stems 20–50 cm, with recurved, long, soft hairs; leaves ovate to lanceolate, rhombic or suborbicular at the base, 20–60 × 10–30 mm, serrate; inflorescence a terminal spike, flowers of both sexes on the same spike; stigmas red; capsule mostly glabrous; seeds brown, pitted. Sandy soil, 3110–7800 ft. EC, C, WC, SE, SC, SW.

ARGYTHAMNIA P. Browne – Silverbush

Annual or perennial herbs or shrubs; appressed or subappressed with malpighiaceous (T-shaped) hairs or rarely glabrous; stems erect, ascending, spreading, or trailing; leaves entire or serrate; inflorescences unisexual or bisexual (pistillate flowers proximal, staminate distal) axillary racemes; glands subtending each bract 0; staminate flowers with 5 sepals, 5 petals, 7–10 stamens; pistillate flowers with 5 sepals, 5 petals, well developed or rudimentary or absent, ovules solitary, styles 3; fruit a schizocarpous capsule; seeds slightly roughened [*Ditaxis* Vahl ex A. Juss.].

1. Leaves sessile…(2)
1'. Leaves petiolate…(3)
2. Plants glabrous or nearly so; pistillate petals evident…**A. cyanophylla**
2'. Plants pubescent; pistillate petals absent or rudimentary…**A. mercurialina**
3. Style branches terete; glands of pistillate flowers triangular…**A. serrata**
3'. Style branches flattened and dilated at the tip; glands of pistillate flowers linear…**A. humilis**

Argythamnia cyanophylla (Wooton & Standl.) J. W. Ingram [*Ditaxis cyanophylla* Wooton & Standl.]—Charleston Mountain silverbush. Herbs, perennial; leaves sparsely hairy, elliptic, obovate, or linear, 1–5 × 0.5–2.2 cm, margins entire; inflorescence unisexual or bisexual, to 4 cm; staminate flower petals 3–5 × 1.5 mm; pistillate flower petals 2.8–3.5 × 0.7–2.8 mm; capsules 3–5.5 mm; seeds 3.5–4 mm. Limestone, piñon-juniper woodlands, 6460–7220 ft. WC, Otero, Rio Arriba, Santa Fe.

Argythamnia humilis (Engelm. & A. Gray) Müll. Arg. [*Ditaxis humilis* (Engelm. & A. Gray) Pax]—Low silverbush. Herbs, perennial; leaves hairy or glabrous, obovate to linear, 0.5–7 × 0.2–2.2 cm, margins entire, sometimes slightly revolute; inflorescences bisexual, rarely unisexual; staminate flower petals 1.5–2.6 × 0.4–0.8 mm; pistillate flower petals 5, 0.3–1.7 × 0.3–0.4 mm; capsules 1.6–4.2 mm; seeds 1.5–2.5 mm. Limestone, sandstone, and clay soils, desert scrub and grasslands, 3200–4800 ft. Chaves, Eddy, Lea.

Argythamnia mercurialina (Nutt.) Müll. Arg. [*Ditaxis mercurialina* (Nutt.) J. M. Coult.]—Tall silverbush. Herbs, perennial; leaves with hairy surfaces, blades lanceolate, elliptic, ovate, or obovate, 1.5–7.5 × 0.6–3.5 cm, margins entire or serrulate; inflorescences bisexual or unisexual, to 12 cm; staminate flower petals 2–2.5 × 0.8–1 mm; pistillate flower petals 0 or 5, < 0.5 mm; capsules 6–14 mm; seeds 3.2–4.5 mm. Sandy soils in shortgrass prairies, 4550–4800 ft. Eddy.

Argythamnia serrata (Torr.) Müll. Arg. [*Ditaxis neomexicana* (Müll. Arg.) A. Heller]—New Mexico silverbush. Herbs; leaves usually densely to sparsely hairy, blades elliptic to ovate, obovate, or narrowly lanceolate, 1–10.5 × 0.2–3 cm, margins serrate, serrulate, or entire; inflorescences bisexual, 0.4–1.5 cm; staminate flower petals 1.5–3.5 × 0.6–2 mm; pistillate flower petals 5, 0.7–3 × 0.4–1.6 mm; capsules 1.8–4.5 mm; seeds 1.5–2.4 mm. Sandy, rocky soils in desert scrub, 3175–5000 ft. SC, SW, Eddy.

BERNARDIA Houst. ex Mill. – Myrtle-croton

Bernardia obovata I. M. Johnst.—Desert myrtlecroton. Shrubs to 0.8 m; leaf blades usually obovate to cuneate, rarely broadly elliptic to suborbiculate, 0.6–3 × 0.5–2.5 cm, margins flat, crenate-serrate, abaxial surface green, sparsely spreading-stellate-pubescent, adaxial surface green, sparsely stellate-pubescent to glabrate; inflorescence a staminate thyrse 5–10 mm; staminate flower petals 0, stamens 3–4(–6), nectary glands claviform; pistillate flower petals 0, pistil 2-carpellate; styles 2, irregularly dissected adaxially; capsules 5 mm, 2-lobed. Chihuahuan desert scrub, 4030–6100 ft. Doña Ana, Eddy.

CROTON L. – Croton

Plants monoecious or dioecious; annual or perennial herbs or shrubs; stellate-pubescent, scaly, or glandular; leaves usually alternate, petiolate, entire to toothed or lobed; flowers in terminal spikes or racemes when both male and female flowers are on the same plant; calyx with 5 or rarely 4 or 6 lobes; petals absent or present; stamens 3+, often 10–20; ovary 2- or 3-celled; capsules 2- or 3-celled, usually 3-seeded, rarely 1-seeded.

1. Leaf blade margins crenate to serrate-dentate, 2 cuplike glands present at junction with petiole…**C. glandulosus**
1'. Leaf blade margins entire, undulate, denticulate, or serrulate, cuplike glands absent at leaf base…(2)
2. Leaf blade abaxial surfaces densely lepidote or stellate-lepidote, often silvery…**C. dioicus**
2'. Leaf blade abaxial surfaces usually stellate-hairy, if stellate-lepidote not markedly silvery…(3)
3. Annual herbs…(4)
3'. Shrubs or perennial herbs…(6)
4. Plants dioecious; inflorescences unisexual; staminate petals 0; capsules verrucose, scurfy, columella 3-winged; plants widespread throughout New Mexico…**C. texensis**

4'. Plants monoecious; inflorescences bisexual or unisexual; staminate petals 3–5; capsules smooth, columella not markedly 3-winged (usually 3-angled or distally 3-lobed)…(5)
5. Leaf blade abaxial surfaces with some stellate hairs with dark brown centers; stamens 3–5; styles 2, terminal segments 4; ovaries 2-locular, only 1 fertile…**C. monanthogynus**
5'. Leaf blade abaxial surfaces lacking stellate hairs with brown centers; stamens 7–16; styles 3, terminal segments 6; ovaries 3-locular…**C. lindheimerianus**
6. Perennial herbs; columella with 3 sharp projections at apex…**C. pottsii**
6'. Shrubs or subshrubs; columella with 3 well-defined, smooth, terminal lobes…**C. fruticulosus**

Croton dioicus Cav. [*C. neomexicanus* Müll. Arg.]—Grassland croton. Subshrubs, 2–5(–9) dm, stellate-lepidote; leaf blades narrowly elliptic-ovate to lanceolate, 1–6.5 × 0.6–2.2 cm, margins entire; staminate flowers with 5 sepals, 1 mm, 0 petals; pistillate flowers with 5 sepals, 1.5–2 mm, 0 petals; ovary 3-locular; styles 3; capsules 5–6 mm diam. Limestone and igneous mountains, canyons, mesas, flats, disturbed areas, 3000–7535 ft. EC, C, WC, SE, SC, SW.

Croton fruticulosus Engelm. ex. Torr. –Bush croton. Shrubs, 2–10 dm, monoecious, stellate-hairy; leaf blades ovate to ovate-lanceolate, 2–8 × 2–4 cm, margins serrulate; staminate flower sepals 5, 0.8–1.2 mm, petals 5; pistillate flower sepals 5, equal, 2.2 mm, petals 0; capsules 5–6 mm diam. Igneous and limestone formations, desert scrub and oak-juniper-piñon communities, 3600–7000 ft. Doña Ana, Eddy, Grant.

Croton glandulosus L.—Sand or tooth-leaved or tropic croton. Herbs, annual, 1–2 dm, stellate-hairy, monoecious; leaf blades mostly > 2 times longer than wide, 1–2(–3) × 0.3–0.8(–1.3) cm, marginal teeth pointed; staminate flower sepals 5, 0.8–1.2 mm, petals 5; pistillate flower sepals 5, subequal, 6–7.5 mm, petals 0 or 5, rudimentary; capsules 3.5–6 × 4–5 mm. New Mexico material belongs to **var. lindheimeri** Müll. Arg. Sandy sites in desert scrub communities, 3250–3650 ft. Eddy, Lea.

Croton lindheimerianus Scheele—Three-seed croton. Herbs, annual, 1–5 dm, monoecious, densely velvety-appressed-tomentose; leaves stellate-hairy, blades suborbiculate, 1–5(–8) × 0.8–2.8(–3.5) cm, margins entire; staminate flower sepals (4)5, 1.5–2 mm, petals 5; pistillate flower sepals 5–6, equal, 3 mm, petals 0; capsules 4–5 × 4–4.5 mm. Our material belongs to **var. lindheimerianus**. Calcareous soils, playa margins, arroyo bottoms, desert scrub, grasslands, 3450–6500 ft. EC, SE, SC, SW, Socorro.

Croton monanthogynus Michx.—One-seed croton, prairie tea. Herbs, annual, 2–5 dm, monoecious, stellate-hairy, leaf blades ovate-oblong to nearly round to narrowly elliptic, 1–3.5 × 0.5–3 cm, margins entire; staminate flower sepals 3–5, 0.7–1 mm, petals 3–5, 0.7–1 mm; pistillate flower sepals 5, subequal, 1.5–2 mm, petals 0; capsules 3.5–4.5 × 1.8–2.2 mm. Desert scrub with scattered juniper; 1 record from Eddy County, 5140 ft.

Croton pottsii (Klotzsch) Müll. Arg.—Leatherweed. Herbs, perennial, 1–6 dm, monoecious or dioecious, stellate-hairy; leaf blades ovate-oblong, 1–6 × 1–3 cm, margins entire; staminate flower sepals (4)5, 1–1.5 mm, petals (4)5; pistillate flower sepals 5, equal, 1.5–3 mm, petals 0; capsules 4–6(–7) × 4–5 mm. Our material belongs to **var. pottsii**. Rocky slopes, desert scrub, juniper-piñon communities, 3285–7500 ft. NE, EC, C, WC, SE, SC, SW.

Croton texensis (Klotzsch) Müll. Arg. [*C. luteovirens* Wooton & Standl.]—Skunkweed, Texas croton. Herbs, annual, 2–7(–9) dm, dioecious, stellate-hairy; leaf blades narrowly ovate-oblong to linear-lanceolate, 1–5 × 0.5–2 cm, margins entire; staminate flower sepals 5, 1–2 mm, petals 0; pistillate flower sepals 5, equal, 1–1.5 mm, petals 0; capsules 5–8 × 4–5.5 mm. Dunes, prairies, creek beds, disturbed areas, 3200–8000 ft. Scattered.

EUPHORBIA L. – Spurge

Herbs, subshrubs, or shrubs, annual, biennial, or perennial, monoecious (sometimes dioecious), indumentum absent or of unbranched hairs; latex milky; stems erect, ascending, or prostrate; leaves alternate, opposite, or whorled, sometimes bractlike and subtending floral structures, stipules absent or present; petioles absent or present, glands absent, blade margins entire or toothed, occasionally revolute, glands absent; inflorescences bisexual (sometimes unisexual), terminal or axillary cymose clusters, capitate glomerules, or solitary; condensed pseudanthia (consisting of a cuplike involucre bearing glands on rim, these sometimes with petaloid appendages, enclosing solitary pistillate flower surrounded by clusters of staminate flowers, entire structure termed the cyathium); staminate flower sepals 0, petals 0, nectary absent, stamen 1; pistillate flower sepals 0, petals 0, nectary absent; pistil 3-carpellate; styles 3; fruit a capsule, seeds globose to ovoid, oblong, cylindrical, deltoid, pyramidal, or bottle-shaped.

Historically, distinctive clades within *Euphorbia* were segregated into a number of satellite genera. In the flora area, these include *Euphorbia*, *Chamaesyce*, and *Poinsettia*. Although these segregate genera are morphologically well-defined, monophyletic assemblages, recent molecular phylogenetic research has demonstrated that they are all nested within a broadly defined *Euphorbia* (Horn et al., 2012; Steinmann & Porter, 2002).

KEY TO SECTIONS AND SUBGENERA OF EUPHORBIA

1. Stems usually prostrate, sometimes erect, ascending, reclining, or decumbent; leaves usually opposite (rarely whorled in *E. fendleri*), blade bases usually asymmetric, stipules interpetiolar (except in *E. acuta*, where they are at base of petiole and deciduous, sometimes appearing absent)...**Euphorbia** sect. **Anisophyllum** [subg. **Chamaesyce**]
1'. Stems erect or ascending, rarely decumbent or prostrate; leaves alternate, opposite, or whorled, blade bases symmetric; stipules at base of petiole or absent...(2)
2. Involucral gland appendages petaloid, occasionally rudimentary; leaf margins entire...**Euphorbia** sect. **Alectoroctonum** [subg. **Chamaesyce**]
2'. Involucral gland appendages not petaloid except in *E. bifurcata*, *E. eriantha*, and *E. exstipulata*; leaf margins entire or toothed...(3)
3. Cyathia in terminal monochasia, dichasia, or condensed pleiochasia; involucral glands shallowly cupped to deeply concave, 1–3 per cyathium (if 4–5 in *E. exstipulata* then involucral gland appendages present; if 4–5 in *E. eriantha* then involucral gland appendages fringed, canescent, and folded over glands); involucral gland appendages petaloid-fringed or absent...**Euphorbia** sect. **Poinsettia** [subg. **Chamaesyce**]
3'. Cyathia in pleiochasia; involucral glands slightly concave, flat, or slightly convex, 4–5 per cyathium, terminal clusters of cyathia never subtended by white to pinkish bracts; involucral gland appendages hornlike or absent...(4)
4. Ovary and capsule not subtended by calyxlike structure; seeds with caruncle; involucral gland appendages hornlike or absent...**Euphorbia** subg. **Esula**
4'. Ovary and capsule subtended by calyxlike structure; seeds without caruncle; involucral gland appendages absent...Euphorbia sect. Nummulariopsis [subg. Euphorbia]

EUPHORBIA L. sect. Alectoroctonum (Schltdl.) Baill.

(Ours) herbs or shrubs, annual or perennial, erect, ascending, decumbent, or prostrate, branched or unbranched; leaves alternate or opposite, glabrous or hairy, blades monomorphic, base symmetric, margins entire, rarely toothed; cyathia solitary or in terminal monochasia, dichasia, or pleiochasia; glands [none or] (2–)5, appendages usually petaloid, occasionally rudimentary; seeds with caruncle present or absent.

1. Shrubs; stems pencil-like, covered with flaky, exfoliating layer of wax...**E. antisyphilitica**
1'. Herbs; stems not as above...(2)
2. Annual herbs with taproots...(3)
2'. Perennial herbs with rootstocks, tubers, or taproots...**E. strictior**
3. Leaves opposite...**E. hexagona**
3'. Leaves mostly alternate (opposite at proximal nodes in *E. bilobata*)...(4)
4. Dichasial bracts with conspicuous white margins...**E. marginata**
4'. Dichasial bracts wholly green or distal one white...(5)
5. Stems 10–35 cm; leaves opposite proximally, alternate distally; dichasial bracts wholly green; involucral glands 5; involucral gland appendages usually bifid...**E. bilobata**
5'. Stems 30–80(–110) cm; leaves usually alternate, sometimes some opposite; distal dichasial bracts often white; involucral glands (1)2–4; involucral gland appendages undivided...**E. graminea**

Euphorbia antisyphilitica Zucc.—Candelilla. Shrubs, erect, stems pencil-like, in age covered with flaky, exfoliating layer of wax; leaves alternate, blades ovate to deltoid-subulate, 2.5–4 × 1 mm, thick, fleshy, margins entire; cyathia in axillary congested cymes, near branch tips or solitary at distal nodes; glands 5, pinkish, 0.3–0.4 × 0.8–1 mm, appendages white to pink; capsules 3.9–4.2 × 3.6–3.9 mm; seeds 2.4–3.1 × 1.4–1.6 mm, rugose-tuberculate, caruncle crescent-shaped. Limestone, desert scrub, White Sands Missile Range, 4400 ft. Doña Ana.

Euphorbia bilobata Engelm.—Blackseed spurge. Herbs, annual, erect, branched; leaves opposite, surfaces usually glabrous, blades linear to narrowly elliptic, 8–52 × 2–7 mm, base attenuate, margins entire; cyathia solitary at distal nodes or in weakly defined cymes or dichasia; glands 5, yellow or pink, 0.2–0.3 × 0.4–0.5 mm, appendages greenish, white, or pink; capsules 0.5–2.6 × 2.1–3.3 mm; seeds 1.3–1.9 × 1–1.4 mm, tuberculate, often with shallow depressions; caruncle absent. Sandy and rocky soils, canyon bottoms, grasslands, juniper-piñon, oak woodlands, ponderosa pine-oak, 3960–8400 ft. WC, SW.

Euphorbia graminea Jacq.—Grassleaf spurge. Herbs, usually annual, rarely perennial; leaves usually alternate, sometimes some opposite, blades ovate, elliptic, linear-elliptic, or oblong, 10–83 × 3–39 mm; cyathia usually terminal, rarely axillary; glands (1)2–4, yellow to greenish; staminate flowers 30–40; pistillate flowers with ovary glabrous; capsules ovoid-oblate, 2.5–3 × 3–3.5 mm, glabrous; seeds gray, brown, or nearly black, 1.5–1.7 × 1.3–1.5 mm, coarsely tuberculate with longitudinal rows of shallow pits. Disturbed sites, ca. 4000 ft. Doña Ana. Introduced.

Euphorbia hexagona Nutt. ex Spreng.—Six-angle spurge. Herbs, annual, erect, unbranched or branched; leaves opposite, surfaces glabrous, blades linear-filiform, linear, or elliptic, 21–40 × 0.9–7.5 mm, base attenuate, margins entire; cyathia solitary in leaf axils or in terminal cymes or dichasia; glands 5, green to deep red, 0.5 × 0.8–1 mm, appendages white to green, tinged red; capsules 4.7–6.5 × 4.9–6.5(–7.1) mm; seeds 3.4 × 2.7 mm, rugose, caruncle absent. Sandy soils, grasslands, stream banks, sandbars, 3380–6200 ft. Eddy, Roosevelt, Sierra, Torrance.

Euphorbia marginata Pursh [*E. bonplandii* Sweet]—Snow-on-the-mountain. Herbs, annual, erect, unbranched or branched; leaves alternate, surfaces glabrous, blades broadly ovate to elliptic, 32–62 (–82) × 18–28(–52) mm, base rounded to cuneate, margins entire, often white on distal leaves; cyathia in terminal pleiochasia, dichasial; glands 4–5, green to greenish yellow, 0.7–1.1 × 1–1.6 mm, appendages white; capsules 3–5 × 3.5–7.5 mm; seeds 3.7–3.9 × 3–3.3 mm, rugose, with 2 transverse ridges, caruncle absent. Grasslands, salt desert scrub, disturbed areas, 3580–6460 ft. NE, NW, EC, Doña Ana, Grant.

Euphorbia strictior Holz.—Panhandle spurge. Herbs, perennial, erect, branched; leaves alternate, persisting, spreading or ascending, blades linear to narrowly oblanceolate, (20–)40–70 × (2–)4–5 mm, margins entire, cyathia in terminal dichasia; glands 5, green, 0.7–0.8 × 1.3–1.6 mm, appendages white; capsules 3.2–4.5 × 4–6.5 mm; seeds 3.8 × 3 mm, shallowly and obscurely pitted; caruncle absent. Open grasslands and uplands, 3770–4600 ft. NE, EC, Bernalillo.

EUPHORBIA L. sect. Anisophyllum Roep.

Herbs, rarely subshrubs or shrubs, annual or perennial, usually prostrate, sometimes erect, ascending, reclining, or decumbent, branched (rarely unbranched); leaves opposite (rarely whorled in *E. fendleri*), surfaces glabrous or hairy, blades monomorphic, base usually asymmetric, margins entire or variously toothed; cyathial arrangement terminal or axillary, solitary or in cymose clusters or capitate glomerules; glands (2–)4; seeds with caruncle absent (except for a carunclelike structure in *E. carunculata*).

1. Ovaries and capsules ± hairy…(2)
1′. Ovaries and capsules glabrous…(14)
2. Cyathia in capitate glomerules (with reduced bractlike leaves subtending cyathia)…(3)
2′. Cyathia solitary or in small, cymose clusters at distal nodes, on congested axillary branches, or at branch tips…(4)
3. Glomerules of cyathia terminal and axillary, axillary glomerules sessile or at tips of elongate, leafless stalks…**E. hirta**
3′. Glomerules of cyathia terminal, on main stems or short, leafy, axillary branches, with reduced, bractlike leaves subtending cyathia…**E. capitellata**

4.   Involucral gland appendages unequal…**E. indivisa**
4'. Involucral gland appendages ± equal in size or rudimentary to absent…(5)
5.   Leaf margins toothed, at least toward the apex…(6)
5'. Leaf margins entire…(10)
6.   Styles unbranched…(7)
6'. Styles bifid…(8)
7.   Stems strigillose; seeds broadly ovoid, 1.2–1.4 × 1–1.1 mm, with 2 well-defined transverse ridges…**E. rayturneri**
7'. Stems pilose to lanate; seeds narrowly oblong-ovoid to ellipsoid, 1–1.5 × 0.5–0.6 mm, almost smooth, rugulose, dimpled, or with short, irregularly interrupted furrows (seeds appearing partially and irregularly few-ridged)…**E. stictospora**
8.   Capsules with pubescence concentrated along keels or toward base, often glabrous between keels…**E. prostrata**
8'. Capsules ± evenly hairy or pubescence at least not concentrated only along keels and base, not glabrous between keels…(9)
9.   Capsules pilose to villous…**E. serpyllifolia** (in part)
9'. Capsules strigose to sericeous   **E. maculata**
10. Involucral gland appendages divided into 3–8 triangular to subulate segments…**E. setiloba**
10'. Involucral gland appendages entire, toothed, or absent…(11)
11. Stems with appressed hairs…(12)
11'. Stems with spreading to erect hairs…(13)
12. Leaf blade apices long-acuminate and spinulose; seeds 2.2–2.6 mm…**E. acuta**
12'. Leaf blade apices acute to obtuse or rounded; seeds 0.8–1.8(–2) mm…**E. lata**
13. Stems and leaves with glistening hairs; stipules (0–)0.1 mm…**E. arizonica**
13'. Stems and leaves without glistening hairs; stipules 0.2–1.6 mm…**E. micromera**
14. Cyathia in capitate glomerules…**E. hypericifolia**
14'. Cyathia solitary or in small cymose clusters at distal nodes or on congested, axillary branches…(15)
15. Leaf blades linear, 5+ times as long as wide, bases symmetric or subsymmetric (sometimes slightly asymmetric in *E. parryi*)…(16)
15'. Leaf blades not linear, 4 times or less as long as wide, bases usually asymmetric, rarely subsymmetric to symmetric…(18)
16. Stipules entire; seeds 0.9–1.4 mm, 4-angled in cross-section…**E. revoluta**
16'. Stipules usually deeply and irregularly fringed, lobed, or lacerate and divided into slender segments, rarely entire; seeds 1.4–2 mm, bluntly 3-angled or rounded-angular in cross-section…(17)
17. Stems erect or ascending; involucral gland appendages 0.4–2.5 × 1.1–1.7 mm…**E. missurica** (in part)
17'. Stems usually prostrate, rarely ascending-erect; involucral gland appendages 0.2–0.6 × 0.3–0.7(–1.1) mm…**E. parryi**
18. Leaf blade margins toothed (at least toward apex or on majority of leaves)…(19)
18'. Leaf blade margins entire…(27)
19. Seeds with prominent transverse ridges interrupting abaxial keel…(20)
19'. Seeds without prominent transverse ridges, or if present, not interrupting abaxial keel…(21)
20. Stems shortly pilose or puberulent proximally (often glabrous distally); capsules 1.3–1.5 × 1.1–1.5 mm…**E. abramsiana**
20'. Stems glabrous; capsules 1.3–1.9 × 1.6–2 mm…**E. glyptosperma**
21. Largest leaf blades > 20 mm…(22)
21'. Largest leaf blades < 20 mm…(23)
22. Stems sparsely to densely pilose or pilose-crinkled proximally, usually glabrous distally; leaf blades glabrous or sparsely pilose toward base (abaxially), glabrous (adaxially); seeds with prominent transverse ridges or coarsely and inconspicuously pitted-reticulate…**E. hyssopifolia**
22'. Stems sparsely to moderately pilose to villous or with short, incurved hairs, pubescence often concentrated at nodes and distally (hairs occasionally in 2 bands along opposite sides of stem); leaf blades usually sparsely to moderately pilose, especially toward base, sometimes glabrous; seeds finely and irregularly wrinkled or with indistinct shallow, rounded cross-ridges…**E. nutans**
23. Leaf blade surfaces papillate; cocci of capsule often elongate and terminating in empty portion…E. villifera (in part)
23'. Leaf blade surfaces not papillate; cocci of capsule not elongate or terminating in empty portion…(24)
24. Stem and leaves glabrous…**E. serpillifolia** (in part)
24'. Stem and leaves usually hairy, rarely glabrate…(25)
25. Leaf blade margins serrate to serrulate, usually with conspicuous teeth at base; capsules 2–2.6 × 3.2–3.7 mm; seeds 1.5–1.8 × 1.1–1.3(–1.5) mm…**E. serrula**
25'. Leaf blade margins usually entire in proximal 1/2 and serrulate in distal 1/2 (rarely some leaves serrulate nearly to base in *E. serpillifolia*); capsules 1.4–1.9 × 1.5–2.1 mm; seeds 1–1.4 × 0.6–0.9 mm…(26)

26. Stems prostrate to ascending, often mat-forming; cyathial glands yellow to pink; seeds smooth to dimpled or rugose, or with faint transverse ridges...**E. serpillifolia** (in part)

26'. Stems prostrate to ascending or erect, not mat-forming; cyathial glands red to reddish green; seeds rugulose and sometimes also with low, transverse ridges...**E. vermiculata**

27. Stipules (at least those of upper side of stem) connate, forming deltate, ligulate, or ovate scale...(28)

27'. Stipules usually distinct, occasionally connate basally, not forming conspicuous deltate, ligulate, or ovate scale...(29)

28. Perennials; involucral glands 0.2–0.5 × (0.2–)0.3–0.8 mm, appendages 0.3–1 × 0.6–1.3 mm; leaf blades often with red blotch in center...**E. albomarginata**

28'. Annuals, rarely short-lived perennials; involucral glands 0.1 × 0.1–0.3 mm, appendages 0–0.2 × 0.1–0.3; leaf blades without red blotch...**E. serpens**

29. Perennial herbs with thickened and often woody rootstocks...(30)

29'. Annual or rarely short-lived perennial herbs with taproots or spreading rootstocks...(32)

30. Seeds 0.8–1.4 mm...E. villifera (in part)

30'. Seeds 1.6–2.4 mm...(31)

31. Stems usually erect, rarely slightly decumbent; leaf blades ovate to lanceolate or oblong- or linear-lanceolate, 3–11 × 0.8–3(–5) mm, apex acute to short-acuminate, base short-tapered, occasionally 1 side rounded...**E. chaetocalyx**

31'. Stems usually prostrate, decumbent or ascending, very rarely erect; leaf blades usually orbiculate to ovate, rarely almost lanceolate, 3–8 × 2.5–7 mm, apex rounded to obtuse, base slightly cordate to rounded or obtuse...**E. fendleri**

32. Capsules 4.7–5.5(–6) mm; seeds (2.8–)4.1–5.2 mm, bottle-shaped, strongly dorsiventrally compressed and weakly 3-angled in cross-section, with linear carunclelike structure...**E. carunculata**

32'. Capsules 1–3.5(–4) mm; seeds 0.7–2.8 mm, not bottle-shaped, weakly dorsiventrally compressed or terete to subterete, 3- or 4-angled in cross-section, without carunclelike structure...(33)

33. Seeds terete or bluntly subangled in cross-section, mostly smooth...(34)

33'. Seeds usually 3–4-angled in cross-section (± weakly angled in *E. villifera*), smooth to rugose or wrinkled, or with transverse ridges...(35)

34. Stems erect or ascending; leaf blades narrowly oblong to narrowly lanceolate-oblong; styles 0.5–1.4 mm; involucral gland appendages present, 0.4–2.5 mm; seeds bluntly angled...**E. missurica** (in part)

34'. Stems prostrate or slightly ascending; leaf blades oblong to oblong-obovate or oblong-elliptic; styles 0.2–0.6 mm; involucral gland appendages rudimentary to absent; seeds terete to bluntly subangled...
E. geyeri

35. Stems erect to ascending...E. villifera (in part)

35'. Stems prostrate or reclining...(36)

36. Seeds 1.3–1.5 mm...**E. golondrina**

36'. Seeds 0.8–1.2 mm...**E. theriaca**

Euphorbia abramsiana L. C. Wheeler [*Chamaesyce abramsiana* (L. C. Wheeler) Koutnik]—Abrams' sandmat. Herbs, annual, prostrate, mat-forming; leaves glabrous, blades ovate, elliptic-oblong, or slightly ovate-cordate, 3–11 × 2–5 mm, margins serrulate at least toward apex, often entire toward base; cyathia solitary at distal nodes of primary stems or at nodes of short, congested axillary branchlets; glands 4, yellowish to pink, 0.1 × 0.1–0.2 mm, appendages absent or white to pink; capsules 1.3–1.5 × 1.1–1.5 mm, seeds 1–1.2 × 0.6–0.7 mm, with 3–5 prominent transverse ridges. Desert scrub, desert grasslands, 3970–4500 ft. Hidalgo, Luna.

Euphorbia acuta Engelm. [*Chamaesyce acuta* Millsp.]—Pointed sandmat. Herbs, perennial, ascending to erect, 5–30 cm; leaf blades ovate to lanceolate, 6–20 × 3–8 mm, base subsymmetric, margins entire, strongly involute; cyathia solitary at distal nodes; glands 4, yellow-green to orange or red, 0.2–0.4 × 0.6–1.5 mm; capsules 2.8–3.7 mm diam.; seeds 2.2–2.6 × 1.1–1.4 mm, smooth to finely reticulate. Limestone, rocky, sandy, or clay soils, desert scrub, grasslands, oak-juniper savanna, 3000–6500 ft. SE, SC.

Euphorbia albomarginata Torr. & A. Gray [*Chamaesyce albomarginata* (Torr. & A. Gray) Small]—Rattlesnake weed or white-margin sandmat. Herbs, perennial, prostrate, occasionally mat-forming; leaf surfaces glabrous, blades ovate, oblong, or orbiculate, 3–8(–15) × 3–7 mm, base strongly asymmetric, margins whitish, entire; cyathia solitary at distal nodes; glands 4, greenish yellow to red, 0.2–0.5 × (0.2–) 0.3–0.8 mm, appendages white to pink; capsules 1.1–2.3 × 1.2–2 mm; seeds 1–1.7 × 0.5–0.8 mm, smooth. Disturbed areas in desert scrub, grasslands, mesquite woodlands, juniper, sagebrush-piñon communities, 3555–7500 ft. Widespread except NE, NC.

Euphorbia arizonica Engelm. [*Chamaesyce arizonica* (Engelm.) Arthur]—Arizona sandmat. Herbs, annual or short-lived perennial, erect to ascending; leaf surfaces pilose with glistening hairs, blades usually ovate, rarely elliptic, 3–11 × 2–7 mm, base asymmetric, margins entire; cyathia solitary at distal nodes; glands 4, dark maroon, 0.2 × 0.2–0.4 mm, appendages white to pink; capsules 1.4–1.8 mm diam.; seeds 0.9–1.1 × 0.5–0.6 mm, rugose. Often in limestone, rocky slopes, dry washes, desert scrub, 4100–4940 ft. Grant, Luna.

Euphorbia capitellata Engelm. [*Chamaesyce capitellata* (Engelm.) Millsp.]—Head spurge, capitate sandmat. Herbs, annual or perennial; leaf surfaces often with red spot in center, glabrous, pilose, or strigillose, blades ovate to narrowly ovate, 4–19 × 2–8 mm, base asymmetric, margins entire or serrulate; cyathia in dense, terminal, capitate glomerules; glands 4, yellow-green to pink or maroon, 0.2–0.4 × 0.2–0.5 mm, appendages white to light pink; capsules 1.3–1.9 × 1.4–2.1 mm; seeds 0.9–1.5 × 0.5–0.7 mm, irregularly dimpled. Desert scrub, oak-juniper woodlands, 4400–5000 ft. Hidalgo, Luna, Otero.

Euphorbia carunculata Waterf. [*Chamaesyce carunculata* (Waterf.) Shinners]—Sand-dune sandmat. Herbs, annual, prostrate, spreading, occasionally mat-forming, ± succulent; leaf surfaces glabrous, blades ovate to elliptic-oblong, 5–26 × 4–12 mm, base subsymmetric to symmetric, margins entire; cyathia solitary at distal nodes; glands 4, yellowish, 0.5–0.7 × 0.5–0.8 mm, appendages white to yellowish; capsules 4.7–5.5(–6) × 3.6–5.1 mm; seeds (2.8–)4.1–5.2 × 1.2–2(–3.4) mm, smooth, with a carunclelike structure. Sand dunes, 3050–4120 ft. Chaves, Lea.

Euphorbia chaetocalyx (Boiss.) Tidestr. [*E. fendleri* Torr. & A. Gray var. *chaetocalyx* Boiss.; *Chamaesyce chaetocalyx* (Boiss.) Wooton & Standl.]—Bristlecup sandmat. Herbs, perennial, usually erect, rarely slightly decumbent; leaf surfaces glabrous, blades ovate to lanceolate or oblong- or linear-lanceolate, 3–11 × 0.8–3(–5) mm, base slightly asymmetric, margins entire; cyathia solitary at distal nodes; glands 4, yellow-brown to reddish, 0.2–0.4 × 0.4–0.6 mm, appendages absent or white; capsules 1.7–2.1 × 1.6–2.4 mm; seeds 1.6–2 × 1–1.2 mm, smooth to slightly wrinkled. Cliff faces, gypsum, sandy sites, wash bottoms, piñon-juniper, ponderosa pine–Gambel oak communities, 3855–8100 ft. Widespread except NE, EC.

Euphorbia fendleri Torr. & A. Gray [*Chamaesyce fendleri* (Torr. & A. Gray) Small]—Fendler's sandmat. Herbs, perennial, usually prostrate, decumbent, or ascending, very rarely erect; leaf surfaces glabrous, blades usually orbiculate to ovate, rarely almost lanceolate, 3–8 × 2.5–7 mm, base slightly asymmetric, margins entire; cyathia solitary at distal nodes; glands 4, yellow-green to reddish, 0.2–0.5 × 0.4–0.9 mm, appendages absent or white, rarely pink; capsules 2–2.4 × 2.2–2.5 mm; seeds 1.7–2 × 1–1.2 mm, smooth to slightly wrinkled. Sand, desert scrub, grasslands, piñon-juniper woodlands, Gambel oak-ponderosa pine, 3465–8655 ft. Widespread.

Euphorbia geyeri Engelm. [*Chamaesyce geyeri* (Engelm.) Small]—Geyer's sandmat. Herbs, annual, stems prostrate or slightly ascending, loosely mat-forming; leaves opposite, glabrous, blades oblong to oblong-obovate or oblong-elliptic, 4–12 × 2–6 mm, base slightly asymmetric, margins entire; cyathia solitary or in small, cymose clusters at distal nodes; peduncle 1–2 mm; glands 4, green to reddish, 0.2–0.4 × 0.2–0.6 mm; appendages rudimentary to absent or white to reddish-tinged; staminate flowers 5–20; pistillate flowers with ovary glabrous; styles 0.2–0.6 mm, bifid nearly 1/2 length; capsules globose-ovoid, 1.5–2 × 1.5–3 mm, glabrous; seeds ashy-white, ovoid, 1.1–1.7 × 0.9–1.2 mm, smooth.

1. Involucral gland appendages present; staminate flowers 5–9; seeds 1.1–1.4(–1.6) mm...**var. geyeri**
1'. Involucral gland appendages absent or rudimentary; staminate flowers 10–20; seeds 1.6–1.7 mm...**var. wheeleriana**

var. geyeri—Geyer's sandmat. Involucral gland appendages present; staminate flowers 5–9; capsules 1.5–2 × 1.5–2.5 mm; seeds 1.1–1.4(–1.6) × 0.9–1.2 mm. Disturbed sandy or gravelly sites, sand barrens, 4100–5250 ft. EC, SE, SC, Hidalgo, San Juan.

var. wheeleriana Warnock & M. C. Johnst. [*Chamaesyce geyeri* (Engelm.) Small var. *wheeleriana* (Warnock & M. C. Johnst.) Mayfield]—Wheeler's sandmat. Involucral gland appendages absent or rudimentary; staminate flowers 10–20; capsules 1.5–2 × 2–3 mm; seeds 1.6–1.7 × 0.9–1.2 mm. Disturbed sandy or gravelly areas, sand barrens, ca. 5000 ft. Doña Ana, Socorro.

Euphorbia glyptosperma Engelm. [*Chamaesyce glyptosperma* (Engelm.) Small]—Ribseed sandmat. Herbs, annual, prostrate, loosely mat-forming; leaf surfaces glabrous, blades narrowly oblong to oblong-obovate or oblong-ovate, 3–15 × 2–7 mm, base asymmetric, margins minutely sparsely serrulate; cyathia solitary or in small, cymose clusters at distal nodes; glands 4, red to purple, 0.1–0.2 × 0.1–0.5 mm, appendages white or pinkish-tinged, capsules 1.3–1.9 × 1.6–2 mm; seeds 1–1.4 × 0.6–0.9 mm, with prominent transverse ridges. Shortgrass prairie, piñon-juniper, ponderosa pine–oak communities, disturbed sites, 3010–8960 ft. NE, NC, NW, C, SE, SC.

Euphorbia golondrina L. C. Wheeler [*Chamaesyce golondrina* (L. C. Wheeler) Shinners]—Canyon spurge. Herbs, annual; stems prostrate, 5–35 cm, glabrous; leaves opposite, glabrous, blades oblong, ovate-oblong, or narrowly elliptic-oblong, 5–11.5 × 1–4 mm, base asymmetric, margins entire, thickened and often revolute on drying; cyathia solitary at distal nodes; glands 4, occasionally rudimentary, red to purple, 0.3–0.4 × 0.3–0.4 mm; appendages white, 0.1–0.3 × 0.5–0.8 mm; staminate flowers 28–40; pistillate flowers with ovary glabrous; styles 0.3–0.4 mm, bifid nearly entire length; capsules 1.7–2 × 1.5–1.6 mm, glabrous; seeds 4-angled in cross-section, 1.3–1.5 × 0.6–0.7 mm, with very faint transverse ridges or wrinkles. Deep, sandy riverbanks, ca. 4000 ft. Doña Ana.

Euphorbia hirta L. [*Chamaesyce hirta* (L.) Millsp.]—Pillpod spurge. Herbs, annual or perennial, erect to ascending, 10–50(–75) cm; leaf blades ovate to rhombic, 7–43 × 3–18 mm, base strongly asymmetric, margins serrulate to double-serrulate; cyathia in dense, axillary and terminal, capitate glomerules; glands 4, greenish to pink, 0.1–0.2 × 0.1–0.2 mm, appendages white to pink; capsules 1–1.3 × 1.1–1.6 mm; seeds 0.7–0.9 × 0.5–0.7 mm, usually rugulose. Hidalgo.

Euphorbia hypericifolia L. [*Chamaesyce glomerifera* Millsp.; *C. hypericifolia* (L.) Millsp.]—Graceful sandmat. Herbs, annual; stems erect to ascending, 15–50 cm, glabrous; leaves opposite, blades obliquely oblong-oblanceolate, 10–35 × 7–15 mm, base asymmetric, margins serrate or serrulate; cyathia in dense, axillary and terminal, capitate glomerules; glands 4, yellow-green to brown, stipitate, subcircular, 0.2 × 0.2 mm, occasionally nearly rudimentary; appendages absent on smaller glands or white to pink; staminate flowers (0–)2–20; pistillate flowers with ovary glabrous; capsules 1.3–1.4 × 1.1–1.5 mm, glabrous; seeds bluntly 4-angled in cross-section, 0.9–1.1 × 0.5 mm, with shallow irregular depressions alternating with low, smooth ridges. Disturbed areas, ca. 4000 ft. Doña Ana.

Euphorbia hyssopifolia L. [*Chamaesyce hyssopifolia* (L.) Small]—Hyssopleaf sandmat. Herbs, annual, erect to ascending; leaf surface glabrous or sparsely pilose, blades lanceolate to oblong or falcate, 8–35 × 7–15 mm, base asymmetric, margins serrulate; cyathia solitary or in small, cymose clusters; glands 4(5) (fifth gland without appendage), yellow-green to maroon, 0.1–0.2 × 0.1–0.3 mm, appendages usually white or turning reddish with age, 0.1–0.3 × 0.2–0.6 mm; capsules 1.5–1.6 × 1.7–1.8 mm; seeds 1–1.4 × 0.7–1.1 mm, inconspicuously pitted-reticulate. Desert grassland, canyon bottoms, lake margins, piñon-juniper, disturbed sites, 3575–8525 ft. EC, C, WC, SC, SW.

Euphorbia indivisa (Engelm.) Tidestr. [*Chamaesyce indivisa* (Engelm.) Millsp.]—Royal sandmat. Herbs, annual or short-lived perennial, prostrate, usually mat-forming; leaf surfaces glabrous or slightly pilose, blades oblong, ovate, or narrowly obovate, 3–10(–12) × 2–6 mm, base strongly asymmetric, margins serrulate; cyathia usually in small cymose clusters on congested, axillary branches; glands 4, yellow to pink, 0.1 × 0.3–0.4(–0.6) mm, appendages pink to reddish; capsules 1.2–1.5 × 1–1.4 mm; seeds 0.8–1 × 0.4–0.5 mm, with low, transverse ridges. Grasslands, oak-juniper-piñon communities, 3850–6410 ft. SC, SW.

Euphorbia lata Engelm. [*Chamaesyce lata* (Engelm.) Small]—Broadleaf spurge. Herbs, perennial, ascending to erect, or prostrate; leaf surfaces strigose to short-sericeous or ± villous, blades narrowly to broadly ovate-deltate, 4–12 × 3–7 mm, base asymmetric, margins entire, often ± revolute; cyathia solitary at distal nodes; glands 4, greenish, 0.2–0.7 × 0.6–1 mm, appendages rudimentary or white; capsules 1.9–2.3 × 2–2.4 mm; seeds 1.5–1.8(–2) × 0.6–0.9 mm, smooth. Sandy or rocky soils, prairies, desert scrub, disturbed sites, 3175–6000 ft. Widespread except N, W.

Euphorbia maculata L. [*Chamaesyce maculata* (L.) Small]—Milk purslane, prostrate spurge. Herbs, annual, usually prostrate, often mat-forming; leaf blades oblong-ovate to ovate-elliptic or oblong-elliptic, 4–18 × 2.5–8 mm, base strongly asymmetric, margins serrulate or subentire; cyathia solitary or in small, cymose clusters at distal nodes or on congested axillary branches; glands 4, green to yellow-green, turning pink with age, 0.1–0.2 × 0.2–0.5 mm, appendages white to reddish-tinged; capsules 1.3–1.5 × 1.2–1.4 mm; seeds 1–1.2 × 0.6–0.9 mm, with 3–4 low, transverse ridges. Disturbed areas, oak-juniper-piñon communities, mixed conifer, 4000–8480 ft. SC, SW, Lea.

Euphorbia micromera Boiss. [*Chamaesyce micromera* (Boiss.) Wooton & Standl.]—Desert spurge. Herbs, annual; stems prostrate, mat-forming, leaves opposite, blades ovate to elliptic, 6–15 × 2–4 mm, base asymmetric; cyathia solitary at distal nodes; glands 4, red, appendages absent; staminate flowers 2–5; pistillate flowers with ovary usually glabrous, rarely pilose; capsules oblong, 1.3–1.5 × 1.1–1.3 mm, usually glabrous, rarely pilose; seeds light gray, 4-angled in cross-section, 0.9–1 × 0.5–0.6 mm, smooth to slightly rugose or with 1–4 faint transverse ridges that do not pass through abaxial keel. Sandy and gravelly areas, desert scrub, grasslands, 3950–7200 ft. NE, NW, EC, SE, Cibola.

Euphorbia missurica Raf. [*Chamaesyce missurica* (Raf.) Shinners]—Prairie or Missouri spurge. Herbs, annual, erect or ascending; leaf surfaces glabrous, blades linear to narrowly oblong or narrowly lanceolate-oblong, (4–)8–30 × 3–7, margins entire, occasionally ± revolute; cyathia solitary or in small, cymose clusters; glands 4, yellowish green, 0.3–0.6 × 0.3–0.7 mm, appendages white or ± pinkish-tinged; capsules 1.9–2.5 × 2–2.5(–3) mm; seeds 1.5–2 × 1.1–1.4 mm, smooth or slightly wrinkled. Sandy and disturbed areas, shin-oak, mesquite-oak, juniper-piñon communities, 3200–7425 ft. NE, NW, EC, C, SE, SW.

Euphorbia nutans Lag. [*Chamaesyce nutans* (Lag.) Small]—Nodding or upright spotted spurge. Herbs, annual, usually ascending, occasionally erect; leaf surfaces usually reddish-mottled or with conspicuous reddish spot, pilose, blades oblong to oblong-lanceolate, 8–40 × 3–12 mm, base asymmetric; cyathia solitary at distal nodes or in small, cymose clusters at branch tips; glands 4, usually green, sometimes reddish purple, 0.2–0.4 × 0.3–0.5 mm, appendages white or pinkish; capsules 1.6–2.3 × 1.5–2.4 mm; seeds 1–1.6 × 0.5–0.8 mm, surface finely and irregularly wrinkled, sometimes faintly so, or with indistinct, shallow, rounded cross ridges. Arroyo margins, juniper–piñon–ponderosa pine communities, 5200–7080 ft. WC, SW, Eddy, San Miguel.

Euphorbia parryi Engelm. [*Chamaesyce parryi* (Engelm.) Rydb.]—Dune spurge. Herbs, annual, usually prostrate, rarely ascending-erect; leaf surfaces glabrous, blades linear to narrowly oblong, (5–)10–25(–30) × 2–5 mm, base usually symmetric, margins entire, occasionally ± revolute; cyathia solitary or in small clusters on short axillary branches at distal nodes; glands 4, reddish pink to greenish yellow, 0.2–0.3 × 0.3–0.5 mm, appendages white; capsules 2–2.3 × 1.5–2.5 mm; seeds 1.4–1.8 × 0.8–1 mm, smooth or only inconspicuously roughened. Sand dunes and other sandy habitats, 3600–8000 ft. NW, C, WC, SE, SC, SW.

Euphorbia prostrata Aiton [*Chamaesyce prostrata* (Aiton) Small]—Prostrate spurge or sandmat. Herbs, annual, prostrate to decumbent, usually not mat-forming; leaf blades broadly elliptic to elliptic-oblong, ovate-spatulate, or ovate, 3–11(–15) × 3–6(–8) mm, base slightly asymmetric, margins serrulate; cyathia solitary or in small, cymose clusters at distal nodes or on congested axillary branches; glands 4, reddish, 0.1 × 0.1–0.2 mm, appendages white to pink, rudimentary; capsules 1.2–2 × 1.4–1.5 mm; seeds 0.8–1.1 × 0.5–0.7 mm, with several narrow, sharp, slightly irregular, transverse ridges. Sandy sites in disturbed areas, fields, gardens, sidewalks, 3110–7200 ft. Scattered.

Euphorbia rayturneri V. W. Steinm. & Jercinovic—Ray Turner's spurge. Herbs, annual, prostrate; leaf blades ovate to elliptic, often slightly falcate, 5–11 × 2–5 mm, base asymmetric, margins sharply serrulate, surfaces often with red spot toward middle; cyathia solitary at distal nodes; glands 4, green, yellow, or light pink, 0.2 × 0.2–0.3 mm, appendages absent or green, yellow, or light pink; capsules 1.7–2 × 2.2–2.7 mm; seeds 1.2–1.4 × 1–1.1 mm, with 2 well-defined transverse ridges. Desert grasslands, 4590–5580 ft. Grant, Hidalgo, Luna. **R & E**

Euphorbia revoluta Engelm. [*Chamaesyce revoluta* (Engelm.) Small]—Threadstem spurge. Herbs, annual, erect; leaf surfaces glabrous, blades 6–27 × 0.6–1.2 mm, base symmetric, margins entire, revolute; cyathia solitary at distal nodes; glands 4, pink to dark purple, 0.1 × 0.1–0.2 mm, appendages white; capsules 1.5–1.8 × 1.6–1.8 mm; seeds 0.9–1.4 × 0.7–1 mm, nearly smooth, rugulose. Rocky soils, desert scrub, sagebrush, juniper-piñon, oak woodlands, ponderosa pine–oak, 3600–8975 ft. NC, NW, C, WC, SE, SC, SW.

Euphorbia serpens Kunth [*Chamaesyce serpens* (Kunth) Small]—Creeping or round-leafed spurge. Herbs, annual, prostrate, frequently mat-forming; leaves glabrous, blades ovate, oblong, or orbiculate, 2–7(–9) × 2–6 mm, base asymmetric, margins entire; cyathia solitary at distal nodes; glands 4, yellow, oblong, 0.1 × 0.2 mm, appendages white to pinkish; capsules 1.3–1.4 × 1.3–1.7 mm; seeds 0.7–1.1 × 0.4–0.7 mm, smooth. Mostly sandy soils, desert scrub, oak-juniper woodlands, dunes, mesquite grasslands, disturbed sites, 3910–6800 ft. Scattered.

Euphorbia serpillifolia Pers. [*Chamaesyce serpillifolia* (Pers.) Small]—Thymeleaf sandmat. Herbs, annual, prostrate to ascending, often mat-forming; leaf blades glabrous, ovate, oblong, elliptic, or obovate, 3–13 × 2–7 mm, base asymmetric, margins usually entire in proximal 1/2 and serrulate in distal 1/2; cyathia solitary or in small, cymose clusters at distal nodes or on congested axillary branches; glands 4, yellow to pink, 0.1 × 0.2–0.3 mm, appendages white to pink; capsules 1.4–1.9 × 1.5–2; seeds 1–1.4 × 0.6–0.9 mm, smooth to dimpled or rugose. Desert scrub, grasslands, oak-pine-juniper, disturbed areas, 4925–8000 ft. Widespread.

Euphorbia serrula Engelm. [*Chamaesyce serrula* (Engelm.) Wooton & Standl.]—Sawtooth sandmat. Herbs, annual, prostrate or ascending; leaf surfaces sparsely pilose to glabrate, blades oblong, ovate, or elliptic, sometimes falcate, 3–11 × 2–5 mm, base asymmetric, margins sharply serrate to serrulate, usually with conspicuous teeth at base of leaf; cyathia solitary at distal nodes; glands 4, greenish yellow, 0.1 × 0.1–0.2 mm; appendages usually white, rarely light pink; capsules 2–2.6 × 3.2–3.7 mm; seeds 1.5–1.8 × 1.1–1.3(–1.5) mm, smooth to minutely rugulose or with scattered small depressions. Desert scrub, grasslands with mesquite, juniper–piñon–ponderosa pine communities, 3115–7200 ft. NC, NW, C, SE, SC, SW.

Euphorbia setiloba Engelm. [*Chamaesyce setiloba* (Engelm.) Millsp.]—Fringed or shaggy spurge. Herbs, annual, prostrate, mat-forming; leaf surfaces villous, blades oblong, ovate, or elliptic, 3–7 × 2–4 mm, base asymmetric, margins entire; cyathia solitary at distal nodes, nodes often congested toward tips of branches; glands 4, red to pink, 0.1–0.2 × 0.2–0.3 mm, appendages white to pink; capsules 1–1.2 mm diam.; seeds 0.8–1 × 0.5–0.6 mm, dimpled or with faint transverse ridges. Sandy areas, desert scrub, grasslands, river valleys, 3785–6200 ft. NC, C, SE, SC, SW.

Euphorbia stictospora Engelm. [*Chamaesyce stictospora* (Engelm.) Small]—Mat or narrow-seeded spurge. Herbs, annual, prostrate, often mat-forming; leaf surfaces sparsely to moderately pilose to lanate, blades usually oblong to oblong-obovate, occasionally nearly circular, 3–10(–15) × 2–5(–10) mm, base asymmetric, margins minutely or conspicuously serrulate at least toward apex; cyathia solitary at leaf nodes or in small, cymose clusters on congested axillary branches; glands 4, reddish, ± unequal, oblong, 0.1 × 0.1–0.3 mm, appendages white to strongly pinkish or reddish-tinged; capsules 1.6–2.3 × 1.4–1.5 mm; seeds usually 1.5 × 0.5–0.6 mm, with irregularly interrupted furrows. Open disturbed sites, 3115–8000 ft. Widespread except NW.

Euphorbia theriaca L. C. Wheeler [*Chamaesyce theriaca* (L. C. Wheeler) Shinners]—Terlingua sandmat. Herbs, annual, prostrate to reclining, not mat-forming; leaf surfaces glabrous, blades ovate, oblong, orbiculate, or obovate, 3–7.1 × 1–3.5 mm, base slightly asymmetric, margins entire, often revolute on drying; cyathia usually solitary at distal nodes, rarely clustered on short, axillary branches; glands 4, yellow-green to red-purple, 0.2–0.5 × 0.2–0.7 mm, appendages absent or white to pink, capsules 1.4–1.6 × 1.5–1.7 mm; seeds 0.8–1.2 × 0.5–0.8 mm, with prominent transverse ridges. New Mexico material belongs to **var. spurca** M. C. Johnst. [*Chamaesyce spurca* (M. C. Johnst.) B. L. Turner]. Limestone or igneous soils, desert scrub, 4410–4600 ft. Doña Ana.

Euphorbia vermiculata Raf. [*Chamaesyce vermiculata* (Raf.) House]—Wormseed spurge. Herbs, annual or short-lived perennial, prostrate to ascending or erect, not mat-forming; leaf surfaces sparsely pilose, villous, or sericeous, blades ovate, oblong, or elliptic, often falcate, 5–18 × 3–9 mm, base asymmetric, margins usually serrulate; cyathia solitary at distal nodes or in small, cymose clusters at branch tips; glands (2–3)4, red to reddish green, 0.1 × 0.1–0.2 mm, appendages absent or white, turning pink with age; capsules 1.4–1.8 × 1.7–2.1 mm, seeds 1.1–1.4 × 0.7–0.8 mm, rugulose and sometimes also with transverse ridges. Clay soils, riparian sites, 4700–6425 ft. WC, SW.

Euphorbia villifera Scheele [*Chamaesyce villifera* (Scheele) Small]—Hairy spurge. Herbs, annual or perennial, erect to ascending, rarely prostrate to decumbent; leaf surfaces usually villous, rarely glabrous, blades ovate, 3–12 × 2–10 mm, base asymmetric, margins entire or serrulate; cyathia solitary at distal nodes; glands 4, pink, 0.1–0.2 × 0.2 mm, appendages white to pink, 0.2–0.4 × 0.2–0.6 mm; capsules 1.5–2 × 2.1–3.1 mm; seeds 1–1.4 × 0.6–0.8 mm, smooth, faintly rugose. Reported for New Mexico, but no specimens are known.

EUPHORBIA L. subg. Esula Pers.

Herbs or shrubs, annual, biennial, or perennial, woody or herbaceous (succulent in *E. myrsinites*), erect or ascending, branched or unbranched; leaves alternate (opposite in *E. lathyris*), surfaces glabrous or hairy, blades monomorphic, herbaceous (fleshy in *E. myrsinites*), base symmetric, margins entire or toothed; cyathia terminal, pleiochasia with (1)2–17 primary branches; glands 4–5, flat or slightly convex, appendages absent or hornlike; seeds with caruncle.

1. Leaves opposite; capsules tardily dehiscent and appearing indehiscent, mesocarp spongy…**E. lathyris**
1'. Leaves alternate; capsules dehiscent, mesocarp not spongy…(2)
2. Cocci verrucose-tuberculate, verrucose, or papillate; involucral gland horns absent…(3)
2'. Cocci smooth; involucral gland horns present or absent…(4)
3. Cocci papillate, papillae raised, 0.2–0.5 mm; montane areas…**E. alta**
3'. Cocci verrucose, verrucae low and round, 0.1–0.2 mm; widespread…**E. spathulata**
4. Annual or biennial herbs with taproots…(5)
4'. Perennial or biennial herbs with rootstocks (taproot in *E. myrsinites*)…(6)
5. Seeds whitish or grayish, 1–1.6 × 0.6–1 mm; annual…**E. peplus**
5'. Seeds cream-and-brown-mottled, 2–2.5 × 1.4–1.7 mm; mostly biennial…**E. crenulata**
6. Stems succulent; leaf blades fleshy, midvein not prominent; involucral gland horns thick, tips rounded, dilated; capsules 5–7 mm…**E. myrsinites**
6'. Stems not succulent; leaf blades not fleshy, midvein prominent; involucral gland horns absent or slender, tips attenuate or rounded; capsules 2–5 mm…(7)
7. Plants with slender, spreading rootstocks; seeds smooth…**E. virgata**
7'. Plants from thick rootstocks; seeds shallowly pitted to almost smooth…(8)
8. Peduncles 1–3 mm; involucral gland horns usually convergent; capsules 4.3–5 mm…**E. chamaesula**
8'. Peduncles 0.3–1 mm; involucral gland horns absent or usually divergent; capsules 2–4 mm…(9)
9. Involucral gland margins usually entire, occasionally slightly crenate to dentate, horns longer than teeth on gland margins…**E brachycera**
9'. Involucral gland margins irregularly to strongly crenate or dentate, horns absent or equaling to slightly longer than teeth or gland margins…**E. lurida**

Euphorbia alta Norton [*Tithymalus altus* (Norton) Wooton & Standl.]—Giant spurge. Herbs, annual or biennial, erect, branched; leaves glabrous, blades oblong-spatulate, 20–50 × 7–18 mm, margins serrulate; cyathia terminal, pleiochasial branches 3, 2–3 times 2-branched; glands 4, 0.3–0.5 × 0.5–0.7 mm; horns absent; capsules 2–3 × 2.5–3.5 mm, 3-lobed; seeds 1.6–2 × 1.3–1.7 mm, reticulate and areolate, caruncle reniform, flat. Pine-oak and mixed conifer forests, disturbed roadsides, logged areas, 7000–8800 ft. EC, C, WC, SE, SC, Quay.

Euphorbia brachycera Engelm. [*E. robusta* (Engelm.) Small]—Horned or shorthorn spurge. Herbs, perennial, erect or ascending, branched; leaves usually glabrous, blades oblong-elliptic, lanceolate, or oblanceolate to broadly ovate, 5–25 × 2–7 mm, margins entire; cyathia terminal, pleiochasial branches 3–5(–8), 1–4+ times 2-branched; glands 4, 0.5–0.8 × 0.7–1.7 mm; horns divergent; capsules 2.8–4 × 3.5–4.5 mm, 3-lobed; seeds 2–2.8(–3) × 1.4–2.2 mm, irregularly shallowly pitted; caruncle sessile to shortly stipitate, conic. Montane areas, canyons, rock crevices, sandy or gravelly slopes, pine-oak woodlands, ponderosa pine and mixed conifer forests, 4200–9570 ft. Widespread except EC.

Euphorbia chamaesula Boiss. [*Tithymalus chamaesula* (Boiss.) Wooton & Standl.]—Mountain spurge. Herbs perennial, erect; leaves glabrous, blades elliptic to oblong, 8–20(–40) × 3–6 mm, margins entire, cyathia terminal, pleiochasial branches 3–5(6), each 3–4 times 2-branched; glands 4, 0.5–0.8 × 1–1.8 mm; horns usually convergent, 0.2–0.8 mm; capsules 4.3–5 × 5–6 mm, 3-lobed; seeds 2.6–3.4 × 2–2.6 mm, shallowly pitted to almost smooth, caruncle conic. Clearings in mixed conifer, oak woodlands, ponderosa pine forests, 5150–8300 ft. NWS, WC, SC, SW.

Euphorbia crenulata Engelm. [*Tithymalus crenulatus* (Engelm.) A. Heller]—Chinese caps. Herbs, usually biennial, occasionally annual, erect, sometimes decumbent at base, unbranched or branched; leaves glabrous, blades obovate-spatulate to oblanceolate, 8–22 × 3–10 mm, margins entire or slightly crisped; cyathia terminal, pleiochasial branches 3, each 2-branched; glands 4, 0.6–1.2 × 1.5–2.3 mm; horns slightly divergent to slightly convergent; capsules 2.5–3 × 3.5–4 mm, 3-lobed; seeds 2–2.5 × 1.4–1.7 mm, usually irregularly vermiculate-ridged and large-pitted, caruncle reniform. One location from McKinley County, Chuska Mountains, ponderosa pine–Gambel oak, ca. 7000 ft.

Euphorbia lathyris L. [*Tithymalus lathyris* (L.) Hill]—Mole or gopher plant. Herbs, annual or biennial, erect, unbranched or branched; leaves opposite, decussate, glabrous, blades linear to oblong-lanceolate, 30–120 × 3–25 mm, base acute, margins entire; cyathia terminal, pleiochasial branches 2–4, each 1–2 times 2-branched; glands 4, elliptic, 0.3–0.6 × 1–1.3 mm; horns divergent; capsules 9–12 × 12–16 mm, deeply 3-lobed, seeds 4.5–6 × 3–4.2 mm, rugose, irregularly reticulate; caruncle hat-shaped. Roadsides and disturbed areas, juniper-piñon; 1 collection from Bernalillo County, 6880 ft.

Euphorbia lurida Engelm. [*E. palmeri* Engelm. ex S. Watson; *Tithymalus luridus* (Engelm.) Wooton & Standl.]—Woodland spurge. Herbs, perennial, with thick rootstock; stems erect or ascending; leaf blades oblanceolate to obovate, 8–20 × 3–7 mm, base truncate or cuneate, margins entire; cyathial arrangement terminal; glands 4, strongly dentate, horns absent or usually divergent or straight, 0.1–0.3 mm; staminate flowers 10–20; pistillate flowers with ovary glabrous or puberulent; capsules ovoid, 3.5–4 × 4–4.5 mm; seeds gray to dark gray, truncate-oblong to truncate-ovoid, 2.8–3 × 1.7–2 mm, irregularly pitted. Open pine-oak forests, 6000–9000 ft. NW, NC, SE.

Euphorbia myrsinites L. [*Tithymalus myrsinites* (L.) Hill]—Myrtle or creeping or blue spurge. Herbs, usually perennial, occasionally biennial, erect or semiprostrate, succulent; leaves glabrous, blades obovate, obovate-oblong, lanceolate, orbiculate, or suborbiculate, 2–30 × 3–17 mm, fleshy; cyathia terminal, pleiochasial branches 2–12, each 1–2 times 2-branched; glands 4, 1–1.5 × 1.5–2.5 mm, horns divergent, thick, tips rounded, capsules 5–7 × 5–6 mm, unlobed; seeds 2.8–4.5 × 2–3.2 mm, vermiculate-rugose, caruncle trapezoidal or mushroom-shaped. One collection from Santa Fe County, 6610 ft. Introduced.

Euphorbia peplus L. [*Esula peplus* (L.) Haw.; *Tithymalus peplus* (L.) Hill]—Petty spurge. Herbs, annual; stems erect, leaf blades obovate, oblong, or suborbiculate, 5–25 × 4–15 mm, base attenuate or cuneate, margins entire; cyathial arrangement terminal; glands 4, horns slightly convergent to divergent, 0.4–0.6 mm; staminate flowers 10–15; pistillate flowers with ovary glabrous; capsules subglobose, 1.3–2 × 1.5–2.2 mm; seeds whitish or grayish, subovoid, 1–1.6 × 0.6–1 mm, abaxial faces regularly large-pitted. Weedy, gardens, roadsides, waste places; 1 record from Doña Ana County, 4150 ft. Introduced.

Euphorbia spathulata Lam. [*E. arkansana* Engelm. & A. Gray]—Warty spurge. Herbs, usually annual, rarely biennial, erect or ascending, unbranched or branched; leaves glabrous, blades oblanceolate, oblong-oblanceolate, spatulate, or cuneate, 10–50 × 6–11 mm, margins finely serrulate; cyathia terminal, pleiochasial branches 3(–5), each 1–3 times 2-branched; capsules 2–3.5 × 4 mm, 3-lobed; seeds 1.3–2.5 × 1.5–1.8 mm, smooth, reticulate, or finely low-ridged, caruncle irregularly reniform to round. Desert scrub, prairies, riparian sites, juniper-piñon, 3100–7500 ft. Widespread except NE, NNC, C, Doña Ana.

Euphorbia virgata Waldst. & Kit.—Leafy spurge. Herbs, perennial, with slender, spreading root-stock; stems erect; leaf blades linear to linear-oblanceolate or linear-oblong, 40–90 × 3–12 mm, base truncate or abruptly attenuate, margins entire, surfaces glabrous; cyathial arrangement terminal; glands 4, crescent-shaped, 0.6–1.5 × 1.3–2.5 mm; horns divergent to convergent, 0.2–0.8 mm; staminate flowers 10–25; pistillate flowers with ovary glabrous; capsules subglobose, 2.5–3.5 × 3–4.5 mm; seeds yellow-brown to gray or mottled, oblong-ellipsoid to oblong-ovoid, 2.2–2.6 × 1.3–1.6 mm, smooth. Open disturbed areas, trails, creek margins, roadsides, 6300–7000 ft. Scattered. Introduced noxious weed.

EUPHORBIA L. sect. **Poinsettia** (Graham) Baill.

Herbs, annual or perennial, erect or ascending, branched; leaves opposite or alternate, surfaces glabrous or variously hairy, blades monomorphic (occasionally polymorphic in *E. cyathophora* and *E. heterophylla*), base symmetric, margins entire or toothed, flat to revolute; cyathia terminal, monochasia, dichasia, or condensed pleiochasia with 1–3 primary branches; caruncle present or absent.

1. Leaf blades linear to linear-elliptic, margins entire or with 2–4 inconspicuous teeth near apex; involucral glands densely canescent, appendages divided into subulate segments, incurved and covering glands, densely canescent; styles unbranched…**E. eriantha**
1′. Leaf blades linear, lanceolate, ovate, oblong, elliptic, or pandurate, margins conspicuously toothed (sometimes subentire in *E. cyathophora*); involucral glands glabrous, appendages absent or entire, undulate, slightly lobed or divided into triangular segments, not incurved and covering glands, glabrous; styles bifid…(2)
2. Branches often arcuate; involucral gland appendages usually present, rarely absent…(3)
2′. Branches ± straight (except occasionally proximal branches arcuate in *E. davidii* and *E. dentata*); involucral gland appendages absent…(4)
3. Leaf blades usually ovate, rarely oblong or elliptic, margins finely serrulate; petioles 15–49 mm; involucral glands 1(–3); ovaries glabrous; caruncles absent or rudimentary…**E. bifurcata**
3′. Leaf blades linear to narrowly elliptic or ovate, margins coarsely serrate; petioles 1–3 mm; involucral glands 4(5); ovaries puberulent on keels; caruncles 0.1 × 0.2 mm…**E. exstipulata**
4. Leaves usually opposite, occasionally alternate distally, blade margins coarsely crenate-dentate or double-crenate; seeds with caruncles; pleiochasial bracts wholly green or with paler green, white, or mauve near base; annual herbs…(5)
4′. Leaves usually alternate, occasionally opposite proximally, blade margins entire, subentire, or glandular-serrulate; seeds usually without caruncles, occasionally caruncles rudimentary; pleiochasial bracts green, often paler green, pink, or red at base, occasionally wholly white, pink, or red; annual or perennial herbs…(7)
5. Ovaries densely pilose; capsules pilose (often sparsely); involucral glands taller than wide, stipitate…**E. dentata** (in part)
5′. Ovaries and capsules glabrous or sparsely strigose; involucral glands shorter than wide, sessile…(6)
6. Hairs of abaxial leaf blade surface stiff, strongly tapered; capsules 4–4.8 mm wide; seeds angular in cross-section, unevenly tuberculate…**E. davidii**
6′. Hairs of abaxial leaf blade surface weak, filiform; capsules 3.5–4 mm wide; seeds rounded in cross-section, evenly tuberculate…**E. dentata** (in part)
7. Pleiochasial bracts wholly green or paler green at base; involucral glands stipitate, opening circular (occasionally flattened from pressing), with annular rim…**E. heterophylla**
7′. Pleiochasial bracts green or purpurescent, often white, pink, or red at base, occasionally wholly white, pink, or red; involucral glands sessile or substipitate, opening oblong (flattened without pressing), without annular rim…**E. cyathophora**

Euphorbia bifurcata Engelm.—Forked spurge. Herbs, annual, erect; leaves usually alternate, occasionally opposite, blades usually ovate, rarely oblong or elliptic, 13–54 × 7–38 mm, margins finely serrulate; cyathia terminal, dichasial branches 2, few-branched; glands 1(–3), greenish, 0.3–0.4 × 0.4–0.8 mm, appendages petaloid, white; capsules 2.8–3.1 × 3.6–4.5 mm; seeds 1.9–2.4 × 1.5–1.8 mm, irregularly and coarsely tuberculate; caruncle absent or rudimentary. Often in riparian areas with cottonwoods and willows; piñon woodlands, pine-oak and mixed conifer forests, 6450–7955 ft. SC, SW, Lincoln.

Euphorbia cyathophora Murray [*Poinsettia cyathophora* (Murray) Bartl.]—Fire on the mountain. Herbs, annual, erect or ascending; leaves usually alternate, occasionally opposite, blades linear, lanceolate, elliptic, or wider, margins subulately glandular-serrulate; cyathia terminal, pleiochasial branches (1–)3, 1–2-branched; gland 1, yellow-green, 1–1.4 × 0.9–1.6 mm, appendages absent; capsules 2.8–3.2 × 4–4.5 mm, 3-lobed; seeds 2.3–3.1 × 1.9–2.5 mm, tuberculate, caruncle absent. Juniper-piñon, ponderosa pine, Douglas-fir, disturbed sites, 5000–8375 ft. Doña Ana, Eddy, Otero.

Euphorbia davidii Subils—Toothed spurge. Herbs, annual, stems erect or ascending; leaves usually opposite, occasionally alternate, strigose, blades usually narrowly to broadly elliptic, occasionally lance-elliptic, 10–100 × 5–35 mm, margins coarsely crenate-dentate; cyathia terminal, pleiochasial branches usually 3, occasionally reduced to congested cyme, 1–2-branched; gland 1, yellow-green, 0.9 × 1.3 mm, appendages absent; capsules 2.9–3.3 × 4–4.8 mm, 3-lobed; seeds 2.4–2.9 × 2.2–2.9 mm, low-tuberculate, caruncle 0.9–1.1 mm. Forests, stream and river banks, prairies, roadsides, open disturbed areas, 3630–7720 ft. Widespread except NW, WC.

Euphorbia dentata Michx. [*E. cuphosperma* (Engelm.) Boiss.; *Poinsettia dentata* (Michx.) Klotzsch & Garcke]—Toothed spurge. Herbs, annual, erect or ascending; leaves usually opposite, occasionally alternate at distal nodes, pilose to glabrate, blades 30–70 × 4–35 mm, narrowly lanceolate to suborbiculate, usually broadest below middle, margins coarsely crenate-dentate or doubly crenate; cyathia terminal, pleiochasial branches usually 3, occasionally reduced to congested cyme, 1–2-branched; glands (1)2, green, 0.7–0.9 × 0.9–1.2 mm; capsules 2.5–2.8 × 3.5–4 mm, 3-lobed; seeds 2.1–2.7 × 1.7–2.1 mm, evenly minute-tuberculate; caruncle 0.4–0.6 mm. Stream and river banks, juniper-piñon, ponderosa pine, disturbed sites, 3400–8100 ft. E, Grant, Otero, San Juan.

Euphorbia eriantha Benth.—Beetle spurge. Herbs, annual or perennial, erect to ascending; leaves alternate, surface usually glabrous, rarely pilose to shortly sericeous, blades linear to linear-elliptic, 20–55 × 1–3 mm, margins entire or with 2–4 inconspicuous teeth; cyathia terminal, cymose branches 1, 1–2-branched; glands (2–)4–5, green to maroon, 0.5–0.6 × 0.5–0.6 mm, appendages petaloid, whitish; capsules 4.4–4.9 × 3.5–4.1 mm; seeds 2.8–4.1 × 2–2.4 mm, irregularly pitted and tuberculate, caruncle 0.4–0.8 × 0.6–1.1 mm. Rocky slopes, washes, desert scrub. Known only from Eddy County; no records seen.

Euphorbia exstipulata Engelm. [*E. exstipulata* var. *lata* Warnock & M. C. Johnst.]—Squareseed or Clark Mountain spurge. Herbs, annual, erect; leaves opposite, blades linear to narrowly elliptic or ovate, 14–42 × 3–28 mm, margins coarsely serrate, occasionally revolute; cyathia terminal, cymose or dichasial branches usually 1–2, occasionally reduced to monochasia, 1–2-branched; glands 4(5), yellow to pink, 0.2 × 0.3–0.4 mm; capsules 2.7–3.3 × 3.1–3.9 mm; seeds 1.9–2.5 × 1.4–1.7 mm, tuberculate, often with 2 transverse ridges; caruncle 0.1 × 0.2 mm. Desert scrub, grasslands, oak-juniper woodlands, piñon-juniper–ponderosa pine communities, 3940–8510 ft. NW, C, SE, SC, SW, Roosevelt.

Euphorbia heterophylla L. [*Poinsettia heterophylla* (L.) Klotzsch & Garcke]—Mexican fireplant. Herbs, annual, erect-ascending; leaves usually alternate, occasionally opposite proximally, blades narrowly lanceolate to elliptic or broadly obovate, 30–200 × 20–140 mm, margins sparsely glandular-serrulate; cyathia terminal, dichasial branches usually 2, occasionally reduced to congested cyme, 1–2-branched; gland 1, yellow-green, 1–1.4 × 1–1.2 mm, appendages absent; capsules 2.8–3.8 × 4–5.3 mm, 3-lobed; seeds 2.4–2.8 × 1.9–2.4 mm, acute-carinate, tuberculate, with broad rounded tubercles, caruncle 0.1 mm. Desert scrub, desert grasslands, disturbed sites, 5260–6270 ft. Grant, Hidalgo.

JATROPHA L. – Jatropha

Jatropha macrorhiza Benth. [*J. arizonica* I. M. Johnst.]—Jirawilla. Herbs, perennial, to 5 dm, monoecious, erect, green, usually sparsely branched, herbaceous, somewhat succulent, glabrous; stipules persistent; leaves glabrous except puberulent on veins, blades cordate in outline, 11–16 × 9.3–11.2 cm, (3–)5–7(–9)-lobed to middle, base cordate, margins coarsely dentate; inflorescences bisexual, terminal; staminate flower sepals lanceolate, 5–7 × 1–2 mm, corolla light pink, often with white striations, petals

8–11.5 × 2.5–4.5 mm; pistillate flowers resembling staminate, but slightly larger, carpels 3; capsules ± spherical, 1.2–1.3 × 1.2–1.3 cm, explosively dehiscent; seeds 8–9 × 6–6.5 mm; caruncle prominent. Hillsides, mesas, sandy washes, desert scrub, 3500–5650 ft. SW, Catron, Otero.

STILLINGIA Garden – Toothleaf

(Ours) herbs or subshrubs, perennial, monoecious, hairs absent, latex white; leaves deciduous, alternate; stipules absent or present, glands absent; blades unlobed, margins dentate, crenate, serrulate, or spinulose-dentate; inflorescences with pistillate flowers proximal, staminate distal, sessile or subsessile; staminate flower sepals 2, petals 0, nectary absent, stamens 2; pistillate flower sepals 3, distinct, petals 0, nectary absent; fruit a capsule, glabrous; seeds ellipsoid or cylindrical, caruncle present (Huft, 2016 and refs. therein).

1. Leaf blades ovate, elliptic, or lanceolate to obovate or oblanceolate, teeth without blackened tips, incurved; capsules 6–12 mm diam.…**S. sylvatica**
1'. Leaf blades linear to linear-lanceolate, teeth with prominent blackened tips, not incurved; capsules 6–8 mm diam. [to be expected]…S. texana

Stillingia sylvatica L. [*S. sylvatica* var. *salicifolia* Torr.]—Queen's delight. Herbs or subshrubs, perennial; stems solitary or fascicled, erect or ascending, mostly unbranched, (1–)2.5–7(–12) dm; leaves alternate, blades ovate, elliptic, lanceolate, obovate, or oblanceolate, 1–10 × 0.5–3 cm, margins serrulate to crenulate, teeth without prominent blackened tips; inflorescences sessile or short-pedunculate; staminate cymules ± crowded, pistillate flowers 3–4, crowded; staminate flower calyx 1 mm; pistillate flower sepals persistent, 3, well developed; capsules 6–12 mm diam.; seeds 4.5 × 3 mm, rugose, caruncle crescent-shaped, 1–1.3 mm. Well-drained sandy soils, 3685–5100 ft. E, Torrance.

Stillingia texana I. M. Johnst.—Texas toothleaf. Herbs or subshrubs, perennial; stems solitary or fascicled, erect, mostly unbranched, 1.5–4.5(–6) dm; leaves alternate, blades linear to linear-lanceolate, (1–)3–6(–7) × 0.3–0.6(–1) cm, margins crenate-dentate, teeth with prominent blackened tips; inflorescences sessile, 3–9 cm; staminate cymules crowded, pistillate flowers 3–4, crowded; staminate flower calyx 1 mm; pistillate flower sepals persistent, 3, well developed; capsules 6–8 mm diam.; seeds 5 × 5 mm, smooth; caruncle broadly crescent-shaped, 1 × 1.3–1.5 mm. Calcareous prairies, 4075–4180 ft. No specimens seen.

TRAGIA L. – Noseburn

Subshrubs (ours), perennial, monoecious, hairy, hairs unbranched, always some stinging, latex absent; leaves deciduous, alternate; stipules present, glands absent; blades unlobed, margins serrate, crenate, dentate, or entire; inflorescences axillary (pistillate flowers proximal, staminate distal); staminate flower sepals 3–5, green or reddish green, petals 0, nectary absent, stamens 3–6(–10); pistillate flower sepals 6, usually green, sometimes red, not petaloid, petals 0, nectary absent, pistil 3-carpellate; capsules usually with 3 carpels maturing; seeds globose to ovoid, caruncle absent (Levin & Gillespie, 2016).

1. Stigmas papillate…**T. nepetifolia**
1'. Stigmas smooth…(2)
2. Leaf blades usually triangular to subhastate, sometimes ovate, base cordate, hastate, or truncate; stems gray-green, apices often flexuous; stigmas undulate to subpapillate; stamens 3–4…**T. amblyodonta**
2'. Leaf blades linear-lanceolate to narrowly ovate, base truncate to weakly cordate; stems dark green to light green, apices rarely flexuous; stigmas smooth to undulate; stamens 3–6(–10)…**T. ramosa**

Tragia amblyodonta (Müll. Arg.) Pax & K. Hoffm.—Dog-tooth or blunt-tooth noseburn. Subshrubs, 1.2–5 dm; stems gray-green, apex often flexuous; leaf blades usually triangular to subhastate, sometimes ovate, 1–4.5 × 0.8–3 cm, base cordate, hastate, or truncate, margins crenate to serrate; inflorescences terminal or axillary, glands absent, staminate flowers 5–16 per raceme, 0.7–1.2 mm, persistent base 0.2–0.8 mm; pistillate flowers 1.5–4 mm in fruit; staminate flower sepals 3–4, green, 0.9–1.2 mm; stamens 3–4; pistillate flower sepals lanceolate, 1–2.5 mm; capsules 7–8 mm wide; seeds brown with tan mottling, 2.5–3.5 mm. Dry, exposed slopes in xerophytic scrublands, 3300–7850 ft. NE, WC, SC, Eddy.

Tragia nepetifolia Cav.—Catnip noseburn. Subshrubs, 1.5–5 dm; stems green to reddish green, apex never flexuous; leaf blades triangular to ovate, sometimes linear, proximal broadly ovate to sometimes suborbiculate, 1.8–5 × 0.9–3.6 cm, often red-green, base truncate to cordate, margins coarsely dentate to coarsely serrate; inflorescences terminal, glands sessile or absent; staminate flowers 8–40 per raceme, distally clustered; staminate bracts 1.3–1.6 mm; staminate flower sepals 3–4, reddish green, 1–2 mm, stamens 3–4; pistillate flower sepals lanceolate [ovate], 1.4–2.3 mm; capsules 6–8 mm wide; seeds 3–4 mm. Pine-oak woodlands, 4100–7500 ft. Widespread.

Tragia ramosa Torr. [*T. stylaris* Müll. Arg.]—Desert noseburn. Subshrubs, 1.2–5 dm; stems dark green to light green, apex rarely flexuous; leaf blades linear-lanceolate to narrowly ovate, 1–4 × 0.5–2 cm, base truncate to weakly cordate, margins serrate; inflorescences terminal, glands few, sessile; staminate flowers 2–20 per raceme; staminate bracts 1.5–2 mm; staminate flower sepals 3–4, green, 1–2.2 mm, stamens 3–6(–10); pistillate flower sepals lanceolate, 0.8–2.5 mm; capsules 6–8 mm wide; seeds 2.5–3.5 mm. Mesquite, desert scrub, pine-juniper and oak woodlands, 3600–8100 ft. Widespread except NC.

FABACEAE – PEA FAMILY

Jason A. Alexander

Annuals to trees; stipules present or absent, pinnately veined; leaves alternate, generally compound; inflorescences solitary, racemose, spicate, umbellate, or headlike; flowers actinomorphic, zygomorphic, or papilionaceous, sepals generally 5, generally fused, petals absent or 1–5, free, fused, or lower 2 ± united into keel, stamens 5 or 10+, filaments free or fused (or with 9 at least partly fused and the uppermost free), pistil 1, ovary superior, 1- or 2-loculed, ovules 1 to many, style and stigma 1; fruit a legume or loment, sessile or stipitate; seeds 1 to many, often ± kidney-shaped, generally hard, with or without a mark or depression called a pleurogram on each side of the seed, the seed coat smooth (Allred, 2020; Barneby, 1989; Isely, 1998; Welsh, 2007).

KEY TO SUBFAMILY GROUPS

1. Flowers actinomorphic, the corolla and calyx inconspicuous; inflorescences generally spicate or in compact heads; stamens 10+, generally long-exserted; seeds with a pleurogram; leaves bipinnately compound or phyllodial...**KEY 2: SUBFAMILY MIMOSOIDEAE**
1'. Flowers zygomorphic, irregular, or reduced to a single petal, the corolla and calyx conspicuous or inconspicuous; inflorescences diverse, generally racemose, paniculate, umbellate, or in compact heads; stamens 5–10, not exserted, or if so, then not exserted beyond the petals; seeds without a pleurogram; leaves variously compound...(2)
2. Corollas irregular or zygomorphic, not differentiated into a banner, keel, and wings; upper petal (banner) inside the lateral petals (wings) in bud; stamens free and visible, not enclosed in petals; leaves pinnately compound or bipinnately compound...**KEY 1: SUBFAMILY "CAESALPINIOIDEAE"** [a traditional but paraphyletic group]
2'. Corollas zygomorphic, differentiated into a banner, keel, and wings (or ± irregular in some *Dalea* and reduced to a single petal in *Amorpha*); upper petal (banner) outside the lateral petals (wings) in bud; stamens connate and enclosed in petals; leaves pinnately or palmately compound, occasionally unifoliate...**KEY 3: SUBFAMILY PAPILIONOIDEAE**

KEY 1: SUBFAMILY "CAESALPINIOIDEAE"

1. Leaves simple; flowers fascicled, appearing before the leaves; a North American native found in cultivation in New Mexico and long-persistent in abandoned cultivated sites, not definitively known to be escaping (no description herein)...**Cercis canadensis**
1'. Leaves compound; flowers not fascicled, appearing after the leaves; native or naturalized taxa documented for New Mexico outside of cultivation...(2)
2. Leaves 1-pinnate...(3)
2'. Leaves 2-pinnate (or both 1- and 2-pinnate in the same plant in some variants of *Gleditsia* and *Parkinsonia*)...(6)
3. Flowers small and the corolla inconspicuous; inflorescence amentiferous...**Gleditsia triacanthos** (in part)
3'. Flowers with a conspicuous corolla; inflorescence not amentiferous...(4)

4. Twigs photosynthetic and greenish; filaments villous…**Parkinsonia aculeata** (in part)
4′. Twigs not photosynthetic and greenish (except for *Senna armata*); filaments glabrous…(5)
5. Fruits elastically dehiscent; stipules conspicuous and persistent…**Chamaecrista**
5′. Fruits inertly dehiscent or indehiscent; stipules inconspicuous or absent…**Senna**
6. Flowers small and the corolla inconspicuous; inflorescence with amentlike branches…**Gleditsia triacanthos** (in part)
6′. Flowers with a conspicuous corolla; inflorescence lacking amentlike branches…(7)
7. Twigs photosynthetic and greenish; filaments villous…**Parkinsonia aculeata** (in part)
7′. Twigs not photosynthetic and greenish; filaments glabrous…(8)
8. Inflorescence axis not stipitate-glandular…**Pomaria jamesii**
8′. Inflorescence axis stipitate-glandular…(9)
9. Shrub or small tree; stems 1–5 m; leaflets 7–11 pairs…**Erythrostemon gilliesii**
9′. Perennial herb; stems 0.6–2 dm; leaflets 4–8 pairs…**Hoffmannseggia**

KEY 2: SUBFAMILY MIMOSOIDEAE

1. Stamens 10 or fewer…(2)
1′. Stamens 10+…(5)
2. Plants armed with nodal spines; fruits indehiscent…**Prosopis**
2′. Plants unarmed or with curved or flattened, internodal prickles; fruits dehiscent…(3)
3. Plants armed with curved or flattened, internodal prickles…**Mimosa**
3′. Plants unarmed…(4)
4. Stipules inconspicuous; plants shrubs or trees…**Leucaena retusa**
4′. Stipules persistent; plants herbaceous perennials or shrubs…**Desmanthus**
5. Filaments fused below, free above; leaves 2-pinnate…(6)
5′. Filaments free; leaves 2-pinnate or simple…(7)
6. Inflorescence axillary spikelike raceme or terminal panicle of headlike racemes; petiole with a proximal or basal gland; fruit valves not recurving…**Albizia julibrissin**
6′. Inflorescence axillary heads; petiole without a gland; fruit dehiscent, valves recurving…**Calliandra**
7. Inflorescence spicate…(8)
7′. Inflorescence of capitate or ovoid heads…(9)
8. Leaves with (4–)6–10 pairs of leaflets…**Mariosousa millefolia**
8′. Leaves with 1–3(4) pairs of leaflets…**Senegalia greggii**
9. Plants armed with prickles or stipular spines…**Vachellia**
9′. Plants without paired nodal spines, unarmed or armed with internodal prickles…(10)
10. Plants herbaceous perennials or subshrubs, suffrutescent only at the base…**Acaciella angustissima**
10′. Plants shrubs or small trees…**Senegalia roemeriana**

KEY 3: SUBFAMILY PAPILIONOIDEAE

1. Plants woody, shrubs or trees…(2)
1′. Plants annuals or perennials, not woody, or if suffrutescent, then the stems herbaceous above and suffrutescent only at the base…(14)
2. Leaves unifoliate, often persistent and inconspicuous or deciduous…(3)
2′. Leaves pinnately or palmately compound, generally persistent…(4)
3. Fruits a legume; herbage glandular-punctate; corolla bluish purple to pinkish purple…**Psorothamnus scoparius**
3′. Fruits a loment; herbage not glandular-punctate; corolla reddish purple…**Alhagi maurorum**
4. Leaves palmately compound…**Lupinus** (in part)
4′. Leaves pinnately compound…(5)
5. Leaves pinnately trifoliate…**Erythrina flabelliformis**
5′. Leaves pinnately compound with > 3 leaflets…(6)
6. Stamens with filaments all free from one another…(7)
6′. Stamens with all filaments fused or 9 fused and 1 free…(8)
7. Flowers lacking a corolla; leaflets involute and filiform…**Parryella filifolia**
7′. Flowers with a corolla of 5 petals, subpapilionaceous; leaves ± flat, broad, coriaceous, ± evergreen…**Dermatophyllum**
8. Corolla consisting of a banner only…**Amorpha** (in part)
8′. Corolla consisting of 5 petals…(9)
9. Herbage conspicuously glandular-punctate…(10)
9′. Herbage not glandular-punctate…(11)
10. Calyx not ribbed; corollas not papilionaceous, the petals ± monomorphic, the upper petal slightly broader and scarcely differentiated from the others; leaves with 6–27 pairs of leaflets, the longest leaf blades > 1 cm…**Eysenhardtia orthocarpa**

10'. Calyx 5-ribbed; corollas papilionaceous, the petals ± differentiated into banner, keel, and wings; leaves with 1–11 pairs of leaflets, the longest leaf blades < 1 cm…**Dalea** (in part)

11. Fruits bladdery-inflated…**Colutea arborescens**

11'. Fruits leathery or ± woody, not bladdery or inflated…(12)

12. Leaves even-pinnate with 4–6 pairs of leaflets; inflorescences in fascicles of 2–4 flowers, each flower solitary on a bracteate peduncle; stipules bristlelike or spiny…**Caragana arborescens**

12'. Leaves odd-pinnate with 2–10 pairs of leaflets; inflorescences in many-flowered axillary racemes; stipules small and herbaceous (or occasionally spinose in *Robinia*)…(13)

13. Hairs of herbage dolabriform; fruits hemispherical and slightly compressed, 3–3.5 mm…**Indigofera sphaerocarpa**

13'. Hairs of herbage basifixed; fruits oblong, laterally compressed, > 10 mm…**Robinia**

14. Leaves with 1–3 leaflets…(15)

14'. Leaves with 4–11 leaflets…(41)

15. Leaflet margins serrate or denticulate…(16)

15'. Leaflet margins entire or broadly lobed…(20)

16. Fruits 1-seeded…(17)

16'. Fruits 2- to many-seeded…(18)

17. Flowering inflorescences in compact spicate racemes, the axis elongating in fruit; corollas 2–3 mm…**Medicago lupulina**

17'. Flowering and fruiting inflorescences in dense heads; corollas 3–11 mm…**Trifolium** (in part)

18. Flowering and fruiting inflorescences in dense heads; fruits straight…**Trifolium** (in part)

18'. Flowering and fruiting inflorescences in elongate or compact racemes; fruits straight, curved, or coiled…(19)

19. Fruits falcate or spirally coiled, unarmed or spiny; leaflets denticulate along the distal 1/3 of the blade; racemes compact…**Medicago**

19'. Fruits straight, unarmed; leaflets serrate or denticulate along the distal 1/2+ of the blade, obscurely denticulate in some species; racemes elongate…**Melilotus**

20. Tendrils present…**Lathyrus** (in part)

20'. Tendrils absent (obscure and poorly developed in *Lathyrus lanzwertii*)…(21)

21. Stamens with filaments all free from one another…**Thermopsis**

21'. Stamens with all or 9 fused filaments…(22)

22. Androecium monadelphous…(23)

22'. Androecium diadelphous…(25)

23. Fruits rounded and inflated…**Crotalaria pumila**

23'. Fruits laterally compressed, not inflated…(24)

24. Fruit a stipitate loment, divided into multiple 1-seeded segments…**Desmodium** (in part)

24'. Fruit not a loment, not divided into multiple segments, seeds 1 to many in each chamber…**Lupinus** (in part)

25. Herbage and fruits glandular-pubescent or glandular-punctate…(26)

25'. Herbage and fruits eglandular…(30)

26. Calyx enlarging in fruit…**Pediomelum** (in part)

26'. Calyx not enlarging in fruit…(27)

27. Plants prostrate, trailing or twining vines…**Rhynchosia senna**

27'. Plants not trailing or twining vines…(28)

28. Fruits elliptic, ± laterally compressed…**Psoralidium tenuiflorum**

28'. Fruits subglobose or ovate, rounded, not compressed…(29)

29. Inflorescence in spikes or spicate racemes; calyx 10-ribbed; fruit ovate with prominent sutures, generally included in the calyx…**Dalea** (in part)

29'. Inflorescence in pseudoracemes bearing 2 to many clusters, each with 2–3 flowers at each node; calyx not conspicuously ribbed; fruit subglobose, exserted from the calyx…**Ladeania lanceolata**

30. Fruit a stipitate loment, divided into multiple 1-seeded segments…**Desmodium** (in part)

30'. Fruit not a loment, not divided into multiple segments, seeds 1 to many in each chamber…(31)

31. Stipules inconspicuous or glandlike…(32)

31'. Stipules herbaceous, bractlike or leaflike, conspicuous…(33)

32. Plants annual; wings ± equal to the keel in length; stigma glabrous…**Acmispon** (in part)

32'. Plants perennial; wings longer than the keel; stigma subtended by a penicillate collar…**Ottleya** (in part)

33. Herbage uncinate-pubescent…**Phaseolus**

33'. Herbage glabrous, glabrescent, or pubescent; the hairs, when present, not uncinate-pubescent…(34)

34. Plants not trailing or twining vines or vinelike herbs…(35)

34'. Plants trailing or twining vines or vinelike herbs with stems sprawling or climbing through vegetation…(37)

35. Flowers 1–2 in leaf axils along the terminal portion of the stem; flowers of 2 forms, the corolla of chasmogamous flowers pinkish purple and white, the cleistogamous flowers apetalous…**Kummerowia striata**

35'. Inflorescences in racemes or pseudoracemes of 2+ flowers; flowers of a single form, all chasmogamous… (36)

36. Inflorescences axillary or pseudoracemes; leaves palmately compound…**Pediomelum** (in part)

36'. Inflorescences axillary racemes, sometimes spicate, subcapitate, or umbellate in appearance; leaves pinnately compound…**Astragalus** (in part)

37. Style straight and glabrous, not bearded or barbellate…(38)

37'. Style straight and bearded (or twisted, terminally hooked, and glabrous with barbellate stigmas in *Macroptilium*)…(39)

38. Flowers solitary or in clusters of 2–3 in leaf axils along the terminal portion of the stem; flowers of 2 forms, the corolla of chasmogamous flowers pinkish purple, the corolla of cleistogamous flowers abortive…**Cologania**

38'. Inflorescences in axillary pseudoracemes; flowers of a single form, all chasmogamous…**Galactia wrightii**

39. Plants rhizomatous; stems angled or winged; style flattened and bearded on the adaxial side…**Lathyrus lanzwertii** (in part)

39'. Plants taprooted annuals or perennials; stems not angled or winged; styles twisted or straight, not flattened…(40)

40. Styles laterally twisted and terminally hooked ca. 180 degrees, glabrous except for the barbellate stigmas; keels coiled 1/2 to a full turn…**Macroptilium gibbosifolium**

40'. Styles straight and bearded on the adaxial side; keels incurved 90–180 degrees but not twisted… **Strophostyles**

41. Corolla consisting of a banner only…**Amorpha** (in part)

41'. Corolla consisting of 5 petals, or in some species of *Dalea*, the readily deciduous keel and wing petals often dropping from the inflorescence, leaving only the banner, therefore appearing to be reduced to only 1 petal…(42)

42. Herbage glandular-pubescent, often punctate…(43)

42'. Herbage glabrous, glabrescent, or variously pubescent, not glandular or glandular-punctate…(45)

43. Fruit body burlike, the valves covered with uncinate prickles; inflorescence axillary; androecium diadelphous…**Glycyrrhiza lepidota**

43'. Fruit body not burlike, unarmed; inflorescence of terminal racemes or spicate racemes; androecium monadelphous or the stamens mostly free and 5-merous…(44)

44. Inflorescence spicate or spicate racemes; flowers sessile or subsessile (pedicels inconspicuous); calyx 10-ribbed; fruit 1–2-seeded…**Dalea** (in part)

44'. Inflorescence racemose; flowers pedicellate, pedicels conspicuous; calyx not ribbed; fruit 1-seeded… **Marina calycosa**

45. Fruit a loment, divided into multiple segments (reduced to a single 1-seeded segment in *Onobrychis*)… (46)

45'. Fruit not a loment, not divided multiple 1–2-seeded segments…(48)

46. Inflorescence umbellate; stems angular; fruit body 4-angled; stipules conspicuously dark-tipped… **Securigera varia**

46'. Inflorescence racemose; stems rounded; fruit body laterally compressed; stipules herbaceous, greenish, not dark-tipped…(47)

47. Fruits a single 1-seeded segment, the valves glabrous and short-prickly around the margin…**Onobrychis viciifolia**

47'. Fruits with 2–8 indehiscent segments, the valves rough and reticulate-veined…**Hedysarum boreale**

48. Stamens with filaments all free…**Vexibia**

48'. Stamens monadelphous or diadelphous…(49)

49. Stipules inconspicuous or glandlike…(50)

49'. Stipules conspicuous, herbaceous, bractlike or leaflike…(52)

50. Plants annual…**Acmispon** (in part)

50'. Plants perennial…(51)

51. Keels obtuse or obscurely rostrate, the wings longer than the keel…**Ottleya** (in part)

51'. Keels prominently rostrate, the wings ± equal to the keel in length…**Lotus corniculatus** (in part)

52. Plants trailing or twining vines or vinelike herbs with stems sprawling or climbing through vegetation… (53)

52'. Plants not trailing or twining vines or vinelike herbs with stems sprawling or climbing through vegetation…(55)

53. Style glabrous; flowers solitary or in clusters of 2–3 in leaf axils along the terminal portion of the stem; flowers of 2 forms, the corolla of chasmogamous flowers pinkish purple, the corolla of cleistogamous flowers abortive…**Cologania angustifolia**

53'. Style bearded; inflorescences of axillary pseudoracemes; flowers of a single form, all chasmogamous…(54)

54. Style bearded adaxially, the beard lateral or apical in orientation, the style apex flattened…**Lathyrus** (in part)

54'. Style bearded distally, the beard tufted or encircling the apex of the style, the style apex rounded...**Vicia**
55. Inflorescence a bracteate umbel, leaflets 5, the lowermost pair stipulelike...**Lotus corniculatus** (in part)
55'. Inflorescence not a bracteate umbel (in some *Astragalus*, the racemes are compressed and umbel-like)...
(56)
56. Stipules spinose...**Peteria scoparia**
56'. Stipules herbaceous, bractlike or leaflike but not spinose...(57)
57. Plants rhizomatous; stems angled or winged; style flattened and bearded on the adaxial side...**Lathyrus lanzwertii** (in part)
57'. Plants not rhizomatous; stems not angled or winged; style glabrous or barbellate, not flattened...(58)
58. Corollas brownish red or orangish red; style barbellate below the stigma, the stigma capitate...**Sphaerophysa salsula**
58'. Corollas various shades of white, purple, pink, or red, but not brownish red or orangish red; style glabrous, the stigma inconspicuous...(59)
59. Androecium monadelphous...**Tephrosia vicioides**
59'. Androecium diadelphous...(60)
60. Keel tip rounded to acute (or beaklike in *A. miser* and *A. humistratus*)...**Astragalus** (in part)
60'. Keel tip conspicuously beaked...**Oxytropis**

ACACIELLA Britton & Rose – Acacia, wattle

Acaciella angustissima (Mill.) Britton & Rose [*Acacia angustissima* (Mill.) Kuntze]—Prairie acacia. Perennial herbs or subshrubs; pubescence basifixed; stems decumbent, erect, or ascending, the herbage glabrous, glabrescent, or strigulose, often more glabrous in age; leaves with 1–17 pairs of pinnae, leaflets 18–40 pairs per pinnule, blades 3–12 mm, elliptic, lanceolate, orbicular, oblong, or ovate; inflorescences headlike, 4–15-flowered; corollas actinomorphic, not papilionaceous, 4- or 5-merous, androecium of separate stamens, stamens 50+; fruits 3–6 × 0.6–1.2 cm, the body slightly curved, the valves thin-papery, glabrescent or strigulose. S, Quay, San Juan.

1. Twigs and herbage glabrous or glabrescent...(2)
1'. Twigs and herbage pubescent...(3)
2. Pinnae 1–8 pairs per leaf...**var. texensis**
2'. Pinnae 9–17 pairs per leaf...**var. angustissima**
3. Twigs, petioles, and rachises densely or sparsely hispid or pilose, the hairs yellowish; pinnae 18–32 pairs per leaf; reported and possibly present, but not documented from New Mexico...var. filicioides (Cav.) L. Rico
3'. Twigs, petioles, or rachises strigulose (often more glabrous in age), the hairs not yellowish; pinnae 9–17 pairs per leaf...**var. angustissima**

var. angustissima [*Acacia angustissima* (Mill.) Kuntze var. *hirta* (Nutt.) B. L. Rob; *A. angustissima* (Mill.) Kuntze var. *suffrutescens* (Rose) Isely]—Herbage and stems glabrous, glabrescent, or strigulose, often more glabrous in age; leaves with 9–17 pairs of pinnae. Hills, canyons, washes, xeric grassland, woodlands, and prairies, often on igneous or limestone substrates but not an obligate calciphile, mostly in alkaline soils, 4500–6200 ft.

var. texensis (Torr. & A. Gray) L. Rico [*Acacia angustissima* (Mill.) Kuntze var. *chisosiana* Isely; *A. angustissima* (Mill.) Kuntze var. *texensis* (Torr. & A. Gray) Isely]—Herbage and stems glabrous or glabrescent; leaves with 1–8 pairs of pinnae. Rocky or sandy sites (often on igneous, limestone, or sandstone substrates but not an obligate calciphile), hills, canyons, scrublands, washes, and xeric woodlands, commonly associated with juniper, piñon, oaks, and creosote bush, 4265–7850 ft.

ACMISPON Raf. – Deervetch, bird's-foot trefoil

Annuals; stems unarmed, pubescence basifixed; stipules inconspicuous, reduced to small glands; leaves odd-pinnate, pinnately compound, petiolate, leaflets 1–7 pairs per leaf, the blades opposite, subopposite, or irregular; inflorescences an umbel or solitary, axillary, peduncle elongate with a foliage leaf or short and the foliage leaf absent, bracts reduced to small glands resembling stipules, persistent; flowers pedicellate, calyx tube not inflated, the teeth acute, corollas papilionaceous, the wings nearly equal to the keel in length, androecium diadelphous; fruits a legume, dehiscent, sessile, the valves papery to leathery, glabrous or pubescent, the beak short.

1. Inflorescences pedunculate, the peduncle not hidden by the bract or leaves; corollas ochroleucous, whitish, or pinkish, often red-striate; fruit body 15–30 mm, the valves glabrous, glabrescent, or hirsute… **A. americanus**
1'. Inflorescences subsessile, the peduncle hidden by the bract and leaves; corollas yellow, turning red in age; fruit body 6–14 mm, the valves villous…**A. brachycarpus**

Acmispon americanus (Nutt.) Rydb. [*Lotus americanus* (Nutt.) Bisch.; *L. purshianus* Clem. & E. G. Clem.]—American bird's-foot trefoil. Annuals; stems erect, ascending, or prostrate, the herbage glabrous, glabrescent, or pubescent, often ± glabrous in age; leaves pinnately trifoliate or the uppermost simple, petiolate or subsessile, leaflets 0–1 pairs per leaf, the blades 5–25 mm, ovate or lanceolate; inflorescences 1-flowered, pedunculate, longer than the subtending, simple bract; calyx tube 1–2 mm, corollas 4–9 mm, ochroleucous or pinkish, often red-striate; fruit 1.5–3 cm, spreading or declined, dehiscent, the body linear, straight or slightly curved, ± rounded or laterally compressed in cross-section, the valves papery, glabrous, glabrescent, or hirsute; seeds 5–7. Foothills, prairies, plains, woodlands, occasionally in ruderal areas and roadsides, 5100–6700 ft. Catron, Grant, Hidalgo.

Acmispon brachycarpus (Benth.) D. D. Sokoloff [*Lotus humistratus* Greene]—Foothill deervetch. Annuals; stems ascending or prostrate, ± mat-forming, the herbage glabrescent or villous, leaves pinnately compound, petiolate, the rachis flattened, leaflets 1–3 pairs per leaf, the blades 5–12 mm, elliptic or obovate; inflorescences 1- or 2-flowered, the peduncle ± hidden by the subtending bract and congested, surrounding leaves; flowers ascending, calyx tube 1–2 mm, corollas 3–7 mm, yellowish, reddish in age; fruit 0.6–1.4 cm, ascending, dehiscent, the body oblong, straight or slightly curved, laterally compressed in cross-section, the valves villous; seeds 2–4. Hillsides, deserts, dunes, washes, flats, occasionally in ruderal areas and roadsides, 3800–5600 ft. SW.

ALBIZIA Durazz. – Silk tree

Albizia julibrissin Durazz.—Persian silk tree. Shrubs or trees, the herbage glabrescent or puberulent; leaves with 5–15 pairs of pinnae, petiolar glands present, rachis glands, if present, restricted to the base of the terminal pinnae, leaflets 20–30 pairs per pinnule, the blades 0.7–1.5 cm, oblong; inflorescences umbellate heads in corymblike racemes, the heads 2.5–4 cm wide, many-flowered; flowers ascending, corollas pinkish; fruit 12–20 × 1.5–2.5 cm, the body oblong, straight, laterally compressed, the valves irregularly constricted between the seeds but not segmented, thin-papery, glabrous. Naturalized, 5000–6250 ft. Bernalillo, Doña Ana, Grant. Introduced.

ALHAGI Gagnebin – Camelthorn, manna tree

Alhagi maurorum Medik.—Camelthorn. Shrubs, rhizomatous; stems armed, ± greenish and photosynthetic, the spines 1–2.5 cm, the herbage glabrous, glabrescent, or strigose; leaves short-petiolate or sessile, leaflets 1 per leaf, the blade 0.7–2 cm, elliptic or obovate; flowers ascending to spreading, the calyx tube 2–3 mm, the teeth reduced and inconspicuous, corollas reddish purple, the petals 8–9 mm; fruit 1–3 cm, short-stipitate, the stipe 2–3 mm, the body linear, curved, ± rounded, the valves glabrous. Naturalized, noxious weed, 4900–7250 ft. Widely scattered. Introduced.

AMORPHA L. – False indigo

Subshrubs, shrubs, or trees; stems unarmed, pubescence basifixed, the herbage glandular-punctate; stipules linear and herbaceous or bristlelike, persistent; leaves odd-pinnate, petiolate, leaflets irregular, opposite or subopposite; inflorescences racemose, compact and spikelike or not, terminal, solitary or clustered, bracts resembling stipules; calyx tube not inflated, the teeth acute, corollas not papilionaceous, the petals purplish, the banner present, keel and wings absent, androecium monadelphous, stamens exserted; fruits a legume, indehiscent or tardily dehiscent, sessile, the body oval or falcate, ± curved, laterally compressed in cross-section, the valves papery to leathery, strongly glandular-punctate or eglandular; seeds 1–2.

1. Fruits 5-9 mm; shrubs or small trees; leaves 10-28 cm, petioles 1-4 cm…**A. fruticosa**
1′. Fruits 3-5 mm; shrubs or subshrubs; leaves 1.5-15 cm, petioles 0.2-1 cm…(2)
2. Inflorescences of solitary racemes or racemes in sparse clusters of 2-5; leaflets conspicuously glandular-punctate, the glands readily visible without magnification; fruit glabrous or glabrescent, the body ± curved…**A. nana**
2′. Inflorescences of racemes in dense clusters of 5-30; leaflets inconspicuously glandular-punctate, the glands obscured by dense pubescence; fruit villous, the body ± straight…**A. canescens**

Amorpha canescens Pursh.—Leadplant. Subshrubs, the herbage glabrescent or tomentose; leaves 2-15 cm, the petioles 0.1-0.5 cm, leaflets 3-7 pairs per leaf, the blades 1-3 cm, elliptic, oblong, or ovate, glandular-punctate, the glands obscured by hairs; inflorescences terminal, in dense clusters of 5-30 racemes, the racemes 5-25 cm, many-flowered; flowers ascending or spreading, calyx tube 3-5 mm, corollas purplish, the banner 4.5-6 mm; fruit 3-4.5 mm, the body oval, ± straight, the valves villous. Sandy, rocky, or gravelly sites in prairies, grasslands, woodlands, hillsides, occasionally in ruderal areas and roadsides, 3950-7875 ft. NE, C.

Amorpha fruticosa L. [*A. fruticosa* var. *angustifolia* Pursh; *A. fruticosa* var. *occidentalis* (Abrams) Kearney & Peebles]—Desert false indigo. Shrubs or small trees, the herbage glabrescent, pilosulous, or puberulent; leaves 10-28 cm, petiolate, the petioles 1-4 cm, leaflets 3-15 pairs per leaf, the blades 1-6 cm, elliptic, oblong, ovate, or lanceolate, glandular-punctate or eglandular, the glands obscured by hairs or inconspicuously glandular-punctate in uncommon forms with glabrous or glabrescent leaflets; inflorescences racemose, terminal, pedunculate, racemes solitary or in clusters of 2-15, the racemes 5-20 cm, many-flowered; calyx tube 2-5 mm, corollas reddish purple or bluish purple, the banner 5-6; fruit 5-9 mm, the body oval, ± curved or falcate, the valves glabrous, glabrescent, or puberulent. Moist meadows, woodlands, wetlands, stream banks, bottomlands, mesic ruderal areas, roadside ditches, 3950-7200 ft. C, WC, SW, Eddy.

Amorpha nana Nutt.—Dwarf false indigo. Shrubs, the herbage glabrous, glabrescent, or strigulose; leaves 1.5-10 cm, subsessile or petiolate, the petioles 0.2-1 cm, leaflets 3-15 pairs per leaf, the blades 0.5-2 cm, elliptic, oblong, obovate, or ovate, the apex acute, emarginate, or obtuse, glandular-punctate, the glands readily visible without magnification; inflorescences racemose, terminal, subsessile or pedunculate, solitary or in sparse clusters of 2-5 racemes, the racemes 2-9 cm; flowers ascending or spreading, calyx tube 1-2.5 mm, corollas purplish, the banner 5-6 mm; fruit 4-5 mm, the body oblong, ± curved, the valves glabrous or glabrescent. In soils derived from various substrates, primarily in prairies, hillsides, and plains, 5575-7550 ft. Lincoln, Otero.

ASTRAGALUS L. – Milkvetch

Annual or perennial herbs, caulescent, acaulescent, or subacaulescent; stems unarmed, pubescence basifixed or dolabriform; stipules herbaceous, persistent; leaves pinnately compound, trifoliate, or simple, petiolate; inflorescences racemose, axillary, pedunculate; flowers pedicellate, calyx tube 5-toothed, the teeth acute, corollas papilionaceous, the wings equal to or longer than the keel, androecium diadelphous; fruits a legume.

Note: Because of the complexity of *Astragalus* morphology, some species appear at two or more locations in the key; fruit shape characters in the keys and descriptions (i.e., ellipsoid, ovate, oblong, etc.) refer to the shape of unpressed pods in longitudinal section. In many species the pods retain a similar shape even when pressed. The degree of compression of the pod in cross-section is a diagnostic fruit character used in the key for many taxa, but it is not discernible in pressed specimens with papery-walled or immature fruits, unfortunately. For specimens in this condition, or if either fruits or flowers are absent, it will be necessary to take both leads in the key.

1. Terminal leaflet continuous with the rachis (in some species, the upper lateral leaflets reduced to phyllodia with a subfiliform rachis)…(2)
1′. Terminal leaflet jointed with the rachis, leaves odd-pinnate…(9)
2. Fruits bladdery-inflated…**A. ceramicus**

2'. Fruits not bladdery-inflated…(3)
3. Stipules at the lowermost stem nodes connate-sheathing opposite the petiole proximally…(4)
3'. Stipules at the lowermost stem nodes free opposite the petiole proximally…(6)
4. Pedicels spreading-declined; keel 5.9–8.4 mm; fruit valves strigulose; introduced in New Mexico…**A. miser** (in part)
4'. Pedicels ascending; keel 3.5–5.4 mm; fruits deflexed, the valves glabrous…(5)
5. Fruits 2.3–3 mm wide; corollas whitish and tinged pale pink…**A. cliffordii**
5'. Fruits 3–4.5 mm wide; corollas pinkish purple with white wing tips or rarely whitish and tinged pinkish purple…**A. wingatanus** (in part)
6. Fruits sessile or substipitate, the stipe obscure and < 0.6 mm…**A. episcopus**
6'. Fruits stipitate, the stipe 4–15 mm…(7)
7. Fruits dorsiventrally compressed; corolla creamy-white to almost white…**A. lonchocarpus**
7'. Fruits strongly laterally compressed; corolla pinkish purple or pale lemon-yellow…(8)
8. Corolla pale lemon-yellow; fruits with stipes 8–15 mm, the valves strigulose…**A. ripleyi**
8'. Corolla pinkish purple; fruits with stipes 4–8 mm, the valves glabrous (rare forms with confluent terminal leaflets)…**A. coltonii** (in part)
9. Plants annual or biennial, pubescence basifixed…(10)
9'. Plants short- to long-lived perennial; pubescence basifixed or dolabriform…(17)
10. Fruits unilocular or subunilocular, the septum incomplete and < 1 mm wide…(11)
10'. Fruits bilocular or semibilocular (and the septum incomplete)…(14)
11. Herbage and fruit valves villosulous or pilosulous, the hairs spreading or ascending, mixed straight and contorted…(12)
11'. Herbage and fruit valves strigulose, the hairs appressed and straight…(13)
12. Fruits inflated but not bladdery; fruits thick-papery or leathery, the body 5–11 mm wide; plants annual… **A. sabulonum**
12'. Fruits bladdery-inflated, papery, the body 12–24 mm wide; plants annual or biennial…**A. wootonii** (in part)
13. Inflorescences 10–20-flowered; plants biennial or short-lived perennial, sometimes flowering the first year and appearing annual…**A. allochrous** (in part)
13'. Inflorescences 2–10-flowered (or up to 15 in rare, robust variants); plants annual or biennial…**A. wootonii** (in part)
14. Fruits deciduous…(15)
14'. Fruits persistent…(16)
15. Fruits 2–3 mm wide, the body not inflated…**A. emoryanus**
15'. Fruits 5–18 mm wide, the body inflated…**A. lentiginosus** (in part)
16. Fruits rounded or dorsiventrally compressed; the body 3.5–5 mm wide…**A. brandegeei** (in part)
16'. Fruits trigonously compressed; the body 2–3.5 mm wide…**A. nuttallianus**
17. Leaflets with apex spinulose, all continuous with the rachis; racemes with 1–3 flowers…**A. kentrophyta**
17'. Leaflets with apex not spinulose, all jointed to the rachis or the terminal continuous with the rachis; racemes 1–50-flowered…(18)
18. Pubescence dolabriform…(19)
18'. Pubescence basifixed…(34)
19. Stipules at the lowermost stem nodes connate-sheathing opposite the petiole…(20)
19'. Stipules at the lowermost stem nodes free opposite the petiole…(26)
20. Fruits bilocular…(21)
20'. Fruits unilocular…(22)
21. Plants taprooted, not woody, caudices superficial and ± at ground surface; flowers with spreading, declined, or deflexed pedicels…**A. laxmannii**
21'. Plants rhizomatous, caudices ± woody and subterranean; flowers with ascending, erect, or spreading pedicels…**A. canadensis**
22. Plants pulvinate-cespitose, cushion- or mat-forming; leaflets 1 pair (trifoliate)…**A. sericoleucus**
22'. Plants not cespitose or cushion- or mat-forming; leaflets 2–10 pairs…(23)
23. Plants seleniferous, the herbage malodorous…(24)
23'. Plants not seleniferous, the herbage not malodorous…(25)
24. Fruits stipitate, the stipe 0.7–3.3 mm…**A. albulus**
24'. Fruits sessile…**A. flavus**
25. Calyx tube 1.2–2.5 mm, the teeth 0.7–1.2 mm; flowers 5–6 mm…**A. kerrii** (in part)
25'. Calyx tube 2.5–4.1 mm, the teeth 1.4–5 mm; flowers 6–12 mm…**A. humistratus**
26. Fruits bilocular or semibilocular…**A. calycosus**
26'. Fruits unilocular or subunilocular…(27)
27. Calyx tube 2–5 mm…(28)
27'. Calyx tube 6–11 mm…(32)

28. Body of fruit 9–37 mm…(29)
28'. Body of fruit 4–7.5 mm…(30)
29. Inflorescences dimorphic; cleistogamous racemes sessile and 1–3-flowered, chasmogamous racemes pedunculate and 5–17-flowered; calyx teeth 2–5.2 mm…**A. lotiflorus**
29'. Inflorescences monomorphic; all racemes chasmogamous and 5–15-flowered; calyx teeth 0.7–1.6 mm… **A. missouriensis** (in part)
30. Racemes 7–26-flowered; leaves with 5–12 pairs of leaflets…**A. gilensis**
30'. Racemes 1–3-flowered; leaves with 2–5 pairs of leaflets…(31)
31. Plants tuft-forming, the caudex clothed with marcescent stipules and leaf bases, the persistent leaf bases becoming spinose; fruits 4–5 mm…**A. humillimus**
31'. Plants pulvinate, the caudex clothed with marcescent stipules and leaf bases, the persistent leaf bases fibrous, not spinose; fruits 5–7.5 mm…**A. siliceus**
32. Racemes 1-flowered; plants acaulescent or subacaulescent, cushion-forming, caudex clothed with marcescent stipules and leaf bases…**A. wittmannii**
32'. Racemes 2–15-flowered; plants short-caulescent or subacaulescent, not cushion-forming, caudex subterranean, superficial, or absent, basal internodes short and obscured by stipules but not clothed with marcescent stipules and leaf bases…(33)
33. Fruits deciduous, usually incurved; caudex, if present, superficial, ± at ground level, not subterranean… **A. amphioxys**
33'. Fruits persistent, straight or slightly incurved; caudex well developed and subterranean…**A. missouriensis** (in part)
34. Stipules at the lowermost stem nodes connate-sheathing opposite the petiole…(35)
34'. Stipules at the lowermost stem nodes free opposite the petiole…(73)
35. Fruits unilocular or subunilocular, the septum < 0.5 mm wide…(36)
35'. Fruits bilocular, subbilocular, or subunilocular, the septum > 0.5 mm wide…(65)
36. Fruits bladdery-inflated, the valves thin and translucent…(37)
36'. Fruits not bladdery-inflated, the valves not translucent…(39)
37. Fruits stipitate…**A. castetteri**
37'. Fruits sessile…(38)
38. Caudex superficial or ± at ground surface…**A. allochrous** (in part)
38'. Caudex subterranean, the stems buried for 1–8 cm…**A. fucatus**
39. Plants acaulescent, subacaulescent, or shortly caulescent, tuft-forming…(40)
39'. Plants caulescent, not tuft-forming…(45)
40. Fruit body laterally compressed; keel apex acute and beaklike…**A. miser** (in part)
40'. Fruit body dorsiventrally compressed; keel apex obtuse…(41)
41. Calyx tube 1.2–2.7 mm; corollas 3–6 mm…(42)
41'. Calyx tube 2.5–13 mm; corollas 6–26 mm…(44)
42. Inflorescence 4–14-flowered…**A. knightii**
42'. Inflorescence 1–4-flowered…(43)
43. Plants subacaulescent, caudices clothed with marcescent leaf bases and peduncles…**A. heilii**
43'. Plants short-caulescent, caudices not clothed with marcescent leaf bases and peduncles…**A. kerrii** (in part)
44. Calyx tube 2.5–4.2 mm; fruits declined or deflexed, the valves hirsute with minute bulbous-based hairs…**A. desperatus**
44'. Calyx tube 6.5–13 mm; fruits ascending and occasionally humistrate, the valves strigulose or villosulous, the hairs not bulbous-based…**A. zionis** (in part)
45. Fruits stipitate; stipes 1–19 mm…(46)
45'. Fruits sessile or substipitate; stipes, if present, < 1 mm…(55)
46. Plants seleniferous, herbage malodorous; fruits dorsiventrally compressed, the ventral surface sulcate in 2 parallel grooves along each side of the prominent sutures…**A. bisulcatus**
46'. Plants not seleniferous, herbage not malodorous; fruits variously compressed, the ventral surface not grooved along each side of the suture…(47)
47. Corollas 12–26 mm…(48)
47'. Corollas 5–12 mm…(50)
48. Calyx tube 2–4 mm; fruit 6–14 mm…**A. alpinus** (in part)
48'. Calyx tube 4–9 mm; fruit 15–30 mm…(49)
49. Fruits with a stipe 3.5–7 mm; calyx teeth 3–10 mm…**A. racemosus**
49'. Fruits with a stipe 1.5–4.5 mm; calyx teeth 0.5–2.5 mm…**A. hallii** (in part)
50. Caudices subterranean, the stems belowground for 1–8 cm…(51)
50'. Caudices ± at ground surface, the stems aboveground (or, at snowline in subalpine areas, the stems aboveground but buried in rocky talus)…(53)
51. Fruits trigonously compressed, the lateral faces ± flat; calyx tube 2–4 mm…**A. alpinus** (in part)

51'. Fruits dorsally or laterally compressed; calyx tube 1.5–2.6 mm…(52)
52. Fruits dorsally compressed, the body 2–3 mm wide…**A. proximus**
52'. Fruits laterally compressed, the body 3–4.5 mm wide…**A. wingatanus** (in part)
53. Calyx tube 2–2.7 mm; pedicels ascending at anthesis, declined in age…**A. tenellus** (in part)
53'. Calyx tube 3–4.5 mm; pedicels spreading or declined at anthesis…(54)
54. Flowers ochroleucous and drying yellowish; leaflets 6–12 pairs; inflorescences 20–45-flowered…**A. altus**
54'. Flowers pinkish purple, bluish purple, or whitish and tinged pinkish purple, the wing petals often white-tipped; leaflets 3–7 pairs; inflorescences 5–25-flowered…**A. robbinsii** (in part)
55. Fruits elevated on a curved, stipelike gynophore to 1 mm; calyx inflated in fruit…**A. oocalycis**
55'. Fruits sessile or substipitate; calyx not inflated or ± inflated in fruit…(56)
56. Flowers large, corollas 12–26 mm…(57)
56'. Flowers small, corollas 5–12 mm…(58)
57. Fruits inflated; stems strigose, the longest hairs 0.2–0.4(–0.6) mm…**A. hallii** (in part)
57'. Fruits not inflated; stems villosulous, the longest hairs 0.4–1.2 mm…**A. puniceus**
58. Caudices subterranean, the stems belowground for 1–20 cm…(59)
58'. Caudices ± at ground surface, the lower stems aboveground…(62)
59. Subterranean stems rooting at the nodes and giving rise to offshoots away from the parent plant; fruits 4–6 mm wide, the body trigonously compressed in cross-section…**A. pictiformis**
59'. Subterranean stems not rooting at the nodes and not giving rise to offshoots away from the parent plant; fruits 2–4.5 mm wide, the body laterally or dorsally compressed in cross-section…(60)
60. Fruits laterally compressed, the body 3–4.5 mm wide…**A. wingatanus** (in part)
60'. Fruits dorsally compressed, the body 2–3 mm wide…(61)
61. Fruits thin- to thick-papery, the body 11–15 mm…**A. flexuosus**
61'. Fruits leathery, the body 4–9 mm…**A. gracilis**
62. Fruit valves hirsutulous, the body 4–5 mm; calyx tube 1.5–2.6 mm…**A. micromerius**
62'. Fruit valves glabrous, glabrescent, pilosulous, or strigulose, the body 5.5–17 mm; calyx tube 2–4 mm…(63)
63. Fruits laterally compressed…**A. tenellus** (in part)
63'. Fruits trigonously compressed…(64)
64. Fruits with a stipelike gynophore 0.1–1 mm; valves strigulose, thin-papery, ± translucent…**A. bodinii**
64'. Fruits sessile; valves pilosulous or strigulose, thick-papery, opaque…**A. chuskanus**
65. Herbage hirsute, the hairs spreading and minutely bulbous-based…**A. drummondii**
65'. Herbage glabrous, glabrescent, or strigulose, the hairs appressed or slightly ascending, not bulbous at the base…(66)
66. Calyx tube 5–8.5 mm…(67)
66'. Calyx tube 1.8–5 mm…(69)
67. Fruits inflated and the valves opaque (not bladdery and translucent); caudex superficial and ± at ground level; stems 3–7 dm; widely introduced and occasionally naturalizing…**A. cicer**
67'. Fruits not inflated and the valves opaque; caudex subterranean for 1–13 cm (or ± at ground level in some forms of *A. agrestis*); stems 1.5–4.5 dm; native…(68)
68. Fruits 7–10 mm, the valves villosulous or tomentose; inflorescences 5–15-flowered, the axis 5–25 mm…**A. agrestis**
68'. Fruits 25–35 mm, the valves glabrous; inflorescences 10–22-flowered, the axis 20–70 mm…**A. scopulorum**
69. Fruits dorsally compressed…(70)
69'. Fruits laterally or trigonously compressed…(71)
70. Caudex superficial and ± at ground level; inflorescence 1–7-flowered…**A. brandegeei** (in part)
70'. Caudex and stems subterranean; inflorescence 8–22-flowered…**A. cobrensis**
71. Herbage glabrous; fruit valves glabrous, thin-papery, ± translucent…**A. egglestonii**
71'. Herbage strigulose, pilosulous, or villosulous; fruit valves glabrescent, strigulose, tomentulose, or villosulous, papery and opaque…(72)
72. Caudices superficial, ± at ground level, taprooted, the branches spreading to ascending, stems diffuse, not mat-forming; fruit valves strigulose, the hairs appressed, the stipe 0.5–2.5 mm…**A. robbinsii** (in part)
72'. Caudices subterranean, rhizomatous, the branches creeping and adventitiously rooting, stems mat-forming; fruit valves villosulous or pilosulous, the hairs spreading or ascending, the stipe 1.5–3 mm…**A. alpinus** (in part)
73. Fruits unilocular or subunilocular (the septum, if present, not > 1.5 mm wide)…(74)
73'. Fruits bilocular, subbilocular, or subunilocular (the septum incomplete and > 1.5 mm wide)…(93)
74. Plants distinctly caulescent, clump-forming in some taxa but not tuft- or mat-forming…(75)
74'. Plants acaulescent, subacaulescent, or shortly caulescent, sometimes dwarf, or tuft- or mat-forming…(85)

75. Plants seleniferous, the herbage malodorous…(76)
75'. Plants not seleniferous, the herbage not malodorous…(79)
76. Fruits sessile or substipitate; stipe, if present, < 0.5 mm…(77)
76'. Fruits stipitate; stipe 1–8 mm…(78)
77. Keel apex acute; calyx tannish or whitish, concolorous with the corolla, the tube 6–8.8 mm…**A. pattersonii**
77'. Keel apex obtuse; calyx greenish or yellowish, occasionally pinkish-tinged, not concolorous with the corolla, the tube 4.4–6.5 mm (to 7.5 mm in robust-flowered forms)…**A. praelongus** (in part)
78. Pedicels at anthesis ascending to spreading; corollas purplish, in some forms the wings white-tipped…**A. preussii**
78'. Pedicels at anthesis deflexed or declined; corollas ochroleucous or whitish, pink-tinged in some forms…**A. praelongus** (in part)
79. Fruits bladdery-inflated, the valves translucent, sessile…(80)
79'. Fruits not inflated, the valves opaque…(83)
80. Calyx tube 1.7–2 mm; herbage villosulous…**A. cerussatus**
80'. Calyx tube 2–3.5 mm; herbage strigulose…(81)
81. Fruits 5–13 mm…**A. thurberi**
81'. Fruits 15–45 mm…(82)
82. Inflorescences 10–30-flowered; plants biennial or short-lived perennial, sometimes flowering the first year and appearing annual…**A. allochrous** (in part)
82'. Inflorescences 2–10-flowered (or up to 15 in rare, robust variants); plants annual or biennial…**A. wootonii** (in part)
83. Fruits strongly laterally compressed…**A. coltonii** (in part)
83'. Fruits not laterally compressed, the body either dorsiventrally compressed or rounded and the compression obscure…(84)
84. Fruits sessile; calyx tube dorsally gibbous, the teeth 2.8–4.4 mm…**A. neomexicanus**
84'. Fruits stipitate, the stipe 6–17 mm; calyx tube not gibbous, the teeth 0.7–2.6 mm…**A. eremiticus**
85. Calyx tube 4.5–7 mm…(86)
85'. Calyx tube 7–14 mm (as short as 6.5 mm in small-flowered forms of *A. zionis*)…(87)
86. Inflorescence axis 2–8.5 cm in fruit; fruit valves pilosulous; caudex not covered in thatch of persistent leaf bases…**A. tephrodes** (in part)
86'. Inflorescence axis 0.5–5 cm in fruit; fruit valves strigulose or villosulous; caudex covered in a thatch of persistent leaf bases and scaly stipules…**A. naturitensis**
87. Mature stem leaves with 1–3 pairs of leaflets…**A. newberryi**
87'. Mature stem leaves with 3–15 pairs of leaflets…(88)
88. Fruits glabrous…(89)
88'. Fruits strigulose, pilosulous, or villosulous…(90)
89. Longest hairs of herbage to 1.4 mm; calyx teeth 2.5–5.5 mm…**A. iodopetalus**
89'. Longest hairs of herbage to 0.9 mm; calyx teeth 1.7–2.8 mm…**A. tephrodes** (in part)
90. Fruit body 25–50 mm…(91)
90'. Fruit body 15–25 mm (to 30 mm in robust forms of *A. tephrodes*)…(92)
91. Longest hairs of herbage to 0.75 mm, at least some hairs flattened and appressed; leaves with 8–14 pairs of leaflets…**A. cyaneus**
91'. Longest hairs of herbage to 1.25 mm, the hairs rounded and ascending or spreading; leaves with 3–9 pairs of leaflets…**A. shortianus**
92. Caudex covered in a thatch of persistent leaf bases and scaly stipules…**A. zionis** (in part)
92'. Caudex not covered in a thatch of persistent leaf bases…**A. tephrodes** (in part)
93. Plants acaulescent, subacaulescent, or shortly caulescent, tuft- or clump-forming in some species…(94)
93'. Plants distinctly caulescent, not tuft- or mat-forming…(99)
94. Inflorescence axis in fruit 0.1–0.6 cm…(95)
94'. Inflorescence axis in fruit 0.6–17 cm…(96)
95. Longest herbage and stem hairs 1–2 mm; corollas 20–23 mm; fruit body 6–8 mm wide…**A. nutriosensis**
95'. Longest herbage and stem hairs 0.3–0.7 mm; corollas 11–17 mm; fruit body 2.6–4.4 mm wide…**A. monumentalis** (in part)
96. Fruits inflated, but not bladdery, rounded in cross-section, not compressed…**A. mollissimus**
96'. Fruits not inflated, trigonously compressed in cross-section, the lateral faces flat or rounded…(97)
97. Corollas 18–23 mm…**A. waterfallii**
97'. Corollas 11–17 mm…(98)
98. Herbage villous or villosulous, the longest hairs 0.7–1.2 mm; fruit trigonously compressed, the lateral faces rounded…**A. feensis**
98'. Herbage and stems strigulose, the longest hairs 0.3–0.7 mm; fruit trigonously compressed, the lateral faces flat…**A. monumentalis** (in part)

99. Plants seleniferous, the herbage malodorous…**A. praelongus** (in part)
99'. Plants not seleniferous, the herbage not malodorous…(100)
100. Fruits trigonously compressed in cross-section, the valves not inflated…(101)
100'. Fruits rounded or dorsally compressed in cross-section, the valves inflated…(103)
101. Fruits stipitate, stipe 6–17 mm…**A. eremiticus** (in part)
101'. Fruits sessile or substipitate, stipe < 1 mm in some forms of *A. nothoxys*…(102)
102. Pedicels declined or deflexed; calyx tube 1.7–2 mm; corollas 3.7–6.2 mm; fruits 6–12 mm…**A. vaccarum**
102'. Pedicels ascending; calyx tube 3.6–5 mm; corollas 8.5–12 mm; fruits 13–22 mm…**A. nothoxys**
103. Fruits persistent; corollas pale yellowish…**A. giganteus**
103'. Fruits deciduous; corollas whitish, greenish, ochroleucous, or pinkish purple, not yellowish…(104)
104. Fruits bladdery-inflated, the walls thin, ± translucent, not succulent, becoming straw-colored to brownish in age, not wrinkled, the interior an open cavity, not alveolate-spongy or pithy…**A. lentiginosus** (in part)
104'. Fruit inflated, but not bladdery, the walls opaque, thick and succulent, becoming brown or blackish and wrinkled in age, the interior alveolate-spongy or pithy…(104)
105. Fruits oblong or elliptic…**A. gypsodes**
105'. Fruits globose or ovate **A. crassicarpus**

Astragalus agrestis Douglas ex G. Don—Field milkvetch. Perennial; stems decumbent or ascending, hairs basifixed, 0.35–0.9 mm; stipules of the lowermost nodes connate-sheathing; leaves with 6–11 pairs of leaflets, the blades 4–20 mm, elliptic or oblong; inflorescences 5–20-flowered, the axis 0.5–2.5 cm in fruit; flowers erect or ascending, calyx tube 4–7.8 mm, corollas 11–20 mm, whitish, pinkish purple, or blue-lavender; fruits 7–10 × 3–4.5 mm, ascending or erect, bilocular, not inflated, sessile or substipitate, persistent, the body ovoid or elliptic, trigonously compressed in cross-section, the lateral faces flat or ± rounded, the valves papery, villous or tomentose. In ± mesic sites in meadows, prairies, stream banks, hills, slopes, or flats, 5000–9700 ft. NE, NC, C.

Astragalus albulus Wooton & Standl.—Cibola milkvetch. Perennial, seleniferous, malodorous; stems erect or ascending, clump-forming, hairs dolabriform, 1–1.7 mm; stipules of the lowermost nodes connate-sheathing opposite the petiole; leaves with 4–11 pairs of leaflets, the blades 4–27 mm, elliptic, linear, or oblanceolate; inflorescences 9–47-flowered; flowers ascending or spreading, calyx tube 5.5–7.5 mm, corollas 13–18 mm, whitish or ochroleucous; fruits 9–12 × 3–5 mm, not inflated, unilocular, stipitate, the body elliptic, lanceolate, or oblong, ± straight or incurved, rounded or trigonously compressed in cross-section, the valves papery, greenish, glabrous, glabrescent, or strigose. Clay soils, badlands, flats, hillsides, talus slopes, bluffs, 5200–7600 ft. NW, NC.

Astragalus allochrous A. Gray—Hassayampa milkvetch. Biennial or short-lived perennial; stems ascending, spreading, or decumbent, the herbage strigulose, the hairs basifixed, 0.4–0.7 mm; stipules of the lowermost nodes free opposite the petiole; leaves with 4–10 pairs of leaflets, the blades 4–20 mm, elliptic, oblanceolate, oblong, or obovate; inflorescences 10–30-flowered; flowers ascending or spreading, calyx tube 2–3.5 mm, corollas 7–9.5 mm, reddish purple or pale pinkish purple; fruits 20–45 × 10–20 mm, bladdery-inflated, unilocular, sessile, deciduous, the valves papery, translucent, strigulose. Sandy or gravelly sites on hillsides, knolls, grasslands, bottomlands, canyon floors, or plains, commonly associated with scrub oak or juniper, 3900–7000 ft. NC, NW, C, WC, SC, SW. Commonly confused with *A. wootonii*, which has been delimited as a variety of this taxon in other major floras.

Astragalus alpinus L.—Alpine milkvetch. Perennial, adventitiously rooting; stems mat-forming, erect, spreading, or ascending, the herbage strigulose, pilosulous, or villosulous, the hairs basifixed, 0.4–0.9 mm; stipules of the lowermost nodes connate-sheathing opposite the petiole; leaves with 7–13 pairs of leaflets, the blades 4–20 mm, ovate, obovate, or elliptic; inflorescences 5–23-flowered; flowers ascending, calyx tube 2–4 mm, corollas 6–14.5 mm, whitish, pinkish purple, or bluish purple; fruits 7–16 × 2.5–4 mm, not inflated, subunilocular, the body elliptic or oblong, the valves thick-papery, glabrescent, villosulous, or pilosulous. New Mexico material belongs to **var. alpinus**. Rocky or sandy sites in soils derived from a diversity of substrates on the shorelines of lakes, stream banks, meadows, flats, forests, and woodlands, commonly associated with willow, aspen, pine, alder, 6500–11,150 ft. NC.

Astragalus altus Wooton & Standl.—Tall milkvetch. Perennial, covered in a thatch of persistent leaf bases; stems 3–7.5 dm, decumbent, erect, or ascending, the herbage glabrous, glabrescent, or strigulose, the hairs basifixed, 0.3–0.5 mm; stipules of the lowermost nodes connate-sheathing opposite the petiole; leaves with 6–12 pairs of leaflets, the blades 3–12 mm, elliptic, lanceolate, orbicular, oblong, or ovate; inflorescences 15–45-flowered; calyx tube 3–4 mm, corollas 9–10 mm, whitish or ochroleucous, yellowish in age; fruits 10–15 × 4–5 mm, not inflated, subunilocular, the body elliptic or oblanceolate, slightly curved, laterally or trigonously compressed in cross-section, the valves thin-papery, strigulose. Open sites in pine forests, 6500–7800 ft. Otero. **R & E**

Astragalus amphioxys A. Gray—Annual or short-lived perennial; stems ascending or decumbent, the herbage strigose, the hairs dolabriform; stipules of the lowermost nodes free opposite the petiole; leaves with 3–10 pairs of leaflets, the blades 4–20 mm, elliptic, oblanceolate, or obovate; inflorescences 5–12-flowered; flowers ascending, calyx tube 5–13 mm, corollas 15–28 mm, pinkish purple; fruits 15–50 × 5–9 mm, not inflated, unilocular or subunilocular, the body lanceolate, straight, incurved, lunate, or coiled to nearly a full circle (± ringlike), laterally or dorsiventrally compressed in cross-section, the valves leathery, strigose.

1. Fruit lunate to coiled nearly to a full circle (± ringlike); corollas 15–24 mm; calyx tube 5.8–10.5 mm…**var. amphioxys**
1'. Fruit straight to slightly incurved; corollas 20–28 mm; calyx tube 8.5–13 mm…**var. vespertinus**

var. amphioxys—Crescent milkvetch. Calyx tube 5.8–10.5 mm, corollas 16–24 mm; fruit 15–50 mm. Sandy sites on soils derived from a variety of substrates on flats, plains, hillsides, badlands, dunes, 4600–7000 ft. NE, NC, NW, C, WC, Doña Ana.

var. vespertinus (E. Sheld.) M. E. Jones—Evening milkvetch. Calyx tube 8.8–13.2 mm, corollas 20–28 mm; fruit 30–40 mm, the body straight or slightly incurved. Sandstone substrates, sandy soils, flats, plains, hillsides, talus slopes, 4600–7000 ft. San Juan.

Astragalus bisulcatus (Hook.) A. Gray—Perennial, caulescent, seleniferous, malodorous; stems clump-forming, ascending, decumbent, or erect, hairs basifixed, 0.3–0.6 mm; stipules of the lowermost nodes connate-sheathing opposite the petiole; leaves with 5–17 pairs of leaflets, the blades 10–32 mm, elliptic, lanceolate, linear, oblanceolate, or obovate; inflorescences 25–80-flowered; calyx tube 3–5.7 mm, gibbous, corollas 7–18 mm, whitish, or ochroleucous and pinkish-tinged; fruits 6–20 × 2–4.5 mm, not inflated, unilocular, the body narrowly elliptic or narrowly oblong, straight or slightly curved, dorsiventrally compressed in cross-section, the ventral surface sulcate in 2 parallel grooves along each side of the prominent and keeled suture, the valves thick-papery or leathery, glabrous, glabrescent, or strigulose.

1. Corolla 10–18 mm; fruit body 8–20 mm, the stipe 3–6 mm…**var. bisulcatus**
1'. Corolla 7–11 mm; fruit body 6–10 mm, the stipe 1.4–3 mm…**var. haydenianus**

var. bisulcatus [*A. bisulcatus* var. *major* (M. E. Jones) S. L. Welsh]—Two-grooved milkvetch. Leaves with 5–15 pairs of leaflets; inflorescence axis 3–18 cm; calyx tube 3.3–5.7 mm, corollas 10–18 mm, whitish, or ochroleucous and pinkish purple–tinged; fruits 8–20 × 2–4.5 mm, the stipe 3–5.2 mm. Plains, prairies, hillsides, knolls, bottomlands, canyon floors (often on clay soils derived from sedimentary substrates), commonly associated with sagebrush, juniper, piñon, 5250–7250 ft. NE, NC.

var. haydenianus (A. Gray) Barneby—Hayden's two-grooved milkvetch. Leaves with 6–17 pairs of leaflets; inflorescence axis 4–25 cm; calyx tube 3–4 mm, corollas 8–11 mm, whitish or ochroleucous; fruits 6–10 × 2–4 mm, stipe 1.4–3 mm. Hillsides, knolls, canyon floors (often on seleniferous soils), commonly associated with sagebrush, juniper, spruce-fir, ponderosa pine, piñon, 5250–8550 ft. NC, NW, C.

Astragalus bodinii E. Sheld.—Bodin's milkvetch. Mostly short-lived perennial; stems spreading or loosely mat-forming, the herbage glabrescent or strigulose, the hairs basifixed; stipules of the lowermost stem nodes connate-sheathing; leaves with 3–9 pairs of leaflets, the blades 2–15 mm, elliptic, oblong, oblanceolate, or ovate; inflorescences 3–16-flowered; flowers ascending or spreading, calyx tube 2.5–4 mm, corollas 8–12 mm, pinkish purple or reddish purple; fruits 4.5–12 × 2.5–4.5 mm, inflated (not bladdery), unilocular, the body elliptic, lanceolate, oblong, or ovate, straight, rounded, or trigonously compressed in cross-section, the valves thin-papery, translucent, strigulose. Commonly in moist meadows, willow thickets, stream banks, alkaline bottomlands, 7550–11,150 ft. San Miguel, Taos, Torrance.

Astragalus brandegeei Porter—Brandegee's milkvetch. Short-lived perennial; stems ascending or decumbent, ± tuft-forming when stems are short, the herbage glabrous, glabrescent, or strigulose, the hairs basifixed, 0.4–0.7 mm; stipules of the lowermost stem nodes shortly connate-sheathing; leaves with 2–7 pairs of leaflets, the blades 4–27 mm, narrowly elliptic, linear, or narrowly oblong; inflorescences 1–7-flowered; flowers ascending, spreading, or declined, calyx tube 1.5–2.5 mm, corollas 4–7 mm, whitish and pinkish purple–tinged; fruits 10–20 × 3.5–5 mm, not inflated, bilocular or semibilocular, body elliptic, obovoid, or oblong, straight or dorsiventrally compressed in cross-section, the valves papery, strigulose. Sandy, sandy-clay, or gravelly sites in washes, intermittent streams, meadows, grasslands, commonly associated with piñon, juniper, yucca, scrub oak, 5250–8000 ft. NW, C, WC, SW.

Astragalus calycosus Torr. ex S. Watson—Perennial, ± covered in a thatch of persistent leaf bases; stems ascending, mat-forming, tuft-forming, or subpulvinate, the herbage strigose, the hairs dolabriform; leaves with 1–6 pairs of leaflets, the blades 5–20 mm, oblanceolate or obovate; inflorescences 1–17-flowered; flowers ascending or spreading, calyx tube 3–6.7 mm, corollas 10–20 mm, ochroleucous, whitish, or pinkish purple; fruits 8–20 × 3–4.5 mm, erect, occasionally humistrate, not inflated, bilocular or semibilocular, the body oblong, straight or slightly curved, laterally or trigonously compressed in cross-section, the valves leathery, strigulose.

1.   Inflorescence axis in fruit 0.2–2 cm; racemes 1–8-flowered…**var. calycosus**
1'. Inflorescence axis in fruit 1–7 cm; racemes 4–17-flowered…**var. scaposus**

var. calycosus—Torrey's milkvetch. Leaves with 1–6 pairs of leaflets (commonly palmately trifoliate); inflorescences 1–8-flowered; corollas whitish or pinkish purple. Sandy, clay, or gravelly sites in washes, plains, hillsides, talus slopes, ridges, commonly associated with juniper, piñon, sagebrush, 5900–7250 ft. NW, WC.

var. scaposus (A. Gray) M. E. Jones—Matted milkvetch. Leaves with 2–6 pairs of leaflets; inflorescences 4–17-flowered, the axis of the raceme usually longer than 2 cm; corollas pinkish purple with white wing tips. Sandy, clay, or gravelly sites in washes, plains, hillsides, talus slopes, ridges, commonly associated with juniper and piñon, 5575–7250 ft. NW, C, WC.

Astragalus canadensis L.—Canada milkvetch. Perennial; stems ascending or erect, herbage strigulose, the hairs dolabriform, 0.3–1 mm; stipules of the lowermost nodes connate-sheathing opposite the petiole; leaves with 4–15 pairs of leaflets, the blades 6–52 mm, elliptic, lanceolate, or oblong; inflorescences 5–25-flowered; flowers spreading or declined, calyx tube 4–8.5 mm, corollas 11–16.5 mm, whitish or ochroleucous; fruits 10–15 × 3–5 mm, not inflated, bilocular, the body elliptic, straight or slightly curved, rounded in cross-section, the valves papery, glabrous or glabrescent. New Mexico material belongs to **var. canadensis**. In ± mesic sites on sandy beaches, shorelines, stream banks, meadows, prairies, hillsides, forests, woodlands, meadows, 4900–10,500 ft. NE, NC.

Astragalus castetteri Barneby—Castetter's milkvetch. Perennial, caulescent, the herbage glabrous, glabrescent, or strigulose, the hairs basifixed, 0.3–0.45 mm; stipules of the lowermost nodes connate-sheathing opposite the petiole; leaves with 5–12 pairs of leaflets, the blades 5–15 mm, elliptic, oblong, or obovate; inflorescences 8–20-flowered; flowers ascending or spreading, declined in fruit, calyx tube 6–7 mm, corollas 14–19 mm, pinkish purple; fruits 22–28 × 11–16 mm, spreading or declined, occasionally humistrate, bladdery-inflated, unilocular, the body elliptic or ovate, rounded in cross-section, the valves papery, translucent, strigulose or villosulous. Gravelly sites in woodlands, commonly associated with piñon, 4900–6600 ft. Doña Ana, Sierra. **R & E**

Astragalus ceramicus E. Sheld.—Painted milkvetch. Perennial; stems ascending or erect, the herbage strigulose, the hairs dolabriform or basifixed, 0.3–1.1 mm; stipules of the lowermost nodes connate-sheathing; leaves with 1–6 pairs of leaflets, the terminal leaflet continuous with the rachis, the blades 3–50 mm, linear or narrowly oblong; inflorescences 2–25-flowered; flowers ascending or spreading, calyx tube 2–3.5 mm, corollas 6–11 mm, whitish, or whitish and pinkish purple; fruits 15–30 × 5–15 mm, bladdery-inflated, unilocular, the body elliptic, ovate, or globose, rounded in cross-section, the valves papery, translucent, glabrous or glabrescent.

1. Pubescence of the leaves dolabriform; calyx tube 1.8–2.5 mm…**var. ceramicus**
1'. Pubescence of the leaves basifixed or obscurely dolabriform; calyx tube 2.5–4 mm…**var. filifolius**

var. ceramicus—Herbage strigulose, the hairs dolabriform; leaves with 1–6 pairs of leaflets; inflorescences 6–25-flowered; calyx tube 2–3.3 mm, the teeth 1–1.8 mm, corollas 6–9.5 mm. Sandy sites in dunes, plains, talus, badlands, washes, 4900–7250 ft. NC, NW, C, Torrance.

var. filifolius (A. Gray) F. J. Herm.—Herbage strigulose, the hairs basifixed or a few hairs incipiently dolabriform; leaves with 1–3 pairs of leaflets; inflorescences 2–7-flowered, the axis 1–5 cm; calyx tube 2.3–3.5 mm, the teeth 1.4–3 mm, corollas 7.4–11 mm. Sandy sites in dunes, plains, sand bars of intermittent streams, 4900–7250 ft. NC, NW.

Astragalus cerussatus E. Sheld.—Powdered milkvetch. Annual, biennial, or short-lived perennial; stems ascending or spreading, the herbage villosulous, the hairs dolabriform, 0.5–0.8 mm; stipules of the lowermost nodes free; leaves with 6–10 pairs of leaflets, the blades 4–18 mm, oblanceolate or oblong; inflorescences 2–7-flowered; flowers ascending or spreading, calyx tube 1.5–2 mm, corollas 5–6 mm, whitish and pinkish purple-tinged; fruits 10–22 × 5–14 mm, spreading or declined, bladdery-inflated, unilocular or subunilocular, the body elliptic or ovate, the valves papery, translucent, strigulose or villosulous. Sandy sites in canyons, washes, stream banks, plains, grasslands, scrublands, 5250–8550 ft. NC.

Astragalus chuskanus Barneby & Spellenb. [*A. chuskanus* var. *spellenbergii* S. L. Welsh & N. D. Atwood]—Chuska milkvetch. Perennial; stems ascending, decumbent, or erect, mat-forming or cushion-forming, the herbage strigulose or pilosulous, the hairs basifixed; stipules of the lowermost nodes connate-sheathing; leaves with 3–7 pairs of leaflets, the blades 1.5–10 mm, elliptic, oblong, or obovate; inflorescences 4–10-flowered; flowers ascending or spreading, calyx tube 2.5–3.8 mm, corollas 7.3–10 mm, whitish or whitish and pinkish purple-tinged; fruits 5.5–6.5 × 2.5–3.5 mm, not inflated, unilocular, the body ovoid, trigonously compressed in cross-section, the valves papery, strigulose or pilosulous. Chuska Sandstone Formation, sandy soil, forests, juniper, oak, ponderosa pine, Douglas-fir, 7350–9050 ft. McKinley, San Juan. **R**

Astragalus cicer L.—Chickpea milkvetch. Perennial; stems prostrate, decumbent, or ascending, the herbage strigose-pilose, the hairs basifixed, 0.4–0.7 mm; stipules of the lowermost nodes connate-sheathing; leaves with 6–15 pairs of leaflets, the blades 5–35 mm, lanceolate, elliptic, or oblong; inflorescences 6–30-flowered; flowers ascending or spreading, calyx tube 5–6 mm, corollas 12–16 mm, ochroleucous; fruits 6–14 × 5–10 mm, bilocular, inflated, ± fleshy, not bladdery, the body ovoid or subglobose, rounded in cross-section, the valves leathery or stiff-papery, pilosulous. Cultivated and escaping, 9185–9850 ft. San Juan. Introduced.

Astragalus cliffordii S. L. Welsh & N. D. Atwood—Clifford's milkvetch. Perennial, the herbage strigulose, the hairs basifixed, 0.2–0.6 mm; stipules of the lowermost nodes connate-sheathing; leaves with 2–4 pairs of leaflets, the terminal leaflet continuous with the rachis, the blades 8–28 mm, narrowly elliptic or linear; inflorescences 5–19-flowered; flowers ascending or spreading, calyx tube 1.2–2 mm, corollas 4–6 mm, whitish and pinkish purple-tinged; fruits 9–12 × 2.5–3 mm, not inflated, unilocular, the body elliptic or oblong, straight or slightly curved, laterally compressed in cross-section, the valves thin-papery, glabrous. Sandy or sandy-clay sites, talus, washes, flats, woodlands, scrublands, 5900–7150 ft. McKinley. **R**

Astragalus cobrensis A. Gray—Copper mine milkvetch. Perennial; stems ascending or decumbent, the herbage strigulose or hirsutulous, the hairs basifixed, 0.25–0.6 mm; stipules of the lowermost nodes connate-sheathing; leaves with 4–11 pairs of leaflets, the blades 3–17 mm, oblong, obovate, ovate, or suborbicular; inflorescences 8–33-flowered; flowers ascending or spreading, calyx tube 2.2–2.7 mm, corollas 6.5–7.8 mm, whitish or whitish and pinkish purple-tinged; fruits 7–15 × 3–6 mm, not inflated, bilocular or semibilocular, the body elliptic or oblong, straight or slightly curved downward, dorsiventrally compressed in cross-section, the valves papery, strigulose.

1. Fruit valves strigulose; the longest hairs on the herbage and stems 0.25–0.6 mm…**var. cobrensis**
1'. Fruit valves villosulous; the longest hairs on the herbage and stems 0.6–0.8 mm…**var. maguirei**

var. cobrensis—Copper mine milkvetch. Herbage and stems strigulose, the hairs 0.25–0.6 mm. In ± mesic sites, hillsides, canyon bottoms, ponderosa pine, juniper, oak, 5575–8200 ft. C, SW.

var. maguirei Kearney—Maguire's milkvetch. Herbage and stems pilosulous, the hairs 0.6–0.8 mm. Dry creek beds, canyon bottoms, stream banks, open slopes, commonly associated with pine, juniper, oak, 5250–6600 ft. Hidalgo. **R**

Astragalus coltonii M. E. Jones [*A. canovirens* (Rydb.) Barneby]—Moab milkvetch. Perennial; stems erect or ascending, the herbage strigulose, the hairs basifixed, 0.4–0.8 mm; stipules of the lowermost nodes free; leaves with 4–9 pairs of leaflets, the blades 4–20 mm, elliptic, oblong, or ovate; inflorescences 6–30-flowered; flowers ascending or spreading; calyx tube 4–6.7 mm, the teeth 0.5–2.3 mm, corollas 12–18.5 mm, bluish purple or pinkish purple; fruits 19–35 × 3–6 mm, not inflated, unilocular, the body narrowly oblong or oblanceolate, straight, slightly decurved, or slightly curved, laterally compressed in cross-section, the valves papery, glabrous. Our material belongs to **var. moabensis** M. E. Jones. Canyon bottoms, talus slopes, ridges, sagebrush, juniper, piñon, 6550–8000 ft. San Juan.

Astragalus crassicarpus Nutt.—Ground-plum. Perennial; stems decumbent or ascending, the herbage strigulose, pilose, or pilosulous, the hairs basifixed, 0.4–1.4 mm; stipules of the lowermost nodes free; leaves with 5–16 pairs of leaflets, the blades 3–24 mm, elliptic, oblong, or ovate; inflorescences 5–35-flowered; flowers ascending or spreading, calyx tube 5–9.5 mm, corollas 16–25 mm, pinkish purple, whitish, ochroleucous, or greenish white; fruits 15–40 × 12–25 mm, ascending or spreading, humistrate, inflated or not, succulent, bilocular, the body globose or ovate, straight, rounded in cross-section, the valves thick, 2–5 mm wide near the sutures, leathery, drying brownish or blackish, glabrous. NE, NC, NW, C, WC, SC, Eddy.

1. Corollas pinkish purple or bluish purple; hairs on the herbage 0.4–1 mm…**var. crassicarpus**
1'. Corollas whitish and pinkish purple-tinged; hairs on the herbage 0.4–0.7 mm…(2)
2. Fruit not inflated, the cavity filled with seeds and pulp, the walls 2–5 mm thick…**var. paysonii**
2'. Fruit inflated, the cavity much larger than the seeds, the walls 1.2–2 mm thick…**var. cavus**

var. cavus Barneby—Hollow ground-plum. Corollas whitish or whitish and pinkish purple-tinged; fruits inflated, the cavity much larger than the seeds and pulp. Sandy or gravelly sites in soils derived from a variety of substrates, on plains, grasslands, washes, flats, meadows, stream banks, talus slopes, commonly associated with juniper or piñon, 2950–7250 ft.

var. crassicarpus—Common ground-plum. Corollas pinkish purple or bluish purple, rarely whitish in albino forms; fruits not inflated, the cavity filled with seeds and pulp. Meadows, prairies, occasionally disturbed sites in old pastures and roadsides, 3300–7250 ft.

var. paysonii (E. H. Kelso) Barneby—Payson's ground-plum. Corollas whitish or whitish and pinkish purple-tinged; fruit not inflated, the cavity filled with seeds and pulp. Hillsides and ridges on soils derived from a variety of substrates, commonly associated with sagebrush, 5900–7000 ft.

Astragalus cyaneus A. Gray—Cyanic milkvetch. Perennial, the herbage strigulose, the hairs basifixed, 0.4–0.8 mm, slightly flattened; stipules of the lowermost nodes free; leaves with 7–15 pairs of leaflets, the blades 4–19 mm, elliptic, oblong, or obovate; inflorescences 12–25-flowered; flowers ascending or spreading, calyx tube 8–11 mm, corollas 18–22 mm, pinkish purple; fruits 25–50 × 7–13 mm, not inflated, unilocular, the body oblong or elliptic, straight, incurved, or lunate, dorsiventrally compressed in cross-section. Sandy or gravelly sites (often on sandstone or granite substrates) on hills or in washes, commonly associated with juniper and piñon, 6900–7300 ft. Rio Arriba, Santa Fe, Taos. **R & E**

Astragalus desperatus M. E. Jones—Rimrock milkvetch. Perennial; stems tuft-forming, the herbage strigulose, the hairs basifixed; stipules of the lowermost nodes connate-sheathing; leaves with 4-8 pairs of leaflets, the blades 2-8 mm, elliptic, oblanceolate, or obovate; inflorescences 3-11-flowered; flowers ascending, spreading, or declined, calyx tube 2.5-4.2 mm, corollas 6-9 mm, whitish and pinkish purple-tinged or pinkish purple; fruits 10-19 × 3-6 mm, not inflated, unilocular, the body elliptic, ovate, or lanceolate, incurved, dorsiventrally compressed in cross-section, the valves thick-papery, hirsute with minute bulbous-based hairs. Our material belongs to **var. desperatus**. Ledges, canyons, washes, commonly associated with sagebrush, piñon, juniper, 5250-5900 ft. San Juan.

Astragalus drummondii Douglas—Drummond's milkvetch. Perennial; stems ascending or erect, the herbage villous or hirsute, the hairs basifixed, minutely bulbous-based, 1-2 mm; stipules of the lowermost nodes connate-sheathing; leaves with 6-16 pairs of leaflets, the blades 4-33 mm, ovate, oblanceolate, or oblong; inflorescences 14-35-flowered; flowers ascending or spreading, calyx tube 4.7-8 mm, corollas 18-26 mm, ochroleucous or whitish and pinkish purple-tinged; fruits 17-32 × 3.5-5.5 mm, not inflated, bilocular or semibilocular, the body narrowly elliptic or narrowly oblong, straight or slightly curved, trigonously compressed in cross-section, the valves thick-papery or leathery, glabrous. Various substrates in hills, plains, shrublands, grasslands, scrublands, commonly associated with ponderosa pine, juniper, oak, 5900-9850 ft. NE, NC.

Astragalus egglestonii (Rydb.) Kearney & Peebles—Eggleston's milkvetch. Perennial; stems erect or ascending, the herbage and stems glabrous, glabrescent, or strigulose, the hairs basifixed, 0.25-0.45 mm; stipules of the lowermost nodes connate-sheathing; leaves with 8-14 pairs of leaflets, the blades 3-12 mm, elliptic, oblanceolate, or oblong; inflorescences 10-30-flowered; flowers spreading, declined, or deflexed, calyx tube 2-2.7 mm, corollas 5.8-8 mm, whitish, ochroleucous, or greenish; fruits 15-35 × 2.5-4 mm, not inflated, bilocular, the body oblanceolate or narrowly oblong, straight or slightly curved, trigonously compressed in cross-section, the valves thin-papery, ± translucent, glabrous. Open, mesic sites (often on soils derived from igneous substrates) in meadows, pine forests, woodlands, 7200-8550 ft. WC, SW, McKinley.

Astragalus emoryanus (Rydb.) Cory—Emory's milkvetch. Annual; stems prostrate, the herbage strigulose, hirsutulous, or villosulous, the hairs basifixed; stipules of the lowermost nodes free; leaves with 3-10 pairs of leaflets, the blades 2-14 mm, oblanceolate, oblong, ovate, obcordate, or elliptic; inflorescences 2-10-flowered; flowers spreading or declined, calyx tube 2-3.5 mm, corollas 6-11 mm, pinkish purple; fruits 12-22 × 2-3 mm, ascending, spreading, or declined, not inflated, bilocular, the body linear, oblong, or oblanceolate, straight, incurved, or lunate, laterally or trigonously compressed in cross-section, the valves papery, glabrous. New Mexico material belongs to **var. emoryanus**. Hillsides, washes, grasslands, flats, commonly associated with yucca, mesquite, creosote brush, piñon, juniper, 3600-7000 ft. NW, C, SE, SC, SW, Bernalillo, Mora.

Astragalus episcopus S. Watson—Bishop's milkvetch. Perennial; stems erect or ascending, subterranean for 1-12 cm, the herbage glabrous, glabrescent, or strigulose, the hairs basifixed, 0.45-75 mm; stipules of the lowermost nodes free; leaves with 1-3 pairs of leaflets, the terminal leaflet continuous with the rachis, the blades 1-15 mm, linear or narrowly elliptic; inflorescences 5-30-flowered; flowers ascending or spreading, spreading or declined in fruit, calyx tube 3.4-6 mm, corollas 11-15.5 mm, pinkish purple or whitish and pinkish purple-tinged; fruits 14-32 × 4-8 mm, not inflated, unilocular, the body elliptic, oblanceolate, or narrowly oblong, straight or slightly curved, laterally compressed in cross-section, the valves thick-papery, glabrous, glabrescent, or strigulose. Sandstone, sandy sites in badlands, flats, hillsides, rocky talus, 4600-5250 ft. San Juan.

Astragalus eremiticus E. Sheld.—Hermit milkvetch. Perennial, the herbage glabrous, glabrescent, or strigulose, the hairs basifixed; stipules of the lowermost nodes free; leaves with 5-11 pairs of leaflets, the blades 2-6 mm, linear, elliptic, or oblong; inflorescences 7-30-flowered; flowers erect or ascending, calyx tube 4.4-8 mm, corollas 12-20 mm, reddish purple or pinkish purple; fruits 12-30 × 3.5-9 mm, ascending or erect, not inflated or scarcely inflated, bilocular or semilocular, stipitate, the body elliptic or oblong, ± straight, trigonously compressed in cross-section, the valves leathery, glabrous. Our material belongs to **var. eremiticus**. Limestone soils, gravelly or rocky sites on hillsides, washes, flats, canyon bottoms, 3600–4600 ft. Hidalgo.

Astragalus feensis M. E. Jones—Santa Fe milkvetch. Perennial, the herbage villosulous, the hairs basifixed; stipules of the lowermost nodes free; leaves with 3–9 pairs of leaflets, the blades 3–13 mm, oblanceolate or obovate; inflorescences 6–15-flowered; flowers ascending, calyx tube 4.5–6.5 mm, corollas 13–16 mm, reddish purple or pinkish purple; fruits 13–30 × 3.5–7 mm, not inflated, bilocular, the body elliptic, oblong, lunate, or incurved, trigonously compressed in cross-section, the valves leathery, strigulose. Gravelly or sandy sites on hillsides, commonly associated with juniper, 4900–6250 ft. C. **R & E**

Astragalus flavus Nutt.—Yellow milkvetch. Perennial, seleniferous, malodorous, caudices covered in a thatch of persistent leaf bases; stems prostrate or decumbent, mat-forming, the herbage and stems glabrescent or strigose, the hairs dolabriform, 0.4–1.1 mm; stipules of the lowermost nodes connate-sheathing; leaves with 4–10 pairs of leaflets, the blades 6–20 mm, elliptic, linear, oblanceolate, or oblong; inflorescences 4–30-flowered; flowers ascending, calyx tube 3–5.2 mm, corollas 9–18 mm, whitish, ochroleucous, or yellowish; fruits 7–13 × 3.5–5.5 mm, not inflated, unilocular, the body elliptic, oblong, or ovoid, straight or slightly curved, dorsiventrally compressed in cross-section, the valves thick-papery or leathery, glabrescent or strigulose.

1. Inflorescence axis at or below the level of the leaves; longest peduncles 3–12 cm…**var. flavus**
1'. Inflorescence axis at or above the level of the leaves; longest peduncles 12–23 cm…**var. higginsii**

var. flavus—Yellow milkvetch. Inflorescence peduncles 3–12 cm, the axis at or below the level of the leaves; calyx 3–5.2 mm. Mostly clay soils, flats, hillsides, badlands, bluffs, 6200–7250 ft. NW.

var. higginsii S. L. Welsh—Higgins's yellow milkvetch. Inflorescence peduncles 12–23 cm, the axis at or above the level of the leaves; calyx 3.7–5.2 mm. Clay soils, flats, hillsides, badlands, bluffs, 6550–7250 ft. McKinley, Rio Arriba.

Astragalus flexuosus (Hook.) Douglas ex G. Don—Perennial; stems ascending or decumbent, herbage strigulose or villosulous, hairs basifixed, 0.2–0.6 mm; stipules of the lowermost nodes connate-sheathing; leaves with 4–12 pairs of leaflets, blades 3–19 mm, linear, obcordate, narrowly obovate, or narrowly oblong; inflorescences 8–30-flowered; flowers ascending, spreading, or declined, calyx tube 2.4–4.3 mm, corollas 7–11 mm, whitish or pinkish purple; fruits 8–24 × 2.7–9 mm, not inflated, unilocular, the body elliptic, oblanceolate, or oblong, straight or slightly curved, dorsiventrally compressed in cross-section, the valves thin- or thick-papery, glabrous, glabrescent, strigulose, or villosulous.

1. Calyx tube 1.9–2.3 mm; fruits sessile, the valves strigulose…**var. diehlii**
1'. Calyx tube 2.7–4.3 mm; fruits sessile or stipitate, the valves glabrous, glabrescent, strigulose, or villosulous…(2)
2. Fruits stipitate, the stipe 0.5–1.3 mm, the valves glabrous, glabrescent, strigulose, or villosulous…**var. flexuosus**
2'. Fruits sessile or substipitate, the stipe 0–0.5 mm, the valves strigulose or villosulous…**var. greenei**

var. diehlii (M. E. Jones) Barneby—Diehl's milkvetch. Leaves with 3–8 pairs of leaflets; inflorescences 12–26-flowered, the axis 2.5–9.5 cm; calyx tube 1.9–2.3 mm, the teeth 0.7–1.4 mm, corollas 7–9 mm; fruits sessile, the valves strigulose. Desert flats, canyons, badlands, washes, 4900–7250 ft. NC.

var. flexuosus—Bent milkvetch. Leaves with 10–12 pairs of leaflets; inflorescences 8–30-flowered, the axis 1–15 cm; calyx tube 2.7–4.3 mm, the teeth 0.5–1.7 mm, corollas 7.4–11 mm; fruits stipitate, the stipe 0.5–1.3 mm, the valves glabrous, glabrescent, strigulose, or villosulous; seeds 14–20. Flats, grasslands, hillsides, meadows, stream banks, washes, 5900–11,150 ft. NC, C, SC.

var. greenei (A. Gray) Barneby—Greene's milkvetch. Leaves with 4–11 pairs of leaflets; inflorescences 8–30-flowered, the axis 1–10 cm; calyx tube 2.8–4.2 mm, the teeth 0.8–2 mm, corollas 9–10.8 mm; fruits sessile or substipitate, the stipe 0–0.5 mm, the valves strigulose or villosulous. Desert flats, badlands, hillsides, commonly associated with piñon, juniper, oak, 4600–8200 ft. NC, NW.

Astragalus fucatus Barneby—Hopi milkvetch. Perennial; stems ascending or decumbent, herbage and stems strigulose, hairs basifixed, 0.2–0.6 mm; stipules of the lowermost nodes connate-sheathing; leaves with 4–9 pairs of leaflets, blades 4–25 mm, narrowly oblanceolate or narrowly oblong; inflorescences 9–27-flowered; flowers ascending or spreading, calyx tube 2.4–3.3 mm, corollas 6.4–8.7 mm, pinkish purple or reddish purple; fruits 17–32 × 12–20 mm, inflated, unilocular, the body ovoid, obovoid, or subglobose, rounded or dorsally compressed in cross-section, the valves thin-papery, strigulose. Sandy sites in dunes, washes, talus below cliff faces, often in soils derived from red sandstones, commonly associated with sagebrush, piñon, juniper, 5200–6250 ft. NW, WC.

Astragalus giganteus S. Watson—Giant milkvetch. Perennial; stems erect or ascending, the herbage pilosulous and tomentulose, hairs basifixed; stipules of the lowermost nodes free; leaves with 8–17 pairs of leaflets, blades 7–55 mm, elliptic, oblong, or ovate; inflorescences 15–65-flowered; flowers ascending or spreading, calyx tube 6–9 mm, corollas 14–22 mm, pale yellowish; fruits 15–25 × 8–13 mm, ± inflated, bilocular, body elliptic or ovate, incurved, rounded or dorsiventrally compressed in cross-section, the valves leathery, glabrous. Woodlands, washes, slopes, gravel bars, roadsides, commonly associated with oak and pine, 5900–9550 ft. SC, Eddy, Hidalgo.

Astragalus gilensis Greene—Gila milkvetch. Perennial, caudices covered in a thatch of persistent leaf bases; stems tuft-forming, the herbage strigulose, hairs dolabriform; stipules of the lowermost nodes free; leaves with 5–11 pairs of leaflets, blades 4–24 mm, narrowly elliptic or narrowly oblong; inflorescences 7–26-flowered; flowers ascending or spreading, calyx tube 2–4 mm, corollas 6–10.5 mm, pinkish purple; fruits 4.5–7.5 × 2–4.2 mm, not inflated, unilocular, body elliptic, oblong, or ovate, ± straight, laterally or trigonously compressed in cross-section, the valves papery, strigulose. Dry sites in woodlands and forests, commonly associated with ponderosa pine, piñon, juniper, 6200–8550 ft. WC, SW, Rio Arriba, Sandoval.

Astragalus gracilis Nutt.—Slender milkvetch. Perennial; stems ascending or erect, herbage glabrous, glabrescent, or strigulose, hairs basifixed, 0.3–0.5 mm; stipules of the lowermost nodes connate-sheathing; leaves with 3–8 pairs of leaflets, blades 5–25 mm, linear, narrowly oblanceolate, or narrowly oblong; inflorescences 12–55-flowered; flowers ascending or spreading, calyx tube 1–2.7 mm, corollas 5–8.5 mm, pinkish purple or bluish purple; fruits 4–9 × 2–3.5 mm, not inflated, unilocular, body ovoid, obovoid, or subglobose, dorsally compressed in cross-section, the valves leathery, strigulose or villosulous. Dunes, washes, woodlands, prairies in soils derived from sandstones, shales, or gypsum, 4600–7250 ft. NE.

Astragalus gypsodes Barneby—Gypsum milkvetch. Perennial; stems ascending, herbage strigulose, hairs basifixed; stipules of the lowermost nodes free; leaves with 5–15 pairs of leaflets, blades 5–20 mm, elliptic, lanceolate, or ovate; inflorescences 10–30-flowered; flowers ascending or spreading, calyx tube 7–10 mm, corollas 16–23 mm, pinkish purple; fruits 25–50 × 8–21 mm, inflated, succulent, the cavity pithy, filled with seeds and pulp, bilocular, body clavate, elliptic, or oblong, straight or slightly curved, laterally or dorsiventrally compressed in cross-section, the valves leathery, strigulose. Gypsum clay soils, hills and washes, 3500–4000 ft. Eddy. **R**

Astragalus hallii A. Gray—Hall's milkvetch. Perennial; stems 1–5 dm, decumbent or prostrate, clump-forming, herbage glabrous, glabrescent, strigulose, or villosulous, the hairs basifixed, 0.2–1 mm; stipules of the lowermost nodes connate-sheathing; leaves with 5–15 pairs of leaflets, blades 3–14 mm, obcordate, oblanceolate, or oblong; inflorescences 10–28-flowered; flowers spreading, declined, or deflexed, calyx tube 5–6.2 mm, corollas 12–18.5 mm, pinkish purple or reddish purple; fruits 12–27 × 3–12 mm, inflated or not, unilocular, body clavate, elliptic, or oblong, rounded or dorsally compressed in cross-section, the valves thick-papery or leathery, glabrous, glabrescent, strigulose, or villosulous.

1. Fruit valves glabrous, glabrescent, or sparsely strigulose, not inflated, the body 4–7 mm wide. (to 8.5 mm wide in rare, robust forms, but still not inflated)...**var. hallii**
1'. Fruit strigulose or villosulous, inflated, the body 7–12 mm wide...**var. fallax**

var. fallax (M. E. Jones) Barneby—Deceptive milkvetch. Herbage and stems strigulose or villosulous, the hairs 0.2–0.6 mm; fruits inflated (not bladdery), unilocular, sessile or stipitate, the stipe 0–4.5 mm, valves strigulose or villosulous. Hillsides and ridges in woodlands, commonly associated with oak, juniper, ponderosa pine, 5500–7900 ft. WC.

var. hallii—Hall's milkvetch. Herbage glabrous, glabrescent, or strigulose, the hairs 0.2–1 mm; fruits 12–27 × 3–7(–8.5) mm, not inflated, unilocular, stipitate, the stipe 1–2 mm, valves glabrous, glabrescent, or strigulose. Hillsides, flats, washes, meadows, commonly associated with sagebrush, 6900–10,500 ft. NE, NC, NW.

Astragalus heilii S. L. Welsh & N. D. Atwood—Heil's milkvetch. Perennial, caudices covered in a thatch of persistent leaf bases; stems tuft-forming, the herbage strigulose, the hairs basifixed; stipules of the lowermost nodes connate-sheathing; leaves with 3–6 pairs of leaflets, blades 1.8–5.5 mm, oblanceolate or ovate; inflorescences 1–4-flowered; flowers ascending or spreading, calyx tube 1.6–1.9 mm, corollas 4–5 mm, whitish and pinkish purple–tinged; fruits 9–9.8 × 4–4.6 mm, inflated (not bladdery), unilocular, body elliptic or oblong, straight, dorsiventrally compressed in cross-section, the valves papery, strigulose. Rocky ledges, primarily Mesozoic sandstone substrates, 6900–7250 ft. McKinley. **R & E**

Astragalus humillimus A. Gray—Mancos milkvetch. Perennial, caudices covered in a thatch of persistent stipules and ± spinose leaf bases; stems tuft-forming, the herbage strigulose, hairs dolabriform; stipules of the lowermost nodes free; leaves with 3–5 pairs of leaflets, the blades 1–5 mm, elliptic, oblong, or ovate; inflorescences 1–3-flowered; flowers ascending, the calyx tube 2–4 mm, corollas 9–12 mm, pinkish purple; fruits 4.5–5.5 × 2–2.5 mm, not inflated, unilocular, body elliptic or oblong, ± straight, laterally compressed in cross-section, the valves thick-papery, strigulose. Cracks or depressions of the Point Lookout Sandstone Formation, ledges and mesa tops, associated with sagebrush, *Rhus*, juniper, 4900–6000 ft. San Juan. **FE**

Astragalus humistratus A. Gray—Perennial; stems decumbent or prostrate, mat-forming or diffuse-spreading, the herbage glabrous, glabrescent, pilose, pilosulous, strigose, strigulose, villous, villosulous, or tomentose, hairs dolabriform, 0.8–1.5 mm; stipules of the lowermost nodes connate-sheathing; leaves with 5–9 pairs of leaflets, all jointed, the blades 4–15 mm, elliptic, oblong, or oblanceolate; inflorescences 3–30-flowered; flowers ascending or spreading, calyx tube 2–4 mm, corollas 7–12 mm, greenish white or ochroleucous, occasionally pinkish purple–tinged; fruits 6–20 × 3.5–6.5 mm, commonly humistrate, not inflated, unilocular, body elliptic, oblong, or broadly lanceolate, straight or slightly curved, rounded in cross-section, the valves thick-papery or leathery, strigulose

1. Herbage villous, villosulous, or tomentose (hairs mostly ascending to spreading and curly or entangled)… **var. crispulus**
1'. Herbage strigose, strigulose, pilose, or pilosulous (hairs mostly ascending to appressed and straight)… (2)
2. Fruits 8–14 mm…(3)
2'. Fruits 13–20 mm…(4)
3. Calyx tube 1.2–2.5 mm, the teeth 0.7–1.2 mm; flowers 5–6.2 mm…(see *A. kerrii*)
3'. Calyx tube 2.5–4.1 mm, the teeth 1.4–5 mm; flowers 7–12 mm…**var. humivagans**
4. Fruits 4–6.5 mm wide, the valves sparsely strigulose, the hairs not obscuring the epidermis…**var. humistratus**
4'. Fruits 3.2–4 mm wide, the valves densely strigulose, the hairs obscuring the epidermis…**var. sonorae**

var. crispulus Barneby—Villous groundcover milkvetch. Herbage villous or tomentose, hairs 1–1.5 mm; leaves with 5–7 pairs of leaflets; inflorescences 3–12-flowered; calyx tube 2.9–4 mm, the teeth 1.2–2.4 mm, corollas 7–9.2 mm; fruits 3–9 × 1.3–3 mm, valves thick-papery, villosulous. Open, dry, sandy sites (on substrates derived from igneous rocks) on hillsides and washes in xeric pine forests, 6900–8200 ft. Catron. **R**

var. humistratus—Groundcover milkvetch. Herbage pilosulous, hairs 0.8–1.5 mm; leaves with 5–9 pairs of leaflets; inflorescences 7–30-flowered, calyx tube 3–3.8 mm, the teeth 2.5–4.5 mm, corollas 9–11.8 mm; fruits 14–18 × 4–6.5 mm, valves thick-papery or leathery, strigulose. Hillsides, forests, woodlands, scrublands, flats, washes, commonly associated with ponderosa pine, piñon, juniper, 5250–9850 ft. NC, NW, C, WC, SC.

var. humivagans (Rydb.) Barneby—Spreading groundcover milkvetch. Herbage strigulose, pilose, or pilosulous, hairs 0.8–1.5 mm; leaves with 4–9 pairs of leaflets; inflorescences 3–22-flowered; calyx tube 2.5–4 mm, the teeth 1.5–5 mm, corollas 7–12 mm; fruits 7–14 × 3.5–5.7 mm, the valves thick-papery, pilosulous or strigulose. Igneous or limestone soils, hillsides, slopes, flats, washes, ponderosa pine at higher elevations and piñon and juniper at lower elevations, 4900–8200 ft. NW, NC.

var. sonorae (A. Gray) M. E. Jones—Sonoran groundcover milkvetch. Herbage strigose, strigulose, pilose, or pilosulous, hairs 0.8–1.5 mm; leaves with 3–8 pairs of leaflets; inflorescences 10–26-flowered; calyx tube 2.5–3.5 mm, the teeth 2–4.5 mm, corollas 8–11 mm; fruits 13–20 × 3.5–4 mm, valves thick-papery, strigulose. Hillsides, grasslands, woodlands, scrublands, washes, commonly associated with yucca, ponderosa pine, piñon, juniper, 4250–8000 ft. WC, SC, SW.

Astragalus iodopetalus (Rydb.) Barneby—Violet milkvetch. Perennial; herbage villous or villosulous, the hairs basifixed; stipules of the lowermost nodes free; leaves with 4–15 pairs of leaflets, blades 3–17 mm, elliptic, oblanceolate, oblong, or obovate; inflorescences 10–25-flowered; flowers ascending, calyx tube 6.8–10.5 mm, corollas 17–24 mm, whitish, reddish purple, or pinkish purple; fruits 17–30 × 7–10 mm, not inflated, unilocular or subunilocular, body ovate, elliptic, or oblong, ± straight or incurved, dorsiventrally compressed in cross-section, the valves leathery or ± woody, glabrous. Open, dry sites on slopes and hillsides, commonly associated with piñon and Gambel oak, 5900–8550 ft. NC. **R**

Astragalus kentrophyta A. Gray—Spiny milkvetch. Perennial; stems erect or decumbent, loosely mat-forming or spreading and diffuse, the herbage glabrous, glabrescent, or strigulose, hairs basifixed or dolabriform, 0.1–0.5 mm; stipules of the lowermost nodes connate-sheathing; leaves with 1–4 pairs of leaflets, all continuous with the rachis and spinose-tipped, the blades 1–17 mm, linear or narrowly elliptic; inflorescences 1–3-flowered; calyx tube 1.2–2.6 mm, subspinose, corollas 3.9–9 mm, whitish and pinkish purple-tinged, or pinkish purple; fruits 4–10 × 1.3–4 mm, declined or spreading, not inflated, unilocular, body elliptic or oblong, straight or slightly curved, laterally compressed in cross-section, the valves thick-papery, strigulose.

1. Calyx teeth 0.7–1.5 mm; fruits 3–4 mm…**var. neomexicanus**
1'. Calyx teeth 1.5–4.2 mm; fruits 4–9 mm…(2)
2. Pubescence dolabriform; stems erect or decumbent, mat-forming or shrublike; fruits 1.5–2 mm wide; seeds 2–4…**var. elatus**
2'. Pubescence basifixed; stems prostrate or decumbent, cushion-forming or mat-forming; fruits 2–2.4 mm wide; seeds 4–8…**var. tegetarius**

var. elatus S. Watson [*A. kentrophyta* var. *impensus* (E. Sheld.) M. E. Jones]—Tall spiny milkvetch. Caudices woody; stems erect or decumbent, mat-forming or shrublike, the herbage glabrous, glabrescent, or strigulose, the hairs dolabriform, 0.1–0.5 mm; leaves with 1–4 pairs of leaflets, the blades 4–17 mm; calyx tube 1.5–2.6 mm, the teeth 1.9–2.5 mm, corollas 3.9–6 mm; fruits 3.5–7 × 1.3–2 mm; seeds 2–4. Hillsides, slopes, talus under cliff faces, rocky washes, commonly associated with piñon at lower elevations, 5250–7550 ft. NW, C.

var. neomexicanus (Barneby) Barneby—New Mexico spiny milkvetch. Caudices woody; stems erect or decumbent, cushion-forming, mat-forming, or shrublike, the herbage strigulose, the hairs dolabriform, 0.3–0.7 mm; leaves with 1–2 pairs of leaflets, the blades 3–13 mm; calyx tube 1.8–2.1 mm, the teeth 0.7–1.5 mm, corollas 4.8–5.2 mm; fruits 3–4 × 1.8–2.4 mm; seeds 2–3. Clay or sandy sites on hillsides, badlands, dunes, 4600–7550 ft. NC, NW, C, Chaves, Guadalupe. A New Mexico endemic. **E**

var. tegetarius (S. Watson) Dorn [*A. kentrophyta* var. *implexus* (Canby ex Porter & J. M. Coult.) Barneby]—Montane spiny milkvetch. Caudices woody; stems prostrate or decumbent, cushion-forming or mat-forming, the herbage glabrous, glabrescent, strigulose, villosulous, or villous, the hairs basifixed, 0.3–1.2 mm; leaves with 1–4 pairs of leaflets, the blades 1–17 mm; calyx tube 1.3–2.8 mm, the teeth

1.5–4.2 mm, corollas 3.9–9.2 mm; fruits 3–9 × 2–2.5 mm; seeds 4–8. Gravelly or rocky sites, slopes, talus under cliff faces, forests, shrublands, rocky washes, 6900–7550. Taos.

Astragalus kerrii P. J. Knight & Cully—Kerr's milkvetch. Perennial; stems tuft-forming, the herbage strigulose, hairs basifixed and obscurely dolabriform; stipules of the lowermost nodes connate-sheathing; leaves with 4–7 pairs of leaflets, blades 2.5–8 mm, elliptic or ovate; inflorescences 1–4-flowered; flowers ascending or spreading, calyx tube 1–2.5 mm, corollas 5–6.2 mm, whitish and pinkish purple–tinged; fruits 8–14 × 4–6 mm, inflated (not bladdery), unilocular, body elliptic, straight, dorsiventrally compressed in cross-section, the valves thin-papery, strigulose. Sandstone substrate, sandy sites in woodlands and forests, commonly associated with juniper, piñon, ponderosa pine, 5400–7500 ft. Lincoln. **R & E**

Astragalus knightii Barneby—Knight's milkvetch. Perennial; stems tuft-forming, the herbage strigulose, the hairs dolabriform; stipules of the lowermost nodes connate-sheathing; leaves with 4–7 pairs of leaflets, blades 2–8 mm, elliptic or ovate; inflorescences 5–14-flowered; flowers ascending or spreading, calyx tube 2–2 7 mm, corollas 5–6 mm, whitish and pinkish purple–tinged; fruits 8–14 × 4–6 mm, inflated (not bladdery), unilocular, body elliptic, oblong, or obovoid, straight, dorsiventrally compressed in cross-section, the valves papery, strigulose. Dakota Sandstone substrates, sandy sites on rocky ledges, juniper and sagebrush, 5700–5900 ft. Sandoval. **R & E**

Astragalus laxmannii Jacq. [*A. adsurgens* Pall. var. *robustior* Hook.]—Laxmann's milkvetch. Perennial; stems clump-forming, herbage glabrous, glabrescent, or pubescent, hairs dolabriform; stipules of the lowermost nodes connate-sheathing; leaves with 4–12 pairs of leaflets, the terminal leaflet continuous with the rachis, blades 8–25 mm, elliptic or linear; inflorescences 5–20-flowered; flowers ascending, calyx tube 4–7 mm, corollas 13–19.5 mm, whitish, pinkish purple, or bluish purple, whitish or ochroleucous in age; fruits 7–14 × 2.3–4 mm, not inflated, bilocular or semibilocular, body elliptic or lanceolate, rounded or trigonously compressed in cross-section, the valves thick-papery, strigulose. New Mexico material belongs to **var. robustior**. Plains, prairies, forests, meadows, hillsides, 5500–9850 ft. NE, NC, Torrance.

Astragalus lentiginosus Douglas—Straggling milkvetch. Annual or perennial, herbage glabrous, glabrescent, or strigulose, hairs basifixed, 0.5–1.2 mm; stipules of the lowermost nodes free; leaves with 5–15 pairs of leaflets; blades 2–25 mm, elliptic or oblong; inflorescences 5–48-flowered; flowers erect, ascending, or spreading, calyx tube 5–7 mm, corollas 7–22 mm, pinkish purple, or pinkish purple with white wing tips; fruits 10–30 × 5–18 mm, bladdery-inflated, bilocular, sessile, deciduous, the body elliptic, oblong, ovate, or subglobose, ± straight or incurved, rounded or dorsiventrally compressed in cross-section, the valves papery or leathery, glabrous, glabrescent, or strigulose.

1. Inflorescence axis in fruit 4–18 cm; the longest hairs of the herbage 0.8–1.2 mm…**var. australis**
1'. Inflorescence axis in fruit 1–4 cm; the longest hairs of the herbage 0.1–0.8 mm…**var. diphysus**

var. australis Barneby—Southern straggling milkvetch. Herbage glabrous, glabrescent, or strigulose, the longest hairs 0.8–1.2 mm; inflorescences 10–33-flowered, the axis 4–18 cm in fruit; fruits 12–22 × 5–15 mm, the valves papery, glabrous, glabrescent, or strigulose. Plains, deserts, hillsides, grasslands, yucca, mesquite, creosote bush, acacia, 3250–4600 ft. SW.

var. diphysus (A. Gray) M. E. Jones—Straggling milkvetch. Herbage glabrous, glabrescent, or strigulose, the longest hairs 0.1–0.8 mm; inflorescences 9–24-flowered, the axis 1–4 cm in fruit; fruits 10–30 × 8–18 mm, the valves papery or leathery, glabrous. Plains, deserts, hillsides, canyon bottoms, yucca, mesquite, piñon, juniper, 3250–8550 ft. NC, NW, C, WC, SE, SC.

Astragalus lonchocarpus Torr.—Rush milkvetch. Perennial herbs or ± subshrubs; stem herbage strigulose, the hairs basifixed, 0.4–0.75 mm; stipules of the lowermost nodes free; leaves with 2–5 pairs of leaflets, lateral leaflets jointed and the terminal leaflet continuous with the rachis, blades 2–36 mm, linear or narrowly elliptic; inflorescences 15–45-flowered, ± secund; flowers spreading or declined, calyx tube 5–8 mm, corollas 13–23 mm, whitish or ochroleucous, orangish in age; fruits 22–47 × 2.5–7.5 mm, pendulous, unilocular, body oblong or oblanceolate, straight or slightly curved, dorsally compressed in cross-section, the valves thinly leathery, glabrous, glabrescent, or strigulose; seeds 12–26. Sandy or gravelly sites in alkaline soils derived from sandstone or limestone; desert scrub, piñon-juniper, sagebrush, 4250–9200 ft. NE, NC, NW, C.

Astragalus lotiflorus Hook.—Lotus milkvetch. Perennial; stems prostrate, the herbage strigulose or villosulous, the hairs dolabriform; stipules of the lowermost nodes free; leaves with 1–9 pairs of leaflets, the blades 4–20 mm, elliptic, oblanceolate, oblong, or ovate; inflorescences dimorphic, cleistogamous racemes sessile and 5–17-flowered; flowers ascending or spreading, calyx tube 3.2–4.5 mm, corollas 8.5–14 mm, whitish, greenish white, ochroleucous, or whitish and pinkish purple–tinged; fruits 12–37 × 5–8 mm, not inflated, unilocular, body elliptic, oblong, or ovate, ± straight, dorsally or trigonously compressed in cross-section, the valves papery, strigulose or villosulous. Often on limestone substrates; ridges, talus slopes, grasslands, ponderosa pine, piñon-juniper, 3300–7550 ft. NE, NC, EC, C.

Astragalus micromerius Barneby—Chaco milkvetch. Perennial; stems decumbent or prostrate, mat-forming, herbage villous and hirsutulous, hairs basifixed, 0.5–0.8 mm; stipules of the lowermost nodes connate-sheathing; leaves with 1–4 pairs of leaflets, ± subpalmate, blades 1–6 mm, elliptic, oblong, or obovate; inflorescences 1–5-flowered; flowers ascending or spreading, calyx tube 1.5–2.5 mm, corollas 5.5–6.5 mm, whitish and pinkish purple–tinged; fruits 4–5 × 2–3.5 mm, not inflated, unilocular, body ovate, slightly curved, rounded or dorsally compressed in cross-section, the valves thin-papery, hirsutulous. Crevices, ledges, talus, sandstone substrates, desert scrub, piñon-juniper, 6550–7400 ft. McKinley, Rio Arriba, San Juan. **R & E**

Astragalus miser Douglas—Weedy milkvetch. Perennial; stems tuft-forming, sterile stems forming basal leafy tuft, fertile stems with 1 to several developed internodes, herbage strigose or pilosulous, hairs basifixed, 0.3–0.75 mm; stipules of the lowermost nodes connate-sheathing; leaves with 3–21 pairs of leaflets, blades 5–25 mm, elliptic, linear, oblong, or ovate; inflorescences 3–25-flowered; calyx tube 2.2–2.9 mm, corollas 5.9–10.2 mm, whitish, ochroleucous, or whitish and pinkish purple–tinged; fruits 12–25 × 2–4 mm, unilocular, body oblanceolate or narrowly oblong, straight, slightly decurved, or slightly incurved, laterally compressed in cross-section, the valves stiff-papery, glabrescent or strigulose. New Mexico material belongs to **var. oblongifolius** (Rydb.) Cronquist. Sagebrush, piñon-juniper, ponderosa pine, mixed conifer communities, 5900–9850 ft. Rio Arriba. Introduced.

Astragalus missouriensis Nutt.—Missouri milkvetch. Perennial; stems tuft-forming, ascending or decumbent, the herbage strigose or strigulose, the hairs dolabriform; stipules of the lowermost nodes free; leaves with 5–11 pairs of leaflets, blades 4–17 mm, elliptic, oblong, or obovate; inflorescences 5–15-flowered; flowers ascending or spreading, calyx tube 4–10 mm, corollas 9.5–24 mm, whitish or pinkish purple; fruits 14–30 × 4–10 mm, not inflated, bilocular, body elliptic or oblong, straight or incurved, rounded or trigonously compressed in cross-section, the valves thick-papery or leathery, glabrous, glabrescent, strigose, or strigulose.

1. Calyx tube 2–5 mm…(2)
1'. Calyx tube 6–11 mm…(3)
2. Corollas pinkish purple or reddish purple; calyx tube 4–4.8 mm…**var. mimetes**
2'. Corollas whitish and pinkish purple–tinged; calyx tube 3.5–4 mm…**var. accumbens**
3. Fruits straight, laterally compressed in cross-section…**var. missouriensis**
3'. Fruits lunately curved, dorsiventrally compressed in cross-section…(4)
4. Racemes 4–8-flowered; fruit valves strigulose…**var. amphibolus**
4'. Racemes 9–12-flowered; fruits valves glabrous or glabrescent…**var. humistratus**

var. accumbens (E. Sheld.) Isely [*A. accumbens* E. Sheld.]—Zuni milkvetch. Calyx tube 3.5–4 mm, corollas 7–8.3 mm, whitish and pinkish purple–tinged; fruits 9–18 × 4–8 mm, persistent, the body ± straight, laterally compressed in cross-section, the valves strigulose. Alkaline soil, piñon-juniper woodlands, 6200–7900 ft. NW, WC. A New Mexico endemic. **R & E**

var. amphibolus Barneby—Missouri milkvetch. Calyx tube 7–10 mm, corollas 15–22 mm, whitish and pinkish purple–tinged or pinkish purple; fruits 11–25 × 7–9 mm, deciduous or persistent, the body

± straight, dorsiventrally compressed in cross-section, the valves strigulose. Hillsides, knolls, washes, piñon-juniper and sagebrush, 4900–6250 ft. NC, NW, C.

var. humistratus Isely—Pagosa milkvetch. Calyx tube 7–10 mm, corollas 15–20.5 mm, whitish and pinkish purple–tinged, pinkish purple, or reddish purple; fruits 17–20 × 7–9 mm, persistent, the body incurved, dorsiventrally compressed in cross-section, the valves glabrous or glabrescent. Clay soils; sagebrush, piñon-juniper, and ponderosa pine communities, 7050–7700 ft. Rio Arriba. **R**

var. mimetes Barneby—Mimic milkvetch. Calyx tube 4–4.8 mm, corollas 9.5–11.8 mm, pinkish purple or reddish purple; fruits 14–24 × 5–9 mm, persistent, the body ± straight, laterally compressed in cross-section, the valves strigulose. Piñon-juniper and ponderosa pine communities, 4550–7250 ft. NC, C, NW. **R**

var. missouriensis—Missouri milkvetch. Calyx tube 6–9.3 mm, corollas 16–24 mm, pinkish purple; fruits 15–30 × 4–9 mm, persistent, the body straight, laterally compressed in cross-section, the valves strigulose. Prairies, grasslands, salt desert scrub, sagebrush, piñon-juniper communities, 3600–8550 ft. N, C, EC, SC.

Astragalus mollissimus Torr.—Woolly milkvetch. Perennial; stems clump-forming, ascending or decumbent, herbage tomentose or villous, the hairs basifixed, 1–3.5 mm; stipules of the lowermost nodes free; leaves with 5–17 pairs of leaflets, blades 4–45 mm, elliptic, orbicular, ovate, or obovate; inflorescences 5–45-flowered; flowers ascending or spreading, calyx tube 4.5–11 mm, corollas 12–24 mm, whitish, ochroleucous, whitish and pinkish purple–tinged, or pinkish purple, fruits 9–25 × 4–13 mm, not inflated, bilocular, body elliptic, lanceolate, oblong, or ovate, straight, incurved, or lunate, rounded or trigonously compressed in cross-section, the valves thick-papery or leathery, glabrous, glabrescent, villosulous, or tomentulose, the epidermis obscured by hairs or not.

1. Fruit valve surface not obscured by hairs, the pubescence sparse…(2)
1'. Fruit valve surface obscured by hairs, the pubescence dense…(3)
2. Corollas 9–13 mm; fruit body 9–14 mm…**var. earlei**
2'. Corollas 14–24 mm; fruit body 14–24 mm…**var. mollissimus**
3. Inflorescence 7–12-flowered…**var. matthewsii**
3'. Inflorescence 12–45-flowered…(4)
4. Fruit body 6–11 mm wide…(5)
4'. Fruit body 4–6 mm wide…(6)
5. Herbage with hairs 2–2.7 mm…**var. bigelovii** (in part)
5'. Herbage with hairs 1–2 mm…**var. thompsoniae**
6. Herbage with hairs 3–3.5 mm; inflorescence axis in fruit 1.6–6 cm…**var. mogollonicus**
6'. Herbage with hairs 2–2.7 mm; inflorescence axis in fruit 5–11 cm…**var. bigelovii** (in part)

var. bigelovii (A. Gray) Barneby [*A. bigelovii* A. Gray]—Bigelow's woolly milkvetch. Hairs basifixed, 2–2.7 mm; leaves with 6–13 pairs of leaflets, the blades 6–25 mm; inflorescences 15–45-flowered, the axis 4–11 cm in fruit; calyx tube 8–10 mm, corollas 16–22 mm, pinkish purple; fruits 10–15 × 4–8 mm, tomentulose or villosulous, the epidermis obscured by hairs. Plains, grasslands, knolls, washes, 3600–7550 ft. EC, C, WC, SC, SW.

var. earlei (Greene ex Rydb.) Tidestr.—Earle's woolly milkvetch. Hairs basifixed, 1–2.5 mm; leaves with 9–17 pairs of leaflets, the blades 5–45 mm; inflorescences 15–36-flowered, the axis 4–14 cm in fruit; calyx tube 4.5–7 mm, corollas 12–17.6 mm, ochroleucous or pinkish purple; fruits 9–14 × 4–9 mm, glabrous, glabrescent, tomentulose, or villosulous, the epidermis not obscured by hairs. Plains, prairies, knolls, piñon-juniper woodlands, 3600–7550 ft. EC, SC.

var. matthewsii (S. Watson) Barneby [*A. bigelovii* A. Gray var. *matthewsii* (S. Watson) M. E. Jones]—Matthews' woolly milkvetch. Hairs basifixed, 2–2.7 mm; leaves with 5–11 pairs of leaflets, blade 3–12 mm; inflorescences 7–12-flowered, axis 1–4 cm in fruit; calyx tube 7–8.6 mm, corollas 18–22 mm, pinkish purple; fruits 12–18 × 7–13 mm, body tomentulose or villosulous, epidermis obscured by hairs. Juniper, piñon, ponderosa pine, 6000–8550 ft. NC, C, WC.

var. mogollonicus (Greene) Barneby [*A. bigelovii* A. Gray var. *mogollonicus* (Greene) Barneby]—Mogollon woolly milkvetch. Hairs basifixed, 3–3.5 mm; leaves with 5–11 pairs of leaflets, the blades 3–13 mm; inflorescences 7–12-flowered, the axis 1–4 cm in fruit; calyx tube 6–8.5 mm, corollas 16–22 mm, pinkish purple; fruits 9–13 × 4.5–6 mm, body tomentulose or villosulous, the epidermis obscured by hairs. Juniper, piñon, ponderosa pine, 5900–8550 ft. WC, SW.

var. mollissimus—Woolly milkvetch. Hairs basifixed, 1–2.5 mm; leaves with 7–16 pairs of leaflets, the blades 5–22 mm; inflorescences 10–40-flowered, the axis 2–14 cm in fruit; calyx tube 6.8–9.5 mm, corollas 16–22 mm, ochroleucous or pinkish purple; fruits 14–24 × 4–7 mm, body glabrous, glabrescent, or villosulous, the epidermis not obscured by hairs. Grasslands, shortgrass prairies, knolls, woodlands, 4250–8200 ft. NE, EC, SE.

var. thompsoniae (S. Watson) Barneby [*A. thompsoniae* S. Watson]—Thompson's woolly milkvetch. Hairs basifixed, 1–2 mm; leaves with 7–14 pairs of leaflets, the blades 2–18 mm; inflorescences 7–25-flowered, the axis 1–18 cm in fruit; calyx tube 7–13 mm, corollas 18–25 mm, pinkish purple; fruits 11–23 × 6–11 mm, the body tomentulose or villosulous, the epidermis obscured by hairs. Gravelly or sandy soil, juniper, sagebrush, piñon, 5250–7250 ft. NC, NW. C.

Astragalus monumentalis Barneby [*A. cottamii* S. L. Welsh]—Cottam's milkvetch. Perennial; stems tuft-forming, the herbage strigulose, the hairs basifixed, or at least a few incipiently dolabriform, 0.35–0.7 mm; stipules of the lowermost nodes free; leaves with 4–10 pairs of leaflets, the blades 2–9 mm, elliptic, oblong, oblanceolate, or obovate; inflorescences 3–9-flowered; flowers ascending or spreading, calyx tube 4–7 mm, corollas 11–17 mm, pinkish purple or bluish purple; fruits 15–37 × 3–4.5 mm, not inflated, bilocular, body linear or lanceolate, lunate, trigonously compressed in cross-section, the lateral faces flat, the valves papery, strigulose. Our material belongs to **var. cottamii** (S. L. Welsh) Isely. Often on ledges and sandstone slickrock; sagebrush, piñon, juniper, 5000–6050 ft. San Juan. Reports of *A. monumentalis* var. *monumentalis* in New Mexico are based on specimens of this variety. **R**

Astragalus naturitensis Payson—Naturita milkvetch. Perennial, caudices covered in a thatch of persistent leaf bases; stems tuft-forming, herbage strigulose, the hairs basifixed; stipules of the lowermost nodes free; leaves with 4–8 pairs of leaflets, blades 2–8 mm, elliptic, oblanceolate, or obovate; inflorescences 3–11-flowered; flowers ascending or spreading, calyx tube 4–6 mm, corollas 11–15.5 mm, whitish and pinkish purple–tinged or pinkish purple, fruits 13–22 × 4–6 mm, not inflated, bilocular, body elliptic, incurved, dorsiventrally compressed in cross-section, the valves thick-papery, strigulose. Ledges, canyons, washes, commonly associated with sagebrush, piñon, juniper, 5350–7250 ft. McKinley, San Juan.

Astragalus neomexicanus Wooton & Standl.—New Mexico milkvetch. Perennial; stem herbage strigulose or pilosulous, the hairs basifixed; stipules of the lowermost nodes free; leaves with 9–21 pairs of leaflets, blades 3–20 mm, elliptic, oblong, lanceolate, or ovate; inflorescences 10–24-flowered; flowers spreading to declined, calyx tube 6–9.4 mm, gibbous, corollas 15–19 mm, reddish purple or pinkish purple; fruits 17–33 × 4–11 mm, not inflated, unilocular, body obliquely oblong-ellipsoid, ± incurved, dorsiventrally compressed in cross-section, the valves fleshy, leathery, strigulose. Hillsides, talus slopes, washes, commonly associated with juniper, piñon, ponderosa pine, 6850–8500 ft. Lincoln, Otero. **R & E**

Astragalus newberryi A. Gray—Newberry's milkvetch. Perennial, caudices covered in a thatch of persistent leaf bases, herbage pilosulous, villosulous, or tomentulose, the hairs basifixed; stipules of the lowermost nodes free; leaves with 2–15 pairs of leaflets, blades 3–20 mm, elliptic, oblanceolate, or obovate; inflorescences 2–8-flowered; flowers ascending, calyx tube 9–12 mm, corollas 14–32 mm, pinkish purple; fruits 13–36 × 7–17 mm, not inflated, unilocular, body obliquely ovoid, curved, dorsiventrally compressed in cross-section, the valves ± fleshy, leathery, strigulose, villosulous, or tomentulose. New Mexico material belongs to **var. newberryi**. Hillsides, washes, grasslands, juniper, piñon, 5900–7550 ft. NC, NW.

Astragalus nothoxys A. Gray—Beaked milkvetch. Perennial, short-lived, sometimes flowering the first year and appearing annual; stems decumbent or prostrate, the herbage strigulose, the hairs basifixed; stipules of the lowermost nodes free; leaves with 3–10 pairs of leaflets, blades 2–12 mm, oblanceo-late, oblong, ovate, obcordate, or elliptic; inflorescences 4–25-flowered; flowers spreading or declined, calyx tube 3.6–5 mm, corollas 8.5–12 mm, pinkish purple or pinkish purple with white wing tips; fruits 13–22 × 2–4 mm, not inflated, bilocular, body linear or lanceolate, straight or incurved, trigonously com-pressed in cross-section, the lateral faces curved, the valves papery, strigulose. Often limestone soil, grasslands, roadsides, mesquite, oak, juniper, piñon, 3900–6250 ft. Grant, Hidalgo.

Astragalus nutriosensis M. J. Sand.—Nutrioso milkvetch. Perennial; stems tuft-forming, ascend-ing or erect, herbage strigose, pilose, or villous, hairs basifixed, 1–2 mm; stipules of the lowermost nodes free; leaves with 5–9 pairs of leaflets, blades 4–8 mm, elliptic or obovate; inflorescences 2–7-flowered; flowers ascending or spreading, calyx tube 6–8 mm, corollas 20–23 mm, whitish and pinkish purple-tinged; fruits 8–10 × 6–8 mm, inflated (not bladdery), bilocular, body elliptic or subglobose, straight, rounded in cross-section, the valves papery, villosulous, the epidermis obscured by hairs. Volcanic silty clay soils, grasslands, piñon-juniper woodlands, 7000–7350 ft. Catron. **R**

Astragalus nuttallianus DC.—Nuttall's milkvetch. Annual; stems strigulose or hirsutulous, the hairs basifixed, 0.6–1.2 mm; stipules of the lowermost nodes free; leaves with 3–10 pairs of leaflets, blades 2–14 mm, linear, obcordate, or elliptic; inflorescences 1–27-flowered, flowers ascending, spread-ing, or declined, calyx tube 1–4 mm, corollas 4–13 mm, whitish or pale pinkish purple; fruits 6–26 × 1–3.5 mm, not inflated, bilocular or subunilocular, body linear, lanceolate, or oblanceolate, straight, incurved, or arched strongly near the base and straight distally or lunate, laterally or trigonously compressed in cross-section, the lateral faces curved, the valves papery, glabrous, glabrescent, strigulose, or villosulous.

1. Keel apex obtuse…(2)
1'. Keel apex subacute…(3)
2. Corollas 8.5–13 mm…**var. macilentus**
2'. Corollas 6–8 mm…**var. micranthiformis**
3. Calyx teeth 1.8–3 mm; herbage with long, spreading hairs, the longest hairs 0.6–1.2 mm…**var. austrinus**
3'. Calyx teeth 1–2 mm; herbage with short, appressed hairs, the longest hairs 0.3–0.7 mm…**var. im-perfectus**

var. austrinus (Small) Barneby—Rio Fronteras milkvetch. Hairs 0.6–1.2 mm; leaves with 3–5 pairs of leaflets, the blades linear or elliptic, all monomorphic, the apex obtuse; inflorescences 2–8-flowered; calyx tube 1–3 mm, corollas 5–7 mm, the keel apex subacute; fruits 13–24 × 1.5–3 mm. Grasslands, desert scrub, roadsides, 3900–7250 ft. NE, NC, EC, C, SC, SW.

var. imperfectus (Rydb.) Barneby—Imperfect milkvetch. Hairs 0.3–0.7 mm; leaves with 3–6 pairs of leaflets, the blades elliptic or ovate, all monomorphic, the apex obtuse or subacute; inflorescences 1–4-flowered; calyx tube 1.9–3 mm, corollas 4–7 mm, the keel apex subacute; fruits 10–21 × 2–3.5 mm. Rocky or gravelly sites, desert scrub, roadsides, 4600–6900 ft. SC, SW.

var. macilentus (Small) Barneby [*A. macilentus* (Small) Cory]—Small flowered milkvetch, turkey-peas milkvetch. Hairs 0.3–0.8 mm; leaves with 3–11 pairs of leaflets, the blades elliptic or ovate, dimor-phic or monomorphic, the apex of all or most leaflets truncate, emarginate, or retuse; inflorescences 2–27-flowered; calyx tube 2–3 mm, corollas 8–13 mm, the keel apex obtuse; fruits 10–25 × 1.8–3 mm. Rocky or gravelly sites, desert scrub, roadsides, 5500–6250 ft. Torrance.

var. micranthiformis Barneby—Montezuma milkvetch. Hairs basifixed, 0.3–0.8 mm; leaves with 3–8 pairs of leaflets, dimorphic, the blades of the lower leaflets oblong or ovate, blades of upper leaves oblanceolate, elliptic, or oblong; inflorescences 2–7-flowered; calyx tube 2–3 mm, corollas 6–8 mm; fruits 12–20 × 2–3.5 mm. Rocky or gravelly sites on hillsides, talus slopes, washes, commonly associated with piñon and juniper, 4600–7900 ft. NE, NC, NW, C, SE, SC.

Astragalus oocalycis M. E. Jones—Arboles milkvetch. Perennial, seleniferous, malodorous; stems clump-forming, herbage glabrescent or strigulose, hairs basifixed, 0.5–0.75 mm; stipules of the lower-most nodes connate-sheathing; leaves with 9–14 pairs of leaflets, blades 10–40 mm, linear or narrowly oblong; inflorescences 35–60-flowered; flowers spreading, declined, or deflexed, calyx tube 8–11 mm, broadly ovoid to subglobose, inflated, thin-papery in age, corollas 14–17 mm, ochroleucous; fruits 6–7.5 × 3–4 mm, declined, not inflated, unilocular or subunilocular, stipelike gynophore 0.5–1 mm, deciduous with the inflated calyx, the body elliptic or oblong, straight or slightly curved, dorsiventrally compressed in cross-section, the valves leathery, glabrous. San Jose Formation, clay soils, piñon-juniper, 5550–7200 ft. Rio Arriba, San Juan. **R**

Astragalus pattersonii A. Gray—Patterson's milkvetch. Perennial, seleniferous, malodorous, herbage glabrous, glabrescent, or strigulose, hairs basifixed; stipules of the lowermost nodes free; leaves with 4–12 pairs of leaflets, blades 6–38 mm, elliptic, obovate, or oblanceolate; inflorescences 6–22-flowered; flowers declined or deflexed, calyx tube 6–8.8 mm, pale tannish or whitish (same color as the petals), corollas 14–24 mm, pale tannish or whitish; fruits 17–35 × 6–10 mm, inflated (not bladdery), unilocular, body elliptic, oblong, or ovoid, ± straight, trigonously compressed in cross-section, the valves leathery, glabrous, glabrescent, or strigulose. Clay, shale, gypsum, hillsides, 3250–7550 ft. NE, NC, NW, Cibola.

Astragalus pictiformis Barneby—Guadalupe milkvetch. Perennial; stems strigulose or villosulous, hairs basifixed, 0.3–0.6 mm; stipules of the lowermost nodes connate-sheathing; leaves with 4–8 pairs of leaflets, blades 2–10 mm, obovate, oblong, oblanceolate, or suborbicular; inflorescences 4–14-flowered; flowers ascending or spreading, calyx tube 2.4–4 mm, corollas 9–11 mm, pale pinkish purple; fruits 9–17 × 4–6 mm, not inflated, unilocular, body elliptic or oblong, straight or slightly curved, trigonously compressed in cross-section, the valves thick-papery, strigulose. Sandy or gravelly sites on soils derived from limestone and sandstone; desert scrub, piñon-juniper woodlands, 4550–7600 ft. NC, C, SE, SC.

Astragalus praelongus E. Sheld.—Stinking milkvetch. Perennial, seleniferous, malodorous, stems glabrous, glabrescent, or strigulose, hairs basifixed; stipules of the lowermost nodes free; leaves with 4–11 pairs of leaflets, blades 3–50 mm, elliptic, oblong, obovate, or oblanceolate; inflorescences 10–33-flowered; flowers declined or deflexed, calyx tube 4–7.5 mm, corollas 14–24 mm, whitish, ochroleucous, or whitish and pinkish purple-tinged; fruits 17–42 × 6–10 mm, inflated (not bladdery), unilocular, body elliptic, oblong, or ovoid, ± straight, rounded in cross-section, the valves ± fleshy, leathery, or ± woody, glabrous, glabrescent, or strigulose.

1. Fruit sessile or substipitate; stipe, if present, < 0.5 mm…**var. praelongus** (in part)
1′. Fruit stipitate; stipe 1–8 mm…(2)
2. Fruit with a stipe 4.5–8 mm…**var. lonchopus**
2′. Fruit with a stipe 1–3 mm…(3)
3. Fruit 6–10 mm wide…**var. ellisiae**
3′. Fruit 10–15 mm wide…**var. praelongus** (in part)

var. ellisiae (Rydb.) Barneby—Ellis's stinking milkvetch. Calyx teeth 2–5.5 mm; fruits 18–34 × 6–11 mm, stipitate, the stipe 1–2.5 mm, the valves glabrous, glabrescent, or strigulose. Grasslands, desert scrub, piñon-juniper woodlands, 3600–7550 ft. NC, NW, EC, C, SC.

var. lonchopus Barneby—Long-stalked stinking milkvetch. Calyx teeth 3–6.5 mm; fruits 20–30 × 6–9 mm, stipitate, the stipe 4.5–8 mm, the valves strigulose. Slickrock surrounded by desert scrub, 5250–5900 ft. San Juan.

var. praelongus—Stinking milkvetch. Calyx teeth 0.3–4.7 mm; fruits 20–42 × 10–25 mm, sessile or stipitate, the stipe 0–2.5 mm, the valves glabrous, glabrescent, or strigulose. Desert scrub, sagebrush, piñon-juniper, ponderosa pine communities, 5900–7550 ft. N, C, WC, SW.

Astragalus preussii A. Gray—Preuss's milkvetch. Perennial, seleniferous, malodorous, herbage glabrous, glabrescent, or strigulose, hairs basifixed; stipules of the lowermost nodes free; leaves with 7–12 pairs of leaflets, blades 6–28 mm, linear, obovate, or oblanceolate; inflorescences 3–22-flowered; flowers erect, ascending, or spreading, calyx tube 5–9.7 mm, corollas 17–24 mm, pinkish purple or whitish and pinkish purple-tinged; fruits 12–34 × 6–13 mm, ascending or erect, inflated (not bladdery), unilocular, body elliptic, oblong, or ovoid, ± straight, trigonously compressed in cross-section, the valves leathery, glabrous. New Mexico material belongs to **var. latus** M. E. Jones. Sandy soil, salt desert scrub, juniper, 5575–6600 ft. McKinley, San Juan.

Astragalus proximus (Rydb.) Wooton & Standl.—Aztec milkvetch. Perennial; stems ascending or erect, herbage glabrescent or strigulose, the hairs basifixed, 0.3–0.6 mm; stipules of the lowermost nodes connate-sheathing; leaves with 3–8 pairs of leaflets, blades 6–22 mm, linear, narrowly oblanceolate, or suborbicular; inflorescences 12–40-flowered; flowers ascending, spreading, or declined, calyx tube 1.8–2.5 mm, corollas 6–7 mm, whitish or whitish and pinkish purple–tinged; fruits 10–15 × 2.3–3.5 mm, not inflated, unilocular, body narrowly elliptic, straight or slightly curved, dorsally compressed and rounded ventrally in cross-section, the valves thin-papery, glabrous. Sandy or sandy-clay sites, salt desert scrub, sagebrush, piñon-juniper, ponderosa pine communities, 5200–8900 ft. Rio Arriba, Sandoval, San Juan.

Astragalus puniceus Osterh.—Trinidad milkvetch. Perennial; stems 1–5 dm, herbage strigulose or villosulous, hairs basifixed, 0.4–1.2 mm; stipules of the lowermost nodes connate-sheathing; leaves with 4–13 pairs of leaflets, blades 5–16 mm, elliptic, oblanceolate, or oblong; inflorescences 5–27-flowered; flowers ascending or spreading, calyx tube 5–8 mm, corollas 13–21 mm, pinkish purple or whitish and pinkish purple–tinged; fruits 15–24 × 4.5–9.5 mm, not inflated, unilocular, body elliptic or oblong, rounded or dorsally compressed in cross-section, the valves leathery, strigulose or villosulous.

1.	Corollas pinkish purple; fruit valves red-mottled, rarely reddish-tinged or greenish…**var. puniceus**
1'.	Corollas whitish and pinkish purple–tinged; fruit valves reddish-tinged or greenish, not red-mottled…**var. gertrudis**

var. gertrudis (Greene) Barneby—Taos milkvetch. Leaves with 6–10 pairs of leaflets; corollas 14.8–18 mm, whitish and pinkish purple–tinged; fruits 15–24 × 6.5–9.5 mm, the body elliptic or ovate, the valves reddish-tinged or greenish, not red-mottled. Hillsides and washes, piñon-juniper woodlands, 5900–7050 ft. Rio Arriba, Taos. **R & E**

var. puniceus—Trinidad milkvetch. Leaves with 4–13 pairs of leaflets; corollas 13–21 mm, pinkish purple; fruits 15–24 × 4.5–8 mm, the body elliptic or oblong, the valves red-mottled, rarely reddish-tinged or greenish. Sandy sites, sagebrush, piñon-juniper woodlands, disturbed sites, 4600–8900 ft. NE, NC, EC.

Astragalus racemosus Pursh [*A. racemosus* var. *longisetus* M. E. Jones]—Alkali milkvetch. Perennial, seleniferous, malodorous; stems clump-forming, herbage glabrous, glabrescent, or strigulose, hairs basifixed, 0.35–1 mm; stipules of the lowermost nodes connate-sheathing; leaves with 5–15 pairs of leaflets, blades 5–35 mm, elliptic, lanceolate, or ovate; inflorescences 10–70-flowered; flowers spreading or declined, calyx tube 5–9 mm, gibbous or not, corollas 16–21 mm, whitish, ochroleucous, or whitish and pinkish purple–tinged; fruits 15–30 × 3–6 mm, not inflated, unilocular or subunilocular, body narrowly elliptic, lanceolate, or narrowly oblong, straight or slightly curved, trigonously compressed in cross-section, valves papery, glabrous, glabrescent, or strigulose. Our material belongs to **var. racemosus**. Grassy hillsides, shortgrass prairies, 5900–9850 ft. NE, NC.

Astragalus ripleyi Barneby—Ripley's milkvetch. Perennial; stems erect or ascending, herbage strigulose, hairs basifixed, 0.3–0.5 mm; stipules of the lowermost nodes free; leaves with 6–10 pairs, the lateral leaflets jointed and the terminal leaflet continuous with the rachis, blades 9–35 mm, linear or narrowly elliptic; inflorescences 15–45-flowered, ± secund; flowers ascending, spreading, declined, or deflexed, calyx tube 5–6.5 mm, corollas 14–17 mm, ochroleucous or pale lemon-yellow; fruits 14–33 × 3.5–6 mm, not inflated, unilocular, body narrowly elliptic, narrowly oblong, or narrowly lanceolate, straight or slightly decurved, laterally compressed in cross-section, the valves papery, strigulose. Sagebrush, piñon-juniper, ponderosa pine communities, 7000–8250 ft. Rio Arriba, Taos. **R**

Astragalus robbinsii (Oakes) A. Gray—Robbins' milkvetch. Perennial; stems strigulose, the hairs basifixed; stipules of the lowermost nodes connate-sheathing; leaves with 3–6 pairs of leaflets, blades 8–30 mm, ovate, obovate, elliptic; inflorescences 5–25-flowered; flowers spreading or declined, calyx tube 2.5–4.5 mm, corollas 7–11.5 mm, pinkish purple or bluish purple; fruits 10–25 × 3.5–5 mm, subunilocular, body elliptic, straight or slightly decurved, rounded or obscurely trigonously compressed in cross-section, the valves thick-papery, strigulose. Our material belongs to **var. minor** (Hook.) Barneby. Montane meadows, stream banks, spruce-fir communities, 8550–10,800 ft. NC, C.

Astragalus sabulonum A. Gray—Gravel milkvetch. Annual; stems villosulous or hirsutulous, hairs basifixed; stipules of the lowermost nodes free; leaves with 4-11 pairs of leaflets, blades 1-5 mm, oblanceolate, oblong, or obovate; inflorescences 2-7-flowered, racemes crowded and headlike in fruit; flowers ascending or spreading, calyx tube 1.8-2.5 mm, corollas 5.2-8 mm, pinkish purple or ochroleucous and pinkish purple–tinged; fruits 9-20 × 5-11 mm, inflated (not bladdery), unilocular, body ovate, obovoid, or subglobose, rounded in cross-section, the valves thick-papery, opaque, villosulous or hirsutulous. Often on dunes and other sandy sites, desert grasslands, desert scrub, piñon-juniper woodlands, 4900-7000 ft. McKinley, San Juan.

Astragalus scopulorum Porter—Rocky Mountain milkvetch. Perennial; stems strigulose, hairs basifixed, 0.5-0.75 mm; stipules of the lowermost nodes connate-sheathing; leaves with 6-17 pairs of leaflets, blades 8-30 mm, elliptic, linear, oblanceolate, oblong, or ovate; inflorescences 10-22-flowered; flowers ascending, spreading, or declined, calyx tube 6.5-8.5 mm, corollas 18-24 mm, ochroleucous; fruits 18-35 × 3.5-6.5 mm, not inflated, bilocular, body narrowly elliptic, narrowly oblong, or oblanceolate, straight, slightly curved, or lunate, trigonously compressed in cross-section, the valves papery, glabrous. Gravelly clay sites, sagebrush, piñon-juniper woodlands, occasionally aspen forests, 6200-12,150 ft. NC, NW, C.

Astragalus sericoleucus A. Gray [*Orophaca sericea* (Nutt.) Britton]—Silky milkvetch. Perennial; stems pulvinate-cespitose, mat-forming or cushion-forming, herbage pilosulous or villosulous, hairs dolabriform; stipules of the lowermost nodes connate-sheathing; leaves with 1 pair of leaflets, ± palmately compound, blades 3-13 mm, oblanceolate, oblong, or ovate; inflorescences 2-5-flowered; flowers ascending or spreading, calyx tube 2-3 mm, corollas 5-8 mm, whitish or pinkish purple; fruits 3-4.5 × 1-2.5 mm, not inflated, unilocular, body elliptic, oblong, or ovate, straight, laterally compressed in cross-section, the valves papery, pilosulous or villosulous. Sandstone outcroppings, bluffs, rocky openings, prairie, 3600-8500 ft. NE, NC.

Astragalus shortianus Nutt. ex Torr. & A. Gray—Short's milkvetch. Perennial, herbage strigulose, hairs basifixed; stipules of the lowermost nodes free; leaves with 3-9 pairs of leaflets, blades 5-25 mm, elliptic or obovate; inflorescences 5-16-flowered; flowers ascending or spreading, calyx tube 7.6-10 mm, corollas 16-22 mm, pinkish purple; fruits 25-50 × 7-13 mm, not inflated, unilocular, body oblong or elliptic, straight or incurved, dorsiventrally compressed in cross-section, the valves thick-fleshy, ± woody, strigulose. Rocky slopes, grasslands, montane meadows, forest openings, 5900-7550 ft. NE, NC.

Astragalus siliceus Barneby—Flint Mountains milkvetch. Perennial, caudices covered in a thatch of persistent stipules and ± spinose leaf bases; stems pulvinate, strigulose, the hairs dolabriform; stipules of the lowermost nodes free; leaves with 1-4 pairs of leaflets, blades 1-5 mm, elliptic, oblanceolate, oblong, or obovate; inflorescences 1-3-flowered; flowers ascending or spreading, calyx tube 3-4.2 mm, corollas 9-12 mm, pinkish purple and the wings white-tipped; fruits 5-7.5 × 2.8-4 mm, not inflated, unilocular, body ovate, ± straight, laterally compressed in cross-section, the valves leathery, strigulose. Calcareous knolls and rocky areas, shortgrass prairie, 6000-6500 ft. Guadalupe, Santa Fe, Torrance. **R & E**

Astragalus tenellus Pursh [*A. multiflorus* (Pursh) A. Gray]—Pulse milkvetch. Perennial; stems glabrous, glabrescent, or strigulose, hairs basifixed; stipules of the lowermost stem nodes connate-sheathing; leaves with 5-9 pairs of leaflets, blades 2-24 mm, linear, elliptic, oblong, oblanceolate, or obovate; inflorescences 3-16-flowered; flowers ascending or spreading, calyx tube 2-2.7 mm, corollas 6-11 mm, whitish or whitish and pinkish purple-tinged; fruits 7-16 × 2.5-4.5 mm, not inflated, unilocular, body elliptic or oblong, straight, laterally compressed in cross-section, the valves papery, glabrescent or strigulose. Montane meadows, sagebrush, high plains, forest openings, 6200-11,000 ft. NC, McKinley.

Astragalus tephrodes A. Gray—Perennial, stems strigulose, pilosulous, or villosulous, hairs basifixed; stipules of the lowermost nodes free; leaves with 5-15 pairs of leaflets, blades 2-27 mm, elliptic, oblanceolate, oblong, orbicular, ovate, or obovate; inflorescences 9-35-flowered; flowers ascending or spreading, calyx tube 3.5-10 mm, corollas 11-24 mm, whitish or pinkish purple; fruits 13-40 × 5-16 mm, ascending or spreading, not inflated, unilocular or subunilocular, sessile, deciduous, the body ovate, elliptic, or oblong, ± straight, incurved, or lunate, dorsiventrally compressed in cross-section, the valves leathery or ± woody, strigulose or pilosulous.

1. Calyx tube 3.5–8 mm; corollas 12–18 mm…**var. tephrodes**
1'. Calyx tube 7–10 mm; corollas 14–24 mm…**var. brachylobus**

var. brachylobus (A. Gray) Barneby—Prescott milkvetch. Calyx tube 7–10 mm, corollas 14–24 mm. Hillsides, washes, desert scrub, mesquite, oak, juniper, piñon, ponderosa pine, 3600–7900 ft. NC, C, WC. Historically, specimen misidentifications that were corrected by Barneby as either *A. waterfallii* or *A. feensis*.

var. tephrodes—Ashen milkvetch. Calyx tube 3.5–8 mm; corollas 12–18 mm. Hillsides, washes, desert scrub, plains, oak, juniper, piñon, ponderosa pine, 3900–6250 ft. C, WC, SC, SW.

Astragalus thurberi A. Gray—Thurber's milkvetch. Perennial, stems strigulose, the hairs basifixed; stipules of the lowermost nodes free, leaves with 4–11 pairs of leaflets, blades 5–18 mm, elliptic, oblanceolate, oblong, or obovate; inflorescences 7–32-flowered, flowers ascending or spreading, calyx tube 2.2–3 mm, corollas 5.7–7 mm, pinkish purple or reddish purple; fruits 5–13 × 6–10 mm, bladdery-inflated, unilocular, body ovate, obovoid, or subglobose, rounded in cross-section, the valves papery, translucent, glabrescent or strigulose. Desert grasslands, creosote, juniper, piñon communities, 4590–7550 ft. WC, SW.

Astragalus vaccarum A. Gray—Cow Spring milkvetch. Perennial; stems strigulose, hairs basifixed; stipules of the lowermost nodes free; leaves with 4–11 pairs of leaflets, blades 4–24 mm, narrowly elliptic or narrowly oblong; inflorescences 15–50-flowered; flowers ascending or spreading, calyx tube 1.7–2 mm, corollas 4–6.2 mm, yellowish and pale pinkish purple–tinged; fruits 6–12 × 1–3 mm, not inflated, bilocular or semibilocular, body elliptic or lanceolate, incurved, trigonously compressed in cross-section, the valves papery, strigulose. Hillsides, washes, desert scrub, scrub oak, piñon, juniper, 3950–4600 ft. Grant, Hidalgo.

Astragalus waterfallii Barneby—Waterfall milkvetch. Perennial; stems strigulose, hairs basifixed; stipules of the lowermost nodes free; leaves with 4–12 pairs of leaflets, blades 3–16 mm, elliptic, oblanceolate, ovate, or obovate; inflorescences 6–18-flowered; flowers ascending or spreading, calyx tube 8–12 mm, corollas 18.5–23 mm, pinkish purple; fruits 17–38 × 5–8 mm, not inflated, body clavate, elliptic, or oblong, ± straight or incurved, trigonously compressed in cross-section, the valves leathery, strigulose. Gravelly or rocky sites, hillsides, desert scrub, desert grasslands, 4200–5900 ft. Chaves, Eddy, Otero.

Astragalus wingatanus S. Watson [*A. wingatanus* var. *dodgianus* (M. E. Jones) M. E. Jones]—Fort Wingate milkvetch. Perennial; stems strigulose, hairs basifixed, 0.2–0.6 mm; stipules of the lowermost nodes connate-sheathing; leaves with 2–8 pairs of leaflets, blades 4–18 mm, narrowly elliptic, linear, narrowly oblanceolate, or narrowly oblong; inflorescences 10–35-flowered; flowers ascending, calyx tube 1.5–2.6 mm, corollas 5.5–8 mm, pinkish purple with white wing tips, or rarely whitish and pinkish purple–tinged; fruits 9–15 × 3–4.5 mm, not inflated, unilocular, body elliptic or oblong, straight or slightly curved, laterally compressed in cross-section, the lateral faces convex, the valves thin-papery, glabrous. Sandy or sandy-clay sites, talus, washes, flats, desert scrub, piñon-juniper woodlands, 5575–7875 ft. NW, Catron.

Astragalus wittmannii Barneby—One flowered milkvetch. Perennial, caudices covered in a thatch of persistent stipules and leaf bases; stems cushion-forming, herbage glabrescent or strigulose, hairs dolabriform; stipules of the lowermost nodes free; leaves with 1–4 pairs of leaflets, blades 2–12 mm, elliptic or oblanceolate; inflorescences 1-flowered; flowers erect or ascending, calyx tube 6–7 mm, corollas 14–21 mm, pinkish purple; fruits 2.5–3.5 × 2.5–3.5 mm, inflated, unilocular, body globose, rounded in cross-section, the valves ± leathery, strigulose. Greenhorn Limestone knolls and hills, shortgrass prairie, 5900–6600 ft. Colfax, Harding, Mora. **R & E**

Astragalus wootonii E. Sheld. [*A. allochrous* A. Gray var. *playanus* (M. E. Jones) Isely]—Wooton's milkvetch. Annual, biennial, or short-lived perennial; stems strigulose, the hairs basifixed, 0.4–0.8 mm;

stipules of the lowermost nodes free; leaves with 3–11 pairs of leaflets, blades 5–20 mm, oblanceolate, oblong, or obovate; inflorescences 2–15-flowered; flowers ascending, spreading, or declined, calyx tube 2–3 mm, corollas 4.5–7.5 mm, whitish and pinkish purple–tinged, pinkish purple, or reddish purple; fruits 10–43 × 10–24 mm, bladdery-inflated, unilocular, body elliptic, ovate, or subglobose, rounded in cross-section, the valves papery, translucent, glabrous, glabrescent, strigulose, or villosulous. New Mexico material belongs to **var. wootonii**. Sandy or gravelly sites, hillsides, knolls, desert grasslands, desert scrub, mesquite, sagebrush, juniper, piñon, 3600–7550 ft. Widespread except E.

Astragalus zionis M. E. Jones—Zion milkvetch. Perennial, caudices covered in a thatch of persistent stipules and leaf bases, mat-forming or tuft-forming, herbage strigulose, pilosulous, or villosulous, the hairs basifixed; stipules of the lowermost nodes usually free; leaves with 6–12 pairs of leaflets, blades 2–16 mm, elliptic, oblanceolate, oblong, orbicular, ovate, or obovate; inflorescences 5–11-flowered; flowers ascending or spreading, calyx tube 6.5–13 mm, corollas 18–26 mm, pinkish purple; fruits 15–28 × 5–9 mm, not inflated, unilocular or subunilocular, body ovate, elliptic, or oblong, ± straight, incurved, or lunate, dorsiventrally compressed in cross-section, the valves leathery or ± woody, strigulose or villosulous. Our material belongs to **var. zionis**. Ledges, talus slopes, sagebrush, piñon-juniper woodlands, 8530–9200 ft. San Juan.

CALLIANDRA Benth. – Fairy duster

Perennial herbs, shrubs, or small trees; stems unarmed, pubescence basifixed; stipules herbaceous, striate, deciduous or persistent; leaves even-pinnate, bipinnately compound, 1 to many pairs of pinnae, petiole glands absent, rachis glands absent; inflorescences axillary, solitary or paired, compact and headlike, pedunculate; flowers pedicellate, corollas actinomorphic, tubular-funnelform, 5-lobed, inconspicuous, stamens numerous, much exserted; fruits a legume.

1. Shrubs, caudex woody; inflorescence heads 2.5–3.5 cm wide…**C. eriophylla**
1'. Perennial herbs, caudex superficial, stems partially subterranean; inflorescence heads 1.5–2.5 cm wide… **C. humilis**

Calliandra eriophylla Benth. [*C. eriophylla* var. *chamaedrys* Isely]—Fairy duster. Shrub, the herbage glabrescent or villosulous, the hairs 0.3–0.6 mm; stipules ± subulate; leaves with 1–4 pairs of pinnae per leaf, leaflets 7–10 pairs per pinnule, the blades 3–6 mm, elliptic or oblong; inflorescences compact and headlike, the heads 2.5–3.5 cm wide, 2–10-flowered, pedunculate; corollas 3.5–4 mm, the petals reddish, at least a few petal lobes involute; fruit 2.5–9 cm, the body oblong or oblanceolate, straight, the valves villosulous. Chihuahuan desert scrub with mesquite and creosote, 3900–5600 ft. NC, SW.

Calliandra humilis Benth.—Dwarf stickpea. Perennial herb, herbage glabrous, glabrescent, or pilose, the hairs 0.3–1 mm; stipules ± striate; leaves with 3–8 pairs of pinnae per leaf, leaflets 6–20 pairs per pinnule, the blades 3–7 mm, oblong; inflorescences headlike, the heads 1.8–2.5 cm wide, 4–10-flowered, subsessile or pedunculate; corollas 4–8 mm, the petals whitish or pinkish; fruit 3–7 cm, the body oblong or oblanceolate, straight, the valves glabrous, glabrescent, or puberulent.

1. Leaves with 3–8 pairs of pinnae, each with 6–20 pairs of leaflets, the blades 3–7 mm…**var. humilis**
1'. Leaves with 1–3 pairs of pinnae, each with 5–8 pairs of leaflets, the blades 7–14 mm…**var. reticulata**

var. humilis—Leaves with 3–8 pairs of pinnae, leaflets 6–20 pairs per pinnule, the blades 3–7 mm, oblong. Juniper, piñon-juniper, oak, ponderosa pine, 5500–8000 ft. NC, C, WC, SW.

var. reticulata (A. Gray) L. D. Benson—Leaves with 1–3 pairs of pinnae, leaflets 5–8 pairs per pinnule, the blades 7–14 mm, 3–6 mm wide, oblong or ovate. Often under trees; piñon-juniper, oak, ponderosa pine, 5500–8000 ft. WC, SW, McKinley.

CARAGANA Fabr. – Pea tree

Caragana arborescens Lam.—Siberian peashrub. Shrubs (ours); stems unarmed or weakly prickly, the herbage glabrous, glabrescent, or puberulent; stipules bristlelike, persistent; leaves petiolate, the petioles 5–10 cm, leaflets 4–6 pairs per leaf, the blades 0.5–2.5 cm, elliptic, oblong, or obovate; inflorescences in fascicles of 2–4; flowers ascending, calyx tube 5–6 mm, corollas 1.5–2 cm, the petals yellowish; fruit ascending, spreading, or deflexed, 4–5.5 cm, the body oblong, the valves ± leathery, glabrous. Naturalized, 6000–7000 ft. Santa Fe. Introduced. In New Mexico, probably no more than a long-persistent waif that might be spreading from abandoned plantings.

CHAMAECRISTA (L.) Moench – Sensitive pea

Annual or perennial herbs, subshrubs, or shrubs; stems armed or unarmed, pubescence basifixed; stipules absent or herbaceous and persistent; leaves even-pinnate, pinnately compound, 1-foliate, or phyllodial, petiole glands present or absent, rachis glands absent, leaflets opposite or alternate; inflorescences axillary, racemose, pedunculate; flowers pedicellate, corollas zygomorphic, 5-merous or the petals absent; stamens in 2 series, 2-, 5-, or 10-merous, extrafloral nectaries present; fruits a legume, dehiscent, the body laterally compressed, the valves coiling and elastically dehiscent.

1. Longest petal 3–8 mm; pedicels 0.5–8 mm…**C. nictitans**
1'. Longest petal 10–20 mm; pedicels 8–45 mm…(2)
2. Leaves with 1–6 pairs of leaflets; calyx 3–6 mm; inflorescence 1- or 2-flowered…**C. serpens**
2'. Leaves with 7–26 pairs of leaflets; calyx 6–14 mm; inflorescence 1–6-flowered…**C. fasciculata**

Chamaecrista fasciculata (Michx.) Greene [*Cassia fasciculata* Michx. var. *rostrata* (Wooton & Standl.) B. L. Turner]—Partridge pea. Annuals, the herbage and stems glabrescent or hirsute; stipules subulate, persistent; leaves petiolate, petiole glands present, leaflets 7–26 pairs per pinna, the blades 0.5–2.5 cm; inflorescences axillary fascicles, 1–6-flowered; flowers pedicellate, calyx 6–14 mm, corollas 10–23 mm, the petals yellowish; fruit 2–8 × 0.3–0.7 cm wide, the body oblong, the valves papery or ± leathery, glabrous, glabrescent, or hirsute. Often grown as an ornamental; woodlands and fields, 3200–4600 ft. Quay.

Chamaecrista nictitans (L.) Moench [*C. nictitans* var. *leptadenia* (Greenm.) Gandhi & S. L. Hatch]—Sensitive partridge pea. Annuals, herbage glabrescent, puberulent, or pilose; stipules subulate, persistent; leaves petiolate, petiole glands present, leaflets 7–26 pairs per pinna, blades 0.5–1.5 cm; inflorescences axillary fascicles, 1- or 2-flowered; flowers pedicellate, calyx 3–6 mm, corollas 4–8 mm, the petals yellowish; fruit 2–5 × 0.3–0.5 cm, the body oblong, laterally compressed, the valves papery or ± leathery, coiling and elastically dehiscent, glabrous, glabrescent, or hirsute. New Mexico material is **var. mensalis** (Greenm.) H. S. Irwin & Barneby. Sandy or rocky sites, grasslands, roadsides, 4200–6600 ft. WC, SC, SW.

Chamaecrista serpens (L.) Greene [*C. wrightii* (A. Gray) Wooton & Standl.]—Slender sensitive pea. Perennial herbs, stems mat-forming, the herbage glabrescent or puberulent; stipules subulate, persistent; leaves petiolate, petiole glands present, leaflets 4–6 pairs per pinna, the blades 0.3–1.2 cm; inflorescences axillary fascicles, 1- or 2-flowered; flowers pedicellate, calyx 3–6 mm, corollas 10–20 mm, the petals yellowish; fruit 1.5–4.5 × 0.3–0.5 cm, the body oblong, the valves papery or ± leathery, glabrous, glabrescent, or puberulent. Our material belongs to **var. wrightii** (A. Gray) H. S. Irwin & Barneby. Sandy, rocky, or gravelly sites in a variety of habitats, occasionally in ruderal areas and roadsides, 3600–4300 ft. Doña Ana. This taxon is included in this treatment based on a single, historical collection from the Mexican Boundary Survey (*Parry 297* determined as *Cassia wrightii* A. Gray) and labeled "on the Rio Mimbres, New Mexico." Isely (1998) found no specimens for New Mexico, and it has apparently not been collected since.

COLOGANIA Kunth – Cologania

Perennial herbs; stems unarmed, pubescence basifixed; stipules inconspicuous, persistent; leaves odd-pinnate, trifoliate, or pinnately compound; inflorescences axillary, solitary or in pairs, the flowers chasmogamous or cleistogamous; flowers pedicellate, calyx with a distinct hypanthial base, the teeth 4, the apex acute, papilionaceous, petals lacking in cleistogamous flowers, androecium diadelphous, reduced to 1 or 2 stamens in cleistogamous flowers; fruits a legume, dehiscent, stipitate, the body straight or curved, ± constricted between the seeds, the valves thick-papery to leathery.

1. Leaflets 5-12 cm, 3-10 times longer than wide, < 1 cm wide, oblong or linear…**C. angustifolia**
1'. Leaflets 1-4 cm, 1-3 times longer than wide, often > 1 cm wide, elliptic…(2)
2. Petioles 1-5(-8) mm, the leaves nearly sessile, not prominently stalked; leaflets 1-2 times longer than wide, the apices rounded to obtuse…**C. obovata**
2'. Petioles 10+ mm, the leaves prominently stalked; leaflets 2-3 times longer than wide, the apices generally acute…**C. pallida**

Cologania angustifolia Kunth—Longleaf cologania. Perennial herbs, caudices ± woody in age, subterranean, the herbage strigose, with some retrorse hairs intermixed; stems ± twining, decumbent or ascending; leaves with 1-2 pairs of leaflets, the blades 5-12 cm, oblong or linear; inflorescences axillary, 1-3-flowered; calyx 9-11 mm, corollas 20-25 mm, the petals pinkish purple or pinkish white; fruit 3-5 cm, the body straight or curved, the valves puberulent. Hillsides, grasslands, washes, talus slopes, oak, juniper-piñon, ponderosa pine, 5575-8200 ft. WC, SW, Quay.

Cologania obovata Schltdl. [*C. humifusa* Hemsl.]—Lemmon's cologania. Plants thickly spreading-hairy; stems 20-50 cm, plants trailing; petioles 1-5(-8) mm; leaflets 3, broadly obovate to suborbicular, 1-2 times longer than wide, short-hairy abaxially; flowers 1-3, calyx ca. 10-12 mm, corollas rose-purple, 2.2-2.5 cm; fruit mostly straight, 3-4.5 cm, densely pubescent. Ponderosa pine forests; 1 record, 7150 ft. Grant.

Cologania pallida Rose—Pale cologania. Perennial herbs, herbage villous with some retrorse hairs intermixed; stems ± twining, decumbent or ascending; leaves with 1 pair of leaflets, the blades 2-4 cm, linear; inflorescences axillary, 1- or 2-flowered; calyx 9-11 mm, corollas 10-25 mm, the petals pinkish purple or purplish; fruit 3-5 cm, the body straight or curved, the valves villous. Juniper, piñon, ponderosa pine-oak communities, 4900-7000 ft. WC, Doña Ana.

COLUTEA L. – Bladder-senna

Colutea arborescens L.—Bladder-senna. Shrubs; stems unarmed, the herbage glabrous, glabrescent, or puberulent; leaves with 3-6 pairs of leaflets, the blades 1.4-2 cm, elliptic or ovate; inflorescences racemose, 2-7-flowered, calyx 6-7 mm, corollas 1.5-2 cm, the petals yellowish or yellowish and gold-tinged, sometimes with darker striations; fruit 5-7 × 2-3 cm, bladdery-inflated, stipitate, the body ovate, lunate, or subglobose, the valves thin-papery, ± translucent, glabrous, glabrescent, or puberulent, 3900-7900 ft. C, Curry. Introduced.

CROTALARIA L. – Rattlepod

Crotalaria pumila Ortega—Low rattlepod. Annual or perennial herbs, herbage glabrous, glabrescent, or puberulent; stems ascending, spreading, or prostrate; leaves with 1 pair of leaflets, ± palmate, blades 0.7-3.5 cm, oblong or obovate; inflorescences racemose, 4-10-flowered, calyx 3-5.5 mm, corollas 7-11 mm, the petals yellowish, keel conspicuously exserted; fruit 10-20 × 4-8 mm, bladdery-inflated, body elliptic, ovate, or subglobose, valves papery or leathery, puberulent. Desert grasslands, desert scrub, juniper, piñon, oak-ponderosa pine, 4250-7000 ft. WC, SW.

DALEA L. – Prairie clover

Annual or perennial herbs, subshrubs, shrubs, or treelike; stems unarmed, pubescence basifixed, glands aromatic; stipules herbaceous or reduced to small glands; leaves odd-pinnate or even-pinnate, pinnately compound, rarely trifoliate, petiolate; inflorescences racemose or spicate, pedunculate, bracts reduced to small glands resembling stipules; flowers pedicellate or sessile, calyx tube 5-toothed, corollas papilionaceous or not, androecium monadelphous, functionally 5-merous; fruits indehiscent, sessile.

1. Plants annuals, or perennials flowering the first year and appearing to be annuals…(2)
1'. Plants biennials, perennial herbs, subshrubs, shrubs, or treelike…(9)

2.   Inflorescence bracts persistent, disjoining from the inflorescence axis only with the fruiting axis in some taxa…(3)
2'.   Inflorescence bracts deciduous, sometimes tardily so, but fully detaching from the calyx and the inflorescence axis…(5)
3.   Androecium deeply cleft, the filaments free…**D. cylindriceps** (in part)
3'.   Androecium shallowly cleft, the filaments free for < 1 mm…(4)
4.   Stems angular-striate; leaflets 1–2 pairs…**D. exigua**
4'.   Stems rounded, not prominently angular; leaflets 2–4 pairs…**D. polygonoides**
5.   Fertile stamens 5; androecium deeply cleft, the filaments free…**D. scariosa**
5'.   Fertile stamens 10, or 5 and 2–4 stamens reduced or sterile; androecium shallowly cleft, the filaments free for < 1 mm…(6)
6.   Leaflets 4–14 pairs…(7)
6'.   Leaflets 1–4 pairs (up to 5 pairs in some forms of *D. brachystachya*)…(8)
7.   Anthers pale or dark bluish…**D. leporina**
7'.   Anthers yellowish…**D. urceolata**
8.   Leaf blades linear; valves of fruit glabrous or glabrescent…**D. filiformis**
8'.   Leaf blades oblanceolate; valves of fruit pilosulous and ± sparsely glandular-punctate…**D. brachystachys** (in part)
9.   Inflorescence racemose, the flowers pedicellate; calyx teeth plumose…(10)
9'.   Inflorescence spicate, the flowers sessile; calyx teeth not plumose…(12)
10.   Fruits 3–3.4 mm…**D. lachnostachys**
10'. Fruits 2.3–3 mm…(11)
11.   Corollas 4–6 mm; perennial herbs…**D. neomexicana**
11'. Corollas 6–11.5 mm; subshrubs…**D. versicolor** (in part)
12.   Inflorescence bracts persistent (falling only with the fruiting calyx)…(13)
12'. Inflorescence bracts deciduous (completely detaching from the axis and calyx in late anthesis in some species)…(22)
13.   Herbage glabrous, glabrescent, or glaucescent…(14)
13'. Herbage pilosulous, tomentulose, or villosulous…(20)
14.   Stems angled-striate…(15)
14'. Stems rounded, not prominently angled…(19)
15.   Petals papilionaceous, keel and wings not deciduous at anthesis…**D. bicolor** (in part)
15'. Petals subhomomorphic and not papilionaceous, keel and wings deciduous at anthesis…(16)
16.   Leaves with 2–3 pairs of leaflets…**D. candida**
16'. Leaves with 3–21 pairs of leaflets…(17)
17.   Leaves with 8–21 pairs of leaflets…**D. grayi**
17'. Leaves with 3–7 pairs of leaflets…(18)
18.   Stamens 5.3–7.7 mm, the filaments and anthers yellowish; fruits 2.5–3 mm, the valves pilosulous…**D. cylindriceps** (in part)
18'. Stamens 3–5.5 mm, the filaments purplish, the anthers yellowish orange; fruits 2–2.6 mm, the valves glabrescent or pilosulous…**D. purpurea**
19.   Corollas whitish; spikes loosely flowered; taproots yellowish…**D. enneandra**
19'. Corollas whitish and pinkish purple–tinged, pinkish purple, or bluish purple; spikes compact but not conelike, ± elongate in fruit; taproots orangish…**D. pogonathera**
20.   Leaflets 7–20 pairs; corollas whitish; androecium deeply cleft, filaments free for 3.5–4.5 mm…**D. albiflora**
20'. Leaflets 2–7 pairs; at least the wings and keel purple or violet, the banner whitish and rubescent in age in some forms; androecium shallowly cleft, filaments free for 1.7–2.9 mm…(21)
21.   Leaflets 2–4 pairs, the blades eglandular; reported but not definitively documented from New Mexico…D. greggii A. Gray
21'. Leaflets 4–7 pairs, the blades glandular-punctate…**D. lanata**
22.   Calyx teeth 4.5–10 mm…(23)
22'. Calyx teeth 0.5–4.2 mm…(27)
23.   Fruits 2.3–3 mm…(24)
23'. Fruits 3–4 mm…(25)
24.   Herbage glabrous or glabrescent; calyx tube 1.5–2 mm…**D. brachystachys** (in part)
24'. Herbage pilose or pilosulous; calyx tube 2.5–3.5 mm…**D. wrightii**
25.   Leaves with 1 pair of leaflets, the leaves palmately 3-foliate…**D. jamesii**
25'. Leaves with 1–5 pairs of leaflets, the leaves pinnately 3–7-foliate…(26)
26   Perennial herbs, the herbage pilose or pilosulous; calyx teeth 3.5–5 mm…**D. aurea**
26'. Subshrub, the herbage glabrous or glabrescent; calyx teeth 4.5–8.5 mm…**D. formosa**
27.   Fruits 2–2.5 mm…(28)

27'. Fruits 2.5–3.5 mm…(29)
28. Calyx teeth 0.5–2 mm; anthers yellowish…**D. bicolor** (in part)
28'. Calyx teeth 2–4.2 mm; anthers whitish…**D. pulchra**
29. Calyx teeth 2.2–4.2 mm…**D. nana**
29'. Calyx teeth 0.8–2 mm…(30)
30. Petals papilionaceous…(31)
30'. Petals subhomomorphic, not papilionaceous…(32)
31. Calyx teeth 2–4 mm; corollas 6–11.5 mm; herbage glabrous, glabrescent, or pilosulous…**D. versicolor** (in part)
31'. Calyx teeth 0.5–1.2 mm; corollas 5–8 mm; herbage glabrous…**D. frutescens**
32. Leaves and stems variously hairy below the inflorescences…**D. villosa**
32'. Leaves and stems glabrous below the inflorescences…(33)
33. Spikes becoming loose, the flowers separated and the rachis visible; calyx teeth as long as or longer than the calyx tube…**D. tenuifolia**
33'. Spikes permanently very dense, the rachis never visible; calyx teeth shorter than the calyx tube…**D. compacta**

Dalea albiflora A. Gray—White flowered prairie clover. Perennial herbs or subshrubs; herbage villosulous, sometimes intermixed with retrorse hairs; leaves with 7–20 pairs of leaflets, the blades 1.5–10 mm, oblanceolate, oblong, obovate, or elliptic, glandular-punctate; inflorescences spicate; calyx tube 2–3.3 mm, deeply cleft behind the banner, the ribs yellowish or brownish, the intervals with 1 row of 3–6 yellow glands; corollas 3.4–6.4 mm, whitish; fruits 2.2–2.8 mm, valves glabrous, glabrescent, or villosulous, glandular-punctate. Grasslands, piñon-juniper, ponderosa pine, 4550–8900 ft. WC, SW.

Dalea aurea Nutt. ex Pursh—Golden prairie clover. Perennial herbs, herbage pilose or pilosulous; leaves with 1–3 pairs of leaflets, blades 3–20 mm, obovate, oblong, or oblanceolate; inflorescences spicate, conelike; calyx tube 2.2–2.8 mm, the ribs stramineous, the intervals with transparent glands, corollas 3–8.5 mm, yellowish or ochroleucous; fruits 3–3.5 mm, the valves glabrous or glabrescent and puberulent at the apex. Sandy soil, shortgrass prairie, 3000–7250 ft. NE, C, EC, SC, Eddy, Sierra.

Dalea bicolor Humb. & Bonpl. ex Willd.—Silver prairie clover. Shrubs or small trees, often thicketforming, suffrutescent perennial herbs, stems many and dense, herbage pilosulous or tomentulose; leaves with 1–11 pairs of leaflets, blades 1.5–9.5 mm, obovate, oblong, elliptic, dorsally glandular-punctate; inflorescences spicate, conelike; calyx tube 2.6–3.2 mm, the ribs thickened, the intervals glandular, corollas 3.2–8 mm, whitish, ochroleucous, pinkish purple to bluish purple; fruits 2.2–2.7 mm, the valves glabrous, glabrescent, or pilosulous, glandular-punctate. New Mexico material belongs to **var. argyrea** (A. Gray) Barneby [*D. argyrea* A. Gray]. Limestone, hills and canyons, Chihuahuan desert scrub, 2950–5250 ft. SE, Lincoln.

Dalea brachystachys A. Gray—Fort Bowie prairie clover. Annual or short-lived perennial herb; stems glandular-punctate, glaucescent, herbage glabrous, glabrescent; leaves with 1–5 pairs of leaflets, bicolored, blades 2–16 mm, oblanceolate to narrowly oblanceolate, glabrous above, glandular-punctate below; inflorescences spicate; calyx tube 1.5–2 mm, the ribs brownish, the intervals eglandular or glandular, corollas 2–6 mm, whitish, ochroleucous, or pinkish brown; fruits 2.3–2.6 mm, the valves pilosulous, sometimes obscurely glandular-punctate. Desert grasslands, prairies, desert scrub communities, 4250–7250 ft. WC, SC, SW, Eddy.

Dalea candida Willd. [*D. occidentalis* (Rydb.) L. Riley; *Petalostemon candidus* (Willd.) Michx. var. *oligophyllus* (Torr.) F. J. Herm.]—White prairie clover. Perennial herb; stems eglandular or glandular-punctate, herbage glabrous or glabrescent; leaves with 2–3 pairs of leaflets, blades 0.6–2.4 cm, obovate, oblong, lanceolate, or elliptic; inflorescences spicate, conelike or amentlike, loosely flowered; calyx tube 1.9–2.7 mm, corollas 3.2–5.7 mm, whitish; fruits 2.6–4.5 mm, the valves glabrous, glabrescent, or pilosulous, glandular-punctate, the glands orangish and blisterlike. New Mexico material belongs to **var. oligophylla** (Torr.) Shinners. Common in rocky or sandy soils, grasslands, desert scrub, piñon-juniper woodland, ponderosa pine, often along roadsides, 4550–9200 ft. Widespread. *Dalea candida* var. *candida* has yet to be documented in New Mexico.

Dalea compacta Spreng. [*Petalostemon compactus* (Spreng.) Swezey]—Compact prairie clover. Perennial with woody taproot; stems several, 4–8 dm, glabrous; leaves numerous, 5–8 cm, leaflets 5–7, oblong to oblong-lanceolate or linear-oblong, short-petiolate, pale green above, gland-dotted and paler beneath, 15–25 mm; spikes cylindrical, dense; calyx 4.5 mm, densely silky-villous with brown hairs, corollas white, blade of the banner 2 mm, claw 1 mm, blades of the other petals 2 mm; pods obovate, 2 mm, pubescent. New Mexico material belongs to **var. pubescens** (A. Gray) Barneby. One record from Socorro County.

Dalea cylindriceps Barneby [*Petalostemon macrostachyus* Torr.]—Andean prairie clover. Perennial herb; stems glandular-punctate, herbage glabrous or glabrescent; leaves with 3–4 pairs of leaflets, blades 12–25 mm, oblanceolate, oblong, or elliptic; inflorescences spicate, conelike; calyx tube 1.9–2.3 mm, the intervals with 1 row of 2–5 pale glands, corollas 2–6.2 mm, whitish or pinkish purple; fruits 2.5–3 mm, the valves pilosulous, glandular-punctate. Sandy sites, dunes, prairies, along rivers or intermittent streams, 4250–7000 ft. NE, NC, NW, C, Doña Ana, Roosevelt.

Dalea enneandra Nutt.—Nine-anther prairie clover. Perennial herb; stems glandular-punctate, ± glaucous, herbage glabrous or glabrescent; leaves with 2–6 pairs of leaflets, the blades 4–12 mm, oblanceolate or elliptic, glabrous or glabrescent, glandular-punctate below; inflorescences spicate, loosely flowered, glands orange or bluish gray and blisterlike; calyx tube 3–3.7 mm, corollas 2.8–7 mm, whitish; fruits 3–3.7 mm, the valves glabrous or glabrescent and puberulent at the apex. Sandy or sandy-clay sites, dunes, prairies, badlands, roadsides, old fields, 3600–5900 ft. NE, EC, C, Grant, Sierra.

Dalea exigua Barneby [*Petalostemon exilis* A. Gray]—Chihuahuan prairie clover. Annual; stems eglandular or glandular-punctate, herbage glabrous; leaves with 1–2 pairs of leaflets, blades 5–40 mm, linear, oblanceolate, elliptic, or oblong, ± bicolored, glandular-punctate below; inflorescences spicate; calyx tube 1.9–2.2 mm, the ribs eglandular, corollas 1.7–4.3 mm, pinkish purple; fruits 2–3 mm, the valves pilosulous. Flats, grasslands, woodlands, commonly associated with pine and oak, 5900–8000 ft. Catron, Grant.

Dalea filiformis A. Gray—Sonoran prairie clover. Annual; stems angled-striate, herbage glabrous or glabrescent; leaves with 1–2 pairs of leaflets, the blades 5–23 mm, linear, glabrous or glabrescent, glandular-punctate below; inflorescences spicate, few-flowered; calyx tube 1.3–2 mm, the ribs filiform, eglandular or glandular, corollas 2.4–3.5 mm, whitish, ochroleucous, pinkish purple, or reddish purple; fruits 2–2.4 mm, the valves glabrous or glabrescent. Sandy soils in grasslands and woodlands, oak, pine, juniper, 5250–8900 ft. WC, SW.

Dalea formosa Torr.—Featherplume. Subshrub, caudices woody, bark papery, exfoliating, fissured in age, herbage glabrous or glabrescent; leaves with 2–7 pairs of leaflets, the blades 0.6–7 mm, obovate or oblanceolate, glabrous above, glandular-punctate below; inflorescences spicate, dense and headlike, 3–9-flowered; calyx tube 3–5 mm, the ribs prominent, glandular, the glands orangish or reddish, blisterlike, corollas 6.5–12 mm, whitish, ochroleucous, pinkish purple, or reddish purple; fruits 3–3.5 mm, the valves glabrous or glabrescent. Grasslands, prairies, desert scrub, piñon-juniper woodland communities, 3250–6600 ft. Widespread except NW, Colfax, Taos.

Dalea frutescens A. Gray—Black prairie clover. Shrubs, forming small thickets; stems glandular-punctate, the epidermis pale brown-furrowed in age, herbage glabrous; leaves with 4–10 pairs of leaflets, blades 1.5–5 mm, obovate or oblanceolate, glandular-punctate below; inflorescences spicate, headlike; calyx tube 2.6–3.5 mm, ribs prominent, glands red or orange and blisterlike, corollas 5–8 mm, whitish or pinkish purple; fruits 2.8–3.5 mm, the valves glabrous, glabrescent, or pilosulous, glandular-punctate or eglandular. Grasslands, mesquite, desert scrub, piñon-juniper, ponderosa pine, mixed conifer communities, 4600–8550 ft. C, SC, Curry.

Dalea grayi (Vail) L. O. Williams—Gray's prairie clover. Perennial herbs or subshrubs; stems angled-striate, eglandular or glandular-punctate, herbage glabrous or glabrescent; leaves with 8–21 pairs of leaflets, blades 1–5.5 mm, elliptic, oblanceolate, or obovate; inflorescences spicate; calyx tube 2.0–2.7 mm, ribs orangish or brownish, the intervals with 1 row of 3–6 small, pale yellow glands, corollas 2–5.7 mm, whitish to greenish; fruits 2.2–2.5 mm, the valves villosulous and finely glandular-punctate. Uncommon; slopes, flats, grasslands, foothills, oak communities, 4250–5900 ft. Grant, Hidalgo.

Dalea jamesii (Torr.) Torr. & A. Gray—James' prairie clover. Perennial herbs; stems tuft-forming, herbage pilose or pilosulous; leaves with 1 pair of leaflets, palmately 3-foliate, blades 5–18 mm, obovate or oblanceolate; inflorescences spicate; calyx tube 2.4–3.5 mm, ribs brownish red, glands transparent, corollas 5.2–8.6 mm, yellowish or ochroleucous, orangish brown in age; fruits 3.5–4 mm, the valves glabrous or glabrescent and puberulent at the apex. Sandy and rocky sites, canyons, washes, hillsides, grasslands, piñon-juniper woodlands, ponderosa pine–oak communities, 3600–7250 ft. Widespread except NW, Cibola.

Dalea lachnostachys A. Gray—Gland-leaf prairie clover. Perennial herbs; stems erect, glandular-punctate, herbage pilose or pilosulous; leaves with 2–5 pairs of leaflets, blades 5–17 mm, obovate or oblong, glandular-punctate; inflorescences racemose; calyx tube 2.6–3.1 mm, ribs prominent, the intervals with 1 row of small yellowish glands, corollas 5–7.6 mm, bluish purple; fruits 3–3.4 mm, the valves pilosulous. Hills, grasslands, desert scrub communities, 4250–5250 ft. SW.

Dalea lanata Spreng.—Woolly prairie clover. Perennial herb; stems glandular-punctate or eglandular, herbage villosulous or pilosulous; leaves with 4–7 pairs of leaflets, blades 3–12 mm, obovate or oblanceolate, glandular-punctate; inflorescences spicate; calyx tube 2.2–2.5 mm, ribs ± prominent, the intervals glandular, corollas 2.8–4.3 mm, reddish purple; fruits 2.5–3.1 mm, the valves pilosulous, glandular-punctate.

1. Calyx tube pilosulous or villosulous, the teeth lanceolate, pilosulous or villosulous…**var. lanata**
1'. Calyx tube glabrous or glabrescent, the teeth deltate or ovate, villosulous…**var. terminalis**

var. lanata—Calyx tube villosulous or pilosulous, the teeth lanceolate, villosulous or pilosulous. Dunes, sandy alluvium, drift-sand plains, 2950–6250 ft. E, SC.

var. terminalis (M. E. Jones) Barneby [*D. glaberrima* S. Watson; *D. terminalis* M. E. Jones]—Calyx tube glabrous or glabrescent, the teeth deltate or ovate, villosulous. Dunes, sandy alluvium, drift-sand plains, 3600–6600 ft. NC, NW, C, SC.

Dalea leporina (Aiton) Bullock [*D. alopecuroides* Willd.]—Foxtail prairie clover. Annual; stems glandular-punctate, with a few striations distally, herbage glabrous; leaves with 4–12 pairs of leaflets, the blades 2–12 mm, oblanceolate, oblong, or obovate; inflorescences spicate; calyx tube 1.7–2.8 mm, ribs filiform, intervals glandular, the glands small, orange, blisterlike, corollas 1.6–6 mm, whitish, ochroleucous, or bluish purple; fruits 2.4–3 mm, tardily dehiscent through the sutures, the valves pilosulous, glandular-punctate or eglandular. A weedy species on disturbed soils in grasslands, desert scrub, piñon-juniper woodlands, oak-ponderosa pine communities, 3900–8200 ft. NC, NW, C, WC, SC, SW.

Dalea nana Torr. ex A. Gray—Dwarf prairie clover. Perennial herb; stems prostrate, angled-striate, eglandular or glandular-punctate, herbage pilose to pilosulous; leaves with 2–3 pairs of leaflets, the blades 3–15 mm, elliptic, oblanceolate, obovate, or oblong, glabrescent or pilose, glandular-punctate below; inflorescences spicate, conelike; calyx tube 2–2.7 mm, rib intervals glandular, the glands transparent, corollas 2.6–5.5 mm, yellowish, brownish, pinkish purple–tinged in age; fruits 2.5–3 mm, the valves pilose or pilosulous.

1. Stems prostrate; inflorescence bracts 2–4 mm wide, the apices acute…**var. nana**
1'. Stems ascending or erect; inflorescence bracts 1–2 mm wide, the apices acuminate…**var. carnescens**

var. carnescens (Rydb.) Kearney & Peebles [*D. rubescens* S. Watson; *D. carnescens* (Rydb.) Bullock]— Stems ascending or erect; inflorescences 10–15 mm wide, bracts ovate or lanceolate, 1–2 mm wide, the apices acuminate. Hillsides and plains, often on gypsum and limestone; desert grasslands, desert scrub, 3250–5900 ft. SE, SC, SW.

var. nana—Stems prostrate; inflorescence bracts ovate or elliptic, 2–4 mm wide, tardily deciduous, the apices acute. Sandy sites, plains, dunes, desert grasslands, mesquite, sometimes becoming weedy in disturbed areas, 3600–5600 ft. NE, EC, C, WC, SE, SC, SW.

Dalea neomexicana (A. Gray) Cory—New Mexico prairie clover. Perennial herbs; stems tuft- or mat-forming, prostrate or ascending, tuberculate-glandular, herbage pilosulous or pilose; leaves with 3–7 pairs of leaflets, the blades 2.5–8 mm, obcordate, glandular, the surface with rows of large glands parallel to the margins and midrib, the dorsal surface with a single large gland; inflorescences racemose; calyx tube 2.1–3.1 mm, ribs glandular, the glands orangish, transparent or not, corollas 4–6 mm, whitish and reddish purple, pinkish purple, or yellow-tinged; fruits 2.3–3 mm, the valves pilose, eglandular. New Mexico material belongs to **var. neomexicana**. Hills, knolls, washes, grasslands, creosote, juniper, 4250–5575 ft. C, Chaves, Luna, Sierra.

Dalea pogonathera A. Gray—Bearded prairie clover. Perennial herbs; stems glandular-punctate, herbage glabrous or glabrescent; leaves with 1–4 pairs of leaflets, blades 1.5–10 mm, oblanceolate, oblong, or elliptic, glandular-punctate below; inflorescences spicate, headlike; calyx tube 2.6–3.4 mm, ribs filiform, the intervals glandular, the glands pale green or light brown, corollas 2.5–7.8 mm, whitish and pinkish purple-tinged, pinkish purple, or bluish purple; fruits 2.8–3.5 mm, the valves glabrous, glabrescent, or pilosulous. Our material belongs to **var. pogonathera**. Hills, plains, creosote, mesquite, 3250–5575 ft. S except Hidalgo, Lea, Luna.

Dalea polygonoides A. Gray—Six-weeks prairie clover. Annual; stems glandular-punctate, herbage glabrous (excluding the inflorescence); leaves with 2–4 pairs of leaflets, blades 5–15 mm, oblong, oblanceolate, glandular-punctate below; inflorescences spicate, 5.5–9 mm wide; calyx tube 1.6–2.1 mm, ± inflated in fruit, rib intervals with 1–2 large orange glands, corollas 2–3.8 mm, pale reddish purple; fruits 2.5–3 mm, the valves pilosulous, eglandular. Grassland and pine forest communities, 4900–8550 ft. Widespread except N, E.

Dalea pulchra Gentry—Indigo bush. Subshrubs or suffrutescent perennial herbs; stems glandular-punctate, herbage pilosulous or tomentulose; leaves with 2–4 pairs of leaflets, blades 1.5–5 mm, obovate or oblanceolate; inflorescences spicate, headlike; calyx tube 2.4–3.2 mm, ribs glandular or not, the glands transparent, corollas 6–10.2 mm, whitish, ochroleucous, or pinkish purple; fruits 2–2.5 mm, the valves glabrous, glabrescent, or pilosulous. Uncommon; desert grasslands, oak, piñon-juniper communities, 4250–4600 ft. Hidalgo.

Dalea purpurea Vent. [*Petalostemon purpureus* (Vent.) Rydb.]—Purple prairie clover. Perennial herbs; stems angled-striate, tuberculate-glandular or eglandular, herbage glabrous, glabrescent, puberulent, pilose, or tomentulose; leaves with 3–7 pairs of leaflets, blades 7–22 mm, linear, oblanceolate, or elliptic, glandular-punctate below; inflorescences spicate, conelike; calyx tube 1.7–2.9 mm, ribs not prominent, intervals heavily brownish red–flecked, eglandular, corollas 3–7.2 mm, reddish purple, pinkish purple, bluish purple, or exceptionally whitish, eglandular; fruit 2.1–2.6 mm, the valves glabrescent or pilosulous, glandular-punctate.

1. Inflorescence spikes 7–8.5 mm wide, peduncles 8–15 cm; herbage glabrous…**var. arenicola**
1'. Inflorescence spikes 9.5–13 mm wide, peduncles 0–8 cm; herbage glabrescent, puberulent, pilose, or tomentulose, rarely glabrous…**var. purpurea**

var. arenicola (Wemple) Barneby [*D. arenicola* (Wemple) B. L. Turner]—Herbage and stems glabrous or glabrescent; inflorescence spikes 7–8.5 mm wide, peduncles 8–15 cm. Sandy sites on bluffs, dunes, flats, stream banks, intermittent streams, dry grasslands, 3900–7000 ft. NE, EC.

var. purpurea—Herbage and stems glabrous, glabrescent, pilose, puberulent, or tomentulose; inflorescence spikes 9.5–13 mm wide, peduncles 0–8 cm. Rocky or gravelly sites (often on sedimentary substrates; petals sometimes more vividly colored on limestone substrates) on bluffs, dunes, woodlands, stream banks, lakeshores, prairies, 4550–8200 ft. NE, NC, EC, C, SE, SW, San Juan.

Dalea scariosa S. Watson [*Petalostemon scariosus* (S. Watson) Wemple]—La Jolla prairie clover.

Annual; stems decumbent, angular-striate, herbage glabrous; leaves with 2–4 pairs of leaflets, blades 3–8 mm, obovate; inflorescences spicate, cylindrical or catkinlike; calyx tube 3–3.8 mm, recessed behind the banner, rib glands golden or reddish and blisterlike, in 1 row of 3–5 or the ventral pair with 2–3 rows of several glands, corollas 4–8 mm, bluish purple or reddish purple; fruit 3–4 mm, the valves glabrous distally except for short beard along the suture, glandular-punctate. Endemic; sandy or gravelly sites, washes, gullies, bluffs, roadsides, 4750–4900 ft. Bernalillo, Sandoval, Socorro, Valencia. **E**

Dalea tenuifolia (A. Gray) Shinners [*Petalostemon tenuifolius* A. Gray]—Slim-leaf prairie clover. Perennial herb; stems angular-striate, herbage glabrous, glabrescent, or pilosulous; leaves with 3–5 pairs of leaflets, blades 1–2.2 cm, linear or oblanceolate, glandular-punctate; inflorescences spicate; calyx tube 1.6–2.4 mm, ribs prominent, the intervals flat, densely brownish red–flecked, eglandular, corollas 4–6.5 mm, reddish purple; fruits 2.8–3.5 mm, the valves pilosulous, glandular-punctate. Limestone, hilltops, bluffs, badlands, prairies, arid grasslands, 3600–8200 ft. NE, EC, C.

Dalea urceolata Greene—Pine-forest prairie clover. Annual; stems with a few small prominent glands and striations distally, herbage glabrous; leaves with 4–14 pairs of leaflets, the blades 2.2–9.5 mm, oblong or oblanceolate; inflorescences spicate; calyx tube 2.4–4.5 mm, ± inflated in fruit, ribs subfiliform, intervals glandular, the glands oblong-elliptic, orange, corollas 1.4–5.5 mm, pale bluish purple or ochroleucous, eglandular; fruits 2–2.5 mm, tardily dehiscent along the sutures, the valves glandular-punctate. New Mexico material belongs to **var. urceolata**. Ponderosa pine–oak forests, 5575–8200 ft. Catron, Grant, Lincoln.

Dalea versicolor Zucc. [*D. sessilis* (A. Gray) Tidestr.; *D. wislizeni* A. Gray var. *sessilis* A. Gray]—Oakwoods prairie clover. Perennial herbs or shrubs, caudices woody; stems glandular-punctate, herbage glabrous, glabrescent, or pilosulous; leaves with 3–9 pairs of leaflets, blades 1–9 mm, oblanceolate, elliptic, or obovate, eglandular above or glandular-punctate throughout; inflorescences spicate or racemose, headlike; calyx tube 2–3 mm, rib intervals glandular, the glands orange or black, corollas 6–11.5 mm, whitish, ochroleucous, or pinkish purple; fruits 2.4–3 mm, the valves pilosulous, sparsely glandular, the glands brownish red or blackish. Our material belongs to **var. sessilis** (A. Gray) Barneby. Hills and canyons, grasslands and woodlands, 4250–7000 ft. Hidalgo.

Dalea villosa (Nutt.) Spreng. [*Petalostemon villosus* Nutt.]—Silky prairie clover. Perennial herb; herbage villosulous or tomentulose; leaves with 4–10 pairs of leaflets, blades 10–11 mm, elliptic or oblanceolate; inflorescences spicate; calyx tube 1.9–2.7 mm, ribs prominent, corollas 2.6–5.6 mm, reddish purple, pinkish purple, or rarely whitish, eglandular; fruits 2.5–3.2 mm, the valves villosulous, glandular-punctate or eglandular. New Mexico material belongs to **var. villosa**. Sandy sites, dunes, talus, bluffs, intermittent streams, prairies, woodlands, 3900–5250 ft. E except Curry, Eddy, Lea.

Dalea wrightii A. Gray—Wright's prairie clover. Perennial herb; herbage pilose or pilosulous; leaves with 1–2 pairs of leaflets, the blades 4–20 mm; inflorescences spicate, dense; calyx tube 2.5–3.5 mm, the rib intervals glandular, the glands transparent, corollas 3–9.5 mm, yellowish or ochroleucous, pinkish or orange-brown in age, eglandular; fruits 2.8–3.5 mm, valves glabrous or glabrescent throughout and ± puberulent apically. Limestone, foothills, desert grasslands, desert scrub, juniper communities, 3400–6250 ft. S, Chaves, Socorro.

DERMATOPHYLLUM Scheele – Mescal bean

Shrubs or trees; stems unarmed, pubescence basifixed; stipules absent or inconspicuous; leaves odd-pinnate, pinnately compound, leaflets 1 to many pairs per leaf, opposite, coriaceous and ± evergreen; inflorescences racemose or paniculate; inflorescence and floral bracts minute or absent, deciduous; petals free, calyx with a distinct hypanthial base, stamens 10, free; fruit a legume, indehiscent, persistent, stipitate; seeds red or yellow.

1. Calyx 9–11 mm; fruits 5–10 cm, laterally compressed in cross-section…**D. gypsophilum**
1'. Calyx 5–9 mm; fruits 4–5 cm, ± rounded in cross-section, not compressed…**D. secundiflorum**

Dermatophyllum gypsophilum (B. L. Turner & A. M. Powell) Vincent [*D. guadalupense* (B. L. Turner & A. M. Powell) B. L. Turner; *Sophora gypsophila* B. L. Turner & A. M. Powell]—Guadalupe mescal bean. Shrubs, the herbage glabrescent, puberulent, or strigose, ± silvery when young; leaves petiolate, leaflets 4-6 pairs, the blades 10-20 mm, elliptic or obovate; inflorescences racemose, compact, termi-nal, 2-8-flowered, pedicels inconspicuous; calyx 9-11 mm, corollas 20-25 mm, purplish or pale pinkish purple; fruit 5-10 × 0.8-1.5 cm, the stipe 4-5 mm, the body straight or curved, laterally compressed in cross-section, irregularly constricted between the seeds, the valves thick-papery to leathery, puberu-lent; seeds 12-20. New Mexico material belongs to **subsp. guadalupense** (B. L. Turner & A. M. Powell) Vincent. Barren clay or gravelly sites on hillsides derived from limestone or gypsum substrates, 5260-6650 ft. Eddy, Otero. **R**

Dermatophyllum secundiflorum (Ortega) Gandhi & Reveal—Texas mountain laurel. Shrubs or small trees, herbage glabrescent or puberulent; leaves petiolate, leaflets 2-5 pairs, blades 25-80 mm, elliptic or obovate; inflorescences racemose, compact, terminal, 5-15-flowered; calyx 8-11 mm, corollas 14-16 mm, whitish, purplish or pale bluish purple in age; fruit 2-10 × 1-2 cm, body straight, ± rounded in cross-section, irregularly constricted between the seeds, the valves thick-papery to leathery, glabres-cent and woody in age. Dunes, washes, 2950-5900 ft. Eddy.

DESMANTHUS Willd. - Bundle-flower

Perennial herbs or subshrubs; stems unarmed, pubescence basifixed; stipules herbaceous; leaves even-pinnate, bipinnately compound; inflorescences axillary, solitary or paired, compact and headlike, pe-dunculate; flowers pedicellate, corollas actinomorphic, tubular-funnelform only at the base or distinct throughout; fruit a legume, dehiscent, sessile, the body laterally compressed, the valves glabrous, ± slightly constricted between the seeds.

1. Leaves with 7-10 pairs of pinnae; fruits oblong, the body 15-25 × 5-6 mm…**D. illinoensis**
1'. Leaves with 2-7 pairs of pinnae (up to 8 pairs in robust individuals of *D. velutinus* and *D. leptolobus*); fruits linear, the body 25-90 × 2-4 mm…(2)
2. Inflorescence with < 10 flowers, the heads 0.5-1 cm diam.…(3)
2'. Inflorescence with > 10 flowers, the heads 1-2 cm diam.…(4)
3. Leaflets 1.5-2.5 mm; peduncles of mature inflorescences 0.5-1 cm…**D. leptolobus**
3'. Leaflets 4-5 mm; peduncles of mature inflorescences 1-4 cm…**D. glandulosus**
4. Fruits 2-3 mm wide; corollas 2-3 mm…**D. obtusus**
4'. Fruits 3-4 mm wide; corollas 3-4 mm…(5)
5. Leaflets villous, glabrescent in age; peduncles of mature inflorescences 4-8 cm…**D. velutinus**
5'. Leaflets glabrous or ciliate; peduncles of mature inflorescences 1-4 cm…**D. cooleyi**

Desmanthus cooleyi (Eaton) Branner & Coville—Cooley's bundleflower. Perennial herbs or sub-shrubs; leaves with 2-6 pairs of pinnae per leaf, petiole glands present, leaflets 2-6 pairs per pinna, blades 2-4 mm, elliptic or oblong, glabrous, glabrescent, or ciliate; inflorescences compact and headlike, 1.5-2 cm wide, 10-20-flowered; corollas 2.5-4 mm, whitish; fruit 4-7 × 0.3-0.4 cm wide, the body linear, straight, falcate, or curved. Foothills, slopes, washes, stream banks, occasionally ruderal areas and road-sides; grasslands, desert scrub, piñon-juniper communities, 3250-7000 ft. Widespread except NC, C.

Desmanthus glandulosus (B. L. Turner) Luckow [*D. virgatus* (L.) Willd. var. *glandulosus* B. L. Turner]—Glandular bundleflower. Perennial herbs or subshrubs; leaves with 2-6 pairs of pinnae per leaf, petiole glands present, leaflets 6-15 pairs per pinna, blades 2-5 mm, elliptic or oblong, glabrous; in-florescences compact and headlike, 4-5 cm wide, 4-10-flowered; corollas 2-3 mm, greenish or whitish; fruit 5-9 × 0.3-0.4 cm, the body linear, straight or curved. Limestone, hillsides, slopes, washes, ruderal areas, roadsides, desert scrub, piñon-juniper communities, 3250-6550 ft. Doña Ana, Eddy.

Desmanthus illinoensis (Michx.) MacMill. ex B. L. Rob. & Fernald—Illinois bundleflower. Perennial herbs; leaves with 2-6 pairs of pinnae per leaf, petiole glands present, leaflets 15-30 pairs per pinna, blades 2-3 mm, elliptic or oblong, glabrous, glabrescent, or puberulent; inflorescences compact and headlike, 1-1.5 cm wide; corollas 1-2 mm, whitish; fruit 1.5-2.5 × 0.5-0.6 cm, the body oblong, falcate or curved. Stream banks, bottomlands, roadsides, grasslands, desert scrub, piñon-juniper communities, 2950-7000 ft. Widely scattered. NC, C, SE, Doña Ana, Luna.

Desmanthus leptolobus Torr. & A. Gray—Prairie bundleflower. Perennial herbs; leaves with 4-8 pairs of pinnae per leaf, petiole glands present, leaflets 10-20 pairs per pinna, blades 1.5-2.5 mm, elliptic or oblong, glabrescent or villous; inflorescences compact and headlike, 0.5-1 cm wide, 4-8-flowered; corollas 2.5-4 mm, whitish; fruit 3.5-6 × 0.2-0.3 cm, the body linear, straight or curved. Naturalized. Apparently collected only a single time in New Mexico in 1983, at Granite Gap at 4500 ft., possibly no more than just a waif. Hidalgo. Introduced.

Desmanthus obtusus S. Watson—Blunt-pod bundleflower. Perennial herbs or subshrubs; leaves with 2-5 pairs of pinnae per leaf, petiole glands present, leaflets 8-15 pairs per pinna, blades 2-4 mm, elliptic or oblong, glabrescent or velutinous; inflorescences compact and headlike, 0.5-1 cm wide, 8-10-flowered; corollas 2-3 mm, ochroleucous or whitish; fruit 2.5-4 × 0.2-0.3 cm wide, the body linear, straight or curved. Limestone, hillsides, slopes, flats, occasionally along roadsides; desert grasslands, desert scrub, 3250-4000 ft. Eddy.

Desmanthus velutinus Scheele—Velvet bundleflower. Perennial herb; leaves with 3-8 pairs of pinnae per leaf, petiole glands present, leaflets 6-14 pairs per pinnae, blades 2-4 mm, elliptic or oblong, glabrescent or villous; inflorescences compact and headlike, 1-1.5 cm wide, 10-15-flowered; corollas 3-4 mm, whitish; fruit 4-6 × 0.2-0.3 cm, the body linear, straight or curved. Limestone and caliche cuestas, hillsides, slopes, flats, roadsides, grasslands, desert scrub, oak-juniper-piñon communities, 3250-6250 ft. Eddy, Lincoln.

DESMODIUM Desv. – Tick-trefoil

Annual or perennial herbs; stems unarmed, pubescence basifixed; stipules herbaceous, deciduous or persistent; leaves pinnately compound, leaflets unifoliate or 1 pair, opposite; inflorescences pseudoracemose, terminal or axillary, with 2 to many clusters, each with 2 to many flowers at each node; flowers papilionaceous, androecium diadelphous or monadelphous, stamens monomorphic; fruit a loment, dehiscent, stipitate or subsessile, the body laterally compressed; seeds 1 per segment.

1. Calyx 1.5-4 mm; corollas 4-9 mm…(2)
1'. Calyx 0.5-1.5 mm; corollas 2-3.5 mm…(5)
2. Fruit articles 3-5 mm…(3)
2'. Fruit articles 5-10 mm…(4)
3. Corollas 5-6 mm; fruit subsessile; stems erect or ascending…**D. arizonicum**
3'. Corollas 7-9 mm; fruit stipitate; stems decumbent…**D. batocaulon**
4. Corollas 6-8 mm; calyx 2.5-3.5 mm…**D. grahamii**
4'. Corollas 4-5 mm; calyx 1.5-2 mm…**D. psilocarpum**
5. Perennial herbs; calyx 1.5-2.5 mm; corollas 4.5-5.5 mm…**D. metcalfei**
5'. Annual or short-lived perennial herbs; calyx 1-1.5 mm; corollas 2.5-3.5 mm…(6)
6. Fruit stipitate, the stipe 1.5-3.5 mm…**D. procumbens** (in part)
6'. Fruit sessile or short-stipitate, the stipe 0-1.5 mm…(7)
7. Corollas 2-3 mm; fruit glabrous or uncinate-pubescent; leaves elliptic or oblong…**D. procumbens** (in part)
7'. Corollas 3-3.5 mm; fruit glabrous; leaves linear or narrowly lanceolate…**D. rosei**

Desmodium arizonicum S. Watson—Arizona tick-trefoil. Perennial herbs; stems erect or ascending, herbage glabrous, glabrescent, or puberulent; leaves 4-6 cm, oblong or linear, subsessile or with petioles 0.1-5 mm; inflorescences simple, pedicels 8-15 mm; calyx 2.5-3.5 mm, corollas 5-6 mm, bluish purple; fruit subsessile, uncinate, the body with 3-5 articles, the articles 4-5 mm wide. Canyons and occasionally ruderal areas and roadsides; oak, juniper, pine, 7200-8200 ft. Catron, Grant.

Desmodium batocaulon A. Gray—San Pedro tick-trefoil. Perennial herbs; stems decumbent, herbage glabrous, glabrescent, or puberulent; leaves 2-6 cm, lanceolate, ovate, or elliptic, petioles > 5 mm; inflorescences axillary or terminal, pedicels 5-10 mm; calyx 2.5-3.5 mm, corollas 7-9 mm, pinkish purple or bluish purple; fruit uncinate or glabrescent, the stipe 1-2 mm, the body with 4-7 articles, the articles

3-5 mm wide. Canyons, waste areas, roadsides, oak, juniper, piñon, ponderosa pine, 4900-7250 ft. Grant, Hidalgo.

Desmodium grahamii A. Gray—Graham's tick-trefoil. Perennial herbs; stems decumbent, mat-forming, the herbage glabrous, glabrescent, strigulose, or puberulent; leaves 1-3 cm, ovate or orbicular, petioles > 5 mm; inflorescences axillary or terminal, pedicels 10-15 mm; calyx 2.5-3.5 mm, corollas 6-8 mm, greenish white, pinkish purple, or bluish purple; fruit uncinate, subsessile or stipitate, the stipe 0-2 mm, the body with 3-6 articles, the articles 5-8 mm wide. Canyons, alluvial deposits, stream banks, disturbed areas, roadsides, grassland, oak, juniper, piñon, ponderosa pine, 4900-8000 ft. WC, SW.

Desmodium metcalfei (Rose & Painter) Kearney & Peebles—Metcalfe's tick-trefoil. Perennial herbs; stems erect or ascending, the herbage glabrous, glabrescent, or strigulose; leaves 4-6 cm, lanceolate or oblong, petioles > 5 mm; inflorescences axillary and terminal, pedicels 10-20 mm; calyx 1.5-2.5 mm, corollas 4.5-5.5 mm, pinkish purple; fruit uncinate, the stipe 1-2 mm, the body with 2-5 articles, the articles 5-6 mm wide. Rocky slopes and canyons, riparian forests, stream banks, grassland, juniper, piñon, 4250-7000 ft. Grant, Sierra. **R**

Desmodium procumbens (Mill.) Hitchc.—Western tick-trefoil. Annual or short-lived perennial herbs; stems erect or ascending, herbage glabrous, glabrescent, or puberulent; leaves 1.5-5 cm, elliptic, lanceolate, ovate, or oblong; inflorescences simple or branched; calyx 1-1.5 mm, corollas 2-3.5 mm, pinkish purple or reddish purple; fruit glabrous, glabrescent, or pubescent, subsessile or stipitate, the stipe 0-3.5 mm, the body with 1-5 articles, the articles 2-4 mm wide. WC, SW.

1. Fruit stipitate, the stipe 1.5-3.5 mm; short-lived perennial herbs…**var. exiguum**
1'. Fruit subsessile or stipitate, the stipe 0-1.5 mm; annual herbs…**var. neomexicanum**

var. exiguum (A. Gray) B. G. Schub. [*D. exiguum* A. Gray]—Western tick-trefoil. Short-lived perennial herbs; leaves lanceolate or oblong; corollas 2.5-3.5 mm; fruit uncinate, the stipe 1.5-3.5 mm, the body with 2-3 articles, the articles 2-3.5 mm wide. Riparian forest, stream banks, oak, walnut, piñon, juniper; 1 record, 5000 ft.

var. neomexicanum (A. Gray) H. Ohashi [*D. neomexicanum* A. Gray]—New Mexico tick-trefoil. Annual herbs; leaves elliptic or oblong; corollas 2-3 mm; fruit glabrous, glabrescent, or pubescent, subsessile or stipitate, the stipe 0-1.5 mm, the body with 1-5 articles, the articles 3-4 mm wide. Canyon bottoms, riparian forests, grasslands, oak, walnut, piñon, juniper, ponderosa pine, 4250-8200 ft.

Desmodium psilocarpum A. Gray—Santa Cruz tick-trefoil. Perennial herbs; stems erect or ascending, the herbage glabrous, glabrescent, pilosulous, or puberulent; leaves 2-10 cm, lanceolate or ovate, petioles > 5 mm; inflorescences branched, pedicels 10-20 mm; calyx 1.5-2 mm, corollas 4-5 mm, pinkish purple or greenish purple; fruit glabrous, glabrescent or uncinate, subsessile or stipitate, the stipe 0-2 mm, the body with 3-6 articles, the articles 6-10 mm wide. Canyons, riparian forests, stream banks, piñon, juniper, 5900-6600 ft. Grant.

Desmodium rosei B. G. Schub.—Rose's tick-trefoil. Annual herbs; stems erect or ascending, the herbage glabrous, glabrescent, or puberulent; leaves 2-7 cm, 7-10+ times longer than wide, linear or narrowly lanceolate, petioles > 5 mm; inflorescences simple or branched, pedicels 15-25 mm; calyx 1-1.5 mm, corollas 3-3.5 mm, pinkish purple; fruit glabrous, the stipe 1-1.5 mm, the body with 2-4 articles, the articles 3-4 mm wide. Canyons, riparian forest, stream banks, grasslands, desert scrub, piñon, juniper, oak, ponderosa pine, 4250-7250 ft. WC, SC, SW.

ERYTHRINA L. - Coral bean

Erythrina flabelliformis Kearney—Chilicote, western coral bean. Shrubs, small trees; stems armed, the herbage glabrous, glabrescent, or puberulent; terminal leaflet 3-7 cm, ovate or elliptic; inflorescences appearing before or at the same times as the leaves; calyx 6-7 mm, corollas 5.5-6.5 cm, reddish; fruit 12-25 × 1.5-2 cm, the body irregularly torulose, the valves glabrous, woody; seeds red with black markings. Canyons, hillsides, woodlands, 4250-6250 ft. Hidalgo, Otero.

ERYTHROSTEMON Klotzsch – Bird-of-paradise

Erythrostemon gilliesii (Hook.) Klotzsch [*Caesalpinia gilliesii* (Hook.) D. Dietr.]—Yellow bird-of-paradise. Shrubs or small trees; stems unarmed, glandular, the herbage glabrous; leaves with 8–12 pairs of pinnae, leaflets 7–11 pairs per pinna, the blades 3–8 mm, elliptic or oblong; inflorescences viscid-glandular, bracts glandular-pubescent, fimbriate-margined, the glands stalked; calyx 1.5–2 cm, pedicels 1.5–3 cm, corollas 2–3.5 cm, yellowish with orange markings, stamens much exserted, 7–8 cm, the filaments reddish; fruit 6–12 × 1.5–2 cm, the body oblong, the valves leathery, punctate-glandular. Naturalized; a common ornamental, 3250–7000 ft. NE, C, WC, SC, SW, Bernalillo, Roosevelt. Introduced.

EYSENHARDTIA Kunth – Kidneywood

Eysenhardtia orthocarpa (A. Gray) S. Watson—Tahitian kidneywood. Shrubs or small trees; stems unarmed, the herbage glabrous, glabrescent, puberulent, or strigose; leaves with 6–27 pairs of leaflets, the blades 5–20 mm, obovate or elliptic; inflorescences pedunculate; calyx 2–3.5 mm, corollas 6–8 mm, whitish, fading reddish, the pedicels spreading or deflexed; fruit spreading or deflexed, 12–15 × 2–4 mm, the body oblong, the valves glabrous, glandular-punctate. Rocky slopes, along waterways; grasslands, piñon-juniper-oak woodlands, 4250–7000 ft. Hidalgo.

GALACTIA P. Browne – Milkpea

Galactia wrightii A. Gray [*G. wrightii* var. *mollissima* Kearney & Peebles]—Wright's milkpea. Perennial herbs; stems vinelike, prostrate and twining, the herbage glabrous, glabrescent, puberulent, or villosulous; leaves trifoliate, the blades 2–5 cm, ovate or elliptic; inflorescences pedunculate; calyx 6–7 mm, corollas 10–14 mm, pinkish purple or whitish; fruit 4–6 × 0.4–0.6 cm, the body straight or ± incurved, the valves villosulous. Canyons, washes, hillsides, rocky sites, riparian forests, grasslands, woodlands, 3250–7000 ft.

GLEDITSIA J. Clayton – Honey locust

Gleditsia triacanthos L.—Honey locust. Trees; stems armed or unarmed, the herbage glabrous or glabrescent; leaves with 1–8 pinnae, the pinnae with 2–14 pairs of leaflets, the blades 1–3 cm, oblong or elliptic; inflorescences solitary in axils or in fascicles from spurs; calyx and corolla 3–5 mm, greenish or whitish; fruit 20–40 × 2.5–4 cm, the body twisted or contorted, the valves ± woody, puberulent, glabrescent in age. Naturalized, 3600–6600 ft. Widely scattered. Introduced.

GLYCYRRHIZA L. – Licorice

Glycyrrhiza lepidota Pursh—American licorice. Perennial herbs; stems glandular, the herbage glabrous, glabrescent, or puberulent; leaves with 4–9 leaflets, the blades 2–4 cm, ovate, lanceolate, or elliptic; inflorescences axillary, pedunculate; calyx 8–10 mm, corollas 9–12 mm, ochroleucous, greenish, or whitish; fruit 12–15 × 8–10 mm, the body elliptic, burlike, the valves covered with uncinate prickles. In ± mesic sites in canyons, washes, prairies, roadside ditches, ruderal areas, 3250–7550 ft. Widespread.

HEDYSARUM L. – Sweetvetch

Hedysarum boreale Nutt. [*H. boreale* var. *cinerascens* (Rydb.) Rollins]—Northern sweetvetch. Perennial herbs; stems erect or ascending, the herbage glabrous, glabrescent, or puberulent; leaves with 3–7 leaflets, blades 1–3.5 cm, oblong or elliptic; inflorescences pedunculate; calyx tube 2.5–3.5 mm, pedicels spreading or ascending, corollas 12–25 mm, reddish purple, pinkish purple, or whitish; fruit spreading or ascending, the body with 2–7 articles, the articles 7–8 mm wide. Canyons, washes, hillsides, roadside ditches, grasslands, sagebrush, piñon, juniper, oak, aspen, 5575–8900 ft. NC, NW, Curry, Quay, Socorro.

HOFFMANNSEGGIA Cav. - Holdback, hog potato

Perennial herbs or subshrubs, caulescent; stems unarmed, pubescence basifixed; stipules herbaceous, deciduous; leaves odd-pinnate, bipinnately compound, 1–13 pairs of pinnae per leaf; inflorescences racemose, axillary, pedunculate; flowers pedicellate, sepals 5, persistent in fruit, corollas zygomorphic, not papilionaceous, petals 5, androecium of separate stamens, stamens 10-merous; fruit a legume, dehiscent or indehiscent, sessile, the body laterally compressed, the valves glandular or eglandular.

1. Stems eglandular; calyx 3–5 mm; corollas 5–6 mm...**H. drepanocarpa**
1'. Stems stipitate-glandular; calyx 6–7 mm; corollas 10–13 mm...**H. glauca**

Hoffmannseggia drepanocarpa A. Gray—Sickle-pod holdback. Perennial herbs; stems unarmed, eglandular, herbage glabrous, glabrescent, puberulent, or villosulous; leaves with 2–5 pairs of pinnae, leaflets 4–8 pairs per pinna, blades 1.5–6 mm, obovate, elliptic, or oblong; inflorescences terminal, 4–10-flowered; calyx 3–5 mm, corollas 5–6 mm, orangish yellow; fruit ascending or spreading, 2–4 × 0.5–0.8 cm, the body oblong, the valves thick-papery, reticulate-veined, reddish brown. Rocky hillsides and flats, grasslands, desert scrub, juniper, piñon communities, 2950–6600 ft. Widespread except N and W.

Hoffmannseggia glauca (Ortega) Eifert—Hog potato. Perennial herbs; roots spreading and bearing spheroid tubers; stems unarmed, spreading to erect, stipitate-glandular, herbage glabrous, glabrescent, or villosulous; leaves with 2–6 pairs of pinnae, leaflets 3–5 pairs per pinna, blades 2.5–6 mm, elliptic or oblong; inflorescences terminal, 4–15-flowered, flowers ascending or spreading, deflexed in fruit; calyx 6–7 mm, corollas 10–13 mm, orangish yellow with red markings; fruit spreading to deflexed, 2–4 × 0.5–0.8 cm, the body oblong, falcate, the valves thick-papery. Sandy sites, plains, roadsides, ruderal areas, 3250–6600 ft. Widespread except NC.

INDIGOFERA L. - Indigo-plant

Indigofera sphaerocarpa A. Gray—Sonoran indigo-plant. Subshrubs or shrubs; stems unarmed, herbage strigose; leaves with 6–9 pairs of leaflets, the blades 20–30 mm, elliptic, obovate, or oblong; inflorescences axillary; calyx 1.5–2 mm, corollas 4–5 mm, orangish or pinkish; fruit spreading or deflexed, 3–3.5 × 2–2.6 cm, indehiscent, the body subglobose, the valves leathery. Rocky hillsides, desert scrub, oak woodland, riparian forest, 3250–6600 ft. Hidalgo.

KUMMEROWIA Schindl. - Japanese-clover

Kummerowia striata (Thunb.) Schindl.—Japanese-clover. Annual, caudices superficial; stems unarmed, prostrate or ascending, the herbage strigose; leaves trifoliate, palmate, the blades 5–14 mm, elliptic, obovate, or oblong; inflorescences axillary; calyx 2.5–3 mm, corollas 3–4 mm, pinkish; fruit covered by the persistent calyx, 3–3.5 mm, the body subglobose, the valves papery. Naturalized, 3250–6600 ft. Catron. Introduced. Apparently collected only once in New Mexico in August 1978 and not persisting (*Hutchins 7800*, UNM). This should be considered a waif.

LADEANIA A. N. Egan & Reveal - Scurfpea

Ladeania lanceolata (Pursh) A. N. Egan & Reveal [*Psoralidium lanceolatum* (Pursh) Rydb.]—Dune scurfpea. Perennial herbs; stems erect, glandular-punctate, the herbage glabrous, glabrescent, or strigose; leaves trifoliate, glandular-punctate, blades 15–40 mm, linear, elliptic, or oblong; inflorescences pedunculate; calyx 2–2.5 mm, corollas 4.5–8 mm, bluish purple or whitish; fruit exserted from the calyx, glandular-punctate, the body subglobose, the valves glabrous, glabrescent, or puberulent. Sandy sites, plains, dunes, washes, bottomlands, grasslands, desert scrub, piñon-juniper, 5250–7250 ft. NE, NC, NW, EC, C, WC, Doña Ana.

LATHYRUS L. - Sweetpea

Annual or perennial herbs; stems angular or winged, unarmed, pubescence basifixed; stipules herbaceous, entire to ± sagittate, persistent; leaves even-pinnate, the rachis terminating in a ± prehensile tendril, petiolate, leaflets 0 or 1 to many pairs per leaf; inflorescences racemose, axillary, pedunculate; flowers pedicellate, calyx tube not inflated, the teeth acute, corollas papilionaceous, the wings not adnate to the keel, androecium diadelphous; style bearded adaxially, the beard lateral or apical in orientation, the style apex flattened; fruits a legume, dehiscent, sessile, the body laterally compressed in cross-section.

1. Leaves with 1 pair of leaflets; naturalized species…(2)
1'. Leaves with 1+ pairs of leaflets; native (or naturalized in *L. venosus*)…(3)
2. Annual; corollas 9-14 mm…**L. hirsutus**
2'. Perennial; corollas 18-25 mm…**L. latifolius**
3. Leaf tendrils absent, reduced to a short bristle, or simple and ± curved, not prehensile…(4)
3'. Leaf tendrils of at least the upper leaves well-developed and prehensile (short-prehensile in *L. eucosmus*)…(7)
4. Corollas 9-14 mm…**L. lanzwertii** (in part)
4'. Corollas 15-30 mm…(5)
5. Leaf tendrils absent or reduced to short bristles; stems conspicuously winged…**L. decaphyllus**
5'. Leaf tendrils poorly developed, either short and straight or those of the uppermost leaves, ± curved; stems not conspicuously winged…(6)
6. Fruit ± sessile or tapered to a substipitate base, the stipe inconspicuous, the fruit body 5-7 mm wide… **L. brachycalyx**
6'. Fruit stipitate, the stipe 3-6 mm, the fruit body 7-12 mm wide…**L. eucosmus** (in part)
7. Corollas 9-14 mm…(8)
7'. Corollas 14-30 mm…(9)
8. Leaf blades 3-12 mm, linear or narrowly oblong, native…**L. graminifolius**
8'. Leaf blades 30-60 mm, ovate or elliptic; naturalized…**L. venosus** (in part)
9. Leaf tendrils poorly developed, short and those of the uppermost leaves short-prehensile; corolla 20-30 mm…**L. eucosmus** (in part)
9'. Leaf tendrils well-developed and prehensile; corollas 14-22 mm…(10)
10. Leaf blade ovate or elliptic; inflorescences 8-20-flowered; naturalized…**L. venosus** (in part)
10'. Leaf blade linear or oblong; inflorescences 2-10-flowered; native…**L. lanzwertii** (in part)

Lathyrus brachycalyx Rydb.—Zion sweetpea. Perennial herbs, rhizomatous; stems erect or prostrate, twining, the herbage glabrous or glabrescent; leaves with 3-6 pairs of leaflets, the tendrils short, the blades 25-80 mm, elliptic or oblong; inflorescences pedunculate, 2-5-flowered; calyx 6-9 mm, corollas 15-25 mm, pinkish purple or whitish; fruit sessile or substipitate, 3-5 × 0.5-0.7 cm, the body oblong, the valves glabrous. New Mexico material belongs to **var. zionis** (C. L. Hitchc.) S. L. Welsh. Sagebrush, piñon-juniper woodlands, 5200-5600 ft. Rio Arriba, San Juan. Often confused with the much more widespread *L. eucosmus*.

Lathyrus decaphyllus Pursh [*L. polymorphus* Nutt. var. *incanus* (J. G. Sm. & Rydb.) Dorn]—Many-stemmed sweetpea. Perennial herbs, rhizomatous; stems erect or prostrate, twining, the herbage glabrescent or villosulous; leaves with 2-5 pairs of leaflets, the tendrils absent or reduced to short bristles, the blades 20-40 mm, linear or oblong; inflorescences pedunculate, 2-8-flowered; calyx 7-11 mm, corollas 20-28 mm, pinkish purple or whitish; fruit 3-6 × 0.7-1.2 cm, the stipe 3-6 mm, the body oblong, the valves glabrous. Our material belongs to **var. incanus** (J. G. Sm. & Rydb.) Broich. Hillsides, badlands, prairies, sagebrush, piñon-juniper communities, 4900-7600 ft. Torrance.

Lathyrus eucosmus Butters & H. St. John [*L. brachycalyx* Rydb. var. *eucosmus* (Butters & H. St. John) S. L. Welsh]—Bush sweetpea. Perennial herbs, rhizomatous; stems erect or prostrate, twining, the herbage glabrous, glabrescent, or villosulous; leaves with 3-6 pairs of leaflets, the tendrils poorly developed, those of the uppermost leaves simple or short-prehensile, the blades 25-80 mm, elliptic, lanceolate, or oblong; inflorescences pedunculate, 2-5-flowered; calyx 9-14 mm, corollas 20-30 mm, pinkish purple or whitish; fruit 3-5 × 0.8-1 cm, the stipe 3-6 mm, the body oblong, the valves glabrous. Washes, roadsides, ruderal areas, sagebrush, piñon, juniper, ponderosa pine, mixed conifer, 4500–9000 ft. N, C, SC, Eddy, Grant, Sierra.

Lathyrus graminifolius (S. Watson) T. G. White—Grass-leaf sweetpea. Perennial herbs, rhizomatous; stems erect or prostrate, twining, the herbage glabrous; leaves with 2–5 pairs of leaflets, the tendrils prehensile, the blades 3–12 mm, linear or oblong; inflorescences pedunculate, 3–10-flowered; calyx 6–9 mm, corollas 10–13 mm, pinkish purple, bluish purple, or whitish, fading orangish in age; fruit 3–5 × 0.5–0.7 cm, the body oblong, the valves glabrous. Hillsides, roadsides, ruderal areas, sagebrush, piñon, juniper, ponderosa pine, mixed conifer, 1700–2700 ft. C, WC, SW, McKinley, San Miguel.

Lathyrus hirsutus L.—Caley pea. Annual, caudices superficial; stems winged, erect or prostrate, twining, the herbage glabrous or glabrescent; leaves with 1 pair of leaflets, the petioles winged, the tendrils prehensile, the blades 20–80 mm, elliptic or oblong; inflorescences pedunculate, 1–3-flowered; calyx 5–7 mm, corollas 9–14 mm, pinkish purple, bluish purple, or whitish; fruit 2–5 × 0.5–0.8 cm, the body oblong, the valves hirsute, the hairs pustulate and 0.5–1 mm. Naturalized, 4000–7500 ft. Colfax, San Miguel, Santa Fe. Introduced.

Lathyrus lanszwertii Kellogg—Mountain sweetpea. Perennial herbs, rhizomatous; stems erect or prostrate, twining, the herbage glabrous, glabrescent, or villosulous; leaves with 1–5 pairs of leaflets, the tendrils absent, reduced to short bristles, or prehensile, the blades 20–80 mm, linear or oblong; inflorescences pedunculate, 2–10-flowered; calyx 5–10 mm, corollas 7–20 mm, pinkish purple, bluish purple, ochroleucous, or whitish; fruit 2–5 × 0.7–0.9 cm, the body oblong, the valves glabrous. NW, C, WC, SC, Hidalgo.

1. Tendrils prehensile; leaves with 2–5 pairs of leaflets; corollas 14–20 mm…**var. laetivirens**
1'. Tendrils absent or reduced to short bristles; leaves with 1–2 pairs of leaflets; corollas 9–14 mm…**var. arizonicus**

var. arizonicus (Britton) S. L. Welsh [*L. arizonicus* Britton; *L. lanszwertii* var. *leucanthus* (Rydb.) Dorn; *L. leucanthus* Rydb.]—Arizona sweetpea. Leaves with 1–2 pairs of leaflets, tendrils absent or reduced to short bristles; corollas 9–14 mm, pinkish purple, ochroleucous, or whitish. Hillsides, sagebrush, piñon-juniper woodlands, ponderosa pine, montane meadows, spruce-bristlecone communities, 6200–10,850 ft.

var. laetivirens (Greene ex Rydb.) S. L. Welsh [*L. laetivirens* Greene ex Rydb.]—Rocky Mountain sweetpea. Leaves with 2–5 pairs of leaflets, the tendrils prehensile; corollas 14–22 mm, ochroleucous or whitish. Hillsides, sagebrush, piñon-juniper woodlands, ponderosa pine, montane meadows, spruce-bristlecone-fir communities, 5575–10,850 ft.

Lathyrus latifolius L.—Perennial peavine. Perennial herbs, caudices superficial; stems winged, erect or prostrate, twining, the herbage glabrous or glabrescent; leaves with 1 pair of leaflets, the petioles winged, the tendrils prehensile, the blades 5–15 cm, elliptic or oblong; inflorescences pedunculate, 2–15-flowered; calyx 10–12 mm, corollas 18–25 mm, pinkish purple, bluish purple, or whitish. Naturalized, 4200–7900 ft. NC, NW, SC, SW, Roosevelt.

Lathyrus venosus Muhl. ex Willd.—Veiny pea. Perennial herbs, caudices superficial; stems erect or prostrate, twining, the herbage glabrous, glabrescent, or puberulent; leaves with 4–7 pairs of leaflets, the tendrils prehensile, the blades 30–60 mm, ovate or elliptic; inflorescences pedunculate, 8–20-flowered; calyx 5–9 mm, corollas 9–20 mm, pinkish purple, bluish purple, or reddish purple; fruit 3–5 × 0.5–0.7 cm, the body oblong, the valves glabrous or glabrescent. Naturalized, 6500–8200 ft. Lincoln, Otero. Introduced.

LEUCAENA Benth. – Leadtree

Leucaena retusa Benth.—Little-leaf leadtree. Shrubs or trees, the herbage glabrescent, hirsutulous, or velutinous; leaves with 2–4 pinnae per leaf, rachis glands knoblike or short-columnar, leaflets 4–8 pairs per pinna, the blades 10–20 mm, elliptic, lanceolate, or obovate; inflorescences compact, the heads 2–2.5 cm wide; calyx and corolla inconspicuous, the petals yellowish gold; fruit 12–25 × 0.8–1.5 cm, the valves leathery, glabrous. Sandy, gravelly, or clay soils derived from igneous or limestone substrates; washes of intermittent streams, flats, oak, juniper, 3600–4600 ft. Eddy.

LOTUS L. – Bird's-foot trefoil

Lotus corniculatus L.—Bird's-foot trefoil. Perennial herbs; stems erect, unarmed, ascending, or prostrate, the herbage glabrous or glabrescent; stipules reduced to small glands; leaves with 2 pairs of leaflets, the lowermost pair stipular in position, the upper 3 palmately trifoliate, the blades 0.5–2.5 cm, ovate or lanceolate; inflorescences 2–8-flowered, the peduncle subtended by a trifoliate bract; calyx tube 2.5–3.5 mm, corollas 10–14 mm, yellowish; fruit 1.3–3.5 × 0.1–0.2 cm, the body oblong or linear, the valves glabrous. Naturalized, 6200–8900 ft. NC, Grant, Lea. Introduced.

LUPINUS L. – Lupine

Annual or perennial herbs, subshrubs, or shrubs; stems unarmed, pubescence basifixed; stipules inconspicuous; leaves odd-pinnate, palmately compound, petiolate; inflorescences racemose, terminal, pedunculate; flowers pedicellate, calyx tube not inflated, 2-lipped, the upper lip 2-toothed, the teeth acute, corollas papilionaceous, androecium monadelphous, anthers dimorphic; fruits a legume, dehiscent, sessile, the body laterally compressed in cross-section.

1. Plants annual…(2)
1'. Plants perennial…(5)
2. Cotyledons petioled, withering and ± deciduous at anthesis; fruits 3–7-seeded…**L. concinnus**
2'. Cotyledons sessile, ± persistent at anthesis; fruits 1- or 2-seeded…(3)
3. Inflorescences subcapitate and dense, the racemes 0.5–2 cm…(4)
3'. Inflorescences elongate and loosely flowered, the racemes 2–10 cm…**L. pusillus**
4. Calyx 5–7 mm; pedicels 1–2 mm…**L. kingii**
4'. Calyx 3–5 mm; pedicels 0.3–1 mm…**L. brevicaulis**
5. Calyx spurs 1–3 mm…**L. caudatus**
5'. Calyx not spurred or the spur ± gibbous and <1 mm…(6)
6. Leaflets pilose, villous or strigose above…(7)
6'. Leaflets glabrous or glabrescent above…(9)
7. Inflorescences included within the leaves; fruit 3–5 mm wide…**L. lepidus**
7'. Inflorescences erect and exserted above the leaves; fruit 5–9 mm wide…(8)
8. Calyx spurred, the spur ± gibbous and <1 mm…**L. argenteus** (in part)
8'. Calyx not spurred…**L. sericeus**
9. Stems glabrous, glabrescent, or inconspicuously strigulose (requiring magnification)…**L. latifolius**
9'. Stems variously pubescent, not glabrous, the hairs conspicuous…(10)
10. Corollas 5–8 mm…**L. argenteus** (in part)
10'. Corollas 8–15 mm…(11)
11. Fruit pubescent with appressed, ± wavy hairs…**L. plattensis**
11'. Fruit pubescent with ascending to erect, ± wavy or straight hairs…(12)
12. Calyx 8–10 mm…(13)
12'. Calyx 4–8 mm…(14)
13. Fruits 3–3.5 cm, the valves hirsute; corollas 10–14 mm…**L. sierrae-blancae**
13'. Fruits 2–2.5 cm, the valves pilose; corollas 8–10 mm…**L. argenteus** (in part)
14. Fruits 2–2.5 cm…(15)
14'. Fruits 2.5–4 cm…(17)
15. Corollas 8–10 mm…**L. argenteus** (in part)
15'. Corollas 10–15 mm…(16)
16. Pedicels 3–4 mm…**L. argenteus** (in part)
16'. Pedicels 4–10 mm…**L. polyphyllus** (in part)
17. Fruits 1.5–2 cm wide, the valves villous…**L. neomexicanus**
17'. Fruits 0.6–0.8 cm wide, the valves pilose…**L. polyphyllus** (in part)

Lupinus argenteus Pursh—Silvery lupine. Perennial herbs, caudices superficial; herbage glabrous, glabrescent, pilose, sericeous, hirsute, or villous; leaves with 6–10 leaflets, the blades 2–13 cm, elliptic or oblanceolate, long-petioled, basal leaves absent or withered at anthesis; inflorescences pedunculate;

calyx 4–14 mm, the tube spurred, the spur ± gibbous and < 1 mm, pedicels 1–4 mm, corollas 5–14 mm, whitish and pinkish purple–tinged, pinkish purple, bluish purple, or reddish purple, the banner glabrous or strigulose, the keel ciliolate; fruit 2–2.5 × 0.5–0.9 cm, the valves pilose.

1. Stem pubescence spreading…**var. palmeri**
1'. Stem pubescence appressed…(2)
2. Corollas 5–8 mm…(3)
2'. Corollas 8–14 mm…(7)
3. Corollas 7–8 mm…**var. argenteus** (in part)
3'. Corollas 5–7 mm…(4)
4. Pedicels 2–4 mm…(5)
4'. Pedicels 1–2 mm…(6)
5. Leaves broadly oblanceolate or broadly elliptic, the blades mostly flat…**var. parviflorus**
5'. Leaves narrowly oblanceolate or narrowly elliptic, the blades mostly folded…**var. hillii**
6. Flowers conspicuously maculate…**var. fulvomaculatus**
6'. Flowers not maculate…**var. myrianthus**
7. Corollas 12–14 mm…**var. moabensis**
7'. Corollas 8–12 mm…(8)
8. Leaves narrowly oblanceolate or narrowly elliptic, the blades mostly folded and pubescent on both sides…
   var. argenteus (in part)
8'. Leaves broadly oblanceolate or broadly elliptic, the blades mostly flat and glabrous above…**var. rubricaulis**

var. argenteus—Stem pubescence appressed; leaves narrowly oblanceolate or narrowly elliptic, the blades mostly folded and pubescent on both sides; pedicels 3–4 mm, corollas 7–12 mm. Washes, flats, hillsides, river valleys, piñon-juniper, ponderosa pine, upper montane meadows, 5500–10,650 ft. McKinley, San Juan.

var. fulvomaculatus (Payson) Barneby—Stem pubescence appressed; pedicels 3–4 mm, corollas 5–8 mm, maculate. Hillsides, piñon-juniper, ponderosa pine, upper montane meadows, mixed conifer, 5500–10,650 ft. NC.

var. hillii (Greene) Barneby—Stem pubescence appressed; leaves narrowly oblanceolate or narrowly elliptic, the blades mostly folded; pedicels 2–4 mm, corollas 5–7 mm. Piñon-juniper, ponderosa pine, 5000–8000 ft. WC.

var. moabensis S. L. Welsh—Stem pubescence appressed; pedicels 3–4 mm, corollas 5–8 mm, maculate. Four Corners; salt desert scrub, piñon-juniper, ponderosa pine, 4000–7450 ft. NW.

var. myrianthus (Greene) Isely—Stem pubescence appressed; pedicels 1–2 mm, corollas 5–7 mm, not maculate. Upper piñon-juniper, ponderosa pine, mixed conifer, 6850–9000 ft. NC.

var. palmeri (S. Watson) Barneby—Stem pubescence spreading; corollas 7–10 mm. Piñon-juniper, ponderosa pine, mixed conifer, 6150–9650 ft. NC.

var. parviflorus (Nutt.) C. L. Hitchc.—Stem pubescence spreading; leaves broadly oblanceolate or broadly elliptic, the blades mostly flat; pedicels 2–4 mm, corollas 5–7 mm. Ponderosa pine, montane meadows, mixed conifer, 8475–10,200 ft. Colfax, Rio Arriba, Taos.

var. rubricaulis (Greene) S. L. Welsh—Stem pubescence spreading; leaves broadly oblanceolate or broadly elliptic, the blades mostly flat and glabrous above; corollas 8–12 mm. Piñon-juniper, ponderosa pine, upper montane meadows, mixed conifer, 6200–9400 ft. NC, NW.

Lupinus brevicaulis S. Watson—Short-stemmed lupine. Annuals, caulescent; stems tuft-forming, cotyledons persistent, the herbage pilose or villous, the longest hairs ± 1 mm; leaves with 3–9 leaflets, the blades 4–20 cm, spatulate or oblanceolate; inflorescences subcapitate; calyx 3–5 mm, pedicels 0.5–1.5 mm, corollas 5–8 mm, whitish and pinkish purple–tinged, pinkish purple, or bluish purple; fruit 0.8–1.2 × 0.4–0.6 cm, the valves pilose. Sandy or gravelly sites, flats, scrublands, woodlands, hillsides, creosote, mesquite, salt desert scrub, piñon-juniper, ponderosa pine, 3900–7550 ft. NW, WC, SW.

Lupinus caudatus Kellogg—Spurred lupine. Perennial herbs; stems erect or ascending, the herbage glabrous, glabrescent, pilose, or villous; leaves with 5-9 leaflets, the blades 10-60 mm, elliptic or oblanceolate; inflorescences pedunculate; calyx 4-14 mm, the tube spurred, the spur gibbous and 1-3 mm, pedicels 2-8 mm, corollas 8-14 mm, whitish and pinkish purple-tinged, pinkish purple, bluish purple, or reddish purple, the banner glabrous or strigulose, the keel ciliolate; fruit 2-3 × 0.8-0.9 cm, the valves pilose.

1. Leaflets broadly oblanceolate, > 9 mm wide…**var. cutleri**
1'. Leaflets narrowly oblanceolate, < 5 mm wide…(2)
2. Basal leaves present at anthesis; leaflets greenish or yellowish above, greenish or grayish below…**var. argophyllus**
2'. Basal leaves deciduous at anthesis; leaflets greenish or grayish above and below…**var. utahensis**

var. argophyllus (A. Gray) S. L. Welsh [*L. argenteus* Pursh var. *argophyllus* (A. Gray) S. Watson]— Spurred lupine. Long-petioled, basal leaves present at anthesis; leaflets narrowly oblanceolate, < 5 mm wide, greenish or yellowish above, greenish or grayish below; corollas 10-12 mm. Piñon-juniper, ponderosa pine, montane meadows, mixed conifer communities, 5500-10,150 ft. NE, NW, C.

var. cutleri (Eastw.) S. L. Welsh—Cutler's spurred lupine. Leaflets broadly oblanceolate, > 9 mm wide. Canyons, hillsides, piñon-juniper and ponderosa pine communities, 5900-8000 ft. NW.

var. utahensis (S. Watson) S. L. Welsh [*L. argenteus* Pursh var. *utahensis* (S. Watson) Barneby]— Utah spurred lupine. Long-petioled, basal leaves absent or withered at anthesis; leaflets narrowly oblanceolate, < 5 mm wide, greenish or grayish above and below; corollas 8-13 mm. Piñon-juniper and ponderosa pine communities, 5700-8800 ft. NW.

Lupinus concinnus J. Agardh [*L. concinnus* var. *orcuttii* (S. Watson) C. P. Sm.]—Bajada lupine. Annuals, caulescent; stems erect, cotyledons stalked, persistent, the herbage pilose or villous, the longest hairs 1-1.5 mm; leaves with 5-9 leaflets, the blades 10-25 mm, obovate, spatulate, or oblanceolate; inflorescences elongate in fruit; calyx 3-5 mm, pedicels 1-2 mm, corollas 6-11 mm, whitish and pinkish purple-tinged, pinkish purple, or bluish purple, eyespot yellowish; fruit 1-2 × 0.4-0.6 cm, the valves villous. Washes, hillsides, roadsides, ruderal areas, desert scrub, piñon-juniper, 3900-7550 ft. WC, SE, SW, McKinley.

Lupinus kingii S. Watson—King's lupine. Annuals, caulescent or subcaulescent; stems erect, cotyledons persistent, the herbage pilose, the longest hairs 1-2 mm; leaves with 4-7 leaflets, the blades 10-25 mm, obovate, spatulate, or oblanceolate; inflorescences elongate in fruit; calyx 5-7 mm, pedicels 1-2 mm, corollas 6-8 mm, whitish and pinkish purple-tinged, pinkish purple, or bluish purple; fruit 1-1.5 × 0.4-0.6 cm, the valves pilose. Sandy sites, washes, dunes, hillsides, woodlands, roadsides, ruderal areas, piñon-juniper, ponderosa pine, 5250-8550 ft. NC, NW, C, WC, SC, SW.

Lupinus latifolius J. Agardh—Broad-leaf lupine. Perennial herbs; stems erect or ascending, the herbage glabrous, glabrescent, or strigulose; leaves with 6-10 leaflets, the blades 10-60 mm, elliptic, oblong, or oblanceolate, long-petioled, basal leaves absent or withered at anthesis; inflorescences pedunculate; calyx 4-10 mm, pedicels 4-10 mm, corollas 10-16 mm, whitish or whitish and pinkish purple-tinged, the keel ciliolate; fruit 2-3 × 0.8-0.9 cm, the valves strigulose. New Mexico material belongs to **var. leucanthus** (Rydb.) Isely. Ponderosa pine communities, 6550-9200 ft. SC, Catron, Rio Arriba.

Lupinus lepidus Douglas ex Lindl.—Utah dwarf lupine. Perennial herbs, caulescent or subcaulescent; stems erect, ascending, or prostrate, mat-forming, the herbage glabrous, glabrescent, sericeous, strigose, or pilose; leaves with 5-9 leaflets, the blades 10-30 mm, oblanceolate, long-petioled, basal leaves present at anthesis; inflorescences pedunculate; calyx 4-7 mm, pedicels 0.5-2 mm, corollas 7-14 mm, whitish and pinkish purple-tinged, pinkish purple, or bluish purple, the keel ciliolate; fruit 1-1.5 × 0.3-0.5 cm, the valves pilosulous. Our material belongs to **var. utahensis** (S. Watson) C. L. Hitchc. Montane meadows and ponderosa pine communities, 6550-9200 ft. San Juan.

Lupinus neomexicanus Greene—New Mexico lupine. Perennial herbs, caulescent; stems erect, ascending, the herbage glabrous, glabrescent, puberulent, hirsutulous, villous, or pilose; leaves with 6–10 leaflets, the blades 20–40 mm, oblanceolate, long-petioled, basal leaves absent or withered at anthesis; inflorescences pedunculate; calyx 5–7 mm, pedicels 5–8 mm, corollas 10–14 mm, pinkish purple or bluish purple, the keel ciliolate; fruit 3–4 × 1–2 cm, the valves villous. Meadows, canyons, woodlands, hillsides, forests, upper piñon-juniper, ponderosa pine, mixed conifer, 5900–9200 ft. NC, C, WC.

Lupinus plattensis S. Watson—Nebraska lupine. Perennial herbs, caulescent, rhizomatous; stems erect or ascending, clump-forming, the herbage glabrous, glabrescent, puberulent, villous, or pilose; leaves with 6–9 leaflets, the blades 20–40 mm, oblanceolate or spatulate; inflorescences pedunculate; calyx 8–9 mm, pedicels 5–8 mm, corollas 10–14 mm, whitish and pinkish purple-tinged, pinkish purple, or bluish purple, the keel ciliolate; fruit 2–4 × 0.8–0.9 cm, the valves appressed-villous, the hairs ± wavy. Sandy sites, meadows, streamsides, prairies, piñon-juniper woodlands, 5900–8550 ft. NE, NC.

Lupinus polyphyllus Lindl.— Perennial herbs, caulescent, rhizomatous; stems erect or ascending, clump-forming, fistulose or not, the herbage glabrous, glabrescent, sericeous, strigulose, hirsute, or pilose; leaves with 3–12 leaflets, the blades 20–110 mm, oblanceolate or spatulate, long-petioled, basal leaves present at anthesis; inflorescences pedunculate; calyx 5–8 mm, pedicels 4–10 mm, corollas 8–14 mm, whitish and pinkish purple-tinged, pinkish purple, or bluish purple, eyespot yellowish, the keel ciliolate; fruit 2–4.5 × 0.4–0.8 cm, the valves pilose. Sandy sites, meadows, prairies, woodlands, streamsides, forest, 4200–9550 ft.

   1. Caudex superficial, plants taprooted; stems pilose…**var. prunophilus**
   1′. Caudex subterranean, plants rhizomatous; stems hirsute…**var. ammophilus**

var. ammophilus (Greene) Barneby [*L. ammophilus* Greene]—Sand lupine. Caudex subterranean, plants rhizomatous; stems hirsute; leaves with 6–10 leaflets, the largest 2.5–6.5 cm. Salt desert scrub, sagebrush, piñon-juniper, 4200–8200 ft. NC, NW.

var. prunophilus (M. E. Jones) L. Ll. Phillips—Many-leaved lupine. Caudex superficial, plants taprooted; stems pilose; leaves with 7–10 leaflets, the largest 3–11 cm. Meadows, prairies, woodlands, streamsides, forests, roadsides, 6550–9550 ft. Rio Arriba.

Lupinus pusillus Pursh—Rusty lupine. Annuals, caulescent or subacaulescent; stems erect, cotyledons sessile, persistent, the herbage pilose; leaves with 5–7 leaflets, the blades 10–40 mm, obovate, elliptic, or oblanceolate; inflorescences elongate in fruit; calyx 4–5 mm, pedicels 1–3.5 mm, corollas 5–12 mm, whitish and pinkish purple-tinged, pinkish purple, or bluish purple; fruit 1–2 × 0.6–0.7 cm, the valves hirsute.

   1. Calyx tube strigulose; corollas 10–12 mm…**var. pusillus**
   1′. Calyx tube glabrous; corollas 7–10 mm…**var. rubens**

var. pusillus—Calyx tube strigulose; corollas 10–12 mm. Sandy sites, washes, prairies, hillsides, flats, desert scrub, sagebrush, piñon-juniper woodlands, ponderosa pine communities, 4900–7550 ft. NE, NC, NW, C.

var. rubens (Rydb.) S. L. Welsh—Calyx tube glabrous; corollas 7–10 mm. Scrublands, washes, hillsides, flats, desert grasslands, sagebrush, piñon-juniper woodlands, 4900–6900 ft. McKinley, Sandoval, San Juan.

Lupinus sericeus Pursh—Silky lupine. Perennial herbs; stems erect or ascending, the herbage glabrous, glabrescent, sericeous, strigulose, pilose, or villous; leaves with 6–9 leaflets, the blades 30–70 mm, elliptic or oblanceolate, long-petioled, basal leaves absent or withered at anthesis; inflorescences pedunculate; calyx 6–10 mm, the tube spurred, the spur gibbous and 1–3 mm, pedicels 2–8 mm, corollas 8–14 mm, whitish and pinkish purple-tinged, pinkish purple, bluish purple, or reddish purple, the banner glabrous or strigulose, the keel ciliolate; fruit 2–3 × 0.8–0.9 cm, the valves pilose. New Mexico material belongs to **var. sericeus**. Piñon-juniper, ponderosa pine, mixed conifer communities, 5250–10,500 ft. NW.

Lupinus sierrae-blancae Wooton & Standl.—White Mountain lupine. Perennial herbs; stems erect or ascending, the herbage glabrous, glabrescent, hirsute, or puberulent; leaves with 7–9 leaflets, the blades 30–70 mm, oblong or oblanceolate, long-petioled, basal leaves absent or withered at anthesis; inflorescences pedunculate; calyx 8–10 mm, pedicels 6–8 mm, corollas 10–14 mm, whitish and bluish purple-tinged, the keel ciliolate; fruit 3–4 × 0.7–0.9 cm, the valves hirsute. Montane meadows, roadsides, ponderosa pine and mixed conifer forests, 5900–10,000. Lincoln, Otero. **R & E**

MACROPTILIUM (Benth.) Urb. – Bushbean

Macroptilium gibbosifolium (Ortega) A. Delgado—Bushbean. Perennial herbs, caudices superficial, woody; stems trailing and vinelike, the herbage strigulose; leaves pinnately trifoliate, the leaflets entire or palmately 1–2-lobed, the blades ovate or elliptic, the terminal leaflets 10–50 mm; inflorescences with 4–10 flowering nodes; calyx 2–3 mm, pedicels ascending, corollas 6–8 mm, reddish or bronzish orange; fruit 1.5–2.5 × 0.7–0.9 cm, the body ± oblong or falcate, the valves strigulose. Canyons, washes, outcrops, ruderal areas, roadsides, Chihuahuan desert scrub, piñon-juniper, 4900–7550 ft. WC, SW.

MARINA Liebm. – Prairie-clover

Marina calycosa (A. Gray) Barneby [*Dalea calycosa* A. Gray]—San Pedro prairie-clover. Perennial herb, caudices subterranean, woody; stems prostrate or decumbent, glandular-punctate, the herbage strigulose; leaves with 4–12 pairs of leaflets, the blades 2–7 mm, grayish, obovate; inflorescences racemose, narrowly oblong; flowers ascending or spreading, deflexed in fruit; calyx tube 5–7 mm, corollas 7–9 mm, whitish and pinkish purple, the petals papilionaceous; fruits 1-seeded. Grasslands, prairies, washes, desert scrub, desert grasslands, 4550–6000 ft. Grant, Hidalgo.

MARIOSOUSA Seigler & Ebinger – Wattle

Mariosousa millefolia (S. Watson) Seigler & Ebinger [*Acacia millefolia* S. Watson]—Fernleaf wattle. Shrubs or trees, the herbage glabrous, glabrescent, or puberulent; leaves with 6–10 pairs of pinnae, leaflets 15–20 pairs, the blades 2–3 mm, oblong; inflorescences axillary, 1–2 per node; flowers whitish or yellowish, short-pedunculate; fruits 5–15 × 1–2 cm, the stipe 4–10 mm, the valves papery, ± straight. Igneous rock, rocky hills, sparse oak, juniper, Mexican piñon; 1 record just north of the Mexican border, 4550–5000 ft. Hidalgo.

MEDICAGO L. – Alfalfa, burclover

Annual or perennial herbs; stems unarmed, pubescence basifixed; stipules partly fused to petiole, papery or membranous, persistent; leaves odd-pinnate, palmately or pinnately compound, trifoliate, petiolate; inflorescences racemose, compact and ± umbel-like, axillary, pedunculate; flowers pedicellate or subsessile, calyx tube not inflated, the teeth acute, corollas papilionaceous, androecium diadelphous; fruits a legume, indehiscent, the body curved or spirally coiled, the valves reticulate-veined or spiny.

1. Fruits 1-seeded, the body curved, unarmed; corollas 2–3 mm…**M. lupulina**
1'. Fruits 2- to many-seeded, the body spirally coiled, spiny or unarmed; corollas 3–11 mm…(2)
2. Calyx 2–2.5 mm; herbage villous…**M. minima**
2'. Calyx 2.5–5.5 mm; herbage glabrous, glabrescent, or puberulent…(3)
3. Fruits spiny; corollas 3–5 mm; calyx 2.5–4 mm…**M. polymorpha**
3'. Fruits unarmed; corollas 8–11 mm; calyx 4–5.5 mm…**M. sativa**

Medicago lupulina L.—Black medic. Annuals, the herbage glabrous, glabrescent, or puberulent; stems prostrate or ascending, 4-angled, glandular or eglandular; leaves pinnately trifoliate, the blades 8–15 mm, obovate; inflorescences compact, ovoid, elongate in fruit; calyx 1–1.5 mm, corollas 2–3 mm, yellowish; fruits 2–2.5 × 1.5–2 mm, curved, unarmed, the valves with concentric nerves, glabrous. Naturalized weed of disturbed areas, 3900–9200 ft. Widespread. Introduced.

Medicago minima (L.) Bartal.—Little burclover. Annuals, the herbage villous; stems prostrate or ascending, rounded; leaves pinnately trifoliate, the blades 4-12 mm, obovate or elliptic; inflorescences compact, headlike; calyx 2-2.5 mm, corollas 3-5 mm, yellowish; fruits 6-10 mm, coiled 2-4 turns, armed, the valves with marginal nerves, leathery. Naturalized weed of disturbed areas, 3900-9200 ft. Doña Ana, Grant. Introduced.

Medicago polymorpha L.—Burclover. Annuals, the herbage glabrous, glabrescent, or puberulent; stems prostrate or ascending, rounded; leaves pinnately trifoliate, the blades 8-20 mm, obovate; inflorescences compact, headlike; calyx 2-3 mm, corollas 3-6 mm, yellowish; fruits 6-12 mm, coiled 2-6 turns, armed, the valves with marginal nerves, leathery. Naturalized weed of disturbed areas, 3900-7250 ft. Scattered. Introduced.

Medicago sativa L.—Alfalfa. Perennial herbs, the herbage glabrous, glabrescent, or puberulent; stems erect or ascending, rounded; leaves pinnately trifoliate, the blades 10-20 mm, obovate or oblanceolate; inflorescences compact, ovate; calyx 4-5.5 mm, corollas 8-11 mm, purplish, greenish blue, greenish yellow, or purplish black; fruits 4-8 mm, coiled 2-3 turns, unarmed, the valves glabrous, glabrescent, or puberulent. Naturalized escape and weed, 3900-8600 ft. Widespread. Introduced.

MELILOTUS (L.) Mill. - Sweetclover

Annual or biennial herbs; stems unarmed, pubescence basifixed; stipules partly fused to petiole, papery or membranous, persistent; leaves odd-pinnate, pinnately compound, trifoliate, petiolate; inflorescences racemose, axillary, pedunculate; flowers pedicellate, calyx tube not inflated, the teeth acute, corollas papilionaceous, androecium diadelphous; fruits a legume, indehiscent, the body straight, the valves reticulate-veined or bumpy; seeds 1 or 2.

1. Corollas 2-3 mm; fruits 2-3 mm…**M. indicus**
1'. Corollas 3-7 mm; fruits 3-7 mm…(2)
2. Corollas whitish, 3-5 mm…**M. albus**
2'. Corollas yellowish, 5-7 mm…**M. officinalis**

Melilotus albus Medik.—White sweetclover. Annuals or biennials, the herbage glabrous, glabrescent, or strigulose; stems erect or ascending; leaves pinnately trifoliate, the blades 10-20 mm, obovate or elliptic; inflorescences elongate; calyx 2-3 mm, corollas 3-5 mm, whitish; fruits 3-4 × 1-2 mm, the valves reticulate-veined. Naturalized; roadsides, weedy and disturbed sites, 3600-9200 ft. Widespread except NE. Introduced.

Melilotus indicus (L.) All.—Sourclover, annual sweetclover. Annuals, the herbage glabrous; stems erect or ascending; leaves pinnately trifoliate, the blades 10-20 mm, oblanceolate, obovate, or elliptic; inflorescences elongate; calyx 1-2 mm, corollas 2-3 mm, yellowish; fruits 2-3 × 1-2 mm, the valves reticulate-veined. Naturalized, weedy, 3600-9200 ft. WC, SC, San Juan, San Miguel. Introduced.

Melilotus officinalis (L.) Lam.—Yellow sweetclover. Biennials, the herbage glabrous, glabrescent, or strigulose; stems erect or ascending; leaves pinnately trifoliate, the blades 10-25 mm, obovate or elliptic; inflorescences elongate; calyx 2-3 mm, corollas 5-7 mm, yellowish; fruits 3-4 × 1-2 mm, the valves reticulate-veined. Naturalized; roadsides, weedy and disturbed sites, 3600-9200 ft. Widespread. Introduced.

MIMOSA L. - Mimosa

Perennial herbs or shrubs; stems unarmed or armed with internodal, nodal, or subnodal prickles, pubescence basifixed; stipules herbaceous or inconspicuous, persistent; leaves alternate or clustered from spurs, even-pinnate, bipinnately compound, petiole prickly or not, leaflets symmetric or asymmetric; inflorescences headlike, spicate or racemose, axillary or terminal, solitary or in clusters, pedunculate, bracts deciduous; flowers sessile or pedicellate, corollas actinomorphic, not papilionaceous, androecium of separate stamens, stamens < 10; fruit a legume, dehiscent, the valves thin-papery, prickly or not, jointed or not, ultimately separating into segments; seeds strongly flattened, the coat with a pleurogram.

1. Inflorescence racemose…**M. dysocarpa**
1'. Inflorescence compact and headlike or spicate…(2)
2. Fruits quadrangular (both dorsiventrally and laterally compressed), the valves not breaking into segments upon maturity…**M. quadrivalvis**
2'. Fruits round and not compressed or laterally compressed, the valves breaking into segments upon maturity (or not segmented and breaking only from the replum in a single unit in *M. grahamii* and *M. aculeaticarpa*)…(3)
3. Leaves with 1 pair of pinnae, each with 1-3 pairs of leaflets…**M. turneri**
3'. Leaves with 1-8 pairs of pinnae, each with 3-15 pairs of leaflets…(4)
4. Calyx 0.8-1.8 mm…(5)
4'. Calyx 0.4-0.8 mm…(6)
5. Fruit margins undulate or conspicuously constricted between the seeds, the valves papery, 3-6 mm wide at maturity…**M. aculeaticarpa**
5'. Fruit margins straight, not constricted between the seeds, the valves leathery, 6-7 mm wide at maturity… **M. grahamii** (in part)
6. Leaves with 6-8 pairs of pinnae; pinnae with 9-15 pairs of leaflets; fruits not segmented…**M. grahamii** (in part)
6'. Leaves with 1-4 pairs of pinnae; pinnae with 3-7 pairs of leaflets; fruits segmented…**M. borealis**

Mimosa aculeaticarpa Ortega [*M. warnockii* B. L. Turner]—Catclaw mimosa. Shrubs, sometimes thicket-forming; stems armed with internodal, nodal, or subnodal, paired or single prickles, the herbage glabrous, glabrescent, or cinereous; leaves with 1-8 pairs of pinnae, leaflets 5-12 pairs per pinna, the blades 1.5-5 mm, elliptic or oblong; inflorescences 1-1.3 cm wide, headlike, solitary and terminal or in fascicles of 2-4 in leaf axils; calyx 1-1.8 mm, corollas 2-3 mm, whitish or pinkish, often fragrant; fruit 2-4 × 0.3-0.6 cm, straight, curved, or contorted, body oblong, glabrous, glabrescent, or puberulent, the replum prickly. New Mexico material belongs to **var. biuncifera** (Benth.) Barneby. Foothills, badlands, washes, prairies, desert grasslands, desert scrub, piñon-juniper woodlands, 4250-7250 ft. EC, SE, SC, SW.

Mimosa borealis A. Gray [*M. texana* (A. Gray) Small]—Fragrant mimosa. Shrubs; stems armed with mostly internodal, single prickles, the herbage glabrous, glabrescent or puberulent; leaves with 1-4 pairs of pinnae per leaf, leaflets 3-7 pairs per pinna, the blades 2-5 mm, elliptic or oblong; inflorescences 0.9-1.5 cm wide, headlike, solitary and terminal or in fascicles; calyx 0.4-0.8 mm, corollas 2-3 mm, reddish purple or pinkish, fragrant; fruit 3-5 × 0.6-0.8 cm, curved or contorted, glabrous, the replum prickly or unarmed. Sandstone and limestone; roadsides and ruderal areas; grasslands, oak, juniper, mesquite, 3250-5900 ft. NE, EC, SE, SC.

Mimosa dysocarpa Benth. [*M. dysocarpa* var. *wrightii* (A. Gray) Kearney & Peebles]—Velvet-pod mimosa, gatuno. Shrubs; stems armed with mostly internodal, single prickles, the herbage glabrous, glabrescent, or puberulent; leaves with 6-8 pairs of pinnae per leaf, leaflets 9-15 pairs per pinna, the blades 3-8 mm, oblong or lanceolate; inflorescences 2-6 × 1-2 cm, elongate, not headlike, solitary or 2 to many per node; calyx 1-2 mm, corollas 2-3 mm, whitish or pinkish; fruit 2-5 × 0.3-0.5 cm, straight or curved, ± sessile, the body oblong, glabrous, the replum prickly or unarmed. Igneous substrates, canyon bottoms, grasslands, oak, juniper, mesquite, 5200-8000 ft. Grant, Hidalgo, Luna.

Mimosa grahamii A. Gray—Graham's mimosa. Shrubs; stems armed with mostly nodal, single or paired prickles, the herbage glabrous, glabrescent, or puberulent; leaves with 1-4 pairs of pinnae, leaflets 7-10 pairs per pinna, the blades 2-5 mm, elliptic or lanceolate; inflorescences 1-1.5 cm wide, headlike, 2 to many per node near the tips of stems; calyx 0.6-1 mm, corollas 2-3 mm, whitish or pinkish; fruit 2-3 × 0.6-0.7 cm, straight or curved, ± sessile, the body oblong or oblanceolate, glabrous, the replum prickly or unarmed. Our material belongs to **var. grahamii**. Hillsides, grasslands, roadsides, oak, juniper, Mexican piñon, 3900-5900 ft. Hidalgo, Luna.

Mimosa quadrivalvis L.—Sensitive-briar. Perennial herbs, caudices superficial, at ground surface, not woody; stems eglandular, armed with mostly internodal, single or paired prickles, the herbage glabrous, glabrescent, or puberulent; leaves with 4–9 pairs of pinnae per leaf, leaflets 10–20 pairs per pinna, the blades 2–9.5 mm, elliptic or oblong, sensitive to touch; inflorescences 1–1.2 cm wide, headlike, in fascicles of 1–3 in leaf axils; calyx 0.3–0.6 mm, corollas 2–4 mm, whitish or pinkish; fruit 8–13 × 1.7–3.5 cm, straight or curved, ± sessile, the body linear or oblong, quadrangulate-compressed, glabrescent or puberulent, the replum prickly. NE, EC, SE, C, SC.

1. Leaf rachis proportionately longer than the petiole; fruit valves 1.7–2.5 mm wide…**var. angustata**
1'. Leaf rachis proportionately equal to or shorter than the petiole; fruit valves 2–3.5 mm wide…**var. occidentalis**

var. angustata (Torr. & A. Gray) Barneby—Little-leaf sensitive-briar. Leaf rachis proportionately longer than the petiole, leaflets 10–20 pairs per pinna, the blades 2–7.5 mm; inflorescences 1–1.2 cm wide; fruit 1.7–2.5 mm wide; seeds 3–5 mm. Igneous and sandy substrates, roadsides, grasslands, desert scrub oak, juniper, piñon communities, 3900–6250 ft. Introduced.

var. occidentalis (Wooton & Standl.) Barneby [*M. rupertiana* B. L. Turner]—Western sensitive-briar. Leaf rachis proportionately shorter than to equal to the petiole, leaflets 10–16 pairs per pinna, the blades 2–9.5 mm; inflorescences 1.2–1.5 cm wide; fruit 2–3.5 mm wide; seeds 5–9 mm. Prairies, dunes, roadsides, juniper, piñon, 3250–6250 ft.

Mimosa turneri Barneby—Western sensitive-briar. Shrubs, caudices woody; stems armed with mostly nodal or subnodal, single or paired prickles, the herbage glabrous, glabrescent, or puberulent; leaves with 1 pair of pinnae per leaf, leaflets 1–3 pairs per pinna, the blades 1.5–4 mm, elliptic or oblong, eglandular; inflorescences 1–1.3 cm wide, headlike, solitary and terminal; calyx 0.4–1 mm, corollas 2–3 mm, whitish or pinkish; fruit 2.5–6 × 0.5–0.8 cm, straight, curved, or contorted, body linear or oblong, glabrous, the replum prickly or unarmed. Limestone, talus, canyons, desert scrub, juniper, 5250–5900 ft. Eddy, Otero.

ONOBRYCHIS Mill. – Sainfoin

Onobrychis viciifolia Scop.—Sainfoin. Perennial herbs, herbage glabrous, glabrescent, or puberulent; leaves with 7–10 pairs of leaflets, the blades 10–25 mm, linear, elliptic, oblanceolate, or obovate, reddish-dotted above; inflorescences 1–4-flowered, peduncles longer than the subtending leaves; calyx 5–7 mm, corollas 8–15 mm, pale pinkish purple; fruit 6–7 mm, the body obovate, exserted from the persistent calyx, the valves leathery, reticulate-veined, prickly around the margin. Naturalized, 5650–8020 ft. NW, Grant, San Miguel. Introduced.

OTTLEYA D. D. Sokoloff – Bird's-foot trefoil

Annual or perennial herbs; stems unarmed, pubescence basifixed; stipules reduced to small glands; leaves odd-pinnate, pinnately compound and petiolate or palmately compound and sessile; inflorescences an umbel or solitary, axillary, peduncle elongate with a foliage leaf, or short and the foliage leaf absent, bracts reduced to small glands resembling stipules, persistent; flowers pedicellate, calyx tube not inflated, the teeth acute, corollas papilionaceous, the wings equal to or longer than the keel, androecium diadelphous, pollen grains with 4–7 apertures; fruits a legume, dehiscent, sessile.

1. Plants subshrubs, the stems woody at the base; internodes 4–9 cm…**O. rigida**
1'. Plants perennial herbs, the stems not woody; internodes < 4 cm…(2)
2. Calyx and fruits sparsely to densely villous, at least some hairs long, soft, spreading, and ± curved…**O. mollis**
2'. Calyx and fruits sparsely to densely strigose, the hairs appressed and straight…(3)
3. Plants with erect to ascending stems; the leaves palmate or subpalmate…**O. wrightii**
3'. Plants with decumbent to ascending stems; the leaves irregularly pinnate or subpalmate…**O. plebeia**

Ottleya mollis (A. Heller) D. D. Sokoloff & Gandhi [*Lotus greenei* (Wooton & Standl.) Ottley]—Greene's bird's-foot trefoil. Perennial herbs, stems ascending or prostrate, ± mat-forming, the herbage villous, villosulose, or tomentulose; leaves pinnately or subpalmately compound, petiolate, leaflets 3–6, the blades 2–15 mm, oblanceolate or obovate; inflorescences 1–5-flowered; calyx tube 3.5–5 mm, corollas 13–20 mm, yellowish, orangish, or ochroleucous, reddish-tinged in age; fruit 1.5–2.5 × 0.2–0.3 cm, dehiscent, sessile, the body oblong, straight, the valves leathery, villous. Hillsides, grasslands, stream banks, washes, flats, canyons, desert scrub, piñon-juniper, ponderosa pine-oak, 3900–8550 ft. C, WC, SC, SW.

Ottleya plebeia (Brandegee) D. D. Sokoloff & Gandhi [*Lotus plebeius* (Brandegee) Barneby]— New Mexico bird's-foot trefoil. Perennial herbs; stems ascending, decumbent, or prostrate, the herbage glabrescent or strigose; leaves pinnately or subpalmately compound, leaflets 3–7, often 3 palmately compound leaflets, the blades 5–15 mm, linear, elliptic, oblanceolate, or obovate; inflorescences 1–4-flowered; calyx tube 2.5–5 mm, corollas 10–18 mm, yellowish or orangish, reddish-tinged in age; fruit 1.5–2.5 × 0.2–0.3 cm, dehiscent, sessile, the body oblong or linear, straight or ± curved, the valves leathery, glabrescent or strigose. Stream banks, washes, canyons, desert scrub, desert grasslands, piñon-juniper, ponderosa pine-oak, 3900–8000 ft. NC, C, WC, SC, SW.

Ottleya rigida (Benth.) D. D. Sokoloff [*Lotus rigidus* (Benth.) Greene]—Shrubby deervetch. Perennial herbs or subshrubs, caudices woody; stems ascending or erect, the herbage glabrescent or strigose; leaves pinnately or subpalmately compound, leaflets 3–7, often 3 palmately compound leaflets, blades 5–15 mm, oblanceolate or obovate; inflorescences 1–4-flowered; calyx tube 4–7 mm, corollas 12–20 mm, yellowish, reddish-tinged in age; fruit 2–4 × 0.3–0.5 cm, persistent, sessile, the body oblong, straight, glabrescent or strigose. Hillsides, grasslands, deserts, washes, flats, canyons, piñon-juniper, ponderosa pine-oak, 6550–7250 ft. Grant.

Ottleya wrightii (A. Gray) D. D. Sokoloff [*Lotus wrightii* (A. Gray) Greene]—Wright's bird's-foot trefoil. Perennial herbs; stems ascending or erect, the herbage glabrescent or strigose; leaves palmately or subpalmately compound, leaflets 3–6, the blades 8–20 mm, linear, elliptic, oblanceolate, or obovate; inflorescences 1- or 2-flowered; calyx tube 4–5 mm, corollas 10–15 mm, yellowish, reddish-tinged in age; fruit 2–3.5 × 0.1–0.3 cm, dehiscent, the body oblong, straight or ± curved, glabrescent or strigose. Washes and canyons, piñon-juniper woodland, ponderosa pine-oak, mixed conifer, 3900–8200 ft. NC, NW, C, WC, SC, SW.

OXYTROPIS DC. – Locoweed

Perennial herbs, acaulescent or subacaulescent; stems unarmed, pubescence basifixed or dolabriform; stipules herbaceous, persistent; leaves pinnately compound, petiolate; inflorescences racemose, terminal, 1- to several-flowered, pedunculate; flowers pedicellate, calyx tube 5-toothed, the teeth acute, corollas papilionaceous, the keel conspicuously beaked, androecium diadelphous; fruits a legume, sessile or stipitate, dehiscent.

1. Fruits deflexed or pendulous; flowers ascending at anthesis, deflexed in age…**O. deflexa**
1'. Fruits and flowers erect, ascending, or spreading…(2)
2. Racemes 1–4-flowered, inflorescences subcapitate and compact, the rachis short and obscured by flowers, 0.5–1 cm in fruit…**O. parryi**
2'. Racemes 5–40-flowered, inflorescences ovate, elliptic, or oblong, the rachis elongate and not obscured by flowers, at least basally, 3–23 cm in fruit…(3)
3. Corollas whitish, yellowish, or ochroleucous…**O. sericea**
3'. Corollas reddish, bluish, pinkish, or purplish…(4)
4. Fruit valves papery; corollas 12–16 mm…**O. splendens**
4'. Fruit valves leathery or woody; corollas 15–25 mm…**O. lambertii**

Oxytropis deflexa (Pall.) DC.—Nodding locoweed. Perennial, caulescent or subacaulescent; stems decumbent, ascending, or erect, the herbage pilose or pilosulous, the hairs basifixed; leaves with 11–20 pairs of leaflets, the blades 3–25 mm, elliptic, linear, oblong, or ovate; inflorescences subcapitate, 3–25-flowered, flowers ascending or spreading, deflexed in fruit; calyx tube 2–4 mm, corollas 6–11 mm, pinkish purple, bluish purple, or whitish and pinkish purple–tinged; fruits 0.8–1.8 × 0.3–0.5 cm, deflexed, ± unilocular, body elliptic or oblong, glabrescent or strigulose. New Mexico plants belongs to **var. seri-**

cea Torr. & A. Gray, silky nodding locoweed. Mixed conifer, wet subalpine meadows, spruce-fir communities, 7850–11,000 ft. Colfax, Rio Arriba, Taos.

Oxytropis lambertii Pursh [*O. lambertii* var. *articulata* (Greene) Barneby; *O. lambertii* var. *bigelovii* A. Gray]—Purple locoweed. Perennial, acaulescent; stems covered in a thatch of decomposed leaf bases; stems decumbent, ascending, or erect, the herbage strigose, the hairs dolabriform; leaves with 3–10 pairs of leaflets, the blades 1–35 mm, elliptic, lanceolate, or oblong; inflorescences subcapitate, elongate in fruit, 3–25-flowered, flowers ascending or spreading; calyx tube 5–9 mm, corollas 15–25 mm, pinkish purple, bluish purple, or whitish and pinkish purple–tinged; fruits 1–2.5 × 0.3–0.5 cm, erect, bilocular, body oblanceolate or oblong, glabrescent, villous, or sericeous. Badlands, dunes, desert scrub, piñon-juniper woodlands, ponderosa pine, subalpine meadows, spruce-fir communities, 5200–10,850 ft. Widespread except EC, SE.

Oxytropis parryi A. Gray—Parry's locoweed. Perennial, acaulescent, caudices ± woody, covered in a thatch of decomposed leaf bases and stipules; stems decumbent, ascending, or erect, mat-forming, the herbage strigose, the hairs dolabriform; leaves with 3–7 pairs of leaflets, the blades 3–12 mm, elliptic, ovate, or oblong; inflorescences 1–3-flowered, flowers erect, ascending, or spreading; calyx tube 3–5 mm, corollas 7–12 mm, pinkish purple, bluish purple, or whitish and pinkish purple–tinged; fruits 1.5–2 × 0.5–0.7 cm, erect, ascending, or spreading, bilocular, persistent, sessile, the body oblanceolate or oblong, ± dorsiventrally compressed in cross-section, the valves leathery, glabrescent, strigulose, or sericeous. Meadows and ridges, often on limestone substrates but not an obligate calciphile, 10,825–11,480 ft. NC.

Oxytropis sericea Nutt.—White locoweed. Perennial, acaulescent; caudices covered in a thatch of decomposed leaf bases; stems decumbent, ascending, or erect, mat-forming, the herbage strigose or sericeous, the hairs basifixed; leaves with 3–7 pairs of leaflets, the blades 5–25 mm, elliptic, ovate, or oblong; inflorescences 1–3-flowered, flowers erect, ascending, or spreading; calyx tube 5–9 mm, corollas 18–25 mm, whitish, yellowish, or ochroleucous, occasionally pinkish-tinged; fruits 1.5–2 × 0.4–0.8 cm, erect, ascending, or spreading, body oblanceolate or oblong, glabrescent, strigulose, or sericeous. New Mexico material belongs to **var. sericea**. Prairies, sagebrush, piñon-juniper, ponderosa pine, wet meadows, mixed conifer, spruce-fir communities, 5350–11,650 ft. NE, NC, C, WC, SC.

Oxytropis splendens Douglas ex Hook. [*O. richardsonii* (Hook.) Wooton & Standl.]—Showy locoweed. Perennial, acaulescent, caudices covered in a thatch of decomposed leaf bases; stems decumbent, ascending, or erect, the herbage puberulent, strigose, or pilose, the hairs basifixed; leaves with 2–7 pairs of leaflets, the blades 3–20 mm, elliptic, lanceolate, or oblong; inflorescences subcapitate, elongate in fruit, 3–25-flowered, flowers erect, ascending, or spreading; calyx tube 5–6 mm, corollas 12–16 mm, pinkish purple, bluish purple, or reddish purple; fruits 1–1.7 × 0.3–0.5 cm, erect, ascending, or spreading, body oblanceolate or oblong, glabrescent, strigulose, or sericeous. Open meadows, forest openings, subalpine slopes, 7500–11,000 ft. NE, NC.

PARKINSONIA L. - Paloverde

Parkinsonia aculeata L.—Retama, Mexican paloverde. Shrubs or small trees; stems armed, the herbage glabrous, glabrescent, or villosulous; leaves solitary or in clusters of 2–3, each with 2–3 pairs of pinnae, leaflets 10–30 pairs per pinna, the blades 2–5 mm, ovate or oblong; inflorescences terminal, 2–15-flowered; calyx 6–7 mm, pedicels ascending to spreading, corollas 15–20 mm, orangish yellow with red markings; fruit 2–10 × 0.4–0.6 cm, indehiscent, the body irregularly torulose, ± rounded in cross-section, the valves leathery. Naturalized in desert scrub communities, 3600–5500 ft. Doña Ana, Grant, Otero. Introduced.

PARRYELLA Torr. & A. Gray - Dunebroom

Parryella filifolia Torr. & A. Gray ex A. Gray—Dunebroom. Shrubs; stems ascending, glandular, the herbage glabrous, glabrescent, or strigose; leaves with 4–20 pairs of leaflets, the blades 0.3–1.5 cm, linear, involute, glandular-punctate; inflorescences racemose; calyx 2.5–3 mm, corollas absent, stamens

9-10, the filaments free distally and fused proximally (within the calyx), the anthers yellow; fruits 5-6 mm, the valves glandular-punctate. Sandy sites, plains, bluffs, badlands, dunes, washes, sagebrush, juniper, piñon-juniper communities, 4550-7250 ft. NW.

PEDIOMELUM Rydb. – Scurf pea

Perennial herbs, caulescent, subacaulescent, or caulescent; stems unarmed, pubescence basifixed, the herbage glandular-punctate or eglandular; leaves odd-pinnate, palmately or pinnately compound, petiolate; inflorescences pseudoracemose, axillary, terminal when acaulescent, with 2 to many clusters, each with 1 to many flowers at each node; flowers pedicellate, calyx ± inflated in fruit, the teeth acute, corollas papilionaceous, androecium diadelphous, anthers dimorphic; fruits a legume, indehiscent, sessile, ± not exserted from the calyx, the valves glandular-punctate or eglandular.

1. Plants caulescent; stems erect or ascending, the main axis > 2 cm, solitary or branched at or above the midpoint, the branches ascending…(2)
1'. Plants acaulescent or subacaulescent; stems < 2 cm, solitary or branched near the base, the lateral branches decumbent…(7)
2. Herbage glabrous or glabrescent; fruit valves glabrous and glandular-punctate…**P. linearifolium**
2'. Herbage strigose or sericeous; fruit valves tomentose, strigose, or sericeous, glandular-punctate (eglandular in *P. argophyllum*)…(3)
3. Fruits tomentose, eglandular…(4)
3'. Fruits strigose, glandular-punctate…(5)
4. Flowers 10-20 mm, clustered into dense, spicate heads…**P. esculentum**
4'. Flowers 5-11 mm, borne in loose racemes or small axillary clusters…**P. argophyllum** (in part)
5. Fruits tomentose, eglandular; calyx 4-5 mm (elongating to 8 mm in fruit)…**P. argophyllum** (in part)
5'. Fruits strigose, glandular-punctuate; calyx 5-15 mm (elongating to 8-15 mm in fruit)…(6)
6. Calyx 5-7 mm; corollas 9-11 mm…**P. digitatum**
6'. Calyx 11-15 mm; corollas 13-17 mm…**P. pentaphyllum** (in part)
7. Racemes 1-4-flowered; inflorescences subcapitate and compact in fruit, the rachis obscured by flowers… **P. hypogaeum**
7'. Racemes 5-40-flowered; inflorescences elongate in fruit, the rachis not obscured by flowers…(8)
8. Leaflet margins undulate…**P. pentaphyllum** (in part)
8'. Leaflet margins entire, not undulate…**P. megalanthum**

Pediomelum argophyllum (Pursh) J. W. Grimes—Silver-leaf breadroot. Perennial herbs, caulescent; stems erect or ascending, the herbage sericeous or tomentose, glandular-punctate; leaves with 3-5 leaflets, the blades 1-5 cm, elliptic or obovate, the margins entire; inflorescences racemose, elongate in fruit, the rachis not obscured by flowers; calyx 4-5 mm, 7-8 mm in fruit, corolla 7-11 mm, pinkish purple or bluish purple; fruits with beaks 3-4 mm, the valves sericeous or tomentose, eglandular. Sandy or rocky soil, prairies, montane meadows, 4900-7900 ft. Colfax, Union.

Pediomelum digitatum (Nutt. ex Torr. & A. Gray) Isely—Palm-leaf breadroot. Perennial herbs, caulescent; stems erect, the herbage strigulose, glandular-punctate; leaves with 3-5 leaflets, the blades 1-4 cm, linear, the margins entire; inflorescences racemose, elongate in fruit, the rachis not obscured by flowers; calyx 5-7 mm, 8-10 mm in fruit, corollas 9-11 mm, pinkish purple or bluish purple; fruits with beaks 2-4 mm, the valves strigose, sparsely glandular-punctate. Sandy or sandy-clay sites, shortgrass prairies; known in New Mexico from a single modern collection from S of Clovis, 3900-4300 ft. Roosevelt.

Pediomelum esculentum (Pursh) Rydb.—Large breadroot. Mostly caulescent herb, to 50 cm, from vertical rhizomes and inflated roots; herbage eglandular and pubescent; leaves mostly 5-foliate, leaflets elliptic to oblanceolate, 2-6 × 1-2.3 cm, adaxially glabrate except on the midvein; flowers 12-20 mm, corolla reddish bluish purple, wings 15-17 mm; pods 4-6 mm, eglandular, hairy. Grasslands, 6300-8120 ft. Colfax.

Pediomelum hypogaeum (Nutt.) Rydb.—Little breadroot. Perennial herbs, acaulescent, subacaulescent, or short-caulescent; stems erect, the herbage strigose, eglandular or glandular-punctate; leaves with 3-5 leaflets, the blades 2-6 cm, rhombic, oblong, or obovate, the margins entire; inflorescences racemose, elongate in fruit, the rachis not obscured by flowers; calyx 8–12 mm, 12–15 mm in fruit, corollas 11–18 mm, pinkish purple or bluish purple; fruits with beaks 8–18 mm, the valves glabrescent or strigose, eglandular. New Mexico material belongs to **var. hypogaeum**. Sandy or sandy-clay sites, bluffs, E plains, 4550–5500 ft. NE, NC, C.

Pediomelum linearifolium (Torr. & A. Gray) J. W. Grimes—Narrow-leaf breadroot. Caulescent; stems erect, the herbage glabrous or glabrescent, glandular-punctate; leaves with 3–4 leaflets, the blades 2–6 cm, linear, the margins entire; inflorescences racemose, elongate in fruit, the rachis not obscured by flowers; calyx 4–6 mm, corollas 9–11 mm, bluish purple; fruits with beaks 3–4 mm, the valves glabrous, glandular-punctate. Rocky or sandy soils, shortgrass prairies; known in New Mexico from a single modern collection from Chaves County, W of Artesia, 2950–3600 ft.

Pediomelum megalanthum (Wooton & Standl.) Rydb.—Breadroot. Acaulescent, subacaulescent, or short-caulescent; stems erect, the herbage strigose, glandular-punctate; leaves with 5–8 leaflets, the blades 0.3–1.5 cm, elliptic, oblong, or obovate, the margins entire; inflorescences racemose, elongate in fruit, the rachis not obscured by flowers; calyx 12–17 mm, corollas 14–20 mm, bluish purple or whitish and bluish purple-tinged; fruits with beaks 5–8 mm, the valves strigose, eglandular. Sandy or sandy-clay sites, bluffs, badlands, canyons, washes, piñon-juniper woodlands, 5250–6550 ft.

Pediomelum pentaphyllum (L.) Rydb.—Chihuahuan scurf pea. Acaulescent, subacaulescent, or short-caulescent; stems erect, the herbage strigose, glandular-punctate; leaves with 5–6 leaflets, palmately or subpinnately arranged, the blades 2–5 cm, elliptic, oblong, or obovate, the margins undulate; inflorescences racemose, elongate in fruit, the rachis not obscured by flowers; calyx 11–15 mm, corollas 13–17 mm, bluish purple or pinkish purple; fruits with beaks 12–15 mm, the valves strigose, glandular-punctate. Sandy or gravelly loam, desert grasslands, desert scrub, 4400–6600 ft. Hidalgo. **R**

PETERIA A. Gray – Spine-vetch

Peteria scoparia A. Gray—Rush spine-vetch. Perennial herbs, caulescent; stems erect, striate, armed, the herbage glabrescent or strigose; leaves with 4–7 pairs of leaflets, the blades 2–10 cm, elliptic or oblong; inflorescences racemose; calyx 7–9 mm, corollas 12–16 mm, whitish, ochroleucous, or ochroleucous and purplish-tinged; fruits 3.5–6 × 0.4–0.5 cm, spreading, the body oblanceolate, the valves glabrous. Limestone, desert grasslands, desert scrub, 4250–6600 ft. SC, SW.

PHASEOLUS L. – Lima bean

Annuals; stems unarmed, pubescence uncinate; stipules striate, persistent; leaves odd-pinnate, pinnately compound, trifoliate, petiolate; inflorescences pseudoracemose, axillary, pedunculate, with 2 to many clusters, each with 1–3 flowers at each node; flowers pedicellate, calyx tube not inflated, 5-lobed, the upper ones ± connate, corollas papilionaceous, the keels twisted 1–3 turns, androecium diadelphous; fruits a legume, dehiscent, sessile or substipitate, the body ± laterally compressed in cross-section.

1. Herbage lacking fine uncinate pubescence; calyx lobes slender, the upper lobe distinct to near the base and as long as the others; fruits hirsute, 3–4 mm wide; often misidentified as a *Phaseolus* species… **Macroptilium**
1′. Herbage with fine uncinate pubescence; calyx lobes broad, the upper lobe partly fused and shorter than the others; fruits glabrous, glabrescent, or variously pubescent, > 4 mm wide…(2)
2. Annuals or short-lived perennials from a slender taproot, not thickened, tuberous, or woody…(3)
2′. Perennial herbs from a rhizome, tuber, or woody taproot…(4)
3. Leaflets symmetric, not lobed; fruits 4–7 cm…**P. acutifolius**
3′. Leaflets asymmetric, 3-lobed; fruits 1–2 cm…**P. filiformis**
4. Flowers 1 or 2; stems erect or spreading, not or weakly twining; caudices subterranean, tuberous, stems arising from vertical rhizomes…**P. parvulus**
4′. Flowers 2 to many; stems trailing or twining; caudices woody, taprooted, not rhizomatous or tuberous… (5)
5. Fruits 3–6 cm; leaflets unlobed and symmetric…**P. maculatus**
5′. Fruits 1–3 cm; leaflets symmetrically 3-lobed, unlobed, or asymmetrically lobed…(6)

6. Fruits 1–2 cm; corollas 6–10 mm; leaflets unlobed or asymmetrically lobed (± arrowhead-shaped)…**P. angustissimus**

6'. Fruits 2.3–3 cm; corollas 12–15 mm; leaflets symmetrically 3-lobed…**P. pedicellatus**

Phaseolus acutifolius A. Gray [*P. acutifolius* var. *tenuifolius* A. Gray]—Tepary bean. Annuals, caulescent; stems erect, trailing, or climbing, the herbage glabrous or glabrescent; leaflets unlobed and ± asymmetric, the blades 3–8 cm, linear or lanceolate; inflorescences with 1–4 flowering nodes, the internodes elongate; calyx 3–4 mm, corollas 6–7 mm, pinkish purple, whitish, ochroleucous, or greenish; fruits 4–7 × 0.5–0.6 cm, the body oblong, the beak 1–2 mm, the valves glabrous or glabrescent. Chihuahuan desert scrub, piñon-juniper woodland, ponderosa pine–oak communities, 3250–8200 ft. WC, SW.

Phaseolus angustissimus A. Gray [*P. angustissimus* var. *laetus* M. E. Jones; *P. dilatatus* Wooton & Standl.]—Narrow-leaf lima bean. Perennial herbs, caulescent; stems trailing or climbing, the herbage glabrous or glabrescent; leaflets unlobed or shallowly 3-lobed, the lobes asymmetric, the blades 2–7 cm, linear or lanceolate (± arrowhead-shaped); inflorescences with 1–7 flowering nodes, the internodes elongate; calyx 2–3 mm, corollas 6–10 mm, pinkish or pinkish purple; fruits 1–2 × 0.6–0.7 cm, the body oblong, the beak 1–2 mm, the valves glabrescent. Desert scrub, piñon-juniper, 4550–7250 ft. NW, C, WC, SC, SW.

Phaseolus filiformis Benth. [*P. wrightii* A. Gray]—Wright's lima bean. Annual or short-lived perennial herbs, caulescent; stems erect, trailing, or climbing, the herbage glabrous or glabrescent; leaflets ± asymmetrically 3-lobed, the blades 1–4 cm, deltate or ovate; inflorescences with 1–3 flowering nodes, the internodes elongate; calyx 1–2 mm, corollas 5–9 mm, pinkish purple or pinkish; fruits 1–2 × 0.5–0.6 cm, the body oblong, the beak 1–2 mm, the valves glabrous, glabrescent, or uncinate-pubescent. Deserts, badlands, hillsides, washes, occasionally ruderal areas and roadsides, 4250–6250 ft. SW.

Phaseolus maculatus Scheele [*P. metcalfei* Wooton & Standl.]—Spotted lima bean. Perennial herbs, caulescent; stems trailing or climbing, the herbage glabrous or glabrescent; leaflets unlobed and ± asymmetric, the blades 4–9 cm, ovate or rhombic; inflorescences with several flowering nodes, the internodes elongate; calyx 4–5 mm, corollas 8–9 mm, pinkish or pinkish purple; fruits 3–6 × 1–1.8 cm, the body oblong, the beak 1–2 mm, the valves glabrescent. Canyons, washes, roadsides, riparian forests, ponderosa pine–oak communities, 4550–7900 ft. WC, SW.

Phaseolus parvulus Greene—Pinos Altos Mountain bean. Perennial herbs, caulescent; stems arising from vertical rhizomes, trailing or climbing, the herbage glabrous or glabrescent; leaflets unlobed, the base asymmetric, the blades 1–5 cm, oblong, elliptic, or lanceolate; inflorescences with 1 or 2 terminal flowers; calyx 6–7 mm, corollas 12–18 mm, pinkish purple; fruits 3–4 × 0.4–0.5 cm, the body oblong, the valves glabrous. Piñon-juniper, ponderosa pine–oak, mixed conifer, 5575–8200 ft. Catron, Grant, Hidalgo.

Phaseolus pedicellatus Benth. [*P. wrightii* A. Gray var. *grayanus* (Wooton & Standl.) Kearney & Peebles; *P. scabrellus* Benth. ex S. Watson]—Gray's lima bean. Perennial herbs, caulescent; stems trailing or climbing, the herbage glabrous or glabrescent; leaflets shallowly or deeply palmately 3-lobed, the lobes symmetric, the blades 2–7 cm; inflorescences with 1–7 flowering nodes, the internodes elongate; calyx 2.5–3.5 mm, corollas 12–15 mm, pinkish or reddish pink; fruits 2.5–3 × 0.4–0.6 cm, the body oblong, the beak 0.5–1 mm, the valves glabrescent or pilosulous. New Mexico material belongs to **var. grayanus** (Wooton & Standl.) A. Delgado ex Isely. Canyons, riparian forests, piñon-juniper, ponderosa pine–oak, mixed conifer, 5250–8200 ft. WC, SC, SW.

POMARIA Cav. – Holdback

Pomaria jamesii (Torr. & A. Gray) Walp. [*Hoffmannseggia jamesii* Torr. & A. Gray]—James' holdback. Perennial herbs, caulescent; stems prostrate, ascending, or erect, striate, the herbage glabrescent or strigose, glandular-punctate; leaves with 2–7 pairs of pinnae, leaflets 5–10 pairs per pinna, the blades 2.5–7 mm, elliptic or oblong; inflorescences axillary and terminal, 16–26-flowered, flowers spreading;

sepals 6-9 mm, corollas 5-9 mm, yellowish with reddish markings; fruits 2-2.5 cm × ca. 7 mm, dehiscent, the body lunate or ovate, slightly curved, the valves thick-papery, stellate-pubescent, glandular-punctate; seeds 2. Rocky or sandy soils, roadsides, open prairie, desert grassland, desert scrub communities, 3900-6550 ft. Widespread except NC, McKinley.

PROSOPIS L. - Mesquite

Shrubs or trees; stems unarmed or armed with nodal spines, pubescence basifixed; stipules spinose or inconspicuous, persistent; leaves even-pinnate, bipinnately compound, 1 or 2 pairs of pinnae per leaf, petiole gland present; inflorescences axillary or terminal, headlike or spicate, solitary or in fascicles in axils, pedunculate; flowers sessile, corollas actinomorphic, not papilionaceous, 5-merous, androecium of separate stamens, stamens < 10; fruit a legume, indehiscent, the valves thick, ± succulent when young, ± woody when mature; seeds strongly flattened, the coat with a pleurogram.

1. Fruits tightly coiled; leaflets 5-8 pairs per pinna…**P. pubescens**
1'. Fruits not tightly coiled; leaflets 10-30 pairs per pinna…(2)
2. Leaflets pubescent; leaves with 1-3 pairs of pinnae…**P. velutina**
2'. Leaflets glabrous or glabrescent; leaves with 1 pair of pinnae…**P. glandulosa**

Prosopis glandulosa Torr.—Honey mesquite. Shrubs or trees; stems armed with 1 or rarely 2 nodal spines, the herbage glabrous, glabrescent, or puberulent; leaves with 1 pair of pinnae, leaflets 10-12 pairs per pinnae, the blades 2.5-6 cm, obovate or oblong; inflorescences axillary and arising from spurs, spicate, 4-8 cm, compact and amentlike; corollas 2-4 mm, yellowish, ochroleucous, or greenish; fruits 10-20 × 1-1.5 cm, the stipe 5-8 mm, ± straight, ± constricted between seeds, the body linear or oblong. Widespread except NW and Taos.

1. Leaflets 2.5-6 cm; twig spines mostly solitary…**var. glandulosa**
1'. Leaflets 1-2.5 cm; twig spines mostly paired…**var. torreyana**

var. glandulosa—Stems mostly armed with 1 nodal spine; leaflets 6-12 pairs per pinna, the blades 2.5-6 cm. Often forming woodlands, on hillsides, stream banks, washes, scrublands, 2950-6600 ft. Widespread except NW, Taos.

var. torreyana (L. D. Benson) M. C. Johnst.—Stems armed with 2 nodal spines; leaflets 10-20 pairs per pinna, the blades 1-2.5 cm. Hillsides, sand dunes, stream banks, washes, woodlands, scrublands, ruderal areas, roadsides, irrigation ditches, 2950-6600 ft. NE, C, SE, SC, SW.

Prosopis pubescens Benth.—Screw bean mesquite. Shrubs or trees; stems armed with 2 nodal spines, the herbage glabrous, glabrescent, or velutinous; leaves with 1-2 pairs of pinnae, leaflets 5-8 pairs per pinna, the blades 0.5-1.5 cm, elliptic or oblong; inflorescences axillary, arising from spurs or solitary, spicate, corollas 2-3 mm, yellowish; fruits 3-5 × 0.4-0.5 cm, the body linear, tightly coiled into an elongate cylinder. Stream banks, washes, woodlands, scrublands, also ruderal areas and irrigation ditches, 3250-5575 ft. C, SC, Cibola, Hidalgo, Sandoval.

Prosopis velutina Wooton—Velvet mesquite. Shrubs or trees; stems armed with 1 or 2 nodal spines, the herbage glabrous, glabrescent, or velutinous; leaves with 1-3 pairs of pinnae, leaflets 15-30 pairs per pinna, the blades 0.5-1.5 cm, elliptic or oblong; inflorescences axillary and arising from spurs, spicate, 5-12 cm, ascending to drooping, compact and amentlike; corollas 4-5 mm, yellowish or greenish; fruits 10-20 × 0.4-0.5 cm, the body linear, ± rounded, not strongly laterally compressed. Hillsides, sand dunes, stream banks, washes, woodlands, scrublands, 3935-5900 ft. Catron, Hidalgo.

PSORALIDIUM Rydb. - Scurf pea

Psoralidium tenuiflorum (Pursh) Rydb. [*Psoralea tenuiflora* Pursh]—Scurf pea. Perennial herbs, caulescent; stems erect or ascending, the herbage glabrous or glabrescent, glandular-punctate; leaves with 3-5 leaflets, the blades 1.5-5 cm, elliptic, oblong, or obovate; inflorescences with crowded or interrupted flower clusters; calyx 1.5-3 mm, corollas 5-7 mm, pinkish purple, bluish purple, or whitish; fruits

7-8 mm, the body laterally compressed, the valves leathery, glabrous, glandular-punctate. Bluffs, road-sides, prairies, piñon-juniper, ponderosa pine-oak communities, 4600-7215 ft. Widespread.

PSOROTHAMNUS Rydb. - Smokebush

Psorothamnus scoparius (A. Gray) Rydb.—Broom smokebush. Shrubs, sparsely leafy or leafless and thatchlike; stems erect or ascending, the herbage strigose, glandular-punctate; leaves with 1-3 leaf-lets, the blades 0.5-2 cm, linear or narrowly oblanceolate; inflorescences subcapitate, 6-15-flowered; calyx 3.5-4.5 mm, corollas 6-9 mm, pinkish purple or bluish purple; fruits 3-4 mm, ± exserted from the calyx, the body ovate, the valves leathery, glandular-punctate. Sandy sites, dunes, roadsides, desert scrub, prairies, scattered piñon-juniper communities, 4550-7250 ft. NW, C, SC, Cíbola, Grant.

RHYNCHOSIA Lour. - Snout-bean

Rhynchosia senna Gillies ex Hook. & Arn. [*R. texana* Torr. & A. Gray]—Texas snout-bean. Perennial herbs, caulescent; stems ascending or twining, the herbage villosulous, glandular-punctate; leaves tri-foliate, the blades 1-5 cm, oblong, ovate, or lanceolate; inflorescences in axillary fascicles, 1-3-flowered; calyx 2.5-3 mm, corollas 5-8 mm, yellowish; fruits 1.2-2.5 × 0.5-0.7 cm, the body lunate, the valves vil-losulous. New Mexico material belongs to **var. texana** (Torr. & A. Gray) M. C. Johnst. Hillsides, rocky outcrops, canyons, desert scrub, piñon-juniper, ponderosa pine communities, 4250-7000 ft. WC, SE, SC, SW, San Juan.

ROBINIA L. - Locust

Shrubs or trees; stems unarmed, pubescence basifixed; stipules herbaceous, persistent; leaves odd-pinnate, pinnately compound, petiolate; inflorescences racemose, axillary, pedunculate; flowers pedi-cellate, calyx tube not inflated, the teeth acute, corollas papilionaceous, aromatic, androecium diadel-phous; fruits a legume, sessile, tardily dehiscent, the body laterally compressed in cross-section.

1. Corollas whitish, the petals 15-20 mm…**R. pseudoacacia**
1'. Corollas pinkish, the petals 20-25 mm…**R. neomexicana**

Robinia neomexicana A. Gray—New Mexico locust. Shrubs or trees; stems armed, the herbage glabrous, glabrescent, or velutinous; leaves with 7-11 pairs of leaflets, the blades 1-4 cm, elliptic or oblong; inflorescences in axillary fascicles; calyx 6-7 mm, corollas 20-25 mm, pinkish; fruits 6-10 × 0.7-1 cm, the body oblong, the valves glabrous, glabrescent, hispid, or sericeous, glandular or eglandular.

1. Fruit valves hispid or glandular-hispid; inflorescence axis glandular-hispid…**var. neomexicana**
1'. Fruit valves sericeous, not glandular-hispid; inflorescence axis not glandular-hispid…**var. rusbyi**

var. neomexicana—Inflorescences glandular-hispid; calyx glandular-hispid; fruit valves hispid or glandular-hispid. Often forms thickets; common along streams; cultivated and escaping along roadsides, near towns and old homesteads; moving into ponderosa pine and mixed conifer communities after fires, 4500-9000 ft. Widespread except EC, SE.

var. rusbyi (Wooton & Standl.) W. C. Martin & C. R. Hutchins ex Peabody—Inflorescences glabrous, glabrescent, or sericeous, eglandular; calyx glabrous, glabrescent, or sericeous, eglandular; fruit valves glabrous, glabrescent, or sericeous, eglandular. Ponderosa pine, mixed conifer communities, 5900-9000 ft. Catron, Grant, Los Alamos, Otero.

Robinia pseudoacacia L.—Black locust. Shrubs or trees; stems armed, the herbage glabrous or glabrescent; leaves with 7-11 pairs of leaflets, the blades 2-4 cm, elliptic or oblong; inflorescences in axillary fascicles; calyx 6-7 mm, corollas 15-20 mm, whitish or whitish and pinkish-tinged; fruits 6-10 × 0.7-1 cm, the body oblong, the valves glabrous, eglandular. Naturalized, widely scattered throughout the state, 3600-7000 ft. Introduced.

SECURIGERA DC. - Crown vetch

Securigera varia (L.) Lassen [*Coronilla varia* L.]—Crown vetch. Perennial herbs, caulescent; stems decumbent or ascending, the herbage glabrous, glabrescent, or puberulent; leaves with 5-8 pairs of leaflets, the blades 0.8-2 cm, obovate or oblong; inflorescences umbellate, 10-15-flowered; calyx 2-3 mm, corollas 8-10 mm, whitish, pinkish-tinged, and purple-striate; fruits 2-5 cm, articles 3-10, the valves glabrous. Planted and escaping, 3900-9500 ft. Widely scattered. Introduced.

SENEGALIA Raf. - Acacia

Shrubs or small trees; stems armed with prickles, stipular spines absent, pubescence basifixed; stipules herbaceous, deciduous or persistent; leaves even-pinnate, bipinnately compound, petiole and rachis glands present; inflorescences racemose or compact and headlike, pedunculate; flowers pedicellate, corollas actinomorphic, inconspicuous, not papilionaceous, 4- or 5-merous, androecium of separate stamens, stamens 50+; fruits a legume, dehiscent or indehiscent; seeds with an aril or not, the coat with a pleurogram.

1.   Inflorescence compact and headlike, 1-1.5 cm diam.; fruits thin-papery…**S. roemeriana**
1'. Inflorescence spicate or racemose, 2.5-5 cm; fruits leathery…**S. greggii**

Senegalia greggii (A. Gray) Britton & Rose [*Acacia greggii* A. Gray]—Catclaw acacia. Shrubs or trees; stems armed, the herbage glabrous, glabrescent, or puberulent; leaves with 1-4 pairs of pinnae, clustered in spurs, leaflets 4-6 pairs per pinnule, the blades 3-6 mm, elliptic, oblong, or ovate; inflorescences spicate, arising from the leaf spurs; fruits 5-10 × 1-1.5 cm, the body oblong, ± constricted between the seeds, the valves leathery, glabrous. Canyons, washes, floodplains, desert scrub with creosote and mesquite, juniper, 2950-5600 ft. C, WC, SE, SW.

Senegalia roemeriana (Scheele) Britton & Rose [*Acacia roemeriana* Scheele]—Roemer's acacia. Shrubs or trees; stems armed, the herbage glabrous, glabrescent, or puberulent; leaves with 1-4 pairs of pinnae, leaflets 3-12 pairs per pinnule, the blades 4-12 mm, elliptic, oblong, or ovate; inflorescences dense, headlike, in clusters of 1-4 in spurs or axillary; fruits 5-10 × 1-3 cm, the body oblong, ± constricted between the seeds, the valves thick-papery, glabrous. Canyons, washes, desert scrub with creosote and mesquite, juniper, 3250-5600 ft. Eddy, Otero.

SENNA Mill. - Senna

Perennial herbs, shrubs, or trees; stems unarmed, pubescence basifixed; stipules absent or inconspicuous and persistent; leaves even-pinnate, pinnately compound, 1-foliate, or phyllodial, petiole glands present or absent, leaflets opposite or alternate; inflorescences axillary, racemose, simple or compound, pedunculate; flowers pedicellate, corollas zygomorphic, distinct beyond the hypanthial rim, not papilionaceous, 5-merous or the petals absent, the uppermost petals not enclosing the others in bud, androecium of separate stamens, stamens 2-10-merous, the anthers monomorphic or dimorphic; fruit a legume, the valves inertly dehiscent or not.

1.   Leaves with 1 pair of leaflets…(2)
1'. Leaves with 2-7 pairs of leaflets…(3)
2.   Flowers 2-5 in clustered, terminal racemes; herbage strigulose…**S. roemeriana**
2'. Flowers 2-3 in axillary racemes; herbage tomentulose or villosulous…**S. bauhinioides**
3.   Leaves with 2-4 pairs of leaflets…(4)
3'. Leaves with 4-7 pairs of leaflets…(5)
4.   Fruits ascending to erect, 2-3.5 cm, the valves papery and villous; plants perennial herbs…**S. covesii**
4'. Fruits deflexed or pendent, 8-13 cm, the valves leathery, glabrous or glabrescent; plants shrubby…**S. wislizeni**
5.   Leaves with 5-7 pairs of leaflets; herbage velutinous; fruit 4-7 cm…**S. lindheimeriana**
5'. Leaves with 4-5 pairs of leaflets; herbage strigose; fruit 8-12 cm…**S. orcuttii**

Senna bauhinioides (A. Gray) H. S. Irwin & Barneby [*Cassia bauhinioides* A. Gray]—Twin-leaf senna. Perennial herbs; stems spreading or ascending, the herbage tomentulose or villosulous; leaves with 1 pair of leaflets, the blades 10–40 mm, elliptic or oblong; inflorescences in axillary racemes, 2–3-flowered; sepals 6–8 mm, corollas 16–18 mm, golden-yellow; fruit 2–5 × 0.4–0.7 cm, dehiscent, the body oblong, the valves thick-papery, villosulous. Rocky hillsides, flats, ruderal areas, roadsides, desert scrub, scattered piñon-juniper, 3250–6600 ft. EC, C, SE, SC, SW, San Miguel.

Senna covesii (A. Gray) H. S. Irwin & Barneby [*Cassia covesii* A. Gray]—Coves' senna, rattlebox. Perennial herbs; stems erect, the herbage tomentulose; leaves with 2–4 pairs of leaflets, the blades 30–40 mm, elliptic, lanceolate, or oblong; inflorescences in axillary racemes, 4–8-flowered; sepals 6–8 mm, corollas 20–28 mm, yellowish; fruit 2–4 × 0.6–0.8 cm, ascending or spreading, dehiscent, the body oblong or oblanceolate, the valves thick-papery, villosulous. Rocky sites, desert scrub, 4550–4950 ft. Grant, Hidalgo.

Senna lindheimeriana (Scheele) H. S. Irwin & Barneby [*Cassia lindheimeriana* Scheele]—Lindheimer's senna, velvet-leaf senna. Perennial herbs, caudices superficial, taproot slender; stems erect or ascending, the herbage velutinous; leaves with 5–7 pairs of leaflets, the blades 10–40 mm, obovate or oblong; inflorescences in axillary and terminal racemes, 5–12-flowered; sepals 6–8 mm, corollas 20–25 mm, yellowish; fruit 4–7 × 0.6–0.7 cm, ascending, dehiscent, the body oblong, the valves thick-papery, glabrescent or puberulent. Rocky ridges, flats, canyons, desert scrub, desert grasslands, juniper communities, 3600–7250 ft. SE, SC, SW, Guadalupe, Lincoln.

Senna orcuttii (Britton & Rose) H. S. Irwin & Barneby [*Cassia orcuttii* (Britton & Rose) B. L. Turner]—Orcutt's senna. Perennial herbs; stems erect or ascending, the herbage glabrescent or strigose; leaves with 4–5 pairs of leaflets, the blades 10–30 mm, elliptic or lanceolate; inflorescences in axillary racemes, 5–15-flowered; sepals 6–7 mm, corollas 16–20 mm, yellowish; fruit 8–12 × 0.3–0.4 cm, ascending, dehiscent, the body linear, the valves thick-papery, glabrescent or strigulose. Rocky ridges, hillsides, canyons, washes, 4250–5900 ft. Otero. Included in this treatment based on a single, historical collection from the Guadalupe Mountains.

Senna roemeriana (Scheele) H. S. Irwin & Barneby [*Cassia roemeriana* Scheele]—Roemer's senna. Perennial herbs; stems erect or ascending, the herbage strigulose; leaves with 1 pair of leaflets, the blades 20–60 mm, lanceolate or oblong; inflorescences in ± clustered, terminal racemes, 2–5-flowered; sepals 6–7 mm, corollas 20–30 mm, yellowish; fruit 2–4 × 0.5–0.7 cm, dehiscent, the body oblong, the valves thick-papery, strigulose. Chihuahuan desert scrub, grasslands, piñon-juniper communities, 3250–6600 ft. NE, EC, C, SE, SC, Hidalgo.

Senna wislizeni (A. Gray) H. S. Irwin & Barneby [*Cassia wislizeni* A. Gray]—Shrubby senna. Shrubs; stems erect, the herbage strigose; leaves with 2–3 pairs of leaflets, mature leaves clustered in spurs, the blades 10–15 mm, elliptic or obovate; inflorescences in axillary racemes, 2–8-flowered; sepals 6–10 mm, corollas 20–50 mm, yellowish; fruit 8–13 × 0.3–0.6 cm, deflexed or pendent, tardily dehiscent, the body linear, the valves leathery, glabrous. Rocky sites, canyons, washes, desert scrub, 4750–4950 ft. Grant, Hidalgo, Luna.

SPHAEROPHYSA DC. – Swainsonpea

Sphaerophysa salsula (Pall.) DC.—Alkali swainsonpea. Perennial herbs, caulescent; stems erect or spreading, herbage glabrous, glabrescent, or strigose; leaves with 7–12 pairs of leaflets, the blades 0.6–2 cm, oblong or elliptic; inflorescences terminal, 5–15-flowered; calyx 4–5 mm, corollas 12–14 mm, reddish orange; fruits 1–2.5 × 1–2 cm, bladdery-inflated, stipitate, the body subglobose, the valves glabrous or glabrescent. Usually in saline soil, roadsides and fields, naturalized, 3600–6600 ft. NW, C, Curry, Doña Ana, Eddy. Introduced.

STROPHOSTYLES Elliott – Woolly bean

Annual or perennial herbs; stems unarmed, pubescence basifixed, the hairs spreading-retrorse; stipules striate, persistent; leaves odd-pinnate, pinnately compound, trifoliate, petiolate, leaflets entire or palmately lobed; inflorescences pseudoracemose, axillary, pedunculate, with 2–6 clusters, each with 1–3 flowers at each node, the nodes ± swollen and functioning as extrafloral nectaries; flowers pedicellate, calyx tube not inflated, 4-lobed, the upper one ± emarginate, corollas papilionaceous, androecium diadelphous; fruits a legume, dehiscent, sessile, the body ± laterally compressed in cross-section; seeds 3–10.

1. Corollas 7–13 mm; fruits ± rounded and not compressed; perennial herbs with thick taproots, ± woody at the base…**S. helvola**
1′. Corollas 3–7 mm; fruits laterally compressed; annuals or short-lived perennials with slender taproots, stems not woody at the base…**S. leiosperma**

Strophostyles helvola (L.) Elliott—Annual woolly bean. Annual, caulescent; stems erect or spreading, angular, the herbage glabrous, glabrescent, or strigose; leaves pinnately trifoliate, the blades 1–8 cm, ovate, orbicular, or lanceolate; inflorescences axillary, angular, with 1–5 flowering nodes; calyx 2–6 mm, corollas 7–13 mm, pinkish; fruits 3–10 × 0.3–0.8 cm, dehiscent, sessile, the body linear, straight or slightly curved, ± rounded in cross-section, the valves glabrous, glabrescent, or sparsely strigose. Naturalized. Socorro. Introduced. Included in this treatment based on a single, historical collection. This taxon is probably best considered a waif.

Strophostyles leiosperma (Torr. & A. Gray) Piper [*S. pauciflora* S. Watson]—Small flowered woolly bean. Annual or short-lived perennial herbs, caulescent; stems erect or spreading, ± round, the herbage glabrous, glabrescent, or sericeous; leaves pinnately trifoliate, the blades 1–6 cm, ovate or lanceolate, entire or shallowly lobed; inflorescences axillary, angular, with 1–4 flowering nodes; calyx 1–4 mm, corollas 3–7 mm, pinkish; fruits 1–4 × 0.2–0.5 cm, dehiscent, sessile, the body linear, straight or slightly curved, valves glabrous, glabrescent, or sparsely strigose. Pastures, irrigation ditches, roadways, 3600–5900 ft. Bernalillo, Doña Ana, Eddy, Valencia.

TEPHROSIA Pers. – Hoary-pea

Tephrosia vicioides Schltdl. [*T. tenella* A. Gray]—Red hoary-pea. Perennial herbs (occasionally flowering the first year and appearing annual), caudices superficial; stems ascending or erect, the herbage glabrescent or strigose; leaves with 3–8 pairs of leaflets, the blades 1–4 cm, elliptic or oblong; inflorescences terminal; calyx 2–3 mm, corollas 6–9 mm, pinkish purple; fruits 3–5 × 0.3–0.4 cm, the body oblong, straight or slightly curved, laterally compressed in cross-section, the valves strigose. Washes and rocky ledges, juniper and oak communities, 4550–5575 ft. Hidalgo, Luna.

THERMOPSIS R. Br. – Golden pea

Perennial herbs, caulescent; stems unarmed, lower nodes leafless, pubescence basifixed; stipules herbaceous, the lowermost ± clasping the stem, persistent; leaves odd-pinnate, palmately compound, petiolate, leaflets trifoliate; inflorescences racemose, terminal, pedunculate, bracts similar to the stipules, ± persistent; flowers pedicellate, calyx tube not inflated, the teeth acute, corollas papilionaceous, the keel rounded, androecium of separate stamens, stamens 10; fruits a legume, tardily dehiscent, stipitate, the body oblong, ± laterally compressed in cross-section; seeds 3–10.

1. Fruits laterally compressed and not conspicuously constricted between the seeds, the body ± straight…(2)
1′. Fruits rounded and conspicuously and irregularly constricted between the seeds (appearing lomentlike), the body ± curved…(3)
2. Corollas 25–29 mm; fruit strigose…**T. divaricarpa** (in part)
2′. Corollas 16–22 mm; fruit villous…**T. montana**
3. Corollas 25–29 mm; calyx 6–8 mm wide; leaflets 3–7 cm, the blades with 7–10 pairs of lateral veins… **T. divaricarpa** (in part)
3′. Corollas 19–21 mm; calyx 4.5–6 mm wide; leaflets 1.7–3.3 cm, the blades with 5–7 pairs of lateral veins… **T. rhombifolia**

Thermopsis divaricarpa A. Nelson—Spreading-fruit golden pea. Perennial herbs; stems ascending or erect, weakly to moderately zigzagged, the herbage glabrescent or strigose; leaves palmately trifoliate, the blades 3–8 cm, elliptic or rhombic, lateral veins 7–10 pairs; inflorescences racemose, 6–15-flowered; calyx 8–11 mm, corollas 25–29 mm, golden-yellow; fruits 6–9 × 0.6–0.8 cm, the body oblong, straight or ± curved, ± rounded or laterally compressed in cross-section, valves strigose. Along streams, montane meadows, upper piñon-juniper, ponderosa pine–oak, mixed conifer communities, 5500–10,850 ft. NE, NC, C.

Thermopsis montana Nutt.—Mountain golden pea. Perennial herbs; stems ascending or erect, weakly to moderately zigzagged, the herbage glabrescent or villous; leaves palmately trifoliate, the blades 3–8 cm, elliptic or obovate, lateral veins 6–11 pairs; inflorescences racemose, 6–25-flowered; calyx 9–11 mm, corollas 16–22 mm, golden-yellow; fruits 4–7 × 0.4–0.6 cm, the body oblong, straight, laterally compressed in cross-section, valves villous. Montane meadows, along streams, upper piñon-juniper, ponderosa pine–oak, mixed conifer communities, 5500–10,850 ft. NC, NW, C, WC, SC, SW.

Thermopsis rhombifolia (Nutt. ex Pursh) Richardson—Prairie golden pea. Perennial herbs; stems ascending or erect, weakly zigzagged, the herbage glabrescent, strigose, or tomentose; leaves palmately trifoliate, the blades 1.7–3.3 cm, elliptic or obovate, lateral veins 5–7 pairs; inflorescences racemose, 5–15-flowered; calyx 6–12 mm, corollas 19–21 mm, golden-yellow; fruits 3–7 × 0.4–0.7 cm, the body oblong, curved, rounded in cross-section, valves glabrescent or tomentose. Plains, piñon-juniper woodlands, ponderosa pine–oak, mixed conifer communities, 5500–9000 ft. NE, NW.

TRIFOLIUM L. – Clover

Annual or perennial herbs; stems unarmed, pubescence basifixed; stipules partly fused to petiole, papery or membranous, persistent; leaves odd-pinnate, palmately or pinnately compound, petiolate, leaflets trifoliate or 4–9-foliate; inflorescences racemose, compact and ± umbel-like, axillary or terminal, subtended by an involucre of bracts or not, pedunculate; flowers pedicellate or subsessile, calyx tube inflated or not, the teeth acute, corollas papilionaceous, the petals ± persistent in fruit, androecium diadelphous; fruits a legume, indehiscent, ± stipitate, the body ± enclosed in the petals.

1. Corollas bright yellow; leaflets often ± pinnately arranged, at least some; plants annual…(2)
1'. Corollas pink-purple, reddish, lavender, or whitish (not yellow); leaflets all palmately arranged; plants annual or perennial…(3)
2. Corollas plainly striate, 4–7 mm; leaflet blades impressed-veined, corrugated abaxially; flower heads usually 20+-flowered…**T. campestre**
2'. Corollas scarcely striate, 3–4 mm; leaflet blades not impressed-veined, smooth abaxially; flower heads usually 2–20-flowered…**T. dubium**
3. Inflorescences subtended by an involucre of bracts (or a psuedoinvolucre of leaves or stipules)…(4)
3'. Inflorescence not subtended by an involucre of bracts…(6)
4. Inflorescences subtended by a psuedoinvolucre of paired, reduced leaves or stipules…**T. pratense**
4'. Inflorescences subtended by an involucre of ± fused bracts…(5)
5. Plants caulescent, with 1+ stem internodes…**T. wormskioldii**
5'. Plants acaulescent or subacaulescent…**T. parryi** (in part)
6. Plants stoloniferous…(7)
6'. Plants not stoloniferous…(8)
7. Calyx in fruit bladdery-inflated; corollas 7–12 mm; fruit 3–5 mm…**T. fragiferum**
7'. Calyx in fruit not inflated; corollas 5–6 mm; fruit 2–3 mm…**T. repens**
8. Plants caulescent, with 1+ stem internodes…(9)
8'. Plants acaulescent or subacaulescent…(11)
9. Calyx 2–4 mm, calyx lobes 1–2.5 mm; pedicels 2–6 mm; plants naturalized…**T. hybridum**
9'. Calyx 4–12 mm, calyx lobes 3–9 mm; pedicels 0–2 mm; plants native…(10)
10. Floral bracts conspicuous, broad, some ± fused; fruits villosulous…**T. attenuatum**
10'. Floral bracts inconspicuous, ± subulate, < 0.5 mm; fruits glabrous or glabrescent, the hairs, if present, restricted to the apex…**T. longipes**
11. Heads 1–4-flowered…**T. nanum**
11'. Heads 5- to many-flowered…(12)
12. Floral bracts membranous, 2–7 mm wide…**T. parryi** (in part)
12'. Floral bracts absent, greenish, or scarious, 0–1.5 mm wide…(13)
13. Calyx lobes 1.5–3 mm…**T. gymnocarpon**
13'. Calyx lobes 3–7 mm…(14)

14. Inflorescence elongating in fruit (± racemose) and the rachis not obscured by flowers; floral bracts absent...**T. brandegeei**
14'. Inflorescence compact in fruits and the rachis obscured; floral bracts greenish and scarious-margined...
T. dasyphyllum

Trifolium attenuatum Greene—Rocky Mountain clover. Perennial herbs, caudices covered in a thatch of decomposed leaf bases; stems clustered, caulescent, herbage villosulous; leaves palmately trifoliate, the blades 0.5-2 cm, ovate, oblanceolate, or oblong, the margins entire; inflorescences head-like, compact, 5- to many-flowered, not subtended by an involucre, floral bracts broad, some ± fused; calyx 6-9 mm, calyx lobes 5-7 mm, corollas 12-18 mm, pinkish purple; fruits 0.5-0.6 cm, villosulous. Hillsides and slopes in mountains, upper montane through subalpine, 9500-13,100 ft. NC, C.

Trifolium brandegeei S. Watson—Brandegee's clover. Perennial herbs, caudices covered in a thatch of decomposed leaf bases; stems clustered, acaulescent, the herbage glabrous, glabrescent, or villous; leaves palmately trifoliate, the blades 1-3 cm, obovate or oblong, the margins serrate or entire; inflorescences headlike, 5- to many-flowered, the rachis not subtended by an involucre, floral bracts absent; calyx 7-9 mm, calyx lobes 3-3.5 mm, corollas 14-16 mm, pinkish purple; fruits 0.5-0.6 cm, glabrescent or villosulous. Often in moist sites, subalpine forests, alpine meadows, 10,000-12,700 ft. Santa Fe, Taos.

Trifolium campestre Schreb.—Hop clover. Annuals; stems erect or ascending, caulescent, the herbage glabrous, glabrescent, or villosulous; leaves pinnately trifoliate, the blades 0.8-1.5 cm, obovate, the margins serrate; inflorescences headlike, compact, not subtended by an involucre, floral bracts filiform, pedicels 0.5-1 mm, at least the lowermost flowers and pedicels recurved in age; calyx 1-2 mm, glabrescent or villosulous, corollas 3-6 mm, yellowish, purplish-tinged in age; fruits 0.2-0.3 cm. Naturalized, disturbed sites, 6550-8000 ft. Colfax, Santa Fe. Introduced.

Trifolium dasyphyllum Torr. & A. Gray—Uinta clover. Perennial herbs, caudices covered in a thatch of decomposed leaf bases; stems subacaulescent, the herbage glabrescent or strigose; leaves palmately trifoliate, the blades 1-3 cm, obovate or oblong, the margins serrate or entire; inflorescences headlike, compact, 5- to many-flowered, not subtended by an involucre, floral bracts greenish and scarious-margined, 0.7-1.5 mm wide, pedicels 0.5-1 mm, the flowers and pedicels spreading to ascending; calyx 5-10 mm, calyx lobes 3-7 mm, corollas 10-16 mm, pinkish purple or whitish and pinkish purple-tinged; fruits 0.4-0.6 cm, glabrescent or strigose. New Mexico material belongs to **var. uintense** (Rydb.) S. L. Welsh. Often in subalpine and alpine meadows, 9200-12,400 ft. Colfax, Mora, Taos.

Trifolium dubium Sibth.—Annuals from slender taproots; caulescent, herbage sparsely short-villous; leaflets 3, mostly pinnately arranged, obovate, 5-11 × 3-6 mm; flowering heads 2-20-flowered, not involucrate; flowers 2-4 mm; calyx sparsely glabrate, corollas bright yellow, obscurely striate; fruits 1-2 mm. Weedy sites, known only from Portales, ca. 4000 ft. Roosevelt. Introduced.

Trifolium fragiferum L.—Strawberry clover. Perennial herbs, caudices stoloniferous; stems tuft-forming, the herbage glabrous, glabrescent, or pilose; leaves palmately trifoliate, the blades 0.5-2 cm, elliptic or oblong, the margins serrate; inflorescences headlike, compact, pilose, the hairs reddish or brown, not subtended by an involucre, floral bracts broad, some ± fused, pedicels 0-0.5 mm, the flowers and pedicels spreading to ascending; calyx 2-4 mm, bladdery-inflated and to 9 mm in fruit, pilose, corollas 5-6 mm, pinkish; fruits 0.2-0.3 cm. Naturalized, common in lawns, 4250-7000 ft. NE, NC, NW. Introduced.

Trifolium gymnocarpon Nutt.—Holly-leaf clover. Perennial herbs, caudices superficial; stems mat-forming or cushion-forming, acaulescent or subacaulescent, the herbage glabrous, glabrescent, pilosulous, villosulous, or strigose; leaves palmately compound, leaflets 3-5, the blades 0.5-2 cm, elliptic, obovate, or oblong, the margins serrate; inflorescences headlike, compact, 5- to many-flowered, not subtended by an involucre, floral bracts scarious, < 1 mm wide, pedicels 1-2 mm, at least the lowermost flowers and pedicels recurved in age; calyx 3-6 mm, calyx lobes 1.5-3 mm, corollas 6-12 mm, ochroleucous or pinkish; fruits 0.4-0.5 cm, villosulous. Dry, sandy slopes, sagebrush communities, 6200-9200 ft. NW, Cibola.

Trifolium hybridum L.—Alsike clover. Short-lived perennial herbs; stems clustered, caulescent, the herbage glabrous or glabrescent; leaves palmately trifoliate, the blades 1–6 cm, elliptic, obovate, or oblong, the margins serrate; inflorescences headlike, compact, 5- to many-flowered, not subtended by an involucre, floral bracts scarious, < 1 mm wide, at least the lowermost flowers and pedicels recurved in age; calyx 2–4 mm, calyx lobes 1–2.5 mm, corollas 6–10 mm, whitish, reddish, or pinkish in age; fruits 0.3–0.4 cm, glabrous. Common in meadows, along streams, disturbed sites, 4900–10,500 ft. NC, NW, C, WC. Introduced.

Trifolium longipes Nutt.—Long-stalked clover. Perennial herbs; stems clustered, caulescent, the herbage glabrous, glabrescent, villous, or villosulous; leaves palmately trifoliate, the blades 1–7 cm, elliptic, lanceolate, or obovate, the margins entire or serrate; inflorescences headlike, compact, 5- to many-flowered, not subtended by an involucre, floral bracts ± subulate and inconspicuous, < 0.5 mm, pedicels 0–2 mm, the flowers and pedicels spreading to ascending; calyx 4–12 mm, calyx lobes 3–9 mm, corollas 9–18 mm, greenish white, ochroleucous, pinkish purple, or bicolored.

1. Plants rhizomatous; corollas greenish white, ochroleucous, or whitish, rarely pinkish purple–tinged…**var. reflexum**
1'. Plants with thickened taproots; corollas pinkish purple or whitish and pinkish purple–tinged or striate…(2)
2. Calyx villous…**var. neurophyllum**
2'. Calyx glabrous, glabrescent, or sparsely villosulous…**var. rusbyi**

var. neurophyllum (Greene) J. S. Martin ex Isely—White Mountain clover. Plants with thickened taproots; calyx villous, corollas pinkish purple or whitish and pinkish purple–tinged or striate. Meadows, springs, riparian areas in montane forests, 6200–9200 ft. Catron. **R**

var. reflexum A. Nelson—Plants rhizomatous; calyx glabrescent, villosulous, or villous, corollas greenish white, ochroleucous, or whitish, rarely pinkish purple–tinged. Meadows, montane forests, subalpine slopes, 8200–11,150 ft. NC, NW.

var. rusbyi (Greene) H. D. Harr.—Plants with thickened taproots; calyx glabrous, glabrescent, or sparsely villosulous, corollas pinkish purple or whitish and pinkish purple–tinged or striate. Meadows, slopes, riparian areas in montane forests, 6550–9200 ft. Rio Arriba.

Trifolium nanum Torr.—Dwarf clover. Perennial herbs, caudices superficial, covered in a thatch of decomposed leaf bases; stems mat-forming or cushion-forming, acaulescent, the herbage glabrous; leaves palmately trifoliate, the blades 0.3–1.5 cm, obovate or oblanceolate, the margins serrate; inflorescences 1–4-flowered, not subtended by an involucre, floral bracts membranous, < 1 mm wide; calyx 4–7 mm, abaxial calyx lobe longest, 1.3–3 mm, corollas 12–19 mm, pinkish purple; fruits 0.8–0.9 cm, glabrous. Common in the alpine, 10,500–13,100 ft. NC.

Trifolium parryi A. Gray—Parry's clover. Perennial herbs, caudices ± rhizomatous; stems mat-forming or cushion-forming, acaulescent or subacaulescent, the herbage glabrous or glabrescent; leaves palmately trifoliate, the blades 1–3 cm, elliptic or oblong, the margins serrate; inflorescences headlike, compact, 5- to many-flowered, not subtended by an involucre or the lowest floral bracts ± fused into an incipient involucre, pedicels 0.3–1 mm, the flowers and pedicels spreading to ascending; calyx 4–8 mm, calyx lobes 1.5–4 mm, corollas 12–16 mm, pinkish purple. Subalpine and alpine meadows, 9800–12,850 ft. Colfax, San Miguel, Taos.

Trifolium pratense L.—Red clover. Short-lived perennial herbs; stems clustered, caulescent, the herbage glabrous, glabrescent, or pilose; leaves palmately trifoliate, the blades 2.5–7 cm, elliptic or ovate, the margins entire or serrate; inflorescences headlike, compact, subtended by a psuedoinvolucre of paired, reduced leaves or stipules, floral bracts absent or ± subulate and inconspicuous, < 0.5 mm, pedicels 0–0.5 mm, at least the lowermost flowers and pedicels recurved in age; calyx 7–10 mm, calyx lobes 4–6 mm, corollas 12–17 mm, reddish purple or whitish; fruits 0.2–0.3 cm. Naturalized, along streams, meadows, fields, pastures, disturbed places, 4250–10,500 ft. NE, NC, NW, C, WC, SC, SW. Introduced.

Trifolium repens L.—White clover. Perennial herbs, caudices ± stoloniferous; stems tuft-forming, caulescent, the herbage glabrous, glabrescent, or villosulous; leaves palmately trifoliate, the blades 0.5–4.5 cm, obovate, the margins obscurely denticulate; inflorescences headlike, compact, not subtended by an involucre, floral bracts white-scarious; calyx 3–6 mm, calyx lobes 1–3.2 mm, corollas 7–12 mm, whitish, pinkish in age; fruits 0.3–0.5 cm, glabrous. Naturalized, 4250–7000 ft. Widespread. Introduced.

Trifolium wormskioldii Lehm. [*T. fendleri* Greene]—Cow clover. Perennial herbs, caudices rhizomatous, ± stoloniferous; stems caulescent, the herbage glabrous or glabrescent; leaves palmately trifoliate, the blades 1–4 cm, elliptic, obovate, or oblong, the margins serrate; inflorescences headlike, compact, subtended by an involucre, the inner floral bracts small and herbaceous or reduced to a scale, pedicels 0.5–1.5 mm; calyx 4–10 mm, calyx lobes 2–7.5 mm, corollas 8–18 mm, ochroleucous, whitish, or pinkish purple; fruits 0.3–0.6 cm, glabrous. W, NCS.

1. Involucre segments united for < 1/3 their length, occasionally free nearly to the base, the lateral margins entire…**var. longicaule**
1'. Involucre segments united for 1/3–1/2 their length, the lateral margins dentate…(2)
2. Corollas 7–9 mm; stolons absent or poorly developed…**var. arizonicum**
2'. Corollas 9–13 mm; stolons present…**var. wormskioldii**

var. arizonicum (Greene) Barneby [*T. mucronatum* Willd. ex Spreng. subsp. *lacerum* (Greene) J. M. Gillett]—Plants not stoloniferous; involucre segments united for 1/3–1/2 their length, the lateral margins dentate; calyx 4–7.3 mm, corollas 7–9 mm. Stream banks and meadows in ponderosa pine–oak and mixed conifer forests, 5000–9000 ft. WC, SW.

var. longicaule (Wooton & Standl.) L. D. Benson [*T. pinetorum* Greene]—Involucre segments united for < 1/3 their length, occasionally free nearly to the base, the lateral margins entire; calyx 4–8 mm, corollas 9–11.5 mm. Meadows in ponderosa pine and mixed conifer forests, 5900–10,850 ft. C, WC.

var. wormskioldii—Plants stoloniferous; involucre segments united for 1/3–1/2 their length, the lateral margins dentate; calyx 6–9.5 mm, corollas 9–13 mm. Moist meadows in forests and woodlands, 5250–8200 ft. NC.

VACHELLIA Wight & Arn. – Wattle

Shrubs or small trees; stems armed with stipular spines, prickles absent; pubescence basifixed; stipules herbaceous, deciduous or persistent; leaves even-pinnate, bipinnately compound, petiole and rachis glands present; inflorescences racemose or compact and headlike, pedunculate; flowers pedicellate, corollas actinomorphic, inconspicuous, not papilionaceous, 4- or 5-merous, androecium of separate stamens, stamens 50+; fruits a legume, dehiscent or indehiscent; seeds with an aril or not, the coat with a pleurogram.

1. Leaves with 4–7 pairs of pinnae, the largest mature leaflets 2–5 mm…**V. constricta**
1'. Leaves with 1–4 pairs of pinnae, the largest mature leaflets 1–3 mm…**V. vernicosa**

Vachellia constricta (Benth.) Seigler & Ebinger [*Acacia constricta* Benth.; *A. constricta* Benth. var. *paucispina* Wooton & Standl.]—Whitethorn acacia. Shrubs or small trees; stems armed, the herbage glabrous, glabrescent, or hirsutulous; leaves from spurs, solitary or 2–5 per cluster, with 4–7 pairs of pinnae, leaflets 6–18 pairs per pinnule, the blades 2–5 mm, ovate or oblong, thick; inflorescences headlike, solitary or 2–3 per cluster; corollas yellow, fragrant; fruits 4–12 × 0.4–0.5 cm. Rocky slopes, washes, stream banks, canyons, desert scrub, 2950–5600 ft. SE, SC, SW, Socorro.

Vachellia vernicosa (Britton & Rose) Seigler & Ebinger [*Acacia constricta* Benth. var. *vernicosa* (Britton & Rose) L. D. Benson; *A. neovernicosa* Isely]—Viscid acacia. Shrubs; stems armed, occasionally unarmed, the herbage glabrous or glabrescent; leaves from spurs, solitary or 2–5 per cluster, with 1–3 pairs of pinnae, leaflets 5–10 pairs per pinnule, the blades 1–3 mm, ovate, thick, ± glutinous; inflorescences headlike, solitary or 2–3 per cluster; corollas yellow, fragrant; fruits 4–12 × 0.3–0.4 cm. Plains, washes, stream banks, canyons, desert scrub, 2950–5600 ft. SE, SC, SW.

Pl. 1

Abronia bigelovii

Abronia bolackii

Abronia nealleyi

Agalinis calycina

Agastache cana

Agastache mearnsii

Agastache pringlei var. verticillata

Aliciella cliffordii

Aliciella formosa

Allium gooddingii

Amsonia fugatei

Amsonia tharpii

Pl. 2

Anticlea mogollonensis

Anulocaulis leiosolenus
var. *gypsogenus*

Anulocaulis leiosolenus
var. *howardii*

Apacheria chiricahuensis

Aquilegia chaplinei

Argemone pinnatisecta

Asclepias sanjuanensis

Asplenium scolopendrium

Astragalus altus

Astragalus castetteri

Astragalus chuskanus

Astragalus cliffordii

Pl. 3

Astragalus cobrensis var. *maguirei*   *Astragalus cyaneus*   *Astragalus feensis*

Astragalus gypsodes   *Astragalus heilii*   *Astragalus humillimus*

Astragalus humistratus
var. *crispulus*   *Astragalus iodopetalus*   *Astragalus kentrophyta*
var. *neomexicana*

Astragalus kerrii   *Astragalus knightii*   *Astragalus micromerius*

Pl. 4

Astragalus missouriensis
var. *acumbens*

Astragalus missouriensis
var. *humistratus*

Astragalus missouriensis
var. *mimetes*

Astragalus monumentalis
var. *cottamii*

Astragalus neomexicanus

Astragalus nutriosensis

Astragalus oocalycis

Astragalus puniceus
var. *gertrudis*

Astragalus ripleyi

Astragalus siliceus

Astragalus wittmannii

Atriplex griffithsii

Pl. 5

Bartlettia scaposa

Berlandiera macvaughii

Boechera perennans subsp. *sanluisensis*

Boechera perennans subsp. *zephyra*

Calochortus gunnisonii var. *perpulcher*

Castilleja organorum

Castilleja ornata

Castilleja tomentosa

Chaetopappa hersheyi

Cirsium vinaceum

Cirsium wrightii

Cleomella multicaulis

Pl. 6

Coryphantha robustispina subsp. scheeri   **Crataegus wootoniana**   **Cuscuta draconella**

Cuscuta warneri   **Cylindropuntia viridiflora**   **Cymopterus davidsonii**

Cymopterus spellenbergii   **Dalea scariosa**   **Delphinium alpestre**

Delphinium novomexicanum   **Delphinium robustum**   **Delphinium sapellonis**

Pl. 7

Dermatophyllum gypsophilum subsp. *guadelupense*

Desmodium metcalfei

Draba abajoensis

Draba heilii

Draba henrici

Draba mogollonica

Draba smithii

Draba standleyi

Echinocereus fendleri var. *kuntzleri*

Ericameria nauseosa var. texensis

Erigeron acomanus

Erigeron hessii

Pl. 8

Erigeron rhizomatus

Erigeron rybius

Erigeron scopulinus

Erigeron sivinskii

Erigeron subglaber

Eriogonum aliquantum

Eriogonum gypsophilum

Eriogonum lachnogynum
var. *colubum*

Eriogonum lachnogynum
var. *sarahiae*

Eriogonum wootonii

Eryngium sparganophyllum

Erythranthe plotocalyx

Pl. 9

Euphorbia rayturneri

Geranium dodecatheoides

Grindelia arizonica
var. *neomexicana*

Grusonia clavata

Hackelia hirsuta

Hedeoma apiculata

Hedeoma pulcherrima

Hedeoma todsenii

Helianthus arizonensis

Helianthus paradoxus

Heterotheca cryptocephala

Heterotheca sierrablancensis

Pl. 10

Heuchera glomerulata

Heuchera pulchella

Heuchera soltisii

Heuchera woodsiaphila

Heuchera wootonii

Hexalectris arizonica

Hexalectris colemanii

Hexalectris nitida

Hieracium brevipilum

Hymenoxys ambigens
var. *neomexicana*

Hymenoxys brachyactis

Hymenoxys vaseyi

Pl. 11

Ionactis elegans   *Ipomoea gilana*   *Ipomopsis sancti-spiritus*

Ipomopsis wrightii   *Justicia wrightii*   *Lechea mensalis*

Lepidospartum burgessii   *Leucosyris blepharophylla*   *Limosella pubiflora*

Linum allredii   *Lorandersonia microcephala*   *Lupinus sierrae-blancae*

Pl. 12

Malaxis abieticola

Mentzelia conspicua

Mentzelia filifolia

Mentzelia humilis var. guadelupensis

Mentzelia perennis

Mentzelia sivinskii

Mentzelia springeri

Mentzelia todiltoensis

Microthelys rubrocallosa

Monarda humilis

Mostacillastrum subauriculatum

Muhlenbergia villiflora var. *villosa*

Pl. 13

Muilla lordsburgana	*Nama xylopoda*	*Nerisyrenia hypercorax*
Oenothera organensis	*Opuntia arenaria*	*Packera cardamine*
Packera cliffordii	*Packera sanguisorboides*	*Packera spellenbergii*
Panicum mohavense	*Paronychia wilkinsonii*	*Pediocactus knowltonii*

Pl. 14

Pediomelum pentaphyllum *Pelecyphora duncanii* *Pelecyphora guadalupensis*

Pelecyphora leei *Pelecyphora orcuttii* *Pelecyphora organensis*

Pelecyphora sandbergii *Pelecyphora sneedii* *Pelecyphora villardii*

Peniocereus greggii var. greggii *Penstemon alamosensis* *Penstemon bleaklyi*

Pl. 15

Penstemon cardinalis
subsp. *cardinalis*

Penstemon cardinalis
subsp. *regalis*

Penstemon crandallii
var. *taosensis*

Penstemon inflatus

Penstemon metcalfei

Penstemon neomexicanus

Perityle cernua

Perityle quinqueflora

Perityle staurophylla
var. *homoflora*

Perityle staurophylla
var. *staurophylla*

Phacelia cloudcroftensis

Phacelia serrata

Pl. 16

Phacelia sivinskii

Phemeranthus brachypodius

Phemeranthus humilis

Phemeranthus rhizomatus

Philadelphus microphyllus
subsp. *argenteus*

Phlox caryophylla

Phlox cluteana

Phlox vermejoensis

Physaria aurea

Physaria gooddingii

Physaria lata

Physaria navajoensis

Pl. 17

Physaria newberryi
subsp. *yesicola*

Physaria pinetorum
subsp. *iveyana*

Physaria pruinosa

Potentilla sierrae-blancae

Proatriplex pleiantha

Puccinellia parishii

Rhinotropis rimulicola
var. *mescalerorum*

Rhinotropis rimulicola
var. *rimulicola*

Rhodiola integrifolia
subsp. *neomexicana*

Ribes mescalerium

Salix arizonica

Salvia summa

Pl. 18

Schoenus nigricans

Sclerocactus mesae-verdae
subsp. *depressus*

Sclerocactus mesae-verdae
subsp. *mesae-verdae*

Sclerocactus papyracanthus

Scrophularia laevis

Scrophularia macrantha

Scrophularia montana

Senecio sacramentanus

Senecio warnockii

Sicyos glaber

Silene plankii

Silene thurberi

Pl. 19

Silene wrightii

Solidago capulinensis

Solidago correllii

Spermolepis organensis

Sphaeralcea wrightii

Stellaria porsildii

Streptanthus platycarpus

Synthyris oblongifolia

Tetradymia filifolia

Townsendia gypsophila

Trifolium longipes
subsp. *neurophyllum*

Valeriana texana

Pl. 20

Viola calcicola   *Yucca baileyi* var. *intermedia*

AUTUMN ON THE ANIMAS RIVER   A welcome day when the Animas was flowing full.

CEDAR SENTINEL   High on the northwestern plateau, this old cedar keeps watch.

DARK CHOCOLATE HILLS   Near Counselors, New Mexico, these richly colored clay formations are dotted with sage and piñon.

ECHO OF MY YOUTH   A favored hiking and painting place in the south La Plata canyon.

FOREST OF LIGHT   Aspens in the Sangre De Cristo mountains near Santa Fe.

HORSE CANYON DAYBREAK   *Sage and scrub oaks in this box canyon in San Juan County.*

MORNING AT TAOS MOUNTAIN   *One of the lovely views of the fabled mountain, in this famous artist community.*

NEW MEXICO CLOUDS   New Mexico truly is the Land of Enchantment.

NORTH BY NORTHWEST   Painted from White Rock, near Los Alamos, the sweeping panorama is full of grandeur.

RIO GRANDE - RIVER OF DREAMS   *A passage for thousands of souls seeking a new life.*

SANTA FE ARROYO   A romantic sand wash in the south part of this high desert city.

SUNFLOWER FIELDS FOREVER   Endless beauty in the northwestern town of Taos.

WATERS OF AVALON   A romantic title for a simple pond at Sunrise Springs, near Santa Fe.

OLD SANTA FE   A typical adobe neighborhood in the old part of Santa Fe.

ARID GRACE   Autumn on the Animas River, during the long drought of 2013.

SCOTTIE'S FIELDS   Fallow fields in Taos, dedicated to my young nephew who lost his life to diabetes.

WINTER RUST   Fresh snow on the Animas River in Farmington, New Mexico.

CHECKERBOARD WATERS   *The famous San Juan River, fisherman's paradise.*

DRAGON'S EVENING   *Late summer monsoon at sunset, features the popular shrub I call Chamisa Amarilla.*

VEXIBIA Raf. – Necklace-pod

Perennial herbs, shrubs, or trees; stems unarmed, pubescence basifixed; stipules inconspicuous or conspicuous at the lower nodes in herbaceous species, persistent; leaves odd-pinnate, pinnately compound, petiolate; inflorescences racemose, axillary, pedunculate; flowers pedicellate, calyx tube not inflated, 5-lobed, the upper 2 connate and ± emarginate, corollas subpapilionaceous, the wings and keels similar, not differentiated, androecium of separate stamens, stamens 10; fruits a legume, indehiscent, stipitate.

1. Leaflets linear; corollas bluish, 18–22 mm; fruits 4–5 cm…**V. stenophylla**
1'. Leaflets oblong, ovate, or oblanceolate; corollas whitish or ochroleucous, 10–16 mm; fruits 5–7 cm…**V. sericea**

Vexibia sericea Raf. [*Sophora sericea* Nutt., non Andrews; *S. nuttalliana* B. L. Turner]—Silky necklace-pod. Perennial herbs, caulescent; stems erect or ascending, the herbage glabrous, glabrescent, or villous; leaves with 3–12 pairs of leaflets, the blades 0.5–1.5 cm, oblong, ovate, or oblanceolate; inflorescences terminal, 6–20-flowered; calyx 5–8 mm, gibbous, corollas 10–16 mm, whitish or ochroleucous; fruits 5–7 × 0.4–0.5 cm, body oblong, constricted between the seeds, the valves glabrous, glabrescent, or villous. Ruderal areas, roadsides, grasslands, desert scrub, piñon-juniper woodlands, ponderosa pine, 3600–7250 ft. Widespread.

Vexibia stenophylla (A. Gray) W. A. Weber [*Sophora stenophylla* A. Gray]—Silvery necklace-pod. Perennial herbs, caulescent; stems erect or ascending, the herbage glabrous, glabrescent, or tomentulose; leaves with 4–7 pairs of leaflets, the blades 1–2 cm, linear; inflorescences terminal, 10–20-flowered; calyx 5–9 mm, corollas 18–22 mm, bluish purple; fruits 4–5 × 0.5–0.6 cm, persistent, body oblong, constricted between the seeds, the valves glabrous, glabrescent, or tomentulose. Sandy sites on dunes and in washes, 4250–6250 ft. NW, EC, C, SE, SC.

VICIA L. – Vetch

Annual or perennial herbs; stems unarmed, pubescence basifixed; stipules herbaceous, entire to ± sagittate, persistent; leaves even-pinnate, the rachis terminating in a ± prehensile tendril, petiolate, leaflets 1 to many pairs per leaf; inflorescences solitary or racemose, the racemes axillary, pedunculate; flowers pedicellate or subsessile, calyx tube not inflated, the teeth acute, corollas papilionaceous, the petals whitish or purplish, the wings adnate to the keel, androecium diadelphous; style bearded distally, the beard tufted or encircling the apex of the style, the style apex rounded; fruits a legume, dehiscent, sessile, the body oblong, the valves eglandular.

1. Inflorescences composed of solitary flowers or clusters of 2 in leaf axils…(2)
1'. Inflorescences composed of 3–10-flowered racemes (10+-flowered in some forms of *V. villosa*)…(4)
2. Calyx 6–15 mm; corollas 10–23 mm…**V. sativa**
2'. Calyx 2.5–5 mm; corollas 4–9 mm…(3)
3. Fruit valves puberulent; corollas whitish or ochroleucous, purple-striate; calyx 4–5 mm…**V. leucophaea**
3'. Fruit valves glabrous; corollas bluish purple, pale pinkish purple, or pinkish; calyx 2.5–3.5 mm…**V. ludoviciana** (in part)
4. Calyx 3–5 mm; corollas 4–10 mm…(5)
4'. Calyx 5–8 mm; corollas 10–25 mm…(6)
5. Annual; corollas bluish purple or pinkish purple…**V. ludoviciana** (in part)
5'. Perennial; corollas whitish…**V. pulchella**
6. Inflorescences not secund, 3–10-flowered…**V. americana**
6'. Inflorescences secund, 12–40-flowered…**V. villosa**

Vicia americana Muhl. ex Willd.—American vetch. Perennial herbs, rhizomatous; stems erect or prostrate, twining, the herbage glabrous, glabrescent, or puberulent; leaves with 2–8 pairs of leaflets, the tendrils simple, prehensile, the blades 10–35 mm, linear or elliptic; inflorescences pedunculate, 3–10-flowered; calyx 6–8 mm, corollas 15–25 mm, pinkish purple, bluish purple, or whitish; fruit 2–3.5 × 0.4–0.7 cm, stipitate, the body oblong, the valves glabrous or puberulent. Common in meadows and along streams, ponderosa pine–oak, mixed conifer forests, 5500–10,850 ft. Widespread except EC, SE.

Vicia leucophaea Greene—Mogollon Mountain vetch. Perennial herbs; stems erect or prostrate, twining, the herbage glabrescent, puberulent, or pilose; leaves with 2-4 pairs of leaflets, the tendrils simple or branched, prehensile or not, the blades 10-20 mm, elliptic; inflorescences pedunculate, 1- or 2-flowered; calyx 4-5 mm, corollas 7-9 mm, ochroleucous or whitish, purple-striate; fruit 2-4 × 0.4-0.6 cm, sessile or stipitate, the body oblong, the valves puberulent. Piñon-juniper, ponderosa pine–oak, mixed conifer, 5900-9200 ft. Catron, Hidalgo, Socorro.

Vicia ludoviciana Nutt. ex Torr. & A. Gray [*V. exigua* Nutt.]—Louisiana vetch. Annuals; stems erect or ascending, twining, the herbage glabrous, glabrescent, or puberulent; leaves with 3-8 pairs of leaflets, the tendrils simple, prehensile or not, the blades 10-20 mm, elliptic, linear, or oblong; inflorescences pedunculate, 1-15-flowered; calyx 3-3.5 mm, corollas 4.5-8 mm, bluish purple or pinkish purple; fruit 1.5-3.5 × 0.4-0.8 cm, stipitate, the body oblong, the valves glabrous. New Mexico material belongs to **var. ludoviciana**. Woodlands, shrublands, meadows, dry hillsides, 3250-7250 ft. NE, NW, EC, C, SE, SC, SW.

Vicia pulchella Kunth [*V. melilotoides* Wooton & Standl.]—Sweetclover vetch. Perennial herbs, caudices superficial; stems erect or prostrate, twining, the herbage glabrous, glabrescent, puberulent; leaves with 4-8 pairs of leaflets, the tendrils branched, prehensile, the blades 10-30 mm, elliptic or linear; inflorescences pedunculate, 5-10-flowered; calyx 3-3.5 mm, corollas 7-9 mm, ochroleucous or whitish, purple-striate; fruit 2-4 × 0.6-0.8 cm, substipitate, the body oblong, the valves glabrous. Piñon-juniper, springs, montane meadows, ponderosa pine–oak, mixed conifer, 5900-10,500 ft. C, WC, SE, SW, Mora, San Miguel.

Vicia sativa L.—Tare, common vetch. Annuals, caudices superficial; stems erect or ascending, twining, the herbage glabrescent or villosulous; leaves with 3-8 pairs of leaflets, the tendrils branched, prehensile, the blades 14-32 mm, elliptic, linear, ovate, or oblong; inflorescences pedunculate, 1- or 2-flowered; calyx 6-15 mm, corollas 10-23 mm, bluish purple, pinkish purple, or whitish; fruit 2-7 × 0.4-1 cm, substipitate, the body linear, the valves glabrous or glabrescent. Our material belongs to **var. nigra** L. Naturalized, roadsides and ditches, 4250-8200 ft. Rio Arriba. Introduced.

Vicia villosa Roth—Hairy vetch. Annuals, caudices superficial; stems erect or ascending, twining, the herbage glabrescent or villosulous; leaves with 5-8 pairs of leaflets, the tendrils branched, prehensile, the blades 13-20 mm, lanceolate, linear, or oblong; inflorescences pedunculate, 12-40-flowered, secund; calyx 5-9 mm, corollas 13-18 mm, bluish purple, pinkish purple, or whitish; fruit 2-4 × 0.8-1 cm, stipitate, the body oblong or elliptic, the valves glabrous or glabrescent. New Mexico material belongs to **subsp. villosa**. Naturalized, along roadsides, sometimes planted in fields, 5200-8200 ft. NC, Curry, Doña Ana. Introduced.

FAGACEAE – BEECH FAMILY

Ross A. McCauley

Trees or shrubs, evergreen or deciduous; leaves alternate, simple; blades lobed or unlobed, pinnately veined, margins serrate, dentate, or entire; inflorescences unisexual or androgynous catkins; staminate and androgynous catkins consisting of few- to many-flowered clusters, bracts present or absent; pistillate catkins with 1 to several cupules bearing pistillate flowers; fruit a 1-seeded nut, subtended or enclosed individually or in groups of 2-3(-15) by a scaly or spiny, multibracteate cupule (Nixon, 1997; Spellenberg, 2001).

QUERCUS L. - Oak

Trees or shrubs, evergreen or winter-deciduous; leaf blades thin or leathery, margins entire, toothed, or awned-toothed; inflorescences in axils of leaves or bud scales, usually clustered at base of new growth;

staminate inflorescences lax, spicate; pistillate inflorescences usually stiff, with terminal cupule and sometimes 1 to several sessile, lateral cupules; fruits (acorns) with annual or biennial maturation, cup covering base of nut, scaly, with scales imbricate or reduced to tubercles, not or weakly reflexed, never hooked, nut 1 per cup.

1. Mature bark smooth or deeply furrowed, not scaly or papery; cup scales flattened, never tuberculate, never embedded in tomentum; leaf blades if lobed then with awned teeth, if entire then often with bristle at apex; endocarp densely tomentose or silky over entire inner surface of nut (sect. **Lobatae**, red or black oaks)…(2)

1'. Mature bark scaly or papery, rarely deeply furrowed; cup scales thickened, tuberculate, or embedded in tomentum; leaf blades if lobed then without awned teeth, if unlobed then without bristle at apex, merely mucronate; endocarp glabrous or somewhat sparsely and irregularly hairy…(3)

2. Leaf blades not conspicuously bicolored, abaxially glabrous expect for tufts of tomentum in vein axils or at base, blades planar…**Q. emoryi**

2'. Leaf blades conspicuously bicolored, adaxially bluish green and abaxially whitish, abaxially uniformly pubescent or tomentose, blades flat to slightly convex…**Q. hypoleucoides**

3. Acorn maturation biennial, nut requiring 2 seasons to mature; cup scales embedded in tawny or glandular tomentum, only scale tips visible; plants evergreen, leaves leathery, abaxially glaucous; pine-oak mountains of the SW portion of the state (sect. **Protobalanus**, intermediate or golden-cap oaks)…
Q. chrysolepis

3'. Acorn maturation annual, nut requiring 1 growing season to mature; cup scales not embedded in tawny or glandular tomentum, although scales may be tomentose; plants variously evergreen or deciduous; widespread (sect. **Quercus**, white oaks)…(4)

4. Leaf blades regularly or irregularly lobed, at least some sinuses extending > 1/3 distance from margin to midrib, or distally 3-lobed (3-dentate); widespread in montane areas…**Q. gambelii**

4'. Leaf blades unlobed or sometimes shallowly lobed or sinuate-lobed, if lobed, sinuses < 1/3 distance to midrib; leaf margins never 3-dentate, usually more localized in various habitats…(5)

5. Leaf blades with 10+ secondary veins on each side, these parallel and ± straight and not branching, margins dentate, regularly toothed with a single tooth for each secondary…**Q. muehlenbergii**

5'. Leaf blades with 2–12 irregular secondary veins on each side, these curved, crooked, and/or branching, or obscure, not noticeably parallel, margins entire or irregularly toothed or spinose, if toothed then not regularly dentate…(6)

6. Leaves and twigs glabrous at maturity, leaf blades unlobed, glaucous and blue-green to gray-green, margins entire to obscurely toothed…**Q. oblongifolia**

6'. Mature leaves pubescent, hairy, or glandular on abaxial surface (visible with ×10 lens), twigs variously hairy or sometimes glabrate, leaf blades greenish to grayish or blue-green…(7)

7. Mature leaf blades abaxially dull or glossy green or yellowish green, glabrous or with few stellate hairs…
Q. toumeyi

7'. Mature leaf blades abaxially with stellate or erect fasciculate hairs distributed ± evenly (hairs sometimes minute or obscured by glandular hairs or wax)…(8)

8. Leaf blades usually convexly cupped, abaxially with prominent raised reticulum formed by ultimate venation; secondary veins often adaxially impressed…(9)

8'. Leaf blades not convexly cupped, without prominent raised reticulum formed by ultimate venation; secondary veins not strongly impressed adaxially…(10)

9. Acorns on axillary peduncles 1–9 cm, usually 3 to several acorns per peduncle; leaf blades broadly obovate to orbiculate…**Q. rugosa**

9'. Acorns subsessile or on axillary peduncles to 3.2 cm, usually 1–2(3) acorns per peduncle; leaf blades elliptic to lanceolate, ovate, or obovate…**Q. arizonica**

10. Rhizomatous shrub restricted to sand dunes; leaves oblong to lanceolate to obovate, margins flat to revolute to undulate, 2–5 rounded teeth on each side…**Q. havardii**

10'. Trees or shrubs, rhizomatous or not, occurring in a variety of habitats, never in deep sand…(11)

11. Leaf blade abaxial surfaces with appressed stellate hairs, the hairs often minute, not felty or velvety to the touch…(12)

11'. Leaf blade abaxial surfaces with erect or semierect stellate or fasciculate hairs, the hairs often curly or wavy, often felty or velvety to the touch…(14)

12. Acorns on axillary peduncles 3–30 mm; leaf margins regularly toothed, teeth often spinose, leaf blades grayish-glaucous or yellowish-glandular…**Q. turbinella**

12'. Acorns subsessile or borne on axillary peduncles to 3 mm; leaf margins entire to irregularly toothed, teeth mucronate, leaf blades green, never glandular…(13)

13. Leaf blades adaxially sandpapery with harsh hairs, leaves elliptic to oblong in outline, 1.5–4 cm wide…
Q. pungens

13'. Leaf blades adaxially smooth, leaves oblong to lanceolate to obovate, 3–5 cm wide…**Q. ×undulata**

14.   Leaf blades often strongly bicolored, abaxially densely gray- or white-tomentose, oblong or elliptic; acorns subsessile or on stalks to 1.5 cm...**Q. mohriana**

14'. Leaf blades weakly bicolored, abaxially densely gray-green or yellowish-tomentose, lanceolate, ovate, or slightly obovate; acorns sessile or on stalks to 1 cm...**Q. grisea**

Quercus arizonica Sarg.—Arizona oak. Evergreen or subevergreen shrub or tree 1–30 m; buds russet-brown, ovoid, ca. 3 mm; leaves broadly linear-oblong or linear-lanceolate to oblanceolate or obovate, 2–12.5 × 0.9–6 cm, leathery, flat or slightly convex, margins entire or irregularly low-sinuate or with 1–6 triangular mucronate teeth on each side, abaxial surface with sparse to moderately dense stellate hairs, secondary veins 8–11 on each side; acorns 11–15 mm, 1–5 in clusters, on stalks to 3.2 cm. Oak and pine woodlands, igneous slopes and arroyos, 4000–8000 ft. C, WC, SC, SW. Known to hybridize with *Q. gambelii*, *Q. grisea*, *Q. oblongifolia*, and *Q. rugosa*.

Quercus chrysolepis Liebm.—Canyon live oak. Evergreen tree or shrub 0.5–10 m; buds conic, 2–8 mm; leaves elliptic to ovate or broadly lanceolate, 1–5 × 1–2.5 cm, thin and stiff, flat or somewhat undulate, margins entire or with 1–4 sharply triangular teeth on each side ending in a stiff spine to 1.5 mm, abaxial surface golden when young from dense amber glandular hairs, turning white with age, secondary veins 12+ on each side; acorns 15–35 mm, solitary or paired, sessile or on stalks to 3 mm. Canyon slopes in piñon-oak woodland, 4500–6000 ft. Grant.

Quercus emoryi Torr.—Emory oak. Evergreen tree or shrub 2–15 m; buds reddish brown, ovoid to subconic, 2.5–6.5 mm; leaves narrowly lanceolate to narrowly ovate, 2–7 × 0.8–3.5 cm, thin and stiff, flat, margins entire or with 1–4 prominent triangular teeth on each side bearing a short mucro or arista, abaxial surface smooth, glabrous except for tufts of hairs in the axils of basal veins, secondary veins 6–11 on each side; acorns annual, 9–10 mm, solitary or paired on stalks to 2 mm. Oak, pine, and juniper woodlands, alluvial washes, 4200–8000 ft. WC, SW.

Quercus gambelii Nutt.—Gambel oak. Deciduous shrub or tree 1–20 m, often clonal; buds brown, ovoid, ca. 3 mm; leaves elliptic to obovate in outline, 6–18 × 3–11 cm, thin and flexible, flat, with 4–6 rounded lobes on each side that reach more than halfway to the midrib, abaxial surface with scattered stellate hairs, secondary veins 4–6 on each side; acorns 12–15 mm, solitary or paired, sessile or on stalks to 3 cm. Montane oak-conifer woodlands, higher margins of piñon-juniper woodlands, 5500–10,700 ft. Widespread except EC. Hybridizes with *Q. grisea* and *Q. turbinella*.

Quercus grisea Liebm.—Gray oak. Deciduous or subevergreen tree or shrub 1–8 m; buds dark red-brown with yellowish stellate hairs, ovoid, 1–2 mm; leaves lanceolate, ovate or slightly obovate, 1.2–3.5 × 0.9–2.1 cm, thin and flexible, flat or undulate, margins entire or with 1–2 blunt teeth on each side near the apex, abaxial surface with scattered to dense stellate hairs, secondary veins 6–8 on each side; acorns 10–12 mm, sessile or paired on stalks 0.6–1 cm. Slopes in oak and juniper woodlands, 3700–8800 ft. Widespread except DeBaca, Roosevelt, San Juan. Known to hybridize with *Q. mohriana*, *Q. muehlenbergii*, *Q. turbinella*, *Q. gambelii*, *Q. pungens*, *Q. toumeyi*, and *Q. arizonica*.

Quercus havardii Rydb.—Havard oak, shinnery oak. Deciduous shrub to 1.5 m, rhizomatous, clonal; buds dark red-brown, subglobose, to 2 mm; leaves polymorphic, oblong to lanceolate to obovate, 5–10 × 2–5 cm, thick and hard, margins flat to revolute to undulate, 2–5 rounded teeth on each side, abaxial surface densely grayish-tomentulose, secondary veins 6–9 on each side; acorns 12–25 mm, solitary or paired, subsessile or on stalks to 1 cm. In deep, shifting, or stabilized sand dunes, 3450–6900 ft. EC, SE. Hybridizes with *Q. gambelii*, *Q. grisea*, and *Q. mohriana*. Our plants are **var. havardii**. Plants assigned to **var. tuckerii** S. L. Welsh [*Q. welshii* R. A. Denham] may enter the state in the far NW corner but have not been definitively confirmed. Such plants would be morphologically identical but represent a distinct genetic grouping isolated to the Navajo Basin.

Quercus hypoleucoides A. Camus—Silverleaf oak. Evergreen tree 3–25 m; buds light brown, ovoid, 2.4–4.5 mm; leaves lanceolate to oblanceolate or elliptic, 4–15 × 0.8–5 cm, leathery and flexible, flat to slightly convex, margins entire or with 1–4 triangular lobes on each side bearing aristae 1–3 mm, abaxial surface densely covered by a tomentum of sessile, overlapping stellate hairs, secondary veins 4–10 on each side; acorns biennial, 12–20 mm, solitary, sessile or on stalks to 15 mm. Woodland slopes, canyon bottoms, 4200–9000 ft. WC, SC, SW.

Quercus mohriana Buckley ex Rydb.—Mohr oak. Evergreen or deciduous shrub or tree 0.5–3 m; buds dark red-brown, round-ovoid, ca. 2 mm; leaves oblong or elliptic, 3–5 × 2–3 cm, leathery, undulate to flat, margins entire, toothed, or denticulate, abaxial surface densely tomentose with curly stellate hairs, secondary veins 8–9 on each side; acorns 8–15 mm, solitary or paired, subsessile or on stalks to 1.5 cm. Hillsides and ravines, particularly on calcareous substrates, 5200–7250 ft. NE, C, EC, SC, SE. Known to hybridize with *Q. gambelii* and *Q. grisea*.

Quercus muehlenbergii Engelm.—Chinkapin oak. Deciduous tree to 30 m; buds brown to red-brown, ovoid, 20–40 mm; leaves obovate to oblanceolate, 5–15 × 4–8 cm, leathery, margins regularly undulate, toothed, or shallow-lobed, teeth or lobes rounded, abaxial surface appearing glabrate; secondary veins usually 10–14 on each side; acorns 15–20 mm, solitary or paired, subsessile or on stalks to 8 mm. North-facing slopes and canyon bottoms, particularly on calcareous substrates, 4600–7550 ft. Eddy, Harding, Lincoln, Mora.

Quercus oblongifolia Torr.—Sonoran blue oak. Evergreen tree 4–13 m; buds reddish brown, subspherical to ovoid, 1–2 mm; leaves oblong to oblong-ovate, 3–6 × 1–3 cm, thin and stiff, flat, margins entire or rarely with 1–3 low lobes or rounded teeth on each side, abaxial surface glabrous or with few stellate hairs, secondary veins 8–10 on each side; acorns 3–18 mm, solitary or paired, sessile or on stalks to 2 cm. Open oak grasslands, hillsides in piñon-juniper woodlands, canyon bottoms, 4300–6000 ft. Grant, Hidalgo. Known to hybridize with *Q. arizonica*, *Q. grisea*, *Q. pungens*, and *Q. toumeyi*.

Quercus pungens Liebm.—Pungent oak, sandpaper oak. Evergreen or subevergreen shrub or tree 1–6 m; buds dark red-brown, ca. 2 mm; leaves elliptic to oblong in outline, 2–9 × 1.5–4 cm, thin and stiff, usually strongly undulate, margins with 3–5 mucronate teeth or shallow lobes on each side, abaxial surface densely covered with minute stellate hairs, secondary veins 5–8 on each side; acorns 9–10 mm, solitary or paired, on stalks to 3 mm. Hillsides in oak, pine, and juniper woodlands, often on igneous or limestone substrates, 4600–8300 ft. NW, WC, SE, SC, SW, San Miguel.

Quercus rugosa Née—Netleaf oak. Evergreen shrub or tree 4–15 m; buds brown, ovoid, 2–4 mm; leaves ovate, 6–11 × 4–8 cm, stiff, usually strongly convex, margins entire or with 2–6 low triangular teeth bearing short mucros on each side, abaxial surface densely covered with vermiform glandular hairs, secondary veins 7–10 on each side; acorns 8–13 mm, 1–8 in clusters on stalks 1–9 cm. Wooded slopes, canyons, ridgelines, 4900–8500 ft. WC, SC, SW.

Quercus toumeyi Sarg.—Toumey oak. Deciduous or subevergreen shrub or tree 4–25 m; buds reddish brown, ovoid, ca. 1 mm; leaves oblong, oval, or obovate, 1–3.5 × 0.5–2.2 cm, stiff, flat or undulate, margins entire or with 1–4 low triangular teeth bearing short mucros on each side, abaxial surface glabrous or with few stellate hairs, secondary veins 6–10 on each side; acorns 13–14 mm, solitary or paired, on stalks 4–8 mm. Oak, pine, and juniper woodlands, 4800–8000 ft. Grant, Hidalgo, Luna. Known to hybridize with *Q. turbinella*.

Quercus turbinella Greene—Sonoran scrub oak. Evergreen or subevergreen shrub or tree 1–4 m; buds brown, round to ovoid, 1–2 mm; leaves broadly elliptic, 1.3–3 × 0.5–2 cm, thick and stiff, flat to slightly undulate, margins entire or with 3–5 triangular teeth on each side bearing a spine 1–1.5 mm, abaxial surface minutely stellate-pubescent and ± golden-brown with glandular amber hairs, secondary veins 4–8 on each side; acorns 10–12 mm, solitary or paired, on stalks to 4 cm. Desert scrublands, riparian woodlands, 4260–8900 ft. Widespread except NC, Cibola. Known to hybridize with *Q. gambelii*, *Q. toumeyi*, and *Q. grisea*.

Quercus ×undulata Torr. [*Q. ×pauciloba* Rydb.]—Wavyleaf oak. Deciduous or subevergreen shrub 1–3 m; buds variable; leaves oblong to lanceolate to obovate, 4–9 × 3–5 cm, stiff, flat to revolute to undulate, margins rarely entire to more commonly lobed with 2–5 mucronate lobes on each side, abaxial surface variable but generally with few scattered stellate hairs, secondary veins 4–9 on each side; acorns

15–18 mm, sessile or paired, on stalks to ca. 2 cm. Oak, pine, and juniper woodlands, open slopes, canyon bottoms, mesa tops, 4390–9500 ft. Widespread except Lea. *Quercus ×undulata* represents a mix of hybrid derivatives, often occurring in mixed stands near one of the parental taxa or on their own. Highly variable morphologically.

FOUQUIERIACEAE – OCOTILLO FAMILY

Kenneth D. Heil

FOUQUIERIA Kunth – Ocotillo

Fouquieria splendens Engelm.—Spiny shrub to small tree with a short trunk, 10–20 cm, bearing numerous usually simple, erect, spiny branches 2–6 m; leaves on new long growth petioled, soon deciduous in dry weather, the petioles in part remaining as a spine when the leaf falls, secondary leaves sessile or nearly so, borne in fascicles on short shoots in axils of spines, leaves 1–2 × 2–8 mm; inflorescences narrow terminal panicles, 5–20 cm; flowers red, reddish orange, yellow, or pinkish purple; sepals 5, free; petals 5, united into a cylindrical tube 6.5–22 mm; stamens 10–15(–23), adnate to corolla tube; pistils 3-carpelled, the ovary superior, 3-lobed; fruits 3-valved, capsules 9. Chihuahuan desert scrub and mesquite grasslands, 3300–7500 ft. EC, C, WC, SE, SC, SW (Henrickson, 1972).

FRANKENIACEAE – ALKALI HEATH FAMILY

Kenneth D. Heil

FRANKENIA L. – Alkali heath

Frankenia jamesii Torr. ex A. Gray—James' seaheath. Salt-tolerant small shrub to 30 cm; stems jointed, puberulent; leaves decussate, appearing whorled, linear to elliptic, somewhat terete, aculeate, somewhat clasping, hispid, margins strongly revolute, base margins ciliate, 3–5 mm, ca. 1 mm wide; flowers sessile, actinomorphic, perfect, rarely unisexual and monoecious or dioecious; calyx tube ribbed, 4–5 mm, lobes toothlike, 5; petals 5, clawed, about twice as long as calyx tube, white; stamens 6, 3 long and 3 short; ovary superior, with 2–4 united carpels, with a single locule and parietal placentation; style with 3 branches; fruit a linear, angled capsule ca. 5 mm. Primarily in desert scrub, usually associated with gypsum deposits, 3700–6500 ft. NW, C, SC (Reeves, 2013a).

GARRYACEAE – SILKTASSEL FAMILY

Kenneth D. Heil

GARRYA Douglas ex Lindl. – Silktassel

Shrubs or trees, evergreen, dioecious; leaves opposite (decussate), simple, blades flat to concave-convex, margins entire, flat, revolute, or strongly undulate; flowers unisexual; perianth epigynous; staminate flower sepals 4, linear to lanceolate-oblong, petals 0, stamens alternate with sepals, anthers basifixed; pistillate flower sepals 2, sometimes rudimentary, petals 0; ovules 2, pistil 1, 1-carpellate, ovary inferior, 1-locular, placentation apical; ovules 1–2 per locule; styles 1–2(3), distinct; stigmas 1–2(3); fruits berries, dark blue to black, drying whitish gray; seeds 1–2 per fruit (Nesom, 2016a).

1. Leaf blade abaxial surfaces glabrous or sparsely strigose…**G. wrightii**
1'. Leaf blade abaxial surfaces persistently densely tomentulose with coiling to recurved hairs…**G. goldmanii**

Garrya goldmanii Wooton & Standl. [*G. ovata* Benth. var. *goldmanii* (Wooton & Standl.) B. L. Turner]—Goldman's silktassel. Shrubs (0.5-)1-2.5 m, branchlets puberulent, glabrescent; leaf blades gray-green, flat to concave-convex, elliptic to broadly lanceolate, mostly 1.6-4 × 0.7-2.5 cm, margins undulate, both surfaces usually persistently densely tomentulose; staminate aments 2-3 cm, pistillate aments loose; berries 4-8 mm diam., glabrous. Limestone, oak scrub, oak-pine-juniper woodlands, 5800-8200 ft. Eddy, Otero, Sierra.

Garrya wrightii Torr.—Wright's silktassel. Shrubs or trees 1-3(-4) m, branchlets sparsely strigose or glabrescent; leaf blades yellowish green, flat, elliptic, mostly 3-5.5 × 1.5-3 cm, abaxial surface glabrous, glabrate, or sparsely strigose, adaxial surface dull, glabrous or glabrate; staminate aments 1-2 cm, pistillate aments loose; berries 5-7 mm diam., sparsely strigose or glabrate. Cliff crevices, bluffs, piñon, ponderosa pine-oak, 4540-8300 ft. C, WC, SE, SC, SW.

GENTIANACEAE – GENTIAN FAMILY

John R. Spence

Annual, biennial, or perennial herbs, cespitose to erect, from taproots, rhizomes, stolons, or fibrous roots; stems erect to decumbent, rounded to angled; leaves simple, opposite, whorled, or rarely alternate, mostly sessile, often basal, bases often sheathing-connate, sometimes white-margined, stipules absent; inflorescence of simple to complex thyrsoid cymes or sometimes flowers solitary; flowers bisexual, actinomorphic, hypogenous, 4- or 5-merous, subtended by bracts; sepals persistent, united to cleft; petals united, tubular to rotate or campanulate, often fringed, sometimes plicate between lobes; nectaries 1-2; stamens alternate, epipetalous; pistil 1, 2-carpellate, 1- or 2-locular, placentation parietal; style obsolete or 1, thick to filiform, stigmas 2, entire to 2-lobed; fruit a 2-valved septicidal capsule, tips often recurved when dry; seeds small, numerous (Mason, 1998).

1. Petals free to near base, distinct, flowers rotate to campanulate...(2)
1'. Petals strongly united into connate tube, with lobes usually shorter than tube, flowers funnelform to tubular...(6)
2. Flowers 5-merous, petals large, showy, 2.5-5 cm, tips erose-dentate...**Eustoma**
2'. Flowers 4- or 5-merous, petals smaller, mostly < 2.5 cm, tips ± entire...(3)
3. Petal lobes with 1-2 distinct fringed or scaled nectaries; plants biennial or perennial...(4)
3'. Petal lobes lacking distinct fringed or scaly nectaries although sometimes fimbriate; plants annual...(5)
4. Flowers 4-merous, petals white to green, rarely tinged with purple, nectaries 1, or if 2 then plants robust with numerous whorls of leaves; plants biennial from taproot...**Frasera**
4'. Flowers predominantly 5-merous, petals blue to blue-purple, rarely white, nectaries 2; plants rhizomatous perennial...**Swertia**
5. Stems conspicuously winged; petals pink to white; stem leaves broadly ovate to ovate-lanceolate, to 4 cm...**Sabatia**
5'. Stems not winged; petals blue to rarely white; stem leaves narrow, lanceolate, < 2 cm...**Lomatogonium**
6. Flowers salverform, anthers coiling spirally after anthesis, petals pink to rose; plants annuals of wet sites...**Zeltnera**
6'. Flowers funnelform to tubular, anthers not coiling spirally after anthesis, petals blue, purple, white, or yellow; plants annual, biennial, or perennial, habitats various...(7)
7. Corolla spurred, yellow, tube somewhat shorter than lobes; plants annual...**Halenia**
7'. Corolla not spurred, corolla blue, purple or white; plants annual, biennial, or perennial...(8)
8. Corolla lobes fringed along margins, flowers 4-merous...**Gentianopsis**
8'. Corolla lobes entire or serrulate but not fringed, flowers 4- or 5-merous...(9)
9. Corolla with conspicuous folded plaits between lobes; sepals with inner membrane connecting lobes; plants annual or perennial...**Gentiana**
9'. Corolla lacking plaits; sepals lacking inner membrane; plants annual or rarely biennial...(10)
10. Flowers on long pedicels longer than internodes; corolla lobes with 2 fringed scales at base...**Comastoma**
10'. Flowers on short pedicels, much shorter than internodes; corolla lobes smooth, somewhat fringed at base or with single fringed scale...**Gentianella**

COMASTOMA (Wettst.) Toyok. – Lapland gentian

Comastoma tenellum (Rottb.) Toyok.—Lapland gentian. Annuals, stems to 20 cm, often with curved-ascending branches; basal leaves elliptic to ovate, often early-deciduous, stem leaves ovate to ovate-lanceolate, to 10 mm, not white-margined; flowers solitary, terminal or axillary, on long pedicels to 6 cm, much longer than branch internodes; sepals lacking inner membrane, lobes cleft nearly to base; perianth tubular, blue to white, to 10 mm, lobes ovate, lacking plaits, bearing 2 fimbriate scales at lobe base; pistil sessile, style absent, stigmas 2; fruit a fusiform capsule, to 15 mm. Wet soil in meadows, wetlands, and along streams in the alpine, 11,000–13,000 ft. Santa Fe, Taos.

EUSTOMA Salisb. – Prairie gentian, catchfly gentian

Annuals, biennials or short-lived perennials, stems to 60 cm, simple or branched, not angled to winged; basal rosette usually present, leaves obovate, stem leaves opposite, strongly sheathing-connate, sessile, oblong to elliptic, not white-margined, sometimes glaucous; inflorescence a several-flowered panicle; flowers 5- or rarely 6-merous, pedicellate, pedicels often longer than the internodes; calyx deeply cleft, tube short or absent, lobes linear-lanceolate, lacking inner membrane; corolla campanulate, tube short to almost absent, lobes elliptic to obovate, large, showy, to 5 cm, sometimes dentate or sometimes weakly lacerate-erose, blue, blue-purple, or white, lacking plaits, nectaries and fimbriae inconspicuous; pistil sessile, style slender, often as long as pistil, stigmas 2; fruit an elliptic capsule.

1. Corolla lobes 3–5 cm, > 2 cm wide…**E. grandiflorum**
1'. Corolla lobes 2–3 cm, mostly < 2 cm wide…**E. exaltatum**

Eustoma exaltatum (L.) Salisb. ex G. Don—Catchfly gentian. Annual, biennial, or short-lived perennial, stems to 70 cm; stem leaves elliptic to oblong, to 9 cm; flowers on long pedicels; calyx lobes to 25 mm, mostly much longer than tube; corolla 2–3 cm, < 2(–2.5) cm wide, lobes mostly twice length of tube, ovate to oblong, lavender or white; capsule ellipsoid, to 15 mm. Wet to damp alkaline soil in meadows and wetlands, 2975–6500 ft. NC, EC, C, SE, SC.

Eustoma grandiflorum (Raf.) Shinners [*E. russellianum* (Hook.) G. Don]—Showy prairie gentian. Annual, biennial, or short-lived perennial, stems to 60 cm; stem leaves elliptic, oblong, or ovate-lanceolate, to 7 cm; flowers on long pedicels; calyx lobes to 20 mm, mostly much longer than tube; corolla (3–)3.5–5 cm, > 2 cm wide, lobes mostly twice length of tube, elliptic to obovate, blue-purple, pink, or white; capsule ellipsoid, to 20 mm. Wet to damp alkaline soil in meadows, prairie, wetlands, 3600–6500 ft. C, EC, SE, SC. *Eustoma grandiflorum* and *E. exaltatum* are very similar to each other, differing primarily in flower size and geographic distribution.

FRASERA Walter – Green gentian, frasera

Biennials or short-lived perennials, stems to 20 dm, simple, not angled to winged; basal rosette leaves persistent or early-deciduous, stem leaves opposite to whorled, sessile, connate-sheathing, sometimes white-margined; inflorescence cymose and branched to ± spicate, many-flowered; flowers predominantly 4-merous, pedicellate, pedicels shorter than the internodes; calyx with short tube or tube nearly absent, lobes long, linear to lanceolate, equal, lacking inner membrane; corolla rotate to campanulate, pale green to white, sometimes streaked with purple or green, tube short or absent, lobes lanceolate to ovate, not fringed-erose to dentate, lacking plaits, bearing 1–2 nectaries at base, fringed by scales or hairs, pistil sessile, style slender, stigmas 2; fruit an ovoid to elliptic capsule.

1. Leaves lacking white margins, in numerous whorls; corolla lobes to 30 mm, with 2 distinct glands; inflorescence spikelike, not branched…**F. speciosa**
1'. Leaves white-margined, opposite in whorls of 4; corolla lobes to 10 mm, with single glands; inflorescence a branched panicle or cyme…(2)
2. Leaves in whorls of 4; nectaries lobed at apex…**F. albomarginata**
2'. Leaves in opposite pairs; nectaries lobed at base…**F. paniculata**

Frasera albomarginata S. Watson—White-margined frasera. Stems to 60 cm, single or sometimes several; leaves in whorls of 4, basal leaves oblanceolate, 3–10 cm, white-margined, somewhat undulate along margins, stem leaves few, smaller; inflorescence a wide-branched panicle; calyx lobes 2–6 mm; corolla lobes 6–10 mm, white, sometimes with green spots, nectaries 1, split above into 2 lobes with white fringes, scales absent, lacking plaits; fruit an ellipsoid capsule to 16 mm. Open sandy to clay soils in sagebrush, desert scrub, piñon-juniper woodlands, 5550–5900 ft. San Juan.

Frasera paniculata Torr. [*F. utahensis* M. E. Jones]—Paniculate frasera. Stems to 100 cm, single or sometimes several; leaves opposite, basal leaves lanceolate, 6–12 cm, white-margined, not or weakly undulate along margins, stem leaves few, smaller; inflorescence a wide triangular panicle; calyx lobes 3–6 mm; corolla lobes 6–10 mm, white or green, often with darker green spots, nectaries 1, broad above, split at the base into 2 lobes with white fringes, scales absent; fruit an ellipsoid capsule to 16 mm. Open sandy, rocky, or clay soils in slickrock, sagebrush, desert scrub, piñon-juniper woodlands, sometimes in springs, 5000–7050 ft. Cibola, Rio Arriba, Sandoval, San Juan.

Frasera speciosa Douglas ex Griseb.—Elk gentian, elkweed. Stems to 2 m, single, unbranched; leaves in whorls, basal leaves numerous, spatulate to oblanceolate, to 50 cm, not white-margined, margins not wavy, leaves gradually becoming smaller along the stem but still numerous; inflorescence a dense, compact, spikelike series of cymes, typically verticillate; calyx lobes 9–25 mm; corolla lobes 9–20 mm, pale green with purple splotches, nectaries 2, unsplit, fringed with long hairs, corona scales present, alternating with nectaries, lacerate; fruit an oblong capsule to 25 mm. Open soils, slopes, forest clearings, meadows in the mountains (5700–)6000–10,600 ft. N, C, WC, SE, SC.

GENTIANA L. – Gentian

Annuals, biennials, or perennials, stems to 60 cm, simple or branched, round or somewhat angled; basal rosette leaves persistent, stem leaves opposite, sessile, connate-sheathing, not or sometimes white-margined; inflorescence of terminal or axillary cymes or flowers sometimes solitary; flowers 4- or 5-merous, sessile or pedicellate, pedicels shorter than the internodes; calyx with thin, membranous inner membrane; corolla tubular, funnelform or rarely fusiform, blue or rarely white or green-purple, 10–50 mm, lobes entire or sometimes weakly crenate along upper margins, shorter than tube, symmetric to asymmetric plaits between the lobes, entire or lacerate, nectaries inconspicuous, at base of ovary, scales and fimbriae lacking; pistil sessile or stipitate, style short or absent, stigmas 2; fruit an elliptic to ovate capsule. *Gentiana* is a large genus that has been shown to be monophyletic, but with its great morphological diversity it has been divided into numerous smaller genera by some authors.

1. Plants small, to 15 cm, annual or rarely biennial; flowers solitary, terminal, mostly 4-merous; corolla narrow, to 20 × 2–5 mm…(2)
1'. Plants larger, (5–)10–60 cm, perennial; flowers solitary or more often several to many, terminal or on secondary branches; corolla mostly > 20 mm, narrow or wide…(3)
2. Fruit included in corolla tube at maturity; leaves and sepals distinctly white-margined; flowers pale blue, white, or green-purple…**G. fremontii**
2'. Fruit exserted on long stipe beyond corolla tube at maturity; leaves and sepals mostly green-margined, rarely narrowly white-margined; flowers deep blue, rarely white…**G. prostrata**
3. Plants short, to 20 cm; flowers sometimes solitary, white to pale yellow, plaits purple, purple to green-spotted; strictly alpine species…**G. algida**
3'. Plants taller, to 60 cm; flowers solitary to several or many, pale to deep blue or purple, rarely white, sometimes with green mottling or streaks; widespread…(4)
4. Flowers 1 to few, mostly terminal, subtended by broadly ovate to boat-shaped, somewhat scarious bracts; calyx tube long, to 20 mm; corolla large, to 5 cm, broadly tubular to campanulate, mouth wide…**G. parryi**
4'. Flowers numerous, in terminal and axillary cymes, bracts linear to lanceolate, predominantly green; calyx tube to 10 mm; corolla to 4 cm, funnelform to fusiform, mouth narrow or closed…**G. affinis**

Gentiana affinis Griseb. [*Pneumonanthe affinis* (Griseb.) Greene; *G. rusbyi* Greene ex Kusn.]—Rocky Mountain bottle gentian. Perennial; stems to 60 cm, erect; basal rosette lacking, cauline leaves lanceolate to elliptic, not white-margined; flowers numerous, in open diffuse inflorescence, subfloral bracts green, shorter than the corolla; calyx tube rarely split to base on 1 side; corolla funnelform to campanulate, blue to purple with green streaks, 20–45 mm, lobes entire, plicae split into 2–3 lobes; fruit an ovoid to elliptic capsule. Wet to damp soils in meadows and along streams, montane to subalpine, (4850–)5700–12,250 ft. NC, NW, C, WC.

Gentiana algida Pall. [*Gentianodes algida* (Pall.) Á. Löve & D. Löve]—Arctic gentian. Perennial; stems to 15 cm, mostly erect, sometimes branched; basal leaves linear-lanceolate, to 10 cm, cauline leaves lanceolate to elliptic, smaller, not white-margined; flowers 1–3, 5-merous, sessile; calyx 15–30 mm, lobes somewhat unequal in length, shorter than tube, not white-margined, inner membrane present; corolla funnelform, white to pale yellow with purple-green spots and blotches, 30–50 mm, lobes short, plicae entire; fruit an elliptic to fusiform capsule, not much exserted past corolla. Wet soil in meadows and along streams in the subalpine and alpine, (9100–)10,500–13,000 ft. NC, Otero.

Gentiana fremontii Torr. [*Chondrophylla fremontii* (Torr.) A. Nelson; *G. aquatica* L.]—Moss gentian. Annual or biennial; stems to 10 cm, often prostrate; basal leaves oblanceolate to obovate, to 12 mm, cauline leaves similar but smaller, distinctly white-margined; flower solitary, 4(5)-merous, pedicellate; calyx 6–15 mm, lobes much shorter than tube, white-margined, inner membrane present; corolla tubular-funnelform, light blue to white, with green to purple spots, 10–22 mm, lobes short, plicae entire; fruit an obovate capsule to 12 mm, ± included in corolla, valves strongly divergent when split, forming cup. Wet soil in meadows, wetlands, and along streams in the mountains up to the alpine, 8300–12,400 ft. NC, Sandoval.

Gentiana parryi Engelm. [*Pneumonanthe parryi* (Engelm.) Greene]—Parry's bottle gentian. Perennial; stems to 50 cm, erect; basal rosette lacking, cauline leaves broadly ovate to elliptic, not white-margined; flowers few, terminal, subfloral bracts shorter than the corolla, broadly obovate to boat-shaped, scarious-margined; calyx tube sometimes split to base on 1 side; corolla broadly funnelform to campanulate, blue to purple, 30–50 mm, lobes entire, plicae split into 2–5 lobes; fruit an ovoid to elliptic capsule. Wet to damp soils in meadows and along streams, montane to alpine, 8450–13,025 ft. NE, NC.

Gentiana prostrata Haenke [*Chondrophylla prostrata* (Haenke) J. P. Anderson]—Pygmy gentian. Annual or biennial; stems to 12 cm, often prostrate; basal leaves ovate to obovate, to 12 mm, cauline leaves similar but smaller, green or faintly white-margined; flower solitary, 4(5)-merous, pedicellate; calyx 6–12 mm, lobes much shorter than tube, equal, inner membrane present; corolla tubular-funnelform, dark blue or rarely white, lacking green to purple spots, 10–20 mm, lobes short, plicae split into 2–3 segments; fruit a slender ellipsoid, fusiform, or obovate capsule, to 14 mm, long-exserted on stipe, valves divergent when split but not forming cup. Wet soil in alpine tundra; 1 record from Big Costilla Peak, 12,535 ft. Taos.

GENTIANELLA Moench – Little gentian, felwort

Annuals, stems to 50 cm, simple or branched, somewhat angled to winged; basal rosette leaves early-deciduous, stem leaves opposite, sessile, ovate-lanceolate to elliptic, not white-margined; inflorescence of terminal or axillary cymes or flowers sometimes solitary; flowers 4(5)-merous, pedicellate, pedicels shorter than the internodes; calyx lacking inner membrane, tube short to almost absent, lobes equal or strongly unequal in shape and length; corolla tubular, funnelform to campanulate, blue or rarely white, pink, or yellow, 10–25 mm, lobes ovate, entire, shorter than tube, lacking plaits, bearing 1 fimbriate scale at base; pistil sessile, style absent, stigmas 2; fruit an ovate capsule.

1. Calyx lobes equal to weakly unequal, lobes not distinctly larger or foliaceous, to 12 mm; corolla lobes 3–5 mm, fimbriae separate to base…**G. acuta**
1'. Calyx lobes distinctly unequal, 2 outer lobes larger and ± foliaceous, to 20 mm; corolla lobes 4–8 mm, fimbriae united at base into scalelike sheath…**G. heterosepala**

Gentianella acuta (Michx.) Hiitonen [*G. amarella* (L.) Börner subsp. *acuta* (Michx.) J. M. Gillett]—Gentianella. Stems to 40 cm, branches erect; basal leaves oblanceolate to elliptic, early-deciduous, stem leaves not cordate at base, lanceolate to ovate, to 6 cm; flowers on short pedicels; calyx 4–12 mm, lacking inner membrane, lobes equal to slightly unequal, 2 lobes somewhat longer; corolla tubular-campanulate, blue to blue-purple, to 17 mm, lobes ovate, lacking plaits, bearing scattered fimbriae at base or rarely absent; fruit a fusiform capsule to 20 mm. Wet soil in meadows, wetlands, and along streams in the mountains up to the alpine, 6700–12,250 ft. NC, NW, C, WC, SC.

Gentianella heterosepala (Engelm.) Holub [*G. amarella* (L.) Börner subsp. *heterosepala* (Engelm.) J. M. Gillett]—Gentianella. Stems to 50 cm, branches open to ascending-erect; basal leaves spatulate to obovate, early-deciduous, stem leaves with slightly cordate base, oblanceolate to ovate, to 6 cm; flowers on short pedicels; calyx 6–20 mm, lacking inner membrane, lobes strongly unequal, with 2 larger outer foliaceous lobes covering smaller inner lobes; corolla tubular-campanulate, blue, white, pink, or rarely yellow, to 25 mm, lobes ovate, about equal to tube, lacking plaits, bearing fimbriate former scale at base; fruit a fusiform capsule to 18 mm. Wet soil in meadows, wetlands, and along streams in the mountains up to the subalpine, 7500–10,600 ft. NW, WC, SC.

GENTIANOPSIS Ma – Fringed gentian

Annual, biennial, or perennial, stems to 60 cm, simple or branched, not angled to winged; basal rosette leaves persistent or early-deciduous, stem leaves opposite, sessile, strongly connate, ovate, elliptic, or oblanceolate, not white-margined; inflorescence of terminal solitary flowers or sometimes axillary flowers below; flowers 4-merous, sessile to pedicellate, pedicels shorter than the internodes; calyx to 25 mm, tube as long as lobes, lobes in 2 unequal series, sometimes with discontinuous inner membrane; corolla tubular, funnelform to campanulate, blue or rarely white, to 80 mm, lobes oblong to obovate, strongly fringed-erose to dentate, shorter than tube, lacking plaits, bearing 1 naked gland at base, lacking fimbriate scales; pistil stipitate, sometimes elongate in fruit, style absent, stigmas 2; fruit a slender fusiform to elliptic capsule.

1. Perennials; flowers sessile or short-pedicelled, subtended by 2 bracts; sepal lobes with discontinuous inner membrane; corolla lobes oblong; anthers to 2.5 mm…**G. barbellata**
1'. Annuals; flowers long-pedicellate, bracts absent; sepals lacking inner membrane; corolla lobes obovate, widest near tip; anthers to 4 mm…**G. thermalis**

Gentianopsis barbellata (Engelm.) Iltis—Fringed gentian. Perennial from fibrous rhizomes, stems to 15 cm, simple or branched above; basal leaves oblanceolate, to 8 cm, strongly connate-sheathing at base, stem leaves few, to 4 cm; flowers solitary or 2–3, sessile or short-pedicellate, subtended by 2 leafy bracts; calyx 10–25 mm, lobes with discontinuous inner membrane, equal to tube; corolla broadly funnelform, to 40 mm, lobes slightly longer than tube, to 25 mm, oblong, anthers < 2.5 mm. Damp to wet soil in meadows, rocky slopes, forest clearings up to the alpine, 10,170–12,250 ft. NC.

Gentianopsis thermalis (Kuntze) Iltis—Fringed gentian. Annual from slender taproot, stems to 60 cm, simple or branched above; basal leaves elliptic to spatulate, to 4 cm, often lacking at anthesis, stem leaves numerous, lanceolate to elliptic, to 7 cm, not or weakly connate-sheathing at base; flowers solitary, terminal, long-pedicellate, lacking bracts; calyx 10–40 mm, lobes lacking inner membrane, longer than tube; corolla broadly funnelform, to 80 mm, lobes shorter than tube, to 20 mm, obovate; anthers 2–4 mm. Damp to wet soil in meadows, rocky slopes, forest clearings up to the alpine, 8000–12,300 ft. NC.

HALENIA Borkh. – Spurred gentian

Halenia recurva (Sm.) C. K. Allen [*H. rothrockii* A. Gray]—Spurred gentian. Annuals, stems to 50 cm, simple; basal leaves elliptic to spatulate, to 3 cm, stem leaves few, linear-lanceolate, to 4 cm, not white-margined; flowers 4-merous, long-pedicellate, pedicels to 3 cm; calyx lobes lanceolate, to 6 mm, often longer than tube, papillate; corolla to 12 mm, lobes longer than tube, yellow, papillate, lacking plaits, with distinct spurs, spurs horizontal to erect-recurved; pistil sessile, style absent, stigmas 2; fruit an ovate-lanceolate capsule, to 12 mm. Wet soil in meadows, wetlands, and along streams in the mountains, 6500–10,000 ft. Catron, Grant, Sierra.

LOMATOGONIUM A. Braun – Marsh-felwort

Lomatogonium rotatum (L.) Fr.—Marsh-felwort. Annuals, stems to 25 cm, simple; basal leaves elliptic to spatulate, often early-deciduous, stem leaves narrowly lanceolate, < 20 mm, not white-margined; flowers 4- or 5-merous, solitary, terminal or sometimes axillary, pedicels short; sepals ± free to base, lacking inner membrane, lobes linear-lanceolate; corolla rotate, tube very short, blue to white, lobes to 20 mm, bearing a few fimbriae at base but lacking distinct nectaries, lobes entire, lacking plaits; pistil sessile, style absent, stigmas 2; fruit a septicidal capsule, to 15 mm. Wet soil in meadows, wetlands, and along streams in the subalpine; Vermejo Park area, ca. 9700 ft. Taos.

SABATIA Adans. – Rose gentian

Sabatia angularis (L.) Pursh—Rose-pink. Annual or biennial, to 80 cm, stems typically 2-branched above, 4-angled, distinctly winged on angles; basal leaves ovate to ovate-lanceolate, to 4 cm, early-deciduous, stem leaves smaller, few, not white-margined; inflorescence strongly branched, peduncles winged; flowers mostly 5-merous, rotate; calyx tube short, lobes lanceolate-linear, to 20 mm, lacking inner membrane; corolla tube ca. 1/3 length of lobes, lobes to 20 mm, spatulate-elliptic, pink or rarely white, with green spot at base, entire, lacking plaits between lobes; pistil sessile, style distinct, slender, stigmas 2; fruit an ovate capsule to 10 mm. Wet sand, soil in saline marshes, wetlands, and lake margins at low elevations; an eastern and southeastern US species adventive in extreme SC New Mexico, ca. 3900 ft. Bernalillo. Introduced.

SWERTIA L. – Star gentian, felwort

Swertia perennis L.—Felwort. Perennials to 60 cm, typically single-stemmed, from fibrous rhizomes; basal leaves oblong to oblanceolate, to 20 cm, petiole distinctly long, stem leaves smaller, sessile, few, not white-margined; flowers 5-merous, solitary, in terminal or sometimes axillary cymes; sepals ± free to base or sometimes short tube present, lacking inner membrane, lobes lanceolate, to 8 mm; corolla ± rotate, tube short, lobes oblong to elliptic, often erose at tips, to 15 mm, blue to purple, purple-spotted, or rarely white, lacking plaits between lobes, with 2 circular fimbriate nectaries; pistil sessile, style absent or nearly so, stigmas 2; fruit an ovate capsule to 12 mm. Wet soil in meadows, wetlands, and along streams in the subalpine and alpine, 7600–11,900 ft. NE, NC, Socorro.

ZELTNERA G. Mans. – Centaury

Annual or biennial, to 60 cm, stems typically freely branched, not 4-angled or winged; basal leaves petiolate, sometimes early-deciduous, stem leaves sessile, opposite, not white-margined; inflorescence cymose or spicate; flowers mostly 4- or 5-merous; calyx tube short, lobes slender, white to green, lacking inner membrane; corolla funnelform to salverform, tube much longer than lobes, pink, rose, or white, lacking plaits between lobes, nectaries inconspicuous, fimbriae absent; anthers spirally twisted after dehiscence; pistil sessile, style distinct, slender, stigmas capitate, flabelliform, or reniform; fruit an elliptic to ovate capsule.

1. Cauline leaves subulate-linear, few, not extending into inflorescence, basal rosette of leaves present… **Z. nudicaulis**
1'. Cauline leaves common, broader, elliptic, ovate, or lanceolate, often extending into inflorescence, basal rosette present or absent…(2)
2. Flowers small, corolla tube < 12 mm, lobes < 5 mm, corolla barely exserted past calyx; basal rosette generally absent…**Z. exaltata**
2'. Flowers large, corolla tube (10–)12–30 mm, lobes mostly > 10 mm, mostly well exserted past calyx; basal rosette present or absent at anthesis…(3)
3. Basal rosette persistent, of numerous leaves; stigma ± capitate; corolla tube green; mostly restricted to gypsiferous soils…**Z. maryanniana**
3'. Basal rosette mostly absent or withered at anthesis; stigmas reniform-flabellate to weakly bilobed; corolla tube white to yellow; on a variety of substrates…(4)

4. Inflorescence a lax, open, racemose cyme; flowers 20–30 mm wide…**Z. arizonica**
4'. Inflorescence a dense, helicoid to corymbose cyme; flowers 10–20 mm wide…**Z. calycosa**

Zeltnera arizonica (A. Gray) G. Mans. [*Centaurium arizonicum* (A. Gray) A. Heller]—Arizona pink, Arizona centaury. Annual or rarely biennial, stems to 40 cm, spreading to erect; basal leaves lanceolate to oblong, to 7 cm, withering at anthesis, stem leaves similar, smaller, extending into inflorescence; inflorescence an open racemose cyme; flowers several, corolla lobes bright pink, 20–30 mm wide at throat, corolla tube 12–25 mm, white-green, well exserted past calyx, lobes 7–10 mm; stigma reniform or weakly bilobed; capsule cylindrical, to 10 mm. Wet saline soil in wetlands, along streams, around springs and seeps, 3500–9000 ft. NC, NW, WC, SW, Eddy, Guadalupe.

Zeltnera calycosa (Buckley) G. Mans. [*Centaurium calycosum* (Buckley) Fernald; *C. breviflorum* (Shinners) B. L. Turner]—Buckley's centaury. Annual or rarely biennial, stems to 70 cm, spreading to erect; basal leaves lanceolate to oblong, to 6 cm, withering at anthesis, stem leaves similar, smaller, extending into inflorescence; inflorescence a dense helicoid to corymbose cyme; flowers several, corolla lobes bright pink to rose or rarely white, 10–20 mm wide at throat, corolla tube 10–20 mm, white-green, well exserted past calyx lobes, lobes 8–12 mm, stigma reniform or weakly bilobed; capsule cylindrical, to 15 mm. Dry to damp, gravelly or sandy saline soil along edges of wetlands, along streams, around springs and seeps, 3100–9000 ft. C, WC, SE, SC.

Zeltnera exaltata (Griseb.) G. Mans. [*Centaurium exaltatum* (Griseb.) W. Wight]—Exalted centaury. Annual or rarely biennial, stems to 40 cm, spreading to erect; basal leaves lanceolate to oblong, to 3.5 cm, withering at anthesis, stem leaves similar, smaller, extending into inflorescence; inflorescence a few-flowered open cyme; flowers 1 to few, corolla lobes pink to rarely white, < 10 mm wide at throat, corolla tube 7–12 mm, white-green, barely exserted past calyx, lobes 3–5 mm, stigma reniform or weakly bilobed; capsule cylindrical, < 10 mm. Wet saline soil in wetlands, along streams, around springs and seeps, 4900–5750 ft. San Juan.

Zeltnera maryanniana (B. L. Turner) G. Mans. [*Centaurium maryannianum* B. L. Turner]—Gypsum centaury. Annual or rarely biennial, stems to 20 cm, spreading to erect, often densely tufted; basal leaves numerous, lanceolate to oblong, to 5 cm, persistent, stem leaves similar, smaller, extending into inflorescence; inflorescence an open racemose cyme; flowers several, corolla lobes pink, 15–25 mm wide at throat, corolla tube 10–20 mm, green, exserted past calyx, lobes 8–12 mm, stigma ± capitate; capsule cylindrical, to 10 mm. Dry to damp, fine gypsiferous soils in arid regions, 3150–5200 ft. EC, SE.

Zeltnera nudicaulis (Engelm.) G. Mans. [*Centaurium nudicaule* (Engelm.) B. L. Rob.]—Naked-stem centaury. Annual or rarely biennial, stems to 20 cm, spreading to erect; basal leaves oblong, to 2 cm, persistent at anthesis, stem leaves few, linear-lanceolate, 3–25 mm, not extending into inflorescence; inflorescence an open cyme; flowers several, corolla lobes bright pink to rarely white, 8–15 mm wide at throat, corolla tube 7–10 mm, white-green, weakly exserted past calyx, lobes 4–5 mm, stigma reniform or weakly bilobed; capsule cylindrical, to 10 mm. Moist soil and gravel along streams, around springs and seeps, 4600–5600 ft. Doña Ana, Otero.

GERANIACEAE – GERANIUM FAMILY

Kenneth D. Heil

Annual or perennial herbs; leaves alternate (often the upper) or opposite (often the lower), simple or compound, petiolate, stipulate; flowers perfect, actinomorphic, solitary or in cymes; sepals 5, distinct; petals 5, distinct, clawed; stamens 5, 10, or 15, the filaments united at the base; pistil 1; style 1; stigmas 5; ovary superior, 5-carpellate; fruit a schizocarp, splitting and coiling along a central beak at maturity (Hanks & Small, 1907; Martin & Hutchins, 1980b; Moore, 2013).

1. Leaves pinnately dissected or tripartite; flowers small, 3–6 mm, rose-purple…**Erodium**
1'. Leaves palmately lobed or divided; flowers 3–25 mm, white to pink-purple or rose-purple…**Geranium**

ERODIUM L'Hér. – Stork's bill

Annuals; leaves in basal rosettes and opposite and cauline, pinnately dissected; stamens 5, alternating with 5 staminodes; fruit with spirally twisted styles along a central beak at maturity.

1. Leaves pinnate; petals 3–6(–7) mm…**E. cicutarium**
1'. Leaves tripartite; petals 6–12 mm…**E. texanum**

Erodium cicutarium (L.) L'Hér. ex Aiton—Filaree. Plants prostrate, 0.5–8+ dm; leaves 1–12 cm; sepals 3–6 mm; petals 5–7 mm, pink to light purple, with darker spots; stylar column 20–40 mm. Common weed in disturbed areas, 3600–9600 ft. Widespread. Introduced.

Erodium texanum A. Gray—Texas stork's bill. Plants decumbent to prostrate, reaching 2.5 dm; leaves 1–3.3 cm; petals 6–12 mm, purplish red; stylar column 45–70 mm. Desert grasslands, Chihuahuan desert scrub, 3150–5500 ft. C, SE, SC, SW.

GERANIUM L. – Geranium, cranesbill

Annuals, biennials, or perennials; leaves alternate or opposite, mostly basal, palmately lobed or divided; stamens 10; fruit with long styles curved or coiling at maturity (Ackerfield, 2015; Correll & Johnston, 1979c).

1. Plants annual, rarely biennial; petals 3–9 mm…(2)
1'. Plants perennial, from a thick, usually branched caudex often covered with old petiole bases and stipules; petals (8–)10–25 mm…(3)
2. Sepals 2–3.7(–4) mm, without a long awn tip but usually with a short callous tip; fertile stamens 5, alternating with 5 sterile ones; seeds smooth…**G. pusillum**
2'. Sepals 4.5–8 mm, awn-tipped; stamens 10, all fertile; seeds finely reticulate…**G. carolinianum**
3. Petals 5–9(–10) mm; sepals 4–8 mm…(4)
3'. Petals 11–26 mm; sepals 7–11 mm…(5)
4. Nectaries glabrous; stigmatic remains 4–5 mm…**G. lentum**
4'. Nectaries hairy on the back surface; stigmatic remains 1.5–3 mm…**G. wislizeni**
5. Petals white with purple veins, occasionally pinkish; nectaries usually hairy; pedicels glandular with purple-tipped glands…**G. richardsonii**
5'. Petals pink or purple with reddish veins, rarely white; nectaries glabrous with a tuft of hair at the apex; pedicels eglandular, or if glandular then the glands yellow- or white-tipped…(6)
6. Flowers nodding, with strongly reflexed petals; petals 3–4 times longer than wide…**G. dodecatheoides**
6'. Flowers erect to horizontal, with spreading petals; petals 1.3–2 times longer than wide…(7)
7. Basal leaves 1.7–7(–8) cm wide, the lobes obtuse or rounded and abruptly acuminate, 3-parted but not sharply toothed along the margins…**G. caespitosum**
7'. Basal leaves 5–15 cm wide, the lobes coarsely and sharply toothed along the margins…**G. viscosissimum**

Geranium caespitosum E. James [*G. atropurpureum* A. Heller; *G. fremontii* Torr. ex A. Gray]— Rocky Mountain geranium. Perennials, 0.7–5(–8) dm; stems usually glandular above with yellow-tipped glands, sometimes eglandular throughout; leaves palmately 3–5-parted, 1.7–7(–8) cm wide; pedicels 1.3–4(–5) cm; sepals 5.5–9(–11) mm with an awn tip 0.2–1.2(–2) mm; petals (8–)10–17 mm, usually pink or purple; stylar column 22–30(–40) mm with a beak 2.5–5.5 mm. Dry rocky slopes in montane meadows, ponderosa pine, mixed conifer, and spruce-fir forests, 5550–11,600 ft. Widespread except EC and Lea.

Geranium carolinianum L.—Carolina geranium. Annuals or biennials, 1.5–5(–7) dm; stems with retrorse hairs, sometimes glandular above; leaves palmately 5(–7)-parted, 3–7 cm wide; pedicels 3–7 (–15) cm, villous and glandular; sepals 4.5–7 mm, with an awn tip 1–2 mm; petals 4.5–7 mm, pink to rose; stylar column 14–22 mm with a beak 1–2.5 mm. Weedy areas, grasslands, juniper-oak, 4800–9200 ft. Doña Ana, Lincoln. Introduced.

Geranium dodecatheoides P. J. Alexander & Aedo—Shooting-star geranium. Perennials, 4–10 dm; stems with eglandular hairs; leaves palmately 3–5-parted, 4.8–5.6 cm wide; pedicels 4.1–5.9 cm; sepals and petals reflexed; petals 11.9–13.8 mm, purplish; stamens 10, all bearing anthers; filaments 10.9–12.8 mm. Among andesitic boulders and outcrops, mixed conifer forest, 7550–9900 ft. Lincoln, Otero. **R & E**

Geranium lentum Wooton & Standl.—Mogollon geranium, white geranium. Perennial herb to 5 dm, with pilose-spreading to retrorse, nonglandular and glandular yellowish- or whitish-tipped hairs; leaves palmately 5-parted, 2.5–5(–7) cm wide; pedicels densely hirsute, nonglandular; sepals 6–8 mm, tip mucronate, 0.5–1 mm; petals 5–10 mm, white with cream to light pinkish veins; stylar column 22–27 mm with a beak 1–2(–3.5) mm. Canyon bottoms, coniferous forests, black sagebrush, aspen, Gambel oak communities, 6000–8100 ft. NW, WC, SW.

Geranium pusillum L.—Small-flower cranesbill. Annual; herbage glandular and eglandular-pubescent; stems 10–60 cm; leaves basal and cauline, blades reniform to orbicular, palmately 7-divided, 2–5 cm wide; pedicels glandular or eglandular-pubescent; sepals 3–5 mm; petals 2–3 mm, pale purple; nectaries glabrous; fruits 9–11 mm. Disturbed sites, vacant lots, garden beds, roadsides, 5100–8500 ft. Bernalillo, Otero. Introduced.

Geranium richardsonii Fisch. & Trautv.—Richardson's geranium. Perennials, 3–7(–9) dm; stems glandular above, the glands purple-tipped; leaves palmately (3–)5–7-parted, (4–)6–12(–15) cm wide; pedicels 1–3.3 cm, glandular with purple-tipped glands; sepals 6–10 mm with an awn tip 1–2.5 mm; petals 11–20(–25) mm, white or light pinkish, with purple veins; stylar column 20–30(–35) mm with a beak 1.7–4 mm. Common in moist meadows, ponderosa pine, mixed conifer, spruce-fir, 5655–12,850 ft. NC, NW, C, WC, Eddy, Otero.

Geranium viscosissimum Fisch. & C. A. Mey. [*G. nervosum* Rydb.]—Sticky purple geranium. Perennials, (2.5–)4–9 dm; stems glandular above, the glands yellow-tipped; leaves palmately 5(–7)-parted, 5–15 cm wide; pedicels 1.5–4(–5) cm, glandular; sepals 7.5–10 mm with an awn tip 1–3 mm; petals 14–25 mm, pink or purple; stylar column 28–40 mm with a beak 4–7 mm. Meadows, aspen forests, mixed conifer, 7000–11,370 ft. NC, NW, Bernalillo.

Geranium wislizeni S. Watson—Huachuca Mountain geranium. Perennials, 4 dm; stems retrorsely villosulous, eglandular; leaves palmately 5-parted, 8 cm wide; sepals 7 mm, mucronate; petals to 1 cm, white; stylar column to 18 mm. Coniferous forests, 5610–10,540 ft. Catron, Grant.

GROSSULARIACEAE – CURRANT OR GOOSEBERRY FAMILY

Kenneth D. Heil

Shrubs; stems with or without spines; leaves mostly alternate, rarely opposite, stipules absent or present, simple and sometimes deeply cleft; inflorescence a raceme or panicle, or of solitary flowers in the upper axils; flowers perfect, rarely unisexual, regular or nearly so, epigynous with a saucer-shaped to tubular hypanthium, sepals 3–9, sometimes larger than the petals, petals 3–9, alternate with the sepals; stamens as many as, and alternate to the petals; nectary disk often present internal to the stamens, gynoecium of 2 or 3(–7) carpels united to form a compound, superior to partly or wholly inferior ovary, ovules numerous; fruit a berry; seeds numerous (Morin, 2009; Taylor, 2013).

RIBES L. – Currant, gooseberry

Shrubs with or without bristles and spines; leaves alternate, stipules none or adnate to the petiole, palmately lobed, crenate, or dentate; inflorescence a raceme or rarely solitary; pedicels subtended by bracts, usually with 2 bractlets about midlength; flowers perfect, hypanthium mostly corollalike, sepals 5, mostly petaloid, petals (4)5, often smaller than the sepals, stamens (4)5(6), ovary inferior, 1-loculed; styles 2, united or distinct.

1. Plants armed with nodal spines, often with internodal bristles…(2)
1'. Plants unarmed…(6)
2. Hypanthium tubular to narrowly campanulate, 1–6 mm; inflorescence a 1–5-flowered raceme or solitary; pedicels not jointed just below flowers…(3)

2'.   Hypanthium shallowly cupped to saucer-shaped, 0.5–1.5 mm; inflorescence a 3–23-flowered raceme; pedicels jointed just below flowers at maturity…(5)

3.   Sepals orange or purple; berries dark purple and densely spiny…**R. pinetorum**

3'.   Sepals green to greenish white, yellowish, pink, or red, rarely orange; berries red to purple or black, glabrous or bristly…(4)

4.   Stamens about equal to petals; berries glabrous, pubescent, or glandular-pubescent, not spiny; anthers 0.7–1.6 mm, not apiculate; styles glabrous, connate nearly full length, short-bifid at apex…**R. leptanthum**

4'.   Stamens exceeding petals; berries with long, stout spines; anthers > 1.6 mm, apiculate…**R. inerme** (in part)

5.   Leaves densely glandular or glandular-pubescent; racemes 3–11-flowered; pedicels 1–5 mm; berries bright red…**R. montigenum**

5'.   Leaves glabrous or sparsely pubescent; racemes 5–23-flowered; pedicels 2.5–7 mm; berries black or dark purple…**R. lacustre**

6.   Hypanthium saucer-shaped to shallowly cup-shaped or turbinate, 0.6–3.5 mm…(7)

6'.   Hypanthium tubular-campanulate to cylindrical, 4–13 mm…(9)

7.   Ovaries and berries glabrous…**R. inerme** (in part)

7'.   Ovaries and berries with stipitate-glandular hairs…(8)

8.   Racemes loosely flowered, floriferous nearly to base; peduncle absent or to 1.5 mm; plants with weakly decumbent stems; floral bracts 1–3.5 mm, < 1/2 as long as pedicels…**R. laxiflorum**

8'.   Racemes densely flowered, obviously pedunculate, at least 2 cm; plants erect; floral bracts conspicuous, 2.8–7 mm, > 1/2 as long as pedicels…**R. wolfii**

9.   Leaves with large, sessile, yellow gland-dots on the lower surface…**R. americanum**

9'.   Leaves not as above…(10)

10.   Flowers bright yellow, sometimes aging orange or pink to red-purple, nearly always glabrous; berries glabrous; anthers without cuplike glands at tip…**R. aureum**

10'.   Flowers white to pinkish white or greenish white, to yellow, pubescent; berries glabrous to glandular-pubescent; anthers tipped by small, cuplike glands…(11)

11.   Petioles, pedicels, and hypanthium stipitate-glandular…**R. mescalerium**

11'.   Petioles, pedicels, and hypanthium glandular or glabrous, not stipitate-glandular…(12)

12.   Hypanthium narrowly tubular, 6–9.5 × 1.5–4 mm, at least twice as long as the sepals; sepals 1.5–3.5 mm, usually remaining recurved after anthesis; pedicels 0.4–3.4 mm; leaf blades 0.5–5 cm wide…**R. cereum**

12'.   Hypanthium broad, usually campanulate, 4.5–7 × 3.5–7.5 mm, about the same length as the sepals; sepals 3.5–7 mm; pedicels 3–17 mm; leaf blades 2–10.5 cm wide…**R. viscosissimum**

Ribes americanum Mill.—American black currant. Shrubs, 0.5–1.5 m; stems erect to spreading, crisply puberulent to villous, glandular throughout with yellow, shiny, sessile, crystalline, round glands, spines at nodes absent; leaves puberulent to villous, blades 3–5-lobed, (1.5–)2–7 cm, surfaces with amber, sessile glands, lobes broadly deltate, margins usually coarsely bicrenate-serrate; inflorescences 6–15-flowered racemes; pedicels jointed, 0.1–2 mm; flower hypanthium green, broadly tubular-campanulate, 3–4.5 mm, glabrous or sparsely villosulous; sepals usually reflexed, cream to greenish white, 4.5–5 mm; petals whitish, 2.5–3 mm; berries black, ovoid, 10 mm, glabrous, without resinous glands. Swamps, stream banks, wet meadows, fens, moist ravines and canyons, open woods, 5100–9130 ft. NE, C, Catron.

Ribes aureum Pursh—Golden currant. Shrubs, 1–3 m, unarmed, branchlets glabrous; leaves with petioles 0.5–2.5(–3) cm, blades (1–)1.6–4.7 × 1–6.7 cm, orbicular, reniform, obovate, cuneate to truncate basally, strongly 3-lobed, the lobes entire or crenate to lobed, glabrous; inflorescence a raceme with (3–)6–9 flowers, bracts 3–12 mm, entire; pedicels to 3 mm; flower hypanthium cylindrical, yellow, or often reddish in age, corollalike; sepals mostly 4–6 mm, yellow; petals ca. 2 mm, yellow, cream, or reddish; berries 8–12 mm, black, red, orange, or translucent-golden, glabrous. Riparian, irrigation canals, greasewood-shadscale, sagebrush, piñon-juniper, ponderosa pine, Douglas-fir communities, 4500–9000 ft. Widely scattered except SE.

Ribes cereum Douglas—Wax or squaw currant. Shrubs, (0.2–)0.5–1.5(–2) m, unarmed, branchlets pilose-villous and stipitate-glandular; leaves with petioles 0.4–2.2(–2.9) cm, blades 0.5–2.5(–3.4) × 0.7–3(–4.4), orbicular, reniform, rarely ovate, with 3–7 shallow lobes; inflorescence a raceme with 2 or 3 flowers, the axis very short; bracts 2–5 mm; flower hypanthium 4–11 mm, pinkish; sepals ca. 2 mm, spreading, whitish or pinkish; petals ca. 1 mm, whitish, ovaries stipitate-glandular; berries 6–8 mm, reddish, sparingly stipitate-glandular, rarely glabrate. Riparian sites, mountain brush, sagebrush, piñon-juniper, ponderosa pine, aspen, spruce-fir communities, 4450–11,390 ft. Widespread except EC, SE.

Ribes inerme Rydb. [*Grossularia inermis* (Rydb.) Coville & Britton]—Whitestem gooseberry. Shrubs, 7.5–20 dm, branchlets often whitish, glabrous, armed at the nodes with 1(–3) spines, or spines lacking; leaf blades (0.8–)1.5–9 cm, orbicular or nearly so, cordate to truncate, with 3–5 main lobes, these again lobed, not glandular, paler beneath than above; inflorescence a 1–4-flowered raceme; bracts 1–2 mm, greenish, glabrous or glandular-ciliolate and puberulent; flower hypanthium 2–3.5 mm, cylindrical to narrowly campanulate, greenish or greenish cream, sometimes purplish-tinged; sepals ca. 3 mm, colored as the hypanthium; petals ca. 1–1.5 mm, white; berries 7–10 mm, reddish or reddish purple, succulent, ± edible. Piñon-juniper, mountain brush, aspen, willow, Douglas-fir, spruce-fir communities, often in mountain meadows, 5700–11,340 ft. NC, C, WC, Otero.

Ribes lacustre (Pers.) Poir. [*R. oxyacanthoides* L. var. *lacustre* Pers.].—Swamp black gooseberry. Shrubs, 7.5–15 dm; branchlets armed with internodal prickles and nodal spines, puberulent, eglandular; leaf blades (0.6–)1.5–5.6 × (1–)2–8 cm, orbicular in outline, usually 5-lobed, the lobes again lobed; inflorescence a raceme, rather loosely 5–15-flowered, stipitate-glandular, puberulent, bracts 2–3 mm; flower hypanthium < 1 mm, saucer-shaped, yellow-green, pinkish, or reddish; sepals 2.5–3 mm, yellow-green, pinkish, or reddish; petals shorter than the sepals, pinkish; berries 6–8 mm, dark purple. Moist sites, often in conifer and aspen woods, 7545–8600 ft. Rio Arriba.

Ribes laxiflorum Pursh [*R. coloradense* Coville]—Western or trailing black currant. Shrubs to ca. 0.7 m, the stems sprawling or ascending, unarmed; branchlets and some older branches puberulent; leaf blades (1–)2–5 × 1.5–6.5 cm, orbicular or nearly so, cordate, with 3–5 primary lobes, these again lobed; inflorescence a 5–10-flowered raceme, stipitate-glandular and puberulent; bracts 1–2 mm, linear or narrowly triangular, greenish; pedicels jointed just below the ovary; flower hypanthium < 1 mm; sepals 2–3 mm, pinkish or purplish; petals ca. 1 mm; berries to ca. 1 cm, blackish, stipitate-glandular. Moist sites, aspen and spruce-fir communities, 8300–10,800 ft. Colfax, Rio Arriba, Sandoval.

Ribes leptanthum A. Gray [*Grossularia leptantha* (A. Gray) Coville & Britton]—Trumpet gooseberry. Shrubs, 0.5–2 m; branchlets armed at the nodes with 1–3 spines, usually lacking internodal bristles, puberulent; leaf blades 0.5–1.6 × 0.7–2 cm, orbicular, mostly 5-lobed, the main lobes again lobed; inflorescence a 1–3-flowered raceme, the axis very short; bracts glabrous except glandular-ciliate or -toothed; flower hypanthium 4–5.5 mm, whitish; sepals 4–6 mm, whitish; petals 2.5–3 mm, whitish; berries mostly 6–10 mm, blackish, glabrous. Piñon-juniper, mountain brush, ponderosa pine, aspen, spruce-fir, mountain meadow communities, 5300–12,530 ft. N, C, W, SC.

Ribes mescalerium Coville—Mescalero currant. Shrubs, 1–2 m; stems erect, glandular-pubescent, glabrescent, spines and prickles absent; leaf blades reniform-orbiculate, shallowly 3–5-lobed, cleft 1/8–1/4 to midrib, 1.5–3 cm, surfaces pubescent, sessile- and stipitate-glandular; inflorescence a 6–10-flowered raceme, 3–5 cm; pedicels jointed, 0.5–1 mm, pubescent, stipitate-glandular; bracts obovate, 3–7 mm, long-stipitate-glandular; flower hypanthium greenish white, tubular, tube evenly wide, 3–5 mm, pubescent, stipitate-glandular; berries black, globose, 5–8 mm, glandular-pubescent. Dry slopes in open montane and ponderosa pine forests; Sacramento and Guadalupe Mountains, 7000–9000 ft. Lincoln, Otero. **R**

Ribes montigenum McClatchie—Red prickly currant. Shrubs, 0.5–2 m; branchlets armed at the nodes with 1–3 spines, usually lacking internodal bristles, puberulent; leaf blades 0.5–1.6 × 0.7–2 mm, orbicular, cordate basally, mostly 5-lobed, the main lobes again shallowly lobed or toothed; inflorescence a 1–3-flowered raceme, the axis very short; bracts glabrous except glandular- or ciliate-toothed; flower hypanthium 4–5.5 mm, whitish; sepals mostly 2.5–4 mm, yellowish green, pink, red, or orange; petals 2.5–3 mm, whitish; berries ca. 6–10 mm, blackish, glabrous. Douglas-fir, spruce-fir, aspen, often in talus and scree slopes, 8700–13,000 ft. NC, C, SC, Cibola.

Ribes pinetorum Greene—Orange gooseberry. Shrubs, 1–2 m; branchlets erect to sprawling, tomentose, stipitate-glandular; spines at nodes 1–3, 5–12 mm; leaf blades roundish to broadly triangular, 3–5-lobed, cleft 1/2 to midrib, 1–3 cm, with long-stalked glands, or glabrous; inflorescence with solitary flowers, 2–3 cm; pedicels not jointed, 2–3 mm; flower hypanthium green, tubular, 6–8 mm, pubescent;

sepals orange, orangish, or purplish, 6–16 mm; petals pale orange, revolute or inrolled, 4–6 mm; berries dark purple, globose, 10–15 mm, densely spiny. Coniferous forests, 6550–11,390 ft. WC, SC, McKinley.

Ribes viscosissimum Pursh—Sticky currant. Shrubs, 1–2 m, unarmed; branchlets pilose-hirsute and stipitate-glandular; leaf blades 0.9–6.6 × 1.3–10 cm, orbicular, rarely ovate, cordate basally, 3–7-lobed, the main lobes crenate or dentate and sometimes again lobed; inflorescence a 4–12-flowered raceme, glandular; bracts 5–8.5 mm, entire to toothed, glandular; flower hypanthium 5–9 mm, whitish or pale green, stipitate-glandular and pilose-hirsute; sepals 3–5.5 mm, white or yellow-green, occasionally pinkish; petals 2–3 mm, whitish; berries 10–13 mm, black, rather dry, stipitate-glandular. Moist canyon bottoms and slopes in aspen, fir, Douglas-fir, lodgepole pine, spruce-fir communities, 8000–10,000 ft. Union.

Ribes wolfii Rothr.—Rothrock's currant. Shrubs, 0.5–3 m, unarmed; branchlets glabrous or puberulent; leaf blades 1.2–5.7 × 1.2–8 cm, orbicular, cordate basally, 3(–5)-lobed, the main lobes again lobed; inflorescence an 8–16 flowered raceme, glandular; bracts 3–6 mm, mostly entire; flower hypanthium 0.7–1.5 mm, green, bowl-shaped, glabrous or puberulent; sepals 2–3 mm, whitish; petals ca. 1–1.5 mm, white; berries 6–10 mm, blackish, not very fleshy, stipitate-glandular. Aspen, Douglas-fir, and spruce-fir communities, usually in shade, 5600–12,530 ft. N, C, WC, SC.

HALORAGACEAE – WATER-MILFOIL FAMILY

C. Barre Hellquist

Rhizomatous, submersed and emersed herbs; leaves simple, opposite, alternate, or whorled; flowers small, submersed or emersed in leaf or bract axils; perfect or imperfect, polypetalous or apetalous; sepals usually developed; petals small or lacking; stamens twice the number of sepals; ovary 1–4-celled; ovule 1 per cell; styles 1–4; fruit indehiscent, nutlike (Scribailo & Alix, 2021).

MYRIOPHYLLUM L. – Water-milfoil

Perennial, aquatic herbs; leaves whorled or opposite, submersed leaves capillary, emersed leaves bract-like; inflorescence sessile, in axils of leaves or bracts, the uppermost staminate; the upper flowers often in an emersed terminal spike, flowers perfect or imperfect; calyx 4-merous; petals 4 or none; stamens 4–8; ovaries inferior, 4-celled; ovule single in each cell; stigmas 4, recurved; fruit hard, nutlike, 4-locular, splitting into 4 mericarps.

1. Leaves whorled and alternate along stem…**M. pinnatum**
1'. Leaves whorled (occasionally alternate in *M. heterophyllum*)…(2)
2. Spikes with large feathery leaves, similar to submersed leaves…**M. aquaticum**
2'. Spikes with reduced bracteate leaves, not feathery…(3)
3. Bracts usually > twice as long as pistillate flowers…(4)
3'. Bracts usually < twice as long as flowers and fruit…(5)
4. Inflorescence bracts pectinate to pinnatifid; clavate winter buds formed in late summer; all leaves whorled…**M. verticillatum**
4'. Inflorescence bracts serrate, slightly pectinate at water line; winter buds not formed; some lower leaves occasionally alternate…**M. heterophyllum**
5. Middle leaves with 11 or fewer segments on each side of the rachis; uppermost leaves rounded at apex… **M. sibiricum**
5'. Middle leaves with 12+ segments on each side of the rachis; many upper leaves truncate at apex…**M. spicatum**

Myriophyllum aquaticum (Vell.) Verdc. [*M. brasiliense* Cambess.]—Parrot-feather. Perennial; stems simple or sparsely branched; leaves all whorled, stiffish, 2–5 cm, puberulent when young; 10+ pinnately arranged linear-filiform divisions on each side, upper divisions 3–6 mm; flowers unisexual in axils of leaves; pistillate flowers 1.5 mm, seen as a tuft of white or pink plumose stigma lobes; staminate flowers not seen. Shallow water of lakes, ponds, and streams where mainly emergent, 3900–7825 ft. C, WC, SC, Colfax, Rio Arriba. Introduced.

Myriophyllum heterophyllum Michx.—Variable-leaved water-milfoil. Stems often swollen on upper portion of plant; leaves in whorls of 4–6, occasionally scattered along lower portion of stem, pinnate leaves 2–5 cm, with 7–10 flaccid capillary divisions on each side; bracts firm, entire or serrate, to 3 × 1 cm; spikes emersed to 40 cm; flowers in whorls of 4–6; fruit subglobose, 1–1.5 × 1–1.5 mm, minutely papillose, carpels 2-ridged on back, rounded on sides, prominently beaked. Ponds, lakes, streams, 6600–6700 ft. Mora.

Myriophyllum pinnatum (Walter) Britton, Sterns & Poggenb. [*M. scabratum* Michx.]—Cut-leaf water-milfoil. Plants submersed or terrestrial, freely branched in mud, elongating in water; leaves in whorls of 3–5, subverticillate and scattered to ca. 3 cm; submersed leaves with 5+ short remote capillary divisions on each side of rachis; emersed leaves linear to oblanceolate, pectinate or serrate to 2 cm; flowers ca. 1 cm, petals purplish, 1.5–2 mm; fruit ovoid, 1.3–1.8 mm, carpels with flat sides and 2 tuberculate dorsal ridges. Muddy shores of shallow waters, 5500–5600 ft. Harding, Quay.

Myriophyllum sibiricum Kom. [*M. exalbescens* Fernale]—Northern water-milfoil. Stems simple or branching, purple to brownish purple, turning white upon drying; leaves in whorls of 3 or 4, ca. 1+ cm apart, midstem whorls to 3 cm apart, simple-pinnate with 11 or fewer segments on each side of the rachis, lower leaves often shorter; inflorescence a spike; flowers in verticils, lower pistillate, upper staminate, in axils of tiny bracts, bracts usually shorter than flowers; fruit subglobose, 2–3 mm, mericarps rounded on back, smooth or rugulose. Lakes, ponds, slow-flowing streams, ditches, 5650–9050 ft. NC, NW, WC.

Myriophyllum spicatum L.—Eurasian water-milfoil. Stems simple or branching, reddish pink; leaves in whorls of 3–5, usually 1+ cm apart, simple-pinnate with 12+ segments on each side of the rachis; tips truncate on some leaves, appearing flattened as if cut across; inflorescence a spike; flowers in whorls of 4 in axils of reduced bracts; all but the lowest bracts entire and shorter than flowers and fruits; petals caducous, ca. 3 mm; stamens 8; fruit subglobose, 4-lobed. Lakes and ponds, 4375–5025 ft. NC, C, WC. Introduced and invasive.

Myriophyllum verticillatum L.—Northern water-milfoil. Stems simple, in late season producing clavate winter buds; leaves in whorls of 4 or 5; submersed leaves 0.8–4.5 cm, with 9–13 opposite or alternate pairs of capillary flaccid divisions; divisions to 28 mm; emersed leaves and bracts smaller, pectinate-pinnate; inflorescence a spike; flowers in whorls of 4–6, petals rudimentary in pistillate flowers, spoon-shaped, obtuse, 2.5 mm; stamens 4 or 8; fruit subglobose, 2–2.5 mm, deeply 4-furrowed, mericarps rounded on back, smooth or slightly tuberculate. Lakes and ponds, 3900–8300 ft. Doña Ana, Catron.

HELIOTROPIACEAE – HELIOTROPE FAMILY

Don Hyder

Herbs, annual or perennial, sometimes shrubby; leaves simple, alternate; flowers perfect, borne in scorpioid cymes, actinomorphic; sepals and petals 5; corolla white, yellow, or purple; stamens 5; single superior pistil; 2 carpels; fruit 2 or 4 1-seeded nutlets. Formerly placed in Boraginaceae (Hasenstab-Lehman, 2017).

1. Plants glabrous, green or glaucous and partially succulent; inflorescence naked; fruit usually separating into 4 1-seeded nutlets…**Heliotropium**
1'. Plants hairy, never glaucous or succulent; inflorescence irregularly leafy-bracteate; nutlets often cohering in pairs…**Euploca**

EUPLOCA Nutt. - Heliotrope

Annual or perennial herbs; leaves all cauline, hirsute; inflorescence of irregularly bracteate scorpioid cymes or few-flowered among leafy bract clusters at the stem tip; calyx hirsute; corolla pubescent on outer surface, limb white, tube and throat often yellowish; fruit 2- or 4-lobed (previously included in *Heliotropium*).

1. Plants perennial, rhizomatous, but dying back to ground level each season…**E. greggii**
1'. Plants annual, not rhizomatous…(2)
2. Corolla 8-15 mm across, the tube long and exserted; style elongate, many times longer than the stigma…
   E. convolvulacea
2'. Corolla 2-4 mm across, the tube included; style short, about as long as the stigma…**H. fruticosa**

Euploca convolvulacea Nutt. [*Heliotropium convolvulaceum* (Nutt.) A. Gray]—Showy heliotrope. Annual herbs, 10-40 cm; stems freely branched, alternate, on short-hairy petioles; blades ovate to lanceolate, 1-4 cm; flowers white, solitary in leaf axils and at stem tips, and often opposite a leaflike bract; calyx 5-lobed; corolla funnelform, white, 10-15 mm; fruits separate into 2 pubescent nutlets, 3 mm. Dry, sandy places, 3050-6000 ft. NE, EC, C, WC, S.

Euploca fruticosa (L.) J. I. M. Melo & Semir [*Heliotropium fruticosum* L.]—Key West heliotrope. Annual herbs, 3-15 cm; stems sparingly branched from the base; leaves alternate and upward-pointing, blades elliptic to oblanceolate, 7-20 mm, margins entire and tightly revolute; flowers white to purple, inconspicuous, and solitary in leaf axils or arranged in leafy-bracteate spikes; sepals unequal, to 1.5 mm, strigose and somewhat pustulate; corollas campanulate, 1-2 mm, covered by finely stiff hairs; fruits nutlets, puberulent with fine white hairs. Dry, sandy, and gravelly soils, 5250 ft. Grant, Hidalgo.

Euploca greggii (Torr.) Halse & Feuillet [*Heliotropium greggii* Torr.]—Gregg's heliotrope. Annual or perennial; stems erect or spreading; leaves linear, flat, 20-25 mm; flowers small, fragrant, in terminal scorpioid cymes; calyx lobes equal; corolla white; anthers minutely bearded at the tip; fruit 4 1-seeded nutlets. Sandy plains, 3000-7500 ft. C, SE, SC.

HELIOTROPIUM L. - Heliotrope

Heliotropium curassavicum L.—Seaside heliotrope. Perennial herbs or rarely annuals, subshrubs, 10-50 cm; stems diffusely branched, semisucculent; leaves nearly sessile, alternate, lanceolate to oblanceolate or obovate, the surfaces glabrous and glaucous, often purplish in age; flowers white, in helicoid spikes, sometimes in clusters of 3 or 4; corollas funnel-shaped, 3-6 mm wide, white with yellow center and fading purplish; fruits depressed-globose, ca. 2 mm diam., separating into 4 ovoid nutlets. Saline soils, low-lying areas that are wet or seasonally wet; greasewood, saltgrass, saltbush communities, 3400-8500 ft. Widespread except EC.

HYACINTHACEAE - HYACINTH FAMILY

Steve L. O'Kane, Jr.

MUSCARI Mill. - Grape-hyacinth

Muscari neglectum Guss. ex Ten. [*Hyacinthus racemosus* L.]—Grape-hyacinth. Herbs perennial, scapose, from brown, tunicate, ovoid bulbs, with or without offsets (bulblets); leaves 3-6, basal; blades linear, glabrous, rather fleshy; scapes terete, equaling leaves; inflorescences terminally racemose, 20-40-flowered, dense, bracteate, elongating in fruit; distal flowers smaller, sterile, differing in color, forming a tuft (coma); bracts minute; flowers fragrant; perianth tube blackish blue, obovoid to oblong-urceolate or cylindrical, 4-6 × 2-3 mm, teeth white; tepals 6, connate most of their length, distal portions distinct, reflexed, short, toothlike; stamens 6, epitepalous, in 2 rows, included; anthers dark blue, dorsifixed, globose; ovary superior, green, 3-locular, inner sepal nectaries present; style 1; stigma 3-lobed; fruits capsular. Potentially escaping in the state, but no authentic material seen. Formerly in Liliaceae. Introduced.

HYDRANGEACEAE – HYDRANGEA FAMILY

Kenneth D. Heil and Craig C. Freeman

Subshrubs, shrubs, or trees, deciduous; leaves opposite, simple, petiolate, blade margins entire; flowers usually perfect, actinomorphic or the marginal ones sterile and zygomorphic; sepals 4–5, distinct or connate basally; petals 4–5, connate basally; nectary usually present; stamens 8–90; pistil 1, (2)3–5-carpellate, ovary < 1/2 inferior, 1/2 inferior, or completely inferior, (2)3–5-locular; styles (2)3–5; stigmas 1–5; fruits capsules, dehiscence septicidal or loculicidal; seeds 1–50 per locule, funicular appendage present or absent (Freeman, 2016).

1. Stamens 13–64...**Philadelphus**
1'. Stamens 8–10...(2)
2. Filament apices 2-lobed, prolonged beyond anthers; seeds (1)2–4(–6) per locule...**Fendlera**
2'. Filament apices not 2-lobed; seeds 1 or 25–50 per locule...(3)
3. Leaf blade margins crenate to dentate; blades ovate, obovate, rhombic, or suborbiculate, venation pinnate; seeds 25–50 per locule...**Jamesia**
3'. Leaf blade margins entire; blades elliptic to lanceolate, oblanceolate, obovate, or linear-oblong; venation acrodromous; seeds 1 per locule...**Fendlerella**

FENDLERA Engelm. & A. Gray – Fendler-bush

Fendlera rupicola Engelm. & A. Gray [*F. falcata* Thornber ex Wooton & Standl.; *F. wrightii* (Engelm. & A. Gray) A. Heller]—Cliff fendler-bush. Shrubs, 5–30 dm; leaf blades usually linear to elliptic, 10–38 × 2–10 mm; inflorescence of solitary flowers or flowers in clusters of 2–3; flower hypanthium and calyx tube 2–3 mm; sepals 4, 4–8.5 × 1.5–4 mm, tomentose; petals 4, white or tinged with pink or red, 8–15 × 5–14 mm, clawed; stamens 8; ovary to 1/2 inferior; capsules 7–15 × 5–8 mm, dehiscence septicidal. Rocky hillsides with oak, piñon, juniper, 4015–10,300 ft. NC, NW, C, WC, SC, SW, Eddy, Mora.

FENDLERELLA (Greene) A. Heller – Yerba desierto

Fendlerella utahensis (S. Watson) A. Heller [*Whipplea utahensis* S. Watson]—Utah fendler-bush. Stems to 10 dm, twigs reddish to orangish, strigose; leaf blades elliptic to lanceolate, oblanceolate, obovate, or linear-oblong, strigose, with some hairs pustulate-based, blades 5–20(–25) × 1–5(–6.5) mm; inflorescences compound cymes, bracteate; flower hypanthium 0.5–2 × 0.9–1.2 mm; sepals (4)5, 0.8–2.2 × 0.2–0.6 mm; petals (4)5, white, 2–4 × 0.8–1.2 mm, clawed; stamens (8–)10; ovary slightly inferior; capsules 4–6.1 × 1.5–1.9 mm, dehiscence septicidal. Rocky cliffs and crevices, usually in limestone and sandstone; piñon-oak-*Rhus*-juniper communities, 4200–8300 ft. WC, SC, SW, Eddy.

JAMESIA Torr. & A. Gray – Cliffbush, waxflower

Jamesia americana Torr. & A. Gray—American cliffbush. Shrubs 5–20(–40) dm, bark exfoliating; leaf blades ovate or broadly ovate to obovate, rhombic, or suborbiculate, 0.7–8 cm, abaxial surface moderately to densely canescent or sericeous; inflorescences cymose, usually 3–35-flowered; flower hypanthium 1.3–2 × 2.5–4.5 mm; sepals 5, mostly 1.5–7 mm; petals 5, white, mostly 5.5–11 mm, clawed; stamens 10; ovary to 1/2 inferior; capsules 3.5–7 mm, dehiscence septicidal. Our material belongs to **var. americana**. Rocky slopes and crevices, stream banks, forested canyon bottoms, 5600–12,300 ft. NE, NC, C, WC, SC.

PHILADELPHUS L. – Mock orange, syringa

Shrubs with tight or exfoliating bark; leaf blades oblong-lanceolate, linear-lanceolate, or elliptic to ovate; flower sepals 4; petals 4, white; stamens 13–64; ovary inferior to 1/2 inferior; capsule dehiscence loculicidal.

1. Hairs on abaxial and adaxial surfaces ± equally moderately strigose; axillary buds exposed; styles 1... **P. mearnsii**
1'. Hairs on abaxial surface longer, denser than those of adaxial surface; axillary buds hidden in pouches; styles 4...**P. microphyllus**

Philadelphus mearnsii W. H. Evans ex Rydb.—Mearns's mock orange. Shrubs, 5–15(–40) dm; stems tan-brown, weathering gray, axillary buds exposed; leaf blades oblong-lanceolate, linear-lanceolate, or elliptic to ovate, 0.5–1.7(–3) × 0.1–0.6(–1.1) cm, margins entire, surfaces ± equally moderately strigose; inflorescences solitary flowers; sepals 2–3.5(–4.2) × 1.3–2.3 mm; petals (5–)7–8.5(–10) × 2.5–4.5(–5.8) mm; stamens 13–18(–24); capsules 3–5.5 × 3.2–5.5 mm. Limestone mountains, oak-piñon zones, 4000–5800 ft. S except Lea.

Philadelphus microphyllus A. Gray—Small-leaf mock orange. Shrubs, 5–12(–20) dm; stems copper to reddish brown, axillary buds hidden in pouches; leaf blades linear-lanceolate, narrowly ovate to ovate, (0.5–)0.8–3(–5.5) × (0.2–)0.3–1.3(–3.3) cm, margins entire, abaxial surface short sericeous-strigose, or sericeous-villous, adaxial surface mostly glabrous, glabrate; inflorescences usually solitary flowers, sometimes 3–5-flowered cymes; sepals (2.5–)4–8.5(–10) × (2.5–)3–4.3(–5) mm, petals (5.8–)7–16(–21) × (5.3–)6–11(–15) mm; stamens 26–64; capsules (3.6–)5–8(–9.5) × (3.5–)4–7(–9.5) mm.

1. Pedicels, hypanthia, and sepals densely tomentose with short, woolly hairs and longer straight hairs; leaf blade abaxial surface with hairs usually erect and chaotically oriented, sometimes appressed; mountains of SC New Mexico...**var. argyrocalyx**
1'. Pedicels, hypanthia, and sepals glabrous or sparsely hairy at the base; leaf blade abaxial surface with hairs appressed or loosely appressed; widespread but mostly not SC New Mexico...**var. microphyllus**

var. argyrocalyx (Wooton) Henrickson [*P. argyrocalyx* Wooton]—New Mexican mock orange. Mountain slopes of mostly sedimentary rock; piñon-juniper woodlands, montane coniferous forests, 6900–8000 ft. Lincoln, Otero. **R**

var. microphyllus [*P. argenteus* Rydb.; *P. microphyllus* var. *argenteus* (Rydb.) Kearney & Peebles]—Small-leaf mock orange. Open slopes, canyons, montane scrub, pine-oak woodlands, ponderosa pine forests, 5270–11,000 ft. Widespread except EC, SE.

HYDROCHARITACEAE – FROG-BIT FAMILY

C. Barre Hellquist

Annual or perennial, aquatic, submersed herb; caulescent, with or without evident stem; submersed and floating leaves or submersed leaves; stems rhizomatous, creeping or rooting at proximal nodes, leafy; leaves basal, alternate, opposite, or whorled, sessile or petiolate, margins entire, spinulose to serrate; inflorescence axillary, terminal, or scapose, 1-flowered or cymose, subtended by spathe or as involucres; flowers unisexual, staminate and pistillate on same or different plants, epigynous, free; mostly 6-parted or with perianth absent; stamens (0–)2 to many; fruit berrylike or achenelike; seeds 1 to many, fusiform, ellipsoid, ovoid, spherical, or straight (Haynes, 2000).

1. Plants with basal leaves with obvious broad lacunae band along midvein...**Vallisneria**
1'. Plants with cauline leaves, lacking obvious lacunae band along midvein...(2)
2. Leaves nearly opposite, appearing whorled, with basal sheaths...**Najas**
2'. Leaves whorled or opposite, lacking basal sheaths...(3)
3. Leaves in whorls of 2–4...**Elodea**
3'. Leaves in whorls of 5+...**Egeria**

EGERIA Planch. – Anacharis

Egeria densa Planch.—Anacharis. Perennial, freshwater; stems rooted in substrate, branched or unbranched, 1–3 mm diam., cauline leaves in whorls of 5–7, submersed, recurved, 1.5–4.5 mm wide, sessile, blades linear; inflorescence 1-flowered, sessile; flowers unisexual, staminate and pistillate on different plants, projected to surface by elongate floral tube base; staminate spathes 2–4-flowered, flowers with distinct filaments; pistillate flowers with 1-locular ovary; fruits ovoid, smooth, dehiscing regularly. Lakes and ponds, 6400–6500 ft. Catron, Grant. Introduced.

ELODEA Michx. – Waterweed

Perennial, freshwater herbs; stems rooted in substrate, branched or unbranched; leaves submersed, cauline whorled, 2–4 at each node, or leaves opposite at base of stems, sessile, blades linear to linear-lanceolate, apex acute, margins with fine serrations; inflorescence solitary, sessile; flowers unisexual, staminate and pistillate on separate plants, rarely bisexual, usually raised to the water surface by elongate floral tube base, ovoid to lance-ellipsoid, smooth; seeds cylindrical to fusiform. Dr. Donald Les (University of Connecticut) has run many DNA samples and notes that many populations are hybrids between *E. canadensis* and *E. nuttallii* and that positive identification can be obtained only with fertile plants or DNA analysis.

1. Leaves usually < 1.7 mm wide; staminate spathes 4 mm or less…**E. nuttallii**
1'. Leaves usually > 2 mm wide; staminate spathes 6+ mm…(2)
2. Staminate pedicels remaining after anthesis; midstem leaves mainly in 2s, opposite, with whorls of 3 toward tip, spreading; pistillate spathes 9–67 mm; seeds 2.8–3 mm…**E. bifoliata**
2'. Staminate pedicels detached before or during anthesis; midstem leaves in whorls of 3, recurved; pistillate spathes 8.3–17.5 mm; seeds 4.5–5.7+ mm…**E. canadensis**

Elodea bifoliata H. St. John—Two-leaf waterweed. Leaves many in 2s, especially on lower stem, linear to narrowly elliptic, 4.7–24.8 × (0.8–)1.8–4.3 mm; inflorescence of staminate spathes, 10.2–42 mm; stamens 7–9, pollen in monads; styles 2.3–3 mm; seeds 2.8–3 mm. Lakes and ponds, 5625–9075 ft. NC, NW, Catron.

Elodea canadensis Michx.—Canadian waterweed. Leaves mostly in 3s, spreading or recurved, linear, oblong, or ovate, 5–13 × 2–5 mm; inflorescence of staminate spathes, 8.2–13.5 mm, peduncles abscising just before or during anthesis; pistillate spathes 8.3–17.7 mm; flowers imperfect; pedicels detaching before anthesis; stamens 7–9(–18), pollen in tetrads; styles 2.6–4 mm; seeds 4.5–5.7+ mm, basal hairs absent. Lakes, ponds, streams, often alkaline, 6725–9775 ft. NC, Grant.

Elodea nuttallii (Planch.) H. St. John—Western waterweed. Leaves mostly in 3s, spreading or recurved, linear to lanceolate, 4–15.5 × 0.9–1.7 mm, margins often folded; inflorescence of staminate spathes, 2.2–4 mm, peduncles abscising in bud; pistillate spathes 8.5–15 mm; flowers imperfect; pedicels briefly attached following anthesis; stamens 9, pollen in tetrads; styles 1.2–2 mm; seeds 4–4.6 mm, base with hairs. Lakes and ponds, 5000–10,125 ft. Rio Arriba.

NAJAS L. – Water-nymph

Annual, aquatic, submersed, stems slender, highly branched, rooting at proximal nodes, some with prickles on internodes; leaves with variously shaped basal sheaths, blades linear, 1-veined; margins serrate to minutely serrulate with 50–100 teeth per side, apex acute to acuminate; inflorescence involucres mostly present in staminate flowers, rare in pistillate flowers; flowers unisexual; staminate flowers subtended by membranous involucre; pistillate flowers sessile; fruits closely enveloping seed; seeds fusiform to obovoid, areola usually arranged into longitudinal rows.

1. Leaves finely serrate (magnification often needed); internodes without prickles…**N. guadalupensis**
1'. Leaves dentate; internodes with prickles…**N. marina**

Najas guadalupensis (Spreng.) Magnus—Southern water-nymph. Stems branched, internodes lacking prickles; leaves lax with age; sheaths 1–3.4 mm wide, rounded to truncate; blades 0.2–2.1 mm wide, margins with 18-100 minute serrations per side; apex rounded or slightly auriculate; flowers 1-3 per axil; staminate in distil axils, 1.5-3 mm; pistillate in proximal axils, 1.5-4 mm; styles 0.3-1.5 mm; seeds 1.2-3.8 mm; areoles of testa regularly arranged in 20-60 longitudinal rows. Alkaline waters, 5800-7775 ft. Doña Ana, Grant, Otero.

Najas marina L.—Spiny naiad. Stems branched, internodes with prickles; leaves stiff with age, sheaths 2-4.4 mm wide, apex acute; blades 0.4-4.5 mm wide; margins with 8-13 coarse serrations per side, apex acute; flowers 1 per axil; staminate and pistillate on separate plants; staminate flowers 1.7-3 mm; pistillate flowers 2.5-5.7 mm; styles 1.2-1.7 mm; seeds 2.2-4.5 mm, areoles of testa irregularly arranged, not in distinct rows. Brackish or waters with high calcium or sulfur content, lakes and ponds, 5450 ft. San Juan.

VALLISNERIA L. – Tape-grass

Vallisneria americana Michx.—Eel-grass. Perennial aquatic of fresh and brackish waters; rhizomes and stolons present; leaves basal, base flattened in cross section, linear, elongate with distinct broad lacunae band along midvein, 0.3-1.5 cm broad, margins entire to serrate; staminate scapes 30 50 mm, submersed; inflorescences cymose, long-pedunculate; flowers unisexual, staminate and pistillate on different plants; pistillate scapes reaching surface by peduncle elongation, peduncle recoiling in fruit; staminate flowers 1-1.5 mm wide; fruit cylindrical to ellipsoid, dehiscent. Fresh to brackish water of ponds, lakes, streams, 5200-5500 ft. Rio Arriba.

HYDROPHYLLACEAE – WATERLEAF FAMILY

Jennifer Ackerfield

Annual, biennial, or perennial herbs; leaves alternate or opposite, simple or pinnately compound or dissected; flowers perfect, actinomorphic; sepals 5, connate at the base or distinct; petals 5, connate; stamens 5, epipetalous, the filaments usually with basal appendages; pistil 1; ovary 2-carpellate, superior, with intrusive parietal placentation giving it the appearance of having 2 locules even though only 1 is present; fruit a capsule.

1.   Leaf margins entire…**Phacelia** (in part)
1'. Leaf margins crenate, or leaves pinnately compound, lobed, or dissected…(2)
2.   Flowers solitary in leaf axils…(3)
2'. Flowers several to many in cymes…(4)
3.   Inflorescence glandular…**Eucrypta**
3'. Inflorescence lacking glandular hairs…**Ellisia**
4.   Flowers in globose, dichotomously branching cymes; plants from fibrous roots…**Hydrophyllum**
4'. Flowers in elongate, helicoid or scorpioid cymes, or few-flowered cymes; plants from a taproot or woody caudex…**Phacelia** (in part)

ELLISIA L. – Ellisia

Ellisia nyctelea (L.) L.—Aunt Lucy. Annuals, 0.5-4 dm, with diffusely branched, retrorsely hispid stems; leaves pinnately divided into lanceolate or linear-oblong segments, lobes acute; inflorescence solitary; calyx 3-4 mm at flowering, 5-7 mm in fruit, hirsute near the margins; corolla campanulate, white to purple; style cleft to 1/2 its length; capsules 5-6 mm wide; seeds ca. 4 per capsule, globose. Moist meadows, open slopes, disturbed places, barely entering N New Mexico, elevation unknown. Union.

EUCRYPTA Nutt. – Hideseed

Eucrypta micrantha (Torr.) A. Heller—Dainty desert hideseed. Annuals, 0.5-2 dm, densely covered in glandular hairs, strongly odoriferous; leaves 1.5-5 cm, with rounded lobes; inflorescence a cyme; calyx 2-5 mm, enclosing the fruit; corolla campanulate, white, pale purple, or lavender, the throat yellow; capsules 2–3 mm wide; seeds 7–15 per capsule, wrinkled, black or dark brown. Dry, rocky soil of the Chihuahuan Desert, 4100–5300 ft. SC, SW.

HYDROPHYLLUM L. – Waterleaf

Hydrophyllum fendleri (A. Gray) A. Heller—Fendler's waterleaf. Perennials or biennials, 2–10 dm, from fibrous roots and rhizomes; leaves pinnately lobed, 6–30 cm, the lobes sharply toothed along the margins; peduncles 3–20 cm, usually longer than the leaves; calyx divided nearly to the base; corolla white to pale lavender, 6–10 mm; seeds 1–3. Our plants are **var. fendleri**. Moist, often shady places, 6500–10,800 ft. NC, C, SC.

PHACELIA Juss. – Phacelia

Annual, biennial, or perennial herbs; leaves usually alternate, entire to pinnately dissected or compound; flowers in helicoid cymes or rarely solitary and terminal; corolla white, pink, blue, purple, or yellow; stamens included or exserted, filaments white, blue, or purple.

1. Perennials from a woody caudex…(2)
1′. Annuals or biennials lacking a woody caudex…(4)
2. Leaf margins entire or a pair of lateral lobes at the base…**P. heterophylla**
2′. Leaf margins pinnatifid to pinnately compound…(3)
3. Corolla purple; seeds 8–12…**P. sericea**
3′. Corolla white; seeds 4…**P. rupestris**
4. Leaf margins entire…(5)
4′. Leaves crenate, toothed, pinnatifid, or pinnately compound…(6)
5. Leaves elliptic to oblong; corolla 3–5 mm, barely surpassing the calyx…**P. cephalotes**
5′. Leaves rounded-ovate to nearly orbicular; corolla 6–11 mm, longer than the calyx…**P. demissa**
6. Ovules numerous; seeds 10–30 per capsule, 1–1.5 mm, reticulate and transversely cross-corrugated…(7)
6′. Ovules 4; seeds 1–4 per capsule, 1.5–4.5 mm, the surface various…(8)
7. Glandular hairs on the stem absent or not black-tipped; inflorescence usually not elevated above the leaves; calyx lobes linear to linear-lanceolate or oblong in fruit…**P. ivesiana**
7′. Glandular hairs on the stem black-tipped; inflorescence usually elevated above the leaves; calyx lobes wider, spatulate in fruit…**P. affinis**
8. Leaf margins irregularly crenate-dentate with shallow, rounded lobes, the lower leaves sometimes few-lobed at the base but in general the sinuses not reaching the midrib…(9)
8′. Leaf margins deeply pinnatifid or bipinnatifid, with most sinuses reaching the midrib…(16)
9. Stems delicate, slender, branched from the base; stamens included or barely exserted < 1 mm past the corolla…**P. caerulea**
9′. Stems robust, branched above or unbranched, or if branched from the base then the stamens exserted > 1 mm past the corolla…(10)
10. Stamens included in the corolla or exserted up to 1.7 mm; corolla white to light blue; mature seeds 3 per capsule…**P. cloudcroftensis**
10′. Stamens exserted > 2 mm beyond the corolla; corolla lavender, purple, or blue (sometimes fading white); mature seeds usually 4 per capsule…(11)
11. Corolla campanulate, blue to blue-violet or purple; anthers usually yellow or pale…(12)
11′. Corolla tubular to tubular-campanulate, pale lavender, light violet, or dull blue-lavender; anthers blue, greenish blue, or bronze…(13)
12. Calyx to 3.5 mm; corolla 5–6 mm; seeds 2.2–2.5 mm; leaves oblong to oblong-ovate in outline…**P. bombycina**
12′. Calyx 4–5.5 mm; corolla 6–10 mm; seeds (2.5–)2.7–4.5 mm; leaves narrowly oblong to lanceolate in outline…**P. crenulata**
13. Cymes usually present from near the bottom of the plant to the top; seeds 2.2–2.7 mm…**P. sivinskii**
13′. Cymes present above the middle of the plant (not throughout); seeds 2.5–4.4 mm…(14)
14. Plants with light-colored stipitate-glandular hairs; corolla 5–8 mm; seeds sometimes with or without corrugations on the ventral surface…**P. integrifolia**
14′. Plants with black-tipped or dark red stipitate-glandular hairs; seeds with corrugations on the ventral surface…(15)
15. Mature seeds black; corolla pale lavender (fading white)…**P. pinkavae**
15′. Mature seeds dark brown; corolla blue to light violet…**P. serrata**

16. Corolla white with a pinkish midvein or light purple with a darker midvein; plants with procumbent or procumbent-ascending reddish stems…**P. arizonica**
16'. Corolla variously colored but lacking a darker midvein; plants with erect stems, these sometimes reddish…(17)
17. Corolla white to pale blue or purple, the lobes evidently erose or denticulate…(18)
17'. Corolla blue, purple, lavender, or rarely white, the lobes not evidently erose or toothed…(20)
18. Stamens included within the corolla…**P. denticulata**
18'. Stamens exserted from the corolla at least 1 mm…(19)
19. Filaments dark purple; seeds 3.2-3.5 mm…**P. neomexicana**
19'. Filaments white or purplish only at the base; seeds 2.5-3.2 mm…**P. alba**
20. Laves glabrous or nearly so, rather thick and succulent…**P. splendens**
20'. Leaves evidently hairy or glandular-hairy, not thick and succulent…(21)
21. Anthers bluish green to green; leaves and sepals with black-tipped glandular-stipitate hairs…**P. bakeri**
21'. Anthers yellow or whitish; leaves and sepals with pale-tipped glandular-stipitate hairs…(22)
22. Stamens included in the corolla; seeds lacking excavations on both sides of the ventral ridge…**P. cryptantha**
22'. Stamens exserted beyond the corolla; seeds with excavations on both sides of the ventral ridge (23)
23. Calyx lobes linear; seeds 2.5-3.2 mm…**P. congesta**
23'. Calyx lobes oblanceolate or oval; seeds 1.8-2 mm…**P. popei**

Phacelia affinis A. Gray—Limestone phacelia. Annuals, aromatic, 0.6-3 dm; stems short-hairy, minutely glandular above; leaves deeply pinnately lobed or compound; calyx 4-5 mm, to 10 mm in fruit; corolla 3-5 mm, white to pale lavender; seeds 15-30, ca. 1 mm, brown, reticulate and transversely corrugated. Dry canyons, limestone slopes, 4450-4550 ft. Grant, Hidalgo.

Phacelia alba Rydb.—White phacelia. Annuals, 1-7 dm; stems with numerous soft, short, spreading hairs and fewer longer, spreading hairs, glandular at the top and in the inflorescence; leaves pinnatifid; calyx 2-4 mm; corolla 4-5 mm, white to pale lavender, the lobes erose-dentate; seeds (2-)4, 2.5-3.2 mm, brown or black, alveolate, not corrugated, deeply excavated on each side of the conspicuous ventral ridge. Rocky and gravelly slopes, meadows, roadsides, 6100-9800 ft. NE, NC, NW, C, SC.

Phacelia arizonica A. Gray—Arizona phacelia. Perennials, 0.2-3 dm, prostrate to ascending, with numerous branches at the base; stems villous and hirsute, glandular above in the inflorescence, often reddish; leaves pinnatifid; calyx 2-4.5 mm; corolla 3-5 mm, the lobes shallowly erose; seeds 4, 1.8-2 mm, ventral surface excavated on both sides of the ventral ridge, dorsal surface alveolate and transversely ridged. Sandy soil, gravelly hillsides, often in oak or juniper communities, 4250-5600 ft. SW.

Phacelia bakeri (Brand) J. F. Macbr.—Baker's phacelia. Annuals, 0.5-4.8 dm; stems pilose to somewhat hirsute with black-tipped multicellular stipitate glands; leaves 2-8 cm, pinnately divided to the midrib, the lobes crenate to dentate; calyx 2-3.5 mm; corolla 7-8 mm, violet to dark blue; seeds 4, 2.7-3 mm, brown, pitted, lacking excavations on either side of the ventral ridge. Talus and scree slopes, 10,500-12,450 ft. Colfax, Taos.

Phacelia bombycina Wooton & Standl.—Mangas Spring phacelia. Annuals, 1-4 dm; stems puberulent to setose, glandular above in the inflorescence; leaves pinnatifid; calyx to 3.5 mm; corolla 5-6 mm, blue to violet; seeds 2.2-2.5 mm, dark brown, ventral surface with corrugated margins and corrugated on 1 side of the ventral ridge, dorsal surface pitted. Sandy, gravelly slopes, rocky hillsides, often in desert shrub communities, 4150-5900 ft. WC, SW.

Phacelia caerulea Greene [*P. intermedia* Wooton]—Skyblue phacelia. Annuals, 1-4 dm; stems 1 to several from the base, hispid and viscid-glandular; leaves pinnately round-lobed, only 1 or 2 basal lobes divided to the midrib; calyx 2-3.5 mm; corolla 3-5 mm, lavender to purple; seeds 4, 2-2.5 mm, ventral surface deeply excavated on both sides of the ventral ridge, with strongly corrugated margins, the ventral ridge corrugated along 1 side. Gravelly, dry, calcareous hillsides, sandy stream banks, rocky ledges, often in desert shrub or creosote communities, 3700-7200 ft. C, WC, SC, SW, Sandoval, Taos.

Phacelia cephalotes A. Gray—Chinle phacelia. Annuals, 0.5-1.3 dm; stems nearly prostrate, densely glandular and hispid; leaves entire, oblong to elliptic, to 1.8 cm; calyx 3-4 mm; corolla 3-5 mm, white to pale lavender with a yellow tube; seeds 8–12, 1–1.5 mm, pitted, angular. Shale slopes, desert shrub communities, 6600–7000 ft. McKinley.

Phacelia cloudcroftensis N. D. Atwood—Cloudcroft phacelia. Annuals, to 5 dm; stems with retrorse, appressed to spreading-hirsute hairs and dark red to black glandular hairs; leaves irregularly lobed to crenate-dentate; calyx 2–3.5 mm; corolla 4–5 mm, white to light blue, the lobes crenate to erose-denticulate; seeds 3, 3.6–4.3 mm, ventral surface pitted, with the ventral ridge usually corrugated on 1 side. Rocky roadcuts near Cloudcroft, 6100–7600 ft. Otero. **R & E**

Phacelia congesta Hook.—Caterpillars. Annuals, 1–10 dm; stems simple or diffusely branched throughout, with hispid and stipitate-glandular hairs; leaves pinnately compound; calyx 3–5 mm; corolla 4–6 mm, blue or rarely white; seeds usually 4, 2.5–3.2 mm, brown, reticulate, the ventral surface excavated on both sides of the ventral ridge. Sandy soil, sandstone outcrops, rocky limestone, usually in desert scrub, 3100–6500 ft. SE, SC, SW, except Hidalgo, Lea.

Phacelia crenulata Torr. ex S. Watson—Heliotrope phacelia. Annuals, 0.5–4(–8) dm; stems densely glandular-stipitate, interspersed with short, spreading, eglandular hairs; leaves crenately pinnatifid with rounded, shallow lobes; calyx 4–5.5 mm; corolla 6–10 mm, blue-lavender to purple, sometimes proximally whitish; seeds usually 4, (2.5–)2.7–4.5 mm, dark brown with paler margins, alveolate-pitted, deeply excavated on either side of the ventral ridge. Sandstone bluffs, sandy soil, clay slopes, gravelly hillsides, mesas, 3800–7550 ft. Our plants are **var. corrugata** (A. Nelson) Brand. NC, NW, C, WC, SC, SW, except Catron. Can cause allergic skin reactions in some people.

Phacelia cryptantha Greene—Hiddenflower phacelia. Annuals, 1–4 dm; stems simple or branched throughout, hispid and glandular-puberulent at least above; leaves pinnately compound; calyx 4–7 mm; corolla 4–7 mm, lavender to blue; seeds 4, 1.5–3 mm, pitted-reticulate, lacking excavations on either side of the ventral ridge. Gravelly and rocky slopes of canyons, often in desert shrub or piñon-juniper communities, 5600–5750 ft. Catron, Hidalgo.

Phacelia demissa A. Gray—Brittle phacelia. Annuals, to 1.5 dm; stems puberulent or villosulous with glandular hairs; leaves entire, rounded to rounded-ovate; calyx 2–5 mm; corolla 6–11 mm, bright to pale lavender limb with a light yellow tube; seeds fewer than or equal to the number of ovules, rarely 4, 1–1.7 mm, pitted-reticulate. Our plants are **var. demissa**. Mancos Shale, clay slopes, badlands, often in desert shrub communities, 4800–6250 ft. McKinley, Sandoval, San Juan.

Phacelia denticulata Osterh.—Rocky Mountain phacelia. Annuals, 0.5–5.4 dm; stems bristly and glandular-stipitate; leaves 1–7.5 cm, pinnately cleft or divided; calyx ca. 2.5 mm; corolla 3.5–4.5 mm, light blue, the lobes denticulate; seeds 4, alveolate, slightly excavated on either side of the curved ventral ridge. Rocky slopes, sandy soil, 8200–9450 ft. Colfax, Union.

Phacelia heterophylla Pursh—Wand phacelia. Perennials, 2–12 dm; stems green or grayish, sometimes silvery, with long, spreading bristles and short, spreading, glandular hairs; leaves entire or with a pair of lateral lobes; corolla 3–6 mm, dull whitish to ochroleucous or sometimes purplish; seeds 1–2(–4), 2–2.5 mm. Mixed conifer forests, meadows, 6550–11,500 ft. Widespread except EC, SE, Doña Ana, Harding.

Phacelia integrifolia Torr.—Gypsum phacelia. Annuals, 1–5 dm; stems densely glandular-stipitate or with short, spreading, sometimes glandular-stipitate hairs; leaves crenate with shallow lobes; calyx 3.5–4.5 mm; corolla 6–8 mm, dull blue-lavender; seeds 4, 3.2–4.4 mm, dark brown to black with paler margins, alveolate, deeply excavated on either side of the curved ventral ridge, with distinct dorsal transverse ridges. Sandy soil, gypsum outcrops, sagebrush flats, desert shrublands, piñon-juniper woodlands, 2500–8000 ft. Two varieties:

1. Seeds lacking corrugations on the ventral ridge or margins…**var. integrifolia**.
1'. Seeds with corrugations on the ventral ridge and/or ventral margins…**var. texana**

var. integrifolia—Gypsum scorpionweed. Seeds lacking corrugations on the ventral ridge or margins. NC, NW, C, WC, SE, SC, SW.

var. texans (J. W. Voss) N. D. Atwood. Seeds with corrugations on the ventral ridge and/or ventral margins. C, EC, SE, SC.

Phacelia ivesiana Torr.—Ives' phacelia. Annuals, to 2.5 dm; stems with spreading-hispid hairs, often finely glandular distally; leaves deeply pinnatifid, the lobes small, slender, few-toothed, or lobed; calyx 2.5–4(–4.5) mm; corolla 2.5–4(–4.5) mm, white, often with a yellowish tube or throat; seeds 10–15, 1–1.5 mm, alveolate-reticulate and cross-corrugated. Sandy or clay soil, dry washes, 5250–6650 ft. NW, C, Chaves, Doña Ana, Eddy.

Phacelia neomexicana Thurb. ex Torr.—New Mexico phacelia. Annuals, 0.8–7 dm; stems erect and sparsely branched, with spreading-hispid hairs and small stipitate-glandular hairs; leaves deeply pinnatifid; calyx 2.7–4.5 mm; corolla 3–4 mm, white, the lobes shallowly erose; seeds 4, 3.2–3.5 mm, alveolate, the ventral surface excavated on both sides of the ventral ridge. Sandy to rocky soil, canyons, mountain slopes, often in pine or oak woodlands, 3900–9600 ft. Widespread except E.

Phacelia pinkavae N. D. Atwood—Pinkava's phacelia. Annuals or biennials, (1.5–)3–5 dm; stems densely covered with short spreading hairs and multicellular stipitate-glandular hairs, leaves shallowly lobed to irregularly crenate-dentate; calyx 2.5–3.3 mm; corolla 4.2–5 mm, pale lavender but fading white; seeds 3–3.5 mm, alveolate, excavated on both sides of the ventral ridge, corrugated faintly on 1 side. Rocky talus slopes, volcanic gravels, pumic soils, grasslands, desert shrub, piñon-juniper communities, 5450–7800 ft. C, Catron, Grant, Socorro.

Phacelia popei Torr. & A. Gray—Pope's phacelia. Annuals, 0.5–4 dm; stems with spreading-hirsute hairs and glandular hairs; leaves pinnately to bipinnately compound, with linear to lanceolate divisions; calyx 2.3–4 mm; corolla 3.5–7 mm, blue to purple; seeds 4, 1.8–2 mm, reticulate, the ventral surface deeply excavated on both sides of the ventral ridge, dorsal surface transversely ridged. Sandy, clay, or rocky soil limestone, desert shrub communities, 3250–7000 ft. NE, EC, SE, SC, SW, Santa Fe.

Phacelia rupestris Greene—Rock phacelia. Perennials, 1–6 dm; stems with spreading-hispid hairs and sometimes with glandular hairs; leaves pinnately compound; calyx 3–5 mm; corolla 2–4 mm, white; seeds 4, 2–3 mm, reticulate, the ventral surface excavated on both sides of the ventral ridge. Sandy soil, gravel bars, cracks of limestone cliffs, canyon ledges, desert shrub, piñon-juniper communities, 4400–6600 ft. C, SE, SC, SW.

Phacelia sericea (Graham ex Hook.) A. Gray—Silky phacelia. Perennials, 1–4 dm; stems thinly strigose, loosely short-hairy, densely sericeous, or loosely woolly, with spreading pubescence in the inflorescence; leaves pinnatifid with entire or cleft lobes; corolla 5–7 mm, purple, pubescent; seeds 8–18, 1–2 mm, pitted-reticulate, with narrow ridges separating the longitudinal rows of alveolae. Spruce-fir forests, subalpine meadows, alpine slopes, 10,000–13,000 ft. Taos.

Phacelia serrata J. W. Voss—Saw phacelia. Annuals, 1–3.5 dm; stems with hirsute and multicellular glandular hairs; leaves shallowly lobed to dentate; calyx 3–4 mm; corolla 4–6 mm, blue to light violet; seeds 4, 2.5–3.4 mm, pitted, ventral ridge corrugated on 1 side, margins corrugated, dorsal surface smooth to faintly pitted. Deep volcanic cinders in ponderosa pine and piñon-juniper woodlands, 5900–7200 ft. Cibola. The calyx length from the original description was misreported as the corolla length by previous authors. **R**

Phacelia sivinskii N. D. Atwood, P. J. Knight & Lowrey—Sivinski's phacelia. Biennials, 2–4 dm; stems densely covered with glandular hairs and intermixed with longer hispid hairs; leaves irregularly crenate-dentate; calyx 3–3.6 mm; corolla (3.7–)4.5–5 mm, light violet; seeds 4, 2.2–2.7 mm, ventral surface pitted, the ventral ridge deeply excavated on both sides, entire to corrugated, dorsal surface cross-corrugated. Gypsum soil in juniper or desert shrub communities, 5650–6250 ft. NC, Cibola. **R & E**

Phacelia splendens Eastw.—Patch phacelia. Annuals, 0.5–2.7 dm; stems puberulent and with scattered glandular-stipitate hairs; leaves 2–7.5 cm, pinnatifid; calyx 2.5–3 mm; corolla 4–8 mm, bright blue with a yellow tube, glabrous to sparsely pubescent; seeds 4, 3–4 mm, finely honeycomb-pitted, excavations on either side of the ventral ridge, the margins mostly revolute. Barren clay or Mancos Shale slopes, 5000–6250 ft. McKinley, San Juan.

HYPERICACEAE – ST. JOHN'S WORT FAMILY

Kenneth D. Heil

Mostly herbs and shrubs; leaves simple, entire, opposite or whorled, often gland-dotted, exstipulate; flowers actinomorphic, perfect or unisexual, sepals 4–5, separate, petals 3–6, distinct, yellow, pink, or white; stamens 4 to many, the filaments distinct or connate into 3–5 groups; pistil superior, of 3–5 united carpels; fruit a capsule (Robson, 2015). Formerly included in Clusiaceae.

HYPERICUM L. – St. John's wort

Annual to perennial herbs and shrubs; leaves sessile to petiolate; leaf blades suborbicular to lanceolate; petals yellow to orange, rarely red-tinged; stamens 10–300, the filaments distinct or connate basally; capsules 2–5-valved.

1. Petals 1.7–3.5 mm; leaf blades bicolored, paler abaxially…**H. mutilum**
1'. Petals 7–13 mm; leaf blades concolored, with equal color abaxially and adaxially…(2)
2. Stems from rooting creeping base, without black glands, rarely with reddish glands; petals sometimes red-tinged…**H. scouleri**
2'. Stems from rooting, not creeping, base, with black glands; petals not red-tinged…**H. perforatum**

Hypericum mutilum L. [*Sarothra mutila* (L.) Y. Kimura]—Dwarf St. John's wort. Herbs annual or perennial, 0.5–8 dm; leaf blades paler abaxially, ovate, elliptic, or lanceolate, 3–27(–40) × 1–15 mm; sepals 2–4.5 × 0.6–1.5 mm, margins sometimes ciliate; petals pale yellow, 1.7–3.5 mm; stamens 5–16, scarcely grouped; capsules 2–5 × 1.6–2.4 mm, usually broadest at or near middle. Ditches, marshes, lake margins, temporary pools; 1 record at a ciénega in Hidalgo County, 5160 ft.

Hypericum perforatum L.—Klamath weed. Herbs erect, with rooting, not creeping, base, 2–12 dm, stems with black glands, usually in lines; leaf blades oblong or elliptic to lanceolate-elliptic or linear, 7–30 × 2–16 mm; sepals 2.5–7 × 0.6–2(–3) mm; petals golden-yellow, 7–13 mm; stamens 40–90; capsules 6–10 × 3.5–5 mm. Dry, rocky meadows and along roadsides and streams, 7130–7840 ft. Cibola. Introduced.

Hypericum scouleri Hook. [*H. formosum* Kunth var. *scouleri* (Hook.) J. M. Coult.; *H. scouleri* subsp. *nortoniae* (M. E. Jones) J. M. Gillett]—Scouler's St. John's wort. Herbs with rooting, creeping, branching base, 0.5–6.6(–8) dm, without black glands; leaf blades oblong-elliptic or elliptic to triangular-ovate, 12–28(–32) × 6–15(–18) mm; sepals 2.5–5.5 × 1–2 mm; petals golden-yellow, sometimes red-tinged, 7–12 mm; stamens 50–90(–109); capsules 6–10 × 3.5–6 mm. Moist meadows, ditches, pond margins, coniferous forests, 6000–10,000 ft. Widespread except E, SW.

HYPOXIDACEAE – STAR-GRASS FAMILY

Don Hyder

HYPOXIS L. – Star-grass

Hypoxis hirsuta (L.) Coville—Common goldstar. Herbs, perennial; stems subterranean, usually vertical, to 3 dm, 8–20 mm wide, fleshy; leaves basal, grasslike, blades linear to setaceous, 3–6, to 30 cm; inflorescences depauperate umbels, borne singly in leaf axils; bracts (1–)2–10(–17) mm; flowers actinomorphic; tepals 6, distinct, often greenish or yellow, 6–15 mm, hairy below; anthers 6, spreading; ovary inferior, usually densely pubescent to pilose, sometimes glabrate; fruits capsular; seeds (5–)10–50 per capsule, ± globose, black, surfaces sharply to bluntly muricate or with rounded pebbling, sometimes with iridescent, membranous coat (Herndon, 2002). Moist to dry woodlands and prairies; 1 record from the Zuni Mountains, 8000 ft. Cibola.

IRIDACEAE – IRIS FAMILY

Kenneth D. Heil

Perennial (ours) or annual herbs, from rhizomes, bulbs, or corms; leaves equitant, mostly basal, long and narrow, 2-ranked; inflorescence a raceme or panicle, subtended by 2 spathelike bracts; flowers perfect, regular (ours) or irregular, showy, 6 petaloid segments (tepals) in 2 whorls, similar or dissimilar; stamens 3, usually partly adnate to the outer petaloid segments; style usually 3-branched, branches often flattened, opposite the stamens and outer petaloid segments; ovary inferior with 3 locules; fruit a loculicidally dehiscent capsule; seeds many, membranous, usually with fleshy endosperm, sometimes arillate (Henderson, 2002).

1. Style branches large, petal-like; flowers large, > 3 cm wide; sepals and petals dissimilar...**Iris**
1'. Style branches small, not petal-like; flowers small, < 3 cm wide; sepals and petals alike...**Sisyrinchium**

IRIS L. – Iris, flag

Perennial, rhizomatous or bulbous; stems erect, simple or branched; leaves mostly basal, linear to swordlike, flat (ours), terete, or quadrangular; inflorescence a spike, raceme, or panicle bearing 1 to several flowers from paired spathelike bracts; flowers showy, brightly colored, petaloid perianth parts differentiated into 2 whorls of 3 each, united below into a tube; stamens 3, alternating with the inner whorl and next to the style branch; style divided into 3 petaloid branches, arched over the outer perianth whorl.

1. Flowers purple, blue-purple, or rarely white...**I. missouriensis**
1'. Flowers yellow...**I. pseudacorus**

Iris missouriensis Nutt.—Western iris. Rhizomatous; stems 20–60 cm, barely surpassing the leaves; leaves linear, flat, erect; flowers typical of the genus, light blue, occasionally white; capsule 3–5 cm, short-cylindrical, 6-ridged. Wetlands such as marshes, moist meadows, stream banks, and pond shores, often in places that dry up as the season progresses, 5250–12,000 ft. Widespread except NE, EC, SE, SW.

Iris pseudacorus L.—Yellow iris. Plants to 10 dm; leaves 4–10 dm × 2–3 cm; flowers yellow, outer tepals 4.5–8 cm; capsules elliptic to cylindrical, 5–8 cm. Known from the Rio Grande and Rio Sapello, with cottonwood, coyote willow, saltgrass, *Ranunculus*, *Anemone*, and horsetail, 4950–7265. Bernalillo. Introduced.

SISYRINCHIUM L. – Blue-eyed grass

Perennial (ours) or annual herbs, often cespitose, rhizomatous or not; stems scapelike or branched, compressed and 2-winged; leaves basal or basal and cauline, alternate, flat, ensiform, glabrous; inflorescence a terminal cluster from a 2-bracted spathe, bracts approximately equal, or greatly unequal; flowers actinomorphic, tepals 6, all alike, widely spreading to reflexed, bluish violet to light blue, white, lavender to pink, magenta, purple, or yellow, stamens symmetrically arranged, styles 3, erect, connate basally; fruit a capsule, ± globose; seeds many, seed coat black, granular to rugulose (Cholewa & Henderson, 2002).

1. Tepals yellow to orange; seeds usually hemispherical with a depression or flattened side or globose...(2)
1'. Tepals purple to light blue; seeds globose to obconic, depression absent...(3)
2. Stems 3.5–8 mm wide; tepals 11–23 mm; capsules 8–19 mm...**S. arizonicum**
2'. Stems 0.6–2.3 mm wide; tepals 7–11 mm; capsules 4.5–9 mm...**S. longipes**
3. Stems branched...(4)
3'. Stems simple...(5)
4. Main stems 2–4.8 mm wide...**S. chilense**
4'. Main stems mostly 2 mm or less wide...**S. demissum**

5. Outer spathes usually at least 16 mm longer than inner; stems mostly 2-3.7 mm wide; keel of inner spathe ± gibbous...**S. idahoense**

5'. Outer spathes not > 16 mm longer than inner; stems mostly 1-2.5 mm wide; keel of inner spathe not gibbous...**S. montanum**

Sisyrinchium arizonicum Rothr.—Arizona blue-eyed grass. Herbs, perennial, cespitose; stems to 6 dm, branched, with 2-4 nodes, 3.5-8 mm wide; spathes green, obviously wider than supporting branch, outer 30-55 mm, equaling or to 3.7 mm longer than inner, tapering evenly toward apex, margins basally connate 5-6 mm; tepals yellow to orange, usually with brownish veins, outer tepals 11-23 mm; capsules black or blackish brown, 8-19 mm. Moist meadows, stream banks, open places in coniferous forest, 7200-10,200 ft. WC.

Sisyrinchium chilense Hook. [*S. ensigerum* E. P. Bicknell; *S. scabrum* Cham. & Schltdl.]—Swordleaf blue-eyed grass. Herbs, perennial; stems to 3.2 dm, branched, with 1-2 nodes, 1.9-4.8 mm wide; spathes green, obviously wider than supporting branch, outer 15-24.5 mm, 1.4 mm shorter to 2.2 mm longer than inner, tapering evenly toward apex, margins basally connate; tepals blue to light bluish violet, bases yellow, outer tepals 8-15 mm; capsules light to dark brown, obovoid to ± globose, 3.8-5.3 mm. Near springs and streams, 4900-8000 ft. Eddy, Otero, Sierra.

Sisyrinchium demissum Greene—Star blue-eyed grass. Perennial herb; stems to 5 dm, branched, with 1 or 2 nodes, with a leaflike bract subtending 2 to several pedunculate spathes; outer spathe bract 11-25 mm, 2.5 mm shorter to 5 mm longer than the inner, basally connate; tepals deep bluish violet, bases yellowish, outer tepals 6-15 mm; capsules tan when mature, 4-8 mm. Moist areas such as wet meadows, springs, seeps, stream banks, 4620-9100 ft. Widespread except EC, SE, SW.

Sisyrinchium idahoense E. P. Bicknell [*S. occidentale* E. P. Bicknell]—Idaho blue-eyed grass. Perennial herb, stems to 45 cm; spathes glabrous, tapering evenly to the apex, outer bract typically 13-16 mm longer than the inner; tepals light to deep bluish violet, occasionally purple, bases yellowish, outer tepals 8-13 mm; capsule beige to light or dark brown when mature, 3-6 mm. Our material belongs to **var. occidentale** (E. P. Bicknell) Douglass M. Hend. Moist montane areas such as wet meadows, springs, seeps, stream banks, 7000-7380 ft. NW, WC, Eddy, Guadalupe, San Miguel.

Sisyrinchium longipes (E. P. Bicknell) Kearney & Peebles [*Hydastylus longipes* E. P. Bicknell]—Timberland blue-eyed grass. Herbs, perennial; stems to 4.6 dm, 0.6-1.7(-2.3) mm wide; spathes green, glabrous, outer 12.5-36 mm, 0-16 mm longer than inner; tepals yellow to orange, usually with brownish or occasionally purplish veins, outer tepals 7-11 mm; capsules dark brown to black, slightly turbinate to ± globose, 4.5-9 mm. Wet to moist meadows, stream banks, moist open areas in forests, 7000-10,200 ft. Grant.

Sisyrinchium montanum Greene—Montana blue-eyed grass. Perennial herb; stems to 50 cm; spathes with outer bract 12-46 mm longer than inner bract; tepals dark bluish violet, bases yellowish; outer tepals 9-14.5 mm; capsule tan to dark brown, globose, 4-7 mm. Our material is **var. montanum**. Moist montane areas such as wet meadows, springs, seeps, stream banks, 4560-11,005 ft. N, C, WC. Doña Ana.

JUGLANDACEAE – WALNUT FAMILY

Kenneth D. Heil

JUGLANS L. – Walnut

Primarily large trees with scaly, hard bark; leaves oddly pinnately compound, leaflets 3-23, margins serrate or entire, gland-dotted, aromatic; flowers unisexual, staminate and pistillate on same plants, subtended by 2-6 bracts; sepals 4; corolla absent; fruit a drupe with fibrous indehiscent exocarp and mesocarp and stony, furrowed endocarp (Stone, 1997).

1. Fruits 1.4–2.3 cm; nuts (stones) 1.1–1.7 cm; leaflets 0.8–1.1(–2.2) cm wide, with capitate-glandular hairs, nonglandular hairs limited to axils of proximal veins on abaxial surface; shrubs or trees, 3–10 m…**J. microcarpa**
1'. Fruits 2–8 cm; nuts (stones) 1.8–4 cm; leaflets 1.5–5.5 cm wide, with both glandular and nonglandular hairs abaxially; trees, 5–50 m…(2)
2. Fruits 2–3.5 cm, nuts longitudinally grooved, surface between grooves smooth; leaflets 9–15, adaxially with capitate-glandular hairs…**J. major**
2'. Fruits 3.5–8 cm, nuts longitudinally grooved, surface between grooves coarsely warty; leaflets (9–)15–19 (–23), adaxially glabrous except for scattered hairs on midrib…**J. nigra**

Juglans major (Torr.) A. Heller—Arizona walnut. Trees, 5–18 m; bark light to medium gray or brownish, divided into narrow checkered plates; leaves 18–38 cm, leaflets 9–15, lanceolate to lance-ovate, 6.5–10.5 × 1.5–3.4 cm, margins serrate, apex narrowly acuminate, surfaces abaxially with capitate-glandular hairs and simple or 2–4-rayed fasciculate hairs, adaxially with capitate-glandular hairs, becoming glabrate except along major veins; staminate catkins 5–8 cm; fruits 1–3, subglobose or short-ovoid, 2–3.5 cm, smooth, densely covered with capitate-glandular hairs and peltate scales; nuts globose to ovoid, 1.8–2.7 cm, deeply longitudinally grooved, surfaces between grooves smooth. New Mexico material belongs to **var. major**. Along streams and rocky canyon sides, 3500–9000(–10,000) ft. C, WC, SE, SC, SW.

Juglans microcarpa Berland. [*J. rupestris* Engelm. ex Torr.]—Little walnut. Shrubs or small trees, to 10 m; bark medium gray, split into ± rough ridges; leaves 12–29 cm, leaflets 17–25, lanceolate or narrowly lanceolate, weakly to strongly falcate, 5.2–6.3(–9.6) × 0.8–1.1(–2.2) cm, margins entire or toothed, apex long-acuminate; surfaces abaxially with capitate-glandular hairs; staminate catkins 3–7 cm; fruits 1–3, globose, 1.4–2.3 cm, smooth, with capitate-glandular hairs; nuts globose to depressed-globose, 1.1–1.7 cm, grooved, surface between grooves smooth. Along creeks and rivers, 3400–8550 ft. C, WC, SE.

Juglans nigra L.—Black walnut. Trees, to 40(–50) m; bark medium to dark gray or brownish, deeply split into narrow rough ridges; leaves 20–60 cm, leaflets (9–)15–19(–23), lanceolate or ovate-lanceolate, margins serrate, apex acuminate; surfaces abaxially with capitate-glandular hairs, adaxially glabrous except for scattered capitate-glandular and fasciculate hairs on midrib; staminate catkins 5–10 cm; fruits 1–2, subglobose to globose, rarely ellipsoid, 3.5–8 cm, warty, with scales and capitate-glandular hairs; nuts subglobose to globose, rarely ellipsoid, 3–4 cm, very deeply longitudinally grooved, surface between grooves coarsely warty. Bandelier National Monument along Frijoles Canyon, ca. 6100 ft. Sandoval. Introduced.

JUNCACEAE – RUSH FAMILY

Max H. Licher and Glenn R. Rink

Perennial or annual grasslike herbs, cespitose or rhizomatous; culms usually round, sometimes compressed; leaves simple, linear, flat to channeled or terete, basal (usually) and/or cauline, in some species reduced to bladeless sheaths (cataphylls) only; sheath open or closed; inflorescence terminal, sometimes appearing lateral, cymose, composed of clusters of flowers or single flowers, sometimes compressed into a compact head, usually subtended by 1+ leaflike bracts; flowers perfect, with or without 2 translucent smaller bracts (bracteoles), the perianth of 6 tepals in 2 ranks, persistent; ovary superior, stigma 1, style 3-branched; stamens 3 or 6, anthers basifixed; fruit a capsule, dehiscent, unilocular to 3-locular; seeds 3 to many.

1. Leaves glabrous, or blades absent; capsules with numerous seeds; bracteoles when present entire; sheaths open…**Juncus**
1'. Leaves ciliate, with hairs at least on the basal margins; capsules with 3 seeds; bracteoles subentire to lacerate or fringed; sheaths closed…**Luzula**

JUNCUS L. - Rush

Plants grasslike, glabrous, perennial or less frequently annual; cespitose to colonial from long rhizomes; culms round, sometimes flattened; leaves clustered at base or alternate up the culm; blades lacking or well developed and linear, flat with face toward the culm (grasslike), flat with edge toward the culm (ensiform, irislike), channeled, or terete, with crosswalls (septa) or without; sheaths open from the base, often auriculate; cataphylls (bladeless basal sheaths) present and diagnostic in some groups/species; inflorescence terminal, sometimes appearing lateral, cymose, diffuse to strongly congested, with 1 to many heads or clusters; primary inflorescence bract shorter to longer than the inflorescence; flowers perfect, with a perianth of 6 tepals in 2 ranks, with or without 2 additional bracteoles, persistent; stamens 3 or 6 (when 3 always under the outer tepals); ovary superior, with 1 style and 3 stigma branches; fruit a capsule, 1–3-locular, containing many minute seeds; seeds sometimes with slender appendages (tails) at 1 or both ends (Brooks & Clemants, 2000; Hermann, 1975; Hurd et al., 1997; Kirschner, 2002a, 2002b; Zika, 2012a).

1. Flowers usually in heads or clusters, rarely single; bracteoles absent at the base of each flower (though 1 to several individual bracts may be present in a cluster); leaf blades with or without septa [subgenus **Juncus**]…(2)
1'. Flowers borne singly (sometimes on very short pedicels in a loose cluster); a pair of bracteoles present on opposite sides at the base of each flower; leaf blades without septa…(15)
2. Leaf blades flat…(3)
2'. Leaf blades channeled to terete…(5)
3. Leaf blades ensiform (flat with edge toward stem, irislike, with the free edges becoming fused upward from the stem), with partial septa…**J. saximontanus**
3'. Leaf blades with face toward stem (grasslike), without septa…(4)
4. Perianth segments (tepals) 1.8–3.5 mm; capsules nearly globose, 1.8–2.9 mm; stamens 3; inflorescence heads 5–200, with 2–10(–20) flowers each; plants cespitose; cataphylls absent; leaves to 5 mm wide… **J. marginatus**
4'. Perianth segments (tepals) 4.5–6 mm; capsules obovoid, 3–5 mm; stamens 6; inflorescence heads 1–8 (–12), with 3–12 flowers each; plants from elongate rhizomes; cataphylls present on at least some culm bases; leaves to 3 mm wide…**J. longistylis**
5. Leaf blades with or without imperfect septa; seeds tailed; small cespitose plants of alpine tundra, or rhizomatous plants of subalpine to alpine habitats…(6)
5'. Leaf blades with septa; seeds not tailed, merely apiculate; plants rhizomatous to at least some degree, of various habitats…(8)
6. Plants rhizomatous; leaves partially septate, inrolled or folded most of the way to the terete tip, 1–2.5 mm wide at midlength; tepals dark brown to purplish, 4.5–7.5 mm; inflorescence generally of > 1 cluster…**J. castaneus**
6'. Plants cespitose; leaves channeled to terete, 0.5 mm diam. at midlength, tepals pale to dark brown, 2.5–5 mm; inflorescence a single terminal cluster of flowers…(7)
7. Inflorescence bract much longer than the inflorescence on at least some culms; capsules retuse…**J. biglumis**
7'. Inflorescence bract shorter than to almost equaling the inflorescence; capsules obtuse to subtruncate… **J. triglumis**
8. Some to most mature heads in an inflorescence spherical to subspherical, with flowers spreading significantly below the horizontal (sometimes barely so in *J. acuminatus*); flowers greenish to tan, sometimes with reddish tinting…(9)
8'. Most mature heads in an inflorescence hemispherical to obpyramidal (flowers in mature heads mostly spreading or ascending to erect, few, if any, definitely reflexed), sometimes subspherical, but then flowers dark brown to blackish; flowers variously colored…(11)
9. Plants cespitose; capsules broadly lanceoloid with a bluntly acute tip…**J. acuminatus**
9'. Plants rhizomatous with tuberous nodes (sometimes not present on herbarium collections); capsules narrowly lanceoloid with a long-tapering apex…(10)
10. Leaves erect to ascending; auricles < 1 mm; perianth 3–4 mm, tepals often reddish, inner tepals equal to or longer than the outer ones; cataphylls sometimes present; anthers 3 or 6; plants low, 10–40 cm… **J. nodosus**
10'. Leaves divaricate; auricles 2–5 mm; perianth 4–5 mm, tepals rarely reddish, inner tepals shorter than the outer ones; cataphylls never present; anthers 6; plants taller, 20–100 cm…**J. torreyi**
11. Inflorescence highly diffuse, 5–20 cm, with 30–70(–130) heads; capsules twice as long as tepals; stamens 3…**J. diffusissimus**

11'. Inflorescence compact to moderately diffuse, 0.5-8 cm, with 1-30(-50) heads; capsules slightly shorter to exserted and up to 1.5 times the length of the tepals; stamens 6...(12)

12. Inflorescence of 5-30(-50) heads; tepals 1.8-3 mm, apex obtuse to acuminate, green to brown...(13)

12'. Inflorescence of 1-11 heads; tepals 2.5-6 mm, apex strongly contracted and mucronate, dark brown to almost black...(14)

13. Stems decumbent to erect; inflorescence branches spreading, sometimes widely so; inner tepals acuminate...**J. articulatus**

13'. Stems erect; inflorescence branches ascending to erect; inner tepals obtuse...**J. alpinoarticulatus**

14. Inflorescence a single head (rarely a second one); tepals 2.3-4.9 mm; anther length 0.25 times to equal to filament length; auricles 0.3-0.6(-1.2) mm; capsules abruptly narrowed to beak to truncate or even retuse; plants with densely branching rhizomes, tending to form loose identifiable clumps...**J. mertensianus**

14'. Inflorescence with (1)2-10 heads; tepals 2.4-6.2 mm; anther length 1-2 times filament length; auricles 1-3.2 mm; capsules more gradually to abruptly narrowed to beak; plants with long rhizomes, tending to express as individual stems in mixed turf composed of other graminoids...**J. nevadensis**

15. Plants annual; inflorescence usually at least 1/2 the height of the plant; roots fine, fibrous; leaf blades generally inrolled and <1 mm wide; plants generally <30 cm...**J. bufonius**

15'. Plants perennial; inflorescence usually <1/2 the height of the plant; roots coarse or rhizomes present; leaf blades narrow or wide, inrolled to flat or cylindrical; mature plants typically larger...(16)

16. Inflorescence arising laterally through what appears to be a slit in the cylindrical stem, or if much reduced and more leaflike, then plants of subalpine to alpine habitats with few flowers and tailed seeds; leaf blades absent or basal, terete...(17)

16'. Inflorescence terminal; seeds not tailed; leaf blades basal and cauline, flat with face toward stem, sometimes inrolled...(21)

17. Rhizomes long; stems scattered or in lines, in loose colonies; inflorescence generally with >5 flowers (except in some depauperate specimens); seeds not tailed; plants generally below subalpine elevations...(18)

17'. Rhizomes short; stems cespitose, in dense tufts like bunchgrass; inflorescence typically with 1-4(-7) flowers; seeds tailed; plants of subalpine to alpine habitats...(19)

18. Blades well developed on the upper sheath of some culms in a colony, >5 cm, stemlike; culms and leaves often compressed and helicoid...**J. mexicanus**

18'. Blades 0, or awn-like and <1 cm, not stemlike; culms and leaves usually not much compressed, less frequently helicoid...**J. balticus**

19. Leaf blades absent, reduced to bristles only; capsules blunt-tipped to slightly retuse...**J. drummondii**

19'. Leaf blades present on at least some stems; capsules either acute or strongly retuse...(20)

20. Capsules acute; inflorescence bract exceeding the inflorescence by 2-4 cm...**J. parryi**

20'. Capsules strongly retuse; inflorescence bract reduced and scarious to leaflike but scarcely exceeding the flowering head...**J. hallii**

21. Inflorescence congested, 2 cm or less; tepals usually with brown stripes marginal to the central green stripe (with immature specimens, these stripes can be very light but are often thickened or have a different texture than the central stripe); mature capsules retuse, 2.5-3.5 cm, with 3 chambers, the locular partitions united almost to tip; anthers 0.3-0.5 mm; >7000 ft....**J. confusus**

21'. Inflorescence open (appearing congested only when immature), 1.5-7 cm; tepals more uniform in color; mature capsules obtuse to truncate, 3-4.7 mm, with single chamber, the locular partitions separated except at base; anthers 0.1-1 mm; diverse habitats, from low to high elevations...(22)

22. Capsules chestnut to dark brown; tepals with obtuse apices, shorter than the capsules...**J. compressus**

22'. Capsules tan to light brown; tepals with acute to acuminate apices, slightly shorter to slightly longer than the capsules...(23)

23. Auricles stiff, thick-margined, plasticlike, rounded, shiny, often yellowish; tepals 4-6 mm; anthers 0.6-1 mm...**J. dudleyi**

23'. Auricles not stiff, thin-margined, rounded to acuminate, dull, white or translucent; tepals 3.3-4.4 mm; anthers 0.1-0.6(-1) mm...(24)

24. Auricles generally rounded, thicker and opaque below, thinner and more translucent above, 0.2-0.6 mm; bracteoles acuminate, sometimes bristle-tipped; stems with 2-6 strong ridges per side; anthers 0.4-0.6(-1) mm; plants often with pinkish bases...**J. interior**

24'. Auricles generally acute to acuminate, ± uniformly translucent, generally 1-8 mm until late in season when generally broken or missing; bracteoles generally acute to blunt; stems with or without strong ridges on side; anthers 0.1-0.2 mm; plants rarely with pinkish bases...**J. tenuis**

Juncus acuminatus Michx.—Sharp-fruited rush. Plants perennial, cespitose; culms terete, 15-90 cm; leaf blades terete, septate, to 2.5 mm wide, auricles 0.4-2.5 mm; inflorescence composed of 3-25

(-50) hemispherical to spherical heads; heads 5-30(-50)-flowered; flowers without bracteoles, tepals 3-4 mm, lanceolate, tip acuminate, often reddish; stamens 3(6), anthers much shorter than filaments; capsules 3-4 mm, narrowly ovoid, shorter than to equaling the tepals; seeds 0.3-0.4 mm, not tailed. Stream banks, lakeshores, wet meadows, 5120-5610(-8200) ft. Cibola, Hidalgo, Otero.

Juncus alpinoarticulatus Chaix [*J. alpinus* Vill.]—Northern green rush. Plants perennial, rhizomatous; culms terete, 5-50 cm; leaf blades terete, septate, to 1.1 mm wide, auricles 0.5-1.2 mm; inflorescence composed of 5-25 obconic heads; heads 2-10-flowered; flowers without bracteoles, tepals 1.8-3 mm, lanceolate to oblong, apex obtuse, mucronate; stamens 6, the anthers 1/2 filament length; capsules 2.3-3.5 mm, oblong to ovoid, usually longer than the tepals; seeds 0.5-0.7 mm, not tailed. Wet meadows, marshy lakeshores; known only from Vermejo Park Ranch, 8250 ft. Colfax.

Juncus articulatus L.—Jointed rush. Plants perennial, ± cespitose; culms terete, (5-)10-60 cm; leaf blades terete, septate, to 1.8 mm wide, auricles 0.5-1 mm; inflorescence terminal, composed of 3-20 (-50) obconic to hemispherical heads; heads 3-5(-10)-flowered; flowers without bracteoles, tepals 1.8-3 mm, ovate to lanceolate, tip broadly acute; stamens 6, anthers equal to the filaments; capsules 2.8-4 mm, narrowly ovoid; seeds 0.4-0.5 mm, not tailed. Wetlands, lake and stream margins, ditches, roadside swales, 5085-9515 ft. Otero, Rio Arriba, San Juan.

Juncus balticus Willd.—Baltic rush. Plants perennial, colonial from stout rhizomes; culms terete to slightly compressed, rarely helicoid, 20-100 cm; leaf blades lacking, sheaths mucronate to awned; inflorescence appearing lateral, compact to open with many branches, with 3-100+ flowers; flowers with bracteoles, tepals (2.5-)3.5-5(-6) mm, lanceolate; stamens 6; capsules (3-)4-6 mm, ovoid, longer than the tepals; seeds 0.6-0.8 mm, not tailed. Our plants are **subsp. ater** (Rydb.) Snogerup [*J. arcticus* Willd. var. *montanus* (Engelm.) S. L. Welsh; *J. balticus* Willd. var. *montanus* Engelm.]. Wet habitats, often in alkaline areas, tolerating soils that dry out seasonally, 4000-12,530 ft. Widespread.

Juncus biglumis L.—Two-flowered rush. Plants perennial, loosely cespitose; culms subterete, 2.5-16 cm, exceeding leaves; leaves basal, blades nearly terete, imperfectly septate, to 1.5 mm wide, auricles 0-0.5 mm; inflorescence 1 upright cluster; 1-2(-4)-flowered, bract upright, longer to much longer than the inflorescence; flowers without bracteoles, tepals 2.5-4 mm, oblong, tip obtuse; stamens 6; capsules 4-5.5 mm, narrowly ovoid, apex retuse, apiculate; seeds short-tailed, body 0.7-0.9 mm. Wet soil or gravel in alpine tundra, slopes, stream banks, mossy pond margins; known from 1 collection in Vermejo Park Ranch, 12,530 ft. Taos.

Juncus bufonius L.—Toad rush. Plants annual, single to many-stemmed; culms terete, 2-30 cm; leaf blades flat or involute with face toward culm, to 1.1 mm wide; inflorescence diffuse, at least 1/2 total height of plant, of 1-20 single flowers spaced along the axis; flowers with bracteoles, tepals 2.4-7(-8.5) mm, lanceolate, tip acuminate; stamens (3)6; capsules (2.7-)3.2-4.2 mm, ellipsoid, shorter than outer tepals; seeds 0.3-0.5 mm, not tailed. Receding pond and lake margins, stream banks, moist soil in washes, ditches, and roadsides, usually in open sites, 3900-9130 ft. Widespread except NE, EC, SE.

Juncus castaneus Sm.—Chestnut rush. Plants perennial, culms solitary from long rhizomes, terete to flattened, 9-40 cm, slightly exceeding leaves; leaf blades channeled, imperfectly septate, to 2 mm wide, auricles absent; inflorescence of 1-3(-5) upright clusters; clusters 2-10-flowered, bract basally inflated, upright, longer than the inflorescence; flowers without bracteoles, tepals 4.5-6.6 mm, lanceolate, tip acute to obtuse; stamens 6; capsules 6.5-8.5 mm, narrowly oblong, apex acute to long-tapering; seeds long-tailed, body 0.6-0.7 mm, tails 0.8-1.1 mm. Wet soil in subalpine to alpine tundra and bogs; found only in the Sangre de Cristo Mountains, 10,500-12,540 ft. NC.

Juncus compressus Jacq.—Roundfruit rush. Plants perennial, cespitose; culms slightly compressed, 10-40(-80) cm; leaf blades flat or canaliculate with face toward culm, to 2 mm wide, auricles 0.3-0.5 mm, scarious to membranous; inflorescence of 5-60 flowers in a compact and congested cyme, proximal bract exceeding the inflorescence; flowers with bracteoles, tepals 1.7-2.7 mm, ovate to oblong, tip obtuse; stamens 6; capsules 2.5-3.5 mm, broadly ellipsoid to obovoid, tip obtuse to truncate with mucro; seeds 0.3-0.6 mm, not tailed. Disturbed soils, ditch banks, roadsides; known from 1 collection, 5930 ft. San Juan. Introduced.

Juncus confusus Coville—Colorado rush. Plants perennial, cespitose; culms terete, 30-50 cm; leaf blades flat or involute with face toward culm, to 1.3 mm wide, auricles 0.2-0.7 mm, membranous; inflorescence of 3-15 flowers in a compact and congested cyme, proximal bract exceeding the inflorescence; flowers with bracteoles, tepals 3-4.5 mm, lanceolate, tip acute; stamens 6; capsule 2.5-3.5 mm, subglobose to broadly obovoid, tip strongly retuse; seeds 0.4-0.5 mm, not tailed. Moist, grassy meadows and stream banks, 8900-10,500 ft. NC, NW.

Juncus diffusissimus Buckley—Slimpod rush. Plants perennial, cespitose; culms terete, 25-65 cm; leaf blades terete to compressed, septate, to 2.4 mm wide, auricles 1-2.1 mm; inflorescence of 30-70(-130) hemispherical to obconic heads; heads (1)2-10-flowered; flowers without bracteoles, tepals (1.8-)2.3-3.2 mm, outer longer than inner, lanceolate, tip acute; stamens 3, anthers 1/3-1/2 as long as filaments; capsules 4-5.2 mm, linear-lanceoloid, strongly exserted; seeds 0.3-0.4 mm, not tailed. Marshy shores, sloughs, ditches, in mucky substrates; known from an irrigation ditch near Farmington, 5560 ft. San Juan. Introduced.

Juncus drummondii E. Mey.—Drummond's rush. Plants perennial, cespitose; culms terete to slightly flattened, 8-40 cm; leaf blades lacking, basal sheaths mucronate to awned; inflorescence appearing lateral near the tip, loosely compact, of (1)2-3(-5) flowers; flowers single or several close together, with bracteoles, tepals 4-7(-8) mm, lanceolate, tip acuminate; stamens 6; capsules 4.5-7 mm, narrowly oblong with a blunt to retuse tip, subequal to the tepals; seeds long-tailed, body and tails each 0.5-0.6 mm. Wet and dry meadows, stream banks, talus slopes, ridges in subalpine to alpine habitats, 10,000-13,000 ft. NC.

Juncus dudleyi Wiegand [*J. tenuis* Willd. var. *dudleyi* (Wiegand) F. J. Herm.]—Dudley's rush. Plants perennial, cespitose; culms terete, 20-90 cm; leaf blades flat or involute with face toward culm, to 1 mm wide, auricles 0.2-0.4 mm, yellowish, leathery to cartilaginous with thick rim, basal sheaths rarely pinkish-tinged; inflorescence composed of (5-)10-80 flowers in an open cyme, sometimes compact, proximal bract exceeding the inflorescence; flowers with bracteoles, tepals 4-6 mm, lanceolate, tip acuminate; stamens 6; capsules 2.2-4.2 mm, ellipsoid, tip obtuse to truncate with mucro; seeds 0.4-0.65 mm, not tailed. Moist areas, 4900-9040 ft. Widespread except EC, SE.

Juncus hallii Engelm.—Hall's rush. Plants perennial, cespitose; culms terete, 20-40 cm; leaf blades terete or channeled, to 1 mm wide, auricles to 0.2 mm; inflorescence compact, cymose, with 2-7 flowers, inflorescence bract reduced, scarious to leaflike and longer than the inflorescence; flowers single or several close together, with bracteoles, tepals 4-5 mm, lanceolate, tip acute; stamens 6; capsules 3.5-5 mm, oblong-ovoid, retuse, equal to or longer than the tepals; seeds short-tailed, body 0.5 mm, tails 0.3 mm. Wet and dry meadows, ponds, stream banks, rocky slopes, montane and alpine habitats, 7950-12,500 ft. Colfax, Mora, Rio Arriba.

Juncus interior Wiegand [*J. arizonicus* Wiegand; *J. interior* var. *neomexicanus* (Wiegand) F. J. Herm.]—Interior rush. Plants perennial, cespitose; culms terete, 20-60 cm; leaf blades flat to involute with face toward culm, to 1.1 mm wide, auricles 0.2-0.6 mm, thickened basally, scarious-margined, basal sheaths often pinkish-tinged; inflorescence of (5-)10-30(-50) flowers in an open cyme, sometimes compact, proximal bract exceeding the inflorescence; flowers with bracteoles, tepals 3-3.8(-4.4) mm, lanceolate, tip acute to acuminate; stamens 6; capsules 2.4-4(-4.7) mm, ellipsoid to subglobose, tip obtuse to truncate with mucro, subequal to slightly longer than the tepals; seeds 0.4-0.6 mm, not tailed. Moist areas around stream banks, ditches, springs, and drier upland areas, 4500-10,800 ft. Widespread except EC.

Juncus longistylis Torr.—Longstyle rush. Plants perennial, stems single or few together from long rhizomes; culms somewhat compressed, 20-60(-70) cm; leaf blades flat with face toward culm, to 3 mm wide, auricles 1–2.5 mm; inflorescence of 1–8(–12) obconic heads on 1–2 short primary branches; heads 3–10(–15)-flowered; flowers without bracteoles, tepals 5–6 mm, lanceolate, tip acute to acuminate; stamens 6; capsules 3–6.5 mm, obovoid, apex bluntly rounded to slightly retuse with a mucro, shorter than the tepals; seeds 0.4–0.6 mm, not tailed. Wet mountain meadows, springs, stream banks, 5000–10,800 ft. N, C, WC, SC, SW.

Juncus marginatus Rostk. [*J. biflorus* Elliott; *J. setosus* (Coville) Small]—Grassleaf rush. Plants perennial, cespitose; culms 30–80(–130) cm; leaf blades flat with face toward culm, to 5 mm wide, auricles 0.2–1(–3) mm; inflorescence of 10–50(–200+) small hemispherical heads, usually diffuse; heads (1–)3–9(–20)-flowered; flowers without bracteoles, tepals (1.5–)2–3(–3.2) mm, broadly ovate-lanceolate, tip obtuse to acute; stamens 3; capsules (1.8–)2–2.5(–2.9) mm, obovoid to nearly globose, shorter than to equaling the tepals; seeds 0.3–0.4(–0.6) mm, not tailed. Moist areas, seasonally dry washes, desert scrub to oak woodland; known only from the Peloncillo Mountains, 5000–5600 ft. Hidalgo.

Juncus mertensianus Bong.—Merten's rush. Plants perennial, loosely to densely clumping from short rhizomes; culms terete, 5–35 cm, slightly exceeding leaf length; leaf blades terete, septate, to 1.2 mm wide, auricles 0.2–0.6(–1.2) mm; inflorescence of 1(2) dark hemispherical to subspherical heads; heads 12–60-flowered; flowers without bracteoles, tepals 3.2–6 mm, lanceolate to lance-ovate, tip acute with subulate mucro; stamens 6; capsules 2.2–2.6 mm, obovoid, apex rounded to retuse with mucro; seeds 0.4–0.5 mm, not tailed. Wet soil in alpine meadows, stream banks, springs, 8900–12,000 ft. NC.

Juncus mexicanus Willd. ex Schult. & Schult. [*J. arcticus* Willd. var. *mexicanus* (Willd. ex Schult. & Schult.) Balslev]—Mexican rush. Plants perennial, colonial from stout rhizomes; culms slightly to greatly compressed, often helicoid, 8–80 cm; leaf blades present on some stems, terete to compressed, not septate, auricles 0–1.2 mm; inflorescence appearing lateral, compact to open with many branches, of 3–60+ flowers; flowers 1 to several close together, with bracteoles, tepals 3.5–5(–5.5) mm, lanceolate, tip acuminate; stamens 6; capsules 2.5–4.2(–4.5) mm, ovoid; seeds 0.6–0.8 mm, not tailed. Wet habitats, often in alkaline areas, tolerating soils that dry out seasonally, 3700–9950 ft. NC, NW, C, WC, SC, SW.

Juncus nevadensis S. Watson. [*J. badius* Suksd.]—Nevada rush, Sierra rush. Plants perennial, culms single or few together from long rhizomes; culms terete, to 70 cm, exceeding most leaves; leaf blades terete to flattened, septate, to 2.3 mm wide, auricles 1–2.6(–3.2) mm; inflorescence of 1–4(–12) obconic to nearly globose heads; heads (3–)20–60-flowered; flowers without bracteoles, tepals 2.4–4 mm, broadly lanceolate, tip acute to awned; stamens 6; capsules 2–3.7 mm, subequal to the tepals, broadly ellipsoid, apex abruptly contracted to a minute beak; seeds 0.4–0.5 mm, not tailed. Wet soil, sometimes in standing water, 8000–10,200 ft. NC, NW, Cibola.

Juncus nodosus L.—Knotted rush. Plants perennial, colonial from long rhizomes with occasional small tuberlike segments; culms terete, (4–)15–35(–70) cm; leaf blades terete to channeled above, septate, to 1.6 mm wide, auricles 0.2–1.7 mm; inflorescence of 3–9(–15) ± globose heads; heads (3–)8–30-flowered; flowers without bracteoles, tepals 2.4–3.5(–4.1) mm, lance-subulate, tip acuminate, often reddish-tinged; stamens 3 or 6; capsules 3.2–4.2(–5) mm, lance-subulate, longer than the tepals; seeds 0.4–0.5 mm, not tailed. Stream banks, lakeshores, wet meadows, swamps, ditches, open wetlands, 5640–9100 ft. N, Cibola.

Juncus parryi Engelm.—Parry's rush. Plants perennial, cespitose; culms terete, 5–30 cm; leaf blades terete to channeled below, to 1 mm wide, basal sheaths tan to reddish brown or pinkish, auricles 0.2–0.3 mm; inflorescence appearing lateral in the upper 1/4 of the culm, loosely compact with 1–3(4) flowers; flowers individual or several close together, with bracteoles, tepals 5.5–8 mm, lanceolate, tip acuminate; stamens 6; capsules 6–9 mm, narrowly oblong, tip acute, usually longer than the tepals; seeds tailed, body 0.6 mm, tails 0.4 mm. Wet and dry meadows, talus slopes, ridges, alpine habitats, 10,800–12,600 ft. NC.

Juncus saximontanus A. Nelson [*J. brunnescens* Rydb.]—Rocky Mountain rush. Plants perennial, colonial from rhizomes; culms compressed, 13–70 cm; leaf blades irislike, with edges toward the stem, fused above the auricles, 3-6 mm wide, imperfectly septate, auricles 0-0.6 mm; inflorescence highly variable, from compact with few branches and few (2-5) many-flowered heads, to elongate with many branches and many (35+) few-flowered heads; heads 2-20-flowered, usually obconic to hemispherical but sometimes subspherical; flowers without bracteoles, tepals 2.7-3.6 mm, lanceolate, tip acuminate; stamens 6; capsules 2.4-4.3 mm, oblong, tip rounded to acute with mucro, subequal to the tepals; seeds 0.4-0.5(-1) mm, not tailed. Wet areas, 4000-11,800 ft. Widespread except EC, SE.

Juncus tenuis Willd. [*J. macer* Gray]—Poverty rush. Plants perennial, cespitose; culms terete, (10-)15-55(-80) cm; leaf blades flat to involute with face toward the culm, to 1.7 mm wide, auricles 1.5-6 mm, thin and membranous throughout, delicate and often broken; inflorescence of (5-)15-45 flowers in a compact to usually more open cyme, proximal bract exceeding the inflorescence; flowers with bracteoles, tepals 3.3-4.4 mm, lanceolate, tip acuminate; stamens 6; capsules 3-3.5(-4.7) mm, ellipsoid, tip rounded to obtuse with mucro, shorter than the tepals; seeds 0.3-0.5 mm, not tailed. Moist areas, 7120-8000 ft. Sandoval.

Juncus torreyi Coville [*J. nodosus* L. var. *megacephalus* Torr.]—Torrey's rush. Plants perennial, colonial from long rhizomes with occasional tuberous nodes; culms terete, (3-)15-90(-100) cm; leaf blades terete to channeled above, septate, to 2.2(-4) mm wide, auricles (1-)2-4.2(-5) mm; inflorescence of (1)2-12(-20) globose heads; heads 25-100-flowered; flowers without bracteoles, tepals 4-5 mm, lance-subulate, tip acuminate, tips often reddish; stamens 6; capsules 4-5.1(-5.7) mm, lance-subulate, subequal to tepals; seeds 0.4-0.5 mm, not tailed. Wet areas, tolerates alkaline conditions, 3500-8730 ft. Widespread.

Juncus triglumis L.—Three-flowered rush. Plants perennial, cespitose; culms subterete, 3-20(-30) cm, exceeding leaves; leaf blades nearly terete, not septate, to 1 mm wide, auricles to 1 mm; inflorescence of 1 upright cluster, 2-3(-5)-flowered, bract equal to or slightly shorter than the inflorescence; flowers without bracteoles, tepals 3-5 mm, oblong-lanceolate; stamens 6; capsules (4-)4.5 mm, equal to or longer than the tepals, ellipsoid, tip subobtuse, mucronate; seeds long-tailed, body 0.5-1 mm, tails 0.6-1 mm. Wet gravel soils in alpine tundra, mossy pond margins, bogs, 10,300-12,540 ft. NC. Known only from the Sangre de Cristo Mountains of NC New Mexico.

1. Tepals blackish to pale brown; capsule exserted, usually > 4.5 mm…**var. triglumis**
1'. Tepals pale brown to straw; capsule scarcely exserted, usually 4-4.5 mm…**var. albescens** (Lange) Hultév

LUZULA DC. - Woodrush

Plants grasslike, perennial, cespitose; culms round in cross-section; leaves clustered at base or alternating and reduced up the culm, blades linear to lanceolate, flat with face toward the culm (grasslike), not septate, sparsely to densely ciliate on the margins, sheaths closed, without auricles, cataphylls (bladeless basal sheaths) absent; inflorescence terminal, cymose, diffuse to strongly congested, sometimes with 1 to many racemose or paniculate clusters (glomerules), primary inflorescence bract shorter to longer than the inflorescence, or sometimes reduced to a scale; flowers perfect, with a perianth of 6 tepals, persistent at fruiting, with 1 bract at the pedicel base and 2 additional subentire to fringed bracteoles immediately subtending the tepals; stamens 6; ovary superior, with 1 style and 3 suberect stigma branches; fruit a capsule, globose to ovoid-trigonous, with a beak formed by persistent style base, unilocular, containing 3 seeds; globose to ovoid (Kirschner, 2002c; Swab, 2000).

1. Flowers borne singly or several together on long, slender branch tips in an open, drooping panicle; leaves 3-13 mm wide, sparsely hairy on the lower margins…**L. parviflora**
1'. Flowers borne in congested spikes, each with 5-20 sessile or nearly sessile flowers; leaves 1-6 mm wide, obviously hairy along the lower margins…(2)
2. Spikes tightly clustered into a single irregularly continuous inflorescence, usually nodding at maturity; leaves 1-4 mm wide; plants of subalpine forests to alpine tundra…**L. spicata**
2'. Spikes widely separated on ascending, unequal, stiff peduncles, culm strictly upright below inflorescence; leaves 3-6 mm wide; plants of mountain forests and meadows…**L. comosa**

Luzula comosa E. Mey.—Pacific woodrush. Plants perennial, cespitose; culms terete, 10–40(–50) cm; basal leaf blades flat, to 5 mm wide, with dense hairs at the throat and long, soft marginal hairs, apex obtuse and thickened; inflorescence of 1(–4) subsessile and 2–6 pedunculate cylindrical clusters of flowers, branches stiffly ascending to spreading, of varying lengths; clusters 5–12-flowered; bracteoles clear, distal margins ciliate, tepals (3.3–)3.5–4 mm, lanceolate; stamens 6; capsules to 2.5–3 mm, oblong-ovoid to globose, subequal to the tepals; seeds 1.1–1.3 mm. Our plants belong to **var. laxa** Buchenau. Meadows and coniferous forests, 8000–10,300 ft. NC, Sierra.

Luzula parviflora (Ehrh.) Desv.—Small-flowered woodrush. Plants perennial, cespitose; culms terete, (8–)30–70(–100) cm; leaf blades flat, to 10(–13) mm wide, mostly glabrous with long, soft marginal hairs proximally, apex acute to acuminate, not thickened; inflorescence of 30–140+ small floral clusters on 1–4 primary, drooping branches; clusters 1–4-flowered; bracteole margins entire to lacerate, tepals 1.8–2.7 mm, broadly lanceolate, tip acute; stamens 6; capsules to 2.7 mm, globose, as long as to longer than the tepals; seeds 1.1–1.3 mm. Meadows and forest glades, wooded slopes, 7870–12,040 ft. NC, WC, SC, Lincoln.

Luzula spicata (L.) DC.—Spiked woodrush. Plants perennial, single or several-stemmed to larger clumps; culms terete, (3–)5–27(–33) cm; basal leaf blades flat, to 4 mm wide, mostly glabrous with dense hairs at the throat and sparse hairs along the lower margins, apex blunt to acute, not thickened; inflorescence a condensed spikelike panicle, often nodding at maturity, of 3–12 overlapping floral clusters on short branches; clusters 3–20-flowered; bracteoles clear with ciliate margins that disintegrate into tangled woolly hairs, tepals 2–3 mm, broadly lanceolate, tip acute to acuminate, awned; stamens 6; capsules to 1.3–2 mm, globose, shorter than the tepals; seeds 0.9–1.1 mm. Alpine tundra and scree slopes to subalpine forests, 9840–13,160 ft. NC.

JUNCAGINACEAE – ARROW-GRASS FAMILY

C. Barre Hellquist

TRIGLOCHIN L. – Arrow-grass

Perennial emersed herbs; roots occasionally with tubers, with stout rhizomes; leaves basal, erect, linear, sessile, sheath longer than the blade, ligulate, auriculate; inflorescence a spikelike raceme, scape shorter or longer than leaves; flowers bisexual, short-pedicellate; tepals 6, in 2 series; stamens 4 or 6; pistils 6, 3 fertile, 3 sterile, or 6 fertile; fruit a schizocarp, mericarps 3 or 6.

1. Carpels and stigmas 3; mature fruit 1 mm wide, ca. 5–7 times as long as broad; stolons present, bearing small bulbs…**T. palustris**
1'. Carpels and stigmas 6; mature fruit 2–3 mm wide, ca. 2 times as long as wide; no stolons formed…**T. maritima**

Triglochin maritima L. [*T. concinna* Burtt Davy]—Seaside arrow-grass. Leaves erect from stem, 2.2–11.5 cm; sheath 0.7–2.5 cm; ligules occasionally hoodlike, 2-lobed, apex obtuse to round; inflorescence scape often purple at base, usually exceeding leaves, pedicels 1–4 mm; flower tepals elliptic, apex acute; pistils 6, all fertile; fruit receptacles lacking wings; schizocarps linear to globose, 2–4.5 mm; mericarp linear to linear-obovate, 1.5–3.5 mm, beak 0.2 mm. Marshes, wet alkaline meadows, 4850–8250 ft. NC, NW, C, WC, SE, SC.

Triglochin palustris L.—Marsh arrow-grass. Leaves erect from sheath, 5–24.5 cm; sheath 3.5–5 cm; ligule not hooded, unlobed; apex acute; inflorescence scape often purple at base, usually exceeding leaves, pedicel 0.4–4.5 mm; flower tepals elliptic, apex round; pistils 6, 3 fertile, 3 sterile; fruit a receptacle with wings; schizocarp linear, 7–8.3 mm; mericarps linear, 6.5–8.5 mm; beak 0.3 mm. Mountain marshes, wet alkaline meadows, 4875–11,425 ft. NC, Otero.

KOEBERLINIACEAE – CRUCIFIXION-THORN FAMILY

Kenneth D. Heil

KOEBERLINIA Zucc. – Crucifixion-thorn

Koeberlinia spinosa Zucc.—Crown of thorns. Plants 0.5–10 m, shrubs or trees, deciduous, thorny, glabrous or glabrate, spreading, globose, compact to somewhat open; stems ± open, terete, green twigs thorn-tipped, 25–70 cm; leaves 0.2–1.5 × 0.5 mm; inflorescences 3–15 mm; pedicels 3–6 mm; flowers 1–2 mm; sepals 4, distinct, greenish white, glabrous; petals 4, greenish white or cream, 3–4.8 × 0.8–1.4 mm, ovary ovoid, 1–1.2 mm; berry ca. 5 mm; seeds 3–3.5 mm. Our material belongs to **var. wivaggii** W. C. Holmes, K. L. Yip & Rushing. Chihuahuan desert scrub, 3100–6800 ft. SE, SC, SW, Socorro.

KRAMERIACEAE – RATANY FAMILY

Kenneth D. Heil

KRAMERIA Loefl. – Ratany

Shrubs and perennial barbs; leaves alternate, simple or trifoliate, sessile, linear to linear-lanceolate, margins entire; inflorescences uniflorous in leaf axils, terminal racemes, or open panicles; flowers zygomorphic, perfect, sepals 4 or 5, showy, petals 4 or 5, small, not showy, stamens 3 or 4, ovary superior; fruit a 1-seeded hard capsule, bearing spines, with retrorse barbs or glochids along the terminal end of the spines; seeds globose, gray-brown, smooth (Simpson & Salywon, 1999).

1. Perennial sprawling herbs; flowers secund; filaments connate 1–2.5 mm beyond the point of insertion; fruit spines appearing without barbs…**K. lanceolata**
1'. Shrubs with upright stems; flowers not secund; filaments distinct beyond the point of insertion; fruit spines mostly with barbs…(2)
2. Petaloid petals connate basally for 1–3 mm of their length, their blades triangular; fruit spines most often with a few to several retrorse barbs near but not at the tip; old stems gray…**K. erecta**
2'. Petaloid petals not connate, their blades oblanceolate; fruit spines with a whorl of recurved barbs at the tip; old stems blue-green…**K. bicolor**

Krameria bicolor S. Watson [*K. grayi* Rose & J. H. Painter]—White ratany. Shrubs, angular, mound-forming, 0.2–1.5 m, young branches canescent, old stems blue-green with spinose tips; leaves mostly sparse, 4–20 × 1–5 mm; flowers solitary, to 25 mm, sepals 5, reflexed, purple or deep red pink, petaloid petals 3, pink or purple at the tip, green basally, stamens 4, ovary 3–4 mm; fruits circular, 5.5–10 mm wide excluding spines, with white or brown trichomes, individual spines mostly 2.5–5 mm, bearing a whorl of amber-colored recurved barbs. Rocky slopes, Chihuahuan desert scrub, oak, juniper, piñon communities, 3500–4500 ft. SW, Eddy.

Krameria erecta Willd. [*K. parvifolia* Benth. var. *imparata* J. F. Macbr.]—Littleleaf ratany. Much-branched shrub, 1–2 m, occasionally bearing glandular trichomes, old stems gray; leaves borne on long and short shoots, mostly 0–12 × 0.5–2 mm, strigose; flowers solitary in the leaf axils, sepals 8–10 mm, pink to light purple, petaloid petals 3, 4–6 mm, stamens 4, ovary ca. 3 mm; fruit cordate in outline, ca. 6 mm wide excluding the spines, the spines 2–5 mm, usually with a few to several retrorse barbs near but not at the tip. Chihuahuan desert scrub, oak savanna, chaparral, 3100–6020 ft. S except Lea.

Krameria lanceolata Torr.—Trailing krameria. Sprawling, decumbent, perennial herbs with branches radiating from a central woody underground stem, the branches to 1 m, tomentose to sparsely strigose; leaves 5–25 × 0.9–4 mm; flowers borne in secund racemes, sepals 5, pink, apricot-colored, or reddish, stamens 4, ovary 3–4 mm; fruits globose in outline, 5.5–8 mm excluding the spines, the spines stout, scattered, yellow, 1.8–5.3 mm, roughened at the distal ends. Grassy areas, often with oak and pine, 3250–7500 ft. Widespread except NC, NW, WC.

LAMIACEAE – MINT FAMILY

Jennifer Ackerfield

Herbs or shrubs, the stems usually square; leaves opposite or whorled, simple or compound, exstipulate; flowers usually perfect; calyx usually 5-lobed, actinomorphic or zygomorphic, usually tubular; corolla 4–5-lobed, bilabiate, unequal but not bilabiate, or nearly actinomorphic; stamens 2 or 4, epipetalous; pistil 1; ovary superior, 2-carpellate but appearing 4-carpellate because of false septa; style usually gynobasic; fruit of 4 nutlets (Ackerfield, 2015; Epling, 1942; Harley & Paton, 1999; Irving, 1980; Sammons, 2011; Sanders, 1987).

1. Shrubs with stems woody throughout…(2)
1'. Herbaceous plants or stems woody only at the base…(4)
2. Leaves palmately compound…**Vitex**
2'. Leaves simple…(3)
3. Leaves narrow, mostly linear; stems and leaves of the current season densely soft-canescent; sepals densely white-pubescent, purple…**Poliomintha**
3'. Leaves broader than linear; plants otherwise not as above in all respects…**Salvia** (in part)
4. Inflorescence a dense terminal head subtended by broad bracts…(5)
4'. Inflorescence not a dense terminal head subtended by broad bracts, or if the inflorescence subtended by broad bracts, then the inflorescence longer than wide…(6)
5. Corolla obscurely bilabiate, with 5 slender, linear to narrowly oblong lobes; stamens 4; leaves 1–3.5 cm… **Monardella**
5'. Corolla conspicuously bilabiate; stamens 2; leaves 2–10 cm…**Monarda**
6. Corolla actinomorphic or nearly so…(7)
6'. Corolla zygomorphic, bilabiate or appearing unilabiate…(10)
7. Stamens 2; flowers and fruit sessile…**Lycopus**
7'. Stamens 4; flowers and fruit pedicellate…(8)
8. Plants glandular-puberulent; flowers solitary or few in leaf axils…**Trichostema**
8'. Plants glabrous to villous, not glandular-puberulent; flowers not as above, flowers in loose clusters or interrupted whorls, sometimes terminal and spikelike…(9)
9. Inflorescence of several whorls at the tips of the stems, these sometimes in axils of well-developed leaves; corolla 4-lobed with one lobe larger than the others; plants strongly aromatic…**Mentha**
9'. Inflorescence of loose flowers in the axils; corolla of 5 ± equal lobes; plants hardly aromatic…**Tetraclea** (*T. brachiatum*)
10. Calyx with an erect, prominent, shieldlike appendage 2–4 mm long on the upper side, inconspicuously nerved; corolla blue to purple or rarely white, the tube straight or more often sigmoidally curved, the upper lip with a rounded and helmetlike central lobe…**Scutellaria**
10'. Calyx without an appendage on the upper side, conspicuously nerved or not; corolla various…(11)
11. Corolla appearing unilabiate, the upper lip deeply cleft and appearing to arise laterally from the margins of the 3-lobed lower lip…**Teucrium**
11'. Corolla conspicuously bilabiate…(12)
12. Stamens 2…(13)
12'. Stamens 4…(14)
13. Anther sacs separated by an elongate connective joined to the filament; stems ridged; calyx glabrous within, not gibbous at the base, the teeth without long, slender bristles at the tips…**Salvia** (in part)
13'. Anther sacs not separated; stems not ridged; calyx with a ring of hairs within the throat, gibbous at the base, the teeth with long, slender bristles at the tips…**Hedeoma**
14. Leaves 3-cleft or palmately cleft or parted; corolla whitish to pink with dark purple spots, the upper lip pubescent on the back; flowers in dense clusters in leaf axils…**Leonurus**
14'. Leaves entire or merely toothed; plants otherwise unlike the above in all respects…(15)
15. Calyx with 10 subulate lobes, the lobes rigid and hooked at the tip; plants white-woolly tomentose throughout…**Marrubium**
15'. Calyx unlike the above; plants glabrous or variously pubescent…(16)
16. Leaves on the upper part of the stem and calyx lobes tipped with a long, slender bristle; corolla barely surpassing the calyx…**Dracocephalum**
16'. Calyx lobes not tipped with long, slender bristles, or if these present then the upper leaves lacking bristle tips; corolla various…(17)
17. Stamens arching above the corolla…**Trichostema**
17'. Stamens not arching above the corolla…(18)
18. Calyx 15-nerved…(19)

18'.   Calyx with fewer than 15 nerves...(20)
19.   Anther sacs parallel; stamens (at least 2) conspicuously exserted from the corolla...**Agastache**
19'.   Anther sacs divaricate; stamens included or scarcely exserted from the corolla...**Nepeta**
20.   Flowers subtended by suborbicular to broadly ovate bracts with long-acuminate tips and ciliate margins; calyx usually purple- or green-tinged with purple on the margins...**Prunella**
20'.   Flowers not subtended by suborbicular bracts; calyx various...(21)
21.   Calyx distinctly bilabiate, 13-nerved...**Clinopodium**
21'.   Calyx actinomorphic, 5-10-nerved...(22)
22.   All or most flowers borne in axillary whorls...**Lamium**
22'.   All or most flowers borne in terminal inflorescences...(23)
23.   Stems glabrous; anther sacs parallel...**Physostegia**
23'.   Stems variously pubescent; anther sacs divaricate...**Stachys**

AGASTACHE J. Clayton ex Gronov. – Giant hyssop

Perennial herbs, leaves simple; inflorescence dense, terminal, spikelike; calyx with 15+ prominent veins, tubular or campanulate, the teeth and upper tube often whitish and scarious or pink or blue; corolla pink to purple or white, bilabiate; stamens 4.

1.   Calyx tube 1.5-3 mm...(2)
1'.   Calyx tube > 3 mm...(3)
2.   Corolla white; inflorescence continuous, the verticils not distinct and widely spaced...**A. micrantha**
2'.   Corolla violet-purple to pale blue; inflorescence interrupted with distinct, widely spaced verticils...**A. wrightii**
3.   Stems with exfoliating bark at the base; corolla tubes 20-30 mm...(4)
3'.   Stems lacking exfoliating bark at the base; corolla tubes 6-18 mm...(5)
4.   Leaves linear to linear-lanceolate, densely tomentose below, with entire or sparsely dentate margins...**A. rupestris**
4'.   Leaves lanceolate to deltoid-ovate, densely puberulent below, with entire to crenate-dentate margins...**A. cana**
5.   Calyx tube 0.5-1.5 mm diam. (sometimes to 2 mm when pressed) at the mouth at anthesis, in fruit the base swelling to a larger diam. than the mouth...(6)
5'.   Calyx tube usually 2-4 mm diam. at the mouth at anthesis, in fruit the base not swelling to a larger diam. than the mouth...(7)
6.   Inflorescence typically interrupted; leaf blades (middle and distal) triangular-lanceolate, 1.5-2.5 times longer than wide; calyx tube usually arching, the veins bowed or curved; upper calyx teeth 3-6 times longer than wide, the secondary veins fusing with primary veins; Organ Mountains...**A. pringlei**
6'.   Inflorescence typically continuous or the longer ones interrupted in the lower 1/2; leaf blades all deltate-ovate to broadly ovate, 1-1.6 times longer than wide; calyx tube rigidly straight, the veins straight; upper calyx teeth 1.5-3.5 times longer than wide, the secondary veins ending free; bootheel...**A. breviflora**
7.   Calyx purplish-tinged with long-subulate teeth; corolla rose-purple, the tube 12-19 mm...**A. mearnsii**
7'.   Calyx lacking long-subulate teeth, or if these present then the calyx green; corolla white to rose-purple, the tube 9-15 mm...**A. pallidiflora**

Agastache breviflora (A. Gray) Epling [*Brittonastrum breviflorum* (A. Gray) Briq.]—Trans-Pecos giant hyssop. Stems 3-8 dm; leaves 1-4 cm, deltate-ovate to broadly ovate, thinly hirtellous to short-pilose, margins crenate-serrate; inflorescence dense, continuous or the longer ones interrupted in the lower 1/2; calyx 4.5-8.5 mm; corolla tube rose to purple, 5-10 mm. Bootheel; riparian sites, moist slopes, pine-oak, 5400-5500 ft. Hidalgo.

Agastache cana (Hook.) Wooton & Standl. [*Cedronella cana* Hook.]—Mosquito plant. Stems 2-7 dm; leaves 0.8-1.2 cm, lanceolate to deltoid-ovate, minutely puberulent, the margins entire to crenate-dentate; inflorescence mostly continuous, with 4-10 verticillasters; calyx 8-11 mm, purplish-tinged; corolla tube 20-25 mm, rose or pink. Granite cliffs and crevices in canyons of C and S mountains, 5350-8750 ft. Doña Ana, Grant, Luna, Sierra. **R**

Agastache mearnsii Wooton & Standl.—San Luis Mountain giant hyssop. Stems 4-10 dm; leaves 1.5-5 cm, deltoid-ovate, puberulent, the margins crenate-dentate; inflorescence continuous, with mostly 4-10 verticillasters; calyx 8-12 mm, purplish-tinged; corolla tube 12-19 mm, pinkish purple. Forested montane slopes, ponderosa pine-oak communities, 5900-8500 ft. Hidalgo. **R**

Agastache micrantha (A. Gray) Wooton & Standl. [*Cedronella micrantha* A. Gray]—White giant hyssop. Stems 3–10 dm; leaves 2–6 cm, cordate-ovate to ovate-lanceolate or lanceolate, puberulent, with apparent glands, the margins crenate-dentate; inflorescence usually continuous, with mostly 15–30 verticillasters; calyx 2.5–4 mm, greenish; corolla tube 2.5–3.5 mm, scarcely if at all exceeding the calyx, white. SW mountains and foothills, riparian areas, granitic outcrops, piñon-juniper, ponderosa pine–oak, mixed conifer communities, 5200–8500 ft. WC, SC, SW, San Miguel.

Agastache pallidiflora (A. Heller) Rydb.—Bill Williams Mountain giant hyssop. Stems 3–10 dm; leaves 1–7 cm, deltoid-ovate to deltoid, puberulent, with obscure glands, the margins crenate-dentate; inflorescence usually continuous, with mostly 4–15 verticillasters; calyx 6.5–13 mm, pinkish-tinged or greenish; corolla tube 8.5–15 mm, white, pinkish purple, bluish purple, or rose-purple. Common in meadows and forest openings throughout W and C mountains, 7000–10,000 ft.

1. Bracts in inflorescence (4–)7–17 mm, conspicuous and often protruding among calyces; upper calyx teeth not or slightly overtopping lower calyx teeth by ca. 0.3 mm…(2)
1'. Bracts in inflorescence 2–8 mm, inconspicuous and hidden in mature inflorescences; upper calyx teeth overtopping lower teeth by 0.5-1 mm…(4)
2. Calyx rose-purple at anthesis over most of its length (rarely green); upper calyx teeth 0.9–1.5 times longer than wide; corollas rose-purple or rarely white, (10–)12–17 mm…**var. gilensis**
2'. Calyx green or tinged purple in the upper 1/2 at anthesis; upper calyx teeth 1.5–3.5 times longer than wide; corolla white to light pinkish purple…(3)
3. Upper calyx teeth mostly 2–3.5 times longer than wide, their bases strongly connate; corolla white or occasionally tinged with pink…**var. pallidiflora**
3'. Upper calyx teeth mostly 1.5–2 times longer than wide, their bases weakly connate; corolla light pinkish purple or occasionally white…**var. greenei**
4. Leaves mostly (1.2–)1.4–1.8 times longer than wide; corolla rose or pink-purple, rarely white…**var. neomexicana**
4'. Leaves mostly 0.9–1.4 times longer than wide; corolla dull bluish purple…**var. havardii**

var. gilensis R. W. Sanders—Pine-oak, mixed conifer communities, 6050–10,700 ft. Gila National Forest, WC.

var. greenei (Briq.) R. W. Sanders—Pine-oak, mixed conifer communities, 7500–9200 ft. Chuska Mountains, NW.

var. havardii (A. Gray) R. W. Sanders—Mixed conifer communities. Single record, Guadalupe Mountains, 6900 ft. SE.

var. neomexicana (Briq.) R. W. Sanders—Pine-oak, mixed conifer, spruce-fir communities, 5685–11,400 ft. W to C New Mexico, NC, C, WC, SC, SW.

var. pallidiflora—Pine-oak, mixed conifer, spruce-fir communities, 6050–10,000 ft. W New Mexico, C, WC.

Agastache pringlei (Briq.) Lint & Epling [*A. verticillata* Wooton & Standl.]—Pringle's giant hyssop. Stems 2–8 dm; leaves 0.5–6 cm, deltoid-ovate or ovate-lanceolate, puberulent, with apparent glands, margins crenate to crenate-dentate; inflorescence usually interrupted, with mostly 6–18 verticillasters; calyx 5–6 mm, purplish-tinged; corolla tube 6.5–8 mm, pale pink to dull rose-purple. Our plants are **var. verticillata** (Wooton & Standl.) R. W. Sanders. Oak and pine woodlands in the Organ Mountains, 5000–7000 ft. Doña Ana. **R & E**

Agastache rupestris (Greene) Standl. [*Cedronella rupestris* Greene]—Threadleaf giant hyssop. Stems 5–10 dm; leaves 2–5 cm, linear to linear-lanceolate, densely tomentose below, with apparent glands, the margins entire to sparsely dentate; inflorescence usually continuous, with mostly 6–15 verticillasters; calyx 6–8 mm, purplish-tinged; corolla tube 20–30 mm, rose-purple to dark pink. Granitic foothills, 4500–7000 ft. Catron, Sierra.

Agastache wrightii (Greenm.) Wooton & Standl. [*Cedronella wrightii* Greenm.]—Sonoran giant hyssop. Stems 4–10 dm; leaves 1–5 cm, deltoid-ovate, puberulent, with obscure glands, the margins crenate-dentate; inflorescence usually interrupted, with mostly 10–25 verticillasters; calyx 3.5–5.5 mm, purplish-tinged; corolla tube 3.5–5 mm, violet-purple to pale blue. Igneous substrates, rocky slopes, canyons, ponderosa pine–oak, 4850–8100 ft. Catron, Grant, Hidalgo.

CLINOPODIUM L. – Calamint

Perennial herbs; leaves simple, entire or few-toothed; inflorescence of cymules or an interrupted spike; calyx 10–13-nerved, weakly zygomorphic to bilabiate; corolla bilabiate, the upper lip entire, emarginate, or 4-toothed, the lower lip 3-lobed; stamens 4, didynamous.

1. Flowers numerous, in dense, headlike clusters; leaves ovate to deltoid-ovate…**C. vulgare**
1'. Flowers few (<10), in loose clusters; leaves linear to linear-oblanceolate…**C. arkansanum**

Clinopodium arkansanum (Nutt.) House [*Satureja arkansana* (Nutt.) Briq.]—Limestone calamint. Stems 1–4 dm; inflorescence of axillary and terminal few-flowered cymules; leaves linear to linear-oblanceolate, to 25 × 5 mm wide, entire or nearly so; calyx to 3 mm, glabrous, sharply toothed; corolla bluish to purple, 8–12 mm. Seeps and springs, Sacramento Mountains, 7800–8000 ft. Otero.

Clinopodium vulgare L. [*Satureja vulgaris* (L.) Fritsch]—Wild basil. Stems 1–5 dm; inflorescence of axillary or terminal, dense cymules in headlike clusters; leaves ovate to deltoid-ovate, 10–40 × 6–15 mm, entire to minutely toothed; calyx 8–9 mm, villous, with subulate teeth; corolla white to purple or rose, 8–12 mm. Moist, disturbed places, ponderosa pine–oak, mixed conifer communities, 7000–8900 ft. NC, C, WC, SC.

DRACOCEPHALUM L. – Dragonhead

Dracocephalum parviflorum Nutt. [*Moldavica parviflora* (Nutt.) Britton]—American dragonhead. Annuals or biennials, 3–8 dm; leaves lanceolate to elliptic, 3–10 × 0.7–3(–5) cm, glabrous above, glabrous to pubescent below, the margins sharply spinose-serrate; inflorescence of dense clusters, bracts spinulose-pectinate; calyx bilabiate, 5-toothed, 8–15 mm with lobes 3–6 mm; corolla blue to purple or sometimes white, bilabiate, just longer than the calyx; stamens 4; nutlets 2–3 mm. Along roadsides, in disturbed areas, common after forest fire, often in moist places, 4600–11,370 ft. Widespread except E, SW.

HEDEOMA Pers. – False pennyroyal

Annual or perennial, aromatic herbs; leaves simple, entire or toothed; flowers borne in leaf axils; calyx bilabiate, the lower 2 teeth longer than the upper 3, 13-nerved, gibbous at the base; corolla bilabiate; stamens 2.

1. Corolla orange-red or rarely yellowish, ca. 30 mm…**H. todsenii**
1'. Corolla white, pink, blue, or lavender, to 20 mm…(2)
2. Plants mat-forming; corolla pink, 19–20 mm; leaves thick and coriaceous, apiculate at the apex…**H. apiculata**
2'. Plants not mat-forming; corolla white, blue, or lavender, 7–20 mm; leaves unlike the above…(3)
3. Calyx teeth convergent at maturity, closing the orifice or nearly so…(4)
3'. Calyx teeth not convergent at maturity, the upper ones at least spreading to reflexed…(6)
4. Calyx tube gibbous but not saccate, not tapered upwardly; calyx teeth only slightly convergent…**H. pulcherrima** (in part)
4'. Calyx tube conspicuously saccate and tapered into a narrow neck; calyx teeth conspicuously convergent…(5)
5. Suffrutescent perennials, smelling strongly of lemon or camphor; corolla white or lavender; leaves gray or dark green…**H. reverchonii**
5'. Perennials but not suffrutescent, smelling of mint; corolla blue; leaves bright green…**H. drummondii**
6. Leaf margins conspicuously toothed or just the basal leaves entire…(7)
6'. Leaf margins entire or nearly so…(10)
7. Underside of leaves with indistinct veins; plants robust, 2–5.5 dm, usually with unbranched shoots…(8)
7'. Underside of leaves with conspicuously elevated veins; plants tufted perennials, 1–4 dm, usually with freely branching shoots…(9)
8. Basal leaves ovate, glabrous, with entire margins…**H. oblongifolia**

8'. Basal leaves elliptic to ovate or rhombic, densely pubescent, with dentate margins…**H. dentata**
9.   Corolla well exserted from the calyx, 10-20 mm; leaves ovate to elliptic…**H. costata**
9'. Corolla 7-10 mm; leaves elliptic to rhombic…**H. plicata**
10. Leaves linear to narrowly lanceolate…**H. hyssopifolia**
10'. Leaves elliptic to ovate, oval, or oblong…(11)
11. Corolla 6-9 mm…**H. nana**
11'. Corolla 10-14 mm…**H. pulcherrima** (in part)

Hedeoma apiculata W. S. Stewart—McKittrick's false pennyroyal. Perennials, suffrutescent, 1-1.5 dm; stems densely covered with minute, retrorsely curling hairs; leaves lanceolate-elliptic to ovate, sparsely hispid, coriaceous, 7-15 mm, entire, sessile; calyx chartaceous, 8.5-9 mm; corolla pink, 19-20 mm, the tube ca. 10 mm, upper lip ca. 5 mm; nutlets 1.5-2 mm. Limestone walls and crevices, Guadalupe Mountains, 5120-8400 ft. Eddy. **R**

Hedeoma costata A. Gray—Ribbed false pennyroyal. Perennials, 1-1.5 dm; stems canescent with retrorsely curling hairs; leaves ovate to elliptic, puberulent to villous, 5-10 mm, coarsely serrate, the nerves distinct but not conspicuously elevated, shortly petiolate; calyx 6.5-8 mm; corolla blue and well exserted from the calyx, 10-20 mm, the tube 8.5-18.5 mm, upper lip ca. 1.5 mm; nutlets ca. 1.2 mm. Ours are **var. pulchella** (Greene) R. S. Irving. Dry, calcareous hills in the S and SE mountains, 3400-8460 ft. SE, SC, Sierra.

Hedeoma dentata Torr.—Dentate false pennyroyal. Perennials, 2.5-4 dm; stems densely hirtellous with spreading or retrorsely curling hairs; leaves ovate to elliptic or rhombic, strigulose or hirtellous, 8-17 mm, coarsely serrate, subsessile; calyx 4.8-5.5 mm; corolla lavender, 8-9 mm, the tube 6.5-7.5 mm, upper lip ca. 1.5 mm; nutlets ca. 1 mm. Juniper-oak woodlands and chaparral in the SE desert, 5000-7200 ft. SC, SW.

Hedeoma drummondii Benth.—Drummond's false pennyroyal. Perennials, 0.5-3 dm; leaves linear to narrowly elliptic, 5-20 mm, punctate, glabrous above and densely pubescent below, entire, subsessile to shortly petiolate; calyx 5-7.5 mm, hirsute, teeth subulate to triangular, 1-2 mm; corolla blue, 7-9 mm; nutlets 1.3-1.6 mm. Dry hills, plains, common throughout the state, 3000-10,500 ft. Widespread.

Hedeoma hyssopifolia A. Gray—Aromatic false pennyroyal. Perennials, 1.5-5 dm; stems puberulent with retrorsely curling hairs; leaves narrowly lanceolate or linear, glabrate, 8-18.5 mm, entire, sessile; calyx 6-8.3 mm; corolla lavender, 11-16 mm, upper lip 2.2-5 mm; nutlets ca. 1.3 mm. Rocky slopes, canyons, understory in pine forests, 5300-8800 ft. WC, SC, SW.

Hedeoma nana (Torr.) Briq.—Dwarf false pennyroyal. Annuals or short-lived perennials, 0.5-3.5 dm; leaves ovate to oval, 5.5-13 mm, glabrous to sparsely pubescent above, hirtellous below, entire or occasionally obscurely crenate, petiolate; calyx 4-5.5 mm; corolla lavender, 6-9 mm; nutlets ca. 1.1 mm. Ours are **subsp. nana**. Dry, rocky slopes of the S mountains, 3000-8700 ft. Widespread except NE, C, EC, Lea, Roosevelt.

Hedeoma oblongifolia (A. Gray) A. Heller—Oblongleaf false pennyroyal. Perennials, 2-5.5 dm; leaves narrowly elliptic to ovate-elliptic, 8.5-14 mm, glabrous to sparsely pubescent above, hirtellous below, denticulate, shortly petiolate; calyx 5.5-7 mm; corolla lavender, 10-11 mm; nutlets ca. 1 mm. Dry slopes in oak woodlands and chaparral, 4000-9200 ft. NW, C, WC, SC, SW, Chaves, Eddy.

Hedeoma plicata Torr.—Veiny false pennyroyal. Perennials, 1.5-4 dm; leaves elliptic to rhombic, 5.5-10 mm, glabrous to sparsely pubescent above, hirtellous below, coarsely serrate, the nerves conspicuously elevated, shortly petiolate; calyx 4.5-6 mm; corolla blue, 7-10 mm; nutlets ca. 1 mm. Pine-oak woodlands in the S mountains, 4000-7500 ft. Doña Ana, Eddy, Lincoln, Otero.

Hedeoma pulcherrima Wooton & Standl.—White Mountain false pennyroyal. Perennials, 1.5-4 dm; leaves elliptic to oblong, 8-18 mm, glabrous above, hirtellous below, entire to obscurely serrate, subsessile to shortly petiolate; calyx 6-7 mm; corolla lavender, 10-14 mm; nutlets ca. 1.2 mm. Endemic to pine and spruce-fir forests in the White Mountains, 5000-9000 ft. Lincoln, Otero. **R & E**

Hedeoma reverchonii A. Gray—Reverchon's false pennyroyal. Perennials, suffrutescent, 1.5–4 dm; leaves elliptic-oblong, 6–10 mm, glabrous to sparsely pubescent above, hirtellous below, entire, subsessile to shortly petiolate; calyx 5–6 mm; corolla white or lavender, 8–10 mm; nutlets ca. 1.2 mm. Ours are **var. serpyllifolia** (Small) R. S. Irving. Calcareous outcrops and roadcuts; known from the Guadalupe Mountains, 4100–8700 ft. Chaves, Eddy, Otero.

Hedeoma todsenii R. S. Irving—Todsen's false pennyroyal. Perennials, suffrutescent, 1–2 dm; leaves lanceolate-elliptic, coriaceous, 8–14 mm, glandular-punctate below, shortly hispid on the margins and midrib, entire, subsessile; calyx 11–12 mm; corolla orange-red, rarely yellowish, ca. 30 mm; nutlets ca. 2 mm. Dry limestone slopes of the S mountains, 6200–7400 ft. Doña Ana, Otero, Sierra. **FE & E**

LAMIUM L. – Deadnettle

Annual or perennial herbs; leaves usually toothed; flowers whorled in axillary or terminal clusters; calyx actinomorphic, 5-toothed, the teeth all equal or the upper one larger; corolla purple, white, or rarely yellow, bilabiate; stamens 4.

1. Bracts and upper leaves sessile and clasping, rarely shortly petiolate, broad-based…**L. amplexicaule**
1'. Bracts and upper leaves petiolate…**L. purpureum**

Lamium amplexicaule L.—Henbit. Annuals, 1–3.5 dm; leaves orbicular to ovate, 5–16 mm, crenate, the upper leaves usually sessile and clasping; calyx actinomorphic, 5–7 mm, densely hairy; corolla bilabiate, pink-purple, 10–20 mm, tube 10–15 mm, upper lip 3–5 mm, lower lip 1.5–2.5 mm; stamens 4; nutlets 1.5–2 mm. Disturbed places, fields, 4500–6500 ft. NC, NW, C, WC, SE, SC, SW. Introduced.

Lamium purpureum L.—Purple deadnettle. Annuals, 1–4 dm; leaves mostly cordate, 1–3 cm, petiolate, those among the flowers purplish; calyx 5–6 mm, hairy on the lobes; corollas 10–15 mm, pink-purple, hairy outside and inside below the filaments, the upper lip with whitish hairs. Moist, weedy sites; 1 record, 6100 ft. Taos. Introduced.

LEONURUS L. – Motherwort

Leonurus cardiaca L.—Motherwort. Perennials, 5–20 dm, glabrous to sparsely hairy; leaves ovate to suborbicular, 3–12 cm, the larger leaves 3–5-cleft with serrate margins; inflorescence of dense, leafy-bracteate, axillary clusters; calyx 3.5–8 mm, 5-nerved, the teeth with spinulose tips; corolla bilabiate, pink or white, 8–12 mm; stamens 4; nutlets 1.5–2.3 mm. Ours are **subsp. cardiaca**. Moist, disturbed places, 5000–8500 ft. C, WC. Introduced.

LYCOPUS L. – Water horehound, bugleweed

Perennial herbs, not much aromatic; leaves simple, toothed or pinnatifid; inflorescence of dense sessile clusters in leaf axils; calyx actinomorphic, 4–5-toothed, campanulate to ovoid, 4–5-nerved; corolla nearly actinomorphic, usually white, 4–5-lobed; stamens 2.

1. At least the lower leaves with pinnatifid margins, the others irregularly sharply serrate…**L. americanus**
1'. Leaf margins merely serrate…**L. asper**

Lycopus americanus Muhl. ex W. P. C. Barton—American water horehound. Plants 3–9 dm; leaves narrowly elliptic to ovate-lanceolate, 1.5–12 cm, irregularly sharply serrate, the lower leaves pinnatifid; calyx 2–3.3 mm; corolla 4-lobed, 2.5–3.5 mm, just longer than the calyx; nutlets 1–1.5 mm. Moist soil, sometimes in standing water, 4500–7000 ft. N, C, WC, SW, except Rio Arriba.

Lycopus asper Greene—Rough bugleweed. Plants 3–13 dm; leaves narrowly elliptic, 5–8 cm, evenly serrate; calyx 2.5–4.5 mm; corolla 4-lobed, 3.5–6 mm, 1–2 mm longer than the calyx; nutlets 1.7–2.5 mm. Moist places, 4500–7500 ft. NC, NW, C, WC.

MARRUBIUM L. – Horehound

Marrubium vulgare L.—Common horehound. Perennials, 3–7 dm; leaves ovate to orbicular, 1.5–5 × 1–4 cm, crenate-serrate, densely woolly; inflorescence of dense axillary clusters; calyx tube 3–5 mm, the teeth alternately long and shorter with hooked apices; corolla bilabiate, 4–6 mm, white; stamens 4; nutlets 1.8–2.3 mm, smooth. Disturbed places, roadsides, old pastures, 4500–7000 ft. Widespread except EC, Lea. Introduced.

MENTHA L. – Mint

Perennial herbs, very aromatic; leaves simple and toothed, punctate; inflorescence of dense axillary clusters or terminal spikes; calyx actinomorphic or obscurely bilabiate, 5-toothed, 10-nerved; corolla 4-lobed, nearly actinomorphic; stamens 4, exserted, equal.

1. Flowers in dense axillary clusters at the nodes…**M. arvensis**
1′. Flowers in terminal spikelike clusters…**M. spicata**

Mentha arvensis L.—Wild mint. Plants 3–5(–9) dm; leaves lanceolate to ovate, 2.5–12 cm, serrate; inflorescence of dense axillary clusters at the nodes; calyx 2.5–3.3 mm; corolla 4.5–6.5 mm, white to lavender, glabrous; nutlets 0.7–1.5 mm, smooth. Moist places, along streams and ditches, 4500–9000 ft. Widespread except EC, SE, SW.

Mentha spicata L.—Spearmint. Plants 3–7 dm; leaves lance-ovate to ovate, 3–9 cm, serrate; inflorescence of terminal spikelike clusters; calyx 1–3 mm; corolla 1.5–3 mm, white to purple. Moist places, along streams and ditches, 4000–6500 ft. NW, CE, C, SC, SW. Introduced.

MONARDA L. – Beebalm

Annual or perennial herbs or shrubs; leaves simple, subentire or usually toothed; inflorescence of dense, headlike clusters; calyx 13–15-nerved, 5-toothed, the teeth nearly equal; corolla white, pink, or purple, strongly bilabiate; stamens 2.

1. Flowers in terminal, solitary, headlike clusters; corollas rose-purple…**M. fistulosa**
1′. Flowers in axillary clusters in addition to a terminal cluster; corollas pink or white…(2)
2. Calyx tips acute…(3)
2′. Calyx tips bristlelike and aristate…(4)
3. Upper lip of corolla white, yellow, or pink, usually with maroon spots; leaves 3–10 cm…**M. punctata**
3′. Upper lip of corolla purple, lacking spots; leaves 2–5 cm…**M. humilis**
4. Inflorescence bracts gradually narrowing to a cuspidate tip; calyx teeth densely villous…**M. pectinata**
4′. Inflorescence bracts abruptly narrowing to a cuspidate tip; calyx teeth sparsely villous…**M. citriodora**

Monarda citriodora Cerv. ex Lag.—Lemon beebalm. Stems to 7.5 dm; leaves lanceolate, 3–8 × 0.3–1 cm, glabrous above, hirsute below, the margins serrate; inflorescence of axillary and terminal headlike clusters, the bracts reflexed from the upper 1/4 or less and green to purple; calyx 7–8.5 mm with teeth 1.2–4.5 mm; corolla whitish or pink, often dotted with purple, giving an overall impression of pink or purplish, the tube 10–18 mm. Grassy meadows, forests, mountain slopes, roadsides, 4500–8500 ft. Widespread.

Monarda fistulosa L.—Wild bergamot. Perennials, 3–12 dm; leaves lanceolate to ovate, 3–10 cm, the margins serrate to nearly entire, with a petiole 2–20 mm; inflorescence of solitary and terminal headlike clusters; calyx tube 5.5–11 mm with teeth ca. 1 mm; corolla purple, pink, or rose-purple, rarely white, 20–35 mm, the tube 15–25 mm. Ours are **var. menthifolia** (Graham) Fernald. Meadows and roadsides, sometimes in moist places, 5000–8500 ft. Widespread except EC, SE.

Monarda humilis (Torr.) Prather & J. A. Keith—Low beebalm. Annuals to 5 dm; leaves lanceolate, 2–5 × 0.3–0.8 cm, puberulent, the margins serrate; inflorescence of axillary and terminal headlike clusters, the bracts green to purple; calyx 4–7.5 mm with teeth 0.5–1 mm; corolla white with maroon margins and spots, the upper lip 4–5.5(–6.5) mm. Piñon-juniper communities, sandy dunes, 5800–10,000 ft. Endemic. NW, WC, Socorro. **E**

Monarda pectinata Nutt.—Plains beebalm. Annuals, 2–5 dm; leaves lanceolate to ovate, 2–4 cm, the margins serrate to subentire; inflorescence of 2+ headlike clusters in an interrupted spike; calyx tube 6–8 mm with teeth 2–4 mm; corolla usually white, sometimes pale pink or rarely purple, 12–20 mm, the tube 8–15 mm and the upper lip helmetlike. Common on open slopes and in meadows, 5000–8500 ft. Widespread.

Monarda punctata L.—Spotted beebalm. Stems to 8.5 dm; leaves lanceolate to ovate, 3–10 cm, the margins serrate; inflorescence of 2+ headlike clusters in an interrupted spike; calyx tube 6–8 mm with teeth 2–4 mm; corolla usually white, sometimes pale pink or rarely purple, 12–20 mm, the tube 8–15 mm, the upper lip helmetlike. Ours are **var. occidentalis** (Epling) E. J. Palmer & Steyerm. Common on open slopes and in meadows, 5000–8500 ft. NW, EC, C, SE, SW.

MONARDELLA Benth. – Monardella

Monardella odoratissima Benth.—Stinking horsemint. Perennials, 2–4 dm, pubescent with downward-curling hairs; leaves lanceolate to oblong or narrowly ovate, 1–3 cm, glabrous to puberulent, the margins entire to denticulate; floral bracts orbicular to ovate, rose purple or whitish, ciliate; calyx 5–10 mm, 13-nerved; corolla pale purple, ca. 15 mm. Canyons, open slopes, roadsides, 7000–10,000 ft. WC, San Juan.

NEPETA L. – Catnip

Nepeta cataria L.—Catnip. Perennials, 3–10 dm; leaves deltoid-ovate, 2–7(–10) cm, crenate to crenate-serrate, petiole 0.5–3(–5) cm; inflorescence a terminal, interrupted spike; calyx 5–7 mm, green; corolla bilabiate, white with purple spots, 7–12 mm; stamens 4; nutlets 1.3–2 mm. Disturbed areas, roadsides, often in moist places, 4000–8000 ft. NC, NW, C, WC, SC, SW. Introduced.

PHYSOSTEGIA Benth. – False dragonhead

Physostegia virginiana (L.) Benth.—Western false dragonhead. Perennials, 2–8 dm; leaves lanceolate to oblong, 4–12 cm, glabrous; bracts lanceolate to ovate; inflorescence a terminal, spikelike raceme; calyx 3–6 mm, to 7 mm in fruit, with triangular teeth ca. 1.5 mm; corolla bilabiate, purple to pink-purple, to 15 mm, the tube 6–9 mm; stamens 4; nutlets 2–3.3 mm. Our plants are **var. arenaria** Shimek. Shores of streams and lakes, moist places, 3000–5500 ft. Eddy.

POLIOMINTHA A. Gray – Rosemary-mint

Poliomintha incana (Torr.) A. Gray—Purple sage. Canescent shrubs, 3–10 dm; leaves linear to narrowly oblong, white-tomentose, 1–3 cm, sessile; inflorescence of small, axillary clusters in the upper leaf axils; calyx 6–7 mm with 5 subulate teeth; corolla bilabiate, pink, purple, or white, with purple spots on the lower lip, 10–15 mm; stamens 2. Sandy, dry deserts and dunes, 4000–5000 ft. NW, SE, SC.

PRUNELLA L. – Selfheal

Perennial herbs; leaves simple, entire or obscurely toothed, petiolate; flowers borne in a dense bracteate terminal spike; calyx 10-nerved, bilabiate; corolla pink, blue-violet, or white, bilabiate; stamens 4, the lower pair longer.

1. Stems and lower leaf surfaces densely hispid-pilose; leaves ovate to ovate-lanceolate, 1–1.5 times longer than wide; upper lip of the corolla hispid on the back...**P. hispida**
1'. Stems and lower leaf surfaces glabrous or sparsely pilose; leaves lanceolate, 1.5–3 times longer than wide; upper lip of the corolla glabrous or nearly so on the back...**P. vulgaris**

Prunella hispida Benth.—Selfheal. Perennial herbs, herbage with conspicuous spreading-hispid to stiff-pilose hairs; leaf blades ovate to ovate-lanceolate, 1.5–3 × 1.5 cm, 1–1.5 times longer than wide, margins undulate-crenate to crenate-serrate; spikes 2–3 cm, the floral bracts 5–6 mm; calyx 8–10 mm, purplish, the teeth spinulose-tipped; corolla dark purple to bluish, 1.5–2 cm. Weedy ground, lawns, elevation unknown. Quay.

Prunella vulgaris L.—Heal-all. Perennials, prostrate to decumbent or erect, the stems to 10 dm; inflorescence a dense, bracteate terminal spike; leaves lanceolate to narrowly ovate or oblong, 2–9 cm, entire to crenate-serrate, with a cuneate to attenuate base; bracts ovate to suborbicular, 5–15 mm, abruptly short-acuminate at the apex; calyx 6–10 mm; corolla bilabiate, pink to blue-violet or white, 10–18 mm; stamens 4; nutlets 1.7–2 mm, carunculate. Moist places, along streams, 5500–9500 ft. Widespread except E, SW.

SALVIA L. – Sage

Annual or perennial herbs or shrubs, usually aromatic; leaves entire or toothed, simple to pinnatifid or pinnately compound; flowers in terminal bracteate racemes or interrupted spikes; calyx strongly 10–15-nerved, bilabiate; corolla strongly bilabiate; stamens 2.

1. Leaves pinnately compound or lobed…(2)
1'. Leaves simple with entire to toothed margins but not deeply pinnately lobed…(4)
2. Corolla 10–13 mm, blue, only slightly exceeding the calyx; annuals…**S. columbariae**
2'. Corolla 25–45 mm, pink, red, or bluish, greatly exceeding the calyx; perennials…(3)
3. At least some leaves truly compound…**S. henryi**
3'. Leaves pinnately lobed, not truly compound…**S. summa**
4. Shrubs…(5)
4'. Perennials, sometimes woody at the base but not throughout…(7)
5. Leaves oblong to elliptic, entire or the upper ones obscurely dentate; flowers in slender open racemes…
   S. lycioides
5'. Leaves ovate to deltoid-ovate, crenate; flowers in dense racemes…(6)
6. Leaf blades strongly bicolored, greenish above, densely whitish canescent below, the whitish hairs obscuring the dark glands; calyx limb green to bluish; flowers blue to lavender…**S. pinguifolia**
6'. Leaf blades not or only obscurely bicolored, glabrous to finely pubescent, the glands not at all obscured, evident and giving the lower surface a rusty or reddish appearance; calyx limb wine-colored; flowers dark pinkish to purplish…**S. vinacea**
7. Annuals…(8)
7'. Perennials…(9)
8. Leaves sharply toothed; calyx with stalked glands…**S. subincisa**
8'. Leaves irregularly toothed or subentire; calyx with short, appressed hairs…**S. reflexa**
9. Calyx with bristly hairs; floral bracts equal to or longer than the calyx…**S. texana**
9'. Calyx lacking bristly hairs; floral bracts rarely as long as the calyx…(10)
10. Calyx and inflorescence densely tomentose…**S. farinacea**
10'. Calyx and inflorescence not densely tomentose…(11)
11. Corolla pinkish red to crimson or occasionally lavender-pink…**S. lemmonii**
11'. Corolla blue, violet-blue, or rarely pink or white…(12)
12. Leaves linear-lanceolate to lanceolate…**S. azurea**
12'. Leaves ovate to ovate-oblong…**S. pratensis**

Salvia azurea Michx. ex Vahl—Blue sage. Perennials, 5–15 dm; leaves simple, lanceolate to linear-lanceolate, 3–7 cm, serrate or minutely toothed; bracts 2–8 mm; calyx 4–10 mm, with acute, triangular teeth; corolla blue or rarely white, 10–25 mm. Ours are **var. grandiflora** Benth. Roadsides, sandy soil of plains, 4000–8000 ft. NC, C, WC, Quay. Introduced.

Salvia columbariae Benth.—Chia. Annuals, 2–5 dm; leaves 1–2-pinnatifid into toothed or irregularly incised divisions, 5–15 cm, cinereous-tomentose, petiolate; flowers borne in capitate verticils subtended by awn-tipped bracts; calyx 8–10 mm; corolla blue, 10–13 mm, slightly exceeding the calyx. Ours are **var. columbariae**. Desert hills; known from the vicinity of Redrock, 3900–4400 ft. Grant.

Salvia farinacea Benth. [*S. earlei* Wooton & Standl.]—Mealycup sage. Perennials, 6–10 dm; leaves simple, linear to narrowly oblong, 3–8 cm, glabrous, petiolate; flowers borne in crowded, often glomerate verticils, inflorescence tomentose; calyx 6–8 mm; corolla blue or violet-purple, 20–25 mm, greatly exceeding the calyx. Plains, grasslands, canyons, 3400–6000 ft. Chaves, Eddy, Otero.

Salvia henryi A. Gray [*S. davidsonii* Greenm.]—Crimson sage. Perennials, 2–5 dm; leaves 3-foliate with a terminal leaflet larger than the laterals or simple, 4–8 cm, petiolate; flowers borne in few-flowered terminal inflorescences, the floral bracts sometimes deciduous; calyx 8–14 mm; corolla pink to red, to 25–45 mm, greatly exceeding the calyx. Rocky hillsides, arroyo banks, limestone cliffs, canyons, mesas, 4000–7700 ft. C, SE, SC, SW, Cibola.

Salvia lemmonii A. Gray [*S. microphylla* Kunth]—Lemmon's sage. Perennials, subshrubs, 3–7 dm; leaves simple, deltoid-ovate to oblong-ovate, to 2.5 cm, petiolate; flowers racemose, mostly 2 per node and opposite; calyx 7–10 mm; corolla lavender-pink to pinkish red or crimson, to 25 mm, greatly exceeding the calyx. Uncommon on dry, rocky slopes and in canyons, 6000–7000 ft. Hidalgo.

Salvia lycioides A. Gray—Canyon sage. Shrubs; leaves simple, oblong to elliptic, 0.8–3 cm, entire or the upper ones obscurely toothed; flowers opposite; calyx 7–10 mm, tinged with purple; corolla blue or white, to 18 mm, exceeding the calyx. Rocky slopes, limestone banks, canyons, 4000–7800 ft. SE, SC, Lincoln.

Salvia pinguifolia (Fernald) Wooton & Standl.—Rock sage. Shrubs, 10.1 dm; leaves simple, deltoid-ovate to ovate, 1.2–3 cm, serrate-crenate, petiolate; flowers solitary or verticillate in the upper nodes; calyx 11–13 mm, canescent; corolla bluish purple, the tube 5–10 mm, exceeding the calyx. Canyons, rocky slopes, foothills, 4000–7000 ft. Catron, Doña Ana, Grant.

Salvia pratensis L.—Meadow clary. Perennials, 3–6 dm; leaves simple, ovate to ovate-oblong, 7–12 cm, crenate-serrate; racemes 10–20 cm, interrupted; calyx 7–11 mm, upper lip minutely 3-toothed and the lower lip with acute to aristate teeth; corolla violet-blue, rarely pink or white, 15–30 mm. Jemez Mountains, 8000–8500 ft. Sandoval. Introduced.

Salvia reflexa Hornem.—Rocky Mountain sage. Annuals, 2–7 dm; leaves simple, lanceolate to narrowly elliptic, 2–6 cm, irregularly toothed to subentire; calyx 4–6 mm, puberulent; corolla blue or sometimes white, 6–9 mm. Rocky slopes, piñon-juniper woodlands, canyon bottoms, 3400–9000 ft. Widespread.

Salvia subincisa Benth.—Sawtooth sage. Annuals, 1.5–4 dm; leaves simple, linear to lanceolate or oblong, 2.5–5 cm, sharply toothed, glabrous; flowers in terminal racemes, usually 3+ per verticil; calyx 6–10 mm, glandular; corolla blue, 10–15 mm, exceeding the calyx. Plains, mesas, rocky slopes, 4500–7500 ft. Widespread except far N, E.

Salvia summa A. Nelson—Great sage. Perennials, to 3 dm; leaves simple but deeply pinnately lobed, the terminal lobe larger than the lateral lobes and toothed, resinous-dotted and finely pubescent; flowers axillary; calyx ca. 10 mm; corolla pink to bluish, 35–45 mm, greatly exceeding the calyx. Limestone cliffs, crevices, ledges, rocky outcrops, 4000–6500 ft. Chaves, Doña Ana, Eddy. **R**

Salvia texana (Scheele) Torr.—Texas sage. Perennials, 1–4 dm; leaves simple, lanceolate, to 6 cm, entire or obscurely toothed; flowers axillary, the floral bracts equal to or longer than the calyx; calyx ca. 8 mm, with spreading hairs; corolla purplish blue, 16–22 mm, exceeding the calyx. Limestone cliffs, piñon-juniper communities; known from Carlsbad Caverns National Park, 7000–7500 ft. Eddy.

Salvia vinacea Wooton & Standl.—Woody shrub, 0.4–2 m, herbage sparingly tomentulose with dark, red to orangish glands; leaves simple, deltate-ovate-orbicular, 1–2.5(–4) cm long and wide, crenate-toothed; flowers 2–4 per node; calyx 8–11 mm in flower, enlarging in fruit, purplish to wine-colored, glandular-punctate; corolla dark pink to purplish, 1.5–2.5 cm, the upper lip with white hairs on the hood. Bajadas, foothills, lower mountain slopes, 5200–6700 ft. SC, SW, Lincoln.

SCUTELLARIA L. – Skullcap

Annual or perennial herbs or low shrubs; leaves simple, sessile or shortly petiolate; flowers 1–3 in leaf axils or sometimes in terminal bracteate racemes or spikes; calyx bilabiate, the lips entire, with an erect scutellum on the upper lip; corolla blue, purple, or white, bilabiate; stamens 4.

1. Leaves crenate-serrate…(2)
1.' Leaves entire or nearly so…(3)
2. Flowers in racemes; corolla 5–8 mm…**S. lateriflora**
2.' Flowers solitary in upper leaf axils; corolla 12–20 mm…**S. galericulata**
3. Annuals; nutlets straw-colored…**S. drummondii**
3.' Perennials, usually from rhizomes; nutlets dark brown or black…(4)
4. Leaves glabrous, or if pubescent then the hairs glandular; corolla 20–30 mm…**S. brittonii**
4.' Leaves with short, retrorse hairs and longer, spreading eglandular hairs; corolla 10–20 mm…**S. potosina**

Scutellaria brittonii Porter—Britton's skullcap. Perennials, (6–)8–25(–33) cm; herbage glabrous or with appressed or curved hairs, sometimes glandular-pubescent; leaves lanceolate to narrowly elliptic, 1.5–3.5 × 0.5–1 cm, entire, petiole 5–10 mm; calyx 4–6 mm, to 6–8 mm in fruit, with sessile glands and villous hairs, the scutellum 3–4 mm in fruit; corolla violet to blue-violet, 20–30 mm. Open forests and canyons in the N mountains, 6500–8000 ft. Colfax, Mora.

Scutellaria drummondii Benth.—Drummond's skullcap. Annuals, 5–25 cm; herbage glandular and densely spreading-villous; leaves oblong-obovate to ovate, 0.8–2 cm, usually entire; calyx 3–5.5 mm, to 7 mm in fruit, glandular and pilose, the scutellum ca. 3 mm in fruit; corolla violet to blue-violet, 5–15 mm. Ours are **var. edwardsiana** B. L. Turner. Desert grasslands, shortgrass prairie, canyons, 3000–6500 ft. SE, SC.

Scutellaria galericulata L.—Marsh skullcap. Perennials, delicate, (10–)20–60(–100) cm; leaves lance-ovate, (2–)3–6(–9) × (0.5–)1–3 cm, crenate-serrate, petiole 0–4 mm; calyx 3–4 mm in flower and 4–6 mm in fruit, with sessile glands and hairs, the scutellum ca. 3 mm in fruit; corolla blue, 12–20 mm. Along streams, floodplains, marshes, lake margins, 5000–9500 ft. NC, NW. Introduced.

Scutellaria lateriflora L.—Blue skullcap. Perennials, delicate, 10–60(–100) cm; leaves ovate, 3–11 × 1.5–5.5 cm, crenate-serrate, petiolate; flowers in axillary racemes; calyx ca. 2 mm in flower and 3 mm in fruit, with coarse, antrorse hairs, the scutellum ca. 2 mm in fruit; corolla blue, 5–8 mm. Ours are **var. lateriflora**. Uncommon along streams and near springs, elevation unknown. Mora.

Scutellaria potosina Brandegee—Mexican skullcap. Perennials, woody at the base, 30–50 cm; herbage with short, retrorse hairs and longer, spreading eglandular hairs; leaves ovate, 1–2.5 cm, entire, petiole 2–3 mm; calyx 3–4.5 mm, to 6 mm in fruit, resinous-dotted and with spreading, often glandular hairs, the scutellum ca. 4.5 mm in fruit; corolla violet with a white palate, 10–20 mm. Ours are **var. tessellata** (Epling) B. L. Turner. Rocky washes, rocky outcrops, arroyos, 4600–6500 ft. EC, C, SE, SW.

STACHYS L. – Hedge-nettle

Perennial herbs; leaves oblong to deltoid or deltoid-ovate, petiolate or sessile, the margins dentate; flowers in clusters at the upper nodes, subtended by foliaceous bracts; calyx actinomorphic, 5-lobed, the lobes equal or nearly so; corolla pink, purple, red, or whitish, bilabiate, the upper lip entire or notched, the lower lip 3-lobed; stamens 4, didynamous.

1. Corolla bright red, conspicuously longer than the calyx; petioles 12–35 mm…**S. coccinea**
1.' Corolla pink, pinkish purple, or whitish, only slightly longer than the calyx; leaves sessile or the petioles < 5 mm…(2)
2. Plants villous-hirsute with spreading hairs; leaves mostly acute…**S. pilosa**
2.' Plants woolly-tomentose with soft, appressed hairs; leaves obtuse…**S. rothrockii**

Stachys coccinea Ortega—Scarlet hedge-nettle. Stems 3–8 dm; leaves deltoid-ovate or deltoid, 2–5 cm, the margins crenate to serrate, puberulent to hirsute, petioles 12–35 mm; calyx 10–15 mm with teeth 2.5–5 mm; corolla bright red, 2–3 cm. Canyons, rocky slopes, cliff bases, SW mountains, 4500–8000 ft. WC, SW.

Stachys pilosa Nutt.—Hairy hedge-nettle. Stems 1.8–6 dm; leaves elliptic-lanceolate to oblong, 1.5–6 cm, the margins crenate-serrate, villous-hirsute with spreading hairs, sessile or on petioles < 5 mm;

calyx 5–9 mm; corolla pinkish purple, 1–1.5 cm, the tube equal to or slightly longer than the calyx. Moist places, wet meadows, usually shady, 6500–9000 ft. NC, NW, C, WC, SC.

Stachys rothrockii A. Gray—Rothrock's hedge-nettle. Stems 1–2 dm; leaves lanceolate to oblong-lanceolate, 1–3 cm, the margins entire to crenate-serrate, woolly-tomentose below with appressed hairs, sessile or on petioles < 5 mm; calyx 5–9 mm; corolla whitish to pink or pinkish purple, 1.2–1.5 cm, the tube 5–9 mm. Meadows, riparian areas, along rivers and lake margins, 6500–8000 ft. NC, NW, WC, Lincoln, Otero.

TETRACLEA A. Gray – Tetraclea

Tetraclea coulteri A. Gray—Coulter's tetraclea. Low perennial herbs branching from a woody base, ascending or spreading, to 0.4 m; stems densely white, with appressed pubescence, not glandular; leaves simple, opposite, petiolate, petioles 4–10 mm, appressed-puberulent, blades ovate, 16–35 × 6–18 mm, sharply acute and mucronulate, irregularly dentate; inflorescence a cyme, usually with 3 flowers, sometimes reduced to 1 or 2, peduncles 3–8 mm; calyx deeply 4- or 5-parted; corollas salverform, cream-colored, tinged with red, limb sprawling, 5-parted; stamens 4; anthers 2-celled; ovary shortly 4-lobed at apex; seeds 4. Rocky hillsides and canyons, desert scrub, grasslands, roadsides, 3300–8515 ft. C, SE, SC, SW, Guadalupe, San Miguel.

TEUCRIUM L. – Germander

Herbs or rarely shrubs; leaves simple, the margins entire to deeply lobed; inflorescence of terminal bracteate spikes or racemes, or sometimes flowers solitary in the axils of upper leaves; calyx 5-lobed with unequal teeth, usually actinomorphic, 10-nerved; corolla unequal but not bilabiate, the upper lip conspicuously shorter than the lower lip, lower lip conspicuous and spreading, with small lateral lobes; stamens 4, usually exserted.

1. Leaves not lobed; corolla pink or pinkish purple…**T. canadense**
1'. Leaves palmately or pinnately lobed; corolla white to pale blue, sometimes with purple lines or spots…(2)
2. Leaves palmately 3-lobed; annuals…**T. cubense**
2'. Leaves pinnately lobed; perennials…**T. laciniatum**

Teucrium canadense L.—Canada germander. Perennials, mostly 1–1.5 m; leaves narrowly elliptic or lanceolate, 4–9 cm, the margins serrate, petioles narrowly winged, 6–10 mm; flowers in a terminal bracteate spike; calyx 5–7 mm, finely canescent; corolla pink or pinkish purple, 8–20 mm. Moist places, floodplains, marshes, 4000–8000 ft. Both varieties are widely scattered.

1. Calyx glandular…**var. occidentale** (A. Gray) E. M. McClintock & Epling
1'. Calyx not glandular…**var. canadense**

Teucrium cubense Jacq. [*T. depressum* Small]—Small coastal germander. Annuals, 1.5–7 dm; leaves obovate to elliptic in outline but shallowly to deeply palmately 3-lobed, the lobes linear to narrowly oblong, 2–4 cm, petioles narrowly winged, 10–20 mm; flowers solitary in the axils of leaflike bracts; calyx tube 2–3 mm, the teeth 3–6 mm; corolla white to pale blue, often marked with purple lines and/or spots, 7–15 mm. Ours are **var. densum** Jeps. Drainage bottoms, stream banks, floodplains, often in sandy soil, 3000–5000 ft. EC, SE, SC, SW.

Teucrium laciniatum Torr.—Lacy germander. Perennials, 0.7–2 dm; leaves deeply pinnately lobed, the lobes linear, 1–5 cm, glabrous to sparsely pubescent; flowers solitary in the axils of leaflike bracts; calyx 10–12 mm, glabrous or nearly so; corolla white, sometimes with purple lines, 9–20 mm. Sandy plains, desert grasslands, roadsides, 4000–8000 ft. Widespread except NW, WC, SW.

TRICHOSTEMA Gronov. – Bluecurls

Annual or perennial herbs, glandular-puberulent or with short, curled hairs; leaves ovate, petiolate or sessile, the margins entire; flowers solitary or axillary; calyx actinomorphic, 5-lobed, the lobes equal;

corolla blue or purple, deeply 5-cleft into oblong segments; stamens 4, didynamous, often well exserted at anthesis.

1. Perennials; corolla 17–12 mm; stamens conspicuously arching above the corolla…**T. arizonicum**
1'. Annuals; corolla 3–6 mm; stamens straight or only slightly arching [reported]…T. brachiatum

Trichostema arizonicum A. Gray—Arizona bluecurls. Perennials, 2–6 dm; leaves oblong to ovate, 1–3 cm, the margins entire to crenate, with retrorsely curled short hairs, sometimes glandular, petiolate or sessile; calyx ca. 6 mm with teeth about as long as the tube, hirtellous; corolla white with the middle lobe of the lower lip blue to purple, 17–12 mm, the tube slightly exserted; stamens arching above the corolla, filaments white to purplish. Desert grassland, canyons, foothills, 4000–6500 ft. SW.

Trichostema brachiatum L.—Fluxweed. Annuals, to 4 dm; leaves lanceolate to elliptic, 2–5 cm, the margins entire to serrate, with curled short hairs, sometimes glandular, petiolate; calyx 3–6 mm with teeth longer than the tube; corolla white with the middle lobe of the lower lip pink to purple, 3–6 mm, slightly exceeding the calyx; stamens straight or slightly arched, filaments blue to purple. Reported for the state but no specimens known.

VITEX L. - Chastetree

Vitex agnus-castus L.—Lilac chastetree. Aromatic shrubs or small trees, to 5 m; leaves palmately compound, mostly 5–9-foliate, leaflets narrowly elliptic, long-pointed at both ends, short-hairy to tomentose below, the central leaflet 4.5–11.5 cm; calyx 2–3 mm; corolla purple to lavender or blue-purple, the tube 6–7 mm; stamens 4. Escaped from cultivation, scattered across the state, 3700–6000 ft. Bernalillo, Doña Ana, Socorro. Introduced.

LENTIBULARIACEAE - BLADDERWORT FAMILY

C. Barre Hellquist

UTRICULARIA L. - Bladderwort

Utricularia vulgaris L. [*U. macrorhiza* Leconte]—Greater bladderwort. Herbs; perennial, submersed aquatic, floating just below surface; stems elongate, 1–3 m, 0.5+ mm thick, plumose branches to 12 cm diam.; leaves elliptic to ovate, 14–20 mm, much dissected, with capillary segment bearing bladderlike traps; scape erect, 10–80 cm, emersed, bearing 5–20 yellow flowers; calyx 2-lobed; corolla 12–18 mm broad, 2-lipped, the lower 3-lobed, spurred at base, palate conspicuous, spur conic; stamens 2; ovary free, style short or lacking; stigma 1–2-lipped; fruiting pedicels reflexed; winter buds 1–2 cm. New Mexico material belongs to **subsp. macrorhiza** (Leconte) R. T. Clausen. Still water, 8150–9100 ft. NC, NW, Catron (Johnson, 1987).

LILIACEAE - LILY FAMILY

Steve L. O'Kane, Jr.

Perennial herbs, usually with bulbs, rhizomatous in some; leaves basal and/or cauline, sheathing, simple, parallel-veined; inflorescence a terminal raceme or a solitary flower; flowers bisexual, actinomorphic; perianth of 2 whorls of 3 tepals or differentiated into 3 sepals and 3 petals, often spotted, striate, or with elaborate glands; stamens 6, distinct, free; anthers dorsifixed or pseudobasifixed, longitudinally dehiscent; ovary superior, carpels 3, locules 3, stigmas 3; placentation axile; fruit a capsule or berry. Here we follow the family circumscription of APG (2016), which excludes many taxa traditionally included in the family but includes species formerly in Calochortaceae. Key modified from Ackerfield (2015) (Simpson, 2019; Utech, 2002).

1. Perianth differentiated into 2 dissimilar whorls; sepals and petals present, the petals broad and with a prominent hairy, colored gland at the base…**Calochortus**
1'. Perianth not differentiated into dissimilar whorls; all parts petaloid and called tepals, ± of the same size and texture, prominent hairy gland lacking…(2)
2. Tepals shades of orange or red, yellow, or purplish brown with yellowish mottling on the inside…(3)
2'. Tepals white, greenish white, or cream-white…(5)
3. Tepals red-orange or red-magenta, sometimes pale orange, with maroon spotting; flowers large, 4.5–9 cm…**Lilium**
3'. Tepals not colored as above; flowers smaller…(4)
4. Cauline leaves present, some whorled or nearly so; tepals yellow or purplish brown with yellowish or white mottling on the inside, < 2.5 cm…Fritillaria
4'. Cauline leaves absent, the 2 leaves basal and subopposite; tepals yellow, showy, 2.5–4 cm…**Erythronium**
5. Leaves linear, mainly basal with stem leaves reduced and grading into bracts; fruit a capsule; > 10,000 ft. elevation…**Gagea**
5'. Leaves elliptic or ovate, stem leaves not reduced; fruit a reddish berry; various elevations…(6)
6. Flowers borne in leaf axils, drying cream-white; pedicels glabrous and sharply geniculate; stems glabrous or coarsely hairy below…**Streptopus**
6'. Flowers borne terminally, appearing to arise from between 2 leaves; pedicels hairy, not sharply geniculate; stems densely short-hairy throughout…**Prosartes**

CALOCHORTUS Pursh – Mariposa lily, sego lily

Herbs, perennial, from bulbs; stems glabrous, often glaucous, straight (twining or straggling in *C. flexuosus*); leaves sessile, linear to narrowly linear-lanceolate; inflorescences 1- to few-flowered, bracteate; perianth globose to broadly campanulate; sepals 3, distinct; petals 3, distinct, longer and broader than sepals, sometimes clawed, usually hairy adaxially, bearing an adaxial gland near base, often spotted to ± patterned; stamens 6; ovary superior; style absent; stigmas 3; fruit a 3-locular, 3-angled, or 3-winged capsule (Fiedler & Zebell, 2002). Key modified from S. L. Welsh et al. (2015).

1. Flowers deep yellow or yellow-orange…**C. aureus**
1'. Flowers white, cream, or lavender to purplish (pale yellow or pale greenish yellow in *C. gunnisonii* var. *perpulcher*)…(2)
2. Stems sprawling and flexuous; flowers in a few-flowered flexuous raceme; petals purplish; gland not depressed or surrounded by a membrane…**C. flexuosus**
2'. Stems erect; flowers solitary or in an umbellate cluster, variously colored; glands depressed and surrounded by a membrane or not…(3)
3. Anthers acute; petal base with a broad purple band above the gland extending across the entire width of the petal to the margins…**C. gunnisonii**
3'. Anthers obtuse; petal base with a purple inverted V shape or crescent above the gland, this typically separated from the gland by a yellow stripe, the purple marking not extending to petal margins (in some flowers the purple V shape or crescent may be suppressed)…(4)
4. Gland of petal circular, shield-shaped, or longitudinally elongate; petal hairs simple, not branched or enlarged apically…**C. nuttallii**
4'. Gland of petal transversely oval; petals hairs bilobed or conspicuously enlarged apically…**C. ambiguus**

Calochortus ambiguus (M. E. Jones) Ownbey—Doubting mariposa lily. Stems usually branching; sepals invested near gland with yellowish, distally enlarged or branching, gland-tipped hairs, base sometimes marked with dull purple, apex acuminate; petals pinkish or bluish gray, infrequently with longitudinal gray stripe with median blotch usually surrounded by yellow zone, ± equaling sepals, base sometimes marked with dull purple; glands round to lunate, depressed, surrounded by narrow, deeply fringed membrane, densely covered with short, distally branching hairs. Rocky slopes, dry meadows, open habitats of woodlands and forests, sandy arroyos, (3900–)4850–8700(–10,750) ft. WC, SW, Sandoval.

Calochortus aureus S. Watson—Golden mariposa lily. Stems usually not branching; sepals with maroon blotch distal to gland, broadly lanceolate; petals lemon-yellow, with maroon blotch distal to gland, broadly obovate, cuneate, ± glabrous or with a few lemon-yellow hairs near gland; glands round, depressed, surrounded by conspicuously fringed membranes, densely covered with short, unbranched or distally branching hairs. Woodlands, savannas, grassy areas, dry clayey or sandy areas, 5700–7900 ft. NW, Cibola.

Calochortus flexuosus S. Watson—Winding mariposa lily. Stems twining or straggling over ground, branching, usually sinuous; sepals lanceolate to lance-ovate, 2–3 cm; petals white with lilac tinge, darker-veined, especially distally, sometimes with purple blotch distal to gland, obovate to cuneate, with sparse or dense, short or long hairs near gland, apex rounded; glands transverse-lunate or wider, densely short-hairy. Dry stony slopes, desert hills, mesas, creosote scrub, sagebrush scrub, 3900–5800 ft. Grant, Hidalgo, San Juan.

Calochortus gunnisonii S. Watson—Gunnison's mariposa lily. Stems not branching; sepals marked similar to petals, lanceolate, usually much shorter, glabrous, apex acute; petals white to purple, greenish adaxially, clawed, often with narrow, transverse purple band distal to gland and purple blotch on claw, obovate, cuneate, usually obtuse and rounded distally; glands transversely oblong, not depressed, densely bearded with distally branching hairs, outermost of which somewhat connate at base to form discontinuous, deeply fringed membranes. Two varieties in the species and in our area:

1. Petals white to purple, occasionally yellow; wide distribution…**var. gunnisonii**
1'. Petals pale yellow to pale greenish yellow, at least adaxially; endemic to mountains of Colfax, Mora, and San Miguel Counties…**var. perpulcher**

var. gunnisonii—Grassy slopes, meadows, grassy openings in coniferous forests, 7400–10,900 ft. N, C, WC.

var. perpulcher Cockerell—Aspen glades, moist meadows, open areas in subalpine forests, 9600–10,800 ft. Colfax, Mora, San Miguel. Ownbey (1940: 512) notes that this is "hardly more than a color form, but locally constant, and easily recognized." **R & E**

Calochortus nuttallii Torr. & A. Gray—Sego lily. Stems usually not branching or twisted; sepals marked similar to petals, usually shorter, lanceolate, glabrous, apex acuminate; petals white, tinged with lilac or infrequently magenta, yellow at base, with reddish-brown or purple band or blotch distal to gland, broadly obovate, cuneate, sparsely invested near gland with slender hairs, apex usually short-acuminate; glands round, depressed, surrounded by conspicuously fringed membrane, densely covered with short, unbranched or distally branching hairs. Piñon-juniper, piñon-juniper-ponderosa pine, open areas in Douglas-fir, grassy slopes, grasslands, 4850–9800 ft. NC, NW, Grant, Hidalgo.

ERYTHRONIUM L. – Fawn-lily, trout-lily

Erythronium grandiflorum Pursh—Avalanche-lily, glacier-lily. Perennial from slender bulbs; leaves 2, 5–20 cm, lanceolate, ± glaucous, gradually narrowed to petiole, margins ± wavy; scape 5–30 cm; inflorescences 1(–5)-flowered; flowers nodding; tepals 6, recurved, bright yellow with narrow paler zone at base, or white to creamy-white with yellow base, narrowly ovate, 20–35 mm, inner usually auriculate at base; stamens 11–18 mm; stigma unlobed or with slender, recurved lobes (1–)2–4 mm. Our plants are **subsp. grandiflorum**. Moist meadows and slopes, moist openings in spruce-fir forests, 10,150–10,500 ft. Rare in Rio Arriba County.

FRITILLARIA L. – Fritillary, missionbells

Fritillaria atropurpurea Nutt.—Spotted-fritillary, missionbells. Herbs, perennial, from a bulb, these with 2–5 large fleshy scales; stem 1–6 dm; leaves in whorls of 2–3, 4–12 cm, usually shorter than inflorescence; blades linear to lanceolate; flowers nodding; perianth widely open; tepals 6, purplish brown, mottled yellow or white, oblong to ± diamond-shaped, 1–2.5 cm, apex not recurved; ovary superior, 6-locular; style branched > 1/2 its length, the branches > 1.5 mm. One report from the Carrizo Mountains, Arizona, just outside New Mexico; to be expected in NW mountains with ponderosa pine-oak, mountain shrub, and aspen groves, 7350–9800 ft.

GAGEA Salisb. - Alp-lily, star-of-Bethlehem

Gagea serotina (L.) Ker Gawl. [*Lloydia serotina* (L.) Rchb.]—Alp-lily. Herbs, perennial, caulescent, glabrous, from short rhizomes with persistent leaf sheaths; basal leaves 2, linear, 4-10(-20) cm × 0.8-1 mm, cauline leaves 2-4, blades lanceolate, 1-5 × 0.1-0.2 cm; inflorescences 1-2-flowered, leafy-bracteate; perianth white with greenish to purplish veins, or yellow; tepals 6, spreading, distinct, equal, transversely corrugate, with small, nectariferous gland above base; ovary superior, 3-locular; style 3-fid. Tundra, subalpine-alpine grasslands and slopes, edges of spruce-fir forest, 9900-13,000 ft. NC, SC, Lincoln. The genus *Lloydia* is now included in a larger *Gagea* (Reveal & Utech, 2002; Zarrei et al., 2009).

LILIUM L. - Lily

Lilium philadelphicum L. [*L. andinum* Nutt.]—Wood lily. Bulbs chunky with 2(-4) years' growth visible; scales 1-2-segmented; stems to 1.2 m, glaucous; leaves ± horizontal and drooping at tips, or ascending in sun, 2.9-10.2 cm, usually narrowly elliptic; inflorescences umbellate, 1-3(-6)-flowered; flowers ± erect, widely campanulate; sepals and petals somewhat recurved 1/4-2/5 along length from base, red-orange or red-magenta, sometimes pale orange, pure red, or rarely yellow, distinctly clawed, nectar guides above claws yellow to orange and spotted maroon, more pronounced on sepals; stamens strongly exserted. Our plants are **var. andinum** (Nutt.) Ker Gawl. Moist canyon bottoms in coniferous zone, marshy areas, moist to wet meadows, 7500-9600 ft. Otero, Sandoval, San Miguel, Santa Fe.

PROSARTES D. Don - Fairybells

Prosartes trachycarpa S. Watson [*Disporum trachycarpum* (S. Watson) Benth. & Hook. f.]— Rough-fruit fairybells. Herbs, from slender, knotty rhizomes with fibrous roots; plants 3-8 dm, crisp-pubescent, becoming glabrate, sparingly branched; leaves 4-12 cm, blades ovate to oblong-lanceolate, subcordate to oblique basally, with 7-9 prominent veins, lower surface and margins moderately pubescent; flowers 1-2(3), narrowly campanulate; tepals creamy to greenish white, narrowly oblanceolate, 8-15 mm; stamens mostly exserted; berries reddish orange to bright red. Rich soils of slopes and drainages in mixed conifer and mixed confer-aspen communities, 7800-9850 ft. NC, NW, WC.

STREPTOPUS Michx. - Twistedstalk

Streptopus amplexifolius (L.) DC.—Clasping twistedstalk. Herbs, perennial, from thick rhizomes; stems branched, stout, 5-12 dm, often with reddish hairs basally; leaves 5-15 × 2.5-6 cm, sessile, ovate-oblong to oblong-lanceolate, base cordate-clasping, apex acuminate; flowers 1-2 per axil; perianth campanulate; tepals spreading, recurved at tips, white to greenish yellow, narrowly oblong-lanceolate, 9-15 mm; style stout, 4-5 mm; stigma unlobed; berries whitish green maturing to yellowish orange or red. Streamsides, lake margins, canyon bottoms, seeps of mixed conifer and aspen communities, 7600-11,000 ft. NC, NW.

LINACEAE - FLAX FAMILY

Kenneth D. Heil

LINUM L. - Flax

Annual (ours) or perennial herbs from a taproot; stems 4-90 cm, often branched; leaves alternate, simple, 3-35 × 0.5-4 mm, appressed-ascending to ascending, stipular glands present or absent; sepals 5, 3.5-9 mm, lanceolate, acuminate, or elliptic, entire, often with a slender awn tip; petals 5, soon falling, yellow, orange, blue, or white, 5-23 mm; filaments united at the base; style 2.5-10 mm, often cleft at the apex; fruit a capsule, ellipsoid or ovoid to globose, 3-8 mm; seeds mucilaginous (Allred, 2020; Heil & Porter, 2013b).

1. Petals blue or rarely white…(2)
1′. Petals yellow or orange…(4)
2. Inner sepals with ciliate margins; stigmas slender; escaped ornamentals…**L. usitatissimum**
2′. Inner sepals entire; stigmas capitate; native…(3)
3. Perennial; petals > 10 mm; styles 4–9 mm…**L. lewisii**
3′. Annual; petals < 10 mm; styles 1–3 mm…**L. pratense**
4. Styles distinct to the base…(5)
4′. Styles united nearly to the apex…(7)
5. Plants annual or biennial; stipular glands absent; SW and SC…**L. neomexicanum**
5′. Plants perennial; stipular glands present; SE…(6)
6. Leaves lanceolate to oblanceolate or broader, some lower ones in whorls of 4…**L. schiedeanum**
6′. Leaves linear, the lower ones alternate or opposite…**L. rupestre**
7. Sepals entire or fringed, not glandular-toothed; flowers few; SE…**L. hudsonioides**
7′. Sepals glandular-toothed; flowers racemose or paniculate; endemic to widespread…(8)
8. Plants a long-lived perennial with a branching woody base and thick lateral roots; gypsophiles; Eddy County…**L. allredii**
8′. Plants annual or perennial, without a woody base or thick lateral roots; gypsophiles or in other substrates; widespread…(9)
9. Stems densely hirtellous with spreading hairs; petals mostly orange to salmon, the bases reddish… **L. puberulum**
9′. Stems glabrous throughout or nearly so…(10)
10. Stipular glands absent; E plains…(11)
10′. Stipular glands present, at least on the lower leaves; widespread…(12)
11. Styles 3–4 mm; petals 6–11 mm…**L. compactum**
11′. Styles 6–11 mm; petals 10–18 mm…**L. rigidum**
12. Styles 2–4 mm; petals 5–10 mm…**L. australe**
12′. Styles 4–9 mm; petals mostly 10–19 mm…(13)
13. False septa incomplete, inner margin terminating in a loose fringe; sepals persistent in fruit…(14)
13′. False septa complete; sepals usually deciduous in fruit…(15)
14. Plants annual; stipular glands usually present; foliage green; petals yellow-orange or salmon, reddish below the middle…**L. vernale**
14′. Plants usually perennial, rarely annual; stipular glands absent; foliage glaucous; petals lemon-yellow, rarely with pale red streaks…**L. subteres**
15. Sepals narrowly lanceolate, acuminate-aristate; plants broomlike and bushy with long, slender, stiffly spreading-ascending, few-flowered branches…**L. aristatum**
15′. Sepals broadly lanceolate to narrowly ovate, acute-aristate; plants not broomlike, few-branched at the base or in the inflorescence…**L. berlandieri**

Linum allredii Sivinski & M. O. Howard—Allred's flax. Long-lived perennial, suffrutescent from a woody branching base; stems to 25 cm, puberulent or glabrescent at the base; leaves mostly 3–10 mm, linear to linear-lanceolate, with stipular glands; flowers 10–13 × 6–7 mm, petals orange or salmon with adaxial dark orange or reddish band across the lower 1/2; fruit oval, 3.7–4 mm; seeds reddish brown. Sandy gypsum, 3900 ft. Eddy. **R**

Linum aristatum Engelm. [*Cathartolinum aristatum* (Engelm.) Small]—Broom-flax. Stems 8–45 cm, usually branched from near the base; leaves 3–15 × 0.5–1 mm, appressed-ascending, stipular glands present; flowers 7–12 mm, petals yellow to yellow-orange; fruit ellipsoid, 3.5–4.5 mm. Sand dunes and other sandy sites in grassland, sand sagebrush, piñon-juniper, and ponderosa pine communities, 3900–6060 ft. Widespread.

Linum australe A. Heller [*Cathartolinum australe* (A. Heller) Small]—Southern flax. Annual, scabrous herbs from taproots; stems 10–40 cm, usually branched above the base; leaves mostly 6–20 × 0.5–1.5 mm, appressed-ascending to ascending, stipular glands present on ours; flowers 5–9 mm, petals yellow to yellow-orange; fruit ovoid, 3.5–4.5 mm. Piñon-juniper, sagebrush, ponderosa pine communities, 6600–9075 ft. Widespread.

Linum berlandieri Hook. [*L. rigidum* Pursh var. *berlandieri* (Hook.) Torr. & A. Gray]—Berlandier's flax. Annual, glabrous; stems 10–40 cm, stout, leafy throughout; leaves 20–30 mm, 3-ribbed, stipular glands present; flowers 12–20 mm, long-aristate, lanceolate, petals yellow, brick-red at the base; fruit ovoid-elliptic, 4 mm. Chihuahuan desert scrub and grassland communities, 3495–5050 ft. SE, Union.

Linum compactum A. Nelson [*Cathartolinum compactum* (A. Nelson) Small]—Wyoming flax. Annual, glabrous; stems 20–50 cm; leaves 1–3 cm, linear, stipular glands none; flowers 4–7 mm, glandular-toothed; petals 6–11 mm, yellow; fruit ellipsoid, 3.5–4.5 mm. Sandy soils of grassland communities, 4780–6440 ft. Colfax, Roosevelt, Union.

Linum hudsonioides Planch. [*Mesynium hudsonioides* (Planch.) W. A. Weber]—Texas flax. Annual; stems 5–30 cm, ascending to erect, hirsutulous on the angles above, otherwise glabrous; leaves 5–10 mm, narrow, opposite near the base, alternate above, imbricate, stipular glands none; flowers 8–12 mm, petals yellow, with or without a brick-red base; fruit broadly ovoid, 2.7–3.5 mm. Grassland communities in sandy, limy, and gravelly soils, 4590–6425 ft. SE, Harding, San Miguel.

Linum lewisii Pursh [*L. perenne* L. subsp. *lewisii* (Pursh) Hultén]—Blue flax. Perennial, glabrous herbs from a caudex and stout taproot; stems 15–80 cm, usually simple; leaves 4–30 × 0.5–4 mm, ascending to spreading-ascending; flowers 12–23 mm, petals blue with a whitish or yellowish base or white; fruit ovoid to globose, 6–8 mm. Frequent along roadsides; sagebrush, mountain brush, piñon-juniper, ponderosa pine communities, 4100–10,500 ft. Widespread.

Linum neomexicanum Greene [*Cathartolinum neomexicanum* (Greene) Small]—New Mexico yellow flax. Perennial, glabrous herbs from a caudex; stems to 40 cm, usually branching above the base; leaves mostly 5–15 × 1–2 mm, appressed-ascending to ascending; flowers 5–8 mm, petals yellow; fruit ovoid, 4–5 mm. Rocky hillsides in piñon-juniper, mixed conifer, and oak woodland communities, 5900–8800 ft. WC, SW, McKinley.

Linum pratense (Norton) Small [*L. lewisii* Pursh var. *pratense* Norton]—Meadow flax. Annual, glabrous; stems commonly branched at the base, 5–40 cm; leaves 1–2 cm, linear to linear-lanceolate, stipular glands none; flowers 5–10 mm, petals blue to pale blue or white; fruit ovoid, 4–6 mm. Sandy substrates in grassland, Chihuahuan desert scrub, oak woodland, juniper, ponderosa pine communities, 3600–7000 ft. Scattered.

Linum puberulum (Engelm.) A. Heller [*Cathartolinum puberulum* (Engelm.) Small]—Plains flax. Annual, hirtellous herbs from a taproot; stems 4–20 cm, usually branched from the base; leaves 3–15 × 0.5–1 mm, appressed-ascending to ascending, stipular glands present in ours; flowers 8–15 mm, petals orange to salmon, with a dark base, 8–15 mm; fruit ovoid, 3–4 mm. Grassland, salt desert scrub, sagebrush, piñon-juniper, oak woodland, mixed conifer communities, 2975–7600 ft. Widespread.

Linum rigidum Pursh [*Cathartolinum rigidum* (Pursh) Small]—Stiffstem flax. Annual, glabrous; stems 10–35 cm; leaves 1–3 cm, linear to linear-lanceolate, alternate, stipular glands none; flowers 10–15 mm, petals yellow; fruit 3.5–4.5 mm, ellipsoid. Grassland, Chihuahuan desert scrub, shinnery oak, piñon-juniper woodland communities, 3285–5560 ft. NE, EC, SE.

Linum rupestre (A. Gray) Engelm. ex A. Gray [*Cathartolinum rupestre* (A. Gray) Small]—Rock flax. Perennial, glabrous or sparsely hairy; stems 20–70 cm; leaves 1–2 cm, linear to linear-lanceolate, alternate throughout or lowermost opposite, stipular glands present; flowers 6–10 mm, petals yellow; fruit 2–3 mm, ovoid. Limestone soils, often in canyon bottoms, desert scrub, piñon-juniper communities, 4150–6500 ft. Chaves, Eddy, Lincoln.

Linum schiedeanum Schltdl. & Cham. [*Cathartolinum schiedeanum* (Schltdl. & Cham.) Small]—Schied's flax. Perennial, glabrous; stems 20–60 cm; leaves 1–2 cm, lanceolate to oblanceolate, the upper alternate, the lower in whorls of 4, stipular glands present; flowers 3–6 mm, petals yellow; fruit to 2.5 mm, broadly ovoid. Limestone substrates, canyon bottoms with Apache plume, bigtooth maple, and oak; known from 1 collection at Longview Spring, 5710–6155 ft. Eddy.

Linum subteres (Trel.) H. J. P. Winkl. [*L. aristatum* Engelm. var. *subteres* Trel.]—Sprucemont flax. Annual or perennial; leaves alternate or opposite proximally, blades lanceolate to oblanceolate, 8–17 × 1–2.3 mm, entire; sepals 4–7 mm, the inner sepals conspicuously toothed, the outer ones very coarsely glandular-toothed; petals lemon-yellow, rarely with interior pale red streaks, 9–15 mm; capsules 3.5–4.5 mm. Rocky, sandy sites, sagebrush and piñon-juniper communities, 5000–6000 ft. Sandoval, San Juan.

Linum usitatissimum L.—Common flax. Annual, glabrous; stems 30–90 cm, commonly branched at the base; leaves 1.5–3.5 cm, linear to narrowly lanceolate, stipular glands none; flowers 10–15 mm, inner sepals fringed, petals blue or white; fruit 6–8 mm, ovoid. Grasslands, water courses in desert scrub up to ponderosa pine communities, 4600–8000 ft. Hidalgo, Luna. Introduced. A sporadic escape from cultivation and native of the Old World.

Linum vernale Wooton [*Cathartolinum vernale* (Wooton) Small]—Chihuahua flax. Annual, glabrous; stems 10–50 cm; leaves 1–1.5 cm, linear, alternate or the lower opposite, stipular glands present in most plants; flowers 10–15 mm, sepals glandular, petals yellow-orange to salmon with a maroon base; fruit 3–4 mm, ovoid. Limestone substrates in Chihuahuan desert scrub, desert grassland, piñon-juniper woodland communities, 3900–7000 ft. C, SC, Eddy.

LINDERNIACEAE – FALSE-PIMPERNEL FAMILY

Kenneth D. Heil

LINDERNIA All. – False-pimpernel

Lindernia dubia (L.) Pennell [*Gratiola dubia* L.; *Ilysanthes dubia* (L.) Barnhart]—Yellowseed false-pimpernel. Annual herbs; leaves cauline, thin, blades spatulate, lanceolate, oblanceolate, elliptic, ovate, or obovate, (1–)5–37 × (0.5–)3–18 mm, margins entire or sometimes coarsely toothed; sepals 0.7–6.1 mm, connate to 1/8 length; corolla tube and adaxial lip lavender or blue to white, abaxial lobes white with purple to blue markings, tube 2.5–8 mm, adaxial lip 1/2 abaxial; stamens 2; staminodes each with appendage and distal segment; capsules 1.4–6.3(–7.5) × 1.2–3.3 mm. Moist to wet disturbed habitats, 5400–5900 ft. Rio Arriba, Sandoval (Lewis et al., 2019). Previously included in Plantaginaceae.

LOASACEAE – STICKLEAF FAMILY

John J. Schenk, Josh M. Brokaw, and Larry Hufford

Herbs or subshrubs with stinging or nonstinging hairs; leaves alternate, simple, stipules absent, petiole present or absent, blade margins entire, toothed, or lobed; inflorescences terminal cymes, thyrses, or solitary flowers; flowers radially symmetric; hypanthium present; sepals 5, distinct or basally connate; petals 5, distinct; nectary disk on dome of ovary or absent; stamens 5 to many, distinct; staminodes present or absent, sometimes petaloid and petals appearing to be > 5; pistil 1, 3–7-carpellate, ovary inferior, 1-locular, placentae parietal or subapical; style 1; stigma 1, 2–7-lobed; fruits capsules or cypselae; seeds 1 to many per fruit (Kartesz, 1999).

1. Sepals longer than petals; stamens 5; fruit a cypsela; seed 1; stinging hairs present...**Cevallia**
1'. Sepals shorter than petals; stamens 10+; fruit a capsule; seeds usually > 1; stinging hairs absent... **Mentzelia**

CEVALLIA Lag. – Stingleaf

Cevallia sinuata Lag.—Stinging serpent. Subshrubs, to 6 dm; leaves ca. 60 × 20 mm, pinnately lobed about halfway to midrib; flowers with 5 linear or lanceolate sepals longer than petals; sepals ca. 8 mm; petals straplike, similar to sepals, inner surface yellowish, densely covered by long, pointed hairs; stamens 5, filaments dorsiventrally flattened and shorter than anthers; gynoecia uniloculate with 1 ovule, stigma covered densely by pointed hairs; cypselae with persistent perianth. Gypsum and limestone hills, gravelly flats of open grassland scrub vegetation, 2950–6575 ft. EC, C, SE, SC, SW.

MENTZELIA L. – Blazingstar, stickleaf

Plants annual, biennial, or perennial herbs or subshrubs; stems woody to herbaceous, turning white, gray, or tan and often exfoliating with age; leaves petiolate or sessile, stinging hairs absent; inflorescences dichasial cymes or solitary flowers; sepals basally connate, shorter than petals; petals alternate to sepal lobes; stamens 10+, inserted on hypanthium, all fertile or some infertile (petal-like) staminodes, filaments all dorsiventrally flattened when staminodial; carpels 3 or 5 (6, 7); ovules 1 to many, oriented perpendicular (horizontal) or parallel (vertical) to long axis of ovary; nectary disk on dome of ovary; fruits capsules that dehisce by apical valves, straight or curved, perianth absent in fruits, prominent costal ridges running lengthwise (in *M. rusbyi*) or absent; seeds 1 to many.

1.   Outermost stamens opposite the sepals petal-like, with or without anthers; seeds with a peripheral wing…(2)
1′. Outermost stamens opposite the sepals filiform or spatulate, all stamens with anthers; seeds without a peripheral wing…(20)
2.   Petals white…(3)
2′. Petals light to golden-yellow…(6)
3.   Anther epidermis papillate…**M. humilis**
3′. Anther epidermis smooth…(4)
4.   Petals 13+ mm wide; androecia white to yellow…**M. decapetala**
4′. Petals < 11 mm wide; androecia white…(5)
5.   Petals 22.6–49 × 3.6–10.3 mm; bracts adnate to or subtending ovary pinnate…**M. nuda**
5′. Petals 14.7–22(–24.4) × 1.9–4.4 mm; bracts adnate to or subtending ovary entire to slightly toothed…
     M. strictissima
6.   Capsules with prominent longitudinal costal ridges…**M. rusbyi**
6′. Capsules without prominent longitudinal costal ridges…(7)
7.   Petals with pubescent abaxial surfaces…**M. cronquistii**
7′. Petals with glabrous abaxial surfaces…(8)
8.   Anther epidermis papillate…(9)
8′. Anther epidermis smooth…(10)
9.   Flowers with > 5 staminodes, the 5 outermost stamens opposite sepal lobes and the second whorl of stamens without anthers…**M. perennis**
9′. Flowers with 5 staminodes, the 5 outermost stamens opposite sepal lobes lacking anthers, the second whorl fertile…**M. todiltoensis**
10. Plants with multiple branches arising from a subterranean branching caudex…**M. springeri**
10′. Plants with a single primary branch, or multiple branches arising from a ground-level caudex…(11)
11. Leaves of primary axis pinnatisect (sometimes becoming pinnate in *M. laciniata*)…(12)
11′. Leaves on primary axis entire, dentate, or serrate to pinnately lobed…(15)
12. Petals > or equal to 30 mm; the 5 outermost stamens opposite sepal lobes > 26 mm…**M. conspicua**
12′. Petals < 26 mm; the 5 outermost stamens opposite sepal lobes < 22 mm…(13)
13. Leaf lobes of primary shoot strongly angled toward leaf apex…**M. holmgreniorum**
13′. Leaf lobes 1/3 of primary shoot perpendicular or slightly angled toward leaf apex…(14)
14. Leaf intersinus distance 1–2.4 mm, lobes perpendicular to leaf axis; seed coat cells with 42–48 central papillae; McKinley and San Juan Counties…**M. filifolia**
14′. Leaf intersinus distance 1.4–4 mm, lobes slightly angled toward leaf apex; seed coat cells with 5–14 central papillae; Rio Arriba, San Juan, and Taos Counties…**M. laciniata**
15. Anticlinal walls of seed coat cells straight…(16)
15′. Anticlinal walls of seed coat cells wavy to sinuate…(17)
16. Capsules generally > 2 times as long as wide; NE New Mexico…**M. reverchonii**
16′. Capsules 2 times as long as wide or less; S New Mexico…**M. longiloba** (in part)
17. Outermost stamens opposite sepal lobes with anthers; NW New Mexico…**M. sivinskii**
17′. Outermost stamens opposite sepal lobes without anthers (sometimes present in *M. longiloba*); NC or S New Mexico…(18)
18. Leaf intersinus distances at widest point no wider than 3.9 mm; petals light yellow…**M. procera**
18′. Leaf intersinus distances at widest point of some leaves > 3.9 mm; petals light to golden-yellow…(20)
19. Seed coat cells with 4–6 or 67–106 papillae per cell; petals light to golden-yellow, (11.4–)13.8–24.4(–26.9) mm; Chihuahuan Desert…**M. longiloba** (in part)
19′. Seed coat cells with 29–48 papillae per cell; petals golden-yellow, 11.3–20.4 mm; N half of New Mexico …**M. multiflora**
20. Petals abaxially pubescent on upper 1/2; fruits erect or recurved downward from base (sometimes slightly

curved in *M. oligosperma*); seeds oblong, oval, or pyriform, dorsiventrally flattened or trigonal and 3-ridged; seed coat testal cells oblong and usually sinuate, much longer than wide…(21)

20′. Petals abaxially glabrous; fruits axillary, curved to 45°-180°; seeds irregularly polygonal, angular, or rounded or triangular prisms; seed coat testal cells polygonal, nearly equal-sided…(22)

21. Plants annual, to 2.5 dm; petals 5-8 mm; stamens nearly all the same length…**M. asperula**

21′. Plants perennial, to 5 dm; petals (6-)8-18.5 mm; outer stamens longer than inner stamens…**M. oligosperma**

22. Basal leaves not persisting; margins of proximal-most remaining leaves (proximal cauline) dentate or entire; leaves to 60 mm…**M. thompsonii**

22′. Basal leaves persisting; proximal-most leaves usually deeply to shallowly lobed, rarely entire; leaves to 130-150 mm…(23)

23. Bracts green with entire margins, or if lobed, lateral lobes not prominent; capsules 8-28(-35) mm (longest capsules usually > 15 mm), axillary curved to 180°…**M. albicaulis**

23′. Bracts with either toothed or lobed margins, or if entire, green with white base, margins usually 3-7-lobed, rarely entire, lateral lobes usually prominent; capsules 6-17(-20) mm, axillary curved to 45°…**M. montana**

Mentzelia albicaulis (Douglas) Douglas ex Torr. & A. Gray [*M. mojavensis* H. J. Thomps. & J. E. Roberts]—White-stem blazingstar. Annual herbs, (0.2-)1-4(-5) dm; leaves to 150 mm, margins entire to deeply lobed; petals abaxially glabrous, orange proximally, yellow distally, 3-7(-8) × 3-6 mm, adaxially; capsules 8-28(-35) × 1.5-3.5(-15+) mm; unwinged seeds usually irregularly polygonal, occasionally triangular prisms proximal to midfruit, coat tuberculate. Sand dunes, gravel fans, washes, 3770-7545 ft. NC, NW, C, WC, SC, SW.

Mentzelia asperula Wooton & Standl.—Mountain stickleaf. Annual herbs, to 2.5 dm, from a taproot; leaves to 150 mm, margins serrate; petals orange, 5-8 × 3-5 mm, hairs limited to apex on abaxial surface, abaxially otherwise glabrous; capsules 12-25 × 3-5 mm, erect; unwinged seeds pyriform, but dorsiventrally flattened to uncommonly trigonal in cross-section. Rocky limestone, igneous slopes, arroyo bottoms in grasslands and oak woodlands, 4200-8000 ft. SC, SW.

Mentzelia conspicua Todsen—Remarkable blazingstar. Biennial herbs, 3.8-11.1 dm, branches from a ground-level caudex; leaves 69-195 mm, margins pinnatisect, lobes slightly angled or perpendicular to leaf axis; petals golden-yellow, 30-42.2 × 5.7-10.8 mm, abaxially glabrous; anther epidermis smooth; capsules 15-26 × 5-7.2 mm, erect; seeds winged, seed coat anticlinal cell walls sinuate. Slopes of piñon-juniper woodlands and grasslands, sparsely vegetated red and brown loam, 5900-7875 ft. Rio Arriba. **R**

Mentzelia cronquistii H. J. Thomps. & Prigge [*M. marginata* (Osterh.) H. J. Thomps. & Prigge var. *cronquistii* (H. J. Thomps. & Prigge) N. H. Holmgren & P. K. Holmgren]—Cronquist's blazingstar. Biennial herbs, 3-8 dm, branches from a ground-level caudex; leaves 21-101 mm, margins dentate to pinnate, lobes perpendicular to leaf axis; petals golden-yellow, 9-16.6 × 2.5-5.1 mm, abaxially pubescent; capsules 5.9-10.6(-11.4) × 5-7.6 mm, erect; seeds winged, seed coat anticlinal cell walls wavy. Washes, roadside banks, steep slopes, sandy and rocky soils, 4700-7550 ft. San Juan.

Mentzelia decapetala (Pursh) Urb.—Ten-petalled western star. Perennial (biennial or short-lived) herbs or subshrubs, 3.1-10 dm, branches from a ground-level caudex; leaves 72-295 mm, margins serrate to pinnate; petals white, 47-75 × 13-22.7 mm, abaxially glabrous; anther epidermis smooth; capsules 30-43 × 12-17 mm, erect; seeds winged, seed coat anticlinal cell walls straight. Rock outcrops, slopes of dry shortgrass prairies, riverbanks, roadsides in loam, limestone, sandy, clay, and gravelly soils, 3735-7875 ft. NC, EC.

Mentzelia filifolia J. J. Schenk & L. Hufford—Narrow-leaved blazingstar. Biennial herbs, 3.3-7.5 dm, branches from a ground-level caudex; leaves 43-94(-115) mm, margins pinnatisect; petals golden-yellow, 13-18.5 × 3.7-6.1 mm, abaxially glabrous; anther epidermis smooth; capsules 11-19 × 5-7.5 mm, erect; seeds winged, seed coat anticlinal cell walls sinuate. Roadcuts and slopes in dark loam and rocky soils, 6890-7200 ft. McKinley, San Juan. **R**

Mentzelia holmgreniorum J. J. Schenk & L. Hufford—Holmgren's blazingstar. Biennial herbs, 2.7–5 dm, branches from a ground-level caudex; leaves 42–89 mm, margins pinnatisect; petals golden-yellow, 13.5–18.8 × 5.2–6.6 mm, abaxially glabrous; anther epidermis smooth; capsules 13.1–14.6 × 5.8–6.9 mm, erect; seeds winged, seed coat anticlinal cell walls sinuate. Dry sandy washes, roadsides other disturbed areas, 4595–7545 ft. Catron.

Mentzelia humilis (Urb. & Gilg) J. Darl. [*Nuttallia gypsea* Wooton & Standl.]—Gypsum blazingstar. Perennial herbs, 1.2–5.7 dm, branches from a ground-level caudex; leaves 25–95 mm, margins entire, dentate, or pinnatisect; petals white, 11.8–21.3(–24.6) × 1.4–4 mm, abaxially glabrous; anther epidermis papillate; capsules 5.3–10.2 × (4.2–)5.2–8.6 mm, erect; seeds winged, seed coat anticlinal cell walls wavy. Sparsely vegetated areas in dry grasslands, knolls, roadsides, level areas or gentle slopes in gravelly, clay, and sandy gypsum substrates.

1. Leaf intersinus distance at widest point generally not > 2.6 mm…**var. humilis**
1'. Leaf intersinus distance at widest point on some leaves > 2.6 mm, to 9.1 mm…**var. guadalupensis**

var. guadalupensis Spellenb.—Guadalupe Mountain blazingstar. Leaves 25–85 mm, margins pinnate; petals 10.3–13(–28.6) × 1.6–2.3(–4) mm; outer whorl of staminodes 8.8–11(–23.3) × 1–1.4(–1.9) mm; fruits 5.3–6.8(–8) × 5.2–6.1 mm; seeds with 6–9 central papillae per cell. 4595–5250 ft. Otero. **R & E**

var. humilis—Gypsum blazingstar. Leaves 25.8–95 mm, margins entire, dentate, or pinnatisect; petals 11.8–21.3(–24.6) × 1.4–3.7 mm; outer whorl of staminodes 10.6–19(–22.3) × 0.7–3.3 mm; fruits 4.8–10.2 × (4.2–)5.2–8.6 mm; seeds with 7–12 central papillae per cell. 2950–5250 ft. EC, SE.

Mentzelia laciniata (Rydb.) J. Darl.—Cut-leaf blazingstar. Biennial herbs, 2.3–8 dm, branches from a ground-level caudex; leaves 52–112 mm, margins (occasionally, especially toward apex) pinnate to pinnatisect; petals golden-yellow, 14–23.8(–26) × 3.8–7.4 mm, abaxially glabrous; capsules 12–20.2 × 4.5–8.1 mm, erect; seeds winged, seed coat anticlinal cell walls sinuate. Dry hillsides, roadcuts, and roadsides in sandy or clay soils, 4595–7545 ft. NC, NW.

Mentzelia longiloba J. Darl.—Dune blazingstar. Biennial or perennial herbs, 1.5–7 dm, branches from a ground-level caudex; leaves 35–112 mm; margins dentate, serrate, or pinnate; petals golden-yellow, 11.3–20.4 × 2.9–7.2(–8.9) mm, abaxially glabrous; anther epidermis smooth; capsules (7.6–)9.1–16.4 × 5.7–9.2 mm, erect; seeds winged. Roadsides, sand dunes, hills, and washes in dry clay or sandy soils.

1. Seed coat cells with anticlinal walls straight, central papillae < 7 per cell; leaves on proximal 1/3 of primary axis with dentate margins…**var. chihuahuaensis**
1'. Seed coat cells with anticlinal walls sinuate, central papillae 10+ per cell; leaves on proximal 1/3 of primary axis generally with pinnate margins…**var. longiloba**

var. chihuahuaensis J. J. Schenk & L. Hufford—Chihuahuan blazingstar. Leaves 35–110 mm, margins dentate (lower) or serrate to pinnate (upper); petals 11.3–16.3 × 3.1–5.1 mm; capsules 9.1–15 × 5.7–8.3 mm; seed coat anticlinal cell walls straight. 2840–5250 ft. Hidalgo.

var. longiloba—Dune blazingstar. Leaves 38–112 mm, margins pinnate; petals 12.2–20.4 × 3.7–7.2 mm; capsules 9.6–16.4 × 6–9.2 mm; seed coat anticlinal cell walls sinuate. 4595–6400 ft. Catron, Doña Ana.

Mentzelia montana (Davidson) Davidson—Mountain blazingstar. Annual herbs, (0.5–)2–4(–5) dm; leaves to 130 mm, margins entire to deeply lobed; petals orange proximally, yellow distally, 2–6(–8) × 2–5 mm; fruits 6–17(–20) × 2–3 mm, axillary curved to 45° at maturity; unwinged seeds irregularly polygonal, occasionally triangular prisms proximal to midfruit; seed coat tuberculate. Open, disturbed slopes or flats, grasslands, sagebrush scrub, coniferous forests, 4500–11,155 ft. NW, WC, SW, Doña Ana.

Mentzelia multiflora (Nutt.) A. Gray [*M. lutea* Greene]—Adonis blazingstar. Biennial herbs, 2.3–11.7 dm, branches from a ground-level caudex; leaves 35.9–125(–146) mm, margins pinnate; petals light to golden-yellow, (11.4–)13.8–24.4(–26.9) × 4–7.5 mm, abaxially glabrous; anther epidermis smooth; capsules 11.2–24.7 × 5.6–8.7 mm, erect; seeds winged, seed coat anticlinal cell walls sinuate. Dry roadsides, hillsides, and washes in clay, rocky, and/or sandy soils, 3935–6890 ft. NE, NC, NW, C, WC.

Mentzelia nuda (Pursh) Torr. & A. Gray [*M. stricta* (Osterh.) G. W. Stevens]—Goodmother. Perennial, biennial, or annual herbs, 2.3–11.7 dm, branches from a ground-level caudex; leaves 37–120 mm, margins serrate; petals white, 22.6–49 × 3.6–10.3 mm, abaxially glabrous; anther epidermis smooth; capsules 14.5–29 × 6.9–12.3 mm, erect; seeds winged, seed coat anticlinal cell walls wavy. Disturbed roadsides, hillsides, and creek banks in sandy and rocky soils, 3060–7545 ft. NE, EC.

Mentzelia oligosperma Nutt.—Stickleaf. Perennial herbs to 5 dm, from a single primary branch; leaves to 100 mm, margins serrate or less commonly crenate (entire); petals orange, (6–)8–18.5 × (3–)4–10.5 mm, pubescent abaxially on upper 1/2; capsules (5–)7–17 × 2–3.5 mm, erect (slightly curved < 45°) or sometimes bent strongly downward from near base; unwinged seeds oval with 1 side flat to concave but otherwise convex. Limestone, gypsum, or sandstone rock outcrops or cliffs in clay or loam flats, 3460–6825 ft. NE, EC, C, SW, Eddy.

Mentzelia perennis Wooton—Perennial blazingstar. Perennial herbs, 1.6–5.5 dm, branches from a ground-level caudex; leaves 25–100 mm, margins entire, dentate, serrate, or pinnatisect; petals light yellow, 11.4–19(–22.7) × 1.8–3.6(–5) mm, abaxially glabrous; anther epidermis papillate; capsules 6.8–12 × (4–)4.8–9.1 mm, erect; seeds winged, seed coat anticlinal cell walls wavy. Roadsides and hillside slopes in gypsum-rich soils, 3935–7215 ft. C, SC. **E**

Mentzelia procera (Wooton & Standl.) J. J. Schenk & L. Hufford [*M. pumila* Torr. & A. Gray var. *procera* (Wooton & Standl.) J. Darl.]—Upright blazingstar. Biennial herbs, 2.7–9.4 dm, branches from a ground-level caudex; leaves 38.3–108 mm, margins pinnate, petals light yellow, 9.5–16.8 × 3.4–5.6 mm, abaxially glabrous; anther epidermis smooth; capsules 9.8–18.8 × 5.2–7.3 mm, erect; seeds winged, seed coat anticlinal cell walls sinuate. Dry hillsides and roadsides in sandy, clayey, or silty soils, 4595–8200 ft. Widespread.

Mentzelia reverchonii (Urb. & Gilg) H. J. Thomps. & Zavort. [*M. hintoniorum* B. L. Turner & A. L. Hempel]—Reverchon's blazingstar. Biennial or perennial herbs, 1.6–8.5 dm, branches from a ground-level caudex; leaves 24.2–54 mm, margins dentate to pinnate; petals golden-yellow, (9.6–)11.1–14.1(–22) × 2.5–4.2(–5.5) mm, abaxially glabrous; anther epidermis smooth, capsules 9.4–22 × 6–9.11 mm, erect; seeds winged, seed coat anticlinal cell walls straight. Grasslands on eroded riverbanks, roadsides, roadcuts, and sparsely vegetated hillsides in sandy, gravelly, clayey, and occasionally gypsum soils, 4000–7000 ft. Union.

Mentzelia rusbyi Wooton—Rusby's blazingstar. Biennial or perennial herbs, 3.4–12.7 dm, branches from a ground-level caudex; leaves 41–123 mm, margins dentate; petals light yellow, 11.8–23.8 × 3–6.9 mm, abaxially glabrous; anther epidermis smooth; capsules (13–)18.8–29 × 7–10.5 mm, erect; seeds winged, seed coat anticlinal cell walls sinuate. Mesic habitats along moist washes, roadsides, and roadcuts, steep to gentle slopes in rocky sand and loam, 5905–10,170 ft. NE, NC, NW, C, WC, SC.

Mentzelia sivinskii J. J. Schenk & L. Hufford [*M. linearifolia* S. L. Welsh & N. D. Atwood]—Sivinski's blazingstar. Biennial herbs, 5.2–7 dm, branches from a ground-level caudex; leaves 32.8–112.2 mm, margins pinnate; petals light to golden-yellow, 9–14.7 × 3.1–6.4 mm, abaxially glabrous; capsules 8.2–12.7 × 5.1–7.7 mm, erect; seeds winged, seed coat anticlinal cell walls sinuate. Knolls, slopes, and grassland roadsides in gypsum or brown clay soils, 4920–6235 ft. NW. **R**

Mentzelia springeri (Standl.) Tidestr.—Springer's blazingstar. Perennial subshrubs, 1.8–4.1 dm, multiple branches from a subterranean branching caudex; leaves 18–56 mm, margins entire (upper) or dentate to pinnate (lower); petals golden-yellow, 8.7–14.2 × 3.1–4.7 mm, abaxially glabrous; anther epidermis smooth; capsules 5.9–10.3 × 3.8–4.8 mm, erect; seeds winged, seed coat anticlinal cell walls wavy. Sparsely vegetated steep talus and pumice slopes, 5250–7215 ft. Los Alamos, Sandoval, Santa Fe. **R & E**

Mentzelia strictissima (Wooton & Standl.) J. Darl.—Grassland blazingstar. Biennial herbs, 2.6–9.8 dm, branches from a ground-level caudex; leaves 28–70 mm, margins (serrate) dentate to pinnate; petals white, 14.7–22(–24.4) × 1.9–4.4 mm, abaxially glabrous; anther epidermis smooth; capsules 10.2–20.1 × 6.7–9.6 mm, erect; seeds winged, seed coat anticlinal cell walls wavy. Arid grasslands, 3060–5250 ft. EC, SE.

Mentzelia thompsonii Glad [*Acrolasia humilis* Osterh.]—Thompson's stickleaf. Annual herbs, 0.5–2 dm from a taproot; leaves to 60 mm, margins entire; petals yellow, 2–4 × 2–4 mm, abaxially glabrous; capsules 5–16(–20) × 2–4 mm, axillary curved to 45° at maturity; unwinged seeds irregularly polygonal, seed coat smooth to minutely tessellate. Barren clay to silt slopes, 4750–6560 ft. San Juan.

Mentzelia todiltoensis N. D. Atwood & S. L. Welsh—Jemez Mountains blazingstar. Biennial or perennial herbs, 1.7–6.3 dm, branches from a ground-level caudex; leaves 41–121 mm, margins entire or serrate to pinnatisect; petals light to golden-yellow, (10.4–)11.7–24.6 × 1.8–5.1 mm, abaxially glabrous; anther epidermis papillate, capsules 6.7–20.2 × 4.5–8.5 mm, erect; seeds winged, seed coat anticlinal cell walls sinuate. Hillside slopes in hard clay soils rich with gypsum, 5250–7215 ft. Bernalillo, Sandoval, Santa Fe. **R & E**

LYTHRACEAE – LOOSESTRIFE FAMILY

Kenneth D. Heil

Herbs, shrubs, or trees; leaves opposite, whorled, or alternate, simple, entire; stipules minute or wanting; flowers perfect, regular or irregular, solitary or clustered, 4–7-parted; calyx tubular to campanulate, 4–6-toothed; petals inserted in the throat of the hypanthium between the lobes or rarely absent; stamens 4 to many, inserted on the hypanthium; style simple or wanting; stigma capitate; fruit a capsule (Correll & Correll, 1972; Graham, 1986, 2021; Martin & Hutchins, 1981a).

1. Plants large shrubs, cultivars; leaves opposite; hypanthium in flower leathery, 2–5 cm across…**Punica**
1'. Plants herbaceous or woody only at the base; leaves opposite or alternate; hypanthium in flower membranous, to 1 cm across…(2)
2. Flowers irregular; plants glandular-hispid…**Cuphea**
2'. Flowers regular or nearly so; plants glabrous…(3)
3. Hypanthium elongate, cylindrical or tubular…**Lythrum**
3'. Hypanthium campanulate or turbinate, becoming hemispherical or globose…**Ammannia**

AMMANNIA L. – Ammannia

Annual herbs, low and inconspicuous; stems mostly 4-angled; leaves opposite, sessile, entire, flowers small, mostly 3- to many-flowered axillary cymes, 4-merous; calyx globose or campanulate, usually with a little horn-shaped appendage in each sinus; petals 4, small, deciduous; stamens 4–8; seeds numerous.

1. Plants perennial, woody at the base; petals 6; stamens 10–14…**A. grayi**
1'. Plants annual; petals 0 or 4; stamens 4–8…(2)
2. Peduncles filiform; capsules surpassing the calyx, ca. 2.5 mm diam.…**A. auriculata**
2'. Peduncles lacking; capsules equal to or shorter than the calyx, ca. 4–5 mm diam.…**A. coccinea**

Ammannia auriculata Willd.—Eared redstem. Stem branches to 8 dm; leaves linear-lanceolate to linear, to 5 cm × 7 mm wide; cymes loosely 3+-flowered on peduncles to 5+ mm or with solitary flowers on pedicels; calyx 1.5–2 mm, in fruit becoming subglobose; petals minute, purple to white, soon dropping; seeds reddish brown. Ditches and pond margins, 4000–4500 ft. Doña Ana.

Ammannia coccinea Rottb.—Valley redstem. Stems to 5 dm; leaves linear-oblong to linear-lanceolate, to 1 dm × 15 mm wide; cymes closely 2–5-flowered, essentially sessile; calyx 2.5–5 mm, in fruit 3–5 mm diam.; petals pink to purple, 1–2 mm; capsule ca. 4 mm. Mud of ditches, ponds, marshes, lakes, streams, 4400–5200 ft. C, SC.

Ammannia grayi S. A. Graham & Gandhi—Gray's redstem. Stems to 9+ dm; leaves linear, to 7 cm × 4 mm wide; flowers paired at the nodes, petals pink-purple; capsules ca. 4 mm. Limestone seeps, elevation unknown. Eddy (probably).

CUPHEA P. Browne – Waxweed

Cuphea wrightii A. Gray—Wright's waxweed. Glandular-hispid annuals, stems to 40 cm, erect, leafy, herbage sparsely hirsute, the hairs spreading, tinged with purple or red, glandular; leaves entire, opposite, ovate to lance-ovate, to 35 mm; flowers axillary, solitary, on pedicels to 6 mm, the upper leaves reduced to small purplish bracts; calyx purple, 12-ribbed, with 6 small teeth; petals 6, purple; stamens 10–12; capsules ovoid, 5–8 mm. Fields, roadsides, edges of streams and lakes, 5000–7500 ft. Grant, Hidalgo.

LYTHRUM L. – Loosestrife

Perennial herbs with basal offshoots, 4-angled stems 1–12 dm; leaves opposite to subopposite, ovate to linear; flowers solitary, paired in the axils or numerous and forming terminal spikes; floral tubes cylindrical, greenish, 8–12-nerved; calyx lobes alternating with appendages, the appendages longer than the lobes; petals 6, ours purple or rose-purple; stamens 6 or 12; ovary 2-locular; seeds many, 1 mm or less.

1. Flowers solitary or paired in the axils; stamens 6(-8)…**L. californicum**
1'. Flowers numerous in showy terminal spikes; stamens 12…**L. salicaria**

Lythrum californicum Torr. & A. Gray—California loosestrife. Perennial, 2–6(–10) dm; leaves linear-oblong to linear-lanceolate, 1–2 cm × 3–8 mm; flowers solitary or paired in the axils; floral tubes 4–7 mm; petals purple, 4–6 mm; stamens 6; thickened ring at the base of the ovary, narrowed on 1 side. Wet soils, ditches, roadsides, lake margins, 3300–8250 ft. WC, SE, SC, SW.

Lythrum salicaria L.—Purple loosestrife. Often pubescent perennial, to 12 dm; leaves opposite or whorled, lanceolate, 2–20 cm × 5–15 mm; flowers in showy terminal spikes; floral tubes 4–6 mm; petals rose-purple; stamens mostly 12; ring at the base of the ovary absent. Wet meadows, marshes, ditches, 5600–6560 ft. Bernalillo, Grant, Rio Arriba, Valencia. Considered a noxious weed in much of its range; however, in New Mexico it is widely scattered. Introduced.

PUNICA L. – Pomegranate

Punica granatum L.—Pomegranate. (Ours) an upright shrub ca. 1 m; leaves opposite, lanceolate to oblanceolate or oblong, shiny green, glabrous, blades 2–9 cm; hypanthium tube fleshy, ca. 2 cm in flower, persistent in fruit; petals 5–9 or many, 15–25 mm; fruit globose, with a leathery skin. Commonly cultivated and escaped, 3900–5200 ft. C, SC, SW. Introduced.

MALPIGHIACEAE – MALPIGHIA FAMILY

Kenneth D. Heil

Shrubs (ours) or trees; leaves opposite, simple, blade margins usually entire, often bearing multicellular glands on margin or abaxial [sometimes adaxial] surface; inflorescences terminal or axillary; flowers bisexual, radially or bilaterally symmetric, sepals 5, usually glandular, petals mostly 5, posterior (flag) petal often different from lateral 4, distinct, mostly clawed; stamens 10, the filaments connate into a tube; pistil single, superior, of 3 united carpels; fruit a schizocarp splitting into winged segments, or a nutlet (Anderson, 2016).

1. Flower petals carrot-yellow; fruit a nutlet…**Aspicarpa**
1'. Flower petals lemon-yellow; fruit a schizocarp, breaking into 3 samaras…**Cottsia**

ASPICARPA Rich. – Asphead

Aspicarpa hirtella Rich. [*A. longipes* A. Gray]—Chaparral asphead. Plants erect, 10–20 cm, or decumbent; leaf blades narrowly to broadly lanceolate or ovate, 15–45 × 6–23 mm, surfaces thinly sericeous or velutinous; chasmogamous flowers mostly in (2–)4(–7)-flowered umbels, petals carrot-yellow; cleistogamous flowers often borne singly, or in clusters; nutlet 3–3.5 mm diam., smooth or rugose. Dry rocky slopes with oak, ponderosa pine, juniper, 4800–6270 ft. Hidalgo.

COTTSIA Dubard & Dop – Janusia

Cottsia gracilis (A. Gray) W. R. Anderson & C. Davis [*Janusia gracilis* A. Gray]—Slender janusia. Vines, twining; leaf blades narrowly lanceolate or elliptic, 12–40(–50) × (1.5–)3–7(–9) mm, surfaces persistently sericeous or sometimes glabrescent; flowers all chasmogamous, 6+ mm diam., showy, with visible petals; calyx glands mostly 8; petals lemon-yellow; fruit a schizocarp, breaking into 3 samaras, 9–15(–17) mm. Open rocky slopes and deserts, 4000–6920 ft. SC, SW.

MALVACEAE – MALLOW FAMILY

Margaret M. Hanes

Herbs, subshrubs, or shrubs (ours), usually stellate-hairy; leaves alternate, usually spiral, sometimes distichous, usually petiolate, sometimes subsessile or sessile, stipulate (usually well developed), simple; blades unlobed or palmately lobed, palmately veined; inflorescences axillary, terminal, or leaf-opposed; flowers bisexual or unisexual; epicalyx present, absent or deciduous, 3–17-bracted (ours); sepals 5; petals 5; style 5–28-branched (ours); stamens 5–10 to many; ovules 1–2 to many per ovary; fruits capsules or schizocarps (ours); seeds glabrous or hairy.

1. Petals clawed; staminodes present, fused into cylindrical or campanulate tube; anthers 3-thecate… **Ayenia**
1′. Petals without clawlike appendage; staminodes absent or much reduced; anthers 1-thecate…(2)
2. Epicalyx absent (sometimes present in *Callirhoe* and *Malvella*)…(3)
2′. Epicalyx present (sometimes absent in *Callirhoe*, *Malvella*, and *Sphaeralcea*)…(11)
3. Stigmas introrsely decurrent…(4)
3′. Stigmas capitate…(5)
4. Staminal columns not divided, filaments all arising equally…**Callirhoe** (in part)
4′. Staminal columns apically divided into inner and outer series of filaments…**Sidalcea**
5. Mericarps 2-celled…**Allowissadula**
5′. Mericarps 1-celled…(6)
6. Calyces enclosing fruits, green, membranous…**Rhynchosida**
6′. Calyces closely subtending but not enclosing or concealing fruits…(7)
7. Fruits inflated, fragile-walled, setose, pendent…**Herissantia**
7′. Fruits not inflated, often indurate, variously hairy, seldom setose, usually erect…(8)
8. Mericarps 3–6-seeded…**Abutilon**
8′. Mericarps 1-seeded…(9)
9. Lateral walls of mericarps evanescent at maturity, dorsal walls usually spurred at dorsal angle (spur rarely absent)…**Anoda**
9′. Lateral walls of mericarps persistent, dorsal spur usually absent…(10)
10. Leaves asymmetric; plants prostrate, often in saline habitats; pubescence sometimes ± lepidote…**Malvella** (in part)
10′. Leaves symmetric; plants usually erect (less often prostrate), not in saline habitats; pubescence never lepidote…**Sida**
11. Fruits capsules; carpels 3–5…(12)
11′. Fruits schizocarps; carpels 6+…(13)
12. Epicalyx bracts 3; seeds comose…**Gossypium**
12′. Epicalyx bracts 4+; seeds not comose…**Hibiscus**
13. Epicalyx bracts 6–9…**Alcea**
13′. Epicalyx bracts 3 or absent…(14)

14. Leaves asymmetric; herbage mixed stellate- and lepidote-hairy…**Malvella** (in part)
14′. Leaves symmetric; herbage usually stellate-hairy or glabrous, seldom lepidote…(15)
15. Stigmas introrsely decurrent…(16)
15′. Stigmas capitate…(17)
16. Mericarps obtusely beaked…**Callirhoe** (in part)
16′. Mericarps not beaked…**Malva**
17. Stigmas obliquely capitate; leaf blades aceriform…**Iliamna**
17′. Stigmas capitate; leaf blades not aceriform…**Sphaeralcea**

ABUTILON Mill. – Indian mallow

Subshrubs, shrubs, or herbs; stems erect, trailing, or ascending, glabrous or pubescent; stipules usually persistent; leaf blades ovate, sometimes shallowly lobed; epicalyx absent; calyx usually not accrescent (ours), not inflated; corolla yellow or orange, less often pinkish, sometimes with dark red center; style 5-25-branched; fruits schizocarps, erect, not inflated, variably hairy; mericarps attached to adjacent mericarps, apex usually acute or acuminate, seeds 3 per mericarp (ours) (Fryxell, 1977, 1983, 1988; Spencer, 1984).

1. Mericarps and style branches 5…(2)
1′. Mericarps and style branches 6-15…(4)
2. Plants trailing or decumbent; leaf surfaces sparsely pubescent; flowers solitary, petals without dark center…**A. parvulum**
2′. Plants erect; leaf surfaces densely pubescent; flowers solitary or in panicles, petals with or without dark center…(3)
3. Leaf margins sharply serrate; sepals 6-8 mm; petals 9-15 mm, yellow, without dark center…**A. malacum**
3′. Leaf margins irregularly crenate-serrate; sepals 3-5 mm; petals 4-6 mm, yellow or pink, with dark center…**A. incanum**
4. Mericarp apices distinctively spinose…**A. theophrasti**
4′. Mericarp apices acuminate or acute (or minutely spinose, subequal to 4 mm)…(5)
5. Plants to 0.5 m; leaves 1.5-4 cm, about as long as wide, discolorous; style 6-9-branched…(6)
5′. Plants 0.5-2 m; leaves 1-10 cm, usually longer than wide, discolorous or not; style 6-10-branched…(8)
6. Plants erect; petals orange or yellow; style 7-10-branched…(7)
6′. Plants procumbent or ascending; petals yellow, style 6-9-branched…**A. wrightii**
7. Petals orange; style 7-branched [to be expected in most SC part of state]…A. pinkavae Fryxell
7′. Petals yellow; style 8-10-branched…**A. mollicomum**
8. Sepals 3-5 mm; petals yellow, 5-10 mm…**A. fruticosum**
8′. Sepals 9-12 mm; petals yellow-orange, 10-12 mm [to be expected in SW]…A. abutiloides (Jacq.) Garcke ex Britt. & Wilson

Abutilon fruticosum Guill. & Perr. [*A. texense* Torr. & A. Gray]—Pelotazo. Subshrubs 1-1.5 m; leaves 1-10 cm, ovate, base cordate, apex acute, margins irregularly serrate, blades concolorous, petioles shorter than blade; flowers solitary or in few-flowered terminal panicles; sepals 2-5 mm, reflexed in fruit, lanceolate to ovate; petals 5-10 mm, spreading, cream to yellow; style 6-9-branched; mericarp apex acute. Open, dry habitats, 3700-6565 ft. SE, SC.

Abutilon incanum (Link) Sweet [*Sida incana* Link]—Pelotazo chico. Subshrubs 1-2 m; leaves to 6 cm, longer than wide, base cordate, apex acute or acuminate, margins irregularly serrulate or crenate, surface densely tomentose; flowers solitary or in open leafy-bracted panicles; sepals 3-5 mm, lanceovate; petals 4-6 mm, reflexed, yellow or pink with dark red center; style 5-branched. Open, dry hills and plains, 3460-5425 ft. S.

Abutilon malacum S. Watson—Yellow Indian mallow. Subshrubs to 1 m, stellate-pubescent throughout; leaves 3-10 cm, ovate, base cordate, margins serrate; flowers in compact panicles; sepals 6-8 mm, lanceolate, apex acute to acuminate; petals 6-15 mm, yellow or orange; style 5-branched. Dry, open habitats, 3900-5575 ft. S.

Abutilon mollicomum (Willd.) Sweet [*A. sonorae* A. Gray]—Sonoran Indian mallow. Shrubs, 1-2 m; stems hirsute; leaves 10-20 cm, broadly ovate to shallowly 3-5-lobed, stellate-hairy; sepals 4-6 mm; petals 5-8 mm, yellow; style 8-10-branched. Dry desert slopes, 4700-5025 ft. SW.

Abutilon parvulum A. Gray—Dwarf Indian mallow. Herbs or subshrubs, perennial, to 0.5 m, trailing; leaves broadly ovate, base cordate, apex acute, margins coarsely dentate, blades concolorous, petioles 1/2 to as long as blade; flowers axillary, solitary; sepals 3-5 mm, reflexed in fruit; petals 4-7 mm, pink, rarely orange; style 5-branched. Dry, open habitats, 3800-6200 ft. NE, NC, S.

Abutilon theophrasti Medik. [*Sida abutilon* L.]—Butterprint, velvetleaf. Annual herbs to 1 m; leaves 8-15 cm, ovate to suborbicular, base cordate, apex acuminate, margins crenulate, surface softly pubescent, blades concolorous, petioles subequal to blade; flowers usually axillary, solitary; sepals 10 mm, nonoverlapping, erect in fruit; petals 8-13 mm, pale yellow; style 13-15-branched; mericarp apex spinose. Disturbed habitats, 3900-4600 ft. C, SE. Introduced.

Abutilon wrightii A. Gray—Wright's Indian mallow. Subshrubs to 0.5 m; stems procumbent or ascending, sometimes purplish; leaves 1.5-4 cm, ovate, ± as long as wide, base deeply cordate, apex acute to obtuse, margins dentate, blades markedly discolorous, petioles often exceeding blade; flowers solitary; sepals 10-15(-20) mm, accrescent, overlapping at base; petals 14-18 mm, pale yellow; staminal column glabrous; style 6-9-branched; mericarp apex acuminate. Dry, open habitats, roadsides, 3880 5200 ft. SC, SE.

ALCEA L. – Hollyhock

Alcea rosea L. [*A. ficifolia* L.; *A. glabrata* Alef.; *Althaea ficifolia* (L.) Cav.; *A. mexicana* Kunze; *A. rosea* (L.) Cav.]—Common hollyhock. Biennial herbs with coarse stellate hairs, 1-3 m, erect; leaves petiolate, 5-7-lobed, 3-15 × 8-30 cm, orbicular, cordate at base, lobed; inflorescences terminal and/or axillary; epicalyx 6-9-bracted; calyx lobes triangular; corolla rotate, white, pink, red, or purple, darker or paler in center; fruit a disk-shaped dry schizocarp. Roadsides, disturbed areas, 5300-7095 ft. Cosmopolitan cultivated ornamental, frequently grown in the American West. Widespread. Introduced.

ALLOWISSADULA D. M. Bates – False Indian mallow

Allowissadula holosericea (Scheele) D. M. Bates [*Abutilon holosericeum* Scheele; *A. marshii* Standl.; *Wissadula holosericea* (Scheele) Garcke; *W. insignis* R. E. Fr.]—Plants 1-2 m, erect or spreading, subshrubs, variably glandular, stellate-hairy, sometimes velvety; leaves ovate, unlobed to 3-lobed, base cordate, margins entire to dentate; inflorescences terminal, open panicles or cymes in 2+ floral units, occasionally solitary; sepals 4-6.5 × 4-5.5 mm, campanulate, equal to or briefly exceeding fruits, ovate to triangular; petals 15-25 mm, orange-yellow; carpels stellate and glandular-hairy. Open habitats, roadsides; Guadalupe Mountains, 3460-5200 ft. SE.

ANODA Cav. – Anoda

Annual herbs or subshrubs; leaves variable in shape, base, and margins; flowers axillary, solitary or in terminal racemes or panicles; epicalyx absent; calyx apex acute or acuminate; corolla yellow, lavender, purple, rarely white; style 5-19-branched; fruit a schizocarp, erect, not inflated, not indurate; mericarps 1-seeded (Fryxell, 1987).

1. Corolla lavender to bluish lavender (rarely white)…(2)
1′. Petals pale to bright yellow…(3)
2. Plants decumbent to sometimes erect; petals exceeding the calyx, 8-30 mm; mericarps with a spur 1.5-4 mm…**A. cristata**
2′. Plants erect; petals barely exceeding the calyx, 4-7 mm; mericarps with a spur to 1 mm…**A. thurberi**
3. Petals bright yellow; pedicels 4-6 mm; sepals 6-9 mm; styles 10-12-branched; mericarp dorsal spine > 1 mm…**A. lanceolata**
3′. Petals pale yellow; pedicels 1.5-3.5 mm; sepals 3-5 mm; styles 5-8-branched; mericarp dorsal spine much < 1 mm…**A. pentaschista**

Anoda cristata (L.) Schltdl. [*Sida cristata* L.; *A. arizonica* A. Gray; *A. hastata* Cav.]—Crested anoda. Herbs to 1 m, usually shorter; leaves ovate to hastate, usually 3-9 cm, base cordate or truncate, apex acute, often with purple blotch on midvein, blades concolorous, petioles 1/2 as long as to equaling blade;

flowers solitary; pedicels 4–12 cm; sepals 5–10 mm (12–20 mm in fruit); petals 8–26 mm, lavender (rarely white); style 10–19-branched; mericarps with dorsal spur 1.5–4 mm. Weedy in disturbed areas, 3500–8200 ft. NW, WC, S.

Anoda lanceolata Hook. & Arn. [*A. wrightii* A. Gray]—Lanceleaf anoda. Herbs or subshrubs 0.5–1 m; leaves 3–7(–12) cm, base truncate to cuneate, margins crenate to subentire, apex acute, blades some-what discolorous, petioles shorter than blade; flowers solitary or in panicles; pedicels 4–6 cm; sepals 6–9 mm, 9 mm in fruit, apex acute; corolla bright yellow, 9–16 mm; style 10–12-branched; mericarps with dorsal spur 1–1.5 mm. Arid, often disturbed habitats in foothills, 5200–7350 ft. Grant.

Anoda pentaschista A. Gray [*Sidanoda pentaschista* (A. Gray) Wooton & Standl.]—Field anoda. Subshrubs 1–2 m; leaves variable in size and form, blade discolorous, margins usually entire, apex usually acute, petioles usually 1/4 length of blade; flowers in panicles; pedicels 1.5–3.5 cm; sepals 3–5 mm; pet-als 1 cm, yellow, sometimes fading reddish; style 5–8-branched; mericarps with dorsal spur to 0.5 mm. Disturbed habitats, 4380–5600 ft. SW.

Anoda thurberi A. Gray—Thurber's anoda. Annual, to 1 m; leaf blades ovate-cordate to hastately 3-lobed or triangular, minutely stellate-hairy, 3–8 cm, margins entire; sepals 3–6 mm, enlarging to 6–8 in fruit; petals 4–7 mm, bluish lavender; style 6–8-branched; mericarps with a dorsal spur to 1 mm. Shrub-lands in the Animas Mountains, ca. 5680 ft. SW.

AYENIA L. – Ayenia

Kenneth D. Heil

Mostly shrubs (ours); leaves alternate, simple, entire, often with stellate hairs; stipules present; epicalyx absent; flowers actinomorphic, perfect or unisexual; sepals 3–5; petals clawed, 5, distinct or connate; style 5-branched; stamens 5 or 10, the outer whorl reduced to staminodes, staminodes fused into cylin-drical or campanulate tube; pistil superior, of 2 to many united carpels; fruit a capsule (ours) (Correll & Johnston, 1979b; Dorr, 2015a). Previously included in Sterculiaceae.

1.   Stems stiffly erect; flowers mostly in fascicles; leaf blades typically lanceolate, often narrowly so; fruit on a slender stipe, 2–3 mm, sometimes more; petals with an appendage on the inner face…**A. filiformis**
1'. Stems with spreading branches; flowers usually solitary or as many as 3; leaf blades typically broadly ovate to ovate-elliptic; fruit on stout stipe, ca. 1.5 mm or less; petals with or without an appendage on the inner face…(2)
2.   Blade of petals notched at apex and with a somewhat clavellate appendage on the upper part of the inner surface and 2 teeth on the lower part; leaves typically ovate-elliptic, their lower surface sparsely coarsely pubescent with simple or once-branched hairs, rarely subglabrous or with scattered stellate hairs…**A. pilosa**
2'. Blade of petals not notched and without an appendage or teeth; leaves typically suborbicular-ovate, their lower surfaces felty with a dense covering of short stellate hairs…**A. microphylla**

Ayenia filiformis S. Watson [*A. cuneata* Brandegee; *A. reflexa* Brandegee]—Trans-Pecos ayenia. Subshrubs, decumbent or erect, 0.2–0.5(–0.9) m; stems hairy; blades of proximal leaves ovate to orbic-ulate, mostly 0.5–2 cm, distal oblong-ovate to ovate-lanceolate or linear, unlobed, mostly 1–4.3 cm, mar-gins serrate to doubly serrate or dentate, surfaces usually stellate-puberulent; cymes axillary; sepals 1–3 mm; petal claws 2–2.5(–6) mm, filiform, pink to red, lamina 1–2.5 × 1–2 mm; capsules oblate, 2–4 × 5 mm, densely stellate-hairy. Limestone and granite soil, steep rocky slopes, canyons, sandy washes, Chihua-huan desert scrub, 4400–5400 ft. NE, S.

Ayenia microphylla A. Gray—Dense or little-leaf ayenia. Shrubs, erect or spreading, 0.2–0.6 m; stems hairy; blades of proximal leaves orbiculate to suborbiculate, 0.5 × 0.4 cm, distal ovate to narrowly ovate, unlobed, mostly 0.5–2.1 cm, margins serrate, stellate-hairy (not ciliate), surfaces minutely, densely stellate-hairy; cymes axillary, borne on short shoots; sepals 2.5–3.2 mm; petal claws 4 mm, red, lamina 1.4–1.6 × 1.5 mm; capsules 4–5 × 4–5 mm, puberulent to densely stellate-pubescent. Dry limestone, ig-neous rocky slopes, arroyos, Chihuahuan desert scrub, 4300–5100 ft. SC, SW.

Ayenia pilosa Cristóbal—Dwarf ayenia. Subshrubs, decumbent, 0.1–0.2(–0.3) m; stems hairy, hairs simple and retrorse, or simple, fasciculate, and stellate; blades of proximal leaves broadly ovate to orbiculate, 0.5–1.3 × 0.5–1.1 cm, ovate to oblong-ovate or oblong-lanceolate, unlobed, 0.5–2(–3.5) × 0.4–1.4 (–1.7) cm, margins serrate to doubly serrate; cymes axillary, not borne on short shoots; sepals 2 mm; petal claws 4 mm, yellow to green, lamina rhombic, 1 × 1 mm; capsules 3–3.5 × 3.5 mm, sparingly stellate-hairy. Rocky soils and sandy loam, mesquite-chaparral and Chihuahuan desert scrub, 5000–6050 ft. SC, SW.

CALLIRHOE Nutt. – Poppy mallow

Perennial herbs (ours); stipules persistent, lanceolate to ovate; leaves unlobed to deeply lobed; inflorescences racemose (ours); flowers bisexual or functionally pistillate; epicalyx present or absent; calyx lobes with prominent midrib; corolla cup-shaped, red, pink, purple, or white; style 10–28-branched; fruit schizocarps; mericarps with or without prominent beaks (Dorr, 1990, 2015b).

1.   Epicalyx present, 3-bracted; mericarp beaks not prominent; petals red purple with white basal spot, or
     with white or pink margins…**C. involucrata**
1'. Epicalyx absent; mericarp beaks prominent; petals evenly pink or white…**C. alcaeoides**

Callirhoe alcaeoides (Michx.) A. Gray [*Sida alcaeoides* Michx.]—Light or pale or plains poppy mallow. Plants erect or ascending, to 8.5 dm; leaves 4–13 × 3–10 cm, triangular, cordate, or ovate, unlobed or shallowly or deeply lobed; petioles 2–20 cm; epicalyx absent; calyx lobes valvate in bud; petals 1.5–2.5 cm (male sterile, 0.7–1.7 cm), evenly pink or white; mericarp beaks prominent. 6070–6500 ft. NE. Introduced.

Callirhoe involucrata (Torr & A. Gray) A. Gray—Purple poppy mallow. Plants decumbent to weakly erect, stems to 8 dm; leaves 2.5–15 mm, suborbiculate to ovate, 3–5-lobed; petioles 0.7–13 cm; epicalyx 3-bracted; calyx not forming a point in bud; petals red-purple with white basal spot, white or pink with white margins; mericarp beaks not prominent. Cultivated. Widespread.

1.   Sinuses between lobes of cauline leaves extending to within 5–15 mm of petiole; epicalyx bracts 6–17.5
     × 1.5–3.5 mm; petals red-purple with white basal spot, rarely entirely white; mericarps hairy…**var.
     involucrata**
1'. Sinuses between lobes of cauline leaves extending to within 2–5 mm of petiole; epicalyx bracts 4.5–10 ×
     0.5–2 mm; petals red-purple with white basal spot, entirely white or dark pink with white margins; mericarps glabrous or hairy…**var. lineariloba**

var. involucrata [*C. involucrata* var. *novomexicana* Baker]—Sinuses between lobes of cauline leaves extending to within 5–15 mm of petiole; epicalyx bracts 6–17.5 × 1.5–3.5 mm; petals red-purple with white basal spot, rarely entirely white; mericarps hairy.

var. lineariloba (Torr. & A. Gray) A. Gray [*Malva involucrata* Torr. & A. Gray var. *lineariloba* Torr. & A. Gray; *C. geranioides* Small; *C. sidalceoides* Standl.]—Sinuses between lobes of cauline leaves extending to within 2–5 mm of petiole; epicalyx bracts 4.5–10 × 0.5–2 mm; petals red-purple with white basal spot, entirely white or dark pink with white margins; mericarps glabrous or hairy.

GOSSYPIUM L. – Cotton

Shrubs with stellate hairs (ours); leaves ovate, base cordate; stipules persistent; flowers sympodially arranged (ours); epicalyx 3-bracted; calyx not accrescent, not inflated; petals cream (ours); fruits capsules, with 3–5 carpels; seeds comose (ours), usually white (Fryxell, 1992).

1.   Involucral bractlets broadly cordate-ovate, margins laciniate; petals 2–5 cm; capsules 2–4 cm; seeds cottony; leaf blades shallowly lobed, lobes broadly ovate…**G. hirsutum**
1'. Involucral bractlets ligulate, margins entire or apically toothed; petals 1.5–2.5 cm; capsules 1–1.5 cm; seeds glabrous or glabrate; leaf blades deeply 3–5-lobed, lobes narrowly lanceolate…**G. thurberi**

Gossypium hirsutum L. [*G. mexicanum* Tod.; *G. punctatum* Schumach. & Thonn.; *G. religiosum* L.]—Upland cotton. Plants 1–2 m, widely branching; leaves 4–10 cm, shallowly 3–5-lobed, apex acuminate to acute, surfaces glabrous to hairy; stipules 5-15 mm, subulate to falcate; flowers sympodial; pedicels 2-4 cm, epicalyx 2-4.5 cm, margins laciniate; calyx 5-6 mm, apex truncate or 5-toothed; petals 2-5 cm, cream, with or without red spot at base; staminal column glabrous; capsules 2-4 cm, ovoid to subglobose, glabrous; seeds 8-10 mm. Cultivated fields, 3820-4500 ft. SE, SC. Introduced.

Gossypium thurberi Tod. [*Thurberia thespesioides* A. Gray]—Desert cotton. Plants 2 m, freely branching; stems 5-angular when young, glabrate; stipules linear, 5-10 mm, leaf blades deeply 3-5-lobed, lobes narrowly lanceolate (4+ times as long as wide), 5-15 cm, surfaces glabrate; flowers sympodial; pedicels 1-3 cm, with 3 prominent nectaries; calyx 3 mm, apex truncate; petals cream, with red spot at base, sometimes spot absent, 1.5-2.5 cm; staminal column 9 mm, glabrous; capsules 3-locular, subglobose to oblong, 1-1.5 cm, punctate, externally glabrous, internally ciliate; seeds 3-4 mm, glabrous or glabrate. Oak woodland scrub, washes. Hidalgo.

HERISSANTIA Medik. – Herissantia

Herissantia crispa (L.) Brizicky [*Sida crispa* L.; *Abutilon crispum* (L.) Medik.]—Bladder mallow. Subshrubs, sometimes trailing or scandent, soft-tomentose, 1-6 dm; leaves 1-8 cm, ovate to sometimes elongate-triangular, apex acuminate, margins dentate; epicalyx absent; sepals 3-7 mm, reflexed in fruit; petals 6-11 mm, white; staminal column 2-2.5 mm; schizocarps papery, dorsally dehiscent. Weedy; dry slopes, canyons, roadsides, 4500-4750 ft. SC, SW.

HIBISCUS L. – Rose-mallow

Annual or perennial herbs, or subshrubs (ours); leaves linear to lanceolate, oblong or ovate (ours), lobed or unlobed, base and margins variable; flowers axillary, solitary (ours), usually lasting for only 1 day; epicalyx 7-17-lobed (ours), rarely minute or absent (*H. denudatus*); calyx persistent, sometimes accrescent; petals white, cream, pink, or light purple (ours); style 5-branched; fruits 5-parted capsules, ovoid or spheroid (Blanchard, 1976; Fryxell, 1980).

1. Annual herbs; leaves 3-5-lobed…**H. trionum**
1'. Subshrubs or perennial herbs; leaves unlobed or shallowly lobed…(2)
2. Pedicels much longer than subtending leaves; capsules ovoid or ellipsoid…**H. coulteri**
2'. Pedicels shorter than subtending leaves; capsules ovoid to subglobose…(3)
3. Subshrubs; leaves mostly 1.2-3 cm; petals 1.3-3 cm; epicalyx sometimes minute or absent…**H. denudatus**
3'. Perennial herbs; leaves mostly 8-20 cm; petals 4-12 cm; epicalyx always present…**H. moscheutos**

Hibiscus coulteri Harv. ex A. Gray—Desert rose-mallow. Subshrubs to 2 m; leaves 1-3.5 × 1-4 cm, ovate, 3-lobed, sometimes unlobed, base cuneate to truncate, margins coarsely dentate or serrate; epicalyx 8-14-lobed; sepals 1.4-2.2 cm, equal to or slightly longer than epicalyx, lanceolate-triangular, divided > 3/4 of length; petals 1.6-4 cm, yellow to cream, usually with dark maroon lines at base; style cream; staminal column yellow or cream; pollen yellow-orange; capsules ovoid or ellipsoid. Rocky slopes, 4600-5700 ft. SC. Otero.

Hibiscus denudatus Benth.—Paleface rose-mallow. Subshrubs to 0.8 m; leaves 1.2-3 × 1-2.5 cm, oblong or ovate, usually not lobed, base cuneate to truncate, margins irregularly and coarsely dentate; epicalyx 7-10-lobed, sometimes absent; sepals 0.7-1.6 cm, narrowly triangular-ovate, divided at least 2/3 of length; petals 1.3-3 cm, pale purple or pink to white, with maroon spot at base; style pale pink; staminal column pink; pollen dark orange; capsules ovoid to subglobose. Rocky, desert soils, 3450-5600 ft. S.

Hibiscus moscheutos L.—Common rose-mallow. Perennial herbs to 2.5 m; leaves 8-20 cm, lanceolate to triangular to ovate, unlobed or 3-lobed, base cordate to cuneate, margins crenate to dentate or serrate; epicalyx (8-)10-14(15)-lobed; sepals 1.5-4 cm, triangular, divided to 1/2-2/3 of length; petals 4-12 cm, white, rarely pink with maroon basal spot (ours); style white; staminal column white or cream; pollen yellow; capsules ovoid to subglobose. Our material belongs to **subsp. lasiocarpos** (Cav.) O. J. Blanch. [*H. lasiocarpos* Cav.]. Near streams, roadside ditches, 3800-4775 ft.

Hibiscus trionum L.—Flower-of-an-hour, bladder ketmia, Venice mallow. Annual herbs to 0.6 m; leaves linear, 1–2.5 cm, 3- or 3–5-lobed, margins ciliate or not; flowers lasting a few hours, erect or ascending; epicalyx 10–15-lobed; sepals 0.8–1.8 cm, campanulate, divided to 1/2 of length, accrescent and inflated, enclosing fruit; petals 1.5–3 cm, yellow or cream with dark purple to maroon spot at base; style white to cream, yellow, or maroon; staminal column dark red to purple; pollen yellow-orange; capsules ellipsoid to ovoid, dark brown. Gardens and open spaces, 4760–5740 ft. N, SC. Introduced.

ILIAMNA Greene – Globe mallow, wild hollyhock

Iliamna grandiflora (Rydb.) Wiggins—Large-flowered wild hollyhock. Perennial herbs, 5–20 dm, stellate-pubescent; leaves petiolate, 6–13 cm wide, palmately 3–7-lobed, margins coarsely dentate; epicalyx 3-bracted; inflorescence an interrupted spicate raceme; epicalyx bracts 8–12 mm, lance-ovate; sepals 5–10 mm, triangular-ovate; petals 2–3.7 cm, pink or lavender. Streamsides and mesic slopes, 7550–10,300 ft. NC.

MALVA L. – Mallow

Annual, biennial, or perennial herbs (ours); stipules persistent (ours); leaves reniform to (sub)orbiculate; flowers axillary, solitary or in 2–6-flowered fascicles (ours); epicalyx 3-bracted, distinct, shorter than calyx (ours); petals white, pink, or purple; style 8–15-branched (ours); fruit a flattened, wheel-shaped schizocarp.

1. Petals deep purple or pink, veins darker, > 2 times longer than calyx…**M. sylvestris**
1′. Petals white to pale purple, up to 2 times longer than calyx…(2)
2. Calyx accrescent; petals 3–4.5 mm; pedicels < 0.5 cm…**M. parviflora**
2′. Calyx not accrescent; petals 5–13 mm; pedicels > 0.5 cm…(3)
3. Plants prostrate to trailing, usually < 0.6 m; leaves small, 1.5–3.5 cm…**M. neglecta**
3′. Plants erect, 0.5–2.5 m; leaves larger, 3–10 cm…**M. verticillata**

Malva neglecta Wallr.—Common mallow, cheeses, mauve négligée. Plants annual or biennial, prostrate-trailing to 1 m; leaves 1.5–3.5 × 1–4 cm, orbicular or reniform, base cordate, unlobed or shallowly lobed, margins crenate to dentate; pedicels 1–5 cm, longer in fruit; epicalyx not adnate to calyx, linear, margins entire; sepals 3–6 mm, longer in fruit; petals 5–11 mm, pale purple to white; style 8–11-branched; mericarps glabrous. Weeds of disturbed sites, 3970–9500 ft. Widespread. Introduced.

Malva parviflora L.—Small-flowered mallow. Plants annual, 2–8 dm, erect or ascending; leaves 2–8 × 2–8 cm, orbiculate-cordate or reniform, base cordate, shallowly 5–7-lobed, margins crenate; flowers solitary or in 2–4-flowered fascicles; pedicels 0.2–0.4 cm; epicalyx not adnate to calyx, linear to filiform, margins entire; sepals 3–4.5 mm, apex acuminate, accrescent; petals 3–4.5 mm, white to pale purple; style 10- or 11-branched; mericarps glabrous or hairy. Weed of disturbed sites, 3900–7950 ft. NE, NC, SC, SW. Introduced.

Malva sylvestris L. [*M. mauritiana* L.]—High or garden mallow. Plants biennial, 0.5–1 m; leaves 5–10 × 5–10 cm, orbiculate-cordate or reniform, base cordate, unlobed or shallowly lobed, margins crenate; flowers solitary or in 2–4-flowered fascicles; pedicels 1–2.5 cm; epicalyx sometimes adnate to calyx; sepals 5–6 mm; petals 2.5–3 times length of calyx, pink, purple, or red-purple; style 10–12-branched. Gardens; unlikely to escape, as this species is not tolerant of hot or arid conditions. 4280–5200 ft. SC. Introduced.

Malva verticillata L. [*M. crispa* (L.) L.]—Whorled or clustered or Chinese or curled mallow. Plants annual, erect, 0.5–2.5 m; leaves 3–10 × 5–10 cm, reniform, base cordate, unlobed or shallowly lobed; flowers in 2–6-flowered fascicles; pedicels 0.5–1 cm; sepals 4–6 mm; petals subequal or up to 2 times longer than calyx, white to pale purple; style 8–11-branched; mericarps glabrous. Uncommon in disturbed sites, 4500–7300 ft. SC, SW. Introduced.

MALVELLA Jaub. & Spach – Alkali mallow

Prostrate perennial herbs; leaves reniform or ovate to triangular, with asymmetric bases, margins dentate, serrate, or entire; stipules persistent; flowers axillary, solitary; epicalyx present or absent; petals white or pale yellow, sometimes fading to rose, asymmetric; style 7–10-branched; fruits schizocarps, seeds 1 per mericarp.

1. Leaves mostly reniform, wider than long; epicalyx usually present, 3-bracted…**M. leprosa**
1'. Leaves triangular to ovate, longer than wide, apex acute, with silvery, lepidote scales; epicalyx usually absent…(2)
2. Leaves narrowly triangular, 3–5 times as long as wide, with few teeth at base, otherwise entire, stellate hairs absent; epicalyx never present…**M. sagittifolia**
2'. Leaves triangular, 1–2 times as long as wide, margins dentate, stellate hairs present; epicalyx sometimes present…**M. lepidota**

Malvella lepidota (A. Gray) Fryxell [*Sida lepidota* A. Gray; *Disella lepidota* (A. Gray) Greene]—Scurfymallow. Plants with stellate hairs and silvery, lepidote scales; leaves 1–2 cm, triangular to ovate, usually 1–2 times longer than wide, base truncate or cuneate, apex acute, margins irregularly dentate, abaxial surface dense with stellate hairs, adaxial surface dense with silvery, lepidote scales; petioles 1/2–2 times as long as blades; pedicels subequal to subtending leaves; epicalyx usually absent, 3-bracted if present; sepals 6–8 mm, with silvery, lepidote scales, apex acuminate; petals 12–15 mm, white or pale yellow, asymmetric, style ca. 7-branched. Roadsides, mudflats, 3850–6000 ft. Mostly S. Less common and less weedy than *M. leprosa*.

Malvella leprosa (Ortega) Krapov. [*Malva leprosa* Ortega; *Sida hederacea* (Douglas) Torr. ex A. Gray; *S. leprosa* (Ortega) K. Schum.]—Dollar-weed, oreja de ratón, scurfy sida. Plants with stellate hairs and sublepidote hairs; leaves 1–4 cm, orbicular or reniform, wider than long, margins serrate to dentate, surfaces dense with stellate hairs and sublepidote hairs; petioles 1/2 as long as to equaling blades; pedicels subequal to subtending petiole; epicalyx 3-bracted, filiform or absent; sepals 4–10 mm, apex acuminate; petals 10–15 mm, pale white, cream, or yellow, may fade to rose, asymmetric; style ca. 7-branched. Disturbed areas, often in saline soils, 3575–6025 ft. Widespread.

Malvella sagittifolia (A. Gray) Fryxell [*Disella sagittifolia* (A. Gray) Greene; *Sida sagittifolia* (A. Gray) Rydb.]—Plants prostrate with silvery, lepidote scales; leaves 1.5–3.5 cm, 3–5 times as long as wide, narrowly triangular, apex acute, base with 2–4 hastate teeth and truncate, surface with sparse silvery, lepidote scales; petioles to 1/3 as long as blades; pedicels shorter than subtending leaves; epicalyx absent; calyx 7–9 mm, with silvery, lepidote scales, apex acuminate; petals 15 mm, white or pale yellow, may fade to rose, asymmetric; style ca. 7- or 8-branched. Roadsides, mudflats, 3150–5500 ft. S. Doña Ana.

RHYNCHOSIDA Fryxell – Rhynchosida

Rhynchosida physocalyx (A. Gray) Fryxell [*Sida physocalyx* A. Gray]—Beaked sida, buff-petal. Perennial herbs trailing to ascending, stellate-hairy; taproots large; leaves 2–5 cm, oblong-ovate, apex obtuse to acute, surfaces coarsely hairy, margins crenate or serrate; pedicels 1–2 cm, slender; flowers solitary in leaf axils; epicalyx absent; calyx lobes apiculate, strongly 5-angled at base; corolla yellow, subequal to calyx, style 8–14-branched; fruit a drooping schizocarp, the mericarps 1-celled and 1-seeded, beaked. Chihuahuan desert shrublands, grasslands, roadsides, fence rows, disturbed habitats, 3200–5500 ft. S.

SIDA L. – Fanpetals, wireweed

Perennial herbs or subshrubs (ours); erect, ascending, or reclining to procumbent; leaves spirally arranged, petiolate or subsessile, usually unlobed and dentate; flowers usually solitary in leaf axils, though sometimes paired or clustered; epicalyx absent; calyx usually 10-ribbed at base; corolla white, cream, yellow, yellow-orange, salmon-pink, red-orange, or red, sometimes with dark red center; schizocarps glabrous or pubescent (Fryxell, 1985; Kearney, 1954).

1. Procumbent herbs; leaves ovate-oblong; flowers solitary in leaf axils, petals white; mericarps and styles 5-branched...**S. abutilifolia**
1'. Multistemmed herbs or subshrubs; leaves oblong-lanceolate; flowers axillary but apically congested on shortened internodes, petals yellow-orange; mericarps and styles 10-12-branched...**S. neomexicana**

Sida abutilifolia Mill. [*S. diffusa* Kunth; *S. filicaulis* Torr. & A. Gray; *S. procumbens* Sw.]—Creeping sida. Procumbent, perennial, stellate-pubescent herbs; leaves distributed evenly along stems, ovate to oblong, stipules inconspicuous, free from petiole; flowers solitary in leaf axils on slender pedicels; corolla white; style 5-branched; mericarps with apical small spines. Arid, sandy plains, disturbed areas, 3200–7120 ft. S. Introduced.

Sida neomexicana A. Gray—New Mexico sida. Multistemmed perennial herbs or subshrubs usually < 0.5 m; leaves narrowly oblong-lanceolate; stipules 5–7 mm, linear; flowers axillary, solitary, though congested apically; corolla yellow-orange to reddish; styles 10–12-branched. Open, arid habitats, rocky slopes and canyons, 4300–7800 ft. Widespread.

SIDALCEA A. Gray – Checkerbloom, checker mallow

Perennial herbs (ours) to 1 m; glabrous to stellate-hairy; stems single to several in mature plants; leaves dimorphic, basal leaves shallowly lobed and cauline leaves palmately divided; petioles long, to 5 times leaf length; inflorescences 20+-flowered, erect or ascending, dense or open; epicalyx absent (ours); petals (6–)10–20 mm, cuplike, white, pink, or purple; style 5–10-branched; fruit a schizocarp.

1. Stems hirsute; stipules persistent; all flowers bisexual, petals white or tinged with pink; anthers blue-pink...**S. neomexicana**
1'. Stems glabrous; stipules deciduous; flowers bisexual and sometimes pistillate, petals white, or pink to purple, bases paler; anthers white or yellow...**S. candida**

Sidalcea candida A. Gray—White checkerbloom. Plants 0.3–1 m with spreading rhizomes; leaves ovate, basal leaf margins crenate; stipules deciduous; petioles 6–18 cm, 1/2 times to as long as blade; flowers in 20+-flowered, erect, dense, spiciform inflorescences; sepals 7–9 mm; petals 10–20 mm, white to pale pink, not overlapping; style 6–9-branched; anthers blue-pink; schizocarps 5–7 mm diam. Stream banks, meadows, seeps, other wet places, (5175–)6600–11,600 ft. NC, NW, C, SC.

Sidalcea neomexicana A. Gray [*S. confinis* Greene; *S. crenulata* A. Nelson]—New Mexico checkerbloom. Plants 0.2–0.8(–1.2) m with taproot or fascicled roots; basal leaves orbiculate, unlobed margins crenate or shallowly lobed; stipules persistent; petioles 10–25 cm, up to 5 times length of blade; inflorescences 20+-flowered, erect or ascending, dense or open; flowers bisexual and occasionally pistillate, plants gynodioecious; sepals 5–8 mm; petals on pistillate flowers 8–12 mm, petals on bisexual flowers 18–20 mm; pink to purple, bases paler; style 8–9-branched; anthers white, rarely yellow; schizocarps 5 mm diam. Stream banks, meadows, seeps, other wet places, 5100–9800 ft. NC, NW, C, WC, SC, SW.

SPHAERALCEA A. St.-Hil. – Globemallow

Perennial herbs or subshrubs, stellate-canescent to silvery-stellate-lepidote; leaves linear, lanceolate, orbiculate, or ovate, margins entire, crenate, or serrate, petiolate or sessile, stipules persistent or deciduous; epicalyx present (absent or deciduous in *S. coccinea*); calyx not accrescent, not inflated, lobes connate for 1/2 length; corolla campanulate, orange, red, red-orange, sometimes pink, purple, or white; fruits schizocarps, erect, not inflated, indurate, hairy; mericarps often remaining attached to fruit axis after maturity by threadlike extension of dorsal vein; 1 or 2 seeds per mericarp (La Duke, 1985, 2015). Species boundaries in *Sphaeralcea* are not always clear because of hybridization, polyploidy, and morphological variation in response to rainfall.

1. Dehiscent part of mericarp narrower and shorter than indehiscent part...(2)
1'. Dehiscent part of mericarp as wide as or wider than indehiscent part...(3)
2. Stem and leaf surfaces with stellate and silvery-lepidote hairs; epicalyx present; mericarp and style branches 7–9...**S. leptophylla**

2'. Stem and leaf surfaces with stellate hairs, lacking silvery-lepidote hairs; epicalyx absent, or if present early-deciduous; mericarp and style branches 9–14…**S. coccinea**
3. Inflorescences usually racemes, sometimes panicles…(4)
3'. Inflorescences usually panicles…(6)
4. Leaf blades 3-lobed to pedately divided; inflorescences racemes…**S. digitata**
4'. Leaf blades unlobed, 3-lobed, or pedately divided; inflorescences racemes or panicles…(5)
5. Leaf blades unlobed; sepals forming beak in bud…**S. hastulata** (in part)
5'. Leaf blades usually pedately divided; sepals forming sphere in bud…**S. wrightii**
6. Inflorescences open panicles…**S. laxa**
6'. Inflorescences crowded panicles…(7)
7. Leaves linear to lanceolate, unlobed or hastately lobed; inflorescence usually with leaves throughout… **S. angustifolia**
7'. Leaves deltate, ovate, ovate-lanceolate, subhastate, lanceolate, or triangular, if lobed, with or without elongate central lobe, or pedately divided; inflorescence usually without leaves throughout…(8)
8. Leaves unlobed, deltate, ovate, ovate-lanceolate, or weakly 3–5-lobed…(9)
8'. Leaves strongly 3–5-lobed or pedately divided…(11)
9. Sepals not forming beak in bud; stems sometimes yellowish…(10)
9'. Sepals forming beak in bud; stems never yellowish…**S. hastulata** (in part)
10. Plants 1.5–4 dm…**S. parvifolia**
10'. Plants 6–18 dm…**S. incana**
11. Plants 1.5–30 dm; leaves 3–5-lobed…(12)
11'. Plants 1–4 dm; leaves deeply lobed or pedately divided, sometimes dissected…(15)
12. Plants 4–7 dm; petals red-orange…**S. fendleri** (in part)
12'. Plants 10–30 dm; petals red-orange, pink, light purple, or white…(13)
13. Stem glabrous or coarsely hairy; leaves ovate or triangular to lanceolate, 3-lobed; inflorescence not leafy at tip…**S. emoryi**
13'. Stem softly pubescent; leaves deltate, ovate or triangular to lanceolate, subhastate to 3-lobed; inflorescence leafy at tip…(14)
14. Mericarps 12–14, 3.5–5.5 mm…**S. polychroma**
14'. Mericarps 10, 3 mm…**S. procera**
15. Plants 1–3 dm; stems green to green-gray…**S. hastulata** (in part)
15'. Plants 4–10 dm; stems green-gray to purple…(16)
16. Leaf surfaces hirsute or soft-pubescent…**S. fendleri** (in part)
16'. Leaf surfaces stellate-pubescent…**S. grossulariifolia**

Sphaeralcea angustifolia (Cav.) G. Don [*Malva angustifolia* Cav.]—Copper or narrow-leaved globemallow. Plants 0.6–2 m; leaves 3.5–10 cm, lanceolate, linear, or trullate, unlobed to hastately lobed; margins dentate to crenate; flowers in panicles; sepals 5–9 mm; petals 6–20 mm, mauve, light purple, red-orange, pink, or white; styles 10–15-branched; anthers yellow or purple; fruit schizocarps, ovoid or a truncated cone; mericarps 3.5–7 × 1.5–2 mm; seeds 2 per mericarp, brown or black, pubescent.

1. Leaves linear-lanceolate, with or without hastate lobes…**var. angustifolia**
1'. Leaves linear to trullate, with broad hastate lobes…**var. oblongifolia**

var. angustifolia [*Malva longifolia* Sessé & Moc.; *S. angustifolia* subsp. *cuspidata* (A. Gray) Kearney]—Leaves linear-lanceolate, with or without hastate lobes. Disturbed roadsides, 2950–7900 ft. Widespread.

var. oblongifolia (A. Gray) Shinners [*S. lobata* Wooton; *S. incana* Torr. ex A. Gray var. *oblongifolia* A. Gray; *S. fendleri* A. Gray subsp. *elongata* Kearney; *S. fendleri* A. Gray var. *elongata* (Kearney) Kearney]—Leaves linear to trullate, with broad lobes. Disturbed roadsides, 3800–6930 ft. Widespread.

Sphaeralcea coccinea (Nutt.) Rydb. [*Malva coccinea* Nutt.; *Malvastrum coccineum* (Nutt.) A. Gray]—Scarlet or common globemallow. Plants subshrubs, rhizomatous, to 0.5 m; leaves deeply 3–5-lobed, lobes toothed or lobed; epicalyx absent or deciduous; calyx 3–10 mm; petals 5–20 mm, red-orange; styles 9–14-branched; anthers yellow or red; fruit schizocarps, flattened spheres; mericarps 3–3.5 × 2.5–3 mm; seeds 1 per mericarp, gray or black.

1. Leaf midlobe ± equal to secondary lobes…**var. coccinea**
1'. Leaf midlobe longer than secondary lobes…**var. elata**

var. coccinea [*Malvastrum coccineum* (Nutt.) A. Gray var. *dissectum* (Nutt. ex Torr. & A. Gray); *M. cockerellii* A. Nelson]—Leaf midlobe ± equal to secondary lobes. Dry plains, disturbed roadsides, 3500–8700 ft. Widespread.

var. elata (Baker f.) Kearney [*Malvastrum coccineum* (Nutt.) A. Gray var. *elatum* Baker f.; *M. elatum* (Baker f.) A. Nelson; *M. micranthum* Wooton & Standl.]—Leaf midlobe longer than secondary lobes. Rocky or sandy soil, slopes, flats, washes, scrublands, 3400–7800 ft. N.

Sphaeralcea digitata (Greene) Rydb. [*Malvastrum digitatum* Greene; *S. digitata* subsp. *tenuipes* (Wooton & Standl.) Kearney; *S. pedata* Torr. ex A. Gray var. *angustiloba* A. Gray]—Juniper globemallow. Plants 3–4 dm; leaves 1.5–2.5 cm, triangular, 3-lobed or pedately divided, base cuneate, margins entire; calyx 3.5–8 mm; petals 8–14 mm, red-orange; styles 9–13-branched; anthers purple; fruits schizocarps, cylindrical to widely conic; mericarps 3–5 × 2–2.5 mm; seeds 1(2) per mericarp, brown or black. Desert scrub, grasslands, piñon-juniper woodlands, ponderosa pine forests, well-drained slopes on diverse substrates, 3800–8650 ft. NW, C, WC, SE, SC, SW.

Sphaeralcea emoryi Torr. ex A. Gray [*S. arida* Rose; *S. emoryi* var. *arida* (Rose) Kearney]—Emory's globemallow. Plants subshrubs to 2 m; leaves ovate-triangular to lanceolate, unlobed to 3-lobed, base cordate to truncate, margins crenate to serrate; flowers in panicles; epicalyx green or tan, rarely red; calyx 6–8 mm; petals 10–12 mm, color highly variable, red-orange, light purple, pink, or white; styles 10–16-branched; anthers yellow; fruits schizocarps, truncate to conic; mericarps 4.5–5 × 2.5 mm; seeds 1 or 2 per mericarp, brown or black, pubescent. Mesquite shrublands, oak-pine woodlands, mixed conifer forests, 4110–8000 ft. SC, SW.

Sphaeralcea fendleri A. Gray—Fendler's globemallow. Plants to 1.5 m, stems usually green, sometimes purple or black; leaves ovate, lanceolate, or triangular, shallowly 3-lobed, base truncate, margins crenate to dentate; calyx 4–6 mm; petals 8–15 mm, red-orange (ours); styles 9–16-branched; anthers yellow; fruits schizocarps, cylindrical to conic; mericarps 4–5.5 × 2–2.5 mm; seeds 2 per mericarp, black, pubescent. Our material belongs to **var. fendleri**. Desert scrub, piñon-juniper woodlands, ponderosa pine forests, 3330–8380 ft.

Sphaeralcea grossulariifolia (Hook. & Arn.) Rydb. [*Sida grossulariifolia* Hook. & Arn. (as *grossular-iaefolia*); *Sphaeralcea grossulariifolia* subsp. *pedata* (Torr. ex A. Gray) Kearney]—Gooseberry-leaf or currant-leaf globemallow. Plants to 1 m, stems green or purple, white-, gray- or yellow-canescent or glabrous; leaves 1.7–3.5 cm, gray-green or yellow-green, triangular, 3-lobed or pedately divided, base truncate to cordate, margins crenate to dentate; calyx 5–8 mm; petals 8–15 mm, red-orange; styles 10–12-branched; anthers yellow; fruits schizocarps, truncate-conic to spherical; mericarps 2.5–3.5 × 2–2.5 mm; seeds 1(2) per mericarp, gray or black. Rocky or sandy soil, salt desert scrub, piñon-juniper woodlands, ponderosa pine forests, 4900–7600 ft.

Sphaeralcea hastulata A. Gray [*S. pumila* Wooton & Standl.; *S. arenaria* Wooton & Standl.; *S. gla-brescens* Wooton & Standl.; *S. martii* Cockerell; *S. subhastata* J. M. Coult.; *S. subhastata* J. M. Coult. var. *martii* (Cockerell) Kearney]—Spear globemallow. Plants to 4 dm, stems ascending to decumbent, canescent; leaves 2–6 cm, ovate-lanceolate, unlobed, lobed, or pedately divided, base cuneate to cordate, margins entire, crenate, or dentate; calyx 4–11 mm, apices forming beak in bud; petals 10–20 mm, red-orange, pink, or purple; styles 10–30-branched; anthers yellow or purple; fruits schizocarps, widely conic; mericarps 3–7 × 1.5–2.5 mm; seeds 1 or 2 per mericarp, black or brown. Desert scrub, piñon-juniper woodlands, ponderosa pine forests, 3150–8300 ft. NW, C, WC, SE, SC, SW.

Sphaeralcea incana Torr. ex A. Gray [*S. incana* subsp. *cuneata* Kearney; *S. incana* var. *cuneata* (Kearney) Kearney]—Gray globemallow. Plants 6–18 dm; stems yellow-green, canescent; leaves 3–5 cm, light green to yellow-green, deltate, unlobed to weakly 3-lobed, base cuneate to cordate, margins entire or undulate; flowers on crowded panicles; calyx 3.5–6.5 mm, forming a sphere in bud; petals 10–17 mm, red-orange; styles 11–15-branched; anthers yellow; fruits schizocarps, hemispherical; mericarps 4–5.5 × 2–2.5 mm; seeds 2 per mericarp, brown. Rocky slopes, 3900–8100 ft. Widespread.

Sphaeralcea laxa Wooton & Standl. [*S. incana* Torr. ex A. Gray var. *dissecta* A. Gray; *S. ribifolia* Wooton & Standl.]—Caliche globemallow. Plants 4–7 dm, gray- or white-tomentose; leaves 1.5–5 cm, gray or white, ovate-deltate, 3–5-lobed, base truncate, margins crenate; flowers in open panicles; epicalyx red to purple; calyx 7–11 mm; petals 12–18 mm, red-orange; styles 11–16-branched; anthers usually purple; fruits schizocarps, truncate-ovoid; mericarps 4–6 × 2–3 mm; seeds 2 or 3 per mericarp, gray, pubescent. Rocky outcrops, juniper woodlands, 4000–6600 ft. WC, SW.

Sphaeralcea leptophylla (A. Gray) Rydb. [*Malvastrum leptophyllum* A. Gray; *S. janeae* (S. L. Welsh) S. L. Welsh]—Plants 1–4(–6) dm, ascending, with silvery-stellate-lepidote hairs; leaves 1–3.5 cm, linear to triangular, mostly unlobed or 3-lobed, with silvery-stellate-lepidote hairs, base truncate to cuneate, margins entire; inflorescences racemose, open; epicalyx with silvery-stellate-lepidote hairs; calyx 4.5–5.5 mm; petals 8–15 mm, red-orange; styles 7–9-branched; anthers yellow, rarely purple; fruits schizocarps, flattened spherical to conic; mericarps 2.5–3.5 × 2–3 mm; seeds 1 per mericarp, brown, usually glabrous. Rocky, dry areas, desert scrub, piñon-juniper woodlands, 4145–6600 ft. C, W.

Sphaeralcea parvifolia A. Nelson [*S. arizonica* A. Heller ex Rydb.]—Small-leaf globemallow. Plants 1.5–4(–10) dm; leaves 1–5.5 cm, ovate, mostly unlobed or weakly 3–5-lobed, base cuneate to cordate, margins entire or crenate; flowers in panicles; epicalyx usually green or tan, rarely red-purple; calyx 6–9 mm; petals 8–14 mm, red-orange, rarely white; styles 12-branched; anthers yellow; fruits schizocarps, ellipsoid; mericarps 3.5–5.5 × 1.5–3 mm; seeds 1 per mericarp, gray or black. Dry slopes, 4800–7000 ft. NW.

Sphaeralcea polychroma La Duke—Hot springs globemallow. Plants 10–20 dm, with white or yellow pubescence; leaves 4–7 cm, variable, deltate with subhastate lobes to deeply trilobed, base cuneate, margins crenate to dentate; flowers in open panicles; calyx 6–7 mm; petals 10–13 mm, red, red-orange, purple, pink, or white; styles 12–14-branched; anthers yellow; fruits schizocarps, urceolate; mericarps 4–5.5 × 2–3 mm; seeds 1 or 2 per mericarp, gray, black, or brown. Desert scrub, piñon-juniper woodlands, scattered ponderosa pine, 4200–8900 ft. C, SC, SW.

Sphaeralcea procera Ced. Porter—Luna County globemallow. Plants 30 dm; leaves 1–5 cm, yellow-pubescent, lanceolate or ovate to triangular, 3-lobed, base cuneate, margins irregularly dentate; flowers in panicles; calyx 5 mm; petals 10–13 mm, rose-purple; styles 10-branched; anthers yellow; fruits schizocarps, helmet-shaped; mericarps 2–3 mm; seeds 1 per mericarp, gray or black, pubescent. Sandy soil; known only from Luna County (type collection), ca. 4593 ft. SW. This taxon may no longer be present, or it may be a form of *S. polychroma* La Duke.

Sphaeralcea wrightii A. Gray—Wright's globemallow. Plants 2–5 dm, canescent; leaves 2–4 cm, ovate, triangular, or orbiculate, most deeply pedately divided, base truncate to cordate, margins entire; flowers in racemes or panicles; calyx 6–7 mm; petals 10–13 mm, light purple, pink, or red-orange; styles 12–15-branched; anthers yellow or purple; fruits schizocarps, hemispherical to truncate-conic; mericarps 4–7 × 2.5–3 mm; seeds 2 per mericarp, black, pubescent. Rocky dry areas, Chihuahuan desert scrub, grasslands, mesquite, 4600–5000 ft. SW. **R & E**

MARTYNIACEAE – SESAME FAMILY

Don Hyder

PROBOSCIDEA Schmidel – Unicorn plant

Herbs, often with sticky hairs; leaves opposite, the upper sometimes alternate; flowers showy, zygomorphic, perfect; sepals 4–5, usually none; corolla somewhat bilabiate, petals 5; stamens the same number as sepals; pistil single, superior, united; fruit a drupaceous capsule with a long, incurved, hooked beak that is longer than the fruit body (Allred, 2020; Hevly, 1969).

1. Plants perennial, arising from a tuberous root; corolla light yellow to bronze; fruit crested on 2 sides...
   P. altheifolia

1'. Plants annual, arising from a slender taproot; corolla purplish, reddish, pinkish, cream-yellow, or whitish; fruit crested on a single side…(2)
2. Sepals united only in basal 1/4 or less; seeds > 3 times longer than wide; flowers nearly hidden within the foliage; corollas purplish…**P. sabulosa**
2'. Sepals united > 1/4 their length; seeds < 3 times longer than wide; flowers mostly conspicuous; corolla lobes various colors…(3)
3. Upper 2 lobes of the corolla each lacking a single large purplish splotch (may have numerous small spots); corollas generally whitish to pale pink…**P. louisianica**
3'. Upper 2 lobes of the corolla each with a single large splotch, sometimes the entire corolla dark; corollas dark pink, magenta, or purple to maroon…(4)
4. Inflorescence nestled among or barely exceeding the foliage; flowers < 10 per inflorescence; corollas 2 cm, white or pale pink to purplish…**P. parviflora**
4'. Inflorescence raised above the foliage; flowers > 15 per inflorescence; corollas to 4 cm, purplish to maroon…**P. fragrans**

Proboscidea altheifolia (Benth.) Decne. [*Martynia altheifolia* Benth.]—Desert unicorn plant. Perennial herbs to 2 dm; leaves simple, blades nearly reniform or suborbicular to broadly ovate, 2–7 × 2–8 cm; flowers 3–16; calyx 1–1.5 cm; corolla yellowish brown externally, yellow to bronze-orange internally, the tube with pale blotches and maroon, reddish-brown, or rust-colored spots; fruit 5–6 cm and ca. 12 mm thick, the horns about twice as long as the body. Sandy soil, dunes, gravelly hills, 3825–8000 ft. WC, SW.

Proboscidea fragrans (Lindl.) Decne. [*Martynia fragrans* Lindl.]—Fragrant unicorn plant. Annuals with decumbent stems; leaf blades reniform to broadly ovate, margins entire to palmately 5-lobed; racemes many-flowered; sepals 1.5–2 cm; corollas 2–4 cm, dark purple to maroon, spotted internally, the upper 2 lobes each with a single large blotch; fruit body to 8 cm, crested on a single suture, the beak to 18 cm, ca. 2 times the length of the body. Sandy to rocky soils, desert scrub, 3660 ft. SE.

Proboscidea louisianica (Mill.) Thell.—Ram's-horn. Sprawling-ascending annual herbs to 0.6 × 2 m; stems covered with glandular hairs, foul-scented; lower leaves opposite, upper leaves sometimes alternate, to 25 cm; inflorescence a raceme; corollas dull white to lavender with purple and yellow spots, 3.5–5.5 cm; fruit 10–20 cm, with 2 long curved beaks at the tip. Open areas with sandy soil, 3600–5800 ft. E.

Proboscidea parviflora (Wooton) Wooton & Standl.—Doubleclaw. Erect or spreading annual herbs to 2.5 × 1 m; leaves simple, blades broadly triangular-ovate to suborbicular-ovate; flowers 5–15 per inflorescence; calyx 1–1.5 cm; corollas reddish purple, pink, or white; fruit 5–10 cm, horns ca. 1–3.5 times as long as the body. Chihuahuan desert, semiriparian shrubland, 3315–6500 ft. NC, NW, C, WC, S.

Proboscidea sabulosa Correll—Sand-dune unicorn plant. Sprawling annual herbs to 40 × 120 cm; leaves opposite along the stems, blades triangular-ovate to broadly subreniform, to 12 × 12 cm; inflorescence a few-flowered raceme; sepals 5, unequal, 10–15 mm; corollas 2 cm, 2-lipped, sparsely glandular, reddish purple, with spots inside the corolla tube; capsules hook-shaped, the claw at the end of the capsule strongly curled. Deep sand, 3000–5280 ft. E, C.

MELANTHIACEAE – DEATH CAMAS FAMILY

Steve L. O'Kane, Jr.

Herbs, perennial, from bulbs or tuberlike rhizomes; leaves simple, alternate, spiral, or in a basal rosette, venation parallel; flowers bisexual, actinomorphic, several to many; perianth of 6 tepals, distinct to basally connate; stamens 6, anthers only 2-locular, confluent apically and appearing peltate, together opening by a single slit across the top (usually and in ours); the single ovary superior to 1/2 inferior, with 3 locules, axile placentation; styles free or connate, stigmas 3; fruit a dry or fleshy ventricidal capsule. The family is here narrowly circumscribed (Christenhusz et al., 2017; Schwartz, 2002; Simpson, 2019; Zomlefer et al., 2001; Zomlefer & Judd, 2002).

1. Plants robust, tall, 1–2.5 m; main leaves cauline, ovate, distal-most lanceolate to lance-linear, 10–25 cm wide…**Veratrum**
1′. Plants smaller and shorter, not > 0.9 m; main leaves ± basal, narrower, < 2 cm wide…(2)
2. Flowers ± sessile, in a condensed spike (or spikelike raceme)…**Schoenocaulon**
2′. Flowers with well-developed pedicels, in typical racemes or panicles…(3)
3. Ovary partly inferior; tepal glands obcordate, bilobed; bulbs slim…**Anticlea**
3′. Ovary superior; tepal glands obovate or obscure, not bilobed; bulbs ovoid…**Toxicoscordion**

ANTICLEA Kunth – Death camas

Plants from bulbs, these slim, narrowly ovoid, with a membranous coat; leaves basal and cauline, linear, weakly to strongly folded, often curved, entire; inflorescence loosely racemose or paniculate, 10–50-flowered (usually < 30-flowered); tepals petal-like, bases cuneate to gradually narrowed (but not clawed), white to cream-yellow (sometimes darker), often tinged green, adaxially with 1 gland near base, this bilobed (obcordate); ovary ± 1/2 inferior. Previously included in *Zigadenus*.

1. Perianth rotate to rotate-campanulate; pedicels erect or ascending at anthesis; tepals cream-colored to greenish…**A elegans**
1′. Perianth campanulate; pedicels recurved to divergent 90° at anthesis; tepals greenish and often tinged with red or purple…(2)
2. Tepals 4–6 mm…**A. virescens**
2′. Tepals 12–16 mm…**A. mogollonensis**

Anticlea elegans (Pursh) Rydb. [*Zigadenus elegans* Pursh]—Mountain death camas. Plants 2–8 dm; proximal leaf blades 10–30 cm × 3–15 mm; flowers ascending to erect; perianth rotate to rotate-campanulate, 15–20 mm diam.; tepals persistent in fruit, cream-colored to greenish, ovate, 7–12 × 4–5 mm, somewhat narrowed at base; pedicels erect at anthesis; bracts often tinged with purple or pink, ovate. Our plants are **subsp. elegans**. Moist areas of montane to alpine meadows, riparian areas, open forests and forest openings, aspen groves, seeps and dripping cliffs, 5800–13,150 ft. NE, NC, NW, C, WC, SC.

Anticlea mogollonensis (W. J. Hess & Sivinski) Zomlefer & Judd [*Zigadenus mogollonensis* W. J. Hess & Sivinski]—Mogollon death camas. Plants 4.5–8.5 dm; proximal leaf blades 25–45 cm × 10–20 mm; flowers nodding; perianth campanulate, 8–20 mm diam.; tepals greenish, often tinged with red-purple, especially the margins, ovate to elliptic, 12–16 × 7–9 mm, somewhat narrowed at base; pedicels recurved at anthesis, erect in fruit; bracts often tinged with purple or red, longer than pedicels, apex acute. Spruce-fir forests and upper ponderosa-aspen forests, in organic duff, 8800–10,000 ft. Catron. **R & E**

Anticlea virescens (Kunth) Rydb. [*Zigadenus virescens* (Kunth) J. F. Macbr.]—Green death camas. Plants 3.5–8.5 dm; proximal leaf blades 25–45 cm × 3–20 mm; flowers nodding; perianth campanulate, 4–8 mm diam.; tepals greenish, sometimes tinged with purple, ovate, 4–6 × 1.5–3 mm, somewhat narrowed at base, apex narrowly obtuse; pedicels recurved at anthesis; bracts often tinged with purple or pink, ovate. Moist areas, mixed conifer, spruce-fir, ponderosa pine, meadows, riparian areas, 6000–11,150 ft. WC, Otero.

SCHOENOCAULON A. Gray – Feathershank

Schoenocaulon texanum Scheele—Texas feathershank. Bulb covered with dark brown to black fibers, these often forming a dense thatch; leaves basal, linear, grasslike; inflorescence unbranched, spicate (to somewhat racemose), dense, 30–250-flowered, bracts small; flowers tiny, sessile to subsessile; tepals free, pale green to yellowish white, with serrulate margin and inconspicuous nectar glands; stamens longer than the tepals; anthers suborbicular. Canyon bottoms, seeps, cliff bases, hillsides, piñon-juniper, desert scrub, 3550–7000 ft. EC, SE, SC.

TOXICOSCORDION Rydb. – Death camas

Plants from ovoid bulbs with a membranous coat; leaves ± basal, linear, generally folded, ± curved, entire; inflorescence loosely racemose or paniculate, 10–80-flowered; tepals petal-like, inner tepal bases conspicuously clawed (sometimes the outer as well), white to pale yellow, sometimes darker, adaxially with 1 gland near base, this obovate or obscure, not bilobed; ovary superior. Previously included in *Zigadenus*.

1. Flowers in panicles; outer tepals usually not clawed…**T. paniculatum**
1'. Flowers in racemes or paniculate with only 1 or 2 basal branches; outer tepals often clawed…**T. venenosum**

Toxicoscordion paniculatum (Nutt.) Rydb. [*Zigadenus paniculatus* (Nutt.) S. Watson]—Foothill death camas, sand-corn. Plants 2–7 dm; proximal leaf blades 15–35 cm × 3–15 mm; inflorescences paniculate, 10–80-flowered, narrow; perianth campanulate, 5–10 mm diam.; tepals cream-colored, ovate, 2–5 × 1–4 mm, outer usually not clawed; pedicels ascending to perpendicular in fruit; bracts green, lanceolate. Piñon-juniper and juniper woodlands, grasslands, sagebrush, 5500–8000 ft. McKinley, Rio Arriba, San Juan.

Toxicoscordion venenosum (S. Watson) Rydb. [*Zigadenus venenosus* S. Watson]—Meadow death camas. Plants 2–7 dm; proximal leaf blades 12–50 cm × 2–10 mm; inflorescences racemose, or if paniculate with only 1 or 2 basal branches, 10–50-flowered; perianth campanulate, 5–10 mm diam.; tepals cream-colored, 2–5 × 1–3 mm, outer often clawed; pedicels usually ascending in fruit, occasionally perpendicular; bracts usually green, sometimes white. Our plants are **var. gramineum** (Rydb.) Brasher (grassy death camas). Sagebrush, piñon-juniper, grassland, shrubby ponderosa pine, 5700–8000 ft. Rio Arriba, San Juan.

VERATRUM L. – False hellebore

Veratrum californicum Durand—California false hellebore. Plants from short, thick, vertical rhizomes and swollen basal bulbs; stems erect, simple, hollow, basally thickened, leafy, 1–2.5 m; leaves coarse, ovate, 20–40 × 15–25 cm, narrower and smaller distally; inflorescences densely paniculate, with spreading to stiffly erect branches; tepals creamy-white, greenish basally, lanceolate to elliptic or oblong-ovate, not or very slightly clawed, 8–17 mm, basal gland 1, green, V-shaped; ovary superior to partly inferior. Our plants are **var. californicum**. Riparian areas, lakeshores, moist to wet meadows, marshy areas, wet openings in coniferous forest, 7250–12,100 ft. NC, NW, C, WC, SC.

MELIACEAE – MAHOGANY FAMILY

Kenneth D. Heil

MELIA L. – Chinaberry tree

Melia azedarach L.—Chinaberry tree. Tree to 15 m; wood-scented, hard; leaves alternate, bipinnate, to 3+ dm, without stipules, leaflets numerous, ovate to elliptic-lanceolate, acuminate, to ca. 6 cm, margins crenate-dentate; inflorescences panicles of small regular flowers; flowers fragrant; sepals 1–2 mm, puberulent; petals white to pale lavender, mostly 1 cm; drupes yellow, subglobose, mostly 15 mm diam. Often cultivated and escaped in S New Mexico, 3500–5900 ft. C, SE, SW. Introduced (Pennington, 1981).

MENYANTHACEAE - BOG BUCKBEAN FAMILY

C. Barre Hellquist

MENYANTHES L. - Buckbean

Menyanthes trifoliata L.—Buckbean. Perennial aquatic or wetland herb; rhizomes thick, creeping, sheathed by membranous bases of the long petioles; leaves basal, emergent, 3-foliate, compound with a sheathing base; leaflets elliptic to obovate, 2-10 cm; inflorescence a bracteate cyme; corolla rotate to funnel-shaped, 1-2 cm across, lobes 5, fringed or hairy, upper surfaces white to light pink, mostly covered by long beard; anthers sagittate; nectaries usually 5, at ovary base; pistil 1, ovary ± superior; stigma 2-lobed; fruit a 2-4-valved capsule; seeds few to many, smooth and shiny. Shallow margins of bogs, ponds, lakes, streams, 4700-4800 ft. NW (rare) (Hellquist, 2013c).

MOLLUGINACEAE - CARPETWEED FAMILY

Steven R. Perkins

MOLLUGO L. - Carpetweed

Herbs, annual or sometimes perennial, glabrous; stems branching from base, prostrate to erect; leaves alternate, opposite, or whorled, basal leaves largest and cauline leaves reduced, margins entire, stipules rudimentary or absent; inflorescences axillary or terminal, umbellate or cymose; flowers pedicellate, perfect, small, actinomorphic, hypogynous, sepals 5 and persistent, petals absent, stamens 3-5, styles 3-5; fruits capsular, 3-valved; seeds flattened laterally, reniform.

1. Plants erect; leaves glaucous; inflorescences stalked, axillary and terminal; seeds finely reticulate…**M. cerviana**
1'. Plants prostrate to ascending; leaves not glaucous; inflorescences sessile and axillary; seeds with curved edges or smooth…**M. verticillata**

Mollugo cerviana (L.) Ser. [*Pharnaceum cerviana* L.]—Threadstem carpetweed. Plants erect, 3-20 cm; leaves glaucous, in whorls of 4-12, linear to spatulate, 3-15 mm, base cuneate, apex acute to obtuse, basal rosette present; sepals pale green, glaucous abaxially, white adaxially, elliptic to obovate; stamens 5 and alternate with sepals; seeds 20-40, finely reticulate. Associated with open woodlands and dry sandy soils, 3815-7000 ft. NW, NC, C, SE, SW. Introduced.

Mollugo verticillata L. [*M. costata* Y. T. Chang & C. F. Wei]—Green carpetweed. Plants prostrate to ascending, 3-45 cm; leaves not glaucous, in whorls of 4-8, linear to elliptic or obovate, 5-40 mm, base cuneate, apex obtuse to rounded or acute, basal rosette present; sepals green abaxially, white adaxially, oblong-elliptic; stamens 3-4 and alternate with sepals; seeds 15-35, dark or reddish brown with parallel curved ridges or smooth. Weedy in fields, gardens, roadsides, 3400-6500 ft. NW, EC, WC, SE, SW.

MONTIACEAE - MINER'S LETTUCE FAMILY

David J. Ferguson and Thomas R. Stoughton

Annual, biennial, or perennial herbs, sometimes woody or cushion-forming shrubs, often succulent; leaves simple, entire, opposite or alternate; flowers radially or bilaterally symmetric; perianth biseriate, usually with 2 sepals and (2-)4-6 petals; stamens most commonly as many as and opposite petals, varying to many; filaments free or sometimes basally adnate to petals; pistil 2- or 3-carpellate; ovary superior, unilocular with free central or basal placentae; style 1 with 3-12 stigma lobes, or seldom with a single (minutely lobed) stigma; fruit usually a loculicidal or circumscissile capsule, seeds mostly lenticular (Ferguson, 2001; Hershkovitz, 2019; Packer, 2003). Previously included in Portulacaceae.

1. Leaves terete, alternate or in a crowded basal clump; capsules mostly basipetally dehiscent into 3 valves, sometimes requiring disturbance to open; seeds numerous with chartaceous pellicle...**Phemeranthus**
1′. Leaves not terete, flat at least dorsally, opposite or in basal rosette or cluster (see 4); seeds with no membranous pellicle...(2)
2. Low-spreading, semiaquatic, succulent herbaceous perennial; stems spreading, branching, rooting at nodes; leaves opposite; inflorescence a short, terminal or axillary raceme or cyme; seeds 1–3...**Montia**
2′. Nonaquatic, terrestrial; stems (when present) arising from common central rootstock; inflorescences various from base of plant; seeds 3+...(3)
3. Leaves thick-succulent, linear, without visible midrib, not obviously petiolate; sepals often dentate to erose; petals usually > 5; capsule basally circumscissile, dehiscent into valves by longitudinal slits; seeds 15+...**Lewisia**
3′. Leaves flat, midrib often visible, petiolate (petiole wide and short in *Claytonia megarhiza*); sepals entire; capsule not basally circumscissile, dehiscent into longitudinal valves...(4)
4. Stem and inflorescence with alternate leaves and bracts; petals 2–4; stigma lobes 2; capsule 2-valved... **Calyptridium**
4′. Stem and inflorescence with opposite leaves and bracts; petals 5; stigma lobes 3; capsule 3-valved...(5)
5. Annual or perennial; stem and inflorescence with 1–2 pairs of leaflike leaves/bracts; stamens 5; seeds 3–6...**Claytonia**
5′. Annual; stem and inflorescence usually with > 2 pairs of leaflike bracts; stamens 5–15; seeds 10–20... **Calandrinia**

CALANDRINIA Kunth – Redmaids, rock-purslane

Thomas R. Stoughton

Calandrinia menziesii (Hook.) Torr. & A. Gray [*C. ciliata* (Ruiz & Pav.) DC.]—Menzies' redmaids. Plants annual, generally prostrate to erect, usually branched from the base, 3–40 cm; stems glabrous; leaves linear to oblanceolate or spoon-shaped, glabrous or not, alternate, sometimes clasping at the base, flattened; inflorescences racemose, pedicels 4–25 mm; flowers actinomorphic; sepals 2.5–8 mm; petals 4–15 mm, bright pink (white); stamens 3–15, usually opposite but not adnate to petals; seeds 1–2.5 mm wide, elliptic. Sandy to loamy soil, grassy or disturbed areas, cultivated fields, 4750–5500 ft. SW.

CALYPTRIDIUM Nutt. – Pussypaws

David J. Ferguson

Calyptridium monandrum Nutt. [*Cistanthe monandra* (Nutt.) Hershk.]—Common pussypaws. Semisucculent annual, often reddish, 1.5–18 cm; root a slender taproot; stems 2+ from each rosette, appearing acaulescent; leaves basal and cauline, crowded, oblanceolate to narrowly spatulate, 1–5 cm; inflorescences racemose or paniculate cymes, often scorpioid, spreading horizontally to decumbent, to 18 cm; sepals 1–2 mm, ovate to deltoid, margins scarious in fruit; petals 2–4, 1–3 mm, white to reddish; stamens 1–3; capsules ovoid to cylindrical, 4–8 mm, 2-valved; seeds compressed-globose, 0.3–0.8 mm, shiny black. Ephemeral after rains, disturbed areas; 1 record, 4025 ft. SW.

CLAYTONIA L. – Spring beauty, miner's lettuce

Thomas R. Stoughton and David J. Ferguson

Annual or perennial, succulent herbs and cushion-forming shrubs, often with thickened, subterranean perennation structures; basal leaves alternate and rosette-forming; cauline leaves subtending inflorescences, opposite (rarely alternate to subopposite), fused partially or completely around aerial stems into a disk, or free; flowers showy; sepals persistent, leaflike, unequal; petals 5, white (pink or yellow), often with pink veins and a yellowish blotch at the base of the blade above the claw; stamens 5; ovary globose, ovules 3 or 6; capsules 3-valved, longitudinally dehiscent from apex; seeds (1–)3–6, black, shiny, smooth to tuberculate (Davis, 1966; Halleck & Wiens, 1966; Miller & Chambers, 1993).

1. Plants annual from weakly thickened taproot; cauline leaf pair fused partially or completely into a perfoliate disk...**C. perfoliata**
1′. Plants perennial from subterranean tuber or strongly thickened caudex; cauline leaf pair free...(2)
2. Plants generally rosette-forming, aerial stems arising from woody caudices; alpine habitat...**C. megarhiza**

2'. Plants not rosette-forming, aerial stems arising from subterranean tubers; subalpine habitat and below…
    (3)
3. Inflorescences multibracteate; cauline leaves 1-veined, blades linear; basal leaves generally present at
   flowering, apices obtuse…**C. rosea**
3'. Inflorescences unibracteate; cauline leaves 3-veined, blades generally lance-ovate to elliptic; basal leaves
    usually absent at flowering, apices acute…**C. lanceolata**

Claytonia lanceolata Pursh—Western spring beauty. Perennials, 1–10 cm, from globose tubers; basal leaves 1–6, often absent at flowering, linear to lanceolate, 5–40 × 0.2–1.6 cm; cauline leaves ovate to narrowly lanceolate, 1–6 × 0.5–2 cm, sessile; inflorescence unibracteate; flowers white, pink, rose, or magenta, 8–14 mm diam. Melting snowbanks in subalpine meadows, moist meadows, and subalpine forests, 7000–12,000 ft. NW, NC.

Claytonia megarhiza (A. Gray) Parry ex S. Watson—Alpine spring beauty. Perennials, 5–25 cm from a stout, woody caudex; basal leaves often forming a tight rosette, numerous, oblanceolate to spatulate, 2–10 × 0.4–3 cm, dilated at the base into a succulent sheath; cauline leaves oblanceolate, 2–10 × 2–5 cm; inflorescence multibracteate; flowers 12-20 mm diam.; petals white, pink, or rose. Rock crevices in the alpine, talus, scree, gravelly slopes, 11,700–13,160 ft. NC.

Claytonia perfoliata Donn ex Willd.—Miner's lettuce. Herbaceous semisucculent annual, usually bearing small rounded subterranean tubers, 5–50 cm, often red or purple; basal leaves in rosette, ovate to broadly rhomboid or deltoid, obtuse, petiolate, 1–5 cm; cauline leaves opposite, united on 1 side or perfoliate, forming a suborbicular disk, 1–5 cm; inflorescence bract(s) mostly unpaired, oblong to ovoid, 0.5–15 mm; flowers 3–10 mm; sepals ±3 mm, rounded-ovate to suborbicular; petals white to pink, 2–5 mm. New Mexico material belongs to **subsp. intermontana** John M. Mill. & K. L. Chambers. Mostly in moist disturbed areas up to 7000 ft. NW.

Claytonia rosea Rydb.—Rocky Mountain spring beauty. Perennials, 2–15 cm from globose tubers; basal leaves linear to narrowly spatulate or sometimes absent, 1–7 × 0.4–2 cm; cauline leaves linear, sessile, 2–5 cm; inflorescence multibracteate (rarely unibracteate), bracts sometimes ephemeral; flowers 8–14 mm diam., pink or rose. Ponderosa pine forests and oak belts (early spring), 5500–8500 ft. W.

LEWISIA Pursh – Bitterroot

David J. Ferguson

Herbaceous succulent perennials; stems from a thick, short, inconspicuous caudex, obscured by densely crowded sessile leaves; leaves (ours) succulent, deciduous at or soon after flowering; inflorescence (ours) lateral, spreading to suberect, single-flowered to short-cymose, few-flowered, mostly little exceeding leaves; bracts (ours) small, mostly in subequal pairs; flowers pedicellate or sessile; sepals 2–4, equal or subequal when paired; petals (5)6–10, white to magenta, often longitudinally striped; stamens 9–20; capsules basally circumscissile and splitting acropetally into valves; seeds brown or black, smooth or minutely sculpted (Davidson, 2000; Matthew, 1989).

1. Flowers sessile; bract and sepal pairs decussate and appearing to form 4-merous calyx; flowers usually >
   3 cm diam.…**L. brachycalyx**
1'. Flowers pedicellate; bracts and sepals not resembling 4-merous calyx; flowers < 3 cm diam.…(2)
2. Flowers usually single, mostly 2+ cm diam., usually slightly zygomorphic (laterally "pinched"); sepal margins mostly entire, sometimes obscurely or irregularly toothed, not glandular, apex acute to subacute;
   petals white (rarely pinkish)…**L. nevadensis**
2'. Flowers usually multiple per inflorescence, mostly < 2 cm diam., regular; sepal margins regularly toothed,
   teeth usually glandular-tipped, apex truncate or sometimes rounded, obtuse, subacute, or apiculate; petals mostly pink, sometimes white…**L. pygmaea**

Lewisia brachycalyx Engelm. ex A. Gray [*Oreobroma brachycalyx* (Engelm. ex A. Gray) Howell]— Mogollon bitterroot. Succulent perennials; leaves oblanceolate, 3–8 cm; inflorescences to 6 cm, 1-flowered; flowers 2.5–6 cm diam.; bracts 2, 5–7 mm; sepals suborbiculate, 3–9 mm; petals white to pink with darker longitudinal stripes; capsules 5–9 mm, base persistent after dehiscence; seeds 40–50, 1.5 mm, shiny. Igneous and sandy gravel substrates, damp areas from snowmelt, mixed conifer, 4000–8000 ft. The presence of this species in New Mexico is uncertain, but it was originally described from the state.

Lewisia nevadensis (A. Gray) B. L. Rob. [*Oreobroma nevadense* (A. Gray) Howell; *L. pygmaea* (A. Gray) B. L. Rob. var. *nevadensis* (A. Gray) Fosberg]—Nevada bitterroot. Succulent perennials; stem a short caudex; leaves narrowly linear to linear-oblanceolate, 2–15 cm; inflorescences to 10 cm, 1(–3)-flowered; flowers often appearing slightly zygomorphic, 1.5–2 cm diam.; sepals broadly ovate, 5–13 mm; petals white or rarely pink; capsules 5–10 mm, persistent after dehiscence; seeds 20–50, 1.3 mm, shiny, muricate. Wet grassy slopes and meadows, ponderosa pine and mixed conifer forests, 7300–8800 ft. Presence in New Mexico is uncertain because of confusion with *L. pygmaea* and difficulty in determining herbarium material.

Lewisia pygmaea (A. Gray) B. L. Rob.—Pygmy bitterroot, alpine bitterroot. Succulent perennials; leaves linear to linear-oblanceolate, to 9 cm; inflorescences to 2–9 cm, 1–7-flowered; flowers 1.5–2 cm diam.; sepals suborbiculate, 2–6 mm; petals white, pink, or magenta, sometimes green at base, often longitudinally striped darker; capsules 4–5 mm, base persistent after dehiscence; seeds 15–24, 1–2 mm, shiny, smooth. Often following edges of melting snowbanks, gravelly or rocky substrates, damp areas, ponderosa pine forests, mixed conifer, subalpine, alpine tundra, 7000–12,600 ft. NC, NW, WC.

MONTIA L. – Miner's lettuce

David J. Ferguson

Montia chamissoi (Ledeb. ex Spreng.) Greene [*Crunocallis chamissoi* (Ledeb. ex Spreng.) Rydb.]—Water miner's lettuce. Perennials, 5–25 cm, appearing tangled and matted; roots slender rhizomes and stolons; basal leaves absent or reduced; cauline leaves opposite, mostly oblanceolate, 1–6 cm; inflorescences short terminal or axillary cymes, pedicels recurved in bud and in fruit; flowers 2–10, sometimes replaced by bulblets; sepals obovate to orbicular, 2–3 mm; petals 4–5, white to pink, 2–8 mm, obovate; stamens 5; capsules obovoid, 1–1.5 mm. Cool, permanently wet areas in mountains, often in shade, 8400–10,650 ft. NC, WC.

PHEMERANTHUS Raf. – Fameflower, rockpink

David J. Ferguson

Succulent perennial herbs with tuberous taproots; stems procumbent to erect, mostly few-branched, inconspicuous, usually obscured by densely crowded leaves; leaves alternate or in a tight stemless cespitose cluster; inflorescence cymose (rarely single-flowered), axillary, terminal, or lateral from stemless leaf cluster; flowers diurnal or nocturnal, opening briefly during 1 day; petals 5–8; stamens 5 to many; stigma lobes 3–6; capsules primarily basipetally dehiscent; seeds many, with or without concentric arcuate ridges, black or appearing gray (Ferguson, 2001; Kiger, 2001; Price & Ferguson, 2012).

1. Plants appearing acaulescent, leaves in tight basal cluster; flowers yellow; seeds with prominent concentric arcuate ridges…(2)
1'. Plants caulescent, leaves distributed along well-developed stems; flowers white to magenta; seeds with or without prominent concentric arcuate ridges…(3)
2. Leaves mostly < 3 cm, narrowed at base and appearing petiolate (narrow portion sometimes hidden below soil level); flowers usually < 8 mm diam.…**P. parvulus**
2'. Leaves mostly > 3 cm, not appearing petiolate; flowers usually > 8 mm diam.…**P. humilis**
3. Plants with a mostly vertical aspect; inflorescences terminal, erect, usually many-flowered; scape erect, slender, usually greatly exceeding leaves; seeds with prominent ridges…(4)
3'. Plants with a horizontal to procumbent aspect; inflorescences appearing axillary, not erect; scape short, thick, and not or little exceeding leaves; seeds without prominent ridges…(6)
4. Flowers mostly > 2 cm across, lavender to magenta, fragrant; pistil exceeding stamens; stamens > 15; sandhill areas of grasslands…**P. calycinus**
4'. Flowers mostly < 1.5 cm across, white, greenish, or pink to magenta, not fragrant; pistil approximately equaling stamens; stamens usually 5–10; locations various…(5)
5. Stamens pink to magenta; capsules appearing lumpy, conforming to seeds inside; seeds with prominent concentric arcuate ridges; on calcareous substrates…**P. longipes**

5'. Stamens greenish to whitish; capsules somewhat trigonous, smooth, not conforming to seeds inside; seeds without prominent ridges; on noncalcareous substrates…**P. confertiflorus**

6. Leaves acute; petals usually magenta (rarely white), acute; sepals acute, exceeding fruit, persistent into fruiting; style mostly erect and central; inflorescence indeterminate and usually with 3-5 flowers (occasionally 1, or > 5); fruit persistent at maturity but very delicate…**P. brevicaulis**

6'. Leaves usually obtuse or blunt; petals usually obtuse; sepals usually obtuse and not exceeding fruit, early-deciduous; style mostly turned downward so as to appear off-center; inflorescence 1-flowered (rarely 2); fruit deciduous upon dehiscing…(7)

7. Plants bearing long, thin rhizomes; leaves usually < 1.7 cm. flowers usually < 2 cm diam., magenta, mostly with 5-6 petals; growing on conglomerate; SW New Mexico…**P. rhizomatus**

7'. Plants without long, thin rhizomes; leaves and flowers various; substrates various; not in SW New Mexico…(8)

8. Leaves usually > 1.7 cm; flowers usually > 2 cm diam., white to pink or magenta, mostly with 6-8 petals; growing on travertine, gypsiferous limestone, and calcareous sandstone…**P. brachypodius**

8'. Leaves usually < 1.7 cm; flowers usually < 2 cm diam., white to magenta (usually magenta in New Mexico), usually with 5-6 petals; noncalcareous (mostly red sandstone) substrates…**P. brevifolius**

Phemeranthus brachypodius (S. Watson) D. J. Ferguson [*Talinum brachypodium* S. Watson]— Laguna fameflower, Laguna rockpink. Stems to 10 cm, usually less; leaves subterete, to 2.5 cm; flowers to 2.8 cm; sepals early-deciduous, oval to orbiculate, to 7 mm; petals 6-8, to 17 mm, white, pink, or magenta; stamens to 35. Soil pockets on ledges, calcareous rock substrates, piñon-juniper woodlands, 5575-5900 ft. Cibola. Very similar to and closely related to *P. brevifolius*, but larger in all proportions. **R & E**

Phemeranthus brevicaulis (S. Watson) Kiger [*Talinum brevicaule* S. Watson; *T. pulchellum* Wooton & Standl.]—Showy fameflower. Stems to 10 cm; leaves subterete, apically pointed, to 2.5 cm; inflorescence cymulose, 1- to multiflowered; flowers to 2.5 cm diam.; sepals persistent, obovate to broadly lanceolate, obtuse to attenuate, to 10 mm, often purplish or pinkish; petals 5-6, to 15 mm, magenta (rarely pink or white); stamens 15-40. Calcareous substrates, mostly shallow soil, desert scrub, grasslands, piñon-juniper woodlands to ponderosa pine forests, 2950-7200 ft. Widespread except far N.

Phemeranthus brevifolius (Torr.) Hershk.—Canyonland fameflower. Stems to 10 cm, usually much less; leaves subterete, apically rounded, to ± 1.5 cm; inflorescences 1-flowered (rarely cymulose, few-flowered); flowers to 2 cm diam.; sepals early-deciduous, oval to orbiculate, to 5 mm; petals 5(6), to 12 mm, white, pink, or ours usually magenta; stamens 10-25. Slickrock; shallow, fine sand pockets overlying noncalcareous, often reddish sandstone; grassland, piñon-juniper woodlands, 3450-7225 ft. NC, W. Published records are badly confused with *P. brevicaulis*.

Phemeranthus calycinus (Engelm.) Kiger [*Talinum calycinum* Engelm.; *Claytonia calycina* (Engelm.) Kuntze]—Sandhills fameflower, sandpink. Stems to 15 cm; leaves subterete, to 5 cm; inflorescences cymose with long, slender scape, much overtopping leaves, to 25 cm; flowers fragrant, "chocolate"-scented, 2-3 cm diam.; sepals persistent, ovate to suborbiculate, 4-6 mm; petals 10-15 mm, pink to magenta; stamens 25-45. Sand, mostly eolian deposits, often somewhat gravelly; grasslands, 2000-5000 ft. Roosevelt. Growth and flowering are generally brief following warm-season rains.

Phemeranthus confertiflorus (Greene) Hershk. [*Talinum confertiflorum* Greene; *T. gracile* Rose & Standl.]—Rocky Mountain fameflower. Stems to 10 cm, usually orange; leaves terete, to 4 cm, often purplish apically; inflorescences cymose with long, slender scape, much overtopping leaves, to 12 cm; sepals persistent, ovate, to 5 mm, apex often purplish; petals white to pink, rarely purplish, to 6 mm; stamens 5(-10). Diverse habitats, often abundant; desert scrub, piñon-juniper woodlands to ponderosa pine forests, 5600-8850 ft. Widespread except SE.

Phemeranthus humilis (Greene) Kiger [*Talinum humile* Greene]—Pinos Altos fameflower. Appearing acaulescent; leaves cespitose, to 8 cm; inflorescences short-spreading cymes to 8 cm, little exceeding leaves; mostly bearing < 10 flowers; sepals ovate, ± 3 mm, deciduous; petals yellow, to 4 mm; stamens usually 5-8. Gravelly clay soils, arid grasslands, piñon, juniper, and oak woodlands, 3900-5900 ft. SW. **R**

Phemeranthus longipes (Wooton & Standl.) Kiger—Tortugas fameflower. Stems to 6 cm, simple or few-branched, somewhat suffrutescent; leaves terete, to 3 cm; inflorescence cymose with long, slender scape, much overtopping leaves, to 10 cm; sepals orbiculate to suborbiculate, 2–3 mm, early-deciduous; petals white to pale pink, 4–5 mm; stamens ± 10. Calcareous substrates, mostly shallow gravelly soils, desert scrub, arid grasslands, scrublands, piñon-juniper woodlands, 3000–9100 ft. NC, C, SC, SE. *Phemeranthus longipes* is often confused with *P. confertiflorus*.

Phemeranthus parvulus (Rose & Standl.) D. J. Ferguson & T. M. Price [*Talinum marginatum* Greene]—Bottle-leaf fameflower. Appearing acaulescent; root a roughly globose, cormlike, tuberous taproot; leaves subterete, mostly 2–4 cm, narrowed basally, appearing roughly bottlelike; inflorescences cymose, mostly ascending, crowded, usually exceeding leaves, to 8 cm; sepals 3 mm, ovate, deciduous; petals yellow, 4 mm; stamens 5–8. Arid grasslands, piñon, juniper, and oak woodlands, ponderosa pine-mixed conifer forests, 3180–9850 ft. SW.

Phemeranthus rhizomatus D. J. Ferguson—Gila fameflower. As in *P. brevifolius*, except producing long, thin, stolonlike underground rhizomes; leaves tending to be more dorsiventrally compressed, obtuse; color averaging darker; flowers magenta. Conglomerate exposures, open piñon, juniper, and oak communities, 5000–6000 ft. Catron, Grant. **R**

MORACEAE – MULBERRY FAMILY

Kenneth D. Heil

Primarily trees and shrubs with milky sap; leaves most often alternate, but also spiral or opposite, stipules present, blades elliptic, lanceolate, ovate, or cordate, occasionally fleshy or prominently veined, margins entire, dentate, serrate, or lobed; inflorescence monoecious or dioecious, sometimes cauliflorous on large branches or trunks, rarely solitary, in catkins, aments, or sometimes on flattened, hollowed, and/or variously thickened receptacles; flowers imperfect, actinomorphic or tubular, small, often minute, perianth parts 4, corolla generally none, stamens 4, rarely 1 or 2, carpels 2, 1 abortive, with persistent style, ovary superior; fruit variable, fleshy, multiple and/or accessory, produced by the receptacle (syconium), occasionally an achene or a drupe (Wunderlin, 1997).

1. Leaves with entire margins, leaf venation pinnate; branches spiny; fruit globose, with a crustlike rind… **Maclura**
1'. Leaves serrate or dentate, undivided or lobed, leaf venation weakly palmate; branches not spiny; fruit a fleshy multiple cluster or a pear-shaped fig…(2)
2. Leaves 12–25 × 10–18 cm, deeply lobed; fruit a pear-shaped fig…**Ficus**
2'. Leaves 4–15 × 3–12 cm, toothed to lobed; fruit a multiple cluster…**Morus**

FICUS L. – Fig

Ficus carica L.—Common fig. Shrubs or small trees, deciduous, to 5 m; sap milky; bark grayish, slightly roughened, branchlets pubescent; leaves alternate, monomorphic, with stipules 1–1.2 cm, leaf blades obovate, nearly orbiculate, or ovate, palmately 3–5-lobed, 15–30 × 15–30 cm, base cordate, margins undulate or irregularly dentate, apex acute to obtuse; inflorescences small, borne on inner walls of fruitlike and fleshy receptacle (syconium); staminate and pistillate flowers on same plant; staminate flowers sessile or pedicellate, calyx of 2–6 sepals, stamens 1–2; pistillate flowers sessile; ovary 1-locular; style unbranched, lateral; cypselae completely embedded in enlarged, fleshy, common receptacle. Disturbed sites, old homesteads, 4190–4870 ft. SC, SW. Introduced.

MACLURA Nutt. – Osage orange

Maclura pomifera (Raf.) C. K. Schneid. [*Toxylon pomiferum* Raf.]—Osage orange. Dioecious trees or shrubs to 12 m, often thicket-forming, spiny, deciduous, dioecious, sap milky; bark furrowed, yellow-brown, thorns to 25 mm, stout; leaves alternate or opposite, stipules minute, blades entire, ovate to elliptic-lanceolate, to 12 cm, base broadly cuneate to subcordate, apex acuminate; staminate inflorescence a loose axillary head or umbel, pistillate a head; flowers imperfect, 4-merous; pistillate flowers sessile, 2–2.5 cm across, staminate flowers pedicelled, 25–35 mm; fruit a globose multiple of achenelike fruits to 15 cm diam.; rind wrinkled, hard. Escaped along rivers and at old homesteads, 3700–7350 ft. Widely scattered throughout the state.

MORUS L. – Mulberry

Deciduous, dioecious or monoecious trees or shrubs; bark scaly and/or thick; leaves with stipules, lanceolate, blades undivided or lobed, serrate or dentate; inflorescence a stalked, axillary, pendent, catkinlike spike or ament, both sexes often present; flowers imperfect; calyx 4-parted, sepals involute, partially enclosing filaments; stamens 4, stigmas 2; fruit an ovoid compressed achene, aggregating into a multiple accessory structure.

1. Mature leaf blades < 7 cm, adaxially harshly scabrous or pubescent, adaxially harshly scabrous; petiole to 1.5 cm…**M. microphylla**
1'. Mature leaf blades usually > 8 cm, adaxially slightly if at all scabrous; petiole 2+ cm…**M. alba**

Morus alba L.—White mulberry. Small trees to 15 m; young branches slightly pubescent, becoming glabrous; leaves with petioles 1–3 cm, ovate to subcordate, to 20 cm, dentate, sometimes lobed, usually glabrous above, hairy in tufts at axils of major veins on underside or glabrous; inflorescence a pendent catkinlike spike; flowers small, greenish or whitish; fruit a fleshy multiple, ovoid to oblong-elliptic, 1–2.5 cm, white, pink, violet, or reddish, edible, sweet and/or tart. Along rivers and acequias or escaped from cultivation near habitations, 3900–6250 ft. NW, NC, C, SC, SW. Introduced.

Morus microphylla Buckley—Texas mulberry. Shrubs or trees to 7.5 m; bark gray, fissured, scaly, branchlets greenish, pubescent; lenticels prominent; leaves 3–5 mm, papery, pubescent, petioles 0.3–0.6(–1.5) cm, pubescent, blades ovate, sometimes 3–5-lobed, 2–7(–9) × 1–4(–7) cm, margins serrate or crenate-serrate, surfaces abaxially harshly scabrous or pubescent, adaxially harshly scabrous; inflorescence a catkin; staminate and pistillate flowers on different plants; staminate flowers with calyx lobes green to reddish; syncarps red, purple, or black, 1–1.5 cm; cypselae yellowish, oval, flattened, ca. 2 mm. Canyons on limestone and igneous slopes, usually along streams, 3650–7000 ft. NE, NC, S.

NAMACEAE – NAMA FAMILY

Steve L. O'Kane, Jr. and Jennifer Ackerfield

NAMA L. – Nama

Annual or perennial herbs; leaves alternate, entire to crenate-dentate, strigose and/or hispid; inflorescence solitary or of terminal cymes, not coiled; calyx lobes united at base or nearly to apex; corolla sympetalous, white to purple or lavender, the tube sometimes yellow; gynoecium bicarpellate, superior, ovary bilocular, placentae narrow, membranous or cartilaginous, completely dividing ovary, style terminal, stylodia 2, distinct to base (and appearing as 2 styles) in ours; stamens included, usually unequal; styles 1–2; capsules ovoid to elliptic.

1. Perennials with a woody base…(2)
1'. Annuals without a woody base…(3)
2. Corolla white to off-white; leaves linear, densely crowded, strongly revolute; plants branched only near the summit; seeds finely alveolate, brown…**N. carnosa**
2'. Corolla purple; leaves oblanceolate, not densely crowded, weakly revolute; plants branched from the base; seeds reticulate, yellow…**N. xylopoda**
3. Plants minutely glandular-puberulent…**N. dichotoma**
3'. Plants hispid, strigose, or hirsute but lacking stalked, glandular hairs…(4)
4. Plants with spreading-hispid hairs and shorter, retrorse hairs underneath; flowers reddish purple, 4–7 mm …**N. retrorsa**

4'. Plants strigose to hirsute, lacking retrorse hairs underneath; flowers pink to purple, 8–15 mm…(5)
5. Leaves linear-oblong to obovate, flat or strongly revolute; seed surface reticulate…**N. hispida**
5'. Leaves linear-lanceolate, strongly revolute; seed surface alveolate…**N. stevensii**

Nama carnosa (Wooton) C. L. Hitchc.—Sand fiddleleaf. Perennials, 2–4 dm, branched near the summit only, sparsely strigose and pustulose; leaves crowded, linear, strongly revolute, 1–3 cm × 1–1.5 mm; flowers in crowded terminal cymes; calyx lobes 7–10 mm; corolla to 7 mm, white to off-white; seeds brown, finely alveolate. Sandy gypsum soil, often in piñon-juniper woodlands, 3250–6650 ft. EC, C, SE, SC.

Nama dichotoma (Ruiz & Pav.) Choisy—Wishbone fiddleleaf. Annuals, 0.5–2 dm, branched dichotomously throughout, minutely glandular-puberulent; leaves linear to spatulate, 1–3 cm × 2–5 mm; flowers solitary or in pairs in the upper branches; calyx lobes 5–10 mm; corolla to 3–8 mm, white to pale lavender or light blue; seeds brown, regularly large-pitted and cross-ridged. Dry, gravelly soil, often in piñon-juniper woodlands, 3700–10,500 ft. Widespread except EC, SE.

Nama hispida A. Gray—Bristly nama. Annuals, 0.5–3(–5) dm, branched from the base and forming rounded mounds, strigose-hispid to hirsute; leaves linear-oblong to obovate, flat or strongly revolute, 1–7 cm × 1–8 mm; flowers solitary or in terminal clusters; calyx lobes 4–7 mm; corolla 8–15 mm, pink to purple; seeds yellowish brown, reticulate. Dry, gravelly or sandy soil of plains, foothills, and mesas, 3200–6900 ft. Widespread except NE.

Nama retrorsa J. T. Howell—Betatakin fiddleleaf. Annuals, 1–3 dm, branched from the base, with spreading-hispid hairs and shorter, retrorse hairs underneath; leaves linear to spatulate, 1.5–5 cm × 2–5 mm; flowers in terminal clusters; calyx lobes to 5 mm; corolla 4–7 mm, reddish purple; seeds brown. Dry, sandy soil of desert scrub communities, 4900–6700 ft. NW.

Nama stevensii C. L. Hitchc.—Stevens' fiddleleaf. Annuals, 0.5–3 dm, branched from the base and forming rounded mounds, strigose; leaves linear-lanceolate, strongly revolute, 1–3 cm × 1–3 mm; flowers in small axillary clusters; calyx lobes 5–8 mm; corolla 8–10 mm, purple; seeds yellow, alveolate. Sandy gypsum outcrops, 3100–5200 ft. SE, Lincoln.

Nama xylopoda (Wooton & Standl.) C. L. Hitchc.—Yellowseed fiddleleaf. Perennials, 0.5–1 dm, branched from the base, strigose to hispid; leaves oblanceolate, weakly revolute, 0.5–1.2 cm × 2–4 mm; flowers solitary or in terminal clusters; calyx lobes 4–5 mm; corolla 6–8 mm, purple; seeds yellow, reticulate. Rocky crevices and limestone cracks of canyons in the Guadalupe and Brokeoff Mountains, 4100–6200 ft. Chaves, Eddy, Otero. **R**

NITRARIACEAE – NITRARIA FAMILY

Kenneth D. Heil

PEGANUM L. – Peganum

Peganum harmala L.—African rue. Mucilaginous shrubs; stems prostrate to ascending, pubescent to glabrous, to 3 dm; leaves alternate, fleshy, petiolate, simple, glabrous to slightly pubescent; sepals 5 usually longer than petals; petals 5, white to yellow, to 15 mm; filaments dilated basally; disk cupuliform; fruit fleshy, a capsule, to 15 mm diam. Disturbed areas in Chihuahuan desert scrub, rarely at higher elevations, 3000–8700 ft. (Correll & Johnston, 1979d). An introduced noxious weed (Watson & Dallwitz, 1992). Sheahan and Chase (1996) recognize this family as belonging to the order Sapindales and not being closely related to Zygophyllaceae.

NYCTAGINACEAE – FOUR-O'CLOCK FAMILY

Kenneth D. Heil

Annual or perennial herbs, subshrubs, or shrubs; stems unarmed (ours), prostrate, procumbent, ascending to erect, often clambering on other plants, sometimes swollen at the nodes, leaves deciduous, sessile to petiolate, opposite or rarely alternate, variably glabrous or pubescent, often somewhat thick and fleshy, sometimes viscid, stipules lacking; inflorescence a terminal or lateral bracteate cyme or raceme, or flowers solitary, bracts sometimes sepaloid; flowers actinomorphic to sometimes zygomorphic, perigynous, perfect or unisexual, showy to inconspicuous; calyx 5-merous, inconspicuous to showy; corolla absent; stamens typically 5, sometimes more, alternate with calyx; nectary disk sometimes present at base of pistil, ovary of 11 carpels, ovules solitary, placentation basal; fruit an anthocarp consisting of an achene, or a utricle enclosed in persistent base of calyx, woody to fleshy, often sticky-glandular, smooth or ribbed to winged; seeds 1 (Allred, 2020; Spellenberg, 2003).

1. Flowers numerous in capitate umbels, subtended by 5–10 broad bracts; stigmas linear…(2)
1'. Flowers single or in few-flowered cymose to umbellate clusters, subtended by 1–5 narrow to broad bracts; stigmas capitate to hemispherical…(3)
2. Plants annual; fruits large, > 1 cm long and wide, with large, conspicuous, papery-translucent wings extending above and below body of fruit…**Tripterocalyx**
2'. Plants perennial or rarely annual; fruits small, < 1 cm, with smaller, less conspicuous, opaque wings that do not extend above or below the fruit…**Abronia**
3. Flowers subtended by foliaceous or broad, translucent, often connate bracts forming an involucre…(4)
3'. Flowers ebracteate or subtended by narrow, small, distinct bracts that do not form an involucre…(7)
4. Involucral bracts connate for at least 40% of their length…**Mirabilis** (in part)
4'. Involucral bracts distinct or connate only at base…(5)
5. Flowers 3 per cluster, highly bilaterally symmetric, blooming simultaneously and resembling single flowers; involucral bracts 3; concave surface of fruit usually bearing 2 rows of teeth, often with glands at tips; fruits dorsoventrally compressed…**Allionia**
5'. Flowers > 3 per cluster, radially symmetric or slightly bilaterally symmetric, if blooming asynchronously or simultaneously, clearly representing several flowers; involucral bracts 5–20; fruit surfaces without glands; fruits radially symmetric…(6)
6. Perianth magenta or creamy-white, campanulate; flowers borne on midvein of involucral bract; fruits ellipsoid…**Mirabilis** (in part)
6'. Perianth orange-red or yellow, funnelform; flowers borne in capitate clusters; fruits turbinate… **Nyctaginia**
7. Fruits with 3–5 thin, scarious, translucent wings; perianth broadly to narrowly funnelform…**Acleisanthes** (in part)
7'. Fruits without wings, or if wings present, then thick and obscurely translucent; perianth campanulate or low-domed to widely funnelform…(8)
8. Inflorescence racemose; fruits clavate, ± gibbous on abaxial (lower) side…**Cyphomeris**
8'. Inflorescence racemose, umbellate, cymose, or capitate, or flowers borne singly; fruits fusiform, biturbinate, oblong or narrowly ellipsoid, obovoid, clavate, or obpyramidal, not gibbous…(9)
9. Fruits clavate, with large sticky glands; inflorescences umbellate…**Commicarpus**
9'. Fruits biturbinate, oblong or narrowly ellipsoid, fusiform, obovoid, clavate, or obpyramidal, eglandular or glandular-puberulent; inflorescences racemose, cymose, umbellate, or capitate, or flowers borne singly…(10)
10. Plants stout; stems erect, with sticky bands on upper internodes; fruits biturbinate or fusiform, often with equatorial wings, obscurely to prominently 10-ribbed…**Anulocaulis**
10'. Plants slender; stems decumbent to erect, with or without sticky bands on upper internodes; fruits oblong or narrowly ellipsoid, clavate, fusiform, obovoid, or obpyramidal, without equatorial wings, prominently 5-angled or 5-ribbed…(11)
11. Flowers usually borne singly, relatively large (> 2 cm); fruits oblong or narrowly ellipsoid…**Acleisanthes** (in part)
11'. Flowers usually in capitate, cymose, umbellate, or racemose clusters, rarely borne singly, relatively small (usually < 1 cm); fruits clavate, fusiform, obovoid, or obpyramidal…**Boerhavia**

ABRONIA Juss. – Sand verbena

Annual or perennial herbs from taproots or rhizomes; stems prostrate or ascending to erect, or plants sometimes acaulescent, glabrous, glandular, or pubescent; leaves commonly basal with fewer cauline leaves; inflorescence axillary or scapose, in capitate cymose to umbellate heads; flowers perfect, funnelform to salverform, tube often constricted above ovary; stamens 5–9, included in tube, styles included, stigmas linear; fruit a symmetric anthocarp, not distinctly ribbed, turbinate or obovate, usually winged. Many species require mature fruits to identify and are highly variable morphologically, suggesting hybridization.

1. Plants perennial, acaulescent or nearly so, usually cespitose, rarely with short-branched aerial caudices; leaves all basal; inflorescences not prominently axillary, appearing scapose; wings on fruit not dilated…(2)
1'. Plants annual or perennial, usually caulescent, infrequently cespitose; stems elongate, rarely with short internodes; leaves basal and cauline (basal leaves sometimes absent on older plants); inflorescences prominently axillary; wings on fruit dilated or not…(3)
2. Leaf blades 5+ times as long as wide…**A. bigelovii**
2'. Leaf blades < 3 times as long as wide…**A. nana**
3. Wings of fruit not dilated distally, or wings absent…**A. fragrans**
3'. Wings of fruit dilated distally and flattened perpendicular to plane of lamina…(4)
4. Plants usually annual, occasionally perennial…**A. angustifolia**
4'. Plants perennial…(5)
5. Perianth limb pale pink to light magenta; inflorescence bracts acute to attenuate at apex; fruits 4–7 mm; calcareous or gypseous soils…**A. nealleyi**
5'. Perianth limb white or greenish white; inflorescence bracts acute to obtuse or rounded at apex; fruits 5–12 mm; usually sandy or gravelly soils, sometimes gypseous…(6)
6. Stems arising from a taproot; fruits 5–12 mm; flowers 25–75 per head; sandy soils…**A. elliptica**
6'. Stems arising from long, cordlike rhizomes; fruits 5–7 mm; flowers 15–25 per head; gypseous sandy or gravelly soils…**A. bolackii**

Abronia angustifolia Greene [*A. angustifolia* var. *arizonica* (Standl.) Kearney & Peebles]—Purple sand verbena. Plants annual (or perennial); stems decumbent to ascending, glandular-pubescent; leaves ovate-oblong to elliptic; 1–5.5 × 0.7–3 cm; bracts lanceolate to oblong-lanceolate, 5–10 × 1–3 mm, papery, glandular-pubescent; perianth 10–20 mm, tube pink, limb bright magenta to pale pink, 6–8 mm diam.; fruits 5–10 × 4–8 mm, broadly obdeltate in profile, scarious, apex tapered to a prominent beak, wings 5. Sandy soils, desert scrub, gypsum flats and dunes, 3500–5375 ft. S.

Abronia bigelovii Heimerl—Tufted sand verbena. Plants perennial, acaulescent or rarely with short branched aerial caudices, cespitose; leaves linear to narrowly oblanceolate, 1.5–8 × 0.3 0.4 cm, 5+ times as long as wide; bracts broadly ovate to ovate-elliptic, 7–11 × 2–8 mm, thinly papery, glabrate to minutely glandular-puberulent; perianth 3–8 mm, tube greenish pink, limb pale pink, 3–4 mm diam.; fruits 6–9 × 5–8 mm, scarious, apex slightly beaked; wings 5. Limestone, Todilto Formation, hills and ridges of gypsum, 5700–7400 ft. NW. **R & E**

Abronia bolackii N. D. Atwood, S. L. Welsh & K. D. Heil—Bolack's sand verbena. Plants perennial, short-caulescent or nearly acaulescent; stems decumbent to erect, minutely and sparsely glandular-pubescent or ± glabrate, arising from cordlike rhizomes; leaves elliptic-oblong to ovate, 1–4 × 0.5–2 cm, minutely puberulent or glabrous; bracts broadly lanceolate, ovate, obovate, or almost round, 5–10 × 3–10 mm, scarious, glandular-puberulent; perianth 7–11 mm, tube greenish, limb white, 3 mm diam.; fruits 5–7 × 3–5 mm, scarious, apex truncate and slightly beaked, wings (3–)5. Ojo Alamo Formation, gypsiferous soil, scattered juniper communities, 5250–5750 ft. NW. Some DNA differences occur between *A. elliptica* and *A. bolackii*, and one could go either way in recognizing *A. bolackii* as a good taxon (Sonia Nosratinia, University of California, Berkeley, pers. comm.). **R & E**

Abronia elliptica A. Nelson [*A. fragrans* Nutt. ex Hook. var. *elliptica* (A. Nelson) M. E. Jones; *A. nana* S. Watson var. *harrisii* S. L. Welsh]—Fragrant white sand verbena. Plants perennial, sometimes nearly acaulescent; stems decumbent to erect, glandular-pubescent, infrequently glabrous; leaves elliptic-oblong to ovate, 1.5–6 × 0.5–3.5 cm, surface glabrous or puberulent, abaxial surface thinly puberulent to pubescent; bracts ovate to obovate, 5-20 × 3-10 mm, scarious, glandular-pubescent to villous; perianth 10-20 mm, tube rose to greenish, limb white, 5-8 mm diam.; fruits 5-12 × 4-8.5 mm, scarious, tapered at both ends; wings (2-)5. Sandy or gravelly soils, desert grasslands, scrub, 4750-6700 ft. NW, C, SC, SW.

Abronia fragrans Nutt. ex Hook.—Fragrant sand verbena. Plants perennial; stems procumbent to semierect, glandular-pubescent; leaves ovate to triangular or lanceolate, 3-12 × 1-8 cm, adaxial surface glandular-pubescent, abaxial surface pubescence denser and longer, or villous; bracts linear-lanceolate to oval-ovate, 7-25 × 2-12 mm, scarious, glandular-puberulent to short-villous; perianth 10-25 mm, tube greenish to reddish purple, limb white, (2-)6-10 mm diam.; fruits 5-12 × 2.5-7 mm, winged or not, fusiform and appearing deeply grooved when wingless; wings 4-5. Dry sandy soils, desert scrub, grasslands, piñon-juniper communities, 3185-7500 ft. Widespread except SW.

Abronia nana S. Watson—Dwarf sand verbena. Plants perennial, acaulescent or nearly so, usually cespitose; leaves elliptic-lanceolate to elliptic-ovate, (0.4-)0.5-2.5 × (0.2-)0.4-1.2 cm, < 3 times as long as wide, surfaces glabrous or glandular-pubescent; bracts lanceolate to ovate, 4-9 × 2-7 mm, scarious-glandular-puberulent; perianth 8-30 mm, tube pale pink, limb white to pink, 6-10 mm diam.; fruits obovate to obcordate, 6-10 × 5-7 mm, scarious, apex low and broadly conic, wings 5. Our material belongs to **var. nana**. Desert scrub and piñon-juniper communities, 5600-6800 ft. NW.

Abronia nealleyi Standl.—Nealley's sand verbena. Plants perennial; stems ascending, infrequently procumbent, often whitish, viscid-puberulent to glandular-pubescent; leaves lanceolate to elliptic-oblong, 2.5-5 × 0.5-3 cm, adaxial surface glabrous, abaxial surface glabrous or puberulent; bracts ovate, 4-8 × 2-5 mm, papery, glandular-pubescent, sometimes villous basally; perianth 8-16 mm, tube greenish, limb pale pink to light magenta, 5-7 mm diam.; fruits 4-17 × 3-5 mm, scarious, apex broadly obtuse and beaked; wings 5. Calcareous, gypseous, clay, or silty soils, shrublands, 3870-4725 ft. Eddy, ?Lincoln. **R**

ACLEISANTHES A. Gray – Trumpets

Plants perennial, finely pubescent; stems prostrate to erect, often clambering through other vegetation, herbaceous or woody; leaves sessile to petiolate, ± thick and succulent; inflorescences terminal or axillary, solitary flowers or few-flowered cymes, bracts persistent, not accrescent, 1-3 beneath each flower, narrowly lanceolate; flowers bisexual, chasmogamous and/or cleistogamous; cleistogamous perianth narrow domelike tube atop basal portion; chasmogamous perianth radially symmetric, short- to elongate-funnelform; stamens 2(3) or 5-6 (2-5 in cleistogamous flowers), exserted; styles exserted beyond anthers; stigmas peltate; fruits oblong or narrowly ellipsoid, smooth, glabrous or minutely puberulent; ribs 5, rounded, often with large, dark, sticky gland near apex, or wings 3-5.

1. Fruits with ridges, not winged…**A. longiflora**
1'. Fruits with thin, hyaline wings…(2)
2. Perianth 4-15 mm, limbs pink or brownish orange; leaf blades ovate or deltate; young stems and leaves with minute, white, T-shaped hairs…**A. chenopodioides**
2'. Perianth 17-52 mm, limbs white, cream, or greenish yellow; leaf blades lanceolate, ovate to ovate-oblong, or lanceolate; young stems and leaves with minute, white hairs, not T-shaped…(3)
3. Pubescence of minute, flat, white hairs only; petioles 0-3 mm; leaf margins entire…**A. lanceolata**
3'. Pubescence of glandular hairs or multicellular conic hairs and minute, flat, white hairs; petioles 1-20 mm; leaf margins undulate…**A. diffusa**

Acleisanthes chenopodioides (A. Gray) R. A. Levin [*Selinocarpus chenopodioides* A. Gray; *Ammocodon chenopodioides* (A. Gray) Standl.]—Goosefoot moonpod. Plants lightly pubescent with minute, white, T-shaped hairs; stems erect or ascending, 15-40 cm; leaves petiolate, those of a pair very unequal, ovate to ovate-oblong or deltate, 15-50 × 6-40 mm, margins entire or undulate; flowers 3-25 in umbellate clusters; perianth 4-6 mm diam.; fruits 5 mm, puberulent with flattened, white hairs; wings 1.5-2 mm wide. Dry, sandy and gravelly areas, rock, gypseous clay, desert scrub, 3900-6400 ft. S.

Acleisanthes diffusa (A. Gray) R. A. Levin [*Selinocarpus diffusus* A. Gray]—Spreading moonpod. Plants with appressed, white, flattened, minute, basally attached hairs; stems erect, decumbent or prostrate, 10-30 cm; leaves petiolate, those of a pair slightly unequal; ovate to ovate-oblong, 10-30 × 4-17 mm, margins undulate; flowers usually 1 in axils of leaves; perianth 35-48 mm, tube pale dull green, limbs greenish white or pale greenish yellow, often with greenish stripes, 15 mm diam.; fruits 5-7 mm, with flattened, white hairs on body; wings 1-2 mm wide. Dry clay and sandy calcareous or gypseous soils, desert scrub, grasslands, scattered juniper, 3200-5700 ft. NC, EC, C, SE, SC.

Acleisanthes lanceolata (Wooton) R. A. Levin [*Selinocarpus maloneanus* B. L. Turner]—Lanceleaf moonpod. Plants herbaceous or basally woody, lightly pubescent, hairs appressed, mostly retrorse, white, minute; stems erect or decumbent, 10-50 cm; leaves grayish green, sessile or petiolate, thick and succulent, lanceolate to broadly ovate, 12-40 × 4-13 mm, margins entire; flowers 1(2) in axils of distal leaves; perianth 30-42 mm, tube pale greenish, limbs cream or pale yellow, 10-15 mm diam.; stamens 5(6); fruits 6-9 mm, puberulent with flattened, white hairs; wings 2-4 mm wide. Gypsum hills and flats, desert scrub, grassland, juniper communities, 3000-6250 ft. NC, EC, C, SE, SC.

Acleisanthes longiflora A. Gray—Angel trumpets. Plants herbaceous, often slightly woody at base, overall pubescence of white, capitate hairs; stems ascending to prostrate or sprawling, profusely branched, to 100 cm; leaves grayish green, petiolate, puberulent to glabrate; blades lanceolate to linear-lanceolate or triangular-lanceolate to deltate, 3-40 × 1-30 mm, margins undulate or crispate; chasmogamous flower perianth with tubes 7-17 cm × 1-2 mm, puberulent to glabrate, limbs 10-20 mm diam., perianth 5-12 mm, puberulent; fruits with 5 hyaline ridges and pair of shallow, parallel grooves between ridges, 6-10 mm, hirtellous to puberulent. Rocky, gravelly, or gypseous soils, desert scrub, grasslands, shrublands, 2900-7000 ft. C, S.

ALLIONIA L. - Windmills

Herbs, annual or perennial, finely pubescent, often viscid; stems procumbent, often clambering; leaves paired, unequal; inflorescences axillary; bracts persistent, 3 forming involucre; flowers bisexual, chasmogamous; perianth strongly bilaterally symmetric, funnelform; stamens 4-7, ± exserted; stigmas capitate; fruits oblong to somewhat obovate, dorsoventrally compressed, surface glabrous.

1. Fruits deeply convex adaxially; lateral ribs developed as curved wings with 0-4 irregular or regular triangular teeth; viscid glands on concave side on stalks usually equaling or shorter than diam. of gland head; mostly perennials...**A. incarnata**
1'. Fruits shallowly convex adaxially; lateral ribs developed as curved wings with 4-8 slender teeth; viscid glands on concave side on stalks usually equaling or longer than diam. of glandular head; mostly annuals ...**A. choisyi**

Allionia choisyi Standl. [*A. incarnata* L. var. *glabra* Choisy]—Annual windmills. Annuals (probably also perennials), glabrate to viscid-pubescent; stems often reddish; leaves ovate to elliptic-ovate, 10-30 × 6-22 mm; perianth pale pink to magenta (nearly white), 2-7 mm; fruits shallowly convex, 3.2-4.3 × 1.8-3.7 mm, lateral ribs with 4-8 teeth, concave side of fruit with 5-7 glands per row. Sandy or gravelly soils, sometimes on clay or gypsum, desert scrub, grassland, piñon-juniper communities, 3000-7000 ft. Widespread.

Allionia incarnata L.—Pink windmills. Perennials or sometimes annuals, glandular-puberulent to spreading-viscid-villous; stems often reddish; leaves ovate, 20-65 × 10-35 mm; perianth deep pink to magenta, 5-15 mm; fruits deeply convex, 2.9-4.7 × 1.5-2.8 mm, lateral ribs with 0-4 teeth, wings entire or with only irregular undulations and incisions, concave side of fruit with 4-7 glands per row. Widespread.

1. Abaxial perianth limb 3-8(-10) mm; flower cluster 5-10(-12) mm diam. in anthesis; fruits 3-5 mm...**var. incarnata**
1'. Abaxial perianth limb 10-15 mm; flower cluster 20-25 mm diam. in anthesis; fruits mostly 4-6 mm...**var. villosa**

var. incarnata—Pink windmills. Peduncle 3-15(-30) mm, sparingly glandular-puberulent, sparsely to densely spreading-villous; flower cluster 5-10(-12) mm diam. in anthesis; abaxial perianth limb 3-8(-10) mm; fruits 3–5 mm. Sandy or gravelly soils, sometimes on gypsum, desert scrub, grasslands, piñon-juniper communities, 3000–7400 ft.

var. villosa (Standl.) Munz—Trailing windmills. Peduncle 15–25(–30) mm, usually glandular-puberulent, densely spreading-villous; flower cluster 20–25 mm diam. in anthesis; abaxial perianth limb 10–15; fruits 4–6 mm. Sandy or gravelly soils, desert scrub, 3800–4950 ft.

ANULOCAULIS Standl. - Ringstem

Anulocaulis leiosolenus (Torr.) Standl. [*Boerhavia leiosolena* Torr.]—Ringstem. Herbs, usually short- to long-lived perennials, pubescent to glabrate; stems 0.5–1.5 m, erect to ascending, unarmed; leaves 1–3 pairs, ± equal in size in each pair, thick and fleshy, broadly ovate to almost round, 5–15 × 4–19 cm, glabrous or sparsely villous with fine hairs; inflorescences terminal; bracts persistent, 1–2 at base of pedicel; flowers bisexual, chasmogamous; perianth radially symmetric or slightly bilaterally symmetric, tubular-funnelform, 25–35 mm, tube greenish bronze, limbs white, pink, or rose-pink; stamens 3 or 5, exserted; fruits biturbinate, 4.2–8 × 2.6–4.6 mm, constricted beyond the base, stiffly coriaceous, smooth or irregularly wrinkled, glabrous, with 10 narrow ribs. SE, SC.

1. Leaf blades glabrous, smooth, pale blue-green; flower buds glabrous externally at apex…**var. gypsogenus**
1'. Leaf blades at least sparsely beset with hairs with pustulate bases, pale grayish green or green; flower buds minutely puberulent or glabrous at apex…(2)
2. Flower buds glabrous at apex; leaf blades dull green…**var. leiosolenus**
2'. Flower buds minutely pubescent at apex; leaf blades pale grayish green to green…**var. howardii**

var. gypsogenus (Waterf.) Spellenb. & Wootten [*A. gypsogenus* Waterf.]—Gypsum ringstem. Leaves pale blue-green, smooth, glabrous; perianths white to very pale pink; buds glabrous; fruits 6–8 mm. Gypsum outcrops, desert scrub, 3300–5500 ft. Chaves, Eddy. **R**

var. howardii Spellenb. & Wootten—Howard's ringstem. Leaves ± pale grayish green, sparsely beset with pustulate-based hairs; perianths deep rose-pink; buds minutely pubescent at apex; fruits 4.2–6.8 mm. Gypsum outcrops, desert scrub, 4400–4800 ft. Guadalupe Mountains. Otero. **R & E**

var. leiosolenus—Ringstem. Leaves dull green, moderately beset with pustulate-based hairs; perianths white to pale pink; buds glabrous at apex; fruits 4.2–6.8 mm. Calcareous clays and shales, sometimes on gypsum, desert scrub, 3480–4200 ft. Doña Ana.

BOERHAVIA L. - Spiderling

Herbs, annual or perennial, often glandular, glabrous or pubescent; stems procumbent, decumbent, ascending, or erect; leaves symmetric to asymmetric, those in each pair unequal in size; inflorescences terminal and axillary; bracts ± persistent and not accrescent, 1–3 beneath each flower, lanceolate, minute, thin; flowers bisexual, chasmogamous; perianth radially or slightly bilaterally symmetric, campanulate or widely funnelform; stamens 2–8, included or exserted; stigmas peltate; fruits stiffly coriaceous, ribs (3–)5, rounded, angular, or winglike, smooth, glabrous or glandular-pubescent.

1. Fruits glandular-pubescent or minutely pubescent; plants perennial…(2)
1'. Fruits glabrous (rarely with some minute pubescence in sulci); plants annual or perennial…(3)
2. Leaves mostly distributed throughout plant; inflorescences axillary or terminal; branches spreading-villous or hispid to minutely and finely pubescent; flowers usually > 5 per cluster; fruits narrowly obovate and tapering at both ends or clavate, apex round or narrowly round-conic…**B. coccinea**
2'. Leaves mostly concentrated in basal 1/2 of plant; inflorescences mostly terminal; branches glabrate or glabrous; flowers usually borne singly or up to 5 per cluster, occasionally more; fruits oblong-clavate or obpyramidal…**B. gracillima** (in part)
3. Plants perennial; fruit ribs rounded or bluntly round-angled…(4)
3'. Plants annual; fruit ribs obtusely to acutely angled, ribs sometimes winglike, rarely bluntly round-angled…(5)
4. Bracts at base of perianths soon deciduous after anthesis; perianths wine-red to brick-red; sulci of fruit usually smooth…**B. gracillima** (in part)

4'. Bracts at base of perianths persistent; perianths red-pink, pink-lavender, pink, or white; sulci of fruit smooth or papillate...**B. linearifolia**
5. Branches of inflorescence densely glandular-villous, rarely minutely pubescent or glabrous, without sticky bands on distal internodes; bracts at base of perianth 1.5–4 mm, ovate (occasionally lance-acuminate in *B. wrightii*), persistent; fruits 4- or 5-ribbed, ribs never winglike...(6)
5'. Branches of inflorescence usually glabrous, sometimes minutely pubescent but not glandular, often with sticky bands on distal internodes; bracts at base of perianth 0.4–1.8 mm, usually lanceolate or narrower, deciduous; fruits (3–)5-ribbed, ribs sometime winglike...(7)
6. Fruits 4(5)-ribbed; inflorescences racemose or spicate, axis 10–35 mm...**B. wrightii**
6'. Fruits 5-ribbed; inflorescences subcapitate or capitate, axis 0–2.5 mm...**B. purpurascens**
7. Terminal portions of inflorescences spicate or racemose...(8)
7'. Terminal portions of inflorescences subracemose, subumbellate, umbellate, or capitate, or flowers borne singly...(10)
8. Fruits broadly obovoid, (length/width usually 1.7–2.1), usually overlapping in inflorescence; sulci and ribs slightly rugose; sulci usually ca. 0.5 times as wide as base of ribs; stems usually glandular and spreading-pilose basally   **B. spicata**
8'. Fruits narrowly obovoid or obpyramidal (length/width usually 2.1–3.1), overlapping in inflorescence or remote; sulci and ribs slightly rugose to smooth; sulci 0.1–1 times as wide as base of ribs; stems puberulent, often sparsely pilose, rarely glandular basally...(9)
9. Epidermal surface of sulci papillose; sulci 0.5–1 times as wide as base of ribs; sides of ribs strongly rugose...**B. torreyana**
9'. Epidermal surface of sulci glabrous; sulci 0.1–0.3 times as wide as base of ribs; sides of ribs smooth or slightly rugose...**B. coulteri**
10. Fruit ribs 3–4(5), ribs acute or winglike...(11)
10'. Fruit ribs (4)5, ribs acute, slightly rugose...(12)
11. Inflorescences of capitate clusters on short peduncles among leaves and terminal on branches; fruit tapering to stipelike base distal to pedicels...**B. pterocarpa**
11'. Inflorescences repeatedly forked, ending in umbels or single flowers, usually well beyond leaves; fruit tapering to pedicels, without stipelike base...**B. triquetra**
12. Terminal flower clusters usually precise umbels, all pedicels attaching at 1 node; occasionally terminal inflorescences 1-flowered; fruits 2–3.2 mm...**B. intermedia**
12'. Terminal flower clusters irregularly umbellate or subracemose, at least some pedicels attaching well below others; terminal inflorescences rarely 1-flowered; fruit 2.7–4 mm...**B. erecta**

Boerhavia coccinea Mill.—Red spiderling, scarlet spiderling. Herbs, perennial; stems prostrate to decumbent, 3–15 dm, minutely pubescent, often glandular; leaves broadly lanceolate or ovate, 20–70 × 10–60 mm, mostly sparsely puberulent; inflorescences axillary or terminal, subumbellate or capitate 5-flowered clusters; perianth maroon or magenta, rarely white, yellow, or pink, 1–3.5 mm; stamens 2–3, slightly exserted; fruits narrowly obovate and tapering at both ends or clavate, 2.6–4 × 0.9–1.2 mm, stipitate-glandular on ribs, ribs 5. Rocky slopes, gravelly outwash fans, arroyos, grasslands, desert scrub, piñon-juniper woodlands, 4100–7800 ft. S.

Boerhavia coulteri (Hook.) S. Watson—Coulter's spiderling. Herbs, annual; stems erect to decumbent-ascending, 2–8(–15) dm, minutely puberulent, often also with long-spreading hairs; leaves mostly in basal 1/2, lanceolate, ovate, or oval to deltate, 10–50 × 6–32 mm, adaxial surface mostly glabrous, sometimes sparsely hirtellous, abaxial surface usually glabrous; inflorescences terminal and axillary, usually with sticky internodal bands, terminating in spicate or racemose flower clusters; perianth white to pale pink, 0.7–2 mm; fruits narrowly obovoid to narrowly obpyramidal, 2–3.6 mm; ribs 5. NC, C, SW.

1. Fruits often overlapping 50%–100% of their length, often ± fasciculate in groups of 2–4, 2.5–3.6 mm, apex truncate, round-truncate, bluntly conic, or rounded...**var. coulteri**
1'. Fruits remote or some overlapping 1%–50% of their length, infrequently 2–3 in a cluster, 2–2.4+ mm, occasionally longer, usually rounded apically...**var. palmeri**

var. coulteri—Coulter's spiderling. Stems 2–8 dm; leaf blades 10–50 × 6–32 mm; perianth 1–2 mm; stamens 2–3(4), slightly exserted. Sandy or loamy soils, arid grasslands, roadsides, desert scrub, 4040–4525 ft. Doña Ana, Hidalgo.

var. palmeri (S. Watson) Spellenb.—Palmer's spiderling. Stems 2–7(–15) dm; leaf blades 15–40 × 10–30 mm; perianth 0.7–1 mm; stamens 1–2, included or barely exserted. Sandy or rocky soils, arid grasslands, desert scrub, 3850–4900 ft. Doña Ana, Hidalgo, Luna.

Boerhavia erecta L.—Erect spiderling. Herbs, annual; stems usually erect, 2-12 dm, minutely puberulent with bent hairs basally; leaves mostly basal, broadly rhombic-ovate, ovate, or lanceolate, 20-50(-80) × 10-45 mm, adaxial and abaxial surfaces usually glabrous, sometimes minutely puberulent; inflorescences terminal, usually with sticky internodal bands, with umbellate or subracemose clusters of flowers; perianth whitish, usually tinged with pink or purple, 1-1.5 mm; fruits narrowly obconic, 2.7-4 × 1.2-1.5 mm, ribs 5. Disturbed sites, roadsides, stream beds, desert scrub, 3900-6100 ft. WC, SW.

Boerhavia gracillima Heimerl [*B. organensis* Standl.]—Slimstock spiderling. Herbs, perennial; stems decumbent to erect, 2-15 dm, usually minutely pubescent; leaves mostly basal, broadly rhombic to elliptic-oblong or ovate, 18-45 × 13-50 mm, adaxial surface glabrous, abaxial surface glabrous or with hairs on veins; inflorescences axillary or terminal; perianth wine-red to brick-red, 2-4.5 mm; fruits oblong-clavate, 2.8-4.2 × 1-1.5 mm, ribs 5. Rocky areas, often along roads, desert scrub, arid grasslands, piñon-juniper woodlands, 4230-6600 ft. C, SC, SW.

Boerhavia intermedia M. E. Jones [*B. erecta* L. var. *intermedia* (M. E. Jones) Kearney & Peebles]—Fivewing spiderling. Herbs, annual; stems usually erect or ascending, 2-6(-8) dm, minutely puberulent with bent hairs basally, glabrous or minutely puberulent distally; leaves mostly basal, broadly ovate or oval to lanceolate, 20-45 × 7-16 mm, adaxial surface usually glabrous or glandular-puberulent, abaxial surface glabrous or glabrate; inflorescences terminal, usually with sticky internodal bands, terminating in umbels or flowers borne singly; perianth whitish to pale pink or purplish, 0.7-1.2(-2) mm; fruits obconic or broadly low-conic, 2-3.2 × 0.7-1.3 mm, ribs 5. Sandy or gravelly areas, desert scrub, arid grasslands, 4000-5700 ft. S.

Boerhavia linearifolia A. Gray [*B. lindheimeri* Standl.]—Narrowleaf spiderling. Herbs, perennial, sometimes woody at base; stems usually erect or ascending, 2-5(-9) dm, hirsute, puberulent or glandular; leaves mostly in basal 2/3 of plant, mostly lanceolate to linear, 15-35 × 1-15 mm, adaxial and abaxial surfaces glabrous or sparsely hispid; inflorescences axillary or terminal, terminating in loose, 1- to few-flowered cymose clusters; perianth purplish pink, 4-7 mm; fruits oblong-clavate, 2.5-3.5 × 1.2-1.5 mm, ribs (4)5. Calcareous soils or rock, arid grasslands, desert scrub, 3275-7000 ft. SE, SC.

Boerhavia pterocarpa S. Watson—Apache Pass spiderling. Herbs, annual; stems procumbent or decumbent to ascending, 1-4 dm, puberulent with bent hairs throughout; leaves throughout, rhombic-ovate to ovate or lanceolate, 15-25 × 9-15 mm; adaxial surface glabrous or sparsely puberulent, abaxial surface glabrous; inflorescences terminal or axillary, with capitate clusters of flowers; perianth white to pale pinkish, 1-1.5 mm; stamens 2, included or barely exserted; fruits broadly obpyramidal, 2.9-3.4 × 2.8-3.2 mm; ribs 3-4, winglike. Sandy loam to clay soils, disturbed areas, weedy; Deming, 4300 ft.

Boerhavia purpurascens A. Gray—Purple spiderling. Herbs annual; stems erect or ascending, 10-60 dm, minutely puberulent to glandular-pubescent; leaves mostly in basal 1/2, oval, ovate-oblong, or lanceolate, 12-37 × 5-20 mm; adaxial surface glabrate, sparsely puberulent or glandular-pubescent, abaxial surface glabrous or glandular-pubescent; inflorescences terminal, with capitate or subcapitate flower clusters; perianth whitish or pale pink, 2.5-4 mm; stamens 3-4(5), well exserted; fruits broadly obovoid, 2.3-3 × 1.3-1.7 mm, ribs 5, smooth. Sandy soils, arid grasslands, desert scrub, piñon, juniper, and oak woodlands, ponderosa pine forests, 4000-7850 ft. WC, SE, SW.

Boerhavia spicata Choisy—Creeping spiderling. Herbs, annual; stems erect or ascending, 30-70 dm, densely glandular-villous, or glandular; leaves mostly in basal 1/2, oval, oblong, ovate, or ± triangular, 18-45 × 13-30 mm, adaxial and abaxial surfaces glandular-pubescent; inflorescences terminal and axillary, with sticky internodal bands, terminating in spicate or racemose flower clusters; perianth white to pale pink, 1-1.3 mm; stamens (2)3, slightly exserted or included; fruits broadly obovoid, 1.9-2.8 × 1.1-1.3 mm, ribs 5. Sandy or rocky soils, arid grasslands, desert scrub, mesquite communities, 3300-6750 ft. Widespread except N.

Boerhavia torreyana (S. Watson) Standl. [*B. spicata* Choisy var. *torreyana* S. Watson]—Torrey's spiderling. Herbs, annual; stems prostrate or decumbent-ascending, usually profusely branched throughout, 10-80 dm, minutely puberulent with flat, spreading hairs; leaves mostly in basal 1/2, oblong-ovate, oval, ovate, or lanceolate, 20–45 × 9–25 mm, adaxial and abaxial surfaces glabrous or sparsely and minutely puberulent; inflorescences terminal and axillary, with sticky internodal bands, terminating in spicate or racemose flower clusters; perianth whitish to purplish pink, 1–1.3 mm; stamens 2, slightly exserted; fruits narrowly obovoid, 2–3.8 × 0.9–1.3 mm, glabrous, ribs 5. Sandy or rocky soils, arid grasslands, desert scrub, 3900–6900 ft. NC, C, EC, SC, SW.

Boerhavia triquetra S. Watson—Slender spiderling. Herbs, annual; stems erect or ascending, occasionally decumbent, 1–5 dm, minutely puberulent with bent hairs basally, glabrous or minutely puberulent; leaves in basal 1/2 of plant, broadly ovate or oval to lanceolate, 10–35 × 6–13 mm, adaxial surface usually glabrous, rarely minutely puberulent, abaxial surface glabrous; inflorescences terminal or axillary, with sticky internodal bands, terminating in umbels or flowers borne singly; perianth whitish to pale pink or purplish, 1 mm; stamens 2–3, included or barely exserted; fruits obpyramidal, 2.5–3.1 × 1.3–1.9 mm, glabrous, ribs 3–4(5). New Mexico material belongs to **var. intermedia** (M. E. Jones) Spellenb. Sandy or gravelly areas, desert scrub, 4000–6000 ft. C, SW.

Boerhavia wrightii A. Gray—Wright's spiderling. Herbs, annual; stems erect or ascending, 20–60 dm, densely glandular-pubescent; leaves mostly in basal 1/2, ovate-triangular, ovate, or broadly lanceolate, 15–55 × 7–35 mm, adaxial surface usually pubescent, often glandular, sometimes glabrate with hairs along midrib, abaxial surface similar; inflorescences terminal, terminating in spicate or racemose flower clusters; perianth whitish to pale pink (rarely golden-yellow), 1.2–1.4 mm; stamens 2–3, included or barely exserted; fruits broadly obovoid, 2.1–2.5 × 1–2 mm, glabrous, ribs 4(5). Sandy and gravelly soils, desert scrub, piñon-juniper and oak woodlands, 3900–6300 ft. C, SC, SW.

COMMICARPUS Standl. – Wartclub

Commicarpus scandens (L.) Standl. [*Boerhavia scandens* L.]—Climbing wartclub. Herbs or shrubs, perennial, suffrutescent, glabrous or glabrate; stems decumbent to erect, often tangled among themselves, 3–20 dm; leaves subequal in each pair, blades ovate, triangular, or rhombic-orbiculate, 15–60 × 10–45 mm; inflorescences terminal and axillary, umbellate, 4–11-flowered; perianths pale greenish yellow, 3–4 mm; stamens 2(–6), exserted; fruits clavate, glabrous or minutely puberulent, with very sticky glands, especially near the apex, 7–10 × 1.5–2 mm. Gravelly areas, often among boulders or shrubs, roadsides, desert scrub, 3000–5230 ft. C, SE, SW.

CYPHOMERIS Standl. – Cyphomeris

Cyphomeris gypsophiloides (M. Martens & Galeotti) Standl. [*Boerhavia gypsophiloides* (M. Martens & Galeotti) J. M. Coult.]—Red cyphomeris. Herbs, perennial, glabrous or sometimes minutely and sparsely pubescent; stems erect to reclining, often clambering through other vegetation, with glutinous bands on internodes, 5–15 dm; leaves unequal in size, broadly oblong-lanceolate or narrowly lanceolate to linear, 10–90 × 1–30 mm, glabrous or glabrate; flowers slightly bilaterally symmetric, forming low domes atop basal portion; perianths deep pink to red-violet, 7–10 mm; stamens 5; fruits clavate, slightly to notably gibbous, 8–14 mm, sometimes weakly warty at least on gibbous side, ribs 10, not well defined, glabrous. Limestone and calcareous soils, rocky soils, washes, roadsides, desert scrub, piñon-juniper and oak woodland, ponderosa pine forests, 3500–8500 ft. C, EC, SC, SW.

MIRABILIS L. – Four-o'clock

Herbs, perennial (or annual), often suffrutescent, glabrous or pubescent, sometimes viscid; stems erect to decumbent; leaves thin to thick and fleshy; inflorescences terminal and axillary, terminating in pedunculate involucres subtending 1–16 flowers, usually cymose; bracts persistent, 5, ± ovate, forming an herbaceous to papery involucre; perianth a small dome atop basal portion, radially symmetric or slightly bilaterally symmetric, campanulate to funnelform; stamens 3–6, exserted; styles exserted beyond stamens; fruits obovoid, ellipsoid, or nearly globose, smooth or tuberculate, glabrous or pubescent, ribs (4)5.

1. Fruits weakly ribbed or bluntly 5-angled, round in cross-section, ovoid or ellipsoid, 2.5–12 mm…(2)
1'. Fruits prominently 5-ribbed, ± obovoid, sometimes narrowly so and tapering at both ends, 3–6 mm…(4)
2. Flowers 6 per involucre; perianth 2.5–6 cm…**M. multiflora**
2'. Flowers 1–3 per involucre; perianth 0.5–17 cm…(3)
3. Fruits bluntly 5-angled in cross-section; perianth 7–17 cm, long-funnelform; flowers 1 per involucre…**M. longiflora**
3'. Fruits circular in cross-section; perianth 0.5–0.9 cm, short-funnelform; flowers 3 per involucre…**M. oxybaphoides**
4. Involucres not or only slightly accrescent after anthesis, greenish at maturity, opaque, moderately veiny, widely bell-shaped, 4–8 mm in fruit; perianth bright red-purple; leaf blades linear…**M. coccinea**
4'. Involucres strongly accrescent after anthesis, becoming tan, translucent, prominently net-veined, very widely bell-shaped or rotate, 4–15 mm in fruit; perianth white, pink, or bright purple-pink; leaf blades linear, linear-lanceolate, or ovate…(5)
5. Fruits usually glabrous, sometimes very lightly puberulent…**M. glabra**
5'. Fruits usually puberulent or pubescent, sometimes sparsely so…(6)
6. Leaf blades linear to linear-lanceolate, mostly 0.1–1 cm wide; tubercles on fruits only slightly elevated above surface, low-rounded to round-angled; ribs on fruit not bearing prominent tubercles…**M. linearis**
6'. Leaf blades linear-lanceolate to ovate, mostly 1–8 cm wide; tubercles on fruits markedly elevated above surface, prominently rounded or ± angular, often paler than background, or only slightly elevated above surface, low-rounded; ribs sometimes bearing prominent tubercles…(7)
7. Involucres green and blushed with dark violet or black; crosswalls of hairs on involucres and peduncles dark purple or black; leaf blades narrowly triangular-ovate to ovate, usually glabrous; perianth bright purple-pink…**M. melanotricha**
7'. Involucres green or blushed with red; crosswalls of hairs on involucres and peduncles usually pale; leaf blades lanceolate to ovate, glabrous or pubescent; perianth white, pale pink, or bright purple-pink…(8)
8. Leaf blades linear-lanceolate to ovate-lanceolate, ovate, or deltate, glabrous or pubescent; ribs of fruit notched, tuberculate, smooth, or somewhat rugose; involucres in axils, loosely clustered on branches or scattered in open, divaricate inflorescences, sparsely to densely pubescent…**M. albida**
8'. Leaf blades ovate-lanceolate to ovate or triangular, glabrous, puberulent, or sparsely hispidulous; ribs of fruit usually irregularly notched, especially toward apex; involucres usually clustered at end of branches, glabrous or glabrate…**M. nyctaginea**

Mirabilis albida (Walter) Heimerl [*M. comata* (Small) Standl.; *M. lanceolata* (Rydb.) Standl.]—White four-o'clock. Plants 0.8–15 dm; stems glabrous to puberulent and often viscid; leaves linear-lanceolate to ovate-lanceolate or ovate, 2–11 × 0.6–6.5 cm, glabrous, puberulent to hirsute, often viscid; bracts widely bell-shaped, 50%–80% connate; flowers 1–3 per involucre; perianth white, pink, or deep red-violet, 0.8–1.5 cm; fruits obovoid to narrowly obovate, 3.5–5.5 mm, pubescent. Rocky slopes, dry meadows, desert scrub, grasslands, piñon-juniper and oak woodlands, ponderosa pine–mixed conifer forests, 3215–11,000 ft. Widespread.

Mirabilis coccinea (Torr.) Benth. & Hook. [*Oxybaphus coccineus* Torr.; *Allionia coccinea* (Torr.) Standl.; *A. gracillima* Standl.]—Scarlet four-o'clock. Plants 1.5–9 dm; stems glabrous or very sparsely puberulent; leaves linear to linear-lanceolate, 4.5–10 × 0.1–0.5 cm, adaxially glabrous or rarely with a few small hairs, abaxially glabrous; bracts narrowly ovate or triangular, 30%–40% connate; perianth bright red-purple, 1.3–2 cm; fruits 4.5–6 mm, ribs round, smooth or slightly warty. Igneous rock, shrubs, piñon-juniper woodlands, ponderosa pine–mixed conifer forests, 4265–9250 ft. WC, SW.

Mirabilis glabra (S. Watson) Standl. [*Oxybaphus glaber* S. Watson; *Allionia carletonii* Standl.; *A. glabra* (S. Watson) Kuntze]—Smooth four-o'clock. Plants 5–20 dm; stems glabrous, glandular-puberulent, or puberulent; leaves linear to narrowly ovate or ovate-oblong, 5–10 × 0.2–7.5 cm, surfaces glabrous to short-pilose; bracts widely bell-shaped, 60%–90% connate; perianth white to pale pink, 0.6–0.9 cm; fruits 4–5.5 mm, usually glabrous, sometimes very lightly puberulent. Sandy soils, grasslands, oak, juniper, mesquite, disturbed areas, 3500–7400 ft. Widespread except NE, EC, SW.

Mirabilis linearis (Pursh) Heimerl [*Allionia linearis* Pursh; *M. hirsuta* (Pursh) MacMill. var. *linearis* (Pursh) B. Boivin; *Oxybaphus linearis* (Pursh) B. L. Rob.]—Narrowleaf four-o'clock. Plants 1–1.3 dm; stems minutely puberulent in 2 lines or hirsute, rarely glabrate or glabrous, usually glandular-puberulent; leaves mostly linear to linear-lanceolate, 3–11.5 × 0.1–1.8 cm, surfaces glabrous, glandular-pubescent, or hirsute; bracts widely bell-shaped, 40%–70% connate; flowers 3 per involucre; perianth white to purple-pink, 0.7–1.1 cm; fruits narrowly obovate, 3–5.5 cm, pubescent. Widespread.

1. Stems hirsute, at least basally…**var. subhispida**
1'. Stems minutely puberulent, glabrate, or glabrous basally…(2)
2. Leaf blades linear, grayish or bluish green; perianth white to deep rose-pink…**var. linearis**
2'. Leaf blades linear-lanceolate, green; perianth pink to deep purple-pink…**var. decipiens**

var. decipiens (Standl.) S. L. Welsh [*Allionia decipiens* Standl.]—Narrowleaf four-o'clock. Stems usually decumbent-ascending, 1.5–13 dm, minutely puberulent, glabrate, or glabrous basally; leaf blades green, linear-lanceolate to lanceolate, 3–11 × 0.4–1.8 cm, surfaces glabrous or glabrate; inflorescences narrowly to widely forked, usually with main axis; fruiting involucres 4–6 mm, crosswalls of peduncle hairs usually dark; perianth pink to deep purple-pink. Gravelly or sandy areas, piñon, juniper, and oak woodlands, mountain brush, ponderosa pine and mixed conifer forests, 4025–10,650 ft.

var. linearis [*Allionia decumbens* (Nutt.) Spreng.; *A. pinetorum* Standl.; *M. decumbens* (Nutt.) Daniels; *Oxybaphus decumbens* (Nutt.) Sweet]—Narrowleaf four-o'clock. Stems erect, ascending, or decumbent-ascending, 1–13 dm, minutely puberulent, glabrate, or glabrous basally; leaf blades grayish or bluish green, linear, 3–10 × 0.1–1.3 cm, surfaces glabrous or glandular-pubescent; inflorescences of single involucres in axils or terminal, well branched with ± well-defined main axis; fruiting involucres 5–10(–15) mm, crosswalls of peduncle hairs usually pale; perianth white to deep rose-pink. Sandy, gravelly, or rocky places, grasslands, piñon, juniper, and oak woodlands, mountain brush, ponderosa pine forests, 3300–9200 ft.

var. subhispida (Heimerl) Spellenb. [*Allionia gausapoides* Standl.; *M. gausapoides* (Standl.) Standl.]—Narrowleaf four-o'clock. Stems usually erect or ascending, 2.5–12 dm, hirsute, at least basally; leaf blades 5–11.5 × 0.1–1 cm, surfaces glabrous or densely hispid; inflorescences single involucres in axils, or terminal, well branched, with ± well-defined main axis; fruiting involucres 5–10 mm, crosswalls of peduncle hairs usually pale; perianth pale to deep pink. Sandy, rocky, or calcareous areas, prairies, piñon, juniper, and ponderosa pine communities, 3900–7800 ft.

Mirabilis longiflora L.—Sweet four-o'clock. Plants 5–15 dm; stems lightly puberulent basally, glandular-puberulent distally; leaves usually cordate, less often deltate, ovate, or ovate-lanceolate, 5–14 × 3–8 cm; inflorescences in dense clusters of flowers, bracts 40%–60%, apex triangular to narrowly triangular; perianth white, tube blushed with green or purple; fruits ovoid to slightly obovoid, 7–12 mm. Rocky canyons, riparian communities, arid grasslands, piñon, juniper, and oak woodlands, ponderosa pine and mixed conifer forests, 3900–8225 ft. NC, C, WC, SW.

Mirabilis melanotricha (Standl.) Spellenb. [*Allionia melanotricha* Standl.]—Mountain four-o'clock. Plants 5–12 dm; stems pubescent basally with minute curved hairs in 2 lines, glandular-pilose distally; leaves bright green, ascending, triangular-ovate to ovate, 3–10 × 0.8–4 cm; inflorescences axillary and terminal, 3 per involucre, bracts widely 40%–50%, widely bell-shaped; perianth bright purple-pink; fruits narrowly obovoid, 3–4 mm, spreading-pilose. Piñon, juniper, and oak woodlands, ponderosa pine and mixed conifer forests, 5600–10,700 ft. NE, NC, C, SC, SW.

Mirabilis multiflora (Torr.) A. Gray [*Oxybaphus multiflorus* Torr.]—Colorado four-o'clock. Plants 4–7 dm; stems forming hemispherical clumps, glabrous or densely pubescent; leaves spreading, ovate to widely ovate, sometimes suborbiculate, 5–10 × 4–8 cm; flowers 6 per involucre; involucres erect or ascending; bracts 5, usually > 50% connate; fruits ovoid or globose, 6–11 mm, smooth to rugulose, glabrous or pubescent. Widespread.

1. Fruits tuberculate, mucilaginous when wetted; involucral bracts obtuse…**var. glandulosa**
1'. Fruits smooth to slightly tuberculate, not mucilaginous when wetted; involucral bracts acute…**var. multiflora**

var. glandulosa (Standl.) J. F. Macbr. [*Quamoclidion multiflorum* (Torr.) Torr. ex A. Gray subsp. *glandulosum* Standl.]—Colorado four-o'clock. Involucral bracts obtuse; fruits dark reddish brown to black, ribs inconspicuous, fruit surface tuberculate, mucilaginous when wetted. Gravelly and sandy soils, sage-brush scrub, shadscale, piñon-juniper woodlands, 5875–6700 ft. San Juan.

var. multiflora—Colorado four-o'clock. Involucral bracts acute; fruits dark brown to black, ribs inconspicuous, fruit surface smooth to slightly tuberculate, not mucilaginous when wetted. Gravelly or sandy soils, piñon-juniper woodlands, ponderosa pine forests, 3950–8500 ft.

Mirabilis nyctaginea (Michx.) MacMill. [*Allionia nyctaginea* Michx.; *Oxybaphus nyctagineus* (Michx.) Sweet]—Heartleaf four-o'clock. Plants 4–15 dm; stems usually glabrous or puberulent in 2 lines, rarely spreading-pubescent or glabrate; leaves triangular to ovate, 3–10 × 2–6.5 cm; bracts 50%–90% connate; flowers 2–5 per involucre; perianth usually pink to reddish purple, rarely white; fruit obovate, 3–5 mm, shaggy-pubescent. Weedy areas, often disturbed sites, 5600–7600 ft. NE, C, SC, SW.

Mirabilis oxybaphoides (A. Gray) A. Gray [*Allionia oxybaphoides* (A. Gray) Kuntze]—Smooth spreading four-o'clock. Plants 2–12 dm; stems decumbent to prostrate, often tangled in other vegetation, puberulent in lines or throughout, glandular or not; leaves ovate to deltate, 1.5–8 × 1–7.5 cm; bracts 50%–70% connate; flowers 3 per involucre; perianth purplish to pale pink (white); fruit obovoid to nearly spherical, 2.5–3.5 mm, glabrous, dark-mottled. Shady and moist areas, piñon-juniper woodlands, ponderosa pine forests, 4600–7500 ft. Widespread except EC, SW.

NYCTAGINIA Choisy – Nyctaginia

Nyctaginia capitata Choisy—Scarlet musk-flower, devil's bouquet. Herbs, perennial, viscid-pubescent; stems erect to spreading, 10–90 cm; leaves subtriangular, rarely ovate, 3–13 × 1–11 cm; inflorescences axillary and terminal; bracts 6–20, 6–15 mm; flowers in clusters of > 3; perianth orange-red, mottled or streaked with yellow or rarely yellow; stamens 5–8, slightly exserted, nearly 2 times length of perianth; stigmas capitate; fruits turbinate, 5–8 × 3.5–4 mm, ribs 10, glabrous, smooth, without glands. Dry sandy or loamy soils, desert scrub, arid grasslands, roadsides, 3000–4920 ft. SE, NW.

TRIPTEROCALYX (Torr.) Hook. – Sand-puffs

Herbs, annual, viscid-pubescent to nearly glabrous; stems decumbent to semierect; leaves ± thick and succulent; inflorescences axillary, capitate clusters; bracts persistent, 5–10, forming an involucre, linear-lanceolate to ovate; flowers bisexual; perianth funnelform or salverform, with 4–5-lobed limb; stamens (3)4–5, included; stigmas linear; fruits fusiform, minutely puberulent or glabrous, wings 2–4, translucent, prominently veined, scarious, extending beyond apex and/or base.

1. Perianth tube 6–18 mm; perianth limb 3–5 mm diam., lobes inconspicuous...**T. micranthus**
1'. Perianth tube 12–30 mm; perianth limb 8–13 mm diam.; lobes > 1.5 mm...(2)
2. Perianth limb pale pink to magenta; fruits usually > 20 mm...**T. carneus**
2'. Perianth limb white adaxially, white to pink abaxially; fruits usually 20 mm or less...**T. wootonii**

Tripterocalyx carneus (Greene) L. A. Galloway [*Abronia carnea* Greene]—Winged sand-puffs. Stems often reddish, at least at nodes, short-glandular-pubescent, viscid; leaves oblong-lanceolate to ovate, 2.5–9.5 × 1–4.5 cm, glabrous or glandular-puberulent; inflorescences 10–35-flowered; bracts lanceolate to ovate, 7–19 × 2–6 mm, papery, glabrous or glabrate; perianth tube pink to magenta, 12–30 mm, limb pale pink to magenta, 8–13 mm diam., lobes showy, > 1.5 mm; fruits oval, 18–30 × 11–25 mm, wings (2)3(4). Sandy soils, desert scrub, 3800–7500 ft. W, San Miguel.

Tripterocalyx micranthus (Torr.) Hook. [*Abronia micrantha* Torr.]—Smallflower sand verbena. Stems reddish, glandular-pubescent, ± viscid; leaves lance-ovate, 1–6 × 0.5–2.5 cm, surfaces short glandular-pubescent, ± viscid; inflorescences 5–15-flowered; bracts lanceolate to ovate, 3–9 × 1–3 mm, green or ± papery, glabrate to lightly glandular-puberulent; perianth tube greenish to pink, 6–18 mm, limb greenish to pink, 3–5 mm diam.; fruits oval to round, 10–20 × 10–20 mm, wings (2)3(4). Sandy soils, desert scrub, arid grasslands, 3850–6520 ft. NW, C, SW.

Tripterocalyx wootonii Standl. [*Abronia wootonii* (Standl.) Tidestr.; *T. carneus* (Greene) L. A. Galloway var. *wootonii* (Standl.) L. A. Galloway]—Wooton's sand-puffs. Stems often reddish, short-glandular-pubescent, viscid; leaves ovate to elliptic-oblong, glabrous or glandular-puberulent, ± glaucous; inflorescences 10–25-flowered; bracts linear-lanceolate to ovate, 5–14 × 1.5–5 mm, papery, glabrous or glabrate; perianth tube pink to pinkish red, 12–25 mm, limb white adaxially, often pinkish abaxially, 8–11 mm diam., lobes showy; fruits oval, 13–25 × 9.5–17.5 mm, wings (2)3(4). Sandy soils, desert scrub, 3850–7185 ft. NW.

NYMPHAEACEAE – WATER-LILY FAMILY

C. Barre Hellquist

Perennial, aquatic, rhizomatous herbs; leaves alternate, floating, submersed, and emersed; stipules present or absent; petioles long; blades lanceolate to ovate, or orbicular with basal sinus, margins entire, spinose, or dentate; flowers solitary, bisexual, diurnal or nocturnal, floating, emersed, occasionally submersed; perianth often persistent; hypogynous, perigynous, or epigynous; sepals usually 4–12; petals numerous, often transitional to numerous stamens; pistil 1, 3–35-carpellate and -locular; ovules numerous; stigma sessile, radiate on stigmatic disk; fruit a many-seeded indehiscent berry (Allred, 2020).

1.   Leaves with rounded basal lobes, venation mainly pinnate; petals all stamenlike; sepals 5–12…**Nuphar**
1′. Leaves with angular basal lobes, venation mainly palmate; inner petals grading into stamens; sepals 4…
     Nymphaea

NUPHAR Sm. – Spatterdock, cow-lily, yellow pond-lily

Nuphar polysepala Engelm.—Rocky Mountain pond-lily. Rhizomes horizontal, branched, 3–8+ cm diam.; leaves floating, submersed, and emersed; blades green, widely ovate, 10–40(–45) × 7–30 cm, primary venation mostly pinnate, lobes divergent to overlapping; petioles terete; flowers 5–10 cm diam.; sepals 6–9(–12), green on undersurface, yellow above, sometimes red-tinged toward base; petals oblong, thick; anthers 3.5–9 mm, shorter than the filaments; fruit green to yellow, cylindrical to ovoid, 4–6(–9) × 3.5–6 cm, strongly ribbed, borne on peduncles; disk green, 20–35 mm diam.; seeds ovoid, 3.5–5 mm, aril absent. Still water of lakes and ponds, 8800–8900 ft. Rio Arriba.

NYMPHAEA L. – Water-lily

Perennial, rhizomatous or stoloniferous; leaves floating and submersed, suborbicular, oval, cleft at base, with angular basal lobes, margins entire, undulate, or toothed, venation mainly palmate; sepals 4, mainly green; petals numerous, grading into stamens, white, pink, yellow, or blue; ovary 12–35-celled; fruit maturing underwater, depressed-globose, usually covered with persistent petal and stamen bases; seeds covered by a saclike aril.

1.   Petals yellow; plants bearing stolons…**N. mexicana**
1′. Petals white or pink; plants rhizomatous…**N. odorata**

Nymphaea mexicana Zucc.—Yellow water-lily. Plants stoloniferous; leaves green above, often with streaks or speckles, purple, brownish, or crimson below, often spotted; petals bright yellow, more elliptic than sepals; stamens 50–60, petaloid stamens 2–2.5 cm; fruit 2–2.5 cm; seeds 4–5 mm diam. Lakes, ponds, slow streams, 5050–6250 ft. Sierra. Introduced.

Nymphaea odorata Aiton—American white water-lily. Plants rhizomatous, with persistent tubers; leaves green above, green, red, or purplish red below; flowers extremely fragrant, petals > 25, white, rarely pink; stamens > 70, petaloid stamens 3–4 cm; fruit 2.5–3 cm diam. Lakes, ponds, slow streams, 3900–6200 ft. Our material belongs to **subsp. odorata**. Doña Ana, Sierra. Introduced.

OLEACEAE – OLIVE FAMILY

Kenneth D. Heil

Trees, shrubs, or subshrubs; leaves opposite (rarely alternate in *Menodora*), simple or pinnately compound, stipulate; inflorescence racemose, paniculate, or thyrsoid; flowers perfect or imperfect; calyx mostly 4-lobed or absent; corolla usually united or with distinct petals, or lacking; stamens 2 or 4, distinct; pistil 1, the ovary superior, 2-carpelled and 2-loculed, style 1 or lacking, stigmas 1 or 2; fruit a berry, drupe, loculicidal capsule, circumscissile capsule, or samara (Ackerfield, 2015; Heil, 2013c).

1. Herbs with a woody base; flowers yellow…**Menodora**
1'. Trees or shrubs; flowers purple or white, or petals absent…(2)
2. Fruit a samara; leaves odd-pinnately compound or simple, broadly ovate to suborbicular with a rounded apex; petals absent…**Fraxinus**
2'. Fruit a drupe, berry, or capsule; leaves simple; flowers purple or white, or petals absent…(3)
3. Leaves ovate to cordate, 3–12 × 2–8 cm; flowers purple or white, in large panicles to 20 cm; fruit a capsule…**Syringa**
3'. Leaves elliptic to ovate-lanceolate, 1.5–6 × 0.5–2 cm; flowers white, in small panicles to 6 cm, sessile or in cymes, or petals absent; fruit a berry or drupe…(4)
4. Petals white; flowers in panicles; fruit a black berry; leaves entire…**Ligustrum**
4'. Petals absent; flowers sessile or in paniclelike cymes; fruit a blue-black drupe; leaves entire to minutely toothed…**Forestiera**

FORESTIERA Poir. – Desert olive

Sprawling shrubs; leaves opposite, often appearing fascicled at the ends of the branches, serrate or entire; flowers very small, dioecious, appearing before the leaves, calyx none or minute, unequally 5–6-cleft, corolla none or 1 or 2 small petals, stamens 2 or 4, ovary 2-loculed, with 2 ovules; fruit a thin-fleshed drupe; seeds bony, 1.

1. Leaf blades 3+ times longer than wide, margins entire or slightly sinuate; Hidalgo County…**F. phillyreoides**
1'. Leaf blades 1–3 times longer than wide, margins crenulate or serrulate, rarely entire; widespread…**F. pubescens**

Forestiera phillyreoides (Benth.) Torr. [*F. shrevei* Standl.]—Desert olive. Shrubs to 4 m; twigs firm, branching, angular; leaves 1.5–2.5 × 0.3–.5 cm, pubescent, elliptic-oblong to ± oblanceolate, somewhat revolute; flowers in clusters of 3–5; anthers purple, stamens 4; drupes 5 × 3 mm, purplish black. Rocky canyon walls and slopes; known only from Guadalupe Canyon, 4600 ft. SW. Hidalgo.

Forestiera pubescens Nutt. [*F. neomexicana* A. Gray]—New Mexico privet, New Mexico olive. Shrubs to 3 m; bark gray with spiny branches; leaves (0.8-)1.5–5.5 × (0.3-)0.5–2 cm, oblanceolate to elliptic, entire to serrulate; inflorescence with flowers in fascicles; staminate flowers sessile, pistillate pedicellate; drupes 5–8 mm, ellipsoid, blue-black. River terraces with Russian olive, saltcedar, skunkbush, and cottonwood, 3500–8965 ft. Widespread except NE, EC, SE.

FRAXINUS L. – Ash

Deciduous trees or shrubs; leaves opposite, pinnately compound (simple in *F. anomala*); inflorescence a panicle; flowers perfect or unisexual, inconspicuous, calyx lacking or 4-cleft, corolla lacking or of 2+ distinct petals, stamens mostly 2, ovary 2-loculed, style 1, stigmas 1 or 2; fruit a samara.

1. Leaves simple or 3–7-foliate; twigs quadrangular…**F. anomala**
1'. Leaflets 3–7; twigs terete or nearly so…(2)
2. Flowers with a corolla, in terminal branches on lateral leafy branchlets of the current year; leaflets mostly 0.5–1.5 cm wide…**F. cuspidata**
2'. Flowers without a corolla, in axillary panicles from separate buds in the axils of leaves of the previous year; leaflets mostly 1.5–3 cm wide…(3)
3. Leaflets mostly 7–9; twigs and petioles mostly glabrous; mostly cultivated and rarely escaping…**F. pennsylvanica**

3'. Leaflets mostly 3-5; twigs and petioles mostly puberulent; indigenous and also widely cultivated...**F. velutina**

Fraxinus anomala Torr. ex S. Watson—Singleleaf ash. Shrub or small tree, mostly 2-5 m; twigs 4-sided; leaves glabrous, mostly simple, occasionally 3-7-foliolate, the blades ovate to oval, entire or crenulate, 1-5 × 1-7.5 cm, acute or subcordate basally, acute to rounded apically; inflorescence a many-flowered panicle, 3-12 cm; flowers mostly perfect; calyx 1-2 mm; petals none; samaras 12-27 × 5-12 mm, obovate-oblanceolate, winged almost to the base. W.

1. Leaves simple or sometimes trifoliate with the terminal leaflet orbicular; W New Mexico...**var. anomala**
1'. Leaves commonly compound with 5 leaflets, may be single; Catron County...**var. lowellii**

var. anomala—Singleleaf ash. Slickrock, ephedra, shadscale, piñon-juniper communities, 5000-6000 ft. Cibola, San Juan.

var. lowellii (Sarg.) Little—Lowell's ash. 5065-6600 ft. SC, SW.

Fraxinus cuspidata Torr. [*F. cuspidata* var. *macropetala* (Eastw.) Rehder]—Fragrant ash. Shrub or small tree to 6 m, trunk to 2 dm diam.; bark gray, smooth, becoming fissured into ridges; leaves petioled, with (3)4-7(-9) leaflets, leaflets lanceolate to broadly ovate, long-pointed and sometimes cuspidate, 3.5-7 cm, entire to coarsely toothed, dark green above, paler and pubescent beneath when young; inflorescence a panicle, 7-10 cm, loose; flowers with 4 petals, ca. 10 × 1 mm, white; samaras including the wing ca. 1.2 cm, oblong-obovate to lanceolate, wing 6 mm wide in upper 1/2 and extending nearly to base of flattened fruit body. Local on rocky slopes of canyons; willow, Gambel oak, piñon-juniper, ponderosa pine communities, 4350-8960 ft. NW, C, WC, SC, SW.

Fraxinus pennsylvanica Marshall—Green ash. Trees to 20 m; leaves pinnately compound, 5-7(-9) leaflets, leaflets 6-15 cm, serrate to subentire; samaras 2-5 cm. Escaping along floodplains and margins of lakes, 4200-8000 ft. Sandoval, San Juan, San Miguel. Introduced.

Fraxinus velutina Torr. [*F. pennsylvanica* Marshall subsp. *velutina* (Torr.) G. N. Mill.]—Velvet ash, Arizona ash. Moderate-size trees; branchlets terete, spreading-hairy to sparingly or nearly glabrous; leaves with 3-5 leaflets (rarely simple), petiolulate, lanceolate to ovate, elliptic, or orbicular, acuminate to rounded apically, nearly entire to serrate, hairy or glabrous on the lower surface; flowers imperfect; calyx persistent, campanulate; corolla none; samaras 16-34 × 4-6, the blade decurrent about halfway along the terete body. Stream courses and floodplains, 3900-8235 ft. C, WC, SE, SC, SW.

LIGUSTRUM L. - Privet

Ligustrum vulgare L.—Common privet. Shrubs to 3 m; twigs cream-colored to brown or gray; leaves opposite, short-petioled, entire, oblong to ovate, 2-6 cm, dark green, with lateral veins ending near the margin and not running to the apex; inflorescence a terminal panicle; corolla lobes ca. 3 mm, white or yellowish white; stamens 2; berries black. An introduced European species planted as hedge-rows, escaped along the Pecos River in San Miguel County, 5500-6900 ft. Introduced.

MENODORA Bonpl. - Menodora

Subshrubs; leaves alternate, lowermost opposite, simple, sessile or subsessile, entire; inflorescence solitary, corymbose, or paniculate; flowers showy, calyx 5-15-lobed, corolla yellow, subrotate, 5-6-lobed, stamens 2, inserted on the corolla tube, ovary superior, 2-loculed, 2-4 ovules per locule, style slender, stigma capitate; fruit a circumscissile capsule.

1. Corolla tube elongate, 2.5-5 cm, salverform, glabrous...**M. longiflora**
1'. Corolla tube < 1 cm, rotate to shortly funnel-shaped, pilose at the opening...**M. scabra**

Menodora longiflora A. Gray—Showy menodora. Suffrutescent perennials, 25-60 cm; leaves mostly opposite, nearly sessile, linear to lanceolate or elliptic to spatulate, 1-5 cm × 1-10 mm, margins entire; corolla yellow, tube 2.5-5 cm, limbs 10-15 mm; capsules 8-10 mm. Limestone, ridges and gravelly plains, 3545-8500 ft. SE, SC.

Menodora scabra A. Gray—Rough menodora. Erect or ascending subshrub, mostly 2-3.5 dm, woody at the base only; leaves 5-30 × 2-5 mm, narrowly elliptic to oblong or lanceolate, glabrous or scaberulous; flower calyx glabrous to scabrous, with 7-11 linear lobes, 4-5 mm; corolla bright yellow, subrotate, lobes 5-9 mm; capsule 5-7 × 8-12 mm; seeds 4-5 × ca. 3 mm. Rocky areas including limestone gypsum soils, ponderosa pine, Rocky Mountain juniper, piñon pine, Gambel oak, desert scrub, desert grassland communities, 3000-7950 ft.

SYRINGA L. - Lilac

Syringa vulgaris L.—Common lilac. Shrubs to 4+ m; branches mostly erect; leaves petiolate, opposite, simple; 3-12 × 1.5-8 cm, ovate to cordate to rounded, truncate or obtuse basally, acute to acuminate apically, glabrous; inflorescence a panicle, 10-20 cm; flowers perfect, campanulate, 4-toothed; corolla tubular, limb 4-lobed, purple, lilac, or white; stamens 2, included; ovary 2-loculed, each locule with usually 2 ovules, style with a 2-lobed stigma; fruit a loculicidal capsule. An ornamental that persists around homesites, 4350-7700 ft. N, WC. Introduced.

ONAGRACEAE - EVENING PRIMROSE FAMILY

Warren L. Wagner, Peter C. Hoch, and Nancy R. Khan

Perennials, annuals, shrubs, or sometimes trees, terrestrial, amphibious, or aquatic; often with epidermal oil cells; erect to decumbent or prostrate; leaves alternate or opposite, sometimes spirally arranged, simple, cauline, sometimes basal and forming rosettes, sessile or subsessile to petiolate, margins entire, sometimes toothed or pinnately lobed; inflorescences axillary, flowers solitary or in leafy spikes, racemes, or panicles; flowers bisexual and usually actinomorphic, (2-)4(-7)-merous, epigynous; sepals green or red, valvate; petals rarely absent, often fading darker with age, imbricate or convolute, sometimes clawed; nectary present; stamens usually 2 times as many as sepals and in 2 series, antisepalous set longer, or as many as sepals; filaments distinct; anthers versatile, sometimes basifixed; placentation axile or parietal; style 1, stigma 1; ovules 1 to numerous per locule; fruit a loculicidal capsule; seeds smooth or sculptured, sometimes with a coma or wings, with straight, oily embryo (Wagner et al., 2007; Wagner & Hoch, 2021).

1. Sepals persistent or tardily caducous after anthesis; flowers 4-5(6)-merous; floral tube absent...**Ludwigia**
1'. Sepals deciduous after anthesis (along with other flower parts); flowers (2-)4-merous; floral tube usually present, often elongate, if absent then petals rose-purple or pink, rarely white...(2)
2. Stipules present and soon deciduous; fruit an indehiscent capsule, burlike, with stiff, hooked hairs... **Circaea**
2'. Stipules absent; fruit a capsule, sometimes indehiscent...(3)
3. Sepals erect or spreading; seeds comose; stigma with dry multicellular papillae, entire or 4-lobed, lobes commissural...(4)
3'. Sepals reflexed; seeds not comose; stigma wet, nonpapillate, entire or (3)4-lobed, noncommissural...(5)
4. Floral tube absent; stamens 8, in 2 subequal whorls; style deflexed at anthesis, later erect, stamens initially erect, later deflexed; leaves spirally arranged, very rarely subopposite proximally...**Chamaenerion**
4'. Floral tube present; stamens 8, in 2 unequal whorls; style and stamens erect, episepalous stamens longer or rarely subequal; leaves opposite, at least near base of stem...**Epilobium**
5. Ovary 2-locular; stems delicate...**Gayophytum**
5'. Ovary (3)4-locular; stems not especially delicate...(6)
6. Styles with peltate indusium at base of stigma, at least at younger stages prior to anthesis; stigma (3)4-lobed, or peltate to discoid or nearly square, entire surface receptive...**Oenothera**
6'. Styles without indusium; stigma subglobose to globose, subcapitate, or capitate, rarely conic-peltate and ± 4-lobed...(7)
7. Leaves mostly basal, blades often pinnately lobed, rarely bipinnately, sometimes unlobed, or lateral lobes greatly reduced or absent, terminal lobe large, abaxial veins with conspicuous brown oil cells; capsules pedicellate; seeds in 2 rows per locule...**Chylismia**
7'. Leaves not predominantly basal, blades not lobed or pinnatifid, without oil cells; capsules sessile or very shortly pedicellate; seeds in 1 row per locule...(8)
8. Petals white, rarely red or tinged red, fading pink or red; flowers vespertine...**Eremothera**
8'. Petals yellow, fading red, often with red dots basally; flowers diurnal...**Camissonia**

CAMISSONIA Link – Suncup

Camissonia parvula (Nutt. ex Torr. & A. Gray) P. H. Raven—Annuals, caulescent, with a taproot; stems 2–15 cm, erect, slender, wiry, often branched, glabrous or densely strigillose proximally, sometimes sparsely glandular-puberulent distally; leaves alternate, cauline, not clustered basally, stipules absent; blades linear or linear-filiform, 1–3 × 0.04–0.1 cm, margins entire or subentire, base attenuate, apex acute; inflorescence a leafy spike, flowers opening during the day; floral tube 1.3–2 mm, glabrate, with basal nectary; sepals 4, reflexed separately; petals 4, 1.5–3.6 mm, yellow without red dots at base; stamens 8, in 2 unequal series; ovary 4-locular; stigma surrounded by anthers; fruit a capsule, subterete or 4-angled, often flexuous or curled, 15–28 × 0.6–1 mm, pedicels 0–2 mm; seeds numerous. Sandy soils, sagebrush scrub to piñon-juniper woodland, 3600–6900 ft. San Juan.

CHAMAENERION Ség. – Fireweed, rosebay willowherb

Chamaenerion angustifolium (L.) Scop. [*Epilobium angustifolium* L.]—Fireweed. Robust herb with a caulescent, woody caudex, colonial; stems 30–200 cm, erect, unbranched, strigillose, usually glabrous proximally; leaves cauline, spirally arranged or very rarely subopposite, petioles 2–7 mm; blades oblong- to elliptic-lanceolate, (6–)9–23 × (0.7–)1.5–3.4 cm, margins entire to denticulate, base subcuneate to attenuate, apex attenuate-acute, surface strigillose on adaxial side and on abaxial midrib, or rarely glabrous; inflorescence an erect raceme, subglabrous, flowers opening laterally; floral tube absent; sepals 4, purplish green, oblong-lanceolate, 9–19 × 1.5–3 mm; petals 4, rose-purple to pale pink, obovate to suborbicular, 14–25 × 7–14 mm, entire or scarcely emarginate; anthers 8, red to rose-purple; ovary 4-locular, surface densely canescent; nectary disk raised on ovary apex; style white or flushed pink, proximally villous; stigma spreading to revolute, deeply 4-lobed; capsules 5–9.5 cm, densely appressed-canescent, pedicels 1–3 cm; seeds with dense coma, 10–17 mm. New Mexico material belongs to **subsp. circumvagum** (Mosquin) Moldenke. Moist, often disturbed places in mountains or lower areas, frequent especially after fires, 5900–12,150 ft. NC, C, WC, SC, SW.

CHYLISMIA (Torr. & A. Gray) Nutt. ex Raim. – Suncup

Annuals, sometimes perennials, caulescent; stems ascending to erect, branched; leaves mostly basal, forming well-developed rosette, cauline often reduced, alternate, long-petiolate, stipules absent, blades often pinnately lobed, sometimes unlobed, or lateral lobes greatly reduced or absent, terminal lobe large, margins irregularly sinuate-dentate, serrulate, or subentire; inflorescence a raceme, erect or nodding, flowers opening at sunrise or sunset; floral tube with basal nectary; sepals 4, reflexed singly; petals 4, yellow or white, often fading orange-red, sometimes lavender or purple, often with 1+ red dots near base; stamens usually 8, in 2 series; ovary 4-locular; stigma entire and capitate; fruit a capsule, straight or slightly curved (never twisted or curled), subterete and clavate or oblong-cylindrical; seeds numerous, finely pitted, in 2 rows per locule (Raven, 1962, 1969).

1. Capsules oblong-cylindrical, 1.2–1.8 mm diam....**C. walkeri**
1'. Capsules distinctly clavate, 1.6–2.6 mm diam....(2)
2. Petals bright yellow, often with red dots near base, 1.7–5 mm; flowers opening at sunrise; stigma surrounded by anthers at anthesis...**C. scapoidea**
2'. Petals white, 3–7.5 mm; flowers opening at sunset; stigma exserted beyond anthers at anthesis...**C. claviformis**

Chylismia claviformis (Torr. & Frém.) A. Heller [*Camissonia claviformis* (Torr. & Frém.) P. H. Raven]—Browneyes. Stems 5–60 cm, branched mostly from base, glandular-puberulent and strigillose, rarely glabrous; leaf petioles 0.7–12 cm, blades 1.5–20 × 0.3–3.5 cm, pinnately lobed, lateral lobes irregular, well developed, terminal lobe narrowly ovate, to 7 × 3 cm, margins irregularly sinuate-dentate; inflorescences nodding, flowers opening at sunset; floral tube 3–5.5 mm, orange-brown inside, villous inside proximally; petals white, often fading purple; stamens subequal, anthers ciliate; capsules ascending to spreading, 8–40 mm, pedicels 4–40 mm. Our material belongs to **subsp. peeblesii** (Munz) W. L. Wagner & Hoch. Flat, sandy plains, washes, 3940–4265 ft. SW. Known from 5 recent collections in Grant and Hidalgo

Counties at higher elevations than previously known; at least 1 collection does not have any glandular hairs, and the relationship to subsp. *aurantiaca* (Munz) P. H. Raven from Arizona, California, and Mexico is under study.

Chylismia scapoidea (Torr. & A. Gray) Nutt. ex Raim. [*Camissonia scapoidea* (Torr. & A. Gray) P. H. Raven]—Paiute suncup. Stems 3–45 cm, unbranched, sometimes branched from base, strigillose, villous, or glandular-puberulent; leaf petioles 0.5–6.5 cm, blades 1–18 × 0.5–3.5 cm, usually unlobed, margins subentire, oil cells on abaxial surface inconspicuous; inflorescences nodding, flowers opening at sunrise; floral tube 1–4 mm, sparsely villous or glabrous inside; petals bright yellow, often with red dots near base, fading pale yellow or yellowish orange; stamens unequal; anthers ciliate or glabrous; capsules ascending, (10–)15–30 mm, pedicels 4–20 mm. Our material belongs to **subsp. scapoidea**. Sandy or clay flats, salt desert scrub, 4985–6235 ft. NW.

Chylismia walkeri A. Nelson [*Camissonia walkeri* (A. Nelson) P. H. Raven]—Walker's suncup. Annual or short-lived perennial; stems 10–60 cm, slender, unbranched or branched from base, villous, densely so proximally, less dense to glabrate distally; leaves primarily cauline, often purple-dotted, petioles 0.4–8 cm, blades 2–22 × 0.4–3.5 cm, with only terminal lobe well developed, 1–5 × 0.5–3.2 cm, margins serrate; inflorescences erect, flowers opening at sunrise, buds individually reflexed; floral tube 0.5–1.3 mm, glabrous or sparsely villous inside; petals bright yellow, fading pale orange or lavender, 1–3 mm; stamens unequal; anthers glabrous; stigma surrounded by anthers; capsules spreading or ascending, 11–45 mm, pedicels 5–15 mm. New Mexico material belongs to **subsp. walkeri**. Loose slides of limestone and other sedimentary rock, sandy washes, salt desert scrub, 4590–4920 ft. San Juan.

CIRCAEA L. – Enchanter's nightshade

Perennials, caulescent, colonial, stolons numerous, often terminated by a tuber; stems erect, unbranched or sparsely branched; leaves cauline, opposite, petiolate; stipules present; blade margins dentate to prominently dentate; inflorescences simple or branched racemes, terminal or at apex of branches; flowers zygomorphic, 2-merous, clustered at apex of raceme on ascending to erect pedicels; floral tube a mere constriction to 0.6 mm, nectary wholly within and filling proximal portion of tube; sepals reflexed to spreading, oblong or ovate to broadly ovate; petals alternate to sepals, emarginate, clawed; anthers basifixed; ovary 1–2-locular, stigma bilobed or obpyramidal, surface wet, minutely papillate; fruit indehiscent, burlike, with stiff, hooked hairs, spreading or slightly reflexed, pedicellate; seed 1 per locule, ellipsoid, glabrous, without appendages (Boufford, 1982; Xie et al., 2009).

1. Stems glabrous at least proximally, sometimes glandular-puberulent distally; leaf blade margins conspicuously dentate, base cordate to subcordate, rarely truncate or rounded...**C. alpina**
1′. Stems pubescent with at least a few recurved, falcate hairs, glandular-puberulent distally; leaf blade margins subentire to minutely denticulate, base rounded to subcordate, rarely cordate...**C. pacifica**

Circaea alpina L.—Small enchanter's nightshade. Stems (3–)5–25(–30) cm, soft, glabrous proximally, glabrous or sometimes glandular-puberulent distally; leaf petioles 0.3–4 cm, blades 1.5–7.5 × 1.5–5.5 cm, base cordate to subcordate, rarely truncate or rounded, margins conspicuously dentate; ovary 1-locular. Our plants belong to **subsp. alpina**. Moist to wet places in cool, temperate or boreal forests, 7545–10,170 ft. NE, C, WC.

Circaea pacifica Asch. & Magnus [*C. alpina* L. var. *pacifica* (Asch. & Magnus) M. E. Jones]—Enchanter's nightshade. Stems 10–40 cm, firm, terete, retrorsely puberulent with some falcate hairs proximally, glandular-puberulent distally; leaf petioles 1.5–5 cm, blades 3–7.5(–11) × 2.5–5.5(–8) cm, base rounded to subcordate, margins subentire to minutely denticulate; ovary 1-locular. Cool, temperate deciduous and mixed forests, forest margins, along streams, 7545–9515 ft. C, WC.

EPILOBIUM L. – Willowherb

Annuals or perennials, sometimes suffrutescent, caulescent; with basal rosettes, turions, soboles, stolons, woody caudex, or taproot; stems erect to ascending or decumbent, simple to well branched distally, strigillose, glandular-puberulent, villous, often mixed, or glabrous, often with raised hairy lines decurrent from leaf axils; leaves cauline, sometimes also basal, subsessile to petiolate, stipules absent, blade margins entire or toothed; inflorescences in racemes, spikes, or panicles, terminal; flowers solitary in leaf axils, actinomorphic; floral tube with nectary at base; sepals 4, spreading individually, green or flushed with red or cream, lanceolate; petals 4, rose-purple to white; ovary 4-locular; stigma entire and clavate to capitate, or deeply 4-lobed, commissural; fruit a capsule, straight or slightly curved, narrowly cylindrical to fusiform, terete, loculicidally dehiscent, splitting to base, rarely only upper 1/2, with intact central column, pedicellate or sessile; seeds numerous (1–100+ per locule), in 1, rarely 2, rows per locule, surface papillose to finely reticulate or longitudinally ridged (Baum et al., 1994; Wagner, 2021).

1. Annual, with taproot; stem epidermis peeling proximally; leaves opposite in proximal pairs, alternate or fasciculate distally…(2)
1'. Perennial, sometimes woody, ± without taproot; stem epidermis not peeling; leaves opposite proximal to inflorescences, alternate distally, not fasciculate…(3)
2. Plants 15–200 cm; floral tube 1–16 × 0.8–2.9 mm; capsules 15–32 mm; seed coma present, often easily or readily detached…**E. brachycarpum**
2'. Plants 1.5–50 cm; floral tube 0.3–1.1 × 0.2–0.8 mm; capsules 4.5–8 mm; seed coma absent…**E. campestre**
3. Floral tube 16–24 mm, bulbous near base; floral tube, sepals, and petals orange-red, very rarely white, slightly zygomorphic, upper petals ± flared 90° to calyx tube…**E. canum**
3'. Floral tube 0.5–2.6 mm, not or very rarely bulbous near base; floral tube and sepals green or flushed reddish green, petals pink, white, or rose-purple, rarely cream, actinomorphic, very rarely slightly zygomorphic…(4)
4. Plants with ± threadlike stolons, with or without terminal turions; stems erect from base, rarely ascending; sometimes with faint raised strigillose lines from margins of petioles…(5)
4'. Plants with rosettes, turions, fleshy soboles, or shoots; stem bases erect or ascending, subglabrous proximal to inflorescence with raised or sparsely strigillose lines decurrent from margins of petioles, ± densely mixed strigillose and glandular-puberulent distally…(6)
5. Stems 15–95 cm, simple to well branched, densely strigillose; leaf blades densely strigillose on both sides; stolons terminating in fleshy turions…**E. leptophyllum**
5'. Stems often matted, 5–30(–40) cm, not or rarely branched, subglabrous; leaf blades subglabrous on both sides; stolons not terminating in turions…**E. oregonense**
6. Plants with small-leafed epigeous or scaly hypogeous soboles; stems ± ascending, rarely erect, ± clumped or matted, rarely with basal scales…(7)
6'. Plants with leafy rosettes or fleshy hypogeous turions; stems ± erect, usually not clumped, with dark basal scales…(9)
7. Stems 3–20(–25) cm; leaf blades (0.5–)0.8–2.5 × 0.3–1 cm, margins subentire to sparsely denticulate; capsules 17–40(–55) mm…**E. anagallidifolium**
7'. Stems 10–50 cm; leaf blades 1.5–5.5 × 0.7–2.9 cm, margins denticulate; capsules 50–100 mm…(8)
8. Petals rose-purple to light pink, rarely white; pedicels 5–15(–25) mm in fruit; capsules 40–65 mm; petioles 3–7 mm proximally, to 0 mm distally, not winged; seed surfaces distinctly papillose…**E. hornemannii**
8'. Petals white, rarely red-veined or flushed light pink; pedicels 15–45 mm in fruit; capsules 50–100 mm; petioles 3–12 mm, often winged; seed surfaces reticulate or sometimes barely rugose…**E. lactiflorum**
9. Seed surfaces ridged; stems (3–)10–120(–190) cm; leaf blades (1–)3–12(–16) cm; petals 2–14 mm, white or pink to rose-purple…**E. ciliatum**
9'. Seed surfaces papillose or reticulate; stems 2–55(–60) cm; leaf blades 0.5–5.5(–6.5) cm; petals 1.6–5.5 (–7) mm, white fading pink or rarely pink…(10)
10. Leaves not clasping, veins inconspicuous; capsules ascending, spreading; pedicels 8–40 mm in fruit; inflorescences ± nodding in bud…**E. hallianum**
10'. Leaves clasping, veins conspicuous; capsules erect, appressed to stem; pedicels 1–5 mm in fruit; inflorescences ± erect…**E. saximontanum**

Epilobium anagallidifolium Lam. [*E. alpinum* L.]—Alpine or pimpernel willowherb. Stems 3–20 (–25) cm, often sigmoidally bent, clumped or mat-forming, simple, subglabrous; leaf petioles 1–6 mm, blades spatulate to oblong proximally, elliptic to narrowly lanceolate or sublinear distally, subglabrous, margins with 2–5 low teeth per side, base attenuate to cuneate, apex obtuse or rounded proximally to subacute distally, bracts reduced and narrow; inflorescences few-flowered racemes, subglabrous to sparsely strigillose and/or glandular-puberulent, flowers suberect, pedicels 1–6(–15) mm; floral tube 0.6–1.2 × 0.8–1.8 mm, with slightly raised subglabrous ring inside mouth; petals pink to rose-purple, (1.7–)2.5–6.5(–8) × 1.6–3.5 mm, apical notch 0.5–1.2 mm; ovary often reddish purple, subglabrous or sparsely strigillose and glandular-puberulent; stigma broadly clavate to subcapitate, entire, surrounded by longer anthers; capsules often reddish purple, surfaces subglabrous or with scattered hairs, pedicels 5–35(–68) mm; seeds narrowly obovoid, surface reticulate, coma persistent, 2–4 mm. Moist flats, stream banks, high montane and alpine meadows and seeps, (5775–)8200–11,810 ft. NC.

Epilobium brachycarpum C. Presl [*E. adenocladum* (Hausskn.) Rydb.]—Tall annual willowherb. Stems 15–200 cm, slender, erect, simple to paniculate-branched, glabrous proximally, strigillose distally; leaf petioles 1–4 mm, blades linear to linear-lanceolate or narrowly elliptic, 1–5.5(–7) × 0.1–0.8 cm, subglabrous, margins with 2–10 teeth per side, base tapered or cuneate, apex acute or acuminate, bracts very reduced; inflorescences erect, open racemes or panicles, glabrous or strigillose, flowers erect; floral tube with ring of spreading hairs inside mouth; petals white to pink or deep rose-purple, 1.5–15(–20) × 1–7.5 mm, apical notch 0.5–6.5 mm; ovary strigillose; stigma clavate to subcapitate, entire to deeply 4-lobed; capsules erect or ascending, pedicels 1–17 mm; seeds obovoid with constriction, surface low-papillose, coma 5–10 mm. Dry or seasonally moist, often disturbed ground in open woods, meadows, prairies, roadsides, stream banks, 6900–9850 ft. NC, C, WC.

Epilobium campestre (Jeps.) Hoch & W. L. Wagner [*Boisduvalia glabella* (Nutt.) Walp.]—Smooth spike-primrose. Stems 1.5–50 cm, suberect, glabrous proximally or throughout; leaves crowded, subsessile, blades lanceolate to narrowly lanceolate or oblong, 0.8–3.5 × 0.2–0.6(–1) cm, strigillose and ± villous, at least along veins and margins, margins with 4–7 teeth per side, base cuneate, apex acute; inflorescences erect spikes, congested, unbranched, densely strigillose and ± villous or subglabrous, flowers erect, often cleistogamous; floral tube with raised ciliate ring inside mouth; sepals reddish green; petals pale pink, fading purplish rose, 0.9–3.5 × 0.7–0.9 mm, apical notch 0.3–1.3 mm; ovary densely villous; stigma clavate, irregularly 4-lobed to subentire, surrounded by longer anthers; capsules with beak 0.8–1 mm, villous, subsessile; seeds 7–14 per tightly packed row, lacking coma. Vernally moist flats, depressions, shores, open fields, clay soils, 8530–10,500 ft. Rio Arriba.

Epilobium canum (Greene) P. H. Raven [*Zauschneria cana* Greene]—Hummingbird trumpet. Stems 10–50(–70) cm, villous and short-glandular-puberulent, mixed especially distally, sometimes predominantly glandular-puberulent, rarely glabrate; leaves lanceolate to ovate or broadly elliptic, 0.8–5 (–6) × 0.4–1.8(–2.2) cm, villous and glandular-puberulent, margins subentire to distinctly denticulate, 4–10 teeth per side, surfaces green to grayish green, rarely silvery-canescent; inflorescence erect spikes or racemes, often branched, glandular-puberulent and sometimes mixed strigillose or villous, flowers subsessile; floral tube 16–24 mm; sepals 6–9 mm; petals 7–9 mm; capsules 8–22 mm, pedicels 0–2 mm; seed coma easily detached, 5.5–7 mm. Our material belongs to **subsp. latifolium** (Hook.) P. H. Raven. Woodland and montane areas, stabilized talus slopes, disturbed ravines, roadsides, granite cliffs, stream banks, stabilized gravel bars in sandy or rocky soils, 4920–5900 ft. Grant, Hidalgo.

Epilobium ciliatum Raf.—Fringed willowherb. Stems erect, ± densely mixed strigillose and glandular-puberulent distally; leaves subglabrous with strigillose margins, margins serrulate, (8–)15–40 irregular teeth per side, base rounded to cuneate or short-attenuate, veins prominent, 4–10 per side; inflorescences erect racemes or panicles, flowers erect, pedicels 2–14(–20) mm; floral tube with ring of spreading hairs inside mouth; apical notch on petals 0.4–2.5 mm; stigma narrowly to broadly clavate or subcapitate, surrounded by anthers; surfaces of inflorescence, ovary, and capsules ± strigillose and glandular-puberulent; seed surface ridged, coma readily detached, 2–8 mm.

1. Leaf blades very narrowly lanceolate to narrowly ovate or elliptic, proximally narrowly obovate to spatulate; bracts very reduced on open inflorescence; petals 2–6(–9) mm, white or sometimes pink; usually with leafy basal rosettes…**subsp. ciliatum**
1'. Leaf blades narrowly ovate to ovate or broadly elliptic, sometimes lanceolate, proximally obovate to broadly elliptic; bracts little reduced on crowded inflorescence; petals 4.5–12(–15) mm, rose-purple to pink, rarely white; usually with condensed, subsessile turions 1–10 cm belowground…**subsp. glandulosum**

subsp. ciliatum—Stems (3–)10–120(–190) cm, well branched; leaf blades (1–)3–12 × (0.3–)0.6–3.7 cm, apex acute to proximally obtuse; inflorescences racemes or panicles; floral tube 0.5–1.8 × 0.9–3 mm; sepals 2–6 × 0.7–1.6 mm; petals 2–6(–9) × 1.3–4 mm; filaments of longer stamens 1.4–4.8 mm, those of shorter ones 0.6–2.6 mm; style 1.1–6.5 mm; capsules (15–)30–100 mm, pedicels 2–15(–40) mm. Disturbed, open, mesic areas, along roadsides, stream banks, lake margins, seeps, 5085–9840 ft. Widespread except E.

subsp. glandulosum (Lehm.) Hoch & P. H. Raven—Stems 20–110(–170) cm, simple or sparsely branched distally; leaf blades 3–10.5(–16) × 1–4.5(–5.5) cm, apex obtuse to subacute; inflorescences racemes, simple or sometimes branched; floral tube 1–2.6 × 1.4–3.5 mm; sepals 4.5–7.5 × 1.2–1.8 mm; petals 4.5–12(–14) × 2.5–6.3 mm; filaments of longer stamens 3–6.5 mm, those of shorter ones 1.4–4.3 mm; style 2.4–8.2 mm; capsules 40–85 mm, pedicels 5–25 mm. Damp banks of streams and lakes, seeps, wet meadows in montane, subalpine, 7200–11,150 ft. N mountains.

Epilobium hallianum Hausskn.—Glandular willowherb. Stems 2–50(–60) cm, erect, rarely branched, glabrescent with strigillose lines or sometimes ± densely long-villous throughout, with condensed fleshy turions 1–5 cm belowground; leaf petioles 1–1.5 mm, blades ovate proximally to lanceolate or narrowly elliptic distally, 0.5–4.7 × 0.2–1.4 cm, margins subentire proximally to denticulate distally, 8–20 teeth per side, base rounded to cuneate, apex obtuse to subacute, surfaces mostly glabrous with strigillose margins, bracts much reduced; inflorescences erect open racemes, mixed strigillose and glandular-puberulent, flowers erect, pedicels 3–8 mm; floral tube 0.5–1.7 × 0.8–1.6 mm, with slightly raised ring of spreading hairs inside mouth; petals white or pink, 1.6–5.5 × 1.2–3 mm, apical notch 0.3–1.2 mm; ovary strigillose and glandular-puberulent or subglabrous; stigma clavate, entire, surrounded by anthers; capsules (15–)24–60 mm, surfaces subglabrous; seed surfaces papillose, coma easily detached, 3–6 mm. Semishaded stream banks, wet grassy slopes and meadows, bogs, seasonally wet sites, vernal pools, 7875–12,150 ft. NC, NW.

Epilobium hornemannii Rchb.—Hornemann's willowherb. Stems 10–45 cm, simple, ± sparsely mixed strigillose and glandular-puberulent distally; leaves 3–7 mm proximally, to 0 mm distally, not winged, blades broadly elliptic to spatulate proximally, ovate to lanceolate distally, 1.5–5.5 × 0.7–2.9 cm, glabrous, margins denticulate distally with 10–25 teeth per side, base attenuate to cuneate or rounded, apex obtuse to subacute, bracts reduced; inflorescences erect or nodding, open racemes, mixed strigillose and glandular-puberulent, flowers erect, pedicels 2–5 mm; floral tube 1–2.2 × 1.3–2.8 mm, with or without sparse ring of hairs inside mouth; petals 3–9 × 2–5.5 mm, apical notch 0.7–2.4 mm; ovary glandular-puberulent; stigma clavate or cylindrical, entire, surrounded by anthers; capsules 40–65 mm; pedicels 5–15(–25) mm in fruit; seed surface papillose, coma readily detached, 6–11 mm. New Mexico material belongs to **subsp. hornemannii**. Banks of montane to alpine streams and lakes, open tussock meadows, willow swales, gravelly ridges, stabilized scree slopes, roadside ditches, 6900–11,750 ft. NE, NC, NW, C, WC.

Epilobium lactiflorum Hausskn.—Milkflower willowherb. Stems 15–50 cm, simple, mixed strigillose and glandular-puberulent distally; leaf petioles 3–12 mm, often winged, blades broadly spatulate to ovate proximally, narrowly ovate to narrowly lanceolate distally, 2–5.5 × 0.8–2.4 cm, glabrous except for strigillose margins, margins denticulate distally with 7–16 teeth per side, base attenuate to cuneate, apex obtuse proximally to subacute distally, bracts reduced and narrow; inflorescences nodding in bud, later erect, ± open racemes, mixed strigillose and glandular-puberulent, flowers suberect, pedicels 5–15 mm; floral tube 1–2.2 × 1–3 mm, inner surface glabrous, without ring; petals 3–8.5 × 1.6–4.5 mm, apical notch 0.7–1.4 mm; ovary glandular-puberulent; stigma clavate, entire, surrounded by anthers; capsules 50–100 mm; pedicels 15–45 mm in fruit; seed surface reticulate to rugose, coma easily detached, 7–14 mm. Montane stream banks, moist crevices and ledges, gravelly roadsides, burned-over woodlands, sandy moraines, subalpine forests, alpine meadows, 5400–11,000 ft. NC, NW, C, WC.

Epilobium leptophyllum Raf.—Bog willowherb. Stems 15–95 cm, erect, simple to loosely clustered; leaves opposite proximally, alternate or rarely fasciculate distally, subsessile, blades linear to very narrowly elliptic or sublanceolate, 2–7.5 × 0.1–0.7 cm, margins subentire, with 4–7 inconspicuous teeth per side, base rounded to subcuneate; inflorescences erect racemes, densely strigillose, often mixed sparsely glandular-puberulent, flowers erect, pedicels 5–12 mm; floral tube 0.8–1.5 × 1.2–1.8 mm, with ring of spreading hairs inside mouth; petals white to light pink, 3.5–7 × 1.6–4 mm, apical notch 1–1.8 mm; ovary densely strigillose; stigma narrowly clavate, entire, surrounded by anthers; capsules 35–80 mm, surfaces densely strigillose, pedicels 10–35 mm; seed surface papillose, coma persistent, 6–8 mm. Marshy ground, bogs, fens, low thickets, seepage areas, damp pastures; rare in New Mexico, 6560–8860 ft. Lincoln.

Epilobium oregonense Hausskn.—Oregon willowherb. Stems 5–30(–40) cm, erect or ascending, often flushed purple distally; leaves subsessile, blades broadly elliptic proximally, narrowly elliptic or lanceolate to sublinear distally, 5–25 × 1–7 mm, margins subentire, base cuneate to rounded, apex obtuse, bracts extremely reduced and linear; inflorescences erect racemes, open, unbranched, sparsely strigillose and glandular-puberulent, flowers suberect or nodding, pedicels 2–7 mm; floral tube 0.8–1.8 × 1–2.1 mm, with faint ring of hairs inside mouth; petals white to pink, 5–8 × 2.8–4 mm, apical notch 0.8–1.5 mm; ovary sparsely strigillose and glandular-puberulent; stigma subcapitate, surrounded by longer anthers; capsules 21–40(–52) mm, subglabrous, pedicels 20–65 mm; seed surface low-papillose, coma persistent, 3–4 mm. Montane to subalpine boggy or mossy areas, wet meadows, semishaded stream banks; rare in New Mexico, 9840–10,170 ft. Rio Arriba.

Epilobium saximontanum Hausskn.—Rocky Mountain willowherb. Stems 4–55 cm, erect, simple or well branched in age, mixed strigillose and glandular-puberulent distally; leaves often ± appressed, subsessile, blades obovate proximally to ovate, lanceolate, or narrowly elliptic distally, 1–5.5(–6.5) × 0.4–2(–2.4) cm, subglabrous with strigillose margins, margins low-denticulate, with 9–30 teeth per side, base rounded or obtuse, apex subacute, veins ± conspicuous, 3–6 per side, bracts much reduced; inflorescences erect racemes, sometimes sparsely branched, flowers erect, pedicels 0–1 mm; floral tube 0.8–1.4 × 0.8–1.9 mm, with ring of sparse spreading hairs inside mouth; petals white, 2.2–5(–7) × 1.7–3.2 mm, apical notch 0.4–1.5 mm; ovary densely strigillose and glandular-puberulent; stigma narrowly to broadly clavate, surrounded by at least longer anthers; capsules 30–55(–70) mm, surfaces mixed strigillose and glandular-puberulent, subsessile; seed surfaces rugose to papillose, coma readily detached, 3–9 mm. Montane semishaded stream banks, damp meadows, mossy seeps, other seasonally damp areas, 4590–11,500 ft. NC, C. WC.

EREMOTHERA (P. H. Raven) W. L. Wagner & Hoch – Suncup

Annuals, caulescent, with a taproot; stems erect, sometimes ascending, well branched from base and distally, with white or reddish-green exfoliating epidermis; leaves alternate, cauline, proximal ones often clustered basally, petiolate to subsessile distally, stipules absent, blade margins denticulate, crenate-dentate, sinuate-toothed, or entire; inflorescence a distal spike, nodding, flowers opening at sunset; floral tube villous in proximal 1/2 inside, with basal nectary; sepals 4, reflexed singly or in pairs; petals 4, white, without spots, fading pinkish; stamens 8 in 2 unequal series; ovary 4-locular; style villous proximally, stigma entire, subglobose; fruit a capsule, straight or much contorted, narrowly cylindrical throughout, terete, regularly but tardily loculicidally dehiscent, sessile; seeds numerous, gray, finely reticulate (Raven, 1969; Wagner et al., 2007).

1. Floral tube 1.5–3 mm; petals 1.8–3 mm; sepals 1.5–2.5 mm; stigma surrounded by anthers at anthesis…
E. chamaenerioides
1′. Floral tube 4–7 mm; petals 3.5–10 mm; sepals 4–6 mm; stigma exserted beyond anthers at anthesis…
E. refracta

Eremothera chamaenerioides (A. Gray) W. L. Wagner & Hoch [*Camissonia chamaenerioides* (A. Gray) P. H. Raven]—Suncup. Stems 8–50 cm, glandular-puberulent and sparsely strigillose distally, especially in inflorescences; leaf petioles 0.1–3.5 cm, blades very narrowly elliptic to narrowly elliptic, (0.7–)2–8(–10) × 0.1–2.5 cm, margins entire or sparsely denticulate; episepalous filaments 0.7–1.5 mm, epipetalous filaments slightly shorter; capsules spreading, straight, 35–60 × 0.7–1 mm. Desert slopes and flats in sandy soil, 4200–5085 ft. C, WC, SC, SW.

Eremothera refracta (S. Watson) W. L. Wagner & Hoch [*Camissonia refracta* (S. Watson) P. H. Raven]—Narrowleaf suncup. Stems 6–45 cm, sparsely strigillose, sometimes also glandular-puberulent, especially in inflorescences; leaf petioles 0–2 cm, blades narrowly lanceolate to narrowly elliptic-lanceolate or narrowly oblanceolate, those distally on stems linear to linear-lanceolate, 2–6(–8) × 0.1–0.8 cm, margins sparsely and weakly denticulate, sometimes sinuate-toothed; episepalous filaments 2–4.5 mm, epipetalous filaments slightly shorter; capsules spreading or reflexed, straight to ± contorted, 20–50 × 0.7–1 mm. Desert slopes and flats in sandy soil; a single collection in 1930 near Lordsburg, 4230 ft. Hidalgo.

GAYOPHYTUM A. Juss. – Groundsmoke

Annuals, caulescent; stems delicate, erect or spreading, densely branched or unbranched; leaves cauline, alternate, sessile or petiolate, stipules absent, blade margins entire; inflorescences erect panicles or racemes; floral tube inconspicuous; sepals 4, reflexed singly or in pairs; petals 4, white with 1 or 2 yellow or greenish-yellow areas at base, fading pink or red; stamens 8, in 2 unequal series; anthers ± basifixed; ovary 2-locular; stigma entire, surrounded by anthers; fruit a capsule, straight or slightly curved, subterete and often constricted between seeds, loculicidally dehiscent, all valves free, pedicellate; seeds 4–50+, in 1 row per locule, ovoid, arranged ± parallel to septum (Lewis & Szweykowski, 1964; Thien, 1969).

1. Stems branched throughout, with 2–8 nodes between branches; capsules with inconspicuous constrictions between seeds; seeds in each locule not crowded or overlapping, with 1 row in each locule, forming 2 even rows in capsule...**G. decipiens**
1'. Stems branched throughout, with 1 or 2 nodes between branches; capsules with inconspicuous to conspicuous constrictions between seeds; seeds in each locule crowded and overlapping, appearing to form 2 irregular rows in each locule or staggered, those in one locule alternating with those in the other, forming a single row in capsule...(2)
2. Petals 0.7–1.2(–1.5) mm; pedicels longer than capsules, (3–)5–12 mm; capsules 3–9 mm; seeds 10–30 per capsule...**G. ramosissimum**
2'. Petals 1.2–3 mm; pedicels shorter than capsules, 2–10(–15) mm; capsules 3–15 mm; seeds (3–)6–18 per capsule...**G. diffusum**

Gayophytum decipiens F. H. Lewis & Szweyk.—Deceptive groundsmoke. Stems 5–50 cm, erect or proximal-most branches decumbent, branching not dichotomous, glabrous or villous to strigillose; leaf petioles 0–5 mm, blades narrowly lanceolate to sublinear, 10–32 × 1–4 mm; inflorescences with flowers arising at first 1–5 nodes from base; sepals reflexed singly; petals 1.1–1.8 mm; capsules ascending, not conspicuously flattened, 6–15 × 0.6–1 mm, valve margin slightly undulate, pedicels 0–5 mm; seeds 10–25, arranged subopposite to those in adjacent locule, forming 2 even rows in capsule, glabrous or densely puberulent. Piñon-juniper woodlands, pine forests, desert ranges mountains bordering desert areas in sandy or gravelly soil, 3935–9515 ft. NW, C.

Gayophytum diffusum Torr. & A. Gray—Spreading groundsmoke. Stems 5–60 cm, erect, distal branching dichotomous or lateral branches shortened, glabrous to strigillose, sometimes villous; leaf petioles 0–10 mm, blades very narrowly lanceolate, 10–60 × 1–5 mm; inflorescences with flowers arising at first 1–20 nodes from base; sepals reflexed singly or in pairs; capsules ascending to reflexed, subterete, 3–15 × 1–1.5 mm, valve margin somewhat undulate, seeds arranged alternate to those in adjacent locule, when crowded often appearing to form 2 irregular rows in each locule, or seeds often staggered, those in one locule alternating with those in the other, forming a single row in capsule, glabrous or puberulent. Our plants belong to **subsp. parviflorum** F. H. Lewis & Szweyk. Open pine forests, sagebrush slopes, dry margins of meadows, 3700–9840 ft. NW, NC.

Gayophytum ramosissimum Torr. & A. Gray—Piñon groundsmoke. Stems 10–50 cm, erect, profusely branched throughout, usually at every other node, branching dichotomous except near base, glabrous or sparsely strigillose distally; leaf petioles 0–3(–10) mm, blades very narrowly lanceolate to sublinear, 10–40 × 1–5 mm; inflorescences with flowers arising at first 5–15 nodes from base; sepals 0.4–0.8 mm, reflexed singly; capsules ascending to reflexed, subterete, 3–9 × 0.8–1.2 mm, valve margins entire or weakly undulate; seeds arranged alternate to those in adjacent locule, crowded and often appearing to form 2 irregular rows in each locule, glabrous. Sagebrush communities, 3700–8200 ft. NW, NC.

LUDWIGIA L. – Primrose-willow, water-purslane, water-primrose

Perennials; glabrous or villous; stems erect to spreading and then rooting at nodes; leaves cauline, petiolate, stipules often deciduous, blades linear to lanceolate, elliptic, oblong, or obovate, margins entire or glandular-serrulate, without oil cells; inflorescences in spikes or racemes; bracteoles 2 at or near base of ovary; flowers 4–5(6)-merous, pedicellate or sessile; sepals spreading to suberect; petals caducous, yellow, sometimes white, margins entire; often with raised or depressed nectary lobes surrounding base of each epipetalous stamen; nectary disk elevated on ovary apex; style glabrous; stigma entire or irregularly lobed; fruit a capsule, spreading to erect; seeds 50–400, 1 to several rows per locule, usually free, sometimes embedded in endocarp (Raven [1963]1964; Wagner et al., 2007).

1.   Stems floating or creeping and ascending to erect; leaves alternate or fascicled; flowers 5(6)-merous; stamens 2 times as many as sepals, in 2 unequal series…**L. peploides**
1'. Stems prostrate or decumbent, erect at tips, sometimes ascending; leaves opposite; flowers 4-merous; stamens as many as sepals, in 1 series…(2)
2.   Petals 0; sepals 1.1–2 mm; anthers 0.2–0.4 mm; capsules (1.6–)2–5 mm, < 2 times as long as broad, walls thin…**L. palustris**
2'. Petals 4, caducous; sepals 1.8–5 mm; anthers 0.4–0.9 mm; capsules 4–10 mm, generally > 2 times as long as broad, walls hard…**L. repens**

Ludwigia palustris (L.) Elliott—Marsh seedbox, water-purslane. Stems 10–50(–70) cm, prostrate or decumbent and ascending at tips, creeping and forming mats, well branched, glabrous; leaves narrowly to broadly elliptic or ovate-elliptic, 0.5–4.5 × 0.3–2.3 cm, margins entire and minutely strigillose, apex subacute; inflorescences in leafy spikes or racemes; flowers 4-merous, sepals ascending, ovate-deltate, margins finely serrulate with minute hairs, apex acuminate; petals 0; stamens translucent; style pale green; stigma subglobose or capitate; capsules oblong-obovoid, 4-angled, (1.6–)2–5 × 1.5–3(–3.5) mm, thin-walled, pedicels 0–0.5 mm; seeds in several rows per locule, free. Roadside ditches, wet meadows, pond margins, swamps, rivers, 3650–5600 ft. Eddy, Guadalupe, Hidalgo.

Ludwigia peploides (Kunth) P. H. Raven—Floating primrose-willow. Emergent aquatic, stems 10–100(–300) cm, glabrous or sparsely villous; leaves narrowly oblong or elliptic to ovate, broadly obovate, or orbiculate, 0.8–4(–8.5) cm, margins entire, apex obtuse or rounded to acute; inflorescences in leafy racemes; flowers 5(6)-merous, sepals spreading, narrowly deltate or lanceolate, apex acute or acuminate; petals yellow, obpyramidal, 7–24 × 4–13 mm, apex mucronate or emarginate; stamens 10(–12), bright yellow; stigma sometimes shallowly or deeply 5-lobed; capsules cylindrical to obscurely 5-angled, straight or curved, 10–17(–25) × 2–3 mm, with thick woody walls, pedicels 10–35 mm; seeds in 1 row per locule, embedded in endocarp. New Mexico material belongs to **subsp. peploides**. Wet places, along slow-moving rivers, streams, ditches, 3800–5175 ft. Doña Ana, Union. Introduced.

Ludwigia repens J. R. Forst.—Creeping primrose-willow, red ludwigia. Stems 30–80 cm, prostrate, ascending to suberect at tips, creeping, sparsely branched, glabrous; leaves narrowly elliptic to broadly lanceolate-elliptic or suborbiculate, 0.8–4.5 × 0.4–2.7 cm, margins entire or sometimes with hydathodal glands, apex acute or apiculate; inflorescences in erect racemes; bracteoles attached in opposite pairs to pedicels; flowers 4-merous, sepals ascending, ovate-deltate to narrowly so, margins minutely strigillose, apex acuminate to elongate-acuminate, surfaces subglabrous; petals oblanceolate to elliptic-oblong, 1.1–3 × 0.4–1.4 mm, base attenuate, apex obtuse; stamens pale yellow; style pale yellow, stigma broadly capitate; capsules elongate-obpyramidal, 4-angled, 4–10 × 2.5–4 mm, hard-walled, pedicels 0.1–3 mm; seeds in several rows per locule, free. Muddy or damp, sandy edges of pools, lakes, swamps, creeks, roadside ditches, 3660–5100 ft. Chaves, Eddy, Guadalupe.

OENOTHERA L. – Evening primrose

Annuals, biennials, or perennials, sometimes suffrutescent, caulescent or acaulescent; usually with taproots; stems erect or ascending, sometimes decumbent, epidermis green or whitish and exfoliating; leaves in basal rosette and cauline, alternate, sessile or petiolate, stipules absent, blade margins toothed to pinnatifid, sometimes subentire; flowers solitary in leaf axils, often forming leafy terminal spikes, erect or nodding, actinomorphic, sometimes zygomorphic and petals positioned in distal 1/2; floral tube glabrous with basal nectary; sepals (3)4, reflexed; petals (3)4, yellow, purple, or white, fading orange, purple, pale yellow, reddish, or whitish; stamens (6-)8, subequal or in 2 unequal series; anthers versatile; ovary (3)4-locular or septa incomplete and 1-locular, ovules numerous or 1-8, style glabrous or pubescent; fruit a capsule, straight or curved, lanceoloid or ovoid, ellipsoid to cylindrical, rhombic-obovoid, or globose, terete or (3)4-angled or -winged, usually loculicidally dehiscent, sometimes an indehiscent, nutlike capsule with woody walls, sessile; seeds numerous (1-160+), in 1 or 2(-4) rows or sometimes reduced to 1-8 (Wagner et al., 2007).

1. Stigma peltate, discoid to quadrangular or obscurely and shallowly 4-lobed...(2)
1'. Stigma deeply divided into (3)4 linear lobes...(8)
2. Sepals with conspicuously keeled midribs; stamens in 2 unequal series, antisepalous filaments 2 times as long as antipetalous filaments...(3)
2'. Sepals without keeled midribs; stamens in subequal series...(5)
3. Petals 5-12(-20) mm; stigma surrounded by anthers at anthesis **O. serrulata**
3'. Petals 6-25 mm; stigma exserted beyond anthers at anthesis...(4)
4. Leaves (0.1-)0.3-0.6 cm wide; stems (10-)25-40 cm, several to many, moderately branched, decumbent to ascending; sepals 4-12 mm; floral tube 5-20 mm; petals 6-25 mm...**O. capillifolia**
4'. Leaves 0.1-0.2 cm wide; stems 15-30(-40) cm, many, branched from base, ascending to erect; sepals 4-6 mm; floral tube 7 mm; petals 15-20 mm...**O. gayleana**
5. Flowers diurnal, opening near sunrise; floral tubes 5-25(-33) mm, funnelform in distal 1/2+...**O. tubicula**
5'. Flowers vespertine, opening in afternoon or near sunset; floral tubes (15-)16-60(-70) mm, funnelform in distal 1/2 or less...(6)
6. Leaf axils with conspicuous fascicles of small leaves, these 0.2-2.5 cm; flower buds with free tips 2-9 (-12) mm; capsules somewhat papery and dehiscent in distal 1/2...**O. toumeyi**
6'. Leaf axils sometimes with axillary fascicles of small leaves, when present 0.1-1.5 cm; flower buds with free tips 0.3-6 mm; capsules hard, dehiscent throughout...(7)
7. Plants not cespitose, stems erect to ascending, 4-60 cm, strigillose, glandular-puberulent, glabrous, hirtellous, or short-pilose...**O. hartwegii**
7'. Plants low, often cespitose, stems spreading to decumbent or ascending, 4-20(-30) cm, densely strigillose throughout, sometimes glandular-puberulent distally...**O. lavandulifolia**
8. Capsules cylindrical to lanceoloid or ovoid, without ridges or wings, sometimes angled...(9)
8'. Capsules ellipsoid to oblong, ovoid, or obovoid, sometimes lanceoloid, cylindrical, pyramidal, fusiform, or globose, angled or winged, or valve margins with ridges or tubercles...(23)
9. Petals white...(10)
9'. Petals pale yellow to yellow...(14)
10. Seeds in 2 rows per locule, ellipsoid to subglobose, surface regularly pitted, pits in longitudinal lines; capsules straight or sometimes curved upward, dehiscent 1/2 their length...(11)
10'. Seeds in 1 row per locule, obovoid, surface minutely alveolate but appearing smooth; capsules straight, curved upward, or contorted, dehiscent 1/2 to nearly throughout their length...(12)
11. Annual, from a taproot; mouth of floral tube glabrous...**O. albicaulis**
11'. Perennial, lateral roots producing adventitious shoots; mouth of floral tube conspicuously pubescent... **O. coronopifolia**
12. Winter annual or short-lived perennial from a taproot; leaves sessile; flower buds without free tips...**O. engelmannii**
12'. Perennial, sometimes annual, from a taproot, also with lateral roots producing adventitious shoots or with long, fleshy roots; leaf petioles 0-2(-4.5) cm; flower buds with free tips 0-4 mm...(13)
13. Capsules erect or strongly ascending, 2-3 mm diam.; style 50-70 mm...**O. neomexicana**
13'. Capsules spreading to reflexed, 1.5-2.5 mm diam.; style 25-55 mm...**O. pallida**
14. Seeds obovoid to oblanceoloid, surfaces coarsely rugose, also with turgid and collapsed papillae; capsules twisted or curved...**O. primiveris**
14'. Seeds prismatic and angled, ellipsoid to subglobose, rarely obovoid and obtusely angled, surfaces reticulate and regularly or irregularly pitted; capsules not twisted...(15)
15. Floral tubes 100-165(-190) mm; perennial...**O. organensis**
15'. Floral tubes (5-)12-50(-160) mm; annual, biennial, or short-lived perennial...(16)
16. Flowers usually many opening per day near sunset; seeds prismatic and angled, surface irregularly pitted ...(17)
16'. Flowers usually 1(2) to several opening per day near sunset; seeds ellipsoid to subglobose, not angled, surface regularly pitted...(20)
17. Stigma exserted beyond anthers at anthesis; petals (28-)30-65 mm...(18)
17'. Stigma surrounded by anthers; petals 7-25(-30) mm...(19)

18. Floral tube 60-135 mm…**O. longissima**
18'. Floral tube (20-)30-45(-50) mm…**O. elata**
19. Inflorescences open (internodes in fruit as long as or longer than capsule), villous with appressed to erect hairs, with pustulate bases, also glandular-puberulent; petals 7-20 mm; sepals 9-18 mm, red-striped or flushed red…**O. villosa**
19'. Inflorescences dense (internodes in fruit shorter than capsule), sparsely to moderately strigillose and glandular-puberulent, sometimes also scattered-villous, hairs erect to appressed, with or without pustulate bases; petals 12-25(-30) mm; sepals 12-22(-28) mm, green or yellowish, rarely red-striped or flushed red…**O. biennis**
20. Young flower buds nodding by recurved floral tube…**O. pubescens**
20'. Young flower buds with floral tube curved upward or straight…(21)
21. Petals rhombic to elliptic or rhombic-ovate…**O. rhombipetala**
21'. Petals shallowly or deeply obcordate…(22)
22. Petals 25-40 mm; styles 40-75 mm; stigma exserted beyond anthers at anthesis…**O. grandis**
22'. Petals 5-22 mm; styles 20-50 mm; stigma surrounded by anthers at anthesis…**O. laciniata**
23. Petals white, pink, or rose-purple, rarely streaked or flecked with red…(24)
23'. Petals yellow…(33)
24. Capsules cylindrical to obtusely angled, with tubercles or an undulate ridge along valve margins…**O. cespitosa**
24'. Capsules angled or winged, without tubercles or ridge along valve margins…(25)
25. Capsules dehiscent, at least in distal portion; seeds clustered in each locule…**O. speciosa**
25'. Capsules indehiscent; seeds in 2-4 rows in locules or reduced to 1-8…(26)
26. Petals pink, rarely white, streaked or flecked with red; flowers actinomorphic; seeds in 2-4 rows per locule…**O. canescens**
26'. Petals white, not streaked or flecked; flowers zygomorphic or nearly actinomorphic; seeds not in rows, reduced to 1-8…(27)
27. Capsules with a slender stipe 2-10 mm…**O. cinerea**
27'. Capsules sessile or with stipe to 2.2 mm…(28)
28. Capsules fusiform…(29)
28'. Capsules ellipsoid, ovoid, or obovoid…(30)
29. Annuals; flowers nearly actinomorphic, sepals 2-3.5 mm; floral tube 1.5-5 mm…**O. curtiflora**
29'. Perennials; flowers zygomorphic, sepals 5-9(-10) mm; floral tube 4-11(-13) mm…**O. suffrutescens**
30. Capsules sharply 4-angled, not winged; stems 30-300 cm…(31)
30'. Capsules winged, furrowed between wings; stems 15-100 cm…(32)
31. Stems glaucous at least in proximal part; sepals 4-6 mm; basal leaves 3-7 cm, margins entire…**O. glaucifolia**
31'. Stems not glaucous, villous and strigillose proximally; sepals 11-15 mm; basal leaves 6-20 cm, margins subentire or repand-denticulate…**O. dodgeniana**
32. Sepals 6-12 mm; floral tube 6-10 mm; capsule stipe 0…**O. podocarpa**
32'. Sepals 11-21 mm; floral tube 10-20 mm; capsule stipe 0.2-2 mm…**O. nealleyi**
33. Capsules ± winged, wings 0-3(-5) mm or walls with corky thickening and wings not developed, then capsule appearing only 4-angled; seeds with erose wing distally, surfaces coarsely rugose…**O. brachycarpa**
33'. Capsules angled and valves with a prominent median ridge, or if winged, then wing oblong to triangular, confined to distal 1/2-2/3, 2-5(-10) mm wide; seeds sometimes with a small wing at distal end and with a ridge or small wing along 1 adaxial margin, surfaces beaded…(34)
34. Petals pale yellow; capsules woody, with broad, triangular wings 5-10 mm wide, these often terminating in a hooked tooth; annual or sometimes biennial…**O. triloba**
34'. Petals bright yellow; capsules leathery, with narrowly oblong wings (2-)3-5(-6) mm wide, without a hooked tooth; perennial, rarely short-lived…**O. flava**

Oenothera albicaulis Pursh [*O. ctenophylla* (Wooton & Standl.) Tidestr.]—White evening primrose. Winter annual; stems 5-30 cm, ascending to decumbent, 1 to several from base, densely strigillose, also sparsely villous; leaf blades oblanceolate to oblong, 1.5-10 × 0.3-2.5 cm, margins subentire or coarsely dentate or pinnatifid; flowers opening near sunset, buds nodding, weakly quadrangular, without free tips; floral tube 15-30 mm; sepals 15-30 mm; petals white, fading pink, obcordate, sometimes obovate, (15-)20-35(-40) mm; stigma exserted beyond anthers; capsules ascending to erect, cylindrical, weakly 4-angled, 20-40 × 3-4 mm. Dry, sandy flats and slopes, 3700-7550 ft. Widespread.

Oenothera biennis L.—Common evening primrose. Biennial; stems 30-200 cm, erect, green or flushed with red on proximal parts, unbranched or with oblique side branches, densely to sparsely strigillose and villous, with somewhat appressed to spreading hairs, these often pustulate; leaf blades narrowly oblanceolate to oblanceolate, basal 10–30 × 2–5 cm, cauline 5–22 × (1–)1.5–5(–6) cm, margins flat, bluntly dentate, teeth widely spaced; inflorescences erect, unbranched or with secondary branches proximal to main one; flowers opening near sunset, buds erect, with free tips, erect or spreading; floral tube (20–)25–40 mm; petals yellow, fading yellowish white and somewhat translucent, very broadly obcordate; stigma surrounded by anthers; capsules erect or slightly spreading, lanceoloid, 20–40 × 4–6 mm, valve tips free. Open, disturbed sites; naturalized in Sandia Mountains, 7550 ft. N, NC, SC. Introduced. This species is a self-pollinating, permanent translocation heterozygote species (PTH) and has only ca. 50% pollen fertility.

Oenothera brachycarpa A. Gray—Shortfruit evening primrose. Perennial; stems (when present) 0–20(–36) cm, ascending, becoming decumbent, densely leafy, strigillose, also hirsute, often with reddish-purple pustulate base, glandular-puberulent distally; leaf petioles (0.8–)2.5–11(–15) cm, blades lanceolate to elliptic or rhombic-obovate, (3.1–)5–21(–34) × (0.3–)1.5–3.5(–5.3) cm, margins irregularly pinnatifid, some sinuses nearly to midrib, with a large terminal lobe, erose; flowers opening near sunset, buds with unequal free tips; floral tube (90–)120–210(–220) mm; sepals 38–55 mm; petals yellow, fading pale orange to pink, broadly rhombic-obovate, (38–)45–58(–62) mm, distal margins erose; stigma exserted beyond anthers; capsules leathery or corky, ovoid to narrowly ellipsoid, (12–)18–40 × 6–10 mm, dehiscent 1/4 length, pedicels 0–3 mm. Rocky sites on limestone, shale, or gypsum, igneous substrates in canyons and slopes in Chihuahuan desert scrub, grasslands, oak-pine-juniper woodlands, open sites in ponderosa pine–Douglas-fir forests, 3935–7215(–8530) ft. C, S.

Oenothera canescens Torr. & Frém.—Spotted evening primrose. Perennial, forming low clumps 10–50 cm diam., lateral roots producing adventitious shoots; stems (10–)15–25(–38) cm, many-branched from base, leafy, densely strigillose throughout; leaves cauline, fascicles of small leaves 0.2–0.6 cm often present in nonflowering axils, petioles 0–0.1 cm, blades lanceolate to linear, (0.3–)0.6–1.5(–2.5) × (0.05–)0.15–0.4(–0.6) cm, base cuneate, apex acute; flowers opening near sunset, buds without free tips; floral tube (8–)10–15(–17) mm; sepals (7–)8–12 mm; petals pink, rarely white, streaked or flecked with red, fading bright purple, obovate, (8–)10–17 mm; anthers often with red longitudinal stripe; stigma exserted beyond anthers; capsules woody, ovoid, narrowly winged, (7–)9–12(–14) × 2–4 mm (excluding wings), abruptly constricted to a conspicuous, sterile beak. Prairie depressions, playas, margins of ditches, temporary wet areas, 4000–5900 ft. NE, EC, C.

Oenothera capillifolia Scheele—Evening primrose. Short-lived perennial or annual, glabrous or strigillose; stout taproot; stems (10–)25–40 cm; leaves cauline, sometimes with fascicles of small leaves to 2 cm present in nonflowering axils, petioles 0–0.6 cm, blades linear to narrowly lanceolate or oblanceolate, 1–4 × (0.1–)0.3–0.6 cm, often folded lengthwise, proximal-most leaves sometimes spatulate, margins subentire or serrate, base attenuate, apex acute; flowers opening at sunrise, buds with free tips; floral tube 5–20 mm; sepals with keeled midrib, 4–12 mm; petals yellow, fading orangish to purplish, 15–20 mm; stigma exserted beyond anthers; capsules hard, 10–35 × 1–2 mm, dehiscent 1/2 length, often tardily dehiscent throughout. New Mexico material belongs to **subsp. berlandieri** (Spach) W. L. Wagner & Hoch. Relatively dry areas, grassy prairies, plains, low hills, in sandy, gravelly, and limestone soil, 2950–3600(–4920) ft. NW, C, EC, SC.

Oenothera cespitosa Nutt.—Tufted evening primrose. Perennial; stems (when present) 0–40 cm, ascending or decumbent, unbranched or branched from near base, hirsute or villous and glandular-puberulent; leaves petiolate, blades oblanceolate to rhombic or spatulate, 1.7–26(–36) × (0.3–)0.5–4.5(–6.5) cm, margins irregularly sinuate-dentate, serrate, pinnatifid, or lobed; flowers 1–4(–6) per stem, opening per day near sunset, buds erect, without free tips; floral tube (20–)40–140(–165) mm; sepals (15–)18–45(–54) mm; petals white, fading rose or rose-pink to dark or deep rose-purple, obovate or obcordate, (16–)20–50(–60) mm; stigma exserted beyond anthers; capsules straight or curved, cylindrical to lanceoloid or ellipsoid, obtusely 4-angled, (10–)13–50(–68) × 4–9 mm, tapering to a sterile beak 6–8 mm, dehiscent 1/3–7/8 length, pedicellate.

1. Capsules oblong-lanceoloid; buds often recurved when young; floral tube (35–)40–70(–80) mm; plants shaggy-villous, sometimes densely so…**subsp. navajoensis**

1'. Capsules cylindrical to lanceoloid-cylindrical; buds erect; floral tube (41–)75–140(–165) mm; plants hirsute, sometimes glabrous…(2)
2. Capsules somewhat curved, valve margins with nearly smooth to irregular, undulate ridges; leaf blades oblanceolate to spatulate, margins dentate…**subsp. macroglottis**
2'. Capsules straight, valve margins with minute to conspicuous tubercles, these sometimes coalesced into a sinuate ridge; leaf blades oblanceolate to narrowly elliptic, rarely lanceolate, margins pinnately lobed to dentate, rarely serrate…**subsp. marginata**

subsp. macroglottis (Rydb.) W. L. Wagner, Stockh. & W. M. Klein—Tufted evening primrose. Stems 4–8 cm, unbranched; leaf petioles (3–)4–11(–14) cm, blades (6.8–)9.5–23(–32) × (1.3–)2.4–4.5(–6.5) cm, margins often undulate; floral tube (45–)75–110(–153) mm; petals (21–)35–43(–50) mm; capsules symmetric throughout, (17–)25–45(–56), valve margins with nearly smooth to irregular, undulate ridges, pedicels 2–7 mm. Open, igneous rocky slopes, talus, roadcuts, open or shaded, sandy or gravelly sites along streams, upper piñon-juniper woodlands, oak scrub, pine forests, 6560–9185 ft. N, C.

subsp. marginata (Nutt. ex Hook. & Arn.) Munz—Tufted evening primrose. Stems 10–40 cm, unbranched, moderately to densely hirsute; leaf petioles (3–)4–11(–14) cm, blades (2.8–)10–26(–36) × (0.6–)1–3(–4.5) cm; floral tube (41–)80–140(–165) mm; petals (24–)35–50(–60) mm; capsules slightly asymmetric, (21–)25–50(–68) × 6–8 mm, valve margins with minute to conspicuous tubercles, these sometimes coalesced into a sinuate ridge, pedicels (0–)1–40(–55) mm. Rocky slopes, cracks in rocks, talus, along gravelly creek beds and arroyos, roadcuts, on sandstone, limestone, volcanic cinder, mostly in foothill communities of piñon-juniper woodlands, sagebrush scrub, chaparral, grasslands, openings in ponderosa pine forests, (4100–)4600–7220(–8860) ft. Widespread.

subsp. navajoensis W. L. Wagner, Stockh. & W. M. Klein—Navajo evening primrose. Stems (0–)10–25 cm, unbranched to few-branched; leaf petioles (1.3–)1.7–10(–12) cm, blades oblanceolate to rhombic-obovate, (3.5–)4–13(–16) × (0.7–)1–3.2 cm, margins often coarsely and irregularly dentate or serrate, sometimes pinnately lobed; petals (25–)28–32(–34) mm; capsules straight, base asymmetric, 13–35(–40) × 5–6 mm, valve margins with a low sinuate ridge to 8–15 low, nearly distinct tubercles, 0.4–1 mm, pedicels 1–3 mm. Colorado Plateau region, on loose or compacted soil derived from clay, shale, fine-textured sandstone, or gypsum, on slopes and along small drainage patterns, often around harvester ant mounds, arroyos in somewhat sandy or gravelly soil, sagebrush scrub, sage-grasslands, 5080–6890 ft. NW, C.

Oenothera cinerea (Wooton & Standl.) W. L. Wagner & Hoch [*Gaura cinerea* Wooton & Standl.]—High-plains beeblossom. Perennial, suffrutescent, deep, twisted, woody rootstock; stems 60–280 cm, erect, several-branched near ground, also branched proximal to inflorescences, densely soft-villous, hairs mostly appressed, also strigillose; leaves sessile, blades narrowly lanceolate to very narrowly elliptic or linear, 0.5–8 × 0.15–2 cm, margins subentire or shallowly sinuate-dentate, often undulate; inflorescences slender, glabrous or sparsely glandular-puberulent, opening near sunset; floral tube 2–5 mm; sepals 6–14 mm; petals white, fading pink to red, 8.5–13 mm, slightly unequal, elliptic, clawed; stamens present in proximal 1/2 of flower; stigma exserted beyond anthers; capsules lanceoloid to narrowly ovoid, 4-winged, 9–19 × 1–3.5 mm. Our plants belong to **subsp. cinerea**. Sandy flats and dunes on high plains and rolling plains, 3445–5575 ft. E.

Oenothera coronopifolia Torr. & A. Gray—Crownleaf evening primrose. Perennial; stems 10–60 cm, ascending to erect, 1 to several from base, unbranched to well branched, strigillose, also hirsute; leaves in a weakly developed basal rosette and cauline, axillary fascicles of reduced leaves often present, blades oblanceolate to oblong, cauline 2–7 × 0.2–1.5 cm, margins pinnatifid; flowers opening near sunset, buds nodding, weakly quadrangular, without free tips; floral tube 10–25 mm, closed with straight, white hairs; sepals 10–20 mm; petals white, fading pink, ovate or shallowly obcordate, 10–15(–20) mm; stigma exserted beyond anthers; capsules ascending to erect, straight, fusiform, weakly 4-angled, 10–20 × 3–5 mm. Dry, open sites, grassy meadows, slopes, drainages, foothills, mountains, (4600–)5900–9840 ft. Widespread except E.

Oenothera curtiflora W. L. Wagner & Hoch [*Gaura mollis* E. James]—Velvetweed. Annual, heavy taproot; stems (20–)30–200(–300) cm, erect, unbranched or many-branched distally, strigillose, glandular-puberulent, and long-villous; basal leaf petioles 0–1.8 cm, blades broadly oblanceolate, 4–15 × 1.5–3 cm, margins sinuate-dentate to dentate; cauline leaf petioles 0–2 cm, blades narrowly elliptic to narrowly ovate, 2–13 × 0.5–5 cm, margins sinuate-dentate to dentate; inflorescences relatively long, dense, nearly actinomorphic, opening near sunset; floral tube 1.5–5 mm; sepals 2–3.5 mm; petals white, fading pale to dark pink, slightly unequal, oblong-obovate to elliptic-oblanceolate, 1.5–3 mm, abruptly clawed; stigma surrounded by anthers; capsules fusiform, terete, angles becoming broad and rounded in proximal part, 5–11 × 1.5–3 mm, tapering abruptly toward base; seeds 3 or 4. Rocky prairie slopes, woodlands, streams, roadsides, disturbed areas, 4265–6900 ft. Widespread.

Oenothera dodgeniana Krakos & W. L. Wagner—Dodgen's evening primrose. Biennial, stout, fleshy taproot; stems 50–120 cm, 1- or few-branched from base, villous and strigillose proximally, glandular-puberulent distally; leaf blades lanceolate to narrowly elliptic, basal 6–20 × 1–3 cm, cauline 5–10 × 1–2.5 cm, margins subentire or repand-denticulate, glabrate or strigillose; opening at sunset; floral tube 10–11 mm, sepals 11–15 mm; petals white, fading pink, rhombic-obovate, 11–14 mm; stigma exserted beyond anthers; capsules ellipsoid or ovoid, deep furrows alternating with angles near apex, ribbed distally, 9–11 × 3–5 mm; seeds 2–4. Mountain meadow openings in coniferous forests, 5900–8860 ft. NW, C, SC.

Oenothera elata Kunth [*O. hookeri* Torr. & A. Gray]—Hooker's evening primrose. Biennial or short-lived perennial; stems 30–250 cm, erect, flushed with red proximally or throughout, unbranched or branched, strigillose, also villous, with appressed or spreading hairs; leaf blades narrowly oblanceolate or oblanceolate to narrowly elliptic or narrowly lanceolate, basal 10–43 × 1.2–4(–6) cm, cauline 4–25 × 1–2.5(–4) cm, margins flat, bluntly dentate or subentire, bracts persistent; inflorescences erect, un-branched, flowers opening near sunset, buds erect with free tips, buds and sepals green to yellowish green, red-striped, or sometimes red throughout; floral tube (20–)30–45(–50) mm; sepals 27–50 mm; petals yellow, fading orange or pale yellow, very broadly obcordate, 30–47(–55) mm; stigma elevated above anthers; capsules erect or slightly spreading, narrowly lanceoloid, 20–65 × 4–7 mm, with free tips. Our material belongs to **subsp. hirsutissima** (A. Gray ex S. Watson) Munz. Montane sites along streams, mesic meadows, roadsides, near permanent or seasonally wet sites, ditch banks, riverbanks, floodplains, fallow agricultural land, 3600–10,000 ft. Widespread except E.

Oenothera engelmannii (Small) Munz—Engelmann's evening primrose. Winter annual; stems 30–50(–80) cm, erect, unbranched or with few, spreading branches, conspicuously villous throughout, also strigillose on leaves and distal parts; leaves with basal rosette weakly developed or absent, sessile, blades lanceolate to oblong-lanceolate, (1–)2–6(–8) × 1–2(–3) cm, margins coarsely repand-dentate or pinnatifid; flowers opening near sunset, buds nodding, without free tips; floral tube 20–30 mm; sepals 13–21 mm; petals white, fading pink, broadly obovate or obcordate, 15–30 mm; stigma exserted be-yond anthers; capsules widely spreading, woody in age, straight or slightly curved, cylindrical, obtusely 4-angled, 30–60 × 2–3 mm, tapering gradually from base to apex. Sandy prairies, dunes, disturbed areas, roadsides, 3280–4265 ft. E.

Oenothera flava (A. Nelson) Garrett—Yellow evening primrose. Perennial; stems (when present) 0–2 cm, ascending, 1 to several, densely leafy, glabrate to moderately strigillose, also glandular-puberulent; leaves primarily in a basal rosette, ± fleshy, petioles (0.2–)2–7(–10) cm, blades oblanceolate to linear, (3.4–)6–30(–36) × (0.5–)1.5–5(–7) cm, margins irregularly and coarsely pinnately lobed, apex acute; flowers opening near sunset, buds with free tips; floral tube (24–)40–200(–265) mm; sepals (8–)11–40(–42) mm; petals yellow, fading pale orange, drying purple, (7–)10–45(–50) mm; stigma exserted beyond or sur-rounded by ring of anthers; capsules leathery in age, surface usually conspicuously reticulate, usually narrowly ovoid or ellipsoid, sometimes ovoid or lanceoloid, winged, wings narrowly oblong, (2–)3–5(–6) mm wide, confined to distal 2/3 of capsule, (10–)20–35(–43) × 4–8 mm (excluding wings), gradually constricted to a short beak, dehiscent 1/4–1/2 length. Swales, desiccating flats and ponds, montane meadows, margins of watercourses, open sites in wet clay to gravelly sand, 3675–11,500 ft. NC, NW, WC, SE, SC, SW.

Oenothera gayleana B. L. Turner & M. J. Moore—Gypsum evening primrose. Perennial, sometimes suffrutescent, stout taproot; stems 15–30(–40) cm, many, branched from base, ascending to erect, strigillose, sometimes glabrous; leaf petioles 0–0.1 cm, blades linear to narrowly linear-lanceolate, 2.5–3.5 × 0.1–0.2 cm, folded lengthwise, margins subentire or serrulate, base long-attenuate, apex acute; flowers opening near sunrise, buds without or with free tips; floral tube ca. 7 mm; sepals with keeled midrib, 4–6 mm; petals yellow, fading yellow to orange, 15–20 mm; stigma discoid to quadrangular, exserted beyond anthers; capsules hard, 18–20 × 2 mm, dehiscent 1/2 length, often tardily dehiscent throughout. Gypsum outcrops, 3770–4265 ft. DeBaca, Eddy.

Oenothera glaucifolia W. L. Wagner & Hoch—False gaura. Probably biennial; stems 30–300 cm, erect, branched or unbranched, glabrous, becoming sparsely to densely glandular-puberulent and short-villous distally, glaucous at least in proximal parts; basal leaves sessile, blades oblong to oblong-lanceolate, 3–7 × 0.5–2 cm, margins entire, base ± auriculate; cauline blades lanceolate to oblong-lanceolate, becoming linear-subulate distally, 3–8(–10) × 0.4–1.8 cm; inflorescences wandlike, unbranched or branched, flowers opening near sunrise; floral tube 6–17 mm; sepals 4–6 mm; petals white, fading off-white or tinged pink, slightly unequal, rhombic, 4–6 mm, abruptly clawed; stigma exserted beyond anthers; capsules ovoid, 4-angled, somewhat flattened, 3–4 × 1.5–2.3 mm, valves with raised midrib and conspicuous lateral veins; seeds 1. Rocky prairie slopes, outcrops, bluffs, along streams, roadsides, usually on limestone, 3700–6800 ft. EC, C, WC, SW.

Oenothera grandis Smyth—Showy evening primrose. Annual; stems 15–60(–100) cm, erect to ascending, often with ascending lateral branches, strigillose and sparsely villous, also glandular-puberulent distally; leaf blades narrowly oblanceolate to narrowly elliptic, basal 5–13 × 1–3 cm, cauline 3–10 × 1.5–3.5 cm, margins lobed or dentate, lobes often dentate, bracts spreading, flat; flowers opening near sunset; buds erect, with free tips, erect or hornlike; floral tube 25–45 mm; sepals 15–30 mm; petals yellow, very broadly obovate or shallowly obcordate, 25–40 mm; stigma exserted beyond anthers; capsules cylindrical, 25–50 × 2–3 mm. Open, sandy sites, 3935–6430 ft. Lea, Roosevelt, San Miguel.

Oenothera hartwegii Benth.—Hartweg's sundrops. Perennial, sometimes suffrutescent, stout taproot; stems 4–60 cm, 1 to many, unbranched to densely branched, erect to ascending, strigillose, glandular-puberulent, glabrous, hirtellous, or short-pilose; leaf petioles 0–0.2 cm, blades elliptic, lanceolate, linear, or filiform to ovate or oblanceolate, not much reduced distally, proximal-most leaves sometimes obovate to spatulate, 0.3–6.5 × 0.04–1.2 cm, margins entire or serrate, often undulate, base attenuate to obtuse, truncate, or subcordate, sometimes clasping, apex acute; flowers opening in afternoon or near sunset; floral tube 16–50(–60) mm; sepals 7–28 mm; petals yellow, fading pale pinkish or pale purple, 10–35 mm; stigma yellow, quadrangular, exserted beyond anthers; capsules 6–40 × 2–4 mm.

1. Base of leaf blades (except proximal-most) truncate or subcordate and clasping; plants densely pubescent with mixture of hair types, but always short-pilose and also hirtellous, sometimes also strigillose, especially on leaves, or glandular-puberulent distally…**subsp. pubescens**
1'. Base of leaf blades attenuate or obtuse; plants glabrous, sparsely strigillose, or glandular-puberulent…(2)
2. Plants glabrous throughout, sometimes glandular-puberulent on distal parts, especially on ovary…**subsp. fendleri**
2'. Plants glandular-puberulent throughout, more densely so on distal parts, sometimes also sparsely strigillose on ovary and leaves…**subsp. filifolia**

subsp. fendleri (A. Gray) W. L. Wagner & Hoch [*Calylophus hartwegii* (Benth.) P. H. Raven subsp. *fendleri* (A. Gray) Towner & P. H. Raven]—Axillary fascicles of small leaves sometimes present, blades linear to oblanceolate or lanceolate, 1–5 × 0.15–1 cm, margins entire or subentire, base attenuate to obtuse; floral tube 30–50 mm; petals 10–30 mm. Grasslands to woodlands on clay or gravelly soil, sometimes calcareous, 3700–7215 ft. C, S.

subsp. filifolia (Eastw.) W. L. Wagner & Hoch [*Calylophus hartwegii* (Benth.) P. H. Raven subsp. *filifolius* (Eastw.) Towner & P. H. Raven]—Axillary fascicles of small leaves often present, blades 0.3–4 × 0.04–0.3(–0.4) cm, margins entire or remotely serrulate, sometimes undulate, base attenuate; floral tube 16–50 mm; petals 12–23 mm. Semiarid gypsum flats, dunes, outcrops, 3280–6235 ft. E, C.

subsp. pubescens (A. Gray) W. L. Wagner & Hoch [*Calylophus hartwegii* (Benth.) P. H. Raven subsp. *pubescens* (A. Gray) Towner & P. H. Raven]—Axillary fascicles of small leaves often absent or much reduced, blades very narrowly elliptic or narrowly lanceolate to ovate, 0.6–4 × 0.15–1.2 cm, margins entire or sparsely serrulate; floral tube 20–50 mm; petals 12–35 mm. Moderately dry, open places, grasslands, plains, hills, sandy to gravelly soil, limestone or gypsum, 4130–6900 ft. NE, EC, C, SE, SC.

Oenothera laciniata Hill—Cutleaf evening primrose. Annual; stems 5–50 cm, erect to ascending, unbranched to much branched, sparsely to moderately strigillose, sometimes also villous and/or glandular-puberulent distally; leaf blades narrowly oblanceolate to narrowly elliptic or narrowly oblong, basal 4–15 × 1–3 cm, cauline 2–10 × 0.5–3.5 cm, margins dentate or deeply lobed, bracts spreading, flat; flowers opening near sunset, buds erect, with free tips; floral tube 12–35 mm; sepals 5–15 mm; petals yellow, fading orange or reddish-tinged, 5–22 mm, broadly obovate or obcordate; capsules cylindrical, 20–50 × 2–4 mm. Open, usually sandy sites, disturbed habitats; known in New Mexico only from Roosevelt County, 3935–4100 ft. This species is a self-pollinating, permanent translocation heterozygote species (PTH) and has only ca. 50% pollen fertility.

Oenothera lavandulifolia Torr. & A. Gray [*Calylophus lavandulifolius* (Torr. & A. Gray) P. H. Raven]—Lavenderleaf sundrops. Perennial, from a stout taproot; stems 4–20(–30) cm, several to many, spreading-decumbent to ascending, branched, often cespitose, densely strigillose throughout, sometimes glandular-puberulent distally; axillary fascicles of small leaves often present, sessile, blades narrowly lanceolate or narrowly oblanceolate, 0.6–5 × 0.08–0.6 cm, margins entire or subentire, base attenuate to truncate, apex acute to obtuse; flowers opening near sunset, buds with free tips; floral tube 25–60 mm; sepals 8–20 mm; petals yellow, fading pale pink or pale purple, 12–28 mm; stigma yellow, quadrangular, exserted beyond anthers; capsules 6–25 × 1–3 mm. High plains or mountains in sandy, rocky, calcareous soil, 4265–8200 ft. N, C, SE, SC. Widespread.

Oenothera longissima Rydb. [*O. clutei* A. Nelson]—Longstem evening primrose. Biennial or short-lived perennial; stems 60–300 cm, erect, flushed with red proximally or sometimes green, unbranched or branched, sparsely strigillose, sometimes also villous and with pustulate hairs near inflorescence; leaf blades flat, narrowly oblanceolate, oblanceolate to narrowly elliptic, or narrowly lanceolate, basal 9–40 × 1.4–5 cm, cauline 5–22 × 0.8–2.5 cm, margins bluntly dentate or subentire, teeth widely spaced, bracts persistent; inflorescences open, erect, unbranched, flowers opening near sunset, buds erect, with free tips; floral tube 60–135 mm; sepals yellowish green, flushed with red, 25–55 mm; petals yellow, fading orange or pale yellow, very broadly obcordate, 28–65 mm; capsules erect or slightly spreading, narrowly lanceoloid, 25–55 × 4–9 mm, valve tips free. Seasonally moist sites, desert washes, streams, seeps, roadsides in sandy or sandy-loam soil, often limestone, 6230–8200 ft. NW.

Oenothera nealleyi (J. M. Coult.) Krakos & W. L. Wagner [*Gaura nealleyi* J. M. Coult.]—Nealley's evening primrose. Annual, stout taproot; stems 20–70(–100) cm, well branched, sparsely villous proximally, glandular-puberulent in distal parts; basal leaves 3.5–9 × 0.5–1.5 cm, blades lyrate; cauline leaves 1.5–7 × 0.1–0.6 cm, blades narrowly lanceolate to linear, margins sinuate-dentate, undulate, glabrate to sparsely villous along veins and on margins; flowers zygomorphic, opening at sunset; petals white, fading pink to red, elliptic to elliptic-obovate, 10–15 mm; stigma exserted beyond anthers; capsules ellipsoid or ovoid, narrowly 4-winged, 4.5–8 × 2–5 mm, stipe 0.2–2.2 mm; seeds 3 or 4(5). Washes, sandy places, grasslands, piñon-juniper or ponderosa pine woodlands, 3280–7220 ft. C, WC, SE, SC.

Oenothera neomexicana (Small) Munz—New Mexico evening primrose. Perennial; stems 30–60 cm, glabrate proximally, strigillose and villous distally; leaves with basal rosette weakly developed or absent, petioles 0–2 cm, blades oblong to lanceolate or narrowly ovate, 3–9 × (0.6–)1–2.5 cm, margins irregularly sinuate-dentate; flowers opening near sunset, buds nodding, weakly quadrangular, with free tips; floral tube 30–50 mm; sepals 20–30 mm; petals white, fading pink, broadly obovate, 20–30 mm; stigma exserted beyond anthers. Capsules not woody, straight or slightly curved, subcylindrical, obtusely 4-angled, 20–30 × 2–3 mm, tapering gradually from base to apex. Coniferous forest openings, stream valleys, roadsides in rocky, sandy clay, or loamy soil, 4920–9850 ft. C, NC, SE, SC.

Oenothera organensis Munz ex S. Emers.—Organ Mountain evening primrose. Perennial, forming clumps 1–1.5 m diam., developing numerous adventitious shoots from taproot and lateral roots, root system appearing fibrous; stems 30–60 cm, weakly erect to ascending, often branched distally, moderately hirsute (hairs often with reddish-purple, pustulate bases), also strigillose and becoming glandular-puberulent distally; leaf petioles 0.5–1.5 cm, blades very narrowly oblanceolate to narrowly elliptic, basal (weakly developed) 9–23 × 1–2.5 cm, cauline 5–11 × 1.5–3.5 cm, margins undulate, remotely and bluntly dentate; flowers opening near sunset, buds with free tips; floral tube 100–165(–190) mm; sepals 25–50 mm; petals yellow, fading deep reddish orange, broadly obovate with truncate apex, or obcordate, 30–55 mm; stigma exserted beyond anthers; capsules erect to slightly spreading at acute angle, cylindrical, 25–35 × 4–5.5 mm, dehiscent at least 3/4 length; seeds in 2 distinct rows per locule. Rhyolite canyons, water courses, eroded basins filled with gravel and rocks, 5900–7550 ft. Doña Ana. Of conservation concern. **R & E**

Oenothera pallida Lindl.—Pale evening primrose. Perennial, sometimes annual; stems 10–50(–70) cm, erect or ascending, single to several from base, unbranched or many-branched, glabrous, strigillose and/or villous, sometimes more villous distally; leaves cauline, petiole 0–2(–4.5) cm, blades lanceolate, oblong, linear-lanceolate, or ovate, 1–5(–7.8) × 0.3–1(–1.5) cm, margins subentire or remotely denticulate, deeply sinuate-dentate, or pinnatifid; flowers opening near sunset, buds nodding, weakly quadrangular, with free tips; floral tube 15–40 mm; sepals 10–30 mm; petals white, fading pink to deep pink, broadly obovate or obcordate, (10–)15–25(–40) mm; stigma exserted beyond anthers. Capsules cylindrical, obtusely 4-angled, 15–60 × 1.5–2.5 mm, tapering slightly from base to apex.

1. Annual, sometimes perennial from a taproot, when perennial, sometimes with lateral roots producing adventitious shoots, strigillose throughout and villous distally, especially on floral parts...**subsp. trichocalyx**
1'. Perennial from a taproot and with lateral roots producing adventitious shoots, glabrous, strigillose, or sparsely villous...(2)
2. Plants glabrous, sometimes strigillose, rarely sparsely villous; leaf blade margins subentire or remotely denticulate, rarely pinnatifid; capsules contorted to curved...**subsp. pallida**
2'. Plants strigillose, rarely villous or glabrous; leaf blade margins shallowly sinuate-dentate or denticulate, or deeply sinuate-dentate to pinnatifid, rarely only dentate; capsules straight or curved, sometimes contorted...(3)
3. Leaf blades (0.4–)0.7–1.5 cm wide, margins shallowly sinuate-dentate or denticulate...**subsp. latifolia**
3'. Leaf blades 0.4–1(–1.5) cm wide, margins deeply sinuate-dentate to pinnatifid, rarely dentate only...**subsp. runcinata**

subsp. latifolia (Rydb.) Munz [*O. latifolia* (Rydb.) Munz]—Perennial; stems several, branched from base, densely strigillose throughout; leaf blades narrowly ovate to oblong-lanceolate or lanceolate, 1–5 (–7) × (0.4–)0.7–1.5 cm, margins shallowly sinuate-dentate or denticulate; floral tube 15–40 mm; sepals 12–30 mm; petals 15–40 mm. Open sites, sandy soil, dunes, rocky sites in grasslands, 6560–7550 ft. NE, NC, EC, SW.

subsp. pallida—Perennial; stems branched throughout, glabrous, sometimes strigillose, rarely sparsely villous; leaf blades lanceolate to linear-lanceolate or oblong, 2–6 × 0.3–0.8(–1) cm, margins subentire or remotely denticulate, rarely pinnatifid; floral tube 20–35 mm; sepals 12–18 mm; petals 12–25 mm. Dunes or disturbed areas in sandy, alkaline soil, 5250–6560 ft. Los Alamos, San Juan.

subsp. runcinata (Engelm.) Munz & W. M. Klein [*O. runcinata* (Engelm.) Munz]—Perennial; stems branched throughout, strigillose, sometimes also sparsely villous, or glabrous; leaf blades oblong to narrowly lanceolate, 2–3.5(–5) × 0.4–1(–1.5) cm; floral tube 15–30 mm; sepals 10–25 mm; petals 10–30 mm. Dunes, disturbed areas, shrublands, open woodlands, in sandy, alkaline soil, 3600–7550 ft. Widespread.

subsp. trichocalyx (Nutt.) Munz & W. M. Klein—Annual, sometimes perennial; stems single to sometimes several from base, unbranched, strigillose throughout and villous distally, especially on floral parts; basal rosette usually present, leaf blades narrowly lanceolate to oblong, 3–5(–7.8) × 0.4–0.8(–1.2) cm, margins pinnatifid or dentate; floral tube 20–30 mm; sepals 10–18 mm; petals 10–20 mm. Piñon-juniper woodlands or shrublands in sandy, silty, or rocky soil, 3600–8200 ft. W.

Oenothera podocarpa (Wooton & Standl.) Krakos & W. L. Wagner [*Gaura hexandra* Ortega]— Harlequin bush. Annual, stout taproot; stems 15–100 cm, ascending to erect, unbranched or well branched, villous proximally, glabrate, strigillose, and/or glandular-puberulent distally; basal leaves 3–15 × 0.5–1 cm, blades lyrate; cauline leaves 1–9 × 0.1–0.8 cm, blades linear to very narrowly elliptic or narrowly lanceolate, margins sinuate-dentate to subentire, densely villous, glabrate in age; flowers opening at sunset; floral tube 6–10 mm; petals white, fading pink to red, narrowly obovate, 5.5–9.5 mm, short-clawed; stigma surrounded by anthers; capsules ellipsoid or narrowly obovoid, narrowly 4-winged, 6–8 × 2–3 mm, narrowed at base; seeds 4. Disturbed sites, sandy washes, slopes, grasslands, meadows, woodlands, volcanic cinders, 4920–9185 ft. WC, SW, Doña Ana.

Oenothera primiveris A. Gray—Desert evening primrose. Winter annual, caulescent to short-caulescent, weakly fleshy taproot; stems (when present) 5–35 cm, unbranched and erect, densely leafy, long-hirsute, hairs often with reddish-purple pustulate bases, also moderately strigillose, and glandular-puberulent distally, often on leaves; leaves usually in a basal rosette, petioles (0.9–)3.5–8(–14) cm, blades oblanceolate to linear-oblanceolate, (1.4–)6–15(–28) × (0.2–)1–3.5(–5.6) cm, pinnatifid or bipinnatifid to shallowly pinnately lobed, margins sinuate-dentate or subentire, apex obtuse; flowers opening 1–2 hours before sunset; floral tube (20–)26–60(–72) mm; sepals (7–)12–25(–30) mm; petals yellow, fading reddish orange to purple, obcordate to obovate, (6–)13–35(–40) mm; stigma exserted beyond anthers or surrounded; capsules woody, lanceoloid to ovoid, 4-angled, 10–45(–60) × 4–8 mm, beaked, dehiscent 1/4–2/3 length; seeds numerous, in 2 rows per locule. Flats, low hills, margins of sand dunes, arroyos, roadsides, desert scrub, grasslands in sandy soil, 4100–5250 ft. S.

Oenothera pubescens Willd. ex Spreng. [*O. amplexicaulis* (Wooton & Standl.) Tidest.]—South American evening primrose. Annual or biennial; stems 5–50(–80) cm, unbranched or with branched central stem and ascending to decumbent lateral branches arising from rosette, densely to sparsely strigillose, sometimes also villous or glandular-puberulent distally; leaf blades narrowly oblanceolate to narrowly elliptic or narrowly oblong, basal 5–14 × 0.5–2.5 cm, cauline 2–8 × 0.5–2.5 cm, margins dentate to deeply lobed, bracts spreading, flat; flowers opening near sunset, buds with free tips; floral tube 15–50 mm; sepals 5–25 mm; petals yellow, fading reddish orange, broadly obovate to obcordate, 5–25(–35) mm; stigma surrounded by or slightly exserted beyond anthers; capsules cylindrical, 20–45 × 2–4 mm. Open sites in montane habitats, 5900–9320 ft. NC, C, WC, SC, Hidalgo. This species is a self-pollinating, permanent translocation heterozygote species (PTH) and has only ca. 50% pollen fertility.

Oenothera rhombipetala Nutt. ex Torr. & A. Gray—Fourpoint evening primrose. Biennial; stems 30–100(–150) cm, sometimes with lateral branches arising obliquely from rosette, densely to sparsely strigillose, sometimes also sparsely glandular-puberulent distally; leaf blades narrowly oblanceolate, gradually narrowly elliptic to narrowly lanceolate, oblanceolate, or ovate distally, basal 6–20 × 0.6–2 cm, cauline 3–15 × 0.8–2.5 cm, margins lobed to remotely dentate or subentire, bracts slightly longer than capsule; inflorescences dense, without lateral branches, mature buds not overtopping spike apex; flowers opening near sunset, buds erect, with free tips; floral tube 30–45 mm; sepals 15–30 mm; petals yellow, 15–35 mm; stigma exserted beyond anthers; capsules narrowly lanceoloid, 13–25 × 2.5–3 mm. Fields or prairies in sandy soil, 3050–4265 ft. Lea, Roosevelt.

Oenothera serrulata Nutt. [*Calylophus serrulatus* (Nutt.) P. H. Raven]—Yellow sundrops. Perennial, stout taproot; stems 10–60(–80) cm, 1 to many, weakly decumbent to erect, unbranched to moderately branched, glabrous or strigillose; fascicles of small leaves to 2 cm sometimes present in nonflowering axils, petioles 0–0.6 cm, leaf blades linear to narrowly lanceolate or oblanceolate, 1–9 × 0.1–1 cm, often folded lengthwise, margins subentire or spinulose-serrate, base attenuate, apex acute; flowers opening near sunrise, buds with free tips; floral tube 2–12(–16) mm; sepals with keeled midrib, 1.5–9 mm; petals yellow, fading dark yellow to orange; stigma discoid to quadrangular, surrounded by anthers; capsules hard, 6–25 × 1–3 mm, dehiscent 1/2 length. Prairies, grassy open areas in woods, rarely in mountains in sandy or rocky soil, 3445–7220 ft. E, NW, C, WC, SW. This species is a self-pollinating, permanent translocation heterozygote species (PTH) and has only ca. 50% pollen fertility.

Oenothera speciosa Nutt.—Pinkladies. Perennial, caulescent, spreading by rhizomes; stems 4–60 cm, many, erect, glabrate to strigillose, also sparsely hirsute; basal leaves 2–9 × 0.3–3.2 cm, blades oblanceolate to obovate; cauline leaves 1–10 × 0.3–3.5 cm, blades narrowly elliptic to ovate, margins subentire or serrulate to sinuate-pinnatifid; inflorescences sharply nodding, buds with free tips; floral tube 12–25 mm; sepals 15–50 mm; petals pink to rose, fading darker, or white, fading pink; stigma exserted beyond anthers; capsules narrowly obovoid to narrowly rhombic-ellipsoid, angled, 10–25 × 3.5–6 mm, apex attenuate to a sterile beak, valve midrib prominent, proximal stipe cylindrical, not tapering to base, (4–)8–15 mm. Grasslands, glades, open woodlands, disturbed places, pastures, railroads, roadsides, in loamy or sandy soil, sometimes clay, 4920–8860 ft. Widely scattered. SW, Bernalillo, Roosevelt, Taos. Introduced. Widely cultivated; New Mexico is outside its natural range. It is not known to be definitely naturalized anywhere because of self-incompatibility, but it tends to persist or become adventive thanks to its aggressive vegetative reproduction.

Oenothera suffrutescens (Ser.) W. L. Wagner & Hoch [*Gaura coccinea* Pursh]—Scarlet beeblossom. Perennial, deep, thick taproot, often with horizontal branching stems that give rise to new plants; stems 10–120 cm, erect or ascending, many-branched, densely strigillose, sometimes also long-villous proximally or glabrate; leaves in a basal rosette (when young) and cauline, blades linear to narrowly elliptic, 0.7–6.5 × 0.1–1.5 cm, margins entire or remotely and coarsely serrate; flowers opening near sunset; floral tube 4–11(–13) mm; sepals 5–9(–10) mm; petals white, fading salmon-pink to scarlet-red, slightly unequal, obovate to elliptic-obovate or elliptic, 3–7(–8) mm, abruptly clawed; stigma exserted beyond anthers; capsules erect, pyramidal in distal 1/2 and abruptly constricted to terete proximal part, pyramidal part angled, 4–9 × (1–)1.5–3 mm; seeds (1–)3 or 4. Desert shrublands to woodlands, grasslands, disturbed areas in sandy or clay soil, often calcareous, 3445–7550 ft. Widespread.

Oenothera toumeyi (Small) Tidestr. [*Calylophus toumeyi* (Small) Towner]—Toumey's sundrops. Perennial or sometimes annual, stout taproot; stems 15–70 cm, 1 to several, ascending to erect, unbranched to densely branched, glabrate to strigillose; leaf blades narrowly lanceolate, 1–3.5 × 0.1–0.7 cm, margins entire or obscurely and sparsely serrulate, base acute-attenuate, apex acute; flowers opening at sunset; floral tube (15–)30–60(–70) mm; sepals 10–25 mm; petals yellow, fading pale pink or pale purple, 10–20 mm; stigma yellow, quadrangular, exserted beyond anthers; capsules 10–50 × 1.5–4 mm. Shaded, rocky slopes, disturbed areas, pine-oak forests, 4920–8530 ft. Catron, Hidalgo.

Oenothera triloba Nutt. [*O. hamata* (Wooton & Standl.) Tidestr.]—Stemless evening primrose. Winter annual, sometimes biennial, acaulescent or very short-caulescent; stems 0–20 cm, 1 to several, ascending, densely leafy, moderately strigillose and glandular-puberulent, sometimes also sparsely hirsute; leaves in a basal rosette, sometimes also cauline, thin, petioles (0.5–)1–8 cm, blades oblanceolate to elliptic, (2.5–)6–25(–32) × (0.6–)1.5–4(–5) cm, margins irregularly pinnatifid, sometimes subentire, apex acute to obtuse or rounded; flowers opening near sunset, buds with subequal free tips; floral tube (20–)28–95(–138) mm; sepals (6–)10–30(–35) mm; petals pale yellow, fading pale orange, drying lavender, (10–)12–30(–38) mm; stigma surrounded by anthers; capsules woody in age, rhombic-obovoid, winged, wings broad, triangular, 5–10 mm wide, these often terminating in a hooked tooth, (10–)15–25(–28) × 4–8 mm (excluding wings), valve surface reticulate, dehiscent 1/8–1/3 length. Playas, floodplains, creek beds, slopes, flats, moist sites, disturbed sites, roadsides, old fields, deserts, prairies, glades in clay, sandy or rocky soil, 3445–6235 ft. E, C.

Oenothera tubicula A. Gray [*Calylophus tubiculus* (A. Gray) P. H. Raven]—Texas sundrops. Short-lived perennial, stout taproot; stems 4–53 cm, 1 to many, decumbent to erect, unbranched to densely branched, glandular-puberulent; fascicles of small leaves 0.2–1.5 cm sometimes present in nonflowering axils, petioles 0–0.2 cm, blades linear to ovate or obovate, 0.7–4.6 × 0.1–1.2 cm, margins entire, base attenuate, apex acute; flower buds with free tips; floral tube 5–25(–33) mm; sepals 3–13 mm; petals yellow, fading pale pink or pale purple, 5–20(–25) mm; stigma yellow, quadrangular, exserted beyond anthers; capsules hard, 8–20 × 1.5–2.5 mm, dehiscent throughout. New Mexico material belongs to **subsp. tubicula**. Flat, arid grasslands in primarily limestone soil, 3115–5905 ft. EC, C, SE.

Oenothera villosa Thunb.—Hairy evening primrose. Biennial; stems 50–200 cm, erect, flushed with red at least proximally, unbranched or with branches obliquely arising from rosette and secondary branches from main stem; leaf blades narrowly oblanceolate, oblanceolate to narrowly elliptic, or narrowly lanceolate, basal 10–30 × 1.2–4(–5) cm, cauline 5–20 × 1–2.5(–4) cm, margins denticulate to subentire, sometimes dentate, bracts persistent; inflorescences erect, unbranched, flowers opening near sunset, buds erect, with free tips; floral tube 23–44 mm; petals yellow, fading orange or pale yellow, very broadly obcordate, 7–20 mm; capsules erect or slightly spreading, lanceoloid, 20–43 × 4–7 mm, valve tips free. Our material belongs to **subsp. strigosa** (Rydb.) W. Dietr. & P. H. Raven. Open, often wet sites, streamsides, fields, roadsides, 5740–9185 ft. N, C, WC, Grant, Otero. This species is a self-pollinating, permanent translocation heterozygote species (PTH) and has only ca. 50% pollen fertility.

ORCHIDACEAE – ORCHID FAMILY

Ronald A. Coleman[†]

Perennial herbs, ours terrestrial, epiphytic or lithophytic elsewhere; strongly mycotrophic; mostly green and photosynthetic, but some genera lacking green color and mycoheterotrophic; leaves 1 to many or reduced to scales; flowers with bilateral symmetry, 1 to many, 3 sepals, 3 petals with 1 petal modified to a lip; flowers usually resupinate, pistil and stamen fused to form a single structure called a column; fruit an anterior ellipsoidal to slightly spherical capsule; seeds minute with no endosperm (Martin & Hutchins, 1980c).

1.   Flowers 1–2(3)…(2)
1′. Flowers > 3…(3)
2.   Lip a pink to white pouch with yellow hairs and markings at pouch opening…**Calypso**
2′. Lip a yellow pouch with or without reddish dots at pouch opening…**Cypripedium**
3.   Flowers with a spur, from small and saccate to long and slender…(4)
3′. Flowers without a spur…(6)
4.   Leaves faded or yellowing at flowering…**Piperia**
4′. Leaves still fresh and green at flowering…(5)
5.   Apex of lip acute…**Platanthera**
5′. Apex of lip 2- or minutely 3-lobed…**Dactylorhiza**
6.   Flowers tubular…(7)
6′. Flowers not tubular…(9)
7.   Flowers spiraled about inflorescence axis…**Spiranthes**
7′. Flowers not spiraled about inflorescence axis…(8)
8.   Lip with green stripes at apex…**Schiedeella**
8′. Lip without green stripe at apex…**Microthelys**
9.   Leaves not present at flowering…(10)
9′. Leaves present at flowering…(11)
10. More than 2 raised ridges along entire length of central lobe of lip…**Hexalectris**
10′. Two or fewer ridges along less than upper 1/2 of central lobe of lip…**Corallorhiza**
11. Leaves in basal rosette…**Goodyera**
11′. Leaves not in basal rosette…(12)
12. Leaves 1…**Malaxis**
12′. Leaves > 1…(13)
13. Leaves 2…**Neottia**
13′. Leaves > 2…**Epipactis**

CALYPSO Salisb. – Fairy slipper

Calypso bulbosa (L.) Oakes—Fairy slipper, calypso. Plants rather succulent; roots fleshy; stems scapose, arising from a slender to stout, fleshy corm, 7–21 cm; sheathing bracts usually 2; leaves solitary, basal, plicate, dark green, ovate, < 5 × 3 cm; flowers 1, terminal, nodding, 2 × 3.5 cm, sepals and petals mostly purplish to pink to whitish, lightly veined, lanceolate, 1.4–2 × 0.25–0.3 cm; lip pouch shaped, with top covered by white to purple lamina and purplish spots, area near orifice of pouch yellow, with 3 ridges bearing yellow hairs, bifurcate tip of pouch extending beyond margin of lamina. Our material belongs to **var. americana** (R. Br.) Luer. Shaded spots in spruce, pine, fir, aspen, and mixed conifer–aspen forests, 8000–10,800 ft. NC, NW, C, WC, Lincoln.

CORALLORHIZA Gagnebin – Coralroot

Stems scapelike, from stout rhizomes; with little or no chlorophyll; leaves reduced to scales; flowers in terminal racemes; perianth inconspicuous; sepals and lateral petals about equal, united to the base of the column, sometimes projecting into a short spur or sac; labellum lobed or unlobed, sometimes with 1–3 nerves but not strongly ridged or crested.

1. Flowers without mentum, lamellae fused...**C. striata**
1'. Flowers with minute to well-defined mentum, 2 free lamellae...(2)
2. Lip without lateral lobes...**C. wisteriana**
2'. Lip with 2 small lateral lobes...(3)
3. Sepals 1-veined...**C. trifida**
3'. Sepals 3-veined...**C. maculata**

Corallorhiza maculata (Raf.) Raf.—Spotted coralroot. Stems leafless, 10–50 cm, with little or no chlorophyll, with coralloid rhizomes; flowers mostly 15–30 × 1.7–1.8 cm, sepals and petals spreading, often spotted, brown, tan, pinkish, reddish, yellow, or purple; sepals 3-veined, to 1.4 × 0.4 cm; petals 1.5 × 0.7 cm; lip white, usually but not always with reddish to purplish spots, 3-lobed with small lateral lobes pointing forward, central lobe either with parallel sides or broadly spreading.

1. Central lobe broadly spreading...**var. occidentalis**
1'. Central lobe with parallel sides...**var. maculata**

var. maculata—Bright light to medium shade in dry coniferous forests, 7200–10,700 ft. NC, NW, WC, SC.

var. occidentalis (Lindl.) Ames—Dry coniferous forests, 6900–10,000 ft. N, C, WC, SC.

Corallorhiza striata Lindl.—Striped coralroot. Stems leafless, glabrous, 10–40 cm, with little or no chlorophyll, with coralloid rhizomes; flowers 0.9–1.2 × 1–1.5 cm, reddish purple, tan, or yellow, usually with stripes on sepals, petals, and lip, lacking mentum; sepals lanceolate, 1.1 × 0.35 cm; petals elliptic-oblanceolate, 1 × 0.4 cm; 2 lamellae fused, often appearing as single bilobed callus. Bright to moderate shade in oak–Douglas-fir–pine forests, 6900–9500 ft. NC, C, WC, SC.

Corallorhiza trifida Châtel.—Early coralroot. Stems leafless, glabrous, 10–25 cm, greenish or yellow, with coralloid rhizomes; flowers 1 cm across, green (also brownish elsewhere) with white lip; sepals green, 1-veined, 3–4 × 1 mm; petals green, oblanceolate, 3 × 1 mm; lip white, usually without dots in our area, often spotted elsewhere, 5 × 3 mm, with margins nearly parallel to slightly expanded on lower 1/2, rounded at apex; small lateral lobes; 2 raised calli in upper 1/3 of lip, margin of lip ruffled. Damp to mesic coniferous forest, 9000–10,500 ft. NC, Otero.

Corallorhiza wisteriana Conrad—Wister's coralroot. Stems leafless, glabrous, 6.5–38 cm, dark brown to tan, purplish, yellow, or greenish, with coralloid rhizomes; flowers 1.8 × 1.6 cm; sepals and petals connivent to spreading, 3-veined; sepals tan, purplish, yellow, lighter with some green toward base, acute to lanceolate and slightly falcate, 1 × 0.2–2.5 cm, with small mentum; petals tan, purplish, yellow, lighter with some green toward base, oblanceolate, 1 × 0.2 cm; lip elliptic to ovate, 0.4–1 × 0.7–0.9 cm, white spotted with fine reddish to purplish dots to pure white, entire, with fine fringes along margins, narrowing down to small claw at column; 2 basal lamellae. Juniper-oak to oak, pine, fir, usually in deep duff, 5500–9800 ft. NC, NW, C, WC, SC.

CYPRIPEDIUM L. – Lady's slipper

Cypripedium parviflorum Salisb.—Yellow lady's slipper. Plants with chlorophyll; stems 16–60 cm, stems, bracts, and leaves pubescent; leaves cauline, 4–6, alternate or subopposite, ovate-lanceolate, 4–9 × 5–14 cm; inflorescence with 1 flower, rarely 2; sepals yellowish green with brownish to reddish stripes that turn to dots near pouch, 4 × 2.2 cm; petals similar in color to sepals, 5.5 × 0.7 cm; lip bright yellow, pouch shaped. Our material belongs to **var. pubescens** (Willd.) O. W. Knight. Moderate shade to nearly full sun in fir, pine, and aspen forests, 6000–9500 ft. NC, NW, WC, Otero.

DACTYLORHIZA Neck. ex Nevski – Long-bracted orchid

Dactylorhiza viridis (L.) R. M. Bateman, Pridgeon & M. W. Chase [*Coeloglossum viride* (L.) Hartm.]—Long-bracted frog orchid. Plants with chlorophyll; stems 15–40 cm, with 3–6 leaves alternating along stem; leaves 12 × 5 cm, lower ones elliptic, becoming elliptic-lanceolate; inflorescence a terminal, spike-like raceme with large bracts with up to 40 flowers; flowers green with yellow and rose highlights on lip, 1–1.8 cm × 0.4–1 cm; sepals dark green, 7–8 × 3–4 mm; petals pale green to whitish green, lanceolate, 4.5 × 1 mm; lip pale green to yellowish green with some rose to reddish shading, 1 × 0.4 cm; spur whitish, saclike, 2.2 × 2 mm. Shade in aspen and fir forest, 9000–10,000 ft. Colfax, San Miguel, Sierra.

EPIPACTIS Zinn – Helleborine

Plants with chlorophyll; leaves cauline, alternate; flowers in a terminal raceme with foliaceous bracts; petals greenish yellow, the lip saccate at the base.

1.   Lip deeply 3-lobed, the epichile elongate; wet places…**E. gigantea**
1′.   Lip not lobed, the epichile blunt, wider than long; mesic to dry places…**E. helleborine**

Epipactis gigantea Hook.—Giant helleborine. Stems glabrous, 20–100 cm; leaves green (to purplish elsewhere), 5–11, narrowly ovate-lanceolate to broadly ovate-lanceolate, 1.5–7 × 11–25 cm, becoming bractlike near flowers; flowers sessile, 3 × 4 cm; lateral sepals ovate-lanceolate, dark green, slightly suffused with rose and faint darker veins, to 2.4 × 0.9 cm; petals ovate, oblique, concave, rose to intense pink, 1.7 × 0.7 cm; lip 2.3 × 2.1 cm, deeply 3-lobed, lateral lobes with reddish raised veinlike ridges on a yellowish to greenish background. Streamsides, seeps, 4300–7500 ft. NC, NW, C, WC, SE, SC.

Epipactis helleborine (L.) Crantz—Broadleaf helleborine. Stems 15–50(100) cm, few to > 50 flowers; leaves green here but white to pink elsewhere, 4–7, elliptic, 10 × 5 cm; flowers greenish white, with rose to brown center on lip, mostly 1.5 × 1.2 cm; lateral sepals green, nearly ovate, 1 × 0.6 cm; petals greenish white, ovate-elliptic, slightly cupped, 0.9 × 0.7 cm; lip 1 × 0.7 cm with cup portion rose to dark brown, covered with nectar. Meadows, mixed forests of willows and cottonwoods. NC, Bernalillo.

GOODYERA R. Br. – Rattlesnake plantain

Plants with chlorophyll; leaves basal; flowers sessile in a terminal spike; petals white to greenish, the dorsal ones connivent and forming a hood covering the lip.

1.   Plants usually < 12 cm; flowers pure white, lateral sepals not recurved at apex…**G. repens**
1′.   Plants usually > 20 cm; flowers with greenish or brownish sepals, lateral sepals recurved at apex…**G. oblongifolia**

Goodyera oblongifolia Raf.—Western rattlesnake plantain. Stems evergreen, 1-sided inflorescence 8–50 cm, pubescent, with few to > 30 small white flowers; leaves 5–7, evergreen in basal rosette, oblong-elliptic, 3.5 × 8 cm, usually with white along central vein; flowers 1 × 0.5 cm, greenish to white, pubescent on ovary and back side of sepals and petals; dorsal sepals and petals connivent, forming a hood over the column and lip; lateral sepals white with trace of green to greenish brown, ovate, 8 × 4 mm; petals white, 9 × 4 mm; lip white. Humus of coniferous forests, in light to deep shade, 5745–10,000 ft. NC, NW, C, WC, Otero.

Goodyera repens (L.) R. Br.—Lesser rattlesnake plantain. Stems mostly 9–12 cm with few to > 20 flowers; leaves evergreen, 4–8 per rosette, ovate to oblong-elliptic, 1.2 × 2.3 cm, solid dark green or very faintly marked with darker veining; flowers white, ca. 3 × 3 mm; lateral sepals white, back covered with fine white hairs, ovate, 4 × 3 mm; petals white, back covered with fine hairs, spatulate, 2–4.1 mm; lip white, deeply saccate with narrow acute apex, 2 × 3 mm. Moderate to heavy shade in mixed fir, spruce, and aspen forest, 8000–10,000 ft. NC, WC, Otero.

HEXALECTRIS Raf. – Crested coralroot

Stems scapelike; leaves reduced to scales; flowers in terminal racemes; sepals and petals similar, free, somewhat spreading, several-nerved; labellum obovate, 3-lobed, with 5 or 6 longitudinal ridges, the middle lobe concave.

1. Lip deeply 3-lobed, fissure > 3 mm deep…**H. colemanii**
1'. Lip shallowly 3-lobed, fissure < 2 mm deep…(2)
2. Midlobe of lip spatulate and entire; lamellae nonbranching…**H. nitida**
2'. Midlobe of lip spreading at least slightly, apex crenulate-crenate; lamellae branching…**H. arizonica**

Hexalectris arizonica (S. Watson) A. H. Kenn. & L. E. Watson—Arizona crested coralroot. Stems leafless, spicate, 2.7–6.5 dm, nearly 1 cm thick at base, 9–20 flowers, dark pink to tan to dark brown; leaves none; flowers 2–2.5 × 1.5–2.5 cm, tan to dark brown, with purple ridges and stripes on lip; lateral sepals tan to brown with faint veining, oblong-elliptic, slightly falcate, 1.7 × 0.7 cm; petals tan to brown with faint veining, oblanceolate, slightly falcate, 2 × 0.6 cm; lip white with purple dots, stripes, and ridges, 3-lobed, 1.6 × 1.4 cm. Partial shade in oak woodlands, mixed oak and conifer forest. Hidalgo, Otero, Sierra. **R**

Hexalectris colemanii (Catling) A. H. Kenn. & L. E. Watson—Coleman's crested coralroot. Stems leafless, spicate, 40–50 cm, with 10–20 flowers, pale pink to rose to tan; leaves none; lateral sepals pale rose-tan with light veining, revolute, with outer 1/3 rolled back to form complete coil, elliptic-lanceolate, oblique, 2 × 0.8 cm; petals pale rose-tan with light veining, elliptic to obovate, slightly falcate, 0.6 × 1.8 cm, revolute, with outer 1/3 rolled back in a full coil; lip 3-lobed, 1.5 × 1.2 cm, white to pale rose-tan, with purple veining on lateral lobes, and purple raised ridges on central lobe. Oak woodlands and canyons, 4500–5200 ft. Hidalgo. **R**

Hexalectris nitida L. O. Williams—Glass Mountain crested coralroot. Stems leafless, spicate, 13–20 cm, tan to dark brown, with 12–20 waxy, shiny flowers; leaves none; flowers 1.2 × 1.2 cm, rose-brown to tan with white and purple on lip, sepals and petals spreading, turned back at apices; lateral sepals rose-brown with faint parallel veining; oblanceolate with rounded apex, slightly falcate, 1.2 × 0.3 cm; petals rose-brown with faint parallel veining, narrowly elliptic, 1.1 × 0.3 cm; lip white with purple markings, elliptic, 3-lobed, 0.7–0.9 × 0.4–0.5 cm. Moderate to heavy shade in oak and juniper woodlands, 4000–5000 ft. Eddy. **R**

MALAXIS Sol. ex Sw. – Adder's-mouth

Stems scapose; leaves 1 or 2, basal, bracts scalelike; flowers inconspicuous, arranged mostly in terminal racemes or spikes; perianth greenish to purple; sepals oblong-lanceolate to ovate; petals linear to lanceolate; lip ovate or lanceolate, cordate, unlobed to 3-lobed; fruit a capsule.

1. Flowers purple…**M. porphyrea**
1'. Flowers green…(2)
2. Flowers closely appressed to inflorescence axis…**M. soulei**
2'. Flowers not closely appressed to inflorescence axis…**M. abieticola**

Malaxis abieticola Salazar & Soto Arenas—Slender-flowered malaxis. Stems to 18 cm, single leaf midway on stem, up to 40 tiny, mostly light green flowers; leaf ovate, mostly 9 × 6.5 cm; flowers green, narrow, 1 × 0.2 cm; lateral sepals green, narrowly lanceolate, 5 × 0.6 mm; petals green, translucent, filiform, 4 × 0.1 mm; lip green with dark green stripes, arrowhead-shaped. Damp mossy and grassy places in spruce-fir forests, 8000–9200 ft. Catron, Otero. **R**

Malaxis porphyrea (Ridl.) Kuntze—Adder's-mouth. Stems to 34 cm, up to 125 purple flowers, single leaf midway on stem; leaf ovate, 7 × 4.5 cm; flowers purple, 5 × 1.5 mm; lateral sepals purple, lanceolate-elliptic, 3 × 0.6 mm; petals purple, filiform, 3 × 0.1 mm; lip purple with cream to yellowish center, arrowhead-shaped, 3 × 2 mm. Mixed oak, fir, and pine forests, 7000–10,200 ft. NC, C, WC, SE, SC, SW.

Malaxis soulei L. O. Williams—Mountain malaxis. Stems to 35 cm, with single leaf; inflorescence spicate with tiny flowers tightly appressed against inflorescence axis; leaf ovate to elliptic to linear-elliptic, to 12 × 4.5 cm; flowers green to yellowish green, 4.5 × 2.5 mm; lateral sepals green to greenish yellow, oblong-ovate, mostly 1 × 2 mm; petals green, filiform, 0.2 × 2 mm, usually curved around behind the dorsal sepal; lip green to yellowish, 1 × 2 mm, 3-lobed, notched at top and bottom. Oak, juniper, pine, and fir forests, 5300–9200 ft. NC, C, WC, SC.

MICROTHELYS Garay – Medusa orchid

Microthelys rubrocallosa (B. L. Rob. & Greenm.) Garay—Green medusa orchid. Stems erect with slender sheaths, to 32 cm, up to 30 flowers; leaves 1 or 2, dark bluish green, narrowly lanceolate, to 10 × 1.5 cm, senescing before or at anthesis; inflorescence a few- to many-flowered spike; flowers tubular, 5 × 2 mm; sepals and petals greenish with white edges forming a tight hood around the lip and column; lip broadly dilated in the middle, narrowing to a tiny claw where it joins the column. Ponderosa pine forests, 8000–9000 ft. Otero. **R**

NEOTTIA Guett. – Twayblade

Perennial herbs with chlorophyll; leaves 2, opposite or subopposite near the middle of the stem; flowers in a terminal raceme; petals green to yellowish green, the lip notched or deeply lobed at the apex; fruit a capsule.

1. Lip bilobed, deeply forked...**N. cordata**
1'. Lip minutely trilobed, not forked...**N. borealis**

Neottia borealis Morong—Northern twayblade. Stems 4–20 cm; leaves 2, opposite, midway on stem, elliptic to ovate-elliptic; flowers few to 20, ca. 15 mm, yellowish to dark green; sepals and petals pale green to dark green, lanceolate to elliptic-lanceolate, reflexed backward; lip < 10 cm, pale green to bluish green with dark green center. Dense shade in cold, wet areas such as stream bottoms with mosses and grasses, 9500–10,600 ft. Taos.

Neottia cordata (L.) R. Br.—Heart-leaved twayblade. Stems mostly to 10 cm in our area, up to 30 flowers; leaves 2, cordate, opposite, midway on stem, mostly 2 × 2 cm; flowers green, perianth star-shaped, with protruding forked lip, 8 × 5 mm; lateral sepals green, lanceolate, 2 × 1 mm; petals green, broadly elliptic, 2 × 1.5 mm; lip green to reddish, linear over first 1/2, deeply forked on lower 1/2, 4 × 2 mm, 2 hornlike appendages near column. Our material belongs to **var. nephrophylla** (Rydb.) Hultén. Damp places in pine, fir, and aspen forests, 9000–10,300 ft. NC.

PIPERIA Rydb. – Piperia

Piperia unalascensis (Spreng.) Rydb.—Alaska piperia. Plants with chlorophyll; stems mostly 15 × 40 cm; 2–6 basal leaves withering at anthesis; leaves oblanceolate to elliptic, 8–10 × 2–3 cm; floral bracts minute, ovate; inflorescence a spikelike raceme with up to 100 tiny flowers; flowers green, with musty aroma, 0.5 × 0.6 cm; sepals green, ovate-lanceolate to ovate-elliptic, 3 × 1.5 mm; petals green, ovate-lanceolate, 3 × 1 mm; lip green, ovate-lanceolate to triangular-ovate, curling upward at tip, 4 × 2.5 mm. Mixed conifer forest, 8000–9000 ft. McKinley.

PLATANTHERA Rich. – Bog orchid

Perennial herbs with chlorophyll; leaves sometimes basal but usually cauline and alternate; flowers in a terminal spike; petals white to green or yellowish green, the lip petal entire or lobed with a saccate or cylindrical spur at the base.

1. Leaves reduced to bracts along the stem...**P. brevifolia**
1'. Leaves well formed...(2)
2. Leaves 1...**P. obtusata**
2'. Leaves > 1...(3)
3. Column 1/2+ as large as hood formed by dorsal sepal and petals...**P. sparsiflora**
3'. Column < 1/2 as large as hood formed by dorsal sepals and petals...(4)

4. Flowers self-pollinating, with pollinia rotated forward and spilling onto stigma…(5)
4'. Flowers not self-pollinating, pollinia within anther sacs unless scattered by pollinator…(6)
5. Lip rhombic to rhombic-lanceolate, yellowish…**P. aquilonis**
5'. Lip dilated at base, acuminate, whitish…**P. huronensis** (in part)
6. Flowers white…**P. dilatata**
6'. Flowers greenish white to yellowish or bluish green…(7)
7. Flowers distinctly whitish green…**P. huronensis** (in part)
7'. Flowers green to yellow-green or bluish green…(8)
8. Spur at least 1.5 times as long as lip…**P. limosa**
8'. Spur much < 1.5 times as long as lip…**P. purpurascens**

Platanthera aquilonis Sheviak—Northern bog orchid. Stems 30–40 cm; leaves 4+, linear-lanceolate, to 10 × 2 cm; flowers yellowish green, self-pollinating; lateral sepals green, broadly lanceolate, folded down along spur, 4.2 × 2 mm; petals yellow-green, narrowly lanceolate, with basal 1/2 broadly triangular, tapering sharply to narrow apex; lip not conspicuously dilated, often held nearly perpendicular to column, 4.5 × 1.5 mm; column about as high as wide, anther sacs diverging toward bottom, pollinia spilling onto stigma on unopened flowers; spur clavate to cylindrical, 3.5 × 1 mm. Moist spruce-fir forests, meadows, streams, 8000–11,200 ft. NC, Lincoln, Otero.

Platanthera brevifolia (Greene) Kraenzl. [*Habenaria brevifolia* Greene]—Blunt-leaved bog orchid. Stems 20–40 cm; leaves reduced to mere bracts, sheathing and scattered along stem, ca. 3.5 × 1 cm; flowers green, dorsal sepals and petals forming hood over column; lateral sepals elliptic-lanceolate, slightly oblique, to 0.3 × 1 cm; petals falcate, 6 × 3 mm; lip lanceolate to slightly elliptic-lanceolate, 0.3 × 1 mm, apex upturned; spur cylindrical, acute, slightly curved, as long as 1.5 times the lip, to 1.5 × 1.2 mm; column 4 × 4 mm. Pine-oak woodlands, coniferous forests, 6500–9000 ft. WC, SC, Mora.

Platanthera dilatata (Pursh) Lindl. ex L. C. Beck [*Habenaria dilatata* (Pursh) Hook.]—Scentbottle, bog candle. Stems (ours) ca. 20–50 cm; flowers loosely to densely packed at top of stem; leaves green, lanceolate, to 35 cm, becoming bractlike near the flowers; flowers white, to 2 cm from the top of the dorsal sepal to the tip of the lip; lateral sepals lanceolate, held out to the side; petals linear-lanceolate, falcate; lip rhombic-lanceolate, widely dilated at the base, to ~0.3 cm wide. Our material belongs to **var. leucostachys** (Lindl.) Luer. Along streams, seeps, wet meadows, mostly 8000–12,500 ft. Rio Arriba, Taos.

Platanthera huronensis (Nutt.) Lindl. [*P. hyperborea* (L.) Lindl. var. *huronensis* (Nutt.) Luer]—Tall green bog orchid. Stems (ours) mostly 20–70 cm; racemes densely flowered with 50–100 flowers; leaves 5 or 6, lanceolate or linear-lanceolate to oblanceolate, to 5–30 cm; flowers whitish green, mostly 1–1.5 × 0.7–1.5 cm; lateral sepals greenish white, spreading, falcate, 5 × 1.8 mm; petals greenish white with 3 faint green veins along length, falcate, 5 × 1.5 mm; lip greenish white, 8 × 2 mm wide; spur about as long as lip, cylindrical, with rounded tip, 6 × 0.9 mm thick; column 5 × 3 mm. Moist forests, meadows, marshes, streams, creeks, 8000–10,200 ft. NC, Catron, Lincoln.

Platanthera limosa Lindl. [*Habenaria limosa* (Lindl.) Hemsl.]—Thurber's bog orchid. Stems 20–100 cm; densely flowered with up to nearly 200 flowers; leaves 5–9, linear-lanceolate, lower > 20–40 × 1–4 cm, uppermost often bractlike; flowers green to slightly yellowish green, 0.9 × 1.1 cm, with musty odor; lateral sepals green, elliptic-lanceolate, 5 × 2 mm; petals yellow-green, ovate-lanceolate, slightly falcate, forming a hood over column with dorsal sepal; lip yellow-green, elliptic-lanceolate, 5.5 × 1.5 mm, with small jagged callus; spur yellow-green, cylindrical, slender, 2–3 times as long as lip, 1.4 × 1 mm; column 2 × 1.5 mm. Open to lightly forested springy marshes, seeps, stream banks, 6300–9150 ft. WC.

Platanthera obtusata (Banks ex Pursh) Lindl. [*Habenaria obtusata* (Banks ex Pursh) Richardson]—Bluntleaved bog orchid. Stems 10–15 cm, laxly flowered; leaf single, basal, linear-oblanceolate, 6–8 cm; flowers few, whitish green to slightly yellowish green; dorsal sepal and petals forming hood over column; petals lanceolate, with central green vein; lip usually < 8 mm, descending, green to yellowish, narrowly lanceolate; spur slender, narrowly conic, as long as lip. Spruce-fir forests and along creeks, 9000–11,500 ft. Colfax, Mora, Taos.

Platanthera purpurascens (Rydb.) Sheviak & W. F. Jenn.—Purple-petal bog orchid. Stems mostly 30–60(–90) cm, laxly to densely flowered with up to 100 flowers; leaves 3–6, lanceolate to slightly elliptic-lanceolate, 10–15 × 1–3 cm; flowers deep green, often with purple or reddish highlights; lateral sepals elliptic-lanceolate, 5 × 3 mm; petals lanceolate, slightly falcate, 3.5 × 2 mm; lip linear-lanceolate, strongly dilated at the base, 6 × 3 mm; spur clavate, shorter than the lip, 3 mm; column small, about as broad as high. Moist meadows and spruce forests, along lakes, streams, marshes, 7000–10,240 ft. NC, NW, WC, Lincoln.

Platanthera sparsiflora (S. Watson) Schltr. [*Habenaria sparsiflora* S. Watson]—Stream bog orchid. Stems 8–60(–100) cm, with 30–100 flowers; leaves 3–5, lanceolate to elliptic-lanceolate, to 11 × 2 cm; flowers green to yellowish green, 1.7 × 0.3 cm; petals and dorsal sepal forming hood over column; dorsal sepal dark, dull green, ovate, 5 × 3 mm; petals pale green, ovate-lanceolate, falcate, 6 × 2 mm; lip pale green to yellowish green, linear to lanceolate, 7 × 1.5 mm; column green, occupying 1/2+ of hood formed by petals and dorsal sepal; pollinia granular, yellow; spur pale green, cylindrical, curved, 1 × 1 mm. Canyons, floodplains, meadows, seeps, floodplains, 5200–8500 ft. NC, NW, C, WC, SC.

SCHIEDEELLA Schltr. – Parasitic lady's tresses

Schiedeella arizonica P. M. Br.—Fallen lady's tresses. Stems leafless in bloom, green to tan or light rose, mostly 10–20(–33) cm, with 3–14 flowers; leaves 2 or 3 in basal rosette appearing after flowering, ovate-elliptic, 1.5 × 6 cm; flowers whitish, rose or tan, 6 × 4 mm; petals and dorsal sepal forming hood over lip; lateral sepals whitish, rose, or tan, 6 × 1.5 mm; petals whitish, rose, or tan, 4 × 1.5 mm; lip whitish, rose, or tan, 5–6 × 4 mm, nearly quadrate, constricted above the middle and near the apex, narrowing to a slight claw near the column, apex crenulate, thickened in the center, with red spot, 3 green stripes running length of lip, central portion finely pubescent. Dry coniferous forest, hillsides, creek canyons, 7120–8900 ft. WC, SC, SW.

SPIRANTHES Rich. – Lady's tresses

Perennial herbs; leaves basal or cauline and alternate; flowers in a dense terminal spike, usually in spirally twisted rows; sepals connivent with the 2 lateral petals and forming a hood enclosing the column and most of the lip (ours); petals white to yellowish.

1. Lip pandurate, the apex dilated; sepals and petals united throughout their length and forming a hood… **S. romanzoffiana**
1'. Lip ovate to oblong, the apex only slightly or not at all dilated; sepals and petals with apices free and spreading…**S. magnicamporum**

Spiranthes magnicamporum Sheviak—Great Plains lady's tresses. Stems 20–50 cm, with 20–50+ flowers in dense multiranked spirals; leaves usually 2–5, basal or nearly so, linear-lanceolate, 19 × 1 cm, may or may not be present at anthesis; flowers white, tubular, 1 × 0.8 cm, strongly scented; petals connivent with dorsal sepal to form hood over column; lateral sepals white, lanceolate, 2 × 0.2 cm, finely pubescent on back side; petals white, linear-lanceolate to slightly elliptic-lanceolate, 1 × 0.3 cm; lip white with cream to pale yellow center, ovate to oblong, with 2 small tubercles at claw with column. Crusty alkaline soils, saturated sandbars in river bottoms, often higher up above the bank in thick, clay loam, 5000–6000 ft. Bernalillo, Guadalupe, Rio Arriba, Santa Fe.

Spiranthes romanzoffiana Cham.—Hooded lady's tresses. Stems mostly 10 × 40(–60) cm; leaves 3–6, lanceolate, 18 × 1.1 cm, mostly basal but scattered on lower stem on largest plants; flowers up to 60 in 3 dense spirals, white, tubular, with a sweet aroma; sepals and petals forming a tight hood over the downward-curving spreading lip; lateral sepals white with green suffusion near base, lanceolate, 1.2 × 0.4 cm; petals white, linear, with 3 faint green stripes, 1 × 0.3 cm; lip white with 5 pale green stripes in center, pandurate (fiddle-shaped), with 2 minor tubercles at base. Full sun in damp to almost wet soil, meadows, streams, seeps, 7400–11,000 ft. NC, Cibola.

OROBANCHACEAE – BROOMRAPE FAMILY

Kenneth D. Heil

Annual or perennial herbs, sometimes fleshy, without chlorophyll, parasitic; stems ascending or erect; leaves alternate, opposite, or basal, stipules absent, petioles present or absent; inflorescences terminal and/or axillary; flowers bisexual, zygomorphic, strongly bilabiate, bracteate; calyx 2–5-lobed, the lobes subequal; petals 5-lobed, tubular; stamens 4, didynamous, subequal or equal; pistil 1; ovary superior, 2-carpellate, unilocular; fruit a capsule (Ackerfield, 2015; Freeman et al., 2019a).

1. Plants without chlorophyll, holoparasitic, purplish to brown or yellow…(2)
1'. Plants with chlorophyll, hemiparasitic, green…(3)
2. Calyx spathelike, deeply cleft below, upper lip large and 4-toothed; plants yellow to brown; stems glabrous; flowers in a dense, bracteate spike; parasitic on *Quercus*, *Pinus*, and *Juniperus*…**Conopholis**
2'. Calyx regular, the lobes equal or nearly so; plants with some purplish tint; stems viscid glandular-hairy; flowers solitary on long pedicels or in a dense spikelike inflorescence; parasitic on herbaceous plants (often Asteraceae species)…**Aphyllon**
3. Corollas bilabiate, adaxial lips not galeate, cucullate, or beaked…(4)
3'. Corollas strongly bilabiate or bilabiate, adaxial lips galeate, cucullate, or beaked…(5)
4. Leaves whorled…**Brachystigma**
4'. Leaves alternate…**Agalinis**
5. Perennials, caudices woody or fleshy…(6)
5'. Annuals, rarely biennials, caudices absent…(7)
6. Pollen sacs equal; adaxial lip of corolla sometimes with an upward or coiled beak…**Pedicularis**
6'. Pollen sacs unequal; adaxial lip of corolla straight, rarely hooked…**Castilleja** (in part)
7. Calyces ovate to suborbiculate, flattened laterally…**Rhinanthus**
7'. Calyces tubular to campanulate, not flattened laterally…(8)
8. Adaxial lip of corolla ± straight, opening directed forward, rarely beaked, bent or hooked at tip and opening directed downward…**Castilleja** (in part)
8'. Adaxial lip of corolla rounded at the apex, sometimes obscurely so, opening directed downward…(9)
9. Sepals 4, calyces tubular…**Orthocarpus**
9'. Sepals 2, calyces spathelike…**Cordylanthus**

AGALINIS Raf. – False foxglove

Annual herbs (ours); hemiparasitic; stems erect or rarely leaning; leaves opposite or nearly so; inflorescence a raceme (ours); pedicels present; sepals 5; petals 5, weakly bilabiate; stamens 4, didynamous (Canne-Hilliker & Hays, 2019).

1. Leaf blades fleshy, adaxial surfaces with sessile, dome-shaped hairs; pedicels 5–9 mm…**A. calycina**
1'. Leaf blades not fleshy, adaxial surfaces scabridulous to scabrous; pedicels 6–25 mm [reported for New Mexico but no specimens seen]…A. tenuifolia (Vahl) Raf.

Agalinis calycina Pennell [*Gerardia calycina* (Pennell) Pennell]—Leoncita false foxglove. Stems branched, 40–62 cm, branches ascending; leaves spreading, blades narrowly linear, 2–4 cm × 1.5 mm, fleshy, margins entire, adaxial surface with siliceous, sessile, dome-shaped hairs; inflorescence a raceme or panicle, flowers 1 or 2 per node; calyx 4.5–6 mm, glabrous; corolla pink with 2 yellow lines and dark red spots in abaxial throat, 17–23 mm, glabrous externally; capsules obovoid to oblong, 10–12 mm. Perennially moist, alkaline-saline-calcareous soils in wet meadows, Bitter Lake National Wildlife Refuge, 3490 ft. Chaves. **R**

APHYLLON Mitch. – Broomrape, cancer root

L. Turner Collins

Annual parasitic herbs lacking chlorophyll, succulent, viscid-glandular-pubescent aboveground; stems erect, fleshy, mostly underground; leaves reduced to scales, appressed or reflexed; inflorescence a terminal spike, raceme, corymb, or paniculate; flowers zygomorphic, 5-merous, perfect, 2-lipped; calyx ± equally 5-cleft, corolla tubular, constricted above ovary, arched, palatal folds in throat, stamens 4, epipetalous; ovary 1-loculed, 2–4 partial placentae; fruit a 2-valved capsule. Root parasites mostly on Aster-aceae (Collins et al., 2019; Schneider, 2016).

1. Pedicels 10–120 mm, much longer than the calyx, often longer than the stem; inflorescence solitary or in short racemes; bracteoles absent; flowers 1 to several; sect. **Aphyllon**…(2)

1'. Pedicels < 10 mm (sometimes longer proximally), shorter than the calyx, always shorter than the stem; inflorescence racemose, corymbose, or paniculate; bracteoles 2; flowers numerous; sect. **Nothaphyllon** …(3)

2. Stems slender, bearing 1–4 flowers; pedicels many times longer than the stem; calyx lobes longer than the calyx tube…**A. purpureum**

2'. Stems stout, bearing 3–15 flowers; distal pedicels shorter than the stem; calyx lobes as long as or shorter than the calyx tube…**A. fasciculatum**

3. Inflorescence paniculate, compactly corymbose, or rarely somewhat racemose; palatal folds glabrous… (4)

3'. Inflorescence an elongate raceme or spike, sometimes branched or clustered; palatal folds pubescent… (5)

4. Inflorescence paniculate; flowers widely spaced, corollas < 20 mm, pale lavender or yellow, inflated distally, palatal folds poorly developed, surface smooth; stem base usually swollen; host *Holodiscus*…**A. pinorum**

4'. Inflorescence a corymb or rarely racemose; flowers crowded, corollas 18–30 mm, lavender, rose, or white, tube gibbous adaxially, palatal folds prominent, surface with blisterlike swellings; stem base usually not swollen; host *Artemisia*…**A. corymbosum**

5. Corolla lobes pointed, triangular, acute or obtuse, adaxial lip 3–6 mm, mostly dark purple…(6)

5'. Corolla lobes rounded, adaxial lip 4–9 mm, purple, lavender, rose-purple, or yellowish…(8)

6. Corolla lobes reflexed or revolute, 4–6 mm, usually with apiculate tooth; stalked glands usually present on anthers; Chihuahuan Desert…**A. cooperi**

6'. Corolla lobes erect, not revolute, apical tooth absent; stalked glands absent from anthers; widespread… (7)

7. Corolla glandular-puberulent, lips dark purple, often glabrate, tube white; host *Gutierrezia*; Colorado Plateau…**A. arizonicum**

7'. Corolla copiously glandular-pubescent, lips lavender or purple, tube lavender or creamy white; hosts *Ambrosia*, *Xanthium*, or *Dicoria*; riparian habitats…**A. riparium**

8. Anthers woolly; corollas 22–35 mm; calyx 14–20 mm; filament base pubescent; plants appearing whitish- or grayish-canescent…**A. multiflorum**

8'. Anthers glabrous or hairy; corollas 14–20 mm; calyx 8–14 mm; filament base glabrous; plants glandular-pubescent, not appearing canescent…**A. ludovicianum**

Aphyllon arizonicum (L. T. Collins) A. C. Schneid. [*Orobanche arizonica* L. T. Collins]—Arizona broomrape. Lavender, purple, or pallid herbs; stems 10–40 cm, simple or rarely branched; inflorescence a dense spikelike raceme, viscid-glandular-pubescent; flowers numerous, pedicels < 5 mm, bracteoles 2; calyx 8–12 mm; corolla 15–20 mm, tube white, glandular-pubescent, upper lip 4–6 mm, the lobes triangular-acute, erect or reflexed, dark purple; filaments glabrous, anthers glabrous or ± hairy, included. Sand dunes, desert scrub, piñon-juniper woodland communities, 4910–6500 ft. NW. Host *Gutierrizia*. Previously treated as *A. cooperi* A. Gray. Often confused with *A. ludovicianum* (Nutt.) A. Gray.

Aphyllon cooperi A. Gray [*Orobanche cooperi* (A. Gray) A. Heller]—Desert broomrape. Stems 5–30 cm, single or clustered, sometimes branched; inflorescence a spikelike raceme, dark purple, glandular-pubescent, bracts strongly reflexed; flowers numerous, pedicels < 10 mm, bracteoles 2; calyx 8–12 mm, lobes reflexed; corolla 15–18(–22) mm, adaxial lobes 4–6 mm, dark purple, reflexed or revolute, palatal folds densely villous; anthers glabrous or tomentulose; stalked glands near connective, often obscure, sometimes absent; stigma bilobed, rhomboid. New Mexico material belongs to **subsp. palmeri** (Munz) A. C. Schneid. Rocky, sandy soils, dry washes, canyons, volcanic mountains, 3000–6000 ft. C, SC. Host *Sidneya* (*Viguiera* s.l.).

Aphyllon corymbosum (Rydb.) A. C. Schneid. [*Orobanche corymbosa* (Rydb.) Ferris]—Flat-topped broomrape. Stems clustered or single, 5–16 cm; inflorescence corymbose or somewhat racemose, glandular-puberulent; flowers numerous, pedicels 2–20 mm; calyx 12–20 mm; corollas 18–30 mm, purple, lavender or pinkish with pink or purple veins, rarely white, slightly ampliate or gibbous, adaxial lip 3–8 mm, palatal folds yellow, glabrous; anthers woolly or sometimes glabrous; fruit a capsule, 8–14 mm.

Sandy shadscale, sagebrush, rabbitbrush, piñon-juniper communities, 5000–6500 ft. NW. Hosts various *Artemisia* species.

1. Inflorescence short, corymbose; calyces (13–)15–24 mm; corollas glandular-pubescent…**subsp. corymbosum**. NW.
1'. Inflorescence elongate, racemose; calyces 12–18 mm; corollas glabrate to slightly glandular-pubescent… **subsp**. **mutabile** (Heckard) A. C. Schneid [*Orobanche corymbosa* (Rydb.) Ferris subsp. *mutabilis* Heckard]. Rare

Aphyllon fasciculatum (Nutt.) Torr. & A. Gray [*Orobanche fasciculata* Nutt.]—Cluster broomrape, cluster cankerroot. Purplish or yellowish herbs; stems solitary or clustered, mostly belowground, 5–17 cm; inflorescence a raceme, yellow or purple-tinged, glandular-puberulent, pedicels 10–120 mm, equal to or slightly longer than the stem, bracteoles 0; flowers 3–12; calyx 7–11 mm, the lobes triangular, 3–5 mm, shorter than the tube; corolla 15–30 mm, purple to pinkish or yellow, the lobes rounded, 2–5 mm; anthers glabrous to woolly. Rocky, sandy, or clay soils in saltbush, greasewood, sagebrush, rabbitbrush, cottonwood, piñon-juniper, mountain brush communities, 5300–7650 ft. Widespread except E. Hosts various *Artemisia* species.

Aphyllon ludovicianum (Nutt.) A. Gray [*Orobanche ludoviciana* Nutt.]—Louisiana broomrape, Louisiana cankerroot. Stems solitary or clustered, 7–40 cm; inflorescence a spikelike raceme, viscid-glandular-pubescent, purplish, rose, or yellowish; flowers sessile or pedicels to 20 mm proximally; calyx 8–14 mm; corolla 14–20 mm, adaxial lips purple, lavender, or pinkish, the tube pale, the lobes rounded or obtuse, 3–6 mm; filaments glabrous, anthers glabrous to sparingly pubescent. Sandy plains, dunes, 3500–7500 ft. NW, Curry, Roosevelt. Hosts *Artemisia filifolia* or *Heterotheca*, occasionally other Asteraceae.

Aphyllon multiflorum (Nutt.) A. Gray [*Orobanche multiflora* Nutt.]—Many-flower broomrape. Stems solitary or clustered, 10–40 cm; inflorescence a dense spike or thyrsoid, viscid-glandular-pubescent, appearing grayish- or whitish-canescent; flowers numerous; calyx 15–21 mm, lobes unequal, pallid externally, purple internally; corolla dark purple, lavender, or pinkish, 22–35 mm, adaxial lip 5–9 mm, lobes broadly rounded at the apex, often with purple veins in lower lobes; filaments pubescent at base, anthers woolly, included. Sandy soils in desert grassland, juniper, piñon-juniper woodland communities, 5500–7500 ft. Widespread except E. Host *Gutierrezia*.

Aphyllon pinorum (Geyer ex Hook.) A. Gray [*Orobanche pinorum* Geyer ex Hook.]—Conifer broomrape, pine broomrape. Stems 10–30 cm, usually single, slender from a swollen base, with globose root mass; inflorescence paniculately branched or simple, ochroleucous, red-brown, purple or purple-streaked, or yellow; flowers numerous, widely spaced or clustered distally; bracts narrowly lanceolate, pedicels 0–2 mm, bracteoles 2; calyx 5–8 mm; corolla 13–19 mm, creamy-white, yellow, reddish brown, or pale lavender, with red-brown or purple veins, glandular-pubescent, palatal folds pale yellow, glabrous; lips 3–4 mm, abaxial lobes spreading, adaxial lip erect, lobe apex rounded; filaments with ring of hairs at base, anthers glabrous or sparsely pubescent, 1 pair exserted. Douglas-fir forests, 6000–8000 ft. Lincoln, Otero. Host *Holodiscus*.

Aphyllon purpureum (A. Heller) Holub [*Orobanche uniflora* L.; *Aphyllon uniflorum* (L.) Torr. & A. Gray]—Naked broomrape, one-flower cankerroot. Small, pale, yellowish herbs; stems usually solitary, mostly belowground, 0.5–5 cm; inflorescence a short raceme, glandular-villous; flowers 1–3, pedicels 3–15 cm, much longer than the stem, bracteoles 0; calyx 6–12 mm, the lobes narrowly triangular-lanceolate, 4–9 mm, longer than the tube; corolla 15–35 mm, ochroleucous to purple, the tube curved, the lobes rounded, 2–7 mm; anthers glabrous or woolly-pubescent. Rocky, moist places in sagebrush, piñon-juniper, mountain brush, ponderosa pine communities, 5300–7600 ft. NC, NW, NC, Roosevelt.

Aphyllon riparium (L. T. Collins) A. C. Schneid. [*Orobanche riparia* L. T. Collins]—River broomrape. Dark purple, lavender, or pallid herbs; stems 5–35 cm, simple or branched; inflorescences slender, open or dense spikelike racemes, viscid-glandular-pubescent; flower pedicels 0–10 mm, bracteoles 2; calyx 7–11(–13) mm; corolla (13–)15–22 mm, glandular-pubescent, often persistent, tube white tinted with purple, adaxial lips 4–6 mm, erect, lavender or dark purple, lobes triangular, apex acute, abaxial lip pale,

3–4 mm, apex acute; filaments glabrous or pubescent at base. Sandy soils, riparian habitats along Rio Grande, San Juan River, and tributaries, 5000–7000 ft. NE, NW, C, SC. Hosts *Ambrosia*, *Xanthium*, *Dicoria*, or *Nicotiana*.

BRACHYSTIGMA Pennell – Desert foxglove

Brachystigma wrightii (A. Gray) Pennell [*Gerardia wrightii* A. Gray; *Agalinis wrightii* (A. Gray) Tidestr.]—Arizona desert foxglove. Perennials; stems 30–50 cm; leaf blades filiform-linear, 30–60 × 2–3 mm, proximal longer than distal; inflorescence a raceme of paired flowers; pedicels 11–30 mm; calyx 3–5 × 4–7 mm; corolla abruptly inflating just beyond the calyx, 24–30 mm diam., externally pubescent; filaments 8–11 mm, anthers included; capsules globular-ovoid, 8–10 × 5–8 mm. Scattered juniper and oak woodlands, 5400–7200 ft. Hidalgo.

CASTILLEJA Mutis ex L. f. – Indian paintbrush

J. Mark Egger

Herbs or subshrubs, annual or perennial, (or biennial); hemiparasitic, caudex woody, with taproot or fibrous roots (or rhizomatous); stems 1 to many, strongly decumbent to erect, often with leafy axillary shoots, glabrous or pubescent, hairs eglandular or stipitate-glandular, unbranched to branched; leaves mostly alternate, reduced proximally, margins entire or divided; inflorescences compact to elongate spikes, floral bracts often brightly colored; flower calyx pale green, ± contrastingly colored distally, tubular, cleft distally into 4 subequal lobes; petals 5, corollas pale proximally, usually becoming more colorful distally, tubular proximally, bilabiate distally with an adaxial beak of 2 lobes joined to the tip and an abaxial lip with 3 highly variable lobes, consisting of either greatly reduced, ± incurved, usually greenish lobes or subpetaloid, contrastingly colored lobes; stamens 4; locules 2, placentation axile; stigma capitate, entire or 2-lobed; fruit a capsule (Egger et al., 2019).

1. Plants annual…(2)
1'. Plants perennial…(4)
2. Leaves and bracts deeply divided; at least the distal portion of bracts usually pink-purple…**C. exserta**
2'. Leaves and bracts entire; distal portion of bracts never pink-purple…(3)
3. Leaves conspicuously wavy-margined; bracts distally white (pale yellow), often aging dull pink or dull red purple…**C. ornata**
3'. Leaves ± plane-margined; bracts distally reddish (yellow)…**C. minor**
4. Abaxial calyx clefts conspicuously (> 2 mm) deeper than adaxials…(5)
4'. Abaxial and adaxial calyx clefts subequal or slightly (< 2 mm) deeper adaxially or abaxially…(8)
5. Middle stem leaves deeply divided; inflorescences clearly secund…**C. patriotica**
5'. Middle stem leaves entire; inflorescences usually not at all secund…(6)
6. Lateral calyx clefts 0(–2.5) mm; bracts 0-lobed, rarely with 1 pair of lateral lobes arising near apex; stems densely short-pubescent [reported for NM but no specimens seen]…C. tenuiflora
6'. Lateral calyx clefts 1.5–7 mm; bracts usually with 1–3 pairs of lateral lobes arising above or below midlength; stems glabrous to sometimes ± pubescent…(7)
7. Abaxial calyx clefts 3–5 times deeper than adaxials; bracts usually 3-lobed; widespread in N counties, usually with sagebrush…**C. linariifolia**
7'. Abaxial calyx clefts 1.2–2 times deeper than adaxials; bracts usually 3–5-lobed; coniferous forests and subalpine of the Sacramento and White Mountains of Lincoln and Otero Counties…**C. wootonii**
8. Middle stem leaves divided…(9)
8'. Middle stem leaves entire…(13)
9. Corolla strongly curved distally, with tube conspicuously exserted from calyx; teeth of lower corolla lip spreading, petaloid, conspicuous…**C. sessiliflora**
9'. Corolla ± straight or only slightly curved, with tube not or only obscurely exserted from calyx; teeth of lower corolla lip erect or incurved, reduced…(10)
10. Inflorescences pale yellow to pale yellow-orange…**C. lineata**
10'. Inflorescences usually purplish or bright red to orange-red…(11)
11. Inflorescences purple to pink-purple; alpine or upper subalpine…**C. haydenii**
11'. Inflorescences usually bright red to orange-red, sometimes yellowish, low-elevation to lower subalpine…(12)
12. Stems decumbent-ascending; proximal-most leaves strongly reduced and scalelike; corollas 25–40(–45)

mm, with beak and lower lip usually exserted; on sandstone substrates, with no preference for sagebrush…**C. scabrida**

12′. Stems ascending to erect; proximal-most leaves only slightly reduced and not at all scalelike; corollas 18–35 mm, with beak and lower lip usually only partially exserted; on a variety of substrates, with a strong preference for sagebrush…**C. chromosa**

13. Stems woolly to tomentose, hairs at least partially obscuring surfaces…(14)

13′. Stems glabrous, glabrate, or variously hairy but never woolly to tomentose, hairs often dense but not obscuring surfaces…(16)

14. Bracts usually entire, sometimes with 1(2) pairs of short lateral lobes usually arising at or above midlength, with the primary lobe broadly rounded distally…**C. integra**

14′. Bracts usually deeply divided, usually with 1 pair of much longer lateral lobes arising from well below midlength, with the lobes acute to rounded distally…(15)

15. Primary calyx lobes truncate, rounded, or emarginate, or sometimes divided 0–5 mm into obtuse to rounded lateral lobes; stems densely woolly, hairs branched or unbranched; widespread…**C. lanata**

15′. Primary calyx lobes divided 5–7 mm into linear to lanceolate lateral lobes, apices ± acute; stems moderately woolly-tomentose, hairs unbranched; limited to S margins of the Animas Valley…**C. tomentosa**

16. Bracts primarily whitish to yellowish or pink-purple to crimson…(17)

16′. Bracts primarily reddish to orange-red…(19)

17. Bracts primarily pink-purplish to purple or crimson, occasionally other colors, but only scattered in mixed-color populations…**C. rhexiifolia**

17′. Bracts primarily whitish to yellowish, sometimes dull reddish to dull brownish purple proximally, especially with age…(18)

18. Stems 0.7–2(–3) dm, usually short-decumbent at base, unbranched; primarily alpine plants extending downward into the upper subalpine…**C. occidentalis**

18′. Stems (1.5–)2.5–5.5(–7) dm, rarely short-decumbent at base, unbranched to often branched distally, primarily montane plants extending upward into the subalpine…**C. septentrionalis**

19. Bract veins usually yellow or white, conspicuously contrasting with distal base color; calyces mostly yellow, with scarlet to red or red-orange apices, the apices matching distal coloration of bracts; plants of the W and SW mountains…**C. nelsonii**

19′. Bract veins not yellow, not conspicuously contrasting with base color; calyces rarely yellow, and if so not contrasting with the apices; plants not limited to the W mountains…(20)

20. Corollas 20–48 mm, beaks usually 14–25 mm; bracts 0–5(–7)-lobed; inflorescences often bearing a thin coat of whitish, powdery exudate; absent from the Organ Mountains…**C. miniata**

20′. Corollas 15–24 mm, beaks 6–10 mm; bracts 0(–3)-lobed, lacking white exudate; limited to canyon slopes in the Organ Mountains…**C. organorum**

Castilleja chromosa A. Nelson—Desert paintbrush. Herbs, perennial, 1.5–3.5(–4.5) dm; stems several to many, ascending to erect, usually unbranched; hairs spreading-erect, stiff, mostly eglandular; leaves (0–)3–5(–7)-lobed; lobes spreading, linear to oblanceolate, ± involute; bracts distally bright red to scarlet or orange-red (yellow, dull orange, pink), (0–)3–7-lobed, lobes spreading, linear to oblong, apices obtuse to rounded, ± widened; calyces (17–)20–27 mm; abaxial cleft 4–10 mm, adaxial 6–12 mm, laterals 1–4 mm; corollas 18–35(–40) mm; beak (9–)10–18 mm; abaxial lip deep green, incurved. Sage slopes and flats, piñon-juniper to open pine forests, 3500–9900 ft. NW. Sometimes called *C. angustifolia* (Nutt.) G. Don var. *dubia* A. Nelson, but the two are not synonymous, and that species is absent from New Mexico.

Castilleja exserta (A. Heller) T. I. Chuang & Heckard [*Orthocarpus purpurascens* Benth.]—Purple owl's-clover. Herbs, annual, 0.1–4.5 dm; stems solitary, erect to ascending, usually branching near base; hairs spreading, mixed with short gland-tipped ones; leaves (0–)3–9(–11)-lobed, lobes filiform to narrowly spatulate, ± involute; bracts pink-purple (white) on lobe apices, (3–)5(–9)-lobed, lobes ascending to spreading, linear to narrowly spatulate, apices rounded to acute; calyces 10–26 mm, abaxial cleft 4–12 mm, adaxial 9–18 mm, laterals 2.5–9 mm; corollas 12–33 mm, beak 5–13 mm, hooked near apex, stigma emerging horizontally; abaxial lip inflated, saccate, ours proximally pink-purple, distally with magenta blotches below yellow apices. Our plants are **var. exserta**. Grasslands, sandy washes, rocky slopes; Peloncillo Mountains, ca. 4300–4400 ft. Hidalgo.

Castilleja haydenii (A. Gray) Cockerell—Hayden's paintbrush. Herbs, perennial, 0.7–2 dm; stems few to many, spreading to ascending, unbranched; glabrate to distally puberulent, hairs ± short, soft, eglandular; leaves (0–)3–7(–9)-lobed; lobes spreading, linear-lanceolate; bracts pink-purple to magenta at least distally, 3-7(-9)-lobed; lobes spreading to erect, linear-lanceolate, central lobe apex rounded to acute, laterals acute; calyces 12-26 mm; abaxial and adaxial clefts 5-10 mm, laterals 0.2-6 mm; corollas 20-25 mm; beak 6-8 mm; abaxial lip greenish at base, becoming white to pink-purple on teeth, reduced, 2-3 mm. Rocky slopes, meadows, fell-fields, upper subalpine to alpine, 10,000-13,150 ft. NC.

Castilleja integra A. Gray [*C. gloriosa* Britton]—Entire-leaved paintbrush. Herbs, perennial, 0.9-5(-10) dm; stems few to several, erect to ascending, unbranched; hairs mostly appressed, dense, ± matted, partially obscuring surface; leaves 0(-3)-lobed, linear to narrowly lanceolate or oblong, involute; bracts red to red-orange (orange, yellow), elliptic to oblong or obovate, 0-3(-5)-lobed; lobes ascending, short, central lobe rounded, laterals acute; calyces (18-)21-35(-38) mm; abaxial and adaxial clefts (6-)9-16(-18) mm, laterals (2-)4-14(-16) mm; corollas (21-)25-45(-50) mm, beak (8-)10-17(-18) mm; abaxial lip deep green, incurved, reduced, 1-2.8 mm. Rocky slopes and flats, grasslands, open forests, ledges, road banks, valleys to subalpine, 4000-10,800 ft. Widespread except SE.

Castilleja lanata A. Gray—Woolly paintbrush. Herbs or subshrubs, perennial, 1.8-9(-10) dm; stems few to many, erect to ascending, usually unbranched, woolly, hairs dense, white to yellowish, branched or unbranched in different populations, eglandular, mostly obscuring surface; leaves 0(-3)-lobed, surfaces woolly, linear to narrowly oblong or lanceolate; bracts distally bright red to orange-red (pinkish, salmon, or yellow), deeply 0-3(-5)-lobed, lobes spreading, oblanceolate or linear, apices usually rounded; calyces 15-29 mm; abaxial and adaxial clefts 6-14 mm, laterals 0-5 mm; corollas 23-35(-42) mm; beak 11-22 mm; abaxial lip dark green, incurved, reduced, 1-4 mm. Dry rocky slopes, flats, valleys to montane, 2900-7600 ft. C, WC, SC, SW.

Castilleja linariifolia Benth.—Wyoming paintbrush. Herbs, perennial, 1.8-10(-20) dm; stems few to many, ascending to erect, often much branched; glabrous or glabrescent, sometimes hairy, eglandular; leaves 0-5-lobed, linear to linear-lanceolate, involute; bracts red to red-orange at least distally (yellow), 3(-5)-lobed; lobes spreading to ascending, linear-lanceolate, apices obtuse to acuminate; calyces 18-30(-35) mm; abaxial cleft 10-20(-22) mm, adaxial 2-6(-12) mm, laterals 1.5-5(-6) mm, lobes curved slightly to adaxial side, narrowly oblong to lanceolate, apices acute; corollas 25-45 mm, often pendent from calyces, beak 9-21(-24) mm; abaxial lip deep green, reduced, incurved, 0.5-3 mm. Sagebrush steppe, grasslands, open forests, lowlands to subalpine, 3500-11,000 ft.

Castilleja lineata Greene—Linear-lobed paintbrush. Herbs, perennial, 1-4 dm; stems few to many, ascending-erect; white-woolly, hairs unbranched, ± appressed, matted, often mixed with shorter stipitate-glandular ones; leaves 3-7 lobed, lobes divergent, spreading-ascending, linear, apices acute to acuminate; bracts green to yellow-green or yellow at least distally (yellow-orange), 3(-7)-lobed, lobes ascending to spreading, linear to oblong, central lobe rounded to obtuse, laterals acute; calyces 15-20 mm; abaxial and adaxial clefts 5.5-8 mm, laterals 5-6 mm, apices acute; corollas 14-22 mm; beak 4-7 mm; abaxial lip green to yellow, reduced, 1-4 mm. Dry to moist slopes and meadows, open forests, montane to subalpine, 6800-12,000 ft. NC, NW, WC.

Castilleja miniata Douglas ex Hook. [*C. confusa* Greene]—Scarlet paintbrush. Herbs, perennial, 1.2-8(-10) dm; stems few to many, erect to ascending (proximally decumbent), branched or unbranched; glabrous, glabrate, or sometimes hairy, hairs eglandular; leaves 0(-5)-lobed, flat to involute; bracts scarlet, red, or red-orange, at least distally, often varying to other colors, surfaces often with white, powdery exudate, 0-5(-7)-lobed, lobes ascending-erect, linear to lanceolate, apices acuminate to rounded; calyces 15-38 mm; abaxial and adaxial clefts 4-24 mm, laterals (1-)3-8(-12) mm, apices acuminate to acute (obtuse); corollas 20-48 mm; beak 9-25 mm; abaxial lip incurved, deep green, reduced, 0.5-3.5 mm. Our plants are **var. miniata**. Moist to dry meadows, stream banks, shores, open forests, rocky slopes, roadsides, mostly montane to subalpine, 5000-12,000 ft. N, C, WC, SC, SW. Sometimes hybridizes with *C. rhexiifolia* and/or *C. septentrionalis*.

Castilleja minor (A. Gray) A. Gray—Seep paintbrush. Herbs, annual, 2-10(-15) dm; stems solitary or few, erect, branched to unbranched; hairs sparse to dense, a mix of longer, eglandular and short, stipitate-glandular ones; leaves 0-lobed, linear to lanceolate, margins plane (± wavy), ± involute; bracts proximally greenish, distally red, red-orange, or pale orange (yellow) on apices, narrowly lanceolate, 0-lobed, apex acuminate; calyces 13–27(–28) mm; abaxial and adaxial clefts 6–15 mm, laterals 0.5–4 mm, apices acute or acuminate; corollas 13–39 mm; beak 5–15(–20) mm; abaxial lip divaricate, red to red-purple or white to greenish yellow, 1–3 mm.

1. Abaxial lips of corollas red to reddish purple; leaves linear or linear-lanceolate, ± soft-textured; plants slender…**var. minor**
1'. Abaxial lips of corollas whitish, pale green, or pale yellowish; leaves linear-lanceolate to lanceolate, ± coarse-textured; plants more robust…**var. exilis**

var. exilis (A. Nelson) J. M. Egger—Alkaline marshes, hot springs, seeps, shores, dune swales, 4000–8000 ft. Bernalillo, San Juan.

var. minor—Wet, sometimes alkaline sites, including marshes, hot springs, seeps, stream banks, 4000–8200 ft. NW, C, SW.

Castilleja nelsonii Eastw. [*C. austromontana* Standl. & Blumer]—Southern mountains paintbrush. Herbs, perennial, 2.5–8(–10) dm; stems few to many, ascending to erect, unbranched to strongly, diffusely branched distally; hairs sparse to dense, becoming puberulent distally, eglandular; leaves 0(–3)-lobed, broadly to narrowly lanceolate, flat to ± involute; bracts distally scarlet to red or orange-red (yellow, crimson), veins yellow, yellow-green, or white, contrasting with base color, lanceolate to obovate, 0–3(–5)-lobed, lobes ascending, ± lanceolate, central lobe obtuse to rounded, laterals acute to obtuse; calyces mostly yellowish, with reddish apical band, 15–27 mm; abaxial cleft (5–)9–11 mm, adaxial 4.5–9.5 mm, laterals 2–4 mm; lobe apices acute to acuminate; corollas 15–35 mm; beak 10–16 mm; abaxial lip green, reduced, incurved, 0.5–1.5 mm. Rocky slopes, meadows, moist ground in open forests, montane to subalpine, 6100–10,000 ft. NW, WC, SW.

Castilleja occidentalis Torr.—Western paintbrush. Herbs, perennial, 0.7–2(–3) dm; stems several to many, erect or ascending, ± short-decumbent, unbranched; hairs spreading, long, soft, mixed with shorter stipitate-glandular ones distally; leaves 0–3-lobed, linear-lanceolate to broadly lanceolate, flat; bracts pale yellowish to white, often dull reddish brown or purplish proximally, broadly lanceolate to ovate, 0–3(–7)-lobed; lobes ascending, triangular to lanceolate, central lobe apex obtuse, laterals usually acute; calyces colored as bracts, 12–20 mm; abaxial and adaxial clefts 5–9(–10) mm, laterals 1–3(–4.5) mm; lobes lanceolate to triangular; corollas 16–25 mm; beak (2.5–)5–9 mm; abaxial lip green, incurved, reduced, 1.5–3 mm. Meadows, talus, ridges, upper subalpine to alpine, 7000–13,150 ft. NC.

Castilleja organorum Standl.—Organ Mountains paintbrush. Herbs or subshrubs, perennial, 2.7–8 dm; stems several to many, erect to sprawling, usually profusely branched; hairs short, dense, retrorse, ± stiff, eglandular; leaves 0-lobed, linear-lanceolate to lanceolate, involute or flat; bracts distally red to reddish orange, broadly lanceolate to oblong, 0(–3)-lobed; lobes ascending, lanceolate, short, apices acute to obtuse; calyces proximally pale yellow-green, distally pale red to red-orange, 12.5–20.5 mm; abaxial and adaxial clefts 6–9 mm, laterals (1.5–)3–4 mm, lobes lanceolate or triangular, apices acute; corollas 15–24 mm; beak 6–10 mm; abaxial lip green, reduced, incurved, 0.5–1.5 mm. Rocky slopes, shaded canyons, open coniferous forests, 4900–8200 ft. Doña Ana. Reports of this species from the Mogollon Rim are based on *C. nelsonii*. **R & E**

Castilleja ornata Eastw.—Ornate paintbrush. Herbs, annual, 1.7–3.5(–5) dm; stems 1 to several from near base, erect or ascending, unbranched distally; hairs spreading to retrorse, soft, eglandular, mixed with shorter stipitate-glandular ones; leaves 0-lobed, linear-lanceolate to oblanceolate, proximal ones clasping, margins wavy or plane, involute; bracts distally white (pale yellow), often aging dull pink or dull red-purple, spatulate, 0-lobed, apex wavy-margined, obtuse to rounded; calyces green throughout or distally white, aging pink, 15–17 mm, abaxial and adaxial clefts 6–14 mm, laterals 0(–0.7) mm, lobes short-triangular, apex acute to rounded; corollas 22–24 mm; beak 5–10 mm; abaxial lip pale greenish, slightly incurved, reduced, 0.5–1.5 mm. Seasonally damp ground, dry or sandy grasslands; S Animas Valley, ca. 5200 ft. Hidalgo. **R**

Castilleja patriotica Fernald—Native paintbrush. Herbs, perennial, 1.7–6 dm; stems few to many, erect or ascending, usually much branched, hairs retrorse, short, ± stiff, eglandular; leaves (0–)3–7-lobed, linear-lanceolate to lanceolate, flat to involute, lateral lobes spreading-divaricate; bracts green or distal-most red to red-orange on apices, (0–)3–5-lobed, lobes spreading-divaricate, apices acute; calyces prox-imally green, distal 2/3 bright red to pale red-orange (yellow), (17–)25–35(–40) mm; abaxial cleft 14–29 mm, adaxial 2.5–7.3 mm, laterals 0–0.3(–2) mm; lobes broadly triangular, entire or shallowly cleft, apices acute to rounded; corollas (24–)27–43 mm, often pendent from calyces, 22–39 mm; abaxial lip deep green, reduced, incurved, 1–2.5 mm. Dry slopes and flats, open pine forests, Animas Mountains, 7200–8300 ft. Hidalgo.

Castilleja rhexiifolia Rydb.—Rhexia-leaved paintbrush. Herbs, perennial, (1–)2.5–6(–8) dm; stems few to several, erect or ascending, unbranched; proximally glabrous to glabrate, distally hairy, hairs spreading, soft, eglandular (stipitate-glandular); leaves 0(–3)-lobed, linear to broadly lanceolate or ob-long, 3–6(–7) cm, flat to ± involute; bracts pink-purple to purple or crimson, at least distally (pale orange, white), broadly lanceolate to ovate or obovate, 0–3(–5)-lobed, lobes ascending to erect, short, central lobe apex obtuse to rounded, lateral ones ± acute; calyces proximally pale, distally colored as bracts, 15–25 mm; abaxial and adaxial clefts 8–12(–15) mm, laterals 2–5(–8) mm, apices acute to rounded; corollas 15–30(–36) mm; beak 7–12 mm; abaxial lip deep green, incurved, reduced, 1.5–3.5 mm. Moist meadows, open forests, slopes and ridges, subalpine to alpine, 6000–13,100 ft. NC, C.

Castilleja scabrida Eastw.—Scabrous paintbrush. Herbs, perennial, 0.7–1.5(–2.2) dm; stems several, decumbent to sprawling, distally ascending, usually unbranched; hairs spreading, whitish, ± stiff, eglandular; leaves 0–3(–5)-lobed, becoming reduced and scalelike near base, linear to lanceolate, flat to invo-lute; bracts bright red to orange-red distally, 3–5(–7)-lobed, lobes spreading, linear-lanceolate, apices ± expanded, acute to obtuse; calyces colored as bracts, 18–33 mm; abaxial cleft 6–12 mm, adaxial 8–15 mm, laterals 2–6 mm, lobes lanceolate to triangular, apices acute; corollas 25–40(–45) mm, beak 10–17 (–20) mm; abaxial lip green, incurved, reduced, 1.5–2.5 mm. Our plants are **var. scabrida**. Sandstone slopes, ledges, washes, sometimes on clay or cryptogamic soils, 4000–9200 ft. NW.

Castilleja septentrionalis Lindl. [*C. sulphurea* Rydb.]—Northern paintbrush. Herbs, perennial, (1.5–) 2.5–5.5(–7) dm; stems few to several, erect to ascending, unbranched or branched, glabrate proximally, distally often with spreading, soft, eglandular to stipitate-glandular hairs; leaves 0(–3)-lobed, linear-lanceolate to broadly lanceolate, flat to involute; bracts white to pale yellow (dull reddish with age), broadly lanceolate to oblong, 0–3(–5)-lobed, sometimes with irregular teeth distally, lobes erect, trian-gular, center lobe obtuse to rounded, laterals acute; calyces colored as bracts, 13–23(–28) mm; abaxial cleft (6–)8–13 mm, adaxial (5–)6–10(–11) mm, laterals 1–4 mm, lobes triangular, apices obtuse to acute; corollas (16–)18–30 mm; beak 6–12 mm; abaxial lip pale green or whitish, incurved, reduced, 1.5–3 mm. Meadows, open forests, riparian areas, rocky slopes, lowlands to subalpine, 4000–10,000 ft. NC, NW, Grant.

Castilleja sessiliflora Pursh—Downy painted-cup. Herbs, perennial, 1–4 dm; stems few to many, ascending to erect, often decumbent near base, unbranched; hairs spreading, sometimes matted, ± soft, eglandular, often mixed with minute glandular ones; leaves (0–)3–5-lobed, linear to narrowly lanceolate, involute, lobes spreading; bracts pale green or whitish to pink-purplish at least distally (salmon, pink, lavender, pale yellow), 3(–5)-lobed; lobes spreading, apices mostly acute; calyces colored as bracts, 20–40 mm; abaxial and adaxial clefts 12–20 mm, laterals 5–15 mm, apices acute; corollas strongly exserted, conspicuously curved distally, 35–55 mm, beak 9–15 mm; abaxial lip green to purple, protruding, shelf-like, 4–8 mm, teeth spreading, petaloid, 3–4 mm. Dry grasslands, sand-sage plains, limestone flats, open forests, 3800–6400 ft. Widespread except NC, NW. Note: the similar *C. mexicana* (Hemsl.) A. Gray was reported from the state but is not verified. It has an annual root, a strongly exserted, usually yellowish corolla, and conspicuously wavy-margined leaves.

Castilleja tenuiflora Benth. [*C. laxa* A. Gray]—Santa Catalina paintbrush. Herbs or subshrubs, peren-nial, 1.8–6 dm; stems few to many, erect or ascending, unbranched or branched, hairs dense, ± reflexed, short, ± stiff, eglandular; leaves 0(-3)-lobed, lanceolate to linear-lanceolate, flat to ± involute; bracts entirely green or reddish distally (pale orange, yellow), lanceolate, 0(-3)-lobed, apices acute; calyces entirely red or proximally green, distally red, red-orange (yellow), 27-35(-40) mm; abaxial cleft 12-18 mm, adaxial 8-14 mm, laterals 0(-2.5) mm, apices emarginate or rounded to acute; corollas 36-47 mm, beak 15-26(-30) mm; abaxial lip incurved, deep green (white), reduced, 1-2 mm. Open pine-oak woodlands, chaparral, 4000-8200 ft. Note: collections attributed to this species from the state are almost all referable to *C. integra*. However, it is likely that *C. tenuiflora* does occur in the mountains adjacent to Arizona in the SW counties, especially Hidalgo County.

Castilleja tomentosa A. Gray—Tomentose paintbrush. Herbs or subshrubs, perennial, 1.3-5 dm. Stems few to many, ascending to erect, unbranched or branched; moderately woolly-tomentose, hairs prostrate to spreading, unbranched, short, eglandular; leaves 0-3(-5)-lobed, linear-lanceolate, (0.8-) 3-5 cm, margins plane, involute; bracts proximally dull brownish to deep greenish purple, distally red to orange, deeply 3(-5)-lobed; lobes ascending, linear to lanceolate, arising below midlength, central lobe rounded to obtuse, laterals acute; calyces colored as bracts, (10-)13-19 mm; abaxial and adaxial clefts 4-8(-11) mm, laterals 5-7 mm, apices acute; corollas 12-20 mm; beak 8-11.5 mm; abaxial lip incurved, green or red-violet, reduced, 1.5-2 mm. Dry grasslands in and near S Animas Valley, 4250-5580 ft. Hidalgo. **R**

Castilleja wootonii Standl.—Wooton's paintbrush. Herbs, perennial, 1.6-6.5 dm; stems few to many, erect, unbranched to much branched, glabrous or hairy proximally and/or distally, hairs sparse to dense, spreading to erect, soft, eglandular; leaves 0(-5)-lobed, lanceolate to linear-lanceolate, sometimes ± wavy, flat to involute; bracts proximally greenish, distally red to orange-red, sometimes with a purplish medial band, lanceolate to broadly lanceolate or ovate, (0-)3-5(-7)-lobed, lobes ascending, linear-lanceolate, apices acuminate to obtuse; calyces proximally green, distally reddish, 20-25 mm; abaxial cleft 11-14(-17) mm, adaxial 8-9 mm, laterals 5-7 mm, apices acute to acuminate; corollas 25-37 mm; beak 11-25 mm; abaxial lip incurved, green or red, reduced, 2 mm. Meadows, rocky slopes, canyons, open forests, montane to subalpine, 6500-12,000 ft. C, SC.

CONOPHOLIS Wallr. – Cancer root

L. Turner Collins

Conopholis alpina Liebm. [*C. alpina* var. *mexicana* (A. Gray ex S. Watson) R. R. Haynes; *C. mexicana* A. Gray ex S. Watson]—Cancer root. Plants yellowish, glabrous, holoparasitic; stems 11-33 cm; leaves triangular to narrowly lanceolate, 12-22 × 3-9 mm, mostly glandular-pubescent; pedicels 0-3(-4) mm; calyx 6-9 mm, lobe apex obtuse; corolla 14-20 mm; anthers sparsely pilose or glabrous; capsules 8-15 × 6-12 mm. Oak woodlands, mixed conifer communities, 4800-10,300 ft. NE, NC, C, WC, SE, SC, SW. Parasitic on oak, pine, and juniper.

CORDYLANTHUS Nutt. ex Benth. – Bird's-beak

Herbs, annual; hemiparasitic; stems erect or ascending; leaves alternate, margins entire or 3-7-lobed; sepals 2, calyx bilaterally symmetric, spathelike; petals 5, corolla bright or pale yellow or lavender-pink with purple markings, bilabiate, adaxial lip galeate, rounded at apex; stamens 4; staminode 0; fruit a capsule; seeds brown (Barringer, 2019a).

1. Leaves puberulent or glandular-puberulent, often glabrescent; corolla bright yellow, pale yellow, lavender, or lavender-pink...**C. wrightii**
1'. Leaves densely pilose; corolla bright yellow...**C. laxiflorus**

Cordylanthus laxiflorus A. Gray—Nodding bird's-beak. Stems erect, 30-90 cm, hirsute to pilose; leaves densely pilose, 5-20 mm, margins entire; inflorescence a 1(-4)-flowered spike, bract 1, 5-7 mm; calyx 10-17 mm, tube 2 mm; corolla bright yellow, 15-20 mm, throat 3-5 mm diam., abaxial lip 7-10 mm; stamens 4, filaments hairy; capsules oblong-lanceoloid, 7-8 mm; seeds 15-20, light brown, 1.5-2 mm. Often on limestone, rocky slopes, mesas, creeks, washes, redrock area, ca. 4200 ft. Grant.

Cordylanthus wrightii A. Gray [*Adenostegia wrightii* (A. Gray) Greene]—Wright's bird's-beak. Stems erect or ascending, 20–90 cm, puberulent and glandular-puberulent or scabrous, often glabrate; leaves puberulent or glandular-puberulent, often glabrescent, 20–35 mm, margins 3–7-lobed; inflorescence a capitate spike, 2–12-flowered, 20–40 mm; bracts 2–10, green or purple distally; calyx 20–25 mm, apex bifid; corolla bright or pale yellow or lavender-pink with purple markings, 20–30 mm; stamens 4, filaments hairy; capsules oblong, 10–15 mm. Mostly sandy soils, sagebrush, piñon-juniper woodlands, scattered ponderosa pine–oak forests, 5000–7800 ft. NC, NW, C, WC, SW.

ORTHOCARPUS Nutt. – Owl clover

Annual, hemiparasitic herbs; leaves alternate, sessile, entire to pinnatifid; flowers in a bracteate spike; calyx 4-lobed; corolla bilabiate, yellow, white, or purple, the upper lip hooded and beaklike, the lower lip saccate; stamens 4, didynamous (Barringer, 2019b).

1. Flowers yellow, 9–12(14) mm; leaves mostly linear and entire with only the uppermost sometimes 3-cleft …**O. luteus**
1'. Flowers purple and white, 15–20 mm; leaves mostly 3-cleft or sometimes only the lowermost linear and entire…**O. purpureoalbus**

Orthocarpus luteus Nutt. [*O. strictus* Benth.]—Yellow owl clover. Annuals, 10–40(–60) cm; stems erect to ascending, pubescent; leaves 15–50 mm, margins entire or 3-lobed; inflorescence a spike, (2–) 5–20 cm; bracts 10–20 mm, margins 3-lobed; calyx 5–8 mm; corolla 10–15 mm, longer than the bracts, golden-yellow, glandular-puberulent, tip minutely hooked; capsules 5–7 mm. Grasslands, sagebrush, mountain meadows, disturbed sites, 6000–11,000 ft. NE, NC, NW, C, WC, SW.

Orthocarpus purpureoalbus A. Gray ex S. Watson—Purple owl clover. Annuals, (5–)15–45 cm; stems erect, scabrous and densely puberulent; leaves 15–35 mm, margins entire, distal 3-lobed; inflorescence a spike, 10–25 cm; bracts 10–20 mm, margins 3-lobed; calyx 5–8 mm; corolla 15–20 mm, longer than the bracts, purple and white, puberulent, tip notably hooked; capsules 6–9 mm. Sagebrush communities, piñon-juniper woodlands, 5800–8750 ft. NC, NW, WC, SW.

PEDICULARIS L. – Lousewort

Perennial herbs; leaves alternate, opposite, or basal, toothed to bipinnatifid; inflorescence a spike or spicate raceme; calyx 2–5-lobed, accrescent, corolla strongly bilabiate, yellow, white, purple, or reddish, the upper lip hooked, the lower lip 3-lobed; stamens 4, didynamous (Ackerfield, 2015; Robart, 2019).

(Key adapted from Jennifer Ackerfield)
1. Upper lip of the corolla (galea) with a prolonged, slender, upcurved beak 7–15 mm (resembling an elephant's head with trunk); flowers 6–8 mm, not including the beak, usually violet to purple, rarely white; calyx 3.5–5.5 mm, with prominent veins…**P. groenlandica**
1'. Galea beakless or the beak smaller; flowers not resembling an elephant's head with trunk; flowers and calyx various in color…(2)
2. Flowers rose, red, or purple throughout…(3)
2'. Flowers yellow, cream, or white, sometimes with a purple or reddish galea, or yellowish with reddish or purplish veins…(5)
3. Leaves simple, not pinnately divided, linear to narrowly lanceolate, with serrate to crenate margins; stems with longitudinal lines of hairs [expected in N mountains]…P. crenulata
3'. Leaves pinnatifid not quite to the midrib; stems glabrous…(4)
4. Stems short, 0.4–0.7(–1) dm, surpassed by the leaves; racemes usually surpassed by the leaves; calyx 15–22 mm; flowers pale violet with dark tips on the galea and lobes of the lower lip, corolla 30–42 mm… **P. centranthera**
4'. Stems longer than the leaves; racemes longer than the leaves; calyx 8–10 mm; flowers lacking dark tips on the galea and lobes of the lower lip, corolla 15–20 mm…**P. sudetica**
5. Leaves crenulate, not pinnatifid; calyx lobes 2; leaves glabrous…(6)
5'. Leaves deeply pinnatifid; calyx lobes 5; leaves various…(7)
6. Upper lip or hood of the corolla (galea) blunt and without a beak; Mogollon Mountains and S…**P. angustifolia**
6'. Upper lip or hood of the corolla (galea) with a long, incurved beak; N mountains…**P. racemosa**

7. Bracts in the inflorescence pinnately 3-7-lobed; galea prolonged into a short beak, 2 mm; calyx with dark longitudinal lines; leaves mostly basal, pinnatifid to the midrib or nearly so…**P. parryi**

7'. Bracts in the inflorescence entire or serrate but not lobed; galea beakless; calyx usually lacking dark longitudinal lines; leaves various but if mostly basal then pinnatifid only to 2/3 of the way to the midrib…(8)

8. Calyx lobes 2; leaves pinnatifid to bipinnatifid with the lobes rather shallow, extending about 1/2-2/3 of the way to the midrib; stems hairy; plants 1-3(-4) dm…**P. canadensis**

8'. Calyx lobes 5; leaves pinnatifid with the lobes extending to the midrib or nearly so, or bipinnatifid; stems glabrous; plants 3-12 dm…(9)

9. Lower lip of the corolla 4-6(-7) mm; calyx 7-10 mm; flowers yellow or cream without reddish or purplish veins…**P. bracteosa**

9'. Lower lip of the corolla 7-12 mm; calyx 10-12(-16) mm; flowers yellow or cream and usually with darker reddish or purplish veins, these usually drying brown and remaining conspicuous…**P. procera**

Pedicularis angustifolia Benth. [*P. angustissima* Greene]—Mogollon Mountain lousewort. Plants 35-55 cm; basal leaves 0, cauline leaves 10-20, blades linear to narrowly lanceolate, 15-70 × 1-6 mm, margins serrate, surfaces glabrous; inflorescence a raceme, paniculate or simple or buds present in cauline leaf axils; calyx 5.5-8.5 mm, glabrous; corolla 12-20 mm, tube yellow, 4-10 mm, galea yellow, 8-11 mm, beakless, abaxial lip yellow, 6-8 mm. Ponderosa pine and mixed conifer forests, 7500-10,200 ft. Catron.

Pedicularis bracteosa Benth.—Bracted or towering lousewort. Plants 20-80 cm; basal leaves 0-10, blades lanceolate, 20-120 × 10-60 mm, 1- or 2-pinnatifid, surfaces glabrous, cauline leaves 4-10, blades lanceolate, 1-27 × 0.8-15 cm, undivided or 1- or 2-pinnatifid, serrate to 2-serrate, surfaces glabrous, hispid, or tomentose; calyx 7-15 mm, tomentose; corolla 14-27 mm, tube yellow, 6-12 mm, galea yellow, 6-15 mm, beakless or beaked, abaxial lip yellow, 4.5-6.5 mm. New Mexico material belongs to **var. paysoniana** (Pennell) Cronquist [*P. paysoniana* Pennell]. Moist spruce-fir forests, grassy meadows, alpine slopes, 10,000-12,500 ft. Mora, Rio Arriba, San Miguel, Taos.

Pedicularis canadensis L. [*P. canadensis* var. *fluviatilis* (A. Heller) J. F. Macbr.]—Canadian lousewort. Plants 4-50 cm; basal leaves 2-20, blades lanceolate, 15-100 × 3-40 mm, 1- or 2-pinnatifid, cauline leaves 1-10, blades lanceolate, 10-70 × 5-20 mm, 1-pinnatifid; inflorescence a simple raceme, 1-5 branches, each 10-40-flowered; calyx 7-12 mm, glabrous, hispid, or tomentose; corolla 18-25 mm, tube yellow, 8-15 mm, galea yellow, yellow with red veins, or red, sometimes purple, 10-14 mm, beakless; abaxial lip expanded, yellow or white, 6-7 mm. Moist forests, moist meadows, marshes, streams, 7200-10,600 ft. NE, NC, Otero.

Pedicularis centranthera A. Gray—Dwarf lousewort. Plants 4-12 cm; basal leaves 6-8, blades elliptic or spatulate, 35-120 × 10-30 mm, undivided or 1- or 2-pinnatifid, surfaces glabrous or with scattered abaxial glands; cauline leaves 0-4, blades elliptic or lanceolate, 20-110 × 5-30 mm, 1- or 2-pinnatifid, surfaces glabrous; inflorescences a simple raceme, 1-4 branches, each 8-14-flowered; calyx 17-22 mm, glabrous; corolla 28-40 mm, tube white or pale purple, 15-30 mm, galea white or pale purple, apically sometimes dark violet to purple, 13-15 mm, beakless, abaxial lip purple, 1-4 mm. Piñon-juniper woodlands, ponderosa pine forests, 6200-10,000 ft. NC, NW, C, WC, SC, SW.

Pedicularis crenulata Benth.—Meadow lousewort. Plants 10-40 cm; basal leaves 8-10, blades narrowly elliptic to linear, 15-40 × 3-6 mm, glabrous, cauline leaves 10-40, blades linear to narrowly oblanceolate, 10-60 × 2-6 mm, surfaces glabrous; inflorescence a simple raceme, 1-10 branches, each 10-50-flowered; calyx 8.5-11 mm, hirsute along veins or glabrous; corolla 20-26 mm, tube light pink, rarely white, 12-15 mm; galea reddish violet, sometimes white, 8-11 mm, beakless, abaxial lip reddish violet, sometimes white, 4-8 mm. Moist meadows and marshes, 7200-10,500 ft. To be looked for in N mountains.

Pedicularis groenlandica Retz. [*Elephantella groenlandica* (Retz.) Rydb.]—Elephant's head. Plants 10-60 cm; basal leaves 5-20, blades lanceolate, 20-150 × 5-25 mm, 1-pinnatifid or slightly 2-pinnatifid, surfaces glabrous; cauline leaves 3 × 31, blades lanceolate, 10-150 × 1-25 mm, 1-pinnatifid; inflorescence a simple raceme, 1-2 branches, 20-75-flowered; calyx 3-5 mm, glabrous or hispid; corolla 5-8 mm, tube purple, rarely white, 3–5 mm, galea pink to purple, rarely white, 1.5–3 mm, beaked, beak coiled, 5–18 mm, abaxial lip purple, rarely white. Common in moist meadows, forested seepage areas, alpine bogs, fens, streams, creeks, 8200–12,600 ft. NC.

Pedicularis parryi A. Gray—Parry's lousewort. Leaves basal, 4–20, blades elliptic or lanceolate, 10–70 × 3–15 mm, 1-pinnatifid, surfaces glabrous; cauline leaves 0–20, blades lanceolate, 10–50 × 2–10 mm, 1-pinnatifid, surfaces glabrous or tomentose; inflorescence a simple raceme, 1–10 branches, each 5–50-flowered; calyx 6–10 mm, glabrous or tomentose; corolla 14–22 mm, tube white, yellowish, or light purple to purple, 7–15 mm; galea white, yellowish, or light purple to purple, 7–10 mm, beaked, beak straight, 5–8 mm, abaxial lip white, yellowish, or light purple to purple, 4–9 mm. Mixed conifer and spruce-fir forests, alpine tundra communities, 8000–13,000 ft. NC, Grant.

Pedicularis procera A. Gray [*P. grayi* A. Nelson]—Giant lousewort. Plants 75–150 cm; basal leaves 2–4, blades lanceolate, 15–25 × 8–12 cm, 2-pinnatifid, surfaces glabrous or hirsute, cauline leaves 4–10, blades triangular to lanceolate, 6–30 × 0.5–9 cm, undivided or 1- or 2-pinnatifid, surfaces glabrous; inflorescence a simple raceme or paniculate, 1–3 branches, each 10–50-flowered; calyx 10–15 mm, hispid to hirsute; corolla 22–30 mm, tube light yellow, greenish yellow, or light pink, 10–15 mm, galea light yellow, greenish yellow, or light pink, with purple to red veins, 9–15 mm, beakless, abaxial lip light yellow or light pink with purple veins, 9–15 mm. Common in aspen groves, mixed conifer forests, along streams, 7200–11,950 ft. Widespread except E and SW.

Pedicularis racemosa Douglas ex Benth.—Sickletop lousewort. Plants 0.5–15 cm; basal leaves 0; cauline leaves 8–25, blades linear to linear-lanceolate, 10–80 × 3–15 mm, 1- or 2-serrate, surfaces glabrous; inflorescence a simple raceme, 1–4 branches, each 3–25-flowered; calyx 4.5–7 mm, glabrous; corolla 10–15 mm, tube white, 6–9 mm, galea white, 4–6 mm, beaked, beak sickle-shaped, 5–8 mm, abaxial lip white, 4–5 mm. New Mexico material belongs to **var. alba** (Pennell) Cronquist. Common in spruce-fir forests and moist meadows, 7000–11,950 ft. NC.

Pedicularis sudetica Willd.—Sudeten lousewort. Plants 2–45 cm; basal leaves 1–20, blades elliptic to lanceolate, 10–110 × 3–26 mm, 1- or 2- pinnatifid, surfaces glabrous; cauline leaves 0–5, blades lanceolate to elliptic, 20–90 × 2–20 mm, 1- or 2-pinnatifid, surfaces glabrous, some hairs along veins on abaxial surface; inflorescence a simple raceme, 1–4 branches, 10–50-flowered; calyx 7–13 mm, glabrous, lanate, or pilose; corolla 16–21 mm, tube purple to magenta, 9–11 mm, galea purple to magenta, 7–12 mm, beakless, abaxial lip purple to magenta, 4–8 mm. Our material belongs to **subsp. scopulorum** (A. Gray) Hultén [*P. scopulorum* A. Gray]. Alpine tundra, bogs, marshes, moist meadows, 10,170–12,150 ft. Colfax, San Miguel, Taos.

RHINANTHUS L. – Yellow rattle

Rhinanthus minor L. [*Alectorolophus minor* (L.) Dumort.]—Little yellow rattle. Stems 5–20(–55) cm, glabrous, sometimes with black streaks; leaves dark green, ovate-oblong to linear-lanceolate, margins crenate-serrate, (2–)5–15 mm wide; bracts longer than to slightly shorter than the calyx, glabrous; corollas 13–15 mm, yellow, teeth of galea bluish or bluish gray, apex rounded; capsules 10–12 mm. Our material belongs to **subsp. minor**. Often on calcareous soils or rocky slopes, wet montane meadows and fens, 8000–9750 ft. NC. Introduced.

OXALIDACEAE – WOODSORREL FAMILY

Kenneth D. Heil

Ours annual or perennial herbs with sour juice, some with creeping rhizomes or bulbs; leaves palmately or pinnately compound or monofoliate, usually with petioles; flowers perfect, regular, sepals 5, distinct; petals 5, distinct, stamens 10, united at the base, pistil 1, superior, ours with 5 carpels, the locules equaling carpels in number, styles 5, distinct; fruit a capsule (Nesom, 2016b; Ornduff & Denton, 1998).

OXALIS L. – Woodsorrel

1. Plants with basal leaves originating from a bulb; orange calli (or black when dry) at apices of sepals and often on leaves; flowers purple, blue, pink, or rarely white, not yellow…(2)
1'. Plants with aerial stems, without a bulb; calli absent; flowers yellow…(5)
2. Flanges (conspicuous membranous margins) at petiole bases extended 0.5-1.5 cm above the bulb; leaflets 3-5, often with 2+ oxalate deposits at the distal end of each lobe…**O. caerulea**
2'. Flanges at petiole bases barely, if at all, extended above the bulb; leaflets 3-11, with oxalate deposits at the notch, randomly distributed along the margins, or absent…(3)
3. Leaflets 5-11 on mature plants, usually 1.2+ times longer than wide; bulb scales usually with 5+ veins… **O. decaphylla**
3'. Leaflets 3 on mature plants, as wide as or wider than long; bulb scales 3-veined…(4)
4. Seeds with longitudinal ridges, rarely with faint transverse ridges, 0.8 mm wide or less; leaflet lobes to 1/5 leaflet length; numerous bulblets formed…**O. metcalfei**
4'. Seeds with longitudinal and transverse ridges, 0.8+ mm wide; leaflet lobes to 1/4+ leaflet length; bulblets absent…**O. latifolia**
5. Leaves pinnately compound, the terminal leaflet on an extended petiolule, the lateral leaflets sessile… **O. frutescens**
5'. Leaves palmately compound, all leaflets sessile…(6)
6. Stipule margins with wide free flanges, apical auricles free; stems prostrate or decumbent, often rooting at the nodes; rhizomes absent…**O. corniculata**
6'. Stipule rudimentary or margins narrowly to very narrowly flanged or without free portions, apical auricles slightly free or absent; stems erect, ascending, decumbent, or prostrate, rooting at the nodes or not; rhizomes present or not…(7)
7. Stems villous, petioles and usually stems with septate and nonseptate hairs; rhizomes present…**O. stricta**
7'. Stems variously pubescent, not villous, petioles and stems glabrous or with nonseptate hairs; rhizomes present or absent…(8)
8. Stems usually strigillose or strigose, hairs straight, antrorsely appressed to closely ascending…**O. dillenii**
8'. Stems moderately strigose, hairs curved or crisped, sometimes straight…(9)
9. Stems puberulent to hirtellous-puberulent, hairs usually antrorsely curved or crisped, sometimes ± straight, longer hairs 0.2-0.3(-0.8) mm…**O. albicans**
9'. Stems sparsely to densely pilose, hairs spreading irregularly to ± deflexed, longer hairs 0.6-1.2 mm…**O. pilosa**

Oxalis albicans Kunth [*O. corniculata* L. subsp. *albicans* (Kunth) Lourteig]—Hairy woodsorrel. Bulbs absent; stems lax, to 40 cm, decumbent, not rooting at the nodes; leaves cauline, leaflets 3-15 cm; flowers yellow or orange, 8-12 mm; fruit 6-18 mm; seeds 1.2-1.7 mm. Riparian sites in canyon bottoms and meadows, piñon pine, ponderosa pine, spruce communities, 4400-8300 ft. Hidalgo.

Oxalis caerulea (Small) R. Knuth—Blue woodsorrel. Bulbs 0.8-1.5 cm, bulblets rarely formed, bulb scales 3-nerved, flanges at petiole bases extending 0.5-1.5 cm above the bulb; stems absent; leaves basal, leaflets 3-5, 5-12 mm and equally wide; flowers rose-red to pinkish lavender, yellow-green at base, 9-14 mm, sepals with 2 calli 0.4-0.8 mm; fruit 3.5-6 mm, glabrous; seeds 1 mm, with both transverse and longitudinal ridges. Rocky hillsides, pine forests, 6560-7010 ft. Catron, Lincoln, San Miguel.

Oxalis corniculata L.—Creeping yellow woodsorrel. Bulbs absent; stems creeping or decumbent, to 40 cm, often rooting at the nodes, hairs pointed in all directions; leaves cauline, leaflets 3, to 2 cm, often purplish; flowers yellow, homostylous, 4-8 mm; fruit 10-20 mm, hairs spreading; seeds 1.2-1.5 mm. Common in lawns, gardens, disturbed sites, greenhouses, occasionally escaping to woodlands and grasslands, 5000-8000 ft. NE, NC, NW, C, WC, SE, SC, SW. Introduced.

Oxalis decaphylla Kunth—Tenleaf woodsorrel. Bulbs 1-2+ cm, bulblets seldom formed, < 10 when present, bulb scales (3-)5-20(-30)-nerved; stems absent, septate hairs on the scapes and petioles; leaves cauline, leaflets mostly 5-13, usually 1.5-13 times longer than wide, lacking calli; flowers pinkish purple, pink, or lavender, tubes yellow or yellow-green, 9-17 mm; fruit 3-11 mm; seeds 0.8-1.2 mm. Disturbed habitats near oak, pine, and fir forests, grasslands, thorn-scrub communities, 6000-8000 ft. NW, C, WC, SC, SW.

Oxalis dillenii Jacq. [*O. corniculata* L. var. *dillenii* (Jacq.) Trel.]—Southern yellow woodsorrel. Bulbs absent; stems erect or decumbent, to 40 cm, hairs antrorse, pointed, nonseptate; leaves cauline, leaflets

3, 4–18 mm; flowers yellow, 5–12 mm; fruit 8–25 mm, strigose; seeds 1–1.5 mm, brown with white spots or lines on transverse ridges. Grassy openings in coniferous forest, 7500–8500 ft. Eddy, San Miguel, Sierra, Union.

Oxalis latifolia Kunth [*O. violacea* L.]—Broadleaf woodsorrel. Bulbs 1–2 cm; leaves shorter than the scape; leaflets 3, 2.5–4 cm; flowers violet, sepals lanceolate to oblong, 4–5 mm; flowers 9–12 mm; fruit 4–12 mm; seeds 1–1.5 mm. Piñon-juniper woodland, ponderosa pine, Douglas-fir, spruce-fir communities, often along drainages and meadows, 6500–9100 ft.

Oxalis metcalfei (Small) R. Knuth [*O. alpina* (Rose) Rose ex R. Knuth]—New Mexico woodsorrel. Bulbs 0.8–2 cm, up to 20 bulblets; leaves basal, 4–30 cm; leaflets 3, 1–3 cm wide, usually wider than long; flowers purple, pink, or white, petals to 27 mm; fruit 5–12 mm; seeds 1–1.5 mm. Moist, rocky places in pine-oak and coniferous forests, 6000–9000 ft.

Oxalis pilosa Nutt [*O. albicans* Kunth subsp. *pilosa* (Nutt.) G. Eiten]—Hairy western woodsorrel. Herbs perennial, caulescent, rhizomes and stolons absent, bulbs absent; stems usually 2–8 from base, 10–40 cm, sparsely to densely pilose, leaves basal and cauline, leaflets 3, glaucous, obcordate, 5–12 mm, surfaces glabrous to loosely strigose to hirsute-villous; sepal apices without tubercles; petals yellow, rarely with red lines proximally, 8–12 mm; capsules angular-cylindrical, abruptly tapering to apex, 12–17 (–20) mm, strigose-hirsute; seeds brown to blackish brown, transverse ridges rarely with whitish lines or spots. Canyon bottoms surrounded by piñon-juniper-oak, 4465–5415 ft. Grant, Hidalgo.

Oxalis stricta L.—Yellow woodsorrel. Bulbs absent, main root not thicker than stem, mature plants with herbaceous underground rhizomes; stems usually erect, unbranched, to 75 cm, with spreading septate trichomes; leaves cauline, leaflets 3, 5–20 mm; flowers yellow, 3.5–11 mm, homostylous or heterostylous; fruit 8–15 mm, with spreading septate hairs; seeds 0.8–1.3 mm. Montane grasslands, 6500–8500 ft. Catron, Hidalgo. Introduced.

PAPAVERACEAE – POPPY FAMILY

Lynn M. Moore

Annual, biennial, or perennial herbs, often with milky or clear juice; leaves simple or compound, basal or alternate, dissected, toothed, or lobed, without stipules; inflorescence a raceme, panicle, or solitary; flowers actinomorphic or zygomorphic, perfect; petals 4–6 (many), usually separate, sometimes with outer 2 petals forming a saccate base, inner 2 petals forming a hood over the stigma and anthers; sepals 2 or 3, separate, sometimes completely enclosing bud, caducous; stamens numerous, distinct, sometimes 6 in 2 bundles of 3 with filaments connate at base; pistil 1, carpels 2 to many, ovary superior, 1 locule, stigma(s) distinct to connate; fruit a dehiscent or indehiscent septicidal or poricidal capsule; seeds 1 to many (Kiger, 1997; Stern, 1997).

1. Corolla zygomorphic; stamens 6 in 2 bundles of 3…**Corydalis**
1'. Corolla actinomorphic; stamens many and equal in length…(2)
2. Sepals connate; receptacle expanded, forming a cup beneath the calyx…**Eschscholzia**
2'. Sepals distinct; receptacle not expanded, cup absent…(3)
3. Leaf blades, sepals, and capsules not prickly; leaves cauline or basal…**Papaver**
3'. Leaf blades, sepals, and capsules stoutly armed with prickles; leaves cauline…**Argemone**

ARGEMONE L. – Prickly poppy

Plants annual or perennial herbs with taproots, sap white to orange, stems leafy and branching; leaves sessile with basal rosette, blades lobed 2/3–4/5+ to midrib (unlobed), surfaces glaucous, unarmed or prickly, glabrous or hispid, margins dentate, each tooth spine-tipped; inflorescence a terminal cyme, bracteate; flowers actinomorphic, perianth 7–12(–16) cm broad, sepals 2 or 3, unarmed or prickly, petals 6 in 2 whorls of 3, ours white, sometimes pale lavender, carpels 3 or 4(5), locule 1, style short, stigma 3- or

4(5)-lobed, stamens 20–250+; fruit an erect capsule, 3–5(–7)-valved, prickly, seeds numerous, subglobose, minutely pitted, 2–2.5 mm.

1. Longest prickles on capsule 8–15 mm, branched…**A. squarrosa**
1'. Longest prickles on capsule 4–10(–12) mm, simple…(2)
2. Leaf surfaces densely crisped-hispid between main veins…**A. hispida**
2'. Leaf surfaces variously prickly, but not crisped-hispid…(3)
3. Upper leaves clasping, leaves at base lobed 2/3 to midrib…**A. polyanthemos**
3'. Upper leaves not clasping, leaves at base lobed 4/5+ to midrib…(4)
4. Bud prickles branched; capsule closely prickly, surface obscured…**A. pleiacantha**
4'. Bud prickles simple; capsule sparingly prickly, surface visible…**A. pinnatisecta**

Argemone hispida A. Gray—Rough prickly poppy. Sap white with faint yellow cast; stems branched toward top, often overtopping main axis, densely crisped-hispid-prickly; distal leaves not clasping, proximal leaves sparsely to densely crisped-hispid between veins, abaxial surface mostly densely prickly on midrib and main veins, adaxial surface less so; buds oblong, 16–20 × 14–18 mm, prickly and hispid; sepal horns 4–7 mm, prickly, apical prickle flattened, indurate; filaments pale yellow; capsules ovoid, 30–40 × 12–18 mm, densely prickly, surface obscured, longest prickle straight or incurved, 5–8 mm, simple. Prairies, slopes, foothills, 5400–8662 ft. NE, NC, EC.

Argemone pinnatisecta (G. B. Ownbey) S. D. Cerv. & C. D. Bailey [*A. pleiacantha* Greene subsp. *pinnatisecta* G. B. Ownbey]—Sacramento prickly poppy. Sap pale lemon to almost white; stems branched from base, sparingly to closely prickly; distal leaves not clasping, proximal leaves often pinnatifid to midrib, midribs and veins with stout yellow spines, abaxial surface sparingly prickly on veins, adaxial surface unarmed or sparingly prickly on veins; buds oblong-ovoid or subglobose, 14–20 × 11–18 mm, bud prickles simple, spreading; sepal horns terete or adaxially flattened, 4–12 mm; filaments pale yellow; capsules ovoid to ellipsoid-lanceoloid, 25–45 × 10–16 mm, sparingly prickly, longest prickle 4–8 mm, simple. Loose, gravelly soils of open disturbed sites, canyon bottoms, slopes, sometimes along roadsides; known only from the Sacramento Mountains, 4200–7100 ft. Otero. **FE & E**

Argemone pleiacantha Greene—Southwestern prickly poppy. Sap bright yellow; stems sparingly branched in the upper 1/2, sparingly to closely prickly; distal leaves not clasping, abaxial surface of proximal leaves sparingly prickly on veins, adaxial surface unarmed or sparingly prickly on veins; buds oblong-ovoid or subglobose, 14–20 × 11–18 mm, bud prickles often branched from base, spreading; sepal horns terete or adaxially flattened, 4–12 mm; filaments pale yellow to red; capsules ovoid to ellipsoid-lanceoloid, 25–45 × 10–16 mm, closely prickly, surface partially obscured, longest prickle 4–8 mm, simple. Our material belongs to **subsp. pleiacantha**. Foothills and adjacent plains, 3500–7600 ft. WC, SW.

Argemone polyanthemos (Fedde) G. B. Ownbey—Crested prickly poppy. Sap bright yellow; stems cymosely branched, sparingly prickly; distal leaves clasping, abaxial surface of proximal leaves with scattered prickles on main veins, adaxial surface unarmed; buds ellipsoid-oblong, 15–22 × 10–15 mm, sparingly prickly; perianth 7–10 cm broad; sepal horns terete, usually unarmed, 6–10(–15) mm; filaments yellow; capsules narrowly to broadly ellipsoid, 35–50 × 10–17 mm, prickly surface clearly visible, prickles widely spaced, longest 4–10(–12) mm, simple, interspersed with a few shorter ones. Prairies, foothills, mesas, 4200–7300 ft. E, NC, NW, EC, WC, Chaves, Curry.

Argemone squarrosa Greene—Hedgehog prickly poppy. Sap yellow; stems widely branching; distal leaves clasping, basal leaves lobed 2/3–4/5 to midrib, lobe apices angular, marginal teeth 3+ mm; buds subglobose to oblong, 16–25 × 15–20 mm, prickly; sepal horn angular in cross-section, 8–14(–18) mm, apical prickle indurate; filaments pale yellow; capsules ellipsoid, oblong, or lance-ovoid, 25–50 × 10–18 mm. Prairies, foothills, 3400–6500 ft. NE, EC, SE, SC.

1. Capsules densely short-prickly as well as long-prickly; abaxial surface of leaves ± densely hispid between the larger veins…**subsp. squarrosa**
1'. Capsules moderately short-prickly as well as long-prickly; abaxial surface of leaves sparingly hispid to nearly glabrous between the larger veins…**subsp. glabrata**

subsp. **glabrata** G. B. Ownbey—Hedgehog prickly poppy. Stems with 40–100 prickles per cm^2; leaves unarmed between veins, abaxial surface sparingly prickly on main veins, adaxial surface unarmed or very sparingly prickly on main veins; capsules moderately short-prickly as well as long-prickly, fewer short prickles between long spinescent prickles, longest prickles 8–12 mm, branched. Prairies, slopes, 3200–5298 ft. Reported from SE New Mexico.

subsp. squarrosa Greene—Hedgehog prickly poppy. Stems with 80–200 prickles per cm^2; leaves prickly on veins and prickly-hispid between main veins, often densely so; capsules densely short-prickly as well as long-prickly, many short prickles between the long spinescent prickles, longest prickle 10–15 mm, branched. Prairies, foothills, 3400–6500 ft.

CORYDALIS DC. – Corydalis, fumewort

Plants winter-annual, biennial, or perennial, erect to prostrate herbs, glaucous, 20 cm to 2 m; leaves basal or cauline, compound, 2–4 times pinnate or pinnate-pinnatifid; flowers zygomorphic, sepals 2, small and bractlike, peltate, not enclosing the bud, persistent or caducous, petals 4, unequal, outer 2 petals connate, forming a saccate base, inner 2 petals fused, forming a hood over the stigma and anthers; capsule linear or ellipsoid, often torulose, seeds few to many, small.

1. Petals white, inner petals red to purple-tipped; perennial; plants tall, to 2 m…**C. caseana**
1'. Petals pale to bright yellow; winter annual or biennial; plants < 50 cm…**C. aurea**

Corydalis aurea Willd.—Scrambled eggs, golden smoke. Plants prostrate or ascending, < 50 cm, roots from a ± branched caudex; stems 10–50, 2–3.5 dm; leaves bipinnate, leaflets pinnatifid; inflorescence racemose; petals pale to bright yellow, seeds smooth, without marginal ring. Widespread.

1. Raceme not longer than leaves, slender or weak; spur < 1/2 the length of the rest of the corolla from the pedicel; capsules pendent or spreading at maturity…**subsp. aurea**
1'. Raceme longer than leaves, stout or robust; spur about equaling the length of the rest of the corolla from the pedicel; capsules erect at maturity, stout…**subsp. occidentalis**

subsp. aurea—Scrambled eggs, golden corydalis, goldensmoke. Raceme not longer than leaves, slender or weak; petals with spur < 1/2 the length of the rest of the corolla from the pedicel; fruit pendent or spreading at maturity, slender, 18–24 mm. Hillsides, roadsides, dry washes in the montane, ponderosa pine, sagebrush, piñon-juniper, 4500–10,800 ft.

subsp. occidentalis (Engelm. ex A. Gray) G. B. Ownbey—Curvepod fumewort, curvepod corydalis. Raceme longer than leaves, stout or robust; petals with spur about equaling the length of the rest of the corolla from the pedicel; fruit erect at maturity, stout, 12–20 mm. Hillsides, sandy terraces, disturbed sites in desert scrub, sagebrush, mountain shrub, piñon-juniper, ponderosa pine, mixed conifer, 3450–10,100 ft.

Corydalis caseana A. Gray—Fitweed, cimarrona. Plants erect, to 2 m, roots large, fleshy; stems 1 to several, 10–15 dm; leaves ca. 5, 2–4 times pinnate, large, 20–30 cm; inflorescence racemes or panicles, with 50+ flowers; petals light pink to white, inner petals red to purple-tipped, sepals 2–3 mm, persistent, ovate-attenuate, irregularly dentate; fruit capsules ellipsoid, reflexed, 10–15 mm; seeds with numerous minute protuberances. Our material belongs to **subsp. brandegeei** (S. Watson) G. B. Ownbey. Wet aspen and spruce-fir forests, 8700–11,800 ft. Rio Arriba.

ESCHSCHOLZIA Cham. – California poppy

Eschscholzia californica Cham.—California poppy. Annual (perennial) herb, with taproots; stems caulescent, erect or spreading, 5–60 cm; leaves 1–4 times pinnately deeply lobed; inflorescence cymose or 1-flowered, buds erect; flowers actinomorphic, perianth with spreading rim of receptacular cup often inconspicuous, petals yellow to orange, often with orange spot at base, sepals 2, connate, deciduous as a unit; capsules 3–9 cm, 2-valved, dehiscing from base along placentas, often explosively; seeds many, globose to ovoid, reticulate. Our material belongs to **var. mexicana** (Greene) N. H. Holmgren & P. K. Holmgren. Roadsides, sandy gravelly substrates, open areas, 4000–6000 ft. NC, C, WC, SC, SW.

PAPAVER L. – Poppy, pavot

Plants annual, biennial, or perennial herbs with taproots, sap white, orange, or red, stems scapose or caulescent; leaves cauline, with basal rosette, distal subsessile to clasping, proximal petiolate, blades lobed (unlobed), margins entire, dentate, crenate, or incised; inflorescence cymose on long scapes or peduncles, bracteate, buds nodding (erect); flowers actinomorphic, sepals 2–3, distinct, petals 4–6, carpels 3–18(–22), locule 1, sometimes multilocular by placental intrusion, style absent, stigmas 3–18(–22), stamens numerous; fruit an erect poricidal or short-valved capsule, 3–18(–22)-valved, valves pubescent, seeds many, minutely pitted, aril absent.

1. Plants perennial; scapose, leaves all basal; petals yellow…**P. coloradense**
1'. Plants annual; caulescent, leaves basal and cauline; petals white, pink, orange, red, or purple…(2)
2. Blades of distal leaves clasping stem, upper blades toothed or lobed, but not deeply dissected…**P. somniferum**
2'. Blades of distal leaves not clasping stem; upper blades deeply dissected…**P. rhoeas**

Papaver coloradense (Fedde) Fedde ex Wooton & Standl.—Alpine poppy, arctic poppy. Perennial herb, 12–15 cm; stems loosely cespitose, leaf bases persistent; leaves gray to blue-green on both sides, 2–8 cm, pinnately lobed into 5 unequal elliptic segments; inflorescence scapose, peduncles with short, light brown hairs; petals yellow (pink-tinged or brick-red), 2 cm broad or less; filaments white or yellow, filiform; anthers yellow; stigmas 4–7, disk convex; capsules ellipsoid-subglobose to oblong-obconic, strigose, hairs light brown. Dry, rocky ridges, 10,000–13,000 ft. Taos (rare).

Papaver rhoeas L.—Corn poppy. Annual herb to 8 dm; stems simple or branching, hispid to setulose; leaves with upper blades 1 or 2 times lobed and deeply dissected, distal often somewhat clustered, not clasping stem, to 15 cm; inflorescence cymes on long scapes or peduncles, peduncles sparsely to moderately hispid throughout; petals white, pink, orange, or red, often with dark basal spot, to 3.5 cm; filaments purple, filiform; anthers bluish; stigmas 5–18, disk ± flat; capsules glabrous, sessile or substipitate, turbinate to subglobose, obscurely ribbed, to 2 cm, < 2 times longer than broad. Disturbed areas; escaped cultivar, 3750–6250 ft. Doña Ana, Hidalgo, Taos. Introduced.

Papaver somniferum L.—Opium poppy. Annual herb to 15 dm; stems caulescent, simple or branching, glabrate and glaucous; upper leaf blades toothed or lobed, sometimes sparsely setose abaxially on midrib, margins usually shallowly to deeply dentate, distal leaves clasping stem, to 30 cm; inflorescence cymes on long scapes or peduncles, peduncles sparsely setose; petals white, pink, red, or purple, often with a dark basal spot, to 6 cm; filaments white, clavate; anthers pale yellow; stigmas 5–18, disk ± flat; capsules stipitate, subglobose, not ribbed, glaucous, to 9 cm. Disturbed areas; escaped cultivar, ca. 3850 ft. Doña Ana. Introduced.

PETIVERIACEAE - PETIVERIA FAMILY

Kenneth D. Heil

RIVINA L. - Rivina

Rivina humilis L. [*R. laevis* L.; *R. portulaccoides* Nutt.]—Rougeplant. Herb, shrub, or vinelike (ours), 0.4–2 m; leaves alternate, simple, entire, glabrous or densely pubescent, lanceolate, elliptic, or oblong to deltate or ovate, petioles 1–11 cm; racemes 4–15 cm, peduncles 1–5 cm, pedicels 2–8 mm; sepals 4, white or green to pink or purplish, 1.5–3.5 mm; stamens 4; carpel 1, ovary superior; berries red to orange or yellow, subglobose, 2.5–5 mm diam.; seed 1, lenticular, 2–3 mm. Washes and roadsides, 3320–5600 ft. S, Eddy, Grant, Hidalgo, Luna (Lee, 2013; Neinaber & Thieret, 2003).

PHRYMACEAE - LOPSEED FAMILY

Guy L. Nesom

ERYTHRANTHE Spach - Monkeyflower

Annual or perennial (rhizomatous) herbs; leaves petiolate or sessile; inflorescences of axillary flowers at medial to distal nodes or at all nodes, rarely solitary; pedicels in fruit usually distinctly longer than calyces; calyx bilaterally or radially symmetric, sometimes inflated and sagittally compressed, with abaxial lobes characteristically upcurving and closing throat or not; corollas yellow, red, orange, or scarlet to pink or purplish, bilabiate to symmetric, sometimes sagittally compressed; ovary 2-locular, placentation axile; capsule dehiscence loculicidal partly to base along both sutures. Segregated from *Mimulus* (Barker et al., 2012).

Erythranthe primuloides (Benth.) G. L. Nesom & N. S. Fraga has been reported from southwestern counties at up to 5500 ft. These are misidentifications. In Arizona it grows at 7600–9500 ft. *Erythranthe cardinalis* (Douglas ex Benth.) Spach has been included in the New Mexico flora based only on speculation that it might occur here.

1. Corollas scarlet to red or orange; tube throat 20–30 mm…**E. eastwoodiae**
1'. Corollas yellow or (in *E. rubella*) sometimes pinkish to violet; tube throat 4–26 mm…(2)
2. Calyx 2-lipped…(3)
2'. Calyx regular…(7)
3. Stems 4-angled; plants without stolons or rhizomes…**E. nasuta**
3'. Stems terete; plants with or without rhizomes…(4)
4. Corolla lobes laciniate to fimbriate; plants procumbent and mat-forming; leaves 3–11 mm; fruiting calyces 4–5 mm…**E. parvula**
4'. Corolla lobes entire; plants erect to procumbent or prostrate; leaves 6–50(–125) mm; fruiting calyces 8–18(–20) mm…(5)
5. Stems prostrate and forming floating mats, sometimes ascending distally; flowers axillary, spaced along the stem…**E. geyeri**
5'. Stems erect to decumbent-ascending, not forming floating mats; flowers usually clustered distally in distinct racemes…(6)
6. Plants rhizomatous; corolla tubes (10–)12–20(–26) mm, exserted 3–5 mm beyond calyx margin, limb expanded 12–24(–25) mm (pressed); styles 15–20 mm, exserted 6–9 mm beyond fruiting calyx margin; stigma above level of anthers…**E. guttata**
6'. Plants without rhizomes but sometimes rooting at proximal nodes; corolla tubes 8–14 mm, exserted 1–3 mm beyond calyx margin, limb expanded 9–14 mm (pressed); styles 7–10 mm, exserted 1–3 mm beyond fruiting calyx margin; stigma at same level as anthers…**E. cordata**
7. Leaves glandular-villous…(8)
7'. Leaves eglandular…(9)

8. Plants erect to ascending-erect; cauline leaves basally attenuate to sessile or subsessile, epetiolate, blades oblanceolate to elliptic or elliptic-lanceolate, primarily palmately 3–5-nerved, sometimes with an additional 1–2 smaller lateral pairs; fruiting pedicels 20–45 mm; corolla tube throats 5–6 mm; fruiting calyces 5–8 mm…**E. plotocalyx**

8'. Plants laxly erect to decumbent, sprawling, or procumbent; cauline leaves abruptly and distinctly petiolate, blades generally ovate with a rounded to truncate or cordate base, mostly pinnately to subpalmately veined; fruiting pedicels 5–15(–26) mm; corolla tube throats (4–)5–10 mm; fruiting calyces 4–7 mm…**E. floribunda**

9. Internodes usually longer than the leaves; fruiting pedicels 2–18 mm; calyx lobes usually ciliate; corollas yellow or pinkish to violet…**E. rubella**

9'. Internodes usually shorter than the leaves; fruiting pedicels 2–10 mm; calyx lobes not ciliate; corollas yellow, often with red markings on lower lip…**E. suksdorfii**

Erythranthe cordata (Greene) G. L. Nesom [*Mimulus cordatus* Greene]—Tinytooter monkeyflower. Annual, fibrous-rooted, stems becoming erect, 12–40(–100) cm, sometimes rooting at proximal nodes and appearing rhizomelike, sparsely stipitate-glandular; leaves basal and cauline, petiolate, cauline broadly ovate to narrowly reniform, auriculate-subclasping, basal and midcauline 15–30(–50) mm, margins shallowly, evenly to unevenly dentate, surfaces glabrous; flowers in loose, bracteate racemes or distally clustered; fruiting pedicels 10–30(–45) mm, minutely stipitate-glandular; fruiting calyces (8–)14–18(–20) mm, inflated, compressed, glabrous or sparsely stipitate-glandular to hirsutulous, throat closing; corollas yellow, weakly bilabiate or regular, tube throat 8–14 mm. Canyon bottoms, riparian areas, lake banks, springs, 5000–5500 ft. NW, WC, SC, SW.

Erythranthe eastwoodiae (Rydb.) G. L. Nesom & N. S. Fraga [*Mimulus eastwoodiae* Rydb.]—Eastwood's monkeyflower. Perennial herbs, stoloniferous, sometimes also rhizomatous; stems scandent to pendent; stems and leaves villous-glandular to stipitate-glandular; leaves cauline, epetiolate, blades flabellate distally to obovate, oblong-oblanceolate, or elliptic, subclasping, (5–)13–40(–55) mm, relatively even-sized, palmately 3-veined, margins coarsely serrate distally; flowers axillary; fruiting pedicels 10–30(–40) mm; corollas scarlet to red or orange, strongly bilabiate, tube throats 20–30 mm. Hanging gardens and sandstone seeps, sometimes in caves; 1 record, 5800 ft. San Juan.

Erythranthe floribunda (Lindl.) G. L. Nesom [*Mimulus floribundus* Lindl.]—Many-flowered monkeyflower. Annual, fibrous-rooted or taprooted; stems erect to decumbent or procumbent-trailing, straight or geniculate at nodes, villous-glandular; leaves cauline, petiolate, blades ovate, 3–35 mm, margins serrate to dentate, surfaces villous-glandular; fruiting pedicels mostly 5–20 mm; fruiting calyces 4–7 mm; corollas yellow, weakly bilabiate, tube throat (4–)5–10 mm. Seeps, hillslope springs, creek banks, pond edges, cliff edges, 5000–7100(–9100) ft. WC, SW.

Erythranthe geyeri (Torr.) G. L. Nesom [*Mimulus geyeri* Torr.; *M. glabratus* Kunth var. *fremontii* (Benth.) A. L. Grant; *M. glabratus* Kunth var. *jamesii* (Torr. & A. Gray ex Benth.) A. Gray]—Geyer's monkeyflower. Stems prostrate and forming floating mats to decumbent-ascending or erect-ascending, not distinctly fistulose; flowers axillary, scattered along the stem or sometimes from distal nodes; fruiting pedicels 18–30 mm; fruiting calyces obtriangular; calyces and pedicels glabrous or sparsely villous-glandular; corolla tube throat 6–8 mm, barely exserted beyond calyx margins. Streamsides, springs, seeps, usually in running water but sometimes at pond and lake edges, sand and gravel bars, 4800–8800 ft. NE, NC, C, SW.

Erythranthe guttata (DC.) G. L. Nesom [*Mimulus guttatus* DC.]—Seep monkeyflower. Perennial, rhizomatous; stems, pedicels, and calyces villous-glandular (S and SW counties) or hirtellous with eglandular hairs (N counties); stems erect to ascending-erect, (6–)15–65(–80) cm, sometimes fistulose, sometimes rooting at proximal nodes; proximal leaves long-petiolate, blades 4–12.5 mm, ovate-elliptic to ovate or suborbicular, reduced in size distally and epetiolate, margins crenate to coarsely dentate; flowers mostly from distal nodes in compact racemes with reduced bracts; fruiting pedicels 15–40(–60) mm; fruiting calyces 11–17(–20) mm, closing; corolla tube throat (10–)12–20 mm, exserted 3–5 mm beyond the calyx margin, expanded 12–24(–24) mm (pressed). Along streams and in intermittent stream beds, among rocks in streams, springs, hillside seeps, wet meadows, 3800–6200 ft. in S counties (glandular form), 7300–11,500 ft. in N counties (hirtellous form). N, C, SW.

Erythranthe nasuta (Greene) G. L. Nesom [*Mimulus nasutus* Greene; *M. guttatus* DC. var. *gracilis* (A. Gray) G. Campb.]—Calyx-nose monkeyflower. Annual, fibrous-rooted; stems 4-angled, glabrous or minutely stipitate-glandular just above the nodes; commonly producing cleistogamous flowers on proximal branches separate from those with larger flowers; leaf blades elliptic-ovate to broadly ovate, suborbicular, or depressed-ovate, margins irregularly dentate to dentate-serrate or nearly lacerate-dentate, commonly double-toothed, distal short-petiolate, hirsute to hirsutulous at least on adaxial surface; fruiting calyces with upper lobe usually beaklike, distinctly longer than the lower and slightly falcate; corolla tube throat (5–)8–12 mm. Around springs and along streams, stream beds, washes, wet cliffs and rock faces, cliff bases, 1000–5700(–6800) ft. Grant, Hidalgo, Luna.

Erythranthe parvula (Wooton & Standl.) G. L. Nesom [*Mimulus parvulus* Wooton & Standl.]—Southwestern mat monkeyflower. Perennial, stems mostly procumbent, often rooting at the nodes and mat-forming, becoming erect in the inflorescence; pedicels and distal stems stipitate-glandular; leaves densely villous-hirsute on both surfaces with thickened and flattened, stiff, whitish, gland-tipped hairs; calyces villous-hirsute; lobes of lower corolla lip laciniate to fimbriate. Wet cliffs, seepage banks in piñon-juniper woodlands, 5000–6400 ft. Catron, Grant.

Erythranthe plotocalyx G. L. Nesom—Floating calyx monkeyflower. Annual, fibrous-rooted; stems erect to ascending-erect; stems, pedicels, and calyces villous-glandular with gland-tipped hairs; leaves cauline, oblanceolate to elliptic, epetiolate, 11–27 mm, margins entire to shallowly toothed; fruiting pedicels ascending-erect, 20–45 mm; fruiting calyces inflated, suburceolate, 5–8 mm; corollas yellow with red-dotted tube floor, exserted 2–4 mm beyond the calyx, tube throat 5–6 mm, limbs essentially radially symmetric, expanded 3–4 mm across (pressed). Around grassland ciénegas and seasonal drainages, wet soil in marshes, wet meadows, along sandy washes, pond and creek edges; known only from the Animas and Peloncillo Mountains, 3000–5500 ft. SW. **R**

Erythranthe rubella (A. Gray) N. S. Fraga [*Mimulus rubellus* A. Gray; *M. gratioloides* Rydb.]—Little redstem monkeyflower. Annual, taprooted; stems 3–32 cm, minutely puberulent; leaves cauline, linear to elliptic, 5–22(–30) mm, minutely puberulent; flowers mostly from distal nodes; fruiting calyces 4–9 mm; corollas weakly bilabiate, tube throat 4–10 mm. Sandy washes, sandbars, creek sides, sandy grassland, gravelly sand, rocky soil, cinder cones, among boulders, desert scrub to piñon-juniper woodlands and ponderosa pine forests, (1500–)4500–7300 ft. NC, NW, WC, SC, SW.

Erythranthe suksdorfii (A. Gray) N. S. Fraga [*Mimulus suksdorfii* A. Gray]—Suksdorf's monkeyflower. Annual, taprooted; stems mostly 1–10 cm, minutely puberulent; leaves cauline, mostly from distal nodes, linear to lanceolate or ovate, 2–20(–25) mm, minutely puberulent; flowers mostly from distal nodes; fruiting calyces 3–6 mm; corollas weakly bilabiate, tube throat 4–6 mm. Sandstone outcrops, canyon bottoms, sandy loam, clay, piñon-juniper, ponderosa pine woodlands, meadows, 6000–8100 ft. NW, WC.

PHYLLANTHACEAE – PHYLLANTHUS FAMILY

Kenneth D. Heil and Paul E. Berry

PHYLLANTHUS L. – Leaf-flower

(Ours) annual or perennial herbs, usually monoecious, rarely dioecious, glabrous or hairy, hairs simple (or branched); branching phyllanthoid or not; stems erect to prostrate; leaves persistent or deciduous, alternate, simple; stipules persistent; blade margins entire; inflorescences unisexual or bisexual, cymules or flowers solitary; pedicels present, pistillate sometimes elongating in fruit; staminate flower sepals 5–6, connate basally, petals 0; nectary extrastaminal, 4–6 glands, stamens 2–3; pistillate flower sepals persistent, 5–6, connate basally, petals 0; nectary annular to cupular or lobed; fruit a capsule; seeds 2 per locule, verrucose or ribbed, caruncle absent (Levin, 2016).

1. Sepals dark reddish purple, medially incurved and distally spreading; stamens 2, filaments distinct; staminate nectary intrastaminal, annular, 4-lobed; capsules 7–9.8 mm diam.; seeds mottled, (4.4–)4.7–6.2(–6.6) mm…**P. warnockii**
1'. Sepals green, greenish yellow, or pale brown, sometimes suffused with red, flat; stamens 3, filaments connate 2/3+ length; staminate nectary extrastaminal, glands 6-lobed; capsules 2–4 mm diam.; seeds uniformly colored, 0.9-1.8 mm…(2)
2. Perennial, with woody caudex; leaves well developed…**P. polygonoides**
2'. Annual, sometimes becoming woody and appearing as a perennial; leaves on main stem scaly…**P. abnormis**

Phyllanthus abnormis Baill.—Drummond's leaf-flower. Herb, annual, sometimes becoming woody and then appearing perennial, monoecious, 1–5 dm; leaves on main stems spiral, scalelike, leaves on ultimate branchlets well developed, glabrous on both surfaces or sparsely to moderately scabridulous, blades elliptic to oblong, 3–10 × 1–4 mm; staminate flower sepals 5–6, pale yellowish green, sometimes suffused with red, flat, 0.5–1 mm, stamens 3; pistillate flower sepals 5–6, green with nearly white margins, flat, (0.5–)0.7–1.1 mm, nectary glands strongly unequal; capsules 2.3–2.7 mm diam.; seeds uniformly brown, 1.1–1.5 mm, longitudinally ribbed. Our material belongs to **var. abnormis**. Sandy soils, open oak woodlands, prairies, barrens, dunes, 3320–4250 ft. SE.

Phyllanthus polygonoides Nutt. ex Spreng.—Smartweed leaf-flower. Herb, perennial with woody caudex, usually monoecious, rarely dioecious, 1–5 dm; leaves spiral, 5–10 × 1.5–5 mm, blades narrowly oblong to obovate; staminate flower sepals (5)6, greenish yellow, sometimes suffused with red, with white margins, flat, 0.7–1.3 mm, stamens 3; pistillate flower sepals (5)6, green with white margins, flat, 1.5–2.5 mm; capsules 2.7–3.2 mm diam.; seeds uniformly brown, irregularly verrucose. Usually in calcareous soils, grasslands, grass-shrublands, 3300–7000 ft. EC, SE, SC, SW, Guadalupe.

Phyllanthus warnockii G. L. Webster [*Reverchonia arenaria* A. Gray]—Sand reverchonia. Herbs, annual, monoecious, 2–5 dm; leaves spiral, blades elliptic to narrowly oblong-elliptic or nearly linear, (15–)20–40(–45) × (1.8–)2.5–8(–9) mm; inflorescences cymules, borne on lateral branches only, bisexual; staminate flower sepals 4, dark reddish purple, central portion sometimes paler, 1.5–2.5 mm, nectary intrastaminal, annular, 4-lobed; stamens 2, filaments distinct; pistillate flower sepals (5)6, mostly dark reddish purple, (1.3–)1.5–2.5(–2.9) mm; capsules 7–9.8 mm diam., smooth; seeds mottled light and dark brown, (4.4–)4.7–6.2(–6.6) mm, 2 surfaces minutely papillate, 1 surface smooth. Dunes, 3000–5900 ft. Lea, Otero, Socorro.

PHYTOLACCACEAE – POKEWEED FAMILY

Steven R. Perkins

PHYTOLACCA L. – Pokeweed

Phytolacca americana L.—American pokeweed. Herbaceous plants to 3(–7) m; leaves lanceolate to ovate, apex acuminate, petioles 1–6 cm; racemes divergent or drooping, 12–30 cm; sepals 5, white to greenish, pinkish, or purplish; stamens 9–12 in 1 whorl; carpels 6–12; berries dark red or purple-black; seeds black, lenticular, 3 mm. Disturbed areas, 3645–4025 ft. SE. Toxic plant native to E United States, occurs in disturbed areas of SE New Mexico (Allred & Ivey, 2012; Nienaber & Thieret, 2003).

PLANTAGINACEAE – PLANTAIN FAMILY

Kenneth D. Heil

Herbs, vines, shrubs, annual or perennial, sometimes biennial; leaves alternate, opposite, or whorled, stipules absent, petioles present or absent, blades fleshy or not, margins entire to subentire, toothed, or lobed; inflorescences axillary or terminal; flowers perfect or imperfect, actinomorphic or zygomorphic; calyx 4–5-lobed; corolla 4–5-lobed or absent; stamens 1, 2, 4, didynamous or with 4 fertile and 1 sterile staminode; pistil 1; style 1; stigma 2-lobed; ovary superior or sometimes inferior, 2-carpellate; fruit a capsule (Freeman et al., 2019b).

1. Corolla lobes 0; stamens 1; sepals 0 or minute rims at summits of ovaries; aquatics…(2)
1'. Corolla lobes 3–5, rarely 0; stamens 2–4(5); sepals 2–5; plants not true aquatics…(3)
2. Leaves whorled; fruit a drupe; ovaries inferior; styles 1; plants emergent aquatics…**Hippuris**
2'. Leaves opposite; fruit a schizocarp; ovaries superior; styles 2; plants submerged or submerged and floating aquatics…**Callitriche**
3. Corolla scarious, thin, dry, transparent; all leaves basal…**Plantago**
3'. Corolla not scarious, thin, dry, transparent; leaves various…(4)
4. Corolla tubes with rounded sacs at bases of median lobes…(5)
4'. Corolla tubes spurred, gibbous, or saccate…(8)
5. Corolla with a small pouch at the base on the upper side…**Collinsia**
5'. Corolla lacking a small pouch at the base…(6)
6. Corolla tube bases spurred…(7)
6'. Corolla tube bases gibbous…(8)
7. Corolla abaxial lips as long as or slightly longer than adaxial ones; filaments usually hairy proximally; capsules 9–12 mm…**Linaria**
7'. Corolla abaxial lips much longer than adaxial ones; filaments glabrous; capsules 2–4.8 mm… **Nuttallanthus**
8. Stamens 2 or 3…(9)
8'. Stamens 4(5)…(12)
9. Ovaries 1-locular; stamens conspicuously exserted; leaves basal and cauline…**Synthyris**
9'. Ovaries 2-locular; stamens included or slightly exserted; leaves cauline, sometimes basal and cauline… (10)
10. Bracts absent; stems prostrate; corolla weakly zygomorphic, regular…**Bacopa** (in part)
10'. Bracts present; stems creeping to decumbent, ascending or erect; corollas zygomorphic, sometimes regular…(11)
11. Corolla lobes 5; corolla bilabiate…**Gratiola**
11'. Corolla lobes 4; corolla regular…**Veronica**
12. Leaves alternate…(13)
12'. Leaves opposite or whorled…(14)
13. Annuals; seed wings present…**Epixiphium**
13'. Perennials; seed wings absent…**Maurandella**
14. Staminodes 1…(15)
14'. Staminodes 0…(16)
15. Calyx mostly connate; flowers cream-white or greenish white; leaves mostly basal, cauline leaves reduced…**Chionophila**

15'.   Calyx lobed to the base or nearly so; flower color various, rarely white; leaves various…**Penstemon**
16.   Leaf blade margins pinnatifid; bracteoles absent…**Schistophragma**
16'.   Leaf blade margins entire or weakly to sharply serrate…(17)
17.   Leaf blade margins entire…**Bacopa** (in part)
17'.   Leaf blade margins weakly to strongly serrate…**Mecardonia**

BACOPA Aubl. – Water hyssop

Bacopa rotundifolia (Michx.) Wettst. [*Monniera rotundifolia* Michx.]—Disk water hyssop. Annual (or perennial) aquatics; stems floating or prostrate, rooting at the lower nodes, 15–60 cm, hairy, sometimes sparsely so, glabrescent; leaves 12–30(–40) × (8–)12–20 mm, hairy, blade base broadly cuneate, margins entire; sepals 5, 3–4.5(–6) mm in fruit; corolla white with yellow throat, 5–10 mm; stamens 2–4; capsules subglobose, 3.5–5.5 mm. Ephemeral pools, marshes, lake and pond margins, 3670–5100 ft. Hidalgo, Lea, Roosevelt (Ahedor, 2019a).

CALLITRICHE L. – Water-starwort

C. Barre Hellquist

Annual or perennial, aquatic and terrestrial herbs; leaves opposite, entire, lacking stipules; flowers unisexual, perianth lacking, each flower subtended by a pair of bracteoles or lacking in some; staminate flowers 1–3 in axils of foliar leaves, styles 2, often longer than the ovary; fruit formed with the splitting of carpels, resulting in 4 achenelike mericarps; mericarp flattened, winged-margined or smooth; 1-seeded (Hellquist, 2013a).

1.   Flowers and fruits lacking bracts at base; leaves all submersed and linear…**C. hermaphroditica**
1'.   Flowers and young fruits with 2 thin, often whitish bracts at base; leaves dimorphic, with linear submersed leaves and floating rosettes of ovate to orbicular leaves…(2)
2.   Fruits longer than broad, narrowed to base, sharply keeled or narrowly winged above…**C. palustris**
2'.   Fruits as long as broad, rounded at base and on edges…**C. heterophylla**

Callitriche hermaphroditica L.—Northern water-starwort. Perennial aquatic herb; leaves submersed, fulvous, uniform, oblong-linear, round to obtuse at apex, 1-veined, 4–12 mm, clasping the stem; flowers lacking bracts at base; fruit 1–2.5 mm wide, about as high as wide; carpels with broad wing, separated by a deep notch; styles long and reflexed, often breaking and leaving a persistent base. Rivers, ponds, lakes, 3950–9050 ft. Rio Arriba, Sandoval, San Juan.

Callitriche heterophylla Pursh—Two-headed starwort. Perennial, aquatic herbs; leaves submersed and floating; submersed leaves linear, notched at apex, 1-veined; floating leaves 3–5-veined, ovate to orbicular, leaves forming rosettes; flowers 2-bracted at base; fruit sessile, 0.6–1.2 mm, the height approximately equaling the width; carpels more broadly rounded at the apex than at the base, so that fruit is slightly heart-shaped, thickest just above the base; margins of carpel wingless or with a narrow wing to apex; styles 1–6 mm, persistent or falling off early. Still water of lakes and ponds, 5050–10,150 ft. N, WC, SW.

Callitriche palustris L. [*C. verna* L.]—Vernal starwort. Leaves submersed and floating; lower submersed leaves linear, clasping at base, 1-veined with a shallow notch at tip; upper submersed leaves spatulate and petioled, 3-veined, 1–3(–5) cm; floating leaves forming a rosette, blades ovate to orbicular, 3–5-veined, rounded or notched at apex; bracteoles conspicuous; flowers 2-bracted at base; fruit sessile, suborbicular, 0.6–1.4 mm wide, height always exceeding width (drying may conceal this); margin of carpel with a scarious wing, widest at the summit. Still water, 8050–10,150 ft. N, WC, SW.

CHIONOPHILA Benth. – Snowlover

Chionophila jamesii Benth.—Perennial herbs; stems 1 or 2(3), (3–)5–15 cm, puberulent or retrorsely hairy, sometimes glabrate; leaves opposite, in a basal rosette, entire, blades lanceolate to spatulate, 12–78 × 2–18 mm, surfaces glabrous or glabrate; inflorescence a raceme, 1–5 cm, verticillasters;

calyx 8–9 mm, glandular-pubescent; corolla greenish white or creamy-white, 10–15 mm, bilabiate, the upper lip erect and retuse, the lower lip bearded, horizontally flattened at the apex, stamens 4, staminode 1. Gravelly slopes, alpine tundra, 11,850–12,800 ft. Taos (Freeman, 2019a).

COLLINSIA Nutt. – Blue-eyed Mary

Collinsia parviflora Lindl. [*C. grandiflora* Lindl. var. *pusilla* A. Gray]—Maiden blue-eyed Mary. Annual herbs, 3–40 cm, glabrous below and usually glandular in the inflorescence; stems erect to ascending; leaves opposite or partly whorled, linear-lanceolate to obovate, 1.5–3(–5) cm, margins subentire; inflorescences finely glandular, proximal nodes 1-flowered, distal 3–5(–7)-flowered; calyx 3.5–5.5 mm; corolla blue, banner whitish or blue-tipped, 4–8 mm, banner length 0.8–1 times wings, banner lobes and wings blue, sometimes purplish, tube and throat white; capsules 4–5.5 mm. Open places in upper piñon-juniper woodlands, ponderosa pine, mixed conifer forests, 6000–8950 ft. NW, Mora.

EPIXIPHIUM (Engelm. ex A. Gray) Munz – Sand snapdragon

Epixiphium wislizeni (Engelm. ex A. Gray) Munz [*Maurandya wislizeni* Engelm. ex A. Gray]—Sand snapdragon. Annual herbs, vines; stems climbing or scrambling, glabrous; leaves alternate, cauline, blades hastate to broadly sagittate, 21–75 × 8–48 mm, glabrous, fleshy, entire, petioles twining; inflorescences axillary, flowers solitary, bracts absent; pedicels present; sepals 5, 13–18 × 2–4; corolla blue to violet, bilabiate, tubular, tube not spurred or gibbous; stamens 4, didynamous; filaments basally hairy; staminode 1; fruit a capsule, 11–15 mm. Active and stabilized siliceous and gypseous dunes and sandy soils, 3000–6800 ft. NC, NW, WC, SE, SW (Elisens, 2019a).

GRATIOLA L. – Hedge hyssop

Gratiola neglecta Torr. [*G. neglecta* var. *glaberrima* (Fernald) Fernald]—Clammy hedge hyssop. Annuals; stems ascending or erect, few- to many-branched, (4–)10–40 cm, glabrous or glandular; leaves opposite, sessile, blades linear, narrowly ovate to elliptic, 10–65 × 3–11(–18) mm, margins with blunt to sharp teeth, surfaces glabrate or glandular-puberulent; flowers solitary or in pairs in the leaf axils, subtended by 2 bracts, pedicellate; calyx 5-lobed nearly to the base, 2.5–6 mm; corolla 7–12 mm, bilabiate, the upper lip shallowly 2-lobed, white with a yellowish tube; capsules ovoid, 2.6–6 × 3–5 mm. Uncommon in moist soil along ditches, stream banks, pond margins, 7500–9350 ft. Cibola, Rio Arriba (Freeman, 2019b).

HIPPURIS L. – Mare's-tail

C. Barre Hellquist

Hippuris vulgaris L.—Mare's-tail. Rhizomatous, aquatic herbs; stems submersed and flaccid, or erect, hollow, simple; leaves 6–12 in a whorl, linear-attenuate, firm and thick, or flaccid; inflorescence forming a long terminal spike; flowers minute along middle to upper stem, pistillate perfect, epigynous; hypanthium enclosing the ovary and bearing perfect flowers at the tip; petals absent; stamen 1, inserted on anterior edge of calyx; ovary inferior; fruit 2–3 mm, 1-celled and 1-seeded. Ponds, lakes, streams, 6050–11,000 ft. N, Otero (Hellquist, 2013b).

LINARIA Mill. – Toadflax

Perennial herbs (ours); leaves alternate or the lowermost sometimes opposite, entire, sessile; flowers in a terminal raceme; calyx of 5 distinct segments; corolla bilabiate, spurred ventrally, yellow, the throat nearly closed by a palate; stamens 4, didynamous (Goñalons, 2019).

1. Leaves ovate to lance-ovate, 10–20(–35) mm wide, sessile and clasping; flowers bright yellow, corolla 14–25 mm excluding the spur…**L. dalmatica**
1'. Leaves linear, 2–6(–8) mm wide, sessile, not clasping; flowers yellow with an orange palate, corolla 10–15(–18) mm excluding the spur…**L. vulgaris**

Linaria dalmatica (L.) Mill.—Dalmatian toadflax. Plants reproducing vegetatively by adventitious buds or stolons; leaf blades ovate to lanceolate, 10-65 × 3-31 mm; racemes 1-35-flowered; calyx (1.9-)3-9.5 × (0.9-)1-3.5 mm in flower; corollas pale to bright yellow, (27-)28-49 mm, tube 6-11 mm wide, spurs straight or curved, 11-30 mm, slightly shorter than, subequal to, or longer than rest of corolla; capsules subglobular, 4-7 × 4-7 mm, glabrous. Meadows, roadsides, disturbed areas, 3850-8400 ft. NW, EC, C, WC, Grant, Otero. Introduced.

Linaria vulgaris Hill—Butter-and-eggs. Reproducing vegetatively by adventitious buds from roots; leaves linear, 2.5-5 cm × 2-6(-8) mm; inflorescence a dense raceme, 11-31(-50)-flowered; calyx 2.5-4.5 × 0.8-1.5 mm in flower; corollas white, pale or bright yellow, sometimes with an orange palate, 27-32 (-33) mm, tube 5.5-9(-10) mm wide, spurs straight or curved, 11-15 mm, slightly shorter than or subequal to rest of corolla; capsules oblong-globular or ovoid, 5-10(-11) × 3-6 mm. Meadows, open forests, roadsides, disturbed areas, 4850-9700(-11,300) ft. N, EC, C, Grant, Otero. Introduced.

MAURANDELLA (A. Gray) Rothm. – Snapdragon vine

Maurandella antirrhiniflora (Humb. & Bonpl. ex Willd.) Rothm. [*Antirrhinum antirrhiniflorum* (Humb. & Bonpl. ex Willd.) Hitchc.]—Snapdragon vine. Perennial vines with a woody caudex, 50-300 cm; stems climbing, glabrous; leaves cauline, alternate, blades hastate to sagittate, 5-30 × 4-35 mm; petioles twining, margins entire; inflorescences axillary, flowers solitary, bracts absent; sepals 5, 9-14 × 2-3 mm, surfaces glabrous; corolla blue to violet, pink, or red, bilabiate, tube base not spurred or gibbous; stamens 4, staminode 1; capsules globular, 7-10 mm, glabrous. Siliceous and gypseous substrates, canyons, rocky slopes, arroyos, desert flats, 3000-7500 ft. NC, EC, C, S (Elisens, 2019b).

MECARDONIA Ruiz & Pav. – Baby jump-up

Mecardonia procumbens (Mill.) Small [*Erinus procumbens* Mill.; *Bacopa procumbens* (Mill.) Greenm.]—Baby jump-up. Herbs, caudex herbaceous or woody, herbage tending to dry black; stems spreading, prostrate, or ascending, 10-60 cm, glandular; leaves glandular, blades 10-25 × 6-12 mm, margins sharply serrate; pedicels 7-25 mm; inflorescences axillary, flowers solitary; sepals 5, lobes equal; corolla lemon-yellow with reddish veins in throat, abaxial lip prominent, tube base not spurred or gibbous; stamens 4, staminode 0; capsules 4-5 mm. Moist rocks in small streams surrounded by oak, juniper, and Mexican piñon pine; Guadalupe Canyon area, 4400-5200 ft. Hidalgo (Ahedor, 2019b).

NUTTALLANTHUS D. A. Sutton – Toadflax

Nuttallanthus texanus (Scheele) D. A. Sutton [*Linaria texana* Scheele]—Texas toadflax. Annual or biennial herbs; stems 17-70 cm; leaves cauline, whorled or alternate, blades linear to narrowly elliptic or obovate, 7-34 × (0.5-)1-3.1 mm; inflorescence a spikelike raceme; calyx (2-)2.5-4.2 × 0.8-1.6 mm, sparsely glandular-pubescent, sometimes glabrous; corolla blue to pale violet, (11-)14-22 mm, spurs curved, sometimes straight, 4.5-11 mm, abaxial lip 5-11 mm, adaxial 3-6 mm; stamens 4, staminode 0 or 1; capsules oblong-ovoid, 2.6-4.8 × 2.5-4 mm. Dry, rocky sites, desert scrub, arid grasslands, scattered piñon pine and juniper, 3800-6800 ft. NE, NC, EC, WC, S.

PENSTEMON Mitch. – Beardtongue, penstemon

Perennial herbs, or rarely shrubs with a taproot; stems decumbent or erect; leaves opposite, simple or rarely serrate; inflorescence axillary, in verticillate, cymose, or thyrsoid panicles; flower sepals 5; corolla zygomorphic, bilabiate, the upper lip 2-lobed and the lower 3-lobed, usually some shade of blue, less commonly purple, red, pink, or white; fertile stamens 4; anthers divided into 2 sacs, these usually dehiscing the entire length of cells and often opening completely (explanate), but sometimes opening only at bases (and across connective), rarely the pair curving into a horseshoe shape; staminode (sterile stamen) 1, conspicuous; fruit a capsule (Ackerfield, 2015; Heflin, 1997; New Mexico Rare Plant Technical Council, 1999).

(Key adapted from Bleakly [1998])

1. Plants shrubby; leaves linear…(2)
1′. Plants herbaceous or woody only at the base; leaves linear or broader…(3)
2. Corolla blue, blue-purple, or reddish purple, 8–14 mm, all lobes spreading, tube gradually expanded; S New Mexico…**P. thurberi**
2′. Corolla pink, white on the face of the lobes, 15–24 mm, upper lobes reflexed, lower lobes projecting, tube narrow and curved; S and E New Mexico…**P. ambiguus**
3. Leaves linear and short, < 35 mm…(4)
3′. Leaves not linear, or if so, > 35 mm…(6)
4. Corolla red, tubular, strongly bilabiate…**P. pinifolius**
4′. Corolla some shade of blue or purplish, bottom of corolla plicate (2-ridged), forming a low palate…(5)
5. Stems and leaves puberulous with flat, appressed, scalelike hairs, especially on lower leaves; leaves scattered on flowering stems; calyx lobes acute or very short-acuminate, scarious-margined almost to tip …**P. linarioides**
5′. Stems and leaves puberulous with fine, erect or retrorse hairs, or leaves glabrous; leaves numerous on flowering stems; calyx lobes long-acuminate, scarious-margined only at the base…**P. crandallii**
6. Upper stem leaves connate-perfoliate; corolla pink to rose, 25–35 mm; leaves usually serrate…(7)
6′. Upper stem leaves sessile or subcordate; corolla various colors (rarely pink to rosé or white); leaf margins various…(8)
7. Corolla expanding gradually, pale pink to rose; staminode glabrous; anthers explanate…**P. pseudospectabilis**
7′. Corolla expanding abruptly, white to pink; staminode bearded; anthers explanate or not…**P. palmeri**
8. Corolla a shade of red (including deep coral-pink), usually tubular or slightly expanding…(9)
8′. Corolla a shade of blue or purple, less frequently white or pink (but not deep coral-pink), of various shapes…(15)
9. Corolla constricted at orifice and with long yellow hairs; staminode bearded near tip…**P. cardinalis**
9′. Corolla not constricted at orifice, with hairs or not; staminode hairs various…(10)
10. Anther sacs dehiscent by a short slit across the connective, the tips remaining closed…**P. rostriflorus**
10′. Anther sacs completely or partially dehiscent, the tips open…(11)
11. Anther sacs explanate…(12)
11′. Anther sacs not explanate…(13)
12. Staminode glabrous; foliage not glaucous; corolla bright red; limestone areas in Sacramento Mountains only…**P. alamosensis**
12′. Staminode bearded; foliage strongly glaucous; corolla orange-pink to scarlet; rocky areas and washes in SW New Mexico…**P. superbus**
13. Corolla strongly bilabiate, lower lobes long, narrow, reflexed, upper lobes projecting; throat usually bearded…**P. barbatus**
13′. Corolla not strongly bilabiate, lower lobes short, rounded, usually spreading; throat glabrous…(14)
14. Corolla barely bilabiate, almost regular; inflorescence glabrous or puberulent; anthers U shaped, opening at tips only, minutely puberulent; staminode glabrous to slightly bearded at the tip; widespread…**P. eatonii**
14′. Corolla definitely bilabiate; inflorescence glandular; anthers opening almost completely, glabrous; staminode glabrous; SC and SW New Mexico…**P. lanceolatus**
15. Foliage glabrous and slightly to heavily glaucous; leaves usually thickened or fleshy; staminode tip expanded…(16)
15′. Foliage glabrous, puberulent, and/or glandular, not glaucous; leaves not thickened; staminode expanded or not…(21)
16. Most of the inflorescence bracts prominent; inflorescence compact, not secund, the very short internodes, pedicels, and peduncles giving the effect of a spike of flowers…(17)
16′. Only the lower inflorescence bracts prominent; inflorescences not spikelike, usually open, secund or not…(19)
17. Bracts lance-ovate to ovate, smaller, usually caudate…**P. angustifolius**
17′. Bracts lance-ovate to orbicular, acute to short-acuminate, large, conspicuous, often overlapping, clasping…(18)
18. Plants mostly 5–10 dm; calyx lobes 7–13 mm; flowers 35–48 mm; inflorescence open; corolla pink, bluish, lavender, or pale blue…**P. grandiflorus**
18′. Plants mostly < 5 dm; calyx lobes < 7 mm; flowers 14–20 mm; inflorescence congested; corolla pale lavender-blue…**P. buckleyi**
19. Inflorescence not secund; bracts broadly ovate with a short, abruptly pointed tip, throat narrow and often somewhat curved, expanded only at the orifice, glabrous at base of lower lobes…**P. fendleri**
19′. Inflorescence ± secund; bracts lanceolate, throat gradually expanded, usually bearded at base of lower lobes…(20)

20. Calyx margins broadly scarious, often pinkish or purplish; pedicels and peduncles usually short...**P. secundiflorus**

20'. Calyx margins narrowly scarious, usually not colored; pedicels and peduncles usually elongate...**P. lentus**

21. Inflorescence and corollas glandular-pubescent...(22)

21'. Inflorescence not glandular (glabrous or puberulous)...(37)

22. Fascicles of small, obscurely toothed leaves present in axils of some stem leaves [uncertain if present in New Mexico; no records seen]...P. campanulatus

22'. Fascicles of leaves absent and not as above (fascicles occasionally present in *P. dasyphyllus*)...(23)

23. Anther sacs explanate...(24)

23'. Anther sacs not explanate...(31)

24. Plants forming loose mats...**P. bleaklyi**

24'. Plants not mat-forming...(25)

25. Corolla dull purple (rarely white), lower lobes projecting 3-5 mm longer than the upper lobes...**P. whippleanus**

25'. Corolla white, pale lavender, violet-blue, or blue-purple, lower lobes not projecting, not noticeably longer than upper lobes...(26)

26. Plants with glabrous stems above and below (glandular-hairy only in the inflorescence); anther sacs glabrous with short hairs on the side opposite dehiscence; plants mostly > 10,000 ft. elevation...**P. hallii**

26'. Plants without the combination of features above...(27)

27. Corolla not bearded at base of lower lobes; staminode sparsely to moderately bearded...(28)

27'. Corolla bearded at base of lower lobes; staminode conspicuously bearded...(29)

28. Corolla 1-2 cm, densely glandular-pubescent externally and internally, mostly white, tube funnelform and moderately inflated...**P. albidus**

28'. Corolla 3.6-5 cm, glandular-pubescent externally, glabrous internally, white or pink to pale-violet, tube abruptly inflated...**P. cobaea**

29. Corolla 5-6 mm wide, orifice as high as or higher than wide; lower lip not glandular within; staminode not or barely exserted; throat not to moderately inflated...**P. breviculus**

29'. Corolla 8-19 mm wide, orifice much wider than high; lower lip glandular within; staminode usually prominently exserted, throat much inflated...(30)

30. Corolla 24-35 × 10-15 mm...**P. jamesii**

30'. Corolla 14-22 × 8-19 mm...**P. ophianthus**

31. Staminode glabrous, not dilated at tip; corolla markedly ampliate; anthers U-shaped, sutures spinescent; leaves occasionally in fascicles...**P. dasyphyllus**

31'. Staminode bearded; corolla narrow to expanded; anthers not U-shaped, sutures not spinescent; leaves never in fascicles...(32)

32. Leaves finely toothed; corolla pale lavender to pale violet; corolla floor narrow, 2-ridged...**P. gracilis**

32'. Leaves entire or undulate (occasionally denticulate in *P. auriberbis*); corolla darker in color; corolla floor ridged or not...(33)

33. Corolla floor with or without ridges on underside; base of lower lobes villous; staminode slightly included to distinctly exserted, densely golden-bearded for most of its length; bracts relatively large...(34)

33'. Corolla floor deeply to moderately ridged on underside; base of lower lobes with a few white or many yellow hairs; staminode usually included, densely bearded for all of its length; bracts always reduced...(35)

34. Stems arising from a simple or branched woody caudex; corolla scarcely bilabiate, corolla floor without ridges; NE...**P. auriberbis**

34'. Stems arising from a slender, creeping, woody rhizome; corolla strongly bilabiate, corolla floor with ridges; Black Range...**P. metcalfei**

35. Corolla floor deeply 2-ridged; base of lower lobes and floor of corolla and staminode densely covered with golden hairs; corolla 17-25 mm; flowers drooping to horizontal...**P. griffinii**

35'. Corolla floor less strongly ridged, glabrous; base of lower lobes with a few white hairs; staminode orange-bearded; corolla 11-27 mm; flowers ascending to drooping...(36)

36. Corolla fairly abruptly but moderately inflated, 17-27 mm; flowers ascending to horizontal; basal rosette usually gone at anthesis; all but most uppermost cauline leaves usually well developed and similar to basal...**P inflatus**

36'. Corolla straight or little inflated, 11-20 mm; basal rosette often present at anthesis; basal leaves best developed; cauline leaves usually smaller than basal...**P. oliganthus**

37. Inflorescence not at all secund; flowers 10-14 mm, in dense fascicles usually separated by long internodes...**P. rydbergii**

37'. Inflorescence at least somewhat secund, often distinctly so; flowers 15-40 mm, not in dense fascicles, or if so, the fascicles not separated by long internodes...(38)

38. Leaves large and broad, lance-ovate or oblong; inflorescence usually broad and compact; corolla 30-40 mm...**P. glaber**

38'. Leaves linear or lanceolate; inflorescence usually narrow and elongate; corolla 15–38 mm…(39)
39. Anthers glabrous, not explanate…(40)
39'. Anthers pubescent (sometimes very sparsely), not opening proximally…(42)
40. Staminode with tuft of golden hairs at tip…**P. deaveri**
40'. Staminode glabrous…(41)
41. Corolla 25–35 × 10–17 mm, usually strongly white-bearded at base of lower lobes; sepals 4–8 mm…
    P. neomexicanus
41'. Corolla 17–28 × 7–10 mm, glabrous or lightly white-bearded at base of lower lobes; sepals 2–4 mm…
    P. virgatus
42. Anthers with flexuous hairs < the length of the sac; staminode short-bearded on distal 1/2; calyx 8–10
    mm, segments usually lanceolate, acuminate, or caudate…**P. strictiformis**
42'. Anthers usually densely villous with hairs > or equaling the length of the sac; staminode glabrous or with
    a few hairs at the tip; calyx 3–6(–8) mm, segments usually ovate or rounded…(43)
43. Inflorescence narrow, cymes 1–2-flowered on short, usually appressed peduncles and pedicels; corolla
    deep blue…**P strictus**
43'. Inflorescence usually broader, cymes often much branched, peduncles and pedicels elongate and divaricate; corolla pale blue to lavender…**P. comarrhenus**

Penstemon alamosensis Pennel & G. T. Nisbet—Alamo penstemon. Perennial; stems solitary or few, glabrous, 3–7 dm; leaves entire, opposite, basal leaves elliptic to lanceolate, stem leaves 2–4 pairs, sessile, lanceolate; inflorescence slightly glandular, secund; calyx lobes slightly glandular, 3–5 mm, narrowly scarious; corolla bright red, 20–25 mm, slightly bilaterally symmetric, lightly glandular outside, throat moderately inflated, lobes 3–6 mm, spreading; stamens included, anthers glabrous, explanate; staminode glabrous, tip not expanded. Limestone, sheltered rock areas, canyon sides and bottoms; Sacramento and San Andres Mountains, 4300–5300 ft. Doña Ana, Lincoln, Otero. **R & E**

Penstemon albidus Nutt.—White penstemon. Perennial; stems 1–5.5 dm, retrorsely hairy below and glandular-hairy near the inflorescence; leaves entire to toothed, lanceolate to obovate, glabrous to scabrous; inflorescence a thyrse, to 30 cm, glandular-hairy; calyx 4–7 mm, glandular-hairy; corolla (12–) 16–20 mm, white to faintly pink, glandular-hairy externally and internally; anthers glabrous, explanate; staminode sparsely to moderately yellow-bearded, included. Shortgrass prairie and E plains, 4475–7705 ft. NE, NC.

Penstemon ambiguus Torr.—Sand beardtongue. Stems 2–6 dm, profusely branched, candelabra-like, woody well above the base, glabrous; leaves entire, 6–30 mm, linear, mucronate, usually inrolled, margins not or remotely and very minutely scabrescent; inflorescence narrow, not secund, not glandular; calyx 2–3 mm, lobes ovate, acute, margins scarious; corolla 15–24 mm, glabrous outside, throat pubescent within, tubular, pale to deep pink, nectar guidelines prominent; stamens included; anthers ultimately explanate, glabrous; staminode included, glabrous, tip not expanded.

1. Stems puberulous…**subsp. ambiguus**
1'. Stems glabrous…**subsp. laevissimus**

subsp. ambiguus—Sand beardtongue. E plains, 3350–5780 ft. EC.

subsp. laevissimus D. D. Keck—Bush beardtongue. Desert scrub, S New Mexico, 4000–7000 ft. Widespread except N and EC.

Penstemon angustifolius Nutt. ex Pursh—Broadbeard penstemon. Stems 1 to several, 2–5 dm, stout, plants glabrous and glaucous; leaves entire, 2–9 × 2–25(–35) mm, fleshy, basal usually shorter and narrower than cauline, lanceolate to oblanceolate; inflorescence not secund, usually elongate, verticillasters usually dense and many-flowered but widely separated; calyx 4–7 mm, lobes lanceolate to narrowly ovate, margins scarious at base; corolla 17–23 mm, usually glabrous, tube gradually expanding, blue or bluish purple to lavender or pinkish; stamens included; anthers completely dehiscent but not explanate, glabrous; staminode bearded with short, deep yellow hairs.

1. Corollas pale blue to bluish purple…**var. caudatus**
1'. Corollas lavender to pinkish…**var. venosus**

var. caudatus (A. Heller) Rydb.—Broadbeard penstemon. S Great Plains, 5000–8580 ft. N, C.

var. venosus (D. D. Keck) N. H. Holmgren—Taperleaf penstemon. Sagebrush and piñon-juniper communities, 4900-7400 ft. NC, NW.

Penstemon auriberbis Pennell—Colorado beardtongue. Perennial, 1-3.5 dm; stems retrorsely short-hairy below, glandular-hairy near the inflorescence; leaves linear to lanceolate, (1.5-)3-6(-10) cm × 2-5(-7) mm, margins entire or toothed; inflorescence with 3-8 verticillasters, somewhat secund; calyx mostly 7-9 mm, densely glandular-hairy; corolla mostly 18-22 mm, pale lilac to purplish blue, glandular-pubescent externally, yellow hairs on lower lip; staminode barely to conspicuously exserted, bearded most of its length with stiff or twisted yellow-orange hairs to 2.5 mm; anthers papillose along sutures. Dry hillsides, shale slopes, grasslands, 4700-6700 ft. Colfax, Rio Arriba, Union.

Penstemon barbatus (Cav.) Roth—Torrey or scarlet beardtongue. Stems 1 to few, 3-11 dm; plants glabrous or puberulous below, glabrous above; leaves entire, 5-10+ cm, basal and lower cauline wider than upper leaves; inflorescence glabrous, ± secund, verticillasters usually widely separated; calyx 3-7 mm, lobes lanceolate to ovate, acute to short-acuminate, usually glabrous; corolla scarlet-red, 25-35 mm, externally glabrous, throat glabrous or sparsely white-bearded, essentially tubular, markedly bilabiate; stamens exserted; anther sacs opening partially from the distal end, not across the connective, glabrous (lightly villous in subsp. *trichander*); staminode included, glabrous.

1. Base of lower corolla lobes bearded with yellow hairs; calyx lobes 6-10 mm...**subsp. barbatus**
1'. Base of lower corolla lobes glabrous or bearded with a few white or yellowish hairs; calyx lobes < 6 mm ...(2)
2. Anthers glabrous...**subsp. torreyi**
2'. Anthers villous...**subsp. trichander**

subsp. barbatus—Scarlet beardtongue. Base of lower lobes bearded with yellow hairs; calyx lobes 6-10 mm. Ponderosa pine, piñon, juniper, 4900-8300 ft. NE, NC, C, SW.

subsp. torreyi (Benth.) D. D. Keck—Torrey's beardtongue. Anthers glabrous; palate glabrous to sparsely villous. Piñon-juniper, oak, ponderosa pine, Douglas-fir communities, 4385-10,845 ft. Widespread except NW, EC, SE.

subsp. trichander (A. Gray) D. D. Keck—Scarlet or beardlip beardtongue. Anthers lightly villous; palate glabrous. Piñon-juniper, Gambel oak, ponderosa pine, Douglas-fir communities, 5800-9850 ft. NC, NW, C, WC.

Penstemon bleaklyi O'Kane & K. D. Heil—Bleakly's penstemon. Plants perennial, forming loose mats from numerous branched and tangled rhizomes; stems ascending-decumbent, 2-8 cm, minutely puberulent; stem leaves opposite, sessile, glabrous or with a few trichomes at the base, entire to serrulate, narrowly lanceolate to linear-lanceolate; inflorescence terminal, secund, ± racemose; sepals lanceolate, 5-8 mm; corolla lavender, 12-25 mm; stamens included, anthers dark purple, finely puberulent with scattered longer trichomes, opening across the connective; staminode bearded at the very tip. Reddish, gneissic granite, alpine scree slopes, 12,500-12,900 ft. Taos. **R & E**

Penstemon breviculus (D. D. Keck) G. T. Nisbet & R. C. Jacks—Narrowmouth beardtongue. Stems 1 to few, 1-3.5 dm, plants puberulent or glabrous below, usually retrorsely puberulent and glandular above; leaves entire (rarely irregularly toothed), 2-8 cm × 4-16 mm, mostly narrowly oblanceolate; inflorescence glandular, subsecund, narrow; calyx glandular, 5-7 mm, lobes lanceolate, acute, scarious at base; corolla 12-20 mm, glandular outside, dark blue to bluish purple with conspicuous dark guidelines, funnel-shaped, palate with long, pale yellow hairs; stamens included; anthers glabrous, explanate; staminode mostly included, densely bearded with threadlike, yellow hairs. Sandy, gravelly, or clay soils in sagebrush and piñon-juniper woodlands, 5000-6500 ft. McKinley, Rio Arriba, San Juan.

Penstemon buckleyi Pennell—Buckley's penstemon. Erect or ascending perennials, 1.5-5.5(-8.1) dm; stems glabrous and glaucous; basal leaves oblanceolate to spatulate, 1.9-11.6(-15) × 0.3-2.2(-3.1) cm, glabrous, thick; cauline leaves lance-ovate to ovate, glabrous, thick, margins entire; inflorescence with mostly 4-20 verticillasters; calyx 3.5-6 mm, glabrous; corolla (12-)14-20 mm, pale pink or lavender to very pale blue, glabrous externally and internally; staminode included, densely to moderately golden-yellow-bearded; anthers minutely papillose. Sandy soils, shortgrass prairie, 4035–6435 ft. E.

Penstemon campanulatus (Cav.) Willd. [*P. pulchellus* Lindl.]—Bellflower beardtongue. Stems to medium height, woody at the base, pubescent; leaves serrate, smooth, oblong to lanceolate, acute, fascicles of smaller leaves in the axils of larger leaves; inflorescence secund, glandular-pubescent; corolla 20–25 mm, glandular, abruptly ampliate, ventricose, violet to purple; anthers fully dehiscent, not explanate; staminode dilated distally, yellow-bearded. Rocky areas at middle elevations; presence in New Mexico is uncertain. Material found near New Mexico is **subsp. chihuahuensis** Straw.

Penstemon cardinalis Wooton & Standl.—Scarlet penstemon. Perennial herb; stems gray-green, glabrous, to 70 cm, branching from the base; leaves opposite, entire, thickened, basal leaves short-petiolate, ovate to elliptic, to 12 cm, cauline leaves smaller, sessile, ovate to lanceolate; inflorescence secund; calyx 3–6 mm, lobes ovate, acute or obtuse; corolla red, 22–30 mm, narrowed at the mouth, lower lobes bearded with soft yellow hairs; staminode bearded near the tip.

1. Leaves slightly thickened; calyx mostly 3 mm…**subsp. cardinalis**
1'. Leaves highly thickened; calyx 4–6 mm…**subsp. regalis**

subsp. cardinalis—Scarlet penstemon. Stems to 70 cm; leaves slightly thickened; calyx mostly 3 mm, lobes ovate, acute or obtuse. Canyon bottoms and rocky slopes in piñon-juniper woodlands and lower montane coniferous forests; Sacramento, Capitan, and Oscura Mountains, 7000–9000 ft. Lincoln, Otero. **R & E**

subsp. regalis (A. Nelson) G. T. Nisbet & R. C. Jacks. [*P. regalis* A. Nelson]—Guadalupe penstemon. Stems to 100 cm; leaves strongly thickened; calyx 4–6 mm, lobes ovate and sharp-tipped. Limestone slopes and canyon bottoms in montane scrub, piñon-juniper woodlands, and lower montane conifer forests; Guadalupe Mountains, 4500–6000 ft. **R & E**

Penstemon cobaea Nutt.—Cobaea beardtongue. Erect perennials, (1.5–)2.5–6.5(–12) dm; stems retrorsely pubescent below, glandular-hairy above; basal leaves lanceolate to spatulate, 3.5–18(–26.4) × 0.8–5.5(–7.6) cm, thick, margins subentire to sharply toothed, cauline leaves lanceolate to ovate; inflorescence with mostly 3–6 verticillasters; calyx (8–)10–16 mm, densely glandular-hairy; corolla 35–55 mm, white or pink to pale violet-purple, glandular-hairy; staminode included or slightly exserted, sparsely bearded with golden-yellow hairs; anthers glabrous. Cobaea beardtongue is native to the E plains and is now spreading to roadsides in New Mexico, 5480–7150 ft. NC, Quay. Introduced.

Penstemon comarrhenus A. Gray—Dusty beardtongue. Stems 1 to few, 3–8(–12) dm; plants glabrous or rarely puberulent below; leaves entire, 3–12 × 2–20+ mm, mostly oblanceolate, lower wider and petiolate, upper narrower to linear and sessile; inflorescence glabrous, usually somewhat secund, open; calyx glabrous, 3–6(–10) mm, ovate, margins scarious; corolla 25–35 mm, glabrous outside and inside, tube pale bluish white, throat markedly expanded, pale pinkish or bluish lavender with dark guidelines; stamens exserted; anthers usually densely villous, mostly dehiscent (opening on distal 3/4+); staminode essentially included, usually glabrous (rarely sparsely bearded). Sagebrush, oak and piñon-juniper woodlands, ponderosa pine forests, 6200–8300 ft. NW, WC.

Penstemon crandallii A. Nelson—Crandall's beardtongue. Stems usually several, 5–25 cm, ascending to decumbent or sometimes prostrate, older, woodier stems often rooting, plants usually puberulent throughout with fine, terete, erect or retrorse hairs; leaves entire, essentially linear, sometimes slightly wider, 10–35 × 1–3 mm, mucronate; inflorescence secund, glandular-puberulent; calyx 5–8 mm, lobes ovate to lanceolate, acuminate to caudate; corolla 14–23 mm, glandular-puberulent outside, palate sparsely bearded, throat plicate, blue to bluish purple or reddish purple; stamens reaching to near orifice; sacs dehiscing completely but not explanate, glabrous; staminode included, densely bearded with golden hairs.

1. Leaves glabrous ventrally…**subsp. glabrescens**
1'. Leaves puberulous with fine, erect or retrorse hairs…**subsp. taosensis**

subsp. glabrescens (Pennell) D. D. Keck—Crandall's penstemon. Dry hillsides in Gambel oak, piñon-juniper woodlands, ponderosa pine, and spruce-fir forests, 6300–8900 ft. NC, NW.

subsp. taosensis (D. D. Keck) Kartesz & Gandhi—Taos penstemon. Gambel oak, piñon-juniper woodlands, sagebrush and ponderosa pine communities, 6850–9670 ft. Taos. **E**

Penstemon dasyphyllus A. Gray—Cochise penstemon. Stems 20–40 cm, ascending, pubescent throughout; leaves 4–12 cm, linear, tapering to a sharp point, glabrous or pubescent; inflorescence secund, glandular; calyx 4–7 mm, lobes oblong to oblong-lanceolate; corolla 25–35 mm, blue or purplish, glandular; stamens included; anthers slightly toothed on the edges; staminode smooth, not dilated. Grassland savanna with oak, pine, lovegrass, grama, and muhly, 3950–8460 ft. Hidalgo, Luna.

Penstemon deaveri Crosswh.—Deaver's penstemon. Stems 16–60 cm, glabrous to puberulent; lower leaves oblanceolate to spatulate, 3.5–11 × 1.5–2 cm, entire, cauline leaves sessile, elliptic-oblong to oblanceolate, 2–9 × 1–1.5 cm; calyx lobes 3–6 mm, lobes glabrous to sparsely puberulent; corolla purple, 16–25 mm, glabrous externally, the throat gradually inflated; anther sacs opposite, dehiscing completely or not; staminode sparsely to moderately hairy; capsules 9–12 mm. Ponderosa pine, mixed conifer, spruce-fir forests, 7200–8800 ft. WC, SW.

Penstemon eatonii A. Gray—Eaton's or firecracker beardtongue, scarlet bugler. Stems few to several, 3–10 dm, plants robust, glabrous or puberulent, not glaucous; leaves entire, sometimes wavy, 3–10+ × 1–5 cm, lower wider, upper sessile, sometimes subcordate and/or clasping; inflorescence glabrous or puberulent, secund; calyx 3–6 mm, lobes ovate, obtuse or acute, margins scarious; corolla cardinal-red, 20–30 mm, glabrous inside and out; stamens included to exserted; anthers horseshoe-shaped, dehiscing at tips only; minutely puberulent; staminode included, glabrous.

1. Herbage glabrous throughout…**var. eatonii**
1'. Herbage puberulent…**var. undosus**

var. eatonii—Firecracker penstemon. Plants glabrous. Piñon-juniper woodlands and ponderosa pine forests, 5200–6200 ft. Rio Arriba, San Juan.

var. undosus M. E. Jones—Scarlet bugler. Plants puberulent. Piñon-juniper woodlands and ponderosa pine forests, 4800–7500 ft. San Juan.

Penstemon fendleri Torr. & A. Gray—Fendler's penstemon. Stems single to several, 20–50 cm; leaves thick, gray-green, acute, basal widest at the base, narrowly ovate or elliptic, cauline leaves smaller, lanceolate, clasping; inflorescence in mostly 3–4 verticillasters; calyx 4–7 mm, lobes ovate, margins scarious; corolla 17–25 mm, violet or blue with dark guidelines, tubes narrow with wide-open throats, a few white hairs on the lower lip; staminode included, bearded with brownish hairs at or near the dilated tip. Desert scrub, blue grama–shin oak and piñon-juniper woodlands, 3800–7300 ft. Widespread.

Penstemon glaber Pursh—Sawsepal penstemon. Erect perennials, (1–)5–6.5(–8) dm; stems glabrous to minutely hairy; basal leaves lanceolate to obovate, 2–8(–15.5) × 0.5–2(–4.5) cm, glabrous or hairy; cauline leaves linear-lanceolate to lanceolate, glabrous or hairy; inflorescence with (5–)8–12 verticillasters, secund; calyx 2–7(–10) mm, glabrous to minutely hairy; corolla 24–40 mm, bluish purple, pink, or deep blue, glabrous externally; staminode included or slightly exserted, glabrous or sparsely bearded; anthers hirsute or rarely glabrous.

1. Staminode notched at the tip; leaves ovate, usually with crisped or undulate margins upon drying; corolla 30–40 mm…**var. brandegeei**
1'. Staminode rounded or obscurely to slightly notched at the tip; leaves mostly lanceolate or ovate, usually lacking crisped or undulate margins; corolla 24–35 mm…**var. alpinus**

var. alpinus (Torr.) A. Gray—Alpine penstemon. Open slopes and meadows, 7000–12,205 ft. NC.

var. brandegeei (Porter) C. C. Freeman [*P. brandegeei* (Porter) Porter ex Rydb.]—Brandegee's penstemon. Mountain meadows, open slopes, 6700–8055 ft. NE, NC.

Penstemon gracilis Nutt.—Lilac penstemon. Erect perennials, (1.5-)2-5 dm; stems with minute retrorse hairs below and glandular hairs; basal leaves lanceolate to ovate or oblanceolate, 2.5-7.5 × 0.4-1.5 cm, cauline leaves linear to lanceolate; inflorescence with 2-7 verticillasters; calyx 4-6 mm, glandular-hairy; corolla 14-22 mm, pale lavender to mauve, glandular-hairy externally, palate bearded with whitish eglandular hairs; staminode reaching orifice or barely exserted, densely bearded; anthers glabrous. Dry grasslands, open forests of Gambel oak, ponderosa pine, and Douglas-fir, 4820-8700 ft. NE.

Penstemon grandiflorus Nutt.—Perennials, 4.5-10 dm; stems glabrous; lower leaves obovate to spatulate, 3-16 × 3-5 cm, cauline leaves sessile, spatulate to orbicular, 2-20 × 3-5 cm, entire, bases clasping; calyx lobes 7-11 mm, glabrous; corollas blue to lavender or pink, bilabiate, 35-48 mm, glabrous, throat abruptly inflated; anther sacs opposite; staminode included to reaching the orifice, sparsely villous; capsules 16-20 mm. Along highways, 5000-5800 ft. Union.

Penstemon griffinii A. Nelson—Griffin's beardtongue. Erect perennials, 2-5 dm; stems minutely hairy at least below; basal leaves mostly oblanceolate to elliptic, 2-6 × 5-10 mm, margins entire, cauline leaves linear-oblanceolate to lance-linear; inflorescence few-flowered, glandular-hairy; calyx 4-5 mm, sparsely glandular-hairy; corolla 15-25 mm, pale blue or purplish, obscurely to moderately glandular-hairy externally, palate densely golden-bearded; staminode not exserted, densely bearded most of its length; anthers glabrous. Sagebrush, mountain mahogany, piñon-juniper woodlands, Gambel oak, ponderosa pine communities, 7300-9670 ft. NC.

Penstemon hallii A. Gray—Hall's beardtongue. Erect perennials, 1-2 dm; stems sparsely glandular-hairy above; leaves oblanceolate to linear-lanceolate, 2-5 cm × 3-8 mm; inflorescence few-flowered to many-flowered, secund; calyx 5-9 mm, glandular-hairy; corolla 18-24 mm, bluish to blue-purple, minutely glandular to glabrous externally, glabrous internally; staminode mostly exserted, short-bearded its entire length, densely so at the apex; anthers minutely hairy. Alpine scree slopes, montane and alpine meadows in gravelly soils; known from 2 records in the San Juan Mountains near the Colorado state line (Barnett and Barnett, 2022: 157; *O'Kane and Heil 8426* (SJNM)). Rio Arriba.

Penstemon inflatus Crosswh.—Inflated beardtongue. Erect perennials, stems 1 or several, 0.4-2.5 dm, minutely hairy; basal leaves lanceolate, often not present at flowering, lower stem leaves lanceolate, narrower near the inflorescence; inflorescence secund, glandular; calyx 3-10 mm, lobes lanceolate to narrowly linear, glandular; corolla 17-27 mm, abruptly inflated, blue, guidelines on the lower side of the throat, glabrous; staminode bearded with yellow-orange hairs for ca. 1/2 the length, not exserted. Sagebrush, piñon-juniper woodlands, Gambel oak, ponderosa pine, Douglas-fir, spruce-fir communities, 6500-10,675 ft. NE, NC, C. **E**

Penstemon jamesii Benth.—James' beardtongue. Erect perennial, 1.4-5.2 dm, stems glabrate to retrorsely puberulent below, glandular-hairy above; leaves linear to lanceolate, 2-11 × (2-)5-15(-30) cm, margins entire to toothed; inflorescence with 2-8 verticillasters, secund; calyx 8-12 mm, glandular-hairy; corolla 24-32 mm, pinkish, or pale lavender to violet-blue, glandular-hairy externally, often sparsely glandular-hairy internally, palate moderately to densely pilose; staminode conspicuously exserted, bearded with hairs to 3.5 mm; anthers glabrous. Shortgrass prairie, oak, mesquite, piñon-juniper, ponderosa pine, montane meadows, 4000-7460 ft. NE, EC, C, SE, SC.

Penstemon lanceolatus Benth. [*P. ramosus* Crosswh.]—Lanceleaf beardtongue. Robust perennial from a woody caudex; stems 3-7.5 dm, puberulent, often branching midway up the stem; leaves of the branches linear, 25 × 1 mm, leaves of the main stem 6-11 × 3-6 mm, linear-lanceolate, puberulent, involute; inflorescence glandular-pubescent, a modified cyme with 1 branch at a node bearing 1-2 flowers; sepals 5-10 mm; corollas straight, red, 28-40 mm; anther sacs opening almost throughout, not explanate; staminodes smooth, included. Rocky areas in desert grasslands and piñon-juniper woodlands, 4000-6000 ft. WC, SW, Eddy.

Penstemon lentus Pennell—Abajo beardtongue. Stems 1 to few, 3-5 dm, plants glabrous and glaucous throughout; leaves entire, fleshy, 2-10 × 1-2 cm, mostly obovate, longer, and wider below, usually lanceolate, shorter, and narrower above; inflorescence glabrous, ± secund, verticillasters usually separated; calyx 4-7 mm, lobes usually ovate to lanceolate, margins scarious, entire or erose; corolla 17-22 mm, glabrous outside, palate sometimes sparsely white-bearded, blue, bluish violet, or white; stamens included, glabrous; staminode essentially included, bearded with short hairs in distal 1/2, tip dilated. Sandy or gravelly soils in sagebrush, piñon-juniper, oak, and ponderosa pine habitats, 5700-7900 ft. NW.

Penstemon linarioides A. Gray—Colorado toadflax beardtongue. Stems few to many, 15-30 cm; sometimes decumbent and rooting at nodes, thus occasionally forming small mats; plants with tiny, retrorsely appressed, scalelike hairs on leaves; stems usually with terete, slender, erect or retrorse hairs; leaves entire, nearly linear or lanceolate to narrowly oblanceolate, 1-2(-3.5) × 1-2.5 mm; inflorescence secund, glandular-puberulent; calyx 5-8 mm, lobes ovate-lanceolate, mostly acuminate, margins scarious; corolla 15-20 mm, glandular-puberulent outside, palate lightly bearded, throat with prominent dark nectar guidelines, reddish lavender to bluish purple; stamens reaching orifice; staminode included, bearded most of length, with longer hairs at slightly widened tip.

1. Principal leaves mostly lanceolate to oblanceolate…**subsp. maguirei**
1′. Principal leaves essentially linear…(2)
2. Staminode sparsely bearded with short hairs, longer golden hairs in apical tuft; base of lower lobes lightly bearded; plants 1-3.5 dm; NW New Mexico…**subsp. coloradoensis**
2′. Staminode densely bearded with long hairs most of its length; base of lower lobes strongly bearded; plants 2-5 dm; SW New Mexico…**subsp. linarioides**

subsp. coloradoensis (A. Nelson) D. D. Keck [*P. coloradoensis* A. Nelson]—Toadflax penstemon. Rocky, dry hillsides, sagebrush communities, piñon-juniper, Gambel oak, ponderosa pine communities, 6500-8760 ft. NC, NW, C.

subsp. linarioides—Toadflax penstemon. Canyons and foothills in piñon-juniper woodlands, mountain brush, oak, ponderosa pine communities, 5270-8500 ft. WC, SW.

subsp. maguirei D. D. Keck—Maguire's penstemon. Limestone cliffs, piñon-juniper woodlands, 6000-6500 ft. Grant.

Penstemon metcalfei Wooton & Standl.—Metcalfe's penstemon. Perennial from a slender, creeping, woody rhizome; stems few, usually erect, to 4 dm, puberulent, glandular; leaves entire, the lower ovate, the upper lanceolate, minutely puberulent, sessile; inflorescence a short interrupted thyrse; calyx to 6 mm, tapering to a sharp tip; corolla 15-25 mm, pale purple, glandular-hairy, with white hairs on the lower lip; staminode bearded with golden hairs, visible but not protruding. Black Range; cliffs or steep N-facing slopes in lower and upper montane coniferous forests, 6000-9500 ft. Grant, Sierra. **R & E**

Penstemon neomexicanus Wooton & Standl.—New Mexico beardtongue. Plants perennial, glabrous; stems stout, 3-7 dm; leaves opposite, basal usually withered at flowering time, lanceolate or oblanceolate, stem leaves narrowly lanceolate or linear; inflorescence secund, not glandular, not congested; calyx lobes lanceolate, 4-8 mm, corolla blue, purple, or violet-blue, 26-34 mm, throat broadly expanded, lower lip expanded; staminode smooth, dilated, often notched at the tip. Wooded slopes and montane meadows in ponderosa pine, spruce, and fir forest, 6000-9000 ft. C, SC. **R**

Penstemon oliganthus Wooton & Standl. [*P. pseudoparvus* Crosswh.]—Apache beardtongue. Perennial, erect, 1.5-6 dm, single or several stems from a basal rosette; leaves glabrous or pubescent, basal leaves ovate, elliptic, or lanceolate with sharp tips, cauline leaves lanceolate or linear; inflorescence a short interrupted thyrse, glandular; calyx 4-6 mm, with broadly lanceolate to lanceolate lobes tapering to a point; corolla 15-26 mm, tubes nearly straight, often drooping, throat bearded with pale golden hairs, glandular; staminode bearded with golden hairs visible at the mouth. Montane meadows in ponderosa pine, Douglas-fir, white pine, spruce communities, 7100-11,400 ft. NC, C, WC.

Penstemon ophianthus Pennell [*P. jamesii* Benth. subsp. *ophianthus* (Pennell) D. D. Keck]—Coiled anther beardtongue. Stems 1 to several, 1-4 dm, plants glabrous or puberulent below, glandular-pubescent above; leaves entire to sometimes sinuate-dentate, 3-12 cm × 6-22 mm; inflorescence glandular,

± secund; calyx 6-10 mm, lobes (narrowly) lanceolate, acute to acuminate; corolla 15-22 × 7-10 mm, glandular-puberulent outside, palate with numerous long, whitish, villous hairs, also sometimes with short, glandular hairs, bluish to pale lavender or blue-violet; stamens included or slightly exserted; staminode clearly exserted, densely bearded with long yellow hairs. Sandy or gravelly soils in sagebrush, piñon-juniper, oak, and ponderosa pine habitats, 5720-8000 ft. W.

Penstemon palmeri A. Gray—Palmer's beardtongue. Stems few to several, robust, 5-14 dm, plants glaucous, glabrous below; leaves usually serrate, fleshy, 5-12 × 2-5 cm, lower petiolate, becoming con-nate-perfoliate above; inflorescence glandular-pubescent, ± secund; calyx 4-6 mm, lobes ovate, acute; corolla sweetly fragrant, 25-35 mm, abruptly expanded, throat inflated, white to pink, dark guidelines conspicuous, palate with sparse white hairs; stamens included, glabrous; staminode exserted, densely bearded with yellow hairs. Washes and roadsides in shrublands, piñon-juniper woodlands, ponderosa pine forests, 4500-6000 ft. Planted in the Southwest for revegetation on roadsides, 3600-7130 ft. Widespread, scattered sites across the state. Introduced.

Penstemon pinifolius Greene—Pine needle beardtongue. Stems woody, forming low mats that send up numerous stems to 2-2.5 dm; leaves dark green, needlelike, 6-20 mm, smooth, scattered on the blooming stalk; inflorescence glandular-hairy, secund; calyx 5-7 mm, lobes lanceolate, acute; corolla 25-32 mm, tubular, scarlet; strongly bilabiate, base of lower lobes with long, flat yellow hairs; anthers explanate, glabrous; staminode bearded most of length with bright yellow hairs. Isolated rocky crags or among pine needles, piñon-juniper, oak, ponderosa pine habitats, 5800-10,000 ft. WC, SW.

Penstemon pseudospectabilis M. E. Jones [*P. connatifolius* A. Nelson; *P. spectabilis* Thurb. ex A. Gray]—Desert beardtongue. Stems several to many, 4-10 dm, from a woody base, plants glabrous (except corolla), glaucous; leaves serrate, petiolate below, becoming connate-perfoliate above, upper perfoliate leaves to 18 × 7.5 cm; inflorescence ± secund, glabrous, of several verticillasters; calyx 5-7 mm, lobes typically ovate, apices acute to acuminate, margins narrowly scarious; corolla pale to deep rose or rose-magenta, with darker guidelines, 22-33 mm, glandular-puberulent externally and internally, palate glabrous; stamens included, anther sacs explanate, glabrous; staminode included, glabrous, narrow. New Mexico material belong to **subsp. connatifolius** (A. Nelson) D. D. Keck. Piñon-juniper woodlands to ponderosa pine forests, 4200-8000 ft. WC, SW.

Penstemon rostriflorus Kellogg [*P. bridgesii* A. Gray]—Bridges' beardtongue. Stems few to several, 3-10 dm, plants glabrous or obscurely puberulent, glandular-pubescent above; leaves entire, mostly cauline, 2-9 × 2-8 mm, lower oblanceolate and petiolate, upper linear or narrow and sessile; inflorescence glandular-pubescent, secund, typically open; calyx glandular-pubescent, 4-6 mm, lobes mostly lanceolate, corolla red to orange-red, 23-33 mm, lightly glandular-pubescent outside, palate pale yellow, glabrous, narrowly funnel-shaped, markedly bilabiate; stamens exserted; anthers horseshoe-shaped, dehiscing near connective only, glabrous; staminode ± included, glabrous, tip not expanded. Desert shrublands, oak and piñon-juniper woodlands, ponderosa pine and montane coniferous forests, 5400-6400 ft. Catron, San Juan.

Penstemon rydbergii A. Nelson—Rydberg's beardtongue. Stems few to several, 2-5 dm, plants usually puberulent; leaves entire, 3-12 × 3-20 mm, basal and lower petiolate, longer, and wider, upper sessile, shorter, and narrower; inflorescence typically puberulent but not glandular, usually in distinctly separated, densely flowered verticillasters; calyx 4-6 mm, lobes ovate, acuminate, margins broadly scarious and erose; corolla 12-20 mm, glabrous outside, palate bearded, throat moderately ampliate, blue to purple; stamens slightly exserted, sacs dehiscing completely including across connective, not explanate, glabrous, blackish; staminode reaching orifice, usually densely yellow-bearded. Our material belongs to **var. rydbergii**. Relatively moist to dry places in meadows, mountain brush, forests, 6200-10,600 ft. N, Catron.

Penstemon secundiflorus Benth.—Sidebells penstemon. Perennials, 1.5-5 dm; stems glabrous, glaucous; basal leaves oblanceolate to spatulate, 2-8(-10.2) cm × 2-25 mm, glabrous and glaucous; cauline leaves lanceolate to ovate, (1.6-)2-7.8 cm × 3-24 mm, glabrous and glaucous; inflorescence a thyrse with (2)3–12 verticillasters, secund; calyx 4–7 mm, glabrous; corolla 15–25 mm, pink to lavender or pale to deep blue, glabrous externally, palate sparsely bearded; staminode densely bearded with pale yellow to golden-yellow hairs. Often in rocky areas, grasslands and meadows in piñon-juniper woodlands, ponderosa pine and montane coniferous forests, 4560–9500 ft. N, DeBaca.

Penstemon strictiformis Rydb. [*P. strictus* Benth. subsp. *strictiformis* (Rydb.) D. D. Keck]—Mancos beardtongue. Stems 1 to few, mostly 2–6 dm, plants glabrous; leaves entire, 2–15 cm × 5–15 mm, mostly narrowly (ob)lanceolate and similar width throughout; basal and lower cauline leaves petiolate and longer, upper leaves sessile, shorter, and often folded; inflorescence glabrous, ± secund, rather narrow or contracted; calyx 5–8 mm, lobes lanceolate, acuminate, or caudate, lower margins scarious, corolla (15–) 20–30 mm, glabrous outside and inside, throat ampliate, pale bluish lavender or bluish purple; stamens exserted; staminode included, glabrous to sparsely villous, tip slightly expanded. Sagebrush, oak and piñon-juniper woodlands, ponderosa pine forests, 5300–10,000 ft. NC, NW.

Penstemon strictus Benth.—Rocky Mountain beardtongue. Stems 1 to few, 3–8 dm, plants glabrous or slightly puberulent in lower parts; leaves entire, 5–15 cm × 4–20 mm, mostly narrowly oblanceolate, basal and lower petiolate, upper sessile, somewhat shorter and narrower, often folded; inflorescence strongly secund, narrow; calyx 2.5–5 mm, lobes ovate, usually obtuse, margins scarious; corolla 23–32 mm, glabrous outside and in, throat expanded, blue to dark blue or blue-purple; stamens exserted, sacs villous, mostly dehiscent; staminode essentially included, sparsely villous or glabrous. Sagebrush, oak and piñon-juniper woodlands, ponderosa pine, Douglas-fir, spruce-fir forests, 6900–10,500 ft. N, C, Catron, Grant.

Penstemon superbus A. Nelson [*P. puniceus* A. Gray]—Superb beardtongue. Plants perennial, glaucous; stems 3–12 dm, margins scarious; leaves grayish, basal petiolate, oblanceolate, spatulate, or elliptic, cauline cordate or connate-perfoliate; inflorescence narrow, glandular or glabrous, flowers in dense fascicles separated by long internodes; calyx 4–5 mm, lobes ovate or elliptic, acute, margins scarious; corolla 17–22 mm, trumpet-shaped, not secund, glandular hairs outside and on the lobes, obscurely bilabiate, coral-pink; stamens included; anthers glabrous, explanate; staminode bearded near the tip with a few short hairs. Rocky canyons, washes, grasslands, sandy or gravelly soil, piñon-juniper woodlands, 3100–5800 ft. Hidalgo.

Penstemon thurberi Torr.—Thurber's penstemon. Stems from a woody base, shrubby, highly branched, 2–4 dm; leaves filiform, smooth, 5–25 mm; inflorescence with flowers mostly single; calyx lobes to 2 mm, ovate, acute, slightly scarious; corolla 10–14 mm, dark pink with a blue cast, slightly downy at the base of the lower lobes, with 2 rows of hairs down the throat; anthers small; staminode narrow and smooth. Often along roadsides in grassland, juniper, and oak communities, 4000–7800 ft. C, WC, SC, SW.

Penstemon virgatus A. Gray—One-side penstemon. Erect perennials, 2–10 dm, stems glabrous; basal leaves narrowly to widely oblanceolate, 2–10 cm × 3–10 mm, glabrous, cauline leaves linear-lanceolate or ovate-lanceolate, glabrous with entire margins; inflorescence a thyrse, secund, glabrous; calyx 3–5 mm, glabrous; corolla 17–26 mm, blue to blue-violet or pinkish, glabrous externally, glabrous to bearded internally; staminode glabrous or with sparse short hairs near apex.

1. Plants puberulous…**subsp. virgatus**
1'. Plants glabrous…**subsp. asa-grayii**

subsp. asa-grayii Crosswh. [*P. unilateralis* Rydb.]—One-side penstemon. Meadows in piñon-juniper and ponderosa pine communities, 5850–9400 ft. NC, C.

subsp. virgatus—One-side penstemon. Montane meadows and hillsides, piñon-juniper, ponderosa pine, and Douglas-fir communities, 6100–9500 ft. NC, C, WC, SW.

Penstemon whippleanus A. Gray—Whipple's or dusky beardtongue. Stems 1 to several, 2–10 dm, plants glabrous below, glandular-pubescent above; leaves entire or dentate, 2–10 × 10–30+ mm; cluster of well-developed basal leaves usually present, lower wider than upper, upper narrower (but not linear) and sessile or clasping, mostly lanceolate throughout; inflorescence glandular-pubescent, secund, cymes usually in dense clusters; calyx 6–11 mm, lobes (narrowly) lanceolate, acuminate, margins scarious at base; corolla 20–30 mm, glandular outside, often very dark (blackish purple), violet, (dark) blue, or lavender to whitish, lower lip projecting well beyond the short upper lip, throat ampliate, palate with long white hairs; stamens included or slightly exserted, sacs glabrous; staminode usually exserted beyond the upper lobe, white, tip with tuft of whitish hairs. Open and forested areas in mountains, subalpine to alpine habitats, 8200–12,500 ft. Widespread except E and SW.

PLANTAGO L. – Plantain

Tina I. Ayers and Kristin Huisinga Harned[†]

Annual or perennial herbs; stems erect; leaves basal, alternate, simple, entire; inflorescence a pedunculate terminal spike; flowers many, crowded; sepals 3–4, free (2 fused in *P. lanceolata*), usually with overlapping scarious margins; corolla actinomorphic or weakly zygomorphic, lobes 4, whitish or scarious, persistent, each with a thickened or colored basal spot; stamens 2 or 4, equal; staminode 0; fruit a circumscissile capsule; seeds 2 to many, mucilaginous when wet (Sivinski, 2001; Shipunov, 2019).

1. Leaves linear to linear-lanceolate, rarely > 1 cm wide; plants annual…(2)
1'. Leaves lanceolate to ovate, rarely < 1 cm wide; plants annual, biennial, or perennial…(6)
2. Seeds 4–8 per capsule…**P. elongata**
2'. Seeds 2 per capsule…(3)
3. Spikes 8–15 mm wide; corolla lobes 3.5–4 mm…**P. helleri**
3'. Spikes 4–8 mm wide; corolla lobes < 3.5 mm…(4)
4. Bracts linear-triangular, 2–16 mm, longer than sepals…**P. patagonica**
4'. Bracts triangular, 2–3.5 mm, shorter than sepals…(5)
5. Leaves silky; spikes interrupted near base at maturity; corolla lobes 1.6–2.3 mm; seeds dark brown to black…**P. argyrea**
5'. Leaves sparsely villous to glabrous; spikes rarely interrupted at maturity; corolla lobes 2.3–3.5 mm; seeds light to dark brown…**P. wrightiana**
6. Corolla lobes erect at maturity…(7)
6'. Corolla lobes spreading or reflexed at maturity…(8)
7. Sepals acute to acuminate, 2.5–3.2 mm; seeds deep red, the inner surface nearly flat…**P. rhodosperma**
7'. Sepals obtuse to acute, 1.5–2.2 mm; seeds yellowish brown, the inner surface deeply concave…**P. virginica**
8. Leaves broadly ovate, abruptly narrowing to a petiole; seeds 6 to many per capsule…**P. major**
8'. Leaves lanceolate, oblanceolate, or elliptic, gradually tapering to a petiole; seeds 2–4 per capsule…(9)
9. Outer pair of sepals (those adjacent to the bract) connate, appearing as a solitary, 2-veined, apically notched or entire sepal; corolla lobes 2–2.5 mm…**P. lanceolata**
9'. Outer pair of sepals (those adjacent to the bract) distinct; corolla lobes 0.6–1.8 mm…(10)
10. Stems densely woolly; corolla lobes 1.2–1.8 mm; seeds 2…**P. eriopoda**
10'. Stems scarcely woolly; corolla lobes 0.6–0.8 mm; seeds 4…**P. tweedyi**

Plantago argyrea E. Morris—Saltmeadow plantain. Plants annual, densely villous; stems to 0–15 mm; leaves without a distinct petiole; blades linear to linear-lanceolate, 2.4–17.4 × 0.2–0.5 cm; peduncles 5–26 cm; spikes 1–9.5 cm, interrupted at maturity; bracts broadly triangular, 2–3 mm; sepals 2.3–3 mm; corolla lobes spreading or reflexed, broadly ovate, 1.6–2.3 mm, stamens 4; seeds 2, ca. 2.5 mm, dark brown to black. Dry piñon-juniper and pine woodlands, 6800–9100 ft. NW, WC.

Plantago elongata Pursh [*P. bigelovii* A. Gray subsp. *californica* (Greene) Bassett]—Prairie plantain, many-seeded plantain. Plants annual, sparsely pubescent; stems 0–5 mm; leaves without a distinct petiole, blades linear, 15–70 × 1–2 mm; peduncles 1–5 cm; spikes 0.5–3 cm, sometimes interrupted at base; bracts ovate, 1.5–2.2 mm; sepals 1.2–2 mm; corolla lobes spreading or reflexed, 0.6–0.8 mm; stamens 2; seeds 6–12, ca. 1.5 mm, dark brown to black. Alkaline clay in sagebrush and piñon-juniper woodlands, 4000–8800 ft. NW, SW.

Plantago eriopoda Torr.—Red-woolly plantain. Plants perennial, sparsely villous or glabrate; stems 0–10 mm; leaves petiolate, blades lanceolate, 5–12 × 1.0–2.5 cm; peduncles 5.5–16 cm; spikes 2–10 cm, interrupted near base; bracts ovate, 2–2.5 mm; sepals ca. 2.5 mm; corolla lobes spreading or reflexed,

1.2–1.8 mm; stamens 4; seeds 2, 2–2.5 mm, light to reddish brown. Alkaline marshes, 7000–8500 ft. McKinley.

Plantago helleri Small—Heller's plantain. Plants annual, densely villous; stems 0–10 mm; leaves without a distinct petiole, blades linear to linear-oblanceolate 50–130 × 5–7 mm; peduncles 8–25 cm; spikes 5–12 cm, not interrupted at maturity; bracts triangular, 4–6.5 mm; sepals 3.5–4 mm; corolla lobes reflexed, 3–3.5 mm; stamens 4; seeds 2, ca. 3.5 mm, dark brown. Dry slopes on limestone, 3150–5700 ft. SE, SC.

Plantago lanceolata L.—Buckhorn plantain. Plants perennial, sparsely pubescent to glabrate; stems 0–20 mm; leaves petiolate, blades lanceolate, 4–25 × 0.5–4.2 cm; peduncles 20–80 cm; spikes 10–30 cm; bracts broadly ovate, (1.5–)2 mm; sepals obovate, 1.5–2.5 mm, the 2 outermost fused; corolla lobes spreading or reflexed, 2–2.5 mm; stamens 4; seeds 2, 1.5–3 mm, brown. Moist soils, 3200–8200 ft. Widespread. Introduced from Eurasia.

Plantago major L.—Common plantain. Plants perennial, sparsely pubescent to glabrate; stems 0–20 mm; leaves petiolate, blades broadly ovate, 3.5–15 × 2–9 cm; peduncles 4–20 cm; spikes 3–24 cm, interrupted near base; bracts ovate, 1–2 mm; sepals broadly ovate to elliptic, 1.2–2.3 mm; corolla lobes spreading or reflexed, 0.7–1.5 mm; stamens 4; seeds 6 to many, ca. 1 mm, olive-green to dark brown. Moist soils, 3235–8200 ft. Widespread. Introduced from Eurasia.

Plantago patagonica Jacq.—Pastoral or woolly plantain. Plants annual, sparsely to densely villous; stems 0–10 mm; leaves without a distinct petiole, blades linear to linear-lanceolate, 1–15 × 0.1–0.7 cm; peduncles 2–22 cm; spikes 3–13 cm; bracts linear-triangular to subulate, 5–16 mm; sepals obovate, 2–5 mm; corolla lobes spreading or reflexed, 1.2–2.5 mm; stamens 4; seeds 2, ca. 2.2 mm, reddish brown to dark brown (rarely black). Deserts and desert grasslands, 3000–8175 ft. Widespread.

Plantago rhodosperma Decne.—Red-seeded plantain. Plants annual, sparsely villous; stems 0–10 mm; leaves petiolate, blades lanceolate, 3–8.5 × 0.8–2.2 cm; peduncles 2–10 cm; spikes 2–15 cm, interrupted near base; bracts subulate to broadly triangular, 2.4–4 mm; sepals 2–3 mm; corolla lobes erect, 2–3.5 mm; stamens 4; seeds 2, ca. 2.5 mm, deep red. Sandy soils, common in shrublands, 3100–5800 ft. EC, SE, SC, SW.

Plantago tweedyi A. Gray—Tweedy's plantain. Plants perennial, sparsely villous to glabrate; stems 0–20 mm; leaves petiolate, blades lanceolate, 3–10 × 1–2.5 cm; peduncles 5–17 cm; spikes 2.5–10 cm, interrupted near base; bracts ovate, 1.4–2 mm; sepals 1.8–2.5 mm; corolla lobes spreading or reflexed, 0.6–0.8 mm, inconspicuous in fruit; stamens 4; seeds 4, ca. 2 mm, brown. Moist meadows, 8600–8700 ft. Mora, San Miguel, Santa Fe.

Plantago virginica L.—Pale-seeded plantain. Plants annual, sparsely villous; stems 0–20 mm; leaves petiolate, blades lanceolate, 5–15 × 0.8–3 cm; peduncles 1–20 cm; spikes 1.5–17 cm; bracts narrowly ovate, 1.8–2.2 mm; sepals ovate, 2.5–3 mm; corolla lobes erect, 2–3 mm; stamens 4; seeds 2, ca. 1.5 mm, yellowish brown. Disturbed areas, 4650–5500 ft. Doña Ana, Eddy.

Plantago wrightiana Decne.—Wright's plantain. Plants annual, sparsely villous to glabrate; stems 0–40 mm; leaves without a distinct petiole, blades linear to linear-lanceolate, 5–22 × 0.1–1 cm; peduncles 9–39 cm; spikes 1–7 cm, rarely interrupted at maturity; bracts broadly triangular, 2.8–3.5 mm; sepals obovate, 3–4 mm, corolla lobes spreading or reflexed, 2.3–3.5 mm; stamens 4; seeds 2, ca. 3 mm, light to dark brown. Moist soils, deserts, desert grasslands, 4000–4300 ft. EC, SE, WC, SW.

SCHISTOPHRAGMA Benth. - Spiralseed

Schistophragma intermedium (A. Gray) Pennell [*Conobea intermedia* A. Gray]—Harlequin spiralseed. Annual herbs; stems erect or ascending, 3–10(–15) cm; leaves cauline, opposite, 10–20 × 3–10 mm, blades ovate to lanceolate, margins pinnatifid (entire); inflorescence axillary, flowers solitary; calyx 2–4 × 0.3–0.5 mm, glabrous or glandular; corolla glabrous, pink or purple, bilabiate, tubular; stamens included,

4; staminode 0; capsules ovoid, 4–6(–8) × 1.5–2.5 mm, slightly upcurved. Sandy soils and scoria, desert scrub, piñon-juniper woodlands, 4150–7000 ft. SC, SW, Cibola (Barringer, 2019c).

SYNTHYRIS Benth. – Kittentail

Perennial herbs, rhizomatous; stems erect (ours); leaves mostly basal, margins with teeth; inflorescences axillary racemes; flowers zygomorphic and bilabiate, in a dense spike; calyx bilaterally symmetric; corolla yellow, green, or blue; stamens 2; staminode 0; ovary 1-locular; capsules flattened. Our species traditionally included in *Besseya* (Hufford, 2019).

1. Corollas blue, purple, lavender, or reddish…**S. alpina**
1'. Corollas pink to white…(2)
2. Corollas 2–3 mm longer than calyces; ovaries puberulent to villous at apex; Sierra Blanca Range…**S. oblongifolia**
2'. Corollas 0–2 mm longer than calyces; ovaries glabrous; N New Mexico…**S. plantaginea**

    Synthyris alpina A. Gray [*Besseya alpina* (A. Gray) Rydb.]—Alpine kittentail. Plants 0.5–2 dm; leaf blades ovate to slightly cordate, 2–5 × 1.5–4 cm, margins crenate; racemes erect, to 10 cm in fruit; sepals 4, 4–6 mm, densely villous; petals (3)4(5), corolla blue, bluish purple, lavender, or reddish, bilabiate; filaments purple; capsules 3–5 mm. Rocky alpine meadows and fell-fields, 10,000–13,165 ft. NC.

    Synthyris oblongifolia (Pennell) L. Hufford & M. McMahon [*Besseya oblongifolia* Pennell]—Eggleaf kittentail. Perennial herb; stems to 30 cm, densely hairy; leaf blades narrowly ovate to ovate or oblong-ovate, 7–11 cm × 25–30 mm, margins crenate, surfaces glabrous or sparsely hairy; sepals 3 or 4; petals 3, corolla pink or white, bilabiate, 5–6 mm; capsules hairy, 4–5 mm. Alpine and subalpine meadows, 11,000–12,000 ft. Lincoln, Otero. **R & E**

    Synthyris plantaginea (E. James) Benth. [*Besseya plantaginea* (E. James) Rydb.]—Foothills kittentail. Perennial herb; stems 1–4 dm; leaf blades narrowly to broadly ovate or elliptic, 5–15 cm, margins crenate, surfaces sparsely hairy to villous; inflorescence a raceme, to 40 cm in fruit; sepals 4, villous; petals (3)4(5), corolla pink to white, bilabiate, 5–8 mm; filaments rose to purple; capsules 3–6 mm. Openings in mixed conifer forests, subalpine meadows, 6575–10,000 ft. NC, NW, Catron.

VERONICA L. – Speedwell

Perennial or annual herbs; stems creeping to erect, glabrous or hairy, leaves opposite, cauline, margins entire, dentate, or serrate; inflorescences terminal and/or axillary racemes, flowers subtended by foliaceous bracts; sepals 4(5), bilaterally symmetric; corolla nearly actinomorphic, purple, blue, pink, or white, 4-lobed; stamens 2; staminode 0; ovary 2-locular (Albach, 2019).

1. Racemes 1–25, axillary…(2)
1'. Racemes 1(–20), terminal, sometimes also with 1–4 axillary racemes, flowers sometimes appearing solitary…(4)
2. Petioles 2–6(–10) mm…**V. americana**
2'. Petioles 0(–5) mm…(3)
3. Corollas white to pale pink; calyx lobe apices obtuse; stamens 5 mm; pedicels equal to or ± shorter than subtending bract; leaf margins entire or subentire…**V. catenata**
3'. Corollas pale lilac to pale blue or lavender (rarely white); calyx lobe apices acute; stamens 2–3.5 mm; pedicels equal to or longer than subtending bract; leaf margins ± serrulate or denticulate…**V. anagallis-aquatica**
4. Perennials; leaf blades usually at least 3 mm longer than the bracts; bracts 1–9(–15) mm…(5)
4'. Annuals; leaf blades usually < 10 mm…(6)
5. Capsules wider than long; stems with scattered eglandular hairs, often with glandular hairs, sometimes glabrate…**V. serpyllifolia**
5'. Capsules about as long as or longer than wide; stems eglandular or glandular-hairy…**V. wormskjoldii**
6. Pedicels 6–30 mm…(7)
6'. Pedicels 1–2 mm…(8)
7. Pedicels 15–30 mm; corollas 8–12 mm across…**V. persica**
7'. Pedicels 6–15 mm; corollas 4–8 mm across…**V. polita**

8. Corollas white or pale pink; leaf blades 3–10 times as long as wide, margins entire or dentate distally…**V. peregrina**
8'. Corollas ± blue; leaf blades 1–2.5 times as long as wide, margins crenate-serrate…**V. arvensis**

Veronica americana Schwein. ex Benth. [*V. beccabunga* L. var. *americana* Raf.]—American speedwell. Perennials; stems decumbent or ascending, 5–50 cm, glabrous; leaves (5–)30–50(–100) × (3–)7–20(–30) mm, 2–4 times as long as wide, margins entire or serrate; inflorescence an axillary raceme; pedicels 3(–12) mm, equal to or ± longer than subtending bract, glabrous; calyx 2.5–4.5(–5.5) mm; corolla blue to violet, 5–10 mm wide; capsules 2–4 × 3–5 mm, glabrous. Along creeks and streams, moist meadows, flowing water, 4450–11,200 ft. Widespread except E.

Veronica anagallis-aquatica L. [*V. salina* Schur]—Water speedwell. Annuals or perennials; stems erect or prostrate basally, (20–)30–100(–170) cm, glabrous; leaves (15–)30–80(–145) × (7–)10–30(–45) mm, 1.5–3 times as long as wide, blades ovate, elliptic, or oblong, margins ± serrulate or denticulate, surfaces glabrous, rarely glandular-hairy; inflorescence an axillary raceme; calyx 3–5.5 mm; corolla blue to violet, 5–10 mm wide; capsules 2.8–4 mm, glabrous or sparsely short-glandular-hairy. Moist places, stream margins, ditches, springs, wet meadows, 3500–9400 ft. Widespread except EC, SE. Introduced.

Veronica arvensis L.—Corn speedwell. Annuals; stems erect to ascending, 1–30(–40) cm; leaves in 3–6 pairs per stem, blades oblong to broadly ovate, (2–)5–14(–35) × (2–)3–10(–18) mm, 2.5 times as long as wide, surfaces mostly sparsely eglandular-hairy; inflorescence a terminal raceme, eglandular- and glandular-hairy; pedicels 0–4 mm, shorter than subtending bract; calyx 3–5 mm; corolla blue to violet, 2–3 mm wide; capsules 2–4 × 2.5–5 mm, glandular-hairy on margins. A weed in disturbed sites, especially lawns and gardens, 3600–7000 ft. Doña Ana, Hidalgo, Luna. Introduced.

Veronica catenata Pennell [*V. catenata* var. *glandulosa* (Farw.) Pennell]—Chain speedwell. Annuals or perennials; stems erect or ascending, 15–60(–80) cm, glabrous or glandular-hairy distally; leaves oblong-ovate to oblong-lanceolate, (5–)25–50(–100) × 4–15(–30) mm, 2.5–5 times as long as wide, margins entire or subentire, glabrous; inflorescence an axillary raceme; pedicels equal to or ± shorter than subtending bract; calyx lobes 2.5–3 mm; corolla white to pale pink with darker veins, 4–5 mm diam.; capsules 2.5–3(–3.5) × 3–4 mm, glabrous or glandular-hairy. Wet places, running water, lakeshores, ditches, stream channels, 5050–8225 ft. N, C, SW.

Veronica peregrina L. [*V. peregrina* var. *xalapensis* (Kunth) Pennell]—Purslane speedwell. Annuals; stems erect or ascending, (2.5–)4–25(–35) cm, glabrous or densely glandular-hairy; leaves 5–28(–35) × 2–6(–10) mm, oblong to oblanceolate, 3–10 times as long as wide; surfaces glabrous or densely glandular-hairy; inflorescence a terminal raceme, glabrous or densely glandular-hairy; calyx 3–6 mm; corolla white, 2–3 mm wide; capsules 2.5–5 × 2.5–6 mm, glabrous. Moist places, creeks, streams, pond shorelines, moist meadows, 3570–11,500 ft. Widespread except EC.

Veronica persica Poir. [*V. persica* var. *aschersoniana* (E. B. J. Lehm.) Drabble & J. E. Little]—Large field or bird's eye speedwell. Annuals; stems creeping to decumbent, 10–50(–60) cm, eglandular-hairy; leaves (6–)9–18(–30) × (5–)8–15(–20) mm, broadly ovate or broadly lanceolate, margins serrate, surfaces sparsely eglandular-hairy; inflorescence a terminal raceme, eglandular-hairy; pedicels 1–2(3) times length of subtending bract; calyx 4.5–8 mm; corolla blue, 8–12 mm wide; capsules 4–6 × 5–9 mm. A weed of lawns and gardens, scattered, 3600–4350 ft. NC, C, SE, SC. Introduced.

Veronica polita Fr.—Gray field speedwell. Annuals; stems 5–25 cm, prostrate or reclining, pilose to short-hirsute; leaves short-petiolate, blades ovate to oval, 5–15 mm, coarsely crenate-serrate; flowers single in the axils or floral bracts, pedicels 6–15 mm; calyx 4–8 mm; corollas blue, 4–8 across; capsules 4–6 mm wide, wider than long, glandular-pubescent. Lawns and gardens. Doña Ana, Otero.

Veronica serpyllifolia L. [*V. humifusa* Dicks.]—Thyme-leaved speedwell. Perennials; stems creeping to ascending, 5–40 cm, with scattered eglandular hairs, often also with glandular hairs, sometimes glabrate; leaves 8–25 × 5–13 mm, 1.5–2.5 times as long as wide, margins subentire or serrulate-crenate; inflorescence a terminal raceme, glandular-hairy; pedicels 2–5 mm, 4–6 mm in fruit, shorter than subtending bract; calyx 2.5–4 mm; corolla blue, 4–8 mm; capsules 2.5–3.5 × 4–5.5 mm, sparsely glandular. Moist places, meadows, along creeks, streams, boggy sites, 5600–12,050 ft. NC, NW, WC, SW. Introduced.

Veronica wormskjoldii Roem. & Schult. [*V. alpina* L. subsp. *wormskjoldii* (Roem. & Schult.) Elenevsky; *V. nutans* Bong.]—American alpine speedwell. Perennials; stems erect or ascending, (3–)8–50 cm, villous-hirsute; 8–40 × 5–20 mm, elliptic to lanceolate or oblong-ovate, margins entire, dentate, or serrate, surfaces sparsely to densely villous-hirsute or glabrous; inflorescence a terminal raceme, glandular; pedicels 2–10(–15) mm, equal to subtending bract; calyx 3.5–5.5 mm; corolla blue, 6–10 mm wide; capsules 4–6(–8) × (2.8–)4–5.5 mm, usually densely glandular-hairy. Along creeks and streams, moist meadows, seep areas, alpine, 7900–12,500 ft. NC.

PLATANACEAE – SYCAMORE FAMILY

Kenneth D. Heil

Trees, bark peeling, scaly; leaves alternate, simple, palmately lobed, deciduous; stipules large, early-deciduous; flowers tiny, wind-pollinated, unisexual; plants monoecious; sepals 3–7; petals 3–7; sepals mostly absent in female flowers; stamens the same number as perianth parts; pistil superior, 3–9 separate carpels; fruit a hairy achene (Kaul, 1997).

PLATANUS L. – Sycamore

Platanus wrightii S. Watson—Arizona sycamore. Tree to 20 m; 1 to several trunks; leaf blades 9–25 × 9–25 cm, 3–7-lobed, margins entire or remotely serrulate, glabrescent; fruiting heads mostly 2–4; inflorescence rachis 20–30 cm. Riparian communities, 3935–8000 ft. WC, SW.

PLUMBAGINACEAE – LEADWORT FAMILY

Don Hyder

LIMONIUM Mill. – Sea lavender

Limonium limbatum Small—Trans-Pecos sea lavender. Herbs perennial; stems erect, leafless; leaves in a basal tuft, thick and leathery; inflorescences large, many-branched panicles 30–60 × 30 cm, flowers clustered at branchlet tips; calyx obconic to slightly funnel-shaped, 4–5 mm; petals 5 per flower, fused at the base and divergent above, petals nearly distinct, blue to nearly white; ovary superior, 1-locular; fruit a utricle. Marshy ground, ciénegas, floodplains, saline wet grasslands, roadside ditches, 3000–6500 ft. C, SC, Cibola, Guadalupe (Reeves, 2013b).

POACEAE (GRAMINEAE) – GRASS FAMILY

Kelly W. Allred

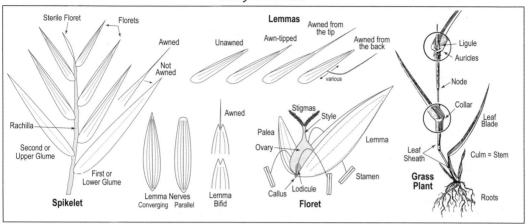

Herbs or less commonly shrubby or even treelike, the stems usually round and with hollow internodes; leaves alternate or basal, simple, differentiated into sheath, ligule, and blade, the sheath margins overlapping or fused; flowers small, inconspicuous, hidden within an inflorescence of spikelets, spikelets composed of basal glumes, lemmas, paleas, and flowers; perianth, if present, modified as tiny lodicules; stamens 1 or 3 (ours); pistil single, superior, of 3 united carpels, the styles usually 2 and the stigmas featherlike; fruit a caryopsis, rarely nutlike or a utricle. This treatment includes all grasses known to grow outdoors in New Mexico, whether cultivated, ornamental, crop, exotic, adventive, or native (Allred, 2005).

MAIN KEY

1. Plants not known to flower in New Mexico; spikelets not produced; blades constricted at the base into a narrow stalklike portion with a tuft of stiff bristles on each side: cultivated ornamentals…**Phyllostachys**
1'. Plants usually flowering each year; spikelets present; blades not constricted at the base into a narrow stalklike portion and without stiff bristles on each side; cultivated or wild grasses…(2)
2. All or some spikelets concealed and hidden from view within modified structures, such as spiny burs, involucres, bony rachis joints, dense fleshy cobs (ears), or detachable clusters of hard bracts…**KEY A**
2'. Spikelets not concealed or hidden within modified structures, but evident and easily seen, sometimes closely subtended by foliage leaves or covered by hairs…(3)
3. One or more bristles (sterile branchlets) borne immediately below the spikelets, the bristles sometimes clustered into a bur or involucre…(4)
3'. Bristles not borne immediately below the spikelets, a bur or involucre absent…(5)
4. Spikelets disarticulating singly, leaving the bristles on the plant…**Setaria**
4'. Spikelets disarticulating with the involucre of bristles, the two falling together…**Pennisetum**
5. Glumes with numerous hooked prickles 1–2 mm…**Tragus**
5'. Glumes lacking hooked prickles…(6)
6. Lemmas with 7–13 awns (rarely 5)…**KEY B**
6'. Lemmas with 1–3 awns or awnless…(7)
7. Flowering shoots 2+ m…**KEY C**
7'. Flowering shoots < 2 m…(8)
8. All or many spikelets sessile and borne on the main axis; inflorescence branches absent, the inflorescence a spike, spicate raceme, or dense headlike cluster of spikelets…**KEY D**
8'. All or most spikelets borne on branches; inflorescence a panicle, or if branches absent then all the spikelets with evident pedicels and few (if any) sessile…(9)
9. Andropogoneae tribe: Glumes mostly hardened (membranous in *Zea* and *Imperata*), completely enclosing the florets, dorsally compressed; disarticulation below the glumes and nearly always in units consisting of a sessile spikelet with attached rachis joint and pedicel (pedicelled spikelet present or absent); spikelets borne in pairs, 1 spikelet sessile or subsessile and 1 spikelet pedicelled (sometimes pedicelled spikelet absent, but pedicel always present); lemmas very thin and translucent, delicate, awned or awnless…**KEY E**
9'. Combination of features other than above…(10)
10. Spikelets with a single floret only…**KEY F**

10'. Spikelets with at least 2 florets (some may be small and poorly developed; look carefully)…(11)
11. Paniceae tribe: Spikelets with 2 florets, the upper bisexual and usually with a hardened lemma at maturity, the lower male or neuter; lemma of the lower floret similar to the second glume in size and texture; disarticulation below the glumes; spikelets dorsally compressed…**KEY G**
11'. Combination of features other than above…(12)
12. Lemmas with 3 nerves, the nerves usually prominent…**KEY H**
12'. Lemmas with 5 to many nerves, at least at the base, or the nerves not discernible…**KEY I**

KEY A: SPIKELETS VARIOUSLY CONCEALED

1. Spikelets enclosed in a bur (involucre) of bristles or stiff spines, the bur falling entire…(2)
1'. Spikelets not enclosed in a bur (involucre) of bristles or spines…(3)
2. Bur of sharp, stiff spines…**Cenchrus**
2'. Bur of bristles, without spines…**Pennisetum**
3. Plants mat- or sod-forming, with stolons or rhizomes…(4)
3'. Plants not mat-forming, without stolons or rhizomes…(5)
4. Sheaths strongly compressed-keeled; spikelets all alike and sunken into 1 side of a corky or succulent, flattened rachis; cultivated lawn grasses…**Stenotaphrum**
4'. Sheaths rounded; spikelets unisexual and different in appearance, the male on spicate, flaglike primary branches raised above the foliage, the female in bony clusters hidden in the foliage; native range grasses, but sometimes also grown as a lawn grass (*Bouteloua* [*Buchloe*] *dactyloides*)…**Bouteloua**
5. Glumes with numerous hooked prickles 1–2 mm…**Tragus**
5'. Glumes lacking hooked prickles…(6)
6. Female spikelets borne singly in hard, whitish beads at the ends of long stalks; ornamental or garden grasses only infrequently grown…**Coix**
6'. Female spikelets borne in cobs, or if beadlike then several borne adjacent to each other; cultivated or wild grasses…(7)
7. Spikelets borne in spicate racemes not > 2 cm; spikelets paired, the sessile one bisexual, grenade-shaped, and covered with square pits, the pedicelled one male and flattened; rare…**Mnesithea**
7'. Spikelets borne in panicles or cobs > 10 cm; spikelets all unisexual, the sexes in different part of the same inflorescence or in separate inflorescences on the same plant…(8)
8. Male spikelets borne in a terminal panicle (tassel); female spikelets borne below in a thick axillary spike (cob) and covered by leaf sheaths, the styles (silk) protruding from the tip; cultivated grasses…**Zea**
8'. Male and female spikelets borne together in the same panicle, the male ones papery and in pairs at the terminal portion of the spicate branches, the female ones bony and at the base of the same branches; wild grasses, but probably not extirpated from the state…**Tripsacum**

KEY B: LEMMAS WITH 7–13 AWNS

1. Awns plumose, feathery, ± equal in length…**Enneapogon**
1'. Awns glabrous to scabrous, not plumose and not equal in length…(2)
2. Glumes 1-nerved…**Pappophorum**
2'. Glumes many-nerved…**Cottea**

KEY C: FLOWERING SHOOTS 2+ METERS

1. Grasses cultivated for ornament, landscaping, or as a harvested crop, occasionally escaping around fields or dwellings…(2)
1'. Grasses wild or weedy, or seeded for range or pasture improvement, but not crop or ornamental plants …(8)
2. Corn: male spikelets borne in a terminal panicle (tassel); female spikelets borne on the stem in a thick axillary spike (cob or ear) covered by leaf sheaths, the styles (silks) protruding from the tip…**Zea**
2'. Plants not as above…(3)
3. Plants growing in large, thick tussocks with numerous flowering shoots; rhizomes lacking…(4)
3'. Plants not in large tussocks, the shoots single, or if clustered then with strong vigorous rhizomes…(6)
4. Blades sharply saw-toothed on the margins; spikelets borne singly on rebranching branches of the inflorescence, with several florets extending beyond the thin glumes…**Cortaderia**
4'. Blades scabrous to smooth on the margins; spikelets borne in pairs on spicate branches, with no florets extending beyond the stiff glumes…(5)
5. Panicle branches breaking apart at the nodes (joints) when mature…**Tripidium**
5'. Panicle branches remaining intact, the spikelets falling separately when mature…**Miscanthus**
6. Plants annual, lacking rhizomes…**Sorghum bicolor**
6'. Plants perennial, with vigorous rhizomes…(7)

7. Panicles plumelike, with very dense silky hairs; plants commonly to 6–7 m…**Arundo**
7'. Panicles slightly pubescent but not plumelike; plants rarely > 3 m…**Sorghum** (in part)
8. Plants tufted, not developing rhizomes…(9)
8'. Plants developing rhizomes…(14)
9. Spikelets subtended by numerous bristles; plants annual…**Setaria magna**
9'. Spikelets not subtended by bristles, but may be pubescent; plants perennial…(10)
10. Inflorescence a spike, no branches developed…**Elymus**
10'. Inflorescence a panicle with branches…(11)
11. Disarticulation above the glumes; spikelets awned…(12)
11'. Disarticulation below the glumes; spikelets awned or awnless; sheaths mostly rounded…(13)
12. Basal sheaths compressed-keeled; spikelets purplish; awns < 1.5 cm…**Muhlenbergia**
12'. Basal sheaths round; spikelets greenish or tawny; awns 2–3 cm…**Eriocoma robusta**
13. Inflorescence branches 2–5 and mostly not rebranched, clustered toward the tip of the shoot…**Andropogon gerardi** (in part)
13'. Inflorescence branches numerous and rebranched, not clustered toward the tip of the shoot…**Panicum** (in part)
14. Disarticulation below the glumes, the spikelets falling entire…(15)
14'. Disarticulation above the glumes, the glumes remaining on the plant and the florets falling…(18)
15. Inflorescence a panicle of 2–5 spicate, unbranched primary branches clustered at the tip of the shoot, sometimes a few branches rebranching…**Andropogon gerardi** (in part)
15'. Inflorescence a rebranched panicle, the numerous primary branches always rebranching…(16)
16. Outer bracts of the spikelet (glumes) membranous, thin, and flexible, not hardened; upper floret hardened at maturity; spikelets awnless…**Panicum** (in part)
16'. Outer bracts of the spikelet (glumes) stiff, hardened; inner floret very thin and delicate, not at all hardened; spikelets awned, at least when young…(17)
17. Spikelets dull, fuzzy-hairy, the hairs standing out from the spikelet; awn persistent through maturity…**Sorghastrum**
17'. Spikelets somewhat shiny, glabrous or slightly pubescent, the hairs pressed against the spikelet; awn early-deciduous…**Sorghum** (in part)
18. Panicles with unbranched spicate branches…**Spartina**
18'. Panicles with rebranched branches…(19)
19. Spikelets with a single floret…**Sporobolus** (in part)
19'. Spikelets with several florets…(20)
20. Glumes nearly equal in length; rachilla glabrous; lemmas long-hairy…**Arundo**
20'. Glumes unequal, the first about 1/2 as long as the second; rachilla beset with long silky hairs; lemmas glabrous…**Phragmites**

KEY D: INFLORESCENCE A SPIKE, SPICATE RACEME, OR DENSE HEADLIKE CLUSTER; ALL OR MANY SPIKELETS SESSILE ON THE MAIN AXIS; BRANCHES ABSENT FROM THE INFLORESCENCE

1. Disarticulation below the glumes, the spikelets falling entire or in clusters, no spikelet parts left on the axis…(2)
1'. Disarticulation above the glumes, the glumes often remaining on the inflorescence…(27)
2. Main axis of the inflorescence breaking apart at maturity…(3)
2'. Main axis of the inflorescence remaining intact…(16)
3. Spikelets borne in pairs of 1 sessile and 1 pedicelled (sometimes only the pedicel present); glumes mostly enclosing the spikelet, the florets mostly not visible (Andropogoneae tribe)…(4)
3'. Spikelets borne other than above; glumes may be longer than, but not enclosing the spikelet, the florets usually visible (Triticeae tribe)…(8)
4. Spikelets awned, the awns at least 5 mm…(5)
4'. Spikelets awnless, or with awns 1–2 mm…(7)
5. Awns 1–2 cm…**Schizachyrium**
5'. Awns 4–12 cm…(6)
6. Racemes 4–8 cm; awns 5–12 cm; main axis (or most of it) breaking apart when mature…**Heteropogon**
6'. Racemes 10–18 cm; awns 4–6 cm; main axis persistent…**Trachypogon** (in part)
7. Racemes < 3 cm, glabrous or only sparsely pubescent; plants annual…**Mnesithea**
7'. Racemes > 4 cm, densely woolly-pubescent; plants perennial…**Elionurus**
8. Spikelets 3 at each node of the main axis, the lateral pair pedicelled, the central spikelet sessile; spikelets with 1 floret…**Hordeum** (in part)
8'. Spikelets mostly 1 or 2 at each node of the main axis, if 3 then not otherwise as above; spikelets with 2 to many florets…(9)

9.   Spikelets mostly 1 at each node of the main axis…(10)
9'. Spikelets mostly 2 at each node of the main axis…(14)
10. Plants annual…(11)
10'. Plants perennial…(12)
11. Spikes 0.6-2 cm…**Eremopyrum** (in part)
11'. Spikes 5-10 cm…**Aegilops**
12. Inflorescence very dense, almost headlike, the rachis obscured and viewed only with difficulty; fertile plants of alpine or subalpine habitats (*Elymus scribneri*)…**Elymus** (in part)
12'. Inflorescence less congested and somewhat elongate, not at all headlike, the rachis easily observed; sterile hybrid plants of low-elevation or midmontane habitats…(13)
13. Awns of the lemma 4-17 mm, usually erect; rachis internodes 2.5-6(-7) mm…these are *Elymus elymoides* × *E. trachycaulus* hybrids [*Elymus saundersii* Vasey; *Agropyron saundersii* (Vasey) Hitchc.]
13'. Awns of the lemma (14-)18-37 mm, spreading to recurved downward; rachis internodes mostly 7-10 mm…these are *Elymus elymoides* × *E. spicata* hybrids
14. Glumes 3-7 mm; anthers 4-5 mm…**Psathyrostachys** (in part)
14'. Glumes 12-100 mm; anthers, when present, ca. 2 mm…(15)
15. Glumes 12-24 mm; sterile hybrid plants…these are *Elymus trachycaulus* × *Hordeum jubatum* hybrids [*Elyhordeum macounii* (Vasey) Barkworth & D. R. Dewey; *Elymus macounii* Vasey]
15'. Glumes 25-100 mm; fertile plants…**Elymus** (in part)
16. Plants strongly rhizomatous or stoloniferous perennials…(17)
16'. Plants tufted annuals or perennials, not stoloniferous or rhizomatous…(19)
17. Wild range grasses, not cultivated in lawns; spikelets falling in clusters of 3…**Hilaria**
17'. Lawn grasses, occasionally escaping in weedy ground in residential areas; spikelets not falling in clusters of 3…(18)
18. Plants mostly stoloniferous; blades fleshy and somewhat succulent; spikelets borne on 1 side of a flattened, succulent main axis…**Stenotaphrum** (in part)
18'. Plants mostly rhizomatous; blades thin and membranous, not at all succulent; spikelets variously disposed on short pedicels around the thin, nonsucculent main axis…**Zoysia**
19. Plants cultivated lawn grasses or weedy in lawns…(20)
19'. Plants of various habitats, but never cultivated or weedy in lawns…(21)
20. Spikelets pointed at the tip and arranged on 1 side of a thickened rachis…**Stenotaphrum** (in part)
20'. Spikelets blunt at the tip and arranged on both sides of the rachis…**Sclerochloa** (in part)
21. First glume with 2 or 3 awns; lower stems angled or flattened somewhat…**Muhlenbergia** (in part)
21'. First glume with a single awn or awnless; lower stems rounded…(22)
22. Awns 4-6 cm…**Trachypogon** (in part)
22'. Awns, if present, < 2 cm…(23)
23. Ligules hairy; sheaths prominently inflated; blades widely spreading to reflexed; inflorescence dense and headlike or spikelike, the base often included in the sheath; much-branched annuals (*Sporobolus alopecuroides* and *S. schoenoides*)…**Sporobolus**
23'. Plants not as above in all respects…(24)
24. Spikelets in pedunculate clusters of 3, usually hanging downward, and falling together…**Aegopogon**
24'. Spikelets not so arranged…(25)
25. Glumes awnless; lemma awned (use a lens)…**Alopecurus**
25'. Glumes awned…(26)
26. Glumes strongly flattened laterally, ciliate on the keeled midnerve…**Phleum** (in part)
26'. Glumes rounded on the back, not keeled, not ciliate on the midnerve but may be pubescent elsewhere…**Polypogon**
27. Spikelets of 2 different kinds, the male spikelets awnless and the female spikelets with awns 9-10 cm; plants mostly dioecious and stoloniferous…**Scleropogon**
27'. Spikelets all similar, awnless or with awns mostly < 6 cm; plants tufted or if stoloniferous then with short awns…(28)
28. Spikelets in very dense, ovoid, woolly or bristly heads, at most 2 times longer than wide, with longer awns conspicuous and protruding (resembling *Polypogon*); plants annual…(29)
28'. Plants not as above in every characteristic…(30)
29. Seed heads stiff-bristly; plants essentially glabrous…**Cynosurus**
29'. Seed heads soft-woolly; plants with markedly pubescent leaves and sheaths…**Lagurus**
30. Lemmas with 3 awns…**Aristida**
30'. Lemmas with 1 awn or awnless…(31)
31. Spikelets with 1 floret only…(32)
31'. Spikelets with > 1 floret, some may be poorly developed, rudimentary, or vestigial…(34)
32. Plants annual; leaves with prominent, clawlike auricles 2-6 mm; awns 50-160 mm…**Hordeum** (in part)

32'. Plants perennial; leaves without auricles, or occasionally with small rounded auricles ca. 1 mm; awns 1–4 mm…(33)

33. Spikelets strongly compressed; glumes flattened, keeled on the midnerve, completely enclosing the floret…**Phleum** (in part)

33'. Spikelets not strongly compressed; glumes rounded on the back, only slightly keeled, not completely enclosing the floret…**Muhlenbergia** (in part)

34. Spikelets in dense, sessile, headlike clusters that are mostly surpassed by and nestled within the foliage …**Munroa**

34'. Spikelets not in dense, headlike clusters, or if so then elevated well above the foliage…(35)

35. Lemmas with 3 conspicuous nerves…(36)

35'. Lemmas with 1 or 5 to several nerves…(37)

36. Lemmas conspicuously pubescent; spikelets with several well-developed florets; blades white-margined …**Erioneuron**

36'. Lemmas glabrous or scabrous; spikelets with 1 well-developed floret and 1–3 rudiments above it; blades not white-margined…**Bouteloua**

37. Plants low annuals; inflorescence not a true spike but the branches very short, with 1–3 spikelets borne on short pedicels nearly on the main axis; lemmas ca. 2 mm, glumes mostly shorter…**Catapodium**

37'. Plants, inflorescence, lemmas, and glumes not as above…(38)

38. Spikelets 2+ per rachis node…(39)

38'. Spikelets mostly 1 per rachis node…(43)

39. Rhizomes present, evident, creeping…**Leymus** (in part)

39'. Rhizomes absent, occasionally short rhizomes developed but the plants still forming dense clumps… (40)

40. Glumes absent or reduced to 1 or 2 minute bristles; spikelets horizontally spreading or ascending at maturity…**Elymus** (in part)

40'. Glumes present; spikelets rarely horizontally spreading…(41)

41. Glumes 2–10 cm…**Elymus** (in part)

41'. Glumes < 1.5 cm…(42)

42. Glumes 2–5-nerved; anthers 1.5–3 mm…**Elymus** (in part)

42'. Glumes 1-nerved; anthers 3–5 mm…**Psathyrostachys** (in part)

43. Spikelets placed edgewise to the rachis, the first glume absent on all but the terminal spikelets… **Lolium**

43'. Spikelets placed flatwise to the rachis; both glumes present on all spikelets…(44)

44. Plants annual…(45)

44'. Plants perennial…(49)

45. Spikes very short, 0.6–2 cm; plants usually < 30 cm…(46)

45'. Spikes longer, mostly 5–15 cm; plants usually much > 30 cm…(47)

46. Inflorescence exserted from the sheath at maturity; glumes and lemmas awn-tipped; blades with small auricles…**Eremopyrum** (in part)

46'. Inflorescence often partially enclosed in the upper sheath; glumes and lemmas blunt-tipped; blades lacking auricles…**Sclerochloa** (in part)

47. Glumes narrow, linear, 1-nerved; spikelets with 2 florets…**Secale**

47'. Glumes broad, oblong to ovate, 3- to several-nerved; spikelets mostly with 3–5 florets…(48)

48. Nerves of the lemma converging at the apex; plants commonly glaucous…**×Triticosecale**

48'. Nerves of the lemma ± parallel, not converging at the apex; plants commonly green and not glaucous… **Triticum**

49. Spikelets borne in pairs, of 1 pedicelled and 1 nearly sessile; glumes awnless; lemmas awned, the awns 4–6 cm…**Trachypogon** (in part)

49'. Spikelets not as above…(50)

50. Glumes linear, needlelike, 1-nerved (occasionally broader at the base and 3-nerved)…**Leymus** (in part)

50'. Glumes lanceolate or broader, usually 3–7-nerved…(51)

51. Spikelets spreading away from the rachis, placed very close together on the main axis; rachis internodes between the spikelets 0.3–3 mm in the middle of the spike…**Agropyron**

51'. Spikelets mostly pressed against the rachis, or curving outward toward the tip of the spikelet; rachis internodes between the spikelets 4–25 mm…(52)

52. Glumes acuminate, symmetric to curved and somewhat sickle-shaped, gradually tapering to an awn-tip; blades somewhat rigid and prominently rigid above; plants rhizomatous, commonly bluish (*P. smithii*)… **Pascopyrum**

52'. Glumes various, blunt to acuminate, symmetric, not curving, not gradually tapering to an awn-tip; blades often lax, not prominently ridged above; plants tufted to rhizomatous, not commonly bluish…**Elymus** (in part)

KEY E: TRIBE ANDROPOGONEAE

1. Spikelets all unisexual, the male and female spikelets conspicuously different in form and borne either separately in the same inflorescence or in separate inflorescences on the same plant; plants monoecious …(2)
1'. Spikelets unisexual or bisexual but usually not conspicuously different in form, borne in pairs and not separated one from the other; plants not monoecious…(4)
2. Female spikelets borne singly in hard, whitish beads at the end of long stalks; domesticated grasses… **Coix**
2'. Female spikelets in cobs, or if beadlike then not borne singly at the end of long stalks but adjacent to other bony spikelets; wild or domesticated grasses…(3)
3. Male spikelets borne in a terminal panicle (tassel); female spikelets in a separate inflorescence and borne below in a thick axillary spike (cob) and covered by leaf sheaths, the styles (silks) protruding from the tip; domesticated grasses…**Zea**
3'. Male and female spikelets borne together in the same panicle, the male ones papery and in pairs at the terminal portions of the spicate branches, the female spikelets bony and at the base of the same branches; wild grasses…**Tripsacum**
4. Each inflorescence a panicle with branches (occasionally a few inflorescences with a single branch), with or without inflated sheaths subtending the inflorescence (spathes)…(5)
4'. Each inflorescence a single unbranched spicate raceme without branches, subtended by a somewhat inflated bladeless sheath (spathe), the flowering shoot usually bearing numerous such inflorescences…(15)
5. Spikelets all similar in appearance and size…(6)
5'. Spikelets not all similar, the pedicelled ones often smaller than the sessile ones or different in appearance…(13)
6. Pedicels without a spikelet borne at the tip…(7)
6'. Pedicels with a spikelet borne at the tip…(8)
7. Flowering shoots mostly with 1 or a few large, terminal panicles 10+ cm…**Sorghastrum**
7'. Flowering shoots with numerous small panicles clustered together, each < 3 cm and each with a subtending spathe…**Andropogon** (in part)
8. Pedicels and rame segments (rachis joints) with a central longitudinal groove or membrane, flattened in cross-section…**Bothriochloa** (in part)
8'. Pedicels and rame segments without a central groove or membrane, nearly round in cross-section…(9)
9. Panicles narrow and spikelike, with soft silky hairs, 8–18 × 1–3 cm, the branches scarcely noticeable at arm's length…**Imperata**
9'. Panicles not as above, usually wider and/or shorter or the branches obvious at arm's length…(10)
10. Panicles with 2–5 primary branches…**Andropogon** (in part)
10'. Panicles with > 10 branches…(11)
11. Hairs at the bases of the spikelets much shorter than the spikelets, < 1 mm; plants grown for crops or adventive in weedy ground…**Sorghum** (in part)
11'. Hairs at the bases of the spikelets nearly equaling or longer than the spikelets, 4–12 mm; plants grown for ornament…(12)
12. Panicle branches breaking apart at the nodes (joints) when mature…**Tripidium**
12'. Panicle branches remaining intact, the spikelets falling separately when mature…**Miscanthus**
13. Pedicels and rame segments (rachis joints) with a central groove or membrane running lengthwise, flattened in cross-section…**Bothriochloa** (in part)
13'. Pedicels and rame segments without a central groove or membrane, nearly round in cross-section, at least at the apex…(14)
14. Inflorescence with numerous (> 5) branches; sessile spikelets ovoid to nearly globose…**Sorghum** (in part)
14'. Inflorescence with 2–5 nearly digitate branches; sessile spikelets lanceolate…**Andropogon gerardi**
15. Spikelets awnless, or with awns 1–2 mm…(16)
15'. Spikelets awned, the awns at least 5 mm…(17)
16. Racemes < 3 cm, glabrous or only sparsely pubescent…**Mnesithea**
16'. Racemes > 4 cm, densely woolly-pubescent…**Elionurus**
17. Awns 0.5–2 cm…**Schizachyrium**
17'. Awns 4–12 cm…(18)
18. Racemes 4–8 cm; awns 5–12 cm, main axis breaking apart at maturity, at least most of it…**Heteropogon**
18'. Racemes 10–18 cm; awns 4–6 cm, the main axis persistent…**Trachypogon**

KEY F: SPIKELETS WITH A SINGLE FLORET

1. Glumes absent; leaf blades strongly saw-toothed on the edges…**Leersia**
1'. Glumes present, at least 1; leaf blades smooth to slightly saw-toothed on the edges…(2)
2. Glumes and lemmas awnless…(3)

2'.   Glumes and/or lemmas awned…(19)

3.   Inflorescence a panicle of evident, unbranched, spicate primary branches…(4)

3'.   Inflorescence a panicle of rebranched branches, or dense and spikelike…(7)

4.   Panicle branches all attached at the tip of the main axis…**Cynodon**

4'.   Panicle branches attached along the length of the main axis, not only at the tip…(5)

5.   Glumes equal in length or nearly so; spikelets nearly round in outline…**Beckmannia** (in part)

5'.   Glumes unequal, the first glume shorter than the second; spikelets lanceolate in outline…(6)

6.   Spikelets widely spaced, rarely overlapping, appearing embedded in the branches; blades spirally twisted …**Muhlenbergia paniculata**

6'.   Spikelets very closely spaced, overlapping, not at all appearing embedded in the branches; blades not spirally twisted…**Spartina** (in part)

7.   Disarticulation below the glumes…(8)

7'.   Disarticulation above the glumes…(11)

8.   Ligules hairy; sheaths prominently inflated; blades widely spreading to reflexed; inflorescence dense and headlike or spikelike, the base often included in the sheath; much-branched annuals (*Sporobolus alope-curoides* and *S. schoenoides*)…**Sporobolus** (in part)

8'.   Plants not as above in all respects…(9)

9.   Spikelets nearly round in outline; glumes somewhat inflated or puffy-looking…**Beckmannia** (in part)

9'.   Spikelets mostly lanceolate in outline; glumes not at all inflated or puffy-looking…(10)

10.   Glumes softly pubescent on the midnerves; inflorescence dense and spikelike, rarely lobed…**Alope-curus** (in part)

10'.   Glumes glabrous to scabrous, not softly pubescent; inflorescence usually lobed at least below…**Poly-pogon** (in part)

11.   Lemma hardened at maturity, enclosing the palea and flower…(12)

11'.   Lemma remaining thin and flexible, not hardened, not enclosing the palea…(14)

12.   Lemma with 1 or 2 slender bracts, bristles, or scales at the base of the floret, these sometimes pubescent and often difficult to see without dissecting carefully…**Phalaris**

12'.   Lemma without any bracts, bristles, or scales at the base of the floret…(13)

13.   Florets dorsally compressed; lemma margins not overlapping, the palea exposed, at least in part…**Pip-tatheropsis** (in part)

13'.   Florets terete; lemma margins slightly overlapping, the palea hidden…**Oryzopsis** (in part)

14.   Lemma with a single nerve; ligule a ring of hairs…(15)

14'.   Lemma with 3+ nerves; ligule a membrane…(16)

15.   Rare turf grasses planted for lawns; first glume absent, the second glume enclosing the floret…**Zoysia**

15'.   Mostly common grasses of numerous habitats, but never lawn grasses…**Sporobolus** (in part)

16.   Sheath margins fused together for 1/2 their length or more…**Catabrosa**

16'.   Sheath margins overlapping most of their length…(17)

17.   Palea about as long as the lemma; body of the glumes (not including awn tips) shorter than the lemma; lemma mostly 3-nerved…**Muhlenbergia** (in part)

17'.   Palea 1/2 or less as long as the lemma; body of the glumes longer than the lemma; lemma obscurely nerved…(18)

18.   Rachilla prolonged beyond the palea as a short bristle to 0.6 mm…**Podagrostis**

18'.   Rachilla not prolonged beyond the palea…**Agrostis** (in part)

19.   Inflorescence a panicle of several evident, unbranched, spicate, primary branches…(20)

19'.   Inflorescence a panicle of rebranched branches, or a raceme, or in some the pedicels and branches poorly developed and the inflorescence spikelike…(22)

20.   Spikelets nearly round in outline; glumes somewhat inflated…**Beckmannia** (in part)

20'.   Spikelets lanceolate in outline; glumes not at all inflated…(21)

21.   Panicle branches all < 2 cm…**Bouteloua**

21'.   Panicle branches mostly > 2 cm…**Spartina** (in part)

22.   Lemma hard at maturity, usually enclosing or clasping the palea and flower, mostly with a well-developed and pointed callus…(23)

22'.   Lemma not hard (somewhat so in *Apera* but then the rachilla prolonged beyond the palea), not enclosing the flower and palea; mostly without a well-developed callus…(32)

23.   Ligule a ring of hairs; lemma terminating in 3 awns, the 2 lateral awns occasionally shortened and inconspicuous…**Aristida**

23'.   Ligule a membrane; lemma terminating in a single awn, this may be deciduous…(24)

24.   Palea hardened, longitudinally grooved and slightly longer than the lemma, protruding from between the lemma margins as a small point; lemma margins involute, fitting into the grooves of the palea… **Piptochaetium**

24'.   Palea usually membranous, not grooved, shorter than or equaling the lemma, not protruding as a small point; lemma margins flat…(25)

25. Lemma margins strongly overlapping; palea < 1/3 the length of the lemma, glabrous, lacking veins...**Nassella**
25'. Lemma margins not or only slightly overlapping; palea 1/3 to equaling the length of the lemma, always pubescent when short, sometimes glabrous when longer, 2-veined...(26)
26. Awns 6–20+ cm; glumes > 1.8 cm...(27)
26'. Awns 0.5–7.5 cm, if > 6 cm then glumes 1–1.5 cm...(28)
27. Membranous ligules of lower leaves densely ciliate, with hairs 0.3–1 mm...**Pappostipa**
27'. Membranous ligules of lower leaves glabrous or at most minutely ciliate...**Hesperostipa**
28. Palea pubescent, the apex flat, the veins terminating below the apex; lemma coriaceous at maturity but not strongly indurate...(29)
28'. Palea glabrous or pubescent, the apex appearing prow-tipped or pinched, the veins extending to the apex; lemma indurate at maturity...(31)
29. Glumes without evident nerves, the apices rounded to acute; plants alpine, growing on mossy hummocks in wet ground...**Ptilagrostis**
29'. Glumes with 1–5 evident nerves and/or the apices attenuate; plants growing in various habitats, but rarely as above...(30)
30. Plants with neither woody nor bamboolike culms 3–6 mm thick below, with mostly 2–3 nodes...**Eriocoma**
30'. Plants with ± woody, bamboolike culms 3–6 mm thick below, with 3–13 nodes...**Pseudoeriocoma**
31. Florets dorsally compressed; lemma margins not overlapping, the palea exposed, at least in part...**Piptatheropsis** (in part)
31'. Florets terete; lemma margins slightly overlapping, the palea hidden...**Oryzopsis** (in part)
32. Inflorescence spikelike or headlike, the branches absent or highly shortened...(33)
32'. Inflorescence a panicle with evident branches...(37)
33. First glume 2-nerved with 2 or 3 awns; lower stems angled or flattened somewhat...**Muhlenbergia** (in part)
33'. First glume 1-nerved with a single awn or awnless; lower stems rounded...(34)
34. Glumes plumose; spikelets in dense ovoid heads, rarely any > 2 times longer than wide; plants annual with markedly pubescent sheaths and blades, grown for ornament and dried bouquets, rarely escaping ...**Lagurus**
34'. Plants not as above in all respects...(35)
35. Glumes awnless...**Alopecurus** (in part)
35'. Glumes awned...(36)
36. Glumes strongly flattened laterally, ciliate on the keeled midnerve...**Phleum** (in part)
36'. Glumes rounded, not keeled, not ciliate on the midnerve, but may be pubescent on the body...**Polypogon** (in part)
37. Disarticulation below the glumes...(38)
37'. Disarticulation above the glumes...(43)
38. First glume with 2 or 3 awns; spikelets falling in pairs...**Muhlenbergia** (in part)
38'. First glume with a single awn or awnless...(39)
39. Spikelets nearly circular in outline; glumes and lemma awnless (glumes with a tiny point, but not awned) ...**Beckmannia** (in part)
39'. Spikelets elongate, not circular in outline; glumes and/or lemma awned...(40)
40. Glumes awnless...(41)
40'. Glumes awned...(42)
41. Panicle loose, the branches at least 5 cm and drooping at maturity...**Cinna**
41'. Panicle cylindrical, dense, the branches very short...**Alopecurus** (in part)
42. Glumes strongly flattened laterally, ciliate on the keeled midnerve...**Phleum** (in part)
42'. Glumes rounded, not keeled, not ciliate on the midnerve, but may be pubescent on the body...**Polypogon** (in part)
43. Glumes strongly flattened laterally, ciliate on the keeled midnerve...**Phleum** (in part)
43'. Glumes rounded, not keeled, not ciliate on the midnerve...(44)
44. Lemma awned from the back, at about the middle or below...(45)
44'. Lemma awned from the apex or just below...(46)
45. Floret with a tuft of hairs at the base; rachilla prolonged beyond the palea as a slender bristle...**Calamagrostis**
45'. Floret without a tuft of hairs at the base; rachilla not prolonged beyond the palea...**Agrostis** (in part)
46. Rachilla prolonged beyond the palea as a slender bristle; plants annual...**Apera**
46'. Rachilla not prolonged beyond the palea; plants annual or perennial...**Muhlenbergia** (in part)

KEY G: TRIBE PANICEAE

1. Spikelets subtended by 1+ bristles or enclosed in an involucre of spines or bristles…(2)
1'. Spikelets not subtended by bristles or spines…(4)
2. Spikelets subtended by 1 to several bristles, these remaining on the plant when the spikelets fall… **Setaria**
2'. Spikelets enclosed in a bowl-like cluster (bur or involucre) of bristles or flattened spines, these falling with the spikelets and not remaining on the plant…(3)
3. Bur of sharp spines, sometimes also with a whorl of bristles…**Cenchrus**
3'. Bur of bristles, without spines…**Pennisetum**
4. Inflorescence spikelike, the spikelets embedded in the side of a somewhat corky rachis…**Stenotaphrum**
4'. Inflorescence a panicle, the spikelets not at all embedded in the rachis…(5)
5. Spikelets covered with long, silky, reddish hairs 2–4 mm…**Melinis**
5'. Spikelets glabrous or pubescent, but any hairs never as above…(6)
6. First glume usually < 0.5 mm, absent or vestigial…(7)
6'. First glume usually > 0.5 mm, well developed, evident…(12)
7. Inflorescence an open rebranched panicle, the spikelets on long pedicels…**Leptoloma**
7'. Inflorescence a panicle of unbranched branches, the spikelets sessile or short-pedicelled…(8)
8. Spikelets with a small cuplike structure at the base (the first glume); lemma of upper floret awn-tipped… **Eriochloa**
8'. Spikelets without a cuplike structure at the base; lemma of upper floret not awn-tipped…(9)
9. Spikelets rounded on one side and flattened on the other, orbicular to ovate in outline; margins of the lemma of the upper floret firm and hard when mature, the apex rounded…**Paspalum**
9'. Spikelets not rounded and flattened as above, lanceolate in outline; margins of the lemma of the upper floret thin and translucent when mature, the apex acute to acuminate…(10)
10. Spikelets glabrous or with short, stiff hairs; plants annual…**Digitaria** (in part)
10'. Spikelets silky-pubescent with long, whitish hairs; plants perennial…(11)
11. Panicles with 3+ nodes, the branches not subdigitate; plants known in the wild, relatively common… **Trichachne**
11'. Panicles with only 1–2 nodes, the branches subdigitate; plants not known in the wild (*Digitaria eriantha*) …**Digitaria** (in part)
12. Ligule absent, the ligular region glabrous; plants annual…**Echinochloa**
12'. Ligule present, the ligular region often pubescent; plants annual or perennial…(13)
13. Lemma of the upper floret with a stiff bristle projecting from the otherwise blunt apex…**Urochloa** (in part)
13'. Lemma of the upper floret without a bristle, the apex rounded to acute…(14)
14. Plants stoloniferous perennials…**Hopia**
14'. Plants tufted annuals or perennials…(15)
15. Inflorescence a panicle of simple or nearly simple spicate branches; spikelets nearly sessile; back of fertile lemma and second glume turned toward the branch axis; plants annual…**Urochloa** (in part)
15'. Inflorescence an open rebranched panicle; spikelets often pedicelled; back of fertile lemma and second glume turned away from the branch axis; plants annual or perennial…(16)
16. Palea of the lower floret inflated, enlarged, obovate, forcing the spikelet to gape open; rare or extirpated plants not known in New Mexico since 1895…**Steinchisma**
16'. Palea of the lower floret not inflated as above, the spikelet closed (except open somewhat during anthesis); including many common grasses…(17)
17. Sheaths keeled; lemmas of fertile florets finely roughened-rugose, dull; bases of culms mostly thickened into bulblike corms…**Zuloagaea**
17'. Sheaths rounded; lemmas of fertile florets smooth and shiny; bases of culms never thickened into bulblike corms…(18)
18. Plants perennial, with 2 distinct growth phases: during the cool season producing a basal rosette of short broad blades and terminal panicles; during the warm season producing much-branched lateral shoots with small axillary panicles; palea of lower floret vestigial…**Dichanthelium**
18'. Plants annual or perennial, with a single growth phase; basal rosettes not produced; flowering during the warm season only; palea of lower floret vestigial to well developed…**Panicum**

KEY H: LEMMAS 3-NERVED; FLORETS > 1

1. Some spikelets (female ones) with long awns 5+ cm; plants stoloniferous, monoecious or dioecious, with awnless male spikelets…**Scleropogon**
1'. All spikelets with awns < 1 cm or awnless; plants stoloniferous or tufted, unisexual in *Bouteloua dactyloides* or bisexual…(2)
2. Spikelets in dense, sessile, headlike clusters closely subtended and mostly surpassed by the leaves…(3)

2′. Spikelets not in dense, sessile, headlike clusters, and/or elevated well above the leaves…(6)

3. Disarticulation below the glumes, the spikelets in bony clusters and falling together; plants strongly sto-loniferous perennials (*Bouteloua dactyloides*)…**Bouteloua** (in part)

3′. Disarticulation above the glumes, the spikelets not falling in bony clusters; plants annual or perennial, stoloniferous or tufted…(4)

4. Plants annual; blades mostly flat (*Munroa squarrosa*)…**Munroa** (in part)

4′. Plants perennial; blades mostly rolled and needlelike…(5)

5. Plants tufted, lacking stolons; lemmas with 3 ciliate awns from the nerves…**Blepharidachne**

5′. Plants producing short stolons; lemmas with a single awn, the lateral nerves extending into lobes (*Munroa pulchella*)…**Munroa** (in part)

6. Inflorescence a panicle of definite and obvious spicate or racemose unbranched primary branches…(7)

6′. Inflorescence a raceme, or a panicle of rebranched primary branches…(17)

7. Spikelets all male, 2-flowered, with orange-red anthers; lemmas awnless (*Bouteloua dactyloides*)…**Bouteloua** (in part)

7′. Combination of features otherwise…(8)

8. Panicle branches all digitate or in whorls near the apex of the main axis   (9)

8′. Panicle branches distributed all along the main axis and most not in whorls, or with a single branch only…(13)

9. Spikelets with 2 to several well-developed, bisexual florets…(10)

9′. Spikelets with 1 well-developed, bisexual floret with 1–4 rudimentary and mostly neuter florets above it…(11)

10. Second glume and some lemmas short-awned or mucronate; rachis projecting as a stiff point beyond the terminal spikelet…**Dactyloctenium**

10′. Second glume and lemmas awnless; rachis not projecting beyond the terminal spikelet…**Eleusine**

11. Spikelets awnless; upper rudimentary floret single and represented by a minute scale…**Cynodon**

11′. Spikelets awned (awnless or mucronate in *Chloris submutica*); upper rudimentary florets 1–4 and obvious…(12)

12. Lemma of the lower floret with 3 awns 8–12 mm…**Leptochloa**

12′. Lemma of the lower floret with a single awn or awnless…**Chloris**

13. Spikelets with a single fertile, well-developed floret and with 1–3 smaller, rudimentary florets above…**Bouteloua** (in part)

13′. Spikelets with usually 3 to many fertile, well-developed florets…(14)

14. Axils of primary panicle branches with tufts of long hairs; spikelets mostly few and widely spaced on each branch…**Eragrostis** (in part)

14′. Axils of primary panicle branches glabrous; spikelets mostly numerous and usually crowded on each branch…(15)

15. Plants perennial…**Disakisperma**

15′. Plants annual…(16)

16. Ligules 2–8 mm, attenuate, not lacerate except by tearing…**Diplachne**

16′. Ligules 1–3 mm, truncate to rounded, often erose or lacerate…**Dinebra**

17. Sheath margins fused together for 1/2+ their length…(18)

17′. Sheath margins overlapping for most of their length…(19)

18. Spikelets < 5 mm…**Catabrosa**

18′. Spikelets usually > 10 mm…**Bromus**

19. Lemmas pubescent on the nerves or at the base (except *Tridens albescens*), the midnerve usually exserted as an awn or short point (except *Poa*)…(20)

19′. Lemmas glabrous on the nerves and at the base, awnless…(27)

20. Ligules membranous; lemma midnerves not exserted as a small point…**Poa**

20′. Ligules a ring of hairs, or if membranous (*Triplasiella eragrostoides*) then the lemma midnerve exserted as a small point…(21)

21. Plants strongly rhizomatous; lemma nerves glabrous…**Redfieldia**

21′. Plants lacking rhizomes; lemma nerves pubescent (except *Tridens albescens*)…(22)

22. Palea densely long-ciliate on the upper 1/2; plants annual…**Triplasis**

22′. Palea not long-ciliate on the upper 1/2; plants perennial…(23)

23. Blades with white margins…**Erioneuron**

23′. Blades not white-margined…(24)

24. Panicles open, loose, the branches spreading to drooping…(25)

24′. Panicles narrow, contracted, the branches erect…(26)

25. Lemmas 2–3 mm, only the midnerve projecting as a short point…**Triplasiella**

25′. Lemmas 3–5 mm, the midnerve and lateral nerves projecting as short points (*Tridens flavus*)…**Tridens** (in part)

26. Nerves of the lemma plainly pubescent…**Tridentopsis**

26'. Nerves of the lemma glabrous or pubescent only at the base (*Tridens albescens*)...**Tridens** (in part)

27. Ligule a membrane...(28)

27'. Ligule a ring of hairs...(31)

28. Spikelets on long pedicels mostly much longer than the spikelets; plants spreading from stolons or rhizomes...**Muhlenbergia**

28'. Spikelets sessile or nearly so, the pedicels much shorter than the spikelets; plants tufted...(29)

29. Lemmas conspicuously awned from the back, the awns 3–6 mm (*K. spicata, K. vaseyi*)...**Koerleria** (in part)

29'. Lemmas awnless or with an awn to 2 mm...(30)

30. Second glume broadened below the middle; lemmas commonly short-awned, the awns 0–2 mm; palea colored, at least on the nerves...**Graphephorum**

30'. Second glume broadened above the middle; lemmas completely awnless; palea colorless, scarious, white (*K. macrantha*)...**Koeleria** (in part)

31. Panicles dense, congested, spikelike, usually light greenish or whitish; lemmas notched at the apex with a minute point; plants perennial (*Tridens albescens*)...**Tridens** (in part)

31'. Panicles usually open, loose, often olive- or dark-colored; lemmas lacking a minute notch and point; plants annual or perennial...(32)

32. Plants with extensive creeping rhizomes; blades very stiff and sharp-pointed...**Kalinia**

32'. Plants lacking rhizomes or with short knotty rhizomes only; blades usually rather lax, not sharp-pointed... **Eragrostis** (in part)

KEY I: LEMMAS WITH 5 TO MANY NERVES; FLORETS > 1

1. Glumes and lemmas stiff-ciliate on the midnerves and keels; spikelets arranged in dense, 1-sided clusters at the branch tips; sheath margins fused together...**Dactylis**

1'. Glumes and lemmas glabrous or variously pubescent but not ciliate on the midnerves and keels; spikelets not so arranged; sheath margins fused or overlapping...(2)

2. Sheath margins fused together for 3/4+ their length...(3)

2'. Sheath margins free from each other, overlapping, or fused only in the lower 1/3 or less...(7)

3. Callus of the floret with a prominent tuft of stiff hairs (otherwise glabrous); lemmas prominently awned... **Schizachne**

3'. Callus of the floret lacking a tuft of hairs and/or lemmas awnless...(4)

4. Lemmas with 7 nerves, these nearly parallel, not converging at the truncate or rounded apex...**Glyceria**

4'. Lemmas with 3–11 nerves, these converging at the obtuse to acute apex, or if parallel then < 7...(5)

5. Spikelets awned, or if awnless then > 15 mm; palea and grain strongly adherent to each other when mature...**Bromus**

5'. Spikelets awnless and < 15 mm; palea and grain free from each other when mature...(6)

6. Spikelets on mostly racemose unbranched primary branches, hanging like flags away from the axis; upper florets empty, inrolled, and represented by a club-shaped rudiment...**Melica**

6'. Spikelets variously arranged, but mostly on rebranched primary branches; upper florets usually not empty nor as above...**Poa** (in part)

7. Disarticulation below the glumes...(8)

7'. Disarticulation above the glumes...(10)

8. Florets 2 per spikelet, the upper with a short, hooked awn, the lower awnless...**Holcus**

8'. Florets 2 to several per spikelet, all either awnless or awned, but the awn never short and hooked...(9)

9. Lemmas mostly awnless; glumes dissimilar in shape, one narrowly lanceolate and the other obovate or spatulate...**Sphenopholis**

9'. Lemmas prominently awned; glumes similar in shape...Trisetum (in part)

10. Spikelets (glumes and/or lemmas) awned...(11)

10'. Spikelets (glumes and lemmas) awnless or at most with an awn tip not > 1 mm...(27)

11. Inflorescence a panicle of unbranched, spicate primary branches all clustered toward the apex of the stalk; plants annual...**Chloris**

11'. Inflorescence a panicle, but the main branches rebranched or the spikelets on obvious pedicels; plants annual or perennial...(12)

12. Florets 3 per spikelet, the lower 2 florets sterile, silky with brownish hairs, and awned, the upper floret fertile, glabrous, awnless, hidden within the sterile florets and appearing as a hardened grain... **Anthoxanthum**

12'. Florets not as above...(13)

13. Florets dissimilar, some awned, some awnless...(14)

13'. All florets alike and awned...(16)

14. Glumes large, > 15 mm...**Avena** (in part)

14'. Glumes small, < 12 mm...(15)

15. Plants perennial, robust, to 1+ m; mountain plants…**Arrhenatherum**
15'. Plants annual, delicate, to 30 cm or so; disturbed ground…**Aira**
16. Glumes not extending beyond the lowermost floret…(17)
16'. Glumes, at least the second, equal to or surpassing the lowermost floret…(20)
17. Spikelets 2(-4)-flowered; awn arising from the back of the lemma or from a deeply cleft apex…Trisetum (in part)
17'. Spikelets mostly 3- to many-flowered; awn arising from an entire apex…(18)
18. Plants annual; flowers with 1 stamen…**Vulpia**
18'. Plants perennial; flowers with 3 stamens…(19)
19. Auricles present; blades mostly wider than 3 mm, flat when fresh…**Schedonorus** (in part)
19'. Auricles absent; blades mostly narrower than 3 mm, rolled and somewhat stiff (but see *Festuca sororia*) …**Festuca** (in part)
20. Lemmas awned from the back or base…(21)
20'. Lemmas awned from an entire or cleft apex, if cleft, the awn arising from the sinus at the tip of the mid-nerve, or lemmas awnless…(25)
21. Spikelets not large, the glumes 2-8 mm…(22)
21'. Spikelets large, the glumes 10-30 mm…(23)
22. Awn of the lemma attached above the middle; lemmas 4-9 mm (sometimes slightly shorter)… Trisetum (in part)
22'. Awn of the lemma attached below the middle; lemmas 1.5-4 mm (sometimes slightly longer)… **Deschampsia**
23. Plants annual; glumes 18-30 mm…**Avena** (in part)
23'. Plants perennial; glumes 10-15 mm…(24)
24. Panicles 2-5 cm; blades rolled, usually pubescent…**Helictotrichon**
24'. Panicles 5-15 cm; blades flat or folded, mostly glabrous…**Avenula**
25. Awns of the lemma minute and nearly obsolete, scarcely visible…**Schismus** (in part)
25'. Awns of the lemma well developed, easily visible…(26)
26. Spikelets mostly 2-flowered, 3.5-6.5 mm; rachilla extending beyond the uppermost floret…Trisetum (in part)
26'. Spikelets 3-7-flowered, 6-15 mm; rachilla not extending beyond the uppermost floret…**Danthonia**
27. Glumes mostly > 2 cm and longer than the florets…**Avena** (in part)
27'. Glumes < 2 cm and/or shorter than the florets…(28)
28. Spikelets appearing 1-flowered, but the large fertile floret subtended by 1 or 2 smaller scales or bristles representing rudimentary florets, these often appressed to the fertile floret and not immediately apparent…**Phalaris**
28'. Spikelets not as above…(29)
29. Glumes and lemmas at maturity stiff, firm, greenish to straw-colored; leaves distichous, the lower ones bladeless as the stems grade into rhizomes; lemmas 7-11-nerved, the nerves obscure; plants strongly rhizomatous, dioecious perennials of alkaline areas and floodplains…**Distichlis**
29'. Glumes and lemmas pliable, thin, often greenish to purplish (stiff in the annual *Catapodium*); leaves not distichous, the lower ones usually with well-developed blades; lemmas generally 5-7-nerved (9-nerved in the annual *Schismus*); plants annual or perennial, of various habitats…(30)
30. Glumes and lemmas spreading at right angles to the rachilla, inflated and papery; florets and spikelets about as wide as long; spikelets on long capillary pedicels, resembling the rattles of a rattlesnake…**Briza**
30'. Glumes, lemmas, florets, and spikelets not all as above…(31)
31. First glume 5-7-nerved; blades threadlike; small tufted annuals of sandy desert areas…**Schismus** (in part)
31'. First glume 1-3-nerved; blades threadlike to much broader; annuals and perennials of various habitats… (32)
32. Glumes, at least the second, equaling or surpassing the lowermost floret…(33)
32'. Glumes, at least 1 but usually both, not extending beyond the lowermost floret…(35)
33. Florets 3, the lower (outer) 2 as large as the upper (middle) 1 but male, their margins prominently ciliate, the upper (middle) floret fertile, somewhat hardened, pubescent at the tip…**Hierochloe**
33'. Florets not as above…(34)
34. Second glume broadened below the middle; palea colored, at least on the nerves…**Graphephorum wolfii**
34'. Second glume broadened above the middle; palea colorless, scarious, white…**Koeleria** (in part)
35. Lemmas awned or narrowing at the apex to an awn-tip…(36)
35'. Lemmas completely awnless, often blunt…(37)
36. Auricles present; blades mostly wider than 3 mm, flat when fresh…**Schedonorus** (in part)
36'. Auricles absent; blades mostly narrower than 3 mm, rolled and somewhat stiff (but see *Festuca sororia*) …**Festuca** (in part)
37. Second glume broadened above the middle; palea colorless, scarious, white; pedicels puberulent… **Koeleria** (in part)

37′. Second glume, palea, and pedicels not all as above…(38)
38. Inflorescence scarcely branched, the spikelets on short stout pedicels ± on the main axis; plants annual …**Catapodium**
38′. Inflorescence noticeably branched, the spikelets not borne as above; plants annual or perennial…(39)
39. Plants rhizomatous and dioecious; glumes hyaline and translucent…**Leucopoa**
39′. Plants not rhizomatous *and* dioecious *and* with translucent glumes…(40)
40. Sheath margins fused at least at the base; nerves of the lemma converging toward the acute apex; base of lemma with or without a tuft of cobwebby hair…**Poa** (in part)
40′. Sheath margins overlapping at the base; nerves of the lemma ± parallel, not converging toward the truncate apex; base of lemma never with a tuft of cobwebby hairs…(41)
41. Nerves of the lemma conspicuous; plants with creeping rhizomes; blades mostly flat, 4–15 mm wide; plants of freshwater habitats…**Torreyochloa**
41′. Nerves of the lemma obscure; plants tufted, lacking rhizomes; blades rolled, or if flat then 1–3(–4) mm wide; plants of usually alkaline or saline habitats…**Puccinellia**

AEGILOPS L. – Goatgrass

Aegilops cylindrica Host [*Triticum cylindricum* (Host) Ces., Pass. & Gibelli]—Goatgrass. Tufted annuals, 15–50 cm, erect to decumbent at the base, usually highly branched basally; sheaths open with hyaline margins; auricles present, ciliate; ligules membranous; blades flat; inflorescence a jointed spike with segments that break apart in maturity; spikes 3–12 cm, ca. 3 mm wide; 1 spikelet per node, spikelets with 3–5 florets, the lower 2–3 fertile; glumes of upper spikelets with awns 3–6 cm; anthers 3. A troublesome weed of crop fields and roadsides, along railroads, disturbed ground; widely distributed throughout the state and expected in every county, 4000–7000 ft. Widespread. Introduced noxious weed.

AEGOPOGON Humb. & Bonpl. ex Willd. – Fragilegrass

Aegopogon tenellus (DC.) Trin.—Fragilegrass. Tufted annual, 2–25 cm; sheaths open; auricles absent; ligules membranous; blades flat, 1–2 mm wide; racemose panicles 2–6 cm; pedicels of the lateral spikelets 1–1.3 mm, those of the central spikelets 0.3–0.6 mm; glumes 1–2 mm, fan-shaped, with short awns; lemmas 3-nerved, 2.5–3.2 mm, with a central awn 3–8 mm. Known only from desert plains and foothills of the bootheel region, in shaded canyons and beneath shrubs and trees, sometimes roadsides; a single record, 6500 ft. Hidalgo. Introduced.

AGROPYRON Gaertn. – Wheatgrass

Tufted perennials; sheaths overlapping; auricles present; ligules membranous; inflorescence a spike with very short internodes between the spikelets; spikelets single per node, longer than the internodes, diverging sharply from the main axis, several-flowered; disarticulation above the glumes, between the florets; glumes shorter than the florets, keeled on the midnerve; lemmas 5–7-nerved; anthers 3. Most species previously placed in *Agropyron* are here treated in *Elymus* or *Eremopyrum*.

1. Lemmas with an awn 1–6 mm; spikelets diverging from the rachis at angles of 30°–95°, often giving the spike a bristly appearance…**A. cristatum**
1′. Lemmas awnless or at most mucronate; spikelets scarcely diverging from the rachis at angles < 30°, the spike not at all bristly…**A. fragile**

Agropyron cristatum (L.) Gaertn. [*A. desertorum* (Fisch. ex Link) Schult.]—Crested wheatgrass. Tufted, but occasionally producing short rhizomes, 25–85+ cm; spikes 2–15 cm, 0.5–2.5 cm wide, sometimes tapering distally; glumes 3–6 mm with awns 1–3 mm; lemmas 5–9 mm with awns 1–6 mm. Widely introduced for rangeland rehabilitation (so-called) and soil stabilization, except in the S desert, 5000–10,400 ft. Widespread except EC, SE, SW. Introduced.

Agropyron fragile (Roth) P. Candargy [*A. sibiricum* (Willd.) P. Beauv.]—Siberian wheatgrass. Tufted, never rhizomatous, 30–70+ cm; spikes 8–15 cm, 0.5–1.3 cm wide; glumes 3–5 mm with awns 1–3 mm; lemmas 5–9 mm, awnless. Old fields, roadsides; known as yet from a only few scattered counties, 5100–7550 ft. Colfax, McKinley, Socorro. Introduced.

AGROSTIS L. – Bentgrass

Perennial, sometimes with rhizomes or stolons; sheaths open; auricles absent; ligules membranous; inflorescence a rebranching panicle; spikelets 1-flowered; glumes exceeding the floret, awnless or rarely with a short awn; lemmas 3-5-nerved, unawned or awned; paleas minute to as long as the lemma; anthers 1 or 3.

1. Palea well developed, 0.5-2 mm, 1/2-3/4 the length of the lemma…(2)
1'. Palea obsolete or a small scale < 0.4 mm, never as much as 1/2 the length of the lemma…(5)
2. Panicle dense, compact, interrupted; spikelets usually disarticulating below the glumes…go to **Polypogon viridis**
2'. Panicle open or closed but not dense or compact; spikelets disarticulating above the glumes…(3)
3. Plants 3-20 cm; anthers 0.5-0.7 mm; rachilla prolonged beyond the floret; alpine and subalpine meadows and boggy ground…go to **Podagrostis humilis**
3'. Plants taller, mostly 40+ cm; anthers 0.8-1.4 mm; rachilla not prolonged beyond the floret; occurring in a wide variety of habitats, and common at lower elevations…(4)
4. Panicles open during anthesis but contracted thereafter and when mature, mostly 1-1.5 cm broad, the branches erect-appressed; plants often stoloniferous and decumbent at the base, if short rhizomes developed then these bearing no more than 3 scale leaves…**A. stolonifera**
4'. Panicles open both during and after anthesis, > 1.5 cm broad, the branches ascending to widely spreading; plants with well-developed rhizomes bearing > 3 scale leaves, not stoloniferous, erect at the base… **A. gigantea**
5. Panicles narrow, contracted, several times longer than broad, at least some branches spikelet-bearing to the base…(6)
5'. Panicles open to diffuse, often < 3 times longer than broad, the branches naked at the base…(7)
6. Stems slender, generally not much > 20 cm; blades mostly not > 1 mm wide…**A. variabilis**
6'. Stems usually stout, mostly much > 20 cm; blades mostly 2-10 mm wide…**A. exarata**
7. Lemmas with a slender, flexuous awn; plants annual; anther 1…**A. elliottiana**
7'. Lemmas awnless or with a straight awn; plants perennial, though they may appear annual; anthers 3…(8)
8. Cauline leaves well developed, the basal ones often withered by anthesis; blades 2-5 mm wide, flat, 6-20 cm…**A. perennans**
8'. Cauline leaves weakly developed, the basal ones usually persistent or at least not withered; blades 1-2 mm wide, rolled to flat, 1-14 cm…(9)
9. Lower panicle branches 1-4 cm; panicles not detaching at maturity; blades 1-7 cm…**A. xidahoensis**
9'. Lower panicle branches 4-12 cm; panicles often detaching at the base at maturity; blades 4-14 cm…**A. scabra**

Agrostis elliottiana Schult. [*A. exigua* Thurb.]—Elliott's bentgrass. Annual, 5-35 cm; basal leaves withered by anthesis; panicles 3-18 cm, open and diffuse when mature, the branches ultimately widely spreading or drooping and rebranched beyond midlength; spikelets 1.5-2 mm; lemmas with a flexuous awn 3-10 mm; paleas absent or minute. Along stream banks and in moist woods of the S desert mountains, uncommon; a single record, 5510 ft. Hidalgo.

Agrostis exarata Trin.—Spike bentgrass. Perennial, 20-80 cm, sometimes decumbent-based or with short rhizomes; blades flat when fresh, 2-7 mm wide; panicles narrow and contracted, 8-28 cm, 1-4 cm wide; spikelets 1.5-3.5 mm; lemmas unawned (ours); paleas < 0.5 mm. Our material belongs to **var. minor** Hook. Widespread in all the mountains and surrounding foothills and plains, moist meadows, stream banks, shady understory, 4770-11,435 ft. Widespread except EC, SE.

Agrostis gigantea Roth [*A. alba* of numerous authors]—Redtop. Plants perennial, 20-100+ cm, with short to rather long (20 cm) rhizomes; leaves mostly cauline; blades 3-8 mm wide; panicles 10-25 cm, 3-12 cm wide, open both during and after anthesis, some branches spikelet-bearing to the base; spikelets green to purple, 1.7-3.2 mm; lemmas awnless (rarely with a short awn); paleas about 1/2 the length of the lemma. Moist pastures, ditches, stream banks, meadows; very widespread and expected in all counties, 4500-10,360 ft. Widespread except EC, SE. Introduced.

Agrostis xidahoensis Nash—Clubbed bentgrass. Perennial, 8-40 cm; leaves mostly basal; panicles 3-13 cm, 1-6 cm wide, open and diffuse, the branches spreading and rebranching above the middle; spikelets 1.5-2.5 mm; lemmas awnless; paleas < 0.2 mm. Wet meadows, seeps, moist ground at high elevations in the N mountains, 6000–11,610 ft. NC, Bernalillo. This is a catch-all name for intermediates (presumably due to hybridization) between *A. scabra* and *A. variabilis* and could be merged with the former.

Agrostis perennans (Walter) Tuck.—Upland bentgrass. Perennial, 20–80 cm; leaves mostly cauline with flat blades 2–5 mm wide; panicles 10–25 cm, to 11 cm wide, the branches ascending to spreading, rebranching at or below the middle; spikelets 1.8–3.2 mm; lemmas awnless (rarely awned to 2 mm); paleas absent or to 0.1 mm. Stream banks, moist meadows, shady roadsides, 5500–8000 ft. Grant, Sandoval, Santa Fe. Not common.

Agrostis scabra Willd. [*A. hiemalis* sensu Wooton & Standl.]—Rough bentgrass. Perennial, 10–80 cm; leaves mostly basal with rolled blades; panicles 10–40 cm, to 20 cm wide, the branches widely spreading when mature and rebranching above the middle; spikelets 1.8–3 mm; lemmas awnless or with an awn to 2 mm; paleas < 0.2 mm. Meadows, grassy slopes, rocky ground, roadsides, foothills to high mountains, 5000–13,025 ft. Widespread except EC, SE.

Agrostis stolonifera L. [*A. alba* of numerous authors; *A. palustris* Huds.]—Redtop. Perennial, mostly 30–60 cm, with stolons 5–50+ cm; blades flat, 2–6 mm wide; panicles narrow and contracted when mature, open during anthesis, mostly 5–16 cm, 1–4 cm wide, the branches spikelet-bearing to the base; spikelets greenish to purplish, 1.5–3 mm; lemmas awnless (rarely with a short awn); paleas about 1/2 the length of the lemma. Moist pastures, ditches, stream banks, meadows, 3540–10,880 ft. Widespread except EC, SE. Introduced.

Agrostis variabilis Rydb.—Alpine bentgrass. Perennial, 5–20(–25) cm, mostly tufted or rarely with short rhizomes to 2 cm; leaves mostly basal in dense tufts; blades 0.5–1.5(–2) mm wide; panicles 2–6 cm, 0.5–1.5 cm wide, narrow and dense, the branches spikelet-bearing to the base; spikelets greenish purple, 1.8–2.5 mm; lemmas awnless, rarely with a short awn; paleas tiny, ca. 0.2 mm. Subalpine and alpine slopes; uncommon in the N mountains, 10,435–12,125 ft. NC.

AIRA L. - Hairgrass

Aira caryophyllea L.—Silver hairgrass. Tufted annual, 5–45 cm; sheaths open; auricles absent; ligules membranous, 1–8 mm; panicles 1–14 cm, open, nearly as wide; pedicels many times longer than the spikelets; spikelets 1.7–3.5 mm; lemma awns 2.1–3.9 mm, protruding past the glumes and visible; anthers 3. Our material belongs to **var. capillaris** (Mert. & W. D. J. Koch) Bluff & Fingerh. [*A. elegans* Roem. & Schult.]. Found once in New Mexico in 1998; weakly adventive in ornamental plantings in Las Cruces, 3900 ft. Not likely persisting. Doña Ana. Introduced.

ALOPECURUS L. - Foxtail

Annual or perennial, usually tufted; sheaths overlapping; auricles absent; ligules membranous; inflorescence a spikelike panicle, the branches much reduced; disarticulation below the glumes; spikelets 1-flowered; glumes equaling or longer than the floret, 3-veined, ciliate on the midnerve; lemmas 3–5-nerved, awned from the back just above the base; paleas absent or very small; anthers 3.

1. Spikelets 5–6 mm…(2)
1'. Spikelets 2–4 mm…(3)
2. Glumes conspicuously ciliate on the keel…**A. pratensis**
2'. Glumes glabrous to scabrous on the keel…**A. myosuroides**
3. Awn slightly exserted beyond the glumes, scarcely visible without magnification…**A. aequalis**
3'. Awn well exserted beyond the lemma, easily visible without magnification…(4)
4. Plants annual; anthers 0.3–0.5 mm…**A. carolinianus**
4'. Plants perennial; anthers 1.2–2 mm…**A. geniculatus**

Alopecurus aequalis Sobol.—Short-awn meadow foxtail. Perennial, 10–70 cm, the culms erect to decumbent; upper sheaths not inflated; glumes 1.8–3.5 mm; lemma awns 0.7–3 mm, slightly exceeding the glumes and scarcely visible without magnification; anthers golden-yellow to orange when mature.

Ponds, ditches, wet ground; widespread in the state from low to high elevations, 5000-11,210 ft. NC, NW, C, WC, SC.

Alopecurus carolinianus Walter—Tufted meadow foxtail. Annual, 5-50 cm, the culms erect to decumbent; upper sheaths not inflated; glumes 2-3 mm; lemma awns 3-6 mm, extending beyond the glumes. Moist ground, ditch banks, irrigated ground, fields, 4700-8585 ft. Hidalgo, Santa Fe, Sierra.

Alopecurus geniculatus L.—Marsh meadow foxtail. Perennial, 15-60 cm, the lower shoots decumbent and often rooting at the nodes; upper sheaths somewhat inflated; glumes 2-3.5 mm; lemma awns 3.5-6 mm, extending beyond the glumes. Moist or wet ground, stream and canal banks, irrigated ground; uncommon, 5500-10,335 ft. WC, Hidalgo, San Juan, Taos. Introduced.

Alopecurus myosuroides Huds.—Slender meadow foxtail. Annual, 30-80 cm; upper sheaths somewhat inflated; glumes 4.5-7.5 mm; lemma awns 7-12 mm, extending beyond the glumes. Known from only a single collection in the late 1800s from a farm in Las Cruces, 3900 ft. Doña Ana. Introduced.

Alopecurus pratensis L.—Field meadow foxtail. Perennial with short rhizomes, 30-100 cm; upper sheaths not or scarcely inflated; glumes 4-6 mm; lemma awns 5-10 mm, extending beyond the glumes. Moist woods and ciénegas; uncommon in the mountains, introduced for erosion control and reseeding, 3540-10,880 ft. NC, Catron, Otero. Introduced.

ANDROPOGON L. - Bluestem

Perennial, tufted or rhizomatous; sheaths overlapping; ligules membranous, sometimes also ciliate; inflorescence a terminal or axillary panicle of 2 to several branches; primary branches (rame) composed of repeating units of sessile and pedicellate spikelets, these units breaking apart at maturity and becoming the dispersal unit; sessile spikelets bisexual and fertile, awned, the 2 glumes enclosing the florets; pedicelled spikelets well developed and staminate to reduced to a single empty glume.

1. Pedicelled spikelets vestigial or absent; sessile spikelets < 4 mm...**A. eremicus**
1'. Pedicelled spikelets present, nearly as large as the sessile ones; sessile spikelets at least 6 mm...**A. gerardi**

Andropogon eremicus Wipff & R. B. Shaw [*A. glomeratus* (Walter) Britton, Sterns & Poggenb var. *scabriglumis* C. S. Campb.]—Bushy bluestem. Tufted, 80-150 cm; shoots highly branched in the distal 1/2; sheaths scabrous; panicles axillary, partially enclosed in inflated terminal sheaths; sessile spikelets 3-4 mm; pedicels densely ciliate; pedicelled spikelets absent. Seasonally wet places, seeps, and springs in desert foothills, 3925-6000 ft. SC, SE, Grant.

Andropogon gerardi Vitman—Big bluestem. Perennial, 1-3 m; rhizomes, when present, with internodes < 2 cm; panicles mostly terminal, with 2-6 rames, often purplish; sessile spikelets 5-12 mm, with awns 0-20 mm; pedicelled spikelets 3.5-12 mm, about the same size and shape as the sessile spikelets; anthers 3. Prairies, plains, sand dunes, wooded slopes, forests. Two subspecies, which intergrade freely when sympatric:

1. Awn of sessile spikelet 0-8 mm; rhizomes well developed; ligules 2-4.5 mm; hairs of rame internodes 3.7-6.6 mm...**subsp. hallii**
1'. Awn of sessile spikelet 8-20 mm; rhizomes absent or well developed; ligules 0.5-2.5 mm; hairs of rame internodes 2.2-4.2 mm...**subsp. gerardi**

subsp. gerardi—Found essentially throughout the state in a variety of communities, but generally absent from arid desert areas, 3400-6800 ft. Widespread.

subsp. hallii (Hack.) Wipff [*A. chrysocomus* Nash; *A. gerardi* var. *paucipilus* (Nash) Fernald]—Mostly on the E plains, but scattered populations elsewhere, 4075-11,000 ft. NE, NC, EC, SE.

ANTHOXANTHUM L. - Vernalgrass

Anthoxanthum odoratum L.—Sweet vernalgrass. Tufted perennial, 25-60+ cm; sheaths open; auricles present to absent; ligules membranous; inflorescence a narrow, sometimes spikelike, poorly developed panicle; panicles 4–14 cm, the branches short; pedicels pubescent; spikelets 8–10 mm; glumes unequal, the lower 3–4 mm, the upper 8–10 mm; awn of first lemma 2–4 mm; awn of second lemma 4–9 mm, equaling or only slightly exceeding the upper glume; upper fertile floret 1–2.5 mm, awnless; palea 1-nerved. Disturbed ground, pastures, meadows, sporadic; known from a 1968 collection in Colfax County, and a 1997 collection in Doña Ana County. Introduced.

APERA L. – Silky bent

Apera interrupta (L.) P. Beauv. [*Agrostis interrupta* L.]—Dense silky bent. Tufted annual, 10–70 cm; sheaths open, often purplish; auricles absent; ligules membranous, 1.5–5 mm; panicles narrow, 3–20 cm, 1–3 cm wide, somewhat interrupted below; spikelets 2–3 mm; glumes somewhat unequal, the lower 1-nerved, the upper 3-nerved; rachilla extended beyond the floret 0.2–0.6 mm; lemmas 1.5–2.5 mm, slightly involute, the awn 4–16 mm; anthers 3. Disturbed moist sites. Lincoln, Torrance. Introduced.

ARISTIDA L. – Threeawn

Annual or perennial, tufted (ours); sheaths overlapping; auricles absent; ligules a ring of hairs or a tiny long-ciliate membrane; inflorescence a panicle, sometimes weakly developed and racemelike; spikelets 1-flowered; disarticulation above the glumes; glumes equal to strongly unequal, 1–3-nerved, thin; lemmas 3-nerved, convolute around the palea and flower, indurate at maturity, usually with 3 awns extending from the nerves but the lateral awns sometimes reduced or absent; callus with a tuft of short stiff hairs; palea shorter than the lemma; anthers (1–)3.

1. Plants annual…(2)
1'. Plants perennial…(3)
2. Awns mostly 1–2 cm; glumes mostly 5–12 mm…**A. adscensionis**
2'. Awns 2–7 cm; glumes mostly 20+ mm…**A. oligantha**
3. Lateral awns shortened, rarely > 3 mm…(4)
3'. Lateral awns > 3 mm, well developed, though often shorter than the central awn…(6)
4. First glume noticeably shorter than the second; inflorescence narrow, contracted, the branches erect (var. *wrightii* f. *brownii*)…**A. purpurea**
4'. First glume equal to or longer than the second; inflorescence open, the branches spreading from axillary swellings at maturity…(5)
5. First glume longer than the second; awn usually bent at a wide angle, the column twisted; blades flat and curling like wood shavings in age; base of blade glabrous (do not confuse with ligule hairs)…**A. schiedeana** (in part)
5'. First glume subequal to the second; awn mostly straight or only slightly bent, the column straight or slightly twisted; blades rolled or flattened at the base, but not curling like wood shavings; base of blade with scattered long hairs (var. *ternipes*)…**A. ternipes**
6. Panicle closed, contracted, the branches erect-appressed…(7)
6'. Panicle open, at least the lower branches spreading…(8)
7. Glumes equal or nearly so; blades usually flat and curling like wood shavings in age…**A. arizonica**
7'. Glumes noticeably unequal; blades usually rolled and not curling like wood shavings, but sometimes arcuate…**A. purpurea** (in part)
8. Primary panicle branches somewhat capillary and curving or drooping under the weight of the spikelets but without axillary swellings; awns mostly (2–)3–8 cm…**A. purpurea** (in part)
8'. Primary panicle branches stiffly divaricate to ascending from axillary swellings; awns mostly 1–2.5 cm…(9)
9. Anthers 0.8–1 mm…(10)
9'. Anthers 1.2–2+ mm…(11)
10. Plants > 25 cm; secondary branchlets present and usually well developed; primary branches 5–13 cm; apex of lemma strongly twisted 4+ turns…**A. divaricata**
10'. Plants < 25 cm; secondary branchlets absent or nearly so; primary branches 2–6 cm; apex of lemma not twisted or twisted only 1 or 2 turns…**A. havardii**
11. Glumes strongly unequal, the first ca. 1/2–2/3 the length of the second (var. *perplexa*)…**A. purpurea**
11'. Glumes equal or nearly so in length…(12)
12. Base of blades with scattered, soft, weak hairs 1.5–3 mm on the upper surface or margin (var. *gentilis*)…**A. ternipes**
12'. Base of blades glabrous to minutely pubescent on the upper surface, lacking long hairs, any hairs present < 0.5 mm (do not confuse with hairs at the collar or summit of the sheath)…(13)

13. Blades flat, loosely curling like wood shavings in age; summit of lemma conspicuously twisted...**A. schiedeana** (in part)
13'. Blades rolled, straight to arcuate but not curling; summit of lemma not or only slightly twisted...**A. pansa**

Aristida adscensionis L.—Sixweeks threeawn. Plants annual, mostly 10-50 cm, sometimes shorter or taller, usually highly branched and often geniculate; sheaths shorter than the internodes; blades flat to involute; panicles 5-18 cm, the branches sparsely flowered and lacking axillary pulvini; glumes 1-nerved, unequal, the lower about 1/2 the length of the floret; lemmas 6-9 mm; awns 7-20 mm, the lateral slightly shorter, not disarticulating. Waste ground, disturbed sites, roadsides, sparsely vegetated ground, 3300-8000 ft. Widespread.

Aristida arizonica Vasey—Arizona threeawn. Tufted perennial, 30-80+ cm, unbranched above the base; leaves mostly basal; blades usually flat, 1-3 mm wide, curling like wood shavings in age; panicles 10-25 cm, 1-3 cm wide, the primary branches appressed and lacking pulvini; glumes subequal, the lower one slightly shorter, 1-2-nerved; lemmas 12-18 mm; awns mostly equal, 20-35 mm. Somewhat dry mountain slopes and forest clearings at medium elevations, especially associated with ponderosa pine forests; widespread in the mountainous regions of the state, 5300-9515 ft. Widespread except NE, EC, SE.

Aristida divaricata Humb. & Bonpl. ex Willd.—Poverty threeawn. Plants perennial, 25-70 cm, not or sparingly branched; sheaths longer than the internodes, glabrous except at the summit, the collars densely pilose; blades flat to loosely rolled; panicles open, 10-30 cm and nearly as wide, the branches 5-13 cm, stiffly spreading to divaricate from axillary pulvini; spikelets infrequently with pulvini; glumes subequal, 1-nerved, 8-12 mm; floret subequal to the glumes, the terminal 2-3 mm with 4+ twists at maturity; awns 8-20 mm, the central erect to curving, the lateral slightly to much shorter. Dry plains and foothills, 4100-8615 ft. Widespread.

Aristida havardii Vasey—Havard threeawn. Plants perennial, 15-40 cm; unbranched; sheaths longer than the internodes, glabrous except for the summit, the collars densely pilose; blades flat to loosely rolled; panicles widely open, 8-18 cm and somewhat less wide, the branches 2-6 cm, stiffly divaricate to reflexed from axillary pulvini; spikelets usually with axillary pulvini; glumes subequal, 1-nerved, 8-12 mm; floret subequal to the glumes, the terminal 2-3 mm straight or with 1-2 twists at maturity; awns 8-20 mm, the lateral mostly thinner and shorter than the central. Dry plains and foothills, 3100-7665 ft. Widespread.

Aristida oligantha Michx.—Oldfield threeawn. Plants annual, 20-45 cm, geniculate at the base, branching at the nodes; sheaths shorter than the internodes; blades flat or loosely rolled; inflorescence spikelike or racemelike, the spikelets borne on very short pedicels on the main axis, branches rarely developed; pulvini in the axils of the spikelets; glumes large, the lower 3-7-nerved, 10-25 mm, with an awn, the upper 1-nerved and 2-3 mm shorter; floret subequal to the upper glume; awns mostly 2-5 cm; anther usually 1 and tiny, rarely 3 and 3-4 mm. Disturbed areas and old fields, 4060-5400 ft. DeBaca, Roosevelt, San Juan. Introduced.

Aristida pansa Wooton & Standl.—Wooton threeawn. Plants perennial, 20-60 cm, unbranched above the base; collars densely pilose with cobwebby hairs often arrayed downward; panicles 10-20 mm, the primary branches stiffly ascending to spreading from axillary pulvini; glumes subequal, 6-12 mm, 1-nerved; lemmas 7-13 mm; awns 6-15 mm, subequal. Dry, sandy plains and mesas of the S regions, 3600-6290 ft. EC, C, WC, SE, SC, SW.

Aristida purpurea Nutt.—Purple threeawn. Densely tufted perennial, 10-100 cm, usually unbranched above the base; leaves basal or cauline; collar hairs straight, not cobwebby; ligules < 0.5 mm; blades mostly involute, usually glabrous; panicles sparingly branched, sometimes racemelike, with or without axillary pulvini; glumes mostly unequal, the lower shorter than the upper, 1-nerved, the lower 4-12 mm, the upper 7-25 mm; lemmas 6-16 mm; awns usually subequal, but sometimes the lateral ones reduced. Dry plains, slopes, foothills, sandy sites, disturbed ground.

1. Panicle branches with axillary swellings, causing the branches to spread abruptly from the main axis…**var. perplexa**
1'. Panicle branches without axillary swellings, the branches erect or drooping, but not spreading abruptly from the main axis…(2)
2. Awns 4-10 cm…(3)
2'. Awns (at least the central) 1-3.5 cm…(4)
3. Summit of lemma 0.1-0.3 mm broad; awns rather delicate, mostly 0.2 mm or less wide at the base, 4-5 cm; second glume mostly < 16 mm…**var. purpurea** (in part)
3'. Summit of lemma 0.3-0.8 mm broad; awns usually stout, > 0.2 mm wide at the base, 4-10 cm; second glume 14-25 mm…**var. longiseta**
4. Summit of lemma mostly < 0.2 mm broad; awns delicate, mostly < 0.2 mm wide at the base…(5)
4'. Summit of lemma mostly > 0.2 mm broad; awns stout, mostly 0.2+ mm wide at the base…(6)
5. Panicle branches and pedicels erect, stiff, occasionally spreading…**var. nealleyi**
5'. Panicle branches and pedicels drooping to flexuous…**var. purpurea** (in part)
6. Mature panicle branches and pedicels capillary and flexuous or drooping…**var. purpurea** (in part)
6'. Mature panicle branches and pedicels moslly stiff and straight…(7)
7. Panicles mostly 3-14 cm; blades mostly basal and < 10 cm…**var. fendleriana**
7'. Panicles mostly 15-30 cm; blades mostly cauline and > 10 cm…**var. wrightii**

var. fendleriana (Steud.) Vasey [*A. fendleriana* Steud.]—Fendler's threeawn. 3500-9515 ft. Widespread.

var. longiseta (Steud.) Vasey [*A. longiseta* Steud.]—Red threeawn. 4100-7565 ft. Widespread.

var. nealleyi (Vasey) Allred [*A. glauca* (Nees) Walp.]—Blue threeawn. 3180-7565 ft. Widespread except WC.

var. perplexa Allred & Valdés-Reyna—Purple threeawn. 3600-5410 ft. C, SE, SC.

var. purpurea [*A. micrantha* (Vasey) Nash; *A. purpurea* var. *laxiflora* Merr.; *A. roemeriana* Scheele]—Purple threeawn. 3050-7300 ft. Widespread.

var. wrightii (Nash) Allred [*A. wrightii* Nash]—Wright's threeawn. 3150-7200 ft. Widespread except NW, WC. Forms with lateral awns reduced and nearly absent have been called **f. brownii** (Warnock) Allred & Valdés-Reyna.

Aristida schiedeana Trin. & Rupr. [*A. orcuttiana* Vasey]—Single threeawn. Plants perennial, 30-120 cm; collar and throat of the sheath usually glabrous; blades mostly flat, coiled or curling at maturity; panicles open, 1-30 cm and nearly as wide, the primary branches stiffly spreading from axillary pulvini; glumes mostly 1-nerved, subequal or more commonly the upper somewhat shorter; floret 10-15 mm, the beak twisted; awns strongly unequal, the central 5-12 mm, the lateral 0-3 mm. New Mexico material belongs to **var. orcuttiana** (Vasey) Allred & Valdés-Reyna. Mountain slopes and foothills in piñon and ponderosa zones of the SW mountains, 4200-8500 ft. WC, SC, SW.

Aristida ternipes Cav.—Spidergrass. Tufted perennial, 25-120 cm, stems mostly unbranched above the base; leaves basal and cauline, the sheaths mostly longer than the internodes; ligules usually < 0.5 mm; blades flat to folded, the adaxial surfaces with scattered hairs 1.5-3 mm near the ligule; panicles open, wiry, 15-40 cm and equally as wide, the primary branches with axillary pulvini; spikelets congested; glumes subequal, 9-15 mm, 1-nerved; lemmas 9-15 mm, smooth to tuberculate-scabrous; awns nearly equal to the lateral completely reduced; anthers 3. Dry plains and mesas, roadsides, 3600-7000 ft. Plants with well-developed lateral awns have been called **var. gentilis** (Henrard) Allred. C, WC, SE, SC, SW, McKinley.

ARRHENATHERUM P. Beauv. - Oatgrass

Arrhenatherum elatius (L.) P. Beauv. ex J. Presl & C. Presl—Tall oatgrass. Tufted to sometimes rhizomatous perennial, 0.5-1+ m, often with bulbous bases (in var. *bulbosum* (Willd.) Spenn.); sheaths open; auricles absent; ligules membranous, sometime ciliate; ligules 1-3 mm; panicles 7-30 cm, spreading before and during anthesis, then contracted; spikelets 7-11 mm; lower glumes 4-7 mm, the upper ca. 3 mm longer; lemmas mostly 6–10 mm; awn of lower lemma 10–20 mm; awn of upper lemma absent or to 5 mm; upper floret 7-nerved, usually awnless; anthers 3. Planted for hay and forage, found escaped in moist, shady places in the mountains, 3900–9600 ft. NC, SE, SC. Introduced.

ARUNDO L. – Giant reed

Arundo donax L.—Giant reed. Large, bamboolike giant grasses; culms to 10 m from vigorous, stout rhizomes > 1 cm thick; ligules < 1 mm; blades 2–9 cm wide, with a wedge-shaped brownish-colored area near the base; panicles 30–60 cm, to 30 cm wide; spikelets with 2–4 florets, 10–15 mm; glumes 3-nerved; florets 8–12 mm, pilose, the hairs 4–9 mm; paleas shorter than the lemmas; anthers 3. Ditches, culverts, roadsides, places where water accumulates, 3805–7200 ft. NC, C, SC, SW, Chaves. Introduced.

AVENA L. – Oats

Plants annual (ours) or perennial; sheaths overlapping, the basal often pilose and the distal glabrous; auricles absent; ligules a membrane; blades usually flat; inflorescence a panicle, sometimes weakly developed; disarticulation above the glumes, above or below the florets; spikelets with few to several florets, ± laterally compressed, the rachilla not prolonged; glumes 3–11-nerved, papery, usually surpassing the florets; lemmas usually indurate and enclosing the grain at maturity, 5–9-nerved, often strigose, the apex bifid into teeth; awns absent or single from the backs of the lemmas, twisted below, 1-geniculate; paleas shorter than the lemmas; anthers 3; caryopsis longitudinally grooved, terete, hairy.

1. Teeth at apex of lemma very thin, elongate, needlelike; pedicels capillary…**A. barbata**
1'. Teeth at apex of lemma acute but not elongate and needlelike; pedicels slender but not capillary…(2)
2. Awns usually absent or short and straight; florets not disarticulating and remaining on the plant, or falling together; when broken apart mechanically a portion of the rachilla remains attached to the glabrous callus…**A. sativa**
2'. Awns usually well developed and bent abruptly; florets separating and falling separately, leaving a circular scar or "sucker-mouth" at the bearded callus…**A. fatua**

Avena barbata Pott ex Link—Slender oats. Plants annual, 60–100+ cm; blades glabrous to variously hairy; panicles open, 15–40 cm, 6–12 cm wide; spikelets with 2–3 florets, each floret falling separately; glumes 15–30 mm, 7–9-nerved; lemmas 15–26 mm, densely strigose below, the apices with needlelike teeth 2–4 mm; awns 30–45 mm, arising about midlength. A weed in fields and along roads; a few collections from Doña Ana County, 3900 ft. Introduced.

Avena fatua L.—Wild oats. Plants annual, 10–120+ cm; blades mostly glabrous; panicles open, 8–35 cm, 5–18 cm wide, the branches nodding; spikelets with mostly 2 florets, each floret falling separately; glumes 18–30 mm, 9–11-nerved; lemmas 14–22 mm, densely strigose below, the apices bifid with teeth to 1.5 mm; awns 22–40 mm, arising in the middle of the lemma. Weed in grain fields and along roads, 4300–9700 ft. NE, NC, NW, WC, SE, SC. Introduced.

Avena sativa L. [*A. fatua* L. var. *sativa* (L.) Hausskn.]—Cultivated oats. Plants annual, 40–150+ cm; blades mostly glabrous; panicles 10–40 cm, 5–15 cm wide, the branches nodding; spikelets mostly with 2 florets, the florets falling together; glumes 20–32 mm, 9–11-nerved; lemmas 14–18 mm, usually glabrous but sometimes strigose, the apices bifid with teeth to 0.5 mm; awns often absent, 15–30 mm when developed, arising in the middle of the lemma. Commonly cultivated, sometimes escaping along fields, 3800–8000 ft. NC, NW, C, WC, SC, SW, Chaves. Introduced.

AVENULA (Dumort.) Dumort. – Little oats

Avenula hookeri (Scribn.) Holub [*Helictotrichon hookeri* (Scribn.) Henrard]—Little oats. Tufted perennial, 10–75 cm; sheaths open; ligules 3–7 mm; blades flat or folded; panicles 5–15 cm, 1–2.5 cm wide; spikelets with 2–7 florets; 12–16 mm, disarticulation above the glumes and between the florets; lower glume 9–13 mm; lemmas 10–12 mm; awns 10–17 mm; anthers 3. Alpine and subalpine slopes and ledges, 9050–10,100 ft. Mora, Taos.

BECKMANNIA Host – Sloughgrass

Beckmannia syzigachne (Steud.) Fernald [*B. eruciformis* (L.) Host]––American sloughgrass. Tufted annual, 30–120 cm; leaves mostly cauline; sheaths open, glabrous; ligules membranous, 5–11 mm, acute; blades glabrous; panicles 8–28 cm, with few secondary branchlets; spikelets 2–3 mm, with 1 floret (sometimes 2); lemma 2.5–3.5 mm; palea about as long as the lemma. New Mexico material belongs to **subsp. baicalensis** (V. A. Kusn.) Hultén. Irrigation ditches, marshes, floodplains, riverbanks, sloughs, 5540–9200 ft. NC, NW, C, WC.

BLEPHARIDACHNE Hack. – Desertgrass

Blepharidachne bigelovii (S. Watson) Hack. [*Eremochloe bigelovii* S. Watson]—Desertgrass. Tufted perennials from a knotty base, often nearly mat-forming, 6–20 cm, the stems branching from the base; basal sheaths shorter than the internodes; ligules to 0.3 mm; blades rolled, 1–2 cm, < 1 mm wide, the lower blades deciduous from the sheath; panicles 1.5–3 cm, sometimes partially enclosed in the subtending sheaths; spikelets 5–7 mm; glumes subequal, shorter than the florets, 5–6 mm, 1-nerved, awnless or nearly so; fertile third lemma 5–6 mm, 3-nerved and 3-lobed, awned to 3 mm; palea of fertile third floret slightly longer than the lemma; anthers 2. Limestone knolls and ledges, 3090–4000 ft. Doña Ana, Eddy.

BOTHRIOCHLOA Kuntze – Bluestem

Plants tufted (ours) perennials; sheaths open; auricles absent; ligules membranous, sometimes with a ciliate fringe; blades usually flat when mature; inflorescence a terminal panicle of several branches; primary branches unbranched, composed of repeating units of sessile and pedicelled spikelets, these breaking apart at maturity and becoming the dispersal unit; internode and pedicels with a longitudinal membranous area (native species) or a longitudinal groove (exotic species); sessile spikelets awned, relatively large and fertile; pedicelled spikelets awnless, staminate or sterile and often much smaller.

1. Pedicelled spikelets well developed, about as large and broad as the sessile ones…(2)
1'. Pedicelled spikelets much shorter and narrower than the sessile ones…(4)
2. Sessile spikelets > 5 mm…**B. wrightii**
2'. Sessile spikelets < 5 mm…(3)
3. Panicle axis longer than the branches…**B. bladhii** (in part)
3'. Panicle axis shorter than the branches…**B. ischaemum**
4. Sessile spikelets < 4.5 mm; awns < 18 mm…(5)
4'. Sessile spikelets > 4.5 mm; awns > 18 mm…(6)
5. Panicle reddish; hairs subtending the sessile spikelet ca. 1/4 the length of the spikelet, sparse, not at all obscuring the spikelet…**B. bladhii** (in part)
5'. Panicle silvery; hairs subtending the sessile spikelet at least 1/2 the length of the spikelet or longer, copious, at least somewhat obscuring the spikelet…**B. torreyana**
6. Panicle axis mostly < 5 cm, with 2–8 branches; rachises and pedicels densely white-long-pubescent; nodes densely white-long-pubescent with spreading hairs…**B. springfieldii**
6'. Panicle axis 5–15 cm, usually with numerous branches; rachises and pedicels long-pubescent but with off-white hairs; nodes bearded with stiff tan or off-white hairs…(7)
7. Panicles of the larger shoots 14–25 cm; stems stout, stiffly erect, little branched above the base, 1.2–2.5 m, bluish-glaucous below the nodes; nodes bearded with spreading hairs 3–6 mm…**B. alta**
7'. Panicles mostly 7–13 cm; stems tending to be bent at the base and much branched in age, mostly 1.2 m or less, not bluish-glaucous below the nodes; nodes bearded with appressed hairs < 3 mm…**B. barbinodis**

Bothriochloa alta (Hitchc.) Henrard—Tall beardgrass. Plants perennial, 1.2–2.5+ m, little branched, glaucous below the nodes, the nodes hirsute with stiff hairs 2–6 mm; panicles 14–25 cm; sessile spikelets 4.5–6 mm, the first glume short-hairy, pitted or not; awns 18–22 mm; pedicelled spikelets 3.8–4.4 mm. Plains in the S region, uncommon, usually along roadways and ditch banks where extra water accumulates, 4310–6235 ft. C, Chaves, Otero.

Bothriochloa barbinodis (Lag.) Herter—Cane bluestem. Plants perennial, 60–120 cm, often branched in age, the nodes hirsute with commonly appressed hairs < 3 mm; panicles 5–14(–20) cm;

sessile spikelets 4.5-7.3 mm, the first glume short-hairy, pitted or not; awns 20-35 mm; pedicelled spikelets 3-4 mm. Arid plains and grasslands, commonly along roadsides, 3425-7500 ft. EC, C, WC, SE, SC, SW, Taos.

Bothriochloa bladhii (Retz.) S. T. Blake—Australian bluestem. Plants perennial, 50-100+ cm, the nodes glabrous or hairy; foliage greenish; panicles 5-15(-20) cm, reddish when mature, the axis longer than the branches, which spread from the main axis from axillary pulvini, the lower branches rebranched; sessile spikelets 3.5-4 mm, the first glume mostly glabrous, a pit present or absent; awns 10-17 mm; pedicelled spikelets about same size and shape as sessile ones to 1/2 as large, staminate or sterile. Planted for range restoration, stabilization of roadsides, and erosion control, 4000-7000 ft. NE, NC, C, WC, Hidalgo, Quay. Introduced.

Bothriochloa ischaemum (L.) Keng—Yellow bluestem. Plants perennial, 30-85 cm, sometimes nearly stoloniferous from geniculate bases, the nodes glabrous or short-hairy; leaves mostly basal; panicles 5-10 cm, fan-shaped, reddish, the branches longer than the main axis and with axillary pulvini; sessile spikelets 3-4.5 mm, the first glume hairy, not pitted; awns 9-17 mm; pedicelled spikelets about as long as the sessile, but narrower. Planted for improving dry-land pastures and roadside stabilization, escaping along roadways, 3450-7700 ft. Widespread. Introduced.

Bothriochloa springfieldii (Gould) Parodi—Springfield's bluestem. Plants perennial, 30-80 cm, little branched, the nodes with conspicuous horizontal white hairs 3-7 mm; panicles 4-9 cm, the branches longer than the main axis; sessile spikelets 5.5-8.5 mm, the first glume densely hairy on the lower 1/2, sometimes pitted; awns 18-26 mm. Rocky to sandy slopes and plains, roadsides, grasslands, woodlands, 3050-8500 ft. Widespread.

Bothriochloa torreyana (Steud.) Scrivanti & Anton [*B. laguroides* (DC.) Herter subsp. *torreyana* (Steud.) Allred & Gould]—Silver bluestem. Plants perennial, 35-100+ cm, the nodes glabrous or short-hairy; panicles 4-12 cm, silvery-white, lacking pulvini; sessile spikelets 2.5-4.5 mm, the first glume glabrous to short-hairy, lacking a pit; awns 8-16 mm. Well-drained soils of grasslands, river valleys, roadsides, watered lawns, cemeteries, 3275-6825 ft. Widespread.

Bothriochloa wrightii (Hack.) Henrard—Wright's bluestem. Plants perennial, 45-70 cm; nodes glabrous or hirsute; foliage glaucous; panicles 5-6 cm, fan-shaped, the axis mostly shorter than the branches; sessile spikelets 5.5-7 mm; first glume glabrous, usually lacking a pit; awns 10-15 mm; pedicelled spikelets staminate, nearly as long as the sessile one. Perhaps extirpated, but to be looked for in rocky, grassy foothills of the piñon zone in the SW mountains, 5200-6000 ft. Grant, Sierra.

BOUTELOUA Lag. - Grama

Plants annual to perennial, tufted to rhizomatous or stoloniferous; sheaths open; auricles absent; ligules hairy, membranous, or a combination; inflorescence a panicle of 1 to many spicate primary branches; disarticulation below the branches or above the glumes, and the florets falling together; spikelets with a single fertile floret and 1-2 smaller staminate or sterile florets above; glumes mostly unequal, 1-nerved; lemmas 3-nerved and usually 3-awned.

1.	Stem internodes (not the sheaths) woolly-pubescent...**B. eriopoda**
1'.	Stem internodes glabrous (distal internodes of *B. breviseta* with a chalky-whitish bloom)...(2)
2.	Plants unisexual, dioecious, stoloniferous, forming low mats often < 10 cm...**B. dactyloides**
2'.	Plants bisexual, tufted or shortly rhizomatous, usually taller...(3)
3.	Inflorescence branches deciduous at maturity; spikelets 1-16 per branch...(4)
3'.	Inflorescence branches and glumes persistent on the plant; spikelets usually 20-60 per branch...(11)
4.	Branches of the inflorescence 15-80 per stem, or if < 15 then the branches (including the spikelets) < 1 cm...(5)
4'.	Branches of the inflorescence 1-13 per stem, or if > 13 then the branches (including the spikelets) 1.5+ cm...(6)
5.	Leaf blades 1-2(-2.5) mm broad; plants not rhizomatous; anthers purple...**B. warnockii**

5′. Leaf blades mostly > 2.5 mm broad; plants with or without rhizomes; anthers red, orange, or yellow… **B. curtipendula**
6. Plants annual…**B. aristidoides**
6′. Plants perennial…(7)
7. Glumes and often the lemmas densely pubescent, the hairs not confined to the midnerves…(8)
7′. Glumes and lemmas glabrous, or scabrous to ciliate on the midnerves only…(9)
8. Inflorescence axis 3-6 cm; spikelet clusters (including awns) mostly < 1 cm…**B. rigidiseta**
8′. Inflorescence axis 7-10 cm; spikelet clusters mostly > 1 cm…**B. eludens**
9. Middle inflorescence branches with 12-20 spikelets; lemma of lower floret 4-6 mm…**B. repens** (in part)
9′. Middle inflorescence branches with 4-16 spikelets; lemma of lower floret 4.5-8 mm…(10)
10. Shoots from hard, stout, rhizomatous bases, the stems thus appearing ± in linear progression and close together; basal sheaths mostly flattened, ribbonlike; middle branches mostly 2-3 cm (excluding awns) …**B. radicosa**
10′. Shoots solitary or several together in somewhat concentric tufts or from weak rhizomes; basal sheaths little flattened, mostly somewhat keeled and not ribbonlike; middle branches mostly 0.7-2 cm (excluding awns)…**B. repens** (in part)
11. Inflorescence reduced to a single branch…(12)
11′. Inflorescence with 2+ branches (*B. barbata* rarely with a single branch)…(14)
12. Plants annual…**B. simplex**
12′. Plants perennial…(13)
13. Primary inflorescence branch extending well beyond the attachment of the terminal spikelet…**B. hirsuta** (in part)
13′. Primary inflorescence branch not extending beyond the attachment of the terminal spikelet…**B. gracilis** (in part)
14. Second glume of some spikelets with stiff, bulbous-based hairs…(15)
14′. Second glume glabrous or pubescent without bulbous-based hairs…(17)
15. Primary branch extending well beyond the attachment of the terminal spikelet…**B. hirsuta** (in part)
15′. Primary branch not extending beyond the attachment of the terminal spikelet…(16)
16. Lemma 2-3(-3.5) mm; inflorescence branches (2)3-6 in number…**B. parryi**
16′. Lemma 4-6 mm; inflorescence branches (1)2(-4) in number…**B. gracilis** (in part)
17. Plants annual…**B. barbata** (in part)
17′. Plants perennial…(18)
18. Inflorescence branches (1)2(-4) in number…(19)
18′. Inflorescence branches 3-30 in number…(20)
19. Stem usually with 2-3 nodes; distal internodes lacking a chalky bloom…**B. gracilis** (in part)
19′. Stem usually with 5+ nodes, the plants somewhat bushy; distal internodes with a chalky bloom…**B. breviseta**
20. Lemma of first floret glabrous…**B. trifida**
20′. Lemma of first floret pubescent at the base…**B. barbata** (in part)

Bouteloua aristidoides (Kunth) Griseb.—Needle grama. Plants annual, tufted, 4-45+ cm; blade margins usually with bulbous-based hairs at the base; panicles 2-12 cm, with 4-15 branches densely pubescent basally, the entire branch somewhat resembling an *Aristida* spikelet; disarticulation at the base of the branch; spikelets appressed to the branch; lemmas and paleas of distal spikelets short-awned. Alluvial plains and uplands, dry mesas, disturbed rangelands.

1. Panicle branches with 2-4 spikelets, mostly 1.6 cm or less to the tip of the terminal spikelet; rachis extended 6-10 mm beyond the point of attachment of the terminal spikelet…**var. aristidoides**
1′. Panicle branches with 6-10 spikelets, 1.5-3.5 cm; rachis extended 2.5(-7) mm beyond the point of attachment of the terminal spikelet…**var. arizonica**

var. aristidoides—Common, with a wide distribution, 3600-8000 ft. NC, SC, SW, Bernalillo, Chaves, San Juan.

var. arizonica M. E. Jones—Uncommon, known only from Hidalgo County, 5500 ft.

Bouteloua barbata Lag.—Sixweeks grama. Plants tufted annuals or short-lived perennials, sometimes stoloniferous, 1-70 cm; panicles 1-22 cm, with 4-10 branches and numerous spikelets on each branch; disarticulation above the glumes; spikelets crowded and pectinate on the branch, with 1 fertile and 2 reduced florets; first glume lacking bulbous-based hairs; lemmas 3-awned; anthers yellow and red. Alluvial flats and slopes, plains, rocky slopes, washes, dry woodlands, roadsides, fields, often disturbed ground, 3045–9900 ft. Perennial plants with erect culms and 15–30 panicle branches are referred to as **var. rothrockii** (Vasey) Gould—Rothrock's grama, 4000–4400 ft. Widespread.

Bouteloua breviseta Vasey—Gypsum grama. Tufted or rhizomatous perennials to 40 cm, some-what bushy, the culms with 4–5+ nodes and slightly woody-based, distal portion of the lower internodes with a chalky bloom; panicles 2–4 cm, with 1–3 branches, each branch ending in a bristlelike spikelet; disarticulation above the glumes; hairs on glumes not bulbous-based; lemmas 3-awned. Gypsum plains, hills, grasslands, 3445–5000 ft. EC, SE, SC.

Bouteloua curtipendula (Michx.) Torr.—Sideoats grama. Tufted to rhizomatous perennials mostly 20–70 cm; blades mostly wider than 2.5 mm; panicles 14–30 cm, with 15–80 short, often hanging branches; disarticulation at the base of the branch; spikelets appressed; lemmas awnless to awned; anthers red, orange, yellow, or purple. Prairies, grasslands, woodlands, forest openings, usually on well-drained soils, 3200–8900 ft. Widespread. Tufted plants, lacking short rhizomes, are referred to as **var. caespitosa** Gould & Kapadia, 4865–7800 ft. Widespread except NW.

Bouteloua dactyloides (Nutt.) Columbus [*Buchloe dactyloides* (Nutt.) Engelm.]—Buffalograss. Dioecious (sometimes monoecious), stoloniferous perennials 1–30 cm, the pistillate shoots much shorter than the staminate; leaves sparsely pilose in the ligule region; staminate panicle usually exceed-ing the foliage, with 1–3 branches, disarticulation above the glumes, the spikelets closely crowded and pectinate, with prominent red to orange anthers; pistillate inflorescence partially hidden in leaf sheaths, the branches burlike with 3–5 enclosed spikelets, disarticulation at the base of the bur. Plains, prairies, grasslands, 3200–7000 ft. Widespread except C corridor.

Bouteloua eludens Griffiths—Santa Rita grama. Tufted perennials 20–60 cm; panicles 6–10 cm, with 10–16 pubescent branches, the axis extended beyond the last spikelet; disarticulation at the base of the branch; glumes and lemmas densely hairy between the veins; lemmas 3-lobed. Dry, rocky slopes, desert grasslands, 4500–5000 ft. Hidalgo.

Bouteloua eriopoda (Torr.) Torr.—Black grama. Plants perennial, regularly producing long stolons, sometimes also with short rhizomes; sheaths glabrous and about 1/2 as long as the woolly-hairy inter-nodes; panicles 2–16 cm with 2–8 branches, a tuft of woolly hairs at the base of the branch; disarticula-tion above the glumes; spikelet hairs not bulbous-based; lemmas mostly with a single awn, the lateral awns absent or very short; anthers yellow to orange. Desert grasslands, dry plains, rocky slopes, 2850–7560 ft. Widespread.

Bouteloua gracilis (Kunth) Lag. ex Griffiths—Blue grama. Densely tufted perennials, 20–70 cm, sometimes with short rhizomes from the base; blades with bulbous-based hairs near the ligule; panicles 2–10 cm, with 1–5 branches, the rachis bearing spikelets to the end; disarticulation above the glumes; spikelets densely crowded on the rachis, pectinate, with 1 fertile and 1 reduced floret; lower lemmas lobed and mostly awned; anthers yellow or purple. Plains, mesas, grasslands, woodlands, forest openings, 3300–11,000 ft. Widespread.

Bouteloua hirsuta Lag.—Hairy grama. Tufted perennials, 15–60 cm; blades with bulbous-based hairs near the ligule; panicles 1–18 cm, with 1–4 branches, the rachis of the branch extending beyond the terminal spikelet as a slender bristle; disarticulation above the glumes; spikelets densely crowded on the rachis, pectinate, with 1 fertile and 1–2 reduced florets; lower lemma lobed and awned; anthers yellow or cream. Plains, rocky slopes, woodlands, 3400–8260 ft. Widespread.

Bouteloua parryi (E. Fourn.) Griffiths—Tufted annuals, 20–55 cm; sheaths with tufts of long hairs at the sides of the collar; blades with bulbous-based hairs on both surfaces; panicles 3–10 cm, with 4–8 branches, the rachis prolonged into a terminal bristle; disarticulation above the glumes; spikelets densely crowded on the branch, pectinate, with 1 fertile floret and 1–2 reduced florets; glumes with bulbous-based hairs; lower lemmas hairy, awned; anthers yellow. Dry sandy plains, 4000–5205 ft. SE, SC, SW.

Bouteloua radicosa (E. Fourn.) Griffiths—Purple grama. Tufted-rhizomatous perennials from hard knotty bases, 40–70 cm; basal sheaths flattened and ribbonlike; panicles 10–15 cm, with 7–12 branches each bearing 8–10 spikelets, the middle branches 2–3 cm (excluding awns); disarticulation at the base of the branch; spikelets appressed, with 1 fertile and 1 reduced floret; glumes glabrous; lemmas 3-lobed, 3-awned. Dry rocky slopes, desert grasslands and woodlands; uncommon, 4200–9000 ft. WC, SW, Eddy.

Bouteloua repens (Kunth) Scribn. & Merr.—Slender grama. Tufted perennials, 15–65 cm, geniculate-based; panicles 4–14 cm, with 6–12 branches (sometimes fewer), the rachis prolonged beyond the terminal spikelet; disarticulation at the base of the branch; spikelets appressed, with 1 fertile and 1 reduced floret; lemmas mostly glabrous, 3-awned; anthers usually orange or yellow. Semiarid rangelands and woodlands, 5015–5510 ft. WC, SW.

Bouteloua rigidiseta (Steud.) Hitchc.—Texas grama. Tufted slender perennials, 10–40 cm, un-branched; panicles 3–6 cm, with 6–8 branches, the rachis prolonged beyond the terminal spikelet; disarticulation at the base of the branch; spikelets appressed-spreading, with 1 fertile and 1–2 reduced florets; glumes and lemmas sparsely hairy on the nerves only; lemmas 3-awned. Shin-oak mottes on the E plains, 4240 ft. Roosevelt.

Bouteloua simplex Lag.—Mat grama. Tufted annuals, the culms decumbent, 3–18(–26) cm; panicles usually with a single branch 1–3 cm, rarely with 2–3 branches, the rachis not prolonged beyond the terminal spikelet; disarticulation above the glumes; spikelets densely crowded, pectinate; glumes glabrous; lemmas hairy on the nerves, 3-awned. Dry rocky plains, mesas, hills, and disturbed ground in the mountains, 4750–9185 ft. Widespread except far E.

Bouteloua trifida Thurb. ex S. Watson—Red grama. Tufted perennials, older plants sometimes with short rhizomes, 8–35 cm; panicles 3–10 cm, with 2–7 branches, the rachis not prolonged beyond the spikelets; spikelets reddish, with 1 fertile and 1 reduced floret; lemmas glabrous, 3-awned; anthers yellow. Calcareous, rocky slopes in the S desert grasslands, infrequently collected, 3315–5200 ft. SC, SW, Eddy, Socorro.

Bouteloua warnockii Gould & Kapadia—Warnock's grama. Tufted perennials with rhizomes, 20–40 cm; blades narrow, 1–2(–2.5) mm wide, involute when dry; panicles 8–20 cm, with 9–20+ branches, the rachis prolonged beyond the terminal spikelet; disarticulation at the base of the branch; spikelets appressed, somewhat greenish; glumes and lemmas mostly glabrous; lemmas 3-awned; anthers dark purple. Dry plains on limestone in desert grasslands, on ledges and outcrops, often on gypsum, 4790–5600 ft. SE, SC.

BRIZA L. – Quaking grass

Plants annual (ours) or perennial, tufted; sheaths open; auricles absent; ligules membranous; blades flat; inflorescence an open panicle, the branches capillary; spikelets erect to drooping, with several florets, awnless, the glumes and lemmas spreading from the rachilla; disarticulation above the glumes and between the florets; lemmas inflated, about as wide as long, with rounded backs, with several indistinct nerves; anthers 3.

1. Spikelets 10–20 mm; panicles bearing < 10 spikelets…**B. maxima**
1'. Spikelets 2–5 mm; panicles bearing numerous (> 20) spikelets…**B. minor**

Briza maxima L.—Big quaking grass. Plants 25–75 cm; sheaths often < 1/2 as long as the inter-nodes; panicles 4–10 cm, 2–6 cm wide; spikelets 10–20 mm, with several florets; lemmas 7–9 mm, gla-brous proximally, becoming villous distally; anthers 1–1.8 mm. Weakly adventive; found recently in Union County, 6500 ft.

Briza minor L.—Little quaking grass. Plants 10–70 cm; sheaths 1/2–3/4 the length of the inter-nodes; panicles 4–15 cm, 3–11 cm wide; spikelets 2–5 mm, with several florets; lemmas 1.5–2.5 mm, mostly glabrous; anthers ca. 0.5 mm. Weakly adventive; collected from Las Cruces, 3900 ft., but not likely persisting. Doña Ana.

BROMUS L. – Brome

Plants annual, biennial, or perennial, tufted to rhizomatous; sheaths fused to near the top, often pubescent; auricles present or absent; ligules membranous; inflorescence a panicle, sometimes racemelike; spikelets with several florets; disarticulation above the glumes and between the florets; lemmas several-nerved, awned or awnless; paleas ciliate on the nerves, adnate to the grain; anthers 3.

1.   Plants perennial…(2)
1'.   Plants annual…(12)
2.   Rhizomes present…(3)
2'.   Rhizomes absent…(4)
3.   Culm nodes usually glabrous; leaves (blades and sheaths) usually glabrous; lemmas mostly glabrous or scabrous; awns 0–3 mm; seeded or disturbed sites, widespread…**B. inermis**
3'.   Culm nodes often pubescent; leaves often pilose; lemmas pubescent; awns 1–8 mm; native plant communities, uncommon in the N mountains…**B. pumpellianus**
4.   Spikelets strongly flattened; lemmas V-shaped in cross-section; second (upper) glume 5–9-nerved…(5)
4'.   Spikelets not strongly flattened, but ± terete; lemmas rounded on the back in cross-section; second (upper) glume 3-nerved…(6)
5.   Lemma awns 0–2.5 mm…**B. catharticus** (in part)
5'.   Lemma awns 3–15 mm (rarely as short as 2 mm)…**B. carinatus** (in part)
6.   First glume 3-nerved…(7)
6'.   First glume 1(2)-nerved…(9)
7.   Glumes mostly glabrous; leaf blades often glaucous…**B. frondosus**
7'.   Glumes mostly pubescent; leaf blades not glaucous…(8)
8.   Pedicels puberulent; blades of the upper 1/2 of the shoot erect, the midrib not narrowed below the collar; auricles absent…**B. porteri**
8'.   Pedicels glabrous; all blades mostly lax or spreading, the midrib mostly narrowed below the collar; auricles frequently present on the lower leaves…**B. anomalus** (in part)
9.   Sheaths densely lanate, the hairs spreading from the sheath but becoming matted at the tips…**B. lanatipes**
9'.   Sheaths glabrous to lightly pilose or hirtellous, if pubescent then not becoming matted…(10)
10.   Midrib of the culm leaves abruptly narrowed below the collar; anthers 2–4 mm; lemmas pubescent across the back as well as on the margins…**B. anomalus** (in part)
10'. Midrib of the culm leaves not narrowed below the collar; anthers 1–2.7 mm; lemmas glabrous to pubescent across the back, pubescent on the margins…(11)
11.   Anthers 1–1.4 mm; upper glumes 7–8 mm; backs of all lemmas glabrous…**B. ciliatus**
11'. Anthers 1.6–2.7 mm; upper glumes 9–11 mm; backs of the upper lemmas in a spikelet with appressed hairs, the backs of the lower lemmas glabrous…**B. richardsonii**
12.   Lemma awns 0–2.5 mm…(13)
12'. Lemma awns > 3 mm…(14)
13.   Lemmas lanceolate, broadest at the base, 9–14 mm; anthers ca. 3–4 mm…**B. catharticus** (in part)
13'. Lemmas inflated, broadest at the middle, 7–9 mm; anthers 1 mm or less…**B. briziformis**
14.   First glumes mostly 3–5+-nerved…(15)
14'. First glumes mostly 1-nerved (sometimes 3-nerved in *B. diandrus*)…(20)
15.   Spikelets strongly flattened; lemmas V-shaped in cross-section…**B. carinatus** (in part)
15'. Spikelets not strongly flattened, but ± terete; lemmas rounded on the back in cross-section…(16)
16.   Awns mostly < 5 mm; lemmas rounded, margins usually rolled around the grain; plants glabrous…**B. secalinus**
16'. Awns mostly > 5 mm; lemmas somewhat flattened, margins not rolled around the grain; plants pubescent…(17)
17.   Panicles dense, compact, 3–8(10) cm, branches stiffly erect…**B. hordeaceus**
17'. Panicles open, 6–20 cm, branches ascending to widely spreading…(18)
18.   Awns arising < 1.5 mm below apex of the lemma…**B. commutatus**
18'. Awns arising 1.5+ mm below apex of the lemma…(19)
19.   Lemmas with hyaline margins 0.3–0.6 mm wide, only slightly curved above the middle toward the apex; panicle branches somewhat drooping, sometimes sinuous, often with more than a single spikelet…**B. japonicus**
19'. Lemmas with hyaline margins 0.6–0.9 mm wide, strongly angled above the middle toward the apex; panicle branches usually not drooping nor sinuous, usually bearing a single spikelet…**B. squarrosus**

20. Panicle dense, compact, ovoid; panicle branches stout, erect, mostly shorter than 2 mm…**B. madritensis**
20'. Panicle loose, open, elongate; panicle branches often spreading or drooping and mostly much longer than 2 mm…(21)
21. Awns mostly 3-6 cm; lemmas 20-35 mm…**B. diandrus**
21'. Awns mostly 1-3 cm; lemmas 9-20 mm…(22)
22. Primary panicle branches mostly with 1(-3) spikelets; awns 15-30 mm; lemmas 14-20 mm…**B. sterilis**
22'. Primary panicle branches mostly with > 3 spikelets, at least on mature shoots; awns 10-18 mm; lemmas 9-12 mm…**B. tectorum**

Bromus anomalus Rupr. ex E. Fourn. [*Bromopsis anomala* (Rupr. ex E. Fourn.) Holub]—Mexican brome. Tufted perennials, 40-90 cm; nodes glabrous to hairy; sheaths glabrous or pilose, the midrib of the culm leaves abruptly narrowed just below the collar; auricles usually present on the lower leaves; blades 2-4 mm wide; panicles 10-20 cm, open, the branches spreading, often drooping; glumes usually pubescent, sometimes glabrous, the lower mostly 1-nerved, the upper 3-nerved; lemmas rounded, the back and margin pubescent; awns 1-3(-5) mm; anthers 2-4 mm. Mountain scrub, oak and piñon-juniper woodlands, ponderosa parklands, aspen groves, mountain meadows, often growing with *B. ciliatus* and *B. richardsonii*, 5250-11,360 ft. NC, NW, C, WC, SC, SW.

Bromus briziformis Fisch. & C. A. Mey.—Rattlesnake chess. Tufted annuals, 20-60 cm; sheaths densely pilose-hispid, the hairs retrorse; auricles absent; blades 2-4 mm wide, pilose; panicles 5-15 cm, open, the branches spreading to reflexed; spikelets laterally flattened; lemmas 5-8 mm wide, obovate, inflated; awns 0-3 mm; anthers 0.7-1 mm. Weedy, dry sites; 1 record, 7000 ft. Santa Fe. Introduced.

Bromus carinatus Hook. & Arn. [*B. arizonicus* (Shear) Stebbins]—California brome. Tufted annuals to short-lived perennials, 50-100 cm; nodes glabrous to hairy; sheaths glabrous to rather densely pilose; auricles absent (?or present); blades 2-10 mm wide, glabrous to hairy, flat to rolled; panicles 8-40 cm, the branches stiffly ascending to divergent; glumes glabrous to pubescent, with several nerves; lemmas keeled on the back, glabrous to pubescent; awns 3-15 mm; anthers 0.4-6 mm. Mountain slopes and forest clearings, 5000-12,500 ft. Widespread except far E.

Bromus catharticus Vahl [*B. willdenowii* Kunth]—Rescuegrass. Annuals or short-lived perennials, 30-100+ cm; nodes glabrous; sheaths densely pilose; auricles absent; blades 3-10 mm wide, glabrous to pubescent, flat; panicles 10-25 cm, the branches ascending to spreading; glumes mostly glabrous; lemmas strongly keeled, usually glabrous; awns 0-3 mm; anthers < 1 mm in cleistogamous florets, 2-4 mm in chasmogamous florets. Disturbed ground, lawns, weedy sites, roadsides, 3135-9400 ft. Widespread. Introduced.

Bromus ciliatus L.—Fringed brome. Tufted perennials, 40-120+ cm; upper nodes mostly pubescent; sheaths glabrous or pilose, the midrib of the culm leaves not abruptly narrowed just below the collar; auricles mostly absent; blades 3-9 mm wide, the upper mostly pilose adaxially; panicles 6-18 cm, the branches spreading to somewhat drooping; glumes glabrous, the lower mostly 1-nerved, the upper 3-nerved, 7-8 mm; lemmas rounded, the back glabrous, the margins ciliate; awns 3-5 mm; anthers 1-1.4 mm. Ponderosa, mixed conifer, spruce-fir forests, mountain meadows, but also extending to lower elevations, 5500-12,000 ft. Widespread except far E.

Bromus diandrus Roth—Ripgut brome. Tufted annuals, 20-80+ cm, the culms puberulent below the panicle; sheaths pilose; auricles absent; blades 1-9 mm wide, pilose on both surfaces; panicles 12-25 cm, the branches ascending to spreading and with only 1-2 spikelets; lemmas rounded on the back; awns 30-65 mm; anthers 0.5-1 mm. Dry, disturbed ground in scattered locales, 4000-8600 ft. WC, SC, SW, Bernalillo. Introduced.

Bromus frondosus (Shear) Wooton & Standl.—Drooping brome. Tufted perennials, 45-100 cm; nodes usually glabrous; sheaths mostly glabrous, sometimes pilose, the midrib often abruptly narrowed just below the collar; auricles absent; blades 3-6 mm wide, usually glabrous; panicles 10-20 cm, the branches ascending to drooping; glumes glabrous, 3-nerved; lemmas rounded, the backs glabrous to pubescent, the margins hairy; awns 2-4 mm; anthers 1.5-3.5 mm. Semidesert mountain scrub and riparian areas, oak and piñon-juniper woodlands up to ponderosa forests, 4175-10,640 ft. C, WC, SC, SW.

Bromus hordeaceus L. [*B. mollis* L.]—Barley brome. Tufted annuals, 10-100+ cm; nodes minutely to densely hairy; sheaths pilose; auricles absent; blades 1-5 mm wide, pilose above, flat; panicles 4-14 cm, the branches very short and often with a single spikelet; lemmas 2-2.5 mm wide, hairy on the back, the nerves raised when mature; awns 4-8 mm; anthers 0.3-1.3 mm. Weedy ground, roadsides, disturbed sites; uncommon, 3900-9175 ft. Doña Ana, Otero, San Juan. Introduced. Often treated as a phase of *B. squarrosus* L.

Bromus inermis Leyss. [*Bromopsis inermis* (Leyss.) Holub]—Smooth brome. Rhizomatous perennials, 50-130 cm; nodes glabrous; sheaths usually glabrous; auricles sometimes present; blades 5-15 mm wide, glabrous, flat; panicles 10-20 cm, the branches mostly ascending; glumes glabrous; lemmas rounded on the back, usually glabrous; awns 0-3 mm; anthers 3.5-6 mm. Pastures, mountain slopes, roadside swales and slopes, 4400-11,370 ft. Widespread except far EC, SE. Introduced.

Bromus japonicus Houtt.—Japanese brome. Tufted annuals, 20-80 cm; nodes glabrous, dark; sheaths densely pilose; auricles absent; blades 1-6 mm wide, densely pilose; panicles 8-26 cm, open, the branches ascending to spreading or sometimes nodding and with 1-6 spikelets; rachilla sometimes visible at maturity; lemmas 1-2 mm wide, the nerves obscure; awns 5-13 mm; anthers 0.6-1 mm. Weedy sites, disturbed ground, roadsides, 3700-9750 ft. Widespread. Introduced. Often treated as a phase of *B. squarrosus* L.

Bromus lanatipes (Shear) Rydb. [*Bromopsis lanatipes* (Shear) Holub]—Woolly brome. Tufted perennials, 35-90 cm; nodes pubescent to occasionally hairy; lower sheaths densely pilose to woolly, the hairs spreading from the sheath but becoming matted at the tips, becoming pilose to glabrous upward, the midrib sometimes abruptly narrowed just below the collar; auricles absent; blades 3-7 mm wide, glabrous; panicles 5-15 cm, the branches erect to ascending or spreading; glumes usually glabrous, sometimes sparsely pilose, mostly 1-nerved; lemmas rounded, the backs pubescent or occasionally glabrous, the margins hairy; awns 2-4 mm; anthers 1.8-4 mm. Semidesert riparian areas, mountain brush, oak and piñon-juniper woodlands, plains, 6500-7600 ft. NC, NW, C, WC, SC, SW.

Bromus madritensis L.—Red brome. Tufted annuals, 15-50+ cm; sheaths pilose to glabrous; auricles absent; blades 1-5 mm wide, pubescent to glabrous, flat; panicles 3-12 cm, dense and compact, reddish, the branches erect and mostly < 2 cm; spikelets pubescent; awns 12-23 mm; anthers 0.5-1 mm. Weedy, dry, disturbed ground, roadsides, old fields. 5000-7000 ft. San Miguel. Introduced.

Bromus porteri (J. M. Coult.) Nash [*Bromopsis porteri* (J. M. Coult.) Holub]—Nodding brome. Tufted perennials, 30-100 cm; sheaths glabrous to pilose, the midrib of the culm leaves not abruptly narrowed just below the collar; auricles absent; blades 2-5 mm wide, noticeably erect rather than spreading or drooping; panicles 7-20 cm, the branches spreading, often drooping; pedicels puberulent; glumes usually pubescent, rarely glabrous, the lower and upper mostly 3-nerved; lemmas rounded, the back and margin pubescent; awns 1-3 mm; anthers mostly 2-3 mm. Ponderosa and spruce-fir forests, aspen groves, often at high elevations, 5400-12,200 ft. NC, NW, C, WC, SC.

Bromus pumpellianus Scribn.—Pumpelly's brome. Rhizomatous perennials (rarely tufted), 50-130 cm; nodes glabrous to pubescent; sheaths mostly pilose to villous, sometimes glabrous; auricles sometimes present on lower leaves; blades 3-9 mm wide, mostly hairy, flat; panicles 10-24 cm, the branches erect to ascending; glumes glabrous to hairy; lemmas rounded on the back, variously hairy; awns 0-7 mm; anthers 3.5-7 mm. Uncommon in forests of the N mountains, 6480-9060 ft. Colfax, Sandoval, San Miguel.

Bromus racemosus L. [*B. commutatus* Schrad.]—Bald brome. Tufted annuals, 20-85+ cm; sheaths densely pubescent with retrorse hairs; auricles absent; blades 1-4 mm wide, pilose, flat; panicles 4-10 cm, the branches erect to ascending and mostly with 1 spikelet per branch; rachilla mostly obscured at maturity; spikelets terete to somewhat laterally flattened; lemmas 1-2 mm wide, the several nerves visible but not conspicuously raised; awns 3.5-9 mm; anthers 1-1.5 mm. Weedy sites, disturbed ground, roadsides; scattered locales and expected elsewhere, 3850-7000 ft. NE, NC, NW, Doña Ana. Often treated as a phase of *B. squarrosus* L.

Bromus richardsonii Link—Richardson's brome. Tufted perennials, 40–110+ cm; upper nodes usually glabrous; sheaths glabrous to pilose, the midrib not abruptly narrowed just below the collar; auricles absent; blades 3–10 mm wide, glabrous adaxially; panicles 10–20 cm, the branches spreading to somewhat drooping; glumes mostly glabrous, the lower mostly 1-nerved, the upper 3-nerved, 9–11 mm; lemmas rounded, the backs of the lower lemmas glabrous, those of the upper lemmas appressed-hairy, the margins ciliate; awns 3–5 mm; anthers 1.6–2.7 mm. Ponderosa, mixed conifer, spruce-fir forests, mountain meadows, 6700–11,800 ft. NC.

Bromus secalinus L.—Rye brome. Tufted annuals, 20–75+ cm; sheaths loosely pilose to glabrous; auricles absent; blades 2–4 mm wide, pilose to glabrous, flat; panicles 5–30 cm, the branches ascending to spreading; spikelets laterally flattened, the rachilla easily visible when mature; lemmas 1.5–2.5 mm wide, rounded on the back, usually glabrous; awns (0–)3–6 mm; anthers 1–2 mm. Disturbed ground, weedy sites, sometimes a weed in crop fields, 4800–7050 ft. NC, NW, C. Introduced. Often treated as a phase of *B. squarrosus* L.

Bromus squarrosus L.—Corn brome. Tufted annuals, 20–60 cm; nodes pubescent; sheaths densely pilose; auricles absent; blades 1.5–3 mm wide, the upper surface densely hairy, flat; panicles 6–10 cm, open and lax, the branches spreading and often with a single spikelet; lemmas 2–3 mm wide, obovate, the back glabrous to densely hairy; awns 8–12 mm; anthers (0.5–)1–1.5 mm. Weedy sites, disturbed ground, roadsides. Colfax. Introduced.

Bromus sterilis L.—Poverty brome. Tufted annuals, 30–80+ cm; sheaths densely pilose; auricles absent; blades 1–6 mm wide, pilose on both surfaces; panicles 10–20 cm, open, the branches mostly drooping and with 1–2(3) spikelets; lemmas 14–20 mm, rounded on the back; awns 15–30 mm; anthers 1–1.5 mm. Dry, disturbed ground, 7200–8650 ft. San Miguel, Taos. Introduced.

Bromus tectorum L.—Cheatgrass. Tufted annuals, 10–90 cm, the culm puberulent below the panicle; sheaths densely pilose, the upper sometimes glabrous; auricles absent; blades 1–6 mm wide, pilose on both surfaces; panicles 5–20 cm, open, the branches drooping and at least some with 4+ spikelets; lemmas 9–12 mm, usually pubescent; awns 10–18 mm; anthers 0.5–1 mm. Dry, disturbed ground, sometimes an indicator of formerly abused rangelands, 3400–10,000 ft. Widespread except SE. Introduced noxious weed.

CALAMAGROSTIS Adans. - Reedgrass

Plants perennial, tufted to commonly rhizomatous; sheaths open; auricles absent; ligules membranous; blades flat to involute; inflorescence a panicle, the primary branches rebranched; spikelets mostly 1-flowered, small, the rachilla prolonged beyond the palea and mostly hairy; disarticulation above the glumes; glumes mostly surpassing the floret, 1–3-nerved; lemmas mostly 3-nerved, the apex toothed; awns arising from the back of the lemma, from the base to near the tip, often obscure; paleas well developed; anthers 3.

1. Plants cultivated ornamentals, not known in the wild…**C. ×acutiflora**
1'. Plants native wild grasses, not known in cultivation…(2)
2. Awns exserted well beyond the glumes, easily visible, 4.5–8 mm; blades usually densely hairy on the upper surface…**C. purpurascens**
2'. Awns scarcely if at all exserted beyond the glumes, < 4.5 mm; blades glabrous or sparsely hairy…(3)
3. Pedicels glabrous or nearly so; panicles contracted, 1–2(–3) cm wide…**C. scopulorum**
3'. Pedicels evidently scabrous; panicles contracted to open…(4)
4. Glumes oblong, the apex abruptly acute and not drawn out to an awn-tip; blades 1–4 mm wide, usually rolled and stiffly ascending; lemmas not translucent on the upper 1/3; callus hairs 1/2–2/3 as long as the lemma…**C. stricta**
4'. Glumes lance-ovate, the apex of especially the first drawn out to an awn-tip; blades 3–10 mm wide, mostly flat and lax; lemmas translucent on the upper 1/3; callus hairs 2/3 up to as long as the lemma…**C. canadensis**

Calamagrostis ×acutiflora (Schrad.) DC.—Feather reedgrass. Densely tufted perennials, with short rhizomes at the base, 1–1.8+ m; panicles 15–30 cm, 1–2 cm wide, greenish purple in anthesis, becoming straw-colored; spikelets 4–5 mm; floret much shorter than the glumes; awns ca. 3.5 mm, attached near the base of the lemma, exserted and distinguished from the long callus hairs. Not known in the wild; this is an attractive ornamental grass that is being planted more and more in the state. No records seen.

Calamagrostis canadensis (Michx.) P. Beauv. [*C. scribneri* Beal]—Canada reedgrass. Loosely tufted perennials with moderate rhizomes, 50–100+ cm; blades 3–10 mm wide, mostly flat and lax, the upper surface scabrous; panicles 8–20 cm, 2–6 cm wide, open when mature, purplish to greenish; spikelets 2–5 mm; glumes lance-ovate, the apex especially of the lower glume drawn out to an awn-tip; rachilla 0.5–1 mm, with hairs 1.5–3.2 mm; lemmas translucent on the upper 1/3, the callus hairs 2/3 as long as to longer than the lemma; awns 1–3 mm, attached on the lower 1/2 of the lemma, difficult to observe. Wet meadows, seeps, marshy ground, 7520–12,200 ft. NC.

Calamagrostis purpurascens R. Br.—Purple reedgrass. Tufted perennials with short rhizomes, 30–80 cm; blades flat to involute, stiff, the upper surface densely hairy; panicles 4–14 cm, 1–2.5 cm wide, purplish; spikelets 4.5–7 mm; rachilla 1–2 mm, with hairs ca. 2 mm; awns 4.5–8 mm, easily seen, attached on the lower 1/3 of the lemma, bent and stout. Open rocky slopes, meadows, alpine plains, 10,700–12,855 ft. Taos.

Calamagrostis scopulorum M. E. Jones—Ditch reedgrass. Loosely cespitose perennials with short rhizomes, 40–90 cm; blades flat, pale bluish green, the upper surface glabrous or sparsely hairy; panicles 6–16 cm, 1–3 cm wide, pale to purplish, straw-colored when mature; spikelets 4–6 mm; rachilla 1–2 mm, with hairs 1.5–2.5 mm; awns 1–2 mm, attached on the upper 1/2 of the lemma and lying against the lemma, not exserted, straight, easily overlooked. Known in the state from a single collection, ca. 6000 ft., along a seep in a hanging garden in a piñon-juniper community. San Juan.

Calamagrostis stricta (Timm) Koeler—Slimstem reedgrass. Tufted perennials with short rhizomes, 30–90+ cm; blades mostly 1–4 mm wide, usually rolled and not lax, the upper surface scabrous; panicles 4–20 cm, 1–3 cm wide, narrow, purplish to pale green; spikelets 2–4.5 mm; glumes oblong, the apex abruptly acute and not drawn out to an awn-tip; rachilla 0.5–1.5 mm, with hairs 1.5–3 mm; lemmas membranous on the upper 1/3, the callus hairs 1/2–2/3 as long as the lemma; awns 1.5–2.5 mm, attached on the lower 1/2 of the lemma. New Mexico material belongs to **subsp. inexpansa** (A. Gray) C. W. Greene. Stream banks, wet meadows, seeps, and marshy or wet ground in the mountains, often in rather open clearings on mesic mountain slopes, 6600–9700 ft. NC, WC, Otero.

CATABROSA P. Beauv. – Brookgrass

Catabrosa aquatica (L.) P. Beauv.—Water whorlgrass. Tufted to stoloniferous perennials, usually decumbent-based and rooting at the nodes, 10–60 cm; sheaths glabrous; auricles absent; ligules membranous, 1–8 mm; panicles 5–28 cm, 2–10 cm wide; spikelets with (1)2(3) florets, 1.5–4 mm; disarticulation above the glumes and beneath the florets; glumes much shorter than the florets, awnless; lemmas 2–3 mm, prominently 3-nerved; anthers 3. Quiet stream banks in the N mountains; a single record, 8530 ft. Colfax.

CATAPODIUM Link – Ferngrass

Catapodium rigidum (L.) C. E. Hubb.—Rigid fescue. Tufted annuals, the culms to 60 cm, but usually much shorter, often procumbent; sheaths open, glabrous, the margins membranous; ligules membranous, 1.5–4 mm; panicles 2–18 cm, the branches spreading to divaricate; spikelets several-flowered, 4–10 mm, awnless; glumes glabrous, keeled; disarticulation above the glumes and between the florets; rachillae not prolonged, puberulent; glumes shorter than the florets; lemmas 5-nerved, 2–3 mm, rounded on the back, obtuse to mucronate; anthers 3. Weakly adventive from horticultural plantings in the S region, but likely to appear almost anywhere in the state; 1 record, 4265 ft. Doña Ana. Introduced.

CENCHRUS L. – Sandbur

Tufted annuals and perennials, the culms often geniculate or decumbent; sheaths open; auricles absent; ligules a ciliate membrane; inflorescence of highly reduced branches clustered into a fascicle or bur, the burs racemose on the main axis; burs of many stiff, spiny, partially fused bristles surrounding and mostly concealing the enclosed spikelets; disarticulation at the base of the bur; spikelets awnless, with 2 florets, the lower usually sterile, the upper fertile; lower glume short, 1-nerved; upper glume and lower lemma subequal, 3–9-nerved; upper floret becoming indurate, flattened on 1 side, enclosing the grain.

1. Burs with a single whorl of flattened spines subtended by 1 to several whorls of bristles…**C. echinatus**
1'. Burs with > 1 whorl of flattened spines, the spines projecting at irregular intervals throughout the body of the bur…(2)
2. Burs mostly with 8–40 spines, the bases of the larger spines frequently 1–2 mm wide; upper floret of the spikelets 3.4–5.8 mm; only 1 margin of the blade of uppermost leaf crinkled near the base…**C. spinifex**
2'. Burs mostly with 45–75 spines, the bases of the larger spines seldom > 1 mm wide; upper floret of the spikelets 5–7.6 mm; both margins of the blade of uppermost leaf conspicuously crinkled near the base… **C. longispinus**

Cenchrus echinatus L.—Southern sandbur. Tufted annuals, 20–100 cm; sheaths compressed; panicles 3–12 cm; burs 5–10 mm, with only a single whorl of inner fused spines subtended by 10–20 shorter unfused outer bristles; spikelets 5–7 mm. Disturbed ground; known from only a single, old collection in Doña Ana County, ca. 4000 ft. Introduced.

Cenchrus longispinus (Hack.) Fernald—Mat sandbur. Tufted annuals or occasionally short-lived perennials from pseudorhizomes, 20–90 cm, the bases geniculate; sheaths compressed-keeled; uppermost blade with both margins crinkled near the base; panicles 2–10 cm; burs 8–12 mm, short-hairy, with 45–75 spines, the inner bristles 0.5–1 mm wide at the base, the outer bristles mostly terete and reflexed; spikelets 4.5–8 mm; upper florets 4–7.6 mm. Disturbed ground, plains, grasslands, 3850–7000 ft. Widespread.

Cenchrus spinifex Cav. [*C. incertus* M. A. Curtis]—Coastal sandbur. Tufted annuals (rarely short-lived perennials), 30–100 cm, the bases geniculate; sheaths compressed-keeled; uppermost blade with only 1 margin crinkled near the base; panicles 3–8 cm; burs 5–10 mm, glabrous or sparsely pilose, with 8–40 spines, the inner bristles 1–2 mm wide at base, the outer bristles usually flattened at base; spikelets 3.5–10 mm; upper florets 3.4–5.8 mm. Disturbed ground, plains, grasslands, 5400–7200 ft. Widespread.

CHLORIS Sw. – Windmill grass

Tufted annuals or perennials; sheaths open, compressed-keeled; auricles absent; ligules a ciliate membrane; inflorescence a panicle of usually several spicate primary branches, these borne at the tip of the peduncle (digitate) or in a few separate whorls below the apex; spikelets with a single fertile floret at the base and 1(–3) staminate or sterile florets above; disarticulation above the glumes all the florets falling together, or sometimes below the glumes and the spikelet falling; glumes unequal, shorter than the florets; lemmas 3-nerved, generally pubescent on the lateral nerves, awned or awnless; paleas present, shorter than the lemmas; anthers 3.

1. Lemma of the lower floret with 3 awns 8–12 mm…go to **Leptochloa crinita**
1'. Lemma of the lower floret with a single awn or awnless…(2)
2. Lowermost lemma awnless or with a short awn < 2 mm…(3)
2'. Lowermost lemma prominently awned, the awn > 3 mm…(5)
3. Upper floret inflated-spheroidal, bowl-shaped, ca. 1 mm wide…**C. cucullata**
3'. Upper floret not inflated, < 0.5 mm wide…(4)
4. Florets with a short awn 0.5–2 mm…**C. ×subdolichostachya** (in part)
4'. Florets awnless or with a short mucro to 0.5 mm…**C. submutica**
5. Panicle branches typically in several whorls along an axis 2+ cm…**C. verticillata**
5'. Panicle branches in a single terminal whorl, or if in several whorls then the axis < 2 cm…(6)
6. Tip of lower lemma with a tuft of spreading hairs to 2 mm; plants annual…**C. virgata**
6'. Tip of lower lemma with short, appressed hairs; plants perennial…**C. ×subdolichostachya** (in part)

Chloris cucullata Bisch.—Hooded windmill grass. Tufted perennials, 15–60 cm; blades glabrous; panicle branches 10–20 in mostly 1–3 whorls, 2–5 cm; disarticulation below the glumes and the spikelets falling entire, or above the glumes and the florets falling together; spikelets with 1 fertile and 1 sterile floret; glumes 1-nerved, hyaline; sterile (upper) floret inflated, spherical, about as wide as long, with awn 0–1.5 mm. Plains, grasslands, roadsides, disturbed ground, 2900–4600 ft. EC, SE, Hidalgo.

Chloris ×subdolichostachya Müll.-Hal.—Plants referred to this name are hybrid derivatives involving *C. andropogonoides* E. Fourn. (not in New Mexico), *C. cucullata*, and *C. verticillata*, not accurately assignable to a biological species, and will have features of all 3 parents: perennial; panicle branches digitate or in very close whorls; and slightly inflated florets with short awns.

Chloris submutica Kunth—Mexican windmill grass. Plants perennial, mostly cespitose but occasionally with short stolons, 30–75 cm; blades sometimes with long basal hairs; panicle branches 5–17 in 1–3 closely spaced whorls, to 7 cm, ascending to reflexed when mature; spikelets with 1 fertile and 1 sterile floret; disarticulation above the glumes; florets awnless or with a short mucro. Disturbed ground, lawns, fields, 4200–4265 ft. Doña Ana, Otero. Introduced.

Chloris verticillata Nutt.—Tumble windmill grass. Tufted perennials, 15–40 cm; blades with basal hairs, otherwise glabrous; panicle branches 10–16 in mostly 2–5 whorls, 5–15 cm; disarticulation at the base of the panicle and concomitantly above the glumes, releasing the florets as the panicle rolls and bounces across terrain; spikelets with 1 fertile and 1 sterile floret; florets with awns 3–9 mm. Plains, grasslands, roadsides, 3570–7720 ft. Widespread except SW.

Chloris virgata Sw.—Feather finger grass. Tufted annuals, sometimes with stolons, 10–80+ cm; blades with long hairs near the base; panicle branches 4–20 in a single terminal whorl, 5–10 cm; spikelets with 1 fertile and 1–2 sterile florets; disarticulation above the glumes, the florets falling together; lower (fertile) floret with a conspicuous tuft of long hairs at the apex; first and second florets with awns 3–15 mm; third floret when present greatly reduced, awnless. Disturbed fields, roadsides, waste areas, 2975–7025 ft. Widespread introduced weed.

CINNA L. – Woodreed

Cinna latifolia (Trevir. ex Göpp.) Griseb.—Drooping woodreed. Tufted perennials, 20–100+ cm, sometimes rhizomatous; sheaths open; auricles absent; ligules thin-membranous, 2–8 mm; blades flat; panicles 5–40 cm, the branches spreading; disarticulation below the glumes; spikelets 1(2), 2–5 mm; rachilla prolonged, 0–1 mm; glumes and lemmas 1.8–3.8 mm, awnless or with an awn to 2.5 mm; paleas 1–2-nerved; anther 1. Moist places in mixed conifer woodlands and forests, 7765–10,500 ft. NC, Catron.

COIX L. – Job's tears

Coix lacryma-jobi L.—Job's tears. Monoecious, tufted annual or perennial, 0.8–3 m; leaves glabrous, blades 2–7 cm wide, the base cordate; inflorescences axillary, many per shoot, composed of 1 staminate and 1 pistillate rame; beads 8–12 mm diam., of varying color; staminate rames 1–3.5 cm, breaking apart when mature; glumes exceeding the florets, with 15+ nerves, 5–9 mm. Occasionally cultivated in flower gardens for the beadlike female involucres; not known in the wild.

CORTADERIA Stapf – Pampas grass

Cortaderia selloana (Schult. & Schult.) Asch. & Graebn.—Pampas grass. Giant tussocky perennials, 2–4 m, often dioecious, sometimes monoecious; sheaths open; auricles absent; ligules a ring of hairs; leaves primarily basal; sheaths glabrous, with a dense tuft of hairs at the collar; blades to 2 m, 3–8 cm wide, the margins sharply serrate; panicles 30–130 cm, only slightly elevated above the foliage; spikelets 15–17 mm; lemmas long-attenuate to an awn 2–5 mm; anthers 3 when present. Introduced as an ornamental landscape plant, with numerous cultivars; not known in the wild in New Mexico.

COTTEA Kunth – Cottagrass

Cottea pappophoroides Kunth—Cottagrass. Tufted perennial from knotty bases, 25–70 cm; foliage softly pilose; sheaths open; auricles absent; ligules hairy; inflorescence a panicle, purplish, 8–15 cm, 2–6 cm wide, the branches ascending; spikelets several-flowered, pilose, the distal reduced; disarticulation above the glumes and between the florets; glumes subequal, 4–5 mm, 7–13-nerved; lemmas 3–4 mm, 9–13-nerved and awned; anthers 3. Rocky volcanic hills and plains of the S desert regions, 4450–5270 ft. Doña Ana, Hidalgo, Sierra.

CYNODON Rich. – Bermudagrass

Cynodon dactylon (L.) Pers.—Bermudagrass. Stoloniferous and rhizomatous perennial, 5–40 cm, often forming a dense turf or sod; sheaths glabrous, the collars usually with long hairs; auricles absent; ligules ca. 0.5 mm; blades flat when mature, 1–10 cm, 2–5 mm wide; panicles with 2–9 branches 2–6 cm; spikelets 2–3.2 mm; glumes subequal, 1.5–2.2 mm; lemmas a bit longer than the glumes. A common grass for lawns and improved pastures, also escaping into gardens, fields, and along roads, 3200–6500 ft. Widespread except NE. Introduced.

CYNOSURUS L. – Dog's-tail grass

Cynosurus echinatus L.—Bristly dog's-tail grass. Tufted annual, 10–100+ cm; sheaths open; auricles absent; ligules membranous, 3–10 mm; panicles 2–8 cm, ovoid to globose; spikelets dimorphic and paired; sterile spikelets persistent, several-flowered, the glumes and lemmas similar and with awns to 8 mm; fertile spikelets 1–5-flowered; glumes shorter than the florets; lemmas with awns 5–20 mm. Known as yet only from moist weedy ground in Bandelier National Monument, 6100 ft. Sandoval. Introduced.

DACTYLIS L. – Orchardgrass

Dactylis glomerata L.—Orchardgrass. Tufted perennial with abundant foliage, 75–150+ cm; internodes hollow, the nodes glabrous; sheaths closed, the margins fused together at least 1/2 their length, mostly longer than the internodes; auricles absent; ligules conspicuous, 3–11 mm; blades 4–10 mm wide; inflorescence a somewhat pyramidal panicle, 4–20 cm, the distal branches ascending to appressed; spikelets 5–8 mm, subsessile; lemmas 4–8 mm; paleas slightly shorter than the lemmas. Widely introduced for meadow and pasture improvement, 3850–11,600 ft. Widespread. Adventive weed.

DACTYLOCTENIUM Willd. – Crowfoot grass

Dactyloctenium aegyptium (L.) Willd.—Egyptian grass. Tufted annual, 10–40 cm, the culms geniculate-based; sheaths open, with bulbous-based hairs distally; ligules to 1.5 mm, ciliate; blades 2–10 mm wide, with bulbous-based hairs, especially proximally; panicle branches mostly 2–6 in number, 2–6 cm; spikelets 2-rowed on the abaxial sides of the branches, with several florets, 3–5 mm; glumes with curved awns 1–2.5 mm; lemmas ovate, with curved awns 0.5–1 mm, the paleas about as long; anthers 3, yellow. An infrequent weed of cultivated fields, moist waste places, and lawns, 4000–5900 ft. Doña Ana, Hidalgo. Introduced.

DANTHONIA DC. – Danthonia, oatgrass

Tufted perennials; sheaths open, with tufts of hair at the edges of the collars; auricles absent; ligules of hairs; inflorescence a weakly developed panicle, raceme, or a single spikelet; spikelets several-flowered; disarticulation above the glumes and between the florets; glumes subequal, usually exceeding the florets or nearly so, 1–7-nerved; lemmas 7–11-nerved, the apex with 2 aristate lobes, densely hairy on the callus and margins; awns from between the lemma lobes, geniculate and twisted; anthers 3; cleistogamous spikelets sometimes developed in the lower sheaths, 1- to several-flowered.

1. Pedicels and branches puberulent…**D. parryi**
1'. Pedicels glabrous…(2)
2. Lemmas 3.5-5 mm; blades curly…**D. spicata**
2'. Lemmas 7-8 mm; blades ± straight…**D. intermedia**

Danthonia intermedia Vasey—Timber oatgrass. Shoots 20-70 cm, the culms not breaking at the nodes; sheaths usually glabrous; uppermost blades erect or only slightly diverging from the culm at maturity; panicles with several spikelets, the branches/pedicels erect and appressed; spikelets 11-18 mm, usually greenish- to purplish-tinged; lemma teeth 1.5-2.5 mm; awns 6-8 mm; cleistogamous spikelets rarely produced. Forest meadows and clearings at high elevations, 8100-13,030 ft. NC. Greenish plants, including the spikelets, and erect blades help identify this species.

Danthonia parryi Scribn.—Parry's oatgrass. Shoots in well-developed tussocks, 30-100 cm, the culms not breaking at the nodes; sheaths glabrous or sparsely pubescent; lower blades breaking at the collar, leaving clumps of basal sheaths; uppermost blades erect or only slightly diverging from the culm at maturity; panicles with few to many spikelets, sometimes racemose, the branches/pedicels ascending to spreading and puberulent; spikelets 16-24 mm, commonly straw-colored when mature; lemma teeth 2.5-8 mm; awns 12-15 mm; cleistogamous spikelets perhaps produced in the lower sheaths. Coniferous forests, mountain meadows and grasslands, 6900-10,600 ft. NC, C, WC.

Danthonia spicata (L.) P. Beauv. ex Roem. & Schult.—Poverty oatgrass. Shoots 10-65 cm, breaking at the nodes when mature; sheaths pilose to glabrous; blades usually curled when mature (sometimes nearly straight in shade plants); panicles with few to numerous spikelets, the branches/pedicels appressed to ascending or spreading; spikelets 7-15 mm, greenish to straw-colored; lemma teeth 0.5-2 mm; awns 5-8 mm. Dry, sandy mineral soil in ponderosa pine forests, 6300-9800 ft. NC, WC.

DESCHAMPSIA P. Beauv. – Hairgrass

Plants annual to perennial; leaves tending to be basal; sheaths open; auricles absent; ligules membranous; inflorescence a panicle; spikelets mostly 2-flowered; disarticulation above the glumes, the florets falling together; glumes exceeding the lower floret; lemmas 5-7-nerved, the apex toothed, awned from the lower back; rachilla prolonged beyond the palea; anthers 3.

1. Plants perennial; blades 1-5 mm wide; panicle loose and open at maturity, the branches spreading…**D. cespitosa**
1'. Plants annual; blades 0.5-1.5 mm wide; panicle narrow at maturity, the branches mostly erect…**D. danthonioides**

Deschampsia cespitosa (L.) P. Beauv.—Tufted hairgrass. Perennial, 35-100+ cm; panicles 10-30 cm, 5-25 cm wide, the branches flexuous; spikelets 3-7 mm, sometimes with 1 or 3 florets; lemmas purplish proximally, pale distally; awns 1-8 mm, straight to bent, often exceeding the glumes slightly; anthers 1-3 mm. Widespread in moist mountain meadows, bogs, grasslands, forest openings, 6000-12,540 ft. NC, WC, C, SC.

Deschampsia danthonioides (Trin.) Munro—Annual hairgrass. Annual, 10-50 cm; panicles 5-20 cm, 2-8 cm wide, the branches rather straight; spikelets 4-9 mm; lemmas pale green to purplish but not bicolored; awns 4-9 mm, strongly geniculate; anthers 0.3-0.5 mm. An infrequent weed of moist waste places; 1 record, 6142 ft. Torrance (?Grant). Introduced.

DICHANTHELIUM (Hitchc. & Chase) Gould – Rosettegrass

Tufted (ours) perennials, producing a winter rosette of broad basal leaves often sharply different from the narrow cauline leaves; sheaths open; auricles absent; ligules mostly a ciliate membrane or ring of hairs, often with a nearly indistinguishable second ring of hairs (pseudoligule) adjacent on the blade; inflorescence a terminal (spring) or axillary (summer-fall) panicle; spikelets dorsally compressed, with 2 florets, the lower staminate or sterile, the upper fertile and indurate; disarticulation below the glumes; lower glume shorter than the spikelet; upper glume as long as the spikelet, 5-11-nerved; lower lemma similar to the upper glume; upper floret forming a hard seedcase, the lemma rounded, shiny, and clasping the flat palea; anthers 3.

1. Basal leaf blades similar in shape to those of the lower cauline leaves, usually erect to ascending; culms branching from near the base in the fall, with 2-4 leaves, only the upper 2-4 internodes elongate; spikelets 2.4-3.4 mm…(2)
1'. Basal leaf blades usually well differentiated from those of the lower cauline leaves, spreading, forming a rosette; culms usually branching from the midculms in the fall, with many leaves, usually all internodes elongate; spikelets 1.4-3.8 mm…(3)
2. Panicles 1-2 cm wide, narrow with appressed spikelets; upper cauline blades 10-20 cm, distinctly longer than those below…**D. perlongum**
2'. Panicles 2-4 cm wide, open with spreading spikelets; upper cauline blades 4-8 cm, similar to those below …**D. wilcoxianum**
3. Spikelets 1.4-2 mm; upper glume lacking an orange or purplish spot at the base…**D. acuminatum**
3'. Spikelets 2.7-3.5 mm; upper glume with an orange or purplish spot at the base…**D. oligosanthes**

Dichanthelium acuminatum (Sw.) Gould & C. A. Clark—Tapered rosettegrass. Plants 20-75+ cm; basal rosettes differentiated; sheaths shorter than the internodes, glabrous to densely pubescent; blades usually densely pubescent; spikelets 1.4-2.1 mm, pubescent; lower glume obtuse to acute; lower floret sterile; upper glume lacking an orange or purplish spot at the base. Moist woodlands, stream banks, shaded canyons, 4650-5900 ft. Eddy, Otero.

Dichanthelium oligosanthes (Schult.) Gould—Scribner's rosettegrass. Plants 20-5 cm; basal rosettes differentiated; sheaths shorter than the internodes, glabrous to pubescent; blades glabrous to pubescent, 6-15 mm wide; spikelets 2.7-3.5 mm, usually glabrous; lower glume acute; upper glume with a prominent orange to purplish spot at the base. New Mexico material belongs to **var. scribnerianum** (Nash) Gould. Moist shaded places along mountain streams and rivers, 5300-8270 ft. NC, NW, EC, WC, Eddy.

Dichanthelium perlongum (Nash) Freckmann—Slimleaf panic grass. Plants 10-50 cm; basal rosettes weakly developed; sheaths longer than the internodes, pilose; blades pubescent; spikelets 2.6-3.4 mm, finely pubescent; lower glume broadly ovate. Moist shaded woodlands and canyon bottoms, 6930-8000 ft. NE, EC, Catron.

Dichanthelium wilcoxianum (Vasey) Freckmann—Fall rosettegrass. Plants 15-35 cm; basal rosettes poorly developed; sheaths mostly longer than the internodes, hirsute; blades stiffly erect, sparsely pilose; spikelets 2.4-3.2 mm, finely pubescent; lower glume triangular. Moist open grassland clearings in the W mountains. No records seen. Catron.

DIGITARIA Haller - Crabgrass

Tufted or stoloniferous annuals to perennials, the culms often rooting at the lower nodes; sheaths open; auricles absent; ligules membranous, sometimes with a ciliate fringe; inflorescence a panicle of spicate, digitate branches, not detaching at maturity, the spikelets appressed in 2 rows along 1 side of the branches; spikelets awnless, glabrous to pubescent, 2-flowered, the lower floret staminate to sterile, the upper floret fertile, the rachilla not elongate between the 2 glumes; disarticulation below the glumes; lower glume absent to not > 1/4 the length of the spikelet; upper glume short to subequal to the spikelet; lower lemma pubescent; upper lemma clasping the palea and forming a seedcase, this membranous to somewhat cartilaginous but not indurate, pale- to dark-colored; anthers 3.

1. Spikelets silky-pubescent with long, whitish hairs; plants perennial…(2)
1'. Spikelets glabrous or with short, stiff hairs; plants annual…(3)
2. Panicles with 3+ nodes, the branches not subdigitate…go to **Trichachne**
2'. Panicles with only 1-2 nodes, the branches subdigitate…**D. eriantha**
3. Blades usually with prominent, stiff, bulbous-based hairs on both surfaces; lower lemma scabrous on the lateral nerves (use ×10 or higher magnification)…**D. sanguinalis**
3'. Blades glabrous, only rarely with scattered hairs; lower lemma smooth on the lateral nerves…(4)
4. Spikelets 1.7-2.3 mm, borne in 3s on the middle portion of the branch; lower glume absent or a nerveless membranous rim < 0.3 mm…**D. ischaemum**

4′. Spikelets 2.8-4.1 mm, borne in 2s on the middle portion of the branch; lower glume 0.2-0.8 mm...**D. ciliaris**

Digitaria ciliaris (Retz.) Koeler—Southern crabgrass. Tufted annual, sometimes longer-lived, 10-60+ cm; sheaths with bulbous-based hairs; blades mostly glabrous except toward the base; panicle branches 2-10, digitate or in 1-2 whorls below the apex, 2-20 cm; spikelets 2.8-4.1 mm; lower glume vestigial, < 0.3 mm; upper glume 2/3 to nearly as long as the spikelet, short-hairy; upper floret 2.5-4 mm, glabrous Weed of moist waste places, 3850-4000 ft. Chaves, Otero. Introduced.

Digitaria eriantha Steud.—Pangolagrass. Tufted, rhizomatous, or stoloniferous perennial, 35-100+ cm; basal sheaths glabrous to densely pubescent, the hairs 4-6 mm; blades 3-6 mm wide; panicle branches 3-15, subdigitate, 5-25 cm; spikelets 2.8-3.5 mm; lower glume 0.3-0.5 mm; upper glume and lower lemma woolly-hairy. Introduced for experimental planting in Quay County at the Tucumcari Historical Research Institute, New Mexico State University, but not known to escape, 4090 ft. Introduced.

Digitaria ischaemum (Schreb.) Muhl.—Smooth crabgrass. Tufted annual, the culms 20-60 cm, decumbent and rooting at the nodes; sheaths glabrous or sparsely pubescent; blades mostly glabrous except toward the base; panicle branches 2-7, subdigitate, 6-17 cm; axillary panicles present in lower sheaths; spikelets 1.7-2.3 mm; lower glume vestigial, at most a nerveless membranous rim; upper glume 3/4 to as long as the spikelet; upper floret 2-3 mm. Weed of lawns and gardens, 3940-5520 ft. Colfax, Doña Ana, Quay, San Juan. Introduced.

Digitaria sanguinalis (L.) Scop.—Hairy crabgrass. Tufted annual, the culms decumbent and rooting at the nodes, 20-70 cm; sheaths sparsely pubescent with bulbous-based hairs; blades usually with bulbous-based hairs on both surfaces; panicle branches 4-13 in number, 3-30 cm, subdigitate, the panicle sometimes with an extended rachis; spikelets 1.7-3.4 mm; lower glume 0.2-0.4 mm; upper glume 1/3-1/2 as long as the spikelet; upper floret 1.7-3 mm, purplish to brown. Weed of gardens and open, moist waste ground, 3850-8400 ft. NE, NW, C, WC, SE, SC, SW. Introduced.

DINEBRA Jacq. - Sprangletop

Tufted annuals; sheaths open; auricles absent; ligules a truncate membrane, sometimes erose; blades flat; inflorescence a panicle of spicate or racemose branches; spikelets several-flowered, awnless (ours); disarticulation above the glumes and between the florets; glumes 1-nerved; lemmas 3-nerved, hairy on the nerves, awned or awnless; anthers 3. Formerly included in a polyphyletic *Leptochloa*.

1. Sheaths sparsely to densely hairy, the hairs bulbous-based; spikelets 2-4 mm; lemmas < 2 mm, awnless ...**D. panicea**
1′. Sheaths glabrous (sometimes hairy near the base); spikelets 4.5-8 mm; lemmas > 2 mm, short-awned... **D. viscida**

Dinebra panicea (Retz.) P. M. Peterson & N. Snow [*Leptochloa brachiata* Steud.; *L. filiformis* (Pers.) P. Beauv.]—Red sprangletop. Tufted annual, 15-100+ cm; sheaths sparsely to densely hairy, the hairs bulbous-based; blades 3-20 mm wide, glabrous to sparsely pilose on both surfaces; panicles with 3-80+ branches arranged along the main axis; branches mostly 4-18 cm; spikelets 2-4 mm; glumes sometimes as long as the spikelet, acute to aristate; lemmas awnless, acute to obtuse at the apex. Our material belongs to **subsp. brachiata** (Steud.) P. M. Peterson & N. Snow. Moist weedy ground, 3850-6000 ft. SC, SW, Chaves, Lincoln.

Dinebra viscida (Scribn.) P. M. Peterson & N. Snow [*Leptochloa viscida* (Scribn.) Beal]—Sticky sprangletop. Tufted annual, the culms prostrate to erect, 5-50 cm; sheaths mostly glabrous, sometimes sticky near the base; blades 1-5 mm wide, glabrous; panicles with 5-20 branches arranged along the main axis; branches 1-3 cm; spikelets 4.5-7.5 mm; glumes shorter than the spikelet, acute; lemmas sticky on the back, with short awns 0.5-1.5 mm. Plains and swales, 4000-4325 ft. SC, SW, Lea, Roosevelt.

DIPLACHNE P. Beauv. - Sprangletop

Diplachne fusca (L.) P. Beauv. ex Roem. & Schult.—Malabar sprangletop. Tufted annuals, sometimes longer-lived, 10–100+ cm; sheaths open, glabrous; auricles absent; ligules an attenuate membrane, lacerate at maturity; blades flat or involute upon drying, glabrous, 2–7 mm wide; inflorescence a panicle of spicate branches, 10–80+ cm; branches numerous, arranged along the main axis, ascending to spreading to 22 cm wide; spikelets several-flowered, to 12 mm, awned to awnless; disarticulation above the glumes and between the florets; glumes 1-nerved; lemmas 3-nerved, hairy on the nerves, awned or awnless; anthers 3. Weedy, moist ground. Formerly included in *Leptochloa*.

1. Uppermost leaf blades exceeding the panicles, which are usually enclosed in the subtending sheaths; mature lemmas whitish with a dark spot on the basal 1/2, acute to awned to 3.5 mm...**subsp. fascicularis**
1'. Uppermost leaf blades exceeded by the panicles, which are usually completely exserted; mature lemmas usually lacking a basal dark spot, truncate to obtuse, awnless...subsp. uninervia

subsp. fascicularis (Lam.) P. M. Peterson & N. Snow [*D. fascicularis* (Lam.) P. Beauv.; *Leptochloa fascicularis* (Lam.) A. Gray]—Bearded sprangletop. Widespread, 3500–7300 ft. NE, NC, EC, C, SE, SC, SW.

subsp. uninervia (J. Presl) P. M. Peterson & N. Snow [*Leptochloa uninervia* (J. Presl) Hitchc. & Chase]—Mexican sprangletop. Uncommon. No records seen.

DISAKISPERMA Steud. - Sprangletop

Disakisperma dubium (Kunth) P. M. Peterson & N. Snow [*Leptochloa dubia* (Kunth) Nees]—Green sprangletop. Perennials, 30–110 cm; sheaths sparsely pilose; the nodes glabrous; sheaths open; auricles absent; ligules membranous, typically 1–1.5 mm, truncate; blades to 35 cm, 2–8 mm wide, flat (drying involute), the midrib prominent; panicle branches 5–15 cm, alternate along the main axis (infrequently subdigitate); cleistogamous spikelets commonly produced in the basal sheaths; spikelets several-flowered, awnless; disarticulation above the glumes and between the florets; glumes 2–6 mm; lemmas 3–5 mm, the green lateral nerves usually prominent against the pale body, pubescent, the apices blunt, notched; paleas subequal to the lemmas; anthers 1–1.6 mm; caryopses 1.5–2.3 mm, to 1 mm wide. Plains, slopes, bajadas, ravines, roadsides, often shady sites, 3320–7500 ft. Widespread except N, NW. This is a segregate genus from the traditional *Leptochloa*, which has been dismembered into the additional genera *Dinebra*, *Diplachne*, and *Disakisperma*.

DISTICHLIS Raf. - Saltgrass

Distichlis spicata (L.) Greene [*D. stricta* (Torr.) Rydb.]—Saltgrass. Mostly rhizomatous perennial, 10–60 cm, usually unisexual; foliage usually exceeding the pistillate panicle but often shorter than the staminate; leaves strongly distichous, glabrous; sheaths open; auricles absent; ligules a short membrane; blades stiff; spikelets several-flowered, awnless, 5–20 mm; glumes unequal, 3–7-nerved; lemmas 9–11-nerved, 3–6 mm; anthers 3. Floodplains, saline soils, swales, salt flats, marshes, 3100–8515 ft. Widespread.

ECHINOCHLOA P. Beauv. - Barnyardgrass

Tufted annuals (ours); sheaths open; auricles absent; ligules absent (ours) or of hairs; blades flat; inflorescence a panicle of spicate branches; disarticulation below the glumes, the spikelets falling entire; spikelets convex, with 2 florets, the lower staminate or sterile, the upper fertile; lower glume much shorter than the spikelet; upper glume similar in size and texture to the lower lemma; upper floret indurate, the lemma clasping the palea; anthers 3.

1. Palea of lower floret absent or vestigial, much < 1/2 as long as the lemma...**E. crus-pavonis**
1'. Palea of lower floret well developed, nearly as long as the lemma...(2)
2. Hairs of the panicle branches and spikelets not bulbous-based; panicle branches simple, usually 2(–3) cm or less; spikelets awnless, 2–3 mm, arranged in 4 rows on the branch...**E. colona**
2'. Hairs of the panicle branches and/or spikelets bulbous-based; panicle branches usually rebranched, the

lower branches usually > 2 cm; spikelets awnless or awned, 2.8–4 mm (excluding the awns), mostly arranged in 2 rows on the panicle branch…(3)

3.   Shiny apical portion of the fertile lemma obtuse or broadly acute, with a line of minute hairs, the tip sharply differentiated and withering; hairs of the panicle branches, at least some, > 3 mm…**E. crus-galli**
3′.   Shiny apical portion of the fertile lemma narrowly acute to acuminate, without a line of minute hairs, with a gradual transition to a membranous, stiff tip; hairs of the panicle branches absent to rarely > 3 mm…**E. muricata**

Echinochloa colona (L.) Link—Jungle rice. Erect to decumbent annual, rooting from the lower nodes, the culms 10–70 cm; blades often with purplish bars; panicles glabrous to hispid but lacking bulbous-based hairs; branches simple, < 3 cm; spikelets 2–3.5 mm, in 4 rows, awnless; fertile floret 2.5–3 mm. Moist disturbed ground, lawns, gardens, 3475–6200 ft. EC, SE, SC, SW, San Juan. Introduced.

Echinochloa crus-galli (L.) P. Beauv.—Barnyardgrass. Decumbent to stiffly erect annual, 30–100+ cm; panicles with bulbous-based hairs to 5 mm; branches 2–10 cm, usually rebranched; spikelets 2.5–4 mm, awnless or with awns to 5 cm; fertile floret 2.5–3.5 mm. Wet ground, muddy places, ditch banks, around stock ponds, 2975–7600 ft. A widespread weed.

Echinochloa crus-pavonis (Kunth) Schult.—Gulf cockspur grass. Annual or short-lived perennial, 30–100+ cm; panicles with bulbous-based hairs; branches 4–12 cm, usually rebranched with shorter second branches; spikelets 2.5–3.5 mm, awnless or with curved awns to 10 mm; lower palea absent or vestigial. New Mexico material belongs to **var. macera** (Wiegand) Gould. Marshy ground and wet disturbed places; uncommon, 3770–8200 ft. Curry, Doña Ana, Quay.

Echinochloa muricata (P. Beauv.) Fernald [*E. muricata* var. *microstachya* Wiegand]—American barnyardgrass. Annual, 75–160 cm; panicles with bulbous-based hairs < 3 mm, or absent; branches 2–8 cm, usually rebranched; spikelets 2.5–5 mm, usually with bulbous-based hairs, awnless or with awns to 16 mm. Moist to wet swales and seeps, disturbed ground, roadsides, 3850–8200 ft. Scattered introduced weed.

ELEUSINE Gaertn. – Goosegrass

Eleusine indica (L.) Gaertn.—Goosegrass. Annuals, (10–)25–65 cm; sheaths open, strongly keeled, hairy near the throat; auricles absent; blades with prominent white midnerve; ligules membranous, ciliate; panicles with several branches whorled at the apex and often with 1 branch attached below; spikelets 4–7 mm, awnless, several-flowered, laterally compressed; disarticulation above the glumes and between the florets; lemmas 3-nerved; anthers 3. Weed of lawns, cultivated fields, moist waste places, 3935–5500 ft. Bernalillo, Doña Ana. Adventive.

ELIONURUS Kunth – Balsamscale

Elionurus barbiculmis Hack.—Woollyspike balsamscale. Tufted perennial, 40–80 cm, densely hirsute beneath the nodes; sheaths open; auricles absent; ligules a ciliate membrane or entirely of hairs; blades mostly involute, to 30 cm, densely pilose on upper surface; inflorescence 5–10 cm, the axis villous, spikelike, unbranched, composed of repeating pairs of sessile and pedicelled spikelets; disarticulation below the spikelet pair, the 2 falling together; spikelets densely villous, the sessile 4.5–8 mm. Rocky, grassy slopes and foothills, 4640–5505 ft. SC, SW, Lincoln.

ELYMUS L. – Wheatgrass, wildrye

Tufted to rhizomatous, sometimes stoloniferous, perennials; sheaths open; auricles often developed; ligules membranous, sometimes ciliate; inflorescence a spike, with 1–5 spikelets per node; spikelets 1- to several-flowered, awned to awnless; disarticulation various, from above the glumes and between the florets to below the spikelet and in the rachis; glumes 2, sometimes highly reduced; lemmas usually 5–7-nerved; anthers 3.

1.   Spikelets mostly solitary at each node of the rachis…(2)
1′.   Spikelets 2+ at each node of the rachis…(16)

2. Spikelets (glumes and lemmas) long-awned, the awns prominent and mostly > 10 mm…(3)
2'. Spikelets (glumes and lemmas) awnless or nearly so, any awns usually < 5 mm…(7)
3. Awns erect-appressed or nearly so, scarcely diverging as much as 15° from vertical; glumes 3/4 to equaling the length of the spikelet…**E. trachycaulus** (in part)
3'. Awns widely spreading to reflexed, diverging at least 30°+ from the vertical; glumes 1/2-2/3 the length of the spikelet…(4)
4. Anthers 4-6 mm; spikelets widely spaced and hardly overlapping…(5)
4'. Anthers 1-2 mm; spikelets at least moderately congested and overlapping…(6)
5. Spikes 15-30 cm, often nodding; blades 4-6 mm wide…**E. arizonicus**
5'. Spikes 8-15 cm, usually erect; blades 1-2 mm wide…**E. spicatus**
6. Spikes 3-7 cm, very dense, the lowermost internodes 3-7 mm; plants 15-45 cm, the bases usually decumbent to prostrate; midculm nodes mostly 0.5-1.5 mm wide; glumes 1-3(-5)-nerved…**E. scribneri**
6'. Spikes 7-20 cm, not especially dense, the lowermost internodes 8-15 mm; plants 30-70 cm, the bases usually erect; midculm nodes mostly 1.5-2.5 mm wide; glumes (3-)5-7-nerved…**E. bakeri**
7. Glumes blunt, nearly truncate, thick and very firm; spikelets awnless; sheaths typically ciliate on at least 1 margin…(8)
7'. Glumes acute to acuminate, thin and membranous to stiff, but not thick; spikelets awned or awnless; sheaths rarely ciliate…(9)
8. Plants with evident, long-creeping rhizomes…**E. hispidus**
8'. Plants densely tufted, lacking evident rhizomes…**E. ponticus**
9. Anthers 1-2 mm…(10)
9'. Anthers 4-16 mm…(12)
10. Glumes 1- or 2(3)-nerved; rachis tending to break apart at maturity; sterile hybrid plants; occurring where the 2 parents grow together…**E. trachycaulus** × **E. elymoides**
10'. Glumes (3-)5-nerved; rachis remaining intact; fertile to sterile plants…(11)
11. Plants mostly with rhizomes…**E. ×pseudorepens**
11'. Plants tufted…**E. trachycaulus** (in part)
12. Plants with evident, long-creeping rhizomes…(13)
12'. Plants lacking evident rhizomes, occasionally rhizomes weakly developed and short…(15)
13. Glumes acuminate, asymmetric or somewhat sickle-shaped, gradually tapering to an awn-tip; blades somewhat rigid and prominently ridged above…go to **Pascopyrum smithii**
13'. Glumes acute to acuminate, symmetric, not gradually tapering to an awn-tip; blades often lax, not prominently ridged above…(14)
14. Blades flat, mostly 5-15 mm wide, dark green, often with a circular constriction toward the tip; anthers (3-)4-7 mm…**E. repens**
14'. Blades rolled or < 4 mm wide when flat, usually glaucous, lacking a circular constriction toward the tip; anthers 3-5 mm…**E. lanceolatus**
15. Spikes 15-30 cm, often nodding; blades 4-6 mm wide…**E. arizonicus**
15'. Spikes 8-15 cm, usually erect; blades 1-2 mm wide…**E. spicatus**
16. Rachis fragile and breaking apart at maturity…(17)
16'. Rachis persistent, not breaking apart at maturity…(19)
17. Glumes 1+ mm wide and conspicuously hardened…**E. virginicus** (in part)
17'. Glumes < 1 mm wide, flexible and not hard…(18)
18. Lemma awns 4-17 mm; rachis internodes 2.5-7 mm…these are *E. elymoides* × *E. trachycaulus* hybrids [*E. ×saundersii* Vasey]
18'. Lemma awns 20-80 mm; rachis internodes mostly 5-12 mm…**E. elymoides** (in part)
19. Glumes absent or reduced to 1 or 2 minute bristles; spikelets horizontally spreading or ascending at maturity…**E. hystrix**
19'. Glumes present and well developed…(20)
20. Glumes nearly subulate, 1-2-nerved…(21)
20'. Glumes narrowly lanceolate and broadened above the base, mostly conspicuously 3-7-nerved…(23)
21. Sheaths villous; glumes short-pilose…**E. villosus**
21'. Sheaths mostly glabrous; glumes glabrous…(22)
22. Glumes 1-3 cm, indurate on the lower portion…**E. interruptus**
22'. Glumes 4-15 cm, rarely slightly shorter, only slightly hardened if at all…**E. elymoides** (in part)
23. Glumes firm and hardened on at least the lower portion, bottom bowed out slightly; lemmas 6-9 mm… **E. virginicus** (in part)
23'. Glumes not hardened or bowed out at the base; lemmas 8-14 mm…(24)
24. At maturity, the spikes erect and the awns erect-appressed; glumes mostly < 20 mm; lemmas glabrous to scaberulous…**E. glaucus**
24'. At maturity, the spikes usually nodding or curved and the awns spreading outward; glumes 20+ mm; lemmas scabrous to short-hairy (rarely glabrous)…**E. canadensis**

Elymus arizonicus (Scribn. & J. G. Sm.) Gould [*Elytrigia arizonica* (Scribn. & J. G. Sm.) D. R. Dewey]—Arizona wheatgrass. Tufted perennial, 45–85 cm; auricles present; blades 4–6 mm wide; spikes 12–30 cm, usually nodding in age, with 1 spikelet per node; spikelets with 4–6 florets, disarticulating above the glumes and between the florets; glumes 3-nerved; lemmas 8–15 mm, with awns 10–25 mm; anthers 3–5 mm. Dry rocky slopes, 5500–9570 ft. NC, SC, SW, McKinley.

Elymus bakeri (E. E. Nelson) Á. Löve—Baker's wheatgrass. Tufted perennials, 25–50 cm; auricles usually present; spikes 8–15(–20) cm, not especially dense, straight to somewhat drooping; glume awns to 8 mm, straight to curving; lemma awns 10–35 mm, curved to divergent. These are sterile hybrids and perhaps partially fertile hybrid derivatives between *E. trachycaulus* and several other species, including *E. scribneri*, *E. elymoides*, and *E. canadensis*, and the features of such plants are diverse and reflect the parentage, 8600–13,030 ft. NC, Cibola, Lincoln.

Elymus canadensis L.—Canada wildrye. Loosely tufted perennial, mostly 50–150+ cm; auricles present; spikes 8–30 cm, with 2(3) spikelets per node, usually nodding; disarticulation above the glumes and between the florets; glumes narrow, 3–5-nerved, with awns 10–25 mm; lemmas variously hairy to glabrous, with divergent awns 10–45 mm. Stream banks, ditch banks, floodplains, moist sandy soil, 3500–10,500 ft. Widespread except E, SW.

Elymus elymoides (Raf.) Swezey [*E. brevifolius* (J. G. Sm.) M. E. Jones; *E. longifolius* (J. G. Sm.) Gould; *Sitanion hystrix* (Nutt.) J. G. Sm. in part]—Squirreltail. Tufted perennial, 15–70 cm, the stems erect to geniculate-based, the nodes mostly concealed by the sheaths; sheaths glabrous to densely villous; auricles to ca. 1 mm; blades 2–6 mm wide, glabrous to villous; spikes 5–20 cm, erect to bent sharply over, with 2–3 spikelets per node; spikelets mostly with 2–5 florets; disarticulation in the rachis of the spike, then each floret falling; glumes 2–13 cm, 1–3-nerved, the awns sometimes split; lemmas 6–12 mm, glabrous to hairy, the awns 2–12 cm, flattish and grooved. Plains, grasslands, woodlands, clearings in forests, roadsides, 3915–11,700 ft. Widespread.

Elymus glaucus Buckley—Blue wildrye. Tufted perennial, sometimes weakly rhizomatous, 30–100+ cm; auricles usually present; spikes 5–20 cm, mostly erect, with usually 2 spikelets per node; disarticulation above the glumes and between the florets; glumes 3–5-nerved, slightly united and overlapping at the base, obscuring the florets, thin and flat, not indurate, with awns 0–6 mm; lemmas 9–15 mm, with erect to flexuous awns 1–30 mm; anthers 1.5–3.5 mm. Open woods, aspen groves, edges of mountain meadows, 6560–11,000 ft. Widespread except E, SW.

Elymus hispidus (Opiz) Melderis [*Thinopyrum intermedium* (Host) Barkworth & D. R. Dewey]—Intermediate wheatgrass. Rhizomatous perennial, often somewhat glaucus, 50–100+ cm; auricles present; blade margins whitish, thicker than the midnerve; spikes 10–20 cm, with 1 spikelet per node; spikelets 11–18 mm, glabrous to pubescent, awnless or rarely with short awns from the lemmas; glumes ca. 1/2 the spikelet length, blunt, thick, and rigid, the midnerve longer than the lateral nerves, offset, acute to mucronate; lemmas 7–10 mm, acute; anthers 5–7 mm. Introduced for range revegetation and erosion control, widespread in forests and foothills, 5610–10,640 ft. N, EC, C, WC, SC. Introduced. Plants with pubescent spikelets are referred to as **subsp. barbulatus** (Schur) Melderis, pubescent wheatgrass.

Elymus hystrix L.—Bottlebrushgrass. Tufted perennial, 50–120+ cm; auricles present; blades glossy green; spikes 8–20 cm, with 2 spikelets per node; spikelets diverging at nearly right angles from the main axis; disarticulation above the glumes and between the florets, the rachis persistent; glumes vestigial, 0.3–3(–5) mm; lemmas 8–11 mm, with mostly straight awns 20–40 mm; anthers 2.5–5 mm. A single collection in 1939. Colfax. Introduced.

Elymus interruptus Buckley—Interrupted wheatgrass. Tufted perennial, 50–100 cm; auricles present or absent; spikes 5–20 cm, with 2 spikelets per node; spikelets ascending to divergent; disarticulation above the glumes and between the florets; glumes subulate, 1–2-nerved, 10–30 mm; lemmas 7–10 mm, with awns 15–22 mm; anthers 2–4.5 mm. Moist canyons and woodlands in rich soil; known from a single collection. Sierra.

Elymus lanceolatus (Scribn. & J. G. Sm.) Gould [*Agropyron dasystachyum* (Hook.) Scribn.]—Streamside wheatgrass. Rhizomatous perennial, 25–100+ cm; leaves often mostly basal; auricles usually present; blades usually rolled; spikes 5–25 cm, usually with 1 spikelet per node; spikelets appressed, nearly awnless; glumes shorter than the adjacent lemma; lemmas 7–12 mm, acute to awned to 2 mm; anthers 3–6 mm. Moist to dry plains and forest clearings, 5610–9580 ft. NC, NW, WC, Otero.

Elymus ponticus (Podp.) N. Snow [*Agropyron elongatum* (Host) P. Beauv.]—Tall wheatgrass. Tufted, sometimes tussocky perennial, 50–200 cm; sheaths ciliate on the lower margins; auricles present; spikes 10–40 cm, with 1 spikelet per node; spikelets glabrous, awnless; glumes ca. 1/2 the length of the spikelet, thick and firm, nearly truncate, the midnerve about same length as lateral nerves; lemmas 9–12 mm. Introduced for range revegetation, pasture improvement, and erosion control, widespread in forests and foothills, 5000–8400 ft. NC, NW, C, WC, SE, SC. Introduced.

Elymus ×pseudorepens (Scribn. & J. G. Sm.) Barkworth & D. R. Dewey—False quackgrass. Mostly rhizomatous perennial, sometimes nearly cespitose, 30–90+ cm; auricles present or absent; spikes 6–14 cm, with 1 spikelet per node; glumes shorter than or equaling the adjacent lemma, 5–9-nerved, slightly curved, sometimes shor.t-awned; lemmas 7–15 mm, mucronate or awned to 3 mm; anthers 1–2 mm. Mountain slopes, grasslands, roadsides, 8500–9200 ft. NC, NW, C, WC, SC, Chaves.

Elymus repens (L.) Gould [*Agropyron repens* (L.) P. Beauv.; *Elytrigia repens* (L.) Desv. ex Nevski]—Quackgrass. Strongly rhizomatous perennial, 50–100 cm; auricles present; spikes 5–15 cm, with 1 spikelet per node; spikelets appressed to ascending; disarticulation above the glumes and between the florets; glumes ca. 1/2 the length of the spikelet, several-nerved, awnless or awned to 3 mm; lemmas 8–12 mm, glabrous, awnless or awned to 6 mm; anthers 4–7 mm. Aggressive weed of moist disturbed ground, gardens, flower beds, 4300–12,040 ft. NC, NW, Cibola, Grant, Otero. Introduced.

Elymus scribneri (Vasey) M. E. Jones—Spreading wheatgrass. Tufted perennial, the culms prostrate to erect, 15–40 cm; auricles usually present; spikes 3–9 cm, very dense, mostly with 1 spikelet per node; disarticulation below the spikelets in the rachis, the florets also eventually falling; spikelets appressed to ascending; glumes 4–8 mm, 1–3(–5)-nerved, with a divergent awn 12–30 mm; lemmas 7–10 mm, with divergent awns 15–30 mm; anthers 1–1.5 mm. Rocky slopes at high elevations, 8000–13,025 ft. NC, SC.

Elymus spicatus (Pursh) Gould [*Agropyron spicatum* (Pursh) Scribn. & J. G. Sm.]—Bluebunch wheatgrass. Tufted perennial, sometimes with short rhizomes, 30–100 cm; auricles present; blades involute when dry; spikes 8–15 cm, with 1 spikelet per node; spikelets scarcely if at all overlapping, appressed to slightly divergent; disarticulation above the glumes and between the florets; glumes ca. 1/2 the length of the spikelet; lemmas 5-nerved, awnless (*inermis* phase) to awned, the awns straight to divergent to 25 mm; anthers 4–8 mm. Sagebrush flats, piñon-juniper foothills, 5000–9500 ft. NC, NW, C, SC, Hidalgo.

Elymus trachycaulus (Link) Gould [*Agropyron trachycaulum* (Link) Malte ex H. F. Lewis]—Wheatgrass. Tufted perennial, sometimes producing weak rhizomes, 30–130 cm; auricles absent or present; blades mostly flat when fresh, 2–6 mm wide; spikes mostly 8–20 cm, with mostly 1 spikelet per node; spikelets well spaced to strongly overlapping; disarticulation above the glumes and between the florets; glumes subequal, nearly as long as the spikelet, green to purplish, the nerves prominent, awned or awnless; lemmas 6–13 mm, awned to awnless; anthers 1–3 mm. Mountain slopes, meadows, roadsides, from foothills to alpine.

1. Lemma awns > 1/2 as long as the lemma body, generally 14–40 mm…**subsp. subsecundus**
1'. Lemma awns absent to < 1/2 as long as the lemma body, generally < 8 mm…(2)
2. Glumes mostly 3-nerved; spikelets strongly overlapping, the rachis internodes 4–5 mm; 9000–12,000 ft. elevation…**subsp. violaceus**
2'. Glumes mostly 5–7-nerved; spikelets scarcely overlapping, the rachis internodes 8–15 mm; 6000–10,000 ft. elevation…**subsp. trachycaulus**

subsp. subsecundus (Link) Á. Löve & D. Löve [*Agropyron subsecundum* (Link) Hitchc.]—Slender wheatgrass. Ponderosa pine to alpine communities, 6600–12,800 ft. NE, NC, SC.

subsp. trachycaulus [*Agropyron novae-angliae* Scribn.]—Widespread, throughout all mountainous regions, 6000-10,000 ft. or sometimes higher. NC, NW, EC, C, WC, SC.

subsp. violaceus (Hornem.) Á. Löve & D. Löve [*Agropyron latiglume* (Scribn. & J. G. Sm.) Rydb.]—Slender wheatgrass. Not common; scattered high-elevation areas, 9000-12,000 ft. NC, SC.

Elymus villosus Muhl. ex Willd.—Hairy wildrye. Tufted perennial, 40-100+ cm; sheaths villous to pilose; auricles present; blades short-villous on the upper surface; spikes 4-12 cm, with usually 2 spikelets per node; spikelets 7-12 mm, overlapping; disarticulation above the glumes and between the florets; glumes awnlike, hirsute, 12-25 mm; lemmas 5-9 mm, usually villous, with awns to 30 mm; anthers 1.5-3 mm. Roadsides. No records seen. SC, SW, Union. Introduced.

Elymus virginicus L.—Virginia wildrye. Tufted perennial, 30-120 cm; foliage mostly glabrous; auricles absent to present; spikes 4-16 cm, with 2 spikelets per node; disarticulation in the rachis, or below the glumes and the rachis remaining intact; glumes 3-5-nerved, the basal portion nearly terete, indurate, and bowed out, gradually narrowing into a straight awn 3-12 mm; lemmas 6-10 mm, glabrous to villous, with straight awns 8-20 mm; anthers 2-4 mm. Moist woods, bottomlands, roadsides, 4900-8790 ft. Bernalillo, Eddy, San Juan, Union.

ENNEAPOGON Desv. ex P. Beauv. – Pappusgrass

Enneapogon desvauxii P. Beauv.—Nineawn pappusgrass. Tufted perennial, 20-45 cm; nodes hairy; sheaths open, shorter than the internodes, hairy; auricles absent; ligules a ring of hairs; panicles 2-10 cm, mostly < 2 cm wide, lead-colored; spikelets with 3-6 florets, often only the lowermost fertile, the upper progressively reduced; disarticulation above the glumes, the florets falling together; glumes subequal, multinerved, 3-5 mm; lower lemma 1.5-2 mm, prominently nerved; awns 3-4 mm, plumose; anthers 3. Plains and alluvial hills in desert or arid grasslands, 3400-6400 ft. Widespread except NE, EC.

ERAGROSTIS Wolf – Lovegrass

Plants annual to perennial, tufted, rhizomatous, or stoloniferous; sheaths open; auricles absent; ligules usually a ciliate membrane; inflorescence a panicle; spikelet with several florets; disarticulation generally above the glumes, the grain and lemma falling, the palea remaining on the rachilla; lemmas 3-nerved, awnless; anthers 2-3.

1. Plants annual…(2)
1′. Plants perennial…(10)
2. Plants with stolons, rooting at the nodes and forming mats…**E. hypnoides**
2′. Plants lacking stolons, not forming mats…(3)
3. Lemma keel (midnerve) with tiny craterlike glands toward the apex; mature spikelets 2-4 mm wide; lemmas with prominent green nerves contrasting sharply with the otherwise whitish body…**E. cilianensis**
3′. Lemma keel lacking craterlike glands; mature spikelets < 2.5 mm wide; lemmas generally colored otherwise…(4)
4. Mature grains with a groove on the side opposite the embryo…(5)
4′. Mature grains lacking a groove (slightly flattened in *E. barrelieri*)…(6)
5. Spikelets with 5-15 florets; rather common and widespread in the state…**E. mexicana**
5′. Spikelets with 3-6 florets; rare or now absent…**E. frankii** (in part)
6. Mature panicles 0.5-2 cm wide; spikelets light yellowish, occasionally purplish…**E. lutescens**
6′. Mature panicles 2-15 cm wide; spikelets generally darkish…(7)
7. Spikelets with 3-6 florets; rare or now absent…**E. frankii** (in part)
7′. Spikelets, at least many of them, with 7-20 florets…(8)
8. Culms with prominent glandular rings below the nodes…**E. barrelieri**
8′. Culms lacking glandular rings, but sometimes with a few glandular pits…(9)
9. Panicle branches usually solitary at the lowest 2 nodes; spikelets 1.2-2.5 mm wide…**E. pectinacea**
9′. Panicle branches usually paired or whorled at the lowest 2 nodes; spikelets 0.6-1.4 mm wide…E. pilosa
10. Plants with extensive creeping rhizomes; blades very stiff and sharp-pointed…go to **Kalinia obtusiflora**
10′. Plants lacking rhizomes or with short knotty rhizomes only; blades usually rather lax, not sharp-pointed…(11)

11. Spikelets 3-10 mm wide, disarticulating below the glumes at maturity and the spikelets falling entire…
    E. superba
11'. Spikelets 1-5 mm wide, disarticulating above the glumes at maturity…(12)
12. Spikelets sessile and borne on divergent unbranched primary branches…**E. sessilispica**
12'. Spikelets pedicelled, at least shortly so, and/or the primary panicle branches rebranched…(13)
13. Lateral (not the terminal) pedicels 2 mm or less…(14)
13'. Lateral (not the terminal) pedicels > 2 mm…(17)
14. Mature spikelets 3-5 mm wide and arranged in overlapping clusters…**E. secundiflora** (in part)
14'. Mature spikelets < 3 mm wide and not arranged in overlapping clusters…(15)
15. Panicle branches gummy, stout, and stiffly spreading…**E. curtipedicellata**
15'. Panicle branches not gummy and stiff, but at least somewhat lax or drooping…(16)
16. Basal sheaths ± glabrous on the back; culms usually geniculate-based; lemmas mostly < 1.8 mm…
    E. lehmanniana
16'. Basal sheaths villous on the back; culms usually erect at the base; lemmas mostly > 2 mm…**E. curvula**
17. Mature spikelets 3-5 mm wide and arranged in dense, overlapping clusters…**E. secundiflora** (in part)
17'. Mature spikelets < 3 mm wide and not arranged in dense, overlapping clusters…(18)
18. Paleas conspicuously ciliate; lateral nerves of lemma prominent; panicle breaking away when mature
    and tumbling before the wind…**E. spectabilis**
18'. Paleas smooth or minutely ciliate; lateral nerves of lemma prominent or obscure; panicle usually not
    breaking away…(19)
19. New basal shoots breaking through the base of the sheaths (extravaginal); stem bases knotty…**E. palmeri**
19'. New basal shoots not breaking through the base of the sheaths, but emerging from the top or off to the
    side; stem bases not knotty…(20)
20. Mature lemmas mostly < 2.2 mm…**E. intermedia**
20'. Mature lemmas mostly > 2.2 mm, usually > 2.4 mm…(21)
21. Grains squarish; lemmas reddish, acuminate with smooth tips; basal nodes and internodes crowded…
    E. trichodes
21'. Grains elongate to elliptic; lemmas greenish, acute with usually fringed tips; basal nodes and internodes
    not crowded…**E. erosa**

Eragrostis barrelieri Daveau—Mediterranean lovegrass. Tufted annual, 10-60 cm, with a yellowish ring of glandular tissue below the nodes; panicles open, with pulvini in the axils, 4-20 cm, 2-10 cm wide, the axils glabrous; spikelets 1-2.2 mm wide, diverging from the branches, lacking glands, with 7-15 florets; disarticulation leaving the paleas on the rachilla but the glumes falling; lemmas 1.4-1.8 mm; grains not grooved. Disturbed sites, flower beds, roadsides, 3150-6900 ft. Widespread except NE, NC, EC. Introduced.

Eragrostis cilianensis (All.) Vignolo ex Janch.—Stinkgrass. Tufted annual, 15-60 cm, sometimes with craterlike glands below the nodes and on the sheath; panicles 5-16 cm, 2-8 cm wide, with pulvini in most axils; spikelets often with numerous florets, 2-4 mm wide, greenish lead-colored to whitish; disarticulation leaving only the rachilla; lemmas with glands on the keels, membranous between the conspicuous nerves; paleas prominently ciliate on the keels; grains not grooved. Disturbed and weedy ground, 3500-8000 ft. Widespread.

Eragrostis curtipedicellata Buckley—Gummy lovegrass. Tufted perennial, much of the plant viscid-sticky, sometimes with short knotty rhizomes, 20-65 cm; panicles 18-35 cm, nearly as wide, with secondary branches from the divergent primary branches; spikelets 1-1.5 mm wide; disarticulation leaving the glumes, but not the rachilla; lemmas glandless; grains not grooved. Sandy or clayey plains and grasslands, 3150-6350 ft. NE, EC, SE.

Eragrostis curvula (Schrad.) Nees [*E. curvula* var. *conferta* Nees]—Weeping lovegrass. Tufted perennial, the culms usually erect basally, 50-100+ cm; ligules 0.6-1.3 mm; blades to 65 cm and forming fountainlike arrays; panicles 16-40 cm, ca. 1/2 as wide, hairy to glabrous in the axils; spikelets appressed, 4-9 mm, 1-2 mm wide, lead-colored; disarticulation leaving the paleas on the rachilla but the glumes falling; lemmas 1.8-3 mm, the nerves somewhat obscure; grains grooved or not. Plains and prairies to foothills and midelevations in the mountains, often along roadsides, 3050-9025 ft. Introduced and widespread.

Eragrostis erosa Scribn. ex Beal—Chihuahuan lovegrass. Tufted perennial, 70–110 cm, lacking glands; panicles 25–45 cm, 10–30 cm wide, hairy to glabrous in the axils; spikelets appressed to divergent, 1–3 mm wide; disarticulation leaving the paleas on the rachilla but the glumes falling; lemmas 2.4–3 mm, greenish, the lateral nerves obscure; grains grooved, 0.8–1.6 mm. Rocky limestone hills and mountain slopes, often in piñon-juniper areas, 4250–9700 ft. NC, SE, SC, Bernalillo, Hidalgo.

Eragrostis frankii (Fisch., C. A. Mey. & Avé-Lall.) C. A. Mey. ex Steud.—Sandbar lovegrass. Tufted annual, 10–50 cm, often with glandular pits below the nodes, on the sheaths, and on the panicle branches; panicles < 1/2 the length of the plant, 2–12 cm wide, glabrous in the axils; spikelets 1–2 mm wide, with 3–6 florets, lead-colored to reddish; disarticulation leaving the paleas on the rachilla, the glumes falling; lemmas 1–1.5 mm, the lateral nerves obscure; grains flat or shallowly grooved. Disturbed ground, moist weedy sites; known only from an 1847 collection by Augustus Fendler at Santa Fe; probably now absent from the state. Adventive.

Eragrostis hypnoides (Lam.) Brltlun, Sterns & Poggenb.—Teal lovegrass. Stoloniferous annual, forming loose mats, 5–20 cm; sheaths short-pilose on the margins and collars; panicles weakly developed, 1–4 cm, mostly open; spikelets 1–1.5 mm wide, greenish yellow to purplish, with 12–35 florets; disarticulation leaving the paleas and the rachilla, the glumes falling; lemmas 1.4–2 mm; grains somewhat translucent, not grooved; anthers 2, 0.2–0.3 mm. Sand and mud bars along slow-moving streams and lakeshores, 4500–6580 ft. Uncommon, in scattered locales. NC, NW, C.

Eragrostis intermedia Hitchc.—Plains lovegrass. Tufted perennial, 30–90+ cm; sheaths lightly hairy on the margins and at the apex; panicles 15–40 cm, 10–30 cm wide, glabrous or hairy in the axils; spikelets 1–1.8 mm wide, with 5–11 florets; disarticulation leaving the paleas on the rachilla but the glumes falling; lemmas 1.6–2.2 mm, the lateral nerves obscure; grains strongly grooved. Our plants belong to **var. intermedia**. Sandy or rocky plains, prairies, mountain slopes, disturbed ground, 4050–9000 ft. NC, C, WC, SE, SC, SW.

Eragrostis lehmanniana Nees—Lehmann lovegrass. Tufted perennial, 30–80 cm, the bases geniculate, sometimes rooting at the nodes and forming surprisingly long stolons; sheaths mostly glabrous basally; ligules 0.3–0.5 mm; blades to 12 cm; panicles 7–18 cm, ca. 1/2 as wide, glabrous in the axils; spikelets generally spreading from the branch (sometimes only slightly), 5–14 mm, 0.8–1.2 mm wide, lead- to straw-colored; disarticulation leaving the paleas on the rachilla but the glumes falling; lemmas 1.5–1.7 mm, the nerves obscure; grains sometimes with a shallow groove. For rangeland rehabilitation and roadside erosion control, 3315–7200 ft. C, SE, SC, SW, Cibola. Adventive.

Eragrostis lutescens Scribn.—Sixweeks lovegrass. Tufted annual, 6–25 cm, with glandular pits (but not rings) below the nodes; sheaths and leaf bases with scattered glandular pits; panicles 4–15 cm, 0.5–2 cm wide, the main axis and branches with glandular pits; spikelets 1.2–2 mm wide, pale yellowish (sometimes purple-mottled), glabrous in the axils; disarticulation leaving the paleas on the rachilla but the glumes falling; lemmas straw-colored with greenish nerves, lacking glands; grains without a groove. Sandy, moist soil, 5600–7200 ft. C, SC.

Eragrostis mexicana (Hornem.) Link—Mexican lovegrass. Tufted annual, 10–95(–110) cm, the culms sometimes with a glandular ring below the nodes; sheaths sometimes with glandular pits; panicles 10–40 cm, 4–18 cm wide, glabrous in the axils; spikelets 0.7–2.4 mm wide, with 5–12 florets, gray-green to purplish green; disarticulation leaving the paleas on the rachilla but the glumes falling; lemmas 1.2–2.4 mm, the lateral nerves evident; grains mostly deeply grooved. Our plants belong to **var. mexicana**. Roadsides, moist disturbed sites in a variety of habitats, often gravelly or rocky sites, 3570–9000 ft. NC, NW, C, WC, SE, SC, SW.

Eragrostis palmeri S. Watson—Palmer's lovegrass. Tufted perennial with knotty bases, the axillary shoots breaking through the base of the sheath (extravaginal), rhizomes lacking, 50–100 cm, lacking glands; sheaths short-villous to glabrous; panicles 12–40 cm, 4–20 cm wide, glabrous to sparsely hairy in the axils; spikelets appressed to divergent, 1–2 mm wide; disarticulation leaving the paleas on the rachilla but the glumes falling; lemmas 2–2.6 mm, the lateral nerves obscure to conspicuous; grains grooved, 0.6–0.8 mm. Rocky plains and mountain slopes, 3000–7050 ft. Eddy.

Eragrostis pectinacea (Michx.) Nees—Carolina lovegrass. Tufted annual, lacking glands, 10–65 cm; panicles 5–25 cm, 3–12 cm wide, usually open, the axils glabrous or sparsely hairy; spikelets 1.2–2.5 mm wide, with 6–20 florets; disarticulation leaving the paleas on the rachilla but the glumes falling; lemmas 1–2.2 mm, the lateral nerves obvious; grains not grooved. Roadsides, fields, alkali flats, sandy plains, disturbed ground, 3600–8000 ft. Widespread except E. Chaves. Plants with spreading to divaricate panicle branches can be referred to as **var. miserrima** (E. Fourn.) Reeder.

Eragrostis pilosa (L.) P. Beauv.—Indian lovegrass. Tufted annual, 8–50 cm; sheaths sometimes with glandular pits; blades sometimes with glands along the midnerve; panicles 4–20 cm, 2–15 cm wide, the branches paired or whorled at the lower 2 nodes, the axils glabrous to hairy; spikelets 0.6–1.4 mm wide, with 5–15 florets; disarticulation leaving the paleas on the rachilla but the glumes falling, and then the paleas eventually falling; lemmas 1.2–2 mm, the lateral nerves obscure; grains not grooved. Roadsides, disturbed ground, gardens, fields. No records seen.

Eragrostis secundiflora J. Presl—Red lovegrass. Tufted perennial, lacking glands, 30–75 cm; panicles 5–30 cm, ca. 1/2 as wide, the branches appressed to ascending, the axils nearly glabrous; spikelets 2.5–5 mm wide, strongly flattened, reddish or somewhat straw-colored, with 10–45 florets; disarticulation dropping the florets and rachilla before the glumes; lemmas 2–6 mm, the lateral nerves conspicuous; grains not grooved. Our plants belong to **subsp. oxylepis** (Torr.) S. D. Koch. Sandy grasslands and prairies, roadsides, 3490–6400 ft. NE, EC, C, SE, SC.

Eragrostis sessilispica Buckley—Tumble lovegrass. Tufted perennial, lacking glands, 30–90 cm; panicles 20–60 cm, ca. 1/2 as wide; branches spicate, unbranched, widely spaced, with hairy axils; spikelets sessile or nearly so on the primary branches, widely spaced and not overlapping, 1.4–3 mm wide, with 3–12 florets; disarticulation leaving the glumes, the florets falling, the panicle also breaking away and tumbling before the wind; lemmas 3–5 mm, indurate; grains not grooved. Sandy hills and prairies, 4100–5905 ft. E.

Eragrostis spectabilis (Pursh) Steud.—Purple lovegrass. Tufted perennial, 30–75 cm, with short knotty rhizomes; sheaths hairy on the margins; blades glabrous to pilose; panicles 20–50 cm, 15–35 cm wide, hairy in the axils; spikelets 1–2 mm wide, spreading or appressed, reddish purple; disarticulation leaving the glumes, the florets falling; lemmas 1.2–2.5 mm, the lateral nerves obvious; grains grooved between 2 ridges. Sandy soil in the NE grasslands, 4040–6755 ft. Harding, Mora, Roosevelt, San Miguel.

Eragrostis superba Peyr.—Wilman's lovegrass. Tufted perennial, lacking glands, 45–95 cm; sheaths hairy on the margins; panicles 10–30 cm, 1–6 cm wide, glabrous in the axils; spikelets 3–10 mm wide, strongly flattened, straw-colored to greenish or reddish; disarticulation below the glumes, the spikelets falling entire; lemmas 3–5 mm long, papery to leathery, the green lateral nerves conspicuous; grains not grooved. Introduced in seeding trials and for erosion control, 3935–4310 ft. Hidalgo, Roosevelt, Sierra. Adventive.

Eragrostis trichodes (Nutt.) Alph. Wood—Sand lovegrass. Tufted perennial, lacking glands, 30–120 cm; sheaths sometimes hairy along the margins; blades sometimes pilose; panicles 30–80 cm, 10–30 cm wide, glabrous to hairy in the axils; spikelets diverging, 1.5–3.6 mm wide, with 4–18 florets; disarticulation leaving the paleas on the rachilla but the glumes falling; lemmas 2.2–3.5 mm, the lateral nerves obvious; grains grooved. Sandy prairies and open woodlands, 4135–7165 ft. Scattered.

EREMOPYRUM (Ledeb.) Jaub. & Spach – Annual wheatgrass

Eremopyrum triticeum (Gaertn.) Nevski [*Agropyron triticeum* Gaertn.]—Annual wheatgrass. Tufted annual, 5–30 cm; culms geniculate-based; sheaths open; auricles present; ligules a truncate membrane; blades flat, short; inflorescence a spike, with 1 spikelet per node; spikes 1–3 cm; disarticulation above the glumes and between the florets, the rachis remaining; spikelets crowded and overlapping, with 2–3 florets; glumes 1-nerved, awn-tipped; lemmas 5–7.5 mm, the lowermost hairy; anthers 3. Dry plains, 5000–6845 ft. NW. Introduced.

ERIOCHLOA Kunth – Cupgrass

Mostly tufted annuals and perennials; sheaths open; auricles absent; ligules a ciliate membrane; inflorescence a panicle with spikelike branches, lacking bristles; spikelets with 2 florets, the lower sterile, the upper fertile, awned to awnless; disarticulation below the glumes (below the cup); lower glume reduced to a cuplike structure; upper and lower lemma similar in shape and texture; upper floret indurate, the lemma clasping the palea and forming a seedcase; anthers 3.

1. Spikelets solitary in the middle of the branches…**E. contracta**
1′. Spikelets in pairs in the middle of the branches…(2)
2. Adaxial blade surface velvety-hairy; lower paleas 1–4 mm…**E. lemmonii**
2′. Adaxial blade surface glabrous to sparsely pilose; lower paleas absent…**E. acuminata**

Eriochloa acuminata (J. Presl) Kunth—Tapertip cupgrass. Tufted annual, 30–100+ cm; sheaths sometimes inflated; blades sparsely pilose; panicles 7–16 cm, the branches 5–20 in number, 1–5 cm, with spikelets in unequally pedicellate pairs in the middle of the branch, solitary distally; spikelets 3–6 mm; upper glume awnless to awned to 1.2 mm; lower palea absent; upper floret awn-tipped 0.1–0.3 mm. Disturbed moist ground, rocky slopes, 5000–8000 ft. EC, SC, SW. Plants with short spikelets (< 4 mm) that are obtuse to acute apically may be referred to as **var. minor** (Vasey) R. B. Shaw. SC, SW.

Eriochloa contracta Hitchc.—Prairie cupgrass. Tufted annual, 20–80+ cm; blades sparsely to densely pubescent on both surfaces; panicles 6–20 cm, the branches 10–28 in number, 1.5–5 cm, with solitary spikelets in the middle of the branch, sometimes paired proximally; spikelets 3–5 mm; upper glume acuminate to awned to 1 mm; lower palea absent; upper floret awned 0.4–1.1 mm. Loamy soil of prairies and swales, 3295–4005 ft. Chaves, Doña Ana, Eddy.

Eriochloa lemmonii Vasey & Scribn.—Canyon cupgrass. Tufted annual, 20–80 cm; blades velutinous on the upper surface; panicles 5–15 cm, with 2–10 branches 1–4 cm, with spikelets in unequally pedicellate pairs in the middle of the branch, solitary distally; spikelets 3–5 mm; upper glume acute, awnless; lower palea 1–4 mm; upper floret rounded to acute, sometimes mucronate, awnless. Rocky, grassy slopes, 4200–5325 ft. Bernalillo, Doña Ana, Hidalgo.

ERIOCOMA Nutt. – Needlegrass

Perennial, ours mostly cespitose; sheaths open; auricles absent; ligules membranous; inflorescence a panicle; spikelets 1-flowered; disarticulation above the glumes; glumes longer than the floret, 1–3-nerved or sometimes more; florets terete, usually fusiform, the lemma convolute around the palea and flower but the margins only slightly overlapping, awned from the tip with a single awn; anthers 3. Species of *Eriocoma* were formerly recognized in the genera *Achnatherum*, *Oryzopsis*, and *Stipa* (Peterson et al., 2019).

1. Lemma densely covered with long hairs; awn short, 3–5 mm, quickly deciduous; panicle widely spreading at maturity, with dichotomous branches…**E. hymenoides**
1′. Lemma glabrous or covered with short appressed hairs; awn > 6 mm, persistent or deciduous; panicle narrow with ascending branches…(2)
2. Basal segment of the once-geniculate awn plumose with long hairs 3–8 mm…go to **Pappostipa**
2′. Basal segment of the awn glabrous or with hairs < 2 mm…(3)
3. Awn ± readily deciduous; blades 1–2 mm wide [to be expected]…E. ×bloomeri
3′. Awn persistent; blades various…(4)
4. Lower segment of the awn (not the lemma tip) with hairs 1–2 mm…**E. curvifolia**
4′. Lower segment of the awn (not the lemma tip) scabrous or with hairs < 1 mm…(5)
5. Awns 3–7.5 cm, obscurely bent, the terminal segment flexuous or curving…(6)
5′. Awns 1–3 cm, usually plainly bent, the terminal segment ± straight…(7)
6. Ligule minute, < 1 mm, hardly visible; panicle narrow, contracted, the main axis obscured…**E. arida**
6′. Ligule 1–2 mm, evident; panicle open when mature, the branches spreading, the main axis visible…go to **Pseudoeriocoma**

7.   Palea approximately 2/3 the length of the lemma…(8)
7'. Palea 1/3-1/2 the length of the lemma…(9)
8.   Hairs at the tip of the palea about the same length as those below; mature stems 60-180 cm, 2-6 mm diam.; blades 4-10 mm wide…**E. robusta**
8'. Hairs at the tip of the palea longer than those below; mature stems 25-80 cm, 1-2 mm diam.; blades 1-2 mm wide…**E. lettermanii**
9.   Hairs at the lemma tip 2.5-3 mm; callus with a pointed extension…**E. scribneri**
9'. Hairs at the lemma tip 1-2.2 mm; callus blunt, without a pointed extension…(10)
10. Apical lemma hairs erect; lemma lobes 0.5-1.2 mm; florets widest about midlength…**E. lobata**
10'. Apical lemma hairs ascending to divergent; lemma lobes 0.2-0.5 mm; florets widest below midlength… (11)
11. Awns mostly 2-3 cm; blades 3-7 mm wide…**E. nelsonii**
11'. Awns mostly 1-2 cm; blades 2-3 mm wide…**E. perplexa**

Eriocoma arida (M. E. Jones) Romasch. [*Achnatherum aridum* (M. F. Jones) Barkworth; *Stipa arida* M. E. Jones]—Mormon needlegrass. Perennial, 30-80 cm; blades flat when fresh, 1-3 mm wide; panicles 5-17 cm, 1-2 cm wide, the bases often enclosed in the subtending sheath at anthesis; glumes subequal to unequal, the lower 8-14 mm, the upper 1-5 mm shorter; florets 4-6.5 mm; awns 4-8 cm, obscurely once-geniculate, the terminal segment flexuous. Desert scrub vegetation, 5000-6000 ft. San Juan.

Eriocoma ×bloomeri (Bol.) Romasch. [*Achnatherum ×bloomeri* (Bol.) Barkworth; *Oryzopsis bloomeri* (Bol.) Ricker]—Not definitely recorded for New Mexico, but to be expected in the Four Corners area. This is a catchall name for hybrids between *E. hymenoides* and various other species of *Eriocoma*, generally with the readily deciduous awns of *E. hymenoides* and the longer, narrow florets and wider blades of the other parent.

Eriocoma curvifolia (Swallen) Romasch. [*Achnatherum curvifolium* (Swallen) Barkworth; *Stipa curvifolia* Swallen]—Guadalupe ricegrass. Perennial, 25-60 cm; basal sheaths puberulent, sometimes tomentose at the base; blades folded to involute, minutely puberulent, moderately to strongly curled in age; panicles 7-11 cm; glumes subequal, 3-nerved, 10-14 mm; florets 6-8 mm; awns 20-50 mm, mostly once-geniculate, the first segment with hairs 1-2 mm. Crevices and rocky ledges and cliffs, limestone substrate, 3850-6600 ft. Doña Ana, Eddy, Otero.

Eriocoma hymenoides (Roem. & Schult.) Rydb. [*Achnatherum hymenoides* (Roem. & Schult.) Barkworth; *Oryzopsis hymenoides* (Roem. & Schult.) Ricker ex Piper]—Indian ricegrass. Perennial, 20-50+ cm; blades convolute, to 1 mm wide; panicles with dichotomous branching, to 20 cm and 15 cm wide; spikelets on long divergent pedicels; glumes subequal, 5-7-nerved; florets obovoid, with hairs 2-6 mm; awns 3-6 mm, straight, early-deciduous. Sandy plains and dunes, 3600-9050 ft. Widespread.

Eriocoma lettermanii (Vasey) Romasch. [*Achnatherum lettermanii* (Vasey) Barkworth; *Stipa lettermanii* Vasey]—Letterman's needlegrass. Perennial, 25-80(-90) cm; collars of the flag leaves glabrous; blades rolled to flat when fresh, 1-2 mm wide; panicles 7-20 cm, to 1 cm wide; glumes subequal, 6-10 mm; florets fusiform, 4-6 mm; paleas 3/4 to nearly as long as the lemma; awns 12-25 mm, twice-geniculate. Sagebrush flats and hills, dry mountain meadows and clearings, from sagebrush to subalpine communities, 5500-11,215 ft. Widespread except E, S.

Eriocoma lobata (Swallen) Romasch. [*Achnatherum lobatum* (Swallen) Barkworth; *Stipa lobata* Swallen]—Littleawn needlegrass. Perennial, to 100 cm; blades flat when fresh, 1-4 mm wide; panicles 12-20(-25) cm, to 2 cm wide; glumes unequal, 1-3-nerved, the lower 10-12 mm, the upper 2-3 mm shorter; florets fusiform, 5-8 mm, with apical lemma lobes 0.5-1.2 mm; awns 10-20 mm, once- to twice-geniculate. Rocky hills and woodlands, 4100-8650 ft. NC, C, WC, SC.

Eriocoma nelsonii (Scribn.) Romasch. [*Achnatherum nelsonii* (Scribn.) Barkworth; *Stipa nelsonii* Scribn.]—Columbia needlegrass. Perennial, 40-100+ cm; blades flat when fresh, 3-7 mm wide (sometimes less); panicles to 30 cm and 2 cm wide; glumes subequal, the lower slightly longer, 6-12 mm; florets fusiform, 4-7 mm; awns 18-30 mm, twice-geniculate. Infrequent and only recently reported (2006) in mixed conifer forests, meadows and clearings in forests, 7550-10,000 ft. Rio Arriba, Sandoval.

Eriocoma perplexa (Hoge & Barkworth) Romasch. [*Achnatherum perplexum* Hoge & Barkworth; *Stipa columbiana* auct.; *S. nelsonii* auct.; *S. perplexa* (Hoge & Barkworth) Wipff & S. D. Jones]—New Mexico needlegrass. Perennial, 35–85 cm; blades flat when fresh, (1-)2–3 mm wide; panicles to 25 cm and 2 cm wide; glumes unequal, the lower 10–15 mm and exceeding the upper by 1–3 mm; florets fusiform, 5–10 mm; awns 10–25 mm, once- to twice-geniculate. Mountain grasslands, clearings, dry slopes in piñon to ponderosa pine communities, 5100–9175 ft. NC, C, WC, SE, SC, SW.

Eriocoma robusta (Vasey) Romasch. [*Achnatherum robustum* (Vasey) Barkworth; *Stipa robusta* (Vasey) Scribn.]—Sleepygrass. Perennial, (60-)80–180+ cm; collars hairy, those of the flag leaves densely so; blades flat when fresh, 4–10 mm wide; panicles 15–30 cm, 1–3.5 cm wide; glumes subequal, 9–12 mm; florets fusiform, 6–9 mm, the apical hairs to 1.5 mm; paleas 2/3–3/4 as long as the lemmas; awns 20–35 mm, twice-geniculate. Mountain grasslands, plains, disturbed pastures, 5310–11,370 ft. Widespread except far E, SW.

Eriocoma scribneri (Vasey) Romasch. [*Achnatherum scribneri* (Vasey) Barkworth; *Stipa scribneri* Vasey]—Scribner needlegrass. Cespitose perennials, 30–90 cm; blades flat or rolled when fresh, 2–4 mm wide when flat; panicles to 24 cm and 1 cm wide; glumes strongly unequal, the lower 10–17 mm and often curving outward, the upper 2–4 mm shorter; florets fusiform, 6–10 mm, the sharp callus extension to 1.5 mm, with apical hairs 2–3 mm; awns 14–25 mm, mostly once-geniculate. Dry rocky hills and woodlands, piñon to ponderosa pine, 5500–12,000 ft. NW, WC, SE, SC, Quay.

ERIONEURON Nash – Tridens, woollygrass

Tufted or stoloniferous perennials; leaves mostly basal; sheaths open; auricles absent; ligules a ring of hairs; blade margins white, cartilaginous; inflorescence a panicle, sometimes racemelike; spikelets several-flowered; disarticulation above the glumes and between the florets; lemmas 3-nerved, the nerves strongly hairy; awned or awnless; anthers 1 or 3.

1. Spikelets arranged in leafy clusters borne down among the pungent, spine-tipped blades; plants often stoloniferous and < 10 cm…go to **Munroa pulchella**
1'. Spikelets borne on an elongate, leafless stalk elevated above the leaves; plants not or rarely stoloniferous and often > 10 cm…(2)
2. Tip of lemma acute or with a notch 0.5 mm or less deep; both glumes shorter than the lowermost floret…**E. pilosum**
2'. Tip of lemma with a notch 1–2.5 mm deep; upper glume equaling or surpassing the lowermost floret…(3)
3. Spikelets of vigorous plants 10–15 mm, usually silvery or only slightly purple-tinged; lemmas copiously pubescent at the base; lateral lemma nerves not extended into a mucro…**E. avenaceum**
3'. Spikelets seldom > 10 mm, usually purplish-tinged or brownish purple; lemmas with some hairs but not copiously pubescent at the base; lateral lemma nerves extended into a mucro to 1 mm…**E. nealleyi**

Erioneuron avenaceum (Kunth) Tateoka [*E. avenaceum* var. *grandiflorum* (Vasey) Gould]—Shortleaf woollygrass. Tufted or infrequently stoloniferous perennial, 7–40 cm; blades sparsely pilose; panicles 2–10 cm; spikelets 6–10 mm, with 4–20 florets; upper glume equaling or surpassing the lower floret; lemmas 4–7 mm, the lobes 1–2 mm, with awns 2–4 mm; anthers 3. Our material belongs to **var. avenaceum**. Limestone hills and rocky outcrops, 3575–6600 ft. C, SE, SC, SW.

Erioneuron nealleyi (Vasey) Tateoka—Nealley's woollygrass. Tufted perennial, 15–65 cm; blades pilose to villous; panicles 5–10, compact; upper glumes surpassing the lower floret; lemmas 4–6 mm, the lobes 1.5–2.5 mm, with awns 1–3.5 mm; anther 1. Limestone hills and rocky outcrops, 4650–6150 ft. C, SC, Eddy.

Erioneuron pilosum (Buckley) Nash—Hairy woollygrass. Tufted perennial, 6–40 cm; blades sparsely pilose to glabrous; panicles 1–6 cm; spikelets 6–15 mm, with 5–20 florets; glumes shorter than the lower floret; lemmas 3–6 mm, the teeth 0.3–0.5 mm, with awns 0.5–2.5 mm; anthers 3. Our material belongs to **var. pilosum**. Limestone hills and rocky outcrops, 3000–8800 ft. Widespread except WC.

FESTUCA L. - Fescue

Tufted perennials, with or without rhizomes in addition, rarely stoloniferous; sheaths open to closed to ca. 3/4 their length, sometimes to near the top; auricles absent; ligules membranous, often ciliate; blades flat to variously folded or rolled; inflorescence a panicle, sometimes racemelike; spikelets with few to several florets; disarticulation above the glumes and between the florets; glumes usually surpassed by the florets, 1-3-nerved; lemmas 5-7-nerved, awned or awnless; anthers 3. The classification of *Festuca* and related groups is problematic and controversial.

1.	Blades mostly > 3 mm wide, usually at least somewhat lax and flat when fresh…(2)
1'.	Blades mostly < 3 mm wide, usually rolled and somewhat stiff…(3)
2.	Spikelets 2-4-flowered, 8-11 mm; auricles absent; panicle branches spreading, at least below…**F. sororia**
2'.	Spikelets (4)5-9-flowered, 10-17 mm; small auricles usually developed; panicle branches usually ascending…go to **Schedonorus**
3.	Glumes (both) equaling or exceeding the upper florets; lemma awns 0-1.3 mm…**F. hallii**
3'.	Glumes distinctly shorter than the upper florets; lemma awns various…(4)
4.	Ligules 2.5-5(-9) mm; lemma awns 0-0.3 mm; nodes usually visible and conspicuous; plants generally > 50 cm…**F. thurberi**
4'.	Ligules < 2 mm; lemma awns usually > 0.5 mm, occasionally shorter; nodes often not visible or conspicuous; plant height various…(5)
5.	Plants usually with short rhizomes, the shoots often loosely tufted; basal sheaths reddish and rapidly separating into threadlike fibers (whitish veins)…(6)
5'.	Plants lacking rhizomes, the shoots loosely to densely tufted; basal sheaths usually not reddish or separating into threadlike fibers (sometimes thus separating in *F. calligera*)…(7)
6.	Anthers 1.8-4.5 mm; ovary apices glabrous…**F. rubra**
6'.	Anthers 0.6-1.4 mm; ovary apices densely pubescent…**F. earlei** (in part)
7.	Anthers 2-4 mm (sometimes shorter in *F. trachyphylla*)…(8)
7'.	Anthers 0.4-1.7 mm, rarely longer…(11)
8.	Blades, especially the older ones, strongly laterally compressed, thickened, and stiff, 0.5-1 mm wide…**F. trachyphylla**
8'.	Blades, even the older ones, at least somewhat terete, not thickened, but threadlike, 0.2-0.4 mm wide…(9)
9.	Peduncle and lower panicle branches densely scaberulous; old basal sheaths conspicuous at the base of the clump, generally 4-12 cm (rarely shorter); body of larger lemmas 5-9 mm, the awn 0.5-2.5 mm; ovary apex pubescent…**F. arizonica**
9'.	Peduncle and lower panicle branches glabrous or nearly so; old basal sheaths conspicuous or not at the base of the clump, 1-3 cm; body of larger lemmas 3-5.5 mm, the awn 1-7 mm; ovary and grain apex glabrous or with a few sparse hairs…(10)
10.	Body of larger lemmas 3.5-5 mm, the awn 1-2.5 mm; lower glume 2.5-3.5 mm; ovary apex with a few sparse hairs at maturity (glabrous when very young); grain 2-3 mm…**F. calligera**
10'.	Body of larger lemmas (4.5-)5-5.5 mm, the awn 2-7 mm; lower glume 3.5-4.5 mm; ovary apex glabrous at maturity and when very young; grain 4-5 mm…**F. idahoensis**
11.	Plants found only as ornamentals and border plants (in New Mexico), never in native habitats; foliage markedly bluish-glaucous in dense hemispherical tufts; ovary and grain apex densely pubescent…**F. glauca**
11'.	Plants not growing as ornamental landscape plants, planted infrequently as a pasture grass, common in native mountain habitats; foliage somewhat glaucous to green, growth form various, but usually not in dense hemispherical tufts; ovary and grain apex glabrous or pubescent…(12)
12.	Plants 3-10 cm…(13)
12'.	Plants > 10 cm, usually 15-50 cm…(14)
13.	Lemma body 2-3 mm, with an awn 0.5-1.5 mm; spikelets with 2, occasionally 3, florets; panicle branches at lowest node usually 2-3; ovary and grain apex pubescent…**F. minutiflora**
13'.	Lemma body 3-5.5 mm, with an awn 2-3.6 mm; spikelets with 3-4 florets, occasionally only 2; panicle branches at lowest node 1; ovary and grain apex glabrous…**F. brachyphylla** (in part)
14.	Basal sheaths reddish and splitting into threadlike fibers (whitish veins) in age; ovary and grain apex pubescent…**F. earlei** (in part)
14'.	Basal sheaths mostly straw-colored to brownish, not splitting into threadlike fibers in age (occasionally so in *F. brachyphylla*); ovary and grain apex glabrous…(15)
15.	Blades soft, striate from the veins showing, somewhat wrinkled in drying, with little or no sclerenchyma

tissue; spikelets and foliage greenish; culms usually < twice the height of the leaves; anthers 0.5-1.3 mm; rachilla internodes of middle florets 0.6-0.8 mm...**F. brachyphylla** (in part)

15'. Blades stiff, terete or sulcate, not striate or wrinkled, the veins generally not visible because of a buildup of sclerenchyma tissue; spikelets and foliage often glaucous; culms usually twice+ the height of the leaves; anthers 1-1.7 mm (rarely longer); rachilla internodes of middle florets 0.9-1.1 mm...**F. saximontana**

Festuca arizonica Vasey—Arizona fescue. Densely tufted perennial, without rhizomes, 40-100+ cm; sheaths persistent, > 3 cm; ligules 0.5-2 mm; peduncle densely scabrous or pubescent below the panicle; panicles 6-20 cm; lemmas 5.5-9 mm; awns 0-3 mm; anthers 2-4 mm; ovary apices densely pubescent. High mountain grasslands throughout the mountain regions of the state; our most common fescue. Widespread in E, far S.

Festuca brachyphylla Schult. & Schult.—Alpine fescue. Densely or loosely tufted perennial, 5-40 cm; sheaths persistent or sometimes splitting into threadlike fibers; ligules 0.1-0.4 mm; panicle 2-5 cm, narrow; lemmas 3-4.5 mm; awns 1-3 mm; anthers 0.5-1 mm; ovary apices glabrous. New Mexico material belongs to **subsp. coloradensis** Fred. Alpine grasslands In the N mountains, 8400-12,960 ft. NC, Lincoln, Otero.

Festuca calligera (Piper) Rydb.—Southwest fescue. Densely tufted perennial, without rhizomes, 15-50 cm; sheaths persistent (rarely splitting in age); ligules 0.3-0.5(-1) mm; panicles 5-15 cm; lemmas 4-6 mm; awns 1-2.5 mm; anthers 2.2-3.5 mm; ovary apices sparsely pubescent. Relatively rare, mostly in the SC mountains (but extending N to Colorado), usually growing with Arizona fescue, 7965-11,400 ft. NC, NW, C, SC.

Festuca earlei Rydb.—Earle's fescue. Loosely tufted perennial, often with short rhizomes, 15-45 cm; sheaths closed ca. 1/2 their length, separating into fibers; ligules 0.1-1 mm; panicles 3-8 cm; lemmas 3-4.5 mm; awns mostly 1-1.5 mm; anthers 0.6-1.4 mm; ovary apices densely pubescent. Subalpine and alpine meadows, grassy slopes at medium to high elevations. NC, Lincoln.

Festuca glauca Vill.—Blue fescue. Perennial; introduced from Europe as an ornamental landscape plant, ideal for borders and accents, with numerous cultivars; not known outside of cultivation. We use this provisional name for several species that have been used in the nursery trade; they are all characterized by dense rounded clumps with markedly bluish foliage and narrow blades. They have most commonly gone by the name F. glauca Vill., with various additional cultivar names, but also by F. arvernensis Auquier, Kerguélen & Markgr.-Dann., and F. ovina L. var. glauca Fr. in various works.

Festuca hallii (Vasey) Piper [*Melica hallii* Vasey]—Hall's fescue. Densely tufted perennial, short rhizomes, 20-85 cm; sheaths open for 1/3+ their length, not separating into fibers, persistent; ligules 0.3-0.6 mm; blades usually folded, to ca. 1.5 mm across, rarely flat to 3 mm wide; peduncles glabrous to sparsely scabrous; panicles 6-16 cm, with 1-2 branches per node, the branches erect to stiffly spreading; glumes as long as the spikelet, exceeding the florets, 7-10 mm; lemmas 5.5-9 mm; awns absent to 1.3 mm; anthers 4-6 mm; ovary apices sparsely pubescent. High-elevation meadows and forest glades in the N mountains; a single record from Taos County, 10,200 ft.

Festuca idahoensis Elmer—Idaho fescue. Densely tufted perennial, without rhizomes, 25-85 cm; sheaths persistent, < 4 cm; ligules 0.2-0.6 mm; peduncles mostly glabrous, occasionally somewhat scabrous; panicles 5-16 cm; lemmas 4.5-5.5 mm; awns 2-7 mm, occasionally shorter; anthers 2.5-4.5 mm; ovary apices glabrous. Mountain grasslands, 7135-11,675 ft. NC.

Festuca minutiflora Rydb.—Smallflower fescue. Loosely or densely tufted perennial, without rhizomes, mostly 3-10 cm in our populations; sheaths persistent; ligules 0.1-0.3 mm; panicles 1-5 cm; spikelets with 2-3 florets; lemmas 2-3.5 mm; awns 0.5-1.5 mm; anthers 0.5-1.2 mm; ovary apices sparsely hairy, rarely glabrous. Alpine grasslands, 11,340-12,960 ft. NC.

Festuca rubra L.—Red fescue. Loosely to somewhat tufted perennial, with short rhizomes, 10-50 cm; sheaths (especially the vegetative ones) closed ca. 3/4 their length, rapidly separating into threadlike fibers (whitish veins); ligules 0.1-0.5 mm; panicles 3-25 cm; lemmas 4-9 mm; awns 0.5-4.5 mm;

anthers 2–4.5 mm; ovary apices glabrous. High mountain grasslands and open clearings, sometimes in lawns, 6000–11,600 ft. NC, NW, WC, SC. Plants in the wild seem to belong to **subsp. rubra**; lawn and turf mixtures might also include **subsp. commutata** Gaudin (Chewing's fescue).

Festuca saximontana Rydb.—Rocky Mountain fescue. Densely to loosely tufted perennial, without rhizomes, 10–50 cm; foliage often glaucous; sheaths persistent, rarely splitting; blades stiff, the veins usually not visible; ligules 0.1–0.5 mm; panicles 3–12 cm; lemmas 3–5 mm; awns 1–2.5 mm; anthers 1–2 mm; ovary apices glabrous. Mountain grasslands and forest clearings, 6900–12,700 ft. NC, Grant, Otero.

Festuca sororia Piper—Ravine fescue. Loosely tufted perennial, without rhizomes, 60–120 cm; sheaths splitting into threadlike fibers in age; ligules 0.3–1.5 mm; blades flat, 3–6+ mm wide; panicles 10–25 cm, the branches lax and spreading below; spikelets 2–4(5)-flowered; lemmas 5–8 mm; awnless or with awns to 2 mm; anthers 1.5–2.5 mm; ovary apices hairy. Moist, shaded slopes and stream banks in the mountains, 7500–11,300 ft. NC, NW, C, WC, SC.

Festuca thurberi Vasey—Thurber's fescue. Densely tufted perennial, without rhizomes, 50–100+ cm; nodes visible and conspicuous; sheaths persistent; ligules 2–9 mm; panicles 10–17 cm, the branches loosely arranged; lemmas 6–10 mm, awnless; anthers 3–4.5 mm; ovary apices densely hairy. High mountain grasslands, 7700–12,855 ft. NC, C, WC, SE.

Festuca trachyphylla (Hack.) R. P. Murray—Hard fescue. Densely tufted perennial, without rhizomes, 15–75 cm; ligules 0.1–0.5 mm; blades strongly folded, thickened and stiff, 0.5–1 mm wide (folded); panicles 3–15 cm; lemmas 3.8–6 mm; awns 0.5–3 mm; anthers 2–3.4 mm; ovary apices glabrous. Introduced for reseeding, erosion control, and rangeland restoration; grassy slopes of the N mountains, 8120–11,500 ft. NC, NW, WC, SC. Adventive.

GLYCERIA R. Br. – Mannagrass

Rhizomatous perennials, rarely annuals; sheaths closed; ligules a scarious membrane, fringed; inflorescence a rebranching panicle; spikelets several-flowered, awnless, the terminal floret sterile; disarticulation above the glumes and between the florets; glumes much smaller than the spikelet, 1-nerved; lemmas several-nerved, the nerves ± parallel and not converging at the apex; anthers 2–3.

1. Spikelets linear, nearly round in cross-section, 9–18 mm, 8–12-flowered; lemmas 3–5.5 mm...**G. borealis**
1'. Spikelets ovate or oblong, somewhat compressed, 2.5–7 mm, 3–6(7)-flowered; lemmas 1.5–3 mm...(2)
2. Apices of lemmas flat; anthers 3; nerves of 1 or both glumes usually extending to the apex of the glume...**G. grandis**
2'. Apices of lemmas prow-shaped; anthers 2; nerves of both glumes ending below the apex of the glume...(3)
3. Blades 6–15 mm wide; anthers 0.5–0.8 mm; culms 2.5–8 mm thick, spongy, 75–150+ cm, often decumbent-based...**G. elata**
3'. Blades 2–6 mm wide; anthers 0.2–0.6 mm; culms 1.5–3.5 mm thick, not or slightly spongy, 20–80 cm, generally erect...**G. striata**

Glyceria borealis (Nash) Batch.—Small floating mannagrass. Perennial, 60–100 cm; culms 1.5–5 mm thick, often decumbent and rooting at the nodes; blades 2–7 mm wide, the adaxial surface of midstem leaves densely papillose; panicles 18–45 cm; spikelets 9–18 mm, cylindrical, 1–2.5 mm wide; first glume 1.2–2.2 mm; lemmas 3–5.5 mm, the apices acute to obtuse and nearly entire; anthers 3, 0.4–1.5 mm. Borders of lakes and ponds in the N mountains, 7950–10,500 ft. NC, NW, Catron.

Glyceria elata (Nash) M. E. Jones—Mannagrass. Perennial, 75–150 cm; culms 2.5–8 mm thick, spongy in the internodes, decumbent and rooting at the lower nodes; blades 6–15 mm wide, lacking papillae but scabrous; panicles 15–30 cm; spikelets 3–6 mm, flattened, 1.5–3 mm wide; lemmas 1.7–2.2 mm, the apices prow-shaped; anthers 2, 0.5–0.8 mm. Mountain springs and marshy ground in subalpine communities, 7100–10,900 ft. NC, WC, SC.

Glyceria grandis S. Watson—American mannagrass. Perennial, 50–150+ cm; culms 8–12 mm thick, erect to decumbent and rooting at the nodes; blades 5–15 mm wide, lacking papillae; panicles 16–40 cm;

spikelets 3.2–10 mm, flattened, 2–3 mm wide; lemmas 1.8–3 mm, the apices flat or nearly so, not prow-shaped; anthers 3, 0.5–1.2 mm. Marshes, swampy ground, irrigation banks, springs, 6549–10,025 ft. NC, NW, WC.

Glyceria striata (Lam.) Hitchc.—Fowl mannagrass. Perennial, 20–80 cm, sometimes taller; culms 1.5–3.5 mm thick, not or slightly spongy, only sometimes rooting at the nodes; blades 2–6 mm wide, lacking papillae but scabrous; panicles 6–25 cm; spikelets 1.8–4 mm, flattened, 1.2–3 mm wide; lemmas 1.2–2 mm, the apices prow-shaped; anthers 2, 0.2–0.6 mm. Marshes, springs, wet ground, stream banks, 5475–10,570 ft. Widespread except E, SW.

GRAPHEPHORUM Desv. – Trisetum

Graphephorum wolfii (Vasey) Vasey ex Coulter [*Trisetum wolfii* Vasey]—Wolf's trisetum. Tufted perennial with short rhizomes, 20–80 cm+ tall; sheaths glabrous to retrose-pilose; ligules 2–6 mm; blades glabrous to pilose; panicles 10–40 cm, congested, usually < 2 cm wide, branches erect-ascending; spikelets 4–8 mm, mostly with 2 florets; glumes subequal and usually exceeding the lower floret; rachillas 1.5–2 mm, with hairs to 1 mm; lowermost lemmas 4–6.5 mm, obscurely bifid with broad points but not setaceous teeth, awnless or an awn coming from the sinus to 2 mm. Marshy ground around seeps and springs at high elevations in the N mountains, 8000–11,740 ft. NC.

HELICTOTRICHON Besser – Oatgrass

Tufted perennials; sheaths open; auricles absent; ligules membranous, short, about as long as wide; blades rolled, the adaxial surface ribbed; inflorescence a weakly developed panicle, racemelike; spikelets with 2–8 florets; disarticulation above the glumes and between the florets; glumes nearly as large as the spikelet, 1–3-nerved; rachillae pilose on all sides; lemmas 3–5-nerved, awned from about midlength; awns geniculate, twisted below; anthers 3.

1. Panicles 2–5 cm; blades rolled, usually pubescent…**H. mortonianum**
1′. Panicles 5–15 cm; blades flat or folded, mostly glabrous…go to **Avenula hookeri**

Helictotrichon mortonianum (Scribn.) Henrard—Tufted perennial, 5–20 cm; blades 1–2 mm wide, rolled or folded, strigose at least on the adaxial surface; panicles 2–5(–8) cm; spikelets 8–12 mm, with (1)2–3 florets, the distal florets awned or awnless; lower lemmas 7–10 mm, 3-nerved; awns 10–16 mm. Alpine slopes and forest edges in the N mountains, 11,450–12,510 ft. Mora, San Miguel, Taos.

HESPEROSTIPA (M. K. Elias) Barkworth – Needle-and-thread

Tufted perennials, lacking rhizomes; sheaths open; auricles absent; ligules membranous, sometimes ciliate; inflorescence a panicle; spikelets with 1 floret; disarticulation above the glumes, the floret falling; glumes large, nearly equaling or exceeding the body of the lemma; lemmas indurate at maturity, the margins overlapping, the distal portion fused into a crown, awned; awns twice-geniculate; paleas equal to the lemmas, indurate at the apex; anthers 3.

1. Terminal segment of awn plumose, with feathery hairs 2–3 mm…**H. neomexicana**
1′. Terminal segment of awn not plumose, any hairs present < 1 mm…(2)
2. Lemmas evenly white-hairy, sometimes glabrous above the callus; lower ligules usually acute, thin, often cut or torn; margins of lower sheaths mostly glabrous…**H. comata**
2′. Lemmas unevenly brownish-hairy, densely hairy on the margins and in lines on the proximal portion, glabrous distally; lower ligules rounded to truncate, thick, not cut or torn; margins of lower sheaths often ciliate…**H. spartea**

Hesperostipa comata (Trin. & Rupr.) Barkworth—Needle-and-thread. Tufted perennial, 25–100+ cm, the nodes hidden to exposed; margins of lower sheaths not ciliate; ligules 1–7 mm, the lower ones usually thin and lacerate; panicles 10–30 cm; glumes 16–35, the upper glume slightly shorter than the lower; florets 7–13 mm, the lemmas evenly white-hairy; awns 6–22 cm, many hairs < 1 mm, the terminal segment straight or curling to flexuous. Plains, prairies, woodland clearings, 3760–9755 ft. NC, NW, WC.

Widespread except SE, SW. Plants with the terminal awn segment straight, the lower panicle branches mostly exserted from the sheath, and the lower nodes exposed may be referred to as **subsp. intermedia** (Scribn. & Tweedy) Barkworth.

Hesperostipa neomexicana (Thurb.) Barkworth—New Mexico feathergrass. Tufted perennial, 40–100 cm; sheaths not ciliate; ligules 0.5–3 mm; panicles 10–30 cm; glumes 30–60 mm, subequal; florets 15–18 mm; awns 12–22 cm, hairs on the terminal segment (1–)2–3 mm. Plains, grassy hills, rocky slopes, usually on limestone, 3900–7800 ft. Widespread.

Hesperostipa spartea (Trin.) Barkworth—Porcupine grass. Tufted perennial, 45–100+ cm; margins of lower sheaths usually ciliate; ligules 0.3–7.5 mm, the lower ones thick and not lacerate; panicles 10–25 cm; glumes 22–45 mm, subequal; florets 15–25 mm, the lemmas with brownish hairs on the margins and in lines proximally, glabrous distally; awns 9–19 cm, any hairs < 1 mm. Plains and prairies, 6235–8040 ft. NC, NW, Bernalillo.

HETEROPOGON Pers. - Tanglehead

Heteropogon contortus (L.) P. Beauv. ex Roem. & Schult.—Tanglehead. Tufted perennial, 20–100+ cm; sheaths open, keeled; auricles absent; ligules membranous, sometimes ciliate; blades flat or folded, blunt, 2–7 mm wide; racemes 3–7 cm; awned sessile spikelets 5–10 mm, with awns 6–10 cm; pedicelled spikelets 6–10 mm, lacking glandular pits. Desert hills, 3695–7500 ft. SE, SC, SW.

HIEROCHLOE R. Br. - Sweetgrass

Hierochloe odorata (L.) P. Beauv. [*Anthoxanthum nitens* (Weber) Y. Schouten & Veldkamp]—Sweetgrass. Loosely tufted perennial, 10–75 cm, with elongate rhizomes; sheaths open; auricles absent; ligules membranous, sometimes ciliate; panicles 3–12 cm, 1.5–7 cm wide; spikelets 3(4), 3–7 mm, tawny-colored, spikelets with 3(4) florets, the lower 2 staminate and appearing on either side of the upper one, the upper (central) one fertile, sometimes a terminal smaller sterile floret present at the tip; disarticulation above the glumes, the florets falling together; glumes nearly concealing the spikelets, 3-nerved, thin, papery, subequal, exceeding the florets; staminate florets 3.5–5 mm, sometimes with an awn to 0.5 mm; fertile florets 2.5–4 mm. Wet high mountain meadows and subalpine to alpine slopes, 7270–12,000 ft. NC, NW, C, WC. Introduced.

HILARIA Kunth - Curly mesquite

Annuals and perennials, tufted, rhizomatous, or stoloniferous; sheaths open; auricles absent; ligules membranous, often ciliate; inflorescence functionally a spike, with 3 spikelets per node, the cluster sitting on a ledgelike indentation on the zigzag rachis; disarticulation at the base of the spikelet cluster, which falls entire; lateral spikelets of the cluster with 1–4 staminate or sterile florets, the glumes deeply lobed and short-awned; central spikelet of the cluster with a single fertile floret, the glumes lobed and awnless. Rhizomatous members of this group are sometimes treated in the genus *Pleuraphis*.

1. Glumes thickened, indurate, and fused at the base; plants stoloniferous and not rhizomatous, rarely > 30 cm...(2)
1'. Glumes papery or membranous throughout, not fused at the base; plants usually rhizomatous, rarely < 30 cm...(3)
2. Glumes of the lateral spikelets pale to purplish, lacking glandular dots or these only at the base, awned from below midlength...**H. belangeri**
2'. Glumes of the lateral spikelets blackish or purplish, evenly covered with glandular dots, awned from above midlength...**H. swallenii**
3. Lower cauline internodes tomentose; known only from Doña Ana County...**H. rigida**
3'. Lower cauline internodes glabrous; widespread...(4)
4. Glumes of the lateral spikelets fan-shaped, the awns not exceeding the apical lobes; cauline nodes short-hairy, sometimes glabrous...**H. mutica**
4'. Glumes of the lateral spikelets lanceolate or parallel-sided, the awns exceeding the apical lobes; cauline nodes long-hairy or glabrous...**H. jamesii**

Hilaria belangeri (Steud.) Nash—Curly mesquite. Tufted perennial, 5-35 cm, usually with stolons; blades sparsely pilose on the adaxial surface with bulbous-based hairs; spikes 2-4 cm; lateral spikelets with 2-3 staminate florets or 1 sterile floret, the glumes indurate and fused at the base, eglandular or the bases spotted with a few dark glands, the awns from below midlength. Desert hills and rocky slopes, 4680-6100 ft. WC, SW, Chaves.

Hilaria jamesii (Torr.) Benth. [*Pleuraphis jamesii* Torr.]—Galleta. Loosely tufted perennial, with strong rhizomes and sometimes stolons, 25-65 cm; spikes 2-6 cm; spikelet cluster 6-8 mm; lateral spikelets with 3 staminate florets, the glumes lanceolate or parallel-sided with dorsal awns exceeding the apices. Plains and foothills, 3700-8700 ft. Widespread except SE, SW.

Hilaria mutica (Buckley) Benth. [*Pleuraphis mutica* Buckley]—Tobosa. Loosely tufted perennial, with often strong rhizomes, 30-60 cm; spikes 4-8 cm; spikelet cluster 5-8 mm; lateral spikelets with 1-2 staminate florets, the glumes fan-shaped, with dorsal awns not exceeding the apices. Flats and swales, gravelly hillsides, 3000-6900 ft. EC, SE, SC, SW.

Hilaria rigida (Thurb.) Benth. ex Scribn. [*Pleuraphis rigida* Thurb.]—Big galleta. Tufted perennial, sometimes rhizomatous, 40-200+ cm, much branched and becoming shrubby; lower internodes tomentose; spikes 4-12 cm; spikelet cluster 6-12 mm; lateral spikelets with 2-4 florets, the lower ones staminate, the upper sterile, the glumes lanceolate or parallel-sided with dorsal awns exceeding the apices. Introduced from California and Arizona for range reseeding trials, without success, but a few plants remain in the test plots of the College Ranch of New Mexico State University, 4700 ft. Doña Ana.

Hilaria swallenii Cory—Swallen's curly mesquite. Loosely tufted perennial, 10-35 cm, with stolons; blades mostly glabrous but also sparsely pilose; spikes 1-4 cm; lateral spikelets with 2 florets, the lower sterile and the upper staminate, the glumes indurate and fused at the base, evenly spotted with dark glands, the awns from above midlength. Desert hills and rocky slopes; 1 record, ca. 4600 ft. Luna.

HOLCUS L. - Velvetgrass

Holcus lanatus L.—Common velvetgrass. Tufted perennial, lacking rhizomes, 35-100 cm; lower internodes densely pilose; sheaths open, densely pubescent, auricles absent; ligules membranous, ciliate; blades 5-10 mm wide, densely soft-hairy; panicles 3-20 cm, the branches and pedicels hairy; spikelets 2(3), 3-6 mm; glumes ciliate on the midnerve, the upper awn-lipped; upper lemma with a hooked awn 1-2 mm; anthers 3. Adventive in cool, moist waste places, 4700-7250 ft. Doña Ana, Grant, Socorro.

HOPIA Zuloaga & Morrone - Vine mesquite

Hopia obtusa (Kunth) Zuloaga & Morrone—Vine mesquite. Stoloniferous perennial sometimes also with short rhizomes, 20-80 cm, the long stolons with conspicuously hairy nodes; lower sheaths pilose; ligules membranous, 0.2-2 mm; panicles 5-15 cm, the branches 2-6 in number; spikelets with 2 florets shaped like a rugby ball, 2.8-4.5 mm, awnless, the lower staminate, the upper fertile; lower glume 3/4 to nearly equaling the spikelet length. Usually heavy soils of swales, playas, flats, and low spots; sometimes planted to control soil erosion, 3300-9500 ft. Widespread.

HORDEUM L. - Barley

Tufted annuals or perennials; sheaths open; auricles present or absent; ligules membranous; inflorescence a spike (spicate raceme) with 3 spikelets per node, the central sessile and fertile, the lateral pedicelled and usually staminate or sterile (fertile in cultivated forms); disarticulation below the spikelets and in the rachis, the 3 spikelets falling together with the rachis internode (above the glumes in cultivated forms); spikelets with a single floret; glumes awnlike; lemmas awned or awnless; anthers 3.

1. Rachis persistent, not breaking apart when mature; plants annual...**H. vulgare**
1'. Rachis breaking apart when mature; plants annual or perennial...(2)

2. Glumes of the central spikelet with conspicuous ciliate margins; auricles usually well developed, mostly 1-8 mm…**H. murinum**
2′. Glumes of the central spikelet without ciliate margins, at most scabrous; auricles usually lacking or weakly developed and < 0.5 mm…(3)
3. Plants perennial…(4)
3′. Plants annual…(6)
4. Glumes of the central spikelet flattened near the base…**H. arizonicum** (in part)
4′. Glumes of the central spikelet terete throughout, not flattened near the base…(5)
5. Glumes 7-20 mm; awns of the lemmas 5-10(-20) mm…**H. brachyantherum**
5′. Glumes 20-150 mm; awns of the lemmas 10-70 mm…**H. jubatum** (in part)
6. Glumes bent outward at the base, strongly divergent when mature…(7)
6′. Glumes erect at the base, ascending to only slightly divergent when mature…(8)
7. Glumes of the central spikelets terete throughout, not flattened near the base, 20-150 mm…**H. jubatum** (in part)
7′. Glumes of the central spikelets flattened near the base, 11-28 mm…**H. arizonicum** (in part)
8. Glumes of lateral spikelets prominently flattened near the base; ligules 0.2-0.8 mm…**H. pusillum**
8′. Glumes of lateral spikelets terete to slightly flattened near the base; ligules 0.6-1.8 mm…**H. arizonicum** (in part)

Hordeum arizonicum Covas—Arizona barley. Tufted annual to sometimes a short-lived perennial, 20-70 cm; auricles absent; ligules 0.6-1.8 mm; spikes 5-12 cm, 6-10 mm wide; glumes 11-28 mm, flattened near the base; lemmas of the central spikelet 5-9 mm. Weedy ground, uncommon, 3890-4100 ft. Doña Ana. Introduced.

Hordeum brachyantherum Nevski—Meadow barley. Tufted perennial, 25-80 cm; auricles absent; spikes 3-8.5 cm; glumes 7-20 mm, ascending to spreading, not or rarely flattened near the base; lemmas of central spikelet 5-10 mm. Moist mountain slopes and grassy hills, 5000-11,130 ft. NC, NW, WC.

Hordeum jubatum L.—Foxtail barley. Tufted perennial, sometimes appearing annual, 20-80 cm; auricles absent; spikes 3-15 cm; glumes 15-85 mm, divergent, sometimes strongly so; lemmas of the central spikelet 4-8.5 mm. Moist ditches, meadows, roadsides, disturbed ground, 3400-11,100 ft. Widespread except SE. Stabilized hybrids between *H. jubatum* and *H. brachyantherum*, with glumes of the central spikelet (including awns) 2-3 cm, and lemmas of the central spikelet (including awns) 2-4 cm, may be recognized as **subsp. intermedium** Bowden.

Hordeum murinum L. [*H. glaucum* Steud.; *H. leporinum* Link]—Mouse barley, wall barley. Loosely tufted annual, 3-100+ cm; ligules 1-4 mm; auricles well developed, 1-8 mm; spikes 3-8 cm, 7-16 mm wide; glumes 10-25 mm, with distinct ciliate margins; lemmas of the central spikelet 8-14 mm, with awns to 40 mm. Weedy ground, 3135-8965 ft. Widespread except NE, NC. Introduced.

Hordeum pusillum Nutt.—Little barley. Loosely tufted annual, 10-40 cm; auricles absent; ligules 0.2-0.8 mm; spikes 2-9 cm, 3-7 mm wide; glumes erect to ascending, not divergent, 8-18 mm; lemmas of central spikelet 5-8.5 mm, with awns to 10 mm. Waste places, 3200-7500 ft. Widespread except WC.

Hordeum vulgare L.—Barley. Loosely tufted annual, to 150 cm; auricles to 1-6 mm; spikes 5-10 cm, 8-20 mm wide, the rachis usually persistent; central and lateral spikelets sessile when both are fertile, the lateral spikelets pedicelled when sterile; glumes 10-30 mm, flattened near the base; lemmas of central spikelet 6-12 mm, awned to awnless. Introduced barley crop also used for erosion control along roads, adventive along fields and roadsides, 4700-10,600 ft. Scattered. Introduced.

IMPERATA Cirillo – Satintail

Imperata brevifolia Vasey—Strongly rhizomatous, erect, mostly unbranched perennial, 50-120 cm; sheaths open; auricles absent; ligules membranous; inflorescence a narrow, plumose panicle; panicle branches (rames) composed of repeating pairs of pedicelled spikelets, panicles 16-34 cm; callus hairs 8-12 mm; disarticulation below the glumes, the pairs of spikelets falling together; spikelets all alike, fertile, awnless, in pairs, 2.7-4 mm, unequally pedicelled, surrounded by long silky hairs from the rachis;

glumes subequal, membranous, with long silky hairs; anthers 1–2, 1.3–2.3 mm. In New Mexico, known only from Doña Ana County along the Rio Grande floodplain; last found in 1939 and no records seen.

KALINIA H. L. Bell & Columbus – Alkali lovegrass

Kalinia obtusiflora (E. Fourn.) H. L. Bell & Columbus [*Eragrostis obtusiflora* (E. Fourn.) Scribn.]— Kalinia grass. Rhizomatous and stoloniferous perennial, 15–40 cm; sheaths open, hairy at the summit; auricles absent; ligules membranous, 0.2–0.4 mm; blades rolled, stiff and sharp-pointed; panicles 6–20 cm, 2–10 cm wide, the upper portion becoming racemelike, with glabrous to pubescent pulvini; disarticulation above the glumes and between the florets, leaving the glumes; spikelets with 5–10 florets, 8–14 mm, 1.4–3 mm wide, appressed along the branch; lemmas 3-nerved (sometimes 4 or 5), 3.8–4.5 mm, awnless; anthers 3. Along dry shores of Playas Lake, 4285 ft. Hidalgo. Formerly included as an anomalous member of *Eragrostis*. Easily distinguished by the rhizomes and puncturing blades.

KOELERIA Pers. – Junegrass

Tufted, sometimes rhizomatous, perennials; sheaths open; auricles absent; ligules membranous; inflorescence a panicle, often dense and spikelike, the rachis and branches velvety-pubescent; spikelets with 2–4 florets, awnless; disarticulation above the glumes and between the florets; glumes subequal to exceeding the lemmas; lemmas thin, 5-nerved, with scarious margins; anthers 3.

1. Florets awnless…**K. macrantha**
1ʹ. Florets conspicuously and definitely awned…(2)
2. Panicles 3–10 cm; spikelets densely congested, the branches mostly < 1 cm and erect-appressed; leaves tending to be basal…**K. spicata**
2ʹ. Panicles 8–24 cm; spikelets somewhat crowded to loosely arranged, the branches (1–)2–6 cm and ascending to somewhat divergent; leaves tending to be cauline…**K. vaseyi**

Koeleria macrantha (Ledeb.) Schult. [*K. cristata* auct.; *Aira macrantha* Ledeb.]—Junegrass. Tufted perennial, 20–85+ cm; leaves mostly basal; panicles 5–35 cm, the primary branches with axillary pulvini, causing them to spread during anthesis; spikelets 2.5–6.5 mm, obovate; lemmas 2.5–6.5 mm, shiny, glabrous, rarely with an awn to 1 mm. Mountain slopes, foothills, plains, 4000–11,620 ft. Widespread except EC.

Koeleria spicata (L.) Barberá, Quintanar, Soreng & P. M. Peterson [*Trisetum spicatum* (L.) Richt.]— Spike trisetum. Tufted perennial, sometimes with short rhizomes, 10–60 cm; sheaths glabrous; ligules ca. 1 mm; blades glabrous to rarely hairy; panicles 3–10 cm, densely congested and spikelike, the branches mostly < 1 cm, the peduncle densely hairy below the panicle, the hairs 0.2–1 mm; spikelets 4.5–6 mm; rachilla 0.8–1 mm, with hairs 0.5–1 mm; glumes subequal or the first at least 3/4 the second; lowermost lemmas 4–5 mm, the awn 4–6 mm; ovary glabrous. Alpine to subalpine ridges, slopes, forest clearings, 7950–13,100 ft. NC, SC.

Koeleria vaseyi Barberá, Quintanar, Soreng & P. M. Peterson [*Trisetum spicatum* (L.) K. Richt. subsp. *montanum* (Vasey) W. A. Weber]—Mountain trisetum. Tufted perennial, 30–70 cm; sheaths glabrous to pilose; ligules ca. 3 mm; blades glabrous to pilose, generally not very canescent; panicles 10–24 cm, lax and somewhat open to contracted, the peduncle glabrous below the panicle or with minute hairs; spikelets 4–6 mm; rachilla ca. 0.8 mm, with hairs < 0.5 mm; glumes unequal, the first ca. 2/3 the second; lowermost lemma 4.5–5.5 mm, the awn 3–6 mm; ovary glabrous. Mountain woodlands and grasslands, clearings, grassy slopes, roadsides, 7380–11,800 ft. NC, C, WC, SC.

LAGURUS L. – Hare's tail

Lagurus ovatus L.—Hare's tail grass. Tufted annual, 10–50 cm, the culms erect to ascending, the foliage pilose to villous; sheaths open, inflated 2–8 cm; auricles absent; ligules membranous, densely villous on the back; blades 3–10 mm wide; inflorescence a dense, ovoid, woolly panicle; panicles 1.5–3 × 1–2 cm; spikelets with a single floret, the rachilla prolonged; disarticulation above the glumes, the floret falling; glumes exceeding the floret, 1-nerved, 6–10 mm, the awn to 3 mm; lemmas 3-nerved, 3–5 mm; central awn 10–22 mm, the laterals much shorter; anthers 3. A recently found adventive, rarely escaping from cultivation for ornament and dried bouquets; 1 record, 4000 ft. Doña Ana.

LEERSIA Sw. - Cutgrass

Leersia oryzoides (L.) Sw.—Rice cutgrass. Rhizomatous perennial, often rooting at the nodes, 35–150 cm; sheaths open, scabrous; auricles absent; ligules membranous, to 1 mm; blades 5–15 mm wide, the margins scabrous; panicles 10–30 cm, the branches spreading, disarticulation below the spikelets; spikelets with 1 floret, usually awnless, elliptic, 4–6 mm, nearly 2 mm wide; glumes absent; lemmas and paleas subequal and hardened somewhat, ciliate on the keels and margins; lemmas 5-nerved; anthers 1, 2, 3, or 6, chasmogamous (1–3 mm) or cleistogamous (to 1 mm). River and stream banks in the S region, often aquatic, 4470–6065 ft. Doña Ana.

LEPTOCHLOA P. Beauv. - Sprangletop

Leptochloa crinita (Lag.) P. M. Peterson & N. Snow [*Trichloris crinita* (Lag.) Parodi]—False Rhodes grass. Tufted perennial, sometimes with short stolons, 45–100+ cm; sheaths open, glabrous to sparsely hirsute; auricles absent; blades 5–10 mm wide; panicle narrow, 8–15 cm; ligules a membrane, truncate to obtuse, somewhat erose or ciliate; panicle branches 6–20 in number, erect, to 15 cm; spikelets 2–5-flowered, with 1 fertile and 1 sterile floret; disarticulation above the glumes, the florets falling together; lemmas 3-nerved, both 3-awned; awns 5–12 mm; anthers 2–3. Disturbed ground, roadsides, fields, drainages in desert grasslands, 3030–4300 ft. C, SC, SW.

LEPTOLOMA Chase - Fall witchgrass

Leptoloma pubiflorum (Vasey) Wipff & R. B. Shaw—Fall witchgrass. Densely tufted to loosely rhizomatous perennial, 20–70 cm; sheaths open; auricles absent; ligules membranous, sometimes with a ciliate fringe; foliage glabrous to densely pubescent; panicles widely open, 5–25 × 6–30 cm, the branches divergent and rebranched; spikelets 2-flowered, on long divergent pedicels, 2.3–3.3 mm, the lower floret staminate to sterile, the upper floret fertile, the rachilla elongate between the 2 glumes, awnless; disarticulation below the glumes; upper glume 2–3 mm, densely pilose between the 3 nerves; lower lemma 5-nerved; upper floret 2–3.1 mm, glabrous, dark when mature; upper lemma clasping the palea and forming a seedcase, this not indurate; anthers 3. S and E plains, 3800–6400 ft. NE, EC, C, SE, SC, SW.

LEUCOPOA Griseb. - Spike fescue

Leucopoa kingii (S. Watson) W. A. Weber [*Hesperochloa kingii* (S. Watson) Rydb.]—Spike fescue. Tufted and rhizomatous perennial, 30–100+ cm, dioecious; sheaths open; auricles absent; ligules membranous; panicles 7–22 cm, with erect to ascending branches; spikelets unisexual with 2–6 florets, 6–12 mm; disarticulation above the glumes and between the florets; glumes mostly hyaline and thinner than the lemmas; lemmas 5-nerved, 4.5–10 mm, awnless or with a mucro; the paleas subequal; anthers 3, 3–6 mm; ovaries in pistillate plants hairy at the apices. Woodlands and brushy hills, 5500–10,000 ft. Sandoval, San Juan.

LEYMUS Hochst. - Wildrye

Tufted to rhizomatous perennials; sheaths open; auricles usually present; ligules membranous; inflorescence a spike with 1–8 spikelets per node, sometimes paniculate; spikelets usually sessile, with 2–12 florets; disarticulation above the glumes and between the florets; glumes often reduced and needlelike; lemmas 5–7-nerved, awned or awnless; anthers 3. All species of *Leymus* derive from *Elymus*, based on a suite of morphological, chromosomal, anatomical, and molecular features.

1. Plants strongly rhizomatous, the rhizomes long and slender, not bunch-forming…(2)
1'. Plants tufted, or with short rhizomes but still bunch-forming…(3)

2. Culms 8-12 mm thick; blades 8-20 mm wide; spikes with 3-8 spikelets per node; glumes 12-25 mm…**L. racemosus**
2'. Culms 1-3 mm thick; blades 3-10 mm wide; spikes with 2 spikelets per node at midspike; glumes 5-16 mm…**L. triticoides**
3. Plants in giant clumps to 2+ m, usually much > 100 cm; blades flat, 5-15 mm wide; spikelets usually 3-6 per node…**L. cinereus**
3'. Plants much smaller, rarely as much as 1 m and usually < 70 cm; blades mostly involute or rarely flat, 2-5 mm wide; spikelets 1-2 per node…(4)
4. Spikelets mostly 1 per rachis node; blades often flat or sometimes involute…**L. salina**
4'. Spikelets mostly 2 per middle rachis node (solitary at the apex and base of the spike); blades almost always involute…**L. ambiguus**

Leymus ambiguus (Vasey & Scribn.) D. R. Dewey [*Elymus ambiguus* Vasey & Scribn.]—Colorado wildrye. Loosely tufted perennial, sometimes with short rhizomes, 60-110 cm, the culms 1-1.5 mm thick; auricles to 1.1 mm; spikes 8-17 cm, 5-10 mm wide, with 2 spikelets per node; glumes needle-shaped; lemmas 8-14 mm, with awns 1-7 mm. Dry, rocky foothills and plains, sometimes mountain slopes with oak brush, 7500-10,000 ft. C.

Leymus cinereus (Scribn. & Merr.) Á. Löve [*Elymus cinereus* Scribn. & Merr.]—Great Basin wildrye. Tussocky perennial, weakly rhizomatous, 70-270 cm, the culms 2-5 mm thick, usually green and not glaucous, populations appearing ash-colored from a distance; auricles to 1.5 mm; spikes 10-30 cm, 8-17 mm wide, with 2-7 spikelets per node; glumes needle-shaped; lemmas 6-12 mm, acute to awned to 3 mm. Known only from Colfax and San Juan Counties, where it appears to be adventive or deliberately planted; 2 records, 5670-6050 ft. Introduced.

Leymus racemosus (Lam.) Tzvelev [*Elymus racemosus* Lam.]—Mammoth wildrye. Strongly rhizomatous perennial, weakly tufted, often glaucous, 50-100+ cm, the culms 8-12 mm thick; spikes 15-35 cm, with 3-8 spikelets per node; glumes tapering to needle-shaped, equal to or exceeding the florets; lemmas 15-20 mm, tapering to an awn 1.5-2.5 mm. Known only from a few collections along weedy roadsides, 7000-8200 ft. Colfax, San Miguel. Introduced.

Leymus salina (M. E. Jones) Á. Löve [*Elymus salina* M. E. Jones]—Salina wildrye. Tufted perennial, sometimes weakly rhizomatous, 35-140 cm, the culms 1.5-3 mm thick; spikes 4-14 cm, with mostly 1(?-3) spikelet per node; glumes needlelike, shorter than the florets; lemmas 7-12 mm, awnless or awned to 2.5 mm. Dry plains, 6100-8200 ft. San Juan.

Leymus triticoides (Buckley) Pilg. [*Elymus simplex* Scribn. & T. A. Williams; *E. triticoides* Buckley]—Creeping wildrye. Strongly rhizomatous perennial, 45-125 cm, the culms 1.8-3 mm thick; auricles to 1 mm; spikes 5-20 cm, 5-15 mm wide, with 2 spikelets per node (varying proximally and distally); glumes needlelike, shorter than the florets; lemmas 5-12 mm, with awns to 3 mm. Some of our plants are from high mountain forest clearings (?introduced), but the species is more common on clay flats and swales at much lower elevations, 7200-11,675 ft. SC, San Juan.

LOLIUM L. – Ryegrass

Tufted annuals and perennials, sometimes short-rhizomatous; sheaths open; auricles present or absent; ligules membranous; blades flat; inflorescence a spike, with 1 spikelet per node, the spikelets attached edgewise to the rachis; spikelets with several florets; disarticulation above the glumes and between the florets; glumes 1, 2 in the terminal spikelet, 3-9-nerved, awnless; lemmas 3-7-nerved; awned or awnless; anthers 3. To avoid undue confusion, we follow here the treatment of the recent *Flora of North America*, vol. 24, which maintains the genera *Festuca*, *Leucopoa*, *Lolium*, *Schedonorus*, and *Vulpia*, while acknowledging the interrelatedness of all 5 species groups (cf. Catalán et al., 2007).

1. Glume exceeding the uppermost floret…**L. temulentum**
1'. Glume shorter than the spikelet, the florets extending beyond the glume…**L. perenne**

Lolium perenne L. [*L. multiflorum* Lam.]—Perennial ryegrass. Tufted perennial, sometimes annual; spikes 5-45 cm; spikelets with 5-22 florets; lemmas 3.5-9 mm, awned or awnless; anthers 2-5 mm.

Lawns, roadsides, and pastures, escaping to moist weedy ground, 3500-11,340 ft. NC, NW, EC, C, WC, SE, SC. Introduced. Plants with awned lemmas have been called **var. aristatum** Willd. [*Lolium multiflorum* Lam.].

Lolium temulentum L.—Tufted annual, 10-120 cm; spikes 2-40 cm; spikelets with 3-20 florets; lemmas 3.5-8.5 mm, awnless or with awns to 23 mm; anthers 1.4-4 mm. Moist weedy ground, known only from Santa Fe Ski Basin, 10,350 ft. Santa Fe. Adventive.

MELICA L. – Melica

Tufted perennials, occasionally producing rhizomes; sheaths closed; auricles sometimes present; ligules thinly membranous; inflorescence a panicle with rebranched branches; spikelets with several fertile florets, terminating in a sterile rudimentary floret or florets; disarticulation above or below the glumes; lemmas 5- to several-nerved, awned or awnless; anthers (2)3.

1. Rudiments at end of rachilla clublike, not resembling the other florets, 2-3 mm…**M. nitens**
1′. Rudiments at end of rachilla resembling the other florets, 2-5 mm…**M. porteri**

Melica nitens (Scribn.) Nutt. ex Piper—Threeflower melicgrass. Shortly rhizomatous perennial, 55-130 cm; panicles 9-26 cm, the branches 3-6 cm and often spreading to reflexed; pedicels sharply bent below the spikelets; disarticulation below the glumes; spikelets 8-12 mm, awnless; lower glumes 5-9 mm; rudiments 2-3 mm, clublike. Calcareous soil and rocky outcrops in the Guadalupe Mountains, 3950-5040 ft. Eddy.

Melica porteri Scribn.—Porter's melicgrass. Shortly rhizomatous perennial, 55-100 cm; panicles 13-25 cm, the branches 1-9 cm and erect-appressed to strongly divergent; pedicels sharply bent below the spikelets; disarticulation below the glumes; spikelets 8-16 mm, awnless; lower glumes 3.5-6 mm; rudiments 1.8-5 mm. Mountain slopes and forest clearings, 6000-11,480 ft. Widespread except NW, NE, EC, SE. Plants with spreading panicle branches are referred to as **var. laxa** Boyle. Lincoln.

MELINIS P. Beauv. – Stinkgrass

Melinis repens (Willd.) Zizka [*Rhynchelytrum repens* (Willd.) C. E. Hubb.]—Natalgrass. Tufted annual or short-lived perennial, 30-100+ cm, the base decumbent and rooting at the nodes; sheaths open; auricles absent; ligules of hairs or membranous and ciliate; leaves glabrous to hairy; panicles 5-22 × 2-12 cm, pinkish to reddish and quite showy; disarticulation below the glumes, the spikelets falling, or sometimes also above the glumes and the upper floret falling; spikelets with 2 florets, 2-6 mm, the lower staminate or sterile, the upper fertile; upper floret forming a leathery seedcase; glumes as long as the spikelet, villous-hairy, the hairs to 7 mm, awnless to awned to 4 mm; upper floret 1.8-2.7 mm, awnless. Known from only 2 localities, 4860-4920 ft. Hidalgo. Introduced. The genus *Rhynchelytrum* has been submerged within *Melinis*.

MISCANTHUS Andersson – Silvergrass

Miscanthus sinensis Andersson—Eulalia. Perennial in large tussocks, with short rhizomes, 60-200+ cm; leaves mostly basal; sheaths open, glabrous; auricles absent, blades flat, to 20 mm wide, with conspicuous midnerves, 1-2 mm wide; panicles 15-25 cm, the branches 8-20+ cm; disarticulation below the glumes; spikelets all alike, with long callus hairs 6-12 mm, awns 6-12 mm, with 2 florets, the lower sterile, the upper fertile; anthers 2-3. Widely used as an ornamental landscape plant.

MNESITHEA Kunth – Pitgrass

Mnesithea granularis (L.) de Koning & Sosef [*Hackelochloa granularis* (L.) Kuntze]—Pitscale grass. Tufted annual, 20-80+ cm, with bulbous-based hairs; sheaths open, shorter than the internodes; auricles absent; ligules membranous, ciliate, 2-3 mm; blades 6-13 mm wide, cordate-based; racemes 5-27 mm, composed of pairs of sessile and pedicelled spikelets; sessile spikelets fertile, spherical, grenadelike, pitted, 1–1.3 mm, awnless; pedicelled spikelets extending beyond the sessile ones, flattened, ovate, 1.6–2.2 mm. Dry desert plains and foothills in the bootheel region (extreme SW); 1 record, 5450 ft. Hidalgo. Introduced. Our species was formerly placed in the genus *Hackelochloa*.

MUHLENBERGIA Schreb. – Muhly

Tufted to rhizomatous (rarely stoloniferous) perennials; sheaths open; auricles absent; ligules membranous, sometimes ciliolate; inflorescence a panicle, sometimes spikelike; spikelets with 1(2) florets; disarticulation above the glumes; glumes shorter than to longer than the floret; lemmas 3-nerved, awned or awnless; anthers usually 3. We have followed Peterson et al. (2010) in expanding *Muhlenbergia* to include the genera *Blepharoneuron*, *Lycurus*, and *Schedonnardus*, whose similarity to *Muhlenbergia* is immediately obvious.

1. Plants annual…(2)
1'. Plants perennial…(12)
2. First glume prominently 2-nerved, usually cleft; panicle branches falling as a unit, bearing 2–3(4) spikelets…(3)
2'. First glume 1-nerved; panicle branches persistent…(4)
3. Glumes ca. 1/2 the length of the floret; spikelets 4–6 mm; lemma awns (5–)10–20 mm…**M. brevis**
3'. Glumes and floret about equal in length; spikelets 2.5–3.5 mm; lemma awns 0.5–5(–10) mm…**M. depauperata**
4. Lemma awns 10–30 mm…(5)
4'. Lemma awns 0–5 mm…(6)
5. Second glume (1)2–3-nerved, the apex truncate to acute, 2–3-toothed…**M. peruviana**
5'. Second glume 1-nerved, the apex acute to acuminate…**M. tenuifolia** (in part)
6. Mature panicles narrow, contracted, the branches appressed to the main axis…**M. filiformis** (in part)
6'. Mature panicles open, the branches spreading…(7)
7. Glumes glabrous or nearly so…(8)
7'. Glumes minutely-pubescent to long-pubescent, at least at the apex (use a lens)…(9)
8. Pedicels 0.3–1 mm, stout, of equal thickness throughout; blades lacking white margins…**M. ramulosa**
8'. Pedicels 2–8 mm, capillary but straight, narrowed downward; blades with thickened white margins…**M. fragilis**
9. Terminal pedicels 2 mm, the lateral ones appressed to the branchlets…**M. eludens**
9'. Terminal pedicels mostly > 5 mm, the lateral ones spreading to flexuous…(10)
10. Pedicels sinuous, often tangled with one another; anthers 0.9–1.4 mm…**M. sinuosa**
10'. Pedicels straight or subflexuous, not tangled; anthers 0.3–0.5 mm…(11)
11. Lemmas awnless, 0.8–1.5 mm…**M. minutissima**
11'. Lemmas usually awned, 1.3–2 mm…**M. texana**
12. Second glume evidently 3-nerved, often 3-toothed; lower sheaths flattened, ribbonlike…(13)
12'. Second glume 1-nerved, entire or fringed; lower sheaths usually not ribbonlike…(15)
13. Sheaths usually becoming coiled and appearing like wood shavings; second glume acute, entire or occasionally toothed, nearly as long as the floret…**M. straminea**
13'. Sheaths not conspicuously coiled; second glume toothed to awned, shorter than the floret…(14)
14. Ligules 2–5 mm; stems and blades very slender and narrow; plants usually 15–30 cm…**M. filiculmis**
14'. Ligules 10–20 mm and the tip often shredded; stems and blades more robust; plants 25–80 cm…**M. montana**
15. Stems stiff, wiry, much branched, the plants bushlike…**M. porteri**
15'. Stems not as above, the plants not bushlike…(16)
16. Plants with evident, slender, creeping rhizomes…(17)
16'. Plants tufted, or sometimes the bases decumbent and spreading, but lacking creeping rhizomes…(31)
17. Callus hairs copious, as long as the body of the lemma…**M. andina**
17'. Callus hairs long-pubescent to glabrous, but the hairs much shorter than the body of the lemma…(18)
18. Awn of the lemma 6–25 mm…(19)
18'. Awn of the lemma 0–3(–5) mm…(21)
19. Blades mostly 2–6 mm wide, mostly flat…**M. mexicana** (in part)
19'. Blades 0.5–2(–2.5) mm wide, mostly rolled…(20)
20. Anthers purple, 1.3–3 mm; lemmas lanceolate, 3.5–5 mm, the awns 4–12(–20) mm; ligules with lateral lobes to 1.5 mm…**M. arsenei** (in part)
20'. Anthers orange, 1.5–2 mm; lemmas elliptic, 2–3.5 mm, the awns 10–25 mm; ligules lacking lateral lobes…**M. polycaulis** (in part)
21. Panicles open, loosely flowered with usually spreading to divergent branches at maturity…(22)

21'. Panicles contracted, narrow and usually densely flowered, the branches mostly erect to appressed…(24)
22. Awns 1–1.5(–2) mm; panicle branches attached in clusters…**M. pungens**
22'. Awns 0–0.3 mm; panicle branches not clustered…(23)
23. Ligules with pointed lateral extensions 1–2 mm; blades with thickened white margins and midribs…**M. arenacea**
23'. Ligules without lateral extensions; blades without thickened white margins or midribs…**M. asperifolia**
24. Blades (2.5–)3–6 mm wide, mostly flat…(25)
24'. Blades 0.5–2(–3) mm wide, rolled…(27)
25. Glumes 2–3.5 mm, subequal to the lemma…**M. mexicana** (in part)
25'. Glumes 4.5–6 mm, the awn-tips much exceeding the lemma…(26)
26. Internodes dull and puberulent, usually terete; culms seldom branched above the base; ligules 0.2–0.6 mm…**M. glomerata**
26'. Internodes polished and glabrous, keeled; culms much branched above the base; ligules 0.6–1.7 mm…**M. racemosa**
27. Lemmas long-pubescent below…(28)
27'. Lemmas glabrous or scabrous only…(30)
28. Blades 4+ cm; glumes acuminate or aristate…**M. glauca**
28'. Blades 2–4(–5) cm; glumes acute…(29)
29. Lemmas 2–2.5 mm; glumes about 1/2 as long as the floret…**M. villiflora**
29'. Lemmas 3–4 mm; glumes shorter than to nearly as long as the floret…**M. thurberi**
30. Inflorescence usually included in the sheath at least below, with 9 nodes or fewer; ligules 0.5–1.5 mm; glumes 1/2 as long as to equaling the floret…**M. repens**
30'. Inflorescence usually well exserted from the sheath, with 11–12 nodes; ligules 1–3 mm; glumes 1/3–1/2 as long as the floret…**M. richardsonis**
31. Panicles of long, strongly divergent, unbranched primary branches bearing widely spaced, sessile, awnless spikelets; blades usually spirally twisted…**M. paniculata**
31'. Panicles and blades not as above…(32)
32. Nerves of lemmas and paleas densely pubescent…**M. tricholepis**
32'. Nerves of lemmas and/or paleas glabrous, scabrous, or short-pubescent, but not densely or noticeably so…(33)
33. Sheaths (at least the lower) compressed-keeled; blades flat or folded…(34)
33'. Sheaths rounded on the back; blades usually becoming rolled…(38)
34. Panicles 20–40 cm; plants 50–100+ cm in large tussocks…**M. emersleyi**
34'. Panicles 5–10 cm; plants 20–60 cm in small tufts…(35)
35. First glume 1-nerved, awnless or with an awn to 1 mm; lemma awns 0.3–1 mm…(36)
35'. First glume 2-nerved, with awns 1–3.5 mm; lemma awns 1.5–3 mm…(37)
36. Ligules < 1 mm; glumes gradually narrowed to a mucro at most 0.3 mm…**M. cuspidata**
36'. Ligules 1–5 mm; glumes abruptly narrowed to awn 0.5–1 mm…**M. wrightii**
37. Blades terminating in a slender, hairlike bristle 3–12 mm; ligules acute to acuminate, 3–10 mm; culms erect…**M. alopecuroides**
37'. Blades acute or with a bristle 1–3 mm; ligules 1.5–3 mm, with lateral acuminate projections on either side; culms erect to ascending, often geniculate…**M. phleoides**
38. Lemma awns 0–4(–5) mm…(39)
38'. Lemma awns 7–40 mm…(46)
39. Glumes, excluding the awn, 3/4+ the length of the floret…(40)
39'. Glumes, excluding the awn, 2/3 or less the length of the floret…(41)
40. Ligules 1–3 mm…**M. rigens**
40'. Ligules 6–20 mm…**M. longiligula**
41. Blades 25–60 cm…(42)
41'. Blades 1–15 cm…(43)
42. Awns (3–)5–10 mm; panicles reddish; glumes 1.5–2 mm…**M. rigida** (in part)
42'. Awns 0–4(–5) mm; panicles greenish; glumes 2–3 mm…**M. dubia**
43. Mature panicles narrow, 0.5–1 cm wide, the primary branches erect to appressed…**M. filiformis** (in part)
43'. Mature panicles open, 4–15 cm wide, at least the primary branches widely spreading…(44)
44. Blades mostly flat, the margins white-cartilaginous…**M. arizonica**
44'. Blades mostly rolled or folded, rarely flat, the margins not white-cartilaginous…(45)
45. Blades strongly arcuate, curving, < 1 mm wide, 1–3(–4) cm; leafy portion 1/8 or less the length of the plant; lateral pedicels commonly longer than the spikelets…**M. torreyi**
45'. Blades rather straight, 1–2 mm wide, 3–15 cm; leafy portion 1/3–1/2 the length of the plant; lateral pedicels commonly shorter than the spikelet…**M. arenicola**
46. Awns 7–10 mm…(47)

46'. Awns 10–40 mm…(50)
47. Blades 20–60 cm…(48)
47'. Blades 1–14 cm; glumes acute to aristate…(49)
48. Glumes awned; panicles 8–20+ cm wide; introduced ornamental plants…**M. capillaris**
48'. Glumes awnless; panicles 2–4 cm wide; native plants in the wild…**M. rigida** (in part)
49. Blades mostly 1–4(–5) cm; glumes acute; lemmas and paleas sparsely but noticeably short-pilose on the lower 1/2; lateral lobes of ligules < 1.5 mm…**M. arsenei** (in part)
49'. Blades mostly 4–14 cm; glumes acuminate to aristate; lemmas and paleas glabrous or minutely scaberulous; lateral lobes of ligules 1.5–3 mm…**M. pauciflora** (in part)
50. Ligules 3–15 mm…(51)
50'. Ligules 0.5–3 mm…(52)
51. Lemmas purple, scaberulous near the apex; glumes 1–1.3 mm…**M. rigida** (in part)
51'. Lemmas straw-colored, smooth and shining; glumes 1.5–2.1 mm…**M. setifolia**
52. Glumes obtuse, 0.5–1 mm; lemma awn 20–40 mm…**M. spiciformis**
52'. Glumes acute to subaristate, 1–2 mm; lemma awn mostly 10–15 mm…(53)
53. Lemmas essentially glabrous, with only a few closely appressed callus hairs; ligules with lateral lobes 1.5–3 mm…**M. pauciflora** (in part)
53'. Lemmas pubescent on the lower 1/2; ligules without lateral lobes…(54)
54. Plant bases usually geniculate and rooting at some of the lowest nodes, sometimes erect, with the lowest sheaths nearly lacking blades; anthers 1.5–2 mm, orange; lemma hairs to 0.5 mm…**M. polycaulis** (in part)
54'. Plant bases usually erect, sometime geniculate, rarely rooting at the lowest nodes; anthers 1–1.5 mm, yellowish; lemma hairs 0.5–1.5 mm…**M. tenuifolia** (in part)

Muhlenbergia alopecuroides (Griseb.) P. M. Peterson & Columbus [*Lycurus setosus* (Nutt.) C. Reeder]—Wolftail. Tufted perennial, 30–60 cm; sheaths compressed-keeled; ligules 2–12 mm, long-acuminate, sometimes with a short cleft on either side; blades with prominent whitish midribs extending as fragile bristles 3–12 mm; panicles 4–10 cm, 5–8 mm wide, with spikelets borne in unequally pedicelled pairs; disarticulation below the spikelet pair; lower glume 2-nerved, awned to 1–3.5 mm; upper glume 1-nerved, awned to 5 mm; lemmas 3–4 mm, with an awn 1.5–3 mm. Dry slopes, plains, woodlands, 3280–9000 ft. Widespread.

Muhlenbergia andina (Nutt.) Hitchc.—Foxtail muhly. Rhizomatous perennial, 25–85 cm; panicles 2–15 cm, 0.5–3 cm wide, contracted and dense, the branches erect-ascending; spikelets 2–4 mm; glumes as long as the floret or longer; lemmas 2–3.5 mm, with long hairs 2–3.5 mm from the callus and base of lemma body; awns 1–10 mm. Mountain meadows, forest clearings, gravelly river beds, 5825–9500 ft. NW, Sandoval, San Miguel.

Muhlenbergia arenacea (Buckley) Hitchc.—Ear muhly. Rhizomatous perennial, 10–35 cm; ligules 0.5–2 mm, with pointed lateral lobes 1–2 mm; blades 1–5 cm, with thickened white margins; panicles 5–15 cm, nearly as wide; spikelets 1.5–2.6 mm, sometimes with 2 florets; glumes slightly shorter than the floret, awnless to mucronate; lemmas 1.5–2.5 mm, awnless or with a mucro to 0.3 mm. Playas and clay flats, often growing with *Scleropogon brevifolius*, 3305–7800 ft. EC, C, WC, SE, SC, SW.

Muhlenbergia arenicola Buckley—Sand muhly. Tufted perennial, 15–65 cm, the base decumbent; blades 4–15 cm, the margins not whitish-thickened; panicles 12–30 cm, 5–20 cm wide, 1/2+ the length of the plant; spikelets 2.5–4.2 mm; glumes shorter than the floret, acute to awn-tipped to 1 mm; lemmas 2.5–4.2 mm, with awns 0.5–4.2 mm. Sandy plains, 3300–6800 ft. Widespread except NC, NW.

Muhlenbergia arizonica Scribn.—Arizona muhly. Tufted perennial, lacking rhizomes, the shoots erect to decumbent, 15–40 cm; blades 4–7 cm, with whitish thickened margins; panicles 4–20 cm, 4–15 cm wide; spikelets 2–3 mm; glumes shorter than the floret, awnless; lemmas 2–3 mm, with short awns 0.5–1 mm. Moist plains and rocky hillsides, 4300–5510 ft. Hidalgo.

Muhlenbergia arsenei Hitchc.—Navajo muhly. Rhizomatous to tufted perennial, 10–50 cm, decumbent-based; ligules with lateral lobes; blades 1–6 cm; panicles 4–13 cm, 1–5 cm wide, loosely contracted; spikelets 3.5–5 mm; glumes shorter than the floret, awnless or with a short awn to 1.5 mm; lemmas 3.5–5 mm, pubescent on the lower 1/3–1/2, with awns 4–12 mm (rarely longer). Limestone rock outcrops, gypsum sands, stream banks, 4600–6500 ft. NC, C, NW. **R** [not pictured in plates]

Muhlenbergia asperifolia (Nees & Meyen ex Trin.) Parodi—Scratchgrass. Rhizomatous perennial, occasionally stoloniferous, 10–100 cm, branching profusely and bushlike; ligules < 1 mm, without lateral extensions; panicles 6–21 cm, 4–16 cm wide, breaking in the peduncle; spikelets 1–2 mm, occasionally with 2–3 florets; glumes shorter than the floret, awnless; lemmas 1–2 mm, glabrous, awnless to mucronate. Damp or wet ground along streams and rivers, floodplains, alkaline meadows, seeps and springs, 3300–11,420 ft. Widespread except far E.

Muhlenbergia brevis C. O. Goodd.—Short muhly. Tufted annual, 3–20 cm; blade margins whitish-thickened; panicles 3–12 cm, the branches bearing the spikelets in pedicellate pairs, which fall together or with other pairs on the short branch; spikelets 2.5–6 mm; glumes shorter than the floret, the lower glume 2-nerved, with teeth or awns to 2 mm, the upper glume 1-nerved, awned to 2 mm; lemmas 3.5–6 mm, with awns 10–20 mm. Grassy slopes and clearings in volcanic soils, 6000–8575 ft. WC, SC, SW.

Muhlenbergia capillaris (Lam.) Trin.—Pink muhly. Tufted perennial in expansive clumps, 60–100+ cm; panicles 20–60 cm, 5–30 cm wide, reddish to pinkish, open and diffuse, the branches and pedicels capillary; spikelets 3–5 mm; glumes shorter than the spikelets, short-awned; lemmas 3–5 mm, with awns 2–13(–15) mm. Introduced as an ornamental landscape plant, not known to escape to the wild.

Muhlenbergia cuspidata (Torr. ex Hook.) Rydb.—Tufted perennial with knotty bases, not rhizomatous, 20–60 cm; sheaths compressed-keeled; ligules 0.2–0.8 mm; panicles 4–14 cm, spikelike, < 1 cm wide; glumes shorter than the floret, 1-nerved, gradually acute, with a mucro to 0.3 mm; lemmas 2.5–3.6 mm, with a mucro to 0.6 mm. Plains and gravelly slopes, 6000–7860 ft. NE, NC, Eddy.

Muhlenbergia depauperata Scribn.—Sixweeks muhly. Tufted annual, 3–15 cm, branching at the nodes; sheaths often longer than the internodes; panicles 2–9 cm, < 1 cm wide; spikelets 2.5–5 mm, in unequally pedicelled pairs, the pair falling as a unit or with other pairs on the short branch; glumes equaling or exceeding the floret, the lower glume 2-nerved and 2-cleft or -awned; lemmas 2.5–4.5 mm, with an awn 6–15 mm. Grassy slopes and clearings in volcanic soils, 3800–9030 ft. NW, C, WC, SE, SC, SW.

Muhlenbergia dubia E. Fourn.—Pine muhly. Densely tufted perennial, forming large clumps, 30–100 cm; ligules 4–10 mm; blades 10–60 cm; panicles 10–40 cm, 0.5–2.5 cm wide; contracted, grayish green; spikelets 3.8–5 mm; glumes shorter than the floret, bluntish, awnless; lemmas 3.8–5 mm, awnless or awned to 6 mm. Woodlands, rocky mountain slopes, canyons, 8600–9500 ft. C, WC, SE, SC, SW, San Miguel.

Muhlenbergia eludens C. Reeder—Gravelbar muhly. Scant, tufted annual, 10–40 cm; sheaths longer than the internodes; panicles 10–20 cm, 3–7 cm wide; spikelets 1.7–3 mm; glumes shorter than the floret, sparsely short-hirsute, 1-nerved, awnless or awned to 1 mm; lemmas (1.7–)2–2.5 mm, with awns 1.2–3.5 mm. Rocky woodlands and forest clearings in the W mountains; 1 record, 8500 ft. Catron.

Muhlenbergia emersleyi Vasey—Bullgrass. Robust, tussocky perennial, lacking rhizomes, 50–150 cm; sheaths compressed-keeled; ligules 10–25 mm; blades 20–50 cm, flat or folded, 2–6 mm wide; panicles 20–45 cm, loosely contracted to open, 3–15 cm wide, the branches not bearing spikelets in the lower parts; spikelets 2.2–3.2 mm; glumes longer than the floret, 1-nerved; lemmas 2–3 mm, short-hairy on lower 3/4, with awns 1–15 mm. Rocky hills and woodlands, 3700–8000 ft. WC, SE, SC, SW.

Muhlenbergia filiculmis Vasey—Slimstem muhly. Tufted perennial, lacking rhizomes, 5–30 cm; sheaths becoming flattened and ribbonlike but not coiling like wood shavings; ligules 2–5(–8) mm; panicles 2–7 cm, 1–2 cm wide; spikelets 2.2–3.5 mm; glumes shorter than the floret, the lower 1-nerved and awn-tipped, the upper 3-nerved and 3-toothed; lemmas 2.2–3.5 mm, appressed short-hairy on lower 1/2, with awns 1–5 mm. Moist, sandy ground in high mountain grasslands and clearings, 2173–3050 ft. NC, NW, Catron.

Muhlenbergia filiformis (Thurb. ex S. Watson) Rydb.—Pull-up muhly. Tufted annual, sometimes appearing perennial, 5–35 cm, erect to geniculate-based and rooting at the lower nodes; sheaths shorter than the internodes; panicles 1–6 cm, < 1 cm wide, spikelike, the branches well-spaced; spikelets 1.5–3.2

mm; glumes shorter than the floret, 1-nerved, rounded to acute; lemmas 1.5–3 mm, appressed-hairy with hairs to 0.3 mm, awnless or with a mucro < 1 mm. Ponderosa–Douglas-fir forests, occasionally higher, 6560–9830 ft. NC, NW, WC.

Muhlenbergia fragilis Swallen—Delicate muhly. Tufted annual, 10–35 cm; ligules 1–3 mm, with long lateral lobes or projections on the sides; blade margins whitish-thickened; panicles 10–24 cm, 4–11 cm wide, open and diffuse; pedicels capillary, 2–8 mm, generally straight, narrowed downward from the spikelet; spikelets tiny, 1–1.2 mm; glumes shorter than or subequal to the floret, glabrous or nearly so, awnless; lemmas 1–1.2 mm, glabrous to hairy on the nerves, awnless. Moist sandy soil and rocky clearings, 4000–10,250 ft. NW, C, WC, SC, SW.

Muhlenbergia glauca (Nees) B. D. Jacks.—Desert muhly. Rhizomatous perennial, 25–60 cm, often decumbent; blades 4–12 cm, 1–2.5 mm wide; panicles 4–17 cm, 0.5–2.5 cm wide, contracted; spikelets 2.4–3.5 mm; glumes shorter than to subequal to the floret, 1-nerved, acuminate, with awns to 1 mm; lemmas 2.4–3.4 mm, hairy on the lower 1/2, short awned to 5 mm. Desert plains, 5320–8600 ft. Doña Ana, Hidalgo.

Muhlenbergia glomerata (Willd.) Trin.—Spike muhly. Rhizomatous perennial, 30–120 cm, seldom branched above the base; internodes dull, puberulent, terete; ligules 0.2–0.6 mm; blades 2–6 mm wide; panicles 2–12 cm, 0.5–2 cm wide, lobed; spikelets 3–8 mm; glumes subequal, much exceeding the floret (including the awn-tip); lemmas 2–3 mm, long-hairy on the lower 1/2, awnless or awn-tipped to 1 mm. Moist shaded ground in coniferous forest; 1 record, 8100 ft. Colfax.

Muhlenbergia longiligula Hitchc.—Longtongue muhly. Tufted, tussocky perennial, 60–130 cm; sheaths rounded; ligules (6–)10–30 mm; panicles 15–55 cm, 1–15 cm wide, loosely contracted; spikelets 2–3.5 mm; glumes mostly longer than the floret, awnless or with a mucro to 0.2 mm; lemmas 2–3 mm, awnless or awned to 2 mm. Canyons and rocky slopes, mostly in the SW region, 5200–8600 ft. WC, SW, Lincoln, Sierra.

Muhlenbergia mexicana (L.) Trin.—Mexican muhly. Rhizomatous perennial, 30–90 cm, much branched above the base; sheaths somewhat keeled; ligules 0.4–1 mm; blades 2–6 mm wide; panicles 3–20 cm, 0.5–3 cm wide, dense, interrupted; spikelets 1.5–3.8 mm; glumes equaling or slightly shorter than the floret, acuminate, awnless or awn-tipped to 2 mm; lemmas 1.5–3.8 mm, hairy on the lower 1/2, awnless or with awns to 10 mm. Moist thickets, woodlands, canyon bottoms, 5000–8400 ft. NE, C, WC, Lincoln.

Muhlenbergia minutissima (Steud.) Swallen—Least muhly. Tufted, annual, 5–40 cm, much branched from the base; ligules 1–2.5 mm, sometimes with lateral projections; blade margins not whitish-thickened; panicles 5–20 cm, 1.5–6 cm wide, open, diffuse; pedicels straight or curving somewhat, but not flexuous; spikelets 0.8–1.5 mm; glumes shorter than the floret, sparsely hairy, 1-nerved, acute; lemmas 0.8–1.5 mm, awnless. Moist, sandy or rocky slopes, 4600–10,650 ft. Widespread except NE, EC, C, SE.

Muhlenbergia montana (Nutt.) Hitchc.—Mountain muhly. Tufted perennial, 20–80 cm; sheaths becoming flattened and ribbonlike, usually not coiled like wood shavings but sometimes so; ligules 4–20 mm; panicles 5–25 cm, 1–6 cm wide; spikelets 3–7 mm; glumes shorter than the floret, the lower 1-nerved and awnless or nearly so, the upper 3-nerved and 3-toothed to shortly 3-awned; lemmas 3–7 mm, appressed-hairy on the lower portion, with awns 6–25 mm. Rocky or grassy slopes, ledges, forest clearings, 5500–11,390 ft. Widespread except EC, SE.

Muhlenbergia paniculata (Nutt.) Columbus [*Schedonnardus paniculatus* (Nutt.) Branner & Coville]—Tumblegrass. Tufted perennial, 10–55 cm; leaves mostly basal; sheaths compressed-keeled; blades usually spirally twisted in age, the margins whitish-thickened; panicles 8–45 cm, breaking at the base and rolling, the spikelike branches widely spaced, 5–16 cm; spikelets mostly sessile and widely spaced, 3–5.5 mm, awnless; glumes a bit shorter than the floret; lemmas 3–5 mm, awnless or awn-tipped. Plains and grasslands, 3600–9000 ft. Widespread.

Muhlenbergia pauciflora Buckley—New Mexico muhly. Tufted perennial, often bushy, sometimes rhizomatous or geniculate and rooting at the lower nodes, 30–70 cm; sheaths shorter than the internodes; ligules 1–5 mm, with lateral projections 1.5–3 mm longer than the central portion; blades 4–15 cm; panicles 3–15 cm, 1–3 cm wide, loosely contracted, the branches spikelet-bearing to the base; spikelets 3.5–5.5 mm, occasionally with 2 florets; glumes shorter than the floret, acuminate to awn-tipped to 2 mm; lemmas 3–5.5 mm, often purplish, glabrous except for the callus, with awns 2–25 mm. Rocky slopes, ledges, mountain outcrops, 4500–9055 ft. Widespread except NE, EC.

Muhlenbergia peruviana (P. Beauv.) Steud. [*M. pulcherrima* Scribn. ex Beal]—Tufted annual, 3–25 cm; ligules 1.5–3 mm; blades 1–5 mm, somewhat stiff and straight, at least when young; panicles 2–8 cm, 0.5–3.5 cm wide, contracted to somewhat open; spikelets 1.5–4.2 mm; glumes shorter than the floret, awnless, awn-tipped, or the upper one 2–3-toothed; lemmas 1.5–4.2 mm, sparsely short-hairy below, with awns 3–10 mm. Mountain meadows and ciénegas; 1 record, 8200 ft. Catron.

Muhlenbergia phleoides (Kunth) Columbus [*Lycurus phleoides* Kunth]—Wolftail. Tufted perennial, 20–50 cm; sheaths compressed-keeled; ligules 1.5–3 mm, acute to acuminate, with pointed lateral projections 1.5–4 mm on either side; blades with prominent whitish midribs extending as a bristle only as much as 3 mm; panicles 4–10 cm, 5–8 mm wide, with spikelets borne in unequally pedicelled pairs; disarticulation below the spikelet pair; lower glume 2-nerved, awned to 3.5 mm; upper glume 1-nerved, awned to 5 mm; lemmas 3–4 mm, with an awn 1.5–3 mm. Dry slopes, plains, woodlands, 3805–9130 ft. C, WC, SC, SW, Curry, Eddy, San Miguel.

Muhlenbergia polycaulis Scribn.—Cliff muhly. Loosely tufted perennial usually with geniculate-decumbent bases and rooting at some lower nodes, sometimes shortly rhizomatous, 15–45 cm; ligules 0.5–2.5 mm, without lateral projections; panicles 2–12 cm, 1–2 cm wide, interrupted; spikelets 2.5–4 mm; glumes shorter than the floret, acute to acuminate, awnless or awn-tipped to 1.5 mm; lemmas 2–3.5 mm, appressed-hairy on the lower portion, the hairs to 0.5 mm, awns 10–25 mm. Shaded ledges and grassy slopes, 5000–8000 ft. WC, SC, SW, Eddy.

Muhlenbergia porteri Scribn.—Bush muhly. Tufted, bushy perennial, lacking rhizomes, 25–100+ cm, the culms wiry, stiff, geniculate, zigzaggy; sheaths shorter than the internodes; ligules 1–4 mm, with short lateral lobes; panicles 4–14 cm, 6–15 cm wide, widely open and diffuse, breaking at the base and blowing in the wind; spikelets 3–4.5 mm; glumes shorter than the floret, acuminate to mucro-tipped; lemmas 3–4.5 mm, appressed-pubescent on the lower portion, with awns 2–13 mm. Dry plains, 2900–7000 ft. Widespread except NE, EC, WC.

Muhlenbergia pungens Thurb. ex A. Gray—Sandhill muhly. Rhizomatous perennial, 15–70 cm; sheaths gray-woolly proximally, glabrous distally; ligules < 1 mm, with short lateral lobes; blades 2–8 cm, stiff and pungent; panicles 8–20 cm, 3–14 cm wide, the primary branches widely spreading, the pedicels often appressed; spikelets 2.5–4.5 mm; glumes shorter than the floret, acuminate to awn-tipped to 1 mm; lemmas 2.5–4.5 mm, with an awn 1–2 mm; paleas 2-awned to 1 mm. Sand dunes and plains, 4000–9680 ft. NE, NC, NW, C, WC, SC.

Muhlenbergia racemosa (Michx.) Britton, Sterns & Poggenb.—Green muhly. Rhizomatous perennial, 30–110 cm, much branched above the middle of the shoot; internodes flattened; sheaths slightly keeled; ligules 0.6–1.7 mm; blades 2–5 mm wide, flat; panicles 2–16 cm, 0.5–2 cm wide, dense, lobed; spikelets 3–8 mm; glumes longer than the lemmas, awned to 5 mm; lemmas 2.2–3.8 mm, pilose on the lower 1/2, the hairs to 1 mm; awnless or awn-tipped to 1 mm. Canyon bottoms, riparian strands, irrigation ditches, moist prairies, roadsides, 3850–9500 ft. Widespread except NE, EC, SE.

Muhlenbergia ramulosa (Kunth) Swallen [*M. wolfii* (Vasey) Rydb.]—Red muhly. Tufted annual, 3–25 cm, much branched; ligules 0.2–0.5 mm, without lateral projections; panicles 1–9 cm, 1–3 cm wide, the primary branches spreading; pedicels tending to be appressed, 0.3–1.1 mm, stout, of equal thickness throughout; spikelets tiny, 0.8–1.3 mm; glumes shorter than the floret, glabrous, awnless; lemmas 0.8–1.3 mm, oval, awnless. Moist soil in forest clearings, 7215–10,460 ft. NW, C, WC, SC.

Muhlenbergia repens (J. Presl) Hitchc.—Creeping muhly. Rhizomatous perennial, 5-40 cm, the shoots decumbent-based, forming mats; ligules 0.2-1.8 mm; blades 1-6 cm; panicles 1-9 cm, < 1 cm wide, with few widely spaced spikelets, usually partially included in the subtending sheath; spikelets 2.6-4.2 mm, occasionally with 2 florets; glumes 1/2 as long as to equaling the floret; lemmas 2.6-4 mm, awnless or with a mucro to 0.3 mm. Flats, roadside swales, moist plains, 3400-8700 ft. Widespread except far E.

Muhlenbergia richardsonis (Trin.) Rydb.—Mat muhly. Rhizomatous perennial, 5-30 cm, shoots decumbent to erect, forming loose mats; ligules 0.8-3 mm; panicles 1-15 cm, < 2 cm wide, exserted, often spikelike; spikelets 1.7-3.1 mm, 1/3-1/2 as long as the floret, awnless; lemmas 1.7-3 mm, awnless to mucronate to 0.5 mm. Mountain meadows and ciénegas, 6000-10,050 ft. NC, NW, C, WC, Lincoln.

Muhlenbergia rigens (Benth.) Hitchc. [*M. mundula* I. M. Johnst.]—Deergrass. Tufted, tussocky perennial, 40 150 cm; sheaths terete basally; ligules 0.5-3 mm; blades 10-50 cm, stiff; panicles 15-60 cm, 0.5-1.5 cm wide, spikelike; spikelets 2.4-4 mm; glumes subequal to the floret, awnless; lemmas 2.4-4 mm, awnless or with a mucro to 1 mm. Dry woodland stream banks, rocky canyons, gullies, 4300-8050 ft. WC, SC, SW, Eddy, San Miguel.

Muhlenbergia rigida (Kunth) Kunth [*M. metcalfei* M. E. Jones]—Purple muhly. Tufted, tussocky perennial, 40-100 cm; lower sheaths terete; ligules 1-15 mm; blades 12-35 cm; panicles 10-35 cm, 2-10 cm wide, narrow but loose and sometimes quite open, purplish; spikelets 3.5-5 mm, purplish; glumes shorter than the floret, awnless; lemmas 3.5-5 mm, purplish, with awns (5-)10-22 mm. Rocky hillsides, canyon slopes, woodlands, 3700-7900 ft. WC, SE, SC, SW, Sandoval, San Miguel.

Muhlenbergia setifolia Vasey—Curlyleaf muhly. Tufted perennial, 30-80 cm; basal sheaths terete; ligules 4-10 mm; blades 5-20 cm, tightly rolled, curly; panicles 10-25 cm, 2-5 cm wide, loosely contracted; spikelets 3.5-5 mm, straw-colored to purplish; glumes shorter than the floret, awnless or with a short mucro, the lower one nerveless; lemmas 3.5-5 mm, smooth and shiny (except for the callus), with flexuous awns 10-30 mm. Dry gravelly plains and hillsides, juniper woodlands, 3300-7200 ft. C, WC, SE, SC.

Muhlenbergia sinuosa Swallen—Marshland muhly. Tufted, branched annual 15-40 cm; ligules 1.5-3 mm, with lateral projections longer than the central portion; blades mostly flat, puberulent adaxially; panicles 10-26 cm, 3-8 cm wide, open and diffuse; pedicels curved or sinuous, often tangled; spikelets 1.4-2 mm; glumes shorter than the floret, conspicuously hairy, awnless; lemmas 1.4-2 mm, awnless. Moist soil of canyon bottoms, riparian habitats, rocky hills, mostly in the C and W mountains. C, WC, SC, SW, Harding, Sandoval.

Muhlenbergia spiciformis Trin.—Long-awn muhly. Tufted perennial, lacking rhizomes, 25-80 cm, erect and wiry; ligules 1-3 mm; panicles 5-20 cm, 0.5-3 cm wide, narrow; spikelets 3-4 mm; glumes much shorter than the floret, awnless; lemmas 2.8-4 mm, slightly appressed-hairy below, with awns 10-40 mm. Canyons and moist woodlands, 6000-8000 ft. Eddy, Lincoln.

Muhlenbergia straminea Hitchc. [*M. virescens* (Kunth) Trin.]—Screwleaf muhly. Tufted perennial, 25-70 cm; sheaths becoming flattened and ribbonlike, coiling like wood shavings in age; ligules 6-20 mm; panicles 8-25 cm, 0.5-3 cm wide, loosely contracted; spikelets 3.5-7 mm; glumes awnless or shortly awn-tipped, the upper equaling or exceeding the floret, 3-nerved and sometimes 3-toothed; lemmas 3.5-6 mm, hairy on the lower 1/2, with awns 12-30 mm. Rocky slopes and clearings, mostly in pine forests, 5900-8930 ft. WC, SW, Sandoval.

Muhlenbergia tenuifolia (Kunth) Kunth [*M. monticola* Buckley]—Mesa muhly. Tufted annual to short-lived perennial, 20-70 cm; sheaths shorter than the internodes; ligules 1.2-5 mm; panicles 7-20 cm, 0.5-3 cm wide, lax, often nodding, interrupted; spikelets 2-4 mm; glumes shorter than the floret, awnless or awn-tipped; lemmas 2.3-4 mm, hairy on the lower 1/2, with an awn 10-30 mm. Rocky ledges and outcrops, canyons, sandy drainages, 3700-8915 ft. C, WC, SE, SC, SW. McKinley.

Muhlenbergia texana Buckley—Texas muhly. Slender tufted annual, 10–35 cm; ligules 1–2.5 mm; panicles 9–21 cm, 2–7 cm wide, loosely diffuse; spikelets 1.3–2 mm; glumes shorter than the floret, sparsely short-hairy, awnless or with a mucro; lemmas 1.3–2 mm, short-appressed-hairy on the lower 1/2, awnless or with an awn to 2 mm. Rocky outcrops, sandy drainages, disturbed ground, 5600 ft. Catron, Grant, Hidalgo.

Muhlenbergia thurberi (Scribn.) Rydb.—Thurber's muhly. Rhizomatous perennial, 12–36 cm, the aerial stems densely clustered at the base of the shoot; sheaths shorter than the internodes; ligules 0.9–1.2 mm; blades 0.5–4 cm, tightly rolled; panicles 1–6 cm, < 1 cm wide; spikelets 2.6–4 mm; glumes shorter than the floret, awnless; lemmas 2.6–4 mm, hairy on the lower 3/4, awnless or with an awn to 1 mm. Dry hills, 5700–8300 ft. NW, WC, Taos.

Muhlenbergia torreyi (Kunth) Hitchc. ex Bush—Ring muhly. Tufted perennial lacking rhizomes, the base decumbent, 10–45 cm; leaves concentrated at the base of the plant, the leafy portion in the lower 1/4 or less; blades 1–5 cm, tightly folded or rolled, strongly curving, the margins not whitish-thickened; panicles 7–21 cm, 3–15 cm wide; spikelets 2–3.5 mm; glumes shorter than the floret, awnless or with an awn to 1 mm; lemmas 2.3–3.5 mm, with awns 0.5–4 mm. Sandy plains, 3850–10,200 ft. Widespread.

Muhlenbergia tricholepis (Torr.) Columbus [*Blepharoneuron tricholepis* (Torr.) Nash]—Pine dropseed. Plants perennial, tufted, 15–65 cm; panicle 3–22 cm, the branches spreading; pedicels capillary, sinuous; spikelets with a single floret, grayish, 1.8–3 mm; disarticulation above the glumes; lemma slightly longer than the glumes, 3-nerved, silky-hairy on the nerves and margins, awnless; palea shorter than the lemma, densely silky-hairy; anthers 3. Widespread on rocky or gravelly slopes in the mountains and foothills, 5340–12,700 ft. Widespread except E.

Muhlenbergia villiflora Hitchc.—Hairy muhly. Rhizomatous perennial, 5–30 cm; sheaths shorter than the internodes; ligules 0.5–1.5 mm; blades tightly rolled; panicles 1–5 cm, < 1 cm wide, not dense, usually exserted from the subtending sheath; spikelets 1.4–2.5 mm; glumes shorter than the floret, glabrous, awnless; lemmas 1.4–2.5 mm, villous nearly throughout, awnless or with a short mucro to 0.5 mm. Our plants belong to **var. villosa**. Dry plains, 3885–4155 ft. Eddy, Otero. **R**

Muhlenbergia wrightii Vasey ex J. M. Coult.—Spike muhly. Tufted perennial, 15–60 cm; sheaths compressed-keeled; ligules 1–5 mm; panicles 5–16 cm, 1 cm wide or less, spikelike, interrupted below; spikelets 2–3 mm; glumes 3/4 or less the length of the lemma, abruptly acuminate to an awn-tip to 1 mm; lemmas 2–3 mm, minutely hairy on the lower 3/4, awnless or with an awn-tip to 1 mm. Plains, grassy hills and slopes, 4000–9400 ft. Widespread except EC, SE.

MUNROA Torr. – False buffalograss

Stoloniferous annuals or perennials, often mat-forming, the stolons ending in fascicles of leaves from which new shoots arise; sheaths open, with a tuft of hair at the throat; auricles absent; ligules a ring of hairs; blades flat or involute; inflorescence a dense, headlike cluster of spikelike branches, the branches hidden in the subtending sheath and bearing a few spikelets; spikelets with few to several florets, the lower fertile, the upper sterile; disarticulation both above the glumes and between the florets or beneath the leaves subtending the panicle; glumes subequal to the adjacent lemmas; lemmas 3-nerved, 2-lobed, pilose along the nerves; paleas subequal to the lemmas; anthers 2–3. We include the genus *Dasyochloa* here.

1. Plants perennial; blades rolled; glumes longer than the lower lemma…**M. pulchella**
1'. Plants annual; blades flat; glumes shorter than the lower lemma…**M. squarrosa**

Munroa pulchella (Kunth) L. D. Amarilla [*Dasyochloa pulchella* (Kunth) Willd. ex Rydb.]—Fluffgrass. Stoloniferous perennial (though sometimes short-lived), 2–15 cm, the peduncle/stolon 3–10 cm; blades 2–6 cm, involute; panicles 1–2 cm, densely white-hairy; glumes 6–9 mm; lemmas 3–6 mm, the apex 2-lobed to ca. 3 mm, awned from the midnerve 2–4 mm. Rocky desert flats and hills, 3045–7000 ft. Widespread except NE, NC.

Munroa squarrosa (Nutt.) Torr.—False buffalograss. Annual, producing short stolons, 3–15(–30) cm; blades 1–5 cm, flat, 1–2 mm wide; spikelets with 3–5 florets; glumes of lower spikelets 2.5–4 mm, acute; lemmas of lower spikelets with tufts of hair on the lateral nerves, the midnerve extended into a mucro or awn-tip 0.5–2 mm. Sandy plains and flats, 6300–9000 ft. Widespread.

NASSELLA (Trin.) E. Desv. – Needlegrass

Mostly tufted perennials, rarely annual or rhizomatous; sheaths open; auricles absent; ligules membranous, sometimes ciliate; inflorescence generally a rebranching panicle; spikelets with 1 floret, the rachilla not prolonged; disarticulation above the glumes; glumes 3–5-nerved, longer than the floret; lemmas 5-nerved, convolute, wrapped around the palea and grain; awns present; paleas to 1/2 the length of the lemma; anthers 1 or 3. Formerly treated in the genus *Stipa*.

1. Awns 4–10 cm, capillary; florets 2–3 mm; summit of sheath glabrous or obscurely pubescent...**N. tenuissima**
1'. Awns 2–3 cm, stout; florets 4–6 mm; summit of sheath with a conspicuous tuft of hair . **N. viridula**

Nassella tenuissima (Trin.) Barkworth—Mexican feathergrass. Tightly tufted perennial, 25–100 cm; sheaths without conspicuous hairs at the summit; ligules 1–5 mm; blades to 60 cm, convolute and very thin, to 1.5 mm wide; panicles 8–50 cm, loosely contracted; glumes 5–13 mm, aristate; florets 2–3 mm; awns 4–10 cm. Rocky slopes and woodlands, 4650–7800 ft. C, WC, SE, SC, SW, Colfax.

Nassella viridula (Trin.) Barkworth—Green needlegrass. Tufted perennial, 35–120 cm; sheath margins ciliate, with a conspicuous tuft at the summit; ligules 0.2–1.2 mm; blades 10–30 cm, 1.5–3 mm wide, flat to rolled; panicles 3–7 cm, loosely contracted; glumes 7–13 mm, apiculate; florets 3.5–6 mm; awns 2–3 cm. Grassy hills, plains, flats, 5550–8250 ft. NE, NC, NW, Cibola.

ORYZOPSIS Michx. – Ricegrass

Oryzopsis a monotypic genus in North America. Former members will now be found in the genera *Eriocoma* and *Piptatheropsis*.

Oryzopsis asperifolia Michx.—Mountain ricegrass. Tufted perennial, 25–65 cm; leaves mostly basal; sheaths open; auricles absent; ligules membranous, < 1 mm; blades of basal leaves 30–90 cm, 4–9 mm wide, the bases twisted so the adaxial surface is uppermost, remaining green through the winter; panicles 3–13 cm; spikelets with 1 floret, the rachilla not prolonged; disarticulation above the glumes; glumes subequal to the floret, 6–10-nerved, 5–7.5 mm, subequal to the lemma; lemmas 5–7 mm; awns straight, deciduous, 7–15 mm; anthers 2–4 mm. Moist wooded sites in the N mountains, usually in the shade, 6840–10,205 ft. NC.

PANICUM L. – Panicum, panicgrass

Annuals and perennials of various habit; basal leaves not forming a basal rosette; sheaths open; auricles absent; ligules membranous, usually ciliate; inflorescence a panicle of various arrangement, bristles absent; spikelets dorsally compressed, awnless, with 2 florets, the lower staminate or sterile, the upper fertile; disarticulation below the glumes and the spikelet falling, sometimes the upper floret falling out; glumes usually unequal, many-nerved, the lower often short but sometimes as along as the spikelet; upper lemma similar in size and texture to the lower, becoming indurate, clasping the edges of the palea, together forming a seedcase; anthers 3.

1. Plants annual...(2)
1'. Plants perennial...(9)
2. Lemma of the upper floret wrinkled; spikelets nearly sessile on simple or nearly simple primary branches ...go to **Urochloa**
2'. Lemma of the upper floret smooth, not wrinkled; spikelets pedicelled in a usually open, freely rebranched panicle...(3)
3. First glume ca. 1/4 as long as the spikelet, obtuse or rounded at the tip; stems as much as 1 m, coarse and often somewhat trailing...**P. dichotomiflorum**

3'. First glume > 1/4 as long as the spikelet, acute to acuminate at the tip; stems various…(4)

4. Spikelets 4.5-5 mm; panicle nodding at maturity…**P. miliaceum**

4'. Spikelets < 4 mm; panicle usually not nodding…(5)

5. Mature panicles 2-3 cm and congested among the leaves, never exceeding the foliage; plants 2-8 cm …**P. mohavense**

5'. Mature panicles longer, exceeding the leaves; plants usually taller…(6)

6. Mature panicles > 1/2 the length of the entire plant; panicle axils pubescent…**P. capillare**

6'. Mature panicles not > 1/3 the length of the entire plant; panicle axils glabrous…(7)

7. Palea of lower floret well developed, as long as the upper floret; first glume 1/3-1/2 the length of the spikelet…P. stramineum

7'. Palea of lower floret 1/2 or less the length of the upper floret; first glume 1/2 to nearly equaling the length of the spikelet…(8)

8. Upper floret ovoid to ellipsoid, not stipitate, lacking thickenings at the base, but with 2 small scars, the base with a cavity when mature, the palea usually bulging outward at the base…**P. hirticaule**

8'. Upper floret obovoid at maturity, shortly stipitate, with 2 fleshy thickenings at the base, the base lacking a cavity, the palea not protruding but even with the lemma…**P. alatum**

9. Terminal spikelet of each branch subtended by 1+ bristles (vestigial branchlets)…go to **Setaria**

9'. Terminal spikelets not subtended by a bristle…(10)

10. First glume about as long as the second; primary panicle branches mostly unbranched; long stolons developed…go to **Hopia**

10'. First glume shorter than the second; primary panicle branches often rebranched; stolons not developed …(11)

11. Spikelets 4-8 mm…(12)

11'. Spikelets < 4 mm…(15)

12. Spikelets 6-8 mm…**P. havardii**

12'. Spikelets 4-5(-6) mm…(13)

13. Panicles narrow, contracted…**P. amarum**

13'. Panicles open, not contracted…(14)

14. Plants with stout scaly rhizomes; blades not curling…**P. virgatum** (in part)

14'. Plants tufted, lacking rhizomes; blades often curling…**P. hallii** (in part)

15. Palea of the lower floret inflated, enlarged, obovate, forcing the spikelet to gape open…go to **Steinchisma**

15'. Palea of the lower floret not inflated as above, the spikelet closed (except open somewhat during anthesis)…(16)

16. Stems hard and somewhat woody in age, becoming much branched above; basal buds silky long-pubescent; spikelets 2.5-3 mm…**P. antidotale**

16'. Stems not hard and woody, or if so, then not much branched above; basal buds not silky long-pubescent …(17)

17. Spikelets appressed and usually closely clustered on simple or nearly simple panicle branches or on short spur branches…(18)

17'. Spikelets not appressed on simple panicle branches, the pedicels and branches spreading and open… (19)

18. Lower floret staminate, producing anthers, which are usually visible; plants usually dark green, the blades rarely curling…**P. coloratum**

18'. Lower floret neuter, anthers not produced; plants usually bluish green, the blades often curling…**P. hallii** (in part)

19. Second glume and lower lemma 5-nerved; sheaths keeled; culms conspicuously swollen and bulblike at the base in many (but not all) populations…go to **Zuloagaea**

19'. Second glume and lower lemma 7-11-nerved; sheaths not keeled; culms not swollen and bulblike at the base, though they may be thickened in *P. virgatum*…(20)

20. Plants with stout, scaly rhizomes; blades usually not curling…**P. virgatum** (in part)

20'. Plants lacking rhizomes; blades often curling…**P. hallii** (in part)

Panicum alatum Zuloaga & Morrone—Winged panicgrass. Glabrous to hispid, tufted annual, 8-65+ cm; ligules of hairs, 0.7-1.8 mm; blades 3-17 mm wide, sparsely pilose; panicles 4-23 cm, 2-11 cm wide, the branches stiffly spreading; pedicels and terminal branchlets tending to be appressed; spikelets 2.4-4.5 mm; lower glume 1/2-3/4 the length of the spikelet; upper floret obovoid when mature, dull and papillose to smooth and shiny, very shortly stipitate, with 2 fleshy thickenings at the base, the base lacking a cavity and the palea not protruding but even with the lemma. Sandy to clayey disturbed ground, roadsides, swales; in the bootheel region, 3850-5500 ft. Doña Ana, Grant, Hidalgo. Plants with the upper floret smooth and shiny may be called **var. minus** (Andersson) Zuloaga & Morrone. Hidalgo.

Panicum amarum Elliott—Bitter panicgrass. Rhizomatous perennial, glabrous and glaucous throughout, 30–200+ cm; ligules 1–5 mm; blades 2–13 mm wide; panicles 10–80 cm, 2–17 cm wide, loosely contracted; spikelets 4–7.7 mm, glabrous; lower glume 1/2 to nearly as long as the spikelet; upper glume and lower lemma exceeding the fertile floret by 1.5–3 mm; lower floret staminate; fertile floret 2.4–4 mm, smooth, shiny, the lemma clasping the palea only at the base. Planted for erosion control near Zuñi, ca. 6300 ft. McKinley. Introduced.

Panicum antidotale Retz.—Blue panicum. Rhizomatous, glaucous perennial, 50–300 cm, becoming somewhat woody and much branched in age; scales of the rhizomes densely brownish-pilose; basal buds silky long-hairy; nodes swollen, sometimes hairy; ligules 0.3–1.5 mm; blades 3–20 mm wide; panicles 10–45 cm, ca. 1/2 as wide; spikelets 2.4–3.4 mm, glabrous; lower glume 1/2 or less the spikelet length; lower floret staminate; fertile floret 1.8–2.8 mm, smooth, shiny. Introduced for range restoration, 3550–4900 ft. Chaves, Doña Ana, Eddy, Luna, Rio Arriba. Adventive.

Panicum capillare L.—Witchgrass. Often densely hirsute or hispid annual, 15–130 cm, the hairs bulbous-based, ; ligules membranous-ciliate, 0.5–1.5 mm; blades 3–18 mm wide; panicles 13–50 cm, 7–24 cm wide, very open and diffuse, usually > 1/2 the length of the plant, the bases often in the subtending sheath, breaking and rolling in the wind when mature, or remaining intact; spikelets 2–4.5 mm; lower glume 1/2 or less the spikelet length; lower floret sterile, the lemma extending beyond the fertile floret by 0.5–1 mm, the palea present or absent; fertile floret sometimes falling out of the spikelet. Roadsides and other disturbed sites, 3700–9210 ft. Widespread.

Panicum coloratum L.—Kleingrass. Tufted perennial, but also with short rhizomes, 50–140 cm; shoots with abundant foliage; sheaths shorter than the internodes; ligules 0.5–2 mm; blades 2–8 mm wide, flat; panicles 5–30 cm, 3–14 cm wide; spikelets 2.5–3.5 mm, glabrous; lower glume ca. 1/3 the length of the spikelet; lower floret staminate, the palea 2–3 mm; fertile floret 2–2.5 mm, smooth, shiny. Introduced for irrigated pastures, escaping along roadsides, 3880–5425 ft. DeBaca, Doña Ana, Eddy. Adventive.

Panicum dichotomiflorum Michx.—Fall panicum. Tufted annual or short-lived perennial under mild conditions, 10–200 cm, geniculate-based and rooting at the lower nodes in wet soil or water; culms often zigzagging and somewhat flattened; sheaths compressed, inflated; ligules 0.5–2 mm; blades 3–25 mm wide, glabrous to pilose; panicles both terminal and axillary, 5–40 cm, diffuse when mature; spikelets 1.8–3.8 mm, glabrous; lower glume very short, ca. 1/4 the spikelet length; lower floret sterile, the palea vestigial to well developed; fertile floret 1.5–2.5 mm, smooth, shiny. Moist stream banks, meadows, roadsides, 4925–6080 ft. NW, Doña Ana, Grant, Hidalgo. Introduced.

Panicum hallii Vasey—Hall's panicgrass. Tufted perennial, 10–100 cm; leaves clustered at the base of the plant; ligules 0.6–2 mm; blades 2–10 mm wide, mostly flat, generally curling in age; panicles 7–30 cm, 3–15 cm wide, the branches ascending, the pedicels appressed (common) to spreading (var. *filipes*); spikelets 2–4.2 mm, glabrous; lower glume 1/2–3/4 the spikelet length; lower floret sterile, the palea nearly as long as its floret; fertile floret 1.5–2.5 mm, smooth, darkening. Plains and rocky slopes, foothills, often on limestone, also clay swales and flats.

1. Spikelets mostly appressed along the primary panicle branches; sheaths mostly papillose-hirsute...**var. hallii**
1'. Spikelets mostly spreading from the panicle branches, the panicle open; sheaths mostly glabrous...**var. filipes**

var. filipes (Scribn.) F. R. Waller—Hall's panicgrass. Dry plains in the SE corner of the state, 3630–6800 ft. Eddy.

var. hallii—Hall's panicgrass. E and S plains, 3300–7545 ft. NE, NC, EC, C, WC, SE, SC.

Panicum havardii Vasey—Havard's panicgrass. Robust rhizomatous perennial, 0.6–1.5 m, with decumbent base, shoots arising singly or in small clusters, glabrous throughout, often glaucous; sheaths longer than the internodes; ligules a ciliate membrane, 2–4 mm; blades 5–10 mm wide; panicles 17–40 cm, nearly as wide, the branches stiffly ascending to spreading; spikelets 6–8 mm; lower glume 1/2–2/3 the spikelet length; upper glume and lower lemma exceeding the fertile floret by 1–2 mm; fertile floret 4.5–5 mm. Sandy plains and dunes, 3050–4000 ft. SE, Lincoln, San Miguel.

Panicum hirticaule J. Presl—Mexican panicgrass. Tufted, branched annual, 15–75+ cm; foliage generally hispid with bulbous-based hairs; sheaths shorter than the internodes; ligules 1–1.5 mm; blades 0.5–2 mm wide; panicles 1/3 or less the length of the plant, 9–20 cm, 5–10 cm wide, glabrous in the axils; pedicels generally appressed along the branchlets; spikelets 1.8–4 mm, glabrous; lower glume 1/2–3/4 the length of the spikelet; lower floret sterile, with a palea 1/2 or less the length of the adjacent fertile floret; fertile floret 1.5–2.4 mm, smooth, shiny to dull, with 2 small scars at the base but lacking fleshy thickenings; palea of fertile floret usually bulging outward at the base. Rocky to sandy slopes, plains, washes, 3770–7745 ft.

1. Fertile floret dull, covered with conspicuous papillae on both lemma and palea...**var. verrucosum**
1′. Fertile floret shiny, smooth, with sparse papillae only at the tip of the palea...**var. hirticaule**

var. hirticaule—Mexican panicgrass. WC, SE, SC, Luna, Quay, San Juan.

var. verrucosum Zuloaga & Morrone—Mexican panicgrass. Hidalgo.

Panicum miliaceum L.—Proso millet. Tufted annual, 20–150+ cm; sheaths densely pilose with bulbous-based hairs; ligules 1–3 mm; blades 7–25 mm wide; panicles 6–20 cm, ca. 1/2 as wide, heavy, often drooping; spikelets 4–6 mm, glabrous; lower glume 1/2–3/4 the spikelet length; lower floret sterile, the palea no more than 1/2 the length of the adjacent fertile floret; fertile floret 3–3.8 mm, smooth to faintly lined, shiny, blackish to straw-colored, persisting in the spikelet or disarticulating singly. Waste places, sometimes under bird feeders, 3900–7400 ft. Scattered. Adventive.

Panicum mohavense Reeder—Mojave panicgrass. Tiny annual, 2–8 cm, the foliage generally hispid with bulbous-based hairs; ligules 0.2–0.4 mm; blades 1–4 cm, 1–3 mm wide, flat; panicles down among the blades, about as wide as long; spikelets 2–2.2 mm, glabrous, plump, obtuse-acute; lower glume ca. 1/3 the spikelet length; lower floret sterile, with a tiny palea; fertile floret 1.4–1.8 mm. Limestone ridges of the Oscura Mountains, 8000 ft. Socorro. **R**

Panicum stramineum Hitchc. & Chase—Sonoran panicgrass. Tufted annual, 10–70+ cm; sheaths glabrous to hirsute with bulbous-based hairs; ligules 1–2.5 mm; blades 3–20 mm wide; panicles 3–30 cm, 4–15 cm wide; spikelets 2.3–3.2 mm; lower glume 1/3–1/2 the length of the spikelet; palea of the lower floret as long as or longer than the adjacent fertile floret; fertile floret 1.5–2.5 mm. Reported but no specimens seen.

Panicum virgatum L.—Switchgrass. Rhizomatous perennial, 40–200+ cm, the culms solitary or more commonly forming dense clumps; sheaths longer than the lower internodes, glabrous or pilose, the margin usually ciliate; ligules 2–6 mm; blades 2–15 mm wide; panicles 10–50 cm, ca. 1/2 as wide; spikelets (3–)4–8 mm, glabrous, acuminate; lower glume 1/2 to nearly as long as the spikelet, acuminate; upper glume and lower lemma extending 1–3 mm beyond the upper floret; lower floret staminate, the palea well developed; fertile floret 2.3–3 mm, smooth, shiny. Moist plains, prairies, meadows, roadsides, 3600–8180 ft. NE, NC, EC, C, WC, SE, SC, SW.

PAPPOPHORUM Schreb. – Pappusgrass

Pappophorum vaginatum Buckley [*Panicum mucronulatum* Mez]—Tufted glabrous perennial, 40–100 cm; sheaths open, with a tuft of hairs at the throat; auricles absent; ligules a ring of hairs; panicles 10–25 cm, whitish or tawny, rarely pinkish; disarticulation above the glumes, all florets falling together; spikelets with 3–5 florets, 1–2 lower fertile florets and 2 reduced florets; glumes 1-nerved, reaching at least to the awns of the florets, 3–4.5 mm, glabrous; lemmas ± 7-nerved, the nerves extending into minutely hairy awns, fertile lemma bodies ca. 3 mm; awns about twice as long as lemma bodies; anthers 3. Infrequent in the S plains and foothills, 3180–3650 ft. Doña Ana, Eddy, Otero.

PAPPOSTIPA (Speg.) Romasch., P. M. Peterson & Soreng – Needlegrass

Pappostipa speciosa (Trin. & Rupr.) Romasch. [*Achnatherum speciosum* (Trin. & Rupr.) Barkworth; *Stipa speciosa* Trin. & Rupr.]—Desert needlegrass. Tufted perennial, 30–60 cm; basal sheaths open, reddish brown, becoming flat and ribbonlike in age; auricles absent; ligules variable, those of the lower leaves densely ciliate with hairs 0.3–1 mm; panicles 10–15 cm, contracted, usually partially included in the subtending sheath; spikelets with a single floret, 16–24 mm (excluding awns); glumes large, subequal, longer than the floret (except the awn); florets fusiform, 6–10 mm; awns 3.5–8 cm, once-geniculate, with hairs 3–8 mm on the basal segment, the terminal segment glabrous; lemmas rolled around the palea, but the margins only slightly overlapping; awns 2–10+ cm, strongly once-geniculate, the basal segment densely pilose with hairs 3–10 mm; paleas 1/2 as long as to subequal to the lemma. Desert canyons and rocky hills; known in New Mexico only from San Juan and Sandoval Counties, 5800–9000 ft.

PASCOPYRUM Á. Löve – Wheatgrass

Pascopyrum smithii (Rydb.) Barkworth & D. R. Dewey [*Agropyron smithii* Rydb.; *Elymus smithii* (Rydb.) Gould; *Elytrigia smithii* (Rydb.) Nevski]—Western wheatgrass. Rhizomatous perennial, 25–100+ cm, leaves tending toward the base; foliage with a bluish cast; sheaths open; auricles present; ligules tiny, ca. 0.2 mm, membranous; blades rigid, prominently veined on the upper surface; disarticulation above the glumes and between the florets; spikes mostly with 1 spikelet per node, 5–18 cm; spikelets with several florets, 12–26 mm, ascending, sometimes curving outward; glumes ca. 1/2 the length of the spikelet, stiff, curving to 1 side; lemmas lanceolate, 6–12 mm, glabrous to hairy, acute to awned to 5 mm; anthers 3, 4–6 mm. Plains, swales, grassy hills and slopes, forming thick stands often with a bluish tint, 3700–11,000 ft. Widespread.

PASPALUM L. – Crowngrass

Tufted, rhizomatous, or stoloniferous annuals and perennials; sheaths open; auricles usually absent (ours), sometimes present; ligules membranous or absent (ours); inflorescence a panicle of spikelike branches, sometimes rebranching shortly; spikelets dorsally compressed, flat on 1 side and rounded on the other, borne on 1 side of the branch, awnless, with 2 florets, the lower sterile, the upper fertile; glumes unequal, the lower vestigial to absent, the upper as long as the spikelet and similar to the lower lemma in size and texture; upper floret fertile, indurate at maturity, the lemma edges clasping the palea.

1. Inflorescence branches 2, attached < 1 cm apart (1 or 2 additional branches occasionally present below) …**P. distichum**
1'. Inflorescences branches 1 to numerous, when 2 then > 1 cm apart…(2)
2. Spikelets 3–4 mm, the margins conspicuously ciliate with soft hairs…**P. dilatatum**
2'. Spikelets 1.5–2.6 mm, the margins glabrous or minutely pubescent…**P. setaceum**

Paspalum dilatatum Poir.—Dallisgrass. Tufted perennial with short rhizomes, 40–125+ cm; sheaths glabrous or pubescent; ligules 1.5–3.8 mm; blades to 16 mm wide; panicles with 2–7 branches, 2–12 cm; spikelets 2.3–4 mm, ovate, pilose; lower glume absent. Introduced as a pasture grass and persisting along roadsides and in old moist fields and waste places, 3665–6400 ft. WC, SE, SC.

Paspalum distichum L.—Knotgrass. Tufted or rhizomatous perennial, 10–65 cm; sheaths glabrous to weakly pilose distally; ligules 1–2 mm; panicles with 2 branches, 2–7 cm, attached < 1 cm apart, a third branch sometimes present below; spikelets 2.4–3.2 mm, elliptic, glabrous; lower glume absent or to 1 mm. Weedy spots along ditch banks, ponds, slow-moving streams, sloughs, 3235–6770 ft. NC, NW, EC, C, WC, SE, SC, SW. Introduced.

Paspalum setaceum Michx.—Thin paspalum. Tufted perennial, sometimes with short rhizomes, 25–110 cm; sheaths glabrous or pubescent; ligules 0.2–0.5 mm; panicles with 1–6 branches, 2–15 cm; spikelets 1.4–2.6 mm, ovate to nearly orbicular, glabrous to minutely hairy; lower glume absent. Sandy plains and dunes, scattered locales, more frequent on the E plains, 3325–6500 ft. NE, NC, EC, C, SE, SC, Hidalgo.

PENNISETUM Rich. – Fountaingrass

Annuals and perennials; sheaths open; auricles absent; ligules a ciliate membrane or a ring of hairs; inflorescence bristly, spikelike, the spikelet clusters in fascicles or burs of reduced panicle branches; burs sessile to stalked, composed of 3–100+ bristles free or fused at the base, not spiny; disarticulation below the bur; spikelets glabrous, hidden within the bur, with 2 florets, the lower staminate or sterile, the upper fertile; fertile floret leathery, the lemma clasping the palea by the edges.

1. Panicles white to tawny, ovoid; longer bristles 3–5 cm…**P. villosum**
1'. Panicles purplish or rosy, generally elongate; longer bristles 1–3 cm…(2)
2. Blades generally reddish or purplish; plants cultivated ornamentals, not known in the wild…P. advena
2'. Blades green; plants escaped to the wild, also known in cultivation…(3)
3. Blades convolute or folded, the midribs noticeably thickened; primary bristles 26–34 mm; panicle rachis hairy proximally…**P. setaceum**
3'. Blades flat, the midribs not thickened; primary bristles 10–23 mm; panicle rachis scabrous but not hairy …**P. ciliare**

Pennisetum advena Wipff & Veldkamp—Purple fountaingrass. Tufted, tussock-forming perennial, 1–1.5 m; leaves burgundy-colored; sheaths glabrous, the margins ciliate; ligules 0.5–0.8 mm; blades to 55 cm, 6–11 mm wide, the midnerve not noticeably thickened; panicle 20–35 cm, 3–6 cm wide, exserted, burgundy-colored, the rachis hairy; burs with 1–3 spikelets; outer bristles 1–18 mm; inner bristles 11–25 mm; primary bristles 21–24 mm. Introduced as an ornamental landscape plant.

Pennisetum ciliare (L.) Link [*Cenchrus ciliaris* L.]—Buffelgrass. Tufted perennial, 20–150 cm, from a hard, knotty base, sometimes with short rhizomes; sheaths glabrous but with ciliate margins; ligules 0.5–3 mm; blades to 50 cm, 13 mm wide, with ciliate margins; panicles 3–20 cm, 4–35 mm wide, exserted, green or brownish to purplish, the rachis scabrous but not hairy; burs with 1–12 spikelets; outer bristles 1–12 mm; inner bristles 4–14 mm; primary bristles 10–23 mm. Adventive in a few places in the S desert and foothills, 3950–5600 ft. Doña Ana, Grant, Lea.

Pennisetum setaceum (Forssk.) Chiov.—Crimson fountaingrass. Tufted, tussock-forming perennial, 45–150 cm; sheaths glabrous but with ciliate margins; ligules 0.5–1.1 mm; blades to 65 cm, 2–3.5 mm wide, convolute or folded, the midnerve noticeably thickened; panicles 8–32 cm, 4–5 cm wide, exserted, pink to burgundy, the rachis hairy; burs with 1–4 spikelets; outer bristles 1–19 mm; inner bristles 8–27 mm; primary bristles 26–34 mm. Ornamental, recently found to escape to the wild in the S desert regions, 3900–4920 ft. Doña Ana. Introduced.

Pennisetum villosum R. Br. ex Fresen.—Feathertop. Tufted perennial, also with short rhizomes, 20–75 cm; sheaths glabrous but with ciliate margins; ligules 1–1.3 mm; blades to 40 cm, 2–4.5 mm wide, flat to folded, glabrous to pubescent, with ciliate margins; panicles ovoid, 4–12 cm, 5–8 cm wide, white to tawny, the rachis hairy proximally; burs with 1–4 spikelets; outer bristles 1–14 mm; inner and primary bristles 13–50 mm. Cultivated as an ornamental landscape grass, but known as an escape from at least 1 site in Doña Ana County, 4000–4025 ft. Introduced.

PHALARIS L. – Canarygrass

Tufted or rhizomatous annuals and perennials; leaves glabrous; sheaths open; auricles absent; ligules membranous; blades usually flat; inflorescence a dense, narrow, often spikelike panicle; spikelets with 1–3 florets, awnless, with 1–2 lower reduced sterile florets or rudiments and 1 terminal well-developed fertile floret; disarticulation below the glumes (ours), the florets falling together; glumes exceeding the florets, keeled and sometimes winged; sterile florets reduced, < 3/4 the length of the fertile floret, sometimes obscure and mistaken for tufts of hair; fertile floret leathery to indurate, shiny, 5-nerved; anthers 3.

1. Plants perennial, with rhizomes…**P. arundinacea**
1'. Plants annual, without rhizomes…(2)
2. Sterile floret (appearing as a scale) solitary, at the base and to 1 side of the large, fertile floret…**P. minor**

2′. Sterile florets (appearing as chaff or bristles) 2, at the base and on both sides of the large, fertile floret…(3)

3. Glumes broadly winged, the wings obvious; sterile florets broad and chaffy, usually at least 1/2 as long as the fertile floret…**P. canariensis**

3′. Glumes wingless or if slightly winged then the wings narrow and obscure; sterile florets needlelike, mostly <1/2 as long as the fertile floret…(4)

4. Sterile florets 1.5-2.5 mm; grain 2-2.3 mm; panicle ovate-lanceolate…**P. caroliniana**

4′. Sterile florets 0.7-1.5 mm; grain 1.4-1.6 mm; panicle narrowly cylindrical…**P. angusta**

Phalaris angusta Nees ex Trin.—Timothy canarygrass. Tufted annual, lacking rhizomes, 10-150+ cm; ligules 4-7 mm; blades 2-12 mm wide; panicles 2-10 cm, 6-15 mm wide, cylindrical; spikelets 2-5.5 mm; glumes with winged keels; sterile florets 2, needlelike, obscure, sparsely hairy; fertile floret 2-3.8 mm wide, hairy. Known from a single old collection at Mangas Springs, ca. 4735 ft. Grant. Introduced.

Phalaris arundinacea L.—Reed canarygrass. Rhizomatous perennial, forming dense stands, 40-200+ cm; ligules 4-10 mm; blades 5-20 mm wide; panicles 5-40 cm, 1-2 cm wide, branched at least basally; spikelets 4-8 mm; glumes not or scarcely winged; sterile florets 2, <1/2 the length of fertile floret, scalelike, hairy; fertile floret 2.5-4.2 mm, shiny when mature. Marshy ground, sloughs, wet meadows, 3400-9800 ft. NC, NW, C, WC, SC, Chaves.

Phalaris canariensis L.—Annual canarygrass. Tufted annual, 30-100 cm; ligules 3-6 mm; blades 2-10 mm wide; panicles 1-5 cm, 1-2 cm wide, ovoid, not lobed, truncate at the base; spikelets 7-10 mm; glumes broadly winged on the keels, enlarging distally; sterile florets 2, broad and chaffy, 1/3+ the length of the fertile floret; fertile floret 4.5-6.8 mm, densely hairy, shiny when mature. Moist weedy ground near human habitation; widely used in birdseed mixes and found around bird feeders, 2900-4200 ft. Bernalillo, Chaves, Doña Ana, San Juan. Introduced.

Phalaris caroliniana Walter—Carolina canarygrass. Tufted annual, lacking rhizomes, 30-150 cm; ligules 1-7 mm; blades 2-11 mm wide; panicles 1-9 cm, 8-20 mm wide, ovoid to cylindrical; spikelets 3.8-8 mm; glumes narrowly to broadly winged on the keels, the lateral nerves prominent; sterile florets needlelike, 1/2+ the length of the fertile floret; fertile floret 3-4.7 mm, hairy, shiny when mature. Moist weedy ground, 3500-4235 ft. WC, SE, SC, SW, Quay.

Phalaris minor Retz.—Littleseed canary grass. Tufted annual, 15-100 cm; ligules 5-12 mm; blades 2-10 mm wide; panicles 1-8 cm, 1-2 cm wide, dense, narrowly ovoid to elliptic, truncate at the base; spikelets 4-6.5 mm; glumes winged on the keels, the lateral nerves prominent; sterile florets single, scalelike, 1/2 or less the length of the fertile floret; fertile floret 2.5-4 mm, hairy, becoming shiny. Adventive weed escaping from agricultural fields, 3850-7350 ft. Doña Ana.

PHLEUM L. – Timothy

Tufted annuals and perennials, sometimes with short rhizomes; sheaths open; auricles absent; ligules membranous, not ciliate; blades usually flat; inflorescence a narrow, dense, spikelike panicle, the branches highly reduced; spikelets strongly laterally compressed, with 1 floret, the rachilla not or sometimes prolonged beyond the floret; disarticulation above the glumes and below the floret, or sometimes below the glumes late in the season; glumes longer than the floret, strongly keeled and ciliate, 3-nerved, abruptly awn-tipped to awned; lemma membranous, 5-7-nerved, awnless; palea shorter than or subequal to the lemma; anthers 3.

1. Panicles several times longer than wide, (3-)4-16 cm and 5-7.5(-10) mm wide; awns of glumes 1-1.5 mm…**P. pratense**

1′. Panicles only 2 or 3 times longer than wide, 1-5(-6) cm and (7-)8-12 mm wide; awns of glumes (1.2-)1.5-2.5 mm…**P. alpinum**

Phleum alpinum L. [*P. commutatum* Gaudin]—Alpine timothy. Tufted perennial, sometimes with short rhizomes, 15-50 cm, often decumbent-based; sheaths subtending the panicles usually inflated; ligules 1-4 mm; panicles 1-6 cm, 5-12 mm wide, 1/2-3 times longer than wide; glumes 2.5-4.5 mm, their junction forming a notch, the keels strongly ciliate, with dark awns 1-3 mm; lemmas ca. 3/4 the length of the glumes. Subalpine meadows, moist grasslands, mossy rivulets and seeps, 6000-13,200 ft. NC, NW, WC, Otero.

Phleum pratense L.—Timothy. Tufted perennial, lacking rhizomes, 30-150 cm, the lower internodes sometimes bulbous; sheaths subtending the panicles not inflated; ligules 2-4 mm; panicles 5-16 cm, 5-10 mm wide, 5-20 times longer than wide; glumes 3-4 mm, their junction not forming a notch but joined straight across, the keels strongly ciliate, with greenish to somewhat darkened awns 1-2 mm; lemmas ca. 1/2 the length of the glumes. Roadsides, fields, mountain meadows, 5500-11,750 ft. NE, NC, NW, C, WC, SC. Introduced.

PHRAGMITES Adans. - Reed

Phragmites australis (Cav.) Trin. ex Steud. [*P. communis* Trin.]—Common reed. Tall, rhizomatous and stoloniferous perennial, 1-4 m, the culms 0.5-1.5 cm thick; sheaths open; auricles absent; ligules ca. 1 mm, with a ciliate membrane; blades flat or folded, 2-4 cm wide, breaking from the sheath in age; panicles plumose, 15-35 × 8-20 cm; spikelets with 3-10 florets, the proximal ones staminate, the middle ones fertile, the distal ones sterile; rachilla 4-10 mm, hairs long-silky; disarticulation above the glumes and below the florets; lemmas 3-nerved, 8-15 mm, narrow, glabrous, awnless; anthers 1. Forming dense thickets and fencerows along streams, rivers, canals, and ditches and in wet ground of springs and seeps, 3500-8100 ft. Widespread except far E. Adventive.

PHYLLOSTACHYS Siebold & Zucc. - Bamboo

Phyllostachys aurea Carrière ex Rivière & C. Rivière—Golden bamboo, fishpole bamboo. Shrubby, rhizomatous perennial to 3 m (ours), the culms 1-4 cm thick; sheaths glabrous except for a basal line of whitish hairs, not glaucous; auricles absent on culm leaves, present on foliage leaves; ligules a ciliate membrane; inflorescence and spikelets never produced in New Mexico. Bamboo is seen sparingly in the state as a landscape ornamental. Not known in the wild in New Mexico.

PIPTATHEROPSIS Romasch., P. M. Peterson & Soreng - Ricegrass

Tufted perennials, sheaths open; auricles absent; ligules membranous; inflorescence a panicle, the re-branching sometimes slight; spikelets with 1 floret; glumes subequal, slightly longer than the floret, 1-3-nerved; florets dorsally compressed to terete; disarticulation above the glumes, the floret falling, the scar circular or elliptic; lemmas leathery, 3-5-nerved, the central nerve obscure, the margins not overlapping, the palea exposed, the awn deciduous. Formerly included in *Piptatherum*, which is considered strictly Eurasian (Romaschenko et al., 2011).

1. Florets 3-4 mm; lemma pubescent; awn 1-2 mm (when present)...**P. pungens**
1'. Florets 1.5-2.5 mm; lemma mostly glabrous (rarely pubescent); awn 4-10 mm (when present)...**P. micrantha**

Piptatheropsis micrantha (Trin. & Rupr.) Romasch., P. M. Peterson & Soreng [*Oryzopsis micrantha* (Trin. & Rupr.) Thurb.; *Piptatherum micranthum* (Trin. & Rupr.) Barkworth]—Littleseed ricegrass. Tufted perennial, 20-85 cm, the basal branching extravaginal; ligules 0.4-2.5 mm; blades 0.5-2.5 mm wide, usually rolled; panicles 5-20 cm, the branchlets and pedicels usually appressed; glumes 2.5-3.5 mm; florets 1.5-2.5 mm, the lemma usually glabrous, the awn 4-10 mm, deciduous. Moist, shaded, often rocky ground in the mountains and foothills, 5165-9125 ft. Widespread except EC, SE, SW.

Piptatheropsis pungens (Torr. ex Spreng.) Romasch., P. M. Peterson & Soreng [*Oryzopsis pungens* (Torr. ex Spreng.) Hitchc.; *Piptatherum pungens* (Torr. ex Spreng.) Dorn]—Mountain ricegrass. Tufted perennial, 10-90 cm, the basal branching intravaginal; ligules 0.5-2.5 mm; blades 0.5-1.8 mm wide, flat to rolled; panicles 4-6 cm, the branchlets and pedicels usually diverging; glumes 3.5-4.5 mm; florets 3-4 mm, the lemma evenly pubescent, the awn 1-2 mm, early-deciduous. Pine forests; as yet known only from Valles Caldera National Preserve, 8500-8745 ft. Sandoval.

PIPTOCHAETIUM J. Presl – Ricegrass

Tufted perennials, lacking rhizomes; sheaths open, the margins glabrous; ligules membranous, some-times ciliate; blades flat to rolled; inflorescence a panicle, often narrow, the branches usually spikelet-bearing on the distal 1/2; spikelets with 1 floret, the rachilla not prolonged; disarticulation above the glumes, the floret falling; glumes subequal, longer than the floret; florets terete; lemmas indurate, the margins involute, not encircling but fitting into the grooved palea, the apex fused into a crown, awned; paleas longer than the lemmas, indurate, grooved; anthers 3.

1. Glumes 3–4 mm; awns 0.5–1 cm…go to **Piptatheropsis micrantha**
1'. Glumes 5–10 mm; awns 1–3 cm…(2)
2. Glumes ca. 5 mm; blades rolled and threadlike, elongate and weeping…**P. fimbriatum**
2'. Glumes ca. 10 mm; blades flat or loosely rolled, firm and somewhat erect…**P. pringlei**

Piptochaetium fimbriatum (Kunth) Hitchc.—Piñon ricegrass. Tufted perennial, 35–100 cm; ligules of upper leaves 1–3 mm; blades usually involute, 0.3–1 mm wide; panicles 7–25 cm, 4–14 cm wide, few-flowered, the spikelets on long pedicels; glumes 4–6.2 mm, 5–7-nerved; florets 3–5.5 mm; lemmas tan to dark brown, hairy when young, becoming glabrous and shiny; awns 10–22 mm. Shaded, moist sites in woodlands, commonly under piñon, 4800–8165 ft. NC, C, WC, SE, SC, SW.

Piptochaetium pringlei (Beal) Parodi—Pringle's speargrass. Tufted perennial, 50–100+ cm; ligules of upper leaves 1–3.5 mm; blades flat or sometimes loosely rolled, 1–3.5 mm wide; panicles 6–20 cm, 3–12 cm wide, the spikelets on long pedicels; glumes 9–12 mm, 5–7-nerved; florets 6.5–10 mm; lemmas brown, sometimes dark, usually stiffly pubescent; awns 19–30 mm. Pine and oak woodlands, 5800–9025 ft. NC, C, WC, SE, SC, SW.

POA L. – Bluegrass

Tufted, rhizomatous, and stoloniferous annuals and perennials; sheaths open to closed; auricles absent; ligules membranous; blades with a groove on each side of the midnerve, the apices often prow-shaped; inflorescence a panicle, infrequently poorly developed and racemelike; spikelets laterally compressed, with several spikelets, awnless; disarticulation above the glumes and between the florets; glumes usu-ally not exceeding the lowermost lemma; callus often with a dense tuft of hairs, sometimes cobwebby; lemmas mostly 5-nerved. Robert J. Soreng assisted with earlier versions of this key.

1. Florets modified and forming small leafy plantlets; stems slightly to strongly bulblike at the base…**P. bulbosa**
1'. Florets not modified into small leafy plantlets; stems rarely somewhat bulblike…(2)
2. Anthers 1 mm or less, nearly all well developed; plants annual or perennial…(3)
2'. Anthers mostly > 1 mm, or vestigial and poorly developed; plants perennial…(8)
3. Callus without a tuft of long hairs (but the nerves of the lemma pubescent)…**P. annua**
3'. Callus with a tuft of long, cobwebby hairs…(4)
4. Panicles narrow, contracted; paleas pubescent on the keels; plants mostly annual or infrequently short-lived perennial…**P. bigelovii**
4'. Panicles open when mature; paleas glabrous or pubescent; plants perennial…(5)
5. Sheath margins fused together 1/5 or less their length; first glume mostly 3-nerved…**P. palustris** (in part)
5'. Sheath margins fused together 1/4–2/3 their length; first glume mostly 1-nerved…(6)
6. Sheaths densely scabrous with downward-pointing hairs, rarely glabrous; panicles (8–)13–40 cm, the internodes of the main axis mostly 4+ cm…**P. occidentalis**
6'. Sheaths glabrous to sparsely scabrous with downward-pointing hairs; panicles mostly < 12 cm, the inter-nodes of the main axis < 3.5 cm…(7)
7. First glume linear-lanceolate, much narrower than the second; paleas glabrous to scabrous on the keels …**P. leptocoma**
7'. First glume about the same shape and width as the second, both broadly lanceolate; paleas short-pubescent on the keels…**P. reflexa**
8. Stems and nodes strongly flattened; plants strongly rhizomatous; sheath margins open to near the base …(9)
8'. Stems and nodes round or nearly so; plants tufted or rhizomatous; sheath margins fused or open…(10)

9. Lemmas 5–6 mm; spikelets unisexual, the plants dioecious with the sexes on separate plants; rare…**P. arachnifera** (in part)

9'. Lemmas 2–3 mm; spikelets bisexual, with both sexes in the same floret…**P. compressa**

10. Callus of the floret with a tuft of cobwebby hairs, these short-kinky to long-sinuous, borne on the back surface of the lemma and distinct from any hairs on the lemma midnerve…(11)

10'. Callus not with cobwebby hairs as above, glabrous or with hairs similar to and continuous with those of the lemma keel, or in *P. secunda* with short, straight hairs around the top of the callus and not restricted to the back side of the lemma…(18)

11. Plants dioecious, with unisexual spikelets and the sexes on separate plants; long, delicate rhizomes developed; panicles oblong, compact, the terminal branches densely flowered from near the base; rare …**P. arachnifera** (in part)

11'. Plants bisexual, or if female then the panicle more open and the branches sparsely flowered; rhizomes present or absent…(12)

12. Sheath margins fused 1/2+ their length; panicles mostly 13–29 cm, the lower internodes of the main axis mostly > 3.5 cm; anthers averaging 2.2 mm…**P. tracyi**

12'. Sheath margins fused 1/2 their length or less; panicles mostly < 13 cm (longer in *P. palustris*), the internodes of the main axis rarely > 3.5 cm; anthers mostly < 1.9 mm…(13)

13. Plants with strong rhizomes; sheath margins fused together 1/4–1/2 their length; panicle branches glabrous to moderately scabrous, round…(14)

13'. Plants tufted, lacking rhizomes (in wet habitats occasionally producing decumbent stems that root at the nodes); sheath margins fused together 1/4 or less their length (to 1/2 in *P. trivialis*); panicle branches distinctly scabrous, mostly angled…(15)

14. Glumes distinctly keeled, scabrous on the nerves, the second glume plainly shorter than the first lemma; panicles often with 4+ branches at the lowermost node (some occasionally vestigial); ligules mostly 1–2 mm…**P. pratensis**

14'. Glumes weakly keeled, nearly glabrous, the second glume subequal to or longer than the first lemma; panicles usually with < 4 branches at the lowermost node; ligules 2–4 mm…**P. arctica** (in part)

15. Ligules 3–10 mm; lemmas sparsely pubescent on the keel near the base and mostly glabrous on the marginal nerves and between the nerves; first glume very narrow, sickle-shaped, 1-nerved…**P. trivialis**

15'. Ligules mostly < 4 mm; lemmas pubescent on the keel and marginal nerves and often between the nerves; first glume narrow to broad, not sickle-shaped, 1–3-nerved…(16)

16. Panicles mostly 10–30 cm, abundantly rebranched; stems often decumbent and rooting at the nodes, stout and leafy well above the middle, 25–120 cm…**P. palustris** (in part)

16'. Panicles mostly < 12 cm, sparingly rebranched if at all; stems never decumbent and rooting at the nodes, leafy or not, mostly < 50 cm…(17)

17. Lemmas glabrous between the nerves; leaves green…**P. interior**

17'. Lemmas mostly pubescent between the nerves; leaves glaucous…**P. glauca** (in part)

18. Plants unisexual, all the spikelets of a plant either male or female…(19)

18'. Plants bisexual, the spikelets with both anthers and pistil in a single floret…(20)

19. Plants rhizomatous; uppermost stem blade well developed; rare in New Mexico…**P. wheeleri**

19'. Plants mostly tufted; uppermost stem blade very reduced; common in New Mexico…**P. fendleriana**

20. Lemmas glabrous to scabrous; sheath margins not fused together…**P. secunda** (in part)

20'. Lemmas prominently pubescent or puberulent; sheath margins fused together or not…(21)

21. Plants rhizomatous…(22)

21'. Plants tufted, not rhizomatous…(23)

22. Sheath margins fused together 1/3–1/2 their length; glumes weakly keeled; plants subalpine to alpine… **P. arctica** (in part)

22'. Sheath margins overlapping most of their length, fused 1/5 or less; glumes strongly keeled; plants of plains and valleys…**P. arida**

23. Stem bases enclosed in persistent, thickened, closely overlapping sheaths; panicle branches widely spreading at maturity; spikelets ovate to subcordate; blades 2–4 mm wide…**P. alpina**

23'. Stem bases not enclosed in persistent sheaths as above; panicle branches not widely spreading; spikelets ovate to more elongate, not at all cordate at the base; blades usually < 2 mm wide…(24)

24. Lemmas keeled on the back, the pubescence on the nerves longer and denser than between the nerves; ligules 1–3 mm (subsp. *rupicola*)…**P. glauca** (in part)

24'. Lemmas rounded on the back, minutely pubescent all across the base, the hairs on nerves and between nerves similar; ligules 2–7 mm…**P. secunda** (in part)

Poa alpina L.—Alpine bluegrass. Tufted perennial, 10–40 cm; leaves mostly basal, the basal sheaths persistent, closely overlapping; upper ligules 4–5 mm; blades 2–4 mm wide; panicles 2–8 cm, open; spikelets ovate to subcordate, with 3–7 florets; lemmas 3–5 mm, the nerves villous, without tuft of cobwebby hairs; anthers 1.3-2.3 mm. Alpine to subalpine slopes, meadows, talus, moist ledges, 10,330–12,960 ft. NC.

Poa annua L.—Annual bluegrass. Densely tufted annual, rarely with short stolons from the base, 2-25 cm; sheaths closed on lower 1/3; ligules 0.5-5 mm; blades flat to folded, 1-5 mm wide; panicles 1-8 cm, ca. 1/2 as wide; spikelets with 2-6 florets; lemmas 2.5-4 mm, the nerves villous, without a tuft of cobwebby hairs at the base; anthers 0.6-1 mm. Lawns, flower beds, moist disturbed ground, 3500–11,590 ft. NC, NW, C, WC, SE, SC. Introduced.

Poa arachnifera Torr.—Texas bluegrass. Dioecious, loosely tufted to rhizomatous perennial, 20-85 cm; sheaths closed on lower 1/3; ligules 1-4 mm; blades 1.5-4.5 mm wide; panicles 5-18 cm, loosely (staminate) to narrowly (pistillate) contracted, with numerous spikelets; spikelets unisexual, with 2-10 florets; staminate lemmas with a weak tuft of cobwebby hairs, the anthers vestigial (to 0.2 mm) or well developed (1.5-2.5 mm); pistillate florets 4-6.5 mm, with copious cobwebby hairs from the base and villous hairs running up the nerves. Known from only a single collection at the Bosque del Apache National Wildlife Refuge, 4540 ft. Socorro. Introduced.

Poa arctica R. Br. [*P. aperta* Scribn. & Merr.; *P. grayana* Vasey]—Arctic bluegrass. Rhizomatous perennial, 10-60 cm, usually with solitary aerial culms; sheaths closed only at the base; ligules 2-7 mm, blades flat to folded, 1-6 mm wide; panicles 2-15 cm, ovoid and to nearly as wide; spikelets with 3-6 florets; lemmas 3-7 mm, the nerves long-villous, with or without a basal tuft of cobwebby hairs; anthers 1.5-2.5 mm. Forests and subalpine and alpine meadows, usually in deep, rich soil, 8500–13,030 ft. NC, Bernalillo, Cibola.

Poa arida Vasey [*P. glaucifolia* Scribn. & T. A. Williams]—Plains bluegrass. Rhizomatous perennial, the culms loosely tufted to solitary, 15-80 cm; sheaths closed only at the base; ligules 1-5 mm; blades flat to folded, 2-5 mm wide; panicles 4-12 cm, narrow and loosely contracted, the branches mostly erect to ascending; spikelets with 2-7 florets; glumes strongly keeled; lemmas 2.5-4.5 mm, the nerves short-hairy, the callus mostly glabrous and without a tuft of cobwebby hairs; anthers 1.3-2.2 mm. Prairies and floodplains, 3545–12,210 ft. NE, NC, EC, C.

Poa bigelovii Vasey & Scribn.—Bigelow's bluegrass. Densely tufted annual, rarely longer-lived, 5-60 cm; sheaths closed 1/2 or less their length, compressed-keeled; ligules 2-6 mm; blades flat, 1.5-5 mm wide; panicles 5-15 cm, contracted-cylindrical, the branches erect; spikelets with 3-7 florets; lemmas 2.5-4.2 mm, the nerves short-villous, usually glabrous between the nerves, with a tuft of cobwebby hairs at the base; anthers 0.2-1 mm. Rocky hills, arroyo bottoms, wooded slopes, 3450-10,000 ft. Widespread except E.

Poa bulbosa L.—Bulbous bluegrass. Densely tufted perennial, 15-60 cm, the bases bulbous; sheaths closed ca. 1/4 their length, the lowest sheaths with inflated bases; ligules 1-3 mm; blades flat, 1-3 mm wide, soon withering; panicles 3-12 cm, loosely contracted, the branches erect to ascending; spikelets with 4-7 florets, most florets modified into leafy shoots; unmodified lemmas 3-4 mm, the callus webbed or glabrous. New Mexico material belongs to **subsp. vivipara** (Koeler) Arcang. Moist hills and slopes in the mountains, 3940-7900 ft. NW, Doña Ana, Hidalgo. Introduced.

Poa compressa L.—Canada bluegrass. Strongly rhizomatous perennial, the aerial culms mostly solitary or few together, 20-60 cm, the bases geniculate, the nodes and internodes flattened; sheaths closed only at the base; ligules 1-3 mm; blades flat, 1.5-4 mm wide; panicles 2-10 cm, loosely contracted, 1/3 or less as wide, the scabrous branches erect-ascending, sometimes spreading; spikelets with 4-7 florets; glumes distinctly keeled; lemmas 2.3-3.5 mm, the nerves densely short-villous, usually with a tuft of cobwebby hairs at the base; anthers 1.3-1.8 mm. Forest clearings, disturbed meadows, roadsides, 5360-10,925 ft. Widespread except E, SW. Introduced.

Poa fendleriana (Steud.) Vasey—Muttongrass. Densely tufted perennial, often with weak short rhizomes, 15-70 cm, erect to decumbent-based, the plants sometimes staminate or pistillate; sheaths closed ca. 1/3 their length; ligules 0.2-18 mm; cauline blades shortened distally, usually rolled, 1-4 mm wide; panicles 4-15 cm, congested, narrow; spikelets with 2-10 florets, ovate, flattened; glumes distinctly keeled; lemmas 3-6 mm, the nerves glabrous to villous, lacking a tuft of cobwebby hairs at the base; anthers vestigial (0.1-0.2 mm) or well developed (2-3 mm). Woodlands, rocky hills, mountain slopes.

1. Lemma keels commonly scabrous to glabrous…**subsp. albescens**
1ʹ. Lemmas commonly pubescent on the keels…(2)
2. Ligules of middle to upper stem leaves 1.8-11 mm…**subsp. longiligula**
2ʹ. Ligules of middle to upper stem leaves 0.2-1(-2) mm…**subsp. fendleriana**

subsp. albescens (Hitchc.) Soreng [*P. albescens* Hitchc.]—Known only from the bootheel region, 6000-7000 ft. Hidalgo.

subsp. fendleriana—Widespread, except for the E plains, 3500-12,855 ft.

subsp. longiligula (Scribn. & T. A. Williams) Soreng [*P. longiligula* Scribn. & T. A. Williams]—Widespread in the NW 1/3 of the state, 5600-8950 ft. NW, WC.

Poa glauca Vahl—Glaucous bluegrass. Densely tufted, usually glaucous perennial, 5-60 cm; sheaths closed only at the base; ligules 1-5 mm; blades flat or folded, 1-2.5 mm wide; panicles 4-20 cm, 3-5 times longer than wide, narrowly pyramidal to contracted; spikelets with 2-5 florets; lemmas 2.5-4 mm, the nerves short-villous, the callus glabrous or with a tuft of cobwebby hairs; anthers 1-2.5 mm. Alpine and subalpine ridges, grassy slopes, meadows, mossy ledges in the mountains.

1. Calluses usually with cobwebby hairs…**subsp. glauca**
1ʹ. Calluses usually without cobwebby hairs…**subsp. rupicola**

subsp. glauca—Alpine, 11,000-13,100 ft. NC, SC.

subsp. rupicola (Nash) W. A. Weber—Montane meadows and alpine, 9500-13,100 ft. NC, SC.

Poa interior Rydb.—Inland bluegrass. Densely tufted, green to glaucous perennial, 10-80 cm; sheaths closed only at the base; ligules 0.5-3 mm; blades flat, 1-3 mm wide; panicles 3-16 cm, ca. 2-4 times longer than wide, the branches often lax and scabrous; spikelets with 1-5 florets; lemmas 2.5-4 mm, the nerves short-villous; with a tuft of cobwebby hairs at the base, this sometimes scant to glabrous; anthers well developed, 1.3-2.5 mm. Alpine and subalpine ledges, meadows, forest clearings, 6250-12,855 ft. NC, Cibola, Lincoln.

Poa leptocoma Trin.—Bog bluegrass. Loosely tufted perennial, occasionally with short rhizomes, 15-100 cm; sheaths closed in the lower 1/4-2/3; ligules 1.5-6 mm; blades flat, 1-4 mm wide; panicles 5-15 cm, lax, open, sparsely flowered, the branches spreading to reflexed, scabrous; spikelets with 2-5 florets; lower glume much narrower than the upper, 1-nerved; lemmas 3-4 mm, the nerves pubescent, with a tuft of cobwebby hairs at the base; paleas glabrous to scabrous on the keels, sometimes ciliate; anthers well developed, 0.2-1 mm. Alpine or subalpine springs, meadows, boggy ground, 7725-12,000 ft. NC, Cibola, Lincoln, Otero.

Poa occidentalis (Vasey) Vasey—New Mexico bluegrass. Densely tufted perennial, 20-110 cm; sheaths closed to 2/3 their length; ligules 3-12 mm; blades flat, lax, 1.5-10 mm wide; panicles 12-40 cm, open, lax, the branches spreading to drooping and densely scabrous; spikelets with 3-7 florets; lemmas 2.6-4.2 mm, the nerves short-villous, with a tuft of long cobwebby hairs at the base; anthers well developed, 0.3-1 mm. Forest clearings and moist woods, 7100-11,800 ft. NC, C, WC, SC, SW.

Poa palustris L.—Fowl bluegrass. Loosely to densely tufted perennial, sometimes with stolons, lacking rhizomes, 25-120 cm, the culms erect to decumbent-based and rooting at the nodes; sheaths closed only at the base; ligules 1-6 mm; blades flat, 2-8 mm wide, held in an ascending position; panicles 10-40 cm, lax, open, ca. 1/2 as wide; spikelets with 2-5 florets; lower glume 1-nerved; lemmas 2-3 mm, keeled, the margins hyaline, the nerves short-villous, with a tuft of cobwebby hairs at the base; anthers 1.3-1.8 mm. Moist meadows, marshy ground, sloughs, 5025-11,570 ft. NC, NW, C, WC, SC, SW.

Poa pratensis L.—Kentucky bluegrass. Strongly rhizomatous perennial, the aerial stems loosely tufted to solitary, 10-100 cm; sheaths closed 1/4-1/2 their length; ligules 1-3 mm; blades flat or folded to rolled, 1-4 mm wide, the apices usually broadly prow-shaped; panicles 3-20 cm, loosely contracted to broadly open and diffuse, the branches often whorled at the lower nodes; spikelets flattened, green and white, with 2-5 florets; glumes keeled; lemmas 2-4.5(-6) mm, keeled, the nerves villous, the lateral nerves prominent, with a tuft of cobwebby hairs at the base; anthers 1-2 mm. Common throughout the state in a wide variety of habitats, generally in the mountains, also disturbed ground along ditches and streams, lawns, moist open fields, and meadows; very widespread and expected in all counties. We have 3 subspecies scattered in the state, 2 of which may represent native populations:

1. Panicle branches scabrous…**subsp. pratensis**
1'. Panicle branches smooth or nearly so…(2)
2. Panicles 4-18 cm; blades flat or folded, soft, adaxial surfaces usually glabrous, sometimes sparsely hairy; plants of alpine areas in the N mountains…**subsp. alpigena**
2'. Panicles 4-8 cm, blades folded or rolled, somewhat firm, adaxial surfaces often sparsely hairy; plants generally from subalpine to lower elevations…**subsp. agassizensis**

subsp. agassizensis (B. Boivin & D. Löve) Roy L. Taylor & MacBryde [*P. agassizensis* B. Boivin & D. Löve]—Many of our upland meadows and dry mountain grasslands, 7000-8030 ft. NC, NW, WC, SW.

subsp. alpigena (Lindm.) Hiitonen—Known in New Mexico from a single collection, Philmont Scout Ranch, 11,200 ft. Colfax.

subsp. pratensis—Kentucky bluegrass. Pastures, meadow reseeding, and lawns, escaping to similar moist sites in natural habitats, 3200-12,000 ft. Widespread. Introduced.

Poa reflexa Vasey & Scribn.—Nodding bluegrass. Densely tufted perennial, 10-60 cm; sheaths closed 1/3-2/3 their length; ligules 1.5-3.5 mm; blades flat, 1-4 mm wide; panicles 4-15 cm, the branches usually spreading to reflexed, flexuous, smooth or scabrous; spikelets with 3-5 florets; glumes similar in shape and width; lemmas 2-3.5 mm, the nerves villous, with a tuft of cobwebby hairs at the base; anthers 0.6-1 mm. Alpine and subalpine meadows, ridges, rocky ledges, 10,600-12,545 ft. NC.

Poa secunda J. Presl—Sandberg's bluegrass. Densely tufted perennial, sometimes glaucous, 15-120 cm, the base erect to slightly decumbent; sheaths closed 1/4 or less their length; ligules 1-8 mm; blades flat or folded to rolled, 0.5-5 mm wide, shortened upward; panicles 4-25 cm, loosely contracted, 2-4 cm wide, the branches erect-ascending, glabrous to densely scabrous; spikelets mostly with 3-5 florets, nearly terete to slightly flattened; lemmas 3.5-6 mm, the nerves glabrous to short villous, the base glabrous or with a short tuft of soft hairs; anthers 1.5-3 mm. Forest clearings, sagebrush plains, meadows, disturbed ground.

1. Lemmas prominently crisp-puberulent on the back toward the base; ligules of axillary shoots usually longer than 2 mm…**subsp. secunda**
1'. Lemmas glabrous to minutely scabrous on the back; ligules of axillary shoots to 2 mm…subsp. juncifolia

subsp. juncifolia (Scribn.) Soreng [*P. juncifolia* Scribn.]—No records seen.

subsp. secunda—11,500-12,600 ft. NC, NW, WC, SW.

Poa tracyi Vasey—Tracy's bluegrass. Loosely tufted perennial with short rhizomes, 30-100+ cm; sheaths closed from 1/2 to nearly all their length, keeled, frequently retrorsely pubescent; ligules 2-4.5 mm; blades lax, flat, 2-5 mm wide; panicles 10-30 cm, open, to 1/2 as wide, the branches spreading to drooping, flexuous, scabrous; spikelets with 2-8 florets; lemmas 2.5-5 mm, the nerves villous, with a tuft of cobwebby hairs at the base; anthers vestigial (0.1-0.2 mm) or well developed (1.5-3 mm). Rich humus and moist loam of forests and woodlands, 6900-11,500 ft. NE, NC, C, SC.

Poa trivialis L.—Rough bluegrass. Loosely to densely tufted perennial, with weak stolons or trailing stems, 25-100+ cm; sheaths closed 1/3-1/2 their length, densely scabrous; ligules 3-10 mm; blades 1-5 mm wide, flat, lax; panicles 8-25 cm, open, loose, the branches ascending to spreading, densely scabrous; spikelets with 2-4 florets; lower glume very narrow, sickle-shaped, 1-nerved; lemmas 2.3-3.5 mm, the marginal nerves glabrous, with a tuft of long cobwebby hairs at the base; anthers 1.3-2 mm. Shaded, moist sites in the mountains, 8000-10,500 ft. Lincoln, Otero. Introduced.

Poa wheeleri Vasey—Wheeler's bluegrass. Tufted perennial with short rhizomes, 35-80 cm, often decumbent-based; sheaths closed 1/3-3/4 their length, some retrorsely pubescent; ligules 0.5-2 mm; blades flat or folded, 2-3.5 mm wide, the uppermost blades well developed; panicles 5-18 cm, loosely contracted to open, the branches ascending to reflexed, scabrous; spikelets with 2-7 florets; lemmas 3-6 mm, the nerves generally glabrous but also short-villous, without a tuft of cobwebby hairs at the base; anthers usually vestigial, 0.1-0.2 mm, but sometimes aborted late in development and up to 2 mm. Subalpine mountain slopes in rich soils, 7560-8365 ft. NC.

PODAGROSTIS (Griseb.) Scribn. & Merr. – Alpine bentgrass

Podagrostis humilis (Vasey) Björkman [*Agrostis humilis* Vasey]—Alpine bentgrass. Tufted to rhizomatous, low perennial, 3-25 cm, sometimes with short rhizomes; sheaths open; auricles absent; ligules membranous; panicles 2-6 × < 2 cm; spikelets with a single floret, purplish, 1.5-2 mm; rachilla prolonged beyond the floret 0.1-0.6 mm, often with a tiny tuft of hairs at the tip; paleas well developed, 1-1.5 mm; anthers 3. Alpine tundra. No records seen but to expected in NC.

POLYPOGON Desf. – Polypogon

Tufted annuals and perennials; sheaths open; auricles absent; ligules membranous; inflorescence a dense panicle; disarticulation below the glumes, the spikelet falling; spikelets with 1 floret, the rachilla not prolonged; glumes exceeding the floret, awned (usually) or awnless; lemmas 1-3-nerved, often awned; paleas arising from the back to terminal, 1/3 to as long as the lemma; anthers 3.

1. Glumes awnless…**P. viridis**
1'. Glumes awned…(2)
2. Awns 1-3(-5) mm; glumes acute and entire to minutely cleft at the tip…**P. interruptus**
2'. Awns 4-12 mm; glumes obtuse to shallowly lobed at the tip…(3)
3. Glumes deeply lobed, the lobes 1/6-1/3 the length of the glume body and evident…**P. maritimus**
3'. Glumes not lobed or only very slightly so…**P. monspeliensis**

Polypogon interruptus Kunth [*P. littoralis* Sm.]—Ditch rabbitsfoot grass. Tufted perennial, 20-80 cm, decumbent-based; ligules 2-6 mm; blades 3-6 mm wide; panicles 3-15 cm, 1-3 cm wide, lobed; glumes 2-3 mm, with awns to 3.2 mm; lemmas 1-1.5 mm, ovoid, with awns 1-3.2 mm. Wet ground, ditches, seeps, springs, 3820-8885 ft. NE, NC, NW, Bernalillo, Sierra. Introduced.

Polypogon maritimus Willd.—Mediterranean polypogon. Tufted annual, 10-50 cm, geniculate-based; uppermost sheaths sometimes inflated; ligules 3-7 mm; blades 1-5 mm wide; panicles 2-12 cm, dense, ellipsoid, sometimes lobed; glumes 1.8-3.2 mm, with awns 4-12 mm; lemmas tiny, 0.5-1.5 mm, awnless or with awns to 1 mm. Disturbed wet places; known in New Mexico from a single collection near Sitting Bull Falls, 3500 ft. Eddy. Introduced.

Polypogon monspeliensis (L.) Desf.—Rabbitsfoot grass. Tufted annual, 10-100 cm, erect to geniculate-based; uppermost sheaths sometimes inflated; ligules 3-16 mm; blades 1-7 mm wide; panicles 2-17 cm, 2-4 cm wide, very dense, occasionally lobed; glumes 1-3 mm, with awns 4-10 mm; lemmas tiny, 0.5-1.5 mm, with awns 1-4 mm. Ditch banks, seeps, wet disturbed ground, 2950-11,945 ft. Widespread. Introduced.

Polypogon viridis (Gouan) Breistr. [*Agrostis semiverticillata* (Forssk.) C. Chr.]—Beardless rabbitsfoot grass. Tufted perennial, 10-90 cm, sometimes decumbent and rooting at the lower nodes; sheaths not inflated; ligules 1-5 mm; blades 1-6 mm wide; panicles 2-10 cm, 1-3 cm wide, dense but lobed and interrupted; glumes 1.5-2 mm, awnless; lemmas 0.8-1.2 mm, awnless. Wet ground of springs, seeps, ponds, ditch banks, 3800-8500 ft. NC, NW, C, SE, SC, SW. Introduced.

PSATHYROSTACHYS Nevski – Wildrye

Psathyrostachys juncea (Fisch.) Nevski [*Elymus junceus* Fisch.]—Russian wildrye. Densely tufted perennial, 25–80+ cm, sometimes with rhizomes or stolons; old sheaths persistent; auricles sometimes present; ligules membranous, 0.2–0.3 mm; blades flat or rolled; spikes 7–12 × 0.5–2 cm; spikelets 2–3 per node, strongly overlapping, the lateral spikelets slightly larger; disarticulation below the spikelets in the main axis; glumes 3–9 mm, 1-nerved, needlelike, shorter than the florets; lemmas 5–7-nerved, 5.5–7.5 mm, glabrous or with short hairs, awn-tipped or with awns to 4 mm; paleas equaling or longer than the lemmas; anthers 3, 2.5–5 mm. Planted for range restoration and erosion control, scattered localities, 5585–10,085 ft. NC, NW, Grant, Otero. Introduced.

PSEUDOERIOCOMA Romasch., P. M. Peterson & Soreng – Needlegrass

Pseudoeriocoma eminens (Cav.) Romasch. [*Achnatherum eminens* (Cav.) Barkworth; *Stipa eminens* Cav.]—Southwestern needlegrass. Perennials, 50–100 cm, with knotty bases, sometimes with short-rhizomatous shoots; blades flat when fresh, 1–3 mm wide; sheaths open, shorter than the internodes, glabrous to hairy; auricles absent; ligules membranous; panicles with drooping branches, 20–40 cm, 3–8 cm wide; ligules membranous; spikelets with 1 floret, the rachilla not prolonged; glumes unequal, longer than the floret, 3–5-nerved, the lower 5–12 mm, the upper 1–4 mm shorter; florets fusiform, 4–7 mm, the callus sharp and 1–2 mm; awns 3.5–7 cm, twice-geniculate, the terminal segment flexuous; lemmas coriaceous, the margins enveloping most of the floret, the awns twice-geniculate, flexuous; paleas 1/3–3/4 as long as the lemma, 2-nerved, the nerves not prolonged; anthers 3. Rocky foothills, upland plains, bajadas, 4200–7500 ft. SE, SC, SW, Socorro.

PTILAGROSTIS Griseb. – False needlegrass

Ptilagrostis porteri (Rydb.) W. A. Weber [*Stipa porteri* Rydb.]—Porter's needlegrass. Tightly tufted perennials, 23–50 cm, leaves mostly basal; sheaths open; auricles absent; ligules membranous, 0.7–1.5 mm; blades rolled-filiform, convolute; panicles 7–12 cm, open or contracted, the branches ascending to spreading, with few spikelets; disarticulation above the glumes and below the floret; spikelets with 1 floret, the rachilla not prolonged; florets fusiform; glumes hyaline, without evident nerves, 4.5–6 mm, broad, acute to obtuse, awnless; florets fusiform, 2.5–4 mm; awns 5–25 mm, the basal segment hairy with hairs 1–2 mm; lemmas awned, the margins not overlapping; anthers 1–3 mm. Mossy hummocks at very high elevations; 1 record, 10,880 ft. Colfax.

PUCCINELLIA Parl. – Alkaligrass

Tufted to stoloniferous annuals to perennials, the culms sometimes decumbent; sheaths open; auricles absent; ligules membranous; inflorescence a rebranching panicle; spikelets with several florets, awnless; disarticulation above the glumes and between the florets; glumes 1–3-nerved, shorter than the lowest lemma; lemmas mostly 5-nerved, the apex rounded to truncate and somewhat erose, the nerves parallel, not reaching the apex and not converging; anthers 3.

1. Plants annual, 3–10(–15) cm…**P. parishii**
1'. Plants perennial (sometimes short-lived), 15+ cm…(2)
2. Lemmas with conspicuous nerves; plants with creeping rhizomes; blades mostly flat, 4–15 mm wide; freshwater habitats…go to **Torreyochloa**
2'. Lemmas with obscure or indistinct nerves; plants tufted, lacking rhizomes; blades rolled, or if flat then 1–3(–4) mm wide; usually alkaline or saline habitats…(3)
3. Plants with yellow-green herbage and erect culms; lower panicle branches erect to divergent at maturity; lemmas 2–3.5 mm; anthers 0.6–2 mm…**P. nuttalliana**
3'. Plants with blue-green herbage and often geniculate-based culms; lower panicle branches divergent to reflexed at maturity; lemmas 1.5–2.2 mm; anthers 0.4–0.8 mm…**P. distans**

Puccinellia distans (Jacq.) Parl.—Weeping alkaligrass. Tufted perennial, 10–60 cm, not forming mats, erect to decumbent; blades 1–7 mm wide; panicles 3–20 cm, open, the branches divergent-horizontal to reflexed; spikelets 2.5–7 mm; lemmas 1.5–2.2 mm; anthers 0.4–0.8 mm. Alkali flats and floodplains, 5100–9580 ft. NC, NW, C. Introduced.

Puccinellia nuttalliana (Schult.) Hitchc. [*P. airoides* S. Watson & J. M. Coult.]—Nuttall's alkaligrass. Tufted perennial, 10–100 cm, not forming mats, erect; blades 1–4 mm wide; panicles 5–30 cm, open, the branches erect to divergent; spikelets 3.5–9 mm; lemmas 2–3.5 mm; anthers 0.6–2 mm. Alkali flats and floodplains, 5040–8800 ft. NC, NW, C, WC.

Puccinellia parishii Hitchc.—Bog alkaligrass. Tufted annual, 3–15 cm; leaves mostly basal; blades 0.2–1.2 mm wide; panicles 1–8 cm, mostly contracted in our populations, sometimes open; spikelets 3.5–5 mm; lemmas 1.8–2.2 mm, the nerves densely hairy on the lower 1/2–3/4; anthers 0.4–0.5 mm. Alkali flats and seeps, 2600–7200 ft. NC, NW, WC, SW. **R**

REDFIELDIA Vasey – Blowout grass

Redfieldia flexuosa (Thurb. ex A. Gray) Vasey [*Muhlenbergia ammophila* P. M. Peterson]—Blowout grass. Strongly rhizomatous perennial, 50–130 cm, rooting at the nodes as the sand covers them; leaves cauline; sheaths open, shorter than the internodes; auricles absent; blades rolled; ligules a ciliate membrane 0.5–1.5 mm; blades to 45 cm, 2–8 mm wide; panicles open, diffuse, rebranching, 20–50 × 8–28 cm, the lower branches ascending, the upper spreading; pedicels longer than the spikelets; spikelets ovate to obovate, olive-green to brownish, awnless, with 2–6 florets, the distal florets sometimes sterile; disarticulation above the glumes and between the florets; glumes 1–3-nerved, acute-attenuate; fertile lemmas narrowly lanceolate to falcate with a tuft of straight soft hairs at the callus; 4.5–6 mm; anthers 2–3.6 mm. Deep sand hills and dunes, blowout areas. Colfax, Roosevelt.

SCHEDONORUS P. Beauv. – Fescue

Plants tufted perennials, rarely rhizomatous; sheaths open; auricles present, clasping; ligules membranous; inflorescence a panicle, the branches weakly rebranched; disarticulation above the glumes and between the florets; spikelets several-flowered; lemmas 3–7-nerved, the apex short-awned to ca. 3 mm; anthers 3.

1. Auricles lacking cilia (×10 or greater); 2 panicle branches borne at the lowermost node, together rarely bearing > 6 spikelets; old sheaths brown, decaying to fibers; blades 3–6(–7) mm wide…**S. pratensis**
1′. Auricles with minute cilia (×10 or greater); 2 or 3 panicle branches borne at the lowermost node, together usually bearing 5–18(–30) spikelets; old sheaths pale straw-colored, often remaining intact; blades 3–12 mm wide…**S. arundinaceus**

Schedonorus arundinaceus (Schreb.) Dumort. [*Festuca arundinacea* Schreb.]—Tall fescue. Plants to 2 m, often shorter; auricles ciliate; blades rolled in young shoots; panicle branches with abundant spikelets (up to 30+); lemmas 5–10 mm, awnless to short-awned to ca. 3 mm. Lawns, improved pastures, revegetation, 4335–9500 ft. Widespread except EC, SE. Introduced.

Schedonorus pratensis (Huds.) P. Beauv. [*Festuca pratensis* Huds.; *Lolium pratense* (Huds.) Darbysh.]—Meadow fescue. Plants to 1.3 m; auricles glabrous; blades folded or rolled in young shoots; lower panicle branches with few spikelets (< 8); lemmas 5–8 mm, awnless or with a very short mucro. Lawns, improved pastures, revegetation, 3850–9600 ft. Widespread except far E. Introduced.

SCHISMUS P. Beauv. – Mediterranean grass

Schismus barbatus (L.) Thell. [*S. arabicus* Nees]—Common Mediterranean grass. Often densely tufted annual, 2–25 cm; sheaths open, usually shorter than the internodes, with tufts of hair at the corners; ligules a ciliate membrane; blades 3–15 cm × 0.3–2 mm, glabrous to hairy adaxially; panicles dense, 1–7 × 1–3 cm, glumes 4–6 mm; lower lemmas 1.5–2.5 mm, variously hairy between the nerves and on the margins; ligules a ciliate membrane; paleas shorter to longer than the lemmas. Dry waste places, fields, roadsides, 3900–5500 ft. SC, SW, Bernalillo, San Juan. Introduced.

SCHIZACHNE Hack. – False melic

Schizachne purpurascens (Torr.) Swallen—False melic. Loosely tufted perennial, 30–80+ cm, often decumbent-based; sheaths closed almost to the summit; ligules membranous, 0.5–1.5 mm; blades folded or loosely rolled, 2–5 mm wide; panicles 7–17 cm, narrow to open; spikelets with 3–6 florets; disarticulation above the glumes and between the florets; glumes unequal, 3–5-nerved, often flushed with purple; lemmas papery, 7–9-nerved, the nerves parallel, 8–12 mm; awns 8–15 mm, divergent; anthers 3, 1.4–2 mm. Moist woods, pine forests, streamsides, meadows, 8200–9500 ft. NC, WC, Guadalupe.

SCHIZACHYRIUM Nees – Bluestem

Tufted to rhizomatous annuals and perennials; sheaths open; auricles absent; ligules membranous; blades of uppermost leaves highly reduced; inflorescence a single spicate raceme (rame) composed of repeating pairs of sessile and pedicelled spikelets subtended by a spathe, generally several of these separate rames on a flowering shoot; disarticulation below the glumes, the pair of spikelets falling together with a section of the rame; rame internode somewhat flattened, without a medial groove; sessile spikelets with 2 florets, the lower sterile, the upper fertile; glumes enclosing the florets and awnless; upper lemma deeply bifid and awned from the sinus; pedicelled spikelets staminate or sterile, shorter than or equaling the sessile spikelet, awned or awnless.

1. First glume of the sessile spikelet pubescent on the back…**S. sanguineum**
1′. First glume of the sessile spikelet glabrous on the back, but this sometimes obscured by subtending hairs…(2)
2. Pedicelled spikelets about the same size as the sessile; internodes and pedicels nearly glabrous or short-ciliate with very short hairs to 1 mm that do not at all obscure the spikelets…**S. cirratum**
2′. Pedicelled spikelets much shorter and narrower than the sessile; internodes and pedicels densely ciliate with hairs 1.5–6 mm that often obscure the spikelets…**S. scoparium**

Schizachyrium cirratum (Hack.) Wooton & Standl. [*Andropogon cirratus* Hack.]—Texas bluestem. Tufted perennial, ours lacking short rhizomes, 30–75 cm, often decumbent-based but not rooting; ligules 1–2.5 mm; blades 2–4 mm wide; rames 4–6 cm, usually exserted from the spathe; internodes straight, with a tuft of short hairs at the base but otherwise nearly glabrous; sessile spikelets 8–9 mm, the subtending hairs < 1/4 their length; lower glumes glabrous; upper lemmas cleft to 1/2 their length; awns 8–12 mm; pedicelled spikelets 6–10 mm, about the same length as the sessile, staminate to sterile, awnless, the pedicel stiffly ciliate on 1 side near the tip. Woodlands and rocky hills, 5000–7980 ft. WC, SW.

Schizachyrium sanguineum (Retz.) Alston [*Andropogon feensis* E. Tourn.]—Crimson bluestem. Tufted perennial, 40–110+ cm, glaucous; ligules 1–2 mm; blades 2–5 mm wide, with long bulbous-based hairs at the base; rames 4–12 cm, usually exserted from the spathe; internodes straight, glabrous to hirsute; sessile spikelets 5–9 mm, the subtending hairs to 1/2 their length; lower glumes hirsute; upper lemmas cleft 2/3 to nearly their length; awns 10–15 mm; pedicelled spikelets 3–5 mm, smaller than the sessile, staminate to sterile, with awns 0.3–5 mm, the pedicels curving outward in age. Our material belongs to **var. hirtiflorum** (Nees) S. L. Hatch. Woodlands and rocky hills, 5335–7500 ft. Doña Ana, Grant, Hidalgo.

Schizachyrium scoparium (Michx.) Nash [*Andropogon scoparius* Michx.]—Little bluestem. Tufted (ours) to rhizomatous perennials, greenish to purplish, 10–100+ cm (taller elsewhere); ligules 0.5–2 mm; blades 2–9 mm wide; rames 3–8 cm, usually exserted from the spathe; internodes sparsely to densely ciliate; sessile spikelets 4–11 mm, the subtending hairs 1/2 to nearly as long as the spikelet; lower glumes glabrous; upper lemmas cleft to 1/2 their length; awns 3–17 mm; pedicelled spikelets 1–10 mm, narrower than the sessile spikelet, sterile to staminate, awnless or awned to 4 mm. Our plants are **var. scoparium**. Hills, plains, woodlands, rocky slopes, 3400–8900 ft. Widespread.

SCLEROCHLOA P. Beauv. – Hardgrass

Sclerochloa dura (L.) P. Beauv.—Common hardgrass. Tufted annual, 2–15 cm, the shoots usually prostrate to sometimes ascending; sheaths open to closed 1/2 their length; auricle absent; foliage overlapping and usually surpassing the foliage; ligules membranous; blades 1–4 mm wide, the apices prow-shaped; racemes 1–5 cm, the base usually still in the subtending sheaths; spikelets 4–10 mm, laterally compressed, glabrous, subsessile on short pedicels to 1 mm, with 2–7 florets, awnless; both glumes much shorter than the lowermost lemma; lemmas 4–7 mm, the nerves scabridulous, the distal lemmas smaller; anthers 3. Adventive in lawns, golf courses, athletic fields, other moist waste places, 3235–6700 ft. Eddy.

SCLEROPOGON Phil. – Burrograss

Scleropogon brevifolius Phil. [*S. longisetus* Beetle]—Burrograss. Plants perennial, 10–20 cm, usually dioecious; stolons rooting at the nodes and producing new plants, the stolon internodes 5–15 cm; basal sheaths hairy; leaves mostly basal; ligules a ring of hairs ca. 1 mm; blades flat or folded, stiff, ca. 1 cm, 1–2 mm wide; inflorescence a spikelike raceme or weakly paniculate, rising above the leaves; staminate spikelets with 5–10 florets, the glumes and lemmas 1–3-nerved, awnless or short-awned, not disarticulating; pistillate spikelets with several florets, the glumes and lemmas 3-nerved, lemma awns 3–15 cm, widely divergent. Grassy plains and clay flats, 3480–8105 ft. NE, EC, C, WC, SE, SC, SW, Sandoval.

SECALE L. – Rye

Secale cereale L.—Cereal rye. Tufted annual, mostly 50–120+ cm; blades 4–12 mm wide; auricles present; ligules membranous; inflorescence a spike, with 1 spikelet per node; spikes 3–15 cm, erect to nodding; disarticulation below the glumes in the rachis or above the glumes and below the florets; spikelets with 2 florets; glumes needlelike, 1-nerved, 8–20 mm, with an awn 1–3 mm; lemmas strongly compressed and keeled, the keel ciliate, 14–18 mm, with awns 7–50 mm; anthers 3, ca. 7 mm. A cultivated crop plant, also widely used for erosion control along roadsides, 4270–8900 ft. NC, NW, WC, SC, SW, Roosevelt. Introduced.

SETARIA P. Beauv. – Bristlegrass

Tufted annuals and perennials, rarely with rhizomes; sheaths open; ligules a ring of hairs, or a ciliate membrane; inflorescence a usually dense, spikelike panicle, the branches sometimes nearly absent but usually discernible, also composed of numerous bristles below the spikelets that represent highly reduced branchlets; disarticulation below the glumes and above the bristles; spikelets with 2 florets, the lower staminate or sterile, the upper fertile, awnless; lower palea usually developed; upper floret indurate at maturity, transversely rugose (rarely smooth); anthers 3.

1. A single bristle usually present at the base of only the terminal spikelet of each branch…**S. reverchonii**
1′. Bristles present below all or nearly all the spikelets…(2)
2. Bristles with downward-pointing barbs, the seed heads thus readily clinging to clothing and to each other…(3)
2′. Bristles with upward-pointing barbs, the seed heads not readily clinging…(4)
3. Margins of the upper sheaths thin and translucent, glabrous, often with a slight auricle at the summit; blades stiff-pubescent on both surfaces…**S. adhaerens**
3′. Margins of the upper sheaths not thin and translucent, pubescent, lacking an auricle at the summit; blades scabrous or stiff-pubescent on the upper surface only…**S. verticillata**
4. Margins of the sheaths glabrous; bristles 4–13 below each spikelet; second glume 1/2–2/3 the length of the adjacent upper lemma…(5)
4′. Margins of the sheaths pubescent (rarely glabrous in *S. leucopila*); bristles 1–3 below each spikelet; second glume 3/4 to equaling the length of the adjacent upper lemma…(6)
5. Plants perennial from hard, knotty, nearly rhizomatous bases, the stems arising singly or in small tufts; spikelets 2–2.8 mm….**S. parviflora**
5′. Plants annual, the stems in large or small tufts; spikelets 2.8–3.4 mm…**S. pumila**
6. Plants annual, though often coarse and robust…(7)
6′. Plants perennial…(10)
7. Panicles contracted, but relatively loose and often lobed or interrupted below, the main axis visible…**S. grisebachii**
7′. Panicles dense, cylindrical and spikelike, lobed and interrupted in *S. italica*, otherwise the main axis obscured…(8)

8. Terminal panicles 18+ cm, as much as 40 cm; shoots 1.2–3 m...**S. magna**
8'. Terminal panicles 3–15 cm; shoots mostly 0.2–0.7 m...(9)
9. Panicles lobed; disarticulation above the glumes, the upper floret falling away from the spikelet...**S. italica**
9'. Panicles not lobed, cylindrical; disarticulation below the glumes...**S. viridis**
10. Palea of the lower floret nearly as long as the adjacent upper palea; spikelets mostly 2–2.3 mm, appearing globose; blades, at least some, 7–15 mm wide...**S. macrostachya**
10'. Palea of the lower floret 1/2–3/4 as long as the adjacent upper palea; spikelets 2.2–3 mm, elliptic; blades typically 2–5(–7) mm wide...**S. leucopila**

Setaria adhaerens (Forssk.) Chiov.—Clinging bristlegrass. Tufted annual, 20–60 cm; sheath margins thin and translucent distally, glabrous; ligules 1–2 mm, a ring of hairs; blades 5–10 mm wide, with bulbous-based hairs on both surfaces; panicles 2–6 cm, the bristles 4–7 mm, retrorsely scabrous, solitary beneath each spikelet; spikelets 2–2.3 mm; lower glume ca. 1/2 the spikelet length; lower palea < 1/2 the spikelet length; upper floret finely transversely rugose. Weedy sites, roadsides, lawns, 3800–6145 ft. SC, SW, Bernalillo, Eddy, Mora. Introduced.

Setaria grisebachii E. Fourn.—Grisebach's bristlegrass. Tufted annual, 30–100 cm; sheath margins ciliate; ligules a ring of hairs; blades flat, 4–20 mm wide, pilose-hispid on both surfaces; panicles 3–18 cm, lobed and interrupted, the main axis visible, often purple, the bristles 1–3 below each spikelet, 5–15 mm; spikelets 1.5–2.2 mm; lower glume ca. 1/3 the spikelet length; lower palea ca. 1/3 the spikelet length; upper floret finely transversely rugose. Canyon bottoms, rocky hills, stream banks, 4385–7600 ft. WC, SC, SW, Lincoln, Mora.

Setaria italica (L.) P. Beauv.—Foxtail millet. Tufted annual, 10–100+ cm; sheaths mostly glabrous, the margins sparsely ciliate; ligules 1–2 mm; blades 1–2 mm wide; panicles 8–30 cm, dense and congested, often drooping, sometimes lobed below, the bristles 1–3 below each spikelet, 3–12 mm; spikelets ca. 3 mm, the fertile floret falling out; lower glume ca. 1/2 the spikelet length; lower palea ca. 1/2 the spikelet length; upper floret appearing smooth, but very finely transversely rugose with magnification. Introduced as a cultivated crop and present in birdseed mixes, escaping but rarely persisting for long, 4500–5500 ft. Bernalillo, Colfax, Roosevelt.

Setaria leucopila (Scribn. & Merr.) K. Schum.—Plains bristlegrass. Tufted perennial, 20–100 cm; sheath margins villous distally; ligules a ciliate membrane 1–2.5 mm; blades 2–5 mm wide; panicles 6–15 cm, congested, the bristles usually solitary, 4–15 mm; spikelets 2.2–2.8(–3) mm, elliptic; lower glume ca. 1/2 the spikelet length; lower palea 1/2–3/4 the length of the adjacent upper palea; upper floret finely transversely rugose. Plains, rocky hills, slopes, 3275–6900 ft. Widespread except NC, NW.

Setaria macrostachya Kunth—Large-spike bristlegrass. Tufted perennial, 60–120 cm; sheath margins with only a few hairs at the throat; ligules 2–4 mm, densely ciliate; blades 7–15 mm wide; panicles 10–30 cm, congested, the bristles usually solitary, 10–20 mm; spikelets 2–2.3 mm, nearly spherical; lower glume 1/3–1/2 the spikelet length; lower palea nearly equaling the adjacent upper palea; upper floret finely transversely rugose. Rocky hills, 3850–6500 ft. WC, SW, Eddy.

Setaria magna Griseb.—Giant foxtail. Tufted annual, 1.5–4+ m; sheath margins villous distally; ligules 1–2 mm, ciliate; blades 1.5–3.5 cm wide; panicles 20–50 cm, 2–5 cm wide, congested, the bristles 1–2 below each spikelet, 10–20 mm; spikelets ca. 2 mm, the upper floret falling out; lower glume 1/3 the spikelet length; lower palea large, equaling the lower lemma and adjacent upper palea; upper floret smooth and shiny brown. Marshy ground at the Bitter Lake National Wildlife Refuge and moist roadsides, 3500–4500 ft. Chaves, Curry, Quay. Introduced.

Setaria parviflora (Poir.) Kerguélen [*S. geniculata* P. Beauv.]—Knotroot bristlegrass. Semitufted perennial with short knotty rhizomes, 30–120 cm; sheath margins glabrous; ligules a short ring of hairs, < 1 mm; blades 2–8 mm wide; panicles 3–10 cm, dense, the bristles 4–12 below each spikelet, 2–12 mm; spikelets 2–2.8 mm; lower glumes ca. 1/3 the spikelet length; lower palea equaling the lower lemma; upper floret distinctly transversely rugose. Open moist habitats in the foothills, 5825–6900 ft. C, Doña Ana, Hidalgo, San Miguel.

Setaria pumila (Poir.) Roem. & Schult. [*S. lutescens* (Weigel ex Stuntz) F. T. Hubb.]—Yellow bristle-grass. Tufted annual, 30-100+ cm; sheath margins glabrous; ligules ciliate; blades 4-10 mm wide, with bulbous-based hairs proximally on the adaxial surface; panicles 3-15 cm, dense, yellowish, the bristles 4-12 below each spikelet, 3-8 mm; spikelets 2-3.4 mm, turgid; lower glumes ca. 1/3 the spikelet length; upper glumes ca. 1/2 the spikelet length; lower palea equaling the lower lemma; upper floret exposed (not covered by the upper glume), strongly rugose. Weedy ground along roads, fields, in lawns, 4230-6600 ft. Widespread. Introduced.

Setaria reverchonii (Vasey) Pilg.—Reverchon's bristlegrass. Rhizomatous perennial, the bases knotty, 30-90 cm; sheath margins ciliate distally; ligules a ring of stiff hairs 1-2 mm; blades 1-7 mm wide, rolled, stiff; panicles 5-20 cm, loose, lobed and interrupted, the branches weakly developed, the bristles 2-8 mm, solitary below usually only the terminal spikelet of the branches, sometimes present below other spikelets as well; spikelets 2-4.5 mm; lower glumes 1/2 the spikelet length; lower paleas absent; upper floret finely transversely rugose. New Mexico material belongs to **subsp. ramiseta** (Scribn.) W. E. Fox. Dry plains and scrublands. Chaves, Eddy.

Setaria verticillata (L.) P. Beauv.—Hooked bristlegrass. Tufted annual, 30-100 cm; sheath margins ciliate distally, not thin and translucent; ligules a densely ciliate membrane 0.3-1 mm; blades 5-15 mm wide, scabrous on the adaxial surface only; panicles 5-15 cm, dense, the bristles solitary, 4-7 mm, re-trorsely scabrous; spikelets 2-2.3 mm; lower glumes ca. 1/3 the spikelet length; lower palea ca. 1/2 the spikelet length; upper floret finely transversely rugose. Weedy ground, known only from a few old collections, 3800-5300 ft. Doña Ana. Introduced.

Setaria viridis (L.) P. Beauv.—Green bristlegrass. Tufted annual, 20-100 cm; sheath margins ciliate distally; ligules a ciliate membrane 1-2 mm; blades 4-25 mm wide, glabrous; panicles 3-20 cm, dense, sometimes the apex nodding, the bristles 1-3 below each spikelet, 5-10 mm; spikelets 1.8-2.2 mm; lower glumes ca. 1/3 the spikelet length; lower palea ca. 1/3 the lower lemma length; upper floret very finely transversely rugose, greenish. Common weed in disturbed ground, 3570-9320 ft. Widespread. Introduced.

SORGHASTRUM Nash – Indiangrass

Sorghastrum nutans (L.) Nash—Indiangrass. Rhizomatous perennial, the aerial shoots sometimes forming small clumps, 50-200+ cm; sheaths open, with thick erect auricles at the collars adjacent to the ligule; ligules 2-6 mm, membranous; blades 1-4 mm wide, usually glabrous; panicles 20-75 cm, yellowish to brownish or copper-colored, loosely contracted, the branches ascending; pedicels and rachises copiously hairy; disarticulation below the spikelets; spikelets sessile, with 2 florets, the lower sterile, the upper fertile, 5-8.7 mm, sparsely long-hairy; awns 10-30 mm; glumes coriaceous, enclosing the florets; upper lemma fertile, bifid, awned from the sinus; pedicels slender, lacking a spikelet at the tip, but sometimes with a vestige of a scale. Grasslands, open woods, prairies, moist rocky hillsides, 4260-8000 ft. Widespread.

SORGHUM Moench – Sorghum

Tufted to rhizomatous annuals and perennials; sheaths open; auricles absent; ligules membranous and ciliate, or a ring of hairs; blades usually broad and flat; inflorescence a rebranching panicle, the terminal branches (rames) composed of repeating pairs of sessile and pedicelled spikelets, the sessile ones fertile, the pedicelled ones staminate or sterile; sessile spikelets with 2 florets, the lower staminate or sterile, the upper fertile, both highly reduced and hidden within the large glumes, awned; pedicelled spikelets usually with a single floret or none, well developed and subequal to the sessile spikelet, awnless.

1. Plants perennial, with strong rhizomes; rame segments breaking apart easily…**S. halepense**
1′. Plants annual, lacking rhizomes; rame segments persistent or breaking apart tardily and inconsistently…
S. bicolor

Sorghum bicolor (L.) Moench—Sorghum, milo. Tufted robust annual, sometimes longer-lived, 50–300+ cm, the culms 1–5 cm thick; blades to 1 m long and 10 cm wide; panicles 10–60 cm, open or contracted; disarticulation not occurring or tardily so; sessile spikelets 3–9 mm, glabrous to various hairy, awnless or with an awn 5–30 mm; pedicelled spikelets 3–6 mm, somewhat shorter than the sessile spikelet, awnless. Grown as a cultivated crop, infrequently escaping along fields but not persisting long, 3800–6400 ft. Bernalillo, Doña Ana, Roosevelt. Introduced.

Sorghum halepense (L.) Pers.—Johnsongrass. Rhizomatous perennial, 50–200+ cm, the culms 0.5–2 cm thick; blades to 90 cm and 4 cm wide; panicles 10–50 cm, generally open, 5–25 cm wide; disarticulation below the spikelet pair; sessile spikelets 4–6.5 mm, pubescent, awnless or with awns to 13 mm; pedicelled spikelets 3.5–5.5 mm, somewhat shorter than the sessile spikelets, awnless. An aggressive weed of fields, ditches, moist waste places, 3100–7550 ft. NE, EC, C, SE, SC, SW. Adventive.

SPARTINA Schreb. – Cordgrass

Tufted to rhizomatous perennials; sheaths open; auricles absent; ligules ciliate from a short membrane; inflorescence a panicle of spikelike branches, the branches scattered along the rachis, not digitate, erect-appressed to divergent; spikelets sessile, with 1 floret, laterally compressed, crowded on 1 side of the branch; disarticulation below the glumes, the spikelet falling; glumes keeled, unequal, the lower usually shorter than the florets and the upper usually longer; lemmas 1–3-nerved, keeled; anthers 3.

1. Plants slender, the shoots 2–4 mm thick; most blades < 5 mm wide; upper (longer) glume only slightly longer than the floret, 6–10 mm, acute to attenuate but not awned…**S. gracilis**
1'. Plants robust, the shoots 3–11 mm thick; most blades > 5 mm wide; upper (longer) glumes nearly twice as long as the floret, 10–25 mm, including the awn…**S. pectinata**

Spartina gracilis Trin.—Alkali cordgrass. Strongly rhizomatous perennial, 40–100 cm, the culms slender, 2–4 mm thick; ligules 0.5–1 mm; blades 2–8 mm wide; panicles 8–25 cm, with 3–12 branches 2–8 cm, with 10–30 spikelets; spikelets 6–11 mm; upper glumes 6–10 mm, subequal to the floret, acute to attenuate, awnless. Marshes and wet prairies; last found in 1945, elevation unknown. San Miguel.

Spartina pectinata Bosc ex Link—Prairie cordgrass. Strongly rhizomatous perennial, 50–150+ cm, the culms robust, 3–11 mm thick; ligules 1–3 mm; blades 5–15 mm wide; panicles 30–90 cm, with 5–50 branches 2–15 cm, with 10–80 spikelets; spikelets 10–25 mm; upper glumes 10–25 mm (including the awn), much exceeding the floret, with an awn 3–8 mm. Marshes and wet prairies, 3500–6000 ft. Chaves, Guadalupe, Roosevelt, Union.

SPHENOPHOLIS Scribn. – Wedgescale

Tufted annuals and perennials; sheaths open; auricles absent; ligules membranous; inflorescence an open to contracted panicle; disarticulation below the glumes, the upper florets sometimes falling prior; spikelets with 2–3 florets, laterally compressed, the rachilla prolonged beyond the terminal floret; glumes unequal and dissimilar, with scarious margins, somewhat wedge-shaped; lemmas 3–5-nerved, but the nerves scarcely visible, awned or awnless; anthers 3.

1. Plants annual; spikelets awned…**S. interrupta**
1'. Plants perennial; spikelets awnless…(2)
2. Second glume rounded to broadly obovate, somewhat hood-shaped, 1/3–1/2 as wide as long; panicles dense, spikelike…**S. obtusata**
2'. Second glume blunt to acute, oblanceolate, not hood-shaped, 1/6–1/3 as wide as long; panicles loose, somewhat open…**S. intermedia**

Sphenopholis intermedia (Rydb.) Rydb.—Slender wedgescale. Tufted perennial, 20–100+ cm; sheaths glabrous to hairy; ligules 1.5–2.5 mm; blades 2–6 mm wide; panicles 5–20 cm, usually nodding, the spikelets loosely arranged; spikelets 2–4 mm; upper glume 2–3 mm, oblanceolate to obovate, not cucullate, the apex acute to rounded; lower lemma 2–3 mm. Moist ground in forests, shaded ground along streams, 7020–8555 ft. NC, Lincoln. Sometimes confused with *Koeleria macrantha*, but that species has (most conspicuously) minutely fuzzy pedicels and panicle branches.

Sphenopholis interrupta (Buckley) Scribn. [*Trisetum interruptum* Buckley]—Prairie false oat. Tufted annual, 5–50 cm; leaves tending to be basal; sheaths glabrous to pilose; ligules 1–2.5 mm; blades glabrous to hairy, the margins ciliate; panicles 2–15 cm, to 1.5 cm wide, congested, the branches short and usually erect; spikelets 3.5–6 mm, with 2–3 florets; disarticulation above and below the glumes; rachillae 0.8–1 mm, with hairs ca. 0.5 mm; glumes subequal, elliptic to oblanceolate, the lower ca. 1/2 as wide as the upper; lowermost lemmas 3–4.5 mm; ovary glabrous to sparsely hairy near the apex. Dry, rocky, desert hills, 4225–9000 ft. EC, SE, SC, SW, San Miguel.

Sphenopholis obtusata (Michx.) Scribn.—Prairie wedgescale. Tufted perennial, 15–100+ cm; sheaths glabrous to hairy; ligules 1–2.5 mm; blades 1–8 mm wide; panicles 5–25 cm, usually erect, the spikelets densely arranged; spikelets 2–3.5 mm; upper glume 1.5–2.5 mm, broadly obovate, somewhat cucullate, the apex rounded to blunt; lower lemma 2–2.8 mm. Moist or wet ground along streams, springs, canals, ditches, 3750–8300 ft. Widespread except EC.

SPOROBOLUS R. Br. – Dropseed

Tufted, rhizomatous, and stoloniferous annuals and perennials; sheaths open, often with tufts of hair at the collar edge or ciliate on the margins; ligules a ring of hairs; inflorescence a panicle, sometimes reduced and racemose; spikelets with 1 floret (rarely 2 or 3), awnless; disarticulation below the glumes, or above the glumes with the lemmas and paleas separating and the seeds falling from the ovary as well; lemmas 1-nerved; paleas well developed, often as long as or longer than the lemma; fruit a utricle, the pericarp becoming free from the seed. Included herein are the genera *Crypsis* and *Calamovilfa* (Peterson et al., 2014).

1.   Plants annual…(2)
1′.   Plants perennial…(6)
2.   Sheaths prominently inflated; blades widely spreading to reflexed; inflorescence dense and headlike or spikelike, the base often included in the sheath…(3)
2′.   Sheaths, blades, and panicles not all as above…(4)
3.   Inflorescence at maturity 5–6 times longer than broad, spikelike, exserted beyond the sheath; spikelets often black-tinged…**S. alopecuroides**
3′.   Inflorescence at maturity 3–4 times longer than broad, headlike, often remaining partially within the sheath; spikelets pale to purple-tinged…**S. schoenoides**
4.   Spikelets all < 2 mm; glumes very unequal; panicles narrow when in flower and open at maturity, the lower branches whorled…**S. pyramidatus** (in part)
4′.   Spikelets, at least some, > 2 mm; glumes equal or nearly so; panicles narrow, the lower branches often included in the subtending sheath…(5)
5.   Florets glabrous…**S. neglectus**
5′.   Florets pubescent…**S. vaginiflorus**
6.   Lemmas with a prominent tuft of hairs at the base…(7)
6′.   Lemmas lacking a tuft of hairs at the base…(8)
7.   Lemmas and paleas long-pubescent along the back above the callus hairs…**S. arenicola**
7′.   Lemmas and paleas glabrous above the callus hairs…**S. rigidus**
8.   Lateral pedicels 5–25 mm…**S. texanus**
8′.   Lateral pedicels 4 mm or less…(9)
9.   Spikelets 1–2(–2.9) mm…(10)
9′.   Spikelets, at least some, 3+ mm…(17)
10.   Panicles dense and spikelike, the branches appressed…(11)
10′. Panicles open, the branches spreading at least from the middle and at the tip, the lower portion often enclosed in the subtending sheath…(12)
11.   Stems robust, 1–2 m, 3–8 mm thick at the base; anthers 0.6–1 mm…**S. giganteus** (in part)
11′. Stems more slender, mostly < 1 m, 1.5–3.5 mm thick at the base; anthers 0.3–0.5 mm…**S. contractus**
12.   Base of the plant knotty, nearly rhizomatous; blades stiff, spreading at right angles; stems mostly < 30 cm…**S. nealleyi**
12′. Base of plant loosely tufted, not knotty; blades erect or ascending; stems often > 30 cm (except *S. pyramidatus*)…(13)
13.   Primary panicle branches with sticky-glandular streaks or patches; lowermost branches in definite whorls; stems 10–60 cm…**S. pyramidatus** (in part)
13′. Primary panicle branches lacking any sticky-glandular patches; lowermost branches whorled or not, often in the sheath; stems often 40–120+ cm…(14)

14.   Sheaths with many long hairs at the summit; plants more slender, the shoots easily pulled from the ground; basal sheaths not shiny, often darkened; roots thin…(15)

14'.   Sheaths glabrous or with only a few long hairs at the summit; plants robust, the shoots difficult to pull from the ground; basal sheaths shiny and cream-colored; roots thick…(16)

15.   Mature panicle branches and pedicels divaricate and flexuous, usually tangled with other branches or other panicles; branch pulvini pubescent; spikelets loosely arranged on the branches…**S. flexuosus**

15'.   Mature panicle branches erect to spreading but not flexuous or tangled; branch pulvini glabrous; spikelets crowded on the branches…**S. cryptandrus**

16.   Panicles 10–45 cm; branchlets naked below, the pedicels 0.5–2 mm, often spreading…**S. airoides**

16'.   Panicles 20–60 cm; branchlets densely flowered to the base, the pedicels < 0.5 mm, appressed to the branchlets…**S. wrightii**

17.   Second glume shorter than the lemma, the floret extending beyond the glume…**S. compositus**

17'.   Second glume equal to or longer than the lemma, the floret not extending beyond the glume, but often surpassed by it…(18)

18.   Panicles usually spikelike; spikelets 2.5–3.5 mm; grain not globe-shaped; blades as much as 10 mm wide …**S. giganteus** (in part)

18'.   Panicles usually loose, the branches spreading; spikelets 4–6 mm; grain globe-shaped; blades 1–2 mm wide…**S. heterolepis**

Sporobolus airoides (Torr.) Torr.—Alkali sacaton. Densely tufted perennial, forming large tussocks to 1+ m across, lacking rhizomes, 35–150 cm; adventitious roots thick, 2–4 mm diam.; lower sheaths shiny, becoming straw-colored; sheath apices glabrous or only sparsely hairy with long hairs; ligules < 0.5 mm; blades to 60 cm, weeping; panicles 15–45 cm, very open and diffuse at maturity, the base often included in the sheath; pulvini glabrous; spikelets 1.3–2.8 mm; glumes unequal, the upper subequal to the floret, the lower 1/2 or less the spikelet length. Sandy, gravelly, clayey plains, flats, mesas, playas, floodplains, 3400–8200 ft. Widespread.

Sporobolus alopecuroides (Piller & Mitterp.) P. M. Peterson [*Crypsis alopecuroides* (Piller & Mitterp.) Schrad.]—Foxtail pricklegrass. Tufted annuals, 5–70 cm, rarely branched above, the shoots prostrate to erect; blades 1–3 mm wide; panicles dense, spikelike, 1.5–6.5 cm, 4–6 mm wide, often purplish, completely exserted from the subtending sheath when fully mature; spikelets 1.8–2.8 mm. Shorelines of ponds and lakes, 4600–7400 ft. Cibola, McKinley, Socorro.

Sporobolus arenicola P. M. Peterson [*Calamovilfa gigantea* (Nutt.) Scribn. & Merr.]—Big sandreed. Perennial with elongate shiny rhizomes, 1–2.4 m; sheaths glabrous or pubescent at the throat; ligules 0.7–2 mm; blades elongate, to 90 cm and 12 mm wide; panicles 25–80 cm, 20–55 cm wide, the branches ascending to divergent; spikelets 7–11 mm; tuft of hairs at base of floret 1/4–3/4 the floret length; lemmas 6–10 mm, pubescent on the back above the callus hairs. Sandy hills and dunes in the E plains, 3060–4145 ft. NE, EC, SE, San Miguel.

Sporobolus compositus (Poir.) Merr. [*S. asper* (P. Beauv.) Kunth]—Composite dropseed. Tufted perennial, without rhizomes (ours), 25–100+ cm; sheath apices sparsely pilose; ligules < 0.5 mm; panicles 8–30 cm, contracted and spikelike, with smaller cleistogamous panicles produced in the axils of the sheaths; spikelets 4–8 mm; glumes 1/2–2/3 the spikelet length; lemmas (2.3–)3–6 mm, extending beyond the upper glume; paleas subequal to the lemmas. Plains and grasslands, sometimes roadsides, 3900–7550 ft. Scattered.

Sporobolus contractus Hitchc.—Spike dropseed. Tufted perennial, lacking rhizomes, 40–100+ cm, the culms 2–5 mm thick; sheath margins densely ciliate, the apices with conspicuous tufts of pilose hairs; ligules 0.4–1 mm; panicles 10–50 cm, 0.3–1 cm wide, spikelike, the base often in the sheath; spikelets 1.7–3.2 mm; glumes unequal, the lower ca. 1/2 or less the length of the upper; lemmas 2–3.2 mm, surpassing both glumes; paleas subequal to the lemmas. Sandy hills and plains, 3815–7400 ft. Widespread.

Sporobolus cryptandrus (Torr.) A. Gray—Sand dropseed. Tufted perennial, lacking rhizomes, 30–120 cm, the culms 1–3.5 mm thick; sheath margins sometimes ciliate, the apices with conspicuous tufts of pilose hairs 2–4 mm; ligules 0.5–1 mm; panicles erect, 15–40 cm, 2–14 cm wide, open in age but often spikelike when young, the branches ascending to reflexed, not entangled; secondary branchlets and spikelets generally appressed along the primary branches; pulvini glabrous; spikelets 1.5–2.7 mm; glumes unequal, the lower 1/2 or less the length of the upper; upper glume 2/3+ the spikelet length; lemmas 1.4–2.7 mm; paleas subequal to the lemmas. Sandy or gravelly plains, mesas, roadsides, waste places, 3475–10,200 ft. Widespread.

Sporobolus flexuosus (Thurb. ex Vasey) Rydb.—Mesa dropseed. Tufted perennial, lacking rhizomes, 30–120 cm, the culms 1–3 mm thick near the base; sheath margins glabrous to ciliate, the apices with tufts of hair 1–3 mm; ligules 0.5–1 mm; panicles often nodding, 10–30 cm, 4–12 cm wide, open and diffuse, often becoming entangled with other branches and panicles, the branches diverging to strongly reflexed; secondary branchlets and spikelets widely spreading; pulvini hairy; spikelets 1.8–2.5 mm; glumes unequal, the lower 1/2 or less the length of the upper; upper glume subequal to the floret; lemmas 1.4–2.5 mm; paleas equal to the lemmas. Sandy plains and mesas, 3750–7500 ft. Widespread except NE, EC.

Sporobolus giganteus Nash—Giant dropseed. Large tufted perennial, lacking rhizomes, 100–200 cm, the culms 4–10 mm thick; sheath margins ciliate distally, the apices with tufts of pilose hairs to 2 mm; ligules 0.5–1.5 mm; panicles 25–75 cm, 1–4 cm wide, spikelike, the base often in the sheath, the branches mostly appressed but sometimes ascending outward; pulvini glabrous; spikelets 2.5–4 mm; glumes unequal, the lower ca. 1/2 the length of the upper; upper glumes subequal to the floret; lemmas 2.5–4 mm; paleas equal to the lemmas. Sandy hills and plains, 3215–7300 ft. Widespread.

Sporobolus heterolepis (A. Gray) A. Gray—Prairie dropseed. Tufted perennial, lacking rhizomes, 30–80 cm; sheath margins glabrous, the apices with sparse hairs; ligules 0.1–0.3 mm; blades 1–2.5 mm wide; panicles 5–25 cm, 1–11 cm wide, contracted to loosely open, the branches erect to ascending-spreading; pulvini glabrous; spikelets 3–6 mm; glumes unequal, the lower ca. 1/2 the upper; upper glume usually surpassing the floret; lemmas 2.7–4.3 mm; paleas equal to longer than the lemmas; fruit globose, indurate, without a loose pericarp. Grasslands and woodlands, 6900–7565 ft. Colfax.

Sporobolus nealleyi Vasey—Gypgrass. Tufted perennial, the bases hard and knotty, 10–50 cm, the culms 0.7–1.2 mm thick; sheath margins villous-ciliate, the apices with a tuft of hairs to 4 mm; ligules 0.2–0.4 mm; blades stiff, spreading at right angles; panicles 3–10 cm, 1–6 cm wide, the branches eventually spreading to divergent, the secondary branchlets spreading to appressed; spikelets 1.4–2.1 mm; glumes unequal, the lower one ca. 1/2 the upper; upper glumes ca. as long as the spikelet; lemmas 1.4–2.1 mm; paleas equal to the lemmas. Sandy, alkaline, mostly gypsiferous plains and flats, 3350–6300 ft. NE, NC, EC, C, SE, SC.

Sporobolus neglectus Nash—Puffsheath dropseed. Tufted delicate annual, 10–40 cm; sheath inflated, the margins glabrous, the apices with small tufts of hairs to 3 mm; ligules 0.1–0.3 mm; panicles 2–5 cm, to 0.5 cm wide, often completely hidden in the sheaths; spikelets 1.6–3 mm; glumes nearly equal, shorter than the floret; lemmas 1.6–2.9 mm, glabrous; paleas equaling the lemmas. Sandy fields, floodplains, stream banks, disturbed ground, 5000–6000 ft. Bernalillo, Chaves, Grant, San Juan, Santa Fe. Introduced.

Sporobolus pyramidatus (Lam.) Hitchc. [*S. pulvinatus* Swallen]—Madagascar dropseed. Tufted annual, sometimes longer-lived, 7–40(–60) cm, erect to decumbent; sheath margins ciliate, the apices with a tuft of pilose hairs to 3 mm; ligules 0.3–1 mm; blades 2–6 mm wide, the margins pectinate-ciliate; panicles 4–18 cm, open, pyramidal, 1–6 cm wide, the lower branches whorled, with elongate glandular patches or streaks; pulvini glabrous; spikelets 1.2–1.8 mm; glumes strongly unequal, the lower < 1 mm, the upper 1.2–1.8 mm, 2/3+ the spikelet length; lemmas 1.2–1.7 mm; paleas equal to the lemmas. Sandy plains, clay flats, disturbed ground, 4050–6815 ft. NC, NW, WC, SC, SW, Bernalillo, Chaves, Roosevelt.

Sporobolus rigidus (Buckley) P. M. Peterson [*Calamovilfa longifolia* (Hook.) Hack. ex Scribn. & Southw.]—Prairie sandreed. Strongly rhizomatous perennial, 0.5–2 m; sheaths glabrous to softly pilose, the margins glabrous to ciliate, the apices often with scattered hairs; blades elongate, to 65 cm, 12 mm wide; panicles 15–70 cm, 4–25 cm wide, the branches ascending to divergent; spikelets 5–8.5 mm; lemmas 4.5-7 mm, glabrous on the back above the callus hairs. Sandy hills and dunes in the E plains; 1 record, 4075 ft. NE, NC.

Sporobolus schoenoides (L.) P. M. Peterson [*Crypsis schoenoides* (L.) Lam.]—Swamp pricklegrass. Tufted annual, the shoots 2-75 cm, usually not branching, but some plants strongly branched above the base; blades 2-6 mm wide; panicles dense, spikelike or headlike, 0.5-7 cm, 5-8(-12) mm wide, the base usually remaining in the subtending sheath at maturity; spikelets 2.7-3.2 mm, strongly laterally compressed; glumes subequal, larger than the floret; lemmas 2.4-3 mm, the palea slightly smaller. Wet ground along ponds and marshes, 4690-6280 ft. McKinley, Socorro.

Sporobolus texanus Vasey—Texas dropseed. Tufted perennial, 20-70 cm; sheath margins glabrous, the apices glabrous or with sparse pilose hairs to 4 mm; ligules 0.2-0.6 mm; panicles 10-35 cm, widely diffuse, 1/2 to nearly as wide, the base usually in the sheath, all branches, branchlets, and pedicels spreading; pedicels 5-25 mm; spikelets 2.3-3 mm; glumes strongly unequal, the lower 1/3 or less the length of the upper, upper glumes 2/3 as long as to slightly longer than the floret; lemmas 1.8-3 mm, glabrous; paleas equal to the lemmas. Low wet plains and swales, 3460-8000 ft. Chaves, Doña Ana, Eddy, Sandoval.

Sporobolus vaginiflorus (Torr. ex A. Gray) Alph. Wood—Poverty dropseed. Tufted delicate annual, 15-60 cm; sheaths often inflated, the margins glabrous, the apices sometimes with small tufts of hair to 3 mm; ligules 0.1-0.3 mm; panicles 1-5 cm, 2-5 mm wide, spikelike, enclosed in the sheath; spikelets 3-6 mm; glumes subequal, narrowly lanceolate, slightly shorter or longer than the floret; lemmas and paleas 3-5.4 mm. Sandy and disturbed ground; Oasis State Park, 4070 ft. Bernalillo, Doña Ana, Roosevelt. Introduced.

Sporobolus wrightii Munro ex Scribn.—Giant sacaton. Large tussocky perennial, 1-2.5 m; sheath margins not ciliate, the apices sparsely hairy with hairs to 6 mm; ligules 1-2 mm; blades to 70 cm, flat, 3-10 mm wide; panicles 20-60 cm, 12-26 cm wide, open, broadly lanceolate, generally fully exserted, the secondary branches, branchlets, and pedicels usually appressed; spikelets 1.5-2.5 mm; glumes unequal, the lower ca. 1/2 the length of the upper; upper glumes 3/4 as long as to subequal to the floret; lemmas 1.2-2.5 mm, glabrous; paleas equal to the lemmas. Swales, playas, ditches, often in hard-packed alkaline soil, 3770-6630 ft. Widespread except NE, NC.

STEINCHISMA Raf. – Gaping grass

Steinchisma hians (Elliott) Nash [*Panicum hians* Elliott]—Gaping grass. Tufted perennial, 20-75 cm, the shoots and sheaths compressed; sheaths open, usually keeled; auricles absent shorter than the internodes, sparsely ciliate at the summit; ligules a minute ciliate membrane, 0.2-0.5 mm; blades 2-5 mm wide, sparsely pilose adaxially at the base; panicles 5-20 cm, ca. 1/2 as wide, open, the branches lax, the secondary branchlets appressed; spikelets elongate, gaping open in age, with 2 florets, the lower staminate or sterile, the upper fertile, 1.8-2.4 mm, often purplish; lower glumes ca. 1/2 the spikelet length; lower palea slightly longer than the lower lemma; fertile floret 1.6-2 mm, dull, minutely papillose, with an acute apex. Collected once in Las Cruces in 1895. Doña Ana.

STENOTAPHRUM Trin. – St. Augustine grass

Stenotaphrum secundatum (Walter) Kuntze—St. Augustine grass. Stoloniferous perennial, 10-30 cm; sheaths open, sparsely pilose, constricted at the summit, shorter than the internodes; ligules a ciliate membrane or ring of hairs ca. 0.5 mm; blades 3-15 cm, 4-10 mm wide, thick, flat; spicate panicles 4-10 cm, < 1 cm wide, the rachis flattened and winged; disarticulation below the glumes; spikelets awnless, 3.5-5 mm; lower glume 1/3 or less the spikelet length; upper lemma papery, weakly clasping the palea. Cultivated as a coarse-textured lawn grass for shaded areas, New Mexico State University campus, 3925 ft. Doña Ana.

TORREYOCHLOA G. L. Church – Mannagrass

Torreyochloa pallida (Torr.) G. L. Church [*Glyceria pauciflora* J. Presl; *T. pauciflora* (J. Presl) G. L. Church]—Rhizomatous perennial, 20-140 cm, erect to decumbent-based; sheaths open; auricles absent; larger ligules membranous, 3-9 mm; blades flat, to 18 mm wide; panicles 5-25 cm, 2-14 cm wide, the branches often lax and flexuous, erect to reflexed; disarticulation above the glumes and between the florets; spikelets with 2-8 florets, 3.6-7 mm; glumes shorter than the lowermost floret; upper glumes 1-1.8 mm; lemmas rounded, with 7-9 parallel nerves, the lateral nerves shorter, 2.2-3.3 mm, truncate to acute. Our plants belong to **var. pauciflora** (J. Presl) J. I. Davis. Wet ground of high-mountain streams and freshwater ponds, 7000-10,500 ft. NC, Catron, Lincoln.

TRACHYPOGON Nees – Crinkleawn grass

Trachypogon spicatus (L. f.) Kuntze—Crinkleawn. Tufted perennial, 60-120 cm, generally unbranched from the base, the nodes appressed-hirsute; sheaths open, sparsely appressed-pilose; auricles absent; ligules membranous, 2-5 mm; racemes 10-18 cm, composed of repeating pairs of subsessile and pedicelled spikelets; disarticulation beneath the pedicelled spikelet; subsessile spikelets staminate or sterile, awnless, otherwise similar to the pedicelled spikelets; pedicelled spikelets 6-8 mm, the glumes pilose, the awns 4-6 cm, with pilose hairs to 2 mm proximally, nearly glabrous distally. Rocky hills and slopes, 4500-5900 ft. Hidalgo.

TRAGUS Haller – Bur grass

Tragus berteronianus Schult.—Spiked bur grass. Tufted annual, the shoots 3-40 cm, erect to prostrate; sheaths open, shorter than the internodes; ligules a ciliate membrane; blades 1-8 cm, margins ciliate; panicles 1-13 cm × 4-8 mm; branches 0.5-3 mm, with 2(3) spikelets, the branches burlike; disarticulation at the base of the branch; spikelets with 1 floret, the lower spikelets fertile, the distal spikelets sterile and smaller; lower (inner) glumes 0.1-0.6 mm; upper (outer) glumes 1.8-4.3 mm, 5-nerved, with 5 longitudinal rows of hooked projections 0.3-1 mm. Disturbed ground in desert plains, mesas, bajadas, 4000-7000 ft. NW, EC, C, WC, SC, SW. Introduced.

TRICHACHNE Nees – Cottontop

Trichachne californica (Benth.) Chase [*Digitaria californica* (Benth.) Henrard]—Arizona cottontop. Tufted perennial, with knotty or spreading rhizomes, 40-100 cm; basal sheaths villous, open; auricles absent; ligules membranous, sometimes with a ciliate fringe; panicles 8-15 cm, densely villous; panicle branches appressed to ascending, not digitate but attached along the main axis; spikelets short-pedicelled, in pairs of 1 sessile and 1 pedicellate, appressed in 2 rows along 1 side of the branch, awnless, silky-pubescent, the hairs generally longer than the spikelet, 2-flowered, the lower floret staminate to sterile, the upper floret fertile, the rachilla elongate between the glumes and florets, the upper floret ± stipitate; disarticulation below the glumes; lower glume absent to no more than 1/4 the length of the spikelet; upper glume and lower lemma densely villous with hairs to 5 mm; anthers 3. Rocky plains, foothills, bajadas, 3450-7000 ft. EC, C, WC, SE, SC, SW, Bernalillo, Santa Fe.

TRIDENS Roem. & Schult. – Tridens

Tufted perennials, sometimes with short rhizomes, lacking stolons; sheaths open, shorter than the internodes; auricles absent; ligules a ciliate membrane or a ring of hairs; inflorescence a panicle, sometimes racemose; spikelets with several florets, sometimes the distal ones sterile; disarticulation above the glumes and between the florets; glumes awnless; lemmas 3-nerved, usually hairy on the nerves, awnless or the midnerve excurrent to a mucro or awn-tip; paleas glabrous, widened or bowed-out below; anthers 3, purplish.

1. Panicles open, loose, the branches spreading to drooping…(2)
1′. Panicles narrow, contracted, the branches erect…(3)

2. Lemmas 2–3 mm, only the midnerve projecting as a short point…go to **Triplasiella**
2'. Lemmas 3–5 mm, the midnerve and lateral nerves projecting as short points…**T. flavus**
3. Nerves of the lemma glabrous or pubescent only at the base…**T. albescens**
3'. Nerves of the lemma plainly pubescent…go to **Tridentopsis**

Tridens albescens (Vasey) Wooton & Standl.—White tridens. Tufted perennial, often with a knotty short-rhizomatous base, 30–100 cm; ligules to 0.5 mm; panicles 8–25 cm, 0.5–1.5 cm wide; glumes about as long as the adjacent lemma, 1-nerved; lemmas 3–5 mm, papery, whitish or purplish, glabrous or with a few short hairs at the base of the lateral nerves; paleas long-hairy. Low swales and ditch banks in the plains, deserts, prairies, 3150–6500 ft. EC, SE, SC, Luna, Sandoval.

Tridens flavus (L.) Hitchc.—Purpletop. Tufted perennial, with short rhizomes from a knotty base, 60–180 cm; ligules to 0.5 mm; panicles 15–40 cm, 2/3 as wide, open, pyramidal, the branches rebranched and drooping, covered with glandular dots; glumes shorter than the adjacent lemmas, 1-nerved; lemmas 3–5 mm, the nerves prominently villous-ciliate and extended as 3 tiny mucros; paleas glabrous, widened or bowed-out below. Prairies and grassy hills; known from a single collection near Clines Corners, ca. 7000 ft. Torrance. Introduced.

TRIDENTOPSIS P. M. Peterson – Tridens, tridentopsis

Tridentopsis mutica (Torr.) P. M. Peterson [*Tridens elongatus* (Buckley) Nash]—Slim tridens. Tufted perennial, with short rhizomes from a knotty base, 20–80 cm; lacking stolons, the nodes often bearded; sheaths open, shorter than the internodes; auricles absent; ligules a ciliate membrane or a ring of hairs, 0.5–1 mm; panicles 7–25 cm, somewhat racemose, to 1 cm wide, the branches racemose; spikelets with several florets, sometimes the distal ones sterile; disarticulation above the glumes and between the florets; glumes glabrous, awnless, shorter than adjacent lemmas, 1–7-nerved; lemmas 3-nerved, usually hairy on the nerves, awnless or the midnerve excurrent to a mucro or awn-tip, 3.5–7 mm, purplish, the nerves pilose above midlength, rarely excurrent; paleas shorter than the lemmas; anthers 3, purplish. Dry, flat hills, outcrops, often on limestone, 3450–5600 ft. EC, SE, SC, SW, Harding. Plants with elongate inflorescences, and second glume 3–7-nerved and 6–8 mm may be referred to as **var. elongata** (Buckley) P. M. Peterson & Romasch.

TRIPIDIUM H. Scholz – Ravennagrass

Tripidium ravennae (L.) H. Scholz [*Saccharum ravennae* (L.) L.]—Ravennagrass. Large, tussocky, tufted perennial, 2–4 m; culms and leaves glabrous; sheaths open; ligules a ciliate membrane, ca. 1 mm; blades flat, to 100 × 1.4 cm; panicles large, often plumose, elevated usually far above the leaves, 40–80 cm and ca. 1/3 as wide; disarticulation below the spikelet pairs, which fall as a unit; spikelets 4–6 mm, each with 2 florets, the silky basal hairs as long as the spikelet; awns 2–5 mm, obscured by the hairs; anthers 3. Increasingly cultivated as an ornamental landscape plant and escaping to the wild in scattered locales, 3410–5000 ft. Bernalillo, Chaves, Doña Ana, San Juan, Santa Fe. Introduced.

TRIPLASIELLA P. M. Peterson & Romasch. – Lovegrass tridens

Triplasiella eragrostoides (Vasey & Scribn.) P. M. Peterson & Romasch. [*Tridens eragrostoides* (Vasey & Scribn.) Nash]—Lovegrass. Tufted perennial, with a knotty short-rhizomatous base, 50–100 cm; sheaths open; ligules a ciliate membrane, 1.2–3 mm; panicles 10–30 × 8–20 cm, the branches somewhat racemose, ascending to reflexed in age; spikelets with several florets, the distal ones sometimes sterile, the rachilla prolonged; disarticulation above the glumes and between the florets; glumes shorter than the lowermost lemma, callus hairy, 1-nerved, purple, dissimilar, the upper one subequal to the adjacent lemma; lemmas 2–3.5 mm, 1-nerved, the nerves puberulent, the midnerve sometimes excurrent as a mucro, otherwise unawned; paleas glabrous or scabrous proximally; anthers 3. Desert plains and bajadas in brushy country; known from a single collection in Luna County, elevation unknown. Introduced. A monotypic genus, most recently included in the genus *Tridens*, but its recognition renders *Tridens* and its segregates monophyletic.

TRIPLASIS P. Beauv. – Sandgrass

Triplasis purpurea (Walter) Chapm.—Purple sandgrass. Tufted annual to short-lived perennial with short rhizome-tillers, 15–100 cm, nodes hairy; auricles absent; sheaths open, inflated late in season with development of axillary cleistogamous panicles; ligules a ring of hairs to 1 mm; blades hairy; panicles 3–7 cm, the base often in the sheath; spikelets 6–9 mm, with 3–4 florets, the distal ones sometimes sterile, the rachilla prolonged; disarticulation above the glumes and between the florets, or the florets falling together, also in the nodes of the culm and the axillary panicles dispersing; glumes shorter than the lowermost lemma, 1-nerved; callus hairy; lemmas 1-nerved, 3–4 mm, the apical lobes < 1 mm; awns straight, < 2 mm; paleas ca. 2.5 mm, the keels conspicuously villous; anthers 3. Sandy flats and plains, disturbed ground, 3410–4225 ft. Chaves, Eddy, Roosevelt.

TRIPSACUM L. – Gamagrass

Tripsacum lanceolatum Rupr. ex E. Fourn.—Mexican gamagrass. Monoecious perennials, rhizomatous, 1–2 m; ligules a ciliate membrane; the staminate and pistillate spikelet separated in the same inflorescence; lower sheaths open, hispid; blades to 1 m × 3 cm; terminal panicle with 3–7 rames; pistillate spikelets 2–3 mm wide, shiny, beadlike; staminate spikelets in pairs of 1 sessile and 1 pedicelled, the glumes 5–10 mm; pedicels 2–5 mm, chaffy, like a corn tassel, awnless. To be looked for in bootheel region.

TRISETUM Pers. – Trisetum

Recent phylogenetic analyses have resulted in the reorganization of species formerly classed within *Trisetum*, which remains a small genus of 1–2 species native to Europe and Asia, none of which occur in New Mexico.

1. Lemmas awnless or more commonly with short awns < 2 mm, scarcely visible…go to **Graphephorum**
1'. Lemmas with awns > 3 mm, easily visible…(2)
2. Plants annual; spikelets eventually disarticulating below the glumes and falling as a unit…go to **Sphenopholis**
2'. Plants perennial; spikelets disarticulating above the glumes and between the florets…go to **Koeleria**

×TRITICOSECALE Wittm. ex A. Camus – Triticale

Tufted, erect annual, to 130 cm; spikes 8–20 cm; spikelets with 2–3 florets, the terminal floret usually reduced; glumes ovate, 9–12 mm, asymmetrically keeled and toothed distally, with awns 3–4 mm; lemmas ovate, 10–15 mm, the nerves converging at the apex, the awns 3–50 mm. A rather common, though nonpersistent, waif of agriculture, more frequent than collections indicate. The name ×*Triticosecale* refers to tetraploids derived by artificial hybridization between wheat (*Triticum*) and rye (*Secale*). There is no valid specific epithet for this agricultural invention. Colfax, Doña Ana.

TRITICUM L. – Wheat

Triticum aestivum L.—Wheat. Tufted, erect annual, 20–150 cm, sometimes branched basally; sheaths open; auricles present; ligules membranous; spikes 4–18 cm; disarticulation in the rachis (wild species) or not disarticulating (domesticated species); spikelets with 2–9 florets, the distal often sterile; 10–15 mm, with 3–9 florets; glumes ovate, thick, leathery, 6–12 mm, prominently keeled in the distal 1/2, toothed or awned to 4 mm; lemmas ovate, 10–15 mm, the nerves prominent and converging at the apex, toothed or awned to 12 mm. A cultivated crop in most regions of the state, found sporadically along roadsides and old fields. Widespread except NE, SW, WC. Introduced.

UROCHLOA P. Beauv. – Signalgrass

Tufted annuals and perennials, sometimes stoloniferous; sheaths open; auricles absent; ligules with a scant basal membranous portion terminated by hairs; blades flat; inflorescence a panicle, the branches spikelike, rarely rebranched; disarticulation beneath the spikelets; spikelets borne on 1 side of the branch, with 2 florets, the lower staminate or sterile, the upper fertile; lower glumes 2/3 or less the spikelet length; upper glumes and lower lemmas similar in size and texture; upper florets perfect, fertile, becoming indurate and forming a seedcase.

1. Spikelets with conspicuous and dense villous hairs (easily visible without magnification) on the second glume and lemma of lower floret; plants perennial with short rhizomes…**U. ciliatissima**
1′. Spikelets glabrous or with short, inconspicuous hairs (hardly visible without magnification); plants annual, lacking rhizomes…(2)
2. Leaf margins noticeably crinkled; lemma of upper floret with a stiff bristle projecting from an otherwise blunt apex…**U. panicoides**
2′. Leaf margins not crinkled, smooth; lemma of the upper floret without a bristle, the apex rounded to acute…(3)
3. Spikelets 5–6 mm; plants often 50+ cm…**U. texana**
3′. Spikelets 2–4 mm; plants rarely > 50 cm and usually much shorter (ours)…(4)
4. Spikelets glabrous or nearly so, mostly 2–3 mm, the base + truncate; upper lemma with deep transverse furrows…**U. fusca**
4′. Spikelets definitely puberulent, mostly 3–4 mm, the base drawn out somewhat and attenuate; upper lemma with minute bumps but lacking obvious transverse furrows…**U. arizonica**

Urochloa arizonica (Scribn. & Merr.) Morrone & Zuloaga—Arizona signalgrass. Tufted annual, 15–65 cm, erect to geniculate-based; sheaths glabrous or pubescent; panicles 6–20 cm, hairy with bulbous-based hairs, open, the branches 3–7 cm; spikelets 3–4 mm, appressed to the branches, generally puberulent; lower glumes 1/2 the spikelet length; upper floret 2.8–3 mm, lacking deep transverse furrows. Disturbed ground and rocky slopes in deserts and woodlands, 4100–6000 ft. SC, SW, Socorro.

Urochloa ciliatissima (Buckley) R. D. Webster—Fringed signalgrass. Rhizomatous or stoloniferous perennial, 10–40 cm; nodes villous; sheaths glabrous or with bulbous-based hairs; panicles 3–6 cm, narrow, the branches erect-appressed, to 2 cm, rarely rebranched; spikelets 3–4.5 mm, with long-villous hairs on the upper glume and lower lemma; lower glume 2/3+ the spikelet length; upper floret 2.4–2.8 mm. Sandy plains and desert grasslands, 3400–4265 ft. Chaves, Eddy.

Urochloa fusca (Sw.) B. F. Hansen & Wunderlin—Browntop signalgrass. Tufted annual, 15–120 cm; nodes glabrous or pilose; sheaths glabrous to hispid, the margins ciliate; panicles 5–15 cm, the branches appressed to divergent and 2–10 cm, scabrous to sparsely pilose; secondary branchlets developed below; spikelets 2–3.4 mm, glabrous or rarely puberulent, usually dark at maturity, lower glume ca. 1/3 the spikelet length; upper glume and lower lemma with evident cross-venation; upper floret 1.8–3 mm. Disturbed ground, 3625–5500 ft. SC, SW, Chaves, Socorro.

Urochloa panicoides P. Beauv.—Liverseed grass. Tufted annual, erect to decumbent, 10–55+ cm, rooting at the lower nodes; sheaths hispid, the margins ciliate; blades with bulbous-based hairs on both surfaces; panicles 3–10 cm, narrow to open, the branches 1–7 cm, erect to spreading, ciliate with bulbous-based hairs; pedicels with a few long hairs below the spikelet; spikelets 2.5–5.5 mm, appressed to the branches, glabrous; lower glume < 1/2 the spikelet length; upper florets 2.6–3.5 mm, short-awned to 1 mm. Weedy ground along sidewalks, flower beds, waste ground, 3925–4000 ft. Doña Ana. Introduced.

Urochloa texana (Buckley) R. D. Webster—Texas signalgrass. Tufted annual, 20–100+ cm, erect or geniculate-based and forming stolons; sheaths hairy with bulbous-based hairs, the margins ciliate; blades pubescent on both surfaces; panicles 8–24 cm, narrow, < 1 cm wide, the branches erect-appressed; spikelets 4.8–6 mm, sparsely puberulent, appressed to the branches; lower glumes 1/2 or less the spikelet length; upper floret 3.6–4.1 mm, acute. Disturbed weedy ground, Las Cruces area, ca. 3420 ft. Doña Ana.

VULPIA C. C. Gmel. - Sixweeks fescue

Tufted annuals, sometimes longer-lived; sheaths open, usually glabrous; auricles absent; ligules membranous; inflorescence a panicle or raceme, open or spikelike; spikelets with 1–17 florets, the distal florets often reduced; disarticulation above the glumes and between the florets; glumes shorter than the adjacent lemmas, awnless or awn-tipped; lemmas membranous, 3-5-nerved, the margins rolled over the edges of the grain, acute to awned; anthers usually 1, sometimes 3 in chasmogamous spikelets. A segregate genus of ca. 25 species, formerly (and sometimes currently) treated in *Festuca*, from which they differ in being annual and having only a single anther in each floret.

1. First glume < 1/2 the length of the second glume, often nearly absent...**V. myuros**
1'. First glume > 1/2 the length of the second glume...(2)
2. Panicle branches 1-2 per node; spikelets with 4-17 florets; rachilla internodes 0.5-0.7 mm; awn of the lowermost lemma 0.3-9 mm; caryopses 1.7-3.7 mm...**V. octoflora**
2'. Panicle branches solitary; spikelets with 1-8 florets; rachilla internodes 0.6-1.2 mm; awn of the lowermost lemma 2-20 mm; caryopses 3.5-6.5 mm...(3)
3. Panicle branches and pedicels erect at maturity, without swellings in the axils...**V. bromoides**
3'. Panicle branches or pedicels spreading or reflexed at maturity, at least below, with swellings usually present in the axils...**V. microstachys**

Vulpia bromoides (L.) Gray [*Festuca bromoides* L.]—Brome fescue. Annual, 5-50 cm; sheaths glabrous or pubescent; panicles 1-15 cm, 0.5-3 cm wide, mostly narrow; spikelets with 4-8 florets, the rachilla internodes 0.6-1.1 mm; lower glumes 1/2 to nearly equaling the length of the upper; upper glumes 4.5-9 mm; lemmas 4-8 mm, scabrous distally, the lower awns 2-13 mm. Dry, disturbed ground, 4400-5500 ft. NC, NW, C, SC, SW. Introduced.

Vulpia microstachys (Nutt.) Munro [*Festuca microstachys* Nutt.]—Small fescue. Annuals, 15-75 cm; panicles 2-24 cm, 1-8 cm wide, the branches or pedicels spreading to reflexed from axillary pulvini when mature, erect when young; spikelets with 1-6 florets, glabrous to pubescent, the rachilla internodes 0.6-1.2 mm; lower glumes 1/2-3/4 the length of the upper; lemmas 3.5-9.5 mm, scabrous to pubescent, the lower awns 3-20 mm. Dry, disturbed ground, 5100-5500 ft. Doña Ana, Hidalgo. Plants with glabrous spikelets are referred to as **var. pauciflora** (Scribn. ex Beal) Lonard & Gould.

Vulpia myuros (L.) C. C. Gmel. [*Festuca megalura* Nutt.]—Vulpia. Annual, 10-75 cm; sheaths glabrous; panicles 3-25 cm, 0.5-2 cm wide, rather dense, sometimes racemose, pulvini absent; spikelets with 3-7 florets, the rachilla internodes 0.7-2 mm; glumes strongly unequal, the lower < 1/2 the length of the upper; lemmas 4.5-7 mm, margins sometimes ciliate, the lower awns 5-22 mm. Dry, disturbed ground, 3500-7100 ft. Chaves, Cibola, Lincoln, Doña Ana, McKinley. Introduced.

Vulpia octoflora (Walter) Rydb. [*Festuca octoflora* Walter]—Sixweeks fescue. Annual, 5-60 cm; panicles 1-20 cm, 0.5-1.5 cm wide, the branches appressed to spreading; spikelets with 4-17 florets evenly spaced like a comb, the rachilla internodes 0.5-0.7 mm; lower glumes 1/2-2/3 the length of the upper; lemmas 2.7-6.5 mm, the lower awns 0.3-9 mm. Dry, disturbed ground, roadsides, rocky slopes and plains, 3150-8500 ft. Widespread.

ZEA L. – Corn, maize

Zea mays L.—Corn. Tufted annual, often with prop roots, 0.5-6 m × 2-5 cm; blades 2-12 cm wide; ears (cobs) 15-40 cm, permanently enclosed by subtending sheaths, not disarticulating at maturity; sheaths open; auricles sometimes present; ligules membranous, ciliate; pistillate spikelets in 8-24+ rows; tassels with a thick rachis and thinner nondisarticulating branches. Our plants belong to **subsp. mays**. Cultivated throughout the state, rarely found along old fields, 3900-6750 ft. SC, Bernalillo, San Juan. Introduced.

ZOYSIA Willd. – Zoysiagrass

Rhizomatous to stoloniferous perennials, turf-forming, densely branched at ground level; sheaths open; auricles absent; ligules short, ciliolate; blades conspicuously distichous, stiff, flat to involute; inflorescence a dense spicate raceme, the spikelets appressed to the axis; disarticulation below the glumes, the spikelets falling entire, the pedicels persistent; spikelets laterally compressed, with 1 floret; lower glume usually absent; upper glume as long as the spikelet, enclosing the floret, leathery, glossy, the apices acute to mucronate; lemma membranous, 1-3-nerved; palea reduced to absent. Introduced worldwide for turf.

1. Pedicels 1.6–3.5 mm; spikelets ovate, 1–1.4 mm wide; culm internodes 2–10 mm; blades ascending…**Z. japonica**
1'. Pedicels 0.6–1.6 mm; spikelets lanceolate, 0.6–1 mm wide; culm internodes 5–40 mm; blades spreading …**Z. matrella**

Zoysia japonica Steud.—Korean lawngrass. Producing long stolons, mat-forming, the culms erect, to 20 cm if unmowed; sheaths glabrous, pilose at summit, the basal sheaths persistent; blades 2–6 cm, 2–4 mm wide, thinly pilose adaxially, the apices pungent; racemes 2–4 cm, rising above the leaves; spikelets 2.5–3.5 mm, 1–1.5 mm wide, the lemma slightly shorter than the glume, 1-nerved, the palea absent, the anthers ca. 1.5 mm. Occasionally planted as a lawn grass; not known outside of cultivation.

Zoysia matrella (L.) Merr.—Manila grass. Producing stolons, mat-forming, the culms erect, to 20 cm if unmowed; sheaths glabrous, pilose at summit; blades 3–8 cm, 1–2.5 mm wide, thinly pilose adaxially, the apices acute; racemes 2–4 cm, rising above the leaves; spikelets 2–3 mm, ca. 1 mm wide, the lemma obscurely 3-nerved, the palea present, 1/2 as long as the lemma, the anthers 1–1.5 mm. Occasionally planted as a lawn grass; not known outside of cultivation.

ZULOAGAEA E. Bess – Panicgrass

Zuloagaea bulbosa (Kunth) E. Bess [*Panicum bulbosum* Kunth]—Bulb panicgrass. Tufted perennial, with or without rhizomes (creeping rootstock), 25–150+ cm, lowest internodes often thickened into hard, bulblike corms; sheaths open, glabrous to pilose, keeled; auricles absent; ligules membranous, ciliate, 0.3–5 mm total; blades flat, to 75 cm, 0.2–1.5 mm wide; inflorescence an open, pyramidal panicle, 10–75 cm, the branches ascending to diverging, bristles absent; spikelets dorsally compressed, awnless, 3–5.5 mm, with 2 florets, the lower staminate or sterile, the upper fertile; disarticulation below the glumes; lower glumes ca. 2/3 as long as the spikelet; upper glume similar in size and texture to the lower lemma; lower palea present, about as long as its companion lemma in the lower floret; upper lemma 2–5 mm, finely rugose, clasping the edges of the palea, together forming a hardened seedcase; anthers 3. Canyon bottoms and moist slopes in the mountains and foothills, 3925–9125 ft. Widespread except E.

POLEMONIACEAE – PHLOX FAMILY

J. Mark Porter

Annuals, perennial herbs, or shrubs, pubescent, rarely glabrous; leaves alternate to opposite, simple or compound; inflorescence cymose, open or congested, the cymes usually arranged in panicles, flowers rarely solitary; flowers perfect, actinomorphic to zygomorphic, hypogynous; sepals usually 5, united, with equal or unequal lobes, the tube usually with herbaceous ribs separated by hyaline membranes; corolla tubular, with 5 lobes; stamens usually 5, epipetalous, alternate with the corolla lobes; filaments equal to unequal in length; ovary usually with 3 locules; style simple, included to exserted, with (2)3 stigmatic lobes; fruit a capsule, with 1 to many ovules and seeds in each locule (Porter & Johnson, 2000).

1. Calyx tube herbaceous nearly throughout or hyaline only in the sinus of calyx lobes, membranous and not ruptured in fruit…(2)
1'. Calyx tube with herbaceous ribs separated by thin, translucent membranes, these usually distended or ruptured in fruit, or if herbaceous or membranous, then the leaves opposite and palmately lobed…(3)
2. Annual; leaves entire to rarely toothed…**Collomia**
2'. Perennial; leaves pinnately compound…**Polemonium**
3. Cauline leaves mostly opposite, upper ones sometimes alternate, either simple or palmately lobed to compound…(4)
3'. Cauline leaves mostly alternate, simple to pinnately lobed or compound…(9)
4. Stamens inserted at different levels on the tube; corolla salverform; leaves simple, linear to linear-oblong or narrowly lanceolate…(5)
4'. Stamens inserted at the same level on the tube; corolla rotate, campanulate, or funnelform, if salverform then the leaves somewhat rigid and spinulose; leaves simple and filiform to palmately compound, the blades or lobes filiform to linear…(6)

5. Annual; corolla tubes 4–8 mm…**Microsteris**
5′. Perennial; corolla tubes 8–30 mm…**Phlox**
6. Perennial, suffrutescent to woody; corolla salverform…(7)
6′. Annual; corolla rotate to campanulate or funnelform…(8)
7. Hyaline membranes separating the calyx lobes narrow and somewhat obscure; uppermost leaves opposite; corolla diurnal and often open at night, corolla tube glandular-pilose externally…**Leptosiphon** (in part)
7′. Hyaline membranes separating the calyx lobes prominent; uppermost leaves often alternate; corolla nocturnal, closed during the day, corolla tube glabrous externally…**Linanthus** (in part)
8. Corolla rotate, the throat orange or maroon…**Leptosiphon** (in part)
8′. Corolla campanulate to funnelform, the throat white to lavender or bluish…**Linanthus** (in part)
9. Leaves deeply palmately to subpalmately lobed, the lobes rigid…(10)
9′. Leaves entire or pinnately lobed to toothed, the lobes usually soft and herbaceous…(11)
10. Corolla white to cream-colored or yellowish, opening nocturnally, usually closed during the day…**Linanthus** (in part)
10′. Corolla bright blue, opening from noon to sundown, closed during the night…**Giliastrum** (in part)
11. Inflorescence capitate to densely corymbose, terminal at tips of ascending to erect branches; flowers subsessile to sessile…(12)
11′. Inflorescence open and diffuse, with 2–5 flowers at the branch tips, or composed of several compact to dense, lateral clusters; flowers short- to long-pedicellate…(14)
12. Calyx lobes unequal; floral bracts pinnately lobed, woolly or with entangled hairs, the apices firm or rigid; corolla usually with blue lobes, yellow throat, and white tube, sometimes pale; leaf lobes at the proximal end of leaf…**Eriastrum**
12′. Calyx lobes equal; floral bracts usually entire to toothed or absent, the apices acute to short-mucronate but not sharply rigid; corolla lobes, throat, and tube various; leaf lobes various…(13)
13. Corolla campanulate to funnelform, blue or with white to bluish lobes…**Gilia** (in part)
13′. Corolla salverform, uniformly white to lavender…**Ipomopsis** (in part)
14. Corolla salverform, white to bluish or red; flowers usually > 7, congested at tips of lateral branches…**Ipomopsis** (in part)
14′. Corolla funnelform to salverform, if salverform then usually with white to bluish lobes and a yellow tube or throat; flowers 2–5(–7), in loose, terminal clusters…(15)
15. Corolla broadly funnelform, rotate, uniformly deep blue; leaves gradually reduced upward, the lobes needlelike…**Giliastrum** (in part)
15′. Corolla narrowly funnelform to salverform, the lobes white, magenta, or bluish, the throat or tube often yellow; leaves abruptly reduced above a basal rosette, the lobes usually acute or mucronate…(16)
16. Glandular hairs on calyx, pedicels, and upper leaves colorless; basal and lower leaves glabrous or mostly glandular; seeds not gelatinous when wet, or inconspicuously so…**Aliciella**
16′. Glandular hairs on calyx, pedicels, and leaves, if present, dark or reddish; basal and lower leaves with short, curled hairs or cobwebby hairs; seeds gelatinous when wet…**Gilia** (in part)

ALICIELLA Brand – Aliciella

Annual or perennial herbs, leafy to more often scapose, mostly stipitate-glandular; leaves basal and alternate, entire to deeply pinnately lobed, lobes confluent with rachis, flat or terete; inflorescence terminal, paniculate, open to congested, composed of 2–7 pedicelled flowers subtended by a single bract; flowers actinomorphic; calyx tube longer than mucronate lobes; corolla funnelform to salverform; stamens equally inserted on the corolla tube or throat; filaments equal or unequal in length; anthers included or exserted; style included or exserted; capsule ovoid to spheroid; seeds 2 to many per locule, not or only slightly gelatinous when wet (Porter, 1998). *Aliciella subnuda* (A. Gray) J. M. Porter has erroneously been reported from New Mexico (Allred et al., 2020; Martin & Hutchins, 1981b); it is restricted to Arizona and Utah.

1. Anthers exserted well beyond the corolla tube, filaments nearly equaling the corolla lobes; basal leaves pinnatifid with narrow rachis…**A. pinnatifida**
1′. Anthers only slightly exserted beyond the corolla tube, filaments much shorter than the corolla lobes; basal leaves entire, dentate or pinnatifid with broad rachis…(2)
2. Seeds 0.5–0.9 mm; plants annuals…(3)
2′. Seeds 1.5–2 mm; plants biennials or short-lived perennials, rarely flowering the first year…(4)
3. Corolla tube somewhat constricted at the orifice, ± salverform, lobes 3-toothed…**A. triodon**
3′. Corolla tube flared at the orifice, funnelform, lobes oblanceolate-acute or truncate and cuspidate but not 3-toothed…**A. leptomeria**

4. Rosette leaves linear, entire, reduced in size but morphologically similar distally on stem…**A. formosa**
4'. Rosette leaves lanceolate, oblanceolate, or spatulate, dentate or irregularly pinnatifid with a broad rachis, rarely entire, reduced in size and transitioning to linear bracts distally on stem…(5)
5. Corolla blue, paling to nearly white when fresh; corolla lobes narrowly lanceolate, 3-6 mm, 1.9-3.5 mm wide; free portion of filaments 0.2-1.5 mm (mean 0.9 mm)…**A. cliffordii**
5'. Corolla magenta when fresh; corolla lobes oval to oblanceolate, slightly wider than above, 3.5-9 mm, 2-4.2 mm wide; free portion of filaments 0.8-2.7 mm (mean 1.5 mm)…**A. haydenii**

Aliciella cliffordii J. M. Porter—Diné star, Clifford's aliciella. Short-lived, perennial herb, 10-90 cm, stems sparsely and coarsely glandular-pubescent, erect and divaricately branching; leaves forming a basal rosette, coarsely toothed or once-pinnatifid, 1.2-5 cm, glandular and crisp-puberulent, lobes and apex cuspidate or mucronate, cauline leaves gradually to abruptly reduced; inflorescence a loosely open cymose panicle; flowers actinomorphic; calyx 3-4.7 mm, glandular; corolla narrowly funnelform-salverform, 6-18 mm, blue, paling to nearly white, ± glabrous externally, lobes narrowly oblanceolate, 3-5.5(-6) × 1.9-3.5 mm; stamens equally inserted at sinuses of corolla lobes, slightly exserted; style well exserted or included; capsule 3-5 mm, ovoid. Sandy or clay badlands, associated with piñon-juniper woodland, ponderosa pine, or Douglas-fir forests, 5000-6500 ft. San Juan. Restricted to the Navajo Nation on Beautiful Mountain and the adjacent E slopes of the Lukachukai and Chuska Mountains. **R**

Aliciella formosa (Greene ex Brand) J. M. Porter [*Gilia formosa* Greene ex Brand]—Aztec aliciella, Aztec gilia. Perennial herb, from a branched, woody caudex, 5-15 cm; stems glandular, erect and ± openly branching above the middle; leaves forming a basal rosette, entire, linear, 1-4.5 cm, glandular and crisp-puberulent, apex mucronate, cauline leaves gradually to abruptly reduced; inflorescence a few-flowered open cymose panicle; flowers actinomorphic; calyx 3.5-6.1 mm, glandular; corolla narrowly funnelform-salverform, 14.7-27 mm, magenta to pink-lavender but drying lead-blue; stamens equally inserted below sinuses of corolla lobes, slightly exserted; stigma maturing above the anthers; capsule 3.5-7 mm, ovoid. Dry, saline clay or sandy-clay badlands, sagebrush-shadscale, piñon-juniper woodlands, 5380-6500 ft. San Juan. **R & E**

Aliciella haydenii (A. Gray) J. M. Porter [*Gilia haydenii* A. Gray]—San Juan aliciella, San Juan gilia. Short-lived perennial herb, 12-50 cm; stems glabrous to glandular-puberulent; leaves forming a basal rosette, gradually to abruptly reduced above; basal and lower sparsely crisp-pubescent above, glabrous beneath, lobed to toothed, the rachis broader than the lobes; cauline leaves usually glabrous, toothed to entire; inflorescence open; flowers actinomorphic; calyx 4-5 mm, the lobes acute, ± mucronate; corolla funnelform, 6-10 mm, magenta; stamens inserted at sinuses of corolla lobes; anthers located at the orifice or slightly exserted; stigma above to below the anthers; capsule 3-5 mm, ovoid.

1. Corolla 10-20 mm, lobes 3.5-6 mm, magenta, drying bluish, tube with a few glands externally…**subsp. haydenii**
1'. Corolla 17-26 mm, lobes 6-9 mm, magenta, drying pink, tube glandular externally…**subsp. crandallii**

subsp. crandallii (Rydb.) J. M. Porter [*Gilia crandallii* Rydb.]—Crandall's aliciella. Rocky soils, clay badlands, piñon-juniper woodland, ponderosa pine communities, 6070-7070 ft. Rio Arriba, Sandoval, San Juan.

subsp. haydenii—Rocky soils, clay badlands, piñon-juniper woodland, saltbush and desert scrub, 5175-5900 ft. McKinley, Rio Arriba, San Juan.

Aliciella leptomeria (A. Gray) J. M. Porter [*Gilia leptomeria* A. Gray]—Sand aliciella, sand gilia. Annual or winter-annual, 6-30 cm, simple to branched; stems usually glandular; leaves glandular, rarely glabrous, abruptly reduced above the basal rosette; basal and lower dentate to shallowly lobed; cauline dentate to entire; inflorescence open, with 1-3 pedicelled flowers at branch tips, the terminal pedicel subsessile or often shorter than those below; calyx 1-3 mm, usually glabrous; corolla funnelform, 3-7 mm, the tube equal to or slightly longer than the calyx; the tube and lobes white, pink, or lavender, the throat yellow; stamens inserted on the upper throat; anthers slightly exserted; stigmas slightly exceeding the anthers; capsule 3-5 mm, narrowly ovoid. Washes, rocky slopes, desert shrublands, woodlands, 4630-5575 ft. McKinley, Sandoval, San Juan.

Aliciella pinnatifida (Nutt. ex A. Gray) J. M. Porter [*Gilia pinnatifida* Nutt. ex A. Gray; *G. calcarea* M. E. Jones; *G. viscida* Wooton & Standl.]—Sticky gilia. Biennial or short-lived perennial, 10–60 cm; stems erect and diffusely branched in flower, glandular-puberulent; leaves 1.4–7 cm; basal leaves oblanceolate, deeply pinnatifid, glandular-puberulent; cauline leaves pinnatifid, transitioning to linear-entire; inflorescence cymose-paniculate; calyx 2.5–5.5 mm, glandular-puberulent; corolla salverform to narrowly campanulate, 5.5–12 mm, white, blue, or lavender, the tube longer than the calyx; stamens inserted on the upper tube, anthers long-exserted; stigma maturing above the anthers; capsule 2.5–5 mm, narrowly ovoid. Sandy soils, sandstone outcrops, montane communities, 6365–10,335 ft. NE, NC, WC.

Aliciella triodon (Eastw.) Brand [*Gilia triodon* Eastw.]—Coyote aliciella, coyote gilia. Annual, 5–15 (–25) mm, simple or branched; stems glandular; leaves spatulate, obovate, or oblanceolate to lanceolate, entire or coarsely dentate, abruptly reduced above basal rosette; cauline usually entire, linear; inflorescence open, with 1–3 short-pedicelled flowers on distal branches; calyx 1–3(–4.5) mm, usually glabrous or glandular-puberulent, the lobes acuminate; corolla salverform or narrowly funnelform, 3–6(–7) mm, the tube exserted, white to pink, the throat light yellow, the lobes 3-toothed, the teeth subequal, white to pink; stamens inserted on the upper throat; anthers slightly exserted; stigma slightly exceeding the anthers; capsule 2–4.5 mm, narrowly ovoid. Sandy to gravelly soils, sagebrush shrubland, piñon-juniper woodland, 3935–6900 ft. San Juan.

COLLOMIA Nutt. – Trumpet, collomia

Collomia linearis Nutt.—Tiny trumpet. Annual; stems 10–50 cm; leaves subsessile, entire, 1–6 cm, the lower cauline leaves linear to lanceolate, the upper leaves narrowly lanceolate; inflorescence terminal, rarely axillary, compact, the bracts ovate to lanceolate, 8–20 × 3–6 mm, the lower margins glandular; calyx lobes acuminate; corollas salverform, 8–15 mm, bluish violet to nearly white; pollen cream to blue; style equal to the corolla tube; capsule 3–5 mm, seeds 1 per locule, gelatinous when wet. Open sites, valleys, hills, meadows, ponderosa pine, spruce, Douglas-fir, aspen forests, 7200–10,415 ft. NE, NC, NW, C, WC.

ERIASTRUM Wooton and Standl. – Woollystar

Eriastrum diffusum (A. Gray) H. Mason—Miniature woollystar. Annual, herbaceous plant 3–35 cm, erect and simple to diffusely branching; stems 1–25 cm, usually wiry and suffused with purple; leaves subglabrous to sparsely floccose, entire or with 1–2 pairs of lobes near the base of the rachis; inflorescence with few to many bracteate heads, bracts pinnatifid; calyx 6–7 mm, lobes unequal to subequal; corolla 6–12 mm, actinomorphic, funnelform, the throat white to yellow or yellow-spotted, the lobes white to pale blue or bluish lavender; stamens inserted on the throat near sinuses, shorter than corolla lobes; filaments slightly unequal in length; style included in the throat; capsule 2.7–4 mm. New Mexico material belongs to **subsp. diffusum**. Open sandy, gravelly, or rocky sites, desert grassland, shrubland, sagebrush, piñon-juniper woodland, 3835–7055 ft. NC, NW, C, WC, SC, SW. Frequently confused with *Ipomopsis pumila* (Nutt.) V. E. Grant in the Colorado Plateau region.

GILIA Ruiz & Pav. – Gilia

Annual or winter annuals; stems simple or branched, leafy or scapose, glabrous, glandular, floccose, or cobwebby; leaves forming a basal rosette, 1–2 times pinnately lobed, flat to terete, cauline leaves alternate, entire to deeply pinnately lobed; inflorescence terminal, paniculate, open to ± congested, composed of 2–7 pedicelled flowers subtended by a single bract, rarely solitary; flowers actinomorphic; calyx tube membranes usually ruptured in fruit; corolla funnelform to salverform, the tube, throat, and lobes often with different hues; stamens equally to subequally inserted on the corolla tube or throat; filaments equal or unequal in length; anthers included or exserted; style included or exserted; capsule ovoid to spheroid; seeds 2 to many per locule, seed coat gelatinous when wet.

1. Flowers many in dense capitate heads; corolla blue to bluish violet; stamens and style exserted…**G. capitata**

1'. Flowers not in dense heads, sometimes glomerulate in bud or early flower; corollas white, pink, lavender, or purplish, sometimes with yellow in throat; stamens and style included or exserted…(2)
2. Basal and lower cauline leaves with white, sharply bent (sometimes branched), nonglandular hairs…**G. stellata**
2'. Basal and lower cauline leaves lacking white, sharply bent (or branched), nonglandular hairs…(3)
3. Basal leaves and stems near base glabrous and glaucous; cauline leaves sessile and clasping…**G. sinuata**
3'. Basal leaves and stems near base densely to sparsely floccose or cobwebby, glandular or not, not glaucous; cauline leaves not clasping, pinnatifid or awl-like…(4)
4. Corolla 7-18 mm, tube purple, violet, or yellow with violet streaks…(5)
4'. Corolla 4-9.3 mm, tube white, rarely yellowish…(6)
5. Corolla tube subequal to the calyx; stamens unequal, longest exserted…**G. ophthalmoides**
5'. Corolla tube longer than the calyx; stamens subequal, slightly exserted…**G. flavocincta**
6. Leaf lobes broad and spreading; capsule globular to broadly ovoid; calyx in fruit translucent except for the often reddish lobe apices…**G. clokeyi**
6'. Leaf lobes narrow and antrorse; capsule narrowly ovoid; calyx in fruit not translucent or with red lobe apices…**G. mexicana**

Gilia capitata Sims—Bluehead gilia. Annual, 10-50 cm, strict or branched throughout; stems glabrous or sparsely floccose, usually glandular; leaves often forming a rosette, sparsely floccose, gradually reduced upward, deeply lobed, once- or twice-pinnatifid, the lobes linear; inflorescence usually globose or capitate, with many subsessile flowers at the branch tips; calyx 4-5 mm, pubescent, the lobes acuminate; corolla broadly funnelform, 7-12 mm, blue to bluish violet, the tube equal to the calyx, the throat exserted; stamens inserted on the throat; anthers exserted; stigmas among or above the anthers; capsule 5-6 mm, ovoid to globose. Canyons, washes, desert shrubland, 5775-6400 ft. San Juan. Probably an escape from cultivation.

Gilia clokeyi H. Mason—Clokey's gilia. Annual, 8-20(-35) cm, branched throughout; stems cobwebby-pubescent below, glandular above; leaves forming a rosette, floccose or cobwebby-pubescent, reduced above the rosette; basal and lower cauline deeply lobed; upper cauline lobed proximally or entire; inflorescence open, with 1-2 flowers at branch tips, each pedicelled; flowers ± actinomorphic; calyx 2-5 mm, the lobes acute, becoming translucent in fruit, only the lobe apices reddish-pigmented; corolla funnelform, 4.3-9.3 mm, the tube equal to the calyx, the throat pale blue to white, the lobes light to deep violet, yellow-spotted at base; stamens inserted on the throat; anthers above the throat; stigma among the anthers; capsule 3-6.5 mm, globose to ovoid. Open sandy, clay, or volcanic soils, shrubland, piñon-juniper woodland, 4600-6400 ft. NW, CW.

Gilia flavocincta A. Nelson [*G. lyndana* Allred]—Lesser yellowthroat gilia. Annual, 6-30(-45) cm, branched throughout; stems cobwebby below, glabrous to sparsely glandular above; leaves sparsely cobwebby, reduced above the basal rosette; basal and lower lobed once or twice; cauline entire or basally lobed; inflorescence congested, with short-pedicelled to subsessile flowers in terminal clusters; calyx 3-5(-8) mm, glabrous to sparsely cobwebby, the lobes acuminate to attenuate; corolla funnelform, 7-27 mm, the tube equal to or longer than the calyx, violet or yellow, the throat yellow, the lobes pink to yellow with violet flecks; stamens inserted on the throat; anthers slightly exserted; stigma exserted or among the anthers; capsule 5-8.5 mm, broadly ovoid. Our material belongs to **subsp. australis** (V. E. Grant) A. G. Day & V. E. Grant. Sandy to gravelly soils, washes, canyons, bajadas, desert shrublands, piñon-juniper woodland, 4265-7220 ft. NW, WC, SC, SW.

Gilia mexicana A. D. Grant & V. E. Grant—El Paso gilia. Annual, 10-35 cm, usually branched from the base; stems slender, cobwebby-pubescent below, sparsely glandular above; leaves floccose or cobwebby-pubescent, reduced above the basal rosette; basal and lower leaves deeply lobed; cauline leaves lobed at the proximal end or entire; inflorescence open, with 1-2 pedicelled flowers at branch tips; flowers actinomorphic; calyx 2.5-5 mm, glabrous, occasionally sparsely cobwebby or glandular; corolla funnelform, 4-8 mm, the tube and throat equal to or slightly exceeding the calyx, white, the throat white with yellow flecks, lobes white to pale blue, sometimes streaked with violet flecks; stamens inserted on the throat; anthers slightly exserted; stigma among the anthers; capsule 3.5-6 mm, oblong-ovoid. Sandy soils, bajadas, canyons, desert shrublands, piñon-juniper, coniferous, or oak woodlands, 3970-7380 ft. WC, SW.

Gilia ophthalmoides Brand—Eyed gilia. Annual, 8–30 cm, branched throughout; stems cobwebby-pubescent below, glandular above; leaves cobwebby-pubescent, reduced above the basal rosette; basal and lower leaves deeply once- or twice-lobed; cauline leaves lobed at proximal end or entire; inflorescence open, with 1–2 pedicelled flowers at the branch tips; flowers actinomorphic; calyx 3–5 mm, glabrous or occasionally sparsely glandular, the lobes acuminate; corolla funnelform, 7–12 mm, the tube usually exserted, purple, the throat yellow, the lobes pink; stamens inserted on the throat; anthers exserted; stigma among the anthers; capsule 4–6 mm, ovoid to subglobose. Open, often sandy soils, shrublands, woodlands, 4200–8100 ft. NC, NW.

Gilia sinuata Douglas ex Benth.—Rosy gilia. Annual, 9–30(–35) cm, simple or branched above rosette, glabrous and glaucous below, glandular above; leaves usually forming a basal rosette, sometimes not, glabrous and glaucous, rarely cobwebby-pubescent on upper surface, abruptly reduced above the basal rosette; basal leaves deeply lobed; cauline leaves clasping, dentate to entire; inflorescence open, with 1–3 short-pedicelled flowers at branch tips; flowers actinomorphic; calyx 3–5 mm, glandular; corolla funnelform, 7–12 mm, the tube exserted, purple and white striate, the throat yellow or purple-tinged below, the lobes white to lavender; stamens inserted on the throat; anthers exserted; stigmas among or slightly above the anthers; capsule 4–7 mm, ovoid. Sandy soils, shrubland, woodland, 4200–8900 ft. NC, NW.

Gilia stellata A. Heller—Star gilia. Annual, 7–70 cm, simple to branched; stems densely pubescent below, with crisp, white trichomes bent to curled and sometimes branched, glandular above; leaves pubescent with curled hairs, reduced above the basal rosette; basal and lower deeply once- or twice-pinnatifid; cauline deeply lobed to entire; inflorescence open, with 2–8 pedicelled flowers on the distal branches; flowers actinomorphic; calyx 3–5 mm, pubescent or glandular, the lobes acuminate; corolla funnelform, 6–10 mm, white, the tube usually 2 times the calyx, the throat yellow with purple spots; stamens inserted on the upper throat; anthers slightly exserted; stigmas slightly above the anthers; capsule 5–7 mm, broadly ovoid. Sandy washes, slopes, desert shrublands and woodlands, oak shrublands at higher elevations, 4500–5900 ft. Catron, Grant, Socorro.

GILIASTRUM (Brand) Rydb. – Bluebowls

Perennial herbs, simple to branched, leafy to subscapose, glabrous or sparsely pubescent; leaves forming a basal rosette or essentially cauline, alternate, gradually reduced upward, pinnately or subpalmately lobed, rarely entire, usually flat; inflorescence terminal, paniculate, open to slightly congested, with 2–5 pedicelled flowers subtended by a single bract, rarely solitary; flowers actinomorphic; calyx tube membranes ruptured in fruit; corolla rotate to broadly funnelform, the tube, throat, and lobes often with the same hues; stamens equally inserted on the lower corolla tube; filaments equal or unequal in length; anthers included or exserted; style included or exserted; capsule ovoid to spheroid; seeds 2 to many per locule, seed coat mucilaginous when wet.

1. Calyx tubes 1–1.8 mm, < 1/2 length of lobes; leaves generally forming a persistent basal rosette; seeds 0.5–0.8 mm…**G. incisum**
1'. Calyx tubes 2–4.5 mm, usually subequal to or slightly < length of lobes; leaves generally not forming a basal rosette; seeds 1–2 mm…**G. acerosum**

Giliastrum acerosum (A. Gray) Rydb. [*Gilia rigidula* Benth. subsp. *acerosa* (A. Gray) Wherry]—Bluebowls. Perennial, 6–15 cm, branched at the woody base; stems 5–12, spreading to erect, woody below, glandular throughout; leaves slightly reduced upward; lower deeply pinnately lobed, the lobes needlelike; upper pinnatifid to subpalmately lobed, the lobes 3–5; inflorescence open, with 1–3 pedicelled flowers at branch tips; flowers actinomorphic; calyx 7–8 mm, the lobes attenuate, acerose or mucronate; corolla broadly funnelform, 7–10 mm, deep blue, the tube usually shorter than the calyx; stamens inserted on the lower tube; anthers above the throat, bright yellow; stigma among the anthers or slightly above; capsule globose-ovoid, 4–6.5 mm, seeds elliptic, sometimes angled, 1.2–2 mm, seed coat mucilaginous when wet. Gravelly soils, rocky slopes, canyons, limestone or gypsum, desert grasslands or shrublands, piñon-juniper woodlands, creosote scrub, 3400-7500 ft. NE, NW, C, WC, SE, SC, SW. Widespread.

Giliastrum incisum (Benth.) J. M. Porter [*Gilia incisa* Benth.]—Splitleaf bluebowls, splitleaf gilia. Perennial herb, sometimes appearing annual, (10-)25-50 cm, branched at the base; stems erect to lax, woody below, glandular-puberulent to villous throughout; leaves forming a rosette, blades spatulate, lyrate-pinnatifid, or pinnate, upper leaves usually pinnatifid, sometimes simple; inflorescence open, with 1-2 pedicelled flowers; flowers actinomorphic; calyx 4-6.5 mm; corolla campanulate-rotate, 5-9(-12) mm, lobes lavender to white, the tube white, yellow at the orifice, usually shorter than the calyx; stamens inserted on the lower tube; anthers above the throat, bright yellow; stigma among the anthers or slightly above; capsule 4-6 mm, ovoid; seeds ovoid to angled, 0.5-0.8 mm, narrowly winged, seed coat mucilaginous when wet. Gravelly soils, rocky slopes, canyons, limestone or gypsum, desert grasslands and shrublands, piñon-juniper woodlands, creosote scrub, 3600-6430 ft. Eddy.

IPOMOPSIS Michx. – Ipomopsis

Annual or perennial herbs, or subshrubs, simple to branched, vesture of white, crisp trichomes and often stipitate-glandular; leaves forming a basal rosette or not, cauline leaves alternate, entire to deeply pinnately lobed, flat to terete, usually linear to oblong; inflorescence terminal, indeterminate racemes, corymbs, glomerulate thyrses, occasionally open, the basic unit composed of 2-7 pedicelled flowers subtended by a single bract, these sometimes arranged along 1 side of the rachis; flowers actinomorphic to slightly or profoundly zygomorphic; calyx tube membranes usually ruptured in fruit; corolla rotate to salverform, white, blue-lavender, purplish, pink, or red; stamens unequally inserted on the corolla tube or throat; filaments equal or unequal in length; anthers included to exserted; style included to exserted; capsule ovoid; seeds 1 to many per locule, seed coat mucilaginous when wet (Porter, 1998, 2010, 2011, and refs. therein).

1. Flowers in terminal, capitate, or glomerulate clusters; corolla tubes 3-8 mm...(2)
1′. Flowers usually in clusters at tips of lateral branches, inflorescence a diffuse or congested panicle (open in *I. longiflora* and *I. laxiflora*); corolla tubes 5-50 mm (5-15 mm in *I. multiflora*)...(7)
2. Perennials, often branched and woody at base; stems 5-60 cm; seeds 1 per locule, sometimes locules empty...(3)
2′. Annuals, usually branching from primary axis; stems 5-30 cm; seeds several per locule (except sometimes in *I. gunnisonii*)...(5)
3. Woody-based perennials (polycarpic); leaves mostly cauline, 8-30 mm; inflorescence of solitary or paniculately arranged heads or in glomerulate clusters...(4)
3′. Monocarpic, rosette-forming perennials; leaves basal and cauline, 20-45(-70) mm; inflorescence spicate, cylindrical, continuous or interrupted...**I. spicata**
4. Cauline leaves gray-hairy and mostly entire; Doña Ana County...**I. wrightii**
4′. Cauline leaves green, nearly glabrous, mostly entire; entire to pinnatifid depending on subspecies; NW...**I. congesta**
5. Lower cauline leaves linear and entire; seeds 1-2 per locule...**I. gunnisonii**
5′. Lower cauline leaves with 3-5 teeth or ovate to oblong lobes; seeds 2-3(4) per locule...(6)
6. Outer inflorescence bracts toothed; stems with short curly hairs; corolla tube 3-5 mm, lobes 1-1.5(-1.8) mm...**I. polycladon**
6′. Outer inflorescence bracts reduced, entire; stems with woolly hairs; corolla tube 4-8 mm, lobes 2-3.5 (-4) mm...**I. pumila**
7. Inflorescence diffusely branched, the flowers pedicelled, solitary or in pairs...(8)
7′. Inflorescence open to narrow, often 1-sided, the flowers short-pedicelled to subsessile, in lateral, pedunculate clusters...(9)
8. Corolla tubes 10-25 mm, lobes broadly elliptic, 4-7 mm...**I. laxiflora**
8′. Corolla tubes 30-50 mm, lobes ovate to broadly obovate, 7-12 mm...**I. longiflora**
9. Corollas scarlet, pink, magenta, or salmon...(10)
9′. Corollas white, lavender, or pale violet to purplish...(11)
10. Corolla tubes 18-25 mm; anthers and style in the throat or strongly exserted...**I. aggregata** (in part)
10′. Corolla tubes 10-15 mm; anthers and style included in the tube...**I. sancti-spiritus**
11. Corolla tubes 4-15(-20) mm...(12)
11′. Corolla tubes 15-50 mm...(13)

12. Corolla tubes sigmoid distal to calyx; stamens inserted unequally on the tube, included…**I. pinnata**
12′. Corolla tubes straight; stamens inserted at the same level on the upper tube or throat, exserted…**I. multiflora**
13. Corolla lobes not reflexed but spreading outward, elliptic, elliptic-ovate, acute, or attenuate, white…**I. aggregata** (in part)
13′. Corolla lobes, at least the upper 2, reflexed, oblong-ovate, obovate, oblong-obovate, obtuse, truncate, apiculate, or attenuate, white, lavender, or pale violet to purplish…(14)
14. Corolla tubes 15-25 mm, bent slightly downward; calyx lobes short-aristate…**I. macombii**
14′. Corolla tubes 25-50 mm, straight; calyx lobes mucronate…(15)
15. Corolla tubes (23-)29-50 mm, throat 2-5 mm, only slightly flaring, lobes ovate to obovate, rounded and obtuse or acute…**I. thurberi**
15′. Corolla tubes (15.5-)21-27 mm, throat 4-6.2 mm, ampliate, lobes ovate to obovate, 5.5-7.5 mm, acuminate, somewhat apiculate…**I. pringlei**

Ipomopsis aggregata (Pursh) V. E. Grant [*Gilia aggregata* (Pursh) Spreng.]—Scarlet gilia, skyrocket. Biennial or short-lived perennial, 20-100 cm, simple or branched at base; stems with short glandular hairs, often with longer, crisped, white, eglandular hairs below; leaves basal in nonpersistent rosette and cauline, subglabrous to short-pilose, deeply pinnatifid to bipinnatifid, cauline leaves gradually reduced in size; inflorescence thyrsoid, diffuse to 1-sided, with subsessile to short-pedicelled flowers on lateral branches; flowers subactinomorphic or somewhat zygomorphic; calyx 3-8 mm, short-glandular-pubescent, the lobes shorter than or equal to the tube and acuminate to triangular; corolla usually scarlet, sometimes white, less often pink, the tube 15-30 mm, the throat 3-6 mm wide, the lobes lanceolate, acuminate, often with dark red flecks; stamens inserted unequally above midtube; filaments unequal; anthers in the throat or exserted; stigma slightly exceeding the anthers; capsule 8-12 mm; seeds 5-10 per locule, mucilaginous when wet.

1. Corolla generally white and fragrant (rarely pink or salmon in introgressed populations); corolla tube 20-40 mm; anthers included (at the orifice in introgressed populations)…**subsp. candida**
1′. Corolla generally red, scarlet, or crimson (rarely pink or salmon); corolla tube 15-25 mm; anthers included or exserted…(2)
2. Anthers all exserted, the longest 3-7 mm beyond the orifice…**subsp. formosissima**
2′. Anthers included or only slightly exserted from the orifice, the longest 2 mm…(3)
3. Calyx lobes 3-5 mm; corolla tube 18-20 mm; anthers slightly exserted to slightly included, the longest 2 mm beyond the orifice…**subsp. aggregata**
3′. Calyx lobes 1-2 mm; corolla tube 20-25 mm; anthers included, sometimes only slightly so, the longest at the orifice…**subsp. collina**

subsp. aggregata—Scarlet gilia, skyrocket. Plants 30-60 cm; inflorescence ± open; calyx lobes acuminate, 3-5 mm, subequal to the tube; corolla scarlet with yellow flecking in the throat, rarely pink, paling to nearly white, corolla tube 18-20 mm, the throat 3-4 mm wide; anthers slightly exserted to slightly included, the longest 2 mm beyond the orifice; flowers generally lacking fragrance. Rocky slopes and openings in sagebrush, mountain grasslands, ponderosa pine forests, riparian habitats, 5545-10,990 ft. NC, NW.

subsp. candida (Rydb.) V. E. Grant & A. D. Grant—White-flowered skyrocket. Plants 20-60 cm; inflorescence somewhat congested, with shorter internodes (< 2 mm); calyx lobes triangular-acute, 1-2 mm, much shorter than the tube; corolla white to pink, rarely red in introgressed populations, corolla tube 20-40 mm, the throat 2-4 mm wide; anthers included; flowers generally fragrant. Sandy clay soils and alluvium of canyons, hills, ridges, roadsides, meadows, streamsides in coniferous forests, 6400-10,335 ft. NE, NC.

subsp. collina (Greene) Wilken & Allard—Hillside scarlet gilia, hillside skyrocket. Plants 30-60 cm; inflorescence ± open; calyx lobes triangular, 1-2 mm, much shorter than the tube; corolla scarlet or pink, sometimes with pink or white flecking in the throat, paling to nearly white in introgressed populations, corolla tube 20-25 mm, the throat 2-4 mm wide; anthers included, the longest at or below the orifice; flowers generally lacking fragrance. Sandy or clay soils, often rocky sites, roadsides, meadows, slopes, juniper-oak and pine-oak woodlands, mixed conifer forests, 5740-12,200 ft. NC.

subsp. formosissima (Greene) Wherry—Beautiful scarlet gilia, beautiful skyrocket. Plants 30–60 cm; inflorescence ± open; calyx lobes acuminate, 2–4 mm, shorter than the tube; corolla scarlet with yellow flecking in the throat, tube 18–25 mm, the throat 4–8 mm wide; anthers and stigma usually exserted; flowers generally lacking fragrance. Open sites, shrublands, piñon-juniper woodland, coniferous forest, 5510–9320 ft. Widespread except E.

Ipomopsis congesta (Hook.) V. E. Grant—Ballhead ipomopsis, ballhead gilia. Perennial herbs, woody at base, 20–60 cm, simple to branched at base; stems with short, white, crisped or floccose, eglandular trichomes; leaves cauline, subglabrous to floccose, linear-entire to pinnately or subpalmately lobed; inflorescence a dense, bracteate, terminal head; flowers subsessile to sessile, actinomorphic; calyx 3–5 mm, the lobes lanceolate, acuminate; corolla white or cream, the tube 3–5 mm, the throat 0.5–1 mm wide, the lobes rounded, sometimes with pinkish flecks; stamens inserted on the throat; filaments subequal; anthers and stigma slightly exserted; capsule 1.5–2 mm; seeds 1–2 per locule. New Mexico material belongs to **subsp. matthewii** J. M. Porter, Matt's ballhead ipomopsis. It has several ascending stems, woody at the base, 4–30 cm, with short, crisped hairs; lower leaves entire, upper leaves pinnatifid with narrow lobes and rachis. Rocky outcrops or badlands, sandy or clay soils associated with salt desert scrub and piñon-juniper woodland, 5740–7840 ft. Sandoval, San Juan.

Ipomopsis gunnisonii (Torr. & A. Gray) V. E. Grant [*Gilia gunnisonii* Torr. & A. Gray]—Gunnison's ipomopsis. Annual, 3–30 cm, simple to branched; stems glabrous or with short glandular hairs; leaves glabrous to sparsely short-pilose, linear-entire or rarely remotely few-toothed; inflorescence with flowers congested in terminal heads, bracteate; flowers subactinomorphic to slightly zygomorphic, subsessile; calyx 3–4.5 mm, glabrous to sparsely glandular; corolla white to pale lavender, the tube 4–7 mm, the throat 0.5–1 mm wide, the lobes rounded; stamens inserted on the upper tube or throat; filaments subequal; anthers and stigma slightly exserted; capsule 3–4 mm; seeds 1–2 per locule. Sandy, clay, or alluvial soils, dunes, badlands, shrublands, piñon-juniper woodland, 4820–6200 ft. Sandoval, San Juan.

Ipomopsis laxiflora (J. M. Coult.) V. E. Grant [*Gilia laxiflora* (J. M. Coult.) Osterh.]—Iron ipomopsis. Annual herb, 9–40 cm, simple to moderately branched; stems glabrous or sparsely villous with white, crisped, eglandular trichomes, occasionally stipitate-glandular; leaves cauline, glabrous to sparsely short-pilose, deeply pinnatifid-lobed, lobes and rachis filamentous, cauline leaves simple or 3-lobed; inflorescence open and diffuse, with 1–3 subsessile to long-pedicelled flowers at tips of branches; flowers ± actinomorphic, calyx 5.5–7 mm, short-glandular-pubescent, the lobes lanceolate, acuminate, mucronate; corolla white to lavender or pinkish, the tube 10–25 mm, the lobes broadly elliptic, 4–7 mm, acute to acuminate; stamens unequally inserted on the upper tube; filaments unequal; anthers included or 1 exserted, occasionally all slightly exserted; stigma included; capsule 8–11 mm, exceeding the calyx; seeds 4–6 per locule, fusiform, mucilaginous when wet. Sandy or rocky soils, mesas, hills, flats, riverbanks, roadsides, associated with grasslands, mesquite, juniper, 3385–7925 ft. Widespread except W, SC.

Ipomopsis longiflora (Torr.) V. E. Grant [*Gilia longiflora* (Torr.) G. Don]—Long-flowered skyrocket. Annual or biennial, sometimes woody-based, 25–120 cm, simple or branched; stems glabrous to sparsely short-pubescent; leaves cauline, glabrous to sparsely short-pilose, deeply pinnatifid-lobed; lobes and rachis narrow and filamentous, cauline leaves trifid or simple; inflorescence an open corymbose panicle, with 1–3 subsessile to long-pedicelled flowers at tips of branches; flowers actinomorphic; calyx 5–11 mm, short-glandular-pubescent, the lobes lanceolate to ovate, acuminate; corolla white to bluish or lavender, tube 30–50 mm, the throat 2–3 mm wide, the lobes 7–12 mm, ovate to broadly obovate, rounded to acuminate; stamens unequally inserted on the tube; filaments unequal; anthers with 2 included and 3 slightly exserted; stigma at the throat; capsule 7–15 mm; seeds 8–15 per locule (Wilken, 2001 and refs. therein).

1. Capsule 7–10 mm, equal to or slightly longer than the calyx in fruit…**subsp. australis**
1'. Capsule 10–15 mm, 1.5–2 times longer than the calyx…(2)
2. Inflorescence branching distributed along the primary axis; base of stem 1–4 mm diam.; cauline leaves generally with 5 lobes…**subsp. neomexicana**
2'. Inflorescence branching in distal 1/2 of the primary axis; base of stem 4–9 mm diam.; cauline leaves generally with 7 lobes…**subsp. longiflora**

subsp. australis R. A. Fletcher & W. L. Wagner—Annual to biennial; adaxial apices of calyx lobes short-pubescent; capsules 7–10 mm. Sandy soils, open sites, arroyos, bajadas, desert shrublands, creosote woodlands, piñon-juniper woodlands, 3935–5085(–6725) ft. C, WC, SW.

subsp. longiflora—Annual to biennial; adaxial apices of calyx lobes glabrous to sparsely short-pubescent; capsules 10–15 mm. Sandy soils, open sites, washes, desert scrub, sagebrush, and grasslands, 3020–4430 ft. NC, NW, C, WC, SC, SW.

subsp. neomexicana Wilken—Annual; adaxial apices of calyx lobes glabrous to sparsely short-pubescent; capsules 10–15 mm. Sandy, gravelly, or clay soils, open sites, washes, desert and sagebrush shrublands, piñon-juniper woodlands, 3840–10,000 ft. Widespread.

Ipomopsis macombii (Torr. ex A. Gray) V. E. Grant [*Gilia macombii* Torr. ex A. Gray; *G. calothyrsa* I. M. Johnst.]—Macomb's ipomopsis. Short-lived perennial, 20–75 cm, simple or branched at base; stems short-villous to floccose, stipitate-glandular distally; leaves basal in nonpersistent rosette, and cauline, glabrous to sparsely short-pilose or glandular, basal leaves few and not persistent, lower cauline leaves once-pinnatifid; inflorescence usually 1-sided, with subsessile flowers crowded near tips of short, lateral branches; flowers zygomorphic; calyx 4–7 mm, short-glandular-pubescent; corolla purplish or white, the tube 13–25 mm, with a somewhat sigmoid bend in the proximal region, the throat 2–4 mm wide, the lobes unequally deflexed, obovate, obtuse-truncate, apiculate, sometimes with white or dark purple flecks; stamens inserted on the tube; filaments unequally inserted on the throat; anthers included to slightly exserted; stigma slightly exceeding the anthers; capsule 3–5(–7.5) mm; seeds 1–2(3) per locule. Sandy to gravelly soils, oak woodland, coniferous forest, 4920–8530 ft. C, WC, SW.

Ipomopsis multiflora (Nutt.) V. E. Grant [*Gilia multiflora* Nutt.]—Many-flowered ipomopsis. Short-lived perennial, 15–70(–80) cm, simple to branched at base; stems with short to long, glandular to eglandular hairs; leaves glabrous to sparsely short-pilose or glandular, the lower deeply lobed, the upper entire to few-lobed; inflorescence glomerulate, ± 1-sided, flowers subsessile; flowers actinomorphic or slightly zygomorphic, calyx 4–8 mm, short-glandular-pubescent, calyx lobes short-aristate; corolla pale violet to purplish, the tube 5–12 mm, the throat 1–2.5 mm wide, the lobes subequal, the lower 3 partly united, often with purple flecks; stamens inserted on the upper tube or throat; filaments unequal; anthers exserted; stigma slightly exceeding the anthers; capsule 4.5–7 mm; seeds 2–8 per locule. Widespread except E.

1. Corolla weakly zygomorphic, usually with 2–3 erect to spreading (adaxial) lobes and 2–3 descending (abaxial) lobes, adaxial lobes with proximal purple streaks or white patches alternating with purple streaks; stamens unequal, the shortest ones straight, the longest declinate...**subsp. whitingii**
1'. Corolla actinomorphic, adaxial lobes lavender to violet, sometimes flecked or proximally streaked; stamens equal to subequal, straight, not declinate...(2)
2. Corolla tube 5–8 mm, included within or slightly exserted from the calyx; stamens 3.5–7 mm, all exserted ...**subsp. brachysiphon**
2'. Corolla tube 7–10 mm, exserted from the calyx; stamens 1–4 mm, included in the throat or the longest exserted...**subsp. multiflora**

subsp. brachysiphon (Wooton & Standl.) Wilken—Short-tubed many-flowered ipomopsis. Calyx 4.5–6 mm, lobes mucronate, mucros 0.9–1.3 mm; corolla actinomorphic, tube 5–8 mm, included within or slightly exserted from the calyx, lobes obovate, lavender to violet, sometimes with proximal purple streaks, equal to subequal, 3–5 × 1–2.5 mm wide, apices rounded; staminal filaments 3.5–7 mm, straight and exserted. Alluvial soils, associated with Chihuahuan desert grassland and scrub, 4790–6920 ft. Doña Ana, Sierra, Socorro.

subsp. multiflora—Many-flowered ipomopsis. Calyx 4–6.5 mm, lobes mucronate, mucros 0.5–0.9 mm; corolla actinomorphic, tube 7–10(–12) mm, exserted from the calyx, lobes oblong to obovate, lavender to violet, rarely purple, sometimes with proximal purple streaks on 2 or 3 lobes, equal to subequal, 4–5 × 1.5–2 mm, apices obtuse to rounded; staminal filaments subequal, 1–5 mm, straight, included in the throat or the longest exserted. Sandy, clay, or gravelly soils of hills, mountain slopes, canyons, arroyos, roadsides, sagebrush shrublands, piñon-juniper woodlands, ponderosa pine forests, 4000–10,200 ft. Widespread.

subsp. whitingii (Kearney & Peebles) Wilken—Whiting's many-flowered ipomopsis. Calyx 4.5–5.5 mm, lobes mucronate, mucros 0.3–0.7 mm; corolla weakly zygomorphic, usually with 2–3 erect to spreading (upper) lobes and 2–3 descending (lower) lobes, tube 6–12 mm, exserted from the calyx, adaxial lobes elliptic to obovate, with purple streaks or proximal white patches alternating with purple streaks, 2.5–5 × 1.5–2.2 mm, abaxial lobes elliptic to narrowly elliptic, 3–4.5 × 1.2–2 mm wide, lavender to violet, rarely with proximal marks; staminal filaments unequal, 3–7.5 mm, exserted, shortest ones straight, longest declinate. Sandy, clay, or gravelly soils of hills, canyons, arroyos, piñon-juniper woodlands, ponderosa pine forests, 6135–7000 ft. Cibola, Grant, McKinley.

Ipomopsis pinnata (Cav.) V. E. Grant [*Phlox pinnata* Cav.; *Gilia campylantha* Wooton & Standl.]— San Luis Mountain ipomopsis. Short-lived perennial, 20–75 cm, simple or branched at base; stems with short to long, glandular and nonglandular hairs; leaves basal in nonpersistent rosette, and cauline, 1.5–3.4 cm, glabrous to sparsely short-pilose or glandular, the lower deeply pinnate-lobed, the lobes narrowly linear, mucronate, the uppermost linear entire to few-lobed; inflorescence thyrsoid, with short-pedicellate flowers crowded on short, lateral branches; calyx 3–4 mm, short-glandular-pubescent; corolla cream tinged with violet or pale violet, the tube 4–12(–20) mm, with a sigmoid bend in the proximal region, the throat 1–2.7 mm wide, the lobes 3.5–5 mm, subequal, the lower 3 partly united, often with purple flecks, upper 2 reflexed; stamens unequally inserted on the lower tube; filaments short; anthers included; stigma included; capsule 2.5–4 mm, subglobose; seeds 1–2 per locule. Limestone-derived soils or loamy rhyolitic and basaltic soils, eroding slopes in Douglas-fir–oak forest and pine-oak woodlands; Animas and Mimbres Mountains, 6560–8530 ft. Hidalgo.

Ipomopsis polycladon (Torr.) V. E. Grant [*Gilia polycladon* Torr.]—Manybranched ipomopsis. Annual, 4–12 cm, usually with ascending to spreading branches; stems with short glandular hairs and some nonglandular crisped trichomes; leaves subglabrous to short-glandular-pubescent, coarsely toothed to lobed, the lobes ovate; inflorescence a congested terminal head, bracteate, the outer bracts leaflike; flowers sessile to subsessile; calyx 3–6 mm, the lobes lanceolate, acuminate; corolla white, the tube 3–5 mm, the throat 0.5–1 mm wide, the lobes ca. 3 mm, rounded; stamens unequally inserted on the tube; filaments subequal; anthers and stigma slightly exserted; capsule 4–5 mm; seeds 1–3 per locule. Sandy, gravelly, or clay soils, washes, desert shrublands, piñon-juniper woodland, 4265–6240 ft. NW, SW.

Ipomopsis pringlei (A. Gray) Henrickson [*Gilia pringlei* A. Gray]—Pringle's skyrocket. Short-lived perennial, 35–100 cm, simple to branched at base; stems often woody at base, crisp-puberulent throughout, glandular above; leaves basal in nonpersistent rosette, and cauline, subglabrous to crisp-puberulent and sparsely glandular, pinnatifid, lobes filiform; inflorescence secund, with subsessile to short-pedicelled flowers on lateral branches; calyx 7.5–9 mm, glandular-puberulent, the lobes attenuate; corolla lavender to bluish purple, the tube (15.5–)21–27 mm, straight to arcuate at calyx orifice, the throat (3.3–)4–6.2 mm wide, ampliate, the lobes ovate to obovate, 5.5–7.5 mm, acuminate, somewhat apiculate, often speckled with darker purple; stamens inserted slightly unequally on the tube; filaments unequal; anthers exserted; stigma slightly exserted; capsule 5.5–7 mm; seeds 3–4 per locule. Sandy to rocky soils, desert shrublands, woodlands, coniferous forest, 5350–8530 ft. WC, SW. Similar to *I. thurberi* (A. Gray) V. E. Grant and *I. macombii* (Torr. ex A. Gray) V. E. Grant, but corolla lobes more attenuate and throat broader than in either of those.

Ipomopsis pumila (Nutt.) V. E. Grant [*Gilia pumila* Nutt.]—Dwarf ipomopsis. Annual, 3–18 cm, simple to branched; stems glandular-puberulent and short-floccose; leaves subglabrous to floccose, often glandular, lower deeply pinnate lobed, upper lobed or entire; inflorescence a congested terminal head, bracteate, the bracts entire to toothed; flowers subsessile; calyx 3–6 mm, the lobes lanceolate, acuminate; corolla lavender to purplish or nearly white, the tube 4–8 mm, the throat 0.5–1 mm wide, the lobes acute to rounded; stamens inserted on the throat between the lobes; filaments subequal; anthers and stigma slightly exserted; capsule 3–5.5 mm; seeds 2–5 per locule. Sandy soils, desert shrublands, piñon-juniper woodland, 5350–8530 ft. Widespread except NE, EC, SE, SW.

Ipomopsis sancti-spiritus Wilken & R. A. Fletcher—Holy Ghost ipomopsis. Short-lived perennial, 35–80 cm, simple to branched at base; stems often woody at base, short-glandular above, short-pilose below; leaves basal in nonpersistent rosette, and cauline, subglabrous to sparsely glandular, deeply pinnately lobed; inflorescence secund, with short-pedicelled flowers on lateral branches; calyx 5–7 mm, short-glandular-pubescent, the lobes attenuate and awned; corolla deep pink to magenta, the tube 15–20 mm, the throat 4–6 mm wide, the lobes ovate to obovate, 6–8 mm, obtuse; stamens inserted unequally on the tube; filaments subequal; anthers included; stigma included; capsule 4–5 mm; seeds 2–4 per locule. Limestone-conglomerate-derived soils, pine-oak forests, 8170–8200 ft. San Miguel. **FE & E** (Wilken et al., 1988).

Ipomopsis spicata (Nutt.) V. E. Grant [*Gilia spicata* Nutt.]—Spiked ipomopsis. Perennial, herbaceous to woody at base, 5–23(–35) cm, simple to branched above; stems with short, white, woolly-villous hairs; leaves subglabrous to villous, entire to trifid, the central lobe longest; inflorescence a dense terminal spicate head, bracteate; flowers subsessile to sessile; calyx 4–6 mm, glandular-puberulent; corolla cream, brownish when dry, the tube 4.5–7 mm, the throat 0.5–1 mm wide, the lobes 1–2.5 mm, ovate to elliptic, acute; stamens equally or unequally inserted on the tube; filaments subequal, short; capsule 3–5 mm; seeds 3–4 per locule. Limestone-derived or sandy soils of hills and windswept flats in shortgrass prairies, 4400–6000 ft. Colfax, Harding, Union.

Ipomopsis thurberi (A. Gray) V. E. Grant [*Gilia thurberi* (A. Gray) A. Gray]—Thurber's ipomopsis. Short-lived perennial, 35–100 cm, simple to branched at base; stems often woody at base, short-glandular above, short-pilose below; leaves subglabrous to sparsely glandular, deeply lobed; inflorescence 1-sided, with subsessile to short-pedicelled flowers on lateral branches; calyx 6–10 mm, short-glandular-puberulent, the lobes attenuate; corolla lavender to bluish purple, the tube (23–)29–50 mm, often pale, usually straight, the throat 2–5 mm wide, the lobes ovate to obovate, rounded and obtuse or acute; stamens inserted unequally on the tube; filaments unequal; anthers exserted; stigma slightly exserted; capsule 8–10 mm; seeds 5–9 per locule. Sandy to rocky soils, canyons, pine-oak woodlands, coniferous forest, 4000–8000 ft. Grant, Hidalgo.

Ipomopsis wrightii (A. Gray) Shinners—Wright's ipomopsis. Subshrubs to 5 dm, with multiple simple branches; leaves mostly undivided, to 25 mm; glomerules few-flowered at branch tips; sepals 4–6 mm, united to about the middle, blades ciliate and awned; corolla salverform, tube 5–7 mm, lobes white to lavender; stamens exserted; seeds 3 or 4 per cell, viscid when wet (Correll & Johnston, 1979e). Chihuahuan desert scrub, 4060 ft. Doña Ana. **R**

LEPTOSIPHON Benth. – Leptosiphon

Annuals or perennials; stems simple to much branched, erect to decumbent, leafy; leaves opposite, rarely alternate above, palmately lobed, the lobes confluent with the rachis, linear, flat to terete, acute to weakly spinulose; inflorescence terminal and compact or axillary, glabrous to pubescent or glandular; flowers pedicelled, actinomorphic; calyx tube membranes usually ruptured or distended in fruit, the lobes equal, linear to attenuate, often mucronate; corolla white to yellow or lavender; stamens equally inserted on the corolla throat or tube; anthers slightly exserted or included; filaments equal in length; style included to exserted; capsule ovoid to oblong; seeds 1 to several per locule.

1. Perennial; corolla salverform, lobes white, internal corolla glabrous, external corolla tube glandular-pilose ...**L. nuttallii**
1.' Annual; corolla funnelform-rotate, lobes yellow, internal corolla with a ring of hairs between the tube and throat, external corolla tube glabrous...**L. chrysanthus**

Leptosiphon chrysanthus J. M. Porter & R. Patt. [*Gilia aurea* Nutt.; *Linanthus aureus* (Nutt.) Greene; *Leptosiphon aureus* (Nutt.) J. M. Porter & L. A. Johnson]—Golden linanthus. Annual, 3–15 cm, usually branched throughout; stems ascending, glabrous to pilose or glandular; leaves 3–5, linear, mucronate, 3–8 mm, glabrous to glandular; inflorescence open, the flowers 1–3, mostly terminal; flowers pedicelled, the filiform pedicels 4–13 mm; calyx glabrous, campanulate, 3–8 mm, the lobes equaling the tube, the hyaline membranes as wide as the herbaceous ribs; corolla diurnal, closed at night, rotate, 6–15 mm, bright yellow, the throat maroon to orange, with a ring of hairs between the tube and the throat; stamens inserted on the throat; style slightly exserted. Our material belongs to **subsp. chrysanthus**. Sandy, gravelly, rocky soils of flats, hills, slopes, and mesas, associated with grasslands, mesquite, piñon-juniper and pine-oak woodlands, 4500–7000 ft. WC, SW.

Leptosiphon nuttallii (A. Gray) J. M. Porter & L. A. Johnson [*Linanthus nuttallii* (A. Gray) Greene ex Milliken; *Linanthastrum nuttallii* (A. Gray) Ewan]—Nuttall's linanthus. Suffrutescent perennial to 3 dm, branching at the base; stems erect to decumbent, glabrous to short-pubescent; leaves palmately lobed, lobes (4)5–9, linear to linear-lanceolate, spinulose, 10–20 mm, short-pubescent; inflorescence compact, the flowers 2–5 in terminal, bracteate clusters; flowers subsessile; calyx glabrous to pubescent, narrowly campanulate, 7–10 mm, the lobes longer than the tube, the membranes mostly herbaceous, the hyaline part narrow, often obscure; corolla diurnal, salverform, 8–15 mm, the tube and lobes white, the throat yellow; stamens inserted on the throat; style slightly exserted. New Mexico material belongs to **subsp. tenuilobus** (R. Patt.) J. M. Porter [*Linanthus nuttallii* (A. Gray) Greene ex Milliken subsp. *tenuilobus* R. Patt.]. Sandy to rocky soils, meadows, coniferous forest, oak woodland, 5250–5415 ft. W except SW, San Juan.

LINANTHUS Benth. – Spiny gilia

Annuals or perennials; stems simple to much branched, erect to decumbent, leafy; leaves mostly opposite, sometimes alternate above, simple, pinnately lobed or deeply palmately lobed, the lobes confluent with the rachis, linear, flat to terete, acute to weakly spinulose; inflorescence terminal and compact or axillary, glabrous to glandular; flowers sessile to short-pedicelled, actinomorphic; calyx tube membranes ruptured or distended in fruit, the lobes equal, linear to attenuate, often mucronate; corolla white, cream, yellowish, or pale yellow-orange, often suffused with purple or lavender, open at dusk or night, usually closed during the day; stamens equally inserted on the corolla throat or tube; anthers exserted or included; filaments equal in length; style included to exserted; capsule ovoid to oblong; seeds 1 to several per locule.

1. Perennial; leaves firm to rigid, sharply acute to spinulose…**L. pungens**
1ʹ. Annual; leaves herbaceous, thin…**L. bigelovii**

Linanthus bigelovii (A. Gray) Greene—Bigelow's linanthus. Annual, 6–30 cm, simple or with 1–5 widely spaced, dichotomous branches above the base, glabrous; leaves mostly simple, sometimes with 3 lobes, linear, 1–3.5 cm, glabrous; inflorescence open, the flowers 1–3, axillary and terminal; flowers sessile to subsessile; calyx glabrous, narrowly campanulate, 8–13 mm, the lobes equaling the tube, the hyaline membranes usually wider than the herbaceous ribs; corolla nocturnal, closed during the day, funnelform, 8–16 mm, white to cream, the lobes tinged with purple; stamens inserted on the upper tube; filaments glabrous; style included. Our plants belong to **subsp. bigelovii**. Sandy, rocky, volcanic, limestone, or igneous soils, decomposed granite, slopes, washes, bajadas, desert shrubland, woodland, 3940–5300 ft. SC, SW.

Linanthus pungens (Torr.) J. M. Porter & L. A. Johnson [*Leptodactylon pungens* (Torr.) Nutt.]—Prickly phlox. Subshrub, 1–4 dm, branching mostly below the middle; stems ascending to erect, glandular and pilose; leaves opposite below, alternate above, palmately to pinnately lobed, the 3–9 lobes linear to narrowly oblong, glabrous to pubescent, spinulose, the upper leaves subtending clusters of short leaves; calyx 6–10 mm, glabrous to sparsely pubescent, the lobes slightly unequal, usually shorter than the tube; corolla nocturnal, closed during the day, salverform, 14–20(–25) mm, cream to pale yellow-orange, the throat often tinged lavender or purple; stamens 5, inserted on the upper tube; stigmas 3; ovary with 3 locules. Our plants belong to **subsp. pungens**. Sandy to rocky soils, shrubland, coniferous forest, woodland, 5250–9250 ft. NC, NW, WC, SW.

MICROSTERIS Greene – Slender phlox

Microsteris gracilis (Hook.) Greene [*Gilia gracilis* Hook.; *Phlox gracilis* (Hook.) Greene]—Slender phlox. Annual with 1–3 erect stems, 3–15 cm, often branched below, the lower stems ascending or spreading; leaves linear to elliptic or narrowly oblanceolate, pubescent and glandular, 1–2(–4) cm × 2–5 mm; flowers 1–2, subsessile to pedicelled at tips of terminal branches, sometimes axillary; pedicels glandular; calyx 3–8 mm; corolla white to bluish lavender, the throat sometimes yellow-tinged, the tube 4–8 mm, the lobes 1–2 mm, obtuse to retuse; stamens inserted on the upper tube and throat; stigmas among the lower stamens. Open, sandy to gravelly sites, 4000–9025 ft. NC, NW, WC, SW.

PHLOX L. – Phlox

Perennial, often cespitose; stems erect to decumbent, often much branched; leaves mostly opposite, often alternate only in the inflorescence, simple, entire, linear to elliptic or lanceolate; inflorescence terminal, the flowers 1–3; flowers pedicelled to sessile; calyx tube membranes ruptured in fruit, the lobes equal; corolla salverform, white to red, blue, or purple; stamens unequally inserted on the tube; filaments short, usually equal in length, glabrous; anthers and style usually included; capsule ovoid to ellipsoid; seeds 1(2–3) per locule, usually not gelatinous when wet (Wherry, 1955).

1. Plants compact to matted, sometimes loosely so, mostly < 1 dm; flowers subsessile to sessile…(2)
1'. Plants open; stems ascending to erect, mostly 1–6(–8) dm; flowers pedicelled…(7)
2. Stems and leaves subglabrous to loosely pilose; calyx membranes ridged to slightly distended below…
   P. austromontana
2'. Stems and leaves short-hirsute to canescent or woolly; calyx membranes flat or transversely wrinkled below…(3)
3. Leaves ciliate, the trichomes coarse and conspicuous…(4)
3'. Leaves not ciliate, or if so, the trichomes fine…(5)
4. Leaves spreading; style 2.5–5.5 mm…**P. pulvinata**
4'. Leaves appressed; style 2–3 mm…**P. condensata**
5. Upper leaves, stems, and calyx glandular-villous; corolla lobes with a pair of purple markings at base…
   P. vermejoensis
5'. Upper leaves, stems, and calyx eglandular, pilose, arachnoid-tomentose, or canescent; corolla lobes lacking pair of purple markings at base…(6)
6. Leaves subulate, 6–12 × 1–2 mm; flowers 1–3 per stem; pedicels 0.5–2.5 mm…**P. canescens**
6'. Leaves linear, (5–)12–25(–30) × 1.5–2.5 mm; flowers 1 per stem; pedicels 3–7 mm…**P. multiflora**
7. Vegetative and floral shoots widely spaced, from long, slender, horizontal rhizomes; corolla lobes erose to emarginate, only slightly if at all notched…**P. cluteana** (in part)
7'. Vegetative and floral shoots arising from a single, thick, ascending rhizome; corolla lobes deeply retuse, truncate, or rounded…(8)
8. Style 1.5–4 mm…(9)
8'. Style 8–18 mm…(12)
9. Corolla lobes deeply retuse, 1–2 mm deep…**P. woodhousei**
9'. Corolla lobes obtuse and entire, erose, or emarginate, but not deeply retuse…(10)
10. Pedicels and calyx eglandular…**P. triovulata**
10'. Pedicels and calyx with some glandular trichomes…(11)
11. Stems with few nodes; leaves linear; corolla tube glabrous to sparsely puberulent…**P. mesoleuca**
11'. Stems with many nodes; leaves narrowly elliptic to lanceolate; corolla tube glandular-puberulent, rarely glabrescent…**P. nana**
12. Calyx membrane plicate-carinate (gibbous basally)…**P. longifolia**
12'. Calyx membrane flat to ± plicate (not gibbous basally)…(13)
13. Calyx puberulent to villous with eglandular trichomes…**P. caryophylla**
13'. Calyx puberulent to villous with glandular trichomes…**P. cluteana** (in part)

Phlox austromontana Coville [*P. austromontana* subsp. *densa* (Brand) Wherry; *P. austromontana* subsp. *prostrata* (E. E. Nelson) Wherry]—Mountain phlox. Compact, matted, taprooted perennial, with 8 to many decumbent to ascending stems; stems 6–10(–15) cm, the lower internodes obscured by the leaves; leaves linear, firm, thick, grayish green, mucronate, glabrous to sparsely pubescent proximally, 8–15(–18) × 1–2 mm; inflorescence reduced, flowers solitary, subsessile; calyx 6–10(–12) mm, the membranes weakly ridged to distended; corolla white to light lavender, the tube 8–18 mm, the lobes 5–8 mm, obovate, obtuse; stamens inserted on the upper tube; stigmas below most of the stamens. Rocky soils, slopes, coniferous forest and woodland, 5300–9200 ft. W except Catron.

Phlox canescens Torr. & A. Gray [*P. hoodii* Richardson subsp. *canescens* (Torr. & A. Gray) Wherry]— Carpet phlox. Compact, matted perennial with many decumbent stems; stems 2–5 cm, the internodes obscured by the leaves, densely villous to woolly; leaves linear, firm, thick, green, subulate, 6–10(–12) × 1–2 mm, the margins and midrib often thick, glabrous to densely cobwebby-floccose below; inflorescence reduced, with flowers solitary, sessile to subsessile; calyx 5–9 mm, the membranes flat or wrinkled, the lobe margins glabrous, glandular, or floccose; corolla white to light lavender, the tube 8–12 mm, lobes 4–7 mm, obtuse; stamens inserted on the upper tube; stigma among the stamens. Sandy, gravelly, or clay soils, gypsum, rocky slopes and ledges, sagebrush, piñon-juniper woodlands, 5575–7220 ft. NW.

Phlox caryophylla Wherry—Pagosa phlox. Perennial with 1–5 erect stems from a short, deep-seated rhizome, rarely suffrutescent; stems 8–25 cm, sparsely glandular, the lower internodes evident; leaves linear-oblong to narrowly elliptic, flat, acute to acuminate, 30–50 × 2–4 mm, the upper leaves eglandular and pilose; inflorescence terminal, composed of 2–3-flowered cymes; pedicels short-pilose; calyx 9–15 mm, the membranes flat or only obscurely carinate-inflated; corolla bright pink to red-purple, the tube 10–17 mm, the lobes 5–10 mm, obtuse or retuse to emarginate; stamens inserted on the upper tube; stigmas usually among the stamens. Open sites, gravelly or clay soils in sagebrush, piñon-juniper woodlands, ponderosa pine forests, 6900–7900 ft. Rio Arriba. **R**

Phlox cluteana A. Nelson—Navajo Mountain phlox. Perennial with 1–2 stems from long, slender rhizomes, forming colonies of scattered flowering and vegetative shoots; stems 8–15(–20) cm, the internodes evident, the upper internodes often longer than the leaves, glabrous to glandular and short-villous; leaves linear, elliptic, or narrowly lanceolate, flat, acute to obtuse, glabrous to ciliate, sometimes glandular, 10–50 × 2–5 mm; inflorescence composed of 2–3-flowered cymes; pedicels glandular; calyx 6–10 mm, the membranes flat; corolla pink to deep red-purple, the tube 15–17(–20) mm, the lobes 8–10 mm, emarginate; stamens inserted on the upper tube; uppermost anthers slightly exserted; stigmas in the upper tube among the upper stamens. Open sites, coniferous forest, 6500–8300 ft. San Juan. **R**

Phlox condensata (A. Gray) E. E. Nelson [*P. caespitosa* Nutt. subsp. *condensata* (A. Gray) Wherry]— Dwarf phlox. Loosely to densely tufted perennial with many erect to decumbent stems; stems 1–3.5 cm, sparsely pilose, glandular, the lower internodes obscured by appressed leaves; leaves linear-subulate, 5–10 × 0.7–1.5 mm, the upper leaves ciliate and short pilose; inflorescence composed of 2–3-flowered cymes; pedicels glandular-pubescent; calyx 4–7 mm, the membranes narrow and flat; corolla white, the tube 6–9 mm, the lobes 3–5 mm, ovate, obtuse; stamens inserted on the upper tube; stigmas usually among the stamens. Open sites, rocky ridges, cliffs, meadows, alpine and subalpine, 10,500–12,500 ft. Cibola, San Miguel, Santa Fe.

Phlox longifolia Nutt. [*P. longifolia* subsp. *compacta* (Brand) Wherry; *P. grayi* Wooton & Standl.; *P. stansburyi* (Torr.) A. Heller]—Longleaf phlox. Taprooted perennial with 3–8(–10) erect to ascending stems, often suffrutescent; stems 1–3(–4) dm, sometimes loosely clumped, the internodes evident, glabrous to glandular and short-villous; leaves linear, lanceolate, or oblong-lanceolate, acuminate, 12–80 (–100) × 1.5–4(–8) mm; inflorescence composed of (1)2–3-flowered cymes; pedicels glandular to short-pilose; calyx 7–14(–15) mm, the membranes carinate to ridged-inflated at the base; corolla white to deep pink, the tube 12–25(–30) mm, the lobes 5–12(–15) mm, obovate, obtuse, rarely emarginate; stamens inserted on the upper tube; stigmas among the stamens. Sandy to rocky soils, open sites, shrublands, woodlands, 4800–8800 ft. Widespread except E. Geographic and ecological variation with respect to habit, leaf size, and floral morphology is complex throughout the range of the species and is in much need of study.

Phlox mesoleuca Greene [*P. nana* Nutt. subsp. *ensifolia* Brand]—Threadleaf phlox. Open perennial with few erect stems; stems 10–40(–60) cm, the internodes evident, glandular-pubescent; leaves linear, the margins and midrib somewhat thick, sparsely glandular-hirsute, somewhat folded at the midrib below, 50–85 × 3–4 mm; inflorescence composed of (1)2–3-flowered cymes; pedicels glandular-pubescent;

calyx 13–18 mm, the membrane flat or slightly plicate; corolla pink, often with a white to yellowish eye at the orifice, rarely white or yellow, the tube 13–18 mm, the lobes 12–16 mm, obtuse, entire to erose; stamens unequally inserted on the lower tube; stigmas below most of the stamens. Open rocky sites, piñon-juniper, pine woodlands, 4800–6800 ft. Widely scattered. C, WC, SC, SW. Closely related to and perhaps conspecific with *P. nana*.

Phlox multiflora A. Nelson—Flowery phlox. Compact, loosely tufted perennial with many decumbent to erect stems; stems 2–5(–10) cm, the internodes ± obscured by the leaves; leaves linear, green, the midrib thick, glabrous to minutely puberulent, 5–25(–30) × 1.5–2.5 mm wide; inflorescence reduced, flowers usually solitary; pedicels short; calyx 7–13 mm, the membrane flat, the lobes linear, somewhat thick, glabrous; corolla white to lavender or pink, the tube 8–15 mm, the lobes 6–9(–11) mm, ovate to orbicular, obtuse; stamens unequally inserted on the upper tube; stigmas among most of the stamens. Open clay, sandy, or rocky sites, coniferous woodlands to alpine tundra, 7240–12,000 ft. Sandoval, San Miguel.

Phlox nana Nutt.—Santa Fe phlox. Taprooted, tufted perennial with 1–7+ ascending to erect stems; stems 1–3 dm, the internodes evident; leaves linear to narrowly elliptic, flat, acute, 25–45 × 1–4(–5) mm, glandular-pubescent, the upper narrower; inflorescence composed of 1–3-flowered cymes; pedicels moderately to densely glandular; calyx 10–18 mm, the membrane flat or slightly carinate, glandular-villous; corolla bright pink or rose to lavender, the tube 11–18 mm, the lobes 9–16 mm, ovate to orbicular, obtuse; stamens unequally inserted midtube; stigmas below most of the anthers; ovary with (1)2–3 ovules per locule. Open rocky slopes, desert shrublands, woodlands, 3800–10,000 ft. NC, C, S.

Phlox pulvinata (Wherry) Cronquist—Cushion phlox. Compact, loose-matted perennial with many decumbent to erect stems; stems 1–3(–5) cm, the internodes obscured by the leaves; leaves linear-subulate, grayish green, the margins and midrib slightly thickened, ciliate, glabrous or pilose to glandular, 5–12(–15) × 1–2.5 mm wide; inflorescence reduced, flowers solitary, subsessile; calyx 5–8 mm, glandular, eglandular, or glabrous, the membrane flat; corolla white to pale bluish, the tube 8–12 mm, the lobes 5–8 mm, obovate, obtuse; stamens unequally inserted on the upper tube; stigmas below most of the stamens. Open, sandy to rocky sites, mixed conifer woodlands and alpine, 10,500–13,200 ft. NC.

Phlox triovulata Thurb. ex Torr.—Threeseed phlox. Suffrutescent perennial with 1–6+ stems; stems 1–4(–6) dm, the internodes evident, pilose to short-villous but eglandular; leaves linear to narrowly lanceolate, flat, 25–90 × 2–6 mm wide, glabrous to sparsely villous; inflorescence composed of 2–3-flowered cymes; pedicels short-pilose; calyx 7–12 mm, the membranes flat; corolla narrowly funnelform, white to lavender, the tube 7–16 mm, the lobes 3–5 mm wide, obtuse to truncate; stamens inserted on the upper tube; stigmas among the anthers. Open to shaded sites, canyons, rocky ravines, shrublands, woodlands, 3115–8730 ft. S. Closely related and perhaps conspecific with *P. nana* Nutt.

Phlox vermejoensis B. S. Legler—Vermejo phlox. Perennial herb from obscure caudex and taproot with numerous branched and tangled rhizomes to 15 cm; stems leafy, tufted, 1–4 cm, glabrous or sometimes glandular-pubescent; leaves with sheathing bases obscuring the stem, leaves spreading to ascending, not appressed, blades elliptic, oblong, or oblanceolate, 6–25 × 3–8 mm, soft and slightly fleshy, glabrous to sparsely glandular-pubescent, acute to obtuse, occasionally apiculate but not pungent; inflorescence terminal, composed of (1–)3-flowered leafy cymes; pedicels short, densely glandular-pubescent; sepals 8–11 mm, densely glandular-pubescent, intercostal membrane flat; corolla white, rarely tinged with pink, tube (7–)8–11 mm, lobes 6–9 mm, entire, not notched at the tip, usually with two purple marks at the proximal end; filaments unequally attached on upper 1/2 of corolla tube; anthers above the style and stigmas, some slightly exserted from the tube. Alpine scree slopes and ridges, 12,370–12,630 ft. Taos. **R** (Legler, 2011).

Phlox woodhousei (A. Gray) E. E. Nelson—Woodhouse's phlox. Perennial with 1–4 stems from a deep-seated rhizome, sometimes suffrutescent; stems 6–15 cm, the lower internodes evident; leaves linear-oblong to narrowly lanceolate, flat, acute to obtuse, glandular and short-villous, 15–40 × 3–5 mm wide; inflorescence composed of 2–3-flowered cymes; pedicels glandular, short-villous; calyx 7–9 mm, the membrane flat; corolla usually bright pink, the tube 10–15 mm, the lobes 6–10 mm, retuse to emarginate; stamens inserted midtube; stigmas well below the stamens. Open sites, shrubland, coniferous forest and woodland, 3935–8040 ft. Catron.

POLEMONIUM L. – Jacob's ladder, sky pilot

Perennial herbs (ours); stems erect to decumbent; leaves alternate, pinnately lobed to compound; leaflets sessile, entire or divided into 2–5 lobes; inflorescence terminal, open to compact and capitate or rarely axillary and then solitary; flowers actinomorphic; calyx herbaceous, becoming membranous, not ruptured in fruit, the lobes rounded to attenuate; corolla rotate to campanulate, funnelform or salverform, white, yellow, blue, or bluish violet; stamens equally inserted on the corolla tube; filaments equal in length, basally pilose; anthers included or exserted; style included to exserted, with 3 stigmatic branches; capsule globose to ovoid, dehiscent; seeds 1–20 per locule, gelatinous or not when wet (Grant, 1989; Rose, 2021).

1. Inflorescence capitate to obovoid; leaflets divided into 3–5 lobes, at least some…(2)
1'. Inflorescence open, cymose to paniculate; leaflets simple and entire…(3)
2. Flowers blue-violet, rarely white; leaflets all divided…**P. viscosum**
2'. Flowers ochroleucous or pale yellow, sometimes suffused with red; some leaflets simple, others divided …**P. brandegeei**
3. Flowering stems 5–10 dm, with 10–15 internodes, solitary or few from woody caudex or herbaceous rhizome; anthers yellow…(4)
3'. Flowering stems < 3 dm, with 3–6 internodes, tufted, from creeping woody rhizome; anthers white…**P. pulcherrimum**
4. Pubescence of the inflorescence floccose, matted; corolla lobes yellow, sometimes with anthocyanic streaks, acuminate, throat yellow…**P. flavum**
4'. Pubescence of the inflorescence of straight hairs; corolla lobes blue (rarely white), acute or more often rounded, throat white…(5)
5. Midstem and below spreading-villous; leaflets of midstem leaves mostly 13–19 per leaf, pubescent on both surfaces…**P. foliosissimum**
5'. Midstem and below glabrous; leaflets of midstem leaves mostly 17–27 per leaf, glabrous on both surfaces …**P. filicinum**

Polemonium brandegeei (A. Gray) Greene [*Gilia brandegeei* A. Gray; *G. mellita* Greene; *P. viscosum* Nutt. subsp. *mellitum* (A. Gray) J. F. Davidson]—Brandegee's Jacob's ladder. Perennial herb with 1–4 stems to 8–25(–30) cm, from an underground, branching caudex; leaves 5–16 cm, glandular-short-pubescent; leaflets 13–39, lanceolate or oblong to elliptic, 2.5–8(–10) × 1–3 mm; inflorescence subcapitate to spicate; flowers subsessile, the pedicels 1–3 mm; calyx 6–14 mm, the lobes shorter than the tube; corolla ochroleucous to yellow, funnelform, 15–24(–31) mm, the lobes rounded, shorter than the tube; anthers included in the throat to slightly exserted; style usually exserted; capsule 4–5 mm; seeds 4–5 per locule. Rare on talus at or above timberline, 9975–12,100 ft. NC.

Polemonium filicinum Greene [*P. grande* Greene]—Perennial with 1–5 leafy stems, 3.9–11.4 dm, glabrous on lower 1/2, pilose or villous on upper 1/2, from an underground, woody caudex; leaves 3–15 cm, glabrous below and sparsely glandular above; leaflets 15–27(–31), lanceolate to elliptic or ovate, 15–28(–36) × 3–6(–11) mm; inflorescence open to congested, branched; pedicels 2–8 mm; calyx 4–11 mm, the tube and lobes equal in length; corolla bluish violet to white or light yellow, campanulate, 10–15 mm, the lobes rounded and longer than the tube; anthers exserted; style exserted, usually exceeding the stamens; capsule 4–6 mm; seeds 3–5 per locule. Montane, riparian areas, 7215–9875 ft. NC, C, SC.

Polemonium flavum Greene [*P. foliosissimum* A. Gray var. *flavum* (Greene) Anway]—Yellow Jacob's ladder. Perennial with 1–5 leafy stems, 2.5–11.6 dm, sparsely or densely glandular-villous; leaves 3–10 × 0.8–2 cm, glabrous to sparsely glandular; leaflets 11–25, lanceolate to elliptic or ovate, 8.6–35 × 2–9(–14) mm; inflorescence open to congested, branched; pedicels 1.5–4.7 mm; calyx 5–10 mm, the tube and lobes equal in length; corolla pale to deep yellow, with reddish, anthocyanic streaks or flecking, campanulate, 10–19 mm, the lobes rounded and longer than the tube; anthers exserted; style exserted, usually exceeding the stamens; capsule 6-7 mm; seeds 3-5 per locule. Sandy, gravelly, or loamy soils along streams, meadows, openings in coniferous forests and aspen groves, 5720-11,350 ft. NC, C, SC.

Polemonium foliosissimum A. Gray—Leafy Jacob's ladder. Perennial with 1-5 leafy stems, 2-11 dm, pilose, villous, or floccose, glandular; leaves (1.2-)3-10.5 cm, densely to sparsely pubescent, glandular; leaflets 13-19(-23), narrowly lanceolate to elliptic or ovate, 4.5-25(-28) × 1.5-6(-10) mm; inflorescence congested to open, branched; pedicels 2-9 mm; calyx 4-9 mm, the tube and lobes equal in length; corolla light to dark bluish violet, campanulate, 9-15(-18) mm, the lobes rounded and longer than the tube; anthers exserted; style exserted, usually exceeding the stamens; capsule 4-6 mm; seeds 3-5 per locule. Wet meadows, canyon bottoms, along streams, moist areas, ponderosa pine, mixed conifer and riparian woodlands (5250-)6465-10,825 ft. NC, C, WC, SC, SW.

Polemonium pulcherrimum Hook.—Jacob's ladder. Perennial from a subrhizomatous caudex, bearing several stems, 5-40 cm, glabrous to sparsely pilose-glandular; leaves 3-20 cm, villous to glandular-pilose; leaflets 7-25, ovate to elliptic, 3-20 × 1-9 mm; inflorescence congested to open, corymbose, sometimes elongate; pedicels 1-8 mm; calyx 4-7 mm, the lobes and tube subequal; corollas 8-14 mm, rotate-campanulate, blue-violet, with a yellow or white throat, the tube 2-5 mm, the lobes ovate to obovate; anthers slightly exserted; style included to slightly exserted, usually equaling the stamens; capsule 3-4 mm; seeds 1-3 per locule. New Mexico material is **subsp. delicatum** (Rydb.) Brand [*P. delicatum* Rydb.], delicate Jacob's ladder. Crevices, rocky slopes, subalpine coniferous forests, talus above timberline, 8500-12,500 ft. NC.

Polemonium viscosum Nutt.—Sky pilot. Perennial herb from a branched caudex, clothed in persistent leaf bases; stems few, 4-20(-30) cm, glandular-puberulent, mephitic; leaves 5-11(-15) cm, glandular-pubescent; leaflets 13-39, each usually divided into 3-5 lobes, pseudoverticillate, 1-7 × 0.5-3 mm; inflorescence dense, capitate to slightly obovoid; flowers subsessile, the pedicels 1-3 mm; calyx 6-12 mm, the lobes shorter than the tube; corolla bluish violet, rarely white, funnelform, 15-30 mm, the lobes rounded, shorter than the tube; anthers included in the throat to slightly exserted; style usually exserted; capsule 4-5 mm; seeds 4-5 per locule. Meadows, talus or rocky slopes near or above timberline, 9085-13,160 ft. San Miguel, Santa Fe, Taos.

POLYGALACEAE - MILKWORT FAMILY

Kenneth D. Heil and Craig C. Freeman

Annual or perennial herbs or low, suffrutescent shrubs; leaves alternate, opposite, or whorled, simple, entire; flowers perfect, irregular, papilionaceous in appearance; sepals 5, inner usually petaloid; petals 3, united at the base, lowest usually keeled and often crested or beaked, the upper petals united to the staminal tube; stamens 6-8, united into a tube; anthers 1-celled; ovary superior, 2-carpellate; fruit a capsule; seeds often carunculate (Abbott, 2011, 2021; Allred & Ivey, 2012; Correll & Johnston, 1979f; Martin & Hutchins, 1980d).

1. Capsules 1-celled, indehiscent; annual…**Monnina**
1'. Capsules 2-celled, dehiscent; perennial (except *Polygala sanguinea*)…(2)
2. Keel petal with a fimbriate crest…**Polygala**
2'. Keel petal without a fimbriate crest…(3)
3. Keel petal with a cylindrical or conic hollow beak…**Rhinotropis**
3'. Keel petal without a beak or crest…**Hebecarpa**

HEBECARPA (Chodat) J. R. Abbott - Prostrate milkwort

Perennial, pubescent herb or dwarf shrub; stems erect to spreading; leaves alternate, blades entire; inflorescences terminal or axillary racemes; sepals 5; flowers purple, some parts greenish yellow or white; keel petal blunt, crestless; capsule 2-celled, narrowly winged or margined. Formerly included within *Polygala*.

1. Leaves and fruits gland-dotted…**H. macradenia**
1'. Leaves and fruits not gland-dotted…(2)
2. Capsules pubescent on the sides…**H. obscura**
2'. Capsules glabrous on the sides but ciliolate…(3)
3. Stems with widely spreading hairs…**H. rectipilis**
3'. Stems with incurved hairs…**H. barbeyana**

Hebecarpa barbeyana (Chodat) J. R. Abbott [*Polygala longa* S. F. Blake; *P. barbeyana* Chodat]—Blue milkwort. Stems suffruticulose below, 1.5-5 dm, densely incurved-griseous-puberulous; lower leaves oblong to oblong-elliptic or linear, 8-17 mm, middle and upper leaves oblong-lanceolate to linear, 17-36 × 1.5-4 mm, sometimes obtuse at apex; racemes loose, 2-10.5 cm; flowers purplish; sepals 2.2-3.5 mm; wings oval to suborbicular, 3.5-5.5 × 2.4-4 mm, rounded at the apex; keel 4-5.7, capsule ca. 7-8 mm. Mostly limestone or igneous, talus, caprock, rocky washes and arroyos, Chihuahuan desert scrub, desert grasslands, oak-juniper-Mexican piñon-ponderosa pine, 3115-6800 ft. EC, C, SE, SC, SW.

Hebecarpa macradenia (A. Gray) J. R. Abbott [*Polygala macradenia* A. Gray]—Glandleaf milkwort. Stems numerous, 3.5-21 cm, fruticulose below, from a thick woody rootstock; leaves linear-oblong to oblong-lanceolate, 2-6.3 × 0.6-1.3 mm, canescently puberulous, thickly gland-dotted; racemes 1- or 2-flowered, flowers purple; sepals gland-dotted, 1.7-2.1 mm; wings 5-5.5 × 2.5, rounded at apex, purple; keel ca. 4.5 mm; capsule canescently incurved-puberulous and gland-dotted, ca. 5-5.5 mm. Rocky outcrops, mostly limestone, Chihuahuan desert scrub, desert grasslands, 3000-5935 ft. SE, SC, SW.

Hebecarpa obscura (Benth.) J. R. Abbott [*Polygala puberula* A. Gray; *P. obscura* Benth.]—Velvetseed milkwort. Many-stemmed, suffruticulose below, 1.2-4 dm, with incurved or rarely slightly spreading hairs; lower leaves oblong to oval-oblong, 12-28 × 4-7 mm, middle and upper leaves oblong to oval-lanceolate to linear, 18-42 × 1.5-12 mm, puberulous; racemes 3.3-3.9 cm; flowers purple; sepals lanceolate; wings 4.5-5.8 × 2-3.5 mm, rounded; keel ca. 4.8-5 mm; capsule mostly 8-9 mm; seeds veil-like. Oak-juniper-piñon-ponderosa pine woodlands, 4000-7000 ft. San Juan, Quay, S.

Hebecarpa rectipilis (S. F. Blake) J. R. Abbott [*Polygala rectipilis* S. F. Blake]—New Mexico milkwort. Perennial herbs, 1.6-2.3 dm, stems with widely spreading hairs; leaves oblong to oblong-oval, 8-14 × 1.5-4 mm; inflorescence a raceme; flowers purple; sepals ca. 2 mm; petals ca. 6 mm; capsules 7.5-8 mm. Dry limestone hills; 1 record from Sierra County, ca. 2 miles S of Hillsboro, ca. 5400-5500 ft. Sierra.

MONNINA Ruiz & Pav. - Monnina

Monnina wrightii A. Gray—Blue pygmyflower. Plants to 3 dm; leaves 20-50 mm; inflorescence a terminal raceme; flowers ca. 3 mm, blue, petals free or nearly so, the keel not united to the staminal tube. Ponderosa pine-piñon-oak-juniper, Douglas-fir communities, 5415-8500 ft. WC, SW.

POLYGALA L. - Milkwort

Annual or perennial herbs or low shrubs; leaves simple, entire, alternate, opposite, or whorled; flowers in terminal racemes, solitary and axillary; lateral sepals much larger than the other 3 and petaloid; petals united at least to the base and to the staminal tube, the keel petal often crested or appendaged to form a beak; stamens 6-8, the filaments united into a tube; ovary 2-carpellate; style simple; capsule 2-celled, compressed at right angles to the partition, sometimes winged; seeds solitary in each cell, often carunculate (Allred, 2020; Correll & Johnston, 1979f; Martin & Hutchins, 1980d).

1. Plants annual; flowers in very dense, cylindrical to capitate racemes…**P. sanguinea**
1'. Plants perennial; flowers mostly in loose terminal or axillary racemes…(2)
2. Fruit not winged…**P. alba**
2'. Fruit winged on the upper cell…(3)
3. Leaves spreading, flexible; inflorescence glabrous…**P. hemipterocarpa**
3'. Leaves erect, rigid; inflorescence puberulent…**P. scoparioides**

Polygala alba Nutt.—White milkwort. Erect perennials with a cluster of short leafy branches at the base, 2-3.5 dm; leaves in 1 or 2 whorls at the base, 4-12 mm, others scattered, linear, acuminate, 8-25 mm; racemes dense; flowers white with a green center, the crest often purple; sepals 1.3-1.5 mm; wings ca. 3 × 1.5 mm, rounded at the apex; keel 3 mm; capsule 0.8-1.5 mm. Rocky sites, rimrock areas, sandy soils, mesquite plains, piñon-juniper, dry washes, 3000-8100 ft. Widespread.

Polygala hemipterocarpa A. Gray—Winged milkwort. Perennial from a fruticulose base, erect, stems glabrous, 12-56 cm; leaves linear, 6-23 × 0.6-1 mm; racemes 3-21 cm; flowers white, greenish-veined; sepals 1.6-2.3 mm; wings obovate, 3.5-4 × 1.6-1.8 mm; keel ca. 3.4 mm, the crest on each side consisting of about three 2-4-lobed processes; capsule 5 × 2.3 mm, winged. Rocky sites, mountain foothills, canyons, desert scrub, 4890-5610 ft. Grant, Hidalgo.

Polygala sanguinea L. [*P. viridescens* L.]—Purple milkwort. Slender annuals, 4 dm or less, simple or branched; leaves alternate, linear to elliptic-linear, 7-39 × 1-4.5 mm, papillose-serrulate on margins; racemes terminal, cylindrical to capitate; flowers greenish, rose, or purplish; sepals 1.3-1.8 mm; wings 4.8-6.3 mm, rounded at apex; keel 2.5-2.7 mm; capsules 1.5-1.7 mm. Clay outcrops, piñon-juniper-*Nolina*; 1 record in Socorro County, 6680 ft.

Polygala scoparioides Chodat—Broom milkwort. Stems numerous, finely incurved-puberulous, 9-30 cm; leaves linear-acicular, 7-14 × 0.6-1.3 mm, glabrous or nearly so; racemes 15-78 mm; flowers white, greenish-veined; sepals 1.3 mm; wings 2.6-3 mm, mucronulate at the obtuse apex; keel 2.8 mm, the crest on each side consisting of 3 weak lobes; capsules 3-3.5 mm, winged. Limestone slopes and igneous canyons, Chihuahuan desert scrub, mesquite, oak-juniper-Mexican piñon, 3200-6235 ft.

RHINOTROPIS (S. F. Blake) J. R. Abbott - Beaked milkwort

Perennial herbs or subshrubs; stems weakly ascending to erect; leaves alternate; inflorescence an axillary or terminal raceme; flowers white, pink, or rose; wings pink to rose; keel with a cylindrical or conic beak at the apex; fruit a 2-celled compressed capsule.

1. Stems woody; branches tending to be thorny; wings of flowers 7-12 mm...**R. subspinosa**
1'. Stems herbaceous or merely woody at the base, branches never thorny; wings of flowers 4-6 mm...(2)
2. Stems 15-40 cm, obviously incurved-pubescent; leaves (4-)9-30+ mm...**R. lindheimeri**
2'. Stems 1-5 cm, appearing glabrous but with scattered incurved hairs; leaves 2-5 mm...**R. rimulicola**

Rhinotropis lindheimeri (A. Gray) J. R. Abbott [*Polygala tweedyi* Britton; *P. parvifolia* (Wheelock) Wooton & Standl.]—Shrubby milkwort. Stems several from a fruticulose base, suberect to decumbent, to 18 cm; lower leaves mostly elliptic to oval or orbicular, 5-13 × 3-12 mm, upper oblong to oblong-lanceolate; racemes 2-8-flowered, 1-3.5 cm; upper sepals 2.8-3.2 mm, lower sepals ca. 2 mm; wings oblong-obovate, 4.5-6 × 2.3 mm, apex rounded; keel 4.2-4.8 mm, the beak 0.6-1.1 mm; capsule ca. 4.8 × 2.4 mm. New Mexico material belongs to **var. parvifolia** (Wheelock) J. R. Abbott. Brushy limestone cliffs, rocky slopes and outcrops in Chihuahuan desert scrub and desert grassland, 5500-6500 ft. Eddy, Grant, SC.

Rhinotropis rimulicola (Steyerm.) J. R. Abbott [*Polygala rimulicola* Steyerm.]—Steyermark's milkwort. Perennial from a woody base, glabrous or minutely puberulent, prostrate or slightly ascending, to 1.5 cm; leaves elliptic-ovate, 1.5-4 × 1-2.5 mm; racemes 1 or 2, terminal, flowers rose-purplish and white, ca. 5 mm; upper sepals ovate, 1.5-2 × 1 mm; lower sepals oblong-obovate, 2-3 × 1 mm; wings 4 × 2.5 mm, obtuse at apex; keel unbeaked or narrowly beaked, greenish yellow, 3 mm; upper petals united to the keel, purplish red, 4-5 × 1 mm; capsule 2 × 1.5-2 mm.

1. Beak of the keel rounded, mostly 0.1-0.3 mm...**var. rimulicola**
1'. Beak of the keel linear, prominent, 0.3-0.7 mm...**var. mescalerorum**

var. mescalerorum T. Wendt & Todsen—Mescalero milkwort. Crevices in sandy limestone cliffs, montane scrub; San Andres Mountains, 5700-6300 ft. Doña Ana. **R & E**

var. rimulicola—Steyermark's milkwort. Crevices in limestone cliffs, montane to lower montane coniferous forest; Guadalupe Mountains, 5000-8000 ft. Eddy. **R**

Rhinotropis subspinosa (S. Watson) J. R. Abbott [*Polygala subspinosa* S. Watson]—Spiny milk-wort. Low, much-branched shrub, 10–15 cm, end of branches spinose; leaves oblanceolate to obovate, 1–2 cm, puberulent; racemes few-flowered, terminal; sepals 4–6 mm, puberulent along the midline; wings glabrous, 8–10 mm; lateral petals 7–8 mm, purple, the keel petal mostly 10 mm, yellow, the beak straight, 2 mm; capsule ca. 8 mm; seeds ellipsoid, pubescent. Salt desert scrub, piñon-juniper, 4830–6240 ft. San Juan.

POLYGONACEAE – BUCKWHEAT FAMILY

Craig C. Freeman

Shrubs, subshrubs, herbs, or vines, annual, biennial, or perennial; stem nodes swollen or not; leaves mostly alternate, sometimes opposite or whorled, simple; fused (at least basally) stipules (ocreae) present or absent; blades simple, margins usually entire, rarely lobed; flowers usually bisexual, sometimes unisexual, actinomorphic; tepals 2–6; stamens 6–9; pistil 1(2) or 3(4)-carpellate, ovary superior, locule 1, placentation basal or free-central, styles 1–3, stigma 1 per style; fruits achenes, seed 1 (Freeman & Reveal, 2005).

1. Ocreae absent; nodes not swollen; involucral bracts present, enclosing or subtending flowers…(2)
1'. Ocreae present; nodes swollen; involucral bracts 0…(5)
2. Involucral bracts 3–12, connate proximally…(3)
2'. Involucral bracts 1 or 2, distinct…(4)
3. Involucral bracts 3–12, involucre tubular; perennials, biennials, or annuals…**Eriogonum**
3'. Involucral bracts 6, in 2 whorls of 3, involucre not tubular; annuals…**Stenogonum**
4. Involucral bract 1, 2-winged; leaf blade margins entire or lobed…**Pterostegia**
4'. Involucral bracts 2, not winged; leaf blade margins entire…**Chorizanthe**
5. Tepals 6…**Rumex**
5'. Tepals 4 or 5…(6)
6. Tepals 4; cypselae lenticular, winged; leaves mostly basal…**Oxyria**
6'. Tepals 4 or 5; cypselae trigonous, biconvex, lenticular, or quadrangular, unwinged or essentially so; leaves cauline or basal and cauline, rarely mostly basal…(7)
7. Stems climbing, scandent, or sprawling…**Fallopia**
7'. Stems prostrate, decumbent, ascending, or erect…(8)
8. Inflorescences terminal, spikelike; stems unbranched; leaves mostly basal, some cauline…**Bistorta**
8'. Inflorescences terminal and axillary or axillary only; stems usually branched; leaves cauline, sometimes also basal…(9)
9. Tepals distinct; cypselae strongly exserted…**Fagopyrum**
9'. Tepals connate to 2/3 their length; cypselae included or exserted…(10)
10. Ocreae often hyaline, silvery, glabrous, 2-lobed distally, often disintegrating into fibers or completely and plants annual herbs; if ocreae chartaceous or coriaceous and brown then plants subshrubs… **Polygonum**
10'. Ocreae chartaceous, usually tan, brown, or reddish, rarely silvery, glabrous or scabrous to variously hairy, never 2-lobed distally, often tearing with age; plants annual or perennial herbs…**Persicaria**

BISTORTA (L.) Scop. – Bistort

Herbs, perennial; stems erect, simple, nodes swollen; leaves mostly basal, some cauline, alternate; ocreae persistent or disintegrating with age and deciduous entirely or distally, chartaceous; blade margins entire; inflorescences terminal, spikelike, involucral bracts 0; perianth white to pink or greenish proximally and white or pink distally, rarely red; tepals 5, connate proximally ca. 1/5 their length; stamens 5–8; styles 3; cypselae included or exserted, trigonous.

1. Inflorescences short-cylindrical to ovoid, 10–50 × 8–25 mm, bulblets absent…**B. bistortoides**
1'. Inflorescences narrowly elongate-cylindrical, 15–90 × 4–10 mm, usually bearing pyriform pink to brown or purple bulblets proximally…**B. vivipara**

Bistorta bistortoides (Pursh) Small [*Polygonum bistortoides* Pursh]—Western or American bistort. Plants (1–)2–7.5 dm; leaf blades elliptic to oblong-lanceolate or oblong-oblanceolate, 35–220 × 8–48 mm, abaxially glabrous or hairy, adaxially glabrous; cauline leaves 2-6; inflorescences short-cylindrical to ovoid, 10-50 × 8-25 mm, bulblets absent; perianth white or pale pink; tepals 4-5 mm; cypselae 3.2-4.2 mm, shiny. Stream banks, moist to wet meadows, alpine slopes, 7000-13,000 ft. NC, WC, SW.

Bistorta vivipara (L.) Delarbre [*Polygonum viviparum* L.]—Alpine bistort. Plants (0.2-)0.8-3(-5) dm; leaf blades linear to lanceolate or oblong-ovate, 10-100 × 5-17(-23) mm, abaxially hairy, adaxially glabrous; cauline leaves 2-4; inflorescences narrowly elongate-cylindrical, 15-90 × 4-10 mm, usually bearing pink to brown or purple pyriform bulblets proximally; perianth greenish proximally and white or pink distally, rarely red, tepals 2.1-4 mm; cypselae rarely produced, 2.2-3.3 mm, dull. Subalpine and alpine meadows, stream banks, 8400-13,000 ft. NC, Otero.

CHORIZANTHE R. Br. ex Benth. – Spineflower

Chorizanthe brevicornu Torr.—Brittle spineflower. Herbs, annual, 0.5-3(-5) dm; stems spreading to erect, much branched, nodes not swollen; leaves basal, alternate; ocreae absent; blades linear-oblanceolate to elliptic or spatulate, 10-40 × 1-10 mm, margins entire, hairy abaxially and adaxially; inflorescences terminal, cymose; involucral bracts 2, distinct, not winged; perianth greenish white, yellowish white, or white; tepals (5)6, connate 3/4 their length, 2-3.5 mm; stamens 3; styles 3; cypselae included, lenticular, 3-4 mm. Our plants belong to **var. brevicornu**. Sandy desert grasslands, desert shrub communities, 4000-4400 ft. Hidalgo.

ERIOGONUM Michx. – Wild-buckwheat

Shrubs, subshrubs, or herbs, perennial, biennial, or annual; stems prostrate to erect, infrequently absent, usually branched, nodes not swollen; leaves basal, cauline, or basal and cauline, usually alternate, sometimes fascicled; ocreae absent; blade margins usually entire, sometimes wavy or crenulate; inflorescences terminal or terminal and axillary, cymose, racemose, simple or compound-umbellate, subcapitate, or capitate; involucre tubular, bracts 2-13; perianth usually white to red or yellow; tepals 6, connate proximally to 1/2 their length; stamens 9; styles 3; cypselae included to exserted, usually trigonous, rarely lenticular, rarely winged or ridged.

1. Annual, rarely biennial…**KEY 1**
1'. Perennial…(2)
2. Cypselae 3-winged at least in distal 1/2…**KEY 2**
2'. Cypselae not 3-winged…(3)
3. Flowers pubescent, pilose, hirsute, or tomentose…**KEY 3**
3'. Flowers glabrous…**KEY 4**

KEY 1: ANNUAL, RARELY BIENNIAL

1. Leaves basal and cauline or cauline only…(2)
1'. Leaves basal, cauline leaves absent or reduced and bractlike…(9)
2. Peduncles 0 mm…(3)
2'. Peduncles 1-60(-70) mm…(5)
3. Leaf blades glabrous on both surfaces…**E. aliquantum** (in part)
3'. Leaf blades puberulent to pilose or tomentose on both surfaces…(4)
4. Leaf blades puberulent to pilose on both surfaces; flowering stems puberulent to pilose…**E. divaricatum**
4'. Leaf blades tomentose on both surfaces; flowering stems tomentose…**E. polycladon**
5. Basal leaf blades linear to linear-oblanceolate, 2-4 mm wide…**E. pharnaceoides**
5'. Basal leaf blades lanceolate to oblanceolate, oblong, obovate, or elliptic, 3-30 mm wide…(6)
6. Stems simple…**E. annuum**
6'. Stems usually branched from bases, sometimes distally…(7)
7. Flowers glandular-puberulent; involucres glandular-puberulent…**E. maculatum**
7'. Flowers glabrous; involucres glabrous or villous-canescent…(8)
8. Basal leaf blades glabrous on both surfaces; involucres glabrous…**E. aliquantum** (in part)
8'. Basal leaf blades villous to tomentose on both surfaces; involucres villous-canescent…**E. abertianum**
9. Flowers pustulose, glandular-puberulent, puberulent, or hirsute…(10)
9'. Flowers glabrous…(13)

10. Involucres glabrous…(11)
10'. Involucres scabrellous or glandular-puberulent…(12)
11. Perianth white to red, rarely yellow; leaf blades reniform to round; margins plane; involucral teeth 5…**E. subreniforme** (in part)
11'. Perianth yellow to greenish yellow with green to red midribs; leaf blades broadly oblong, margins wavy; involucral teeth 4(5)…**E. trichopes**
12. Involucres scabrellous; flowers pustulose…**E. scabrellum**
12'. Involucres glandular-puberulent; flowers glandular-puberulent…**E. thurberi**
13. Stems floccose to tomentose; peduncles 0 mm…**E. palmerianum**
13'. Stems glabrous or sparsely hispid, floccose, or villous proximally; peduncles (0-)1-30(-40) mm…(14)
14. Peduncles or involucres (if peduncles absent) usually deflexed, sometimes some spreading, ascending, cernuous, or erect…(15)
14'. Peduncles erect…(17)
15. Perianths yellow to reddish yellow; involucres campanulate to hemispherical; peduncles 0 mm…**E. hookeri**
15'. Perianths white to pink, sometimes with green to red midveins, turning pink or red; involucres turbinate; peduncles (0-)1-25 mm…(16)
16. Outer tepals pandurate; peduncles (0-)1-25 mm…**E. cernuum**
16'. Outer tepals cordate to ovate; peduncles (0-)2-5 mm…**E. deflexum**
17. Involucral teeth 4; cypselae lenticular, 0.6-1 mm…**E. wetherillii**
17'. Involucral teeth 5; cypselae trigonous, 1.3-2.5 mm…(18)
18. Tepals monomorphic, lanceolate to spatulate, oblong, elliptic, or ovate…(19)
18'. Tepals dimorphic, outer pandurate or flabellate…(20)
19. Leaf blades sparsely villous to hirsute or glabrescent abaxially; involucres campanulate; cypselae 2-2.5 mm…**E. gordonii**
19'. Leaf blades densely tomentose abaxially; involucres turbinate; cypselae 1.7-2 mm…**E. subreniforme** (in part)
20. Leaf blades sparsely villous to hirsute abaxially; peduncles 10-30 mm; cypselae 1.3-1.6 mm…**E. capillare**
20'. Leaf blades densely tomentose abaxially; peduncles 3-15 mm; cypselae 1.5-2 mm…**E. rotundifolium**

KEY 2: PERENNIAL; CYPSELAE 3-WINGED AT LEAST IN DISTAL 1/2

1. Leaves basal; stems often fistulose; perianths purple to red or maroon…**E. atrorubens**
1'. Leaves basal and cauline; stems never fistulose; perianths yellow to yellowish green, usually turning red …(2)
2. Flowers pilose; cypselae winged in distal 1/2…**E. hieracifolium**
2'. Flowers glabrous; cypselae winged entire length…**E. alatum**

KEY 3: PERENNIAL; CYPSELAE NOT 3-WINGED; FLOWERS PUBESCENT, PILOSE, HIRSUTE, OR TOMENTOSE

1. Flowers with stipelike bases…(2)
1'. Flowers without stipelike bases…(5)
2. Plants 3-20 dm; cypselae tomentose…**E. longifolium**
2'. Plants 0.2-5 dm; cypselae glabrous except for pubescent beak…(3)
3. Perianths white to cream…**E. jamesii**
3'. Perianths yellow…(4)
4. Plants matted, 0.2-2.5 dm; leaf blades 10-30 mm…**E. arcuatum**
4'. Plants not matted, 4-5 dm; leaf blades 25-65 cm…**E. wootonii**
5. Cypselae villous to tomentose…(6)
5'. Cypselae glabrous…(7)
6. Perianths yellow; leaf blades 4-25(-30) mm; inflorescences capitate, subcapitate, umbellate-cymose, or cymose…**E. lachnogynum**
6'. Perianths white to rose or yellow; leaf blades 2-8(-12) mm; inflorescences capitate…**E. shockleyi**
7. Involucres tomentose; leaf blades densely tomentose on both surfaces…**E. havardii**
7'. Involucres glabrous; leaf blades hirsute on both surfaces, sometimes glabrous adaxially…**E. inflatum**

KEY 4: PERENNIAL; CYPSELAE NOT 3-WINGED; FLOWERS GLABROUS

1. Herbs…(2)
1'. Shrubs or subshrubs…(8)
2. Perianths yellow…(3)

2'. Perianths white, pink, rose, red, maroon, or purple, sometimes with green or red midribs…(4)
3. Leaf blades glabrous abaxially except for hairs on margins and veins…**E. gypsophilum**
3'. Leaf blades lanate abaxially…**E. umbellatum**
4. Inflorescences capitate; plants 0.2-3 dm…**E. ovalifolium**
4'. Inflorescences cymose or racemiform; plants 1-12 dm…(5)
5. Inflorescences racemiform; involucres tomentose to floccose; leaf blades 10-40(-50) mm wide…**E. racemosum**
5'. Inflorescences cymose; involucres glabrous (sometimes tomentose adaxially); leaf blades 2-30 mm wide…(6)
6. Perianths purple to red or maroon; cypselae 2-5 mm…**E. atrorubens**
6'. Perianths white to pink, or white with green to red midveins; cypselae 2-3 mm…(7)
7. Leaves basal or cauline on proximal 1/2 of stem; blades lanceolate to oblanceolate or elliptic, 15-70(-90) mm…**E. lonchophyllum** (in part)
7'. Leaves basal; blades elliptic, 3-15 mm…**E. tenellum**
8. Inflorescences racemiform or spiciform; peduncles 0 mm…(9)
8'. Inflorescences cymose or umbellate; peduncles 0-24 mm…(10)
9. Leaves fascicled at nodes, occasionally 1 per node…**E. wrightii**
9'. Leaves 1 per node…**E. leptocladon** (in part)
10. Leaves basal or cauline on proximal 1/2 of stem…**E. lonchophyllum** (in part)
10'. Leaves cauline…(11)
11. Involucres glabrous…(12)
11'. Involucres tomentose to floccose or glabrate…(14)
12. Leaf blade margins usually plane, infrequently revolute; peduncles 0 mm…**E. leptocladon** (in part)
12'. Leaf blade margins revolute; peduncles (0-)0.5-8 mm…(13)
13. Leaf blades 10-15(-20) mm; tepals dimorphic, outer ovate to flabellate; peduncles 1.5-8 mm…**E. clavellatum**
13'. Leaf blades (5-)20-60 mm; tepals ± monomorphic, oblong to obovate; peduncles (0-)0.5-2 mm…**E. leptophyllum**
14. Leaf blades 10-35 mm wide; peduncles 0 mm…**E. corymbosum**
14'. Leaf blades 1-7 mm wide; peduncles (0-)3-24 mm…(15)
15. Leaf blades 2-7 mm wide, margins plane…**E. effusum**
15'. Leaf blades 1-2 mm wide, margins revolute…**E. microtheca**

Eriogonum abertianum Torr.—Abert's wild-buckwheat. Herbs, annual, 0.5-6(-7) dm; leaves basal and cauline; basal blades oblong to obovate, 10-40 × 10-30 mm, villous to tomentose on both surfaces; inflorescences cymose; involucres campanulate, 2-3 mm, villous-canescent; flowers 3-4.5 mm, glabrous, stipelike base absent; perianth white to pale yellow, turning reddish or rose; cypselae lenticular, 0.6-1 mm, glabrous. Sandy, gravelly, or clayey flats, washes, slopes, grasslands, desert shrublands, oak and coniferous woodlands, 3675-8200 ft. NC, NW, S. Widespread.

Eriogonum alatum Torr.—Winged wild-buckwheat. Herbs, perennial, 5-20(-25) dm; leaves basal and cauline; basal blades linear-lanceolate or lanceolate to oblanceolate to spatulate, (30-)50-200 × 3-20 mm, strigose or glabrescent on both surfaces; inflorescences cymose; involucres turbinate to campanulate, 2-4(-4.5) mm, strigose or glabrous; flowers 1.5-2.5 mm, glabrous, stipelike base absent; perianth yellow to yellowish green, rarely red, turning red or maroon; cypselae 3-winged entire length, trigonous, 5-9 mm, glabrous.

1. Flowering stems and inflorescence branches strigose; peduncles ± strigose…**var. alatum**
1'. Flowering stems and inflorescence branches ± glabrous; peduncles usually glabrous, sometimes strigose…**var. glabriusculum**

var. alatum [*E. triste* S. Watson]—Sandy to gravelly flats and slopes, grasslands, saltbush and sagebrush shrublands, oak, piñon, juniper, and coniferous woodlands, 5000-10,900 ft. Widespread except E.

var. glabriusculum Torr.—Sandy to gravelly flats and slopes, grasslands, saltbush and mesquite shrublands, oak woodlands, 4300-4600 ft. Curry.

Eriogonum aliquantum Reveal—Cimarron wild-buckwheat. Herbs, annual, 1.5-3.5 dm; leaves basal and cauline; basal blades elliptic, 15-20 × 10-15 mm, glabrous on both surfaces; inflorescences cymose; involucres turbinate-campanulate, 1-1.3 mm, glabrous; flowers 1.2-2 mm, glabrous, stipelike base absent; perianth yellow with darker midribs; cypselae trigonous, 1.7–2.3 mm, glabrous. Shaly saltbush and sagebrush shrublands, clayey grasslands, 6000–6700 ft. Colfax. **R**

Eriogonum annuum Nutt.—Annual wild-buckwheat. Herbs, annual or biennial, 5–20 dm; leaves basal and cauline or cauline, sometimes fascicled; blades oblanceolate to oblong, 10–70 × 3–15 mm, densely tomentose abaxially, floccose adaxially; inflorescences cymose; involucres turbinate to campanulate, 2.5–4 mm, tomentose to floccose abaxially, glabrous adaxially; flowers 1–2.5 mm, glabrous, stipelike base absent; perianth white or cream to rose or reddish brown; cypselae trigonous, 1.5–2 mm, glabrous. Sandy flats and slopes, sand dunes, sagebrush shrublands, 3500–7300 ft. Widespread.

Eriogonum arcuatum Greene—Baker's wild-buckwheat. Herbs, perennial, matted, 0.2–2.5 dm; leaves basal; blades oblanceolate to elliptic, (5–)10–30 × 5–15 mm, densely tomentose abaxially, floccose adaxially; inflorescences umbellate or compound-umbellate, rarely capitate; involucres campanulate, 3–7 mm, floccose; flowers (4–)5–8 mm, densely pubescent, stipelike base 0.7–2 mm; perianth yellow; cypselae trigonous, 4–5 mm, glabrous except for pubescent beak.

1. Involucres 3–8 mm wide; inflorescences umbellate to compound-umbellate, rarely capitate; flowers (4–)5–8 mm...**var. arcuatum**
1'. Involucres 3–5 mm wide; inflorescences capitate; flowers 6–7 mm...**var. xanthum**

var. arcuatum [*E. jamesii* Benth. var. *flavescens* S. Watson]—Baker's wild-buckwheat. Sandy, clayey, gravelly, or rocky flats and slopes, ledges, cliffs, grasslands, desert shrublands, piñon, juniper, and coniferous woodlands, 6500–7000 ft. NC, NW.

var. xanthum (Small) Reveal—Ivy League wild-buckwheat. Gravelly to sandy slopes, subalpine and alpine meadows and woodlands, 12,100–12,600 ft. Taos.

Eriogonum atrorubens Engelm.—Red wild-buckwheat. Herbs, perennial, 5–12 dm; leaves basal; blades oblanceolate to oblong or elliptic, 40–80 × 10–30 mm, strigose on both surfaces, sometimes less so adaxially; inflorescences cymose; involucres turbinate to campanulate, 1.5–4(–4.5) mm, glabrous; flowers (1.5–)2–2.5 mm, glabrous, stipelike base absent; perianth purple to red or maroon; cypselae obscurely 3-winged distally, trigonous, 2–5 mm, glabrous. Our plants are **var. atrorubens**. Rocky to gravelly slopes in grassland, oak, oak-juniper woodlands, 4300–5300 ft. Hidalgo.

Eriogonum capillare Small—San Carlos wild-buckwheat. Herbs, annual, (1–)2–4 dm; leaves basal; blades obovate to round, 10–30 × 10–30 mm, sparsely villous to hirsute on both surfaces; inflorescences cymose; involucres campanulate, 1–1.5 mm, glabrous, teeth 5; flowers 1–1.6 mm, glabrous, stipelike base absent; perianth white with greenish or reddish midribs, turning pink to rose; cypselae trigonous, 1.3–1.6 mm, glabrous. Rocky desert scrub communities, ca. 4500 ft. Hidalgo.

Eriogonum cernuum Nutt.—Nodding wild-buckwheat. Herbs, annual, 0.5–6 dm; leaves basal; blades round-ovate to round, (5–)10–20(–25) × (5–)10–20(–25) mm, tomentose abaxially, tomentose to floccose or glabrate adaxially; inflorescences cymose; involucres turbinate, (1–)1.5–2 mm, glabrous; flowers 1–2 mm, glabrous, stipelike base absent; perianth white to pink, turning red; cypselae trigonous, 1.5–2 mm, glabrous. Clayey, sandy, and gravelly flats and slopes, grassland, desert shrublands, oak, piñon-juniper, and coniferous woodlands, 4000–8500 ft. Widespread except E.

Eriogonum clavellatum Small—Comb Wash wild-buckwheat. Subshrubs, perennial, 1–2.5 dm; leaves cauline, sometimes fascicled; blades oblanceolate, 10–15(–20) × 1–2 mm, densely tomentose abaxially, sparsely tomentose or rarely glabrous adaxially; inflorescences umbellate to cymose; involucres turbinate-campanulate, (3–)3.5–4.5(–5) mm, glabrous; flowers 2.5–3.5 mm, glabrous, stipelike base absent; perianth white; cypselae trigonous, 3–3.5 mm, glabrous. Sandy to clayey washes and slopes, saltbush shrublands, 4400–5700 ft. San Juan.

Eriogonum corymbosum Benth.—Crispleaf wild-buckwheat. Shrubs or subshrubs, perennial, 5–10 dm; leaves cauline; blades elliptic-oblong to oblong, ovate, or round, rarely cordate, 10–35 × 10–35 mm, tomentose to floccose abaxially, glabrous or glabrate, sparsely to densely lanate, or floccose adaxially;

inflorescences cymose; involucres turbinate, 2–3.5 mm, tomentose to floccose; flowers 2–3 mm, glabrous, stipelike base absent; perianth white to cream; cypselae trigonous, 2–3 mm, glabrous.

1. Leaf blades broadly ovate to round, usually glabrous or glabrate adaxially, sometimes tomentose to floccose…**var. orbiculatum**
1′. Leaf blades elliptic to ovate, rarely cordate, sparsely to densely lanate to floccose adaxially…**var. velutinum**

var. orbiculatum (S. Stokes) Reveal & Brotherson—Orbiculate-leaf wild-buckwheat. Sandy, gravelly, or clayey flats and slopes in desert shrublands and piñon-juniper woodlands, 6500–7300 ft. NW.

var. velutinum Reveal—Velvety wild-buckwheat. Gravelly or clayey flats and washes, grassland, desert shrublands, juniper and piñon-juniper woodlands, 5200–7200 ft. NC, NW, C.

Eriogonum deflexum Torr.—Flat-top skeleton-weed. Herbs, annual, (0.5–)1–5(–20) dm; leaves basal; blades cordate to reniform or nearly round, 10–25(–40) × 20–40(–50) mm, densely tomentose abaxially, sparsely tomentose to floccose or glabrate adaxially; inflorescences cymose; involucres turbinate, 1.5–2.5 mm, glabrous; flowers 1–2 mm, glabrous, stipelike base absent; perianth white to pink with green to red midribs, turning pink to red; cypselae trigonous, 1.5–3 mm, glabrous. Our plants are **var. deflexum**. Sandy to gravelly washes, flats, and slopes, saltbush, creosote bush, greasewood, and sagebrush shrublands, 6000–7000 ft. Hidalgo.

Eriogonum divaricatum Hook.—Divergent wild-buckwheat. Herbs, annual, 1–2(–3) dm; leaves basal and cauline; basal blades elliptic-oblong to round, 10–20(–25) × 10–20(–25) mm, puberulent to pilose on both surfaces; inflorescences cymose; involucres somewhat appressed to branches, campanulate, 1–2 mm, pilose; flowers 1–2 mm, glandular-hispidulous, stipelike base absent; perianth yellow; cypselae trigonous, 1.5–2 mm, glabrous. Clayey to sandy flats and slopes, badlands, desert shrublands, piñon-juniper woodlands, 4700–7100 ft. NC, NW.

Eriogonum effusum Nutt.—Spreading wild-buckwheat. Shrubs, perennial, (1.5–)2–5(–7) dm; leaves cauline; blades oblanceolate to oblong or obovate, 10–30 × 2–7 mm, densely tomentose abaxially, floccose to glabrate adaxially; inflorescences cymose; involucres turbinate, 1.5–2.5(–3) mm, tomentose to floccose, teeth 5; flowers 2–4 mm, glabrous, stipelike base absent; perianth white to pinkish white or pale yellow; cypselae trigonous, 2–2.5 mm, glabrous. Sandy to rocky slopes and flats, grasslands, sagebrush shrublands, juniper woodlands, 4500–6500 ft. NC, NW, C.

Eriogonum gordonii Benth.—Gordon's wild-buckwheat. Herbs, annual, (0.5–)1–5(–7) dm; leaves basal; blades obovate to reniform or round, 10–50 × 10–50 mm, sparsely villous to hirsute or glabrescent on both surfaces; inflorescences cymose; involucres campanulate, 0.6–1.5 mm, glabrous, teeth 5; flowers 1–2.5 mm, glabrous, stipelike base absent; perianth white with green or red midribs, rarely yellow, turning pink to rose; cypselae trigonous, 2–2.5 mm, glabrous. Sandy to clayey flats and slopes, desert shrublands, piñon, piñon-juniper, and montane coniferous woodlands, 4800–6300 ft. San Juan.

Eriogonum gypsophilum Wooton & Standl.—Gypsum wild-buckwheat. Herbs, perennial, 1.2–2 dm; leaves basal; blades cordate to truncate, rarely reniform, 10–25 × 15–25(–30) mm, glabrous except for hairs on margins and veins; inflorescences cymose; involucres campanulate, 1–1.5 mm, glabrous; flowers 1–2 mm, glabrous except for hairs along midrib, stipelike base absent; perianth yellow; cypselae trigonous, 1.5–2 mm, glabrous. Eroded gypsum hills and fans, creosote bush shrublands, 3200–3600 ft. Eddy. **FT & E**

Eriogonum havardii S. Watson—Havard's wild-buckwheat. Herbs, perennial, 2–6 dm; leaves basal; blades oblanceolate to elliptic, 10–30(–50) × 2–13 mm, densely tomentose on both surfaces; inflorescences cymose; involucres campanulate, 1.5–2.5 mm, tomentose; flowers 2.5–3 mm, densely pubescent, stipelike base absent; perianth yellow; cypselae trigonous, 2–2.5 mm, glabrous. Gravelly to sandy flats and outcrops, grassland, shrublands, juniper woodlands, 3300–7400 ft. Widespread except N, W.

Eriogonum hieracifolium Benth.—Hawkweed wild-buckwheat. Herbs, perennial, 4–7 dm; leaves basal and cauline; basal blades oblanceolate to spatulate, 30–150 × 5–20(–25) mm, sparsely to densely strigose on both surfaces; inflorescences cymose; involucres turbinate to campanulate, 2.5–4 mm, hirsute to strigose; flowers 1.5–2.5 mm, pilose, stipelike base absent; perianth yellow, turning red; cypselae 3-winged in distal 1/2, trigonous, 4.5–6 mm, strigose. Sandy to gravelly flats and slopes, grasslands, desert shrublands, oak, juniper, and montane coniferous woodlands, 4800–8100 ft. Widely scattered in S.

Eriogonum hookeri S. Watson—Hooker's wild-buckwheat. Herbs, annual, 1–6 dm; leaves basal; blades cordate to reniform, (10–)20–50 × 20–60 mm, densely tomentose abaxially, tomentose adaxially; inflorescences cymose; involucres deflexed, campanulate to hemispherical, 1–2 mm, glabrous; flowers 1.5–2 mm, glabrous, stipelike base absent; perianth yellow to reddish yellow; cypselae trigonous, 2–2.5 mm, glabrous. Silty to clayey badlands and arroyos, 5000–5500 ft. San Juan.

Eriogonum inflatum Torr. & Frém.—Desert trumpet. Herbs, perennial, 1–10(–15) dm; leaves basal; blades oblong-ovate to oblong, reniform, or round, (5–)10–25(–30) × (5–)10–20(–25) mm, hirsute on both surfaces or glabrous adaxially; inflorescences cymose; involucres turbinate, 1–1.5 mm, glabrous; flowers (1–)2–3(–4) mm, hirsute, stipelike base absent; perianth yellow with green or red midribs; cypselae lenticular or trigonous, 2–2.5 mm, glabrous. Sandy to gravelly washes, flats, and slopes, mixed grassland, saltbush, creosote bush, mesquite, and sagebrush communities, piñon and/or juniper woodlands, 5000–5500 ft. San Juan.

Eriogonum jamesii Benth.—Antelope-sage. Herbs or subshrubs, perennial, matted, 0.5–2.5 dm; leaves basal; blades narrowly elliptic, 10–30 × 3–10 mm, densely tomentose abaxially, sparsely tomentose, floccose, or glabrous adaxially; inflorescences compound-umbellate; involucres turbinate, 1.5–7 mm, tomentose to floccose, 5–8-toothed; flowers 3–8 mm, densely pubescent, stipelike base 0.7–2 mm; perianth white to cream; cypselae trigonous, 4–5 mm, glabrous except for pubescent beak.

1. Leaf blade margins undulate, frequently crisped; flowers 3–5(–6) mm…**var. undulatum**
1'. Leaf blade margins plane, not crisped; flowers 4–8 mm…**var. jamesii**

var. jamesii—James' antelope-sage. Sandy to gravelly flats and slopes, grasslands, desert shrublands, sagebrush shrublands, oak, piñon, juniper, and montane coniferous woodlands, 4300–12,200 ft. Widespread except EC, SE.

var. undulatum (Benth.) S. Stokes ex M. E. Jones—Wavy-margined antelope-sage. Sandy, gravelly, and rocky flats and slopes, grasslands, desert shrublands, oak and coniferous woodlands, 5800–8500 ft. WC, SW.

Eriogonum lachnogynum Torr. ex Benth.—Herbs, perennial, sometimes cespitose and matted, (0.5–)1–4 dm; leaves basal; blades lanceolate or oblanceolate to elliptic, 4–25(–30) × 1–5(–8) mm, densely tomentose abaxially, densely sericeous adaxially; inflorescences capitate, subcapitate, umbellate-cymose, or cymose; involucres broadly campanulate, 2–4 mm, tomentose; flowers 2.5–6 mm, densely pubescent, stipelike base absent; perianth yellow; tepals monomorphic, lanceolate; cypselae trigonous, 3–4 mm, villous to tomentose.

1. Plants not matted; inflorescences subcapitate, umbellate-cymose, or cymose; flowering stems 1–2 dm at anthesis; leaf blades 10–25(–30) × 3–5(–8) mm…**var. lachnogynum**
1'. Plants cespitose and matted; inflorescences capitate; flowering stems 0.01–0.7 dm at anthesis; leaf blades 4–12 × 1–4 mm…(2)
2. Flowering stems 0.1–0.5(–0.7) dm, exceeding leaves…**var. sarahiae**
2'. Flowering stems (0.01–)0.02–0.05(–0.12) dm, shorter than or barely exceeding leaves…**var. colobum**

var. colobum Reveal & A. Clifford—Clipped wild-buckwheat. Sandy or gypseous limestone flats and slopes in piñon-juniper woodlands, 6800–7800 ft. McKinley, Taos. **R & E**

var. lachnogynum—Woolly-cup wild-buckwheat. Clayey, sandy, and gravelly flats and washes, grasslands, desert shrublands, piñon and piñon-juniper woodlands, 5200–6500 ft. NE, NC.

var. sarahiae (N. D. Atwood & A. Clifford) Reveal—Sarah's wild-buckwheat. Sandy limestone ridges and edges of mesas in piñon-juniper woodlands, 5900–7500 ft. McKinley, San Juan. **R**

Eriogonum leptocladon Torr. & A. Gray—Sand wild-buckwheat. Shrubs, perennial, 2–10 dm; leaves cauline; blades linear-lanceolate to oblong or elliptic, 15–40 × 2–12 mm, densely tomentose abaxially, sparsely tomentose adaxially; inflorescences racemiform or spiciform; involucres turbinate to turbinate-campanulate, 1.5–3 mm, tomentose to floccose or glabrous; flowers 2–3.5 mm, glabrous, stipelike base absent; perianth white; cypselae trigonous, 2.5–3.5 mm, glabrous except for papillate beak.

1. Flowering stems glabrous…**var. papiliunculi**
1'. Flowering stems tomentose to floccose…**var. ramosissimum**

var. papiliunculi Reveal—Butterfly wild-buckwheat. Windblown sands on flats, washes, and slopes, grasslands, sagebrush shrublands, piñon-juniper woodlands, 3900–7000 ft. San Juan.

var. ramosissimum (Eastw.) Reveal—San Juan wild-buckwheat. Windblown sand on flats, washes, and slopes, grasslands, desert shrublands, sagebrush shrublands, piñon-juniper woodlands, 5900–7200 ft. NW, C.

Eriogonum leptophyllum (Torr.) Wooton & Standl.—Slender-leaf wild-buckwheat. Shrubs or subshrubs, perennial, (0.5–)2–8(–13) dm; leaves cauline, sometimes fascicled; blades linear to linear-oblanceolate, (5–)20–60 × 1–3 mm, densely to sparsely tomentose abaxially, sparsely tomentose or glabrous adaxially; inflorescences cymose; involucres narrowly turbinate, 2–4(–4.5) mm, glabrous; flowers 2.5–4 mm, glabrous, stipelike base absent; perianth white; cypselae trigonous, 2.5–4 mm, glabrous. Clayey flats and slopes, rock outcrops, grasslands, sagebrush shrublands, piñon-juniper woodlands, 5000–8300 ft. NC, NW, C, WC.

Eriogonum lonchophyllum Torr. & A. Gray [*E. nudicaule* (Torr.) Small]—Spearleaf wild-buckwheat. Shrubs, subshrubs, or herbs, perennial, 1–5 dm; leaves basal or cauline on proximal 1/2 of stem; blades lanceolate to oblanceolate or elliptic, 15–70(–90) × 2–20 mm, densely tomentose abaxially, sparsely tomentose to floccose or glabrous adaxially; inflorescences cymose; involucres turbinate to turbinate-campanulate, 2.5–4 mm, glabrous; flowers 2–3.5(–4) mm, glabrous, stipelike base absent; perianth white; cypselae trigonous, 2–3 mm, glabrous. Clayey, sandy, gravelly, and rocky grasslands, desert shrublands, piñon-juniper and montane coniferous woodlands, 5400–8500 ft. N.

Eriogonum longifolium Nutt.—Longleaf wild-buckwheat. Herbs, perennial, 3–20 dm; leaves basal or cauline, sometimes fascicled; blades lanceolate or oblanceolate to oblong, 5–200 × 3–30 mm, tomentose abaxially, slightly tomentose to floccose or glabrous adaxially; inflorescences cymose; involucres turbinate to campanulate, 4–6 mm, tomentose; flowers 5–11 mm, densely tomentose, stipelike base 0.5–2.5 mm; perianth yellow; cypselae trigonous, 4–6 mm, tomentose. Our plants are **var. longifolium**. Sandy to gravelly grasslands, limestone outcrops, oak and coniferous woodlands, 3500–6500 ft. Eddy.

Eriogonum maculatum A. Heller—Spotted wild-buckwheat. Herbs, annual, 1–2(–3) dm; leaves basal and cauline; basal blades lanceolate to obovate, 10–30(–40) × 10–20 mm, tomentose abaxially, sparsely floccose to glabrate adaxially; inflorescences cymose; involucres campanulate, 1–1.5(–2) mm, glandular-puberulent; flowers 1–2.5 mm, glandular-puberulent, stipelike base absent; perianth white to yellow, turning pink or red with purple spot on each outer tepal; cypselae trigonous, 1–1.5 mm, glabrous. Sandy, gravelly, and clayey flats and slopes, grasslands, desert shrublands, 4000–4200 ft. SW.

Eriogonum microtheca Nutt.—Simpson's wild-buckwheat. Shrubs, perennial, (1–)2–15 dm; leaves cauline, sometimes fascicled; blades narrowly elliptic, 5–18(–25) × 1–2 mm, densely tomentose abaxially, floccose adaxially, margins revolute; inflorescences cymose; involucres turbinate, 2–3 mm, tomentose to floccose or glabrate; flowers 2–3 mm, glabrous, stipelike base absent; perianth white to pink or rose; cypselae trigonous, 2–3 mm, glabrous. Our plants are **var. simpsonii** (Benth.) Reveal. Clayey, sandy, and gravelly washes, flats, and slopes, grassland, desert shrublands, piñon-juniper and montane coniferous woodlands, 6000–7000 ft. NC, NW, C, WC.

Eriogonum ovalifolium Nutt.—Cushion wild-buckwheat. Herbs, perennial, pulvinate to cespitose, 0.2–3 dm; leaves basal; blades spatulate to oblong, obovate, or oval, 2–60 × 5–20 mm, lanate to tomentose or floccose, sometimes less so adaxially; inflorescences capitate; involucres turbinate to campanulate, 4–5 mm, tomentose; flowers 4–5 mm, glabrous, stipelike base absent; perianth white to rose or purple; cypselae trigonous, 2–3 mm, glabrous. Our plants are **var. purpureum** (Nutt.) Durand. Sandy to gravelly flats, washes, slopes, and ridges, grasslands, desert shrublands, sagebrush communities, piñon and piñon-juniper woodlands, 4500–6700 ft. Rio Arriba, San Juan.

Eriogonum palmerianum Reveal—Palmer's wild-buckwheat. Herbs, annual, (0.5–)1–3 dm; leaves basal; blades nearly round to cordate, 5–15(–18) × 5–20 mm, densely tomentose abaxially, sparsely tomentose or glabrate adaxially; inflorescences cymose, distally 1-parous due to suppression of secondary branches; involucres appressed to branches, campanulate, 1.5–2 mm, floccose to tomentose; flowers 1.5–2 mm, glabrous, stipelike base absent; perianth white to pink or pale yellow, turning pink to red; cypselae trigonous, 1.5–1.8 mm, glabrous. Sandy to gravelly washes, flats, and slopes, desert shrublands, sagebrush shrublands, piñon and piñon-juniper woodlands, 4000–4800 ft. Grant, Hidalgo.

Eriogonum pharnaceoides Torr.—Wirestem wild-buckwheat. Herbs, annual, 1–5 dm; leaves basal and cauline; basal blades linear-lanceolate to linear-oblanceolate, 20–40 × 2–4 mm, lanate abaxially, villous adaxially, inflorescences cymose; involucres campanulate, 1–2 mm, villous; flowers 1–3 mm, glabrous, stipelike base absent; perianth white to rose, sometimes turning red; cypselae lenticular, 1.8–2 mm, glabrous. Our plants are **var. pharnaceoides**. Sandy or gravelly slopes, arroyos, sagebrush shrublands, oak, piñon-juniper, and coniferous woodlands, 5800–8000 ft. WC, SW, Taos.

Eriogonum polycladon Benth. [*E. densum* Greene]—Sorrel wild-buckwheat. Herbs, annual, (0.5–)1–6 dm; leaves cauline; blades oblanceolate to elliptic, (7–)10–30 × 5–15 mm, tomentose on both surfaces; inflorescences narrowly cymose; involucres appressed to branches, turbinate, 1.5–2.5 mm, usually tomentose, rarely glabrous; flowers 1–2 mm, glabrous, stipelike base absent; perianth white, turning pink or red; cypselae trigonous, 1–1.3 mm, glabrous. Sandy to gravelly arroyos, flats, and slopes, desert shrublands, sagebrush shrublands, oak, piñon, and juniper woodlands, 4600–7600 ft. C, SC, W.

Eriogonum racemosum Nutt.—Redroot wild-buckwheat. Herbs, perennial, 3–8(–10) dm; leaves basal; blades elliptic to ovate or oval to nearly round, (15–)20–60(–100) × 10–40(–50) mm, lanate to sparsely tomentose abaxially, floccose or glabrous adaxially; inflorescences racemiform; involucres turbinate to campanulate, (2–)3–5 mm, tomentose to floccose; flowers 2–5 mm, glabrous, stipelike base absent; perianth white to pink; cypselae trigonous, 3–4 mm, glabrous. Sandy to gravelly flats and slopes, grasslands, sagebrush shrublands, mountain mahogany, scrub oak, piñon, juniper, and coniferous woodlands, 4500–10,500 ft. NC, NW, C, WC, SW.

Eriogonum rotundifolium Benth.—Round-leaf wild-buckwheat. Herbs, annual, 0.5–4 dm; leaves basal; blades cordate to round, 10–20(–30) × 10–25(–30) mm, densely tomentose abaxially, floccose or glabrate adaxially; inflorescences cymose; involucres turbinate to campanulate, 1–2 mm, glabrous, teeth 5; flowers 1–2.5 mm, glabrous, stipelike base absent; perianth white to pink with green to red midribs, turning rose to red; cypselae trigonous, 1.5–2 mm, glabrous. Sandy to gravelly flats and slopes, grasslands, desert shrublands, piñon-juniper woodlands, 3700–7300 ft. Widespread except E, NC.

Eriogonum scabrellum Reveal—Westwater wild-buckwheat. Herbs, annual, 1–3(–5) dm; leaves basal; blades cordate, 10–30(–40) × 10–30(–40) mm, densely tomentose abaxially, sparsely floccose adaxially; inflorescences cymose; involucres spreading, turbinate, 1.5–2.5 mm, scabrellous; flowers 1–1.5 mm, pustulose, stipelike base absent; perianth white with green midribs, turning pink to red; cypselae trigonous, 2 mm, glabrous. Clayey to gravelly arroyos, flats, and slopes, desert shrublands, sagebrush shrublands, piñon-juniper woodlands, 5000–5500 ft. NW.

Eriogonum shockleyi S. Watson—Shockley's wild-buckwheat. Herbs, perennial, cespitose, matted, 0.3–0.5(–0.7) dm; leaves basal; blades oblanceolate to elliptic or spatulate, (2–)3–8(–12) × 2–4(–6) mm, tomentose to floccose on both surfaces; inflorescences capitate; involucres campanulate, (2–)2.5–5(–6) mm, tomentose; flowers 2.5–4 mm, densely pilose, stipelike base absent; perianth white to rose or yellow;

cypselae trigonous, 2.5–3 mm, tomentose. Gravelly, clayey, or sandy flats, washes, and slopes, desert shrublands, piñon-juniper woodlands, 5500–7400 ft. NW.

Eriogonum subreniforme S. Watson—Kidney-shaped wild-buckwheat. Herbs, annual, 1–5(–7) dm; leaves basal; blades reniform to round, (3–)10–35(–40) × (3–)10–40(–50) mm, densely tomentose abaxially, hirsute to floccose or glabrous adaxially; inflorescences cymose; involucres turbinate, 0.5–1 mm, glabrous, teeth 5; flowers 0.6–1.6(–2) mm, glabrous or puberulent to hirsute, stipelike base absent; perianth white to red, rarely yellow; cypselae trigonous, 1.7–2 mm, glabrous. Sandy to clayey flats and slopes, desert shrublands, oak and piñon-juniper woodlands, 5000–7300 ft. NW.

Eriogonum tenellum Torr.—Tall wild-buckwheat. Herbs, perennial, 1–5 dm; leaves basal; blades elliptic, 3–15 × 3–10 mm, densely tomentose on both surfaces; inflorescences cymose; peduncles erect or spreading, 6–60 mm, glabrous; bracts 3, 1–3 mm; involucres turbinate to campanulate, 2–4 mm, glabrous but tomentose adaxially; flowers 1.5–3.5 mm, glabrous, stipelike base absent; perianth white to pink with green to red midribs, turning pink or orange-brown to red; cypselae trigonous, 2–3 mm, glabrous. Our plants are **var. tenellum**. Sand dunes, sandy to gravelly flats, slopes, and bluffs, grasslands, desert shrublands, sagebrush shrublands, juniper-oak and piñon-juniper woodlands, 3800–4000 ft. NE, EC, C, SE, SC.

Eriogonum thurberi Torr.—Thurber's wild-buckwheat. Herbs, annual, 0.5–4 dm; leaves basal; blades oblong to ovate, 8–45 × 5–30 mm, densely tomentose abaxially, floccose or glabrous adaxially; inflorescences cymose; involucres turbinate, 1.8–2 mm, glandular-puberulent; flowers 1–1.7 mm, glandular-puberulent, stipelike base absent; perianth white with green or red midribs, turning red; cypselae usually lenticular, 0.6–0.8 mm, glabrous. Sandy arroyos, flats, and slopes, desert shrublands, oak, piñon, and piñon-juniper woodlands, 4000–9300 ft. Grant.

Eriogonum trichopes Torr.—Little desert trumpet. Herbs, annual, 1–4.5(–6) dm; leaves basal; blades broadly oblong, (5–)10–25(–40) × (5–)10–20(–30) mm, hirsute on both surfaces; inflorescences cymose; involucres turbinate, 0.7–1 mm, glabrous, teeth 4(5); flowers 1–2 mm, hirsute, stipelike base absent; perianth yellow to greenish yellow with green to red midribs; cypselae lenticular or trigonous, 1–1.5 mm, glabrous. Clayey, sandy, or gravelly flats, washes, and slopes, grasslands, desert saltbush, piñon and juniper woodlands, 4000–6000 ft. C, SC, SW, Cibola.

Eriogonum umbellatum Torr.—Sulphur flower. Herbs, perennial, cespitose, usually matted, 1–3.5 dm; leaves basal; blades elliptic to ovate, 10–25(–30) × 5–15(–18) mm, lanate abaxially, lanate to floccose or glabrous adaxially; inflorescences umbellate; involucres turbinate to campanulate, 1–6 mm, tomentose to floccose or glabrous, 6–12-toothed; flowers 4–8 mm, glabrous, stipelike base (0.7–)1.3–2 mm; perianth yellow; cypselae trigonous, 2–7 mm, glabrous except for pubescent beak. Our plants are **var. umbellatum**. Sandy to gravelly flats and slopes, grasslands, sagebrush shrublands, scrub oak and montane coniferous woodlands, 5500–8300 ft. Rio Arriba.

Eriogonum wetherillii Eastw.—Wetherill wild-buckwheat. Herbs, annual, 0.5–2.5 dm; leaves basal; blades oblong to round, (5–)10–40 × (5–)10–30 mm, densely tomentose abaxially, floccose to glabrate adaxially; inflorescences cymose; involucres turbinate, (0.3–)0.5–1 mm, glabrous, teeth 4; flowers 1–1.5 mm, glabrous, stipelike base absent; perianth yellow to red, turning pink to rose; cypselae lenticular, 0.6–1 mm, glabrous. Sandy to clayey flats, washes, and slopes, desert shrublands, oak, piñon, and piñon-juniper woodlands, 5300–5700 ft. San Juan.

Eriogonum wootonii (Reveal) Reveal [*E. jamesii* Benth. var. *wootonii* Reveal]—Wooton wild-buckwheat. Herbs, perennial, not matted, 4–5 dm; leaves basal; blades elliptic, 25–65 × 15–30 mm, densely tomentose abaxially, sparsely tomentose to floccose adaxially; inflorescences compound-umbellate; involucres campanulate, 3–6 mm, floccose; flowers 4–7(–9) mm, pubescent, stipelike base 1–1.5 mm; perianth yellow; cypselae trigonous, 4–8 mm, glabrous except for pubescent beak. Sandy to gravelly slopes in sagebrush shrublands and montane coniferous woodlands, 6000–11,500 ft. Lincoln, Otero. **R & E**

Eriogonum wrightii Torr. ex Benth.—Bastard-sage. Shrubs or subshrubs, perennial, (1-)1.5-5(-7) dm; leaves basal or cauline, sometimes fascicled; blades oblanceolate to elliptic, 5-15 × 2-7 mm, tomentose to floccose on both surfaces, sometimes glabrate or glabrous adaxially; inflorescences racemiform; involucres turbinate to campanulate, 2-2.5 mm, tomentose to floccose or glabrous; flowers 2.5-3.5 mm, glabrous, stipelike base absent; perianth white to pink or rose; cypselae trigonous, 2.5-3 mm, glabrous. Our plants are **var**. **wrightii**. Gravelly to rocky slopes, grasslands, desert shrublands, oak, piñon, and piñon-juniper woodlands, 4000-8000 ft. Widespread except E, NC.

FAGOPYRUM Mill. - Buckwheat

Fagopyrum esculentum Moench [*F. sagittatum* Gilib.]—Common buckwheat. Herbs, annual, (0.7-)1.5-9 dm; stems ascending to erect, branched, nodes swollen; leaves cauline, alternate; ocreae persistent or deciduous, chartaceous; petioles 1.5-6(-9) cm; blades hastate-triangular, sagittate-triangular, or cordate, 25-80 × 20-80 mm, margins cillolate; inflorescences terminal and axillary, paniclelike, 1-4 cm; involucral bracts 0; perianth creamy-white to pale pink; tepals 5, distinct, 2.5-5 mm; stamens 8; styles 3; cypselae strongly exserted, sharply trigonous, 4-6 mm. Cultivated as crop plant, waif along roadsides, fields, and waste places, ca. 7700 ft. Bernalillo, Rio Arriba.

FALLOPIA Adans. - False-buckwheat

Vines or herbs, annual or perennial; stems climbing, scandent, or sprawling, branched, nodes swollen; leaves cauline, alternate; ocreae persistent or deciduous, chartaceous or hyaline; blade margins entire or wavy; inflorescences terminal and spikelike, or terminal and axillary and paniclelike, involucral bracts 0; perianth pale green or white to pink; tepals 5, connate nearly completely or only basally, outer 3 winged or keeled, larger than inner 2; stamens 6-8; styles 3; cypselae included, trigonous.

1. Perennials; stems woody, climbing…**F. baldschuanica**
1'. Annuals; stems herbaceous, scandent or sprawling…**F. convolvulus**

Fallopia baldschuanica (Regel) Holub—Bukhara fleeceflower. Vines, perennial, 30-100 dm; stems climbing, woody, glabrous; leaf blades narrowly ovate to ovate-oblong, 30-100 × 10-50 mm, abaxial face glabrous or scabrid along midvein, adaxial face glabrous; inflorescences axillary and terminal, spreading or drooping, paniclelike, 3-15 cm; perianth greenish white with white wings or mostly pink, sometimes pink in fruit, 5-8 mm; cypselae 2-4 mm, smooth; fruiting perianth wings 2-4 mm wide. Disturbed urban areas, roadsides, fencerows, 4200-6400 ft. Grant, Luna, Rio Arriba, Roosevelt.

Fallopia convolvulus (L.) Á. Löve [*Polygonum convolvulus* L.]—Black bindweed. Herbs, annual, 5-10 dm; stems scandent or sprawling, herbaceous, puberulent, sometimes mealy; leaf blades cordate-ovate, cordate-hastate, or sagittate, 20-60(-150) × 20-50(-100) mm, abaxial face usually mealy, adaxial face glabrous; inflorescences axillary, erect or spreading, spikelike, 2-10(-15) cm; perianth greenish white, often with pink or purple base, 3-5 mm; cypselae 4-6 mm, minutely granular-tuberculate; fruiting perianth wings 0-0.9 mm wide. Sandy grasslands and meadows, sandbars, sandy washes, cultivated ground, waste places, 4400-8100 ft. Widespread except EC, SE.

OXYRIA Hill - Mountain-sorrel

Oxyria digyna (L.) Hill—Alpine mountain-sorrel. Herbs, perennial, 0.3-5 cm; nodes swollen; leaves basal, rarely also cauline, alternate; ocreae sometimes deciduous, chartaceous; blades reniform to round-cordate, 5-65 × 5-60 mm; inflorescences terminal, paniclelike or racemelike; involucral bracts 0; perianth greenish to reddish brown; tepals 4, distinct, outer 2 spreading in fruit, 1.2-1.7 mm, inner 2 elliptic to round or obovate, 1.4-2.5 mm, appressed in fruit; stamens (2-)6; styles 2; cypselae exserted, lenticular, 3-4.5 mm, 2-winged. Snow beds and zones of snow accumulation, gravel bars, mudflats, tundra, scree slopes, crevices in rock outcrops, talus slopes, 9000-12,700 ft. NC, SC.

PERSICARIA (L.) Mill. – Smartweed

Herbs, perennial or annual; stems prostrate to erect, simple or branched, nodes swollen; leaves cauline, alternate; ocreae persistent or disintegrating with age and deciduous entirely or distally, chartaceous or partially to entirely foliaceous; blades lanceolate or ovate, margins entire; inflorescences terminal or terminal and axillary, spikelike or paniclelike; involucral bracts 0; perianth white, greenish white, pink, red, or purple, sometimes glandular-punctate; tepals 4 or 5, connate 1/4–2/3 their length, outer not winged or keeled; stamens 5–8; styles 2 or 3; cypselae included or exserted, discoid, biconvex, or trigonous.

1.   Some or all ocreae foliaceous and green distally…**P. amphibia** (in part)
1'.   All ocreae chartaceous and hyaline, tan, brown, or reddish brown throughout…(2)
2.   Perianths glandular-punctate…(3)
2'.   Perianths not glandular-punctate…(6)
3.   Cypselae minutely roughened, dull; axillary inflorescences sometimes enclosed in ocreae…**P. hydropiper**
3'.   Cypselae smooth, shiny; axillary inflorescences never enclosed in ocreae…(4)
4.   Outer tepals with anchor-shaped veins; cypselae discoid…**P. lapathifolia** (in part)
4'.   Outer tepals without anchor-shaped veins; cypselae trigonous or biconvex…(5)
5.   Perianth puncta on tubes and inner tepals…**P. hydropiperoides** (in part)
5'.   Perianth puncta uniformly distributed…**P. punctata**
6.   Peduncles stipitate-glandular…(7)
6'.   Peduncles not stipitate-glandular…(10)
7.   Perennials; inflorescences terminal…**P. amphibia** (in part)
7'.   Annuals; inflorescences terminal and axillary…(8)
8.   Outer tepals with anchor-shaped veins; tepals 4(5); inflorescences mostly arching or nodding…**P. lapathifolia** (in part)
8'.   Outer tepals without anchor-shaped veins; tepals 5; inflorescences erect or, rarely, nodding…(9)
9.   Flowers homostylous; cypselae without central hump on 1 side…**P. pensylvanica** (in part)
9'.   Flowers heterostylous; cypselae usually with central hump on 1 side…**P. bicornis** (in part)
10.   Perennials…(11)
10'. Annuals…(12)
11.   Cypselae biconvex; perianths pink to red…**P. amphibia** (in part)
11'. Cypselae trigonous; perianths white or greenish white to pink…**P. hydropiperoides** (in part)
12.   Margins of ocreae with bristles 0.2–1.3(–2) mm; ocreolae not overlapping proximally; cypselae discoid, biconvex, or trigonous…**P. maculosa**
12'. Margins of ocreae eciliate or with bristles to 1 mm; ocreolae mostly overlapping; cypselae discoid, rarely trigonous…(13)
13.   Outer tepals with anchor-shaped veins; tepals 4(5); inflorescences mostly arching or nodding…**P. lapathifolia** (in part)
13'. Outer tepals without anchor-shaped veins; tepals 5; inflorescences mostly erect, rarely nodding…(14)
14.   Flowers homostylous; cypselae without central hump on 1 side…**P. pensylvanica** (in part)
14'. Flowers heterostylous; cypselae usually with central hump on 1 side…**P. bicornis** (in part)

Persicaria amphibia (L.) Delarbre [*Polygonum amphibium* L. var. *stipulaceum* N. Coleman]— Water smartweed. Perennials, 2–12 dm on land, to 30 dm in water; ocreae brown and chartaceous, sometimes green and foliaceous distally; leaf blades lanceolate to elliptic or oblong, 20–150(–230) × 10–80 mm, glabrous or strigose; inflorescences terminal, ascending to erect, 1–15 cm; flower perianth pink to red, not glandular-punctate; tepals 5, 4–6 mm; cypselae biconvex, 2–3 mm, shiny or dull, smooth or minutely granular. Shallow water, shorelines of ponds and lakes, stream banks, moist meadows, 3500–9500 ft. Widespread except SE.

Persicaria bicornis (Raf.) Nieuwl. [*Polygonum longistylum* Small]—Pink smartweed. Annuals, 2–18 dm; ocreae brownish, cylindrical, chartaceous; leaf blades linear-lanceolate to ovate-lanceolate, 23–130(–180) × (4–)10–23 mm, glabrous or appressed-pubescent along midvein; inflorescences terminal and axillary, erect, 1–6 cm; flowers heterostylous; flower perianth pink, not glandular-punctate; tepals 5, obovate to elliptic, 3–4.6 mm; cypselae discoid, rarely trigonous, 1 side usually slightly concave and other with central hump, 2–2.9 mm, shiny, smooth. Moist, disturbed places, wetlands, ditches, cultivated fields, shorelines of ponds and reservoirs, 4000–7300 ft. Scattered. NE, Bernalillo, Catron, Eddy, Roosevelt.

Persicaria hydropiper (L.) Spach [*Polygonum hydropiper* L.]—Marsh-pepper smartweed. Annuals, 2-8(-10) dm; ocreae brown, chartaceous; leaf blades lanceolate to rhombic, (15-)40-100(-150) × 4-25 mm, glabrous or scabrous along midvein; inflorescences terminal and axillary, erect or nodding; flower perianth greenish proximally, white or pink distally, glandular-punctate with puncta ± uniformly distributed; tepals 4 or 5, 2-3.5 mm; cypselae biconvex or trigonous, 1.9-3 mm, dull, minutely roughened. Shorelines of lake and ponds, stream banks, pastures, occasionally waste ground, 6500-7600 ft. Colfax. Introduced.

Persicaria hydropiperoides (Michx.) Small [*Polygonum hydropiperoides* Michx.]—Swamp smartweed. Perennials, 1.5-10 dm; ocreae brown, chartaceous; leaf blades lanceolate to linear-lanceolate, 50-250 × 4-37 mm, glabrous or appressed-pubescent; inflorescences terminal, sometimes also axillary, erect, 3-8 cm; flower perianth white to greenish white or pink, not glandular-punctate or glandular-punctate on tubes and inner tepals; tepals 5, 1.5-4 mm, veins not anchor-shaped; cypselae trigonous, 1.5-3 mm, shiny, smooth. Wetlands, stream banks, shallow water, ditches, 4000-7000 ft. McKinley. Known only from an 1871 collection from Zuni that appears to be this species.

Persicaria lapathifolia (L.) Delarbre [*Polygonum lapathifolium* L.]—Pale smartweed. Annuals, 0.5-10 dm; ocreae brown, chartaceous; leaf blades lanceolate, 40-120(-220) × 3-60 mm, strigose on main vein, glabrous or tomentose abaxially; inflorescences terminal, sometimes also axillary, mostly arching or nodding, 3-8 cm; flower perianth greenish white to pink; tepals 4(5), 2.5-3 mm, veins of 2 or 3 outer tepals prominently bifurcate distally, anchor-shaped; cypselae discoid, rarely trigonous, 1.5-3.2 mm, shiny or dull, smooth. Moist sites, wetlands, roadsides, waste places, cultivated fields, 3000-10,700 ft. Widespread.

Persicaria maculosa Gray [*Polygonum persicaria* L.]—Spotted lady's-thumb. Annuals, 0.5-7(-13) dm; ocreae brown, chartaceous; leaf blades lanceolate to ovate, (10-)50-100(-180) × (2-)10-25(-40) mm, glabrous or strigose; inflorescences terminal and axillary, erect, 1-6 cm; flower perianth pink, sometimes greenish white proximally, not glandular-punctate; tepals 4 or 5, 2-3.5 mm; cypselae discoid or biconvex to trigonous, 1.9-2.7 mm, shiny, smooth. Moist, disturbed sites, 4500-8200 ft. N, C, WC, SC, SW.

Persicaria pensylvanica (L.) M. Gómez [*Polygonum pensylvanicum* L.]—Pennsylvania smartweed. Annuals, 1-20 dm; ocreae brown, chartaceous; leaf blades lanceolate, 40-170(-230) × 5-48 mm, glabrous or appressed-pubescent; inflorescences terminal and axillary, erect or nodding, 0.5-5 cm; flowers homostylous; perianth greenish white to pink, not glandular-punctate; tepals 5, 2.5-5 mm; cypselae discoid, rarely trigonous, without central hump on 1 side, 2.1-3.4 mm, shiny, smooth. Shorelines of ponds and lakes, moist, disturbed places, stream banks, cultivated fields, 3800-7500 ft. NE, NW, EC, C, SW.

Persicaria punctata (Elliott) Small [*Polygonum punctatum* Elliott]—Dotted smartweed. Annuals or perennials, 1.5-12 dm; ocreae brown, chartaceous; leaf blades lanceolate to ovate, 40-100(-150) × 6-24 mm, glabrous or scabrous along midveins; inflorescences mostly terminal, sometimes also axillary, erect, 5-20 cm; flower perianth greenish white, rarely tinged pink, uniformly glandular-punctate; tepals 5, 3-3.5 mm, veins not anchor-shaped; cypselae trigonous, rarely biconvex, 1.8-2.2 mm, shiny, smooth. Wetlands, shallow water, shorelines of lakes and ponds, 5100-5800 ft. WC, SW.

POLYGONUM L. - Knotweed

Annual herbs, rarely perennial subshrubs; stems prostrate to erect, simple or branched, nodes swollen; leaves cauline, alternate; ocreae persistent or distal part disintegrating into fibers, white or silvery, rarely brown, hyaline, rarely coriaceous; blades linear to subround, margins entire; inflorescences axillary, axillary and terminal, or terminal, spikelike, racemelike, or flowers solitary; involucral bracts 0; perianth white or greenish white to pink; tepals 5, connate 3%-70% of their length, rarely distinct, outer sometimes keeled and cucullate distally; stamens 3-8; styles (2)3; cypselae included or exserted, trigonous, rarely quadrangular (includes *Polygonella* Michx.).

1. Subshrubs, perennial; tepals distinct, outer 2 sharply reflexed, inner 3 ascending to erect; ocreae brown…
   P. americanum
1′. Herbs, annual; tepals connate proximally, ascending to erect; ocreae hyaline, white, or silvery…(2)
2. Stems 4-angled, especially near nodes, ribs obscure or absent; leaf blade venation parallel, secondary veins inconspicuous; anthers pink to purple…(3)
2′. Stems ± round, distinctly ribbed; leaf blade venation pinnate, secondary veins conspicuous; anthers yellow …(6)
3. Pedicels reflexed…**P. douglasii**
3′. Pedicels erect or erect to spreading…(4)
4. Apices of tepals acute to acuminate; cypselae yellow to greenish brown or brown; stamens 3…**P. polygaloides**
4′. Apices of tepals rounded; cypselae black; stamens 3-8…(5)
5. Ocreae 4-10 mm, distal part lacerate or disintegrating into fibers; perianths (2.5-)3-3.5 mm; cypselae 2.5-3 mm…**P. sawatchense**
5′. Ocreae 1-4 mm, distal part entire or lacerate; perianths 1.8-2.5 mm; cypselae 1.8-2.3 mm…**P. minimum**
6. Stems erect, branched distally; margins of tepals greenish yellow, yellow, or whitish green, rarely white or pink…(7)
6′. Stems prostrate to decumbent, ascending, or erect, branched mostly from base; margins of tepals white, pink, or red…(8)
7. Leaf blades elliptic to obovate; distal part of ocreae persistent, usually entire or lacerate; cypselae dull, striate-tubercled…**P. erectum**
7′. Leaf blades elliptic to lanceolate or oblanceolate; distal part of ocreae disintegrating into fibers; cypselae shiny or dull, smooth to roughened or obscurely tubercled…**P. ramosissimum**
8. Cypselae shiny, smooth, 1.3-2.3 mm…**P. argyrocoleon**
8′. Cypselae dull, usually striate-tubercled, sometimes obscurely tubercled, 1.2-5 mm…**P. aviculare**

Polygonum americanum (Fisch. & C. A. Mey.) T. M. Schust. & Reveal [*Polygonella americana* (Fisch. & C. A. Mey.) Small]—American jointweed. Subshrubs, perennial, 5.5–9 dm; stems erect, branched from near base, ± round, distinctly ribbed; ocreae 3–10 mm, usually persistent, brown; leaf blades linear to spatulate, 4–19 × 0.5–1.2 mm, venation parallel, secondary veins inconspicuous; inflorescences terminal, racemelike; pedicels spreading, 0.4–2.3 mm; perianth 1.2–2.9 mm; tepals distinct, outer 2 sharply re-flexed, inner 3 ascending to erect, ovate to ± round, white to pink; cypselae brown, 2.5–4 mm. Sandy roadsides, sand dunes, sand fields, 3900–4300 ft. DeBaca, Roosevelt.

Polygonum argyrocoleon Steud. ex Kunze—Silversheath knotweed. Herbs, annual, 1.5–10 dm; stems decumbent to ascending or erect, branched mostly from base, ± round, distinctly ribbed; ocreae 4–8 mm, distal part disintegrating into fibers, silvery; leaf blades lanceolate or linear-lanceolate, 15–50 × 2–8 mm, venation pinnate, secondary veins conspicuous; inflorescences axillary and terminal, spikelike; pedicels spreading to erect, 1–2 mm; perianth 1.8–2.4 mm; tepals connate proximally, ascending to erect, green or white with pink, red, or white margins; cypselae brown, 1.3–2.3 mm. Fields, gardens, roadsides, disturbed sites, 6900–9200 ft. McKinley, Socorro, Taos.

Polygonum aviculare L.—Prostrate knotweed, doorweed. Herbs, annual, 0.5–20 dm; stems pros-trate to ascending or erect, branched mostly from base, ± round, distinctly ribbed; ocreae 3–15 mm, disintegrating into persistent fibers or completely deciduous, silvery; leaf blades elliptic to lanceolate, obovate, or spatulate, 6–60 × 0.5–22 mm, venation pinnate, secondary veins conspicuous; inflores-cences axillary; pedicels spreading to erect, 1.5–5 mm; perianth 1.8–5.5 mm; tepals connate proximally, ascending to erect, green or reddish brown with white, pink, or red margins; cypselae brown, 1.2–5 mm, dull, usually coarsely striate-tubercled, sometimes obscurely tubercled. Many specimens of *P. aviculare* are not determined to subspecies; the following four subspecies appear to be represented in New Mex-ico based on examination of a limited number of specimens. Widespread except EC.

1. Perianth tubes 40%-57% of perianth length…(2)
1′. Perianth tubes (15-)20%-40(-42)% of perianth length…(3)
2. Tepals green or reddish brown, margins white, veins unbranched…**subsp. depressum**
2′. Tepals green, margins usually pink or red, rarely white, veins branched…**subsp. neglectum** (in part)
3. Perianths 3.3-5.5 mm; cypselae 2.5-4.2 mm…**subsp. aviculare** (in part)
3′. Perianths 1.9-3.6 mm; cypselae 1.2-2.8(-3) mm…(4)

4. Ocreae with distal parts relatively persistent, silvery; perianths 0.9–1.3(–1.5) times as long as wide, outer tepals pouched at base...**subsp. buxiforme**
4'. Ocreae soon disintegrating into fibers or with no fibrous remains; perianths 1.5–2.9 times as long as wide; outer tepals not pouched at base...(5)
5. Leaf blades (6–)10–20 mm wide, 2–4.5 times as long as wide; cymes 3–8-flowered, aggregated at tips of stems and branches; cypselae enclosed in or barely exserted from perianth...**subsp. aviculare** (in part)
5'. Leaf blades 1.5–6.8(–8) mm wide, (3.4–)4.2–9.2 times as long as wide; cymes 1–3(–5)-flowered, uniformly distributed along stems and branches; cypselae exserted from perianth...**subsp. neglectum** (in part)

subsp. aviculare—Common knotweed. Fields, roadsides, waste places, disturbed sites, 4000–10,000 ft.

subsp. buxiforme (Small) Costea & Tardif—American knotweed. Fields, roadsides, waste places, disturbed sites, 5000–7500 ft.

subsp. depressum (Meisn.) Arcang.—Oval-leaf knotweed. Fields, roadsides, waste places, disturbed sites, 6000–8500 ft.

subsp. neglectum (Besser) Arcang.—Narrow-leaf knotweed. Fields, roadsides, waste places, disturbed sites, 4000–7000 ft.

Polygonum douglasii Greene [*P. montanum* (Small) Greene]—Douglas's knotweed. Herbs, annual, 0.5–8 dm; stems erect, simple or branched from base, 4-angled, especially near nodes, ribs obscure or absent; ocreae 6–12 mm, distal part lacerate, hyaline; leaf blades linear to narrowly oblong or oblanceolate, 15–55 × 2–10 mm, venation parallel, secondary veins inconspicuous; inflorescences axillary and terminal, spikelike; pedicels reflexed, 2–6 mm; perianth 3–4.5 mm; tepals connate proximally, ascending to erect, green to tan with white or pink margins; cypselae black, 3–4.5 mm, shiny or dull. Montane meadows, 6000–12,000 ft. Widespread except EC, SE, SC.

Polygonum erectum L.—Erect knotweed. Herbs, annual, 1.5–7.5 dm; stems ascending to erect, branched distally, ± round, distinctly ribbed; ocreae 7–12 mm, distal part persistent, usually entire or lacerate, silvery; leaf blades elliptic to obovate, 30–80 × 8–25 mm; inflorescences axillary, venation pinnate, secondary veins conspicuous; pedicels spreading to erect, 3–7 mm; perianth 2.8–3.8(–4.2) mm; tepals connate proximally, ascending to erect, green with yellow or whitish-green margins; cypselae brown, 2.3–5 mm, dull. Waste ground, disturbed sites, 3700–8700 ft. NE, NW, C, WC, SE.

Polygonum minimum S. Watson—Broadleaf knotweed. Herbs, annual, 0.3–3 dm; stems prostrate to erect, simple or branched from base, 4-angled, especially near nodes, ribs obscure or absent; ocreae 1–4 mm, distal part entire or lacerate, silvery; leaf blades elliptic to ovate or obovate, 6–27 × 3–8 mm, venation parallel, secondary veins inconspicuous; inflorescences axillary; pedicels spreading to erect, 2–3 mm; perianth 1.8–2.5 mm; tepals connate proximally, ascending to erect, green with white or pink margins; cypselae black, 1.8–2.3 mm, shiny, smooth. Alpine and subalpine meadows, 10,600–10,700 ft. San Juan.

Polygonum polygaloides Meisn.—Polygala knotweed. Herbs, annual, 0.3–1.5 dm; stems erect, branched from base, rarely simple, 4-angled, especially near nodes, ribs obscure or absent; ocreae 4–8 mm, distal part lacerate, silvery; leaf blades linear, 10–40 × 1–2.5 mm, venation parallel, secondary veins inconspicuous; inflorescences axillary and terminal, spikelike; pedicels erect, 0.1–2 mm; perianth 1.5–3 mm; tepals connate proximally, ascending to erect, white, pink, or red; cypselae yellow to greenish brown or brown, 1.3–2 mm, shiny or dull.

1. Margins of bracts green, not scarious...**subsp. kelloggii**
1'. Margins of bracts white, scarious border 0.2–0.4 mm wide...**subsp. confertiflorum**

subsp. confertiflorum (Nutt. ex Piper) J. C. Hickman [*P. confertiflorum* Nutt. ex Piper; *P. watsonii* Small]—Fruitleaf knotweed, white-margined knotweed. Wet meadows, 9400–9700 ft. Rio Arriba.

subsp. kelloggii (Greene) J. C. Hickman [*P. kelloggii* Greene]—Kellogg's knotweed. Mountain meadows, dry montane slopes, 6900–9000 ft. Rio Arriba, San Juan.

Polygonum ramosissimum Michx.—Bushy knotweed. Herbs, annual, 1-10(-20) dm; stems erect, branched distally, ± round, distinctly ribbed; ocreae 6-15 mm, distal part disintegrating into fibers, silvery; leaf blades elliptic to lanceolate or oblanceolate, 8-70 × 4-30 mm, venation pinnate, secondary veins conspicuous; inflorescences axillary or axillary and terminal, spikelike; pedicels spreading to erect, 1-6 mm; perianth 2-4 mm; tepals connate proximally, ascending to erect, greenish yellow with greenish-yellow or yellow margins, or rarely pink or white; cypselae brown, 1.6-3.5 mm, shiny or dull. NE, NC, NW, C, WC, SC, SW.

1. Plants heterophyllous, yellowish green when fresh or dried; apices of leaf blades acute to acuminate; pedicels 2.5-6 mm…**subsp. ramosissimum**
1'. Plants homophyllous, bluish green when fresh, dark brown or black after drying; apices of leaf blades rounded or obtuse; pedicels 1-2 mm…**subsp. prolificum**

subsp. prolificum (Small) Costea & Tardif—Wet slopes, salt marshes, sandy to clayey shorelines, 3900-9500 ft.

subsp. ramosissimum—Disturbed sites, sandy floodplains, shorelines, roadsides, playas, 3700-9700 ft.

Polygonum sawatchense Small—Sawatch knotweed. Herbs, annual, 0.4-5 dm; stems erect, branched from base, 4-angled, especially near nodes, ribs obscure or absent; ocreae 4-10 mm, distal part lacerate or disintegrating into fibers, silvery; leaf blades lanceolate to oblong or oblanceolate, 15-45 × 2-12 mm, venation parallel, secondary veins inconspicuous; inflorescences axillary or axillary and terminal, racemes; pedicels erect, 1-4 mm; perianth (2.5-)3-3.5 mm; tepals connate proximally, ascending to erect, greenish or reddish with white or pink margins; cypselae black, 2.5-3 mm, shiny, smooth. Our plants are **subsp. sawatchense**. Montane meadows, pastures, 5100-10,000 ft. Widespread except EC, SE, SW.

PTEROSTEGIA Fisch. & C. A. Mey. – Pterostegia

Pterostegia drymarioides Fisch. & C. A. Mey.—Woodland threadstem. Herbs, annual, 1-10 dm diam.; stems sprawling and spreading, much branched, nodes not swollen; leaves cauline, opposite; ocreae absent; blades elliptic to flabellate, 3-20 × 5-30 mm; inflorescences terminal, cymose; involucral bract 1, 2-winged, reticulately veined, lobed or notched; perianth pale yellow to pink or rose; tepals (5)6, connate for ca. 1/3 their length, 0.9-1.2 mm; cypselae included, winged, globose, 1.2-1.5 mm. Sandy, disturbed sites, 4400-5700 ft. Luna.

RUMEX L. – Dock, sorrel

Herbs, perennial, biennial, or annual; stems prostrate to ascending or erect, simple or branched, nodes swollen; leaves basal or basal and cauline, alternate; ocreae persistent or partially deciduous, greenish white to green or brown, hyaline, rarely coriaceous; blades linear to round, margins usually entire, rarely lobed; inflorescences usually terminal, sometimes terminal and axillary, paniclelike, rarely simple; involucral bracts 0; perianth green, yellowish green, pink, or red; tepals 6, connate proximally, outer 3 remaining small, inner 3 usually enlarging, sometimes with central vein enlarging into tubercle; stamens 6; styles 3; cypselae included, trigonous.

1. Flowers unisexual; some leaf blade bases usually lobed…(2)
1'. Flowers usually bisexual, sometimes bisexual and unisexual within same inflorescence; leaf blade bases never lobed…(3)
2. Pedicels jointed near base of tepals…**R. acetosella**
2'. Pedicels jointed in proximal 1/2…**R. hastatulus**
3. Leaves cauline; inflorescences terminal and axillary…(4)
3'. Leaves basal and cauline; inflorescences terminal…(8)
4. Inner tepals 13-18(-20) mm; pedicels 8-16 mm…**R. venosus**
4'. Inner tepals 2.5-6 mm; pedicels (2-)3-8 mm…(5)
5. Inner tepal margins denticulate proximally, tubercles 0 or 1 and < 0.5 times width of tepal…**R. californicus**

5′. Inner tepal margins entire or erose, tubercles (2)3, or if 1 then 0.5 times width of tepal…(6)
6. Leaf blades ovate-lanceolate, elliptic-lanceolate, or lanceolate, 30-55 mm wide; inner tepals 4.5-6 mm …**R. altissimus**
6′. Leaf blades linear-lanceolate to lanceolate, 10-40(-50) mm wide; inner tepals (2-)2.5-4.5(-5) mm…(7)
7. Inner tepals 3.5-4.5(-5) mm; cypselae 2-3 mm…**R. mexicanus**
7′. Inner tepals (2-)2.5-3.5(-3.8) mm; cypselae 1.7-2 mm…**R. triangulivalvis**
8. Tubercles 0…(9)
8′. Tubercles 1-3…(13)
9. Inner tepals 11-16 mm; cypselae 4-7 mm…**R. hymenosepalus**
9′. Inner tepals 5-10(-12) mm; cypselae 2.5-4.8 mm…(10)
10. Plants with creeping or fusiform rhizomes; inner tepal entire, erose, or denticulate proximally…(11)
10′. Plants with vertical or oblique and fusiform rootstocks; inner tepal margins ± entire…(12)
11. Leaf blade apices obtuse to acute, lateral veins unequal in length, long veins alternating with short veins; inner tepal bases subcordate…**R. densiflorus**
11′. Leaf blade apices acute to acuminate, lateral veins ± equal in length; inner tepal bases truncate…**R. orthoneurus**
12. Pedicels 12-20 mm; inner tepals 4-5(-6) × 3-4(-5) mm…**R. nematopodus**
12′. Pedicels 5-17 mm; inner tepals 5-10(-12) × 5-8(-11) mm…**R. occidentalis**
13. Inner tepal margins entire or obscurely erose…(14)
13′. Inner tepal margins usually denticulate to dentate, rarely ± entire, teeth 2-10, triangular to subulate or bristlelike…(15)
14. Leaf blades linear-lanceolate to lanceolate, 150-300(-350) × 20-60 mm, margins undulate or crisped; inner tepals 3.5-6 × 3-5 mm…**R. crispus**
14′. Leaf blades ovate-lanceolate or oblong-lanceolate, 300-450(-500) × 100-150 mm, margins plane or weakly undulate; inner tepals (5-)5.5-8(-10) × 5-9(-10) mm…**R. patientia**
15. Inner tepals 1.5-2.5 mm wide (excluding teeth), marginal teeth bristlelike; cypselae 1-1.5 mm; flowers 15-30 per whorl…**R. fueginus**
15′. Inner tepals 2.5-6 mm wide (excluding teeth), marginal teeth triangular to subulate; cypselae 1.8-3 mm; flowers 10-25 per whorl…(16)
16. Inner tepal marginal teeth denticulate, 0.2-0.5 mm, 2 or 3; annuals or biennials…**R. violascens**
16′. Inner tepal marginal teeth dentate, 0.2-3(-5) mm, 2-10; annuals, biennials, or perennials…(17)
17. Leaf blades 100-150 mm wide, base usually cordate, sometimes rounded or truncate…**R. obtusifolius**
17′. Leaf blades 20-70 mm wide, base cuneate, truncate, rounded, or weakly cordate…(18)
18. Annuals, rarely biennials…**R. dentatus**
18′. Perennials…(19)
19. Inflorescence open, broadly paniculate, branches spreading to ascending, whorls mostly distinct…**R. pulcher**
19′. Inflorescences dense, narrowly paniculate, branches ascending, whorls mostly not distinct…**R. stenophyllus**

Rumex acetosella L.—Common sheep-sorrel. Perennials, with vertical rootstock and/or creeping rhizomes; stems 1-4(-4.5) dm; leaves basal and cauline, blades obovate-oblong, ovate-lanceolate, lanceolate-elliptic, or lanceolate, 20-60 × 3-20 mm, base usually hastate, occasionally unlobed and base broadly cuneate, margins plane, apex acute or obtuse; inflorescences terminal; pedicels jointed near base of tepals, 1-3 mm; flowers unisexual, (3-)5-8(-10) per whorl; inner tepals not or slightly enlarged, 1.2-1.7(-2) × 0.5-1.3 mm, base cuneate, margins entire, tubercles 0; cypselae 0.9-1.5 mm. Our plants appear to be **subsp. pyrenaicus** (Pourr. ex Lapeyr.) Akeroyd. Roadsides, waste places, disturbed areas, lawns, sandy to gravelly meadows, 5000-12,700 ft. NC, NW, SC, WC, SW.

Rumex altissimus Alph. Wood [*R. ellipticus* Greene]—Tall dock. Perennials, with vertical rootstock; stems 5-9(-12) dm; leaves cauline, blades ovate-lanceolate, elliptic-lanceolate, or lanceolate, 100-150 × 30-55 mm, base broadly cuneate, rarely rounded, margins plane, apex acute or attenuate; inflorescences terminal and axillary; pedicels jointed in proximal 1/3, (2-)3-7(-8) mm; flowers bisexual, 12-20 per whorl; inner tepals broadly triangular to ovate-triangular, 4.5-6 × 3-5 mm, base truncate or indistinctly cordate, margins entire, tubercles (2)3; cypselae 2.5-3.5 mm. Marshes, wet depressions, shorelines of rivers and streams, 3800-8800 ft. NE, NC, NW, C, WC, EC, SW. Widespread.

Rumex californicus Rech. f. [*R. salicifolius* Weinm. var. *denticulatus* Torr.]—California willow dock. Perennials, with vertical rootstock; stems 3-6 dm; leaves cauline, blades linear-lanceolate to linear-oblanceolate, 50–100 × 10–30 mm, base cuneate, margins plane or undulate proximally, apex acute or attenuate; inflorescences terminal and axillary; pedicels jointed in proximal 1/3, 3–8 mm; flowers bisexual, 10–15(–20) per whorl; inner tepals triangular, 2.5–3.5 × 2.2–3.3 mm, base truncate, margins denticulate proximally, tubercles 0 or 1 and < 0.5 times width of tepal; cypselae 2 × 1.3 mm. Shorelines of rivers and streams, moist sites, 5800–7800 ft. Grant, Rio Arriba, Taos.

Rumex crispus L.—Curly dock. Perennials, occasionally biennials, with vertical and fusiform rootstock; stems 4–10(–15) dm; leaves basal and cauline, blades linear-lanceolate to lanceolate, 150–300 (–350) × 20–60 mm, base cuneate, truncate, or weakly cordate, margins undulate or crisped, apex acute; inflorescences terminal; pedicels jointed in proximal 1/3, 3–8 mm; flowers bisexual, 10–25 per whorl; inner tepals ovate to ovate-deltate, 3.5–6 × 3–5 mm, base truncate to cordate, margins entire or obscurely erose, tubercles (1–)3, unequal; cypselae 2–3 mm. Disturbed sites, waste places, cultivated fields, roadsides, meadows, shorelines, 3800–12,000 ft. Widespread.

Rumex densiflorus Osterh.—Dense-flower dock. Perennials, with creeping rhizomes; stems 5–10 dm; leaves basal and cauline, blades oblong or oblong-lanceolate, 300–400(–500) × 100–120 mm, base cuneate, truncate, or weakly cordate, margins plane or slightly undulate, apex obtuse to acute, lateral veins unequal in length, long veins alternating with short veins; inflorescences terminal; pedicels jointed in proximal 1/3, 6–16 mm; flowers bisexual, 10–20 per whorl; inner tepals ovate-triangular to cordate, 5–6 × 4.5–6 mm, base subcordate, margins entire, erose, or denticulate proximally, tubercles 0; cypselae 2.5–4.5 mm. Stream banks, moist meadows, 7300–11,400 ft. NC.

Rumex dentatus L.—Toothed dock. Annuals, rarely biennials, with vertical and fusiform rootstock; stems 2–7(–8) dm; leaves basal and cauline, blades oblong, elliptic-lanceolate, or ovate-elliptic, 30–80 (–120) × 20–50 mm, base usually truncate, sometimes weakly cordate, margins weakly undulate or crisped, apex obtuse or subacute; inflorescences terminal; pedicels jointed in proximal 1/3, 2–5 mm; flowers bisexual, 10–20 per whorl; inner tepals ovate-triangular, 3–5.5(–6) × 2–3 mm (excluding teeth), base truncate, margins usually dentate, rarely ± entire, teeth 2–4(5) per margin, triangular, 1–3(–5) mm; tubercles (1–)3, equal or subequal; cypselae 2–2.8 mm. Waste places, wet sites, shorelines of reservoirs and ponds, stream banks, 3700–5800 ft.

Rumex fueginus Phil.—Tierra del Fuego dock. Annuals or biennials, with vertical and fusiform rootstock; stems (0.4–)1.5–6(–7) dm; leaves basal and cauline, blades linear-lanceolate to lanceolate or oblong-lanceolate, 30–250(–300) × 10–40 mm, base truncate to cordate, margins undulate or crisped, apex acute; inflorescences terminal; pedicels jointed in proximal 1/3, 3–9 mm; flowers bisexual, 15–30 per whorl; inner tepals triangular to rhombic-triangular, 1.5–2.5 × 0.7–0.9(–1.2) mm (excluding teeth), base truncate to cuneate, margins usually dentate, rarely ± entire, teeth 2 or 3 per margin, bristlelike, 1–3 mm, tubercles 3, equal or subequal; cypselae 1–1.5 mm. Shorelines of reservoirs and ponds, stream banks, 4500–9000 ft. NE, NC, NW, WC, SW.

Rumex hastatulus Baldwin—Heartwing sorrel. Annuals or perennials, with vertical rootstock; stems 1–4(–4.5) dm; leaves basal and cauline, blades obovate-oblong, ovate-lanceolate, oblong-lanceolate, or lanceolate; 20–60(–100) × 5–20 mm, base hastate, auriculate, or occasionally unlobed, margins plane, apex obtuse to acute; inflorescences terminal; pedicels jointed in proximal 1/2, 1.5–3 mm; flowers unisexual, 3–6(–8) per whorl; inner tepals round or broadly ovate, 2.5–3.2 × 2.7–3.2 mm, base broadly cordate or rounded, margins entire, tubercles 0; cypselae 0.9–1.2 mm. Sandy disturbed sites, waste places, shorelines of streams and lakes, 5600–9500 ft. Catron, Grant, Sandoval, Taos.

Rumex hymenosepalus Torr.—Canaigre dock. Perennials, with tuberous roots and creeping rhizomes; stems 2.5–10 dm; leaves basal and cauline, blades oblong, oblong-elliptic, or obovate-lanceolate, (50–)80–300 × 20–80(–120) mm, base cuneate, margins plane or indistinctly crisped, apex acute or acuminate; inflorescences terminal; pedicels jointed in proximal 1/2, 5–15(–20) mm; flowers bisexual, 5–20 per whorl; inner tepals oblong-cordate or round-cordate, 11–16 × 9.5–14 mm, base sinuate or emarginate, margins entire, rarely with few denticles proximally, tubercles 0; cypselae 4–7 mm. Sandy to rocky meadows and grasslands, pine-juniper and oak woodlands, stream banks, roadsides, disturbed sites, 3300–9500 ft. Widespread except NC, EC, C, WC.

Rumex mexicanus Meisn.—Mexican dock. Perennials, with vertical rootstock and sometimes creeping rhizomes; stems 3–6(–9) dm; leaves cauline, blades linear-lanceolate to lanceolate, 60–140 × 10–40 mm, base cuneate, margins plane or undulate, apex acute to attenuate; inflorescences terminal and axillary; pedicels jointed in proximal 1/3, 4–7 mm; flowers bisexual, 10–20 per whorl; inner tepals ovate-triangular to triangular, 3.5–4.5(–5) × 3.5–4(–5) mm, base truncate to cordate, margins entire or obscurely erose, tubercles 3, equal or subequal; cypselae 2–3 mm. Stream banks, wet meadows, roadsides, 3800–10,700 ft. Widespread except E.

Rumex nematopodus Rech. f.—Arizona dock. Perennials, with vertical and fusiform rootstock; stems 4–8.5(–10) dm; leaves basal and cauline, blades lanceolate to oblong-lanceolate, 200–350 × 50–120 mm, base cordate to cuneate, margins entire, apex acute to attenuate; inflorescences terminal; pedicels jointed in proximal 1/3, 12–20 mm; flowers bisexual, 10–20 per whorl; inner tepals ovate-triangular, 4–5(–6) × 3–4(–5) mm, base truncate to subcordate, margins entire, tubercles 0; cypselae 3 mm. Stream banks, wet meadows, 4100–8000 ft. Cibola, Lincoln, San Miguel, Torrance.

Rumex obtusifolius L.—Bitter dock. Perennials, with vertical and fusiform rootstock; stems 6–12 (–15) dm; leaves basal and cauline, blades oblong to ovate, 200–400 × 100–150 mm, base usually cordate, sometimes rounded or truncate, margins plane, rarely undulate or crisped, apex obtuse or acute; inflorescences terminal; pedicels jointed in proximal 1/3(–1/2), 2.5–10 mm; flowers bisexual, 10–25 per whorl; inner tepals ovate-triangular to triangular, 3–6 × 2–3.5 mm (excluding teeth), base truncate, margins usually dentate, rarely ± entire, teeth 2–5 per margin, triangular to subulate, 0.5–1.8 mm, tubercle usually 1, if 3 then 2 much smaller; cypselae 2–2.7 mm. Stream banks, wet meadows, roadsides, disturbed sites, 3900–8000 ft. NW, C, WC, SW.

Rumex occidentalis S. Watson—Western dock. Perennials, with vertical or oblique and fusiform rootstock; stems 5–10(–14) cm; leaves basal and cauline, blades ovate-triangular to ovate-lanceolate or oblong-lanceolate, 100–350 × 50–120 mm, base cordate, truncate, or rounded, margins plane to undulate or indistinctly crisped, apex acute or acute; inflorescences terminal; pedicels jointed in proximal 1/3, 5–17 mm; flowers bisexual, 10–25 per whorl; inner tepals round to ovate or ovate-triangular, 5–10(–12) × 5–8(–11) mm, base truncate to weakly cordate, margins entire or subentire, tubercles 0; cypselae 3–4.8 mm. NE, NC, NW, C, SW.

1. Plants glabrous or papillose...**var. occidentalis**
1'. Plants densely tomentose...**var. tomentellus**

var. occidentalis—Wet meadows, montane marshes, 5100–12,900 ft.

var. tomentellus (Rech. f.) Reveal [*R. tomentellus* Rech. f.]—Mogollon dock. Stream banks; in New Mexico known only from the 1928 type collection from Catron County, ca. 7600 ft. **R** [not pictured in plates]

Rumex orthoneurus Rech. f.—Chiricahua Mountain dock. Perennials, with creeping or fusiform rhizomes; stems 6–10 dm; leaves basal and cauline, blades oblong or oblong-lanceolate, 200–400(–500) × 80–150(–180) mm, base cuneate, obtuse, or weakly cordate, margins plane, apex acute to acuminate, lateral veins ± equal in length; inflorescences terminal; pedicels jointed in proximal 1/2, (5–)12–17 mm; flowers bisexual, 10–20 per whorl; inner tepals ovate-deltate, 4.5–7 × 3.5–7 mm, base truncate, margins erose or obscurely denticulate proximally, tubercles 0; cypselae 2.5–4 mm. Stream banks, wet meadows, 3800–12,000 ft. WC, C, SW.

Rumex patientia L.—Patience dock. Perennials, with vertical and fusiform rootstock; stems 8–15 (–20) dm; leaves basal and cauline, blades ovate-lanceolate or oblong-lanceolate, 300–450(–500) × 100–150 mm, base truncate, cuneate, or weakly cordate, margins plane or weakly undulate, apex acute; inflorescences terminal; pedicels jointed in proximal 1/3, 5–13(–17) mm; flowers bisexual, 10–20(–25) per whorl; inner tepals ovate to round, (5-)5.5-8(-10) × 5-9(-10) mm, base cordate, margins entire or obscurely erose, tubercles usually 1, if 3 then 2 much smaller; cypselae 3-3.5 mm. Waste places, roadsides, old fields, disturbed meadows, 5600-7800 ft. Sandoval, San Miguel, Santa Fe.

Rumex pulcher L.—Fiddle dock. Perennials, with vertical and fusiform rootstock; stems 2-6(-7) dm; leaves basal and cauline, blades lanceolate to oblong or ovate-oblong, 40-100(-150) × (20-)30-50 mm, base truncate or weakly cordate, sometimes rounded, margins plane or undulate to slightly crisped, apex obtuse to acute; inflorescences terminal, open, broadly paniculate, branches spreading to ascending, whorls mostly distinct; pedicels jointed in proximal 1/3(-1/2), 2-6 mm; flowers bisexual, 10-20 per whorl; inner tepals ovate-triangular to oblong-triangular, 3-6 × 2-3 mm (excluding teeth), base truncate, margins usually dentate, rarely ± entire, teeth 2-5(-9) per margin, narrowly triangular, 0.3-2.5 mm, tubercles (1-)3, equal or not equal; cypselae 2-2.8 mm. Waste places, disturbed sites, roadsides, fields, meadows, 3900-6700 ft. SC, SW, San Juan, Taos.

Rumex stenophyllus Ledeb.—Narrow-leaf dock. Perennials, with vertical and fusiform rootstock; stems 4-8(-13) dm; leaves basal and cauline, blades oblong-lanceolate to lanceolate, 150-250(-300) × 20-70 mm, base cuneate or truncate, margins usually crisped or undulate, sometimes plane, apex acute to obtuse; inflorescences terminal, dense, narrowly paniculate, branches ascending, whorls mostly not distinct; pedicels jointed in proximal 1/3, 3-8 mm; flowers bisexual, 20-25 per whorl; inner tepals ovate to ovate-triangular, 3.5-5 × 3-5 mm, base truncate to cordate, margins denticulate, each with 4-10 teeth, triangular, 0.2-1.5 mm, tubercles usually 3, equal or subequal; cypselae 2-3 mm. Waste places, roadsides, fields, meadows, marshes, shorelines, 5000-6800 ft. Lincoln, Sandoval, Union. Introduced.

Rumex triangulivalvis (Danser) Rech. f.—Willow dock. Perennials, with vertical rootstock and sometimes creeping rhizomes; stems (3-)4-10 dm; leaves cauline, blades linear-lanceolate, 60-170 × 10-40(-50) mm, base cuneate, margins plane or undulate, apex acute; inflorescences terminal and axillary; pedicels jointed in proximal 1/3, 4-8 mm; flowers bisexual, 10-25 per whorl; inner tepals triangular, (2-)2.5-3.5(-3.8) × (2-)2.5-3(-3.5) mm, base truncate to rounded, margins entire or obscurely erose proximally, tubercles 3, rarely 1 and then 0.5 times width of tepal; cypselae 1.7-2.2 mm. Ruderal sites, roadsides, fields, stream banks, wet meadows, 3800-10,700 ft. Widespread except EC, SE.

Rumex venosus Pursh—Veiny dock. Perennials, with creeping rhizomes; stems (1-)1.5-3(-4) dm; leaves cauline, blades ovate-elliptic, obovate-elliptic, or ovate-lanceolate, (20-)40-120(-150) × 10-50 (-60) mm, base cuneate, margins plane to slightly undulate, apex acute to acuminate; inflorescences terminal and axillary; pedicels jointed near middle, 8-16 mm; flowers bisexual, 5-15 per whorl; inner tepals reniform to round, 13-18(-20) × (20-)23-30 mm, base cordate, margins entire, tubercles 0; cypselae 5-7 mm. Sand dunes, arroyos, sandy riverbanks and grasslands, 4600-6000 ft. San Juan, Union.

Rumex violascens Rech. f.—Violet dock. Annuals or biennials, with ± vertical and fusiform rootstock; stems 2.5-7.5 dm; leaves basal and cauline, blades oblong-lanceolate to obovate-elliptic, 60-120 (-150) × (20-)30-40(-50) mm, base cuneate or rounded, margins plane or weakly undulate, apex obtuse to acute; inflorescences terminal; pedicels jointed in proximal 1/3, 3-8 mm; flowers bisexual, 10-20 per whorl; inner tepals triangular, 2.5-3.7(-4) × 2-2.8(-3) mm (excluding teeth), base truncate, margins denticulate, rarely ± entire, teeth 2 or 3 per margin, triangular, 0.2-0.5 mm, tubercles 3, usually unequal, sometimes subequal; cypselae 1.8-2.3 mm. Stream banks, wet meadows, 3800-6800 ft. Doña Ana.

STENOGONUM Nutt. – Two-whorl buckwheat

Herbs, annual; stems spreading to erect, branched, nodes not swollen; leaves basal or basal and cauline, alternate; ocreae absent; blade margins entire; inflorescences terminal, cymose; involucre not tubular, bracts 6, in 2 whorls of 3, connate proximally; perianth yellow to reddish yellow; tepals 6, connate 1/3-1/2 their length; stamens 9; styles 3; cypselae usually included, trigonous, not winged.

1. Leaves basal; stems erect, sparsely glandular...**S. flexum**
1'. Leaves basal and cauline; stems spreading, glabrous...**S. salsuginosum**

Stenogonum flexum (M. E. Jones) Reveal & J. T. Howell—Bent two-whorl buckwheat. Plants 0.5–3 dm; stems erect, sparsely glandular; leaves basal; blades round, 5–20 × 5–20 mm, minutely strigose, glabrescent; inflorescences erect to spreading, 5–25 cm; bracts 0.5–2 mm, glabrous or sparsely glandular; involucral lobes 2–3 × 2–4 mm, glabrous or sparsely glandular; perianth yellow to reddish yellow, 1.5–3.5 mm; cypselae 2–2.5 mm. Clayey grassland and shrub communities, 5600–5700 ft. San Juan.

Stenogonum salsuginosum Nutt. [*Eriogonum salsuginosum* (Nutt.) Hook.]—Smooth two-whorl buckwheat. Plants 0.5–2 dm; stems spreading, glabrous; leaves basal and cauline, basal blades spatulate, 10–40 × 5–25 mm, glabrous; cauline blades lanceolate to oblanceolate, 5–45 × 2–10 mm; inflorescences spreading, 5–30 cm; bracts 0.5–4 mm, glabrous; involucral lobes 2–8 × 2–3 mm; perianth yellow, 1.5–3 mm; cypselae 2–2.5 mm. Clayey grassland and shrub communities, 5000–7500 ft. NW.

PONTEDERIACEAE – PICKEREL-WEED FAMILY

C. Barre Hellquist

Perennial or annual herbs, aquatic, rooted or free-floating; stems vegetative and flowering; leaves sessile and petiolate; dead stipules retained, associated with petiolate leaves; sessile leaves submersed, rarely emersed; blades linear, occasionally oblanceolate, base sheathing, margins entire; petiolate leaves floating or emersed, blades cordate, reniform, or ovate; inflorescences paniculate, spicate, umbellate, or 1-flowered; flowers sessile, 3-merous; perianth yellow, blue, mauve, or white; stamens 3 or 6; pistils 3-locular; fruit a capsule or utricle; seeds with longitudinal wings (Horn, 2002).

HETERANTHERA Ruiz & Pav. – Mud plantain

Aquatic herbs, submersed, floating, or rooted in mud; perianth with a slender tube; stamens in the throat, usually unequal; anthers erect; capsule 1-locular or incompletely 3-locular; 1- to few-flowered spathes from the sheathing side or base of petiole; stamens 3, equal or unequal; anthers ovate to sagittate; capsule many-seeded.

1. Flowers yellow, inflorescences 1-flowered; sessile leaf blades linear [reported]…H. dubia
1'. Flowers blue, mauve, or white; inflorescences 1–24; sessile leaf blades linear to occasionally oblanceolate or absent…(2)
2. Inflorescence 1-flowered…**H. limosa**
2'. Inflorescence 2–24-flowered…**H. rotundifolia**

Heteranthera dubia (Jacq.) MacMill. [*Zosterella dubia* (Jacq.) Small]—Water star-grass. Perennial; submersed grasslike herb; slender branching stems often rooting at nodes; leaves linear, ribbonlike, sessile, finely parallel-veined, lacking a distinct midvein; spathe terminal, rarely > 2 cm; exposed above water, or forming on emersed plants; perianth yellow; parietal placentae capsule 1-celled with 3 filaments, dilated below, yellow; stamens 3, equal in size; seeds oblong-ovoid, finely cross-lined with prominent raised longitudinal ribs. Streams, ponds, quiet waters. Reported in New Mexico but no specimens seen.

Heteranthera limosa (Sw.) Willd.—Blue mud plantain. Annual; plants rooted in mud, forming rosettes, elongate and creeping, rooted at nodes; leaf blades ovate to elliptic or elliptic-lanceolate, to 10 cm; spathe conspicuously peduncled, peduncles 2–24 cm; flowering stems 1-flowered; spathes 0.9–4.5 cm, glabrous; flowers opening after dawn, wilting by midday; perianth white to purplish blue; lateral stamens 2.3–7.8 mm, central stamens 3.3–7.2 mm; seeds 9–14-winged. Wet shorelines of ponds, ditches, and steams, 4175–5275 ft. Scattered.

Heteranthera rotundifolia (Kunth) Griseb.—Roundleaf mud plantain. Annual; vegetative stems submersed, with elongate internodes; flowering stems 2–12 cm, distal internode 1–6 cm; sessile leaves with floating basal rosettes, blades linear to oblanceolate, thickened; petiolate leaves floating or emersed, petioles 3–11 cm, blades round to oblong; inflorescence 1-flowered; spathes 1–2.8 mm, glabrous; flowers opening after daybreak and wilting by noon; perianth blue or white, tube 11–29 mm; lateral stamens 2.8–8 mm, central stamens 3.9–8.5 mm; style glabrous; seeds 8–15-winged. Shallow water or exposed shores, 4525–5450 ft. Scattered.

PORTULACACEAE – PURSLANE FAMILY

David J. Ferguson

PORTULACA L. – Purslane

Annual or perennial herbs, usually fleshy; stems succulent, prostrate to erect, usually branched, sometimes suffrutescent, usually glabrous; leaves succulent, alternate or opposite, blades terete, subterete, or flattened, simple, entire; inflorescence terminal, multiflowered, with flowers tightly crowded; flowers mostly diurnal, perfect, actinomorphic; sepals usually 2, distinct; petals mostly 5–7, distinct or basally connate, stamens 4 to many, usually the same number as the petals, opposite the petals; pistil 1; ovary 1/2 inferior, with free-central placentation, 2–9-carpellate; fruit a circumscissile capsule (Carolin, 1987; Packer, 2003).

1. Leaves linear, terete or nearly so; axils pilose…(2)
1'. Leaves flat, lanceolate, obovate, or spatulate; axils not or inconspicuously pilose…(6)
2. Flowers < 16 mm across…(3)
2'. Flowers > 16 mm across…(5)
3. Flowers usually < 5 mm across, yellow to slightly orange…**P. halimoides**
3'. Flowers usually > 5 mm across, usually pink to magenta…(4)
4. Taproot thickened, tuberous; plants mostly short-lived perennials with upright single to cespitose stems; seeds usually appearing silvery-gray (black if pellicle rubbed off), with prominent peglike tubercles; native in shallow soils in rocky areas, most often on calcareous substrates; mostly E of the Rio Grande…**P. mundula**
4'. Taproot usually slender and not distinctly tuberous; plants mostly annuals with horizontal spreading; seeds appearing black, mostly smooth or with low, rounded elevations; apparently introduced, uncommon and mostly in urban or agricultural areas…**P. pilosa**
5. Taproot tuberous; plants mostly with upright single to cespitose stems; flowers yellow to red (usually orange), 16–25 mm across; native in igneous rocky and gravelly areas in S New Mexico…**P. suffrutescens**
5'. Taproot usually slender and nontuberous; plants mostly spreading; flowers various colors, often double, usually > 25 mm across; introduced ornamental, sometimes escaping…**P. grandiflora**
6. Fruit with a winglike encircling rim at top margin, with "cap," convex-domed to conic, not crested…**P. umbraticola**
6'. Fruit without a winglike rim or "cap," laterally compressed and ridged to crested…(7)
7. Flowers typically with 4 or 5 yellow petals < 3 mm; stigma lobes 3–4; seeds often appearing gray or silvery, strongly "echinate"-tubercled, mostly < 0.8 mm; usually in arid or semiarid situations and growing after summer rains; abundant and widespread…**P. retusa**
7'. Flowers typically with 5 yellow petals > 3 mm; stigma lobes 5–6; seeds black, not or weakly tubercled, mostly > 0.8 mm; disturbed urban and agricultural areas where introduced…**P. oleracea**

Portulaca grandiflora Hook.—Moss-rose. Annual herb; roots fibrous; stems succulent, prostrate to suberect, well branched; trichomes conspicuous at nodes and in inflorescence, to 30 cm; leaves terete, 5–30 × 1–5 mm; flowers 25–60 mm diam.; petals obovate, 15–25 mm, white or yellow to red or magenta, often 2+ colors in cross bands; capsule broadly ellipsoid to turbinate, 3.5–6.5 mm diam.; seeds gray, often iridescent. Escaped ornamental; 3 records, elevation unknown.

Portulaca halimoides L. [*P. parvula* A. Gray]—Sand moss-rose. Succulent annual herb; roots fibrous; stems prostrate to suberect, mostly 3–20 cm; nodes and inflorescence moderately villous, this more conspicuous as the plant ages; leave terete, to 2–14(–20) × 0.4–2(–3) mm; flowers diurnal, 3–8 mm diam.; petals yellow, rarely orange to red or marked reddish, obovate, to 4 mm; capsule ovoid, 1.1–2 mm diam.; seeds gray to nearly black. Sandy soils in mostly open, semiarid to arid habitats, 3300–8400 ft. Widespread except SE.

Portulaca mundula I. M. Johnst. [*P. pilosa* L. var. *mundula* (I. M. Johnst.) D. Legrand]—Rock moss-rose. Succulent annual or short-lived perennial herb with taproot; stems suberect to erect, to 15 cm; leaves terete to 15(-20) mm; flowers diurnal, 4-15 mm diam.; petals pink, magenta, or purple, rarely white, obovate, to 8 mm; capsule 1.5-6 mm; seeds gray to lustrous/metallic. Mostly shallow soils overlying calcareous rock, 4000-8000 ft. EC, SC, SW, San Juan. Included by some under a broad concept of *P. pilosa* L. Evidence for or against is conflicting, but the two are clearly distinct as found in New Mexico.

Portulaca oleracea L. [*P. neglecta* Mack. & Bush]—Common purslane. Annual herbs (ours); roots fibrous or with thick tuberous taproot; stems prostrate to upright, mostly > 15 cm diam.; leaves obovate or spatulate, flattened, 4-30 × 2-15 mm; flowers diurnal, 3-10 mm diam.; petals yellow (ours), oblong, 1.5-4 mm; capsule 4-9 mm, with rooflike ridge or crest dorsally; seeds somewhat flattened, 0.4-1.5 mm diam., ± smooth to tuberculate. Weedy, riparian and disturbed waste places, fields, yards, 2975-8730 ft. Widespread.

Portulaca pilosa L.—Purple moss-rose. Annual herb; roots fibrous or with slightly swollen fleshy taproot; stems spreading, branched, prostrate to sometimes procumbent; trichomes at nodes and in inflorescence; leaves linear to oblong-lanceolate, terete to hemispherical, 5-20 × 1-3 mm; flowers 5-12 mm diam., petals dark pink to purple (white), obovate, to 7 mm; capsules ovoid, 1.5-4 mm diam.; seeds black, ± 0.6 mm, nearly smooth to weakly tuberculate. Moist disturbed sites, cultivated areas, 2975-8000 ft. Widespread except far N. Introduced weed.

Portulaca retusa Engelm.—Desert purslane. Annual herbs (ours); roots fibrous or with thick tuberous taproot; stems prostrate to upright; stems prostrate to upright, mostly < 15 cm diam., commonly reddish; leaves obovate or spatulate, flattened, 4-30 × 2-15 mm, often emarginate; flowers diurnal, 3-10 mm diam.; petals yellow (ours), oblong, 1.5-4 mm; capsule 4-9 mm, with rooflike ridge or crest dorsally; seeds somewhat flattened, 0.4-1.5 mm diam., ± smooth to tuberculate. Arid grasslands, desert scrub, piñon-juniper woodlands, ponderosa pine forests, 4000-8000 ft. Widespread.

Portulaca suffrutescens Engelm.—Copper moss-rose. Succulent short-lived perennial herb; roots thickened, tuberous; stems erect, sometimes slightly suffrutescent basally, to 30 cm, often with reddish nodes; trichomes conspicuous at nodes and in inflorescence; leaves terete, to 20(-30) mm; flowers diurnal, ± 25 mm diam.; petals orange or dull yellow-orange to copper or near red, often darker basally, usually emarginate, to 13 mm; capsule 2.5-5 mm diam.; seeds grayish-lustrous, 0.5-0.65 mm, tuberculate. Mostly on rhyolite- or granite-derived substrates, grasslands, piñon-juniper-oak woodlands, 4000-6700 ft. SC, SW.

Portulaca umbraticola Kunth—Wing-pod purslane. Annual herb; roots fibrous; stems prostrate to suberect, 5-20 cm, glabrous; trichomes sparse at nodes and in inflorescence; leaves flat, obovate, spatulate, or sometimes lanceolate, to 35 mm; flowers diurnal, 5-12 mm diam.; petals yellow or yellow tipped with red or copper, spatulate or obovate, to 8 mm; capsule 3-5 mm diam., with encircling membranous wing 0.5-1.5 mm wide proximal to suture; seeds gray, 0.5-1 mm, with prominent central tubercle. New Mexico material belongs to **subsp. lanceolata** J. F. Matthews & Ketron. Sandy and rocky sites, arroyos, grasslands, piñon-juniper woodlands, 3100-6000 ft. Hidalgo, Luna, Sierra.

POTAMOGETONACEAE – PONDWEED FAMILY

C. Barre Hellquist

Perennial or annual herbs; rhizomatous or nonrhizomatous; caulescent; turions absent or present; leaves stipulate, simple, alternate, or opposite, submersed, or submersed and floating, sessile or petiolate; inflorescence a spike, or panicle of axillary or terminal spikes, submersed or emersed; flowers of 4 tepals in 1 series, stamens (2-)4, epipetalous in 1 series; anthers dehiscing vertically, borne in spikes or axillary clusters (in *Zannichellia*), or some enclosed in leaf sheaths at time of anthesis, submersed or emersed or both; pistils 1-4, mostly not stipitate; fruit a drupe (Haynes & Hellquist, 2000b; Wiegleb & Kaplan, 1998).

1. Submersed leaves opposite or subopposite; floating leaves absent; flowers unisexual, in axillary clusters; fruits stalked, dentate or papillate on 1 side…**Zannichellia**
1'. Submersed leaves alternate; floating leaves present or absent; flowers bisexual, in axillary or terminal spikes; fruits sessile, sometimes with a smooth or undulate lateral keel, but not dentate or papillate…(2)
2. Submersed leaves with stipular sheaths adnate to leaf base for 2/3+ stipule length, submersed leaves opaque, channeled, septate; peduncles flexible; inflorescences, if reaching water surface, floating on surface…**Stuckenia**
2'. Submersed leaves with stipulate sheaths free from leaf blade base, or adnate for < 1/2 stipule length; submersed leaves translucent, not channeled or septate; peduncles stiff; inflorescences, if reaching surface, emergent above the water…**Potamogeton**

POTAMOGETON L. – Pondweed

Perennial or annual herbs; turions formed on some species; stem nodes with oil glands present or absent; leaves submersed or submersed and floating, alternate, stipules connate or convolute, sheathing the stem, submersed leaves sessile, petiolate, or perfoliate, apex subulate to obtuse; stipules either free or adnate to the leaf base for < 1/2 the length of the stipule, floating leaves petiolate, rarely sessile, blades elliptic to ovate, leathery; inflorescence a spike, submersed and/or emersed, peduncles often projecting above the water surface. Unnamed hybrids listed have been DNA determined.

1. Plants producing floating leaves…(2)
1'. Plants lacking floating leaves…(8)
2. Stipule sheaths fused to leaf base; fruits with distinct coil and sharply keeled…**P. diversifolius** (in part)
2'. Stipule sheaths free from leaf base; fruits lacking distinct coil and not sharply keeled…(3)
3. Submersed leaves linear, 0.8-2 mm wide; petiole of floating leaves pale at junction with blade…**P. natans** (in part)
3'. Submersed leaves broadly linear-oblong, lanceolate to elliptic, or ovate to oblanceolate, 10-58 mm wide; petiole of floating leaves not pale at junction with blade…(4)
4. Submersed leaves petiolate; fruits 2.7-5.7 mm…(5)
4'. Submersed leaves sessile; fruits 1.9-3.5 mm…(7)
5. Petiole of submersed leaves 0-2.1(-6) cm; apex mucronate…**P. illinoensis** (in part)
5'. Petiole of submersed leaves mainly > 6 cm; apex acute to obtuse…(6)
6. Floating leaves 21-51-veined; submersed leaves arcuate, 21-41-veined, 2.5-7.5 cm wide, margins entire; fruits 3.9-5.2(-5.7) mm; stems often rusty-spotted…P. amplifolius
6'. Floating leaves 11-29-veined; submersed leaves not arcuate, 7-21-veined, 11-38 mm wide, margins minutely denticulate, rarely crisped, not arcuate; fruits 2.7-4.1(-4.3) mm; stems lacking spots…**P. nodosus** (in part)
7. Stems branched; submersed leaves green to brownish, margins denticulate, tip acute to obtuse; fruits sessile, 1.9-2.3 mm…**P. gramineus** (in part)
7'. Stems unbranched; submersed leaves usually reddish, margins entire, tip blunt; fruits pedicellate, (2.5-)3-3.5 mm…**P. alpinus** (in part)
8. Leaf margins distinctly serrate; fruit beaks 2-3 mm…**P. crispus**
8'. Leaf margins entire or minutely denticulate; fruit beaks lacking or < 2 mm…(9)
9. Submersed leaves linear, 2 mm or less wide…(10)
9'. Submersed leaves broadly linear-oblong to lanceolate to elliptic, 10-75 mm wide…(11)
10. Submersed leaves phyllodial, 0.8-2 mm wide (young plants), lacking distinct veins…**P. natans** (in part)
10'. Submersed leaves flattened, < 2 mm wide, with 1-5 distinct veins…(14)
11. Stipule sheaths fused to leaf base, fruits with distinct coil and sharply keeled; leaves 1-3-veined…**P. diversifolius** (in part)
11'. Stipule sheaths free from leaf base, fruits lacking distinct coil and not sharply keeled; leaves 3-5-veined…(12)
12. Nodal glands present at base of leaves; peduncles terminal or terminal and axillary, straight, (0.6-)1-4.5(-8) cm…(13)
12'. Nodal glands absent at base of leaves; peduncles axillary, often slightly recurved, thickened to tip, 0.3-1.1(-3.7) cm…**P. foliosus**
13. Stipules mostly connate (fused, surrounding stem); peduncles 1.5-8 cm, usually terminal; inflorescences usually of 3-5 distinct, interrupted whorls of flowers/fruits…**P. pusillus**
13'. Stipules convolute (wrapped around stem); peduncles 0.35-3 cm, terminal and/or axillary; inflorescences with flowers/fruits in a crowded spike…**P. berchtoldii**
14. Submersed leaves clasping the stem…(15)
14'. Submersed leaves sessile (not clasping) or petiolate…(16)

15. Rhizomes spotted rusty-red; stems with a zigzag appearance, changing direction at each node; leaf tips boat-shaped, usually splitting when pressed…**P. praelongus**

15′. Rhizomes white, unspotted; stems straight, lacking a zigzag appearance; leaf tips flat, not splitting when pressed…**P. richardsonii**

16. Petiole of submersed leaves 0–2.1(–6) cm; apex mucronate…**P. illinoensis** (in part)

16′. Petiole of submersed leaves mainly > 6 cm; apex acute to obtuse…(17)

17. Upper submersed leaves folded along the midvein, falcate in outline…P. amplifolius

17′. Upper submersed leaves not folded along the midvein, not falcate in outline…(18)

18. Submersed leaves petiolate; fruits 3.5–4.3 mm…**P. nodosus** (in part)

18′. Submersed leaves sessile; fruits 1.9–3.5 mm…(19)

19. Leaf apex obtuse to somewhat acute, leaves with reticulate portion along midrib; plant branched; fruits stalked…**P. alpinus** (in part)

19′. Leaf apex sharply acute or awl-shaped, leaves lacking reticulate portion along midrib; plant unbranched; fruits lacking a stalk…**P. gramineus** (in part)

Potamogeton alpinus Balb.—Alpine pondweed. Perennial; leaves submersed and floating or only submersed, often transitionirıg into floating leaves; submersed leaves sessile, reddish green, 0.5–2 cm wide, apex obtuse to acute, stipules persistent, convolute; floating leaves reddish green, 1–2.5(–4) cm, base gradually tapering into petiole; apex obtuse or acute, veins (7–)9–13(–15); fruits olive-green, (2.5–)3–3.7 mm, beak curved, 0.5–0.9 mm. Cold, neutral to alkaline waters, 7700–11,300 ft. NC. Hybrids with *P. gramineus* = *P.* ×*nericus* Hagstr.; with *P. nodosus* = *P.* ×*subobtusus* Hagstr.

Potamogeton amplifolius Tuck.—Broad-leaved pondweed. Perennial; stems often with rusty-colored spots; leaves submersed and floating or only submersed; submersed leaves petiolate, with petioles 0.9–11.7 cm, leaves ovate to oblanceolate, arcuate, 2.5–7.5 cm wide, apex acute to round-apiculate; fruits reddish brown, 3.9–5.2(–5.7) mm; beak erect, 0.5–0.8 mm. Acid to alkaline waters, 8650–8750 ft. Hybrids with *P. illinoensis* = *P.* ×*luxuriaris* Z. Kaplan. No records seen.

Potamogeton berchtoldii Fieber [*P. pusillus* L. subsp. *tenuissimus* (Mert & W. D. J. Koch) R. R. Haynes & Hellq.]—Berchtold's pondweed. Annual; stems with nodal glands to 0.5 mm diam.; turions common, terminating branches; leaves submersed, stipules convolute; inflorescences submersed or emersed; peduncles axillary or terminal, erect, rarely recurved, filiform to slightly clavate, 0.35–3 cm, 1–3 per plant; fruits green to brown, 2–2.5 mm; beak erect, 0.1–0.6; embryo with < 1 full spiral. Acid to alkaline waters, 2850–6720 ft. Colfax, Rio Arriba.

Potamogeton crispus L.—Curly-leaved pondweed. Annual; turions common, axillary and terminal, hard; leaves submersed, linear, 4–10 mm wide, apex round to acute, margins serrate, lacunae in 2–5 rows on each side of midvein; inflorescences emersed, peduncles terminal, rarely axillary, 2.5–4 cm; spikes 10–15 mm; fruits 4–6 mm, beak 2–3 mm. Alkaline or polluted waters, 5115–6300 ft. NW, WC, Eddy, Hidalgo, Union. Different species that form hybrids with *P. crispus* = *P.* ×*undulatus* Wolfg.

Potamogeton diversifolius Raf.—Water-thread pondweed. Perennial; leaves submersed, or submersed and floating; submersed leaves sessile, (0.2–)0.4–1.5 mm wide, stipules adnate, 0.8–1.6 cm; floating leaves 3–17(–20) mm wide, (3–)5–17-veined; inflorescence peduncles dimorphic, submersed, recurved, clavate, 0.1–0.8 cm, or emersed, axillary or terminal; spikes submersed, capitate, 3–3.2 cm; fruits 1–2 mm, dorsal keels with sharp points, beak present, erect, 0.1 mm, embryo with > 1 full coil. Acid to alkaline waters, 5315–7760 ft. Hidalgo, San Juan, San Miguel.

Potamogeton foliosus Raf. [*P. foliosus* var. *macellus* Fernald]—Leafy pondweed. Annual, rarely perennial; stems slightly compressed, glands usually absent, turions produced; leaves submersed, linear, 0.3–2.3 mm wide, apex usually acute, rarely apiculate, lacunae 0–2 on each side of the midvein, veins 1–3(–5), stipules 0.2–2.2 cm; inflorescence often with recurved peduncles, clavate, 0.3–1.1(–3.7) cm, spikes 1.5–7 mm; fruits olive-green or brown, 1.5–2.7 mm, keel 0.2 mm, undulate, beak 0.2–0.6 mm. Neutral to alkaline waters, 3685–11,540 ft. NE, NC, NW, WC, SE, SW, Lincoln. An unnamed hybrid is *P. foliosus* × *P. pusillus*.

Potamogeton gramineus L.—Variable pondweed. Perennial from rhizomes; stems branching; leaves submersed and floating; submersed leaves sessile, blades elliptic, lacunae in 1–2 rows on either side of the midvein, 3–9 veins, apex acute to obtuse, margins minutely apiculate; floating leaves 1.6–2 cm wide, apex acuminate, veins 11–13; petioles 3–4.5 cm, stipules convolute; inflorescences emersed, peduncles erect to ascending, 3.2–7.7 cm, spikes 1.5–3.5 cm; fruits greenish brown, 1.9–2.3 mm, beak 0.3–0.5 mm. Acid to alkaline waters, 3115–10,480 ft. NC, NW.

Potamogeton illinoensis Morong—Illinois pondweed. Perennial; leaves both submersed and floating; submersed leaves sessile or petiolate, petioles, if present, 0.5–4 cm, blades elliptic to lanceolate or rarely linear, often arcuate, margins entire to minutely denticulate, apex acute-mucronate, lacunae in 2–5 rows on each side of midrib, veins 7–19; floating leaf apex round-mucronate, veins 13–29; inflorescences 4–13 cm; fruit 2.7–3.6 mm, beak 0.5–0.8 mm. Alkaline waters, 2845–8500 ft. Colfax, Eddy. Hybrids with *P. gramineus* = *P.* ×*deminutus* Hagstr.; with *P. nodosus* = *P.* ×*faxonii* Morong.

Potamogeton natans L.—Floating-leaf pondweed. Perennial; leaves submersed and floating; submersed leaves 0.7–2.2 mm wide, apex obtuse, stipules persistent, free from blades 4.5–1 cm; floating leaves with petiole a lighter color at junction with leaf blades, blades elliptic to ovate, 3.5–11 cm wide, veins 17–37; fruits sessile, unkeeled, 3.5–5 mm, beak 0.4–0.8 mm. Acid to alkaline waters, 6235–9450 ft. NC, NW, WC, SW. Unnamed hybrid with *P. illinoensis*.

Potamogeton nodosus Poir.—Long-leaf pondweed. Perennial; leaves submersed and floating; submersed leaves with petioles 2–13 cm, blades linear-lanceolate to lance-elliptic, 1–3.5 cm wide, apex acute, margins minutely denticulate, 2–5 rows of lacunae on each side of midvein; floating leaf petioles 3.5–26 cm, blades elliptic, 1.5–4.5 cm wide, apex acute to rounded, veins 9–21; fruits 2.7–4.3 mm, beak erect. Acid to alkaline waters, 5350–8565 ft.

Potamogeton praelongus Wulfen—White-stemmed pondweed. Perennial; stems white; submersed leaves linear-lanceolate, base clasping, apex hooded or boat-shaped, obtuse, splitting when pressed, lacunae absent, 1.1–4.6 cm wide, stipules persistent, conspicuous, convolute, 3–8.1 cm, fibrous, shredding at apex; inflorescences emersed, spikes cylindrical, 3.4–7.5 cm; fruits greenish brown, obovoid, keeled, 4–5.7 mm, beak 0.5–1 mm. Neutral to moderately alkaline waters, 3120–10,950 ft. Taos. Unnamed hybrid with *P. amplifolius*.

Potamogeton pusillus L. [*P. panormitanus* Biv.]—Small pondweed. Mostly annual; turions common, lateral and terminal; stems with nodal glands present on most nodes, 0.5 mm diam.; submersed leaves 0.5–1.9 mm wide; veins (1–)3, apex acute, rarely apiculate, lacunae 0–2 on each side of the midvein, stipules connate; inflorescence with 1–3 peduncles per plant, spikes mostly terminal, interrupted, 1.5–10.1 cm, in 3–5 distinct whorls; fruits 1.9–2.8 mm, beak 0.1–0.6 mm. Neutral to alkaline waters, 5280–10,750 ft. NC, McKinley, Roosevelt.

Potamogeton richardsonii (A. Benn.) Rydb. [*P. perfoliatus* L. var. *richardsonii* A. Benn.]—Clasping-leaved pondweed. Perennial; stems terete; leaves submersed, clasping, ovate-lanceolate to narrowly lanceolate, apex acute to obtuse, 0.5–2.8 cm wide, veins 3–35, margins minutely denticulate, lacunae absent, stipules disintegrating to persistent fibers; inflorescence with terminal or axillary peduncles, 1.5–14.8 cm; fruits keeled, 2.2–4.2 mm, beak 0.4–0.7 mm. Alkaline waters, 6850–9950 ft. Rio Arriba, Sandoval. *P. richardsonii* × *P. gramineus* = *P.* ×*hagstroemii* A. Benn.

STUCKENIA Börner – Sheath-leaved pondweed

Perennial from rhizomes, often with tubers; leaves submersed, alternate, filiform, channeled, turgid, margins entire, apex acute, obtuse, veins 1–5; stipules adnate to leaf base for 2/3+ stipule length, terminating in a free ligule; inflorescence a capitate or cylindrical spike, submersed, peduncles flexible, floating on the water surface; flowers with 4 pistils; fruit a drupe, beaked or unbeaked. Hybrids frequent in flowing water.

1. Leaf apex acute, apiculate, cuspidate, rarely rounded; proximal stipular sheaths not inflated, convolute; fruits distinctly beaked…(2)
1'. Leaf apex notched, obtuse, or rounded; proximal stipular sheath often inflated; fruits without beak…(3)

2. Midstem leaves < twice the width of leaves on branches; stipules 10–70 mm; roots forming apical tubers; fruits 3–4.7(–5.1) mm, beak to 1.3 mm…**S. pectinata**

2′. Midstem leaves > twice the width of leaves on branches; stipules 12–34 mm; roots not forming apical tubers; fruits 3–3.9 mm, beak < 0.5 mm…**S. striata**

3. Ligule extending beyond stipule sheath distinct, to 20 mm; summits of midstem stipulate sheaths only slightly inflated, < twice the width of stem; fruits 2–3 mm…**S. filiformis**

3′. Ligule barely formed, extending beyond stipule sheath 1 mm or less; summits of midstem stipular sheaths distinctly inflated, twice the diameter of stem; fruits 3–3.5 mm…**S. vaginata**

Stuckenia filiformis (Pers.) Börner [*Potamogeton filiformis* Pers. var. *alpinus* (Blytt) Asch. & Graebn.; *S. filiformis* subsp. *occidentalis* (J. W. Robbins) R. R. Haynes, Les & M. Král.]—Threadleaf pondweed. Leaves with lower stipular sheaths often inflated, 1–4.5(–9.5) cm, apex notched, blunt, or short-apiculate, veins 1–3; inflorescence with spikes 5–55 mm, 2–6(–9) whorls; fruits dark brown-greenish, 2–3 mm, beak obscure. Calcareous or alkaline still or flowing water, 6500–12,180 ft. Cibola, Colfax, Doña Ana, San Juan. This species forms hybrids with *S. pectinata* – *S.* ×*suecica* (K. Richt.) Holub; with *S. vaginata* = *S.* ×*fennica* (Hagstr.) Holub.

Stuckenia pectinata (L.) Börner [*Potamogeton pectinatus* L.]—Sago pondweed. Leaves with stipular sheaths not inflated, ligule to 0.8 mm, blades 0.2–1.1 mm wide, apex acute, mucronate or blunt on young stems, veins 1–3; inflorescence peduncles terminal or axillary, 4.5–11.4 cm, spikes 14–22 mm, 3–5 whorls; fruits yellow-brown to brown, 3.8–4.5 mm, beak erect, 0.5–1.1 mm. Calcareous, saline, and alkaline waters, lakes and rivers, 2980–8250 ft. NC, NW, WC, SE, SC, SW. Chaves. Hybrids with *S. filiformis* = *S.* ×*suecica* (K. Richt.) Holub; with *S. vaginata* = *S.* ×*fennica* (Hagstr) Holub.

Stuckenia striata (Ruiz & Pav.) Holub [*Potamogeton latifolius* (J. W. Robbins) Morong]—Nevada pondweed. Leaves on stems 2+ times as broad as those on branches, stipular sheaths not inflated, 1.2–3.4 cm, ligule 0.2–1.1 cm, blades linear, 0.4–5.1(–8.5) mm wide, apex apiculate, cuspidate, rarely rounded, veins 3–5; inflorescence with axillary peduncles, rarely terminal, erect to ascending, cylindrical, 1.2–5.2 cm; spikes 13–45 mm, 4–9 whorls; fruits brown to reddish brown, 3–3.9 mm. Calcareous and alkaline-loving or still waters, 2650–5275 ft. Eddy, San Juan.

Stuckenia vaginata (Turcz.) Holub [*Potamogeton vaginatus* Turcz.]—Big-sheath pondweed. Leaves 0.2–2.9 mm wide, apex rounded, obtuse, or slightly notched, veins 1(–3), stipular sheaths inflated 3–5 times the stem thickness, 2–9 cm, ligule absent or to 0.2 mm; inflorescence with peduncles 3–15 cm; spikes 10–80 mm, 3–12 whorls; fruits brown, 3–3.8 mm, beak small. Still, calcareous or alkaline waters, 9045 ft. Chuska Mountains. San Juan. This species forms hybrids with *S. filiformis* = *S.* ×*fennica* (Hagstr.) Holub; with *S. pectinata* = *S.* ×*bottnica* (Hagstr.) Holub.

ZANNICHELLIA L. - Horned pondweed

Zannichellia palustris L.—Horned pondweed. Submersed annual herbs with turions; leaves opposite, 3.5–4.2 cm, apex acute, entire; staminate flowers with filaments 1.5–2 mm, pistillate flowers with 4 or 5 pistils, style 0.4–0.7 mm; fruits 1.7–2.8 mm, rostrum 0.7–2 mm, podogyne 0.1–1.5 mm, pedicel 0.3–1.2 mm. Brackish and freshwater ponds, lakes, streams, and ditches, 4000–9100 ft. Widespread.

PRIMULACEAE – PRIMROSE FAMILY

Anita F. Cholewa

Perennial or annual herbs (occasionally woody), sometimes rhizomatous or stoloniferous; leaves in basal rosettes; inflorescences scapose, terminal umbels or solitary flowers; flowers radially symmetric, hypogynous; sepals 4–5, united basally; petals 4–5, united basally; corolla campanulate to salverform or tubular, lobes reflexed; stamens 4–5, opposite and united to petals, distinct or united basally; staminodes absent; carpels 5 and united, ovary superior, 1-loculed with free-central placentation; fruit a capsule (Cholewa & Kelso, 2009).

1. Corolla reflexed; stamens exserted, with connivent anthers…**Primula** (in part)
1'. Corolla salverform; stamens included, anthers not connivent…(2)
2. Corolla white or fading to pink, limb to 5 mm diam.; flowers homostylous…**Androsace**
2'. Corolla rarely white, limb 10+ mm diam.; flowers heterostylous or homostylous…**Primula** (in part)

ANDROSACE L. – Rock jasmine

Annual or perennial herbs, sometimes mat-forming, taprooted or fibrous-rooted; leaf blades lanceolate to spatulate or cuneate, surfaces usually pubescent with simple or forked hairs; inflorescences umbels, pedicels often elongating in fruit; flowers homostylous, calyx broadly campanulate or hemispherical, angled but not strongly keeled, lobes shorter than or equal to tube, corollas white with yellow bases (fading to pink or reddish), campanulate to salverform, constricted at throat, 5-parted, lobes shorter than tube, apex emarginate to entire; filaments distinct, short, included within corolla rube, anthers not connivent; capsules valvate; seeds 3–4 in perennial species, 20–50 in annual species, angled or trigonous.

1. Plants perennial, mat-forming; scapes, pedicels, and calyces villous…**A. chamaejasme**
1'. Plants annual, not mat-forming; scapes, pedicels, and calyces strigillose or glabrous…(2)
2. Involucral bracts broadly lanceolate to ovate…**A. occidentalis**
2'. Involucral bracts narrowly lanceolate to linear-lanceolate…**A. septentrionalis**

Androsace chamaejasme Wulfen [*A. lehmanniana* Spreng.]—Sweetflower rock jasmine. Plants perennial, forming open or dense mats with multiple rosettes; leaf blades 3–15 × 3–4 mm; scapes usually 1, villous; inflorescence 3–6-flowered, bracts broadly lanceolate to ovate, villous; calyx villous, obscurely angled. Our material belongs to **subsp. lehmanniana** (Spreng.) Hultén. Alpine tundra, 10,500–13,050 ft. NC.

Androsace occidentalis Pursh [*A. arizonica* A. Gray; *A. platysepala* Wooton & Standl.]—Western rock jasmine. Plants annual or biennial, with a single rosette; leaf blades 5–30 × 4–9 mm; scapes 1–15, strigillose with simple or forked hairs; inflorescence 3–15-flowered, bracts broadly lanceolate to ovate, strigillose; calyx sparsely strigillose at least on lobes, angled. Open meadows, sandy or gravelly soils, often disturbed sites, 4000–8855 ft. Widespread except E.

Androsace septentrionalis L. [*A. pinetorum* Greene; *A. puberulenta* Rydb.]—Pygmy rock jasmine, pygmyflower. Plants annual or biennial, with a single rosette, highly variable in terms of height, pubescence, and size of floral parts; leaf blades 5–30 × 3–10 mm; scapes 1–5, strigillose with simple or forked hairs; inflorescence 5–20-flowered, bracts narrowly lanceolate to linear-lanceolate, strigillose to glabrous; calyx glabrous or sparsely strigillose, prominently angled. Rocky open habitats, 5700–13,000 ft. Widespread except E and far SW.

PRIMULA L. – Primrose

Perennial herbs (rarely semiwoody), rhizomatous, sometimes mat-forming, sometimes stoloniferous, roots fibrous; leaf blades linear, lanceolate, oblanceolate, oblong, or elliptic to cuneate, surfaces usually glabrous, base tapering into usually winged petiole; inflorescences umbels, bracts present, pedicels erect, spreading, nodding, or recurved; flowers usually homo- or heterostylous, calyx broadly campanulate to cylindrical, angled or keeled, lobes subequal to tube, corollas white, lavender, pink, magenta (often drying bluish), violet, or yellow, often yellow at throat, salverform, apex rounded and often emarginate, lobes strongly reflexed in some; filaments distinct or united, included or exserted, anthers connivent; capsules valvate; seeds 10–100, globose to ovoid to oblong. Species previously included in *Dodecatheon* are included here (Mast & Reveal, 2007).

1. Corolla reflexed; stamens exserted, with connivent anthers…(2)
1'. Corolla salverform; stamens included, anthers not connivent…(3)
2. Corolla lobes usually white; leaves lacking glands, blade abruptly transitioning to petiole; inflorescence 1–6-flowered…**P. standleyana**
2'. Corolla lavender to magenta; leaves with reddish glands, blade gradually transitioning to petiole; inflorescence 2–15-flowered…**P. pauciflora**

3. Calyces with prominent farinose stripes; leaf margins generally denticulate, serrulate, or spiculose…**P. rusbyi**

3'. Calyces efarinose or rarely sparsely farinose; leaf margins generally entire…(4)

4. Plants small, to 8 cm; corolla limb 10-15 mm diam. (dried)…**P. angustifolia**

4'. Plants taller, at least 10 cm; corolla limb 14-25 mm diam. (dried)…**P. parryi**

Primula angustifolia Torr. [*P. angustifolia* var. *helenae* Pollard & Cockerell]—Alpine primrose. Plants 0.5-8 cm at flowering, efarinose, not aromatic; leaves 1-1.7 × 0.3-1 cm, blades linear-lanceolate to oblanceolate, margins entire or rarely sparsely denticulate, usually slightly thickened; inflorescence 1-2-flowered; calyx 5-8 mm; corolla bright rose-pink, magenta, or rarely white, tube 5-8 mm, limb 8-15 mm diam. Upper subalpine spruce-fir krummholz to windy, exposed alpine sites, 11,000-13,010 ft. NC.

Primula parryi A. Gray—Parry's primrose. Plants 10-50 cm at flowering, efarinose, aromatic (skunky); leaves 1-33 × 1.5-7 cm, blades oblong-obovate, broadly lanceolate, or oblanceolate, margins usually entire or rarely denticulate distally; inflorescence 5-25-flowered; calyx 8-15 mm; corolla rose to magenta, tube 5-20 mm, limb 14-25 mm diam. Wet meadows, streamsides, rocky crevices, subalpine streams and snowmelt seeps, 10,500-12,855 ft. NC.

Primula pauciflora (Greene) A. R. Mast & Reveal [*Dodecatheon pulchellum* (Raf.) Merr.; *D. pauciflorum* Greene; *D. radicatum* Greene]—Dark-throated shootingstar. Plants 10-40 cm; leaf blades narrowly to broadly lanceolate to narrowly ovate, base gradually transitioning to petiole, margins usually entire, surfaces glabrous or with sessile or sunken glands appearing as reddish dots or streaks; inflorescence 2-15-flowered, bracts 2-15 mm; calyx usually with reddish sessile glands, tube 1.5-4 mm; corolla 7-20 mm, lobes magenta to lavender with thin red or magenta throat ring, filaments united, 0.7-3.6 mm, yellow, maroon, purple, or nearly black; capsules 5-14 mm, tan to light brown, often reddish apically. New Mexico material belongs to **var. pauciflora**. Moist open slopes and meadows, sometimes wet shrubby marshes, sometimes saline or alkaline flats, 7500-12,200 ft. NC, SC, Catron.

Primula rusbyi Greene [*P. ellisiae* Pollard & Cockerell]—Rusby's primrose. Plants 5-25 cm at flowering, farinose on bracts, pedicels, and in stripes on calyx, not aromatic; leaves 4-15 × 0.5-2.5 cm, blades lanceolate to spatulate, margins generally denticulate, serrulate, or spiculose; inflorescence 3-12-flowered; calyx 4-8 mm; corolla rose-magenta, tube 5-15 mm, limb 10-25 mm diam. Moist alpine meadows, ridges, rocky edges, 6700-12,000 ft. NC, C, WC, SC, Grant.

Primula standleyana A. R. Mast & Reveal [*Dodecatheon ellisiae* Standl.; *D. dentatum* Hook. var. *ellisiae* (Standl.) N. H. Holmgren]—Ellis's shootingstar. Plants 10-30 cm; leaf blades broadly elliptic to ovate, base abruptly transitioning to petiole, margins sinuate to dentate, surfaces glabrous; inflorescence 1-6-flowered, bracts 3-8 mm; calyx glabrous, tube 2-3 mm; corolla 13-20 mm, lobes white with thin red throat ring; filaments distinct, 0.6-1 mm, yellow or cream; capsules 5-13 mm, tan to light brown. Moist shady slopes in oak and coniferous woodlands, 7875-10,170 ft. Torrance.

RANUNCULACEAE – CROWCUP/BUTTERCUP FAMILY

Kenneth D. Heil and Alan T. Whittemore

Perennial or annual herbs, sometimes woody or herbaceous climbers; leaf blades entire, toothed, or lobed, simple or variously compound, basal and/or cauline, alternate or opposite; inflorescences terminal or axillary, racemes, cymes, umbels, panicles, or spikes, or flowers solitary; flowers actinomorphic or zygomorphic, bisexual (sometimes unisexual in *Thalictrum* and *Clematis*), inconspicuous or showy; sepals 3-6(-20), often petaloid and colored, occasionally spurred; petals 0-26, greenish, white, blue, purple, yellow, pink, or red, stamens 5 to many, ovary superior; fruit an aggregate of cypselae or follicles (rarely utricles, capsules, or solitary berry) (Parfitt & Whittemore, 2013).

1. Flowers zygomorphic, blue to white; pistils 3(-5) per flower (1 in *Consolida*); fruits several-seeded follicles…(2)

1.   Flowers actinomorphic, blue, white, yellow, green, or red; pistils 1 or 4 to many per flower (2–6 in *Thalic-trum alpinum*); fruit an aggregate of 1-seeded cypselae or utricles, or several-seeded follicles or a berry …(4)

2.   Perennials; petals 2 or 4; pistils 3(–5)…(3)

2'.   Annuals; petals 2; pistils 1…**Consolida**

3.   Upper (adaxial) sepal hood- or helmet-shaped; petals completely hidden by colored calyx…**Aconitum**

3'.   Upper (adaxial) sepal spurred; petals at least partly exserted from colored calyx…**Delphinium**

4.   Flowers in congested, leafless racemes; pistil 1 per flower; fruit (ours) a berry…**Actaea**

4'.   Flowers in leafy cymes, umbels, panicles, or corymbs, or flowers solitary (open leafy raceme in some *Thalictrum* spp.); pistils 2 to many per flower; fruit an aggregate of achenes, utricles, or follicles…(5)

5.   Plants annual, scapose, from a slender taproot; flowers solitary, relatively inconspicuous; fruit an aggregate of achenes, many (> 20) per flower…(6)

5'.   Plants usually perennial, aerial (or aquatic) stems elongate, leafy (or if scapose or nearly so and perennial, then inflorescence a raceme or flowers showy with yellow petals); flowers few to many per stem, or if solitary then showy; fruits achenes, utricles, follicles, or a berry, 1 to many per flower…(7)

6.   Plants gray-tomentose; leaves ternate to biternate with linear lobes; cypselae in a globose to thick-cylindrical aggregate…**Ceratocephalus**

6'.   Plants glabrous; leaves entire, linear or very narrowly oblanceolate; cypselae in a long, tapered, slender aggregate…**Myosurus**

7.   Petals prominent, spurred; fruit an aggregate of follicles, 5 per flower…**Aquilegia**

7'.   Petals (if present) planar, sometimes with a cupped nectary near the base; fruit an aggregate of achenes, utricles, or follicles…(8)

8.   Fruit a several-seeded aggregate of follicles, 4–15 per flower…**Caltha**

8'.   Fruit an aggregate of 1-seeded cypselae or utricles, usually many per flower (2–6 in *Thalictrum alpinum*; 15–16 in *Trautvetteria*)…(9)

9.   Sepals valvate, 4; leaves all cauline and opposite; stems ± woody vines, or if not woody then sepals petaloid, violet-blue or pink (rarely white), > 25 mm…**Clematis**

9'.   Sepals imbricate, 3–9, if sepals 4 then < 15 mm and/or not violet-blue or pink; leaves basal, or basal and cauline, alternate (leaflike involucral bracts alternate, opposite, or whorled); stems herbaceous…(10)

10.   Plants without whorled or paired involucral bracts; cauline leaves (if present) alternate (rarely a pair of opposite, unlobed leaves in *Ranunculus alismifolius* and *R. flammula*, with green sepals and showy yellow petals)…(13)

10'.   Plants with whorled or paired (opposite) involucral bracts, these leaflike; leaves basal; (sepals petaloid, white, violet-blue, or rarely white)…(11)

11.   Sepals 20–40 mm, usually lavender or blue-purple; persistent style of the achenes plumose, 20–40 mm; stamens numerous, 150–200 per flower; plants softly villous throughout…**Pulsatilla**

11'.   Sepals 5–20(25) mm, white, green, yellow, blue, purple, or pink; persistent style of the achenes glabrous, 0.5–6 mm; stamens 40–100 per flower; plants variously hairy or glabrous…(12)

12.   Stem or involucre (subtending the inflorescence) leaves sessile, not narrowed to a petiole; achenes winged, strigose to glabrous…**Anemonastrum**

12'.   Stem or involucre leaves petiolate or narrowed to a widened petiole at the base (the petiole may be very short); achenes not winged, woolly to tomentose or villous…**Anemone**

13.   Petals absent; inflorescences panicles, racemes, or corymbs…(14)

13'.   Petals present, white or yellow; inflorescences simple or compound cymes or flowers solitary…(15)

14.   Leaves pinnately or ternately compound, leaflets < 3 cm wide; flowers commonly unisexual (bisexual in scapose, racemose *Thalictrum alpinum*)…**Thalictrum**

14'.   Leaves simple, 8–30(–40) cm wide; flowers bisexual…**Trautvetteria**

15.   Achene wall thick and firm, smooth (with coarse transverse ridges in *R. aquatilis*)…**Ranunculus**

15'.   Achene wall papery, longitudinally nerved…(16)

16.   Stems erect, without stolons; leaves biternately or bipinnately compound; cypselae cylindrical, not flattened…**Cyrtorhyncha**

16'.   Stems dimorphic, with erect or ascending flowering stems and prostrate stolons; leaves simple and undivided, margins crenate or crenate-serrate; cypselae discoid to flattened-lenticular…**Halerpestes**

ACONITUM L. – Monkshood

Aconitum columbianum Nutt.—Columbian monkshood. Perennial herbs; stems 2–30 dm; leaf blades 5–20 cm wide, deeply 3–5(–7)-divided; inflorescence terminal, sometimes also axillary, 2–32+-flowered racemes or panicles; flowers zygomorphic, commonly blue, sometimes white, 18–50 mm, pendent sepals 5, 6–16 mm, upper sepal (hood) conic-hemispherical, hemispherical, or crescent-shaped, 11–34 mm from receptacle to top of hood; petals 2–5, blue to whitish; fruit 12–20 mm. Ours are **subsp.**

columbianum. Moist places in riparian areas, meadows, forests of alder-willow, pine–Douglas-fir, oak, aspen, spruce-fir, 7500–12,960 ft. Widespread except E, SW.

ACTAEA L. – Baneberry

Actaea rubra (Aiton) Willd.—Red baneberry. Perennial herbs; leaves compound, with leaflets abaxially glabrous or pubescent, blades broadly ovate to reniform in outline, 1–3 times ternate or 1–3 times pinnate; inflorescences usually terminal, 20–50-flowered racemes; flowers actinomorphic; sepals 3–5, whitish green; petals acute to obtuse at apex, 4–10, cream-colored; stigma nearly sessile, 0.7–1.2 mm diam. during anthesis, much narrower than ovary; fruit a red or white berry, widely ellipsoid to spherical, 5–11 mm; seeds 2.9–3.6 mm. Moist, shaded areas in forests of aspen, ponderosa pine, Douglas-fir, spruce, spruce-fir, riparian areas, 6950–12,120 ft. Widespread except E, SW.

ANEMONASTRUM Holub – Anemone

Anemonastrum canadense (L.) Mosyakin [*Anemone canadensis* L.]—Canada anemone. Aerial shoots (15–)20–80 cm, from caudices on rhizomes; leaves (basal) 1–5, simple, deeply divided, blades orbiculate, margins serrate and incised on distal 1/3–1/2, apex acuminate, surfaces puberulous, lateral segments again lobed or parted; inflorescences 1(–3+)-flowered, rarely cymes; involucral leaves 3; flower sepals (4)5(6), white; fruit a head of achenes, spherical to ovoid. Damp thickets, meadows, wet prairies, lakeshores, streamsides, clearings, occasionally swampy areas, 5855–9500 ft. NC, C (Ackerfield, 2022).

ANEMONE L. – Windflower

Perennial herbs with rhizomes, caudices, or tubers; leaves basal, simple or compound, blades reniform to obtriangular or lanceolate in outline; inflorescence a terminal 2–9-flowered cyme or an umbel, or flower solitary; flowers actinomorphic, bisexual, sepals 4–20(–27), white, purple, blue, green, yellow, pink, or red, petals usually absent (reduced petals in *A. patens*); fruit an aggregate, sessile or stalked head of achenes.

1. Sepals (18–)20–40 mm; involucral leaves connate, sessile; achene beak plumose, 20–40 mm (on mature fruit)…Go to **Pulsatilla**
1′. Sepals 5–12(–15) mm; involucral leaves distinct, petioled or sessile; achene beak glabrous, 0.3–6 mm (on mature fruit)…(2)
2. Basal leaves simple (often divided)…Go to **Anemonastrum**
2′. Basal leaves compound…(3)
3. Aerial shoots 10–30(–40) cm, from caudexlike tubers, tubers ascending to vertical; S New Mexico, < 6200 ft.…**A. tuberosa**
3′. Aerial shoots 10–70(–80) cm, from caudices, rarely with short-ascending rhizomes, caudices ascending to vertical; N New Mexico, White and Sacramento Mountains (*A. cylindrica*), > 6500 ft.…(4)
4. Involucral leaves clearly petioled, petioles terete, often with an adaxial groove, uninterrupted portion of blades (4–)6+ mm wide…**A. cylindrica**
4′. Involucral leaves sessile, or if petioled, then petioles flattened or winged at least distally, uninterrupted portion of blades 1.5–3(–4.3) mm wide or less…**A. multifida**

Anemone cylindrica A. Gray—Long-headed anemone. Stems (14–)28–70(–80) cm; leaves (basal) (2–)5–10(–13), ternate; terminal leaflet broadly rhombic to oblanceolate, deeply incised on distal 1/2, strigose, lateral leaflets 1–2 times parted and lobed; inflorescence a 2–8-flowered cyme or flower solitary, sometimes appearing umbel-like; involucral leaves 3–7(–9), distinct, ± similar to basal leaves; flower sepals 4–5(6), green to whitish, 5–12(–15) × 3–6 mm; petals absent; fruit a head of cylindrical achenes, beak usually recurved. Montane forest, spruce-fir, 6560–9000 ft. NE, NC, C, Grant.

Anemone multifida Poir. [*A. globosa* Nutt. ex A. Heller]—Cut-leaved anemone. Stems 3–70 cm; leaves (basal) 3–8(–10), 1–2-ternate, terminal leaflet broadly and irregularly rhombic to obovate in outline, deeply divided into narrow lobes, lateral leaflets (2)3 times parted; inflorescence a 2–7-flowered cyme or flower solitary; involucral leaves usually 3–5, distinct, ± similar to basal leaves; flower sepals 5–9, usually purple, red, or yellow and red, sometimes green to yellow, white, or possibly blue, 5–12 × (3.5–)

5-7(-9) mm; petals absent; fruit a spherical head of achenes. Rock outcrops, gravelly hills and ridges, montane meadows, aspen, spruce-fir, alpine, 7910-12,500 ft. NC.

Anemone tuberosa Rydb.—Desert anemone. Aerial shoots 10-30(-40) cm, from caudexlike tubers, tubers ascending to vertical; leaves (basal) 1-3(-5), 12-ternate, terminal leaflet mostly sessile, irregularly oblanceolate, (1.5-)2-3(-3.5) × 1-2(-2.5) cm, lateral leaflets 1-2 times parted and/or lobed; inflorescences 2-3(-5)-flowered cymes or flowers solitary; involucral leaves primarily 3, (1)2-tiered, simple, ± similar to basal leaves; flower sepals 8-10, pink to white, 10-14(-20) × (2-)3-5(-6) mm, sparsely hairy; fruit a head of achenes. Rocky slopes, streamsides, 2935-6200 ft. SC, SW, Eddy.

AQUILEGIA L. - Columbine

Perennial herbs with slender woody rhizomes; leaves basal and cauline, alternate, blades broadly ovate to reniform in outline; inflorescence an open, terminal, 1-10-flowered cyme; flowers actinomorphic; sepals 5, white to blue, yellow, or red; petals 5, white to blue, yellow, or red, each basally forming a backward-pointing tubular spur; fruit an aggregate, sessile, cylindrical follicle.

1. Sepals and spurs blue, white, or sometimes pink, no trace of yellow; flowers erect; spurs slender…**A. coerulea**
1'. Sepals and spurs yellow or red; flowers erect or nodding; spur shape various…(2)
2. Sepals red (at least proximally); spurs red (at least proximally), stout, abruptly narrowed near middle, 12-23 mm; flowers nodding…(3)
2'. Sepals and spurs yellow; spurs slender, evenly tapered from base, 30-65 mm; flowers erect or nodding…(4)
3. Sepals red proximally, yellow-green distally, not much longer than petal blades; stamens 8-14 mm…**A. elegantula**
3'. Sepals red or apex yellow-green, ca. 2 times length of petal blades; stamens 14-19 mm…**A. desertorum**
4. Spurs 42-65 mm; flowers erect…**A. chrysantha**
4'. Spurs 30-40 mm; flowers suberect to inclined…**A. chaplinei**

Aquilegia chaplinei Standl. ex Payson [*A. chrysantha* A. Gray var. *chaplinei* (Standl. ex Payson) E. J. Lott]—Chaplin's columbine. Stems 20-50 cm; leaves (basal) 2-3 times ternately compound, 7-25 cm, much shorter than stems, leaflets to 9-19 mm, not viscid; flowers suberect to inclined; sepals pale yellow, broadly lanceolate, 9-19 × 4-6 mm; petal spurs yellow, straight, ± parallel or divergent, 30-40 mm, slender, evenly tapered from base; stamens 10-19 mm; follicles 18-22 mm. Limestone seeps in montane scrub and riparian canyon bottoms; Guadalupe and S Sacramento Mountains, 4700-5500 ft. **R**

Aquilegia chrysantha A. Gray—Golden columbine. Stems 30-120 cm; leaves (basal) 2-3 times ternately compound, 9-45 cm, much shorter than stems, not viscid; flowers erect; sepals yellow, 20-36 × 5-10 mm; petal spurs yellow, straight, ± parallel or divergent, 42-65 mm, slender, evenly tapered from base; follicles 18-30 mm. Damp places in canyons, 4000-9000 ft. NC, C, WC, SE, SC, SW.

Aquilegia coerulea E. James—Colorado blue columbine, Rocky Mountain columbine. Stems 15-80 cm; leaves (basal) 2-3-ternate, 9-37 cm, much shorter than stems; leaflets to 13-42(-61) mm, not viscid; flowers erect; sepals white to medium or deep blue or pink, 28-51 × 8-26 mm; petal spurs white, blue, or sometimes pink, (25-)34-70 mm, slender, evenly tapered from base; stamens 13-19 mm; fruit 20-30 mm.

1. Sepals medium to deep blue; petal spurs 34-48 mm (population mean 39-45 mm); stamens 13-19 mm… **var. coerulea**
1'. Sepals white, pale blue, or pink; petal spurs 45-70 mm (population mean 50-58 mm); stamens 18-24 mm…**var. pinetorum**

var. coerulea—Colorado blue columbine. Leaves 2-ternate; sepals medium or deep blue, 28-43 mm; petal spurs 34-48 mm, petal blades (17-)20-24 mm; stamens 13-19 mm. Rocky slopes or near streams, in open woodland or herbland, Gambel oak–ponderosa pine, aspen, Douglas-fir, spruce-fir, subalpine to alpine, 7300-12,960 ft. NC, NW, Grant.

var. pinetorum (Tidestr.) Payson ex Kearney & Peebles—Colorado blue columbine. Leaves 2-3-ternate; sepals white, pale blue, or pink, (22-)29-51 mm; petal spurs 43-72 mm, petal blades 20-28 mm; stamens 17-24 mm. Riparian, aspen-fir forest, 8400-8810 ft. NC, NW.

Aquilegia desertorum (M. E. Jones) Cockerell ex A. Heller [*A. formosa* Fisch. ex DC. var. *desertorum* M. E. Jones; *A. triternata* Payson]—Desert columbine. Stems 15–60 cm; leaves (basal) 2–3 times ternately compound, 7–30 cm, much shorter than stems, leaflets to 9–26(–32) mm, not viscid; flowers nodding, red or apex yellow-green, 7–20 × 3–8 mm; petal spurs red, straight, ± parallel, 16–32 mm, stout proximally, blades yellow or red and yellow, 4–12 × 3–8 mm, stamens 14–19 mm; follicles 15–30 mm. Open rocky places, 5300–10,605 ft. C, WC, SC, SW.

Aquilegia elegantula Greene—Western red columbine. Stems 10–60 cm; leaves (basal) 2-ternate, 7–30 cm, usually shorter than stems; leaflets green adaxially, to 11–33 mm, not viscid; glabrous or pilose; flowers pendent; sepals erect, red proximally, yellow-green distally, 7–11 × 4–5 mm; petal spurs red, straight, 16–23 mm, stout, abruptly narrowed near the middle; terminal blade yellow-green, 6–8 × 3–4 mm; stamens 8–14 mm; fruit 13–20 mm. Moist coniferous forests, especially along streams, ponderosa pine and subalpine spruce-fir forests, 6900–12,500 ft. Widespread except E, SC, Hidalgo.

CALTHA L. – Marsh-marigold

Caltha leptosepala DC.—White marsh-marigold. Perennial herbs with thick caudices, plants 2–30 × 0.5–2 cm; leaves usually all basal, blades oblong-ovate to orbiculate-reniform, largest 1.5–11.5(–15) × 1–13 cm, margins entire or crenate to dentate; inflorescence a terminal or axillary 2–6-flowered cyme, or flower solitary; flowers actinomorphic, 15–40 mm diam.; sepals 5–12, adaxially white, abaxially bluish, 8.5–23 mm; petals and nectary absent; fruit a follicle. Open riparian areas, wet meadows, marshy edges of lakes, montane to alpine, 8925–12,960 ft. NC.

CERATOCEPHALUS Moench – Butterwort

Ceratocephalus testiculatus (Crantz) Roth [*C. orthoceras* DC; *Ranunculus testiculatus* Crantz]—Bur-buttercup. Stems erect or ascending, not rooting nodally, villous, not bulbous-based; tuberous roots absent; basal leaf blades broadly spatulate in outline, 1–2 times dissected, 0.9–3.8 × 0.5–1.5 cm, segments linear, margins entire, apex obtuse to acuminate; flower receptacles glabrous; sepals spreading, 3–6 × 1–2 mm, villous; petals yellow, 3–5 × 1–3 mm; heads of cypselae cylindrical, 9–16(–27) × 8–10 mm; cypselae 1.6–2 × 1.8–2 mm, tomentose; beak persistent, lanceolate, 3.5–4.5 mm. Disturbed areas, especially in grassland and sagebrush communities, 5225–8700 ft. NW, NC. Introduced.

CLEMATIS L. – Clematis

Perennial herbs or ± woody vines, sometimes woody only at base; stems erect or prostrate to climbing by means of tendril-like petioles; leaves opposite, simple or compound; inflorescence an axillary and/or terminal 2- to many-flowered cyme or panicle, or flowers solitary or in a fascicle; flowers actinomorphic, bisexual or unisexual; sepals 4(5), white, blue, violet, red, yellow, or greenish; petals absent; stamens many; fruit an achene.

1. Sepals ± thick, leathery, connivent proximally and usually much of length; perianth bell- to urn-shaped, blue to violet...(2)
1′. Sepals thin, spreading, not connivent; perianth widely bell-shaped to rotate, violet blue, greenish yellow, or white to cream...(4)
2. Leaf blades 1–2-pinnate and many simple...**C. pitcheri**
2′. Larger leaf blades (1)2–3-pinnate, ternate, or deeply dissected...(3)
3. Leaflets usually < 1.5 cm wide, mostly > 2.5 times as long as wide, mostly unlobed; blades abaxially sparsely to densely hirsute; beak plumose...**C. hirsutissima**
3′. Leaflets usually > 1.5 cm wide and/or < 2.5 times as long as wide; blades glabrous or near so; beak glabrous or inconspicuously appressed-pubescent...**C. bigelovii**
4. Staminate flowers with petaloid staminodes between stamens and sepals; perianth widely bell-shaped or tardily rotate...**C. columbiana**
4′. Staminate flowers without staminodes between stamens and sepals; perianth rotate, sepals wide-spreading or recurved at least toward the tip...(5)
5. Sepals greenish yellow, ascending or wide-spreading and recurved...**C. orientalis**
5′. Sepals white to cream, wide-spreading, not recurved...(6)

6. Leaflets deltate to ovate, strongly 3-parted to 3-sect, segments ovate, deltate, or linear; achene beak 4–9 cm...**C. drummondii**
6'. Leaflets ovate to lanceolate, variously lobed or toothed, but without narrow segments; achene beak 3–3.5 cm...**C. ligusticifolia**

Clematis bigelovii Torr.—Bigelow's clematis. Stems erect or sprawling; leaf blades 1–2(3)-pinnate, primary leaflets 7–11; inflorescences terminal, 1-flowered; flowers broadly urn- to bell-shaped; sepals purple, 1.5–3 cm, margins narrowly expanded distally to ca. 1 mm wide; cypselae appressed-long-pubescent, beak 2–3 cm. Mountain slopes, moist sites in canyons, 5400–8850 ft. C, SC, SW, Sandoval.

Clematis columbiana (Nutt.) Torr. & A. Gray [*C. pseudoalpina* (Kuntze) A. Nelson]—Rock clematis. Stems ± woody, viny, climbing or trailing; leaf blades 2–3-ternate, leaflets usually very deeply 2–3-lobed; inflorescence terminal on shoots, 1-flowered; flowers bisexual, ± nodding, perianth widely bell-shaped to rotate; sepals not connivent or connivent only in proximal 1/4, violet-blue (rarely white), 25–60 mm; petaloid staminodes present between stamens and sepals; cypselae flattened, beak 3.5–5 cm. Our material belongs to **var. columbiana**. Riparian areas, piñon-juniper, ponderosa pine, oak, Douglas-fir, aspen, spruce-fir, 6235–10,900 ft. Widespread except E, SW.

Clematis drummondii Torr. & A. Gray [*C. nervata* Benth.]—Drummond's clematis. Stems scrambling to climbing, with tendril-like petioles and leaf rachises, 4–5+ m; leaf blades odd-pinnate, usually 5-foliate; inflorescences usually axillary, 3–12-flowered simple cymes or compound with central axis or flowers solitary; flowers unisexual, sepals wide-spreading, not recurved, white to cream; cypselae elliptic to ovate, beak 4–9 cm. Chaparral, xeric scrub, oak and grasslands, often along streams, 3300–6700 ft. WC, S.

Clematis hirsutissima Pursh [*C. hirsutissima* var. *arizonica* (A. Heller) R. O. Erickson]—Hairy clematis. Herbaceous perennials; stems erect, not branched or viny; leaf blades 2–3-pinnate; leaflets often deeply 2- to several-lobed; inflorescence terminal, 1-flowered; flowers bisexual, nodding, perianth broadly cylindrical to urn-shaped; sepals thick, usually leathery, connivent at least proximally and usually much of length, very dark violet-blue or rarely pink or white; cypselae flattened, beak 4–9 cm.

1. Leaflets and lobes linear to narrowly lanceolate, 0.5–9(–10) mm wide...**var. hirsutissima**
1'. Leaflets and lobes narrowly to broadly lanceolate to ovate, 5–15 mm wide...**var. scottii**

var. hirsutissima—Hairy clematis. Stems generally simple, erect; primary leaflets 3–7 or not distinctly differentiated, leaflets and larger lobes narrowly linear to narrowly lanceolate, 1–6 cm × 0.5–6 (–10) mm; surfaces nearly glabrous to densely silky-hirsute. Moist mountain meadows, prairies, open woods and thickets, 5700–10,900 ft. NC, NW, C, SC, Grant.

var. scottii (Porter) R. O. Erickson [*C. scottii* Porter]—Scott's clematis. Stems simple or branched, erect or ± sprawling; primary leaflets 7–13, leaflets and larger lobes narrowly to broadly lanceolate or ovate, 1–6 cm × 5–15 mm, surfaces nearly glabrous to sparsely hirsute. Dry to moist mountain meadows, thickets, rocky slopes, 6900–8555 ft. NE, NC.

Clematis ligusticifolia Nutt.—Virgin's-bower. Stems woody, viny, clambering or climbing; leaf blades pinnately 5-foliate, 2-ternately 9-foliate, or bipinnately 8–15-foliate; inflorescence axillary on current year's stems, usually 7–30(–65)-flowered compound cymes; flowers unisexual, staminate and pistillate on different plants, not nodding; sepals wide-spreading, not recurved, not connivent, white to cream; cypselae flattened or nearly terete, beak 3–4.5 cm. Piñon-juniper, montane or riparian woodland and forest edges, canyon bottoms, hanging gardens, 3900–9565 ft. Widespread except EC, SE.

Clematis orientalis L.—Oriental virgin's-bower. Stems climbing, 2–8 m; leaf blades pinnately 5–7-foliate, proximal leaflets sometimes 3-foliate; inflorescences axillary, sometimes terminal, 3- to many-flowered cymes or flowers solitary; flowers bisexual; sepals wide-spreading and recurved, greenish yellow; staminodes absent; cypselae turgid, beak 2–5 cm. Roadsides, other secondary habitats, open woods, 6400–6600 ft. San Miguel.

Clematis pitcheri Torr. & A. Gray—Bellflower clematis. Stems viny, to 4 m, very sparsely short-pilose, sometimes nearly glabrous; leaf blades usually 1-pinnate, distal-most leaf occasionally simple; primary leaflets 2–8 plus additional tendril-like terminal leaflet; inflorescences axillary, 1–7-flowered; sepals pale to dark bluish or reddish purple, sometimes whitish toward tip, mostly 4–11 × 1.5–6 cm; cypselae appressed-pubescent, beak 1–3 cm. Limestone outcrops in dry to moist woods and thickets, disturbed sites, 3800–8500 ft. SE, SC except Lea.

CONSOLIDA (DC.) Gray – Knight's-spur

Consolida ajacis (L.) Schur [*Delphinium ajacis* L.]—Doubtful knight's-spur. Herbs, annual; stems 3–8(-10) dm, glabrous to sparsely puberulent; leaves all cauline, alternate, leaf blades palmately finely dissected, ± semicircular with orbiculate lobes glabrous to puberulent, lobes < 1.5 mm wide; inflorescences 6–30(75)-flowered, simple or with 3 or fewer branches; sepals 5, the upper one spurred, 4–20 mm, blue to purple, rarely pink or white; petals 2, connate, of same color as sepals or whiter; follicles 12–25 mm. Drainage ditches and roadsides, 4000–4400 ft. Widely scattered. Colfax, Curry, Eddy, Grant, Roosevelt.

CYRTORHYNCHA Nutt. – Tadpole buttercup

Cyrtorhyncha ranunculina Nutt. [*Ranunculus ranunculinus* (Nutt.) Rydb.]—Tadpole buttercup. Stems erect, never rooting nodally; basal and lower cauline leaves ovate to semicircular in outline, biternately or bipinnately compound; upper cauline leaves twice compound; sepals 3–6 × 1–3 mm, glabrous; petals (0-)5–6, yellow, 3–8 × 1–3 mm; style present; heads of cypselae hemispherical to globose, 4–5 × 6–7 mm; cypselae cylindrical, not flattened, wall papery, longitudinally nerved, beak filiform, strongly reflexed from base, 0.8–1.5 mm, brittle, often broken. Open, grassy or brushy slopes in the mountains, 6765–9200 ft. Colfax, Los Alamos, San Miguel, Santa Fe.

DELPHINIUM L. – Larkspur

Perennial herbs; leaves basal and/or cauline, alternate; blades round to pentagonal or reniform in outline, deeply palmately divided; inflorescence terminal, 2–100+-flowered racemes or few-branched panicles; flowers zygomorphic; sepals 5, usually blue to purplish (rarely white to pink); lateral and lower sepals 4, upper sepal 1, spurred; petals 5, upper petals 2, often white, spurred, bearing a nectary concealed inside sepal-spur; lower petals 2, often colored as the sepals; fruit a cylindrical follicle. Roots are needed for definitive identification.

1. Roots very easily separable from stems; if not extracted, roots typically breaking cleanly from stem and often absent from herbarium specimens…(2)
1ʹ. Roots not easily separable from stems; roots typically breaking raggedly, leaving portion attached to lower stem…(3)
2. Primary root segments ± succulent, brittle, roots with single primary segment (cormlike); N New Mexico …**D. nuttallianum**
2ʹ. Primary root segments dry, braided, tough, major root branches at least 1 cm from stem attachment; W New Mexico…**D. scaposum**
3. Seeds with transverse wavy ridges visible without magnification; pedicels appressed-ascending…**D. wootonii**
3ʹ. Seeds lacking transverse wavy ridges visible without magnification; pedicels rarely appressed-ascending …(4)
4. Proximal internodes much shorter than those of midstem; leaves largest near base of stem (but sometimes absent at anthesis), often abruptly reduced upward…(5)
4ʹ. Proximal internodes similar in length to those of midstem; leaves largest near midstem, gradually reduced upward…(6)
5. Midstems and leaf blades pubescent; sepals dark blue to purple…**D. geraniifolium**
5ʹ. Midstems and leaf blades glabrous to subglabrous; sepals bright dark blue…**D. scopulorum**
6. Leaves present on proximal 1/5 of stem at anthesis…(7)
6ʹ. Leaves absent from proximal 1/5 of stem at anthesis…(8)
7. Stems < 30 cm…**D. alpestre**

7'.   Stems > (45–)70 cm…**D. ramosum** (in part)
8.   Sepals brownish, yellowish, or purple…(9)
8'.   Sepals blue or purple, rarely white or pink, not brownish or yellowish (although sometimes so in press)…(10)
9.   Sepals (in bud) yellowish or brownish purple…**D. sapellonis**
9'.   Sepals (in bud) purple to lavender…**D. novomexicanum**
10.   Hairs in inflorescence gland-based…**D. barbeyi**
10'. Hairs in inflorescence (if present) not gland-based…(11)
11.   Stems finely, evenly puberulent throughout…**D. ramosum** (in part)
11'. Stems glabrous…**D. robustum**

Delphinium alpestre Rydb. [*D. ramosum* Rydb. var. *alpestre* (Rydb.) W. A. Weber]—Alpine larkspur. Stems 5–25 cm, base green, puberulent; leaves cauline, 5–20, on proximal 1/5 of stem at anthesis, blades round to pentagonal; inflorescences 2–8-flowered, puberulent; flower sepals dark blue, puberulent, spurs straight, 8–12 mm; fruits 7–12 mm. Exposed talus slopes on high peaks, alpine tundra, subalpine meadows, 11,500–13,000 ft. Colfax, Taos. **R**

Delphinium barbeyi (Huth) Huth—Subalpine larkspur, tall larkspur, Barbey's larkspur. Roots twisted, fibrous, dry; stems 50–150 cm; base green, glabrous or sparsely pubescent; leaves cauline, usually absent from lower 1/5 of stem at anthesis; midstem blades lacerate, 7–15 cm broad; inflorescence a few-branched panicle or raceme; pedicels densely golden-velutinous with finely glandular hairs; flower sepals dark bluish purple, sparsely puberulent; fruit 22 mm. Subalpine and alpine, wet soils, riparian areas, ponderosa pine, Douglas-fir, spruce-fir, tundra, scree slopes, 8450–12,500 ft. NC, C.

Delphinium geraniifolium Rydb.—Mogollon larkspur. Stems 60–100 cm, base reddish, puberulent, midstems pubescent; leaves mostly on proximal 1/3 of stem, basal leaves (0–)2–7 at anthesis, cauline leaves 12–20 at anthesis; inflorescences 20–90-flowered, dense, cylindrical; flower sepals dark blue to purple, puberulent, 10–14 × 3–5 mm, spurs 12–15 mm; fruits 13–18 mm. Heavy clay soil, dry meadows in coniferous woods, 5900–9500 ft. NC, SC, SW, San Juan.

Delphinium novomexicanum Wooton [*D. sierrae-blancae* Wooton]—New Mexico larkspur. Stems 90–180(–250) cm; leaves cauline, 12–20, absent from proximal 1/5 of stem at anthesis, leaf blades round to pentagonal, 5–10 × 8–18 cm, nearly glabrous; inflorescences (20–)30–70(–140)-flowered; flower sepals (in bud) purple to lavender, fading brownish, puberulent, spurs 7–11 mm; fruits 12–16 mm, puberulent. Meadows in coniferous forests; Sacramento, White, and Capitan Mountains, 7200–11,200 ft. Lincoln, Otero. **R & E**

Delphinium nuttallianum Pritz.—Two-lobed larkspur. Roots fascicled, tuberous; stems 11–45(–70) cm, usually long-tapered to the deep root crown; leaves basal and cauline, present on lower 1/5 of stem at anthesis, midstem blades dissected; inflorescence a raceme; flower sepals usually bluish purple (rarely white) to pink, puberulent; fruit 7–17 mm. Spruce-fir, aspen, Ponderosa pine, Gambel oak, meadow edges, piñon-juniper, sage and desert scrub, 5645–10,900 ft. Widespread in N, W.

Delphinium ramosum Rydb.—Mountain larkspur. Stems (45–)70–100 cm; base sometimes reddish, puberulent; leaves cauline, 8–24, absent from or present on proximal 1/5 of stem at anthesis, blades round to pentagonal, 2–8 × 4–14 cm; inflorescences (10–)15–40(–120)-flowered; flower sepals bright dark blue, puberulent, spurs 9–13 mm; fruits 11–17 mm, puberulent. Spruce-fir forests, aspen woodlands, montane meadows, ponderosa pine woodlands, sagebrush scrub, 6450–10,850 ft. NE, NC. *Delphinium ramosum* hybridizes with *D. barbeyi*.

Delphinium robustum Rydb.—Robust larkspur. Stems 100–200(–250) cm; base sometimes reddish, glabrous, glaucous; leaves cauline, 12–22, absent from proximal 1/5 of stem at anthesis, blades round to pentagonal, 7–12 × 10–20 cm, nearly glabrous; inflorescences 40–90(–180)-flowered; flower sepals bluish purple to pale lavender, nearly glabrous, spurs 10–13 mm; fruits 13–18 mm, puberulent. Riparian woodlands, subalpine meadows, 7200–11,200 ft. NC. **R**

Delphinium sapellonis Tidestr.—Sapello Canyon larkspur. Stems (50–)100–180(–220) cm; base sometimes reddish, glabrous, sometimes glaucous; leaves cauline, 10–20, absent from proximal 1/5 of stem at anthesis; blades round to pentagonal, 6–10 × 8–16 cm, nearly glabrous; inflorescences (12–)30–80(–120)-flowered; flower sepals (in bud) yellowish or brownish purple, becoming browner or yellower with age, glandular-puberulent, spurs 8–11 mm; fruits 12–18 mm. Subalpine meadows and open coniferous forest, 7300–12,075 ft. NC, C. **E**

Delphinium scaposum Greene [*D. andersonii* A. Gray var. *scaposum* (Greene) S. L. Welsh]—Tall mountain larkspur. Roots (at least the main one) elongate, fibrous, dry; stems 17–61 cm, glabrous; leaves all basal or nearly so; leaves present on lower 1/5 of stem (or less) at anthesis, midstem blades absent; inflorescence a raceme (panicled in 1 individual); flower sepals bright dark blue (rarely ± pale), glabrous to very sparsely hairy; fruit 12–16 mm. Desert scrub, sagebrush, grassland, piñon-juniper, ponderosa-piñon woodland, 3900–7500 ft. W.

Delphinium scopulorum A. Gray [*D. macrophyllum* Wooton]—Rocky Mountain larkspur. Stems 50–120 cm, base often reddish, puberulent, midstems glabrous to subglabrous; leaves mostly on proximal 1/3 of stem, basal leaves (0–)3–7 at anthesis, cauline leaves 6–15 at anthesis, blades round to pentagonal, 1.5–10 × 2–16 cm, nearly glabrous; inflorescences 10–30-flowered; flower sepals bright dark blue, nearly glabrous, spurs 15–20 mm; fruits 16–20 mm, nearly glabrous. Riparian forests and open woodlands, 6380–9800 ft. NC.

Delphinium wootonii Rydb. [*D. virescens* Nutt. subsp. *wootonii* (Rydb.) Ewan]—Wooton's larkspur. Stems (15–)30–50(–60) cm; base sometimes reddish, pubescent; leaves mostly on proximal 1/4 of stem at anthesis, basal leaves 0–10 at anthesis, cauline leaves 3–10 at anthesis, blades reniform to fan-shaped, 1.5–3 × 2.5–4 cm, puberulent; inflorescences 15–30(–49)-flowered; flower sepals white to lavender, nearly glabrous, spur (8–)13–20(–25) mm; fruits (10–)11–20(–24) mm. Oak woods, grasslands, desert scrub, 3300–7350 ft. NE, NC, C, SE, SC, SW.

HALERPESTES Greene – Alkali buttercup

Halerpestes cymbalaria (Pursh) Greene [*Ranunculus cymbalaria* Pursh]—Alkali buttercup. Perennial herbs; stems dimorphic, flowering stems erect or ascending, stolons prostrate, rooting nodally; basal leaves oblong to cordate or circular, simple and undivided, bases rounded to cordate, margins crenate or crenate-serrate, apices rounded; upper cauline leaves small and undivided; sepals 5, yellow-green, 2.5–6 × 1.5–3 mm, glabrous; petals 5, yellow, 2–7 × 1–3 mm; style present; heads of cypselae long-ovoid or cylindrical, 6–12 × 4–5(–9) mm; cypselae discoid to flattened-lenticular, oblong to obovate in outline, moderately flattened, wall papery, longitudinally nerved, beak conic, straight, 0.1–0.2 mm. Bogs, marshes, ditches, lake margins, often in saline sites, 4770–9860 ft. Widespread except EC, SE.

MYOSURUS L. – Mousetail

Annual herbs, scapose, with a slender taproot; stems absent; leaves all basal, simple, blades linear or very narrowly oblanceolate, tapering to a filiform base, entire, glabrous; inflorescence a solitary flower at the end of a scape without an involucre; flowers actinomorphic, bisexual; sepals (3–)5(–8), green or with scarious margins, lanceolate or oblanceolate, 1.5–4 mm; petals 0–5, white, 1–2.5 mm, glabrous, nectary present; fruit with sessile prismatic achenes, exposed face of achene forming a planar outer surface, sides faceted or curved by compression against adjacent achenes, 0.05–1.8 mm.

1. Outer face of achene circular to square or broadly rhombic, 1–1.3 times higher than wide; petal claw 3 times longer than blade…(2)
1'. Outer face of achene elliptic or oblong to linear, 1.5–5 times higher than wide; petal claw 1–2 times longer than blade…(3)
2. Outer face of achene bordered by prominent ridge…**M. cupulatus**
2'. Outer face of achene not bordered…**M. nitidus**
3. Beak of achene 0.05–0.4 mm, parallel to outer face of achene (strongly appressed), head of cypselae thus appearing smooth…**M. minimus**
3'. Beak of achene 0.6–1.8 mm, diverging at a narrow angle, head of cypselae thus conspicuously roughened with projecting achene beaks…**M. apetalus**

Myosurus apetalus Gay [*M. minimus* L. subsp. *montanus* G. R. Campb.]—Bristly mousetail. Plants 1.5–12.5 cm; leaves linear or narrowly oblanceolate, 0.9–4 cm; inflorescence scape 0.9–10.5 cm; flower sepals faintly 3-nerved; petal claw 1–2 times longer than blade; cypselae 11–33 × 1.5–2.5 mm, achene outer face narrowly rhombic, 1–2.2 × 0.4–1 mm, 2–5 times higher than wide, beak 0.6–1.4 mm, 0.4–1 times longer than achene body, diverging from outer face of achene, head of cypselae thus strongly roughened by projecting achene beaks. Ours are **var. montanus** (G. R. Campb.) Whittem. Wet meadows, margins of lakes and intermittent ponds, mudflats, moist, sandy soils, 4900–9600 ft. NW, WC.

Myosurus cupulatus S. Watson—Arizona mousetail. Herbs, 3.3–16 cm; leaf blades linear or very narrowly oblanceolate, 1.8–9.5 cm; inflorescence scape 2.2–12 cm; flower sepals faintly 3-veined; petal claw 1–3 times as long as blade; cypselae 13–42 × 2–3 mm, long-exserted from leaves, 0.8–1.2 × 0.6–1 mm, outer face orbiculate or sometimes square, 0.8–1.2 times as high as wide, bordered by prominent ridge, beak 0.6–1.2 mm, 0.6–1.2 times as long as achene body, heads of cypselae roughened by projecting achene beaks. Dry hillsides or canyon bottoms in shrubland communities, 3985–6600 ft. WC, SW.

Myosurus minimus L.—Tiny mousetail. Plants 4–16.5 cm; leaves narrowly oblanceolate or linear, 2.2–11.5 cm; inflorescence scape 1.8–12.8 cm; flower sepals faintly or distinctly 3–5-nerved; petal claw 1–2 times longer than blade; fruit a head of cypselae 16–50 × 1–3 mm, achene outer face narrowly rhombic to elliptic or oblong, 0.8–1.4 × 0.2–0.6 mm, 1.5–5 times higher than wide, beak 0.05–0.4 mm, 0.05–0.3 times longer than achene body, head of cypselae appearing smooth. Muddy margins of lakes, stock ponds, dry washes, greasewood, piñon-juniper communities, 3750–9560 ft. NC, NW, C, WC, SW.

Myosurus nitidus Eastw. [*M. egglestonii* Wooton & Standl.]—Western mousetail. Plants 1–4 cm; leaves linear or very narrowly oblanceolate, 0.6–2.4 cm; inflorescence scape 0.5–2.4 cm; flower sepals sometimes strongly nerved, scarious margin narrow or broad; petal claw 3 times longer than blade; fruit a head of cypselae 8–17 × 2–3 mm; achene outer face circular to rhombic, 0.8–1 × 0.6–0.9 mm, 1–1.3 times higher than wide, beak 0.7–1 mm, 0.8–1.2 times longer than outer face of achene, head of cypselae ± roughened by projecting achene beaks. Ephemeral wet places under sagebrush or in shallow depressions in bedrock, woodlands with ponderosa pine and Gambel oak, 5400–7500 ft. Cibola, Los Alamos, Rio Arriba, Sandoval, San Juan.

PULSATILLA Mill. – Pasque flower

Pulsatilla nuttalliana (DC.) Spreng. [*Pulsatilla patens* (L.) Mill.]—Pasqueflower. Stems, including inflorescence, 5–40(–60) cm; leaves (basal) (1–)3–8(–10), primarily 3-foliate, terminal leaflet obovate in outline, deeply divided into very narrow lobes, hairy (rarely glabrous), lateral leaflets 3–4 times parted; inflorescence 1-flowered, involucral leaves usually 3, connate-clasping, dissimilar to basal leaves, 1-tiered, simple, divided almost to the base, sessile; flower sepals 5–8, blue or purple to rarely nearly white, (18–)20–40 × (8–)10–16 mm; petals 1.5–2 mm; fruit a spherical to ovoid head of achenes. Meadows and clearings, sagebrush, Gambel oak, ponderosa pine, 6500–10,600 ft. NC, C.

RANUNCULUS L. – Buttercup, crowsfoot

Annual or perennial herbs; leaves simple or compound, basal, cauline, or both, if cauline then alternate; inflorescence an open terminal (seldom axillary) cyme or flower solitary; flowers actinomorphic; sepals 3–5, green or purple; petals 3–16 (usually 5), distinct, yellow (white in *R. aquatilis*), nectary basal, covered by a scale; fruit a head of achenes, achene wall thick and firm, smooth (with coarse transverse ridges in *R. aquatilis*).

1. All leaves simple and unlobed…(2)
1'. Some or all leaves deeply lobed or compound…(6)
2. Sepals covered with dense brown tomentum; distal leaves and bracts 3-crenate or shallowly 3-lobed apically, otherwise entire…**R. macauleyi**
2'. Sepals glabrous or with colorless hairs; leaves not as above…(3)
3. Stems never rooting nodally, erect or ascending…(4)
3'. Stems rooting nodally, erect to prostrate…(5)

4. Petals 2–3 mm broad; heads of cypselae 3–7 mm…**R. alismifolius**
4'. Petals 5–12 mm broad; heads of cypselae 7–12(–20) mm…**R. glaberrimus** (in part)
5. Leaves ovate to broadly ovate, 8–19 mm wide, bases rounded to weakly cordate; beaks of cypselae 0.4–1 mm…**R. hydrocharoides**
5'. Leaves lance-elliptic to lanceolate or linear, 2–8 mm wide, bases acute to filiform; beaks of cypselae 0.1–0.6 mm…**R. flammula**
6. Stems creeping and rooting at nodes, or floating in water (then rootless)…(7)
6'. Stems upright, or if decumbent then rooting only at the base…(10)
7. Basal leaves 3-foliate; terrestrial, sometimes emergent from shallow water…(8)
7'. Basal leaves simple; floating in water or reclining on very wet soil…(9)
8. Petals 6–18 mm, much longer than sepals; achene beak curved…**R. repens**
8'. Petals 4–6 mm, about as long as sepals; achene beak straight or nearly so…**R. macounii** (in part)
9. Petals white or white with yellow claws; cypselae with strong transverse ridges; stems floating in water; leaves completely dissected into filiform segments…**R. aquatilis**
9'. Petals yellow; cypselae smooth; stems reclining or sometimes floating; leaves 3-parted, segments again 1–3 times lobed or dissected…**R. gmelinii**
10. Style absent, stigma sessile; achene margins thick and corky; emergent aquatic, sometimes also found on very wet soil…**R. sceleratus**
10'. Style present; achene margins ridges or narrow wings, or else differentiated margins not evident; wet or dry soil (*R. macounii* sometimes emergent from shallow water)…(11)
11. Cypselae strongly flattened, discoid, 3–15 times as wide as thick; basal leaves deeply 3–5-parted or 3-foliate…(12)
11'. Cypselae weakly compressed, thick-lenticular, no more than as wide as thick; basal leaves simple, lobed, or divided (sometimes compound in *R. eschscholtzii*)…(16)
12. Petals 8–21 mm…(13)
12'. Petals 2–6 mm…(14)
13. Petals 11–16; basal leaves 3(–5)-foliate, ovate to deltoid in outline…R. fasciculatus
13'. Petals 5; basal leaves deeply 3–5-parted but not compound, pentagonal in outline…**R. acris**
14. Achene beak 1.2–2.5 mm, strongly curved; basal leaves 3-parted (rarely 3-foliate)…**R. uncinatus**
14'. Achene beak 0.6–1.2 mm, straight; basal leaves 3-foliate…(15)
15. Petals 2–4 mm, 1–2.5 mm wide; heads of cypselae cylindrical, 5–7 mm wide…**R. pensylvanicus**
15'. Petals 4–6 mm, 3.5–5 mm wide; heads of cypselae globose to ovoid, 7–10 mm wide…**R. macounii** (in part)
16. All basal leaves deeply lobed or compound…**R. eschscholtzii**
16'. Some or all basal leaves simple and unlobed…(17)
17. Heads of cypselae globose; stems prostrate or ascending; basal leaves entire (rarely with 3 broad, shallow crenae)…**R. glaberrimus** (in part)
17'. Heads of cypselae ovoid or cylindrical; stems erect or nearly so; basal leaves crenate (with > 3 crenae), the innermost leaves sometimes lobed or parted…(18)
18. Basal leaves all (except sometimes innermost) shed before anthesis, leaving a dense brush of fibers… **R. arizonicus**
18'. Basal leaves present at anthesis…(19)
19. Sepals glabrous; petals 1.5–3.5 × 1–2 mm…**R. abortivus**
19'. Sepals pilose; petals 4–13 × 2–13 mm…(20)
20. Sepals 3–5 × 2–3 mm; leaf base acute to rounded…**R. inamoenus**
20'. Sepals 5–8 × 3–7 mm; leaf base cordate to broadly obtuse…**R. cardiophyllus**

Ranunculus abortivus L.—Littleleaf buttercup. Perennial, 10–60 cm; stems glabrous; basal leaves undivided or the innermost 3-parted, reniform to orbiculate, 1.4–4.2 × 2–5.2 cm, crenate; cauline leaves 3–5-cleft into linear-lanceolate, entire segments; sepals 2.5–4 mm, glabrous; petals 5, yellow, 1.5–3.5 mm; cypselae 1.4–1.6 × 1–1.5 mm, with a curved beak 0.1–0.2, glabrous. Riparian sites, streams, seeps, forests, 7960–8850 ft. NC.

Ranunculus acris L.—Tall buttercup. Stems erect, never rooting nodally; basal leaves pentagonal in outline, deeply 3–5-parted, segments once or twice lobed or parted; upper cauline leaves deeply parted or compound; sepals 4–6(–9) × 2–5 mm, hispid; petals 5, yellow, 8–11(–17) × 7–13 mm; style present; heads of cypselae globose, 5–7(–10) mm wide, cypselae discoid, much thinner than broad, wall thick, firm, smooth, beak deltoid and usually with a short or long, straight or curved subulate tip, 0.2–1 mm. Meadows, stream banks; known only from the Glenwood Fish Hatchery along stream banks and pond margins, 4740 ft. Catron.

Ranunculus alismifolius Geyer ex Benth.—Plantainleaf buttercup. Stems erect or ascending, never rooting nodally; lower cauline leaves lanceolate to narrowly ovate or elliptic, undivided, bases acute to rounded-obtuse, margins entire (rarely serrulate), apices obtuse to acuminate; upper cauline leaves undivided; sepals 2-6 × 1-4 mm, glabrous or hirsute; petals 5-12, yellow, 5-9 × 2-3 mm; style present; heads of cypselae hemispherical to globose, 3-7 × 4-8 mm; cypselae globose-lenticular to globose, not or weakly flattened, wall thick, firm, smooth, beak lance-subulate, straight or weakly curved, 0.4-1.2 mm. Our material belongs to **var. montanus** S. Watson. Wet meadows or stream banks in Douglas-fir and spruce-fir communities, 8000-11,500 ft. Rio Arriba, San Juan.

Ranunculus aquatilis L.—White water crowfoot. Stems creeping or floating, rooting nodally; basal leaves none; leaves completely dissected into filiform segments; sepals 2-4 × 1-2 mm, glabrous; petals 5, white or white with yellow claws, 4-7 × 1-5 mm; style present; heads of cypselae hemispherical to ovoid, 2-4 × 2-5 mm; cypselae ellipsoidal or flattened-ellipsoidal, not or weakly flattened, wall thick, firm, with coarse transverse ridges, beak filiform, 0.2-1.2 mm. New Mexico material belongs to **var. diffusus** With. Ponds, lakes, streams, ditches, edges of rivers, 4300-9955 ft. Widespread except E, SW.

Ranunculus arizonicus Lemmon ex A. Gray—Arizona buttercup. Stems erect, never rooting nodally; basal leaves all (except sometimes innermost) shed before anthesis, leaving a dense brush of fibers, semicircular to reniform, outer leaves undivided, inner 5-7-parted, bases obtuse to cordate, apices acute to rounded; upper cauline leaves lobed; sepals 3-5 × 1-3 mm, glabrous or sparsely pilose; petals 5(-11), yellow, 7-15 × 2-6 mm; style present; heads of cypselae ovoid or cylindrical, 4-10 × 4-6 mm; cypselae thick-lenticular to compressed-globose, not or weakly flattened, wall thick, firm, smooth, beak subulate, straight, 0.8-1.2 mm but brittle and often broken. Dry, rocky slopes, intermittent stream beds, 4920-7875 ft. Grant.

Ranunculus cardiophyllus Hook.—Heartleaf buttercup. Stems erect, never rooting nodally; basal leaves ovate or elliptic, crenate but undivided or the innermost 3-5-parted, bases cordate to broadly obtuse, apices rounded to broadly acute; upper cauline leaves lobed; sepals 5-8 × 3-7 mm, pilose; petals 5-10, yellow, 6-13 × 4-13 mm; style present; heads of cypselae ovoid or cylindrical, 5-16 × 5-9 mm; cypselae thick-lenticular to compressed-globose, not or weakly flattened, wall thick, firm, smooth, beak subulate, curved or straight, 0.6-1.2 mm. Wet or dry meadows in ponderosa pine, Douglas-fir, spruce-fir communities, 7500-11,365 ft.

Ranunculus eschscholtzii Schltdl.—Eschscholtz's buttercup. Stems erect or decumbent from prominent caudices, never rooting nodally; basal leaves reniform or cordate, 3-parted with at least the lateral segments again lobed, bases truncate or cordate; upper cauline leaves lobed; sepals 4-8 × 2-6 mm, glabrous or pilose; petals 5-8, yellow, 6-12 × 4-16 mm; style present; heads of cypselae cylindrical or ovoid, 5-10 × 4-7 mm; cypselae thick-lenticular to compressed-globose, not or weakly flattened, wall thick, firm, smooth, beak lanceolate or subulate, straight (sometimes curved when immature), 0.6-1.8 mm. New Mexico material belongs to **var. eschscholtzii**. Open rocky slopes and meadows, usually alpine, often along the margins of melting snowbanks, 9000-12,800 ft. Rio Arriba, Taos.

Ranunculus fasciculatus Sessé & Moc.—Latin American buttercup. Stems erect or decumbent, not rooting nodally; basal leaves ovate to deltoid in outline, 3(-5)-foliate, leaflets once or twice lobed or parted; upper cauline leaves deeply parted or compound; sepals 5-10 × 3-5 mm, hispid; petals 11-16, yellow, 8-21 × 2-5 mm; style present; heads of cypselae globose to ovoid, 6-13 × 7-9 mm; cypselae discoid, much thinner than broad, wall thick, firm, smooth, beak filiform from a deltoid base, straight, 1.8-2.5 mm, filiform tip deciduous, leaving a 1-1.2 mm deltoid beak. Stream banks, lakeshores, marshes. This taxon is reported for New Mexico; however, it is found in the E United States and is unlikely to be found here.

Ranunculus flammula L.—Greater creeping spearwort. Stems erect to prostrate or sometimes ascending, usually rooting nodally; lower cauline leaves lance-elliptic to lanceolate or linear, undivided, bases acute to filiform, margins entire or serrulate, apices acute to filiform; upper cauline leaves undivided; sepals 2-3 × 1-2 mm, glabrous or appressed-hispid; petals 5-6, yellow, 3-5 × 2-3 mm; style present; heads of cypselae globose or hemispherical, 2–4 × 3–4 mm; cypselae globose-lenticular to globose, not or weakly flattened, wall thick, firm, smooth, beak lanceolate to linear, straight or curved, 0.1–0.6 mm. Muddy ground or shallow water in piñon-juniper, ponderosa pine, Douglas-fir, spruce-fir communities, 5350–10,065 ft. NW, NC. Our material belongs to **var. ovalis** (J. M. Bigelow) L. D. Benson.

Ranunculus glaberrimus Hook.—Sagebrush buttercup. Stems prostrate or ascending, never rooting nodally; basal leaves elliptic to oblong or reniform, entire to deeply 3-crenate, bases obtuse to truncate, apices rounded; upper cauline leaves deeply 3-crenate to unlobed; sepals 5–8 × 3–7 mm, glabrous or sparsely pilose; petals 5–10, yellow, 8–13 × 5–12 mm; style present; heads of cypselae globose, 7–12(–20) × 6–11(–20) mm; cypselae thick-lenticular to compressed-globose, not or weakly flattened, wall thick, firm, smooth, beak subulate or lance-subulate, straight or curved, 0.4–1 mm. New Mexico material belongs to **var. ellipticus** (Greene) Greene. Moist seepy slopes, sagebrush, ponderosa pine, Gambel oak communities, 6225–10,380 ft. NW.

Ranunculus gmelinii DC.—Gmelin's buttercup. Stems reclining or sometimes floating, rooting nodally; basal leaves none; leaves reniform to circular, blades 3-parted, segments again 1–3 times lobed to dissected, bases cordate; sepals 2–5 × 2–4 mm, glabrous or sparsely pilose; petals 4–14, yellow, 3–7 × 2–5 mm; style present; heads of cypselae globose or ovoid, 3–7 × 2–5 mm; cypselae thick-lenticular or compressed-ellipsoidal to discoid, not or weakly flattened, wall thick, firm, smooth, beak narrowly lanceolate or filiform, 0.4–0.8 mm. Shallow water or drying mud in the mountains, 6500–9800 ft. N.

Ranunculus hydrocharoides A. Gray—Frogbit buttercup. Stems erect to prostrate, usually rooting nodally; lower cauline leaves ovate to broadly ovate, undivided, bases rounded to weakly cordate, margins entire or dentate, apices rounded or obtuse; upper cauline leaves undivided; sepals 1.5–3 × 1–2 mm, glabrous; petals 5–6, yellow, 3–5 × 1–2 mm; style present; heads of cypselae hemispherical or globose, 2–4 × 3–4 mm; cypselae globose-lenticular to globose, not or weakly flattened, wall thick, firm, smooth, beak lanceolate to lance-filiform, straight or curved, 0.4–1 mm. Wet soil or shallow water in marshes, edges of streams and lakes, 5350–8920 ft. WC, SW.

Ranunculus inamoenus Greene—Graceful buttercup. Stems erect, never rooting nodally; basal leaves ovate, obovate, or orbicular (rarely reniform), crenate and undivided or the innermost with 2 clefts or partings near apex, bases acute to rounded, apices rounded; upper cauline leaves lobed; sepals 3–5 × 2–3 mm, pilose; petals 5, yellow, 4–9 × 2–5 mm; style present; heads of cypselae cylindrical, 7–17 × 5–8 mm; cypselae thick-lenticular to compressed-globose, not or weakly flattened, wall thick, firm, smooth, beak subulate, straight or hooked, 0.4–0.9 mm. Our material belongs to **var. inamoenus**. Meadows, riparian communities, rocky slopes, piñon-juniper, ponderosa pine–Gambel oak, white fir, spruce-fir communities, 4670–12,960 ft. Widespread except E, SW, McKinley.

Ranunculus macauleyi A. Gray—Rocky Mountain buttercup. Stems erect from short caudices, never rooting nodally; basal leaves narrowly elliptic to lanceolate or oblanceolate, undivided and entire except for apex, bases acute or long-attenuate, apices truncate or rounded and 3(–5)-toothed; upper cauline leaves simple or apically lobed; sepals 6–12 × 2.5–8 mm, densely brown-pilose; petals 5(–8), yellow, 10–19 × 6–17 mm; style present; heads of cypselae ovoid or cylindrical, 5–10 × 4–5.5 mm; cypselae thick-lenticular to compressed-globose, not or weakly flattened, wall thick, firm, smooth, beak slender, straight or recurved, 0.5–1.5(–2.2) mm. Sunny open soil of alpine meadows and slopes, 10,500–13,000 ft. NC, Lincoln, Otero.

Ranunculus macounii Britton—Macoun's buttercup. Stems prostrate to suberect, often rooting nodally; basal leaves cordate to reniform in outline, 3-foliate, leaflets 3-lobed or -parted; upper cauline leaves deeply parted or compound; sepals 4–6 × 1.5–3 mm, glabrous or hirsute; petals 5, yellow, 4–6 × 3.5–5 mm; style present; heads of cypselae globose or ovoid, 7–11 × 7–10 mm; cypselae discoid, much thinner than broad, wall thick, firm, smooth, beak lanceolate to broadly lanceolate, straight or nearly so, 1–1.2 mm. Meadows, ditches, edges of streams and ponds, wet soil or emergent from shallow water; piñon-juniper, ponderosa pine–Gambel oak, white fir, spruce-fir communities, 4500–10,025 ft. Widespread except E, SW, Sierra, Socorro.

Ranunculus pensylvanicus L. f.—Pennsylvania buttercup. Stems erect, never rooting nodally; basal leaves broadly cordate in outline, 3-foliate, leaflets cleft (usually deeply so); upper cauline leaves deeply parted or compound; sepals 3–5 × 1.5–2 mm, ± hispid; petals 5, yellow, 2–3 × 1–2.5 mm; style present; heads of cypselae cylindrical, 9–12 × 5–7 mm; cypselae discoid, much thinner than broad, wall thick, firm, smooth, beak broadly lanceolate or subdeltoid, straight or nearly so, 0.6–0.8 mm. Moist woodlands in *Acer*, juniper, Douglas-fir, white fir communities, 5660–8545 ft. NC, WC, SW.

Ranunculus repens L.—Creeping buttercup. Stems decumbent or creeping, rooting nodally; basal leaves ovate to reniform in outline, 3-foliate, leaflets lobed, parted, or parted and again lobed; upper cauline leaves deeply parted or compound; sepals 4–7(–10) × 1.5–3(–4) mm, hispid or sometimes glabrous; petals 5, yellow, 6–18 × 5–12 mm; style present; heads of cypselae globose or ovoid, 5–10 × 5–8 mm; cypselae discoid, much thinner than broad, wall thick, firm, smooth, beak lanceolate to lance-filiform, curved, 0.8–1.2 mm. Meadows, borders of marshes, stream margins, ponderosa pine, Douglas-fir, and white fir communities, 7130–8865 ft. Los Alamos, Sandoval, San Miguel.

Ranunculus sceleratus L.—Cursed buttercup. Stems erect, only very rarely rooting at lower nodes; basal and lower cauline leaves reniform to semicircular in outline, blades 3-parted (often deeply so), segments again lobed or parted, bases truncate to cordate; upper cauline leaves deeply parted; sepals 2–5 × 1–3 mm, glabrous or sparsely hirsute; petals 3–5, yellow, 2–5 × 1–3 mm; style none; heads of cypselae ellipsoidal or cylindrical, 2–5 × 1–3 mm; cypselae thick-lenticular or compressed-ellipsoidal to discoid, not or weakly flattened, wall thick, firm, smooth, beak triangular, usually straight, 0.1 mm. New Mexico material belongs to **var. multifidus** Nutt. Wet ground or shallow water in bogs, ponds, shores of lakes and rivers; desert scrub, piñon-juniper, ponderosa pine–Gambel oak communities, 3915–9800 ft. Widespread except EC, SE, SC.

Ranunculus uncinatus D. Don—Woodland buttercup. Stems erect, never rooting nodally; basal leaves cordate to reniform in outline, 3-parted or sometimes 3-foliate, segments again lobed; upper cauline leaves deeply parted or compound; sepals 2–3.5 × 1–2 mm, pubescent; petals 5, yellow, 2–4(–6) × 1–2(–3) mm; style present; heads of cypselae globose or hemispherical, 4–7 × 4–7 mm; cypselae discoid, much thinner than broad, wall thick, firm, smooth, beak lanceolate, curved and hooked, 1.2–2.5 mm. Moist meadows or woods, often along streams, piñon-juniper, ponderosa pine–Gambel oak, aspen, spruce-fir, 4800–10,140 ft. NC, C, Catron.

THALICTRUM L. - Meadow-rue

Perennial herbs with woody rhizomes, caudices, or tuberous roots; stems erect or ascending, usually branched in large species, unbranched in *T. alpinum*; leaves basal and cauline, alternate; blades ovate to reniform in outline, 1–4 times ternately or pinnately compound; inflorescence a terminal, sometimes also axillary, (1)2–200-flowered panicle, raceme, corymb, or umbel; flowers actinomorphic, all bisexual, bisexual and unisexual on same plant, or all unisexual on monoecious or dioecious plants; sepals 4–10, whitish to greenish yellow or purplish; petals absent; fruit sessile or stipitate, aggregate achenes.

1. Flowers bisexual…**T. alpinum**
1'. Flowers unisexual (very rarely also bisexual)…(2)
2. Leaflets apically 3–12-lobed, lobe margins crenate; filaments variously colored, not white…(3)
2'. Leaflets undivided or apically 3-lobed, lobe margins entire; filaments usually white…(4)
3. Cypselae laterally compressed…**T. fendleri**
3'. Cypselae terete to slightly flattened…**T. venulosum**
4. Achenes, peduncles, abaxial surfaces of leaflets, and/or petioles and rachises with stipitate glands…**T. amphibolum** (in part)
4'. Achenes, peduncles, abaxial surfaces of leaflets, and/or petioles and rachises without stipitate glands…(5)
5. Leaflet length 0.9–5.25 times width; filaments 2.5–7.8 mm; anthers (0.7–)1.2–2.7(–3) mm; stipe 0.2–1.7 mm…**T. amphibolum** (in part)
5'. Leaflet length 0.9–2.6 times width; filaments 2–6.5 mm; anthers 1–3.6(–4) mm; stipe 0–1.1 mm…**T. dasycarpum**

Thalictrum alpinum L.—Arctic meadow-rue. Stems erect, scapose or nearly so, 6–20(–30) cm, with very slender rhizomes; leaves all basal or a single cauline leaf near the base, leaf blades twice ternately or twice pinnately compound, apically 3–5-lobed; inflorescence an elongate raceme, few-flowered; flowers bisexual; sepals 5, filaments filiform; fruits 2–6 per flower, 2–3.5 mm, stigmatic beak ca. 0.7 mm. Wet meadows, cold (often calcareous) bogs in willow-sedge and spruce-fir outcrops, subalpine and alpine, 9580–13,010 ft. Colfax, Mora, Taos.

Thalictrum amphibolum Greene [*T. revolutum* DC.]—Wax-leaved meadow-rue. Stems erect, coarse, 50–150 cm; leaves cauline, proximal leaves petiolate, distal sessile; petioles and rachises stipitate-glandular to glabrous; leaf blades 1–4 times ternately compound; inflorescences racemes to panicles; flowers usually unisexual, staminate and pistillate on different plants; sepals 4(–6), whitish, filaments white; cypselae 8–16. Dry open woods, brushy banks, thickets, barrens, prairies, 6500–7300 ft. Sandoval, San Miguel, Taos

Thalictrum dasycarpum Fisch. & Avé Lall—Purple meadow-rue. Stems erect, stout, 40–150 (–200) cm; leaves chiefly cauline, petioles and rachises glabrous or occasionally pubescent and/or stipitate-glandular, basal and proximal cauline leaf blades 3–5 times ternately compound, inflorescences panicles, apically ± acutely pyramidal, many-flowered; flowers usually unisexual, staminate and pistillate on different plants, sepals 4(–6), whitish, filaments white to purplish, filiform; cypselae numerous. Riparian woods, damp thickets, wet meadows, prairies; in an irrigated field near Mora, 7130 ft. Colfax, Mora, San Miguel.

Thalictrum fendleri Engelm. ex A. Gray—Fendler's meadow-rue. Stems mostly erect, sometimes reclining, leafy, (20–)30–60(–160) cm, from rhizomes or branched caudices; leaves mainly cauline, leaf blades (2)3–4 times ternately compound, glandular or glabrous; leaflets (5–)10–20 mm; inflorescence a leafy panicle, many-flowered; flowers mostly unisexual (plants dioecious); sepals 4(–6), filaments filiform; fruits 7–11(–14) per flower. Gambel oak, ponderosa pine, Douglas-fir, aspen, spruce-fir communities, 5310–12,055 ft. Widespread except E.

Thalictrum venulosum Trel.—Veiny meadow-rue. Stems erect, 20–50 cm, glabrous, from rhizomes; leaves basal and cauline, cauline 1–3, leaf blades 3–4 times ternately compound; leaflets obovate to orbiculate, apically 3–5-lobed, 5–20 mm, lobe margins crenate; inflorescences terminal panicles, narrow and dense, many-flowered; flower sepals greenish white, filaments colored, not white; cypselae 5–17, nearly terete to slightly flattened. Prairies, riparian woods, and coniferous, deciduous, and mixed forests, 8000–12,140 ft. Colfax.

TRAUTVETTERIA Fisch. & C. A. Mey – False bugbane

Trautvetteria caroliniensis (Walter) Vail—False bugbane, tassel rue. Perennial herbs with short, slender rhizomes; plants 0.5–1.5 m; stems 1 to several, erect, usually unbranched below distal-most bract; leaves (basal) with petiole 7–45 cm, blades palmately 5–11-lobed, 8–30(–40) cm wide; inflorescence peduncle 10–80 cm; flowers actinomorphic, bisexual; sepals 3–5(–7), greenish white; petals and nectaries absent; stamens white, 5–10 mm; fruit papery, veins prominent along angles and on 2 adaxial faces. Riparian and other wet areas in montane spruce-fir forests to subalpine and alpine meadows, 8020–13,000 ft. NC, Catron, Socorro.

RESEDACEAE – MIGNONETTE FAMILY

Kenneth D. Heil

OLIGOMERIS Cambess. – Whitepuff

Oligomeris linifolia (Vahl) J. F. Macbr.—Lineleaf whitepuff. Herbs; annual, succulent, erect, glabrous, to 35 cm; leaves alternate with glandlike stipules, numerous, linear, often fascicled, to 25 mm;

spike densely, often loosely flowered, to 15 cm; flowers perfect, zygomorphic, ca. 2 mm; sepals 4, white-marginate; petals 2, white, with irregular apical margins; stamens 3; pistil 1; ovary superior, (2)3-6-carpellate; style absent; capsule depressed-globose to ca. 2.5 × 3.4 mm thick, 4-beaked. Salt and clay flats, about boulders and on gravel bars along streams, 4015-5880 ft. Doña Ana, Otero, Sierra.

RHAMNACEAE - BUCKTHORN FAMILY

John R. Spence

Shrubs to small trees, deciduous to evergreen, armed or unarmed; leaves simple, alternate, fascicled, or opposite, petiole present or rarely absent, margins smooth to variously serrate or dentate, sometimes spinose, venation pinnate to conspicuously palmately 3-veined from leaf base, stipules present although sometimes deciduous; inflorescence unisexual or bisexual, axillary or terminal, sometimes flowers solitary; flowers bisexual or unisexual, epigynous or perigynous; sepals 4-5, distinct; petals (0-)4-5, distinct, hypanthium usually present, fused to ovary base, nectaries present, often fleshy; stamens 4-6, opposite petals; pistil 1, 2-4-carpellate, ovary superior or inferior, (1)2-4-locular, ovules 1 per locule; styles 1-4, connate proximally or fused; fruits capsules or drupes; seeds 1 per locule.

1. Plants with obvious thorns, or branch tips strongly thorn-tipped…(2)
1'. Plants lacking thorns or thorns obscure…(5)
2. Leaves deciduous, sometimes absent when plants in flower, pinnately veined…(3)
2'. Leaves evergreen, distinctly palmately 3-veined from base…**Ceanothus** (in part)
3. Leaves alternate, fascicled or clustered terminally; fruit a drupe; at least some leaves present when plants in flower…(4)
3'. Leaves mostly opposite or subopposite; fruit a capsule; plants essentially leafless, leaves largely absent when plants in flower…**Adolphia**
4. Leaves of short shoots 2-3; petals absent or present, style 1; drupes beaked…**Condalia**
4'. Leaves of short shoots mostly single; petals present, styles 2-4; drupes not beaked…**Ziziphus**
5. Leaves opposite or subopposite; fruit a drupe or capsule…(6)
5'. Leaves alternate or predominantly so; fruit a drupe…(7)
6. Leaves thin, shiny bright green, pinnately veined; fruit a drupe…**Sageretia**
6'. Leaves thick and dull-colored, or if somewhat thinner and green then distinctly palmately 3-veined from base; fruit a capsule…**Ceanothus** (in part)
7. Flowers mostly 5-merous, bisexual, styles 1; bud scales absent, buds hairy; secondary veins of leaves ± straight, parallel; seeds beaked, smooth…**Frangula**
7'. Flowers mostly 4-merous, unisexual, styles 2-4; bud scales present, buds glabrous; secondary veins of leaves arching, not parallel; seeds furrowed, not beaked…**Rhamnus**

ADOLPHIA Meisn. - Spinebush

Adolphia infesta (Kunth) Meisn. [*Ceanothus infestus* Kunth; *Colubrina infesta* (Kunth) Schltdl.]— Texas spinebush. Shrubs, 0.2-1(-2) m, early-deciduous, armed with thorns, branches green, rigid; leaves opposite or subopposite, petiolate, pinnately veined, blades oblanceolate to linear, glabrous, 3-10 × 1-3 mm, margins usually entire, stipules present, not spinelike; inflorescence axillary, of cymes or sometimes flowers solitary; flowers bisexual; sepals (4)5, spreading, whitish green; petals (4)5, white to yellow, clawed; stamens (4)5; nectary 5-angled; ovary superior, 3-locular; style 1; fruits capsules, beaked. Rocky places, outcrops, washes. Upper Guadalupe Canyon, no elevation given. Hidalgo. This species is characterized by usually being leafless and having rigid green photosynthetic stems.

CEANOTHUS L. - Buckbrush, deerbrush, mountain lilac

Shrubs to 3 m, armed or unarmed, branches various, generally not green, flexible to rigid, bud scales present; leaves evergreen or deciduous, alternate or opposite, petiolate, thin to coriaceous, pinnately veined to distinctly palmately 3-veined from leaf base, stipules deciduous or persistent, not spinelike; inflorescence of axillary or terminal cymose panicles; flowers bisexual; sepals 5, white or blue-purple;

petals 5, white to pale blue or purple; stamens 6; nectary cupulate; ovary partially inferior, 3-locular; styles 3; fruits capsules, each locule sometimes apically beaked.

1. Stipules persistent, fleshy; leaves opposite, thick, coriaceous, pinnately veined; plants unarmed...**C. pauciflorus**
1'. Stipules deciduous, thin; leaves alternate, thin to coriaceous, distinctly palmately 3-veined; plants armed or unarmed...(2)
2. Plants armed, branchlets rigid, thorn-tipped; leaves evergreen, coriaceous, resinous...**C. fendleri**
2'. Plants unarmed, branchlets mostly flexible, not thorn-tipped; leaves deciduous, thin, not resinous...(3)
3. Leaf blades entire; flowers white, blue, or pink; inflorescence axillary or terminal, paniculate to racemose, 5-25 cm...**C. integerrimus**
3'. Leaf blades distinctly serrulate to serrate; flowers white; inflorescence terminal, flat-topped to umbellate, 4-8 cm...**C. herbaceus**

Ceanothus fendleri A. Gray—Fendler's ceanothus. Shrubs 0.5-1.5 m, evergreen, branchlets rigid, spine-tipped; leaves coriaceous, resinous, elliptic, ovate, or orbiculate, blades 8-30 × 3-12 mm, margins entire to rarely serrulate; inflorescence terminal or axillary, umbel-like to racemose, 1-3 cm; sepals, petals, and hypanthium white or pink; capsule 4-6 mm, lobed, not or rarely beaked. Rocky slopes, shrublands, oak-pine woodlands, forests, 4600-10,500 ft. Widespread except E, Luna.

Ceanothus herbaceus Raf.—New Jersey tea. Shrubs 0.5-1 m, deciduous, unarmed, branches flexible; leaves thin, herbaceous, not resinous, elliptic-lanceolate to ovate, blades 20-70 × 10-33 mm, margins serrate; inflorescence terminal, umbel-like, 4-8 cm; sepals, petals, and hypanthium white; capsule 3-5 mm, lobed, not or rarely beaked. Our material is referred to **var. pubescens** (S. Watson) Shinners. Dunes and limestone bluffs, grasslands, woodlands. Eddy.

Ceanothus integerrimus Hook. & Arn.—Deerbrush. Shrubs 1-3 m, deciduous, unarmed, branches flexible; leaves thin, herbaceous, not resinous, oblong-elliptic or lanceolate to ovate, blades 20-80 × 10-40 mm, margins entire to distally denticulate; inflorescence terminal or axillary, racemose to paniculate, 5-25 cm; sepals, petals, and hypanthium white, pale blue, violet, or pink; capsule 4-5 mm, lobed, not or rarely beaked. Rocky slopes, shrublands, scrub, woodlands, forests, 5500-8000 ft. Grant, Socorro. Following Nesom (2016c), varieties are not recognized.

Ceanothus pauciflorus DC. [*C. greggii* A. Gray; *C. vestitus* Greene]—Desert ceanothus. Shrubs 0.2-2 m, evergreen, branches rigid, spine-tipped; leaves coriaceous, not resinous, oblong or obovate to suborbicular, blades 5-20 × 13-15 mm, margins entire to distally denticulate; inflorescence mostly axillary, 1-3 cm; sepals, petals, and hypanthium white, pale blue, cream, or lavender; capsule 3-6 mm, not or weakly lobed, distinctly beaked. Rocky slopes and ridges, granite or metamorphic rock, piñon-juniper woodlands to montane forests, usually 4200-6600 ft. Catron, S except Lea, Luna.

CONDALIA Cav. – Snakewood

Shrubs to 3 m, armed, branches various, generally not green, flexible to rigid, thorn-tipped, bud scales absent; leaves deciduous, alternate, sessile to subsessile, in fascicles on short shoots, ± coriaceous, not glandular, pinnately veined, stipules deciduous to persistent, not spinelike; inflorescence of axillary cymes; flowers bisexual; sepals 5, yellow-green; petals 0-5, yellow if present; stamens 5; nectary absent or very thin; ovary superior, 2-locular becoming 1-locular; styles 3; fruits drupes, beaked, stones 1.

1. Leaf blades linear, ≤ 1 mm wide; petals 5, yellow...**C. ericoides**
1'. Leaf blades obovate, spatulate, or elliptic, mostly 2-5 mm wide; petals 0...(2)
2. Leaf blades spatulate, villous to hispidulous, abaxial venation conspicuous...**C. warnockii**
2'. Leaf blades elliptic to obovate, glabrous to hirtellous, abaxial venation obscure...**C. correllii**

Condalia correllii M. C. Johnst.—Correll snakewood. Shrubs 1-3 m, branches thorn-tipped or not; leaves with petiole 1-3 mm, blades green, elliptic to obovate, 7-20 × 4-6(-10) mm, sparsely hispidulous to glabrous, margins plane; inflorescence of 1-4 flowers; petals 0, sessile or very short-pedicellate; drupes globose to somewhat elongate, usually not flattened, 5-8 mm, stones elongate, 2-seeded. Dry slopes, canyons, shrublands, scrub, 4400-5020 ft. Hidalgo.

Condalia ericoides (A. Gray) M. C. Johnst.—Javelina bush. Shrubs 1–2.5 m, branches thorn-tipped; leaves sessile or with petiole to 0.1 mm, blades green or gray-green, linear, 2–12 × 1 mm, glabrous, margins revolute or appearing so; inflorescence of 1–6 flowers; petals 5, yellow; drupes elliptic to fusiform, not flattened, 7–12 mm, stones 1–2-seeded. Clay, gypsum, or limestone soils, dry slopes, canyons, shrublands, grasslands, scrub, 3000–5400 ft. C, SE, Curry.

Condalia warnockii M. C. Johnst.—Warnock snakewood. Shrubs 0.5–3 m, branches mostly thorn-tipped; leaves sessile or with petiole to 2 mm, blades yellow, brown, or gray-green adaxially, spatulate, 3–10 × 1–5 mm, hispidulous to villous, often densely so, margins plane; inflorescence of 1–2 flowers; petals 0, pedicels 0.5–3 mm; drupes globose to fusiform, often flattened, 4–6 mm, stones elongate, 1–2-seeded. Our material is referred to as **var. warnockii**. Dry slopes, canyons, shrublands, scrub, sandy plains, 3800–6000 ft. C, SE, SC, SW.

FRANGULA Mill. – Buckthorn

Shrubs or small trees to 5 m, unarmed, branches various colors, generally not green, mostly flexible, bud scales absent, buds pubescent; leaves evergreen or deciduous, thin to thickened-coriaceous, alternate, petiolate, blades glabrous or pubescent, pinnately veined, secondary veins ± straight, parallel, stipules present, not spinelike; inflorescence axillary, obscured by leaves, umbellate or solitary; flowers bisexual; sepals (4)5, white or yellow to green; petals (4)5, cream to yellow; stamens (4)5; nectary thin, attached to hypanthium; ovary superior, (2)3-locular; style 1; fruits drupes.

1. Leaves evergreen; blades thick, coriaceous, white-tomentose abaxially…**F. californica**
1′. Leaves deciduous or predominantly so; blades thin to coriaceous, glabrescent to sparsely hirtellous…(2)
2. Some leaves often persistent, not fully deciduous; blades thick, coriaceous, broadly ovate, < 2 times longer than wide; plants of the San Juan River basin [to be expected in the Four Corners area]…F. obovata Blume
2′. Leaves fully deciduous; blades thin, herbaceous, narrowly ovate, mostly > 2 times longer than wide; plants of S and C New Mexico…**F. betulifolia**

Frangula betulifolia (Greene) Grubov—Birchleaf buckthorn. Shrubs 1–4 m, predominantly deciduous; leaves thin, ± herbaceous, blades glabrescent to hirtellous, yellow-green abaxially, green adaxially, oblong-elliptic to ovate, 2–10 × 2–6 cm, 1.5–2.5 times as long as wide, margins serrate to crenate, tips not glandular; inflorescence of 2–30(–35) flowers; drupes black, globose, 5–10 mm; stones 2–4. Streamsides, wet banks, seepy cliff faces and ledges, often in shaded locations, 4800–9200 ft. Catron, Doña Ana, Grant, Hidalgo, Sierra.

Frangula californica (Eschsch.) A. Gray—California coffeeberry. Shrubs 0.5–5 m, predominantly evergreen or sometimes partly deciduous; leaves thick, coriaceous, blades glabrous to tomentose abaxially, green to yellow-green abaxially, green adaxially, elliptic to ovate, 3–8 × 1.5–4.5 cm, > 2 times as long as wide, margins serrate to serrulate with gland-tipped teeth; inflorescence of 5–60 flowers; drupes black, globose to elliptic, 10–15 mm; stones 2. Our material is referred to as **subsp. ursina** (Greene) Kartesz & Gandhi. Woodlands, chaparral, riparian zones, 4375–6700 ft. WC, SC, SW.

RHAMNUS L. – Buckthorn

Shrubs or small trees to 4 m, unarmed, branches various colors, generally not green, mostly flexible, bud scales present, buds glabrous; leaves deciduous, thin to somewhat coriaceous, alternate or rarely sub-opposite, petiolate, blades glabrous or pubescent, pinnately veined, secondary veins ± arched-curved, not parallel, stipules present, not spinelike; inflorescence axillary, cymose, fascicled or solitary, obscured by leaves; flowers unisexual; sepals 4, white or yellow to green; petals (0–)4, cream to yellow; stamens 4 (rudimentary in pistillate flowers); nectary thin, attached to hypanthium; ovary superior, (2–)4-locular; styles 2–4; fruits drupes. The European species *R. cathartica* L. (common buckthorn) is cultivated in the United States and sometimes escapes. It can be distinguished by the presence of thorns, opposite leaves, and 5-merous flowers.

1. Leaf blades elliptic-oblong, 2–5 cm, minutely hirtellous on both sides; bud scales thick, coriaceous, dark red to red-brown…**R. serrata**
1'. Leaf blades lanceolate to narrowly elliptic or oblong-lanceolate, 3–8 cm, glabrous; bud scales thin, pale golden-yellow…**R. smithii**

Rhamnus serrata Schult. [*Endotropis serrata* (Humb. & Bonpl. ex Willd.) Hauenschild]—Sawleaf buckthorn. Shrubs 0.5–2.5 m, bud scales thick, coriaceous, dark red to red-brown, dull; leaves herbaceous, minutely hirtellous on both sides, petioles 3–4 mm, blades yellow-brown to green abaxially, green adaxially, not distinctly glossy, elliptic-lanceolate to elliptic-oblong, 1.5–5.5 cm, tips obtuse. Rocky ledges, often on limestone, canyons, streamsides, mesic to coniferous forests, 5000–8600 ft. C, SE, SC, Hidalgo.

Rhamnus smithii Greene [*Endotropis smithii* (Greene) Hauenschild]—Smith's buckthorn. Shrubs 1–3 m, bud scales thin, pale golden-yellow, glossy; leaves herbaceous to subcoriaceous, glabrous on both sides, petioles 3–8 mm, blades green abaxially, glossy yellow-green adaxially, elliptic-lanceolate to lanceolate, 3–8 cm, tips acute or obtuse. Dry rocky slopes, meadows, woodlands, forests, 4600–8600 ft. C, SC, SW, Rio Arriba.

SAGERETIA Brongn. – Mock buckthorn

Sageretia wrightii S. Watson—Wright's mock buckthorn. Shrubs, 1–4 m, ± unarmed to obscurely thorny, branches intricately branched, erect to spreading-curved, generally not green, flexible, bud scales present; leaves evergreen, thin, shiny green, opposite to subopposite, blades 5–30 × 5–20 mm, margins entire or rarely distally serrulate, tomentose becoming glabrescent, obscurely pinnately veined, stipules present, not spinelike; inflorescence terminal or axillary toward branch tips, extending beyond leaves, spicate; flowers bisexual; sepals 5, erect, yellow-green, fleshy; petals 5, white, cream to yellow, hooded, 3–5 mm, white to pale blue or purple; stamens 5; ovary superior, 2–3-locular; styles 1; hypanthium yellow to orange, nectary cupulate; fruit a drupe, black, ± globose, 5–10 mm wide, not beaked. Dry rocky slopes, canyons, open oak woodlands, 4800–5450 ft. Catron, Hidalgo.

ZIZIPHUS Mill. – Graythorn

Shrubs or small trees to 4 m, strongly armed with thorns, branches various, generally gray-green to brown, thorn-tipped, flexible to rigid, bud scales absent; leaves deciduous, alternate or fascicled, petiolate, small, ovate to linear, gray-green, 1-veined, stipules present, spinelike; inflorescence axillary, often within foliage, of thyrses, 2–30 flowers, sometimes flowers solitary; flowers bisexual; sepals 5, green, yellow, or orange; petals 5, white to yellow; stamens 5; hypanthium yellow or orange, nectary 5–10-lobed; ovary superior, 2–4-locular; styles 1–4; fruits drupes.

1. Secondary branches short-hirtellous to villous; leaf blades hirtellous to villous, thin, herbaceous; hypanthium hirtellous; inflorescences with (5–)10–30 flowers…**Z. divaricata**
1'. Secondary branches mostly glabrous; leaf blades mostly glabrous, thickened-coriaceous; hypanthium strigose with loose curved hairs; inflorescences with 1–6 flowers…**Z. obtusifolia**

Ziziphus divaricata (A. Nelson) Davidson & Moxley [*Z. obtusifolia* (Hook. ex Torr. & A. Gray) A. Gray var. *canescens* (A. Gray) M. C. Johnst.]—Graythorn. Shrubs or small trees 1–4 m, secondary branches hirtellous to villous; leaves small, blades 3–25 mm, ovate to oblong or elliptic, thin and herbaceous, hirtellous to villous on both sides; inflorescence of pedunculate thyrses, 10–20 mm, with (5–)10–30 flowers, peduncles as long as or longer than leaf petioles; hypanthium densely strigose; drupes black to black-purple or blue-black, 7–10 mm, globose to slightly elongate. Floodplains, washes, mesquite bosques; 1 record reported near Tularosa, 5200 ft.

Ziziphus obtusifolia (Hook. ex Torr. & A. Gray) A. Gray [*Sarcomphalus obtusifolius* (Hook. ex Torr. & A. Gray) Hauenschild]—Chihuahuan graythorn, Texas buckthorn. Shrubs 1–2.5 m, secondary branches glabrous or sometimes sparsely pilose; leaves small, blades 5–40 mm, ovate to oblong, elliptic, or linear, thickened, subcoriaceous, glabrous, sometimes shortly villous to strigose on abaxial surface; inflorescence of pedunculate thyrses, or flowers solitary, with 1–6 flowers, peduncles shorter than leaf petioles; hypanthium densely hirtellous; drupes black to black-purple or black-blue, 5–10 mm, globose to slightly elongate. Often on calcareous and gypsum substrates, washes, floodplains, gypsum flats, grasslands, mesquite bosques, 3275–6160 ft.

ROSACEAE – ROSE FAMILY

Kenneth D. Heil

Herbs (annual or perennial), shrubs, or trees; stems simple or branched; leaves persistent or deciduous, basal and/or cauline, usually alternate, rarely opposite, simple or compound; stipules usually present, inflorescences terminal, sometimes axillary, panicles, 1-flowered, glomerules, fascicles, spikes, racemes, corymbs, umbels, or cymes; flowers usually bisexual, rarely unisexual, perianth and androecium perigynous or epigynous; epicalyx bractlet sometimes present, hypanthium flat to hemispherical, or cylindrical to funnelform or urceolate, sepals (0–)4 or 5(–10), distinct, free, petals (0–)4 or 5(–12), distinct, free, stamens 0–130(–220), distinct, free, anthers usually longitudinally dehiscent, pistils 1–250(–450), distinct or ± connate, free or ± adnate to hypanthium, ovary superior or inferior, stigmas usually capitate, ovules 1 or 2(–5+); fruits achenes, follicles, drupes, nutlets, pomes, or capsules (Heil et al., 2013; Phipps, 2014).

1. Plants herbs, sometimes woody at the very base…(2)
1'. Plants shrubs or small trees…(10)
2. Petals lacking; inflorescence a dense spike, flowers numerous; leaves pinnately compound…**Poterium**
2'. Petals present (minute in *Sibbaldia*); inflorescence either not a dense spike or flowers not numerous; leaves various…(3)
3. Leaves simple; hypanthium densely hairy…**Alchemilla**
3'. Leaves compound or variously divided; hypanthium not densely hairy…(4)
4. Flowers yellow (or dark red in *Potentilla thurberi*)…(5)
4'. Flowers white or pink…(6)
5. Petals minute, narrow; leaflets toothed at apex; plants prostrate or mat-forming; plants of high elevations…**Sibbaldia**
5'. Petals showy, not narrow; leaflets entire, lobed, or serrate; plants not prostrate or mat-forming; plants not limited to high elevations…**Potentilla** (in part)
6. Leaves trifoliate; plants with well-developed stolons; receptacle ripening into an accessory fruit…**Fragaria**
6'. Leaves usually with > 3 leaflets; plants lacking stolons; receptacle not ripening into an accessory fruit…(7)
7. Flowers in a narrow raceme; upper 1/2 of mature hypanthium covered with hooked bristles; rare in our area…**Agrimonia**
7'. Flowers in a branched inflorescence; hypanthium without hooks; common to rare…(8)
8. Leaves pinnately lobed or compound or more usually lyrate-pinnatifid; styles at maturity elongate and conspicuous…**Geum**
8'. Leaves palmately or pinnately lobed or compound, not lyrate-pinnatifid; styles at maturity not elongate and conspicuous…(9)
9. Style attached near the top of the ovary; leaves either palmately compound or pinnately compound with narrow leaflets…**Potentilla** (in part)
9'. Style attached near the base of the ovary; leaves pinnately compound with broadly oval leaflets… **Drymocallis**
10. Leaves compound…(11)
10'. Leaves simple…(14)
11. Flowers yellow; leaves pinnately compound, crowded and appearing to be palmate…**Dasiphora**
11'. Flowers rose, white, or pink; leaves distinctly pinnate…(12)
12. Leaflets 11–15; small trees with orange berries (pomes); stems and/or leaves lacking prickles…**Sorbus**
12'. Leaflets 5–7; shrubs without orange berries; stems and/or leaves with prickles…(13)
13. Leaves glaucous beneath; flowers white; fruit an aggregate of drupelets (a raspberry)…**Rubus** (in part)
13'. Leaves green, often pale; flowers rose or rarely white; fruit a hip…**Rosa**
14. Low mat-forming shrubs; flowers in dense spikes on leafless or merely bracteate scapes…**Petrophytum**
14'. Shrubs or small trees, never mat-forming; flowers various but not scapose or subscapose…(15)
15. Ovary or ovaries superior, free from the hypanthium…(16)
15'. Ovary inferior, 2–5 carpels fused and the fruit a pome…(23)

16. Petals lacking, flowers inconspicuous; leaves entire, evergreen (except *Cercocarpus montanus*)... **Cercocarpus**
16'. Petals present, flowers showy or small; leaves mostly toothed or lobed, not evergreen...(17)
17. Pistil 1; fruit a drupe; leaves usually with glands at the base of the blade or on the petiole...**Prunus**
17'. Pistils 1 to many; fruit not a drupe; leaves without glands...(18)
18. Flowers large, 2-5 cm across, in few-flowered cymes; leaves 2-15 cm wide; fruit an aggregate...**Rubus** (in part)
18'. Flowers < 2 cm across, solitary or in corymbs; leaves to 2 cm wide; fruit a follicle or an achene, plumose-tailed or not...(19)
19. Fruit a follicle with several seeds...(20)
19'. Fruit an achene with a single seed...(21)
20. Leaves palmately veined and lobed...**Physocarpus**
20'. Leaves pinnately veined, toothed but not lobed...**Vauquelinia**
21. Flower petals white, in terminal, many-flowered pyramidal clusters; usually on talus slopes and cliff faces ...**Holodiscus**
21'. Flower petals white, cream, or pale yellowish, borne singly or in few-flowered cymes; various habitats (22)
22. Pistils numerous; petals white; leaf lobes tightly revolute; often in wash bottoms and roadsides...**Fallugia**
22'. Pistils 1-5; petals white to cream or pale yellow; leaf lobes not tightly revolute; various habitats...**Purshia**
23. Stems armed with thorns...(24)
23'. Stems unarmed (rarely so in *Malus*)...(25)
24. Leaves evergreen; petals < 4 mm...**Pyracantha**
24'. Leaves deciduous, petals > 4 mm...**Crataegus**
25. Leaves narrowly elliptic, entire or indistinctly toothed...**Peraphyllum**
25'. Leaves serrate to doubly serrate...(26)
26. Flowers white, in racemes; plants indigenous, rarely cultivated; leaves toothed at the apex... **Amelanchier**
26'. Flowers white, creamy, or otherwise, in corymbs or short panicles; plants cultivated, sometimes escaping; leaves toothed or lobed...(27)
27. Styles united below into a column; fruit subglobose, the persistent sepals sunken in a depression... **Malus**
27'. Styles free to the base; fruit pear-shaped, broader at the end opposite the stalk, the persistent sepals not sunken in a depression...**Pyrus**

AGRIMONIA L. – Agrimony

Perennial herbs; stems unbranched below inflorescences; leaves pinnately compound, leaflet margins serrate, stipules persistent, resembling smallest leaflets; inflorescence of racemes terminating stems; flowers perfect, yellow, small; hypanthium turbinate to hemispherical, throat constricted, outer rim with numerous hooked bristles that elongate and become indurate in fruit, sepals 5, petals small, ovary superior; fruit an accessory.

1. Stems with glistening glandular hairs, ± hirsute; major leaflets 5-7...**A. gryposepala**
1'. Stems with glandular hairs, ± glistening, pubescent to villous or hirsute; major leaflets 5-13...(2)
2. Major leaflets 5-7 on midcauline leaves; widespread...**A. striata**
2'. Major leaflets 9-13 on midcauline leaves; rare or absent in New Mexico...**A. parviflora**

Agrimonia gryposepala Wallr. [*A. eupatoria* L. var. *parviflora* Hook.]—Common agrimony. Herbs, 3.5-15 dm; stems with glistening stipitate-glandular hairs, hirsute; leaves midcauline, major leaflets 5-7, blades ovate to elliptic or rhombic, 2.4-10.5 × 1.4-5.6 cm, margins serrate, abaxial surface usually with glistening stipitate-glandular hairs, hirsute; flowers usually ± alternate; fruiting hypanthia turbinate to broadly campanulate, 2.3-5.8 × 2.8-6.2 mm, with hooked bristles in rows, with glistening stipitate-glandular hairs, sparsely hirsute. Moist mixed and riparian communities, 6300-8600 ft. NC, WC, SC, Grant.

Agrimonia parviflora Aiton [*A. polyphylla* Urb.]—Harvestlice. Herbs 5-20 dm; stems with glistening glandular hairs, pubescent to villous or hirsute; leaves midcauline, margins dentate, major leaflets 9-13 blades lanceolate to narrowly elliptic, 3.4-8.5 × 1.2-2.4 cm, margins serrate to dentate, abaxial surface usually glistening with sessile-glandular hairs, pubescent to villous or hirsute; flowers usually ± subopposite; fruiting hypanthia mostly broadly campanulate to broadly turbinate, 1.3-3 × 1.7-3.8 mm, with hooked bristles, usually with glistening sessile-glandular hairs. A single record in Santa Fe, ca. 7200 ft. Perhaps no longer in New Mexico.

Agrimonia striata Michx. [*A. brittoniana* E. P. Bicknell]—Agrimony. Herbaceous perennial with rhizomes; stems 30–100 cm, hirsute and somewhat glandular; leaves with 5–7 leaflets, 3–12 cm, crenate-serrate, resinous-glandular beneath and sparingly hirsute only on the veins; inflorescence a raceme, 20–40 cm; flower calyx tube in fruit mostly 5 mm, petals obovate, ca. 5 mm, yellow, stamens 5–15, pistils 1 or 2, stigmas 2-lobed; fruit 1 or 2 achenes. Ponderosa pine and Douglas-fir communities, 5570–10,500 ft. NE, NC, NW, C, WC, Grant, Otero.

ALCHEMILLA L. – Lady's mantle

Alchemilla monticola Opiz [*A. pastoralis* Buser]—Lady's mantle. Perennial herbs to 40 cm; stems densely spreading-hairy; leaves reniform to reniform-orbiculate, shallowly 7–9-lobed, surfaces densely pubescent; inflorescences terminal, multiflowered cymes, with primary branches densely hairy; flowers green, often becoming reddish, hypanthium urceolate, densely pubescent, sepals 4; fruit an achene, ovoid. Riparian community with scattered spruce; Wheeler Peak area, 9375–10,500 ft. Taos. Introduced.

AMELANCHIER Medik. – Serviceberry

Deciduous, unarmed shrubs or small trees; stems woody, bark smooth, pale, with shallow longitudinal fissures; leaves deciduous, alternate, simple, petiolate, toothed, pinnately veined; stipules linear; inflorescence racemose, leafy, appearing with the leaves; flowers perfect, white, showy, large nectar ring present, hypanthium campanulate to urceolate, sepals 5, petals 5, narrow, not clawed, stamens usually 20, carpels 2–5, partially adnate to hypanthium, ovary ca. 1/2 inferior; fruit a fleshy pome, purple, orange to yellow, or nearly white.

1. Larger leaf blades (2–)2.5–6 cm, sparsely pubescent to glabrous with age; styles (4)5, united into a column below; petals 7–13 mm; pome becoming fleshy and purplish at maturity…**A. alnifolia**
1'. Leaf blades 1–3 cm, permanently pubescent; styles 2 or 3 (4 or 5), usually distinct to the base; petals 5.5–9 (–10) mm; pome ± dry or mealy, orangish, yellowish, or whitish, but sometimes with a purplish tinge…**A. utahensis**

Amelanchier alnifolia (Nutt.) Nutt. ex M. Roem.—Western serviceberry. Shrub 1–4.5 m, bark of the young stems smooth, reddish brown, eventually becoming gray, herbage glabrous to sparsely puberulent when young; leaves with petioles 0.7–1.8(–2.8) cm, blades obovate to suborbicular, (2–)2.5–6 × 1.5–4.5 cm, serrate or coarsely toothed; inflorescence a short, erect, 5–15-flowered raceme; flowers fragrant, hypanthium brownish within, sepals 1.8–4(–5.5) mm, persistent, petals 7–18 × 2.5–6.7 mm, white; fruit a globose pome, 7–10 mm thick, becoming fleshy and purplish at maturity; seeds 3.5–5 mm, dark brown. NE, NC, NW except Union.

1. Herbage and inflorescence pubescent, glabrescent with age, the top of the ovary remaining densely hairy through development into a pome…**var. alnifolia**
1'. Herbage and inflorescence glabrous, including the top of the ovary (pome)…**var. pumila**

var. alnifolia [*Pyrus alnifolia* (Nutt.) Lindl. ex Ser.; *A. bakeri* Greene]—Gambel oak, mountain mahogany, ponderosa pine, Douglas-fir, aspen, spruce-fir communities, 7000–9000 ft.

var. pumila (Torr. & A. Gray) C. K. Schneid. [*A. pumila* Nutt ex Torr. & A. Gray; *A. polycarpa* Greene]—Dwarf serviceberry. Gambel oak, mountain mahogany, ponderosa pine, Douglas-fir, aspen, spruce-fir communities, 7500–11,000 ft.

Amelanchier utahensis Koehne [*A. oreophila* A. Nelson]—Utah serviceberry. Shrub to 3 dm, often much branched, bark reddish when young, soon becoming gray; leaf blades obovate to suborbicular, (1–)1.5–3 × 0.8–2(–2.3) cm, usually prominently tomentose on both surfaces; inflorescence a 3–6(–10)-flowered raceme; flower sepals 1.5–4.2 mm, petals 5.5–9(–10) × 1.7–3.7(–4.2) mm, white; fruit a pome, 5–10 mm thick, dry or at least not juicy at maturity. Streamsides, foothills, mountain slopes, grassland, mountain mahogany, Gambel oak, mountain brush, piñon-juniper woodland, aspen, ponderosa pine communities, 4500–8065 ft. NC, NW, C, WC, SC, SW.

CERCOCARPUS Kunth – Mountain mahogany

Evergreen or deciduous shrubs or small trees, unarmed or spinescent; stems woody, bark smooth, gray to reddish brown; leaves simple, clustered at tips of short shoots, entire or dentate, usually thick or coriaceous; inflorescence of solitary flowers or few-flowered clusters terminating short shoots, bracts small; flowers perfect, sessile, not showy, hypanthium of 2 parts, lower part narrowly tubular, persistent, upper part cup-shaped, deciduous, petals absent, sepals 5, small, stamens 10–45, carpel 1, free from hypanthium; fruit an achene.

1. Leaves linear; highly and intricately branched low shrub; slickrock sites and cliffs; San Juan County…**C. intricatus**
1'. Leaves broader; tall, open-branched shrub; various sites; widespread…(2)
2. Leaves usually > 1.8 cm, thin, winter-deciduous, mostly ovate to obovate-orbicular, coarsely crenate to serrate-dentate on the apical 1/2; flowers and fruits relatively large…**C. montanus**
2'. Leaves usually < 2.5 cm, moderately thick, evergreen, usually elliptic to oblanceolate or obovate, entire or shallowly cuneate-dentate near the apex; flowers and fruits relatively small…**C. breviflorus**

Cercocarpus breviflorus A. Gray [*C. breviflorus* var. *eximius* C. K. Schneid.]—Eastern mountain mahogany. Shrubs, (5–)10–30(–60) dm, sparsely to moderately branched; stems sericeous, pilose-hirsute or villous, glabrate; leaves persistent or drought-deciduous, blades narrowly oblanceolate or oblong-oblanceolate to obovate, sometimes ovate, (3–)7–35(–52) × (1.3–)4–15(–28) mm, margins serrulate to crenate in distal 1/3–1/2; flowers 1–3(–7) per short shoot, sepals 5, 1–2 mm, stamens (15–)20–40; achenes 6–9(–10.7) × 1.3–2 mm. Yellow pine, fir forests, piñon, juniper, oak woodlands, chaparral, grasslands, desert mountains, rocky, limestone, sandstone, or rhyolite substrates, 4300–8900 ft. Widespread except far E and N.

Cercocarpus intricatus S. Watson [*C. ledifolius* Nutt. var. *intricatus* (S. Watson) M. E. Jones]—Littleleaf mountain mahogany. Evergreen, small shrub, intricately branched, spinescent, to 2.5 m, young branches reddish brown, finely villous with crinkly white hairs or glabrous, older stems ashy-gray; leaves 5–15 × 0.5–2 mm, blades linear, mucronulate-tipped, entire, revolute-margined; inflorescence of 1–3 flowers terminating short, lateral branches; flowers subsessile, hypanthium tube persistent, 4–6 mm, sepals recurved, 0.7–1.2 mm, stamens 10–15, anthers 0.8–0.9 mm; cypselae ca. 6 mm, spirally coiled, plumose. Cracks and crevices of sandstone outcrops, shallow rocky soils, desert scrub, piñon-juniper woodland communities, 5100–10,900 ft. San Juan.

Cercocarpus montanus Raf. [*C. montanus* var. *argenteus* (Rydb.) F. L. Martin]—Alderleaf mountain mahogany, birchleaf mountain mahogany. Deciduous shrub, to 4 m, young branches reddish brown, older bark gray to brown, smooth; leaves light brown, blades ovate, rhombic, 12–50 × 8–20 mm, margins crenate-serrate, plane or slightly revolute, dark green, pale beneath; flowers with a persistent hypanthium tube, 6.5–11 mm, sepals 1–1.8 mm, rounded, recurved, stamens 22–44; cypselae 9–12 mm, spirally coiled, densely plumose. Sagebrush, piñon-juniper woodland, Gambel oak, ponderosa pine, Douglas-fir communities, 5000–7500 ft. Widespread except EC, SE.

CRATAEGUS L. – Hawthorn

Small trees and shrubs, usually armed with stout thorns; stems woody, young bark smooth and often reddish, becoming gray, rough, and scaly or checked; leaves deciduous, alternate, petiolate, stipules often glandular-toothed, small and caducous, blades simple, entire to variously toothed or lobed; inflorescence a corymb, cyme, or panicle; flowers perfect, white, showy, hypanthium cup-shaped to urceolate, sepals 5, petals ± circular, deciduous, stamens 5–25, ovary inferior; fruit a fleshy pome containing pyrenes.

1. Thorns mostly 1–4 cm; mature pome purplish black; leaves mostly unlobed, 2 times longer than wide…
   C. rivularis

1'. Thorns mostly 2.5–6 cm; mature pome dark red or reddish orange to dark brown; leaves with coarse toothing and shallow lobing, 1.5 times longer than wide…(2)
2. Pedicels villous at anthesis…(3)
2'. Pedicels glabrous at anthesis…(4)
3. First-year twigs golden-green to tan; autumn leaves yellow…**C. chrysocarpa**
3'. First-year twigs dark purple-bronze; autumn leaves brown…**C. macracantha**
4. Thorns on 2-year-old twigs grayish to brown, dull; fruit ellipsoid…**C. wootoniana**
4'. Thorns on 2-year-old twigs blackish to deep reddish purple, glossy; fruit orbicular…**C. erythropoda**

Crataegus chrysocarpa Ashe [*C. coloradoides* Ramaley]—Red hawthorn. Shrubs, 20–30 dm, twigs with new growth usually appressed-pubescent, 1-year-old twigs usually dull yellowish to greenish brown or gray-brown to light or dark tan, thorns straight to slightly recurved, 1-year-old thorns shiny, dark mahogany or black, ± slender to ± stout, 3–6 cm; leaf blades ovate to rhombic-ovate; inflorescences with branches sparsely to densely villous; flowers 15–20 mm diam., hypanthium densely villous, stamens 10; pomes red (darkened to burgundy only if dried and shriveled), suborbicular to broadly ellipsoid, 8–10 mm diam., hairy. Rocky slopes and canyon bottoms, with Gambel oak, New Mexico locust, and aspen, 7350–7770 ft. Colfax.

Crataegus erythropoda Ashe—Rocky Mountain hawthorn. Shrubs or trees, 50 dm; twigs with new growth greenish, glabrous, 1-year-old twigs dark reddish mahogany, thorns on twigs straight or slightly recurved, 2-year-old thorns black, shiny, moderately stout, 2–4 cm; leaf blades rhombic-elliptic, 3–5 cm, length 1.6 times width, ± coriaceous-shining, lobes 3 or 4 per side, margins serrate; inflorescences 5–10-flowered, branches glandular-punctate; flowers 14–18 mm diam., sepals narrowly triangular, 3–4 mm; pomes deep red to purple when mature, orbicular, 10 mm diam. Along streams and canyons in piñon-juniper, ponderosa pine, Douglas-fir, aspen communities, 5500–9500 ft. NC, NW, Catron.

Crataegus macracantha Lodd. ex Loudon—Big-spine hawthorn. Shrub or small tree to 7 m; twigs with new growth reddish green, usually glabrous, 1-year-old twigs dark, shiny reddish brown, becoming dark gray at 2–3 years, older ± paler gray, thorns on twigs usually numerous, recurved, shiny, dark blackish brown, older gray, stout, 3–7(–11) cm; leaves glabrous, blades mostly narrowly rhombic-elliptic to broadly rhombic-ovate, 4–7(–10) cm; inflorescence a flat-topped corymb; flower sepals narrowly lanceolate, 4–6 mm, petals 6–8 mm, white; fruit a globose to subglobose pome, ca. 7–11 mm diam., bright red to purplish, pubescent to glabrous. Our material belongs to **var. occidentalis** (Britton) Eggl. Wet to dry meadows and canyons, 6560–8900 ft. NC, Bernalillo, Catron.

Crataegus rivularis Nutt. [*C. douglasii* (Lindl.) var. *rivularis* (Nutt.) Sarg.]—Mountain river hawthorn. Shrubs or trees, 30–50 dm; twigs with new growth greenish, sparsely pubescent, at 1–2 years old often red-purple, thorns on twigs straight or ± recurved, at 2 years old black or purple-black, glossy, fine, 1.5–4 cm; leaf blades elliptic to narrowly elliptic, 3–8 cm, length 2+ times width, lobes 0 or small, margins serrate; inflorescences 6–12-flowered, branches glandular-punctate; flowers 14–17 mm diam., sepals broadly deltate, 6–8 mm, stamens 10; pomes deep red (mid-August), black or blackish purple when mature, 10 mm diam. Mostly in canyons, along streams in piñon-juniper woodlands and ponderosa pine communities, 6000–8300 ft. NW, Catron, Grant.

Crataegus wootoniana Eggl.—Wooton's hawthorn. Shrubs, 20 dm; twigs branched, thorny, glabrous, 1-year-old twigs reddish brown, older gray, thorns on 1-year-old twigs reddish brown, ± stout, 4 cm; leaf blades ovate, 3–4(–5) cm, lobes 3 or 4 per side; inflorescences 5–10-flowered, branches glabrous; flowers 10–12 mm diam., hypanthium glabrous, sepals 2–3 mm, stamens 10; pomes shiny red, ellipsoid, 8 mm. Canyon bottoms, near streams, forest understory in lower montane coniferous forest, 6500–8000 ft. C, WC, SC, Grant. **R & E**

DASIPHORA Raf. – Shrubby cinquefoil

Dasiphora fruticosa (L.) Rydb. [*Potentilla fruticosa* L.; *Pentaphylloides floribunda* (Pursh) Á. Löve; *P. fruticosa* (L.) O. Schwarz]—Shrubby cinquefoil. Deciduous shrub, mostly 1–15 dm, bark shiny brown, current year's growth villous, the stem becoming glabrous with shreddy, reddish-brown bark; leaves sericeous beneath; leaves with stipules, petioles to 15 mm; blades pinnate, mostly 5-foliate, leaflets ser-rate, narrowly elliptic, 0.8–1.8 cm; inflorescence cymose or of solitary axillary flowers; flowers perfect, 4–6 mm; sepals deltate-ovate, 4.5–6.5 mm; petals 7–12 mm, yellow; stamens 20–25; fruit numerous achenes. Moist meadows, mountain streams, 6000–13,030 ft. NC, NW, WC, SC, Grant.

DRYMOCALLIS Fourr. ex Rydb. – Sticky cinquefoil

Ours perennial herbs to 10 dm; herbage brownish-viscid, some hairs glandular; leaves pinnate with a well-defined terminal leaflet; flower petals white to cream, style fusiform, basally attached.

1. Basal leaves with (4)5–6(–10) pairs of leaflets, cauline with 4–6(–10) pairs; mostly E of the Continental Divide…**D. fissa**
1'. Basal leaves with (1)2–5 pairs of leaflets, cauline with 1–4 pairs; W of the Continental Divide…(2)
2. Inflorescence not leafy, congested; petals cream-white…**D. arguta**
2'. Inflorescence leafy, widely branched; petals usually yellow, sometimes cream-white…**D. glandulosa**

Drymocallis arguta (Pursh) Rydb. [*Potentilla arguta* Pursh]—Tall cinquefoil. Stems 5–10 dm, herbage brownish-viscid-villous, pubescence on stems and rachis of leaves shaggy, mostly unbranched below the inflorescence; leaves pinnate, 7–9-foliate, 1–2.5 dm; upper cauline leaves few and reduced upward; inflorescence a narrow, mostly flat-topped cyme, viscid-villous; flower bractlets lanceolate, 3–5 mm, sepals ovate, 5–10 mm; petals 4–8 mm, white, cream, or pale yellow, stamens 25–30; cypselae 0.8–1.5 mm, glabrous. Ponderosa pine, Gambel oak, aspen, spruce-fir communities, 8085–9900 ft. NE, NC.

Drymocallis fissa (Nutt.) Rydb. [*Potentilla fissa* Nutt.]—Leafy drymocallis. Stems mostly 1.5–3.5 dm; ± densely septate-glandular; leaves sparsely to moderately hairy, basal (3–)7–19 cm, leaflet pairs (4)5–6(–10), additional reduced leaflets sometimes interspersed; inflorescences 5–15-flowered, leafy, congested to ± open, 1/6–1/2 of stem, narrow to wide; flowers opening widely, bractlets linear-oblanceolate, 3–7 × 1–2 mm, sepals 6–10 mm, apex acute to acuminate, petals spreading, yellow, 7–11 × 5–11 mm; achenes light brown, 1 mm. Sagebrush slopes, open forests, stream banks, often in rocky or moderately disturbed sites, 7890–10,215. Colfax, San Miguel.

Drymocallis glandulosa (Lindl.) Rydb. [*Potentilla glandulosa* Lindl.]—Sticky drymocallis. Stems 2–6 dm; moderately to densely septate-glandular; leaves sparsely to moderately hairy, basal mostly 10–25 cm, leaflet pairs 2–3, cauline 1–3, developed or reduced; inflorescences (2–)5–40-flowered, leafy or not, open or of congested flowers opening widely, 2–6 × 0.5–2 mm, sepals 4.5–10 mm, apex usually broadly obtuse with mucronate tip, sometimes acute, petals spreading or reflexed, cream-white to yellow, narrowly to broadly obovate or ovate to nearly round, 3.5–5 × 3–4 mm; achenes reddish brown, 1–1.4 mm. Open slopes, stream banks, road banks, shrublands, open woodlands, 7800–9600 ft. DeBaca, Socorro.

FALLUGIA Endl. – Apache plume

Fallugia paradoxa (D. Don) Endl. ex Torr.—Apache plume. Deciduous or semievergreen shrubs to 2 m, unarmed; stems woody, much branched, erect, bark exfoliating, light gray with reddish fissures; leaves with small stipules, caducous, the larger ones 7–20 mm, pinnately 3–5(–9)-lobed, the larger lobes sometimes divided again, blades deeply pinnatifid, often in fascicles; inflorescence of solitary flowers or few-flowered loose cymes terminating stems; flowers mostly imperfect and the plants functionally dioecious, white, showy, hypanthium hemispherical, sepals 5, petals 5, subcircular, deciduous; stamens many, carpels many, free from each other and the hypanthium, the ovaries superior; fruit an achene with a greatly elongated plumose style. Desert scrub in sandy to rocky drainages; rocky uplands in grasslands, chaparral, piñon-juniper, oak, and ponderosa pine communities, 3150–9500 ft. Widespread except Curry, Lea, Quay.

FRAGARIA L. – Strawberry

Perennial herbs; stems scaly crowns terminating underground rhizomes, producing stolons that root at nodes; leaves in rosettes, trifoliate stipules adnate to bases of petiole, persistent, forming scales on crowns; leaflets coarsely serrate; inflorescence few- to several-flowered cymes or short racemes; flowers perfect or partially to wholly imperfect, white, showy; hypanthium saucer-shaped, sepals 5, petals 5, stamens 20–35; carpels many, the ovaries superior; fruit accessory, with many cypselae on surface of enlarged, fleshy, fragrant red receptacle.

1. Terminal tooth of the leaflets well developed, projecting beyond the adjacent lateral pair; inflorescence equal to or exserted beyond the leaves; leaves green or yellowish green on the upper surface, veiny…**F. vesca**
1'. Terminal tooth of the leaflets relatively small, usually shorter than the adjacent lateral pair; inflorescence usually surpassed by the leaves; leaves glaucous on the upper surface, not so veiny…**F. virginiana**

Fragaria vesca L.—Woodland strawberry. Stoloniferous, scapose herb from a branched caudex; pubescence of the scape, petiole, and pedicels spreading to ascending-pilose; leaves 3-foliate, lanceolate to broadly lanceolate, acuminate; leaflets ovate, mostly 3–6 × 2–4.5 cm, terminal leaflet the largest, coarsely crenate-serrate in the upper 2/3, terminal tooth projecting beyond the adjacent lateral teeth, all leaflets glaucous beneath and green to yellowish green above, veins well developed; inflorescence a cyme, exceeding the leaves, 0.5–2 dm, 3–15-flowered; flowers mostly 4.5–7 mm, sepals 4.5–8 mm, petals mostly 5–8 mm, rounded, white with some pinkish tinge; fruit a receptacle, becoming fleshy, to 1 cm broad; cypselae 1.3–1.4 mm. Shady and moist coniferous woodland and aspen communities, 5100–12,855 ft. Widespread except E and far S.

Fragaria virginiana Mill. [*F. glauca* (S. Watson) Rydb.]—Mountain strawberry. Stoloniferous, scapose herb from a branched caudex, scapes 0.2–2.5 dm, usually surpassed by the leaves, stolons and scapes sparsely to abundantly silky-villous beneath; leaves 3-foliate, leaflets narrowly to broadly obovate, 2.2–7 × 1.3–4.5 cm, coarsely crenate-serrate in the upper 1/3–2/3, the terminal tooth reduced and often surpassed by the adjacent lateral teeth; inflorescence a cyme, 2–15-flowered; flower sepals 4–8 mm, petals 5–12 mm, white or sometimes pinkish; fruit a fleshy receptacle, mostly 1 cm broad; cypselae 1.2–1.4 mm. Our material belongs to **var. glauca** S. Watson. Moist soils, meadows and coniferous forests, 6500–12,960 ft. Widespread except E and far S.

GEUM L. – Avens

Perennial herbs with rosettes terminating short vertical rootstocks or elongate horizontal rhizomes; stems leafy or scapose; leaves basal and sometimes also cauline, pinnately compound, leaflets toothed, the terminal one larger than lower ones; inflorescence several- to many-flowered open cymes or corymbs; flowers white to pink or yellow, showy, sepals 5, petals 5, stamens 20 to many, ovaries superior, styles terminal, of 2 types, entire and wholly persistent and becoming plumose on the fruit, or jointed and geniculate near the middle, with the apical part deciduous and leaving a hooked beak on the fruit; fruit an aggregate of achenes.

1. Style jointed, the lower part persistent, with a terminal hook; leaves pinnately compound, with few unequal leaflets…(2)
1'. Style continuous, without a hook; leaves with many narrow segments…(4)
2. Sepals purple; petals violet…**G. rivale**
2'. Sepals green; petals yellow…(3)
3. Portion of style below the hook glabrous or minutely pubescent, not glandular; terminal leaf segment not enlarged…**G. aleppicum**
3'. Portion of style below the hook with stalked glands; terminal leaf segment greatly enlarged…**G. macrophyllum**
4. Flowers nodding; petals pinkish…**G. triflorum**
4'. Flowers erect; petals yellow…**G. rossii**

Geum aleppicum Jacq.—Yellow avens. Perennial herb from a thick, fleshy crown; herbage and inflorescence hirsute, the hairs bulbous-based; stems few, erect, 4–12 dm, leafy; basal leaves well developed, 15–28 cm, pinnately divided, blades obovate, terminal leaflet 3–5-lobed, cauline leaves 3–5-foliate with stipules to 2 cm; inflorescence a few- to several-flowered cyme on long pedicels; flowers erect, sepals reflexed, 3–7.5 mm, petals spreading, 4–7 mm, yellow, stamens numerous, pistils 200–250, hairy;

cypselae 3.5–4 mm, mature style strongly geniculate and jointed, hooked at the apex. Wet meadows and along streams, 5000–10,805 ft. NC, C, WC, Otero, Grant.

Geum macrophyllum Willd.—Large-leaved avens. Perennial herb from a thick, scaly crown, herbage and inflorescence hirsute, stems partly glandular-puberulent; stems few, erect, to 12 dm; basal leaves well developed, mostly 10–25 cm, petiolate, pinnately divided, 3–5-lobed, the lobes coarsely toothed, cauline leaves alternate, smaller, 3(–7)-foliate, terminal leaflets 2.5–13 cm, base cordate or rounded; inflorescence a divergently branched 4–9-flowered cyme with long pedicels; flowers erect, sepals mostly 3–6 mm, petals spreading, 3.5–6.5 mm, yellow; stamens numerous, pistils ca. 250; cypselae 2.2–3.2 mm, mature style geniculate and jointed, glandular-puberulent, hooked at the apex. Our material belongs to **var. perincisum** (Rydb.) Raup. Wet meadows and stream banks, ponderosa pine, Douglas-fir, aspen, white fir, spruce-fir communities, 5500–11,210 ft. NC, NW, C, WC, Grant, Otero.

Geum rivale L.—Purple avens. Perennial; stems few, 25–60 cm; hirsute and glandular-pilose; basal leaves lyrate-pinnate, leaflets obovate, 2–10 cm, doubly serrate to deeply incised, cauline leaves ternate; inflorescence a 1–4-flowered cyme; flower sepals 8–12 mm, densely pilose, purplish, petals 6–10 mm, flesh-colored or yellow-tinged, purple-veined; receptacle short-hirsute, ± stalked in fruit, mature style geniculate, the lower segment ca. 7–9 mm, hirsute below and ± glandular-pubescent, articulated to lower by a curved and hooked joint; cypselae mostly 4 mm. Meadows in Douglas-fir zone and subalpine, 7800–10,620 ft. NC.

Geum rossii (R. Br.) Ser.—Alpine avens. Perennial herb from a thick caudex forming dense clumps to 3+ dm broad, herbage glabrous to sericeous or villous; stems erect, subscapose, 0.5–2.5 dm, simple or branched; basal leaves well developed, 3–14 cm, pinnately divided, the lateral leaflets mostly 7–15 mm, cleft 1/2 their length into narrow segments, cauline leaves reduced; inflorescence a 1–4-flowered cyme; flower hypanthium turbinate, 2.5–6 mm, purple-tinged, sepals erect, 3.5–5.5 mm, petals spreading, 5–9 mm, yellow, stamens numerous, receptacle short, rounded, pistils few to many; cypselae 2.5–4 mm. New Mexico material belongs to **var. turbinatum** (Rydb.) C. L. Hitchc. Alpine and subalpine meadows, 9000–13,165 ft. NC, C, SC.

Geum triflorum Pursh [*G. ciliatum* Pursh]—Old man's whiskers. Perennial herb from a thick caudex, branching to form clumps to 3+ dm broad, herbage hirsute or pilose; stems erect, 1.5–4 dm, subscapose; basal leaves well developed, mostly 6–15 cm, oblanceolate in outline, pinnately lobed, becoming somewhat pinnatifid above, the lateral leaflets 13–22 mm; inflorescence a 1–3(–9)-flowered cyme, the flowers nodding on long pedicels; flower hypanthium hemispherical, sepals erect, 7–13 mm, petals erect, 8–15 mm, white to pinkish, stamens numerous, receptacle short-clavate, to 2 mm, short-pubescent, pistils numerous; cypselae mostly 3 mm, pyriform, the mature style slightly geniculate and jointed, purplish, not hooked at the apex. Our material belongs to **var. ciliatum** (Pursh) Fassett. Mountain meadows, grassy and rocky slopes, open coniferous and aspen woodlands, mountain brush communities, 7150–12,400 ft. NC, NW, WC, SW.

HOLODISCUS (K. Koch) Maxim. – Oceanspray

Holodiscus discolor (Pursh) Maxim. [*Spiraea dumosa* Nutt. ex Hook.; *H. dumosus* (S. Watson) A. Heller]—Oceanspray. Shrub to 2 m, the terminal branches often becoming weakly spinescent; stems of the current year reddish tan; leaf blades oblanceolate to obovate, 2.5–3.5 × mostly 0.5–1.2 cm, toothed above the middle, pilose above, lower surface less pubescent and with sessile, glandular hairs; inflorescence a diffuse panicle, flowers subtended by bractlets; flower sepals ovate, 1.2–2 mm, petals spatulate, 1.7–2.5 mm, white to cream, hypanthium discoid to shallowly cupulate, 1.8–2.3 mm across, stamens 20, cypselae 1.5–2 mm, densely pilose with spreading hairs. Rocky sites, often steep canyon cliff faces, piñon-juniper woodland, ponderosa pine communities, 5750–12,205 ft. Widespread except E.

MALUS Mill. – Apple

Deciduous trees, unarmed or with thorns; stems woody, bark smooth with horizontal lenticels or with scaly plates; leaves deciduous, simple, toothed or lobed; stipules small, deciduous; inflorescence of corymbs, panicles, or umbels; flowers perfect, white to pink, showy, large nectar ring present, hypanthium saucer- to cup-shaped, sepals 5, petals circular or broadly elliptic, deciduous, stamens 15–30; carpels 3–5, ovary inferior; fruit a large, fleshy pome, red to yellow or green.

1. Leaves on elongate shoots lobed or notched…**M. ioensis**
1'. Leaves on elongate shoots neither lobed nor notched…**M. pumila**

Malus ioensis (Alph. Wood) Britton [*Pyrus ioensis* (Alph. Wood) L. H. Bailey]—Prairie crab apple. Small tree to 9 m, branchlets tomentose, sometimes spinescent; leaves ovate to oblong, 2.5–10 cm, tomentose on both sides; flower petals white to pink, 12–25 mm; fruit 2–3 cm diam. Cultivated tree that persists and escapes, 5100–8710 ft. Rio Arriba, San Miguel, Taos. Introduced.

Malus pumila Mill. [*M. domestica* (Suckow) Borkh.]—Common apple. Tree to 10 m, branchlets tomentose when young, becoming glabrous in age; leaves ovate to oblong or elliptic, 1.5–10 cm, tomentose on 1 or both sides; flower petals mostly white, often pink dorsally, 12–25 mm; fruit mostly 2.5–12 cm diam., red, reddish purple, or yellow. Widely cultivated, rarely escapes, 5300–8500 ft. NC, NW, C, Catron, Grant, Otero.

PERAPHYLLUM Nutt. – Peraphyllum

Peraphyllum ramosissimum Nutt.—Wild crab apple. Shrub 0.5–2 m, bark reddish on young growth, becoming gray, herbage appressed-pubescent; leaves simple, with petioles 1.5–4.5 mm, blades narrowly elliptic or narrowly oblanceolate, 2–5 × 0.4–1.2 cm; inflorescence a fascicle of 1, 2, or 3 flowers; flowers fragrant, hypanthium campanulate or cylindrical, sepals 2–5.5 mm, white-tomentose inside, petals 5–10 mm, white to pinkish, stamens 15–20; fruit a globose pome, 8–10(–12) mm diam., yellowish to reddish, with a bitter taste. Sagebrush, serviceberry, piñon-juniper woodland, Gambel oak communities, 5575–7700 ft. Rio Arriba, San Juan.

PETROPHYTUM (Nutt.) Rydb. – Rock-spiraea

Petrophytum caespitosum (Nutt.) Rydb. [*Spiraea caespitosa* Nutt.]—Rock-spiraea. Mat-forming shrub, herbage sericeous; stems much branched, to 25 cm; leaves simple, sessile, linear-spatulate to oblanceolate, 3–30 × mostly 2–4 mm apically; inflorescence a dense cylindrical to globular raceme, 2–4 cm wide; flowers turbinate or cupulate, 0.5–1 mm deep, sericeous, sepals 1–1.6 mm, petals 1.5–2 mm, white, stamens mostly 20; follicles 1.2–1.5 mm, pilose. Steep, exposed andesite porphyry and sandstone surfaces, rooting in cracks in the rock, 5900–8300 ft. C, WC, SE, SC, SW.

PHYSOCARPUS (Camb.) – Ninebark

Physocarpus monogynus (Torr.) J. M. Coult.—Mountain ninebark. Unarmed shrubs 4–20 dm; bark exfoliating; twigs glabrous to stellate-hairy; leaves alternate, ovate to orbicular-ovate, 3–6 cm, crenate or double-crenate on the margins; inflorescence a 15–20-flowered corymb; sepals 5, 2.5–3.2 mm; petals 5, 5–6 mm, white or cream; stamens 20–40; pistils 1–5; fruit a follicle, 2.5–4.5 mm, densely stellate-hairy. Along streams, steep wooded slopes, canyons, rocky crevices, (5400)6000–10,500 ft. NC, NE, WC, C, SC.

POTENTILLA L. – Cinquefoil

Perennial or rarely annual herbs; stems ascending to erect or spreading, rarely decumbent, often with long rhizomes with persistent leaf bases, plants in rosettes, cespitose, or with runners rooting at nodes; leaves cauline and basal, palmately or pinnately compound with 3–15 leaflets, leaflets entire, lobed, or serrate; inflorescence a narrow or open cyme, 1- to several-flowered, or flowers solitary at the nodes; flowers perfect, mostly pale to bright yellow, rarely dark red, showy, hypanthium saucer-shaped, sepals 5, petals 5, stamens 10–30, often 20; carpels many, ovaries superior; fruit accessory with many cypselae on a dry receptacle. A taxonomically diverse and difficult genus. Segregate genera recognized here are *Dasiphora* and *Drymocallis*.

1. Annual, biennial, or short-lived perennial; mostly weedy plants…(2)
1'. Perennial; native plants…(4)
2. Achene with a protuberance on the ventral suture…**P. supina**
2'. Achene lacking a protuberance…(3)
3. Leaves pinnate to subpalmate or palmate, leaflets 5–7; stem hairs not stiff, not tubercle-based…**P. rivalis**
3'. Leaves ternate or palmate, leaflets 3–5; stem hairs ± stiff, tubercle-based…**P. norvegica**
4. Style < 1.2 mm, mostly thick and conic, thick just below the stigma, shorter than the achene…(5)
4'. Style > 1.2 mm, usually much longer, usually thin just below the stigma and longer than the achene…(8)
5. Stems from centers of ephemeral basal rosettes, primary leaves cauline; primarily ± disturbed habitats …**P. recta**
5'. Stems lateral to persistent basal rosettes, primary leaves basal, mostly alpine and prairie habitats …(6)
6. Leaves ternate or palmate; alpine…**P. nivea**
6'. Leaves pinnate to subpalmate; other plant communities…(7)
7. Basal leaves pinnate, short hairs usually abundant to dense; leaflets mostly 3–7 per side, cottony and crisped hairs absent; ponderosa pine to subalpine…**P. pensylvanica**
7'. Basal leaves subpinnate to subpalmate, short hairs absent or sparse; leaflets mostly (1)2–3 per side, usually cottony hairs ± dense; shortgrass prairie, sagebrush, and disturbed sites…**P. bipinnatifida**
8. Plants densely glandular-puberulent and sparsely hirsute, 0.4–0.8 dm; leaves subpinnately compound with 5–7 leaflets, green above and below…**P. subviscosa**
8'. Plants unlike the above in all respects…(9)
9. Leaves pinnate with (5–)7+ leaflets…(10)
9'. Leaves digitate or subdigitate, or occasionally 5-digitate above with 1 or 2 extra leaflets separated by a long rachis internode…(17)
10. Plants stoloniferous; flowers solitary at the nodes…**P. anserina**
10'. Plants not stoloniferous; flowers few to several in cymes…(11)
11. Leaves subdigitate, with 7 leaflets…(12)
11'. Leaves definitely pinnate, with 9+ leaflets…(13)
12. Leaves with 3 terminal leaflets and 2 lower pairs, leaflets narrowly toothed; mostly alpine…**P. subjuga** (in part)
12'. Leaves highly variable, some pinnate and some digitate; upper leaflets mostly 5; lower montane to higher montane…**P. pulcherrima** (in part)
13. Stems decumbent to sometimes ascending, becoming prostrate; leaflets incised 3/4 to midvein…**P. plattensis**
13'. Stems not as above; leaflets not incised 3/4 to midvein…(14)
14. Leaves folded, 3-toothed in the upper 1/3, narrow, silvery-hirsute…**P. crinita** (in part)
14'. Leaves not folded, several-toothed throughout their length, broader, densely tomentose below, often bicolored…(15)
15. Leaves densely tomentose on lower surface…**P. hippiana**
15'. Leaves green on upper and lower surface, merely strigose…(16)
16. Stems 6–7 dm; leaflets coarsely serrate, not folded…**P. ambigens**
16'. Stems 1–4 dm; leaflets toothed at apex, folded…**P. crinita** (in part)
17. Leaves strictly digitate, no rachis between the leaflets; bicolored or green on both surfaces…(18)
17'. Leaves subdigitate, with at least a short rachis visible between the leaflets on some leaves; surfaces various…(25)
18. Leaflets green on both surfaces…**P. pulcherrima** (in part)
18'. Leaflets green above and white-tomentose beneath…(19)
19. Stems 2–8 dm; leaflets mostly 7…(20)
19'. Stems 0.2–2.8 dm; leaflets mostly 5…(23)
20. Glands in the inflorescence usually conspicuous, red-tipped…**P. pucherrima** (in part)
20'. Glands in the inflorescence absent or inconspicuous and not red-tipped…(21)
21. Leaflets blue-green and glaucous or less often greenish, 0.5–2.5(–4) cm, usually toothed in the upper 1/2 only, with 1–3(–5) teeth per side; anthers 0.35–0.7 mm…**P. glaucophylla**
21'. Leaflets green or grayish green, not glaucous, 2–5(–10) cm, toothed most of their length with (2–)5–10 teeth per side; anthers 0.7–1.4…(22)
22. Leaflets narrowly elliptic, incised 1/4–1/2 the distance to the midrib, the teeth 1–2 mm…**P. townsendii**
22'. Leaflets elliptic to obovate, incised 1/2–3/4+ the distance to the midrib, the teeth > 2 mm…**P. gracilis**
23. Leaflets toothed to lobed or incised for most of their length…**P. concinna**

23'. Leaflets entire or only 2-3-toothed…(24)
24. Leaflets white-tomentose beneath…**P. bicrenata**
24'. Leaflets green (but may be sparsely villous) beneath…**P. sierra-blancae**
25. Petals rose-red or dark red…**P. thurberi**
25'. Petals yellow…(26)
26. Leaves with 3 terminal leaflets and 2 lower pairs, leaflets narrowly toothed; mostly alpine…**P. subjuga** (in part)
26'. Leaves highly variable, some pinnate and some digitate; upper leaflets mostly 5; lower montane to higher montane…**P. pulcherrima** (in part)

Potentilla ambigens Greene—Silkyleaf cinquefoil. Perennial; stems stout, silky-villous, to 7 dm; basal leaves pinnate, to 20 cm, leaflets 9–15-foliate, larger leaflets 3–6 cm, coarsely serrate, silky-villous below, ± glabrous above, color contrast not striking; inflorescence a narrow cyme; flower bractlets as long as or longer than the sepals, sepals mostly 6–7 mm, lanceolate, strigose, petals ca. 8 mm, yellow; style filiform near the apex; cypselae glabrous. Dry meadows in coniferous woodlands, 6900–10,090 ft. NC, C, WC, Grant, Lincoln, Otero, Union.

Potentilla anserina L.—Silverweed. Stoloniferous perennial, the stolons to 50 cm and producing a rosette of leaves at each node; leaves pinnate, 8–20 cm, 9–25-foliate, larger leaflets oblanceolate to cuneate-obovate, 1–4 cm, toothed or lobed, sparsely appressed-villous to glabrate above, white-tomentose beneath; flowers solitary at the stolon nodes, silky-tomentose, bractlets lanceolate or ovate-elliptic, 4–7 mm; flower sepals 3–5 mm, petals obovate, 5.5–11 mm, yellow, stamens 20–25, style 1.5–2.5 mm; cypselae ca. 2 mm, light brown. Floodplains, lakeshores, meadows, other wet areas, 5000–10,300 ft. Widespread except E, S.

Potentilla bicrenata Rydb. [*P. concinna* Richardson var. *bicrenata* (Rydb.) S. L. Welsh & B. C. Johnst.]—Elegant cinquefoil. Perennial herb from a branched caudex; stems ascending at anthesis, prostrate in fruit, 0.2–0.8 dm; leaves mostly basal, digitate, 5–7(–9)-foliate, leaflets oblanceolate, 10–40 mm, the terminal one the longest, entire or 2- or 3-toothed at the apex, green above and white-tomentose beneath; inflorescence mostly a 2- or 3-flowered cyme; flower sepals dentate-lanceolate, 2.5–5.5 mm, petals obovate, 4–6 mm, yellow; stamens 20, style 1.5–2.5 mm; cypselae 1.7–2 mm. Ponderosa pine and sagebrush communities, 6775–9200 ft. NC, W, WC.

Potentilla bipinnatifida Douglas—Tansy cinquefoil. Perennial, 1–5 dm; leaves 2–6 × 10 cm, pinnately compound, 5–7 leaflets, deeply pinnatifid, with linear, rounded segments 6–15 mm, green above, grayish-tomentulose below, margins revolute; flower sepals 3–6 mm, petals yellow, 3–5 mm, style 1–1.2 mm; cypselae 1–1.2 mm. Meadows, dry grasslands, sagebrush communities; Capitan and Sacramento Mountains, 7300–10,500 ft. Lincoln, Taos.

Potentilla concinna Richardson—Elegant cinquefoil. Perennial herb from a branched caudex; stems numerous, ascending in flower, prostrate in fruit, 0.5–3 dm; leaves mostly basal, digitate, 5–7(–9)-foliate, sometimes subpinnate, leaflets 7–20 mm, the terminal one longest, toothed to shallowly lobed, oblanceolate to obovate, often folded; inflorescence a 1–7-flowered cyme, divaricately branched; flower sepals 3.5–5 mm, petals obovate to oblanceolate, 4–6.5 mm, yellow, stamens 20, style 1.3–2.5 mm; cypselae 1.7–2 mm. Spruce-fir communities to alpine, 8950–12,935 ft. NC, WC, Sierra.

Potentilla crinita A. Gray—Bearded cinquefoil. Perennial herb, 2–5 dm from a branched caudex; stems ascending to erect; leaves strigose-sericeous, mostly basal, pinnate, mostly 11–13-foliate, 4–20 cm, leaflets oblanceolate to narrowly oblanceolate, 10–30 mm, shallowly toothed beyond the middle or at the apex only; inflorescence a branched cyme, strigose-sericeous; flower sepals mostly 4–6 mm, petals mostly 4.5–7.5 mm, yellow, stamens 20, styles ca. 1.8–2.5 mm; cypselae 1.5–1.7 mm, smooth to weakly rugulose. Meadows, piñon-juniper woodland, ponderosa pine, aspen, Gambel oak communities, 6230–8800 ft. NC, WC, SW.

Potentilla glaucophylla Lehm. [*P. diversifolia* Lehm.; *P. diversifolia* Lehm. var. *glaucophylla* (Lehm.) Lehm.]—Blueleaf cinquefoil. Stems ± ascending, to 4.5 dm; basal leaves often 2-ranked, palmate, sometimes subpalmate, 2–20 cm, leaflets mostly 5–7, largest oblanceolate or cuneate to obovate, 1–4(–6) × 0.5–1.5(–2) cm, ± evenly incised halfway or nearly to midvein, ± hairy, blue-green, usually glaucous, cauline hairs (0)1–2(3) mm; flower sepals mostly 2.5–4.5 mm, petals short-acuminate, (4–)5–10(–12) × 4–9(–10) mm, styles filiform above papillate-swollen base, 1.5–2.5(–3) mm; cypselae 1.2–1.6 mm. Meadows on the edge of spruce-fir forests, alpine tundra, 8640–12,700 ft. NC, NW.

Potentilla gracilis Douglas ex Hook.—Slender cinquefoil. Perennial, 2–9 dm, lacking glands or glands lacking red tips; leaves usually palmately compound with 5–9 leaflets, the leaflets mostly 2.5–10 cm, shallowly toothed to pinnatifid with linear segments, sparsely to densely hairy; flower sepals 4–9 mm, petals yellow, 5–9 mm, style 1.6–2.5 mm, filiform; cypselae 1.1–1.4 mm, smooth. NC.

1. Leaflets distinctly bicolored, white below, the margins deeply dissected 2/3+ the distance to the midrib… **var. elmeri**
1′. Leaflets usually about equally grayish green above and below, the margins dissected ca. 1/4–1/2 the distance to the midrib…**var. fastigiata**

var. elmeri (Rydb.) Jeps.—Combleaf cinquefoil. Valley bottoms and meadows, 6900–10,280 ft.

var. fastigiata (Nutt.) S. Watson—Slender cinquefoil. Meadows in montane and subalpine communities, 8200–12,435 ft.

Potentilla hippiana Lehm.—Woolly cinquefoil. Perennial herb; stems 0.6–5 dm, ascending to erect, sericeous; leaves mostly basal, pinnate, 5–13-foliate, 3–14 cm, leaflets densely white-tomentose beneath, 12–40 mm, the upper 3 the largest, oblanceolate, deeply cleft to nearly halfway to the midrib or less, cauline leaves few, ovate to lanceolate; inflorescence an open cyme; flower sepals 4–6.5 mm, petals 4–6.2 mm, slightly longer than the sepals, yellow, stamens ca. 20, style 1.7–2.3 mm; cypselae ca. 1.5 mm, smooth. Meadows, ponderosa pine, Douglas-fir, aspen, spruce-fir woodland communities, 7870–11,800 ft. N, C, WC, SW, Otero. Often hybridizes with *P. pulcherrima*.

Potentilla nivea L.—Snow cinquefoil. Mat-forming perennial herb; stems arachnoid-lanate or spreading-hirsute, decumbent to ascending, 0.2–2.5 dm; leaves green above, white-tomentose below, mostly basal, 3-foliate, leaflets obovate to flabellate, 5–15 mm, coarsely toothed; inflorescence a 1- to few-flowered cyme; flower petals 4–5.5 mm, yellow, stamens ca. 20, style 0.8–1.1 mm, thickened at the base; cypselae ca. 1.5 mm, smooth. Rocky slopes and ridges, alpine communities, 11,600–12,800 ft. Taos.

Potentilla norvegica L.—Norwegian cinquefoil. Erect annual, biennial, or short-lived perennial herb, 1–7 dm, from a slender taproot, herbage hirsute with stiff spreading hairs; stems leafy, few to several, usually branched in the upper part; leaves mostly cauline, reduced upward, 3-foliate, leaflets oblanceolate to obovate, 1.5–4 cm, shallowly lobed to toothed or coarsely serrate; inflorescence a leafy cyme, flowers inconspicuous; flower sepals 3.5–8.5 mm, petals mostly shorter than the sepals, 2–4 mm, yellow, stamens 15–20, style 0.7–0.8 mm, thickened below the middle, subterminally attached; cypselae 0.8–1.1 mm, often becoming rugulose at maturity, brown. Lakeshores, irrigation ditches, meadows, other moist sites, 4500–10,440 ft. N, C, WC, Doña Ana, Lincoln.

Potentilla pensylvanica L. [*P. strigosa* (Pursh) Pall. ex Tratt.]—Pennsylvania cinquefoil. Perennial herb; stems few to several, erect to decumbent, 0.5–3 dm, spreading-villous and substrigose, grayish green; leaves mostly basal, 5–15 cm, with 2–6(7) lateral pairs of leaflets, green, strigose above and grayish-tomentulose beneath, pinnatifid, cleft halfway to the midrib, margins revolute; inflorescence a several-flowered glomerulate or open cyme; flower sepals 3–6 mm, petals 3–6 mm, pale yellow, stamens mostly 20, style 0.8–1.2 mm, thickened at base; cypselae 1.1–1.4 mm. Rocky sites, ponderosa pine, black sage, aspen, spruce-fir communities, 6320–11,550 ft. N, C, WC, Doña Ana.

Potentilla plattensis Nutt.—Platte cinquefoil. Plants rosetted to tufted; stems initially decumbent to sometimes ascending, becoming prostrate, 0.3–4.5 dm, basal leaves pinnate, straight hairs common, tightly appressed, primary lateral leaflets 3–8 per side, largest ones obovate, 0.5–2 × 0.5–1.3 cm, margins pinnately incised 3/4+ to midvein, surfaces green to grayish green, not glaucous; inflorescences mostly 3–15-flowered, loosely cymose; flower sepals 3–6 mm, petals 4–7 × 3–6 mm, style 1.5–2.5 mm; achenes

(1.3–)1.5–1.9 mm, smooth. Moist meadows, streamsides, reservoir margins, 8600–10,320 ft. Colfax, Santa Fe, Taos.

Potentilla pulcherrima Lehm. [*P. diversifolia* Lehm.; *P. gracilis* Douglas ex Hook. var. *pulcherrima* (Lehm.) Fernald]—Beautiful cinquefoil. Perennial herb from a branched caudex; stems erect or ascending, 2–8 dm, sparsely to densely pubescent with spreading or appressed hairs; leaves mostly basal, digitate, 5–9-foliate, leaflets oblanceolate to obovate, mostly 2.5–6 cm, shallowly toothed to pinnatifid, green above and whitish-lanate beneath; inflorescence a 2–3(–8)-flowered cyme; flower sepals mostly 2.6–5 mm, petals 4–6.5 mm, yellow, stamens 20, style 1.5–2.5 mm; cypselae 1.7–2.1 mm. Moist meadows and open woods to alpine communities, 6970–12,590 ft. NC, C, WC, SC, SW.

Potentilla recta L. [*P. sulphurea* Lam.]—Sulphur cinquefoil. Perennials, (2–)3–5(–8) dm, hirsute; leaves palmately compound with 5–7 leaflets, 1.5–10 × 0.5–3.5 cm, long hairs common; flower sepals 5–9(–12) mm, petals yellow, 7–11(–13) mm, style thickened, warty; cypselae 1.2–1.8 mm. Meadows and along streams, 6675–7300 ft. San Miguel, Santa Fe.

Potentilla rivalis Nutt.—Brook or river cinquefoil. Stems decumbent to erect, sometimes prostrate, 0.5–7 dm, hairs at base not stiff; leaves ternate, palmate, or subpalmate, basal mostly 3–15 cm, leaflets 3–5(–7), largest ones oblanceolate-elliptic to obovate, 1–5 × 0.5–2(–2.5) cm, 1/2–3/4 of margin evenly to unevenly incised 1/3–1/2 to midvein, surfaces moderately to abundantly hairy; inflorescences 20–100+-flowered; flower sepals 3–5 mm, petals 1.5–2 × 1 mm, pale yellow to yellow, stamens 5–15, style 0.5–0.6 mm; cypselae yellowish, 0.7–0.9 mm, ± smooth. Moist meadows, stream banks, lakeshores, gravel bars in floodplains, drying marshes, open areas in river-bottom forests, 3850–9710 ft. NC, WC, SW, McKinley.

Potentilla sierrae-blancae Wooton & Rydb.—Stems 0.2–1 dm; basal leaves palmate, 2–8 cm; straight hairs sparse to abundant, sometimes absent, ± appressed, usually ± stiff, leaflets (3)4–5, margins incised 1/2–3/4 to midvein, green, straight hairs mostly absent (except on margins); inflorescences solitary flowers or 2–3-flowered; flower sepals 3–5 mm, petals 5–7 × 4–6 mm, style 2 mm; achenes 1.5 mm, smooth. Windswept barren ridges, subalpine grasslands, rock outcrops, 8000–12,000 ft. Lincoln, Otero. **E**

Potentilla subjuga Rydb.—Colorado cinquefoil. Perennial from a woody caudex; stems 1–3 dm, tufted, silky-villous; leaves digitate, (3–)5-foliate with an additional smaller pair of leaflets on what appears to be the petiole, leaflets 1–4 cm, oblong or oblanceolate to obovate, deeply incised, silky, denser below; flower sepals 5–6 mm, petals larger than the sepals, yellow, style filiform, > 1.2 mm; cypselae glabrous. Alpine communities, 10,000–12,850 ft. NC.

Potentilla subviscosa Greene—Navajo cinquefoil. Perennials, stems to 1.1 dm; leaves palmately or subpinnately compound, with 5(–7) leaflets, leaflets 1.5–2.5(–5) cm, lobed, lobes obovate to oblanceolate, glandular, hirsute, green above and below; flower sepals 2.8–5 mm, petals yellow, (3–)4–8 mm, style 2–3 mm; cypselae 1.5–1.6 mm. Piñon-juniper woodlands to mixed conifer forests, 6525–9100 ft. NC, WC, Otero.

Potentilla supina L. [*P. paradoxa* Nutt.]—Contrary cinquefoil. Annual or short-lived perennial herb from a slender taproot, herbage villous-hirsute with spreading hairs; stems leafy, spreading to ascending, 1–5 dm, often branched in the upper part, divaricate; leaves mostly cauline, reduced upward, pinnate, 5–9-foliate, 4–10 cm, leaflets lanceolate to ovate, 10–27 mm; inflorescence a divaricately branched cyme; flower sepals 2.8–6 mm, petals 2.5–3 mm, yellow, stamens 20+, style 0.5–0.6 mm, thickened at the base; cypselae 0.7–1 mm, brown, rugose, with a protuberance on the ventral suture. Shorelines of lakes and reservoirs, 3850–10,580 ft. NC, C, Doña Ana, Grant, San Juan.

Potentilla thurberi A. Gray—Thurber's cinquefoil. Perennial herb with horizontal rhizomes; stems 2–6 dm, erect or nearly so, sparsely villous; basal leaves palmate, (2–)4–15(–30) cm, with 5–7 leaflets 1–5 cm, oblong to oblanceolate, toothed, abaxially pale green to white, straight hairs sparse to abundant, adaxially green; inflorescence (4–)10–35-flowered; flower bractlets 2–2.5 mm, sepals villous, 4–5 mm, petals broad, longer than the sepals, dark red or red, style 2.5-3.5 mm; achenes 1.5 mm, ± rugose. C, WC, SC, SW. McKinley, San Miguel.

Potentilla townsendii Rydb.—Townsend's cinquefoil. Perennials; stems 2-6 dm, leaves palmately compound, 4-20 cm, hairy, leaflets 5-7, largest 2-6 cm, elliptic-oblanceolate, green; flower sepals 4-9 mm, petals yellow, slightly longer than the sepals. Mixed conifer, spruce-fir, alpine communities, 9800-12,600 ft. NC, Otero.

POTERIUM L. - Burnet

Poterium sanguisorba L.—Burnet. Perennial from a branched caudex; stems simple or branched above, to 7 dm; herbage glabrous or sparsely pilose with moniliform hairs; leaves mostly basal, pinnate, basal and lower cauline leaves mostly 10-15 cm, 11-21-foliate, blades ovate to ovate-flabellate; leaflets 10-25 mm, crenate-toothed; inflorescence a globose to ovoid spike, 10-25 × 7-12 mm; flower sepals 4, 4-5 mm, greenish brown, often red-tinged; petals none; stamens mostly 12, fruiting hypanthium pyriform, 3-5 mm. Open and disturbed sites, often in piñon-juniper woodland, sagebrush and ponderosa pine communities, 4650-9400 ft. New Mexico material belongs to **subsp. muricatum** Bonnier & Layens. Introduced.

PRUNUS L. - Cherry, plum

Deciduous (in ours) trees or shrubs, unarmed or with thorns; stems woody, bark often reddish, thin, smooth, with elongate horizontal lenticels; leaves simple, usually serrate, with prominent gland on petioles or base of blades; stipules small, caducous; inflorescences of umbels, corymbs, racemes, or solitary flowers; flowers perfect, white to pink, showy, large nectar disk present, hypanthium mostly cup-shaped, free from carpels, the ovary superior, sepals 5, deciduous; petals spreading, quickly falling; stamens usually 15-20; carpel 1; fruit a drupe, red to purple or black, rarely yellow; pits very hard, indehiscent, smooth or variously textured.

1. Inflorescence consisting of 15+ flowers in an elongate raceme; a common native species…(2)
1.' Inflorescence consisting of 15 or fewer flowers in a corymbose raceme, or umbel, or solitary; a rare native or cultivated species…(3)
2. Leaf margins crenulate-serrulate to serrate, teeth incurved or appressed; lateral veins 15-30 per side; sepal margins usually entire…**P. serotina**
2.' Leaf margins serrulate to serrate, teeth ascending to spreading; lateral veins 6-18 per side; sepal margins usually toothed, erose…**P. virginiana**
3. Drupes hairy…(4)
3.' Drupes glabrous…(5)
4. Leaf blades broadly ovate to suborbiculate, abaxial surface with tufts of hairs in vein axils; stones not pitted…**P. armeniaca**
4.' Leaf blades lanceolate to oblong, abaxial surface glabrous; stones pitted…**P. persica**
5. Twigs with terminal end buds; plants not thorny; stones not or slightly flattened…(6)
5.' Twigs with axillary end buds; plants sometimes thorny; stones slightly or strongly flattened…(8)
6. Leaf apices usually rounded to obtuse, rarely acute…**P. emarginata**
6.' Leaf apices acute to abruptly acuminate…(7)
7. Leaf abaxial surfaces moderately hairy, especially along midribs and veins, blades mostly 7-14 cm; petioles mostly 20-40 mm, with 1-3 glands distally and/or glands on margins at bases of blades…**P. avium**
7.' Leaf abaxial surfaces glabrous or glabrate, blades mostly 4.4-6 cm; petioles 10-24 mm, usually eglandular, sometimes with glands on margins at bases of blades…**P. cerasus**
8. Leaf blade margins with sharp teeth, usually eglandular, sometimes with glands, especially near base of blade…(9)
8.' Leaf blade margins with blunt teeth, usually glandular…(10)
9. Abaxial leaf surfaces glabrous or sparsely hairy along main veins; usually hairy adaxially, rarely glabrous, usually eglandular; widespread…**P. americana**
9.' Abaxial leaf surfaces moderately to densely hairy; usually glandular distally; plains…**P. gracilis**
10. Trees 10-50 dm, thorny; leaf blades usually lanceolate to narrowly elliptic; apex acute…**P. angustifolia**
10.' Trees 40-80 dm, not or slightly thorny; leaf blades ovate, elliptic, or obovate; apex obtuse to acute [a cultivar]…P. cerasifera

Prunus americana Marshall—American plum. Mostly shrubs, rarely treelike, to 5 m, glabrous; leaf blades to 7 × 0.5-3 cm, elliptic to ovate or lanceolate, serrate, apex acuminate to long-attenuate, acute to obtuse basally; inflorescence of 1-4 sessile or subsessile umbels, pedicels to 12 mm; flower sepals puberulent, spreading, petals white, 5-7 × 2.5-3 mm; fruit a yellow to red plum, glabrous. Cultivated and escaped, 5670-8000 ft. N, NC, Grant, Sierra.

Prunus angustifolia Marshall—Chickasaw plum. Shrubs or trees, often suckering, 10-50 dm, thorny; twigs with axillary end buds, glabrous; leaf blades lanceolate to narrowly elliptic, 1.5-6 × 0.8-2 cm, margins crenulate-serrulate, teeth blunt, glandular, glands reddish orange, apex acute; inflorescences 2-4-flowered, umbellate fascicles, pedicels 3-10 mm, glabrous; flower sepals ovate, 1-2 mm, margins entire, petals white, 3-6 mm; drupes red to yellow, 15-20 mm, glabrous, mesocarps fleshy, stones ovoid, ± flattened. Thickets, upland sandy soil, open woods, sand dunes, fencerows, pastures, roadsides, stream bottoms, 4165-6930 ft. Curry, Roosevelt.

Prunus armeniaca L.—Apricot. Small trees to 8 m, branchlets green to brown; leaf blades 1.5-7 × 1-6 cm, cordate-ovate to (ob)ovate, serrate, attenuate, obtuse to cordate basally, glabrous; inflorescence a solitary flower, pedicel short; flower sepals glandular, petals white to pink, 8-12 mm; fruit pubescent, fleshy, the stone longitudinally furrowed, compressed. Cultivated and escaped, 5500-7380 ft. Scattered; Colfax, Lincoln, San Juan, Santa Fe, Taos.

Prunus avium (L.) L.—Sweet cherry. Trees to 8 m, branchlets brown; leaf blades to 15 × 2.5-8 cm, oblanceolate to obovate, serrate or doubly serrate, attenuate, obtuse to rounded basally, glabrous to long-hairy; inflorescence of 2-4 flowers per bud; flower sepals glabrous, petals white to pink, 8-14 mm; fruit glabrous, red to black, the stone globose or ovoid. Cultivated and escaped, 4900-7000 ft. San Juan.

Prunus cerasifera Ehrh.—Cherry plum. Trees, sometimes suckering, 40-80 dm, not or slightly thorny; twigs with axillary end buds, glabrous; leaf blades ovate, elliptic, or obovate, 3-7 × 1.5-3.5 cm, glabrous except for a few hairs on adaxial surface, eglandular, margins singly to doubly crenate-serrate, teeth blunt, glandular; inflorescences usually solitary flowers, sometimes 2-flowered fascicles; flower sepals oblong-ovate, 2-4 mm, petals white, 7-14 mm; drupes purple-red to yellow, ovoid, ellipsoid, or globose, 15-30 mm, glabrous; mesocarps fleshy; stones ellipsoid to ovoid, ± to strongly flattened. A cultivar that rarely escapes, 5000-6500 ft.

Prunus cerasus L.—Sour or pie cherry. Shrubs or trees, suckering, 30-50(-80) dm, not thorny; twigs with terminal end buds, glabrous; leaf blades broadly elliptic to ovate or obovate, 4.4-6(-8) × 2.8-4(-6) cm, glabrous, usually eglandular, margins doubly crenate-serrate, teeth blunt, glandular, apex acute to abruptly acuminate; inflorescences 1-4-flowered, umbellate fascicles; flower sepals oblong, 4-7 mm, petals white, suborbiculate, 10-14 mm; drupes bright red, globose, 13-20 mm, glabrous; mesocarps fleshy; stones subglobose, not flattened. Roadsides, thickets, woodland borders, abandoned fields, 4900-6600 ft. McKinley, San Juan.

Prunus emarginata (Dougl.) D. Dietr.—Bitter cherry. Shrub to 6 m, the branches erect to spreading, the young stems with smooth reddish bark, herbage glabrous; leaves in fascicles on short lateral branches or spread along elongating stems, blades broadly oblanceolate to obovate, 1.7-4.5 × 0.5-2.5 cm; inflorescence a 6-10(-12)-flowered corymbose raceme; flower sepals ovate-rounded, 1.5-2.5 mm, petals 3-4.5 mm; drupes 7-8 mm thick, bright red with pulpy exocarp, glabrous. Gravelly or sandy soil along streams, rocky mountain slopes, subalpine, thickets on exposed sites, cutover and burned areas, understory of coniferous and oak forests, 6035-9400 ft. NE, NC, WC, Grant, Otero, Sandoval.

Prunus gracilis Engelm. & A. Gray—Oklahoma plum. Shrubs, suckering, 3-15 dm, not thorny; twigs with axillary end buds, densely hairy; leaf blades usually ovate or elliptic, rarely obovate, 2-5(-7) × 1-2.5 (-3.6) cm, margins singly or doubly serrulate, teeth sharp, usually eglandular, apex acute to obtuse, abaxial surface densely hairy, adaxial sparsely hairy; inflorescences 2-4(-6)-flowered, umbellate fascicles; flower sepals ovate-oblong, 1.5-2.5 mm, surfaces hairy, petals white, oblong to obovate, 4-7 mm; drupes yellow, orange, or red, globose to ellipsoid, 9-18 mm, glabrous; mesocarps fleshy; stones subglobose to ellipsoid, ± flattened. Sandy roadsides, upland thickets, open woods, waste places, 3500-4100 ft. Chaves, Quay.

Prunus persica (L.) Batsch—Peach. Trees to 4 m, branchlets green, aging to gray, glabrous; leaf blades to 15 × 0.5-5.5 cm, oblong-lanceolate to narrowly (ob)lanceolate, serrate, attenuate, obtuse to acute basally, glabrous; inflorescence solitary; flower sepals villous at the margin, petals white to pink or red; fruit pubescent, fleshy at maturity, the stone usually furrowed on the margin, somewhat compressed. Cultivated and escaped, 5200-8165 ft. Scattered; SC, Bernalillo, Colfax, Grant, Roosevelt, Santa Fe.

Prunus serotina Ehrh.—Black or rum cherry. Shrubs or trees, not suckering, 40-400 dm, not thorny; twigs with terminal end buds, glabrous or hairy; leaf blades usually narrowly elliptic, oblong-elliptic, or obovate, sometimes lanceolate, rarely ovate, 2-13.5 × 1.1-6.5 cm, margins crenulate-serrulate to serrate, inflorescences 18-55(-90)-flowered racemes; flower sepals 0.5-1.5 mm, petals white, 2-4 mm; drupes dark purple to nearly black, globose, stones subglobose, not flattened. New Mexico material belongs to **var. rufula** (Wooton & Standl.) McVaugh [*P. serotina* Ehrh. var. *virens* (Wooton & Standl.) McVaugh]. Along streams, moist slopes in canyons, mixed oak-pine-juniper woodlands, 4335-9000 ft. NE, NC, NW, C, WC, SC, SW.

Prunus virginiana L.—Chokecherry. Shrub or small tree to 8 m, young stems puberulent, greenish at first, soon becoming glabrous, reddish, older stems ashy-gray with a reddish-brown undertone; leaf petioles 1-2.5 cm, with a pair of distal reddish glands, blades (ob)ovate or elliptic, 4-12 × 1.5-4.5(-7.5) cm, serrulate with ascending teeth, dark green above and pale green beneath; inflorescence an elongate raceme, 5-14(-20) cm, with numerous flowers; flower sepals 0.6-1.3(-2) mm, petals 3-5(-6.5) mm, white; drupes 6-8 mm diam., becoming dark red or bluish purple to nearly black. J. R. Rohrer states, "I believe that fruit color and hairiness are worthless taxonomic characters in the chokecherry (*P. virginiana*). Thus, I treated the western plants as one taxon. Almost all of the New Mexican plants are the western variety, *Prunus virginiana* var. *demissa*." The eastern variety is only in the northwest corner of the state adjacent to the Oklahoma Panhandle (pers. comm., March 1, 2016).

1. Racemes (18-)40-70(-95) mm; petals (2-)2.5-4 mm; leaf blades usually obovate, length of larger leaves < 2 times width…**var. virginiana**
1'. Racemes (30-)60-110(-130) mm; petals 4-5(-7) mm; leaf blades elliptic to oblanceolate, length of larger leaves at least 2 times width…**var. demissa**

var. demissa (Nutt.) Torr. [*P. virginiana* var. *melanoarpa* (A. Nelson) Sarg.]—Western chokecherry. Leaf blades elliptic to oblanceolate, length of larger leaves at least 2 times width; inflorescences racemes, (30-)60-110(-130) mm; flower hypanthium 2-3 mm, sepals 0.7-1.4 mm, petals 4-5(-7) mm. Canyons, gullies, mountain slopes, stream banks and terraces, chaparral, pine-oak forests, thickets, grassy rocky slopes, bluffs, forest edges, roadsides, 5500-11,200 ft. NC, C.

var. virginiana—Eastern chokecherry. Leaf blades usually obovate, sometimes elliptic to ovate, length of larger leaves < 2 times width; inflorescences racemes, (18-)40-70(-95) mm; flower hypanthium 1.5-2 mm, sepals 0.7-1 mm, petals (2-)2.5-4 mm. Stream banks, fencerows, roadsides, hillsides, 4800 ft. N, C, WC, SE, SC, Grant.

PURSHIA DC. ex Poir. – Bitterbrush

Fragrant shrubs or small trees, mostly deciduous; stems much branched, woody, bark thin and gray to brown or shredding; leaves simple, pinnatifid or apically 3-lobed; inflorescence of solitary flowers terminating short shoots; flowers mostly perfect, white to cream or yellow, fragrant, showy, sepals 5, petals 5, stamens numerous, ovaries superior; fruit an achene, either large with a short, nonplumose style or smaller with a long, plumose style.

1. Carpels several; style long, plumose…**P. stansburyana**
1'. Carpel 1(2-3); style short, not plumose…**P. tridentata**

Purshia stansburyana (Torr.) Henrickson [*Cowania stansburyana* Torr.; *C. mexicana* D. Don var. *stansburyana* (Torr.) Jeps.]—Cliffrose, cowania. Shrubs to 7.5 m, with ascending branches, reddish-brown bark becoming gray and exfoliating in age; leaves fasciculate, revolute-margined, the larger ones 15 mm, pinnately (3-)5-7-lobed, green, finely pubescent along the midrib above, white-tomentose beneath; inflorescence of solitary flowers, pedicels and hypanthium stipitate-glandular; flowers perfect or rarely staminate or pistillate, sepals 3-5 mm, tomentose on the margins, petals 7-14 mm, white, cream, or pale yellow; cypselae 4-12, densely plumose. Rocky foothills, along washes and canyon bottoms, salt desert scrub, piñon-juniper woodland, ponderosa pine communities, 4300-8000 ft. *Purshia stansburyana* often hybridizes with *P. tridentata*. NW, NC, SW, Colfax.

Purshia tridentata (Pursh) DC.—Antelope bitterbrush. Shrub, upright or low and prostrate with trailing branches, gray bark; leaves fasciculate, 10-23 mm, 3-lobed apically, margins revolute, veins impressed above and prominent beneath, blades thinly pubescent and green above and greenish-white-tomentose beneath; inflorescence of solitary flowers; flowers fragrant, perfect, sepals ovate, 1.8-3 mm, petals 4-10 mm, yellow; cypselae 7-11 mm, nonplumose. Often in sandy soils, sagebrush, juniper, piñon-juniper woodland, ponderosa pine communities, 5000-9000 ft. NW, NC, Colfax.

PYRACANTHA M. Roem. – Pyracantha

Pyracantha coccinea M. Roem.—Scarlet firethorn. Shrubs 10-60 dm; stems 1-3+, erect or divergent, bark grayish, thorns present; leaves simple, blades elliptic or ovate to lanceolate or oblanceolate, 2-4 × 0.7-1.5 cm, margins finely crenulate-serrulate; inflorescences terminal, 6-40-flowered, panicles 3-4 cm diam.; flowers 6-8 mm diam., sepals triangular, 1.5-2 mm, petals 3-5 mm; pomes bright red, 5-8 mm diam. Introduced and rarely escapes; 1 collection from the foothills of the Sandia Mountains, 6725 ft. Bernalillo, San Juan.

PYRUS L. – Pear

Pyrus communis L.—Common pear. Trees to 5.5 m; sometimes armed with thorns; leaf blades pubescent when young, becoming glabrous, broadly ovate to elliptic, 2-7 cm; flowers 25-35 mm wide; sepals 5-9 mm; petals white, 12-15 mm; fruit a pome. Orchard plant that is rare in the wild, 5600-8200 ft. Widely scattered.

ROSA L. – Rose

Walter H. Lewis and Barbara Ertter

Deciduous shrubs armed with prickles; stems erect to spreading, climbing, or trailing; prickles straight, curved, or hooked, often paired; leaves alternate, petiolate, stipules conspicuous, paired, entire to pinnatifid, pricklets common to rachis and petiole; leaflets uniserrate, multiserrate, or a combination of both, often gland-tipped; inflorescence a 1- to many-flowered determinate corymb or panicle; flowers perfect, 5-merous, strongly showy, often aromatic, sepals 5, petals 5, light pink to deep rose, often fading with age, or white, stamens numerous; fruit accessory (hips), fleshy or pulpy, usually red or orange, containing few or numerous bony achenes.

1. Flowers yellow…**R. ×harisonii**
1'. Flowers red to pink to white…(2)
2. Flowers mostly white, rarely pink; sepals lobed or fringed…(3)
2'. Flowers mostly pink to rose or red; sepals fringed or not…(6)
3. Stems arching and sprawling to 5 m, forming dense, impenetrable clumps; petals white, very small (7-10 mm); styles connate and exserted beyond hypanthium (hip); introduced species, rare in New Mexico… **R. multiflora**
3'. Stems upright, short to infrequently 3 m; petals pink to rose, > 10 mm; styles free and ± inserted in the hypanthium (hip); native species…(4)
4. Hip densely bristly; sepals mostly lobed; leaflets wedge-shaped…**R. stellata**
4'. Hip not bristly; sepals lobed or not; leaflets not wedge-shaped…(5)
5. Leaflets mostly hairy and glandular abaxially…**R. obtusifolia**

5′. Leaflets glabrous and eglandular abaxially…**R. canina**
6. Flowering stems densely bristly between the nodes, usually without stout broad-based prickles…(7)
6′. Flowering stems with stout broad-based prickles at the nodes, occasionally with some bristles…(8)
7. Leaflets mostly 5-7, rarely 9, the teeth on the margins often gland-tipped; flowers solitary or in clusters of
   2-3…**R. acicularis**
7′. Leaflets mostly 9-11, rarely 7 or fewer, the teeth on the margins rarely gland-tipped; flowers usually in
   clusters of 3+…**R. arkansana**
8. Flowers mostly solitary, rarely 2-3, petals 25-40 mm; hip 12-20 mm diam.…**R. nutkana**
8′. Flowers mostly 2-3(-5), petals 15-25 mm; hip 6-12 mm diam.…**R. woodsii**

Rosa acicularis Lindl. [*R. sayi* Schwein.]—Prickly rose. Stems 1 m or commonly shorter, armed to apices of floral branches with aciculae or rarely with aciculae and small round prickles, occasionally with few bristles; leaves with stipules edged with few or numerous glands, blades (3-)5-7-foliate, leaflets elliptic to oval, large, 25-40(-50) × 12-21(-30) mm, abaxially mostly puberulent to pubescent, frequently slightly glandular, uniserrate or biserrate; inflorescence solitary or rarely 2-3-flowered; flowers 35-50 mm across; hypanthium often with a distinct neck, 2.5-4 mm across, eglandular; sepals 19 × 3 mm basally, petals pink to rose, ± 20 × 20 mm; fruit orange-red at maturity, 13-18+ × 9-12 mm, eglandular. Our material belongs to **subsp. sayi** (Schwein.) W. H. Lewis. Edges of spruce forest communities above 9000 ft. NE, NC, Catron.

Rosa arkansana Porter—Stems densely bristly with straight, slender, unequal prickles; leaflets (7-)9-11 × 1-5(6), serrate, teeth rarely gland-tipped, glabrous to villous below; flowers in clusters of 3+; sepals 1.5-2(-3) cm, glandular-stipitate, petals white to pink or rose, 1.5-3 cm. Prairie communities, edges of piñon-juniper woodlands, creek margins, 5525-10,200 ft. NE, NC, EC, SE.

Rosa canina L.—Dog rose. Shrubs, arching, not rhizomatous; stems usually erect to sprawling, 10-25(-50) dm; distal branches arching, infrastipular prickles paired, curved or appressed; leaves 6-11 cm, with pricklets, glabrous, eglandular, leaflets 5-7, blades ovate, obovate, or elliptic, 15-40 × 12-20 mm, abaxial surfaces glabrous, rarely pubescent; inflorescence a panicle, sometimes a corymb, solitary; flowers 3.5-5 cm diam., petals rose, pink, or white; hips red, globose. Foothills, woodlands, abandoned homesteads; a single collection from Vermejo Park Ranch, 7930 ft. Colfax.

Rosa ×harisonii Rivers—Harison's yellow rose. Stems with stout, flattened, strongly curved or hooked thorns, sometimes with slender bristlelike prickles intermixed; leaves with 5 or 7 leaflets, 1-2 cm, serrate with gland-tipped teeth, glandular-stipitate on the midrib; flowers usually solitary, petals yellow, 2.5-3 cm, usually double. Cultivated and persisting near old homesteads, 4700-7000 ft. Colfax.

Rosa multiflora Thunb.—Japanese rose. Stems few to many, forming dense, impenetrable clumps reaching 10 m across, erect and arching to trailing or sprawling, 5-3(-5) m, armed with stout, curved prickles; leaf blades (5-)7-9(-11)-foliate, 10-35 × 8-20 mm, abaxially glabrous to usually pubescent; inflorescence of few to usually many flowers in a conic corymb; flowers small, 15-20 mm across, petals white, rarely pink; fruit small, ovoid to globose, < 7 mm diam., red. Escape, disturbed sites, 5600-7020 ft. Los Alamos, Rio Arriba, San Juan.

Rosa nutkana C. Presl—Nootka rose. Stems stout, few to many from base, much branched, 0.5-3 m, prickly throughout or unarmed, infrastipular prickles with scattered prickles internodally, straight and/or curved to hooked; leaves glabrous or pubescent, eglandular or glandular, pricklets common, 10-18 mm, blades 5-7(-9)-foliate, leaflets elliptic to oval, 25-35 × (6-)15-20 mm, glabrous or pubescent; inflorescence solitary or few-flowered; flowers large, to 50-60 mm across, flowering at ends of lateral branches, petals pink to deep rose; fruit large, 12-20(-24) × 10-18 mm, red to purplish. Our material belongs to **subsp. melina** (Greene) W. L. Lewis & Ertter. Wooded regions or open areas from sea level to moderate or high mountainous elevations, higher montane in aspen, fir, spruce, pine forests, 8450-12,005 ft. NC.

Rosa obtusifolia Desv.—Round-leaved dog rose. Differs from *R. canina* in having hairy leaflets that are usually glandular abaxially. S of Montezuma in Lime Canyon below Peterson Dam, 6700 ft. San Miguel.

Rosa stellata Wooton—Desertberry. Shrubs forming dense, low thickets; stems erect, rarely arching, 2.5–8(–15) dm, distal branches densely pubescent, usually with stellate hairs, rarely without, rarely glabrous, sometimes densely or sparsely sessile- or stipitate-glandular; infrastipular prickles paired or single, erect, rarely curved; leaves 2–3 cm, leaflets 3–5, terminal blades obovate or deltate, 8–18 × 5–13 mm, margins broadly 1- or rarely multi-crenate, sometimes 1-serrate, glandular, abaxial surfaces pubescent or tomentulose, rarely glabrous, adaxial dull to lustrous, glabrous or tomentulose; inflorescences 1(2)-flowered; flowers 4–5 cm diam., petals pink or dark pink, with a strong almond fragrance, 15–25 × 14–25 mm; hips dull red and darkening with age.

1. Leaflets (3–)5; distal branches usually glabrous, rarely with stellate hairs; petioles and rachises glabrous or puberulent; SE New Mexico…**subsp. mirifica**
1′. Leaflets 3; distal branches tomentose-woolly, with stellate hairs; petioles and rachises pubescent or puberulent, sometimes glabrous; SC New Mexico…**subsp. stellata**

subsp. mirifica (Greene) W. H. Lewis—Wonderful rose. Stems erect, 10–15 dm; distal branches usually glabrous, rarely with stellate hairs, usually stipitate-glandular, infrastipular prickles paired, 10–12 × 3–4 mm; leaves with petioles and rachises glabrous or puberulent, leaflets (3–)5, blades deltate, margins 1-serrate, abaxial surfaces glabrous or tomentulose on midveins and elsewhere, adaxial dull; flower petals pink, 4 cm diam. Igneous or limestone cliffs, arroyos, edges of piñon pine woods, along streams, roads, openings in pine-juniper woods, 5575–7545 ft. Eddy.

subsp. stellata—Desert or gooseberry or star rose. Stems erect, 4–15 dm; distal branches tomentose-woolly, with stellate hairs, sessile- and stipitate-glandular, infrastipular prickles single or paired, 11–13 × 2.5–6 mm; leaves with petioles and rachises pubescent, sometimes glabrous, leaflets 3, blades obovate, margins broadly 1- or multi-crenate, abaxial surfaces tomentulose on midveins, otherwise glabrous; flower petals pink, 4.5 cm diam. Streams, roadsides, openings in pine-juniper woods, dry rocky hillsides and slopes, overhangs in rocky canyons, 5200–8500 ft. SE, SC, SW.

Rosa woodsii Lindl.—Woods' rose. Stems much branched, from 2–3 dm to 3(–4) m, usually forming thickets, commonly with infrastipular prickles, differing from those of internodes where prickles are often dense, variable, and with aciculae mostly straight; leaves with petioles glabrous or pubescent, eglandular or glandular, blades 5–7(–9)-foliate, leaflets elliptic to oval or obovate, variable in size, 10–30 × 6–20 mm, mostly uniserrate, with glandular tips, abaxial surface glabrous to pubescent, eglandular to glandular; inflorescence 1- to few-flowered or a corymb with 5–10+ flowers; flowers borne mostly at ends of lateral branches, 30–35 mm across, petals 20–25 × 9–17 mm, pink to deep rose; fruit fleshy, 7–12 × 7–10 mm.

1. Stems frequently tall, 1–2(–3) m, not compact or highly branched; internodes long, prickles mostly curved/hooked; terminal leaflets mostly elliptic; low riparian and wetland areas…(2)
1′. Stems short to infrequently tall, 0.2–1(–2) m, compact and markedly branched; internodes short, prickles straight or curved/hooked; terminal leaflets (some to many) obovate; various habitats…(3)
2. Sepals usually eglandular abaxially; prickles usually falcate, sometimes erect, declined, or introrse; terminal leaflets 20–40 mm…**subsp. puberulenta**
2′. Sepals usually densely stipitate-glandular abaxially; prickles usually strongly curved or hooked, sometimes introrse; terminal leaflets 10–30 mm…**subsp. arizonica**
3. Prickles straight; prairies and plains to limited high montane regions, mostly C Rocky Mountains…**subsp. woodsii**
3′. Prickles mostly curved/hooked; high montane regions, mostly S Rocky Mountains…**subsp. manca**

subsp. arizonica (Rydb.) W. H. Lewis & Ertter [*R. arizonica* Rydb.]—Arizona rose. Along creeks and other riparian habitats, 4730–7875 ft. Catron, McKinley, San Juan.

subsp. manca (Greene) W. H. Lewis & Ertter—Mancos rose. Montane meadows, hillsides, and other openings, edges of aspen groves and pine-spruce-fir forests, 6890–10,800 ft. McKinley, San Juan, Taos.

subsp. puberulenta (Rydb.) W. H. Lewis & Ertter—Plateau rose. Edges of streams, flats, riparian woodlands dominated by cottonwood and pine, 4265–7875 ft. Santa Fe.

subsp. woodsii—Wood's rose. Prairies and high plains to montane regions, often in wetland and riparian habitats, 5300–10,000 ft. Widespread except EC, Lea, Luna.

RUBUS L. – Blackberry, raspberry

Mostly erect to trailing shrubs or scramblers from rootstocks or creeping stems, usually abundantly armed with prickles, sometimes unarmed; stems mostly biennial; leaves simple and then palmately lobed or compound with 3-7 leaflets, the leaflets arranged palmately or pinnately; inflorescences simple or compound cymes, racemes, or panicles, or flowers solitary; flowers perfect, white to pink or rose-purple, showy, sepals 5, petals 5, stamens many; carpels many, the ovaries superior, inserted on a cone-like to nearly flat receptacle; fruit red to purple or black, an aggregate of many drupelets, drupelets fleshy or rarely nearly dry.

1. Leaves simple; stems unarmed…(2)
1'. Leaves compound, 3-5-foliate; stems armed with prickles…(4)
2. Leaf blades (5-)10-25(-30) cm wide; style glabrous…**R. parviflorus**
2'. Leaf blades (2.5-)3-5.5(-8) cm wide; style villous…(3)
3. Leaf blades orbiculate to reniform; fruits dark purple…**R. deliciosus**
3'. Leaf blades cordate to broadly ovate; fruits red or deep red…**R. neomexicanus**
4. Stems strongly pruinose; abaxial surfaces of leaf blades white-tomentose…(5)
4'. Stems not or slightly pruinose; abaxial surfaces of leaf blades white-tomentose or not…(6)
5. Leaves pinnately compound, lateral leaflets sessile or nearly so…**R. idaeus**
5'. Leaves palmately compound or ternate, lateral leaflets stalked…**R. leucodermis**
6. Leaflet abaxial surface white-tomentose and short-velutinous to tomentose…**R. bifrons**
6'. Leaflet abaxial surface glabrous or sparsely to moderately hairy, not whitened…**R. flagellaris**

Rubus bifrons Vest—Himalayan blackberry. Woody shrub to 3 m, much branched, erect, arching, sprawling, or creeping, usually eglandular with stellate hairs; prickles sharp, stout, broad-based, laterally flattened, slightly to moderately recurved; leaves ± evergreen, leaflets usually ovate to obovate, margins serrate to doubly serrate, 12-20 × 9-14 cm, pinnately to pedately 3-foliate or palmately 5-foliate; inflorescence bracteate, compound or simple terminal cymes with 11 to numerous flowers; flower petals white to pink, 9-16 mm; fruit very large, drupelets black, glabrous to slightly pubescent, fleshy. Cultivated, escaped, and naturalized in shady riparian areas, 4700-6200 ft. Catron, Doña Ana, Sierra. Introduced.

Rubus deliciosus Torr. [*R. medius* Kuntze]—Delicious raspberry. Shrubs, 5-15(-20) dm, unarmed; stems decumbent to erect, glabrous or sparsely short-hairy, eglandular or sparsely stipitate-glandular; leaves simple, blades orbiculate to reniform, (1.5-)2-4(-5) × (2.5-)3-4.5(-7) cm, margins doubly dentate, abaxial surfaces sparsely hairy, mostly along veins, sparsely stipitate-glandular; inflorescences 1-flowered; flower petals white; fruits dark purple, drupelets 10-40. Rocky canyons, outcrops, stream banks, 6000-9000 ft. NE, NC, C, WC, Doña Ana.

Rubus flagellaris Willd. [*R. arizonicus* (Greene) Rydb.]—Dewberry. Shrubs, to 3 dm, armed; stems usually creeping, sometimes low-arching and then creeping, glabrous or densely hairy; prickles sparse to dense, hooked, 1-4 mm, broad-based; leaves ternate or palmately compound, leaflets 3-5, terminal ovate or elliptic to suborbiculate, 3-11 × 2-7.5 cm, margins moderately to coarsely serrate to doubly serrate or serrate-dentate; flowers bisexual, petals white, 8-20 mm; fruits black, sometimes dark red, globose to cylindrical, 1-2 cm. Shady sites along mountain streams and canyons, 5900-8500 ft. Cibola, Socorro.

Rubus idaeus L. [*R. strigosus* Michx.]—American red raspberry, wild raspberry. Prickly shrub; stems to 2 m, bark yellow to yellowish brown, exfoliating, prickles numerous, slender, narrow-based, terete to subterete, straight or nearly so; leaves 3-foliate, the terminal leaflet largest, lanceolate to ovate, 3-5 × 1.5-3.2 cm, coarsely double-serrate, gray-tomentose beneath; inflorescence a raceme, a few-flowered thyrse, or 1 or 2 flowers in the leaf axils; flower petals 3-7 mm, white, stamens numerous; fruit an aggregate, red. Moist slopes and canyon bottoms, ponderosa pine, Douglas-fir, white fir, spruce-fir communities, 6700-13,025 ft. Widespread except E, S.

Rubus leucodermis Douglas ex Torr. & A. Gray—Western black raspberry, whitebark raspberry. Prickly shrub; stems arched, 1-3 m, bark yellowish at first, becoming reddish brown, often with a whitish bloom, prickles stout, on the stems, petioles, and some leaf veins, stems and petioles with a whitish bloom; leaves 3- or 5-foliate, the 5-foliate ones with 3 petiolulate leaflets and 2 sessile, the terminal leaflet largest, lanceolate to ovate, 3–5 × 1.5–4.5 cm, sharp-serrate, double-serrate, whitish-tomentose below; inflorescence a 2–7(–10)-flowered corymb, sometimes reduced to solitary flowers in the upper leaf axils; flower petals 5–6 mm, white; fruit an aggregate, drupelets densely tomentose basally, dark red to purple. Slopes, stream banks, moist canyons, 5000–8300 ft. WC, SW, Eddy.

Rubus neomexicanus A. Gray—New Mexico raspberry. Unarmed shrub, erect; stems to 2 m, becoming woody, bark reddish or orange-brown; leaves simple, petiolate, blades 2.7–8.5 cm and about as wide, palmately 3-lobed, sometimes 5–7-lobed, double-serrate, sparsely pilose above, pale, soft pubescence below; inflorescence of 1 or 2 flowers in the leaf axils or terminating branches; flower petals (8–)11–28 mm, white, stamens numerous; fruit an aggregate, red, ca. 15 mm diam., ± dry. Rocky slopes, canyons, and along streams in ponderosa pine communities, 6400–8775 ft. NE, NC, WC, SW, McKinley.

Rubus parviflorus Nutt.—Thimbleberry. Unarmed shrub, erect, to 3 m, gray, flaky bark; herbage glandular-pubescent; leaves simple, petiolate, petiole 2–11 cm, blades mostly 10–20 cm and about as wide, palmately (3–)5(–7)-lobed, irregularly serrate; inflorescence a (1)2–4(–7)-flowered raceme; flowers with 5(6–7) sepals and petals, petals 8–20 mm, white, sometimes pinkish-tinged, stamens numerous; fruit an aggregate (of coherent drupelets), thimble-shaped. Moist places, often near streams, ponderosa pine, Douglas-fir, white fir, spruce-fir communities, 7000–10,880 ft. NC, NW, C, WC, Otero.

SIBBALDIA L. – Sibbaldia

Sibbaldia procumbens L.—Creeping sibbaldia. Low, perennial herbs, mat- or cushion-forming; stems arising from a branched caudex covered in persistent leaf bases, rhizomatous, herbage appressed-pilose; leaves basal, 3-foliate, leaflets 1–2.5 cm, the terminal one slightly larger than the lateral pair, (2)3(–5)-toothed apically, teeth rounded, petiole 1–5.5 cm; inflorescence a few-flowered cyme, usually glandular-pubescent; sepals mostly 2.5–4 mm, petals 1.3–1.6 mm, pale yellow, stamens 5; fruit an achene. Rocky slopes and ridges, subalpine and alpine communities, 9600–13,030 ft. NC.

SORBUS L. – Mountain ash

Sorbus scopulina Greene—Rocky Mountain ash. Shrub or small tree, 1–4 m, bark yellowish to reddish purple, becoming grayish red, herbage strigose-pilose, the stems becoming sparsely so, leaves becoming glabrate; winter buds glutinous-brown, glossy, glabrous to sparsely whitish-pubescent; leaves 11–15-pinnate, 14–21 cm, the rachis grooved above and glandular-pubescent near the nodes, leaflets broadly lanceolate or elliptic, 4–7.5(–8.5) × 1.5–2.5 cm, finely serrate; inflorescence a corymb, ± flat-topped, 80–200-flowered; flower sepals 0.7–1.8 mm, petals 3–5(–6) mm, white or cream, stamens 20; fruit a globose pome, bright glossy orange-red, drying purplish. Mountain slopes, ponderosa pine, Douglas-fir, white fir, aspen, spruce-fir communities, 7500–10,600 ft. NC, NW, C, WC, Eddy.

VAUQUELINIA Corrêa ex Bonpl. – Rosewood

Vauquelinia californica (Torr.) Sarg.—Arizona rosewood. Shrubs or trees, (10–)15–80(–100) dm; stems 1–10+, bark gray to dark gray, unarmed, tomentulose to villous-canescent; young stems loosely tomentulose, tardily glabrescent; leaves with petioles (1.5–)4–16(–22) mm, blades green or yellow-green, lanceolate or oblong-lanceolate to elliptic or oblong-elliptic, sometimes oblong-ovate, (2.2–)3–7.5(–9) × (0.6–)0.8–1.4(–2) cm, surfaces glabrate or puberulent along midveins; inflorescence a corymb; flower sepals 1.1–2.2 × 1.4–2 mm, abaxially puberulent to glabrate, petals white, 3.4–5.4 × 2.4–3.4 mm; capsules (4.5–)5–6 × 3.5–4(–4.5) mm; fruit a capsule. New Mexico material belongs to **subsp. pauciflora** (Standl.) W. J. Hess & Henrickson. Limestone substrates in arid chaparral–desert scrub, 4490–5690 ft. Hidalgo.

RUBIACEAE - MADDER FAMILY

Craig C. Freeman

Shrubs, subshrubs, or herbs, annual or perennial; leaves opposite, whorled, or sometimes appearing whorled by means of leaflike interpetiolar stipules, simple; stipules interpetiolar, sometimes leaflike; blades simple, margins entire; inflorescences usually cymose, sometimes paniclelike or 1-flowered; flowers usually bisexual, rarely unisexual, homostylous or heterostylous, actinomorphic; sepals 4 or 5, connate; corolla lobes (3)4 or 5; stamens (3)4 or 5, alternate with corolla lobes, adnate to corolla tube; pistil 1, 1(3–5)-carpellate, ovary inferior or 1/2 inferior, locules 2, placentation axile; styles 1 or 2; stigmas 1 or 2; fruits capsules, seeds 2–50, or schizocarps, seed 1 per mericarp.

1. Leaves whorled or apparently whorled, blades never acicular...(2)
1'. Leaves opposite, sometimes fasciculate, if whorled then blades acicular...(3)
2. Corollas red or orange-red; fruits capsules; leaves 3- or 4-whorled; style 1...**Bouvardia**
2'. Corollas white, cream-colored, yellow, pink, or red; fruits schizocarps; leaves 4–12-whorled by means of leaflike interpetiolar stipules; styles 2...**Galium**
3. Fruits schizocarps; seeds 1 per mericarp...(4)
3'. Fruits capsules; seeds 2–50...(6)
4. Pedicels 9–60(–100) mm; inflorescences terminal cymes, open...**Kelloggia**
4'. Pedicels 0 mm; inflorescences terminal and axillary cymes, congested, or axillary and 1(2)-flowered...(5)
5. Inflorescences congested cymes, to 30-flowered; 2 calyx lobes at right angle to carpels often reduced or absent...**Crusea**
5'. Inflorescences 1(2)-flowered; 1 or 2 calyx lobes sometimes shorter than others...**Hexasepalum**
6. Shrubs; leaf blades linear, 0.3–1 mm wide...**Arcytophyllum**
6'. Herbs or subshrubs; leaf blades oblanceolate, elliptic, or linear, 0.1–6 mm wide...(7)
7. Perennials or annuals; if annuals then fruiting pedicels recurved; corolla tubes 1.5–36 mm...**Houstonia**
7'. Annuals; fruiting pedicels erect; corolla tubes 1–3.5 mm...**Stenotis**

ARCYTOPHYLLUM Schult. & Schult. f. – Star-violet

Arcytophyllum fasciculatum (A. Gray) Terrell & H. Rob. [*Houstonia fasciculata* A. Gray; *Hedyotis intricata* Fosberg]—Tangled star-violet. Shrubs, perennial, 2–10 dm; stems prostrate to erect, puberulent, pubescent, or glabrate; leaves opposite, fasciculate, blades linear, 2–14 × 0.3–1 mm; inflorescences terminal, cymose; flowers bisexual; hypanthium cupulate; calyx lobes 4(5); corolla white, funnelform, tube 0.5–2.5 mm, lobes 4(5), ovate to lanceolate, 1–3 mm; fruits capsules, oblong or ellipsoid, 1.6–3 mm, puberulent or glabrate. Rocky or gravelly slopes, arroyos, limestone ledges, cliff crevices, deserts, semideserts, pine-oak, pine-juniper, 4600–5800 ft. Doña Ana, Otero, Socorro.

BOUVARDIA Salisb. – Bouvardia

Bouvardia ternifolia (Cav.) Schltdl. [*B. glaberrima* Engelm.]—Firecracker bush. Subshrubs or shrubs, perennial, 6–15 dm; stems spreading to erect, hispidulous when young; leaves 3- or 4-whorled, blades usually elliptic-lanceolate, sometimes linear, ovate, or obovate, (1–)4–75(–110) × 2–19(–31) mm; inflorescences terminal, cymose, 3–40-flowered; flowers bisexual; hypanthium funnelform; calyx lobes 4; corolla red or orange-red, tubular to salverform, tube 10–32 mm, lobes 4, ovate to oblong, 1.5–3.5(–5) mm; fruits capsules, subglobose, 4.5–9 mm, glabrous or hispid. Rocky slopes, canyon bottoms, pine-oak and juniper-oak woodlands, 5000–7200 ft. Grant, Hidalgo.

CRUSEA Cham. & Schltdl. – Saucerflower

Crusea diversifolia (Kunth) W. R. Anderson [*C. subulata* (DC.) A. Gray]—Mountain saucerflower. Herbs, annual, 0.3–3 dm; stems erect, glabrous or papillose to hispidulous; leaves opposite, blades subulate to narrowly elliptic, 6–27 × 1–4 mm; inflorescences terminal or axillary congested cymes, to 30-flowered; flowers bisexual; hypanthium funnelform; calyx lobes 4, 2 at right angle to carpels often reduced or absent; corolla white, salverform, tube (2.5–)3–4.3(–4.8) mm, lobes 4, triangular, 1–2 mm; fruits schizocarps, 1.2–3.5 mm, glabrous or scabrous to antrorsely aculeolate. Open grassy sites, roadsides, stream banks, woodlands, rocky pastures, 5300–8500 ft. C, WC, SW, McKinley.

GALIUM L. – Bedstraw

Herbs or subshrubs, annual or perennial, 0.5–9 dm, usually synoecious, sometimes dioecious or polygamous; stems clambering to prostrate, spreading, ascending, or erect; leaves opposite but appearing 4–12-whorled by means of foliaceous interpetiolar stipules; inflorescences terminal and/or axillary, cymose, paniclelike, or 1-flowered; flowers bisexual or functionally unisexual; hypanthium ellipsoid to spherical; calyx lobes absent or reduced to a minute rim; corolla usually rotate, rarely campanulate or cupulate, 0.4–1.5 mm, lobes (3)4; stamens (3)4; styles 2; fruits schizocarps, 1–3.5 mm, glabrous or hairy; mericarps 1 or 2, ellipsoid to reniform or spherical.

1. Annuals…(2)
1′. Perennials…(4)
2. Leaves 6–8(–10) per node…**G. aparine**
2′. Leaves 4 per node, sometimes 2 per node distally…(3)
3. Fruiting pedicels (3–)5–30 mm; corolla lobes 3; stems glabrous…**G. bifolium**
3′. Fruiting pedicels 0–0.5 mm; corolla lobes 4; stems glabrous or scabrous to hispidulous…**G. proliferum**
4. Schizocarps glabrous…(5)
4′. Schizocarps hispidulous, hirsute, pilose, uncinate-hairy, or with curved hairs…(6)
5. Fruiting pedicels erect, 0–0.5 mm; leaf blade apices acute, apiculate…**G. microphyllum**
5′. Fruiting pedicels 2–9 mm; leaf blade apices obtuse to acute, not apiculate…**G. trifidum**
6. Schizocarps aculeolate or uncinate-hairy…(7)
6′. Schizocarps hispidulous, hirsute, or pilose, hairs straight or curved but not aculeolate or uncinate…(10)
7. Leaves 4 per node…(8)
7′. Leaves (4)5–12 per node…(9)
8. Leaf blades 3-nerved…**G. boreale** (in part)
8′. Leaf blades 1-nerved [record in doubt]…G. pilosum
9. Hairs of schizocarps 0.1–0.4 mm; cymes 2–5(–7)-flowered; peduncles ascending to spreading or divergent…**G. mexicanum**
9′. Hairs of schizocarps 0.5–1 mm; cymes (1)2–3(–5)-flowered; peduncles divergent to divaricate…**G. triflorum**
10. Subshrubs, polygamous; flowers bisexual and unisexual…(11)
10′. Herbs or subshrubs, synoecious or dioecious; flowers bisexual or unisexual…(12)
11. Corollas usually red or pink, rarely yellow; corolla lobe apices acuminate…**G. wrightii**
11′. Corollas greenish yellow or yellow; corolla lobe apices obtuse to acute…**G. fendleri**
12. Plants synoecious; flowers bisexual; stems pilosulous, sometimes ± glabrous except near nodes…**G. boreale** (in part)
12′. Plants dioecious; flowers unisexual; stems glabrous or scabridulous…**G. multiflorum**

Galium aparine L. [*G. aparine* var. *echinospermum* (Wallr.) T. Durand]—Cleavers. Herbs, annual, 1–10 dm, synoecious; stems clambering or prostrate to ascending, retrorsely aculeolate; leaves 6–8(–10) per node, blades linear-oblanceolate or oblanceolate, 8–70 × 1–6 mm, 1-nerved; inflorescences axillary cymes, 1–5-flowered; fruiting pedicels spreading, 1–10 mm; flowers bisexual; corolla white, yellowish white, or greenish white, lobes 4, ovate; schizocarps greenish black to greenish brown, 1.5–3 mm, sparsely to densely uncinate-hairy or hispid. Stream banks, shaded slopes, forest edges, roadsides, 4400–9800 ft. N, C, SW, Cibola, Doña Ana.

Galium bifolium S. Watson—Twinleaf bedstraw. Herbs, annual, 0.5–2 dm, synoecious; stems erect, glabrous; leaves mostly 4 per node proximally, sometimes 2 per node distally, blades lanceolate to elliptic, 10–25 × 1–5 mm, 1-nerved, surfaces glabrous; inflorescences axillary cymes, 1-flowered; fruiting pedicels spreading to slightly reflexed, sharply reflexed proximal to fruit, (3–)5–30 mm; flowers bisexual; corolla white, 0.8–1.3 mm diam., lobes 3, ovate; schizocarps black, 2.4–3.5 mm, uncinate-hirsute. Moist or wet places in coniferous forest, gravelly slopes, stream banks, meadows, 6500–9200 ft. Sandoval, San Juan, Taos.

Galium boreale L.—Northern bedstraw. Herbs, perennial, 3–7(–9) dm; stems ascending to erect, pilosulous, sometimes ± glabrous except near nodes; leaves 4 per node, blades linear to lanceolate or oblong, 5–45 × 1–5(–10) mm, 3-nerved, surfaces glabrous or abaxially scabridulous to hirsutulous along veins; inflorescences terminal and axillary cymes, 2–9-flowered; fruiting pedicels spreading, 1–4 mm; flowers bisexual; corolla white to cream-colored, 3.5–6 mm diam., lobes 4; schizocarps dark brown to black, 1.2–1.7 mm, densely hispidulous or with curved hairs, rarely glabrate or ± uncinate-hairy. Our plants are **subsp. septentrionale** (Roem. & Schult.) H. Hara. Meadows, stream banks, moist shaded places, 5300–11,200 ft. N, C, Grant, McKinley, Otero.

Galium fendleri A. Gray—Fendler's bedstraw. Subshrubs, perennial, 1.5–4.5 dm, polygamous; stems ascending to erect, puberulent, sometimes only proximal to nodes; leaves 4 per node, blades linear to lanceolate or narrowly oblanceolate, 10–25 × 1–3 mm, 1-nerved, surfaces glabrous or sparsely antrorsely hairy, especially along abaxial midvein; inflorescences terminal and axillary cymes, 3–8-flowered; fruiting pedicels spreading to ascending, 1–4 mm; flowers bisexual and unisexual; corolla greenish yellow or yellow, 1.4–2 mm diam., lobes 4; schizocarps dark brown, 1.4–2 mm, hirsute. Moist shaded forests, talus slopes, rock crevices, hillsides, 4700–10,500 ft. Widespread except E.

Galium mexicanum Kunth—Mexican bedstraw. Herbs, perennial, 2–8 dm; stems ascending to erect, retrorsely aculeolate; leaves 5–12 per node, blades linear to lanceolate, oblanceolate, or elliptic, (8–)10–45 × 1–10(–12) mm, 1-nerved, surfaces glabrous, glabrate, or sparsely hirsute, abaxial midvein retrorsely aculeolate; inflorescences terminal and axillary cymes, (1)2–5(–7)-flowered; fruiting pedicels ascending to spreading or divergent, 2–10(–16) mm; flowers bisexual; corollas white, pink, or red, 2.5–5 mm diam., lobes 4, deltate to ovate; schizocarps brown to dark brown, 1.8–2.1 mm, moderately to densely aculeolate or uncinate-hairy. Our plants are **subsp. asperrimum** (A. Gray) Dempster. Stream banks, moist slopes, meadows, mixed conifer forests, 5000–10,700 ft. NE, NC, C, WC, SC, SW.

Galium microphyllum A. Gray—Bracted bedstraw. Herbs or subshrubs, perennial, 1–4 dm; stems decumbent to ascending or erect, glabrate or sparsely hirsutulous; leaves 4 per node, blades linear to narrowly elliptic or oblong, 5–15 × 1–2.5 mm, apex acute, apiculate, surfaces glabrous; inflorescences axillary cymes, 1–3-flowered; flowering pedicels erect, 0–0.5 mm, glabrous; flowers bisexual; corolla greenish white or white, 2–2.5 mm diam., lobes 4, ovate; schizocarps purple, 1.6–2 mm, glabrous. Stream banks, moist places in pine-oak forest, grasslands, rocky hillsides, 3900–5000 ft. S except SE.

Galium multiflorum Kellogg—Herbs or subshrubs, perennial, (0.8–)1.3–4 dm, dioecious; stems spreading to erect, glabrous or scabridulous; leaves 4 per node, blades linear to lanceolate, ovate, or nearly orbiculate, 4–20 × 1–8 mm, 1- or 3-nerved, surfaces glabrous or scabridulous; true leaves often 3-nerved; inflorescences terminal and axillary cymes, 2–9-flowered; fruiting pedicels recurved, 2–15 mm; flowers unisexual; corolla white, cream-colored, or yellow, sometimes tinged pink, 2–3 mm diam., lobes 4; schizocarps brown to dark brown, 2.5–3 mm, pilose to hirsute.

1. Leaf blades linear to lanceolate, usually 5–12 times longer than wide…**var. coloradoense**
1'. Leaf blades ovate to nearly orbiculate, usually 2–6 times longer than wide…**var. multiflorum**

var. coloradoense (W. Wight) Cronquist—Colorado bedstraw. Shaded areas among rocks, crevices of sandstone ledges and cliffs, pine-juniper woodlands, 5000–9200 ft. San Juan.

var. multiflorum—Shrubby bedstraw. Sandy or gravelly talus, N- or E-facing canyon slopes or cliffs, sagebrush, juniper-pine, aspen, 5400–6200 ft. San Juan.

Galium pilosum Aiton—Hairy bedstraw. Herbs, perennial, 1.5–9 dm; stems usually ascending to erect, sometimes decumbent, sparsely to densely pilose; leaves 4 per node, blades elliptic to ovate or obovate, 2–28 × 2–14 mm, 1-nerved, apex rounded, usually apiculate, surfaces hirsute to pilose; inflorescences terminal and axillary, (1)2–5(–7)-flowered; fruiting pedicels spreading, straight or sharply recurved proximal to fruit, 4–12 mm; flowers bisexual; corolla yellowish red or brownish red, 2.8–4 mm diam., lobes 4; schizocarps green or brown, 2.5–3 mm, densely uncinate-hairy. Rocky sites in pine and spruce-pine forests, 7400–10,700 ft. Record doubtful.

Galium proliferum A. Gray—Limestone bedstraw. Herbs, annual, 1–3 cm; stems decumbent to ascending or erect, glabrous or scabrous to hispidulous; leaves 4 per node, blades linear-oblong to narrowly ovate or obovate, 3–9 × 1–4 mm, 1-nerved, surfaces glabrous or scabrous to hispidulous; inflorescences axillary cymes, 1- or 2-flowered; fruiting pedicels recurved, 0–0.5 mm; flowers bisexual; corolla white to pale yellow, 2.5–3 mm diam., lobes 4, ovate; schizocarps greenish brown to brown, 1.5–2 mm, uncinate-hairy. Moist places along streams, washes, moist slopes, grassy areas, under shrubs or trees, 3400–4900 ft. Lincoln, S except E.

Galium trifidum L. [*G. brandegeei* A. Gray]—Three-petal bedstraw. Herbs, perennial, 0.7–3.5 dm; stems spreading to ascending, glabrous or retrorsely scabrous; leaves 4(5) per node, blades linear to narrowly oblanceolate or oblong, 8–15 × 1–2.5(–3) mm, 1-nerved, surfaces glabrous or scabrous; inflorescences terminal and axillary, 1- or 2-flowered; fruiting pedicels arcuate, 2–9 mm; flowers bisexual; corolla white to greenish white, 1.2–1.8(–2) mm diam., lobes 3 or 4, ovate to deltate; schizocarps dark brown to black, 1–1.8 mm, glabrous. Our plants are **subsp. subbiflorum** (Wiegand) Puff. Lakeshores, wet banks of mountain rivulets, wet depressions, 7000–10,600 ft. Widespread except E, C.

Galium triflorum Michx.—Fragrant bedstraw. Herbs, perennial, 2–7 dm; stems clambering to decumbent, spreading, or ascending, glabrous or retrorsely scabrous; leaves (4)5 or 6 per node, blades elliptic to ovate-obovate, 6–40(–60) × 2–10(–15) mm, 1-nerved, surfaces glabrous except for retrorsely scabrous abaxial midvein; inflorescences terminal and axillary, cymes (1)2–3(–5)-flowered; peduncles divergent to divaricate; fruiting pedicels divergent to divaricate, 1–10 mm; flowers bisexual; corolla white, 2–4 mm diam., lobes 4, ovate; schizocarps dark brown, 1.6–2.3 mm, densely uncinate-hairy. Moist to dry upland forests, stream banks, moist disturbed sites, 7000–9700 ft. NW, C, WC, Colfax, Otero.

Galium wrightii A. Gray [*G. rothrockii* A. Gray]—Wright's bedstraw. Subshrubs, perennial, 1.5–5 dm; stems ascending to erect, glabrous or scabrous; leaves 4 per node, blades linear to oblanceolate, rarely elliptic, 7–20 × 0.5–2 mm, 1-nerved, surfaces glabrate, scabrous, or hispid; inflorescences terminal and axillary cymes, (1)2–5-flowered; fruiting pedicels spreading to divergent, 1–6 mm; flowers bisexual and unisexual; corollas usually red or pink, rarely yellow, 1.5–3 mm diam., lobes 4; schizocarps greenish brown to brown, 1–1.5 mm, densely hirsute. Moist banks, shaded and sheltered slopes in pine, Douglas-fir, aspen forests, 5000–8900 ft. NW, C, WC, SC, SW.

HEXASEPALUM Bartl. ex DC. – Buttonweed

Hexasepalum teres (Walter) J. H. Kirkbr. [*Diodia teres* Walter]—Buttonweed. Herbs, annual, (0.2–)0.5–6 dm; stems ascending to erect, glabrous or hirtellous to hirsute; leaves opposite, blades elliptic, 15–45 × (2–)3–7 mm; inflorescences axillary, 1(2)-flowered; flowers bisexual; hypanthium turbinate; calyx lobes 4, 1 or 2 shorter than others; corolla purple, pink, or white with purple lobes, funnelform, tube 3–6 mm, lobes 4, ovate, 1.5–2.4 mm; fruits schizocarps, 2.8–4.6 mm, hirsute. Sandy to gravelly slopes and washes, meadows, stream banks, juniper-oak, juniper-pine, and pine-oak woodlands, pine forests, 5300–8500 ft. WC, SW.

HOUSTONIA L. – Bluet

Herbs or subshrubs, perennial or annual, 0.1–6.2 dm; stems prostrate to spreading or erect, scaberulous, puberulent, hirsute, glabrate, or glabrous; leaves opposite or 3- or 4-whorled, sometimes fasciculate, blades linear to lanceolate, oblong, oblanceolate, or elliptic; inflorescences terminal or axillary, cymose or 1-flowered; pedicels 0–22 mm; flowers bisexual; calyx lobes 4; corolla white, blue, purple, violet, lavender, pink, or rose, salverform or funnelform, lobes 4; stamens 4; style 1; capsules subglobose to turbinate or ellipsoid, glabrate or scabrous, puberulent, hirsutulous, or hirsute.

1. Annuals; calyx lobes linear to narrowly lanceolate…**H. humifusa**
1'. Perennials; calyx lobes linear to lanceolate or deltate…(2)
2. Leaf blades acicular, 0.3–1.3 mm wide, opposite or 3- or 4-whorled, usually fasciculate…**H. acerosa**
2'. Leaf blades linear, lanceolate, oblong, oblanceolate, or elliptic, 0.3–8 mm wide, opposite, usually not fasciculate…(3)
3. Cymes 1-flowered; corolla tubes 6–36 mm…**H. rubra**
3'. Cymes (1)2–10-flowered; corolla tubes 1.5–6 mm…(4)

4. Fruiting pedicels erect; capsules turbinate to ellipsoid; leaves opposite, often fasciculate…**H. nigricans**
4'. Fruiting pedicels recurved; capsules subglobose; leaves opposite…**H. wrightii**

Houstonia acerosa (A. Gray) Benth. & Hook. f. [*Hedyotis acerosa* A. Gray]—Needleleaf bluet. Herbs or subshrubs, perennials, 0.2–3 dm; stems ascending to erect, hirsute or puberulent to glabrate; leaves cauline, opposite or 3- or 4-whorled, usually fasciculate, blades 3–15 × 0.1–1.3 mm, acicular, surfaces glabrate to puberulent or hirsute; cymes 1–3-flowered; fruiting pedicels erect, 1–22 mm; corolla white, pink, rose, or purple, salverform or funnelform, tube 3–12 mm; capsules subglobose, 1.5–4 mm, hirsute or glabrate. Our plants are **var. polypremoides** (A. Gray) Terrell. Rocky or gravelly slopes, mesas, arroyos, desert grasslands, desert scrublands, 3000–9000 ft. C, EC, SE, SC.

Houstonia humifusa (A. Gray) A. Gray [*Hedyotis humifusa* A. Gray]—Matted bluet. Herbs, annual, 0.1–1.5 dm; stems prostrate to ascending or erect, scaberulous to pubescent; leaves essentially cauline, opposite, blades linear to narrowly elliptic or oblanceolate, 5–45 × 0.5–6 mm, surfaces glabrous, scaberulous, or hirsutulous; cymes 1–3-flowered; fruiting pedicels recurved, 2–5 mm; corolla pink, white, lavender, violet, or purplish, funnelform, tube 2–5 mm, lobes ovate or elliptic, 1.5–5 mm; capsules subglobose, 1–3 mm, glabrate or puberulent. Desert grasslands, desert shrublands, sandhills, 3000–7100 ft. EC, SF, Doña Ana, Socorro, Torrance.

Houstonia nigricans (Lam.) Fernald [*Hedyotis nigricans* (Lam.) Fosberg]—Diamond flowers. Herbs or subshrubs, perennial, 1–6.2 dm; stems spreading or erect, glabrous, scaberulous, papillose, or pubescent; leaves cauline, opposite, often fasciculate, blades linear, lanceolate, or narrowly oblong, 7–40 × 0.3–8 mm, surfaces glabrous, scaberulous, or puberulent; cymes 3–10-flowered; fruiting pedicels erect, 0–10 mm; corolla white, pink, lavender, or purple, salverform or funnelform, tube 1.5–5.5 mm; capsules turbinate to ellipsoid, 1.5–4.5, glabrous or hirsutulous. Our plants belong to **var. nigricans**. Rocky to gravelly slopes, grasslands, desert shrublands, pine, oak, and juniper woodlands, 3900–9100 ft. Widespread except NC, NW, W.

Houstonia rubra Cav. [*Hedyotis rubra* (Cav.) A. Gray]—Red bluet. Herbs, perennial, 0.1–1 dm; stems spreading to erect, puberulent or glabrous; leaves essentially cauline, opposite, linear to oblanceolate or elliptic, 5–30 × 0.5–3 mm, surfaces glabrous or scaberulous; cymes 1-flowered; fruiting pedicels recurved in fruit, 0.1–4 mm; corolla pink, purplish, reddish, rose, or white, salverform, tube 6–36 mm; capsules subglobose, 2–3.5 mm, glabrous or scabrous to puberulent. Rocky, sandy, silty, or gypsum soils in desert flats and arroyos, piñon-juniper and pine-oak woodlands, 4200–7000 ft. Widespread except N, E.

Houstonia wrightii A. Gray [*Hedyotis pygmaea* Roem. & Schult.]—Pygmy bluet. Herbs, perennial, 0.2–1.8(–3.2) dm; stems prostrate to spreading or erect, puberulent or glabrous; leaves basal and cauline, basal sometimes absent, opposite; cauline 5–15(–30) × (0.5–)1–4 mm, linear, elliptic, or oblanceolate, glabrous or scabrous; cymes (1)2–3(4)-flowered; fruiting pedicels recurved, 0–6 mm; corolla purple, violet, pink, or white, funnelform or salverform, tube 2–6 mm; capsules subglobose, 1–3 mm, glabrous or scabrous. Rocky slopes, meadows, grasslands, ponderosa pine, oak, juniper, subalpine forests, 4000–8800 ft. NW, WC, SW.

KELLOGGIA Torr. ex Hook. f. – Kelloggia

Kelloggia galioides Torr.—Milk kelloggia. Herbs, perennial, 1–6 dm; stems erect, glabrous; leaves opposite, blades linear-lanceolate to lanceolate or narrowly ovate, (10–)13–55 × 2–15 mm; inflorescences terminal cymes, open, 2–6(–8)-flowered; pedicels 9–60(–100) mm; flowers bisexual; calyx lobes 4 or 5; corolla white or pink, funnelform, tube 1.5–3 mm, lobes 4 or 5, deltate to linear-lanceolate or lanceolate, 2–2.5 mm; stamens 4 or 5; style 1; fruits schizocarps, 2–3.8 mm, uncinate-hairy. Piñon-juniper and mountain mahogany woodlands, sagebrush shrublands, pine, pine-spruce, and aspen forests, stream banks, 7000–8700 ft. McKinley, San Juan.

STENOTIS Terrell – Star-violet

Stenotis greenei (A. Gray) Terrell & H. Rob. [*Hedyotis greenei* (A. Gray) W. H. Lewis; *Houstonia greenei* (A. Gray) Terrell]—Greene's star-violet. Herbs, annual, 0.3–1.6 dm; stems erect, glabrous or scaberulous; leaves opposite, blades linear to narrowly oblanceolate or narrowly elliptic, 5–32 × 0.5–5 mm; inflorescences terminal and axillary, cymose; fruiting pedicels 0–16 mm, erect; flowers bisexual; calyx lobes 4; corolla white, funnelform, tube 1–3.5 mm, lobes 4, ovate, 1–2 mm; capsules subglobose, 2–4 mm, glabrous. Rocky to gravelly slopes, rock outcrops, oak-juniper woodlands, pine-oak forests, 4800–7000 ft. WC, SW.

RUPPIACEAE – DITCH-GRASS FAMILY

C. Barre Hellquist

RUPPIA L. – Ditch-grass, widgeon-grass

Ruppia cirrhosa (Petagna) Grande—Spiral ditch-grass. Annual, rarely perennial submerged aquatic herbs usually of brackish or saline water, rhizomes lacking, rooting at proximal nodes; stems 0.1–0.3 mm wide; leaves alternate or subopposite, submersed, sessile, veins 1, blades linear, 3.2–45.1 cm, 0.2–0.5 mm wide, minutely serrulate distally, apex obtuse to acute; inflorescence terminal, capitate, with subtending spathe, peduncles with 5–30 coils, 30–300 × 0.5 mm; perianth absent, stamens 2; pistils 4–16; fruit drupaceous, 1.5–1.2 × 1.1–1.5 mm; gynophore 2–3.5 cm; beak 0.5–1 mm; seeds 1. Brackish waters, or waters of high sulfur and/or calcium content, in ditches, ponds, lakes, 3175–9075 ft. Scattered.

RUSCACEAE – BUTCHER'S BROOM FAMILY

Steve L. O'Kane, Jr.

Perennial herbs (usually) to shrubs and small trees, typically rhizomatous; leaves simple, entire to serrate in some, usually spiral or 2-ranked, cauline, but often basal, venation parallel; flowers bisexual, actinomorphic, hypogynous; tepals 6, petaloid, ± distinct to connate and then the perianth urn-, bell-, or wheel-shaped, not spotted; stamens 6, filaments usually distinct, typically adnate to tepals, at least at the very base; ovary superior, 1(2) or 3 locules with axile placentation, stigma 1, capitate or 3-lobed; fruit indehiscent, a berry or capsular and dry. Includes Convallariaceae and Nolinaceae. Recognized as subfamily Nolinoideae of a large Asparagaceae s.l. in APG (2016) (Judd et al., 2016; Woodland, 2009).

1. Plants woody (shrubby) at least at base, stem > 15 cm wide at maturity; leaves leathery, serrate (sometimes ± entire) or with prominent hooked marginal prickles, usually > 30 cm…(2)
1'. Plants herbaceous; leaves soft, not leathery, leaf margins entire, < 25 cm…(3)
2. Leaf margins with prominent hooked marginal prickles; ovary 1-locular with 1 seed…**Dasylirion**
2'. Leaf margins entire to serrulate; ovary 3-locular with multiple seeds…**Nolina**
3. Perianth forming a fused tube; flowers pendent on long peduncles from axils of leaves…**Polygonatum**
3'. Perianth not fused or fused only at the very base; flowers borne in terminal racemes or panicles… **Maianthemum**

DASYLIRION Zucc. – Sotol

Plants shrubby, typically with an erect or reclining trunk; leaves persistent, in dense rosettes, rigid, long-linear, fibrous, glabrous, sometimes waxy-glaucous, bases expanded and spoon-shaped, margins with sharp, curved prickles, apex fibrous; dense inflorescences to 5 m including the long, woody stalks; flowers small, functionally unisexual, plants male or female; tepals distinct, whitish, greenish, or purple, obovate, margins denticulate; stamens rudimentary in pistillate flowers; ovary superior, 3-angled, abortive in staminate flowers; fruits capsular, 1-locular, dry, indehiscent, 3-winged. Traditionally placed in a large, nonmonophyletic Liliaceae or in Agavaceae (Bogler, 2002).

1. Leaves whitish or bluish green, waxy-glaucous, papillose, dull; marginal prickles pointing forward; fascicles of flowers on pendent branches...**D. wheeleri**
1'. Leaves bright green, glabrous, not waxy, smooth, shiny; marginal prickles pointing backward; fascicles of flowers on mostly erect branches...**D. leiophyllum**

Dasylirion leiophyllum Engelm. ex Trel.—Green sotol. Plants with trunks to 1 m, erect or reclining; leaf blades bright green, 90–110 cm, usually smooth and shiny, glabrous, not waxy; prickles mostly retrorse; inflorescences 2.5–5 m; fascicle branches mostly basal, erect; tepals whitish or greenish, 1.8–2 × 1 mm. Acacia communities, desert scrub, bajadas, canyon bottoms, gravelly slopes, 3350–5300 ft. Chaves, Eddy, Lea.

Dasylirion wheeleri S. Watson—Common sotol. Plants with trunks to 1.5 m, usually reclining; leaf blades whitish or bluish green, 35–100 cm, densely waxy-glaucous, papillose, dull; prickles all antrorse; inflorescences often massive, to 5 m; stalk 3–6 diam. at base; fascicle branches lateral, pendent in fruit; tepals sometimes tinged purple, 2.4 × 1–1.5 mm. Rocky slopes, grassy areas, lava flows, acacia communities, desert scrub, 4000–7200 ft. C, WC, SE, SC, SW.

MAIANTHEMUM F. H. Wigg. – False Solomon's-seal

Herbs from spreading, filiform rhizomes; stems simple, arching or erect; leaves 2–15, cauline, distichous, sessile, clasping or short-petiolate; blades usually ovate, base rounded or cordiform, denticulate or entire, apex acute or caudate; inflorescences terminally paniculate or racemose, 5–250-flowered; flowers 3-merous (6 tepals, 6 stamens) or, by reduction, 2-merous (4 tepals, 4 stamens); perianth spreading; tepals distinct, white, ovate or triangular, 0.5–5 mm; ovary 2- or 3-carpellate; fruit a berry, bright red at maturity, globose to somewhat lobed. Traditionally placed in a large, nonmonophyletic Liliaceae or in Convallariaceae. Species traditionally included in the genus *Smilacina* (LaFrankie, 2002).

1. Inflorescences paniculate, 70–250-flowered, branches well developed; rhizomes 8–14 mm wide; tepals ≤1 mm (yes, the species is misnamed!)...**M. racemosum**
1'. Inflorescences racemose, 6–15-flowered, simple; rhizomes 1–4.5 mm wide; tepals >1 mm...**M. stellatum**

Maianthemum racemosum (L.) Link [*Smilacina racemosa* (L.) Desf.]—Feathery false Solomon's-seal. Stems erect or arching, 7.5–12.5 dm × 7–9 mm; leaf blades elliptic to ovate, 9–17 × 5–8 cm; inflorescences paniculate, 70–250-flowered, branches well developed, pyramidal; tepals inconspicuous, 0.5–1 × 0.5 mm; style 0.1–0.3 mm; berries green with copper spots when young, maturing to deep translucent red. Our plants are **subsp. amplexicaule** (Nutt.) LaFrankie [*M. amplexicaule* (Nutt.) W. A. Weber; *Smilacina racemosa* (L.) Desf. var. *amplexicaulis* (Nutt.) S. Watson]. Mixed conifers, openings in spruce-fir forests, riparian areas, 5000–10,850 ft. Widespread except EC, SE.

Maianthemum stellatum (L.) Link [*Smilacina stellata* (L.) Desf.]—Starry false Solomon's-seal. Stems erect, 2.5–5 dm × 2–3.5 mm; leaf blades ovate-elliptic to lanceolate, 5–6 × 2.5–3.5 cm; inflorescences racemose, simple, 6–15-flowered; tepals conspicuous, 4–5 × 1.5–2 mm; style 1.3–1.8 mm; berries green with black stripes along median carpel vein when young, maturing to red. Piñon-juniper woodlands, ponderosa pine-oak forests, oak scrub, mixed conifer, spruce-fir forests, riparian areas, 5700–11,700 ft. Widespread except EC, SE, SW.

NOLINA Michx. – Bear-grass

Plants cespitose and acaulescent (ours), scapose, from branched, woody caudices; usually forming colonies of rosettes; stems to 2.5 m; leaf blades linear, wiry, not glaucous or rigid or fibrous, bases broadly expanding, margins serrulate or entire; scape 0.5–25 dm; inflorescences paniculate, 3–18 dm; bracts caducous or persistent; flowers functionally unisexual, pistillate flowers with staminodes, staminate flowers with reduced pistils; tepals white to cream or yellow-green, 1.3–5 mm, apex glandular; pedicel jointed near middle; fruits capsular, 3-locular, 3-lobed, often inflated. Traditionally placed in a large, nonmonophyletic Liliaceae or in Agavaceae. Key modified from Jercinovic (2012) (Hess, 2002).

1. Bracts of the inflorescence caducous, rarely persistent; inflorescence much exceeding the leaves…**N. microcarpa**
1'. Bracts of the inflorescence persistent or mainly so; inflorescence not or only slightly exceeding the leaves …(2)
2. Inflorescence usually conspicuously tinged purple, diffuse; main rachis and division slender and flexible; fruiting pedicels jointed near the middle, not noticeably dilated, 1-2.5 mm…**N. micrantha**
2'. Inflorescence not purplish or only rarely so, dense; main rachis and divisions thick and rigid; fruiting pedicels jointed near the base, noticeably dilated, 4-7 mm…(3)
3. Leaf margins remotely serrulate with close-set, cartilaginous teeth (sometimes entire); leaf blades (4-)5-8 mm wide, concavo-convex in cross-section; seeds becoming coppery-colored…**N. greenei**
3'. Leaf margins entire (rarely remotely serrate with widely separated, noncartilaginous teeth); leaf blades 2-4(-7) mm wide, triangular in cross-section (except toward base); seeds not becoming coppery-colored …**N. texana**

Nolina greenei S. Watson ex Trel.—Woodland bear-grass. Leaf blades stiff or slightly lax, concavo-convex, 45-90(-110) cm × (4-)5-8 mm, margins remotely serrulate with close-set, cartilaginous teeth, or sometimes entire; scape curling, 0.5-2 dm; inflorescences 3-6.5 dm; bracts persistent; tepals white, sometimes with purple midveins, 2.2-3.3 mm, margins hyaline; fruiting pedicel erect, proximal to joint 1-1.5 mm, distal to joint 2.5-4.5 mm; capsules hyaline, thin-walled, inflated, 2.1-3.8 mm, distinctly notched distally; seeds becoming coppery. Gravelly or rocky slopes, limestone hillsides, volcanic flows, piñon-juniper and pine-oak woodlands and adjacent grasslands, 4550-7600 ft. NE, NC, EC, C, SC.

Nolina micrantha I. M. Johnst.—Chaparral bear-grass. Leaf blades stiff, concavo-convex, 80-130 cm × 4-6 mm, margins entire or remotely serrulate with widely separated, noncartilaginous teeth; scape 0.5-2 dm; inflorescences usually conspicuously tinged purple, 3.5-7.5 dm, held partially within rosettes; bracts mostly persistent; tepals white, 1.9-3.2 mm; fruiting pedicel erect, slender, articulating near middle, not noticeably dilated, proximal to joint to 1.5 mm, distal to joint to 2.5 mm; capsules firm-walled, inflated, 3-4 mm. Rocky limestone slopes or sandy soils of grasslands, desert scrub, adjacent to mixed conifers, 3600-7000 ft. EC, C, WC, SE, SC.

Nolina microcarpa S. Watson—Sacahuista bear-grass. Leaf blades lax, concavo-convex, 80-130 cm × 5-12 mm, margins serrulate, with close-set, cartilaginous teeth; scape 3-15 dm; inflorescences 4-12 dm, surpassing leaves; bracts caducous, rarely persistent; tepals white, 1.5-3.3 mm; fruiting pedicel erect, proximal to joint 1-2 mm, distal to joint 3-6 mm; capsules hyaline, thin-walled, inflated, 4.2-6 mm, indistinctly notched at apex. Hillsides in coarse, often rocky soil, desert grasslands, oak, piñon-juniper, desert scrub, 4350-8750 ft. EC, C, SE, SC, SW, Colfax.

Nolina texana S. Watson—Texas bear-brass. Leaf blades stiff, triangular, slightly concavo-convex toward base, 40-90 cm × 2-4(-7) mm, margins entire, rarely remotely serrate with widely separated, noncartilaginous teeth; scape curling distally, 0.5-2 dm; inflorescences rarely purple, 2.5-7 dm, held completely or partially within rosettes; bracts persistent; tepals yellow-green, 2.5-3.5 mm; fruiting pedicel ascending, thick, articulate near base, noticeably dilated into perianth, proximal to joint 2.5-6 mm, distal to joint 1.5-2 mm; capsules thin-walled, inflated, 3-4 mm. Rocky hillsides, limestone, granite, desert grasslands, desert scrub, mesquite-juniper, oak-juniper, piñon-juniper, juniper woodlands, (3350-)4400-7500 ft. NE, NC, EC, C, WC, SE, SC, SW.

POLYGONATUM P. Mill. - Solomon's-seal

Polygonatum biflorum (Walter) Elliott—Solomon's-seal. Herbs from knotty, creeping rhizomes; stems simple, erect to arching, 5-20 dm; leaves simple, alternate, sessile to clasping, narrowly lanceolate to broadly elliptic, 5-25 cm, glabrous, margins entire; inflorescences with 2-10(-15) flowers on pendent axillary peduncles; tepals whitish to greenish yellow, connate; perianth tube (13-)17-22 mm, distinct tips gently spreading, 4-6.5 mm; stamens inserted near middle of tube; berries dark blue to black or red, globose. Montane forests, spruce-fir forests, riparian areas, canyon bottoms, seepy areas, 5500-10,500 ft. NC, WC, Otero. Traditionally placed in a large, nonmonophyletic Liliaceae or in Convallariaceae.

RUTACEAE – CITRUS FAMILY

Kenneth D. Heil

Perennials or small trees; leaves alternate or opposite, simple, pinnately or palmately compound, often aromatic with oil glands (obscure in *Ptelea*); inflorescence of racemes, cymes, terminal panicles, or corymbiform; sepals mostly 4 or 5(6); petals 4 or 5(6); stamens 1–3 times as many as the petals; ovary and fruit superior (Correll & Johnson, 1979h).

1. Leaves opposite or subopposite…**Choisya**
1ʹ. Leaves alternate…(2)
2. Herbs or subshrubs, usually < 3 dm…**Thamnosma**
2ʹ. Shrubs or small trees, usually > 3 dm…**Ptelea**

CHOISYA Kunth – Mexican orange, zorrillo

Choisya dumosa (Torr. & A. Gray) A. Gray—Shrubs, mostly 1–2 m; leaves opposite or subopposite, palmately compound; leaflets 5–10, linear, 1–5 cm × 1–3 mm, with gland-tipped teeth; flowers perfect, in axillary corymbiform clusters; sepals 4 or 5, 4–5 mm; petals 4 or 5, white; fruits of 4 or 5 coriaceous 2-valved carpels. Rocky canyon slopes of limestone and clay loam, desert scrub and juniper communities, 4200–7220 ft. SE, SC.

PTELEA L. – Hoptree

Ptelea trifoliata L.—Common hoptree. Deciduous shrubs or small trees with whitish bark; leaves alternate, trifoliate; leaflets variable in shape, mostly 2–6 cm; flowers in terminal panicles, greenish white; sepals 4 or 5(6), 1–2 mm; petals 4 or 5(6), mostly 4–5 mm; stamens 4 or 5(6), alternate to the petals; fruit an indehiscent samara. Protected canyons, rocky areas, 4000–9000 ft. Widespread except NW, NE, EC, Lea. Several weakly distinguished and intergrading infraspecific taxa have been described that are unrealistic.

THAMNOSMA Torr. & Frém. – Dutchman's breeches

Thamnosma texana (A. Gray) Torr.—Ruda del monte (rue of the mountains). Perennial herb or subshrub, mostly < 3 dm; leaves alternate, simple, linear, entire, strongly gland-dotted, 5–15 mm, somewhat fleshy, aromatic when bruised; flowers in racemes or racemose cymes; sepals 4; petals 4; stamens 8; ovary 2-celled, 2-lobed; fruit a leathery 2-celled and 2-lobed capsule, 3–7 mm. Chihuahuan desert scrub, desert grassland with scattered juniper, 3080–6500 ft. S, Quay, San Miguel.

SALICACEAE – WILLOW FAMILY

Robert D. Dorn

Trees or shrubs; leaves simple, alternate (opposite), entire or toothed, rarely lobed; inflorescence a catkin; flowers unisexual, the plants dioecious or very rarely monoecious; perianth none or vestigial, each flower subtended by a bract and/or cupular disk or 1 or 2 nectar glands; stamens (1)2–70; pistil 1, 2–4-carpellate; ovary superior, sessile or stipitate; locule 1; style 1 (rarely none); stigmas 2–4; fruit a 2–4-valved capsule; seeds usually many, long-hairy for wind dispersal (Argus, 2010; Dorn, 2010; Eckenwalder, 2010). Hybrids should be expected but should not be overemphasized as some authors have done.

1. Bud scales > 1; bracts subtending flowers usually fringed (often deciduous); catkins mostly hanging downward; stamens 6–70; trees…**Populus**
1ʹ. Bud scale 1; bracts subtending flowers usually entire, rarely toothed or slightly lobed (sometimes deciduous); catkins mostly erect to spreading or drooping; stamens 2–8(–12); trees or shrubs…**Salix**

POPULUS L. – Cottonwood, aspen, poplar

Trees; bud scales several, often resinous; leaves mostly toothed, rarely lobed; inflorescence a catkin, mostly pendulous; flower usually subtended by a fringed bract as well as a cup-shaped disk; stamens 6–70; fruit a 2–4-valved capsule. Two taxa are commonly planted but are not documented as escaping cultivation. Carolina poplar (*P. canadensis* Moench), common in cities and towns, is similar to *P. deltoides* but the leaves are not nearly so deltoid, with the base tapering slightly forward; the teeth are finer, averaging ca. 2 mm deep; and the stamens are fewer (15–30). Lombardy poplar (*P. nigra* L. var. *italica* Du Roi) has suberect branches forming a narrowly conic growth form. It is otherwise similar to *P. deltoides*, but the leaves are not nearly so deltoid and have finer teeth, and the stamens are fewer (12–30). Additional hybrids are sold in the horticultural trade.

1. Leaves usually white-tomentose on underside, coarsely toothed to deeply 3–5-lobed; stamens 6–10… **P. alba**
1'. Leaves usually glabrous or glabrate on underside, mostly toothed; stamens 6–70…(2)
2. Leaves mostly suborbicular or cordate to deltoid; petioles strongly laterally flattened just below blade…(3)
2'. Leaves mostly lanceolate or ovate; petioles usually not flattened…(5)
3. Leaves mostly suborbicular or cordate; bark smooth, whitish green to nearly white…**P. tremuloides**
3'. Leaves mostly deltoid or nearly so; bark rough, usually dark…(4)
4. Capsules ovoid, with saucer-shaped disks mostly 1.5–3.5 mm wide; stipes 1–15 mm; young branchlets glabrous or sparsely hairy; expanded blades of later-developing leaves often wider than long…**P. deltoides**
4'. Capsules mostly globose or ellipsoid, with cup-shaped disks mostly 3.5–9 mm wide; stipes 1–5(–7) mm; young branchlets often densely hairy; expanded blades of later-developing leaves about as wide as long …**P. fremontii**
5. Petioles mostly < 1/3 the blade length; leaves mostly lanceolate or lance-ovate (ovate); stamens 10–20… **P. angustifolia**
5'. Petioles mostly > 1/3 the blade length; leaves mostly ovate; stamens 25–40…**P. ×acuminata**

Populus ×acuminata Rydb.—Lanceleaf cottonwood. Tree to 25 m, bark furrowed; leaf blades mostly ovate, crenate-serrate, the expanded 5–9(–13) × 3–7 cm; catkins 6.5–9(–16) cm; stamens 25–40; capsules 5–7 mm, ovoid, 2- or 3-valved; stipes 1–2(–4) mm. Stream banks, shores, washes, 4200–8500 ft. Widely scattered from N to S. Usually treated as a hybrid between *P. angustifolia* and *P. deltoides* but often occurring in the absence of one or both reputed parents. Best treated as a hybrid-derived species.

Populus alba L.—Silver poplar. Tree to 30 m, bark furrowed below; leaf blades ovate, undulate-toothed to deeply 3–5-lobed, usually white-tomentose on underside, the expanded 3–5(–7) × 2–4(–6) cm; catkins 2–9 cm; stamens 6–10; capsules 2–5 mm, ovoid, 2- or 3-valved; stipes 0.5–1(–2) mm. Introduced from Europe and occasionally becoming naturalized, mostly along roadsides, 3400–7700 ft. Widely scattered from N to S.

Populus angustifolia E. James—Narrowleaf cottonwood. Tree to 20 m, bark furrowed; leaf blades lanceolate or lance-ovate (ovate), finely crenate-serrate, the expanded 2–7(–10) × 0.8–2.5(–4) cm; catkins 3–8 cm; stamens 10–20; capsules 3–8 mm, ovoid or orbiculoid, 2-valved; stipes 0.5–1.5(–3) mm. Stream banks, seeps, mountain slopes, 4500–9700 ft. Widespread except E.

Populus deltoides W. Bartram ex Marshall—Cottonwood. Tree to 20 m, bark deeply furrowed; branchlets usually glabrous or nearly so; leaf blades deltoid to deltoid-ovate, coarsely toothed to almost lobed, the expanded 3–8 cm, usually slightly wider, acuminate or acute; catkins 5–13(–21) cm; stamens 30–40(–55); capsules (4–)8–16 mm, ovoid, 3- or 4-valved; floral disk 1–3.5(–4) mm wide, saucer-shaped (shallowly cup-shaped); stipes 1–15 mm. Stream banks, shores, washes, 2900–7800 ft. Two varieties gradually intergrade in the C part of the state.

1. Stipes 1–6(–8) mm; leaves generally with tips slender, acuminate, needlelike, with concave lateral sides, sometimes with 2+ glands on upper side of blade near tip of petiole or on petiole near blade…**var. occidentalis**
1'. Stipes (5–)8–15 mm; leaves generally with tips broadly acute, with relatively flat lateral sides, lacking glands near tip of petiole…**var. wislizeni**

var. occidentalis Rydb. [*P. deltoides* W. Bartram ex Marshall subsp. *monilifera* (Aiton) Eckenw.]—Plains cottonwood. NE, EC, SC.

var. wislizeni (S. Watson) Dorn—Basin cottonwood. Widespread.

Populus fremontii S. Watson—Fremont cottonwood. Tree to 30 m, bark deeply furrowed; branchlets of year and year-old branchlets mostly pubescent; leaf blades deltoid or deltoid-cordate (deltoid-ovate), mostly coarsely rounded-toothed to almost lobed, the expanded (4-)5-11(-14) cm and about as wide, acute to short-acuminate (long-acuminate); catkins 4-13 cm; stamens 30-70; capsules 6-10 mm, globose or ellipsoid (ovoid), usually (3)4-valved, floral disk 3.5-9 mm wide, deeply cup-shaped; stipes 1-5(-7) mm. Stream banks, shores, washes, 3600-7400 ft. SE, SC, SW. Our plants are **var. fremontii**. There is apparent intergrading with *P. deltoides*, and the plants are sometimes treated as *P. deltoides* var. *fremontii* (S. Watson) Cronquist.

Populus tremuloides Michx.—Quaking aspen. Tree to 30 m but usually 1/2 that, forming clones, bark mostly smooth and whitish green to nearly white; leaf blades cordate (ovate) to suborbicular, finely crenate-serrate, the expanded 2-7 cm and about as wide; catkins 2-10 cm; stamens 6-12; capsules 4-7 mm, lanceoloid, 2-valved; stipes 1-2 mm. Moist areas in the mountains and in steep shaded canyons at lower elevations, 5700-11,900 ft. Widespread except EC, SE.

SALIX L. – Willow

Lowland trees to tiny, creeping Arctic-alpine shrubs; bud scales solitary, nonresinous; leaves entire or toothed, usually stipulate but the stipules often deciduous; inflorescence a mostly spreading or drooping to erect catkin, the catkins sessile or on floriferous branchlets (peduncles) that are usually leafy; flower subtended by an entire or rarely toothed to erose bract and 1 or 2 nectaries; stamens (1)2-8(-12); fruit a 2-valved capsule. Globe willow (*S. matsudana* Koidz. Cultivar f. *umbraculifera*) is often planted in yards. It is a small tree with a graceful, spherical crown and narrowly lanceolate, glaucous leaves. White willow (*S. alba* L.) is sparingly planted. It is similar to *S. fragilis* but differs by its persistently pubescent leaves and shorter styles to 0.2 mm. The first key below is for pistillate material, the second for staminate material, and the third for vegetative material. Avoid sucker shoots with vigorous growth, severely browsed shoots, and young catkins and leaves in early stages of development. When obtaining specimens, include current year, 1-year-old, and 2-year-old branchlets.

PISTILLATE KEY

1. Capsules hairy…(2)
1'. Capsules glabrous…(13)
2. Plants creeping shrubs 1-10 cm with branchlets usually rooting; near or above timberline…(3)
2'. Plants erect shrubs or trees mostly well > 10 cm; only occasionally above timberline…(4)
3. Catkins pseudoterminal opposite the terminal leaf, peduncles usually without leaves; styles < 0.5 mm; leaves mostly oval to suborbicular, the tip usually rounded or obtuse…**S. reticulata**
3'. Catkins often borne laterally, peduncles usually leafy; styles 0.5-2 mm; leaves mostly elliptic, the tip usually pointed…**S. arctica**
4. Flower bracts yellowish, greenish, whitish, or tawny, deciduous in fruit; leaves linear to narrowly lanceolate or narrowly elliptic (oblong), nonglaucous…(5)
4'. Flower bracts only occasionally yellowish, greenish, whitish, or tawny, usually brown or black, persistent in fruit; leaves often broader, glaucous or glaucescent on underside…(7)
5. Bud scales split down the side toward branch, with the free margins overlapping; leaves mostly closely serrulate or serrate; petioles 3-10 mm…**S. gooddingii** (in part)
5'. Bud scales caplike, not split down the side; leaves remotely denticulate or serrulate to entire; petioles 0-6 mm…(6)
6. Catkins (1-)1.5-10 cm; stipes 0-1.8(-2) mm; expanded leaf blades mostly (3-)4-17 × 0.3-1.2(-2.3) cm; mostly shrubs with slender trunks; statewide…**S. exigua** (in part)
6'. Catkins 0.4-2.3 cm; stipes 0-0.3 mm; expanded leaf blades 0.5-3.5(-4.2) × 0.1-0.5 cm; mostly treelike with 1 to several trunks; SW counties…**S. exilifolia**
7. Branchlets of previous year, and sometimes those of current season, glaucous, sometimes apparent only at nodes, especially behind buds…(8)

7′. Branchlets not glaucous…(9)

8. Catkins 1.5–6(–11) cm, sessile or nearly so, densely flowered; leaves often densely silvery-hairy on underside, glabrous or glabrate on upper side; flower bracts usually dark…**S. drummondiana**

8′. Catkins 0.6–2(–2.5) cm, on leafy branchlets 0.1–1.8 cm, loosely flowered; leaves sparsely to moderately sericeous on 1 or both sides; flower bracts tawny or light brown in fruit…**S. geyeriana**

9. Stipes mostly 2–5 mm; styles 0.4 mm or less; flower bracts light brown or tawny; buds often with depressed margins…**S. bebbiana**

9′. Stipes 2 mm or less, or if as long as 3 mm, the styles often > 0.4 mm; flower bracts tawny or dark; buds without depressed margins…(10)

10. Catkins appearing with the leaves on leafy branchlets 0.2–3.5 cm; leaves and branchlets often conspicuously hairy; plants mostly to 1.5 m (rarely to 4 m)…(11)

10′. Catkins appearing mostly before the leaves, sessile or nearly so or sometimes on branchlets to 1.3 cm; leaves and branchlets often sparsely hairy to glabrous; plants sometimes > 1.5 m…(12)

11. Catkins mostly < twice as long as wide, 0.5–2(–3) cm; stipes < 0.5 mm; petioles mostly 1–3 mm; expanded leaf blades 2–3(–4) × 0.6–1.6 cm…**S. brachycarpa**

11′. Catkins mostly > twice as long as wide, (2–)3–4(–6.5) cm; stipes 0.2–1.8 mm; petioles mostly 3–16 mm; expanded leaf blades 3–6(–8) × 0.7–3.5 cm…**S. glauca**

12. Stipes 0–1 mm; stigmas usually < 0.5 mm; leaves mostly elliptic, usually very shiny on upper surface; year-old branchlets often reddish and shiny; plants of wet places…**S. planifolia**

12′. Stipes 0.8–2.8 mm; stigmas usually > 0.5 mm; leaves mostly oblanceolate to obovate (elliptic), not very shiny; year-old branchlets mostly yellowish to reddish brown, dull; plants of drier upland areas…**S. scouleriana**

13. Flower bracts yellowish, greenish, whitish, or tawny, deciduous in fruit (sometimes tardily so); catkins usually on leafy floriferous branchlets; styles mostly < 1 mm…(14)

13′. Flower bracts often blackish or brown, persistent in fruit; catkins sessile or on leafy floriferous branchlets; styles 0–2 mm…(20)

14. Bud scales split down the side toward branch, with the free margins overlapping; native trees at lower elevations…(15)

14′. Bud scales caplike, not split down the side; mostly shrubs or an introduced tree with branchlets of previous 3 years each very brittle and easily broken off at base…(18)

15. Leaves glaucous on underside…(16)

15′. Leaves not glaucous on underside…(17)

16. Year-old branchlets mostly yellowish to grayish; branchlets of year mostly glabrous; later-expanded leaves mostly lanceolate to ovate with acuminate tips; widespread…**S. amygdaloides**

16′. Year-old branchlets mostly reddish, purplish, or brownish; branchlets of year sometimes hairy; later-expanded leaves lanceolate or lance-linear to narrowly elliptic with mostly acute (acuminate) tips; W counties…**S. bonplandiana**

17. Year-old branchlets usually gray, yellowish, or pale yellowish brown; widespread…**S. gooddingii** (in part)

17′. Year-old branchlets mostly reddish brown to reddish yellow; E edge of state…**S. nigra**

18. Leaves mostly linear to narrowly elliptic, remotely denticulate or serrulate to entire, usually hairy; petioles 0–6 mm; catkins sometimes 2 to several in a terminal cluster in addition to lateral solitary catkins …**S. exigua** (in part)

18′. Leaves predominantly lanceolate to ovate or broadly elliptic, mostly closely serrate or serrulate, often glabrous; petioles 5–30 mm; catkins all lateral and solitary…(19)

19. Catkins mostly slender and elongate, mostly 1 cm or less wide; plants naturalized trees; branchlets of previous 3 years each very brittle and easily broken off at base…**S. fragilis**

19′. Catkins often thick and not especially elongate, mostly 1+ cm wide; plants native shrubs; branchlets not especially brittle…**S. lasiandra**

20. Leaves glaucous on underside (catkins sometimes appearing before the leaves)…(21)

20′. Leaves not glaucous on underside…(25)

21. Year-old branchlets, and sometimes branchlets of current season, glaucous, sometimes apparent only at nodes, especially behind buds; flower bracts mostly obovate to suborbicular and densely fringed with long, relatively straight hairs; catkins appearing mostly before the leaves, sessile or on branchlets to 0.5 cm; expanded leaves glabrous or essentially so…**S. irrorata**

21′. Year-old branchlets and branchlets of current season not glaucous; flower bracts sometimes narrower, with often sparser, shorter, or curly hairs or glabrate; catkins appearing before or with the leaves, sessile or on branchlets to 1.7 cm; expanded leaves glabrous or hairy…(22)

22. Flower bracts densely fringed with relatively straight untangled hairs; catkins appearing mostly before the leaves, sessile or on branchlets to 0.7 cm; typical buds short and plump, barely longer than wide; expanded leaves oblong, narrowly elliptic, or oblanceolate (obovate), usually hairy at least on underside …**S. lasiolepis**

22'. Flower bracts with often sparse hairs or with curly or tangled hairs; catkins appearing before or with the leaves, often on prominent branchlets; buds often elongate; expanded leaves mostly lance-linear or lanceolate to elliptic or ovate (obovate to oblong), sometimes glabrous or glabrate…(23)

23. Styles averaging 0.7+ mm; hairs of floral bracts long and straight or curly; leaf blades tending to be < 3 times as long as wide…**S. monticola**

23'. Styles averaging 0.7 mm or less; hairs of floral bracts short or crinkly and tangled mostly toward base of bract or sometimes bracts glabrate; leaf blades tending to be > 3 times as long as wide…(24)

24. Stipes 0.5–2(–2.5) mm; plants native shrubs with spreading to ascending branchlets; expanded leaves lanceolate to elliptic or occasionally oblong to oblanceolate, the base rounded, entire or serrulate…**S. eriocephala**

24'. Stipes 0–0.3 mm; plants introduced trees with long, pendulous branchlets; expanded leaves narrowly lanceolate or lance-linear, the base acute, spinulose-serrate…**S. babylonica**

25. Expanded leaf blades mostly 1.5–2.4 times as long as wide, 1–5 × 0.5–2.8(–3.1) cm, cordate or subcordate (rounded) at base, the margins of at least the young leaves or those of the flowering branchlets with prominent glands that stand out from the margin…**S. arizonica**

25'. Expanded leaf blades usually not as above…(26)

26. Catkins 0.8–2(–3) cm; stipes 0–0.8 mm; leaf blades entire, mostly 2–4(–6) × 0.5–1.5(2) cm, persistently hairy on both sides; shrubs mostly < 1 m, occasionally to 2 m…**S. wolfii**

26'. Catkins 1.5–5(–7) cm; stipes 0.5–2(–2.5) mm; leaf blades serrulate to entire, (1–)2–8(–10) × (0.4–)0.8–2.5 (–3.5) cm, often becoming glabrous or glabrate; shrubs to 6 m…**S. boothii**

STAMINATE KEY

1. Stamens 3–8 per flower…(2)

1'. Stamens 2 per flower…(6)

2. Bud scales caplike, not split down the side; catkins generally thick and stiff, mostly 1+ cm wide; shrubs…**S. lasiandra**

2'. Bud scales split down the side toward branch, with the free margins overlapping; catkins generally slender and lax, often 1 cm or less wide; trees at lower elevations…(3)

3. Leaves glaucous on underside…(4)

3'. Leaves not glaucous…(5)

4. Year-old branchlets mostly yellowish to grayish; branchlets of current year mostly glabrous; later-expanded leaves mostly lanceolate to ovate with acuminate tips; widespread…**S. amygdaloides**

4'. Year-old branchlets mostly reddish, purplish, or brownish; branchlets of current year sometimes hairy; later-expanded leaves lanceolate or lance-linear to narrowly elliptic with mostly acute (acuminate) tips; W counties…**S. bonplandiana**

5. Year-old branchlets usually gray, yellowish, or pale yellowish brown; widespread…**S. gooddingii**

5'. Year-old branchlets mostly reddish brown to reddish yellow; E edge of state…**S. nigra**

6. Plants creeping shrubs to 10 cm with branchlets usually rooting; near or above timberline…(7)

6'. Plants upright trees or shrubs mostly well > 10 cm; only occasionally above timberline…(8)

7. Catkins pseudoterminal opposite the terminal leaf, the prominent peduncle usually without leaves; leaves mostly oval to suborbicular with rounded or obtuse tips…**S. reticulata**

7'. Catkins often borne laterally, the peduncle usually leafy; leaves mostly elliptic with pointed tips…**S. arctica**

8. Leaves glaucous on underside (catkins sometimes appearing before the leaves)…(9)

8'. Leaves not glaucous…(21)

9. Year-old branchlets glaucous, sometimes apparent only behind buds, branchlets of current year sometimes also glaucous…(10)

9'. Year-old branchlets and branchlets of current year not glaucous…(12)

10. Catkins 0.6–2(–2.5) cm, appearing mostly with the leaves, borne on leafy branchlets 0.1–1.8 cm; flower bracts mostly tawny or light brown…**S. geyeriana**

10'. Catkins 1.5–6(–11) cm, appearing before the leaves, mostly sessile or subsessile; flower bracts usually dark…(11)

11. Flower bracts mostly obovate (oblong), dark brown, with rather dense, long, relatively straight hairs conspicuously exceeding bract tip; leaves soon becoming glabrous or nearly so…**S. irrorata**

11'. Flower bracts often narrower or darker or less hairy; leaves remaining mostly densely silvery-hairy on underside…**S. drummondiana**

12. Plants naturalized trees either with branchlets of the previous 3 years each very brittle and easily broken off at base or with long, pendulous branchlets; flower bracts yellowish, greenish, whitish, or tawny; expanded leaves mostly lance-elliptic, lance-linear, or lanceolate and glabrous or glabrate…(13)

12'. Plants usually shrubs, the branchlets usually not very brittle or long and pendulous; flower bracts often blackish or brownish, occasionally lighter; expanded leaves variously shaped, sometimes hairy…(14)

13. Catkins 1–3.5(–4) cm; branchlets long and pendulous…**S. babylonica**
13'. Catkins (2–)4–8 cm; branchlets spreading to ascending…**S. fragilis**
14. Flower bracts mostly brown to black, obovate to oval (oblong), densely fringed with straight, untangled hairs; leaves hairy at least on underside; typical buds mostly short and plump, barely longer than wide; anthers mostly 0.8 mm or less; mostly lowlands…**S. lasiolepis**
14'. Flower bracts sometimes lighter, mostly narrower or pointed at tip, with often sparse hairs or with curly or tangled hairs; leaves glabrous or hairy; buds variable; anthers sometimes > 0.8 mm; mostly mountains…(15)
15. Nectaries generally 2+, both dorsal and ventral; catkins appearing with the leaves, terminating leafy branchlets; plants mostly to 1.5 m (rarely to 4 m); mostly high mountains…(16)
15'. Nectaries generally 1, ventral; catkins sometimes appearing before the leaves and mostly sessile or sub-sessile; plants often well > 1.5 m; sometimes lower in mountains or in lowlands…(17)
16. Catkins mostly < twice as long as wide, 0.5–2(–3) cm; petioles mostly 1–3 mm; expanded leaf blades 2–3(–4) × 0.6–1.6 cm…**S. brachycarpa**
16'. Catkins mostly > twice as long as wide, (2–)3–4(6.5–) cm; petioles mostly 3–16 mm; expanded leaf blades generally 3–6(–8) × 0.7–3.5 cm…**S. glauca**
17. Young emerging leaves with some reddish hairs, especially on margins or at tip; flower bracts often with tawny hairs; leaves mostly elliptic, usually very shiny on upper surface; year-old branchlets often reddish and shiny; mostly wet sites…**S. planifolia**
17'. Young emerging leaves with white hairs, or if with some reddish, then flower bracts with white hairs; leaves variable; year-old branchlets rarely both reddish and shiny; sometimes in drier upland sites…(18)
18. Bracts at base of catkin densely fringed with long hairs, often tawny; catkins appearing before the leaves; filaments often 8+ mm; leaves mostly oblanceolate to obovate; plants of drier uplands…**S. scouleriana**
18'. Bracts at base of catkin becoming greenish and glabrous or glabrate, at least at tip, rarely lacking; catkins appearing before or with the leaves; filaments rarely as much as 8 mm; leaves often mostly ovate or lanceolate to elliptic, rarely obovate or oblanceolate; mostly wet sites…(19)
19. Flower bracts tawny (greenish yellow); buds generally with depressed margins; leaves elliptic or sometimes ovate or obovate; 2-year-old branchlets usually with cracked bark, giving a white-streaked appearance; branchlets of current year usually red-purple and appressed-hairy…**S. bebbiana**
19'. Flower bracts usually brown or black but sometimes paler; buds and branchlets usually not as above…(20)
20. Flower bracts generally dark and prominent, with long, straight or curly hairs; leaf blades tending to be < 3 times as long as wide…**S. monticola**
20'. Flower bracts often brownish and inconspicuous, with short, tangled, crinkly hairs mostly toward base of bract or sometimes bracts glabrate (hairs on catkin rachis may be long and tangled); leaf blades tending to be > 3 times as long as wide…**S. eriocephala**
21. Flower bracts yellowish, whitish, greenish, or tawny; catkins occasionally 2 to several in a terminal cluster in addition to lateral solitary catkins; petioles 0–6 mm; expanded leaves predominantly linear or linear-elliptic, remotely denticulate or serrulate to entire…(22)
21'. Flower bracts blackish or brownish; catkins all lateral and solitary; petioles 2–17 mm; expanded leaves lanceolate to elliptic or oblanceolate, or ovate to obovate, entire or closely toothed…(23)
22. Catkins (1–)1.5–10 cm; expanded leaf blades mostly (3–)4–17 × 0.3–1.2(–2.3) cm; mostly shrubs with slender trunks; statewide…**S. exigua**
22'. Catkins 0.4–2.3 cm; expanded leaf blades 0.5–3.5(–4.2) × 0.1–0.5 cm; mostly treelike with 1 to several trunks; SW counties…**S. exilifolia**
23. Expanded leaf blades mostly 1.5–2.4 times as long as wide, 1–5 × 0.5–2.8(–3.1) cm, cordate or subcordate (rounded) at base, the margins of at least the young leaves or those of the flowering branchlets with prominent glands that stand out from the margin…**S. arizonica**
23'. Expanded leaf blades usually not as above…(24)
24. Catkins 0.8–2(–3) cm; leaf blades entire, mostly 2–4(–6) × 0.5–1.5(–2) cm, persistently hairy on both sides; shrubs mostly < 1 m, occasionally to 2 m…**S. wolfii**
24'. Catkins 1.5–5(–7) cm; leaf blades serrulate to entire, (1–)2–8(–10) × (0.4–)0.8–2.5(–3.5) cm, often becoming glabrous or glabrate; shrubs to 6 m…**S. boothii**

VEGETATIVE KEY

1. Leaves glaucous on underside, or rarely underside much lighter from dense hairs that obscure leaf surface…(2)
1'. Leaves not glaucous…(22)
2. Plants creeping shrubs to 10 cm with usually rooting branchlets; near or above timberline…(3)
2'. Plants upright trees or shrubs usually well > 10 cm; only occasionally above timberline…(4)

3. Leaf blades somewhat leathery, predominantly oval to suborbicular with prominent reticulate venation on underside, the tips mostly rounded to obtuse...**S. reticulata**
3'. Leaf blades not leathery, predominantly elliptic with reticulate venation not especially prominent, the tips usually pointed...**S. arctica**
4. Bud scales split down the side toward branch, with the free margins overlapping; native lowland trees... (5)
4'. Bud scales caplike, not split down the side; mostly shrubs or naturalized trees...(6)
5. Year-old branchlets mostly yellowish to grayish; branchlets of current year mostly glabrous; expanded leaves mostly lanceolate to ovate and acuminate; widespread...**S. amygdaloides**
5'. Year-old branchlets mostly reddish, purplish, or brownish; branchlets of current year sometimes hairy; expanded leaves mostly lanceolate or lance-linear to narrowly elliptic and acute (acuminate); W counties ...**S. bonplandiana**
6. Plants naturalized trees either with branchlets of previous 3 years each very brittle and easily broken off at base or with long, pendulous branchlets; expanded leaves predominantly lance-linear or lanceolate to lance-elliptic and glabrous or glabrate; year-old branchlets mostly yellow or yellow-brown...(7)
6'. Plants usually shrubs, the branchlets not especially brittle or long and pendulous; expanded leaves variable; year-old branchlets sometimes more reddish or dark brown...(8)
7. Branchlets long and pendulous; expanded leaf blades 0.9–1.8 cm wide...**S. babylonica**
7'. Branchlets spreading to ascending; expanded leaf blades 1.7–3.5 cm wide...**S. fragilis**
8. Year-old branchlets, and sometimes branchlets of current year, glaucous, sometimes apparent only behind buds...(9)
8'. Year-old branchlets and branchlets of current year not glaucous...(11)
9. Expanded leaves glabrous or essentially so...**S. irrorata**
9'. Expanded leaves hairy at least on underside...(10)
10. Expanded leaves with silver hairs on underside that are often so dense that they obscure the leaf surface, the upper side glabrous or sparsely hairy; leaf margins slightly revolute, the blades 1–2.6 cm wide...**S. drummondiana**
10'. Expanded leaves sparsely to moderately sericeous at least on underside, often about equally hairy on both sides; leaf margins not revolute, the blades 0.6–1.5 cm wide...**S. geyeriana**
11. Branchlets of current year usually red-purple and appressed-hairy; bark of 2-year-old branchlets cracked, giving a white-streaked appearance; buds often with depressed margins...**S. bebbiana**
11'. Branchlets and buds not as above...(12)
12. Plants with mostly oblanceolate to obovate leaves; freshly stripped bark of living branchlets of previous year usually with a "skunky" odor; large shrubs of drier upland sites...**S. scouleriana**
12'. Plants not as above...(13)
13. Most leaves entire or nearly so, rarely serrulate only toward base...(14)
13'. Most leaves toothed...(18)
14. Year-old branchlets usually reddish and shiny; upper side of leaf shiny, leaf base acute...**S. planifolia** (in part)
14'. Year-old branchlets mostly reddish brown and dull; upper side of leaf dull, leaf base often rounded...(15)
15. Expanded leaves glabrous or nearly so...**S. eriocephala** (in part)
15'. Expanded leaves usually obviously hairy, rarely glabrate (rarely glabrous in high-mountain species)...(16)
16. Leaf blades mostly oblong or narrowly elliptic to oblanceolate (obovate), (2.5–)3.5–12.5(–22) cm; mostly lowland shrubs...**S. lasiolepis** (in part)
16'. Leaf blades mostly elliptic to oblanceolate or elliptic-obovate, oblong, or oval, 2–6(–8) cm; mostly high-mountain shrubs...(17)
17. Petioles mostly 3–16 mm; expanded leaf blades 3–6(–8) × 0.7–3.5 cm, usually glabrous to moderately hairy...**S. glauca**
17'. Petioles mostly 1–3 mm; expanded leaf blades 2–3(–4) × 0.6–1.6 cm, usually densely hairy...**S. brachycarpa**
18. Leaves predominantly elliptic, dark green and shiny on upper side; year-old branchlets usually reddish and shiny...**S. planifolia** (in part)
18'. Leaves mostly lanceolate, oblong, or oblanceolate to ovate or obovate; if elliptic, the leaves not shiny on upper side and the year-old branchlets not reddish and shiny...(19)
19. Petioles usually with glands near base of leaf blade; leaf tips mostly acuminate...**S. lasiandra** (in part)
19'. Petioles usually lacking glands; leaf tips mostly acute to obtuse...(20)
20. Petioles often reddish; leaf blades often ovate, obovate, or broadly elliptic, tending to be < 3 times as long as wide; branchlets of current year hairy; mountain plants...**S. monticola**
20'. Petioles usually green; leaf blades often lanceolate, elliptic, oblanceolate, or oblong, tending to be > 3 times as long as wide; branchlets of current year often glabrous; mountains or lowlands...(21)
21. Leaves oblong or narrowly elliptic to oblanceolate (obovate), usually hairy at least on underside; typical buds mostly short and plump, barely longer than wide...**S. lasiolepis** (in part)

21'. Leaves predominantly lanceolate to elliptic, occasionally oblong, rarely oblanceolate, glabrous or glabrate when expanded; buds often obviously longer than wide…**S. eriocephala** (in part)
22. Bud scales split down the side toward branch, with the free margins overlapping; lower-elevation trees …(23)
22'. Bud scales caplike, not split down the side; mostly shrubs…(24)
23. Year-old branchlets usually gray, yellowish, or pale yellowish brown; widespread…**S. gooddingii**
23'. Year-old branchlets mostly reddish brown to reddish yellow; E edge of state…**S. nigra**
24. Expanded leaves mostly linear or linear-elliptic, remotely denticulate or serrulate to entire; petioles 0–6 mm…(25)
24'. Expanded leaves lanceolate to elliptic or oblanceolate, or ovate to obovate, entire or closely toothed; petioles 2–30 mm…(26)
25. Expanded leaf blades mostly (3–)4–17 × 0.3–1.2(–2.3) cm; mostly shrubs with several slender trunks; statewide…**S. exigua**
25'. Expanded leaf blades 0.5–3.5(–4.2) × 0.1–0.5 cm; mostly treelike with 1 to several stoutish trunks; SW counties…**S. exilifolia**
26. Petioles usually with glands near base of leaf blade; leaves usually glabrous when expanded, the tips mostly acuminate…**S. lasiandra** (in part)
26'. Petioles usually lacking glands; leaves often persistently hairy, the tips mostly acute to obtuse…(27)
27. Expanded leaf blades mostly 1.5–2.4 times as long as wide, 1–5 × 0.5–2.8(–3.1) cm, cordate or subcordate (rounded) at base, the margins of at least the young leaves often with prominent glands that stand out from the margin…**S. arizonica**
27'. Expanded leaf blades usually not as above…(28)
28. Leaf blades entire, mostly 2–4(–6) × 0.5–1.5(–2) cm, persistently hairy on both sides; shrubs mostly < 1 m, occasionally to 2 m…**S. wolfii**
28'. Leaf blades serrulate to entire, (1–)2–8(–10) × (0.4–)0.8–2.5(–3.5) cm, often becoming glabrous or glabrate; shrubs to 6 m…**S. boothii**

Salix amygdaloides Andersson—Peachleaf willow. Tree to 15(–30) m; leaf blades mostly lanceolate to ovate, the expanded 5.5–13 × (0.7–)1–3.7 cm, base rounded to acute, tip acuminate, serrulate, glaucous on underside, glabrous or becoming so; catkins coetaneous, 2.5–11 cm, on leafy branchlets 0.4–3.5 (–6) cm; capsules glabrous, 3–5.5 mm; styles 0.3–0.6 mm; stipes 1.2–3.2 mm. Stream banks, floodplains, shores, marshes, seeps, 2900–7900 ft. N, C, Eddy, Otero. Hybridizes with *S. gooddingii* (*S.* ×*wrightii* Andersson).

Salix arctica Pall.—Arctic willow. Creeping shrub < 10 cm; leaf blades mostly elliptic (oval), the expanded (1–)1.5–4 × 0.4–1.5(–2) cm, base acute to obtuse, tip acute to obtuse, entire, glaucous on underside, either glabrous or pubescent mostly on underside; catkins coetaneous, (0.7–)1–5.5 cm, on leafy branchlets (0.5–)1–3(–5.5) cm; capsules pubescent, 3–6(–7) mm; styles 0.5–2 mm; stipes 0–0.8 mm. Our plants are **var. petraea** (Andersson) Bebb [*S. petrophila* Rydb.]. Alpine and subalpine, 11,200–13,100 ft. Mora, Rio Arriba, Santa Fe, Taos.

Salix arizonica Dorn—Arizona willow. Shrub to 1.5(–2.5) m; leaf blades ovate to obovate, the expanded 1–5 × 0.5–2.8(–3.1) cm, base cordate or subcordate (rounded), tip acute to obtuse, serrulate, nonglaucous, pubescent to sometimes glabrous; catkins precocious to coetaneous, (0.5–)1–4 cm, sessile or on leafy branchlets to 1.2 cm; capsules glabrous, 3–5 mm; styles 0.5–1.5 mm; stipes 0.2–1.5 mm. Wet meadows, low-gradient stream banks, 10,100–10,700 ft. Mora, Rio Arriba, Taos. **R**

Salix babylonica L.—Weeping willow. Introduced tree to 20 m; leaf blades narrowly lanceolate or lance-linear, the expanded 7–14 × 0.9–1.8 cm, base acute, tip acuminate, spinulose-serrate, glaucous on underside, glabrous or glabrate; catkins coetaneous, 1–3.5(–4) cm, on leafy branchlets 0.2–1.5 cm; capsules glabrous, (1.5–)2.8–3.8 mm; styles 0–0.5 mm; stipes 0–0.3 mm. Introduced and rarely escaping along streams and ditch banks, 4000–6500 ft. Grant, Guadalupe, Otero, San Juan. Some authors treat this as a hybrid, *S.* ×*pendulina* Wender. or *S.* ×*sepulcralis* Simonk.

Salix bebbiana Sarg.—Bebb willow. Shrub or small tree to 10 m; leaf blades elliptic or sometimes ovate to obovate, the expanded (2–)4–8 × 1–3.3 cm, base acute to obtuse or rounded, tip acute or rarely obtuse, crenate or irregularly serrate to entire, glaucous on underside, pubescent to glabrate; catkins coetaneous or subprecocious, 0.6–6(–8) cm, on leafy branchlets 0.1–6 cm; capsules pubescent, strongly beaked, 5–9 mm; styles 0.1–0.4 mm; stipes 2–5 mm. Swamp edges, moist woods, stream banks, meadows, 5500–10,800 ft. NC, NW, C, WC, SC, Grant.

Salix bonplandiana Kunth—Bonpland willow. Tree to 15(–20) m; leaf blades narrowly elliptic (obovate) to lanceolate or lance-linear, the expanded 5–17(–19) × 0.7–3(–4) cm, base acute to subcordate, tip acute or acuminate, serrulate or serrate-crenate to entire, glaucous on underside, glabrous or becoming so; catkins coetaneous or sometimes serotinous, the latter usually appearing in summer or early fall in axils of leaves and sometimes sessile or nearly so, the former on leafy (leafless) branchlets (0.4–)0.5–4 cm or rarely sessile, 1.2–7.5(–9) cm, often not very lax; capsules glabrous, 3–6 mm; styles 0–0.4 mm; stipes 0.8–2.8(–3.5) mm. Stream banks, washes, shores, seeps, ditch banks, 4800–6800 ft.

1. Flowering branchlets (sometimes leafless) 0–1.2(–3) cm; leaves, especially those on middle 1/3 of branchlets, generally (4)5+ times as long as wide; spring and sometimes flowering in summer and early fall; SW counties…**var. bonplandiana**
1′. Flowering branchlets mostly (0.4–)0.5–4 cm; leaves, especially those of flowering branchlets and those on middle 1/3 of vegetative branchlets, generally 4(5) times or less as long as wide; flowering in spring; NW counties…**var. laevigata**

var. bonplandiana—Bonpland willow. Grant. A specimen could not be located (*Felger 09-101*), but photographs of the living tree suggest that it is this variety (Felger, 2010; Felger & Kindscher, 2010; Kleinman & Felger, 2009). Other reports were all either misidentified or from old Mexico just across the border.

var. laevigata (Bebb) Dorn [*S. laevigata* Bebb]—Red willow. McKinley, San Juan.

Salix boothii Dorn—Booth willow. Shrub to 6 m; leaf blades lanceolate to elliptic or oblanceolate (ovate), the expanded (1–)2–8(–10) × (0.4–)0.8–2.5(–3.5) cm, base rounded to acute, tip acute or short-acuminate, serrulate to entire, nonglaucous, glabrous or pubescent; catkins coetaneous or subprecocious, 1.5–5(–7) cm, on leafy branchlets 0.1–1(–1.5) cm; capsules glabrous, (2.5–)3–6 mm; styles 0.2–1.2(–1.5) mm; stipes 0.5–2(–2.5) mm. Stream banks, swamps, seeps, 7000–11,500 ft. NC except Santa Fe, Taos.

Salix brachycarpa Nutt.—Shortfruit willow. Shrub to 1.5(–3) m; leaf blades elliptic to elliptic-obovate, oblong, or oval, the expanded 2–3(–4) × 0.6–1.6 cm, base rounded to acute, tip acute to obtuse, entire or with a few glands near base, glaucous on underside, pubescent on both sides; catkins coetaneous, 0.5–2(–3) cm, on leafy branchlets 0.2–2 cm; capsules pubescent, (3–)5–7 mm; styles 0.1–0.8(–1.5) mm; stipes 0–0.5 mm. Our plants are **var. brachycarpa**. Meadows, slopes, bogs, wet alkaline barrens, stream banks, 8200–13,100 ft. NC.

Salix drummondiana Barratt ex Hook.—Drummond willow. Shrub to 6 m; leaf blades elliptic or oblong (lanceolate) to elliptic-obovate, the expanded 4–11 × 1–2.6 cm, base and tip acute, entire to rarely crenate, glaucous on underside, glabrous or sparsely pubescent on upper side, silver-sericeous to rarely glabrate on underside; catkins precocious (subprecocious), 1.5–6(–11) cm, sessile or subsessile; capsules pubescent, 3–5.6 mm; styles 0.4–1.8 mm; stipes 0.1–2 mm. Stream banks, swamps, thickets, 7400–12,000 ft. NC.

Salix eriocephala Michx.—Heartleaf willow. Shrub, or rarely treelike, to 6 m; leaf blades lanceolate to elliptic, occasionally oblong, rarely oblanceolate, the blades 5–10(–12) × 1–2.5(–3.5) cm, base mostly rounded, tip acute, serrulate to entire, glaucous on underside, glabrous or glabrate; catkins mostly coetaneous, 2–6 cm, sessile or on leafy branchlets to 0.9 cm; capsules glabrous, 3.5–6 mm; styles 0.1–0.7 mm; stipes 0.5–2(–2.5) mm. Our plants are **var. ligulifolia** (C. R. Ball) Dorn [*S. ligulifolia* C. R. Ball ex C. K. Schneid.], strapleaf willow. Stream banks, shores, swamps, other moderately wet areas, 5600–9200 ft. NC, NW, C, WC, Grant.

Salix exigua Nutt.—Sandbar willow. Shrub to 5(–10) m; leaf blades mostly linear (narrowly elliptic), the expanded (3–)4–17 × 0.3–1.2(–2.3) cm, base and tip mostly acute, entire or serrulate (denticulate) with teeth widely spaced, nonglaucous, pubescent or sometimes glabrous; catkins coetaneous or serotinous, (1–)1.5–10 cm, on leafy branchlets 0.5–18 cm; capsules glabrous or pubescent, 3–8(–10) mm; styles

0-0.2 mm; stipes 0-1.8(-2) mm. Stream banks, floodplains, washes, shores, ditch banks, 2900-10,200 ft. (subsp. *exigua*), 2900-4600 ft. (subsp. *interior*). Two subspecies occur in the state. Widespread.

1. Leaves often persistently hairy even when fully expanded, sometimes entire or nearly so, often not very veiny; capsules 3-5(-7) mm; statewide…**subsp. exigua**
1'. Leaves often glabrous or glabrate when expanded, regularly serrulate, often spinulosely so, prominently veiny; capsules (4-)5-8(-10) mm; E edge of state…**subsp. interior**

subsp. exigua—Coyote willow. 2900-10,200 ft.

subsp. interior (Rowlee) Cronquist [*S. interior* Rowlee]—Sandbar willow. 2900-4600 ft.

Salix exilifolia Dorn [*S. taxifolia* auct. non Kunth]—Slenderleaf willow. Small tree (occasionally shrublike) to 10(-16) m; leaf blades mostly linear (narrowly elliptic or oblanceolate), the expanded 0.5-3.5(-4.2) cm × 1-5 mm, base and tip acute, entire or sometimes with widely spaced glands or teeth, nonglaucous (very rarely glaucescent), pubescent or glabrate; catkins coetaneous or serotinous, 0.4-2.3 cm, on leafy branchlets 0.1-29 cm; capsules pubescent to glabrate, 3-6 mm; styles 0-0.3 mm; stipes 0-0.3 mm. Floodplains, washes, seeps, 4800-6500 ft. Catron, Grant, Hidalgo. The specimens of *S. taxifolia* reported from San Juan County, New Mexico (*Clark 8249, 8940* UNM), were actually collected in Chiricahua National Monument, Cochise County, Arizona (J. Mygatt, pers. comm.).

Salix fragilis L.—Crack willow. Introduced tree to 15(-25) m; leaf blades lanceolate or lance-elliptic, the expanded (7-)10-17 × 1.7-3.5 cm, base acute to obtuse, tip acute or acuminate, serrate, glaucous on underside, glabrous; catkins coetaneous, (2-)4-8 cm, on leafy branchlets 1-5 cm; capsules glabrous, 4-5.5 mm; styles 0.3-0.8 mm; stipes 0.5-1 mm. Introduced from Eurasia and occasionally escaping on stream banks, ditch banks, and shores, 5100-7700 ft. McKinley, Mora, San Miguel, Taos. Some authors treat *S. fragilis* as a hybrid. Hybrids with *S. alba* are sometimes encountered. These have only slightly brittle branchlets.

Salix geyeriana Andersson—Geyer willow. Shrub to 7 m; leaf blades mostly lance-elliptic or elliptic, the expanded 2-8 × 0.6-1.5 cm, base and tip mostly acute, entire or nearly so, mostly glaucous or glaucescent on underside, usually pubescent on both sides; catkins coetaneous, 0.6-2(-2.5) cm, on leafy branchlets 0.1-1.2(-1.8) cm; capsules pubescent, 3-6 mm; styles 0.1-0.8 mm; stipes 1-3 mm. Edges of swamps, moist meadows, stream banks, 7000-10,700 ft. Catron.

Salix glauca L.—Gray willow. Shrub to 1.5(-4) m; leaf blades mostly elliptic to oblanceolate (obovate), the expanded 3-6(-8) × 0.7-3.5 cm, base acute to rounded, tip acute to rounded, entire or sometimes serrulate toward base, glaucous on underside, pubescent to glabrous; catkins coetaneous, (2-)3-4(-6.5) cm, on leafy branchlets 0.5-3.5 cm; capsules pubescent, 4-9 mm; styles 0.3-1(-1.4) mm; stipes 0.2-1.8 mm. Our plants are **var. villosa** Andersson. Stream banks and subalpine slopes, 9700-13,000 ft. NC. Sometimes appearing to intergrade with *S. brachycarpa*.

Salix gooddingii C. R. Ball—Goodding willow. Tree to 15(-30) m; leaf blades linear or oblong to lance-linear or narrowly elliptic, the expanded 6-13 × 0.8-1.6 cm, base acute, tip acuminate, serrulate or serrate, nonglaucous, glabrous or becoming so; catkins coetaneous, 2.2-8 cm, rather lax, on leafy branchlets 0.4-3 cm; capsules glabrous or pubescent, 3-7 mm; styles 0-0.4 mm; stipes 1-3.2 mm. Stream banks, shores, floodplains, washes, seeps, 3000-7000 ft. Widespread except NE, EC.

Salix irrorata Andersson—Bluestem willow. Shrub to 7 m; leaf blades oblong, narrowly elliptic, or oblanceolate, the expanded 4.7-12 × (0.5-)0.8-2.2 cm, base acute, tip acute to obtuse, entire to serrate or crenate, glaucous on underside, glabrous or rarely sparsely pubescent; catkins precocious or subprecocious, 1.5-4.2 cm, sessile or on leafy branchlets to 0.5 cm; capsules glabrous, 3-5 mm; styles 0.2-0.9 mm; stipes 0.3-1.2 mm. Rocky stream banks and washes, 4000-10,500 ft. Widespread except E, SW.

Salix lasiandra Benth.—Longleaf willow. Shrub (ours) or tree to 15 m; leaf blades lanceolate or sometimes elliptic, the expanded 2.4-17(-20) × 0.9-4.3 cm, base acute to rounded, tip acute or acuminate, serrate or serrulate, glaucous or not on underside, glabrous to occasionally pilose; catkins coetaneous,

1.7–10 cm, on leafy branchlets 0.8–6.5 cm; capsules glabrous, 4–7(–11) mm; styles 0.2–1 mm; stipes 0.5–2(–4) mm. Stream banks, shores, wet meadows, seeps, 5400–10,400 ft. NC, NW, C. Two varieties occur in the state, which are sometimes treated as subspecies of *S. lucida* Muhl., but definite intergrading with that E species has not been demonstrated.

1. Leaves glaucous on underside…**var. lasiandra**
1'. Leaves not glaucous on underside…**var. caudata**

var. caudata (Nutt.) Sudw.—Whiplash willow.

var. lasiandra—Longleaf willow.

Salix lasiolepis Benth.—Arroyo willow. Shrub (ours) or tree to 12 m; leaf blades oblong, narrowly elliptic, or oblanceolate to obovate, the expanded (2.5–)3.5–12.5(–22) × 0.6–3.2(–4.5) cm, base acute, tip acute to obtuse, entire to irregularly serrate or crenate, glaucous on underside, densely pubescent to glabrate on both sides, or becoming glabrous on upper side (and lower side); catkins precocious or sub-precocious, 1.5–7(–9) cm, sessile or on leafy branchlets to 0.7 cm; capsules glabrous, 2.5–5.5 mm; styles 0.2–0.8(–1) mm; stipes 0.5–1.8(–2.2) mm. Stream banks, washes, seeps, shores, 4800–8100 ft. Widespread except NW, WC, SC, SW. Some specimens suggest possible gene exchange with *S. irrorata*.

Salix monticola Bebb—Mountain willow. Shrub to 6 m; leaf blades ovate, lanceolate, or elliptic to obovate, the expanded 3–8(–9.5) × (1–)1.5–3.5 cm, base acute to rounded, tip acute, crenate or serrate or serrulate (subentire), glaucous on underside, glabrous; catkins precocious to coetaneous, 1–5(–6) cm, subsessile or on leafy branchlets to 0.8(–1.7) cm; capsules glabrous, 3–6 mm; styles 0.6–1.5(–1.8) mm; stipes 0.3–1.5(–2) mm. Stream banks and wet meadows in or near mountains, 5900–11,800 ft. NC, NW, Catron.

Salix nigra Marshall—Black willow. Tree to 20+ m; leaf blades narrowly lanceolate or elliptic (linear), the expanded (5–)7–13(–19) × (0.4–)0.8–1.2(–2.3) cm, base acute to rounded, tip acuminate, serrulate, nonglaucous, glabrous to sparsely pilose; catkins coetaneous, 1.7–9 cm, on leafy branchlets 0.4–2.3 (–4.5) cm; capsules glabrous, 3–6 mm; styles 0.1–0.4 mm; stipes 0.5–1.5(–2) mm. Stream banks, shores, floodplains, swamps, other moist or wet places, 2900–3900 ft. Quay, Union.

Salix planifolia Pursh—Planeleaf willow. Shrub to 5 m; leaf blades elliptic (lance-elliptic to obovate), the expanded (2–)3.5–5(–8) × 0.9–1.5(–2.2) cm, base and tip acute, entire to sometimes crenate or serrate, shiny and glabrous or glabrate on upper side, sparsely pubescent often with some reddish hairs (glabrous) and glaucous on underside; catkins precocious or subprecocious, (1–)1.5–6 cm, sessile or sub-sessile; capsules pubescent, (3.5–)5–6 mm; styles 0.4–1.8 mm; stipes 0–1 mm. Wet meadows, subalpine slopes, fens, willow swamps, 8300–13,100 ft. NC.

Salix reticulata L.—Net willow. Creeping shrub < 10 cm; leaf blades mostly oval to suborbicular (elliptic), the expanded 0.4–2.5(–3.6) × 0.3–1.5(–2.3) cm, base obtuse to rounded (subcordate), tip obtuse to rounded (retuse), entire, slightly revolute, glaucous on underside, glabrous or with sparse, long hairs mostly on underside; catkins serotinous, 0.5–2(–3) cm, on naked branchlets 0.2–2 cm; capsules pubescent, 1.5–4 mm; styles 0.1–0.4 mm; stipes 0–0.5 mm. Our plants are **var. nana** Andersson [*S. nivalis* Hook.], snow willow. Alpine and subalpine, 11,000–13,100 ft.

Salix scouleriana Barratt ex Hook.—Scouler willow. Shrub or small tree to 15(–20) m; leaf blades elliptic to obovate, the expanded 3–8(–10) × (1.3–)2–3.5 cm, base acute, tip acute to rounded, entire to irregularly serrulate or somewhat crenate, glaucous on underside, pubescent at least along midrib on underside; catkins precocious, 1.5–5(–7) cm, sessile or on leafy branchlets to 1.3 cm; capsules pubescent, strongly beaked, 4.5–11 mm; styles 0.2–1.1 mm; stipes 0.8–2.8 mm. Woods, slopes, meadows, rarely shores, 7300–11,000 ft. N, C, NC, SC, Grant.

Salix wolfii Bebb—Wolf willow. Shrub to 1(–2) m; leaf blades mostly elliptic, lanceolate, or oblanceolate, the expanded 2–6 × 0.5–1.5(–2) cm, base rounded or obtuse (acute), tip acute (obtuse), entire, nonglaucous, pubescent on both sides; catkins coetaneous, 0.8–2(–3) cm, subsessile or on leafy branchlets to 1.2 cm; capsules glabrous, 3.5–5 mm; styles 0.2–1.3 mm; stipes 0–0.8 mm. Our plants are **var. wolfii**. Wet meadows and stream banks, 9600–11,700 ft. Rio Arriba, San Miguel.

SAPINDACEAE – SOAPBERRY FAMILY

Kenneth D. Heil

(Ours) trees, shrubs, or herbs; leaves alternate or opposite, petiolate, often gland-dotted, stipulate or exstipulate; flowers imperfect or perfect; sepals 4-5, usually distinct; petals 4-5, usually united; stamens 8-10, in 2 whorls, distinct; pistil 1; ovary superior, (2)3(-8)-carpellate; fruit a capsule, berry, drupe, nut, or samara (Ackerfield, 2015; Correll & Johnston, 1979i).

1. Leaves opposite, palmately lobed or ternately compound (occasionally with 5-7 leaflets); fruit a schizo-carp of 2 winged, 1-seeded samaras…**Acer**
1'. Leaves alternate, pinnately compound; fruit a globose berry or woody capsule…(2)
2. Leaflets entire; fruit a berry…**Sapindus**
2'. Leaflets toothed; fruit a capsule…**Ungnadia**

ACER L. – Maple

Shrubs or trees; leaves opposite, simple, palmately lobed or ternately compound; fruit a schizocarp of 2 winged 1-seeded samaras.

1. Leaves ternately compound with 3 leaflets or 5-7 leaflets, the terminal leaflet with an evident petiolule; nectary disk absent…**A. negundo**
1'. Leaves simple, palmately lobed or ternately compound with sessile leaflets; nectary disk present and well developed…(2)
2. Leaves glabrous, palmately 3-5-lobed or ternately compound, with numerous sharp teeth along the margins; sepals distinct…**A. glabrum**
2'. Leaves hairy below, usually 5-lobed, never ternately compound, with a few blunt teeth; sepals irregularly connate…**A. grandidentatum**

Acer glabrum Torr.—Rocky Mountain maple. Shrubs or small trees to 8 m; leaves palmately 3-5-lobed or ternately compound, mostly 2-8 cm wide, sharply toothed along the margins, glabrous; sepals 3-5 mm, greenish, distinct; samaras glabrous, 2-3 cm.

1. Leaves lobed, not deeply parted or divided into leaflets…**var. glabrum**
1'. Leaves deeply parted and/or divided into leaflets…**var. neomexicanum**

var. glabrum—Rocky Mountain maple. Along streams, canyon bottoms, ravines, coniferous forests with ponderosa pine, Douglas-fir, aspen, Gambel oak, spruce, fir, 6300-10,660 ft. EC, SE, SW.

var. neomexicanum (Greene) Kearney & Peebles—New Mexico maple. Stream and creek bottoms, canyons, coniferous forests with ponderosa pine, white pine, Douglas-fir, aspen, spruce, fir, 6500-10,750 ft. NC, C, WC, Otero.

Acer grandidentatum Nutt.—Big-tooth maple. Small trees, 4-8 m; leaves mostly palmately lobed, mostly 2.5-10(-13) cm wide, with a few blunt teeth along the margins, hairy below; sepals 3-5 mm, greenish, connate to near the middle or above; samaras hairy. Often on rocky soils, canyon bottoms, stream and creek beds, coniferous forests with ponderosa pine, Gambel oak, alligator juniper, Douglas-fir, white fir, aspen, 4200-9040 ft. C, WC, SC, SW.

Acer negundo L. [*A. negundo* var. *arizonicum* Sarg.; *A. negundo* var. *interius* (Britton) Sarg.; *A. negundo* var. *texanum* Pax; *Negundo aceroides* Moench]—Box elder. Trees to 12 m; leaves ternately compound, occasionally with 5 or 7 leaflets, the terminal leaflet on a petiolule, coarsely toothed, hairy or glabrous; sepals 1-2 mm; samaras glabrous or hairy, 2.5-4 cm. Widespread except E.

SAPINDUS L. – Soapberry

Sapindus drummondii Hook. & Arn. [*S. saponaria* L.]—Western soapberry. Trees to ca. 10-15 m; bark grayish or tan; leaves alternate, pinnate, leaflets as many as 18, elliptic-lanceolate to narrowly lance-olate, to 1 dm × 4 cm, entire; flowers 4-5 mm broad, white, in large dense terminal panicles; sepals and petals 4 or 5; stamens 8 or 10 with long hairs on the filaments; fruits globose, mostly ca. 13 mm diam.

Canyons, ravines, sandy washes, sand hills with shinnery oak, mesquite, *Yucca elata*, 3275–6800 ft. SE, EC, WC, S except Otero.

UNGNADIA Endl. – Texas buckeye, Mexican buckeye

Ungnadia speciosa Endl.—Monilla. Shrub or small tree, to 10 m, trunk to 2 dm diam., bark light gray; leaves alternate, odd-pinnate, exstipulate, leaflets 3–7, ovate to ovate-lanceolate, serrate, to 12 × 4 cm, pubescent beneath when young, soon glabrate; flowers in lateral fascicles, pink to purplish pink, fragrant; calyx 5-lobed; petals 4 or 5, with a pilose claw; stamens 7–10; fruit a woody, 3-lobed, smooth pod, pale green, 3.5–5 cm thick. Limestone, rocky areas of canyons, slopes, and ridges, with creosote, sotol, scrub oak, 3880–6500 ft. Doña Ana, Eddy, Otero.

SAPOTACEAE – SAPODILLA FAMILY

Kenneth D. Heil

SIDEROXYLON L. – Bully

Sideroxylon lanuginosum Michx. [*Bumelia lanuginosa* (Michx.) Pers.]—Gum bully. Shrubs or trees, with milky sap, to 15 m; stems ± thorny; leaves 2–14 mm, alternate, simple, blade margins entire, villous or sparsely hairy to glabrate; blades oblong or oblanceolate to spatulate, 15–97 × 7–40 mm, abaxial surface usually villous, rarely glabrate, adaxial surface glabrate; calyx 1.9–3.2 mm diam.; sepals (4)5, 1.8–2.8 × 1.4–1.9 mm; petals (4)5(6), white, 0.8–1.8 mm; ovary superior, (1–)3–12(–15)-locular; styles 1; stigmas 1; berries purplish black, 7–12 mm, glabrate. Sandy and rocky soils, arroyos and margins of creek beds; desert grasslands, Chihuahuan desert scrub, 4400–5950 ft. Cibola, Grant, Hidalgo (Elisens et al., 2009).

SARCOBATACEAE – GREASEWOOD FAMILY

Ross A. McCauley

SARCOBATUS Nees – Greasewood

Sarcobatus vermiculatus (Hook.) Torr.—Greasewood. Monoecious shrubs with axillary thorns, 1–2(–5) m; stems erect, much branched, branches becoming spine-tipped; leaves alternate or in fascicles, succulent, blades terete, linear, fleshy, (0.3–)1.5–5 cm, usually glabrous; staminate flowers lacking sepals, with (1)2–4(5) stamens, each arising from near a stalked, peltate, scarious scale, in a terminal spike; pistillate flowers with 2 stigmas and a cuplike, shallowly lobed or entire perianth that develops into a membranous wing in fruit; fruit a winged achene. Alkaline flats in desert scrub, piñon-juniper, sagebrush communities, 4700–7900 ft. NC, NW, C, WC, SW, Chaves, Otero (Behnke, 1997).

SAURURACEAE – LIZARD-TAIL FAMILY

Kenneth D. Heil

ANEMOPSIS Hook. & Arn. – Yerba mansa

Anemopsis californica (Nutt.) Hook. & Arn. [*A. californica* var. *subglabra* Kelso]—Yerba mansa. Perennial herbs, rhizomatous or stoloniferous, ± aromatic; stems simple or branched, often scapelike; plants 1–5 dm, glabrous or pubescent; leaves basal and cauline, alternate, basal leaves petioled, stipules adnate to the petiole, the blades 3.5-10 cm, elliptic-oblong with a truncate or cordate base, entire, cauline leaves sessile and clasping; inflorescence a terminal, congested, conic spike of 75-100(-150) flowers, subtended by 4-9 white to reddish petaloid bracts, 0.5-3.5 cm; flowers each subtended by an adnate, white, ovate, clawed bract, flowers small, perfect, perianth lacking; stamens (3-)6-8; pistil 1, of 3-5(-7) carpels, styles and stigmas distinct; fruit a capsule. Alkaline soils of wet meadows, along streams (Heil, 2013d). NC, C, WC, SC, SW.

SAXIFRAGACEAE – SAXIFRAGE FAMILY

Kenneth D. Heil and Jennifer Ackerfield

Annual or mostly perennial herbs; leaves usually basal or sometimes alternate or opposite, simple, entire to pinnately or palmately compound, the margins entire to crenate, serrate, or dentate; flowers usually perfect, usually actinomorphic, with a hypanthium, sepals (4)5, distinct, petals 5 or absent, distinct, lobed or unlobed, stamens 10 or sometimes 5, rarely 4 or 8, distinct, pistil 1, sometimes appearing 2-3, ovary superior to inferior, 2-carpellate or rarely 3-carpellate; fruit a capsule (Wells & Elvander, 2009).

1. Leaves all basal or with very reduced, small, scalelike stem leaves…(2)
1'. Leaves both basal and cauline, the stem leaves usually reduced distally but not scalelike…(5)
2. Flowers in a narrow, loose, simple raceme or spike…(3)
2'. Flowers in a panicle or thyrse sometimes densely crowded into a dense, terminal, conic or cylindrical, headlike thyrse…(4)
3. Petals entire or 3-parted, alternate with the stamens; leaves scarcely lobed, teeth very shallow and blunt… **Ozomelis**
3'. Petals pinnatifid, 5-11-parted, opposite the stamens; leaves lobed, coarsely toothed…**Pectiantia**
4. Stamens 10; leaves oblanceolate, triangular, or ovate to elliptic with a cuneate base, or orbiculate with a cordate to truncate base and regularly dentate margins (not double); plants from thick, fleshy rhizomes… **Micranthes**
4'. Stamens 5; leaves reniform to orbiculate or broadly ovate, with a cordate or truncate base, the margins doubly crenate to dentate; plants from a stout, often branched caudex or rhizome…**Heuchera** (in part)
5. Petals deeply palmately lobed, white or pink; leaves deeply palmately lobed or compound… **Lithophragma**
5'. Petals entire, variously colored; leaves not deeply palmately lobed or compound…(6)
6. Petals violet-purple to red-purple or bright pink, (5-)7-11 mm; plants densely stipitate-glandular-hairy… **Telesonix**
6'. Petals green, white, or yellow, sometimes with reddish-pink spots, if purple then the petals only 1.5-4.5 mm; plants densely stipitate-glandular or not…(7)
7. Flowers green or greenish yellow, in a dense, one-sided thyrse; ovary unilocular with parietal placentation …**Heuchera** (in part)
7'. Flowers white, yellow, rose, or purple, solitary or loosely arranged in an open cyme, thyrse, or panicle, ovary 2-locular with axile or marginal placentation…**Saxifraga**

HEUCHERA L. – Alumroot

Perennial herbs; leaves in a basal rosette and cauline, the cauline leaves very reduced and scalelike; flowers actinomorphic or zygomorphic, sepals 5, petals 5 or sometimes absent or minute, stamens 5, opposite sepals, pistil 1, 2-carpellate; ovary 1/2 inferior (Folk & Freudenstein, 2014).

1. Styles to 1.5 mm; hypanthia usually short-stipitate-glandular; stamens included…(2)
1'. Styles 0.2-7 mm; hypanthia short- to long-stipitate-glandular, rarely glabrous; stamens exserted or barely included…(8)
2. Hypanthia flat, saucer-shaped; sepals reflexed…(3)
2'. Hypanthia campanulate, broadly campanulate or basally turbinate; sepals erect or incurved…(4)
3. Petioles long-stipitate-glandular; hypanthia green…**H. wootonii**
3'. Petioles glabrate or short-stipitate-glandular; hypanthia greenish or cream to yellow…**H. parvifolia**
4. Sepals < 2 mm…(5)
4'. Sepals 2.5-4 mm…(7)

5. Petioles and flowering stems glandular-puberulent and glandular-villous…**H. nova-mexicana**
5'. Petioles and flowering stems only glandular-puberulent…(6)
6. Sepals much longer than the petals; nectary disk absent; hypanthium white or pinkish white, petals the same color…**H. glomerulata**
6'. Sepals equaling or slightly shorter than the petals; nectary disk present (yellow); hypanthium yellowish or greenish white, petals white…**H. soltisii**
7. Petals absent…**H woodsiaphila**
7'. Petals present…**H. hallii**
8. Hypanthia dark pink to red; stamens included, 1.5–3 mm…**H. sanguinea**
8'. Hypanthia pink or rose to red or purple; stamens to 2 mm included, to 3 mm exserted…(9)
9. Petals relatively narrow…**H. rubescens**
9'. Petals relatively wide…**H. pulchella**

Heuchera glomerulata Rosend., Butters & Lakela—Chiricahua alumroot. Plants 2.5–4 dm; leaves ovate to orbiculate, shallowly 5-lobed, 2.5–9 cm, with dentate margins, stipitate-glandular; inflorescence dense, interrupted; hypanthium campanulate, 0.8–1.2 mm; sepals erect or incurved at tip, 1.5–2 mm; petals erect, white, 1.5–1.8 mm, entire. Shaded, rocky slopes, 5700–7900 ft. Catron, Grant, Hidalgo. Nearly **R**

Heuchera hallii A. Gray—Front Range alumroot. Plants 1–3 dm; leaves rounded-reniform, deeply 5–7-lobed, 1–3.8 cm, with dentate margins, glabrous or short-glandular-stipitate; inflorescence dense, occasionally secund; hypanthium broadly campanulate, 4–5.5 mm; sepals erect, 1.2–2 mm; petals spreading, white, 1.8–2.5 mm, entire. Rocky crevices and cliffsides, 8200–8800 ft. Colfax.

Heuchera nova-mexicana Wheelock—Range alumroot. Plants 3–5 dm; leaves reniform or rounded-cordate, shallowly 5-lobed, 2–6 cm, with dentate margins, stipitate-glandular on veins abaxially, glabrous or short-stipitate-glandular adaxially; inflorescence dense; hypanthium campanulate, 3–5 mm; sepals erect or incurved, 0.6–1.2 mm; petals usually erect, white, 0.5–1 mm, entire. Shaded rocky ledges and outcrops, 6950–9900 ft. Catron, Grant, Sierra.

Heuchera parvifolia Nutt. ex Torr. & A. Gray—Common alumroot. Plants 0.4–7.1 dm; leaves orbiculate or reniform to broadly cordate, shallowly to deeply 5–7-lobed, 1–8 cm, with dentate margins, glabrate or short- stipitate-glandular, occasionally long-stipitate below; inflorescence dense at anthesis, spreading in fruit; hypanthium flat, saucer-shaped, 2.5–5 mm; sepals reflexed, 0.5–1.5 mm; petals reflexed, white or yellowish, 0.7–3 mm, entire. Common on rocky outcroppings, among boulders, in meadows, sometimes with sagebrush or piñon-juniper, 7000–11,700 ft. N, C, SC.

Heuchera pulchella Wooton & Standl.—Sandia alumroot. Plants 0.7–1.5 dm; leaves orbiculate, deeply 5-lobed, 0.5–2 cm, with dentate margins, stipitate-glandular and sparsely long-stipitate-glandular on veins; inflorescence dense, secund; hypanthium campanulate, 4–5.5 mm; sepals spreading, 0.8–1.8 mm; petals spreading, pink, 2–2.5 mm, margins entire. Limestone canyons, mountain slopes and cliffs, 8000–10,700 ft. Bernalillo, Sandoval, Torrance. **R & E**

Heuchera rubescens Torr.—Red alumroot. Plants (0.6–)1–3.2(–5) dm; leaves suborbiculate or broadly ovate, shallowly 3–7-lobed, 0.6–4.5 cm, with dentate margins, glabrous or short- to long-stipitate-glandular; inflorescence dense and often secund, spreading in fruit; hypanthium narrowly campanulate, becoming urceolate, 3–6.5 mm; sepals erect, 1–3.3 mm; petals spreading, pink to rose, 2–3(–6) mm, entire. Rock crevices and cliffsides, 5750–11,800 ft. Widespread except E.

Heuchera sanguinea Engelm. [*H. sanguinea* var. *pulchra* (Rydb.) Rosend.]—Coralbells. Plants 2–4 dm; leaves reniform to orbiculate, shallowly 5–7-lobed, 2–5.5 cm, with dentate margins, stipitate-glandular on veins, glabrous or sparsely long-stipitate-glandular adaxially; inflorescences moderately dense to diffuse; hypanthium broadly campanulate or urceolate, 4–8 mm; sepals spreading, 2–3 mm; petals spreading, pink or cream, 1.2–1.8 mm, entire. Moist, shaded rocks, 4935–8000 ft. Hidalgo.

Heuchera soltisii R. A. Folk & P. J. Alexander—Plants 3–7 dm; basal leaves variegated green and white, ebracteate or with 1–2 reduced yellowish bracts; inflorescence racemose, 5–15 cm; flowers 3.5–

4.5 mm, hypanthium campanulate with a yellowish nectary disk, basal adnate portion green, greenish yellow, or pinkish; petals white; stamens included. Doña Ana, Luna, Socorro. **R**

Heuchera woodsiaphila P. J. Alexander—Capitan Peak alumroot. Plants with an elongate branching caudex; leaves subreniform to broadly ovate, shallowly 5-lobed, 2–3.5 cm, with dentate margins, glandular-puberulent; inflorescence 1-sided or nearly so; hypanthium 1–1.5 mm; sepals greenish white to cream, densely glandular-puberulent, 2.75–3.75 mm. Stable granitic talus in montane coniferous forests, 8370–9510 ft. Lincoln. **R & E**

Heuchera wootonii Rydb.—White Mountain alumroot. Plants 0.4–6 dm; leaves orbiculate, shallowly 5–7-lobed, 4–7 cm, with dentate margins, stipitate-glandular along veins, glabrous adaxially; inflorescence dense; hypanthium saucer-shaped, 2–3 mm; sepals reflexed, green-tipped, 0.5 mm. Rock outcrops, talus slopes from piñon-juniper to alpine communities, 7000–12,000 ft. Lincoln, Otero. **R & E**

LITHOPHRAGMA (Nutt.) Torr. & A. Gray – Woodland star

Rhizomatous perennial herbs with underground bulbils; leaves in a basal rosette and the cauline leaves usually alternate, palmately dissected; flowers actinomorphic; sepals 5; petals 5, white or pink, the limb palmately or pinnately 3–7-cleft or -divided, sometimes entire; stamens 10; ovary superior to almost completely inferior.

1. Reddish-purple bulbils usually present in the leaf axils and often replacing flowers; basal leaves usually glabrous below or sometimes sparsely hairy; stigma apex with numerous round, short bumps (papillae) ...**L. glabrum**
1'. Bulbils usually absent from the leaf axils and not replacing flowers; basal leaves usually sparsely to densely hairy below; stigma glabrous at the apex, with a subapical narrow band of papillae below...(2)
2. Hypanthium rounded at the base, campanulate or hemispherical; petals palmately 5–7-lobed; stem leaves deeply 3-lobed and appearing pinnatifid, ultimately with very narrow lobes and appearing markedly different from the basal leaves...**L. tenellum**
2'. Hypanthium acute at the base and gradually tapering to the pedicel, obconic-elongate; petals palmately 3-lobed; stem leaves palmately 3-lobed, similar to the basal leaves except with longer lobes...**L. parviflorum**

Lithophragma glabrum Nutt. [*L. bulbiferum* Rydb.]—Bulbous woodland star. Plants 0.8–3.5 dm; leaves round, trilobed, or trifoliate, the ultimate segments 1–4-lobed, glabrous or sparsely hairy; inflorescence a 2–5(–7)-flowered raceme; hypanthium narrowly campanulate; sepals erect in bud, spreading in flower, triangular; petals widely spreading, pink or rarely white, 3.5–7 mm, 3–5-lobed. Meadows, aspen groves, sagebrush; 1 record, 8025 ft. Rio Arriba.

Lithophragma parviflorum (Hook.) Nutt.—Smallflower woodland star. Plants 2–3 dm; leaves trilobed or trifoliate, the cauline leaves reduced, glabrous or sparsely to densely hairy; inflorescence a 4–14-flowered raceme; hypanthium obconic-elongate at anthesis, elongate in fruit; sepals erect, triangular; petals widely spreading, white or pink, 7–16 mm, 3-lobed. Moist, rich soil along streams, in meadows, sometimes in sagebrush, 7370–7550 ft. Rio Arriba.

Lithophragma tenellum Nutt.—Slender woodland star. Plants 0.8–3 dm; leaves trilobed, the cauline leaves reduced; inflorescence a 3–12-flowered raceme; hypanthium hemispherical to campanulate; petals pink or sometimes white, 3–7 mm, 5–7-lobed. Rich, often moist soil of meadows, forest openings, sagebrush slopes, 6240–9800 ft. N, WC.

MICRANTHES Haw. – Saxifrage

Annual or perennial herbs; leaves basal; flowers actinomorphic; sepals 5; petals 5 or absent; stamens 10; pistils 2(3), 2(3)-carpellate; ovary superior or inferior (Elvander, 1984).

1. Ovaries ± superior to at most 1/2 inferior; sepals reflexed...**M. odontoloma**
1'. Ovaries ca. 1/2 inferior to inferior (sometimes appearing more superior in fruit); sepals ascending to erect ...(2)

2. Inflorescence lax, open…**M. eriophora**
2'. Inflorescence ± congested…**M. rhomboidea**

Micranthes eriophora (S. Watson) Small [*Saxifraga eriophora* S. Watson]—Redfuzz saxifrage. Perennials, solitary or in groups; leaves ovate to elliptic, 1–2 cm, base attenuate, margins sharply serrate, sparsely to densely ciliate, some glandular-tipped, surfaces densely tangled, reddish-brown-hairy abaxially, glabrous adaxially; inflorescence 10+-flowered, flowers stipitate-glandular; sepals erect; petals white, sometimes purplish-tipped, not spotted, 4–7 mm. Rocky slopes and ledges, 6300–7500 ft. Doña Ana.

Micranthes odontoloma (Piper) A. Heller [*Saxifraga odontoloma* Piper]—Brook saxifrage. Perennials, (1.3–)2–7 dm, glandular-hairy above; leaves round, cordate to truncate at the base, the margins dentate and eciliate, glabrous above, sparsely brownish-hairy below; inflorescence 10–30+-flowered, glabrous below and purple-tipped subsessile-glandular above; sepals reflexed; petals white with 2 basal yellow spots, 3–4.5 mm. Common in moist soil along streams and around lakes, 6000–12,000 ft. NC, WC.

Micranthes rhomboidea (Greene) Small [*Saxifraga rhomboidea* Greene]—Diamondleaf saxifrage. Perennials, 0.25–3 dm, densely glandular above; leaves broadly ovate to triangular, 1–6 cm, the margins coarsely serrate and ciliate, glabrous above, sparsely to moderately tangled, reddish-brown-hairy below; inflorescence (5–)10–40-flowered, usually densely yellow-glandular-stipitate; sepals ascending; petals white, not spotted, 2–4 mm. Common from the foothills to the alpine, 8420–12,960 ft. NC, NW, Cibola, Socorro.

OZOMELIS Raf. – Miterwort

Ozomelis stauropetala (Piper) Rydb. [*Mitella stauropetala* Piper]—Side-flowered miterwort. Plants 1–4.5 dm, glandular-hairy; leaves orbicular to ovate or reniform, shallowly 1–2 times crenate, hirsute or glabrous; inflorescence (7–)10–35-flowered, secund; hypanthium short-campanulate, 1.5–3 mm diam.; sepals 1–2.2 mm, erect or recurved; petals 1.5–4 mm, with 3 linear lobes, white. Our material belongs to **var. stenopetala** (Piper) Rosend. Common along streams, in moist meadows, and in forests, 9750–10,850 ft. Rio Arriba, Taos (Folk & Freudenstein, 2014; Rosendahl, 1914; Soltis & Freeman, 2009).

PECTIANTIA Raf. – Miterwort

Pectiantia pentandra (Hook.) Rydb. [*Mitella pentandra* Hook.]—Five-star miterwort. Plants (0.8–)1–4 dm, glandular above; leaves ovate-cordate, 2.5–7 cm, doubly crenate-dentate, sparsely hirsute or glabrous, inflorescence 6–20(–25)-flowered; hypanthium strongly flattened, 2.4–3 mm diam.; sepals 0.5–0.6 mm, spreading or reflexed; petals 1.5–3 mm, pectinate-pinnatifid with 5–11 narrow spreading divisions, greenish yellow. Common along streams, in shady forests, and in moist meadows, 9750–10,250 ft. Rio Arriba (Folk & Freudenstein, 2014; Rosendahl, 1914; Soltis & Freeman, 2009).

SAXIFRAGA L. – Saxifrage

Perennial herbs; leaves in a basal rosette and cauline, the cauline leaves reduced distally, stipules absent; flowers actinomorphic or zygomorphic; sepals 5; petals 5 or absent; stamens 10; pistil 1, 2-carpellate; ovary superior or inferior (Brouillet & Elvander, 2009).

1. Leaves unlobed, the margins entire; flowers yellow, or white with purple or reddish spots…(2)
1'. Leaves lobed or toothed; flowers white (sometimes greenish- or yellowish-tinged at the tips) to purple, not spotted…(5)
2. Flowers white with purple to reddish spots on the lower part of the petals and yellowish spots on the upper part; leaves with a white-spinulose top and stiff, white-ciliate margins; plants mat-forming with trailing stems…**S. bronchialis**
2'. Flowers yellow or orange-yellow; leaves various; plants mat-forming or not but lacking trailing stems…(3)
3. Plants with slender stolons; sepals erect, glandular-stipitate-hairy; basal leaves stiffly ciliate…**S. flagellaris**

3'. Plants lacking stolons; sepals spreading to reflexed, glabrous to glandular-stipitate-hairy; basal leaves not stiffly ciliate…(4)
4. Flowering stems sparsely to densely reddish-brown-villous below the inflorescence; sepals glabrous or with reddish-brown-ciliate margins…**S. hirculus**
4'. Flowering stems sparsely purplish-glandular-stipitate below the inflorescence; sepals glandular-stipitate with glandular-ciliate margins…**S. chrysantha**
5. Leaves oblanceolate to obovate or spatulate, 2-5-lobed or toothed at the apex…(6)
5'. Leaves round, reniform, or orbiculate, 3-7-lobed…(7)
6. Leaves shallowly toothed into ovate segments; plants solitary or tufted with 2 stems but not strongly cushion-forming or cespitose [to be looked for]…S. adscendens
6'. Leaves deeply lobed into linear segments; plants usually with > 2 stems, strongly cespitose and densely or loosely cushion- or mat-forming…**S. cespitosa**
7. Bulbils present in the leaf axils; petals 5-12 mm; sepal margins glandular-ciliate…**S. cernua**
7'. Bulbils absent from the leaf axils; petals (1.5-)2-5(-6.2) mm; sepal margins with or without glandular-ciliate hairs…**S. debilis**

Saxifraga adscendens L. [*Muscaria adscendens* (L.) Small]—Wedgeleaf saxifrage. Plants solitary or tufted, not stoloniferous, with a caudex; leaves oblanceolate to obovate, (2)3(-5)-toothed or shallowly lobed, (2-)4-15 mm, the margins entire, glabrate to glandular-stipitate; inflorescence a (2-)6-15(-40)-flowered thyrse, densely purple-tipped glandular-stipitate; sepals erect; petals white, not spotted, 3-6 mm. Uncommon along streams in alpine meadows and on scree slopes. Potential habitat in N. No records seen.

Saxifraga bronchialis L. [*Ciliaria austromontana* (Wiegand) W. A. Weber]—Spotted saxifrage. Plants mat-forming, not stoloniferous; leaves linear or linear-lanceolate, unlobed, 3-15 mm, the margins entire and often white- or glandular-ciliate, glabrous, occasionally sparsely glandular above; inflorescence a 2-15-flowered cyme or thyrse, sparsely short-stipitate-glandular; sepals erect, purplish; petals yellowish-spotted proximally, purple- or red-spotted distally, 3-7 mm. New Mexico material belongs to **var. austromontana** (Wiegand) M. Peck. Common on rocky outcrops, spruce-fir forests, alpine communities, 7500-13,000 ft. NC, C, Cibola, Otero, Sierra.

Saxifraga cernua L.—Nodding saxifrage. Plants solitary or in tufts, not stoloniferous; leaves round to reniform, 3-7-lobed, (3-)5-20 mm, the margins entire and eciliate, sometimes sparsely glandular-ciliate, glabrous or sparsely glandular-stipitate; inflorescence a 2-5-flowered paniculate or racemelike thyrse, glandular-stipitate; sepals erect; petals white, not spotted, 5-12 mm. Alpine fell-fields and talus slopes and along creeks, 12,000-12,700 ft. Rio Arriba, Santa Fe, Taos.

Saxifraga cespitosa L. [*Muscaria delicatula* Small]—Tufted alpine saxifrage. Plants densely cushion-to loosely mat-forming, not stoloniferous or rhizomatous; leaves elliptic to obovate or spatulate, 3(-5)-lobed apically, 4-16(-25) mm, the margins entire and eciliate, or sometimes sparsely glandular-ciliate, glabrous or sparsely glandular-stipitate; inflorescence a 2-5-flowered paniculate or racemelike thyrse, glandular-stipitate; sepals erect; petals white, not spotted, 5-12 mm. Alpine fell-fields, talus and scree slopes, 12,325-12,550 ft. Taos.

Saxifraga chrysantha A. Gray [*Hirculus serpyllifolius* (Pursh) W. A. Weber]—Golden saxifrage. Plants mat-forming, slenderly rhizomatous, not stoloniferous; leaves linear-oblanceolate to narrowly spatulate, unlobed, (2-)3-12 mm, the margins entire and eciliate, sometimes sparsely glandular-ciliate, glabrous or occasionally sparsely glandular-stipitate; inflorescence a 2-3-flowered cyme, sparsely glandular-stipitate; sepals strongly reflexed; petals golden-yellow, sometime orange-tipped proximally, 4-8 mm. Alpine meadows and scree slopes, 12,000-13,000 ft. Mora, Santa Fe, Taos.

Saxifraga debilis Engelm. ex A. Gray [*S. rivularis* L. var. *debilis* (Engelm. ex A. Gray) Dorn]—Saxifrage. Plants usually densely tufted, not stoloniferous, not rhizomatous; leaves round or reniform, (3-)5-7-lobed, (3-)4.5-6.7(-10.3) mm, margins entire, eciliate, surface glabrous; inflorescence a 2-3(-5)-flowered capitate cyme, nonglandular-hairy; sepals erect; petals white to pale purple, not spotted, (1.7-)3-4.4(-6.2) mm. Talus, ravines, cliffs, alpine meadows, seepage areas, stream and lake margins, 11,500-13,000 ft. Rio Arriba, Taos.

Saxifraga flagellaris Willd. [*Hirculus platysepalus* (Trautv.) W. A. Weber subsp. *crandallii* (Gand.) W. A. Weber]—Stoloniferous saxifrage. Plants in solitary clumps, stoloniferous, slenderly rhizomatous; leaves oblong-lanceolate or elliptic to obovate, unlobed, 5–20 mm, the margins entire and sparsely to densely glandular-ciliate, glabrous; inflorescence a 2–3(–5)-flowered lax cyme, densely purplish-glandular-stipitate; sepals erect; petals yellow, not spotted, 4–9(–10) mm. Our material belongs to **subsp. crandallii** (Gand.) Hultén. Alpine meadows and scree slopes, 11,700–13,000 ft. NC.

Saxifraga hirculus L. [*Hirculus prorepens* (Fish. ex Sternb.) Á. Löve & D. Löve]—Yellow marsh saxifrage. Plants loosely tufted, rhizomatous, sometimes shortly stoloniferous; leaves linear to spatulate, unlobed, (5–)10–30 mm, the margins entire and eciliate, or sparsely reddish-brown-ciliate, glabrous or sparsely reddish-brown-villous; inflorescence a 2(–4)-flowered cyme, flowers initially nodding, sparsely to densely reddish-brown-villous; sepals ascending to spreading; petals yellow, often proximally orange-spotted, usually drying cream, 6–18 mm. Uncommon in moist meadows, bogs, fens, and along creeks and lakeshores, 10,750–11,000 ft. Colfax.

TELESONIX Raf. – Brookfoam

Telesonix jamesii (Torr.) Raf. [*Boykinia jamesii* (Torr.) Engl.; *Saxifraga jamesii* Torr.]—Alumroot, brookfoam. Perennial herbs; plants 1–1.5 dm, glandular-hairy; leaves in basal rosettes and cauline, the cauline leaves reduced, reniform to orbiculate, 2–3.5 cm wide, long-petiolate with crenate margins; inflorescence a dense paniculate cyme; sepals 5, 4–5 mm; petals 5, violet-purple, red-purple, or bright pink, (5–)7–11 mm; stamens 10; pistils 1, 2-carpellate; ovary 1/2 inferior, 2-locular. Granite outcrops, 9000–12,000 ft. San Miguel, Santa Fe.

SCROPHULARIACEAE – FIGWORT FAMILY

Kenneth D. Heil

Trees, shrubs, or herbs; leaves alternate or opposite, simple, exstipulate; flowers perfect, 5-merous, nearly actinomorphic or zygomorphic; calyx 5-lobed; corolla 5-lobed; stamens 4 with a staminode or 5 all fertile, epipetalous, alternate with the corolla lobes; pistil 1; ovary superior, 2-carpellate; fruit a capsule (Ackerfield, 2015; Correll & Johnston, 1979j; Martin & Hutchins, 1980e).

1. Aquatic or semiaquatic acaulescent herbs; leaves simple, entire, on long petioles, mostly oblong-elliptic; flowers solitary, corolla white or pinkish…**Limosella**
1'. Plants unlike the above…(2)
2. Plants well-developed shrubs…(3)
2'. Plants herbaceous or woody at the base; leaves variously pubescent or glabrous…(4)
3. Leaves silvery-canescent…**Leucophyllum**
3'. Leaves grayish green…**Buddleja**
4. Leaves opposite, petiolate; corolla strongly zygomorphic and bilabiate; stamens 4 with a fifth sterile staminode; stem 4-angled…**Scrophularia**
4'. Leaves alternate or in a basal rosette, sessile; corolla nearly actinomorphic to zygomorphic, not bilabiate; stamens 5, all fertile, stem round…**Verbascum**

BUDDLEJA L. – Butterfly-bush

Shrubs with glandular-canescent or -tomentose indumenta; leaves opposite, sessile (ours), entire to crenate or dentate; inflorescences racemose or crowded into capitate clusters; calyx campanulate; corolla tubular to rotate-campanulate; anthers sessile or nearly so in throat or tube of corolla; pods globose to ellipsoid, bivalved.

1. Flowers purplish; leaves alternate; introduced; San Miguel County…**B. alternifolia**
1'. Flowers yellowish; leaves opposite; native; Eddy County…**B. scordioides**

Buddleja alternifolia Maxim.—Alternate-leaved butterfly-bush. Shrubs to 8 m; stems slender, much branched; leaves alternate or opposite, lanceolate, entire to denticulate, 4–10 × 0.6 cm, whitish-stellate-tomentose abaxially, mostly glabrous adaxially; inflorescence a tight verticil at the leafy nodes; corolla purple, lilac, or violet with an orange throat, tube 6–10 mm; fruit a capsule, ellipsoid, mostly 5 × 2 mm. A single collection known from San Miguel County, 7300 ft. Probably an escape.

Buddleja scordioides Kunth—Butterfly-bush. Densely branched aromatic shrub to 12 dm, ferruginous-tomentose throughout; leaves sessile, narrowly oblong to linear-cuneate, coarsely crenate, to 45 mm × 1 cm, usually much smaller; flowers minute, greenish yellow to yellowish in dense clusters that are sessile in the axils of the uppermost leaves. Limestone soil and rocky areas, 3300–4000 ft. Eddy.

LEUCOPHYLLUM Bonpl. – Barometerbush, cenizo

Much-branched shrubs, mostly densely scurfy-tomentose with mostly silvery branched woolly hairs; leaves alternate, entire, subsessile to short-petioled; flowers showy, solitary on short bractless pedicels in the leaf axils; calyx 5-cleft; corolla funnelform-campanulate, usually purplish or violet, with 5 rounded lobes, subequal; stamens 4 (Correll & Johnston, 1979j).

1. Corolla purple, throat relatively inflated, 1–1.5 cm, lower lobes hairy within; leaves elliptic-obovate, closely pubescent, to 25 mm, midrib evident [to be expected]…L. frutescens
1'. Corolla violet, throat narrow, < 1 cm, all lobes somewhat hairy within; leaves obovate-spatulate, finely pubescent, < 20 mm, midrib obscure…**L. minus**

Leucophyllum frutescens (Berland.) I. M. Johnst. [*L. texanum* Benth.]—Cenizo, purple-sage, Texas silver-leaf. Shrub to ca. 2.5 m, densely stellate-tomentose throughout; leaves sessile or nearly so, elliptic-obovate, to ca. 2.5 cm; calyx lobes oblong-lanceolate; corolla mostly campanulate, delicately soft-villous within. In cultivation throughout S New Mexico and not known to escape to the wild and potential habitat in SW communities.

Leucophyllum minus A. Gray—Big Bend silver-leaf. Low shrub to ca. 1 m, finely stellate-tomentose throughout, silvery-white with branched woolly hairs; leaves alternate, entire, silvery-canescent, obovate-spatulate, to 15 mm; flowers showy, solitary; calyx 5-cleft; corolla lavender or purplish, 18–25 mm, with narrow and funnelform tube, sparsely pubescent; stamens 4. Rocky hills in Chihuahuan desert scrub, 3900–5500 ft. Eddy, Otero.

LIMOSELLA L. – Mudwort

C. Barre Hellquist

Annual, wetland plant; small-rosulate; leaves basal, erect, rarely cauline; flowers solitary on 1-flowered scapes, white, pink, or pale blue; calyx and corolla campanulate; upper surface of petals minutely papillate; stamens 4; style terminal or subterminal; capsule globose to ellipsoid, many-seeded.

1. Leaves 2 mm or less wide; styles 0.5–1 mm…**L. acaulis**
1'. Leaves 2–8 mm wide; styles 0.2–06 mm…(2)
2. Inner surface of corolla lobes with finely evident pubescence; styles ca. 0.6 mm; capsule 3.5–4 mm; seeds grayish brown and a little longer than wide; rare in SW New Mexico…**L. pubiflora**
2'. Inner surface of corolla lobes sparsely papillate; styles < 0.5 mm; capsule 3 mm; seeds brown, several times longer than wide; widespread in New Mexico…**L. aquatica**

Limosella acaulis Sessé & Moc.—Owyhee mudwort. Cespitose stoloniferous plants forming mats; leaves flat, linear to linear-spatulate, 1–6 cm; flowers solitary on erect scapes; petals sparsely papillate on inner surface; stamens 4; style 0.5–1 mm; capsule to 3 mm, ridged and reticulate. Margins and shallow water of lakes, ponds, and streams, 4950–7825 ft. Catron, Hidalgo, Sierra.

Limosella aquatica L.—Water mudwort. Soloniferous herbs; leaves with long, slender petioles, 3–10 cm; blades linear-spatulate to broadly oblong-elliptic, 1–3 cm × 3–12 mm; peduncles shorter than leaves; corolla slightly longer than calyx, oblong, acute, slightly papillate within, white or pink; stamens 4; style 0.2–0.4 mm; seeds brown, several times longer than broad, Shallow water of lakes, ponds, and streams, 6900–10,250 ft. NC, NW, C, WC.

Limosella pubiflora Pennell—Chiricahua mudwort. Possibly stoloniferous; leaves 1–6 cm, 3–5 mm wide, oblanceolate and glandular-spotted; calyx 3–3.5 mm, glabrous, lobes ovate-acute, ca. 1 mm; corolla lobe inner surface with fine pubescence, ca. 8 mm, white; stamens 4; style 0.6 mm, long-ribbed, slightly longer than wide, grayish brown, ridges sharp; capsule 3.5–4 mm; seeds grayish brown, ridges sharp. Pond margins, 4730 ft. Hidalgo. **R**

SCROPHULARIA L. – Figwort

Perennial herbs with 4-angled stems, to 1 m; leaves opposite or whorled, ovate to lanceolate, petiolate, toothed (ours); flowers in a terminal panicle; corolla strongly bilabiate, greenish yellow to brownish, dull red, or crimson-red; stamens 4, slightly didynamous, with a fifth sterile staminode; capsule ovoid, 2-valved.

1. Corolla crimson-red, 15–20 mm…**S. macrantha**
1'. Corolla dull greenish, greenish yellow, greenish brown, or dull red…(2)
2. Branches of the inflorescence ascending, forming a dense, narrow panicle…(3)
2'. Branches of the inflorescence spreading, the panicle open and relatively few-branched…(4)
3. Stems rounded to angled or 4-sided; calyx margins scarious and slightly erose…**S. lanceolata**
3'. Stems > 4-angled; calyx margins green and entire…**S. montana**
4. Plants glabrous or glabrate throughout; petioles about 1/2 as long as the blades…**S. laevis**
4'. Plants densely puberulent at least on the stems; petioles not > 1/3 as long as the blades…**S. parviflora**

Scrophularia laevis Wooton & Standl.—Smooth figwort. Stems 4–10(–12) dm, glabrous; leaves lanceolate to ovate, obtuse to cordate at base, coarsely laciniate-serrate, mostly glabrous, blades 5–7 × 2–3.5 cm; flowers dull red to greenish brown, 7–12 mm; capsules 8–11 mm. Moist canyons on quartz monzonite in piñon-juniper woodlands and Rocky Mountain montane coniferous forests, 6900–8500 ft. Doña Ana. **R & E**

Scrophularia lanceolata Pursh—Lanceleaf figwort. Plants 8–15 dm, glandular-hairy; leaves lanceolate to narrowly ovate, with a truncate or cordate base, 8–13 × 3–5(–7) cm, coarsely toothed; calyx 2–4 mm; corolla 8–12 mm, dull greenish brown; capsules 6–8 mm. Coniferous forests and along creeks, mixed conifer, spruce, fir, 7500–10,525 ft. NC, NW.

Scrophularia macrantha Greene ex Stiefelh. [*S. coccinea* A. Gray]—Mimbres figwort. Plants to ca. 4–11 dm, herbage glabrous; leaves opposite or whorled, lanceolate to ovate, 6–8 cm, coarsely serrate; flowers showy, bright red, 13–22 mm, glandular-pubescent; capsules 8–11 mm. Igneous cliffs and talus slopes, occasionally canyon bottoms; piñon-juniper woodlands, lower montane coniferous forest, 6500–8200 ft. Catron, Grant. **R & E**

Scrophularia montana Wooton—Mountain figwort. Plants often > 10 dm, puberulent, branched above; leaves lanceolate, 8–15 cm, serrate but not coarsely; calyx 3–4 mm; corolla 6–10 mm, dull purplish or greenish; capsules 10–15 mm. Montane meadows, ponderosa pine, mixed conifer, spruce-fir communities, 5760–10,700 ft. Scattered. NC, C, WC, SC, SW. **E**

Scrophularia parviflora Wooton & Standl.—Pineland figwort. Plants ca. 10+ dm, simple or sometimes few-branched; puberulent toward the summit; leaves lanceolate or triangular-lanceolate, 5–10 cm, laciniate-serrate; calyx ca. 3–4 mm; corolla 6–8 mm, greenish purple. Ponderosa pine, Gambel oak, mixed conifer, 5510–9600 ft. SW.

VERBASCUM L. – Mullein

Biennial or perennial herbs; leaves alternate or in a basal rosette, simple, sessile, clasping or somewhat decurrent on the stem; flowers in a raceme or spike; corolla yellow or white, nearly actinomorphic; stamens 5, all fertile.

1. Leaves glabrous…**V. blattaria**
1'. Leaves puberulent, hispid, or densely woolly…(2)
2. Leaves densely woolly…**V. thapsus**
2'. Leaves puberulent and hispid…**V. virgatum**

Verbascum blattaria L.—Moth mullein. Stems 4-15 dm; leaves oblanceolate to narrowly ovate, margins toothed to lobed or pinnatifid, 5-20 × 1-3 cm, glabrous; inflorescence a simple raceme, glandular; corolla white or rarely yellow, 2.5-3 cm diam.; stamens with purple hairy filaments. Disturbed sites along roads, grasslands, gardens, 4200-5190 ft. Hidalgo. Introduced

Verbascum thapsus L.—Woolly mullein. Stems 3-20 dm; leaves oblanceolate to obovate or ovate, margins entire to shallowly toothed, to 5 dm, tomentose, decurrent; inflorescence a dense spikelike panicle; corolla yellow, 1.2-2(-3.5) cm diam.; stamens mostly with yellow filaments. Meadows, roadsides, open slopes, disturbed areas, 4000-10,500 ft. Widespread except EC, Doña Ana, Hidalgo, Lea. Introduced.

Verbascum virgatum Stokes—Wand mullein. Stems to 10 dm; leaves of the first year in a basal rosette, the cauline leaves elliptic to ovate, 2-12 cm, crenate, clasping at the base; inflorescence a loose raceme; corolla yellow or white, 2.5-3 cm diam.; capsules 6-8 mm. Disturbed sites in fields and along roadsides, 4465-5150 ft. Grant, Hidalgo, Roosevelt. Introduced.

SIMAROUBACEAE - QUASSIA FAMILY

Steven R. Perkins

AILANTHUS Desf. - Tree of heaven

Ailanthus altissimus (Mill.) Swingle [*A. glandulosus* Desf.]—Tree of heaven. Slender tree or large shrub with gray bark, to 20 m, often occurring in thickets due to rhizomes; stems green, glandular-puberulent; leaves deciduous, alternate, odd-pinnately compound, 10-100 cm, leaflets 9-31, 5-15 cm, elliptic-oblong to lanceolate, acuminate; stipules absent; inflorescence a large panicle, 10-40 cm; flowers often pungent, small, greenish, whitish, or yellowish; sepals 5-6, < 1 mm; petals 5-6, 2-3 mm; fruit samaralike, 3-5 × 3-7 mm. Weedy species often associated with developed, disturbed, and riparian areas, 3900-6640 ft. Widespread except NE, NC (Heil, 2013e).

SIMMONDSIACEAE - JOJOBA FAMILY

Kenneth D. Heil

SIMMONDSIA Nutt. - Jojoba

Simmondsia chinensis (Link) C. K. Schneid. [*Buxus chinensis* Link; *S. californica* Nutt.]—Jojoba. Evergreen shrubs to 2(-3) m, dioecious; leaves opposite, simple, stipules absent, dull pale green or gray-green, blades elliptic or oblong-ovate, (1.5-)2-5.5 × (0.5-)1-3 cm, thick-leathery, often becoming wrinkled when dry, margins entire; staminate flowers greenish yellow; sepals 2-4 mm, densely hairy abaxially; stamen filaments 0.1-0.3 mm; anthers 1.5-2 mm; pistillate flowers pale green, sepals (in fruit) triangular-ovate, surfaces densely hairy; capsules shiny greenish brown, 15-25 × 9-11 mm; seeds reddish brown or brown, 13-17 mm. Rocky, gravelly, or sandy slopes, desert scrub; Peloncillo Mountains, up to 4900 ft. Hidalgo (Gillespie, 2006).

SOLANACEAE - POTATO FAMILY

Jennifer Ackerfield

Herbs, shrubs, trees, or vines; leaves alternate or alternate below and becoming opposite above, simple to compound, exstipulate; flowers perfect, actinomorphic or zygomorphic; calyx usually 5-lobed, usually persistent and sometimes accrescent or inflated in fruit; corolla 5-lobed; stamens 5, epipetalous, alternate with the corolla lobes; pistil 1; ovary superior, 2-carpellate or falsely 3- or 5-carpellate, with axile placentation; fruit a berry or capsule (Ackerfield, 2015; Felger & Rutman, 2016).

1. Shrubs or small trees…(2)
1'. Herbs or vines…(3)
2. Flowers bright yellow; stems lacking spines; leaves 3–6 cm wide; fruit a capsule…**Nicotiana** (in part)
2'. Flowers white, purple, or greenish yellow; stems often spiny on older growth; leaves 0.5–1.5 cm wide; fruit a berry…**Lycium**
3. Fruit a circumscissile capsule surrounded by a persistent, urn-shaped calyx; flowers in a dense 1-sided spike or raceme with leafy bracts, cream to greenish yellow with purple veins, zygomorphic…**Hyoscyamus**
3'. Fruit a berry, or if a capsule then not circumscissile; flowers unlike the above in all respects…(4)
4. Corolla narrowly long-tubular or funnelform; fruit a capsule…(5)
4'. Corolla campanulate, rotate, or urceolate; fruit a dry or fleshy berry…(7)
5. Corolla 5–20 cm; fruit a globose, often spiny, capsule…**Datura**
5'. Corolla 0.5–4 cm; fruit a berry, or if a capsule then lacking spines or prickles…(6)
6. Corolla 0.5–0.7 cm, purple with a whitish tube; plants low, spreading…**Calibrachoa**
6'. Corolla 1.8–4 cm, white or greenish white; plants erect…**Nicotiana** (in part)
7. Stamens connivent around the style, the anthers opening by terminal pores or rarely also with longitudinal slits; flowers usually 2+ in a cymose or racemose inflorescence; calyx not becoming dry and inflated and completely enclosing the berry; stems and leaves sometimes armed with spines or prickles…**Solanum**
7'. Stamens not connivent around the style, the anthers opening by longitudinal slits; flowers solitary on axillary peduncles; calyx becoming dry and inflated and completely enclosing the berry or not; stems and leaves unarmed…(8)
8. Calyx not becoming dry and bladdery-inflated and completely enclosing the berry; leaves mostly linear to lanceolate with pinnatifid margins; flowers white, yellow, or greenish yellow…**Chamaesaracha**
8'. Calyx becoming dry, bladdery-inflated, and papery, completely enclosing the berry; leaves unlike the above; flowers various…(9)
9. Flowers yellow, or if purple then the corolla urceolate; plants with glandular or eglandular hairs, or glabrous; seeds glossy…**Physalis**
9'. Flowers purple or rarely white, rotate, erect; plants with flat, spherical white hairs; seeds dull…**Quincula**

CALIBRACHOA Cerv. – Petunia

Calibrachoa parviflora (Juss.) D'Arcy—Seaside petunia. Annuals, spreading, glandular-viscid; leaves alternate to subopposite, simple and entire, 4–12 mm; flowers actinomorphic, purple, solitary in leaf axils; calyx with narrow lobes, 2–3 mm in flower and 5–8 mm in fruit; corolla funnelform, 5–7 mm; fruit an ovoid capsule, 3–4 mm. Along streams, sandy river shores, lakeshores, irrigation ditches, 3300–6800 ft. EC, WC, Colfax, Eddy, Santa Fe.

CHAMAESARACHA (A. Gray) Benth. & Hook. f. – Five eyes

Perennial herbs; leaves alternate, simple, entire to pinnatifid; flowers actinomorphic, white, ochroleucous, or greenish yellow; fruit a berry, enclosed in a persistent, herbaceous calyx (Turner, 2015).

1. Leaves glabrous or nearly so; leaf blades 1–3 times longer than wide; (glabrous phase)…**C. pallida** (in part)
1'. Leaves pubescent with branched, stellate, simple, or glandular hairs; leaves other than above…(2)
2. Herbage eglandular, pubescent with stellate, branched, or scurfy-like hairs…(3)
2'. Herbage predominantly with glandular hairs, these usually intermixed with eglandular hairs…(4)
3. Leaves linear to linear-lanceolate, 4–6 times longer than wide, sinuate to lobate; sparsely pubescent with branched hairs…**C. arida**
3'. Leaves rhombic, broadly lanceolate, or oblanceolate; midstem leaves mostly entire; pubescent with mostly branched hairs…**C. pallida** (in part)
4. Leaf margins pinnately or irregularly lobed or toothed; calyx villous…**C. coniodes**
4'. Leaf margins entire to undulate; calyx densely glandular-pubescent…**C. sordida**

Chamaesaracha arida Henrickson—Greenleaf five eyes. Perennials, rhizomatous, with branched or stellate hairs, these often intermixed with longer, tangled hairs; leaves alternate, simple, linear, lanceolate, narrowly elliptic, or oblanceolate, 1.5–8 cm; flowers cream to greenish white, solitary in leaf axils; calyx campanulate, densely stellate-pubescent, 2.5–6 mm; corolla rotate, 10–15 mm diam.; fruit a whitish berry, 4–8 mm diam. Disturbed areas, deserts, dry grasslands, 3500–7500 ft. EC, WC, SW, Colfax, Eddy, Los Alamos, Santa Fe.

Chamaesaracha coniodes (Moric. ex Dunal) Benth. & Hook. f. ex B. D. Jacks.—Gray five eyes. Perennials, stipitate-glandular-pubescent with simple, eglandular hairs intermixed, rarely with a few stellate hairs; leaves alternate, simple, broadly lanceolate, 2–6 cm; flowers cream, sometimes purple-tinged, solitary in leaf axils; calyx campanulate, villous, to 4 mm; corolla rotate, 10–15 mm diam.; fruit a pale yellow to whitish berry, ca. 7 mm diam. Dry hillsides, woodlands, 3100–7500 ft. NE, EC, SE.

Chamaesaracha pallida Averett [*C. edwardsiana* Averett of New Mexico reports]—Pale five eyes. Perennials, densely pubescent with branched or stellate hairs; leaves alternate, simple, ovate-lanceolate to ovate or oblanceolate, 2–4 cm; flowers white to cream or greenish white, solitary in leaf axils; calyx campanulate, hispid, 3–4 mm; corolla rotate, 10–15 mm diam.; fruit a whitish berry, 5–7 mm diam. Pastures, roadsides, dry slopes, 4000–6000 ft. Eddy, Otero, Roosevelt.

Chamaesaracha sordida (Dunal) A. Gray—Hairy five eyes. Perennials, rhizomatous, densely glandular-pubescent mixed with longer, simple hairs; leaves alternate, simple, lanceolate to oblanceolate, elliptic, or rhombic, 1–4 cm; flowers cream to yellowish, solitary in leaf axils; calyx campanulate, densely glandular-pubescent, 3–5 mm; corolla rotate, ca. 10 mm diam.; fruit a whitish berry, 4–8 mm diam. Limestone hillsides, piñon-juniper woodlands, grasslands, dry washes, 2900–7500 ft. S.

DATURA L. – Jimsonweed

Perennial or annual herbs; stems erect or spreading; leaves alternate, simple, entire to toothed; flowers nocturnal, fragrant, solitary, actinomorphic, large, to 20 cm, in the forks of branching stems; calyx tubular, 5-toothed; corolla funnelform, white, purplish, or pinkish; stamens 5, the anthers basifixed; fruit a capsule, spiny or not, with numerous seeds.

1. Fruit smooth, lacking spines; calyx split on 1 side; plants growing in shallow, temporary pools and playas …**D. ceratocaula**
1'. Fruit spiny; calyx not split; plants not growing in water…(2)
2. Calyx 6–14 cm; corolla tube 12–26 cm; capsules nodding…(3)
2'. Calyx 2–5 cm; corolla tube 4–10 cm; capsules erect…(4)
3. Stems white-puberulent, lacking glandular hairs; corolla 14–26 cm…**D. wrightii**
3'. Stems densely villous or glandular-villous; corolla 12–16 cm…**D. inoxia**
4. Capsules armed with few, stout, unequal prickles; sharp-pointed apex of the corolla lobes mostly 2 mm; leaves mostly deeply cleft…**D. quercifolia**
4'. Capsules armed with numerous slender prickles; sharp-pointed apex of the corolla lobes 3–8 mm; leaves mostly shallowly lobed…**D. stramonium**

Datura ceratocaula Ortega—Latin thorn-apple. Annuals to 15 dm; leaves narrowly ovate, 6–15 × 3–5 cm, shallowly to deeply pinnately lobed, canescent below; calyx 6.5–8 cm; corolla white or purple, the tube 8.5–15 cm; capsule ovoid, 2.5–3.5 cm, smooth and unarmed. Playas and temporary pools, 4500 ft. Hidalgo.

Datura inoxia Mill.—Pricklyburr, angel's trumpet. Annuals or short-lived perennials to 15 dm; leaves ovate, 5–25 × 5–12 cm, entire to sinuate-dentate, canescent, becoming glabrous in age except for the primary veins; calyx 6–10 cm; corolla white, the tube 12–19 cm; capsule ovoid, 3–4 cm, armed with numerous prickles ca. 10 mm. Dry desert, plains, 4000–6500 ft. Catron, Lincoln, Otero.

Datura quercifolia Kunth—Chinese thorn-apple. Annuals to 15 dm; leaves ovate to elliptic, 6–20 × 3–12 cm, sinuate-dentate to pinnately lobed, moderately hairy; calyx 2–3 cm; corolla white, violet, or purple, the tube 4–7 cm; capsules ovoid, 3–4.5 cm, armed with unequal prickles 0.5–3.5 mm. Sandy soil, floodplains, roadsides, canyon slopes, 3400–7600 ft. Widely scattered except NW.

Datura stramonium L.—Jimsonweed. Annuals to 12 dm; leaves lance-ovate, ovate, or elliptic, 5–25 × 4–18 cm, sinuate-dentate, glabrous to slightly hairy; calyx 3.5–6 cm; corolla white or sometimes tinged with purple, the tube 6–10 cm; capsules ovoid, 3.5–5 cm, smooth to sparsely hairy, with numerous prickles 3–9 mm. Disturbed areas, railroads, fields, 4600–6500 ft. Scattered except E. Introduced.

Datura wrightii Regel—Indian apple. Perennials from thick caudices, to 10 dm; leaves ovate, 4–25 × 5–12 cm, entire to shallowly sinuate-dentate, softly cinereous-pubescent; calyx 6–14 cm; corolla white to purplish, the tube 15–26 cm; capsules ovoid, armed with numerous prickles 5–12 mm. Fields, disturbed areas, deserts, sometimes planted, 3200–7000 ft. Widespread except NE, NC, Torrance.

HYOSCYAMUS L. – Henbane

Hyoscyamus niger L.—Black henbane. Biennials, glandular-viscid; leaves alternate, simple, shallowly dentate to deeply pinnately lobed or parted, 4–10 × 2–6 cm; flowers zygomorphic, pale yellow with purple veins, solitary in leaf axils; calyx tubular-campanulate to urceolate, becoming enlarged in fruit, with needlelike lobes, 1–1.5 cm; corolla campanulate, 2–3 cm; fruit an ovoid circumscissile capsule, ca. 1.5 cm. Roadsides, grasslands, disturbed areas, 6500–9800 ft. NC, NW.

LYCIUM L. – Wolfberry, desert-thorn

Vines or thorny shrubs; leaves alternate, usually fascicled, simple, entire, sometimes fleshy; flowers solitary or in clusters, actinomorphic; calyx campanulate to tubular; corolla white, yellow, greenish, or purple, campanulate, funnelform to tubular; fruit a berry, subtended by a persistent calyx; seeds 2 to many.

1. Clambering vines or shrubs with slender, arched and recurved stems…**L. barbarum**
1'. Shrubs with stout, erect or spreading stems…(2)
2. At least some leaves < 3 mm wide; calyx 1–3 mm…(3)
2'. Leaves mostly > 3 mm wide, to 15 mm wide; calyx 2.5–8 mm…(4)
3. Stems usually tan or gray; flowers longer than wide; corolla tube narrow, not conspicuously expanded above; filaments sparsely hairy to glabrous at the base of the free portion…**L. andersonii**
3'. Stems usually dark reddish purple; flowers as wide as or wider than long; corolla tube conspicuously expanded above; filaments densely hairy at the base of the free portion…**L. berlandieri**
4. Corolla 15–25 mm…**L. pallidum**
4'. Corolla 8–15 mm…**L. torreyi**

Lycium andersonii A. Gray—Red-berry desert-thorn. Shrubs 0.5–3 m, armed with thorns; leaves linear-spatulate, 3–35 × 1–8 mm; flowers white, greenish white, or pale purple, solitary in leaf axils; calyx 1.5–3 mm; corolla tubular to narrowly funnelform, 6–16 mm; fruit red or orange-red, 3–9 mm. Dry slopes, deserts, washes, 4000–5300 ft. Hidalgo, Luna, Otero, Socorro.

Lycium barbarum L.—Matrimony vine. Clambering vines or shrubs 0.8–2 m, thorny; leaves lanceolate to narrowly elliptic, 20–30 × 3–6 mm; flowers purple, solitary in leaf axils; calyx 4–5 mm; corolla funnelform, 8–10 mm; fruit red, ca. 1 mm diam. Roadsides, near abandoned homes, 3900–8400 ft. Colfax, San Miguel, Santa Fe, Taos.

Lycium berlandieri Dunal—Silver desert-thorn. Shrubs 0.7–3 m, with few thorns at the ends of branches; leaves linear to spatulate-obovate, 1.5–15 × 1–2.5(–4.5) mm; flowers white to pale purple, solitary in leaf axils; calyx 1–3 mm; corolla funnelform, 4–9 mm; fruit red or orange-red, ca. 5 mm diam. Dry slopes, desert scrub, limestone slopes, canyon bottoms, washes, 3300–8100 ft. EC, S.

Lycium pallidum Miers—Frutilla. Shrubs 1–2.5 m, armed with thorns; leaves oblanceolate, spatulate, or elliptic, 10–15 × 3–15 mm; flowers white to purple or greenish with purple veins, solitary in leaf axils; calyx 5–8 mm; corolla funnelform, 15–25 mm; fruit red or reddish blue, ca. 10 mm diam. Desert grasslands, piñon-juniper woodlands, washes, canyon bottoms, 3300–10,200 ft. Widespread except far E.

Lycium torreyi A. Gray—Squawthorn. Shrubs 1–3 m, armed with thorns; leaves oblanceolate, spatulate, or obovate, 10–35(–50) × 3–15 mm; flowers greenish lavender or white, solitary in leaf axils; calyx campanulate to tubular, 2.5–6 mm; corolla tubular to narrowly funnelform, 8–15 mm; fruit red, 6–12 mm diam. Washes, canyon bottoms, alluvial flats, 3400–7500 ft. C, WC, SC, SW.

NICOTIANA L. – Tobacco

Heavy-scented annual or perennial herbs, shrubs, or small trees; leaves alternate, simple, entire, glabrous to glandular and/or pubescent; inflorescence of terminal panicles or racemes; calyx not inflated in fruit; corolla funnelform, salverform, or nearly tubular, the limb usually shallowly 5-lobed and spreading, white to cream; fruit a septicidal capsule, 2- or 4-valved at summit; seeds many, small, ovoid to reniform.

1. Flowers yellow; shrubs or small trees…**N. glauca**
1'. Flowers white to greenish white; plants herbaceous…(2)
2. Cauline leaves petiolate; corolla glabrous to sparsely hairy externally; annuals…**N. attenuata**
2'. Cauline leaves sessile and clasping the stem at the base; corolla densely hairy externally; perennials, often woody at the base…**N. obtusifolia**

Nicotiana attenuata Torr. ex S. Watson—Coyote tobacco. Annuals, glabrate to glandular; leaves lance-ovate to ovate, cauline leaves petiolate, 5–15 cm; flowers white, in racemose-paniculate inflorescences; calyx ovoid-campanulate, 5–8 mm; corolla funnelform, the tube 18–20 mm; fruit 8–10 mm. Washes, sandy slopes, fields, roadsides, piñon-juniper, 4300–8500 ft. NW, C, WC, SW.

Nicotiana glauca Graham—Juan loco, tree tobacco. Shrub to small tree; leaves lanceolate to ovate, thick and rubbery, 5–20 cm; flowers yellow; calyx unequally toothed, ca. 10 mm; corolla funnelform, the tube 30–40 mm; fruit 7–15 mm. Uncommon in disturbed areas, fields, pastures, 4300–4500 ft. Doña Ana, Sierra.

Nicotiana obtusifolia M. Martens & Galeotti—Desert tobacco. Perennials, glandular; leaves ovate to obovate, cauline leaves sessile and clasping at the base, 2–10 cm; flowers white to greenish white, in racemose-paniculate inflorescences; calyx ovoid-campanulate, 10–15 mm; corolla funnelform, the tube 15–26 mm; fruit 8–10 mm. Gypsum hillsides, cliff faces, washes, 3000–7500 ft. S, Cibola, Guadalupe, Sandoval, San Miguel.

PHYSALIS L. – Groundcherry

Annual or perennial herbs, glabrous or pubescent; leaves alternate to subopposite, simple, entire to toothed, sinuate or shallowly lobed; inflorescence usually solitary or in 2s or 3s; calyx campanulate to cup-shaped, shallowly 5-lobed, shorter than the corolla at anthesis and becoming greatly inflated, dry, membranous, and completely enclosing the fruit; corolla usually yellowish, sometimes white or bluish, often darker-spotted at the base; stamens 5; anthers yellow or sometimes blue; fruit a berry enclosed in an enlarged, persistent calyx (Pretz and Deanna, 2020; Rydberg, 1895; Seithe & Sullivan, 1990).

1. Corolla urceolate, constricted toward the apex…**P. solanaceus**
1'. Corolla rotate or campanulate…(2)
2. Plants with stellate or branched hairs…(3)
2'. Plants glabrous or with simple, glandular or nonglandular hairs…(4)
3. Calyx 10-ribbed; pubescence of short, branched hairs…**P. fendleri**
3'. Calyx not 10-ribbed; pubescence of stellate or branched hairs…**P. cinerascens**
4. Plants with glandular hairs, these sometimes intermixed with nonglandular hairs, often viscid…(5)
4'. Plants glabrous or with nonglandular hairs…(7)
5. Annuals from taproots; corolla 0.6–0.7 cm wide…**P. neomexicana**
5'. Perennials from somewhat woody or deeply buried caudices or with rhizomes; corolla 1.1–1.8 cm wide… (6)
6. Leaves smaller, 2.5–4 × 1–3.5 cm; calyx 1.4–2.7 cm in fruit…**P. hederifolia** (in part)
6'. Leaves larger, 5–10 × 3.5–6 cm; calyx 2.5–3 cm in fruit…**P. heterophylla**
7. Pubescence partially or wholly of long-villous or multicellular hairs 3–4.5 mm…(8)
7'. Plants glabrous, or the pubescence of short, often antrorse hairs to 1.5 mm…(10)
8. Annuals from taproots; corolla 0.6–0.7 cm wide…**P. pubescens**
8'. Perennials from somewhat woody or deeply buried caudices or with rhizomes; corolla 1.1–2.1 cm wide… (9)
9. Leaves 2–4 times as long as wide…**P. caudella**
9'. Leaves 1–2.5 times as long as wide…**P. hederifolia** (in part)

10.   Anthers blue, twisting after dehiscence; fruiting calyx filled with and often burst by the mature berry...
       P. ixocarpa
10'.   Anthers yellow, or if bluish then not twisting after dehiscence; fruiting calyx not filled or burst by the
       mature berry...(11)
11.   Corolla white to cream with a yellow center; leaves distinctly and evenly dentate to serrate; annuals...
       (12)
11'.   Corolla yellow or yellowish green with a darkened, purplish center; leaves entire to sinuate-dentate; pe-
       rennials...(13)
12.   Corolla 1–2.2 cm wide...**P. acutifolia**
12'.   Corolla 0.4–1 cm wide...**P. angulata**
13.   Calyx tube with 10 lines of short, ascending hairs or nearly glabrous; leaves glabrous or minutely hairy on
       the main veins...**P. longifolia**
13'.   Calyx tube uniformly hairy on its entire surface; leaves sparsely hairy, the hairs not confined to the main
       veins...(14)
14.   Leaf margins with stiff, spreading hairs ca. 1 mm...**P. hispida**
14'.   Leaf margins glabrous or with recurved hairs < 1 mm...**P. virginiana**

Physalis acutifolia (Miers) Sandwith [*P. wrightii* A. Gray]—Sharpleaf groundcherry. Annuals, sparsely antrorsely pubescent; leaves lanceolate, elliptic, or ovate, dentate to serrate or incised, 4–12.5 × 1–5 cm; calyx 0.3–0.5 cm in flower, 1.2–2.3 cm in fruit; corolla white to cream with a yellow center, rotate, 0.7–1.2 × 1–2.2 cm; berry 0.5–1.3 cm diam. Fields, sandy washes, roadsides, disturbed areas, 3700–5500 ft. Doña Ana, Hidalgo, Luna.

Physalis angulata L.—Cutleaf groundcherry. Annuals, sparsely pubescent; leaves elliptic, dentate to serrate, 2.5–8 × 0.5–2.5 cm; calyx 0.2–0.5 cm in flower, ca. 1 cm in fruit; corolla white to cream with a yellowish center, campanulate, to 1 × 0.4–1 cm; berry 0.5–1 cm diam. Introduced from planters, gardens, 3700–5300 ft. Doña Ana.

Physalis caudella Standl.—Southwestern groundcherry. Perennials, villous to pubescent, with mul-ticellular hairs to 4.5 mm; leaves lanceolate, rhomboid, or elliptic, entire to sinuate or coarsely toothed, 2.5–9.5 × 1–5 cm; calyx 0.7–2 cm in flower, 2–5 cm in fruit; corolla yellow with a darkened center, cam-panulate, 0.8–1.1 × 1.1–2.1 cm; berry 0.4–1.5 cm diam. Cliffs; known from the Mogollon Mountains, 7500 ft. Catron.

Physalis cinerascens (Dunal) Hitchc.—Smallflower groundcherry. Perennials with stellate or branched hairs; leaves linear-lanceolate to ovate, entire to undulate or coarsely toothed; calyx 0.5–1 cm in flower, 1.5–3.5 cm in fruit; corolla yellow with a darkened center, campanulate, 0.8–2 × 0.8–2 cm. Grasslands, prairie, disturbed sites, 3600–7000 ft. SE, Harding.

Physalis fendleri A. Gray [*P. hederifolia* A. Gray var. *fendleri* (A. Gray) Cronquist]—Fendler's groundcherry. Perennials with short, branched hairs; leaves ovate to reniform, coarsely dentate to ser-rate, 2.5–4 × 1–3.5 cm; calyx 0.5–1.1 cm in flower, 1.4–2.5 cm in fruit; corolla yellow with a darkened cen-ter, campanulate, 0.8–1.1 × 1.1–1.5 cm; berry 0.8–2 cm diam. Piñon-juniper woodlands, desert grasslands, sandy stream beds, canyon bottoms, 3800–8300 ft. NC, NW, C, WC, SC, SW.

Physalis hederifolia A. Gray—Ivyleaf groundcherry. Perennials with long, multicellular hairs or short, glandular hairs; leaves ovate to reniform, coarsely dentate to serrate, 2.5–4 × 1–3.5 cm; calyx 0.5–1.1 cm in flower, 1.4–2.7 cm in fruit; corolla yellow with a darkened center, campanulate, 0.8–1.1 × 1.1–1.5 cm; berry 0.8–2 cm diam. Shortgrass prairie, desert scrub, rocky canyons, ponderosa pine forests, piñon-juniper woodlands, 3000–9300 ft. NE, NC, C, SE, SC, SW.

1.   Pubescence of short, glandular hairs...**var. comata**
1'.   Pubescence of long, multicellular hairs...**var. hederifolia**

var. comata (Rydb.) Waterf.—Ivyleaf groundcherry. Pubescence of short, glandular hairs. Prairie, desert scrub, piñon-juniper woodlands, ponderosa pine forests, often in disturbed locations, 3850–7300 ft. NC, NE, C, SC, Eddy.

var. hederifolia—Ivyleaf groundcherry. Pubescence of long, multicellular hairs. Short grass prairie, piñon-juniper woodlands, ponderosa pine forests, often in disturbed locations, 3000–9300 ft. Scattered.

Physalis heterophylla Nees—Clammy groundcherry. Perennials, villous with spreading hairs and viscid-glandular, rarely glandular hairs absent; leaves ovate to rhombic, shallowly sinuate-dentate or entire, 5–10 × 3.5–6 cm; calyx 0.7–1.2 cm in flower, 2.5–3 cm in fruit; corolla yellow with a darkened center, campanulate, 1–2 × 1.2–1.8 cm; berry 1–1.2 cm diam. Meadows, ponderosa pine forests, 6500–8500 ft. SW.

Physalis hispida (Waterf.) Cronquist—Prairie groundcherry. Perennials, sparsely pubescent with short, antrorse hairs, leaf margins and stems with stiff, spreading hairs ca. 1 mm; leaves lanceolate to elliptic-lanceolate, sinuate-dentate or entire; calyx 0.4–1.2 cm in flower, 2–4 cm in fruit; corolla yellow with a darkened center, campanulate, 1–1.5 cm; berry 0.8–1.5 cm diam. Grasslands, shortgrass prairie, sandy hills, 4200–7300 ft. NE, NC.

Physalis ixocarpa Brot. ex Hornem. [*P. philadelphica* Lam. subsp. *ixocarpa* (Brot. ex Hornem.) Sobr.-Vesp. & Sanz-Elorza]—Tomatillo. Annuals, glabrous to sparsely pubescent; leaves ovate to ovate-lanceolate, entire to dentate or sinuate, 2–7 cm; calyx 1.5–3 cm in fruit; corolla yellow with five blue-tinged spots at the base, campanulate, 0.8–1.5 cm; berry usually filling or even bursting the fruiting calyx. Cultivated, sometimes escaping gardens or in old homesteads, 5900–6100 ft. Cibola, Santa Fe.

Physalis longifolia Nutt. [*P. virginiana* Mill. var. *subglabrata* (Mack. & Bush) Waterf.]—Longleaf groundcherry. Perennials, glabrous to sparsely pubescent; leaves lanceolate to elliptic, entire to repand-sinuate, 4–11 × 1–4 cm; calyx 0.8–1.2 cm in flower, 1.5–3.5 cm in fruit; corolla cream to greenish yellow with a darkened center, campanulate, 1.1–1.6 cm; berry 0.6–1.2 cm diam. Shortgrass prairie, canyon bottoms, dry washes, fields, 3600–8200 ft. Scattered. N, C, WC, SW, Chaves, Cibola, Eddy.

1. Anthers yellow…**var. longifolia**
1′. Anthers light blue or tinged with blue…**var. subglabrata**

var. longifolia—Longleaf groundcherry. Anthers yellow. Grasslands, canyon bottoms, dry washes, fields, 3800–8200 ft. NW, NE, C, SW, SC.

var. subglabrata (Mack. & Bush) Cronquist—Longleaf groundcherry. Anthers blue or tinged with blue. Arroyos, river margins, and piñon-juniper woodlands, 6000–7200 ft. Sandoval, Sierra.

Physalis neomexicana Rydb. [*P. subulata* Rydb. var. *neomexicana* (Rydb.) Waterf. ex Kartesz & Gandhi]—New Mexican groundcherry. Annuals, densely pubescent with multicellular hairs and glandular hairs; leaves elliptic to ovate, sinuate to coarsely dentate, 1.1–5.5 × 1–5 cm; calyx 0.4–0.7 cm in flower, 1.2–2.4 cm in fruit; corolla yellow with a darkened center, campanulate, 0.5–0.7 × 0.6–0.7 cm; berry 0.7–1.5 cm diam. Piñon-juniper woodlands, Douglas-fir forests, streams, canyon bottoms, 5400–9500 ft. Widespread except E, NW.

Physalis pubescens L.—Husk tomato. Annuals, densely pubescent with villous hairs; leaves ovate, entire to sparsely toothed, 1–8 × 1–8 cm; calyx 0.4–0.7 cm in flower, 1.7–3.5 cm in fruit; corolla yellow with a darkened center, campanulate, 0.5–0.7 × 0.6–0.7 cm; berry 0.7–1.7 cm diam. Canyons, riparian areas, piñon-juniper woodlands, 5500–7200 ft. Grant.

Physalis solanaceus (Schltdl.) Axelius [*Margaranthus solanaceus* Schltdl.]—Netted groundcherry. Annuals, sparsely pubescent on older growth, densely strigose on young growth; leaves ovate-lanceolate to ovate, entire to repand-sinuate, 2.5–7 × 1–3.5 cm; calyx ca. 0.2 cm in flower, 1–2 cm in fruit; corolla purple or yellow with a purple base, urceolate, 0.3–0.5 × 0.2–0.4 cm; berry 0.4–0.7 cm diam. Piñon-juniper woodlands, limestone slopes, canyon bottoms, sandy washes, 4000–7500 ft. S, Catron, Lincoln.

Physalis virginiana Mill.—Virginia groundcherry. Perennials, glabrous to pubescent with short, antrorse hairs; leaves lanceolate to ovate, sinuate-dentate or entire, 1.5–7 × 0.8–2.5 cm; calyx 0.8–1.1 cm in flower, 2.5–5 cm in fruit; corolla yellow with a darkened center, campanulate, 1.5–2.5 cm. Piñon-juniper woodlands, floodplains, sandy washes, canyon bottoms, 3500–8000 ft. Mora, San Miguel.

QUINCULA Raf. – Chinese lanterns

Quincula lobata (Torr.) Raf. [*Physalis lobata* Torr.]—Purple groundcherry. Perennials, decumbent to spreading, sparsely whitish-scurfy; leaves alternate, oblong to ovate, simple, entire to lobed, 1–10 cm; flowers actinomorphic, purple or rarely white, usually in pairs from leaf axils; calyx becoming enlarged in fruit, 3–4.5 mm in flower, 15–20 mm in fruit; corolla rotate, 1–2 cm diam.; fruit a berry, greenish yellow, 5–8(–10) mm diam. Shortgrass prairie, fields, roadsides, 3000–7500 ft. Widely scattered, E, NC, NW, C, WC, SC, SW.

SOLANUM L. – Nightshade

Annual or perennial herbs or vines, glabrous to pubescent or tomentose, often glandular, sometimes prickly; leaves alternate, simple, or pinnately compound, the margins entire to pinnatifid; inflorescence commonly umbels or cymes, often 1-sided; calyx 5-cleft or -toothed, ± bell-shaped; corolla rotate to pentagonal, white to purple or yellow; stamens 5, inserted on the corolla tube; fruit a berry, sometimes nearly dry; seeds many, generally reniform (Basset & Munro, 1984; Bates et al., 2009; Schilling, 1981).

1. Plants with sharp prickles on the leaves and stem…(2)
1'. Plants lacking sharp prickles…(6)
2. Corolla yellow…**S. rostratum**
2'. Corolla purple, bluish purple, or rarely white…(3)
3. Leaves pinnatifid to pinnately compound; anthers dissimilar, one purple, beaked, and conspicuously longer than the others; berries tightly enclosed in the burlike accrescent tube of the calyx…**S. heterodoxum**
3'. Leaves simple, often with sinuate or lobed margins; anthers all similar; berries not enclosed in a persistent calyx…(4)
4. Leaves silvery-gray-canescent; stellate hairs with rays fused at the center, scalelike…**S. elaeagnifolium**
4'. Leaves green; stellate hairs not fused at the center…(5)
5. Corolla white to pale purple; fruit 8–20 mm diam.…**S. carolinense**
5'. Corolla purple or bluish purple; fruit 25–30 mm diam.…**S. dimidiatum**
6. Stems climbing or clambering, vinelike; leaves 3-lobed or hastate; berries red…**S. dulcamara**
6'. Stems erect; leaves unlike the above; berries white, green, orangish brown, or blackish…(7)
7. Leaves deeply pinnatifid or pinnately compound…(8)
7'. Leaves simple, the margins entire or toothed but not lobed…(10)
8. Leaves deeply pinnatifid; plants lacking stolons or tubers…**S. triflorum**
8'. Leaves pinnately compound; plants with stolons or tubers…(9)
9. Corolla deeply lobed nearly to the base, white…**S. jamesii**
9'. Corolla shallowly lobed to about the middle, purple or blue, or white with purple tinges…**S. stoloniferum**
10. Plants densely glandular-pubescent, usually viscid…**S. sarrachoides**
10'. Plants glabrous to pubescent, not glandular…(11)
11. Anthers 2.5–4 mm; style as long as the anthers or to 3 mm longer…**S. douglasii**
11'. Anthers 1–2 mm; style as long as or only slightly longer than the anthers (the remaining taxa belong to the *S. nigrum* complex and are difficult to distinguish)…(12)
12. Sclerotic granules (small, hard inclusions ca. 0.5 mm diam. present in the fleshy matrix of berries) absent; inflorescence usually racemiform…S. nigrum L.
12'. Sclerotic granules (small, hard inclusions ca. 0.5 mm diam. present in the fleshy matrix of berries) present; inflorescence usually umbelliform…(13)
13. Sclerotic granules 6–15 per berry…**S. ptychanthum**
13'. Sclerotic granules 5 or fewer per berry…**S. interius**

Solanum carolinense L.—Carolina horsenettle. Perennials, rhizomatous, to 10 dm, lacking tubers or stolons, armed with prickles to 7 mm, stellate-pubescent; leaves simple, ovate, 12 × 8 cm; calyx ca. 10 mm; corolla white to pale purple, ca. 30 mm wide; berry 8–20 mm diam., green with pale green to grayish markings when immature, yellow at maturity, not enclosed in an accrescent calyx. Cultivated and sometimes escaping, along roadsides, 3900–7000 ft. Widespread, NW, EC, C, SC, SW.

Solanum dimidiatum Raf.—Western horsenettle. Perennials, rhizomatous, 0.3-1 dm, lacking tubers or stolons, armed with prickles to 7 mm, stellate-pubescent; leaves simple, ovate to elliptic-lanceolate, lobed or sinuate, 6-15 × 5-10 cm; calyx 8-13 mm; corolla purple or bluish purple, 30-50 mm wide; berry 25-30 mm diam., pale yellow, not enclosed in an accrescent calyx. Disturbed areas, roadsides, 3700-7700 ft. Catron, Eddy, Lea.

Solanum douglasii Dunal—Greenspot nightshade. Annuals or short-lived perennials, < 10 dm, lacking tubers or stolons, unarmed, glabrous to strigose-pubescent; leaves simple, ovate to lanceolate-elliptic, entire to coarsely toothed, 2-9 × 1-5 cm; calyx 1.5-2.5 mm; corolla white or tinged with purple, sometimes gland-dotted, 8-25 mm wide; berry 6-12 mm diam., green, orangish brown, or blackish at maturity, not enclosed in an accrescent calyx. Piñon-juniper woodlands, canyons, dry stream beds, 3600-8000 ft. WC, SW, Colfax, Mora, Torrance.

Solanum dulcamara L.—Climbing nightshade. Perennials, rhizomatous, stems climbing or clambering, lacking stolons or tubers, unarmed, glabrous to pubescent, rarely stellate; leaves simple, 3-lobed, or hastate, 7-12 × 4-9 cm; calyx 10-15 mm; corolla purple or bluish purple, lobes reflexed, 6-9 mm; berry 8-12 mm diam., red, not enclosed in an accrescent calyx. Disturbed areas, irrigation ditches, fencerows, 5300-7200 ft. NW.

Solanum elaeagnifolium Cav.—Silverleaf nightshade. Perennials, rhizomatous, to 10 dm, lacking tubers or stolons, armed with prickles to 5 mm, stellate-pubescent; leaves simple, linear to oblong-lanceolate, entire, sinuate-repand, or lobed, to 10 × 2.5 cm; calyx ca. 10 mm; corolla purple or rarely white, to 35 mm wide; berry to 20 mm diam., green with pale green to grayish markings when immature, yellow at maturity, not enclosed in an accrescent calyx. Piñon-juniper woodlands, disturbed areas, sandhills, roadsides, 3000-8000 ft. Widespread.

Solanum heterodoxum Dunal—Melonleaf nightshade. Annuals, 3-7 dm, lacking tubers or stolons, armed with prickles 2-9 mm, with simple and/or gland-tipped hairs and stellate hairs on lower leaf surfaces; leaves bipinnatifid to compound, broadly ovate to obovate or deltoid, 4-11 cm; calyx 1.5-2.2 mm; corolla violet to blue, 10-17 mm wide; berry 9-12 mm diam., tightly enclosed in the burlike accrescent tube of the calyx. Gravelly slopes, playas, dry stream beds, canyon bottoms, 3800-7500 ft.

1. Prickles on the stem scattered, not dense…**var. novomexicanum**
1'. Prickles on the stem densely packed, with 30+ prickles per cm of stem…**var. setigeroides**

var. novomexicanum Bartlett—NE, NC, Roosevelt.

var. setigeroides Whalen—WC, SW.

Solanum interius Rydb.—Deadly nightshade. Annuals or short-lived perennials, 3-8 dm, lacking tubers or stolons, unarmed, strigose; leaves simple, ovate, ovate-lanceolate, or rhombic, entire, undulate, or sinuate-dentate, 3-9 cm; calyx 1.5-2 mm; corolla white or tinged with purple, lobes reflexed, 3-4 mm wide; berry 7-10 mm diam., purplish black at maturity, not enclosed in an accrescent calyx. Canyons, rocky slopes, 3900-8100 ft. C, SC, Grant.

Solanum jamesii Torr.—Wild potato. Perennials, 0.1-0.5 dm, with tubers, unarmed, glabrous or with gland-tipped hairs and stellate hairs on lower leaf surfaces; leaves pinnately compound, 7-15 × 4-9 cm; calyx to 8 mm; corolla white, 20-35 mm wide; berry 9-10 mm diam., green, not enclosed in an accrescent calyx. Stream banks, canyon bottoms, piñon-juniper woodlands, 3600-9200 ft. Widespread except E.

Solanum ptychanthum Dunal—West Indian nightshade. Annuals, 3-8 dm, lacking tubers or stolons, unarmed, glabrous or sparsely strigose; leaves simple, ovate, ovate-lanceolate, triangular-ovate, or elliptic-lanceolate, entire, undulate, or sinuate-dentate, 5-10(-17) cm; calyx 1-1.5 mm; corolla white, often with a yellow center, lobes 1-1.5 mm; berry 5-9 mm diam., purplish black at maturity, not enclosed in an accrescent calyx. Disturbed areas, 4900-6200 ft. Doña Ana.

Solanum rostratum Dunal—Buffalobur. Annuals, to 7 dm, lacking tubers or stolons, armed with prickles to 10 mm, stellate-pubescent; leaves pinnatifid to bipinnatifid, elliptic to broadly ovate, lobed, to 12 cm; calyx 5-12 mm; corolla yellow, 15-25 mm wide; berry 9-12 mm diam., tightly enclosed in the bur-like accrescent tube of the calyx. Piñon-juniper woodlands, canyon bottoms, prairie, roadsides, disturbed areas, 3300-7500 ft. Widespread.

Solanum sarrachoides Sendtn.—Hairy nightshade. Annuals, 1-8 dm, lacking tubers or stolons, unarmed, densely glandular; leaves simple, ovate, entire, repand, or sinuate-dentate, 5-12 × 3-6 cm; calyx 1-1.5 mm; corolla white, 4-8(-11) mm wide; berry 6-9 mm diam., green to nearly black, calyx enlarged in fruit and covering about 1/2 of the fruit. Disturbed areas, roadsides, rocky slopes, 3900-8700 ft. W, C, SC, Colfax, Los Alamos, Mora, Taos.

Solanum stoloniferum Schltdl.—Creeping nightshade. Perennials, 0.1-1.5(-5) dm, with stolons and tubers, unarmed, pilose; leaves pinnately compound, ovate to elliptic, 7-22 × 3.5-8 cm; calyx 4-7 (-30) mm; corolla purple, blue, or white with purple tinges, 18-35 mm wide; berry 9-17 mm diam., white to green, not enclosed in an accrescent calyx. Canyons, shaded woodlands, 5500-10,000 ft. WC, SW, San Miguel.

Solanum triflorum Nutt.—Cut-leaf nightshade. Annuals, 1-5 dm, lacking stolons or tubers, unarmed, sparsely to densely appressed-pubescent; leaves deeply pinnatifid, ovate to oblong, 2-5 × 1-2 cm; calyx 2.5-5 mm; corolla white, drying yellow, to 8 mm wide; berry 9-14 mm diam., green, not enclosed in an accrescent calyx. Disturbed areas, roadsides, stream banks, 4200-9600 ft. Widespread except E.

TALINACEAE – FAMEFLOWER FAMILY

David J. Ferguson

TALINUM Adans. – Fameflower

Succulent, perennial herbs (± woody basally in *T. polygaloides*); roots tuberous; stems prostrate to erect; leaves succulent or semisucculent, flat with midvein evident, sessile or short-petiolate, alternate or sub-opposite, 1-7 cm wide, margins entire; inflorescences paniculate, racemose, or cymose, few- to many-flowered; flowers diurnal (ours); sepals 2; petals 5-6 (rarely more); stamens 10-45; stigma lobes 2-6; fruit subglobose to somewhat trigonous, a pendent capsule, dehiscing upward from base; seeds smooth, tubercled, or with concentric ridges. Previously included in Portulacaceae (Brown-Carter & Murdy, 1985; Ferguson, 1995, 2001; Kiger, 2001; Mueller, 1933; Packer, 2003; Price & Ferguson, 2012).

1. Scape and peduncles triangular in cross-section; inflorescence terminal, cymose…**T. fruticosum**
1'. Scape and peduncles roughly terete in cross-section, sometimes with low longitudinal ridges or wings; inflorescence various…(2)
2. Inflorescence a terminal panicle of cymes; flowers usually < 7 mm across; seeds approximately lenticular, nearly smooth to tubercled, without concentric ridges…(3)
2'. Inflorescences axillary, single-flowered or short few-flowered cymes; flowers > 7 mm across; seeds globose with raised concentric arcuate ridges…(5)
3. Flowers yellow (rarely whitish or pale peach-pink)…**T. spathulatum**
3'. Flowers rose-pink to magenta…(4)
4. Cauline leaves mostly spatulate, mostly truncate to rounded apically; seeds nearly smooth, not tuberculate…**T. sarmentosum**
4'. Cauline leaves mostly elliptic to obovate, obtuse to acute; seeds tubercled…**T. paniculatum**
5. Stems perennial, becoming suffrutescent, slender; leaves thick and rather succulent; flowers mostly < 1.5 cm wide, yellow…**T. polygaloides**
5'. Stems strictly annual, herbaceous, succulent; leaves thin, revolute only in drought; flowers mostly > 1.5 cm wide, usually orange…(6)
6. Leaves linear; sepals membranous, not prominently ridged, mostly early-deciduous…**T. lineare**
6'. Leaves broadly linear to broadly elliptic or obovate; sepals foliaceous, prominently 3-5-ribbed, persistent until fruit matures…(7)

7. Stems usually < 2 dm, simple or few-branched, often becoming slightly suffrutescent basally; leaves mostly broadly linear; inflorescence 1-flowered (very rarely more), short, usually < 1 cm, with peduncle much shorter than pedicel…**T. aurantiacum**

7'. Stems usually reaching > 2 dm, several lateral branches normally present, not suffrutescent; leaves mostly broad, elliptic to obovate; inflorescence usually 3-flowered (1-5), mostly well > 1 cm, with peduncles equaling or longer than pedicels…**T. whitei**

Talinum aurantiacum Engelm. [*Phemeranthus aurantiacus* (Engelm.) Kiger]—Orange fameflower. Plants to 5 dm; stems erect, sometimes suffrutescent; leaves linear to narrowly lanceolate, rarely oblanceolate, to 6 cm; inflorescences single, rarely 2; sepals 5-10 mm; petals yellow or orange, sometimes reddish or pinkish, 9-25 mm; capsules ovoid to globose, 4-7 mm; seeds with arcuate ridges, 1.2-1.7 mm. Rocky slopes, desert scrub, arid grasslands, piñon, juniper, and oak woodlands, ponderosa pine forests, 3050-7200 ft. S, Guadalupe, Quay.

Talinum fruticosum (L.) Juss. [*Portulaca fruticosa* L.; *P. triangularis* Jacq.]—Ceylon spinach. Annual or perennial herb to subshrub, erect, to 5 dm; roots fibrous to tuberous; stems succulent, thickened in age; leaves elliptic to obovate, to 10 cm; inflorescence an erect panicle of lateral short cymes along a triangular central scape; flowers open briefly (usually) in afternoon; sepals ovate, usually green, foliaceous, 4-8 mm; petals rounded, elliptic to narrowly obovate, pink to magenta, 3-6 mm; capsule subglobose, 4-6 mm, dehiscing explosively. Widely cultivated and occasional greenhouse weed, occasionally escaping near places of cultivation.

Talinum lineare Kunth—Varicolor flameflower. Herbaceous succulent perennial herb; roots tuberous with fibrous lateral roots; stems ascending to erect, to 4 dm; leaves narrowly linear, to 7 cm; inflorescences few-flowered, rarely exceeding leaves in length; flowers showy, 15-20 mm diam.; petals varied in color. Mostly on gravelly igneous-derived substrates, grasslands, piñon, juniper, and oak woodlands, to 5900 ft. Many botanists feel the name *T. lineare* applies to a species from central Mexico and has been misapplied to some collections of *Phemeranthus aurantiacus* (Engelm.) Kiger.

Talinum paniculatum (Jacq.) Gaertn. [*T. patens* (L.) Willd.]—Jewels of Opar. Perennial herb, mostly erect, to 1 m; roots often enlarged-tuberous; stems succulent; leaves elliptic to obovate, to 12 cm; inflorescence an erect panicle; flowers open briefly in afternoon; sepals ovate to suborbiculate, often brown or red, 2.5-4 mm; petals ovate to obovate, white to yellow, orange, pink, or magenta, 6-18 mm; capsule subglobose to obtusely trigonous, 2.5-5 mm, dehiscing upward from circumscissile base, tuberculate. Widely cultivated and occasional greenhouse weed, occasionally escaping near places of cultivation, 4000-5250 ft. Doña Ana, Hidalgo.

Talinum polygaloides Gillies ex Arn. [*T. angustissimum* (Engelm.) Wooton & Standl.]—Yellow fameflower. Perennial suffrutescent herb or subshrub, to 5 dm; roots tuberous with mostly fibrous lateral roots; stems single to few, ascending to erect, suffrutescent with maturity, becoming woody with age; leaves narrowly linear, to 7 cm; inflorescences few-flowered, rarely exceeding leaves in length; sepals membranous, deciduous; flowers showy, 15-20 mm diam., petals varied in color. Mostly on gravelly igneous-derived substrates in grassland or open woodland, to 6000 ft. Hidalgo, Lea, Otero, Sierra. This taxon is typically recognized as being included in *Phemeranthus aurantiacus* (Engelm.) Kiger.

Talinum sarmentosum Engelm.—Pink baby's breath. Perennial herb, mostly erect, to 1 m; roots often enlarged-tuberous; stems succulent; leaves elliptic to obovate, to 12 cm; inflorescence an erect panicle; flowers open briefly in afternoon; sepals ovate to suborbiculate, often brown or red, 2.5-4 mm; petals ovate to obovate, pink to magenta, 4-10 mm; capsule subglobose to obtusely trigonous, 2.5-5 mm, dehiscing upward from circumscissile base, nearly smooth, not tuberculate. Widely cultivated and occasional. Association of our species with the name *T. sarmentosum* Engelm. is based on distinctive morphological traits cited in the original description of that species. Typically included as a synonym of *T. paniculatum* (Jacq.) Gaertn., but distinctly different.

Talinum spathulatum Engelm. ex A. Gray [*T. chrysanthum* Rose & Standl.]—Yellow baby's breath. Perennial herb, mostly 15 dm; roots often tuberous; stems succulent; leaves elliptic to obovate, to 12 cm, reduced gradually upward and usually extending upward well into inflorescence; inflorescence an erect panicle; flowers open briefly in afternoon; sepals ovate to suborbiculate; flowers small, yellow (sometimes near white or tinged peach-pink); capsule subglobose to obtusely trigonous, 2.5-5 mm, dehiscing upward from circumscissile base, nearly tuberculate. Widely cultivated and occasional. Favors canyon bottoms and other sheltered areas with ample moisture, often among trees and shrubs in arid and semi-arid regions. SE, SC. Often listed as a synonym of *T. paniculatum* (Jacq.) Gaertn., but distinctly different.

Talinum whitei I. M. Johnst.—Santa Eulalia fameflower. Plants to 5 dm; stems erect, not becoming basally suffrutescent; leaves broadly elliptic to obovate, to 8 cm; inflorescences mostly 3-flowered; sepals 5-10 mm; petals mostly orange (yellow, red, or magenta); capsules ovoid to globose, 4-7 mm. Mostly in deep soil, grasslands. No records seen.

TAMARICACEAE - TAMARISK FAMILY

Kenneth D. Heil

TAMARIX L. - Tamarisk

Shrubs or trees (subshrubs), usually halophytes, rheophytes, or xerophytes; leaves alternate, scalelike, small; stipules absent; flowers tiny, actinomorphic; sepals 4-5; petals 4-5; stamens 5-10, attached to fleshy nectar disk; pistil (2)3-4(5)-carpellate; ovary 1-locular, sometimes almost plurilocular, ovules 2+ per placenta; placentation parietal, basal, or parietal-basal; styles (2)3-4(5) (or absent, stigmas sessile); fruits capsular, dehiscence loculicidal; seeds comose at 1 end (or hairy overall); embryo straight; endosperm absent (Gaskin, 2015).

1. Leaves sheathing [cultivar]…T. aphylla
1'. Leaves sessile or amplexicaul…(2)
2. Flowers 4-merous…**T. parviflora**
2'. Flowers 5-merous…(3)
3. Lobes of the star-shaped nectary disk attenuate, gradually passing into the filaments; plants not common in New Mexico…**T. gallica**
3'. Lobes of the nectary disk truncate-emarginate, the filaments abruptly extending from between the lobes; plants very common in New Mexico…**T. chinensis**

Tamarix aphylla (L.) H. Karst. [*Thuja aphylla* L.]—Athel tamarisk. Trees, to 10+ m; leaves sheathing, blades abruptly pointed, 2 mm; inflorescences 3-6 cm × 4-5 mm; bracts exceeding pedicels, not reaching calyx tip; flowers 5-merous; sepals 1-1.5 mm, margins entire; petals oblong to elliptic, 2-2.5 mm; antisepalous stamens 5, filaments alternate with nectar disk lobes, all originating from edge of disk. Sandy soil, lakeshores, riverways, 2885-4000 ft. Cultivar.

Tamarix chinensis Lour. [*T. juniperina* Bunge]—Saltcedar, five-stamen tamarisk. Shrubs or trees, to 8 m; leaf blades lanceolate to ovate-lanceolate, 1.5-3 mm; inflorescences 2-6 cm × 5-7 mm; bracts reaching or exceeding pedicels, not exceeding calyx tip; flowers 5-merous; sepals 0.5-1.5 mm, margins entire; petals elliptic to ovate, 1.5-2 mm; antisepalous stamens 5, filaments alternate with nectar disk lobes, some or all originating from below disk. Riverways, lakeshores, arroyos, 300-8175 ft. Widespread except Curry, Roosevelt.

Tamarix gallica L.—French tamarisk. Shrubs or trees, to 5 m; leaf blades lanceolate, 1.5-2 mm; inflorescences 2-5 cm × 4-5 mm; bracts exceeding pedicels, not reaching calyx tip; flowers 5-merous; sepals 0.5-1.5 mm, margins entire or subentire; petals elliptic to ovate, 1.5-2 mm; antisepalous stamens 5, filaments confluent with nectar disk lobes, all originating from edge of disk. Sandy soil, riverways, 3490-6300 ft. Doña Ana, Lincoln.

Tamarix parviflora DC.—Small-flower tamarisk. Shrubs or trees, to 5 m; leaf blades lanceolate, 2-2.5 mm; inflorescences 1.5-4 cm × 3-5 mm; bracts exceeding pedicels, not reaching calyx tip; flowers

4-merous; sepals 1–1.5 mm, margins entire or denticulate; petals oblong to ovate, 2 mm; antisepalous stamens 4, filaments confluent with nectar disk lobes, all originating from edge of disk. Riverways, lake-shores, 3940–6230 ft. Bernalillo, Doña Ana, Luna, Sandoval.

THEMIDACEAE – BRODIAEA FAMILY

Steve L. O'Kane, Jr.

Perennial herbs from corms, new cormlets formed at base of corms or on short stolons, not smelling of onion; leaves basal, 1–10, linear to narrow-lanceolate; inflorescence scapose, a terminal umbel or flowers solitary; scape erect, generally 1(2); flower bracts 2–4(–10), not enclosing flower buds; perianth parts 6 in 2 petal-like whorls, free or fused below into a tube; stamens 3 or 6, free or fused to perianth, append-aged in some; ovary superior, locules 3; fruit a loculicidal capsule (Pires & Preston, 2019; Utech, 2002).

1. Perianth segments ± free at the base, not forming a discernible tube (< 0.5 mm)…**Muilla**
1'. Perianth segments fused at the base, forming an obvious tube…(2)
2. Flowers solitary, sessile but appearing pedicellate because of the long, slender perianth tube; perianth tube greatly exceeding the lobes; bracts 4…**Milla**
2'. Flowers in umbels, pedicellate; perianth tube shorter than or not much exceeding the lobes; bracts 2+…(3)
3. Staminal filaments fused into a tube; perianth blue to purple…**Androstephium**
3'. Staminal filaments free; perianth white to greenish, with purple veins…**Dipterostemon**

ANDROSTEPHIUM Torr. – Funnel-lily

Androstephium breviflorum S. Watson [*Brodiaea breviflora* (S. Watson) J. F. Macbr.]—Funnel-lily. Leaves 10–30 cm × 1.5–2 mm; scape 1–3.5 dm, scabrous basally; inflorescences 3–12-flowered; perianth white to light violet-purple, 1.5–2 cm, lobes longer than tube; capsules 1–1.5 cm. Clay hills and sandy slopes, piñon-juniper woodlands, arid grasslands, 4600–7100 ft. McKinley, Rio Arriba, San Juan.

DIPTEROSTEMON Rydb. – Snakelily, bluedicks

Dipterostemon capitatus (Benth.) Rydb. [*Dichelostemma capitatum* (Benth.) Alph. Wood]—Bluedicks. Leaves 2–3, 10–70 cm; scape to 65 cm, smooth; umbel dense, 2–15-flowered; perianth blue to pinkish purple, or white, tube narrowly cylindrical to short-campanulate, 3–12 mm; perianth appendages slightly reflexed distally, leaning toward anthers to form corona, white, apex deeply notched; stamens 6, smaller 3 on outer tepals alternating with larger 3 on inner tepals. Our material is **subsp. pauciflorus** (Torr.) R. E. Preston. Habitat variable, but arid, 5050–6800(–8900) ft. Catron, Grant, Hidalgo.

MILLA Cav. – Mexican-star

Milla biflora Cav.—Mexican-star. Leaves 2–7(–10), 1 mm wide, blades channeled; scape 4–55 cm, scabrous on proximal veins; umbels 1–9-flowered (count undeveloped buds); perianth 4.5–18 cm (ap-pearing 2.5–4 cm due to pseudopedicel), white with green abaxial stripe, 3–5-veined, persisting in fruit; ovary proximally adnate to perianth tube, ovoid to obovoid, 1 cm, stipe 3–16 cm; style exserted; stigma capitate; capsules ovoid, 1.5–2 cm. Gravelly or sandy soil, oak woodlands, grassy areas, open areas, 5250–8500 ft. Hidalgo.

MUILLA S. Watson ex Benth. & Hook. – Goldenstar, muilla

Muilla lordsburgana P. J. Alexander—Lordsburg muilla. Leaves 1–2, basal, withering soon after an-thesis, aboveground portion 10–19 cm × 1.5–2 mm, drying to ± 1 mm wide, ± succulent, weakly chan-neled; scape (1–)2.5–7(–9) cm; umbels 2–6-flowered; perianth white to pale lavender adaxially, abaxially white to pale lavender with a prominent dark green midvein, margins of the midvein often purple, the purple color rarely extending outward 1/2+ of the distance to the margins of the tepals, tube 0.2–0.5 mm, lobes 5.5–7 mm; filaments dilated their entire length, distinct but their margins appressed to slightly overlapping and forming a tube, apex rounded, truncate, or retuse, with a narrow introrse acuminate extension holding the anther; ovary superior, broadly ovate, 1.6–2.2 mm; style and stigma 2.5–3.8 mm; capsule loculicidal, globose, 3-lobed, 8.5–10.5 mm. The report of *Muilla coronata* Greene from Hidalgo County is almost certainly *M. lordsburgana*. These taxa can be distinguished as follows (Alexander, 2020). **R & E**.

1. Style + stigma ≤ 2 mm; anthers ≤ 0.9 mm; capsule ≤ 7 mm; seeds ≤ 3 mm…M. coronata
1'. Style + stigma ≥ 2.5 mm; anthers ≥ 1.3 mm; capsule ≥ 8.5 mm; seeds > 4 mm…**M. lordsburgana**

TYPHACEAE – CATTAIL FAMILY

C. Barre Hellquist

Perennial, monoecious, rhizomatous, emergent or floating; leaves basal and cauline, mostly 2-ranked; sheaths open, tapering into blade, auriculate or nonauriculate; blades flat or keeled, apex acute, obtuse, or retuse; inflorescence 1, terminal, erect, emergent or floating, spikelike or globose; staminate flowers above pistillate flowers, deciduous, with a persistent aril; flowers unisexual; staminate flowers stipitate or sessile, stamens 1 to several; pistillate flowers hypogynous, sessile to stipitate, pistil 1; fruit an achene-like drupe, or follicle.

1. Fruits in cylindrical, spikelike heads, brown-cinnamon; stems flattened, twisted, erect…**Typha**
1'. Fruits in globose, burlike heads, green-brown; stems keeled or flattened, floating or erect…**Sparganium**

SPARGANIUM L. – Bur-reed

Herbaceous, monoecious, freshwater, emergent or floating, rhizomatous; leaves flat, plano-convex or keeled, spongy; inflorescences single, terminal, erect or floating; heads globose, sessile or peduncled; pistillate heads subtended by axillary or supra-axillary bracts; staminate flowers sessile, whitish; stamens 2–8, exceeding tepals; pistillate flowers sessile to stipitate, pistil 1, exceeding tepals; stigmas 1–2, white to greenish, linear, ovate, or subcapitate; fruits achenelike drupes, often constricted at or near center, sessile or stipitate; tepals persistent, attached at base in most species; seeds 1–2(3), slender-ovoid.

1. Stigmas 2…**S. eurycarpum**
1'. Stigma 1…(2)
2. Staminate head 1; pistillate heads 8–12 mm diam.…**S. natans**
2'. Staminate heads 2+; pistillate heads 10–35 mm diam.…(3)
3. Stems submersed and floating; leaves flat, not keeled; fruit beak 1.5–2.2 mm…**S. angustifolium**
3'. Stems erect; leaves keeled; fruit beak 2–4.5 mm…**S. emersum**

Sparganium angustifolium Michx.—Narrow-leaf bur-reed. Stems slender, floating, to 2 m; leaves flat or plano-convex, 2–5(–10) mm wide, middle and upper leaves dilated with subinflated base; inflorescence unbranched with 1+ of the 1–3 pistillate heads supra-axillary, 1–3 cm diam., staminate heads 1–4 (–6); flowers with tepals, stigma 1, lance-ovate; fruits usually reddish at base, 3–7 mm, 1.2–1.7 mm thick, constricted at center, beak 1.5–2.2 mm. Acid waters of ditches, lakes, ponds, rivers, 3750–11,275 ft. NC, NW, C.

Sparganium emersum Rehmann—European bur-reed. Stems robust to slender, erect, to 0.8(–2) m, some floating; leaves flat or slightly keeled, slightly dilated at base, 2–12 mm wide; pistillate heads 1–6, supra-axillary, sessile, 16–35 mm diam., the lowest borne 16–65 cm above the base; staminate heads 3–7(–10), contiguous or not; flowers with tepals, stigma 1, linear-lanceolate; fruits green to reddish brown, lustrous, 3–4 mm. Neutral to alkaline waters of ditches, pond, lakes, river shores, 6300–9000 ft. NC, NW, SC.

Sparganium eurycarpum Engelm.—Broad-fruit bur-reed. Stems to 2.5 m; leaves erect, keeled to 2.5 m; pistillate heads mainly 2–6, axillary, peduncled on main rachis, sessile on branches; staminate heads 10–40, on main rachis and branches; flowers with 2 stigmas; fruits straw-colored, darkening with age, body 3-7-sided, 5-10 mm, often nearly as wide; beak straight, 2-4 mm. Marshes, shores of neutral-alkaline or brackish waters, 6125-8300 ft. Otero, Santa Fe.

Sparganium natans L. [*S. minimum* (Hartm.) Wallr.]—Northern or small bur-reed. Stems slender, floating or suberect when stranded, to 60 cm; leaves flat, unkeeled, 1.5-8 mm wide; inflorescence un-ranked, with 1-3(4) axillary; pistillate heads 8-12 mm diam.; staminate head 1, terminal; flowers with 1 stigma, lance-ovate; fruit dark green or brown, subsessile, body ellipsoid to obovoid, slightly constricted at center, 3-3.5 mm. Acid, neutral, and slightly alkaline waters of ponds, lakes, ditches, 7250 ft. San Juan.

TYPHA L. - Cattail

Herbs in fresh, brackish, or slightly saline water, emergent from horizontal rhizomes; leaves linear, per-sistent, twisted, slightly oblanceolate, plano-convex or concavo-convex; inflorescence with staminate scales shorter than or exceeding flowers; pistillate spikes brown, orange-brown, or cinnamon at spike surface; flowers with pistillate hairs exceeded by the stigmas.

1. Staminate heads usually contiguous with pistillate heads; pistillate heads dark brown, 24-36 mm thick…
   T. latifolia
1'. Staminate heads separated from pistillate heads; pistillate heads medium brown to cinnamon-brown, 13-25 mm thick…(2)
2. Pistillate spikes medium brown; leaf sheath with terminal membranous auricles; flowering stems 2-3 mm thick at inflorescence…**T. angustifolia**
2'. Pistillate spikes cinnamon to orange-brown; leaves tapered to blades, occasionally with membranous auricles; flowering stems 3-4 mm thick at inflorescence…**T. domingensis**

Typha angustifolia L.—Narrowleaf cattail. Stems 0.75-1.5(-3) m, flowering shoots 2-3 mm near the spike; leaves 3-8 mm broad, slightly convex on back, leaf sheath with terminal membranous auri-cles; staminate spikes 3-15 cm, pistillate spikes 6-20 cm, 5-6 mm thick, medium brown, separated by 1-8(-12) cm, usually above the height of the leaves; pollen single or in clumps. Fresh, brackish, or slightly saline wetlands, 3150-5850 ft. Hybrids with *T. latifolia* = *T.* ×*glauca* Godr. NE, NC, NW, EC, C, WC, SE.

Typha domingensis Pers.—Southern cattail. Stems 1.5-4 m, flowering shoot 3-4 cm near the spike; leaves 6-18 mm broad, flat, leaf sheath with persistent membranous auricles; staminate spikes 6-35 cm, 5-6 mm thick, cinnamon-brown; pistillate spikes separated by (0-)1-8 cm, usually above the height of the leaves; pollen grains single; pistillate flowers 2 mm. Fresh or brackish wetlands, 3000-6650 ft. NC, NW, EC, C, WC, SE, SC, SW.

Typha latifolia L.—Broadleaf cattail. Stems stout, 1-3 m, flowering shoots 3-7 mm thick near the spike; leaves 6-23 mm broad, flat; staminate spikes 7-13 cm, pistillate spikes 2.5-20 cm, 12-35 mm thick, dark brown, 24-36 mm thick in fruit, usually contiguous, rarely with 4(-8) cm separation, usually at the same height as the leaves; pollen in tetrads. Fresh to slightly brackish wetlands, 3100-7325 ft. Hybrids with *T. angustifolia* = *T.* ×*glauca* Godr. Widespread except far E.

ULMACEAE - ELM FAMILY

Don Hyder

ULMUS L. - Elm

Trees, less often shrubs; bark smooth to deeply fissured or platy and flaky; leaves sometimes tardily de-ciduous, blades ovate to obovate or elliptic, margins serrate to doubly serrate, venation pinnate; flowers small, bisexual; calyx 3-9-lobed, persistent; petals absent; stamens 3-9; ovary superior, styles persistent, deeply 2-lobed; fruit samaras, usually flattened (Sherman-Broyles et al., 1987).

1. Leaves twice-serrate, mostly 7-15 cm…**U. americana**
1'. Leaves once-serrate, mostly 2-5 cm…(2)

   2. Flowers and often fruit appearing in the spring or with the leaves; bark furrowed...**U. pumila**
   2'. Flowers and fruit appearing in the fall, after the leaves are quite mature; bark platy, lacy...**U. parvifolia**

Ulmus americana L.—American elm. Trees, 21–35 m; bark light brown to gray, deeply fissured or split into plates; leaves glabrous to pubescent, blades oval to oblong-obovate, 7–14 × 3–7 cm, margins doubly serrate; inflorescence a fascicle, < 2.5 cm, flowers and fruits drooping on elongate pedicels; flower calyx shallowly lobed, slightly asymmetric, lobes 7–9; samaras yellow-cream when mature, ovate, ca. 1 cm, narrowly winged; seeds thickened, not inflated; cultivated shade trees, perhaps escaping and persisting (most have been killed by Dutch elm disease), 4000–6000 ft. San Juan.

Ulmus parvifolia L.—Chinese elm. Trees, to 25 m; bark olive-green to gray, shedding in irregular, tan to orange plates, distinctly lacy; leaves glabrous or sparsely pubescent with short hairs, blades ellip-tic to ovate-obovate, (3.5–)4–5(–6) × 1.5–2.5 cm, margins mostly singly serrate (some doubly serrate); inflorescence a fascicle; flower calyx reddish brown, deeply lobed, lobes (3)4–5, glabrous; samaras green to light brown, elliptic to ovate, ca. 1 cm, not winged; seeds thickened, nearly filling samara. Commonly cultivated as an ornamental but not found in the wild; 1 record in Aztec, 5650 ft. San Juan.

Ulmus pumila L.—Siberian elm. Trees, 15–30 m; bark gray to brown, deeply furrowed with interlac-ing ridges, wood brittle; leaves glabrous, blades narrowly elliptic to lanceolate, 2–6.5 × 2–3.5 cm, margins singly serrate; inflorescences tightly clustered fascicles, 0.5 cm, flowers and fruits not pendulous, sessile; flower calyx shallowly lobed, lobes 4–5, glabrous; samaras yellow-cream, orbiculate, 10–14 mm diam., broadly winged; seeds thickened, not inflated; widespread throughout the state along roads, fencerows, and other disturbed sites, 3275–8700 ft. Widespread.

URTICACEAE – NETTLE FAMILY

David E. Boufford

Herbs, annual or perennial, or subshrubs, dioecious, monoecious, or hermaphroditic, usually pubescent, some species with stinging hairs, deciduous; leaves opposite or alternate, simple, petiolate; blades with linear or rounded cystoliths; inflorescences axillary or terminal, paniculately or racemosely arranged cymes or spikelike; staminate flowers usually pedicellate, tepals 4 or 5, white or green; stamens 4 or 5, equaling tepals in number; filaments inflexed in bud, reflexing suddenly at maturity to eject pollen ex-plosively; anthers basifixed, dehiscing by longitudinal slits; pistillode 1; pistillate flowers usually sessile, tepals 2–4, hypogynous, greenish or reddish, distinct or connate; staminodes present or absent, pistil 1, 1-locular; ovule 1; style present or stigma sessile; bisexual flower tepals 4, stamens 4, pistil 1; fruit achenes, free or surrounded by persistent accrescent perianth (Boufford, 1992, 1997).

   1. Plants with stinging hairs; tepals of pistillate flowers 4, distinct, inner 2 equal to achene, outer 2 smaller...
      Urtica
   1'. Plants without stinging hairs; tepals of bisexual and pistillate flowers 3 or 4, distinct, equal in length...(2)
   2. Leaves opposite, rarely alternate, margins dentate; inflorescences spikelike, usually > 2 cm...**Boehmeria**
   2'. Leaves alternate, margins entire; inflorescences axillary clusters, < 1 cm...**Parietaria**

BOEHMERIA Jacq. – False nettle

Boehmeria cylindrica (L.) Sw. [*B. drummondiana* Wedd.; *B. cylindrica* var. *drummondiana* (Wedd.) Wedd.]—False nettle. Perennials, 0.1–1.6 m, without stinging hairs; leaves opposite or nearly opposite, rarely alternate; blades elliptic or lanceolate to broadly ovate, 5–18 × 2–10 cm, margins dentate, nearly glabrous on both surfaces or abaxially densely short-pilose or puberulent, adaxially scabrous; inflores-cences axillary, spikelike, often with tufts of reduced leaves at apex; flowers in remote or crowded clusters of 1 to few staminate and several pistillate flowers, rarely staminate and pistillate flowers on different plants. Wet to moist swamps, bogs, marshes, wet meadows, ditches, 3800–6060 ft. Chaves, Taos.

PARIETARIA L. - Pellitory

Annuals or perennials, sparsely to densely pubescent with hooked and straight, nonstinging hairs on all parts, stinging hairs absent; stems often branched from base, erect, ascending, or decumbent; leaves alternate, deltate or orbiculate to narrowly elliptic or lanceolate, margins entire, stipules absent; cystoliths rounded; inflorescences axillary, loose clusters of flowers, < 1 cm; flowers bisexual, staminate, or pistillate; proximal flowers usually bisexual and staminate; distal flowers pistillate; involucral bracts linear to lanceolate, without hooked hairs; tepals 4, distinct, ascending, lacking hooked hairs; stamens 4; style persistent or not; stigma tufted, deciduous; cypselae stipitate, ovoid, acute or mucronate (style base sometimes persisting as apical or subapical mucro), loosely enclosed by tepals.

1. Leaf blades narrowly to broadly ovate, oblong, orbiculate, or reniform, base rounded; lowest pair of lateral veins arising at junction of blade and petiole; involucral bracts usually > 2 times length of achene...**P. hespera**
1'. Leaf blades narrowly to broadly elliptic, lanceolate, oblong, or ovate, base narrowly cuneate; lowest pair of lateral veins arising above junction of blade and petiole; involucral bracts usually < 2 times length of achene...**P. pensylvanica**

Parietaria hespera B. D. Hinton—Rillita pellitory. Annuals, 30-55 cm; leaf blades narrowly to broadly ovate, less frequently oblong or nearly orbiculate, 10-30 × 5-15 mm, longer than wide, base broadly cuneate, rounded, or truncate, apex distally rounded or acuminate to acute or obtuse; involucral bracts 1-4.5 mm, flower tepals erect, loosely connivent at maturity, ca. 2-2.8 mm, apex acute. New Mexico material belongs to **var. hespera**. Chaparral, deserts, roadsides, sand dunes, often in shaded and moist places, 4220-6500 ft. SC, SW.

Parietaria pensylvanica Muhl. ex Willd.—Pennsylvania pellitory. Annuals, 40-60 cm; stems simple or freely branched, decumbent, ascending, or erect; leaf blades narrowly to broadly elliptic, lanceolate, oblong, or ovate, (1-)2-9 × 0.4-3 cm, base narrowly cuneate, apex acuminate to long-attenuate or obtuse to rounded; proximal pair of lateral veins arising above junction of blade and petiole; involucral bracts 1.8-5 mm, usually < 2 times length of achene, flower tepals 1.5-2 mm, shorter than bracts; cypselae light brown, symmetric, 0.9-1.2 × 0.6-0.9 mm, apex obtuse, mucro apical; stipe straight, short-cylindrical, centered, basally dilated. Dry ledges, talus slopes, waste and shaded places, primarily in neutral to basic soils, 3300-8700 ft. WC, C, SE, SC, SW, Sandoval, San Miguel.

URTICA L. - Stinging nettle

Urtica dioica L.—Stinging nettle. Perennial, rhizomatous, 0.5-3 m, with stinging hairs; stems simple or branched, erect or sprawling; leaf blades elliptic, lanceolate, or narrowly to broadly ovate, 6-20 × 2-13 cm, base rounded to cordate, margins coarsely serrate, sometimes doubly serrate, apex acute or acuminate; cystoliths rounded; inflorescences paniculate, pedunculate, elongate; flowers unisexual, staminate and pistillate on same or different plants, staminate ascending, pistillate lax or recurved; pistillate flower outer tepals linear to narrowly spatulate or lanceolate, 0.8-1.2 mm, inner tepals ovate to broadly ovate, 1.4-1.8 × 1.1-1.3 mm; cypselae ovoid to broadly ovoid, 1-1.3(-1.4) × 0.7-0.9 mm. Disturbed sites, upper piñon pine to mixed conifer, 5165-10,800 ft.

1. Stems glabrous or strigose, with few stinging hairs; abaxial surface of leaf blade glabrous or puberulent...**subsp. gracilis**
1'. Stems softly pubescent, with stinging hairs; abaxial surface of leaf blade tomentose to moderately strigose...**subsp. holosericea**

subsp. gracilis (Aiton) Selander—Stems glabrous or strigose, with few stinging hairs; abaxial surface of leaves glabrous or puberulent, with few stinging hairs. Widespread except EC, SW.

subsp. holosericea (Nutt.) Thorne—Stems soft-pubescent, with stinging hairs; abaxial surface of leaves sparsely to densely tomentose to moderately strigose, soft to the touch, with stinging hairs. WC, C.

VERBENACEAE - VERVAIN FAMILY

Susan C. Barber

Plants annuals or perennials; herbs or shrubs; aromatic or not; stems 4-sided or terete; leaves simple, opposite or whorled, venation pinnate, entire or toothed to pinnatifid, stipules absent; inflorescence of spikes, cymes, panicles, or racemes, terminal or axillary, bracts present; flowers perfect; corollas zygomorphic, petals 5 or rarely 4; sepals 5 or 4, fused; stamens 4; pistils 1, compound, carpels 2, ovaries superior, locules 4 or 2; fruit a nutlet or drupe; seeds 4, 2, or 1 (Allred et al., 2020; Nesom, 2010b; Marx et al., 2010).

1. Plants woody…(2)
1′. Plants herbaceous…(4)
2. Sepal lobes indistinct; ovary drupelike, usually fleshy, 2-celled, not splitting at maturity…**Lantana**
2′. Sepals 2-4-lobed; ovaries of 2 nutlets, splitting at maturity…(3)
3. Flowers on elongate flowering rachis; leaf blades without glands…**Aloysia**
3′. Flowers imbricate, not elongating in fruit; leaf blades with inconspicuous glands…**Lippia**
4. Inflorescence racemose; plants annual…**Bouchea**
4′. Inflorescence spicate; plants perennial…(5)
5. Fruit of 2 nutlets; corolla 4-lobed, distinctly 2-lipped…**Phyla**
5′. Fruit of 4 nutlets; corolla 5-lobed, slightly bilateral…(6)
6. Spikes, including bracts, 10-20 mm wide; calyx 2 times longer than nutlets, nutlets completely enclosed within closed calyx; styles 6-20 mm…**Glandularia**
6′. Spikes, including bracts, 3-8 mm wide; calyx equal to length of nutlets, nutlets partially enclosed within open calyx; styles 2-3 mm…**Verbena**

ALOYSIA Paláu - Beebrush

Branched shrubs, aromatic; leaves opposite or whorled with small fascicles, blades oblong, elliptic, ovate, lanceolate, obovate, or suborbicular, entire or toothed, adaxial surfaces strigillous-strigose to hirsutulous-strigose; inflorescence in elongate spikes or racemes; bractlets present, lanceolate to acuminate; corollas white or tinged with violet or pale blue; calyx of 4 sepals, villous or hirsute; fruit a 2-celled schizocarp; seeds 2.

1. Leaf blades ovate or rounded, teeth rounded, abaxial surfaces tomentose, crenate-serrate; leaves opposite; plants to 1.5 m…**A. wrightii**
1′. Leaf blades narrowly oblong or elliptic to lanceolate-oblong, abaxial surfaces strigulose, entire or denticulate; leaves opposite, sometimes whorled with fascicles of small leaves; plants to 3 m…**A. gratissima**

Aloysia gratissima (Gillies & Hook.) Tronc.—Whitebrush. Plants to 3 m, branches numerous, erect; stems grayish-puberulent, often with spiny tips; leaves sessile or subsessile, blades lanceolate to narrowly oblong or elliptic, entire or denticulate, sometimes emarginate, abaxial surfaces very densely puberulent and resinous-punctate; flowers crowded; calyx densely pubescent, glandular, 2-4 mm; corolla limbs 3 mm wide, white or tinged with violet to bluish, throat loosely villous. Sandy or limestone soils, arroyos and rocky slopes. Otero. No records seen.

Aloysia wrightii (A. Gray ex Torr.) A. Heller—Wright's beebrush. Plants to 1.5 m, branches numerous; stems densely grayish-puberulent, becoming glabrous with age; leaves petiolate or subsessile, blades ovate to ovate-orbicular, crenate to crenate-serrate; abaxial surfaces tomentose; flowers crowded; calyx 2-3 mm; corollas white or bluish, limbs 2 mm wide, puberulent on the outside. Rocky or gravelly slopes in arroyos and canyons, 3150-8500 ft. C, WC, SE, SC, SW.

BOUCHEA Cham. - Bouchea

Plants annual, densely pubescent, canescent, puberulent, or glabrous; leaves opposite, often with several smaller ones in axils, entire or serrate, sessile or petiolate; inflorescence a raceme; flowers pink, rose, lavender, purple, or blue; corolla tubes 8.5-15 mm, glabrous; fruit a schizocarp, linear, beaked, splitting at maturity; seeds 2.

1. Leaves petiolate, serrate, elliptic; corolla tubes 8.5–10 mm…**B. prismatica**
1'. Leaves sessile or subsessile, entire, linear to lanceolate; corolla tubes ca. 15 mm…B. linifolia

Bouchea linifolia A. Gray ex Torr.—Groovestem bouchea. Plants low and shrubby, < 1 m; stems conspicuously ridged, glabrate; leaves sessile, blades linear to narrowly lanceolate, 0.7–4.4 × 0.15–0.4 cm, entire, often drying revolute; inflorescence 5–12 × 0.7–1.5 cm; flowers loose, peduncles short or obsolete; bractlets setaceous-subulate, 1–2 mm; calyx slightly curved; corolla lavender, tubes 15 mm. Plains. Otero. No records seen; supposedly last collected in New Mexico in 1853 and likely collected in Texas or Chihuahua (Standley, 1910).

Bouchea prismatica (L.) Kuntze—Prism bouchea. Sprawling herbs to 0.4 m, ascending; stems puberulent and often canescent; leaves petiolate, 11–25 mm, blades mostly elliptic, 10–40 × 8–30 mm, serrate, broad teeth below the middle; inflorescence 5–15 cm, to 1 cm wide; flowers closely appressed to rachis, short-pedunculate; bractlets lanceolate-subulate, 4–6 mm; calyx slightly curved; corolla pink, rose, lavender, purple, or blue, tubes 8.5–10 mm. Rhyolitic soils, grassland with scattered juniper, 1 record, 5060 ft. Hidalgo.

GLANDULARIA J. F. Gmel. – Mock vervain

Herbs, annual or perennial; stems often semiwoody, sometimes rooting at nodes, variously pubescent; leaves opposite, simple, entire, serrate, lobed, or pinnatifid, pubescence stiff, often glandular; inflorescence a spike, corymbose at anthesis, terminal; flowers perfect, zygomorphic; calyx tubular, 5-toothed by vein extension; corollas 5-lobed, blue, purple, lavender, pink, or white, fused, tubes extending beyond the calyx; fruit a schizocarp, separating into 4 mericarps/nutlets.

1. Ultimate segments of the leaf lobes 0.3–1 mm wide; herbage finely appressed-hairy; weedy…**G. aristigera**
1'. Ultimate segments of the leaf lobes > 1 mm wide; herbage with spreading hairs; native…(2)
2. Calyx 5–6 mm; leaves 1–3 cm; corolla limb 3–5 mm wide…**G. pumila**
2'. Calyx 8.5–13 mm; leaves 2.5–9 cm; corolla limb 8–15 mm wide…(3)
3. Corolla tube slightly surpassing the calyx…**G. gooddingii**
3'. Corolla tube 1.5–3 times length of calyx…(4)
4. Calyx eglandular to sparsely glandular with sessile to subsessile glands…(5)
4'. Calyx densely stipitate-glandular…(8)
5. Floral bracts shorter to longer than the calyx; corolla tube 9–14 mm, limb 9–15 mm diam. [to be expected in NE]…G. bipinnatifida
5'. Floral bracts shorter than the calyx; corolla tube 7–13 mm, limb 6–10(–12) mm diam.…(6)
6. Stems ascending to erect, 20–80 cm; calyx 8–10 mm; corolla mostly pink to purplish pink, rarely violet; tube 10–13 mm, limb 7–11(–12) mm diam.…**G. chiricahensis**
6'. Stems decumbent to ascending, or ascending-erect, 12–50 cm; calyx 5–13 mm; corolla rose-pink, purplish to blue, or lavender; tube 7–20 mm, limb 6–15 mm diam.…(7)
7. Calyx stipitate-glandular, 9–13 mm, spreading-hairy; corolla rose-pink to purplish, lavender, or violet, tube 15–20 mm, limb 11–15 mm diam.…**G. canadensis**
7'. Calyx eglandular or sparsely sessile/subsessile-glandular, 5–7 mm, hirsute and strigose; corolla purplish to blue or lavender, tube 7–11, limb 6–9 mm diam.…**G. latilobata**
8. Stems densely hirsute to pilose or hirsutulous, some hairs deflexed; corolla pink to purplish, drying pink, tube 12–15 mm, limb 8–12 diam.…**G. pubera**
8'. Stems stiffly hirsute, spreading at right angles; corolla bright pinkish when fresh, drying purple; tube 8–12 mm, limb 7–12 diam.…**G. wrightii**

Glandularia aristigera (S. Moore) Tronc. [*G. tenuisecta* (Briq.) Small; *Verbena aristigera* S. Moore]—Mock vervain. Annual or perennial, procumbent to decumbent or ascending, 10–30 cm, pilose to strigose or glabrate; leaves 2–4 cm, sparsely hairy, blades 2-pinnatifid, lobes narrowly linear to 1 mm wide; calyx lobes 7–9 mm; corolla rose-purple to violet or white, tube 9–13 mm, limb 7–11 mm. Disturbed sites, 3900–4300 ft. Doña Ana, Hidalgo. Introduced.

Glandularia bipinnatifida (Nutt.) Nutt. [*Verbena bipinnatifida* Nutt.]—Dakota mock vervain. Annual or perennial, prostrate, decumbent to ascending, often rooting at the lower nodes, branched from base, pubescent, not glandular, 5–60 cm; stems hispid-hirsute; leaves petiolate, deeply incised to bipinnatifid, lobes linear to oblong, margins often revolute, hirsute with appressed hairs, 2–6 cm × 1–6 mm; inflorescence a pedunculate spike, short at anthesis, elongating at maturity; calyx 7–10 mm, hispid-hirsute; corolla pink to lavender or purple, limb 7–10 mm wide; 4 nutlets, 3 mm. Sandy soils in desert scrub, grasslands, piñon-juniper to ponderosa pine communities. Potential habitat in NE.

Glandularia canadensis (L.) Small [*Verbena canadensis* (L.) Britton]—Rose mock vervain. Perennial, decumbent to ascending, rooting at the lower nodes; stems glabrate or irregularly spreading-hirsute; leaves variable, ovate to ovate-oblong, 2.5–9 × 1.5–4 cm, incised to incised-pinnatifid to 3-cleft, appressed-hirsute or glabrate; inflorescence a spike, elongating at maturity; ciliate bractlets shorter than or equaling the calyx; calyx 10–13 mm, glandular-hirsute; corolla pink, rose, magenta, blue, lavender, or purple, tube 2 times longer than the calyx, limb 7–15 mm wide. Hillsides and forest openings, 5000–6000 ft. Bernalillo, Grant, Guadalupe. Introduced.

Glandularia chiricahensis Umber—Chiricahua Mountain mock vervain. Perennials, ascending or erect, 20–80 cm; stems with stiff white hairs 1–2 mm; leaves petiolate, 3.5–7 cm, 3-cleft to pinnatifid, hairs short, white, appressed; inflorescence a spike, barely elongating in fruit; bractlets slightly shorter than the calyx, with stiff white hairs; calyx 8–10 mm, eglandular, rarely with glands; corolla various shades of purple, limb 3 times longer than the calyx 8–10 mm wide; 4 nutlets, black, 2.5 mm. Juniper-oak woodlands, mixed conifer communities, 5575–8500 ft. Grant, Hidalgo.

Glandularia gooddingii (Briq.) Solbrig [*Verbena gooddingii* Briq.]—Southwestern mock vervain. Plants erect or decumbent-ascending, often forming mats; stems densely white-villous, often glandular; leaves 3–5 cm, tapering to a short petiole, 3-cleft, divisions coarsely toothed, villous-hirsute; inflorescence a spike, pedunculate, somewhat elongating at maturity; bractlets slightly shorter than the calyx, lanceolate or acuminate, villous-hirsute; calyx 8.5–11 mm, glandular; corolla blue, pink, or lavender, tube slightly surpassing the calyx, limb 8–12 mm wide. Grasslands, desert scrub, piñon-juniper to ponderosa pine communities, 4000–6655 ft. Hidalgo.

Glandularia latilobata (L. M. Perry) G. L. Nesom—Vervain. Perennials; stems decumbent to erect, 12–40 cm, pubescent; leaves 2–4 cm, blades 1–2-pinnatifid, pubescent; calyx 5–7 mm, hirsute or strigose; corollas purple to lavender, tube 7–11 mm, limb 6–9 mm wide. Grasslands, piñon-oak to ponderosa pine forests, 5300–7750 ft. SC, SW.

Glandularia pubera (Greene) G. L. Nesom [*G. bipinnatifida* (Nutt.) Nutt. var. *brevispicata* Umber]—Mock vervain. Perennials; stems ascending to erect, 15–40 cm, hirsute to pilose-hirsute; leaves 4–5 cm, blades 1–3-pinnatifid, eglandular; calyx 8–10 cm, with soft hairs; corolla pink to purplish pink, tube 12–15 mm, limb 8–12 mm wide. Grasslands, piñon-juniper woodlands, ponderosa pine forests, disturbed sites, 4200–8000(–8800) ft. NC, C, WC, SC, SW. Eddy.

Glandularia pumila (Rydb.) Umber [*Verbena pumila* Rydb.]—Pink mock vervain. Stems decumbent-ascending, hirsute, often finely glandular; leaves 1.5–3 cm, triangular, truncate at base, cuneate, 3-parted to -lobed, appressed-hirsute; inflorescence pedunculate, somewhat compact, not elongating; bractlets almost as long as the calyx; calyx ca. 6 mm, with various pubescence, sometimes glandular; corolla inconspicuous, pink or lavender to blue, tube slightly surpassing the calyx, 8–10 mm, limb 3–5 mm wide. Chihuahuan desert scrub, desert grasslands, piñon-juniper woodlands, 3285–7200 ft. NE, SW.

Glandularia wrightii (A. Gray) Umber [*Verbena wrightii* A. Gray]—Wright's vervain. Perennials; stems decumbent to erect, (15–)20–60 cm, hirsute to hirtellous; leaves 2–5 cm, blades 1–2-pinnatifid; calyx 6–8 mm, sparsely hairy; corollas pink to purple or lavender, tube 9–12 mm, limb 7–12 mm wide. Desert scrub, desert grasslands, piñon-juniper woodlands, ponderosa pine, mixed conifer forests, 3050–8750 ft. Widespread except NW, WC, SW.

LANTANA L. – Lantana

Lantana camara L.—Common lantana. Shrubs; stems hairy; leaf blades broadly lanceolate to ovate, 3–8(–16) cm, margins serrate to crenate or dentate, surfaces with short hairs; inflorescence hemispherical; corollas yellow to reddish, aging reddish orange, corolla tube 4-12 mm. Washes, disturbed sites; cultivar, 1 escaped plant, ca. 3900 ft. Doña Ana.

LIPPIA L. – Lippia

Lippia graveolens Kunth—Redbrush lippia. (Ours) slender aromatic shrubs to 3 m; stems slender, villous, densely resinous-punctate; petioles 2-20 mm, blades oblong to ovate-oblong or elliptic, 1.6-5 × 5-30 mm, crenate, sometimes prolonged into the petiole, adaxial surfaces reticulate-rugose, abaxial surfaces puberulent and resinous-glandular; bractlets 4-ranked; corollas yellowish or white with yellow eyes, tubes 3-6 mm. Rocky hills in desert scrub. No records seen since 1851.

PHYLA Lour. – Frog-fruit

Perennial herbs, not aromatic, prostrate or procumbent, often rooting at the nodes; stems glabrate to appressed-strigose with grayish malpighian hairs; leaves opposite, 15-75 × 2-30 mm, variously dentate, at least apically, surfaces strigose; inflorescence axillary, pedunculate, a dense subglobose to cylindrical spike; flowers sessile, small, borne singly in axils of cuneate-obovate bractlets; corollas 4-lobed, distinctly 2-lipped; fruit of 2 nutlets.

1. Leaf blades widest below middle, toothed below middle to apex…(2)
1'. Leaf blades widest toward apex, toothed only at apex…(3)
2. Blades 2-4 times longer than wide, margins serrate with teeth pointed toward the apex, veins not impressed adaxially; floral bracts 2.7-3.2 mm; calyx with scattered appressed hairs…**P. lanceolata**
2'. Blades 1.5-2 times longer than wide, margins dentate with teeth pointed outward at 90° to slightly antrorse, veins shallowly impressed adaxially; floral bracts 1.8-2.5 mm; calyx with hooked hairs…**P. fruticosa**
3. Leaf blades with 1-4 remote teeth above the middle; spikes elongating to 2 cm × 7-12 mm; bractlets 5 mm, acuminate, scarious…**P. cuneifolia**
3'. Leaf blades with numerous pairs of teeth; spikes elongating to 2.5 cm × 6-9 mm; bractlets 2-3 mm, mucronate to acuminate, not scarious…**P. nodiflora**

Phyla cuneifolia (Torr.) Greene [*Lippia cuneifolia* (Torr.) Steud.]—Wedgeleaf frog-fruit. Herbs procumbent, to 1 m, often rooting at the nodes; leaves sessile, rigid, thick-textured, often with fascicles, blades 1.5-2 cm × 2-8 mm, teeth above the middle, with sparsely appressed hairs; inflorescence globose early, elongating to cylindrical, to 2 × 0.8-1.2 cm; peduncles 8-50 mm; bractlets obovate, long-acuminate, 5 mm; corolla white or purple, tube 4-5 mm, limb 2-4.5 mm wide. Lake margins, stream beds, wetlands, playas, 3600-7215 ft. E, NC, C, SC, SW. McKinley.

Phyla fruticosa (Mill.) K. Kenn. ex Wunderlin & B. F. Hansen—Diamond-leaf frog-fruit. Procumbent herbs, 20-60 cm, nodal branches erect; leaves triangular-ovate to rhomboid, 1.5-7 cm × 8-20 mm, margins coarsely dentate to serrate, 5-9 teeth per side; inflorescence 8-16 mm; calyx with hooked hairs; corolla white to pale purple, throat yellow or purple, limb 1.5-2.5 mm across. Ponds, playas, ditches, 3150-5000 ft. Eddy, San Miguel, Socorro. Introduced.

Phyla lanceolata (Michx.) Greene [*Lippia lanceolata* Michx.]—Northern frog-fruit. Stems procumbent or ascending, to 0.6 m, often rooting at the nodes; leaves narrowed to a cuneate base, blades 1.8-7.5 cm × 5-30 mm, venation conspicuous, appressed-strigillose; inflorescence globose early, elongating to cylindrical, to 35 mm, 40-90 × 5-7 mm; bractlets obovate, acute, imbricate, 3 mm; corolla pale blue to purplish or white. Moist soils of river bottoms and lake margins, 3135-4920 ft. Chaves, Eddy, Socorro.

Phyla nodiflora (L.) Greene [*Lippia nodiflora* (L.) Michx.]—Common frog-fruit. Stems prostrate or ascending, rooting at the nodes; leaves tapering into a petiole, blades 1.7-2 cm × 6-25 mm, spatulate or obovate, sharply serrate above the middle, glabrous or strigillose-puberulent; inflorescence globose early, elongating to cylindrical and becoming greatly elongate in fruit, to 12 cm; corolla rose to purple or white. Moist soils, lake margins, river bottoms, pastures, ditches, 3000-7500 ft. Widely scattered.

VERBENA L. – Vervain

Herbs, annuals or perennials; prostrate, procumbent, ascending, or erect; stems glabrous or variously pubescent; leaves opposite, dentate or variously lobed, incised, or pinnatifid; inflorescence a terminal spike, usually densely flowered, but sometimes greatly elongate and sparsely flowered; flowers perfect, zygomorphic, 5-lobed, in bract axils, corolla salverform or funnelform, blue, purple, lavender, pink, or white, fused; calyx tubular, 5-angled, 5-ribbed; stamens 4, didynamous, inserted in the upper 1/2 of the corolla tube, usually included; anthers ovate; fruits dry, schizocarps splitting into 4, 1-seeded nutlets.

1. Leaves linear, usually unlobed, margins not toothed…**V. perennis**
1'. Leaves pinnatifid or margins deeply toothed to regularly coarsely crenate to serrate…(2)
2. Leaves not lobed or deeply toothed, margins regularly coarsely crenate to serrate…(3)
2'. Leaves pinnatifid to deeply toothed…(9)
3. Stems and sepals glandular…(4)
3'. Stems and sepals eglandular…(5)
4. Stems sparsely hirsute-pilose or bristly-hirsute to bristly-pilose, eglandular to very sparsely stipitate-glandular; leaves evenly distributed along the stems; corolla limb 4–6 mm diam.…**V. livermorensis**
4'. Stems densely hirsutulous to hirtellous and minutely stipitate-glandular; leaves often clustered at base; corolla limb (5–)6–9 mm diam.…**V. hirtella**
5. Spikes slender, elongate with loose flowers; fruits not overlapping; bractlet 1/2 length of calyx…(6)
5'. Spikes with dense flowers; fruits overlapping; bractlet > 1/2 length of calyx…(7)
6. Stems glabrous with occasional hairs; leaves strigillose, middle and lower leaves cleft or incised…**V. halei**
6'. Stems hispidulous; leaves scabrous, middle and lower leaves serrate-dentate…**V. scabra**
7. Leaves lanceolate to ovate or oblong, gradually becoming acuminate, glabrous or slightly pubescent, serrate or incised, sometimes hastately 3-lobed at base; bractlet shorter than the calyx; calyx 2.5–3 mm; corolla purple to blue, limb 3.5–4 mm wide…**V. hastata**
7'. Leaves oblong to broadly elliptic, or broadly ovate or ovate-orbicular, hirsute or spreading-hairy, serrate-dentate, incised, or laciniate, shorter or longer than the calyx; calyx 3.5–6 mm; corolla deep purple or sometimes pinkish or white; limb 5–10 mm wide…(8)
8. Leaves short-petiolate, the blades narrowly ovate or narrowly elliptic; floral bracts equaling or slightly longer than the calyx; corolla limb 4–6 mm wide…**V. macdougalii**
8'. Leaves sessile, the blades broadly ovate or broadly elliptic to ovate-orbicular; floral bracts slightly shorter than the calyx; corolla limb 5–10 mm wide…V. stricta
9. Midstem leaves mostly 0.5–1.5 cm…**V. gracilis**
9'. Midstem leaves mostly 2–8 cm…(10)
10. Fruiting spikes dense with overlapping fruits; floral bracts ± leaflike and longer than the fruits…**V. bracteata**
10'. Fruiting spikes and floral bracts not as above…(11)
11. Basal and proximal cauline leaves persistent and present at flowering; midstem and distal cauline leaves reduced in number and size…(12)
11'. Basal and proximal cauline leaves deciduous at flowering; midstem and distal cauline leaves evenly distributed and mostly even-sized…(14)
12. Stems decumbent-ascending; leaves plicate, veins whitish beneath…**V. plicata**
12'. Stems erect; leaves not plicate, veins mostly greenish gray (possibly whitish in *V. xylopoda*)…(13)
13. Lower and midstem leaves sessile nor nearly so; SE…**V. canescens**
13'. Lower and midstem leaves petiolate; SW…**V. xylopoda** (in part)
14. Stems spreading-hairy, eglandular…**V. menthifolia**
14'. Stems hispidulous, hispid, or hirsute to villous, glandular…(15)
15. Stems hirsute, stipitate-glandular; corolla tube 2.5–4 mm, limb 1.5–2.5 diam.…**V. neomexicana**
15'. Stems hispidulous to hirsutulous, sessile to short-stipitate-glandular; corolla tube 4–5 mm, limb 4–8 mm diam.…**V. xylopoda** (in part)

Verbena bracteata Lag. & Rodr. [*V. bracteosa* Michx.]—Bigbract verbena. Perennial herbs, prostrate or decumbent (can flower as an annual), 9–55 cm; stems coarsely hirsute; leaves hirsute, usually 3-lobed, the central lobe cuneate-obovate and larger than the other 2, rarely pinnately incised, narrowed into a petiole 2–20 mm; inflorescences a terminal spike, sessile, 5–15 cm × 2–10 mm; bractlets 8–15 mm, conspicuous, much longer than the calyx, recurved at maturity, coarsely hirsute; calyx 3–4 mm; corolla blue to lavender or purple, inconspicuous. Weedy waste ground and roadsides, 3030–9550 ft. Widespread.

Verbena canescens Kunth [*V. roemeriana* Scheele; *V. neei* Moldenke]—Gray vervain. Perennial herbs, decumbent-ascending to erect, 15–35 cm, densely hirsutulous to hirsute-villous, minutely stipitate-glandular; leaves oblanceolate to elliptic-oblanceolate, midstem blades 2–5(–6) cm × 2–7(–15) mm, margins coarsely serrate to pinnately lobed with 1–5(–7) pairs of teeth or lobes; spikes 1–3(–5); floral bracts ovate-lanceolate, 4–8 mm; calyx 2.1–3(–4) mm, hirtellous to hirsutulous, minutely stipitate-glandular; corolla blue to purple, tube 2.5–4(–5) mm, 0.5–1(–1.5) longer than the calyx, limb 2.5–4(–6) mm diam. Calcareous substrates, desert scrub, oak-juniper-woodlands. Eddy.

Verbena gracilis Desf. [*V. remota* Benth.; *V. arizonica* A. Gray]—Fort Huachuca vervain. Perennial herbs, prostrate to ascending-erect, 10–25 cm, hirsute with bristly hairs, minutely stipitate-glandular; leaves ovate to ovate-lanceolate, lobed or deeply toothed, midstem blades 5–15(–25) × 3–12 mm; spikes 1–3(–5); floral bracts linear-lanceolate to narrowly lanceolate, 3–5(–8 proximally) mm; calyx 2–2.5 mm, hirsute and hispidulous, minutely stipitate-glandular at least near lobes; corolla purplish to pink, tube 3–3.5 mm, 0.5–1 mm longer than the calyx, limb 2–2.5 mm diam. Rocky sites, desert scrub, desert grassland, pine-oak, 4100–5700 ft. Hidalgo.

Verbena halei Small—Texas vervain. Erect, annuals or weak perennials, 22–90 cm; stems glabrous or nearly so; leaves narrowed into a petiole 3–40 mm, hirsute, especially along the veins, uppermost leaves sparingly dentate to entire, basal and lower leaves oblong to ovate, blades irregularly dentate to incised, 1–2-pinnatifid; inflorescence a paniculate elongate spike; bractlets 1/2 the length of the calyx; calyx 1–4 mm, hirsute and/or glandular; corolla blue. Bitter Lake National Wildlife Refuge, 3500 ft. Chaves.

Verbena hastata L.—Swamp verbena. Erect perennial, 33–140 cm; stems sparsely pubescent; leaves petiolate to 30 mm, blades 40–110 × 15–60 mm, sparsely pubescent; inflorescence a spike, 8–90 cm × 1–3 mm; bractlets 1 mm, usually slightly shorter than the calyx; calyx 1–3 mm, sparsely pubescent; corolla purplish blue. Moist areas, stream and river banks, pastures, roadsides, 3300–9100 ft. Scattered.

Verbena hirtella (L. M. Perry) G. L. Nesom [*V. neomexicana* Small var. *hirtella* L. M. Perry]—Hirsute verbena. Plants perennial, stems erect, 15–60 cm, densely hirsutulous to hirtellous and minutely stipitate-glandular; leaves often clustered at base, narrowly oblanceolate to oblanceolate or narrowly obovate, cauline similar to basal, blades 2–4 cm (basal) to 1.5–3 cm (midstem) × 6–10 cm, margins coarsely crenate to serrate with (2–)4–8 pairs of teeth, densely hirsutulous to hirtellous and minutely stipitate-glandular; spikes 1 or 2–3; floral bracts ovate-acuminate to ovate-lanceolate, 3–4 mm, shorter than the calyx; calyx 3.5–4.5(–5) mm, densely hirsutulous to hirtellous and minutely stipitate-glandular; corolla blue to purple or lavender, tube (3.5–)4.5–5.5 mm, 1–1.5(–2) mm longer than the calyx, limbs (5–)6–9 mm diam. Rocky slopes, mesquite desert scrub, 3800–4800 ft. Sierra, Socorro.

Verbena livermorensis B. L. Turner & G. L. Nesom—Davis Mountains verbena. Plants perennial, stems erect to ascending-erect, 35–65 cm, sparsely hirsute-pilose, bristly-hirsute, or bristly-pilose; leaves evenly distributed along stems, narrowly oblanceolate to oblong-oblanceolate, midstem 1.5–5 cm × 3–8(–12) mm, margins coarsely serrate with (2)3–8 antrorse teeth per side; spikes 1–3(–5); floral bracts ovate-lanceolate, slightly shorter than calyx; rachis stipitate-glandular; calyx 3.5–4 mm, hirsute-hirtellous, stipitate-glandular; corolla violet to blue, tube 4–4.5 mm, 1.5–2 mm longer than calyx, limb 4–6 mm diam. Ponderosa pine–oak and mixed conifer forests, 6300–8700 ft. Lincoln, Otero.

Verbena macdougalii A. Heller—MacDougal verbena. Erect perennial, to 1 m; stems villous-hirsute with spreading hairs; leaves short-petiolate or narrowed into a subpetiolar base, blades oblong-elliptic to ovate, 6–10 mm, coarsely and prominently rugose, irregularly serrate-dentate; inflorescence a spike, typically solitary, but sometimes in 3s, 7–10 mm wide, compact, hirsute; bractlets usually longer than the calyx; calyx 4–5 mm, densely pubescent, obtuse lobes with subulate teeth; corolla deep purple, barely exserted. Roadsides, montane meadows, ponderosa pine communities, 5000–9900 ft. Widespread except E, SW.

Verbena menthifolia Benth.—Mint vervain. Plants perennial herbs; stems erect to ascending-erect, 25–60(–120) cm, very sparsely hirsute-strigose, eglandular or sparsely stipitate-glandular; leaves ovate to lanceolate or oblong-lanceolate, 2.5–4(–5) cm × 5–17(–25) mm, margins deeply and irregularly toothed to lobed, not revolute or only slightly so, sometimes 3-lobed or pinnatifid; spikes 1 or 3–7, rachis eglandular or rarely sparsely stipitate-glandular; floral bracts ovate to ovate-lanceolate, slightly shorter than the calyx, glabrous to sparsely strigillose, eglandular to sparsely stipitate-glandular; calyx 2.2–3 mm, hirsute to hirsutulous, densely minutely stipitate-glandular; corolla blue to purplish pink, tube 2.5–3 mm, 0.5–1 mm longer than the calyx, limb 1.5–3(–5) mm diam. Springs, canyon bottoms, seeps, 4200–6060 ft. Grant, Hidalgo, Otero.

Verbena neomexicana Small—Hillside vervain. Erect, weak perennial, to 1 m; stems hirsute; leaves 7–60 × 2–30 mm, usually with clefts to 2 mm, hirsute; inflorescence a spike, 4–29 cm × 2–7 mm; bractlets 2–5 mm, hirsute; calyx 2–5 mm, hirsute; corolla blue, lavender, or purple. Piñon-juniper woodlands, ponderosa pine forests, 4800–7500 ft. Otero, SC, SW except Doña Ana, Luna.

Verbena perennis Wooton—Pinleaf vervain. Plants erect, 20–40 cm, somewhat woody at the base; stems antrorsely hispidulous and glandular or glabrate; leaves 1–4 cm, linear or linear-oblong, entire or a few lower leaves with a few teeth or lobes, strongly revolute; inflorescence a slender and interrupted spike; bractlets shorter than the calyx; calyx ca. 4 mm, hispidulous; corolla blue or purple. Rocky slopes, grasslands, oak-juniper woodlands, 3550–7500 ft. C, SE, SC, SW.

Verbena plicata Greene—Fanleaf vervain. Stems ascending to erect, sometimes decumbent, 30–50 cm, hirtellous; leaves narrowing to a petiole, blades 10–40 mm, incised-toothed to pinnatifid or 3-cleft, usually plicate, lower leaves elliptic-ovate; inflorescence a loose, slender, elongate spike; bractlets much longer than the calyx, 5–7 mm; calyx 3–4 mm, glandular-hirsute; corolla white, blue, lavender, or purple. Grasslands and prairies, piñon-juniper woodlands, ponderosa pine, mixed conifer, subalpine communities, 3000–11,925 ft. NE, EC, SE, SC, SW except Doña Ana, Luna.

Verbena scabra Vahl—Sandpaper vervain. Stems erect, 4–10 dm, solitary or branched near apices, strigillose; leaves short-petiolate, oblong to ovate-lanceolate or ovate, 3–13 × to 5 cm, serrate, scabrous; inflorescence of several to numerous slender spikes; bractlets ovate-lanceolate, hispid, shorter than the calyx; calyx 2–3 mm, hispidulous; corolla white, blue, or violet. Damp ground, shorelines of lakes, rivers, and streams, 3235–3660 ft. Eddy.

Verbena stricta Vent.—Hoary verbena. Perennials, 2.9–9 dm; stems tomentose; leaves 3–10 cm, densely hirsute-tomentose; inflorescence a spike, 4–43 cm × 3–9 mm; bractlets 3–6 mm, hirsute; calyx 3–6 mm, hirsute; corolla blue or purple. Grasslands, oak-juniper woodlands, 7400–8300 ft. Reported specimens are actually *V. macdougalii*.

Verbena xylopoda (L. M. Perry) G. L. Nesom—Verbena. Plants perennial, stems erect, 15–60 cm, moderately hirsute to densely hispidulous, sparsely to very sparsely stipitate-glandular; leaves often clustered at base, narrowly oblanceolate to oblanceolate or narrowly obovate, cauline similar to basal, blades 2–4 cm (basal) to 1.5–3 cm (at midstem) × 6–10 cm, sometimes 3-lobed, margins coarsely crenate-serrate or coarsely toothed with (1)2–3(4) pairs of teeth, densely hirsutulous to hirtellous and minutely stipitate-glandular; spikes 1 or 2–3; floral bracts ovate-acuminate to ovate-lanceolate, 3–4 mm, shorter than the calyx; calyx 3.5–4.5(–5) mm, densely hirsutulous to hirtellous and minutely stipitate-glandular; corolla blue to purple or lavender, tube (3.5–)4.5–5.5 mm, 1–1.5(–2) mm longer than the calyx, limb 4–7 (–8) mm diam. Desert scrub, canyon bottoms, ca. 4200 ft. Hidalgo.

VIBURNACEAE – VIBURNUM FAMILY

Kenneth D. Heil and Jennifer Ackerfield

Herbs, shrubs, or small trees; leaves opposite, simple to compound; inflorescence determinate, often umbellate; flowers perfect, actinomorphic, small; sepals connate, petals (4)5, connate, usually with a short tube and well-developed lobes; fruit a fleshy drupe with 1, 3, or 5 stones or small pits (Donoghue et al., 1992, 2016; Judd et al., 1994; Olmstead et al., 1993). Includes Adoxaceae.

1. Small herbaceous plants, 5-15 cm; rocky scree slopes and stream banks in higher mountains; leaves once or twice ternately compound or parted…**Adoxa**
1'. Shrubs or small trees, mostly 1+ m; various habitats; leaves pinnately compound or simple…(2)
2. Leaves pinnately compound…**Sambucus**
2'. Leaves simple…**Viburnum**

ADOXA L. – Adoxa

Adoxa moschatellina L.—Muskroot. Perennial herbs from rhizomes, 5-15 cm; leaves 3-parted or ternately compound, cauline leaves in 1 pair above midpoint of stem, basal leaves variable in size, mostly averaging 8 cm, leaflets round-toothed and mucronate; calyx 2-4-lobed, persistent in fruit; corolla 4-6-lobed, lobes 1.5-3 mm. Rocky scree slopes and along forested stream banks in midelevation to higher mountains, 8400-12,500 ft. NC.

SAMBUCUS L. – Elderberry

Shrubs or small trees; leaves pinnately to bipinnately compound with serrate leaflets; flowers small, in broad, terminal compound cymes; sepals 5, generally small and inconspicuous; petals white to yellowish; fruit a drupe with 3-5 stones.

1. Inflorescence a pyramidal cyme, as long as or longer than wide; berries red or blackish, lacking a bloom… **S. racemosa**
1'. Inflorescence a flat-topped or broadly rounded cyme, wider than long; berries blue, purplish black, or bluish black, with a whitish bluish bloom…**S. cerulea**

Sambucus cerulea Raf.—Blue elderberry. Shrubs or small trees to 7 m; leaflets 5-9, 3-16 cm, lanceolate, long-attenuate, or nearly ovate; cyme corymbose; fruit blue to nearly black, with a whitish bluish bloom. NC, NW, C, WC, SE, SC, SW.

1. Inflorescences 7-15 cm across; few-stemmed shrub or often treelike; leaflets 3-7 cm; mostly S…**var. mexicana**
1'. Inflorescences 12-30 cm across; many-stemmed shrub; leaflets 6-16 cm; widespread…**var. cerulea**

var. cerulea [*S cerulea* var. *neomexicana* (Wooton) Rehder]—Blue elderberry. Leaflets narrowly to broadly lanceolate, long-attenuate at the apices, flat and not folded. Along creeks and open slopes, ponderosa pine, mixed conifer, spruce-fir communities, 5400-11,400 ft.

var. mexicana (C. Presl ex DC.) L. D. Benson—Mexican elderberry. Leaflets lanceolate-oblong to nearly ovate, abruptly acuminate at the apex, commonly folded along the midrib. Along streams and rivers, 4920-8400 ft.

Sambucus racemosa L.—Red elderberry. Shrubs, 0.5-2 m; leaflets 5-7, lanceolate to elliptic, serrate; cyme paniculate, 3-10 cm; fruit black or red.

1. Fruit purple or black…**var. melanocarpa**
1'. Fruit red or yellowish…**var. microbotrys**

var. melanocarpa (A. Gray) McMinn [*S. melanocarpa* A. Gray]—Black elderberry. Along streams from piñon-juniper to spruce-fir communities, 5650-12,000 ft. C.

var. microbotrys (Rydb.) Kearney & Peebles—Red elderberry. Common along streams, moist slopes, aspen forests, 6200-12,850 ft. N, C, WC, SC, SW.

VIBURNUM L. - Viburnum

Viburnum lantana L.—Wayfaring-tree. Shrubs to 3 m; leaves simple, ovate to obovate, serrate, unlobed, stellate-hairy below, 5–10 cm; flowers in terminal, flat-topped cymes, flowers 4 mm; sepals 5; petals white; fruit red, 8–10 mm diam. Near Los Alamos along canyon bottoms, ca. 7000 ft. Los Alamos. Introduced.

VIOLACEAE - VIOLET FAMILY

Ross A. McCauley

Herbs (ours), shrubs, lianas, and trees; leaves cauline or basal, simple or compound; flowers bisexual, stamens 5, alternate with petals, surrounding ovary; pistil 1, (2)3(–5)-carpellate; ovary superior, 1-locular, placentation parietal; style (0)1, usually enlarged distally, solid or hollow; stigma 1(3–5); fruit capsular [berry or nut], 3-valved, dehiscence loculicidal. Descriptions closely follow those in Little and McKinney (2015).

1. Flowers inconspicuous, nodding on axillary pedicels; sepals not auriculate; upper 2 and lateral 2 petals not showy, 0.5–5 mm; lowest petal showy, narrowed at middle; stamens connate, lowest 2 filaments not spurred with nectary; seeds 6...**Pombalia**
1′. Flowers conspicuous and showy on long peduncles; sepals auriculate; upper 2 and lateral 2 petals showy, 5+ mm; lowest petal showy, not narrowed at middle; stamens connivent but distinct, lower 2 filaments spurred with nectary that protrudes into petal spur; seeds 6–75...**Viola**

POMBALIA Vand. - Baby-slippers

Pombalia verticillata (Ortega) Paula-Souza [*Hybanthus verticillatus* (Ortega) Baill.]—Baby-slippers, nodding green-violet. Plants subshrubs, from a ligneous rhizome; stems erect, 10–40 cm, glabrous or strigose to pilose; leaves proximally opposite to distally alternate; stipules linear-subulate and minute to leaflike, 3–40 mm; petioles 0–1 mm; blades linear to oblanceolate, 1–5.5(–6) × 0.1–0.8(–1.1) cm, margins usually entire, base attenuate, apex acute to acuminate, surfaces glabrous, strigose, or pilose; inflorescences 1-flowered; peduncles usually pendent at anthesis; upper petals greenish white, cream, or yellowish, with purple tips, 2–2.5 mm, glabrous; lowest petal greenish white, cream, or yellowish, sometimes tinged purplish, 2.5–6 mm; capsules ovoid to globose, 4–7 mm; seeds 6, dark brown to black. Piñon-juniper woodlands, grasslands, desert shrublands, arroyos, 3280–6900 ft. NE, C, WC, SE, SC, SW.

VIOLA L. - Violet

Annual or perennial herbs, with or without distinct aboveground stems; leaves alternate or basal; stipules conspicuous; often exhibiting both chasmogamous and cleistogamous flowers; chasmogamous flowers zygomorphic, perfect; sepals 5; petals 5, the lower one prolonged into a spur; stamens 5, connivent, forming a cone around the ovary; pistil superior, of 3 united carpels; fruit a capsule; seeds often arillate with an elaiosome.

1. Plants with branched stems (caulescent); petioles arising from lower and upper stem nodes...(2)
1′. Plants with unbranched stems (acaulescent); petioles arising only from rootstalk or rhizome...(5)
2. Plants annual; petals of multiple colors (upper violet, lateral white, and lowest yellow or white)...**V. tricolor**
2′. Plants perennial; petals all a similar color...(3)
3. Petals violet or blue...**V. adunca**
3′. Petals white or yellow...(4)
4. Petals white; leaf blades ovate to broadly ovate or ovate-reniform...**V. canadensis**
4′. Petals yellow; leaf blades linear to linear-lanceolate or linear-oblong...**V. nuttallii**
5. Leaves 5–9-lobed...**V. pedatifida**
5′. Leaves entire (may be infrequently lobed)...(6)
6. Plants mat-forming, with stolons; petals white; in wet bog habitats...**V. macloskeyi**
6′. Plants not mat-forming, without stolons; petals blue to violet; in a variety of habitats...(7)

7. Plants from a short vertical rhizome; petals white to medium purple; specialist in cracks of limestone; restricted to Guadalupe Mountains…**V. calcicola**

7'. Plants from a horizontal rhizome; petals bluish violet; in moist soil along water courses and gravel alluvium…(8)

8. Spurred petal lightly bearded; leaf blade apex distinctly acute; in gravel alluvium on the E plains…**V. retusa**

8'. Spurred petal thickly bearded; leaf blade apex broadly acute to obtuse or rounded; in wet habitats generally in the mountains…**V. nephrophylla**

Viola adunca Sm.—Western dog violet, hooked violet. Plants perennial, caulescent, not stoloniferous, 1.8–30(–35) cm; stems erect, ascending, or decumbent; leaves basal and cauline; stipules linear to linear-lanceolate, margins entire or laciniate with gland-tipped projections, apex acute to acuminate; blades usually ovate or ovate-deltate to ovate-orbiculate, sometimes ± reniform or oblong, 0.5–6.9 × 0.4–5 cm, margins crenate to crenulate or entire, ciliate or eciliate, base cordate, subcordate, truncate, or attenuate, apex acute to obtuse; chasmogamous flowers with peduncles 1.7–13.8 cm, sepals lanceolate, margins ciliate or eciliate, auricles not enlarged in fruit, 0.5–2 mm; petals light- to deep- to lavender-violet on both surfaces, rarely white, lower 3 usually white basally, dark violet-veined, lateral 2 (and sometimes upper 2) bearded, lowest 7–17(–23) mm, spur purple to violet or white, elongate, 5–7 mm, tip straight or pointed, curved up or lateral; cleistogamous flowers axillary; capsules short-ovoid, 6–11 mm, glabrous; seeds dark brown to olive-black, 1.5–2 mm.

1. Plants (4–)4.5–30(–35) cm; basal blades 1.3–6.9 × 1.2–5 cm; in shaded understory vegetation, 6800–10,500 ft….**var. adunca**

1'. Plants 1.8–4.5(–6.5) cm; basal blades 0.5–1.7 × 0.4–1.4 cm; in subalpine meadows and alpine tundra >10,000 ft….**var. bellidifolia**

var. adunca—Plants erect, decumbent, or prostrate, (4–)4.5–30(–35) cm; leaf blades usually ovate or ovate-deltate, sometimes ± reniform or oblong, 1.3–6.9 × 1.2–5 cm. Understory of mixed conifer and aspen woodlands, meadows, 6800–10,500 ft. N, C, WC, SW.

var. bellidifolia (Greene) H. D. Harr.—Plants erect, usually appearing small and tufted, 1.8–4.5(–6.5) cm; leaf blades ovate to ovate-orbiculate, 0.5–1.7 × 0.4–1.4 cm. Subalpine meadows and alpine tundra, >10,000 ft. C.

Viola calcicola R. A. McCauley & H. E. Ballard [*V. missouriensis* Greene in part; *V. palmata* L. in part]—Limestone violet. Plants perennial, acaulescent from a short vertical rhizome, 3–9(–14) cm; leaves all unlobed or rarely shallowly trilobate; stipules narrowly triangular to linear-lanceolate; blades cordate or triangular-cordate to deltoid, 1–3(–4.5) × 0.8–3(–4.5) cm, margins rounded-serrate; chasmogamous flowers with peduncles 3–6(–11) cm; sepals ovate to lanceolate, petals white to medium purple, lateral petals with few inconspicuous purple veins and sparsely bearded, lowest with prominent and extensive nectar guides 10–13(–15) mm, spur white; cleistogamous flowers on erect peduncles; capsules ellipsoid, (6.5–)7–9 mm, glabrous, seeds yellow-brown, ca. 1.75 mm. Cracks of limestone in sheltered canyons and springs; Guadalupe Mountains, 5000–6000 ft. Eddy. **R**

Viola canadensis L.—Canada violet, tall white violet. Plants perennial, caulescent, not stoloniferous, 3–46(–60) cm; stems erect to ascending; leaves basal and cauline; stipules ± oblong, ovate, lanceolate, or deltate, margins entire to laciniate; blades ovate to broadly ovate or ovate-reniform, 0.7–12.4 × 0.9–11.1(–12.3) cm, margins serrate, base cordate to truncate, apex acute to acuminate; chasmogamous flowers with peduncles of 1–6.1 cm, sepals lanceolate, auricles 0.5–1.3 mm, petals white, usually with yellow patch basally, lower 3 usually purple-veined, lateral 2 bearded, lowest 5.5–20 mm, spur white, gibbous, 1–2 mm; cleistogamous flowers axillary or absent; capsules ovoid to ellipsoid, 2.5–10 mm, glabrous or puberulent; seeds brown to dark brown or purplish black, 1.5–2.5 mm. NC, NW, C, WC, SC, SW.

1. Plants with branched rhizomes…**var. rugulosa**

1'. Plants with unbranched rhizomes…**var. scopulorum**

var. rugulosa (Greene) C. L. Hitchc.—Tall white or rugose violet. Plants with branched rhizomes. Moist woodlands, forest edges, riparian areas, 5300–12,300 ft.

var. scopulorum A. Gray [*V. scopulorum* (A. Gray) Greene]—Plants with unbranched rhizomes. Moist woodlands, forest edges, riparian areas, 5300–12,300 ft.

Viola macloskeyi F. E. Lloyd—Northern white violet. Plants perennial, acaulescent, stoloniferous, 2–10 cm; leaves basal; stipules ovate to linear-lanceolate, margins entire or glandular-toothed, apex acute; blades unlobed, reniform to ovate, 1–6.5 × 1–5.5 cm, margins ± entire to shallowly crenate, base cordate, apex rounded to acute; chasmogamous flowers with peduncles 2.5–11(–21) cm, sepals lanceolate to ovate, auricles 0.5–2 mm, petals white, lower 3 purple-veined, lateral 2 bearded or rarely beardless, lowest 6–12 mm, spur white, gibbous, 1–2.5 mm; cleistogamous flowers on erect peduncles; capsules ovoid, 5–9 mm, glabrous; seeds beige to bronze, 1–1.5 mm. Bogs, fens, wet meadows, usually among mosses, 8530–10,500 ft. NC.

Viola nephrophylla Greene—Northern bog violet. Plants perennial, acaulescent, not stoloniferous, 5–15 cm; leaves basal; stipules lanceolate, margins entire or fimbriate, apex acute; blades usually grayish green or purplish green abaxially, unlobed, ovate, reniform, or broadly reniform to orbiculate, 1–7 × 1–7 cm, usually somewhat fleshy, margins crenate to serrate, base broadly cordate or reniform to ± truncate, apex broadly acute to obtuse or rounded; chasmogamous flowers with peduncles 3–25 cm, sepals ovate, auricles 1–2 mm, petals deep bluish violet, lower 3 white basally and darker violet-veined, lateral 2 bearded, upper 2 sometimes sparsely bearded, lowest densely bearded or beardless, 10–28 mm, spur same color as petals, gibbous, 2–3 mm; cleistogamous flowers on erect peduncles; capsules ovoid, 5–10 mm, glabrous; seeds beige to brown or dark brown, 1.5–2.5 mm. Usually in saturated soil of bogs, fens, sedge meadows, lake and stream edges, occasionally in moist but not saturated soil, 5120–10,400 ft. NC, C, WC, SC, SW.

Viola nuttallii Pursh—Nuttall's violet, yellow prairie violet. Plants perennial, caulescent, not stoloniferous, 2–27 cm; stems ascending to erect, ca. 1/2 subterranean, puberulent; leaves basal and cauline; stipules adnate to petiole (basal leaves) to adnate or free (cauline leaves), linear to linear-lanceolate or linear-oblong, margins entire, apex acute to acuminate; blades lanceolate, ovate, or elliptic, 1–9 × 0.6–2.5 cm, margins entire or serrulate, base attenuate, apex acute to obtuse, mucronulate; chasmogamous flowers with peduncles 3–13 cm, sepals linear-lanceolate, auricles 0.5–1 mm, petals deep lemon-yellow adaxially or on both surfaces, upper 2 petals often brownish purple abaxially, lower 3 veined with dark brown to brownish purple, lateral 2 sparsely bearded, lowest 6–13 mm, spur yellow, gibbous, 0.5–1.5 mm; cleistogamous flowers axillary; capsules subglobose to ovoid, 4–10 mm, usually glabrous; seeds medium brown, 2–3.2 mm. Sagebrush flats, prairie grasslands, dry stream banks, piñon-juniper woodlands, scree slopes, 6150–8400 ft. NE, NC, NW, San Miguel.

Viola pedatifida G. Don—Crowfoot or larkspur violet. Plants perennial, acaulescent, not stoloniferous, 5–30 cm; leaves basal, 5–9-lobed; stipules linear-lanceolate, margins entire, apex acute, lobes lanceolate, spatulate, falcate, or linear, 1–7 × 2–8 cm; leaflet margins entire, base truncate to reniform, apex acute to obtuse, mucronulate; chasmogamous flowers with peduncles 5–18 cm, sepals lanceolate to ovate, auricles 1–2 mm, petals light to soft reddish violet on both surfaces, lower 3 white basally, dark violet-veined, lateral 2 and lowest usually bearded, lowest 10–25 mm, spur same color as petals, gibbous, 2–3 mm; cleistogamous flowers on erect peduncles; capsules ellipsoid, 10–15 mm, glabrous; seeds beige, mottled to bronze, 1.5–2.5 mm. Woodland meadows, slopes, canyon bottoms, grasslands, 7300–9600 ft. NE, NC, Cibola.

Viola retusa Greene [*V. missouriensis* Greene in part; *V. nephrophylla* Greene in part]—Plants perennial, acaulescent, not stoloniferous, 6–15 cm; leaves basal, stipules linear-lanceolate, margins entire, apex acute to pinnatifid; blades broadly cordate-deltoid,(1–)2–5.5 × 1.8–4 cm, margins crenate to serrate, base broadly cordate, apex usually acute to obtuse or rarely rounded; chasmogamous flowers with peduncles 3–10(–25) cm, sepals oblong, auricles 1–2 mm, petals deep bluish violet, lower 3 white basally and darker violet-veined and bearded, lowest 10–28 mm, spur same color as petals, gibbous, 2–3 mm; cleistogamous flowers on erect peduncles; capsules ellipsoid, 5–10 mm, glabrous; seeds brown, ca. 2 mm. Gravel alluvium and riparian zones, 5100–6100 ft. San Miguel, Union.

Viola tricolor L.—Johnny-jump-up. Plants usually annual, rarely perennial, caulescent, not stoloniferous, 3–45 cm; stems prostrate to erect; leaves cauline; stipules palmately lobed or pinnatifid, margins entire or usually crenate, apex acute to obtuse; blades ovate to ± oblong (distal blades lanceolate), 1.5–2.8 × 0.8–2.3 cm, margins coarsely crenate-serrate, base cordate, truncate, or attenuate, apex acute to obtuse; chasmogamous flowers with peduncles 3–12.5 cm, sepals deltate-lanceolate to linear-lanceolate, auricles 2–4 mm, upper 2 petals commonly violet, lateral petals white, lowest petal yellow or white on both surfaces with darker yellow patch basally, lateral 2 petals densely bearded, lowest and lateral 2 veined with dark violet to brownish purple, lateral 2 longer than sepals, lowest 10–15 mm, spur usually yellow to violet, rarely white, elongate, 4–5 mm; cleistogamous flowers absent; capsules subglobose to ovoid, 6–8 mm, glabrous; seeds tan, 1.5–2 mm. Our plants are **var. tricolor**. Introduced, escaped from cultivation in a variety of habitats. Rio Arriba.

VISCACEAE – MISTLETOE FAMILY

John R. Spence and Robert L. Mathiasen

Herbaceous to woody perennials, aerial hemiparasites, evergreen, stems often brittle, variously colored brown, olive, green, orange, or dull red, roots forming haustorial connections with host; leaves opposite, reduced to scales or simple, sessile or petiolate, connate at base or free, stipules absent; inflorescence of axillary unisexual or bisexual spicate thyrses; flowers unisexual, free or embedded in inflorescence axis, epigynous, sepals 0, petals 4–7; stamens 2–6, opposite petals, adnate to petals, 1–2-locular; pistil 1, ovary inferior, 1–3-carpellate, 1-locular; stigma 1, 2-lobed; fruit a fleshy berry, seeds 1. Viscaceae is treated here as a distinct family following *Flora of North America*.

1. Leaves scalelike; ± herbaceous plants to 30 cm; flowers free, not in cavities or grooves on inflorescence axis; anthers 1-locular; berries bicolored, lacking persistent petal, on recurved pedicels, dehiscing explosively…**Arceuthobium**
1'. Leaves scalelike or well developed with blades; semiwoody to woody plants to 1 m; flowers borne in cavities or grooves on inflorescence axis; anthers 2-locular; berries unicolored, with persistent petal remnants, sessile, not dehiscing explosively…**Phoradendron**

ARCEUTHOBIUM M. Bieb. – Dwarf mistletoe

Aerial, parasitic, dioecious herbs to rarely subshrubs, shoots to 30 cm, stems mostly erect or rarely pendent, brittle, disarticulating at nodes, round to sometimes square or angled, branching dichotomous or flabellate, variously colored red-brown, olive, green, orange, or yellow, glabrous, nodes swollen; leaves reduced to opposite connate scales; inflorescence of axillary bracteate spikes; flowers 3- or 4-merous, sessile or pedicellate; perianth of staminate flowers (2)3–4(–7), with inconspicuous persistent tepals, often reduced to teeth, anthers sessile, adnate to tepal, ovary of carpellate flowers fused with perianth, forming a bilobed cap; fruit a pedicellate, fleshy, bicolored berry, pale tan to brown above, glaucous-blue or blue-purple below, capped by persistent petals, dehiscing explosively, pedicels erect to recurved when mature, seeds 1, mucilaginous (Hawksworth & Wiens, 1993; Nickrent, 2016b).

1. Parasitic on *Pinus*…(2)
1'. Parasitic on other Pinaceae, including *Abies*, *Picea*, and *Pseudotsuga*…(5)
2. Parasitic on piñon pines (*P. cembroides*, *P. discolor*, *P. edulis*); flowering early August to late September; shoots 6–13 cm, olive-green to dull brown…**A. divaricatum**
2'. Parasitic on other species of pine; flowering times various but often summer–fall; shoots variable, short to long, 3–25 cm, color various, often orange or yellow…(3)
3. Flowering July–October; parasitic on southwestern white pine (*P. strobiformis*); stems short, < 6(–10) cm …**A. apachecum**
3'. Flowering March–June; parasitic on various pines, rarely on *P. strobiformis*; stem length variable, often to 25 cm…(4)
4. Stems orange to red-brown or yellow, female and male plants similar; parasitic on various pines, commonly *P. arizonica*, *P. engelmannii*, or *P. scopulorum*, rarely *P. aristata*, *P. flexilis*, or *P. strobiformis*…**A. vaginatum**

4'. Stems dull green to brown or rarely yellow, male and female stems dimorphic; parasitic on *P. leiophylla* …**A. gillii**
5. Parasitic on Douglas-fir (*Pseudotsuga menziesii*), rarely on species of fir or spruce (*Abies, Picea*) in vicinity; flowering mid April to May; forming systemic and often large witches' brooms; berries olive to purple proximally; shoots short, mostly ≤ 2 cm; third internode 2-6 mm…**A. douglasii**
5'. Parasitic on spruce (*Picea* spp.), rarely on subalpine fir (*Abies arizonica*) in vicinity; flowering late July to late September; forming localized infections and small round witches' brooms; berries purple to green proximally; shoots to 9(–15) cm; third internode 6-15 mm…**A. microcarpum**

Arceuthobium apachecum Hawksw. & Wiens—Southwestern white pine dwarf mistletoe. Plants occasionally forming systemic witches' brooms, stems mostly < 6(–10) cm, central ones 1-2 mm thick at base, yellow-green or orange to red, third branch internode 3-9 mm, 1-2 mm at base; staminate flower tepals yellow-green; fruit a berry, brown to yellow-orange distally, yellow-green to olive proximally, 4 × 2.5 mm; flowering July–October. Found on southwestern white pine, *Pinus strobiformis*; montane mixed conifer forests, 6900-9920 ft. C, WC.

Arceuthobium divaricatum Engelm.—Piñon dwarf mistletoe. Plants form small, dense witches' brooms and rarely producing systemic witches' brooms, stems 5-10 cm, central ones 1.5-4 mm thick at base, dark green to olive-green or brown, lacking yellow tints, third branch internode 6-15 mm at base; staminate flower tepals pale yellow-green; fruit a berry, olive-brown and glaucous distally, olive, gray, or blue-green proximally, 3-4 × 2 mm; flowering early August to late September. Found on piñon pines, including *P. cembroides*, *P. discolor*, and *P. edulis*; piñon-juniper and mixed oak-pine woodlands, 4500-9800 ft. Widespread except E, Hidalgo.

Arceuthobium douglasii Engelm.—Douglas-fir dwarf mistletoe. Plants forming systemic witches' brooms; stems 2(–8) cm, central ones 1-1.5 mm thick at base, orange or yellow-green to olive-green, third branch internode 2-6 mm; staminate flower tepals red to purple; fruit a berry, yellow-brown to orange distally, green to purple proximally, 3-5 × 1.5-2 mm; flowering mid April to May. Found on *Pseudotsuga menziesii*; mixed conifer forests, 6500-10,700 ft. Widespread except E, Doña Ana, Sierra.

Arceuthobium gillii Hawksw. & Wiens—Chihuahua pine dwarf mistletoe. Plants not forming systemic witches' brooms; stems 8-25 cm, central ones 2.5-8 mm thick at base, yellow-green to olive-green, third branch internode 5-18 mm; staminate flower tepals tan or brown to red-brown; fruit a berry, green to yellow-brown distally, olive-green and distinctly glaucous with blue tints proximally, 4-5 × 2-3 mm; flowering March to May. Found on *Pinus leiophylla*; oak-pine and mixed conifer forests; Animas Mountains, 7000 ft. Hidalgo.

Arceuthobium microcarpum (Engelm.) Hawksw. & Wiens—Spruce dwarf mistletoe. Plants not forming systemic witches' brooms; stems 5-11 cm, central ones 1.5-3 mm thick at base, yellow-green, orange, red, or purple, third branch internode 6-15 mm; staminate flower tepals yellow-green; fruit a berry, brown to yellow-orange distally, yellow-green to olive proximally, 4 × 2.5 mm, flowering July–September. Found on *Picea engelmannii* and *P. pungens*, rarely on *Abies arizonica* in vicinity; mixed conifer and spruce-fir forests, 7800-9500 ft. Catron, Grant, Otero.

Arceuthobium vaginatum (Humb. & Bonpl. ex Willd.) J. Presl—Southwestern dwarf mistletoe. Commonly forming systemic witches' brooms, stems 5-25 cm, yellow-orange, dull reddish brown, or sometimes nearly black; flowers of staminate inflorescences 2-3 mm across, staminate flower tepals yellow-green to pink; fruit a berry, brown or tan distally, glaucous olive-green proximally, 4-6 × 2-3 mm; flowering April–June. Our material belongs to **subsp. cryptopodum** (Engelm.) Hawksw. & Wiens. Found on *Pinus arizonica*, *P. engelmannii*, or *P. scopulorum*, rarely *P. aristata*, *P. flexilis*, or *P. strobiformis*; pine-oak and mixed conifer forests, 5500-9350 ft. Widespread except E, Hidalgo, Luna.

PHORADENDRON Nutt. - Mistletoe

Aerial parasitic dioecious shrubs on branches of host, stems erect to pendent, sometimes brittle and disarticulating at nodes, round to sometimes square or angled, branching dichotomously, variously colored red-brown, green, or yellow, glabrous or pubescent when young, nodes swollen; leaves well developed to scalelike, free or connate at base; inflorescence of axillary bracteate spikes, carpellate spikes with 2 flowers per fertile internode, internodes elongating in fruit, staminate spikes with 2 to many flowers; flowers 3- or 4-merous, sessile; perianth of 3-4 inconspicuous persistent tepals, often reduced to teeth, anthers sessile, adnate to tepal; ovary fused with perianth; fruit a brightly colored, sessile, fleshy, drupelike berry, white, white-pink, yellow, or red, capped by persistent petals, on straight or recurved pedicel, not dehiscing explosively; seeds 1.

1. Plants appearing ± leafless, leaves reduced to scalelike bracts, < 2(-3) mm…(2)
1'. Plants with well-developed leaves, mostly > 7 mm…(3)
2. Plants green to orange, glabrous; berries white to pink; pistillate spikes with a single fertile internode; parasitic on various junipers (*Juniperus*)…**P. juniperinum**
2'. Plants green to red-green, densely pubescent although becoming glabrate with age; berries mostly red; pistillate spikes with 2+ fertile internodes; parasitic on various leguminous shrubs and trees, rarely on other families…**P. californicum**
3. Leaf blades obovate or ovate to suborbicular, mostly > 10 mm wide; parasitic on dicot shrubs and trees…(4)
3'. Leaf blades narrowly spatulate to lanceolate or linear, < 5 mm wide; parasitic on junipers (*Juniperus*)…(5)
4. Parasitic on hardwood trees (mostly *Alnus, Fraxinus, Juglans, Platanus, Populus*, and *Salix*); flowering from fall through spring; berries glabrous; stem internodes > 6 mm…**P. macrophyllum**
4'. Parasitic on oaks (*Quercus*); flowering from summer to fall; berries puberulent proximally; stem internodes < 4 mm…**P. coryae**
5. Leaves and stems densely stellate-pubescent; basal leaf pair of lateral branches in same plane as branching; flowering in winter; parasitic on several juniper species…**P. capitellatum**
5'. Leaves glabrous to sparsely hirtellous with simple hairs; basal leaf pair of lateral branches at right angles to branching; flowering in summer; parasitic on *Juniperus monosperma* or *J. pinchotii*…**P. hawksworthii**

Phoradendron californicum Nutt.—Mesquite mistletoe. Stems 5-20 cm, pendent, gray-green to red-green, densely silver-white-pubescent, internodes > 10 mm; leaves 1.5-3 mm, triangular; pistillate inflorescence 5-10 mm, fertile internodes (1)2-4(-6), each with 2 flowers, staminate inflorescence 5-25 mm, fertile internodes 2-5, each with 6-14 flowers; fruit a translucent white, yellow, pink, or orange-red berry, globose, 3-6 mm. Flowering November–March. Found primarily on *Acacia* and *Prosopis*, but also *Condalia, Dalea, Larrea* (rare), *Olneya*, and *Parkinsonia*; desert scrub, dry washes, bosques, 4400-4750 ft. Hidalgo.

Phoradendron capitellatum Torr. ex Trel.—Downy mistletoe. Stems 30-60 cm, erect to pendent, green, densely stellate-pubescent; leaves green, pubescent, somewhat fleshy, 5-15 × 1-2 mm, elliptic to spatulate; pistillate inflorescence 3-5 mm, fertile internodes 1-2, each with 2 flowers, staminate inflorescence 3-5 mm, fertile internodes 1-2, each with 6 flowers; fruit a pink or white berry, globose, 3-4 mm. Flowering November–February. Found on junipers, including *Juniperus coahuilensis, J. monosperma, J. deppeana*, and *J. osteosperma*; piñon-juniper and mixed oak-pine woodlands, pine forests, 4000-5800 ft. SC, SW.

Phoradendron coryae Trel. [*P. villosum* (Nutt.) Nutt. ex Engelm. subsp. *coryae* (Trel.) Wiens]—Cory's mistletoe. Stems to 1 m, erect, green to olive-green or gray-green, pubescent in clusters and scattered along stem, internodes < 4 mm; leaves green to olive-green, pubescent with stellate hairs, 15-40 × 9-20 mm, ovate or elliptic to orbicular; pistillate inflorescence 10-80 mm, fertile internodes 2-3, each with 6-12 flowers, staminate inflorescence 10-80 mm, fertile internodes 2-5, each with 15-40 flowers; fruit a pink or white berry, globose to oblong, 3 mm. Flowering July–September. Found on *Quercus* species, especially live oak (*Q. turbinella*); oak-pine and riparian woodlands and forests, 4500-8600 ft. C, WC, SW, Eddy.

Phoradendron hawksworthii Wiens [*P. bolleanum* (Seem.) Eichler]—Hawksworth mistletoe. Stems to 1 m, erect, green to brown, glabrous or sparsely hairy, internodes 2 cm; leaves olive to green, glabrous or sparsely hairy, 5-25 × 2-4 mm, narrowly elliptic to lanceolate; pistillate inflorescence 2-4 mm, fertile internodes 1-2, each with 2 flowers, staminate inflorescence 2-4 mm, fertile internodes 1(2), each with 6-20 flowers; fruit a pink or white berry, globose to oblong, 4-5 mm. Flowering April–August.

Found on juniper species, especially on *J. monosperma* and *J. pinchotii*; piñon-juniper and pine-oak woodlands, 4660-6300 ft. C, S except E.

Phoradendron juniperinum A. Gray—Juniper mistletoe. Stems 5-20 cm, erect, green, olive, or yellow-green, glabrous, internodes > 5 mm; leaves 0.5-1.5 mm, triangular, acute; pistillate inflorescence 3-5 mm, reduced to 1 segment with 2 flowers, staminate inflorescence 3-5 mm, fertile internode 1(2), with 6 flowers; fruit a shiny, somewhat translucent white, pink, or pale red berry, 4-5 mm, elliptic to globose. Flowering June–September. Found on most *Juniperus* species; piñon-juniper woodlands and pine forests, 5000-8900 ft. Widespread except E.

Phoradendron macrophyllum Spreng. [*P. leucarpum* (Raf.) Reveal & M. C. Johnst. subsp. *macrophyllum* (Engelm.) J. R. Abbott & R. L. Thomps.]—Bigleaf mistletoe. Stems to 1 m, erect, green, sparsely hairy, becoming glabrous with age, internodes > 6 mm; leaves bright green to yellow-green, glabrous or sparsely hairy, 30-40 × 15-22 mm, ovate or elliptic to spatulate; pistillate inflorescence 20-80 mm, fertile internodes 2-6, each with 6-20 flowers, staminate inflorescence 20-80 mm, fertile internodes 2-7, each with 20-30 flowers; fruit a pink or white berry, globose to oblong, 4-5 mm. Flowering October–March. Found on numerous hardwood shrubs and trees, especially species of *Alnus*, *Fraxinus*, *Juglans*, *Platanus*, *Populus*, and *Salix*; riparian areas, 3800-6400 ft. Sandoval, WC, S except Lea, Luna.

VITACEAE – GRAPE FAMILY

John R. Spence

Vines or lianas, clambering or climbing, bark often exfoliating, tendrils usually present; leaves alternate, simple to palmately or pinnately compound, stipules present, petioles present, leaf blades simple to palmately lobed, margins dentate to serrate; inflorescence of axillary or terminal cymes or thyrses; flowers bisexual or unisexual, actinomorphic; sepals (3)4-5(-9), connate most or all of their length; petals (3)4-5(-9), distinct or connate distally; stamens (3)4-5(-9), opposite petals; pistil 1, 2(3)-carpellate, ovary superior, 2(3)-locular, placentation axile, ovules 2 per locule, style 1, stigma 1; fruit a berry; seeds 1-4 per fruit (Moore & Wen, 2016).

1. Leaves simple, palmately lobed; bark exfoliating, pith brown; petals connate distally…**Vitis**
1'. Leaves palmately compound, rarely simple and 3-lobed; bark adherent, pith white; petals distinct…(2)
2. Leaves 3-foliate, rarely simple, blades ± succulent; petals and stamens 4; young branches succulent tendrils unbranched; berries obovoid…**Cissus**
2'. Leaves 4-7-foliate, blades not succulent; petals and stamens 5; young branches not succulent; tendrils branched; berries ± globose…**Parthenocissus**

CISSUS L. – Possum grape

Cissus trifoliata (L.) L. [*C. incisa* Des Moul.]—Marine vine, sorrel ivy. Lianas, scrambling over shrubs and other low vegetation, to 10 m, arising from a large woody tuber, branchlets becoming woody with age, tendrils unbranched, lacking adhesive disks; leaves deciduous to evergreen, mostly 3-foliate, leaflet margins coarsely and irregularly toothed; sepals with 4 small lobes, greenish; petals 4, greenish white, white, or purple to yellow; stamens 4; berries obovoid, black, 6-12 mm, 1-4-seeded. Rocky slopes and ridges, waste areas, 4000-6000 ft. SE, SC, SW.

PARTHENOCISSUS Planch. – Thicket creeper, woodbine

Lianas, clambering or climbing, arising from fibrous roots, sometimes producing adventitious roots, tendrils with 3-12 branches, with or without adhesive disks; leaves palmately 4-7-compound, turning red in fall; inflorescence bisexual, axillary, opposite leaves or terminal, cymose, often compound; flowers bisexual or functionally unisexual; sepals 5, connate, campanulate, 5-lobed, greenish; petals 5, distinct, greenish; stamens 5; style short-conic; berries globose to subglobose, blue-black, 2-4-seeded.

1. Inflorescence dichotomously branched, lacking distinct central axis; berries 8-12 mm; tendrils 2-5-branched, adhesive disks lacking or rarely present; leaflets shiny adaxially...**P. quinquefolia**
1'. Inflorescence divergently branched, with distinct central axis; berries 5-8 mm; tendrils 4-12-branched, with adhesive disks; leaflets dull adaxially...**P. vitacea**

Parthenocissus quinquefolia (L.) Planch. [*P. inserta* (A. Kern.) Fritsch]—Virginia creeper. Lianas, scrambling over shrubs and other low vegetation; branches becoming flaky and resinous with age, tendrils strongly branched, with adhesive disks; leaves dull green above, glabrous below, 3-7-foliate, leaflets coarsely serrate distally, central leaflet 4-14 × 3-7 cm; flowers functionally unisexual, inconspicuous, green; berries globose, fleshy, black. Cultivated in our area, rarely escaping, waste areas, 4880-7200 ft. Catron, McKinley, San Juan. Introduced.

Parthenocissus vitacea (Knerr) Hitchc.—Woodbine, thicket creeper. Lianas, scrambling over shrubs and other low vegetation; branches becoming flaky and resinous with age, tendrils weakly branched, lacking adhesive disks although swollen at tips; leaves shiny dark green above, glabrous to pubescent below, 3-7-foliate, leaflets coarsely serrate distally, central leaflet 4-15 × 3-8 cm; flowers functionally unisexual, inconspicuous, green; berries globose, fleshy, purple to black, 5-10 mm. Damp ravines and wooded canyons, streams, springs, 3600-8000 ft. Widespread except EC. Native.

VITIS L. – Grape

Lianas, clambering or climbing, bark often exfoliating, interrupted by nodal diaphragms; tendrils usually present, 2-3-branched, lacking adhesive disks; leaves simple, palmately lobed, margins dentate to serrate; flowers functionally unisexual, plants dioecious; sepals minute, reduced to a 5-toothed to entire rim, greenish; petals (3-)5(-9), greenish; nectary free; stamens (3-)5(-9), connate distally, often 0 in pistillate flowers; style conic, short; berries purple to black; seeds 1-4 per fruit. The wine grape, *V. vinifera* L., is often found persisting where planted but does not usually escape cultivation.

1. Branchlet growing tips not enveloped by unfolding leaves; nodal diaphragms 1.5-3 mm; stipules 1-3 mm...**V. arizonica**
1'. Branchlet growing tips enveloped by unfolding leaves; nodal diaphragms 0.5(-1) mm; stipules 3-6 mm...(2)
2. Plants sprawling, clambering, or rarely climbing, strongly branched; tendrils deciduous if not attached; branchlets glabrous or with arachnoid pubescence, growing tips pubescent; inflorescences to 7(-9) cm ...**V. acerifolia**
2'. Plants often high-climbing, not much branched; tendrils persistent; branchlets glabrous, or if pubescent then hirtellous, growing tips glabrous or sparsely pubescent; inflorescences to 9-12 cm...**V. riparia**

Vitis acerifolia Raf.—Bush grape, panhandle grape. Lianas, strongly branched, scrambling over shrubs and other low vegetation, young branches arachnoid-tomentose, becoming glabrous with age; tendrils rare, stipules 3-6 mm; leaves somewhat coriaceous, blades cordate, suborbicular to triangular-ovate, 3-lobed, tips acute to long-acuminate, often wider than long, abaxial surface sparsely arachnoid to glabrate, not glaucous, adaxial surface arachnoid to glabrate; berries black, distinctly glaucous, globose, 8-11 mm. Hillslopes, stream banks, ravines, woodlands, rocky outcrops, 4560-6000 ft. NE.

Vitis arizonica Engelm. [*V. treleasei* Munson ex L. H. Bailey]—Arizona grape, canyon grape. Lianas, weakly to strongly branched, scrambling over shrubs and other low vegetation, often climbing into tree canopies, young branches arachnoid-tomentose, becoming glabrous with age, tendrils rare, stipules 1.5-3 mm; leaves thin, blades cordate to cordate-ovate, not or weakly 3-lobed, tips acute to acuminate, abaxial surface sparsely arachnoid, not glaucous, adaxial surface sparsely arachnoid to glabrate; berries black, not or rarely somewhat glaucous, globose, 6-10 mm. Stream banks, damp seepy low places, springs, 3800-8500 ft. Widespread except E.

Vitis riparia Michx. [*V. cordifolia* Michx. var. *riparia* (Michx.) A. Gray; *V. vulpina* L.]—Riverbank or frost grape. Lianas, weakly branched, scrambling to mostly strongly climbing, young branches glabrate to hirtellous, glabrous to sparsely pubescent, tendrils persistent, branched, stipules 3-5 mm; leaves thin, blades cordate, ± 3-lobed, tips acute, abaxial surface glabrate, not glaucous, adaxial surface glabrate;

berries black, distinctly glaucous, globose, 8–12 mm. Hillslopes, stream banks, ravines, thickets, road-sides, fencerows, pond margins, 5540–7000 ft. Colfax, San Miguel.

ZYGOPHYLLACEAE – CALTROP FAMILY

Kenneth D. Heil

(Ours) herbs or shrubs, annual or perennial; leaves opposite or alternate, pinnately compound or simple, stipulate, margins entire; inflorescences terminal, flowers solitary or in 2-flowered clusters (cymes); flowers perfect, actinomorphic; sepals 4–5, usually distinct; petals 4–5, distinct, white, yellow, orange, or red; stamens (5–)10 in 2 whorls; pistil 1, ovary superior, (2–)5–10-locular; ovules (1)2–10 per locule; style 1; stigma 1; fruit a capsule or schizocarp, splitting into 5 or 10 mericarps (Porter, 2016).

1. Leaflets 2…(2)
1′. Leaflets 6–16(–20)…(3)
2. Leaflets distinct…**Zygophyllum**
2′. Leaflets connate basally, leaves appearing simple and 2-lobed…**Larrea**
3. Ovaries 5-lobed, 5-locular; fruits 5-angled, spiny, breaking into mostly 5 mericarps; petals yellow, base darker; nectary of 10 glands in 2 whorls…**Tribulus**
3′. Ovaries 10-lobed, 10-locular; fruits 10-lobed, not spiny, breaking into mostly 10 mericarps; petals white to bright orange or green to red; nectary of 5 glands at bases of filaments opposite the petals…**Kallstroemia**

KALLSTROEMIA Scop. – Caltrop

Herbs, usually annual, sometimes perennial; stems prostrate to decumbent or ascending, diffusely branched, 1(–1.5) m, somewhat succulent, densely hairy to glabrate; leaves opposite, even-pinnate, leaf-lets 6–16(–20), opposite, distinct, elliptic to broadly oblong, ovate, or obovate, surfaces hairy to glabrate; flowers solitary, regular; sepals 5, distinct, green; petals 5, white to bright orange, base white to bright orange or green to red; nectary of 5, 2-lobed glands at bases of filaments opposite sepals; stamens 10; ovary sessile, 10-lobed, 10-locular, glabrous or hairy; ovules 1 per locule; fruit a schizocarp; seeds 1 per mericarp.

1. Leaves obovate in outline, terminal pairs of leaflets largest…(2)
1′. Leaves elliptic in outline, middle pairs of leaflets largest…(3)
2. Sepals usually deciduous; mericarps tuberculate, 4–5 tubercles oblong, 1–5 mm, other tubercles rounded, much <1 mm…**K. californica** (in part)
2′. Sepals persistent; mericarps tuberculate, cross-ridged or slightly keeled, tubercles if present all rounded, to 1 mm…**K. hirsutissima**
3. Flowers 20–60 mm diam.; petals 10–34 × 7–22 mm, 2-colored; pedicels in fruit 3–10.5 cm…**K. grandiflora**
3′. Flowers 8–25 mm diam.; petals 3–12 × 2.5–5 mm, 1-colored; pedicels in fruit 0.8–4 cm…(4)
4. Pedicels shorter than the subtending leaves; sepals usually deciduous; petals yellow, 3–6 × 2.5–3 mm…**K. californica** (in part)
4′. Pedicels usually longer than or equaling subtending leaves; sepals persistent; petals orange, 5–12 × 3.5–5 mm…**K. parviflora**

Kallstroemia californica (S. Watson) Vail [*K. californica* var. *brachystylis* (Vail) Kearney & Peebles]—California caltrop. Stems appressed-pubescent; leaflets 6–12(–14), elliptic to oblong, 4–17 × 1.5–9 mm, terminal pairs sometimes largest; sepals persistent or falling before the fruit matures, 6–8 mm; petals orange-yellow, 4–6 mm; schizocarps ovoid, to 4 × 3–5 mm (including tubercles); beak 2–4 mm, strongly conic at the base, not exceeding fruit body in length. Sandy sites, disturbed areas in Chihuahuan desert scrub and semiarid grasslands, 2900–7400 ft. Widespread except SE.

Kallstroemia grandiflora Torr. ex A. Gray—Arizona poppy. Stems hispid-hirsute or short-pilose; leaflets 8–16(–20), elliptic to slightly obovate, 8–25 × 2–3 mm, the middle pairs largest; sepals 6–16 mm; petals orange, 10–34 mm; schizocarps ovoid, 4–5 mm diam., strigose; beak cylindrical, 6–18 mm, 3 times as long as fruit body. Sandy sites, Chihuahuan desert scrub, 3210–5600 ft. C, WC, SE, SC, SW.

Kallstroemia hirsutissima Vail ex Small—Carpet weed. Stems spreading-hirsute; leaflets 6-8, broadly elliptic to oblong-ovate or broadly ovate, 12-19 × 5-11 mm, terminal pair largest, surfaces densely hirsute; sepals persistent, 2.5-4 mm; petals yellow, 2-4 mm; schizocarps broadly ovoid, 6-8 mm diam., strigillose; beak conic, 1-4 mm, shorter than fruit body. Chihuahuan desert scrub and adjacent semiarid grasslands, 3300-6050 ft. C, SE, SC, SW, Quay.

Kallstroemia parviflora Norton [*K. intermedia* Rydb.]—Carpet weed. Stems coarsely hirsute and sericeous with white or gray antrorse hairs, becoming glabrate; leaflets 6-10(-12), elliptic to oblong, 8-19 × 3.5-9 mm, middle pairs largest, surfaces appressed-hirsute; sepals persistent, 4-7 mm; petals orange, drying white to yellow, 5-12 mm; schizocarps ovoid, 4-6 mm diam., strigose; beak cylindrical, 3-9 mm, as long as to 3 times as long as fruit body. Roadsides, disturbed areas, grasslands, 3250-6900(-7900) ft. Widespread except NE, NC.

LARREA Cav. – Creosote bush

Larrea tridentata (DC.) Coville [*Zygophyllum tridentatum* DC.]—Creosote bush, gobernadora, hediondilla. Shrubs, divaricately branched, multistemmed, strong-scented, resinous; stems reddish when young, becoming gray or black; leaves 1-4 mm, fleshy, resinous, leaflets green to olive-brown, 4-18 × 1-8.5 mm, inequilateral; flowers to 3 cm diam.; sepals 5-8 × 3-4.5 mm, appressed-pubescent; petals twisted, 7-11 × 2.5-5.5 mm, claw brownish; stamens 5-9 mm; ovary 2-5 mm; schizocarps 4.5 mm diam., pilose-woolly, hairs white, turning reddish brown with age. Creosote bush scrub, 2940-6000(-7000) ft. EC, C, WC, SE, SC, SW.

TRIBULUS L. – Puncture vine

Tribulus terrestris L.—Puncture vine, goat head. Herbs, annual, herbage hairy (whitish), becoming glabrate; stems prostrate, green to reddish, to 1 m, ± hirsute; leaves 2-4.5 × 1 cm, leaflets (largest) 4-11 × 2-4 mm, densely sericeous, younger parts silvery, becoming glabrate; flowers 5(-10) mm diam.; sepals 2-4 × 1.5-2 mm, minutely ciliate, hirsute; petals 2.5-5 × 1-3 mm; outer whorl of nectary glands yellowish, inner whorl distinct, yellow, triangular, 0.2 mm; schizocarps (7-)10-15 mm diam. excluding 4-12 mm spines; mericarps bearing 2 conic, spreading 3-7 mm dorsal spines and sometimes 2 smaller retrorse spines near base. Often spread by the spiny mericarps sticking to bicycle or automobile tires and to the feet or coats of livestock. Agricultural lands, roadsides, railways, other disturbed areas, 3280-7100 ft. Widespread.

ZYGOPHYLLUM L. – Bean-caper

Zygophyllum fabago L.—Syrian bean-caper. Herbs or subshrubs; leaves 2-6 cm, leaflets 1-4.5 × 0.6-3 cm, linear or lanceolate; flowers 6-7 mm diam.; sepals ovate to elliptic, 5-7 × 3.5-5.5 mm, margins white; petals obovate, 7-8 mm; stamens 11-12 mm, filaments red-orange, ± linear, basal scales red-orange, apex notched; anthers red-orange; capsules 1-3.5 × 0.4-0.5 cm. Weedy sites in Chihuahuan desert scrub, 3700-4400 ft. Doña Ana.

PHOTO CREDITS FOR RARE, ENDANGERED, AND ENDEMIC PLANTS OF NEW MEXICO

All species in New Mexico considered rare, endangered, or endemic are included in the color plates. Photos are attributed to their source in the table below. KDH = Kenneth D. Heil; SOK = Steve L. O'Kane, Jr. Photographs of herbarium specimens are listed by herbarium and accession number and were obtained and modified mainly from public domain materials available online. We appreciate the photographers who have allowed us to reproduce their work.

Photograph	Photographer
Abronia bigelovii	SOK
Abronia bolackii	SOK
Abronia nealleyi	Chuck Sexton
Agalinis calycina	KDH
Agastache cana	KDH
Agastache mearnsii	COLO #417912 (specimen)
Agastache pringlei var. *verticillata*	Natures Images
Aliciella cliffordii	SOK
Aliciella formosa	SOK
Allium gooddingii	KDH
Amsonia fugatei	SOK
Amsonia tharpii	KDH
Anticlea mogollonensis	KDH
Anulocaulis leiosolenus var. *gypsogenus*	SOK/inset KDH
Anulocaulis leiosolenus var. *howardii*	SOK
Apacheria chiricahuensis	SOK
Aquilegia chaplinei	SOK
Argemone pinnatisecta	SOK
Asclepias sanjuanensis	KDH
Asplenium scolopendrium	Laura Baumann
Astragalus altus	SOK
Astragalus castetteri	KDH
Astragalus chuskanus	SOK
Astragalus cliffordii	SOK
Astragalus cobrensis var. *maguirei*	SOK
Astragalus cyaneus	SOK
Astragalus feensis	SOK
Astragalus gypsodes	KDH
Astragalus heilii	SOK
Astragalus humillimus	SOK
Astragalus humistratus var. *crispulus*	KDH

Astragalus iodopetalus	Teresa Burkhart
Astragalus kentrophyta var. *neomexicanus*	KDH
Astragalus kerrii	SOK
Astragalus knightii	SOK
Astragalus micromerius	KDH
Astragalus missouriensis var. *accumbens*	SOK
Astragalus missouriensis var. *humistratus*	SOK
Astragalus missouriensis var. *mimetes*	SOK
Astragalus monumentalis var. *cottamii*	SOK
Astragalus neomexicanus	SOK
Astragalus nutriosensis	SOK
Astragalus oocalycis	SOK
Astragalus puniceus var. *gertrudis*	SOK
Astragalus ripleyi	SOK
Astragalus siliceus	KDH
Astragalus wittmannii	KDH
Atriplex griffithsii	SOK
Bartlettia scaposa	KDH
Berlandiera macvaughii	UNM #78130 (specimen)
Boechera perennans subsp. *sanluisensis*	Patrick Alexander
Boechera perennans subsp. *zephyra*	Patrick Alexander
Calochortus gunnisonii var. *perpulcher*	SOK
Castilleja organorum	SOK
Castilleja ornata	KDH
Castilleja tomentosa	SOK
Chaetopappa hersheyi	KDH
Cirsium vinaceum	SOK
Cirsium wrightii	KDH
Cleomella multicaulis	SOK
Coryphantha robustispina subsp. *scheeri*	KDH
Crataegus wootoniana	SOK
Cuscuta draconella	Mihai Costea
Cuscuta warneri	Mihai Costea
Cylindropuntia viridiflora	SOK
Cymopterus davidsonii	SOK
Cymopterus spellenbergii	SOK
Dalea scariosa	KDH
Delphinium alpestre	SOK
Delphinium novomexicanum	SOK
Delphinium robustum	SOK
Delphinium sapellonis	SOK
Dermatophyllum gypsophilum subsp. *guadalupense*	KDH
Desmodium metcalfei	KDH
Draba abajoensis	SOK
Draba heilii	SOK
Draba henrici	SOK
Draba mogollonica	SOK
Draba smithii	SOK
Draba standleyi	SOK
Echinocereus fendleri var. *kuenzleri*	SOK
Ericameria nauseosa var. *texensis*	KDH
Erigeron acomanus	SOK

Erigeron hessii	Jeremy McClain
Erigeron rhizomatus	SOK
Erigeron rybius	KDH
Erigeron scopulinus	SOK
Erigeron sivinskii	SOK
Erigeron subglaber	KDH
Eriogonum aliquantum	KDH
Eriogonum gypsophilum	SOK
Eriogonum lachnogynum var. *colobum*	SOK
Eriogonum lachnogynum var. *sarahiae*	SOK
Eriogonum wootonii	SOK
Eryngium sparganophyllum	KDH
Erythranthe plotocalyx	Sue Carnahan
Euphorbia rayturneri	Patrick Alexander
Geranium dodecatheoides	KDH
Grindelia arizonica var. *neomexicana*	KDH
Grusonia clavata	SOK
Hackelia hirsuta	SOK
Hedeoma apiculata	KDH
Hedeoma pulcherrima	SOK
Hedeoma todsenii	KDH
Helianthus arizonensis	SOK
Helianthus paradoxus subsp. *paradoxus*	KDH
Heterotheca cryptocephala	ASU #243384 (specimen)
Heterotheca sierrablancensis	SOK
Heuchera glomerulata	Patrick Alexander
Heuchera pulchella	SOK
Heuchera soltisii	Patrick Alexander
Heuchera woodsiaphila	Patrick Alexander
Heuchera wootonii	SOK
Hexalectris arizonica	SOK
Hexalectris colemanii	SOK
Hexalectris nitida	Ron Coleman
Hieracium brevipilum	KDH
Hymenoxys ambigens var. *neomexicana*	SOK
Hymenoxys brachyactis	KDH
Hymenoxys vaseyi	SOK
Ionactis elegans	SOK
Ipomoea gilana	Daniella Roth
Ipomopsis sancti-spiritus	SOK
Ipomopsis wrightii	LL #00330879 (specimen)
Justicia wrightii	SOK
Lechea mensalis	BRIT #473939 (specimen)
Lepidospartum burgessii	KDH
Leucosyris blepharophylla	SOK
Limosella pubiflora	KDH
Linum allredii	SOK
Lorandersonia microcephala	KDH
Lupinus sierrae-blancae	SOK
Malaxis abieticola	Jim Fowler
Mentzelia conspicua	KDH
Mentzelia filifolia	KDH

Mentzelia humilis var. *guadalupensis*	SOK
Mentzelia perennis	SOK
Mentzelia sivinskii	KDH
Mentzelia springeri	KDH
Mentzelia todiltoensis	KDH
Microthelys rubrocallosa	Ron Coleman
Monarda humilis	KDH
Mostacillastrum subauriculatum	KDH
Muhlenbergia villiflora var. *villosa*	KDH
Muilla lordsburgana	Patrick Alexander
Nama xylopoda	SOK
Nerisyrenia hypercorax	Patrick Alexander
Oenothera organensis	KDH
Opuntia arenaria	KDH/inset Natures Images
Packera cardamine	SOK
Packera cliffordii	SOK
Packera sanguisorboides	SOK
Packera spellenbergii	SOK
Panicum mohavense	Janet Mygatt
Paronychia wilkinsonii	Patrick Alexander
Pediocactus knowltonii	SOK
Pediomelum pentaphyllum	KDH
Pelecyphora duncanii	KDH
Pelecyphora guadalupensis	SOK
Pelecyphora leei	KDH
Pelecyphora orcuttii	KDH
Pelecyphora organensis	SOK
Pelecyphora sandbergii	KDH
Pelecyphora sneedii	KDH
Pelecyphora villardii	KDH
Peniocereus greggii var. *greggii*	SOK
Penstemon alamosensis	KDH
Penstemon bleaklyi	SOK
Penstemon cardinalis subsp. *cardinalis*	KDH
Penstemon cardinalis subsp. *regalis*	SOK
Penstemon crandallii subsp. *taosensis*	SOK
Penstemon inflatus	KDH
Penstemon metcalfei	KDH
Penstemon neomexicanus	SOK
Perityle cernua	SOK
Perityle quinqueflora	SOK
Perityle staurophylla var. *homoflora*	SOK
Perityle staurophylla var. *staurophylla*	SOK
Phacelia cloudcroftensis	SOK
Phacelia serrata	KDH
Phacelia sivinskii	KDH
Phemeranthus brachypodius	SOK
Phemeranthus humilis	SOK
Phemeranthus rhizomatus	SOK
Philadelphus microphyllus subsp. *argyrocalyx*	SOK
Phlox caryophylla	SOK
Phlox cluteana	SOK

Phlox vermejoensis	Ben Legler
Physaria aurea	SOK/inset KDH
Physaria gooddingii	SOK
Physaria lata	SOK/inset KDH
Physaria navajoensis	SOK
Physaria newberryi subsp. *yesicola*	SOK
Physaria pinetorum subsp. *iveyana*	SOK
Physaria pruinosa	SOK
Potentilla sierrae-blancae	SOK
Proatriplex pleiantha	KDH
Puccinellia parishii	SOK
Rhinotropis rimulicola var. *mescalerorum*	Gregory Penn
Rhinotropis rimulicola var. *rimulicola*	KDH
Rhodiola integrifolia subsp. *neomexicana*	SOK
Ribes mescalerium	SOK
Salix arizonica	SOK
Salvia summa	SOK
Schoenus nigricans	"Ruff tuff cream puff"
Sclerocactus mesae-verdae subsp. *depressus*	KDH
Sclerocactus mesae-verdae subsp. *mesae-verdae*	KDH
Sclerocactus papyracanthus	SOK (plant)/David Schleser (flower)
Scrophularia laevis	KDH
Scrophularia macrantha	SOK
Scrophularia montana	SOK
Senecio sacramentanus	KDH
Senecio warnockii	Roth
Sicyos glaber	Richard Spellenberg
Silene plankii	KDH
Silene thurberi	SOK
Silene wrightii	KDH
Solidago capulinensis	KDH
Solidago correllii	SOK
Spermolepis organensis	Patrick Alexander
Sphaeralcea wrightii	KDH
Stellaria porsildii	SOK
Streptanthus platycarpus	SOK
Synthyris oblongifolia	SOK
Tetradymia filifolia	SOK
Townsendia gypsophila	SOK
Trifolium longipes subsp. *neurophyllum*	SOK
Valeriana texana	SOK
Viola calcicola	KDH
Yucca baileyi var. *intermedia*	SOK

APPENDIX I
A BRIEF HISTORY OF WHITE SANDS MISSILE RANGE (WSMR), NEW MEXICO

Summarized from Eckles[1] by Kenneth D. Heil

Without the financial support of White Sands Missile Range (WSMR), this project would not have seen the light of day. In thanks, a brief history of WSMR is given here.

INTRODUCTION

White Sands Missile Range in southern New Mexico is 3200 square miles, as large as Rhode Island and Delaware combined. Since 1942 most of the land has been restricted by the military. The shape of the WSMR boundary is basically a large rectangle (see map), and within the rectangle are White Sands National Park and the San Andres National Wildlife Refuge. The main post employs several thousand people.

West of the main post are the Organ Mountains and the impressive Organ Needle and Sugarloaf Peak. North of the Organ Mountains, the land slopes into the Tularosa Basin. Near the main post are mounds of sand dunes anchored with mesquite. In the Tularosa Basin is White Sands National Park. The dunes of gypsum originate from Lake Otero, a shallow lake that at one time covered several hundred square miles. Farther north are the El Malpais lava beds, which are several thousand years old, and "mound springs," small hills with craters filled with water and riparian vegetation.

Near the northeast boundary, the Oscura Mountains run south to north. The west side is steep and covered with desert scrub while the east side is covered with piñon-juniper. The Trinity Site is a few miles west of the Oscura Mountains. At the northeast boundary is Stallion Gate, near where Range Road #7 connects with U.S. 380. From the junction of U.S. 70 to Stallion Gate it is 118 miles. At the northwest boundary is a lava flow that is three-quarters of a million years old and approximately 170 square miles. Windblown sand covers several lava tubes, which contain a large population of bats. West of the lava tubes is Jornada del Muerto—"Route of the Dead Man"—a name given by the Spanish conquistadors to the 100-mile dry stretch between Las Cruces and Socorro.

South of Stallion Gate is Mockingbird Gap and the Mockingbird Mountains; this is where the Oscura Mountains separate from the San Andres Mountains. The San Andres Mountains run south to north for 80 miles with a steep west slope and a gentler east slope. Salinas Peak is the highest point at 8967 feet and includes a bench with a small ponderosa pine community. Cutting through the San Andres Mountains is Rhodes Canyon, a major west-to-east drainage with thick stands of piñon pine and juniper. This drainage was the location of the old NM Road 52, a major route to Truth or Consequences (Hot Springs).

South of Rhodes Canyon is the Hembrillo Basin and Hembrillo Canyon. In this canyon are several locations for Apache and Mogollon rock art. Within the basin is the Hembrillo battlefield, where the last battle in the United States occurred between Victorio's Warm Springs Apaches and the "Buffalo Soldiers." Also within the basin is Victorio Peak and its legendary 100 tons of gold bars. South of the Hembrillo area

[1] Eckles, Jim. 2013. Pocketful of Rockets: History and Stories behind White Sands Missile Range. Fiddlebike Partnership, Las Cruces, New Mexico.

is San Andres National Wildlife Refuge and Ash Canyon. Ash Canyon contains perennial water, including small pools and waterfalls surrounded by ash trees, yellow columbine, and maidenhair fern. West of the refuge is San Augustin Pass, and to the south, the Organ Mountains.

PREHISTORIC NATIVE AMERICANS AND SPANISH

Evidence of hunter-gatherers is found in their pictographs and petroglyphs. Large amounts of pottery are found around water sources, and corn has been found in shelters that dates back 3000 years. Near the southeast boundary and the Orogrande Gate, several pit houses have been found dating from 1100 to 1500 years ago.

The Jornada branch of the Mogollon culture dates back to around 1000 years ago. By 700 years ago there were adobe pueblos, and throughout the range numerous arrowheads, beads, pendants, and bracelets have been found. Evidence of the Mogollon culture is found along the Organ and San Andres Mountains as well as at Lake Lucero.

After the Mogollon culture came the Apache, dating back to the 1500s and earlier as evidenced by Apache pictographs and petroglyphs found in Hembrillo Canyon. Also in the canyon are agave-roasting pits. Here the Apache roasted agave bulbs, mostly from *Agave palmeri*.

The first Europeans to enter WSMR might have been the Spanish expedition with Cabeza de Vaca in the 1530s. However, there is no evidence that supports Cabeza de Vaca passing through the Tularosa Basin or San Augustin Pass. The Spanish probably arrived during the seventeenth and eighteenth centuries. Their Camino Real (Royal Road) was just a few miles west of the missile range. Some of the prospect holes in the San Andres Mountains may be Spanish in origin.

THE COX FAMILY AND THE SAN AUGUSTINE RANCH

The Wheeler Mapping Expedition was the first to survey the Tularosa Basin, and once the area was mapped, Texans moved in with their cattle. One of the first families to move into the region was the Coxes, and they are still at the San Augustine Ranch, having occupied it for over 100 years (from Pat Garrett to missiles and rockets).

W. W. Cox arrived in the Tularosa Basin in 1888 and first camped below Black Mountain in the San Andres Mountains. The family lived in a dugout next to the San Augustine Ranch. Eventually W. W. was able to purchase the ranch. It was appealing to W. W. because of the natural springs and the thunderstorms over the Organ Mountains. These water sources provided a dependable water supply; however, drought would still prove to be a major problem over the years. W. W. started by raising sheep but quickly switched to Hereford cattle. Over time, the ranch grew to around 150,000 acres.

Hal Cox and his wife, Alyce Lee, eventually took over the ranch after the death of W. W. in 1926. Hal brought electricity to the ranch. In the 1940s Jim Cox and his wife, Fannie, took over the San Augustine Ranch. At this time, he was forced to sell 90 percent of the ranch to the U.S. Army, and that land later became the White Sands Proving Ground. After the war, land was also taken from the ranch to form the Alamogordo Bombing Range. In 1976, Rob and Murnie Cox took over the ranch, and the relationship between White Sands and the Coxes grew into a warm friendship.

COLORFUL CHARACTERS OF THE OLD WEST

Billy the Kid and Pat Garrett

In late 1880 Sheriff Pat Garrett captured Billy the Kid in northeast New Mexico. The Kid was charged with killing William Brady, the sheriff of Lincoln County. Eventually Garrett and the Kid ended up in Mesilla in 1881 for a trial. The lawyer for the Kid was Albert Fountain, and the Kid was found guilty and was transported to Lincoln to be hanged. On the way to Lincoln the party traveled along the old wagon road over San Augustin Pass near present-day WSMR and the Cox Ranch. On April 28, 1881, Billy the Kid escaped jail and eventually became a hero while Pat Garrett's reputation sank.

Pat Garrett and the Albert Fountain Case

In 1896 Albert Fountain and his eight-year-old son disappeared. Pat Garrett, the Doña Ana County sheriff, was elected to investigate. The disappearance of the two Fountains is one of the great murder mysteries of New Mexico. Fountain had driven a buckboard to Lincoln to get indictments against well-known individuals for cattle rustling. On his return trip Fountain and his son disappeared at Chalk Hill (now in the heart of WSMR). A posse later found an empty buckboard, empty cartridges, and blood-soaked sand. The posse followed tracks to the east toward the Oliver Lee Ranch and then lost them.

Garrett focused his search on Oliver Lee, Jim Gililland, and Bill McNew as suspects. However, Oliver Lee was very popular, powerful, and well connected politically. Eventually, the three suspects were tried in Hillsboro, represented by Albert Fall. After 18 days the case went to the jury and the three suspects were found not guilty. The mystery of the Fountains' disappearance has never been solved and the two bodies have never been found.

The End of Pat Garrett

The Garrett Ranch was east of San Augustin Pass and north of Mineral Hill on WSMR. Garrett added more property north of his ranch in nearby Bear Canyon. By 1908 Garrett was down on his luck and quite poor and was hoping to make some money off his northern property in Bear Canyon. However, because of confusion with his son Poe, the place was leased to Wayne Brazel. Brazel showed up with goats instead of cattle, and this made Garrett furious. After arguing, Garrett threatened him. Later, a solution looked to be made when a party came forth to take over the Bear Canyon lease, buy the goats, and run cattle instead of goats. On February 29, Garrett and Carl Adamson took a buckboard from the Garrett home toward Las Cruces and met Wayne Brazel. Garrett and Brazel started arguing again. Garrett ended up shot in the back of the head and stomach. He died on the spot.

There are several versions of what happened. One is that Garrett was so angry that he reached for his shotgun and Brazel shot him. A jury found Brazel not guilty. Another version is that Jim Miller, a hired assassin, shot him with a rifle.

RANCHING

The military moved into the Tularosa Basin in 1942, and by 1950, more than 80 ranches like the Coxes' were affected by the military takeover of the area. Most ranches ran cattle or goats or rarely sheep. Besides the land they owned, most ranchers leased federal and state lands. To be a successful rancher in the Chihuahuan Desert required quite a bit of land. The "carrying capacity" for most cattle ranches was 5 or 6 cows per 640 acres or a little over 12 cow-calf pairs per square mile, or five times that many goats.

Without potable water for humans and livestock, there would be no ranching in the Tularosa Basin. Near White Sands National Park just underground is a sea of mineral-laden water, which in many places is near the surface. Digging a hole a couple of feet deep caused a puddle of water to form; water near the surface was not as salty as the rest. Therefore, livestock could build up a tolerance for it. Other methods were used to capture water in the Tularosa Basin. Water in the foothills and mountains was excellent, and wells drilled in these areas provided water for both livestock and humans. Often ranchers would store water in tanks for gardening, family use, and livestock.

In late May 1942 and during World War II, the U.S. Army arrived and formed the Alamogordo Bombing Range (later renamed Holloman Air Force Base). Before B-17s could fly, humans and livestock needed to be removed from the area. Therefore, the Army Corps of Engineers visited ranches to tell the owners they had to move out of the area, at least until the end of the war. Most ranchers had to move within two weeks and received no funds from the government for their moving expenses.

In 1944 Germany was launching V-2 rockets against England. It became apparent to the military that America needed a land-based range for development of rocket technology. In 1945 the Corps of Engineers issued a real estate directive that the Alamogordo Bombing Range was the area that best met the requirements for testing rockets. Therefore, ranchers were never getting their land back! Construction of buildings and roads began on June 25. Eventually boundaries were set, and by 1974 Congress authorized the purchase of private ranchlands and about 200 mining claims. Most ranchers sold voluntarily.

THE TRINITY SITE

The Trinity Site (nps.gov/whsa/learn/historyculture/trinity-site.htm), where the first atomic bomb was tested on July 16, 1945, is open to the public twice a year. The bomb was placed atop a 100-foot steel tower at what was called "Ground Zero." From Ground Zero, equipment and observation points were established at varying distances. The observation points were constructed of concrete and earthen barricades, with the nearest point 5.7 miles from Ground Zero. "Gadget," the nuclear device, was detonated at 5:30 a.m. A multicolored cloud rose 38,000 feet into the air, and where the tower had been, there was a crater about half a mile across and eight feet deep. Sand in the crater was fused into a glasslike solid the color of green jade, which was given the name trinitite.

The flash of light and shock wave made an impression over a radius of around 160 miles. On August 6, 1945, the atomic bomb code-named "Little Boy" was exploded over Hiroshima, Japan, and three days later a third bomb, "Fat Man," exploded over Nagasaki.

In 1965, army officials erected a monument at Ground Zero. In 1975, the National Park Service designated Trinity Site as a National Historic Landmark. The landmark includes the base camp, where the scientists and support group lived; the McDonald ranch house, where the plutonium core was assembled; and Ground Zero.

V-2 Rocket

The V-2 rocket (https://wsmrmuseum.com/wp-content/uploads/2020/08/v-2-handout.pdf) was the world's first large-scale rocket. It was developed by Germany during World War II and used as a weapon. The rocket could deliver a 2000-pound warhead at supersonic speed to target areas 150 miles away.

The architect of the rocket was Dr. Wernher von Braun, who based much of his design on the pioneering work done by Robert Goddard during the 1930s near Roswell, New Mexico. In late 1945, V-2 components captured in Europe arrived at WSMR, and from 1946 to 1952, 67 V-2 rockets were fired. This launched the United States into the Space Age.

The V-2 provided the United States with valuable experience in the assembly, preflight testing, handling, fueling, launching, and tracking of large missiles. The scientific experiments conducted aboard the V-2 gave us new information about the upper atmosphere, our first photographic look back at the earth from space, and our first large two-stage rocket. The first biological experiments were also carried into space. The V-2 led to the Redstone and Jupiter missiles, which led to the Saturn series and the capability to send humans to the moon.

WSMR SUMMARY

WSMR (https://www.army-technology.com/projects/white-sands-range/) has grown to be the largest military installation in the United States and now occupies 3200 square miles. It has facilities for testing, research and development, evaluation, and training activities conducted by the U.S. Army, U.S. Navy, U.S. Air Force, other government agencies, and private industry.

After the V-2 missile was launched in April 1946, the U.S. Navy's Viking No. 1 rocket was launched to an altitude of 51.5 miles from White Sands Proving Grounds in May 1949. The U.S.S. Desert Ship, a concrete blockhouse with shipboard condition simulations, was built in 1958.

Now the White Sands Test Facility tests and evaluates hazardous materials, spacecraft components, and rocket propulsion systems. Other organizations located at the missile range are the 46th Test Group, Second Engineer Battalion, the 746th Test Squadron, Army Research Laboratory, Center for Countermeasures, and the National Geospatial-Intelligence Agency.

At the WSMR main post are family housing units, an aquatic center, auto skills center, fire department, gymnasium, restaurants, bowling center, elementary school, golf course, and childcare and youth programs. Much has changed since 1945.

APPENDIX II
CHECKLIST* OF THE FLORA OF
WHITE SANDS MISSILE RANGE (WSMR),
NEW MEXICO

Compilation Committee:
David Anderson
Kenneth D. Heil
Steve L. O'Kane, Jr.
Gregory Silsby

FERNS AND FERN ALLIES

Aspleniaceae
Asplenium trichomanes L.

Dryopteridaceae
Woodsia neomexicana Windham
Woodsia phillipsii Windham
Woodsia plummerae Lemmon

Pteridaceae
Astrolepis cochisensis (Goodd.) D. M. Benham & Windham subsp. cochisensis
Astrolepis sinuata (Lag. ex Sw.) D. M. Benham & Windham subsp. sinuata
Cheilanthes alabamensis (Buckley) Kunze
Cheilanthes fendleri Hook.
Cheilanthes tomentosa Link
Notholaena standleyi Maxon
Pellaea intermedia Mett. ex Kuhn

GYMNOSPERMS

Cupressaceae
Cupressus arizonica Greene
Juniperus coahuilensis (Martínez) Gaussen ex R. P. Adams var. arizonica R. P. Adams
Juniperus deppeana Steud.

Ephedraceae
Ephedra torreyana S. Watson
Ephedra trifurca Torr. ex S. Watson
Ephedra viridis Coville

Pinaceae
Pinus edulis Engelm.
Pinus scopulorum (Engelm.) Lemmon

ANGIOSPERMS

Acanthaceae
Carlowrightia linearifolia (Torr.) A. Gray
Justicia pilosella (Nees) Hilsenb.

---

* The scientific names currently used at the White Sands Missile Range Herbarium may not be the same as those found in the *Vascular Plants of New Mexico*.

Ruellia parryi A. Gray
Stenandrium barbatum Torr. & A. Gray

Adoxaceae

Sambucus caerulea Raf. mexicana (C. Presl ex DC.) L. Benson
Sambucus caerulea Raf. neomexicana (Wooton) Rehder

Aizoaceae

Trianthema portulacastrum L.

Amaranthaceae

Acanthochiton wrightii Torr.
Allenrolfea occidentalis (S. Watson) Kuntze
Amaranthus arenicola I. M. Johnst.
Amaranthus blitoides S. Watson
Amaranthus fimbriatus (Torr.) Benth. ex S. Watson
Amaranthus hybridus L.
Amaranthus palmeri S. Watson
Amaranthus powellii S. Watson
Amaranthus retroflexus L.
Amaranthus torreyi (A. Gray) Benth. ex S. Watson
Atriplex canescens (Pursh) Nutt.
Atriplex elegans (Moq.) D. Dietr. var. elegans
Atriplex rosea L.
Bassia hyssopifolia (Pall.) Kuntze
Chenopodium album L. var. album
Chenopodium album L. lanceolatum (Muhl. ex Willd.) Coss. & Germ.
Chenopodium berlandieri Moq. var. zschackei (Murr) Murr ex Graebn.
Chenopodium desiccatum A. Nelson var. desiccatum
Chenopodium fremontii S. Watson
Chenopodium incanum (S. Watson) A. Heller
Chenopodium leptophyllum (Moq.) Nutt. ex S. Watson
Chenopodium rubrum L.
Chenopodium watsonii A. Nelson
Corispermum americanum (Nutt.) Nutt.
Dysphania botrys (L.) Mosyakin & Clemants
Dysphania graveolens (Willd.) Mosyakin & Clemants
Froelichia arizonica Thornber ex Standl.
Froelichia floridana (Nutt.) Moq. campestris (Small) Fernald
Froelichia gracilis (Hook.) Moq.
Gomphrena caespitosa Torr.
Guilleminea densa (Humb. & Bonpl. ex Schult.) Moq. var. aggregata Uline & W. L. Bray
Halogeton glomeratus (M. Bieb.) C. A. Mey.
Kochia scoparia (L.) Schrad.
Krascheninnikovia lanata (Pursh) A. Meeuse & A. Smit
Salsola tragus L.
Sarcocornia utahensis (Tidestr.) A. J. Scott
Suaeda nigra (Raf.) J. F. Macbr.
Tidestromia lanuginosa (Nutt.) Standl. subsp. lanuginosa
Tidestromia suffruticosa (Torr.) Standl.

Amaryllidaceae

Allium cernuum Roth
Allium geyeri S. Watson var. geyeri
Allium kunthii G. Don
Allium macropetalum Rydb.

Anacardiaceae

Rhus microphylla Engelm.
Rhus trilobata Nutt. var. pilossima Engl.
Rhus trilobata Nutt. var. trilobata

Apiaceae

Aletes acaulis (Torr.) J. M. Coult. & Rose
Cymopterus lemmonii (J. M. Coult. & Rose) Dorn

Apocynaceae
Amsonia arenaria Standl.
Apocynum cannabinum L.
Asclepias arenaria Torr.
Asclepias asperula (Decne.) Woodson
Asclepias brachystephana Engelm. ex Torr.
Asclepias latifolia (Torr.) Raf.
Asclepias subverticillata (A. Gray) Vail

Asparagaceae
Agave gracilipes Trel.
Agave neomexicana Wooton & Standl.
Agave parryi Engelm. var. neomexicana (Wooton & Standl.) McKechnie
Asparagus officinalis L.
Dasylirion wheeleri S. Watson
Nolina micrantha I. M. Johnst.
Nolina microcarpa S. Watson
Nolina texana S. Watson
Yucca baccata Torr. var. baccata
Yucca elata (Engelm.) Engelm.
Yucca treculeana Carrière

Asteraceae
Achillea millefolium L.
Acourtia nana (A. Gray) Reveal & R. M. King
Acourtia wrightii (A. Gray) Reveal & R. M. King
Acroptilon repens (L.) DC.
Ageratina herbacea (A. Gray) R. M. King & H. Rob.
Ageratina wrightii (A. Gray) R. M. King & H. Rob.
Aldama cordifolia (A. Gray) E. E. Schill. & Panero
Amauriopsis dissecta (A. Gray) Rydb.
Ambrosia acanthicarpa Hook.
Ambrosia monogyra (Torr. & A. Gray) Strother & B. G. Baldwin
Ambrosia psilostachya DC.
Aphanostephus ramosissimus DC. var. humilis (Benth.) B. L. Turner & Birdsong
Arida parviflora (A. Gray) Morgan & R. L. Hartm.
Artemisia bigelovii A. Gray
Artemisia carruthii Alph. Wood ex J. H. Carruth
Artemisia dracunculus L.
Artemisia filifolia Torr.
Artemisia frigida Willd.
Artemisia ludoviciana Nutt. subsp. albula (Wooton) D. D. Keck
Artemisia ludoviciana Nutt. subsp. mexicana (Willd. ex Spreng.) D. D. Keck
Artemisia ludoviciana Nutt. subsp. redolens (A. Gray) D. D. Keck
Artemisia ludoviciana Nutt. subsp. sulcata (Rydb.) D. D. Keck
Baccharis emoryi A. Gray
Baccharis pteronioides DC.
Baccharis salicifolia (Ruiz & Pav.) Pers.
Baccharis salicina Torr. & A. Gray
Baccharis sarothroides A. Gray
Bahia absinthifolia Benth. dealbata (A. Gray) A. Gray
Bahia biternata A. Gray
Bahia pedata A. Gray
Baileya multiradiata Harv. & A. Gray
Bebbia juncea (Benth.) Greene var. aspera Greene
Berlandiera lyrata Benth.
Bidens bigelovii A. Gray
Bidens heterosperma A. Gray
Bidens laevis (L.) Britton, Sterns & Poggenb.
Bidens leptocephala Sherff
Bidens tenuisecta A. Gray
Brickellia brachyphylla (A. Gray) A. Gray
Brickellia californica (Torr. & A. Gray) A. Gray

Brickellia eupatorioides (L.) Shinners
Brickellia grandiflora (Hook.) Nutt.
Brickellia laciniata A. Gray
Brickellia venosa (Wooton & Standl.) B. L. Rob.
Calycoseris wrightii A. Gray
Calyptocarpus vialis Less.
Carduus nutans L.
Carphochaete bigelovii A. Gray
Centaurea melitensis L.
Chaenactis stevioides Hook. & Arn.
Chaetopappa ericoides (Torr.) G. L. Nesom
Chloracantha spinosa (Benth.) G. L. Nesom
Chrysactinia mexicana A. Gray
Cirsium arizonicum (A. Gray) Petr.
Cirsium calcareum (M. E. Jones) Wooton & Standl.
Cirsium neomexicanum A. Gray
Cirsium ochrocentrum A. Gray
Cirsium undulatum (Nutt.) Spreng.
Cirsium vulgare (Savi) Ten.
Conoclinium dissectum A. Gray
Conyza bonariensis (L.) Cronquist
Conyza canadensis (L.) Cronquist var. canadensis
Conyza canadensis (L.) Cronquist var. glabrata (A. Gray) Cronquist
Conyza ramosissima Cronquist
Diaperia verna (Raf.) Morefield
Dicranocarpus parviflorus A. Gray
Dieteria asteroides Torr. var. asteroides
Dieteria bigelovii (A. Gray) D. R. Morgan & R. L. Hartm.
Dieteria canescens (Pursh) Nutt. var. glabra (A. Gray) D. R. Morgan & R. L. Hartm.
Dyssodia papposa (Vent.) Hitchc.
Engelmannia peristenia (Raf.) Goodman & C. A. Lawson
Ericameria laricifolia (A. Gray) Shinners
Ericameria nauseosa (Pursh) G. L. Nesom & G. I. Baird var. bigelovii (A. Gray) G. L. Nesom & G. I. Baird
Ericameria nauseosa (Pursh) G. L. Nesom & G. I. Baird var. graveolens (Nutt.) Reveal & Schuyler
Ericameria nauseosa (Pursh) G. L. Nesom & G. I. Baird var. latisquamea (A. Gray) G. L. Nesom & G. I. Baird
Ericameria nauseosa (Pursh) G. L. Nesom & G. I. Baird var. oreophila (A. Nelson) G. L. Nesom & G. I. Baird
Erigeron bellidiastrum Nutt. var. arenarius (Greene) G. L. Nesom
Erigeron bellidiastrum Nutt. var. bellidiastrum
Erigeron bigelovii A. Gray
Erigeron divergens Torr. & A. Gray
Erigeron flagellaris A. Gray
Erigeron modestus A. Gray
Erigeron tracyi Greene
Flaveria trinervia (Spreng.) C. Mohr
Flourensia cernua DC.
Gaillardia pinnatifida Torr.
Gaillardia pulchella Foug.
Grindelia nuda Wood var. aphanactis (Rydb.) G. L. Nesom
Grindelia squarrosa (Pursh) Dunal
Gutierrezia microcephala (DC.) A. Gray
Gutierrezia sarothrae (Pursh) Britton & Rusby
Gutierrezia sphaerocephala A. Gray
Gutierrezia texana (DC.) Torr. & A. Gray
Gymnosperma glutinosum (Spreng.) Less.
Haploësthes greggii A. Gray var. texana (J. M. Coult.) I. M. Johnst.
Hedosyne ambrosiifolia (A. Gray) Strother
Helianthus annuus L.
Helianthus ciliaris DC.
Helianthus laciniatus (L.) E. H. L. Krause
Helianthus petiolaris Nutt. var. canescens A. Gray
Helianthus petiolaris Nutt. var. fallax (Heiser) B. L. Turner
Heliomeris multiflora Nutt. var. multiflora

Heterosperma pinnatum Cav.
Heterotheca fulcrata (Greene) Shinners var. fulcrata
Heterotheca subaxillaris (Lam.) Britton & Rusby var. latifolia (Buckley) Gandhi & R. D. Thomas
Heterotheca villosa (Pursh) Shinners var. minor (Hook.) Semple
Heterotheca villosa (Pursh) Shinners var. nana (A. Gray) Semple
Heterotheca viscida (A. Gray) V. L. Harms
Hieracium fendleri Sch. Bip.
Hymenopappus biennis B. L. Turner
Hymenopappus filifolius Hook. var. cinereus (Rydb.) I. M. Johnst.
Hymenopappus flavescens A. Gray var. canotomentosus A. Gray
Hymenopappus flavescens A. Gray var. flavescens
Hymenothrix wislizeni A. Gray
Hymenothrix wrightii A. Gray
Hymenoxys odorata DC.
Hymenoxys richardsonii (Hook.) Cockerell var. floribunda (A. Gray) K. F. Parker
Hymenoxys vaseyi (A. Gray) Cockerell
Isocoma pluriflora (Torr. & A. Gray) Greene
Isocoma wrightii (A. Gray) Rydb.
Jefea brevifolia (A. Gray) Strother
Koanophyllon solidaginifolium (A. Gray) R. M. King & H. Rob.
Lactuca serriola L.
Laennecia coulteri (A. Gray) G. L. Nesom
Leuciva dealbata (A. Gray) Rydb.
Liatris punctata Hook. var. punctata
Lorandersonia baileyi (Wooton & Standl.) Urbatsch, R. P. Roberts & Neubig
Lorandersonia pulchella (A. Gray) Urbatsch, R. P. Roberts & Neubig
Lorandersonia spathulata (L. C. Anderson) Urbatsch, R. P. Roberts & Neubig
Machaeranthera tanacetifolia (Kunth) Nees
Malacothrix fendleri A. Gray
Malacothrix stebbinsii W. S. Davis & P. H. Raven
Melampodium leucanthum Torr. & A. Gray
Packera neomexicana (A. Gray) W. A. Weber & Á. Löve var. neomexicana
Packera thurberi (A. Gray) B. L. Turner
Palafoxia sphacelata (Nutt. ex Torr.) Cory
Parthenium confertum A. Gray var. lyratum (A. Gray) Rollins
Parthenium incanum Kunth
Pectis angustifolia Torr.
Pectis filipes Harv. & A. Gray var. subnuda Fernald
Pectis papposa Harv. & A. Gray var. papposa
Pectis prostrata Cav.
Pentzia incana (Thunb.) Kuntze
Pericome caudata A. Gray
Perityle staurophylla (Barneby) Shinners var. staurophylla
Porophyllum scoparium A. Gray
Psilostrophe tagetina (Nutt.) Greene var. tagetina
Sanvitalia abertii A. Gray
Senecio flaccidus Less. var. douglasii (DC.) B. L. Turner & T. M. Barkley
Senecio flaccidus Less. var. flaccidus
Senecio neomexicanus A. Gray var. neomexicanus
Solidago wrightii A. Gray var. adenophora S. F. Blake
Stephanomeria exigua Nutt.
Stephanomeria pauciflora (Torr.) A. Nelson
Taraxacum officinale F. H. Wigg.
Tetradymia filifolia Greene
Tetraneuris scaposa (DC.) Greene var. linearis (Nutt.) & K. F. Parker
Thymophylla acerosa (DC.) Strother
Thymophylla pentachaeta (DC.) Small var. belenidium (DC.) Strother
Tragopogon dubius Scop.
Trixis californica Kellogg
Verbesina encelioides (Cav.) Benth. & Hook. f. ex A. Gray var. exauriculata B. L. Rob. & Greenm.
Viguiera dentata (Cav.) Spreng.
Viguiera multiflora (Nutt.) S. F. Blake

Viguiera stenoloba S. F. Blake
Xanthisma gracile (Nutt.) D. R. Morgan & R. L. Hartm.
Xanthisma spinulosum (Pursh) D. R. Morgan & R. L. Hartm. var. spinulosum
Zinnia grandiflora Nutt.

Berberidaceae

Berberis haematocarpa Wooton

Bignoniaceae

Chilopsis linearis (Cav.) Sweet subsp. linearis

Boraginaceae

Cryptantha cinerea (Greene) Cronquist var. cinerea
Cryptantha micrantha (Torr.) I. M. Johnst.
Cryptantha pterocarya (Torr.) Greene var. cycloptera (Greene) J. F. Macbr.
Hackelia pinetorum (Greene ex A. Gray) I. M. Johnst. var. pinetorum
Lappula occidentalis (S. Watson) Greene var. occidentalis
Lithospermum incisum Lehm.
Lithospermum multiflorum Torr. ex A. Gray
Lithospermum viride Greene
Nama hispida A. Gray
Phacelia crenulata Torr. ex S. Watson
Phacelia integrifolia Torr. var. integrifolia
Phacelia pinkavae N. D. Atwood
Phacelia rupestris Greene
Tiquilia canescens (A. DC.) A. T. Richardson var. canescens

Brassicaceae

Boechera fendleri (S. Watson) W. A. Weber
Descurainia pinnata (Walter) Britton subsp. ochroleuca (Wooton) Detling
Draba cuneifolia Nutt. ex Torr. & A. Gray
Erysimum capitatum (Douglas ex Hook.) Greene var. capitatum
Erysimum capitatum (Douglas ex Hook.) Greene var. purshii (Durand) Rollins
Halimolobos diffusa (A. Gray) O. E. Schulz
Hesperidanthus linearifolius (A. Gray) Rydb.
Lepidium alyssoides A. Gray
Lepidium lasiocarpum Nutt. ex Torr. & A. Gray var. wrightii
Physaria fendleri (A. Gray) O'Kane & Al-Shehbaz
Physaria gordonii (A. Gray) O'Kane & Al-Shehbaz
Physaria purpurea (A. Gray) O'Kane & Al-Shehbaz
Schoenocrambe linearifolia (A. Gray) Rollins
Thelypodiopsis linearifolia (A. Gray) Al-Shehbaz
Thelypodium wrightii A. Gray

Cactaceae

Cylindropuntia imbricata (Haw.) F. M. Knuth var. imbricata
Cylindropuntia leptocaulis (DC.) F. M. Knuth
Echinocactus horizonthalonius Lem.
Echinocereus coccineus Engelm.
Echinocereus fendleri (Engelm.) Sencke ex J. N. Haage
Echinocereus stramineus (Engelm.) F. Seitz
Echinocereus triglochidiatus Engelm. var. triglochidiatus
Echinocereus viridiflorus Engelm. var. cylindricus (Engelm.) Rümpler
Echinomastus intertextus (Engelm.) Britton & Rose subsp. dasyacanthus (U. Guzmán)
Escobaria sandbergii Castetter, P. Pierce & K. H. Schwer.
Escobaria strobiliformis (Poselger) Scheer ex Boed.
Escobaria vivipara (Nutt.) Buxb.
Glandulicactus uncinatus (Galeotti ex Pfeiff.) Backeb. var. wrightii (Engelm.) Backeb.
Mammillaria grahamii Engelm.
Mammillaria meiacantha Engelm.
Opuntia engelmannii Salm-Dyck ex Engelm. var. engelmannii
Opuntia macrocentra Engelm.
Opuntia phaeacantha Engelm.

Cannabaceae

Celtis reticulata Torr.

Caprifoliaceae
Symphoricarpos rotundifolius A. Gray

Caryophyllaceae
Paronychia jamesii Torr. & A. Gray
Silene antirrhina L.
Silene plankii C. L. Hitchc. & Maguire
Silene scouleri Hook.

Cleomaceae
Polanisia uniglandulosa (Cav.) DC.

Commelinaceae
Commelina dianthifolia Delile var. dianthifolia
Commelina erecta L. var. angustifolia (Michx.) Fernald
Tradescantia occidentalis (Britton) Smyth var. occidentalis
Tradescantia occidentalis (Britton) Smyth var. scopulorum (Rose) E. S. Anderson & Woodson

Convolvulaceae
Convolvulus equitans Benth.
Cuscuta umbellata Kunth
Evolvulus nuttallianus Schult.
Ipomoea costellata Torr.
Ipomoea cristulata Hallier f.

Crassulaceae
Sedum wrightii A. Gray var. priscum (R. T. Clausen) H. Ohba

Crossosomataceae
Apacheria chiricahuensis C. T. Mason

Cucurbitaceae
Cucurbita foetidissima Kunth

Cyperaceae
Bolboschoenus maritimus (L.) Palla subsp. paludosus (A. Nelson) T. Koyama
Carex geophila Mack.
Cyperus fendlerianus Boeckeler
Eleocharis atropurpurea (Retz.) Kunth
Schoenoplectus americanus (Pers.) Volkart ex Schinz & R. Keller

Euphorbiaceae
Acalypha neomexicana Müll. Arg.
Croton dioicus Cav.
Euphorbia acuta Engelm.
Euphorbia brachycera Engelm.
Euphorbia davidii Subils
Euphorbia fendleri Torr. & A. Gray var. chaetocalyx Boiss.
Euphorbia fendleri Torr. & A. Gray var. fendleri
Euphorbia glyptosperma Engelm.
Euphorbia hyssopifolia L.
Euphorbia lata Engelm.
Euphorbia revoluta Engelm.
Euphorbia serpyllifolia Pers.
Euphorbia stictospora Engelm.
Tragia ramosa Torr.

Fabaceae
Acaciella angustissima (Mill.) Britton & Rose
Astragalus allochrous A. Gray
Astragalus castetteri Barneby
Astragalus flexuosus (Hook.) Douglas ex G. Don var. flexuosus
Astragalus missouriensis Nutt. var. missouriensis
Astragalus mollissimus Torr. var. bigelovii (A. Gray) Barneby
Astragalus nuttallianus DC. var. austrinus (Small) Barneby
Astragalus praelongus E. Sheld. var. ellisiae (Rydb.) Barneby
Caesalpinia gilliesii (Wall ex Hook.) D. Dietr.
Calliandra humilis Benth. var. humilis
Dalea candida Willd. var. oligophylla (Torr.) Shinners

Dalea formosa Torr.
Dalea pogonathera A. Gray
Dalea wrightii A. Gray
Hoffmannseggia glauca (Ortega) Eifert
Lotus plebeius (Brandegee) Barneby
Lupinus concinnus J. Agardh var. concinnus
Melilotus albus Medik.
Prosopis glandulosa Torr. var. torreyana (L. D. Benson) M. C. Johnst.
Robinia neomexicana A. Gray var. neomexicana
Senna bauhinioides (A. Gray) H. S. Irwin & Barneby
Senna lindheimeriana (Scheele) H. S. Irwin & Barneby
Vicia ludoviciana Nutt. ex Torr. & A. Gray subsp. ludoviciana

Fagaceae
Quercus gambelii Nutt.
Quercus grisea Liebm.
Quercus turbinella Greene

Fouquieriaceae
Fouquieria splendens Engelm.

Garryaceae
Garrya ovata Benth. subsp. goldmanii (Wooton & Standl.) B. L. Turner
Garrya wrightii Torr.

Gentianaceae
Zeltnera arizonica (A. Gray) G. Mans.

Geraniaceae
Erodium cicutarium (L.) L'Hér. ex Aiton

Grossulariaceae
Ribes cereum Douglas
Ribes leptanthum A. Gray
Ribes mescalerium Coville

Hydrangeaceae
Fendlera rupicola Engelm. & A. Gray
Jamesia americana Torr. & A. Gray var. americana
Philadelphus microphyllus A. Gray subsp. argenteus (Rydb.) C. L. Hitchc.

Koeberliniaceae
Koeberlinia spinosa Zucc.

Lamiaceae
Hedeoma drummondii Benth.
Hedeoma nana (Torr.) Briq.
Marrubium vulgare L.
Salvia henryi A. Gray
Salvia pinguifolia (Fernald) Wooton & Standl.
Salvia reflexa Hornem.

Linaceae
Linum aristatum Engelm.
Linum lewisii Pursh
Linum puberulum (Engelm.) A. Heller

Loasaceae
Cevallia sinuata Lag.
Mentzelia albicaulis (Douglas) Douglas ex Torr. & A. Gray
Mentzelia humilis (Urb. & Gilg) J. Darl.
Mentzelia procera (Wooton & Standl.) J. J. Schenk & L. Hufford
Mentzelia pumila Torr & A. Gray var. pumila

Lythraceae
Lythrum californicum Torr. & A. Gray

Malpighiaceae
Janusia gracilis A. Gray

Malvaceae
Abutilon parvulum A. Gray

Hibiscus denudatus Benth.
Sphaeralcea angustifolia (Cav.) G. Don
Sphaeralcea digitata (Greene) Rydb.
Sphaeralcea hastulata A. Gray
Sphaeralcea incana Torr. ex A. Gray var. incana
Sphaeralcea polychroma La Duke

Moraceae
Morus microphylla Buckley

Nitrariaceae
Peganum harmala L.

Nyctaginaceae
Abronia angustifolia Greene
Acleisanthes chenopodioides (A. Gray) R. A. Levin
Acleisanthes longiflora A. Gray
Allionia incarnata L.
Boerhavia coulteri (Hook.) S. Watson var. coulteri
Boerhavia gracillima Heimerl
Boerhavia spicata Choisy
Boerhavia triquetra S. Watson var. intermedia (M. E. Jones) Spellenb.
Cyphomeris gypsophiloides (M. Martens & Galeotti) Standl.
Mirabilis albida (Walter) Heimerl
Mirabilis linearis (Pursh) Heimerl var. linearis
Mirabilis multiflora (Torr.) A. Gray
Mirabilis oxybaphoides (A. Gray) A. Gray

Oleaceae
Forestiera pubescens Nutt.
Fraxinus velutina Torr.
Menodora scabra A. Gray

Onagraceae
Gaura coccinea Nutt. ex Pursh
Oenothera albicaulis Pursh
Oenothera brachycarpa A. Gray
Oenothera cespitosa Nutt. subsp. marginata (Nutt. ex Hook. & Arn.) Munz
Oenothera cespitosa Nutt. subsp. navajoensis W. L. Wagner, Stockh. & W. M. Klein
Oenothera curtiflora W. L. Wagner & Hoch
Oenothera pallida Lindl.
Oenothera suffrutescens (Ser.) W. L. Wagner & Hoch

Orobanchaceae
Castilleja integra A. Gray
Castilleja lanata A. Gray

Oxalidaceae
Oxalis decaphylla Kunth
Oxalis latifolia Kunth

Papaveraceae
Argemone polyanthemos (Fedde) G. B. Ownbey
Corydalis aurea Willd. subsp. occidentalis (Engelm. ex A. Gray) G. B. Ownbey
Eschscholzia californica Cham. subsp. mexicana (Greene) C. Clark

Pedaliaceae
Proboscidea parviflora subsp. parviflora (Wooton) Wooton & Standl.

Phrymaceae
Mimulus guttatus DC.

Plantaginaceae
Maurandya antirrhiniflora Humb. & Bonpl. ex Willd.
Penstemon barbatus (Cav.) Roth subsp. torreyi (Benth.) D. D. Keck
Penstemon fendleri Torr. & A. Gray
Penstemon jamesii Benth.
Penstemon linarioides A. Gray subsp. linarioides
Plantago patagonica Jacq.

Plumbaginaceae

Limonium limbatum Small

Poaceae

Achnatherum eminens (Cav.) Barkworth
Achnatherum lobatum (Swallen) Barkworth
Achnatherum perplexum Hoge & Barkworth
Achnatherum robustum (Vasey) Barkworth
Agropyron spicatum (Pursh) Scribn. & J. G. Sm.
Aristida divaricata Humb. & Bonpl. ex Willd.
Aristida longiseta Steud.
Aristida purpurea Nutt. var. fendleriana (Steud.) Vasey
Aristida purpurea Nutt. var. nealleyi (Vasey) Allred
Aristida purpurea Nutt. var. purpurea
Aristida ternipes (Cav.) var. ternipes
Blepharoneuron tricholepis (Torr.) Nash
Bothriochloa barbinodis (Lag.) Herter
Bothriochloa ischaemum (L.) Keng var. ischaemum
Bothriochloa laguroides (DC.) Herter subsp. torreyana (Steud.) Allred & Gould
Bothriochloa springfieldii (Gould) Parodi
Bouteloua aristidoides (Knuth) Griseb. var. aristidoides
Bouteloua barbata Lag. var. barbata
Bouteloua curtipendula (Michx.) Torr. var. curtipendula
Bouteloua eriopoda (Torr.) Torr.
Bouteloua gracilis (Kunth) Lag. ex Griffiths
Bouteloua hirsuta Lag.
Bromus anomalus Rupr. ex E. Fourn.
Bromus ciliatus L.
Bromus frondosus (Shear) Wooton & Standl.
Bromus lanatipes (Shear) Rydb.
Bromus rubens L.
Bromus willdenowii Kunth
Dasyochloa pulchella (Kunth) Willd. ex Rydb.
Digitaria californica (Benth.) Henrard
Digitaria pubiflora (Vasey) Wipff
Distichlis spicata (L.) Greene
Elymus arizonicus (Scribn. & J. G. Sm.) Gould
Elymus longifolius (J. G. Sm.) Gould
Elymus spicatus (Pursh) Gould
Eragrostis cilianensis (All.) Vignolo ex Janch.
Eragrostis intermedia Hitchc.
Eragrostis pilosa (L.) P. Beauv.
Erioneuron nealleyi (Vasey) Tateoka
Erioneuron pilosum (Buckley) Nash
Festuca arizonica Vasey
Hesperostipa neomexicana (Thurb.) Barkworth
Heteropogon contortus (L.) P. Beauv. ex Roem. & Schult.
Koeleria macrantha (Ledeb.) Schult.
Leptochloa dubia (Kunth) Nees
Lycurus setosus (Nutt.) C. Reeder
Melica porteri Scribn. var. porteri
Muhlenbergia arenacea (Buckley) Hitchc.
Muhlenbergia dubia E. Fourn.
Muhlenbergia montana (Nutt.) Hitchc.
Muhlenbergia monticola Buckley
Muhlenbergia pauciflora Buckley
Muhlenbergia porteri Scribn.
Muhlenbergia setifolia Vasey
Muhlenbergia straminea Hitchc.
Muhlenbergia tenuifolia (Kunth) Kunth
Muhlenbergia wrightii Vasey ex J. M. Coult.
Panicum hallii Vasey var. hallii

Panicum mohavense Reeder
Panicum obtusum Kunth
Pleuraphis jamesii Torr.
Pleuraphis mutica Buckley
Poa bigelovii Vasey & Scribn.
Poa fendleriana (Steud.) Vasey subsp. fendleriana
Polypogon monspeliensis (L.) Desf.
Polypogon viridis (Gouan) Breistr.
Schizachyrium scoparium (Michx.) Nash
Setaria leucopila (Scribn. & Merr.) K. Schum.
Sorghum halepense (L.) Pers.
Sphenopholis obtusata (Michx.) Scribn.
Sporobolus airoides (Torr.) Torr.
Sporobolus cryptandrus (Torr.) A. Gray
Tridens muticus (Torr.) Nash var. muticus
Vulpia octoflora (Walter) Rydb. var. octoflora

Polemoniaceae
Gilia sinuata Douglas ex Benth.
Ipomopsis aggregata (Pursh) V. E. Grant subsp. formosissima (Greene) Wherry
Ipomopsis multiflora (Nutt.) V. E. Grant
Microsteris gracilis (Hook.) Greene
Phlox triovulata Thurb. ex Torr.

Polygalaceae
Polygala obscura Benth.
Polygala scoparioides Chodat

Polygonaceae
Eriogonum abertianum Torr.
Eriogonum jamesii Benth. var. jamesii
Eriogonum polycladon Benth.
Eriogonum rotundifolium Benth.
Eriogonum wrightii Torr. ex Benth.
Rumex hymenosepalus Torr.

Portulacaceae
Phemeranthus confertiflorus (Greene) Hershk.
Phemeranthus longipes (Wooton & Standl.) Kiger
Portulaca halimoides L.
Portulaca oleracea L.
Talinum aurantiacum Engelm.

Pottiaceae
Crossidium aberrans Holz. & E. B. Bartram

Primulaceae
Samolus ebracteatus Kunth var. cuneatus (Small) Henrickson

Ranunculaceae
Clematis bigelovii Torr.
Thalictrum fendleri Engelm. ex A. Gray

Rhamnaceae
Condalia warnockii M. C. Johnst. var. warnockii
Rhamnus serrata Schult.

Rosaceae
Cercocarpus breviflorus A. Gray var. breviflorus
Fallugia paradoxa (D. Don) Endl. ex Torr.
Holodiscus dumosus (S. Watson) A. Heller
Prunus serotina Ehrh. var. virens (Wooton & Standl.) McVaugh
Rosa stellata Wooton var. mirifica (Greene) W. H. Lewis
Rosa stellata Wooton var. stellata
Spiraea vanhouttei (Briot) Carrière

Rubiaceae
Galium fendleri A. Gray

Galium microphyllum A. Gray

Hedyotis nigricans (Lam.) Fosberg var. nigricans

Rutaceae

Ptelea trifoliata L.

Salicaceae

Populus deltoides W. Bartram ex Marshall var. wislizeni (S. Watson) Dorn

Populus tremuloides Michx.

Salix exigua Nutt.

Salix gooddingii C. R. Ball

Salix lucida Muhl.

Santalaceae

Arceuthobium divaricatum Engelm.

Arceuthobium vaginatum (Humb. & Bonpl. ex Willd.) J. Presl subsp. cryptopodum (Engelm.) Hawksw. & Wiens

Comandra umbellata (L.) Nutt. var. pallida (A. DC.) M. E. Jones

Phoradendron juniperinum A. Gray

Phoradendron serotinum (Raf.) M. C. Johnst. subsp. tomentosum (DC.) Kuijt

Saxifragaceae

Heuchera nova-mexicana Wheelock

Heuchera rubescens Torr.

Solanaceae

Chamaesaracha arida Henrickson

Chamaesaracha coniodes (Moric. ex Dunal) Benth. & Hook. f. ex B. D. Jacks.

Chamaesaracha coronopus (Dunal) A. Gray

Datura wrightii Regel

Lycium pallidum Miers

Nicotiana trigonophylla Dunal

Physalis hederifolia A. Gray var. fendleri (A. Gray) Cronquist

Physalis hederifolia A. Gray var. hederifolia

Solanum elaeagnifolium Cav.

Tamaricaceae

Tamarix chinensis Lour.

Ulmaceae

Ulmus pumila L.

Urticaceae

Urtica gracilenta Greene

Verbenaceae

Aloysia wrightii (A. Gray ex Torr.) A. Heller

Glandularia bipinnatifida (Nutt.) Nutt. var. ciliata (Benth.) B. L. Turner

Zygophyllaceae

Kallstroemia californica (S. Watson) Vail

Kallstroemia parviflora Norton

Larrea tridentata (DC.) Coville var. tridentata

Tribulus terrestris L.

GLOSSARY

Abaxial – The side away from the axis, usually the underside.

Abortive – Failing to mature, undeveloped.

Acaulescent – Without an obvious stem; plants with basal leaves only are considered acaulescent.

Accessory fruit – Fruit derived from parts other than the ovary, often the receptacle, as in *Fragaria*. Most of the fruit, especially the exocarp, may include some accessory-derived structures.

Accrescent – A structure that becomes larger with age, such as the calyx in some flowers.

Accumbent – Cotyledons lying with one edge against the radicle (see cotyledon illustration).

Acerose – Needle-shaped, as in the leaves of conifers.

Achene – Small, dry, one-seeded, indehiscent fruit developing from a superior ovary, as in *Carex*.

Achlorophyllous – Without chlorophyll, as in nongreen parasitic plants.

Acicular – Needle-shaped.

Acicula (pl. aciculae) – Short, needle-shaped prickles.

Acrodromous – A kind of leaf venation in which two or more primary or strongly developed secondary veins arch upward from the midvein.

Actinomorphic – Flowers with whorls of parts radially symmetric, all parts in each whorl alike; typically based on petals unless otherwise stated.

Aculeate – Sharply pointed; beset with prickles or spines.

Acuminate – Gradually tapering to a point with slightly concave margins.

Acute – Forming an angle of less than 90°.

Adaxial – The side toward the axis, usually the upper side.

Adnate – Union or fusion of two or more unlike parts, as stamens adnate to petals.

Adventitious – Structures or organs developing in unusual places, like roots developing on a stem.

Adventive – Not native; introduced.

Aggregate fruit – Fruit formed from a cluster of pistils that were distinct in a single flower, as in *Fragaria*.

Alete – Forming single spores, and therefore not having a line where individual spores from a tetrad do.

Alkaline – With a basic pH caused by abundant alkaline-metal salts.

Alluvial – Referring to materials (alluvium) deposited by moving water such as mud, silt, sand, gravel, or cobbles.

Alpine – The area above tree line; the elevation of tree line depends on exposure, local climate, latitude, and other factors.

Alpine tundra – The treeless plant community of matted, tufted, and prostrate cushion plants found above timberline.

Alternate – Having one leaf at a node, with successive leaves on opposing sides of the stem.

Alveolate – Like a honeycomb, with depressions separated by thin partitions or ridges.

Ament – Spikelike, commonly pendulous inflorescence of highly reduced unisexual flowers, also known as a catkin.

Amphibious – Plants that live both in water and on land, often shore dwellers.

Amplexicaul – Referring to leaf bases that clasp the stem.

Anastomosing – Interwoven, netlike.

Androecium – All the stamens in a flower.

Androgynous – Containing staminate and pistillate flowers, with the staminate flowers above the pistillate ones.

Angiosperm – A plant producing flowers and bearing an ovary with ovules.

Angustiseptate – Referring to the compression of some fruits of the Brassicaceae where the compression is perpendicular to the replum and the (false) septum. In angustiseptate fruits, the width of the septum is narrower than the widest dimension of the fruit.

Annual – Plants that grow from seed, flower, produce seed, and die in the same year.

Annulus – A ring-shaped structure; a row of specialized, thick-walled cells along one side of a fern sporangium that aids in spore dispersal.

Anther – The portion of the stamen that produces and releases pollen.

Antheridium – The male reproductive structure in moss and fern gametophytes.

Anthesis - Flowering period.

Anthocarp – A fruit that has a flower part other than the pericarp persisting (e.g., the perianth on the fruit of Nyctaginaceae).

Anthocyanic – With a reddish or purplish tinge due to pigments called anthocyanins.

Antipetalous (or antepetalous) – Standing opposite or facing the petals, e.g., antipetalous stamens.

Antisepalous (or antesepalous) – Standing opposite or facing the sepals, e.g., antisepalous stamens.

Antrorse - Forward or upward in direction.

Apetalous - Without petals.

Apex - Tip of a leaf blade or other structure.

Aphyllopodic – Basal leaves reduced to a leaf sheath without a blade; generally referring to graminoids.

Apiculate - With a short, flexible tip.

Appressed – Lying close and flattened to the surface of another organ.

Aquatic – Growing in water, either emergent, submerged, or floating.

Arachnoid - With cobwebby, tangled hairs.

Archegonium – The structure in plants that produces and contains the female gamete; a vase-shaped structure in bryophytes, ferns, and fern allies that produces and contains the egg on the gametophyte.

Arcuate – Curved into an arch; bowed.

Arenophilous – Sand-loving.

Areole – The node of a cactus that bears spines, flowers, or both and may produce new stems.

Aril – A seed appendage at or near the hilum; the fleshy seed coat of some gymnosperms.

Aristate - With an elongate, bristlelike tip.

Armed – Bearing spines, prickles, sharp trichomes, or barbs.

Ascending - Growing upright, often in a curving fashion.

Asymmetric - Not divisible into equal halves.

Attenuate - Gradually narrowed and tapering to a point.

Auriculate - Having an auricle.

Awl-shaped - Narrowly triangular, sharp-pointed.

Awn – A bristlelike appendage.

Axile placentation – Ovules attached to the central axis of an ovary with two or more locules.

Axillary - Located in an axil.

Baccate - Berrylike; soft.

Banner – Upper petal of the flower in legumes.

Barbed – Bearing sharp, rigid, reflexed points like those on a fishhook.

Barbellate - Bearing short, stiff hairs.

Basal placentation – Ovules positioned at the base of a single-chambered ovary.

Basifixed – Hairs or scales fixed at the base.

Beak – A prolonged, usually narrowed tip of a thicker organ, as in some fruits and petals.

Bearded – Bearing tufts of long hairs.

Berry – Fleshy, indehiscent, many-seeded fruit with no true stone (pit) or core, embedded in pulp.

Bicarinate - Two-keeled.

Bicarpellate - With two carpels.

Biennial – Plants that live two years, usually flowering and fruiting the second year. The first year is typically represented by a rosette of leaves.

Bifid - Split into two parts or lobes, as in some styles.

Bifurcate - Forked; divided into two branches.

Bilabiate - With two lips, as in many zygomorphic flowers.

Bipinnate - Pinnately compound with each segment also divided pinnately; twice compound in a featherlike fashion.

Bipinnatifid - Cleft or deeply lobed bipinnately.

Biserrate – Doubly serrate; the individual teeth are also serrate.

Bisexual – With both stamens and pistil(s).

Bladder – A thin-walled, inflated structure.

Blade – The expanded portion of a leaf or petal.

Bloom – A whitish, waxy, powdery coating.

Bract – A leaflike structure usually associated with inflorescences.

Bracteole – An especially small bract or secondary bract.

Bristle – A short, stiff, hairlike structure.

Bud – An undeveloped leaf or flower shoot typically enclosed by bud scales.

Bulb – An underground bud with a flat discoid stem, roots, and thickened storage leaves (as in an onion).

Bulbil – An aerial bulb or reproductive stem.

Bulblets – Small to tiny bulbs.

Burlike – Spiny or prickly and often rounded.

Caducous – Early-deciduous.

Callous – Hardened or thickened.

Callus – The basal extension of the lemma in grasses.

Calyculus – Small bracts surrounding the calyx, appearing as a second calyx whorl.

Calyx – Outermost whorl of flower parts; sepals.

Campanulate – Bell-shaped.

Canaliculate – Having longitudinal channels or grooves.

Canescent – Densely covered with short, fine, gray or white hairs.

Capillary – Hairlike, very fine, slender.

Capitate – Head-shaped or in a headlike inflorescence.

Capsule – Dry, dehiscent fruit of more than one carpel and opening on more than one line of dehiscence.

Carinate – Keeled with at least one longitudinal ridge.

Carpel – A simple pistil; megasporophyll.

Carpophore – A thin extension of the receptacle forming a central axis between the carpels, as in the Apiaceae and Geraniaceae.

Cartilaginous – Tough and firm; flexible.

Caruncle – A protuberance or appendage near the hilum of a seed; often conspicuous in *Euphorbia* subgenus *Esula* and in some *Croton* species.

Caryopsis (grain) – A dry fruit typical of the Poaceae, usually with the ovary wall fused to the seed layers.

Catkin – Spikelike, commonly pendulous inflorescence of highly reduced unisexual flowers, also known as an ament.

Caudate – Bearing a tail or tail-like appendage.

Caudex – The woody and often enlarged base of a perennial.

Caulescent – With an obvious stem.

Cauline – On the stem.

Central spine – One of the innermost spines of an areole.

Cespitose (also caespitose) – Growth in tufts or dense clumps from a common point.

Chaff – Thin, dry scales or bracts, as on the receptacles of heads in the Asteraceae.

Channeled – Possessing one or more deep longitudinal grooves over the length of the structure.

Chartaceous – With a stiff, papery texture.

Chasmogamous – Flowers that open before fertilization and are generally cross-pollinated.

Ciénega – Low to midelevation area characterized by saturated, reducing soils with a reliable water supply via seepage.

Ciliate – With marginal hairs or scales; thinly and finely fringed.

Ciliolate – With short marginal hairs or scales; thinly and finely fringed.

Cinereous – Ash-colored; grayish due to a covering of short hairs.

Circinate – Coiled from the tip down, as in new fern fronds.

Circular – Orbicular, round; flat and round in outline.

Circumscissile – A fruit, sporangium, or anther that opens by dehiscence around the top, forming a lidlike structure.

Cladode (cladophyll) – An expanded stem with the form and function of a leaf.

Clambering – Growing in a vining manner on the substrate.

Clasping – Grasping or partially surrounding; attached on one side; usually a leaf grasping a stem.

Clathrate – Having a regular latticelike appearance.

Clavate – Club-shaped; widening toward the apex.

Claw – A narrow petal base, often petiolelike.

Cleft – Incompletely divided or split nearly to the middle.

Cleistogamous – Flowers that do not open and generally self-fertilize.

Coherent – Sticking together of like parts.

Colleters – Multicellular glandular hairs found in groups near the base of petioles, on stipules, and on sepals.

Columella – An axis to which a carpel of a compound pistil may be attached, which may remain open in the open dry fruit.

Column – The united filament(s) and style of the Orchidaceae; the union of staminal filaments, as in the Malvaceae.

Coma – A tuft of hairs.

Commissure – The face by which two carpels join one another, as in the Apiaceae.

Comose – With a coma or comalike.

Compound leaves – Composed of separate leaflike parts called leaflets.

Concave – Forming an inward-curving space on a structure, such as a seed or floral part.

Concolored – Parts having a single color.

Conduplicate – Folded together lengthwise with the upper surface within, used especially to describe cotyledons within a seed (see cotyledon illustration).

Cone – A usually woody strobilus in nonflowering plants; a dense cluster of sporophylls on an axis.

Confluent – The blending of one part into another.

Conic – Cone-shaped.

Connate – Union of like adjacent structures.

Connivent – Converging, stuck together but not anatomically fused or united.

Constricted – Narrowed or drawn together.

Convergent – Meeting together.

Convex – Bulging out, as in seeds or floral structures.

Convolute – Parts overlapping like roof shingles; rolled up longitudinally.

Cordate – Heart-shaped, with the base rounded and prominently notched, and the sides tapering to a narrowed apex.

Coriaceous – Leathery in texture.

Corm – A short, vertical underground stem covered with small, thin, papery leaves.

Corniculate – With small, hornlike projections.

Corolla – Second from outside whorl of flower parts; the collective name for the petals.

Corona – Petal-like crown structure such as in some *Asclepias* or *Nasturtium* flowers.

Corymb – Nearly flat-topped, indeterminate inflorescence with the lower and outer pedicels longest; simple or compound.

Corymbiform – An inflorescence with the general appearance of a corymb, but not necessarily the form.

Costa – A rib or prominent midvein.

Cotyledon – A seed leaf of a plant embryo, often containing nutrients for the growth of the young plant.

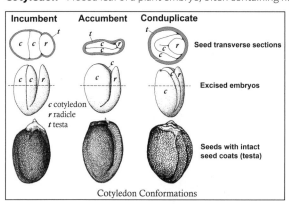

Common conformations of cotyledons, especially in seeds of the Brassicaceae. Drawings modified from Appel & Al-Shehbaz (2003) and Al-Shehbaz & Salariato (2012).

Crenate – With broad, rounded teeth; scalloped.

Crested – With an elevated ridge on a surface.

Crisped – Curled, wavy, or crinkled.

Crown – The perennial stem base of a herbaceous plant; the leaf top of a tree; pappus of an Asteraceae flower that is short, scalelike, and usually continuous around the corolla.

Cruciform – In the form of a cross, as in species of *Marsilea* or flowers of the Brassicaceae.

Cucullate – Hooded or hood-shaped.

Culm – A stem, often hollow or pithy, especially in the Poaceae, Juncaceae, or Cyperaceae.

Cuneate – Wedge-shaped or triangular, with the point of attachment at the narrowest part.

Cuspidate – Tipped with a short, sharp, abrupt point.

Cyathium – Inflorescence of some members of the Euphorbiaceae consisting of a female flower reduced to only a single pistil, and several male flowers reduced to a single stamen each, surrounded by a cuplike involucre. This reduced assemblage of several flowers looks like a single flower.

Cylindrical – Elongate and round in cross-section.

Cyme – Determinate inflorescence with central flowers opening first.

Cypsela – Small, dry, one-seeded, indehiscent fruit developing from an inferior ovary, as in the Asteraceae.

Cystolith – Enlarged cells containing crystals of calcium carbonate (usually) forming outgrowths of the epidermal cell wall.

Deciduous – Plant parts that are shed or fall off either seasonally or after a certain stage of development.

Decompound – More than once compound; subdivided leaflets.

Decumbent – Flat or resting on the ground, but with growth tips ascending.

Decurrent – Extending downward to the point of insertion.

Decussate – Opposite pairs, usually leaves, that alternate perpendicular to the previous and following pair.

Deflexed – Turned abruptly downward.

Dehiscent – Opening at maturity, as in dry fruits and anthers.

Deltoid – Equilaterally triangular.

Dendritic – Hairs with a branching pattern like the branches of a tree (with a central axis and shorter branches coming off both sides).

Dentate – With pointed teeth facing outward.

Denticulate – Minutely dentate or toothed.

Depauperate – A specimen that is poorly developed.

Descending – Pointing downward at an angle.

Determinate – An inflorescence where the terminal (most distal) flower opens first.

Diadelphous – Stamens united into two sets by their filaments.

Dichotomous – Forked in equal branching pairs.

Dicot (dicotyledon) – A paraphyletic group of angiosperms that possess two cotyledons (food storage leaves) in the seed and young seedling.

Didymous – Found in pairs.

Didynamous – With two pairs of stamens of unequal length.

Diffuse – Finely and widely spread out, as in an inflorescence.

Digitate – Divided or lobed from a common point.

Dilated – Broadened but remaining flat.

Dimorphic – Existing in two forms or sizes.

Dioecious – With staminate and pistillate flowers on different plants.

Disarticulate – Separating at maturity.

Discoid – Resembling a disk; with disk flowers, as in the inflorescence of a member of the Asteraceae that lacks ray flowers.

Discolorous – The two sides of an organ, such as a leaf, having different colors.

Disk flower – An actinomorphic flower in an Asteraceae head.

Dissected – Leaf surface deeply cut into numerous fine divisions.

Distal – Toward the tip.

Distinct – Not united with adjacent similar parts; separate.

Divaricate – Widely diverging or spreading, as in some stigmas.

Dolabriform – Hairs attached near the middle. See Malpighian hair (trichome).

Drupe – Fleshy, indehiscent fruit with a stony pit (*Prunus*).

Drupelet – A small drupelike fruit.

Echinate – With stout, sometimes blunt, prickles or spines.

Eciliate – Not ciliate, lacking ciliate trichomes.

Ecotype – Population adapted to a particular set of local environmental conditions.

Edaphic – Pertaining to the soil or substrate.

Eglandular – Without glands.

Elaiosome – A structure attached to a seed, usually high in lipids, that facilitates seed dispersal, especially by ants.

Elliptic – Longer than broad, tapering equally at both ends and widest in the middle.

Emarginate – Noticeably notched, as in the apex of some leaves.

Emersed – Rising from and reaching above the water.

Endemic – Found only in a specific geographic area or ecological situation.

Endocarp – Inner layer of a fruit pericarp, sometimes stony, as in some *Prunus* species.

Entire – Smooth, not toothed, indented, or lobed.

Ephemeral – Not long-lasting, usually referring to pools or streams; annual plants that germinate and flower whenever sufficient moisture is present in any season.

Epicalyx – An involucre of bracts resembling an outer calyx, as in some Malvaceae.

Epigynous – Attached at the top of the ovary when the ovary is completely inferior.

Epipetalous – On top of or attached to the petals.

Epiphyte – A plant that lives on another plant, using it for support and not rooted in the soil, typically with aerial roots.

Episepalous – On top of or attached to the sepals.

Epistemonous – Attached to the stamens.

Equitant – Alternate leaves with overlapping leaf blades that are folded and flattened against the stem axis, partly enclosing the leaf above, causing the leaves of a stem to appear fan-shaped, e.g., in Iridaceae and some Orchidaceae.

Erose (lacerate) – Irregularly indented, appearing gnawed or torn.

Evaporite – A sedimentary rock formed of material deposited from solution by evaporation of water.

Evergreen – Having green leaves year-round.

Excurrent – Projecting beyond the main structure, e.g., the midrib of some leaves.

Exfoliate – Peeling off in flakes or plates, as in the bark of some trees.

Explanate – Flat; spreading.

Exserted – Protruding, not included within the main structure.

Exstipulate – Without stipules.

Falcate – Sickle-shaped.

Farinose – Covered with meal-like particles.

Fascicle – Cluster.

Fen – A calcareous, marshy wetland.

Ferruginous – Rust-colored.

Fertile – Capable of bearing seeds.

Filament – A threadlike structure; the stalk of an anther.

Filiferous – Producing threads.

Filiform – Threadlike.

Fimbriate – A fringelike structure or projection, such as from a leaf or petal.

Fistulose – Hollow and cylindrical.

Flabellate – Fan-shaped.

Flaccid – Limp; not turgid.

Fleshy – Thickened and at least somewhat succulent.

Flexuous – With curves or bends; zigzag.

Floccose – Bearing tufts of long, soft, tangled hairs.

Floral tube – An elongate tube primarily of the perianth and, often, of the stamens.

Floret – A small flower, usually in a group or head of many, such as in the Asteraceae; a reduced flower in the Poaceae consisting of the flower proper and the lemma and palea.

Foliaceous – Leaflike in form, color, and texture; pertaining to leaves; bearing leaves.

Follicle – Dry, one-carpellate fruit opening along one longitudinal suture, as in *Asclepias*.

Forb – Herbaceous plant that is not a grass or grasslike.

Fornix (pl. fornices) – A small crest or scale where the corolla limb meets the throat, typical of many Boraginaceae.

Fovea – A pit on a surface.

Free-central placentation – In a unilocular ovary, ovules attached to a free-standing central structure.

Fringed – With hairs or bristles along the margins.

Frond – A fern leaf.

Fruit – A ripened ovary with enclosed seeds.

Funnelform – In the shape of a funnel.

Fusiform – Spindle-shaped, thicker in the middle, tapering at either end.

Galea – The hoodlike upper lip of a two-lipped corolla that often encloses the style and stigma.

Galeate – Having a galea, as in some flowers of the Lamiaceae or other families.

Geniculate – Possessing angular bends and joints.

Gibbous – Enlarged on one side.

Glabrate – Becoming glabrous in age.

Glabrous – Completely hairless.

Gland – A depression, cavity, protuberance, trichome, or other structure that secretes oil, wax, nectar, or sticky fluid.

Glaucous – With a waxy covering, often appearing waxy.

Globose – Spherical or nearly so.

Glochid – Small, fine spine that often easily detaches from the areole, as in *Opuntia*.

Glomerule – A dense, headlike, often axillary cyme, as in some Asteraceae and Lamiaceae; a dense cluster.

Glume – One of the pair of basal bracts on a spikelet in the Poaceae; a chaffy bract in the Cyperaceae.

Glutinous – Sticky or covered with a gummy exudate.

Grain – A seedlike structure, as seen on the fruit of some *Rumex* species.

Graminoid – A grass or grasslike plant in the Poaceae, Cyperaceae, Juncaceae, etc.

Gymnosperm – A plant producing seeds not borne in an ovary; usually cone-producing.

Gynecandrous (also gynaecandrous) – Containing both staminate and pistillate flowers, with the pistillate flowers above the staminate ones.

Gynobase – A slight enlargement and elongation of the receptacle, especially that forming between the nutlets in the Boraginaceae.

Gynobasic – The attachment of the style directly to the gynobase rather than directly to the carpel apices.

Gynophore – A pedicel-like structure, sometimes elongate, that supports the ovary in some species. Synonymous with one sense of the word "stipe."

Gynostegium – A structure formed from the fusion of the androecium and gynoecium, as in *Asclepias*.

Gypsophile – A plant that grows on gypsum deposits (magnesium sulfate).

Habit – The general appearance or mode of growth of a plant.

Half inferior – A hypanthium that is fused to the lower half of the ovary; the flower parts appear to arise from the equator of the ovary.

Halophyte – A plant that grows in a salty or hypertonic environment.

Hastate – With basal lobes spreading.

Haustoria – Fungal connections with roots.

Head – Dense cluster of sessile or subsessile flowers on an expanded peduncle or receptacle.

Helicoid – Coiled in a helix or spiral, as in one-sided cymose inflorescences in the Boraginaceae.

Hemiparasite – A partial parasite, photosynthetic from green leaves but simultaneously parasitic on a host.

Herb – A plant lacking true wood, typically soft and green, but occasionally with some hardened nonwoody parts.

Herbaceous – Not woody.

Heterosporous – An individual with two or more morphologically different spores, as in some selaginellas and ferns.

Heterostylous – Having styles of different lengths.

Hilum – A scar on a seed that indicates previous attachment to a placenta within the ovary.

Hirsute – Hairy with coarse, stiff hairs.

Hispid – Rough, with bristly hairs.

Hispidulous - Shortly rough with fine, bristly hairs.

Homostylous – Styles all of one length.

Hood – A hollow, arched covering.

Horn – Projection from flower parts or fruit as in *Proboscidea*; part of the corolla in *Asclepias*; long or short hornlike extensions from the gland ends of some species of *Euphorbia* subgenus *Esula*.

Humistrate - Lying on the ground.

Hyaline – Translucent; letting light through, but not transparent.

Hypanthium – A cuplike structure on which floral parts are inserted, formed from the fusion of the calyx, corolla, and stamens.

Hypogeous - Belowground.

Hypogynous - With stamens, petals, and sepals attached below the ovary, with the ovary superior.

Imbricate – With overlapping units, such as petals.

Imperfect – Having flowers that are either male or female.

Incised – Irregularly cut with sharp teeth.

Included – Not projecting beyond surrounding structures, such as stamens within a corolla tube.

Incomplete – Lacking one or more floral whorls.

Incumbent - Cotyledons lying flat against the radicle with the back of one against it (see cotyledon illustration).

Indehiscent – Not opening along suture lines or pores when mature, as in some dry fruits.

Indeterminate – An inflorescence that may continue to flower at the distal end of a stem or axis.

Indurate - Hardened.

Inferior ovary – An ovary that is beneath other parts of the flower.

Inflated – Bladdery, swollen, or expanded.

Inflorescence - The collection of flowers on a plant.

Infrastipular prickles – Single or paired prickles proximal to stipules. See internodal prickles.

Infructescence - The collection of fruits on a plant.

Internodal prickles – Prickles located on the stem between nodes but not proximal to stipules. See infrastipular prickles.

Internode - The portion of the stem between nodes.

Introduced - Not native.

Involucel – A secondary involucre, as in the bracts of the secondary umbels in the Apiaceae.

Involucre – A subtending whorl of bracts (or phyllaries) subtending an inflorescence.

Involute - With margins in-rolled, as in some leaves.

Irregular – Usually synonymous with zygomorphic, but technically meaning lacking symmetry (not dividable into planes of symmetry).

Joint - The node on a stem; a stem unit in certain Cactaceae, especially chollas.

Keel - A prominent dorsal ridge; the two lower united petals of a papilionaceous (pealike) corolla.

Krummholz – Alpine-subalpine timberline trees that are stunted and contorted due to extreme weather conditions.

Labellum – A liplike lower petal, often modified as a landing platform for pollinators or for other functions.

Lacerate (lacinate, laciniate) – Appearing torn; irregularly cut.

Lactiferous – With milky sap.

Lacustrine – Pertaining to lakes; of a lake; a sedimentary deposit typical of a lake.

Lamella (pl. lamellae) – In orchids, a thin flat plate or laterally flattened ridge on the lip.

Lamina - The blade or expanded portion of a structure, often referring to a petal.

Lanate – Woolly; densely covered with long, tangled hairs.

Lanceolate - At least three times longer than wide, tapering to an apex, widest below the middle.

Lateral - Located at the side of a structure.

Lateral spines – Spines on the side or margin of an areole, usually thick, without a bulbous base but often similar in color (dark) to central spines.

Latiseptate - Referring to the compression of some fruits of the Brassicaceae where the compression is parallel to the replum and the (false) septum. In latiseptate fruits, the width of the septum is ± equal to the widest dimension of the fruit.

Leaflet – A segment or leaflike portion of a compound leaf.

Leathery – With a leatherlike, thick, but flexible texture, usually referring to leaves.

Legume – A fruit, the product of a simple pistil, that dehisces along two sutures, typical of the Fabaceae.

Lemma – The lower of the two bracts that subtend a grass flower.

Lenticular – Lens-shaped.

Lepidote – Covered with small, scurfy or peltate scales.

Liana – A woody, long-stemmed, climbing "vine."

Ligulate – Strap-shaped.

Ligule – The flattened part of a ray floret corolla in the Asteraceae; a membranous structure arising from the inner surface of a leaf at the junction with the leaf sheath in the Poaceae and many sedges.

Linear – Long and narrow, with parallel sides, as in many grass leaves; a straight, narrow line.

Lobed – Indented, parted, divided; a leaf may be shallowly, deeply, pinnately, or palmately lobed.

Locule – The cavity of an organ, primarily pistils and anthers.

Loculicidal – Longitudinally dehiscent through the middle back, into the cavity, usually of a capsule or other hollow, dry fruit.

Lodicules – Paired vestigial or rudimentary scales at the base of an ovary in grass florets.

Loment – A specialized legume fruit that has constrictions between the seeds and that dehisces as one-seeded units.

Lumen (pl. luminae) – The inside space of a structure.

Lunate – Moon- or crescent-shaped.

Lyrate – Lyre-shaped; pinnatifid with the terminal lobe large and the lower lobes much smaller or shallower.

Maculate – Spotted or dotted.

Malodorous – Having a foul or unpleasant odor.

Malpighian hair (trichome) – T-shaped hair with a short stalk attached near the middle of the upper part and tapering toward the ends; shaped like a rock pick. Usually considered synonymous with dolabriform hair.

Marcescent – Withering but persistent, as in some basal leaves.

Margin – The edge, such as the edge of a leaf blade or petal.

Marginal placentation – Ovules attached to the margins of a simple pistil.

Mealy – Covered with meal-like particles or secretions, as in goosefoot (*Chenopodium*).

Megasporangium – A megaspore-producing structure.

Membranous – Thin, soft, and pliable or papery.

Mentum – A saclike or tubelike projection formed by the sepals and the extended base of the column in some orchids.

Mericarp – Unicarpellate segment of a schizocarp.

Microphyll – A small, narrow, single-veined leaf of lower vascular plants, such as *Lycopodium*.

Microsporangium – A microspore-producing structure.

Midrib – The central vein of a leaf or other organ.

Monad – A single, free, individual structure; usually referring to pollen grains that are shed singly.

Monadelphous – Stamens with their filaments united to form a tube around the style.

Moniliform – Cylindrical but constricted at regular intervals like beads on a string.

Monochasium – A cymose inflorescence with a single main axis.

Monocot – Plant that has only one cotyledon or seed leaf, generally with parallel veins in the leaves and the flower parts in threes or sixes; includes grasses, orchids, palms, sedges, lilies, etc.

Monoecious – With staminate and pistillate flowers on the same plant.

Monolete – Spores with a single line indicating the splitting axis.

Monotypic – Having only one subordinate taxon, like a family with only one genus, or a genus with only one species.

Mottled – With spots or blotches.

Mucilaginous – Having a mucuslike substance.

Mucro – A short, abrupt, firm tip.

Mucronate – With a short, abrupt, firm tip.

Multiple – Fruit formed from closely clustered ovaries of many separate flowers, as in *Morus* or *Ficus*.

Muricate – Possessing a small, sharp projection.

Nectary – A tissue or group of cells that secretes nectar.

Needle – A slender, needle-shaped leaf found in the Pinaceae.

Net-veined – Reticulate.

Neutral – Lacking functional stamens or pistils.

Node – A stem joint; where buds, leaves, inflorescences, and new stems arise.

Nut – A one-seeded, indehiscent fruit with a hard or woody pericarp, as in acorns of *Quercus*.

Nutlet – A small nut; a section of a mature ovary in the Boraginaceae, Verbenaceae, and Lamiaceae.

Obcompressed – Compressed such that a structure is flattened dorsiventrally while other structures are flattened laterally.

Obconic – Cone-shaped, with the attachment at the narrow end.

Obcordate – Similar to cordate, but with the attachment at the narrowed end and a notch at the apex.

Oblanceolate – Similar to lanceolate, but tapering to the base and widest above the middle.

Oblate – Generally spheroid but flattened at the poles.

Oblique – Asymmetric by having unequal sides.

Oblong – Longer than broad, with margins parallel and ends rounded.

Obovate – Similar to ovate, but tapering to the base and widest above the middle.

Obsolete – Rudimentary or vestigial.

Obtuse – Forming an angle of greater than 90°.

Ochroleucous – Off-white, buff.

Ocrea (pl. ocreae) A sheath around the stem formed by stipules, as in some Polygonaceae.

Ocreola – Minute, stipular sheath around the secondary inflorescence divisions in some Polygonaceae.

Oil tube – Narrow duct in the fruit of many Apiaceae that contains volatile oils.

Opposite – Arranged with two opposing structures, as when two leaves are at a node on opposing sides; one part in front of another, as when sepals and petals align with one another.

Orbicular – Circular in shape.

Oval – Broadly elliptic, widest in the middle.

Ovary – The part of the pistil containing the ovules; develops into the fruit.

Ovate – Egg-shaped in outline, less than three times as long as wide, tapering to the apex and widest below the middle.

Ovule – Unfertilized female gametophyte, which when fertilized develops into a seed. Loosely, an immature seed.

Palate – A raised structure on the lower lip of a corolla that almost completely obscures the throat, as in some *Penstemon* flowers.

Palea – Uppermost bract in a grass floret.

Palmate – Arranged in a palmlike fashion, spreading.

Palustrine – Growing in wet meadows or marshes.

Panicle – Compound or branched raceme.

Papilionaceous – The bilateral flower form of, for example, a garden pea; having an upper banner, two lateral wings, and a keel formed from the partial fusion of the two lower petals.

Papillae – Pimplelike or blisterlike structures, usually on the epidermis.

Papillate – Having papillae; papilliform (see papillae).

Pappus – Modified calyx on florets of the Asteraceae. These may be awns, scales, or bristles at the achene or cypsela apex.

Parasite – An organism that obtains its food and water from another organism.

Parietal – Attached to the outer wall of the ovary, as in parietal placentation, where the ovary has only one locule and the placentae are arranged around the outside of the locule.

Pectinate – Comblike; with closely spaced appendages or hairs often in a single row.

Pedate – Palmately divided, with lateral lobes two-cleft.

Pedicel – The stalk of a single flower in an inflorescence.

Peduncle – The stalk subtending an inflorescence.

Pellicle – A thin, membranous covering, as on seeds of the Montiaceae.

Peltate – With a stalk attached to the lower surface of a leaf blade or other structure well inside the margin.

Pendulous – Hanging downward, usually somewhat loosely.

Penicillate – With a brushlike tuft of short hairs.

Pepo – A type of berry with a hard or leathery rind (exocarp), many seeds, and a usually fleshy (sometimes fibrous) mesocarp, usually one-locular; typical of the Cucurbitaceae.

Perennial – A plant that lives for three or more years.

Perfect (bisexual, hermaphrodite) – Having both pistil(s) and stamen(s) on the same flower.

Perfoliate – With the margins (usually of a leaf) surrounding the stem, giving the impression of the stem passing through the leaf.

Perianth – Collectively, the calyx and corolla.

Pericarp – The wall of a fruit, usually derived from ovary tissue, but sometimes derived from an accessory structure, often with exocarp, mesocarp, and endocarp.

Perigynium – The sac enclosing an achene in some members of the Cyperaceae.

Perigynous – A flower with a tubular hypanthium on which occur stamens, petals, and sepals. The calyx tube surrounds, but is not attached to, the superior ovary.

Persistent – Remaining attached, as in a style on a fruit.

Petal – An individual segment of the corolla, the second whorl (moving inward) of the flower perianth.

Petaloid – Petal-like; may refer to a stamen, sepal (as in many Cactaceae), style (as in *Iris*), other flower part, an entire flower (as in a ray floret in the Asteraceae), or a bract (as in *Poinsettia*).

Petiole – The stem or stalk attached to the blade of a leaf.

Petiolule – The stem or stalk attached to a leaflet.

Phyllary – A bract of the involucre, as in the Asteraceae.

Phyllodium (pl. phyllodia) – A leaf reduced to just a midrib, this sometimes dilated into the form of a blade.

Phyllopodic – Basal leaves with well-developed blades present; generally referring to graminoids.

Pilose – With soft, distinct, straight hairs that are ascending or spreading.

Pinna (pl. pinnae) – A lobe of a pinnatifid leaf, petal, etc.

Pinnate – Arranged in a featherlike or fernlike fashion.

Pinnatifid – Pinnately lobed almost to the point of being pinnately compound.

Pinnatisect – Pinnately cleft to the midrib of a leaf.

Pinnule – The ultimate division of the blade in a bipinnately or more compound leaf.

Pistil – The gynoecium, female parts of the flower: ovary, stigma, and style.

Pistillode – A sterile vestigial pistil.

Pithy – With spongy tissue in the central area of roots and stems, as in some Poaceae.

Placenta (pl. placentae) – The attachment place of the ovule.

Pleiochasial – A cymose inflorescence having more than one branch off the main axis.

Pleurogram – A mark or depression on both sides of the seeds of some Fabaceae. It may be closed or open.

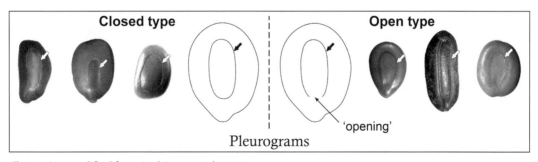

Illustration modified from Rodrigues et al. (2021).

Plicate – Plaited, pleated, or folded like a fan, as in the leaf blades of *Calypso bulbosa* or the corolla in many Gentianaceae and Solanaceae.

Plumose – Plumelike; feathery with hairs or fine bristles on both sides of the main axis.

Podogyne – See stipe and gynophore.

Pollen – Powdery mass shed from anthers (of angiosperms) or microsporangia (of gymnosperms).

Pollinarium (pl. pollinaria) – A structure in the Orchidaceae and Apocynaceae that contains pollinia and other structures that facilitate pollen transport by pollinating insects, birds, etc.

Polyploid – With more than two of the basic sets of chromosomes in the nucleus.

Prickle – Hard, pointed outgrowth from the surface (epidermis) of a plant.

Procumbent – Lying on the ground, but not rooting in it.

Propagule – A structure capable of producing a new plant; includes seeds, spores, bulbils, etc.

Prostrate – Growing or lying flat on the ground or substrate.

Proximal – Toward the base or axis.

Pruinose – Conspicuously glaucous with a waxy, whitish, grayish, or bluish bloom (coating) on the surface.

Puberulent – Minutely pubescent.

Puberulous – Pubescent; bearing hairs.

Pulvinate – Cushionlike or forming a dense mat.

Punctate – Having puncta, i.e., dotted with small holes, depressions, spots, or translucent glands.

Pustulose – With small blisters or pustules, often at the base of a hair, as in many species of *Cryptantha*.

Pyramidal – Of a plant's form, pyramid-shaped.

Pyriform – Pear-shaped.

Quadrate – Squared.

Raceme – Indeterminate inflorescence bearing several pedicelled solitary flowers along a central axis.

Rachilla – A small rachis, usually the axis of a grass spikelet or sedge inflorescence.

Rachis – The axis or stem of a spike, raceme, or compound leaf.

Radial spine – One of the outermost spines of an areole, often radiating or appressed, usually fine and lighter-colored than the central spines.

Radiate – With parts spreading from a single point; when some of the flowers in a head are ligulate, especially in the outer whorl as in the Asteraceae.

Rame – In grasses, a type of raceme with repeating pairs of sessile and pedicellate spikelets.

Ray – The straplike part of a ray or ligulate floret or the flower itself in the Asteraceae; a branch of an umbel, especially in the Apiaceae; the arm of a stellate trichome.

Recurved – Bent or curved backward or downward.

Reflexed – Bent downward.

Reniform – Kidney-shaped.

Replum – Hardened rim around the septum between the two locules of siliques or silicles.

Reticulate – Net-veined; with a network, or netlike markings or structure.

Retrorse – Directed downward or backward.

Retuse – Shallowly notched.

Revolute – Rolled backward from both margins toward the underside.

Rhizome – An underground, often creeping, stem that produces rootlets and vertical stems or leaves.

Riparian – Vegetation adjacent to a river or stream; sometimes referring to seasonally wet vegetation of otherwise dry washes.

Rootstock – The part of a root from which the main stem arises.

Rotate – Wheel-shaped; usually referring to the limb of a corolla.

Rugose – Wrinkled or furrowed.

Rugulose – Minutely wrinkled or furrowed.

Runcinate – Sharply pinnatifid or cleft, with the segments pointing downward.

Saccate – With a sac; bag-shaped.

Sagittate – Shaped like an arrow, with two backward-directed lobes.

Salverform – With a narrow tube, usually of a corolla, abruptly expanding to a flattened portion.

Samara – Dry, indehiscent, winged fruit, as in *Acer* or *Ulmus*.

Saprophyte – A heterotrophic plant that obtains its nutrients and energy from decaying organic matter, often in association with a fungus.

Scabrous – Rough to the touch.

Scape – A leafless peduncle arising from the ground in acaulescent plants.

Scapose – Bearing a scape.

Scarious – Thin, dry, membranous; not green.

Schizocarp – Dry fruit splitting into one- or few-seeded segments, as in many Apiaceae.

Scorpioid – Shaped like a scorpion's tail, as in the coiled cyme of some plants in the Boraginaceae and Hydrangeaceae.

Scree – Loose rock debris covering a slope or at the base of a cliff or steep incline.

Scrotiform – Scrotumlike in appearance; having a drooping pouch or sac.

Scurfy – Covered with small scales.

Secund – Arranged on one side of a rachis or axis, as in some species of *Penstemon*.

Seed – A ripened ovule.

Sensu lato (s.l.) – In the broad sense (Latin).

Sensu stricto (s.s.) – In the narrow sense (Latin).

Sepal – A segment of the calyx, the outer whorl of the perianth.

Septate-nodulose – Divided by small transverse knobs or nodules; generally referring to the leaves of graminoids.

Septum – Partition separating cavities, as in an ovary.

Sericeous – Silky with dense, appressed hairs.

Serrate – With pointed teeth facing toward the apex.

Serrulate – Minutely serrate.

Sessile – Attached directly by the base, not stalked.

Setose – Covered with bristles.

Sheath – Tubular basal portion of a leaf that encloses a stem in the Poaceae and Cyperaceae; in *Opuntia* and *Cylindropuntia* (Cactaceae), a papery, tubular, epidermis-derived layer that encloses a spine.

Shoot – Usually the aerial part of plant; a stem including its dependent parts, leaves, flowers, etc.

Shrub – A woody perennial plant with several or more stems.

Silicle – A dry fruit less than three times longer than broad, dehiscing by two valves, with seeds attached to a central, persistent rim (replum) usually spanned by a membrane, as in the Brassicaceae.

Silique – A dry fruit more than three times longer than broad, dehiscing by two valves, with seeds attached to a central, persistent rim (replum) usually spanned by a membrane, as in the Brassicaceae.

Silky – With soft, distinct, straight hairs that are closely appressed.

Simple – Unbranched; without leaflets or not compound.

Sinuate – Wavy on the margins only, not three-dimensional.

Sinus – A cleft between two lobes of an organ such as a leaf (or petal).

Sobole – A perennating shoot.

Spathe – A large, often colorful or showy bract subtending and often surrounding a usually spikelike inflorescence, as in the Araceae.

Spatulate – Oblanceolate, but with the apex broadly rounded, tapering at the base, like a spatula.

Spherical – Round in outline; three-dimensional.

Spiciform – Resembling a spike.

Spike – Indeterminate inflorescence bearing sessile or subsessile solitary flowers along a central axis.

Spikelet – A section of a larger inflorescence similar in conformation to a spike; the ultimate flower cluster in grasses that consists of two glumes and one or more florets; flower cluster in sedges.

Spine – Modified leaf or leaf parts, including stipules, as in the Cactaceae.

Spinescent – Bearing a spine or spines.

Spinulose – Bearing spines.

Sporangiasters – Sterile sporangia with dark bulbous heads.

Sporophore – The fertile, sporangium-bearing segment of some ferns, especially *Botrychium*.

Spur – An enclosed, often nectar-containing extension from a corolla or calyx.

Stamen (pl. stamens) – The male pollen-bearing portion of a flower; the microsporophyll; composed of anther and filament.

Staminode – A sterile stamen, sometimes petaloid, resembling another structure, or otherwise modified.

Stellate – Star-shaped, with rays extending from a central point, for example a type of hair.

Stigma – The receptive, sticky or feathery part of a pistil on which pollen is deposited and germinates.

Stipe – In ferns, the stalk of a frond; a small stalk, usually synonymous with gynophore.

Stipitate – With a stipe or gynophore.

Stipulate – Bearing stipules; loosely, with small stipulelike projections.

Stipule – One of the pair of leaflike appendages at the base of some leaves.

Stolon – A horizontal, aboveground stem that can produce adventitious roots and daughter plants at its nodes.

Stoloniferous – Bearing a stolon or stolons.

Striate – Striped with longitudinal lines or furrows.

Strigose – With stiff hairs that are closely appressed.

Strophiole – A crest, tubercle, or appendage on the hilum of a seed.

Style – The portion of the pistil between the ovary and the stigma.

Submersed – Growing entirely underwater.

Subspecies – A grouping within a species, usually used for geographically isolated and morphologically distinct entities. Its taxonomic rank is between species and variety.

Subtending – Immediately below and close to another structure.

Subulate – Long, narrow, tapering gradually to a rigid apex; awl-shaped.

Succulent – Juicy or fleshy, as in the Cactaceae or Crassulaceae.

Suffrutescent – Shrublike; slightly woody.

Sulcate – With grooves or furrows, usually longitudinal.

Superior ovary – An ovary that is free from or above other organs in the flower.

Synoecious – Having male and female parts in the same flower.

Synonym – Name of a plant family, genus, species, etc., which, for various reasons, is no longer valid but generally represents the equivalent taxon.

Talus – A layer of broken rock, individually small cobbles to boulders, usually on a slope or below a cliff.

Taproot – A thick, fleshy, main or primary root that functions in food and water storage.

Tawny – Dull yellowish brown.

Tendril – Part or all of a stem, leaf, or petiole modified to form a threadlike structure, which may coil around objects for support.

Tepal – Sterile, sepaloid or petaloid structure of a flower in which the perianth parts are not differentiated into calyx and corolla, as in many monocots and many cactus flowers.

Terete – Round in cross-section.

Ternate – A leaf divided into three leaflets; usually used when it is difficult to determine whether it is pinnately compound or palmately compound.

Terrestrial – Growing in or on the ground; not aquatic, epiphytic, or lithophytic.

Tetradynamous – Flowers with four long and two short stamens, as in the Brassicaceae.

Tetraploid – Possessing an extra complete diploid set of chromosomes.

Thallus – A flat, leaflike structure that is not differentiated into stem and leaves.

Thorn – Pointed, sharp, modified stem, as in *Condalia* or *Elaeagnus*.

Throat – The portion of the corolla between the limb and the tube; the opening or orifice in a zygomorphic corolla; the upper margin of the leaf sheath in grasses.

Thyrse – A panicle that is more compact and oval than usual and has an indeterminate main axis.

Tomentose – With rather short, densely matted, soft, whitish, woolly hairs.

Tomentum – Woolly, short, soft covering of hairs.

Tortuous – Twisted.

Torulose – Slightly constricted at intervals, with a generally cylindrical form.

Translucent – Thin, transmitting light.

Transverse – Horizontal; crosswise.

Travertine – A type of limestone deposited in watery environments by bacteria, typically cyanobacteria.

Tree – A woody perennial with one or a few main trunks.

Trichome – A hairlike, scaly, or prickly outgrowth of the epidermis; loosely, a plant "hair."

Trichotomous – Three-forked.

Tridentate – Three-toothed.

Trigonate (trigonous) – Three-angled.

Trilete – Spores with three lines indicating the splitting axes.

Trophophore – The sterile leaflike segment of some ferns, especially *Botrychium*.

Trullate – Trowel-shaped.

Truncate – Cut off squarely.

Tube – A hollow, cylindrical structure, as in the corolla of some flowers.

Tuber – A short, usually thickened, underground storage stem with buds at the nodes, as in potatoes.

Tubercle – Short, usually rounded, nodule or projection, as in the Cactaceae.

Turbinate – Top-shaped.

Turf – Densely packed grasses or grasslike plants.

Turgid – Swollen or inflated.

Turion – A small shoot, typically budlike, that often overwinters, as in some Cyperaceae and *Epilobium*.

Twining – Coiling around a support and climbing.

Ultimate – The final or top section of a structure.

Umbel – Indeterminate inflorescence with pedicels usually of similar length arising from one point, sometimes flat-topped; can be simple or compound.

Umbellate – Umbel-like.

Umbo – A blunt protuberance, as on the ends of some pine cones.

Uncinate – Hooked at the tip; clawlike.

Undulate – Wavy, as in a leaf surface; three-dimensional.

Unilocular – With a single locule or compartment, as in some ovaries.

Uniseriate – In a single row or series.

Unisexual – A flower with either pistils or stamens.

Urceolate – Pitcherlike; hollow, but contracted near the mouth like an urn.

Utricle – A small, thin-walled, one-seeded, inflated dry fruit.

Valvate – Opening by sutures that separate the sections of a fruit at maturity.

Valve – One of the segments into which a dehiscent capsule or legume separates.

Variety – A unit of classification below the species and subspecies levels.

Velutinous – Velvety.

Venation – The arrangement of veins in a leaf or petal.

Ventral – Lower surface or underside.

Ventricose – Inflated or swollen on one side, as in some salvias or penstemons.

Verrucose – Covered with wartlike structures.

Versatile – Attached near the middle instead of at one end; mobile anthers of this construction, as in many *Lilium*.

Verticil – A whorl; an arrangement of similar parts around a central axis.

Verticillaster – A pair of axillary cymes arising from opposite leaves or bracts, forming a false whorl.

Villous – With long, soft, unmatted hairs.

Virgate – Erect, straight, and slender; usually long and wandlike.

Viscid – Sticky, as in the leaves of *Proboscidea*.

Whorled – With three or more leaves or branches at a node.

Wing – A thin membranous extension of an organ; the lateral petals of a papilionaceous Fabaceae flower.

Woolly – With soft, matted, interwoven hairs.

Zygomorphic – Bilaterally symmetric.

LITERATURE CITED

Abbott, J. R. 2011. Notes on the disintegration of *Polygala* (Polygonaceae) with four new genera for the Flora of North America. J. Bot. Res. Inst. Texas 5: 125–137.

Abbott, J. R. 2021. Polygalaceae. *In* Flora of North America Editorial Committee, eds., Flora of North America North of Mexico, Vol. 10, pp. 380–414. Oxford University Press, New York.

Ackerfield, J. R. 2015 [2018]. Flora of Colorado, 2nd ed. BRIT Press, Fort Worth.

Ackerfield, J. R., D. J. Keil, W. C. Hodgson, M. P. Simmons, S. D. Fehlberg, and V. A. Funk. 2020. Thistle be a mess: Untangling the taxonomy of *Cirsium* (Cardueae: Compositae) in North America. J. Syst. Evol. 58(6): 881–912.

Adams, R. P. 1993. *Juniperus*. *In* Flora of North America Editorial Committee, eds., Flora of North America North of Mexico, Vol. 2, pp. 412–420. Oxford University Press, New York.

Ahedor, A. J. 2019a. *Bacopa*. *In* Flora of North America Editorial Committee, eds., Flora of North America North of Mexico, Vol. 17, pp. 260–263. Oxford University Press, New York.

Ahedor, A. J. 2019b. *Mecardonia*. *In* Flora of North America Editorial Committee, eds., Flora of North America North of Mexico, Vol. 17, pp. 273–275. Oxford University Press, New York.

Aiken, S. G. 1981. A conspectus of *Myriophyllum* (Haloragaceae) in North America. Brittonia 33: 57–69.

Albach, D. C. 2019. *Veronica*. *In* Flora of North America Editorial Committee, eds., Flora of North America North of Mexico, Vol. 17, pp. 305–322. Oxford University Press, New York.

Alexander, P. J. 2020. *Muilla lordsburgana* (Asparagaceae: Brodiaeoideae), a new species found north of Lordsburg, southwestern New Mexico. J. Semiarid Environm. 1: 1–10.

Alford, É. R., J. M. Vivanco, and M. W. Paschke. 2009. The effects of flavonoid allelochemicals from knapweeds on legume–rhizobia candidates for restoration. Restor. Ecol. 17: 506–514.

Allen, G. A. 1985. The hybrid origin of *Aster ascendens* (Asteraceae). Amer. J. Bot. 72: 268–277.

Allen, G. A., and K. R. Robertson. 2002. *Erythronium*. *In* Flora of North America Editorial Committee, eds., Flora of North America North of Mexico, Vol. 26, pp. 153–164. Oxford University Press, New York.

Allison J. W. 2006. *Dicranocarpus*. *In* Flora of North America Editorial Committee, eds., Flora of North America North of Mexico, Vol. 21, pp. 219–220. Oxford University Press, New York.

Allred, K. W. 1999. New plant distribution records [*Lappula squarrosa*]. New Mexico Botanist 13: 7.

Allred, K. W. 2005. A Field Guide to the Grasses of New Mexico, 3rd ed. Agricultural Experiment Station, New Mexico State University, Las Cruces.

Allred, K. W. 2020. Flora Neomexicana I: Annotated Checklist, 3rd ed. Lulu Press.

Allred, K. W., and R. D. Ivey, eds. 2012. Flora Neomexicana III: An Illustrated Identification Manual. Lulu Press.

Allred, K. W., E. M. Jercinovic, and R. D. Ivey, eds. 2020. Flora Neomexicana III: An Illustrated Identification Manual. Part 2: Dicotyledonous Plants, 2nd ed. Lulu Press.

Al-Shehbaz, I. A. 2012a. A generic and tribal synopsis of the Brassicaceae (Cruciferae). Taxon 61: 931–954.

Al-Shehbaz, I. A. 2012b. Notes on miscellaneous species of the tribe Thelypodieae (Brassicaceae). Harvard Pap. Bot. 17: 3–10.

Al-Shehbaz, I. A. 2013. *Draba henrici* (Brassicaceae), a new species from northern New Mexico. Harvard Pap. Bot. 18: 91–93.

Al-Shehbaz, I. A., and S. L. O'Kane, Jr. 2002. *Lesquerella* is united with *Physaria* (Brassicaceae). Novon 12: 319–329.

Al-Shehbaz, I. A., and D. L. Salariato. 2012. Brassicaceae. *In* Anton, A. M., and F. O. Zuloaga, eds., Flora Argentina, Vol. 8. Graficamente Ediciones, Córdoba.

Al-Shehbaz, I. A., M. D. Windham, and R. Elven. 2010. *Draba*. *In* Flora of North America Editorial Committee, eds., Flora of North America North of Mexico, Vol. 7, pp. 269–347. Oxford University Press, New York.

Anderson, E. F. 1986. A revision of the genus *Neolloydia* B. & R. (Cactaceae). Bradleya 4: 1–28.

Anderson, E. F. 2001. The Cactus Family. Timber Press, Portland, Oregon.

Anderson, J. L. 2006. Vascular plants of Arizona: Anacardiaceae. Canota 3: 13–22.

Anderson W. R. 2016. Malpighiaceae. *In* Flora of North America Editorial Committee, eds., Flora of North America North of Mexico, Vol. 12, pp. 354–364. Oxford University Press, New York.

APG (Angiosperm Phylogeny Group). 2016. An update of the Angiosperm Phylogeny Group classification for the orders and families of flowering plants: APG IV. Bot. J. Linn. Soc. 181: 1–20.

Appel, O., and I. A. Al-Shehbaz. 2003. Cruciferae. *In* Kubitzki, K., and C. Bayer, eds., The Families and Genera of Vascular Plants, Vol. 5, Flowering Plants—Dicotyledons Malvales, Capparales and Non-Betalain Caryophyllales, pp. 75–174. Springer, Berlin.

Argus, G. W. 2010. *Salix. In* Flora of North America Editorial Committee, eds., Flora of North America North of Mexico, Vol. 7, pp. 23–162. Oxford University Press, New York.

Atwood, N. D. 1975. A revision of the *Phacelia* Crenulatae group (Hydrophyllaceae) for North America. Great Basin Naturalist 35: 127–190.

Atwood, N. D. 2007. Six new species of *Phacelia* (Hydrophyllaceae) from Arizona and New Mexico. Novon 17: 403–416.

Atwood, N. D., and S. L. Welsh. 2005. New species of *Mentzelia* (Loasaceae) and *Phacelia* (Hydrophyllaceae) from New Mexico. W. N. Amer. Naturalist 65: 365–370.

Austin, D. F. 1998. Convolvulaceae: Morning glory family. J. Arizona-Nevada Acad. Sci. 30: 61–83.

Bailey, R. 1983. Delineation of ecosystem regions. Environm. Managem. 7: 365–373.

Baird, G. I. 2006. *Agoseris. In* Flora of North America Editorial Committee, eds., Flora of North America North of Mexico, Vol. 19, pp. 323–335. Oxford University Press, New York.

Baird, G. I. 2013. *Oxytenia. In* Heil, K. D., S. L. O'Kane, Jr., L. M. Reeves, and A. Clifford, eds., Flora of the Four Corners Region: Vascular Plants of the San Juan River Drainage; Arizona, Colorado, New Mexico, and Utah, p. 264. Missouri Botanical Garden Press, St. Louis.

Baker, M. A., B. D. Parfitt, D. J. Pinkava, and A. D. Zimmerman. 2009. Chromosome numbers in some cacti of western North America—VIII. Haseltonia 15: 117–134.

Baldwin, B. G., S. J. Bainbridge, and J. L. Strother. 2006. *Layia. In* Flora of North America Editorial Committee, eds., Flora of North America North of Mexico, Vol. 21, pp. 262–269. Oxford University Press, New York.

Ball, P. W. 2016. *Parnassia. In* Flora of North America Editorial Committee, eds., Flora of North America North of Mexico, Vol. 12, pp. 113–117. Oxford University Press, New York.

Ball, P. W., and A. A. Reznicek. 2002. *Carex. In* Flora of North America Editorial Committee, eds., Flora of North America North of Mexico, Vol. 23, pp. 254–572. Oxford University Press, New York.

Ball, P. W., A. A. Reznicek, and D. F. Murray. 2002. Cyperaceae. *In* Flora of North America Editorial Committee, eds., Flora of North America North of Mexico, Vol. 23, pp. 3–573. Oxford University Press, New York.

Barker, W. R., G. L. Nesom, P. M. Beardsley, and N. S. Fraga. 2012. A taxonomic conspectus of Phrymaceae: A narrowed circumscriptions for *Mimulus*, new and resurrected genera, and new names and combinations. Phytoneuron 2012–39: 1–60.

Barkley, T. M. 2006a. *Barkleyanthus. In* Flora of North America Editorial Committee, eds., Flora of North America North of Mexico, Vol. 20, p. 614. Oxford University Press, New York.

Barkley, T. M. 2006b. *Psacalium. In* Flora of North America Editorial Committee, eds., Flora of North America North of Mexico, Vol. 20, pp. 621–622. Oxford University Press, New York.

Barkley, T. M. 2006c. *Senecio. In* Flora of North America Editorial Committee, eds., Flora of North America North of Mexico, Vol. 20, pp. 544–570. Oxford University Press, New York.

Barkley, T. M., L. Brouillet, and J. L. Strother. 2006. Asteraceae. *In* Flora of North America Editorial Committee, eds., Flora of North America North of Mexico, Vol. 19, pp. 13–69. Oxford University Press, New York.

Barlow-Irick, P. L. 2002. Chapter 4: Redelimitation of the *Cirsium arizonicum* complex. *In* Biosystematic Analysis of the *Cirsium arizonicum* Complex of the Southwestern United States. Ph.D. Dissertation, University of New Mexico, Albuquerque. ProQuest Dissertations Publishing.

Barneby, R. C. 1989. Fabales. *In* Cronquist, A., A. H. Holmgren, N. H. Holmgren, J. L. Reveal, and P. K. Holmgren, eds., Intermountain Flora, Vol. 3, Part B. New York Botanical Garden, Bronx.

Barnett, D., and D. Barnett. 2016. *Cylindropuntia anasaziensis* sp. nov. *In* The Cactus of Colorado, pp. 247–260. Colorado Cactus and Succulent Society, Denver.

Barnett, D., and D. Barnett. 2022. *Penstemon* of Southeastern Colorado. Ethical Desert, Pueblo, Colorado.

Barrie, F. R. 2013. Caprifoliaceae. *In* Heil, K. D., S. L. O'Kane, Jr., L. M. Reeves, and A. Clifford, eds., Flora of the Four Corners Region: Vascular Plants of the San Juan River Drainage; Arizona, Colorado, New Mexico, and Utah, pp. 400–403. Missouri Botanical Garden Press, St. Louis.

Barringer, K. A. 2019a. *Cordylanthus. In* Flora of North America Editorial Committee, eds., Flora of North America North of Mexico, Vol. 17, pp. 669–678. Oxford University Press, New York.

Barringer, K. A. 2019b. *Orthocarpus. In* Flora of North America Editorial Committee, eds., Flora of North America North of Mexico, Vol. 17, pp. 680–684. Oxford University Press, New York.

Barringer, K. A. 2019c. *Schistophragma. In* Flora of North America Editorial Committee, eds., Flora of North America North of Mexico, Vol. 17, p. 275. Oxford University Press, New York.

Barringer, K. A., and A. T. Whittemore. 1997. Aristolochiaceae. *In* Flora of North America Editorial Committee, eds., Flora of North America North of Mexico, Vol. 3, pp. 44–58. Oxford University Press, New York.

Basset, I. J., and D. B. Munro. 1984. The biology of Canadian weeds. 67. *Solanum ptycanthum* Dun., *S. nigrum* L. and *S. sarrachoides* Sendt. Canad. J. Pl. Sci. 65: 401–414.

Bates, S. T., F. Farruggia, E. Gilbert, R. Gutierrez, D. Jenke, E. Makings, E. Manton, et al. 2009. Solanaceae part two: Key to the genera and *Solanum* L. Cantonia 5: 1–16.

Baum, D. A., K. J. Sytsma, and P. C. Hoch. 1994. A phylogenetic analysis of *Epilobium* (Onagraceae) based on nuclear ribosomal DNA sequences. Syst. Bot. 19: 363–388.

Bauters, K., I. Larridon, M. Reynders, P. Asselman, A. Vrijdagas, A. Muasya, D. Simpson, et al. 2014. A new classification for *Lipocarpha* and *Volkiella* as infrageneric taxa of *Cyperus* s.l. (Cypereae, Cyperoideae, Cyperaceae): Insights from species tree reconstruction supplemented with morphological and floral developmental data. Phytotaxa 166: 33–48.

Bayer, C., and K. Kubitzki. 2003. Malvaceae. *In* Kubitzki, K., and C. Bayer, eds., The Families and Genera of Vascular Plants, Vol. 5, pp. 225–311. Springer, Berlin.

Bayer, R. J. 2006. *Antennaria*. *In* Flora of North America Editorial Committee, eds., Flora of North America North of Mexico, Vol. 19, pp. 388–415. Oxford University Press, New York.

Behnke, H. D. 1997. Sarcobataceae—A new family of Caryophyllales. Taxon 46: 495–507.

Benet-Pierce, N., and M. G. Simpson. 2014. The taxonomy of *Chenopodium desiccatum* and *C. nitens*, sp. nov. J. Torrey Bot. Soc. 141: 161–172.

Benet-Pierce, N., and M. G. Simpson. 2017. Taxonomic recovery of the species in the *Chenopodium neomexicanum* (Chenopodiaceae) complex and description of *Chenopodium sonorense* sp. nov. J. Torrey Bot. Soc. 144: 339–356.

Benson, L. D. 1982. The Cacti of the United States and Canada. Stanford University Press, Redwood City, California.

Biddulph, S. F. 1944. A revision of the genus *Gaillardia*. Res. Stud. State Coll. Wash. 12: 195–256.

Bierner, M. W. 2006a. *Helenium*. *In* Flora of North America Editorial Committee, eds., Flora of North America North of Mexico, Vol. 21, pp. 426–435. Oxford University Press, New York.

Bierner, M. W. 2006b. *Hymenoxys*. *In* Flora of North America Editorial Committee, eds., Flora of North America North of Mexico, Vol. 21, pp. 435–443. Oxford University Press, New York.

Bierner, M. W., and B. L. Turner. 2006. *Tetraneuris*. *In* Flora of North America Editorial Committee, eds., Flora of North America North of Mexico. Vol. 20, pp. 447–453. Oxford University Press, New York.

Blanchard, O. J. 1976. A Revision of Species Segregated from *Hibiscus* Sect. *Trionum* (Medicus) de Candolle Sensu Lato (Malvaceae). Ph.D. Dissertation, Cornell University, Ithaca, New York.

Bleakly, D. 1998. A key to the *Penstemon* of New Mexico. New Mexico Botanist 9: 1–6.

Boedeker, F. 1933. Mammillarien-Vergleichs-Schluessel, 17.

Bogin, C. 1955. Revision of the genus *Sagittaria* (Alismataceae). Mem. New York Bot. Gard. 9: 179–233.

Bogler, D. J. 2002. *Dasylirion*. *In* Flora of North America Editorial Committee, eds., Flora of North America North of Mexico, Vol. 26, pp. 422–423. Oxford University Press, New York.

Bogler, D. J. 2006a. *Crepis*. *In* Flora of North America Editorial Committee, eds., Flora of North America North of Mexico, Vol. 19, pp. 222–239. Oxford University Press, New York.

Bogler, D. J. 2006b. *Hypochaeris*. *In* Flora of North America Editorial Committee, eds., Flora of North America North of Mexico, Vol. 19, pp. 297–299. Oxford University Press, New York.

Bogler, D. J. 2006c. *Leontodon*. *In* Flora of North America Editorial Committee, eds., Flora of North America North of Mexico, Vol. 19, pp. 294–296. Oxford University Press, New York.

Bogler, D. J. 2006d. *Lygodesmia*. *In* Flora of North America Editorial Committee, eds., Flora of North America North of Mexico, Vol. 19, pp. 369–373. Oxford University Press, New York.

Bogler, D. J. 2006e. *Pinaropappus*. *In* Flora of North America Editorial Committee, eds., Flora of North America North of Mexico, Vol. 19, pp. 374–376. Oxford University Press, New York.

Boissevain, C. H., and C. Davidson. 1940. Colorado Cacti. Abbey Garden Press, Pasadena, California.

Boraginales Working Group (BWG). 2016. Familial classification of the Boraginales. Taxon 65: 502–522.

Boufford, D. E. 1982. The systematics and evolution of *Circaea* (Onagraceae). Ann. Missouri Bot. Gard. 69: 804–994.

Boufford, D. E. 1992. Urticaceae: Nettle family. J. Arizona-Nevada Acad. Sci. 26: 42–49.

Boufford, D. E. 1997. Urticaceae. *In* Flora of North America Editorial Committee, eds., Flora of North America North of Mexico, Vol. 3, pp. 400–413. Oxford University Press, New York.

Brack, S., and K. D. Heil, Cact. Succ. J. (Los Angeles) 58: 165(–166), fig. 1986.

Brainerd, R. E., N. Otting, D. Lytjen, B. Newhouse, B. L. Wilson, and the Carex Working Group. 2014. *Kobresia*. *In* Field Guide to the Sedges of the Pacific Northwest, 2nd ed., pp. 406–409. Oregon State University Press, Corvallis.

Britton, N. L., and J. N. Rose. 1919. The Cactaceae, Vol. 1. Publication 248. Carnegie Institution, Washington, DC.

Brooks, R. E., and S. E. Clemants. 2000. *Juncus*. *In* Flora of North America Editorial Committee, eds., Flora North America North of Mexico, Vol. 22, pp. 211–255. Oxford University Press, New York.

Brouillet, L. 2006. *Taraxacum*. *In* Flora of North America Editorial Committee, eds., Flora of North America North of Mexico, Vol. 19, pp. 239–252. Oxford University Press, New York.

Brouillet, L., and P. E. Elvander. 2009. *In* Flora of North America Editorial Committee, eds., Flora of North America North of Mexico, Vol. 8, pp. 132–146. Oxford University Press, New York.

Brouillet, L., J. C. Semple, G. A. Allen, K. L. Chambers, and S. D. Sundberg. 2006. *Symphyotrichum*. *In* Flora of North America Editorial Committee, eds., Flora of North America North of Mexico, Vol. 20, pp. 465–539. Oxford University Press, New York.

Brown-Carter, M. E., and W. H. Murdy. 1985. Systematics of *Talinum parviflorum* Nutt. and the origin of *T. teretifolium* Pursh (Portulacaceae). Rhodora 87(850): 121–130.

Brummitt, R. K. 2012. *Calystegia*. *In* Baldwin, B. G., and D. H. Goldman, eds., The Jepson Manual: Vascular Plants of California. University of California Press, Berkeley.

Canne-Hilliker, J. M., and J. F. Hays. 2019. *Agalinis*. *In* Flora of North America Editorial Committee, eds., Flora of North America North of Mexico, Vol. 17, pp. 534–555. Oxford University Press, New York.

Carolin, R. 1987. A review of the family Portulacaceae. Austral. J. Bot. 35: 385–412.

Castetter, E. F., P. Pierce, and K. H. Schwerin. 1975a. Cact. Succ. J. (Los Angeles) 47: 62(-64), figs. 1–4.

Castetter, E. F., P. Pierce, and K. H. Schwerin, 1975b. Cact. Succ. J. (Los Angeles) 47: 64(-66), f.

Catalán, P., P. Torrecilla, J. A. López-Rodríguez, J. Müller, and C. A. Stace. 2007. A systematic approach to subtribe Loliinae (Poaceae: Pooideae) based on phylogenetic evidence. Aliso 23: 380–405.

Cather, S. M. 2004. Laramide orogeny in central and northern New Mexico and southern Colorado. *In* Mack, G., H., and K. A. Giles, eds., The Geology of New Mexico: A Geologic History, pp. 203–248. New Mexico Geological Society Special Publication 11, Socorro, New Mexico.

Chambers, K. L. 1964. Nomenclature of *Microseris lindleyi*. Leafl. W. Bot. 10: 106–108.

Chambers, K. L. 2006a. *Nothocalaïs*. *In* Flora of North America Editorial Committee, eds., Flora of North America North of Mexico, Vol. 19, pp. 335–337. Oxford University Press, New York.

Chambers, K. L. 2006b. *Prenanthella*. *In* Flora of North America Editorial Committee, eds., Flora of North America North of Mexico, Vol. 19, pp. 359–360. Oxford University Press, New York.

Chambers, K. L., and R. J. O'Kennon. 2006. *Krigia*. *In* Flora of North America Editorial Committee, eds., Flora of North America North of Mexico, Vol. 19, pp. 362–367. Oxford University Press, New York.

Chan, R., and R. Ornduff. 2006. *Lasthenia*. *In* Flora of North America Editorial Committee, eds., Flora of North America North of Mexico, Vol. 21, pp. 336–347. Oxford University Press, New York.

Chapin, C. E., W. C. McIntosh, and R. M. Chamberlin. 2004. The late Eocene-Oligocene peak of Cenozoic volcanism in southwestern New Mexico. *In* Mack, G. H., and K. A. Giles, eds., The Geology of New Mexico: A Geologic History, pp. 271–293. New Mexico Geological Society Special Publication 11, Socorro, New Mexico.

Cholewa, A. F., and D. M. Henderson. 2002. *Sisyrinchium*. *In* Flora of North America Editorial Committee, eds., Flora of North America North of Mexico, Vol. 26, pp. 351–371. Oxford University Press, New York.

Cholewa, A. F., and S. Kelso. 2009. Primulaceae. *In* Flora of North America Editorial Committee, eds., Flora of North America North of Mexico, Vol. 8, pp. 257–301. Oxford University Press, New York.

Christenhusz, J. M., M. F. Fay, and M. W. Chase. 2017. Plants of the World. Key Publishing, Stamford, United Kingdom; University of Chicago Press.

Clark, C. 2006. *Encelia*. *In* Flora of North America Editorial Committee, eds., Flora of North America North of Mexico, Vol. 21, pp. 118–122. Oxford University Press, New York.

Clevinger, J. A. 2006. *Silphium*. *In* Flora of North America Editorial Committee, eds., Flora of North America North of Mexico, Vol. 21, pp. 77–82. Oxford University Press, New York.

Cohen, J. I., and J. I. Davis. 2009. Nomenclatural changes in *Lithospermum* (Boraginaceae) and related taxa following a reassessment of phylogenetic relationships. Brittonia 61: 101–111.

Cohen, K. M., S. C. Finney, P. L. Gibbard, and J.-X. Fan. 2021 (2013; updated). The ICS International Chronostratigraphic Chart, ver. 10. Episodes 36: 199–204.

Coleman, R. 2002. The Wild Orchids of Arizona and New Mexico. Cornell University Press, Ithaca, New York.

Collins, L. T., A. E. L. Colwell, and G. Yatskievych. 2019. *Orobanche*. *In* Flora of North America Editorial Committee, eds., Flora of North America North of Mexico, Vol. 17, pp. 467–488. Oxford University Press, New York.

Constance, L. 1949. A revision of *Phacelia* subgenus *Cosmanthus* (Hydrophyllaceae). Contr. Gray Herb. 168: 3–48.

Cook, C. D. K., and M. S. Nicholls. 1986. A monographic study of the genus *Sparganium* (Sparganiaceae). Part 1. Subgenus *Xanthosparganium* Holmberg. Bot. Helv. 96: 213–167.

Cook, C. D. K., and M. S. Nichols. 1987. A monographic study of the genus *Sparganium* (Sparganiaceae). Part 2. Subgenus *Sparganium*. Bot. Helv. 97: 1–44.

Coop, J. D., and T. J. Givnish. 2007. Gradient analysis of reversed treelines and grasslands of the Valles Caldera, New Mexico. J. Veg. Sci. 18: 43–54.

Correll, D. S. 1943. The genus *Habenaria* in western North America. Leafl. W. Bot. 3(11): 233–247.

Correll, D. S., and H. B. Correll. 1972. Lythraceae. *In* Aquatic and Wetland Plants of Southwestern United States, pp. 1154–1168. Environmental Protection Agency, Washington, D.C.

Correll, D. S., M. C. Johnston, and collaborators. 1979a. Aceraceae. *In* Manual of the Vascular Plants of Texas, pp. 1001–1003. The University of Texas, Dallas.

Correll, D. S., M. C. Johnston, and collaborators. 1979b. *Ayenia*. *In* Manual of the Vascular Plants of Texas, pp. 1055–1057. The University of Texas, Dallas.

Correll, D. S., M. C. Johnston, and collaborators. 1979c. Geraniaceae. *In* Manual of the Vascular Plants of Texas, pp. 890–893. The University of Texas, Dallas.

Correll, D. S., M. C. Johnston, and collaborators. 1979d. *Peganum*. *In* Manual of the Vascular Plants of Texas, pp. 902–903. The University of Texas, Dallas.

Correll, D. S., M. C. Johnston, and collaborators. 1979e. Polemoniaceae. *In* Manual of the Vascular Plants of Texas, p. 263. The University of Texas, Dallas.

Correll, D. S., M. C. Johnston, and collaborators. 1979f. Polygalaceae. *In* Manual of the Vascular Plants of Texas, pp. 915–923. The University of Texas, Dallas.

Correll, D. S., M. C. Johnston, and collaborators. 1979g. Resedaceae. *In* Manual of the Vascular Plants of Texas, p. 712. The University of Texas, Dallas.

Correll, D. S., M. C. Johnston, and collaborators. 1979h. Rutaceae. *In* Manual of the Vascular Plants of Texas, pp. 906–911. The University of Texas, Dallas.

Correll, D. S., M. C. Johnston, and collaborators. 1979i. Sapindaceae. *In* Manual of the Vascular Plants of Texas, pp. 1005–1008. The University of Texas, Dallas.

Correll, D. S., M. C. Johnston, and collaborators. 1979j. Scrophulariaceae. *In* Manual of the Vascular Plants of Texas, pp. 1406–1442. The University of Texas, Dallas.

Costea, M., M. A. García, and S. Stefanović. 2015. A phylogenetically based infrageneric classification of the parasitic plant genus *Cuscuta* (dodders, Convolvulaceae). Syst. Bot. 40: 269–285.

Cronquist, A., A. H. Holmgren, N. H. Holmgren, J. L. Reveal, and P. K. Holmgren, eds. 1984. Valerianaceae. *In* Intermountain Flora, Vol. 4, pp. 546–548. New York Botanical Garden, Bronx.

Crook, R., and R. Mottram, 1995–2005. Opuntia index part 1–11. Bradleya 13: 88–118.

Dai, L., G. C. Tucker, and D. A. Simpson. 2010. *Cyperus*. *In* Wu, Z. Y., P. H. Raven, D. Y. Hong, eds., Flora of China, Vol. 23, pp. 219–241. Science Press, Beijing; Missouri Botanical Garden Press, St. Louis.

Daubenmire, R. 1942. Soil temperature versus drought as a factor determining the lower altitudinal limits of trees in the Rocky Mountains. Bot. Gaz. 105: 1–13.

Daubenmire, R. 1943. Vegetation zonation in the Rocky Mountains. Bot. Rev. 9: 325–393.

Dauphin, B., J. R. Grant, D. R. Farrar, and C. J. Rothfels. 2018. Rapid allopolyploid radiation of moonwort ferns (*Botrychium*; Ophioglossaceae) revealed by PacBio sequencing of homologous and homeologous nuclear regions. Molec. Phylogen. Evol. 120: 342–353.

Davidson, B. L. 2000. Lewisias. Timber Press, Portland, Oregon.

Davis, R. G. 1966. The North American perennial species of *Claytonia*. Brittonia 18: 285–303.

Davis, W. S. 2006. *Malacothrix*. *In* Flora of North America Editorial Committee, eds., Flora of North America North of Mexico, Vol. 19, pp. 310–321. Oxford University Press, New York.

De Groot, S. J. 2016. *Tomus nominum Eriastri*: The nomenclature and taxonomy of *Eriastrum* (Polemoniaceae: Loeselieae). Aliso 34(2): 25–152. https://scholarship.claremont.edu/aliso/vol34/iss2/1, accessed 25 October 2023.

de Paula-Souza, J., and H. E. Ballard, Jr. 2014. Re-establishment of the name *Pombalia*, and new combinations from the polyphyletic *Hybanthus* (Violaceae). Phytotaxa 183: 1–15.

Dickerman, C. 1985. Mid-nineteenth-century botanical exploration in New Mexico. New Mexico Historical Review 60(2): 159–171. https://digitalrepository.unm.edu/nmhr/vol60/iss2/3, accessed 25 October 2023.

Dick-Peddie, W. A. 1993. New Mexico Vegetation. University of New Mexico Press, Albuquerque.

Dillenberger, M. S., and J. W. Kadereit. 2014. Maximum polyphyly: Multiple origins and delimitation with plesiomorphic characters require a new circumscription of *Minuartia* (Caryophyllaceae). Taxon 63: 64–88.

Donoghue, M. J., R. G. Olmstead, J. F. Smith, and J. D. Palmer. 1992. Phylogenetic relationships of Dipsacales based on *rbc* sequences. Ann. Missouri Bot. Gard. 79: 333–345.

Dorn, R. D. 2010. The Genus *Salix* in North America North of Mexico. https://www.lulu.com/shop/robert-dorn/the-genus-salix-in-north-america-north-of-mexico/ebook/product-6517733, accessed 25 October 2023.

Dorr, L. J. 1990. A revision of the North American genus *Callirhoe* (Malvaceae). Mem. New York Bot. Gard. 56: 1–75.

Dorr, L. J. 2015a. *Ayenia*. *In* Flora of North America Editorial Committee, eds., Flora of North America North of Mexico, Vol. 6, pp. 202–207. Oxford University Press, New York.

Dorr, L. J. 2015b. *Callirhoe*. *In* Flora of North America Editorial Committee, eds., Flora of North America North of Mexico, Vol. 6, pp. 240–245. Oxford University Press, New York.

Eckenwalder, J. 2010. *Populus*. *In* Flora of North America Editorial Committee, eds., Flora of North America North of Mexico, Vol. 7, pp. 5–22. Oxford University Press, New York.

Egger, J. M., P. F. Zika, B. L. Wilson, R. E. Brainerd, and N. Otting. 2019. *Castilleja*. *In* Flora of North America Editorial Committee, eds., Flora of North America North of Mexico, Vol. 17, pp. 565–665. Oxford University Press, New York.

Elisens, W. J. 2019a. *Epixiphium*. *In* Flora of North America Editorial Committee, eds., Flora of North America North of Mexico, Vol. 17, pp. 20–21. Oxford University Press, New York.

Elisens, W. J. 2019b. *Maurandella*. *In* Flora of North America Editorial Committee, eds., Flora of North America North of Mexico, Vol. 17, pp. 35–36. Oxford University Press, New York.

Elisens, W. J., R. D. Whetstone, and R. P. Wunderlin. 2009. Sapotaceae. *In* Flora of North America Editorial Committee, eds., Flora of North America North of Mexico, Vol. 8, pp. 232–246. Oxford University Press, New York.

Ellison, W. L. 1964. A systematic study of the genus *Bahia* (Compositae). Rhodora 66: 67–86, 177–215.

Elvander, P. E. 1984. The taxonomy of *Saxifraga* (Saxifragaceae) section *Borophila*, subsection *Integrifoliae* in western North America. Syst. Bot. Monogr. 3: 1–44.

Engelmann, G. 1856. *M. vivipara*. Proc. Amer. Acad. Arts 3: 269.

Engelmann, G. 1876. *M. arizonica*. *In* Brewer, W. H., and S. Watson, Bot. California, Vol. 1, p. 244. Welch, Bigelow, & Co., University Press, Cambridge, Massachusetts.

Epling, C. 1942. The American species of *Scutellaria*. Univ. Calif. Publ. Bot. 20: 1–137.

Evrard, C., and C. Van Hove. 2004. Taxonomy of the American *Azolla* species (Azollaceae): A critical review. Syst. & Geogr. Pl. 74: 301–318.

Ewan, J. 1950. Rocky Mountain Naturalists. University of Denver Press, Denver, Colorado.

Faden, R. B. 2000. Commelinaceae. *In* Flora of North America Editorial Committee, eds., Flora of North America North of Mexico, Vol. 22, pp. 170–197. Oxford University Press, New York.

Felger, R. 2010. Plant distribution reports. New Mexico Botanist 51: 2.

Felger, R., and K. Kindscher. 2010. Trees of the Gila forest region, New Mexico. New Mexico Botanist Special Issue 2: 62.

Felger, R. S., and S. Rutman. 2016. Ajo Peak to Tinajas Altas: A flora of southwestern Arizona, Part 20. Eudicots: Solanaceae to Zygophyllaceae. Phytoneuron 52: 1–66.

Fenstermacher, J. 2016. Club chollas of the Big Bend 3.0: Refining species concepts and distributions for the *Corynopuntia* [*Opuntia*, *Grusonia*] *schotii* complex of western Texas via new chromosome counts, pollen stainability, and morphologic data. Phytoneuron 2016-1: 1–58.

Ferguson, D. J. 1987. *Opuntia cymochila* Engelm. & J. M. Bigelow, a species lost in the shuffle. Cact. Succ. J. (Los Angeles) 59: 256–260.

Ferguson, D. J. 1988. *Opuntia macrocentra* Engelm. and *Opuntia chlorotica* Engelm. & J. M. Bigelow. Cact. Succ. J. (Los Angeles) 60: 155–160.

Ferguson, D. J. 1995. Fameflowers, the genus *Talinum*. Rock Gard. Quart. 53: 83–96, 118–124.

Ferguson, D. J. 1999. Rare plant report. *Cylindropuntia viridiflora*. New Mexico Rare Plant Technical Council, Albuquerque.

Ferguson, D. J. 2001. *Phemeranthus* and *Talinum* (Portulacaceae) in New Mexico. New Mexico Botanist 20: 1–7.

Fiedler, P. L., and R. K. Zebell. 2002. *Calochortus*. *In* Flora of North America Editorial Committee, eds., Flora of North America North of Mexico, Vol. 26, pp. 119–141. Oxford University Press, New York.

Field, A. R., W. Testo, P. D. Bostock, J. A. M. Holtum, and M. Waycott. 2016. Molecular phylogenetics and the morphology of the Lycopodiaceae subfamily Huperzioideae supports three genera: *Huperzia*, *Phlegmariurus*, and *Phylloglossum*. Molec. Phylogen. Evol. 94: 635–657.

Fishbein, M., and A. McDonnell. 2023. *In* Flora of North America Editorial Committee, eds., Flora of North America North of Mexico, Vol. 14, pp. 239–256. Oxford University Press, New York.

Flagg, R. O., G. L. Smith, and A. W. Meerow. 2010. New combinations in *Habranthus* (Amaryllidaceae) in Mexico and southwestern U.S.A. Novon 20: 33–34.

Folk, R. A., and J. V. Freudenstein. 2014. Phylogenetic relationships and character evolution in *Heuchera* (Saxifragaceae) on the basis of multiple nuclear loci. Amer. J. Bot. 101: 1532–1550.

Freeman, C. C. 2009a. *Chimaphila*. *In* Flora of North America Editorial Committee, eds., Flora of North America North of Mexico, Vol. 8, pp. 377–379. Oxford University Press, New York.

Freeman, C. C. 2009b. *Orthilia*. *In* Flora of North America Editorial Committee, eds., Flora of North America North of Mexico, Vol. 8, pp. 385–386. Oxford University Press, New York.

Freeman, C. C. 2009c. *Pyrola*. *In* Flora of North America Editorial Committee, eds., Flora of North America North of Mexico, Vol. 8, pp. 377–379. Oxford University Press, New York.

Freeman, C. C. 2016. Hydrangeaceae. *In* Flora of North America Editorial Committee, eds., Flora of North America North of Mexico, Vol. 12, pp. 462–490. Oxford University Press, New York.

Freeman, C. C. 2019a. *Chionophila*. *In* Flora of North America Editorial Committee, eds., Flora of North America North of Mexico, Vol. 17, pp. 61–62. Oxford University Press, New York.

Freeman, C. C. 2019b. *Gratiola*. *In* Flora of North America Editorial Committee, eds., Flora of North America North of Mexico, Vol. 17, pp. 264–269. Oxford University Press, New York.

Freeman, C. C. 2019c. *Nuttallanthus*. *In* Flora of North America Editorial Committee, eds., Flora of North America North of Mexico, Vol. 17, pp. 40–42. Oxford University Press, New York.

Freeman, C. C. 2019d. *Penstemon*. *In* Flora of North America Editorial Committee, eds., Flora of North America North of Mexico, Vol. 17, pp. 82–255. Oxford University Press, New York.

Freeman, C. C., and J. L. Reveal. 2005. Polygonaceae. *In* Flora of North America Editorial Committee, eds., Flora of North America North of Mexico, Vol. 5, pp. 216–601. Oxford University Press, New York.

Freeman, C. C., R. K. Rabeler, and W. J. Elisens. 2019a. Orobanchaceae. *In* Flora of North America Editorial Committee, eds., Flora of North America North of Mexico, Vol. 17, pp. 456–687. Oxford University Press, New York.

Freeman, C. C., R. K. Rabeler, and W. J. Elisens. 2019b. Plantaginaceae. *In* Flora of North America Editorial Committee, eds., Flora of North America North of Mexico, Vol. 17, pp. 11–323. Oxford University Press, New York.

Fryxell, J. E. 1983. A revision of *Abutilon* sect. Oligocarpae (Malvaceae), including a new species from Mexico. Madroño 30: 84–92.

Fryxell, P. A. 1977. New species of Malvaceae from Mexico and Brazil. Phytologia (USA): 285–288.

Fryxell, P. A. 1980. A revision of the American species of *Hibiscus* section *Bombicella* (Malvaceae). U.S.D.A. Technical Bulletin 1624. Washington, DC.

Fryxell, P. A. 1985. *Sidus sidarum* V. The North and Central American species of *Sida*. Sida 11: 62–91.

Fryxell, P. A. 1987. Revision of the genus *Anoda* (Malvaceae). Aliso 11: 485–522.

Fryxell, P. A. 1988. Malvaceae of Mexico. Syst. Bot. Monogr. 25: 24–68.

Fryxell, P. A. 1992. A revised taxonomic interpretation of *Gossypium* L. (Malvaceae). Rheedea 2: 108–165.

Fryxell, P. A. 1997. The American genera of Malvaceae—II. Brittonia 49: 204–269.

Fuentes-Bazán, S., P. Uotila, and T. Borsch. 2012. A novel phylogeny-based generic classification for *Chenopodium* sensu lato, and a tribal rearrangement of Chenopodioideae (Chenopodiaceae). Willdenowia 42: 5–24.

Funston, A. M. 2006. *Roldana*. *In* Flora of North America Editorial Committee, eds., Flora of North America North of Mexico, Vol. 20, pp. 620–621. Oxford University Press, New York.

Gaskin, J. F. 2015. Tamaricaceae. *In* Flora of North America Editorial Committee, eds., Flora of North America North of Mexico, Vol. 6, pp. 413–417. Oxford University Press, New York.

Gillespie, L. J. 2016. Simmondsiaceae. *In* Flora of North America Editorial Committee, eds., Flora of North America North of Mexico, Vol. 12, pp. 441–442. Oxford University Press, New York.

Gillett, H. J., and K. S. Walter. 1998. 1997 IUCN Red List of threatened plants. International Union for the Conservation of Nature.

Goetghebeur, P., and A. Van den Borre. 1999. *Lipocarpha*. *In* Yatskievych, G., ed., Flora of Missouri, Vol. 1, pp. 405–407. Missouri Botanical Garden Press, St. Louis.

Goff, F., and J. N. Gardner. 2004. Late Cenozoic geochronology of volcanism and mineralization in the Jemez Mountains and Valles Caldera, north central New Mexico. *In* Mack, G. H., and K. A. Giles, eds., The Geology of New Mexico: A Geologic History, pp. 295–312. New Mexico Geological Society Special Publication 11, Socorro, New Mexico.

Goñalons, L. S. 2019. *Linaria*. *In* Flora of North America Editorial Committee, eds., Flora of North America North of Mexico, Vol. 17, pp. 27–33. Oxford University Press, New York.

Goodrich, S. 2013. Cyperaceae. *In* Heil, K. D., S. L. O'Kane, Jr., L. M. Reeves, and A. Clifford, eds., Flora of the Four Corners Region: Vascular Plants of the San Juan River Drainage; Arizona, Colorado, New Mexico, and Utah, pp. 458–491. Missouri Botanical Garden Press, St. Louis.

Goodson, B. E., and I. A. Al-Shehbaz. 2010. *Descurainia*. *In* Flora of North America Editorial Committee, eds., Flora of North America North of Mexico, Vol. 6, pp. 518–529. Oxford University Press, New York.

Gottlieb, L. D. 2006a. *Calycoseris*. *In* Flora of North America Editorial Committee, eds., Flora of North America North of Mexico, Vol. 19, pp. 307–308. Oxford University Press, New York.

Gottlieb, L. D. 2006b. *Rafinesquia*. *In* Flora of North America Editorial Committee, eds., Flora of North America North of Mexico, Vol. 19, pp. 346–349. Oxford University Press, New York.

Gottlieb, L. D. 2006c. *Stephanomeria*. *In* Flora of North America Editorial Committee, eds., Flora of North America North of Mexico, Vol. 19, pp. 350–359. Oxford University Press, New York.

Graham, S. A. 1986. Lythraceae. *In* the Great Plains Flora Association, eds., Flora of the Great Plains, pp. 494–497. University of Kansas, Lawrence, Kansas.

Graham, S. A. 2021. Lythraceae. *In* Flora of North America Editorial Committee, eds., Flora of North America North of Mexico, Vol. 10, pp. 42–66. Oxford University Press, New York.

Grant, V. 1989. Taxonomy of the tufted alpine and subalpine Polemoniums (Polemoniaceae). Bot. Gaz. 150: 158–169.

Griffith, G. E., J. M Omernik, M. M. McGraw, G. Z. Jacobi, C. M. Canavan, T. S. Schrader, D. Mercer, et al. 2006. Ecoregions of New Mexico [color poster with map, descriptive text, summary tables, and photographs; map scale 1:1,400,000]. U.S. Geological Survey, Reston, Virginia.

Griffiths, D. 1908–1911. Illustrated studies in *Opuntia*, I–IV. Rep. (Annual) Missouri Bot. Gard. 1918: 259–272; 1909: 81–95; 1910: 165–174; 1911: 25–36.

Griffiths, D. 1914–1916. New species of *Opuntia*. Proc. Biol. Soc. Washington 27: 23–28; 29: 9–16.

Grusz, A. L., and M. D. Windham. 2013. Toward a monophyletic *Cheilanthes*: The resurrection and recircumscription of *Myriopteris* (Pteridaceae). PhytoKeys 32: 49–64.

Halleck, D. K., and D. Wiens. 1966. Taxonomic status of *Claytonia rosea* and *C. lanceolata* (Portulacaceae). Ann. Missouri Bot. Gard. 53: 205–212.

Hanes, M. M. 2015. Malvaceae. *In* Flora of North America Editorial Committee, eds., Flora of North America North of Mexico, Vol. 6, pp. 187–219. Oxford University Press, New York.

Hanks, L. T., and J. K. Small. 1907. Geraniaceae. N. Amer. Fl., 25(1): 3–24.

Harley, R., and A. Paton. 1999. Notes on New World *Scutellaria*. Kew Bulletin 54: 221–225.

Harms, L. J. 1972. Cytotaxonomy of the *Eleocharis tenuis* complex. Amer. J. Bot. 59: 483–487.

Harms, V. L. 1963. Variation in the *Heterotheca* (*Chrysopsis*) *villosa* Complex East of the Rocky Mountains. Ph.D. Dissertation, University of Kansas, Lawrence.

Harms, V. L. 1974. Chromosome numbers in *Heterotheca*, including *Chrysopsis* (Compositae: Astereae), with phylogenetic interpretations. Brittonia 26: 61–69.

Harnik, P. G., C. Simpson, and J. L. Payne. 2012. Long-term differences in extinction risk among the seven forms of rarity. Proc. Roy. Soc. London, Ser. B, Biol. Sci. 279: 4969–4976.

Harrison, H. K. 1972. Contributions to the study of the genus *Eriastrum*. II. Notes concerning the type specimens and descriptions of the species. Brigham Young Univ. Sci. Bull., Biol. Ser. 16: 1–26.

Hartman, R. L. 1973. New plant records for New Mexico. Southw. Naturalist 18: 241–242. [*Eleocharis compressa*]

Hartman, R. L. 2006. *Xanthisma*. *In* Flora of North America Editorial Committee, eds., Flora of North America North of Mexico, Vol. 20, pp. 382–393. Oxford University Press, New York.

Hartman, R. L., and D. J. Bogler. 2006. *Arida*. *In* Flora of North America Editorial Committee, eds., Flora of North America North of Mexico, Vol. 20, pp. 401–405. Oxford University Press, New York.

Hartman, R. L., and R. K. Rabeler. 2013. Caryophyllaceae. *In* Heil, K. D, S. L. O'Kane, Jr., L. M. Reeves, and A. Clifford, eds., Flora of the Four Corners Region: Vascular Plants of the San Juan River Drainage; Arizona, Colorado, New Mexico, and Utah, pp. 403–419. Missouri Botanical Garden Press, St. Louis.

Hartman, R. L., S. Goodrich, and K. D. Heil. 2013. Apiaceae. *In* Heil, K. D, S. L. O'Kane, Jr., L. M. Reeves, and A. Clifford, eds., Flora of the Four Corners Region: Vascular Plants of the San Juan River Drainage; Arizona, Colorado, New Mexico, and Utah, pp. 108–129. Missouri Botanical Garden Press, St. Louis.

Hasenstab-Lehman, K. E. 2017. Phylogenetics of the borage family: Delimiting Boraginales and assessing closest relatives. Aliso 35: 41–43.

Hasenstab-Lehman, K. E., and M. G. Simpson. 2012. Cat's eyes and popcorn flowers: Phylogenetic systematics of the genus *Cryptantha* s.l. (Boraginaceae). Syst. Bot. 37: 738–757.

Hawksworth, F. G., and D. Wiens. 1993. Viscaceae. *In* Hickman, J. C., ed., The Jepson Manual: Higher Plants of California, pp. 1092–1097. University of California Press, Berkeley.

Haynes, R. R. 2000. Hydrocharitaceae. *In* Flora of North America Editorial Committee, eds., Flora of North America North of Mexico, Vol. 22, pp. 26–28. Oxford University Press, New York.

Haynes, R. R., and C. B. Hellquist. 2000a. Alismataceae, *In* Flora of North America Editorial Committee, eds., Flora of North America North of Mexico, Vol. 22, pp. 7–25. Oxford University Press, New York.

Haynes, R. R., and C. B. Hellquist. 2000b. Potamogetonaceae. *In* Flora of North America Editorial Committee, eds., Flora of North America North of Mexico, Vol. 22, pp. 47–74. Oxford University Press, New York.

Heflin, Jean. 1997. Penstemons: The Beautiful Beardtongues of New Mexico. Jackrabbit Press, Albuquerque.

Heil, K. D. 2013a. Betulaceae. *In* Heil, K. D., S. L. O'Kane, Jr., L. M. Reeves, and A. Clifford, eds., Flora of the Four Corners Region: Vascular Plants of the San Juan River Drainage; Arizona, Colorado, New Mexico, and Utah, pp. 318–319. Missouri Botanical Garden Press, St. Louis.

Heil, K. D. 2013b. Elaeagnaceae. *In* Heil, K. D., S. L. O'Kane, Jr., L. M. Reeves, and A. Clifford, eds., Flora of the Four Corners Region: Vascular Plants of the San Juan River Drainage; Arizona, Colorado, New Mexico, and Utah, pp. 492–493. Missouri Botanical Garden Press, St. Louis.

Heil, K. D. 2013c. Oleaceae. *In* Heil, K. D., S. L. O'Kane, Jr., L. M. Reeves, and A. Clifford, eds., 2013. Flora of the Four Corners Region: Vascular Plants of the San Juan River Drainage; Arizona, Colorado, New Mexico, and Utah, pp. 657–659. Missouri Botanical Garden Press, St. Louis.

Heil, K. D. 2013d. Saururaceae. *In* Heil, K. D., S. L. O'Kane, Jr., L. M. Reeves, and A. Clifford, eds., Flora of the Four Corners Region: Vascular Plants of the San Juan River Drainage; Arizona, Colorado, New Mexico, and Utah, p. 940. Missouri Botanical Garden Press, St. Louis.

Heil, K. D. 2013e. Simaroubaceae. *In* Heil, K. D., S. L. O'Kane, Jr., L. M. Reeves, and A. Clifford, eds., Flora of the Four Corners Region: Vascular Plants of the San Juan River Drainage; Arizona, Colorado, New Mexico, and Utah, pp. 948–949. Missouri Botanical Garden Press, St. Louis.

Heil, K. D., and J. M. Porter. 1994. *Sclerocactus* (Cactaceae): A revision. Haseltonia 2: 20–46.

Heil, K. D., and J. M Porter. 2013a. Cornaceae. *In* Heil, K. D., S. L. O'Kane, Jr., L. M. Reeves, and A. Clifford, eds., Flora of the Four Corners Region: Vascular Plants of the San Juan River Drainage; Arizona, Colorado, New Mexico, and Utah, pp. 452–453. Missouri Botanical Garden Press, St. Louis.

Heil, K. D., and J. M. Porter. 2013b. Linaceae. *In* Heil, K. D., S. L. O'Kane, Jr., L. M. Reeves, and A. Clifford, eds., Flora of the Four Corners Region: Vascular Plants of the San Juan River Drainage; Arizona, Colorado, New Mexico, and Utah, pp. 635–637. Missouri Botanical Garden Press, St. Louis.

Heil, K. D., B. Armstrong, and D. Schleser. 1981. A review of the genus *Pediocactus*. Cact. Succ. J. (Los Angeles) 53: 17–39.

Heil, K. D., S. L. O'Kane, Jr., L. M. Reeves, and A. Clifford, eds. 2013. Flora of the Four Corners Region: Vascular Plants of the San Juan River Drainage; Arizona, Colorado, New Mexico, and Utah. Missouri Botanical Garden Press, St. Louis.

Hellquist, C. B. 2013a. *Callitriche*. *In* Heil, K. D., S. L. O'Kane, Jr., L. M. Reeves, and A. Clifford, eds., Flora of the Four Corners Region: Vascular Plants of the San Juan River Drainage; Arizona, Colorado, New Mexico, and Utah, p. 711. Missouri Botanical Garden Press, St. Louis.

Hellquist, C. B. 2013b. *Hippuris*. *In* Heil, K. D., S. L. O'Kane, Jr., L. M. Reeves, and A. Clifford, eds., Flora of the Four Corners Region: Vascular Plants of the San Juan River Drainage; Arizona, Colorado, New Mexico, and Utah, p. 731. Missouri Botanical Garden Press, St. Louis.

Hellquist, C. B. 2013c. Menyanthaceae. *In* Heil, K. D., S. L. O'Kane, Jr., L. M. Reeves, and A. Clifford, eds., Flora of the Four Corners Region: Vascular Plants of the San Juan River Drainage; Arizona, Colorado, New Mexico, and Utah, pp. 647–648. Missouri Botanical Garden Press, St. Louis.

Henderson, N. C. 2002. *Iris*. *In* Flora of North America Editorial Committee, eds., Flora of North America North of Mexico, Vol. 26, pp. 371–395. Oxford University Press, New York.

Henrickson, J. 1972. A taxonomic revision of Fouquieriaceae. Aliso 7: 439–537.

Hermann, F. J. 1974. Manual of the genus *Carex* in Mexico and Central America. U.S.D.A. Agricultural Handbook No. 467, Washington, D.C.

Hermann, F. J. 1975. Manual of the rushes (*Juncus* spp.) of the Rocky Mountains and Colorado Basin. U.S.D.A. Forest Service General Technical Report RM-18, Washington, D.C.

Herndon, A. 2002. *Hypoxis*. *In* Flora of North America Editorial Committee, eds., Flora of North America North of Mexico, Vol. 26, pp. 201–204. Oxford University Press, New York.

Hershkovitz, M. A. 1993. Revised circumscriptions and subgeneric taxonomies of *Calandrinia* and *Montiopsis* (Portulacaceae) with notes on phylogeny of the Portulacaceous Alliance. Ann. Missouri Bot. Gard. 80: 333–396.

Hershkovitz, M. A. 2019. Systematics, evolution, and phylogeography of Montiaceae (Portulacineae). Phytoneuron 27: 1-77.

Hess, W. J. 2002. *Nolina*. *In* Flora of North America Editorial Committee, eds., Flora of North America North of Mexico, Vol. 26, pp. 415–421. Oxford University Press, New York.

Hess, W. J., and R. L. Robbins. 2002. *Yucca*. *In* Flora of North America Editorial Committee, eds., Flora of North America North of Mexico, Vol. 26, pp. 423–439. Oxford University Press, New York.

Hevly, R. H. 1969. Nomenclatural history and typification of *Martynia* and *Proboscidea* (Martyniaceae). Taxon 18(5): 527–534.

Hoch, P. C. 2021. *Epilobium*. *In* Flora of North America Editorial Committee, eds., Flora of North America North of Mexico, Vol. 10, pp. 112–159. Oxford University Press, New York.

Hodgdon, A. R. 1938. A taxonomic study of *Lechea*. Rhodora 40: 29–69, 87–131.

Holmes, W. C., K. L. Yip, and A. E. Rushing. 2008. Taxonomy of *Koeberlinia* (Koeberliniaceae). Brittonia 60: 171–184.

Holmgren, N. H. 2005. Brassicaceae. *In* Holmgren, N. H., P. K. Holmgren, and A. Cronquist, eds., Intermountain Flora, Vol. 2, Part B, Subclass Dilleniidae. New York Botanical Garden Press, Bronx.

Horn, C. 2002. Pontederiaceae. *In* Flora of North America Editorial Committee, eds., Flora of North America North of Mexico, Vol. 26, pp. 37–46. Oxford University Press, New York.

Horn, J. W., B. W. van Ee, J. J. Morawetz, R. Riina, V. W. Steinmann, and P. E, Berry. 2012. Phylogenetics and the evolution of major structural characters in the *Euphorbia* L. (Euphorbiaceae). Molec. Phylogen. Evol. 63: 305–326.

Hufford, L. D. 2019. *Synthyris*. *In* Flora of North America Editorial Committee, eds., Flora of North America North of Mexico, Vol. 17, pp. 296–304. Oxford University Press, New York.

Huft, M. J. 2016. *Stillingia*. *In* Flora of North America Editorial Committee, eds., Flora of North America North of Mexico, Vol. 12, p. 233. Oxford University Press, New York.

Hurd, E. G., S. Goodrich, and N. L. Shaw. 1997. Field guide to intermountain rushes. U.S.D.A. Forest Service General Technical Report INT-306, Washington D.C.

Hutchins, C. R. 1974. A flora of the White Mountains area, southern Lincoln and northern Otero Counties, New Mexico. Privately published.

Hyatt, Philip. 2006. *Sonchus*. *In* Flora of North America Editorial Committee, eds., Flora of North America North of Mexico, Vol. 19, pp. 273–276. Oxford University Press, New York.

Irving, R. S. 1980. The systematics of *Hedeoma* (Labiatae). Sida 8(3): 218–295.

Isely, D. 1998. Native and Naturalized Leguminosae (Fabaceae) of the United States. Monte L. Bean Life Museum, Brigham Young University, Provo, Utah.

Jacobsen, T. D., and D. W. McNeal, Jr. 2002. *Nothoscordum*. *In* Flora of North America Editorial Committee, eds., Flora of North America North of Mexico, Vol. 26, pp. 276–277. Oxford University Press, New York.

Jercinovic, E. M. 2012. *Nolina*. *In* Allred, K. W., and R. D. Ivey, eds., Flora Neomexicana III: An Illustrated Identification Manual. Lulu Press.

Johnson, D. E., and J. S. Mooring. 2006. *Eriophyllum*. *In* Flora of North America Editorial Committee, eds., Flora of North America North of Mexico, Vol. 21, pp. 353–362. Oxford University Press, New York.

Johnson, L. A., and R. L. Johnson. 2006. Morphological delimitation and molecular evidence for allopolyploidy in *Collomia wilkenii* (Polemoniaceae), a new species from northern Nevada. Syst. Bot. 2: 349–360.

Johnson, W. T. 1987. Bladderwort, Arizona's carnivorous wildflower. Desert Plants 8: 140–141.

Johnson-Fulton, S. B., and L. E. Watson. 2017. Phylogenetic systematics of Cochlospermaceae (Malvales) based on molecular and morphological evidence. Syst. Bot. 42(2): 271–282.

Jones, A. G. 1978. The taxonomy of *Aster* section *Multiflori* (Asteraceae). I. Nomenclatural review and formal presentation of taxa. Rhodora 80: 319–357.

Jones, G. N. 1940. A monograph of the genus *Symphoricarpos*. J. Arnold Arbor. 21: 201–252.

Judd, W. S., R. W. Sanders, and M. J. Donoghue. 1994. Angiosperm family pairs: Preliminary cladistic analyses. Harvard Pap. Bot. 5:1–51.

Judd, W. S., and S. R. Manchester. 1997. Circumscription of Malvaceae (Malvales) as determined by a preliminary cladistic analysis of morphological, anatomical, palynological, and chemical characters. Brittonia 49: 348–405.

Judd, W. S., C. S. Campbell, E. A. Kellogg, P. F. Stevens, and M. J. Donoghue. 2016. Plant Systematics: A Phylogenetic Approach, 4th ed. Sinauer Associates, Sunderland, Massachusetts.

Kadereit, G., and H. Freitag. 2011. Molecular phylogeny of Camphorosmeae (Camphorosmoideae, Chenopodiaceae): Implications for biogeography, evolution of C_4-photosynthesis and taxonomy. Taxon 60: 51–78.

Karlstrom, K. E., J. M. Amato, M. L. Williams, M. Heizler, C. A. Shaw, A. S. Read, and P. Bauer. 2004. Proterozoic tectonic evolution of the New Mexico region: A synthesis. *In* Mack, G. H., and K. A. Giles, eds., The Geology of New Mexico: A Geologic History, pp. 1–34. New Mexico Geological Society Special Publication 11, Socorro, New Mexico.

Kartesz, J. T. 1999. A Synonymized Checklist of the Vascular Flora of the United States, Canada, and Greenland. Timber Press, Portland, Oregon.

Kaul, R. B. 1997. Platanaceae. *In* Flora of North America Editorial Committee, eds., Flora of North America North of Mexico, Vol. 3, pp. 358–361. Oxford University Press, New York.

Kearney, T. H. 1954. A tentative key to the North American species of *Sida* L. Leafl. W. Bot. 7: 138–150.

Keil, D. J. 2006a. *Arctium*. *In* Flora of North America Editorial Committee, eds., Flora of North America North of Mexico, Vol. 19, pp. 168–171. Oxford University Press, New York.

Keil, D. J. 2006b. *Carminatia*. *In* Flora of North America Editorial Committee, eds., Flora of North America North of Mexico, Vol. 21, pp. 511–512. Oxford University Press, New York.

Keil, D. J. 2006c. *Carphochaete*. *In* Flora of North America Editorial Committee, eds., Flora of North America North of Mexico, Vol. 21, pp. 486–487. Oxford University Press, New York.

Keil, D. J. 2006d. *Carthamus*. *In* Flora of North America Editorial Committee, eds., Flora of North America North of Mexico, Vol. 21, pp. 178–181. Oxford University Press, New York.

Keil, D. J. 2006e. *Cirsium*. *In* Flora of North America Editorial Committee, eds., Flora of North America North of Mexico, Vol. 19, pp. 95–164. Oxford University Press, New York.

Keil, D. J. 2006f. *Engelmannia*. *In* Flora of North America Editorial Committee, eds., Flora of North America North of Mexico, Vol. 21, p. 87. Oxford University Press, New York.

Keil, D. J. 2006g. *Onopordum*. *In* Flora of North America Editorial Committee, eds., Flora of North America North of Mexico, Vol. 19, pp. 87–88. Oxford University Press, New York.

Keil, D. J. 2006h. *Pectis*. *In* Flora of North America Editorial Committee, eds., Flora of North America North of Mexico, Vol. 21, pp. 222–230. Oxford University Press, New York.

Keil, D. J. 2006i. *Pentzia*. *In* Flora of North America Editorial Committee, eds., Flora of North America North of Mexico, Vol. 19, p. 543. Oxford University Press, New York.

Keil, D. J. 2006j. *Plectocephalus*. *In* Flora of North America Editorial Committee, eds., Flora of North America North of Mexico, Vol. 19, pp. 175–177. Oxford University Press, New York.

Keil, D. J. 2006k. *Silybum*. *In* Flora of North America Editorial Committee, eds., Flora of North America North of Mexico, Vol. 19, p. 164. Oxford University Press, New York.

Keil, D. J. 2006l. *Trixis*. *In* Flora of North America Editorial Committee, eds., Flora of North America North of Mexico, Vol. 19, pp. 75–76. Oxford University Press, New York.

Keil, D. J. 2012. *Porophyllum*. *In* Jepson Flora Project, eds., Jepson eFlora. https://ucjeps.berkeley.edu/eflora/, accessed on 29 May 2023.

Keil, D. J., and J. Ochsmann. 2006. *Centaurea*. *In* Flora of North America Editorial Committee, eds., Flora of North America North of Mexico, Vol. 19, pp. 181–194. Oxford University Press, New York.

Keller, C. 1999. Solidagos: A preliminary look at a difficult problem, with a tentative key. New Mexico Botanist 11: 1–10.

Keller, C. F., C. T. Martin, T. S. Foxx, and N. R. Greiner. 2017. Additional species for the Jemez Mountains, New Mexico, U.S.A. J. Bot. Res. Inst. Texas 11(2): 513–522.

Kiger, R. W. 1997. Papaveraceae. *In* Flora of North America Editorial Committee, eds., Flora of North America North of Mexico, Vol. 3, pp. 300–339. Oxford University Press, New York.

Kiger, R. W. 2001. New combinations in *Phemeranthus* Raf. (Portulacaceae). Novon 11: 319–321.

Kiger, R. W. 2006. *Cosmos*. *In* Flora of North America Editorial Committee, eds., Flora of North America North of Mexico, Vol. 21, pp. 203–205. Oxford University Press, New York.

Kirschner, J. 2002a. Species Plantarum, Part 7. Juncaceae 2: *Juncus* subg. Juncus. Australian Biological Resources Study, Canberra.

Kirschner, J. 2002b. Species Plantarum, Part 8. Juncaceae 3: *Juncus* subg. Agathryon. Australian Biological Resources Study, Canberra.

Kirschner, J. 2002c. *Luzula*. *In* Species Plantarum, Part 6. Juncaceae 1, pp. 18–228. Australian Biological Resources Study, Canberra.

Kleinman, R., and R. Felger. 2009. Vascular plants of the Gila wilderness. Western New Mexico University Department of Natural Sciences. www.wnmu.edu/academic/nspages/gilaflora/salix_bonplandiana.html, accessed 25 October 2023.

Kral, R. 1971. A treatment of *Abildgaardia*, *Bulbostylis* and *Fimbristylis* (Cyperaceae) for North America. Sida 4: 57–227.

Krings, A., M. Fishbein, and D. Lemke. 2023. Apocynaceae. *In* Flora of North America Editorial Committee, eds., Flora of North America North of Mexico, Vol. 14, pp. 103–268. Oxford University Press, New York.

Kron, S. A., W. S. Judd, P. F. Stevens, D. M. Crayn, A. A. Anderberg, P. A. Gadek, C. J. Quinn, et al. 2002. Phylogenetic classification of Ericaceae: Molecular and morphological evidence. Bot. Rev. (Lancaster) 68: 335–423.

Küchler, A. W. 1980. International Bibliography of Vegetation Maps, 2nd ed. University of Kansas Libraries, University of Kansas, Lawrence.

Kues, B. S., and K. A. Giles. 2004. The late Paleozoic ancestral Rocky Mountain system in New Mexico. *In* Mack, G. H., and K. A. Giles, eds., The Geology of New Mexico: A Geologic History, pp. 95–136. New Mexico Geological Society Special Publication 11, Socorro, New Mexico.

La Duke, J. C. 1985. A new species of *Sphaeralcea* (Malvaceae). Southw. Naturalist 30: 433–436.

La Duke, J. C. 2015. *Sphaeralcea*. *In* Flora of North America Editorial Committee, eds., Flora of North America North of Mexico, Vol. 6, pp. 357–369. Oxford University Press, New York.

LaFrankie, J. V. 2002. *Maianthemum. In* Flora of North America Editorial Committee, eds., Flora of North America North of Mexico, Vol. 26, pp. 206–210. Oxford University Press, New York.

Lamont, Eric E. 2006. *Eutrochium. In* Flora of North America Editorial Committee, eds., Flora of North America North of Mexico, Vol. 21, pp. 474–478. Oxford University Press, New York.

Landolt, E. 1986. The family of Lemnaceae—A monographic study, Vol. 1. Veröffentlichungen des Geobotanischen Institutes der Eidgenössischen Technischen Hochschule, Stiftung Rübel, Zurich.

Landolt, E. 2000. Lemnaceae. *In* Flora of North America Editorial Committee, eds., Flora of North America North of Mexico, Vol. 22, pp. 143–151. Oxford University Press, New York.

Landolt, E., and R. Kandeler. 1987. The family of Lemnaceae—A monographic study, Vol. 2. Veröffentlichungen des Geobotanischen Institutes der Eidgenössischen Technischen Hochschule, Stiftung Rübel, Zurich.

Larridon, I., K. Bauters, M. Reynders, W. Huygh, and P. Goetghebeur. 2014. Taxonomic changes in C4 *Cyperus* (Cypereae, Cyperoideae, Cyperaceae): Combining the sedge genera *Ascolepis*, *Kyllinga* and *Pycreus* into *Cyperus* s.l. Phytotaxa 166: 1–32.

Larson, B. M. H., and P. M. Catling. 1996. The separation of *Eleocharis obtusa* and *Eleocharis ovata* (Cyperaceae) in eastern Canada. Canad. J. Bot. 74: 238–242.

Lee, J. S. Y., S. H. P. Kim, and M. A. Ali. 2013. Molecular phylogenetic relationships among members of the family Phytolaccaceae sensu lato inferred from internal transcribed spacer sequences of nuclear ribosomal DNA. Genet. Mol. Res. 12: 4515–4525.

Legler, B. S. 2011. *Phlox vermejoensis* (Polemoniaceae), a new species from northern New Mexico, U.S.A. J. Bot. Res. Inst. Texas 5(2):397–403.

Les, D. H. 1997. *Ceratophyllum. In* Flora of North America Editorial Committee, eds., Flora of North America North of Mexico, Vol. 3, pp. 82–84. Oxford University Press, New York.

Levin, G. A. 2016. Phyllanthaceae. *In* Flora of North America Editorial Committee, eds., Flora of North America North of Mexico, Vol. 12, pp. 318–347. Oxford University Press, New York.

Levin, G. A., and L. J. Gillespie. 2016. Euphorbiaceae. *In* Flora of North America Editorial Committee, eds., Flora of North America North of Mexico, Vol. 12, pp. 156–324. Oxford University Press, New York.

Lewis, D. Q., R. K. Rabeler, C. C. Freeman, and W. J. Elisens. 2019. Linderniaceae. *In* Flora of North America Editorial Committee, eds., Flora of North America North of Mexico, Vol. 17, pp. 352–359. Oxford University Press, New York.

Lewis, H., and J. Szweykowski. 1964. The genus *Gayophytum* (Onagraceae). Brittonia 16: 343–391.

Little, R. J., and L. E. McKinney. 2015. Violaceae. *In* Flora of North America Editorial Committee, eds., Flora of North America North of Mexico, Vol. 6, pp. 106–164. Oxford University Press, New York.

Lowrey, T. 2020. Asteraceae. *In* Allred, K. W., E. M. Jercinovic, and R. D. Ivey, eds., Flora Neomexicana III: An Illustrated Identification Manual. Part 2: Dicotyledonous Plants, 2nd ed., pp. 77–238. Lulu Press.

Lucas, S. G. 2004. The Triassic and Jurassic systems in New Mexico. *In* Mack, G. H., and K. A. Giles, eds., The Geology of New Mexico: A Geologic History, pp. 137–152. New Mexico Geological Society Special Publication 11, Socorro, New Mexico.

Lucas, S. G., and A. B. Heckert, eds. 1995. Early Permian footprints and facies. New Mexico Museum of Natural History and Science, Bulletin 6, Albuquerque.

Luther, H. E., and G. K. Brown. 2000. *Tillandsia. In* Flora of North America Editorial Committee, eds., Flora of North America, Vol. 22, pp. 288–296. Oxford University Press, New York.

Ma, J., P. W. Ball, and G. A. Levin. 2016. Celastraceae. *In* Flora of North America Editorial Committee, eds., Flora of North America North of Mexico, Vol. 12, pp. 111–132. Oxford University Press, New York.

Mack, G. H. 2004. The Cambro-Ordovician Bliss and Lower Ordovician El Paso Formations, southwestern New Mexico and west Texas. *In* Mack, G. H., and K. A. Giles, eds., The Geology of New Mexico: A Geologic History, pp. 35–44. New Mexico Geological Society Special Publication 11, Socorro, New Mexico

Mack, G. H., and K. A. Giles, eds. 2004 The Geology of New Mexico: A Geologic History. New Mexico Geological Society Special Publication 11, Socorro, New Mexico.

Madhani, H., R. Rabeler, A. Pirani, B. Oxelman, G. Heubl, and S. Zarre. 2018. Untangling taxonomic confusion in the carnation tribe (Caryophyllaceae: Caryophylleae) with special focus on generic boundaries. Taxon 63: 83–112.

Manson, G. 2004. A new classification of the polyphyletic genus *Centaurium* Hill (Chironiinae, Gentianaceae): Description of the New World endemic *Zeltnera*, and reinstatement of *Gyrandra* Griseb. and *Schenkia* Griseb. Taxon 53: 719–740.

Martin, W. C., and C. R. Hutchins. 1980–1981. A Flora of New Mexico, Vols. 1 and 2. J. Cramer, Germany.

Martin, W. C., and C. R. Hutchins. 1980a. Cyperaceae. *In* A Flora of New Mexico, Vol. 1, pp. 317–376. J. Cramer, Germany.

Martin, W. C., and C. R. Hutchins. 1980b. Geraniaceae. *In* A Flora of New Mexico, Vol. 1, pp. 1118–1125. J. Cramer, Germany.

Martin, W. C., and C. R. Hutchins. 1980c. Orchidaceae. *In* A Flora of New Mexico, Vol. 1, pp. 450–469. J. Cramer, Germany.

Martin, W. C., and C. R. Hutchins. 1980d. Polygalaceae. *In* A Flora of New Mexico, Vol. 1, pp. 1150–1158. J. Cramer, Germany.

Martin, W. C., and C. R. Hutchins. 1980e. Scrophulariaceae. *In* A Flora of New Mexico, Vol. 2, pp. 1776–1853. J. Cramer, Germany.

Martin, W. C., and C. R. Hutchins. 1981a. Lythraceae. *In* A Flora of New Mexico, Vol. 2, pp. 1360–1362. J. Cramer, Germany.

Martin, W. C., and C. R. Hutchins. 1981b. Polemoniaceae. *In* A Flora of New Mexico, Vol. 2, pp. 1569–1606. J. Cramer, Germany.

Martin, W. C., and C. R. Hutchins. 1981c. Umbelliferae. *In* A Flora of New Mexico, Vol. 2, pp. 1406–1449. J. Cramer, Germany.

Marx, H. E., N. O'Leary, Y. W. Yuan, P. Lu-Irving, D. C. Tank, M. E. Mulgura, and R. G. Olmstead. 2010. A molecular phylogeny and classification of Verbenaceae. Amer. J. Bot. 97: 1647–1663.

Mason, C. T., Jr. 1992. Crossosomataceae: Crossosoma family. J. Arizona-Nevada Acad. Sci. 26: 7–9.

Mason, C. T., Jr. 1998. Gentianaceae: Gentian family. J. Arizona-Nevada Acad. Sci. 30: 84–95.

Mast, A. R., and J. L. Reveal. 2007. Transfer of *Dodecatheon* to *Primula* (Primulaceae). Brittonia 59: 79–82.

Matthew, B. 1989. The Genus *Lewisia*. Kew Magazine Monograph. Timber Press, Portland, Oregon.

McGrath, J., M. Licher, W. R. Norris, and G. Rink. 2015. A review of *Carex* in New Mexico: Initial findings. New Mexico Botanist 63: 1–8.

McNeal, D. W., Jr., and T. D. Jacobson. 2002. *Allium*. *In* Flora of North America Editorial Committee, eds., Flora of North America North of Mexico, Vol. 26, pp. 297–308. Oxford University Press, New York.

McVaugh, R. 1936. Studies in the taxonomy and distribution of eastern North American species of *Lobelia*. Rhodora 38: 241–298.

McVaugh, R. 1945. The genus *Triodanis* Raf., and its relationships to *Specularia* and *Campanula*. Wrightia 1: 13–52.

Mickel, J. T., and A. R. Smith. 2004. The Pteridophytes of Mexico. Mem. New York Bot. Gard. 88: 1–1055 + xxiv.

Miller, J. M., and K. L. Chambers. 1993. Nomenclatural changes and new taxa in *Claytonia* (Portulacaceae) in western North America. Novon 3: 268–273.

Moir, W. H. 1967. The subalpine tall grass, *Festuca thurberi*, community of Sierra Blanca, New Mexico. Southw. Naturalist 102: 217–331.

Moore, L. M. 2013. Geraniaceae. *In* Heil, K. D, S. L. O'Kane, Jr., L. M. Reeves, and A. Clifford, eds. Flora of the Four Corners Region: Vascular Plants of the San Juan River Drainage; Arizona, Colorado, New Mexico, and Utah, pp. 585–588. Missouri Botanical Garden Press, St. Louis.

Moore, M. O., and J. Wen. 2016. Vitaceae. *In* Flora of North America Editorial Committee, eds., Flora of North America North of Mexico, Vol. 12, pp. 3–23. Oxford University Press, New York.

Moran, R. V. 2009. *Crassula*. *In* Flora of North America Editorial Committee, eds., Flora of North America North of Mexico, Vol. 8, pp. 150–155. Oxford University Press, New York.

Morefield J. D. 2006a. *Chaenactis*. *In* Flora of North America Editorial Committee, eds., Flora of North America North of Mexico, Vol. 21, pp. 400–414. Oxford University Press, New York.

Morefield, J. D. 2006b. *Chamaechaenactis*. *In* Flora of North America Editorial Committee, eds., Flora of North America North of Mexico, Vol. 21, p. 395. Oxford University Press, New York.

Morefield, J. D. 2006c. *Diaperia*. *In* Flora of North America Editorial Committee, eds., Flora of North America North of Mexico, Vol. 19, pp. 460–463. Oxford University Press, New York.

Morefield, J. D. 2006d. *Logfia*. *In* Flora of North America Editorial Committee, eds., Flora of North America North of Mexico, Vol. 19, pp. 443–447. Oxford University Press, New York.

Morefield, J. D. 2006e. *Stylocline*. *In* Flora of North America Editorial Committee, eds., Flora of North America North of Mexico, Vol. 19, pp. 450–453. Oxford University Press, New York.

Morgan, D. R. 2006a. *Dieteria*. *In* Flora of North America Editorial Committee, eds., Flora of North America North of Mexico, Vol. 20, pp. 395–401. Oxford University Press, New York.

Morgan, D. R. 2006b. *Psilactis*. *In* Flora of North America Editorial Committee, eds., Flora of North America North of Mexico, Vol. 20, pp. 462–465. Oxford University Press, New York.

Morgan, D. R., and R. L. Hartman. 2006. *Machaeranthera*. *In* Flora of North America Editorial Committee, eds., Flora of North America North of Mexico, Vol. 20, pp. 394–395. Oxford University Press, New York.

Morin, N. R. 2009. Grossulariaceae. *In* Flora of North America Editorial Committee, eds., Flora of North America North of Mexico, Vol. 8, pp. 8–42. Oxford University Press, New York.

Morse, C. A. 2006. *Stenotus*. *In* Flora of North America Editorial Committee, eds., Flora of North America North of Mexico, Vol. 20, pp. 174–177. Oxford University Press, New York.

Morse, C. A. 2006. *Tonestus*. *In* Flora of North America Editorial Committee, eds., Flora of North America North of Mexico, Vol. 20, pp.181–184. Oxford University Press, New York.

Morton, J. K. 2005a. *Cerastium*. *In* Flora of North America Editorial Committee, eds., Flora of North America North of Mexico, Vol. 5, pp. 74–93. Oxford University Press, New York.

Morton, J. K. 2005b. *Silene*. *In* Flora of North America Editorial Committee, eds., Flora of North America North of Mexico, vol. 5, pp. 166–214. Oxford University Press, New York.

Morton, J. K. 2005c. *Stellaria*. *In* Flora of North America Editorial Committee, eds., Flora of North America North of Mexico, Vol. 5, pp. 96–114. Oxford University Press, New York.

Mueller, C. H. 1933. A new species of *Talinum* from Trans-Pecos Texas. Torreya 33: 148–149.

Mulligan, G. A., and L. J. Bassett. 1959. *Achillea millefolium* complex in Canada and portions of the United States. Canad. J. Bot. 37: 73–79.

Nazaire, M., and L. Hufford. 2014. Phylogenetic systematics of the genus *Mertensia*. Syst. Bot. 39: 268–303.

Nazaire, M. 2023. *Mertensia. In* Flora of North America Editorial Committee, eds., Flora of North America North of Mexico, Vol. 15. Oxford University Press, New York.

Neal, J. T. 1975. Playas and Dried Lakes. Dowden, Hutchinson & Ross, Stroudsburg, Pennsylvania.

Neinaber, W. A., and J. W. Thieret. 2003. Phytolaccaceae. *In* Flora of North America Editorial Committee, eds., Flora of North America North of Mexico, Vol 4, pp. 3–11. Oxford University Press, New York.

Nesom, G. L. 1990. Taxonomy of the genus *Laënnecia* (Asteraceae: Astereae). Phytologia 68: 205–228.

Nesom, G. L. 2006a. *Adenophyllum. In* Flora of North America Editorial Committee, eds., Flora of North America North of Mexico, Vol. 21, pp. 237–239. Oxford University Press, New York.

Nesom, G. L. 2006b. *Ageratina. In* Flora of North America Editorial Committee, eds., Flora of North America North of Mexico, Vol. 21, pp. 547–553. Oxford University Press, New York.

Nesom, G. L. 2006c. *Ageratum. In* Flora of North America Editorial Committee, eds., Flora of North America North of Mexico, Vol. 21, pp. 481–483. Oxford University Press, New York.

Nesom, G. L. 2006d. *Amphiachyris. In* Flora of North America Editorial Committee, eds., Flora of North America North of Mexico, Vol. 20, pp. 87–88. Oxford University Press, New York.

Nesom, G. L. 2006e. *Aphanostephus. In* Flora of North America Editorial Committee, eds., Flora of North America North of Mexico, Vol. 20, pp. 351–353. Oxford University Press, New York.

Nesom, G. L. 2006f. *Asanthus. In* Flora of North America Editorial Committee, eds., Flora of North America North of Mexico, Vol. 21, p. 509. Oxford University Press, New York.

Nesom, G. L. 2006g. *Chaptalia. In* Flora of North America Editorial Committee, eds., Flora of North America North of Mexico, Vol. 19, pp. 78–80. Oxford University Press, New York.

Nesom, G. L. 2006h. *Chloracantha. In* Flora of North America Editorial Committee, eds., Flora of North America North of Mexico, Vol. 20, p. 358. Oxford University Press, New York.

Nesom, G. L. 2006i. *Erigeron. In* Flora of North America Editorial Committee, eds., Flora of North America North of Mexico, Vol. 20, pp. 256–348. Oxford University Press, New York.

Nesom, G. L. 2006j. *Fleischmannia. In* Flora of North America Editorial Committee, eds., Flora of North America North of Mexico, Vol. 21, pp. 540–541. Oxford University Press, New York.

Nesom, G. L. 2006k. *Gamochaeta. In* Flora of North America Editorial Committee, eds., Flora of North America North of Mexico, Vol. 19, pp. 431–438. Oxford University Press, New York.

Nesom, G. L. 2006l. *Gutierrezia. In* Flora of North America Editorial Committee, eds., Flora of North America North of Mexico, Vol. 20, pp. 88–94. Oxford University Press, New York.

Nesom G. L. 2006m. *Gymnosperma. In* Flora of North America Editorial Committee, eds., Flora of North America North of Mexico, Vol. 20, pp. 94–95. Oxford University Press, New York.

Nesom, G. L. 2006n. *Ionactis. In* Flora of North America Editorial Committee, eds., Flora of North America North of Mexico, Vol. 20, pp. 82–84. Oxford University Press, New York.

Nesom, G. L. 2006o. *Isocoma. In* Flora of North America Editorial Committee, eds., Flora of North America North of Mexico, Vol. 20, pp. 439–445. Oxford University Press, New York.

Nesom, G. L. 2006p. *Koanophyllon. In* Flora of North America Editorial Committee, eds., Flora of North America North of Mexico, Vol. 21, pp. 542–543. Oxford University Press, New York.

Nesom, G. L. 2006q. *Leibnitzia. In* Flora of North America Editorial Committee, eds., Flora of North America North of Mexico, Vol. 19, pp. 80–81. Oxford University Press, New York.

Nesom, G. L. 2006r. *Liatris. In* Flora of North America Editorial Committee, eds., Flora of North America North of Mexico, Vol. 21, pp. 512–535. Oxford University Press, New York.

Nesom, G. L. 2006s. *Oreochrysum. In* Flora of North America Editorial Committee, eds., Flora of North America North of Mexico, Vol. 20, pp. 166–167. Oxford University Press, New York.

Nesom, G. L. 2006t. *Pluchea. In* Flora of North America Editorial Committee, eds., Flora of North America North of Mexico, Vol. 19, pp. 478–484. Oxford University Press, New York.

Nesom, G. L. 2006u. *Pseudognaphalium. In* Flora of North America Editorial Committee, eds., Flora of North America North of Mexico, Vol. 19, pp. 415–425. Oxford University Press, New York.

Nesom, G. L. 2006v. *Rayjacksonia. In* Flora of North America Editorial Committee, eds., Flora of North America North of Mexico, Vol. 20, pp. 437–439. Oxford University Press, New York.

Nesom, G. L. 2006w. *Stevia. In* Flora of North America Editorial Committee, eds., Flora of North America North of Mexico, Vol. 21, pp. 483–486. Oxford University Press, New York.

Nesom, G. L. 2010a. *Chaetopappa. In* Flora of North America Editorial Committee, eds., Flora of North America North of Mexico, Vol. 20, pp. 206–209. Oxford University Press, New York.

Nesom, G. L. 2010b. Revision of *Verbena* ser. Tricesimae (Verbenaceae). Phytoneuron 2010-35: 1–38.

Nesom, G. L. 2010c. Taxonomy of the *Glandularia bipinnatifida* group (Verbenaceae) in the USA. Phytoneuron 2010-46: 1–20.

Nesom, G. L. 2015. Cucurbitaceae. *In* Flora of North America Editorial Committee, eds., Flora of North America North of Mexico, Vol. 6, pp. 3–58. Oxford University Press, New York.

Nesom, G. L. 2016a. Garryaceae. *In* Flora of North America Editorial Committee, eds., Flora of North America North of Mexico, Vol. 16, pp. 548–554. Oxford University Press, New York.

Nesom, G. L. 2016b. Oxalidaceae. *In* Flora of North America Editorial Committee, eds., Flora of North America North of Mexico, Vol. 12, pp. 133–153. Oxford University Press, New York.

Nesom, G. L. 2016c. Rhamnaceae. *In* Flora of North America Editorial Committee, eds., Flora of North America North of Mexico, Vol. 12, pp. 43–110. Oxford University Press, New York.

Nesom, G. L. 2019. Taxonomic summary of *Heterotheca* (Asteraceae: Astereae): Part 1, sects. *Heterotheca* and *Ammodia*. Phytoneuron 2019-64: 1–44.

Nesom, G. L. 2020a. Geography of *Erigeron abajoensis* (Asteraceae). Phytoneuron 2020-52: 1–8.

Nesom, G. L. 2020b. Taxonomic summary of *Heterotheca* (Asteraceae: Astereae): Sect. *Chrysanthe*. Phytoneuron 2020-68: 1–359.

Ness, F. 2002. *Fritillaria*. *In* Flora of North America Editorial Committee, eds., Flora of North America North of Mexico, Vol. 26, pp. 164–171. Oxford University Press, New York.

New Mexico Rare Plant Technical Council. 1999. New Mexico Rare Plants. http://nmrareplants.unm.edu, accessed 29 May 2023.

Nickrent, D. L. 2016a. Comandraceae. *In* Flora of North America Editorial Committee, eds., Flora of North America North of Mexico, Vol. 12, pp. 408–412. Oxford University Press, New York.

Nickrent, D. L. 2016b. Viscaceae. *In* Flora of North America Editorial Committee, eds., Flora of North America North of Mexico, Vol. 12, pp. 422–440. Oxford University Press, New York.

Nienaber, M. A., and J. W. Thieret. 2018. Phytolaccaceae. *In* Flora of North America Editorial Committee, eds., Flora of North America North of Mexico, Vol. 4, pp. 3 11. Oxford University Press, New York.

Nilsson, Ö. 1971. Studies in *Montia* L., *Claytonia* L. and allied genera. VI. The genus *Montiastrum* (A. Gray) Rydb. Bot. Notes 124: 87–121.

Nixon, K. C. 1997. Fagaceae. *In* Flora of North America Editorial Committee, eds., Flora of North America North of Mexico, Vol. 3, pp. 436–506. Oxford University Press, New York.

Nummedal, D. 2004. Tectonic and eustatic controls on Upper Cretaceous stratigraphy of northern New Mexico. *In* Mack, G. H., and K. A. Giles, eds., The Geology of New Mexico: A Geologic History, pp. 16–182. New Mexico Geological Society Special Publication 11, Socorro, New Mexico.

Nyffeler, R., and U. Eggli. 2010. Disintegrating Portulacaceae: A new familial classification of the suborder Portulacineae (Caryophyllales) based on molecular and morphological data. Taxon 59: 232–233.

O'Kane, S. L., Jr. 2010. *Physaria*. *In* Flora of North America Editorial Committee, eds., Flora of North America North of Mexico, Vol. 7, pp. 616–665. Oxford University Press, New York.

O'Kane, S. L., Jr. 2013. Brassicaceae. *In* Heil, K. D., S. L. O'Kane, Jr., L. M. Reeves, and A. Clifford, eds., Flora of the Four Corners Region: Vascular Plants of the San Juan River Drainage; Arizona, Colorado, New Mexico, and Utah, pp. 337–387. Missouri Botanical Garden Press, St. Louis.

O'Kane, S. L., Jr., and K. D. Heil. 2022. *Descurainia kenheilii* (Brassicaceae): Revised description and new records from Colorado and New Mexico. Phytologia 104(2): 4–7.

O'Kennon, R. J., and K. N. Taylor. 2013. *Eleocharis microformis* (Cyperaceae): Rediscovered in North America from the Edwards Plateau and Trans-Pecos regions of Texas. J. Bot. Res. Inst. Texas 7: 587–593.

Olmstead, R. G., B. Bremer, K. M. Scott, and J. D. Palmer. 1993. A parsimony analysis of the Asteridae sensu lato based on *rbc*L sequences. Ann. Missouri Bot. Gard. 80: 700–722.

Omernik, J. M. 1987. Ecoregions of the conterminous United States [map, scale 1:7,500,000]. Ann. Am. Assoc. Geogr. 77: 118–125.

Omernik, J. M. 1995. Ecoregions: A spatial framework for environmental management. *In* Davis, W. S., and T. P. Simon, eds., Biological Assessment and Criteria: Tools for Water Resource Planning and Decision Making, pp. 49–62. Lewis Publishers, Boca Raton, Florida.

Omernik, J. M. 2003. Level III Ecoregions of the Continental United States National Health and Environmental Effects Research Laboratory, U.S. Environmental Protection Agency.

Omernik, J. M., S. S. Chapman, R. A. Lillie, and R. T. Dumke. 2000. Ecoregions of Wisconsin. Trans. Wisconsin Acad. Sci. 88: 77–103.

Orcutt, C. R. 1926. *Coryphantha bisbeeana*. Cactography: 3.

Ornduff, R., and M. Denton. 1998. Oxalidaceae: Oxalis family. J. Arizona-Nevada Acad. Sci. 30: 115–119.

Ownbey, M. S. 1940. A monograph of the genus *Calochortus*. Ann. Missouri Bot. Gard. 27: 371–560.

Packer, J. G. 2003. Portulacaceae. *In* Flora of North America Editorial Committee, eds., Flora of North America North of Mexico, Vol. 4, pp. 457–504. Oxford University Press, New York.

Parfitt, B. D., and A. C. Gibson. 2003. Cactaceae. *In* Flora of North America Editorial Committee, eds., Flora of North America North of Mexico, Vol. 4, pp. 99–257. Oxford University Press, New York.

Parfitt, B. D., and A. T. Whittemore. 2013. Ranunculaceae. *In* Heil, K. D, S. L. O'Kane, Jr., L. M. Reeves, and A. Clifford, eds., Flora of the Four Corners Region: Vascular Plants of the San Juan River Drainage; Arizona, Colorado, New Mexico, and Utah, pp. 877–895. Missouri Botanical Garden Press, St. Louis.

Park, M. S. 2019. *Collinsia*. *In* Flora of North America Editorial Committee, eds., Flora of North America North of Mexico, Vol. 17, pp. 62–74. Oxford University Press, New York.

Patterson, T. F., and G. L. Nesom. 2006. *Conoclinium*. *In* Flora of North America Editorial Committee, eds., Flora of North America North of Mexico, Vol. 21, pp. 478–480. Oxford University Press, New York.

Pennington, T. D. 1981. Meliaceae. *In* Luteyn, J. L., and S. A. Mori, eds., Flora Neotropica Monograph no. 28, pp. 1–472. New York Botanical Garden, New York.

Peterson, P. M., K. Romaschenko, and G. Johnson. 2010. A phylogeny and classification of the Muhlenbergiinae (Poaceae: Chloridoideae: Cynodonteae) based on plastid and nuclear DNA sequences. Amer. J. Bot. 97: 1532–1554.

Peterson, P. M., K. Romaschenko, N. Snow, and G. P. Johnson. 2012. A molecular phylogeny and classification of Leptochloa (Poaceae: Chloridoideae: Chlorideae) sensu lato and related genera. Ann. Bot. (London) 109: 1317–1330.

Peterson, P. M., K. Romaschenko, Y. Herrera Arrieta, and J. M. Saarela. 2014. A molecular phylogeny and new subgeneric classification of *Sporobolus* (Poaceae: Chloridoideae: Sporobolinae). Taxon 63: 1212–1243.

Peterson, P. M., K. Romaschenko, and Y. Herrera Arrieta. 2015. A molecular phylogeny and classification of the Eleusininae with a new genus, *Micrachne* (Poaceae: Chloridoideae: Cynodonteae). Taxon 64: 445–467.

Peterson, P. M., K. Romaschenko, R. J. Soreng, and J. Valdes Reyna. 2019. A key to the North American genera of Stipeae (Poaceae, Pooideae) with descriptions and taxonomic names for species of *Eriocoma*, *Neotrinia*, *Oloptum*, and five new genera: *Barkworthia*, ×*Eriosella*, *Pseudoeriocoma*, *Ptilagrostiella*, and *Thorneochloa*. PhytoKeys 126: 89–125.

Phipps, James B. 2014. Rosaceae. *In* Flora of North America Editorial Committee, eds., Flora of North America North of Mexico, Vol. 9, pp. 18–662. Oxford University Press, New York.

Pinkava, D. J. 1999. Vascular Plants of Arizona: Cactaceae. Part 4, *Grusonia*. J. Arizona-Nevada Acad. Sci. 32(1): 48–52.

Pinkava, D. J. 2003. Vascular Plants of Arizona: Cactaceae. Part 6: *Opuntia*. J. Arizona-Nevada Acad. Sci. 35(2): 137–150.

Pinkava, D. J. 2006. *Berlandiera*. *In* Flora of North America Editorial Committee, eds., Flora of North America North of Mexico, Vol. 21, pp. 83–87. Oxford University Press, New York.

Pires, J. C., and R. E. Preston. 2019. Themidaceae. *In* Jepson Flora Project, eds., Jepson eFlora. https://ucjeps.berkeley.edu/eflora/, accessed on 29 May 2023.

Porter, D. M. 2016. Zygophyllaceae. *In* Flora of North America Editorial Committee, eds., Flora of North America North of Mexico, Vol. 12, pp. 12–42. Oxford University Press, New York.

Porter, J. M. 1998. Nomenclatural changes in Polemoniaceae. Aliso 17: 83–85.

Porter, J. M. 2010. Phylogenetic systematics of *Ipomopsis* (Polemoniaceae): Relationships and divergence times estimated from chloroplast and nuclear DNA sequences. Syst. Bot. 35(1) 181–200.

Porter, J. M. 2011. Two new *Aliciella* species and a new subspecies in *Ipomopsis* (Polemoniaceae) from the western United States of America (*Ipomopsis congesta matthesii*). Phytotaxa 15: 15–25.

Porter, J. M., and A. Clifford. 2013. *Artemisia*. *In* Heil, K. D., S. L. O'Kane, Jr., L. M. Reeves, and A. Clifford, eds., Flora of the Four Corners Region: Vascular Plants of the San Juan River Drainage; Arizona, Colorado, New Mexico, and Utah, pp. 170–178. Missouri Botanical Garden Press, St. Louis.

Porter, J. M., and L. A. Johnson. 2000. A phylogenetic classification of Polemoniaceae. Aliso 19: 55–91.

Powell, A. M., and J. F. Weedin. 2004. Cacti of the Trans-Pecos and Adjacent Areas. Texas Tech University Press, Lubbock.

Powell, A. M., and R. D. Worthington. 2018. Flowering plants of Trans-Pecos Texas and adjacent areas. Sida, Bot. Misc. 49: 7–25.

Pretz, C., and R. Deanna. 2020. Typifications and nomenclatural notes in *Physalis* (Solanaceae) from the United States. Taxon 69: 170–192.

Price, T. M., and D. J. Ferguson. 2012. A new combination in *Phemeranthus* (Montiaceae) and notes on the circumscription of *Phemeranthus* and *Talinum* (Talinaceae) from the southwestern United States and northern Mexico. Novon 22: 67–69.

Pruski, J. F., and R. L. Hartman. 2012. Synopsis of *Leucosyris*, including synonymous *Arida* (Compositae: Asteraceae). Phytoneuron 2012-98: 1–15.

Pteridophyte Phylogeny Group. 2016. A community-derived classification for extant lycophytes and ferns. J. Syst. Evol. 54: 563–603.

Rabeler, R. K., and R. L. Hartman. 2005. Caryophyllaceae. *In*: Flora of North America Editorial Committee, eds., Flora of North America North of Mexico, Vol. 5, pp. 3–215. Oxford University Press, New York.

Randle, C. P. 2019. *Brachystigma*. *In* Flora of North America Editorial Committee, eds., Flora of North America North of Mexico, Vol. 17, p. 530. Oxford University Press, New York.

Raven, P. H. 1962. The systematics of *Oenothera* subgenus *Chylismia*. Univ. Calif. Publ. Bot. 34: 1–122.

Raven, P. H. 1963. The Old World species of *Ludwigia* (including *Jussiaea*), with a synopsis of the genus (Onagraceae). Reinwardtia 6: 327–427.

Raven, P. H. 1969. A revision of the genus *Camissonia* (Onagraceae). Contr. U.S. Natl. Herb. 37: 161–396.

Razifard, H., G. C. Tucker, and D. H. Les. 2016. Elatinaceae. *In* Flora of North America Editorial Committee, eds., Flora of North America North of Mexico, Vol. 12, pp. 348–353. Oxford University Press, New York.

Reeves, L. M. 2013a. Frankeniaceae. *In* Heil, K. D., S. L. O'Kane, Jr., L. M. Reeves, and A. Clifford, eds., Flora of the Four Corners Region: Vascular Plants of the San Juan River Drainage; Arizona, Colorado, New Mexico, and Utah, pp. 575–576. Missouri Botanical Garden Press, St. Louis.

Reeves, L. M. 2013b. Plumbaginaceae. *In* Heil, K. D., S. L. O'Kane, Jr., L. M. Reeves, and A. Clifford, eds., Flora of the Four

Corners Region: Vascular Plants of the San Juan River Drainage; Arizona, Colorado, New Mexico, and Utah, pp. 730–731. Missouri Botanical Garden Press, St. Louis.

Reveal, J. L., and W. C. Hodgson. 2002. *Agave*. *In* Flora of North America Editorial Committee, eds., Flora of North America North of Mexico, Vol. 26, pp. 413–465. Oxford University Press, New York.

Reveal, J. L., and F. H. Utech. 2002. *Lloydia*. *In* Flora of North America Editorial Committee, eds., Flora of North America North of Mexico, Vol. 26, p. 198. Oxford University Press, New York.

Rhodes, A. M., W. P. Bemis, T. W. Whitaker, and S. G. Carmer. 1968. A numerical taxonomic study of *Cucurbita*. Brittonia 20: 251–266.

Richardson, A. T. 1977. Monograph of the genus *Tiquilia* (*Coldenia*, sensu lato), Boraginaceae: Ehretioideae. Rhodora 79: 467–572.

Rink, G., and M. Licher. 2015. Cyperaceae sedge family. Part 1: Family description, key to the genera, and *Carex* L. Canotia 11: 1–97.

Roalson, E. H., J. C. Hall, J. P. Riser, II, W. M. Cardinal-McTeague, T. S. Cochrane, and K. J. Sytsma. 2015. A revision of generic boundaries and nomenclature in the North American cleomoid clade (Cleomaceae). Phytotaxa 205: 129–144.

Robart, B. W. 2019. *Pedicularis*. *In* Flora of North America Editorial Committee, eds., Flora of North America North of Mexico, Vol. 17, pp. 510–534. Oxford University Press, New York.

Robertson, K. R., and S. E. Clemants. 2003. Amaranthaceae. *In* Flora of North America Editorial Committee, eds., Flora of North America North of Mexico, Vol. 4, pp. 405–456. Oxford University Press, New York.

Robson, N. K. B. 2015. Hypericaceae. *In* Flora of North America Editorial Committee, eds., Flora of North America North of Mexico, Vol. 6, pp. 71–105. Oxford University Press, New York.

Rodrigues, A. G., Jr., C. C. Baskin, J. M. Baskin, and O. C. De-Paula. 2021. The pleurogram, an under-investigated functional trait in seeds. Ann. Bot. 127: 167–174.

Rollins, R. C. 1993. The Cruciferae of Continental North America: Systematics of the Mustard Family from the Arctic to Panama. Stanford University Press, Palo Alto, California.

Romaschenko, K., P. M. Peterson, R. J. Soreng, O. Futorna, and A. Susanna. 2011. Phylogenetics of *Piptatherum* s.l. (Poaceae: Stipeae): Evidence for a new genus, *Piptatheropsis*, and resurrection of *Patis*. Taxon 60: 1703–1716.

Romero-González, G. A., G. Carnevali Fernández-Concha, R. L. Dressler, L. K. Magrath, and G. W. Argus. 2002. Orchidaceae. *In* Flora of North America Editorial Committee, eds., Flora of North America North of Mexico, Vol. 26, pp. 490–651. Oxford University Press, New York.

Rose, J. 2021. Taxonomy and relationships within *Polemonium foliosissimum* (Polemoniaceae): Untangling a clade of colorful and gynodioecious herbs. Syst. Bot. 46: 519–537.

Rosendahl, C. O. 1914. A revision of the genus *Mitella* with a discussion of geographical distribution and relationships. Bot. Jahrb. Syst. 50: 375–397.

Rothrock, P. E. 2009a. *Bolboschoenus*. *In* Sedges of Indiana and the Adjacent States: The Non-*Carex* Species, pp. 60–63. Indiana Academy of Science, Indianapolis.

Rothrock, P. E. 2009b. *Eleocharis*. *In* Sedges of Indiana and the adjacent states: The non-*Carex* species, pp. 101–142. Indiana Academy of Science, Indianapolis.

Rothrock, P. E. 2009c. *Schoenoplectus*. *In* Sedges of Indiana and the adjacent states: The non-*Carex* species. pp. 180–199. Indiana Academy of Science, Indianapolis.

Rothrock, P. E. 2009d. *Scirpus*. *In* Sedges of Indiana and the adjacent states: The non-*Carex* species, pp. 200–219. Indiana Academy of Science, Indianapolis.

Rydberg, P. A. 1895. The North American species of *Physalis* and related genera. Mem. Torrey Bot. Club 4: 297–372.

Rydberg, P. A. 1922. Flora of the Rocky Mountains and Adjacent Plains. Hafner, New York.

Salls, K. A., and K. S. Bannister. 2014. Allelopathic characteristics of *Artemisia tridentata* and *Purshia tridentata* and implications for invasive species management. Nevada State Undergraduate Research Journal V1:I1 Fall 2014. https://doi.org/10.15629/6.7.8.7.5_1-1_F-2014_1

Sammons, N. 2011. Morphometric Analysis and Monograph of *Monarda* Subgenus *Cheilyctis* (Lamiaceae). Ph.D. Dissertation, Michigan State University, East Lansing.

Sanders, R. W. 1987. Taxonomy of *Agastache* section *Brittonastrum* (Lamiaceae-Nepeteae). Syst. Bot. Monogr. 15: 1–92.

Schilling, E. 1981. Systematics of *Solanum* sect. *Solanum* (Solanaceae) in North America. Syst. Bot. 6: 172–185.

Schilling, E. E. 2006a. *Helianthus*. *In* Flora of North America Editorial Committee, eds., Flora of North America North of Mexico, Vol. 21, pp. 141–169. Oxford University Press, New York.

Schilling, E. E. 2006b. *Heliomeris*. *In* Flora of North America Editorial Committee, eds., Flora of North America North of Mexico, Vol. 21, pp. 169–172. Oxford University Press, New York.

Schilling, E. E. 2006c. *Viguiera*. *In* Flora of North America Editorial Committee, eds., Flora of North America North of Mexico, Vol. 21, pp. 172–174. Oxford University Press, New York.

Schilling, E. E., and J. L. Panero. 2011. A revised classification of subtribe Helianthinae (Asteraceae) II. Derived lineages. Bot. J. Linn. Soc. 167: 311–331.

Schimpf, D. J., J. A. Henderson, and J. A. MacMahon. 1980. Some aspects of succession in the spruce-fir forest zone of northern Utah. Great Basin Naturalist 40: 1–26.

Schneider, A. C. 2016. Resurrection of the genus *Aphyllon* for New World broomrapes (*Orobanche* s.l., Orobanchaceae). PhytoKeys 75: 107–118.

Schwartz, F. C. 2002. *Zigadenus. In* Flora of North America Editorial Committee, eds., Flora of North America North of Mexico, Vol. 26, pp. 81–88. Oxford University Press, New York.

Scott, R. W. 2006a. *Brickellia. In* Flora of North America Editorial Committee, eds., Flora of North America North of Mexico, Vol. 21, pp. 491–507. Oxford University Press, New York.

Scott, R. W. 2006b. *Brickelliastrum. In* Flora of North America Editorial Committee, eds., Flora of North America North of Mexico, Vol. 21, p. 507. Oxford University Press, New York.

Scribailo, R. W., and M. S. Alix. 2021. Haloragaceae. *In* Flora of North America Editorial Committee, eds., Flora of North America North of Mexico, Vol. 10, pp. 12–31. Oxford University Press, New York.

Seithe, A., and J. R. Sullivan. 1990. Hair morphology and systematics of *Physalis* (Solanaceae). Pl. Syst. Evol. 170: 193–204.

Semple, J. C. 1996. A revision of *Heterotheca* sect. *Phyllotheca* (Nutt.) Harms (Compositae: Astereae): The prairie and montane goldenasters of North America. Univ. Waterloo Biol. Ser. 37: 1–164.

Semple, J. C. 2006. *Heterotheca* (Asteraceae). *In* Flora of North America Editorial Committee, eds., Flora of North America North of Mexico, Vol. 20, pp. 230–256. Oxford University Press, New York.

Semple, J. C. 2008. Cytotaxonomy and cytogeography of the goldenaster genus *Heterotheca* (Asteraceae: Astereae). Botany 86: 886–900.

Semple, J. C., and R. E. Cook. 2006. *Solidago. In* Flora of North America Editorial Committee, eds., Flora of North America North of Mexico, Vol. 20, pp. 107–166. Oxford University Press, New York.

Sharples, M. T., and E. A. Tripp. 2019. Phylogenetic relationships within and delimitation of the cosmopolitan flowering plant genus *Stellaria* L. (Caryophyllaceae): Core stars and fallen stars. Syst. Bot. 44: 857–876.

Sheahan, M. C., and M. W. Chase. 1996. A phylogenetic analysis of Zygophyllaceae based on morphological, anatomical and rbcL DNA sequence data. Bot. J. Linn. Soc. 122: 279–300.

Sherff, E. E. 1937. The genus *Bidens*. Publ. Field Mus. Nat. Hist., Bot. Ser. 16: 1–709.

Sherman-Broyles, S. L., W. T. Barker, and L. M. Shultz. 1997. Ulmaceae. *In* Flora of North America Editorial Committee, eds., Flora of North America North of Mexico, Vol. 3, pp. 368–380. Oxford University Press, New York.

Shipunov, A. 2019. *Plantago. In* Flora of North America Editorial Committee, eds., Flora of North America North of Mexico, Vol. 17, pp. 281–293. Oxford University Press, New York.

Shultz, L. M. 2006. *Artemisia. In* Flora of North America Editorial Committee, eds., Flora of North America North of Mexico, Vol. 19, pp. 503–534. Oxford University Press, New York.

Shultz, L. M., and W. A. Varga. 2021. Elaeagnaceae. *In* Flora of North America Editorial Committee, eds., Flora of North America North of Mexico, Vol. 10, pp. 415–421. Oxford University Press, New York.

Simpson, B. B. 2006. *Acourtia. In* Flora of North America Editorial Committee, eds., Flora of North America North of Mexico, Vol. 19, pp. 72–74. Oxford University Press, New York.

Simpson, B. B., and A. Salywon. 1999. Krameriaceae. J. Arizona-Nevada Acad. Sci. 32: 57–61.

Simpson, M. 2019. Plant Systematics, 3rd ed. Academic Press, Massachusetts.

Sivinski, R. C. 1998. Annotated checklist of the genus *Allium* (Liliaceae) in New Mexico. New Mexico Naturalist's Notes 1: 43–56 [reprint 2003, New Mexico Botanist 47: 1–6].

Sivinski, R. C. 2001. The genus *Plantago* (Plantaginaceae) in New Mexico. New Mexico Botanist 18: 1–3.

Sivinski, R. C. 2008. Some observations on the dry, dehiscent-fruited yuccas in New Mexico. New Mexico Botanist 43: 1–4.

Sivinski, R. C. 2013. Boraginaceae. *In* Heil, K. D., S. L. O'Kane, Jr., L. M. Reeves, and A. Clifford, eds., Flora of the Four Corners Region: Vascular Plants of the San Juan River Drainage; Arizona, Colorado, New Mexico, and Utah, pp. 321–337. Missouri Botanical Garden Press, St. Louis.

Sivinski, R. C. 2016. New Mexico Thistle Identification Guide. Native Plant Society of New Mexico. https://www.npsnm.org/thistle-identification-booklet/, accessed 25 October 2023.

Sivinski, R. C. 2023. *Oreocarya worthingtonii* (Boraginaceae): A new species from southeastern New Mexico, U.S.A. J. Bot. Res. Inst. Texas 17(1): 1–8.

Sivinski, R. C., T. Lowrey, and R. Peterson. 1994. Additions to the native and adventive flora of New Mexico. Phytologia 76(6): 473–479. [*Antiphytum floribundum*]

Sivinski, R. C., T. Lowrey, and C. Keller. 1995. Additions to the floras of Colorado and New Mexico. Phytologia 79: 319–324. [*Eleocharis bella*]

Skinner, M. W. 2002. *Lilium. In* Flora of North America Editorial Committee, eds., Flora of North America North of Mexico, Vol. 26, pp. 172–197. Oxford University Press, New York.

Smith, A. R. 1993. Pteridophytes. *In* Flora of North America Editorial Committee, eds., Flora of North America North of Mexico, Vol. 2, pp. 9–342. Oxford University Press, New York.

Smith, A. R. 2006a. *Heliopsis. In* Flora of North America Editorial Committee, eds., Flora of North America North of Mexico, Vol. 21, pp. 67–70. Oxford University Press, New York.

Smith, A. R. 2006b. *Zinnia. In* Flora of North America Editorial Committee, eds., Flora of North America North of Mexico, Vol. 21, pp. 71–74. Oxford University Press, New York.

Smith, S. G. 2001. Taxonomic innovations in North American *Eleocharis* (Cyperaceae). Novon 11: 241–257. [*Eleocharis acicularis* var. *porcata*]

Smith, S. G., J. J. Bruhl, M. S. Gonzalez-Elizondo, and F. J. Menapace. 2002. *Eleocharis. In* Flora of North America Editorial Committee, eds., Flora of North America North of Mexico, Vol. 23, pp. 60–120. Oxford University Press, New York.

Soltis, D. E., and C. C. Freeman. 2009. *Mitella*. *In* Flora of North America Editorial Committee, eds., Flora of North America North of Mexico, Vol. 8, pp. 108–114. Oxford University Press, New York.

Soltis, P. M. 2006. *Tragopogon*. *In* Flora of North America Editorial Committee, eds., Flora of North America North of Mexico, Vol. 19, pp. 303–306. Oxford University Press, New York.

Sørensen, P. D. 2009. *Arbutus*. *In* Flora of North America Editorial Committee, eds., Flora of North America North of Mexico, Vol. 8, pp. 398–400. Oxford University Press, New York.

Sørensen, P. D. 2009. *Arbutus xalapensis*. *In* Flora of North America Editorial Committee, eds., Flora of North America North of Mexico, Vol. 8, p. 400. Oxford University Press, New York.

Sørensen, P. D., C. E. Totten, and D. M. Piatak. 1978. Alkane chemotaxonomy of *Arbutus*. Biochem. Syst. Ecol. 6: 109–111.

Sosa, V., and M. W. Chase. 2003. Phylogenetics of Crossosomataceae based on rbcL sequence data. Syst. Bot. 28: 96–105.

Spellenberg, R. 2001. Oaks of La Frontera and taxonomic overview of the oaks of La Frontera. *In* Webster, G. L., and C. J. Bahre, eds., Changing Plant Life of La Frontera, pp. 176–186, 195–211. University of New Mexico Press, Albuquerque.

Spellenberg, R. W. 2003. Nyctaginaceae. *In* Flora of North America Editorial Committee, eds., Flora of North America North of Mexico, Vol. 4, pp. 14–74. Oxford University Press, New York.

Spencer, N. R. 1984. Velvetleaf, *Abutilon theophrasti* (Malvaceae), history and economic impact in the United States. Econ. Bot. 38: 407–416.

Spooner, D. M. 2006. *Simsia*. *In* Flora of North America Editorial Committee, eds., Flora of North America North of Mexico, Vol. 21, pp. 140–141. Oxford University Press, New York.

Standley, P. C. 1910. The type localities of plants first described from New Mexico. Contr. U.S. Natl. Herbarium 13(6) 143–228.

Steinmann, V. W., and J. M. Porter. 2002. Phylogenetic relationships in Euphorbieae (Euphorbiaceae) based on ITS and *ndbF* sequence data. Ann. Missouri Bot. Gard. 89: 453–490.

Stern, K. R. 1997. Fumariaceae. *In* Flora of North America Editorial Committee, eds., Flora of North America North of Mexico, Vol. 3, pp. 340–357. Oxford University Press, New York.

Stevenson, Dennis Wm. 1993. Ephedraceae. *In* Flora of North America Editorial Committee, eds., Flora of North America North of Mexico, Vol. 2, pp. 428–434. Oxford University Press, New York.

Steyermark, J. A. 1937. Studies in *Grindelia* III. Ann. Missouri Bot. Gard. 24: 225–262.

Stone, D. E. 1997. Juglandaceae. *In* Flora of North America Editorial Committee, eds., Flora of North America North of Mexico, Vol. 3, pp. 416–428. Oxford University Press, New York.

Stoutamire, W. 1977. Chromosome races of *Gaillardia pulchella* (Asteraceae). Brittonia 26: 297–309.

Straley, G. B., and F. H. Utech. 2002. *Asparagus*. *In* Flora of North America Editorial Committee, eds., Flora of North America North of Mexico, Vol. 26, pp. 213–214. Oxford University Press, New York.

Strother, J. L. 2006a. *Ambrosia*. *In* Flora of North America Editorial Committee, eds., Flora of North America North of Mexico, Vol. 21, pp. 10–18. Oxford University Press, New York.

Strother, J. L. 2006b. *Bartlettia*. *In* Flora of North America Editorial Committee, eds., Flora of North America North of Mexico, Vol. 21, pp. 378–379. Oxford University Press, New York.

Strother, J. L. 2006c. *Calyptocarpus*. *In* Flora of North America Editorial Committee, eds., Flora of North America North of Mexico, Vol. 21, p. 65. Oxford University Press, New York.

Strother, J. L. 2006d. *Chrysactinia*. *In* Flora of North America Editorial Committee, eds., Flora of North America North of Mexico, Vol. 21, p. 222. Oxford University Press, New York.

Strother, J. L. 2006e. *Coreopsis*. *In* Flora of North America Editorial Committee, eds., Flora of North America North of Mexico, Vol. 21, pp. 185–199. Oxford University Press, New York.

Strother, J. L. 2006f. *Cyclachaena*. *In* Flora of North America Editorial Committee, eds., Flora of North America North of Mexico, Vol. 21, p. 28. Oxford University Press, New York.

Strother, J. L. 2006g. *Dicoria*. *In* Flora of North America Editorial Committee, eds., Flora of North America North of Mexico, Vol. 21, pp. 24–25. Oxford University Press, New York.

Strother, J. L. 2006h. *Eclipta*. *In* Flora of North America Editorial Committee, eds., Flora of North America North of Mexico, Vol. 21, p. 222. Oxford University Press, New York.

Strother, J. L. 2006i. *Flourensia*. *In* Flora of North America Editorial Committee, eds., Flora of North America North of Mexico, Vol. 21, p. 118. Oxford University Press, New York.

Strother, J. L. 2006j. *Gaillardia*. *In* Flora of North America Editorial Committee, eds., Flora of North America North of Mexico, Vol. 21, pp. 421–426. Oxford University Press, New York.

Strother, J. L. 2006k. *Haploësthes*. *In* Flora of North America Editorial Committee, eds., Flora of North America North of Mexico, Vol. 21, p. 243. Oxford University Press, New York.

Strother, J. L. 2006l. *Hedosyne*. *In* Flora of North America Editorial Committee, eds., Flora of North America North of Mexico, Vol. 21, p. 30. Oxford University Press, New York.

Strother, J. L. 2006m. *Hedypnois*. *In* Flora of North America Editorial Committee, eds., Flora of North America North of Mexico, Vol. 21, p. 302. Oxford University Press, New York.

Strother, J. L. 2006n. *Hieracium*. *In* Flora of North America Editorial Committee, eds., Flora of North America North of Mexico, Vol. 19, pp. 278–294. Oxford University Press, New York.

Strother, J. L. 2006o. *Hymenopappus. In* Flora of North America Editorial Committee, eds., Flora of North America North of Mexico, Vol. 21, pp. 309–316. Oxford University Press, New York.

Strother, J. L. 2006p. *Hymenothrix. In* Flora of North America Editorial Committee, eds., Flora of North America North of Mexico, Vol. 21, pp. 387–388. Oxford University Press, New York.

Strother, J. L. 2006q. *Iva. In* Flora of North America Editorial Committee, eds., Flora of North America North of Mexico, Vol. 21, pp. 25–28. Oxford University Press, New York.

Strother, J. L. 2006r. *Jefea. In* Flora of North America Editorial Committee, eds., Flora of North America North of Mexico, Vol. 21, p. 128. Oxford University Press, New York.

Strother, J. L. 2006s. *Lactuca. In* Flora of North America Editorial Committee, eds., Flora of North America North of Mexico, Vol. 19, pp. 259–263. Oxford University Press, New York.

Strother, J. L. 2006t. *Lasianthaea. In* Flora of North America Editorial Committee, eds., Flora of North America North of Mexico, Vol. 21, p. 134. Oxford University Press, New York.

Strother, J. L. 2006u. *Lepidospartum. In* Flora of North America Editorial Committee, eds., Flora of North America North of Mexico, Vol. 20, pp. 632–634. Oxford University Press, New York.

Strother, J. L. 2006v. *Leuciva. In* Flora of North America Editorial Committee, eds., Flora of North America North of Mexico, Vol. 21, pp. 29–30. Oxford University Press, New York.

Strother, J. L. 2006w. *Melampodium. In* Flora of North America Editorial Committee, eds., Flora of North America North of Mexico, Vol. 21, pp. 34–36. Oxford University Press, New York.

Strother, J. L. 2006x. *Osteospermum. In* Flora of North America Editorial Committee, eds., Flora of North America North of Mexico, Vol. 19, pp. 382–383. Oxford University Press, New York.

Strother, J. L. 2006y. *Palafoxia. In* Flora of North America Editorial Committee, eds., Flora of North America North of Mexico, Vol. 21, pp. 388–391. Oxford University Press, New York.

Strother, J. L. 2006z. *Parthenium. In* Flora of North America Editorial Committee, eds., Flora of North America North of Mexico, Vol. 21, pp. 20–22. Oxford University Press, New York.

Strother, J. L. 2006aa. *Porophyllum. In* Flora of North America Editorial Committee, eds., Flora of North America North of Mexico, Vol. 21, pp. 233–235. Oxford University Press, New York.

Strother, J. L. 2006bb. *Psathyrotopsis. In* Flora of North America Editorial Committee, eds., Flora of North America North of Mexico, Vol. 21, pp. 380–381. Oxford University Press, New York.

Strother, J. L. 2006cc. *Pseudoclappia. In* Flora of North America Editorial Committee, eds., Flora of North America North of Mexico, Vol. 21, pp. 252–253. Oxford University Press, New York.

Strother, J. L. 2006dd. *Psilostrophe. In* Flora of North America Editorial Committee, eds., Flora of North America North of Mexico, Vol. 21, pp. 453–455. Oxford University Press, New York.

Strother, J. L. 2006ee. *Pyrrhopappus. In* Flora of North America Editorial Committee, eds., Flora of North America North of Mexico, Vol. 21, pp. 375–378. Oxford University Press, New York.

Strother, J. L. 2006ff. *Sanvitalia. In* Flora of North America Editorial Committee, eds., Flora of North America North of Mexico, Vol. 21, pp. 70–71. Oxford University Press, New York.

Strother, J. L. 2006gg. *Sartwellia. In* Flora of North America Editorial Committee, eds., Flora of North America North of Mexico, Vol. 21, pp. 243–247. Oxford University Press, New York.

Strother, J. L. 2006hh. *Schkuhria. In* Flora of North America Editorial Committee, eds., Flora of North America North of Mexico, Vol. 21, pp. 381–383. Oxford University Press, New York.

Strother, J. L. 2006ii. *Scorzonera. In* Flora of North America Editorial Committee, eds., Flora of North America North of Mexico, Vol. 19, pp. 306–307. Oxford University Press, New York.

Strother, J. L. 2006jj. *Tagetes. In* Flora of North America Editorial Committee, eds., Flora of North America North of Mexico, Vol. 21, pp. 235–236. Oxford University Press, New York.

Strother, J. L. 2006kk. *Tetradymia. In* Flora of North America Editorial Committee, eds., Flora of North America North of Mexico, Vol. 20, pp. 629–632. Oxford University Press, New York.

Strother, J. L. 2006ll. *Thelesperma. In* Flora of North America Editorial Committee, eds., Flora of North America North of Mexico, Vol. 21, pp. 199–203. Oxford University Press, New York.

Strother, J. L. 2006mm. *Thymophylla. In* Flora of North America Editorial Committee, eds., Flora of North America North of Mexico, Vol. 21, pp. 239–245. Oxford University Press, New York.

Strother, J. L. 2006nn. *Townsendia. In* Flora of North America Editorial Committee, eds., Flora of North America North of Mexico, Vol. 20, pp. 191–203. Oxford University Press, New York.

Strother, J. L. 2006oo. *Uropappus. In* Flora of North America Editorial Committee, eds., Flora of North America North of Mexico, Vol. 19, pp. 296–297. Oxford University Press, New York.

Strother, J. L. 2006pp. *Verbesina. In* Flora of North America Editorial Committee, eds., Flora of North America North of Mexico, Vol. 21, pp. 106–111. Oxford University Press, New York.

Strother, J. L. 2006qq. *Vernonia. In* Flora of North America Editorial Committee, eds., Flora of North America North of Mexico, Vol. 19, pp. 206–213. Oxford University Press, New York.

Strother, J. L. 2006rr. *Zaluzania. In* Flora of North America Editorial Committee, eds., Flora of North America North of Mexico, Vol. 21, pp. 63–64. Oxford University Press, New York.

Strother, J. L., and R. R. Weedon. 2006. *Bidens. In* Flora of North America Editorial Committee, eds., Flora of North America North of Mexico, Vol. 21, pp. 205–218. Oxford University Press, New York.

Strother, J. L., and M. A. Wetter. 2006. *Grindelia*. *In* Flora of North America Editorial Committee, eds., Flora of North America North of Mexico, Vol. 20, pp. 424–436. Oxford University Press, New York.

Stuart, B. L., A. G. Rhodin, L. L. Grismer, and T. Hansel. 2006. Scientific description can imperil species. Science 312(5777): 1137.

Stuessy, T. F., S. Irving, and W. L. Ellison. 1973. Hybridization and evolution in *Picradeniopsis* (Compositae). Brittonia 25: 40–56.

Sundberg, S. D., and D. J. Bogler. 2006. *Baccharis*. *In* Flora of North America Editorial Committee, eds., Flora of North America North of Mexico, Vol. 20, pp. 23–34. Oxford University Press, New York.

Swab, J. C. 2000. *Luzula*. *In* Flora of North America Editorial Committee, eds., Flora of North America North of Mexico, Vol. 22, pp. 255–267. Oxford University Press, New York.

Taylor, C. E., and R. J. Taylor. 1983. New species, new combinations and notes on the goldenrods (*Euthamia* and *Solidago*-Asteraceae). Sida 10: 176–183.

Taylor, N. P. 1979. Notes on *Ferocactus* Britton & Rose. Cact. Succ. J. Gr. Brit. 41: 88–94.

Taylor, N. P. 1985. The Genus *Echinocereus*. Kew Magazine Monograph. Timber Press, Portland, Oregon.

Taylor, R. 2013. Grossulariaceae. *In* Heil, K. D, S. L. O'Kane, Jr., L. M. Reeves, and A. Clifford, eds., Flora of the Four Corners Region: Vascular Plants of the San Juan River Drainage; Arizona, Colorado, New Mexico, and Utah, pp. 588–592. Missouri Botanical Garden Press, St. Louis.

Thien, L. B. 1969. Translocations in *Gayophytum* (Onagraceae). Evolution 23: 456–465.

Thieret, J. W. 1993. Pinaceae. *In* Flora of North America Editorial Committee, eds., Flora of North America North of Mexico, Vol. 3, pp. 356–357. Oxford University Press, New York.

Thomas, D. 1983. *Carlowrightia* (Acanthaceae). Flora Neotropica 34: 1–115.

Tidestrom, I., and Sister T. Kittell. 1941. A Flora of Arizona and New Mexico. Catholic University of America Press, Washington, D.C.

Trock, D. K. 2006. *Packera*. *In* Flora of North America Editorial Committee, eds., Flora of North America North of Mexico, Vol. 20, pp. 541–548. Oxford University Press, New York.

Tryl, R. J. 1975. Origin and distribution of polyploid *Achillea* (Compositae) in western North America. Brittonia 27: 187–196.

Tucker, G. C. 1994. Revision of the Mexican species of *Cyperus* (Cyperaceae). Syst. Bot. Monogr. 43: 1–23.

Tucker, G. C. 2009. Ericaceae. *In* Flora of North America Editorial Committee, eds., Flora of North America North of Mexico, Vol. 8, pp. 364–504. Oxford University Press, New York.

Tucker, G. C., and B. M. Daugherty. 2019. *Rhinanthus*. *In* Flora of North America Editorial Committee, eds., Flora of North America North of Mexico, Vol. 17, pp. 504–506. Oxford University Press, New York.

Tucker, G. C., and S. S. Vanderpool. 2010. Cleomaceae. *In* Flora of North America Editorial Committee, eds., Flora of North America North of Mexico, Vol. 7, pp. 199–223. Oxford University Press, New York.

Turner, B. L. 2015. Taxonomy of *Chamaesaracha* (Solanaceae). Phytologia 97: 226–245.

Turner, B. L., and M. Whalen. 1975. Taxonomic study of *Gaillardia pulchella* (Asteraceae-Heliantheae). Wrightia 5: 189–192.

Turner, M. W. 2006. *Baileya*. *In* Flora of North America Editorial Committee, eds., Flora of North America North of Mexico, Vol. 21, pp. 444–447. Oxford University Press, New York.

Urbatsch, L. E., and P. B. Cox. 2006a. *Ratibida*. *In* Flora of North America Editorial Committee, eds., Flora of North America North of Mexico, Vol. 21, pp. 60–63. Oxford University Press, New York.

Urbatsch, L. E., and P. B. Cox. 2006b. *Rudbeckia*. *In* Flora of North America Editorial Committee, eds., Flora of North America North of Mexico, Vol. 21, pp. 44–63. Oxford University Press, New York.

Urbatsch, L. E., and K. M. Neubig. 2006. *Lorandersonia*. *In* Flora of North America Editorial Committee, eds., Flora of North America North of Mexico, Vol. 20, pp. 177–181. Oxford University Press, New York.

Urbatsch, L. E., R. P. Roberts, and K. M. Neubig. 2006a. *Chrysothamnus*. *In* Flora of North America Editorial Committee, eds., Flora of North America North of Mexico, Vol. 20, pp. 187–193. Oxford University Press, New York.

Urbatsch, L. E., L. C. Anderson, R. P. Roberts, and K. M. Neubig. 2006b. *Ericameria*. *In* Flora of North America Editorial Committee, eds., Flora of North America North of Mexico, Vol. 20, pp. 50–77. Oxford University Press, New York.

Urbatsch, L. E., R. P. Roberts, and R. M. Neubig. 2006c. *Petradoria*. *In* Flora of North America Editorial Committee, eds., Flora of North America North of Mexico, Vol. 20, pp. 171–172. Oxford University Press, New York.

USNVC (U.S. National Vegetation Classification). 2019. U.S. National Vegetation Classification Database, V2.03. Federal Geographic Data Committee, Vegetation Subcommittee, Washington, DC. usnvc.org. Accessed January 25, 2021.

Utech, F. H. 2002. *Liliaceae*. *In* Flora of North America Editorial Committee, eds., Flora of North America North of Mexico, Vol. 26, pp. 50–347. Oxford University Press, New York.

Veno, B. A. 1979. A Revision of the Genus *Pectocarya* (Boraginaceae), Including Reduction to Synonymy of the Genus *Harpagonella* (Boraginaceae). Ph.D. Dissertation, University of California, Los Angeles.

Vincent, M. A. 2003. *Mollugo*. *In* Flora of North America Editorial Committee, eds., Flora of North America North of Mexico, Vol. 4, pp. 510–511. Oxford University Press, New York.

Vivrette, N. J., J. E. Bleck, and W. R. Ferren, Jr. 2003. Aizoaceae. *In* Flora of North America Editorial Committee, eds., Flora of North America North of Mexico, Vol. 4, pp. 75–91. Oxford University Press, New York.

Voss, J. W. 1937. A revision of the *Phacelia crenulata* group for North America. Bull. Torrey Bot. Club 64: 81–96.

Wagner, W. L. 2021. *Oenothera*. *In* Flora of North America Editorial Committee, eds., Flora of North America North of Mexico, Vol. 10, pp. 243–336. Oxford University Press, New York.

Wagner, W. L., and P. C. Hoch. 2021. Onagraceae. *In* Flora of North America Editorial Committee, eds., Flora of North America North of Mexico, Vol. 10, pp. 67–336. Oxford University Press, New York.

Wagner, W. L., P. C. Hoch, and P. H. Raven. 2007. Revised classification of Onagraceae. Syst. Bot. Monogr. 83: 1–222.

Watson, L., and M. J. Dallwitz. 1992 onward. The families of flowering plants: Descriptions, illustrations, identification, and information retrieval. Version: June 1, 2007. https://www.delta-intkey.com, accessed 25 October 2023.

Watson, L. E. 2006. *Cotula*. *In* Flora of North America Editorial Committee, eds., Flora of North America North of Mexico, Vol. 19, pp. 543–544. Oxford University Press, New York.

Weakley, A. S. 2014. *Poteridium*. *In* Flora of North America Editorial Committee, eds., Flora of North America North of Mexico, Vol. 9, pp. 319–320. Oxford University Press, New York.

Weber, W. A. 2006. *Helianthella*. *In* Flora of North America Editorial Committee, eds., Flora of North America North of Mexico, Vol. 21, pp. 114–117. Oxford University Press, New York.

Weber, W. A., and R. C. Whittman. 2001. Colorado Flora: Eastern Slope. University Press of Colorado, Boulder.

Welch, S. L., C. W. Crompton, and S. E. Clemants. 2003. Chenopodiaceae. *In* Flora of North America Editorial Committee, eds., Flora of North America North of Mexico, Vol. 4, pp. 258–404. Oxford University Press, New York.

Wells, E. F., and P. E. Elvander. 2009. Saxifragaceae. *In* Flora of North America Editorial Committee, eds., Flora of North America North of Mexico, Vol. 8, pp. 43–146. Oxford University Press, New York.

Welsh, S. L. 2007. North American Species of *Astragalus* L. (Leguminosae). Monte L. Bean Life Science Museum, Brigham Young University, Provo, Utah.

Welsh, S. L. 2016. Compositae. *In* Welsh, S. L., N. D. Atwood, S. Goodrich, and L. C. Higgins, eds., A Utah Flora, 5th ed., pp. 137–273. Brigham Young University Press, Provo, Utah.

Welsh, S. L., and D. Atwood. 2013. Chenopodiaceae. *In* Heil, K. D., S. L. O'Kane, Jr., L. M. Reeves, and A. Clifford, eds., Flora of the Four Corners Region: Vascular Plants of the San Juan River Drainage; Arizona, Colorado, New Mexico, and Utah, pp. 421–442. Missouri Botanical Garden Press, St. Louis.

Welsh, S. L., N. D. Atwood, S. Goodrich, and L. C. Higgins. 2015 [2016]. A Utah Flora, 5th ed. Brigham Young University, Provo, Utah.

Weniger, Del. 1969. Cacti of the Southwest: Texas, New Mexico, Oklahoma, Arkansas, and Louisiana. University of Texas Press, Austin.

Whalen, M. 2006. *Bebbia*. *In* Flora of North America Editorial Committee, eds., Flora of North America North of Mexico, Vol. 21, p. 177. Oxford University Press, New York.

Wherry, E. 1944. Review of the genera *Collomia* and *Gymnosteris*. Amer. Midl. Naturalist 31: 216–231.

Wherry, E. 1955. The Genus *Phlox*. Morris Arboretum Monographs, 3. Associates of the Morris Arboretum, Philadelphia.

Whetstone, R. D. 1997. *Berberis*. *In* Flora of North America Editorial Committee, eds., Flora of North America North of Mexico, Vol. 3, pp. 356–357. Oxford University Press, New York.

Whittemore, A. T. 2012. Cannabaceae and *Celtis*. *In* Baldwin, B. G., D. Goldman, D. J. Keil, R. Patterson, T. J. Rosatti, and D. Wilken, eds., The Jepson Manual: Vascular Plants of California, 2nd ed., pp. 598–600. University of California Press, Berkeley.

Whittemore, A. T., and A. E. Schuyler. 2002. *Scirpus*. *In* Flora of North America Editorial Committee, eds., Flora of North America North of Mexico, Vol. 23, pp. 8–21. Oxford University Press, New York.

Wiegleb, G., and Z. Kaplan. 1998. An account of the species of *Potamogeton* L. (Potamogetonaceae). Folia Geobot. 33: 241–316.

Wilken, D. H. 2001. A new *Ipomopsis* (Polemoniaceae) from the southwest USA and adjacent Mexico. Madroño 48(2): 116–122.

Wilken, D. H., and R. Fletcher 1988. *Ipomopsis sancti-spiritus* (Polemoniaceae) a new species from northern New Mexico. Brittonia 40(1) 48–51.

Williamson, T. E. 1996. The Beginning of the Age of Mammals in the San Juan Basin, New Mexico: Biostratigraphy and Evolution of Paleocene Mammals of the Nacimiento Formation. New Mexico Museum of Natural History and Science, Bulletin 8, Albuquerque.

Wilson, B. L., R. Brainerd, D. Lytjen, B. Nehouse, and N. Otting. 2014. Field Guide to the Sedges of the Pacific Northwest. Oregon State University Press, Corvalis.

Wingate, J. L. 2017. Sedges of Colorado. PDI Publication Design, Wheat Ridge, Colorado.

Woodland, D. W. 2009. Contemporary Plant Systematics, 4th ed. Andrews University Press, Berrien Springs, Michigan.

Wooton, E. O. 1911. Cacti in New Mexico. New Mexico Agricultural Experiment Station Bulletin 78. The New Mexico Printing Company, Santa Fe.

Wooton, E. O., and P. C. Standley. 1915. Flora of New Mexico. Contr. U.S. Natl. Herb. 19: 1–794.

Wunderlin, R. P. 1997. Moraceae. *In* Flora of North America Editorial Committee, eds., Flora of North America North of Mexico, Vol. 3, pp. 388–399. Oxford University Press, New York.

Xie, L., W. L. Wagner, R. H. Ree, P. E. Berry, and J. Wen. 2009. Molecular phylogeny, divergence time estimates, and historical biogeography of *Circaea* (Onagraceae). Molec. Phylogen. Evol. 53: 995–1009.

Yarborough, S. C., and A. M. Powell. 2006a. *Flaveria*. *In* Flora of North America Editorial Committee, eds., Flora of North America North of Mexico, Vol. 21, pp. 247–250. Oxford University Press, New York.

Yarborough, S. C., and A. M. Powell. 2006b. *Pericome*. *In* Flora of North America Editorial Committee, eds., Flora of North America North of Mexico, Vol. 21, pp. 334–335. Oxford University Press, New York.

Yarborough, S. C., and A. M. Powell. 2006c. *Perityle*. *In* Flora of North America Editorial Committee, eds., Flora of North America North of Mexico, Vol. 21, pp. 317–334. Oxford University Press, New York.

Yatskievych, G. 2015. Apodanthaceae. *In* Flora of North America Editorial Committee, eds., Flora of North America North of Mexico, Vol. 6, pp. 183–184. Oxford University Press, New York.

Yatskievych, G., and P. Fischer. 1984. The Acanthaceae of the southwestern United States. Desert Plants 15: 16–179.

Yu, C.-C., and K.-F. Chung. 2017. Why *Mahonia*? Molecular recircumscription of *Berberis* s.l., with the description of two new genera, *Alloberberis* and *Moranothamnus*. Taxon 66: 1371–1392.

Yuncker, T. G. 1932. The genus *Cuscuta*. Mem. Torrey Bot. Club 18: 113–331.

Zacharias, E. H., and B. G. Baldwin. 2010. A molecular phylogeny of North American Atripliceae (Chenopodiaceae), with implications for floral and photosynthetic pathway evolution. Syst. Bot. 35: 839–857.

Zarrei, M., P. Wilkin, M. F. Fay, M. J. Ingrouille, S. Zarre, and M. W. Chase. 2009. Molecular systematics of *Gagea* and *Lloydia* (Liliaceae; Liliales): Implications of analyses of nuclear ribosomal and plastid DNA sequences for infrageneric classification. Ann. Bot. 104: 125–142.

Zika, P. F. 2012a. *Juncus*. *In* Baldwin, B. G., D. Goldman, D. J. Keil, R. Patterson, T. J. Rosatti, and D. Wilken, eds., The Jepson Manual: Vascular Plants of California, 2nd ed., pp. 1361–1374. University of California Press, Berkeley.

Zika, P. F. 2012b. *Luzula*. *In* Baldwin, B. G., D. Goldman, D. J. Keil, R. Patterson, T. J. Rosatti, and D. Wilken, eds., The Jepson Manual: Vascular Plants of California, 2nd ed., pp. 1374–1375. University of California Press, Berkeley.

Zimmerman, A. D. 1972. Cact. Succ. J. (Los Angeles) 44: 114, figs. 1–4.

Zimmerman, A. D., and B. D. Parfitt. 2003. *Thelocactus*. *In* Flora of North America Editorial Committee, eds., Flora of North America North of Mexico, Vol. 4, pp. 216–218. Oxford University Press, New York.

Zomlefer, W. B., and W. S. Judd. 2002. Resurrection of segregates of the polyphyletic genus *Zigadenus* s.l. (Liliales: Melanthiaceae) and resulting new combinations. Novon 12: 299–308.

Zomlefer, W. B., N. H. Williams, W. M. Whitten, and W. S. Judd. 2001. Generic circumscription and relationships in the tribe Melanthieae (Liliales: Melanthiaceae), with emphasis in Zigadenus: Evidence from ITS and trnL-F sequence data. Amer. J. Bot. 88(9): 1657–1669.

INDEX OF BOTANICAL
AND COMMON NAMES

All scientific names (genera and species) are in *italics*.
Family names are in UPPERCASE. Common names are in roman type.

In memory of Kenneth D. Heil: December 6, 1941–January 16, 2024. This photo of Ken and Steve was taken on the last day they worked together (photo courtesy of Arnold Clifford).